Clinical Biochemistry of Domestic Animals

FIFTH EDITION

Clinical Biochemistry of Domestic Animals

FIFTH EDITION

Edited by

J. Jerry Kaneko
Department of Pathology, Microbiology, and Immunology
School of Veterinary Medicine
University of California, Davis
Davis, California

John W. Harvey
Department of Physiological Sciences
College of Veterinary Medicine
University of Florida
Gainesville, Florida

Michael L. Bruss
Department of Anatomy, Physiology, and Cell Biology
School of Veterinary Medicine
University of California, Davis
Davis, California

Academic Press

San Diego London Boston New York Sydney Tokyo Toronto

This book is printed on acid-free paper. 

Academic Press
a division of Harcourt Brace & Company
525 B Street, Suite 1900, San Diego, California 92101-4495, USA
http://www.apnet.com

Academic Press Limited
24-28 Oval Road, London NW1 7DX, UK
http://www.hbuk.co.uk/ap/

Library of Congress Cataloging-in-Publication Data

Clinical biochemistry of domestic animals / edited by J. Jerry Kaneko,
John W. Harvey, Michael L. Bruss.
p. cm.
Includes bibliographical references and index.
ISBN 0-12-396305-2 (alk. paper)
1. Veterinary clinical biochemistry. I. Kaneko, Jiro J.
II. Harvey, John W. III. Bruss, Michael.
SF769.3.C45 1997
636.089'60756--dc21 97-25865
CIP

PRINTED IN THE UNITED STATES OF AMERICA
97 98 99 00 01 02 MM 9 8 7 6 5 4 3 2 1

Contents

CHAPTER

12

Clinical Enzymology

JOHN W. KRAMER AND WALTER E. HOFFMANN

CHAPTER

13

Hepatic Function

BUD C. TENNANT

CHAPTER

14

Pancreatic Function

DUANE F. BROBST

CHAPTER

15

Gastrointestinal Function

WILLIAM E. HORNBUCKLE AND BUD C. TENNANT

CHAPTER

16

Skeletal Muscle Function

GEORGE H. CARDINET III

CHAPTER

17

Kidney Function

DELMAR R. FINCO

CHAPTER

18

Fluid, Electrolyte, and Acid–Base Balance

GARY P. CARLSON

CHAPTER

19

Pituitary Function

JAN A. MOL AND AD RIJNBERK

CHAPTER

20

Adrenocortical Function

AD RIJNBERK AND JAN A. MOL

CHAPTER

21

Thyroid Function

J. JERRY KANEKO

CHAPTER

22

Clinical Reproductive Endocrinology

LARS-ERIK EDQVIST AND MATS FORSBERG

CHAPTER

23

Calcium-Regulating Hormones and Diseases of Abnormal Mineral (Calcium, Phosphorus, Magnesium) Metabolism

THOMAS J. ROSOL AND CHARLES C. CAPEN

CHAPTER

24

The Vitamins

ROBERT B. RUCKER AND JAMES G. MORRIS

CHAPTER

25

Lysosomal Storage Diseases

MARK HASKINS AND URS GIGER

CHAPTER

26

Tumor Markers

BRUCE R. MADEWELL

CHAPTER

27

Cerebrospinal Fluid

CLETA SUE BAILEY AND WILLIAM VERNAU

CHAPTER

28

Clinical Biochemistry in Toxicology

JAMES R. TURK AND STAN W. CASTEEL

CHAPTER

29

Clinical Biochemistry of Laboratory Rodents and Rabbits

WALTER F. LOEB

CHAPTER

30

Avian Clinical Biochemistry

J. T. LUMEIJ

Appendixes

J. JERRY KANEKO, JOHN W. HARVEY, AND MICHAEL L. BRUSS

Contributors

Numbers in parentheses indicate the pages on which the authors' contributions begin.

Cleta Sue Bailey (785) Department of Surgical and Radiological Sciences, School of Veterinary Medicine, University of California, Davis, Davis, California 95616

Duane F. Brobst (353) Department of Veterinary Clinical Sciences, College of Veterinary Medicine, Washington State University, Pullman, Washington 99164

Michael L. Bruss (83, 885) Department of Anatomy, Physiology, and Cell Biology, School of Veterinary Medicine, University of California, Davis, Davis, California 95616

Charles C. Capen (619) Department of Veterinary Biosciences, The Ohio State University, Columbus, Ohio 43210

George H. Cardinet III (407) Department of Veterinary Anatomy, Physiology, and Cell Biology, School of Veterinary Medicine, University of California, Davis, Davis, California 95616

Gary P. Carlson (485) Department of Medicine and Epidemiology, School of Veterinary Medicine, University of California, Davis, Davis, California 95616

Stan W. Casteel (829) Veterinary Medical Diagnostic Laboratory, College of Veterinary Medicine, University of Missouri, Columbia, Missouri 65205

W. Jean Dodds (241) Hemopet, Santa Monica, California 90403

Lars-Erik Edqvist (589) Department of Clinical Chemistry, National Veterinary Institute, S-75007 Uppsala, Sweden

Thomas B. Farver (1) Department of Population Health and Reproduction, School of Veterinary Medicine, University of California, Davis, Davis, California 95616

Delmar R. Finco (441) Department of Physiology and Pharmacology, College of Veterinary Medicine, University of Georgia, Athens, Georgia 30602

Mats Forsberg (589) Department of Clinical Chemistry, College of Veterinary Medicine, Swedish University of Agricultural Sciences, Uppsala, Sweden

Laurel J. Gershwin (139) Department of Pathology, Microbiology, and Immunology, School of Veterinary Medicine, University of California, Davis, Davis, California 95616

Urs Giger (741) Medical Genetics, Veterinary Hospital, University of Pennsylvania, Philadelphia, Pennsylvania 19104

John W. Harvey (157, 885) Department of Physiological Sciences, College of Veterinary Medicine, University of Florida, Gainesville, Florida 32610

Mark Haskins (741) Pathology and Medical Genetics, School of Veterinary Medicine, University of Pennsylvania, Philadelphia, Pennsylvania 19104

Jens G. Hauge (21) Department of Biochemistry, Physiology, and Nutrition, Norwegian College of Veterinary Medicine, N-0033 Oslo, Norway

Walter E. Hoffmann (303) Department of Veterinary Pathobiology, College of Veterinary Medicine, University of Illinois, Urbana, Illinois 61801

William E. Hornbuckle (367) Department of Clinical Sciences, College of Veterinary Medicine, Cornell University, Ithaca, New York 14853

Mahendra B. Kabbur (285) Department of Pathology, Microbiology, and Immunology, School of Veterinary Medicine, University of California, Davis, Davis, California 95616

J. Jerry Kaneko (45, 117, 205, 571, 885) Department of Pathology, Microbiology, and Immunology, School of Veterinary Medicine, University of California, Davis, Davis, California 95616

John W. Kramer (303) Department of Clinical Medicine and Surgery, College of Veterinary Medicine, Washington State University, Pullman, Washington 99164

Walter F. Loeb (845) Ani Lytics, Inc., Gaithersburg, Maryland 20877

J. T. Lumeij (857) Division of Avian and Exotic Animal Medicine, Department of Clinical Sciences of Companion Animals, Utrecht University, 3584 CM Utrecht, The Netherlands

Bruce R. Madewell (761) Department of Veterinary Surgical and Radiological Sciences, University of California, Davis, Davis, California 95616

Jan A. Mol (517, 553) Department of Clinical Sciences of Companion Animals, Faculty of Veterinary Medicine, Utrecht University, 3584 CM Utrecht, The Netherlands

James G. Morris (703) Department of Molecular Biosciences, School of Veterinary Medicine, University of California, Davis, Davis, California 95616

Ad Rijnberk (517, 553) Department of Clinical Sciences of Companion Animals, Utrecht University, Utrecht, The Netherlands

Thomas J. Rosol (619) Department of Veterinary Pathobiology, Department of Veterinary Biosciences, The Ohio State University, Columbus, Ohio 43210

Robert B. Rucker (703) Department of Nutrition, College of Agriculture and Environmental Sciences, and Department of Biological Chemistry, School of Medicine, University of California, Davis, Davis, California 95616

Joseph E. Smith (223) Department of Diagnostic Medicine/Pathology, Kansas Veterinary Medical Center, Kansas State University, Manhattan, Kansas 66506

Bud C. Tennant (327, 367) Department of Clinical Sciences, College of Veterinary Medicine, Cornell University, Ithaca, New York 14853

James R. Turk (829) Veterinary Medical Diagnostic Laboratory, College of Veterinary Medicine, University of Missouri, Columbia, Missouri 65205

William Vernau (785) Consolidated Veterinary Diagnostics, Inc., West Sacramento, California 95605

Joseph G. Zinkl (285) Department of Pathology, Microbiology, and Immunology, School of Veterinary Medicine, University of California, Davis, Davis, California 95616

Preface to the 5th Edition

The first edition of *Clinical Biochemistry of Domestic Animals* was published in 1963; now, 34 years later, this fifth edition is appearing. We can all appreciate that change is a constant in all human endeavor and that science in particular is changing at an ever more rapid pace. In recognition of this rapid progress, this fifth edition constitutes yet another major change from the previous edition. In recognition of the breadth, magnitude, and complexity of this change, two associate editors, Dr. John W. Harvey of the University of Florida at Gainesville and Dr. Michael L. Bruss of the University of California at Davis, joined a team to guide the preparation of this fifth edition. Many new contributors joined this effort, major revisions of previous contributions were undertaken, and four new chapters were added: Tumor Markers, Lysosomal Storage Diseases, Clinical Biochemistry in Toxicology, and Avian Clinical Biochemistry.

It is also fitting that this volume be dedicated to the memories of Dr. Charles E. Cornelius, co-editor of the first two editions and contributor to all previous editions, and of Dr. George H. Stabenfeldt, contributor to the third and fourth editions. Both were great personal friends and outstanding scientists, administrators, and teachers and are sorely missed.

The acceptance of the Système International d'Unités (SI units) continues to be resisted for many reasons, although its use is expanding, albeit slowly, throughout the world. The concept of the SI unit is retained in this fifth edition as it was in the fourth with the inclusion of both SI and conventional units throughout the text and the appendices. The appendices in this edition have also been expanded to include new analyte reference ranges for the variety of species with which this volume is concerned.

We take this opportunity to acknowledge our gratitude to the many contributors who through their conscientiousness have eased the incredibly formidable task of assembling a multiauthored volume of this magnitude. All contributors have been extremely enthusiastic and cooperative with us as well as with Academic Press in their efforts to move production of this volume forward in a timely fashion. From the outset, Academic Press, ably represented by editor Chuck Crumly and his staff, has been extremely diligent, providing constant encouragement, guidance, and assistance for which we are most appreciative.

This fifth edition represents a new departure with increased editorship and a host of new and emerging subject areas in the field of clinical biochemistry of domestic animals. It recognizes and welcomes the changing nature of the field as new knowledge is uncovered and, as a corollary, the need for the wider view that associate editors bring to this effort. All contributors can justly point with pride to this fifth edition.

Finally, we express our thanks to our families, who have given us incredible support in this undertaking and have persevered without complaint through our single-minded devotion to this effort.

J. Jerry Kaneko
John W. Harvey
Michael L. Bruss

C H A P T E R

1

Concepts of Normality in Clinical Biochemistry

THOMAS B. FARVER

I. POPULATIONS AND THEIR DISTRIBUTIONS

A population is a collection of individuals or items having something in common. For example, one could say that the population of healthy dogs consists of all dogs that are free of disease. Whether a given dog belongs to the population of healthy dogs depends on someone's ability to determine if the dog is or is not free of disease. Populations may be finite or infinite in size.

A population can be described by quantifiable characteristics frequently called *observations* or *measures.* If it were possible to record an observation for all members in the population, one most likely would demonstrate that not all members of the population have the same value for the given observation. This reflects the inherent variability in populations. For a given measure, the list of possible values that can be assumed with the corresponding frequency with which each value appears in the population relative to the total number of elements in the population is referred to as the *distribution* of the measure or observation in the population. Distributions can be displayed in tabular or graphical form or summarized in mathematical expressions. Distributions are classified as discrete distributions or continuous distributions on the basis of values that the measure can assume. Measures with a continuous distribution can assume essentially an infinite number of values over some defined range of values, whereas those with a discrete distribution can assume only a relatively few values within a given range, such as only integer values.

Each population distribution can be described by quantities known as *parameters.* One set of parameters of a population distribution provides information on the center of the distribution or value(s) of the measure that seem to be assumed by a preponderance of the elements in the population. The mean, median, and mode are three members of the class of parameters describing the center of the distribution. Another class of parameters provides information on the spread of the distribution. The spread of the distribution has to do with whether most of the values assumed in the population are close to the center of the distribution or whether a wider range of values is assumed. The standard deviation, variance, and range are examples of parameters that provide information on the spread of the distribution. The shape of the distribution is very important. Some distributions are symmetric about their center, whereas other distributions are asymmetric, being skewed (having a heavier tail) either to the right or to the left.

II. REFERENCE INTERVAL DETERMINATION AND USE

One task of clinicians is determining whether an animal that enters the clinic has blood and urine analyte values that are in the normal interval. The conventional method of establishing normalcy for a particular analyte is based on the assumption that the distribution of the analyte in the population of normal animals is the "normal" or Gaussian distribution. To avoid confusion resulting from the use of a single word having two different meanings, the "normal" distribution henceforth is referred to as the Gaussian distribution.

A. The Gaussian Distribution

Understanding the conventional method for establishing normalcy requires an understanding of the properties of the Gaussian distribution. Theoretically, a Gaussian distribution is defined by the equation:

$$y = \frac{1}{\sqrt{2\pi}\sigma} e^{-(x - \mu)^2/2\sigma^2},$$

where x is any value that a given measurement can assume, y is the relative frequency of x, μ is the center of the distribution, σ is the standard deviation of the distribution, π is the constant 3.1416, and e is the constant 2.7183.

Theoretically, x can take on any value from $-\infty$ to $+\infty$. Figure 1.1 gives an example of a Gaussian distribution and demonstrates that the distribution is symmetric around μ and is bell shaped. Figure 1.1 also shows that 68% of the distribution is accounted for by measurements of x that have a value within 1 standard deviation of the mean, and 95% of the distribution includes those values of x that are within 2 standard deviations of the mean. Nearly all of the distribution (97.75%) is contained by the bound of 3 standard deviations of the mean.

Most analytes cannot take on negative values and so, strictly speaking, cannot have Gaussian distributions. However, the distribution of many analyte values is approximated well by the Gaussian distribution because virtually all of the values that can be assumed by the analyte are within 4 standard deviations of the mean and, for this range of values, the frequency distribution is Gaussian. Figure 1.2 gives an example. The figure, adapted from the printout of BMDP2D (detailed data description including frequencies; Engleman, 1992), gives an example of the distribution of glucose values given in Table 1.1 for a sample of 168 dogs from a presumably healthy population. Though not perfectly Gaussian, the distribution is reasonably well approximated by the Gaussian distribution. Support for this claim is that the distribution has the characteristic bell shape and appears to be symmetric about the mean. Also, the mean [estimated to be 96.4 mg/dl (5.34 mmol/liter)] of this distribution is nearly equal to the median [estimated to be 95.0 mg/dl (5.27 mmol/liter)], which is characteristic of the Gaussian distribution. The estimates of the skewness and kurtosis coefficients are close to zero, also characteristic of a Gaussian distribution (Engleman, 1992; Remington and Schork, 1985; Snedecor and Cochran, 1989).

B. Evaluating Probabilities Using a Gaussian Distribution

All Gaussian distributions can be standardized to the reference Gaussian distribution, which is called the *standard Gaussian distribution.* Standardization in general is accomplished by subtracting the center of the distribution from a given element in the distribution and dividing the result by the standard deviation of the distribution. The distribution of a standardized Gaussian distribution, that is, a Gaussian distribution that has its elements standardized in this form, has its center at zero and has a variance of unity. The elements of the standard Gaussian distribution are traditionally designated by the letter z so that it can be said that z is $N(0,1)$. That all Gaussian distributions can be transformed to the standard Gaussian distribution is convenient in that just a single table is required to summarize the probability structure of the infinite number of Gaussian distributions. Table 1.2 provides an example of such a table and gives the percentiles of the standard Gaussian distribution.

Example 1 Suppose the underlying population of elements is $N(4,16)$ and one element from this population is selected. We want to find the probability that the selected element has a value less than 3.0 or greater than 6.1. In solving this problem, the relevant distribution is specified: x is $N(4,16)$. The probability of observ-

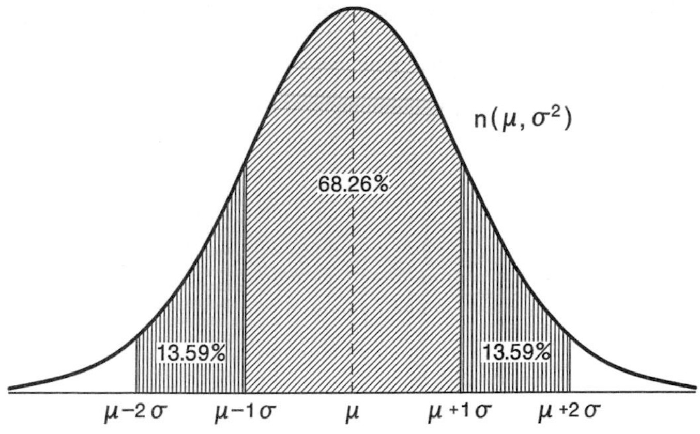

FIGURE 1.1 The Gaussian distribution.

ing $x < 3.0$ in the distribution of x is equivalent to the probability of observing $z < (3.0 - 4)/4 = -.25$ in the standard Gaussian distribution. Going to Table 1.2, $z = .25$ is approximately the 60th percentile of the standard Gaussian distribution and by symmetry $z = -.25$ is approximately the 40th percentile. Thus, the probability of observing a z value less than or equal to $-.25$ is approximately .40. The probability of observing $x > 6.1$ is equivalent to the probability of observing $z > (6.1 - 4)/(4) = +.525$. Table 1.2 gives the probability of observing a $z < .525$ as approximately .70, so the probability of observing a $z > .525$ approximately equals $1 - .70$ or .30. The desired probability of observing a sample observation less than 3.0 or greater than 6.1 is the sum of .40 and .30, which is .7 or 7 chances in 10.

C. Conventional Method for Determining Reference Intervals

The first step in establishing a normal interval by the conventional method involves determining the mean and standard deviation of the distribution of the analyte. This can be accomplished by taking a representative sample (using a sampling design that

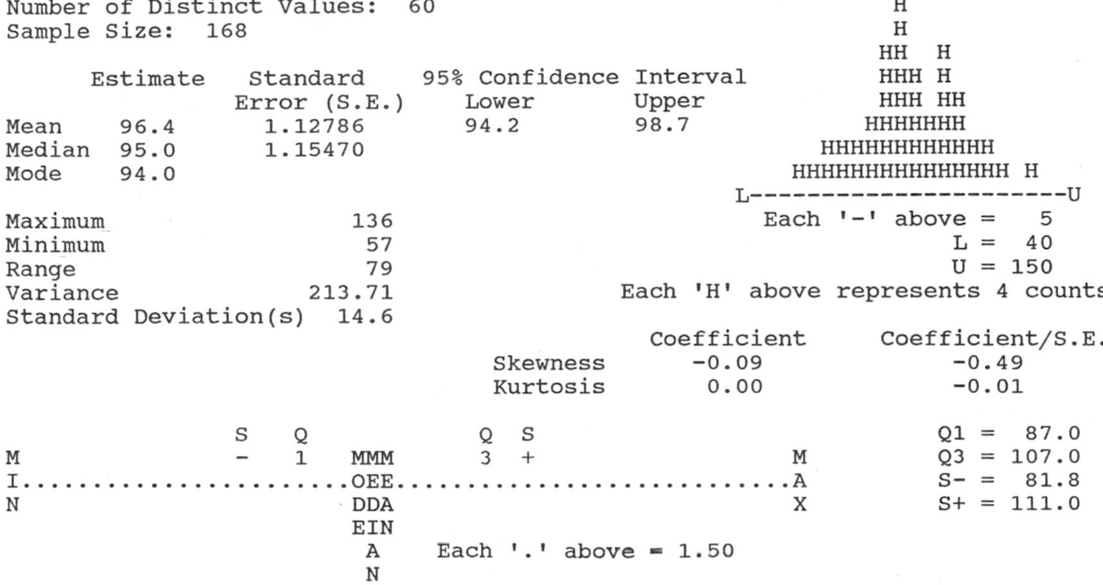

```
Number of Distinct Values:   60                                 H
Sample Size:   168                                              H
                                                              HH  H
      Estimate    Standard      95% Confidence Interval       HHH H
                 Error (S.E.)     Lower       Upper           HHH HH
Mean     96.4      1.12786        94.2        98.7           HHHHHHH
Median   95.0      1.15470                                 HHHHHHHHHHHHH
Mode     94.0                                            HHHHHHHHHHHHHHHH H
                                                       L-----------------------U
Maximum                   136                            Each '-' above =    5
Minimum                    57                                         L =   40
Range                      79                                         U =  150
Variance               213.71                   Each 'H' above represents 4 counts
Standard Deviation(s)   14.6
                                                  Coefficient      Coefficient/S.E.
                                  Skewness          -0.09               -0.49
                                  Kurtosis           0.00               -0.01

                  S   Q               Q  S                                 Q1 =  87.0
M                 -   1     MMM       3  +                      M          Q3 = 107.0
I...................... OEE.............................A          S- =  81.8
N                           DDA                                 X          S+ = 111.0
                            EIN
                             A     Each '.' above = 1.50
                             N
```

FIGURE 1.2 Distribution and summary statistics for the sample of canine glucose values (mg/dl) in Table 1.1. Information adapted from printout of BMDP2D (Engelman, 1992).

TABLE 1.1 Glucose (Glu, mg/dl) and Alanine Aminotransferase (ALT, U/liter) for a Sample of 168 Dogs from the Population of Healthy Dogs[a]

Cow	Glu	ALT	Cow	Glu	ALT	Cow	Glu	ALT	Cow	Glu	ALT
1	88	60	43	86	53	85	110	54	127	108	105
2	104	79	44	86	50	86	78	54	128	90	32
3	89	138	45	115	72	87	95	37	129	100	25
4	99	58	46	98	59	88	111	25	130	96	46
5	63	34	47	98	80	89	116	115	131	86	95
6	97	43	48	99	42	90	108	60	132	100	99
7	94	47	49	94	42	91	76	36	133	122	115
8	105	77	50	104	116	92	111	102	134	109	60
9	86	102	51	107	98	93	86	62	135	77	67
10	124	34	52	107	78	94	101	43	136	88	83
11	118	64	53	119	56	95	106	73	137	94	118
12	112	184	54	114	38	96	92	99	138	92	44
13	85	82	55	94	50	97	67	50	139	121	64
14	109	35	56	109	47	98	75	24	140	86	19
15	96	46	57	110	32	99	127	110	141	84	68
16	72	29	58	99	53	100	87	65	142	86	74
17	91	117	59	105	97	101	136	44	143	105	86
18	94	132	60	102	97	102	94	40	144	91	47
19	90	68	61	100	54	103	89	18	145	92	56
20	68	50	62	83	36	104	72	30	146	89	49
21	84	95	63	83	32	105	87	75	147	123	78
22	94	140	64	108	111	106	96	66	148	109	93
23	91	38	65	114	63	107	85	113	149	117	46
24	90	146	66	105	58	108	95	63	150	115	31
25	72	68	67	74	24	109	96	61	151	83	65
26	87	42	68	92	96	110	117	62	152	94	55
27	94	43	69	97	42	111	106	33	153	92	52
28	97	84	70	85	101	112	113	99	154	109	64
29	103	44	71	83	46	113	107	97	155	92	59
30	70	84	72	86	58	114	96	131	156	93	49
31	91	108	73	110	29	115	94	44	157	92	29
32	58	28	74	121	115	116	100	68	158	101	66
33	89	75	75	87	62	117	127	37	159	113	53
34	81	38	76	88	40	118	106	52	160	92	79
35	106	38	77	114	78	119	93	113	161	110	47
36	94	26	78	96	83	120	99	142	162	116	46
37	57	89	79	107	26	121	94	45	163	111	137
38	67	35	80	101	19	122	82	80	164	111	57
39	93	69	81	90	105	123	130	53	165	70	49
40	89	44	82	100	133	124	76	87	166	94	80
41	80	47	83	65	56	125	99	36	167	106	53
42	112	41	84	95	70	126	81	31	168	102	128

[a] These data were provided by Dr. J. J. Kaneko, Department of Pathology, Microbiology and Immunology, School of Veterinary Medicine, University of California, Davis.

has a random component such as simple random sampling) from the population of normal animals and computing the mean and standard deviation of the sample.

Once these estimates of μ and σ are obtained, an animal coming into the clinic in the future is classified as being normal for a particular analyte if its value for the analyte is within the bound of some multiple of the standard deviation below the mean and some multiple of the standard deviation above the mean. The multiple is determined by the degree of certainty that is desired to be placed on the classification scheme. For example, if the multiple chosen is 2, which is the conventional choice, any animal entering the clinic with an analyte value within 2 standard deviations of the mean would be classified as normal, whereas all animals with a value of the analyte outside this boundary would be classified as abnormal. Because 95% of

TABLE 1.2 Percentiles of the Standard Gaussian (z) Distribution[a,b]

$z_{0.50} = 0$	$z_{0.90} = 1.282$	$z_{0.975} = 1.960$
$z_{0.55} = 0.126$	$z_{0.91} = 1.341$	$z_{0.98} = 2.054$
$z_{0.60} = 0.253$	$z_{0.92} = 1.405$	$z_{0.99} = 2.326$
$z_{0.65} = 0.385$	$z_{0.93} = 1.476$	$z_{0.995} = 2.576$
$z_{0.70} = 0.524$	$z_{0.94} = 1.555$	$z_{0.999} = 3.090$
$z_{0.75} = 0.674$	$z_{0.95} = 1.645$	$z_{0.9999} = 3.719$
$z_{0.80} = 0.842$	$z_{0.96} = 1.751$	$z_{0.99999} = 4.265$
$z_{0.85} = 1.036$	$z_{0.97} = 1.881$	

[a] This table is adapted, with kind permission of the authors and publisher, from Dixon, W. J., and Massey, F. J. (1983). "Introduction to Statistical Analysis," 4th ed., Table A-4, p. 511. McGraw-Hill, New York.

[b] Example: The 75th percentile, or the z value below which is 75% of the Gaussian distribution, equals 0.674, $z_{0.75} = 0.674$. Percentiles smaller than the 50th percentile can be found by noting that the Gaussian distribution is symmetric about zero so that, for example, $z_{0.30} = -0.524$.

the Gaussian distribution is located within 1.96 or approximately 2 standard deviations of the mean, with this classification scheme, 2.5% of the normal animals would have a value of the analyte that would be below 2 standard deviations below the mean, and 2.5% of the animals would have an analyte value above 2 standard deviations above the mean. So with this classification scheme, there is a 5% chance that a true normal animal would be classified as being abnormal. Clinicians, by choosing 2 as the multiple, are willing to designate normal animals with extreme values of a particular analyte as being abnormal as the trade-off for not accepting too many abnormal animals as normals. With this methodology, no consideration is given to the distribution of abnormal animals because in fact there would be multiple distributions corresponding to the many types of abnormalities. The assumption is that for those cases where an analyte would be useful in identification of abnormal animals, the value of the analyte would be sufficiently above or below the center of the distribution of the analyte for normal animals. The reference interval for glucose based on the distribution from the sample of 168 normal dogs is 96.42857 mg/dl ± (1.96 × 14.61873 mg/dl) or 67.8 mg/dl (3.76 mmol/liter) to 125.1 mg/dl (6.94 mmol/liter).

Solberg (1994) gives $1/\alpha$ as the theoretical minimum sample size for estimation of the 100α and $100(1\text{-}\alpha)$ percentiles. Thus a minimum of 40 animals is required to estimate the 2.5th and 97.5th percentiles but many more than 40 is recommended.

D. Methods for Determining Reference Intervals for Analytes Not Having the Gaussian Distribution

The conventional procedure for assessing normalcy works quite well provided the distribution of the analyte is approximately Gaussian. Unfortunately, for many analytes a Gaussian distribution is not a good assumption. For example, Figure 1.3 describes the distribution of alanine aminotransferase (ALT) values given in Table 1.1 for the same sample of 168 normal dogs. This distribution is visibly asymmetric. The distribution has a longer tail to the right and is said to be skewed to the right or positively skewed. The skewness value (.91) exceeds the approximate 99th percentile of the distribution for this coefficient for random samples from a population having a Gaussian distribution. That the distribution is not symmetric and hence not Gaussian is also evidenced by the lack of agreement between the mean, median, and mode as shown in Fig. 1.3. Application of the conventional procedure for computing reference intervals [$\bar{x} \pm (1.96 \times \text{SD})$] reveals a reference interval of 4.4 to 127.7 U/liter so that all the low values of the distribution fall above the value, which is 2 standard deviations below the mean of the distribution and more than 2.5% of the high values fall above the value, which is 2 standard deviations above the mean. The following sections gives two approaches that can be followed in such a situation to obtain reference intervals.

1. *Use of Transformations*

Frequently, some transformation (such as the logarithmic or square root transformation) of the analyte values will make the distribution more Gaussian (Kleinbaum *et al.*, 1988; Neter *et al.*, 1990; Zar, 1984). The boundaries for the reference values are two standard deviations above and below the mean for the distribution of the transformed analyte values. These boundaries then can be expressed in terms of the original analyte values by retransformation. Figure 1.4 describes the distribution of the ALT analyte values after transformation with natural logarithms. The reference boundaries in logarithmic units are equal to 4.08013 ± (1.96 × 0.47591) or (3.14734, 5.01292), which correspond to (23.3, 150.3 U/liter) in the original units of the analyte.

2. *Use of Percentiles*

The second approach that can be followed in the situation where an assumption of a Gaussian distribution is not tenable is to choose percentiles as boundaries (Feinstein, 1977; Herrera, 1958; Mainland, 1963; Mas-

```
Number of Distinct Values:  86                          H
Sample Size:  168                                       H
                                                       HHH
      Estimate   Standard    95% Confidence Interval   HHHH
                Error (S.E.)    Lower       Upper      HHHHH
Mean    66.0      2.42617       61.3        70.8      HHHHHH H H
Median  58.5      3.17543                             HHHHHHHHHH H
Mode    Not Unique                                   HHHHHHHHHHHHHHH   H
                                                   L-----------------------U
Maximum                 184                          Each '-' above =  10
Minimum                  18                                        L =   0
Range                   166                                        U = 220
Variance             988.90                Each 'H' above represents 4 counts
Standard Deviation(s)  31.4
                                             Coefficient     Coefficient/S.E.
                                  Skewness       0.91              4.83
                                  Kurtosis       0.46              1.21

     S  Q             Q   S                                      Q1 = 43.0
M    -  1M   M  M     3   +                            M         Q3 = 83.5
I........O...E..E......................................A         S- = 34.6
N        D   D  A                                      X         S+ = 97.5
         E   I  N
             A            Each '.' above = 3.00
             N
```

FIGURE 1.3 Distribution and summary statistics for the sample of canine alanine aminotransferase values (U/liter) in Table 1.1. Information adapted from printout of BMDP2D (Engelman, 1992).

sod, 1977; Reed *et al.*, 1971; Solberg, 1994). For example, if we wanted to misclassify only 5% of normal animals as being abnormal, the 2.5th and 97.5th percentiles could be chosen as the reference boundaries. Thus, animals would be classified as abnormal when having analyte values either below the value of the analyte below which are 2.5% of all normal analyte values or above the value of the analyte below which are 97.5%

```
Number of Distinct Values:  86                               H
Sample Size:  168                                            HHH
                                                             HHH H
      Estimate   Standard    95% Confidence Interval        HHHHHHHH
                Error (S.E.)   Lower       Upper           HHHHHHHHH
Mean   4.08013    0.03672      4.00764     4.15262        HHHHHHHHHHHH
Median 4.06899    0.05444                                HHHHHHHHHHHH
Mode   Not Unique                                      H HHHHHHHHHHHHHH
                                                   L-----------------------U
Maximum              5.21494                         Each '-' above = 0.15
Minimum              2.89037                                       L = 2.4
Range                2.32457                                       U = 5.7
Variance             0.22649               Each 'H' above represents 3 counts
Standard Deviation(s) 0.47591
                                             Coefficient     Coefficient/S.E.
                                  Skewness      -0.08             -0.41
                                  Kurtosis      -0.53             -1.39

                  S   Q               Q   S                      Q1 = 3.76120
M                 -   1      MM       3   +            M         Q3 = 4.42483
I............................EE........................A         S- = 3.60422
N                     M      DA                        X         S+ = 4.55605
                      O      IN
                      D      A    Each '.' above = 0.04
                      E      N
```

FIGURE 1.4 Distribution and summary statistics for the natural logarithm of the sample of canine ALT values (U/liter) in Table 1.1. Information adapted from printout of BMDP2D (Engelman, 1992).

of all normal analyte values. This method is attractive because percentiles are reflective of the distribution involved.

The 97.5th percentile is estimated as the value of the analyte corresponding to the $(n + 1) \times 0.975$th observation in an ascending array of the analyte values for a sample of n normal animals (Dunn, 1977; Ryan *et al.*, 1985; Snedecor and Cochran, 1989). For the ALT values from the sample of $n = 168$ animals, $(n + 1) \times 0.975 = 169 \times 0.975 = 164.775$. Since there is no 164.775th observation, the 97.5th percentile is found by interpolating between the ALT values corresponding to the 164th and 165th observation in the ascending array commonly referred to as the 164th and 165th order statistics (Ryan *et al.*, 1985; Snedecor and Cochran, 1989). The 164th order statistic is 138 U/liter and the 165th order statistic is 140 U/liter and the interpolation is $138 + .775(140 - 138) = 139.5$ U/liter. The 2.5th percentile is estimated similarly as the $(n + 1) \times 0.025$th order statistic, which is the 4.225th order statistic for the sample of ALT values. In this case, the 4th and 5th order statistics are the same, 24 U/liter, which is the estimate of the 2.5th percentile. Note that there is reasonable agreement between this reference interval and that obtained using the logarithmic transformation. This method of using percentiles as reference values can also be used for analytes having a Gaussian distribution. The 2.5th and 97.5th percentiles for the sample of glucose values are 65.4 mg/dl (3.63 mmol/liter) and 126.3 mg/dl (7.01 mmol/liter), respectively. This interval agrees very well with that calculated earlier using the conventional method.

E. Sensitivity and Specificity of a Decision Based on a Reference Interval

As alluded to earlier, in addition to the "normal" or healthy population, several diseased populations may be involved, each with its own distribution. Figure 1.5 depicts the distributions of one analyte for a single diseased population and for a normal healthy, nondiseased population. Note that there will be some overlap of these distributions. Little overlap may occur when the disease has a major impact on the level of the analyte, whereas extensive overlap could occur if the level of the analyte is unchanged by the disease.

Using a reference value based on values of the analyte for normal animals could lead to two types of mistakes in diagnosis of patients. First, normal patients with values beyond the normal interval would be classified incorrectly as diseased and would be the false positives. Second, diseased patients having values within the normal interval would be classified as non-

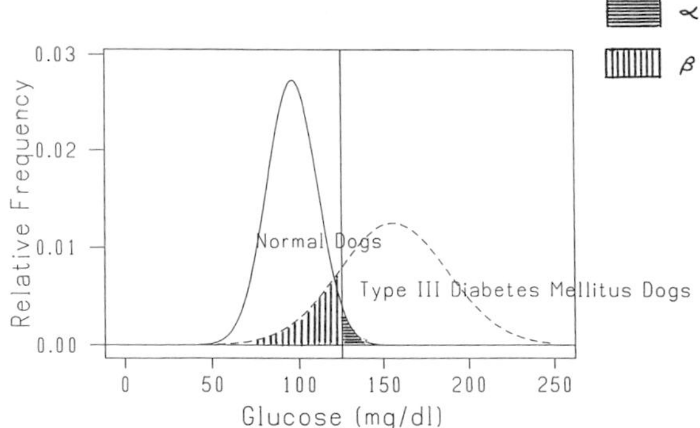

FIGURE 1.5 Overlapping Gaussian distributions of one analyte for a diseased dog population and a healthy, nondiseased dog population with upper limit of reference interval, α, and β error rates shown.

diseased, the false negatives. The two kinds of mistakes in classifying patients on the basis of analyte values are called the Type I error (saying a normal animal is diseased) and the Type II error (saying an animal is normal when in fact the animal is diseased). The probabilities of making these two errors, the error rates, are α and β, respectively. The size of these error rates is determined by the reference values, which are the decision points. We call $1 - \alpha$ the *sensitivity* of the diagnostic or decision process using reference values (the probability of deciding that a truly normal animal is normal on the basis of the given reference value) and $1 - \beta$ is the specificity of the decision process (the probability of deciding that a truly diseased animal is diseased). It is possible to change the reference values to reduce the size of α and in so doing increase the sensitivity of the test. However, such an action will also result in an increase in the size of β and therefore a reduction in the specificity of the test.

Example 2 Type III diabetic dogs have the chemical form of diabetes mellitus generally regarded as the first level of development of the disease offering the highest likelihood "for successful oral hypoglycemic therapy and/or dietary therapy" (Kaneko, 1977). Thus it would be useful to distinguish Type III diabetic dogs from normal dogs. Using the sample mean [155.6 mg/dl (8.63 mmol/liter)] and standard deviation [32.0 mg/dl (1.77 mmol/liter)] of the plasma glucose values given by Kaneko (1977) for five dogs with Type III diabetes mellitus as reasonable estimates of the corresponding parameters for the population of dogs with Type III diabetes mellitus, and assuming that this population distribution is approximately Gaussian, a comparison of this distribution of glucose values can be made with that for the population of normal dogs described by the approximately Gaussian distribution with parameter estimates given in Fig. 1.2 [μ_x = 96.4 mg/dl (5.35 mmol/liter) and σ_x = 14.6 mg/dl (0.81 mmol/liter)]. These two distributions are shown in Fig. 1.5; they have reasonably good separation with moderate overlap. Based on this information, a diagnostic procedure is proposed whereby a dog entering the clinic with a glucose value above 125.1 mg/dl (6.94 mmol/liter), the upper limit of the normal reference interval, will be flagged as possibly having Type III diabetes mellitus, thereby indicating the need for more follow-up. (*Note:* This is an oversimplification of actual practice because a diagnostic decision of this type would be based on additional information, such as the animal's glucose tolerance and insulin response, making the decision rule and subsequent error calculations more complex.) This is an example of a one-sided diagnostic procedure because dogs with a glucose value below the lower limit of the reference interval would not be considered as having Type III diabetes mellitus. Thus the Type I error is concentrated at the upper end of the distribution of glucose values for normal dogs and its magnitude (α) is the area to the right of a glucose value of 125.1 mg/dl in the distribution of glucose values for normal dogs or the area to the right of the corresponding z value, $z = (125.1 - 96.4)/14.6 \sim 1.96$, for the standard Gaussian distribution (see Section II.B); $\alpha = 2.5\%$.

The clinician may be interested in determining the sensitivity and the specificity of the diagnostic procedure. The sensitivity is $1 - \alpha = 97.5\%$. A dog that actually has Type III diabetes mellitus but has a glucose value less than 125.1 mg/dl would be incorrectly classified by the proposed diagnostic procedure as being a normal dog. This is a Type II error and the probability of making this type of error is β, which is the area to the left of a glucose value of 125.1 mg/dl in the distribution of glucose values for dogs with Type III diabetes mellitus or the area to the left of the corresponding z value, $z = (125.1 - 155.6)/32.0 \sim -.953$, for the standard Gaussian distribution. Using Table A-4 from Dixon and Massey (1983), $\beta = 17.62\%$ (or 17.14% by linear interpolation of the information given in Table 1.2), and the specificity of the diagnostic procedure is $1 - \beta = 82.38\%$.

F. Predictive Value of a Decision Based on a Reference Interval

A useful quantity is the probability that a patient having a reference value outside the normal interval actually has the disease. This is known as the predictive value of a positive diagnosis, Prob(D|+). Interest could also be in determining the probability that a patient having a reference value within the normal interval is actually nondiseased or the predictive value of a negative diagnosis, Prob(D|−). The predictive value depends on the sensitivity, specificity, and prevalence (p) of the disease as shown in the following equations:

$$\text{Prob}(D|+) = \frac{p \times \text{Sensitivity}}{p \times \text{Sensitivity} + (1 - p) \times (1 - \text{Specificity})}$$

$$\text{Prob}(D|-) = \frac{(1 - p) \times \text{Specificity}}{(1 - p) \times \text{Specificity} + p \times (1 - \text{Sensitivity})}$$

Fig. 1.6 demonstrates the extent to which the predictive value of a positive diagnosis changes with the prevalence. In general, larger changes are seen in the predictive value of a positive diagnosis for smaller changes in the prevalence for diseases with low prevalence, and smaller changes are seen in the predictive value for larger changes in the prevalence for diseases with high prevalence.

In the example of the diagnostic procedure given in the previous section, assuming the prevalence of Type III diabetes mellitus in the dog population was 2%,

$$\begin{aligned}\text{Prob}(D|+) &= (0.02 \times 0.975)/[(0.02 \times 0.975) + (0.98 \times 0.1762)] \\ &= 0.10 \text{ or } 10\%, \text{ and} \\ \text{Prob}(D|-) &= (0.98 \times 0.8238)/[(0.98 \times 0.8238) + (0.02 \times 0.025)] \\ &= 0.999 \text{ or } 99.9\%.\end{aligned}$$

To demonstrate how specificity and hence the predictive value of a positive test improve with greater separation of the populations, Kaneko (1977) gives estimates (based on a sample of 11

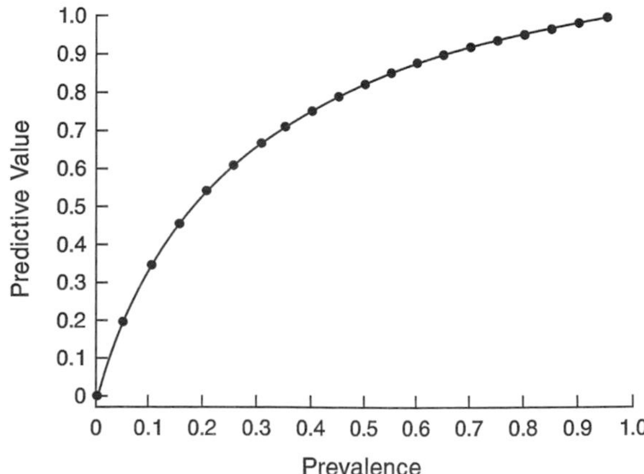

FIGURE 1.6 Impact of disease prevalence on the predictive value of a positive laboratory test having 95% sensitivity and 80% specificity.

dogs) of the mean and standard deviation of the plasma glucose values of the population of dogs with Type I diabetes mellitus (the juvenile or childhood form) as $\hat{\mu}$ = 415.1 mg/dl (23.02 mmol/liter) and $\hat{\sigma}$ = 114.3 mg/dl (6.34 mmol/liter). If we use these values in the preceding calculations with the diagnostic value remaining at 125.1 mg/dl, the specificity improves to 99.44% and the predictive value of a positive test increases to 78%.

III. ACCURACY IN ANALYTE MEASUREMENTS

Accuracy has to do with the conformity of the actual value being measured to the intended true or target value. An analytical procedure having a high level of accuracy produces measurements that on average are close to the target value. An analytical procedure having a low level of accuracy produces measurements that on average are a distance from the target value. Such a procedure in effect measures something other than is intended and is said to be *biased*. Failure of analytical procedures to produce values that on average conform to the target values is due to unresolved problems, either known or unknown, in the assay.

The degree of accuracy of an analytical procedure has been difficult to quantify due to the fact that the target value is unknown. It is now possible for laboratories to compare their assay results with definitive results obtained by the use of isotope dilution-mass spectrometry (Shultz, 1994). Shultz (1994) reports the results of two large surveys of laboratories in the United States (Gilbert, 1978) and Sweden (Björkhem *et al.*, 1981) in which samples from large serum pools were analyzed for frequently tested analytes (calcium, chloride, iron, magnesium, potassium, sodium, cholesterol, glucose, urea-nitrogen, urate, and creatinine). The laboratory averages were compared with the target value obtained using definitive methods, and the results of these surveys indicated that, with the exception of creatinine, all averages expressed as a percentage of the target value were within the accuracy goals published by Gilbert (1975). Results from individual laboratories naturally would vary about the average, and many of these laboratories would not have met the accuracy goal.

IV. PRECISION IN ANALYTE MEASUREMENTS

Precision has to do with how much variability there is about the actual value being measured when the assay is replicated. If in a given laboratory a particular assay is run repeatedly on the same sample and the results obtained have little variability, the assay is said to have high precision. Large variability in the observed results is an indication of low assay precision. Note that precision is defined in reference to what is actually being measured and not to the target value. Clinical analysts have always had a goal of achieving the highest possible level of precision for a particular assay within a laboratory. Emphasis is presently placed on meeting an "average laboratory" level of precision (Shultz, 1994).

The level of precision is stated quantitatively in terms of the coefficient of variation (*cv*). The *cv* is the ratio of the standard deviation to the average of a series of replicated assays and its magnitude depends on the concentration of the analyte. Elevitch (1977) and Harris (1988) provide the guidelines on the desired level of precision in terms of the *cv*. In the case where the analytical results are intended to assist in the diagnostic process or to assist in monitoring a patient's response to treatment, the level of laboratory precision of a given analyte in terms of the *cv* (cv_a) need only be a function of the within day and day-to-day variability or intrasubject variation of healthy subjects. Specifically,

$$cv_a < \tfrac{1}{2}cv_{\text{intrasubject}}.$$

In the case where analytical test results are to be used to screen a population, the laboratory precision goal in terms of the *cv* should be a function of the variability in response among healthy subjects or intersubject variation. Specifically,

$$cv_a < \tfrac{1}{2}cv_{\text{intersubject}}.$$

Use of intrasubject variability as a goal for precision has appeal because this source of variability would be considered in decision processes relating to patients. Unfortunately, a given analysis reflects not only this intrasubject variability, but also imprecision in the assay. Shultz (1994) summarizes the results of a large national survey of laboratory precision. With the exception of high-density lipoprotein and thyroxine (T_4), the precision of the assay for the analytes evaluated from the "average" laboratory met or nearly met the precision goals based on the intrasubject variability. This result has to be regarded as encouraging, no doubt reflecting the tremendous emphasis that has been placed on quality control by laboratories as well as the use of automation in analytical work. On the other hand, there were some analytes for which the assay precision for the "average" laboratory was above the precision goal. It also must be remembered that there would be many individual laboratories that would not have assay precision profiles as good as the "average" laboratory. Assay precision in excess of the precision goal based on physiological variability makes it nearly impossible to rule out the possibility that very large changes in the level of an analyte are a reflection of assay imprecision.

V. INFERENCE FROM SAMPLES

The basis for everything that has been discussed to this point is probability and distributional theory. No other theory is relevant unless one is operating at the level where inference is to be made on the basis of a sample from the underlying population. Most standard statistical theory assumes that the sample was obtained by simple random sampling.

A. Simple Random Sampling

Simple random sampling (SRS) is a method of sampling whereby, at each step of the sampling process, the elements available for selection have an equally likely chance of being selected. In most applications, it is assumed that the elements are selected without replacement, although the elements could be selected with replacement. If the number of elements to be selected is small relative to the number of elements in the population, then it is rather unlikely that an element will be selected more than a single time with replacement sampling, so that in such situations sampling with replacement produces essentially the same results as sampling without replacement. It is only when a small finite population is being sampled that differences may be noted between the two methods.

Three steps are used to select a sample by SRS without replacement: All elements in the population must first be identified by a unique number from 1 to N, the population size. Then n numbers are selected from a table of random numbers or selected by a random number generator, which gives the numbers 1 to N in random order. Numbers appearing more than once are ignored after their first use. Finally, those elements having numbers corresponding to the n numbers selected constitute the sample. There are other probability-based sampling procedures that should be considered in practice; these methods are found in texts on sampling (Cochran, 1977; Jessen, 1978; Levy and Lemeshow, 1991; Murthy, 1967; Raj, 1968,1972; Scheaffer *et al.*, 1996).

B. Descriptive Statistics

Once the data have been collected, so-called "descriptive statistics" can be computed. As the name suggests, these statistics are useful in describing the underlying populations. For example, because complete information for the entire population is not available, it is not possible to know the population mean, $\mu = \Sigma x_i / N$. (Here x_i designates the value of the ith element in the population and Σ indicates summation. Thus the population mean μ is found by summing the values of all N elements in the population and then dividing the sum by N.) However, a sample mean based on the sample can be computed as $\bar{x} = \Sigma x_i / n$, the sum of the values of all n elements in the sample divided by n. If the sample has been selected in a manner that results in a small bias, $\bar{x}$ should be a reasonably good estimate of the population mean, μ, and will be a better estimate as the sample size increases. Other estimates of the measures of central tendency of the population can be obtained from the sample, such as the sample median and the sample mode. Also, sample-based estimates of the measures of dispersion or spread for the population can be obtained. The sample variance is computed as $s^2 = \Sigma(x_i - \bar{x})^2/(n - 1)$, and the sample standard deviation, s, is obtained by taking the square root of s^2. The descriptive statistics are called *point estimates* of the parameters and represent good approximations of the parameters. An alternative to the point estimate is the *interval estimate*, which takes into account the underlying probability distribution of the point estimate called the sampling distribution of the statistic.

C. Sampling Distributions

In actual practice, only a single sample is taken from a population and, on the basis of this sample, a single point estimate of the unknown population parameter is computed. If time and resources permitted repeated sampling of the population in the same manner, that is, with the same probability-based sampling design, one point estimate would be obtained for each sample obtained. The estimates would not be the same because the sample would contain different elements of the population. As the number of such repeated sampling operations increases, a more detailed description emerges of the distribution of possible point estimates that could be obtained by sampling the population. This is the sampling distribution of the statistic.

Some fundamental facts relating to the sampling distribution of the sample mean follow: (1) The center of the sampling distribution of $\bar{x}$ is equal to μ, the center of the underlying distribution of elements in the population. (2) The spread of the sampling distribution of $\bar{x}$ is smaller than σ^2, the spread of the underlying distribution of elements in the population. Specifically, the variance of the sampling distribution of $\bar{x}$ (denoted $\sigma^2_{\bar{x}}$) *equals* σ^2/n, where n is the sample size. So increasing the sample size serves to increase the likelihood of obtaining an $\bar{x}$ close to the center of the distribution because the spread of the sampling distribution is being reduced. (3) The central limit theorem (Daniel, 1995; Remington and Schork, 1985; Zar, 1996) states that regardless of the underlying distribution of the population of elements from which the sample mean is based, if the sample size is reasonably large ($n > 30$), the sampling distribution of $\bar{x}$ is approximated well by the Gaussian distribution. So $\bar{x}$ drawn from any distribution has a sampling distribution that is approximately $N(\mu, \sigma^2/n)$ for $n > 30$. If the distribution of the underlying population of elements is Gaussian or approximated well by a Gaussian distribution, the sampling distribution of $\bar{x}$ will be approximated well by the Gaussian distribution regardless of the sample size on which $\bar{x}$ is based.

Probabilities of the sampling distribution of $\bar{x}$, $N(\mu, \sigma^2/n)$, can be evaluated using the method described in Section II.B.

Example 3 Suppose the underlying population of elements is $N(4,16)$ and a sample of size $n = 9$ is drawn from this population using SRS. We want to find the probability of observing a *sample mean* less than 3.1 or greater than 6.2. In solving this problem, the relevant sampling distribution is specified: $\bar{x}$ is $N(4, 16/9)$. Note that the sampling distribution of $\bar{x}$ is Gaussian because the problem stated that the underlying population was Gaussian. (Otherwise the stated sample size would needed to have been 30 or larger to invoke the central limit theorem.) The probability of observing $\bar{x} < 3.1$ in the distribution of $\bar{x}$ is equivalent to the probability of observing $z < (3.1 - 4)/(4/3) = -0.675$ in the standard Gaussian distribution. Going to Table 1.2, $z = 0.675$ is approximately the 75th percentile of the standard Gaussian distribution and by symmetry $z = -0.675$ is approximately the 25th percentile. Thus, the probability of observing a z value less than or equal to -0.675 is approximately 0.25. The probability of observing a sample mean greater than 6.2 is equivalent to the probability of observing $z > (6.2 - 4)/(4/3) = +1.65$. Table 1.2 gives the probability of observing a $z < 1.65$ as approximately 0.95, so the probability of observing a $z > 1.65$ equals $1 - 0.95$ or 0.05. The desired probability of observing a sample mean less than 3.1 or greater than 6.2 is the sum of 0.25 and 0.05, which is 0.3 or 3 chances in 10.

D. Constructing an Interval Estimate of the Population Mean, μ

This brief exposure to sampling distributions and their standardized forms provides the framework for generating an interval estimate for μ. Consider the probability statement $\text{Prob}(-2 < z < +2) = 0.9544$. Because $z = (\bar{x} - \mu)/(\sigma/\sqrt{n})$, this probability statement is equivalent to the statement $\text{Prob}(-2 < (\bar{x} - \mu)/(\sigma/\sqrt{n}) < +2) = 0.9544$. Some standard algebraic manipulation of the inequality within the parentheses gives $\text{Prob}(\bar{x} + (-2\sigma/\sqrt{n}) < \mu < \bar{x} + (-2\sigma/\sqrt{n})) = 0.9544$. This is the form of the confidence statement about the unknown parameter μ. With repeated sampling of the underlying population, 95.44% of the intervals constructed by adding and subtracting $2\sigma/\sqrt{n}$ to and from the sample mean would be expected to cover the true unknown value of μ. The quantities of $\bar{x} - 2\sigma/\sqrt{n}$ and $\bar{x} + 2\sigma/\sqrt{n}$ are called the *lower* and *upper confidence limits*, respectively, and the interval bounded below by $\bar{x} - 2\sigma/\sqrt{n}$ and above by $\bar{x} + 2\sigma/\sqrt{n}$, that is, $(\bar{x} - 2\sigma/\sqrt{n}, \bar{x} + 2\sigma/\sqrt{n})$, is the 95.44% confidence interval for μ. In practice, only one sample is taken from the population and thus there is a 95.44% chance that the one interval estimate obtained will cover the true value of μ. The value of 95.44% or 0.9544 is called the *confidence level*.

The degree of confidence that is to be had is determined by the amount of error that is to be tolerated in the estimation procedure. For a 0.9544 level of confidence, the error rate is $1 - 0.9544 = 0.0456$. The error rate is designated by alpha, α. The size of α determines the magnitude of the value of z that is multiplied by $\sigma/\sqrt{n}$. The convention is to apportion half of α to the lower end and half of α to the upper end of the sampling distribution of $\bar{x}$ so that the relevant values of z are (1) the z value that has $\alpha/2$ area to its left, $z_{\alpha/2}$, and (2) the z value that has $\alpha/2$ to its right or, equivalently (to conform to Table 1.2), the z value that has $1 - (\alpha/2)$ area to its left, $z_{1-(\alpha/2)}$. Therefore, the most general form of the interval estimate statement is $(\bar{x} - (z_{\alpha/2}\sigma)/\sqrt{n}, \bar{x} + (z_{1-(\alpha/2)}\sigma)/\sqrt{n})$ or $\bar{x} \pm (z_{1-(\alpha/2)}\sigma)/\sqrt{n}$ because of the symmetry of the Gaussian distribution.

Example 4 Assuming the distribution in Example 3, construct a 90% confidence interval for μ. A 90% level of confidence implies a tolerated error rate of 10% and the relevant z values are (1) that which has 5% of the distribution of z to its left or $z_{0.05} = -1.645$ and (2) that which has 95% of the distribution to its left or $z_{0.95} = 1.645$. The 90% confidence interval for μ is therefore $(\bar{x} \pm (1.645 \times 4)/3)$ or $(\bar{x} \pm 2.1933)$ where $\bar{x}$ is the sample mean obtained by taking a sample of size 9 from the population.

The form of the interval estimate given earlier assumes that σ, the standard deviation of the underlying population, is also known. Frequently this parameter, like μ, is unknown and must be estimated from the sample drawn from the population using the estimator $s^2 = \Sigma(x_i - \bar{x})^2/(n - 1)$. For small samples ($n < 30$), the standardized form of $\bar{x}$, $(\bar{x} - \mu)/(s/\sqrt{n})$, does not have a standardized Gaussian distribution, $N(0,1)$, but rather has the t distribution corresponding to the effective sample size, $n - 1$, the degrees of freedom. Table 1.3 gives the percentiles of several t distributions. A given row of Table 1.3 pertains to the t distribution with the indicated effective sample size or degrees of freedom. The entries in the row are percentiles or those points of the given t distribution that have the indicated area to the left. For example, $t_{0.95,10}$ is 1.812, meaning that the 95th percentile of the t distribution with 10 degrees of freedom is 1.812, which is equivalent to saying that 95% of the t distribution with 10 degrees of freedom lies to the left of the t value 1.812. Note

TABLE 1.3 Percentiles of Student's t Distribution[a,b]

df	$t_{0.55}$	$t_{0.65}$	$t_{0.75}$	$t_{0.85}$	$t_{0.90}$	$t_{0.95}$	$t_{0.975}$	$t_{0.99}$	$t_{0.995}$	$t_{0.9995}$
1	0.158	0.510	1.000	1.963	3.078	6.314	12.706	31.821	63.657	636.619
2	0.142	0.445	0.816	1.386	1.886	2.920	4.303	6.965	9.925	31.599
3	0.137	0.424	0.765	1.250	1.638	2.353	3.182	4.541	5.841	12.924
4	0.134	0.414	0.741	1.190	1.533	2.132	2.776	3.747	4.604	8.610
5	0.132	0.408	0.727	1.156	1.476	2.015	2.571	3.365	4.032	6.869
6	0.131	0.404	0.718	1.134	1.440	1.943	2.447	3.143	3.707	5.959
7	0.130	0.402	0.711	1.119	1.415	1.895	2.365	2.998	3.499	5.408
8	0.130	0.399	0.706	1.108	1.397	1.860	2.306	2.896	3.355	5.041
9	0.129	0.398	0.703	1.100	1.383	1.833	2.262	2.821	3.250	4.781
10	0.129	0.397	0.700	1.093	1.372	1.812	2.228	2.764	3.169	4.587
11	0.129	0.396	0.697	1.088	1.363	1.796	2.201	2.718	3.106	4.437
12	0.128	0.395	0.695	1.083	1.356	1.782	2.179	2.681	3.055	4.318
13	0.128	0.394	0.694	1.079	1.350	1.771	2.160	2.650	3.012	4.221
14	0.128	0.393	0.692	1.076	1.345	1.761	2.145	2.624	2.977	4.140
15	0.128	0.393	0.691	1.074	1.341	1.753	2.131	2.602	2.947	4.073
16	0.128	0.392	0.690	1.071	1.337	1.746	2.120	2.583	2.921	4.015
17	0.128	0.392	0.689	1.069	1.333	1.740	2.110	2.567	2.898	3.965
18	0.127	0.392	0.688	1.067	1.330	1.734	2.101	2.552	2.878	3.922
19	0.127	0.391	0.688	1.066	1.328	1.729	2.093	2.539	2.861	3.883
20	0.127	0.391	0.687	1.064	1.325	1.725	2.086	2.528	2.845	3.850
21	0.127	0.391	0.686	1.063	1.323	1.721	2.080	2.518	2.831	3.819
22	0.127	0.390	0.686	1.061	1.321	1.717	2.074	2.508	2.819	3.792
23	0.127	0.390	0.685	1.060	1.319	1.714	2.069	2.500	2.807	3.768
24	0.127	0.390	0.685	1.059	1.318	1.711	2.064	2.492	2.797	3.745
25	0.127	0.390	0.684	1.058	1.316	1.708	2.060	2.485	2.787	3.725
26	0.127	0.390	0.684	1.058	1.315	1.706	2.056	2.479	2.779	3.707
27	0.127	0.389	0.684	1.057	1.314	1.703	2.052	2.473	2.771	3.690
28	0.127	0.389	0.683	1.056	1.313	1.701	2.048	2.467	2.763	3.674
29	0.127	0.389	0.683	1.055	1.311	1.699	2.045	2.462	2.756	3.659
30	0.127	0.389	0.683	1.055	1.310	1.697	2.042	2.457	2.750	3.646
35	0.127	0.388	0.682	1.052	1.306	1.690	2.030	2.438	2.724	3.591
40	0.126	0.388	0.681	1.050	1.303	1.684	2.021	2.423	2.704	3.551
45	0.126	0.388	0.680	1.049	1.301	1.679	2.014	2.412	2.690	3.520
50	0.126	0.388	0.679	1.047	1.299	1.676	2.009	2.403	2.678	3.496
60	0.126	0.387	0.679	1.045	1.296	1.671	2.000	2.390	2.660	3.460
70	0.126	0.387	0.678	1.044	1.294	1.667	1.994	2.381	2.648	3.435
80	0.126	0.387	0.678	1.043	1.292	1.664	1.990	2.374	2.639	3.416
90	0.126	0.387	0.677	1.042	1.291	1.662	1.987	2.368	2.632	3.402
100	0.126	0.386	0.667	1.042	1.290	1.660	1.984	2.364	2.626	3.390
120	0.126	0.386	0.677	1.041	1.289	1.658	1.980	2.358	2.617	3.373
140	0.126	0.386	0.676	1.040	1.288	1.656	1.977	2.353	2.611	3.361
160	0.126	0.386	0.676	1.040	1.287	1.654	1.975	2.350	2.607	3.352
180	0.126	0.386	0.676	1.039	1.286	1.653	1.973	2.347	2.603	3.345
200	0.126	0.386	0.676	1.039	1.286	1.653	1.972	2.345	2.601	3.340
∞	0.126	0.385	0.674	1.036	1.282	1.645	1.960	2.326	2.576	3.291

[a] This table is reprinted, with kind permission of the authors and publisher, from Kleinbaum, D. G., and Kupper, L. L. (1988). "Applied Regressing Analysis and Other Multivariable Methods," 2nd ed., Table A-2, p. 647. PWS-Kent, Boston.

[b] Example: The 75th percentile of a t distribution with 11 degrees of freedom or the t value below which is 75% of the t distribution with 11 degrees of freedom equals 0.697, $t_{0.75,11} = 0.697$.

that the 5th percentile of the t distribution with 10 degrees of freedom is -1.812 because the t distributions, like the standard Gaussian distribution, are symmetric about zero. Thus, for smaller sample sizes when the value of σ is unknown, the form of a confidence interval for μ is $\bar{x} \pm [(t_{1-\alpha/2;n-1}s)/\sqrt{n}]$.

As with the mean, interest also centers around estimating the variance, σ^2, and standard deviation, σ, of the population. The estimates for these parameters are s^2 and s, respectively. One might also be interested in constructing confidence intervals for these parameters. Because space does not permit further elaboration, the interested reader is referred to several introductory statistics books for the relevant formulas and their derivations (Daniel, 1995; Dixon and Massey, 1983; Dunn, 1977; Remington and Schork, 1985).

E. Comparing the Mean Response of Two Populations

1. *Independent Samples*

The presentation thus far has focused on estimation of parameters from a single population. Frequently, interest lies in two populations. For example, in a clinical trial, one group of animals might receive some treatment (t) while a second group of animals serves as a control receiving no treatment (c). Among the several points of interest could be that of estimating the difference in central response for the two populations, that is, $\mu_t - \mu_c$, where the subscripts designate the groups. The point estimate of $\mu_t - \mu_c$ is $\bar{x}_t - \bar{x}_c$. If we assume that the variances of the two populations, that is, σ_t^2 and σ_c^2, are unknown but equal and the common variance is designated as σ^2, then the estimate of σ^2 is $s_p^2 = [(n_{t-1})s_t^2 + (n_c - 1)s_c^2]/(n_t + n_c - 2)$ (called the *pooled variance*) and $[(\bar{x}_t - \bar{x}_c) - (\mu_t - \mu_c)]/s_p(1/n_t + 1/n_c)^{\frac{1}{2}}$ has a t distribution with $n_t + n_c - 2$ degrees of freedom. A 100 $(1 - \alpha)$% confidence interval for $\mu_t - \mu_c$ is $(\bar{x}_t - \bar{x}_c) \pm t_{1-\alpha/2;n_t+n_c-2}s_p(1/n_t + 1/n_c)^{\frac{1}{2}}$. If the interval so constructed covers zero, it may be that there is no difference between the central responses for the two distributions; otherwise, it could be concluded that the central responses differ significantly.

2. *Nonindependent Samples*

The procedure just discussed, in addition to assuming that the variances are homogeneous, also assumes that the two samples are drawn independently. An alternative design for comparing two responses involves using each subject as its own control. For example, a pretreatment response in an individual might be compared with a posttreatment response. Clearly in this design, the pretreatment and posttreatment responses may not be and most likely are not independent so the procedure given above for comparing two groups would not be appropriate. Rather, the differences in response (pretreatment minus posttreatment response) are formed and the population of interest is the single population of differences, having as one of its parameters the mean difference, μ_d. The quantity μ_d is estimated by the mean difference for the sample of n differences, $\bar{d} = \Sigma d_i/n$, and an interval estimate is formed by $\bar{d} \pm (t_{1-\alpha/2;n-1}s_d)/\sqrt{n}$ where s_d is an estimate of σ_d. If the interval thus constructed covers zero, then it may be that there is no difference between the mean pretreatment and posttreatment values.

F. Comparing the Mean Response of Three Populations

Comparisons of more than two groups is the natural progression from the methodology discussed to this point. Consider the comparison of three independent groups. The approach that immediately comes to mind is that of estimating the three groups' means and standard deviations and then constructing three sets of confidence intervals (the first group versus the second group, the first group versus the third group, and the second group versus the third group) using the approach described earlier for two independent groups. However, some modifications need to be made. First, because we are assuming that all three groups have equal variances, pooling of the variances for the three groups provides a better estimate of the common variance than does pooling of just the variances for the two groups being compared. The form of the pooled variance is the natural extension of that for two groups, namely: $s_p^2 = [(n_1 - 1)s_1^2 + (n_2 - 1)s_2^2 + (n_3 - 1)s_3^2]/(n_1 + n_2 + n_3 - 3)$. When the group simple sizes are equal, that is, $n_1 = n_2 = n_3$, $s_p^2 = \Sigma s_i^2/3$. The quantity s_p^2 is used for all three interval estimates.

Second, the error rate of each comparison has to be adjusted so that the error rate over all three comparisons will be α. This is required because theoretically it turns out that the error rate over all three comparisons is larger than that for a single comparison. Several approaches are suggested in the literature for circumventing this problem. One such approach attributed to Fisher (1966) is called the *least significance difference* (LSD) procedure (Kleinbaum and Kupper, 1988; Steel and Torrie, 1980). The LSD procedure involves making each single comparison with an error rate of α/m where m is the total number of comparisons to be made, which in the present example is 3. This approach gives an error rate over all comparisons of α. In general, the form of the interval estimate is $(\bar{x}_i - \bar{x}_j) \pm [(t_{1-\alpha/2m;n_1+n_2+...+n_k-k})(S_p)(1/n_i + 1/n_j)^{\frac{1}{2}}]$ where m is the total number of comparisons to be made and k is the total number of groups. Intervals covering zero would indicate no difference in the central value of the groups being compared.

G. Analysis of Variance and Its Uses

1. Comparing Population Means

The process of deciding whether or not there are differences between groups in the central value of the response being evaluated can be approached using the method of analysis of variance (ANOVA). ANOVA involves decomposing the total variability in a given set of data into parts reflective of the amount of variability attributable to various sources. One source of variability is that within the group. Because the groups are assumed to have the same spread, this source of variability is estimated as the pooling of the estimated variances for the k groups considered and is equal to s_p^2 as defined earlier. The second source of variability results from the variability among groups means. If there is no difference in the groups means, then the k samples can be thought of as being k independent samples from a common population, and the k means, therefore, represent a sample of size k from the sampling distribution of $\bar{x}$. Their variance represents an estimate of the variance of the sampling distribution of $\bar{x}$, that is, $\hat{\sigma}_{\bar{x}}^2 = \Sigma[\bar{x}_i - \Sigma(\bar{x}_i/k)]^2/(k - 1)$, where ˆ denotes an estimate. Because $\hat{\sigma}_{\bar{x}}^2 = \hat{\sigma}^2/n$, where n is the common sample size, inflation of $\hat{\sigma}_{\bar{x}}^2$ by n will produce a second estimate of the variance of the underlying population assuming no differences among the group means. These two estimates of σ^2 should be about equal and their ratio close to 1.

If there is significant separation among some or all of the group means, the second variance estimate, when computed as described, will be larger than the first estimate, indicating that a component of variance is being estimated beyond that embodied only in an estimate of the within-group variability. This additional variance component being estimated is the variance among group means. ANOVA in this example involves generating the two estimates of σ^2 (called *mean squares*) under the hypothesis that all of the group means are equal. The hypothesis is then tested by forming the ratio of the two mean squares (the second mean square divided by the first mean square) called the *F statistic*. The F statistic has a probability distribution called the F distribution. If the computed F value is greater than the table distributional value, then the hypothesis is rejected and subsequently confidence intervals are constructed as described earlier to determine which group means are different. If the hypothesis cannot be rejected, the process stops or perhaps, at most, the data from all the groups might be pooled together and the parameters of this single population estimated. Tables of the percentiles of the F distributions can be found in all introductory texts on statistics (Daniel, 1995; Dixon and Massey, 1983; Dunn, 1977; Remington and Schork, 1985), which also provide instruction on how to read the tables. Also, these texts give the generalization of the among-group mean square when the sample size is not constant for all groups.

The results of an ANOVA are traditionally summarized in a table called the ANOVA table of which Table 1.4 is such an example. The first column of Table 1.4 shows the sources of variability into which the total variability is decomposed. In the present example, these sources are due to the variability within groups and that which is reflective of variability among group means. Column 4 gives the two independent (under a hypothesis of no difference in response among groups) estimates of σ^2 or mean squares, mean square among group means (MS_A) and mean square within groups (MS_W). Columns 3 and 2 give, respectively, the numerator (called the sum of squares) and the denominator (the effective sample size or degrees of freedom) of the corresponding mean square. The sums of squares provide a check on calculations when generating an ANOVA table because the total of the sum of squares attributable to the various sources of variability is equal to the sum of squared deviations of each observation across the k samples from the grand mean of all the observations in the k samples, called the *total sum of squares*. In other words, the total sum of squares is the numerator for estimating the total variability in the data ignoring group membership. Similarly, the degrees of freedom for the sources of variability sum to the effective sample size for estimating the total variability in the data ignoring group membership of the observations [here, $kn - 1 = (k - 1) + k(n - 1)$].

2. Identifying Significant Sources of Variability in Experimental Designs

a. Experimental Design 1: Two Factors Crossed

The ANOVA does not represent a superior method to that of the LSD procedure in the context in which it was presented. In fact, if a significant group effect was noted by the ANOVA, the LSD procedure or any other multiple comparison procedure would be used to identify which group differences were contributing to the overall group effect. The reason analysis of variance has been introduced is that it is a very convenient method for assessing the importance of the various sources of variability encountered in the complex designs of clinical laboratories.

TABLE 1.4 Analysis of Variance Table for Classification of Responses on Basis of One Factor, Equal Responses for Each Class[a]

Source of variation	Degrees of freedom	Sum of squares	Mean square	*F* value
Among group means	$k - 1$	$SS_A = (k - 1)MS_A$	$MS_A = \frac{n \sum_{i=1}^{k} [\bar{x}_i - \sum_{i=1}^{k} (\bar{x}_i / k)]^2}{k - 1}$	MS_A / MS_W
Within group	$k(n - 1)$	$SS_W = k(n - 1)MS_W$	$MS_W = \frac{\sum_{i=1}^{k} s_i^2}{k}$	
Total	$kn - 1$	$SS_T = SS_A + SS_W$		

[a] k is the number of classes or groups, n is the number of responses per group, constant over all groups, s_i^2 is the variability of the responses within the ith group.

As an example of this kind of application of ANOVA, suppose interest centers around an evaluation of the plasma cortisol secretory patterns in dogs with pituitary-dependent hyperadrenocorticism. A part of this type of assessment traditionally involves a determination of the magnitude of intra-animal and interanimal variances. One experiment that would provide estimates of these variances is as follows: Twenty dogs are randomly selected to participate in the experiment. On three randomly selected days of the study period, plasma cortisol samples are obtained at 11:00 A.M. from all 20 dogs. The assay for plasma cortisol is replicated for each sample. This experimental design involves two factors, "animals" and "days." These factors are said to be crossed in that each level of factor "animals" (i.e., each animal) has been evaluated at each level of factor "days" (i.e., at each of the three days on which plasma cortisol samples were drawn). The total variability in this experiment can be decomposed by the method of ANOVA into four parts: that which will identify significant among-animals variability, significant among-days variability, and significant variability attributable to a lack of parallelism among the animals in their response across the three days samples were drawn (called the interaction between the factors "animals" and "days"), and that which will quantify the residual variability embodying all other sources of variability related to the assay.

Table 1.5 lists data that shall be assumed to have been generated from an experiment using the design given in the preceding paragraph. Table 1.6 gives the ANOVA table from these data. In addition to providing the standard entries of an ANOVA table (sum of squares, degrees of freedom, mean square, and *F* statistics with corresponding *p* values), Table 1.6 also gives the expected mean squares, which are the quantities being estimated by the corresponding mean squares in the table. The expected mean squares are functions of the unknown population parameters. Note that each expected mean square involves one or more of the variance components of the design. The expected mean squares indicate which mean squares are appropriate as the denominator for computing the *F* statistic to test the hypothesis for a given effect. For example, to test the hypothesis that the variance component giving the magnitude of the variability attributable to the interaction between the factors "animals" and "days" is not

TABLE 1.5 Plasma Cortisol Levels (mg/dl) in Dogs with Pituitary-Dependent Hyperadrenocorticism[a]

Dog	Day 1	Day 2	Day 3
1	0.8, 0.8[b]	1.5, 1.4	4.0, 4.2
2	2.2, 2.5	4.7, 4.5	3.4, 3.5
3	2.3, 2.4	2.7, 2.7	2.8, 2.7
4	3.1, 3.2	2.0, 2.1	2.7, 2.8
5	4.2, 4.3	3.5, 3.5	2.3, 2.4
6	3.8, 3.7	4.0, 4.0	2.7, 2.9
7	2.6, 2.5	3.9, 4.1	2.7, 2.7
8	2.4, 2.3	4.1, 4.1	2.9, 2.9
9	2.9, 2.9	3.3, 3.5	2.7, 2.8
10	2.2, 2.3	2.0, 2.2	1.6, 1.6
11	3.1, 2.9	3.0, 2.8	2.3, 2.1
12	5.3, 5.2	4.4, 4.2	3.9, 4.0
13	3.2, 3.3	2.7, 2.5	2.7, 2.9
14	4.4, 4.3	4.8, 4.8	3.8, 3.6
15	4.2, 4.1	4.3, 4.2	3.6, 3.7
16	2.9, 3.1	3.7, 3.9	4.6, 4.8
17	3.8, 3.7	3.6, 3.4	3.5, 3.3
18	5.0, 5.2	5.1, 5.1	4.8, 4.7
19	3.7, 3.6	1.4, 1.4	2.2, 2.1
20	2.9, 3.0	2.7, 2.9	3.1, 3.1

[a] These data were adapted from data provided by Dr. E. C. Feldman, Department of Medicine and Epidemiology, School of Veterinary Medicine, University of California, Davis.

[b] Assay replicated for each sample.

TABLE 1.6 Analysis of Variance of Canine Plasma Cortisol Level Data Assuming Experimental Design 1 ("Days" Crossed with "Animals")

Source of variation	df	Sum of squares	Mean square	F value	Expected mean square
Among animals	19	70.8316	3.7280	3.32	$\sigma_e^2 + nn_d\sigma_a^2 + n\sigma_{ad}^2$
Among days	2	1.1546	0.5773	0.51	$\sigma_e^2 + nn_a\sigma_d^2 + n\sigma_{ad}^2$
Animals × days	38	42.6787	1.1231	117	$\sigma_e^2 + n\sigma_{ad}^2$
Error	60	0.5750	0.0096		σ_e^2
Total	119	115.2399			

significant, that is, to test $H_0: \sigma_{ad}^2 = 0$, the test statistic is $F_{ad} = MS_{ad}/MS_e$, where MS_e is the correct denominator for this test statistic because if the hypothesis is true and $\sigma_{ad}^2 = 0$, the expected mean square for the interaction between "animals" and "days" reduces to σ_e^2 so that both MS_{ad} and MS_e would be estimating σ_e^2 and it would be expected that the ratio of MS_{ad} to MS_e (F_{ad}) would not differ significantly from 1. This ratio differing significantly from 1 would be an indication that MS_{ad} was estimating something additional to σ_e^2, namely, $n\sigma_{ad}^2$ where n is the number of replications per sample. Likewise, it can be determined by considering the expected mean squares of Table 1.6 that the tests for no significant variability among animals or among days ($H_0: \sigma_a^2 = 0$ or $H_0: \sigma_d^2 = 0$) are made using MS_{ad} in the denominator of the F statistic.

The results of Table 1.6 show that the hypotheses regarding σ_a^2 and σ_{ad}^2 were significant but that regarding σ_d^2 was not significant. This means that σ_a^2 and σ_{ad}^2 are not equal to zero, indicating significant sources of variability among animals that are attributable to the interaction of animals and days in the plasma cortisol levels recorded in the experiment. The quantity σ_d^2, on the other hand, may be equal to zero.

Figure 1.7, a graphical presentation of the results obtained from the analysis of variance, gives each animal's responses (mean of the two determinations) for the three days during which responses were recorded. We can see that for some dogs the response decreased over time, for some the response increased over time, for some the response at day 2 was depressed compared to that at day 1 but then at day 3 was elevated over that observed at day 2, and for others the response at day 2 was elevated compared to that at day 1 but then at day 3 was depressed compared to that observed at day 2. Differing degrees of elevations and depressions were observed with these basic patterns. Figure 1.7 shows that there was a large degree of variability among dogs in the plasma cortisol values on any given day, and that the plasma cortisol profiles across the three days that samples were taken were not parallel across the dogs. This lack of parallelism was identified by the significant interaction in the ANOVA; significance indicated that the observed departures from parallelism were larger than would have been expected as being a result of chance. The mean plasma cortisol levels recorded on the three days were 3.25 *μg/dl* (89.7 *nmol/liter*), 3.36 *μg/dl* (9.27 *nmol/liter*), and 3.12 *μg/dl* (86.1 *nmol/liter*) for a range of 0.24 *μg/dl* (6.6 *nmol/liter*). This range is very small relative to the range of mean response of 3.00 *μg/dl* (82.8 *nmol/liter*) observed for the 20 dogs ($\bar{x}$ max = 4.98 *μg/dl* or 137.4 *nmol/liter* and $\bar{x}$ min = 1.98 *μg/dl* or 54.6 *nmol/liter*). Thus, although the responses of some dogs increased while those of other dogs decreased from one day to the next day, there was practically no change in the overall variability from day to day. This is reflected in the ANOVA by a nonsignificant variance component attributable to the factor "days." The large variability among animals in mean plasma cortisol is identified in the ANOVA by the rejection of the hypothesis that $\sigma_a^2 = 0$.

b. Estimating Variance Components

Once significance has been established for one or more of the variance components in an experimental design, interest focuses on estimating the variance component(s). Estimates are readily obtainable using

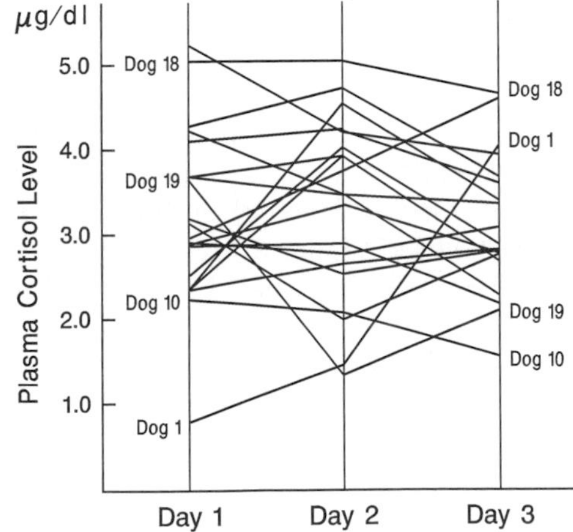

FIGURE 1.7 Interaction between "animal" and "day" factors in the canine plasma cortisol level (μg/dl) based on data from 20 dogs in Table 1.5.

the appropriate expected mean squares in conjunction with the mean squares obtained from the data. For example, Table 1.6 shows that an estimate of σ_e^2 is MS_e (0.0096). An estimate of σ_{ad}^2 can be obtained by noting that MS_{ad} estimates $n\sigma_{ad}^2 + \sigma_e^2$; solving for σ_{ad}^2 followed by substitution of MS_e for σ_e^2 yields $(MS_{ad} - MS_e)/n_e$ as the estimate. Based on the data used in the example, $\hat{\sigma}_{ad}^2 = (1.1231 - 0.0096)/2 = 0.5567$. Estimates of σ_a^2 and σ_d^2, obtained in a similar manner, are $(MS_a - MS_{ad})/n_d n$ (0.4342) and $(MS_d - MS_{ad})/n_a n$ (−0.0136 or zero), respectively, where n_a is the number of animals sampled (here 20) and n_d is the number of days samples were taken (3). Interval estimates for these variance components can be obtained using these point estimates by methods described elsewhere (Dunn and Clark, 1987; Harter and Lum, 1955; Satterthwaite, 1941).

c. Estimating the Variance of the Grand Mean Response

Another way to visualize the importance of these variance components is to analyze their impact on the estimate of the variance of interest. In some applications, there would be interest in estimating the grand mean (μ) of the response. In the present example, this would involve estimating the mean plasma cortisol level taking into account any random animal effect, day effect, and the interaction between the animal and day effects. If it is assumed that the main effects, "days" and "animals," are independent of their interaction in the response model and other distributional assumptions hold (Dunn and Clark, 1987), then the variance of $\hat{\mu}$, Var($\hat{\mu}$), is given as $\sigma_a^2/n_a + \sigma_d^2/n_d + \sigma_{ad}^2/n_a n_d + \sigma_e^2/n_a n_d n$ (Dunn and Clark, 1987). In the present example, Var($\hat{\mu}$) is estimated as $0.4342/20 + 0.0/3 + 0.5567/60 + 0.0096/120 = 0.0217 + 0.0000 + 0.0093 + 0.0001 = 0.0311$, and by far the greatest contribution to this variance is that due to the variability among animals in their response followed by the contribution of interaction to the response.

d. Estimating the Total Variability of a Single Response

In other applications, interest centers about the total variability (σ_{total}^2) associated with a single response. A single response is a linear combination of the terms in the response model and, using the assumption of the independence of terms in the model, has a variance equal simply to the sum of the variance components. Specifically, $\sigma_{total}^2 = \sigma_a^2 + \sigma_d^2 + \sigma_{ad}^2 + \sigma_e^2$, which in this example is estimated as $0.4342 + 0.0000 + 0.5567 + 0.0096 = 1.0005$. In this example, the total variability of a single response is determined by two variance components: (1) that associated with the animal variability and (2) that resident in the interaction between "animals" and "days."

e. Experimental Design 2: One Factor Nested within a Second Factor

A second experimental design that would provide estimates of intra-animal and interanimal variances in the plasma cortisol level is as follows: The 20 dogs are randomly selected over a period of time from the population of dogs identified by a clinic as having pituitary-dependent hyperadrenocorticism. At the time the dog is selected, three days within the next two weeks are randomly selected for plasma cortisol samples to be taken at 11:00 A.M. from the dog. Again plasma cortisol assays are replicated for each sample. Because the animals are sampled across time, it is rather unlikely that many and perhaps any of the three days that samples are drawn will be common for the 20 dogs. If the 20 dogs had no sampling days in common, samples would be taken on 60 (20 × 3) different days. Thus, the factor "days" in this design is not crossed with the factor "animals" but is said to be nested within "animals." The total variability in this experiment can be decomposed by the method of ANOVA into three parts, that which will identify significant variability among animals and significant variability among days (nested within animals), and that which quantifies the residual variability.

The data in Table 1.5 were used again assuming that it had been generated from the nested design described. Table 1.7, the ANOVA table, shows that two hypotheses can be tested. The mean square error is the denominator for testing that the variance component for "days nested within animals" (*d*/*a*) is equal to zero, that is, $\sigma_{d/a}^2 = 0$. This hypothesis is rejected based on the data. In this design, $\sigma_{d/a}^2$ is the true intradog variability for this response. The test result that the day effect is significant means that this intradog variability is larger than zero. The second test is that the variance component for "animals" is equal to zero or $\sigma_a^2 = 0$. The expected mean squares given in Table 1.7 show that the denominator for this test is the mean square for days nested within animals, $MS_{d/a}$. This test is also significant. The interpretation of a significant σ_a^2 is the same as that noted for design 1.

Once significance has been established, estimates of these variance components can be computed using the method described earlier for design 1. These estimates are $\hat{\sigma}_a^2 = 0.4387$, $\hat{\sigma}_{d/a}^2 = 0.5431$ and $\hat{\sigma}_e^2 = 0.0096$. The term $\hat{\sigma}_a^2$ is the estimate of the interanimal variability, whereas $\hat{\sigma}_{d/a}^2$ estimates the intra-animal variability in plasma cortisol level for the underlying population of dogs. For those interested in estimating the grand mean plasma cortisol level, the importance of each of these components is obtained by observing their impact on the estimate of the variance of the estimate of the grand

TABLE 1.7 Analysis of Variance of Canine Plasma Cortisol Level Data Assuming Experimental Design 2 ("Days" Nested within "Animals")

Source of variation	df	Sum of squares	Mean square	F value	Expected mean square
Among animals	19	70.8316	3.7280	389	$\sigma_e^2 + nn_d\sigma_a^2 + n\sigma_{d/a}^2$
Days/animals	40	43.8333	1.0958	114	$\sigma_e^2 + n\sigma_{d/a}^2$
Error	60	0.5750	0.0096		σ_e^2
Total	119	115.2399			

mean, Var($\hat{\mu}$). The expression for this variance is $\text{Var}(\hat{\mu}) = \sigma_a^2/n_a + \sigma_{d/a}^2/n_a n_d + \sigma_e^2/n_a n_d n$ (Little *et al.*, 1991; Neter *et al.*, 1996), which in the present is estimated as $0.0219 + 0.0091 + 0.0001 = 0.0311$. Again by far the greatest contribution to this variance is that due to the variability among animals in their response. The intradog variance component, although slightly larger than the interdog variance component, makes a considerably smaller impact on $\widehat{\text{Var}}(\hat{\mu})$. For those interested in estimating the total variability of a single response, $\sigma_{\text{total}}^2 = \sigma_a^2 + \sigma_{d/a}^2 + \sigma_e^2$ (Kringle, 1994), which in the present example is estimated as $0.4387 + 0.5431 + 0.0096 = 0.9914$. Here the total variability of a single response is divided nearly equally between "animals" and "days nested within animals." Other possible designs could be considered. What has been demonstrated is that the method of analysis of variance in conjunction with experimental design can be useful in answering a variety of questions. Design 1 allowed an assessment of how parallel the plasma cortisol profiles of the dog population are across time. It also allowed a determination of whether there are major shifts in the daily overall variability in the plasma cortisol level in light of changes in individual animal responses across time. This information is not provided by design 2; rather design 2 provides an estimate of the true intradog variability in the plasma cortisol level. Both designs provide an estimate of the interdog variability in the response.

Design 2 is frequently used to assess sources of variability in an assay. For example, several laboratories could be involved in doing a particular assay, with several autoanalyzers in each laboratory and multiple technicians running these autoanalyzers. Inference in this context centers around being able to identify if there are significant sources of variation among the laboratories, among autoanalyzers within a given laboratory, and among technicians operating a given autoanalyzer. The goal of analyses of this sort is to identify large sources of variability. Once the larger sources of variability have been identified, changes are made in the system in an effort to reduce the variability associated with each source. The long-term objective is to have an assay with sources of variability that are as small as possible. Clinical analysts conventionally divide the square root of the estimates of the variance components (the sample standard deviations) resulting from such assay experiments by the grand mean to obtain coefficients of variation for each source of variability (Kringle, 1994). These coefficients of variability should be much smaller than those derived as intra-animal and interanimal variability. Interested readers are strongly encouraged to consult texts written on experimental design and ANOVA (Dunn and Clark, 1987; Neter *et al.*, 1996).

References

Björkhem, I., Bergman, A., and Falk, O. (1981). *Clin. Chem.* **27,** 733–735.

Cochran, W. C. (1977). "Sampling Techniques," 3rd ed. Wiley, New York.

Daniel, W. W. (1995). "Biostatistics: A Foundation for Analysis in the Health Sciences," 6th ed. Wiley, New York.

Dixon, W. J., and Massey, F. J. (1983). "Introduction to Statistical Analysis," 4th ed. McGraw-Hill, New York.

Dunn, O. J. (1977). "Basic Statistics: A Primer for the Biomedical Sciences," 2nd ed. Wiley, New York.

Dunn, O. J., and Clark, V. A. (1987). "Applied Statistics: Analysis of Variance and Regression," 2nd ed. Wiley, New York.

Elevitch, F. R. (1977). *Proc. 1976 Conf. Analytical Goals in Clinical Chemistry,* College of American Pathologists, Skokie, Illinois.

Engleman, L. (1992). *In* "BMDP Statistical Software Manual" (W. J. Dixon, ed.), Vol. 1, pp. 141–153. University of California Press, Berkeley.

Feinstein, A. R. (1977). "Clinical Biostatistics." C.V. Mosby, St. Louis, Missouri.

Fisher, R. A. (1966). "The Design of Experiments," 8th ed. Oliver and Boyd, Edinburgh.

Gilbert, R. K. (1975). *Am. J. Clin. Pathol.* **63,** 960–973.

Gilbert, R. K. (1978). *Am. J. Clin. Pathol.* **70,** 450–470.

Harris, E. K. (1988). *Arch. Pathol. Lab. Med.* **112,** 416–420.

Harter, H. L., and Lum, M. D. (1955). "Partially Hierarchical Models in the Analysis of Variance." Wright Air Development Center Technical Report No. 55–33, Wright-Patterson Air Force Base, Ohio.

Herrera, L. (1958). *J. Lab. Clin. Med.* **52,** 34–42.

Jessen, R. J. (1978). "Statistical Survey Techniques." Wiley, New York.

Kaneko, J. J. (1977). *J. Small Anim. Pract.* **18,** 85–94.

Kleinbaum, D. G., Kupper, L. L., and Muller, K. E. (1988). "Applied Regression Analysis and Other Multivariable Methods," 2nd ed. PWS-Kent, Boston.

Kringle, R. O. (1994). *In* "Tietz Textbook of Clinical Chemistry" (C. A. Burtis and E. R. Ashwood, eds.), 2nd ed., pp. 384–453. Saunders, Philadelphia, Pennsylvania.

Levy, P. S., and Lemeshow, S. (1991). "Sampling of Populations: Methods and Applications." Wiley, New York.

Little, R. C., Freund, R. J., and Spector, P. C. (1991). "SAS System for Linear Models," 3rd ed. SAS Institute, Cary, North Carolina.

Mainland, D. (1963). "Elementary Medical Statistics," 2nd ed. Saunders, Philadelphia, Pennsylvania.

Massod, M. F. (1977). *Am. J. Med. Technol.* **43,** 243–252.

Murthy, M. N. (1967). "Sampling Theory and Methods." Statistical Publication Society, Calcutta.

Neter, J., Wasserman, W., and Kutner, M. H. (1996). "Applied Linear Statistical Models," 4th ed. Richard D. Irwin, Chicago, Illinois.

Raj, D. (1968). "Sampling Theory." McGraw-Hill, New York.

Raj, D. (1972). "The Design of Sampling Surveys." McGraw-Hill, New York.

Reed, A. H., Henry, R. J., and Mason, W. B. (1971). *Clin. Chem.* **17**(4), 275–284.

Remington, R. D. and Schork, M. A. (1985). "Statistics and Applications to the Biological and Health Sciences," 2nd ed. Prentice Hall, Englewood Cliffs, New Jersey.

Ryan, B. F., Joiner, B. L., and Ryan, T. A. (1985). "Minitab Handbook." PWS-Kent, Boston.

Sattherthwaite, F. E. (1941). *Psychometrika* **6,** 309–316.

Scheaffer, R. L., Mendenhall, W., and Ott, L. (1996). "Elementary Survey Sampling," 5th ed. Duxbury Press, Belmont, California.

Shultz, E. K. (1994). *In* "Tietz Textbook of Clinical Chemistry" (C. A. Burtis and E. R. Ashwood, eds.), 2nd ed., pp. 485–507. Saunders, Philadelphia, Pennsylvania.

Snedecor, G. W., and Cochran, W. G. (1989). "Statistical Methods," 8th ed. The Iowa State University Press, Ames.

Solberg, H. E. (1994). *In* "Tietz Textbook of Clinical Chemistry," (C. A. Burtis and E. R. Ashwood, eds.), 2nd ed., pp 454–484. Saunders, Philadelphia, Pennsylvania.

Steel, R. G. D., and Torrie, J. H. (1980). "Principles and Procedures of Statistics," 2nd ed. McGraw-Hill, New York.

Zar, J. H. (1996). "Biostatistical Analysis," 3rd ed. Prentice Hall, Upper Saddle River, New Jersey.

CHAPTER

2

DNA Technology in Diagnosis, Breeding, and Therapy

JENS G. HAUGE

I. INTRODUCTION

During the last 20 years an immense expansion has taken place in our understanding of the structure, organization, and regulation of genes. This expansion has to a large extent been the result of the development of new methods to manipulate, multiply, and study DNA fragments. The same new methods have a number of important applications in human medicine and, gradually, also in veterinary medicine. For good introductions, see Old and Primrose (1990) and Watson *et al.*, (1992). An agricultural perspective is given in the book edited by Hansel and Weir (1990). These surveys also discuss the large impact on microbiological diagnosis and vaccine production, as well as hormone production, aspects that are not covered here.

The development of gene technology rests on the foundations of molecular biology laid during the preceding 40 years, with the discovery of the double helical structure of DNA and the breaking of the genetic code as the major milestones. This chapter starts, therefore, with a brief review of the molecular biological basis of gene technology.

II. MOLECULAR GENETICS

A. Structure and Replication of Genes

The notion that heredity was linked to matter, to certain cellular structures, appeared for the first time in 1903 when the chromosomal theory of heredity was formulated. Chromosomes were recognized to have properties that could accommodate the hereditary units discovered by Mendel in his experiments. A bridge was built between cytology and genetics. But the nature of the molecules that carried the hereditary information was not thereby identified. Many felt that chromosomal protein was the most likely candidate.

In 1944, however, Avery and coworkers showed convincingly that genes in pneumococcal bacteria are molecules of deoxyribonucleic acid (DNA). It soon became apparent that pneumococci were no exception in this respect. In animals and plants, as well as in bacteria, hereditary information is carried by DNA.

A section of a DNA molecule has the appearance shown in Fig. 2.1a. DNA is built in the form of a chain, with phosphoric acid and the sugar deoxyribose alternating as the links. Phosphoric acid always connects the 5′-OH group in one deoxyribose to the 3′-OH in the next deoxyribose. The heterocyclic bases adenine, guanine, cytosine, or thymine are bound to the 1′ carbons in deoxyribose. The base–deoxyribose moiety is a nucleoside, and the repeating unit in the chain, the combination base–deoxyribonucleoside–phosphoric acid, a nucleotide.

It is impractical to write DNA structures with all the atoms, so simplified writings have come into use. If the bases from the top in Fig. 2.1a are adenine, thymine, and guanine, the most used abbreviation is simply ATG. The first letter in the name of the base is used to designate the corresponding nucleotide.

That such a chainlike DNA molecule could carry hereditary information, genes, was not difficult to conceive. Most genes determine the construction of proteins, which are also chainlike molecules, with amino acids making up the links in the chain. One needs only to postulate a one-to-one relationship between an amino acid and a group of DNA bases. The other fundamental property of the hereditary material, that it is accurately duplicated and distributed on both daughter cells when cells divide, was harder to visualize on the basis of the DNA structure, as given in Fig. 2.1a. Watson felt that the solution to the problem perhaps lay hidden in the three-dimensional structure of DNA, and together with Crick he set out to elucidate this structure. The result was the discovery in 1953 of the double-helix structure of DNA. The double helix is built as shown schematically in Fig. 2.1b. A, T, G, and C represent the bases, which point toward each other and are bound to each other through hydrogen bonds. The DNA strands of the double helix have opposite directions (a DNA chain has a 5′-end and a 3′-end, see Fig. 2.1a). The size of the bases and their hydrogen bonding properties make only two pairings possible, namely, A with T and G with C. One turn of the standard double helix has 10 base pairs (bp) and covers a length of 3.4×10^{-6} mm. The diameter of the double helix is 2×10^{-6} mm. It is so slim a structure

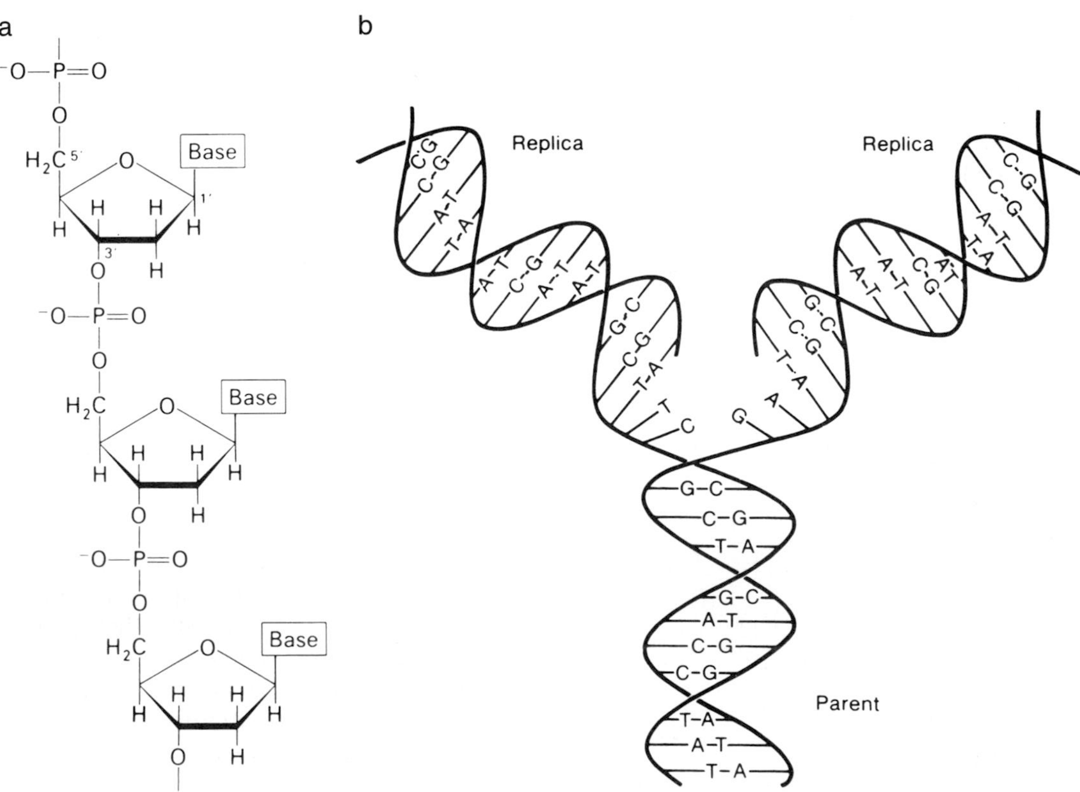

FIGURE 2.1 (a) Structure of part of a DNA chain. (b) DNA double helix during replication. Reproduced from Stryer (1981) and Freifelder (1983), with permission.

that if a DNA molecule could reach from the Earth to the moon, it still would weigh only 1 mg!

The model for the DNA molecule at which Watson and Crick (1953) arrived can in a simple way explain how DNA molecules are duplicated, that is, the process of DNA replication. When the double helix is considered (Fig. 2.1b), we can see that the two strands correspond to each other. The base pairing rules determine which base in one strand must be opposite a given base in the other strand, and the strands are said to be *complementary*. When replication starts, the hydrogen bonds between the two strands are gradually broken (Fig. 2.1b). New DNA chains are constructed from the four deoxynucleoside triphosphates with the help of the enzyme DNA polymerase, one molecule of pyrophosphate being eliminated per nucleotide laid down. Step by step, two new strands, complementary to the original strands, are synthesized. The original strands serve as templates in the process, and the result is two new double helixes that are identical to each other and to the original DNA molecule.

The basic mechanism for DNA replication is simple and elegant. The practical execution of this process in the cells, however, involves a large number of proteins, including some additional enzymes. The unwinding of the DNA molecules causes problems, as does the fact that DNA polymerase lays down a new strand from the 5′-end toward the 3′-end only. Furthermore, for this synthesis to start, a primer nucleic acid is needed for the strand to grow. One important auxiliary enzyme is DNA ligase, which links DNA molecules.

If DNA is heated to 80–90°C, the molecule will denature, and the strands will separate. If the temperature is lowered somewhat, the complementary strands will find each other again and reform the double helix. This property is widely exploited in gene technology (Section III.B.1).

B. From Gene to Gene Product

When the information stored in DNA is used, organisms utilize another type of nucleic acid, namely, ribonucleic acid, RNA. RNA differs from DNA in that it contains ribose instead of deoxyribose, and uracil instead of thymine. Uracil has the same base pairing properties as thymine. RNA does not normally exist in double-helix form, but RNA may form short stretches of double helix when the RNA strand folds back on itself locally. A DNA strand and a complementary RNA strand will form a double helix.

RNA is made by transcription from one of the DNA strands in a double helix (Fig. 2.2). The DNA strand is used as template in a process somewhat similar to DNA replication. The DNA double helix is opened locally, and an RNA strand complementary to the template is synthesized with the help of the enzyme RNA polymerase, ribonucleoside triphosphates being used as activated building blocks. Special sequences of bases in front of the genes (promoters) are recognized by the polymerase, causing it to start at the right place and on the right DNA strand. At the other end of the genes, there are termination signals.

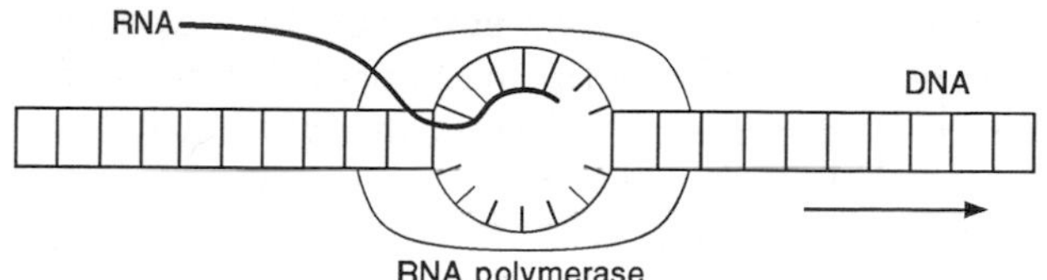

FIGURE 2.2 RNA synthesis (transcription).

In all cells there are three main types of RNA: messenger RNA, ribosomal RNA and transfer RNA. Messenger RNA (mRNA) carries the information specifying the amino acid sequences in the proteins. Ribosomal RNA (rRNA) is a group of RNA molecules that forms part of the structure of the protein-synthesizing machinery of the cells, the ribosomes. Transfer RNA (tRNA) participates in protein synthesis (translation) by carrying amino acids to the ribosomes.

The mRNA type is of special importance. For each kind of protein chain the cell makes, there is a corresponding mRNA. The genetic coding system that biochemists discovered between 1960 and 1965 showed that organisms use groups of three nucleotides, triplets, to code for amino acids. It is, therefore, the sequence of triplets in mRNA that directly determines the sequence of amino acids in the corresponding polypeptide chain. The coding group could not have consisted of only two nucleotides. That would have given only $4^2 = 16$ combinations, and 20 amino acids need coding. But $4^3 = 64$ gives more than enough. Nature has chosen this most economical solution, and done it in a way so that 61 of these 64 triplets code for some amino acid. Almost all the amino acids, therefore, have more than one codon.

When ribosomes directed by mRNA link together the amino acids, it is crucial for the result that the ribosomes start at the correct position on the mRNA. Starting one nucleotide too early or late will give a completely different protein. A correct start is secured by having all protein synthesis start with the triplet AUG, which codes for methionine. This means that all proteins initially are synthesized with methionine as their first amino acid. In posttranslational processing, some proteins lose this initial methionine, however. Because methionine also occurs internally in proteins, something more is needed for a correct start. In bacteria, the ribosomes recognize start AUG by the presence of a short stretch of purines a limited distance before the AUG. In animals and plants, the ribosomes choose the AUG that is closest to the 5′-end of the mRNA. The coding triplets UAA, UAG, and UGA do not code for any amino acid, and these codons are used as termination signals.

The amino acids do not by themselves have the ability to recognize their codons on mRNA. They must, furthermore, be activated in order to form peptide bonds with each other. Both of these problems are solved by each amino acid becoming bound to a tRNA by means of a specific activating enzyme, utilizing ATP. tRNA has in its structure an anticodon, a set of three nucleotides that forms base pairs with the codon on mRNA for the corresponding amino acid. On the ribosome surface the amino acids leave their tRNA carriers as they are linked together to form polypeptide chains. Briefly then, the information flow is as illustrated here for the start of a hypothetical gene:

```
3′ . . . TAC  AAA  AGT  GAC  TGT  CCG . . . 5′  DNA
5′ . . . ATG  TTT  TCA  CTG  ACA  GGC . . . 3′
                        ↓ transcription
5′ . . . AUG  UUU  UCA  CUG  ACA  GGC . . . 3′  mRNA
                        ↓ translation
         Met  Phe  Ser  Leu  Thr  Gly           protein
```

III. GENE MANIPULATION

A. Basic Features

The core of the new gene manipulation techniques is that one can join two pieces of DNA to form a recombinant DNA molecule, which then is multiplied, or cloned, in a suitable host organism so that a sufficient amount of the piece of DNA desired is obtained. The one piece of DNA is often used as a *plasmid,* a ring-shaped, extrachromosomal DNA molecule that occurs in many bacteria and carries genes, for example, for antibiotic resistance. Figure 2.3 illustrates molecular cloning of a piece of DNA with the help of a plasmid vector. The plasmid vector is opened with the help of a special endonuclease and joined to the foreign DNA with the help of another enzyme. Under proper conditions, some of the bacteria will take in the recombinant DNA molecule, which will be replicated as the bacteria grow and multiply. Most of the bacteria will not have taken in any plasmid. A selection process is therefore necessary, usually based on the plasmid carrying an antibiotic resistance gene. From the bacteria, the plasmid and the foreign DNA piece can be isolated in such

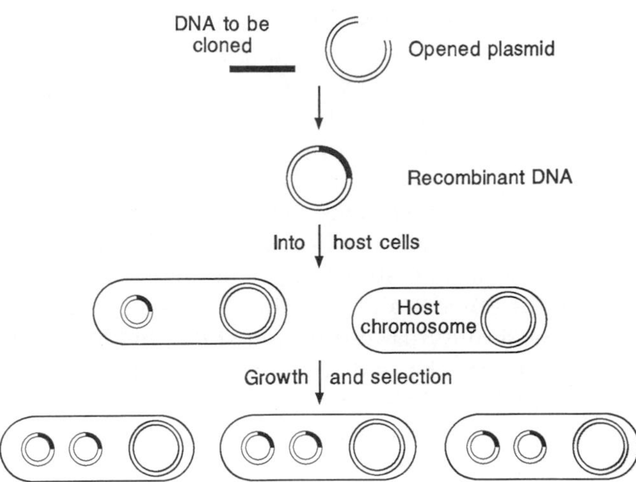

FIGURE 2.3 Basic steps in cloning of DNA in a plasmid.

amounts that it is possible to study the DNA in the electron microscope, determine its base sequence, inject it in fertilized ova, or use it as a diagnostic probe, to mention some important uses.

As is often the case, when a homogeneous DNA preparation is not available as starting material for cloning, but rather a mixture of many types of DNA pieces, special selection techniques are required afterward to find the interesting clone or clones. Plasmids are but one type of vector for DNA cloning. Viruses from bacteria or animals are also used quite often.

B. Methods and Tools

1. Nucleic Acid Hybridization

Nucleic acid hybridization is fundamental both in basic and applied gene technology. This is the process whereby sequence-specific DNA : DNA or RNA : DNA duplexes are formed from single-stranded nucleic acids. For most uses, one of the components in the hybridization is radioactively labeled, or labeled by other means, so that the hybrid can be identified. The labeled molecule is referred to as a *probe* and is used to probe nucleic acid mixtures for the presence of complementary sequences (Fig. 2.4).

The factors affecting the stability of the hybrid and the rate of the hybridization reaction are fairly well known (Marmur and Doty, 1961; Nygaard and Hall, 1964; Young and Anderson, 1985). The stability depends on the temperature, the ionic strength of the medium (high ionic strength reduces the repulsion between the negatively charged phosphate groups and thus stabilizes the duplex), the presence of formamide (which breaks hydrogen bonds), the G + C content of

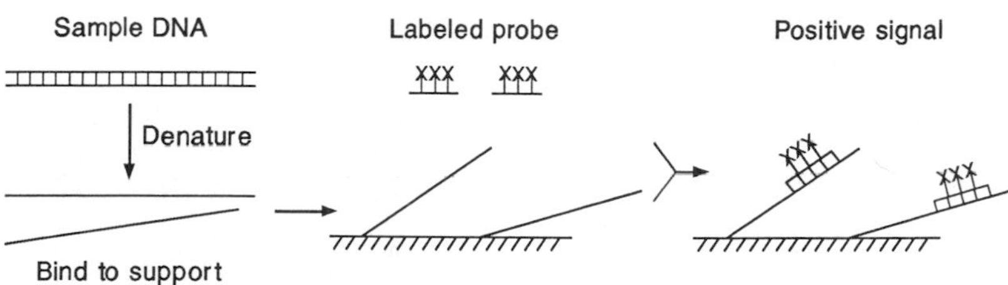

FIGURE 2.4 Hybridization.

the probe (G and C rich probes are bound more tightly because there are three hydrogen bonds linking G to C, compared to two bonds between A and T), and the length of the probe. The temperature at which the duplex is 50% denatured is called its T_m, which can be calculated from the following formula:

$$T_m = 81.5°C + 16.6 \log M + 0.41(\%G + \%C) - 500/n - 0.61\ (\%\ \text{formamide}).$$

This formula assumes perfect complementarity between the hybridizing strands. With mismatches present, the T_m is reduced by about 1°C per 1% mismatch.

The factors affecting stability also affect the rate of hybridization. The rate is optimal about 25°C below T_m. For a reaction in solution, the rate is proportional to the concentrations of the reacting species. When the DNA or RNA to be investigated is immobilized on a membrane, which is often the case, the reaction is slower by a factor of 7 to 10, but the general effects of the factors listed earlier are the same.

The reaction conditions are manipulated to give hybridizations at a desired level of stringency. High stringency, that is, high temperature and/or low ionic strength, will allow only the most stable hybrids, those having perfect complementarity, to form. In certain situations, one may want to probe for a family of related DNA molecules with some sequence differences, and hybridization at a lower stringency will be chosen. The same may be necessary when using a probe from one species to look for a homologous gene in another species.

Several methods are in use for labeling a probe with radioactivity (Arrand, 1985). In our experience, the random primer method (Feinberg and Vogelstein, 1983) is a convenient way of obtaining high specific activity, better than the so-called nick translation method. The probe DNA is denatured, and both strands are used as template for DNA synthesis, using short random oligonucleotides as primers and one or more of the triphosphates being labeled with ^{32}P.

Nonradioactive labeling is also used because it has certain advantages for routine, diagnostic purposes. Attaching biotin to one of the bases (Leary *et al.*, 1983) has received particular attention. The biotinylated probe is localized with streptavidin and a biotin-coupled enzyme for which a substrate is used that gives a colored precipitate. Nonradioactive probes have the drawback of multistep detection procedures and a somewhat lower detection sensitivity than radioactive probes (Syvanen, 1986).

2. *Restriction Endonucleases*

Restriction endonucleases of type II, usually referred to as restriction enzymes, are a major tool in gene analysis and gene manipulation. These enzymes bind to particular sequences on DNA and cut both strands within this DNA site. The majority of the enzymes have sites of tetra-, penta-, hexa-, or hepta-nucleotides which have an axis of rotational symmetry (Fig. 2.5a). Enzymes corresponding to more than 150 different sites have been discovered so far. The names of the enzymes have been constructed from names of the bacteria and the particular substrain from which they have been isolated. The arrows in Fig. 2.5a indicate where the DNA strands are cut by these five enzymes. Treatment of a DNA molecule with a restriction enzyme yields a reproducible set of fragments. An enzyme with a recognition site of four nucleotide pairs will cut relatively often, statistically once every 4^4 (i.e., 256) nucleotide pairs, whereas an enzyme with a hexanucleotide recognition site will cut less frequently, once every 4^6 (i.e., 4096) nucleotide pairs. Figure 2.5b shows the pattern of fragments into which three different restriction enzymes cut the SV40 virus chromosome. After enzyme treatment, the fragments were separated by electrophoresis in an agarose gel and made visible by binding of ethidium bromide, which fluoresces in ultraviolet light. The size of the fragments can be found by running molecular weight markers in the same gel. Short fragments move farther, and long fragments move a shorter distance.

The fragments produced by the restriction enzymes will, for most enzymes, have single-stranded, complementary end sections, because the cuts are not made straight across (Fig. 2.5a), and the fragments are said to have sticky ends. If a plasmid is opened with the same restriction enzyme that has been used to produce

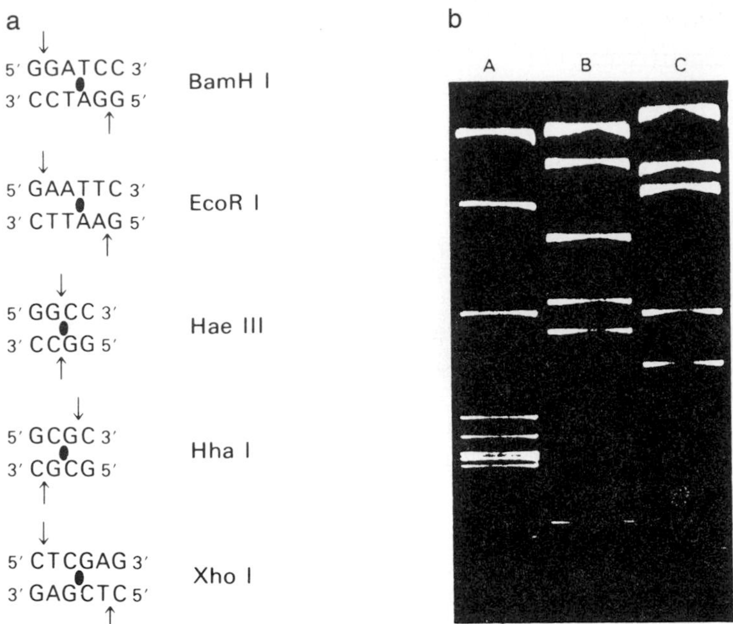

FIGURE 2.5 (a) Specificities of some restriction enzymes. (b) Gel electrophoresis pattern showing SV40 DNA fragments produced with each of three restriction enzymes. Reproduced from Stryer (1981), with permission.

the DNA fragments to be cloned, this DNA will have sticky ends of a type that allows its insertion in the plasmid. Fig. 2.6a shows what the situation would have been in Fig. 2.3 if the enzyme *Eco*RI had been used.

3. *DNA Ligase*

To form a stable, recombinant DNA molecule from two molecules, the strands of these two molecules must be joined. This is effected by the enzyme DNA ligase, which requires for its action a 5′-phosphate and a 3′-OH group. The enzyme from phage T4 uses ATP as energy donor, whereas the *Escherichia coli* enzyme uses NAD. The ligation reaction is carried out at a low temperature, 4–15°C, in order to increase the stickiness of the sticky ends.

4. *Terminal Deoxynucleotidyltransferase*

Terminal Deoxynucleotidyltransferase, or briefly, terminal transferase, is useful when the object is to insert in a plasmid a DNA molecule that has not been produced with the help of a restriction enzyme, a frequently occurring situation. With this enzyme, tails of identical nucleotides can be attached to the 3′-end of DNA strands. The plasmid is opened with a suitable

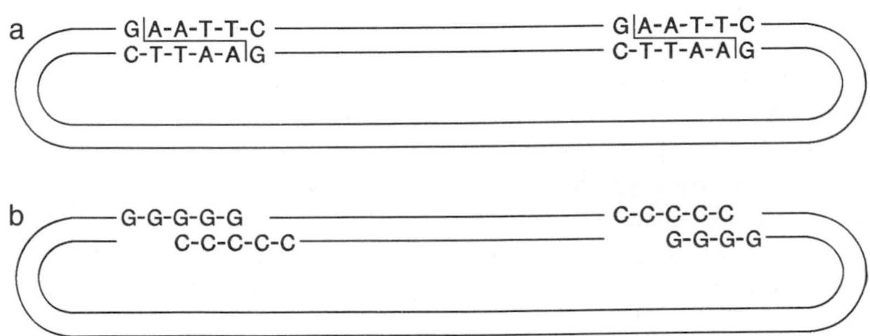

FIGURE 2.6 (a) Fitting a piece of DNA into an *Eco*RI site of a plasmid. (b) Fitting a piece of DNA into a plasmid after tailing the plasmid with G and the insert with C by using terminal transferase.

restriction enzyme and furnished with a tail of about 20 G residues, whereas the DNA to be inserted is furnished with a C tail (Fig. 2.6b).

5. *Adaptors*

Use of an adaptor DNA containing a restriction site is a common alternative strategy when a DNA molecule is to be inserted in a plasmid opened with the corresponding restriction enzyme. A piece of DNA, 6 to 10 base pairs long and containing the desired restriction site, is synthesized or purchased. The ends of the DNA to be inserted are first made blunt by filling in with DNA polymerase. All 5′-ends are phosphorylated with the enzyme, polynucleotide kinase, and the adaptor is attached to both ends of the insert DNA with DNA ligase. Treatment with the restriction enzyme then produces the proper sticky ends for insertion in the plasmid. This method illustrates well the role of both chemistry and enzymology in recombinant DNA technology.

C. Vectors

Through stepwise modifications of natural *E. coli* plasmids, new plasmids have been constructed that are safe and convenient for different types of gene manipulation. One much used plasmid is pBR322 (Fig. 2.7a). This plasmid carries two antibiotic resistance genes, one for tetracycline and one for ampicillin. Within these two genes, there are several restriction sites for enzymes that have only these sites in the pBR322 molecule. If one uses the *Bam*HI site to insert a foreign DNA, the tetracycline resistance will be eliminated. It is then possible to select for bacteria that have taken up the plasmid by using ampicillin in the medium, and then finding those that have an insert in the *Bam*HI seat by investigating which bacterial colonies have lost tetracycline resistance. This approach is useful since there can be some regeneration of the original plasmid in the ligation step.

Lambda virus DNA has also been modified so that it is suitable as a vector. Three of its five *Eco*RI sites have been removed. The DNA between the two remaining sites is not necessary for the infection process and can be exchanged with a foreign piece of DNA of 10–20 kilobase pairs (kbp). The recombinant DNA is packed *in vitro* using lambda coat and tail proteins. These phage particles enter *E. coli* bacteria very efficiently and reproduce as in a normal lytic infection.

For other bacteria, yeast, and animal and plant cells, other vectors exist. Shuttle vectors, which have specificity for two different hosts, have also been constructed.

D. Studies of Gene Structure

Comprehensive solutions to a wide range of biological problems require knowledge of the structure of the gene and regulatory sequences that govern the particular biological phenomenon.

1. *cDNA Cloning*

For studies of genes of viruses and bacteria, cloning can start with fragments of their chromosomal DNA. With genes from animals and plants, the situation is different. Here the genome is so large and contains so many noncoding sections that often one must start with mRNA molecules and preform what is called a copy DNA (cDNA) cloning. Total RNA is isolated with chemical methods from a tissue of interest, and mRNA adsorbed on a column of poly(T)-cellulose. mRNA is bound due to the 3′-tail of poly(A), which mRNA possesses in higher organisms. mRNA is eluted and used as the template for RNA-directed DNA polymerase, so-called "reverse transcriptase." This enzyme occurs in retroviruses, where its role during viral infection is to produce double-stranded DNA corresponding to the viral RNA genome. In principle, the same reaction can take place *in vitro,* and a mixture of cDNA is formed, corresponding to the mixture of mRNA isolated.

After attachment of sticky ends, the mixture of cDNA molecules can be inserted in a plasmid. If the bacteria to be transformed have been made competent by a special treatment, 10^5 to 10^7 transformed bacteria per microgram of recombinant plasmid can be expected. The clones from these bacteria make up a cDNA library for that particular tissue.

The library will have cDNA clones corresponding to different mRNA in the tissue. Clones for the particular gene of interest can be found by various means. The simplest situation exists if something is known about the amino acid sequence of the protein for which the gene codes. Then, a DNA probe of 17 to 19 bases, corresponding to a suitable hexapeptide, can be synthesized. This probe will be long enough that the search will be specific at high stringency. The probe preparation will have to be a mixture of DNA molecules, however, since most amino acids have more than one codon. An imprint of the bacterial colonies is made on a nitrocellulose filter, the cells are opened, and their DNA is denatured. If a radioactive probe is used, it will hybridize to a few colonies, those containing cDNA corresponding to the gene of interest. Plasmid DNA is purified, treated with the proper restriction enzyme, and the cDNA isolated from a hybridizing band on an agarose gel after electrophoresis.

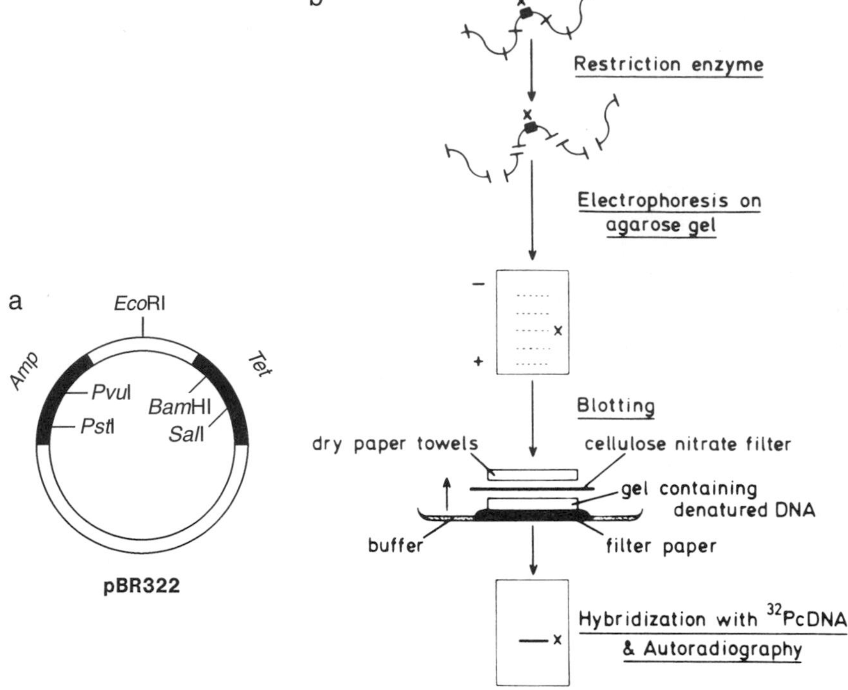

FIGURE 2.7 (a) The cloning vector pBR322 (4362 bp). *Amp* is the gene for ampicillin resistance, and *Tet* is the gene for tetracycline resistance. (b) The Southern blot technique. The symbol x indicates the gene or sequence of interest. Reproduced from Emery (1984), with permission.

2. Gene Cloning

A collection of clones, which includes all DNA in an organism, is called a *genomic library* and is usually made in a lambda vector. DNA from the organism is subjected to incomplete digestion with a frequently cutting restriction enzyme, so that fragments of about 20 kbp are formed. Vector arms are ligated to both sides of these fragments, and the resulting DNA molecules packed into lambda proteins, as described in Section III.C. The lambda clones that result from the infection will, if one has close to a million different clones, represent all the DNA and genes of a mammal. If one has a cDNA clone for a gene of interest, this cDNA can be radioactively labeled and used as a probe to find the lambda clones that contain the gene or parts of it.

3. Splitting of Eukaryotic Genes

One of the big surprises in molecular biology research was a discovery made in 1977 on cloned eukaryotic genes. The chicken gene for ovalbumin was found to be 7700 bp long, while only 2000 bp would be required to code for the protein. Closer investigation of the cloned DNA with restriction enzymes and DNA sequencing revealed that the coding part of the gene was divided into eight pieces, *exons,* separated by larger pieces of noncoding DNA, *introns*.

4. Restriction Maps, Southern Blots, and DNA Sequencing

Characterization of gene structure is done in broad outline by restriction mapping and in detail by establishing the DNA sequence. A restriction map identifies the positions of sites for a set of restriction enzymes. The correct order for the fragments produced by a given enzyme is determined in several ways, for example, through comparison of DNA fragment lengths corresponding to different degrees of partial digestion with the enzyme. The approximate distances between the cut sites are obtained by including molecular weight markers in the electrophoresis. Accurate distances require DNA sequencing.

If one has a cDNA probe for a gene, it is possible to obtain some information on the structure of the gene without having cloned it by using the technique developed by Southern (1979) (Fig. 2.7b). After digestion with a restriction enzyme and separation by agarose electrophoresis, the DNA fragments are denatured in alkali and transferred to a nitrocellulose or nylon membrane with a flow of buffer. A replica of the gel pattern of DNA is thereby produced on the membrane. When the membrane is then treated with the radioactive probe, it will hybridize to those bands that have DNA sequences complementary to the probe and become visible on X-ray film. In this way, information can be obtained about restriction sites in the gene region corresponding to the cDNA.

DNA sequencing is carried out by means of a chemical method developed by Maxam and Gilbert (1977) or by an enzymatic method developed by Sanger *et al.* (1979). The chemical method starts with a cloned DNA fragment that has a few hundred bases and the 5′-end radioactively labeled. From this cloned DNA, a spectrum of radioactive fragments is produced in which all sizes—from the full length down to zero—are represented. In principle, this procedure is done in four reactions, one reaction each for fragments ending with a G, C, T, or A. The fragments are separated by gel electrophoresis at a voltage of about 3000 V, and an autoradiograph is made from the gel (Fig. 2.8a). The sequence can be read directly from the autoradiograph, as indicated in the figure. The enzymatic method gives a similar end result. In this method, DNA polymerase synthesizes DNA in four reactions in the presence of small amounts of 2′,3′-dideoxynucleoside triphosphates, which cause chain termination when incorporated.

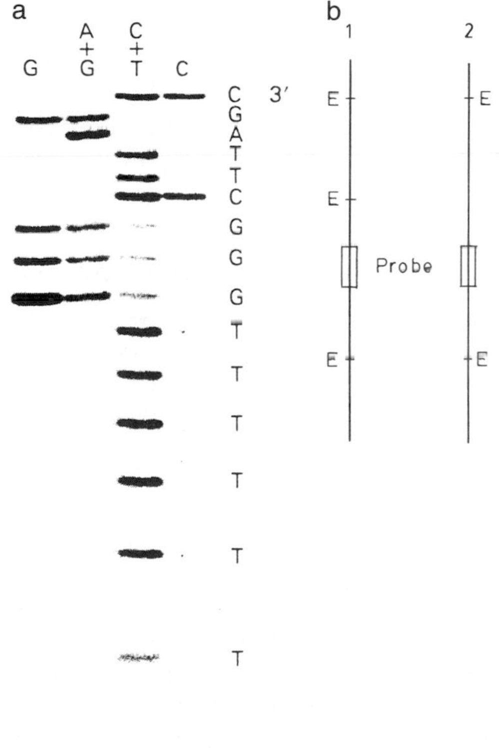

FIGURE 2.8 (a) Autoradiograph of a gel showing labeled fragments obtained in the Maxam and Gilbert DNA sequencing method. The reaction used for adenine affects both adenine and guanine, and the reaction used for cytosine affects both cytosine and thymine. Reproduced from Stryer (1981), with permission. (b) A restriction fragment length polymorphism caused by the presence or absence of a site for restriction enzyme E.

DNA sequencing work has yielded a wealth of information on gene structure and chromosome organization in viruses and bacterial and eukaryotic cells. The following are important examples. Sequencing of the 5375-bp ϕX174 virus showed how the DNA could code for proteins of 2000 amino acids in aggregate by having an overlap between some of the genes. The transition between exons and introns was shown to have certain characteristic bases, common to all genes. The promoter region in front of the genes, the site for attachment of RNA polymerase, was also found to have a common base pattern. DNA sequencing has help elucidate the structure and the function of immunoglobulin genes, histocompatibility genes, and oncogenes.

The sequencing technique has developed quickly. By 1982, the full sequence of lambda DNA with its 48,502 bp was published (Sanger *et al.*, 1982). Now under way is a gigantic international project that calls for sequencing of the total human genome with its 3×10^9 base pairs by the year 2005.

5. *Polymerase Chain Reaction*

An important use of sequence data is in enzymatic amplification of segments of DNA in the polymerase chain reaction (PCR). PCR amplification involves two oligonucleotide primers that flank the DNA segment to be amplified and repeated cycles of heat denaturation, annealing of the primers to the complementary sequences, and extension of the annealed primers with DNA polymerase. The primers hybridize to opposite strands of the target DNA and are oriented so that DNA is synthesized for the region between the primers. Since the extension products are also complementary and capable of binding the primers, each cycle doubles the amount of DNA resulting from the previous cycle.

The basic idea of this method was described in 1971 by Kleppe *et al.* and later developed by Saiki *et al.* (1985). Its full potential appeared when Saiki *et al.* (1988) introduced the thermostable DNA polymerase from *Thermus aquaticus* (*Taq*), thus eliminating the need to add fresh polymerase during each cycle. This modification not only simplified the procedure, making it amenable to automation, but also substantially increased its specificity, sensitivity, yield, and the length of targets that could be amplified. Amplification of 10 million times was demonstrated, and segments up to 2000 bp were readily amplified, although longer segments have reduced yield.

The PCR technique has found wide application (Erlich, 1989; Erlich *et al.,* 1991b). Isolation of cloned segments from their vectors is simplified by PCR amplification of these, using vector-specific primers that flank the insertion site. RNA and cDNA can be amplified from the products of reverse transcription (Todd *et al.,* 1987). DNA sequencing with the enzymatic method can be done directly on the PCR product without ligation into a sequencing vector. For this purpose, however, it is advantageous to purify the sequencing template by incorporating biotin in one of the primers. After immobilization of the PCR product on streptavidin-coated magnetic beads, the unbiotinylated DNA strand is removed by alkali (Hultman *et al.,* 1989). In diagnostic work and genetic mapping, PCR methods are used extensively as will be apparent in later sections. For some applications, it is important to be aware that the *Taq* enzyme allows incorporation of 1 incorrect nucleotide for about 20,000 nucleotides incorporated.

IV. STUDIES OF METABOLIC AND GENE REGULATION

Students of metabolic regulation now have new tools at their disposal through the advent of recombinant DNA technology. The impact has been the largest in the area of analysis of gene expression and its regulation by hormones due to the availability of cDNA probes for a broad variety of genes of metabolic interest. These probes make possible quantification of the level of specific messengers. A number of such studies, covering enzymes in glycolysis, gluconeogenesis, lipid metabolism, and amino acid metabolism, are described by Goodridge and Hanson (1986).

A. Tissue Culture Studies

Several of the corresponding genes have been isolated, and their promoter–regulator regions are being identified and characterized. One example is the rat gene for cytosolic phosphoenolpyruvate carboxykinase (PEPCK), studied by Hod *et al.* (1986). The promoter–regulator region had been localized to be within a 621-bp *Bam*H1/*Bgl*II fragment present at the 5′-end of the gene. The hormonal control region was studied by linking the 621-bp fragment to the structural segment of the herpes simplex thymidine kinase (TK) gene in a plasmid vector. This construct was used to transfect cells of a TK-deficient rat hepatoma cell line. The addition of dibutyryl-cAMP to the cells resulted in a fourfold to sixfold induction of both TK activity and the level of its mRNA. To define the sequences in the PEPCK promoter that are necessary for this cAMP effect, a series of graded deletions was constructed. The transfected chimeric genes retained their responsiveness to cAMP as long as they included the sequence from −61 to −108. Deletion of this 47-bp fragment completely eliminated cAMP inducibility.

Further studies of this and other cAMP-regulated genes have shown that a palindromic DNA sequence, TGACGTCA, termed *CRE* (cAMP response element) is sufficient for the effect. Other hormones, such as glucocorticoids, androgens, mineralocorticoids, estrogen, thyroxin, and retinoic acid, have their specific response segments, where the hormone–receptor complexes bind (Lucas and Granner, 1992). cAMP normally exerts its effect through activation of protein kinase A (PKA), and such is the case here as well. A CRE-binding protein (CREB) was discovered that is phosphorylated by PKA and thereby activated (Yamamoto *et al.,* 1988).

PEPCK is actually regulated by several hormones and the dietary state. Further transfection studies have uncovered eight protein-binding domains between −460 and +73 and identified proteins binding to these response elements (Park *et al.,* 1993). The interaction of these partially tissue-specific transcription factors with the basic transcriptional complex is currently being studied.

B. Regulation in Transgenic Animals

Although much can be learned from studies with cell cultures, including how they react when recombinant DNA is introduced, this approach has its limitations. One fundamental set of regulatory problems is connected to the development of an animal from a fertilized ovum. What sort of mechanisms are responsible for tissue- and time-specific transcription of genes? Such questions are addressed through the use of transgenic organisms, particularly mice (Westphal, 1987; Grosveld and Kollias, 1992). This important technique was introduced by Gordon *et al.* (1980). The DNA to be investigated has been injected into the male pronucleus of a fertilized ovum, and the ovum implanted. Some of the ova develop into mice with the extra DNA integrated into their chromosomes, usually with many copies arranged head to tail.

The first successful experiments dealing with tissue-specific regulation were reported by Chada *et al.* (1985). These investigators showed that a human β-globin gene, in which the front end had been replaced by the corresponding mouse gene, together with 1200 bp of flanking DNA, was expressed in transgenic mice exclusively in erythroid cells. They also showed that the hybrid β-globin gene was expressed at the proper time during development (Magram *et al.*, 1985). The 5′-flanking region must have been responsible for this specificity, along with corresponding tissue-specific regulatory proteins.

Similar results have been found for a number of other genes and tissues (Palmiter and Brinster, 1986). A particularly striking example is the work on elastase I (Swift *et al.*, 1984). Mice were made transgenic with the rat elastase I gene and its flanking regions. A marked tissue specificity was observed, rat elastase mRNA being 500,000 times more abundant in the pancreas than in the kidneys. Experiments with trimmed 5′-flanking DNA showed that 205 bp were sufficient to uphold the pancreas-specific expression (Ornitz *et al.*, 1985). When this regulatory region was joined to the human growth hormone gene, human growth hormone was found in pancreas and not in other tissues. Furthermore, immunofluorescence analysis demonstrated that the growth hormone was present in the acinar cells, not in the endocrine or connective tissue cells of the pancreas.

The faithful tissue specificity observed with this kind of construct has been exploited in the study of oncogene expression (Section V.E). In one such study, Hanahan (1985) directed the expression of the SV40 oncogene to pancreatic β cells using an insulin gene regulator sequence.

This technique has also been used, for instance, to further understand the complex and tissue-specific regulation of the PEPCK gene. McGrane *et al.* (1988) fused the promoter–regulatory sequence of the PEPCK gene to the bovine growth hormone gene as a reporter gene. The transgene was expressed only in liver and kidney and showed dietary and hormonal responsiveness. Further, in order to establish their regulatory roles, mutations in specific regulatory domains were produced in transgenic mice (Patel *et al.*, 1994).

Another method for producing transgenic animals uses embryonic stem (ES) cells. These ES cells are derived from an early embryo, cultivated, and infected with a retrovirus construct (Robertson *et al.*, 1986) or transfected through the phosphate/DNA precipitation technique (Lovell-Badge *et al.*, 1985). When they are reintroduced into the blastocyst, the ES cells contribute to the production of a chimeric animal. In the next generation, some pure transgenic animals appear. This approach has the advantage that somatic cell genetic techniques can be used to modify and to select cells with a desired potential.

V. DIAGNOSIS OF GENETIC DISEASE

A. Gene Changes in Hereditary Disease

Mutations could affect the production of a given gene product in many different ways. Studies of hemoglobin synthesis in humans have found examples of most of these possibilities. Replacement of a single base with another base can have the effect that an amino acid is replaced by another amino acid. One hundred and eighty-nine such structural variants are known for the human β-globin chain (Little, 1981). Not all of these variants cause disease, however. The mechanism of polypeptide chain termination is the source of other disturbances. In individuals with hemoglobin Constant Spring, the stop codon TAA for α-globin is replaced by CAA. In a type of β^0-thalassemia (no production of β chains), the seventh codon, AAG, is replaced by TAG, terminating the chain prematurely.

Single base mutations may also affect transcription and RNA processing, which is often observed in β-thalassemia (low production of β chains). Some mutations in the promoter region reduce transcription markedly. Changes at exon–intron junctions lead to errors in the removal of introns from the primary transcript, with unusable mRNA as the result. Mutations can also generate new, false exon–intron border sequences, again with defective mRNA as the result.

Deletions and insertions are other common types of mutations. Deletions or insertions of one, two, or four bases are the cause of several β^0-thalassemias. Such a mutation destroys the reading frame completely because the wrong set of triplets is read by the ribosomes. Deletions of $n \times 3$ bases can also occur, and the reading frame is not affected. Mutants with β-chain shortening from one to five amino acids are known. Larger deletions also occur; for hemoglobin genes, deletions from 0.6 kbp up to about 20 kbp are known.

Studies have uncovered a similar mutational heterogeneity for many other genes. In addition, a new type of mutation has been discovered for myotonic dystrophy (DM) and six other diseases (Miawa, 1994). The DM gene contains the triplet repetition $(GCT)_n$, with n varying somewhat in the population. For affected individuals n is increased and appears to correlate with the severity of the disease.

B. Detection of Gene Changes with DNA Probes

Recombinant DNA technology makes possible a direct approach to the study and diagnosis of a genetic disease, instead of being limited to study of its phenotypic expression. For animal owners and animal breeders, it is worthwhile to be able to detect carriers of recessive disease genes in order not to propagate the disease gene or to avoid mating the animal to another carrier of the same defect. Furthermore, it is not always the case that the homozygous state for a recessive gene, or the heterozygous state for a dominant gene, is expressed phenotypically. The disease may be dependant on an environmental factor for its expression or have a late onset in the life of the animal. In this case, DNA analysis could provide an early diagnosis.

Application of recombinant DNA methods in diagnosis and gene defect studies have made great advances in the human sector and are well under way in the veterinary sector also. Common to the methods is the need to have cloned or synthetic DNA probes for the disease gene or its neighborhood and / or sequence knowledge that permits the use of PCR methods.

1. Mutations That Create or Destroy a Restriction Site

The ideal situation exists when the mutation either creates a new restriction enzyme site or removes one that was present in the wild type. This is the case for sickle cell anemia in man. In sickle cell anemia, the sixth amino acid in the β chain of hemoglobin, glutamate, has been replaced by valine because an A in the corresponding codon has been replaced by T. The restriction enzyme *Dde*I for normal individuals cuts three places in the β-globin gene, whenever the sequence CTNAG occurs (N can be any base). The mutation has the effect that the second of these sites is destroyed. Sickle cell patients thus will have one longer DNA fragment hybridizing with a β-globin probe, whereas normal persons will have two shorter fragments, which is revealed by Southern blotting as described in Section III.D.4. A heterozygote would be revealed by all three bands being present.

Each restriction digest requires 5^{-10} μg of patient DNA, which is usually obtained from the buffy coat of peripheral blood samples. The leukocytes present in 10–20 ml of human blood can yield as much as 200 μg of DNA after treatment with proteinase K and extraction with phenol and chloroform.

There are many other examples of diagnosis based on change in a restriction site, for example, in the *ras* family of oncogenes and for induced H-*ras* mutations in the rat (Zarbl *et al.*, 1985). A more convenient version of this test amplifies the region around the mutation with PCR and subjects the PCR product to restriction enzyme treatment. Sufficient DNA is often present to allow direct detection of the restriction fragments after electrophoresis using ethidium bromide and UV light.

2. Deletion/Insertion Mutations

A favorable situation exists when deletions or insertions of a few hundred base pairs or more are the cause of the disease. These mutations will directly affect the length of the DNA fragment between two given restriction sites. Deletions occur in human globin genes, the gene causing Duchenne muscular dystrophy, the gene for coagulation factor VIII and others.

3. Single Base Replacement in a Known DNA Sequence

Base replacement mutations that do not affect any restriction site may still be diagnosed if one knows the DNA sequence around the mutation site. One can then synthesize a wild-type allele-specific oligonucleotide (ASO) of 18 to 20 bases, label it with radioactivity, and observe its hybridization with DNA fragments on a membrane. Under stringent hybridization conditions, it is possible to detect a mismatch of only one base between the probe and the DNA investigated. Additionally, it is prudent to make a probe corresponding to the mutant and demonstrate its reduced hybridization to normal DNA. This method is used in diagnosis of α_1-antitrypsin deficiency (Kidd *et al.*, 1984), H-*ras* mutations in rats (Zarbl *et al.*, 1985), and several other mutations.

The method is more reliable and convenient when carried out on a PCR-amplified segment containing the mutation. The sensitivity thus reached makes possible the use of nonradioactive, enzyme-labeled ASO probes in dot blot hybridization (Scharf *et al.*, 1991). The occurrence of many different mutations in the same gene in the population can be handled using a "reverse dot blot" method. Instead of having numerous membranes with the PCR product applied, each to hybridize with a different ASO probe, a membrane that contains an immobilized array of ASO probes is hybridized to the PCR product (Erlich *et al.*, 1991a).

4. Restriction Fragment Length Polymorphisms Linked to the Gene of Interest

Earlier, the most widely used and general diagnostic method was that of restriction fragment length polymorphism (RFLP) linked to the gene that was investigated. With this technique, one needs no knowledge of the structure of the gene. It is based on the fact that the heterozygosity at the level of DNA is large. If the base sequences of two homologous chromosomes are compared in humans, a difference for each 200 to 400 bp is found (Cooper and Schmidtke, 1984). The majority of these differences are neutral mutations. A few of them reside in the coding triplets, but most of them are in introns and in the region between the genes.

Some of these mutations will affect known restriction enzymes and be within or close to the gene of interest, so that they can be observed as RFLP with a probe from the same region (Fig. 2.8b). Such an RFLP can be used as a marker for the gene studied. The procedure is essentially the same as in Fig. 2.7b. Presence and absence of the restriction site will give two different restriction fragment lengths that move different distances on the gel.

a. Study of Linked RFLP in Families

The use of RFLP for diagnosis has one drawback. An answer cannot be obtained by investigating the DNA of a single individual, as was possible with the methods in Sections V.B.1 through V.B.3. The chromosome with the mutant form of the gene will not always have the same restriction site allele, for instance, presence of the restriction site, that may have been the original constellation. However, only a fraction of chromosomes with this restriction site need have the gene mutation. By recombination, the mutant gene may also have landed on a chromosome with this restriction site absent. If the distance between the restriction site of concern and the gene studied is small, it is probable that the constellation will remain the same within a given family.

The use of this method is illustrated in the following example from a family with β-thalassemia. Both parents were healthy, but a child was affected. During the following three pregnancies, prenatal diagnosis was performed. Using the restriction enzyme *Ava*II, Southern blotting, and a labeled β-globin probe, the band pattern shown in Fig. 2.9 was observed. Both parents were RFLP +−, one chromosome having the restriction site, the other not. The affected child was ++; therefore, it must thus have been the + chromosome from both parents that carried the disease gene. For the fetuses, one could conclude that they were noncarrier, affected, and carrier, respectively. The probe cross hybridizes with δ-globin sequences common to all.

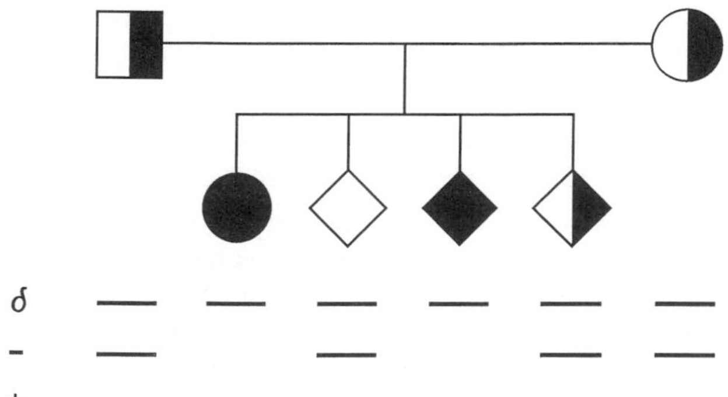

FIGURE 2.9 Prenatal diagnosis of β-thalassemia. Adapted with permission from Francomano and Kazazian (1986).

b. Usefulness of Having More Than One RFLP

The example just described was a favorable situation. It would only have been possible to diagnose 50% of the offspring correctly if the thalassemia gene in one parent had been on an RFLP chromosome. The situation might have been helped if one had another RFLP close by. With two RFLPs, one would have four combinations, so-called "haplotypes." This circumstance is equivalent to having four alleles for the marker locus, which gives a favorable degree of heterozygosity.

If the distance between the gene and the RFLP site is large, for example, some thousand kilobase pairs, it is important to have a second RFLP situated on the opposite side of the gene. A crossing over during meiosis between the first RFLP and the gene would then be detected since a double crossing over is unlikely. Such a flanking RFLP is also useful in work attempting to localize and study the gene if it is not yet characterized.

c. Searching for RFLP

The search for RFLP does not have to be a random trial-and-error test of the large battery of commercially available enzymes. Cooper and Schmidtke (1984), Wijsman (1984), and Feder *et al.* (1985) describe rational approaches and give lists of relative efficiency for many enzymes. Enzymes containing the dinucleotide CG in their recognition sequences detect more variation of the base replacement type than other enzymes, probably because of mutations from methylated CG to TG. Larger probes have been found to be more efficient than small probes.

Screening should be limited, as suggested by Skolnick and White (1982), to a panel of about 10 unrelated individuals in order not to pick up rare polymorphisms. These have low degrees of heterozygosity and therefore are inferior in diagnostic work.

When multiple restriction fragments are found, it must be shown that they represent alleles at the same locus. A probe might recognize nonallelic fragments due to sequence homology with a different chromosomal region. If the fragments are alleles, they will segregate in a Mendelian fashion.

d. Minisatellites

Studies of gene structure have uncovered situations in which a single restriction enzyme reveals a multiallele polymorphism, a polymorphism that does not affect sites for the enzyme. Between two fixed restriction sites, one finds a region consisting of tandem repetition of 15- to 100-bp units, and the number of these repeats varies considerably in the population. Such variable tandem repeat (VNTR) regions, also called minisatellites, have been found in humans close to the insulin gene (Bell *et al.*, 1982), α-related globin genes (Proudfoot *et al.*, 1982), and the c-H-*ras* oncogene (Capon *et al.*, 1983), but similar VNTR regions appear widely dispersed (Jeffreys *et al.*, 1985b). The presence near a disease gene of a VNTR region is very useful, because the multiallelic nature of the corresponding RFLP leads to high degrees of heterozygosity. A method for a systematic search for such RFLP has been reported (Nakamura *et al.*, 1987).

e. Microsatellites

Weber and May (1989) reported the existence of a large set of highly polymorphic microsatellites, consisting of dinucleotide repeats, that could be typed using PCR. Such microsatellites have become the most important source of markers for high-resolution genetic maps. A human map was presented in 1992 with 814 microsatellites (Weissenbach *et al.*, 1992). Two years later the map included 2066 $(AC)_n$ markers, flanked by unique sequences for which PCR primers could be synthesized (Gyapay *et al.*, 1994). The average distance between the markers was 2.9 cM, which means that there are now polymorphic microsatellite markers close enough to most disease genes for diagnostic purposes. These markers are also starting points for work to identify the genes and their mutations. This latter task is being made easier by the parallel development of physical maps of yeast artificial chromosome clones (Cohen *et al.*, 1993) that have been ordered using sequence-tagged sites (Olson *et al.*, 1989).

5. Disease Gene Identification

Rapid progress has been made in the identification and characterization of disease genes in humans, a prerequisite for application of the direct methods of diagnosis described in Sections V.B.1 through V.B.3. Many genes have been identified by functional cloning, using information about the protein product (sequence or antibodies). Mapping of the gene then followed cloning. Starting in 1986, another approach, positional cloning, came into use. In this approach, mapping to a precise location on a chromosome is the basis for the cloning effort. Some 20 disease genes have been identified in this way (Collins, 1992). As the Human Genome Project produces high-resolution genetic maps and physical maps based on overlapping clones, positional cloning will be simplified.

In other instances, map position has been important in order to verify gene identification, but the availability of a plausible candidate protein and its cDNA has made it possible to avoid exhaustive cloning in the specified chromosome region. This positional candidate approach will probably become increasingly common (Ballabio, 1993). In all, about 400 human disease genes can now be diagnosed directly.

6. Genetic Screening

Carrier screening programs have been attempted for some human genetic diseases even before DNA tests became available, notably sickle cell anemia, thalassemia, and Tay–Sachs disease. The thalassemia program in Cyprus for testing couples at marriage has been very successful. Pilot programs that test pregnant women for the cystic fibrosis gene and the partner if the test is positive have taken place in Britain and the United States (Williamson, 1993). Although there are legitimate concerns that screening data may be used to coerce families to make reproductive decisions that meet economic or societal objectives rather than personal wishes, the availability of a screening service is generally welcomed.

For human diseases, there is, however, often the problem of genetic heterogeneity. The number of different cystic fibrosis mutations is now greater than 260. Although most of these are rare, 11 mutations making up 91.5% of the mutations in Northwest England (Ferec *et al.*, 1992), a screening program for cystic fibrosis will clearly leave a small residual risk for unexpected development of the disease. For animal genetic diseases, there is less heterogeneity, making screening an effective and widespread tool (Section V.D).

C. DNA Probes and Disease Susceptibility

For the first time recombinant DNA technology has also made it possible to begin analyzing polygenic diseases, that is, diseases where a combination of unfavorable genes may predispose to the development of disease.

1. HLA Genes

Genes determining susceptibility or resistance to a broad variety of diseases in humans have been localized to the HLA region, the major histocompatibility complex in humans (Vogel and Motulsky, 1986; Kostyu, 1991). Among these diseases are ankylosing spondylitis, celiac disease, dermatitis herpatiformis, and insulin-dependent diabetes mellitus (IDDM). The latter three are associated with the HLA serologic specificity DR3, and IDDM also with DR4. Seventy-five percent of IDDM patients have HLA-DR4, whereas only 32% of controls do. This outcome gives a DR4 individual a relative IDDM risk of 6.4, and the risk is increased in DR3,4 heterozygotes (Knip *et al.*, 1986). The association of HLA-DR genes with IDDM susceptibility is related to the fact that IDDM pathogenesis often involves autoimmune phenomena (Nepom and Erlich, 1991).

With recombinant DNA technology, it became possible to ask whether there might be DNA differences, detectable as restriction fragment length differences, between patients and controls of the same DR3 and/or DR4 specificity. This was indeed found to be the case by Owerbach *et al.* (1983). Using a DQβ probe, these investigators found a significantly decreased frequency of a *Bam*H1 3.7-kbp fragment among DR4 IDDM patients. These studies have been confirmed and extended by others, including Michelsen and Lernmark (1987).

The full impact of DNA technology came, however, with PCR amplification of cDNA from the first exons of α and β chains, the exons containing most of the polymorphisms, followed by sequencing (Todd *et al.*, 1987). This procedure has identified a large number of α- and β-chain alleles, denoted by a four-digit code. Typing of DNA sequence polymorphisms at HLA loci is conveniently done by PCR amplification of genomic DNA with appropriate primers, followed by probing with sequence-specific oligonucleotide probes (Erlich *et al.*, 1991a). Such studies have shown that molecules of DQ(α1*0301,β1*0302), associated with DR4, and DQ(α1*0501,β1*0201), associated with DR3 give a relative IDDM risk of 6 to 18 for Caucasians. In heterozygotes, two further DQ molecules may form by transcombination of the subunits (Thorsby and Rønningen, 1993; Nepom and Erlich, 1991). Results of this nature have been reported for malaria, where DRB1*1302–DQB1*0501 gives some protection (Hill *et al.*, 1991), and for papillomavirus-associated cervical carcinoma (Apple *et al.*, 1994), where DRB1*1501–DQB1*0602 gave increased risk (relative risk of 4.8).

2. *Insulin Gene 5′-VNTR Alleles*

As mentioned in Section V.B.4.d, there is a VNTR associated with the human insulin gene on chromosome 11. The variable region falls in three size classes (class I averaging 570 bp, class II averaging 1200 bp, and class III averaging 2200 bp). A strikingly higher frequency of class I alleles was observed for Caucasian individuals with IDDM (Bell *et al.*, 1984). By sequencing the insulin gene locus from patients and controls, Lucassen *et al.* (1993) have identified the region of association with IDDM to be 4.1 kbp. Ten polymorphisms, including the VNTR, are in strong linkage disequilibrium with each other and show a relative risk for IDDM of 3.8 to 4.5. Analysis of other ethnic groups may reveal which exact polymorphism(s) confer susceptibility. It is possible that the VNTR length directly affects transcription.

The insulin gene associated polymorphism and the HLA variation are not sufficient to account for the development of IDDM. Extensive mapping of the disease in the nonobese diabetic mouse, using microsatellite markers, has revealed eight additional loci that make a contribution to this polygenic disease (Ghosh *et al.*, 1993). One of these codes for interleukin-2, which has a role in autoimmunity, and different sequences were found in nonobese diabetic and normal mice.

3. *Atherosclerosis*

The most plausible link between genes and atherosclerosis is found in the structure and regulation of genes for lipoproteins, the enzymes that are involved in their metabolism, and lipoprotein receptors. Genes for the apolipoproteins and the low-density lipoprotein (LDL) receptor in humans have been cloned and localized on chromosomes. One of every 500 persons is heterozygous for a mutant allele of the LDL receptor, leading to two to three times normal levels of cholesterol. Hobbs *et al.* (1992) have unraveled the mechanisms involved and characterized different types of mutations in the gene. Mapping of the defects in a population can start with a determination of mutation-associated RFLP haplotypes. This information is useful for diagnosis within families (Rødningen *et al.*, 1992). To establish direct gene tests, DNA from affected individuals was screened for mutations (Leren *et al.*,, 1993).

LDL is bound to its receptor via $apoB_{100}$, and hypercholesterolemia may be caused by mutations in the $apoB_{100}$ gene as well (Innerarity *et al.*, 1990). Mutations are known also in the *apoA-I* gene and cause low levels of high-density lipoprotein (HDL) accompanied by atherosclerosis. Type III hyperlipoproteinemia patients have a particular allele for the *apoE* gene. With the knowledge of DNA structure, oligonucleotide probes that differentiate between normal and mutant genes can be made, allowing screening of people at risk.

D. Application to Domestic Animals

The application of recombinant DNA techniques to diagnosis in domestic animals is now quite widespread. The availability of DNA clones, or knowledge of primer sequences from humans and small experimental animals, has enhanced this development.

1. *Diagnosis of Disease Mutations*

a. Citrullinemia in Cattle

The first disease to benefit from a DNA test was citrullinemia, which is widespread in Australian Friesian cattle. Three conditions were helpful for the establishment of a DNA test for the mutation in 1989: Bovine cDNA libraries were available, sequenced cDNA for the enzyme (argininosuccinate synthetase) for humans and rats was available, and the protein is fairly small, only 412 amino acids.

Normal bovine cDNA for the enzyme was isolated with a rat probe and sequenced by Dennis *et al.* (1989). To identify the mutation, total mRNA was prepared from the liver of an affected animal and cDNA synthesized. The product was amplified with PCR using primers corresponding to sequences immediately before and after the coding region of the cDNA. These primers had been extended with *Bam*HI linkers, so the amplification product could be ligated into the *Bam*HI site of the DNA sequencing vector. In clones from the mutant, a C to T transition in the first position of codon 86 was found that generates a TGA termination codon leading to a truncated and inactive protein.

The change in codon 86 leads to the disappearance of an *Ava*II restriction site. This deficiency was utilized by Dennis *et al.* (1989) to design a simple, PCR-based gene test on genomic DNA. A 194-bp DNA could be amplified from within the exon containing the mutation. A portion of the product was digested with *Ava*II, and both digested and undigested DNA were analyzed by electrophoresis. The normal calf has two bands of 72 and 118 bp, the homozygous (affected) calf has one band of 194 bp, and a carrier has all three bands.

b. Malignant Hyperthermia in Pigs

Work with malignant hyperthermia (MH) in pigs had started earlier. We isolated clones for the glucose-phosphate isomerase (GPI) gene, a closely linked gene (Davies *et al.*, 1987). With this DNA as probe, a multiallelic RFLP was detected that was confirmed to be tightly linked to the MH gene (Davies *et al.*, 1988), so that it could serve as a marker. A further search for the MH mutation was based on reports that calcium is more easily released from the sarcoplasmic reticulum of MH pigs. Meissner (1986) had shown that the alkaloid ryanodin interacts directly with the calcium release channel (CRC) of rabbits. Using tritiated ryanodin as a label, Lai *et al.* (1988) purified the channel.

This work prepared the way for the cloning of the rabbit CRC cDNA by Takeshima *et al.* (1989), a 15,100-bp-long cDNA. With a small rabbit cDNA clone as a probe, we isolated a porcine cDNA clone and showed it hybridized *in situ* to metaphase chromosomes in the same chromosomal band that hybridized with GPI DNA (Harbitz *et al.*, 1990). This result confirmed CRC as a good candidate for the product of the MH locus. Sequencing CRC cDNA from a normal and an affected pig, Fujii *et al.* (1991) identified the mutation, a substitution of T for C at nucleotide 1843, which changes an arginine to cysteine. The mutation was found to be the same in five Canadian pig breeds and in British Landrace (Otsu *et al.*, 1991) and Norwegian Landrace pigs (Harbitz *et al.*, 1992).

The mutation destroys a *Hin*P site and creates a *Hgi*AI site. Fujii *et al.* (1991) amplified a 74-bp piece from the exon containing the mutation and made restriction site changes in the product the basis for their diagnosis. A more convenient and reliable method became possible after sequencing of genomic DNA flanking the exon.

c. Leukocyte Adhesion Deficiency in Cattle

Leukocyte adhesion deficiency (LAD) in humans is an autosomal recessive disease characterized by greatly reduced expression of the heterodimeric beta-2 integrin adhesion glycoproteins on leukocytes. Without this protein, neutrophils are unable to enter tissues to destroy invading pathogens. All human cases have been traced to the integrin beta-subunit, CD18. A granulocytopathy syndrome had been described in Holstein cattle, and Schuster *et al.* (1992a) investigated whether it was also caused by a CD18 mutation. These investigators isolated bovine CD18-encoding cDNA by screening a lymphocyte library with a murine CD18 probe and sequenced the clone. Leucocyte RNA was then isolated from a calf with LAD symptoms and from a normal cow (Schuster *et al.*, 1992b). Northern blot analysis revealed a CD18 transcript at normal level and size from the affected calf, ruling out a large deletion or a transcription defect. mRNA was reverse transcribed, and the cDNA was amplified using bovine CD18-specific primers. To sequence the amplified product, single-stranded DNA was generated with a single primer and sequenced directly. A mutation was detected at nucleotide 383, replacing aspartic acid with glycine (D128G). The mutation occurs in a row of 26 consecutive amino acids that are identical in normal bovine, human, and murine CD18.

This mutation introduces a *Hae*III site and eliminates a *Taq*I site. Schuster has utilized this in an assay that PCR amplifies a 58-bp DNA from the mutated exon. Fifteen other LAD calves were found to be homozygous for the D128G mutation. To determine the prevalence of this allele, 2025 U.S. Holstein bulls used as semen donors were screened. The carrier frequency was 14.1% before selection against the disease began. In addition, cows (1559 Holsteins) were screened, and

5.8% were found to be carriers. The LAD incidence in the United States in calves at birth should therefore be about 0.2%. All the LAD calves had pedigrees leading back over five generations to a certain bull, Osborndale Ivanhoe, born in 1952. The carriers with pedigree information also descended from this bull or, in a few cases, his immediate relatives. This bull was demonstrated to be a carrier of D128G by testing frozen semen.

Recently, an improved detection method has been presented by Batt *et al.* (1994), using a thermostable ligase in an oligonucleotide ligation assay, combined with PCR (Barany, 1991). Ligation/amplification fails when a single base mismatch is present. Using biotin in one oligonucleotide, for product capture, and a nonisotopic reporter group in the other, the method can be automated.

d. Hyperkalemic Periodic Paralysis in Quarter Horses

Hyperkalemic periodic paralysis (HYPP) is a genetic disease observed among quarter horses. The disease causes attacks of paralysis, inducible by ingestion of potassium. In humans, HYPP is caused by a single base substitution within the skeletal muscle sodium channel gene. Rudolph *et al.* (1992a) studied a quarter horse pedigree to see if the sodium channel gene was also involved in the horse. Starting with reverse-transcribed horse mRNA, horse sodium channel cDNA was produced by PCR, using rat primer sequences. This procedure revealed a polymorphism that was used to investigate whether the horse sodium channel gene was genetically linked to HYPP, which was found to be the case.

To identify the putative mutation, they used single-strand conformational polymorphism (SSCP) analysis (Orita *et al.*, 1989) of reverse-transcribed mRNA PCR products from muscle biopsies of normal and affected horses (Rudolph *et al.*, 1992b). This method is capable of detecting existing two-allele polymorphic markers (Neibergs *et al.*, 1993). Rat and human sodium channel sequences were used to design primers. A single-strand specific for an affected horse was seen in the PCR product from amino acid region 1348–1524. This region was cloned and sequenced.

In the affected horse, there was a C to G change, replacing phenylalanine with leucine in transmembrane region IV of the channel. All nine normal sodium channel sequences sequenced to date have phenylalanine in this position, indicating its importance for activity. To test whether the leucine substitution cosegregated with the disease, allele-specific hybridization and/or *Taq*I restriction enzyme analysis of PCR-amplified genomic DNA sequences was used to study an extended pedigree. All affected horses in the pedigree showed the leucine substitution, whereas all unaffected horses had phenylalanine. It was theoretically possible still that this change was in strong linkage disequilibrium with a separate disease-causing mutation. To check this possibility, 176 unaffected horses from a variety of breeds were tested. All were homozygous for the normal allele (Phe/Phe), which strongly argues against linkage disequilibrium with some other mutation.

Although this disease gene is dominant, the developed test is important because affected horses commonly do not exhibit signs of disease until they are mature. They may then have episodes of weakness and collapse, posing a hazard to their riders and themselves. Current physiological tests do not have the required sensitivity.

e. Bovine Uridine Monophosphate Synthase Deficiency

Deficiency of uridine monophosphate synthase (DUMPS) is an autosomal recessive disorder in Holstein and Red Holstein cattle that results in early embryonic death of homozygous offspring. Heterozygotes are phenotypically normal, but show about one-half the normal enzyme level and elevated levels of orotic acid. Heterozygotes may have a higher genetic merit in milk and protein production. With accurate DNA-based genotyping, it would be possible to preserve the mutation in the population while avoiding matings between carriers.

Schøber *et al.* (1992) isolated and sequenced wild-type bovine UMPS cDNA using a human UMPS-specific cDNA to screen the library. To identify the mutation responsible for DUMPS, liver RNA from DUMPS heterozygotes was reverse transcribed. Amplification of cDNA with sequence-specific primers and subsequent sequencing of the PCR products revealed a C to T transition in codon 405, resulting in a premature stop codon and a truncated protein. The mutation led to the loss of an *Ava*I site. A genomic DNA-based PCR test was developed that looks for presence of the *Ava*I site in a 108-nucleotide segment. Complete concurrence between low levels of UMPS and presence of the point mutation was found in a large Holstein pedigree, confirming this mutation as the basic defect in DUMPS cattle.

f. Bovine Maple Syrup Urine Disease

Maple syrup urine disease (MSUD) results from a deficiency of the branched chain α-ketoacid dehydrogenase, a mitochondrial multisubunit complex of enzymes. The E1 component catalyzes the oxidative decarboxylation. MSUD has been identified both in humans and in cattle and is relatively common in the Polled Hereford breed of cattle in Australia.

Hu *et al.* (1988) isolated and sequenced cDNA for the bovine E1α subunit, screening an expression library with antiserum to bovine E1. To define the mutation responsible for MSUD, Zhang *et al.* (1990) produced cDNA from MSUD calf fibroblasts by reverse transcription of RNA followed by PCR. These investigators chose primers that amplified the complete coding region of the E1α subunit. Subcloning of the product and sequencing revealed replacement of a CAG glutamine triplet with TAG, a stop codon, in the partial coding for the leader sequence that directs the protein to mitochondria. The mutation was verified by hybridization of amplified cDNA from calves with two radioactive 15-*mer* allele-specific oligonucleotides.

MSUD also occurs in the Polled Shorthorn breed. Amplifying DNA with PCR across the glutamine codon and testing for allele-specific oligonucleotide hybridization revealed, however, that the mutation in the Polled Shorthorn must be in a different position from that in the Polled Hereford (Healy and Dennis, 1994).

2. *Diagnosis of Disease Susceptibility*

Association between diseases and alleles of genes in the major histocompatibility complex (MHC) is also known for domestic animals. This is particularly well documented for Marek's disease of chickens, which is caused by a herpes virus (Hanson *et al.*, 1967; Hepkema *et al.*, 1993). Resistance to bovine leukosis, caused by the BLV retrovirus, also has a clear MHC association (Lewin, 1994), and Mejdell *et al.* (1994) found an influence of the bovine MHC on resistance to mastitis. A highly significant association was found between the bovine MHC class I antigen BoLA-A8 and chronic posterior spinal paresis, a form of ankylosing spondylitis, in Holstein bulls (Park *et al.*, 1993). The relative risk was 34.6. The special interest in developing RFLP to establish haplotypes for the MHC genes in domestic animals must be viewed in this light. MHC RFLP have been found with human probes in a large number of species (Andersson *et al.*, 1986; Juul-Madsen *et al.*, 1993). More informative studies are now being conducted based on sequencing of the MHC polymorphic exons. Such data were presented for cattle (Andersson *et al.*, 1991), sheep (Fabb *et al.*, 1993), chickens (Moon Sung *et al.*, 1993), horses (Szalai *et al.*, 1993), and pigs (Våge *et al.*, 1994).

E. Gene Changes in Cancer

Cancer is, like genetic diseases, a result of changes in the hereditary material, but in the latter case, mainly in somatic cells. A great deal of light has been thrown on the phenomenon of neoplastic growth via the use of recombinant DNA techniques. The molecular structures of a long series of viral oncogenes and cellular proto-oncogenes, often coding for proteins participating in signal transduction from the cell surface to the nucleus, have been elucidated, and the mechanisms for the activation of proto-oncogenes to oncogenes to a large extent clarified (Land *et al.*, 1983; Van de Woude *et al.*, 1984). For the *ras* family of oncogenes, this activation takes place by a base replacement, for others, for example, *myc, myb,* and *ets-1,* by translocation and/or amplification. Oncogene probes are available for such studies. Translocations, and sometimes mutations, result in restriction fragment size changes. Amplification leads to stronger bands of hybridizing DNA. Amplification, or stimulated transcription, leads to increased hybridization to mRNA.

These developments have clinical relevance. DNA methods are becoming useful in tumor diagnosis, and they give new information as to the stage of development of the tumor. Oncogenes are highly conserved in evolution, so it is likely that human and murine oncogene probes can be used for investigation of cancer development in domestic animals. Human probes for c-*myc, myb, HER-2/neu,* H-*ras,* K-*ras,* and N-*ras* thus were found to hybridize to DNA fragments from dogs (Hauge *et al.*, 1988).

While activated oncogenes act in a dominant manner, mutations in tumor suppressor genes are recessive. Their gene products inhibit cell division, and thus balance the stimulatory properties of the proto-oncogene products. Loss of function of the suppressor gene on both chromosomes leads to increased cell division. Because the first mutation is often transmitted from a parent, whereas the second arises somatically, such tumors are found to be hereditary. Well-studied examples are the genes for retinoblastoma and p53, as well as several suppressor genes involved in colorectal cancer (Weinberg, 1991). Gene tests for these mutations are of obvious value for early diagnosis and treatment.

VI. THERAPY OF GENETIC DISEASES

In the foreseeable future, gene therapy may not be of practical value in regular veterinary medicine because of the cost involved. But work with gene therapy in animals will nonetheless be important as preparation for such therapy in humans (Anderson, 1984). What needs to be shown in animals is (1) that the gene is integrated in the target cells, (2) that it is expressed on a suitable level, and (3) that the gene does not harm the cells or the animal. The ideal would be to exchange the mutated gene with a wild-type gene. Techniques for accomplishing this exchange by homologous recombination in ES cells have been developed (Smithies *et al.*, 1985; Thomas and Capecchi, 1986).

Gene therapy can be conceived at different levels: introduction of DNA into the fertilized ovum or early embryo; into cells or tissues that are taken out, modified, and reimplanted; and into the whole animal with the help of an infective vector.

A. Germ Line Gene Therapy

Several successful experiments of this type have been reported. Hammer *et al.* (1984) describes how injection of the rat growth hormone (GH) gene, linked to the mouse metallothionein promoter, into fertilized ova from dwarf mice with low GH levels yielded mice with normal growth. Le Meur *et al.* (1985) restored the immune response to a tripeptide in mice from a line with a defective MHC class II E gene, by injection of DNA containing this gene. β-Thalassemia has also been successfully corrected by injection of β-globin DNA in a murine model of human β-thalassemia (Costantini *et al.*, 1986).

B. Somatic Gene Therapy

Human gene therapy is coming of age (Miller, 1992; Friedman, 1993; Tolstoshev and Anderson, 1993). Protocols for treatment of 14 diseases have been approved (Wivel, 1993) based on results obtained with animals. The first disease treated was adenosine deaminase (ADA) deficiency, which causes immunodeficiency. A retrovirus vector was used to introduce the normal cDNA in lymphocytes. Among the diseases recently attacked are cystic fibrosis and hypercholesterolemia.

Cystic fibrosis affects about 1 of every 2000 Caucasians. Mouse models for cystic fibrosis have been generated by homologous recombination in embryonic stem cells, in one case by replacement of part of the cystic fibrosis transmembrane conductance regulator (CFTR) gene with a mutated gene section (Snouwaert *et al.*, 1992), in the other case by insertion of a mutated gene in the host gene (Dorin *et al.*, 1992). The first type is a null mutation, whereas the latter has a small degree of leakiness. Hyde *et al.* (1993) showed that CFTR cDNA could be delivered to the lungs of the replacement mutant by direct instillation of a cDNA-liposome cocktail, resulting in correction of the ion conductance defect. Other workers have used adenovirus as a vector. Unlike retrovirus, adenovirus is capable of infecting terminally differentiated cell types and is naturally drawn to airway epithelial cells. Good expression of human CFTR cDNA, carried on replication-deficient adenovirus, was observed after intratracheal introduction in cotton rats. Adenovirus is not integrated into host DNA. Studies have therefore been performed on the safety and efficiency of repeated CFTR cDNA transfer in cotton rats and primates (Zabner *et al.*, 1994).

Another cDNA that has been carried on adenovirus into rat lung epithelium and expressed is cDNA for α_1-antitrypsin (Rosenfeld *et al.*, 1991). Adenovirus can also be used for muscle diseases. Vincent *et al.* (1993) have observed long-term correction of dystrophic degeneration in a mouse model for Duchenne muscular dystrophy after intramuscular injection of adenovirus carrying a human dystrophin minigene.

Familial hypercholesterolemia (FH) affects individuals heterozygous for LDL receptor mutations (Section V.C.3) and is very severe in those who are homozygous. A strain of rabbits genetically deficient in LDL receptors was used to demonstrate the potential efficacy of *ex vivo* gene therapy. Part of the liver was resected and perfused with collagenase to free hepatocytes. The isolated hepatocytes were transduced with LDL receptor retrovirus and infused into the rabbit liver with good effect (Chowdhury *et al.*, 1991). The experiment was followed up using dogs and baboons. The first results in humans have been reported (Grossman *et al.*, 1994). A 29-year-old woman, homozygous for the disease, underwent the same procedure, and her LDL/HDL ratio declined from 10 to 13 before treatment to 5 to 8 following gene therapy.

VII. USE OF RECOMBINANT DNA METHODS TO IMPROVE DOMESTIC ANIMALS

A. Markers and Maps

One important use of DNA methods for the improvement of domestic animals has been discussed already: marker-assisted selection against animals that carry a disease gene that has been characterized at the molecular level. On the whole, however, specific genetic disorders are less important than traits affecting growth, reproduction, disease resistance, and the quality of the end product. These traits are usually determined by several genes, by quantitative trait loci (QTL). To enable marker-assisted selection for a QTL, it must be fairly precisely localized on a chromosome. This localization requires genetic and physical maps of high resolution. Much effort has gone into establishing such maps.

The work is furthest advanced for pigs. At the First Pig Gene-Mapping Workshop, the number of mapped loci was 170 (Andersson *et al.*, 1993). Sixty of these loci were anonymous DNA segments, mostly microsatellites, and the rest were polymorphic blood groups and protein polymorphisms. Using a two-generation reference population, Rohrer *et al.* (1994) established a genetic linkage map with 376 microsatellite and 7 RFLP loci. The average distance between adjacent markers was 5.5 cM. Ellegren *et al.* (1994) reported a genetic linkage map for a cross between the European wild boar and a domestic breed (large white) that had a somewhat lower resolution (11-cM average spacing), but had 60 reference markers for comparative mapping and 47 markers physically assigned by *in situ* hybridization. This material has been analyzed with respect to some quantitative traits, and evidence for QTL with large effects on growth, length of the small intestine, and fat deposition was found on chromosome 4 (Andersson *et al.*, 1994). Comparative gene mapping for pig chromosome 4 could indicate candidates for these QTL because clusters of neighboring genes tend to be preserved between species.

An example of the comparative approach was the focus on the long arm of human chromosome 19 in the search for the MH locus. It was known that the GPI gene was located here, and that GPI and MH were closely linked in pigs. Another striking example is the observation of linkage for the human secreted phosphoprotein 1 (SSP1) locus to the sheep Booroola fecundity (FecB) gene, and the subsequent finding of linkage to human epidermal growth factor and complement I genes, near SSP1 in 4q in man. The latter genes define a candidate region on sheep chromosome 6 (Montgomery *et al.*, 1993).

Good progress has also been achieved for genome mapping in cattle. Barendse *et al.* (1994) mapped 171 loci, with an average distance between markers of 15 cM. Fifty-six loci represent DNA information from other species. An American–Swiss collaboration produced a bovine map with 313 polymorphic markers, with an average spacing of 8 cM (Bishop *et al.*, 1994). An extensive comparison between cattle, cats, mice, and humans has been presented by O'Brien *et al.* (1993). A good foundation now appears to exist for the isolation of bovine QTL, as well as for the study of mammalian genome evolution.

B. Parentage Identification

Errors may occur in marking semen portions used in artificial insemination. Piglets may jump from one pen to another and be falsely marked when marking takes place. And it may, at times, be a temptation for a breeder to report a false father or mother. These problems can be addressed using core sequences for minisatellites, which hybridize with the VNTR of many genes, producing a DNA fingerprint (Jeffreys *et al.*, 1985a). Because each band represents a minisatellite locus, the probability that two individuals will have the same pattern is nearly zero except for monozygotic twins. A multilocus microsatellite probe, such as $(GTG)_5$, yields a similar result, with better coverage of the genome (Mørsch and Leibenguth, 1994). Alternatively, the PCR results for a set of highly polymorphic microsatellites can be combined (Marklund *et al.*, 1994).

C. Transgenic Animals

When improving livestock by traditional breeding, one is limited to an exploitation of the alleles that are already present in the animal population. The results 14 years ago of introduction of extra DNA into pronuclei of fertilized mouse eggs suggested the possibility of a quicker and larger improvement of production animals. Experiments with injection of growth hormone (GH) in lactating cattle had demonstrated a significant boost in milk yields. It was natural to ask whether the same result could be obtained with extra gene copies for the hormone or whether pigs with extra GH genes could reach slaughter weight sooner and with lower feed consumption.

The first report on transgenic livestock animals appeared in 1985 (Hammer *et al.*). In the following years, these initial experiments were extended. In a review, Pursel and Rexroad (1993) listed 24 reports on transgenic pigs, 11 on sheep, 6 on cattle, and 2 on goats. The average frequency of success (transgenic animals/eggs transferred) was 0.8–0.9% for these species, compared to 2–5% routinely obtained with mice. This result puts a severe constraint on the adoption of this technology. For sheep and cattle there is, however, a promising development in that viable embryos can be obtained by *in vitro* maturation and fertilization of oocytes removed from slaughterhouse ovaries (Lu *et al.*, 1987).

1. Improving Productivity Traits

Transfer of genes affecting feed conversion, rate of gain, reduction of fat, and improved quality of meat, milk, and wool would be of great importance for both producer and consumer. These traits are, however, usually determined by several genes, by QTL, of which few are presently known. The mapping efforts discussed earlier will gradually change this situation.

Many of the pig and sheep experiments have involved GH and GH-releasing factor genes from various species. The promoter has often been that of metallothionein from mouse and other species, but other promoters have also been used (mouse albumin, Moloney leukemia virus, rat PEPCK).

Pursel *et al.* (1989) found concentrations of GH in expressing founder pigs to be in the range of

14–4000 ng/ml for human GH. The variation could reflect the influence of the chromosomal position for the inserted gene. To obtain a reliable assessment of the effect of the GH transgene on growth, the daily weight gain and feed efficiency were measured for two generations of transgenic and control pigs. The combined transgenic progeny showed 11% faster weight gain than the controls, and there was a 16–18% increased feed efficiency. Equally dramatic was the effect on subcutaneous fat accretion. Mean back fat thickness was reduced from 21 to 7.5 mm. In a follow-up study with bovine GH transgenic pigs, Solomon *et al.* (1994) studied whole-carcass ground tissue lipid composition. At 92 kg, these pigs contained 85% less carcass fat than control pigs, and the reduction in polyunsaturated fatty acids was somewhat lower.

These favorable traits were, however, offset by considerable deleterious side effects. Many of these could be the results of the lasting, high concentrations of GH. There was a high incidence of joint pathology, gastric ulcers, and infertility. What appears necessary is to achieve a tightly controlled GH gene expression, so that it can be confined to 1–2 months during the period of rapid growth. It should be possible to devise a promoter construct that is turned on by a substance added in the feed.

2. *Disease Resistance*

One route to specific disease resistance is introduction of one or more immunoglobulin genes that produce antibodies against a particular pathogen. Model experiments with mice have shown that this could work (Rusconi and Kohler, 1985). Weidle *et al.* (1991) produced transgenic pigs that harbored mouse lambda heavy chain and kappa light chain transgenes from antibodies against the hapten, 4-hydroxy-3-nitrophenylacetate. Titers up to 1 mg/ml were obtained in one founder and its progeny.

An interesting suggestion for reducing mastitis is to fuse the lysostaphin gene with regulatory elements from the betalactoglobulin (BLG) gene before injection in fertilized bovine eggs. The transgenic cow would then produce lysostaphin in its udder. Lysostaphin hydrolyzes the cell wall of *Staphylococcus aureus*. The idea is being tested in mice (Clark *et al.*, 1992).

In chickens, Salter and Crittenden (1989) have described a transgenic line carrying a defective ALV retrovirus genome that expresses only the envelope proteins of the virus. This expression leads to resistance since the produced viral protein competes for the virus receptor binding sites on the cell surface. This approach is similar to a successful technique for making plants virus resistant. The technique has been used to make sheep resistant to visna virus, which causes ovine progressive pneumonia (Pursel and Rexroad, 1993).

3. *Milk Modification*

Various ways of modifying milk protein composition can be envisaged (Clark *et al.*, 1992). Caseins could be altered in their phosphorylation sites and, thereby, in micelle properties. Reduction in the amount of BLG could reduce allergies and inhibit the synthesis of lactose, a disaccharide not tolerated by 90% of the adult world population. It would also seem useful to try to introduce into cows the gene for human lactoferrin, an antimicrobial agent and iron transporter, in order to make bovine milk better suited for human infants. Experiments have been initiated to achieve this (Krimpenfort *et al.*, 1991).

4. *Pharmaceutical Proteins*

The lack of α-1-antitrypsin (AAT) is one of the most common human genetic disorders. It leads to life-threatening emphysema, requiring repeated administrations of intact enzyme (200 g per patient per year). The source has hitherto been human blood plasma with its limitations and dangers. Wright *et al.* (1991) reported that high levels of AAT expression had been obtained in milk of transgenic ewes, in which a sheep BLG regulatory sequence was ligated to a human AAT genomic sequence. One sheep produced AAT at 30 g/liter, 50% of the milk protein. Earlier investigations using cDNA of AAT or human factor IX resulted in much lower yields. Several lines of sheep have now produced AAT stably during several lactations (Carver *et al.*, 1993).

Human protein C (hPC), which has an important role in hemostasis, is produced in the milk of transgenic pigs, the construct here consisting of hPC cDNA inserted into the first exon of the mouse whey acidic protein gene (Velander *et al.*, 1992).

Milk is not the only possible vehicle for medically important proteins from transgenic animals. Swanson *et al.* (1992) have succeeded in making pigs transgenic for human globin genes, expressed in pig erythrocytes. Human hemoglobin could be separated from pig hemoglobin by ion exchange chromatography. Its use would be in crisis treatment. It lacks red cell antigenic components and does not require refrigeration.

5. *Use of Embryonic Stem Cells*

A major limitation in the production of transgenic animals by microinjection of DNA in fertilized ova is that there is no control as to where in the genome the extra DNA finds it place. Sometimes it will disrupt a gene, introducing defects in development or function of the animal, or it may land in a region where it is poorly expressed. Work with mice has shown that for ES cells in culture, one can achieve a site-specific integration of the new DNA with the help of homologous recombination. Cells that are shown to have incorporated the gene properly are then introduced into the blastocyst. Chimeric animals result, but in the next generation, animals with the new gene in their germ cells can be found. For domestic livestock, the ES technique has met with some problems. However, Notarianni *et al.* (1990) described the derivation of apparently pluripotent cell lines from porcine and ovine blastocysts.

In cattle and sheep (but not in mice), young have been born following transfer of nuclei from cells of the inner cell mass of blastocysts to enucleated oocytes (Marx, 1988). It is possible, therefore, if gene-modified ES cells can be established from these species, that we could transfer nuclei from them and avoid the chimeric stage.

References

Anderson, W. F. (1984). *Science* **226,** 401.

Andersson, L., Boehme, J., Peterson, P. A., and Rask, L. (1986). *Anim. Genet.* **17,** 295.

Andersson, L., Sigurdardottir S., Borsch, C., and Gustafsson, K. (1991). *Immunogenetics* **33,** 188.

Andersson, L., Archibald, A. L., Gellin, J., and Schook, L. B. (1993). *Anim. Genet.* **24,** 205.

Andersson, L., Haley, C. S., Ellegren, H., Knott, S. A., Johansson, M., Andersson, K., Andersson-Eklund, L., Edfors-Lilja, I., Fredholm, M., Hansson, I., Håkansson, J., and Lundstrøm, K. (1994). *Science* **263,** 1771.

Apple, R. J., Erlich, H. A., Klitz, W., Manos, M. M., Becker, T. M., and Wheeler, C. M. (1994). *Nature Genet.* **6,** 157.

Arrand, J. E. (1985). *In* "Nucleic Acid Hybridization—a Practical Approach" (B. D. Hames and S. J. Higgins, eds.), pp. 17–45. IRL Press, Washington, D.C.

Avery, O. T., MacLeod, C. M., and McCarty, M. (1944). *J. Exp. Med.* **79,** 137.

Ballabio, A. (1993). *Nature Genet.* **3,** 277.

Barany, F. (1991). *Proc. Nat. Acad. Sci. USA* **88,** 189.

Barendse, W., Armitage, S. M., Kossarek, L. M., Shalom, A., Kirkpatrick, B. W., Ryan, A. M., Clayton, D., Li, L., Neibergs, H. L., Zhang, N., Grosse, W. M., Weiss, J., Creighton, P., McCarthy, F., Ron, M., Teale, A. J., Fries, R., McGraw, R. A., Moore, S. S., Georges, M., Soller, M., Womack, J. E., and Hetzel, D. J. S. (1994). *Nature Genet.* **6,** 227.

Batt, C. A., Wagner, P., Wiedmann, M., Luo, J., and Gilbert, R. (1994). *Anim. Genet.* **25,** 95.

Bell, G. I., Selby, M. J., and Rutter, W. J. (1982). *Nature* **295,** 31.

Bell, G. I., Horita, S., and Karam, J. H. (1984). *Diabetes* **33,** 176.

Bishop, M. D., Kappes, S. M., Keele, J. W., Stone, R. T., Sunden, S. L. F., Hawkins, G. A., Toldo, S. S., Fries, R., Grosz, M. D., Yoo, J., and Beattie, C. W. (1994). *Genetics* **136,** 619.

Capon, D. J., Chen, E. Y., Levinson, A. D., Seeburg, P. H., and Goeddel, D. V. (1983). *Nature* **302,** 33.

Carver, A. S., Dalrymple, M. A., Wright, G., Cottom, D. S., Reeves, D. B., Gibson, Y. H., Keenan J. L., Barrass, J. D., Scott, A. R., Colman, A., and Garner, I. (1993). *Bio/Technol.* **11,** 1263.

Chada, K., Magram, J., Raphael, K., Radice, G., Lacy, E., and Costantini, F. (1985). *Nature* **314,** 377.

Chowdhury, J. R., Grossman, M., Gupta, S., Chowdhury, N. R., Baker, J. R., and Wilson, J. M. (1991). *Science* **254,** 1802.

Clark, A. J., Simons, J. P., and Wilmut, I. (1992). *In* "Transgenic Animals" (F. Grosveld and G. Kollias, eds.), pp. 249–270. Academic Press, New York.

Cohen, D., Chumakov, I., and Weissenbach, J. (1993). *Nature* **366,** 698.

Collins, F. S. (1992). *Nature Genet.* **1,** 3.

Cooper, D. N., and Schmidtke, J. (1984). *Hum. Genet.* **66,** 11.

Costantini, F., Chada, K., and Magram, J. (1986). *Science* **233,** 1192.

Davies, W., Harbitz, I., and Hauge, J. G. (1987). *Anim. Genet.* **18,** 233.

Davies, W., Harbitz, I., Fries, R., Stranzinger G., and Hauge J. G. (1988). *Anim. Genet.* **19,** 203.

Dennis, J. A., Healy, P. J., Beaudet, A. L., and O'Brien, W. E. (1989). *Proc. Nat. Acad. Sci. USA* **86,** 7947.

Dorin, J. R., Dickinson, P., Alton, E. W. F. W., Smith, S. N., Geddes, D. M., Stevenson, B. J., Kimber, W. L., Fleming, S., Clarke, A. R., Hooper, M. L., Anderson, L., Beddington, R. S. P., and Porteous, D. J. (1992). *Nature* **359,** 211.

Ellegren, H., Chowdhary, B. P., Johansson, M., Marklund, L., Fredholm, M., Gustavsson, I., and Andersson, L. (1994). *Genetics* **137,** 1089.

Emery, A. E. H. (1984). "An Introduction to Recombinant DNA." Wiley, Chichester.

Erlich, H. A. (1989). "PCR Technology. Principles and Applications for DNA Amplification." Stockton Press, New York.

Erlich, H., Bugwan, T., Begovich, A. B., Scharf, S., Griffith, R., Saiki, R., Higuchi, R., and Walsh, P. S. (1991a). *Eur. J. Immunogen.* **18,** 33.

Erlich, H. A., Gelfand, D., and Sninsky, J. J. (1991b). *Science* **252,** 1643.

Fabb, S. A., Maddox, J. F., Gogolin-Ewens, K. J., Baker, L., Wu, M-J., and Brandon, M. R. (1993). *Anim. Genet.* **24,** 249.

Feder, J., Yen, L., Wijsman, E., Wang, L., Wilkins, L., Schroder, J., Spurr, N., Cann, H., Blumenberg, M., and Cavalli-Sforza, L. L. (1985). *Am. J. Hum. Genet.* **37,** 635.

Feinberg, A. P., and Vogelstein, B. A. (1983). *Anal. Biochem.* **132,** 6.

Ferec, C., Audrezet, M. P., Mercier, B., Guillermit, H., Moullier, P., Quere, I., and Verlingue, C. (1992). *Nature Genet.* **1,** 188.

Francomano, C. A., and Kazazian, H. H. (1986). *Ann. Rev. Med.* **37,** 377.

Freifelder, D. (1983). "Molecular Biology." Jones and Bartlett, Boston.

Friedman, T. (1993). *In* "Molecular Genetic Medicine" (T. Friedman, ed.), pp. 1–32. Academic Press, New York.

Fujii, J., Otsu, K., Zorzato, F., De Leon, S., Khanna, V. K., Weiler, J. E., O'Brien, P. J., and MacLennan, D. H. (1991). *Science* **253,** 448.

Ghosh, S., Palmer, S. M., Rodrigues, N. R., Cordell, H. J., Hearne, C. M., Cornall, R. J., Prins, J.-B., McShane, P., Lathrop, G. M., Peterson, L. B., Wicker, L. S., and Todd, J. A. (1993). *Nature Genet.* **4,** 404.

Goodridge, A. G., and Hanson, R. W. (1986). "Metabolic Regulation: Application of Recombinant DNA Techniques." *New York Acad. Sci.* **478.**

Gordon, J. W., Scangos, G. A., Plotkin, D. J., Barbos, J. A., and Ruddle, F. H. (1980). *Proc. Nat. Acad. Sci. USA* **77,** 7380.

Grossman, M., Raper, S. E., Kozarsky, S. E., Stein, E. A., Engelhardt, J. F., Muller, D., Lupien, P. J., and Wilson, J. M. (1994). *Nature Genet.* **6,** 335.

Grosveld, F., and Kollias, G. (1992). "Transgenic Animals." Academic Press, New York.

Gyapay, G., Morisette, J., Vignal, A., Dib, C., Fizames, C., Milasseau, P., Marc, S., Bernardi, G., Lathrop, M., and Weissenbach, J. (1994). *Nature Genet.* **7,** 246.

Hammer, R. E., Palmiter, R. D., and Brinster, R. L. (1984). *Nature* **316,** 65.

Hammer, R. E., Pursel, V. G., Rexroad, C. E., Wall, R. J., Bolt, D. J., Ebert, K. M., Palmiter, R. D., and Brinster, R. L. (1985). *Nature* **315,** 680.

Hanahan, D. (1985). *Nature* **315,** 1152.

Hansel, W., and Weir, B. J. (1990). *In* "Genetic Engineering of Animals: Proceedings for the Second Symposium on Genetic Engineering of Animals" (W. Hansel and B. J. Weir, eds.). *J. Reprod. Fertil. Suppl.* **41.**

Hanson, M. P., van Zandt, J. N., and Law, G. R. (1967). *Poult. Sci.* **46,** 1268.

Harbitz, I., Chowdhary, B., Thomsen, P. D., Davies, W., Kaufmann, U., Kran, S., Gustavsson, I., Christensen, K., and Hauge, J. G. (1990). *Genomics* **8,** 243.

Harbitz, I., Kristensen, T., Bosnes, M., Kran, S., and Davies, W. (1992). *Anim. Genet.* **23,** 395.

Hauge, (1988). Unpublished data.

Healy, P. J., and Dennis, J. A. (1994). *Anim. Genet.* **25,** 329.

Hepkema. B. G., Blankert. J. J., Albers, G. A. A., Tilanus, M. G. J., Egberts, E., van der Zijpp, S. A. J., and Hensen, E. J. (1993). *Anim. Genet.* **24,** 283.

Hill, A. V. S., Alsopp, C. E. M., Kwiatkowski, D., Anstey, N. M., Twumasi, P., Rowe, P. A., Bennett, S., Brewster, D., McMichael, A. J., and Greenwood, B. M. (1991). *Nature* **352,** 595.

Hobbs, H. H., Brown, M. S., and Goldstein, J. L. (1992). *Hum. Mut.* **1,** 445.

Hod, Y., Cook, J. S., Weldon, S. L., Short, J. M., Wynshaw-Boris, A., and Hanson, R. W. (1986). *In* "Metabolic Regulation: Application of Recombinant DNA Techniques" (A. G. Goodridge and R. W. Hanson, eds.), pp. 31–45. *New York Acad. Sci.* **478.**

Hu, C-W. C., Lau, K. S., Griffin, T. A., Chuang, J. L., Fischer, C. W., Cox, R. P., and Chuang, D. T. (1988). *J. Biol. Chem.* **263,** 9007.

Hultman, T., Ståhl, S., Hornes E., and Uhlen, M. (1989). *Nucl. Acids Res.* **17,** 4937.

Hyde, S. C., Gill, D. R., Higgins, C. F., Trezise, A. E. O., MacVinish, L. J., Cuthbert, A. W., Ratcliff, R., Evans, M. J., and Colledge, W. H. (1993). *Nature* **362,** 250.

Innerarity, T. L., Mahley, R. W., Weisgraber, K. H., Bersot, T. P., Krauss, R. M., Vega, G. I., Grundy, S. M., Friedl, W., Davignon, J., and McCarthy, B. J. (1990). *J. Lip. Res.* **8,** 1337.

Jeffreys, A. J., Wilson, V., and Thein, S. L. (1985a). *Nature* **314,** 67.

Jeffreys, A. J., Wilson, V., and Thein, S. L. (1985b). *Nature* **316,** 76.

Juul-Madsen, H. R., Hedemand, J. E., Salomonsen, J., and Simonsen, M. (1993). *Anim. Genet.* **24,** 243.

Kidd, V. J., Golbus, M. S., Wallace, R. B., Itakura, K., and Woo, S. L. C. (1984). *N. Engl. J. Med.* **310,** 639.

Kleppe, K., Ohtsuka, E., Kleppe R., Molineux, I., and Khorana, H. G. (1971). *J. Mol. Biol.* **56,** 341.

Knip, M., Ilonen, J., Mustonen, A., and Åkerblom, H. K. (1986). *Diabetologia* **29,** 347.

Kostyu, D. D. (1991). *Curr. Opin. Genet. Devel.* **1,** 40.

Krimpenfort, P. A., Rademakers, A., Eyestone, W., Van de Schans, A., Van den Broek, S., Koiman, P., Kootwijk, E., Platenburg G., Pieper, F., Strijker, R., and de Boer, H. (1991). *Bio/Technol.* **9,** 844.

Lai, F. A., Erickson, H. P., Roussseau, E., Liu, Q.-Y., and Meissner, G. (1988). *Nature* **331,** 315.

Land, H., Parada, L. F., and Weinberg, R. A. (1983). *Science* **222,** 771.

Le Meur, M., Gerlinger, P., Benois, C., and Mathis, D. (1985). *Nature* **316,** 38.

Leary, J. J., Brigati, D. J., and Ward, D. C. (1983). *Proc. Nat. Acad. Sci. USA* **80,** 4045.

Leren, T. P., Solberg, K., Rødningen, O. K., Røsby, O., Tonstad, S., Ose, L., and Berg, K. (1993). *Hum. Genet.* **92,** 6.

Lewin, H. A. (1994). *Anim. Biotech.* **5,** 183.

Little, P. F. R. (1981). *In* "Genetic Engineering 1" (R. Williamson, ed.), pp. 61–102. Academic Press, New York.

Lovell-Badge, R. H., Bygrave, A. E., Bradley, A., Robertson, E., Evans, M. J., and Cheah, K. S. E. (1985). *Cold Spring Harb. Symp. Quant. Biol.* **50,** 707.

Lu, K. H., Gordon, I., Gallagher, M., and McGovern, H. (1987). *Vet. Rec.* **121,** 259.

Lucas, P. C., and Granner, D. K. (1992). *Ann. Rev. Biochem.* **61,** 1131.

Lucassen, A. M., Julier, C., Beressi, J. P., Beitard, C., Froguel, P., Lathrop, M., and Bell, J. I. (1993). *Nature Genet.* **4,** 305.

McGrane, M. M., deVente, J., Yun, J., Bloom, J., Park, E., Wynshaw-Boris, A., Wagner, T., Rottman, F. M., and Hanson, R. W. (1988). *J. Biol. Chem.* **263,** 11443.

Magram, J., Chada, K., and Costantini, F. (1985). *Nature* **315,** 338.

Marklund, S., Ellegren, H., Eriksson, S., Sandberg, K., and Andersson, L. (1994). *Anim. Genet.* **25,** 19.

Marmur, J., and Doty, P. (1961). *J. Mol. Biol.* **3,** 585.

Marx, J. L. (1988). *Science* **239,** 463.

Maxam, A. M., and Gilbert, W. (1977). *Proc. Nat. Acad. Sci. USA* **74,** 560.

Meissner, G. (1986). *J. Biol. Chem.* **261,** 6300.

Mejdell, C. M., Lie, Ø., Solbu, H., Arnet, E. F., and Spooner, R. L. (1994). *Anim. Genet.* **25,** 99.

Miawa, S. (1994). *Nature Genet.* **6,** 3.

Michelsen, B., and Lernmark, Å. (1987). *J. Clin. Invest.* **79,** 1144.

Miller, A. D. (1992). *Science* **357,** 455.

Montgomery, G. W., Crawford, A. M., Penty, J. M., Dodds, K. G., Ede, J., Henry, H. M., Person, C. A., Lord, E. A., Galloway, S. M., Schmack, A. E., Sise, J. A., Swarbrick, P. A., Hanraham, V., Buchanan, F. C., and Hill, D. F. (1993). *Nature Genet.* **4,** 410.

Moon Sung, A., Nordskog, A. W., Lamont, S. J., and Warner, C. M. (1993). *Anim. Genet.* **24,** 227.

Mørsch, G., and Leibenguth, F. (1994). *Anim. Genet.* **25,** 25.

Nakamura, Y., Leppert, M., O'Connel, P., Wolf, R., Holm, T., Culver, M., Martin, C., Fujimoto, E., Hoff, M., Kumlin, E., and White, R. (1987). *Science* **235,** 1616.

Neibergs, H. L., Dietz, A. B., and Womack, J. E. (1993). *Anim. Genet.* **24,** 81.

Nepom, G. T., and Erlich, H. (1991). *Ann. Rev. Immunol.* **9,** 493.

Notarianni, E., Galli, C., Laurie, S., Moor, R. M., and Evans, M. J. (1990). *Proc. 4th World Cong. Genet. Appl. Livest. Prod.* **9,** 58.

Nygaard, A. P., and Hall, B. D. (1964). *J. Mol. Biol.* **9,** 125.

O'Brien, S. J., Womack, J. E., Lyons, L. A., Moore, K. J., Jenkins, N. A., and Copeland, N. G. (1993). *Nature Genet.* **3,** 103.

Old, R. W., and Primrose, S. B. (1990), "Principles of Gene Manipulation." Blackwell, Boston.

Olson, M., Hood, L., Cantor, C., and Botstein, D. (1989). *Science* **245,** 1434.

Orita, M., Suzuki, Y., Sekiya, T., and Hayashi, K. (1989). *Genomics* **5,** 874.

Ornitz, D. M., Palmiter, R. E., Hammer, R. E., Brinster, R. L., Swift, G. E., and MacDonald, R. J. (1985). *Nature* **313,** 600.

Otsu, K., Khanna, V. K., Archibald, A. L., and MacLennan, D. H. (1991). *Genomics* **11,** 744.

Owerbach, D. A., Lernmark, A., Platz, P., Ryder, L. P., Rask, L., Peterson, P. A., and Ludvigsson, J. (1983). *Nature* **303,** 815.

Palmiter, R. D., and Brinster, R. L. (1986). *Ann. Rev. Genet.* **20,** 465.

Park, C. A., Hines, H. C., Monke, D. R., and Threlfall, W. T. (1993). *Anim. Genet.* **24,** 53.

Park, E. A., Gurney, A. L., Nizielski, S. E., Hakimi, P., Cao, Z., Moorman, A., Hanson, R. W. (1993). *J. Biol. Chem.* **268,** 613.

Patel, Y. M., Yun, J. S., Liu, J., McGrane, M. M., and Hanson, R. W. (1994). *J. Biol. Chem.* **269,** 5619.

Proudfoot, N. J., Gil, A., and Maniatis, T. (1982). *Cell* **31,** 553.

Pursel, V. G., and Rexroad, C. E. (1993). *J. Anim. Sci.* **71,** 10.

Pursel, V. G., Pinkert, C. A., Miller, K. F., Bolt, D. J., Campbell, R. G., Palmiter, R. D., Brinster, R. L., and Hammer, R. E. (1989). *Science* **244,** 1281.

Robertson, E., Bradley, A., Kuehn, M., and Evans, M. (1986). *Nature* **323,** 445.

Rødningen, O. K., Leren, T. P., Røsby, O., Tonstad, D., Ose, L., and Berg, K. (1992). *Clin. Genet.* **41,** 288.

Rohrer, G. A., Alexander, L. J., Keele, J. W., Smith, T. P., and Beattie, C. W. (1994). *Genetics* **136,** 231.

Rosenfeld, M. A., Siegfried, W., Yoshimura, K., Yoneyama, K., Fukayama, M., Stier, L. E., Paakko, P. K., Gilardi, P., Stratford–Perricaudet, L. D., Perricaudet, M., Jallat, S., Pavirani, A., Lelcocq, J. P., and Crystal, R. G. (1991). *Science* **252,** 431.

Rudolph, J. A., Spier, S. J., Byrns, G., and Hoffman, E. P. (1992a). *Anim. Genet.* **23,** 241.

Rudolph, J. A., Spier, S. J., Byrns, G., Rojas, C. V., Bernoco, D., and Hoffman, E. P. (1992b). *Nature Genet.* **2,** 144.

Rusconi, S., and Køhler, G. (1985). *Nature* **314,** 330.

Saiki, R., Scharf, S., Faloona, F., Mullis, K. B., Horn, G. T., Erlich, H. A., and Arnheim, N. (1985). *Science* **230,** 1350.

Saiki, R. K., Gelfand, D. H., Stoffel, S., Scharf, S. J., Higuchi, R., Horn, G. T., Mullis, K. B., and Erlich, H. A. (1988). *Science* **239,** 487.

Salter, D. W., and Crittenden, L. B. (1989). *Theor. Appl. Genet.* **77,** 457.

Sanger, F., Nicklen, S., and Coulsen, A. R. (1979). *Proc. Nat. Acad. Sci USA* **74,** 5463.

Sanger, F., Coulson, A. R., Hong, G. F., Hill, D. F., and Petersen, G. B. (1982). *J. Molec. Biol.* **162,** 729.

Scharf, S. J., Griffith, R. L., and Erlich, H. A. (1991). *Hum. Immun.* **30,** 190.

Schøber, S., Simon, D., and Schwenger, B. (1992). *Gene* **124,** 307.

Schuster, D. E., Bosworth, B. T., and Kehrli, M. E. (1992a). *Gene* **114,** 267.

Schuster, D. E., Kehrli, M. E., Ackerman, M. R., and Gilbert, R. O. (1992b). *Proc. Nat. Acad. Sci. USA* **89,** 9225.

Skolnick, M. H., and White, R. (1982). *Cytogenet. Cell Genet.* **32,** 58.

Smithies, O., Gregg, R. C., Boggs, S. S., Koralewski, M. A., and Kucherlapati, R. S. (1985). *Nature* **317,** 230.

Snouwaert, J. N., Brigman, K. K., Latour, A. M., Malouf, N. N., Boucher, R. C., Smithies, O., and Koller, B. H. (1992). *Science* **257,** 1083.

Solomon, M. B., Pursel, V. G., Paroczay, E. W., and Bolt, D. J. (1994). *J. Anim. Sci.* **72,** 1242.

Southern, E. M. (1979). *Meth. Enzymol.* **68,** 152.

Stryer, L. (1981). "Biochemistry." Freeman, San Francisco.

Swanson, M. E., Martin, M. J., O'Donnel, J. K., Hoover, K., Lago, W., Huntress, V., Parsons, C. T., Pinkert, C. A., Pilder, S., and Logan, J. S. (1992). *Biotech.* **10,** 557.

Swift, G. H., Hammer, R. E., MacDonald, R. J., and Brinster, R. L. (1984). *Cell* **38,** 639.

Syvanen, A. C. (1986). *Med. Biol.* **64,** 313.

Szalai, G., Bailey, E., Gerber, H., and Lazary, S. (1993). *Anim. Genet.* **24,** 187.

Takeshima, H., Nishimura, S., Matsumoto, T., Ishida, H., Kanagawa, K., Minamino, N., Matsua, H., Ueda, M., Hanaoka, M., Hirose, T., and Numa, S. (1989). *Nature* **339,** 439.

Thomas, K. R., and Capecchi, M. R. (1986). *Nature* **324,** 34.

Thorsby, E., and Rønningen, K. S. (1993). *Diabetologia* **36,** 371.

Todd, J. A., Bell, J. I., and McDevitt, H. O. (1987). *Nature* **329,** 599.

Tolstoshev, P., and Anderson, W. F. (1993). *In* "Genome Research in Molecular Medicine and Virology" (K. W. Adolph, ed.), pp. 35–50. Academic Press, New York.

Våge, D. I., Olsaker, I., Lingaas, F., and Lie, Ø. (1994). *Anim. Genet.* **25,** 73.

Van de Woude, G. F., Levine, A. J., Topp, W. C., and Watson, J. D. (1984). "Oncogenes and Viral Genes," Cold Spring Harbor Press, New York.

Velander, W. H., Johnson, J. L., Page, R. L., Russel, C. G., Subramanian, A., Wilkins, T. D., Gwazdauskas, F. C., Pittius, C., and Drohan, W. N. (1992). *Proc. Nat. Acad. Sci. USA* **89,** 12003.

Vincent, N., Ragot, T., Gilgenkrantz, H., Couton, D., Chafey, P., Gregoire, A., Briand, H., Kaplan, J. C., Kahn, A., and Perri-Caudet, M. (1993). *Nature Genet.* **5,** 130.

Vogel, F., and Motulsky, A. G. (1986). "Human Genetics." Springer, New York.

Watson, J. D., and Crick, F. H. C. (1953). *Nature* **171,** 737.

Watson, J. D., Gilman, M., Wirkowsky, J., and Zoller, M. (1992). "Recombinant DNA." Freeman, New York.

Weber, J. L., and May, P. E. (1989). *Am. J. Hum. Genet.* **44,** 388.

Weidle, U. H., Lenz, H., and Brem, G. (1991). *Gene* **98,** 185.

Weinberg, R. A. (1991). *Science* **254,** 1138.

Weissenbach, J., Gyapay, G., Dib, C., Vignal, A., Morisette, J., Milasseau, P., Vaysseix, G., and Lathrop, M. (1992). *Nature* **359,** 794.

Westphal, H. (1987). *Bio Essays* **6,** 73.

Wijsman, E. M. (1984). *Nucl. Acids Res.* **12,** 9209.

Williamson, R. (1993). *Nature Genet.* **3,** 195.

Wivel, N. A. (1993). *Trends Biotech.* **11,** 189.

Wright, G., Carver, A., Cottom, D., Reeves, D., Scott, A., Simons, P., Wilmut, I., Garner, I., and Colman, A. (1991). *Biotech.* **9,** 830.

Yamamoto, K. K., Gonzales, G. A., Biggs, W. H., and Montminy, M. R. (1988). *Nature* **334,** 494.

Young, B. D., and Anderson, M. L. H. (1985). *In* "Nucleic Acid Hybridization—A Practical Approach" (B. D. Hames and S. J. Higgins, eds.), pp. 47–111. IRL Press, Washington, D.C.

Zabner, J., Petersen, D. M., Puga, A. P., Graham, S. M., Couture, L. A., Keyes, L. D., Lukason, M. J., St. George, J. A., Gregory, R. J., Smith, A. E., and Welsh, M. J. (1994). *Nature Genet.* **6,** 75.

Zarbl, H., Saraswati, S., Arthur, A. V., Dionisio, M. Z., and Barbacid, M. (1985). *Nature* **315,** 382.

Zhang, B., Healy, P. J., Crabb, D. W., and Harris, R. A. (1990). *J. Biol. Chem.* **265,** 2425.

C H A P T E R

3

Carbohydrate Metabolism and Its Diseases

J. JERRY KANEKO

I. INTRODUCTION

The biochemical mechanisms by which the chemical energy contained in foodstuffs is made available to the animal are collectively described as metabolism. Thus, the description of the metabolism of a foodstuff encompasses the biochemical events that occur from the moment of ingestion to its final breakdown and excretion. Classically, these biochemical events have been divided into the metabolism of the three major constituents of food: carbohydrates, proteins, and lipids. The metabolism of the lipids and proteins is discussed in other chapters.

The major function of ingested carbohydrates is to serve as energy sources and their storage function is relatively minor. Carbohydrates also function as precursors of essential intermediates for use in synthetic processes. When the metabolic machinery of an animal is disrupted, a disease state prevails, for example, diabetes. The literature of the biochemistry of metabolism and disease continues to expand as the intricate details of individual and overall reaction mechanisms are continually refined and elucidated. Additionally, modern molecular approaches have significantly increased our understanding of disease mechanisms and remain fertile fields for investigations into the disease processes. This chapter is presented as a basis for the better understanding of the biochemical mechanisms underlying those diseases associated with carbohydrate metabolism. Exhaustive treatment of carbohydrate metabolism in health and disease is beyond the scope of this chapter.

II. DIGESTION

The digestion of carbohydrates in the animal begins with the initial contact of these carbohydrates with the enzymes of salivary juice. Starch of plant foods and glycogen of meat are split into their constituent monosaccharides by the action of amylase and maltase. This activity ceases as the food matter passes into the stomach, where the enzymatic action is destroyed by the hydrochloric acid. Within the stomach, acid hydrolysis may occur, but the stomach empties too rapidly for complete hydrolysis to take place. Thus, only a small portion of the ingested carbohydrate is hydrolyzed prior to passage into the small intestine. In the small intestine, digestion of carbohydrate takes place quickly by the carbohydrate splitting enzymes contained in the copious quantities of pancreatic juice and in the succus entericus. Starch and glycogen are hydrolyzed to glucose by amylase and maltase; lactose to glucose and galactose by lactase; and sucrose to glucose and fructose by sucrase (sucrose-α-glucosidase). The monosaccharide products of enzymatic hydrolysis of carbohydrates, glucose, fructose, and galactose, are the principal forms in which absorption occurs in the monogastric animal.

III. ABSORPTION

The monosaccharides are almost completely absorbed through the mucosa of the small intestine and appear in the portal circulation as the free sugars. Absorption occurs by two methods: (1) facilitated diffusion and (2) sodium-dependent active transport. Glucose and galactose are absorbed rapidly and by both methods. Fructose is absorbed at about half the rate of glucose with a portion being converted to glucose in the process. Other monosaccharides, for example, mannose, are absorbed slowly at a rate consistent with a diffusion process. The active absorption of glucose across the intestinal mucosa is thought to be by phosphorylation in the mucosal cell. The phosphorylated sugars are then transferred across the mucosal cell, rehydrolyzed, and free glucose appears in the portal circulation for transport to the liver.

The possible role of glucose transporters in this event is not yet elucidated but they are very likely to be involved. Glucose transporters are known to be involved in many tissues, including brain, erythrocytes, kidney, liver, pancreas, skeletal muscle, heart muscle, and fat cells.

IV. METABOLISM OF ABSORBED CARBOHYDRATE

A. General

Liver cells are readily permeable to the absorbed glucose. This process is facilitated by glucose transporter (GLUT) proteins within the plasma membrane, in particular, GLUT-2 is the transporter in the liver cell plasma membrane (Thorens *et al.*, 1988). Within the liver, there are several pathways by which the immediate fate of the absorbed hexose is determined. Glucose, fructose, and galactose first enter the general metabolic scheme through a series of complex reactions to form glucose phosphates (Fig. 3.1). The enzyme, galactose-1-P uridyl transferase, which catalyzes the reaction

$$\text{galactose-1-P} + \text{UDP-glucose} \rightarrow \text{UDP-galactose} + \text{glucose-1-P},$$

is blocked or deficient in congenital galactosemia of humans. The glucose phosphates are then converted

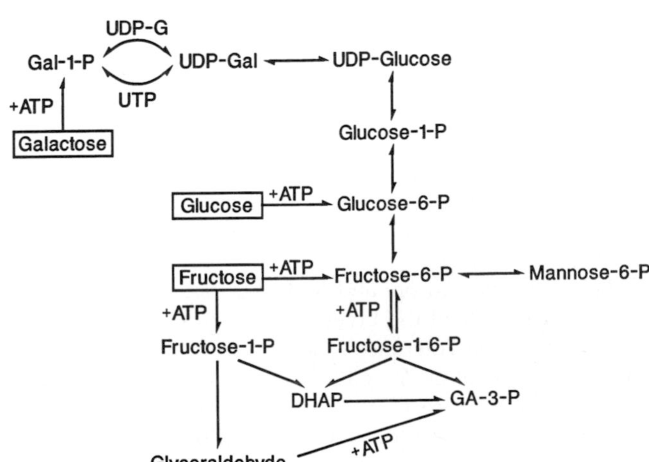

FIGURE 3.1 Pathways for hexose metabolism. ATP, adenosine triphosphate; UTP, uridine triphosphate; UDP-G, uridine diphosphate glucose; DHAP, dihydroxy acetone phosphate; GA-3-P, glyceraldehyde-3-phosphate.

TABLE 3.1 Liver Glycogen Content of Animals

Species	Glycogen in liver (%)	Reference
Dog	6.1	Lusk (1928)
Sheep	3.8	Roderick *et al.* (1933)
Cow (lactating)	1.0	Kronfeld *et al.* (1960)
Cow (nonlactating)	3.0	Kronfeld *et al.* (1960)
Baby pig	5.2	Morrill (1952)
Baby pig (newborn)	14.8	Swiatek *et al.* (1968)

to and stored as glycogen, catabolized to CO_2 and water or, as free glucose, returned to the general circulation. Essentially, intermediate carbohydrate metabolism of animals evolves about the metabolism of glucose, and the liver is the organ of prime importance.

B. Storage as Glycogen

Glycogen is the chief storage form of carbohydrate in animals and is analogous to the storage of starch by plants. It is found primarily in liver and in muscle, where it occurs at about 3–6% and about 0.5%, respectively (Table 3.1). Glycogen is comprised solely of α-D-glucose units linked together through carbon atoms 1 and 4 or 1 and 6. Straight chains of glucose units are formed by the 1–4 links and these are cross-linked by the 1–6 links. The result is a complex ramification of chains of glucosyl units with branch points at the site of the 1–6 links (Fig. 3.2). The internal chains of the glycogen molecule have an average length of four glucosyl units. The external chains beyond the last 1–6 link are longer and contain between 7 to 10 glucose units. The molecular weight of glycogen may be as high as 4×10^6 Mr and contain about 20,000 glucosyl units.

In Table 3.2, the amount of carbohydrate available to meet the theoretical requirements of a hypothetical dog is shown. The amount present is sufficient for about half a day. It is apparent that the needs of the body which must be continually met are satisfied by alternate means and not solely dependent on continuous ingestion of carbohydrates. During and after feeding (postprandial), absorbed hexoses are converted to glucose by the liver and enter the general circulation. Excesses are stored as glycogen or as fat. During the fasting or postabsorptive state, glucose is supplied by the conversion of protein (gluconeogenesis) and by the breakdown of glycogen (glycogenolysis). The continued rapid synthesis and breakdown of glycogen that

TABLE 3.2 Carbohydrate Content of a Dog[a]

Muscle glycogen (0.5%)	25.0 g
Liver glycogen (6%)	18.0 g
Carbohydrate in fluids 5.5 mmol/liter (100 mg/dl)	2.2 g
	45.2 g

Caloric value (45.2×4 kcal/g) = 181 kcal

Caloric requirement ($70\ \text{kg}^{\frac{3}{4}} = 70 \times 5.6$) = 392 kcal/day

$$\frac{181}{392} \times 24 \text{ hours} = 11 \text{ hours}$$

[a] Body weight, 10 kg; liver weight, 300 g; muscle weight, 5 kg; volume of blood and extracellular fluid, 2.2 liters.

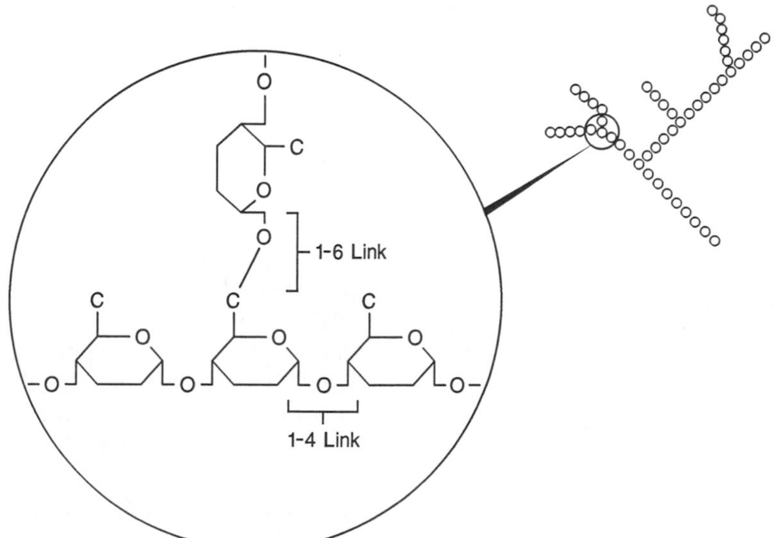

FIGURE 3.2 Glycogen structure. Note that hydrolysis of a 1–6 link by the debrancher enzyme yields a mole of free glucose.

is, turnover, is well illustrated by the biological half-time of glycogen, which is about a day.

C. Glycogen Metabolism

1. *Glycogenesis*

The initial reaction required for the entrance of glucose into the series of metabolic reactions that culminates in the synthesis of glycogen is the phosphorylation of glucose at the C-6 position. Glucose is phosphorylated with adenosine triphosphate (ATP) in liver by an irreversible enzymatic reaction catalyzed by a specific glucokinase (GK):

$$\text{glucose} + \text{ATP} \underset{\text{(HK-IV)}}{\overset{\text{GK}}{\longrightarrow}} \text{glucose-6-P} + \text{ADP}. \quad (3.1)$$

Glucokinase (GK) [also called hexokinase-IV (HK-IV)] is one of the four hexokinase isoenzymes that occurs in all tissues. Glucokinase or HK-IV, which is glucose specific, is the predominant isoenzyme found in liver. The nonspecific hexokinase-I (HK-I) is the isoenzyme found in red cells, brain, and nerve tissue.

Liver contains both GK (HK-IV) and HK-I but GK is the predominant isoenzyme. GK has a high Michaelis constant ($Km = 2 \times 10^{-2}$ mol G/liter) indicating a low affinity for glucose. The rate of the phosphorylation reaction catalyzed by GK is therefore controlled by the glucose concentration. The activity of GK is increased by glucose feeding and by insulin and is decreased during fasting and in insulin lack (i.e., diabetes). In this regard, GK is an inducible enzyme whose activity is increased by glucose or by insulin. The nonspecific HK-I is found in all tissues including liver, brain, and erythrocytes and has a low Michaelis constant ($Km = 5 \times 10^{-5}$ mol G/liter), indicating a high affinity for glucose. HK-I catalyzed phosphorylation in all tissues, therefore, is not controlled by glucose concentration. The activity of HK-I is not affected by fasting or carbohydrate feeding, by diabetes, or by insulin. Therefore, in contrast to GK, HK-I is not an inducible enzyme.

The initial unidirectional phosphorylation reaction permits the accumulation of glucose in the liver cells because phosphorylated sugars do not pass freely into and out of the cell in contrast to the readily transported free sugars. The glucose-6-phosphate (G-6-P) accumulated in the cell next undergoes a mutation in which the phosphate group is transferred to the C-1 position of the glucose molecule. This reaction is catalyzed by the enzyme, phosphoglucomutase (PGM) and involves glucose-1-6-diphosphate as an intermediate:

$$\text{Glucose-6-P} \rightarrow \text{glucose-1-P}. \quad (3.2)$$

Glycogen is synthesized from this glucose-1-phosphate (G-1-P) through a series of reactions involving the formation of uridine derivatives. Uridine diphosphoglucose (UDPG) is synthesized by the transfer of glucose from G-1-P to uridine triphosphate (UTP). This reaction is catalyzed by the enzyme UDP-G-pyrophosphorylase (UDP-G-PPase):

$$\text{UTP} + \text{G-1-P} \rightarrow \text{UDP-G} + \text{PP} \quad (3.3)$$

In the presence of a polysaccharide primer and the enzyme glycogen synthase (glucosyl transferase), the glucose moiety of UDP-G is linked to the polysaccharide chain by an α-1-4 link:

$$\text{UDP-G} + (\text{glucose 1–4})_n \xrightarrow[\text{synthase}]{\text{glycogen}} (\text{glucose 1–4})_{n+1} + \text{UDP}. \quad (3.4)$$

Through repeated transfers of glucose, the polysaccharide chain is lengthened. When the chain length of the polysaccharide reaches a critical level between 11 and 16 glucosyl units, the brancher enzyme, α-glucan glycosyl 4:6 transferase, transfers the terminal 7 residue portion from an α-1-4 linkage to an α-1-6 linkage. The newly established 1–6 linkage thus becomes a branch point in the expanding glycogen molecule. The remaining stub can again be lengthened by the action of glycogen synthase. Approximately 7% of the glucose units of the glycogen molecule are involved in these branch points.

2. *Glycogenolysis*

The breakdown of liver glycogen to glucose (glycogenolysis) takes place via a separate pathway. The key initiating and regulating factor in glycogenolysis is the action of epinephrine on liver and muscle glycogen and of glucagon on liver glycogen only. The mechanism of action of glucagon and epinephrine is through a series of reactions that culminates in the phosphorolytic cleavage of the 1–4 glucosyl links of glycogen. In the liver cell, glucagon and epinephrine stimulate the enzyme adenylate cyclase to form 3′-5′ cyclic adenosine monophosphate (cAMP) from ATP. In turn, cAMP activates a protein kinase, which in its turn activates liver phosphorylase (LP), the phosphorolytic enzyme. As with many enzymes, LP is present in an inactive form, dephospho-liver phosphorylase (dLP), which is converted to its active form, LP (Cherrington and Exton, 1976) by the protein kinase, phosphorylase kinase.

The action of the LP is to cleave the 1–4 glucosyl links of glycogen by the addition of orthophosphate in a manner analogous to a hydrolytic cleavage with water, hence the analogous term *phosphorolysis*. Phosphate is added to the C-1 position of the glucose moiety while H^+ is added to the C-4 position of the other.

cAMP is also a key regulating factor in cellular processes in addition to LP activation. It is required for the conversion of inactive muscle phosphorylase b to active muscle phosphorylase a, again via phosphorylase b kinase. The actions of other hormones known to be mediated by activating adenylate cyclase and cAMP include ACTH, LH, TSH, MSH, T_3, and insulin. From these findings, a general concept of hormone action has evolved in which the hormone elaborated by the endocrine organ is described as the first messenger and cAMP within the target cell is the second messenger.

Glucagon acts only on liver glycogen, whereas epinephrine acts on both liver and muscle glycogen. In liver, glucagon promotes the formation and release of glucose by increasing glycogenolysis and decreasing glycogenesis. In liver, the hydrolysis of G-6-P is catalyzed by the enzyme glucose-6-phosphatase (G-6-Pase) to release free glucose, thus promoting hyperglycemia. Additionally, glucagon promotes hyperglycemia by stimulation of hepatic gluconeogenesis and, thus, glucagon is a potent hyperglycemic factor. With muscle glycogen, however, since the enzyme G-6-Pase is absent from muscle, glycogen breakdown in muscle results in the production and release of pyruvate and lactate rather than glucose. Mainly lactate and some pyruvate is transported to the liver where glucose is resynthesized via reverse glycolysis (Cori cycle; Section IV.D).

The continued action of LP on the 1–4 linkages results in the sequential release of G-1-P units until a branch point in the glycogen molecule is reached. The residue is a limit dextrin. The debrancher enzyme, amylo-1-6-glucosidase, then cleaves the 1–6 linkage, releasing free glucose. The remaining 1–4 linked chain of the molecule is again open to attack by LP until another limit dextrin is formed. Thus, by the combined action of LP and the debrancher enzyme, the glycogen molecule is successively reduced to G-1-P and free glucose units.

G-1-P is converted to G-6-P by the reversible reaction catalyzed by phosphoglucomutase (PGM, Section IV.C.1, Reaction 3.2). The G-6-P is then irreversibly cleaved to free glucose and phosphate by the enzyme G-6-Pase, which is found in liver and kidney. The free glucose formed can, unlike its phosphorylated intermediates, be transported out of the hepatic cell and enter the general circulation, thereby contributing directly to the blood glucose pool. In muscle tissue, there is no G-6-Pase and muscle glycogen cannot supply glucose directly to the circulation by glycogenolysis. Muscle glycogen contributes to blood glucose indirectly via the lactate or Cori cycle (Section IV.D). The series of reactions described is illustrated schematically in Fig. 3.3.

3. *Hormonal Influences on Glycogen Metabolism*

The biochemical basis of the glycogenolytic and hyperglycemic action of glucagon and epinephrine has been discussed in Section IV.C.2. These hormone actions are the bases for the epinephrine and glucagon stimulation tests that are used to assess the availability of liver glycogen and the sensitivity of the carbohydrate regulatory mechanisms to these hormones. Many other hormones influence carbohydrate metabolism to a greater or lesser degree in keeping with the concept that carbohydrate metabolism is a totally integrated metabolic mechanism.

One of the results of successful insulin therapy is a restoration of the depleted glycogen reserve. The mechanism of insulin action on carbohydrate metabolism continues to be a subject for intense study and is discussed more fully in Section VI. Briefly, the primary role of insulin is to promote glucose entry into peripheral cells, mainly muscle and fat cells, and to enhance glucose utilization by liver cells by its effect on enzyme systems at control points in the glycolytic pathways. In the presence of insulin, glucose removal from the blood is enhanced by shifting the direction of glucose metabolism toward utilization by increasing glycogen synthesis and glucose uptake oxidation. The result is a hypoglycemia.

Promotion of liver glycogen storage is also one of the effects of the glucocorticoids. This effect may be attributed to their enhancement of gluconeogenesis, hyperglycemia, decreased glycogenolysis, and decreased glucose oxidation. A tendency toward a mild hyperglycemia is also present in hyperthyroid states, as the result of an overall increase in carbohydrate metabolism. Thyroxine is thought to increase the sensitivity of the liver cell to the action of epinephrine, thereby increasing glycogenolysis and promoting hyperglycemia. Increased glycogenolysis, gluconeogenesis, and the hyperglycemia may also be the compensatory result of an increased rate of tissue metabolism. In rats made hyperthyroid, hepatic G-6-Pase activities are increased, which would enhance hepatic glucose production and hyperglycemia in the hyperthyroid states. An additional factor contributing to the overall tendency for hyperglycemia is the stimulation of glucose absorption by the gastrointestinal tract by thyroxine.

4. *Glycogen in Disease*

In systemic disease, changes in glycogen concentrations in tissues or organs are generally observed as decreases. Depletion of liver glycogen stores is seen in diabetes mellitus, starvation, bovine ketosis, ovine pregancy toxemia or in any condition with nutritional

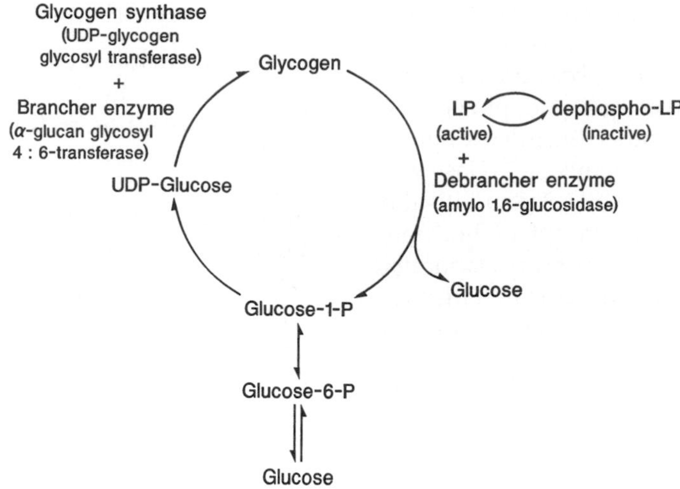

FIGURE 3.3 Summary of liver glycogen metabolism. In muscle, phosphorylase a is the active form and phosphorylase b is the inactive form. UDP, uridine diphosphate; LP, liver phosphorylase.

carbohydrate deficiency or increased carbohydrate turnover. Pathological increases in liver glycogen occur in the rare glycogen storage diseases (GSDs) and are described in Section IX.D.

D. Catabolism of Glucose

Carbohydrate in the form of glucose is the principal source of energy for the life processes of the mammalian cell. All cells require a constant supply of this indispensable nutrient and only relatively small changes are tolerated without adverse effects on the health of the animal. Glucose is not oxidized directly to CO_2 and H_2O but rather through a series of stepwise reactions involving phosphorylated intermediates. The chemical energy of glucose is "stored" through the synthesis of "high-energy" phosphate bonds during the course of these reactions and used in other metabolic reactions. The details of the individual reactions in the pathways of glucose catabolism have been largely elucidated but emphasis here is being placed on the interrelationships of the pathways rather than on the details of the individual reactions.

1. Pathways of Glucose-6-Phosphate Metabolism

The fundamental conversion required to initiate the oxidation of glucose by a cell is its phosphorylation to form G-6-P. This reaction has been described in Section IV.C.1. The G-6-P formed as a result of the GK (HK-IV) catalyzed reaction is central to glucose catabolism. At least five different pathways can be followed by G-6-P: the free glucose, glycogenesis, glycolysis, hexose monophosphate, and glucuronate pathways.

a. Free Glucose Pathway

The simplest direction for G-6-P is a reversal of phosphophorylation by a separately enzyme catalyzed reaction in which G-6-P is hydrolyzed to form free glucose and inorganic phosphate. This reaction is catalyzed by the enzyme G-6-Pase:

$$\text{Glucose-6-PO}_4 \xrightarrow{\text{G-6-Pase}} \text{glucose} + \text{P}_i.$$

This is an irreversible reaction that opposes the previously described unidirectional GK (HK-IV) reaction. These two opposing and independently catalyzed enzyme reactions are the site of metabolic control for glucose because the balance of these enzyme activities regulates the net direction of the reaction. Significant amounts of G-6-Pase are found only in liver and to a lesser extent in the kidney. This is in accord with the well-known function of the liver as the principal source of supply of glucose for the maintenance of blood glucose concentration. The G-6-Pase activity is generally higher than the GK activity for most of a 24-hour day except for a few hours after each meal. This means that for most of the day, the liver is supplying glucose rather than using glucose.

Muscle G-6-P, however, because of the absence of G-6-Pase, does not contribute glucose from its glycogen to blood directly. Muscle G-6-P does, however, contribute glucose to blood indirectly via the lactate or Cori cycle. Lactate formed in muscle by glycolysis is transported to the liver, where it is resynthesized to glucose and its precursors as outlined in Fig. 3.4.

b. Glycogenesis Pathway

This pathway for G-6-Pase leading to the synthesis of glycogen has been discussed in Section IV.C.1.

c. Anaerobic Glycolysis

One of the three oxidative pathways of G-6-P is the classic anaerobic glycolytic or Embden–Meyerhof pathway (EMP). The intermediate steps involved in this pathway of breakdown of G-6-P to three-carbon compounds are summarized in Fig. 3.5. A mole of ATP is used to phosphorylate fructose-6-phosphate (F-6-P) to form fructose-1,6-diphosphate (F-1,6-P). This phosphorylation reaction is also irreversible and catalyzed by a specific kinase, phosphofructokinase (PFK). The opposing unidirectional reaction is catalyzed by a specific phosphatase, fructose-1,6-diphosphatase (F-1,6-Pase). The opposing PFK and F-1,6-Pase catalyzed reactions are a second site of metabolic control regulated by the activities of these two highly specific enzymes. At this point in the process, starting from glucose, a total of two high-energy phosphates from ATP have been donated to form a mole of F-1,6-P.

F-1,6-P is next cleaved to form two three-carbon compounds as shown in Fig. 3.5. The next step is an oxidative step catalyzed by the enzyme glyceraldehyde-3-phosphate dehydrogenase (GA-3-PD) with oxidized nicotinamide adenine dinucleotide (NAD^+) as the hydrogen acceptor. During the process, the molecule is phosphorylated. In the succeeding steps, the molecule is dephosphorylated at the points indicated, and a mole of ATP is generated at each point.

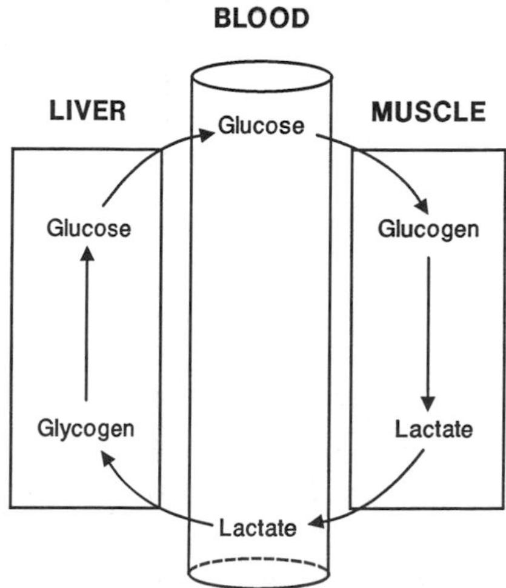

FIGURE 3.4 The lactate or Cori cycle. Muscle cells are devoid of glucose-6-phosphatase, therefore muscle glycogen contributes indirectly to blood glucose by this pathway.

A third site of control of glycolysis is the irreversible formation of pyruvate catalyzed by the enzyme pyruvate kinase (PK). In the reverse direction, two enzymatic reactions operate. Pyruvate carboxylase (PC) first catalyzes the carboxylation of pyruvate to oxaloacetate (OAA) and the OAA is then converted to phosphoenolpyruvate (PEP) by the enzyme PEP carboxykinase (PEPCK) (Figs. 3.5 and 3.8).

Thus, the overall conversion of a mole of glucose to 2 moles of pyruvate requires 2 moles of ATP for the initial phosphorylations. A total of 4 moles of ATP is generated in the subsequent dephosphorylations. This net gain of 2 moles of ATP represents the useful energy of anaerobic glycolysis.

For repeated function of the glycolytic pathway, a supply of NAD^+ must be available for use in the oxidative (GA-3-PD) step. Normally in the presence of molecular O_2, that is, aerobic glycolysis, reduced NADH is reoxidized via the cytochrome system

$$H^+ + NADH + \tfrac{1}{2}O_2 \xrightarrow[\text{(system)}]{\text{(cytochrome)}} NAD^+ + H_2O,$$

which provides a continuous source of NAD^+.

In the absence of O_2, that is, anaerobic glycolysis, NADH is reoxidized to NAD^+ in the reaction catalyzed by lactate dehydrogenase (LDH) where pyruvate is reduced to lactate and the NADH is the H^+ donor. Therefore, by this "coupling" of the LDH system to the GA-3-PD system, anaerobic breakdown of glucose to lactate proceeds in the absence of O_2. As noted earlier, this anaerobic system generates only 2 moles of ATP and when compared to the 36 moles of ATP generated in aerobic glycolysis, anaerobic glycolysis is not very efficient.

d. Hexose Monophosphate Pathway

This alternate route of G-6-P oxidation has been variously referred to as the pentose phosphate pathway (PPP), direct oxidative pathway, Warburg–Dickens scheme, the hexose monophosphate pathway (HMP), or the hexose monophosphate shunt. The initial step of the shunt pathway involves the oxidation of G-6-P at the C-1 position to form 6-phosphogluconate (6-PG) as summarized in Fig. 3.6. The reaction is catalyzed by glucose-6-phosphate dehydrogenase (G-6-PD) and in this pathway, oxidized nicotinamide adenine dinucleotide phosphate ($NADP^+$) serves as the hydrogen acceptor. In the second oxidative step, 6-P-G is oxidatively decarboxylated by 6-phosphogluconate dehydrogenase (6-P-GD) to yield a pentose phosphate, ribulose-5-phosphate (Rib-5-P) again in the presence of $NADP^+$. Thus, in the initial reactions, which are essentially irreversible, 2 moles of NADPH are formed.

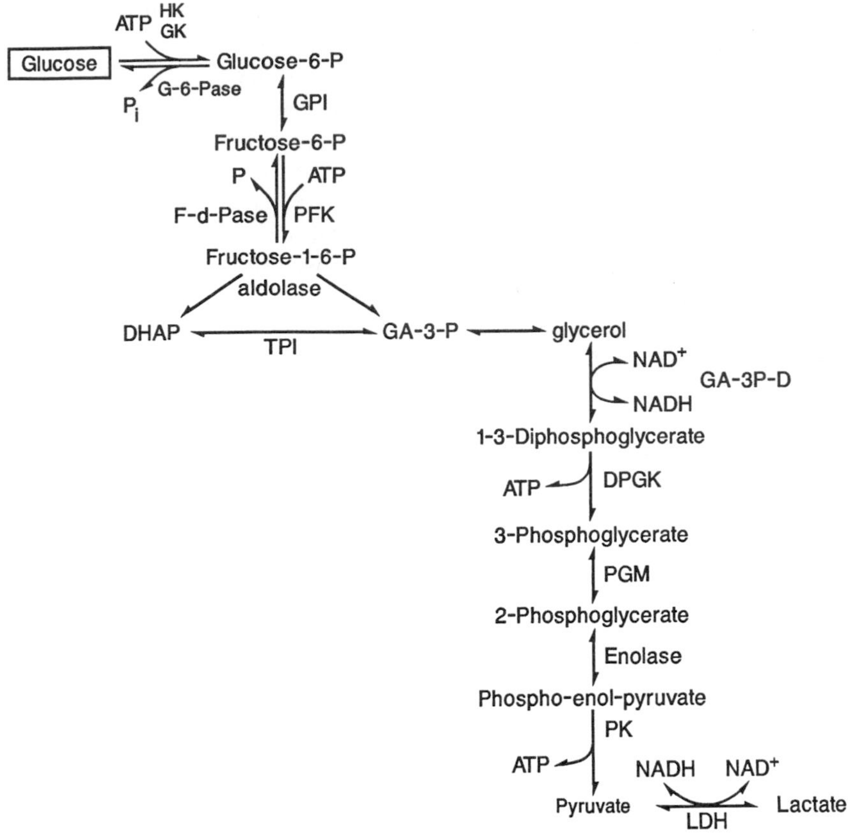

FIGURE 3.5 The glycolytic or classic Embden–Meyerhof pathway. Note that 2 moles of ATP are used and 4 moles of ATP are generated. ATP, adenosine triphosphate; DHAP, dihydroxy acetone phosphate; GA-3-P, glyceraldehyde-3-phosphate; NAD^+, nicotinamide adenine dinucleotide; Pi, inorganic phosphate.

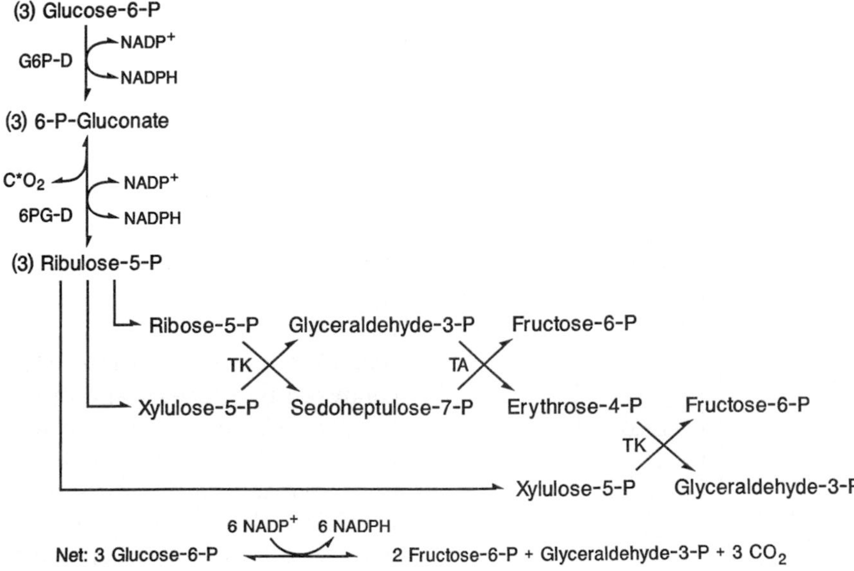

FIGURE 3.6 The pentose phosphate pathway (PPP) or the hexose monophosphate pathway (HMP). $NADP^+$, nicotinamide adenine dinucleotide phosphate; TK, transketolase; TA, transketolase; TA, transaldolase; C^*O_2, is derived from C_1 of glucose.

In this pathway only the C-1 carbon atom of the glucose molecule is evolved as CO_2. By contrast, glucose catabolism via the glycolytic scheme results in the loss of both the C-1 and C-6 carbon atoms as CO_2 when pyruvate is oxidatively decarboxylated to form acetyl-CoA. This difference in CO_2 evolution is used to study partitioning of glucose metabolism through the glycolytic (EMP) pathway and the HMP shunt pathway in domestic animals. The subsequent metabolism of the Rib-5-P in the HMP shunt is also shown in Fig. 3.6. As a result of the series of transformations, F-6-P and GA-3-P are formed and serve as recycling links into the glycolytic pathway.

For continued functioning of the HMP shunt pathway, a supply of $NADP^+$ must be available to act as the hydrogen acceptor. Oxidized $NADP^+$ is regenerated from NADPH via the cytochrome system in the presence of O_2 so the HMP pathway is an aerobic pathway of glucose oxidation. Reduced NADPH is also required as a hydrogen donor in the synthesis of fatty acids. Through generation of NADPH, the HMP shunt route of carbohydrate metabolism is linked to that of fat synthesis. Accordingly, glucose oxidation through the HMP shunt pathway is essential for the synthesis of fat. In general, the HMP pathway is the major source of the NADPH, which maintains the reductive environment for all biosynthetic processes using NADPH as a cofactor.

e. Glucuronate Pathway

This is an alternate pathway of G-6-P oxidation, which has been named the uronate pathway, glucuronate pathway, or the C_6 oxidative pathway. This pathway is shown in Fig. 3.7. The initial steps of this pathway involve the formation of UDPG, which as noted earlier, is an intermediate in glycogen synthesis. G-6-P is first converted to G-1-P, which then reacts with UTP to form UDPG. This product is then oxidized at the C_6 position of the glucose moiety in contrast to the C_1 position, which is oxidized in the HMP shunt pathway. This reaction requires NAD^+ as a cofactor and the products of the reaction are uridine diphosphoglucuronic acid (UDPGA) and NADH. This UDPGA is involved in a large number of important conjugation reactions in animals, for example, bilirubin glucuronide formation, synthesis of mucopolysaccharides (chondroitin sulfate), which contain glucuronic acid, and generally in detoxification reactions. UDPGA is cleaved to release D-glucuronate and UDP.

D-Glucuronate is next reduced to L-gulonate in a reaction catalyzed by the enzyme gulonate dehydrogenase (GUD), with NADPH as the hydrogen donor. The L-gulonate may be converted to a pentose, L-xylulose, or to vitamin C. When converted to L-xylulose, the C-6 carbon of L-gulonate is oxidatively decarboxylated and evolved as CO_2. The L-xylulose is then reduced to xylitol, catalyzed by the enzyme L-xylulose reductase. This is the enzyme that is deficient in pentosuria of humans. As shown in Fig. 3.7, xylitol is converted to D-xylulose, which is then phosphorylated to D-xylulose-5-P, and a cyclical pathway involving the HMP shunt pathway may occur. L-Gulonate is also converted by enzyme catalyzed reactions to L-ascorbate in those species that can synthesize their own vitamin C, that is, all domestic animals. The enzyme, L-gulonolactone oxidase (GLO), is lacking in humans, nonhuman primates, and guinea pigs and therefore vitamin C must be supplied in their diets. The enzyme is present only in the liver of the mouse, rat, pig, cow, and dog. In the dog, the liver GLO activity is low and the ascorbate hydrolytic activity is high so dogs may have additional needs for vitamin C during

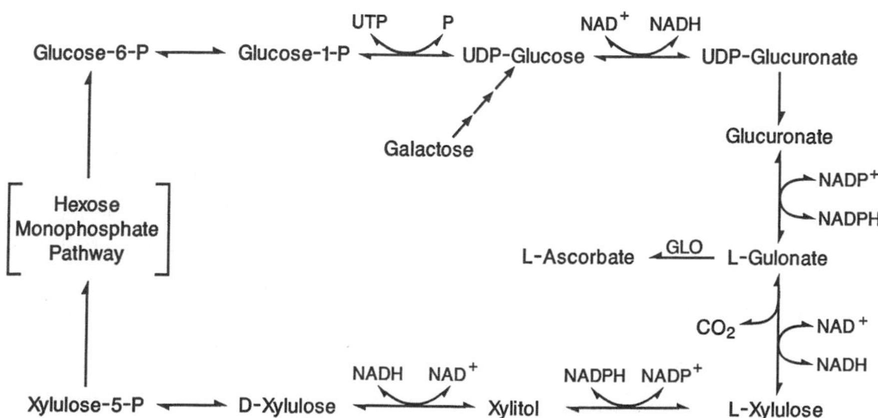

FIGURE 3.7 Glucuronate pathway or the C_6 oxidation pathway. Note that vitamin C is synthesized via this pathway. UTP, uridine triphosphate; NAD^+, Nicotinamide adenine dinucleotide; $NADP^+$, nicotinamide adenine dinucleotide phosphate; P, phosphate.

stress, for example, wound healing or postsurgical stress. For vitamin C synthesis, D-galactose may be an even better precursor than D-glucose. This pathway is also included in Fig. 3.7.

2. *Terminal Oxidation—Aerobic Glycolysis*

The metabolic pathways described thus far are those of the carbohydrates. In analogous fashion, the breakdown of fats and of proteins also follows independent pathways leading to the formation of organic acids. Among the organic acids formed from lipids are acetyl-CoA (AcCoA), acetoacetate (AcAc) and 3-OH-butyrate (3-OH-B) from the β-oxidation of fatty acids. From proteins, pyruvate, OAA, and α-ketoglutarate (α-KG) form from transamination of their corresponding α-amino acids. Direct deamination of amino acids is also a route of formation of organic acids. These organic acid intermediate metabolites are indistinguishable in their subsequent interconversions. Thus, the breakdown of the three major dietary constituents converges into a final common pathway, which also serves as a pathway for the interconversions between them.

a. Pyruvate Metabolism

The pathway for breakdown of glucose to pyruvate has been described in Section IV.D.1. Pyruvate, if it is not reduced to lactate, is oxidatively decarboxylated in a complex enzymatic system requiring the presence of lipoic acid, thiamine pyrophosphate (TPP), coenzyme A (CoA), NAD^+, and pyruvate dehydrogenase (PD) to form AcCoA and NADH. Pyruvate may follow a number of pathways as outlined in Fig. 3.8. The conversion of pyruvate to lactate has already been described in Section IV.D.1. By the mechanism of transamination or amino group transfer, pyruvate may be reversibly converted to alanine. The general reaction for an amino group transfer is

$$\underset{\alpha\text{-keto acid}}{R_1\text{-}\underset{\|}{\overset{}{C}}(=O)\text{-COO-}} + \underset{\alpha\text{-amino acid}}{R_2\text{-}C(NH_2)\text{-COO-}} \xrightarrow{\text{transferase}} \underset{\alpha\text{-amino acid}}{R_1\text{-}C(NH_2)\text{-COO-}} + \underset{\alpha\text{-keto acid}}{R_2\text{-}C(=O)\text{-COO-}}$$

where the amino group of an amino acid is transferred to the α position of an α-keto acid and, as a result, the amino acid is converted to its corresponding α-keto acid. This reaction requires the presence of vitamin B_6 as pyridoxal phosphate and is catalyzed by a specific transferase, in this case alanine aminotransferase (ALT). Serum levels of several of these transferases, for example, ALT and aspartate aminotransferase (AST) have been particularly useful in the diagnosis and evaluation of liver and muscle disorders, respectively. These aspects are discussed in the individual chapters on liver and muscle function.

The energetics of the reaction from PEP to form pyruvate and catalyzed by PK are such that this is an irreversible reaction, as is the PD catalyzed conversion of pyruvate to AcCoA. A two-step separate pathway to reverse this process is present at this step so this is a fourth site of directional metabolic control. Through a CO_2 fixation reaction in the presence of $NADP^+$-linked malate dehydrogenase (MD), malate is formed from pyruvate. Malate is then oxidized to OAA in the presence of NAD^+-linked MD. OAA may also be formed directly from pyruvate by the reaction catalyzed by pyruvate carboxylase (PC). OAA formed by either route may then be phosphorylated and decarboxylated to form PEP in a reaction catalyzed by PEPCK. Thus, a pathway in the reverse direction of the PK reaction is present for gluconeogenesis from lower intermediates. These pathways for pyruvate metabolism are outlined in Fig. 3.8, which includes the dicarboxylic acid cycle.

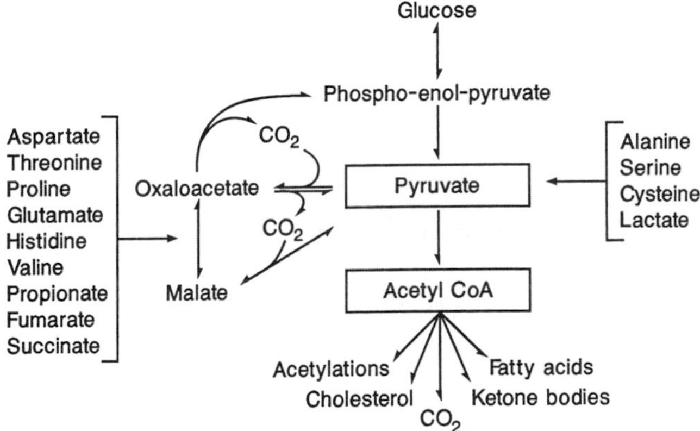

FIGURE 3.8 Pathways of acetate and pyruvate metabolism.

b. Tricarboxylic Acid Cycle

AcCoA formed by the oxidative decarboxylation of pyruvate also has a number of metabolic routes available. AcCoA occupies a central position in synthetic and in oxidative pathways as shown in Fig. 3.8. The oxidative pathway leading to the breakdown of AcCoA to CO_2 and H_2O follows a cyclical pathway, which is the tricarboxylic acid (TCA) cycle, citric acid cycle, or the Kreb's cycle. The major steps involved are given in Fig. 3.9. In a single turn of the cycle, 1 mole of AcCoA enters, 2 moles of CO_2 are evolved, and 1 mole of OAA is regenerated. The regenerated OAA may then condense with another mole of AcCoA, and the cycle continues. Citric acid is a symmet-

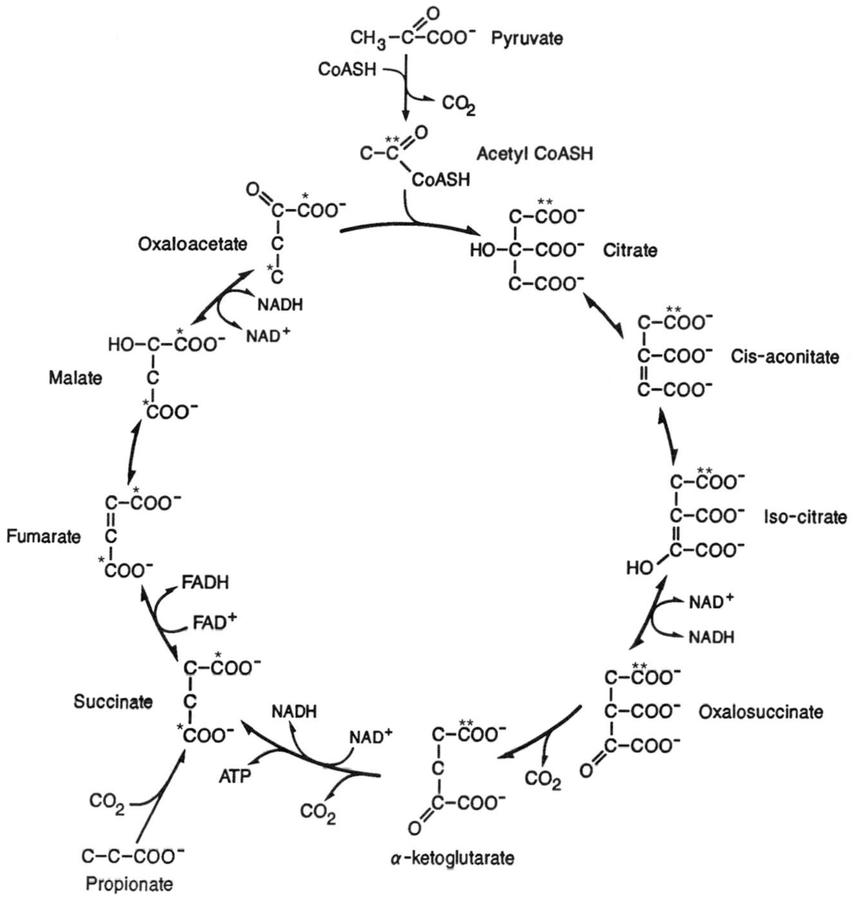

FIGURE 3.9 Tricarboxylic acid cycle. The pathway for the entry of propionate into the metabolic scheme is also included. The asterisks give the distribution of carbon in a single turn of the cycle starting with acetyl-CoA. Note the randomization of carbon atoms at the succinate step.

rical molecule that behaves asymmetrically as shown in Fig. 3.9. Also, the CO_2 that evolves is derived from that portion of the molecule contributed by OAA during each turn of the cycle. The expected distribution of carbon atoms from AcCoA in one turn of the cycle is also given in Fig. 3.9. During one turn of the cycle, a randomization of carbon atoms occurs at the succinate level such that CO_2 derived from the carboxyl group of acetate will be evolved during the next turn of the cycle.

In the process, 3 moles of NAD^+ and 1 mole of a flavin nucleotide (FAD) are reduced and 1 mole of ATP generated as noted in Fig. 3.9. In animal tissues, there is a cytoplasmic $NADP^+$-linked isocitric dehydrogenase (ICD), which is not associated with the mitochondrial NAD^+-linked ICD or other enzymes of the TCA cycle. The $NADP^+$-ICD is another enzyme used as an aid to diagnosis of liver disease.

3. Carbon Dioxide Fixation in Animals

According to Fig. 3.9, the TCA cycle is a repetitive process based on the regeneration of OAA at each turn. In addition, other metabolic pathways are available for intermediates in the cycle. Reversal of the transamination reactions previously described to form aspartate or glutamate would result in a withdrawal of OAA and α-KG, respectively, from the cycle. By decarboxylation, OAA may also be withdrawn to form PEP, and malate may form pyruvate and thence other glycolytic intermediates as shown in Fig. 3.8. Continued losses of these intermediates into other metabolic pathways would theoretically result in a decrease in the rate of operation of the cycle. A number of metabolic pathways are known whereby the losses of cycle intermediates may be balanced by replacement from other sources and are shown in Fig. 3.8. The amino acids, aspartate and glutamate, may function as sources of supply as well as routes for withdrawal. The CO_2 fixation reactions, which are the reversal of the reactions previously described:

Phosphoenolpyruvate + CO_2 ⟶ oxaloacetate,
Pyruvate + CO_2 ⟶ malate,
Pyruvate + CO_2 ⟶ oxaloacetate,

may also function as important sources of supply. A fourth CO_2-fixing reaction,

Propionate + CO_2 ⟶ succinate,

is especially important in ruminants because propionate is a major product of rumen fermentation and is a major supplier of intermediates for the TCA cycle. Propionate is one of the three major fatty acids, with acetate and butyrate, involved in ruminant metabolism.

4. Energy Relationships in Carbohydrate Metabolism

The energy of carbohydrate breakdown must be converted to high-energy phosphate compounds to be useful to the organism, otherwise the energy is dissipated as heat. The total available chemical energy in the reaction

Glucose ⟶ 2 Lactate

is about 50 kcal/mol or about 7% of the 690 kcal/mol that is available from the complete oxidation of glucose to CO_2 and water. The useful energy of anaerobic glycolysis is represented by the net gain of 2 moles of ATP and the available energy of each is about 7 kcal. Thus, the efficiency of glycolytic breakdown of glucose to pyruvate is 14 kcal or 28% of the available 50 kcal or only 2% of the total available 690 kcal in glucose.

The major portion of the energy of glucose is generated in the further aerobic oxidation of pyruvate to CO_2 and H_2O. In the oxidative or dehydrogenation steps, NADH or NADPH (FAD in the succinate step) is formed. In the presence of molecular O_2, these compounds are reoxidized to NAD^+ or $NADP^+$ in the cytochrome system. In the sequence of reactions of this system, 3 moles of ATP are formed per mole of NADH or NADPH oxidized to NAD^+ or $NADP^+$. This transfer of energy to ATP is known as *oxidative phosphorylation* or *ox-phos*. The yield of high-energy phosphate bonds in the form of ATP in the system per atom of oxygen consumed ($\frac{1}{2}O_2$) is conventionally referred to as the P:O ratio, which in this case, is 3.

In Table 3.3, a balance sheet of the ATPs formed in the various steps is given and 36 of the total 38 ATPs are generated in aerobic glycolysis. The complete oxidation of 1 mole of glucose to CO_2 and water yields 690 kcal and, therefore, the net gain of 38 ATPs in anaerobic plus aerobic glycolysis represents 266 kcal for an overall efficiency of 38%. In comparison, the efficiency of the modern internal combustion engine is about 20%.

TABLE 3.3 ATP Yield in Glucose Oxidation

Glucose		
↓ ATP (2×)		−2
Fructose-1-6-diphosphate		
→ NADH →	3 ATP (2×)	+6
→ ATP (4×)		+4
2 Pyruvate		
→ NADH →	3 ATP (2×)	+6
2 Acetyl-CoA		
→ NADH →	3 ATP (6×)	+18
→ ATP (2×)		+2
→ FADH →	2 ATP (2×)	+4
4 CO_2		
Net: Glucose ⟶ 6 CO_2		+38 ATP

V. INTERRELATIONSHIPS OF CARBOHYDRATE, LIPID, AND PROTEIN METABOLISM

The pathways by which the breakdown products of lipids and proteins enter the common metabolic pathway have been described in previous sections. The principal points at which carbohydrate carbon may be interconverted between amino acids and fatty acids are outlined in Fig. 3.10. Thus, certain amino acids (glycogenic) can serve as precursors of carbohydrate through the transamination reactions and by reversal of these transaminations, carbohydrates can serve as precursors of amino acids.

The relationship between carbohydrate and lipid metabolism deserves special mention because the carbohydrate economy and the status of glucose oxidation strongly influences lipid metabolism. A brief description of lipid metabolism follows, and greater detail may be found in the chapter on lipid metabolism.

A. Lipid Metabolism

1. Oxidation of Fatty Acids

Intracellular fatty acids are either synthesized in the cytoplasm or taken up as free fatty acids. Fatty acid oxidation begins in the cytoplasm with the activation

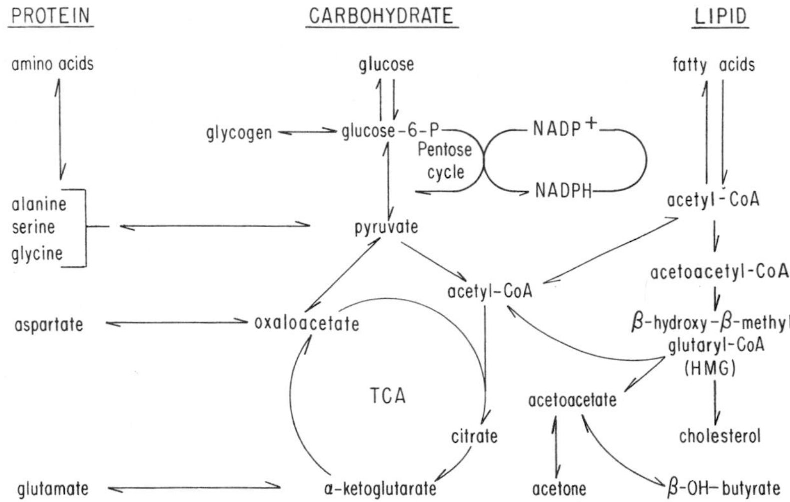

FIGURE 3.10 Interrelationships of carbohydrate, protein, and lipid metabolism.

of fatty acids to form fatty acyl-CoA. The activated fatty acyl-CoA is bound to carnitine for transport into the mitochondria where fatty acyl-CoA is released for intramitochondrial oxidation.

The classical β-oxidation scheme for the breakdown of fatty acids whereby two-carbon units are successively removed is a repetitive process involving four successive reactions. After the initial activation to form a CoA derivative, there is (1) a dehydrogenation, (2) a hydration, (3) a second dehydrogenation, and (4) a cleavage of a two-carbon unit. The result is thc formation of AcCoA and a fatty acid residue shorter by two carbon atoms. The residue can then recycle to form successive AcCoA molecules until final breakdown is achieved In the case of odd-chain fatty acids, propionyl-CoA is formed in the final cleavage reaction. The hydrogen acceptors in the oxidative steps are NAD^+ and FAD. The further oxidation of AcCoA to CO_2 and water proceeds in the common pathway of the TCA cycle. In the process, 2 moles of CO_2 are evolved per mole of AcCoA entering the cycle. Therefore, fatty acids could not theoretically lead to a net synthesis of carbohydrate. Net synthesis of carbohydrate from fatty acids would require the direct conversion of AcCoA into some glucose precursor, that is, pyruvate. The reaction

$$\text{Pyruvate} \longrightarrow \text{acetyl CoA} + CO_2,$$

however, is irreversible and the only route by which fatty acid carbon could theoretically appear in carbohydrate is through the TCA cycle intermediates and this occurs without a net synthesis.

2. *Synthesis of Fatty Acids*

The pathway for fatty acid synthesis is separate from that of the β-oxidation mechanism for fatty acid breakdown. Malonyl-CoA is first formed by the addition of CO_2. Subsequently, two-carbon units from malonyl-CoA are sequentially added to the growing chain with a loss of CO_2 at each addition. At each step, there is also a reduction, dehydration, and a final reduction to form a fatty acid that is two carbons longer than the previous one.

The synthesis of fatty acids also requires NADPH as the hydrogen donor rather than NADH or FADH. The major source of NADPH is during the oxidation of glucose in the HMP shunt pathway. NADPH concentration is also high in the cytoplasm of liver and adipose cells where HMP shunt activity is also high. The availability of this NADPH is the basis for the linkage of carbohydrate oxidation to lipid synthesis.

3. *Synthesis of Cholesterol and Ketone Bodies*

AcCoA is also the precursor of cholesterol and the ketone bodies, AcAc, 3-OH-B, and acetone. The synthesis of cholesterol proceeds through a series of reactions beginning with the stepwise condensation of 3 moles of AcCoA to form β-hydroxy-β-methyl glutaryl-CoA (HMG-CoA). As shown in Fig. 3.10, HMG-CoA is a common intermediate for the synthesis of cholesterol and ketone bodies in the liver cell. In liver, a deacylating enzyme is present that cleaves HMG-CoA to yield AcCoA and free AcAc. This is the HMG-CoA cycle. The free AcAc then diffuses out of the cell and enters the general circulation. For further oxidations to occur, AcAc is "reactivated" with CoA in extrahepatic tissues (muscle) by the transfer of CoA from succinyl-CoA to form AcAcCoA. Increased ketogenesis and ketonemia are the net result of alterations in metabolic pathways and/or enzymes that favor the accumulation of AcAcCoA. Prime examples are diabetes mellitus and bovine ketosis.

The increased mobilization and utilization of fatty acids is a well-known requisite for ketogenesis under conditions of starvation and diabetes. Under these same conditions, lipid synthesis from AcCoA is also depressed. The net effect of either or both of these alterations favors the accumulation of AcCoA and thus ketogenesis.

Increased ketogenesis is always associated with an increased rate of gluconeogenesis in association with an increased activity of the key gluconeogenic enzyme, PEPCK. The increased rate of gluconeogenesis in turn depletes OAA. There is an increase in the NADH/NAD ratio, which would promote the conversion of OAA to malate, thereby depleting OAA. With the depletion of OAA and OAA deficiency, there is insufficient condensing partner for AcCoA for the Kreb's cycle. The AcCoA is then be readily diverted to ketone bodies.

Hepatic ketogenesis is regulated by the rate-limiting transfer of FFA across the mitochondrial membrane. Carnitine acyl transferase, the enzyme system responsible for the mitochondrial uptake of FFA, is increased in diabetes and contributes to the ketogenesis.

B. The Influence of Glucose Oxidation on Lipid Metabolism

In addition to the separation of the biochemical pathways for lipid oxidation and lipid synthesis, an anatomical separation of lipid metabolism is also present. The liver is the major site of fatty acid oxidation and the adipose tissue is the major site of lipid synthesis. Adipose tissue, *in vitro,* converts glucose carbons to fatty acids faster than does liver tissue.

It is well known that, with excessive carbohydrate feeding, fat depots in the body increase. Fasting, on the other hand, depresses the respiratory quotient (RQ) indicating that the animal is now using body fat as the energy source. During fasting, plasma FFAs also increase and when carbohydrate is supplied, they decrease. The presence of glucose both stimulates lipogenesis and spares fatty acid from oxidation. In diseases with an inability to utilize glucose, for example, diabetes, depression of lipogenesis is a characteristic finding. When there is adequate glucose oxidation, for example, successful insulin therapy in diabetes, lipid synthesis is restored and the animal regains its weight.

In those conditions with decreased glucose use or availability, such as diabetes, starvation, and ruminant ketosis, there is an increased release of glucose precursors (amino acids) from muscle and FFA from adipose tissues mediated by activated hormone sensitive lipases (HSLs) (Khoo *et al.,* 1973). The amino acids and FFA are transported to the liver where the amino acids follow gluconeogenic pathways. Fatty acids follow pathways toward oxidation and ketogenesis and, additionally, glucagon promotes hepatic ketogenesis. There is also an underutilzation of ketones in the peripheral tissues of dogs (Balasse and Havel, 1971; McGarry and Foster, 1976). The net result is an overproduction of glucose and ketones in liver and an underutilization of both in the peripheral tissues.

VI. INSULIN AND CARBOHYDRATE METABOLISM

The internal secretions of the anterior pituitary, adrenal cortex and medulla, and the pancreas are closely associated with carbohydrate metabolism. The pituitary and adrenal factors have previously been discussed in Section IV.C together with glucagon. More detailed information is available in the chapters on pituitary and adrenal function. After the successful extraction of insulin by Banting and Best in 1921, a vast amount of literature has accumulated on its role in carbohydrate metabolism and continues to this day. The fine details of insulin action are still being studied and a basic understanding of the major biochemical events that occur in animals with and without insulin has evolved.

A. Proinsulin and Insulin

The elucidation of the insulin structure by Sanger in 1959 was soon followed by the discovery of its precursor, proinsulin, and its structure was quickly known. It has been the subject of many reviews (Kitabchi, 1977; Schade and Eaton, 1985; Raptis and Dimitriadis, 1985; Taylor, 1995). Proinsulin is a single-chain looped polypeptide linked by disulfide bridges (Fig. 3.11). It varies in length from 78 amino acid residues in the dog to 86 for the human, horse, and rat. Its molecular weight is near 9,000 Da. Proinsulin is synthesized in the pancreatic β cells on the rough endoplas-

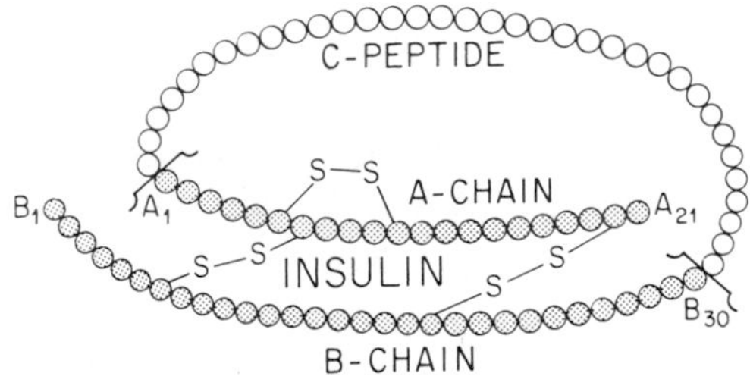

FIGURE 3.11 Insulin and proinsulin. Proinsulin is the coiled polypeptide. When the connecting C-peptide (open circle) is removed, the insulin molecule (solid circle) is released.

mic reticulum (rER) and transported and stored in the secretory granules on the Golgi apparatus. There, the central connecting polypeptide or C-peptide is cleaved from the chain by proteolytic enzymes and the two linked end fragments are the monomeric insulin molecule. C-peptide has a molecular weight of 3600 Da and is devoid of biological activity.

Insulin and C-peptide are released into the circulation in equimolar amounts, but C-peptide's circulatory concentration is higher than that of insulin because of its slower clearance half-time of 20 minutes as compared to 5–10 minutes for insulin. C-peptide is primarily degraded by the kidney and a portion is excreted in the urine (Duckworth and Kitabchi, 1981).

In the pancreatic cells, as the insulin moiety is cleaved from the proinsulin, it crystallizes with zinc for storage in the β-cell granules. The dense central inclusions of these insulin secretory granules consist mainly of crystalline insulin. Insulin release is stimulated by glucose, amino acids, hormones (glucagon, gastrin, secretin, pancreozymin), and drugs (sulfonyl ureas, isoproterenol). Insulin release is inhibited by hypoglycemia, somatostatin, and many drugs, such as dilantin and phenothiazines (Pfeifer *et al.*, 1981). The liver is the primary site of insulin degradation and the kidney is a secondary site. The half-life of insulin in the circulation is between 5 to 10 min (Steiner, 1977).

The A chain of insulin consists of 21 amino acids and the B chain of 30 amino acid residues (Fig. 3.11). The molecular weight of the insulin monomer is 6000 Da and is the smallest unit possessing biological activity. Under physiological conditions, four molecules of insulin are linked to form a tetramer, the active molecule. Insulin obtained from various species differs in amino acid composition in chain A or chain B or both (Table 3.4). Differences occur within species also since rats and mice (Markussen, 1971) have two nonallelic insulins. These structural differences among the various species of animals are not located at critical sites, however, because they do not affect their biological activity. They do, however, affect their immunologic behavior.

The amount of insulin stored in the pancreata of various species also differs. The dog stores about 3.3 units per gram of pancreas, which amounts to about 75 IU in a 10-kg dog. This amount, if suddenly released, would be fatal.

Insulin release is effected by glucose, mannose, leucine, other amino acids, ketone bodies, and fatty acids. This release is mediated by glucagon, a hormone that increases cAMP and potentiates the insulin response. The sulfonylureas are effective as pharmacologic agents to release insulin, the basis for their therapeutic use.

TABLE 3.4 Species Variation in Amino Acid Sequences of Insulin[a]

	Position[b]						
	A chain				B chain		
Species	A-4	A-8	A-9	A-10	B-3	B-29	B-30
Human	Glu	Thr	Ser	Ileu	Asp	Lys	Thr
Monkey	Glu	Thr	Ser	Ileu	Asp	Lys	Thr
Dog	Glu	Thr	Ser	Ileu	Asp	Lys	Ala
Pig	Glu	Thr	Ser	Ileu	Asp	Lys	Ala
Sperm whale	Glu	Thr	Ser	Ileu	Asp	Lys	Ala
Rabbit	Glu	Thr	Ser	Ileu	Asp	Lys	Ser
Horse	Glu	Thr	Gly	Ileu	Asp	Lys	Ala
Cow	Glu	Ala	Ser	Val	Asp	Lys	Ala
Sheep	Glu	Ala	Gly	Val	Asp	Lys	Ala
Sei whale	Glu	Ala	Ser	Thr	Asp	Lys	Ala
Rat 1	Asp	Thr	Ser	Ileu	Lys	Lys	Ser
Rat 2	Asp	Thr	Ser	Ileu	Lys	Met	Ser

[a] From Renold and Cahill Jr. (1966) and Naithani *et al.* (1984).
[b] These are the sites of variation on the A chains and the B chains. Ala, alanine; Asp, aspartic acid; Glu, glutamic acid; Ileu, isoleucine; Met, methionine; Ser, serine; Thr, threonine.

Blood glucose is the primary regulator of both insulin release and its biosynthesis. This is a highly selective process and only insulin, C-peptide, and proinsulin are released and released rapidly. The insulin response curve to a glucose load exhibits two peaks in humans, the early 5-minute peak representing release and the second 10–30 minute peak representing de novo insulin synthesis and release. This bimodal curve is not clear in dogs (Fig. 3.14, shown later) but it is likely to occur.

During proinsulin hydrolysis, C-peptide also accumulates in the granules. Therefore, when the granule contents are released by glucose stimulation, insulin, C-peptide, and proinsulin all appear in plasma and each can be measured by radioimmunoassay (RIA). Whereas studies in humans have focused on all three, in animals, the focus has been on insulin and little is known of proinsulin or C-peptide in health or disease.

The influence of the various gastrointestinal hormones on insulin secretion is of considerable interest because plasma insulin levels are higher at a given plasma glucose level after an oral glucose load as compared to an intravenous load. The oral glucose tolerance test (OGTT) is known to elicit larger total insulin response than the intravenous glucose tolerance test (IVGTT). A number of GI hormones are known to influence insulin secretion to varying degrees and be sufficient to form an enteroinsular axis (Buchanan, 1975). The hormones implicated are secretin, cholecystokinin-pancreozymin (CCK-PZ), gastrin, glucagon-like activity (GLA) of the gut, and gastric inhibitory peptide (GIP). GIP is a powerful stimulator of insulin secretion in humans and dogs and this is associated with a rise in blood glucose (Ross *et al.*, 1974). Thus, GIP is central to the enteroinsular axis.

B. Insulin Transport

Insulin is transported in the cirulation bound to a β-globulin. At a tissue, insulin binds to receptors on the cell membrane. The insulin receptor is a very large glycoprotein on the surface of virtually all cells, including liver, kidney, fat, muscle, erythrocytes, and monocytes. The receptor is a posttranslational derivative of a gene product and is a tetramer of two α and two β subunits. The internal β subunit of the receptor anchors the receptor to the membrane. As a result, insulin moves through the plasma membrane and into the cytoplasmic compartment but the mechanism is unclear. All cells, in particular liver and kidney, are able to inactivate insulin by reductive cleavage of the disulfide bonds. Liver inactivates about 50% of the total insulin.

C. Glucose Transport

Insulin binding also activates receptors both on the plasma membrane surface and in the cytoplasm. This activation induces a variety of reactions, for example, phosphorylations, but the details and their implications are not yet known (Taylor, 1995). However, the end result of these interactions—glucose transport across the membrane and into the cell—is defined. Glucose transport proteins [glucose transporters (GLUT-1 through GLUT-4)] are characterized. They are small membrane proteins, 40–50 kDa, and the different transporters are distributed in different cells; GLUT-1 (brain, RBC, placenta, kidney), GLUT-2 (liver, pancreatic cell), GLUT-3 (brain), and GLUT-4 (skeletal muscle, heart muscle, fat). GLUT-4 has been studied extensively because it is found in those tissues where glucose transport into the cells is regulated by insulin. Insulin mobilizes GLUT-4 to the membrane, thereby facilitating glucose transport into the cell. GLUT-2 is the primary transporter in liver cells (Thorens *et al.*, 1988) Recently, glucose transport activity was studied in the erythrocytes of trained and untrained race horses (Arai *et al.*, 1994). Horses in training had glucose transport activities 2 to 3.5 times greater than those of untrained horses. The specific glucose transporter was not identified but presumably is GLUT-1 as in other animals.

D. Insulin Action on Biochemical Systems

The principal sites of insulin action are in the initial phases of glucose metabolism. Insulin first binds to insulin receptors of the target cell plasma membranes and then facilitates glucose entry into cells such as muscle and fat by activation of glucose transporters, in this case GLUT-4. There is also a high degree of stereospecificity because D-glucose is transported but L-glucose is not. With increased accumulation of glucose in the cells, the movement of glucose into the metabolic scheme is enhanced and glucose utilization is increased.

Insulin influences the metabolism of glucose by the liver cells, the central organ of glucose homeostasis, but with a slightly different focus. GLUT-2 is not significantly regulated by insulin so the liver cell is freely permeable to glucose. Therefore, the major action of insulin in liver is after the initial transport step. The principal step is the first phosphorylation of glucose to form G-6-P in the reaction catalized by GK. This GK reaction is rate limiting and GK activity is influenced by insulin. Additionally, the effect of insulin on other key unidirectional phosphorylative steps directs glucose metabolism toward utilization and FA synthesis. An important effect of insulin is to increase the activity of the pyruvate dehydrogenase (PD) system, which increases AcCoA, thereby promoting increased FA synthesis and oxidation to CO_2 via the Kreb's TCA cycle. These and other reactions are described in Section VII.C. Thus, there are two major roles for insulin, promoting (1) glucose transport across the membranes of muscle and fat cells and (2) glucose utilization by increasing enzyme catalyzed reactions in liver cells.

In nerve cells, insulin binds to receptors and promotes membrane transport of glucose but in this case, the membrane transport system itself appears to be the limiting factor. Thus, even though the HK system is operating maximally [$Km(G) = 5 \times 10^{-5}$], the limited glucose transport of about 1.5 mmols/liter (27 mg glucose/dl) induces the symptoms of hypoglycemia—incoordination, disorientation, weakness—when there is insufficient glucose to compensate by mass action.

In other cells such as the red blood cell, which also has the HK system, insulin does not affect glucose metabolism or limit transport. The HK system is operating maximally and glucose utilization is sufficient to meet the needs of the blood cell at all times.

E. Physiological Effects of Insulin

The principal effects of insulin administration to an animal are summarized in Table 3.5. The most characteristic finding following insulin administration is a hypoglycemia. This occurs regardless of the nutritional state, age, etc., of the animal and is a net result of the increased removal of glucose from the plasma into the tissues. The RQ increases toward unity, indicating that

TABLE 3.5 Effects of Insulin on Animals

Tissue	Increase	Decrease
Whole animal	Anabolism Food intake Respiratory quotient	
Blood		Glucose Ketones Fatty acids Phosphate Potassium Amino acids Ketone bodies
Enzymes	Glucokinase Phosphofructokinase Pyruvate kinase Lipoprotein lipase Acetyl-CoA carboxylase Glycogen synthase	Glucose-6-phosphatase Fructose-1,6-diphosphatase Pyruvate carboxylase PEP-carboxykinase Carnitine acyltransferase Serine dehydratase Hormone sensitive lipase
Liver	Glucose oxidation Glycogen synthesis Lipid synthesis Protein synthesis	Glucose production Ketogenesis
Muscle (skeletal/heart)	Glucose uptake Glucose oxidation Glycogen synthesis Amino acid uptake Protein synthesis Potassium uptake	
Adipose	Glucose uptake Glucose oxidation Lipid synthesis Potassium uptake	

the animal is primarily utilizing carbohydrate. The consequences of this increased utilization of glucose follow a pattern of an increase in those constituents derived from glucose and a decrease in those influenced by increased glucose oxidation. The conversion of glucose to glycogen, fat, and protein is enhanced while gluconeogenesis and ketogenesis are inhibited. The decreases in serum phosphate and potassium levels that parallel those of blood glucose are presumably due to their involvement in the phosphorylating mechanisms.

F. Other Pancreatic Islet Hormones

Numerous hormones oppose the action of insulin and by doing so, prevent or correct the hypoglycemic effects of insulin. Hypoglycemia stimulates a number of counterregulatory hormones including glucagon, epinephrine, and growth hormone. Norepinephrine and cortisol are less responsive to hypoglycemia than the three mentioned.

1. Glucagon

Glucagon is a polypeptide hormone (molecular weight = 3485 Da) secreted by the α (A) cells of the islets. Release of glucagon is stimulated by hypoglycemia. Glucagon acts only in the liver where it stimulates glycogenolysis and gluconeogenesis, thereby increasing blood glucose. The most important physiological role of glucagon is to promote hyperglycemia in response to a hypoglycemia. Glucagon does not act on muscle glycogen, unlike epinephrine, which acts on both liver and muscle glycogen. Like most hormones, glucagon is first bound to surface receptors on a cell, in this case, the hepatocyte. Acting through these receptors, adenylate cyclase is activated, which in turn increases the amount of cAMP. cAMP then activates a phosphorylase kinase, which activates phosphorylase A, which in turn hydrolyzes glycogen. Additionally, glucagon is an insulin secretagogue second only to glucose in the magnitude of the insulin response it elicits. This insulin-releasing action of glucagon is the basis for the glucagon stimulation test (GST), which has been used for the evaluation of diabetes in cats (Kirk *et al.*, 1993).

2. Somatostatin

Somatostatin is secreted by many cells, including the hypothalamus, but its major source is the pancreatic δ (D) cells. Somatostatin has broad inhibitory effects on the release of many hormones, including growth hormone, glucagon, and insulin. Therefore it has a modulating effect on the actions of these two hormones. Administration of somatostatin blocks the secretion of glucagon and, in this way, somatostatin exacerbates an insulin-induced hypoglycemia.

VII. BLOOD GLUCOSE AND ITS REGULATION

A. General

The blood glucose concentration depends on a wide variety of factors and its concentration at any time is the net result of an equilibrium between the rates of entry and of removal of glucose in the circulation. As such, all the factors that exert influence upon entry or removal become of importance in the regulation of blood glucose concentration. Furthermore, when the renal reabsorptive capacity for glucose is exceeded (renal threshold), urinary loss of glucose becomes an additional factor influencing the maintenance of the blood glucose concentration. The blood glucose levels at which this occurs vary between species and are listed in Table 3.6.

B. Glucose Supply and Removal

Glucose is supplied by intestinal absorption of dietary glucose or by hepatic glucose production from its precursors, for example, carbohydrates (glycogen, fructose, galactose) and amino acids (gluconeogenesis). The dietary sources of supply of carbohydrates are especially variable among the various species. The absorptive process varies with the degree of systemic hormonal activity (e.g., thyroid) and gastrointestinal hormone activity (e.g., secretin). All conditions affecting gastrointestinal digestive processes (e.g., gastrointestinal acidity, digestive enzymes, disease) substantially affect absorption of glucose. Hence, it is important to evaluate the blood glucose in virtually all diseases.

In the postabsorptive state, hepatic production is the major source of supply for maintaining blood glucose. The hormones epinephrine and glucagon promote the release of glucose from glycogen as described in Section IV.C.2. The glucocorticoids promote gluconeogenesis and oppose the hypoglycemic action of insulin.

Removal of glucose is governed by a variety of factors, most of which ultimately relate to the rate of utilization of glucose. All tissues constantly utilize glucose either for energy purposes or for conversion into other products (glycogen, pentoses, lipids, amino acids). Therefore, an outflow of glucose from the circulation that is governed by the rate of utilization of glucose by a tissue occurs at all times. The level of blood glucose itself partially governs the rate of utilization and therefore, in a sense, is autoregulatory. At high levels, the rate of glucose uptake by tissues such as muscle and liver increases due to mass action. The presence of insulin increases the rate of glucose utilization, either by increased transport (muscle, fat) or increased phosphorylation (liver). The action of insulin is opposed by the diabetogenic factors, growth hormone, glucagon, cortisol, and epinephrine.

TABLE 3.6 Renal Thresholds for Glucose in Domestic Animals

Species	Thresholds mg/dl	Thresholds mmol/liter	Reference
Dog	180–220	10.0–12.2	Shannon *et al.* (1941)
Horse	180–200	10.0–11.1	Stewart and Homan (1940)
Cow	98–102	5.4–5.7	Bell and Jones (1945)
Sheep	160–200	8.9–11.1	McCandless *et al.* (1948)
Goat	70–130	3.9–7.2	Cutler (1934)

The liver occupies a central position in the regulatory mechanism of blood glucose concentration because it supplies and also removes glucose from the system. The major direction of liver glucose metabolism is directed toward supplying rather than using glucose. When liver takes up glucose, 25% is oxidized to lactate or CO_2 and the remainder forms glycogen. This glycogen is the source of the glucose supplied by the liver to the system during the better part of a day. Muscle, on the other hand, does not contain G-6-Pase, cannot provide free glucose, and is therefore primarily a glucose utilizing tissue.

C. The Role of the Liver

The glucose transporter system across the membrane is rate limiting in peripheral tissues that are sensitive to insulin (muscle, fat). In the liver, however, glucose moves freely across the plasma membrane so this process is not rate limiting at this point. At a blood glucose level of approximately 8.33 mmol/liter (150 mg/dl), the liver does not take up or supply glucose to the circulation. This level is termed the *steady state* or the *glucostatic level* at which the mechanisms of normal supply and removal of glucose are operating at equal rates. Above 8.33 mmol/liter (150 mg/dl), glucose removal is greater than supply; below 8.33 mmol/liter (150 mg/dl), glucose supply is greater than removal. But the fasting blood glucose level in most animals is about 5 mmol/liter (90 mg/dl). This means that the liver supplies glucose throughout most of a day except for the few periods during the day when blood glucose is greater than the steady-state level of 8.33 mmol/liter (150 mg/dl). These periods are the few hours after each meal during a day.

Insulin decreases liver glucose production, output, and glycogenolysis while increasing liver glucose utilization. The net result is an increase in glucose uptake by the liver with increased glucose oxidation, glycogenesis, and hypoglycemia. This directional control is due to the action of insulin on key enzymes of glucose metabolism.

Directional control for glucose production or utilization is governed by coupled sets of opposing and irreversible enzyme reactions at three control points of glucose metabolism. These "key enzyme" couples are GK/G-6-Pase, PFK/F-1-6-Pase, and the PK/PEPCK, PC. The kinases direct metabolism toward glycolysis utilization because they are phosphorylating enzymes and the opposing enzymes reverse the direction so they are gluconeogenic.

The insulin sensitivity of the rate-limiting GK reaction in liver promotes glucose utilization. The opposing G-6-Pase reaction increases during fasting or starvation, which favors liver glucose production. In diabetes mellitus, even though there is a hyperglycemia >8.33 mmol/liter (150 mg/dl), G-6-Pase is increased. Increases in the other key enzymes of gluconeogenesis, F-1-6-Pase, PEP-CK, and PC, are also observed in diabetes. Increases in activity of these gluconeogenic enzymes in insulin deficiency direct metabolic pathways toward excessive production of glucose by the diabetic liver.

The amelioration of diabetes in an experimental animal by hypophysectomy (Houssay animal) is well established. The pituitary factor, which opposes the action of insulin, is growth hormone. The glucocorticoids increase gluconeogenesis and intracellular G-6-P and, by their insulin opposing effect, increase free glucose. An increase also results from the glycogenolytic action of epinephrine and glucagon, and the equilibrium is shifted to favor glucose production. Therefore, it is the balance of hormones that directly (insulin) or indirectly (epinephrine, growth hormone, glucagon, cortisol) affects glucose metabolism, which sets the "steady-state blood glucose" at which the liver uses glucose or produces glucose.

D. Glucose Tolerance

The regulatory events that occur in response to changes in blood glucose concentration are best summarized by a description of the events following ingestion of a test dose of glucose. When administered orally to a normal animal, a typical change in blood glucose concentration with time is observed as shown in Fig. 3.12. During the absorptive phase, phase I, the rate of entry of glucose into the circulation exceeds that of removal and the blood glucose rises. As the blood glucose rises, hepatic glucose output is inhibited and the release of insulin from the pancreas is stimulated by the rising blood glucose. This release of insulin is also influenced by the insulin releasing effect of the GI hormones, secretin, CCK-PZ, and gastrin, and by pancreatic glucagon. In 30–60 minutes, the peak level of blood glucose is reached, after which it begins to fall. During this phase of falling blood glucose, phase II, the rates of removal now exceed those of entry and the regulatory mechanisms directed toward removal of glucose are operating maximally. At the same time, hepatic glucose output decreases and the blood glucose falls rapidly. When the blood glucose reaches its baseline level, it continues to fall below the original level for a short time and then returns to its baseline level. This hypoglycemic phase (phase III) is due to the inertia of the regulatory mechanisms because, in general, the higher the glycemia, the greater the subsequent hypoglycemia. Clinically, this postinsulin hypoglycemia can be marked if there is a defect in the secretion of glucagon.

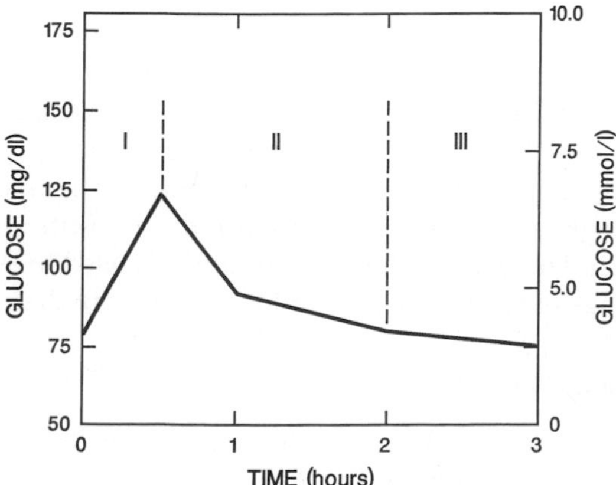

FIGURE 3.12 Oral glucose tolerance in the dog. I, II, and III are phases of the curve.

VIII. METHODOLOGY

A large number of tests have been devised to evaluate the status of the carbohydrate economy of animals but the principal focus continues to lie with the determination of blood glucose levels. The *o*-toluidine, hexokinase, glucose dehdrogenase, and glucose oxidase methods are the four most widely used methods for blood glucose and are used in manual, automated, and point-of-care testing modules.

A. Blood Glucose

1. *Methods*

The *o*-toluidine (*o*-T) method is still used in some blood chemistry kit systems and in some automated analyzers. The *o*-T condenses with glucose in glacial acetic acid to form a blue green color. It is specific for aldose sugars, galactose, and mannose, and because blood contains no significant amounts of aldoses other than glucose, the values are taken as "true glucose."

Three glucose-specific enzyme methods are in use: the glucose oxidase (GO), HK, and GD methods. The GO method is coupled with peroxidase and a dye. GO catalyzes the conversion of glucose to gluconic acid:

$$\text{Glucose} \xrightarrow{\text{GO}} \text{gluconic acid} + H_2O_2.$$

The hydrogen peroxide, with peroxidase, oxidizes a dye to form a colored product. This principle is also used in the glucose-specific paper strips for urine glucose.

In the HK method, HK catalyzes the phosphorylation of glucose and the reaction is coupled to a reaction such as G-6-PD for assay:

$$\text{Glucose} + \text{ATP} \xrightarrow{\text{HK}} \text{G-6-P} + \text{ADP},$$

$$\text{G-6-P} + \text{NADP} \xrightarrow{\text{G-6-PD}} \text{6-PG} + \text{NADPH}.$$

Either NADP or NADPH is measured spectrophotometrically.

In the GD method, GD catalyzes:

$$\text{Glucose} + \text{NAD} \xrightarrow{\text{GD}} \text{gluconolactone} + \text{NADH}$$

and NAD or NADH is measured spectrophotometrically.

Of these enzymatic methods, the method of Banauch *et al.* (1975) was found to be best for the quantitative assay of urine glucose (Kaneko *et al.*, 1978a).

No matter how accurate the method for blood glucose, it cannot compensate for loss of glucose in an improperly handled blood sample. Glucose breakdown (i.e., glycolysis) by red blood cells takes place very rapidly, about 10% per hour loss, at room temperature and is even more rapid if the sample is contaminated by microorganisms. For these reasons, the plasma or serum must be separated from the RBCs as quickly as possible, within one-half hour; otherwise, the glucose in the blood sample must be protected from glycolysis. This is best accomplished by refrigeration or by the use of sodium fluoride (NaF) (10 mg/ml blood). The NaF acts both as an anticoagulant and a glucose preservative. The NaF can also be added to a blood sample vial containing an anticoagulant.

2. *Blood Glucose in Animals*

The normal ranges for blood glucose are given in Table 3.7 and in Appendices VIII, IX, X, and XI. A standard sampling procedure must be used to obtain optimum results and to minimize variations in blood glucose, especially those due to diet. This is best accomplished in the nonruminant and in the young ruminant by a standard overnight (12–16 h) fast prior to sampling. This is not necessary in the mature ruminant because feeding elicits no blood glucose response. Methods for establishing statistically valid reference ranges for analytes such as blood glucose, with examples, are given in Chapter 1.

TABLE 3.7 Blood Glucose Levels in Domestic Animals[a]

Species	Glucose (Reference Range and Mean ± SD) mmol/liter	mg/dl
Horse	4.2–6.4 (5.3 ± 0.4)	75–115 (95 ± 8)
Cow	2.5–4.2 (3.2 ± 0.4)	45–75 (57 ± 7)
Sheep	2.8–4.4 (3.8 ± 0.3)	50–80 (68 ± 6)
Goat	2.8–4.2 (3.5 ± 0.4)	50–75 (63 ± 7)
Pig	4.7–8.3 (6.6 ± 0.9)	85–150 (119 ± 17)
Dog	3.6–6.5 (5.0 ± 0.4)	65–118 (90 ± 8)
Cat	2.8–4.2 (3.5 ± 0.4)	50–75 (63 ± 7)
Monkey (*Macaca sp.*)	4.7–7.3 (5.9 ± 0.7)	85–130 (107 ± 13)
Llama	5.7–8.9 (7.1 ± 0.9)	103–160 (128 ± 16)
Rabbit	2.8–5.2 (4.1 ± 0.5)	50–93 (73 ± 10)

[a] Plasma or serum, glucose oxidase method, adult animals.

B. Indirect Monitoring of Blood Glucose

The phenomenon of glucose molecules irreversibly binding to proteins is widespread in biological systems and the products are known as *glycated proteins.* The glucose molecules are covalently bound to free amino groups of a protein (i.e., lysine and valine) by a nonenzymatic glycation mechanism. The glycated intermediate in the reaction is unstable and immediately undergoes a classic Amadori rearrangement to form a stable ketoamine. The carbon backbone of this ketoamine is identical to fructose. When the protein of the protein–ketoamine complex is hemoglobin (Hb), the product is called hemoglobin Alc (HbAlc) because it was first identified as a fast moving minor Hb component by electrophoresis. When the protein of the complex is albumin or total serum protein, the product is called fructosamine (FrAm) (Armbruster, 1987). When the albumin–ketoamine is specifically measured, the product is sometimes called glycoalbumin (Galb).

The binding of glucose to proteins occurs firmly and constantly over the lifespan of a particular protein. Therefore, these glycated proteins reflect the average blood glucose concentration over the half-life of the protein. Thus, they offer a means and are used to evaluate long-term average blood glucose levels in diabetics.

1. *Hemoglobin Alc*

In the case of canine HbAlc where the canine erythrocyte has a lifespan of 100 days and a half-life of about 60 days, HbAlc reflects the average blood glucose over the previous 2 months prior to sampling. In the cat, with an erythrocyte lifespan of 70 days and a half-life of about 40 days, HbAlc could be used as a measure of the average blood glucose over the previous 6 weeks. This means that bimonthly samplings for the dog and 6 weekly intervals in the cat for HbAlc could be used to monitor long-term blood glucose control. The use of HbAlc is a well-established means for monitoring long-term average blood glucose in human diabetics. The techniques for measuring HbAlc, however, have not been applicable to animals. The methods are complex as well as labor and equipment intensive. The methods all rely on the structural properties of HbAlc for their separation so that methods for human HbAlc are not directly applicable to animal HbAlc. Methods used for human HbAlc are affinity and ion-exchange high-performance liquid chromatography (HPLC), electrophoresis, immunoassay, and colorimetry to measure 5- hydroxymethylfurfural-thiobarbituric acid (HMF-TBA). Of these, the HPLC method is most widely used in humans. A report evaluating a number of methods for canine HbAlc indicated that the HMF-TBA method is the most promising (Hooghuis *et al.*, 1995).

Because HbAlc is directly related to the amount of red cells, anemias or polycythemias must be ruled out. Also, bimonthly samplings may not detect the long-term changes in glucose in a timely manner.

2. *Fructosamine*

The total serum proteins or albumin have half-lives of 2–3 weeks and 7–9 days, respectively. This means that FrAm or Galb could be used as indicators of the average blood glucose over the previous 2 weeks. This biweekly time interval has the advantage of detecting changes in glucose control more quickly than HbAlc and allows for timely clinical intervention. Furthermore, the FrAm assay is a colorimetric assay based on its reducing properties so it is an assay readily performed in any clinical laboratory. An improved version of the original kit is now available from the manufacturer (Roche Diagnostic Systems, Inc., Rahway, NJ). Using this improved version, Jensen and Aaes (1992) report a reference range for FrAm in dogs of 259–344 μmol/liter (301 ± 21.3 μmol/liter). This result is lower than that originally reported for dogs by Kawamoto *et al.* (1992) of 1.7–3.38 mmol/liter (2.54 ± 0.42 mmol/liter) by a factor of about 10. This difference is attributable to the different substrates used in the assays. The reference range for cats as reported by Kaneko *et al.* (1992) is 2.19–3.47 mmol/liter (2.83 ± 0.32 mmol/liter).

C. Tolerance Tests

1. *Glucose Tolerance Tests*

Glucose tolerance (GT) in its original definition referred to the amount of glucose that could be ingested by an animal without producing a glucosuria, hence, tolerance for glucose. Since in the normal animal, the absence of a glucosuria indicates only a limited rise in blood glucose where the renal threshold is not exceeded, GT now refers to the blood glucose curve following glucose administration. Accordingly, an animal with an increased glucose tolerance is one that has a limited rise and rapid fall in blood glucose, that is, can tolerate extra glucose. The animal with a decreased tolerance has an excessive rise and a slow return to its baseline level, that is, cannot tolerate extra glucose. This is the typical diabetic type of GT curve.

It is important to ascertain the nature of the animal's diet, especially in omnivores and carnivores, prior to performance of this test. A carbohydrate-only diet favors a minimum rise in the tolerance curve, whereas a carbohydrate-free diet (meat only) favors a high or diabetic type of glucose tolerance curve. Therefore, for optimum results, the diet must be standardized by placing the dog on a standard diet of 100–200 g carbohydrate plus fat and protein per day for 3 days prior to performance of the test. The GT curve is also affected by the status of the intestinal absorptive process, that is, inflammation, increased motility, thyroxine. Furthermore, the variations due to absorption and the excitement attending intubation or tranquilization can be avoided by use of the intravenous test.

a. Oral Glucose Tolerance Test

In Section VII.D the blood glucose curve following the oral administration of a test dose of glucose was described. The oral glucose tolerance test is ineffective in the ruminant because the ingested carbohydrate is almost totally fermented by the rumen microflora. The OGTT has been used in dogs by feeding of a test meal consisting of 4 g glucose/kg body weight mixed with a few grams of horse meat. A fasting blood sample is taken, the test meal given and blood samples are taken at 30-minute intervals for 3 hours. The OGTT curves in dogs receiving a standard daily diet of either glucose or galactose with meat had normal curves as described in Section VII.D. The maximum level, 6.6–7.7 mmol/liter (120–140 mg/dl) was reached at 1 hour and returned to the fasting level, 3.6–5.3 mmol/liter (65–95 mg/dl), in 2–3 hours. The OGTT may be simplified by taking a single sample at 2 hours after giving the glucose, that is, 2 hours postprandial glucose. A normal blood glucose level at 2 hours postprandially indicates that diabetes is unlikely. A hyperglycemia at 2 hours is indicative of a diabetic curve and should be confirmed with the complete GTT. The insulin response curve during the OGTT can also be evaluated In the OGTT, for a given level of blood glucose, the total insulin secretion (TIS) is greater than in the

IVGTT. This is most likely due to the insulin releasing effect of the gut hormones (Section VII.D).

b. Intravenous Glucose Tolerance Test and the Insulin Response

The intravenous glucose tolerance test in animals must also be standardized for best results (Kaneko *et al.*, 1978a). This is necessary because glucose clearance half-times ($T_{\frac{1}{2}}$) and urinary glucose losses are directly proportional to the glucose dose. The recommended method gives optimal results because (1) it does not overload the animal with glucose, (2) the infusion can be given easily within the time limits, (3) the blood glucose level is high enough to give a maximal insulin response, and (4) urinary loss of glucose is minimal. After a standard overnight (12–16 hour) fast (except for an adult ruminant) a zero-time heparinized blood sample is taken. Next, 0.5 g glucose/kg body weight is infused IV as a sterile 50% solution in 30 seconds. Timing of the test is begun at the midpoint or at 15 seconds after start of the injection. In large animals, the glucose is given within 2–3 minutes or more quickly if possible. Subsequent blood samples are taken at 5, 15, 25, 35, 45, and 60 minutes. The results are plotted on semilogarithmic coordinates from which the time required for the glucose concentration to fall by one-half, the $T_{\frac{1}{2}}$, is graphically estimated between 15- and 45-minutes postinfusion. From the $T_{\frac{1}{2}}$, the fractional turnover rate, k, can also be calculated:

$$k = \frac{0.693}{T_{\frac{1}{2}}} \times 100 = \%/\text{minute}.$$

The fractional turnover rate, k, can also be calculated without graphing the data and using the relationships:

$$k = \frac{\ln 1 - \ln 2}{T_2 - T_1} \times 100 = \%/\text{minute}.$$

From the k value, the $T_{\frac{1}{2}}$ may be calculated:

$$T_{\frac{1}{2}} = \frac{0.693}{k} \times 100 = \text{minutes}.$$

The fractional turnover rate has been variously expressed as the glucose turnover rate, the glucose disappearance rate, the glucose disappearance coefficient or simply as the k value. The normal $T_{\frac{1}{2}}$ and k in dogs are 25 ± 8 minutes and $2.76 \pm 0.91\%$/minute, respectively (Kaneko *et al.*, 1977). The diabetic animal with a decreased glucose tolerance has a longer $T_{\frac{1}{2}}$ and lower k.

The method is equally applicable to and the only practical method in large animals. The k value in a spontaneously diabetic cow was 0.38%/minute ($T_{\frac{1}{2}}$ = 182 minutes) as compared to a normal of 1.98%/minute ($T_{\frac{1}{2}}$ = 35 minutes) and was comparable to the k values obtained using ^{14}C glucose.

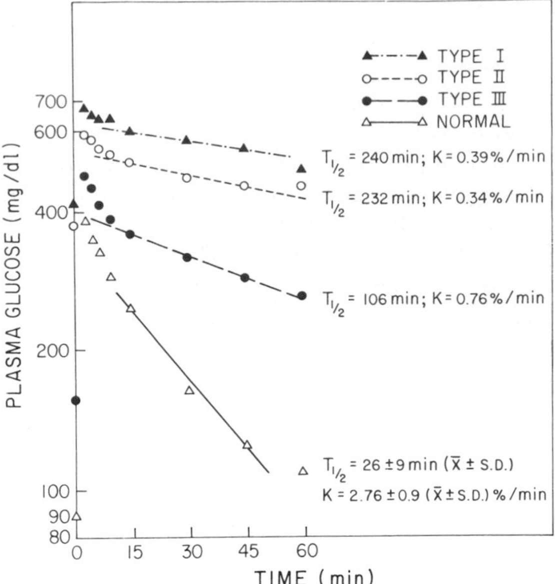

FIGURE 3.13 The IVGTT in normal dogs and in dogs with various types of diabetes mellitus. From Kaneko *et al.*, (1977).

Standardization of the IVGTT as described also has the advantages that an adequate insulin response is provoked, the influence of urinary glucose loss is minimized, and reproducible clearance values are obtained (Fig. 3.13). Other areas of the IVGTT with diagnostic significance for diabetes are the 5-minute peak, which is inordinately high and the 60-minute glucose level, which has not returned to the preinfusion level.

The insulin response curve to the glucose load is obtained from the same samples as is done for glucose (Fig. 3.14). In the normal response curve to a glucose load, the peak insulin response occurs at 5 minutes

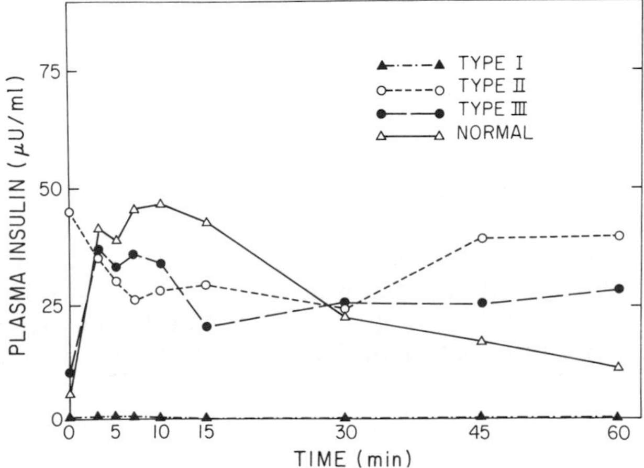

FIGURE 3.14 The insulin response during the IVGTT in normal dogs and in dogs with various types of diabetes mellitus. From Kaneko *et al.*, (1977).

followed by a return to normal at 60 minutes (Kaneko *et al.*, 1977). The early 5-minute peak is due to the stimulation of release of stored insulin by the β cells by glucose. In humans, a second peak is seen at 20–30 minutes, which is attributed to the de novo synthesis of insulin by the β cells. This peak has not been experimentally discernible in dogs (Kaneko *et al.*, 1978b).

The IVGTT and the insulin response are of greatest value in the diagnosis of diabetes, particularly those cases with a mild hyperglycemia and without persistent glucosuria. Furthermore, the insulin response to the IVGTT allows for the accurate differentiation of the different types of diabetes. This has great significance because the Type II diabetic can potentially respond to oral hypoglycemic therapy. Decreased tolerance is also observed, though less consistently, in hyperthyroidism, hyperadrenalism, hyperpituitarism, and in severe liver disease. An increased tolerance is observed in hypofunction of the thyroids, adrenals, pituitary, and in hyperinsulinism.

2. *Insulin Tolerance Test*

The blood glucose response of a normal animal after the administration of a test dose of insulin exhibits a characteristic response as shown in Fig. 3.15. After obtaining a fasting blood sample, 0.1 unit of crystalline zinc insulin per kilogram body weight is injected intramuscularly or subcutaneously and blood samples are taken every 30 minutes for 3 hours. The test measures (1) the sensitivity of the blood glucose level to a test dose of insulin and (2) the response of the animal to insulin-induced hypoglycemia. Normally, the blood glucose level falls to 50% of its fasting level in 20–30 minutes and returns to its fasting level in 1.5–2 hours. Two types of abnormal responses are seen. If the blood glucose level does not fall by 50% or requires longer than 30 minutes to reach the maximum hypoglycemic level, the response is described as *insulin insensitive* or *insulin resistant*. Insulin resistance is found, though inconsistently, in hyperfunction of the pituitary and adrenals. This test has not been used in the Type II diabetic dog or cat but may be of value in evaluating insulin resistance in this type.

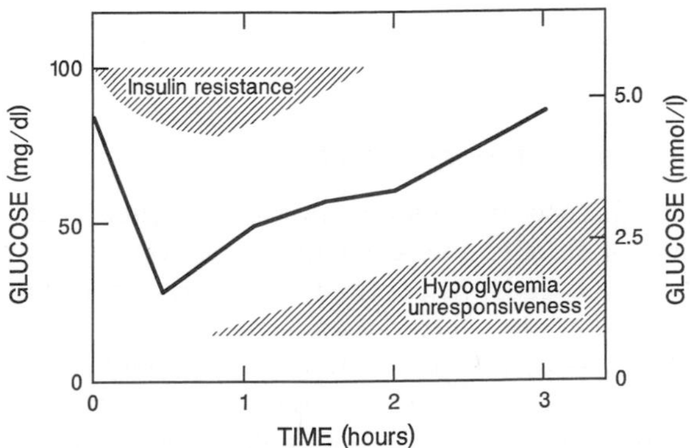

FIGURE 3.15 Insulin tolerance in the dog. Curves falling in the shaded areas are described as noted.

If the hypoglycemia is prolonged and fails to return to the fasting level in 2 hours, the response is described as *hypoglycemia unresponsiveness*. This type of response may be observed in hyperinsulinism, hypopituitarism, and hypoadrenalism and is most often used in suspected cases of the latter two diseases. An inability to secrete sufficient glucagon by the pancreatic δ cells may also be a significant factor in the unresponsiveness. In carrying out this test, because a hypoglycemia is being induced, a glucose solution should be readily available for injection.

3. *Glucagon Stimulation Test*

Glucagon via hepatic glycogenolysis and gluconeogenesis has a hyperglycemic effect that in turn evokes an insulin response. In addition, glucagon is an insulin secretagogue second only to glucose. These are the bases for the glucagon stimulation test (GST) and has been used for the diagnosis of diabetes in cats (Kirk *et al.*, 1993). The test is performed by the IV injection of 30 μg glucagon/kg body weight. Samples for blood glucose and insulin are obtained before injection (0 time) and at 5, 10, 15, 30, 45, and 60 minutes after injection. In cats, the peak insulin response was observed at 15 minutes followed by a rapid decline to baseline levels at 60 minutes. The insulin response curve was flat in the Type I diabetic cats, whereas controls, obese, and Type II diabetic cats had comparable 15-minute peaks and declines to baseline at 60 minutes.

Samplings at half hourly intervals can be continued for up to 3 hours in suspected hyperinsulinism. An exaggerated 15-minute insulin response followed by a marked hypoglycemia at 2–3 hours or longer are characteristic of pancreatic islet cell tumors (Johnson and Atkins, 1977). However, Kruth *et al.* (1982) found that the GST was not diagnostic for these cases.

4. *Epinephrine Tolerance Test*

Epinephrine also has a postinjection hyperglycemic effect via hepatic glycogenolysis. The blood glucose level rises to a maximum of 50% above the fasting level in 40–60 minutes and returns to the original level in 1.5–2 hours. The test is performed by obtaining a fasting blood sample (0 time), injecting 1 ml of 1:1000 epinephrine-HCL (in the dog) intramuscularly and obtaining blood samples every 30 minutes for 3 hours.

The characteristic increase in blood glucose is used as an index of the availability of liver glycogen for the production of blood glucose. On the basis of a lowered response to epinephrine, liver glycogen can indirectly be shown to be depleted in bovine ketosis. This can be confirmed directly by measurement of glycogen in biopsy samples. A lowered glycemic response is also a characteristic response of the glycogen storage diseases where glycogenolysis is inhibited by enzyme deficiencies.

5. *Leucine-Induced Hypoglycemia*

The oral administration of L-leucine induces a marked and persistent hypoglycemia in hyperinsulinism due to pancreatic islet cell tumors. The hypoglycemia is associated with a rise in plasma insulin due to increased release of insulin by the tumorous islet cells. The test is performed by the oral administration of 150 mg L-leucine/kg body weight as an aqueous suspension to the fasting dog. A fasting blood glucose sample is taken before administration (0 time) and every 30 minutes for 6 hours. A hypoglycemic effect is seen quickly at 0.5–1 hour and may persist for as long as 6 hours in hyperinsulinism. The normal dog exhibits no hypoglycemic effect.

6. *Tolbutamide Test*

The intravenous administration of tolbutamide, an oral hypoglycemic agent, induces the release of insulin from the pancreas and is used as a test of the availability of insulin from the pancreas. The blood glucose curve during the test parallels the insulin tolerance test. This test has not been used in animals.

D. Ketone Bodies

The methodology and role of ketone bodies in the carbohydrate economy of animals in health and disease are discussed in the chapter on lipid metabolism. The major ketone bodies are acetone, AcAc, and 3-OH-butyrate (3-OH-B). The 3-OH-B is the precursor of acetone and AcAc so that the measurement of any or all in body fluids is a standard method to evaluate ketosis and ketoacidosis. Additionally, 3-OH-B constitutes one-half or more of the total ketone bodies. Recently, a point-of-care enzymatic and colorimetric method has been developed for assay of plasma 3-OH-B (GDS Diagnostics, Elkhart, IN). The method is based on the enzyme 3-OH-B dehydrogenase (3-OH-BD) and nitroblue tetrazolium (NBT):

$$\text{3-OH-butyrate} + \text{NAD}^+ \xrightarrow{\text{3-OH-BD}} \text{acetoacetate} + \text{NADH} + \text{H}^+$$

$$\text{NADH} + \text{NBT (ox)} \xrightarrow{\text{diaphorase}} \text{NAD}^+ + \text{NBT (red)}.$$

A rapid, reliable method would have decided advantages in the management of diabetic ketoacidosis, bovine ketosis, and ovine pregnancy toxemia.

IX. DISORDERS OF CARBOHYDRATE METABOLISM

Although alterations in blood glucose levels occur in a wide variety of disease states, they are of particular importance in the endocrine disorders. Normal blood glucose levels are the result of a finely balanced system of hormonal interaction affecting the mechanisms of supply and removal from the circulation. When a hormonal imbalance occurs, a new equilibrium is established. Whether this equilibrium is clinically evident as a persistent hypoglycemia or hyperglycemia depends on the total interaction of the hormonal influences on carbohydrate metabolism. Further discussions concerning the disorders of the pituitary, adrenals, and the thyroids are given in their respective chapters. The following sections discuss the conditions in which the principal manifestations are closely related to derangements in carbohydrate metabolisms.

A. Diabetes Mellitus

Although diabetes mellitus has been reported in virtually all laboratory animals (gerbils, guinea pigs, hamsters, mice, rats, nonhuman primates) and in horses, cattle, sheep, and pigs, it is most frequently found in dogs and cats. Estimates of the incidence of diabetes range as high as 1:66 (1.52%) for dogs and 1:800 for cats. Diabetes mellitus in animals has been frequently reviewed (Cotton *et al.*, 1971; Foster, 1975; Ling *et al.*, 1977; Engerman and Kramer, 1982; Kaneko and Howard, 1989).

1. *Natural History of Diabetes*

The disease in dogs occurs most frequently in the mature or older female, often in association with estrus, and in all breeds. In contrast, male cats appear to be more commonly affected than females. In the dog, diabetes is frequently associated with obesity and it is now known that obesity is the single most important contributing factor to the development of diabetes (Mattheeuws *et al.*, 1984). In the obese cat, the GTT is significantly impaired, suggesting that obesity also predisposes cats to diabetes (Nelson *et al.*, 1990). The obese cat also has a GST response like that of the Type II diabetic (Kirk *et al.*, 1993). Little is known of the genetic aspects of diabetes in animals as compared to humans in which the hereditary predisposition is well known. Insulin genes are identified, cloned, and human insulin produced by biotechnology that is now widely used. Diabetes has, however, been reported in the offspring of diabetic dogs (Gershwin, 1975; Kaneko *et al.*, 1977). Kramer (1977, 1981) and Kramer *et al.* (1980) reported their observations on hereditary diabetes in a family of keeshonds and Williams *et al.*, (1981) reported hereditary diabetes in golden retrievers.

On the basis of serum insulin (I) response patterns during the IVGTT, diabetes mellitus of dogs can be divided into at least three types, Types I, II, and III (Table 3.8; Kaneko *et al.*, 1977, 1979). Type I dogs are characterized by no or very low initial I (I^0) level and no I response to the glucose load similar to juvenile diabetes (Type I) in children. Type II is characterized by a normal or high I^0 and, again, no increment of I response to the glucose, which are features of the maturity onset form (Type II) of diabetes in humans. Type III is characterized by a normal I^0, a normal or delayed I response to the glucose, and a delayed return of I to normal at 60 minutes as seen in chemical diabetes (Type III). Types II and III were each later further subdivided into the obese and nonobese types (Mattheeuws *et al.*, 1984).

All diabetics are glucose intolerant so their separation into the various types depends on their insulin response patterns in the IVGTT. The importance of defining their type is the likelihood that Types II and III obese diabetics are the most likely subjects for successful dietary therapy and/or oral hypoglycemic therapy. Nelson *et al.* (1993) have successfully treated some diabetic cats with oral sulfonylureas.

It is also likely that in the natural history of diabetes, Type III (chemical) diabetes precedes the development of Types I and II depending on the nature of the insulin deficiency, whether absolute (Type I) due to islet cell absence or relative (Type II) due to insulin antibodies, receptor defects, or deficiencies.

2. *The Etiology of Diabetes*

The most frequent contributory factors to the onset of diabetes are pancreatitis, obesity, infection, stress, and estrum. The possibility of a viral etiology has also been reviewed (Steinke and Taylor, 1974), and Yoon *et al.* (1979) isolated a virus from a patient who died and the virus produced diabetes in mice.

Currently, autoimmunity is considered to be the fundamental cause of Type I diabetes, possibly as an aftermath to the viral infection. Autoimmunity is evidenced by the lymphocytic infiltration associated with immune processes, and lymphocytic infiltration is found in diabetes of cattle and humans. It is also observed in cattle or rabbits immunized with bovine insulin. Eisenbarth (1986) reviewed the evidence that Type I diabetes in the biobreeding (BB) rat and the nonobese diabetic (NOD) mouse and Type I diabetes of humans is an autoimmune disease. Taniyama *et al.* (1995) found lymphocytic infiltration in all pancreata of four diabetic cattle and their studies were suggestive of Type I diabetes as an autoimmune disease.

In the cat, several studies indicate that there is a strong correlation between pancreatic insular amyloidosis and diabetes although the amyloid does not appear to be the primary cause (O'Brien *et al.*, 1985).

The high estimates of the incidence of diabetes is an indication of its importance as a clinical disease entity. Furthermore, the similarities of the clinical picture of diabetes with other wasting diseases characterized by polyuria and polydypsia attest to the importance of laboratory examinations for the early and accurate diagnosis of diabetes. In no other disease is an understanding of the metabolic alternations so important in diagnosis and proper treatment.

The fundamental defect in diabetes mellitus is an absolute (due to absence of pancreatic β cells) or relative (insulin resistance due to insulin antibodies, receptor defect, or deficiency) lack of insulin resulting in an

TABLE 3.8 Diagnostic Criteria for Types of Diabetes Mellitus in Dogs

	IVGTT			Insulin[a]		
Diabetes type	Fasting glucose G_0 (mg/dl)	$T_{\frac{1}{2}}$ (m)	k (%/min)	Fasting insulin I_0 (μU/ml)	Peak insulin I_0 (μU/ml)	$\Delta I/\Delta G$
I	>200	>70	<1.0	L	L	L
II	>200	>70	<1.0	N, H	L	L
III	100–200	>45	<1.5	N, H	N, D	L, N
Normal	70–110	15–45	>1.5	N	N	N

[a] N, normal; L, low; H, high; D, delayed.

inability to utilize glucose. The lack of insulin using the bioassay technique was demonstrated in the early 1960s in the diabetic cow by the failure of the IVGTT to elicit a serum insulin response. The development of the RIA technique has since provided the impetus for modern advances in diabetology. Using the IVGTT and the serum insulin response, diabetes mellitus in dogs has been classified into Types I, II, and III. In all types, the inability to utilize glucose or glucose intolerance is evident in the IVGTT. In the absence of insulin, the inability of the diabetic animal to utilize glucose is clearly shown in its inability to convert glucose-^{14}C to $^{14}CO_2$ (Phillips *et al.*, 1971). This inability is corrected by insulin. The inability to utilize glucose is reflected in the clinical signs of diabetes, loss of weight, polyuria, polydypsia, and in the advanced stages, ketoacidosis.

Several reports have suggested that the development of diabetes mellitus is the result of the interaction of several hormones, principally insulin and glucagon (Unger and Orci, 1975, 1976). Although excess glucagon is seen in diabetes and it can induce glucose intolerance or changes in diabetic control, it can do so only as long as insulin deficiency is present and pharmacological levels of glucagon are induced (Felig *et al.*, 1976). Thus, insulin deficiency is sine qua non the cause of diabetes and that while glucagon may modify the consequences, it is neither necessary nor sufficient for the development of diabetes (Felig *et al.*, 1976). The modifying action of glucagon on diabetes, however, has important implications because excess glucagon will tend to exacerbate the posthypoglycemic hyperglycemia, for example, the Somogyi effect. A deficiency of glucagon will tend to increase the effect of insulin and contribute to a prolonged hypoglycemia after treatment. Thus glucagon plays a significant role in the variability of diabetes.

3. *Hyperglycemia*

A persistent fasting hyperglycemia is the single most important diagnostic criteria of diabetes mellitus. In the normal animal, the homeostatic level of blood glucose is maintained by the equilibrium between glucose supply and removal, which in turn is based on the endocrine balance. The effect of insulin tends to lower blood glucose, whereas the opposing effects of growth hormone, glucagon, and adrenal cortical hormones tend to raise it. In the diabetic animal with an absolute or relative lack of insulin, the equilibrium is shifted to a higher level of blood glucose. Glucose utilization in the peripheral tissues is decreased while at the same time, hepatic glucose production is increased due to increases in their gluconeogenic enzyme activities.

In the diabetic, the hyperglycemia itself tends to compensate in part for the decrease in peripheral utilization. This occurs as a mass action effect, which promotes the flow of glucose into the peripheral tissues. In this way, the diabetic can continue to use some glucose when insulin is decreased but only at the expense of increased glucose production and hyperglycemia. As the deficiency of insulin progressively becomes more severe, the equilibrium level of blood glucose is established at higher and higher levels and equilibrium is never established without therapeutic intervention. Blood glucose levels in canine diabetics have reached 70 mmol/liter (1260 mg/dl). When the renal threshold of 11.1 mmol/liter (200 mg/dl) for glucose is exceeded, the diabetic is faced with excessive loss of glucose in the urine. It is evident that the blood glucose level is exquisitely sensitive to insulin and, conversely, the blood glucose level is the mainstay for monitoring the success of diabetes therapy. However, a marked post-hypoglycemic hyperglycemia or the Somogyi effect has long been known to occur in humans after insulin therapy (Bolli *et al.*, 1984), which indicates that glucose regulation by insulin is not complete. This hyperglycemic effect is thought to be due to an excess of glucagon, growth hormone, adrenal cortical hormones, or epinephrine. This phenomenon has been seen in diabetic cats given an inadvertent insulin overdose (McMillan and Feldman, 1986), which points to the need for effective monitoring of diabetes.

4. *Glycated Proteins*

The biochemical and physiological bases for using glycated proteins as a monitor for long-term glucose control are now commonplace in human diabetolgy (Section VIII.B). Successful management of diabetes depends on the reliable evaluation of blood glucose levels, and any blood glucose sample only reflects the blood glucose level at the moment of sampling. An effective method for estimating the average blood glucose over an extended time period offers a way of evaluating successful insulin therapy. This can be done by use of the glycated proteins, HbA1c or FrAm. Of these, FrAm offers the most cost-effective method for evaluating the average blood glucose over the preceding 2 weeks.

a. Hemoglobin A1c

The glycated hemoglobin, HbA1c, is known to reflect the average blood glucose level over the preceding 60 days and is widely used to monitor human diabetics (Gabbay *et al.*, 1977; Bunn *et al.*, 1978; Nathan *et al.*, 1984). Several studies in diabetic dogs (Wood and Smith, 1980; Cornelius and Mahaffey, 1981) have also shown that HbA1c is potentially useful for monitoring purposes. Although the reference values for % HbA1c differed in the two studies, 2.29% and 6.43%, the means for the diabetics were increased in each to 4.97% and 9.63%, respectively. Hooghuis *et al.* (1995), using thiobarbituric acid colorimetry (HMF-TBA), report a reference range of 1.4–3.2% HbA1c (2.3 ± 1.96 SD). Previous assays for HbA1c have been time, labor, and equipment intensive as well as giving variable results. The colorimetric HMF-TBA method shows promise of being a clinically viable method.

b. Fructosamine

The fructosamines (FrAm) reflect the average blood glucose over the preceding 2 weeks in a manner analgous to HbAlc. This means that FrAm could be used to monitor the average blood glucose on a biweekly interval. This has the advantage that changes in blood glucose can be detected more quickly than with HbAlc and allows for timely clinical intervention. Furthermore, the FrAm assay is a colorimetric assay that can be readily performed in any clinical laboratory. An improved version of the original kit is now available from the manufacturer (Roche Diagnostic Systems, Inc., Rahway, NJ). Using this improved version, Jensen and Aaes (1992) report a reference range for FrAm for dogs of 259–344 μmol / liter (301 ± 21.3 SD). This result is 10-fold lower than that originally reported by Kawamoto *et al.* (1992). The reference range for cats as reported by Kaneko *et al.* (1992) is 2.19–3.47 mmol/ liter (2.83 ± 0.32 SD). In all cases, FrAm was shown to be significantly elevated in diabetes, indicating that they can be of clinical value to monitor glucose control in treated diabetics. On occasion, especially in cats, hyperglycemia and/or glucosuria is seen on initial presentation and without other indications of diabetes. A FrAm sample taken at this time can be used to differentiate a transient from a persistent hyperglycemia.

5. Glucose Tolerance and the Insulin Response

The GTT is the most important test of carbohydrate function and is of particular value in those cases of diabetes in which the fasting blood glucose is only moderately elevated and the diagnosis is equivocal (Section VIII.C). The diabetic oral GGT curve is high and relatively flat, indicating a decreased tolerance for glucose (Fig. 3.12). The nature of the diabetic curve can be quantitated by using the intravenous GTT. The diabetic curve is characterized by a long $T_{\frac{1}{2}}$ or low k value, which reflects the inability of the animal to use the test dose of glucose. The insulin response curve in Type I (absolute insulin deficiency) diabetes clearly demonstrates the inability of the pancreas to release insulin in response to the glucose load. It is the absence of an insulin response that is responsible for the failure of the diabetic to utilize the added glucose and the prolonged hyperglycemia occurs. An important factor adding to the hyperglycemia is the overproduction of glucose by the liver. The test dose of glucose is in effect added to the already existing oversupply of glucose. Because the steady-state level at which the liver ceases to supply or remove glucose is elevated in diabetes, the liver continues to oversupply glucose, which contributes to the slow return of the tolerance curve to its original level.

In Types II and III diabetes (discussed later), there is also glucose intolerance but this occurs in the presence of a normal to elevated insulin. This would mean that the insulin in the plasma of these types is ineffective, that is, relative deficiency, due to a number of factors including insufficient receptors, receptor blockage, abnormal receptor structure, or antibody binding all of which lead to the glucose intolerance and the phenomenon of insulin resistance. Therefore, glucose intolerance is seen in all types of diabetes whether there is an absolute (Type I) or relative (Types II and III) deficiency of insulin. The insulin response must be evaluated in order to establish the type of diabetes.

6. Insulin and the Insulin Response

Serum insulin is characteristically very low or absent in Type I diabetes, whereas it is normal or even very high in Types II or III. Type I diabetes can be readily differentiated from the other types by an absent or low fasting insulin level. On the other hand, about 40% of diabetics have normal to very high insulins. The classification of these diabetic types is based on the nature of the insulin response curve during the IVGTT. Type II has a normal to high insulin with no increment of insulin response to the glucose load. Type III also has a normal to high insulin; the insulin response is inadequate and there is a delayed return to preinjection levels (Kaneko *et al.*, 1977). Types II and III have been further subdivided on the basis of obesity or nonobesity (Mattheeuws *et al.*, 1984) and their insulin levels are given in Table 3.9.

The classification of diabetes into types has important therapeutic and prognostic implications. Insulin replacement therapy is the only effective treatment for the Type I and Type II nonobese diabetic. Type II obese and the Type III dogs with even a small insulin reserve would be the most likely subjects for successful oral hypoglycemic therapy. The early detection of diabetes and being able to treat these patients using oral drugs would have obvious advantages. Nelson *et al.* (1993)

TABLE 3.9 Insulin Concentrations in the Various Types of Canine Diabetes

Classification	Serum insulin (μU/ml)
Normal	5–20
Type I	0–5
Type II, nonobese	5–20
Type II, obese	20–130
Type III, nonobese	5–20
Type III, obese	8–60

have successfully treated cats with diabetes using oral hypoglycemic drugs. Prognostically, the severity of the diabetes can be assessed by the degree of glucose intolerance and the nature of the insulin response.

Atkins *et al.* (1979) identified diabetes in dogs less than 1 year of age and Atkins and Chin (1983) examined their insulin responses to glucose loading. All dogs were glucose intolerant but could mount a minimal insulin response somewhat akin to the Type II diabetic dogs. It could also be that these young diabetic dogs were identified during the early stages of their natural history of progression of their diabetes to Types I or II.

7. Glucagon Stimulation and the Insulin Response

The GST has been used in humans and cats to differentiate Type I from Type II diabetes. Type I diabetic cats have a minimal or no insulin response to glucagon. Type II diabetic cats have a significant insulin response in the GST. Nondiabetic obese cats also have an insulin response that is similar to the Type II diabetic cats. Thus, obesity is predisposing to the development of diabetes in animals as well as in humans. Type II diabetes is known to be characterized by various forms of insulin resistance (Section VIII.C.2).

8. Ketonemia and Lipemia

As the utilization of glucose progressively decreases in the diabetic, the utilization of fatty acids for energy purposes progressively increases in order to compensate. The supply of fatty acids for hepatic utilization is obtained by mobilization from the body fat depots. Mobilization progressively increases as insulin deficiency becomes more severe due to increases in hormone sensitive lipase. This enzyme is separate and distinct from the hepatic lipoprotein lipase.

In severe diabetes, hyperlipemias are often so marked that the blood appears as "tomato soup." A cream layer may separate out on storage in the cold due to hyperchylomicronemia. The plasma is turbid due to the presence of very low density lipoproteins (VLDLs). On chemical analysis, total triglycerides and cholesterol are elevated (Rogers *et al.*, 1975). Diabetic hyperlipemia appears to be caused by impaired lipolysis of chylomicra secondary to a deficiency of hepatic lipoprotein lipase rather than to an overproduction of VLDLs.

Concurrently with increased fatty acid oxidation in liver, a progressive decrease in fatty acid synthesis occurs. The net effect of the alterations in hepatic fatty acid metabolism is that AcCoA is generated in excess by the liver due to the increased rate of fatty acid β-oxidation catalyzed by the increased activity of the enzyme carnitine acyltransferase. Fatty acyl-CoA from fat mobilization is also a strong inhibitor of citrate synthase, which removes another route for disposal of AcCoA. The accumulated AcCoA units are then diverted into alternate pathways as described in Section V.B and, with the activation of ketogenic mechanisms, excessive synthesis of ketone bodies (Kreisberg, 1978) and cholesterol results. In the peripheral tissues, there is an underutilization of ketone bodies in the diabetic dog (Balasse and Havel, 1971). Ketosis is thus the result of an overproduction of ketone bodies by the liver and an underutilization by the peripheral (muscle) tissues. The Type I diabetic has a greater tendency than the Type II to develop ketoacidosis.

It has been suggested that the development of ketosis requires both a deficiency of insulin and an excess of glucagon (McGarry *et al.*, 1975; McGarry and Foster, 1976). Dobbs *et al.* (1975) and Unger and Orci (1975) proposed that diabetes develops due to a bihormonal interaction of insulin and glucagon because glucagon levels are high in insulin deficiency. The excess glucagon is thought to be caused by an abnormality in the alpha cell. There is also an excessive secretion of glucagon after protein ingestion or amino acid infusions (Unger, 1981). The excess glucagon may then exacerbate the insulin deficiency and lead to the ketoacidosis.

In the ketoacidotic state, marked cholesterolemias as high as 18 mmol/liter (700 mg/dl) have been observed in clinical diabetes of the dog. Net gluconeogenesis from fatty acid does not occur and the precursors for gluconeogenesis are the proteins. Excesses of glucagon, cortisol, and growth hormone in the diabetic also contribute to protein catabolism and gluconeogenesis. The cofactors that provide the reductive environment required for gluconeogenesis can be provided by the increased production of reduced cofactors during the increased fatty acid oxidation. This increase in the reductive environment of the cell is the mechanism that stimulates gluconeogenesis, which is corollary to the development of ketoacidosis.

9. *Electrolyte Balance and Ketoacidosis*

A mild glucosuria with only a few grams of glucose loss per day does not in itself precipitate the acidotic state because some compensation occurs. The liver increases its production and output of glucose even though there is a hyperglycemia so glucose metabolism continues. However, with continued and severe loss of glucose, all the attendant phenomena of attempts to compensate are exaggerated. Liver glycogen stores are depleted but liver glucose production continues to be increased due to increased protein breakdown and gluconeogenesis. The oxidation of fatty acids is accelerated and, with it, the overproduction of the acidic ketone bodies (AcAc, 3-OH-B, and acetone) occurs. The vapor pressure of acetone (bp 56.5°C) is high at body temperature and, thus, this volatile compound is often detected in the breath of the severely ketotic animal. AcAc and 3-OH-B are acidic anions that increase the "anion gap" and reduce the concentrations of HCO_3^-, Cl^-, Na^+, and K^+. Acidosis develops as the HCO_3^- is reduced and respiratory compensation is inadequate. In addition, there is an underutilization of ketone bodies in starvation (Garber *et al.*, 1974) and a similar underutilization of ketone bodies occurs in diabetes (Sherwin *et al.*, 1976). A rapid point-of-care method for quantifying 3-OH-B is now available and is expected to be very usefull in managing ketoacidosis (Section VIII.D).

In hyperketonemia, large amounts of ketones are wasted in the urine with large losses of water and HCO_3^-. The acidic ketones are buffered by ammonium ions derived from glutamine in the renal tubules, but large amounts of ketones are ultimately lost with Na^+ and K^+ in the urine. Even without ketonuria, the loss of electrolytes in the polyuria of diabetes may be considerable. Thus, the acidosis of the diabetic is a primary base deficit fundamentally related to the ketonemia and to the loss of ketones and HCO_3^- in the urine.

Excess glucose in the glomerular filtrate provokes an osmotic diuresis leading to loss of water and dehydration. The progressively severe loss of water and electrolytes and the dehydration and ketoacidosis ultimately lead to collapse, coma, and death. The condition is aggravated by renal impairment, which fortunately is not a common finding in diabetes of the dog. Not all the extracellular sodium deficit is due to urinary loss, however, because as H^+ increases, it enters the cells. In exchange, K^+ leaves the intracellular compartment and some Na^+ enters the cells. As the dehydration progresses, extracellular K^+ concentration may be very high even though there may be a total body deficit. This is an important consideration in the insulin, fluid, and electrolyte replacement therapy of diabetic ketoacidosis. The electrolyte replacement must include K^+ because correction of the acidosis and the rapid expansion of the extracellular fluid compartment leads to the reverse exchange of K^+ and this results in hypokalemia.

10. *Urinalysis*

The renal threshold for glucose in the dog is about 11.1 mmol/liter (200 mg/dl) so that the detection of even trace amounts of glucose in the urine is an important finding and warrants further consideration. In virtually all cases of diabetes suspected on the basis of persistent glycosuria alone, the diagnosis can be later confirmed. Renal diabetes, that is, a low renal threshold for glucose, is an extremely rare occurrence. If it does occur, it can be detected by the normal blood glucose in the presence of the glucosuria. Transient glucosurias may occur for 1–1.5 hours after a heavy carbohydrate meal, but a 2-hour postprandial glucosuria is a strong indication of diabetes.

Currently, detection of glucosuria using the urinalysis sticks is the most common method of point-of-care evaluation of the clinical success of insulin therapy. The disadvantages of this system include owner difficulties, inconsistencies, and inaccuracies. The FrAm method, whereby only biweekly blood samplings need be taken, can have decided advantages in following the course of insulin therapy.

An elevated urinary specific gravity (SG) has in the past been considered to be a good indicator of glucosuria and, hence, of diabetes. SG is a measure of the concentration of solutes in the urine, principally the cations (Na^+, K^+, NH_4^+), anions (PO_4^{2-}, SO_4^{2-}, HCO_3^-, Cl^-), and urea. The observed SG of urine is the result of the additive effect of the contributions of all these solutes. It is for this reason that the osmolality of any fluid, urine or plasma, can be estimated by simply adding the major anions and cations expressed in millimoles per liter (see the chapter on acid–base). Albumin in urine increases the SG by 0.003 units for each 10 g/liter (1 g/dl), whereas glucose increases it by 0.004 units for each 55 mmol/liter (1 g/dl). Even though the presence of glucose does increase the SG linearly, a 4+ reaction, 140 mmol/liter (2.5 g/dl) would increase the SG by only 0.010 unit, an insignificant value on the refractometer. Therefore, although SG is a valuable measure of renal function, it is of no value with respect to the glucosuria of diabetes or to proteinuria. Conversely, by subtracting the contributions of protein and glucose from the observed SG, a more accurate measure of renal function in diabetes may be obtained.

Proteinuria is a common sign of renal disease and is often observed in diabetes in dogs. There is doubt whether this is associated with chronic nephritis or whether it is due to renal failure as an aftermath of diabetes.

Diabetic nephropathies due to microangiopathies of the glomerular tufts and basement membrane injuries are frequent and serious complications of the chronic, poorly controlled, human diabetic. A degree of renal arteriosclerosis is common in diabetic dogs but this lesion is not comparable to the Kimmelstiel–Wilson lesion seen in humans. Also, only 1 of 10 diabetic dogs at necropsy had a significant renal lesion although most had some degree of nephritis (Cotton *et al.*, 1971). In renal function studies of experimental streptozotocin diabetes (Kaneko *et al.*, 1978b) and in spontaneous diabetes (Kaneko *et al.*, 1979), the urea, creatinine, and phosphate clearances were normal. The blood urea and creatinine concentrations were only slightly elevated and it was concluded that renal disease is not a significant complication in the dog.

The ketone bodies are very low renal threshold substances and their appearance in the urine is an early and significant sign of developing ketonemia and acidosis. They are not, however, diagnostic of diabetes because ketonuria is observed in starvation or any form of increased fat catabolism. Ketonuria is also absent in mild diabetes but ketonurias of varying degrees are common in the advanced diabetic state. Therefore, ketonurias can be useful for prognostication. Generally, the Type I diabetic is prone to ketonuria because there is an absence of insulin. In the Type II diabetic, ketonuria occurs less frequently because there is ample insulin and ketonuria is seen only when the diabetes has advanced to the point of complete failure of production. Urine pH is of little value in detecting acidosis, because only in extreme cases does the pH reflect acidosis.

11. *Summary*

The alterations in blood plasma that have been described are summarized in Fig. 3.16. In the diabetic state, the uptake and hence utilization of glucose by muscle and adipose tissue are depressed. In these tissues, protein and lipid breakdown are enhanced, and increased amounts of their constituent amino acids and fatty acids are released to the circulation and carried to the liver. Increased hepatic urea production results from the catabolism of these amino acids. Increases in the key gluconeogenic enzymes of the liver, G-6-Pase, PEPCK, and PC, direct glucose metabolism toward an overproduction of glucose. Simultaneously, lipogenesis is suppressed and, with the increased mobilization of fatty acids, AcCoA accumulates and is followed by increased cholesterolgenesis and ketogenesis. In the

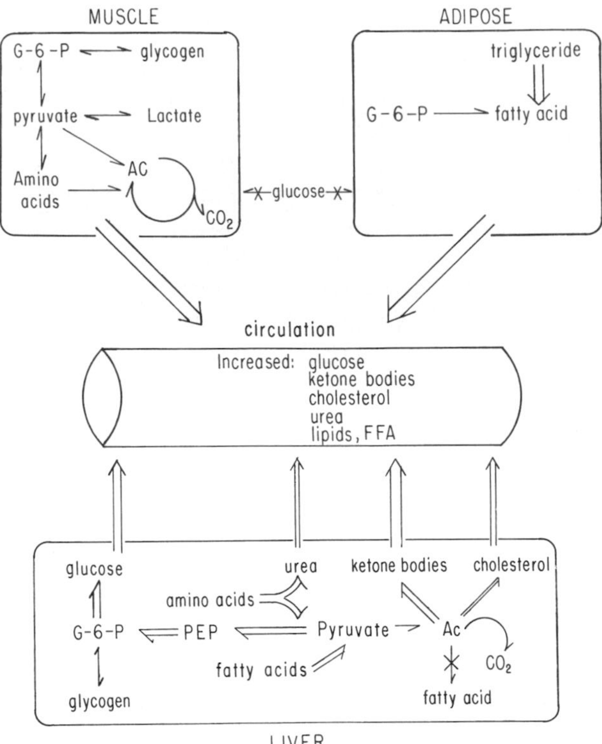

FIGURE 3.16 Summary of metabolic alterations in tissues of major importance in the diabetic animal. Increased flow in the metabolic pathways is noted by larger arrows.

peripheral tissues there is an underutilization of ketones, all of which results in a net increase in blood ketones and subsequent ketoacidosis. Thus, diabetes mellitus is characterized by a fundamental overproduction of and an underutilization of both glucose and ketones as the result of the absolute or relative deficiency of insulin.

B. Hyperinsulinism

After the discovery of insulin, a clinical state with marked similarities to insulin overdosage was recognized as a disease entity in humans and named *hyperinsulinism*. The disease is now known to be due to a persistent hyperactivity of the pancreas as the result of insulin-secreting islet cell tumors. Excess insulin can be extracted from metastatic foci in liver as well as from the pancreatic tumor. There are many reports on this disease in dogs (Teunissen *et al.*, 1970; Spieth, 1973; Hill *et al.*, 1974; Mattheeuws *et al.*, 1976). Priester (1974) and Kruth *et al.* (1982) reviewed pancreatic islet cell tumors in animals in the United States and Canada.

Hyperinsulinism is characterized by a persistent hypoglycemia with periods of weakness, apathy, fainting, and during hypoglycemic crises, convulsions, and coma. A history relating the attacks to periods after fasting or exercise provides a clinical basis for further investigations. Establishment of the diagnosis depends on finding a hypoglycemia of <3mmol/liter (<55 mg/dl) at the time of symptoms and a hyperinsulinemia, usually >20 μU/ml. The symptoms are also relieved by glucose administration. In mild cases, the fasting glucose level may be normal, in which case, diagnostic hypoglycemia may be provoked by sequentially (1) placing on a low-carbohydrate diet (meat only) with frequent feedings for 1 week, (2) placing on a 24-hour fast, and finally (3) by moderate to stressful exercise. Blood glucose is determined at the end of each step and if hypoglycemia is seen at any step, the provocation should be terminated. Serum insulin is determined at this time and a hyperinsulinemia is generally diagnostic of insulinoma. Calculations of ratios—insulin/glucose; glucose/insulin; amended insulin/glucose—do not offer any advantages over the individual insulin and glucose values.

The glucose tolerance curve is generally characteristic of an increased tolerance if the test is modified as follows: (1) The dog is on a standard carbohydrate diet for 3 days, (2) the intravenous test is used, and, most important, (3) blood sampling is continued for 6–8 hours. A prolongation of the hypoglycemic phase (phase III, Fig. 3.12) is the most significant portion of the curve.

A dog with a tendency toward persistent hypoglycemia is likely to have an abnormal response in the insulin tolerance test but this test is not a reliable test of insulinoma. The tolerance curve may have a minimal drop in blood glucose and remain below the original level for a prolonged length of time. Therefore, the curve has "insulin resistance" and "hypoglycemia unresponsiveness." Use of this test carries some risk so a glucose solution for intravenous administration should be at hand. Similarly, the glucagon stimulation test has not been a reliable test for hyperinsulinism.

The hypoglycemia that follows oral administration of leucine in children has been used in human patients with islet cell tumors. Marked hypoglycemia occurs within 30–60 minutes after L-leucine administration. Leucine-induced hypoglycemia is also associated with a rise in plasma insulin. In patients with islet cell tumors, leucine sensitivity disappeared after surgical excision of the tumor, which indicates that the tumorous islet cells alone were being stimulated by the leucine. This test has been used successfully in pancreatic islet cell tumors of dogs.

Currently, the most useful tests are the serum insulin and the fasting plasma glucose taken as described earlier. There is an inappropriately high level of insulin (>20 μU/liter) with a hypoglycemia of <3 mmol/liter (<55 mg/dl).

C. Hypoglycemia of Baby Pigs

Hypoglycemia of baby pigs occurs during the first few days of life and is characterized by hypoglycemias of <2.2 mmol/liter (<40 mg/dl), apathy, weakness, convulsions, coma, and finally death.

The newborn baby pig is particularly susceptible to hypoglycemia. At birth, the blood glucose level is >6 mmol/liter (>110 mg/dl) and, unless the pig is fed, its blood glucose drops rapidly to hypoglycemic levels within 24–36 hours. The liver glycogen, which is high (14.8%) at birth, is almost totally absent at death. In contrast, newborn lambs, calves, and foals are able to resist starvation hypoglycemia for more than a week. If the baby pig suckles, its ability to withstand starvation progressively increases from the day of birth. A 10-day-old baby pig can be starved up to 3 weeks before symptoms of hypoglycemia occur.

Gluconeogenic mechanisms are undeveloped in the newborn pig, which indicates that the gluconeogenic enzymes of the baby pig are inadequate at birth. This also indicates that these enzymes need to be induced by feeding so they can reach their maximal activities within 1 or 2 weeks after birth. The precise hepatic gluconeogenic enzymes and their induceability by feeding have not yet been identified.

The association of baby pig hypoglycemia with complete or partial starvation is shown by the findings that their stomachs are empty at necropsy and the syndrome itself is indistinguishable from experimental starvation of the newborn baby pig. Starvation of the newborn pig under natural conditions can occur due to factors relating to the sow (agalactia, metritis, etc.) or to the health of the baby pig (anemia, infections, etc.), either case resulting in inadequate food intake. The requirement for feeding to induce the hepatic gluconeogenic mechanisms in the newborn baby pig explains its inability to withstand starvation in contrast to the newborn lamb, calf, or foal, which is born with fully functioning hepatic gluconeogenesis.

D. Glycogen Storage Diseases

The glycogen storage diseases are characterized by the pathological accumulation of glycogen in tissues. Based on their patterns of glycogen accumulation, their clinical pathological findings, their enzymes of glycogen metabolism, and the structural analyses of their glycogen, the GSDs in humans have been classified into eight types (Howell and Williams, 1983). All have an autosomal recessive mode of inheritance except for GSD VIII, which is sex linked. Their glycogen structures are normal except for Types III and IV.

Type I or classical von Gierke's disease is characterized by increased liver glycogen leading to a marked hepatomegaly. There is a marked hypoglycemia and the blood glucose response to epinephrine or glucagon is minimal or absent. The liver glycogen structure is normal. The defect in this disease is a deficiency of the enzyme G-6-Pase. Type II or Pompe's disease is a generalized glycogenosis with lysosomal accumulation of glycogen and early death. The defect in this disease is a deficiency of acid-α-glucosidase (AAGase). In Type III or Cori's disease, the debrancher enzyme is deficient, which leads to the accumulation of glycogen of abnormal structure. The branches are abnormally short and there are an increased number of branch points; it is a limit dextrin and the disease is sometimes called a *limit dextrinosis*. There is a variable hypoglycemia, little or no response to epinephrine or glucagon, hepatomegaly, cardiomegaly, and early death. In Type IV or Andersen's disease the brancher enzyme is deficient, which leads to a glycogen with abnormally long branches and few branch points. It is clinically similar to Type III. In Type V or McArdle's disease muscle phosphorylase (MPase) is deficient, whereas in Type VI, it is liver phosphorylase (LPase), which is deficient. Type VII or Tarui's disease is characterized by a deficiency of muscle phosphofructokinase (PFK) with accumulation of glycogen in muscle, and Type VIII is deficient in leukocyte or hepatic phosphorylase b kinase (PBK). This disease is uniquely sex linked.

Of these eight types in humans, only Types I, II, III, and VIII are found in animals. Other forms of glycogen storage in animals are described as GSD-like based on their pathological patterns of glycogen accumulation. GSD in animals has been reviewed by Walvoort (1983).

There is an inherited PFK deficiency in the springer spaniel dog but unlike human Type VII GSD, there is no muscle pathology or glycogen accumulation in muscle. The deficiency in the dog is expressed as a hemolytic anemia caused by a deficiency of the PFK isoenzyme in the erythrocytes and is rightly considered to be an inherited erthrocyte enzyme deficiency rather than a GSD (Giger *et al.*, 1985). Mammalian PFK is present in tissues as tetramers composed of combinations of three different subunits: PFK-M (muscle), PFK-L (liver), and PFK-P (platelets). Human and dog muscle and liver have homogenous tetrameric PFK-M_4 and PFK-L_4, respectively. Human erythrocyte PFK is a mixed tetramer, PFK-L_2/PFK-M_2, whereas the dog erythrocyte PFK is a mixed tetramer, PFK-M_2/PFK-P_2 (Vora *et al.*, 1985). In PFK-M subunit deficiency in the dog erythrocyte, PFK-L replaces PFK-M; PFK-L_2/PFK-P_2. In the human erythrocyte, PFK-P replaces PFK-M; PFK-L_2/PFK-P_2. Although the substituted PFK in the erythrocyte is the same in dog or human, the deficiency in the human is expressed as a GSD, whereas in the dog it is expressed as an exertional hemolytic anemia. The anemia occurs after heavy exertional respiratory stress such as is experienced in vigorous hunting or exercise. Hyperventilation induces a respiratory alkalosis, which in turn increases the fragility of the erythrocyte and the hemolytic anemia occurs (Giger *et al.*, 1985).

A radiation-induced Type I GSD occurs as an autosomal recessive condition in the C3H mouse and is characterized by hypoglycemia, early death, and a deficiency of liver G-6-Pase (Gluecksohn-Welch, 1979).

Type II GSD has been described in Brahman cattle (O'Sullivan *et al.*, 1981), the Lapland dog (Walvoort *et al.*, 1982), and in the Japanese quail (Murakami *et al.*, 1980). In the Brahman cattle, Type II is characterized by early death, generalized glycogen deposition, and a marked decrease in AAGase activity. It is inherited as an autosomal recessive. In the Lapland dog, there is also early death, generalized glycogen deposition, hepatomegaly, and cardiomegaly. There is also a marked decrease in heart and liver AAGase. The Japanese quail with Type II is also characterized by early death, glycogen deposition in the heart, liver, and muscles, and decreased AAGase.

Type III occurs in the German shepherd dog and is characterized by early death, little or no response to epinephrine or glucagon, hepatomegaly, and cardiomegaly with glycogen accumulation. The glycogen has a limit dextrin structure and there is a very low debrancher enzyme activity in liver and muscle (Svenkerud and Hauge, 1978; Ceh *et al.*, 1976).

Type VIII is seen in the rat and the mouse. In the rat, the disease is inherited as an autosomal recessive, appears healthy but is hypoglycemic, has hepatomegaly due to glycogen accumulation in the liver, and also has a very low liver phosphorylase kinase activity (Clark *et al.*, 1980). The affected mouse is apparently healthy but has increased glycogen accumulation in the muscle with a very low muscle PBK. The inheritance is sex linked (Gross, 1975).

X. DISORDERS OF RUMINANTS ASSOCIATED WITH HYPOGLYCEMIA

A. General

The principal disorders of domestic ruminants in which hypoglycemia is a salient feature are bovine ketosis and ovine pregnancy toxemia. Pregnancy toxemia characteristically is a widespread disease of high mortality occurring in the pregnant ewe just prior to term, the time when carbohydrate demands are highest, especially in those ewes carrying more than one fetus. Bovine ketosis, on the other hand, occurs in the high-producing dairy cow characteristically during the early stages of lactation, when milk production is generally the highest. Abnormally high levels of the ketone bodies, acetone, AcAc, 3-OH-B, and isopropanol appear in blood, urine, and the milk. These alterations are accompanied by the clinical signs of ketosis: loss of appetite, weight loss, decrease in milk production, and nervous disturbances.

The energy metabolism of the ruminant is focused on the utilization of the volatile fatty acids produced by rumen fermentation rather than on carbohydrates as in the nonruminant. The carbohydrate economy of the ruminant is significantly different from that of the nonruminant and an appreciation of these differences is important to the understanding of these metabolic disorders of the ruminant.

B. Carbohydrate Balance

1. Glucose Requirements

The heavy demands for glucose in early lactation and in late pregnancy are well known. In the late 1950s, Kleiber calculated that about 60% of the lactating cows' daily glucose requirement is for the production of milk. The balance sheet (Table 3.10) indicates a total daily glucose requirement of 1140 g of which 700 g appear in the milk. For sheep in late pregnancy, about one-third to one-half of the daily glucose turnover of 100 g is utilized by the fetus.

A good approach to assess the glucose requirements of an animal is to measure its turnover rate or the rate at which glucose enters or leaves the circulation. This is best measured by the use of isotopically labeled glucose and has been used in lactating cows. It has been estimated to be 1440 g/day (60 g/hour) in cows and about 144 g/day (6 g/hour) in normal pregnant ewes just prior to term.

2. Glucose Sources

The large amounts of indigestible carbohydrates ingested by ruminants are fermented to volatile fatty acids by the rumen microflora. Little, if any, of the digestible carbohydrates (starch, glucose) in the diet escapes this fermentation so that glucose absorption from the digestive tract accounts for virtually none of the daily glucose requirement of ruminants. However, if any glucose escapes rumen fermentation, for example, in gastrointestinal disease, it is readily absorbed.

An indirect source of blood glucose is ruminal lactic acid. Lactic acid is a product of many fermentation reactions and ruminal lactate can be absorbed. The blood lactate can be a source of blood glucose via the lactic acid cycle (Fig. 3.4). However, the principal source of blood lactate is the breakdown of muscle glycogen. Therefore, some of the ruminant's glucose requirement may be met by lactate but this is minimal because excess lactic acid in the rumen is toxic.

TABLE 3.10 Carbohydrate Balance Sheet[a]

	Carbohydrate carbon
A. Cow's daily glucose flux	
1. *In 12.5 kg milk:*	
610 g lactose	257 g C/day
462 g milk fat with 58 g glycerol	23 g C/day
Carbohydrate carbon in milk/day	280 g C/day
2. Daily glucose catabolism:	
Cow produced daily 3288 liters CO_2 = 1762 g C	
Transfer quotient plasma gluocse → CO_2 is 0.1	
Thus glucose to CO_2/day =	176 g C/day
1 + 2 = daily flux of glucose	456 g C/day
$\frac{180}{72} \times 456 = 1140$ g glucose/day	
B. Cow's glucose sources	
Cow secreted daily in urine 34 g N, indicating catabolism of 213 g protein =	100 g C/day
less c in urea =	14 g C/day
Maximum available for glucose synthesis from protein =	96 g C/day
Glucose flow in milk and respiration =	456 g C/day
Thus glucose flow from nonprotein sources =	360 g C/day
$\frac{180}{72} \times 360 = 900$ g glucose daily must have been supplied from a non-protein source	

[a] From Kleiber (1959).

The carbohydrate balance sheet (Table 3.10) provides the contribution of protein as a source of carbohydrate for the lactating cow. Because glucose absorption in the ruminant is minimal, the balance sheet also illustrates the importance of an alternate nonprotein source of carbohydrate carbon. These sources are the ruminal volatile fatty acids. The principal products of rumen fermentation are the volatile fatty acids, acetate, propionate, and butyrate. These acids are absorbed across the rumen wall and are the major source of nutriment for the ruminant. Various authors have used a variety of techniques to estimate the amounts of production and absorption of these acids. These fatty acids are found in blood in proportions of about acetate, 65; propionate, 20; and butyrate, 10. Further details of fatty acid production and absorption by the ruminant may be found in the chapter on lipid metabolism. In general, carbon atoms of acetate, although they appear in carbohydrate (blood glucose, milk lactose) through the mechanism of rearrangement in the TCA cycle (Fig. 3.9), cannot theoretically contribute to the net synthesis of carbohydrate. Thus, acetate is not a glucogenic compound. The large amounts of acetate provided by rumen fermentation are utilized for energy purposes and for the synthesis of fat. A possible mechanism for the direct incorporation of acetate into a glucose precursor is the so-called "glyoxylate pathway," which occurs in plants but not in animals.

Propionate, on the other hand, is a well-known precursor of carbohydrate. The pathway leading to a net synthesis of glucose from propionate is available via the reaction

$$\text{Propionate} + CO_2 \rightarrow \text{succinate},$$

as shown in Fig. 3.9. According to the scheme, 2 moles of propionate are required for the synthesis of 1 mole of glucose so 1 g of propionate theoretically can provide 1.23 g of glucose. The amounts of propionate available from rumen fermentation can theoretically supply the glucose requirements not accounted for by protein sources.

Butyrate, the third major fatty acid of rumen fermentation, influences glucogenesis but does not contribute carbon directly to glucose. Butyrate stimulates glucose production by liver by increasing phosphorylases and gluconeogenesis. The AcCoA derived from β-oxidation of butyrate also activates pyruvate carboxylase, a key gluconeogenic enzyme that further promotes gluconeogenesis.

3. *Utilization of Glucose*

The overall utilization of glucose by the ruminant has significant differences from that of other animals. Acetate oxidation rather than glucose plays the important role in energy metabolism of the ruminant. Only about 10% of the respiratory CO_2 arises from glucose oxidation, which is considerably less than the 25–60% for the rat, dog, and human. The glucose tolerance of the cow, however, is the same as in other animals. The plasma clearance $T_{1/2}$ of 33 minutes in the cow is similar to that of dogs (Kaneko *et al.*, 1977) and humans.

About 60% of the glucose oxidized in the mammary gland of the lactating cow occurs via the HMP pathway (Fig. 3.6), the same as in the rat mammary gland. HMP pathway activity in the ruminant mammary gland is also evidenced by the high activities of the HMP enzymes, G-6-PD and 6-P-GD in sheep and cows' mammary glands. Thus, even though overall glucose utilization is lower in ruminants, their pathways of glucose catabolism are the same as in other animals. As in other animals, the HMP pathway is the major provider of the reductive atmosphere for the synthetic processes of the mammary gland.

Through the TCA cycle pathway, carbons from acetate, from whatever source, appear in milk products (Fig. 3.17). Glucose carbon atoms may be given off as CO_2, appear in the amino acids of milk protein via transamination of oxaloacetate and α-ketoglutarate, or appear in milk fat. The short-chain fatty acids of butterfat are synthesized from acetate in the mammary gland, whereas the long-chain acids of butterfat are derived from blood lipids. The synthetic pathway for fatty acids in the gland is the same as that in other animal tissues (see Section IX).

The major portion of glucose uptake by the mammary gland provides for the biosynthesis of milk. The glucose and galactose moieties of lactose are derived from blood glucose. The rate of lactose synthesis is

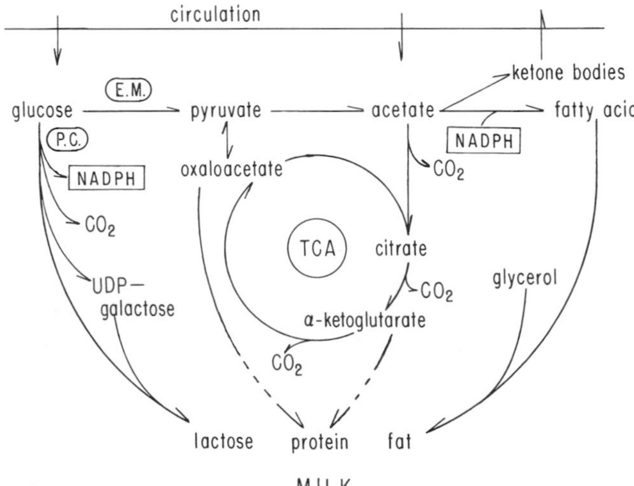

FIGURE 3.17 Summary of some metabolic pathways in the mammary gland.

also constant over a wide range of blood glucose concentrations, 1.1–4.4 mmol/liter (20–80 mg/dl), which indicates that lactose synthesis is maximal even under hypoglycemic conditions. The mammary gland, therefore, is a glucose utilizing tissue, principally for biosynthesis and considerably less is oxidized. The principal metabolic pathways involved are summarized in Fig. 3.17.

Ruminant nervous tissue, that is, brain, is also similar to that of other animals in being an obligatory glucose-utilizing tissue. The HK activity of sheep brain, however, is significantly lower than that of rat brain. This means that even though there is the same obligatory glucose requirement between the ruminant and nonruminant, glucose utilization by ruminant nervous tissue is lower than in the nonruminant. Similarly, ruminant intestine and muscle use less glucose than nonruminants.

With regard to organ distribution of gluconeogenic enzymes, the highest G-6-Pase activities are found in ruminant livers as compared to other organs of ruminants and are generally equal to or slightly lower than the activities found in nonruminant livers. During early lactation, the period when a cow's glucose requirement is highest, hepatic G-6-Pase does not increase. Similarly, cow liver PEPCK, a key gluconeogenic enzyme, is already very high in comparison to that of rat liver. All of this is in keeping with the concept that liver is primarily a glucose producing tissue. This also means that the high-producing dairy cow that has been genetically selected for these qualities is already synthesizing glucose maximally under normal conditions. It follows that any additional demands for glucose from physical stress, disease, etc., are unlikely to be met by increased glucose production. This glucose shortage leads to ketosis, the primary form from excess milk production or secondary form from the stress of a disease.

To summarize, the ruminant appears to be an animal well adapted to a carbohydrate economy based on the endogenous synthesis of glucose from noncarbohydrate sources (gluconeogenesis). The enzymatic mechanisms for gluconeogenesis are already operating at near maximal levels in the high-producing dairy cow. Glucose oxidation by individual tissues as well as by the intact animal is lower in ruminants than in nonruminants. Although overall partitioning of glucose oxidation may be different in ruminants, the pathways by which this oxidation is accomplished are similar to those of other animals (Fig. 3.17). The endocrine relationships of ruminants are also qualitatively similar to those of nonruminants so that the normally low blood glucose concentrations of ruminants are a reflection of their degree of influence or balance rather than their type of action.

C. Biochemical Alterations in Body Fluids

1. *Hypoglycemia*

Hypoglycemia is such a consistent finding in bovine ketosis and in ovine pregnancy toxemia that *hypoglycemia* has been suggested as another name for bovine ketosis. This hypoglycemia has played an important role in ketosis, as a rationale for therapy and as a basis for the concept of ketosis and pregnancy toxemia as manifestations of a carbohydrate deficiency that occurs under conditions of excessive and insurmountable demands.

2. *Ketonemia*

Ketonemia is a consistent feature of bovine ketosis and ovine pregnancy toxemia from which the name *ketosis* is derived. The ketone bodies are the same as those previously mentioned (Section V.3), AcAc, 3-OH-B, and acetone. A fourth compound, isopropanol, is included for the ruminant and interconversions can occur between these ketone bodies. The fundamental mechanism of ketosis is covered in a separate chapter and a brief outline is presented here.

a. Site of Ketone Body Production

Increased ketogenesis occurs under conditions that favor the accumulation of acetate. In the nonruminant, the liver is the sole source of ketone bodies and the ketone bodies appear in the body fluids when production exceeds the capacity for utilization. They are low renal threshold substances as well as being highly volatile so they readily appear in urine, milk, and in the breath. In the ruminant, the rumen epithelium and mammary gland are also sources of ketone bodies. The extent of their contribution to the ketone body pool, however, is uncertain but it could be considerable in the ketotic animal.

b. Hyperketonemia

Hyperketonemia is influenced by a number of conditions, all of which relate to the carbohydrate economy of the ruminant. Starvation is the most well known cause of ketosis. Mild to moderate ketonemia is often seen without detrimental effects during early lactation, late pregnancy, underfeeding, and with high fat diets. In all instances, the constant demands of the body for glucose are not adequately met.

Hyperketonemia is a consistent finding in bovine ketosis and pregnancy toxemia, though the degree of ketonemia does not necessarily parallel the severity of the clinical signs. Normally, total blood ketones in sheep or cows are less than 0.10 mol/liter (0.6 mg/dl).

D. Ruminant Ketosis

The finely balanced carbohydrate economy of the ruminant plays a central role in the development of ketosis in cows and sheep. In the cow, large amounts of glucose must be produced by the liver to meet the heavy demands for lactose, particularly in early lactation, when the demand is highest. Similarly, in the pregnant ewe, especially when carrying twins, there is a progressive and large obligatory demand for hexoses (fructose), which is maximal near term. The precise mechanisms whereby internal metabolic imbalances occur in trying to meet these demands and manifest as ketosis, however, are uncertain. The fundamental imbalance is an inability of hepatic gluconeogenesis to respond to meet the demands of lactation or pregnancy and a hypoglycemia occurs, followed by hyperketonemia, and clinical ketosis develops.

References

Arai, T., Washizu, T., Hamada, S., Sako, T., Takagi, S., Yashiki, K., and Motoyoshi, S. (1994). *Vet. Res. Comm.* **18,** 417.

Armbruster, D.A. (1987). *Clin. Chem.* **33,** 2153.

Ashmore, J., and Weber, G. (1968). *In* "Carbohydrate Metabolism and Its Disorders" (F. Dickens, P. J. Randle, and W. J. Whelan, eds.), p. 336. Academic Press, New York.

Atkins, C. E., and Chin, H. P. (1983). *Am. J. Vet. Res.* **44,** 596.

Atkins, C. E., Hill, J. R., and Johnson, R. K. (1979). *J. Am. Vet. Med. Assoc.* **175,** 362.

Balasse, E. O., and Havel, R. J. (1971). *J. Clin. Invest.* **50,** 801.

Banauch, D., Brummer, W., Ebeling, W., Metz, H., Rinfrey, H., Leybold, K., and Rick, W. (1975). *Z. Klin. Chem.* **13,** 101.

Bell, R. R., and Jones, E. R. (1945). *J. Comp. Pathol. Therap.* **55,** 117.

Bolli, G. B., Gottesman, I. S., Campbell, P. J., Haymond, M. W., Cryer, P. E., and Gerich, J. E. (1984). *N. Engl. J. Med.* **311,** 1214.

Buchanan, K. D. (1975). "Diabetes: Its Physiological and Biochemical Bases," p. 63. MTP Press, Lancaster, UK.

Bunn, H. F., Gabbay, K. H., and Gallop, P. M. (1978). *Science* **200,** 21.

Ceh, L., Hauge, J. G., Svenkerud, R., and Strande, A. (1976). *Acta. Vet. Scand.* **17,** 210.

Cherrington, A. D., and Exton, J. H. (1976). *Metabolism* **25,** 1351.

Clark, D. G., Topping, D. L., Illman, R. J., Trimble, R. P., and Malthus, R. S. (1980). *Metabolism* **29,** 415.

Cornelius, L. M., and Mahaffey, E. A. (1981). *J. Am. Vet. Med. Assoc.* **180,** 635.

Cotton, R. B., Cornelius, L. M., and Theran, P. (1971). *J. Am. Vet. Med. Assoc.* **159,** 863.

Cutler, J. T. (1934). *J. Biol. Chem.* **106,** 653.

Dobbs, R., Sakurai, H., Sakai, H., Faloona, G., Valverde, I., Baetens, D., Orci, L., and Unger, R. (1975). *Science* **187,** 544.

Duckworth, W. C., and Kitabchi, A. E. (1981). *Endocrin. Rev.* **2,** 210.

Eisenbarth, G. S. (1986). *N. Engl. J. Med.* **314,** 1360.

Engerman, R. L., and Kramer, J. W. (1982). *J. Am. Diabetes Assoc.* **31,** 26.

Felig, P., Wahren, J., Sherwin, R., and Hendler, R. (1976). *Diabetes* **25,** 1091.

Foster, S. J. (1975). *J. Sm. Anim. Pract.* **16,** 295.

Gabbay, K. H., Hasty, K., Breslow, J. L., Ellison, R. C., Bunn, H. F., and Gallop, P. M. (1977). *J. Clin. Endocrinol. Metab.* **44,** 859.

Garber, A. J., Menzel, P. H., Boden, G., and Owen, O. E. (1974). *J. Clin. Invest.* **54,** 981.

Gershwin, L. J. (1975). *J. Am. Vet. Med. Assoc.* **167,** 479.

Giger, U., Harvey, J. W., Yamaguchi, R. A., McNulty, P. K., Chiapella, A., and Beutler, E. (1985). *Blood* **65,** 345.

Gluecksohn-Waelsch, S. (1979). *Cell* **16,** 225.

Gross, S. R. (1975). *West. J. Med.* **123,** 194.

Hill, F. W. G., Pearson, H., Kelly, D. F., and Weaver, B. M. Q. (1974). *J. Small Anim. Pract.* **15,** 119.

Hooghuis, H., Rodriguez, M., and Castaño, M. (1995). *Vet. Clin. Pathol.* **23,** 110.

Howell, R. R., and Williams, J. C. (1983). *In* "The Metabolic Basis of Inherited Disease" (J. B. Stanbury, J. B. Wymgaarden, D. S. Frederickson, Goldstein, J. L., and Brown, M. S., eds.) 5th ed., p. 141. McGraw-Hill, New York.

Jensen, A.L., and Aaes, H. (1992). *Vet. Res. Commun.* **16,** 317.

Johnson, R. K., and Atkins, C. E. (1977). *In* "Current Veterinary Therapy" (R. W. Kirk, ed.), Vol. 6, p. 1010. Saunders, Philadelphia, Pennsylvania.

Kaneko, J. J., and Howard, C. F., Jr. (1989). *In* "Clinical Chemistry of Laboratory Animals" (W. F. Loeb and F. W. Quinby, eds.), p. 73. Pergamon Press, New York.

Kaneko, J. J., Mattheeuws, D., Rottiers, R. P., and Vermeulen, A. (1977). *J. Small Anim. Pract.* **18,** 85.

Kaneko, J. J., Mattheeuws, D., Rottiers, R. P., Van Der Stock, J., and Vermeulen, A. (1978a). *Acta Endocrinol.* **87,** 113.

Kaneko, J. J., Mattheeuws, D., Rottiers, R. P., and Vermeulen, A. (1978b). *Am. J. Vet. Res.* **39,** 807.

Kaneko, J. J., Mattheeuws, D., Rottiers, R. P., and Vermeulen, A. (1979). *Cornell Vet.* **69,** 375.

Kaneko, J. J., Kawamoto, M., Heusner, A. A., Feldman, E. C., and Koizumi, I. (1992). *Am. J. Vet. Res.* **53,** 1797.

Kawamoto, M., Kaneko, J. J., Heusner, A. A., Feldman, E. C., and Koizumi, I. (1992). *Am. J. Vet. Res.* **53,** 851.

Khoo, J. C., Steinberg, D., Thompson, D., and Mayer, S. E. (1973). *J. Biol. Chem.* **248,** 3823.

Kirk, C. C., Feldman, E. C., and Nelson, R. W. (1993). *Am. J. Vet. Res.* **54,** 463.

Kitabchi, A. E. (1977). *Metabolism* **26,** 547.

Kleiber, M. (1959). *Proc. Interam. Symp. Peaceful Appl. Nucl. Energy, 2nd, Buenos Aires,* p. 161.

Kramer, J. W. (1977). *Fed. Am. Soc. Exp. Biol.* **36,** 279.

Kramer, J. W. (1981). *Am. J. Pathol.* **105,** 194.

Kramer, J. W., Nottingham, S., Robinette, J., Leaz, G., Sylvester, S., and Dessouky, M. (1980). *Diabetes* **29,** 558.

Kreisberg, R. A. (1978). *Ann. Intern. Med.* **88,** 681.

Kronfeld, D. S., Simesen, M. G., and Dungworth, D. L. (1960). *Res. Vet. Sci.* **1,** 242.

Kruth, S. A., Feldman, E. C., and Kennedy, P. C. (1982). *J. Am. Vet. Med. Assoc.* **181,** 54.

Ling, G. C., Lowenstine, L. J., Pulley, L. T., and Kaneko, J. J. (1977). *J. Am. Vet. Med. Assoc.* **170,** 521.

Lusk, G. (1928). "The Elements of the Science of Nutrition," 4th ed., p. 321. Saunders, Philadelphia, Pennsylvania.

McCandless, E. L., Woodward, B. A., and Dye, J. A. (1948). *Am. J. Physiol.* **154,** 94.

McGarry, J. D., and Foster, D. W. (1976). *Am. J. Med.* **61,** 9.

McGarry, J. D., Wright, P. H., and Foster, D. W. (1975). *J. Clin. Invest.* **55,** 1202.

McMillan, F. D., and Feldman, E. C. (1986). *J. Am. Vet. Med. Assoc.* **188,** 1426.

Markussen, J. (1971). *Int. J. Protein Res.* **3,** 149.

Mattheeuws, D., Rottiers, R., Deijcke, J., DeRick, A., and DeSchepper, J. (1976). *J. Small Anim. Pract.* **7,** 313.
Mattheeuws, D., Rottiers, R., Kaneko, J. J., and Vermeulen, A. (1984). *Am. J. Vet. Res.* **45,** 98.
Morrill, C. C. (1952). *Am. J. Vet. Res.* **13,** 164.
Murakami, H., Takagi, A., Nonaka, S., Ishiura, S., Sugita, H., and Mizutani, M. (1980). *Exp. Anim.* **29,** 475.
Naithani, G., Steffans, G., Tager, H. S., Buse, G., Rubenstein, A. F., and Steiner, D. F. (1984). *Hoppe-Seyler's Z. Physiol. Chem.* **365,** 571.
Nathan, D. M., Singer, D. E., Hurxthal, K., and Goodson, J. D. (1984). *N. Engl. J. Med.* **310,** 341.
Nelson, R. W., Himsel, C. A., and Feldman, E. C. (1990). *Am J. Vet Res.* **51,** 1357.
Nelson, R. W., Feldman, E. C., and Ford, S. L. (1993). *J. Am. Vet. Med. Assoc.* **203,** 821.
O'Brien, T. D., Hayden, D. W., Johnson, K. W., and Stevens, J. B. (1985). *Vet. Pathol.* **22,** 250.
O'Sullivan, B. M., Healy, P. J., Fraser, I. R., Nieper, R. E., Whittle, R. J., and Sewell, C. A. (1981). *Aust. Vet. J.* **57,** 227.
Pfeifer, M. A., Halter, J. B., and Porte, D. (1981). *Am. J. Med.* **70,** 579.
Phillips, R. W., Knox, K. L., Pierson, R. E., and Tasker, J. B. (1971). *Cornell Vet.* **61,** 114.
Priester, W. A. (1974). *J. Nat. Cancer Inst.* **53,** 227.
Rafaguzzaman, M., Svenkerud, R., Strande, A., and Hauge, J. G. (1976). *Acta Vet. Scand.* **17,** 196.
Raptis, S., and Dimitriadis, G. (1985). *Clin. Physiol. Biochem.* **3,** 29.
Renold, A. E., and Cahill, G. F., Jr. (1966). *In* "The Metabolic Basis of Inherited Disease" (J. B. Stanbury, J. B. Wyngaarden, and D. S. Fredrickson, eds.), p. 69. McGraw-Hill, New York.
Roderick, L. M., Harshfield, G. S., and Merchant, W. R. (1933). *Cornell Vet.* **23,** 348.
Rogers, W. A., Donovan, E. F., and Koclba, G. J. (1975). *J. Am. Vet. Med. Assoc.* **166,** 1092.
Ross, S. A., Brown, J. C., and Dupre, J. (1974). *Diabetologia* **10,** 384.
Schade, D. S., and Eaton, R. P. (1985). *N. Engl. J. Med.* **312,** 1120.
Shannon, J. A., Farber, S., and Troast, L. (1941). *Am. J. Physiol.* **133,** 752.
Sherwin, R. S., Fisher, M., Hendler, R., and Felig, P. (1976). *N. Engl. J. Med.* **294,** 455.
Spieth, K. (1973). *Prakt. Thierzt.* **54,** 292.
Steiner, D. E. (1977). *Diabetes* **26,** 332.
Steinke, J., and Taylor, K. W. (1974). *Diabetes,* **23,** 631.
Stewart, J., and Holman, H. H. (1940). *Vet. Rec.* **52,** 157.
Svenkerud, R., and Hauge, J. G. (1978). *Comp. Pathol. Bull.* **10,** 2.
Swiatek, K. R., Kipnis, D. M., Mason, G., Chao, K., and Cornblath, M. (1968). *Am. J. Physiol.* **214,** 400.
Taniyama, H., Ushiki, T., Tajima, M., Kurosawa, T., Kitamura, N., Takahashi, K., Matsukawa, K., and Itokura, C. (1995). *Vet. Pathol.* **32,** 221.
Taylor, S. I. (1995). In "The Metabolic and Molecular Bases of Inherited Disease," (C. R. Scriver, A. L. Beaudette, W. S. Sly, and D. Valle, eds.), 7th ed., Vol. 2, p. 843. McGraw-Hill, New York.
Teunissen, G. H. B., Rijnberk, A., Schotman, W., and Hackeng, W. H. L. (1970). *Kleint. Prax.* **15,** 29.
Thorens, B., Sakhar, H. K., Kaback, H. R., and Lodish, H. F. (1988). *Cell* **55,** 281.
Unger, R. H. (1981). *Diabetologia* **20,** 1.
Unger, R. H., and Orci, L. (1975). *Lancet* **1,** 14.
Unger, R. H., and Orci, L. (1976). *Physiol. Rev.* **56,** 778.
Vora, S., Giger, U., Turchen, S., and Harvey, J. W. (1985). *Proc. Nat. Acad. Sci. USA.* **82,** 8109.
Walvoort, H. C. (1983). *J. Inher. Metab. Dis.* **6,** 3.
Walvoort, H. C., Slee, R. G., and Koster, J. F. (1982). *Biochim. Biophys. Acta* **715,** 63.
Wilder, R. M., Allan, F. N., Power, M. H., and Robertson, H. E. (1927). *J. Am. Med. Assoc.* **89,** 348.
Williams, M., Gregory, R., Schall, W., Gossain, V., Bull, R., and Padgett, G. (1981). *Fed. Proc.* **40,** 740.
Wood, P. A., and Smith, J. E. (1980). *J. Am. Vet. Med. Assoc.* **176,** 1267.
Yoon, J.-W., Marshall, A., Onodera, T., and Notkins, A. L. (1979). *N. Engl. J. Med.* **300,** 1173.

CHAPTER

4

Lipids and Ketones

MICHAEL L. BRUSS

I. INTRODUCTION

This chapter covers the biochemistry and clinical chemistry of long-chain fatty acids, triacylglycerols, phospholipids, cholesterol, and ketones, a list that includes the majority of lipids found in vertebrates. The only remaining major classes are sphingolipids and waxes, which are not discussed. Although lipids have many functions, two of the most important are energy storage and membrane structure. Triacylglycerols are by far the most important lipid with regard to energy storage, and phospholipids and cholesterol are the most important lipoid membrane constituents. Lipids serve other functions, including being precursors for steroids and bile acids (cholesterol), thermal insulation (triacylglycerols), and electrical insulation (various lipids). Virtually all lipids are insoluble in water, which greatly complicates their handling in the body. Because of their insolubility, lipids must rely on proteins for transport for any significant distance in the body and various proteins have evolved to provide this function. The insolubility of lipids is an asset as well as a liability.

Because of their insolubility, lipids generate no osmotic force, so large amounts of triacylglycerol can be stored in adipose without the weight gain due to water that would accompany it if it were soluble. The insolubility of lipids is vital to many of their functions in membranes.

II. LONG-CHAIN FATTY ACIDS

A. Structure, Properties, and Assay of Long-Chain Fatty Acids

Long-chain fatty acids (LCFAs), frequently called free fatty acids or nonesterified fatty acids, are straight-chain fatty acids containing 12 or more carbon atoms. Since LCFAs are usually synthesized in animals or plants from acetyl-CoA and are then degraded two carbons at a time via β-oxidation in animals, the LCFAs found most commonly in animals have an even number of carbon atoms. LCFAs having carbon chain lengths of 16 and 18 comprise the greatest bulk of fatty acids in animal tissues and most animal diets. The saturated 16-carbon LCFAs is palmitic acid, and the saturated 18-carbon LCFA is stearic acid. Unsaturated 18-carbon LCFA are common, with double bonds occurring at C_9-C_{10} (oleic acid), at C_9-C_{10} and C_{12}-C_{13} (linoleic acid), and at C_9-C_{10}, C_{12}-C_{13}, and C_{15}-C_{16} (linolenic acid). The double bonds found in fatty acids in nature are mostly of the *cis* configuration. Ruminant fat contains more *trans*-LCFAs than that of nonruminants because rumen microbes isomerize some plant *cis*-LCFAs to *trans* isomers. Unsaturated LCFAs have a lower melting point than saturated LCFAs with the same number of carbons and are more susceptible to spontaneous oxidation (Gurr and Harwood, 1991). The 20-carbon polyunsaturated fatty acids, arachidonic acid (double bonds at C_5-C_6, C_8-C_9, C_{11}-C_{12}, C_{14}-C_{15}) and eicosapentaenoic acid (also called timnodonic acid), which is arachidonic acid with an additional double bond at C_{17}-C_{18}, are the precursors of the eicosanoids (prostaglandins, leukotrienes, thromboxanes).

Long-chain fatty acids are relatively insoluble in water at physiological pH. They dissolve readily in highly alkaline solutions, forming soaps. LCFAs are amphiphilic, being quite polar (hydrophilic) at their carboxyl end and quite nonpolar (hydrophobic) at the methyl end. All LCFAs must bind to proteins in order to be transported for any significant distance, and albumin is the primary transport protein in plasma (Gurr and Harwood, 1991).

Plasma LCFA concentrations can be determined spectrophotometrically with either a procedure involving extraction and chemical reaction or a procedure involving direct enzymatic reaction. The extraction/chemical reaction method involves causing formation of cobalt or copper salts of LCFAs, which are extracted with organic solvents followed by colorimetric assay of cobalt or copper (Duncombe, 1964; Elphick, 1968; Demacker *et al.*, 1982). The enzymatic method involves direct reaction of plasma LCFA to form LCFA-CoA. Then, LCFA-CoA is oxidized using LCFA-CoA oxidase, which produces hydrogen peroxide. The hydrogen peroxide is used to produce a colored product under the catalysis of peroxidase (Shimizu *et al.*, 1980; Demacker *et al.*, 1982). Generally, the enzymatic method is more satisfactory because it is more rapid, requires less operator skill, and is available commercially as a kit. If a sample contains triacylglycerol and lipase, which is not uncommon, LCFA may be released if the sample is allowed to stand. Falsely high LCFAs may be avoided by centrifuging blood samples and freezing the plasma immediately after collection or by adding paraoxon, a lipase inhibitor (Degen and Van der Vies, 1985).

B. Synthesis of Long-Chain Fatty Acids

LCFAs can be synthesized in most tissues, but only liver, adipose, or mammary tissue do it on a large scale. Synthesis occurs in the cytosol from acetyl-CoA. The precursor of the acetyl-CoA used for LCFA synthesis is usually acetate or glucose, with the former being important in ruminants and the latter being important in nonruminant mammals. When acetate is the acetyl-CoA precursor, it is formed from plasma acetate in the cytosol, the same cellular location as the enzymatic machinery needed to manufacture the LCFAs. However, when glucose is the precursor, it must go through glycolysis, which has its terminal enzyme, pyruvate dehydrogenase, located in the mitochondria. Thus, the acetyl-CoA is produced in the mitochondria, which is a problem if it is to be used for LCFA synthesis because the inner mitochondrial membrane is relatively impermeable to acetyl-CoA (Goodridge, 1991).

This problem has been solved by a mechanism known as the *citrate shuttle,* which is shown in Fig. 4.1. Acetyl-CoA in the mitochondria combines with oxaloacetate under the catalysis of citrate synthase to form citrate. Citrate is translocated across the mitochondrial membrane where it is cleaved into acetyl-CoA and oxaloacetate by ATP-citrate lyase. Thus, acetyl-CoA has been effectively transported from mitochondrion to cytosol. What remains is for the oxaloacetate to reenter the mitochondria to complete the cycle. However, the inner mitochondrial membrane is also impermeable to oxaloacetate, so it is first converted to malate by malate dehydrogenase or aspartate by aspartate aminotransferase in the cytosol. The malate

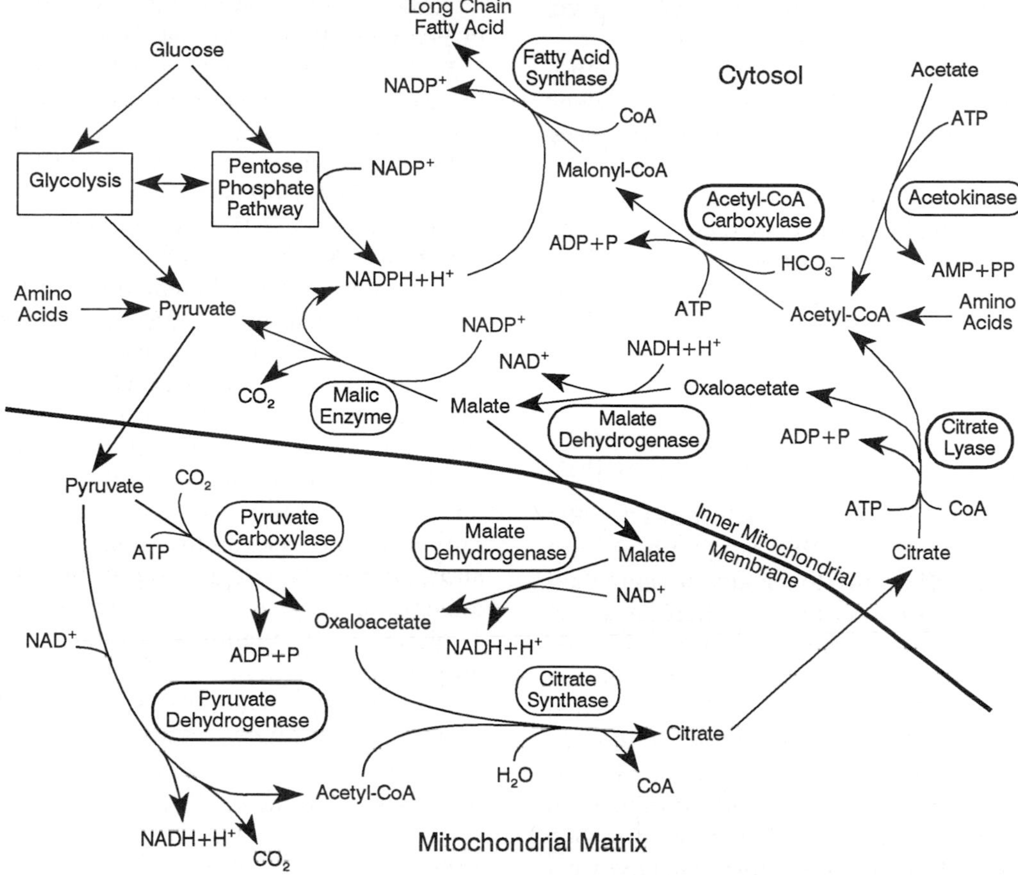

FIGURE 4.1 Fatty acid synthesis. Acetyl-CoA is generated in the mitochondria from pyruvate but cannot penetrate the mitochondrial membrane to reach fatty acid synthesizing enzymes in the cytosol. Citrate is formed from acetyl-CoA and oxaloacetate and migrates to the cytosol where it is cleaved to regenerate acetyl-CoA and oxaloacetate. The acetyl-CoA is converted into malonyl-CoA and used for fatty acid synthesis. The oxaloacetate cannot penetrate the mitochondrial membrane but must be converted to malate or pyruvate, which can penetrate the membrane and be converted back to oxaloacetate in the mitochondria. NADPH needed for fatty acid synthesis is generated by the pentose phosphate pathway and malic enzyme.

or aspartate are translocated into the mitochondrion where they can be converted back to oxaloacetate by reversal of the reactions that occurred in the cytosol. Alternately, malate in the cytosol can be converted to pyruvate by malic enzyme, and the pyruvate can enter the mitochondrion and be converted to oxaloacetate by pyruvate carboxylase (Goodridge, 1991).

Once acetyl-CoA reaches or has been formed in the cytosol, it must be carboxylated to produce malonyl-CoA via acetyl-CoA carboxylase if it is to be used for LCFA synthesis. This biotin-containing enzyme catalyzes the following reaction:

$$CH_3CO\text{-}CoA + CO_2 + ATP \xrightarrow{\text{Acetyl-CoA carboxylase}} {}^{-}OOCCH_2CO\text{-}CoA + H^+ + ADP + P_i.$$

Acetyl-CoA carboxylase is the main regulatory site in the synthesis of LCFA, which makes sense because the cell has little use for malonyl-CoA other than for the synthesis of LCFA. The enzyme is activated by citrate, which is again logical because citrate will be abundant only when there is a plentiful supply of mitochondrial acetyl-CoA. In addition, acetyl-CoA carboxylase is directly inhibited by LCFA-CoA, which can be derived from the synthetic process itself or from uptake and activation of plasma LCFAs. Acetyl-CoA carboxylase is also regulated by hormones via phosphorylation of the enzyme itself. Glucagon and LCFA-CoA stimulate phosphorylation, which inhibits the enzyme. Insulin activates the enzyme quickly, probably by stimulating dephosphorylation (Goodridge, 1991). These controls make sense in that a fasting or exercising animal will have its capacity for LCFA synthesis suppressed by increased plasma glucagon and LCFA levels, decreased plasma insulin, and increased intracellular LCFA-CoA. Conversely, in a recently fed animal, these controls will all be reversed to promote LCFA synthesis.

Malonyl-CoA is used as the building block for LCFA in the cytosol by a large, complex, multiple-unit enzyme called *fatty acid synthase.* Fatty acid synthase uses malonyl-CoA to add two carbon units at a time to a growing LCFA chain that is attached to the enzyme itself, and uses NADPH to reduce the oxygen that was attached to what was the end carbon of the old LCFA chain. The reaction proceeds in a series of distinct steps, all of which occur on the same enzyme complex. The overall reaction is

$$CH_3\text{-}(CH_2)_n\text{-CO-enzyme} + {}^{-}OOCCH_2CO\text{-CoA} + 2\ NADPH + 3\ H^+ \rightarrow CH_3\text{-}(CH_2)_{n+2}\text{-CO-enzyme} + CO_2 + H_2O + 2\ NADPH^+.$$

The subscript n in the structural formula for the growing LCFA is an even number ranging from zero (i.e., the starting acetyl group) to usually no more than eight (stearate). The process begins when an acetyl group binds to the enzyme complex and usually ends when a palmityl (16-carbon) group has been formed on the enzyme, at which point the LCFA is detached from the enzyme. New carbons are added to the carboxyl end, not the methyl end, of the growing LCFA. The carbon atom in the carbon dioxide produced in the fatty acid synthase reaction is the same carbon atom in the carbon dioxide used to form malonyl-CoA from acetyl-CoA.

Cellular synthesis of the enzymes directly involved in LCFA synthesis (acetyl-CoA carboxylase and fatty acid synthase) and the enzymes involved in the generation of NADPH and acetyl-CoA translocation is stimulated by diets that are high in carbohydrate and low in fat and suppressed by fasting, high-fat/low-carbohydrate diets, and diabetes. These changes appear to be brought about, in part, by alterations in plasma insulin and glucagon that accompany diet changes or diabetes (Goodridge, 1991; Gurr and Harwood, 1991).

Fatty acid synthesis is expensive energetically. To add a single acetyl-CoA to a growing LCFA chain, one ATP is used directly and six more are used indirectly (each of the two NADPH is equivalent to three ATP). Because fatty acid synthesis occurs in the cytosol and requires NADPH, there must be a generous source of that cofactor when fatty acid synthesis is active. The main source of NADPH for fatty acid synthesis is the pentose phosphate (i.e., hexose monophosphate) pathway in the cytosol. This pathway utilizes plasma glucose in the case of adipose or mammary tissue, whereas in the liver, it can use plasma glucose, glycogen, or gluconeogenesis as the hexose source. Another source of NADPH in the cytosol is the malic enzyme reaction. These sources of NADPH are illustrated in Fig. 4.1.

Although the most common length for nascent LCFAs when they are released from fatty acid synthase is 16 carbons, they can be as long as 18 carbons or, in the case of fat synthesis in the mammary gland, as short as 4 carbons. When LCFAs are detached from fatty acid synthase, they are rapidly thioesterified to CoA by LCFA-CoA synthetase, an enzyme found in the endoplasmic reticulum and outer mitochondrial membrane. Most of the palmitate produced by fatty acid synthase will be elongated to produce stearate by fatty acid elongase, an enzyme found mainly in the endoplasmic reticulum but also in mitochondria. This enzyme adds two new carbons at the carboxyl end of the existing LCFA. Fatty acid elongase uses the same substrates (malonyl-CoA and NADPH) as fatty acid synthase, but is located in a different part of the cell and prefers palmityl-CoA as its substrate. However, fatty acid elongase can use longer LCFA-CoA as substrates to a limited degree to produce LCFA-CoA with a length of as great as 24 carbons (Cook, 1991).

Nonruminant mammals synthesize LCFA in liver, adipose, and mammary tissue. Ruminants synthesize LCFA primarily in adipose and mammary tissue with acetate being the most important precursor. Ruminants generally have a low capacity for LCFA synthesis in liver, but after eating large amounts of high starch diets, they may synthesize some LCFA in the liver from acetate and propionate (Hanson and Ballard, 1967; Ingle, 1972a, 1972b; Liepa *et al.*, 1978).

C. Catabolism of Long-Chain Fatty Acids

1. *Desaturation*

Most animals are capable of desaturating LCFAs only at the Δ^4, Δ^5, Δ^6, and Δ^9 positions (counting from the carboxyl end). Animals are able to desaturate palmityl-CoA and stearyl-CoA between C_9 and C_{10} by means of the Δ^9 desaturase system located in the endoplasmic reticulum to produce palmitoleyl-CoA and oleyl-CoA, respectively. However, animals are not able to create additional double bonds beyond C_9 in these products to any significant extent, so linoleic and linolenic acids must be absorbed from the intestinal tract (Cook, 1991). By a combination of the actions of the LCFA elongase and Δ^4, Δ^5, and Δ^6 desaturase systems, the livers of most mammals can synthesize arachidonic acid and eicosapentaenoic acid from linoleic and linolenic acids, respectively. However, the cat has very low levels of Δ^6 desaturase in its liver and must have arachidonic acid in its diet (MacDonald *et al.*, 1984).

2. *β-Oxidation*

The main catabolic route for LCFAs is β-oxidation. Most tissues can perform β-oxidation (erythrocytes are an exception), but those most adept at it are liver, skeletal muscle, and heart. In addition, the liver can partially oxidize LCFAs to ketones, an important process that is discussed extensively later. Before LCFAs can be subjected to β-oxidation, they must be esterified to CoA, which is accomplished by the following reaction:

$$\text{LCFA} + \text{ATP} + \text{CoA} \leftrightarrow \text{LCFA-CoA} + \text{AMP} + \text{PP}.$$

The reaction is catalyzed by LCFA-CoA synthetase, an enzyme bound to the endoplasmic reticulum and the outer mitochondrial membrane. The pyrophosphate (PP) is rapidly hydrolyzed, so the reaction effectively consumes two ATP. The activation of LCFA is not rate limiting for β-oxidation (Pande, 1971).

In order for LCFA-CoA to be catabolized, it must pass into the mitochondrion, which presents a problem because the inner mitochondrial membrane is impermeable to it. The CoA must be exchanged for a carnitine moiety, a reaction catalyzed outside the mitochondrion by carnitine acyltransferase I (see Fig. 4.2), as follows:

$$\text{LCFA-CoA} + \text{carnitine} \leftrightarrow \text{LCFA-carnitine} + \text{CoA}.$$

LCFA-carnitine passes readily through the inner mitochondrial membrane and is acted on by carnitine acyltransferase II, which converts the LCFA-carnitine back to LCFA-CoA (Kopec and Fritz, 1973).

Carnitine acyltransferase I appears to be controlled by inhibition by malonyl-CoA (McGarry *et al.*, 1977), and it is logical that when lipogenesis is stimulated, the LCFAs that are produced should be prevented from entering the mitochondrion where they will be catabolized.

In the mitochondrion, the process of β-oxidation per se cleaves the LCFA into acetyl-CoA units. The reaction sequence is

Acyl-CoA dehydrogenase

$$\text{R-CH}_2\text{-CH}_2\text{-CO-CoA} + \text{FAD} \rightarrow \text{R-CH=CH-CO-CoA} + \text{FADH}_2,$$

Δ^2-Enoyl-CoA hydratase

$$\text{R-CH=CH-CO-CoA} + \text{H}_2\text{O} \rightarrow \text{R-C(OH)H-CH}_2\text{-CO-CoA},$$

L(+)-3-Hydoxyacyl-CoA dehydrogenase

$$\text{R-C(OH)H-CH}_2\text{-CO-CoA} + \text{NAD}^+ \rightarrow \text{R-CO-CH}_2\text{-CO-CoA} + \text{NADH} + \text{H}^+,$$

Thiolase

$$\text{R-CO-CH}_2\text{-CO-CoA} + \text{CoA} \rightarrow \text{R-CO-CoA} + \text{CH}_3\text{-CO-CoA}.$$

The resulting acyl-CoA is two carbons shorter and can recycle through the pathway. Each trip of an acyl-CoA through the pathway generates one $FADH_2$ and one NADH + H^+, which can generate 5 ATP via oxidative phosphorylation. If the LCFA has an odd number of carbons, which is rare, the final product of β-oxidation will be propionyl-CoA. The double bond produced by the acyl-CoA dehydrogenase reaction has a *trans* configuration, not the *cis* configuration occurring in unsaturated LCFA found free or esterified to glycerol.

Unsaturated LCFA can proceed through β-oxidation to within three carbons of the double bond. As this point, Δ^2-enoyl-CoA hydratase cannot act because it requires a *trans*, rather than a *cis*, configuration in its substrates, and it requires that the double bond be between C_2 and C_3 rather than between C_3 and C_4. At this point, Δ^3, Δ^2-enoyl-CoA isomerase will convert the Δ^3-*cis* double bond to a Δ^2-*trans* double bond, which will allow β-oxidation to proceed (Kilponen *et al.*, 1991). Polyunsaturated LCFAs require an additional enzyme, 2,4-dienoyl-CoA reductase, because after enoyl-CoA isomerase acts, the new *trans* double bond will still have the second *cis* double bond in proximity, which will prevent Δ^2-enoyl-CoA hydratase from acting. 2,4-Dienoyl-CoA reductase effectively eliminates the second double bond by reducing it with NADPH (Roe *et al.*, 1990).

β-Oxidation in the mitochondria appears to be controlled mainly by substrate availability. The acetyl-CoA units can be oxidized in the citric acid cycle provided there is sufficient oxaloacetate to condense with them to form citrate. Alternatively, acetyl-CoA units can be recondensed to form ketones, which will occur when there is not sufficient oxaloacetate for citrate formation or when citrate synthase is inhibited by high levels of citrate. Although β-oxidation occurs mostly in mitochondria, the process occurs to a minor extent in peroxisomes as well (Wanders *et al.*, 1992).

Although the main catabolic route for LCFA is β-oxidation, there are two quantitatively minor alternatives. α-Oxidation, in which carbons are removed one at a time from the carboxyl end of the LCFA, is used by brain tissue to produce LCFAs of varying lengths for synthesis of complex lipids. ω-Oxidation, which is oxidation that occurs at the methyl, rather than at the carboxyl, end is conducted to a limited extent by the cytochrome P450 system in the endoplasmic reticulum of liver. The resulting dicarboxylic acid can then undergo β-oxidation to a chain length of six carbons (adipate) most of which will be excreted in the urine (Mortensen, 1990; Gurr and Harwood, 1991).

III. TRIACYLGLYCEROL

A. Structure, Properties, and Assay of Triacylglycerol

The main storage form of LCFAs is the triacylglycerols (also called triglycerides), in which three LCFAs are esterified to glycerol. Triacylglycerols are even less

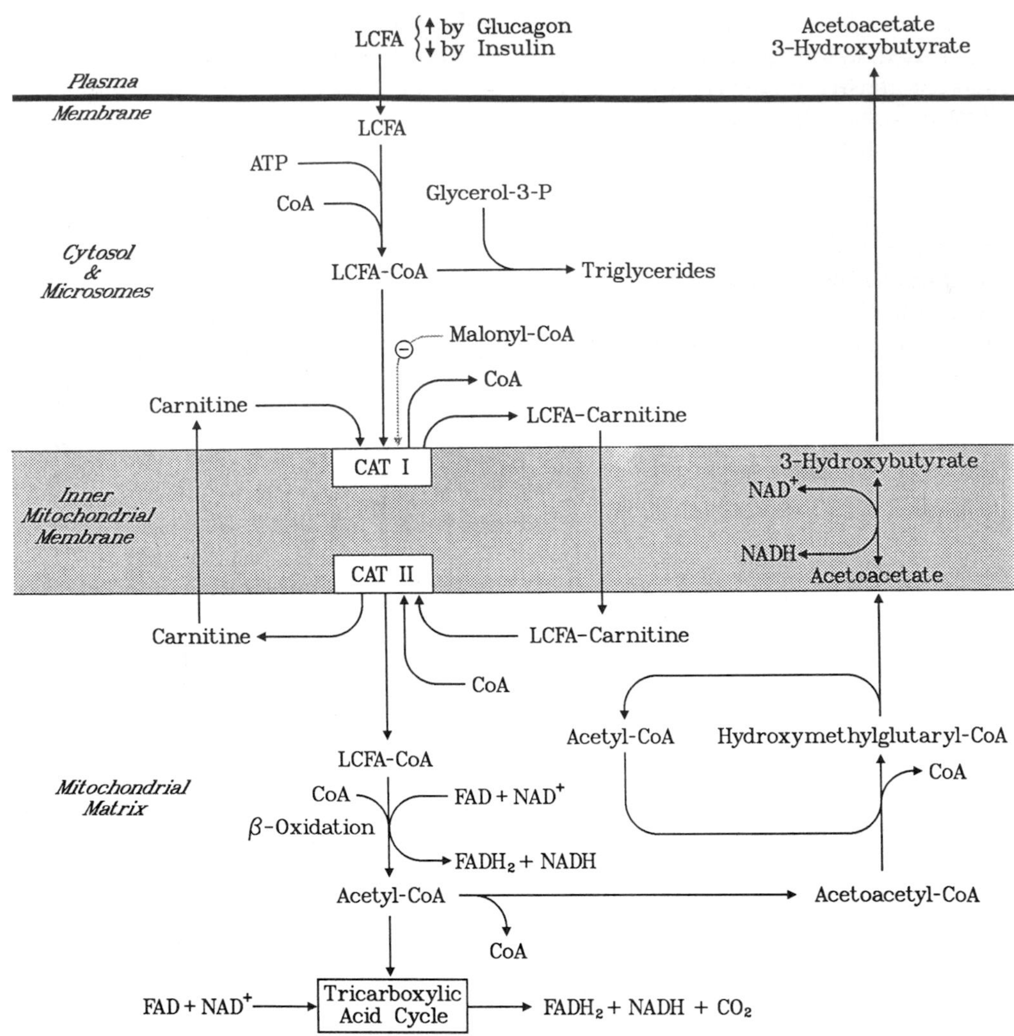

FIGURE 4.2 LCFA oxidation and ketogenesis in the liver. Abbreviations: CAT, carnitine acyltransferase.

soluble than LCFAs, and also must be bound to proteins in complexes called *lipoproteins* for transport through plasma.

Assay of triacylglycerol in plasma or serum is best accomplished by enzymatic hydrolysis using lipase followed by enzymatic determination of the released glycerol (McGowan *et al.*, 1983; Klotzsch and McNamara, 1990). If high plasma glycerol levels are likely, as can occur in animals that have not eaten lately, a plasma blank must be run. Older methods that use alkaline hydrolysis require caustic reagents, consume more time, and may assay phospholipids plus triacylglycerol. Glycerol, which has been used occasionally to lubricate stoppers of blood collection tubes, and soap, which may contain glycerol or triacylglycerols, contamination of samples will lead to falsely elevated values. If the sample contains lipase, which is not uncommon, triacylglycerol levels will decrease if the sample is allowed to stand. Prompt centrifugation of blood samples followed by rapid analysis or freezing of the plasma will prevent falsely low triacylglycerol levels.

B. Synthesis of Triacylglycerol

Although most cells can synthesize triacylglycerols, liver, adipose, mammary gland, and small intestine are particularly adept at it. The LCFA-CoA are the building blocks for triacylglycerol synthesis. There are two sources of LCFA-CoA for triacylglycerol synthesis: LCFAs in the plasma and LCFAs synthesized locally. Generally, physiological or pathological circumstances, such as starvation or diabetes, that promote high plasma levels of LCFAs suppress LCFA synthesis. Physiological circumstances that promote LCFA synthesis, such as eating a carbohydrate meal, also inhibit lipolysis in adipose, so plasma LCFA levels are not elevated.

To form triacylglycerols, the LCFA-CoA are esterified to glycerol-3-P. Glycerol-3-P can be produced in the liver from glycerol, which is absorbed from the plasma, and ATP in a reaction catalyzed by glycerol kinase as follows:

$$\text{Glycerol} + \text{ATP} \rightarrow \text{glycerol-3-P} + \text{ADP}.$$

Glycerol is normally plentiful in plasma only when there is active lipolysis occurring in adipose tissue.

When glucose is plentiful in the plasma and LCFAs are being synthesized from glucose via acetyl-CoA, glycerol-3-P is also synthesized from glucose in liver, mammary gland, and adipose. This process occurs via glycolysis to dihydroxyacetone-P followed by a reduction catalyzed by glycerol-3-P dehydrogenase:

$$\text{Dihydroxyacetone-P} + \text{NADH} + \text{H}^+ \leftrightarrow \text{glycerol-3-P} + \text{NAD}^+.$$

LCFA-CoA is esterified to glycerol-3-P by glycerol-P acyltransferase as follows:

$$\text{glycerol-3-P} + \text{LCFA-CoA} \rightarrow \text{1-acyl-glycerol-3-P} + \text{CoA}.$$

This reaction occurs in both mitochondria and smooth endoplasmic reticulum, but the smooth endoplasmic reticulum enzyme is more plentiful and most important in triacylglycerol synthesis. Next, another LCFA-CoA is esterified by the enzyme, acylglycerol-P acyltransferase, which is located in the smooth endoplasmic reticulum:

$$\text{1-Acyl-glycerol-3-P} + \text{LCFA-CoA} \rightarrow \text{phosphatidate} + \text{CoA}.$$

Phosphatidate (the ionized form of phosphatidic acid) is 1,2-diacyl-glycerol-3-P. Next, the phosphate is hydrolyzed from phosphatidate by phosphatidate phosphohydrolase to produce a diacylglycerol:

$$\text{Phosphatidate} \rightarrow \text{diacylglycerol} + \text{P}.$$

This reaction occurs in the smooth endoplasmic reticulum and cytosol. Finally, a last LCFA-CoA is esterified by the enzyme diacylglycerol acyltransferase, an enzyme located in the smooth endoplasmic reticulum (Brindley, 1991):

$$\text{Diacylglycerol} + \text{LCFA-CoA} \rightarrow \text{triacylglycerol} + \text{CoA}.$$

If the triacylglycerol has been synthesized in adipose, it will migrate into the large storage vesicle that each adipocyte possesses. Most of the triacylglycerol synthesized in liver will normally be incorporated into and exported from the liver as part of very low density lipoproteins (VLDLs). However, if triacylglycerol synthesis exceeds hepatic export capacity, triacylglycerol will accumulate in vesicles in hepatocytes, leading to fatty liver. If the triacylglycerol has been synthesized in mammary gland, the resulting triacylglycerols will accumulate in vesicles of secretory cells and the vesicles will be extruded into the lumina of the gland acini.

The regulation of triacylglycerol synthesis is not fully understood and differs among tissues. In small intestine, substrate availability is most important because triacylglycerol synthesis in that organ is an integral part of triacylglycerol absorption. In mammary gland, substrate availability and the hormones that support lactation regulate triacylglycerol synthesis.

In liver, the limiting enzyme in the pathway appears to be phosphatidate phosphohydrolase. This enzyme is subject to an interesting control mechanism in which it is switched between a less active and more active state by the enzyme itself being translocated between the cytosol and endoplasmic reticulum, respectively. Intracellular cAMP, which increases with high plasma glucagon or low plasma insulin levels (e.g., during fasting or with diabetes), inhibits binding of the enzyme to the endoplasmic reticulum, whereas LCFA or LCFA-CoA promote binding of the enzyme to the endoplasmic reticulum (Brindley, 1991; Gurr and Harwood, 1991). The role of LCFA and LCFA-CoA in promoting synthesis of triacylglycerols in the liver is important and explains how fat synthesis and fatty liver can occur in the fasting state when hormonal changes would oppose triacylglycerol synthesis.

In adipose tissue, the synthesis of triacylglycerol is very much regulated by hormones, especially glucagon, catecholamines, and insulin. The first two hormones increase intracellular cAMP and the latter tends to decrease it, although insulin probably has effects independent of cAMP. In conditions in which glucagon would be elevated and insulin would be decreased (e.g., during fasting), hormone sensitive lipase will be activated and lipolysis will occur. It is important that fat synthesis not be operative during lipolysis, so as not to waste energy. Low insulin and elevated catecholamine or glucagon levels decrease the level of lipoprotein lipase (LPL) in adipose tissue. Fat cells need LPL in order to hydrolyze plasma triacylglycerol so that the resulting LCFAs can be absorbed and used for triacylglycerol synthesis. Decreased plasma insulin levels will decrease entry of glucose into adipocytes, which will result in less glycerolphosphate being synthesized. Increased intracellular cAMP in adipose tissue decreases the activity of several key enzymes in fat synthesis, including fatty acyl-CoA synthetase, glycerolphosphate acyltransferase, phosphatidate transferase, and diacylglycerol acyltransferase; however, the mechanism of inhibition is uncertain (Saggerson, 1988).

C. Catabolism of Triacylglycerol

Catabolism of triacylglycerol involves the action of lipases, which are specialized esterases that hydrolyze glyceride bonds. The major lipases are pancreatic lipase, hepatic lipase, hormone sensitive lipase of adipose, LPL found on endothelial cells, and lysosomal lipases contained in most cells. Pancreatic lipase is the essential lipase for digestion of triacylglycerol in the GI tract and is discussed later. Hepatic lipase is synthesized in hepatocytes from where it migrates to the surface of hepatic endothelial cells. Hepatic lipase primarily attacks triacylglycerol in the plasma, which are part of VLDL remnants to produce low-density lipoproteins (LDLs), and it attacks triacylglycerol in high-density lipoproteins (HDLs) as well.

Lipoprotein lipase attacks triacylglycerol in chylomicrons and VLDL in plasma and is found on the endothelium of many organs and tissues, but is in greatest quantity in adipose, heart, skeletal muscle, and mammary gland. Lipoprotein lipase is synthesized by the underlying tissue and migrates to the capillary endothelium where is it anchored on the cell surfaces to glycoproteins, which have polysaccharide chains structurally similar to heparin. If heparin is injected into an animal, LPL can switch its attachment from cell surface glycoproteins to the free injected heparin and, thus, appears in the plasma. If the animal had a lipemia prior to injecting the heparin, the large amount of lipoprotein lipase released into the plasma will clear the lipemia. Phospholipids and apolipoprotein C-II must be present for lipoprotein lipase to have full activity (Fielding and Fielding, 1991).

IV. PHOSPHOLIPIDS

A. Structure and Properties of Phospholipids

Most of the phospholipids found in the body consist of a core of glycerol that has LCFA esterified to its 1 and 2 carbons and phosphate esterified to its 3 carbon, a compound called phosphatidate. In addition, the phosphate is often esterified to a hydroxyamino compound such as choline, ethanolamine, or serine to produce phosphatidylcholine (also called lecithin), phosphatidylethanolamine, and phosphatidylserine, respectively. Inositol may be esterified to the phosphate to produce phosphatidylinositol. Because of the phosphate group, phospholipids are very polar on one end, but nonpolar on the other end and still must be part of lipoproteins for transport through the plasma. Phospholipids are constituents of all cellular membranes, lipoproteins, and bile micelles. The fatty acid portion of the molecule is oriented toward the center of the membrane or micelle, and the phosphatidyl group is oriented toward the outer surface (i.e., toward the aqueous medium). In micellar structures, like lipoproteins and bile micelles, the surface coating of the polar ends of constituent phospholipids provides a surface charge that helps to keep the micelles in suspension.

B. Synthesis of Phospholipids

Phospholipids are synthesized either from phosphatidate (e.g., phosphatidylinositol) or diacylglycerol (e.g., phosphatidylcholine and phosphatidylethanolamine), both of which are intermediates in the synthesis of triacylglycerol. In all cases, cytidine triphosphate (CTP), a high-energy organophosphate that derives it phosphates from ATP, plays an important role. In the case of phosphatidylinositol, CTP reacts with phosphatidate to form CDP-diacylglycerol, which then reacts with inositol to form phosphatidylinositol and CMP. In the case of choline or ethanolamine, they must first be phosphorylated by reaction with ATP. Then the phosphocholine or phosphoethanolamine reacts with CTP to form CDP-choline or CDP-ethanolamine, respectively, which then react with diacylglycerol to produce phosphatidylcholine and phosphatidylethanolamine, respectively. Phosphatidylserine is formed by serine replacing ethanolamine in phosphatidylethanolamine. In the endoplasmic reticulum of the liver, a methyl group from S-adenosylmethionine can be transferred to phosphatidylethanolamine to produce phosphatidylcholine (Vance, 1991). Fig. 4.3 illustrates the synthesis of phospholipids.

The enzymes that synthesize CDP-choline and CDP-ethanolamine (cytidylyltransferases) appear to be rate limiting for the synthesis of phosphatidylcholine and phosphatidylethanolamine, respectively. Phosphocholine cytidylyltransferase is subject to regulation similar to that of phosphatidate phosphohydrolase, the control enzyme in triacylglycerol synthesis. When phosphocholine cytidylyltransferase is bound to the endoplasmic reticulum it is relatively active, but when it is free in the cytosol, it is relatively inactive. Factors that increase binding of the enzyme to the endoplasmic reticulum are decreased levels of phosphatidylcholine, increased levels of diacylglycerol or LCFA, and dephosphorylation of the enzyme. Opposite changes in these factors inhibit binding of the enzyme to the endoplasmic reticulum, forcing it to remain inactive (Vance, 1991).

C. Catabolism of Phospholipids

Phospholipids are hydrolyzed by phospholipases, which can be found in the lysosomes of most tissues and in pancreatic secretion. Mammalian phospholi-

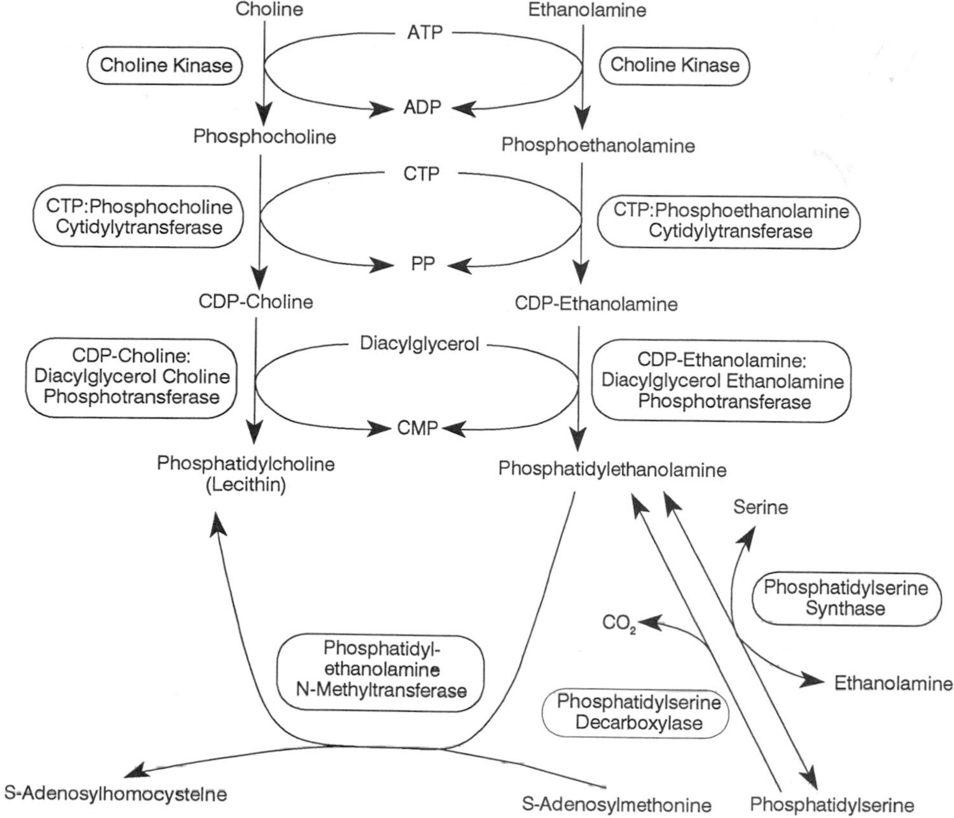

FIGURE 4.3 Synthesis of phospholipids. Diacylglycerol is the lipid to which organic bases and phosphate are transferred via CDP derivatives. Abbreviations: CTP, CDP, CMP, cytidine tri-, di-, and monophosphate, respectively.

pases are primarily of the A type, meaning that they hydrolyze the glycerol-LCFA ester bond at either position 1 (A_1 type) or 2 (A_2 type), but not both (Gurr and Harwood, 1991; Waite, 1991). Phospholipase types B, C, and D, which hydrolyze at other locations in the molecule, exist in mammalian tissues, but with lower activities.

V. CHOLESTEROL

A. Structure, Properties, and Assay of Cholesterol

Structurally, cholesterol is composed of a core of phenanthrene to which a cyclopentane ring is attached, and there is an eight-carbon side chain attached to the cyclopentane ring (see Fig. 4.4). Cholesterol is found only in animals and is not present in plants or microorganisms. Cholesterol is the precursor of steroid hormones, vitamin D, and the bile acids and is a constituent of cell membranes and bile micelles. Cholesterol can be obtained from the diet if it contains animal products or it can be synthesized. The chief synthetic and catabolic organ for cholesterol is the liver. Steroidogenic endocrine organs (adrenal cortex, testis, ovary, placenta) can synthesize small amounts of cholesterol; however, these organs utilize hepatically synthesized cholesterol for most of their steroid synthesis (Pedersen, 1988). Pure cholesterol and cholesterol esters are insoluble waxy white solids and must be transported through plasma as part of lipoproteins.

Enzymatic methods are used almost universally for assay of cholesterol (Stein and Meyers, 1994). Older nonenzymatic methods used harsh reagents and lack specificity. The key enzymes in the assay are cholesterol esterase, which hydrolyses cholesterol esters, and cholesterol oxidase. The latter enzyme is of microbial origin and has an action analogous to that of glucose oxidase, that is, it uses dissolved oxygen to oxidize cholesterol to produce cholest-4-ene-3-one and hydrogen peroxide. In the presence of added peroxidase, hydrogen peroxide will oxidize an added organic dye (e.g., dianisidine, ABTS, 4-aminoantipyrine plus phenol) to generate a colored product that can be quantified spectrophotometrically. If cholesterol esterase is included in the reagent, then total cholesterol will be determined. If cholesterol esterase is omitted from the reagent, then only nonesterified (i.e., free) cholesterol

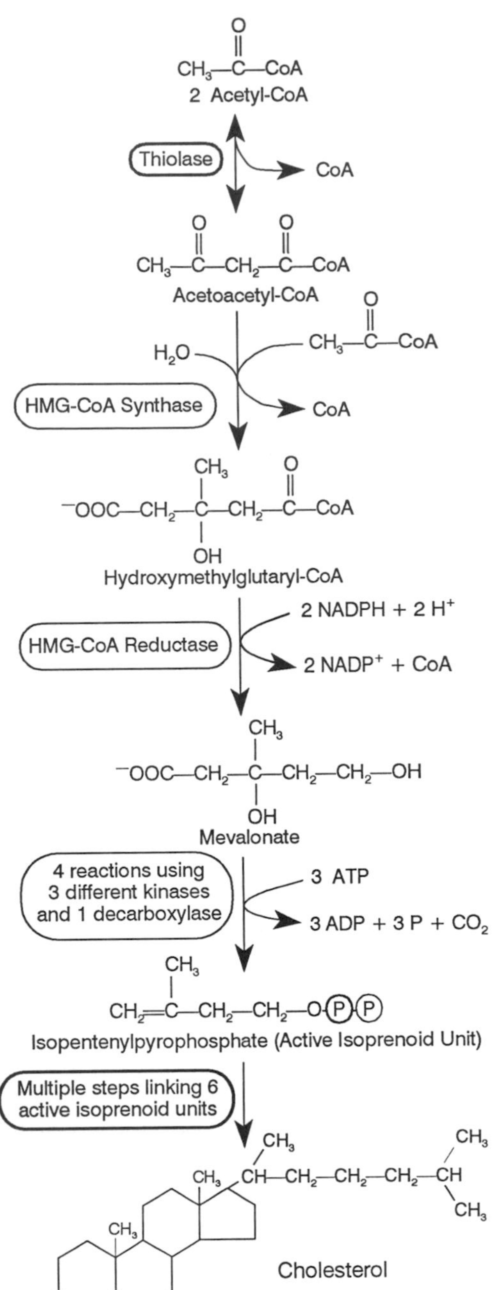

FIGURE 4.4 Synthesis of cholesterol. The first two reactions occur in the cytosol and the remainder in the smooth endoplasmic reticulum.

will be determined. If the assay is done with and without cholesterol esterase, then cholesterol ester concentration can be determined by subtraction.

Because virtually all of the cholesterol and cholesterol esters in plasma are part of lipoproteins, they must be liberated before they can be acted on by the enzymes of the reagent. This liberation can be accomplished by extracting cholesterol and its esters with an organic solvent prior to the assay or, more conveniently, by including small amounts of detergents (bile acids or artificial detergents) in the reagent (Stein and Meyers, 1994).

B. Metabolism of Cholesterol

As is the case for LCFA and ketones, the substrate for cholesterol synthesis is acetyl-CoA. The beginning site of cholesterol synthesis is in the cytosol, so acetyl-CoA, which is generated primarily in the mitochondria, must be transferred to the cytosol via the citrate shuttle mechanism discussed earlier. The process of cholesterol synthesis is shown diagrammatically in Fig. 4.4. In the cytosol, the first two steps of cholesterol synthesis are identical to the first two steps of ketone synthesis except that the process occurs in the cytosol rather than in the mitochondria. The enzymes that catalyze the first two steps are acetyl-CoA: acetoacetyl-CoA thiolase and hydroxymethylglutaryl-CoA (HMG-CoA) synthase:

$$\text{2 Acetyl-CoA} \leftrightarrow \text{acetoacetyl-CoA} + \text{CoA},$$
$$\text{Acetoacetyl-CoA} + \text{acetyl-CoA} \rightarrow \text{HMG-CoA} + \text{CoA}.$$

The remaining enzymes of cholesterol synthesis are located in the endoplasmic reticulum, perhaps because of decreasing solubility of succeeding products formed in the pathway. Next, HMG-CoA is reduced to mevalonate under the catalysis of HMG-CoA reductase as follows:

$$\text{HMG-CoA} + 2\,\text{NADPH} + 2\,\text{H}^+ \rightarrow \text{mevalonate} + 2\,\text{NADP}^+ + \text{CoA}.$$

HMG-CoA reductase is the primary control point for cholesterol synthesis; its control mechanisms are discussed later. Next, via three steps, isopentenylpyrophosphate is formed. Six of these molecules, often called the active isoprenoid units, are linked to form cholesterol in a long and complex pathway that is only partially understood (Faust *et al.*, 1988; Edwards, 1991).

The control of HMG-CoA reductase is complex and not completely understood. Artificially increasing plasma cholesterol levels *in vivo* decreases the activity of the enzyme in liver. However, cholesterol does not inhibit the enzyme directly, but represses synthesis of the enzyme mRNA (Edwards, 1991). Thus, if the amount of cholesterol consumed in the diet increases, the amount synthesized by the liver will decrease. This reciprocal relationship between cholesterol consumed and hepatic synthesis limits the extent to which plasma cholesterol levels can be decreased by restricting the amount of cholesterol in the diet.

Hepatic HMG-CoA reductase is inhibited by phosphorylation of the enzyme and reactivated by dephosphorylation. The protein kinase system responsible for the phosphorylation of HMG-CoA reductase is stimulated by intracellular cAMP (Edwards, 1991). Hepatic intracellular cAMP levels are controlled in part by plasma glucagon, which increases it, and by insulin, which decreases it. Thus, conditions that increase insulin (e.g., eating) will increase cholesterol synthesis. Conditions that decrease insulin (e.g., diabetes) or increase glucagon (e.g., fasting) will decrease cholesterol synthesis. Other hormones that affect hepatic HMG-CoA reductase activity, but probably not by altering intracellular cAMP levels, are thyroid hormones (increase HMG-CoA reductase activity) and glucocorticoids (decrease HMG-CoA reductase activity). Some drugs, such as lovastatin and mevastatin, used in humans to decrease plasma cholesterol levels operate by inhibiting HMG-CoA reductase (Brown and Goldstein, 1990).

Once cholesterol has been synthesized in the hepatocyte, it can be secreted into the plasma as part of lipoproteins (mostly in VLDL), it can be secreted into the canaliculi and become part of bile micelles, it can be degraded to bile acids, or it can be esterified to a LCFA by acyl-CoA : cholesterol acyltransferase (ACAT), which is located in the smooth endoplasmic reticulum. Cholesterol esters are even less soluble than cholesterol and are found in membranes and micelles wherever cholesterol itself is found. Cholesterol ester can be exported as part of lipoproteins or it can be converted back to cholesterol plus LCFA by cholesterol ester hydrolases, which are found in the cytosol, endoplasmic reticulum, and lysosomes. Deesterification is mandatory before cholesterol can be catabolized to bile acids. Because enzymes for the final steps of cholesterol synthesis and the first steps of its degradation are colocated in the endoplasmic reticulum, it might seem that most of newly synthesized cholesterol would be immediately degraded. However, the negative feedback of bile acids on cholesterol degradation keeps this process in check.

HDL contains lecithin : cholesterol acyltransferase (LCAT), which esterifies cholesterol by transferring a LCFA moiety from lecithin (phosphatidylcholine). The cholesterol to be esterified by LCAT can be that secreted with HDL at the time of its synthesis or it can be cholesterol from other lipoproteins or cell membranes that come in contact with HDL at a later time.

VI. LIPOPROTEINS

A. Structure, Properties, and Assay of Lipoproteins

Lipoproteins are very large noncellular conglomerations (micelles) of lipids and proteins that are suspended in plasma or lymph. Their main function is to transport most lipids (steroid hormones and LCFAs being notable exceptions) among tissues. Another function of lipoproteins is the esterification of cholesterol. Lipoproteins have a micellar structure in which the least polar molecules (triacylglycerol and cholesterol) occupy the center and more polar molecules (proteins and phospholipids) coat the exterior. Lipoproteins are synthesized almost exclusively by liver and the small intestine.

The main classes of lipoprotein are defined by their density as determined by ultracentrifugation and are chylomicrons ($d < 0.94$ g/ml), VLDLs ($d = 0.94–1.006$ g/ml), LDLs ($d = 1.006–1.063$ g/ml), and HDLs ($d = 1.063–1.21$ g/ml). Less commonly considered are very high density lipoproteins (VHDL, $d > 1.21$ g/ml), which are usually very low in concentration in plasma. It is common to designate the lighter LDL ($d = 1.006–1.019$ g/ml) as intermediate-density lipoproteins (IDL), and some schemes subdivide the HDL into HDL_1, HDL_2, and HDL_3. (Chapman, 1986; Gotto *et al.*, 1986).

The lipid component of lipoproteins is less dense than the protein component, but the lipids have similar densities and the proteins have similar densities. Therefore, the density of a lipoprotein is almost entirely dependent on its ratio of lipid to protein, with the chylomicrons having the highest ratio and, on the other end of the spectrum, the HDL having the lowest ratio (see Table 4.1). More than one-half of the lipid in chylomicrons and VLDL is triacylglycerol, whereas in LDL and HDL the majority of the lipids are not triacylglycerol (see Table 4.1). In domestic species, HDL is normally the most abundant plasma lipoprotein in the fasting state.

Chylomicrons and VLDL particles are large enough to refract light significantly, so they make plasma appear turbid or creamy if in high enough concentration (lipemic plasma). The chylomicrons have a low enough density that they will rise to the top of an undisturbed refrigerated plasma sample in 6–12 hours. This phenomenon is the basis of the *chylomicron test*, in which a milky plasma sample is placed in the refrigerator overnight. If a "cream layer" has formed at the top, then hyperchylomicronemia is present, and if the bottom portion of the plasma is turbid, then elevated levels of VLDL are present.

Because of the expense, time, and complexity involved with ultracentrifugation, electrophoresis in an alkaline medium has been used as an alternative method of lipoprotein classification. A variety of electrophoretic supports, ranging from paper to acrylamide gels, have been used. The sample is applied at the cathode end of the support, voltage is applied for a variable time, and the proteins are fixed and stained

TABLE 4.1 Composition of Lipoproteins of Domestic Animals[a]

	Triacylglycerol (wt %)	Free cholesterol (wt %)	Cholesterol esters (wt %)	Phospholipid (wt %)	Protein (wt %)
Cattle:					
Chylomicrons	87	4	2	4	3
VLDL	60	5	4	25	6
LDL	1	5	35	36	23
HDL	4	4	30	20	42
Dogs:					
VLDL	68	6	2	10	14
LDL	27	5	25	22	21
HDL	1	5	23	33	38
Horses:					
VLDL	57	5	6	18	14
LDL	6	8	36	23	27
HDL	0	2	20	28	50

[a] References: Cattle: Ferreri and Elbein, 1982 (chylomicrons); Stead and Welch, 1975 (other lipoproteins). Dogs: Mahley and Weisgraber, 1974; Blomhoff *et al.*, 1978. Horses: Watson *et al.*, 1993; Le Goff *et al.*, 1989.

with a lipid stain such as oil red O. A densitometer is used to quantify the lipoprotein fractions on the stained electrophoretogram.

Typically, three to five bands of lipoproteins can be discerned; however, additional bands may be present depending on the species of animal, electrophoretic technique, and presence of abnormal lipoproteins. The fastest moving band is HDL, which is designated as α-lipoprotein. The next fastest moving band is VLDL, which is designated pre-β-lipoprotein followed by the LDL band, which is designated as β-lipoprotein. The slowest moving band, which is still at the origin and seen primarily in the postprandial period, is composed of chylomicrons. With some electrophoresis systems, a separate IDL band, designated as slow pre-β-lipoproteins, can be discerned between the VLDL and LDL bands, and sometimes subbands of the HDL can be discerned. The correlation of electrophoretic and ultracentrifuge fractions established for humans does not always apply to animals. For example, bovine LDL can appear as α or β bands on electrophoresis (Puppione, 1983). Usually, two HDL bands can be discerned for dog plasma (Rogers, 1977). Figure 4.5 illustrates the distribution of lipoproteins in dog plasma.

Although easier and cheaper to perform than ultracentrifugation, lipoprotein electrophoresis still requires considerable time and expense. Consequently, methods have been developed that involve precipitation of one or more lipoprotein classes followed by analysis of a particular lipid, usually cholesterol, in the remaining supernatant. For example, chylomicrons can be removed by low-speed centrifugation (they rise to the top), and then precipitation of VLDL and LDL in human plasma can be accomplished by treatment with magnesium and dextran sulfate. The main lipoprotein remaining in the supernatant will be HDL, and if cholesterol is determined, it will mostly be HDL cholesterol (Stein and Meyers, 1994). Such empirical methods may be species specific. For example, the preceding method, though valid for human plasma, does not work for dog plasma (Rhodes *et al.*, 1992).

B. Apolipoproteins

The protein components of lipoproteins are called *apolipoproteins*. Some apolipoproteins are found in only one class of lipoproteins, whereas others can be found in multiple classes. Although there are species variations in the amino acid sequences of apolipoproteins, individual apolipoproteins in the domestic species are quite similar. The main classes of apolipoproteins are designated with a letter (A through E), sometimes fol-

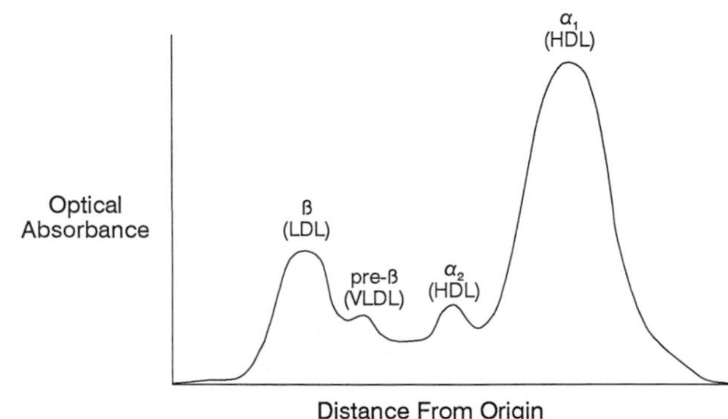

FIGURE 4.5 Densitometric scan of an electrophoretogram of canine plasma lipoproteins. The scan is typical of a fasted dog. In a fed dog, an additional peak due to chylomicrons would be present at the origin. Abbreviations: HDL, LDL, and VLDL, high, low, and very low density lipoproteins, respectively.

lowed by a number to indicate a distinct subclass. The main classes and subclasses of apolipoproteins found in domestic animals are A-I, A-II, A-IV, B_{48}, B_{100}, C-I, C-II, C-III, C-IV, and E. Characteristics of these apolipoproteins are listed in Table 4.2.

The B_{100} apolipoprotein, synthesized in the liver and part of VLDL, is one the largest polypeptide chains in mammals, having a molecular weight of 527,000 in horses (Watson *et al.*, 1991). The B_{48} apolipoprotein is about one-half the size of B_{100} and contains a subset of the B_{100} amino acid sequence (i.e., they are probably coded by the same gene); however, B_{48} is synthesized in the small intestine and is part of chylomicrons. The origin of the "48" and "100" designations stems from the fact that human B_{48} is exactly 48% of the mass of human B_{100}. Both B_{48} and B_{100} are glycoproteins and have a variety of carbohydrates attached to them (Chapman, 1986).

In addition to apolipoproteins, HDL contains an additional protein in the form of the enzyme LCAT, which esterifies cholesterol esters by transferring a LCFA moiety from phosphatidylcholine (lecithin) to cholesterol. LCAT is activated by lipoprotein A-I.

C. Digestion of Fat and Formation of Chylomicrons

The largest lipoproteins are the chylomicrons. To understand their formation, the digestion of triacylglycerol must be discussed. The main site of digestion and absorption of triacylglycerol is the small intestine, and the chief enzyme involved is pancreatic lipase. The pancreas not only supplies lipase to attack triacylglycerol, but also supplies cholesterol esterase to hydrolyze cholesterol esters and phospholipase A_2 to attack phospholipids. If any of these enzymes is to be effective, the lipids in food must first be emulsified with bile. Bile contains micelles composed mostly of bile acids, phospholipids, and cholesterol. Fats in food become part of these micelles, and then the enzymes can attack them on the outer surface of the micelles. The fatty acids, monoacylglycerols, and cholesterol resulting from the attack of the enzymes become part of the lipids of the brush border of the intestinal cells. The intestinal cells then use the monoacylglycerols and fatty acids to resynthesize triacylglycerol. Globules of triacylglycerol coated with protein are extruded from the basolateral membranes into the interstitium as chylomicrons. The lymphatic capillaries of the microvilli are called *lacteals* and have many large openings between the endothelial cells that line them. Consequently, the chylomicrons can enter the lymphatics, but not the blood capillaries. From the small intestine, the lymph flows to larger abdominal ducts to the thoracic duct and enters the right atrium. Thus, unlike most other nutrients, most of the absorbed fat bypasses the portal system and liver (Brindley, 1991).

The main apolipoproteins in chylomicrons are A series, B_{48}, C series, and E. The A series and B_{48} apolipoproteins are added by the small intestine, but the C series and E apolipoproteins, which are synthesized in the liver, appear to transfer from HDL to nascent chylomicrons soon after they are released into the circulation.

Chylomicrons are attacked by LPL, which resides on the surface of endothelial cells and hydrolyses tri-

TABLE 4.2 Apolipoproteins of Domestic Animals[a]

Apolipoprotein	Molecular weight	Major synthetic organ	Constituent of lipoprotein	Special properties
A-I	28 kDa (all species)	Small intestine	Chylomicrons, HDL	Activates lecithin: cholesterol acyltransferase
A-II (monomer)	8.5 kDa (cattle, dog) 6.5 kDa (horse)	Small intestine	Chylomicrons, HDL	Exists mostly as a dimer
A-IV	43 kDa (cattle, dog)	Small intestine	Nascent chylomicrons, HDL	Activates lecithin: cholesterol acyltransferase?
B_{48}	270 kDa (cattle, horse)	Small intestine	Chylomicrons	
B_{100}	534 kDa (cattle, horse) 325 kDa (swine)	Liver	VLDL, LDL	Binds to LDL receptor
C-I	8.0 kDa (cattle)	Liver	VLDL, LDL, HDL, chylomicrons	
C-II	9.5 kDa (cattle, pig) 14 kDa (horse)	Liver	VLDL, LDL, HDL, chylomicrons	Activator of lipoprotein lipase
C-III	8.0 kDa (cattle) 12 kDa (horse)	Liver	VLDL, LDL, HDL	
C-IV	10 kDa (cattle)	Liver	VLDL, LDL, HDL, chylomicrons	
E	37 kDa (all species)	Liver	VLDL, LDL, HDL, chylomicrons	Binds to LDL and chylomicron remnant receptor

[a] References: Chapman, 1986; Demacker *et al.*, 1987; Watson *et al.*, 1991; Bauchart, 1993; Watson and Barrie, 1993.

acylglycerol. Most of the resulting LCFA are absorbed by the tissue cells. As the chylomicron diminishes in size, some of the apolipoproteins, mostly A series and C series, transfer to HDL. Finally, a much diminished chylomicron remnant is left and will attach to an apolipoprotein E receptor on hepatocytes. The remnant will be absorbed, and its components hydrolyzed within the hepatocytes (Brindley, 1991; Schneider, 1991). The transport and metabolism of chylomicrons is illustrated in Fig. 4.6.

D. Very-Low-Density Lipoproteins: Synthesis, Export, and Metabolism

Secretion of VLDLs into the plasma is the main method by which hepatocytes export triacylglycerol. Its main apolipoproteins are B_{100}, C series, and E, but some A series is present as well. The A-series apolipoproteins, which are synthesized in the small intestine, transfer from HDL to VLDL soon after its secretion. Some C-series and E apolipoprotein may transfer from HDL to newly secreted VLDL as well. Like chylomicrons, the main lipid component of VLDL is triacylglycerol (see Table 4.1). VLDL and chylomicrons both serve as a means to distribute triacylglycerol to tissues. In the case of chylomicrons, the triacylglycerol is a product of fat digestion, whereas in the case of VLDL, the triacylglycerol is synthesized in the liver.

The assembly process for VLDL is complex (Fig. 4.7). Final steps in the synthesis of triacylglycerol, phospholipid, and cholesterol occur in the smooth endoplasmic reticulum. Microdroplets containing these three lipids and cholesterol esters move toward the confluence of the rough and smooth endoplasmic reticulums where they are joined by apolipoproteins synthesized on the rough endoplasmic reticulum to form the nascent VLDL.

The nascent VLDL particles move through microtubular membranes to the Golgi apparatus where the apolipoproteins are glycosylated. In the Golgi apparatus, the nascent VLDL particles reach final composition and are surrounded by membranes to form secretory granules. The secretory granules merge with the plasma membrane and spill their contents into the plasma (Alexander *et al.*, 1976; Davis, 1991).

The capacity of the liver to synthesize the protein components of VLDL is stimulated by a diet high in carbohydrate. It has been hypothesized that this stimu-

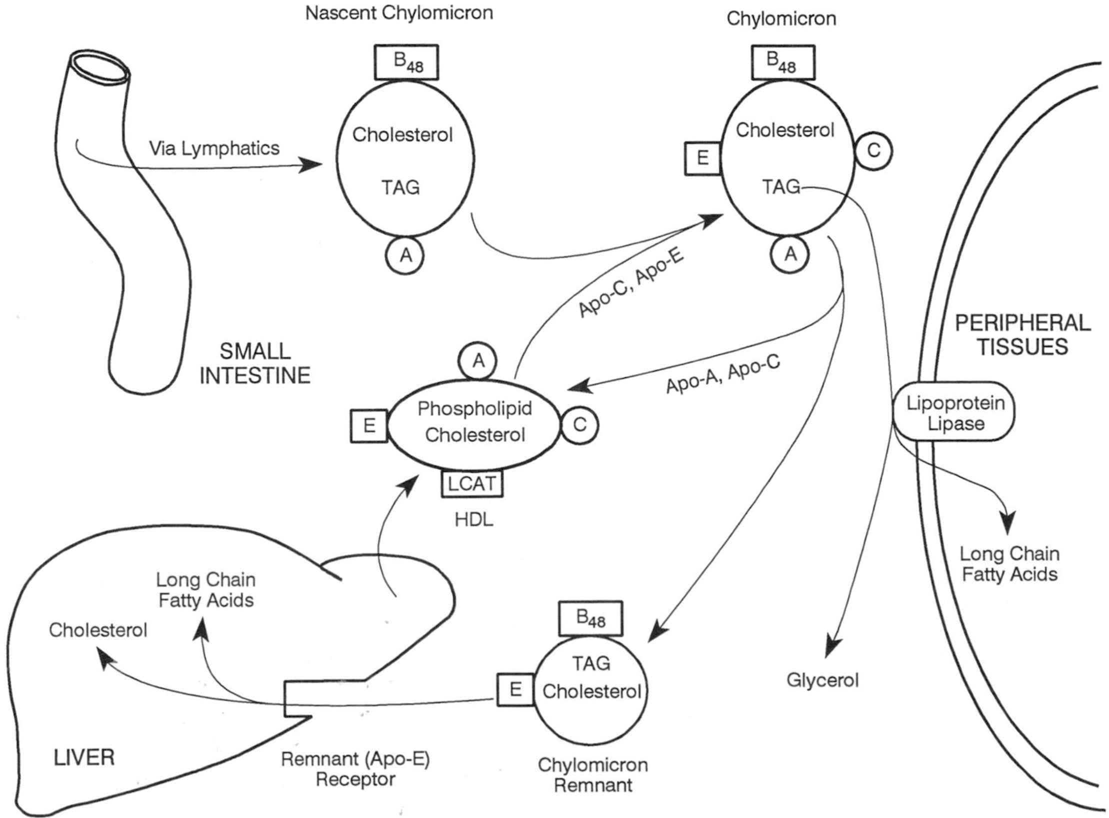

FIGURE 4.6 Metabolism and transport of chylomicrons. Abbreviations: A, B_{48}, C, E, apolipoproteins; HDL, high-density lipoprotein; TAG, triacylglycerol.

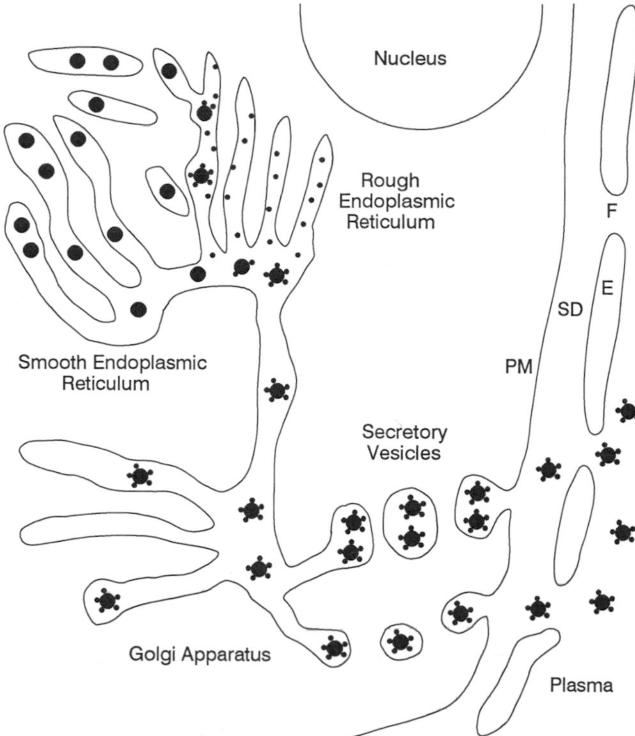

FIGURE 4.7 Synthesis of VLDL in liver. Triacylglycerol and phospholipid synthesis occurs in the smooth endoplasmic reticulum to generate lipid particles (large dots), which acquire small amounts of cholesterol and its esters as well. Apolipoproteins (small dots) are synthesized on the rough endoplasmic reticulum. Lipid particles acquire apolipoproteins at the convergence of the rough and smooth endoplasmic reticulum or by merging of sections of the two organelles. The nascent VLDL move though tubular membranes to the Golgi apparatus where apolipoproteins are glycosylated, and the nascent VLDL are collected in secretory vesicles. The secretory vesicles migrate to and merge with the plasm membrane (PM) and spill VLDL into the space of Dissé (SD). The VLDL migrate through the fenestrae (F) between endothelial cells (E) to enter the plasma in hepatic sinusoids.

lation is due to increased insulin and decreased glucagon levels in plasma. Most studies have shown that glucagon partially inhibits hepatic VLDL secretion, whereas insulin stimulates it (Gibbons, 1990). Estrogens (Crook and Seed, 1990; Sacks and Walsh, 1994; Haffner and Valdez, 1995) and glucocorticoids (Gibbons, 1990; Martin-Sanz *et al.*, 1990) stimulate VLDL secretion.

The inherent capacity of the liver to synthesize the lipid components exceeds its inherent capacity to synthesize the protein components, a fundamental factor in the development of fatty liver. In addition, phosphatidylcholine is essential for lipoprotein assembly, so animals having a deficiency of choline tend to develop fatty livers (Davis, 1991).

Triacylglycerol in plasma VLDL is hydrolyzed by LPL just like triacylglycerol of chylomicrons, and most of the released LCFAs are absorbed by the underlying tissue cells. As the VLDLs shrink some of the apolipoproteins (C series and E) transfer to HDL. Finally, the shrinking VLDL becomes an IDL and then a LDL. The LDL will attach to an apoprotein B_{100} or E receptor on hepatocytes or extrahepatic tissues and be taken into the cell where its component parts will be hydrolyzed. The transport and metabolism of VLDL are illustrated in Fig. 4.8.

E. Metabolism of High-Density Lipoproteins

HDLs are synthesized by both liver and small intestine. Nascent HDL produced in the small intestine has only A-series apolipoproteins and gains C-series and E apolipoproteins and LCAT, which are synthesized in the liver, from other lipoproteins after it enters the circulation. Nascent HDL produced in the liver gains its A-series apolipoprotein, which is synthesized in the small intestine, from other lipoproteins after it enters the circulation. HDL serves two main functions. It is a repository for A-series, C-series, and E apolipoproteins, and it transports cholesterol from peripheral tissues to liver. LCAT is important in this latter function. The conversion of cholesterol to cholesterol ester within HDL creates a favorable concentration gradient from tissue cell to HDL, which promotes migration of cholesterol from tissue cells to HDL (Fielding and Fielding, 1991; Gurr and Harwood, 1991).

HDL is removed from the circulation primarily by the liver, and its component parts can be metabolized within the hepatocyte or some of its lipid components can be incorporated into VLDL and enter the plasma again. Cholesterol can migrate from HDL into hepatocytes without the entire HDL being removed and, as mentioned earlier, apolipoproteins can migrate from HDL to chylomicrons, VLDL, and other HDL.

In summary, chylomicrons and VLDL distribute triacylglycerol, cholesterol and phospholipids from the small intestine and liver, respectively, to other tissues. IDL and LDL are effectively remnants of VLDL. HDL is a reservoir of some apolipoproteins and transports cholesterol from peripheral tissues to liver.

VII. HYPERLIPIDEMIA

A. Introduction

Hyperlipidemia refers to increased plasma levels of cholesterol (hypercholesterolemia) and triacylglycerols (hypertriacylglycerolemia or hypertriglyceridemia). Note that increased plasma levels of LCFA alone

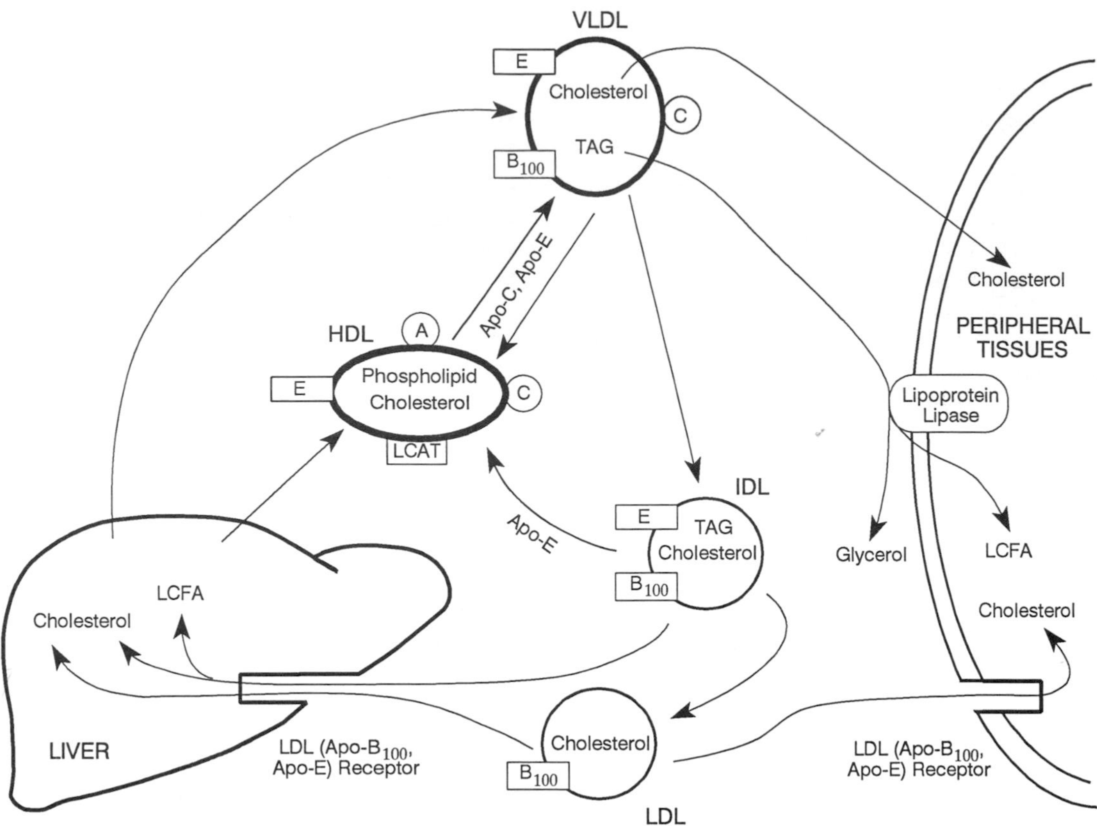

FIGURE 4.8 Metabolism and transport of VLDL. Abbreviations: A, B_{100}, C, E, apolipoproteins; HDL, IDL, and LDL, high-, intermediate-, and low-density lipoproteins, respectively; LCAT, lecithin : cholesterol acyltransferase; LCFA, long-chain fatty acid; TAG, triacylglycerol.

do not constitute hyperlipidemia. Because cholesterol and triacylglycerols must reside within lipoproteins in plasma, hyperlipidemia is synonymous with hyperlipoproteinemia. *Lipemia* is a term denoting that hyperlipidemia is severe enough that the plasma looks milky (i.e., lactescent). If lipemia is marked, whole blood may have a light red color or "tomato soup" appearance. The most common form of hyperlipidemia is postprandial hyperlipidemia, which is observed after an animal consumes a meal containing fat, and is primarily a result of increased chylomicron levels. For evaluation of possible abnormalities in lipid metabolism, it is important that blood samples be taken from fasting animals to avoid confusion caused by postprandial hyperlipidemia. One exception is adult ruminants, which are usually on a very low fat diet and, in addition, because of the volume of the rumen and fermentative nature of digestion there, have absorption spread over a considerable time period.

B. Canine Fasting Hyperlipidemias

Healthy dogs normally do not develop significant hyperlipidemia on fasting. Therefore, fasting hyperlipidemia in a dog usually is an abnormal sign with potential causes being hypothyroidism, diabetes, pancreatitis, hyperadrenocorticism, hepatic disease, nephrotic syndrome, and inherited defects in lipid metabolism. Hyperlipidemia is commonly observed in dogs with hypothyroidism, whether of congenital or acquired origin (Manning *et al.*, 1973; Liu *et al.*, 1986; DeBowes, 1987; Medaille *et al.*, 1988; Barrie *et al.*, 1993; Watson and Barrie, 1993). The main lipid that is increased is cholesterol, but triacylglycerol can be increased too. Most of the increased lipid is in LDL and HDL, but some animals have increased VLDL or chylomicron levels as well (Rogers, 1977; Whitney, 1992). The mechanism by which hypothyroidism causes hyperlipidemia in dogs is unknown; however, a similar phenomenon occurs in humans, and in that species, it appears that hypothyroidism decreases lipoprotein lipase and hepatic lipase activities (Valdemarsson *et al.*, 1983). The prolonged hypercholesterolemia associated with chronic hypothyroidism in dogs may lead to atherosclerosis (Patterson *et al.*, 1985; Liu *et al.*, 1986) although other factors may be involved as well.

Dogs with naturally occurring pancreatitis frequently have hyperlipidemia (Whitney *et al.*, 1987; Hardy, 1992). Because the pancreatitis in some of these animals causes diabetes, the hyperlipidemia in those individuals may be a result of the diabetes. Plasma lipid levels of dogs with pancreatitis induced by injecting bile into or ligating the pancreatic duct are comparable to control dogs (Zieve, 1968; Bass *et al.*, 1976; Whitney *et al.*, 1987). In some cases, hyperlipidemia may play a role in the pathogenesis of pancreatitis rather than being a result of pancreatitis. This proposition is supported by the fact that humans with some forms of hyperlipidemia have increased risk of pancreatitis (Greenberger, 1973; Cameron *et al.*, 1974). One proposed mechanism is that increased lipids, especially chylomicrons, entering the pancreatic capillaries will be hydrolyzed by pancreatic lipase, and the resultant LCFA may injure endothelial or acinar cells (Havel, 1969). Once the initial damage occurs, there is a positive feedback in which more lipase enters the circulation and hydrolyses more triacylglycerol leading to more LCFA release and more damage. In support of this theory, when dog pancreata were perfused with a medium containing high levels of triacylglycerol or LCFA, they became edematous and hemorrhagic and released large amounts of amylase compared to pancreata perfused without these additions (Saharia *et al.*, 1977). In a related proposed mechanism, large amounts of chylomicrons or VLDL may impede the microcirculation of the pancreas, leading to partial stasis, which allows blood lipids and their hydrolysis products more contact with pancreatic cells (Hardy, 1992).

Dogs with uncontrolled diabetes frequently have hyperlipidemia (Rogers, 1977; Medaille *et al.*, 1988; Whitney, 1992). In naturally occurring cases, plasma triacylglycerol levels are increased with concomitant increases in VLDL levels and often hyperchylomicronemia is present as well (Rogers *et al.*, 1975b; Rogers, 1977; DeBowes, 1987; Ford, 1995). The increase in VLDL is due in part to increased mobilization of LCFA from adipose. The liver removes LCFA from plasma and reissues some of them to the plasma as triacylglycerol in VLDL. In addition, synthesis of lipoprotein lipase by peripheral tissues is partially dependent on insulin, so less of this enzyme is available to remove triacylglycerol from the circulation (Brown and Goldstein, 1994).

Hyperlipidemia with increases in plasma triacylglycerol and cholesterol levels has been noted in dogs with cholestasis (Meyer and Chiapella, 1985; Bauer *et al.*, 1989). The increase in cholesterol can be explained in part by the inability of the liver to remove and catabolize cholesterol. However, there is evidence of production of an abnormal LDL, called lipoprotein-X, which is rich in cholesterol (Danielsson *et al.*, 1977; Blomhoff *et al.*, 1978; Meyer and Chiapella, 1985; Bauer *et al.*, 1989).

Dogs with hyperadrenocorticism (Cushing's disease) often have hyperlipidemia with increased total plasma cholesterol levels (Ling *et al.*, 1979; Scott, 1979; Medaille *et al.*, 1988; Barrie *et al.*, 1993; Feldman, 1995). Most of the increased plasma cholesterol is associated with LDL, and although the mechanism of the hyperlipidemia is unclear, it may be related to a decrease in activity of hepatic LDL receptors (Barrie *et al.*, 1993).

Dogs with nephrotic syndrome often have hyperlipidemia (McCullagh, 1978; Lewis and Center, 1984; Medaille *et al.*, 1988; Ford, 1995). Hypercholesterolemia is present most commonly, but hypertriacylglycerolemia may also be present, especially in more severe cases (McCullagh, 1978). In humans with nephrotic syndrome, the hyperlipidemia appears to be related to the loss of albumin or regulatory factors in the urine, and infusion of albumin or dextran into afflicted patients lowers lipid levels (Glassock *et al.*, 1991). Albumin or regulatory factors may inhibit VLDL production by the liver; without this inhibition, more VLDL will be released to the plasma increasing VLDL and LDL levels (Glassock *et al.*, 1991).

Idiopathic hyperlipidemia, which is probably inherited, occurs in some miniature schnauzers (Rogers *et al.*, 1975a; Richardson, 1989; Ford, 1993). Animals present with abdominal pain, diarrhea, and vomiting, and sometimes with seizures and pancreatitis. Affected animals have hypertriacylglycerolemia, hypercholesterolemia, and increased chylomicron levels and often have increased levels of other lipoproteins as well. It has been proposed that these animals may have low levels of LPL or perhaps deficient apolipoprotein C-II, the activator of LPL. However, some dogs have shown clearing of the plasma following heparin injection, so the mechanism remains unknown. The primary treatment is to place the animal on a low-fat diet. A similar syndrome has been reported in mixed breed dogs (Baum *et al.*, 1969; Rogers *et al.*, 1975a) and in Brittany spaniels (Hubert *et al.*, 1987).

C. Feline Fasting Hyperlipidemias

Not surprisingly, some of the same diseases that are associated with hyperlipidemia in dogs are associated with hyperlipidemia in domestic cats, including diabetes and nephrotic syndrome (McCullagh, 1978; Watson and Barrie, 1993; Jones, 1995). Some cats, however, have a well-characterized familial hyperlipidemia due to LPL deficiency (Jones *et al.*, 1983; Brooks, 1989; Watson *et al.*, 1992b; Whitney, 1992; Jones, 1993, 1995). There is lipemia with hyperchylomicronemia and increases in plasma levels of cholesterol and triacylglycerol (Whitney, 1992; Jones, 1993, 1995). The high plasma levels of lipids lead to deposition in tissues (xanthoma formation) in the skin, nerve sheaths, and other locations (Whitney, 1992; Jones, 1993). Pressure on spinal or other nerves from xanthomas or subsequent granulomas may lead to peripheral neuropathy (Jones *et al.*, 1986). The disease appears to be autosomal recessive, and homozygotes apparently manufacture a defective LPL and do not have a defective or missing apolipoprotein C-II activator (Peritz *et al.*, 1990; Watson *et al.*, 1992b).

D. Equine Fasting Hyperlipidemia

The phenomenon of equine hyperlipidemia was reported in horses with maxillary myositis (Hadlow, 1962) and equine infectious anemia (Gainer *et al.*, 1966). It is likely that the hyperlipidemia described in these early reports was due to anorexia, and it has been shown that fasting alone causes hyperlipidemia in horses and that pregnancy, lactation, and obesity accentuate the effect (Schotman and Kroneman, 1969; Schotman and Wagenaar, 1969; Eriksen and Simesen, 1970; Schotman and Wensing, 1977). Total plasma triacylglycerol may increase from a normal fed value of less than 500 mg/liter to more than 2000 mg/liter (Morris *et al.*, 1972; Naylor *et al.*, 1980) and, in severe cases, may exceed 10,000 mg/liter (Schotman and Wensing, 1977; Naylor *et al.*, 1980; Freestone *et al.*, 1991). Although most horses are susceptible to this effect of calorie deprivation, it is harmless for the majority and only becomes pathological for a few. Ponies are more susceptible to the pathological syndrome, which can be fatal. Survival rates are inversely proportional to plasma triacylglycerol concentrations (Schotman and Wagenaar, 1969), and severe fatty liver and increased plasma levels of liver enzymes have been reported (Schotman and Wagenaar, 1969). Like in other mammals, fasting increases the plasma levels of LCFA in horses and ponies (Baetz and Pearson, 1972; Naylor *et al.*, 1980; Watson *et al.*, 1992a), and the hyperlipidemia in horses and ponies is due to increased VLDL levels (Morris *et al.*, 1972; Bauer, 1983; Watson *et al.*, 1992a). Presumably, horse liver is removing LCFAs from plasma and reesterifying them into triacylglycerol, which is released to the plasma as VLDL. It appears that the liver of fasting horses has a high capacity for reesterification of LCFAs into triacylglycerol, which is exported as VLDL. Horses do have increased plasma ketone levels when fasting (Rose and Sampson, 1982), so some of the LCFAs removed from plasma by liver are converted to ketones, but additional LCFAs are reesterified to glycerol and are recycled to the plasma as triacylglycerol in VLDL.

The mechanism of fasting equine lipemia is uncertain; it could be increased secretion or decreased uptake of VLDL or a combination thereof. In one study, fasting horses were injected with Triton WR 1339, a compound shown to inhibit LPL in rats, and the rate of increase in plasma triacylglycerol levels was observed (Morris *et al.*, 1972) and compared with the preinjection level of triacylglycerol. The authors claimed that the rate of increase was not related to the level of triacylglycerol concentration. However, only four horses were used, which may not have been enough given the high variation in fasting triacylglycerol levels observed among them. On close examination of the data, it is apparent that three horses showed a perfect rank correlation of fasting triacylglycerol level and rate of triacylglycerol increase after Triton injection, with the fourth being a considerable outlier to the trend. In addition, the dose of Triton was such that all the horses developed anemia due to intravascular hemolysis. Thus, elucidation of the mechanism of fasting equine lipemia will have to await additional kinetic or enzyme studies. There are reports of apparent therapeutic success with intravenous glucose and oral carbohydrates (Watson and Love, 1994; Mogg and Palmer, 1995). This therapy makes sense in that increased plasma glucose levels should lead to increased insulin and decreased glucagon levels, which should inhibit lipolysis in adipose that is generating the plasma LCFAs used for triglyceride synthesis. In addition, the hormonal changes may stimulate LPL activity. Although supplemental insulin has been used with carbohydrate therapy, its efficacy and safety have not been adequately evaluated.

VIII. KETOGENESIS AND KETOSIS

A. Introduction

The ketones or ketone bodies, which are composed of acetoacetic acid, 3-hydroxybutyric acid (also known as β-hydroxybutyric acid), and acetone, are important compounds in the metabolism of birds and mammals. Ketosis simply means that ketones are present in body fluids in elevated concentrations. Ketones are important clinically and have a rather sinister reputation because of the ketoacidosis that is often present when their plasma levels are high. In recent years though, the survival value of ketogenesis has become clearer, and although increased levels of ketones in biological fluids will continue to be regarded as a pathological sign in many situations, perhaps the beneficial aspects of ketogenesis will be more widely appreciated.

B. Chemistry of Ketones

1. Structure and Properties

The ketones, acetone, 3-hydroxybutyric acid, and acetoacetic acid, are relatively simple chemical structures. Of the three, only 3-hydroxybutyric acid can exist as stereoisomers, having L (+) and D (−) forms. Only the D (−) form is produced in a free state in intermediary metabolism. The L (+) form exists only as its CoA thioester produced and destroyed in β-oxidation (Newsholme and Leach, 1983). Acetone is relatively volatile, whereas the other two ketones are not. Acetone has a characteristic organic solvent odor, which may be detectable in the exhaled breath of animals with elevated blood ketone levels. Anecdotal evidence indicates that people vary greatly in their olfactory sensitivity for acetone.

Acetone does not ionize appreciably, whereas 3-hydroxybutyric acid and acetoacetic acid do readily ionize. Acetoacetic acid has a pK_a of 3.58, and 3-hydroxybutyric acid has a pK_a of 4.41 (Dean, 1985). Consequently, at normal plasma pH of 7.40, 99.9% of either compound exists in its ionized form. Therefore, the compounds will usually be referenced by the names of their ions whenever their metabolism is discussed. Acetoacetic and 3-hydroxybutyric acids are more powerful acids than the volatile fatty acids (VFAs; acetic, propionic and butyric acids), which have pK_a's of 4.76–4.87 (Dean, 1985). Acetoacetic acid is more powerful and 3-hydroxybutyric acid is less powerful as an acid than lactic acid, which has a pK_a of 3.86 (Dean, 1985).

Acetone and acetoacetic acid are miscible in water in all proportions, and 3-hydroxybutyric acid is exceedingly soluble, but not in all proportions (Dean, 1985). The common metallic salts of acetoacetic acid and 3-hydroxybutyric acid are soluble in water. Acetone and 3-hydroxybutyric acid and its salts are relatively stable compounds. Acetoacetic acid spontaneously decomposes to acetone and carbon dioxide. This reaction occurs readily without catalysis, and its rate is accelerated by increased temperature and hydrogen ion concentration. Apparently, some nonspecific catalysis of acetoacetate decarboxylation by cellular proteins can occur (Williamson, 1978). The lithium, sodium, and potassium salts of acetoacetic acid are relatively stable if stored in dry form below 0°C.

2. Detection and Assay

a. Qualitative

The most common qualitative test for ketones is the alkaline nitroprusside test, which is also known as the Rothera test (Rothera, 1908). This test has been used for decades in clinical practice and is still exceedingly useful today. The test relies on the reaction of nitroprusside with acetone or acetoacetate to produce a purple chromogen. The nitroprusside test has been used for virtually every body fluid imaginable including whole blood, serum, plasma, urine, and milk. The test is most sensitive for acetoacetate (0.5 mmol/liter can be detected), gives only a slight response to acetone, and is completely insensitive to 3-hydroxybutyrate.

The nitroprusside test is available commercially in the form of strips, tablets, and powders. The maximum sensitivity of all three forms is approximately 0.5 mmol/liter, although specific formulations may have a sensitivity less than this value. The strip form is commonly used for urine. The powder form and strips are both commonly used for milk. The tablet form is used for serum, plasma, and whole blood and can be used for milk and urine as well. The test is often used in a semiquantitative manner with the result expressed in adjectival form (negative, weak, strong) or as a series of pluses (−, +, ++, etc.).

A number of drugs or other substances may appear in urine and give a false positive with the nitroprusside test. Some compounds react with nitroprusside to yield a purple or near purple color. Included in this group are phenylketones, levodopa, methyldopa, acetaldehyde, paraldehyde (Caraway and Kammeyer, 1972), cysteine, cysteamine, penicillamine, and mesna (Csako, 1987). In general, substances with keto, aldehyde, or sulfhydryl groups have the potential for reacting with nitroprusside. Since the nitroprusside test is performed in an alkaline medium, some substances, like sulfobromophthalein and phenolsulfonphthalein, which may exist in urine and are otherwise colorless, may yield a purple or near purple color simply due to the alkaline pH (Caraway and Kammeyer, 1972).

b. Quantitative

Commonly used means of quantitative assay for ketone concentrations in biological fluids include microdiffusion methods, used primarily for assay of acetone, and enzymatic methods, used primarily for assay of acetoacetate and 3-hydroxybutyrate. Regardless of the method to be used for analysis, proper handling of the samples prior to analysis is crucial for obtaining representative results. In particular, the volatility of acetone and instability of acetoacetate must be respected.

Blood samples should be cooled immediately after collection. Ketones can be determined on whole blood or plasma. Serum is not recommended because of losses, particularly of acetoacetate, that may occur during the time required for clotting. Any of the common anticoagulants (heparin, fluoride, oxalate, citrate, or EDTA) may be used. If whole blood is to be used, it should be mixed with perchloric acid immediately after collection to precipitate proteins. The tube should be chilled on ice until centrifuged, which should be performed within a few hours. The supernatant should be frozen until analyzed. If plasma is to be used, the red cells should be spun down within a few hours, and the plasma proteins precipitated with perchloric acid. The supernatant should be frozen until analyzed.

The microdiffusion method can be used to determine the concentration of acetone or acetone plus acetoacetate in any biological fluid. The reagents are relatively simple and inexpensive although rather corrosive. The diffusion step requires specialized, but inexpensive, apparatus and adds to the complexity and time to complete the assay. The method relies on the reaction of acetone with vanillin (Henry *et al.*, 1974) or salicylaldehyde (Nadeau, 1952) to produce a colored product that can be quantified in a spectrophotometer. In the author's experience, vanillin provides more sensitivity than salicylaldehyde, but variability in the purity of vanillin batches from commercial sources makes salicylaldehyde the reagent of choice. Salicaldehyde must be stored under nitrogen or argon to preserve its purity.

The method as described by Henry *et al.* (1974) was shown to determine acetone, and there was speculation that it would also detect acetoacetate simultaneously. In fact, the method as described by Henry *et al.* (1974) is specific for acetone. It has been found in the author's laboratory that to use the method for acetone plus acetoacetate, it is necessary to preincubate the sample with an equal volume of 10 *N* sulfuric acid for 4 hours at 50°C in a sealed container to decarboxylate all of the acetoacetate. The method can be adapted to measure 3-hydroxybutyrate as well by introducing a step in which 3-hydroxybutyrate is oxidized to acetoacetate with potassium dichromate (Procos, 1961). However, if the primary interest is the determination of acetoacetate or 3-hydroxybutyrate, rather than acetone, the enzymatic method described later should be used.

The enzymatic method for assay of acetoacetate or 3-hydroxybutyrate in biological fluids is accurate and precise (Williamson *et al.*, 1962) and is probably the most common method used for quantitative assay of ketone concentrations. The method has been successfully adapted to a variety of automated analysis systems (Ozand *et al.*, 1975; Työppönen and Kauppinen, 1980; Harano *et al.*, 1985) and is a relatively straightforward spectrophotometric or fluorometric method. For a detailed step-by-step practical description of the method, see Mellanby and Williamson (1974) and Williamson and Mellanby (1974).

The method relies on the reversible reaction catalyzed by 3-hydroxybutyrate dehydrogenase:

$$\text{3-Hydroxybutyrate} + NAD^+ \leftrightarrow \text{acetoacetate} + NADH + H^+.$$

The reaction is run in the forward direction by including an excess of NAD^+ in the reaction mixture to assay 3-hydroxybutyrate and in the backward direction by including an excess of NADH in the reaction mixture to assay acetoacetate. The equilibrium constant of the reaction is 1.42×10^{-9} and, therefore, is highly favorable toward the reduction of acetoacetate at pH 7.0 (Krebs *et al.*, 1962). To force the reaction to completion in the direction of oxidizing 3-hydroxybutyrate, hydrazine is used as a trapping agent to remove acetoacetate as it is formed, and the reaction mixture is buffered at an alkaline pH. The change in NADH concentration is measured by the change in absorbance at 340 nm in either case. Alternately, a fluorometer can be used to measure the change in NADH concentration. To avoid interference from lactate or pyruvate in the sample, the 3-hydroxybutyrate dehydrogenase should be free of lactate dehydrogenase; alternatively, the lactate dehydrogenase inhibitor, oxamic acid, can be added to the reaction mixture (Harano *et al.*, 1985).

Table 4.3 lists normal blood and plasma ketone concentrations for several domestic species. The values are for healthy fed animals. Plasma and blood ketone concentrations are assumed to be similar because of the generally high permeability of cell membranes to ketones and lack of protein binding of ketones; however, reports of definitive studies on this problem are not apparent in the literature. For clinical purposes, there is no lower normal limit for ketone concentrations.

C. Synthesis of Ketones

Ketones are primarily products of intermediary metabolism. Only under unusual circumstances would more than trace amounts be absorbed from the contents of the gastrointestinal tract. The real source of ketones is fatty acids including those with short (1–4 carbons), medium (5–11 carbons), and long (>11 carbons) chains. Of course, any compound (glucose, lactate, glycerol, amino acids, etc.) that can be converted to fatty acids can be considered as a source of ketones, but for the purposes of this discussion, the origin of ketones will be considered to be fatty acids, either esterified or nonesterified.

The predominant source of ketones in healthy animals is LCFAs released during lipolysis in adipose tissue. When plasma insulin levels decrease and plasma glucagon levels increase, cAMP levels in adipose cells increase. Increased cAMP levels lead to activation of hormone sensitive lipase, which hydrolyses triacylglycerols to LCFA and glycerol. The LCFAs bind to plasma albumin for transport to other tissues, whereas glycerol freely dissolves in plasma water (Spector and Fletcher, 1978; McGarry, 1979; Newsholme and Leach, 1983).

TABLE 4.3 Blood and Plasma Ketone Concentrations of Domestic Animals[a]

Species	3-Hydroxy-butyrate (mmol/liter)	Acetoacetate (mmol/liter)	Sample	References
Cow (lactating)	0.41 ± 0.03	0.043 ± 0.00	Plasma	Hibbit *et al.*, 1969
	0.95 ± 0.18	0.13 ± 0.03	Blood	Gröhn, 1985
Cow (nonlactating)	0.27 ± 0.04	0.011 ± 0.003	Plasma	Baird *et al.*, 1968
Dog	0.033 ± 0.015	0.018 ± 0.010	Blood	Balasse, 1970
	0.030 ± 0.006	—	Plasma	Lammerant *et al.*, 1985
Goose	0.042 ± 0.015	0.023 ± 0.003	Plasma	Maho *et al.*, 1981
Horse	0.11 ± 0.01	—	Blood	Snow and Mackenzie, 1977
	0.064 ± 0.006	0.029 ± 0.003	Plasma	Rose *et al.*, 1980
Sheep	0.27 ± 0.04	0.051 ± 0.005	Blood	Brockman, 1976
	0.55 ± 0.04	0.030 ± 0.002	Blood	Heitman *et al.*, 1986

[a] Values are means ± standard errors for healthy fed animals. Ketone concentrations were determined by the method of Williamson *et al.* (1962) or a modification thereof.

1. *Ketogenesis by Liver*

The liver has an enormous capacity to remove LCFAs from plasma. LCFAs unbind from albumin, diffuse through the hepatocyte plasma membrane, and bind to fatty acid binding protein in the cytosol (Burnett *et al.*, 1979). In the cytosol, LCFAs are converted to LCFA-CoA as discussed earlier. The LCFA-CoA can be used to synthesize triacylglycerol or can go through β-oxidation to acetyl-CoA in the mitochondrion, also discussed earlier. Mitochondrial acetyl-CoA can have a number of fates, but under circumstances that elevate plasma LCFA levels, the two main fates are combustion in the TCA cycle or conversion to ketones. Two acetyl-CoA units can be recondensed to form ketones, which will occur when there is insufficient oxaloacetate for citrate formation or when citrate synthase is inhibited by high levels of citrate.

Four enzymes are involved in ketogenesis from acetyl-CoA: acetoacetyl-CoA thiolase, hydroxymethylglutaryl-CoA (HMG-CoA) synthase, HMG-CoA lyase, and D-3-hydroxybutyrate dehydrogenase. These four enzymes catalyze the following four reactions, respectively:

$$\text{2 Acetyl-CoA} \leftrightarrow \text{acetoacetyl-CoA} + \text{CoA},$$
$$\text{Acetoacetyl-CoA} + \text{acetyl-CoA} \rightarrow \text{HMG-CoA} + \text{CoA},$$
$$\text{HMG-CoA} \rightarrow \text{acetoacetate} + \text{acetyl-CoA},$$
$$\text{Acetoacetate} + \text{NADH} + \text{H}^+ \leftrightarrow \text{3-hydroxybutyrate} + \text{NAD}^+$$

Thiolase occurs in both cytosol and mitochondria, whereas the other three enzymes are mainly restricted to the mitochondrion. However, there is some HMG-CoA synthase in the cytosol that is involved with cholesterol synthesis. The first three enzymes are in the mitochondrial matrix, whereas 3-hydroxybutyrate dehydrogenase is in the inner membrane of the mitochondrion, and membrane lipids are required for full activity of the enzyme. Interestingly, livers of ruminants have lower apparent activities of 3-hydroxybutyrate dehydrogenase than the livers of other species (Nielsen and Fleischer, 1969; Watson and Lindsay, 1972), a situation that has not been explained adequately. Because ruminant liver is continuously presented with 3-hydroxybutyrate synthesized by the rumen epithelium, the low hepatic activity of 3-hydroxybutyrate dehydrogenase may be beneficial because more of the compound will reach peripheral tissues in an unoxidized state.

Acetoacetyl-CoA thiolase and 3-hydroxybutyrate dehydrogenase catalyze reactions that are at or near equilibrium; that is, the rates of these reactions are controlled by the concentrations of the substrates and products. HMG-CoA synthase and lyase catalyze reactions that are far removed from equilibrium, and these enzymes may be subject to regulatory controls other than the concentrations of substrates and products. The synthase is considered to be the rate-limiting enzyme and appears to be restricted almost exclusively to the liver (McGarry and Foster, 1969; Duee *et al.*, 1994; Valera *et al.*, 1994). Physiological controls of the enzyme are not completely understood; however, the concentration of enzyme molecules in the mitochondria is increased by cAMP, so fasting and diabetes increase it and refeeding decreases it (Serra *et al.*, 1993). In addition, succinyl-CoA inhibits the enzyme (Quant *et al.*, 1990). Glucagon usually decreases mitochondrial succinyl-CoA, whereas an abundance of glucose or glucose precursors, like propionate, increases it. A summary of ketogenesis in the liver is depicted diagrammatically in Fig. 4.2.

Ketogenesis can occur from VFAs and medium-chain fatty acids. Medium-chain fatty acids are normally in quite low concentration in the diet or in triacylglycerols of mammals and, therefore, are not usually quantitatively important in ketogenesis. Except in unusual circumstances, nonherbivores do not absorb large quantities of VFAs from the gastrointestinal tract. Among the herbivores, the metabolism of VFAs has been studied most thoroughly in ruminants. Propionate is the major gluconeogenic precursor and is not an important precursor of ketones; in fact, propionate inhibits ketogenesis in ruminant liver (Faulkner and Pollock, 1991). The propionate inhibition probably is due to inhibition of carnitine acyltransferase I in ruminant liver by methylmalonyl-CoA, a metabolite of propionate (Brindle *et al.*, 1985). Without active carnitine acyltransferase I, LCFAs cannot enter mitochondria and be oxidized to ketones.

Butyrate is converted to 3-hydroxybutyrate by the rumen epithelium and is discussed later. Acetate must be covalently bound to CoA under the catalysis of acetokinase before it can be catabolized further. Acetokinase is found in the cytosol and mitochondria of most cells in most organs. In ruminants, the liver has a relatively low concentration of acetokinase, and most absorbed acetate passes through the liver and is removed from the plasma by other tissues, particularly heart, skeletal muscle, kidney, and mammary gland (Cook *et al.*, 1969; Bauman and Davis, 1975).

2. *Ketogenesis by the Alimentary Tract*

Butyrate produced during fermentation of feedstuffs in the rumen is readily absorbed by the rumen wall (Stevens, 1970). The rumen epithelial cells possess high activities of butyryl-CoA synthetase, which can convert butyrate to butyryl-CoA (Cook *et al.*, 1969). By β-oxidation, butyryl-CoA is converted to L-3-hydroxybutyryl-CoA, which is oxidized to acetoacetyl-CoA followed by cleavage of the CoA and reduction of the resulting acetoacetate to 3-hydroxybutyrate (Emmanuel, 1980). Rumen epithelium does have HMG-CoA synthase, HMG-CoA lyase, and 3-hydroxybutyrate dehydrogenase activities, although in lesser concentration than in liver (Baird *et al.*, 1970; Emmanuel, 1980). It is possible that rumen epithelium can cleave CoA from acetoacetyl-CoA directly because it contains acetoacetyl-CoA deacylase (Bush and Milligan, 1971). In addition, rumen epithelium possesses 3-ketoacid CoA-transferase, an enzyme to be discussed later when ketone oxidation is discussed (Bush and Milligan, 1971). This enzyme can catalyze the transfer of CoA from acetoacetyl-CoA to succinate, thus liberating acetoacetate. This latter route may be the predominant pathway in rumen epithelium (Bush and Milligan, 1971).

3-Hydroxybutyrate appears in portal blood (Stevens, 1970; Katz and Bergman, 1969). At least 50% of absorbed butyrate is oxidized to ketones in the rumen wall, and of the butyrate that does appear in portal blood, nearly all of it is removed on the first pass through the liver (Ramsey and Davis, 1965; Bergman *et al.*, 1965; Bergman and Wolfe, 1971; Fell and Weekes, 1975). Ruminal production of 3-hydroxybutyrate is probably the main reason why fed ruminants normally have a higher plasma concentration of this compound than fed nonruminants (see Table 4.3).

3. *Ketogenesis by Other Organs*

It has been claimed that mammary gland may synthesize appreciable ketones in ketotic dairy cows; however, the evidence is weak. Arteriovenous concentration differences and mammary blood flow have been used to estimate mammary ketone production and uptake in dairy cows (Kronfeld *et al.*, 1968). It was found that the mammary gland utilized small quantities of acetoacetate and larger quantities of 3-hydroxybutyrate in healthy cows, whereas the mammary gland of ketotic cows produced large quantities of acetoacetate. The increased uptake of 3-hydroxybutyrate by mammary in ketotic cows equaled almost exactly mammary production of acetoacetate. There was no significant difference in mammary uptake of acetate between healthy and ketotic cows.

In yet another study on ketotic cows (Schwalm *et al.*, 1969), arteriovenous (AV) concentration differences across the mammary glands of acetoacetate and 3-hydroxybutyrate were observed. A positive AV difference was noted for 3-hydroxybutyrate, which was almost equal in magnitude to the negative AV difference noted for acetoacetate. The foregoing results point toward mammary conversion of 3-hydroxybutyrate to acetoacetate, which increases in ketosis. This process cannot really be called ketogenesis; perhaps *ketoconversion* would be the appropriate term.

D. Catabolism of Ketones

1. *Reduction and Oxidation*

Reduction is a possibility for acetoacetate and, of course, the reduction product is 3-hydroxybutyrate. 3-Hydroxybutyrate is a metabolic cul de sac because it can be metabolized only by being reconverted to acetoacetate. In comparison to acetoacetate, 3-hydroxybutyrate should be viewed as a means by which the liver can export reducing power (hydrogen) to the peripheral tissues for combustion and energy generation there.

Both acetoacetate and 3-hydroxybutyrate can be reduced by being converted to LCFA. This fate is more likely to occur in mammary gland than in adipose tissue. Plasma 3-hydroxybutyrate has been shown to be a milk fat precursor in cows (Palmquist *et al.*, 1969), goats (Linzell *et al.*, 1967), and rabbits (Jones and Parker, 1978). A substantial portion of 3-hydroxybutyrate used for milk fat synthesis in ruminants is incorporated as a four-carbon unit (Palmquist *et al.*, 1969; Kinsella, 1970).

Ultimately, the fate of most 3-hydroxybutyrate and acetoacetate is oxidation in the peripheral tissues. Once 3-hydroxybutyrate has been oxidized to acetoacetate, the acetoacetate is converted to acetoacetyl-CoA by the following reaction:

$$\text{Acetoacetate} + \text{succinyl-CoA} \leftrightarrow \text{acetoacetyl-CoA} + \text{succinate}.$$

This reaction is catalyzed by 3-ketoacid CoA-transferase and, viewed from the point of converting succinyl-CoA to succinate, it effectively bypasses the succinyl-CoA synthetase reaction of the citric acid cycle. Because the reaction catalyzed by succinyl-CoA synthetase produces one GTP from GDP, the 3-ketoacid CoA-transferase reaction effectively consumes 1 mol of ATP. The reaction also pushes succinyl-CoA toward oxaloacetate in the citric acid cycle, and oxaloacetate will be needed to form citrate from the acetyl-CoA derived from the acetoacetyl-CoA.

3-Ketoacid CoA-transferase is found in spleen, skeletal muscle, brain, adipose, heart, lung, and kidney of rodents and sheep although the activities are very low in sheep heart and brain (Williamson *et al.*, 1971). In general, though, the highest activities are in the heart and kidney. Activity of 3-ketoacid CoA-transferase is absent in liver (Williamson *et al.*, 1971). The absence of the enzyme from liver is logical because if there were a need for NADH for combustion in the liver, it could be obtained directly from acetyl-CoA in the citric acid cycle rather than shunting the acetyl-CoA units into ketones and back again.

Acetyl-CoA is produced from acetoacetyl-CoA via the acetoacetyl-CoA thiolase reaction, which was discussed under ketogenesis. All tissues have thiolase, and it is in greatest activity in heart, liver, and kidney (Williamson *et al.*, 1971). Heart and brain of sheep have significantly less thiolase than in rodents. The activities of 2-ketoacid CoA-transferase and acetoacetyl-CoA thiolase are relatively stable in fasting, fed state, high-fat diet, and diabetes except that in rodents, thiolase increases on feeding a high-fat diet (Williamson *et al.*, 1971). In general, it appears that the ketone utilizing capacity of the body is relatively constant and ketone availability controls ketone oxidation.

For many years, acetone was viewed as a metabolic dead-end, a substance destined to be excreted in the urine or exhaled in the breath. Although much acetone does indeed wind up in the breath and urine, evidence has accumulated that indicates that some acetone is metabolized (Luick *et al.*, 1967; Owen *et al.*, 1982; Vander Jagt *et al.*, 1992; Kalapos *et al.*, 1994). Furthermore, evidence for the catabolic pathway for acetone metabolism indicates that it is metabolized to pyruvate, apparently via hydroxyacetone and pyruvaldehyde (Vander Jagt *et al.*, 1992). Thus, by this mechanism, it is possible that small amounts of fat can be converted to glucose.

Labeling patterns of milk glutamate carbons following injection of 2-^{14}C-acetone into cows indicated that acetone was metabolized via pyruvate (Luick *et al.*, 1967; Black *et al.*, 1972). Labeling patterns of glucose in humans injected with radiolabeled acetone also indicate metabolism via pyruvate (Owen *et al.*, 1981). In rats, however, labeling patterns of glucose following radiolabeled acetone injection indicate that acetone can be metabolized via pyruvate and acetate, but that the latter pathway predominates (Kosugi *et al.*, 1986). Thus, there appear to be real species differences in acetone metabolism. In humans, at least, the fraction of acetone that is metabolized versus the amount excreted varies inversely with acetone concentration (Owen *et al.*, 1982), so it appears that the catabolizing pathways for acetone are not capable of handling large quantities.

2. *Renal Metabolism and Excretion*

The kidney cannot synthesize ketones to any appreciable extent (Lynen *et al.*, 1958; Weidman and Krebs, 1969), but is a voracious consumer of ketones as an energy source in ruminants (Kaufman and Bergman, 1971, 1974) and nonruminants (Weidman and Krebs, 1969; Baverel *et al.*, 1982). It is interesting though that in fasting sheep, the kidney removes from the plasma and catabolizes both acetoacetate and 3-hydroxybutyrate (Kaufman and Bergman, 1974), whereas in fasting humans, there is substantial removal of 3-hydroxybutyrate and a slight production of acetoacetate (Owen *et al.*, 1969).

Ketones are freely filterable in the glomerulus. There appears to be in humans and dogs, at least, a direct or indirect energy-consuming tubular transport system for acetoacetate and 3-hydroxybutyrate that approaches saturation at relatively low plasma concentrations of ketones such as those encountered in the fed state or after a one-day fast. Neither ketone is excreted at these lower concentrations, but they do begin to appear in the urine as plasma levels begin to rise (Visscher, 1945; Schwab and Lotspeich, 1954; Sapir and Owen, 1975; Wildenhoff, 1977). However, as the ketone concentrations increase in the glomerular filtrate, the primary mode of reabsorption is by diffusion down a concentration gradient as water is reabsorbed from the tubular lumen. Acetone begins to appear in the urine as soon as it begins to appear in the plasma (Widmark, 1920); presumably, this effect is due to great lipid solubility of acetone, which allows it to penetrate cell membranes with relative ease.

Renal excretion and reabsorption of ketones is approximately proportional to their filtration rates (or plasma concentrations if glomerular filtration rate remains constant) at concentrations found after more than a 1-day fast in humans and in ruminants (Wildenhoff, 1977; Kaufman and Bergman, 1974). At least some parts of the nephron, probably beyond the proximal tubule, are less permeable to ketones than to water because when plasma ketone levels are substantially increased, the urinary concentration exceeds the plasma concentration.

The dual mode of ketone reabsorption has an advantage in that none of this valuable energy source is lost at lower plasma concentrations; however, there is no transport maximum for the kidney as a whole, so 80–90% of filtered ketones are reabsorbed regardless of how concentrated ketones become in the plasma during pathological conditions or prolonged starvation. Mammals presumably could have evolved a greater activity of the energy-consuming ketone transport system. However, the energy cost of continuously maintaining the system at a higher activity probably outweighed the survival value of having the system available during rare periods of prolonged starvation.

E. Pathophysiology of Ketonemia

As discussed earlier, the acetoacetate and 3-hydroxybutyrate are more powerful acids than the VFA and, in the case of acetoacetate, more powerful than lactic acid. Not surprisingly, then, a high concentration of ketones in the plasma results in a metabolic acidosis known as *ketoacidosis*. The most significant ketoacidoses commonly encountered in domestic animals are in diabetes mellitus and ovine pregnancy toxemia. The ketoacidosis encountered in these syndromes may cause plasma bicarbonate to be below 10 mmol/liter (Reid, 1968; Ling *et al.*, 1977) and is a chief contributor to mortality.

The ketoacidosis in diabetes of dogs and cats can be severe with blood pH being 7.2 or less (Schaer, 1976; Ling *et al.*, 1977; Edwards, 1982). Because plasma ketone concentrations in diabetic dogs have been reported to average 3.2 mmol/liter with some individuals having levels of 7–8 mmol/liter (Balasse *et al.*, 1985), the base deficit in extracellular fluids would be greater than that concentration for two reasons. First, the distribution space of the ketones is greater than that of extracellular fluid and, secondly, some acetoacetate and 3-hydroxybutyrate anions may have been lost in the urine without equal losses of hydrogen ion. (A mineral ion, such as sodium or potassium would have balanced the electrical charge.) Base deficits of more than 15 mmol/liter have been reported in spontaneously diabetic dogs (Ling *et al.*, 1977; Edwards, 1982).

As the metabolic acidosis of diabetes progresses in dogs, increased catabolism of muscle protein occurs (Balasse *et al.*, 1985). Much of the nitrogen from protein degradation is diverted into ammonia rather than urea, and it is ammonium ion that balances most of the electrical charge on excreted acetoacetate and 3-hydroxybutyrate.

Ketones are really an alternate form of lipid, comparable to triacylglycerols, LCFAs or VFAs, and should be considered as such in caloric balance (Williamson, 1971). VFAs and ketones are effectively water-soluble forms of lipids; however, only the ketones can be produced in large quantities in tissue metabolism.

In fed animals, only a nominal caloric production is derived from oxidation of ketones; however, in fasted animals or in some pathological conditions, ketone oxidation accounts for a substantial quantity of expended calories. For example, only 3–4% of expired carbon dioxide is derived from 3-hydroxybutyrate in fed cows (Palmquist *et al.*, 1969), whereas 30% of expired carbon dioxide is derived from ketones in fasted pregnant ewes (Pethick and Lindsay, 1982).

It has been demonstrated in canine perfused liver (Shaw and Wolfe, 1984) and *in vivo* in humans (Mebane and Madison, 1964; Binkiwicz *et al.*, 1974; Miles *et al.*, 1981) and dogs (Paul *et al.*, 1966; Balasse *et al.*, 1967) by infusing acetoacetate or 3-hydroxybutyrate that both ketones inhibit gluconeogenesis. In most of these experiments, evidence was seen of increased plasma insulin concentrations, which could account for the diminution in plasma glucose concentration. The survival value of having ketones inhibit gluconeogenesis is that in starvation, as ketones concentrations increase and become available for tissue energy needs, the rate at which body protein must be catabolized to supply glucose precursors can decrease.

Not surprisingly in view of the increased insulin levels usually observed, decreased levels of LCFAs were noted during ketone infusions in some of the experiments mentioned earlier. Thus, increased ketone levels may serve as a negative feedback on rate of lipolysis in adipose and, therefore, on the plasma levels of ketones themselves.

F. Fasting Ketosis

During fasting, hormonal changes occur that promote lipolysis. Most important, as less glucose is available from the gut or from gluconeogenesis in the liver, plasma glucose concentrations will decrease. Responding to the hypoglycemia, pancreatic islet cells will release less insulin and more glucagon, so that plasma insulin concentrations will decrease and plasma glucagon concentrations will increase. These hormonal changes will increase cAMP concentrations in adipose cells, which leads to the activation of hormone sensitive lipase.

Through the action of hormone sensitive lipase, triacylglycerols are hydrolyzed with release of LCFAs and glycerol. LCFAs are utilized directly by tissues for energy but are also taken up by the liver in proportion to their plasma concentration. During fasting, hepatic concentrations of malonyl-CoA and methylmalonyl-CoA are relatively low, so carnitine acyltransferase I activity is relatively high, and LCFA-CoA are quickly converted to LCFA-carnitine, which is translocated into the mitochondrion (McGarry *et al.*, 1977). Once in the mitochondrion, LCFA-carnitine is converted to LCFA-CoA again.

Following β-oxidation of ketones, some acetyl-CoA is combusted in the citric acid cycle. However, during fasting, gluconeogenesis is quite active in the liver, and much of the mitochondrial oxaloacetate is used for that purpose and is unavailable for citrate formation with acetyl-CoA; consequently, large quantities of acetyl-CoA are shunted into ketogenesis.

Acetoacetate and 3-hydroxybutyrate can be utilized by most extrahepatic tissues. Because peripheral tissues can also use LCFAs, the utility of hepatic production of ketones from LCFAs was not clear originally. However, many tissues have as great or greater capacity for utilizing plasma ketones as for utilizing plasma LCFAs. Among these tissues are heart and kidney (Hall, 1961; Williamson and Krebs, 1961; Owen *et al.*, 1969; Little *et al.*, 1970). In some species, such as the rat (Hawkins *et al.*, 1971) and human (Owen *et al.*, 1967), ketones constitute a major energy source for the brain during fasting. In some other species though, it appears that the brain prefers glucose and utilizes only small quantities of ketones in the fed or fasted state in sheep (Jones *et al.*, 1975; Lindsay and Setchell, 1976; Pell and Bergman, 1983), dog (Wiener *et al.*, 1971), and pig (Tildon and Sevdalian, 1972). Resting skeletal muscle utilizes ketones preferentially as a fuel during short-term starvation (Owen and Reichard, 1971); however, LCFAs are preferred during long-term starvation (Owen and Reichard, 1971) or exercise (Hagenfeldt and Wahren, 1968a, 1968b).

Ketones are quite soluble, require no protein carrier, and diffuse (in their un-ionized form) or are transported rapidly through biological membranes including the blood–brain barrier (Hawkins *et al.*, 1971; Persson *et al.*, 1972). The liver has an advantage over other tissues regarding uptake of LCFAs from plasma albumin because of its unique sinusoidal vascular system. Therefore, the liver can be regarded as a machine that can rapidly remove LCFAs from plasma and convert them to a form, the ketones, that other tissues can utilize rapidly.

Because they must be bound to albumin if they are to be nontoxic, the maximum safe plasma concentration of LCFA is fixed by the albumin concentration. Furthermore, in prolonged fasting, albumin concentration decreases, which lessens the number of LCFA carriers. Generally, LCFA concentrations do not rise above 2 mmol/liter in fasting, whereas ketone concentrations can increase to 3-4 mmol/liter or more. Thus, ketones can have a greater concentration gradient to allow their entry into the cell.

Although the acid nature of ketones has received much attention in the clinical literature, less well recognized is the toxic potential of LCFAs. If LCFAs are released into the plasma in excess of hepatic uptake, albumin binding capacity will be exceeded (Spector and Fletcher, 1978). Unbound fatty acids may damage endothelial cells, perhaps due to detergent action, oxidation of unsaturated LCFAs or changes in cell metabolism (Ramasamy *et al.*, 1991). Such damage to endothelial cells has been proposed as a mechanism in the development of atherosclerotic plaque (Zilversmit, 1973). There is some evidence in humans and guinea pigs that high levels of LCFAs within the heart may predispose it to arrhythmias (Oliver *et al.*, 1968; Cowen and Vaughn-Williams, 1977). The possible role of LCFAs in causing some cases of pancreatitis was discussed earlier.

Ketogenesis in fasting should be viewed as an evolved mechanism with specific survival value for peripheral tissues and not a burden that the liver is placing on the rest of the body. It is important to remember that fasting animals should be expected to have a degree of ketonemia, ketonuria, and ketolactia. Thus, any disease condition that causes anorexia will usually be accompanied by increased ketone levels in body fluids that have no significance other than the fact that the animal has a subnormal caloric intake.

G. Diabetic Ketosis

Although diabetes mellitus is covered in more depth elsewhere in this book, no discussion of ketones would be complete without a mention of this disease. Diabetes is diagnosed more frequently in dogs and cats than other domestic species, and the ketoacidosis that occurs can be fatally severe and was discussed earlier under acid–base balance.

In experimental diabetes in dogs, plasma total ketone concentrations are 3.2 mmol/liter as compared with 0.1 mmol/liter in healthy dogs (Balasse *et al.*,

1985). Diabetes is accompanied by hyperglycemia, whereas most other ketotic syndromes occurring in domestic animals are usually accompanied by normoglycemia or hypoglycemia. The ketonemia in diabetes is due to increased lipolysis in adipose plus accelerated hepatic gluconeogenesis, both brought about by a lack of insulin. Thus, there are abundant plasma LCFAs as ketogenic substrates and metabolic conditions in the liver that favor ketone synthesis.

H. Ketosis Associated with Pregnancy and Lactation

These ketoses are most commonly observed in ruminants although they have been documented in dogs and humans. Before specific syndromes are discussed, a general picture of ketogenesis in pregnancy and lactation is presented.

Fetal demands for glucose are high, and the placenta can transport glucose from maternal to fetal plasma (Setchell *et al.*, 1972; Warnes *et al.*, 1977). When an imbalance occurs between the maternal ability to synthesize or absorb glucose and fetal consumption, hypoglycemia results. Under these circumstances, hypoglycemia will lead to lipolysis in adipose tissue and release of LCFAs as discussed earlier. The LCFAs will be taken up by the liver and converted to ketones with resulting ketosis.

Ketosis in lactation is somewhat more complex than ketosis occurring during pregnancy. The volume of milk produced is almost totally dependent on the rate of lactose synthesis by the mammary gland because milk volume formation is an osmotic phenomenon, and lactose is the predominant molecular species in milk (Peaker, 1977). There is virtually only one precursor of lactose and that precursor is plasma glucose (Kleiber *et al.*, 1955; Bickerstaffe *et al.*, 1974). Therefore, a female that is in heavy lactation will have a heavy drain on plasma glucose. The two sources of plasma glucose are absorption from the gut and gluconeogenesis.

In ruminants, little glucose is absorbed from the gut, so the overwhelming bulk of it is synthesized (Lindsay, 1959; Otchere *et al.*, 1974). Most (approximately 90%) of this synthesis occurs in the liver with the remainder occurring in the kidney (Bergman, 1982). The chief substrates are propionate and amino acids, with the former being most important in animals on a high-grain diet. Other precursors are branched-chain VFAs and lactate absorbed from the rumen and glycerol released during lipolysis (Bergman, 1975). If there is a mismatch between mammary drain of glucose for lactose synthesis and gluconeogenesis in the liver, hypoglycemia will result. Under these circumstances, hypoglycemia will lead to ketosis as explained in the discussion on fasting ketosis.

1. *Bovine Ketosis*

Bovine ketosis is actually at least three different syndromes that occur in cows during lactation (Kronfeld, 1980; Kronfeld *et al.*, 1983). The syndromes are characterized by anorexia, depression (usually), ketonemia, ketolactia, ketonuria, hypoglycemia, and decreased milk production. The three syndromes are underfeeding ketosis, alimentary ketosis, and spontaneous ketosis.

Underfeeding ketosis occurs when a dairy cow receives insufficient calories to meet lactational demands plus body maintenance. This version of ketosis can be conveniently divided into nutritional underfeeding ketosis and secondary (or complicated) ketosis. The former occurs when the cow has a normal appetite but is given an insufficient quantity of feed or a diet with low metabolic energy density. The latter occurs when a cow has some other disease, such as hypocalcemia, mastitis, or metritis, which suppresses appetite and causes the cow to consume insufficient nutrients. In most respects, underfeeding ketosis resembles starvation ketosis as explained earlier, except that there is the additional caloric and glycemic burden of milk production.

Alimentary ketosis occurs when cattle have been fed spoiled silage that contains excessive amounts of butyric acid (Brouwer and Kijkstra, 1938; Adler *et al.*, 1958). As discussed previously, the rumen epithelium has a high capacity to activate butyrate to acetoacetate and 3-hydroxybutyrate. Under conditions where excessive butyrate is presented to the rumen epithelium, large amounts of 3-hydroxybutyrate will be produced and released to the circulation with resulting ketosis. Alimentary ketosis then is really butyrate toxicosis.

Spontaneous ketosis is probably the most common, the most researched, the most controversial, and the least understood form of bovine ketosis. It occurs in high-producing dairy cows that are near the peak of lactation, that have access to abundant high-quality feed, and that have no other disease (Kronfeld, 1980; Baird, 1982). The disease is not accompanied by severe acidosis (Sykes *et al.*, 1941) and spontaneous recovery is common although there is a large decrease in milk production (Kronfeld, 1980; Baird, 1982). Several schemes have been proposed for the molecular pathogenesis of the syndrome. As these schemes are discussed, we will see that they are not necessarily mutually exclusive, and more than one of them may be correct and present simultaneously in the same animal.

The most widely accepted theory of bovine ketosis is the hypoglycemia theory (Baird, 1982). In this theory, hypoglycemia is the driving force in the syndrome and ultimately causes the ketonemia. Dairy cows are selected for remaining in the herd more for milk production than for any other factor. Thus, dairy cows have been selected for many generations to have a metabolically aggressive mammary gland. This selection criterion has dictated that the mammary produce a maximum amount of milk with secondary regard for the metabolic consequences for the rest of the animal. It is not surprising, therefore, that occasionally the mammary gland might withdraw glucose from the plasma more rapidly than the liver can resupply it, which leads to hypoglycemia even in a well-fed animal. The hypoglycemia will lead to ketonemia by mechanisms discussed earlier and below. The hypoglycemia and ketonemia may cause the cow to be ill enough that she will decrease her feed intake. At this point, the syndrome will resemble underfeeding ketosis.

As explained earlier, high milk production equates to a high rate of plasma glucose utilization by the mammary gland, which equates to a high rate of hepatic gluconeogenesis. In a lactating cow, plasma glucose concentration represents the balance point between hepatic glucose production and peripheral glucose utilization, with the mammary gland being the chief user. If peripheral glucose utilization should leap ahead of hepatic glucose production, hypoglycemia will result. In theory, hypoglycemia under these circumstances should lead to a decrease in plasma insulin and an increase in plasma glucagon levels. Lower plasma insulin and higher plasma glucagon should increase the activity of hormone sensitive lipase in adipose tissue, which will lead to increased plasma levels of LCFAs. Consequently, more LCFAs will reach the liver and exceed its capacity to oxidize them completely or to reesterify them, and increased ketogenesis will result.

What evidence supports this theory? First, the vast majority of cows with clinical spontaneous ketosis are indeed hypoglycemic (Baird *et al.*, 1968; Schwalm and Schultz, 1976; Gröhn *et al.*, 1983). Second, cows with spontaneous ketosis usually are hypoinsulinemic (Hove, 1974; Schwalm and Schultz, 1976). Third, compared to the prelactation period, postparturient dairy cows have been found to have elevated levels of plasma immunoreactive glucagon (Manns, 1972; De Boer *et al.*, 1985), which is even greater in cows with ketosis (Sakai *et al.*, 1993). Fourth, ketotic cows have elevated levels of plasma LCFAs (Baird *et al.*, 1968; Ballard *et al.*, 1968; Schwalm and Schultz, 1976).

Some investigation of molecular mechanisms of ketogenesis in the liver ketotic cows has been performed (Baird *et al.*, 1968; Ballard *et al.*, 1968). In particular, there has been interest in hepatic mitochondrial oxaloacetate levels. In the discussion of ketogenesis presented earlier, it was noted that when increased levels of plasma LCFAs occur, the liver can reesterify them or can oxidize them to acetyl-CoA. The acetyl-CoA can be oxidized to carbon dioxide provided there is sufficient oxaloacetate to permit entry into the citric acid cycle as citrate. In order for the citric acid cycle to operate, there must also be a sufficient amount of ADP available for phosphorylation as well or accumulation of NADH will slow the cycle. If acetyl-CoA accumulates, the excess will be diverted into ketogenesis.

Two studies have attempted to investigate oxaloacetate concentrations in the livers of ketotic cows (Baird *et al.*, 1968; Ballard *et al.*, 1968). Different methodologies were used to estimate oxaloacetate concentrations, and one study (Ballard *et al.*, 1968) concluded that there was no change in oxaloacetate concentration during ketosis, and the other concluded that oxaloacetate concentrations were lower in ketotic than in healthy cows (Baird *et al.*, 1968). Actually, both studies measured total hepatic oxaloacetate rather than mitochondrial oxaloacetate, which may be critical in ketogenic control. However, there has been no evidence to indicate that the ruminant liver should be any different from the nonruminant liver with regard to the concept that if the liver is presented with sufficient LCFA, ketogenesis will result. Research on the control of lipolysis in adipose in ruminants is insufficient. In particular, research has been insufficient in differences in plasma levels of lipogenic and lipolytic hormones and sensitivity of adipose to these hormones in cow populations that are susceptible and nonsusceptible to ketosis. No matter how low mitochondrial oxaloacetate levels might be in the liver, ketogenesis will not occur at a significant rate without sufficient precursor in the form of LCFAs and, conversely, ketogenesis could occur with normal oxaloacetate levels if the liver were presented with a sufficiently high concentration of LCFA.

Dairy cattle, however, can become ketonemic without the presence of significant hypoglycemia (Ballard *et al.*, 1968; Gröhn *et al.*, 1983). This is often the case with subclinical ketosis in which ketonemia exists without other signs of ketosis. It has been postulated that there is a lipolytic signal of unknown identity for lipolysis to meet mammary demand for LCFAs that is independent of plasma glucose concentration (Kronfeld, 1982; Kronfeld *et al.*, 1983). The increased plasma LCFAs lead directly to increased hepatic ketogenesis.

When it was first observed that glucocorticoids appeared to be an effective treatment for spontaneous ketosis, the hypothesis was made that the disease was due to adrenal cortical insufficiency (Shaw, 1956). This theory has fallen into disfavor because ketotic cows have been shown to have higher plasma levels of glucocorticoids than healthy cows (Robertson *et al.*, 1957). Glucocorticoids are efficacious and probably have their effect by stimulating proteolysis and inhibiting glucose use in muscle, thereby providing gluconeogenic precursors and glucose (Ryan and Carver, 1963; Bassett *et al.*, 1966; Robertson, 1966; Braun *et al.*, 1970; Reilly and Black, 1973).

The efficacy of glucose or glucose precursors as ketosis treatments favors the hypoglycemic theory. Parenteral glucose provides nearly immediate relief although relapses are common (Kronfeld, 1980). Gluconeogenic precursors, such as propylene glycol, glycerol, and sodium propionate, have been shown to be efficacious (Schultz, 1952; Simesen, 1956; Emery *et al.*, 1964; Kauppinen and Gröhn, 1984). Treatment of cows with bovine somatotropin in one lactation appears to decrease the likelihood of ketosis in the next lactation (Lean *et al.*, 1994). Cows treated with somatotropin appear to have less body fat and more skeletal muscle, so after calving, there is less fat to mobilize to LCFAs and more protein to mobilize as glucose precursor. Therefore, hypoglycemia and subsequent fatty acidemia and ketonemia are less likely to occur.

2. *Ovine Pregnancy Toxemia*

This syndrome occurs in pregnant ewes that are carrying more than one fetus and that have been subjected to caloric deprivation or stress. Because of intense genetic selection for twinning, the syndrome is, to a large extent, a man-made disease. Susceptibility increases as ewes approach term because fetal glucose demands increase with increasing body size. The ovine placenta is capable of extracting glucose from maternal plasma at concentrations below 1 mmol/liter and readily does so. It might seem biologically useless for the fetuses to cause a fatal hypoglycemia in the ewe, which will also lead to their own demise, but the fetuses are highly dependent on glucose as a caloric and synthetic source and would expire without it anyway.

Fetal lambs normally maintain a very low plasma glucose concentration of approximately 0.6 mmol/liter compared to 2.7 mmol/liter in a ewe (Warnes *et al.*, 1977). Thus, the transplacental glucose gradient greatly favors movement from dam to fetus. Curiously, the most concentrated carbohydrate in fetal sheep plasma is fructose (5.1 mmol/liter), which is synthesized from glucose in the placenta by reducing glucose to sorbitol followed by oxidation to fructose (Hers, 1960; Warnes *et al.*, 1977). Despite the abundance of fructose in the plasma of the fetal sheep, glucose constitutes its primary energy supply (Warnes *et al.*, 1977; Lindsay and Pethick, 1983), and the fetuses normally consumed 60–70% of maternal glucose production (Setchell *et al.*, 1972; Prior and Christenson, 1978).

The ovine placenta appears to have a low permeability for acetoacetate. When acetoacetate loads have been infused into pregnant sheep, the concentrations in fetal blood have remained low. Further, *in vitro* experiments with perfused sheep placenta have also demonstrated a low permeability for acetoacetate (Alexander *et al.*, 1966, 1969). Thus, it appears that maternal acetoacetate, and perhaps 3-hydroxybutyrate, cannot be a major energy source for the ovine fetus.

The disease is characterized by depression and weakness in the ewes, which is associated with hypoglycemia, ketonemia, and ketonuria (Reid, 1968). The ketonemia is severe enough to cause acidosis, which can be severe (Holm, 1958; Reid, 1968). There is also considerable fatty deposition in the liver to the extent that it may interfere with liver function (Snook, 1939; Cornelius *et al.*, 1958). Eventually, the ewes are unable to rise, become comatose, and die if untreated.

Mild cases respond to intravenous glucose, glucocorticoids, glucose precursors such propylene glycol, or glycerol coupled with removal of stress and improved nutrition (McClymont and Setchell, 1955a, 1955b; Thompson, 1956). Severe cases, in which the ewes are unable to rise, usually respond only to delivery of the lambs and even then a high mortality will occur (Holm, 1958; Reid, 1968).

3. *Syndromes in Other Species*

Ketosis associated with lactation can occur in dairy goats (Morand-Fehr *et al.*, 1984). The syndrome has also been reported in beef cows with caloric deprivation and nursing two calves (Khan *et al.*, 1986). Pregnancy toxemia has been reported in goats carrying multiple fetuses (Rindsig, 1980; East, 1983; Thedford, 1983; Morand-Fehr *et al.*, 1984). The syndrome can be produced with calorie deprivation, particularly if coupled with stress, and almost always occurs in does carrying more that one fetus. Obesity also may be a predisposing factor in does (Thedford, 1983; Morand-Fehr *et al.*, 1984). Generally, the syndrome in does appears entirely similar to that in ewes.

Pregnancy toxemia has been reported in beef cows in the last two months of gestation (Sampson *et al.*, 1945; Kingrey *et al.*, 1957; Caple *et al.*, 1977; Tyler *et al.*, 1994). The disease occurs predominantly in cows that are carrying twins. The cows may be in good or even obese body condition, but sudden food deprivation or decrease in quality or imposition of stress such as water deprivation may precipitate the syndrome. The disease resembles pregnancy toxemia in sheep in most respects. Pregnancy toxemia has been reported in pregnant bitches (Irvine, 1964; Jackson *et al.*, 1980) and appears similar to the disease in sheep. Hypoglycemia is severe in canine cases, and the animals respond readily to intravenous glucose. If the animals will eat a carbohydrate-containing diet, a relapse is unlikely; otherwise; removal of the fetuses is required for a cure. Pregnancy toxemia occurs in pregnant guinea pigs and, like in pregnant ewes, the syndrome can be precipitated by inadequate calories and stress (Bergman and Sellers, 1960; Wagner, 1976). The syndrome in guinea pigs is quite similar to that in sheep. There is marked ketonemia and acidosis, and the animals become weak and depressed with eventual coma (Wagner, 1976).

I. Postexercise Ketosis

Postexercise ketosis, which was first documented in 1909 (Forssner, 1909), has been investigated most extensively in humans and rats. Neither trained nor untrained humans or rats show much increase in ketones during exercise, but only untrained individuals exhibit a significant ketonemia and ketonuria after exercise (Johnson *et al.*, 1969; Winder *et al.*, 1975; Koeslag, 1982). The experiments of Winder *et al.* (1975) demonstrated a greater enzymatic capacity of muscles of trained rats to catabolize ketones. It also appears that trained athletes have a greater capacity to oxidize LCFAs in muscle than nonathletes (Johnson *et al.*, 1969). A high-carbohydrate diet in conjunction with training also decreases the magnitude of postexercise ketosis (Koeslag *et al.*, 1980).

From the foregoing, it appears that a number of factors are involved in postexercise ketosis. During exercise all forms of fuel, including LCFAs, ketones, and glucose, are oxidized. Postexercise there is a diminution of LCFA release, although plasma LCFA concentrations decrease little at first because of an even greater diminution in LCFA oxidation, and more LCFA may be converted to ketones. Ketone oxidation by muscle is decreased postexercise, which will allow ketones to accumulate. In the postexercise period, there is gluconeogenesis as lactate is cycled back into glucose and glycogen, which may lead to decreased mitochondrial oxaloacetate levels and increased ketogenesis. Finally, compared to the exercise period, in the postexercise period, relatively more of the cardiac output will flow through the portal system, and the rate at which LCFAs are presented to the liver may increase.

Postexercise ketosis undoubtedly occurs in most mammalian species but, among the domestic species, has been best documented in dogs and horses. Postexercise increases in plasma levels of ketones have been observed in racing sled dogs (Hammel *et al.*, 1977). Postexercise ketosis has been reported several times in the horse (Dybdal *et al.*, 1980; Lucke and Hall, 1980; Rose and Sampson, 1982). In these studies, horses were subjected to endurance rides of 80 to 160 km. Plasma 3-hydroxybutyrate concentrations increased two- to threefold 5 to 60 minutes postexercise compared to preexercise levels. Plasma LCFA concentrations increase fivefold or more in horses during exercise (Lucke and Hall, 1980; Rose and Sampson, 1982) and decrease little during 30 to 60 minutes postexercise. Thus, abundant LCFAs are available to the liver postexercise when muscle utilization of LCFAs and ketones is decreased, a situation that results in ketonemia.

References

Adler, J. A., Roberts, S. J., and Dye, J. A. (1958). *Am. J. Vet. Res.* **19,** 314–318.

Alexander, C. A., Hamilton, R. L., and Havel, R. J. (1976). *J. Cell Biol.* **69,** 241–263.

Alexander, D. P., Britton, H. G., and Nixon, D. A. (1966). *J. Physiol.* **186,** 100P–101P.

Alexander, D. P., Britton, H. G., Cohen, N. M., and Nixon, D. A. (1969). *In* "Foetal Autonomy" (C. E. W. Wolstenholme and M. O'Connor, eds.), pp. 95–116. J & A Churchill, London.

Baetz, A. L., and Pearson, J. E. (1972). *Am. J. Vet. Res.* **33,** 1941–1946.

Baird, G. D. (1982). *J. Dairy Sci.* **65,** 1–10.

Baird, G. D., Hibbit, K. G., Hunter, G. D., Lund, P., Stubbs, M., and Krebs, H. A. (1968). *Biochem. J.* **107,** 683–689.

Baird, G. D., Hibbit, K. G., and Lee, J. (1970). *Biochem. J.* **117,** 703–709.

Balasse, E. O. (1970). *Europ. J. Clin. Invest.* **1,** 155–160.

Balasse, E., Courturier, E., and Franckson, J. (1967). *Diabetologia* **3,** 488–493.

Balasse, E. O., De Graef, J., and Neef, M. A. (1985). *Horm. Metabol. Res.* **17,** 554–558.

Ballard, F. J., Hanson, R. W., Kronfeld, D. S., and Raggi, F. (1968). *J. Nutr.* **95,** 160–172.

Barrie, J., Watson, T. D. G., Stear, M. J., and Nash, A. S. (1993). *J. Small. Anim. Pract.* **34,** 507–512.

Bass, V. D., Hoffmann, W. E., and Dorner, J. L. (1976). *Am. J. Vet. Res.* **37,** 1355–1357.

Bassett, J. M., Mills, S. C., and Reid, R. L. (1966) *Metab.* **15,** 922–932.

Bauchart, D. (1993). *J. Dairy Sci.* **76,** 3864–3881.

Bauer, J. E. (1983). *Am. J. Vet. Res.* **44,** 379–384.

Bauer, J. E., Meyer, D. J., Goring, R. L., Beauchamp, C. H., and Jones, J. (1989) *In* "Nutrition of the Dog and Cat" (I. H. Burger and J. P. W. Rivers, eds.), pp. 343–352. Cambridge University Press, New York.

Baum, D., Schweid, A. I., Porte, D., and Bierman, E. L. (1969). *Proc. Soc. Exp. Biol. Med.* **131,** 183–185.

Bauman, C. E., and Davis, C. L. (1975). *In* "Digestion and Metabolism in the Ruminant" (I. W. McDonald and A. C. I. Warner, eds.), pp. 496–509. University of New England Publishing Unit, Armidale, New South Wales.

Baverel, G., Forissier, M., and Pellet, M. (1982). *In* "Biochemistry of Kidney Functions" (F. Morel, ed.), pp. 177–185. Elsevier, Amsterdam.

Bergman, E. N. (1975). *In* "Digestion and Metabolism in the Ruminant" (I. W. McDonald and A. C. I. Warner, eds.), pp. 292–305.

University of New England Publishing Unit, Armidale, New South Wales.

Bergman, E. N. (1982). *In* "Comparative Aspects of Nutritional and Metabolic Diseases" (J. C. Woodard and M. Bruss, eds.), pp. 1–23. CRC Press, Boca Raton, Florida.

Bergman, E. N., and Sellers, A. F. (1960). *Am. J. Physiol.* **198,** 1083–1086.

Bergman, E. N., and Wolfe, J. E. (1971). *Am. J. Physiol.* **221,** 586–592.

Bergman, E. N., Reid, R. S., Murray, M. G., Brockway, J. M., and Whitelaw, F. G. (1965). *Biochem. J.* **97,** 53–58.

Bickerstaffe, R., Annison, E. F., and Linzell, J. L. (1974). *J. Agric. Sci. Camb.* **82,** 71–85.

Binkiwicz, A., Sadeghi-Nejad, A., Hochman, H., Loridan, L., and Senior, B. (1974). *J. Pediatr.* **84,** 226–231.

Black, A. L., Luick, J. R., Lee, S. L., and Knox, K. (1972). *Am J. Physiol.* **222,** 1575–1580.

Blomhoff, J. P., Holme, R., and Östrem, J. (1978). *Scand. J. Gastroenterol.* **13,** 693–702.

Braun, R. K., Bergman, E. N., and Albert, T. F. (1970). *J. Am. Vet. Med. Assoc.* **157,** 941–946.

Brindle, N. P. J., Zammit, V. A., and Pogson, C. I. (1985). *Biochem. J.* **232,** 177–182.

Brindley, D. N. (1991). *In* "Biochemistry of Lipids, Lipoproteins and Membranes" (D. E. Vance and J. Vance, eds.), pp. 171–203. Elsevier, New York.

Brockman, R. P. (1976). *Can. J. Comp. Med.* **40,** 166–170.

Brooks, K. D. (1989). *Compan. Anim. Pract.* **19,** 5–9.

Brouwer, E., and Kijkstra, N. D. (1938). *J. Agric. Sci.* **28,** 695.

Brown, M. S., and Goldstein, J. L. (1990). *In* "Goodman and Gilman's The Pharmacological Basis of Therapeutics" (A. G. Gilman, T. W. Rall, A. S. Nies, and P. Taylor, eds.), 8th ed., pp. 874–896. Pergamon, New York.

Brown, M. S., and Goldstein, J. L. (1994). *In* "Harrison's Principles of Internal Medicine" (K. J. Isselbacher, E. Braunwald, J. D. Wilson, J. B. Martin, A. S. Fauci, and D. L. Kaspar, eds.), 13th ed., Vol. 2, pp. 2058–2069. McGraw-Hill, New York.

Burnett, D. A., Lysenko, N., Manning, J. A., and Ockner, R. K. (1979). *Gastroenterology* **77,** 241–249.

Bush, R. S., and Milligan, L. P. (1971). *Can. J. Anim. Sci.* **51,** 129–133.

Cameron, J. L., Capuzzi, D. M., Zudema, G. D., and Margolis, S. (1974). *Am. J. Med.* **56,** 482–487.

Caple, I. W., Pemberton, D. H., Harrison, M. A., and Halpein, C. G. (1977). *Austral. Vet. J.* **53,** 289–291.

Caraway, W. T., and Kammeyer, C. W. (1972). *Clin. Chim. Acta* **41,** 395–434.

Chapman, M. J. (1986). *In* "Methods in Enzymology. Plasma Lipoproteins. Part A. Preparation, Structure, and Molecular Biology" (J. P. Segrest and J. J. Albers, eds.), Vol. 128, pp. 70–143. Academic Press, San Diego.

Cook, H. W. (1991). *In* "Biochemistry of Lipids, Lipoproteins and Membranes" (D. E. Vance and J. Vance, eds.), pp. 141–169. Elsevier, New York.

Cook, R. M., Liu, S. C., and Quraishi, S. (1969). *Biochem.* **8,** 2966–2969.

Cornelius, C. E., Holm, L. W., and Jasper, D. E. (1958). *Cornell Vet.* **48,** 305–312.

Cowen, P. J., and Vaughn-Williams, E. N. (1977). *J. Molec. Cell. Cardiol.* **9,** 327–342.

Crook, D., and Seed, M. (1990). *Bailliere's Clin. Endocrinol. Metab.* **4,** 851–875.

Csako, G. (1987). *Clin. Chem.* **33,** 289–292.

Danielsson, B., Ekman, R., Johansson, B. G., and Petersson, B. G. (1977). *Clin. Chim. Acta* **80,** 157–170.

Davis, R. A. (1991). *In* "Biochemistry of Lipids, Lipoproteins and Membranes" (D. E. Vance and J. Vance, eds.), pp. 401–426. Elsevier, New York.

De Boer, G., Trenkle, A., and Young, J. W. (1985). *J. Dairy Sci.* **68,** 326–337.

Dean, J. A. (1985). "Lange's Handbook of Chemistry," 13th ed., Chap. 7, pp. 84, 443. McGraw-Hill, New York.

DeBowes, L. J. (1987). *Compend. Contin. Educ. Pract. Vet.* **9,** 727–736.

Degen, A. J. M., and Van der Vies, J. (1985). *Scan J. Clin. Invest.* **45,** 283–285.

Demacker, P. N., Hijmans, A. G., and Jansen, A. P. (1982). *Clin. Chem.* **28,** 1765–1768.

Demacker, P. N. M., van Heijst, P. J., Hak-Lemmers, H. L. M., and Stalenhoef, A. F. H. (1987). *Atherosclerosis* **66,** 113–123.

Duee, P. H., Pegorier, J. P., Quant, P. A., Herbin, C., Kohl, C., and Girar, J. (1994). *Biochem. J.* **298,** 207–212.

Duncombe, W. G. (1964). *Clin. Chim. Acta* **9,** 122–125.

Dybdal, N. O., Gribble, D., Madigan, J. E., and Stabenfeldt, G. H. (1980). *Equine Vet. J.* **12,** 137–140.

East, N. (1983). *Vet. Clin. N. Am. Large Anim. Pract.* **5,** 601–618.

Edwards, D. F. (1982). *J. Am. Vet. Med. Assoc.* **180,** 68–70.

Edwards, P. A. (1991). *In* "Biochemistry of Lipids, Lipoproteins and Membranes" (D. E. Vance and J. Vance, eds.), pp. 383–401. Elsevier, New York.

Elphick, M. C. (1968). *J. Clin. Path.* **21,** 567–570.

Emery, R. S., Burg, N., Brown, L. D., and Blank, G. N. (1964). *J. Dairy Sci.* **47,** 1074–1079.

Emmanuel, B. (1980). *Comp. Biochem. Physiol.* **65B,** 699–704.

Eriksen, L., and Simesen, M. G. (1970). *Nord. Vet.-Med.* **22,** 273–284.

Faulkner, A., and Pollock, H. T. (1991). *Comp. Biochem. Physiol.* **98B,** 283–286.

Faust, J. R., Trzaskos, J. M., and Gaylor, J. L. (1988). *In* "Biology of Cholesterol" (P. L. Yeagle, ed.), pp. 39–69. CRC Press, Boca Raton, Florida.

Feldman, E. C. (1995). *In* "Textbook of Veterinary Internal Medicine" (S. J. Ettinger and E. C. Feldman, eds.), 4th ed., Vol. 2, pp. 1538–1578. Saunders, Philadelphia, Pennsylvania.

Fell, B. F., and Weekes, T. E. C. (1975). *In* "Digestion and Metabolism in the Ruminant" (I. W. McDonald and A. C. I. Warner, eds.), pp. 101–118. University of New England Publishing Unit, Armidale, New South Wales.

Ferreri, L. F., and Elbein, R. C. (1982). *J. Dairy Sci.* **65,** 1912–1920.

Fielding, P. E., and Fielding, C. J. (1991). *In* "Biochemistry of Lipids, Lipoproteins and Membranes" (D. E. Vance and J. Vance, eds.), pp. 427–459. Elsevier, New York.

Ford, R. B. (1993). *J. Small Anim. Pract.* **34,** 488–492.

Ford, R. B. (1995). *In* "Textbook of Veterinary Internal Medicine" (S. J. Ettinger and E. C. Feldman, eds.), 4th ed., Vol. 2, pp. 1414–1419. Saunders, Philadelphia, Pennsylvania.

Forssner, G. (1909). *Skan. Arch. Physiol.* **22,** 393–405.

Free, H. M., Smeby, R. R., Cook, M. H., and Free, A. H. (1958). *Clin. Chem.* **4,** 323–330.

Freestone, J. F., Wolfsheimer, K. J., Ford, R. B., Church, G., and Bessin, R. (1991). *J. Vet. Int. Med.* **5,** 15–22.

Gainer, J. H., Amster, R. L., Needham, J. W., and Schilling, K. F. (1966). *Am. J. Vet. Res.* **27,** 1611–1621.

Gibbons, G. F. (1990). *Biochem. J.* **268,** 1–13.

Glassock, R. J., Adler, S. G., Ward, H. J., and Cohen, A. H. (1991). *In* "The Kidney" (B. M. Brenner and G. V. Rector, eds.), 4th ed., Vol. 1, pp. 1182–1279. Saunders, Philadelphia, Pennsylvania.

Goodridge, A. G. (1991). *In* "Biochemistry of Lipids, Lipoproteins and Membranes" (D. E. Vance and J. Vance, eds.), pp. 111–139. Elsevier, New York.

Gotto, A. M., Pownall, H. J., and Havel, R. J. (1986). *In* "Methods in Enzymology. Plasma Lipoproteins. Part A. Preparation, Structure, and Molecular Biology" (J. P. Segrest and J. J. Albers, eds.), Vol. 128, pp. 3–41. Academic Press, San Diego.

Greenberger, N. J. (1973). *New Engl. J. Med.* **289,** 586–587.
Gröhn, Y. (1985). *Res. Vet. Sci.* **39,** 24–28.
Gröhn, Y., Lindberg, L.-A., Bruss, M. L., and Farver, T. B. (1983). *J. Dairy Sci.* **66,** 2320–2328.
Gurr, M. I., and Harwood, J. L. (1991). "Lipid Biochemistry, An Introduction," 4th ed., pp. 23–33, 69–95, 148–157, 199–220, 307–315. Chapman and Hall, New York.
Hadlow, W. J. (1962). *In* "Comparative Neuropathology" (J. R. M. Innes and L. Z. Saunders, eds.), pp. 147–243. Academic Press, New York.
Haffner, S. M., and Valdez, R. A. (1995). *Am. J. Med.* **98,** 40S–47S.
Hagenfeldt, L., and Wahren, J. (1968a). *Scand. J. Clin. Lab Invest.* **21,** 263–276.
Hagenfeldt, L., and Wahren, J. (1968b). *Scand. J. Clin. Lab Invest.* **21,** 314–320.
Hall, L. M. (1961). *Biochem. Biophys. Res. Comm.* **6,** 177–179.
Hammel, E. P., Kronfeld, D. S., Ganjam, V. K., and Dunlap, H. L. (1977). *Am. J. Clin. Nutr.* **30,** 409–418.
Hanson, R. W., and Ballard, F. J. (1967). *Biochem. J.* **105,** 529–536.
Harano, Y., Ohtsuki, M., Ida, M., Kojima, H., Harada, M., Okanishi, T., Kashiwagi, A., Ochi, Y., Uno, S., and Shigeta, Y. (1985). *Clin. Chim. Acta* **151,** 177–183.
Hardy, R. M. (1992). *In* "Veterinary Gastroenterology" (N. V. Anderson, ed.), 2nd ed., pp. 275–292. Lea & Febiger, Philadelphia, Pennsylvania.
Havel, R. J. (1969). *Adv. Intern. Med.* **15,** 117–154.
Hawkins, R. A., Williamson, D. H., and Krebs, H. A. (1971). *Biochem. J.* **122,** 13–18.
Heitman, R. N., Sensenig, S. C., Reynolds, C. K., Fernandez, J. M., and Dawes, D. J. (1986). *J. Nutr.* **116,** 2516–2524.
Henry, R. J., Cannon, D. C., and Winkelman, J. W. (1974). "Clinical Chemistry: Principles and Techniques" 2nd ed. pp. 1354–1369. Harper and Row, Hagerstown, Maryland.
Hers, H. G. (1960). *Biochim. Biophys. Acta* **37,** 127–138.
Hibbitt, K. G., Neill, D., and Radford, R. (1969). *Res. Vet. Sci.* **10,** 245–253.
Holm, L. W. (1958). *Cornell Vet.* **48,** 348–357.
Hove, K. (1974). *Acta Endocrin.* **76,** 513–524.
Hubert, B., de La Farge, F., Braun, J. P., and Magnol, J. P. (1987). *Companion Anim. Pract.* **1,** 33–35.
Ingle, D. L., Bauman, D. E., and Garrigus, U. S. (1972a). *J. Nutr.* **102,** 609–616.
Ingle, D. L., Bauman, D. E., and Garrigus, U. S. (1972b). *J. Nutr.* **102,** 617–624.
Irvine, C. H. G. (1964). *New Zealand Vet. J.* **12,** 140–143.
Jackson, R. F., Bruss, M. L., Growney, P. J., and Seymour, W. G. (1980). *J. Am. Vet. Med. Assoc.* **177,** 1123–1127.
Johnson, R. H., Walton, J. L., Krebs, H. A., and Williamson, D. H. (1969). *Lancet* **2,** 1383–1385.
Jones, B. R. (1993). *J. Small Anim. Pract.* **34,** 493–499.
Jones, B. R. (1995). *In* "Textbook of Veterinary Internal Medicine" (S. J. Ettinger and E. C. Feldman, eds.), 4th ed., Vol. 2, pp. 1410–1414. Saunders, Philadelphia, Pennsylvania.
Jones, B. R., Wallace, A., Harding, D. R. K., Hancock, W. S., and Campbell, C. H. (1983). *Vet. Rec.* **112,** 543–547.
Jones, B. R., Johnstone, A. C., Cahill, J. I., and Hancock, W. S. (1986). *Vet. Rec.* **119,** 268–272.
Jones, C. S., and Parker, D. S. (1978). *Biochem. J.* **174,** 291–296.
Jones, M. D., Burd, L. I., Makowski, E. L., Meschia, G., and Battaglia, F. C. (1975). *Am. J. Physiol.* **229,** 235–239.
Kalapos, M. P., Mandl, J., Banhegyi, G., Antoni, F., and Garzo, T. (1994). *Int. J. Biochem.* **26,** 1069–1079.
Katz, M. L., and Bergman, E. N. (1969). *Am. J. Physiol.* **216,** 953–960.
Kaufman, C. F., and Bergman, E. N. (1971). *Am. J. Physiol.* **221,** 967–972.
Kaufman, C. F., and Bergman, E. N. (1974). *Am. J. Physiol.* **226,** 827–832.
Kauppinen, K., and Gröhn, Y. (1984). *Acta Vet. Scand.* **25,** 467–479.
Khan, M. A. S., Topps, J. H., Broadbent, P. J., and Stephen, N. H. (1986). *Proc. 6th Int. Conf. Prod. Dis. Farm Anim.*, pp. 39–42.
Kilponen, J. M., Palosaari, P. M., Sormunen, R. T., Vihinen, M., and Hiltunen, J. K. (1991). *In* "New Developments in Fatty Acid Oxidation" (P. M. Coates and K. Tanaka, eds.), pp. 33–40. Wiley-Liss, New York.
Kingrey, B. W., Ladwig, V. D., Monlux, W. S., and Ramsey, F. K. (1957). *N. Am. Vet.* **38,** 321–328.
Kinsella, J. E. (1970). *Biochim. Biophys. Acta* **210,** 28–38.
Kleiber, M., Black, A. L., Brown, M. A., Baxter, C. F., Luick, J. L., and Stadtman, F. H. (1955). *Biochim. Biophys. Acta* **17,** 252–260.
Klotzsch, S. G., and McNamara, J. R. (1990). *Clin. Chem.* **36,** 1605–1613.
Koeslag, J. H. (1982). *Med. Sci. Sports Exercise* **14,** 327–334.
Koeslag, J. H., Noakes, T. D., and Sloan, A. W. (1980) *J. Physiol. (Lond).* **301,** 79–90.
Kopec, B., and Fritz, I. B. (1973). *J. Biol. Chem.* **248,** 4069–4074.
Kosugi, K. Scofield, R. F., Chandramouli, V., Kumaran, K., Schumann, W. C., and Landau, B. R. (1986). *J. Biol. Chem.* **261,** 3952–3957.
Krebs, H. A., Mellanby, J., and Williamson, D. H. (1962). *Biochem J.* **82,** 96–98.
Kronfeld, D. S. (1980). *In* "Bovine Medicine and Surgery" (H. E. Amstutz, ed.), Vol. 1, pp. 537–592. American Veterinary Publications, Santa Barbara, California.
Kronfeld, D. S. (1982). *J. Dairy Sci.* **65,** 2204–2212.
Kronfeld, D. S., Raggi, F., and Ramberg, C. F., Jr. (1968). *Am. J. Physiol.* **215,** 218–227.
Kronfeld, D. S., Donoghue, S., Naylor, J. M., Johnson, K., and Bradley, C. A. (1980). *J. Dairy Sci.* **63,** 545–552.
Kronfeld, D. S., Chalupa, L., and Sklan, D. (1983). *Anim. Nutr. Health* **38**(6), 28–33.
Lammerant, J., Huynh-Thu, T., and Kolanowski, J. (1985). *J. Molec. Cell. Cardiol.* **17,** 421–433.
Le Goff, D., Pastier, D., Hannan, Y., Petit, E., Ayrault-Jarrier, M., and Nouvelot, A. (1989). *Comp. Biochem. Physiol.* **93B,** 371–377.
Lean, I. J., Bruss, M. L., Troutt, H. F., Galland, J. C., Farver, T. B., Rostami, J., Holmberg, C. A., and Weaver, L. D. (1994). *Res. Vet. Sci.* **57,** 200–209.
Lewis, R. M., and Center, S. A. (1984). *In* "Canine Nephrology" (K. C. Bovée, ed.), pp. 461–479. Harwal Publishing, Media, Pennsylvania.
Liepa, G. U., Beitz, D. C., and Linder, J. R. (1978). *J. Nutr.* **108,** 1733–1739.
Lindsay, D. B. (1959). *Vet Rev. Annot.* **5,** 103–128.
Lindsay, D. B., and Pethick, D. W. (1983). *In* "Dynamic Biochemistry of Animal Production" (P. M. Riis, ed.), pp. 431–480. Elsevier, New York.
Lindsay, D. B., and Setchell, B. P. (1976). *J. Physiol.* **259,** 801–823.
Ling, G. V., Lowenstine, L. J., Pulley, T., and Kaneko, J. J. (1977). *J. Am. Vet. Med. Assoc.* **170,** 521–530.
Ling, G. V., Stabenfeldt, G. H., Comer, K. M., Gribble, D. H., and Schechter, R. D. (1979). *J. Am. Vet. Med. Assoc.* **174,** 1211–1215.
Linzell, D. B., Annison, E. F., Fazakerley, S., and Leng, R. A. (1967). *Biochem. J.* **104,** 34–42.
Little, J. R., Goto, M., and Spitzer, J. J. (1970). *Am. J. Physiol.* **219,** 1458–1463.
Liu, S., Tilley, L. P., Tappe, J. P., and Fox, P. R. (1986). *J. Am. Vet. Med. Assoc.* **189,** 227–232.
Lucke, J. N., and Hall, G. M. (1980). *Vet. Rec.* **106,** 405–407.
Luick, J. R., Black, A. L., Simesen, M. G., and Kronfeld, D. S. (1967). *J. Dairy Sci.* **50,** 544–549.

Lynen, F., Henning, U., Bublitz, C., Sorbö, B., and Kröplin-Rueff, L. (1958). *Biochem. Z.* **330,** 269–295.
McClymont, G. L., and Setchell, B. P. (1955a). *Austral. Vet. J.* **31,** 53–68.
McClymont, G. L., and Setchell, B. P. (1955b). *Austral. Vet. J.* **31,** 170–174.
McCullagh, K. G. (1978). *Vet. Ann.* **18,** 41–50.
MacDonald, M. L., Anderson, B. D., Rogers, Q. R., Buffington, C. A., and Morris, J. G. (1984). *Am. J. Vet. Res.* **45,** 1310–1317.
McGarry, J. D. (1979). *Diabetes* **28,** 517–523.
McGarry, J. D., and Foster, D. W. (1969). *J. Biol. Chem.* **254,** 4251–4256.
McGarry, J. D., Mannaerts, G. P., and Foster, D. W. (1977). *J. Clin. Invest.* **60,** 265–270.
McGowan, M. W., Artiss, J. D., Standbergh, D. R., and Zak, B. (1983). *Clin. Chem.* **29,** 538–542.
Mahley, R. W., and Weisgraber, K. H. (1974). *Circ. Res.* **35,** 713–721.
Maho, Y. L., Kha, H. V. V., Koubi, H. Dewasmes, G., Girard, J., Ferrè, P., and Cagnard, M. (1981). *Am. J. Physiol.* **241,** E342–E354.
Manning, P. J., Corwin, L. A., and Middleton, C. C. (1973). *Exp. Molec. Path.* **19,** 378–388.
Manns, J. G. (1972). *Can. Vet. J.* **13,** 151–155.
Martin-Sanz, P., Vance, J. E., and Brindley, D. N. (1990). *Biochem. J.* **271,** 575–583.
Mebane, D., and Madison, L. L. (1964). *J. Lab. Clin. Med.* **63,** 177–192.
Medaille, C., de La Farge, F., Braun, J. P., Valdiguie, P., and Rico, A. G. (1988). *In* "Animal Clinical Biochemistry: The Future" (D. J. Blackmore, ed.), pp. 287–290. Cambridge University Press, New York.
Mellanby, J., and Williamson, D. H. (1974). *In* "Methods of Enzymatic Analysis" (H. U. Bergmeyer, ed.), 2nd English ed., Vol. 4, pp. 1840–1843. Academic Press, Orlando, Florida.
Meyer, D. J., and Chiapella, A. M. (1985). *Vet. Clin. N. Am. Small Anim. Pract.* **15,** 215–227.
Miles, J. M., Haymond, M. W., and Gerich, J. E. (1981). *J. Clin. Endo. Metab.* **52,** 34–37.
Mogg, T. D., and Palmer, J. E. (1995). *J. Am. Vet. Med. Assoc.* **207,** 604–607.
Morand-Fehr, P., Bas, P., Hervieu, J., and Sauvant, D. (1984). *In* "Les Maladies de la Chèvre" (P. Yvore and G. Perrin, eds.), pp. 379–391. Institut National de la Recherche Agronomique, Versailles.
Morris, M. D., Zilversmit, D. B., and Hintz, H. F. (1972). *J. Lipid Res.* **13,** 383–389.
Mortensen, P. B. (1990). *In* "Fatty Acid Oxidation: Clinical, Biochemical, and Molecular Aspects" (K. Tanaka and P. M. Coates, eds.), pp. 249–264. Alan R. Liss, New York.
Nadeau, G. (1952). *Can. Med. Assoc. J.* **67,** 158–159.
Naylor, J. M., Kronfeld, D. S., and Acland, H. (1980). *Am. J. Vet. Res.* **41,** 899–905.
Newsholme, E. A., and Leach, A. R. (1983). "Biochemistry for the Medical Sciences" pp. 259–284. John Wiley & Sons, New York.
Nielsen, N. C., and Fleischer, S. (1969). *Science* **166,** 1017–1019.
Oliver, M. F., Kurien, V. A., and Greenwood, T. W. (1968). *Lancet* **1,** 710–715.
Otchere, E. O., McGilliard, A. D., and Young, J. D. (1974). *J. Dairy Sci.* **57,** 1189–1195.
Owen, O. E., and Reichard, G. A. (1971). *J. Clin. Invest.* **50,** 1536–1545.
Owen, O. E., Morgan, A. P., Kemp, H. G., Sullivan, J. M., Herrera, M. G., and Cahill, G. F. (1967). *J. Clin. Invest.* **46,** 1589–1595.
Owen, O. E., Felig, A. P., Morgan, J., and Cahill, G. F., Jr. (1969). *J. Clin. Invest.* **48,** 547–553.
Owen, O. E., Reichard, G. A., Trapp, V. E., Skutches, C. L., Mozzoli, M. A., Hoeldtke, R. D., and Boden, G. (1981) *Diabetes* **30,** 62A.
Owen, O. E., Trapp, V. E., Skutches, C. L., Mozzoli, M. A., Hoeldtke, R. D., Boden, G. and Reichard, G. A., (1982). *Diabetes* **31,** 242–248.
Ozand, P. T., Hawkins, R. L., Collins, R. M., Tildon, J. T., and Cornblath, M. (1975). *Biochem. Med.* **14,** 170–183.
Palmquist, D. L., Davis, C. L., Brown, R. E., and Sachan, D. S. (1969). *J. Dairy Sci.* **52,** 633–638.
Pande, S. V. (1971). *J. Biol. Chem.* **246,** 5384–5390.
Patterson, J. S., Rusley, M. S., and Zachary, J. F. (1985). *J. Am. Vet. Med. Assoc.* **186,** 499–503.
Paul, P., Issekutz, B., and Miller, H. I. (1966). *Am. J. Physiol.* **211,** 1313–1320.
Peaker, M. (1977). *Symp. Zool. Soc. Lond.* **41,** 113–134.
Pedersen, R. C. (1988). *In* "Biology of Cholesterol" (P. L. Yeagle, ed.), pp. 39–69. CRC Press, Boca Raton, Florida.
Pell, J. M., and Bergman, E. N. (1983). *Am. J. Physiol.* **244,** E282–E289.
Peritz, L. N., Brunzell, J. D., Harvey-Clarke, C., Pritchard, P. H., Jones, B. R., and Hayden, M. R. (1990). *Clin. Invest. Med.* **13,** 259–293.
Persson, B., Settergren, G., and Dahlquist, G. (1972). *Acta Paediatr. Scand.* **61,** 273–278.
Pethick, D. W., and Lindsay, D. B. (1982). *Brit. J. Nutr.* **48,** 549–563.
Prior, R. L., and Christenson, R. K. (1978). *J. Anim. Sci.* **46,** 201–209.
Procos, J. (1961). *Clin. Chem.* **7,** 97–106.
Puppione, D. L. (1983). *In* "Handbook of Electrophoresis" (L. A. Lewis and H. K. Naito, eds.), Vol. 4, p. 185. CRC Press, Boca Raton, Florida.
Quant, P. A., Tubbs, P. K., and Brand, M. D. (1990). *Europ. J. Biochem.* **187,** 169–174.
Ramasamy, S., Boissonneaudt, G. A., Decker, E. A., and Henning, B. (1991). *J. Biochem. Tox.* **6,** 29–35.
Ramsey, H. A., and Davis, C. L. (1965). *J. Dairy Sci.* **48,** 381–390.
Reid, R. L. (1968). *Adv. Vet. Sci.* **12,** 163–237.
Reilly, P. E. B., and Black, A. L. (1973). *Am. J. Physiol.* **225,** 689–695.
Rhodes, D. C., Meyer, D. J., Fish, C. J., and Kerns, W. D. (1992). *Vet. Clin. Path.* **21,** 34.
Richardson, M. (1989). *Canine Pract.* **19,** 33–37.
Rindsig, R. B. (1980). *Int. Goat Sheep Res.* **1,** 113–117.
Robertson, J. M. (1966). *J. Am. Vet. Med. Assoc.* **149,** 1620–1623.
Robertson, W. G., Lennon, H. D., Bailey, W. W., and Mixner, J. P. (1957). *J. Dairy Sci.* **40,** 732–738.
Roe, C. R., Millington, D. S., Norwood, D. L., Kodo, N., Sprecher, H., Mohammed, B. S., Nada, M., Schulz, H., McVie, R. (1990). *J. Clin. Invest.* **85,** 1703–1707.
Rogers, W. A. (1977). *Vet. Clin. N. Am.* **7,** 637–647.
Rogers, W. A., Donovan, E. F., and Kociba, G. J. (1975a). *J. Am. Vet. Med. Assoc.* **166,** 1087–1091.
Rogers, W. A., Donovan, E. F., and Kociba, G. J. (1975b). *J. Am. Vet. Med. Assoc.* **166,** 1092–1100.
Rose, R. J., and Sampson, D. (1982). *Res. Vet. Sci.* **32,** 198–202.
Rose, R. J., Ilkiw, J. E., Sampson, D., and Backhouse, J. W. (1980). *Res. Vet. Sci.* **28,** 393–395.
Rothera, A. C. H. (1908). *J. Physiol.* **37,** 491–494.
Ryan, W. L., and Carver, M. J. (1963). *Proc. Soc. Exp. Biol. Med.* **114,** 816–819.
Sacks, F. M., and Walsh, B. W. (1994). *Curr. Opin. Lipidol.* **5,** 236–240.
Saggerson, E. D. (1988). *In* "Phosphatidate Phosphohydrolase" (D. N. Brindley, ed.), Vol. 1, pp. 79–124. CRC Press, Boca Raton, Florida.
Saharia, P., Margolis, S., Zuidema, G. D., and Cameron, J. L. (1977). *Surgery* **82,** 60–67.
Sakai, T., Hayakawa, T., Hamakawa, M., Ogura, K., and Kubo, S. (1993). *J. Dairy Sci.* **76,** 109–114.
Sampson, J., Morrill, C. C., and Alberts, J. O. (1945). *Cornell Vet.* **35,** 365–369.
Sapir, D. G., and Owen, O. E. (1975). *Metabolism* **24,** 23–33.
Schaer, M. (1976). *Vet. Clin. N. Am.* **6,** 453–461.

Schneider, W. J. (1991). *In* "Biochemistry of Lipids, Lipoproteins and Membranes" (D. E. Vance and J. Vance, eds.), pp. 461–487. Elsevier, New York.

Schotman, A. J. H., and Kroneman, J. (1969). *Neth. J. Vet. Sci.* **2,** 60–64.

Schotman, A. J. H., and Wagenaar, G. (1969). *Zbl. Vet. Med. A* **16,** 1–7.

Schotman, A. J. H, and Wensing, T. (1977). *Vet. Sci. Commun.* **1,** 337–347.

Schultz, L. H. (1952). *Corn. Vet.* **42,** 148–155.

Schwab, L., and Lotspeich, W. D. (1954). *Am. J. Physiol.* **176,** 195–200.

Schwalm, J. W., and Schultz, L. H. (1976). *J. Dairy Sci.* **59,** 255–261.

Schwalm, J. W., Waterman, R., Shook, G. E., and Schultz, L. H. (1969). *J. Dairy Sci.* **52,** 915.

Scott, D. W. (1979). *Vet. Clin. N. Am.* **9,** 3–28.

Serra, D., Casals, N., Asins, G., Royo, T., Ciudad, C. J., and Hegardt, F. G. (1993). *Arch. Biochem. Biophys.* **307,** 40–45.

Setchell, B. P, Bassett, J. M., Hinks, N. T., and Graham, N. M. (1972). *Quart. J. Exp. Physiol.* **57,** 257–266.

Shaw, J. C. (1956). *J. Dairy Sci.* **39,** 402–434.

Shaw, J. H. F., and Wolfe, R. R. (1984). *J. Surg. Res.* **37,** 437–442.

Shimizu, S., Tani, Y., Yamada, H., Tabata, M., and Murachi, T. (1980). *Anal. Biochem.* **107,** 193–198.

Simesen, M. G. (1956). *Nord. Veterinarmotet.* **8,** 1.

Snook, L. C. (1939). *J. Physiol.* **97,** 238–249.

Snow, D. H., and Mackenzie, G. (1977). *Equine Vet J.* **9,** 226–230.

Spector, A. A., and Fletcher, J. E. (1978). *In* "Disturbances in Lipid and Lipoprotein Metabolism" (J. M. Dietschy, A. M. Gotto, and J. A. Ontko, eds.), pp. 229–250. American Physiological Society, Bethesda, Maryland.

Stead, D., and Welch, V. A. (1975). *J. Dairy Sci.* **58,** 122.

Stein, E. A., and Meyers, G. L. (1994). *In* "Tietz Textbook of Clinical Chemistry" (C. A. Burtis and E. R. Ashwood, eds.), 2nd ed., pp. 1002–1093. Saunders, Philadelphia, Pennsylvania.

Stevens, C. E. (1970). *In* "Physiology of Digestion and Metabolism in the Ruminant" (A. T. Phillipson, ed.), pp. 101–112. Oriel Press, Newcastle upon Tyne, England.

Sykes, J. F., Duncan, C. W., and Huffman, C. F. (1941). *J. Dairy Sci.* **24,** 193–197.

Thedford, T. R. (1983). "Goat Health Handbook," pp. 63–65. Winrock International, Morrilton, Arkansas.

Thompson, G. G. (1956). *New Zeal. Vet. J.* **4,** 136–144.

Tildon, J. T., and Sevdalian, D. A. (1972). *Arch. Biochem. Biophys.* **148,** 382–390.

Tyler, J. W., Dowling, P. M., Spano, J. S., McKnight, A. L., and Wolfe, D. F. (1994). *J. Am. Vet. Med. Assoc.* **204,** 1665–1667.

Työppönen, J., and Kauppinen, K. (1980). *Acta Vet. Scand.* **21,** 55–61.

Valdemarsson, S., Hansson, P., Hedner, P., and Nilsson-Ehle, P. (1983). *Acta Endocrinol.* **104,** 50–56.

Valera, A., Pelegrin, M., Asins, G., Fillat, C., Sabater, J., Pujol, A., Hegardt, P. G., and Bosch, F. (1994). *J. Biol. Chem.* **269,** 6267–6270.

Vance, D. E. (1991). *In* "Biochemistry of Lipids, Lipoproteins and Membranes" (D. E. Vance and J. Vance, eds.), pp. 205–240. Elsevier, New York.

Vander Jagt, D. L., Robinson, B., Taylor, K. K., and Hunsaker, L. A. (1992). *J. Biol. Chem.* **267,** 4364–4369.

Visscher, F. E. (1945). *Proc. Soc. Exp. Biol. Med.* **60,** 296–297.

Wagner, J. E. (1976). *In* "The Biology of the Guinea Pig" (J. E. Wagner and P. J. Manning, eds.), pp. 252–267. Academic Press, Orlando, Florida.

Waite, M. (1991). *In* "Biochemistry of Lipids, Lipoproteins and Membranes" (D. E. Vance and J. Vance, eds.), pp. 269–295. Elsevier, New York.

Wanders, R. J. A., van Roermund, C. W. T., Denis, S., Schutgens, R. B. H., Ijlst, L., and Tager, J. M. (1992). *In* "New Developments in Fatty Acid Oxidation" (P. M. Coates and K. Tanaka, eds.), pp. 507–519. Wiley-Liss, New York.

Warnes, D. M., Seamark, R. F., and Ballard, F. J. (1977). *Biochem. J.* **162,** 617–626.

Watson, H. R., and Lindsay, D. B. (1972). *Biochem. J.* **128,** 53–57.

Watson, T. D. G., and Barrie, J. (1993). *J. Small Anim. Pract.* **34,** 479–487.

Watson, T. D. G., and Love, S. (1994). *Compend. Contin. Educ. Pract. Vet.* **16,** 91–98.

Watson, T. D. G., Burns, L., Love, S., Packard, C. J., and Shepherd, J. (1991). *Equine Vet. J.* **23,** 353–359.

Watson, T. D. G., Burns, L., Love, S., Packard, C. J., and Shepherd, J. (1992a). *Equine Vet. J.* **24,** 341–346.

Watson, T. D. G., Gaffney, D., Mooney, C. T., Thompson, H., Packard, C. J., and Shepherd, J. (1992b). *J. Small Anim. Pract.* **33,** 207–212.

Watson, T. D. G., Packard, C. J., and Shepherd, J. (1993). *Comp. Biochem. Physiol. B* **106,** 27–34.

Weidman, M. J., and Krebs, H. A. (1969). *Biochem. J.* **112,** 149–166.

Whitney, M. S. (1992). *Sem. Vet. Med. Surg. Small Anim.* **7,** 292–300.

Whitney, M. S., Boon, G. D., Rebar, A. H., and Ford, R. B. (1987). *Am. J. Vet. Res.* **48,** 1492–1497.

Widmark, E. M. P. (1920). *Biochem. J.* **14,** 364–378.

Wiener, R., Hirsch, H. J., and Spitzer, J. J. (1971). *Am. J. Physiol.* **220,** 1542–1546.

Wildenhoff, K. E. (1977). *Acta Med. Scand.* **201,** 63–67.

Williamson, D. H. (1961). *Biochem. J.* **80,** 540–547.

Williamson, D. H. (1971). *Postgrad. Med. J.*, June Suppl., 371–375.

Williamson, D. H. (1978). *In* "Biochemical and Clinical Aspects of Ketone Body Metabolism" (H.-D. Söling and C.-D. Seufert, eds.), p. 190. Georg Thieme, Stuttgart.

Williamson, D. H., and Krebs, H. A. (1961). *Biochem. J.* **80,** 540–547.

Williamson, D. H., and Mellanby, J. (1974). *In* "Methods of Enzymatic Analysis" (H. U. Bergmeyer, ed.), 2nd English ed., Vol. 4, pp. 1836–1839. Academic Press, Orlando, Florida.

Williamson, D. H., Mellanby, J., and Krebs, H. A. (1962). *Biochem. J.* **82,** 90–96.

Williamson, D. H., Bates, M. W., Page, M. A., and Krebs, H. A. (1971). *Biochem. J.* **121,** 41–47.

Winder, W. W., Baldwin, K. M., and Holloszy, J. O. (1975) *Can. J. Physiol. Pharmacol.* **53,** 86–91.

Zieve, L. (1968). *Med. Clin. N. Am.* **52,** 1493–1501.

Zilversmit, D. B. (1973). *Circ. Res.* **33,** 633–638.

CHAPTER

5

Serum Proteins and the Dysproteinemias

J. JERRY KANEKO

I. INTRODUCTION

The nitrogenous compounds of blood plasma encompass all those organic and inorganic nitrogen-containing compounds of blood. These include the organic macromolecular compounds such as the proteins and nucleic acids; the smaller molecular weight compounds such as glutathione, urea, and creatinine; and the inorganic compounds such as nitrate and ammonia. The nonprotein nitrogen (NPN) compounds are those grouped together as the fraction of the N-containing compounds of plasma that are not removed by the common protein precipitating agents such as trichloracetic acid (TCA). The principal components of the NPN fraction are urea (50%) and amino acids (25%), which total about 35.7 mmol N/liter (50 mg N/dl) plasma in contrast to protein nitrogen, which totals more than 714 mmol N/liter (1 g N/dl) plasma (or 6.25 g protein/dl). This chapter deals primarily with this latter fraction, which is classed collectively as the plasma or serum proteins. The diagnostic and interpretive aspects of the various individual fractions of this pool of proteins are the focus of this chapter.

II. CLASSIFICATION OF PROTEINS

A. Structural Classification

Proteins are most frequently classified according to their form and composition:

Primary structure: the amino acid structure of a single polypeptide chain.
Secondary structure: the helical structure of the chain (α-helix).

Tertiary structure: the folding of the helices into different shapes; globular (myoglobin), linear (fibrinogen), etc. These are monomers.
Quaternary structure: the combination of two or more monomers; dimers (creatine kinase), tetramers (hemoglobin), etc.

B. Chemical Classification

Proteins are also classed according to their chemical composition:

1. Simple proteins contain the basic elements of the amino acids: carbon, hydrogen, oxygen, nitrogen, sulfur.
2. Conjugated proteins are those bound to elements, prosthetic groups, or to other compounds:
   a. Metalloproteins—ferritin (iron)
   b. Phosphoproteins—casein (phosphate)
   c. Lipoproteins—high-density lipoproteins (triglyceride, cholesterol, cholesterol ester)
   d. Glycoproteins—glycohemoglobin, fructosamine
   e. Nucleoproteins—ribosomal proteins.

C. Physical Classification

Proteins have also been classified by their physical behavior: (1) water and salt solubility, (2) density, and (3) molecular weight.

III. METABOLISM OF PROTEINS

A. General

The intake of nitrogenous compounds and the maintenance of nitrogen balance in animals essentially evolve about the intake and metabolism of the amino acids of the ingested proteins. Very few free amino acids and little ammonia are present in the diet. In the adult, excretory losses are balanced by intake so that a nitrogen equilibrium is maintained. Major routes of loss are lactation, illness with cell breakdown, and urinary or gut losses, all of which can result in a negative balance unless nitrogen intake is increased. Positive nitrogen balance occurs during pregnancy, growth, and recovery from disease.

B. Synthesis of Proteins

The fundamental units of protein structure in nature are the 20 natural amino acids. The essential amino acids of this group are those not synthesized by animals and hence must be supplied in the diet (Table 5.1). The nonessential amino acids are those that are synthesized through transamination reactions from the carbon skeletons shown in the figures of Chapter 3. Thus, α-keto-glutarate of the TCA cycle is transaminated to become glutamic acid by accepting the α-amino group from alanine, which then becomes pyruvate. The enzyme catalyzing this reaction is widely used in clinical biochemistry as an index of liver function. The enzyme is named alanine amino transferase (ALT; EC 2.6.1.2) formerly known as glutamic-pyruvate transaminase (GPT).

The α-amino acids are linked together by peptide bonds by the protein synthesizing mechanisms within cells on the rough endoplasmic reticulum (rER). The intricacies of the process by which genetic information for replication is transmitted (transcription) and the mechanism by which this information is expressed in protein structure (translation) are summarized in the genetic code and the mechanisms of protein synthesis. Briefly, in the nucleus, a precise sequence of nucleotides of DNA (chromosomal material) forms a template on which a complementary messenger RNA (mRNA) is replicated (transcription). This transcription is catalyzed by DNA-directed RNA polymerases. The newly formed mRNAs now move to the cytoplasm to become bound with the ribosomes. Meanwhile, transfer RNAs (tRNA) or triplet nucleotides have bound to their specific amino acids also in the cytoplasm. The tRNA (the anticodon) with its amino acid next moves to its complementary triplet segment (the codon) of the mRNA. Succeeding amino acids are brought to the mRNA and sequential peptide linkages are formed until a terminator sequence on the mRNA is reached, at which time the completed polypeptide chain is released. This is the molecular basis of the inheritance of protein structure as represented by DNA transcription and RNA translation into a protein identical in structure to its progenitor. Many replicates of the protein molecule can be synthesized on a single mRNA. When a cell divides, the genetic information is again transmitted from mother to daughter cells via DNA so that identi-

TABLE 5.1 Natural Amino Acids

Essential amino acids:		
Histidine	Lysine	Threonine
Isoleucine	Methionine	Tryptophan
Leucine	Phenylalanine	Valine
Nonessential amino acids:		
Alanine	Cysteine	Proline
Arginine	Glycine	Serine
Asparagine	Glutamate	Tyrosine
Aspartate	Glutamine	

cal proteins are synthesized by succeeding generations of cells. In certain disease states, a single cell or a family of genetically homogenous cells (clone) might exhibit uncontrolled proliferation, for example, plasma cells in multiple myeloma, and excessively produce a single discreet species of protein, immunoglobulin M.

C. Catabolism of Proteins

1. *Turnover of Proteins*

Tissue and plasma proteins are constantly being degraded to their constituent amino acids and the amino acids in turn are sources of energy as well as sources of carbohydrate and fat carbon. The interrelationships linking these major foodstuffs were discussed in Chapter 3. Generally, amino acid carbons, after conversion to fat (lipogenesis) or to carbohydrate intermediates or glucose (gluconeogenesis), become sources of energy after their oxidation to CO_2 and H_2O. Carnivores derive as much as 40–50% of their energy requirements from dietary proteins and their amino acids, whereas omnivores and herbivores derive from less than 10% to about 20%.

The rate of degradation of the plasma proteins is expressed as their turnover, fractional clearance, or as their half-times. Clearance half-times may range from a few hours (some enzyme proteins) to as long as 160 days for hemoglobin in cow red cells (Kaneko, 1983). The clearance half-times of most plasma protein range between 1 to 3 weeks.

2. *The Urea Cycle*

Proteins, as the major dietary source of nitrogen, are hydrolyzed in the gut to their constituent amino acids and absorbed by the intestinal mucosal cells. Gut bacteria may also degrade the amino acids so that ammonia itself may be absorbed. This is an important consideration in the management of liver disease. Sterilization of the gut by the use of antibiotics is one of the management strategies designed to reduce the NH_3 load in liver diseases. The amino acids and ammonia are transported to the liver via the portal circulation and then on to other protein metabolizing tissues.

About 75% of the amino acids or ammonia is transported across the liver cell membrane to be taken up by the liver cell and on into the mitochondria. In the mitochondria, a number of mechanisms of deamination to generate ammonium ion are present, such as glutamate dehydrogenase and glutaminase. In the mitochondria, ammonium ion from any source, bicarbonate, and ornithine form citrulline. Citrulline moves to the cytoplasm where arginine and then ornithine are formed with the release of urea. This cyclical process is known as the Krebs–Henseleit or urea cycle and is given in Fig. 5.1. The compartmentalization between mitochondria and cytosol coordinates the deamination reactions within the mitochondria with the aspartate and TCA cycle within the cytosol. Urea excretion by the kidney tubules is the major route of nitrogen excretion and most others routes, for example, uric acid and nucleic acids, are relatively minor.

All animals are quite intolerant of free ammonia in their cell or body fluids and it is the free ammonia form that is highly toxic. Fortunately, at the pH of blood, 99% of the total ammonia is in the form of ammonium ion:

$$NH_3 + H^+ \rightarrow NH_4^+.$$

The ammonium ion does not readily transfer across cell membranes whereas the NH_3 form readily moves across cell membranes into cells where it is reconverted to the NH_4^+ form. Ammonia is particularly toxic to brain cells where it is thought to reduce their metabolic activity by reducing Krebs cycle activity. Glutamate is often increased and α-ketoglutarate decreased in the presence of increased ammonia. Also, ammonia itself may be toxic by decreasing neurotransmitters. Ammonia is clearly associated with hepatic encephalopathy of humans, dogs (Strombeck *et al.*, 1975), and horses (Divers *et al.*, 1983) but the mechanism is still unknown. It is often found in portasystemic shunts of young dogs.

Disposal of ammonia by the kidney tubules is a second important route of nitrogen excretion. In the

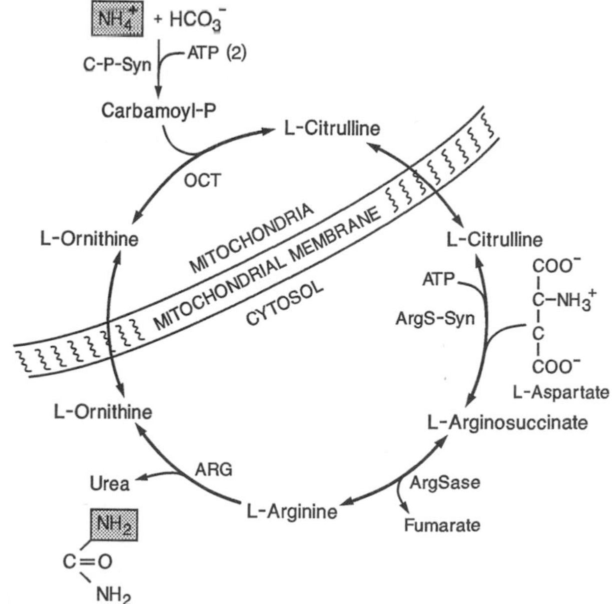

FIGURE 5.1 The Krebs–Henseleit urea cycle. Note the partitioning of the reactions of the cycle between the cytosolic and the mitochondrial compartments.

tubules, glutamine is deaminated to glutamate and ammonia by glutaminase. Ammonia in the tubular lumen is converted to ammonium by binding to H^+ and is therefore an equally important mechanism for removal of excess H^+ for the maintenance of acid–base balance.

IV. THE PLASMA PROTEINS

A. Sites of Synthesis

The major site of synthesis of the plasma proteins is the liver. The second major site is the immune system consisting of the monocyte–macrophage system, lymphoid, and plasma cells. Structural, functional, and enzyme proteins that are synthesized in all body cells and tissues are present in plasma in minor quantities as a result of cell turnover. In general, plasma contains about 5–7% (50–70 g/liter or 5–7 g/dl) protein. If hemoglobin is included, whole blood is composed of about 20% or more of protein.

B. Functions of the Plasma Proteins

The functions of proteins in the body are innumerable. They form the basis of structure of cells, organs, and tissues; they maintain colloid osmotic pressure; they are catalysts (enzymes) in biochemical reactions; they are buffers to maintain acid–base balance, they are regulators (hormones); they function in blood coagulation and in body defenses (antibodies); they are nutritive; and they are transport and carrier compounds for most of the constituents of plasma. The biological activity of proteins and polypeptides for these various functions depends on their structure, from the primary amino acid sequence of the polypeptide hormones to the macromolecular fibers of the fibrin polymers that participate in clot formation.

C. Factors Influencing the Plasma Proteins

1. Age

At birth, plasma proteins of most animals are quite low due to the minimal quantities of immunoglobulins and low albumin. As the newborn animal ingests colostrum, a rapid rise in the immunoglobulins occurs as a result of the absorbed maternal immunoglobulins. As the maternal antibodies decline due to normal turnover, the neonatal animal rapidly gains immunocompetence and begins to synthesize its own immunoglobulins. Upon reaching young adulthood, adult levels of the albumins and globulins are reached.

With increasing age, the plasma proteins are seen to increase above the normal adult levels as a result of a small decrease in albumin and a progressive increase in the globulins.

2. Hormones, Pregnancy, and Lactation

During gestation, total plasma protein decreases due to an albumin decrease even though there is a slight increase in the globulins. Near term, there is a sharp rise in the γ-globulins and a corresponding rise in total plasma protein. In lactation the total plasma protein again decreases due to an albumin decrease.

Some hormones (testosterone, estrogens, growth hormones) effect an increase in total plasma protein due to their anabolic effects, whereas others (thyroxine, cortisol) tend to decrease the total plasma protein due to their catabolic effects.

D. Handling and Identification of Proteins

The process of protein denaturation is the net effect of the alteration of the biological, chemical, and physical properties of the protein by disruption of its structure. It is, therefore, of great importance that proteins be handled in a way to prevent this structural change. Heat, ultraviolet light, surfactant detergents, and chemicals all have some effect on protein structure. If this effect is significant, the protein is denatured, its biological and physical activity is irreversibly lost, and it cannot be accurately measured. A corollary to measurement of any protein is, therefore, the avoidance of denaturation by gentle handling and careful preservation at the appropriate temperature. Assuming proper collection, timely centrifugation, and proper handling of the serum or plasma, Thoresen *et al.* (1995) report that the reliability of the total protein assay is unaffected by storage for up to 240 days either at −20°C or −70°C. Proper handling and storage is of particular importance when enzyme activity needs to be determined because it cannot be assayed if its activity is lost.

Albumin is a water-soluble, globular protein that is usually identifiable as a single discreet molecular species. The globulins are also globular proteins but precipitate in water so they require some salt to maintain their solubility. In contrast to albumin, a major feature of globulins is that they are not a single discreet species of protein but rather a mix of proteins of various types. By virtue of their structure, they migrate in groups in an electric field or precipitate together as rather large families of proteins that are identifiable as the α-, β-, or γ-globulins. The use of cellulose acetate or agarose gel as the support medium for the electrophoretic separation of the various protein fractions is a common first step in the investigation of protein dyscrasias. Agarose gel or "high resolution" is gradually replacing cellulose acetate as the medium of choice. Serum protein electrophoresis has evolved into an extremely useful technique because aberrations are observed in many disease states. There are, however, only a few diseases for which the electrophoretic pattern can be considered pathognomonic.

V. METHODOLOGY

A. Total Serum Protein

A number of methods are available for the determination of total protein in the serum or other body fluids. Classically, the N in the protein has been determined by the Kjeldahl analysis and multiplied by the factor 6.25 (16% N in protein). The technique is rather cumbersome and, for many reasons, is not amenable for use in the clinical laboratory. Other chemical or physical techniques are used.

1. *Chemical Methods*

a. Biuret

This method continues to be the most widely used colorimetric method for the determination of total protein in serum. The biuret reaction is highly specific for protein and is based on the formation of the blue peptide–Cu complex in alkaline solution.

The method is highly accurate for the range of total protein likely to be found in serum (10–100 g/liter; 1–10 g/dl), but it is not sensitive enough for the very low levels found normally in other body fluids, for example, cerebrospinal fluid (CSF). It is the chemical method of choice for total serum or plasma protein in the clinical laboratory because of its simplicity, accuracy, and precision. It is widely adapted for automated chemical analyzers as well as for point-of-care chemical analyzers.

b. Phenol–Folin–Ciocalteau

The Phenol–Folin–Ciocalteau method is very sensitive (Lowry *et al.*, 1951) and is the method of choice for the dilute solutions such as CSF. It is based on the reaction of the phenolic groups of tryptophan and tyrosine of proteins with the reagent to form a blue color. Automated as well as manual techniques are available.

c. Bicinchoninic Acid

The bicinchoninic acid (BCA) method is a relatively new modification of the Lowry method and is more sensitive than the original Lowry method. It is often the method of choice for dilute solutions (Smith *et al.*, 1985).

d. Precipitation Methods

Depending on their charges, proteins can be brought to their isoelectric points, where they will precipitate, by the addition of either anions or cations. Anionic precipitants such as TCA, sulfosalicylic acid, and tungstic acid combine with cationic proteins to cause precipitation. Barium and zinc ions are common cationic precipitants. These reagents are more generally used to prepare protein-free filtrates in the clinical laboratory but have been used for total protein, for example, the sulfosalicylic acid method for urine protein.

2. *Physical Methods*

a. Refractometric

Proteins in solution cause a change in refractive index of the solution that is proportional to the concentration of the protein. Properly controlled and used, this method can be quite accurate at the levels of total protein found in serum or plasma. It is currently in widespread use as a screening method, for the rapid determination of protein in serum, plasma, or other body fluids. Because of its rapidity and simplicity, it is well suited for the office, point-of-care, or emergency clinical laboratory. Due to its dependence on the transmission of light, it is important that the method be used only for clear, nonturbid, and nonlipemic sera and fluids. A moderate degree of hemolysis or icterus does not interfere. The calibration of the refractometer should be frequently checked because this has been found to be a frequent source of error (Booij, 1972). The hand-held Goldberg refractometer is the most useful and versatile refractometer for the veterinary clinical biochemical laboratory because it is scaled to read both total serum protein and urine specific gravity, its scale has finer divisions, and it is temperature controlled.

b. Fibrinogen

Fibrinogen (Fbn) is a large protein of 340,000 Da, which constitutes about 5% of the total proteins of plasma. It is most simply and rapidly estimated by the heat precipitation–refractometer method (Kaneko and Smith, 1967), which takes advantage of the fact that Fbn precipitates on heating to 56°C. The total protein in a sample of plasma is first determined by the refractometer. A sample of the plasma in a microhematocrit tube is heated at 56°C for 3 minutes in a waterbath. The tube is then centrifuged and the total protein in the clear supernate (serum) is determined. The difference between the serum and the plasma protein gives an estimation of the plasma Fbn. In a variation of this method, the height of the precipitated Fbn column can be measured or the amount of precipitate formed can be weighed. The heat precipitation method is now extensively used as a routine screening method but, for more accurate results, fibrinogen is determined as the protein content of the clot (Davey *et al.*, 1972) or as the thrombin time (see the chapter on hemostasis).

B. Fractionation of the Serum Proteins

To determine the amounts of albumins and globulins that make up the total serum protein, various fractionation techniques have been designed to separate and quantitate the serum proteins. All quantitative fractionation techniques first require the determination of the total serum protein, after which the concentrations of the individual fractions are determined by calculations depending on the fractionation method. In its simplest form, if either the albumin or globulin were determined, the other can be obtained by subtraction from the total serum protein. Ultimately, the accuracy of any of the clinical chemical methods is based on comparisons with serum protein electrophoresis, the "gold" standard of reference for serum protein fractionations.

1. Salt Fractionation

Salts added to protein solutions dehydrate the proteins, causing them to precipitate and the solution to become turbid. The most commonly used salts are those of sodium or ammonium sulfate. The salt fractionation technique is based on the differences in solubility of the various protein fractions at different salt concentrations. Albumin is soluble in water and the various globulins precipitate independently in different concentrations of salt. The principle of the technique has been applied for many years in clinical biochemistry in the form of the serum flocculation tests for liver function. Numerous tests have been devised and include the zinc sulfate, sodium sulfate, and ammonium sulfate turbidity tests. These are turbidimetric tests in which the amounts of salts used are adjusted for the normal concentrations of γ-globulins in human sera. In the presence of increased amounts of γ-globulins, flocculation occurs. Because the concentrations of globulins differ in animals, these tests are not directly applicable to animal sera. They can, however, be adapted to animal sera if the concentrations of salt are adjusted to the normal concentrations of γ-globulins in a particular species. This has been done in the calf (Pfeiffer *et al.*, 1977) and foal (Rumbaugh *et al.*, 1978a), where detection of suckling was evaluated using $ZnSO_4$ turbidity. The test successfully detected colostral immunoglobulin transfer to newborn foals when compared to electrophoresis and was deemed better than other rapid tests such as the sodium sulfite precipitation test. Another precipitation test for detection of hypogammaglobulinemia in calves is the glutaraldehyde coagulation test, which also appears to have practical value in identifying hypogammaglobulinemic calves (Tennant *et al.*, 1979).

2. Dye Binding

The acid dye, 2-(4′-hydroxyazobenzene)-benzoic acid (HABA), was widely used in the past for the determination of serum albumin because of its adaptation for automated systems. However, the dye binds so poorly to the albumins of domestic animals that the method is unacceptable. The HABA dye method has now been replaced by bromcresol green (BCG). The BCG method, in comparison with electrophoresis, is quite accurate for albumin within the reference ranges of albumin concentration in animals. However, its accuracy becomes progressively less outside the reference ranges and is usually unacceptable at very low or very high levels. This discrepancy between the BCG and electrophoresis is a common finding in hypoalbuminemic sera. A dye binding method for albumin with properties similar to BCG is the bromcresol purple (BCP) method.

The Coomassie blue dye binding method (Bradford, 1976) for albumin is another innovation that is sensitive at low concentrations of proteins but suffers from variability.

3. Colorimetric Determination of Globulin

The reaction of glyoxylic acid with globulins forms a purple color complex, which is proportional to the amount globulin. The test has not gained acceptance in clinical biochemistry because of its variability and is infrequently used.

C. Electrophoretic Fractionation of the Serum Proteins

The electrophoretic technique is the current standard of reference for the fractionation of the serum proteins (SPE) in clinical biochemistry. The marked advances in technology of the past decade have made this previously elaborate technique into a widely used, routine clinical biochemical test procedure. Its current widespread use is commensurate with its reflection of a variety of changes in serum protein patterns in disease. Although only a few changes in pattern can be considered as diagnostic of a specific disease, the results of electrophoresis, properly interpreted, can be one of the most useful diagnostic aids available to the clinician. There are a large number of methods for serum protein electrophoresis that basically differ only in the type of support media used. Cellulose acetate (CA) continues to be used as a support medium but it is being supplanted by agarose gel (AG) as the medium of choice in clinical laboratories.

1. *Principle*

The principle of the electrophoretic separation of serum proteins is based on the migration of charged protein particles in an electric field. The direction and rate of migration of the particles is based on the type of charge (+ or −) on the protein, size of the protein, intensity of the electric field, and the support medium through which the protein particles are induced to migrate. It follows that it is of importance that the support medium, pH, buffer, and the electric current be described fully in order to be able to compare results of SPE.

When a support medium is used for electrophoresis, the process is called *zone electrophoresis* (as compared to free). Historically, the most commonly used support medium was CA but AG is now widely used in clinical biochemistry for SPE. Many other support media such as agar gel, starch gel, or polyacrylamide gel are also available. Many of the latter yield greater separability of the serum proteins than does CA or AG, but in the clinical laboratory CA and AG are the most useful methods. Recall that the globulins are comprised of a myriad of separate protein moieties—enzymes, carriers, antibodies, clotting factors—and that a specific serum protein can only be identified by additional special techniques.

2. *Cellulose Acetate Electrophoresis*

Depending on factors already noted, particularly the charge and the medium, the protein particles move toward the anode or the cathode of the electric field. Most serum proteins are negatively charged and, on CA in barbital buffer, pH 8.6, they migrate toward the anode. Albumin, with the strongest negativity, migrates farthest toward the anode. The globulins, with weaker negativities, migrate more slowly toward the anode, while some γ-globulins may not move or may even move toward the cathode. The globulins move in groups classed as the α-, β-, and γ-globulins. Depending on the species, there may normally be one or two α, one or two β, and one or two γ fractions (Figs. 5.2 and 5.3).

After the CA membrane is stained for protein, the CA membrane can be cleared and the staining densities of the individual proteins determined using a recording densitometer. The recording provides a visual display of the relative amounts of each protein class in the serum and is the electrophoretogram. When combined with an integrating unit, the intensities of staining of the different classes or bands on the electrophoretogram are translated into a percentage of the total staining intensities. Modern densitometers incorporate a computer that automatically multiplies this percentage times the total serum protein concentration and the amount of each of the separate protein fractions is obtained. This arithmetic calculation can be done manually as well.

A frequent difficulty in dealing with the sera of domestic animals is the difference in the normal pattern between the species. This has created some confusion in the identification of a particular fraction of the electrophoretogram as α, β, or γ. As a first step toward standardization, a useful resolution is to identify the midpoint of the electrophoretogram, which lies between or very nearly between the α_2 and β_1 peaks (Fig. 5.2). This plus the knowledge of the normal number of peaks in a particular species usually suffices to identify the individual peaks with confidence.

3. *Gel Electrophoresis*

Agarose gel electrophoresis (AGE), while retaining most of the limitations of cellulose acetate electrophoresis (CAE), has the same ease of performance and is capable of resolving proteins of serum, CSF, or urine into from 9 to 15 fractions depending on the species. Hence, AGE has been called *high-resolution protein electrophoresis* (HRPE). Because there are so many fractions, it is difficult to quantitate the individual fractions so most often the electrophoretogram is interpreted subjectively by visual examination. Kristensen and Barsanti (1977) using AGE were able to separate cat sera into 10 fractions. Keay (1982) reported similar results in cats, and Keay and Doxey (1982) separated dog, horse, sheep, and cow serum proteins into 7, 8, 9, and 10 fractions, respectively.

Polyacrylamide gel electrophoresis (PAGE) has not been used extensively for the study of animal serum proteins. Furukawa and Sugiyama (1986), using immunoelectrophoresis (IEP), were able to identify 21 feline plasma proteins.

Of the almost infinite number of proteins in serum, only about 200 have been studied; of these, clinical information relative to their interpretation in disease

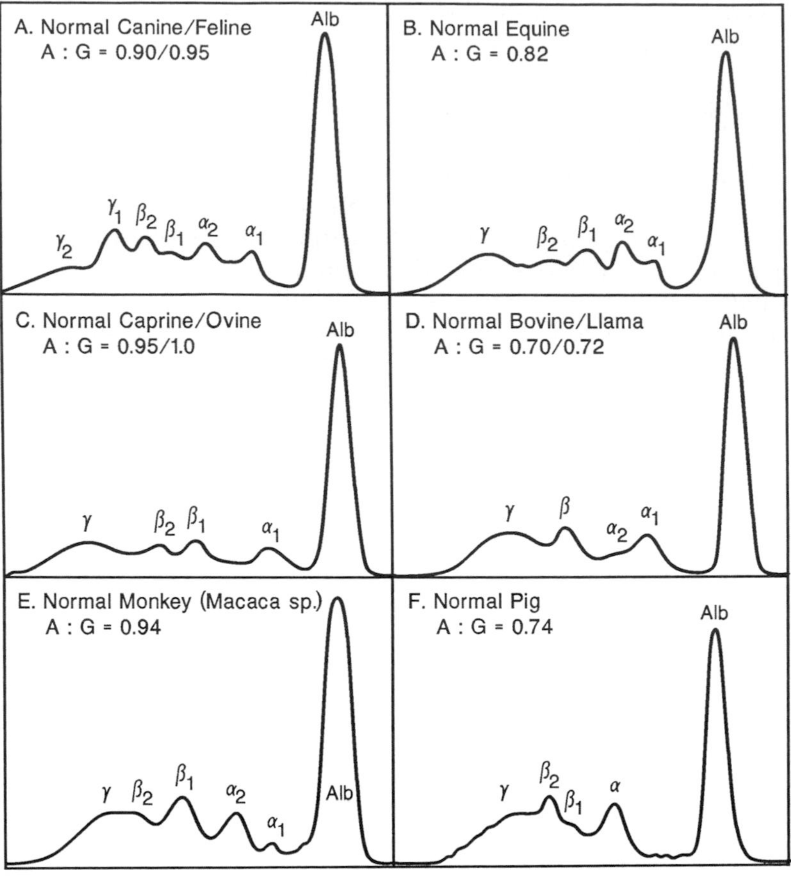

FIGURE 5.2 Cellulose acetate electrophoretograms of serum proteins of some normal mature animals. The albumin-to-globulin (A : G) ratios are within the normal ranges for these animals.

is available in only about 50. Furthermore, present clinically effective separation methods can only resolve a maximum of about 12.

Interpretation of a 5-7 band CAE scanned and quantitated by densitometry remains a very useful technique. Most densitometric methods, however, are unable to resolve fully all of the visible protein bands because of limitations of instrument sensitivity and variations in staining. Thus, it is best to use the densitometric results in conjunction with a visual examinination of the electrophoretogram.

4. *Immunoelectrophoresis*

IEP technique combines electrophoresis in one dimension with immune diffusion in a second dimension to obtain a more precise identification of the various serum proteins. Quantitation of the specific immunoglobulins G, M, A, and E has also been performed in those species of animals in which specific antisera have been available. This field of immunology has experienced unprecedented growth in the last two decades and the scope of its importance continues to increase. This is a highly specialized area and is more fully described in the chapter on clinical immunology.

VI. NORMAL SERUM PROTEINS

More than 200 plasma proteins have been described and quantitated in man and animals. Many of these plasma proteins change markedly in disease and many change only subtley or not at all. Because the proteins of an individual or of a species are synthesized under genetic control, it is to be expected that variations in proteins would occur between individuals and between species. These variations are reflected in the species differences of the normal SPE patterns. Thus,

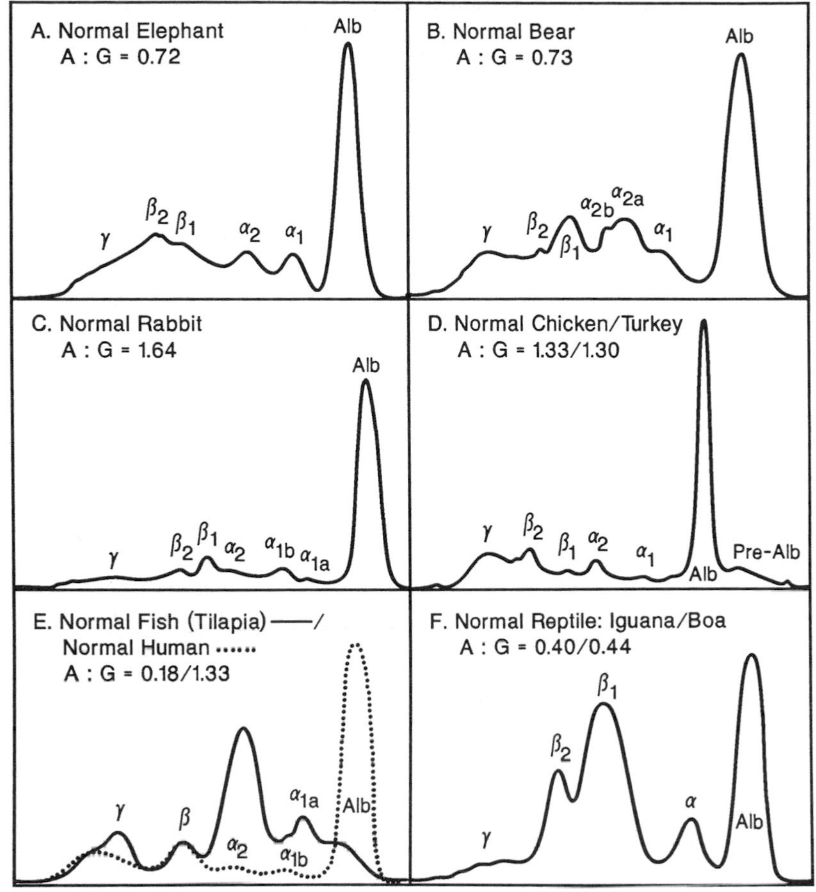

FIGURE 5.3 Cellulose acetate electrophoretograms of serum proteins of some normal mature animals of different genera. E is redrawn from Kaneko (1983).

in ruminants such as the cow, the normal SPE pattern has an albumin, one α, one β, and one γ fraction. Table 5.2 lists some of the important serum proteins with their principal functions and conditions under which alterations may be found. Because there are significant differences in fractionation depending on the method used, reference is made here only to CAE. Reference values for total serum protein and its fractions are given in Appendixes VIII, IX, X, and XI.

A. Prealbumin

Prealbumin is the most rapidly migrating fraction in human serum, is usually not visualized, and may not exist in any domestic animal, but does exist in some birds. However, a protein with the characteristics of prealbumin migrates in the α_2-globulin region of the SPE of dogs (Fex *et al.*, 1977). The only known function of prealbumin is thyroxine binding and transport (Oppenheimer *et al.*, 1965).

B. Albumin

1. Structure

Albumin is the most prominent of the serum proteins on SPE and in animals constitutes 35–50% of the total serum proteins in contrast to humans and nonhuman primates in which albumin accounts for 60–67% of the total. Its tertiary structure is globoid or ellipsoid and it is the most homogenous fraction discernible on SPE. It is the only discreet protein species that can be detected by SPE. However, other methods such as starch gel allow for the separation and detection of genetic polymorphisms of albumin.

It is also important to observe the sharpness of the albumin peak as a measure of the quality of the SPE

TABLE 5.2 Common Serum Proteins, Their Function, and Changes in Disease

Protein	M_r^a (Da)	Function	Change in disease
Prealbumin	54,400	Thyroxine transport	Increase: nephrotic syndrome Decrease: liver disease, protein deficiency
Albumin	69,000	Osmotic pressure regulation, general transport	Increase: dehydration Decrease: liver, kidney, gastrointestinal disease, malnutrition, blood and plasma loss
α-Globulins (α_1 and α_2)			
Thyroxine-binding globulin (TBG)	54,000	Thyroxine transport	Increase: pregnancy
α_1-Fetoprotein	65,000	Unknown	Increase: hepatoma, pregnancy
α_1-Antitrypsin	45,000	Trypsin inhibitor	Increase: acute inflammatory disease Decrease: liver disease, chronic pulmonary disease
α_1-Antichymotrypsin	68,000	Chymotrypsin inhibitor	Increase: acute inflammatory disease
α_1-Acid glycoprotein (orosomucoid, seromucoid)	44,000	Unknown	Increase: acute inflammatory disease Decrease: liver disease, nephrotic syndrome, malnutrition
α_1-Antithrombin III	65,000	Thrombin inhibitor	Increase: disseminated intravascular coagulation, liver disease
α_1-Lipoprotein (HDL, α-lipoprotein)	200,000	Lipid transport	
α_2-Lipoprotein (VLDL, pre-β lipoprotein)	1,000,000	Lipid transport	Increase: nephrotic syndrome, diabetes mellitus, hypothyroidism, steroid therapy
α_2-Macroglobulin	820,000	Insulin binding, trypsin inhibitor	Increase: nephrotic syndrome, chronic active liver disease, acute inflammatory disease
α_2-Globulin	54,000	Thyroxine transport in dogs	
Ceruloplasmin	151,000	Copper transport, ferrioxidase	Increase: acute inflammatory disease
Haptoglobin	100,000	Hemoglobin binding	Increase: acute inflammatory disease
Protein C	62,000	Protease, anticoagulant	Increase: acute inflammatory disease
β-Globulins (β_1 and β_2)			
β_2-Lipoprotein (LDL, β-lipoprotein)	2,750,000	Lipid transport	Increase: nephrotic syndrome, hypothyroidism, hepatocanalicular disease
Transferrin	76,000	Iron transport	Increase: anemias, iron deficiency, pregnancy, acute liver disease, nephrotic syndrome Decrease: iron storage disease, chronic liver disease, acute inflammatory disease
Ferritin	465,000	Iron transport	Increase: iron storage disease, acute inflammatory disease Decrease: iron deficiency
Hemopexin	80,000	Heme transport	Decrease: hemolytic anemia, chronic active liver disease
C3 complement	75,000	Complement C3 factor	Increase: acute inflammatory disease, atopic dermatitis Decrease: autoimmune disease
C-Reactive protein	140,000	Activates complement	Increase: acute inflammatory disease
C4 complement		Complement C4 factor	Increase: acute inflammatory disease Decrease: autoimmune disease
Plasminogen		Proenzyme of plasmin, fibrinolysis	Increase: disseminated intravascular coagulation
Fibrinogen	340,000	Fibrin precursor, coagulation	Increase: acute inflammatory disease Decrease: disseminated intravascular coagulation, afibrinogenemia
γ-Globulins (γ_1 and γ_2)			
Immunoglobulin G (IgG)	150,000	Major antibody formed in response to infectious agents, toxins	Increase: infectious disease, connective tissue disease, liver disease, myelomas and other tumors of the reticuloendothelial system Decrease: fetuses, newborn animals before intake of colostrum, immune deficiency diseases, agammaglobulinemia
Immunoglobulin A (IgA)	150,000	Secretory antibodies in the fluids of the respiratory, gastrointestinal, and the genitourinary tracts	Increase: infectious disease, connective tissue disease, liver disease, myelomas and other tumors of the reticuloendothelial system Decrease: fetuses, newborn animals before intake of colostrum, immune deficiency diseases, agammaglobulinemia

(continues)

TABLE 5.2 (*continued*)

Protein	M_r^a (Da)	Function	Change in disease
Immunoglobulin E, IgE	200,000	Antibodies in allergy	Increase: allergies, anaphylaxis Decrease: aglobulinemia
Immunoglobulin M, IgM	900,000	Cold agglutinin, initiator	Increase: chronic disease, primary cell reactions macroglobulinemia (Waldenstrom's)
Immunoglobulin D, IgD	160,000	Unknown, not seen in animals	
Light chains (Bence-Jones protein)	30,000	Part of the immunoglobulin molecule	Increase: myeloma

[a] M_r, Relative molecular mass.

fractionation. This sharpness or slope of the albumin peak is a useful guide for differentiating the sharp monoclonal globulin peaks from the polyclonal peaks. Additionally, the horse frequently exhibits a minor postalbumin fraction, which appears as a shoulder on the cathodal side of the albumin peak. This shoulder becomes progressively more prominent with the severity of the hypoalbuminemia or with acute inflammatory disease. It is consistently observed in chronic liver disease in horses. A thyroxine binding function of this postalbumin fraction has also been observed in horses.

2. *Function*

Albumin is synthesized by the liver as are all plasma proteins except for the immunoglobulins and it is catabolized by all metabolically active tissues. Its rate of metabolism varies among species and this is reflected in the half-times for clearance (Table 5.3). There appears to be a direct correlation between albumin turnover and body size because clinically significant hypoalbuminemic edemas occur only in the larger animals. This would suggest that edema develops in large animals because of slow replacement of albumin. Hypoalbuminemic edemas do occur in small animals but usually in the presence of posterior vena caval hypertension.

TABLE 5.3 Albumin Turnover in Animals

Species	$T_{1/2}$ (days)	Reference
Mouse	1.90	Allison (1960)
Rat	2.50	Allison (1960)
Guinea pig	2.80	Allison (1960)
Rabbit	5.70	Dixon *et al.* (1953)
Pig	8.20	Dick and Nielsen (1963)
Dog	8.20	Dixon *et al.* (1953)
Sheep	14.28	Campbell *et al.* (1961)
Human	15.00	Dixon *et al.* (1953)
Baboon	16.00	Cohen (1956)
Cow	16.50	Cornelius *et al.* (1962)
Horse	19.40	Mattheeuws *et al.* (1966)

Albumin is a major labile storage reservoir of proteins and as well as being a transporter of its constituent amino acids. It is the most osmotically active plasma protein due to its abundance and small size and accounts for about 75% of the osmotic activity of plasma. A major metabolic function of albumin is its role as a general binding and transport protein. Virtually all constituents of plasma not bound and transported by a specific transport protein and even many that are (e.g., thyroxine) are transported by albumin. Albumin binding solubilizes substances in plasma that are otherwise lipid soluble or only sparingly soluble. Albumin binding permits the effective transport of these substances in the aqueous plasma. Binding to albumin also prevents the loss of plasma chemical constituents through the kidneys. The binding of unconjugated bilirubin or of fatty acids by albumin is an example of this function.

C. Globulins

1. *α-Globulins*

The α fraction is the most rapidly migrating of all the globulins and, in most species except the ruminant, it migrates as an α_1 (fast) and an α_2 (slow) fraction. Most of the globulins of this fraction are synthesized by the liver, except for the α_1-fetoprotein, which is synthesized by fetal liver cells. In general, the α_1-globulins are smaller than the α_2, but there appears to be no functional separation between the two fractions. Important proteins of this fraction are the α-lipoproteins (HDL), which migrate as α_1, and the pre-β-lipoproteins (VLDL), which migrate in the α_2 position.

The β-lipoprotein (LDL), so named because it migrates in the β region on paper, also migrates in the α_2 region on CA. These latter two lipoproteins together with α_2-macroglobulin account for the increase in α_2-globulins seen in the nephrotic syndrome. The α_2-macroglobulin, haptoglobin, ceruloplasmin, and amyloid A (SAA) are also diagnostically important acute phase proteins (Table 5.4).

2. β-Globulins

β-Globulins trail the β_2 and similarly migrate as β_1 (fast) and β_2 (slow) fractions in most domestic animals except ruminants. Important proteins of this fraction are complement (C_3, C_4), hemopexin, transferrin, ferritin, and C-reactive protein (CRP). Fibrinogen slightly trails the β_2 and is another important acute phase protein. CRP and SAA are now considered the most diagnostically important of the acute phase proteins in humans (Maurey, 1985) and a recent study (Caspi *et al.*, 1987) indicates it may be equally valuable in dogs. Some immunoglobulins, IgM and IgA, extend from the β_2 to the γ_2 regions. Therefore, in response to the antigenic stimulus of infectious agents, or in plasma cell malignancies, immunoglobulin can rise in the β_2 zone as well as in the γ_1 and γ_2 zones.

3. γ-Globulins

In most animals, the γ fraction is also observed as two fractions, a γ_1 (fast) and a γ_2 (slow). Of the immunoglobulins observed in animals, IgA, IgM, and IgE are found primarily in the γ_1 region and IgG is found primarily in the γ_2 region. The specific identification and quantitation of the immunoglobulins require the use of immunochemical techniques (Iammarino, 1972; Mancini *et al.*, 1965). Immunochemical assays are widely used in protein research and now in the clinical laboratory. A brief description of immunological principles is given here as a basis for understanding the interpretation of dysproteinemias visualized on SPE. More thorough coverage is given in the chapter on clinical immunology.

TABLE 5.4 Acute Phase Proteins: Markers of Acute Inflammatory Disease

Positive Acute Phase Proteins:
α_1-Globulins
α_1-Antitrypsin
α_1-Acid glycoprotein (orosomucoid, seromucoid)
α_2-Globulins
α_2-Macroglobulin
Ceruloplasmin
Serum amyloid A
Haptoglobin
β-Globulins
Fibrinogen
Complement, C3, C4
Protein C
C-reactive protein
Ferritin
Amyloid A
Negative Acute Phase Proteins:
Prealbumin
Albumin
Transferrin

a. Source

Antigens employed in the immunochemical tests are of two types. complete antigens, which induce formation of specific antibodies and incomplete antigens (haptens), which, though they react with antibodies, do not elicit an immune response. Complete antigens usually have a molecular weight greater than 5000 Da and are proteins, glycoproteins, complex carbohydrates, or nucleic acids that are recognized as foreign by the host and elicit an immunologic response from the host. Haptens are low-molecular-weight compounds that, if coupled to a larger molecular weight compound such as a protein, elicit an antibody response. Most antibodies used in radioimmunoassay (RIA) are produced in response to haptens coupled to albumin, for example, T_4 antibody for T_4 RIA.

Antibodies produced in response to antigens are highly specific if only one antigenic determinant is involved. In nature, however, multiple antigenic determinants are usually involved. Natural haptens are usually multiple and elicit antibodies with multiple specificities or cross reactivities.

The lymphocytic cell line is now known to play the central role in the immune system. There are two subpopulations, the B lymphocytes (bursa) and the T lymphocytes (thymus), which can be identified by special immunologic means.

The T cells are found in blood and in lymph nodes in the deep cortical areas and paracortical sinuses. They are associated with cell-mediated immunity. The B cells were originally identified in the bursa of Fabricius of the chicken and are now thought to be differentiated in fetal liver cells. In the adult, they are found in the blood and in the germinal centers of lymph nodes. The B cells respond to antigenic stimuli with the proliferation of plasma cells that produce the specific antibody or immunoglobulin. Of the five known immunoglobulins, IgG, IgA, IgM, IgD, and IgE, four have been identified in dogs (IgG, IgA, IgM, and IgA), cats (Gorham, 1971), and horses (McGuire *et al.*, 1975a). Under certain conditions, an excess of a portion of the immunoglobulin molecule, that is, light chains, might also be produced. These light-chain fragments appear in the plasma and because of their small size appear in the urine as the Bence–Jones protein.

A specific plasma cell population of defined genetic origin—a clone—produces a specific immunoglobulin. Uncontrolled growth of a single B-cell clone (malignancy) results in the overproduction of a single chemical species of immunoglobulin, which appears as a sharp "monoclonal" spike or monoclonal gammopathy on an electrophoretogram. Occasionally, a "biclonal" or "triclonal" gammopathy can be identified. A group of clones, each of a different genetic origin, can also overproduce a heterogenous mix of immunoglobulins, which appears as a diffuse or broad hyperglobulinemic region on the electrophoretogram. This region is described as a "polyclonal" gammopathy.

b. Immunoglobulins

The immunoglobulins are glycoproteins whose basic structure is a monomer comprised of two heavy (H) and two light (L) chains linked by disulfide bridges. Each H chain consists of 446 amino acids and each L chain consists of 214 amino acids (Edelman, 1973). The structure of the H chain governs the class of immunoglobulin and is named by corresponding Greek letters: $\gamma = G$, $\mu = M$, $\alpha = A$, $\Sigma = E$, and $\delta = D$. The structure of the L chain is either kappa (κ) or lambda (λ) and denotes type. Structural variations in the variable regions of H or L chains provide a basis for further subdivision into subtypes and subclasses. To date, four subclasses of IgG have been identified: IgG_1, IgG_2, IgG_3, IgG_4, and two subclasses of IgA: IgA_1 and IgA_2.

IgG, IgD, and IgE are monomers; IgA is a dimer; and IgM is a pentamer. Most viral, bacterial, and toxin antibodies are of the IgG type and are present in all animals. IgD has been reported only in humans, where its function is unknown. IgE is involved in allergic and anaphylactic reactions and is present in dogs. IgA is a dimer of two basic units joined by a secretory piece. IgA is found in the secretions of the respiratory, genitourinary, and gastrointestinal tracts. IgM is a cyclic pentamer of five basic units that forms a high-molecular-weight unit. These are the macroglobulins or "M" components.

The Bence–Jones proteins are light-chain units and their presence reflects the asynchronous synthesis of H chains so that excess L chains appear. When they appear in urine, they are called Bence–Jones proteins. They are not detected on SPE, but when immunochemical techniques are used, they are often found to accompany gammopathies (Solomon, 1976), especially multiple myelomas.

VII. INTERPRETATION OF SERUM PROTEIN PROFILES

The determination of serum proteins and of their SPE profiles have evolved into important diagnostic aids in clinical biochemistry. This has occurred even though a specific diagnosis can seldom be made with SPE. Abnormal serum protein profiles can be identified with general types of disease processes and in this way provide the rationale for further definitive studies of the patient. A variety of electrophoretograms illustrating some common types of disease in different species are given in Figs. 5.2 through 5.6. Groulade (1985) gives a series of similar SPE profiles in chronic diseases of the dog. Foulon *et al.* (1992), in their study of feline retroviral diseases, observed both an increase in β-globulins and a decrease in γ-globulins in FeLV but only a decrease in γ-globulins in FIV. They suggest that the SPE can be used to differentiate between FeLV and FIV. An added impetus for the use of SPE has been the inclusion of total protein and albumin in automated systems, thereby providing the albumin-to-globulin ratio (A:G). A change in the A:G ratio is often the first signal of a protein dyscrasia, which leads to the further study of the proteins by SPE. Appendixes VIII, IX, X, and XI give reference values for total serum protein and its fractions in animals and birds.

A. Physiological Influences

Abnormalities of SPE must be interpreted in light of the many influences unassociated with disease. Normal physiological variations within an individual are relatively constant over a considerable period of time, therefore even minor changes in the SPE profile can be of significance and warrant close scrutiny.

1. Influence of Age and Development

In the fetus, the concentration of total protein and albumin progressively increases with little change in globulins and an absence of γ. However, γ-globulin was detected in 25% of bovine fetal serum samples (Kniazeff *et al.*, 1967), which they attributed to transplacental transfer. After nursing and up to 24 hours, baby pigs had large amounts of γ-globulin, which progressively decreased to 5% of the total serum protein by 4 weeks of age (Rutqvist, 1958; Rook *et al.*, 1951). In the calf, precolostral serum normally contains no γ-globulin (Pierce, 1955), but within a few hours after ingestion of colostrum, γ-globulin appears in serum and absorption continues for up to 48 hours after birth (Ebel, 1953), after which gut permeability ceases. In colostrum-deprived calves, immunoglobulin increases only minimally. In the developing foal from birth to 12 months of age, progressive increases in albumin, globulins, and total proteins are seen (Sato *et al.*, 1979; Rumbaugh and Adamson, 1983; Bauer *et al.*, 1985).

In all animals, there is a general increase in total protein, a decrease in albumin, and an increase in globulins with advancing age (Forstner, 1968; Tumbleson

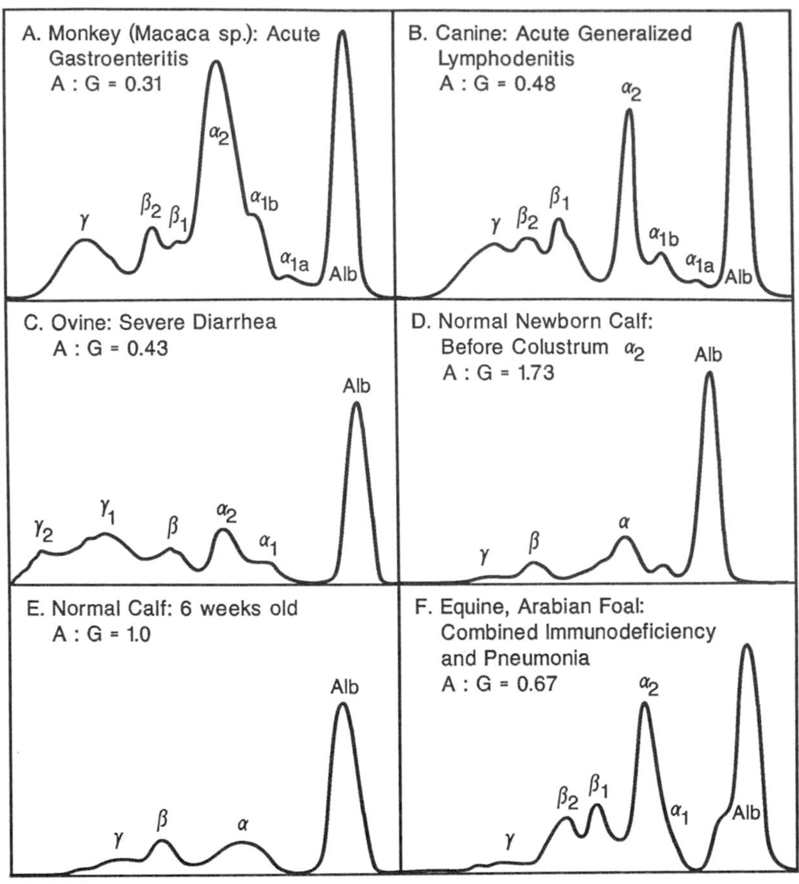

FIGURE 5.4 Cellulose acetate serum protein electrophoretograms in (A, B, F) acute inflammatory disease with acute phase protein increases, (C) gastrointestinal loss of albumin, (D, E) newborn and neonatal calves, and (F) immunodeficiency in a foal.

et al., 1972); and in the very old, the total proteins again decline. Similar changes due to age have also been clearly shown in chickens by Tanaka and Aoki (1963), who periodically followed serum protein changes up to 210 days of age. Thus, age is an important consideration in the interpretation of the SPE.

2. *Hormonal and Sexual Influences*

Hormonal effects on serum protein can be either anabolic or catabolic. Testosterone and estrogens are generally anabolic in all species. Diethylstilbestrol (DES) given to male calves and to chickens (Perk and Loebl, 1960) was found to increase total protein, decrease albumin, and increase globulins. Growth hormone is another well-known anabolic hormone with similar effects. On the other hand, thyroxine decreases total serum protein, most likely due to its catabolic effect (Sturkie, 1951). The glucocorticoids are characterized by gluconeogenic activity, but their effects on serum protein are not marked, because only small decreases in γ-globulin have been observed (Bjorneboe *et al.*, 1952). Thus, in general, hormonal effects of serum proteins are slight even though their effects on weight gains or body composition may be quite marked.

3. *Pregnancy and Lactation*

During gestation, albumins decrease and the globulins increase. In ewes, albumin decreases to a minimum at midgestation and returns to near normal at term. Globulins and the total serum protein progressively decrease throughout gestation (Dunlap and Dickson, 1955). In cows, the total serum protein, γ_1- and β_2-globulins begin to increase at 2 months before term, reach a maximum at 1 month, and then rapidly decline toward term (Larson and Kendall, 1957). These data indicate that the immunoglobulins rapidly leave the plasma during the last month of gestation when colostrum is being formed in the mammary gland. Lactation

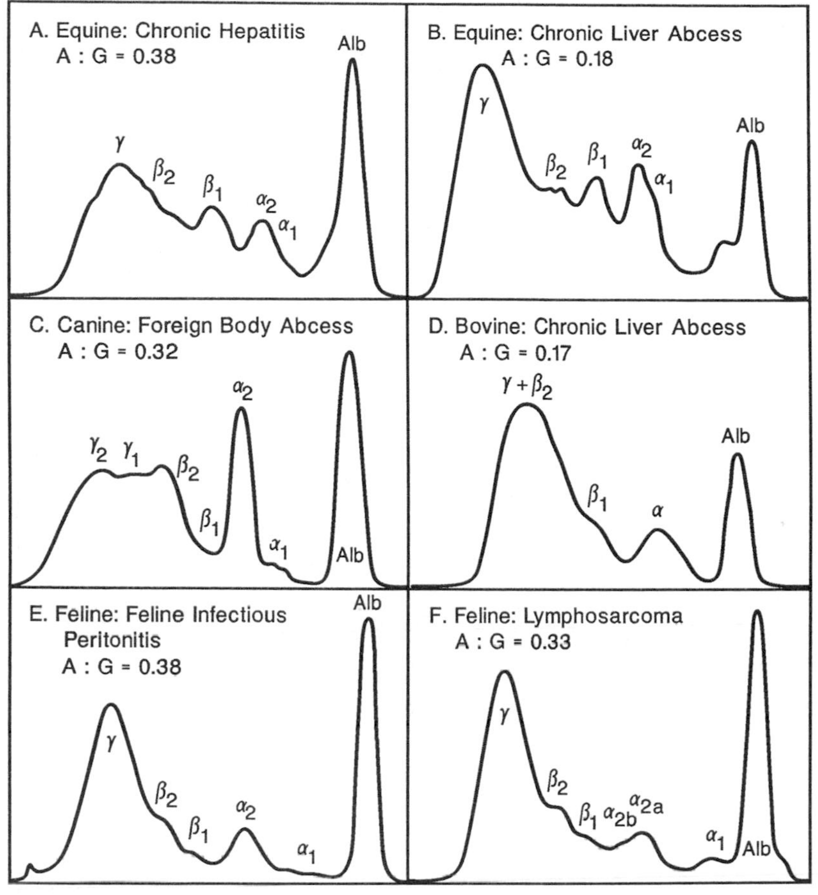

FIGURE 5.5 Cellulose acetate serum protein electrophoretograms of some polyclonal gammopathies in animals. Note that the major globulin peaks are broader and wider than the albumin peaks.

and egg production impose further stresses on protein reserves and metabolism and changes similar to pregnancy also occur.

4. Nutritional Influences

The plasma proteins are sensitive to nutritional influences but the changes are often subtle and difficult to detect and interpret. A direct relationship between vitamin A and lowered albumin has been observed in cows (Erwin *et al.*, 1959), which was corrected by the administration of carotene. Severe dietary protein deficiency induces a hypoproteinemia and hypoalbuminemia in rats (Weimer, 1961), chickens (Leveille and Sauberlich, 1961), and dogs (Allison, 1957). In humans, kwashiorkor and marasmus, diseases of severe protein-calorie malnutrition, are also characterized by hypoproteinemia and hypoalbuminemia. Dietary protein deficiency results in a decreased turnover of serum albumin in rats (Jeffay and Winzler, 1958). Immunoglobulins are affected only on severe protein restriction (Benditt *et al.*, 1949), but the effects are reversible on protein repletion (Wissler *et al.*, 1946).

5. Stress and Fluid Loss

Temperature stress, either febrile or hypothermia, is associated with nitrogen loss, increased adrenal activity, and increased protein turnover. These stresses cause a decrease in total serum protein, decrease in albumin, and often an increase in α_2-globulin. Similar findings are observed in crushing injuries, bone fractures, and extensive surgery. Tissue repair calls on protein reserves, and the increased protein turnover results in decreased albumin and increased α_2-globulin (Hoch-Ligeti *et al.*, 1953). In the inflammatory process, fluids and proteins move into the tissue fluids, inducing edema and contributing to a decrease in albumin. Hemorrhage or massive exudation with large external losses of plasma is followed by a rapid movement

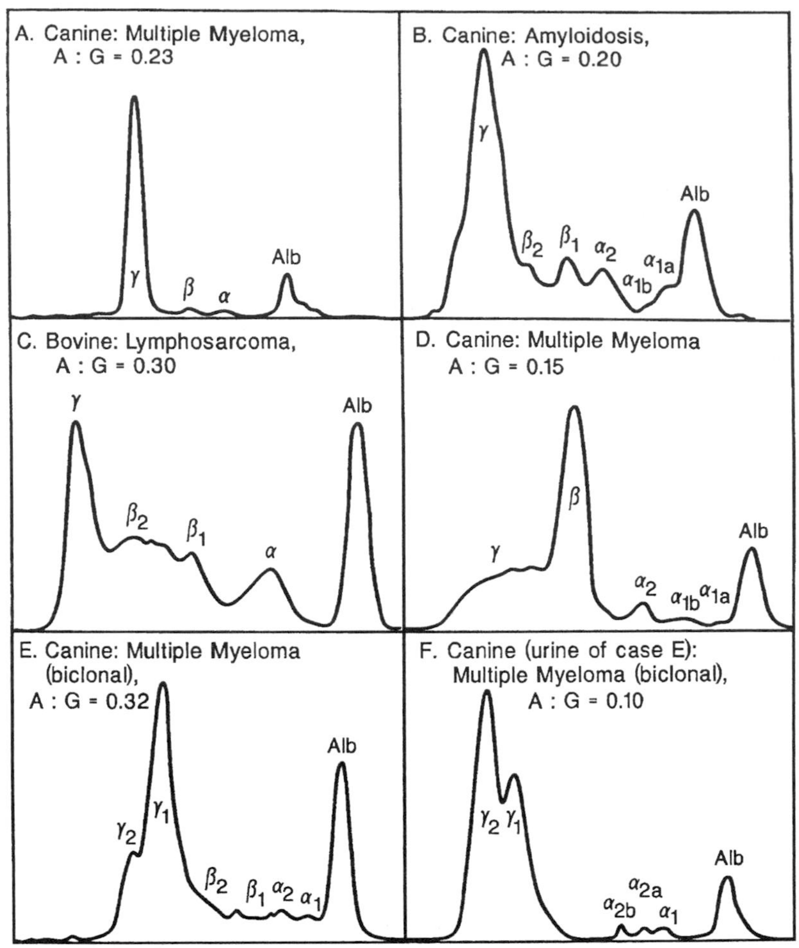

FIGURE 5.6 Cellulose acetate serum protein electrophoretograms of some monoclonal gammopathies in animals. Note that the major globulin peaks are narrow and as sharp or sharper than the albumin peaks. Biclonal gammopathies (E, F) are occasionally seen.

of interstitial fluid (without protein) into the plasma compartment to induce an acute hypoproteinemia. Conversely, dehydration leads to hemoconcentration through reduction in fluid volume and consequent hyperproteinemia. During splenic contraction in the horse, a large mass of erythrocytes moves into the circulation with little or no change in the serum protein.

B. Effects of Inflammation

It has long been known that changes in a select group of constitutive plasma proteins occur in the early stages of the inflammatory process in all animals. In recognition of this phenomenon, this group has been classified as the acute phase proteins (APPs) and the response of the APPs to inflammation is currently the subject of heightened interest. This interest is generated by the potential for use of APPs to provide an early and reliable signal to the clinician of the presence of any form of inflammatory disease or inflammation, including malignancies.

1. Mechanism of the Response

The inflammatory response to tissue injury is a fundamental mechanism by which the host mounts a defense against further injury and initiates a series of chemical events to begin the healing process. This series of events is mediated by the release of cytokines from the leukocytes that gather at the site of injury. The cytokines are polypeptides that may act locally to compound the leukocyte response or distantly to induce a febrile response. Interleukin-1 (IL-1), interleukin-6 (IL-6), and tumor necrosis factor-α (TNF-α) are important cytokines in that they are released into the circulation to reach the liver where they induce the synthesis and release of the APPs. Most of the APPs are globulins but very few are normally in high enough concentrations as to be readily determined. Their rates of release and their clearance half-times are also important factors in their seum levels in response to inflammation and, of course, their clinical diagnostic usefulness. A very important factor in their clinical usefulness is the magnitude of the response, which is minimal in some instances and can be several 100-fold in others. Further, some of the APPs may be reduced (e.g., albumin and transferrin) and these are the negative APPs. All of these factors are known to operate differently in different species so that significant species differences are seen and must be considered.

2. *Functions*

For many years, fibrinogen (Fbn) was the only well known and readily assayable APP. Fibrinogen, as part of the hemostatic process generates fibrin and the accumulation of fibrin is vital to localizing the injury and initiating healing. Haptoglobin (Hp), another important APP, binds to hemoglobin (Hb) and in this way may limit the scope of the injury by conserving the Hb released by local hemolysis, conserve its iron, or make iron unavailable to microorganisms. CRP binds to many products, including phosphorylcholine, lipoproteins, complement, or bacteria, and may play a similar protective role. SAA is a high-density lipoprotein (HDL) apolipoprotein and increased amounts of SAA may serve to increase HDL transport out of the plasma. In general, these are all defensive and healing responses to injury.

Other hepatic protein responses, notably Alb and Tf synthesis, are downregulated in inflammation. The common function of these negative APPs is transport of metabolites and among these Alb is a particularly well-known general metabolic transport protein. Therefore, any decrease in Alb may well enhance the protective and healing function of inflammation by focusing the animals' metabolic activities toward synthesis of protective proteins and away from catabolism of metabolites.

3. *Species Differences*

Species variations in the APPs occur because of inherent variations in synthesis and in rates of entry and exit of these APPs from the circulation as is true of any other plasma proteins such as the enzyme proteins, transport proteins, or coagulation factors. These variations may take the form of increased synthesis, release, or catabolism as expressed in their clearance half-times, but little is known of the impact of these variations on the species-specific response. As a result, widespread differences occur in the major APP responses among the various animal species. Fibrinogen, previously the most widely used APP for clinical purposes because of its ease and historical use, is a low to moderate responder in most species (Hayes, 1994). Although it has been very useful as a screening tool, it is not a major responder in any of the species studied. Therefore, current interest lies with other APPs. In dogs, the CRP is the major responder and there is intense interest in this APP (Yamamoto *et al.*, 1993, Burton *et al.*, 1994). Additionally, Hp (Conner *et al.*, 1988) and α-1-acid-glycoprotein (AGP) (Ohwada and Tamura, 1995) have been studied in the dog as a potentially useful APP. In rabbits, CRP is also the major responder (Kushner and Feldman, 1978). AGP was also increased in hemolymphatic tumors of dogs (Ogilvie *et al.*, 1993). In cats, TNF-α is a cytokine that responds to feline leukemia virus infection (Khan *et al.*, 1992). In ruminants, mainly cattle, Hp is the major APP responder (Skinner *et al.*, 1991) and, frequently, Hp or even Fbn is the better predictor of inflammation than the leukogram. In the horse, SAA is the major responder (Hayes, 1994). These are the species in which most information is available and important APPs in other species are given in Tables 5.4 and 5.5.

4. *Methodology*

A method based on heat precipitation of Fbn and the use of refractometric measurement of the total plasma proteins before and after heating has been widely used as a screening method for Fbn (see Section V.A.2.b). The common method for Hp is based on its hemoglobin binding property (Skinner *et al.*, 1991) as well as on single radial immunodiffusion (RID) (Morimatsu *et al.*, 1992). The AGP is assayed on the basis of its solubility (Ohwada and Tamura, 1995; Tamura *et al.*, 1989). A variety of methods have been used for CRP; RID (Conner *et al.*, 1988), enzyme-linked imunosorbent assay (ELISA) (Eckersall *et al.*, 1989), immunoturbidometry

TABLE 5.5 Acute Phase Protein: Major Responders in Various Animal Species

Species	Major APP
Cat	TNF-α
Cow	Hp, SAA
Dog	CRP, SAA
Horse	SAA
Mouse	SAA, AGP
Pig	CRP
Rabbit	CRP
Rat	α_2-Macroglobulin, AGP

(Eckersall *et al.*, 1991), and latex agglutination (Yamamoto *et al.*, 1993). The major APP in the horse, SAA, has been assayed based on the lipid binding properties of this lipoprotein (Pepys *et al.*, 1989).

C. The Dysproteinemias

The first approach and method of choice for the overall evaluation of protein status is by the use of SPE. The SPE profile and the absolute values of the individual fractions provide an excellent basis for presumptive diagnoses and for additional studies of the patient. The A:G ratio derived from chemistry panels or from the SPE is the basis on which the SPE can be interpreted.

A classification of the SPE profile in conjunction with the A:G provides for a sytematic approach to the interpretation of protein dyscrasias. Table 5.6 gives just such a classification of the SPE based on the A:G ratio and the nature of the profile. This table provides a useful vehicle for alerting the clinical biochemist and the clinician to the underlying significance of the specific dysproteinemia.

1. Normal A:G—Normal Profile

a. Hyperproteinemia

Simple dehydration with water loss is essentially the only instance when a simple hyperproteinemia without change in profile or A:G occurs. In this case, all protein fractions increase proportionately, including albumin, because only water has been removed from the system.

b. Hypoproteinemia

Overhydration through vigorous fluid therapy or excess water intake is a common cause of simple hypoproteinemia. This is simply a dilution of the system. In other instances, for example, after acute blood loss, interstitial fluid moves rapidly into the plasma compartment, thus diluting the system. This dilution may be further intensified by the ingestion of water to satisfy the thirst commonly seen in acute blood loss. Similarly, after acute plasma loss, whether internal or external, by exudation or extravasation, simple hypoproteinemia occurs because the water losses are rapidly replaced by movement of interstitial water into the plasma compartment.

2. Decreased A:G—Abnormal Profile

a. Decreased Albumin

Decreases in albumin are a common form of dysproteinemia. Fundamentally, the decreases can be attributed to either albumin loss or failure of albumin synthesis. Depending on the stage of the disease, it can be associated with either slight hyperproteinemia (acute stage), normoproteinemia (progressive stage), or, in its advanced stages, hypoproteinemia. Therefore, the total serum protein is rarely a reliable index of albumin status and albumin must be determined.

Due to its small size and osmotic sensitivity to fluid movements, albumin is selectively lost in renal disease (Osborne and Vernier, 1973), gut disease (Meuten *et al.*, 1978; Kaneko *et al.*, 1965), and in intestinal parasitism (Dobson, 1965). The hypoalbuminemia of intestinal parasitism is aggravated by increased albumin catabolism (Cornelius *et al.*, 1962; Holliday *et al.*, 1968; Holmes *et al.*, 1968). Furthermore, due to the sensitivity of albumin synthesis to protein and nitrogen loss such as that occurring in any form of diarrhea, albumin loss impairs albumin synthesis and further compounds the hypoalbuminemia. Due to this same sensitivity of albumin synthesis to protein and nitrogen availability, decreased albumin precedes the development of generalized hypoproteinemia in dietary protein deficiencies. Classic human protein-calorie malnutrition, kwashiorkor, is characterized by hypoalbuminemia and hypoproteinemia (see Section VII.A.4).

The liver is the only site of albumin synthesis and hypoalbuminemia is an important feature of chronic liver disease (Figs. 5.4, 5.5, and 5.6). In the horse, a unique postalbumin shoulder (Fig. 5.5) with or without a hypoalbuminemia is virtually pathognomonic of liver disease. Additionally, albumin is a negative APP (Table 5.4) and extensive inflammation accompanying any of the aforementioned conditions may compound the hypoalbuminemia.

b. Increased Globulins

i. α-Globulins α_1-Globulin but mainly α_2-globulin increases are commonly found and are of diagnostic significance (Fig. 5.4). Many APPs migrate in the α_1- and α_2-globulin regions (Table 5.3) so that increases in these globulins are a common finding in acute inflammatory diseases. A rise in α_2-macroglobulin is also seen in acute inflammatory diseases. The specific APPs that account for the γ-globulin increase, however, must be specifically identified by other means (see Section VII.B.4). In the nephrotic syndrome, α_2-globulins increase due in part to the α_2-macroglobulin and the lipoproteins. The triad of azotemia, hypoalbuminemia, and cholesterolemia is a characteristic of the nephrotic syndrome.

ii. β-Globulins Increases in β-globulins alone are infrequent and found in association only with active liver disease, suppurative dermatopathies, and in the nephrotic syndrome. Transferrin appears to be the major component that rises in active liver disease together

TABLE 5.6 Classification of the Dysproteinemias Based on the Albumin-to-Globulin Ratio and the Serum Protein Electrophoretic Profile

A. Normal A : G—normal SPE profile
  1. Hyperproteinemia: dehydration
  2. Hypoproteinemia
    a. Overhydration
    b. Acute blood loss
    c. External plasma loss: extravasation from burns, abrasions, exudative lesions, exudative dermatopathies, external parasites; gastrointestinal disease, diarrhea
    d. Internal plasma loss: gastrointestinal disease, internal parasites

B. Decreased A : G—abnormal SPE profile
  1. Decreased albumin
    a. Selective loss of albumin: glomerulonephritis, nephrosis, nephrotic syndrome, gastrointestinal disease, internal parasites
    b. Decreased synthesis of albumin: chronic liver disease, malnutrition, chronic inflammatory disease
  2. Increased globulins
    a. Increased α_1-globulin
      i. Acute inflammatory disease: $\alpha_1$pantitrypsin, α_1-acid glycoprotein (orosomucoid, seromucoid)
    b. Increased α_2-globulin
      i. Acute inflammatory disease: α_2-macroglobulin, ceruloplasmin, haptoglobin
      ii. Severe active hepatitis: α_2-macroglobulin
      iii. Acute nephritis: α_2-macroglobulin
      iv. Nephrotic syndrome: α_2-macroglobulin, α_2-lipoprotein (VLDL)
    c. Increased β-globulin
      i. Acute hepatitis: transferrin, hemopexin
      ii. Nephrotic syndrome: β_2-lipoprotein (LDL), transferrin
      iii. Suppurative dermatopathies: IgM, C3
    d. β-γ Bridging
      i. Chronic active hepatitis: IgA, IgM
    e. Increased γ-globulin (broad increases)—polyclonal gammopathies: IgG, IgM, IgA
      i. Chronic inflammatory disease, infectious disease, collagen disease
      ii. Chronic hepatitis
      iii. Hepatic abscess
      iv. Suppurative disease: feline infectious dermatitis, suppurative dermatitis, tuberculosis
      v. Immune-mediated disease: autoimmune hemolytic anemia, autoimmune thrombocytopenia, Aleutian disease of mink, equine infectious anemia, systemic lupus erythematosus, autoimmune polyarthritis, autoimmune glomerulonephritis, autoimmune dermatitis, allergies
      vi. Tumors of the reticuloendothelial system (RES): lymphosarcoma
    f. Increased γ-globulin (sharp increases)—monoclonal gammopathies: IgG, IgM, IgA
      i. Tumors of the reticuloendothelial system (RES): lymphosarcoma
      ii. Plasma cell dyscrasias: multiple myeloma, Aleutian disease of mink
      iii. Macroglobulinemia
      iv. Canine ehrlichiosis
      v. Benign

C. Increased A : G—abnormal profile
  1. Increased albumin: does not occur except in dehydration
  2. Decreased globulins
    a. Fetal serum
    b. Precolostral neonate
    c. Combined immunodeficiency of Arabian foals
    d. Aglobulinemia

with hemopexin and complement. IgM can also rise in active liver disease in response to the antigenic stimulus of infectious agents. In the suppurative dermatopathies, a similar antigenic stimulus is thought to account for the IgM and complement increases in the β fraction. In the nephrotic syndrome, increases in β-globulins are associated with increases in Tf. Most increases in β-globulins are polyclonal, and only occasionally the sharp monoclonal spikes of multiple myeloma, Waldenstrom's macroglobulinemia, or lymphosarcoma are seen (Hurvitz *et al.*, 1977; MacEwen *et al.*, 1977).

iii. β-γ Bridging The phenomenon of β-γ bridging is virtually pathognomonic of chronic active hepatitis (Fig. 5.5). In this case, there is no clear separation between the β_2 and γ_1 fraction, which results from an increase of IgA, IgM, or both. Rarely does a low-grade gammopathy of lymphosarcoma result in a β-γ bridge.

iv. Increased γ-Globulin (Broad Increase): Polyclonal Gammopathy The diffuse or broad increases in the γ-globulins that characterize polyclonal gammopathies (Fig. 5.5) are a result of the heterogeneity of the clones of plasma cells, which produce a heterogenous mix of immunoglobulins. Any one or all of the immunoglobulins IgM, IgG, or IgA can be present, but a preponderance of one is usually seen.

The chronic inflammatory disease profile may be manifested by a variety of disease states such as chronic infections, collagen diseases, and malignancies in general. There is a concomitant decrease in albumin as a result of decreased synthesis.

Chronic hepatitis, hepatic abscesses, and suppurative disease processes also exhibit changes characteristic of chronic disease (Rumbaugh *et al.*, 1978b). In these cases, the polyclonal increase is more marked and the hypoalbuminemia more severe than in chronic inflammatory disease. This phenomenon may be a reflection of the severity of the disease process and the more intense antigenic response generated.

Immunologically mediated disease processes are also characterized by polyclonal increases. These may be immune processes directed against "self," that is, autoimmune disease, or against external antigenic stimuli. In either case, a multiple immunologic response is elicited, one or more organs may be affected, and polyclonal increases are observed on the electrophoretogram. Immune complexes trapped in the glomeruli and reacting with antigens are thought to be involved in the glomerulonephritis often seen with heartworms, pyometra, and systemic lupus erythematosus (SLE). SLE is a multifaceted disease in the dog often found in association with autoimmune hemolytic anemia (AIHA), thrombocytopenia (AITP), glomerulonephritis (GN), and rheumatoid polyarthritis. The basic defect in SLE is the LE factor, an anti-DNA antibody, and the widespread dissemination of the DNA–anti DNA complex throughout highly vascular structures. The LE cell is a granulocyte that has phagocytized the DNA–anti DNA complex.

AIHA is characterized by acute erythrocyte destruction, accelerated bone marrow response to the anemia, and the presence of autoantibodies against the patient's own erythrocytes. The antibodies may be of the warm or cold type but usually are warm. A prevailing view holds that the erythrocyte membrane proteins or prosthetic groups of these proteins are altered or mutated to become a "foreign" antigen, which the body's immune mechanism now does not recognize as "self." Antierythrocyte antibodies are formed, coat the erythrocytes, and in the presence of complement, hemolysis, fragmentation, or phagocytosis occurs. Diagnosis of AIHA is made by demonstrating a positive direct Coombs' test.

AITP is another form of autoimmune disease in which the platelets have either absorbed viral or drug antigens on their surfaces and become coated with antibody or have adsorbed antigen–antibody complexes directly. These antibody-coated platelets are rapidly removed from the circulation (Wilkins *et al.*, 1973). Definitive diagnosis is made by use of the PF-3 test (Karpatkin and Siskind, 1969).

Rheumatoid arthritis (RA) is characterized by the development of rheumatoid factor (RF), an autoantibody against IgG, and the immune complexes formed in the joints induce the chronic inflammatory lesion.

Tumors of the monocyte–macrophage system exemplified by lymphosarcoma can elicit either a poly- or monoclonal response. The hyperglobulinemic peaks can occur anywhere between the β_1 to γ_2 regions and range from very broad, diffuse peaks to very sharp, monoclonal spikes. The polyclonal peaks of lymphosarcoma are thought to be the result of a tumorous group of distantly related clones in contrast to the single discreet clones, which give rise to the monoclonal spikes.

v. Increased γ-Globulin (Sharp Increases): Monoclonal Gammopathy The monoclonal forms are characterized by sharp spikes of immunoglobulin and are frequently limited to the γ region (Fig. 5.6). A useful guideline for interpretation is to compare visually the sharpness of these spikes to the albumin peak. One or the other slope of the monoclonal spike is as steep or steeper than one of the slopes of the albumin peak. The monoclonal spike is the result of a single clone producing a single class of immunoglobulin, usually abnormal in nature. Thus, the monoclonal immunoglobulins are of identical structure and move as one on the SPE. These immunoglobulins have been described as "paraproteins" or as the "M" components because of the frequent occurrence of IgM. Waldenstrom's type macroglobulinemia with hyperviscosity and IgM monoclonal spikes have been reported in dogs (Hurvitz *et al.*, 1970; MacEwen *et al.*, 1977). In multiple myelomas, Bence–Jones proteins (light chains) are detected in approximately 50% of the cases in humans (Civantos *et al.*, 1973; Ritzmann *et al.*, 1972), but infrequently in dogs (Hurvitz, 1975). The characteristic monoclonal spike in the γ region, Bence–Jones proteinuria, and plasma cell tumors have been general findings in multiple myelomas reported in the horse (Cornelius *et al.*, 1959), dog (Osborne *et al.*, 1968; Shepard *et al.*, 1972), cat (Farrow and Penny, 1971), and rabbit (Pascal, 1961). Aleutian disease (AD) of mink, a valuable model of immunologic disease, is also characterized by plasma cell infiltration, hypoalbuminemia, hyperproteinemia, and hyper-γ-globuinemia, frequently with a monoclonal spike (Leader *et al.*, 1963; Thompson and Aliferis, 1964). Light-chain disease, in which the M component was identified as L chains (Bence–Jones proteins) in the plasma and urine, has been observed in dogs (Hurvitz *et al.*, 1971). An IgA monoclonal gammopathy in a dog without Bence–Jones proteinuria or plasma cell infiltration has also been observed (Dewhirst *et al.*, 1977).

Tumors of the monocyte–macrophage system (e.g., lymphosarcoma) frequently present with monoclonal spikes depending on the degree of cloning of the tumor cells. The dominant M component was identified as IgM in a case of lymphocytic leukemia in a dog (Braund *et al.*, 1978). A biclonal gammopathy has been observed in a dog with a combined myeloma and cutaneous lymphoma (Jacobs *et al.*, 1986) (Fig. 5.6). Even a triclonal gammopathy can be observed.

Monoclonal gammopathies are also seen in canine amyloidosis (Schwartzman, 1984) (Fig. 5.6) and in canine ehrlichiosis in association with large numbers of reactive plasma cells in the bone marrow (Breitschwerdt *et al.*, 1987).

Generally, the clinical characteristics of the monoclonal gammopathies are referable to the magnitude of plasma cell proliferation, the extent of organ infiltration and production of abnormal protein. Thus, immunologically associated diseases tend to be multifaceted and can present with bleeding tendencies, glomerulonephritis, polyarthritis, arteritis, hepatitis, and SLE. Treatment by cytotoxic drugs is largely ineffective, but symptomatic treatment by plasmapheresis is often followed by a brief period of clinical improvement. Infection is a common sequela and a frequent cause of death because of the immunosuppression. All monoclonal gammopathies are not necessarily pathologic, however, because they can be benign. In consequence, care must be exercised in the final evaluation of the monoclonal gammopathy.

3. *Increased A : G—Abnormal Profile*

a. Increased Albumin

True overproduction of albumin does not occur in any animal. Therefore, any rise in albumin is only a relative hyperalbuminemia due to hemoconcentration as a result of water loss and dehydration.

b. Decreased Globulins

The absence of γ-globulins in fetal serum or in serum from precolostral or colostrum-deprived neonatal animals can be readily demonstrated on SPE (Fig. 5.1). Hypo-γ-globulinemia occurred in about 15% of foals less than 2 weeks of age and was attributed to failure of colostral transfer even though most had nursed (McGuire *et al.*, 1975b). Combined immunodeficiency of Arabian foals is thought to be an inherited autosomal recessive disease. The disease is characterized by lymphopenia failure to synthesize IgG, IgM, and IgA and early death (McGuire *et al.*, 1975a). A selective IgM deficiency has been reported (Perryman *et al.*, 1977).

References

Allison, A. C. (1960). *Nature* **188,** 37.

Allison, J. B. (1957). *J. Am. Med. Assoc.* **164,** 283.

Bauer, J. E., Harvey, J. W., Asquith, R. L., McNulty, P. K., and Kivipelto, J. (1985). *Vet. Clin. Path.* **14,** 14.

Benditt, E. P., Wissler, R. W., Woolridge, R. L., Rowley, D. A., and Steffee, C. H., Jr. (1949). *Proc. Soc. Exp. Biol. Med.* **70,** 240.

Bjorneboe, M., Raaschon, F., and Sondergard, T. (1952). *Acta. Endocrinol.* **9,** 318.

Blackmore, D. Personal communication.

Booij, J. (1972). *Clin. Chem. Acta.* **38,** 355.

Bradford, M. M. (1976). *Anal. Bioch.* **72,** 248.

Braund, K. G., Everett, R. M., and Albert, R. A. (1978). *J. Am. Vet. Med. Assoc.* **172,** 1407.

Breitschwerdt, E. B., Woody, B. J., Zerbe, C. A., DeBuysscher, E. V., and Barta, O. (1987). *J. Vet. Int. Med.* **1,** 2.

Burton, S. A., Honor, D. J., Mackenzie, A. L., Eckersall, P. D., Markham, R. J. F., and Horney, B. S. (1994). *Am. J. Vet. Res.* **55,** 613.

Campbell, R. M., Cuthbertson, D. P., Mackie, W., McFarlane, A. S., Phillipson, A. T., and Sudsaneh, S. (1961). *J. Physiol.* **158,** 113.

Caspi, D., Snel, F. W., Batt, R. M., Bennett, D., Rutteman, G. R., Hartman, E. G., Baltz, M. L., Gruys, E., and Pepys, M. B. (1987). *Am. J. Vet. Res.* **48,** 919.

Civantos, F., Dominguez, C. J., Rywlin, A. M., and DiBella, J. (1973). *Lab. Med.* **4,** 373.

Cohen, S. (1956). *Biochem.* **64,** 286.

Conner, J. G., Echersall, P. D., Ferguson, J., and Douglas, T. A. (1988). *Res. Vet. Sci.* **45,** 107.

Cornelius, C. E., Goodbary, R. F., and Kennedy, P. C. (1959). *Cornell Vet.* **49,** 478.

Cornelius, C. E., Baker, N. F., Kaneko, J. J., and Douglas, J. R. (1962). *Am. J. Vet. Res.* **23,** 837.

Cornelius, C. E., Gazmuri, G., Gronwall, R., and Rhode, E. A. (1965). *Cornell Vet.* **55,** 110.

Davey, F. R., Carrington, C. E., and Nelson, D. A. (1972). *Clin. Chem.* **18,** 1360.

Dewhirst, M. W., Stamp, G. L., and Hurvitz, A. I. (1977). *J. Am. Vet. Med. Assoc.* **170,** 1313.

Dick, J., and Nielson, J. (1963). *J. Comp. Med. Vet. Sci.* **27,** 269.

Dimopoullos, G. T. (1961). *Am. J. Vet. Res.* **22,** 986.

Divers, T. J., Warner, A., Vaala, W. E., Whitlock, R. H., Acland, H. A., Mansmann, R. A., and Palmer, J. A. (1983). *J. Am. Vet. Med. Assoc.* **183,** 1407.

Dixon, F. J., Maurer, P. H., and Deechmiller, M. P. (1953). *Proc. Soc. Exp. Biol. Med.* **83,** 287.

Dobson, C. (1965). *Nature* **207,** 1304.

Dunlap, J. S., and Dickson, W. M. (1955). *Am. J. Vet. Res.* **58,** 91.

Ebel, K. H. (1953). *Zentr. Veterinarmed.* **AI,** 76.

Eckersall, P. D., Conner, J. G., and Parton, H. (1989). *Vet. Rec.* **124,** 490.

Eckersall, P. D. , Conner, J. G., and Harvie, J. (1991). *Vet. Res. Comm.* **15,** 17.

Edelman, G. M. (1973). *Science* **180,** 830.

Erwin, E. S., Varnell, T. R., and Page, H. M. (1959). *Proc. Soc. Exp. Biol. Med.* **100,** 373.

Farrow, B. R. H., and Penny, R. (1971). *J. Am. Vet. Med. Assoc.* **158,** 606.

Fex, G., Laurell, C. B., and Thulin, E. (1977). *Eur. J. Biochem.* **75,** 181.

Forstner, M. J. (1968). *Zentr. Veterinarmed.* **AI,** 76.

Foulon, T., Groslambert, P., and Groulade, P. (1992). *Bull. Acad. Vet. de France* **65,** 279.

Furukawa, T., and Sugiyama, F. (1986). *Jpn. J. Vet. Sci.* **48,** 643.

Gorham, J. H. (1971). *J. Am. Vet. Med. Assoc.* **158,** 846.

Groulade, P. (1985). *Prat. Med. Chir. Anim. Compagnie* **20,** 569.

Hayes, M. A. (1994). *Proc. VI Congr. ISACB* (J. H. Lumsden, ed.), p. 1.

Henry, R. J. (1964). "Clinical Chemistry," p. 197. Harper and Row, New York.

Hoch-Ligeti, C., Irvine, K., and Sprinkle, E. P. (1953). *Proc. Soc. Exp. Biol. Med.* **84,** 707.

Holliday, G. J., Mulligan, W., and Dalton, R. G. (1968). *Res. Vet. Sci.* **9,** 224.

Holmes, P. H., Dargie, J. D., Maclean, J. M., and Mulligan, W. (1968). *Vet. Record* **83,** 227.

Hurvitz, A. I. (1975). *In* "Textbook of Veterinary Internal Medicine" (S. J. Ettinger, ed.), p. 1707. Saunders, Philadelphia, Pennsylvania.

Hurvitz, A. I., Haskins, S. C., and Fischer, C. A. (1970). *J. Am. Vet. Med. Assoc.* **157,** 455.

Hurvitz, A. I., Kehoe, J. M., Capra, J. D., and Prata, R. (1971). *J. Am. Vet. Med. Assoc.* **159,** 1112.

Hurvitz, A. I., MacEwen, E. G., Middaugh, C. R., and Litman, G. W. (1977). *J. Am. Vet. Med. Assoc.* **170,** 511.

Iammarino, R. M. (1972). *In* "Standard Methods of Clinical Chemistry" (G. R. Cooper, ed.), Vol. 7, p. 185. Academic Press, New York.

Jacobs, R. R., Couto, C. G., and Wellman, M. L., (1986). *Vet. Pathol.* **23,** 211.

Jeffay, H., and Winzler, R. J. (1958). *J. Biol. Chem.* **231,** 111.

Johansson, B. G., Kindmark, C. O., Trell, E. Y., and Wollheim, F. A. (1972). *Scand. J. Clin. Lab. Invest.* **29,** Suppl. 124, 117.

Kaneko, J. (1983). Thesis. University of Hawaii, Honolulu, Hawaii.

Kaneko, J. J. (1963). *Ann. N. Y. Acad. Sci.* **104,** 689.

Kaneko, J. J., and Smith, R. (1967). *Calif. Vet.* **21,** 25.

Kaneko, J. J., Moulton, J. E., Broedy, R. S., and Perryman, V. D. (1965). *J. Am. Vet. Med. Assoc.* **146,** 463.

Karpatkin, S., and Siskind, G. W. (1969). *Blood* **33,** 795.

Kawai, T. (1973). "Clinical Aspects of the Plasma Proteins." Lippincott, Philadelphia, Pennsylvania.

Keay, G. (1982). *Res. Vet. Sci.* **33,** 343.

Keay, G., and Doxey, D. L. (1982). *Vet. Res. Comm.* **5,** 263.

Khan, K. N. W., Kociba, G. J., Wellman, M. L., and Reiter, J. A. (1992). *Comp. Hematol. Int.* **2,** 133.

Kniazeff, A. J., Rimer, V., and Gaeta, L. (1967). *Nature* **214,** 805.

Kristensen, F., and Barsanti, J. A. (1977). *Am. J. Vet. Res.* **38,** 399.

Kushner, I., and Feldman, G. (1978). *Res. Vet. Sci.* **148,** 466.

Larson, B. L., and Kendall, K. A. (1957). *J. Dairy Sci.* **40,** 659.

Laurell, C. B. (1972). *Scand. J. Clin. Lab. Invest.* **29,** Suppl. 124, 71.

Leader, R. W., Wagner, G. M., Henson, J. B., and Gorham, J. H. (1963). *Am. J. Path.* **43,** 33.

Leveille, G. A., and Sauberlich, H. E. (1961). *J. Nutr.* **74,** 500.

Lowry, O. H., Rosebrough, N. J., Farr, A. C., and Randall, R. J. (1951). *J. Biol. Chem.* **193,** 265.

MacEwen, E. G., Hurvitz, A. I., and Hayes, A. (1977). *J. Am. Vet. Med. Assoc.* **170,** 1309.

McGuire, T. C., Banks, K. L., and Poppie, M. J. (1975a). *Am. J. Vet. Path.* **80,** 551.

McGuire, T. C., Poppie, M. J., and Banks, K. L. (1975b). *J. Am. Vet. Med. Assoc.* **166,** 71.

Mancini, G., Carbonara, A. O., and Heremans, J. R. (1965). *Immunochemistry* **2,** 235.

Mattheeuws, D. R. G., Kaneko, J. J., Loy, R. G., Cornelius, C. E., and Wheat, J. D. (1966). *Am. J. Vet. Res.* **27,** 699.

Maurey, C. P. (1985). *Clin. Sci.* **68,** 233.

Meuten, D. J., Butler, D. G., Thomson, G. W., and Lumsden, J. H. (1978). *J. Am. Vet. Med. Assoc.* **172,** 326.

Morimatsu, M., Sarikaputi, M., Syuto, B., Yamamoto, S., and Naiki, M. (1992). *Vet. Immunol. Immunopathol.* **33,** 365.

Ogilvie, G. K., Walters, L. W., Greely, S. G., Henkel, S. E., and Salman, H. D. (1993). *J. Am. Vet. Med. Assoc.* **203,** 1144.

Ohwada, K., and Tamura, K. (1995). *Exp. Anim.* **42,** 627.

Oppenheimer, J. H., Surks, M. I., Smith, J. C., and Squef, R. (1965). *J. Biol. Chem.* **240,** 173.

Osborne, C. A., and Vernier, R. L. (1973). *J. Am. Anim. Hosp. Assoc.* **9,** 101.

Osborne, C. A., Perman, V., Sautter, J. H., Stevens, J., and Hanlon, G. F. (1968). *J. Am. Vet. Med. Assoc.* **153,** 1300.

Pascal, R. R. (1961). *Cornell Vet.* **51,** 528.

Pepys, M. B., Baltz, M. L., Tennent, G. A., Kent, J., Ousey, J., and Rossdale, P. D. (1989). *Equine Vet. J.* **21,** 106.

Perk, K., and Loebl, K. (1960). *Refuah Vet.* **17,** 46.

Perryman, L. E., McGuire, T. C., and Hilbert, B. J. (1977). *J. Am. Vet. Med. Assoc.* **170,** 212.

Pfeiffer, N. E., McGuire, T. C., Bendel, R. B., and Weikel, J. M. (1977). *Am. J. Vet. Res.* **38,** 693.

Pierce, A. E. (1955). *J. Hyg.* **53,** 247.

Ritzmann, S. E., Daniels, J. C., and Lawrence, M. C. (1972). *Tex. Med.* **68,** 91.

Rook, J. A. F., Moustgaard, J., and Jacobsen, P. E. (1951). *Biol. Abstr.* **27,** 27237.

Rumbaugh, G. E., and Adamson, P. J. W. (1983). *J. Am. Vet. Med. Assoc.* **183,** 769.

Rumbaugh, G. E., Ardans, A. A., Ginno, D., and Trommershausen-Smith, A. (1978a). *J. Am. Vet. Med. Assoc.* **172,** 321.

Rumbaugh, G. E., Smith, B. P., and Carlson, G. P. (1978b). *J. Am. Vet. Med. Assoc.* **172,** 304.

Rutqvist, L. (1958). *Am. J. Vet. Res.* **19,** 25.

Sato, T., Oda, K., and Kubo, M. (1979). *Cornell Vet.* **69,** 3.

Schwartzman, R. M. (1984). *J. Am. Vet. Med. Assoc.* **185,** 102.

Shepard, V. J., Dodds-Laffin, W. J., and Laffin, R. J. (1972). *J. Am. Vet. Med. Assoc.* **160,** 1121.

Skinner, J. G., Brown, R. A. L., and Roberts, L. (1991). *Vet. Rec.* **128,** 147.

Smith, P. K., Krohn, R. I., Hermanson, G. T., Mallia, A. K., Gartner, F. H., Provenzano, M. D., Fujimoto, E. K., Goeke, N. M., Olson, B. J., and Klenk, D. C. (1985). *Anal. Biochem.* **150,** 76.

Solomon, A. (1976). *N. Engl. J. Med.* **294,** 17.

Stockl, W., and Zacherl, M. K. (1953). *Z. Physiol. Chem.* **293,** 278.

Strombeck, D. R., Weiser, M. G., and Kaneko, J. J. (1975). *J. Am. Vet. Med. Assoc.* **166,** 1105.

Sturkie, P. D. (1951). *Endocrinology* **49,** 565.

Tamura, K., Yatsu, T., Itoh, H., and Motoi, Y. (1989). *Jpn. J. Vet. Sci.* **51,** 987.

Tanaka, K., and Aoki, S. (1963). *Natl. Inst. Anim. Health Quart.* **5,** 49.

Tennant, B., Baldwin, B. H., Braun, R. K., Norcross, N. L., and Sandholm, M. (1979). *J. Am. Vet. Med. Assoc.* **174,** 848.

Thompson, G. R., and Aliferis, P. (1964). *Arthritis Rheum.* **7,** 521.

Thoresen, H. E., Tverdal, A., Havre, G., and Morberg, H. (1995). *Vet. Clin. Pathol.* **24,** 129.

Tumbleson, M. E., Burks, M. F., and Wingfield, W. E. (1972). *Cornell Vet.* **63,** 65.

Vaermans, J. P., and Heremans, J. F. (1969). *Immunochemistry* **6,** 779.

Weimer, H. E. (1961). *Ann. N. Y. Acad. Sci.* **94,** 225.

Wilkins, R. J., Hurvitz, A. I., and Dodds-Laffin, W. J. (1973). *J. Am. Vet. Med. Assoc.* **163,** 277.

Wissler, R. W., Woolridge, R. L., Steffee, C. H., Jr., and Cannon, P. R. C. (1946). *J. Immunol.* **52,** 267.

Yamamoto, S., Shida, T., Miyaji, S., Santsuka, H., Fujise, H., Mukawa, K., Fukukawa, E., Nagae, T., and Naiki, M. (1993). *Vet. Re. Comm.* **17,** 85.

C H A P T E R

6

Clinical Immunology

LAUREL J. GERSHWIN

I. INTRODUCTION

A healthy immune system is critical to survival. Clinical immunology is directed at evaluation of the immune system of patients and the ability of the immune system to respond to antigenic stimuli. Assays developed to target specific parts of the immune system enable the clinician not only to determine if a patient has normal immune responsiveness, but also to target those parts of the immune system that are suspect of inadequate function. Serology has historically been used to determine retrospectively if a patient was infected with a particular disease agent; antibody titers continue to have importance in diagnostics. In the fourth edition of this book, it was stated that veterinary clinical immunology is a rapidly growing field that has lagged somewhat behind its human counterpart due to a lack of specific assays and reagents. In the past six years the field of veterinary immunology has seen immense growth. The widespread use of molecular cloning techniques and the polymerase chain reaction have provided tools for development of probes and expressed genes that allow veterinary immunologists to measure cytokines, adhesion molecules, and a variety of other cell surface molecules. Awareness on the part of practicing veterinarians of some less common immunological diseases has increased and some diseases with unknown etiology have been found to have a significant immune component.

II. THE IMMUNE RESPONSE

A. Innate Immune Responses

Several inherited defects of innate immunity have emerged during the inevitable inbreeding inherent in certain animal breeding programs. These defects demonstrate the importance of the particular function of the defective component. For example, a recently described absence of adhesion molecules on bovine neutrophils was manifested in severe bacterial infections in the presence of ever increasing neutrophil counts. Holstein calves lacking the CD18 glycoprotein on neutrophil cell membranes were unable to deal with normally easily stopped bacterial infections because the neutrophils were unable to marginate along blood vessels. They were thus unable to undergo diapedesis and migrate to the site of a bacterial infection. This syndrome, called *bovine leukocyte adhesion deficiency* (BLAD) can be detected by use of a polymerase chain reaction (PCR) to detect the presence of the defective gene (Kehrli *et al.*, 1990).

Several other inherited defects of innate immunity focus on slightly different functions of the neutrophil. For example, in canine granulopathy syndrome the neutrophils are able to marginate and undergo diapedesis, but their ability to kill organisms once they have arrived at the site of infection is severely compromised (Renshaw and Davis, 1979). This is also true for the Chediak–Higaski syndrome seen in Persian cats. Assays are described later in this chapter for the evaluation of neutrophil function.

B. Acquired Immune Responses: Antigens

The ability of a patient to develop an immune response to a specific antigen depends on the presence of B lymphocytes, helper T lymphocytes, and generally an antigen presenting cell, such as a macrophage. The first exposure to antigen elicits primarily an IgM response, but subsequent exposures trigger the production of IgG. Mucosal sites have a preference for production of secretory immune responses, such as IgA production. IgE production is elicited by certain antigens, *allergens,* in individuals that are genetically high responders, *atopic,* and by parasite antigens.

This chapter addresses the various components of the immune system and how to evaluate whether or not their function is normal. Information is included on the variety of immunological tests that are available for detecting immune response to certain pathogens. Some have importance in the historical development of serological techniques, whereas others are quite newly developed.

An antigen is broadly defined as something foreign. Indeed foreignness is the most important characteristic of an antigen. As an illustration, consider a simple serum protein such as albumin. If bovine serum albumin is injected into a rabbit, the rabbit's immune system recognizes it as foreign and develops an immune response to it. If, however, that same bovine serum albumin is injected into a cow, it will not be recognized as foreign and no immune response will occur. Another important characteristic of an antigen is its size. Most effective antigens are greater than 10,000 Da. Other factors such as chemical complexity and molecular rigidity are important determinants of antigenicity.

A complete antigen is a substance that is able to induce an immune response and can then react with the products of that response, namely, antibodies and sensitized T lymphocytes. An incomplete antigen, also known as a hapten, cannot induce an immune response by itself, but requires prior binding to a carrier molecule, usually a protein. The hapten can then bind to the induced antibodies without attachment to a carrier. There are several examples of drug moieties, such as penicillinoyl, that act as haptens and bind to host proteins or cells to induce an immune response. Once induced, these antibodies can induce immunological disease.

Although an individual protein can be an antigen, in reality the immune response is not directed to the entire molecule, but rather to one or more antigenic determinants or epitopes present on the protein. In a normal (polyclonal) immune response, some antibodies are produced specific for each of these epitopes. This is the basis for the phenomenon of cross reactivity. Related but different antigens will usually react with antibodies induced by immunization with each other. This occurs because they share at least one epitope. Cross reactivity is important because it is essential to have specific antisera as reagents for use in diagnostic assays. One way to obtain such reagents is to absorb with the offending organism antisera that cross react, thereby removing those antibodies that recognize the cross-reactive epitopes and leaving only those that are specific for the immunogen used to prepare the antiserum. Another way to prepare specific antisera is to develop monoclonal antibodies against the agent, as described later in this chapter.

III. ANTIBODIES

Electrophoretic separation of serum proteins creates four broad categories: albumin, α-globulins, β-globulins, and γ-globulins. The antibody activity is present in the γ-globulin fraction, with a slight amount in the β fraction. These immunoglobulins are heterogeneous, having different molecular weights and functional properties.

The five classes (isotypes) of immunoglobulins are IgG, IgM, IgA, IgE, and IgD. They share a basic structure, which consists of four polypeptide chains bound by disulfide bonds. Two of these chains are called light chains, because with a molecular weight of about 22,000 Da each. They are lighter than the other two heavy chains (approximately 55,000 Da each). At the nitrogen terminal of the polypeptide chains on all four chains is a portion of variable amino acid sequences. This is the antigen binding end of the immunoglobulin. The hinge region of the immunoglobulin provides for flexibility of the molecule, so that the two "arms" containing the antigen binding portion can move apart. Treatment of an IgG molecule with papain causes it to break into three pieces, two that contain the antigen binding sites (Fab), and a third consisting solely of heavy chain, and called the Fc (for crystallizable) fragment. Sometimes antibody reagents are described as being *Fc specific*, which means that there is no cross reactivity between isotypes due to recognition of light chains.

A. Immunoglobulin G

In the serum IgG is the antibody class with the greatest concentration, approximately 1–2 g/100 ml, with some species differences (Tizard, 1992). Subclasses of IgG are recognized in most species. IgG has a four polypeptide chain structure with a total molecular weight of 180,000 Da. The heavy chains in IgG are called gamma chains and are unique to IgG. Immunoglobulin G is important in host defense because it can exit the vascular system and distribute throughout the extravascular tissue fluid where it has many protective functions. For example, IgG can agglutinate bacteria, causing them to clump; it can opsonize bacteria, by binding to the bacteria by the Fab fragment and to the phagocyte by receptors for the Fc fragment, thereby facilitating engulfment of the bacteria by the phagocyte. Complement (a series of serum proteins to be discussed later in this chapter) can be fixed by IgG molecules and target cells can be lysed by this mechanism. In addition, IgG can participate with lymphocytes called null cells in antibody-dependent cellular cytotoxicity. This mechanism allows destruction of virus-infected cells after recognition by the antibody bound to the null cell. IgG can neutralize toxins, such as those produced by *Clostridium tetani.*

B. Immunoglobulin M

Immunoglobulin M (IgM) is the first antibody to be synthesized in response to an immunogenic stimulus and is the first antibody seen in ontogeny. In serum IgM is present in the second greatest concentration, generally between 100 and 400 mg/100 ml (species dependent). The structure of IgM consists of five of the basic four polypeptide units held together by a J chain. The large size of IgM (900,000 Da) keeps it confined to the intravascular space. There are a total of 10 potential antigen binding sites on IgM. Even though in reality, due to steric hindrance, only five to seven of the antigen binding sites are functionally active, this large capacity to bind antigen makes IgM a very efficient antibody at agglutination, precipitation, opsonization, complement fixation, and virus neutralization.

C. Immunoglobulin A

Immunoglobulin A exists primarily in two forms, as a monomer (160,000 Da) in the blood-vascular compartment, and in a dimeric secretory form (390,000 Da). Less commonly, polymers of greater number occur. The dimeric form consists of two monomers, each containing a heavy chain (alpha) and a light chain. These are held together by a J chain and include an additional component called the *secretory piece.* The secretory piece is produced by mucosal epithelial cells and functions to assist in transport of IgA dimers from the lamina propria of the intestine through to the lumen where it then protects the IgA dimer from proteolysis by intestinal enzymes. In domestic animals IgA is important as a secretory antibody both within the intestinal tract and the lung. It is capable of neutralizing virus and preventing adherence of bacterial pathogens to target tissues. It does not function as an opsonin and is unable to fix complement.

D. Immunoglobulin E

Immunoglobulin E is recognized and characterized in dogs, cattle, sheep, pigs, and horses and is functionally recognized in cats. The existence of IgE in chickens is still undocumented. IgE occurs normally in very small amounts in the serum (nanogram quantities). In allergic or parasitized individuals the serum concentration of IgE is greatly increased. The basic four polypeptide chain structure of IgE, with epsilon heavy chains that contain one additional domain, has a molecular weight of 196,000 Da. Functions of IgE are mediated through its ability to bind via the Fc portion to receptors on tissue mast cells and blood basophils. When antigen cross-links these cell-bound antibodies the cell degranulates, releasing vasoactive amines, stimulating leukotriene synthesis, and resulting in potent pharmacologic effects.

E. Immunoglobulin D

A complete discussion of immunoglobulin isotypes should include IgD, the antibody that has been demonstrated on lymphocyte surfaces. In mouse and human it has been shown to be an important antigen receptor during the differentiation of B lymphocytes. Although IgD has been demonstrated in pigs and in chickens, information is lacking in other domestic animal species. IgD exists in the four polypeptide chain configuration (heavy chains are called delta) with a molecular weight of 180,000 Da.

F. Epitopes and Specificity of Immunoglobulins

Most antigens contain several different antigenic determinants or epitopes. Larger molecules usually have more epitopes than small molecules. Consider, for example, the protein bovine serum albumin, with a molecular weight of about 60,000 Da. Ten to 12 different epitopes could easily be present on such a molecule. When we immunize an animal with an antigen and then analyze the antibody response, we measure the sum of the immune responses to each of these epitopes. This is called a *polyclonal response.* Each immature B lymphocyte with specificity for the antigen recognizes only one epitope. Consequently, the progeny of that cell, the resulting clone, produces antibody that reacts with a single epitope. However, within the animal are many such cells and the resultant antibodies are mixed together in the serum (a polyclonal antiserum). Recent advances in biotechnology have created a new type of antibody reagent, monoclonal antibodies, produced by hybridomas. Monoclonal antibodies are a characteristic of patients with multiple myeloma. These plasma cell tumors originate from single antibody forming cells. Neoplastic proliferation of these cells results in a high level of circulating antibody all specific for a single epitope and all with identical isotype. Kohler and Milstein (1975) took advantage of this type of tumor and produced a hybrid cell, between a mouse spleen-derived B lymphocyte and a mouse myeloma cell line. These hybrids have the antigenic specificity of the B cell and the ability to grow continuously in the tissue culture from the myeloma cell line. This technique of fusing spleen cells from immunized mice with myeloma cells is now a commonly used technique to produce antibody reagents that have specificity for a single epitope. Additional advantages of these reagents include their ability to grow *in vitro* thus guaranteeing availability of large amounts of standardized antibody reagents for use in different laboratories.

IV. COMPONENTS OF THE IMMUNE RESPONSE

Antibody formation in response to antigen requires the cooperation of several cell types. The B lymphocyte is the most important cell in humoral immunity because it possesses the receptors for antigen recognition and ultimately gives rise to the antibody producing cell. However, the role of the T helper lymphocyte is critical for a response to most antigens. Antigen must also be presented to the T cells, and while B cells can present antigen in some cases, the majority of external antigens, such as bacterial pathogens, first require some processing by a macrophage, which acts as the presenting cell to T helper cells.

A. B Lymphocytes

Although lymphocytes are indistinguishable from each other when observed with a microscope and a simple Wright's or Giemsa stain, there are actually several very distinct populations. B lymphocytes are the cells that differentiate into antibody forming plasma cells following exposure to antigen. B cells develop in the bone marrow of mammals or bursa of Fabricius in fowl. They then seed specific areas of the secondary lymphoid organs, such as germinal centers in lymph nodes. Antigen receptors, present on the B-cell membrane, are actually immunoglobulin molecules, with an additional piece of polypeptide chain that anchors the Fc fragment into the cell membrane. Each B-cell receptor has a unique variable region that allows it to react with the appropriate antigen. The specificity of the antibodies that result from antigenic stimulation of a B-cell clone is identical to that of the antigen receptors on the original B cell that was stimulated. During development of a B-cell clone it displays first IgM on the cell surface. This isotype is then joined by IgD, and finally other isotype receptors, such as

IgG or IgA, are displayed. The final differentiation step creates the antibody forming cell, the plasma cell.

B. Antigen Presenting Cells

As stated earlier, antigen is presented to T cells on a cell membrane surface, frequently a macrophage, dendritic cell, Langerhans' cell, or other cell carrying major histocompatibility complex (MHC) antigen class II. The presence of this molecule is imperative for T helper presentation because the cell surface MHC class II glycoprotein cradles the processed antigen within a specific groove and displays it in such a way that the T-cell receptor can recognize it. Great detail on this phenomenon is available in several reference texts listed at the end of this chapter.

C. T Cells and Cytokines

Progress in the area of antigen-specific T-cell activation and cytokine production has been immense in recent years. Recognition of a variety of chemokines, interleukins and other cell-stimulatory factors has allowed for a better understanding of how the immune response is regulated. For example, the profile of cytokines secreted by a T cell after antigen recognition is dependent on the type of T cell. A subset of T cells, referred to as T helper type 1 cells, which secrete gamma interferon and interleukin 2 (IL-2), promotes delayed-type hypersensitivity reactions; whereas a different subset, T helper type 2 cells, directs B cells to produce antibodies, and thus predisposes for a humoral response (Salgame *et al.*, 1991). The major cytokine involved in this response is interleukin 4 (IL-4), produced by CD4+ T cells. Besides stimulation of B-cell proliferation and differentiation (especially into IgE producing cells), IL-4 also increases expression of MHC class II antigens and CD23 (low affinity IgE receptor) expression on appropriate cells. The discovery of interleukin 13 (IL-13) has added another cytokine with similar activity as IL-4. Interleukin 13 also induces CD23 expression on B cells and upregulates MHC class II expression. Unlike IL-4, IL-13 is produced by activated cells of both the CD4+ and CD8+ phenotypes. Production of IL-13 occurs earlier after cell stimulation than IL-4 and its effect lasts longer. The effects of IL-4 on production of gamma interferon are downregulatory, whereas the effects of IL-13 are stimulatory for gamma interferon production. Although IL-13 behaves like IL-4 and induces a B-cell switch to production of IgG4 and IgE in humans, thus facilitating an allergic response, the same does not appear to be true in mice. Thus there will most likely be some species-specific differences noted among the domestic animals.

Along with the recognition of T-cell subsets that produce different cytokine patterns, a series of disease paradigms has developed: activation of T helper type 2 cells causes the production of IL-4, IL-5, IL-6, and IL-10, which leads to the selective differentiation of B cells into plasma cells that produce IgE, IgA, and some classes of IgG. Activation of T helper type 1 cells causes production of IL-2, gamma interferon, and tumor necrosis factor (TNF), with some selective IgM and IgG induction as well (Mosmann and Coffman, 1989; Romagnani, 1991). This type of response facilitates the delayed-type hypersensitivity reaction. Certain diseases, such as leprosy in humans, can exist in several forms, which can be directly linked to a particular T-cell profile. When immunity is focused on production of a cellular response, there is greater healing as compared to the humoral response, which results in disease exacerbation.

Another group of cytokines is of particular importance in response to disease; these are the inflammatory cytokines, TNF, IL-1, IL-6, and IL-8. The inflammatory cytokines are produced early in disease and their production is independent of antigen. Bacterial lipopolysaccharide derived from gram-negative bacteria is a potent inducer of these cytokines. Chemotaxis of neutrophils is stimulated by IL-8 (Arai *et al.*, 1990).

V. PRIMARY AND SECONDARY IMMUNE RESPONSES

The primary immune response occurs after the animal's immune system is exposed to an antigen for the first time. In the case of young animals with high levels of maternal immunoglobulin specific for the antigen, the primary immune response may be delayed until there are sufficiently low levels of maternal antibody for persistence of antigen to stimulate immune cells in the animal. In the absence of passive antibody, the antigen will stimulate the B and T cells, as described earlier, resulting in the production of IgM antibody for a short period of time. The IgM response takes 10–14 days to develop and is sometimes followed with low levels of IgG, which do not persist long in the circulation.

Subsequent exposures to antigen result in a secondary immune response, a quicker and greater magnitude response than that seen with initial exposure. Also of note is the preponderance of IgG in the secondary response. The concentration of antibody is greater and the titer stays high longer. Further exposure to antigens in the future results in increasing amounts (greater titer) of antibody. This information is extremely useful in determining the result of serological assays for infectious agents. For example, if a horse has a pectoral abscess, which resembles a *Corynebacterium pseudotuberculosis* lesion, and a blood sample obtained for testing has a titer of 16, the result on its own is relatively meaningless. However, if a second sample taken 3 weeks later shows an increase in titer to 128, then the response can be attributed to infection with the *C. pseudotuberculosis* organism.

To illustrate how the isotype-specific immune response can be used in diagnostics, an example can be cited: In the clinical immunology laboratory, serological tests for infection with canine distemper virus can be very confusing, primarily because the presence of maternal antibody and the frequency of vaccination generally create dogs with some level of antibody. However, performance of a serological test that determines an IgM titer separately from an IgG titer is often helpful in determining the likelihood of recent exposure to the virus.

VI. EVALUATION OF IMMUNE RESPONSES: DETECTION OF ANTIGENS AND ANTIBODIES

A. General Principles

Immunological tests to detect antigens or antibodies are based on several properties of antigen–antibody reactions. An antibody binds specifically with the antigen that induced its formation. Antibodies of the IgG and IgM isotypes precipitate with soluble antigen or agglutinate particulate antigen. These reactions can be used to secondary and/or qualitatively observe an antigen–antibody reaction. Antibody to a virus can neutralize its infectivity and antibody to a toxin can neutralize its toxigenic effect. These properties have been used in assays to measure the antibody response. Regardless of what type of assay is performed, if it is based on an antigen–antibody reaction it is important that there are not any cross-reacting antigens or antibodies that could give an erroneous result. Specificity of an immunological test is best achieved with well-defined reagents. Detection of antibody isotypes that do not precipitate with antigen, such as IgE, require either *in vivo* assays that make use of the mast cell degranulation function, such as intradermal skin tests, or assays that use a second labeled antibody to indicate the presence of antigen-bound IgE, such as the radioallergosorbent test (RAST) or enzyme-linked immunosorbent test (ELISA).

A second important concept to remember when evaluating immunological tests is the sensitivity of the test. The most sensitive tests, that is, those that detect the smallest amount of protein, are referred to as primary binding assays, radioimmunoassay, and ELISA. These can detect nanogram amounts of protein. Virus neutralization is also very sensitive. Other techniques such as agar gel diffusion (a precipitin test) are much less sensitive, requiring greater than 10 $\mu g/ml$ before protein antigen is detectable. In immunological tests in which serum is diluted serially, the results are usually expressed as a titer. The titer by convention is the inverse of the last dilution that produces a positive reaction (end point). In studies involving infectious disease, it is desirable to obtain two serum samples, one during the acute phase of the illness and another after convalescence. A rise in titer, as illustrated earlier, of two doubling dilutions or more is suggestive of a positive role for the particular pathogen in the disease process.

B. Serological Tests

1. Precipitin Reactions

Soluble antigen reacts with specific antibody to form immune complexes. These immune complexes can be small or large depending on the proportion of antigen to antibody within the complex. When there is either antigen or antibody excess, complexes are soluble. However when there is an optimal proportion of antigen to antibody, complexes are insoluble and form a visible precipitate. Initially the solution containing the complexes becomes cloudy, then after some time they settle out. This is because a lattice structure is formed between the multivalent antigen and the bivalent antibody. Precipitation is the basis for several immunological assays. There are several ways to perform secondary precipitin reactions. Tube precipitin tests can quantitatively determine the amount of antigen that reacts with a given antiserum at equivalence and thus provide an estimation of the "strength" of a particular antiserum. Laser nephelometry can be used to detect immune complexes in solution. This technique is receiving some application in the clinical immunology laboratory for rapid quantitation of immunoglobulins. Precipitin reactions can be performed in a semisolid matrix, such as agar gel. When the reactants reach the point of optimum proportion, a line of precipitation develops. Ouchterlony double diffusion in gel makes use of this principle. Wells are cut in an agarose slab and are filled with either antigen or antibody. After incubation for 24–48 hours, visible lines are present between antigen and antibody wells, indicating the presence of immune complexes. In this technique several types of reactions are identifiable: identity, nonidentity, and partial identity. Unknown reactants can be identified by this test by demonstration of an arc of identity with a known antigen. When two separate antigens in adjacent wells react with different antibodies in the same antibody well, the precipitin lines cross each other and the reaction is termed *nonidentity*. Partial identity occurs when two adjacent wells contain antigen with a cross-reacting determinant. Partial identity is recognizable as a spur. These principles are useful in several diagnostic tests. The agar gel diffusion test for diagnosis of equine infectious anemia (Coggin's test) is an Ouchterlony double immunodiffusion test in which viral antigen is placed in a central well and positive control sera are alternated with test sera in the peripheral wells. A line of identity between positive and test serum is indicative of a positive test (Fig. 6.1). A similar principle is used in the bovine leukemia test.

Another type of agar gel immunodiffusion test is widely used to quantitate secondary levels of specific immunoglobulin isotypes. In single radial diffusion, antiserum specific for the reactant being measured is incorporated into the agar. Wells are punched and filled with serum to be assayed. During incubation the antigens in the serum diffuse radially into the agarose and cause formation of a precipitate ring around the well at the point where an optimal proportion of antigen to antibody exists for lattice formation. The sizes of the precipitate rings are proportional to the amount of antigen present in the wells. Wells containing known standards are used to draw a standard curve from which quantitative values for the unknown serums are obtained. Total serum IgG, IgM, or IgA is readily measured for most species using this method. Test kits are available from several commercial companies. These kits provide agarose plates containing the appropriate antibody and a set of standards (Fig. 6.2).

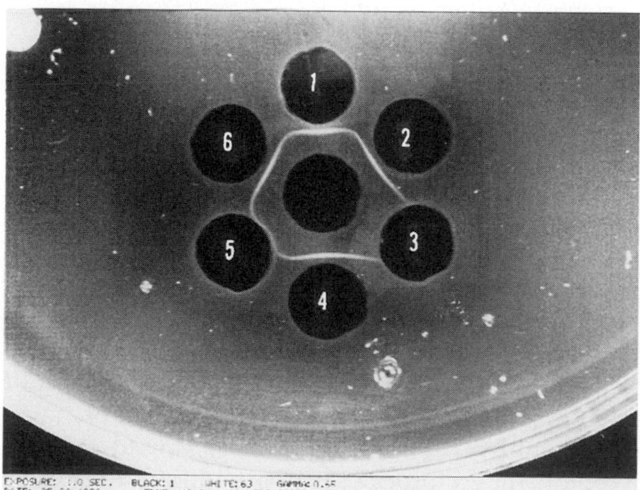

FIGURE 6.1 Coggin's test for equine infectious anemia. Viral antigen is in the center well. Lines of identity between the positive controls are in wells 2, 4, and 6. The sample in well 1 shows positive, the sample in well 5 is weakly positive, and well 3 is negative.

Immunoelectrophoresis is another technique that utilizes the principle of precipitation and provides a useful test in the clinical immunology laboratory (Caron and Penn, 1992). In this test antigens, usually proteins in the patient's serum, are separated by electrical charge in an agarose gel. After protein separation is complete, a trough is removed from the agarose and antiserum specific for components in the serum is added. Wherever antigen and antibody meet at optimal proportions, precipitin arcs develop. In the clinical immunology laboratory a patient's serum is compared with a control serum on opposite sides of each antiserum trough. This technique is a useful adjunct to the diagnosis of hypogammaglobulinemia, paraproteinemia, and isotype-specific immunoglobulin deficiency (Fig. 6.3).

2. *Agglutination*

Agglutination reactions have important applications in diagnostic immunology. Although a relatively quick and easy assay, agglutination has greater sensitivity than precipitin-type reactions. The process of agglutination occurs when particulate antigens such as bacterial cells or erythrocytes are incubated with a source of antibody that recognizes determinants on these cell surfaces. This type of assay has traditionally been performed to detect serum agglutinating antibodies specific for *Salmonella* H and O antigens (Olitzki, 1972). A slide test is performed by mixing a drop of specific antiserum with a drop of suspended bacteria. Rapid development of clumping indicates a positive test. This process was used to define bacterial serotypes. Agglutination tests have found use in test kits, some of which utilize the concept of passive agglutination, in which an antigen is chemically fixed to a solid particle, such as a latex bead or an erythrocyte. The patient's serum is mixed with the solid substrate and observed for agglutination. It provides a quick positive or negative result when performed in this manner. A commonly used test kit for rheumatoid factor in canine serum is based on this principle. A more sensitive and quantitative assay is the passive agglutination test used to determine antibody to *Toxoplasma gondii*. This test kit provides latex beads coated with *Toxoplasma* antigen. The beads are mixed with serial dilutions of serum and are plated into 96-well microtitration plates. After incubation wells are read for agglutination of the beads and a titer is obtained.

3. *Coombs' (Antiglobulin) Test*

The antiglobulin test is performed to detect the presence of incomplete antibody bound to erythrocyte membranes. Its use in the diagnosis of autoimmune

FIGURE 6.2 Single radial immunodiffusion test for measuring the concentrations of bovine IgM. The diameters of the standards (a) are used to plot a curve (b) from which sample concentrations are determined. The Y axis is the amount of IgM and the X axis is the diameter of the diffusion reaction.

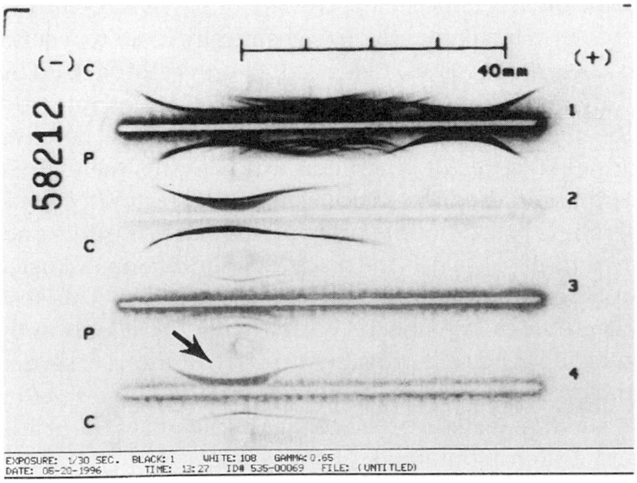

hemolytic anemia is described later in this chapter. The basis of the test is agglutination.

4. Hemagglutination and Hemagglutination Inhibition

Some viruses have receptors for erythrocytes and when incubated in their presence cause them to agglutinate. This phenomenon is called *hemagglutination.* The specifics of the erythrocyte source, mammalian or avian, and optimum temperature and time for reaction vary depending on the virus of interest. Myxoviruses, paramyxoviruses, enteroviruses, and adenoviruses are several virus groups with members of veterinary interest that are capable of hemagglutination (Davis *et al.*, 1990). The hemagglutination procedure is itself of little immunological interest. However, the ability of antiserum to inhibit the hemagglutination caused by virus receptors for erythrocytes has been utilized to develop a serological test called *hemagglutination inhibition* (HI). The antibodies bind to receptor sites for erythrocytes and thus block the hemagglutination reaction. The test

FIGURE 6.3 This immunoelectrophoresis profile is indicative of the presence of an IgA myeloma protein (*arrow*). Trough contents: (1) anti-canine serum; (2) anti-canine IgG (h and l); (3) anti-canine IgM; (4) anti-canine IgA. Well contents: (C) normal control dog; (P) patient.

is used to measure antibody titers to the virus. Alternatively, with a known source of antiserum, one can use the HI test as a preliminary viral identification step.

To perform the HI test, serial twofold dilutions of heat-inactivated serum are prepared in saline. A 0.25-ml aliquot of each dilution is then mixed with a similar amount of viral suspension that contains four hemagglutinating units. These are mixed and incubated. Next 0.25 ml of a 1% erythrocyte suspension is added and the tubes are mixed again and incubated at the appropriate temperature and time for the virus of interest. The agglutination or absence thereof is read and the HI titer of the serum is assigned as the reciprocal of the highest serum dilution that completely prevents hemagglutination. Alternatively, one can perform the test by making serial dilutions of the virus suspension and using a standard amount of serum. Test sera are then compared with known negative and positive sera. The former HI test is called the α procedure and the latter, the β procedure. Appropriate controls must be included in either procedure, particularly to prevent false-positive results from the presence of hemagglutinating substances in test sera.

5. Virus Neutralization Assays

Detection of antibodies specific for viral epitopes is an important indicator of exposure to the virus. However, antibodies detected by precipitin tests or primary binding tests are not necessarily protective. It is therefore often worthwhile to measure the titer of functional antibody (i.e., antibody capable of neutralizing the infectivity of a virus), and which is specific for epitopes responsible for virus adherence to and entry into cells. The two types of virus neutralization are the α and the β procedures. In the α procedure, the virus concentration is varied and the serum dilution is held constant, whereas in the β procedure the serum is diluted and the virus concentration is held constant. The β procedure provides a more quantitative test and wastes less serum. This procedure is performed by incubation of serum dilutions with virus and then observing whether or not the virus is still able to infect cells and produce cytopathic effect (CPE). It is therefore necessary to know the dose of virus that produces CPE in 50% of the cells ($TCID_{50}$). After mixing serum dilutions with virus, the virus is used to inoculate the cell culture. After the appropriate incubation period, the cultures are read to determine the end point: A reduction in CPE is expected for each dilution until the last dilution of serum that produces CPE in 50% of the cells is determined. The inverse of that dilution is the titer (Rovozzo and Burke, 1973).

6. Complement Fixation Tests

No discussion on veterinary immunological tests would be complete without the mention of the complement fixation (CF) test. Once used as a primary means of diagnosis for several important infectious diseases, the CF test has been replaced in many laboratories by more modern primary binding assays, such as ELISA. The principle of the CF test is simple. Dilutions of the patient's serum are mixed with antigen and then a measured amount of complement is added. If antibodies are present in the patient's serum specific for the antigen, then the complement will be fixed. If no antibodies are present, the complement remains to react with an indicator system (erythrocytes sensitized with anti-erythrocyte serum). Fixation of complement by the indicator system results in hemolysis. Hence, an absence of hemolysis is indicative of the presence of antibody in the patient's serum. Titers are obtained by observing tubes for the presence or absence of hemolysis. It is obvious that the performance of the CF test requires very accurate titration of each reactant. Protocols are available for setting up the CF test in the referenced texts (Rovozzo and Burke, 1973).

C. Primary Binding Assays

Each of the assays discussed earlier requires more than antigen–antibody binding for a result to be achieved. A secondary phase of the reaction is required: forming an insoluble lattice (precipitin), settling out of cross-linked particulate antigen (agglutination), or binding to complement (CF test). Generally these kinds of assays are less sensitive than those that require simply the binding of antigen to antibody.

1. Immunofluorescence-Based Assays

The coupling of an antibody to a fluoresceinated compound has been a popular method for detecting antigen or antibody for a number of years. The technique was originally used on tissue sections to demonstrate the presence of antigens and/or antibodies. However, recent advances in technology have given rise to a number of other uses, such as detection of individual cells in suspension using a fluorescence-activated cell sorter. Fluorochromes are compounds that absorb light of one wavelength, and when stimulated with the appropriate wavelength of light, give off light in the fluorescent spectrum. Using specialized optical systems, one can then visualize the fluorescence. When the fluorochrome is coupled to an antibody or antigen, it can effectively demonstrate the location of that antibody or antigen. Fluorescein isothiocyanate, a common fluorochrome conjugate, gives off an apple-green color. It has an absorption / emission spectrum of 495 nm / 517 nm (Goldman, 1986). Another fluorochrome, rhodomine isothiocyanate, fluoresces with a reddish-orange color; its absorption emission spectra are 550 nm / 580 nm. It is possible to combine two or more fluorochromes to perform localization of two substances in one single tissue section or other sample.

Examples of indirect immunofluorescence tests used in the clinical immunology laboratory are the antinuclear antibody test (described later in this chapter), canine distemper antibody titer, feline infectious peritonitis antibody titer, and feline immunodeficiency virus titer. Characteristic of the indirect FA test is that the patient's serum serves as the primary antibody, which is first incubated with the antigen-containing cells. After incubation and washing steps a fluorescein-conjugated secondary antibody is added (e.g., for canine serum a rabbit anti-canine IgG-FITC would be added). These tests are performed on cells that are grown in Lab-Tek slides and fixed *in situ*. Canine distemper antibody assays use a mink kidney cell line infected with canine distemper virus as antigen on slides. Slides are fixed in acetone to allow the cell membrane to become permeable to antibody. A positive serum sample reveals intracellular fluorescence where virus antigen has been bound by antibody. Similarly cat serum is evaluated for antibodies to feline infectious peritonitis (FIP) virus using cells infected with FIP virus using an indirect fluorescent antibody test (Fig. 6.4).

Direct immunofluorescence is used for detection of antigen in tissue sections or in cells obtained from patients. For example, kidney biopsy samples, preserved for frozen section, are incubated with fluorescein-conjugated anti-immunoglobulin serum for the detection of immune complex deposition in glomeruli. Conjunctival smears from distemper suspect dogs are fixed and stained with conjugated antisera specific for distemper virus to demonstrate the presence of distemper virus in infected dogs that have viral antigen in the conjunctiva (Fig. 6.5).

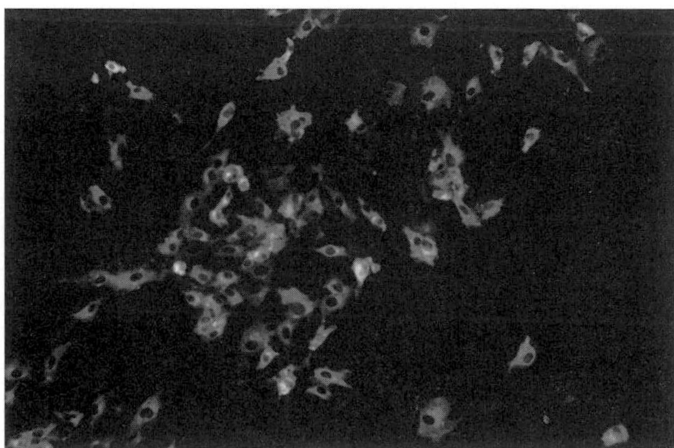

FIGURE 6.4 Positive feline serum reacting with feline infectious peritonitis virus in feline cells as indicated by indirect immunofluorescence.

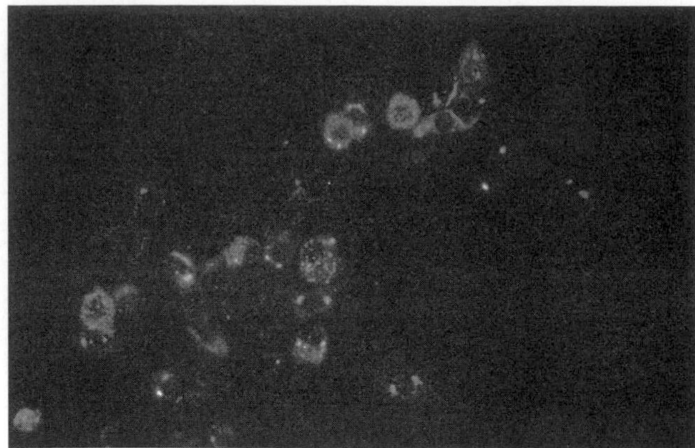

FIGURE 6.5 Canine distemper virus is demonstrated in conjunctival cells by direct immunofluorescence.

2. *Enzyme-Linked Immunosorbent Assay (ELISA)*

The ELISA is a highly sensitive assay that can be used to detect either antigen or antibody. Applications of ELISA include diagnostics for noninfectious diseases involving hormones, drugs, serum components, oncofetal proteins, or autoimmune diseases as well as diagnostics for infectious diseases caused by bacterial, viral, mycotic, or parasitic diseases (Voller *et al.*, 1981). Similarities between the ELISA, fluorescent antibody assays, and radioimmunoassays (RIAs) are notable. All three types of assay take advantage of an indicator that is attached to antibody (sometimes antigen), allowing bound antibodies to be detected by presence of the indicator (i.e., enzymatic digestion of a chromogenic substrate, fluorescence, or radioactivity respectively). ELISA is as sensitive as RIA, but has the advantage that it does not require the use of isotopes with the accompanying concerns for containment and monitoring.

To perform ELISA, antigen or antibody is attached to a solid phase. This makes it possible for reactants that bind immunologically to be separated from the unbound substances during washing procedures. Tubes, beads, and disks can be used, but most commonly 96-well microtitration plates made from polystyrene are used. The plastic will bind noncovalently, but very strongly, protein that is added in an alkaline buffer solution (e.g., carbonate buffer pH 9.6). The amount of antigen that is used to sensitize the plates must be predetermined by checkerboard titration using reference reagents. Generally small amounts of protein antigen are required for optimum sensitivity, 1–10 μg/ml. Sensitization can be achieved using one of several protocols; overnight at 4°C or several hours at 37°C is commonly used. In some systems it is necessary to block any sites on the plastic that have not adsorbed antigen. This is accomplished using a solution of unrelated protein, such as bovine or rabbit serum albumin or gelatin. Washing steps and sample dilution are usually performed using phosphate buffered saline solution to which is added the detergent Tween 20. This assists in prevention of nonspecific binding. When assaying for antibodies, a single solution of test serum can be used to assay and compare to a standard curve. Enzyme is conjugated to antibody using a coupling agent such as glutaraldehyde or sodium periodate. Horseradish peroxidase and alkaline phosphatase are the most frequently used enzymes, but others such as β-galactosidase can also be used. The substrate used is usually colorless and develops into a colored reaction product after enzymatic activation. The colored reaction product is then read visually for a positive/negative assessment or the absorbance value is read on a spectrophotometer for quantitative results. Specially designed ELISA readers are available for this purpose. Direct connection to a computer system allows for easy data manipulation and standard curve development.

Several methods are used for setting up the ELISA, depending on whether antigen or antibody is to be measured. These are similar to the strategies described for fluorescence immunoassay, that is, indirect, double antibody sandwich, and competitive. Commercial kits have become available for use by veterinary practitioners. These take a variety of forms: Some contain presensitized plastic wells accompanied by appropriate positive and negative controls, often utilizing one monoclonal antibody specific for a virus protein as sensitizing antibody and a monoclonal specific for another determinant as conjugate. Test wells are read visually for color development and are compared with positive and negative controls. Other test kits have wandlike devices that are dipped into a succession of reagents to achieve a colored result. Still other tests have evolved as a "dot blot" type of assay in which the patient's serum sample is compared to positive and negative control dots on a membrane-type apparatus. Aside from these test kits, an immense variety of ELISAs is available, both commercially and in research laboratories, for the detection and/or quantitation of antibodies/antigens important in numerous diseases. This technique is both sensitive and quantitative.

Enzyme immunoassay has an additional application in immunohistochemistry. Enzyme conjugates can be used to stain specific antigens in fixed tissue sections just as fluorescent-labeled antibodies are used. Enzyme techniques have several advantages: The test can be performed on paraffin-embedded tissue and the slides can be permanently retained (Mason and Woolston, 1982), and there is no requirement for special adaptation of the microscope for illumination and viewing as there is for fluorescence. This technique has had wide application for diagnosis of infectious and autoimmune disease.

3. *Western Blot*

The technique of Western blotting has developed to allow further differentiation of antibody reactivity than a simple solid-phase ELISA or other antibody assay will provide. In this technique the antigen preparation, which is composed of a number of different proteins, is subjected to electrophoretic separation using SDS-polyacrylamide gel (Heegaard and Bjerrum, 1988). Proteins separated by size are then transferred to a nylon membrane. The membrane is reacted with a source of antibody (such as patient serum), washed, and then reacted with an antibody reagent conjugated to an indicator, either an enzyme or an isotope. The membrane is ultimately developed with substrate or by autoradiography (if a radioactive indicator was used) and the proteins to which antibody bound are visualized as bands on the membrane. This can be compared with an unblotted protein stained gel of the antigen so that the size and number of proteins to which antibody is directed can be determined. This type of assay is very sensitive and can be useful in differentiating antibody titers obtained by vaccination from those obtained by infection if there are differences in the number of proteins present in the vaccine and field isolates of the organism. This is particularly effective when subunit vaccines are used. Western blot is used in diagnosis of equine protozoal myelitis.

D. Detection and Quantitation of Lymphocyte Populations Using Monoclonal Antibodies Specific for Cell Surface Markers and Fluorescence Analysis

The increase in availability of monoclonal antibodies recognizing specific determinants characterizing subpopulations of lymphocytes has enabled researchers to enumerate B-cell, T-cell, and CD4+ helper T-cell subsets, CD8+ cytotoxic cell subpopulations, gamma/

delta T cells, MHC class II expression, interleukin-2 (IL-2) receptor expression, and expression of other markers for memory B cells. A standard protocol for cell purification, staining, and analysis is as follows.

Evaluation of cell populations using monoclonal antibodies and fluorescence analysis can be done on samples from peripheral blood, lymph, bronchoalveolar lavage fluid, and single cell suspensions from tissue. The lymphocytes are purified by centrifugation through a Ficoll–hypaque gradient. Cells are allocated into tubes at 10^7 cells per tube. These cells are resuspended in the primary antiserum and incubated for 30 minutes on ice. They are then washed and resuspended in 100 μl of fluoresceinated antibody and incubated for another 30 minutes on ice. Finally, cells are pelleted, resuspended, and washed twice more in phosphate buffered saline solution. Cells are resuspended and are assayed for fluorescence using a fluorescence-activated cell scanner (FACS) or FACSTAR within 24 hours. Analysis on FACS provides information on the percentage of the cell population that stains with the particular antibody of interest. Normal data on several species are becoming available. Analysis of cell populations by FACS utilizes size and fluorescence intensity as the parameters for selection of the population to be examined. There are several ways to present data obtained by FACS, including the contour graph and the histogram (Fig. 6.6). Ultimately the percentage of positive cells is of greatest interest. One direct application of such data is the calculation of a ratio between CD4+ and CD8+ T lymphocytes.

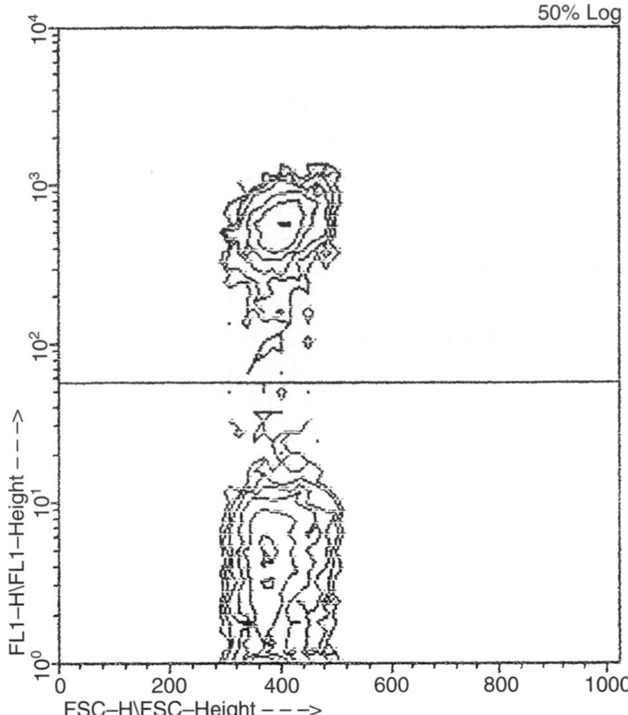

FIGURE 6.6 Contour plot demonstrating CD4+ T cells (*top*) and unstained CD4- T cells (*bottom*) using the fluorescence-activated cell scanner (Becton–Dickinson).

E. Detection and Quantification of Cytokines

In species for which genes coding for individual cytokines have been cloned and expressed, antibodies are available for use in ELISA or other primary binding assays to detect cytokines in serum or secretions. For those species in which there are no immunoassays for cytokine quantitation, detection of the mRNA is one method that is frequently used. Performance of reverse transcriptase–polymerase chain reaction (RT-PCR) on RNA derived from cells is used with appropriate primers and the product is then viewed on an agarose gel and its density is compared to a standard "housekeeping gene" such as actin. The RT-PCR involves steps that first convert the RNA into cDNA. Using a thermal cycler, the DNA is heated so that the two strands separate. Next a set of primers (small sequences of DNA that are complementary to sites on the desired gene) is added and the primers bind in an appropriate spot on the cDNA at the annealing temperature. Addition of nucleotides to the primers occurs with the help of the DNA polymerase enzyme, which is thermostable and able to resist the temperature changes that occur in the thermal cycler. This extends the new DNA strands. Once extended, the strands are again broken apart at a high temperature and the cycle starts once again. After multiple cycles a segment of the desired cDNA has been amplified sufficiently for visualization on an agarose gel.

The third type of assay commonly used to detect cytokines is the functional assay. This type of assay usually involves *in vitro* cell stimulation followed by harvesting of the cell supernatants and using them to stimulate a cell line, which is responsive to the particular cytokine. For example, IL-2 concentration in cell supernatant can be assayed using an IL-2–dependent cell line (see later discussion).

1. Gamma Interferon

There is a commercially available test (Idexx) for the detection of bovine gamma interferon. The test is marketed as a kit for diagnosis of *Mycobacterium paratuberculosis* infection. A facultative intracellular bacteria, the presence of *M. paratuberculosis* is best demonstrated by a cell-mediated immune response. Production of gamma interferon is such an indicator. The test can be adapted for use with other antigen/infectious agent systems in which the presence of gamma interferon is thought to be important. The assay itself is an ELISA, based on an antibody to gamma interferon, but the test sample consists of supernatant obtained from culturing the patient's lymphocytes with the antigen.

2. *Interleukin-2*

Interleukin-2 has been measured in several species using a biological assay long before the gene was cloned in several species. Generally IL-2–dependent cell lines are incubated in the presence of supernatant thought to contain IL-2. The presence of IL-2 is then demonstrated by the ability or lack thereof to sustain growth of the dependent cell line. For example, bovine IL-2 stimulates proliferation of BT-2T-lymphoblast cells and murine IL-2 is able to support growth of CTLL 2 cells. Protocols are available for these biological assays (Coligan *et al.*, 1991).

3. *Interleukin-4*

With the exception of murine and human species there is currently no readily available ELISA to measure IL-4. Biological assays, as described earlier, and RT-PCR are the current methodologies used for detection and semiquantitation of this cytokine.

F. Testing for Autoimmune Disease

1. *Antinuclear Antibodies*

Generally a diagnosis of autoimmune disease is indicated by the presence of autoantibodies, such as antinuclear antibodies. In domestic animals the generalized autoimmune disease systemic lupus erythematosus (SLE), occurs and is accompanied by the presence of a positive ANA test. The ANA test, an indirect immunofluorescence test performed using a human cell line (HEP-2), indicates the presence of antibodies that react with DNA, RNA, or nuclear proteins. To perform the ANA test, a slide containing wells with fixed cells is incubated with serial dilutions of test serum and positive and negative control sera. After incubation and washing, an appropriate dilution of fluorescein-conjugated antiserum (species specific) is applied. The slide is incubated again, washed, and observed under the fluorescence-equipped microscope. Several patterns of fluorescence are described, including speckled, peripheral, homogeneous. A titer as well as a pattern are reported when the sample shows positive fluorescence (Fig. 6.7). Adaptations of the test for veterinary species are detailed by Lewis and Picut (1989).

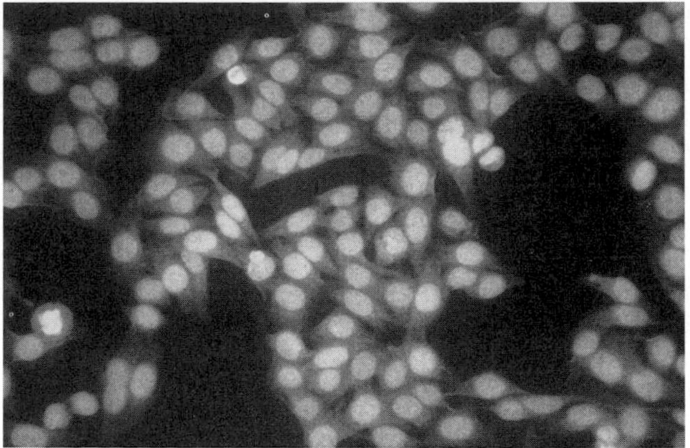

FIGURE 6.7 A positive anti-nuclear antibody test demonstrating a homogeneous pattern of fluorescence.

In addition to the ANA test, which measures a variety of antibodies reactive with nuclear components, some laboratories perform a test for native DNA. This test uses a protozoan parasite, *Crithidia lucia,* which has a structural body called a kinetoplast that contains only native DNA. Staining of the kinetoplast is indicative of antibodies reactive with native DNA (Ballou, 1992).

The presence of LE cells is pathognomonic for SLE. Although seen infrequently in SLE cases, these characteristic neutrophils, containing large homogeneous eosinophilic nuclei and a flattened cellular nucleus, indicate that nuclear material is being opsonized with antinuclear antibodies prior to engulfment by the neutrophil.

2. *Organ-Specific Autoantibody Detection*

Tissue or organ-specific autoimmune diseases are characterized by the presence of particular antibodies reactive with antigens present only in the affected organ. The variety of organ systems involved include the hematopoietic system (autoimmune hemolytic anemia, autoimmune thrombocytopenia), the endocrine system (thyroid, pancreas, adrenal gland), and the neuromuscular system (myasthenia gravis). Assays that detect these are referable to the body system involved. For example, for signs referable to hemolytic anemia of autoimmune etiology, the Coombs' test for the presence of anti-erythrocyte antibodies is performed. In this test the patient's cells are incubated with anti-γ-globulin (species specific). If there is autoantibody on the red blood cell surface, then the added antibodies will bind to those already on the cell surface and cause agglutination. Absence of antibodies on the erythrocyte surface will result in no agglutination. The assay, as described earlier, is a direct Coombs' test. Also performed is the indirect Coombs' test in which the patient's serum is examined for the presence of anti-erythrocyte antibodies by incubation with normal erythrocytes. Agglutination of the cells after addition of the antiglobulin reagent indicates that the patient's serum contains anti-erythrocyte antibodies.

Other antigen–antibody systems involved in autoimmune disease include thrombocytes/megakaryocytes in autoimmune thrombocytopenia, anti-thyroglobulin antibodies in autoimmune thyroiditis, anti-pancreatic islet cell antibodies in autoimmune diabetes mellitus, and anti-acetylcholinesterase antibodies in myasthenia gravis. Many of these autoantibodies are detectable using direct or indirect immunofluorescence on the appropriate tissue section. The IFA test for autoimmune thrombocytopenia is performed on a bone marrow aspirate. The slide is incubated with a fluorescein-conjugated anti-IgG antibody specific for the same species as the patient. The presence of fluorescence on megakaryocytes is indicative of autoantibody on the cells, whereas a lack of fluorescence means that there is no autoantibody present.

Among the several skin diseases that have autoimmune etiology, pemphigus complex shows antibody bound to intercellular cement substance within the epidermis. Systemic lupus erythematosus or discoid lupus show antibody bound to the basement membrane at the dermal–epidermal junction. Direct immunofluorescence is a useful technique to demonstrate both of these lesions in frozen tissue sections taken by biopsy. Isotype specificity of the autoantibody is determined by using isotype-specific fluoresceinated antibodies in the test; use of antiserum specific for the third component of complement (C3) will demonstrate complement deposition in tissue if it has occurred. Preservation of biopsy material is possible for frozen section/FA by placing the sample in Mischel's media, which contains NH_4SO_4 and prevents solubilization of deposited complexes. The fluorescent antibody technique for detection of antibody and/or C3 in tissues is detailed by Valenzuela and Deodhar (1992) as originally developed for human patients. Adaptations for veterinary species are detailed by Lewis and Picut (1989).

VII. DETECTION OF IMMUNE DEFICIENCY

Evaluation of the immune system requires that each element of the immune system be considered separately: humoral, cellular, innate, and complement. Generally elements of the clinical history, age, and breed/species of the animal and the type of agent causing infections will be helpful in formulating a tentative diagnosis. For example, a Holstein heifer 3 weeks of age with recurrent apurulent bacterial infections is a likely candidate for bovine adhesion deficiency (a defect in innate immunity), whereas a 3-month-old Arabian foal with adenovirus pneumonia may have severe combined immunodeficiency.

A. Humoral Immunity

Diagnosis of defects in humoral immunity is accomplished using some of the previously described assays such as the radial immunodiffusion test for antibody. This agar gel test is isotype specific and is quantitative. The exception for this choice is when it is necessary to evaluate a neonate for colostrum consumption within the first day after birth. Although less accurate than SRD the quick tests for IgG will provide a faster result than the SRD test, thereby allowing colostrum supplementation to be performed prior to intestinal closure. Of the available tests, the zinc sulfate test is acceptable, and the more expensive CITE tests are also popular and trustworthy if done according to instructions. Diagnosis of severe combined immunodeficiency generally requires that two measurements be made: lymphocyte count (absolute and percentage) and IgM concentration. Both will be negligible if the foal has severe combined immunodeficiency (SCID).

Selective immunoglobulin deficiencies are present as inherited defects in several species. Those currently recognized are reviewed (Tizard, 1992). For example, in dogs several breeds have inherited IgA deficiencies. For these breeds (Shar-pei, German shepherd, Dalmatian), agar gel diffusion (SRD) is the preferable method to quantitate the amounts of a specific isotype such as IgA in serum. In these breeds, one usually sees levels less than normal, rather than total absence of the antibody. Thus only the truly quantitative method is acceptable.

B. *In Vitro* Assays for Cell-Mediated Immune Responses

1. *Lymphocyte Stimulation*

Unlike serology, there are not many methods available for *in vitro* evaluation of cellular immune responses. The most commonly used assay is one referred to as *lymphocyte stimulation/lymphocyte transformation;* it has been used extensively in several species (Fletcher *et al.,* 1992). Lymphocyte stimulation is performed by incubation of peripheral blood lymphocytes with plant mitogen in 96-well microliter plates. Concanavalin A and phytohemagglutinin are the commonly used T-cell mitogen, while pokeweed mitogen stimulates both T and B lymphocytes. From 2 to 3 days after stimulation, tritiated thymidine is added to the cultures and allowed to incorporate into the DNA of dividing cells. At about 18 hours later the cells are harvested onto filter paper disks, which are read for radioactivity in a scintillation counter. Calculation of the stimulation index by dividing the counts per minute from stimulated wells by the values from the unstimulated controls provides a number indicative of the cellular responsiveness. When performing this test a normal species-matched control is always tested in parallel. It is possible to measure antigen-specific responsiveness by stimulating with antigen instead of or in addition to the mitogen. Stimulation indices are always much lower for antigen-specific systems, because only a small percentage of cells is responsible to an individual antigen, whereas all T cells respond to antigen.

2. *Innate Immunity: Neutrophil Function*

Evaluation of neutrophil function is particularly important in cases of recurrent or persistent bacterial infection with organisms that are generally avirulent or mildly pathogenic. Infection of mucosal surfaces, including the lung and skin, are particularly common in animals with decreased neutrophil function. To evaluate fully the function of neutrophils it is necessary to be aware of each step required for a neutrophil to kill an organism. First, having a sufficient supply of neutrophils is mandatory. In cyclic neutropenia of gray collies, for example, at cyclic periods the bone marrow shuts down its production of neutrophils, creating a peripheral neutropenia. During these times the dogs are at increased risk for developing bacterial infections (Jones *et al.*, 1975). Assuming there are a sufficient number of neutrophils in the peripheral blood, response to an invasion with bacteria requires that the cells be able to marginate (adhere to vascular endothelial cells), undergo diapedesis, and move out of the blood vessel. In BLAD and canine leukocyte adhesion deficiency, the cells are unable to exit the blood–vascular system because they lack an adhesion molecule (CD11b/CD18) necessary for margination to occur; affected individuals are unable to make pus. Once out of the blood–vascular system the neutrophils must be able to locate and travel toward the bacteria. This requires chemotactic ability. Defective chemotaxis can result in a depressed neutrophil response to infection, as seen in Pelger–Huet anomaly in several dog breeds. When the neutrophil reaches the bacteria, it must then engulf by phagocytosis the organism and then kill it using an oxidative and nonoxidative enzyme system. This requires fusion of the phagosome with the lysosomes so that the enzymes used in killing can be combined with the organisms in the phagosome. Inability to kill can occur by several mechanisms, including abnormal lysosomes, as seen in the Chediak–Higashi syndrome in certain breeds of cattle, mink, and Persian cats (Padgett *et al.*, 1964).

Assays to examine each of these neutrophil functions are available. A simple leukocyte count and differential will provide information on the number of circulating neutrophils. In cyclic neutropenia of gray collie dogs, numbers will be very low (Jones *et al.*, 1975). In cattle it is now possible to determine if the genetic defect responsible for BLAD is present by using the PCR. Chemotaxis can be evaluated using one of several assays (Coates *et al.*, 1992). Phagocytic ability can be evaluated by calculation of the phagocytic index; this test requires incubation of the patient's cells with bacteria and observation of the numbers of organisms engulfed during a finite period of time. To evaluate the enzymes of the respiratory burst a nitroblue tetrazolium blue test can be performed, and finally to evaluate overall killing ability the bactericidal test is superior (Quie and Herron, 1992).

3. *Evaluation of Complement*

Proteins that comprise the complement system are important accessory molecules in the immune system. Complement-mediated killing of bacteria, viruses, and parasites is important in innate and acquired immune responses. The absence or subnormal concentration of one or more components of complement may affect the phagocytosis (opsonins), chemotaxis (chemotactic factors), and killing (lysis by C6–C9). While absence of C1, C4, and C2 can have subclinical effect, a relative or absolute absence of C3 will have major implications on the ability of the patient's immune system to eliminate pathogens. In domestic animal species it is most common to test for overall complement function by performing a test called hemolytic complement/50 (Wolfe and Halliwell, 1980; Barta and Oyekan, 1981). In this assay, the ability of the patient's plasma to serve as a lytic complement source is tested for and is compared with that of plasma from a normal animal of the same breed. To analyze individual components of complement, single radial immunodiffusion can be used providing there is a species-specific source of antisera specific for the complement component of interest available (Ruddy, 1992).

VIII. TYPING FOR MAJOR HISTOCOMPATIBILITY ANTIGENS

The major histocompatibility complex (MHC) is a segment of DNA that codes for the histocompatibility antigens, polymorphic cell surface proteins that are inherited in a codominant manner. Class I MHC antigens are present on all nucleated cells, whereas class II antigens are present on only a select population of cells, primarily B cells, macrophages, and dendritic cells. Both class I and class II antigens play important roles in the immune response.

A. Class I MHC Antigens: Microcytotoxicity

Class I antigens are detected with typing antisera, using a test called the *microcytotoxicity test.* Typing sera, specific for the various class I determinants, are added to cells from the patient to be typed in microliter well plates. Complement is added and then trypan blue dye. Cells are observed to determine if they remain uncolored or if they have taken up the blue dye. Cells that have the surface MHC antigen for which the antibody is specific bind antibody and are permeabilized when the complement is fixed. Wells in which antisera that is not specific for the MHC determinants on the cells remain viable and do not take up the blue dye. By observing which wells contain blue cells and which contain unstained cells, one can determine which MHC determinants are present on the patient's cells.

B. Class II MHC Antigens: Mixed Lymphocyte Reaction

The mixed lymphocyte reaction (MLR) is designed for typing of MHC class II molecules on the surface of lymphocytes for the purpose of tissue transplantation. In this assay typing cells with known MHC specificities are rendered metabolically paralyzed and then are mixed with the patient's cells. If, however, it sees foreign determinants, then it will react with blastogenesis. Blastogenesis, as mentioned earlier, can be measured by incorporation of radioactive nucleotides. Mixed lymphocyte reactions have been used to assist in characterization of MHC class II specificities in several species. If the patient shares specificities with the typing cells, it will not react.

C. Restriction Fragment Length Polymorphisms and Southern Blotting

Differences in the MHC genes can be detected by digestion of DNA with restriction endonucleases and then comparing the fragments by polyacrylamide gel electrophoresis. When the gel is blotted onto nylon (as in the Western blot) and probed with a labeled piece of DNA from a particular MHC allele, the size of the fragment can be determined. Some class II specificities have been identified using restriction fragment length polymorphisms (Davies *et al.*, 1992).

IX. SUMMARY

Advances in molecular biology and immunology are gradually being incorporated into the clinical veterinary immunology laboratory. Since the last edition of this book, use of the fluorescence-activated cell scanner to determine cell populations based on markers recognized by monoclonal antibodies has moved from a purely research-based endeavor to diagnostic applications. Similarly the ability to measure cytokines and/or their message has received some application to diagnostics; the gamma interferon assay for diagnosis of Johne's disease is a prime example. In the near future we expect to see increasing movement of methodology from the research laboratory to the clinical laboratory.

References

Arai, K., Lee, F., Miyajima, A., Miyatake, S., Arai, N., and Yokota, T. (1990). *Annu. Rev. Biochem.* **59,** 783.

Ballou, S. P. (1992). *In* "Manual of Clinical Laboratory Immunology" (N. R. Rose, E. C. DeMacario, J. L. Fahey, H. Friedman, and G. M. Penn, eds.), 4th ed., p. 84. American Society for Microbiology, Washington, D.C.

Barta, O., and Oyekan, P. P. (1981). *Am. J. Vet. Res.* **421,** 378.

Caron, J., and Penn, G. M. (1992). *In* "Manual of Clinical Laboratory Immunology" (N. R. Rose, E. C. DeMacario, J. L. Fahey, H. Friedman, and G. M. Penn, eds.), 4th ed., p. 84. American Society for Microbiology, Washington, D.C.

Coates, T. D., Beyer, L. L., and Baehner, R. L. (1992). *In* "Manual of Clinical Laboratory Immunology" (N. R. Rose, E. C. DeMacario, J. L. Fahey, H. Friedman, and G. M. Penn, eds.), 4th ed., p. 84. American Society for Microbiology, Washington, D.C.

Coligan, J. E., Kruisbeek, A. M., Margulies, D. H., Shevach, E. M., and Strober, W. (1991). "Current Protocols in Immunology," Greene Publishing Associates and Wiley-Interscience, New York.

Davies, C. J. Joosten, I., Andersson, L. (1992). *Polymorphism of Bovine MHC class II genes: Joint Report of the Fifth Int. Bovine Lymphocyte Antigen (BoLA) Workshop,* Interlaken, Switzerland.

Davis, B. D., Dulbecco, R., Eisen, H., and H. S. Ginsberg. (1990). "Microbiology," 4th ed., p. 786. Lippincott, Philadelphia, Pennsylvania.

Fletcher, M. A., Klimas, N., Morgan, R., and Gjerset, G. (1992). *In* "Manual of Clinical Laboratory Immunology" (N. R. Rose, E. C. DeMacario, J. L. Fahey, H. Friedman, and G. M. Penn, eds.), 4th ed., p. 84. American Society for Microbiology, Washington, D.C.

Goldman, M. (1986). "Fluorescent Antibody Methods." Academic Press, New York.

Heegaard, N. H. H. and Bjerrum, O. J. (1988). "Handbook of Immunoblotting of Proteins," Vol. I, Technical Descriptions. CRC Press, Boca Raton, Florida.

Jones, J. B., Lange, R. D., and Jones, E. S. (1975). *J. Am. Vet. Med. Assoc.* **166,** 365.

Kehrli, M. E., Schmalstieg, F. C., Anderson, D. C., et al. (1990). *Am J. Vet. Res.* **51,** 1826.

Kohler, G., and Milstein, C. J. (1975). *Nature* **256,** p 495.

Lewis, R. M. and Picut, C. A. (1989). "Veterinary Clinical Immunology, from classroom to clinics." (1989). Lea and Febiger, Philadelphia, Pennsylvania.

Mason, D. Y., and Woolston, R. E. (1982). *In* Techniques in Immunocytochemistry (G. J. R. Bullock and P. Petrusz, eds.), Vol. 1. Academic Press, London.

Mosmann, T. R., and Coffman, R. L. (1989). *Adv. Immunol.* **46,** 111.
Olitzki, A. (1972). "Enteric Fevers," p. 330. Karger, Basel.
Padgett, G. A., Leader, R. W., Gorham, J. R. and O'Mary, C. C. (1964). *Genetics* **49,** 505.
Quie, P. J., and Herron, J. (1992). *In* "Manual of Clinical Laboratory Immunology" (N. R. Rose, E. C. DeMacario, J. L. Fahey, H. Friedman, and G. M. Penn, eds.), 4th ed., p. 84. American Society for Microbiology, Washington, D.C.
Renshaw, H. W., and Davis, W. C. (1979). *Am J. Pathol* **95,** 731.
Romagnani, J. S. (1991). *Immunol. Today* **12,** 256.
Rovozzo, G. C., and Burke, C. N. (1973). In "A Manual of Basic Virological Techniques," p. 103. Prentice Hall, Englewood Cliffs, New Jersey.
Ruddy, S. (1992). *In* "Manual of Clinical Laboratory Immunology" (N. R. Rose, E. C. DeMacario, J. L. Fahey, H. Friedman, and G. M. Penn, eds.), 4th ed., p. 84. American Society for Microbiology, Washington, D.C.
Salgame, P. R., Abrams, J. S., Clayberger, C., Goldstein, H. Convit, J., Modlin, R. L., and Bloom, B. R. (1991). *Science* **254,** 279.
Tizard, I. (1992). "Veterinary Immunology, An Introduction," 4th ed. Saunders, Philadelphia, Pennsylvania.
Valenzuela, R., and Deodhar, S. D. (1992). *In* "Manual of Clinical Laboratory Immunology" (N. R. Rose, E. C. DeMacario, J. L. Fahey, H. Friedman, and G. M. Penn, eds.), 4th ed., p. 84. American Society for Microbiology, Washington, D.C.
Voller, A., and deSavigny, D. (1981). *In* "Techniques in Clinical Immunology" (R. A. Thompson, ed.). Blackwell Scientific Publications, London.
Wolfe, J. H., and Halliwell, R. E. M. (1980). *Vet. Immunol. Immunopathol.* **1,** 287.

CHAPTER

7

The Erythrocyte: Physiology, Metabolism, and Biochemical Disorders

JOHN W. HARVEY

I. INTRODUCTION

Mammalian erythrocytes or red blood cells (RBCs)[1] are anucleate cells that normally circulate for several months in blood despite limited synthetic capacities and repeated exposures to mechanical and metabolic insults. Their primary purpose is to carry hemoglobin

[1] Abbreviations used in tables, figures, and text: EMP, Embden–Meyerhof pathway; PPP, pentose phosphate pathway; HK, hexokinase; GPI, glucose phosphate isomerase; PFK, phosphofructokinase; Aldo, aldolase; TPI, triosephosphate isomerase; GAPD, glyceraldehyde-3-phosphate dehydrogenase; PGK, phosphoglycerate kinase: MPGM, monophosphoglycerate mutase; DPGM, diphosphoglycerate mutase; DPGP, diphosphoglycerate phosphatase; Enol, enolase; PK, pyruvate kinase; G6PD, glucose-6-phosphate dehydrogenase; 6PGD, 6-phosphogluconate dehydrogenase; LDH, lactate dehydrogenase; GOT, glutamic oxaloacetic transaminase; GR, glutathione reductase; GPx, glutathione peroxidase; GST, glutathione S-transferase; MR, methemoglobin reductase; Diaph, diaphorase; TK, transketolase; TA, transaldolase; GSSG, oxidized glutathione; PNP, purine nucleoside phosphorylase; ATPase, adenosine triphosphatase; AST, aspartate aminotransferase; Gluc, glucose; G6P, glucose 6-phosphate; F6P, fructose 6-phosphate; FDP, fructose 1,6-diphosphate; DHAP, dihydroxyacetone phosphate; GAP, glyceraldehyde 3-phosphate; 1,3DPG, 1,3-diphosphoglycerate; 2,3DPG, 2,3-diphosphoglycerate; 3PG, 3-phosphoglycerate; 2PG, 2-phosphoglycerate; PEP, phosphoenolpyruvate; Pyr, pyruvate; Lac,

(Hb), a heme-containing protein that accounts for 95% of the total protein in RBCs. The benefits of having Hb contained within cells, as opposed to free in plasma, include the much slower turnover in blood (free Hb has a half-life of only a few hours), the metabolic capability of RBCs to maintain iron in Hb in the functional ferrous state, and the ability to control Hb oxygen affinity by altering the concentrations of organic phosphates (especially 2,3DPG). In addition, the presence of free Hb in plasma in concentrations normally found in whole blood would exert an osmotic pressure several times greater than that normally exerted by plasma proteins, profoundly affecting the movement of fluid between the vascular system and tissues.

A. Species Differences in Erythrocyte Shape

Most RBCs in normal dogs, cats, horses, cattle, and sheep occur in the shape of biconcave disks (discocytes). The degree of biconcavity is most pronounced in dogs and less so in cats and horses (Jain, 1986). RBCs from goats generally have a flat surface with little surface depression; a variety of irregularly shaped RBCs (poikilocytes) may be present in clinically normal goats (Jain, 1986). The apparent benefit of the biconcave shape is that it gives RBCs high surface area to volume ratios and allows for deformations that must take place as they circulate.

Marked acanthocytosis is reported to occur in young goats (Holman and Drew, 1964) and some young cattle (McGillivray *et al.*, 1985; Sato and Mizuno, 1982). Acanthocytosis of young goats occurs as a result of the presence of HbC at this stage of development (Jain *et al.*, 1980). Normal adult angora goats (Jain and Kono, 1977; Jain *et al.*, 1980) and some breeds of British sheep (Evans, 1968) have variable numbers of fusiform or spindle-shaped RBCs that resemble sickle cells (drepanocytes) in normal deer and people with sickle cell anemia (Taylor, 1983). Drepanocyte formation in deer depends on the Hb types present. It is an *in vitro* phenomenon that occurs when oxygen tension is high and pH is between 7.6 and 7.8 (Taylor, 1983). The proportion of fusiform cells in angora goats varies depending on the individual and *in vitro* alterations in temperature, pH, and oxygenation, but the tendency to form fusiform cells could not be attributed to differences in Hb type (Jain and Kono, 1977). Echinocytosis is a consistent artifact in stained blood films from pigs (Jain, 1986).

B. Functions of Erythrocytes

The RBC functions of oxygen transport, carbon dioxide transport, and hydrogen ion buffering are interrelated. Each Hb tetramer can bind four molecules of oxygen when fully saturated. Assuming a normal arterial pO_2 of 100 mm Hg and a Hb concentration of 15 g/dl in blood, the presence of Hb-containing RBCs increases the oxygen carrying capacity of blood to approximately 70 times more than that which could be transported dissolved in plasma (West, 1985).

Approximately 10% of CO_2 is transported dissolved in blood, 30% is transported bound to amine groups of blood proteins, and 60% is transported in the form of bicarbonate in normal individuals (West, 1985). Carbonic acid is formed when dissolved CO_2 combines with water. This reaction occurs nonenzymatically but is accelerated by the presence of the carbonic anhydrase (CA, carbonate dehydratase) enzyme in RBCs. Bicarbonate is formed by the rapid spontaneous dissociation of carbonic acid as shown:

$$H_2O + CO_2 \overset{CA}{\leftrightharpoons} H_2CO_3 \leftrightharpoons H^+ + HCO_3^-.$$

Hb potentiates the formation of bicarbonate by buffering hydrogen ions and shifting the equilibrium of the reaction to the right. Carbamino groups are formed by the combination of CO_2 with the terminal groups of proteins. The globin of Hb is the most important blood protein in this regard. The reaction can be represented as follows:

$$\text{Hb-NH}_2 + CO_2 \leftrightharpoons \text{Hb-NHCOOH} \leftrightharpoons \text{Hb-NHCOO}^- + H^+.$$

The transportation of CO_2 from the tissues to the lungs as carbamino groups is potentiated because DeoxyHb binds twice as much CO_2 as OxyHb.

Hb is the most important protein buffer in blood because it occurs in high concentration, has a relatively low molecular weight, and has a large number of histidine residues with pK_a values sufficiently close to 7.4 to function as effective buffers. It has about six times the buffering capacity of the plasma proteins. An addi-

lactate; AMP, adenosine monophosphate; ADP, adenosine diphosphate; ATP, adenosine triphosphate; NAD, nicotinamide adenine dinucleotide; NADH, reduced nicotinamide adenine dinucleotide; NADP, nicotinamide adenine dinucleotide phosphate; NADPH, reduced nicotinamide adenine dinucleotide phosphate; Hb, hemoglobin; GSH, reduced glutathione; P_i, inorganic phosphate; RBC, red blood cell; MetHb, methemoglobin; Ep, erythropoietin; ALA, delta-aminolevulinic acid; MB, methylene blue; LMB, leukomethylene blue; BFU-E, burst-forming unit-erythroid; CFU-E, colony-forming unit–erythroid; SOD, superoxide dismutase; Cb_5R, cytochrome-b_5-reductase; OxyHb, oxyhemoglobin; DeoxyHb, deoxyhemoglobin; HzB, Heinz body or bodies; PCV, packed cell volume; MCV, mean cell volume; GCS, gamma-glutamylcysteine synthetase; MCHC, mean cell hemoglobin concentration; HK^+, high potassium; LK^+, low potassium.

tional factor of importance that contributes to the effectiveness of Hb as a blood buffer is the fact that DeoxyHb is a weaker acid than OxyHb. As a result, most of the H^+ produced in the tissues under normal conditions is buffered as a direct result of the H^+ uptake by DeoxyHb owing to an increase in the effective pK_a of Hb following release of oxygen to the tissues (West, 1985).

II. ERYTHROPOIESIS

A. Stem Cells and Progenitor Cells

Throughout life, mammalian blood cells are produced continuously from primitive stem cells within extravascular spaces of the bone marrow. Stem cells are capable of proliferation, self-renewal, and differentiation. They are mononuclear cells that cannot be distinguished morphologically from lymphocytes. The most primitive hematopoietic stem cell can give rise to lymphoid stem cells as well as a pluripotent myeloid stem cell that supports the production of all nonlymphoid blood cells. The myeloid stem cell gives rise to a series of progressively more differentiated progenitor cells with little or no self-renewal capabilities (Quesenberry, 1995). When measured in an *in vitro* cell culture assay, these progenitor cells are referred to as colony forming units (CFUs). The colony-forming unit–granulocyte, erythrocyte, macrophage, megakaryocyte (CFU-GEMM) is an oligopotential progenitor cell that can give rise to multiple cell lines.

The CFU-GEMM, and possibly other oligopotential progenitor cells, produce the burst-forming unit–erythroid (BFU-E), the earliest unipotential progenitor cell committed to the erythroid cell line. The BFU-E gives rise in culture to multiple subcolonies of a more differentiated progenitor cell, the colony-forming unit–erythroid (CFU-E). The CFU-E gives rise to the first morphologically recognizable erythroid cell, the rubriblast (Quesenberry, 1995). Although discussed earlier as discrete cell compartments, the transition from one cell type to another is gradual; consequently, considerable heterogeneity exists within stem cell and progenitor cell compartments.

B. Hematopoietic Microenvironment

Blood cell production occurs in bone marrow of adult animals because of the unique microenvironment present there. The hematopoietic microenvironment is a complex meshwork composed of various stromal cells, accessory cells, glycoprotein growth factors, and extracellular matrix that profoundly affects stem cell and progenitor cell survival, proliferation, and differentiation (Campbell and Wicha, 1988). Stromal cells (endothelial cells, fibroblast-like reticular cells, and macrophages) and accessory cells (subsets of lymphocytes and natural killer cells) produce a variety of positive and negative growth factors (Moore, 1991; Pantel and Nakeff, 1993). Stromal cells also produce components of the extracellular matrix (Long, 1992; Tavassoli and Hardy, 1990).

The extracellular matrix (ECM) consists of broad categories of macromolecules: collagen, proteoglycans, and glycoproteins. In addition to providing structural support (collagen) the ECM is important in the binding of hematopoietic cells and soluble growth factors to stromal cells so that optimal proliferation and differentiation can occur. Adhesion molecules (primarily integrins) on progenitor cells bind to ECM glycoproteins such as hemonectin, fibronectin, thrombospondin, and vascular cell adhesion molecule-1 (Campbell, 1992; Long and Dixit, 1990; Papayannopoulou and Brice, 1992; Yanai *et al.*, 1994). The spectrum of expression of adhesion molecules varies with the type and maturity of hematopoietic cells. Proteoglycans with their glycosaminoglycan moieties such as chondroitin sulfate and heparan sulfate bind growth factors and at least strengthen the bond between progenitor cells and stromal cells (Minguell, 1993; Tavassoli and Hardy, 1990). Growth factors such as stem cell factor (KIT ligand) may also be involved in the adhesion of hematopoietic cells by binding to both the extracellular matrix and to specific receptors on hematopoietic cells (Sieff, 1991).

Proliferation of stem cells and progenitor cells cannot occur spontaneously, but requires the presence of specific hematopoietic growth factors (HGFs). Hematopoietic cells coexpress receptors for more than one HGF on their surface, with the number of each receptor type present depending on the stage of cell differentiation (Metcalf, 1993). Binding of an HGF to its receptor can stimulate proliferation and/or modulation of HGF receptors on the surface of the cell that is acted on. The binding of an HGF generally results in downmodulation of its own receptor and upmodulation (transactivation) of receptors for distal HGFs that act primarily on more differentiated cell types (Testa *et al.*, 1993).

Stem cell factor (KIT ligand) can maintain survival of stem cells, but the presence of additional factors such as interleukin-3 (IL-3), interleukin-6 (IL-6) and granulocyte macrophage colony-stimulating factor (GM-CSF) are needed for proliferation to occur (Hoffman *et al.*, 1993; Shull *et al.*, 1992). CFU-GEMM cells are stimulated to proliferate and differentiate into BFU-E by IL-3 and GM-CSF in the presence of erythropoietin

(EPO). BFU-E proliferation and differentiation into CFU-E results from the presence of these same factors and may be further potentiated by additional factors such as interleukin-9. EPO is the primary growth factor involved in the proliferation and differentiation of CFU-E into rubriblasts (Erickson and Quesenberry, 1992; Metcalf, 1993).

Marrow macrophages appear to be an important component of the hematopoietic microenvironment involved with erythropoiesis. Both early and late stages of erythroid development occur with intimate membrane apposition to central macrophages in so-called erythroblastic islands (Bernard, 1991). These central macrophages may regulate RBC development by producing both positive factors such as burst-promoting activity and EPO (Rich, 1986, 1991; Vogt *et al.*, 1989) and negative factors such as IL-1α, TNFα, interferon and transforming growth factor β (Khan *et al.*, 1992; Wang *et al.*, 1992). Macrophages may be important in the local or basal regulation of erythropoiesis, but humoral regulation is also important, with EPO production primarily occurring in the kidney and various inhibitory cytokines being produced during inflammation throughout the body. In addition to negative factors already listed, leukotrienes B_4 and C_4 are reported to inhibit BFU-E and CFU-E growth (Estrov *et al.*, 1988).

C. Erythropoietin

EPO is a 34-kDa glycoprotein hormone that exhibits a high degree of sequence homology among mammals (Wen *et al.*, 1993). It is the principal HGF that promotes the viability, proliferation, and differentiation of erythroid progenitor cells that express specific cell surface EPO receptors (EPO-Rs). Early BFU-E cells do not express EPO-Rs, but more mature BFU-E cells express EPO-Rs and are weakly responsive to EPO. EPO-R copies on cell surfaces increase to maximum values in CFU-E cells and then decline in rubriblasts and are lost in later stages of development (Porter and Goldberg, 1993; Youssoufian *et al.*, 1993). Because of their EPO-R density, CFU-E cells readily respond to EPO. After binding to its receptor, EPO is internalized and promotes the proliferation, differentiation, and transformation of CFU-E cells into rubriblasts, the first morphologically recognizable erythroid cell type.

EPO stimulation promotes increased heme synthetic enzyme activities and represses heme degradation enzyme activity. The resultant increase in intracellular heme may promote erythroid differentiation. In addition, increased heme may be important in stimulating erythropoiesis by the upmodulation of EPO receptors (Abraham, 1991; Mayeux *et al.*, 1986)

High titers of EPO may accelerate rubriblast entry into the first mitotic division, shortening the marrow transit time and resulting in the early release of stress reticulocytes (Erslev and Beutler, 1995). Although EPO has been assumed to act as a mitogen because of its capacity to amplify RBC production, other studies have suggested it acts primarily as a survival factor preventing apoptosis and permitting cells to proceed with programmed proliferation and maturation (Erslev and Beutler, 1995; Koury and Bondurant, 1988).

In the presence of EPO, other hormones including androgens, thyroid hormones, and growth hormone can enhance the growth of erythroid precursor cells *in vitro* (Fisher *et al.*, 1971; Peschle, 1988; Popovic *et al.*, 1977). These substances apparently increase the number of erythroid colonies formed *in vitro* at suboptimal EPO concentrations, possibly by modulating the fraction of CFU-E cells that will enter terminal differentiation at a given EPO concentration. Both positive and negative effects of glucocorticosteriods have been reported. Hydrocortisone promotes BFU-E growth in culture, but inhibits T-lymphocyte production of HGFs that stimulate BFU-E proliferation (Morra *et al.*, 1989).

The kidney is the major site of EPO production in adult mammals, at least when anemia is present (Oishi *et al.*, 1995). Most studies indicate that EPO is produced by peritubular interstitial cells located within the inner cortex and outer medulla of the kidney (Erslev, 1990; Porter and Goldberg, 1993). The liver is an extrarenal source of EPO in adults and the major site of EPO production in the mammalian fetus. Bone marrow macrophages have also been shown to produce EPO, suggesting the possibility of short-range regulation of erythropoiesis (Rich, 1991).

EPO production is stimulated by tissue hypoxia. The oxygen sensor has not been defined, but there is evidence that a heme protein may be involved (Porter and Goldberg, 1993) Tissue oxygen tension is determined by the oxygen consumption by the tissues and the oxygen-delivering capacity of the blood. Oxygen-delivering capacity depends on cardiovascular integrity, oxygen content in arterial blood, and Hb oxygen affinity. Low oxygen content in the blood can result from low partial pressure of oxygen (pO_2) in blood, such as that occurring with high altitudes or with congenital heart defects where some of the blood flow bypasses the pulmonary circulation. Low oxygen content in blood can also occur with normal pO_2, such as that occurring with anemia and methemoglobinemia. An increased oxygen affinity of Hb within RBCs results in a decreased tendency to release oxygen to the tissues (Erslev, 1990).

III. DEVELOPING ERYTHROID CELLS

A. Morphologic and Metabolic Changes

Rubriblasts are continuously generated from progenitor cells in the extravascular space of the bone marrow. The division of a rubriblast initiates a series of approximately four divisions over a period of 3 or 4 days to produce 16 metarubricytes that are no longer capable of division. These divisions are called *maturational divisions* because there is a progressive maturation of the nucleus and cytoplasm concomitant with the divisions. Each division yields a smaller cell with greater nuclear condensation and increased Hb synthesis. An immature RBC, termed a reticulocyte, is formed following extrusion of the nucleus.

Early precursors have intensely blue cytoplasm, when stained with Romanowsky-type blood stains, owing to the presence of many basophilic ribosomes and polyribosomes that are actively synthesizing globin chains and smaller amounts of other proteins. Because the cells are nonsecretory, rough endoplasmic reticulum is scant and limited to early erythroid precursors (Bessis, 1973). Hb progressively accumulates in these cells, imparting a red coloration to the cytoplasm. Cells with both red and blue coloration are described as having polychromatophilic cytoplasm. Kinetics of erythroid cells and changes in biochemical and metabolic pathways are depicted in Fig. 7.1; time intervals were determined for cattle (Rudolph and Kaneko, 1971).

B. Iron Metabolism

1. Transferrin

Developing erythroid cells generally extract 70–95% of the iron circulating in plasma (Bush *et al.*, 1956; Gillis and Mitchell, 1974; Kaneko and Mattheeuws, 1966; Obara and Nakajima, 1961; Weissman *et al.*, 1960). Plasma iron is bound to transferrin, a β-globulin that can maximally bind two atoms of ferric iron per molecule. Transferrin molecules transport iron to erythroid cells and bind to transferrin receptors on cell surfaces. Transferrin-iron-receptor complexes are internalized in a clathrin-coated vesicle, which matures into an ATP-dependent proton-pumping endosome. Ferric iron is released following acidification and leaves the endosome by an undefined mechanism that appears to require its reduction to the ferrous state (Aisen, 1994). Following iron release, apotransferrin-receptor complexes are returned to the surface where at neutral pH apotransferrin is released and returns to plasma.

The proportion of apo-, mono-, and diferric forms of transferrin present in serum depends on the percent saturation of transferrin with iron. Diferric transferrin is more efficient than monoferric transferrin in delivering iron to cells, because it binds with higher affinity to receptors and can deliver twice the iron per molecule of transferrin incorporated (Huebers *et al.*, 1985).

2. Intracellular Iron Transport

Within the cytoplasm iron appears to be complexed to ATP (Weaver and Pollack, 1989) and possibly other low-molecular-weight compounds for transport to mitochondria, the site of iron incorporation into protoporphyrin to form heme. Entry of iron into mitochondria is not well characterized but is energy dependent and involves the reduction of ferric iron to the ferrous state during transport into mitochondria (Romslo, 1980).

3. Ferritin

Iron not required for heme synthesis is stored as ferritin. Each ferritin molecule is composed of a protein shell of 24 apoferritin subunits surrounding a central core of up to 4500 iron atoms as ferric oxyhydroxide (Leibold and Guo, 1992). Individual ferritin molecules can be visualized by electron microscopy, and large aggregates of ferritin molecules can be visualized by light microscopy when stained for iron. When membrane bound, ferritin aggregates have been called siderosomes (Bessis, 1973).

4. Iron Regulation of Transferrin Receptor and Ferritin Expression

Transferrin receptor and ferritin expression are regulated in response to iron availability and demand for heme synthesis. When increased iron is needed, as occurs in iron-deficient states, transferrin receptor synthesis is increased and ferritin synthesis is decreased. These effects are accomplished by changes in the structure of an iron regulatory factor (IRF) protein, which binds with high affinity to iron responsive elements (IREs) of transferrin receptor mRNA and ferritin mRNA (Melefors and Hentze, 1993).

Free iron is toxic because it forms reactive hydroxyls that can cause oxidation of cellular components. When excess iron is present, transferrin receptor synthesis is decreased and ferritin synthesis is increased, minimizing free iron concentration. These effects occur when IRF binding to IREs on mRNAs is reduced (Melefors and Hentze, 1993).

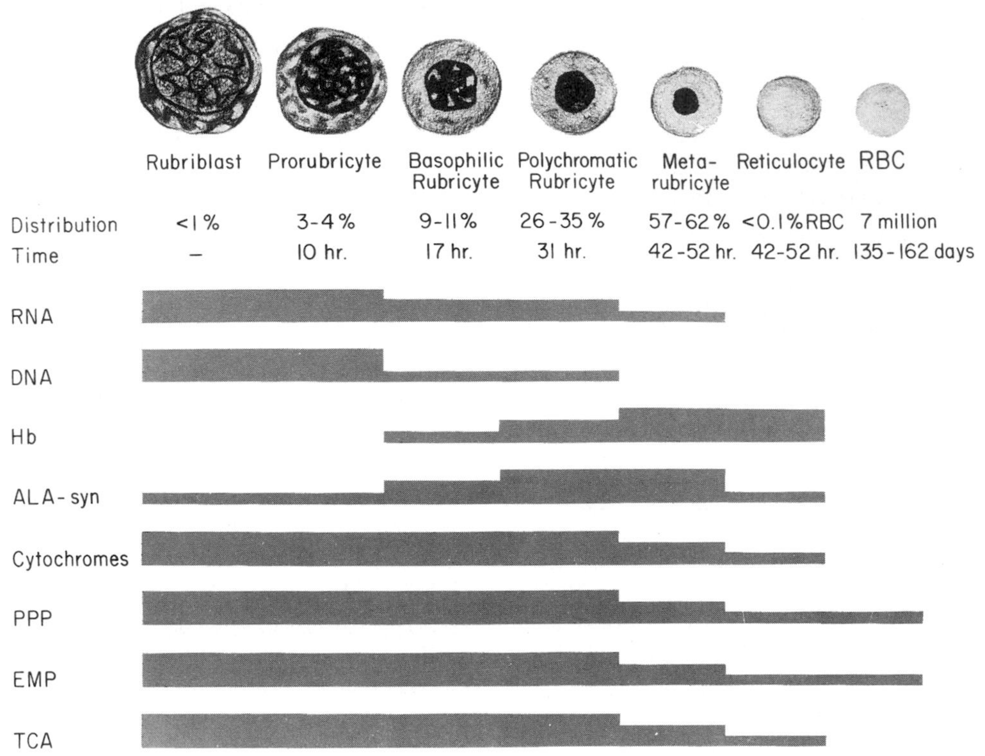

FIGURE 7.1 Summary of metabolic activities of the erythroid series. Maturation progresses from left to right. The time intervals indicated are for cattle. From Kaneko (1980) with permission.

5. Siderocytes and Sideroblasts

Nucleated erythroid cells with iron-staining inclusions are called *sideroblasts,* and anucleated cells are called *siderocytes.* The inclusions occur in erythroid precursors in bone marrow of normal people (Cartwright and Deiss, 1975), dogs (Feldman *et al.,* 1981), and pigs (Deiss *et al.,* 1966). Siderotic inclusions are rare or absent in circulating RBCs of normal animals, but may occur with lead poisoning (George and Duncan, 1979; Goodman and Dallman, 1969), hemolytic anemia, dyserythropoiesis, myeloproliferative diseases (Jain, 1986), chloramphenicol therapy (Harvey *et al.,* 1985), and in pigs fed a pyridoxine-deficient diet (Deiss *et al.,* 1966).

Siderotic inclusions in normal animals and people are composed of ferritin. Mitochondria can accumulate iron during certain pathologic states to a point at which they can be visualized as siderotic inclusions by light microscopy (Cartwright and Deiss, 1975). Mitochondrial iron overload appears when heme synthesis is impaired (Cartwright and Deiss, 1975). Pathologic sideroblasts and siderocytes with iron-loaded mitochondria have been demonstrated in a dog treated with chloramphenicol (Harvey *et al.,* 1985), a dog with dyserythropoiesis of unknown etiology (Canfield *et al.,* 1987), and in pigs fed a pyridoxine-deficient diet (Hammond *et al.,* 1969). Congenital sideroblastic anemias have been reported in people, but not in animals (Bottomley, 1991).

C. Hemoglobin Synthesis

Hb is a tetrameric protein consisting of four polypeptide globin chains, each of which contains a heme prosthetic group within a hydrophobic pocket. In adults the molecule consists of two identical alpha and two non-alpha chains that are generally classified as beta chains.

1. Heme Synthesis and Metabolism

Heme is a planar molecule composed of the tetrapyrrole protoporphyrin IX, containing a central ferrous molecule. The initial rate-controlling step in heme synthesis, the delta-aminolevulinic acid (ALA) synthetase (5-aminolevulinate synthase) reaction, occurs within mitochondria (see Chapter 8). Glycine and the Krebs cycle intermediate succinyl-CoA are utilized as substrates, and vitamin B_6, as pyridoxal phosphate, is required as a cofactor. The ALA formed is transported to the cytoplasm where a series of reactions results in the formation of coproporphyrinogen III, which must enter the mitochondria for the final steps in heme synthesis. The final reaction, heme synthetase, involves the insertion of ferrous iron into protoporphyrin IX.

Following synthesis, heme must be transferred from mitochondria to the cytoplasm for combination with globin chains to complete the synthesis of Hb. Free heme is poorly soluble in water and can bind to and damage cellular components (Hebbel and Eaton, 1989). It apparently is bound to cytosolic proteins for transport to sites of globin chain synthesis (Harvey and Beutler, 1982). Sometime after its synthesis, the iron moiety of heme is oxidized (presumably spontaneously) to the ferric state and is then more specifically called *ferriheme* (Schulman *et al.*, 1974).

Heme affects erythroid cell metabolism in different ways depending on the stage of maturation. It stimulates iron uptake, ferritin synthesis, and heme synthesis in early erythroid cells, but it inhibits iron uptake and heme synthesis in reticulocytes (Abraham, 1991; Battistini *et al.*, 1991). Excess heme accumulation presumably does not occur in early erythroid cells because of the presence of heme oxygenase activity, which decreases during maturation (Abraham, 1991).

2. *Globin Synthesis*

The synthesis of polypeptide globin monomers occurs in association with ribosomes and polyribosomes in the cytoplasm. Ferriheme is inserted into each globin chain, either during translation or shortly thereafter. Following synthesis, a ferriheme-containing alpha chain spontaneously combines with a ferriheme-containing beta chain to form a stable alpha-beta dimer. Two like alpha-beta dimers combine in a readily reversible manner to form the Hb tetramers (Bunn, 1987). Because it is ferriheme that combines with globin, newly synthesized Hb is in the form of MetHb (Schulman *et al.*, 1974).

3. *Control of Hemoglobin Synthesis*

The synthesis of heme and globin chains are coordinated such that minimal amounts of free heme or globin monomers exist in the cytoplasm (Bunn, 1987). Heme plays a central role because it governs the initiation step of translation in globin chain synthesis (Traugh, 1989). Increased amounts of free heme increase globin synthesis, and the lack of free heme inhibits globin synthesis. In addition, α and β chain synthesis is coordinated; surplus α chains inhibit their own synthesis but stimulate β chain synthesis, and surplus β chains inhibit their own synthesis (Jandl, 1987).

4. *Hemoglobin Types in Animals*

Hemoglobin types are different in animal and human embryos than in fetuses or in adults. Embryonal Hbs are composed of either one or two pairs of peptide chains not found in adult Hbs (Kitchen and Brett, 1974). In ruminants and humans, embryonal Hbs are replaced by fetal Hbs composed of two α and two γ chains. Most fetal Hb is replaced by adult Hb types in ruminants during the first month(s) after birth (Aufderheide *et al.*, 1980; Blunt, 1972; Huisman *et al.*, 1969; Kitchen and Brett, 1974; Lee *et al.*, 1971). This switch from production of fetal Hb to the production of adult Hb appears to result from an inherent programming of hematopoietic stem cells (Wood *et al.*, 1985). In cats, dogs, horses, and pigs, embryonal Hbs are replaced by adult Hb types during the fetal period. Hb types present in fetuses are identical to those found in adults (Bunn and Kitchen, 1973; Kitchen and Brett, 1974).

Considerable heterogeneity of Hb types occurs in adult animals. With the possible exception of pigs, two or more types are reported to occur in domestic animal species (Braend, 1988; Kitchen, 1969). Most polymorphism of animal Hbs is determined genetically and usually caused by multiple amino acid interchanges (Kitchen, 1974). Nongenetic alterations in Hb structure can also contribute to apparent Hb heterogeneity. Examples include the *N*-acetylation of β chains in cat HbB (Taketa *et al.*, 1972) and glycosylation of Hb, a function of intracellular glucose concentration and RBC lifespan (Higgins *et al.*, 1982; Rendell *et al.*, 1985). Increased glycosylation of Hb has been reported in diabetic dogs (Mahaffey and Cornelius, 1982; Wood and Smith, 1980).

A unique occurrence in sheep and goats is the synthesis of HbC in response to anemia (Huisman and Kitchen, 1968). This Hb switching from synthesis of HbA in sheep and HbA and HbB in goats to HbC is mediated by EPO (Barker *et al.*, 1980). Carbon dioxide decreases oxygen affinity for HbC more than it does for normal adult Hbs (Huisman and Kitchen, 1968; Winslow *et al.*, 1989).

D. Reticulocytes

1. *Formation*

Although reticulocytes may be formed by denucleation as metarubricytes pass through endothelial cells into vascular sinuses of bone marrow, most are formed within the extravascular space of the bone marrow by a process of nuclear extrusion that requires functional microtubules (Chasis *et al.*, 1989) and is likened to mitosis (Bessis, 1973; Simpson and Kling, 1967). Extruded nuclei are bound and phagocytosed by a novel receptor on the surface of bone marrow macrophages (Qui *et al.*, 1995). Early reticulocytes have polylobulated surfaces. Their cytoplasm contains ribosomes, polyribosomes, and mitochondria necessary for the completion of Hb synthesis. Reticulocytes derive their name from a network or reticulum that appears when stained with basic dyes such as methylene blue and brilliant cresyl green. That network is not preexisting but is an artifact formed by the precipitation of ribosomal ribonucleic acids and proteins. As reticulocytes mature, the amount of ribosomal material decreases until only a few basophilic specks can be visualized with reticulocyte staining procedures. These mature reticulocytes have been referred to as type IV (Houwen, 1992) or punctate reticulocytes (Alsaker *et al.*, 1977; Perkins and Grindem, 1995). To reduce the chance that a staining artifact would result in misclassifying a mature RBC as a punctate reticulocyte using a reticulocyte stain, the cell in question should have two or more discreet blue granules that are visible without requiring fine focus adjustment of the cell being evaluated to be classified as a punctate reticulocyte.

2. *Metabolism*

Reticulocyte metabolism and maturation have been reviewed by Rapoport (1986) and are summarized here. Immature reticulocytes continue to synthesize protein (primarily globin chains) with residual mRNA, tRNA, and rRNA formed prior to denucleation. Synthesis of fatty acids is minimal, but phospholipids are synthesized from preformed fatty acids. Substrates for protein and lipid synthesis and for energy metabolism are provided from endogenous sources (breakdown of mitochondria and ribosomes) as well as from plasma. The reticulocyte can synthesize adenine and guanine nucleotides *de novo*.

Most ATP is generated in reticulocytes by oxidative phosphorylation in mitochondria. Glucose is the major substrate, but amino acids and fatty acids can also be utilized for energy (Rapoport, 1986).

3. *Maturation into Erythrocytes*

Reticulocyte maturation into mature RBCs is a gradual process that requires a variable number of days depending on the species involved. Consequently the morphologic and physiologic properties of reticulocytes vary with the stage of maturation. The cell surface undergoes extensive remodeling with loss of membrane material and ultimately the formation of the biconcave shape of mature RBCs (Bessis, 1973). The loss of membrane protein and lipid components appears to require ATP (Weigensberg and Blostein, 1983) and involve formation of intracellular multivesicular bodies that fuse with the plasma membrane releasing vesicles (exosomes) extracellularly (Johnstone, 1992). This is a highly selective process where some proteins (e.g., transferrin receptor and fibronectin receptor) are lost and cytoskeletal proteins (e.g., spectrin) and firmly bound transmembrane proteins (e.g., the anion transporter and glycophorin A) are retained and concentrated (Johnstone, 1992; Rapoport, 1986).

Some membrane proteins such as the nucleoside transporter, glucose transporter, Na,K-ATPase, insulin receptor and adrenergic receptors decrease to variable degrees depending on the species involved (Chasis *et al.*, 1989; Johnstone, 1992). Examples where reticulocytes from a species exhibit a complete or nearly complete loss of a protein, which is retained in mature RBCs from other species, include the adenosine transporter in sheep (Jarvis and Young, 1982), the glucose transporter in pigs (Zeidler and Kim, 1982), and Na,K-ATPase in dogs (Maede and Inaba, 1985).

Although loss of membrane components accounts for much of the change in membrane protein composition during reticulocyte maturation, certain proteins such as protein 4.1 and glycophorin C increase because they are still being synthesized in reticulocytes (Chasis *et al.*, 1989).

Mitochondria undergo degenerative changes owing to lipoxygenase attack and subsequent ATP-dependent proteolysis (Rapoport, 1986). Degenerating mitochondria are either digested or extruded following entrapment in structures resembling autophagic vacuoles (Simpson and Kling, 1968). The polysomes separate into monosomes and decrease in number and disappear as reticulocytes mature into RBCs. The degradation of ribosomes appears to be energy dependent; it presumably involves proteases and RNAases (Rapoport, 1986).

4. *Species Differences in Marrow Release*

Reticulocyte maturation begins in the bone marrow and is completed in the peripheral blood and spleen in dogs, cats, and pigs. As reticulocytes mature, they lose the surface receptors needed to adhere to fibronectin and thrombospondin components of the extracellular matrix, thereby facilitating their release from the bone marrow (Long and Dixit, 1990; Patel *et al.*, 1985; Vuillet-Gaugler *et al.*, 1990). Residual adhesion molecule receptors on newly released reticulocytes may explain their tendency to concentrate in the reticular meshwork of the spleen (Patel *et al.*, 1985).

Reticulocytes become progressively more deformable as they mature, a characteristic that also facilitates their release from the marrow (Waugh, 1991). To exit the extravascular space of the marrow, reticulocytes press against the abluminal surfaces of endothelial cells that make up the sinus wall. Cytoplasm thins and small pores (0.5–2 μm) develop in endothelial cells that allow reticulocytes to be pushed through by a small pressure gradient across the sinus wall (Lichtman and Santillo, 1986; Waugh, 1991). Pores apparently close after cell passage.

Relatively immature aggregate-type reticulocytes are released from canine bone marrow; consequently, most of these cells appear polychromatophilic when viewed following routine blood film staining procedures (Laber *et al.*, 1974). Absolute reticulocyte counts oscillate with a periodicity of approximately 14 days in some dogs, suggesting that canine erythropoiesis may have a homeostatically controlled physiologic rhythm (Morley and Stohlman, 1969). Reticulocytes are normally not released from feline bone marrow until they mature to punctate-type reticulocytes; consequently, few or no aggregate reticulocytes (<0.4%), but up to 10% punctate reticulocytes, are found in blood from normal adult cats (Cramer and Lewis, 1972). The high percentage of punctate reticulocytes results from a long maturation time (Fan *et al.*, 1978) with delayed degradation of organelles. Reticulocytes are generally absent in peripheral blood of healthy adult cattle and goats, but a small number of punctate types (0.5%) may occur in adult sheep (Jain, 1986). Equine reticulocytes are absent from blood normally and are not released in response to anemia.

5. *"Stress" Reticulocytes*

Except for horses, increased numbers of reticulocytes are released in response to anemia, with better responses to hemolytic anemias than to hemorrhage. When the degree of anemia is severe, basophilic macroreticulocytes or so-called stress reticulocytes may be released into blood. It is proposed that a generation in the maturation sequence is skipped, and immature reticulocytes that are twice the normal size are released (Rapoport, 1986). Increased EPO results in a diminution in the adventitial cell and endothelial cell barrier separating marrow hematopoietic cells from the sinus, thereby potentiating the premature release of stress reticulocytes from the marrow (Chamberlain *et al.*, 1975). Although a portion of these macroreticulocytes apparently is rapidly removed from the circulation (Noble *et al.*, 1990), it is clear from studies in cats that some can mature into macrocytic RBCs with relatively normal lifespans (Weiser and Kociba, 1982).

E. Abnormalities in Erythroid Development

1. *Ineffective Erythropoiesis*

Ineffective erythropoiesis is used to describe the destruction of developing erythroid cells in marrow. Normally, minimal cells die within the marrow (Odartchenko *et al.*, 1971), but ineffective erythropoiesis is prominent in disorders of nucleic acid, heme, or globin synthesis. Examples include folate deficiency, iron deficiency, vitamin B_6 deficiency, lead poisoning, and thalassemia in humans (Jandl, 1987). Ineffective erythropoiesis also occurs in association with myeloproliferative and myelodysplastic disorders (Durando *et al.*, 1994; Jain, 1986) and congenital dyserythropoiesis (Holland *et al.*, 1991; Steffen *et al.*, 1992).

2. *Vitamin and Mineral Deficiencies*

Folate is required for normal DNA synthesis. Folate deficiency impairs the activity of the folate-requiring enzyme thymidylate synthase (Jandl, 1987). Not only is deoxythymidylate triphosphate (dTTP) synthesis decreased, but deoxyuridylate triphosphate (dUTP) accumulates secondarily in the cell such that some becomes incorporated into DNA in place of dTTP. Cycles of excision and attempts to repair these copy errors, with limited thymidine available, result in chromosomal breaks and malformations and slowing of the S phase in the cell cycle. Consequently, erythroid precursors are often large with deranged appearing nuclear chromatin; such cells are classified as megaloblastic cells. Folate deficiency in people causes macrocytic anemia because fewer divisions occur as a result of retarded nucleic acid synthesis in the presence of normal protein synthesis (Jandl, 1987). Possible causes of folate deficiency include dietary deficiency, impaired absorption, and drugs that interfere with folate metabolism.

Macrocytic anemias resulting from folate deficiency are rarely reported in animals. A possible case was reported in a dog on anticonvulsant therapy (Lewis and Rebar; 1979), but serum folate was not measured. Megaloblastic precursors are present in bone marrow of cats with experimental dietary folate deficiency, but packed cell volumes (PCVs) and mean cell volumes (MCVs) remained normal (Thenen and Rasmussen, 1978). Macrocytic anemia occurs in folate-deficient pigs (Bush *et al.*, 1956) but not lambs (Stokstad, 1968).

Vitamin B_{12} (cobalamin) deficiency in people causes hematologic abnormalities similar to folate deficiency because vitamin B_{12} is necessary for normal folate metabolism in humans (Chanarin *et al.*, 1985). In contrast, vitamin B_{12} deficiency does not cause macrocytic anemia in any animal species (Chanarin *et al.*, 1985). Anemia has been reported in some experimental animal studies, but RBCs were of normal size (Stokstad, 1968; Underwood, 1977), although slight increases in MCV have been reported in B_{12}-deficient goats fed diets deficient in cobalt (Mgongo *et al.*, 1981). Cobalamin deficiency has been reported secondary to an inherited malabsorption of cobalamin in giant schnauzer dogs (Fyfe *et al.*, 1989, 1991). Affected animals have normocytic, nonregenerative anemia with increased anisocytosis and poikilocytosis, neutropenia with hypersegmented neutrophils and giant platelets. Megaloblastic changes in the bone marrow were particularly evident in the myeloid cell line. The malabsorption of cobalamin in these dogs apparently results from the absence of an intrinsic factor-cobalamin receptor in the apical brush border of the ileum (Fyfe *et al.*, 1991). No blood or bone marrow abnormalities were recognized in kittens fed a B_{12}-deficient diet for several months (Morris, 1977), but a normocytic nonregenerative anemia was present in a cobalamin-deficient cat that probably had an inherited disorder of cobalamin absorption (Vaden *et al.*, 1992).

A number of disorders exhibit macrocytic anemias with megaloblastic abnormalities in the marrow that mimic findings in human folate or B_{12} deficiency, but have had normal serum levels of these vitamins when measured. Examples include cats infected with the feline leukemia virus (Dunn *et al.*, 1984; Hirsch and Dunn, 1983; Weiser and Kociba, 1983a), cattle with congenital dyserythropoiesis (Steffen *et al.*, 1992), and myelodysplastic syndrome in a horse (Durando *et al.*, 1994). In addition, some miniature and toy poodles exhibit macrocytosis without anemia and variable megaloblastic abnormalities in the bone marrow with normal serum folate and B_{12} values (Canfield and Watson, 1989; Schalm, 1976).

Abnormalities in heme or globin synthesis can result in the formation of microcytic hypochromic RBCs. Cellular division is normal, but Hb synthesis is delayed; consequently, one or more extra divisions occur in RBC development, resulting in smaller cells than normal.

Pyridoxine, vitamin B_6, is required for the first step in heme synthesis. Although natural cases of pyridoxine deficiency have not been documented in domestic animals, microcytic anemias with high serum iron values have been produced experimentally in dogs (McKibbin *et al.*, 1942), cats (Bai *et al.*, 1989; Carvalho da Silva *et al.*, 1959), and pigs (Deiss *et al.*, 1966) with dietary pyridoxine deficiency.

With the exception of young growing animals, iron deficiency in domestic animals usually results from blood loss. Milk contains little iron; consequently, nursing animals can easily deplete body iron store as they grow (Furugouri, 1972; Harvey *et al.*, 1987; Holter *et al.*, 1991; Siimes *et al.*, 1980). Microcytic RBCs are produced in response to iron deficiency (Holman and Drew, 1964, 1966; Holter *et al.*, 1991; Reece *et al.*, 1984) but a low MCV may not develop postnatally in species where the MCV is above adult values at birth (Weiser and Kociba, 1983b). The potential for development of severe iron deficiency in young animals appears to be less in species that begin to eat food at an early age.

Chronic iron deficiency anemia with microcytic RBCs is common in adult dogs in areas where hookworm and flea infestations are severe (Harvey *et al.*, 1982; Weiser and O'Grady, 1983). Severe iron deficiency appears to be rare in adult cats (French *et al.*, 1987; Fulton *et al.*, 1988) and horses (Smith *et al.*, 1986), but it occurs frequently in ruminants that are heavily parasitized with blood-sucking parasites such as *Haemonchus contortus.*

Prolonged copper deficiency generally results in anemia in mammals (Brewer, 1987; Lahey *et al.*, 1952), although it was not a feature of experimental copper deficiency in the cat (Doong *et al.*, 1983). The anemia is generally microcytic hypochromic; however, normocytic anemia has been reported in experimental studies in dogs, and normocytic or macrocytic anemias have been reported in cattle and adult sheep (Brewer, 1987). Copper deficiency results in impaired iron metabolism in at least two ways. In experimental studies in pigs, serum iron concentration is low in early copper deficiency when iron stores are normal (Lahey *et al.*, 1952). Functional iron deficiency results from inadequate iron absorption and mobilization of iron stores caused by a decreased concentration of circulating ceruloplasmin (Lee *et al.*, 1968). Ceruloplasmin is the major copper-containing protein in plasma, and its ferroxidase activity appears to be important in the oxidation of ferrous iron (released from ferritin stores) to the ferric state for binding to transferrin (Frieden, 1983).

If experimental copper deficiency is prolonged, hyperferremia occurs and bone marrow sideroblasts increase (Lee *et al.*, 1968). Reticulocyte mitochondria from copper-deficient pigs are unable to synthesize heme at the normal rate using ferric iron (Williams *et al.*, 1976). A deficiency in copper-containing cytochrome oxidase within mitochondria may slow the reduction of ferric to ferrous iron within mitochondria. That would, in turn, limit heme synthesis, which requires iron in the ferrous state (Porra and Jones, 1963).

3. *Deficiencies in Globin Synthesis*

Hereditary deficiencies in synthesis of the globin α chain (α-thalassemia) and β chain (β-thalassemia) cause microcytic hypochromic anemias in humans with variable degrees of poikilocytosis (Jandl, 1987). Both α- and β-thalassemia occur in mice, but hereditary hemoglobinopathies have not been reported in domestic animals (Kaneko, 1987).

4. *Aplastic Anemia*

Aplastic anemia is generally used to describe anemias where granulocytic, megakaryocytic, and erythrocytic cell lines are markedly reduced in the bone marrow. When only the erythroid cell line is reduced or absent, terms such as *pure red cell aplasia* or *selective erythroid aplasia* or *hypoplasia* are used. These anemias can result from insufficient numbers of stem cells, abnormalities in the hematopoietic microenvironment, or abnormal humoral or cellular control of hematopoiesis (Appelbaum and Fefer, 1981; Juneja and Gardner, 1985). The factors mentioned are interrelated and the exact defect in a given disorder is usually unknown.

Drug-induced causes of aplastic anemia or generalized marrow hypoplasia in animals include estrogen toxicity in dogs (Crafts, 1948; Gaunt and Pierce, 1986; Sherding *et al.*, 1981) and ferrets (Bernard *et al.*, 1983; Kociba and Caputo, 1981), phenylbutazone toxicity in dogs (Schalm, 1979; Watson *et al.*, 1980; Weiss and Klausner, 1990), trimethoprim-sulfadiazine administration in dogs (Fox *et al.*, 1993; Weiss and Klausner, 1990), bracken fern poisoning in cattle and sheep (Parker and McCrea, 1965; Sippel, 1952; Tustin *et al.*, 1968), trichloroethylene-extracted soybean meal in cattle (Strafuss and Sautter, 1967), and various cancer chemotherapeutic agents. Thiacetarsamide, meclofenamic acid, and quinidine (Watson, 1979; Weiss and Klausner, 1990) have also been incriminated as potential causes of aplastic anemia in dogs, as has griseofulvin in a cat (Rottman *et al.*, 1991).

Human parvovirus, hepatitis viruses, Epstein–Barr virus, dengue, cytomegalovirus, and human immunodeficiency virus 1 have, in low percentages of infections, resulted in aplastic anemia in people (Rosenfeld and Young, 1991). Parvovirus infections can cause severe erythroid hypoplasia, as well as myeloid hypoplasia in canine pups (Robinson *et al.*, 1980), but animals usually do not become anemic because of the long lifespans of RBCs. Either affected pups die acutely or the bone marrow returns rapidly to normal before anemia can develop. In contrast to its effects in pups, parvovirus is reported to have a minimal effect on erythroid progenitors in adult dogs (Brock *et al.*, 1989). Parvovirus inhibits colony formation of both myeloid and erythroid progenitor cells in feline bone marrow cultures (Kurtzman *et al.*, 1989), but only myeloid hypoplasia was reported during histologic examination of bone marrow from viremic cats (Larsen *et al.*, 1976).

Although some degree of marrow hypoplasia and/or dysplasia often occurs in cats with feline leukemia virus (FeLV) infections, true aplastic anemia is not a well-documented sequela (Rojko and Olsen, 1984). Anemia is a common finding in ill cats infected with feline immunodeficiency virus (FIV). It does not appear to result from a direct action(s) of the virus on erythroid progenitor cells, but generally occurs secondary to inflammation, dysplasia, or neoplasia (Linenberger *et al.*, 1991; Shelton *et al.*, 1990). Dogs that enter the chronic stage of ehrlichiosis generally have some degree of marrow hypoplasia and in severe cases can develop aplastic anemias (Buhles, *et al.*, 1975).

Idiopathic aplastic anemias have also been reported in dogs (Eldor *et al.*, 1978; Weiss and Christopher, 1985) and horses (Berggren, 1981; Lavoie *et al.*, 1987). CFU-Es were not detected in bone marrow culture from a dog with an idiopathic aplastic anemia (Weiss and Christopher, 1985). One case of erythroid and myeloid aplasia, with normal megakaryocyte numbers, has been reported in a horse, the etiology of which was unknown (Ward *et al.*, 1980).

5. Selective Erythroid Aplasia

Selective erythroid aplasia (pure red cell aplasia) occurs as either a congenital or acquired disorder in people (Dessypris, 1991). Acquired erythroid aplasia is often associated with abnormalities of the immune system. Erythroid aplasia may also occur secondary to disorders including infections, malignancy, and drug or chemical toxicities.

Acquired erythroid hypoplasia or aplasia occurs in dogs (Weiss, 1986; Weiss *et al.*, 1982). Some cases have immune-mediated etiologies based on positive responses to immunosuppressive therapy and the presence of antibodies that inhibit CFU-E development in marrow cultures (Weiss, 1986). Erythroid hypoplasia or dysplasia is reported to be a rare sequela to vaccination against parvovirus in dogs (Dodds, 1983). High doses of chloramphenicol cause reversible erythroid hypoplasia in some dogs (Watson, 1977) and erythroid aplasia in cats (Watson and Middleton, 1978). A dog has been reported to have congenital erythroid aplasia based on histopathologic examination of bone marrow at necropsy, but the M:E ratio was normal when aspirate smears were examined several days prior to euthanasia (Hotston Moore *et al.*, 1993).

Selective erythroid aplasia occurs in cats infected with subgroup C FeLV, but not in cats infected only with subgroups A or B (Onions *et al.*, 1982; Riedel *et al.*, 1986). Although both myeloid and erythroid progenitor cells are infected with virus, only erythroid progenitor cell numbers are decreased in cat marrow. Recent studies suggest that CFU-E numbers are markedly decreased because the envelope glycoprotein gp70 of subgroup C FeLV is uniquely cytopathic for BFU-E or impairs differentiation of BFU-E into CFU-E (Abkowitz, 1991; Dean *et al.*, 1992; Dornsife *et al.*, 1989).

IV. THE MATURE ERYTHROCYTE

Values for RBC glucose utilization, ion concentrations, and survival times in normal animals are given in Table 7.1. Enzyme activities are given in Tables 7.2 and 7.3, and chemical constituents in RBCs are given in Tables 7.4 and 7.5. These are not, however, comprehensive lists. Other values are provided by Friedemann and Rapoport (1974) and in various chapters of a reference book edited by Agar and Board (1983a). Anemia induced by phlebotomy or by hemolytic drugs produces changes in many of the given values owing to the influx of young RBCs into the circulation in response to the anemia (Agar and Board, 1983a). Methods for enzyme assays vary considerably; consequently, each laboratory will need to establish its own normal ranges if enzyme studies are to be done.

A. Membrane Structure

The RBC membrane is composed of a hydrophobic lipid bilayer with a protein skeletal meshwork attached to its inner surface by binding to integral (transmembrane) proteins (Fig. 7.2). Membrane proteins from RBCs have been numbered by their migration location (Smith, 1987) on sodium dodecyl sulfate–polyacrylamide gel electrophoresis (SDS–PAGE); some have also been given one or more names. Electrophoretic patterns of membrane proteins on SDS–PAGE are species variable (Gillis and Anastassiadis, 1985; Smith *et al.*, 1983a; Kobylka *et al.*, 1972; Whitfield *et al.*, 1983).

1. Lipids

The lipid bilayer and associated transmembrane proteins chemically isolate and regulate the cell interior. The bilayer consists of phospholipids arranged with hydrophobic hydrocarbon chains of fatty acids to the center of the bilayer and the polar ends of the molecules in contact with both intracellular and extracellular aqueous environments. Molecules of unesterified cholesterol are intercalated between fatty acid chains in molar concentrations approximately equal to the sum of the molar concentrations of phospholipids. Phospholipids are asymmetrically arranged, with anionic amino-containing phospholipids (phosphatidylserine and most of the phosphatidylethanolamine) located in the inner layer of the bilaminar membrane. These phospholipids are shuttled across the membrane by an ATP-dependent aminophospholipid-specific transporter (Zwaal *et al.*, 1993). Most of the cationic choline-containing phospholipids, phosphatidylcholine (lecithin) and sphingomyelin, are located in the outer layer (Kuypers *et al.*, 1993). These choline-containing phospholipids are readily exchangeable with plasma phospholipids, whereas the aminophospholipids are not (Reed, 1968). Species vary in RBC membrane phospholipid compositions (Engen and Clark, 1990; Garnier *et al.*, 1984; Nelson, 1967; Wessels and Veerkamp, 1973).

TABLE 7.1 Erythrocyte Glucose Utilization, Ion Concentrations, and Survival Times of Various Mammals[a]

Species	RBC glucose utilization (μmol/hr/ml)	RBC Na^+ (mmol/liter)	RBC K^+ (mmol/liter)	RBC survival (days)
Human	1.48 ± 0.11 (1)	6.2 ± 0.8 (7)	102 ± 3.9 (7)	120 (11)
Dog	1.33 ± 0.12 (1)	92.8 ± 11.1 (8)	5.7 ± 1.0 (8)	100 (11)
Cat	0.94 ± 0.09 (1)	105.8 ± 14.4 (8)	5.9 ± 1.9 (8)	72 (11)
Horse	0.64 ± 0.10 (1)	10.4 ± 1.8 (9)	120 ± 11.1 (9)	143 (11)
Cattle	0.56 ± 0.05 (2)	79.1 ± 14.6 (8)	22.0 ± 4.5 (8)	130 (11)
Sheep	0.69 ± 0.19 (3)	HK^+, 17.1 (10) LK^+, 73.7 (10)	HK^+, 98.7 (10) LK^+, 39.4 (10)	135 (11)
Goat	1.94 (4)	13.4 (4)	76.1 (4)	115 (11)
Pig	0.09 (5)	15.6 ± 1.8 (8)	105.9 ± 12.7 (8)	67 (11)
Rabbit	2.26 ± 0.30 (2)	16.8 ± 6.3 (8)	110.1 ± 6.0 (8)	57 (11)
Guinea pig	1.44 (6)	24.4 ± 5.4 (8)	107.2 ± 10.1 (8)	80 (11)
Mouse	2.85 ± 0.20 (2)			43 (11)
Hamster		17.2 (12)	92.0 (12)	50 (11)
Rat	2.38 ± 0.20 (2)	33.5 ± 3.5 (8)	104.7 ± 15.4 (8)	56 (11)

[a] Mean values have been recalculated at times to permit direct comparisons between species. Standard deviation values are given where indicated. Figures in parentheses are the references cited below. Abbreviations used are given in the footnote at the beginning of the chapter. References: (1) Harvey and Kaneko (1976b), (2) Magnani *et al.* (1980), (3) Leng and Annison (1962), (4) Harkness *et al.* (1970), (5) Kim and McManus (1971), (6) Laris (1958), (7) Beutler (1995), (8) Coldman and Good (1967), (9) Contreras *et al.* (1986), (10) Tucker and Ellory (1971), (11) Vacha (1983), (12) Miseta *et al.* (1983).

TABLE 7.2 Erythrocyte Enzymes of Various Animal Species[a]

Enzyme	Human	Dog	Cat	Horse
HK	1.78 ± 0.38 (1)	2.05 ± 0.86 (2)	5.04 ± 0.84 (2)	3.75 ± 0.42 (2)
GPI	24.1 ± 1.0 (3)	16.3 ± 1.8 (3)	49.0 ± 8.3 (3)	27.3 ± 5.9 (3)
PFK	11.0 ± 2.3 (1)	9.7 ± 1.3 (2)	2.2 ± 0.7 (2)	8.4 ± 1.8 (2)
Aldol	3.19 ± 0.86 (1)	2.36 ± 0.32 (4)		2.71 ± 0.98 (9)
TPI	2111 ± 397 (1)	436 ± 70 (40)		
GAPD	226 ± 42 (1)	54.4 ± 2.8 (4)	59.0 (6)	57.2 ± 14.2 (9)
PGK	320 ± 36 (1)	89.6 ± 9.4 (4)	29.2 (6)	69.1 ± 19.8 (9)
MPGM	37.7 ± 5.6 (1)	4.04 ± 1.19 (4)	6.35 (6)	5.7 (6)
Enol	5.39 ± 0.83 (1)	0.84 ± 0.16 (4)		13.6 ± 3.9 (9)
PK	15.0 ± 2.0 (1)	8.4 ± 2.0 (2)	22.9 ± 5.5 (2)	1.3 ± 0.7 (2)
LDH	200 ± 26 (1)	52.2 ± 5.0 (4)	15.1 ± 2.2 (8)	32.3 ± 3.6 (9)
AST	3.02 ± 0.67 (1)	3.14 ± 1.10 (2)		1.53 ± 0.27 (7)
DPGM	2.00 (6)	1.02 (6)	0.08 (6)	0.6 (6)
DPGP	0.021 (6)	0.010 (6)	0.005 (6)	0.006 (6)
G6PD	8.34 ± 1.59 (1)	11.33 ± 1.95 (2)	15.16 ± 3.15 (2)	19.19 ± 3.00 (2)
6PGD	8.78 ± 0.78 (1)	6.73 ± 1.30 (2)	7.00 ± 1.19 (2)	2.69 ± 0.66 (2)
GR	3.37 ± 0.69 (3)	1.38 ± 0.15 (3)	3.69 ± 1.02 (3)	0.89 ± 0.06 (3)
GPx	30.8 ± 4.6 (1)	82.3 ± 24.0 (2)	168.7 ± 24.0 (2)	42.6 ± 17.9 (2)
GST	6.7 ± 1.8 (1)	3.2 ± 0.9 (2)	35.6 ± 8.5 (2)	6.4 ± 1.7 (2)
SOD	2352 (5)	2118 (5)	2885 (5)	
Catalase (× 10^3)	153 ± 24 (1)	9 ± 4 (2)	161 ± 43 (2)	105 ± 21 (2)
NADH-MR	19.2 ± 3.8 (1)	22.2 ± 2.8 (2)	20.0 ± 4.6 (2)	25.9 ± 5.5 (2)
NADPH-Diaph	0.51 ± 0.12 (3)	0.33 ± 0.06 (3)	0.39 ± 0.06 (3)	0.70 ± 0.06 (3)
Na^+, K^+-ATPase	8 ± 2 (10)	Nil (10)	Nil (10)	5 ± 2 (10)

[a] All enzyme units are IU/g Hb, except SOD given in units/g Hb and ATPase given in μmoles phosphorus liberated/g Hb/hr. Mean values have been recalculated at times to permit direct comparisons between species. Standard deviation values are given where indicated. Figures in parentheses are the references cited below. Abbreviations used are given in the footnote at the beginning of the chapter. Temperatures above 25°C are included in reference citations. References: (1) Beutler (1984) (at 37°C), (2) Harvey (1994) (at 37°C), (3) Harvey and Kaneko (1975a), (4) Maede and Inaba (1987) (at 37°C), (5) Kurata *et al.* (1993), (6) Harkness *et al.* (1969), (7) Franken and Schotman (1977), (8) Schechter *et al.* (1973), (9) Smith *et al.* (1972), (10) Gupta *et al.* (1974) (at 44°C).

Early studies indicated that ruminant RBCs lack phosphatidylcholine, a finding at odds with a more recent study (Engen and Clark, 1990). Although RBC cholesterol/phospholipid ratios remain relatively constant, differences occur in membrane phospholipid composition between neonate and adult RBCs (Marin *et al.*, 1990).

A small amount of glycolipid is also located in the outer layer. Species differ in the dominant glycolipids of RBCs (Eberlein and Gercken, 1971; Yamakawa, 1983). Based on studies of human blood group antigens, it is assumed that many animal blood group antigens are also glycolipids and that specificity resides in the carbohydrate moieties.

2. *Integral Membrane Proteins*

Integral membrane proteins penetrate the lipid bilayer. These glycoproteins express carbohydrate residues on the outside surface of the cell. They contribute negative charge to the cell surface, function as receptors or transport proteins and carry RBC antigens (Chasis and Mohandas, 1992; Mohandas and Chasis, 1993; Schrier, 1985). Band 3 (protein 3) is the major integral protein accounting for approximately one-fourth of the total membrane protein (Schrier, 1985). It is important as an anion transporter and provides a site for binding of the cytoskeleton internally. Additional transmembrane glycoproteins called *glycophorins* also help anchor and stabilize the cytoskeleton (Chasis and Mohandas, 1992).

3. *Membrane Skeletal Proteins*

The membrane skeleton appears as a dense sweater-like meshwork bound to the inner surface of the lipid bilayer. It is a major determining factor of membrane shape, deformability, and durability (Mohandas and Chasis, 1993). Major skeletal proteins include spectrin, actin, protein 4.1, ankyrin, and adducin (Fig. 7.2). Spectrin is a heterodimer of long, flexible α and β chains twisted around one another. Heterodimers are bound together by self-association at their head ends to form tetramers. Multiple spectrin tail ends (average 6) are joined by binding to common short filaments of actin to form an anastomosed meshwork of polygons (Fig.

TABLE 7.3 Erythrocyte Enzymes of Various Animal Species[a]

Enzyme	Cattle	Sheep	Goat	Pig
HK	0.36 ± 0.14 (1)	0.58 ± 0.07 (3)	0.52 ± 0.15 (4)	0.18 ± 0.12 (5)
GPI	17.0 ± 1.0 (1)	19.8 ± 2.6 (3)	79.1 ± 8.5 (4)	90.9 ± 16.3 (5)
PFK	2.43 ± 1.26 (1)	1.53 ± 0.14 (3)	2.33 ± 0.46 (4)	1.23 ± 0.85 (5)
Aldol	1.46 (2)	1.28 ± 0.13 (3)	1.44 ± 0.71 (4)	0.78 ± 0.67 (5)
TPI		300 ± 17 (3)	589 ± 122 (4)	719 ± 211 (5)
GAPD	43.3 (2)	57.2 ± 5.2 (3)	73.7 ± 10.2 (4)	40.6 (5)
PGK	16.2 ± 7.3 (1)	49.6 ± 2.6 (3)	87.8 ± 18.3 (4)	94.2 ± 63.7 (7)
MPGM	10.0 ± 4.5 (1)	18.7 ± 1.5 (3)	19.3 ± 6.7 (4)	27.3 ± 12.6 (5)
Enol	3.15 ± 0.98 (1)	8.15 ± 0.51 (3)	13.9 ± 2.0 (4)	8.58 ± 6.45 (5)
PK	5.72 ± 1.13 (1)	2.82 ± 0.22 (3)	5.00 ± 1.23 (4)	12.5 ± 4.8 (5)
LDH	23 (2)	33.6 ± 3.5 (3)	17.5 ± 3.05 (4)	28.7 ± 5.9 (5)
DPGM	0.42 (6)		0.04 (6)	0.58 (6)
DPGP	0.03 (6)		0.01 (6)	0.01 (6)
G6PD	5.40 ± 0.49 (1)	0.76 ± 0.13 (3)	2.06 ± 0.46 (4)	17.3 ± 1.2 (5)
6PGD	0.84 ± 0.19 (1)		0.58 ± 0.24 (4)	3.3 ± 1.4 (5)
GR	0.69 ± 0.47 (7)	2.60 ± 0.22 (3)	3.87 ± 1.57 (4)	2.6 ± 0.8 (7)
GPx	165 (8)	164 ± 21 (3)	179 ± 55 (4)	
GST	4.7 ± 0.5 (9)	7.8 ± 1.0 (9)	8.0 (13)	1.4 ± 0.2 (9)
SOD	2060 ± 75 (10)	1910 ± 100 (10)		1240 ± 100 (10)
Catalase (× 10^3)	81 (8)	16.6 ± 1.5 (12)		
NADH-MR	1.83 ± 0.24 (1)	2.00 ± 0.12 (3)		
Na^+, K^+-ATPase	Nil (11)	HK^+, 3 ± 1 (11) LK^+ Nil		10 ± 2 (11)

[a] All enzyme units are IU/g Hb, except SOD given in units/g HB and ATPase given in μmoles phosphorus liberated/g Hb/hr. Mean values have been recalculated at times to permit direct comparisons between species. Standard deviation values are given where indicated. Figures in parentheses are the references cited below. Abbreviations used are given in the footnote at the beginning of the chapter. Temperatures above 25°C are included in reference citations. References: (1) Zinkl and Kaneko (1973a), (2) Smith *et al.* (1972), (3) Agar and Smith (1973, 1974), (5) McManus (1967), (6) Harkness *et al.* (1969), (7) Agar *et al.* (1974), (8) Kurata *et al.* (1993), (9) Del Boccio *et al.* (1986), (10) Maral *et al.* (1977), (11) Gupta *et al.* (1974) (at 44°C), (12) Suzuki and Agar (1983), (13) Board and Agar (1983).

TABLE 7.4 Erythrocyte Chemical Constituents of Various Species[a]

Analyte	Human	Dog	Cat	Horse
G6P	27.8 ± 7.5 (1)	17.1 ± 1.8 (3)		9.1 ± 2.6 (4)
F6P	9.3 ± 2.0 (1)	5.4 ± 0.5 (3)		3.5 ± 1.1 (4)
FDP	1.9 ± 0.6 (1)	1.4 ± 0.2 (3)		3.9 ± 1.8 (4)
DHAP	9.4 ± 2.8 (1)	6.7 ± 1.0 (3)		7.8 ± 4.5 (4)
3PG	44.9 ± 5.1 (1)	48.8 ± 4.2 (3)		22.4 ± 6.6 (4)
2PG	7.3 ± 2.5 (1)	17.6 ± 4.6 (3)		24.4 ± 4.5 (4)
PEP	12.2 ± 2.2 (1)	20.7 ± 4.1 (3)		5.3 ± 1.3 (4)
Pyruvate	53 ± 33 (1)	24 ± 9 (6)		
Lactate	932 ± 211 (1)	940 ± 517 (6)		600 ± 100 (8)
AMP	21.2 ± 3.4 (1)	35 ± 6 (3)		2.3 ± 1.3 (4)
ADP	216 ± 36 (1)	211 ± 50 (3)	82 (5)	16 ± 3 (4)
ATP	1438 ± 99 (1)	639 ± 140 (2)	529 ± 176 (2)	370 ± 74 (2)
2,3DPG	4171 ± 636 (1)	5989 ± 632 (2)	874 ± 317 (2)	6220 ± 1071 (2)
Pi	480 (5)	350 (5)	260 (5)	210 (5)
GSH	2234 ± 354 (1)	2171 ± 344 (2)	2106 ± 264 (2)	2552 ± 382 (2)
GSSG	4.2 ± 1.5 (1)	6.9 ± 1.7 (7)		

[a] Given in nmole/ml RBC, except lactate and pyruvate which are given in nmole/ml whole blood. Mean values have been recalculated at times to permit direct comparisons between species. Standard deviation values are given where indicated. Figures in parentheses are the references cited below. Abbreviations used are given in the footnote at the beginning of the chapter. References: (1) Beutler (1984), (2) Harvey (1994), (3) Harvey *et al.* (1992b), (4) Smith and Agar (1976), (5) Harkness *et al.* (1969), (6) Maede and Inaba (1987), (7) Maede *et al.* (1982), (8) Snow and Martin (1990).

TABLE 7.5 Erythrocyte Chemical Constituents of Various Species[a]

Analyte	Cattle	Sheep	Goats	Pigs
G6P	52 ± 15 (1)	28 ± 10 (2)	3.2 ± 0.8 (3)	11 (4)
F6P	28 ± 11 (1)	11 ± 2 (2)	1.2 ± 0.7 (3)	4.5 (4)
FDP	16 ± 12 (1)	25 ± 10 (2)		2 (4)
DHAP	35 ± 11 (1)	10 ± 3 (2)	1.0 ± 0.5 (3)	1 (4)
3PG	32 ± 12 (1)	11 ± 21 (2)		53 (4)
2PG	9 ± 4 (1)	14 ± 10 (2)		12 (4)
PEP	19 ± 7 (2)	19 ± 14 (2)	1.0 ± 0.4 (3)	8 (4)
Pyruvate	54 ± 24 (1)	87 ± 21 (2)		22 (4)
Lactate	1989 ± 758 (1)	1623 ± 1203 (2)		14800 (4)
AMP		40 ± 23 (2)	8 ± 2 (3)	250 (4)
ADP	73 ± 22 (1)	138 ± 31 (2)	17 ± 4 (3)	500 (4)
ATP	633 ± 115 (1)	532 ± 126 (8)	363 ± 52 (3)	1670 (4)
2,3DPG	289 (5)	21 ± 16 (8)	59 ± 28 (3)	9500 (4)
Pi	400 (6)	666 ± 206 (9)	850 (6)	870 (6)
GSH	2490 ± 350 (7)	2257 ± 130 (10)	2500 ± 360 (7)	
GSSG		<5 (10)		

[a] Given in nmole/ml RBC, except lactate and pyruvate which are given in nmole/ml whole blood. Mean values have been recalculated at times to permit direct comparisons between species. Standard deviation values are given where indicated. Figures in parentheses are the references cited below. Abbreviations used are given in the footnote at the beginning of the chapter. References: (1) Zinkl and Kaneko (1973), (2) Noble *et al.* (1983), (3) Agar and Smith (1974), (4) Magnani *et al.* (1983), (5) Agar *et al.* (1983), (6) Harkness *et al.* (1969), (7) Agar *et al.* (1974), (8) Travis *et al.* (1985), (9) Battaglia *et al.* (1970), (10) Srivastava and Beutler (1969).

7.3). These junctional complexes with actin are stabilized by other proteins including protein 4.1 and adducin. The meshwork is bound to transmembrane protein 3 by ankyrin in regions of spectrin self-association (Mohandas and Chasis, 1993). Protein 4.1 also binds the meshwork to one or more integral membrane glycophorins (Chasis and Mohandas, 1992).

The cytoskeletal meshwork assumes a condensed configuration in nondeformed discocytes, with spectrin tetramer fibers existing in folded or coiled states (Fig. 7.3). When fully extended (accomplished by removal of lipid from the membrane), the cytoskeleton meshwork extends over an area several times the normal RBC surface area (Liu and Derick, 1992; Palek and Sahr, 1992). Because of the spectrin structure, membrane skeletons have extensional elasticity and can be stretched more than twice the normal RBC diameter without rupture.

B. Shape and Deformability

The normal biconcave shape of most mammalian RBCs (discocytes) represents the resting unstressed geometry of the cell. The biconcave shape results in a large surface area-to-volume ratio compared to that of a sphere, allowing RBCs to undergo marked deformation while maintaining a constant surface area (Lenard, 1974). This is important because an increase of 3–4% of surface area results in cell lysis (Mohandas and Chasis, 1993).

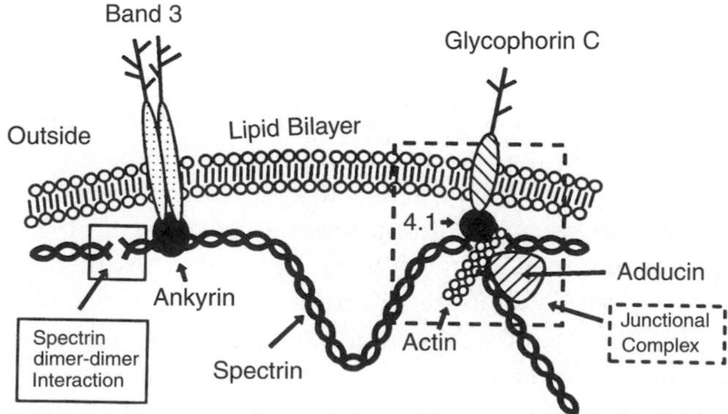

FIGURE 7.2 Schematic model of the organization of the RBC membrane skeleton.

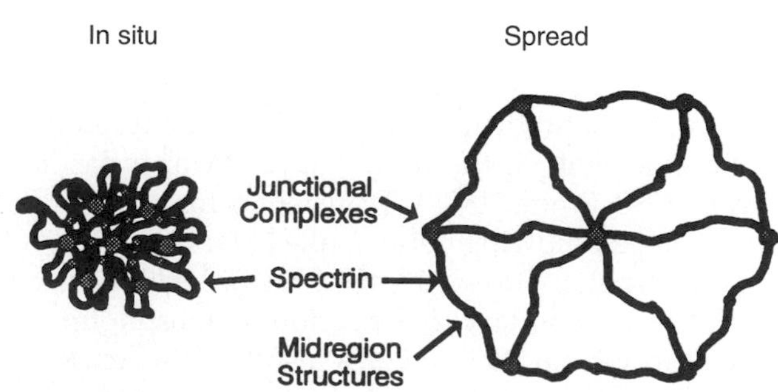

FIGURE 7.3 Schematic model of a hexagon of condensed and fully extended (spread) cytoskeletal meshwork. Spectrin dimer–dimer interactions and binding to band 3 occur in the midregions. See Figure 7.2 for components of junctional complexes.

The RBC spends little time in a discoid shape in the microcirculation. Except for goats, RBCs from domestic animals have diameters (Jain, 1986) greater than those of capillaries (approximately 4 μm) (Henquell *et al.*, 1976; Sobin and Tremer, 1972); consequently, they must be deformable to flow through capillaries. RBCs must pass through even smaller spaces in the sinus wall of the spleen (Chen and Weiss, 1973). The biconcave shape is generally more pronounced in species with larger RBCs (Jain, 1986), presumably because the degree of deformation required to flow through capillaries is greater. RBC deformability also reduces the bulk viscosity of blood in large vessels (Smith, 1991).

RBC deformability is a function of surface-to-volume ratio, viscosity of intracellular contents (determined primarily by intracellular Hb concentration), and viscoelastic properties of the membrane (Mohandas and Chasis, 1993). RBC shape and viscoelastic properties result from interactions between the fluid lipid bilayer and the underlying cytoskeleton, which stabilizes the lipid bilayer and provides both rigid support and elasticity. Membrane lipid fluidity varies with cholesterol composition, concentration of phospholipids present, and degree of saturation of fatty acids and length of acyl chains (Yawata *et al.*, 1984). Comparisons of RBC deformability among various animal species using ektacytometry have been reported (Smith *et al.*, 1979). Large RBCs are generally more deformable than small ones. RBCs from species in the family Camelidae are flat, thin, and not deformable; they apparently flow through vessels by orienting to the direction of flow.

In addition to mechanically-induced deformations, a wide variety of chemical perturbations, genetic defects, and oxidative injury can result in shape changes (Mohandas and Chasis, 1993; Smith, 1987). Small changes in the surface areas of the inner or outer lipid monolayers can result in transformations of discocytes into echinocytes or stomatocytes. Echinocytes are spiculated RBCs in which the spicules are relatively evenly spaced and of similar size. They form when the surface area of the outer lipid monolayer increases relative to the inner monolayer. Echinocytic transformation occurs in the presence of fatty acids, lysophospholipids, and amphiphatic drugs that distribute preferentially in the outer half of the lipid bilayer (Mohandas and Chasis, 1993; Smith, 1987). Transient echinocytosis occurs in dogs following coral snake (Marks *et al.*, 1990) and rattlesnake (Brown *et al.*, 1994a) envenomation, presumably secondary to the action of phospholipases present in venom. Echinocytes also form when RBCs are dehydrated (Weiss and Geor, 1993), pH is increased, RBC ATP is depleted (Jacob *et al.*, 1973), and intracellular calcium is increased (Smith, 1987). Echinocytes and other shape abnormalities have been recognized in dogs with RBC pyruvate kinase deficiency, which results in a decreased ability to generate ATP (Chandler *et al.*, 1975; Schaer *et al.*, 1992; Muller-Soyano *et al.*, 1986). Echinocytosis occurs in horses where total body depletion of cations has occurred. Increased numbers have been reported in horses during endurance exercise (Boucher *et al.*, 1981), following furosemide-induced electrolyte depletion (Weiss *et al.*, 1992), and in ill horses with systemic electrolyte depletion and hyponatremia (Geor *et al.*, 1993).

Although the mechanism(s) is unknown, ATP appears to be required in some manner for maintenance of normal shape and deformability of RBCs (Jacob *et al.*, 1973). Because ATP concentrations must be depleted for a number of hours to demonstrate changes in shape and deformability *in vitro*, its concentration does not directly control these properties; rather, the shapes of cells are altered by processes occurring secondary to ATP depletion (Feo and Mohandas, 1977).

ATP is required for a number of reactions involving the RBC membrane (Cohen and Gascard, 1992). It is used as the phosphoryl donor in a wide variety of phosphorylation reactions involving membrane proteins and for the phosphorylation of membrane phosphoinositides. It provides the energy needed to pump Ca^{2+} out of cells. Increased Ca^{2+} activates neutral proteases (calpains), which can degrade membrane skeletal proteins, and phospholipase C, which cleaves phosphoinositides. ATP is required for the transport of aminophospholipids to the inner half of the lipid bilayer, presumably assisting in the maintenance of the asymmetry of membrane phospholipids. The relative importance of each of these ATP-dependent reactions to the maintenance of RBC shape and deformability remains to be determined (Cohen and Gascard, 1992).

Cup-shaped stomatocytes form when amphiphatic drugs are present that distribute preferentially in the inner half of the lipid bilayer (Smith, 1987). Stomatocytes also form when RBC water content is increased as occurs in hereditary stomatocytosis in dogs (Giger *et al.*, 1988a; Pinkerton *et al.*, 1974).

Acanthocytes are RBCs with irregularly spaced, variably sized spicules. They form when RBC membranes contain excess cholesterol compared to phospholipids. If cholesterol and phospholipids are increased to a similar degree, codocyte formation is more likely than acanthocyte formation (Cooper *et al.*, 1972). Alterations in RBC membrane lipids can result from increasd blood cholesterol content (Cooper *et al.*, 1980) or the presence of abnormal plasma lipoprotein composition (Ulibarrena *et al.*, 1994). Another possible contributing factor is the defective repair (acylation of lysophospholipids) of oxidant-damaged RBC phospholipids reported in human patients with cirrhosis and acanthocytosis (Allen and Manning, 1994). Acanthocytes have been recognized in animals with liver disease, possibly due to alterations in plasma lipid composition that can alter RBC lipid composition (Christopher and Lee, 1994; Shull *et al.*, 1978). They have also been reported in dogs with disorders such as hemangiosarcoma and disseminated intravascular coagulation that result in RBC fragmentation (Weiss *et al.*, 1993).

Poikilocytes can form when oxidant injury results in Heinz body formation and/or membrane injury. One or more blunt RBC surface projections may form as the membrane adheres to Heinz bodies bound to its internal surface. Eccentrocytes are RBCs in which the Hb is localized to part of the cell, leaving a Hb-poor area visible in the remaining part of the cell. Other terms used to refer to eccentrocytes include hemighosts, irregularly contracted cells, double-colored cells, and cross-bonded RBCs (Arese and De Flora, 1990; Chan *et al.*, 1982). They are formed by the adhesion of opposing areas of the cytoplasmic face of the RBC membrane (Fischer *et al.*, 1985; Fischer, 1986, 1988). Denatured spectrin is believed to be of primary importance in the cross bonding of membranes (Fischer, 1988; Arese and De Flora, 1990). Eccentrocytes have been seen in animals ingesting or receiving oxidants (Harvey and Rackear, 1985; Harvey *et al.*, 1986; Reagan *et al.*, 1994) and in a horse with G6PD deficiency (Stockham *et al.*, 1994).

Genetic defects in a variety of membrane proteins result in spherocyte or elliptocyte formation in people (Palek and Sahr, 1992). Hereditary elliptocytosis has been reported in a dog with protein 4.1 deficiency (Smith *et al.*, 1983a).

C. Blood Group Isoantigens

Large numbers of protein and complex carbohydrate antigens occur on the external surface of RBCs. Some antigens are present on RBCs from all members of a species and others (including blood group isoantigens) segregate genetically, appearing in some but not all members of a species. Blood group isoantigens are detected serologically on the surface of RBCs using agglutination and/or hemolysis tests. With detailed genetic studies, these isoantigens can be placed into blood groups (RBC isoantigen systems). Blood groups have individual chromosomal loci and each loci has from two to many allelic genes. Most blood groups (such as the ABO system in man) derive their antigenicity from the carbohydrate composition of membrane-associated glycolipids and glycoproteins. The amino acid sequence of membrane proteins accounts for the antigenic determinants in other blood groups, such as the complex Rh system in man (Agre and Cartron, 1991). Most isoantigens are produced by erythroid cells but some, such as the J group in cattle, the DEA-7 (Tr) group in dogs, the R group in sheep, and the A group in pigs (Tizard, 1992), are produced by other tissues and absorbed from plasma.

Blood groups in domestic animals have been reviewed (Bell, 1983; Smith, 1991; Tizard, 1992). They have been most extensively characterized in horses and cattle, where blood typing is routinely used for animal identification and parentage testing. Cattle have 11 blood groups with multiple alleles per group. Based on all the different combinations of factors that can occur, greater than 2 trillion distinctly different blood type profiles are possible in cattle.

1. Blood Group Isoantigens of Clinical Significance

Isoantigens vary in their potential to cause transfusion reactions when mismatched blood is given. Many isoantigens are weak (do not induce antibodies of high titer) or induce antibodies that do not act at normal body temperature. Fortunately only a few isoantigens appear to be important in producing hemolytic disease in animals. Although dogs have at least 12 blood groups, only the DEA-1.1 isoantigen regularly generates hemolysins in high enough titer to cause significant transfusion reactions (Giger *et al.*, 1995a; Stormont, 1982).

Incompatibilities in the AB blood group of cats have been recognized to cause transfusion reactions and neonatal isoerythrolysis (Auer and Bell, 1983; Giger and Bücheler, 1991; Giger and Akol, 1990; Hubler *et al.*, 1987). It has been proposed that the A and B isoantigens (blood types) result from the action of two different alleles at the same gene locus and that A is dominant over B (Giger *et al.*, 1991a). Type A cat RBCs have glycolipids with terminal *N*-glycolyneuraminic acid (NeuGc) on their surface, whereas type B cat RBCs have glycolipids with terminal *N*-acetylneuraminic acid (NeuAc) on their surface (Andrews *et al.*, 1992). Because the enzyme CMP-*N*-acetylneuraminic hydroxylase converts NeuAc to NeuGc, it has been proposed that type B cats lack this enzyme. The frequency of blood types varies with location and breed of cat. Fewer than 1% of domestic short- and long-hair cats in the United States are type B, but 15–59% of purebred cats in some breeds are type B (Giger *et al.*, 1991a).

Horse RBC isoantigens are recognized to occur at 7 blood group loci. The frequency of expression of RBC isoantigens varies by breed of horse (Bowling and Clark, 1985). Aa and Qa isoantigens are responsible for most cases of neonatal isoerythrolysis in horses, where mares negative for these factors develop antibodies against them and transfer these antibodies to their foals through colostrum (Bailey, 1982). Neonatal isoerythrolysis has been reported in mule foals because of an RBC antigen not found in horses, but present in some donkeys and mules (McClure *et al.*, 1994).

Pig RBC isoantigens are recognized to occur in 15 blood groups numbered A to O. A-negative pigs exhibit intravascular hemolysis when transfused with A-positive blood and neonatal isoerythrolysis has been recognized in pigs, with antibodies usually directed against isoantigens of the E blood group (Tizard, 1992).

Naturally occurring neonatal isoerythrolysis has not been reported in cattle, but it occurs in some calves born to cows previously vaccinated for anaplasmosis or other bovine origin vaccines containing RBC membranes (Dimmock and Bell, 1970; Luther *et al.*, 1985). Several blood group isoantigens have been incriminated, but the most important blood group isoantigens involved in this disorder are uncertain (Dimmock and Bell, 1970). Based on experimentally produced disease, the B isoantigen group appears to generate potent hemolysins (Dimmock *et al.*, 1976).

2. *Natural Antibodies*

Some blood group systems, such as the ABO group in man, the AB group in cats, and the A group in pigs, are characterized by "naturally occurring" antibodies; that is, antibodies occur in plasma in the absence of prior exposure to blood from another individual (Tizard, 1992; Bücheler and Giger, 1993). In other blood groups, such as the Rh system in man and most blood groups in animals, antibody formation results from prior exposure to different RBC isoantigens via transfusion, pregnancy, or vaccination with products containing blood group antigens (Stormont, 1982). Fortunately, naturally occurring antibodies of clinical significance seldom occur in animals; consequently, adverse transfusion reactions to unmatched RBCs generally do not occur at the time of the first blood transfusion. However, exceptions may occur as in the case of the AB group in cats where B-positive cats have naturally occurring anti-A antibodies with high hemolytic titer (Bücheler and Giger, 1993).

D. Membrane Transport

The lipid bilayer is impermeable to most molecules. Consequently, various membrane protein transport systems are utilized for movement of molecules into and out of RBCs.

1. *Anions and Water*

Band 3 appears to function as an aqueous pore or channel for the movement of anions (e.g., bicarbonate and chloride), water, certain nonelectrolytes, and probably cations to some extent (Solomon *et al.*, 1983). Gruber and Deuticke (1973) studied phosphate as a model for anion exchange in RBCs from several species. There was a positive correlation between phosphate influx and the proportion of phosphatidylcholine in membrane phospholipids (Gruber and Deuticke, 1973).

2. *Sodium and Potassium*

Major interspecies, and in some cases intraspecies, differences occur in cation transport and subsequently in intracellular Na^+ and K^+ concentrations (Ellory and Tucker, 1983). There is a strong positive correlation between the intracellular K^+/Na^+ ratio and ATP concentration when RBCs of different species are compared. The cause of this relationship is unknown, but it is not related to differences in glucose utilization (Miseta *et al.*, 1993).

Like man, those animal species with high intracellular K^+ concentrations, horse, pig, and some ruminants, have an active Na^+, K^+-pump that exchanges intracellular Na^+ for extracellular K^+ with the hydrolysis of ATP. This Na^+, K^+-activated ATPase activity is often used as a measure of Na^+, K^+-pump activity. In addition to individuals with high potassium (HK^+) RBCs, some sheep, goats, buffalo, and most cattle have relatively low potassium (LK) and, consequently, high sodium RBCs. These LK^+ RBCs have low Na^+, K^+-pump activity and high passive K^+ permeability (Tosteson and Hoffman, 1960). Studies in sheep and goats have demonstrated the HK^+/LK^+ polymorphism is determined by a single autosomal genetic locus with two alleles, the LK^+ allele being dominant (Tunon *et al.*, 1987; Xu *et al.*, 1994). In sheep, this polymorphism is associated with an M and L blood group antigen polymorphism. The HK^+ cells exhibit M antigens and the homozygous LK^+ cells exhibit L antigens, with heterozygous LK_+ sheep exhibiting M and L antigens. The L_p-antigen in LK^+ RBCs appears to be an endogenous inhibitor of the Na^+, K^+-pump (Xu *et al.*, 1993).

RBCs from cats and most dogs do not have Na^+, K^+-pump activity and have Na^+ and K^+ concentrations near, but not at, those predicted for the Donnan equilibrium with plasma (Parker, 1977). Some clinically normal Japanese Akita dogs (Degen, 1987) and mongrel dogs from Japan (Inaba and Maede, 1984) have HK^+ RBCs. RBCs from these mongrel dogs have substantial Na^+, K^+-ATPase activity and altered amino acid metabolism, as discussed later. Dog, ferret, and bear RBCs have a unique Na^+-Ca^{++} countertransport system that can remove sodium (Parker, 1992). The calcium that enters the cell is subsequently pumped out by an ATP-dependent calcium pump.

Other pathways of sodium and potassium transport occur to variable degrees in certain species. These pathways include passive diffusion, Na^+, K^+, Cl^- cotransport, $Na^+ \times Na^+$ countertransport, band 3 anion transport as $NaCO_3^-$, Na^+-dependent amino acid transport, K^+, Cl^- cotransport, Na^+, H^+ exchange, and a calcium-dependent K^+ channel (Contreras *et al.*, 1986; Ellory and Tucker, 1983; Haas, 1989). A negative linear correlation between internal sodium concentration and membrane protein to lipid ratio was found by comparison of RBC values from nine mammalian species, the significance of which is unknown (Garnier *et al.*, 1984).

RBC volumes influence cation fluxes. Sodium flux increases when cells are shrunken, and potassium flux increases when cells are swollen. The Na^+, K^+, Cl^- cotransport and Na^+-H^+ exchange are activated by cell shrinkage and K^+-Cl^- cotransport is activated by cell swelling (Haas, 1989). Volume changes are believed to be detected from alterations in cytoplasmic macromolecules (Parker, 1992).

Early nucleated erythroid precursors in dog bone marrow have HK^+ content, whereas mature RBCs are of the LK^+ type. The switch from HK^+ to LK^+ content occurs during the maturation from early to late nucleated erythroid cells (Kirk *et al.*, 1983). When erythropoiesis is dramatically stimulated in response to a hemolytic anemia, much of the HK^+ to LK^+ transition does not occur until after denucleation. Consequently, stress reticulocytes, produced in response to anemia, have potassium contents much higher than reticulocytes normally released into blood. The high potassium concentration in canine stress reticulocytes results from membrane Na^+, K^+-ATPase activity that is lost during maturation into RBCs, possibly by ATP-dependent proteolysis (Inaba and Maede, 1986). Stress reticulocytes produced by LK^+-type ruminants (Israel *et al.*, 1972; Tucker and Ellory, 1971; Kim *et al.*, 1980) also have high potassium concentrations as a result of high Na^+, K^+-ATPase activities (Blostein and Grafova, 1990). The decline in the number of Na^+, K^+ pumps on LK^+ sheep reticulocytes during maturation is modulated by the L_p antigen (Xu *et al.*, 1994). Fetal and neonatal RBCs examined from mammals with LK^+ RBCs have higher potassium concentrations than adult RBCs, but the difference in dogs was not as dramatic as that in ruminants (Coulter and Small, 1973; Ellory and Tucker, 1983).

3. *Calcium*

Excessive intracellular Ca^{2+} is deleterious to RBCs; consequently, they actively extrude Ca^{2+} using a calcium pump having Ca^{2+}-activated, Mg^{2+}-dependent ATPase activity. The calcium pump is activated by a calcium-binding protein called calmodulin (Bababunmi *et al.*, 1991; Hinds and Vincenzi, 1986). This pump working in conjunction with a Na^+-Ca^{++} countertransport system appears to be important in RBC volume regulation in dogs (Parker, 1992).

4. *Amino Acids*

Amino acid transport in RBCs provides amino acids for synthesis of reduced glutathione (GSH). In addition, amino acid transporters may be responsible for efflux of amino acids during reticulocyte maturation (Tunnicliff, 1994).

Eight different amino acid transport systems have been recognized in mammalian RBCs, each with its own characteristic species distribution, ion requirements, and substrate specificity (Young, 1983; Fincham *et al.*, 1987). In addition, the band 3 anion transporter can transport glycine and some other amino acids (Fincham *et al.*, 1987). Amino acid transport in sheep appears to be mediated largely by the Na^+-independent C amino acid transporter, a carrier whose optimal substrates are small neutral amino acids and dibasic amino acids (Young, 1983). Sheep deficient in this transporter have low RBC GSH, because of impaired cysteine transport (Tucker *et al.*, 1981). A similar asc transport system occurs in most equine RBCs and deficient horses are predisposed to GSH deficiency (Fincham *et al.*, 1987; Fincham *et al.*, 1988). Dog and cat RBCs have a Na^+-dependent acidic amino acid transporter that optimally transports glutamate and aspartate. The transport of 1 glutamate into dog RBCs is accompanied by 2 Na^+ and by the counter transport of 1 K^+ and 1 anion (Sato *et al.*, 1994). Cat RBCs have an Ly^+ system for the transport of dibasic amino acids (Young, 1983).

5. *Glucose*

Species vary in their permeability to glucose, with human RBCs being very permeable and pig RBCs being poorly permeable (McManus, 1967). RBCs of other domestic animals appear to be intermediate between these extremes (Arai *et al.*, 1992; Bolis, 1973; Widdas, 1955). Glucose is transported into RBCs by passive diffusion. Both band 3 (Bosman and Kay, 1990; Langdon and Holman, 1988) and an integral membrane protein that migrates in the band 4.5 region on SDS–PAGE (Craik *et al.*, 1988) are believed to be involved in glucose transportation. Glucose movement into RBCs is not regulated by insulin (Baldwin, 1993). RBCs from adult pigs lack a functional glucose transporter (Craik *et al.*, 1988) and, therefore, have limited ability to utilize glucose for energy (Kim and McManus, 1971; Magnani *et al.*, 1983). In contrast, RBCs from neonatal piglets and pig reticulocytes have the transporter

(Craik *et al.*, 1988) and, consequently, exhibit substantial glucose transport (Kim and Luthra, 1977). With the exception of cats, fetal and neonatal RBCs studied from man and animals have higher glucose transport than RBCs from adults (Mooney and Young, 1978; Widdas, 1955).

6. *Adenine, Adenosine, and Inosine*

RBC membranes from most animal species have a nucleoside transporter (Young, 1983). The adenosine transporter from human and pig RBCs migrates in the band 4.5 region on SDS-PAGE (Kwong *et al.*, 1986). Rabbit, pig, and human RBCs exhibit substantially more adenosine uptake than those of other species studied (Van Belle, 1969). RBCs from dogs exhibit more adenosine uptake than cats, goats, or cattle, and RBCs from horses and most sheep appear to be nearly impermeable to adenosine. A low percentage of sheep have RBCs with a high affinity nucleoside transport system with a broad specificity for both purine and pyrimidine nucleosides (Young, 1983). While dog RBCs are permeable to adenosine, they are impermeable to inosine (Duhm, 1974). Dog and cat RBCs exhibit adenine uptake and incorporation into nucleotides, but values are much lower than those of human, rabbit, or rodent RBCs (Lalanne and Willemot, 1980).

E. Metabolism of Adenine Nucleotides

Adenine nucleotides in RBCs contain adenine, ribose, and one or more phosphate groups. Mature RBCs cannot synthesize adenine nucleotides *de novo* but can produce these compounds utilizing so-called salvage pathways (Brewer, 1974; Eaton and Brewer, 1974). AMP can be synthesized from adenine or from adenosine, both of which may be supplied to RBCs as they pass through the liver. One molecule of ATP interacts with one molecule of AMP to generate two molecules of ADP in the adenylate kinase reaction. ATP is generated from ADP in glycolysis (see later discussion).

AMP is synthesized from adenine and phosphoribosyl pyrophosphate (PRPP), utilizing the adenine phosphoribosyltransferase enzyme. Adenine is converted to ATP at a slower rate in dog and cat RBCs than in those of man, rodents, or rabbits (Lalanne and Willemot, 1980). AMP degradation to inosine monophosphate and ammonia is catalyzed by AMP-deaminase. The activity of this enzyme is generally lower in mammalian RBCs compared to nucleated RBCs from birds, reptiles, amphibians, and fish (Kruckeberg and Chilson, 1973).

Adenosine can be phosphorylated to AMP using ATP in the adenosine kinase reaction. A competing reaction, adenosine deaminase, converts adenosine to inosine, which cannot be incorporated into AMP. The uptake and/or deamination of adenosine varies considerably by species (Van Belle, 1969). Not only are dog, cat, and cattle RBCs poorly permeable to inosine, but inosine produced by adenosine deamination cannot be readily used for energy in these species because of low purine nucleoside phosphorylase activity, which converts inosine to ribose 1-phosphate and hypoxanthine (Duhm, 1974).

NAD and NADP can apparently be synthesized from nicotinate by way of a series of reactions in RBCs (Eaton and Brewer, 1974). In addition to ATP, PRPP and NH^3, or glutamine, are required. Comparative studies of the synthesis of these compounds in domestic animals have not been reported.

F. Carbohydrate Metabolism

RBCs require energy in the form of ATP for maintenance of shape and deformability, phosphorylation of membrane phospholipids and proteins, active membrane transport of various molecules, partial synthesis of purine and pyrimidine nucleotides, and synthesis of GSH (Nakao, 1974; Reimann *et al.*, 1981). Reducing potential in the form of NADH and NADPH is needed to counteract oxidative processes (see later discussion). Although substrates such as ribose, fructose, mannose, galactose, dihydroxyacetone, glyceraldehyde, adenosine, and inosine may be metabolized to some extent, depending on the species, glucose is the primary substrate for energy needs of RBCs from all species except the pig (Agar and Board, 1983b; Kim, 1983).

RBCs from adult pigs utilize glucose at lower rates than other species (Magnani *et al.*, 1983) because they lack a functional glucose transporter (Craik *et al.*, 1988; Zeidler and Kim, 1982). Inosine appears to be the major substrate for pig RBCs; its production by the liver is sufficient to meet their energy requirements (Young *et al.*, 1985; Zeidler *et al.*, 1985). Inosine can be used because nucleoside phosphorylase converts it to ribose 1-phosphate and hypoxanthine (Sandberg *et al.*, 1955). Ribose 1-phosphate is converted to ribose 5-phosphate, an intermediate of the pentose phosphate pathway (PPP), by phosphoribomutase (Brewer, 1974).

Glucose utilization rates of RBCs vary by species (Table 7.1). Factors such as pH, phosphate concentration, temperature, and leukocyte and platelet contamination of RBC incubations can have substantial effects on glucose utilization rates measured *in vitro*. Consequently, species comparisons of values determined in different laboratories may be misleading. Harvey and Kaneko (1976a) approximated physiologic conditions *in vitro* and measured mean glycolytic rates of 0.64,

0.94, 1.33, and 1.48 μmol/ml RBC for horse, cat, dog, and man, respectively. Once glucose enters the cell, it is phosphorylated to glucose 6-phosphate (G6P) utilizing the hexokinase (HK) enzyme. The G6P is then metabolized through either the Embden–Meyerhof pathway (EMP) or the pentose phosphate pathway (PPP) as shown (in Fig. 7.4).

G. Embden–Meyerhof Pathway

Most of the species variations in glucose utilization appear to result from variations in EMP metabolism, with PPP metabolism being relatively constant when not stimulated by oxidants (Harvey and Kaneko, 1976a). In addition to the phosphorylation of glucose, one molecule of ATP is used to phosphorylate fructose 6-phosphate, and one molecule of ATP is generated for each three-carbon molecule metabolized through the phosphoglycerate kinase (PGK) and pyruvate kinase (PK) reactions (Fig. 7.4). Consequently a net of two molecules of ATP is produced for each molecule of glucose metabolized to two molecules of lactate in the EMP. Since mature RBCs lack mitochondria, the EMP is the only source of ATP production in these cells.

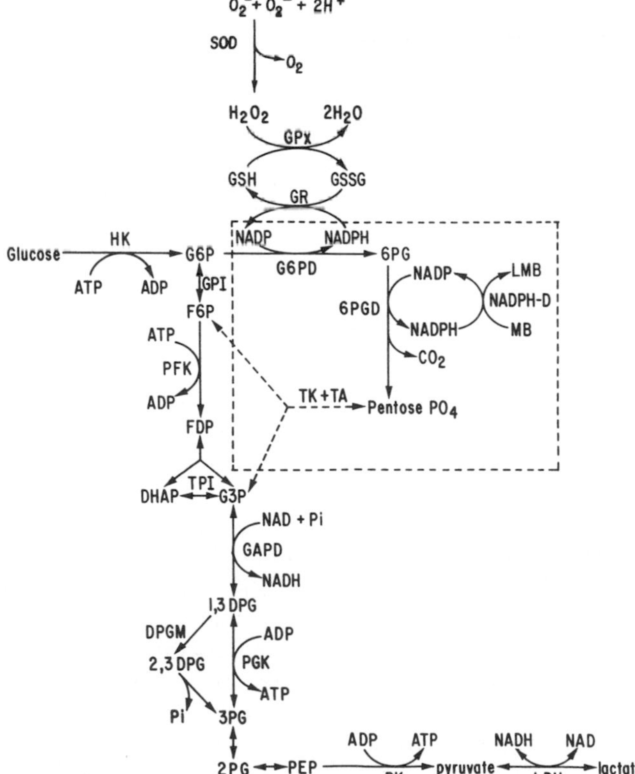

FIGURE 7.4 Metabolic pathways of the mature RBC. For abbreviations see footnote at beginning of chapter.

In human RBCs, reactions catalyzed by HK, phosphofructokinase (PFK), and PK appear to be rate-limiting steps in glycolysis, because these reactions are far displaced from equilibrium (Minakami and Yoshikawa, 1966). Under physiologic steady-state conditions, the PFK enzyme reaction controls glycolysis through the EMP (Rapoport, 1968). Its activity is influenced by a variety of effectors, with ATP being the most important inhibitor and AMP and inorganic phosphate (P_i) being the most influential activators. ADP is a less potent activator (Jacobasch *et al.*, 1974). Most of the adenine nucleotide pool in RBCs is maintained as ATP under normal conditions, with less as ADP and even less as AMP (Tables 7.4 and 7.5). Glycolysis is ultimately controlled by the demand for production of ATP. As ATP is utilized, concentrations of ADP, AMP, and P_i increase. These changes result in the activation of PFK and increased EMP metabolism. Other potential activators of PFK include ammonium ions (Debski and Rynca, 1985; Shimizu *et al.*, 1988), glucose 1,6-bisphosphate (Harvey *et al.*, 1992b; Accorsi *et al.*, 1985), and fructose 2,6-bisphosphate (Gallego and Carreras, 1990; Sobrino *et al.*, 1987; Harvey *et al.*, 1992b).

When PFK is activated, HK is activated secondarily because the concentration of G6P is reduced, and G6P competitively inhibits HK (Rapoport *et al.*, 1976). Although the HK reaction is normally not a rate-controlling step in glycolysis of human RBCs, conceivably it is more important in animal species with lower HK activity (Rapoport, 1968). The glycolytic rate is correlated with HK activity when RBCs of various species are compared at pH values above 8.0. Comparisons may be less reliable when measurements are made at pH 7.4 (Harvey and Kaneko, 1975a, 1976a). In human RBCs, the PK reaction becomes limiting when the PFK reaction is markedly stimulated, that is, at pH values greater than 7.6 (Jacobasch *et al.*, 1974). The pH values listed earlier are external values. The pH within RBCs is generally about 0.2 units lower than the external pH (Waddell and Bates, 1969).

In addition to ATP/ADP and ADP/AMP ratios, various other factors influence EMP metabolism. Alterations in pH of plasma or *in vitro* buffers affect glycolysis. As pH is increased above 7.2, PFK is activated and glucose utilization and EMP metabolism increase (Burr, 1972; Rapoport, 1968).

At physiologic pH values, high concentrations of P_i stimulate glycolysis through the EMP by reducing the ATP inhibition of PFK. Conversely, glycolysis is inhibited by short-term phosphate deficiency, primarily by decreasing intracellular P_i for glyceraldehyde phosphate dehydrogenase (GAPD) (Jacobasch *et al.*, 1974; Ogawa *et al.*, 1989; Wang *et al.*, 1985). Decreased glycolytic rates result in decreased RBC ATP concentrations and hemolytic anemia in experimental dogs made severely hypophosphatemic by hyperalimentation (Yawata *et al.*, 1974; Jacob *et al.*, 1973). Hemolytic anemia associated with hypophosphatemia has also been reported in diabetic cats and a diabetic dog following insulin therapy (Adams *et al.*, 1993; Perman and Schall, 1983; Willard *et al.*, 1987), in a cat with hepatic lipidosis (Adams *et al.*, 1993), and in postparturient cattle in which decreased RBC ATP concentrations have been measured (Ogawa *et al.*, 1987, 1989). In addition to low ATP concentrations, dog RBCs might hemolyze as a result of decreased RBC 2,3DPG concentration, because dog RBCs with low 2,3DPG are more alkaline fragile than those of normal dogs and may hemolyze at physiologic pH values (Harvey *et al.*, 1988).

Several glycolytic enzymes including PFK, GAPD, and aldolase bind to the highly acidic region of the cytoplasmic domain of band 3 (Harris and Winzor, 1990). These enzymes are inhibited by binding to band 3. It is suggested that tyrosine phosphorylation of band 3 by protein tyrosine kinases may enhance glycolysis by preventing the binding and inhibition of one or more glycolytic enzymes (Harrison *et al.*, 1991).

There is a strong positive correlation between intracellular Mg^{2+} and ATP concentrations in RBCs from various species due to the presence of the Mg^{2+}–ATP complex within cells (Miseta *et al.*, 1993). RBCs of rats and dogs with short-term magnesium deficiency have lowered glycolytic rates, because adenine nucleotide substrates in four glycolytic kinase reactions (HK, PFK, PGK, and PK) must be complexed with Mg^{2+} (Rapoport, 1968). Dogs and rats on magnesium-deficient diets become anemic (Elin and Alling, 1978; Kruse *et al.*, 1933), owing to shortened RBC life spans.

The saturation of Hb with oxygen has an effect on glucose utilization. Human RBCs utilize more glucose when incubated anaerobically under nitrogen than under aerobic conditions (Asakura *et al.*, 1966). OxyHb is a stronger acid than DeoxyHb; consequently, the intracellular pH of human RBCs is lower in oxygenated blood than in deoxygenated blood (Takano *et al.*, 1976). The PFK reaction is inhibited as blood is oxygenated, due to the pH effect. In human RBCs, 2,3DPG is bound to DeoxyHb and released on oxygenation. Based on studies of glycolytic intermediates, the increased unbound 2,3DPG in oxygenated RBCs may have additional inhibitory effects on glycolysis (Hamasaki *et al.*, 1970). The effect of oxygenation on glycolysis of RBCs from domestic animals remains to be determined. It may not be important in ruminants because oxygenation results in insignificant decreases in intracellular pH values (Takano *et al.*, 1976).

2,3DPG inhibits glycolysis in part because of its reduction of intracellular pH as a consequence of the Donnan effect of this nonpenetrating anion (Duhm, 1975). 2,3DPG also inhibits glycolysis by inhibiting HK, PFK, and PK in a manner different from its pH effect on these enzymes (Duhm, 1975; Jacobasch *et al.*, 1974). In addition, 2,3DPG inhibits 6-phosphofructokinase-2-kinase, the enzyme responsible for the synthesis of the positive PFK effector fructose-2,6-bisphosphate (Sobrino *et al.*, 1987).

Glycolysis increases in human RBCs with increasing temperature to a maximum at 45°C, with a Q_{10} of 2 (Rapoport, 1968). Based on measurements of glycolytic intermediates, the major effect of temperature appears to be on the PFK reaction (Jacobasch *et al.*, 1974).

H. Diphosphoglycerate Pathway

Molecules of 1,3-diphosphoglycerate (1,3DPG), produced by the GAPD reaction, may be utilized by the PGK reaction in the EMP or may be converted to 2,3DPG by the diphosphoglycerate mutase (DPGM) reaction (Fig. 7.4). 2,3DPG degradation to 3-phosphoglycerate (3PG) is catalyzed by diphosphoglycerate phosphatase activity (DPGP). A single protein is responsible for both the DPGM and DPGP activities (Sasaki *et al.*, 1977). The DPG pathway or shunt (Rapoport–Luebering cycle) bypasses the ATP-generating PGK step in glycolysis; consequently, no net ATP is generated when glucose is metabolized through this pathway (Brewer, 1974).

Normally, from 10 to 30% of triose phosphate metabolism in human RBCs is shunted through this pathway (Jacobasch *et al.*, 1974; Oxley *et al.*, 1984). The proportion of 1,3DPG metabolized by PGK and DPGM is determined mostly by the concentration of ADP (Rapoport, 1968). ATP and 2,3DPG influence the relative amount of flow through each route by product inhibition of their own synthesis.

The absolute flow is also determined by the overall glycolytic rate. 2,3DPG inhibits HK, PFK, and GAPD in human RBCs (Srivastava and Beutler, 1972). The formation of 2,3DPG is stimulated by increased P_i concentration and increased pH, which stimulate glycolysis by activating PFK greater than PK (Jacobasch *et al.*, 1974).

The concentration of 2,3DPG can be affected by a decrease in PK activity. When PK activity is reduced relative to PFK activity, phosphorylated intermediates between the PK and GAPD reactions increase in concentration (Mueggler and Black, 1982). The synthesis of 2,3DPG is increased as a result of the increased concentration of the 1,3DPG substrate. Thyroid hormones can increase 2,3DPG in human (Snyder *et al.*, 1970) and sheep (Studzinski *et al.*, 1982) RBCs, possibly by a direct effect on the DPGM enzyme.

RBCs of dogs, horses, pigs, and man normally contain high concentrations of 2,3DPG, whereas those of cats and domestic ruminants have low concentrations (Tables 7.4 and 7.5). Based on results from a large number of mammalian species, only cats, hyenas and civets among carnivores, and deer, giraffe, antelope, and the cattle family among artiodactyls have low RBC 2,3DPG concentrations (Bunn, 1981; Bunn *et al.*, 1974). Low concentrations of 2,3DPG in cat, goat, and sheep RBCs result primarily from low RBC DPGM activities, whereas RBCs of cattle have relatively high DPGP activity in association with moderately low DPGM activity (Chemtob *et al.*, 1980; Pons *et al.*, 1985; Harkness *et al.*, 1969).

I. Hemoglobin Oxygen Affinity

1. Oxygen Dissociation Curve

The initial binding of a molecule of O_2 to a monomer of tetrameric, deoxygenated Hb facilitates further binding of O_2 to the Hb molecule. Since the O_2 binding of one heme group influences the affinity of other heme groups for O_2, this characteristic has been called the *heme–heme interaction.* The changing oxygen affinity of Hb with oxygenation results in a sigmoid oxygen dissociation curve (Fig. 7.5) when the percent saturation of Hb with oxygen is plotted against the partial pressure of oxygen (pO_2). The pO_2 at which Hb is 50% saturated is the P_{50}. The steepness of the middle portion of the curve is of great physiologic significance because it covers the range of oxygen tensions present in tissues. Consequently, relatively small decreases in oxygen tension result in substantial oxygen release from Hb (Benesch *et al.*, 1975).

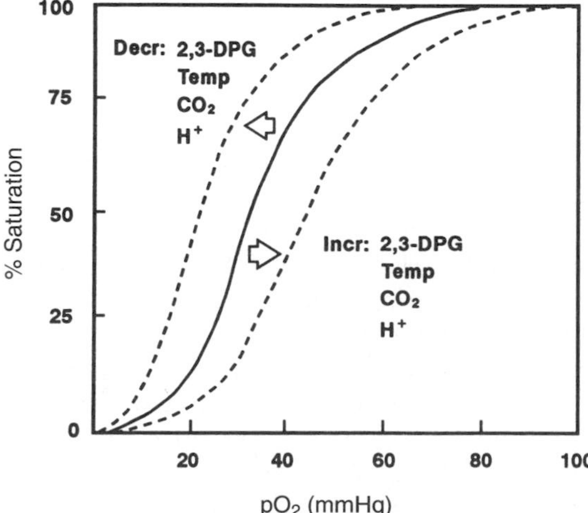

FIGURE 7.5 The hemoglobin–oxygen dissociation curve and factors influencing the position of the curve.

2. Effects of H^+, CO_2, and Temperature

The oxygen affinity of Hb is influenced by H^+ in a manner termed the Bohr effect. In capillaries of metabolizing tissues, CO_2 enters RBCs where it is rapidly converted to H_2CO_3 by carbonic anhydrase (carbonate dehydratase). This carbonic acid spontaneously ionizes to H^+ and HCO_3^-. The increased H^+ concentration decreases the oxygen affinity of Hb and facilitates oxygen delivery to the tissues. DeoxyHb is a weaker acid than OxyHb; therefore, DeoxyHb binds the excess H^+ and limits the decrease in pH. The increased HCO_3^- diffuses out of the cell down a concentration gradient and $C1^-$ moves in (chloride shift) to maintain electrical neutrality. These processes are reversed at the lungs.

The addition of H^+ to a suspension of RBCs results in an increase in P_{50} and a shift of the oxygen dissociation curve to the right (Fig. 7.5). The magnitude of the Bohr effect is defined numerically $\Delta \log P_{50}/\Delta$ pH. The relationship between the magnitude of the Bohr effect and average body size of various animal species is an inverse one (Riggs, 1960). The direct binding of CO_2 to Hb in carbamino groups also lowers oxygen affinity, but this effect is considered to be minor. Increased temperature decreases the oxygen affinity of Hb, a response that appears physiologically appropriate considering that increased heat production accompanies increased oxygen consumption in tissues (Benesch *et al.*, 1975).

The P_{50} for greyhound RBCs in whole blood is lower than that for mongrel dogs, yet the groups have similar 2,3DPG concentrations (Sullivan *et al.*, 1994). The cause of this difference remains to be determined, but it is suggested that the higher PCV found in greyhound dogs may represent a compensatory response to a higher oxygen affinity of Hb in this species.

3. Effects of 2,3DPG

In RBCs from most mammalian species, 2,3DPG decreases the oxygen affinity of Hb, resulting in an increase in P_{50} (Bunn *et al.*, 1974). 2,3DPG reacts with Hb in a ratio of one molecule per Hb tetramer. Negatively charged groups of 2,3DPG bind to specific positively charged groups in the N-terminal region of Hb beta chains. There is a marked preference for binding to DeoxyHb as compared to OxyHb due to differences in the conformation of the molecules. The interaction of 2,3DPG with Hb is represented as follows:

$$\text{HbDPG} + O_2 \leftrightharpoons \text{HbO}_2 + \text{DPG}.$$

When 2,3DPG is increased, the reaction is displaced to the left, and when pO_2 is increased, the reaction is displaced to the right. ATP has a similar effect on Hb oxygen affinity, but is generally much less important than 2,3DPG, because it usually occurs in lower concentration and is complexed with Mg^{2+} (Bunn, 1971).

When the oxygen affinity of Hb is studied in hemolysates dialyzed to remove 2,3DPG and ATP, the "stripped" Hbs from species with low 2,3DPG RBCs have considerably lower oxygen affinities than stripped Hb from species with high 2,3DPG RBCs (Bunn, 1971; Bunn *et al.*, 1974). Furthermore, the oxygen affinity of stripped Hb from these low 2,3DPG RBCs is minimally affected by added 2,3DPG, because alterations are present in the β-chain-binding area that prevent or diminish 2,3DPG binding (Bunn, 1981). However, the Hb oxygen affinity in cattle RBCs (a low 2,3DPG species) is modulated by chloride ions and to a lesser extent phosphate ions (Fronticelli, 1990; Gustin *et al.*, 1994). Because stripped Hbs from species with high 2,3DPG RBCs have high oxygen affinities, 2,3DPG is needed within RBCs of these species to maintain Hb oxygen affinity within a physiologically useful range (Benesch *et al.*, 1975).

When blood from many mammalian species is studied, an inverse linear correlation is recognized between the P_{50} of whole blood and the log of body weight (Scott *et al.*, 1977). The relationship between metabolic rate (oxygen consumption per gram of tissue) and body weight (Kleiber, 1961) is an inverse one. Consequently, the higher P_{50} in smaller animals should be beneficial in meeting tissue oxygen requirements associated with their higher metabolic rates. Oxygen affinity of stripped Hb from various mammals does not correlate with body weight (Nakashima *et al.*, 1985). The maintenance of 2,3DPG is energetically expensive because the ATP-generating PGK reaction is bypassed. 2,3DPG apparently allows for an evolutionary adaption of blood Hb oxygen affinity to metabolic rate.

Potentially, animals with high 2,3DPG RBCs can alter their Hb oxygen affinity to meet metabolic needs. The significance of (and in some cases the appropriateness of) alterations in 2,3DPG in disease states is not always clear. RBC 2,3DPG increases in some anemic animals (Agar *et al.*, 1977; King *et al.*, 1992; Studzinski *et al.*, 1978). In the case of anemias the resultant increase in P_{50} would seem to be beneficial, but in the case of severe hypoxic hypoxemia the change might be detrimental, because Hb could not be fully saturated. Various studies in dogs indicate that cardiac output and microcirculation adjustments are much more important than changes in Hb oxygen affinity in adapting to hypoxia (Liard and Kunert, 1993; Schumacker *et al.*, 1985; Zachara *et al.*, 1981). However, a reduction in Hb oxygen affinity secondary to increased 2,3DPG can be beneficial, because it is far less energy demanding than is an increase in cardiac output (Liard and Kunert, 1993; Mairbäurl, 1994; Teisseire *et al.*, 1985). RBC 2,3DPG increases in hibernating mammals, but the effect of the decrease in body temperature, associated with hibernation, on Hb oxygen affinity *in vivo* would more than offset the effect of increased 2,3DPG (Bunn, 1981). Because pH has a substantial effect on Hb oxygen affinity, changes in RBC 2,3DPG concentration, in response to acidosis and alkalosis, produce effects on Hb oxygen affinity that counteract alterations induced by the respective changes in pH (Bellingham *et al.*, 1971).

4. *Maternal–Fetal Oxygen Transport*

Except for the domestic cat, the oxygen affinity of Hb in fetal blood is higher than that of maternal blood (Bunn and Kitchen, 1973; Novy and Parer, 1969). This difference in oxygen affinity probably enhances the transport of oxygen across the placenta to the fetus (Comline and Silver, 1974; Hebbel *et al.*, 1980). Oxygen is adequately transported to cat fetuses, without the advantage of increased oxygen affinity, due to the nature of the placentation and a countercurrent arrangement of blood flows (Novy and Parer, 1969).

During late gestation, mean umbilical venous pO_2 values of 48, 38, 35, and 30 mm Hg have been reported for horses, cattle, sheep, and pigs, respectively (Comline and Silver, 1974). Fetal systemic arterial pO_2 values may be even lower, especially in species such as the pig and horse which lack a ductus venosus, requiring that all returning umbilical venous blood pass through the liver. These pO_2 values are considerably lower than normal adult arterial pO_2 values of about 100 mm Hg. Because Hb in adult RBCs of these species is only partially saturated at these low arterial pO_2 values (Schmidt-Neilsen and Larimer, 1958), the increased oxygen affinity of fetal blood would result in a greater saturation of Hb and, therefore, a greater oxygen-carrying capacity of blood than would otherwise be present.

Fetal RBCs maintain Hb oxygen affinities higher than those of the mother by one of three mechanisms (Bunn and Kitchen, 1973): (1) Ruminants have structurally distinct fetal Hbs that have higher oxygen affinities than adult Hbs in the absence of organic phosphates (Battaglia *et al.*, 1970; Blunt *et al.*, 1971). Interactions between 2,3DPG and both fetal and adult Hbs in ruminants are weak (Bunn and Kitchen, 1973). (2) A structurally distinct fetal Hb (HbF) also occurs in man, but the oxygen affinity of stripped HbF and adult human Hb is about the same. Differences in whole blood oxygen affinity occur primarily because HbF interacts weakly with 2,3DPG, in contrast to the strong interaction with adult Hbs (Oski and Gottlieb, 1971). (3) Animals without structurally distinct fetal Hbs, such as dog, horse, and pig, have much lower 2,3DPG concentrations in fetal RBCs than in adult RBCs, thereby creating higher oxygen affinities in fetal RBCs (Bunn and Kitchen, 1973; Comline and Silver, 1974; Dhindsa *et al.*, 1972; Tweeddale, 1973). The actual *in vivo* difference in oxygen affinity between fetal and maternal RBCs is less than suggested by differences in P_{50} values measured *in vitro* at pH 7.4, because fetal blood pH is lower than that of maternal blood (Comline and Silver, 1974).

5. *Postnatal Changes in 2,3DPG and Oxygen Affinity*

Although the higher oxygen affinity of Hb in fetal RBCs may be beneficial in the uterus, it would seem to be disadvantageous to the newborn animal breathing air. Whole blood P_{50} values and RBC 2,3DPG concentrations (Fig. 7.6) increase after birth in most domestic animals.

RBC 2,3DPG concentrations are higher in fetuses than in adult ruminants (Aufderheide *et al.*, 1980; Zinkl and Kaneko, 1973a), and increase markedly within a few days after birth (Aufderheide *et al.*, 1980; King and Mifsud, 1981; Zinkl and Kaneko, 1973a). Since neither fetal nor adult Hbs from ruminants bind 2,3DPG to any extent, 2,3DPG decreases oxygen affinity primarily by lowering intracellular pH (Aufderheide *et al.*, 1980; Battaglia *et al.*, 1970; Blunt, 1972). Several mechanisms are involved in the 10-fold increase in RBC 2,3DPG during the first 5 days of life in lambs (Noble *et al.*, 1983). (1) Plasma glucose increases from 40 to 100 mg/dl during the first 2 days of life and allows for an increased consumption of glucose by the glucose-permeable neonatal RBC. (2) The blood pH increases during the first day of life and activates the PFK enzyme as evidenced by changes in RBC intermediates. (3) Plasma P_i concentration increases to a level sufficient to increase GAPD activity at 3 days of age. (4) DPGM activity in neonatal RBCs is 12-fold higher than that of adults. (5) RBC PFK activity is still above adult values, but PK activity has decreased to adult values by birth.

The decline in 2,3DPG in postnatal ruminant RBCs is more gradual, requiring 1–2 months to reach adult values. The whole blood P_{50} is maintained, however, because of concomitant decreases in HbF and increases in adult Hb types (Aufderheide *et al.*, 1980; Blunt *et al.*, 1971; Lee *et al.*, 1971; Zinkl and Kaneko, 1973a). In goat RBCs HbC replaces most of the HbF initially, but after 2 months other adult Hbs begin to replace HbC (Huisman *et al.*, 1969). Only a small percentage of HbC is present in lambs up to 2 months of age. The percentage of HbF at birth varies from 70 to 100%. The signal to switch production of HbF to HbA in sheep is regulated by a developmental clock inherent to hematopoietic stem cells, rather than a change in the hematopoietic inductive microenvironment (Wood *et al.*, 1985).

Rapid, but modest, increases in 2,3DPG and P_{50} occur after birth in horse RBCs (Bunn and Kitchen, 1973). As in ruminants, blood pH increases significantly within 1 hour after birth (Rose *et al.*, 1982). Plasma P_i concentration also increases during the first 2 weeks of life (Bauer *et al.*, 1984). Gradual, but large, increases in 2,3DPG and P_{50} occur postnatally in blood of dogs (Dhindsa *et al.*, 1972; Mueggler *et al.*, 1980; Harvey and Reddy, 1989) and pigs (Kim and Duhm, 1974; Watts and Kim, 1984). A decreasing activity of PK has been reported to account for the increasing 2,3DPG concentration during the first 60 days of life in dogs (Mueggler and Black, 1982), but activation of PFK may also play a role in this increase (Harvey, 1994).

In man, 2,3DPG increases slightly after birth, but most of the increase in P_{50} that occurs during the first 6 months of life results from the replacement of HbF with adult Hbs (Oski and Gottlieb, 1971). Although P_{50} values changed slightly in kittens after birth, RBC

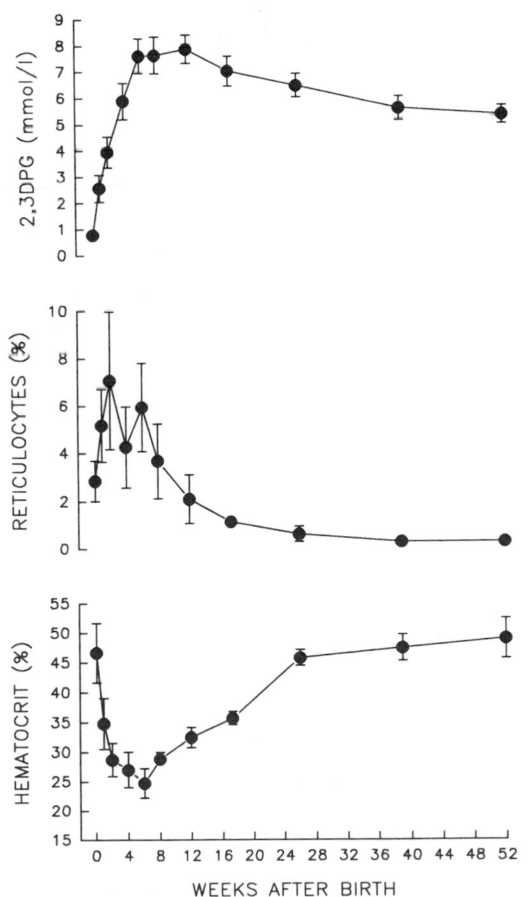

FIGURE 7.6 Changes in hematocrit, uncorrected reticulocyte count, and RBC 2,3DPG content in dogs following birth. Values are mean ± standard deviation (Harvey, 1994).

2,3DPG values remained in the normal adult range (Dhindsa and Metcalfe, 1974). Data concerning changes in these and other species have been compiled by Isaacks and Harkness (1983).

J. Pentose Phosphate Pathway

The pentose phosphate pathway (PPP) generates NADPH, the major source of reducing equivalents in the protection of RBCs against oxidative injury. This pathway also produces ribose 5-phosphate (R5P), which is required for adenine nucleotide synthesis (Eaton and Brewer, 1974). The PPP competes with the EMP for the G6P substrate (Fig. 7.4). Normally only about 5–13% of glucose metabolized by RBCs flows through the PPP (Harvey and Kaneko, 1976a), but this flow can be accelerated markedly by oxidants (Harvey and Kaneko, 1977).

The first step in the metabolism of glucose through the PPP generates NADPH from the oxidation of G6P in the glucose-6-phosphate dehydrogenase (G6PD) reaction. An additional NADPH is generated from the oxidative decarboxylation of 6-phosphogluconate (6PG) to ribulose 5-phosphate in the 6-phosphogluconate dehydrogenase (6PGD) reaction. This is the only known reaction producing CO_2 in mature RBCs. The remainder of the reactions in the PPP are nonoxidative and freely reversible. R5P is produced from ribulose 5-phosphate by the R5P isomerase reaction. The net effect of the metabolism of three molecules of G6P through the PPP is as follows (Eaton and Brewer, 1974):

$$3\ \mathrm{G6P} + 6\ \mathrm{NADP}^+ \rightarrow 3\ \mathrm{CO}_2 + 2\ \mathrm{F6P} + \mathrm{G3P} + 6\ \mathrm{NADPH} + 6\ \mathrm{H}^+.$$

G6PD is the rate-limiting reaction in the PPP under physiologic conditions. Normally, the G6PD reaction in intact human RBCs operates at only 0.1–0.2% of the maximal enzyme activity as determined in hemolysates under optimal conditions. The low rate of this reaction in RBCs occurs because of limited substrate availability (especially $NADP^+$) and because G6PD is strongly inhibited by NADPH and ATP at physiologic concentrations (Yoshida, 1973). The maximal G6PD activities measured in hemolysates from goat and sheep RBCs are much lower than those of humans or of other domestic animals (Tables 7.2 and 7.3). However, this comparatively low enzyme activity does not render sheep RBCs unduly susceptible to the hemolytic effects of oxidant drugs (Maronpot, 1972; Smith, 1968), in part because G6PD is not inhibited by ATP in this species (Smith and Anwer, 1971).

About 91% of total NADP is in the reduced form in horse RBCs (Stockham *et al.*, 1994) and 92-99% of total NADP is NADPH in human RBCs (Kirkman and Gaetani, 1986; Zerez *et al.*, 1987). NADPH is utilized to reduce oxidized glutathione to GSH, the substrate for the glutathione peroxidase reaction, and it is bound to catalase, preventing and reversing the accumulation of an inactive form of catalase that is generated when catalase is exposed to H_2O_2 (Kirkman *et al.*, 1987). In the presence of oxidants, NADPH is oxidized and the PPP is stimulated because the activities of G6PD and 6PGD are directly related to the concentration of NADP and inversely related to that of NADPH (Yoshida, 1973). Glutathione metabolism affects PPP activity via the glutathione reductase (GR) enzyme, which generates NADP as a result of the reduction of GSSG with NADPH (Fig. 7.4).

K. Metabolic Protection Against Oxidants

1. *Nature of Oxidants in Biology*

From a thermodynamic standpoint, oxygen is a strong oxidant, but its reactivity is limited by virtue of its unusual electronic configuration (Green and Hill, 1984). Although relatively unreactive, oxygen can be metabolized *in vivo* to form highly reactive derivatives. A single-electron reduction of O_2 yields the superoxide (O_2^-) free radical. A free radical is defined as any species with one or more unpaired electrons. There are a wide variety of sources of O_2^- production *in vivo;* examples include the ubiquinone-cytochrome b region in mitochondria, uncoupled cytochrome P450 reactions, autoxidation of adrenaline, certain flavins, Hb, and SH groups, enzyme reactions such as xanthine oxidase and tryptophan deoxygenase, and the oxidant burst of activated neutrophils and mononuclear phagocytes (Freeman and Crapo, 1982). Superoxide undergoes spontaneous and enzyme catalyzed dismutation to H_2O_2. In the presence of a metal ion such as iron or copper, O_2^- and H_2O_2 interact to form the highly reactive hydroxyl radical ($OH^\cdot$).

A wide variety of drugs, metabolic intermediates, and environmental agents either exist as free radicals or can be converted to free radicals by cellular metabolic processes (Freeman and Crapo, 1982; Mason, 1982). These free radicals can be more damaging than the partially reduced oxygen species listed earlier. In some situations, the superoxide radical may be beneficial by acting as a free radical scavenger (French *et al.*, 1978).

2. *Superoxide Dismutase*

Superoxide dismutase (SOD) is a copper- and zinc-containing enzyme that was first isolated from cattle blood (McCord and Fridovich, 1969). It dismutates two O_2^- molecules to H_2O_2 and O_2 (Fig. 7.4). The activity of

SOD in RBCs of domestic animals is about the same as or higher than that in man (Harvey and Kaneko, 1977; Suzuki *et al.*, 1984b). RBC SOD activity is reduced in animals fed diets deficient in copper (Andrewartha and Caple, 1980; Williams *et al.*, 1975). Zinc is also needed for optimal activity (Hirose *et al.*, 1992) and, consequently, SOD activity may also be low in zinc-deficient animals (Hirose *et al.*, 1992). The importance of SOD as an oxidant defense in RBCs is unclear. Under conditions where H_2O_2 catabolism is compromised, SOD can increase oxidant injury due to accelerated H_2O_2 generation (Scott *et al.*, 1989).

3. *Glutathione*

Glutathione is a tripeptide of glutamic acid, cysteine, and glycine that occurs in approximately 2 m*M* concentrations in RBCs. It is synthesized *de novo* in RBCs of humans and animals from constituent amino acids via two ATP-requiring reactions, utilizing gamma-glutamylcysteine synthetase and glutathione synthetase (Beutler, 1989). GSH has a highly reactive (easily oxidizable) sulfhydryl (SH) group that, like other thiols, may act nonenzymatically as a free radical acceptor to counteract oxidant damage (Prins and Loos, 1969). GSH also binds free hemin, thereby reducing its potential for membrane injury (Shviro and Shaklai, 1987). Enzyme-mediated reactions involving GSH are discussed later. While GSH is constantly being oxidized to a disulfide (GSSG) to some extent within RBCs, most glutathione is maintained in its reduced form by the GR reaction. GSSG accounts for only about 0.2% of the total glutathione in human RBCs (Beutler, 1984). RBC membranes are not permeable to GSH, but GSSG is transported from RBCs by an active process (Srivastava and Beutler, 1969). The half-life of glutathione in dog and rabbit RBCs (2–5 days) is similar to that in human RBCs (4 days), but longer times (10–12 days) are reported in sheep RBCs. The GSSG transport rate may be the main determinant of glutathione turnover in RBCs (Smith, 1974).

4. *Glutathione Reductase*

Excepting the cat, RBCs of domestic animals have lower GR activity than those of man (Agar *et al.*, 1974; Harvey and Kaneko, 1975b). GR requires flavin adenine dinucleotide (FAD) as a cofactor for activity. Therefore, riboflavin metabolism affects the availability of FAD and the fraction of the protein that exhibits activity (Beutler, 1989; Harvey and Kaneko, 1975b). Horse RBCs are much slower than other species in their ability to regenerate GSH after it has been oxidized *in vitro* (Agar *et al.*, 1974; Mahaffey and Smith, 1975). The nature of the rate-limiting reaction has not been determined. Horse RBCs also appear less able to protect themselves against oxidative injury induced by incubation with high levels of ascorbate, which stimulates the GR reaction by the oxidation of GSH (Harvey and Kaneko, 1977).

5. *Glutathione S-Transferase*

Glutathione S-transferase (GST) catalyzes the formation of glutathione S-conjugates between GSH and certain electrophilic substrates. The same Mg^{+2}–ATPase transport system appears to transport glutathione S-conjugates and GSSG out of RBCs (Beutler, 1989). GST activity is present in RBCs of all mammalian species studied thus far, but natural electrophilic substrates and the potential involvement of these glutathione S-conjugates in glutathione turnover in normal animals are unknown (Board and Agar, 1983). The GST-mediated conjugation of various carcinogens and other electrophilic drugs in the liver is important in the protection of the body against these agents (Chasseaud, 1979), but the importance of this activity in protecting RBCs against xenobiotics is unknown. GST can bind free hemin that is released during Hb oxidation, presumably reducing damage to RBC membranes (Harvey and Beutler, 1982). A direct correlation between GST activity and GSH concentration has been reported in sheep and dog RBCs (Goto *et al.*, 1992). This is probably related to the fact that GST is stabilized by GSH.

6. *Thioltransferase*

Thioltransferase enzymes in human and animal RBCs can catalyze the reduction of not only low-molecular-weight disulfides and intramolecular protein disulfides, but also mixed disulfides (Terada *et al.*, 1992). Using GSH, thioltransferases can restore thiol groups in oxidatively damaged proteins.

7. *Glutathione Peroxidase*

Low levels of H_2O_2 are produced in the course of normal cellular events and higher levels may be generated by exogenously administered redox active compounds (Saltman, 1989). Hydrogen peroxide is catabolized by both glutathione peroxidase (GPx) and catalase. GPx catalyzes the conversion of H_2O_2 to H_2O (Fig. 7.4). In response to the GSSG produced by this reaction, RBCs increase PPP metabolism to provide the NADPH necessary for the regeneration of GSH by the GR reaction (Eaton and Brewer, 1974). The selective oxidation of a renewable thiol helps limit irreversible damage to RBCs that would otherwise occur.

GPx activity in RBCs correlates directly with blood selenium concentration in ruminants, horses, and rats but not in pigs or higher primates (Anderson *et al.*, 1978; Beilstein and Whanger, 1983; Caple *et al.*, 1978; Thompson *et al.*, 1976). Selenium deficiency results in low GPx activity in RBCs because selenium is an essential cofactor that is incorporated into GPx as the enzyme is formed (Rotruck *et al.*, 1973). Although selenium deficiency can be diagnosed by measuring RBC GPx activity in some species (Anderson *et al.*, 1978; Caple *et al.*, 1978), caution is indicated in using this activity as a direct indicator of selenium status because polymorphism in GPx activity may be present, as occurs in Finn sheep (Sankari and Atroshi, 1983). A wide variety of abnormalities or lesions (most notably in skeletal and cardiac muscle) have been described in association with selenium deficiency in animals (Shamberger, 1986), but hemolytic anemia rarely, if ever, occurs in the absence of external oxidant stress. Heinz body hemolytic anemia has been reported during the summer months in selenium-deficient cattle grazing on St. Augustine grass growing on peaty muck soils in South Florida (Morris *et al.*, 1984). Although Heinz body formation was greatly reduced by selenium supplementation, it is likely that these animals were also exposed to a seasonal oxidant, possibly incorporated in the grass.

Fujii *et al.* (1984) have provided evidence for a GSH-dependent protein in RBCs that provides protection against lipid peroxidation. The nature of this presumed enzyme is unknown, but it does not have GST or GPx activity as normally measured. It may be similar to a selenium-containing membrane-bound enzyme found in other tissues that has been called phospholipid hydroperoxide glutathione peroxidase (Ursini *et al.*, 1985).

8. Catalase

Catalase is a heme-containing enzyme that also destroys H_2O_2 by conversion to H_2O and O_2. Recent *in vitro* studies suggests that catalase is of equal (Gaetani *et al.*, 1994) or greater (Scott *et al.*, 1993) importance than GPx in the defense of human RBCs against H_2O_2-generating reactions. However, the low levels of endogenous H_2O_2 produced *in vivo* are more likely cleared by GPx, which has a much higher affinity for substrate (Eaton, 1991).

Except for dogs, mammalian RBCs generally have high catalase activities (Allison *et al.*, 1957; Paniker and Iyer, 1965; Suzuki *et al.*, 1984a). Catalase is linked metabolically to the PPP because NADPH is tightly bound to catalase (Kirkman *et al.*, 1987). The binding of NADPH prevents and reverses the accumulation of an inactive form of catalase that is generated when catalase is exposed to H_2O_2. In addition to its benefit to RBCs, the presence of catalase in RBCs may help protect somatic cells exposed to high levels of H_2O_2, such as in sites of active inflammation (Agar *et al.*, 1986).

9. Vitamin E

Vitamin E deficiency increases the susceptibility of RBCs to peroxidative hemolysis (Duthie *et al.*, 1989; Pillai *et al.*, 1992; Rotruck *et al.*, 1972). Vitamin E (α-tocopherol) is a lipid-soluble vitamin that acts as a free radical scavenger within the membrane. It donates reducing equivalents to lipid peroxy radicals, thereby limiting chain reaction oxidations of polyunsaturated lipids and thus protecting membranes (Hebbel, 1986). Its location in the membrane provides little protection against cytosolic oxidative injury (Rotruck *et al.*, 1972).

10. Methemoglobin Reduction

MetHb differs from Hb only in that the iron moiety of the heme groups is in the ferric rather than the ferrous state. MetHb forms *in vivo* at low levels normally and at much higher levels in the presence of oxidative compounds (Bodansky, 1951).

MetHb is unable to bind oxygen and must be reduced to Hb to be functional. MetHb reduction is more corrective than protective. It is reduced to Hb primarily by NADH-methemoglobin reductase (NADH-MR) or cytochrome-b_5 reductase (Cb_5R). In the reaction, ferricytochrome b_5 is first reduced enzymatically with NADH; then the resulting ferrocytochrome b_5 reduces MetHb nonenzymatically to Hb (Hultquist *et al.*, 1978).

RBCs contain another enzyme, NADPH diaphorase (NADPH methemoglobin reductase, NADPH dehydrogenase), that is capable of MetHb reduction when appropriate electron carriers are present. In addition to redox dyes, such as methylene blue, various flavins may function as substrates for reduction by NADPH, prompting Yubisui *et al.* (1980) to classify the enzyme as a NADPH flavin reductase. The contribution of this enzyme to MetHb reduction in human RBCs is believed to be insignificant because flavin concentrations are normally low (Hultquist *et al.*, 1993). The recent finding that this protein also binds heme, porphyrins, and fatty acids raises additional questions about its possibly physiologic role in RBCs (Xu *et al.*, 1992).

Following the oxidation of Hb to MetHb with nitrite *in vitro*, horse RBCs reduce MetHb at a slower rate than those of other domestic animals (except pigs) when glucose is added as the substrate for energy (Robin and Harley, 1966). RBCs from adult pigs cannot reduce MetHb with glucose as the substrate because they lack a membrane glucose transporter (Kwong *et al.*, 1986). Interestingly horse RBCs reduce partially oxidized

Hb (25–40% MetHb) at a rapid rate but act on more fully oxidized Hb (80–90% MetHb) at a slower rate (Medeiros *et al.*, 1984). In contrast to NADP, only about half of the total NAD is normally present in the reduced (NADH) form (Zerez *et al.*, 1987). Horse (Robin and Harley, 1967; Medeiros *et al.*, 1984) and pig (Rivkin and Simon, 1965) RBCs utilize lactate better than glucose to generate NADH (by the LDH reaction) for the reduction of MetHb. Because lactate occurs in blood and easily diffuses into RBCs, it may be an important substrate of MetHb reduction *in vivo*.

Methylene blue (MB) is used to treat toxic methemoglobinemia because it causes MetHb to be reduced faster than occurs by the relatively slow Cb_5R reaction. MB is reduced to leukomethylene blue (LMB) by the NADPH-dependent diaphorase discussed earlier (Fig. 7.4), and LMB reacts spontaneously with MetHb, reducing it to Hb and regenerating MB (Sass *et al.*, 1969). MB was suggested to be of limited value in the treatment of methemoglobinemia in horses, since it had a limited ability to stimulate MetHb reduction of horse RBCs *in vitro* with glucose as the substrate (Robin and Harley, 1966). Dixon and McPherson (1977), however, found intravenous MB to be effective in reducing MetHb in a horse with persistent methemoglobinemia.

Being a redox dye, MB can participate in various oxidative reactions on entering RBCs. It is important that it be used judiciously lest it potentiate Heinz body formation and hemolysis that might result from the original oxidative insult (Harvey and Keitt, 1983).

V. DETERMINANTS OF ERYTHROCYTE SURVIVAL

A wide variety of conditions, including immune-mediated hemolytic anemias and infectious diseases, result in shortened RBC survival. Readers are referred to *Schalm's Veterinary Hematology* (Jain, 1986) for information concerning conditions beyond the scope of this chapter.

A. Oxidative Injury

Oxidants produce different patterns of intracellular and membrane damage which may be related to differences in lipid solubility, redox potentials, reactivity with SH groups, binding to heme, and the source or site of oxidant generation. Extracellularly produced oxidants can damage the membrane before reaching the cytosolic protective mechanisms. Oxidants that are generated intracellularly in coupled reactions with OxyHb tend to produce more Hb injury than membrane injury. Hb has been described as a "frustrated" oxidase, because it potentiates the generation of oxidants by a variety of drugs (Carrell *et al.*, 1977).

1. Methemoglobin Formation

Although MetHb formation is reversible and does not enhance RBC destruction per se, it is a component of oxidative injury to RBCs and is generally involved as a step in Heinz body (HzB) formation. Clinical signs associated with methemoglobinemia are the result of hypoxia because MetHb cannot bind O_2. Lethargy, ataxia, and semistupor do not become apparent until MetHb content reaches 50%, with a comalike state and death ensuing when it reaches 80% (Bodansky, 1951). The cyanotic appearance of mucous membranes associated with this condition may not be easily recognized in heavily pigmented animals. MetHb content is quantified spectrophotometrically, but a spot test can be used to determine if clinically significant levels of MetHb are present. One drop of blood from the patient is placed on a piece of absorbent white paper and a drop of normal control blood is placed next to it. If the MetHb content is 10% or greater, the patient's blood should have a noticeably brown coloration.

Approximately 3.0% of Hb is spontaneously oxidized to MetHb each day in normal people (Mansouri and Lurie, 1993) and dogs (Harvey *et al.*, 1991). This formation results from spontaneous autoxidation of OxyHb and possibly secondarily in response to oxidants produced in normal metabolic reactions.

While the iron moiety of DeoxyHb is in the ferrous state, in OxyHb it exists in (or near) the ferric state, with an electron being transferred to the O_2 molecule to give a bound superoxide (O_2^-) ion (Mansouri and Lurie, 1993). During deoxygenation, the electron returns to the iron moiety and O_2 is released. Autoxidation to MetHb with the release of O_2^- occurs when the bound O_2^- is replaced by a nucleophile such as Cl^- (Wallace *et al.*, 1974).

Rather than extracting electrons, oxidant drugs often produce MetHb by donating electrons to OxyHb (Carrell *et al.*, 1977) as is shown here for phenylhydroxylamine (PHA):

$$Hb^{3+}\cdot\cdot O_2^- + PHA \rightarrow Hb^{3+}\cdot\cdot O_2^{2-} + PHA^{\cdot},$$
$$Hb^{3+}\cdot\cdot O_2^{2-} + 2H^+ \rightarrow Hb^{3+} + H_2O_2.$$

Drug free radicals, such as $PHA^{\cdot}$, that are generated can donate the unpaired electron to molecular oxygen to form superoxide or to another OxyHb molecule to form more MetHb and H_2O_2. To achieve greater stability, free radicals may also extract electrons by oxidizing SH groups of Hb, enzymes, membrane proteins, or GSH, by oxidizing membrane unsaturated fatty acids, by oxidizing NADPH or NADH, and possibly by extracting an electron from DeoxyHb to form MetHb.

Oxidants such as copper appear to oxidize Hb by extracting electrons. Cupric ions oxidize the reactive SH group on the beta chains of human and animal Hbs (Taketa and Antholine, 1982), and then an electron is transferred from heme iron to regenerate the SH group. Cupric ion is regenerated by interaction with oxygen. The series of reactions by which copper catalyzes Hb oxidation to MetHb (Carrell *et al.*, 1978) is as follows:

$$Cu^{2+} + Fe^{2+}S^{-} \rightarrow Cu^{1+} + Fe^{2+}S^{\cdot},$$
$$Fe^{2+}S^{\cdot} \rightarrow Fe^{3+}S^{-},$$
$$Cu^{1+} + O_2 \rightarrow Cu^{2+} + O_2^{-}.$$

These reactions occur rapidly but continue only until about 50% of total hemes are oxidized because α chains are resistant to oxidation (Taketa and Antholine, 1982). Additional oxidative reactions involving copper can denature Hb and damage membranes (Hochstein *et al.*, 1978).

Nitrite produces MetHb without HzB formation or anemia in animals. Methemoglobinemia occurs in ruminants eating nitrate-accumulating plants, especially when the plants have been fertilized with nitrogenous compounds. Nitrate is relatively nontoxic, but it is reduced to nitrite by ruminal microorganisms (Burrows, 1980). The reactions involved in MetHb formation in the presence of nitrite are complex; superoxide appears to be involved (Doyle *et al.*, 1982; Tomoda *et al.*, 1981). Methemoglobinemia can occur in animals following the topical application of benzocaine-containing products to skin (Harvey *et al.*, 1979; Wilkie and Kirby, 1988), or in laryngeal (Krake *et al.*, 1985) or nasopharyngeal sprays (Davis *et al.*, 1993; Lagutchik *et al.*, 1992). Benzocaine appears to have only a limited ability to produce other forms of RBC injury.

Many compounds can produce MetHb (Bodansky, 1951), and most of them also produce variable degrees of Hb denaturation and membrane injury. Following exposure to oxidants, MetHb forms within minutes, but HzB take hours to form. If the oxidant is rapidly metabolized, MetHb content will generally be reduced to values approaching normal within 24 hours (Harvey and Keitt, 1983).

2. *Heinz Body Formation*

Heinz bodies are composed of oxidized denatured Hb. They are often not recognized on routinely stained blood films, because they either are not stained or stain similarly to the remaining intact Hb. If they are of sufficient size (1–2 μm), they may appear as pale inclusions within RBCs or as nipple-like projections from the surface of RBCs. Heinz bodies can be visualized as dark, refractile inclusions in new methylene blue "wet" preparations and as light blue inclusions with reticulocyte stains (Jain, 1986).

The following sequence of biochemical events leading to HzB formation is proposed (Allen and Jandl, 1961; Chiu and Lubin, 1989; Hebbel and Eaton, 1989; Low, 1989): (1) Ferrohemes are oxidized to ferrihemes (MetHb); (2) SH groups of Hb are oxidized, causing conformational changes in globin chains and resultant hemichrome formation (hemichromes have both the ferric iron's fifth and sixth coordinate positions occupied by a ligand provided by the globin chain); (3) hemichromes bind to band 3 and to a lesser degree other membrane components, forming clusters of copolymers; and (4) the precipitation of these denatured Hb molecules results in HzB formation. Precipitation is potentiated by the dissociation of the ferriheme (hemin) moieties from the hemichromes because the resultant free globin chains are unstable.

Hemolytic anemias associated with HzB formation in domestic animals have resulted from a variety of compounds. Dietary causes include consumption of onions by cattle (Lincoln *et al.*, 1992), sheep (Kirk and Bulgin, 1979; Verhoeff *et al.*, 1985), horses (Pierce *et al.*, 1972), cats (Kobayashi, 1981), and dogs (Harvey and Rackear, 1985; Ogawa *et al.*, 1986) and consumption of kale and other *Brassica* species by ruminants (Greenhalgh *et al.*, 1969; Smith, 1980; Suttle *et al.*, 1987). Three novel thiosulfates have been identified as causative factors in onion extracts (Yamato *et al.*, 1994). In the case of *Brassica* species, the hemolytic factor is reported to be dimethyl disulfide, produced by the action of rumen microbes on *S*-methylcysteine sulfoxide contained within the plants (Smith, 1980).

Heinz bodies hemolytic anemia occurs in Florida in cattle grazing on lush rye (*Secale cereale*) pastures in the winter (Simpson and Anderson, 1980) and in selenium-deficient cattle grazing on St. Augustine grass pastures in the summer (Morris *et al.*, 1984). The nature of oxidants involved is unknown. Methemoglobinemia, HzB formation, severe intravascular hemolysis, and death have followed the consumption of red maple leaves by horses (George *et al.*, 1982; Tennant *et al.*, 1981). Dried leaves are toxic, but freshly harvested leaves are not.

Postparturient hemoglobinuria with HzB formation occurs in cattle in New Zealand grazing primarily on perennial ryegrass (*Lolium perenne*) (Martinovich and Woodhouse, 1971). Postparturient cattle may be more susceptible to the development of anemia because increased food consumption associated with lactation could increase exposure to an unidentified dietary oxidant. Both hypocuprosis (Gardner *et al.*, 1976) and hypophosphatemia (Jubb *et al.*, 1990) have been considered to contribute to the severity of the anemia in these cattle. An apparently different syndrome of postparturient hemoglobinuria has also been reported in hypophosphatemic cattle of North America (MacWilliams *et al.*, 1982). Heinz bodies have not been reported in affected animals, and the anemia appears to develop because affected animals have decreased RBC ATP concentrations (Ogawa *et al.*, 1987, 1989).

Methemoglobinemia and HzB hemolytic anemia occur acutely when large amounts of copper are released from the liver of ruminants that have accumulated excessive amounts of liver copper secondary to increased dietary intake (Brewer, 1987; Kerr and McGavin, 1991; Soli and Froslie, 1977). Zinc toxicity has primarily resulted from the consumption and retention of zinc-containing objects within the stomach of dogs. Sources of zinc include U.S. pennies minted after 1982, metallic hardware, and ointment containing zinc oxide (Breitschwerdt *et al.*, 1986). The mechanism(s) by which zinc produces hemolytic anemia is unclear, but HzB have been recognized in some clinical cases (Houston and Myers, 1993; Luttgen *et al.*, 1990).

Clinical cases of HzB hemolytic anemias have occurred following the administration of a variety of drugs including methylene blue in cats (Schechter *et al.*, 1973) and dogs (Fingeroth and Smeak, 1988; Osuna *et al.*, 1990), phenazopyridine in a cat (Harvey and Kornick, 1976), acetaminophen in cats (Finco *et al.*, 1975; Gaunt *et al.*, 1981; Hjelle and Grauer, 1986) and dogs (Harvey *et al.*, 1986; Houston and Myers, 1993), methionine in cats (Maede *et al.*, 1987), menadione (vitamin K_3) in dogs (Fernandez *et al.*, 1984), and phenothiazine in horses (McSherry *et al.*, 1966). Of these drugs, acetaminophen and phenazopyridine also produce clinically significant methemoglobinemia. A variety of additional oxidants produce HzB hemolytic anemias experimentally in animals (Fertman and Fertman, 1955).

Cats are generally recognized as the species most susceptible to HzB formation. Although species differences in metabolism and excretion of various drugs may partially account for this increased susceptibility, cat Hb generally appears more susceptible to oxidative denaturation than do Hbs of other species (Harvey and Kaneko, 1976b). The presence of 8–10 reactive SH groups per Hb tetramer in cat Hb may render it more susceptible to oxidation (Mauk and Taketa, 1972); apparently no other species has more than 4 reactive SH groups per tetramer (Snow, 1962).

Heinz bodies are rarely recognized in RBCs from most species, but are frequently present in cat RBCs because of the susceptibility of cat Hb to form HzB, combined with a poor ability of the cat spleen to remove HzB from RBCs (Jain, 1986) Even normal cats may have low numbers of HzB (<5%), and increased HzB numbers have been seen in kittens fed fish-based diets (Hickman *et al.*, 1990) and in cats fed commercial soft-moist diets containing propylene glycol (Christopher *et al.*, 1989; Hickman *et al.*, 1990). Increased HzB numbers have also been documented in cats with diabetes mellitus, hyperthyroidism and lymphoma (Christopher, 1989). Diabetic cats with ketoacidosis have more HzB and lower PCV than do nonketotic diabetic cats (Christopher *et al.*, 1995). Although RBC survival tends to be shortened, anemia is either absent or mild in the conditions mentioned. The oxidative reactions involved in these conditions are not clearly defined.

3. Membrane Injury

Various types of oxidant-induced membrane injury have been recognized. These include oxidation of membrane SH groups, lipid peroxidation, cross-linking of spectrin, inhibition of enzymes and membrane transport systems, and band 3 clustering. Damaged cells exhibit increased membrane rigidity, decreased deformability, and impaired ability to maintain ion gradients (Chiu and Lubin, 1989). The relative importance of HzB formation, hemin release, and the generation of various free radicals remains to be clarified.

The binding of HzB to the inner surface of the RBC membrane and clustering of band 3 and other components alter the normal membrane organization. Potential negative effects include weakening or disruption of the cytoskeleton, altered distribution of membrane phospholipids, altered cell surface charges, and formation of abnormal external cell surface antigens that can be recognized by autologous antibodies (Low, 1989). Heinz bodies cause focal membrane rigidity but do not affect global cellular deformability until they nearly cover the entire internal surface of RBCs (Reinhart *et al.*, 1986).

Hemin released during HzB formation binds to and causes damage to RBC membranes. Evidence has been presented that hemin can mediate the dissociation of RBC cytoskeletal proteins, impair the ability of the RBC membrane to maintain ionic gradients, oxidize membrane sulfhydryl proteins, and potentiate peroxide-induced membrane lipid peroxidation (Chiu and Lubin, 1989; Hebbel and Eaton, 1989; Jarolim *et al.*, 1990). The molecular mechanism(s) has not been clearly defined, but hemin may exert its toxic effects via catalysis of the formation of reactive oxygen species (Vincent, 1989).

Oxidative injury to RBC membranes is sometimes recognized by the appearance of eccentrocytes in stained blood films. These cells have been reported to occur in horses (Reagan *et al.*, 1994) and dogs (Ham *et al.*, 1973; Harvey *et al.*, 1986; Harvey and Rackear, 1985) exposed to exogenous oxidants and in a G6PD-deficient horse with inadequate metabolic protection against endogenous oxidants (Stockham *et al.*, 1994). Denatured spectrin dimers and tetramers are believed to be of primary importance in this cross bonding of membranes (Fischer, 1988; Arese and De Flora, 1990).

4. *Erythrocyte Destruction*

Following extreme oxidative injury, RBCs may lyse within the circulation. Hemoglobinuria occurs if intravascular hemolysis is of sufficient magnitude to saturate the haptoglobin-binding capacity of plasma and exceed the ability of renal tubules to reabsorb filtered Hb. In most cases, however, enhanced RBC destruction results primarily from increased phagocytosis of injured RBCs by macrophages of the spleen, liver, and bone marrow.

Of the organs of the mononuclear phagocyte system, the spleen is most adept at recognizing and removing damaged RBCs. In most species, RBCs must pass through narrow slits between endothelial cells lining venous sinus walls of the spleen to reenter the general circulation. As a consequence, oxidant-induced decreased RBC deformability tends to result in sequestration of RBC in the splenic reticular meshwork, thereby enhancing the likelihood of phagocytosis by resident macrophages (Baerlocher *et al.*, 1994; Weiss, 1984).

When HzB are larger than openings in walls of splenic venous sinuses, they are retained within the trabecular meshwork of the spleen. The whole cell may be phagocytosed, or the HzB and closely associated membrane may be removed. The remainder of the RBC reseals and passes through the sinus wall. This removal of HzB is the so-called pitting function of the spleen (Weiss, 1984). Cat spleens have poor pitting capabilities (Jain, 1986) owing to the presence of large openings in venous sinus walls (Blue and Weiss, 1981).

Macrophages can recognize damaged RBCs by antibody-dependent and by antibody-independent mechanisms; however, the relative importance of the different mechanisms remains to be defined. Macrophages do not recognize less-deformable cells per se, but phagocytose them at a higher rate because of their slower transit time or entrapment in the spleen as discussed earlier (Baerlocher *et al.*, 1994).

The cross-linking of band 3 molecules by hemichromes during oxidative denaturation of Hb results in clustering of band 3 molecules and greatly enhanced binding of autologous IgG (Low, 1991). Because macrophages contain Fc receptors on their surfaces, the damaged RBCs would presumably become bound to the surface of macrophages once a sufficient number of clusters of band 3 had formed. Sheep with chronic copper toxicity may become direct Coombs' test positive, indicating a possible *in vivo* role of antibody-mediated removal of oxidatively injured RBCs (Wilhelmsen, 1979).

Macrophages contain antibody-independent scavenger receptors that can bind to oxidized lipoproteins, anionic phospholipids, polysaccharides, and polyribonucleotides (Sambrano *et al.*, 1994). Evidence has been presented to indicate that at least two different scavenger receptors may be involved in the recognition and phagocytosis of oxidatively damaged RBCs. Macrophages can bind to altered membrane glycoproteins via a sialosaccharide receptor (Beppu *et al.*, 1994) and to altered lipid-protein conjugates via a receptor with specificity for oxidized low-density lipoproteins (Sambrano *et al.*, 1994).

B. Erythrocyte Aging and Normal Lifespans

Most RBCs circulate in blood for a finite period (survival time or lifespan) that varies from 2 to 5 months in domestic animals, depending on the species (Table 7.1). Methods for measurement of RBC lifespans and results of lifespan determinations from many species have been compiled by Vacha (1983). RBC lifespans are related to body weight (and consequently metabolic rate) with the smallest animals (highest metabolic rate) having the shortest RBC lifespans (Vacha, 1983). The RBC lifespan can be prolonged by reducing the metabolic rate within an individual animal. Examples include thyroidectomy in rats, hibernation in hibernating mammals, and reduced ambient temperatures in poikilotherms (Landaw, 1988). Various changes occur in RBCs as they get older (Kosower, 1993; Low, 1991), but the nature of the factor(s) that initiates age-related changes and the mechanism(s) of removal of senescent RBCs from the circulation require further clarification. Decreased membrane phospholipid asymmetry, modified membrane carbohydrate residues (e.g., desialation of sialoglycoproteins), and/or modified membrane proteins (e.g., partially degraded Band 3) are possible signals for removal (Kosower, 1993).

1. *Metabolic Impairment*

Because RBCs cannot synthesize new enzymes, one theory is that one or more critical enzymes (e.g., hexokinase), involved in generating ATP or in the protection against oxidants, may decrease to a point that metabolic impairment results in irreversible RBC membrane changes (Piomelli and Seaman, 1993). Although the inability to maintain normal ATP and NADPH concentrations appears to account for premature destruction of RBCs in patients with hereditary enzyme deficiencies of the EMP and PPP, respectively, the theory that metabolic impairment occurs as a result of age-related decreases in critical enzymes has been questioned (Beutler, 1985; Zimran *et al.*, 1990). Much of the decline in activities of enzymes, considered in the past to be correlated with RBC age, occurs as reticulocytes mature into RBCs. Even if metabolic impairment is not the primary factor responsible for RBC senescence, it may render the aged RBC vulnerable to events in the circulation that require a burst of metabolic activity (Piomelli and Seaman, 1993).

2. *Cumulative Oxidant Damage*

Current evidence indicates that cumulative oxidative injury is probably responsible for normal RBC aging and removal (Low, 1991; Seppi *et al.*, 1991). Even the decrease in RBC enzymes may be the result of oxidative damage (Stadtman, 1992). The presence of HzB in RBCs from splenectomized humans, horses, and dogs and in RBCs from nonsplenectomized cats (Jain, 1986; Low, 1989) provides evidence for ongoing oxidant injury *in vivo*. The inverse correlation discussed earlier between RBC lifespan and metabolic rate may result from differences in endogenous oxidant generation, but other factors such as differing amounts of mechanical stress should also be considered as a possible cause of this relationship (Landaw, 1988).

3. *Senescent Cell Antigen*

Although multiple mechanisms may be involved in the recognition and removal of aged RBCs, current evidence suggests that removal of senescent human RBCs is largely immune mediated, following the appearance of a senescent cell antigen (Kay, 1989). This senescent cell antigen is derived from the band 3 anion transporter. The specific alteration required for band 3 to become antigenic remains to be clarified, but oxidative mechanisms are probably involved. Low (1991) has demonstrated copolymerization between hemichromes and band 3 and has suggested that the senescent antigen represents clusters of band 3. In contrast, studies by Kay *et al.* (1989) suggest that the senescent cell antigen results from band 3 degradation rather than clustering.

A natural antibody against the senescent cell antigen is present in human plasma. This antibody binds to senescent cell antigens on the surface of aged cells and, together with bound complement, promotes the phagocytosis of aged RBCs by macrophages that exhibit Fc and C3b surface receptors (Lutz *et al.*, 1991). Increased autologous immunoglobulin binding also occurs on dog RBCs aged *in vivo*, indicating that a similar process likely occurs in this species (Christian *et al.*, 1993). The possibility that senescent RBCs can also be removed via binding to macrophage scavenger receptors requires further study (Sambrano *et al.*, 1994).

C. Anemia of the Newborn

Animals are generally born with PCVs near values for adults. Following birth, there is a rapid decrease in PCV that is followed by a gradual increase to adult values (Jain, 1986). Factors involved to variable degrees in the development of the anemia of the newborn include the following: (1) absorption of colostral proteins during the first day of life, which increases plasma volume through an osmotic effect (Harvey *et al.*, 1987; Möllerberg *et al.*, 1975); (2) decreased RBC production during the early neonatal period; (3) shortened lifespan of RBCs formed *in utero* (Kim and Luthra, 1977; Lee *et al.*, 1976; Mueggler *et al.*, 1979; Landaw, 1988); and (4) rapid growth with hemodilution resulting from expansion of total plasma volumes more rapidly than total RBC mass (Mueggler *et al.*, 1979).

In some species, production of RBCs is decreased because of low EPO concentrations at birth (Halvorsen and Halvorsen, 1974; Huisman *et al.*, 1969; Meberg, 1980; Meberg *et al.*, 1980; Schwartz and Gill, 1983). The decreased stimulus for EPO production may occur as a result of a placental blood transfusion that increases RBC mass immediately after birth (Rossdale and Ricketts, 1980), a rapid increase in pO_2 associated with breathing air, and an increase in P_{50} as discussed previously.

Much of the postnatal anemia of dogs occurs as a physiological response to increased RBC 2,3DPG and subsequent improved oxygen transport (Mueggler *et al.*, 1981). The "anemia" of childhood in humans is also associated with RBC 2,3DPG above adult values. In children this increase appears to occur secondarily to increased plasma P_i concentrations (Card and Brain, 1973). Serum P_i values are also above adult values in young dogs (Pickrell *et al.*, 1974).

Although not involved in the early, rapid decrease in PCV, iron availability may limit the response to anemia in some rapidly growing animals (Chausow and Czarnecki-Maulden, 1987; Dhindsa *et al.*, 1971; Harvey *et al.*, 1987; Holman and Drew, 1966; Mollerberg *et al.*, 1975; Siimes *et al.*, 1980; Weiser and Kociba, 1983b).

VI. INHERITED DISORDERS OF ERYTHROCYTES

Many hereditary disorders of RBCs have been described in humans (Dacie, 1985). A limited number of inherited RBC disorders have been identified in laboratory and domestic animals (Kaneko, 1987; Smith, 1981). Congenital porphyrias are discussed elsewhere in this volume (Chapter 8).

A. Cytosolic Enzyme Deficiencies

1. Phosphofructokinase Deficiency in Dogs

Autosomal recessive inherited PFK deficiency occurs in English springer spaniel (Giger *et al.*, 1985; Giger and Harvey, 1987) and American cocker spaniel (Giger *et al.*, 1992) dogs. Canine PFK is genetically controlled by three separate loci. They code for muscle- (M), liver- (L), and platelet- (P) type subunits (Vora *et al.*, 1985). Random tetramerization of the subunits produces various isozymes. PFK in normal dog RBCs consists of 86% M-type, 2% L-type, and 12% P-type subunits and normal dog muscle is composed exclusively of M-type subunits (Mhaskar *et al.*, 1992).

A point mutation is reported to occur in the M-type gene of deficient dogs, causing a loss of amino acids from the carboxyl terminus of the polypeptide (Giger *et al.*, 1991c). Studies of brain and RBCs from homozygous deficient dogs indicated that native M-type subunits were not present, but small amounts of a structurally unstable truncated M-type subunit were found (Mhaskar *et al.*, 1991, 1992). As would be expected from the subunit composition of normal tissues, total RBC and muscle PFK activities are markedly reduced in affected dogs (Giger and Harvey, 1987; Vora *et al.*, 1985). Changes in concentrations of glycolytic intermediates in muscle and RBCs reflect the block at the PFK step (Harvey *et al.*, 1992a, 1992b). RBCs from affected dogs also exhibit altered enzyme kinetic properties due to the loss of the M-type subunit (Harvey *et al.*, 1992b).

Homozygously affected dogs have persistent compensated hemolytic anemias and sporadic episodes of intravascular hemolysis with hemoglobinuria (Giger *et al.*, 1985; Giger and Harvey, 1987). RBC mean cell volumes are usually between 80 and 90 fl. Reticulocyte counts are generally between 10 and 30%, with hematocrit values between 30 and 40% (Harvey and Smith, 1994), except during hemolytic crises when the hematocrit may decrease to 15% or less. Lethargy, weakness, pale or icteric mucous membranes, mild hepatosplenomegaly, muscle wasting, and fever as high as 41°C may occur during hemolytic crises (Giger and Harvey, 1987).

Hemolytic crises occur secondarily to hyperventilation-induced alkalemia *in vivo*, and PFK-deficient dog RBCs are extremely alkaline fragile *in vitro* (Giger and Harvey, 1987). For unknown reasons, normal dog RBCs are more alkaline fragile than those of humans and other mammals studied (Iampietro *et al.*, 1967; Waddell, 1956). The even greater alkaline fragility of PFK-deficient dog RBCs results from decreased 2,3DPG, which is formed below the PFK reaction (Harvey *et al.*, 1988). Because 2,3DPG is the major impermeant anion in dog RBCs, a substantial decrease in its concentration results in a higher intracellular pH (Hladky and Rink, 1977) and thereby greater alkaline fragility than normal dog RBCs. As expected, the low 2,3DPG concentration also results in an increased oxygen affinity of Hb in affected dog RBCs (Giger and Harvey, 1987).

Hematologic parameters of affected dogs are similar to normal dogs at birth, because all newborn dogs have RBC PFK activities about three times that of normal adult dogs (Harvey and Reddy, 1989). This high PFK activity results from the presence of the L-type subunit of PFK, which is negligible in normal adult canine RBCs (Harvey and Reddy, 1989; Mhaskar *et al.*, 1992). Both total PFK activities and the amounts of L-type subunit present decrease dramatically during the first 6 to 8 weeks of life. The M-type subunit is low at birth, but increases as the L-type decreases in normal dogs. These changes result from the replacement of RBCs formed in the fetus with those formed after birth.

Deficient dogs generally exhibit less evidence of myopathy than is observed in PFK-deficient people, probably because canine skeletal muscle is less dependent on anaerobic glycolysis than human skeletal muscle, owing to a lack of the classical fast-twitch glycolytic (type IIB) fibers in dogs (Snow *et al.*, 1982). Affected dogs appear to tire more easily than normal, and *in vivo* muscle studies of PFK-deficient dogs indicate altered muscle function in these animals (Brechue *et al.*, 1994; Giger *et al.*, 1988b). A severe progressive myopathy with associated abnormal polysaccharide deposits in skeletal muscle has been recognized in an aged PFK-deficient dog (Harvey *et al.*, 1990a). In contrast to PK deficiency, myelofibrosis and liver failure have not been recognized in dogs with PFK deficiency.

Homozygous affected animals over 3 months of age can easily be identified by measuring RBC PFK activity. Heterozygous carrier dogs have approximately one-half normal enzyme activities in RBCs (Harvey and Reddy, 1989). A DNA test using polymerase chain reaction technology has been developed that can clearly differentiate normal, carrier, and affected dogs regardless of age (Giger *et al.*, 1995b).

2. *Pyruvate Kinase Deficiency in Dogs and Cats*

PK deficiency occurs in basenji (Giger and Noble, 1991; Searcy *et al.*, 1971, 1979), beagle (Giger *et al.*, 1991b; Harvey *et al.*, 1977; Prasse *et al.*, 1975), West Highland white terrier (Chapman and Giger, 1990), cairn terrier (Schaer *et al.*, 1992), and American Eskimo dogs (Harvey, 1994), and Abyssinian cats (Ford *et al.*, 1992). PK deficiency is transmitted as an autosomal recessive trait. Homozygously affected animals have decreased exercise tolerance, pale mucous membranes, tachycardia, and splenomegaly.

Affected animals have mild to severe anemia with marked reticulocytosis. An unexplained feature of the disease in dogs is the progressive development of myelofibrosis and osteosclerosis. Affected dogs generally die by 3 years of age because of bone marrow failure and/or liver disease with hemochromatosis and subsequent cirrhosis (Searcy *et al.*, 1979; Weiden *et al.*, 1981).

RBCs of affected dogs lack the normal adult R isozyme of PK but have a persistence of an M_2 isozyme that normally predominates in fetal RBCs (Black *et al.*, 1978; Whitney *et al.*, 1994). Consequently, many affected dogs have normal or increased PK activity, making it difficult to diagnose this defect based solely on total RBC PK activity. In contrast, total RBC PK activity is markedly reduced in cats, with no evidence of a persistent M_2 isozyme (Ford *et al.*, 1992). Heterozygous animals have approximately 50% of normal RBC PK activity.

The enzyme activity in hemolysates of affected dogs is unstable and decreases rapidly when samples are kept at room temperature (Standerfer *et al.*, 1974). If the M_2-isozyme is unstable *in vivo*, as it is *in vitro*, its rapid loss of activity would explain the dramatically shortened lifespan of RBCs in this disorder (Dhindsa *et al.*, 1976).

Additional assays (an enzyme heat stability test, measurement of RBC glycolytic intermediates, electrophoresis of isozymes, and enzyme immunoprecipitation) may be used to reach a diagnosis of PK deficiency in dogs where the total enzyme activity is not decreased (Giger and Noble, 1991; Harvey *et al.*, 1990b; Schaer *et al.*, 1992). The defect in basenji dogs is the result of a single nucleotide deletion in the R-type PK gene (Whitney *et al.*, 1994). A DNA diagnostic test has recently been developed for basenji dogs (Whitney and Lothrop, 1995). Thus far, this test has not been valid in other dog breeds, indicating that the defect is not identical in all breeds.

Because the defect in glycolysis occurs below the diphosphoglycerate shunt, RBCs from PK-deficient dogs have increased concentrations of 2,3DPG. As a consequence, the whole blood P_{50} is higher than that of normal dogs (Dhindsa *et al.*, 1976).

3. *Glucose-6-Phosphate Dehydrogenase Deficiency in a Dog and Horse*

G6PD deficiency is a very common X-linked inherited defect of human RBCs, affecting millions of people worldwide (Beutler, 1994). Smith *et al.* (1976) screened more than 3000 dogs for G6PD activity and found one male dog to have approximately 44% of normal activity. The enzyme was partially purified and characterized, and it was found to be similar to that of normal dogs. The deficient dog was not anemic and exhibited no clinical signs; studies were not done to determine if his RBCs were more sensitive to oxidant damage than normal.

In contrast, a persistent hemolytic anemia and hyperbilirubinemia has been described in an American saddlebred colt with <1% of normal G6PD activity (Stockham *et al.*, 1994). Morphologic abnormalities of RBCs included eccentrocytosis, pyknocytosis, increased anisocytosis, and increased Howell–Jolly bodies. The presence of eccentrocytes in the absence of exposure to external oxidants indicated that the deficient RBCs did not have adequate metabolic capabilities to defend themselves against endogenous oxidants. Biochemical abnormalities in RBCs included low GSH, markedly reduced NADPH, and increased $NADP^+$. RBC catalase activity was normal even though NADPH concentration was <1% of normal. It was suggested that catalase activity may have been maintained by the action of NADH. Polymerase chain reaction amplification of segments of the G6PD gene of the affected colt revealed a G to A mutation, converting an arginine codon to a histidine codon (Nonneman *et al.*, 1993).

4. *Cytochrome-b_5 Reductase (NADH-Methemoglobin Reductase) Deficiency in Dogs and a Cat*

Persistent methemoglobinemia associated with Cb_5R deficiency has been recognized in a Chihuahua, a borzoi, an English setter, a terrier mix, a cockapoo, a Welsh corgi, miniature poodles, pomeranians, and in toy American Eskimo dogs (Atkins *et al.*, 1981; Harvey *et al.*, 1974, 1991; Letchworth *et al.*, 1977), and in a domestic short-haired cat (Harvey *et al.*, 1994). Several types of enzymatic hereditary methemoglobinemia have been recognized in people (Jaffe, 1986), but the exact nature of the defect(s) and the possibility that cell types in addition to RBCs might be deficient in activity have not been reported in animals. The deficiency is presumed to be a hereditary disorder, but family studies have not been reported.

Affected animals have cyanotic-appearing mucous membranes and may exhibit lethargy or exercise intolerance at times, but they usually have no clinical signs of disease. Blood samples appear dark, suggesting hypoxemia, but arterial pO_2 values are normal. MetHb content in dogs with Cb_5R deficiency varies from 13 to 41%. The MetHb content in the deficient cat was 50%. An inverse correlation appears to exist between RBC enzyme activity and MetHb content (Harvey *et al.*, 1991). The PCV is sometimes slightly increased secondary to the chronic methemoglobinemia. Animals with Cb_5R deficiency do not require treatment and have normal life expectancy.

5. Gamma-Glutamylcysteine Synthetase Deficiency in Sheep

An autosomal dominant inherited deficiency in RBC GSH in Corriedale and Merino sheep results from low levels of gamma-glutamylcysteine synthetase (GCS), the first enzyme involved in GSH synthesis (Fisher *et al.*, 1986; Smith *et al.*, 1973). Although the specific activity of the enzyme is low, the molecular weight, K_m values for glutamate and cysteine, K_i for GSH, and other characteristics of purified GCS from deficient sheep are remarkably similar to those from normal sheep (Board *et al.*, 1980). Deficient sheep have only about 20–30% of normal RBC GSH but exhibit no clinical signs, are not anemic, and have normal RBC lifespans (Smith *et al.*, 1973). Although RBCs from these low-GSH sheep do not appear to have greater susceptibility to injury by superoxide or hydrogen peroxide than do RBCs from high-GSH sheep (Eaton *et al.*, 1989), they do exhibit greater HzB formation when exposed to acetylphenylhydrazine *in vitro* (Goto *et al.*, 1993) and kale feeding *in vivo* (Tucker *et al.*, 1981).

B. Membrane Abnormalities

1. Amino Acid Transport Deficiency in Sheep and Horses

A second type of GSH deficiency, inherited as an autosomal recessive trait, occurs in Finnish Landrace sheep (Tucker and Kilgour, 1970). Although affected animals are not anemic, the lifespan of deficient RBCs is shortened (Tucker, 1974), possibly from increased oxidant injury as evidenced by the presence of HzB (Tucker *et al.*, 1981). These sheep are more likely to become anemic following the administration of oxidants *in vivo* (Tucker *et al.*, 1981).

The amino acid transporter normally responsible for cysteine transport (system C) into RBCs is defective (Young *et al.*, 1975), thereby limiting cysteine uptake and restricting GSH synthesis. As a consequence, GSH concentrations in RBCs are about 30% of normal (Young *et al.*, 1975). Intracellular Na^+ and K^+ are decreased, because dibasic and other amino acids accumulate in this disorder (Ellory *et al.*, 1972). The transport deficiency appears to develop during reticulocyte maturation, and intracellular amino acids appearing in these cells are believed to come from protein degradation during reticulocyte maturation (Tucker and Young, 1980). When RBCs are separated by age, using density gradients, intracellular GSH decreases and HzB numbers increase in older RBCs from system C-deficient sheep but not from normal or GCS-deficient sheep (Tucker and Young, 1980). This decreasing GSH with RBC age may explain why system C-deficient sheep are more susceptible to oxidants than are GCS-deficient ones, even though they have similar whole blood GSH levels.

About 30% of thoroughbred horses and 3% of ponies are deficient in a similar amino acid transporter. The lesion results in increased amino acid levels and GSH deficiency in some cases (Fincham *et al.*, 1985).

2. High Membrane Na^+, K^+-ATPase Activity in Dogs

Although dog reticulocytes have considerable membrane Na^+, K^+-ATPase (Na^+, K^+-pump) activity, it is rapidly lost during maturation into mature RBCs in most dogs (Maede and Inaba, 1985). Consequently, RBCs from most dogs have low potassium concentrations owing to the absence Na^+, K^+-pump activity (Parker, 1977). However, some Japanese mongrel dogs, some Japanese Akita dogs, and some Japanese Shiba dogs have HK^+ RBCs, because the Na^+, K^+-pump is retained in mature RBCs (Degen, 1987; Maede *et al.*, 1983, 1991). This trait is inherited in an autosomal recessive manner. RBCs from these dogs also have high GSH, glutamate, glutamine, and aspartate concentrations secondary to increased glutamate and aspartate uptake. GSH concentration is increased 5–7 times normal because the feedback inhibition of GCS by GSH is released by the approximately 90 times normal glutamate concentration (Maede *et al.*, 1982).

Dog RBCs have a high affinity Na^+-dependent transport system for glutamate and aspartate (Young, 1983). The increased transport of these amino acids into RBCs of affected dogs apparently occurs as a consequence of the Na^+ and K^+ concentration gradients produced by the presence of a Na^+, K^+-ATPase activity three times higher than that of human RBCs (Inaba and Maede, 1984). The glycolytic rate of HK^+ dog RBCs is about twice that of LK^+ cells, because greater ATP production is required to provide energy for active cation transport by the Na^+, K^+-pump and for increased GSH synthesis (Maede and Inaba, 1987).

HK^+ dog RBCs have increased osmotic fragility, increased MCV, decreased MCHC, and normal MCH values, suggesting an increase in cell water (Maede *et al.*, 1983). Although dogs with HK^+ RBCs are not anemic, their RBCs have shortened lifespans (Maede and Inaba, 1987), and some dogs have slightly increased reticulocyte counts (Maede *et al.*, 1983). Clinically these dogs appear normal.

The high GSH concentration in the HK^+ dog RBCs provides increased protection against oxidative damage induced by acetylphenylhydrazine (Ogawa *et al.*, 1992), but increased susceptibility to oxidative damage induced by onions (Yamoto and Maede, 1992) and 4-aminophenyl disulfide (Maede *et al.*, 1989). Evidence suggests that the increased GSH concentration potentiates the generation of superoxide through its redox reaction with the aromatic disulfide. It is not known whether a similar redox reaction occurs between GSH and the thiosulfates present in onions (Yamato *et al.*, 1994).

A variant of this HK^+ RBC disorder has been reported that lacks GSH accumulation (Fujise *et al.*, 1993). It was suggested that these dogs had a defect in amino acid metabolism in addition to the persistence of the Na^+, K^+-pump.

3. *Hereditary Stomatocytosis in Dogs*

Stomatocytes are cup-shaped RBCs that have slitlike areas of central pallor on stained blood films. Stomatocytosis is recognized in association with three different inherited syndromes in dogs. All disorders appear to be transmitted as autosomal recessive traits.

No clinical signs occur in miniature schnauzers with stomatocytosis (Brown *et al.*, 1994b; Giger *et al.*, 1988a). Chondrodysplasia (short-limbed dwarfism) occurs along with stomatocytosis in Alaskan malamutes (Fletch *et al.*, 1975; Pinkerton *et al.*, 1974). This disorder is deforming but not life threatening. The syndrome in Drentse patrijshond dogs has been termed familial stomatocytosis–hypertrophic gastritis (Slappendel *et al.*, 1991). Affected animals have polysystemic disease with growth retardation, diarrhea, polyuria/polydipsia, hind limb weakness, pale and/or icteric mucous membranes, and a somnolent mental state. Pathologic findings include hypertrophic gastritis, progressive liver disease, polyneuropathy, and renal cysts. Affected Drentse patrijshond dogs are usually euthanized by the time they reach young adulthood because of a progressive deterioration in clinical condition.

Hb values and RBC counts are slightly reduced, but PCVs are normal in malamutes (Fletch *et al.*, 1975; Pinkerton *et al.*, 1974) and schnauzers (Brown *et al.*, 1994b; Giger *et al.*, 1988a). The MCV is markedly increased and MCHC moderately decreased even though reticulocyte counts are normal or only slightly increased. Affected Drentse patrijshond dogs have lower PCVs and higher reticulocyte counts than those found in the other breeds (Slappendel *et al.*, 1991, 1994). The MCHC is moderately decreased, but the MCV is normal or only slightly increased. RBCs from all breeds have increased osmotic fragility and shortened RBC survival.

The pathogenesis of stomatocyte formation in malamutes and schnauzers is attributed to an increase in monovalent cation, and consequently water content of RBCs caused by abnormal membrane permeability (Giger *et al.*, 1988a; Pinkerton *et al.*, 1974). The swelling associated with water accumulation accounts for the markedly increased MCV and decreased MCHC values. The GSH content in affected RBCs is about 50–60% of normal (Giger *et al.*, 1988a; Pinkerton *et al.*, 1974). This GSH deficiency appears to occur from increased catabolism, but the mechanism responsible is unknown (Smith *et al.*, 1983b).

In contrast to malamutes and schnauzers with stomatocytosis, RBCs from affected Drentse patrijshond dogs do not have increased total monovalent cations and cell water is only slightly increased; consequently stomatocyte formation appears to be caused by a different mechanism (Slappendel *et al.*, 1994). The composition of phospholipids and cholesterol in plasma and RBC membranes is abnormal in these dogs (Slappendel *et al.*, 1994). The authors suggest that a defect in lipid metabolism results in altered membrane lipid composition and a loss or contracture of membrane components.

Hereditary stomatocytosis appears to be a heterogeneous group of disorders in people as well (Pinkerton *et al.*, 1974). A deficiency of a specific band 7 integral membrane protein has been reported in people with hereditary stomatocytois associated with high RBC cation content. This band 7 protein may function in closing a latent ion channel (Eber *et al.*, 1989; Stewart, 1993).

4. *Hereditary Elliptocytosis in Dogs*

Persistent elliptocytosis and microcytosis has been described in a crossbred dog that lacked RBC membrane band 4.1 (Smith *et al.*, 1983a). Although the animal was not anemic, the reticulocyte count was about twice normal in compensation for a shortened RBC lifespan. This dog was an offspring of a father–daughter mating. Both parents had decreased band 4.1 and some elliptocytes. This disorder in dogs resembles that in human patients with band 4.1 deficiency (Dacie, 1985).

C. Miscellaneous Abnormalities

1. *Familial Nonspherocytic Hemolytic Anemia in Poodles*

A nonspherocytic hemolytic anemia occurs in poodles (Randolph *et al.*, 1986). An autosomal dominant with incomplete penetrance mode of transmission was suggested. Affected animals had severe persistent macrocytic hypochromic anemia (PCVs 13–31%) with marked reticulocytosis. As in dogs with PK deficiency, myelofibrosis, osteosclerosis, and excess iron deposition in hepatocytes and mononuclear phagocytes were observed at necropsy. One dog died when 2.5 years old with liver failure. Despite extensive studies, the defect in this disorder could not be determined.

2. *Hereditary Nonspherocytic Hemolytic Anemia in Beagles*

A mild hemolytic anemia with reticulocytosis, slightly increased RBC osmotic fragility, shortened RBC lifespan and normal RBC morphology has been reported in beagle dogs (Maggio-Price *et al.*, 1988). Studies of RBC enzymes, membrane protein electrophoresis and Hb failed to identify a defect. Decreased calcium pump ATPase activity (Hinds *et al.*, 1989) and accelerated RBC swelling under osmotic stress (Pekow *et al.*, 1992) have been reported in RBCs from anemic dogs, but these abnormalities may be the consequence of an unknown membrane defect, rather than representing primary abnormalities. Although the etiology remains elusive, this abnormality appears to be transmitted as an autosomal recessive trait.

3. *Familial Methemoglobinemia and Hemolytic Anemia in Horses*

Methemoglobinemia and mild to moderate hemolytic anemia have been reported in a trotter mare and her dam (Dixon and McPherson, 1977). The animals were examined because of poor performance. GSH values were about 50% of normal and GR activities were 25–50% that of normal RBCs. The GR deficiency was not the result of riboflavin deficiency. It is unclear how these abnormalities in GSH and GR are related to the pathogenesis of this disorder. In the absence of exogenous oxidants, the deficiencies do not seem severe enough to account for the hematologic abnormalities. No abnormal Hb types were present, and G6PD, GPx, and Cb_5R activities were normal.

4. *Familial Erythrocytosis in Cattle*

Marked erythrocytosis (PCVs 60–80%) has been described in calves from a highly inbred Jersey herd (Tennant *et al.*, 1967, 1969). Affected calves had normal Hb types and arterial blood gas values. As with normal calves, PCVs were within the adult range at birth, fell during the first month of life, and then increased during the next 2 months. In contrast to normal calves, however, the PCV continued to increase in affected calves until 6 to 7 months of age. The majority of affected calves died during this time. PCVs of surviving animals returned slowly to normal by maturity. Serum of affected calves lacked measurable EPO but contained a growth factor that appeared to enhance EPO activity *in vitro* (Van Dyke *et al.*, 1968). The pathogenesis of this disorder may involve an abnormally controlled increased RBC production in response to the anemia of the neonate (Tennant *et al.*, 1969). A variety of familial and congenital erythrocytosis syndromes have been described in people (Emanuel *et al.*, 1992). Most are associated with Hb mutations that result in increased oxygen affinity, but some appear to be associated with abnormal responses to EPO.

References

Abkowitz, J. L. (1991). *Blood* **77,** 1442–1451.

Abraham, N. G. (1991). *Blood Rev.* **5,** 19–28.

Accorsi, A., Fazi, A., Ninfali, P., Piatti, E., Palma, F., Piacentini, M. P., and Fornaini, G. (1985). *Comp. Biochem. Physiol. B* **80B,** 839–842.

Adams, L. G., Hardy, R. M., Weiss, D. J., and Bartges, J. W. (1993). *J. Vet. Intern. Med.* **7,** 266–271.

Agar, N. S., and Board, P. G. (1983a). "Red Blood Cells of Domestic Animals," Elsevier, Amsterdam.

Agar, N. S., and Board, P. G. (1983b). *In* "Red Blood Cells of Domestic Mammals" (N. S. Agar and P. G. Board, eds.), pp. 227–251. Elsevier, Amsterdam.

Agar, N. S., and Smith, J. E. (1973). *Anim. Blood Groups Biochem. Genet.* **4,** 133–140.

Agar, N. S., and Smith, J. E. (1974). *Int. J. Biochem.* **5,** 49–52.

Agar, N. S., Gruca, M., and Harley, J. D. (1974). *Aust. J. Exp. Biol. Med. Sci.* **52,** 607–614.

Agar, N. S., Harley, J. D., Gruca, M. A., and Roberts, J. (1977). *Experientia* **33,** 275–277.

Agar, N. S., Suzuki, T., Roberts, J., and Evans, J. V. (1983). *Comp. Biochem. Physiol. B* **75B,** 445–449.

Agar, N. S., Sadrzadeh, S. M. H., Hallaway, P. E., and Eaton, J. W. (1986). *J. Clin. Invest.* **77,** 319–321.

Agre, P., and Cartron, J. P. (1991). *Blood* **78,** 551–563.

Aisen, P. (1994). *Adv. Exp. Med. Biol.* **356,** 31–40.

Allen, D. W., and Jandl, J. H. (1961). *J. Clin. Invest.* **40,** 454–475.

Allen, D. W., and Manning, N. (1994). *Blood* **84,** 1283–1287.

Allison, A. C., Reese, W. A., and Burn, G. P. (1957). *Nature* **180,** 649–650.

Alsaker, R. D., Laber, J., Stevens, J. B., and Perman, V. (1977). *J. Am. Vet. Med. Assoc.* **170,** 39–41.

Anderson, P. H., Berrett, S., and Patterson, D. S. (1978). *J. Comp. Pathol.* **88,** 181–189.

Andrewartha, K. A., and Caple, I. W. (1980). *Res. Vet. Sci.* **28,** 101–104.

Andrews, G. A., Chavey, P. S., Smith, J. E., and Rich, L. (1992). *Blood* **79,** 2485–2491.

Appelbaum, F. R., and Fefer, A. (1981). *Semin. Hematol.* **18,** 241–257.

Arai, T., Washizu, T., Sako, T., Sasaki, M., and Motoyoshi, S. (1992). *Comp. Biochem. Physiol. A* **102A,** 285–287.

Arese, P., and De Flora, A. (1990). *Semin. Hematol.* **27,** 1–40.

Asakura, T., Sato, Y., Minakami, S., and Yoshikawa, H. (1966). *J. Biochem.* **5,** 524–526.

Atkins, C. E., Kaneko, J. J., and Congdon, L. L. (1981). *J. Am. Anim. Hosp. Assoc.* **17,** 829–832.

Auer, L., and Bell, K. (1983). *Res. Vet. Sci.* **35,** 145–152.

Aufderheide, W. M., Parker, H. R., and Kaneko, J. J. (1980). *Comp. Biochem. Physiol. A* **65A,** 393–398.

Bababunmi, E. A., Olorunsogo, O. O., and Bewaji, C. O. (1991). *World Rev. Nutr. Diet.* **64,** 109–138.
Baerlocher, G. M., Schlappritzi, E., Straub, P. W., and Reinhart, W. H. (1994). *Br. J. Haematol.* **86,** 629–634.
Bai, S. C., Sampson, D. A., Morris, J. G., and Rogers, Q. R. (1989). *J. Nutr.* **119,** 1020-1027.
Bailey, E. (1982). *Am. J. Vet. Res.* **43,** 1917–1921.
Baldwin, S. A. (1993). *Biochim. Biophys. Acta* **1154,** 17–49.
Barker, J. E., Pierce, J. E., and Nienhuis, A. W. (1980). *Blood* **56,** 488–494.
Battaglia, F. C., McGaughey, H., Makowski, E. L., and Meschia, G. (1970). *Am. J. Physiol.* **219,** 217–221.
Battistini, A., Coccia, E.-M., Marziali, G., Bulgarini, D., Scalzo, S., Fiorucci, G., Romeo, G., Affabris, E., Testa, U., Rossi, G. B., and Peschle, C. (1991). *Blood* **78,** 2098–2103.
Bauer, J. E., Harvey, J. W., Asquith, R. L., McNulty, P. K., and Kivipelto, J. (1984). *Equine Vet. J.* **16,** 361–363.
Beilstein, M. A., and Whanger, P. D. (1983). *J. Nutr.* **113,** 2138–2146.
Bell, K. (1983). *In* "Red Blood Cells of Domestic Mammals" (N. S. Agar and P. G. Board, eds.), pp. 133–164. Elsevier, Amsterdam.
Bellingham, A. J., Detter, J. C., and Lenfant, C. (1971). *J. Clin. Invest.* **50,** 700–706.
Benesch, R., Benesch, R. E., and Bauer, C. (1975). *In* "The Red Blood Cell" (D. M. Surgenor, ed.), 2nd ed., pp. 825–839. Academic Press, New York.
Beppu, M., Takahashi, T., Hayashi, T., and Kikugawa, K. (1994). *Biochim. Biophys. Acta Mol. Cell Res.* **1223,** 47–56.
Berggren, P. C. (1981). *J. Am. Vet. Med. Assoc.* **179,** 1400–1402.
Bernard, J. (1991). *Blood Cells* **17,** 5–14.
Bernard, S. L., Leathers, C. W., Brobst, D. F., and Gorham, J. R. (1983). *Am. J. Vet. Res.* **44,** 657–661.
Bessis, M. (1973). "Living Blood Cells and Their Ultrastructure," Springer-Verlag, New York.
Beutler, E. (1984). "Red Cell Metabolism. A Manual of Biochemical Methods," 3rd ed. Grune & Stratton, Orlando, Florida.
Beutler, E. (1985). *Br. J. Haematol.* **61,** 377–384.
Beutler, E. (1989). *Ann. Rev. Nutr.* **9,** 287–302.
Beutler, E. (1994). *Blood* **84,** 3613–3636.
Beutler, E. (1995). *In* "Williams Hematology" (E. Beutler, M. A. Lichtman, B. S. Coller, and T. J. Kipps, eds.), 5th ed., pp. 364–369. McGraw-Hill, New York.
Black, J. A., Rittenberg, M. B., Standerfer, R. J., and Peterson, J. S. (1978). *Prog. Clin. Biol. Res.* **21,** 275–290.
Blostein, R., and Grafova, E. (1990). *Am. J. Physiol. Cell Physiol.* **259,** C241–C250.
Blue, J., and Weiss, L. (1981). *Am. J. Anat.* **161,** 135–168.
Blunt, M. H. (1972). *Anim. Blood Groups Biochem. Genet.* **3,** 211–217.
Blunt, M. H., Kitchens, J. L., Mayson, S. M., and Huisman, T. H. J. (1971). *Proc. Soc. Exp. Biol. Med.* **138,** 800–803.
Board, P. G., and Agar, N. S. (1983). *In* "Red Blood Cells of Domestic Mammals" (N. S. Agar and P. G. Board, eds.), pp. 253–270. Elsevier, Amsterdam.
Board, P. G., Smith, J. E., Moore, K., and Ou, D. (1980). *Biochim. Biophys. Acta* **613,** 534–541.
Bodansky, O. (1951). *Pharmacol. Rev.* **3,** 144–196.
Bolis, L. (1973). *In* "Comparative Physiology" (L. Bolis, K. Schmidt-Neilsen, and S. H. P. Maddrell, eds.), pp. 583–590. North-Holland Publishing, Amsterdam.
Bosman, G. J. C. G. M., and Kay, M. M. B. (1990). *Biochem. Cell Biol.* **68,** 1419–1427.
Bottomley, S. S. (1991). *Hosp. Pract.* **26,** suppl 3,37–40.
Boucher, J. H., Ferguson, E. W., Wilhelmsen, C. L., Statham, N., and McMeekin, R. R. (1981). *J. Appl. Physiol.* **51,** 131–134.
Bowling, A. T., and Clark, R. S. (1985). *Anim. Blood Groups Biochem. Genet.* **16,** 93–108.
Braend, M. (1988). *J. Hered.* **79,** 211–212.
Brechue, W. F., Gropp, K. E., Ameredes, B. T., O'Drobinak, D. M., Stainsby, W. N., and Harvey, J. W. (1994). *J. Appl. Physiol.* **77,** 2456–2467.
Breitschwerdt, E. B., Armstrong, P. J., Robinette, C. L., Dillman, R. C., and Karl, M. L. (1986). *Vet. Hum. Toxicol.* **28,** 109–117.
Brewer, G. J. (1974). "The Red Cell" (D. M. Surgenor, ed.), 2nd ed., pp. 387–433. Academic Press, New York.
Brewer, N. R. (1987). *J. Am. Vet. Med. Assoc.* **190,** 654–658.
Brock, K. V., Jones, J. B., Shull, R. M., and Potgieter, L. N. D. (1989). *Am. J. Vet. Res.* **50,** 965–969.
Brown, D. E., Meyer, D. J., Wingfield, W. E., and Walton, R. M. (1994a). *Vet. Pathol.* **31,** 654–657.
Brown, D. E., Weiser, M. G., Thrall, M. A., Giger, U., and Just, C. A. (1994b). *Vet. Pathol.* **31,** 247–250.
Bücheler, J., and Giger, U. (1993). *Vet. Immunol. Immunopathol.* **38,** 283–295.
Buhles, W. C., Jr., Huxsoll, D. L., and Hildebrandt, P. K. (1975). *J. Comp. Pathol.* **85,** 511–521.
Bunn, H. F. (1971). *Science* **172,** 1049–1050.
Bunn, H. F. (1981). *Blood* **58,** 189–197.
Bunn, H. F. (1987). *Blood* **69,** 1–6.
Bunn, H. F., and Kitchen, H. (1973). *Blood* **42,** 471–479.
Bunn, H. F., Seal, U. S., and Scott, A. F. (1974). *Ann. NY Acad. Sci.* **241,** 498–512.
Burr, M. J. (1972). *Comp. Biochem. Physiol. B* **41B,** 687–694.
Burrows, G. E. (1980). *J. Am. Vet. Med. Assoc.* **177,** 82–83.
Bush, J. A., Jensen, W. N., Ashenbrucker, H., Cartwright, G. E., and Wintrobe, M. M. (1956). *J. Exp. Med.* **103,** 161–171.
Campbell, A. D. (1992). *Hematol. Pathol.* **6,** 51–60.
Campbell, A. D., and Wicha, M. S. (1988). *J. Lab. Clin. Med.* **112,** 140–146.
Canfield, P. J., and Watson, A. D. J. (1989). *J. Comp. Pathol.* **101,** 269–278.
Canfield, P. J., Watson, A. D. J., and Ratcliffe, R. C. C. (1987). *Vet. Clin. Pathol.* **16(1),** 21–28.
Caple, I. W., Edwards, S. J. A., Forsyth, W. M., Whiteley, P., Selth, R. H., and Fulton, L. J. (1978). *Aust. Vet. J.* **54,** 57–60.
Card, R. T., and Brain, M. C. (1973). *N. Engl. J. Med.* **288,** 388–392.
Carrell, R. W., Winterbourn, C. C., and French, J. K. (1977). *Haemoglobin* **1,** 815–827.
Carrell, R. W., Krishnamoorthy, R., and Winterbourn, C. C. (1978). *Prog. Clin. Biol. Res.* **21,** 687–695.
Cartwright, G. E., and Deiss, A. (1975). *N. Engl. J. Med.* **292,** 185–193.
Carvalho da Silva, A., Fajer, A. B., De Angelis, R. C., Pontes, M. A., Giesbrecht, A. M., and Fried, R. (1959). *J. Nutr.* **68,** 213–229.
Chamberlain, J. K., Weiss, L., and Weed, R. I. (1975). *Blood* **46,** 91–102.
Chan, T. K., Chan, W. C., and Weed, R. I. (1982). *Br. J. Haematol.* **50,** 575–582.
Chanarin, I., Deacon, R., Lumb, M., Muir, M., and Perry, J. (1985). *Blood* **66,** 479–489.
Chandler, F. W., Prasse, K. W., and Callaway, C. S. (1975). *Am. J. Vet. Res.* **36,** 1477–1480.
Chapman, B. L., and Giger, U. (1990). *J. Small. Anim. Pract.* **31,** 610–616.
Chasis, J. A., and Mohandas, N. (1992). *Blood* **80,** 1869–1879.
Chasis, J. A., Prenant, M., Leung, A., and Mohandas, N. (1989). *Blood* **74,** 1112–1120.
Chasseaud, L. F. (1979). *Adv. Cancer Res.* **29,** 175–274.
Chausow, D. G., and Czarnecki-Maulden, G.L. (1987). *J. Nutr.* **117,** 928–932.
Chemtob, S., Gibb, W., and Bard, H. (1980). *Biol. Neonate* **38,** 36–39.

Chen, L.-T., and Weiss, L. (1973). *Blood* **41,** 529–537.
Chiu, D., and Lubin, B. (1989). *Semin. Hematol.* **26,** 128–135.
Christian, J. A., Rebar, A. H., Boon, G. D., and Low, P. S. (1993). *Blood* **82,** 3469–3473.
Christopher, M. M. (1989). *J. Am. Vet. Med. Assoc.* **194,** 1089–1095.
Christopher, M. M., and Lee, S. E. (1994). *Vet. Clin. Pathol.* **23,** 7–12.
Christopher, M. M., Perman, V., and Eaton, J. W. (1989). *J. Am. Vet. Med. Assoc.* **194,** 1045–1056.
Christopher, M. M., Broussard, J. D., and Peterson, M. E. (1995). *J. Vet. Intern. Med.* **9,** 24–31.
Cohen, C. M., and Gascard, P. (1992). *Semin. Hematol.* **29,** 244–292.
Coldman, M. F., and Good, W. (1967). *Comp. Biochem. Physiol.* **21,** 201-206.
Comline, R. S., and Silver, M. (1974). *J. Physiol.* **242,** 805–826.
Contreras, A., Martinez, R., Deves, R., and Marusic, E. T. (1986). *Biochim. Biophys. Acta* **856,** 388–391.
Cooper, R. A., Diloy-Puray, M., Lando, P., and Greenberg, M. S. (1972). *J. Clin. Invest.* **51,** 3182–3192.
Cooper, R. A., Leslie, M. H., Knight, D., and Detweiler, D. K. (1980). *J. Lipid Res.* **21,** 1082–1089.
Coulter, D. B., and Small, L. L. (1973). *Cornell Vet.* **63,** 462–468.
Crafts, R. C. (1948). *Blood* **3,** 276–285.
Craik, J. D., Good, A. H., Gottschalk, R., Jarvis, S. M., Paterson, A. R., and Cass, C. E. (1988). *Biochem. Cell Biol.* **66,** 839–852.
Cramer, D. V., and Lewis, R. M. (1972). *J. Am. Vet. Med. Assoc.* **160,** 61–67.
Dacie, J. (1985). "The Haemolytic Anaemias. Vol 1. The Hereditary Haemolytic Anaemias," 3rd ed. Churchill Livingston, Edinburgh.
Davis, J. A., Greenfield, R. E., and Brewer, T. G. (1993). *Am. J. Vet. Res.* **54,** 1322–1326.
Dean, G. A., Groshek, P. M., Mullins, J. I., and Hoover, E. A. (1992). *J. Virol.* **66,** 5561–5568.
Debski, B., and Rynca, J. (1985). *Acta Physiol. Pol.* **36,** 345–351.
Degen, M. (1987). *J. Am. Vet. Med. Assoc.* **190,** 541–543.
Deiss, A., Kurth, D., Cartwright, G. E., and Wintrobe, M. M. (1966). *J. Clin. Invest.* **45,** 353–364.
Del Boccio, G., Casalone, E., Sacchetta, P., Pennelli, A., and Di Ilio, C. (1986). *Biochem. Med. Metab. Biol.* **36,** 306–312.
Dessypris, E. N. (1991). *Semin. Hematol.* **28,** 275–284.
Dhindsa, D. S., and Metcalfe, J. (1974). *Resp. Physiol.* **21,** 37–46.
Dhindsa, D. S., Hoversland, A. S., Neill, W. A., and Metcalfe, J. (1971). *Resp. Physiol.* **11,** 346–353.
Dhindsa, D. S., Hoversland, A. S., and Templeton, J. W. (1972). *Biol. Neonate* **20,** 226–235.
Dhindsa, D. S., Black, J. A., Koler, R. D., Rigas, D. A., Templeton, J. W., and Metcalfe, J. (1976). *Resp. Physiol.* **26,** 65–75.
Dimmock, C. K., and Bell, K. (1970). *Aust. Vet. J.* **46,** 44–47.
Dimmock, C. K., Clark, I. A., and Hill, M. W. M. (1976). *Res. Vet. Sci.* **20,** 244–248.
Dixon, P. M., and McPherson, E. A. (1977). *Equine Vet. J.* **9,** 198–201.
Dodds, W. J. (1983). *Adv. Vet. Sci. Comp. Med.* **27,** 163–196.
Doong, G., Keen, C. L., Rogers, Q. R., Morris, J. G., and Rucker, R. B. (1983). *J. Nutr.* **113,** 1963–1971.
Dornsife, R. E., Gasper, P. W., Mullins, J. I., and Hoover, E. A. (1989). *Exp. Hematol.* **17,** 138–144.
Doyle, M. P., Pickering, R. A., Dykstra, R. L., Nelson, C. L., and Boyer, R. F. (1982). *Biochem. Biophys. Res. Commun.* **105,** 127–132.
Duhm, J. (1974). *Biochim. Biophys. Acta* **343,** 89–100.
Duhm, J. (1975). *Biochim. Biophys. Acta* **385,** 68–80.
Dunn, J. K., Hirsch, V. M., and Searcy, G. P. (1984). *J. Am. Anim. Hosp. Assoc.* **20,** 999–1002.
Durando, M. M., Alleman, A. R., and Harvey, J. W. (1994). *Equine Vet. J.* **26,** 83–85.
Duthie, G. G., Arthur, J. R., Nicol, F., and Walker, M. (1989). *Res. Vet. Sci.* **46,** 226–230.
Eaton, J. W. (1991). *J. Lab. Clin. Med.* **118,** 3–4.
Eaton, J. W., and Brewer, G. J. (1974). *In* "The Red Blood Cell" (D. M. Surgenor, ed.), 2nd ed., pp. 436–471. Academic Press, New York.
Eaton, J. W., Hallaway, P. E., and Agar, N. S. (1989). *Prog. Clin. Biol. Res.* **319,** 23–38.
Eber, S. W., Lande, W. M., Iarocci, T. A., Mentzer, W. C., Höhn, P., Wiley, J. S., and Schröter, W. (1989). *Br. J. Haematol.* **72,** 452–455.
Eberlein, K., and Gercken, G. (1971). *J. Chromatogr.* **61,** 285–293.
Eldor, A., Hershko, C., and Bruchim, A. (1978). *J. Am. Vet. Med. Assoc.* **173,** 304–305.
Elin, R. J., and Alling, D. W. (1978). *J. Lab. Clin. Med.* **91,** 666–672.
Ellory, J. C., Tucker, E. M., and Deverson, E. V. (1972). *Biochim. Biophys. Acta* **279,** 481–483.
Ellory, J. C., and Tucker, E. M. (1983). *In* "Red Blood Cells of Domestic Mammals" (N. S. Agar and P. G. Board, eds.), pp. 291–314. Elsevier, Amsterdam.
Emanuel, P. D., Eaves, C. J., Broudy, V. C., Papayannopoulou, T., Moore, M. R., D'Andrea, A. D., Prchal, J. F., Eaves, A. C., and Prchal, J. T. (1992). *Blood* **79,** 3019–3030.
Engen, R. L., and Clark, C. L. (1990). *Am. J. Vet. Res.* **51,** 577–580.
Erickson, N., and Quesenberry, P. J. (1992). *Med. Clin. North Am.* **76,** 745–755.
Erslev, A. J. (1990). *Leuk. Res.* **14,** 683–688.
Erslev, A. J., and Beutler, E. (1995). *In* "Williams Hematology" (E. Beutler, M. A. Lichtman, B. S. Coller, and T. J. Kipps, eds.), 5th ed., pp. 425–441. McGraw-Hill, New York.
Estrov, Z., Halperin, D. S., Coceani, F., and Freedman, M. H. (1988). *Br. J. Haematol.* **69,** 321–327.
Evans, E. T. R. (1968). *Nature* **217,** 74–75.
Fan, L. C., Dorner, J. L., and Hoffman, W. E. (1978). *J. Am. Anim. Hosp. Assoc.* **14,** 219–224.
Feldman, B. F., Kaneko, J. J., and Farver, T. B. (1981). *Am. J. Vet. Res.* **42,** 1109–1113.
Feo, C., and Mohandas, N. (1977). *Nature* **265,** 166–168.
Fernandez, F. R., Davies, A. P., Teachout, D. J., Krake, A., Christopher, M. M., and Perman, V. (1984). *J. Am. Anim. Hosp. Assoc.* **20,** 711–720.
Fertman, M. H., and Fertman, M. B. (1955). *Medicine* **34,** 131–192.
Fincham, D. A., Young, J. D., Mason, D. K., Collins, E. A., and Snow, D. H. (1985). *Res. Vet. Sci.* **38,** 346–351.
Fincham, D. A., Mason, D. K., Paterson, J. Y., and Young, J. D. (1987). *J. Physiol. (London)* **389,** 385–409.
Fincham, D. A., Mason, D. K., and Young, J. D. (1988). *Biochim. Biophys. Acta* **937,** 184–194.
Finco, D. R., Duncan, J. R., Schall, W. D., and Prasse, K. W. (1975). *J. Am. Vet. Med. Assoc.* **166,** 469–472.
Fingeroth, J. M., and Smeak, D. D. (1988). *J. Am. Anim. Hosp. Assoc.* **24,** 175–182.
Fischer, T. M. (1986). *Biochim. Biophys. Acta* **861,** 277–286.
Fischer, T. M. (1988). *Blood Cells* **13,** 377–394.
Fischer, T. M., Meloni, T., Pescarmona, G. P., and Arese, P. (1985). *Br. J. Haematol.* **59,** 159–169.
Fisher, J. W., Samuels, A. I., and Malgor, L. A. (1971). *Israel J. Med. Sci.* **7,** 892–900.
Fisher, T. J., Tucker, E. M., and Young, J. D. (1986). *Biochim. Biophys. Acta* **884,** 211–214.
Fletch, S. M., Pinkerton, P. H., and Brueckner, P. J. (1975). *J. Am. Anim. Hosp. Assoc.* **11,** 353–361.
Ford, S., Giger, U., Duesberg, C., Beutler, E., and Wang, P. (1992). *J. Vet. Intern. Med.* **6,** 123.

Fox, L. E., Ford, S., Alleman, A. R., Homer, B. L., and Harvey, J. W. (1993). *Vet. Clin. Pathol.* **22,** 89–92.

Franken, P., and Schotman, J. H. (1977). *Tijdschr. Diergeneeskd.* **102,** 1197–1203.

Freeman, B. A., and Crapo, J. D. (1982). *Lab. Invest.* **47,** 412–426.

French, J. K., Winterbourn, C. C., and Carrell, R. W. (1978). *Biochem. J.* **173,** 19–26.

French, T. W., Fox, L. E., Randolph, J. F., and Dodds, W. J. (1987). *J. Am. Vet. Med. Assoc.* **190,** 437–439.

Friedemann, H., and Rapoport, S. M. (1974). *In* "Cellular and Molecular Biology of Erythrocytes" (H. Yoshikawa and S. M. Rapoport, eds.), pp. 181–249. University Park Press, Baltimore.

Frieden, E. (1983). *Semin. Hematol.* **20,** 114–117.

Fronticelli, C. (1990). *Biophys. Chem.* **37,** 141–146.

Fujii, S., Dale, G. L., and Beutler, E. (1984). *Blood* **63,** 1096–1101.

Fujise, H., Mori, M., Ogawa, E., and Maede, Y. (1993). *Am. J. Vet. Res.* **54,** 602–606.

Fulton, R., Weiser, M. G., Freshman, J. L., Gasper, P. W., and Fettman, M. J. (1988). *Vet. Pathol.* **25,** 521–523.

Furugouri, K. (1972). *J. Anim. Sci.* **34,** 421–426.

Fyfe, J. C., Jezyk, P. F., Giger, U., and Patterson, D. F. (1989). *J. Am. Anim. Hosp. Assoc.* **25,** 533–539.

Fyfe, J. C., Giger, U., Hall, C. A., Jezyk, P. F., Klumpp, S. A., Levine, J. S., and Patterson, D. F. (1991). *Pediatr. Res.* **29,** 24–31.

Gaetani, G. F., Kirkman, H. N., Mangerini, R., and Ferraris, A. M. (1994). *Blood* **84,** 325–330.

Gallego, C., and Carreras, J. (1990). *Mol. Cell. Biochem.* **99,** 21–24.

Gardner, D. E., Martinovich, D., and Woodhouse, D. A. (1976). *N. Z. Vet. J.* **24,** 117–122.

Garnier, M., de Preville, G., Pilardeau, P., and Boudia, D. (1984). *Comp. Biochem. Physiol. A* **77A,** 315–317.

Gaunt, S. D., and Pierce, K. R. (1986). *Am. J. Vet. Res.* **47,** 906–909.

Gaunt, S. D., Baker, D. C., and Green, R. A. (1981). *Am. J. Vet. Res.* **42,** 1982–1984.

Geor, R. J., Lund, E. M., and Weiss, D. J. (1993). *J. Am. Vet. Med. Assoc.* **202,** 976–980.

George, J. W., and Duncan, J. R. (1979). *Vet. Clin. Pathol.* **8,** 23–30.

George, L. W., Divers, T. J., Mahaffey, E. A., and Suarez, J. H. (1982). *Vet. Pathol.* **19,** 521–533.

Giger, U., and Akol, K. G. (1990). *J. Vet. Intern. Med.* **4,** 315–316.

Giger, U., and Bücheler, J. (1991). *J. Am. Vet. Med. Assoc.* **198,** 411–418.

Giger, U., and Harvey, J. W. (1987). *J. Am. Vet. Med. Assoc.* **191,** 453–459.

Giger, U., and Noble, N. A. (1991). *J. Am. Vet. Med. Assoc.* **198,** 1755–1761.

Giger, U., Harvey, J. W., Yamaguchi, R. A., McNulty, P. K., Chiapella, A., and Beutler, E. (1985). *Blood* **65,** 345–351.

Giger, U., Amador, A., Meyers-Wallen, V., and Patterson, D. F. (1988a). *Proc. ACVIM Forum,* p. 754.

Giger, U., Argov, Z., Schnall, M., Bank, W. J., and Chance, B. (1988b). *Muscle Nerve* **11,** 1260–1265.

Giger, U., Bucheler, J., and Patterson, D. F. (1991a). *J. Hered.* **82,** 15–20.

Giger, U., Mason, G. D., and Wang, P. (1991b). *Vet. Clin. Pathol.* **20,** 83–86.

Giger, U., Smith, B., Griot-Wenk, M., Raipurohit, Y., McCully, K., Haskins, M., and Stedman, H. (1991c). *Blood* **78,** Suppl 1, 365a.

Giger, U., Smith, B. F., Woods, C. B., Patterson, D. F., and Stedman, H. (1992). *J. Am. Vet. Med. Assoc.* **201,** 1569–1571.

Giger, U., Gelens, C. J., Callan, M. B., and Oakley, D. A. (1995a). *J. Am. Vet. Med. Assoc.* **206,** 1358–1362.

Giger, U., Smith, B. F., and Rajpurohit, Y. (1995b). *Proc. ACVIM Forum,* p. 1002.

Gillis, D. B., and Mitchell, R. A. (1974). *Am. J. Vet. Res.* **35,** 31–33.

Gillis, G. H., and Anastassiadis, P. A. (1985). *Comp. Biochem. Physiol. B* **81B,** 131–135.

Goodman, J. R., and Dallman, P. R. (1969). *Blood* **34,** 747–753.

Goto, I., Agar, N. S., and Maede, Y. (1992). *Jpn. J. Vet. Res.* **40,** 99–104.

Goto, I., Agar, N. S., and Maede, Y. (1993). *Am. J. Vet. Res.* **54,** 622–626.

Green, M. J., and Hill, H. A. (1984). *In* "Methods in Enzymology" (L. Packer, ed.), pp. 3–22. Academic Press, New York.

Greenhalgh, J. F. D., Sharman, G. A. M., and Aitken, J. N. (1969). *Res. Vet. Sci.* **10,** 64–72.

Gruber, W., and Deuticke, B. (1973). *J. Membr. Biol.* **13,** 19–36.

Gupta, J. D., Peterson, V. J., and Harley, J. D. (1974). *Comp. Biochem. Physiol. A* **47A,** 1123–1126.

Gustin, P., Detry, B., Cao, M. L., Chenut, F., Robert, A., Ansay, M., Frans, A., and Clerbaux, T. (1994). *J. Appl. Physiol.* **77,** 202–208.

Haas, M. (1989). *In* "Red Blood Cell Membranes" (P. Agre and J. C. Parker, eds.), pp. 663–690. Marcel Dekker, New York.

Halvorsen, K., and Halvorsen, S. (1974). *Pediatr. Res.* **8,** 176–183.

Ham, T. H., Grauel, J. A., Dunn, R. F., Murphy, J. R., White, J. G., and Kellermeyer, R. W. (1973). *J. Lab. Clin. Med.* **82,** 898–910.

Hamasaki, N., Asakura, T., and Minakami, S. (1970). *J. Biochem.* **68,** 157–161.

Hammond, E., Deiss, A., Carnes, W. H., and Cartwright, G. E. (1969). *Lab. Invest.* **21,** 292–297.

Harkness, D., Osta, S., Roth, S., and Grayson, V. (1970). *Int. J. Biochem.* **1,** 575–581.

Harkness, D. R., Ponce, J., and Grayson, V. (1969). *Comp. Biochem. Physiol.* **28,** 129–138.

Harris, S. J., and Winzor, D. J. (1990). *Biochim. Biophys. Acta Protein Struct. Mol. Enzymol.* **1038,** 306–314.

Harrison, M. L., Rathinavelu, P., Arese, P., Geahlen, R. L., and Low, P. S. (1991). *J. Biol. Chem.* **266,** 4106–4111.

Harvey J. W. (1994). Unpublished studies.

Harvey, J. W., and Beutler, E. (1982). *Blood* **60,** 1227–1230.

Harvey, J. W., and Kaneko, J. J. (1975a). *Comp. Biochem. Physiol. B* **52B,** 507–510.

Harvey, J. W., and Kaneko, J. J. (1975b). *Am. J. Vet. Res.* **36,** 1511–1513.

Harvey, J. W., and Kaneko, J. J. (1976a). *Br. J. Haematol.* **32,** 199–209.

Harvey, J. W., and Kaneko, J. J. (1976b). *J. Cell Physiol.* **89,** 219–224.

Harvey, J. W., and Kaneko, J. J. (1977). *Toxicol. Appl. Pharmacol.* **42,** 253–261.

Harvey, J. W., and Keitt, A. S. (1983). *Br. J. Haematol.* **54,** 29–41.

Harvey, J. W., and Kornick, H. P. (1976). *J. Am. Vet. Med. Assoc.* **169,** 327–331.

Harvey, J. W., and Rackear, D. (1985). *Vet. Pathol.* **22,** 387–392.

Harvey, J. W., and Reddy, G. R. (1989). *Blood* **74,** 2556–2561.

Harvey, J. W., and Smith, J. E. (1994). *Comp. Haematol. Int.* **4,** 70–74.

Harvey, J. W., Ling, G. V., and Kaneko, J. J. (1974). *J. Am. Vet. Med. Assoc.* **164,** 1030–1033.

Harvey, J. W., Kaneko, J. J., and Hudson, E. B. (1977). *Vet. Clin. Pathol.* **6,** 13–17.

Harvey, J. W., Sameck, J. H., and Burgard, F. J. (1979). *J. Am. Vet. Med. Assoc.* **175,** 1171–1175.

Harvey, J. W., French, T. W., and Meyer, D. J. (1982). *J. Am. Anim. Hosp. Assoc.* **18,** 946–960.

Harvey, J. W., Wolfsheimer, K. J., Simpson, C. F., and French, T. W. (1985). *Vet. Clin. Pathol.* **14**(1), 36–42.

Harvey, J. W., French, T. W., and Senior, D. F. (1986). *J. Am. Vet. Med. Assoc.* **189,** 1334–1335.

Harvey, J. W., Asquith, R. L., Sussman, W. A., and Kivipelto, J. (1987). *Am. J. Vet. Res.* **48,** 1348–1352.

Harvey, J. W., Sussman, W. A., and Pate, M. G. (1988). *Comp. Biochem. Physiol. B* **89B,** 105–107.

Harvey, J. W., Calderwood Mays, M. B., Gropp, K. E., and Denaro, F. J. (1990a). *Vet. Pathol.* **27,** 1–8.

Harvey, J. W., Peteya, D. J., and Kociba, G. J. (1990b). *Vet. Clin. Pathol.* **19,** 55–58.

Harvey, J. W., King, R. R., Berry, C. R., and Blue, J. T. (1991). *Comp. Haematol. Int.* **1,** 55–59.

Harvey, J. W., Gropp, K. E., and Bellah, J. R. (1992a). *In* "State of Art in Animal Clinical Biochemistry" (A. Ubaldi, ed.), pp. 79, Boehringer Mannheim, Parma.

Harvey, J. W., Pate, M. G., Mhaskar, Y., and Dunaway, G. A. (1992b). *J. Inher. Metab. Dis.* **15,** 747–759.

Harvey, J. W., Dahl, M., and High, M.E. (1994). *J. Am. Vet. Med. Assoc.* **205,** 1290–1291.

Hebbel, R. P. (1986). *J. Lab. Clin. Med.* **107,** 401–404.

Hebbel, R. P., and Eaton, J. W. (1989). *Semin. Hematol.* **26,** 136–149.

Hebbel R. P., Berger E. M., and Eaton J. W. (1980). *Blood* **55,** 969–974.

Henquell, L., LaCelle, P. L., and Honig, C. R. (1976). *Microvasc. Res.* **12,** 259–274.

Hickman, M. A., Rogers, Q. R., and Morris, J. G. (1990). *Am. J. Vet. Res.* **51,** 475–478.

Higgins, P. J., Garuck, R. L., and Bunn, H. F. (1982). *Diabetes* **31,** 743–748.

Hinds, T. R., and Vincenzi, F. F. (1986). *Proc. Soc. Exp. Biol. Med.* **181,** 542–549.

Hinds, T. R., Hammond, W. P., Maggio-Price, L., Dodson, R. A., and Vincenzi, F. F. (1989). *Blood Cells* **15,** 407–420.

Hirose, J., Kano, H., Kidani, Y., Iwamoto, H., and Hiromi, K. (1992). *Chem. Pharm. Bull. (Tokyo)* **40,** 506–508.

Hirsch, V., and Dunn, J. (1983). *J. Am. Anim. Hosp. Assoc.* **19,** 873–880.

Hjelle, J. J., and Grauer, G. F. (1986). *J. Am. Vet. Med. Assoc.* **188,** 742–746.

Hladky, S. B., and Rink, T. J. (1977). *In* "Membrane Transport in Red Cells" (V. L. Lew and J. C. Ellory, eds.), pp. 115–135. Academic Press, New York.

Hochstein, P., Kumar, K. S., and Forman, S. J. (1978). *Prog. Clin. Biol. Res.* **21,** 669–681.

Hoffman, R., Tong, J., Brandt, J., Traycoff, C., Bruno, E., McGuire, B. W., Gordon, M. S., McNiece, I., and Srour, E.F. (1993). *Stem. Cells Dayt.* **11,** suppl 2, 76–82.

Holland, C. T., Canfield, P. J., Watson, A. D. J., and Allan, G. S. (1991). *J. Vet. Intern. Med.* **5,** 151–159.

Holman, H. H., and Drew, S. M. (1964). *Res. Vet. Sci.* **5,** 274–285.

Holman, H. H., and Drew, S. M. (1966). *Vet. Rec.* **78,** 772–776.

Holter, P. H., Framstad, T., Aulie, A., Refsum, H. E., and Sjaastad, O. V. (1991). *Pediatr. Hematol. Oncol.* **8,** 1–11.

Hotston Moore, A., Day, M. J., and Graham, M. W. A. (1993). *Vet. Rec.* **132,** 414–415.

Houston, D. M., and Myers, S. L. (1993). *Vet. Hum. Toxicol.* **35,** 158–161.

Houwen, B. (1992). *Blood Cells* **18,** 167–186.

Hubler, M., Kaelin, S., Hagen, A., Fairburn, A., Canfield, P., and Ruesch, P. (1987). *J. Small. Anim. Pract.* **28,** 833–838.

Huebers, H., Csiba, E., Huebers, E., and Finch, C. A. (1985). *Proc. Natl. Acad. Sci. USA* **179,** 222–226.

Huisman, T. H. J., and Kitchen, J. (1968). *Am. J. Physiol.* **215,** 140–146.

Huisman, T. H. J., Lewis, J. P., Blunt, M. H., Adams, H. R., Miller, A., Dozy, A. M., and Boyd, E. M. (1969). *Pediatr. Res.* **3,** 189–198.

Hultquist, D. E., Slaughter, S.R., Douglas, R.H., Sannes, L.J., and Sahagian, G.G. (1978). *In* "Proceedings of the Fourth International Congress on Red Cell Metabolism and Function " (G. J. Brewer, ed.), pp. 199–211. Alan R. Liss, New York.

Hultquist, D. E., Xu, F., Quandt, K. S., Shlafer, M., Mack, C. P., Till, G. O., Seekamp, A., Betz, A. L., and Ennis, S. R. (1993). *Am. J. Hematol.* **42,** 13–18.

Iampietro, P. F., Burr, M. J., Fiorica, V., McKenzie, J. M., and Higgins, E. A. (1967). *J. Appl. Physiol.* **23,** 505–510.

Inaba, M., and Maede, Y. (1984). *J. Biol. Chem.* **259,** 312–317.

Inaba, M., and Maede, Y. (1986). *J. Biol. Chem.* **261,** 16099–16105.

Isaacks, R. E., and Harkness, D. R. (1983). *In* "Red Blood Cells of Domestic Mammals," (N. S. Agar and P. G. Board, eds.), pp. 315–337. Elsevier, Amsterdam.

Israel, Y., MacDonald, A., Bernstein, J., and Rosenmann, E. (1972). *J. Gen. Physiol.* **59,** 270–284.

Jacob, H. S., Yawata, Y., Craddock, P., Hebbel, R., Howe, R., and Silvis, S. (1973). *Trans. Assoc. Am. Phys.* **86,** 143–153.

Jacobasch, G., Minakami, S., and Rapoport, S. M. (1974). *In* "Cellular and Molecular Biology of Erythrocytes" (H. Yoshikawa and S. M. Rapoport, eds.), pp. 55–92. University Park Press, Baltimore.

Jaffe, E. R. (1986). *Blood Cells* **12,** 81–90.

Jain, N. C. (1986). "Schalm's Veterinary Hematology," 4th ed. Lea & Febiger, Philadelphia, Pennsylvania.

Jain, N. C., and Kono, C. S. (1977). *Am. J. Vet. Res.* **38,** 983–990.

Jain, N. C., Kono, C. S., Myers, A., and Bottomly, K. (1980). *Res. Vet. Sci.* **28,** 25–35.

Jandl, J. H. (1987). "Blood: Textbook of Hematology," Little Brown and Company, Boston.

Jarolim, P., Lahav, M., Liu, S.-C., and Palek, J. (1990). *Blood* **76,** 2125–2131.

Jarvis, S. M., and Young, J. D. (1982). *J. Physiol. (London)* **324,** 47–66.

Johnstone, R. M. (1992). *Biochem. Cell Biol.* **70,** 179–190.

Jubb, T. F., Jerrett, I. V., Browning, J. W., and Thomas, K. W. (1990). *Aust. Vet. J.* **67,** 86–89.

Juneja, H. S., and Gardner, F. H. (1985). *Exp. Hematol.* **13,** 194–199.

Kaneko, J. J. (1980). *In* "Clinical Biochemistry of Domestic Animals" (J. J. Kaneko, ed.), 3rd ed., pp. 119–174. Academic Press, New York.

Kaneko, J. J. (1987). *Clin. Chim. Acta* **165,** 1–19.

Kaneko, J. J., and Mattheeuws, D. R. (1966). *Am. J. Vet. Res.* **27,** 923–929.

Kay, M. M. B. (1989). *Prog. Clin. Biol. Res.* **319,** 199–217.

Kay, M. M. B., Flowers, N., Goodman, J., and Bosman, G. (1989). *Proc. Natl. Acad. Sci. USA* **86,** 5834–5838.

Kerr, L. A., and McGavin, H. D. (1991). *J. Am. Vet. Med. Assoc.* **198,** 99–101.

Khan, K. N. M., Kociba, G. J., Wellman, M. L., and Reiter, J. A. (1992). *Exp. Hematol.* **20,** 900–903.

Kim, H. D. (1983). *In* "Red Blood Cells of Domestic Mammals" (N. S. Agar and P. G. Board, eds.), pp. 339–355. Elsevier, Amsterdam.

Kim, H. D., and Duhm, J. (1974). *Am. J. Physiol.* **226,** 1001–1006.

Kim, H. D., and Luthra, M. G. (1977). *J. Gen. Physiol.* **70,** 171–185.

Kim, H. D., and McManus, T. J. (1971). *Biochim. Biophys. Acta* **230,** 1–11.

Kim, H. D., Theg, B. E., and Lauf, P. K. (1980). *J. Gen. Physiol.* **76,** 109–121.

King, L. G., Giger, U., Diserens, D., and Nagode, L. A. (1992). *J. Vet. Intern. Med.* **6,** 264–270.

King, M. E., and Mifsud, C. V. (1981). *Res. Vet. Sci.* **31,** 37–39.

Kirk, J. H., and Bulgin, M. S. (1979). *Am. J. Vet. Res.* **40,** 397–399.

Kirk, R. G., Andrews, S. B., and Lee, P. (1983). *J. Membr. Biol.* **76,** 281–286.

Kirkman, H. N., and Gaetani, G. F. (1986). *J. Biol. Chem.* **261,** 4033–4038.

Kirkman, H. N., Galiano, S., and Gaetani, G. F. (1987). *J. Biol. Chem* **262,** 660–666.

Kitchen, H. (1969). *Adv. Vet. Sci. Comp. Med.* **13,** 247–329.

Kitchen, H. (1974). *Ann. N. Y. Acad. Sci.* **241,**12–24.

Kitchen, H., and Brett, I. (1974). *Ann. N. Y. Acad. Sci.* **241,** 653–671.

Kleiber, M. (1961). "The Fire of Life," Wiley, New York.

Kobayashi, K. (1981). *Feline Pract.* **11,** 22–27.

Kobylka, D., Khettry, A., Shin, B. C., and Carraway, K. L. (1972). *Arch. Biochem. Biophys.* **148,** 475–487.

Kociba, G. J., and Caputo, C. A. (1981). *J. Am. Vet. Med. Assoc.* **178,** 1293–1294.

Kosower, N. S. (1993). *Am. J. Hematol.* **42,** 241–247.

Koury, M. J., and Bondurant, M. C. (1988). *J. Cell Physiol.* **137,** 65–74.

Krake, A. C., Arendt, T. D., Teachout, D. J., Raffe, M. R., Christopher, M. M., Stowe, C. M., and Perman, V. (1985). *J. Am. Anim. Hosp. Assoc.* **21,** 527–534.

Kruckeberg, W. C., and Chilson, O. P. (1973). *Comp. Biochem. Physiol. B* **46B,** 653–660.

Kruse, H. D., Orent, E. R., and McCollum, E.V. (1933). *J. Biol. Chem.* **100,** 603–643.

Kurata, M., Suzuki, M., and Agar, N. S. (1993). *Comp. Biochem. Physiol. B* **106B,** 477–487.

Kurtzman, G. J., Platanias, L., Lustig, L., Frickhofen, N., and Young, N. S. (1989). *Blood* **74,** 71–81.

Kuypers, F. A., Lubin, B. H., Yee, M., Agre, P., Devaux, P. F., and Geldwerth, D. (1993). *Blood* **81,** 1051–1057.

Kwong, F. Y., Baldwin, S. A., Scudder, P. R., Jarvis, S. M., Choy, M. Y., and Young, J. D. (1986). *Biochem. J.* **240,** 349–356.

Laber, J., Perman, V., and Stevens, J. B. (1974). *J. Am. Anim. Hosp. Assoc.* **10,** 399–406.

Lagutchik, M. S., Mundie, T. G., and Martin, D .G. (1992). *J. Am. Vet. Med. Assoc.* **201,** 1407–1410.

Lahey, M. E., Gubler, C. J., Chase, M. S., Cartwright, G. E., and Wintrobe, M. M. (1952). *Blood* **7,** 1053–1074.

Lalanne, M., and Willemot, J. (1980). *Comp. Biochem. Physiol. B* **66B,** 367–372.

Landaw, S. A. (1988). *Blood Cells* **14,** 47–59.

Langdon, R. G., and Holman, V .P. (1988). *Biochim. Biophys. Acta* **945,** 23–32.

Laris, P. G. (1958). *J. Cell. Comp. Physiol.* **51,** 273–306.

Larsen, S., Flagstad, A., and Aalbaek, B. (1976). *Vet. Pathol.* **13,** 216–240.

Lavoie, J. P., Morris, D. D., Zinkl, J. G., Lloyd, K., and Divers, T. J. (1987). *J. Am. Vet. Med. Assoc.* **191,** 1462–1464.

Lee, C. K., Odell, G. V., Eliot, F. P., Anderson, I. L., and Jones, E. W. (1971). *Am. J. Vet. Res.* **32,** 1039–1044.

Lee, G. R., Cartwright, G. E., and Wintrobe, M. M. (1968). *Proc. Soc. Exp. Biol. Med.* **127,** 977–981.

Lee, P., Brown, M. E., and Hutzler, P. T. (1976). *Am. J. Vet. Res.* **37,** 561–565.

Leibold, E. A., and Guo, B. (1992). *Annu. Rev. Nutr.* **12,** 345–368.

Lenard, J. G. (1974). *Bull. Math. Biol.* **36,** 55–58.

Leng, R. A., and Annison, E. F. (1962). *Aust. J. Agric. Res.* **13,** 31–43.

Letchworth, G. J., Bentinck-Smith, J., Bolton, G. R., Wootton, J. F., and Family, L. (1977). *J. Am. Anim. Hosp. Assoc.* **13,** 75–79.

Lewis, H. B., and Rebar, A. H. (1979). "Bone Marrow Evaluation in Veterinary Practice," Ralston Purina Co., St. Louis.

Liard, J. F., and Kunert, M. P. (1993). *Am. J. Physiol. Regul. Integr. Comp. Physiol.* **264,** R396–R401.

Lichtman, M. A., and Santillo, P. (1986). *Blood Cells* **12,** 11–19.

Lincoln, S. D., Howell, M. E., Combs, J. J., and Hinman, D. D. (1992). *J. Am. Vet. Med. Assoc.* **200,** 1090–1094.

Linenberger, M. L., Shelton, G. H., Persik, M. T., and Abkowitz, J. L. (1991). *Blood* **78,** 1963–1968.

Liu, S.-C., and Derick, L. H. (1992). *Semin. Hematol.* **29,** 231–243.

Long, M. W. (1992). *Exp. Hematol.* **20,** 288–301.

Long, M. W., and Dixit, V. M. (1990). *Blood* **75,** 2311–2318.

Low, P. S. (1989). *In* "Red Blood Cell Membranes" (P. Agre and J. C. Parker, eds.), pp. 237–260, Marcel Dekker, Inc., New York.

Low, P. S. (1991). *Adv. Exp. Med. Biol.* **307,** 173–183.

Luther, D. G., Cox, H. U., and Nelson, W. O. (1985). *Am. J. Vet. Res.* **46,** 107–1079.

Luttgen, P. J. , Whitney, M. S., Wolf, A. M., and Scruggs, D .W. (1990). *J. Am. Vet. Med. Assoc.* **197,** 1347–1350.

Lutz, H. U., Stammler, P., Kock, D., and Taylor, R. P. (1991). *Adv. Exp. Med. Biol.* **307,** 367–376.

McClure, J. J., Koch, C., and Traub Dargatz, J. (1994). *Anim. Genet.* **25,** 119–120.

McCord, J. M., and Fridovich, I. (1969). *J. Biol. Chem.* **244,** 6049–6055.

McGillivray, S. R., Searcy, G. P., and Hirsch, V. M. (1985). *Can. J. Comp. Med.* **49,** 286–290.

McKibbin, J. M., Schaefer, A. E., Frost, D.V ., and Elvehjem, C. A. (1942). *J. Biol. Chem.* **142,** 77–84.

McManus, T. J. (1967). *Fed. Proc.* **26,** 1821–1826.

McSherry, B. J., Roe, C. K., and Milne, F. J. (1966). *Can. Vet. J.* **7,** 3–12.

MacWilliams, P. S., Searcy, G. P., and Bellamy, J. E. C. (1982). *Can. Vet. J.* **23,** 309–312.

Maede, Y., and Inaba, M. (1985). *J. Biol. Chem.* **260,** 3337–3343.

Maede, Y., and Inaba, M. (1987). *Am. J. Vet. Res.* **48,** 114–118.

Maede, Y., Kasai, N., and Taniguchi, N. (1982). *Blood* **59,** 883–889.

Maede, Y., Inaba, M., and Taniguchi, N. (1983). *Blood* **61,** 493–499.

Maede, Y., Hoshino, T., Inaba, M., and Namioka, S. (1987). *Am. J. Vet. Res.* **48,** 289–292.

Maede, Y., Kuwabara, M., Sasaki, A., Inaba, M., and Hiraoka, W. (1989). *Blood* **73,** 312–317.

Maede, Y., Amano, Y., Nishida, A., Murase, T., Sasaki, A., and Inaba, M. (1991). *Res. Vet. Sci.* **50,** 123–125.

Maggio-Price, L., Emerson, C. L., Hinds, T. R., Vincenzi, F. F., and Hammond, W. R. (1988). *Am. J. Vet. Res.* **49,** 1020–1025.

Magnani, M., Piatti, E., Dacha, M., and Fornaini, G. (1980). *Comp. Biochem. Physiol. B* **67B,** 139–142.

Magnani, M., Stocchi, V., Serafini, N., Piatti, E., Dacha, M., and Fornaini, G. (1983). *Arch. Biochem. Biophys.* **226,** 377–387.

Mahaffey, E., and Smith, J. E. (1975). *Int. J. Biochem.* **6,** 853–854.

Mahaffey, E .A., and Cornelius, L. M. (1982). *J. Am. Vet. Med. Assoc.* **180,** 635–637.

Mairbäurl, H. (1994). *Int. J. Sports Med.* **15,** 51–63.

Mansouri, A., and Lurie, A. A. (1993). *Am. J. Hematol.* **42,** 7–12.

Maral, J., Puget, K., and Michelson, A. M. (1977). *Biochem. Biophys. Res. Commun.* **77,** 1525–1535.

Marin, M. S., Fernandez, A., Sanchez-Yagüe, J., Cabezas, J. A., and Llanillo, M. (1990). *Biochimie* **72,** 745–750.

Marks, S. L., Mannella, C., and Schaer, M. (1990). *J. Am. Anim. Hosp. Assoc.* **26,** 629–634.

Maronpot, R. R. (1972). *Can. J. Comp. Med. Vet. Sci.* **36,** 55–60.

Martinovich, D., and Woodhouse, D. A. (1971). *N. Z. Vet. J.* **19,** 259–263.

Mason, R. P. (1982). *In* "Free Radicals in Biology" (W. A. Pryor, ed.), pp. 183–218. Academic Press, New York.

Mauk, A. G., and Taketa, F. (1972). *Arch. Biochem. Biophys.* **150,** 376–381.

Mayeux, P., Felix, J. M., Billat, C., and Jacquot, R. (1986). *Exp. Hematol.* **14,** 801–808.

Meberg, A. (1980). *Exp. Hematol.* **8,** 615–619.

Meberg, A., Haga, P., and Johansen, M. (1980). *Br. J. Haematol.* **45,** 569–574.

Medeiros, L. O., Nurmberger, R., Jr., and Medeiros, L. F. (1984). *Comp. Biochem. Physiol. B* **78B,** 869–871.

Melefors, Ö., and Hentze, M. W. (1993). *Blood Rev.* **7,** 251–258.

Metcalf, D. (1993). *Blood* **82,** 3515–3523.

Mgongo, F. O. K., Gombe, S., and Ogaa, J. S. (1981). *Vet. Rec.* **109,** 51–53.

Mhaskar, Y., Giger, U., and Dunaway, G. A. (1991). *Enzyme* **45,** 137–144.

Mhaskar, Y., Harvey, J. W., and Dunaway, G. A. (1992). *Comp. Biochem. Physiol. B* **101B,** 303–307.
Minakami, S., and Yoshikawa, H. (1966). *J. Biochem.* **59,** 139–144.
Minguell, J. J. (1993). *Exp. Hematol.* **21,** 7–8.
Miseta, A., Bogner, P., Berenyi, E., Kellermayer, M., Galambos, C., Wheatley, D. N., and Cameron, I. L. (1993). *Biochim. Biophys. Acta* **1175,** 133–139.
Mohandas, N., and Chasis, J. A. (1993). *Semin. Hematol.* **30,** 171–192.
Mollerberg, L., Ekman, L., and Jacobsson, S. O. (1975). *Acta Vet. Scand.* **16,** 205-217.
Mooney, N. A., and Young, J. D. (1978). *J. Physiol. (London)* **284,** 229–239.
Moore, M. A. S. (1991). *Blood* **78,** 1–19.
Morley, A., and Stohlman, F. (1969). *Science* **165,** 1025–1056.
Morra, L., Ponassi, A., Moccia, F., Mela, G. S., Ponassi, I., and Bessone, G. (1989). *Acta Haematol.* **82,** 16–21.
Morris, J. G. (1977). *In* "Kal Kan Symposium for the Treatment of Dog and Cat Diseases" pp. 15–18.
Morris, J. G., Cripe, W. S., Chapman, H. L., Walker, D. F., Armstrong, J. B., Alexander, J. D., Jr., Miranda, R., Sanchez, A., Jr., Sanchez, B., Blair-West, J. R., and Denton, D. A. (1984). *Science* **223,** 491–493.
Mueggler, P. A., and Black, J. A. (1982). *Am. J. Physiol.* **242,** H500–H506.
Mueggler, P. A., Peterson, J. S., Koler, R. D., Metcalfe, J., and Black, J. A. (1979). *Am. J. Physiol.* **237,** H71–H75.
Mueggler, P. A., Jones, G., Peterson, J. S., Bissonnette, J. M., Koler, R. D., Metcalfe, J., Jones, R. T., and Black, J. A. (1980). *Am. J. Physiol.* **238,** H73–H79.
Mueggler, P. A., Black, J. A., Carpenter, S., Koler, R. D., and Metcalfe, J. (1981). *Resp. Physiol.* **43,** 189–196.
Muller-Soyano, A., Platt, O., and Glader, B. E. (1986). *Am. J. Hematol.* **23,** 217–221.
Nakao, M. (1974). *In* "Cellular and Molecular Biology of Erythrocytes" (H. Yoshikawa and S. M. Rapoport, eds.), pp. 35–54. University Park Press, Baltimore.
Nakashima, M., Noda, H., Hasegaea, M., and Ikai, A. (1985). *Comp. Biochem. Physiol. A* **82A,** 583–589.
Nelson, G. J. (1967). *Biochim. Biophys. Acta* **144,** 221–232.
Noble, N. A., Jansen, C. A., Nathanielsz, P. W., and Tanaka, K. R. (1983). *Blood* **61,** 920–924.
Noble, N.A., Xu, Q.-P., and Hoge, L. L. (1990). *Blood* **75,** 1877–1882.
Nonneman, D., Stockham, S. L., Shibuya, H., Messer, N. T., and Johnson, G. S. (1993). *Blood* **82** Suppl 1, 466a.
Novy, M. J., and Parer, T. J. (1969). *Resp. Physiol.* **6,** 144–150.
Obara, J., and Nakajima, H. (1961). *Jpn. J. Vet. Sci.* **23,** 247–252.
Odartchenko, N., Cottier, H., and Bond, V. P. (1971). *Cell Tissue Kinet.* **4,** 107–112.
Ogawa, E., Shinoki, T., Akahori, F., and Masaoka, T. (1986). *Jpn. J. Vet. Sci.* **48,** 685–691.
Ogawa, E., Kobayashi, K., Yoshiura, N., and Mukai, J. (1987). *Am. J. Vet. Res.* **48,** 1300–1303.
Ogawa, E., Kobayashi, K., Yoshiura, N., and Mukai, J. (1989). *Am. J. Vet. Res.* **50,** 388–392.
Ogawa, E., Kawakami, A., Yagi, T., Amaya, T., Fujise, H., and Takahashi, R. (1992). *J. Vet. Med. Sci.* **54,** 57–62.
Oishi, A., Sakamoto, H., Shimizu, R. (1995). *J. Vet. Med. Sci.* **57,** 747–749.
Onions, D., Jarrett, O., Testa, N., Frassoni, F., and Toth, S. (1982). *Nature* **296,** 156–158.
Oski, F. A., and Gottlieb, A. J. (1971). *Prog. Hematol.* **7,** 33–67.
Osuna, D. J., Armstrong, P. J., Duncan, D. E., and Breitschwerdt, E. B. (1990). *J. Am. Anim. Hosp. Assoc.* **26,** 410–412.
Oxley, S. T., Porteous, R., Brindle, K. M., Boyd, J., and Campbell, I. D. (1984). *Biochim. Biophys. Acta* **805,** 19–24.
Palek, J., and Sahr, K. E. (1992). *Blood* **80,** 308–330.
Paniker, N. V., and Iyer, G. Y. (1965). *Can. J. Biochem.* **43,** 1029–1039.
Pantel, K., and Nakeff, A. (1993). *Exp. Hematol.* **21,** 738–742.
Papayannopoulou, T., and Brice, M. (1992). *Blood* **79,** 1686–1694.
Parker, J. C. (1977). *In* "Membrane Transport in Red Cells" (J. C. Ellory and V. L. Lew, eds.), pp. 427–465. Academic Press, New York.
Parker, J. C. (1992). *Comp. Biochem. Physiol. A* **102A,** 615–618.
Parker, W. H., and McCrea, C. T. (1965). *Vet. Rec.* **77,** 861–865.
Patel, V. P., Ciechanover, A., Platt, O., and Lodish, H. F. (1985). *Proc. Natl. Acad. Sci. USA* **82,** 440–444.
Pekow, C. A., Hinds, T. R., Maggio-Price, L., Hammond, W. P., and Vincenzi, F. F. (1992). *Am. J. Vet. Res.* **53,** 1457–1461.
Perkins, P. C., and Grindem, C. B. (1995). *Vet. Clin. Pathol.* **24,** 37–43.
Perman, V., and Schall, W. D. (1983). *In* "Textbook of Veterinary Internal Medicine: Diseases of the Dog and Cat" (S. Ettinger, ed.), 2nd ed., pp. 1938–2000. Saunders, Philadelphia, Pennsylvania.
Peschle, C. (1988). *Annu. Rev. Med.* **31,** 303–314.
Pickrell, J. A., Schluter, S. J., Belasich, J. J., Stewart, E. V., Meyer, J., Hobbs, C. H., and Jones, R. K. (1974). *Am. J. Vet. Res.* **35,** 897–903.
Pierce, K. R., Joyce, J. R., England, R. B., and Jones, L. P. (1972). *J. Am. Vet. Med. Assoc.* **160,** 323–327.
Pillai, S. R., Steiss, J. E., Traber, M. G., Kayden, H. J., and Wright, J. C. (1992). *J. Comp. Pathol.* **107,** 399–410.
Pinkerton, P. H., Fletch, S. M., Brueckner, P. J., and Miller, D. R. (1974). *Blood* **44,** 557–567.
Piomelli, S., and Seaman, C. (1993). *Am. J. Hematol.* **42,** 46–52.
Pons, G., Berrocal, F., Tauler, A., and Carreras, J. (1985). *Comp. Biochem. Physiol. B* **80B,** 551–556.
Popovic, W. J., Brown, J. E., and Adamson, J. W. (1977). *J. Clin. Invest.* **60,** 907–913.
Porra, R. J., and Jones, O. T. G. (1963). *Biochem. J.* **87,** 181–185.
Porter, D. L., and Goldberg, M. A. (1993). *Exp. Hematol.* **21,** 399–404.
Prasse, K. W., Crouser, D., Beutler, E., Walker, M., and Schall, W. D. (1975). *J. Am. Vet. Med. Assoc.* **166,** 1170–1175.
Prins, H. K., and Loos, J. A. (1969). *In* "Biochemical Methods in Red Cell Genetics" (J. J. Yunis, ed.), pp. 1–49. Academic Press, New York.
Quesenberry, P. J. (1995). *In* "Williams Hematology" (E. Beutler, M. A. Lichtman, B. S. Coller and T. J. Kipps, eds.), 5th ed., pp. 211–228. McGraw-Hill, New York.
Qui, L. B., Dickson, H., Hajibagheri, N., and Crocker, P. R. (1995). *Blood* **85,** 1630–1639.
Randolph, J .F., Center, S. A., Kallfelz, F. A., Blue, J. T., Dodds, W. J., Harvey, J. W., Paglia, D. E., Walsh, K. M., and Shelly, S. M. (1986). *Am. J. Vet. Res.* **47,** 687–695.
Rapoport, S. (1968). *Essays Biochem.* **4,** 69–103.
Rapoport, S. M. (1986). "The Reticulocyte," CRC Press, Boca Raton, Florida.
Rapoport, T. A., Heinrich, R., and Rapoport, S. M. (1976). *Biochem. J.* **154,** 449–469.
Reagan, W. J., Carter, C., and Turek, J. (1994). *Vet. Clin. Pathol.* **23,** 123–127.
Reece, W. O., Self, H. L., and Hotchkiss, D. K. (1984). *Am. J. Vet. Res.* **45,** 2119–2121.
Reed, C. F. (1968). *J. Clin. Invest.* **47,** 749–759.
Reimann, B., Klatt, D., Tsamaloukas, A. G., and Maretzki, D. (1981). *Acta Biol. Med. German.* **40,** 487–493.
Reinhart, W. H., Sung, L. A., and Chien, S. (1986). *Blood* **68,** 1676–1383.
Rendell, M., Stephen, P. M., Paulsen, R., Valentine, J. L., Rasbold, K., Hestorff, T., Eastberg, S., and Shint, D. C. (1985). *Comp. Biochem. Physiol. B* **81B,** 819–822.
Rich, I. N. (1986). *Exp Hematol* **14,** 746–751.
Rich, I. N. (1991). *Exp. Hematol.* **19,** 985–990.

Riedel, N., Hoover, E. A., Gasper, P. W., Nicolson, M. O., and Mullins, J. I. (1986). *J. Virol.* **60,** 242–250.

Riggs, A. (1960). *J. Gen. Physiol.* **43,** 737–752.

Rivkin, S. E., and Simon, E. R. (1965). *J. Cell Physiol.* **66,** 49–56.

Robin, H., and Harley, J. D. (1966). *Aust. J. Exp. Biol. Med. Sci.* **44,** 519–526.

Robin, H., and Harley, J. D. (1967). *Aust. J. Exp. Biol. Med. Sci.* **45,** 77–88.

Robinson, W. F., Wilcox, G. E., and Fowler, R .L. P. (1980). *Vet. Pathol.* **17,** 589–599.

Rojko, J. L., and Olsen, R. G. (1984). *Vet. Immunol. Immunopathol.* **6,** 107–165.

Romslo, I. (1980). *In* "Iron in Biochemistry and Medicine II" (A. Jacobs and M. Worwood, eds.), pp. 325–362. Academic Press, New York.

Rose, R. J., Rossdale, P. D., and Leadon, D. P. (1982). *J. Reprod. Fert. Suppl.* **32,** 521–528.

Rosenfeld, S. J., and Young, N. S. (1991). *Blood Rev.* **5,** 71–77.

Rossdale, P. D., and Ricketts, S. W. (1980). "Equine Stud Farm Medicine" 2nd ed. Lea & Febiger, Philadelphia, Pennsylvania.

Rotruck, J. T., Pope, A. L., Ganther, H. E., and Hoekstra, W. G. (1972). *J. Nutr.* **102,** 689–696.

Rotruck, J. T., Pope, A. L., Ganther, H. E., Swanson, A. B., Hafeman, D. G., and Hoekstra, W. G. (1973). *Science* **179,** 588–590.

Rottman, J. B., English, R. V., Breitschwerdt, E. B., and Duncan, D. E. (1991). *J. Am. Vet. Med. Assoc.* **198,** 429–431.

Rudolph, W. G., and Kaneko, J. J. (1971). *Acta Haematol.* **45,** 330–335.

Saltman, P. (1989). *Semin. Hematol.* **26,** 249–256.

Sambrano, G. R., Parthasarathy, S., and Steinberg, D. (1994). *Proc. Natl. Acad. Sci. USA* **91,** 3265–3269.

Sandberg, A. A., Lee, G. R., Cartwright, G. E., and Wintrobe, M. M. (1955). *J. Clin. Invest.* **34,** 1823–1829.

Sankari, S., and Atroshi, F. (1983). *Zentralbl. Veterinarmed. A* **30,** 452–458.

Sasaki, R., Ikura, K., Narita, H., and Chiba, H. (1977). *Acta Biol. Med. German.* **36,** 669–680.

Sass, M. D., Caruso, C. J., and Axelrod, D. R. (1969). *Clin. Chim. Acta* **24,** 77–85.

Sato, K., Inaba, M., and Maede, Y. (1994). *Biochim. Biophys. Acta Bio-Membr.* **1195,** 211–217.

Sato, T., and Mizuno, M. (1982). *Nippon. Juigaku. Zasshi.* **44,** 801–805.

Schaer, M., Harvey, J. W., Calderwood Mays, M. B., and Giger, U. (1992). *J. Am. Anim. Hosp. Assoc.* **28,** 233–239.

Schalm, O. W. (1976). *Can. Pract.* **3**(6), 55–57.

Schalm, O. W. (1979). *Can. Pract.* **6,** 47–50.

Schechter, R. D., Schalm, O. W., and Kaneko, J. J. (1973). *J. Am. Vet. Med. Assoc.* **162,** 37–44.

Schmidt-Neilsen, K., and Larimer, J. L. (1958). *Am. J. Physiol.* **195,** 424–428.

Schrier, S. L. (1985). *Clin. Haematol.* **14,** 1–12.

Schulman, H. M., Martinez-Medellin, J., and Sidloi, R. (1974). *Biochem. Biophys. Res. Commun.* **56,** 220–226.

Schumacker, P. T., Suggett, A. J., Wagner, P. D., and West, J. B. (1985). *J. Appl. Physiol.* **59,** 749–757.

Schwartz, E., and Gill, F. M. (1983). *In* "Hematology" (W. J. Williams, E. Beutler, A. J. Erslev, and M. A. Lichtman, eds.), 3rd ed., pp. 37–47. McGraw-Hill, New York.

Scott, A. F., Bunn, H. F., and Brush, A. H. (1977). *J. Exp. Zool.* **201,** 269–288.

Scott, M. D., Eaton, J. W., Kuypers, F. A., Chiu, D. T.-Y., and Lubin, B. H. (1989). *Blood* **74,** 2542–2549.

Scott, M. D., Wagner, T. C., and Chiu, D. T.-Y. (1993). *Biochim. Biophys. Acta Mol. Basis Dis.* **1181,** 163–168.

Searcy, G. P., Miller, D. R., and Tasker, J. B. (1971). *Can. J. Comp. Med.* **35,** 67–70.

Searcy, G. P., Tasker, J. B., and Miller, D. R. (1979). *Am. J. Physiol.* **94,** 689–692.

Seppi, C., Castellana, M. A., Minetti, G., Piccinini, G., Balduini, C., and Brovelli, A. (1991). *Mech. Aging Dev.* **57,** 247–258.

Shamberger, R. J. (1986). *Clin. Physiol. Biochem.* **4,** 42–49.

Shelton, G. H., Linenberger, M. L., Grant, C. K., and Abkowitz, J. L. (1990). *Blood* **76,** 1104–1109.

Sherding, R. G., Wilson, G. P., and Kociba, G. J. (1981). *J. Am. Vet. Med. Assoc.* **178,** 497–501.

Shimizu, T., Kono, N., Kiyokawa, H., Yamada, Y., Hara, N., Mineo, I., Kawachi, M., Nakajima, H., Wang, Y. L., and Tarui, S. (1988). *Blood* **71,** 1130–1134.

Shull, R. M., Bunch, S. E., Maribei, J., and Spaulding, G. L. (1978). *J. Am. Vet. Med. Assoc.* **173,** 978–982.

Shull, R. M., Suggs, S. V., Langley, K. E., Okino, K. H., Jacobsen, F. W., and Martin, F. H. (1992). *Exp. Hematol.* **20,** 1118–1124.

Shviro, Y., and Shaklai, N. (1987). *Biochem. Pharmacol.* **36,** 3801–3807.

Sieff, C. A. (1991). *Exp. Hematol.* **19,** 857–860.

Siimes, M. A., Refino, C., and Dallman, P. R. (1980). *Am. J. Clin. Nutr.* **33,** 2601–2608.

Simpson, C. F., and Anderson, B. (1980). *Florida Vet. J.* **9,**26–27.

Simpson, C. F., and Kling, J. M. (1967). *J. Cell Biol.* **35,** 237–245.

Simpson, C. F., and Kling, J.M. (1968). *J. Cell Biol.* **36,** 103–109.

Sippel, W. L. (1952). *J. Am. Vet. Med. Assoc.* **121,** 9–13.

Slappendel, R. J., Van der Gaag, I., Van Nes, J. J., Van den Ingh, T. S. G. A. M., and Happé, R. P. (1991). *Vet. Q.* **13,** 30–40.

Slappendel, R. J., Renooij, W., and De Bruijne, J. J. (1994). *Blood* **84,** 904–909.

Smith, J. E. (1968). *J. Lab. Clin. Med.* **71,** 826–833.

Smith, J. E. (1974). *J. Lab. Clin. Med.* **83,** 444–450.

Smith, J. E. (1981). *Clin. Haematol.* **10,** 239–251.

Smith, J. E. (1987). *Vet. Pathol.* **24,** 471–476.

Smith, J. E. (1991). *Adv. Vet. Sci. Comp. Med.* **36,** 9–55.

Smith, J. E., and Agar, N. S. (1976). *Equine Vet. J.* **8,** 34–37.

Smith, J. E., and Anwer, M. S. (1971). *Experientia* **27,** 835–836.

Smith, J. E., McCants, M., Parks, P., and Jones, E. W. (1972). *Comp. Biochem. Physiol. B* **41B,** 551–558.

Smith, J. E., Lee, M. S., and Mia, A. S. (1973). *J. Lab. Clin. Med.* **82,** 713–718.

Smith, J. E., Ryer, K., and Wallace, L. (1976). *Enzyme* **21,** 379–382.

Smith, J. E., Mohandas, N., and Shohet, S. B. (1979). *Am. J. Physiol.* **236,** H725–H730.

Smith, J. E., Moore, K., Arens, M., Rinderknecht, G. A., and Ledet, A. (1983a). *Blood* **61,** 373–377.

Smith, J. E., Moore, K., Boyington, D., and Potter, K. A. (1983b). *J. Lab. Clin. Med.* **101,** 611–616.

Smith, J. E., Cipriano, J. E., DeBowes, R., and Moore, K. (1986). *J. Am. Vet. Med. Assoc.* **188,** 285–287.

Smith, R. H. (1980). *Vet. Rec.* **107,** 12–15.

Snow, D. H., and Martin, V. (1990). *Res. Vet. Sci.* **49,** 77–81.

Snow, D. H., Billeter, R., Mascarello, F., Carpene, E., Rowlerson, A., and Jenny, E. (1982). *Histochemistry* **75,** 53–65.

Snow, N. S. (1962). *Biochem. J.* **84,** 360–364.

Snyder, L. M., Reddy, W. J., and Kurjan, L. (1970). *J. Clin. Invest.* **49,** 1993–1998.

Sobin, S. S., and Tremer, H. M. (1972). *Microvasc. Res.* **4,** 330–331.

Sobrino, F., Rider, M. H., Gualberto, A., and Hue, L. (1987). *Biochem. J.* **244,** 235–238.

Soli, N. E., and Froslie, A. (1977). *Acta Pharmacol. Toxicol. Copenh.* **40,** 169–177.

Solomon, A. K., Chasan, B., Dix, J. A., Lukacovic, M. F., Toon, M. R., and Verkman, A. S. (1983). *Ann. N. Y. Acad. Sci.* **414,** 97–124.

Srivastava, S. K., and Beutler, E. (1969). *Biochem. J.* **114,** 833–837.
Srivastava, S. K., and Beutler, E. (1972). *Arch. Biochem. Biophys.* **148,** 249–255.
Stadtman, E. R. (1992). *Science* **257,** 1220–1224.
Standerfer, R. J., Templeton, J. W., and Black, J. A. (1974). *Am. J. Vet. Res.* **35,** 1541–1543.
Steffen, D. J., Elliott, G. S., Leipold, H. W., and Smith, J. E. (1992). *J. Vet. Diagn. Invest.* **4,** 31–37.
Stewart, G. W. (1993). *Baillieres Clin. Haematol.* **6,** 371–399.
Stockham, S. L., Harvey, J. W., and Kinden, D. A. (1994). *Vet. Pathol.* **31,** 518–527.
Stokstad, E. L. R. (1968). *Vitamins Hormones* **26,** 443–463.
Stormont, C. J. (1982). *J. Am. Vet. Med. Assoc.* **181,** 1120–1124.
Strafuss, A. C., and Sautter, J. H. (1967). *Am. J. Vet. Res.* **28,** 25–37.
Studzinski, T., Gluszak, A., and Owczarski, K. (1978). *Acta Physiol. Pol.* **29,** 335–341.
Studzinski, T., Czarnecki, A., and Gluszak, A. (1982). *Acta Physiol. Pol.* **33,** 129–137.
Sullivan, P. S., Evans, H. L., and McDonald, T. P. (1994). *J. Am. Vet. Med. Assoc.* **205,** 838–841.
Suttle, N. F., Jones, D. G., Woolliams, C., and Woolliams, J. A. (1987). *Br. J. Nutr.* **58,** 539–548.
Suzuki, T., and Agar, N. S. (1983). *Experientia* **39,** 103–104.
Suzuki, T., Agar, N. S., and Suzuki, M. (1984a). *Comp. Biochem. Physiol. B* **79B,** 515–520.
Suzuki, T., Agar, N. S., and Suzuki, M. (1984b). *Comp. Biochem. Physiol. B* **79B,** 515–520.
Takano, N., Hayashi, E., and Matsue, K. (1976). *Pflugers Arch.* **366,** 285–288.
Taketa, F., and Antholine, W. E. (1982). *J. Inorg. Biochem.* **17,** 109–120.
Taketa, F., Attermeier, M. H., and Mauk, A. G. (1972). *J. Biol. Chem.* **247,** 33–35.
Tavassoli, M., and Hardy, C. L. (1990). *Blood* **76,** 1059–1070.
Taylor, W. J. (1983). *Adv. Vet. Sci. Comp. Med.* **27,** 77–98.
Teisseire, B. P., Ropars, C., Vallez, M. O., Herigault, R. A., and Nicolau, C. (1985). *J. Appl. Physiol.* **58,** 1810–1817.
Tennant, B., Asbury, A. C., Laben, R. C., Richards, W. P. C., Kaneko, J. J., and Cupps, P. T. (1967). *J. Am. Vet. Med. Assoc.* **150,** 1493–1509.
Tennant, B., Harrold, D., Reina-Guerra, M., and Laben, R. C. (1969). *Cornell Vet.* **59,** 594–604.
Tennant, B., Dill, S. G., Glickman, L. T., Mirro, E. J., King, J. M., Polak, D. M., Smith, M. C., and Kradel, D. C. (1981). *J. Am. Vet. Med. Assoc.* **179,** 143–150.
Terada, T., Oshida, T., Nishimura, M., Maeda, H., Hara, T., Hosomi, S., Mizoguchi, T., and Nishihara, T. (1992). *J. Biochem. (Tokyo)* **111,** 688–692.
Testa, U., Pelosi, E., Gabbianelli, M., Fossati, C., Campisi, S., Isacchi, G., and Peschle, C. (1993). *Blood* **81,** 1442–1456.
Thenen, S. W., and Rasmussen, S. D. (1978). *Am. J. Vet. Res.* **39,** 1205–1207.
Thompson, R. H., McMurray, C. H., and Blanchflower, W. J. (1976). *Res. Vet. Sci.* **20,** 229–231.
Tizard, I. (1992). "Veterinary Immunology. An Introduction" 4th ed. Saunders, Philadelphia, Pennsylvania.
Tomoda, A., Tsuji, A., and Yoneyama, Y. (1981). *Biochem. J.* **193,** 169–179.
Tosteson, D. C., and Hoffman, J. F. (1960). *J. Gen. Physiol.* **44,** 169–194.
Traugh, J. A. (1989). *Semin. Hematol.* **26,** 54–62.
Travis, S. F., Wagerle, L. C., De Alvarado, C. M., Rose, G., and Delivoria-Papadopoulos, M. (1985). *Pediatr. Res.* **19,** 272–277.
Tucker, E. M. (1974). *Res. Vet. Sci.* **16,** 19–22.
Tucker, E. M., and Ellory, J. C. (1971). *Res. Vet. Sci.* **12,** 600–602.
Tucker, E. M., and Kilgour, L. (1970). *Experientia* **26,** 203–204.
Tucker, E. M., and Young, J. D. (1980). *Biochem. J.* **192,** 33–39.
Tucker, E. M., Young, J. D., and Crowley, C. (1981). *Br. J. Haematol.* **48,** 403–415.
Tunnicliff, G. (1994). *Comp. Biochem. Physiol. A* **108A,** 471–478.
Tunon, M. J., Gonzalez, P., and Vallejo, M. (1987). *Anim. Genet.* **18,** 371–375.
Tustin, R. C., Adelaar, T. F., and Meldal-Johnsen, C. M. (1968). *J. S. Afr. Vet. Med. Assoc.* **39,** 91–99.
Tweeddale, P. M. (1973). *Resp. Physiol.* **19,** 12–18.
Ulibarrena, C., Vecino, A., and Cesar, J. M. (1994). *Br. J. Haematol.* **87,** 614–616.
Underwood, E. J. (1977). "Trace Elements in Human and Animal Nutrition," 4th ed. Academic Press, New York.
Ursini, F., Maiorino, M., and Gregolin, C. (1985). *Biochim. Biophys. Acta* **839,** 62–70.
Vacha, J. (1983). *In* "Red Blood Cells of Domestic Mammals" (N. S. Agar and P. G. Board, eds.), pp. 67–132. Elsevier, Amsterdam.
Vaden, S. L., Wood, P. A., Ledley, F. D., Cornwell, P. E., Miller, R. T., and Page, R. (1992). *J. Am. Vet. Med. Assoc.* **200,** 1101–1103.
Van Belle, H. (1969). *Biochim. Biophys. Acta* **192,** 124–132.
Van Dyke, D., Johns, L. E., and Tennant, B. (1968). *Nature* **217,** 1027–1028.
Verhoeff, J., Hajer, R., and Van den Ingh, T. S. G. A. M. (1985). *Vet. Rec.* **117,** 497–498.
Vincent, S. H. (1989). *Semin. Hematol.* **26,** 105–113.
Vogt, C., Pentz, S., and Rich, I. N. (1989). *Exp. Hematol.* **17,** 391–397.
Vora, S., Giger, U., Turchen, S., and Harvey, J. W. (1985). *Proc. Natl. Acad. Sci. USA* **82,** 8109–8113.
Vuillet-Gaugler, M.H., Breton-Gorius, J., Vainchenker, W., Guichard, J., Leroy, C., Tchernia, G., and Coulombel, L. (1990). *Blood* **75,** 865–873.
Waddell, W. J. (1956). *Am. J. Physiol.* **186,** 339–342.
Waddell, W. J., and Bates, R. G. (1969). *Physiol. Rev.* **49,** 285–329.
Wallace, W. J., Maxwell, J. C., and Caughey, W. S. (1974). *Biochem. Biophys. Res. Commun.* **57,** 1104–1110.
Wang, C. Q., Udupa, K. B., and Lipschitz, D. A. (1992). *Blood* **80,** 1702–1709.
Wang, X. L., Gallagher, C. H., McClure, T. J., Reeve, V. E., and Canfield, P. J. (1985). *Res. Vet. Sci.* **39,** 333–339.
Ward, M. V., Mountan, P. C., and Dodds, W. J. (1980). *Calif. Vet.* **12,** 19–22.
Watson, A. D. (1977). *Res. Vet. Sci.* **23,** 66–69.
Watson, A. D., and Middleton, D. J. (1978). *Am. J. Vet. Res.* **39,** 1199–1203.
Watson, A. D., Wilson, J. T., Turner, D. M., and Culvenor, J. A. (1980). *Vet. Rec.* **107,** 239–241.
Watson, A. D. J. (1979). *J. Small Anim. Pract.* **20,** 681–690.
Watts, R. P., and Kim, H. D. (1984). *Biol. Neonate* **45,** 280–288.
Waugh, R. E. (1991). *Blood* **78,** 3037–3042.
Weaver, J., and Pollack, S. (1989). *Biochem. J.* **261,** 787–792.
Weiden, P. L., Hackman, R. C., Deeg, J., Graham, C., Thomas, E. D., and Strob, R. (1981). *Blood* **57,** 66–70.
Weigensberg, A. M., and Blostein, R. (1983). *Proc. Natl. Acad. Sci. USA* **80,** 4978–4982.
Weiser, G., and O'Grady, M. (1983). *Vet. Pathol.* **20,** 230–241.
Weiser, M. G., and Kociba, G. J. (1982). *Blood* **60,** 295–303.
Weiser, M. G., and Kociba, G. J. (1983a). *Vet. Pathol.* **20,** 687–697.
Weiser, M. G., and Kociba, G. J. (1983b). *Vet. Pathol.* **20,** 1–12.
Weiss, D. J. (1986). *Am. J. Vet. Res.* **47,** 2646-2648.
Weiss, D. J., and Christopher, M. M. (1985). *Vet. Clin. Pathol.* **14**(2), 23–25.
Weiss, D. J., and Geor, R. J. (1993). *Comp. Haematol. Int.* **3,** 185–189.
Weiss, D. J., and Klausner, J. S. (1990). *J. Am. Vet. Med. Assoc.* **196,** 472–475.

Weiss, D. J., Stockham, S. L., Willard, M. D., and Schirmer, R. G. (1982). *J. Am. Anim. Hosp. Assoc.* **18,** 353–359.
Weiss, D. J., Geor, R., Smith, C. M., II, and McClay, C. B. (1992). *Am. J. Vet. Res.* **53,** 1769–1772.
Weiss, D. J., Kristensen, A., and Papenfuss, N. (1993). *Vet. Clin. Pathol.* **22,** 117–121.
Weiss, L. (1984). "The Blood Cells and Hematopoietic Tissues," Elsevier, New York.
Weissman, S. M., Waldmann, T. A., and Berlin, N. I. (1960). *Am. J. Physiol.* **8,** 183–186.
Wen, D., Boissel, J. P. R., Tracy, T. E., Gruninger, R. H., Mulcahy, L. S., Czelusniak, J., Goodman, M., and Bunn, H. F. (1993). *Blood* **82,** 1507–1516.
Wessels, J. M., and Veerkamp, J. H. (1973). *Biochim. Biophys. Acta* **291,** 190–196.
West, J. B. (1985). *In* "Best and Taylor's Physiologic Basis of Medical Practice" (J. B. West, ed.), 11th ed., pp. 516–532, 546–571. Williams & Wilkins, Baltimore.
Whitfield, C. F., Mylin, L. M., and Goodman, S. R. (1983). *Blood* **61,** 500–506.
Whitney, K. M., and Lothrop, C. D., Jr. (1995). *J. Am. Vet. Med. Assoc.* **207,** 918–921.
Whitney, K. M., Goodman, S. A., Bailey, E. M., and Lothrop, C. D., Jr. (1994). *Exp. Hematol.* **22,** 866–874.
Widdas, W. F. (1955). *J. Pathol.* **127,** 318–327.
Wilhelmsen, C. L. (1979). *Cornell Vet.* **69,** 225–232.
Wilkie, D. A., and Kirby, R. (1988). *J. Am. Vet. Med. Assoc.* **192,** 85–86.
Willard, M. D., Zerbe, C. A., Schall, W. D., Johnson, C., Crow, S. E., and Jones, R. (1987). *J. Am. Vet. Med. Assoc.* **190,** 1007–1010.
Williams, D. M., Lynch, R. E., Lee, G. R., and Cartwright, G. E. (1975). *Proc. Soc. Exp. Biol. Med.* **149,** 534–536.
Williams, D. M., Loukopoulos, D., Lee, G. R., and Cartwright, G. E. (1976). *Blood* **48,** 77–85.
Winslow, R. M., Swenberg, M.-L., Benson, J., Perrella, M., and Benazzi, L. (1989). *J. Biol. Chem.* **264,** 4812–4817.
Wood, P. A., and Smith, J. E. (1980). *J. Am. Vet. Med. Assoc.* **176,** 1267–1268.
Wood, W. G., Bunch, C., Kelly, S., Gunn, Y., and Breckon, G. (1985). *Nature* **313,** 320–323.
Xu, F., Quandt, K. S., and Hultquist, D. E. (1992). *Proc. Natl. Acad. Sci. USA* **89,** 2130–2134.
Xu, Z.-C., Dunham, P. B., Dyer, B., and Blostein, R. (1993). *Am. J. Physiol. Cell Physiol.* **265,** C99–C105.
Xu, Z.-C., Dunham, P. B., Dyer, B., and Blostein, R. (1994). *Am. J. Physiol. Cell Physiol.* **266,** C1173–C1181.
Yamakawa, T. (1983). *In* "Red Blood Cells of Domestic Mammals" (N. S. Agar and P. G. Board, eds.), pp. 37–53. Elsevier, Amsterdam.
Yamato, O., Yoshihara, T., Ichihara, A., and Maede, Y. (1994). *Biosci. Biotechnol. Biochem.* **58,** 221–222.
Yamoto, O., and Maede, Y. (1992). *Am. J. Vet. Res.* **53,** 134–137.
Yanai, N., Sekine, C., Yagita, H., and Obinata, M. (1994). *Blood* **83,** 2844–2850.
Yawata, Y., Hebbel, R. P., Silvis, S., Howe, R., and Jacob, H. (1974). *J. Lab. Clin. Med.* **84,** 643–653.
Yawata, Y., Sugihara, T., Mori, M., Nakashima, S., and Nozawa, Y. (1984). *Blood* **64,** 1129–1134.
Yoshida, A. (1973). *Science* **179,** 532–537.
Young, J. D. (1983). *In* "Red Blood Cells of Domestic Mammals" (N. S. Agar and P. G. Board, eds.), pp. 271–290. Elsevier, Amsterdam.
Young, J. D., Ellory, J. C., and Tucker, E. M. (1975). *Nature* **254,** 156–157.
Young, J. D., Paterson, A. R., and Henderson, J. F. (1985). *Biochim. Biophys. Acta* **842,** 214–224.
Youssoufian, H., Longmore, G., Neumann, D., Yoshimura, A., and Lodish, H. F. (1993). *Blood* **81,** 2223–2236.
Yubisui, T., Takeshita, M., and Yoneyama, Y. (1980). *J. Biochem.* **87,** 1715–1720.
Zachara, B., Zakrzewska, I., Maziarz, Z., Gaszynski, W., and Wachowicz, N. (1981). *Haematologia. Budap.* **14,** 285–291.
Zeidler, R. B., and Kim, H. D. (1982). *J. Cell Physiol.* **112,** 360–366.
Zeidler, R. B., Metzler, M. H., Moran, J. B., and Kim, H. D. (1985). *Biochim. Biophys. Acta* **838,** 321–328.
Zerez, C. R., Lee, S. J., and Tanaka, K. R. (1987). *Anal. Biochem.* **164,** 367–373.
Zimran, A., Forman, L., Suzuki, T., Dale, G. L., and Beutler, E. (1990). *Am. J. Hematol.* **33,** 249–254.
Zinkl, J., and Kaneko, J. J. (1973a). *Comp. Biochem. Physiol. A* **45A,** 463–476.
Zinkl, J., and Kaneko, J. J. (1973b). *Comp. Biochem. Physiol. A* **45A,** 699–704.
Zwaal, R. F., Comfurius, P., and Bevers, E. M. (1993). *Biochem. Soc. Trans.* **21,** 248–253.

CHAPTER

8

Porphyrins and the Porphyrias

J. JERRY KANEKO

I. INTRODUCTION

The metal-porphyrin complexes are found widespread in nature as constituents of compounds of fundamental importance in the metabolic processes of life. The photosynthetic pigment of plants, chlorophyll, is a magnesium porphyrin. The iron-porphyrin complexes are found as prosthetic groups of proteins, including the hemoglobins, myoglobins, and the heme enzymes such as the cytochromes, catalase, and peroxidase. The porphyrins also exist in nature in their free or uncombined state or as zinc complexes and it is this group that is associated with the porphyrias and the porphyrinurias.

Present knowledge of the porphyrins has its basis in the classic studies of the German physician and chemist, Hans Fischer, whose work on the porphyrins dates back to 1915. More recently, the development of more elegant methods of detection and identification of porphyrins and sophisticated enzymological techniques have resulted in an extremely clear understanding of the mechanisms of porphyrin biosynthesis and the biochemical and molecular bases for the disorders of porphyrin metabolism.

II. THE PORPHYRINS

A. Structure of the Porphyrins

The parent nucleus of the porphyrins is a cyclic tetrapyrrole, which consists of four pyrrole nuclei with their α (adjacent to the β) carbon atoms linked together by methene (–C=) bridges. This compound is called *porphin* and is shown in Fig. 8.1. The various synthetic and naturally occurring porphyrins are derivatives of porphin, distinguished from each other by the type and position of the radicals substituted for the hydrogen atoms at positions 1 through 8. For convenience in discussing the substitutions, we use the simplified representation of the porphin nucleus as shown in Fig. 8.1.

The classification of the porphyrins is based on the synthetic porphyrin, etioporphyrin (ETIO), in which two different radicals are substituted at positions 1 through 8. The substituted radicals are four methyl (M) and four ethyl (E) groups. The number of structural isomers possible with these eight substituted radicals are the four shown at the top of Fig. 8.2. The naturally occurring porphyrins are only those in which the positioning of their substituted radicals corresponds to iso-

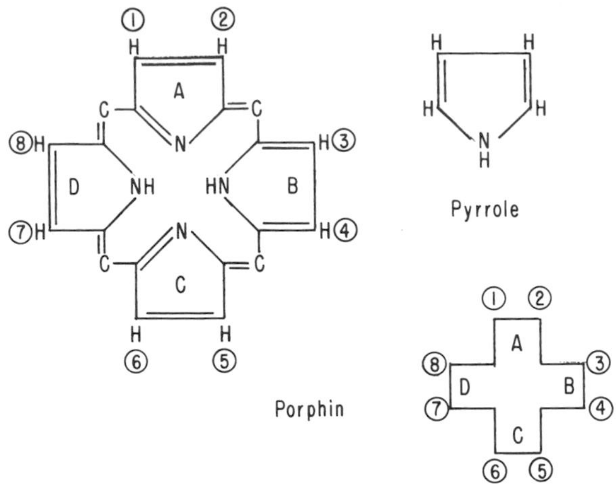

FIGURE 8.1 The precursor pyrrole and the parent porphin nucleus of porphyrins. Sites of isomeric substitutions are given as circled numbers and the pyrrole rings as letters. A schematic representation is also given.

mers I or III of etioporphyrin, ETIO I and ETIO III. This observation led Fischer to speak of a "dualism" of porphyrins in nature, which is in essential agreement with present knowledge of the biosynthesis of the porphyrin isomers as proceeding along parallel and independent paths.

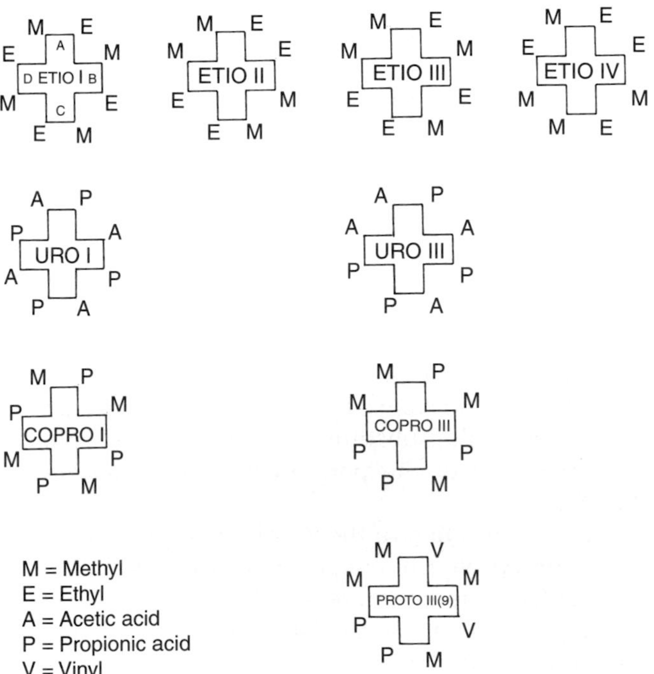

FIGURE 8.2 The isomeric porphyrins. The nomenclature of the porphyrins, URO, COPRO, and PROTO, is based on the isomeric structure of the etioporphyrins. Note that there are only the type I and type III isomers.

The uroporphyrins also contain two different radicals, acetic (A) and propionic (P) acids, and four each of these are arranged to correspond to either isomer ETIO I or ETIO III (Fig. 8.2). In this case, A corresponds to M and P corresponds to E. Therefore, these are designated uroporphyrin I (URO I) or uroporphyrin III (URO III). Similarly, the coproporphyrins contain four M and four P groups and are designated coproporphyrin I (COPRO I) and coproporphyrin III (COPRO III). The protoporphyrin of heme (iron-protoporphyrin, the prosthetic group of hemoglobin) corresponds to the series III isomer. In this case, however, three different radicals instead of two are substituted. These consist of four M, two P, and two vinyl (V) radicals. With three different radicals, a total of 15 isomers is possible, but the protoporphyrin of heme is the only naturally occurring isomer known. This isomer was designated protoporphyrin IX because it was the ninth in the series of protoporphyrin isomers synthesized by Fischer. The arrangement of the methyl groups of this isomer as shown in Fig. 8.2 corresponds to that of a type III etioporphyrin isomer and more properly should be called protoporphyrin III. However, by convention, the name protoporphyrin IX (PROTO IX) is the designation for this porphyrin.

An interesting isomer with only one vinyl group located on the A ring occurs and is known as Harderoporphyrinogen, which is thought to be an intermediate in the synthesis of protoporphyrinogen III. Other naturally occurring or chemically synthesized porphyrins are derivatives of PROTO IX. If the two vinyl groups are hydrogenated to ethyl groups, the product is mesoporphyrin IX. If the two vinyl groups are converted to hydroxy-ethyl groups, the product is hematoporphyrin IX. If the two vinyl groups are replaced by hydrogen atoms, the product is deuteroporphyrin IX. Protoporphyrin and deuteroporphyrin normally occur in feces but these are primarily products of intestestinal bacterial degradation.

B. Synthesis of Porphyrins and Heme

The pathway for heme biosynthesis has its basis in the demonstration by Shemin and Rittenberg (1946a) that the nitrogen atom of glycine is incorporated into the heme of hemoglobin. The concentration of isotopic nitrogen (^{15}N) in heme indicated that the nitrogen atoms of glycine were incorporated as the nitrogen atoms of the porphyrin ring. These findings led to the rapid elucidation of the enzymatic mechanisms of heme synthesis. Porphyrin metabolism and the porphyrias have been periodically reviewed (Meyer and Schmid, 1978; Elder, 1982; Hindmarsh, 1986; Kappas *et al.*, 1995).

In addition to contributing the nitrogen atoms, the methyl carbon atom (C-2) of glycine is also incorporated into the porphyrin ring. These carbons supply 8 of the 34 carbon atoms of protoporphyrin: one for each of the four methene bridges and one for each of the pyrroles (Fig 8.3). The carboxyl carbon atom of glycine is given off as CO_2 and is not incorporated into the protoporhyrin molecule (Fig. 8.3). The direct incorporation of the nitrogen atom and the methyl carbon atom of glycine into the heme of hemoglobin has been the basis for a useful technique to label the erythrocyte and to measure its survival time. In their original studies, Shemin and Rittenberg (1946b) observed that after administering ^{15}N-glycine, the concentration of ^{15}N in heme rose rapidly, remained constant for a time, and then fell. Analysis of the data indicated a survival time of 120 days for the human erythrocyte. On a similar basis, methyl carbon labeled glycine has been used to determine the lifespan of the erythrocytes of a number of domestic animals (Table 8.1).

The remaining carbon atoms of protoporphyrin are supplied by the tricarboxylic acid (TCA) cycle intermediate, succinyl-CoA. An outline of the pathway of heme and porphyrin synthesis is given in Fig. 8.4, which depicts the compartmentalization of their pathways between the mitochondria and cytosol. The nomenclature of the enzymes of porphyrin synthesis varies among authors so a guide to the nomenclature of the enzymes is provided in Table 8.2.

1. δ-Aminolevulinic Acid

The initial step in the pathway of δ-aminolevulinic acid (ALA) synthesis occurs in the mitochondria and involves the enzymatic condensation of glycine with succinyl-CoA to form ALA. This reaction requires vitamin B^6 as pyridoxal phosphate and the pyridoxal–

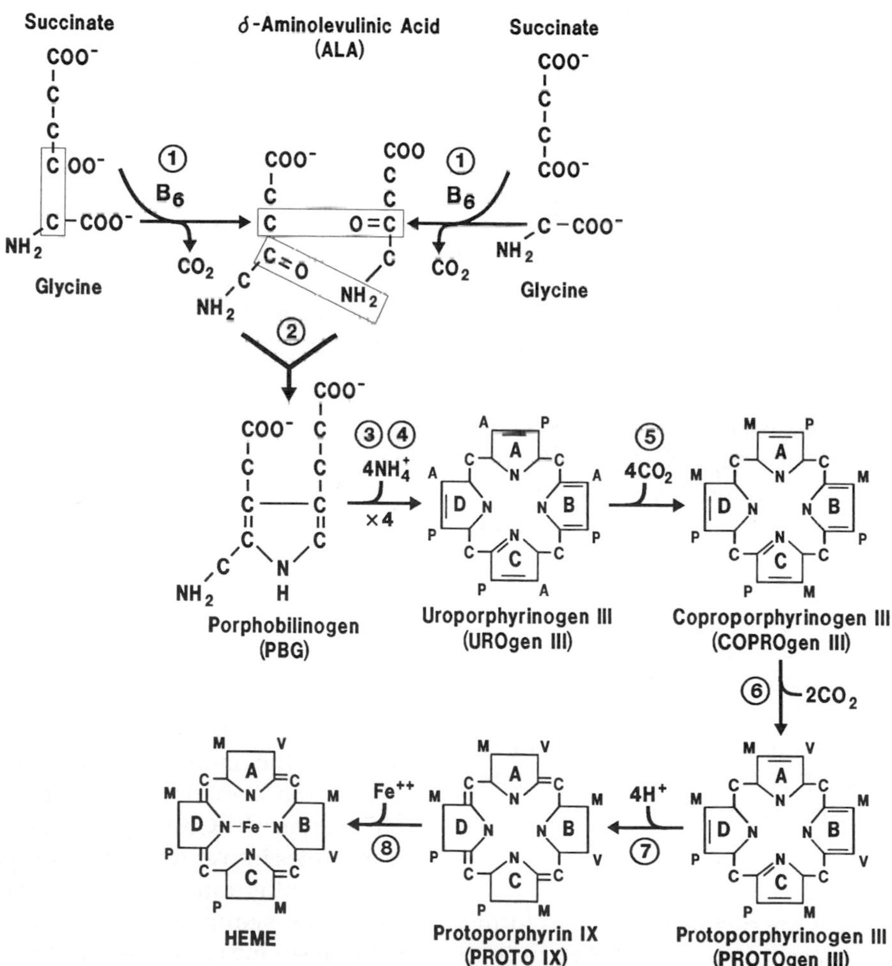

FIGURE 8.3 The synthetic pathway for protoporphyrin and heme. Note that two enzymes are required to catalyze the formation of UROgenIII. The circled numbers correspond to the enzymes listed in Table 8.2.

TABLE 8.1 Erythrocyte Lifespan of Animals Determined by the Cohort Labeling of Heme

Animal	Label	Lifespan (days)	Reference
Antelope	^{14}C	80	Cornelius *et al.* (1959)
Cat	^{15}N	77	Valentine *et al.* (1951)
Cat	^{14}C	66–79	Kaneko *et al.* (1966)
Cat	^{59}Fe	68	Brown and Eadie (1953)
Cat	^{59}Fe	36–66	Liddle *et al.* (1984)
Chicken	^{14}C	20	Brace and Altland (1956)
Cow	^{14}C	135–162	Kaneko (1963)
Mule deer	^{14}C	95	Cornelius *et al.* (1959)
Dog	^{14}C	86–106	Cline and Berlin (1963)
Dog	^{59}Fe	95–109	Finch *et al.* (1949)
Duck	^{14}C	39	Brace and Altland (1956)
Goat	^{14}C	125	Kaneko and Cornelius (1962)
Tahr goat	^{14}C	160–165	Kaneko and Cornelius (1962)
Guanaco	^{14}C	225	Cornelius and Kaneko (1962)
Guinea pig	^{59}Fe	83	Everett and Yoffey (1959)
Horse	^{14}C	140–150	Cornelius *et al.* (1960)
Human	^{14}C	120	Berlin *et al.* (1957)
Human	^{15}N	127	Shemin and Rittenberg (1946b)
Mouse	^{59}Fe	20–30	Burwell *et al.* (1953)
Pig	^{59}Fe	63	Jensen *et al.* (1956)
Pig	^{14}C	62	Bush *et al.* (1955)
Rabbit	^{15}N	65–70	Neuberger and Niven (1951)
Rabbit	^{59}Fe	57	Gower and Davidson (1963)
Rabbit	^{14}C	50	Gower and Davidson (1963)
Rat	^{14}C	64	Berlin and Lotz (1951)
Rat	^{14}C	68	Berlin *et al.* (1951)
Rat	^{59}Fe	45–50	Burwell *et al.* (1953)
Sheep	^{59}Fe	70–153	Tucker (1963)
Sheep	^{14}C	64–118	Kaneko *et al.* (1961)
Bighorn sheep	^{14}C	147	Kaneko *et al.* (1961)
Karakul sheep	^{14}C	130	Kaneko *et al.* (1961)
Aoudad sheep	^{14}C	60 and 170	Cornelius *et al.* (1959)

phosphate–glycine complex condenses with succinyl-CoA (Gibson *et al.*, 1958, Kikuchi *et al.*, 1958). The requirement for pyridoxine explains the pyridoxine responsive anemia of pyridoxine deficiency because in the absence of pyridoxine, the condensation cannot occur. The condensing reaction is catalyzed by the enzyme ALA synthase (ALA-Syn). ALA-Syn is the rate-controlling enzyme for heme synthesis (Granick, 1966). ALA-Syn is induced by heme and is also suppressed by negative feedback inhibition by heme. Thus, the end product, heme, controls its own synthesis (Granick and Levere, 1964). The ALA is next transferred into the cytosol (Sano and Granick, 1961).

2. *Porphobilinogen*

Two molecules of ALA are next condensed to form the precursor pyrrole, porphobilinogen (PBG), in the cytosol (Cookson and Rimington, 1953). This reaction is catalyzed by the enzyme ALA-dehydrase (ALA-D), an enzyme that is strongly inhibited by lead. The activity of this enzyme is commonly assayed in lead poisoning and a reduced activity is generally regarded as presumptive evidence of exposure to lead.

3. *Uroporphrinogen*

Next, two enzymes, uroporphrinogen I synthase (UROgenI-Syn)and uroporphrinogen III cosynthase (UROgenIII-Cosyn) act together to condense 4 moles of PBG into the cyclic tetrapyrrole, uroporphrinogen III (UROgenIII). The mechanism of cyclization continues to escape elucidation. UROgenI-Syn appears to initially catalyze the formation of a symmetrical linear tetrapyrrole. UROgenIII-Cosyn next flips the D ring and closes the pyrroles into an asymmetrical porphyrin ring of the type III configuration. In the absence of the UROgenIII-Cosyn, the symmetrical linear tetrapyrrole spontaneously closes into a symmetrical porphyrin of the type I configuration. Normally, there is a great excess of UROgenIII-Cosyn, and UROgenIII is synthesized. Both enzymes have been isolated from the spleens of anemic

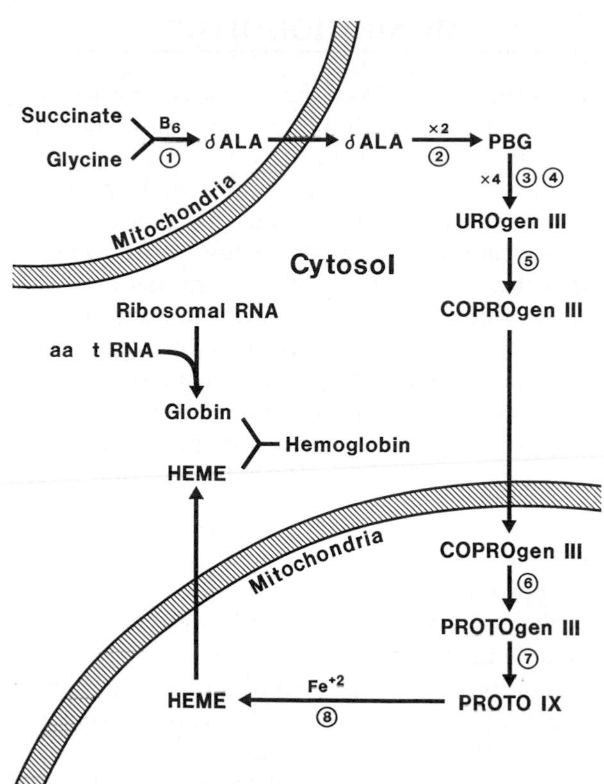

FIGURE 8.4 The synthetic pathway for protoporphyrin and heme. Note the partitioning of the heme synthetic pathway between the mitochondria and the cytosol. The circled numbers correspond to the enzymes listed in Table 8.2.

mice and *in vitro*, in the presence of both enzymes, UROgenIII was produced (Levin and Coleman, 1967).

4. *Coproporphyrinogen*

The eight-carboxyl UROgens I or III are next progressively decarboxylated into the four-carboxyl coproporphyrinogens (COPROgen) I or III with the decarboxylations catalyzed by the enzyme uroporphyrinogen decarboxylase (UROgen-D). UROgen is nonspecific so it catalyzes the decarboxylation of either UROgenI or UROgenIII. The COPROgens now move back into the mitochondria (Fig. 8.4).

TABLE 8.2 Nomenclature for Enzymes of Porphyrin and Heme Synthesis and Their Synonyms

Abbreviations	Nomenclature
1. ALA-Syn	δ-Aminolevulinate synthase (synthetase)
2. ALA-D	δ-Aminolevulinate dehydrase (dehydratase); porphobilinogen synthase
3. UROgenI-Syn	Uroporphyrinogen I synthase (synthetase); porphobilinogen deaminase
4. UROgenIII-Cosyn	Uroporphyrinogen III cosynthase (cosynthetase)
5. UROgen-D	Uroporphyrinogen decarboxylase
6. COPROgenIII-Ox	Coproporphyringen III oxidase
7. PROTOgen-Ox	Protoporphyrinogen oxidase
8. FER-Ch	Ferrochelatase; heme synthase (synthetase)

5. *Protoporphyrinogen*

Within the mitochondria, coproporphyrinogen III oxidase (COPROgenIII-Ox) catalyzes the decarboxylation of the two propionic acid groups on the A and B pyrrole rings of COPROgenIII to vinyl groups and the resulting product is protoporphyrinogen III (PROTOgenIII). COPROgenIII-Ox is highly specific for COPROgenIII and this explains the presence of only type III porphyrin isomers in nature. This also means that COPROgenI is a terminal intermediate that is oxidized to coproporphyrin I, the end product of this path (Fig. 8.5). COPROgenIII, when in excess, is also oxidized to coproporphyrin III. Similarly, the UROgens I and III can be oxidized to their end products, the uroporphyrins. We can see from (Fig. 8.5) that each of the -gen forms can be oxidized to their free forms, which are forms usually found in the circulatory system. These free porphyrins and protoporphyrin are

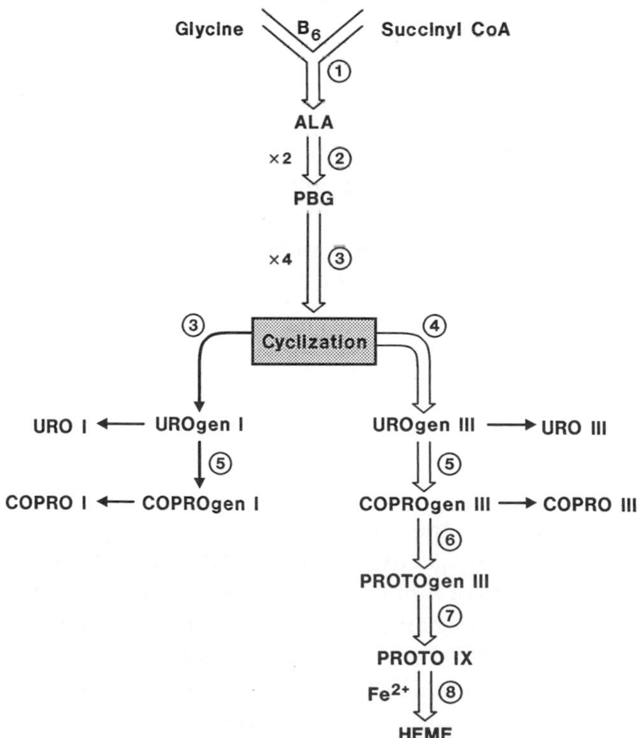

FIGURE 8.5 Alternate pathways for porphyrin synthesis. Normally, enzymes 3 and 4 function together in a coordinated manner to form heme. In the absence of enzyme 4, the alternative and terminal pathway to form the I isomers is taken. The circled numbers correspond to the enzymes listed in Table 8.2.

photoreactive and are the causative agents of the photosensitivity in the porphyrias.

6. Protoporphyrin

PROTOgenIII, the two-carboxyl porphyrinogen, is next oxidized at the carbon bridges to form the methene bridges connecting the pyrroles and is catalyzed by protoporphyrinogen III oxidase (PROTOgenIII-Ox). The resulting product is protoporphyrin IX (PROTO IX).

7. Heme

Within the mitochondria, ferrous iron (Fe^{2+}) is chelated with PROTO IX to form heme and is catalyzed by the enzyme ferrochelatase (FER-Ch). Iron can also be incorporated with relative ease by a nonenzymic method but the enzymic iron incorporation is more than 10 times that of the nonenzymic route (Labbe and Hubbard, 1961). Conditions that help to maintain iron in its ferrous form including the presence of reducing agents (ascorbic acid, cysteine, glutathione) enhance both enzymic and nonenzymic iron incorporation. Because iron is incorporated into heme as an integral part of the molecule, labeled iron can also used as a cohort label to determine the lifespan of the erythrocyte (Table 8.1).

8. Hemoglobin

Heme next moves into the cytosol where it is linked to the heme pocket of globin in a precise and stable position that permits binding of oxygen to the heme. Hemoglobin consists of 4 moles of this heme–globin moiety linked together as a globular tetramer. It is this globular tetrameric form of the hemoglobin molecule that permits the cooperative intearction of oxygen binding, which gives the familiar sigmoid oxygen–hemoglobin saturation curve.

9. Summary

In summary, the synthesis of porphyrins, heme, and globin can only occur in those respiring cells with full complements of mitochondrial and cytosolic enzymes. The TCA cycle is an aerobic cycle and therefore a lack of oxygen would preclude synthesis of succinyl-CoA and hence of heme. PROTO IX formation and the chelation of iron to form heme are also oxygen-requiring systems. Therefore, the reticulocyte with its residual complement of enzymes can synthesize hemoglobin but the mature erythrocyte, which is devoid of mitochondrial enzymes, cannot.

III. METHODOLOGY

The principal method now employed for the detection of porphyrins in biological materials in the clinical laboratory is based on the characteristic red fluorescence observed when acidic solutions of the porphyrins are exposed to ultraviolet light. The color of the fluorescence cannot be used to distinguish between the uroporphyrins and the coproporphyrins and, therefore, these must be separated prior to examination for fluorescence. The separation procedures take advantage of the solubility differences of the porphyrins in various organic solvents. In general, the following solubility properties are principally employed in the separation of the uroporphyrins from the coproprophyrins:

1. The coproporphyrins are soluble in diethyl ether, but the uroporphyrins are not; therefore, uroporphyrins remain in the aqueous phase.
2. Both uroporphyrin and coproporphyrin are soluble in strong acid, 1.5 *N* HCl. Coproporphyrins are therefore extracted from the organic phase with 1.5 *N* HCl. The uroporphyrins in the aqueous phase are absorbed with aluminum trioxide and subsequently eluted with 1.5 *N* HCl.

The acidic solutions of the porphyrins are then observed visually for fluorescence or examined in a sensitive fluorometer. The most suitable condition for the excitation of fluorescence is the use of ultraviolet light in the near visible range using aqueous solutions of the porphyrins at pH 1–2. Further means of identification include spectrophotometric examination, melting points of their methyl esters, and high-performance liquid chromatography. The methods for the quantitative determination of the porphyrins are given by Labbe and Lamon (1986). The following screening procedures are guides for further laboratory examinations.

A. Urinary Porphyrins

The urine for porphyrin examination must be alkalinized because porphyrins readily precipitate in acid urine. Addition of 0.5 g of sodium carbonate to the collecting bottle for each 100 ml of urine will keep the urine alkaline. The alkalinized urine can be stored at 4°C for 2–3 days prior to analysis. All contact of the urine with metal must be avoided and unfiltered urine is used. The following is a simplified screening procedure:

1. Place 5 ml urine in a 250-ml separatory funnel and add 5 ml acetate buffer (4 parts glacial acetic

acid : 1 part saturated sodium acetate) and adjust pH to 4.6–5.0.
2. Add 15 ml cold water.
3. Extract the mixture with two 50-ml aliquots of diethyl ether (or until the ether phases show no fluorescence under UV light) and pool the aliquots. The coproporphyrins will enter the ether phase.
4. Most of the porphyrins in urine are in the form of their nonfluorescent precursors. Storage of the urine for 24 hours in a refrigerator will enhance their conversion to the fluorescent pigments. If fresh urine is used, the ether phase is gently shaken with 5 ml of fresh 0.005% iodine solution (dilute 0.5 ml of a stock 1% iodine in ethanol solution to 100 ml with water) to convert the precursors to the porphyrins.
5. Extract the pooled ether phases with 20 ml 5% HCl (1.5 *N*) and examine for fluorescence under UV light. Fluorescence indicates the presence of coproporphyrins.
6. Uroporphyrins are insoluble in ether. Therefore, fluorescence in the aqueous phase urinary phase indicates the presence of uroporphyrins.

B. Fecal Porphyrins

The screening test as described for urine can also be used with fecal samples after prior extraction with strong acid. Five grams of a fecal sample is emulsified with 10 ml 95% ethanol. Twenty-five milliliters of concentrated HCl and 25 ml water are added to the emulsion and the mixture is kept overnight at room temperature. The mixture is next diluted to 200 ml with water, filtered, and the filtrate examined for porphyrins as described for urine.

C. Blood Porphyrins

Almost all porphyrins in blood are present in the erythrocytes and only trace amounts are in plasma. Therefore, it is important to use whole blood in the test and to understand that the test is a measure of the porphyrin content of the erythrocytes. Mix 2 ml heparinized whole blood and 6 ml ethyl acetate : glacial acetic acid (4 : 1) in a glass centrifuge tube. Stir thoroughly, centrifuge, and pour off the supernate into a second centrifuge tube. Add 1 ml 3 *N* HCl, vortex, and allow the phases to separate. Fluorescence under UV light in the aqueous layer indicates the presence of porphyrins in the erythrocytes. Normally, there will only be a trace of fluorescence so that more than a trace is indicative of an abnormal concentration.

D. Porphobilinogen

The Watson and Schwartz (1941) test is a reliable screening test because false positives are not commonly seen. Mix 3 ml fresh urine and 3 ml Ehrlich's aldehyde reagent (0.7 *g* p-dimethylaminobenzaldehyde, 150 ml concentrated HCl, and 100 ml water), the reagent commonly used for the urine urobilinogen test. Next, 5 ml chloroform is added, shaken vigorously in a separatory flask, and allowed to separate. The porphobilinogen aldehyde formed in the test is insoluble in chloroform and will remain in the lower aqueous phase. If the pink color is due to urobilinogen, it will be extracted into the chloroform phase. Porphobilinogen is found characteristically in the urine of patients with hepatic forms of porphyria, forms that have not been reported in animals.

IV. THE PORPHYRIAS

A. Classification

By convention, the term *porphyria* is used to define those disease states having a hereditary basis and having increased urinary and/or fecal excretion of the uroporphyrins and coproporphyrins. Depending on the fundamental biochemical defect, the porphyrias can be broadly classified on the basis of their tissue of origin, the erythropoietic system, or the liver. The term porphyrinuria is used to define those acquired conditions in which the principal, if not the sole, porphyrins being excreted are the coproporphyrins. Excess coproporphyrin excretion is seen in a wide variety of conditions including infections, hemolytic anemias, liver disease, and lead poisoning. The screening test for coproporphyrinuria has been especially useful for detection of exposure to lead.

Many systems of classification of the porphyrias have been used, most of which are based on the defect in the tissue of origin, erythropoietic, and the liver. There is general agreement on the classification of the erythropoietic forms but there is still some confusion as to the classification of the hepatic forms. These have been classified on the basis of their clinical manifestations, heredity, porphyrins excreted, and on the basis of their enzymatic defects (Eales, 1961; Tschudy, 1965; Elder, 1982; Hindmarsh, 1986; Kappas *et al.*, 1995). A useful system of classification is given in Table 8.3.

Methods are also available for the experimental production of the two major types of porphyria. In lead or phenylhydrazine poisoning, a type of porphyrinuria is seen that has some of the characteristics of erythropoietic porphyria of humans and cattle (Schwartz *et*

TABLE 8.3 Classification of the Porphyrias[a]

Porphyria type	Inheritance	Enzyme deficiency
Erythropoietic porphyrias		
Congenital erythropoietic porphyria	AR	UROgenIII-Cosyn
Erythropoietic protoporphyria	AD	FER-Ch
Hepatic porphyrias		
ALA-D deficiency porphyria	AR	ALA-D
Acute intermittent porphyria	AD	UROgenI-Syn (PBG-D)
Porphyria cutanea tarda	AD	UROgen-D
Hepatoerythropoietic porphyria	AR	UROgen-D
Harderoporphyria	AR	COPROgenIII-Ox
Hereditary coproporphyria	AD	COPROgenIII-Ox
Variegate porphyria	AD	PROTOgen-Ox

[a] Inheritance: A, autosomal; R, recessive; D, dominant. See Table 8.2 for enzyme abbreviations.

al., 1952). A hepatic form can be produced with Sedormid (allylisopropylacetylcarbamide) (Schmid and Schwartz, 1952), dihydrocollidine (Granick and Urata, 1963), or hexachlorobenzene.

B. Erythropoietic Porphyrias

1. *Bovine Congenital Erythropoietic Porphyria*

a. Introduction

One of the characteristic findings in bovine congenital erythropoietic porphyria (CEP) is a reddish brown discoloration of the teeth and bones. Discolorations of this type have been observed in cattle at slaughter since the turn of the century and these cattle are presumed to have had the disease. The first living cases were encountered in South Africa in a herd of grade Shorthorn cattle and were described by Fourie (1936) and Rimington (1936). Since then, CEP has been reported in cattle in Denmark (Jorgensen and With, 1955), the United Kingdom (Amoroso *et al.*, 1957), the United States (Ellis *et al.*, 1958; Rhode and Cornelius, 1958), and Jamaica (Nestel, 1958). The disease has been seen primarily in Holsteins in addition to the cases in the Shorthorns and Jamaican cattle.

The simple Mendelian autosomal recessive heredity of the disease was established by study of the geneology of the affected cattle and by breeding experiments (Fourie, 1939). The affected homozygotes are characterized by discoloration of the teeth and urine, photosensitivity of the light areas of the skin, and generalized lack of condition and weakness. The condition is present at birth and severely affected calves must be protected from sunlight if a state of health is to be maintained.

The predominant symptoms of teeth and urine discoloration and the photosensitization of the severely affected animal are readily apparent and a tentative diagnosis can be confirmed by the orange-red fluorescence of the teeth and urine when examined with near-UV light or a Woods lamp. The symptomatology of affected animals, however, may vary from minimal to severe and with age and time of year (Fourie and Roets, 1939). The discoloration of the teeth may vary in the same animal, being more pronounced in the young and less apparent in the older animals. Porphyrin deposits are heavily concentrated in the dentine so the occlusal surfaces should also be examined. If porphyrins are present, the discoloration and the fluorescence of the dentine will be readily detected.

The degree of photosensitization will vary with the amount of porphyrin deposition in the dermis, coat color, density of the coat and extent of exposure to sunlight. The photosensitization may be so slight as to escape detection. At times, loss of condition may be the only outward symptom for which the veterinarian is called. Marked variations in the urinary excretion of the porphyrins also occur. These may range from minimal to thousands of micrograms in the same animal so that the urine color may vary widely. The variations observed in this disease provide an indication of the dynamic state of flux of porphyrin metabolism in the living animal and the porphyrin deposits constitute a part of this dynamic state.

b. Distribution of Porphyrins

Some normal values for the porphyrins in animals are given in Table 8.4. These values are to be considered best approximations obtained from a relatively few animals. The figures, however, provide an indication

TABLE 8.4 Normal Values of Porphyrin Concentrations[a,b]

	Urine (μg/dl)		Feces (μg/g)		Erythrocytes (μg/dl cells)		Plasma (μg/dl)		Bone marrow (μg/dl cells)		
Species	URO	COPRO	COPRO	PROTO	COPRO	PROTO	COPRO	PROTO	URO	COPRO	PROTO
Cow	1.09 ± 0.92	4.06 ± 1.96	3.12 ± 0.96	0.75 ± 0.30	Trace	Trace	Trace	Trace	1	5	100
	0.80 ± 1.60	2.05 ± 6.15	1.11 ± 4.28	0.15 ± 1.25							
Pig		104[c]				118					
Rabbit	25[c]	41	25[c]		2.6	83.3			Trace	4.5	87.5
Dog	50[c]			50[c]		35.0					

[a] References: cow, Jorgensen (1961b), Watson *et al.* (1959), Kaneko (1969); pig, Cartwright and Wintrobe (1948); rabbit, Schmid *et al.* (1952); dog, Schwartz *et al.* (1960).
[b] URO, uroporphyrin; COPRO, coproporphyrin; PROTO, protoporphyrin.
[c] Micrograms per day.

of the very low concentrations of the free porphyrins normally found in the body. Thus, the finding of porphyrins in greater than trace amounts is always significant. Porphyrin concentrations in porphyric cows and calves are given in Table 8.5 (Kaneko and Mills, 1969a, 1969b).

i. Urine It is known that porphyrin excretion varies over wide limits. Jorgensen (1961) in 52 cases found values for urinary uroporphyrins between 6.3–3900 μg/dl (0.076–46.96 μmol/liter) and coproporphyrins between 2.1–8300 μg/dl (0.032–126.74 μmol/liter). At concentrations of 100 μg/dl (1.53 μmol/liter) or more, a reddish discoloration is discernible in the urine. At 1000 μg/dl (15.27 μmol/liter) or more, an intense red fluorescence of the urine is readily observed when examined in the dark with a Woods light. The principal porphyrins excreted are URO I and COPRO I and only a small fraction is of the type III isomers. The percentage of each appearing in the urine is also variable. There is usually a greater excretion of URO I than COPRO I but Jorgensen (1961) observed a greater excretion of COPRO I than URO I.

TABLE 8.5 Porphyrins in Blood and Excreta of Normal, Porphyric, and Porphyria Carrier Cattle[a,b]

	Erythrocytes		Plasma		Urine		Feces	
Animals[c]	COPRO	PROTO	COPRO	PROTO	COPRO	PROTO	COPRO	PROTO
Normal mature cows								
(N = 10)	Trace	Trace	Trace	Trace	2.05–6.15	0.80–1.60	111–428	15–125
					(4.06)	(1.09)	(312 ± 96)	(75 ± 30)
Mature porphyric cows								
1184	3.0	61	15.3	1.8	410	378	5670	46
652 (N = 3)	3.4	64	4.5	1.6	313	336	1900	12
2026 (N = 2)	3.1	457	8.9	1.5	498	487	2090	62
718	89.7	36			1450	1280		
Mature porphyria carrier cows								
1140	Trace	Trace	Trace	Trace	Trace	Trace	292	88
1141	Trace	Trace	Trace	Trace	Trace	Trace	273	92
Porphyric calves, 2–6 months old								
1857 (N = 3)	2.9	109	1.5	Trace	13	70	796	144
1801 (N = 3)	7.8	104	38.4	2.6	1430	1144	12	72
1959	18.6	288	Trace	Trace	480	265	22	99
Porphyria carrier calf, 5 months old								
1802	Trace	Trace	Trace	Trace	Trace	Trace	495	40

[a] Values are means given in micrograms per deciliter or micrograms per 100 grams; values in parentheses are means plus or minus a single standard deviation (Kaneko and Mills, 1969a).
[b] COPRO, coproporphyrins; PROTO, protoporphyrins; URO, uroporphyrins.
[c] N denotes number of animals or number of determinations per animal.

Porphobilinogen is not usually found in bovine CEP urine and earlier reports of its presence (Ellis *et al.*, 1958; Jorgensen and With, 1955) have not been confirmed (Jorgensen, 1961b). Normally colored, nonfluorescent urine of a CEP cow has consistently given a definite pink Ehrlich reaction but unlike porphobilinogen aldehyde, the pigment is soluble in chloroform (Kaneko, 1969). The nature of the pigment is unknown. On heating in a steam bath for an hour or standing at room temperature for several days, a red fluorescence is apparent on exposure to UV light. Quantitative porphyrin determinations of this urine yielded values of 135 μg/dl (1.63 μmol/liter) and 87 μg/dl (1.33 μmol/liter) for uroporphyrin and coproporphyrin, respectively. Watson *et al.*, (1959) also described a similar experience with bovine CEP urine.

ii. Bile and Feces Bovine fecal porphyrins may be derived from two sources: the bile and from chlorophyll of the food. The porphyrins derived from chlorophyll are excluded by the usual analytical method. Essentially, the only porphyrin found in the bile and feces of CEP cattle is COPRO I, and its concentration varies over wide limits (Tables 8.4 and 8.5). Fecal coproporphyrin varied between 1.9–11,800 μg/g (0.003–18.0 μmol/g) and biliary coproporphyrin between 320–13,600 μg/dl (4.88–207.67 μmol/liter) (Jorgensen, 1961). Only small amounts of COPRO III have been observed in feces. This preponderance of COPRO I in feces was also observed by Watson *et al.* (1959) who also reported the presence of small amounts of URO I.

iii. Plasma and Erythrocytes Only traces of free porphyrins are normally present in the plasma and in the erythrocytes. In bovine CEP plasma, Watson *et al.* (1959) observed variable amounts of porphyrins that were in general equally URO I (1–27 μg/dl; 0.012–0.33 μmol/liter) and COPRO I (4.2–25 μg/dl; 0.064–0.38 μmol/liter). A striking difference as compared to the human disease was the high level of free protoporphyrin in the erythrocytes of the CEP cow. The significance of this high level of protoporphyrin was unclear but was most likely related to the severe hemolytic anemia in the CEP cow. Excess PROTO IX is commonly found in iron deficiency, hemolytic anemia, and in lead poisoning. In iron deficiency, PROTO IX accumulates because of failure to form hemoglobin, but in CEP, serum iron is normal or elevated (Watson *et al.*, 1959; Kaneko, 1963; Kaneko and Mattheeuws, 1966). Similarly, PROTO IX accumulates in lead poisoning due to inhibition of ferrochelatase and a subsequent inability to insert iron into hemoglobin. An unlikely possibility is that the protoporphyrin in CEP is a type I isomer (Schmid, 1966).

iv. Tissue The range of concentrations of porphyrins in the tissues of CEP cattle is given in Table 8.6. The deposition of porphyrins throughout the bones and soft tissues is readily apparent at postmortem of severe cases by the generalized discoloration. A reddish brown discoloration is most apparent in the teeth, bones, and bone marrow. The greatest amount of discoloration of soft tissues occurs in the lungs and spleen in which characteristic fluorescence may be observed with UV light. The high concentration of porphyrins in the spleen is consistent with the hemolytic type of anemias in CEP. Discoloration of skin, muscle, heart, liver, and kidney is also observed but only a part is due to porphyrins. The discoloration is likely due to other porphyrin derivatives and to hemoglobin staining.

c. Hematology

The hematologic picture of the majority of reported cases is one of a responsive hemolytic anemia. In general, the degree of response is correlated with the severity of the hemolytic anemia. The anemia in mild cases of bovine CEP is normocytic and in the more severe cases is macrocytic. In the severe cases of bovine CEP, there is reticulocytosis, polychromasia, anisocytosis, basophilic stippling, and an increase in nucleated erythrocytes. A consistent monocytosis has been observed (Rhode and Cornelius, 1958; Kaneko, 1963) but remains unexplained. There is a markedly decreased M:E ratio in the presence of the anemia indicating a marked bone marrow hyperplasia. Bone marrow is also a principal site of porphyrin deposition, and Watson *et al.* (1959) found high concentrations of uroporphyrins in the bone marrow of a CEP cow.

The presence of porphyrins in the nucleated erythrocytes is clearly evident by examination of unfixed and unstained bone marrow smears with a fluorescent microscope. These fluorescent cells have been called fluorocytes. This phenomenon was originally observed by Schmid *et al.* (1955) in the bone marrow of a human patient and was an important contribution to the localization of the metabolic lesion in the erythropoietic tissue. They also reported that the fluorescence was seen only in morphologically abnormal nucleated erythrocytes that contained abnormal nuclear inclusions. Similar nuclear abnormalities were observed in bovine CEP bone marrow (Watson *et al.*, 1959). Schmid *et al.* (1955) concluded that there were two populations of erythrocytes, one normal and one containing free porphyrins.

The presence of two populations of erythrocytes was reported in humans but this was attributed to the intermittent hemolytic crises that occurred (Gray *et al.*, 1950). Tschudy (1965) has also pointed out that a single

TABLE 8.6 Tissue Porphyrins in Bovine Congenital Erythropoietic Porphyria[a]

Tissue	Uroporphyrins	Coproporphyrins	Protoporphyrins	Total Porphyrins
Bone marrow	Tr–162	Tr–1890	Tr–394	Tr–2,396
Bones	6,000	Tr	Tr	6,000
Teeth	18,550	Tr	Tr	18,550
Spleen	0–10	Tr–342	Tr–60	7–400
Liver	0–Tr	16–340	42–65	66–403
Lung	0–79	0–37	Tr–20	20–130
Kidney	0	Tr–117	5–16	5–133
Lymph node	0	0–40	1–7	1–49
Intestine	0	Tr–65	7–77	18–104
Stomach	0	Tr–58	12–82	12–111
Bile	0–690	112–12,205	0–856	112–13,750
Adrenal	0–6	Tr–202	19–170	19–378
Ovary	0	65	1	66
Testes	0	0	9–14	9–14
Skin	0	0	0	0
Muscle	0	10	30	40
Brain and spinal cord	0	23	57	80
	0			

[a] Given in micrograms per 100 grams of tissue.

population of erythrocytes is more likely to be present rather than two separate lines of erythrocytes. Runge and Watson (1969) studying fluorescing bovine CEP bone marrow cells after bleeding also concluded that there was only a single population of erythrocytes in bovine CEP.

The hematology of newborn CEP calves also exhibits striking differences when compared to that of older CEP calves and cows (Kaneko and Mills, 1969a). There is an intense erythrogenic response in the neo-natal CEP calf that persists for the first 3 weeks of life. Nucleated erythrocyte counts during the first 24 hours of life ranged from 5000 to 63,500/μl. Reticulocyte counts were lower than expected (6.4%) and increased to a peak of only 12.5% at 4 days of age. The persistent reticulocytosis is thought to be due to a delay in maturation of the reticulocytes (Smith and Kaneko, 1966; Rudolph and Kaneko, 1971). This delay in maturation of the reticulocyte, which is proportional to the degree of anemia, is now a well-established phenomenon during the reticulocyte response to a blood loss or hemolytic anemia. In essence, this delay represents the increase in survival time of the reticulocyte beyond its normal 1-day survival time. This increased survival time is the now commonly used maturation correction factor (MCF) for estimating the reticulocyte production index (RPI) when evaluating the response to an anemia.

d. Mechanism of the Anemia

A responsive hemolytic anemia is a well-established occurrence in CEP. Erythrocyte porphyrins are high in CEP and if these erythrocytes with high porphyrin concentrations were more susceptible to destruction, a shortening of their lifespan would be expected. Erythrocyte lifespan is shortened in bovine (Kaneko, 1963) and in human (Gray *et al.*, 1950) CEP. There is general agreement that this shortening of lifespan is associated with the hemolytic process but the mechanism of the hemolysis remains obscure. It has been shown that erythrocyte survival in bovine CEP is inversely correlated with erythrocyte coproporphyrin concentration (Kaneko *et al.*, 1971). The shortest erythrocyte survival time of 27 days (normal = 150 days) was associated with the highest erythrocyte coproporphyrin concentration, which is a reasonable expectation. The porphyrins through their lipid solubility are presumed to damage the erythrocyte membrane leading to the hemolysis. *In vivo* ^{59}Fe metabolic studies were completely compatible with a hemolytic type of anemia, and ineffective erythropoiesis, that is, bone marrow hemolysis, was also demonstrated (Kaneko, 1963; Kaneko and Mattheeuws, 1966). Plasma iron turnover and transfer rates, erythrocyte iron uptake, and organ uptakes were increased as expected in a hemolytic process.

The mechanism of cell damage has also been studied in reticulocytes and in nucleated erythrocytes. A biochemical defect in the bovine CEP reticulocyte *in vitro* was expressed as an increase in porphyrin synthesis, a marked decrease in heme synthesis and a delay in the maturation time of the reticulocyte (Smith and Kaneko, 1966). The $T_{1/2}$ for the maturation of the reticulocyte was 50 hours compared to a normal of 3–10 hours. This delay in reticulocyte maturation is thought to be the direct result of the defect in heme synthesis since the rate of heme synthesis controls the rate of maturation of the reticulocyte (Schulman, 1968). This means that the increase in the reticulocyte survival time is inversely proportional to the degree of anemia.

A similar delay in the maturation of the metarubricyte to the reticulocyte was observed in the bone marrow cells of CEP cows (Rudolph and Kaneko, 1971) but there was no effect on the earlier nucleated erythrocytes. Therefore, the more mature erythrocytic cells are the cells most noticeably affected by the high porphyrin content. This is not surprising because heme and hemoglobin synthesis are most active in the later stages of erythrocytic cell development. Ultimately, the accumulation of porphyrins in these cells, whether in bone marrow or in blood, induces hemolysis. On exposure of surface capillaries to sunlight, photohemolysis of the type observed in erythropoietic protoporphyria (Harber *et al.*, 1964) would further aggravate the hemolysis.

This hemolytic mechanism might also explain the striking erythrogenic response seen in the neonatal porphyric calf. Because most of the porphyrins are within the fetal erythrocytes and these would not normally cross the placenta, the porphyrin containing erythrocytes would accumulate in the fetus and a profound hemolysis would occur *in utero.* This hemolysis in turn would induce a marked erythrogenic response in the fetus and this is seen at birth. At birth, porphyrins are high but they fall to their steady-state level in about 3 weeks in CEP calves. This is comparable to the rate of clearance of ^{14}C-porphyrin into urine, which fell to 0.1% of the initial concentration in 3 weeks (Kaneko, 1963). Furthermore, 3 weeks is also the time at which the erythrogenic response is stabilized at a steady-state level in the neonatal calf (Kaneko and Mills, 1969c).

In summary, as a result of the heme synthetic defect in erythropoietic porphyria, there is excess porphyrin accumulation in the mature and developing erythrocytes that induces their hemolysis in the circulation or in the bone marrow. There is a corresponding shortening of erythrocyte lifespan. In addition, the decrease in heme synthesis induces an increase in the survival time of the reticulocyte by inhibiting the maturation of the developing erythrocytes, which further aggravates the anemia. This biochemical defect in heme synthesis is morphologically expressed in the fluorocytes and in the evidence of erythrogenic response in the blood and bone marrow, the degree of which is directly related to the severity of the enzymatic defect of the porphyria.

e. Detection of the Carrier State

Bovine CEP is inherited as an autosomal recessive trait. Previously, carrier animals were detectable only by the occurrence of the disease in their progeny. Levin (1968a) developed an assay for UROgenIII-Cosyn and found that its activity was considerably less in CEP cattle than in normals (Levin, 1968b). Heterozygous cattle, which are the clinically unaffected carriers of the porphyria gene, had UROgenIII-Cosyn activities intermediate between porphyrics and normals (Romeo *et al.*, 1970). Similarly low UROgenIII-Cosyn activity is found in human CEP but carriers are less readily detectable in humans than in cattle (Romeo and Levin, 1969). These findings are also in keeping with the concept that the genetic defect in CEP is a deficiency of UROgenIII-cosyn. Romeo (1977) has reviewed the genetic aspects of all forms of hereditary porphyrias and concludes that the weight of evidence is conclusive for a UROgenIII-Cosyn deficiency in CEP.

f. Metabolic Basis of Bovine Congenital Erythropoietic Porphyria

The mechanisms for heme biosynthesis and the biochemical nature of the porphyrins in the tissues and excreta provide an explanation for the metabolic defect in CEP. Certain features of the biosynthetic mechanism are particularly important in an explanation of the clinical and metabolic manifestations of CEP, which has its ultimate pathogenesis in a hereditary deficiency of UROgenIII-Cosyn. These can be summarized as follows:

1. There is a compartmentation of the enzymes of heme biosynthesis between the mitochondria and the cytosol.
2. Mitochondrial systems are involved in the synthesis of ALA, PROTO IX, and heme.
3. Cytosolic systems catalyze the formation of PBG, UROgenIII or I, and COPROgenIII or I.
4. Mitochondria are present only in the immature nucleated erythropoietic cells and in the reticulocytes. The most active heme and hemoglobin synthesis occurs in the metarubricyte and secondly in the reticulocyte.
5. There is no heme or hemoglobin synthesis in the mature erythrocyte.

The enzymatic deficiency in CEP is localized in the erythropoietic tissue within the developing erythropoietic cells, which are the mitochondria-containing cells. Normally, the combined action of UROgenI-Syn and UROgenIII-Cosyn catalyzes the formation of the normal type III porphyrin isomer, UROgenIII, leading to the formation of heme. In the absence of UROgenIII-Cosyn, the type I isomer, UROgenI, and then COPROgenI is formed. Thus, the relative activities of these enzymes govern the extent as to which of these pathways is traversed. The type I isomers that are formed in the deficiency state cannot be converted into PRO-

TOgenI and into a type I heme. This is because there is no coproporhyrinogen I oxidase and the COPROgenIIIOx is highly specific only for the type III isomer. The type UROgenI and COPROgenI isomers are oxidized to their corresponding uroporphyrins and coproporphyrins. These oxidized free porphyrins accumulate in the erythropoietic tissues, developing erythrocytic cells, and in the mature erythrocytes where they induce the hemolysis characteristic of CEP. In addition, the porphyrins are released into the circulation and are widely distributed throughout the body in all body fluids and are readily excreted in the feces and urine. They are deposited in all tissues, most notably in the teeth, bones, and skin. When exposed to ultraviolet light, the porphyrins in the skin are excited by absorption of the ultraviolet light energy into an unstable higher level energy state. The excitation energy is then be emitted when the excited molecule returns to its ground state. The energy can be emitted as fluorescence or transferred to molecular oxygen to form singlet oxygen. Singlet oxygen is a powerful oxidant for many forms of biologically important compounds including the peroxidation of membrane lipids, membrane and cellular proteins, cell enzymes, and cell organelles. Peroxidation appears to be the primary event in the photosensitivity and photodermatitis seen in the porphyrias (Meyer and Schmid, 1978; Poh-Fitzpatrick, 1982).

Total deficiency of UROgenIII-Cosyn is obviously incompatible with life so that surviving cases of CEP have only a partial deficiency of UROgeIII-Cosyn. Also, there is a wide variation in the severity of the disease commensurate with the degree of enzyme deficiency as well as with the conditions of husbandry. The severity of the disease, however, is quite constant in each bovine if it is kept under standard controlled conditions. The metabolic basis for bovine CEP is summarized in Fig. 8.6 in which the central theme is the genetically controlled UROgenIII-Cosyn deficiency with the resultant type I porphyrin accumulation and a failure in heme synthesis.

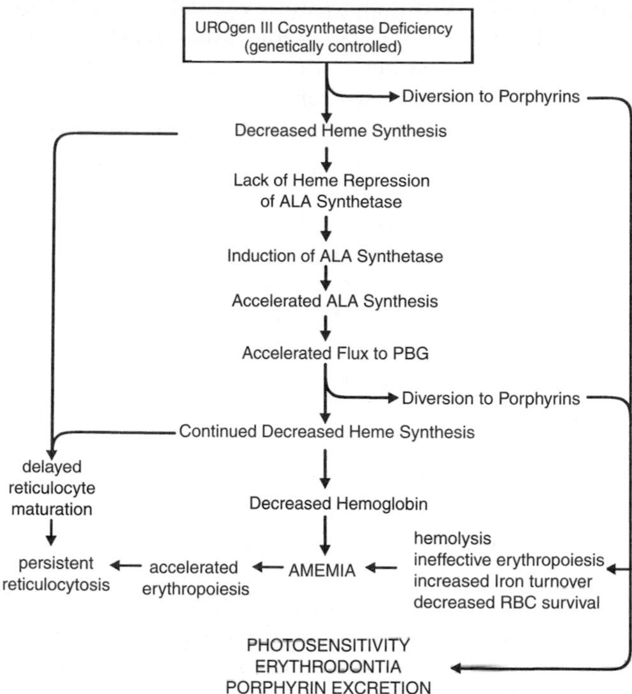

FIGURE 8.6 Metabolic basis of bovine congenital erythropoietic porphyria. The fundamental defect is a deficiency of UROgenIII-Cosyn leading to the accumulation of type I uroporphyrins and coproporphyrins. These type I porphyrins account for the clinical, hematologic, and biochemical features of this disease.

2. *Bovine Erythropoietic Protoporphyria*

This disorder of porphyrin metabolism occurs in humans and in cattle. Erythropoietic protoporphyria (EPP) was first reported in 1961 (Magnus *et al.*, 1961) and is now well recognized in humans (Redeker, 1963; Harber *et al.*, 1964). It is inherited as a dominant autosomal trait (Romeo, 1977). Patients do not have the major signs of CEP such as anemia, porphyrinuria, or discolored teeth. Photosensitivity of the skin is the only significant clinical manifestation of the disease and this is associated with a high plasma protoporphyrin concentration. In the laboratory, the most striking findings are the high concentrations of PROTO IX in the erythrocytes and feces.

In cattle, EPP has a pattern of recessive inheritance in contrast to humans and may be sex linked because to date it has only been seen in females. The photosensitivity also seems to diminish in adult life. Affected cattle also do not have anemia, porphyrinuria, or discoloration of the teeth. Erythrocyte and fecal protoporphyrins are very high in comparison to normal cows (Ruth *et al.*, 1977).

The fundamental enzymatic defect in bovine EPP has been shown to be a deficiency of FER-Ch (Ruth *et al.*, 1977), which results in the accumulation of PROTO IX. Low FER-Ch was found in all tissues of EPP calves so that the defect is a total body defect.

3. *Porphyria of Swine*

Porphyria in swine was first recognized in New Zealand by Clare and Stephens (1944). Later it was recognized in Denmark and a number of studies with these swine have been published (Jorgensen and With, 1955; Jorgensen, 1959). Porphyria in swine is inherited as a dominant characteristic. Except for the very severe cases, there appears to be little or no effect on the general health of the pig. Photosensitivity is not seen even in the white pigs. The predominant feature in the affected pig is a characteristic reddish discoloration of the teeth, which fluoresces upon exposure to ultraviolet light. Porphyrin deposition in the teeth of the newborn pig is virtually pathognomonic of porhyria in swine. Occasionally, darkly discolored teeth may not fluoresce but porphyrins may be extracted from these teeth with 0.5 *N* HCl (With, 1955). Similar, though less apparent, deposition occurs in the bones. The porphyrins are principally URO I and have been found in concentrations of up to 200 μg/g of teeth or bones. The liver, spleen, lungs, kidneys, bones, and teeth are also discolored by another dark pigment, the nature of which is unknown (Jorgensen, 1959).

The urine of the affected pig is discolored only in the more severely affected pig. The 24-hour urinary excretion of uroporphyrins ranged beween 100–10,000 μg and for coproporphyrin, only 50 μg. These were both the type I isomers. PBG is absent in the urine. Close similarities in this pattern of porphyrin excretion to that found in bovine CEP are apparent but the localization of the defect in the erythropoietic tissue has not been established. This disease has not been seen in pigs since the original occurrences.

4. *Porphyria of Cats*

Porphyria in cats was first reported in a male kitten (Tobias, 1964). One of its three littermates and kittens from a previous litter were also reported to have had the same unusually discolored teeth. The kitten's teeth were brown and under ultraviolet light, they fluoresced red. Its urine was amber colored and was qualitatively positive for uroporphyrin, coproporphyrin, and porphobilinogen. There was no evidence of anemia or photosensitization. These cats had been kept indoors all their lives. One of these cats was the propositus for a porphyric cat colony (Glenn *et al.*, 1968). Study of the inheritance of the porphyria in these cats indicated it to be of a simple Mendelian autosomal dominant trait analagous to that seen in swine.

A detailed study of porphyria in a family of Siamese cats has been reported (Giddens *et al.*, 1975) in which excessive accumulation of URO I, COPRO I, and PROTO IX were observed in erythrocytes, urine, feces, and tissues. These cats had photosensitivity, severe anemia, and severe renal disease. They concluded that the principal defect in these cats was a deficiency of UROgenIII-Cosyn similar to CEP of humans and cows.

5. *Normal Porphyrias*

All fox squirrels (*Sciurus niger*) have red bones and this is due to the accumulation of URO I and COPRO I (Flyger and Levin, 1977). The fox squirrel porphyria resembles the disease in humans, cows, and cats by having a deficiency of UROgenIII-Cosyn, type I porphyrins in their urine and feces, and discolored bones, teeth, and tissues that fluoresce on exposure to ultraviolet light. There is increased erythropoiesis but no apparent hemolytic anemia, and no photosensitivity or any other deleterious effects. These relatively benign effects are most likely due to their thick hair coats and nocturnal living habits. It is interesting that an enzyme deficiency with serious health effects in other animals should have evolved as a "normal" characteristic in the fox squirrel. This is understandable when one appreciates that CEP cattle that have always been kept indoors and protected from sunlight thrive and reproduce normally.

The UROgenIII-Cosyn deficiency is found only in the fox squirrel and not in the closely related gray squirrel (*Sciurus carolinensis*). Urine porphyrin excretion in the fox squirrel is 10-fold greater than in the gray squirrel and is markedly increased when erythropoiesis is stimulated by bleeding. The UROgenIII-Cosyn of fox squirrel erythrocytes is very heat sensitive and this may indicate that its CEP is due to an increased lability of the enzyme.

In the feathers of certain brightly colored birds, for example, Touracos and in certain lower animals and microorganisms, porphyrins accumulate but these appear to be normal phenomena.

C. Hepatic Porphyrias

This group of diseases is seen in humans only as naturally occurring diseases. They constitute the most common group of porphyrias seen in humans. The salient features of this group of porphyrias are also summarized in Table 8.2. As the name of this group implies, the predominant site of the metabolic defect is localized in the liver and the group is further subdivided on the bases of their principal clinical manifestations. Specific enzyme deficiencies have been identified for all forms of hereditary porphyria (Table 8.2).

1. *δ-Aminolevulinic Acid Dehydratase Porphyria*

This rare hepatic form of porphyria has a marked deficiency of the enzyme δ-aminolevulinic acid dehydratase (ALA-D) in the homozygous state (Doss *et al.*, 1979; Brandt and Doss, 1981) and is referred to as ALA-D porphyria (ADP). It is characterized by neurologic symptoms without skin photosensitivity. It is inherited as an autosomal recessive trait. A similar though less marked ALA-D deficiency that was without symptoms was reported (Bird *et al.*, 1979) and this is thought to be the heterozygous state of the deficiency.

2. *Acute Intermittent Porphyria*

Acute intermittent porphyria (AIP) is a common form in humans characterized by acute abdominal attacks and neurologic symptoms. Photosensitivity is not a feature of this form. Most patients are not clinically affected unless some form of aggravating factor is present. The attacks are precipitated by a large number of causative factors, the principal ones being barbiturates, sulfonamides, estrogens, and alcohol. The disease occurs more commonly in the adult female than in the male. The principal urinary finding is the excretion of large amounts of ALA and PBG. This is in keeping with the hereditary deficiency of UROgenI-Syn (PBG-D) in this disease (Strand, 1970). AIP is the major autosomal dominant form of hepatic porphyria.

3. *Porphyria Cutanea Tarda*

Porhyria cutanea tarda (PCT) is caused by a deficiency of UROgen-D and presents as both a sporadic form and a familial form. The sporadic form is the acquired form of PCT and is the most common of all forms of human porphyria. The familial form is inherited as an autosomal dominant. As the name implies, the characteristic clinical signs of PCT are the photosensitive lesions of the skin. The disease occurs in mid to late adult life and common precipitating causes of this disease are alcohol and estrogens. The disease can be successfully treated by avoidance of alcohol and estrogens. There is a decrease in hepatic UROgen-D in both forms but the enzyme deficiency is found in extra-hepatic tissues only in the familial form (Pimstone, 1982). In the sporadic form, erythrocyte UROgen-D activity is normal whereas in the familial form, erythrocyte UROgen-D is less than 50% of normal (McManus *et al.*, 1988). The erythrocyte URO-D assay is difficult and not readily available; therefore, indirect means of distinguishing the sporadic form from the familial are used. One of the most simple indirect methods has been to assay the plasma γ-glutamyltransferase activity (GGT). Sporadic PCT has a significant increase in GGT, whereas the familial form has normal GGT activity (Badcock *et al.*, 1993). Furthermore, the ratio of fecal COPRO III : COPRO I when combined with the plasma GGT was found to give an even more accurate differentiation of the sporadic form from the familial form (Badcock *et al.*, 1995).

4. *Hepatoerythropoietic Porphyria*

Hepatoerythropoietic porphyria (HEP) is a form which clinically resembles CEP but there is a severe deficiency of UROgen-D as in PCT. It is thought to be the homozygous form of familial PCT. HEP is characterized by a very severe photosensitivity but there is no liver involvement.

5. *Harderoporphyria*

This is a rare form of porphyria in which the propionate group on the A ring only is converted to a vinyl group. The normal next step of B-ring conversion is somehow disrupted. There is a deficiency of COPROgenIII-Ox but the mechanism explaining why the groups on both rings are not oxidized is unknown.

6. *Hereditary Coproporphyria*

Hereditary coproporphyria (HCP) is clinically similar to PCT with a mild cutaneous photosensitivity and it may also have neurologic symptoms as in AIP. Like AIP, HCP is commonly precipitated by drugs and alcohol. As in Harderoporphyria, COPROgenIII-Ox is the deficient enzyme.

7. *Variegate Porphyria*

The symptoms of variegate porphyria (VP) are generally more variable than the other forms but in most cases, acute abdominal attacks and photosensitivity are seen. VP is most common among the South African white population. VP is inherited as an autosomal dominant. There is a deficiency of PROTOgenIII-Ox, which can be observed in cultured fibroblasts and in leukocytes of VP patients.

D. Acquired Toxic Porphyrias

1. *Chemical Porphyrias*

The two major forms of the acquired toxic porphyrias are those due to organic chemical intoxication and to heavy metal poisonings, mainly lead. Experimentally, hexachlorobenzene (HCB) (Elder *et al.*, 1976), 2,3,7,8-tetrachlorodibenzo-*p*-dioxin (TCDD) (Elder and Sheppard, 1982), allylisopropylacetylcarbamide (Sedormid) (Schmid and Schwartz, 1952), or dihydrocollidine (DHC) (Granick and Urata, 1963) has been used to produce the hepatic forms of porphyria.

2. *Lead Poisoning*

Lead poisoning occurs in all domestic animals and is a significant clinical problem, particularly in the dog. In the dog as in other animals, the principal clinical features are related to the gastrointestinal and the nervous systems. Anemia is usually seen only in the longstanding chronic lead toxicities. The anemia has certain features that are suggestive of lead poisoning but are not diagnostic. The anemia is a mild to moderate normocytic normochromic anemia with basophilic stippling and nucleated erythrocytes (NRBC) out of proportion to the degree of anemia. Zook *et al.* (1970) consider that more than 15 stippled cells per 10,000 erythrocytes is suggestive of and that more than 40 stippled cells per 10,000 erythrocytes is diagnostic of lead poisoning in the dog. Stippling is thought to be accumulated ribosomal RNA aggregates that have not been normally degraded to their nucleotides and subsequently dephosphorylated by pyrimidine-5′-nucleotidase (P5NT). Lead has been shown to decrease the activity of the dephosphorylating enzyme P5NT in humans (Valentine *et al.*, 1976) and in calves (George and Duncan, 1982).

Lead is known to have widespread toxic effects on sulfhydryl-, carboxyl-, and imidazole-containing proteins, which would include enzymes, cell proteins, globins, and membrane proteins (Fell, 1984). However, only a few are altered specifically and significantly to be of diagnostic value. Globin synthesis and therefore hemoglobin synthesis is disrupted and this is the major mechanism of the anemia of lead poisoning. The anemia, however, occurs late in chronic lead poisoning and its nonspecific nature makes it of less diagnostic importance than is usually attributed to it.

A major focus is on the enzyme systems of heme synthesis because several of the enzymes are very sensitive in early exposure to small quantities of lead. The most sensitive are ALA-D and FER-Ch and these enzymes and their accumulated substrates are widely used as screening tests for lead exposure. Erythrocyte ALA-D is strongly inhibited by lead and, as a result, ALA rises in plasma and is excreted in the urine. Measurement of ALA is difficult and the results lack sensitivity for low-level lead exposure; therefore, instead of its substrate ALA, erythrocyte ALA-D is more commonly assayed. Farant and Wigfield (1982) using the ratio of ALA-D activities assayed at two different pH levels demonstrated ALA-D inhibition at blood lead concentrations of 10–15 μg/dl (0.50–0.70 μmol/liter). They found this to be a highly sensitive and reliable index of the blood lead concentration.

Coproporphyrin also rises in plasma and is excreted in urine and, like ALA, it is also difficult to measure and lacks sensitivity. Hence, this method is not used as an index of lead poisoning.

FER-Ch is the second major enzyme that is strongly inhibited by lead and, as a result, PROTO IX accumulates in the erythrocytes. This PROTO IX is zinc PROTO IX instead of the "free" PROTO IX as it is commonly called. Piomelli *et al.* (1982) found that erythgrocyte zinc PROTO IX increased when blood lead concentrations were at 15–18 μg/dl (0.75–0.85 μmol/liter). It is not quite as sensitive an index of blood lead concentration as ALA-D but is well below the diagnostic criteria for lead poisoning. George and Duncan (1981) found marked elevations in erythrocyte PROTO IX in experimental lead poisoning in calves. Modern fluorometers specifically designed to measure porphyrins have greatly simplified the assay. For these reasons, the current test of choice to monitor lead exposure is the blood zinc PROTO IX concentration.

The final diagnosis of lead poisoning ultimately rests on the measurement of blood lead concentration and this is best done using flame atomic absorption spectrophotometry. In children, a blood lead concentration of <30 μg/dl (1.45 μmol/liter) is currently regarded as normal but it has been shown that zinc PROTO IX rises at blood lead levels of one-half that amount (Piomelli *et al.*, 1982). It is clear that the heme synthetic pathway is affected at blood lead concentrations well below that considered normal. Zook *et al.* (1970) reported a normal range for blood lead in the dog of 10–50 μg/dl (0.48–2.41 μmol/liter) and considered a blood lead concentration of >60 μg/dl (2.90 μmol/liter) diagnostic of lead poisoning. In the light of current knowledge, this action level is probably too high. In the domestic rabbit, the blood lead concentration is reported to be 2–27 μg/dl (0.10–1.30 μmol/liter) (Gerken and Swartout, 1986). Therefore, lead concentrations of >30 μg/dl (1.45 μmol/liter) should be considered diagnostic of lead poisoning in the dog as well as in all animals.

References

Amoroso, E. C., Loosmore, R. M., Rimington, C. and Tooth, B. E. (1957). *Nature (London)* **180,** 230.

Badcock, N. R., O'Reilly, D. A., Zoanetti, G. D., Robertson, E. F., and Parker, C. J. (1993). *Clin.Chem.* **39,** 1334.

Badcock, N. R., Szep, D. A., Zoanetti, G. D. and Lewis, B. D. (1995). *Clin. Chem.* **41,** 1315.

Berlin, N. I., and Lotz, C. I. (1951). *Proc. Soc. Exp. Biol. Med.* **78,** 788.

Berlin, N. I., Meyer, L. M., and Lazarus, M. (1951). *Am. J. Physiol.* **166,** 565.

Berlin, N. I., Beechmans, M., Elmlinger, P. J., and Lawrence, J. H. (1957). *J. Lab. Clin. Med.* **50,** 558.

Bird, T. D., Hamernyik, P., Butter, J. Y., and Labbe, R. F. (1979). *Am. J. Hum. Gen.* **31,** 662.

Brace, K. C., and Altland, P. D. (1956). *Proc. Soc. Exp. Biol. Med.* **92,** 615.

Brandt, A., and Doss, M. (1981). *Hum. Genet.* **58,** 194.

Brown, I. W., Jr., and Eadie, G. S. (1953). *J. Gen. Physiol.* **36,** 327.

Burwell, E. L., Brickley, B. A., and Finch, C. A. (1953). *Am. J. Physiol.* **172,** 18.

Bush, J. A., Berlin, N. I., Jensen, W. N., Brill, A. B., Cartwright, G. E., and Wintrobe, M. M. (1955). *J. Exp. Med.* **101,** 451.

Cartwright, G. E., and Wintrobe, M. M. (1948). *J. Biol. Chem.* **172,** 557.

Clare, H. T., and Stephens, E. H. (1944). *Nature (London)* **153,** 252.

Cline, H. T., and Berlin, N. I. (1963). *Am. J. Physiol.* **204,** 415.

Cookson, G. H., and Rimington, C. (1953). *Nature (London)* **171,** 875.

Cornelius, C. E., and Kaneko, J. J. (1962). *Science* **137,** 673.

Cornelius, C. E., Kaneko, J. J., and Benson, D. C. (1959). *Am. J. Vet. Res.* **20,** 917.

Cornelius, C. E., Kaneko, J. J., Benson, D. C., and Wheat, J. D. (1960). *Am. J. Vet. Res.* **21,** 1123.

Doss, M., von Tiepermann, R., Schneider, J., and Schmid, H. (1979). *Klin. Wochenschr.* **57,** 1123.

Eales, L. (1961). *Ann. Rev. Med.* **12,** 251.

Elder, G. H. (1982). *Semin Liver Dis.* **2,** 87.

Elder, G.H., and Sheppard, D. M. (1982). *Biochem. Biophys. Res. Commun.* **109,** 113.

Elder, G. H., Evans, J. O., and Matlin, S. A. (1976). *Clin. Sci. Mol. Med.* **51,** 71.

Ellis, D. J., Barner, R. D., Madden, D., Melcer, I. and Orten, J. M. (1958). *Mich. State Univ. Vet.* **18,** 89.

Everett, N. B., and Yoffey, J. M. (1959). *Proc. Soc. Exp. Biol. Med.* **101,** 318.

Farant, J. P., and Wigfield, D. C. (1982). *Int. Arch. Occup. Environ. Health* **51,** 15.

Fell, G. S. (1984). *Ann. Clin. Biochem.* **21,** 453.

Finch, C. A., Wolff, J. A., Rath, C. E., and Fluharty, R. G. (1949). *J. Lab. Clin. Med.* **34,** 1480.

Flyger, V., and Levin, E. Y. (1977). *Am. J. Pathol.* **87,** 269.

Fourie, P. J. J. (1936). *Onderstepoort J. Vet. Sci. Anim. Ind.* **7,** 535.

Fourie, P. J. J. (1939). *Onderstepoort J. Vet. Sci. Anim. Ind.* **13,** 383.

Fourie, P. J. J., and Roets, G. C. S. (1939). *Onderstepoort J. Vet. Sci. Anim. Ind.* **13,** 369.

George, J. W., and Duncan, J. R. (1981). *Am. J. Vet. Res.* **42,** 1630.

George, J. W., and Duncan, J. R. (1982). *Am. J. Vet. Res.* **43,** 17.

Gerken, D. F., and Swartout, M. S. (1986). *Am. J. Vet. Res.* **47,** 2674.

Gibson, K. D., Laver, W. G., and Neuberger, A. (1958). *Biochem. J.* **70,** 71.

Giddens, W. E., Jr., Labbe, R. F., Swango, L. J., and Padgett, G. A. (1975). *Am. J. Pathol.* **80,** 367.

Glenn, B. L. Glenn, H. G., and Omtvedt, I. T. (1968). *Am. J. Vet. Res.* **29,** 1653.

Gower, D. B., and Davidson, W. M. (1963). *Brit. J. Haematol.* **9,** 132.

Granick, S. (1966). *J. Biol. Chem.* **241,** 1359.

Granick, S., and Levere, R. D. (1964). *Prog. Hematol.* **4,** 1.

Granick, S., and Urata, G. (1963). *J. Biol. Chem.* **238,** 821.

Gray, C. H., Muir, I. M. H., and Neuberger, A. (1950). *Biochem. J.* **47,** 542.

Harber, L. C., Fleischer, A. S., and Baer, R. L. (1964). *J. Am. Med. Assoc.* **189,** 191.

Hindmarsh, J. T. (1986). *Clin. Chem.* **32,** 1255.

Jensen, W. N., Bush, J. A., Ashenbrucker, H., Cartwright, G. E., and Wintrobe, M. M. (1956). *J. Exp. Med.* **103,** 145.

Jorgensen, S. K. (1959). *Br. Vet. J.* **115,** 160.

Jorgensen, S. K. (1961). *Br. Vet. J.* **117,** 61.

Jorgensen, S. K., and With, T. K. (1955). *Nature (London)* **176,** 156.

Kaneko, J. J. (1963). *Ann. N. Y. Acad. Sci.* **104,** 689.

Kaneko, J. J., and Cornelius, C. E. (1962). *Am. J. Vet. Res.* **23,** 913.

Kaneko, J. J., and Mattheeuws, D. R. G. (1966). *Am. J. Vet. Res.* **27,** 923.

Kaneko, J. J., and Mills, R. (1969a). *Am. J. Vet. Res.* **30,** 1805.

Kaneko, J. J., and Mills, R. (1969b). *Fed. Proc. Fed. Am. Soc. Exp. Biol.* **23,** 453.

Kaneko, J. J., and Mills, R. (1969c). *Cornell Vet.* **60,** 52.

Kaneko, J. J., Cornelius, C. E., and Heuschele, W. P. (1961). *Am. J. Vet. Res.* **22,** 683.

Kaneko, J. J., Green, R. A., and Mia, A. S. (1966). *Proc. Soc. Exp. Biol. Med.* **123,** 783.

Kaneko, J. J., Zinkl, J. G., and Keeton, K. S. (1971). *Am. J. Vet. Res.* **32,** 1981.

Kappas, A., Sassa, S., Galbraith, R.A., and Nordmann, Y. (1995). *In* "The Metabolic and Molecular Bases of Inherited Disease" (C. R. Scriver, A. L. Beaudet, W. S. Sly, and D. Valle, eds.), 7th ed., Vol II p. 2103, McGraw-Hill, New York.

Kikuchi, G., Kumar, A., Talmage, P., and Shemin, D. (1958). *J. Biol. Chem.* **233,** 1214.

Labbe, R. F., and Hubbard, H. (1961). *Biochim. Biophys. Acta* **52,** 130.

Labbe, R. F., and Lamon, J. M. (1986). *In* "Textbook of Clinical Chemistry" (N. W. Tietz, ed.), p. 1589, Saunders, Philadelphia, Pennsylvania.

Levin, E. Y. (1968a). *Biochemistry* **7,** 3781.

Levin, E. Y. (1968b). *Science* **161,** 907.

Levin, E. Y., and Coleman, D. L. (1967). *J. Biol. Chem.* **242,** 4248.

Liddle, C. G., Putnam, J. P., Berman, E., and Fisher, H. (1984). *Lab. Anim. Sci.* **34,** 365.

McManus, J., Blake, D., and Ratnaike, S. (1988). Clin.Chem. **34,** 2355.

Magnus, I. A., Jarrett, A., Prankerd, T. A. J., and Rimington, C. (1961). *Lancet* **2,** 448.

Meyer, U. A., and Schmid, R. (1978). *In* "The Metabolic Basis of Inherited Disease" (J. B. Stanbury, J. B. Wyngaarden, and D. S. Fredrickson, eds.) 4th ed., p. 1166. McGraw-Hill, New York.

Nestel, B. L. (1958). *Cornell Vet.* **48,** 430.

Neuberger, A., and Niven, J. S. F. (1951). *J. Physiol.* **112,** 292.

Pimstone, N. R. (1982). *Semin. Liver Dis.* **11,** 132.

Piomelli, S., Seaman, C., Zullow, D., Curran, A., and Davidow, B. (1982). *Proc. Natl. Acad. Sci. USA* **79,** 3335.

Poh-Fitzpatrick, M. B. (1982). *Semin. Liver Dis.* **2,** 164.

Redeker, A. G. (1963). *J. Lab. Clin. Med.* **9,** 235.

Rhode, E. A., and Cornelius, C. E. (1958). *J. Am. Vet. Med. Assoc.* **132,** 112.

Rimington, C. (1936). *Onderstepoort J. Vet. Sci. Anim. Ind.* **7,** 567.

Romeo, E. Y., and Levin, E. Y. (1969). *Proc. Natl. Acad. Sci. USA* **63,** 856.

Romeo, G. (1977). *Hum. Genet.* **39,** 261.

Romeo, G., Glenn, B. C., and Levin, E. Y. (1970). *Biochem. Genet.* **4,** 719.

Rudolph, W. G., and Kaneko, J. J. (1971). *Acta Haematol.* **45,** 330.

Runge, W., and Watson, C. J. (1969). *Blood* **32,** 119.

Ruth, G. R., Schwartz, S., and Stephenson, B. (1977). *Science* **198,** 199.

Sano, S., and Granick, S. (1961). *J. Biol. Chem.* **236,** 1173.

Schmid, R. (1966). *In* "The Metabolic Basis of Inherited Disease" (J. B. Stanbury, J. B. Wyngaarden, and D. S., Fredrickson, eds.), p. 939. McGraw-Hill, New York.

Schmid, R., and Schwartz, S. (1952). *Proc. Soc. Exp. Biol. Med.* **81,** 685.

Schmid, R., Hanson, B., and Schwartz, S. (1952). *Proc. Soc. Exp. Biol. Med.* **79,** 459.

Schmid, R., Schwartz, S., and Sundberg, R. D. (1955). *Blood* **10,** 416.

Schulman, H. M. (1968). *Biochim. Biophys. Acta* **155,** 253.

Schwartz, S., Keprios, M., and Schmid, R. (1952). *Proc. Soc. Exp. Biol. Med.* **79,** 463.

Schwartz, S., Berg, M. H., Bossenmaier, I., and Dinsmore, H. (1960). *Meth. Biochem. Anal.* **8,** 221.

Shemin, D., and Rittenberg, D. (1946a). *J. Biol. Chem.* **166,** 621.

Shemin, D., and Rittenberg, D. (1946b). *J. Biol. Chem.* **166,** 627.

Smith, J. E., and Kaneko, J. J. (1966). *Am. J. Vet. Res.* **27,** 931.

Strand. J. L., Felsher, B. F., Redeker, A. G., and Mawer, H. S. (1970). *Proc. Natl. Acad. Sci. USA* **67,** 1315.

Tobias, G. (1964). *J. Am. Vet. Med. Assoc.* **145,** 462.

Tschudy, D. P. (1965). *J. Am. Med. Assoc.* **191,** 718.

Valentine, W. N., Pearce, M. L., Riley, R. F., Richter, E., and Lawrence, J. S. (1951). *Proc. Soc. Exp. Biol. Med.* **77,** 244.

Valentine, W. N., Paglia, D. E., Fink, K., and Madokoro, G. (1976). *J. Clin. Invest.* **58,** 926.

Watson, C. J., and Schwartz, S. (1941). *Proc. Soc. Exp. Biol. Med.* **47,** 393.

Watson, C. J., Perman, V., Spurell, F. A., Hoyt, H. H., and Schwartz, S. (1959). *AMA Arch. Intern. Med.* **103,** 436.

With, T. K. (1955). *Biochem. J.* **60,** 703.

Zook, B. C., McConnell, G., and Gilmore, C. E. (1970). *J. Am. Vet. Med. Assoc.* **157,** 2092.

CHAPTER

9

Iron Metabolism and Its Disorders

JOSEPH E. SMITH

I. INTRODUCTION

Iron is the second most abundant metal and the fourth most common element. Unfortunately, it is chemically unstable and is easily oxidized to an insoluble ferric form, which is the form found in rocks and soils. Ferric iron is unavailable for most biological systems.

All living organisms, except possibly *Lactobacillus,* require iron. It is important for biological processes because it can exist in two oxidation states: Fe^{2+} and Fe^{3+}, which can be interconverted easily. Because iron is so inaccessible, organisms use a variety of mechanisms to get iron from their environments. Microorganisms have a high-affinity system consisting of three components: (1) a low-molecular-weight, secreted compound with a high affinity for iron, called a *siderophore;* (2) a membrane receptor that attracts the iron-loaded siderophore and transports the ferric chelate across the microbial membrane; and (3) a system to release the iron from the siderophore (Crichton and Charlotteaux-Wauters, 1987).

Because free iron can catalyze free radicals from molecular oxygen and hydrogen ions, it can have disastrous consequences for biological materials. So, intracellular iron is bound to or incorporated into various proteins or other chelates to reduce its toxicity. Those proteins or chelates are responsible for the absorption, storage, and biological activity of iron. Any study of iron metabolism involves a study of the physiological compounds associated with it.

II. IRON COMPARTMENTS

Iron exists in the following compartments (Fig. 9.1) in mammals: hemoglobin, storage, myoglobin, labile iron, tissue iron, and transport (Fairbanks and Beutler, 1983) (Table 9.1).

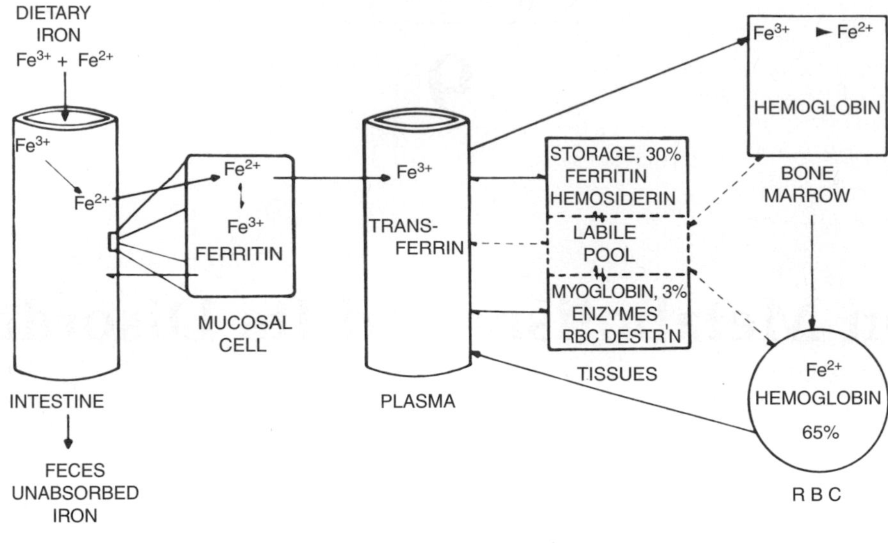

FIGURE 9.1 Pathways of iron metabolism. Single arrows show a unidirectional flow of iron to emphasize the limited excretion of iron and its "closed" cycle in the body. From Kaneko (1964), with permission.

A. Hemoglobin

Most iron in animals is located in erythrocytes as hemoglobin. Each hemoglobin molecule contains four atoms of iron and is 0.34% iron by weight. Each milliliter of erythrocytes contains 1.1 mg of iron, so the exact amount of iron will depend on the animal's packed cell volume and blood volume. Because packed cell volume is constant, but blood volume increases linearly with weight, the total amount of iron in hemoglobin is normally related to body weight (Stahl, 1967).

B. Storage

Iron is stored in various tissues as either a diffuse, soluble, mobile fraction (ferritin) or as insoluble, aggregated deposits (hemosiderin). Storage iron concentration is the major factor affecting the relative distribution of iron between ferritin and hemosiderin in mammals. At low storage levels, more iron is stored as ferritin than as hemosiderin (Torrance and Bothwell, 1980). As the amount of iron increases, the hemosiderin proportion increases. Both hemosiderin and ferritin iron are available to the body. The liver and spleen usually have the highest storage iron concentrations, followed by the kidney, heart, skeletal muscles, and brain (Underwood, 1977).

1. Ferritin

Ferritin consists of protein and iron (Crichton and Charlotteaux-Wauters, 1987). The protein portion, apoferritin, contains 24 monomers of at least two subunit types, designated H and L. The H subunit is predominant in heart ferritin and is larger (21,000 Da) than the L subunit. The L subunit occurs in the liver

TABLE 9.1 Distribution of Iron in Heme and Nonheme[a] Compounds

	Heme iron		Nonheme iron		
Species (wt)	g	%	g	%	Reference
Dog (10 kg)	0.43	55	0.36	45	Lintzel and Radeff (1931)
Horse (400 kg)	11.5	67	5.6	33	Obara and Nakajima (1961a)
Cow (386 kg)	9.2	55	7.7	45	Kaneko (1963)
Man (70 kg)	3.19	76	1.01	24	Drabkin (1951)

[a]Approximate nonheme iron distribution is as follows: 12, hemosiderin; 13%, ferritin; 3%, myoglobin; 1%, transferrin, cytochromes, peroxidase, catalase; 4%, unknown.

and spleen and has a molecular weight of 19,000 Da. Other tissues have ferritins made up of various ratios of H and L subunits. The H/L ratio is species and tissue specific and varies from 1 : 9 in horse spleen to 8.5 : 1 in horse heart ferritin. Both subunits are encoded by a family of genes. The active gene has introns, but several pseudogenes without introns exist and are located on different chromosomes.

Each subunit is shaped like a short rod and interacts with other subunits to form a hollow sphere (Fig. 9.2) (Harrison *et al.*, 1986). Pores are formed in the exterior surface between apoferritin monomers. These pores allow iron to enter or leave the interior cavity of the ferritin molecule (Harrison, 1977).

The mineral core of ferritin is hydrated ferric oxide with some phosphate. The apoferritin shell has a molecular weight of 441,000 Da and, if maximally saturated, can hold 4500 iron atoms. The apoferritin and iron together would weigh about 800,000 Da and be 31% iron by weight. Because ferritin more commonly is 620,000 Da and 18% iron, maximal saturation probably is rare.

Acute administration of iron will induce liver ferritin synthesis. The level of chelatable iron within the cell regulates ferritin synthesis by redistributing the messenger RNA between the free mRNA pool and the polyribosomes (Rogers and Munro, 1987). Apparently, iron passes through the pores in the apoferritin coat as the ferrous ion. Once inside the shell, it must be oxidized to the ferric form, hydrolyzed, and polymerized to the ferric oxyhydroxide polymer. Iron can exit ferritin by reversing the process. The physiological compounds involved in reducing the iron back to the ferrous form have not been determined (Crichton and Charlotteaux-Wauters, 1987).

2. Hemosiderin

Hemosiderin appears similar in structure to ferritin, but it has a higher iron to protein ratio. It may be formed from soluble cytosol ferritin by lysosomal action. The following sequence of events has been proposed (Richter, 1984):

Cytosol ferritin
↓
Lysosomal-siderosome ferritin
↓
Ferritin protein is denatured
↓
Ferritin protein is decomposed by proteases
↓
FeO·OH core is divested
↓
FE core is disintegrated
↓
Amorphous hemosiderin

Because hemosiderin is insoluble in water, it remains in tissues processed for histological examination. It

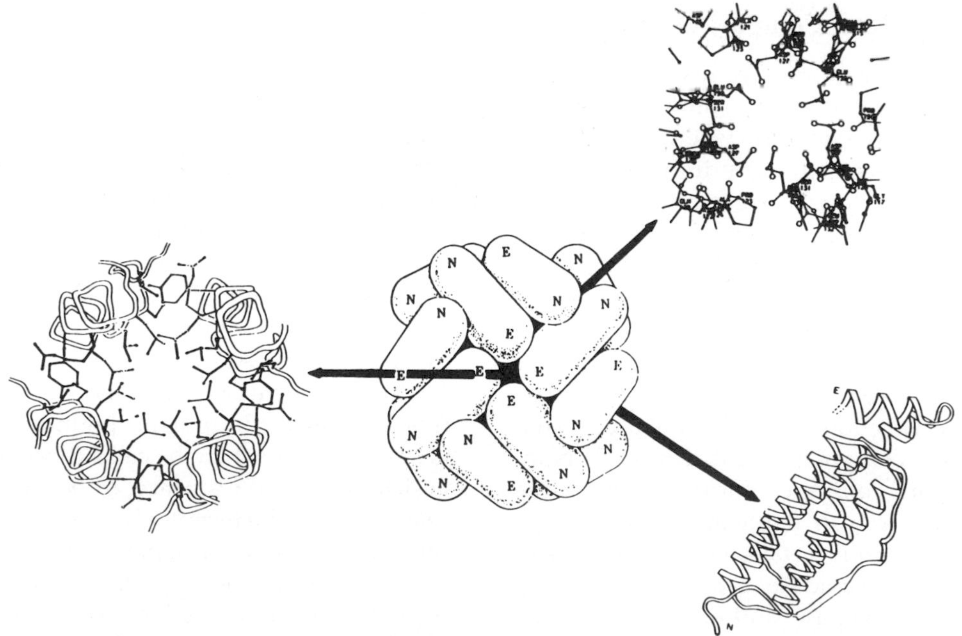

FIGURE 9.2 Structural features of ferritin. The gross structure of the assembled molecule is illustrated in the middle, details of the two types of channels are shown on the left (surrounded by four chains) and upper right (three chains), and the individual subunit in the lower right. From Harrison *et al.* (1986) with permission. N, amino terminal of the polypeptide; E, carboxyterminal of the polypeptide.

appears in unstained sections as clumps or granules of golden refractile pigment and stains readily with the Prussian blue reaction.

C. Myoglobin

Myoglobin is similar to one subunit of hemoglobin and is found mostly in muscle. It serves as a reservoir for oxygen and can temporarily provide oxygen during anaerobic conditions. Each myoglobin molecule contains one atom of iron (0.34% iron by weight). Myoglobin is species related and dependent on muscular activity. For example, the muscle myoglobin content of race horses (7.4 mg/kg) is much higher than that of man (1.2 mg/g) (Kolb, 1963). This iron pool varies between species, because the amount of myoglobin varies among muscles within a species and among species.

D. Labile Iron Pool

When radioiron disappearance curves are analyzed using a multicompartment model, an iron pool is found that is in dynamic equilibrium with the plasma iron pool (Pollycove and Mortimer, 1961). Iron disappears from the plasma into this pool, but can reflux back to the plasma. Although the exact physical nature of the pool is unknown, it may be an intermediate between the plasma, storage, and hemoglobin pools (Fig. 9.1).

E. Tissue Iron Compartment

Although the amount of iron in this compartment is small, it is extremely important. Almost half of the enzymes of the tricarboxylic acid cycle either contain iron or require it as a cofactor. The iron compounds can be classified into four categories (Dallman *et al.*,1978): (1) Heme-containing compounds are structurally similar to hemoglobin and include myoglobin, catalase, peroxidase, and cytochromes. Cytochromes *a, b,* and *c* are located in the mitochondria and are important in oxidative phosphorylation. Other cytochromes (such as cytochrome P450) are located in the endoplasmic reticulum and function in the oxidative degradation of endogenous compounds and drugs. (2) Nonheme iron-containing enzymes form another large group that includes iron in nonheme such as iron–sulphur compounds and metalloflavoproteins. This group includes xanthine oxidase, cytochrome c reductase, succinate dehydrogenase, and nicotinamide adenine dinucleotide dehydrogenase, which contain more iron in mitochondria than in cytochromes. (3) Enzymes requiring iron or heme as a cofactor include aconitase and tryptophan pyrrolase. (4) Enzymes containing iron in an unknown form include ribonucleotide reductase and α-glycerophosphate.

F. Transport Compartment

Iron is transported between some compartments by plasma transferrin. Transferrin is a single polypeptide chain of about 700 amino acids (molecular weight: ~80,000 Da). Transferrin is a glycoprotein, as are most plasma proteins. It contains two branched oligosaccharide chains that are attached to asparagine residues (Crichton and Charlotteaux-Wauters, 1987). It has a dilobal structure; the N- and C-terminal halves form separate globular lobes connected by a short α-helix. Each half carries one iron-binding site that requires concomitant binding of one carbonate or bicarbonate ion with each atom of iron. Iron as a ferric ion is bound tightly at neutral pH, but is dissociated when the pH goes below 5.5.

In domestic animals, plasma transferrin type is a polymorphic trait and can be used for parental exclusion. The genetically related, electrophoretic variability resides in both the polypeptide chain and the oligosaccharide side chains (Stratil and Glasnak, 1981; Maeda *et al.*, 1984).

The transfer of iron from transferrin to cells requires internalization of the iron-laden transferrin molecule (Fig. 9.3). Ferrotransferrin binds to a membrane receptor at neutral pH. The receptor-ferrotransferrin complex collects in specialized sites on the plasma membrane called *coated pits.* These pits are coated on the inner surface with a fibrous protein, clathrin. The coated-pit region of the membrane then can invaginate to form a vesicle, with the cell's outer plasma membrane and receptor–ferrotransferrin complex as the inner surface and the clathrin coat as the outer surface. The vesicles are converted into structures with the acronym CURL (compartment of uncoupling of receptor and ligand). An enzyme in the endostomal membrane exploits the energy stored in adenosine triphosphate to pump protons into the CURL lumen and, thus, lower the internal pH. At the lower pH, transferrin's affinity for iron is decreased, and the iron is released. Iron-free apotransferrin remains bound to its receptor and is transported back to the cell surface. When the coated vesicle reaches the surface, iron-free apotransferrin is released from the receptor because the affinity between the two is low at neutral pH. The iron-free apotransferrin can enter the blood plasma to bind more iron, and the receptor can be used to bind more ferrotransferrin. The free iron must escape from the CURL, cross the mitochondrial membrane, and be incorporated into heme. The exact mechanism for free iron movement remains unknown. The entire process requires about

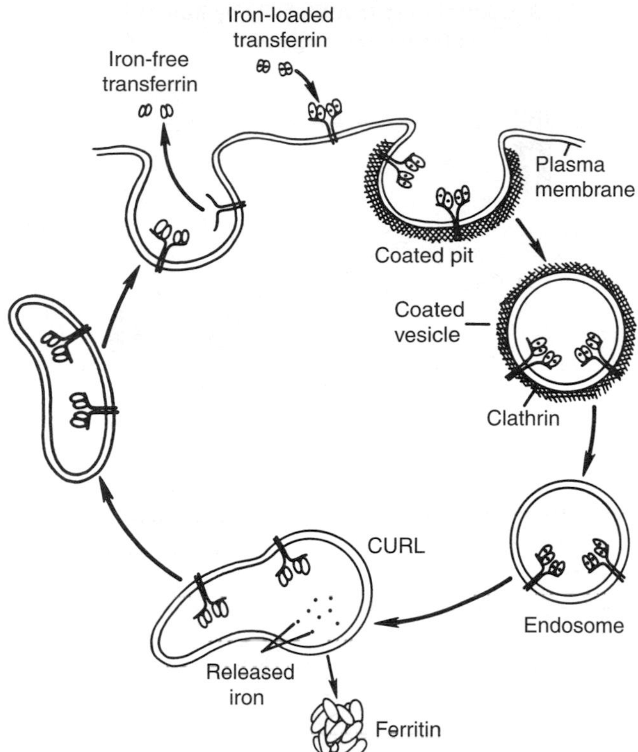

FIGURE 9.3 Transferrin cycle to deliver iron into cells. Iron-loaded transferrin binds to a specific receptor. The receptor–transferrin complex moves to a coated pit and is internalized in a coated vesicle. The coated vesicle is converted to an endosome by losing its clathrin coat and is transformed to the compartment of uncoupling of receptor and ligand (CURL). The iron dissociates from the ferritin–receptor complex at the CURL's low pH and is added to cytoplasmic ferritin. The iron-free transferrin–receptor complex recycles to the surface, where the iron-free transferrin is released, and the receptor is made available to bind more iron-loaded transferrin (Dautry-Varsat and Lodish, 1984).

16 minutes: 4 minutes for a transferrin receptor to bind a ferrotransferrin, 5 minutes for the receptor–ferrotransferrin to be internalized, 7 minutes for iron to dissociate from transferrin and return to the cell surface, and 16 seconds for the apotransferrin to be released (Dautry-Varsat and Lodish, 1984; Dautry-Varsat *et al.*, 1983).

The amount of iron delivered to immature erythrocytes depends on the plasma iron concentration, percent iron saturation, and the number of membrane receptors (Finch and Huebers, 1985). When the percentage of transferrin containing iron is low, most transferrin is monoferric; as the percentage increases, the diferric form increases. That increases the amount of iron delivered in two ways because (1) the payload of the diferric molecules is twice that of monoferric transferrin and (2) diferric transferrin has a higher affinity for the membrane receptor than monoferric transferrin and, thus, is preferentially bound to available receptors (Huebers *et al.*, 1985). The receptor number probably increases when erythropoiesis is increased.

The Belgrade rat is unable to release iron within the CURL. So, these rats have an anemia resembling iron deficiency, despite being hyperferremic (Edwards *et al.*, 1986).

Copper plays an important role in transporting iron across membranes (Rosen *et al.*, 1995). When pigs are copper deficient, they have hypoceruloplasminemia and signs of a functional iron deficiency. Iron accumulates in the liver, macrophage–phagocyte system, and enterocytes. Presumably, this "iron deficiency" results from the inability to mobilize the iron from these sites (Lee *et al.*, 1968). Most of the circulating copper in plasma is attached to the serum glycoprotein, ceruloplasmin. Ceruloplasmin has ferroxidase activity and may be required to deliver iron into the circulation. In man, a genetic defect in the ceruloplasmin gene causes aceruloplasminemia and systemic hemosiderosis (Harris *et al.*, 1995).

Plasma iron in laying hens is high (greater than 500 μg/dl), and about two-thirds of it is bound to a specific phosphoprotein, phosvitin. It is responsible for transporting iron to ovocytes and egg yolks. The average egg contains about 1 mg of iron (Kolb, 1963).

G. Genetic Control of Iron Proteins

Several genes critical to the uptake, storage, and use of iron are regulated posttranslationally. The mRNAs of these proteins contain iron-responsive elements (IREs) in the form of stem-loop structures in untranslated regions (Fig. 9.4). A single IRE in the 5′ untranslated region is involved in controlling translation of the mRNA message. A cluster of IREs in the 3′ untranslated region confers iron-regulated control of mRNA stability (Basilion *et al.*, 1994). IREs are not the definitive iron sensors, but require a cytosolic protein called an *iron regulatory protein* (IRP). The binding of IRP to IREs depends on iron availability. If iron is abundant, the affinity of IRP for IREs is low (Klausner and Harford, 1989; Beinert and Kennedy, 1993).

Two IRPs have been described. When IRP1 occurs in iron-replete cells, it is a cytosolic aconitase. When iron is limiting, IRP1 exists in an alternative form that is devoid of a cubane Fe-S cluster and has a high affinity for IREs (Klausner and Harford, 1989). IRP2 is the product of a second gene, but its mRNA abundance and tissue distribution differ from those of IRP1. IRP2

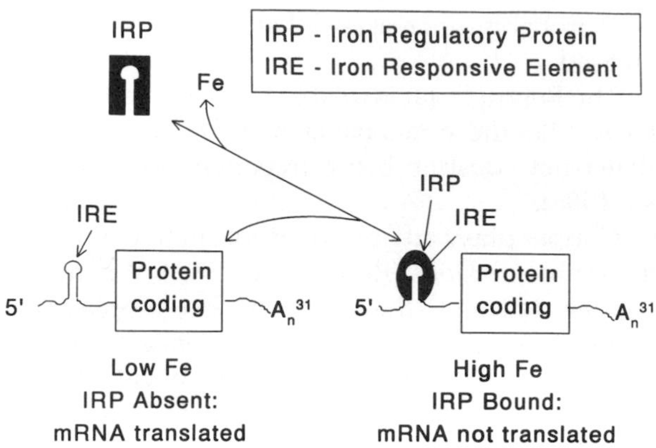

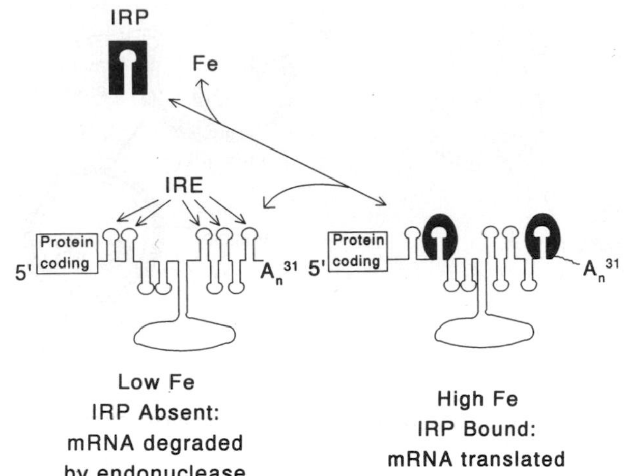

FIGURE 9.4 Posttranslational control of iron proteins. Iron regulatory protein (IRP) binds to the iron responsive element (IRE) of mRNA only when cells are iron replete. If IRP is bound to an IRE on the 5′ end of mRNA, mRNA is not translated (left panel). When it binds to the mRNA's 3′ end, mRNA degradation by endonuclease is inhibited (right panel).

does not have aconitase activity in an iron-adequate environment (Samaniego *et al.*, 1994).

When iron is low, an IRP binds to the 5′ IRE with high affinity and prevents the translation of ferritin mRNA and erythroid 5-aminolevulinate synthase mRNA. Similarly, when IRP binds to 3′IRE, it stabilizes the short-lived transferrin receptor mRNA by blocking its degradation by endonuclease. Thus, IRP exerts an iron-dependent, dual, and reciprocal control of ferritin concentrations and transferrin receptors (Klausner and Harford, 1989).

III. IRON ABSORPTION

A. Iron Requirements

Iron requirements for domestic animals are influenced by age, growth rate, availability of a dietary iron source, and the criteria of adequacy (NRC, 1978b). Definite iron requirements for most domestic species have not been determined. Most recommendations (Table 9.2) are estimates (Harrison, 1977; Rogers and Munro, 1987; NRC, 1975, 1976, 1978a, 1978b, 1979, 1981, 1984, 1985). The intake for mature animals can be adequately met from the usual dietary sources. Because milk is low in iron and growth rates of neonatal animals are high relative to their weights, the young of most species can become iron deficient. Anemia should not be used to determine iron adequacy, because use of dietary iron for hemoglobin synthesis can take precedence over demands for other iron compounds (Nathanson and McLaren, 1984).

B. Dietary Iron

The amount of iron available for absorption by the intestine depends on the amount of iron in the diet and its bioavailability. Iron content of various foods and feeds can be divided into high (greater than 50 ppm), intermediate (10 to 50 mg iron), and low (less than 10 ppm) levels. Those with high iron content include organ meats such as heart and liver, brewer's yeast, wheat germ, egg yolks, oysters, and certain dried beans and fruits. Those with intermediate iron content include most muscle meats, fish and fowl, most green

TABLE 9.2 Estimated Dietary Iron Requirements for Domestic Animals

Species	Iron (ppm)	Reference
Pigs, mature	80	NCR (1979)
neonatal	150	
Cattle, adult	50	NRC (1978b, 1984)
neonatal	100	
Horses, adult	40	NRC (1978)
neonatal	50	
Sheep, adult	30–50	NRC (1975)
Dogs and cats	80	Chausow and Czarnecki-Maulden (1987)

vegetables, and most cereals. Foods low in iron include milk and milk products and most nongreen vegetables (Finch, 1980).

Iron is absorbed as either heme or nonheme compounds. Heme is absorbed readily and independently of the composition of the diet. Nonheme iron is largely unavailable, and its absorption is affected by other ingredients in the diet. Tannates and phosphates inhibit but meat and ascorbate enhance nonheme iron absorption (Finch and Cook, 1984).

C. Mechanisms of Absorption

Despite many investigations over several years, the exact mechanism of intestinal absorption of iron remains unknown. It must be a finely tuned mechanism, to maintain homeostasis without a major excretory mechanism. Because free iron is inherently unstable and can catalyze free-radical-mediated membrane damage, it is unlikely that iron crosses the intestinal cell wall in a free, soluble form. Iron probably is bound to a ligand that is involved in regulating its uptake. Originally, apoferritin was implicated as a transport and regulatory protein. Apoferritin may be important in keeping iron out of the body or in excreting iron. Apoferritin is made as mucosal cells are formed in crypts. The iron trapped in apoferritin is retained in those cells as they migrate to the tips of intestinal villi, and it is lost when the mucosal cells are sloughed. The most recent theory of iron absorption involves mucin and an intestine protein designated mobilferrin. According to this theory, gastric juice stabilizes dietary inorganic iron and prevents iron from being precipitated as insoluble ferric hydroxide. At acid pH, iron combines with mucins. Intestinal mucin delivers inorganic iron to intestinal absorptive cells in a form that is acceptable for absorption. In the small intestine, iron is transferred from mucin to the membrane integrins of the intestinal absorptive cells. Membrane integrins help the transfer of iron through the cell membrane for binding to mobilferrin. Mobilferrin serves as the shuttle protein within the absorptive cell. If an excess of iron occurs in the cell, ferritin synthesis is stimulated, and iron is deposited in ferritin to prevent oxidative damage to the cell from ionic iron. Transferrin receptors located on the basolateral membranes of the absorptive cell act to permit iron to enter the cell from plasma similar to cells in other organs (Conrad and Umbreit, 1993).

Iron uptake is influenced by previous dietary exposure, the amount of storage iron within the body, and erythropoietic activity (Finch and Huebers, 1985). After one dose of iron is given orally, a second dose will be absorbed more slowly (Hahn, 1943). Iron absorption increases 6- to 15-fold in dogs with chronic blood loss anemia and with increased erythropoiesis (Weiden *et al.*, 1981).

The "mucosal blockage" of iron absorption has limited capacity, because higher doses of iron can overcome the blocking mechanism. Several children are poisoned each year because they eat too many sugar-coated iron tablets. In domestic animals, iron toxicity can occur when they are exposed to high iron levels in feed or water.

IV. TESTS FOR EVALUATING IRON METABOLISM

A. Hematology

With certain constraints, hematological examination can be used to evaluate iron adequacy. Blood can be obtained easily and it contains the largest iron pool. Unfortunately, iron is preferentially shunted from other iron pools to hemoglobin (Nathanson and McLaren, 1984). Thus, hemoglobin may be the last pool to show the effects of iron inadequacy.

A high concentration of hemoglobin is a signal to stop cell division and extrude the nucleus (Stohlman, *et al.*, 1963). During severe iron deficiency, hemoglobin synthesis in immature erythrocytes is slowed, and nucleated erythrocytes continue to divide. Erythrocytes released into the circulation are small, contain occasional nuclei, and show increased central pallor because of low hemoglobin content. Hematology has been used to test biological availability of iron in diets or to determine the iron requirements of various species, in spite of the drawbacks. If the erythrocyte number, packed cell volume, and hemoglobin are determined, three erythrocyte indices can be calculated:

Mean corpuscular hemoglobin (pg)

$$= \frac{\text{Hb (g/dl)}}{\text{RBC }(10^6/\mu\text{l})} \times 10,$$

$$\text{Mean corpuscular volume (fl)} = \frac{\text{PCV (\%)}}{\text{RBC }(10^6/\mu\text{l})} \times 10,$$

Mean corpuscular hemoglobin concentration (%)

$$= \frac{\text{Hb (g/dl)}}{\text{PCV (\%)}} \times 100.$$

Classically, all three indices decrease in iron deficiency anemia. However, similar changes can occur in any deficiency or disease that inhibits hemoglobin synthesis, such as pyridoxine-responsive anemia or copper deficiency.

B. Serum Iron

Serum iron can be measured to assess the transport compartment. Conditions of sample collection are particularly important. Most evacuated containers used for routine serum collection are satisfactory. However, if zinc is to be determined, special trace element tubes are required, because most stoppers are contaminated with zinc (Handy, 1979). Although plasma may be used, the anticoagulant should be tested for iron content. Samples must be handled carefully to avoid post-sampling contamination. Glassware must be cleaned carefully with acid solution. Disposable plastic containers and tubes offer an acceptable alterative and usually do not require acid cleansing.

Iron can be determined by atomic absorption, colorimetry, and coulometry (Henry *et al.*, 1974; Smith *et al.*, 1981). Direct measurement of serum iron by atomic absorption may give unreliable results because the sensitivity is limited, matrix interference occurs, and hemoglobin iron cannot be distinguished from transferrin iron. Colorimetric determination requires that the iron be separated from transferrin by lowering the pH and then reduced to the ferrous ion. The ferrous iron can be determined with reagents such as thiocyanate, *o*-phenanthroline, or 2,2′-dipyridyl. Although most colorimetric methods are not elevated by mild hemolysis, other serum constituents such as bilirubin may interfere. Coulometric methods can determine iron in small samples. They are not affected seriously by mild hemolysis and are free of interference from other compounds such as bilirubin, lipids, and cholesterol in the serum (Smith *et al.*, 1981).

Serum iron declines in severe iron deficiency, acute-phase inflammatory reactions, hypoproteinemia, hypothyroidism, renal disease, and chronic inflammation. It may be elevated in hemolytic anemia, refractory anemia, iron overload, and liver disease (Kaneko, 1980).

C. Serum Total Iron Binding Capacity

Total transferrin can be measured by immunological methods, but the technique is not used commonly. Transferrin usually is measured in terms of iron content after it has been saturated with iron. When transferrin is saturated with iron, the iron content is called the total iron-binding capacity (TIBC). Because transferrin can bind more iron than is normally present, the TIBC is greater than the serum iron and the difference between them is the unsaturated iron-binding capacity (UIBC). Thus, serum iron can be expressed as a percentage of the TIBC and reported as the percent saturation (Henry *et al.*, 1974).

Several strategies are used to measure serum TIBC or UIBC (Henry *et al.*, 1974). (1) Excess iron can be added to the serum, and the increase in "salmon-pink" color can be measured colorimetrically. (2) Iron can be added in excess, and the unbound excess can be measured colorimetrically. (3) Excess iron can be added to saturate transferrin and then removed by adsorption onto a resin. The bound iron remaining in the supernatant can be determined by a typical serum iron assay. (4) A resin that is partially saturated with iron can be incubated with serum. Under proper incubation conditions, transferrin competes successfully for resin-bound iron, but resin-bound iron will not leave without transferrin. If serum iron exceeds the transferrin-binding capacity, the resin can adsorb the excess iron. The supernatant iron can be determined by any serum iron assay (Smith *et al.*, 1981).

D. Serum Ferritin

Although ferritin functions *in vivo* as an iron storage compound and is primarily intracellular, it can be detected in serum (Table 9.3). In man, serum ferritin concentrations correlate with total body iron stores. They are low in iron deficiency and high in iron overload (Addison *et al.*, 1972; Jacobs *et al.*, 1972; Lipschitz *et al.*, 1974; Siimes *et al.*, 1974; Walters *et al.*, 1973).

Serum ferritin is assayed in antibody-driven reactions such as radioimmunoassay and enzyme-linked immunosorbent assay. Unfortunately, the antibodies against ferritin are usually species specific (Richter, 1967); that is, an antibody against human ferritin will not cross-react with horse ferritin. Thus, for each new species, ferritin must be isolated, an antibody must be made, and the assay conditions must be developed. Despite this difficulty, serum ferritin has been measured in horses (Smith *et al.*, 1983), rats (Hunter, 1978; Ward *et al.*, 1977), pigs (Smith *et al.*, 1984), dogs (Weeks *et al.*, 1988b), cats (Andrews *et al.*, 1994), and cows (Miyata *et al.*, 1984). In dogs (Weeks *et al.*, 1989), cats (Andrews *et al.*, 1994), horses (Smith *et al.*, 1983), and pigs (Smith *et al.*, 1984), but not rats (Hunter, 1978; Ward *et al.*, 1977), serum ferritin correlates significantly with nonheme iron in the liver and spleen. In calves, serum ferritin increases after iron therapy (Miyata *et al.*, 1984).

Serum ferritin can be increased in several conditions that are not related to body stores of iron. It is an acute-phase protein and increases during serious inflammatory reactions when interleukin-1 is produced. Clinically, serum ferritin should be monitored relative to other acute-phase proteins such as haptoglobin, fibrinogen, or C-reactive protein, because it is possible for the low serum ferritin associated with iron defi-

TABLE 9.3 Serum Iron Analytes by Domestic Animals

Species	Serum analyte[a]			Reference
	Iron (μg/dl)	TIBC (μg/dl)	Ferritin (ng/ml)	
Horse	50–198	231–455	43–261	Smith *et al.* (1986)
Cattle	39–155	186–270	33–55	Kolb (1963), Furugouri (1984)
Sheep	179–207	298–370		Kolb (1963)
Pig	55–187	241–393	20–125	Kolb (1963), Smith *et al.* (1984)
Dog	33–147	282–386	80–800	Weeks *et al.* (1988)
Cat	33–135	169–325	32–123	Andrews *et al.* (1994)

[a] Reference ranges.

ciency to be elevated into the normal range by a concurrent infection (Lorier *et al.*, 1985). Serum ferritin also can be increased during liver disease, hemolytic diseases, and some neoplastic disorders (Newlands *et al.*, 1994). Serum ferritin increases and transferrin decreases with malnutrition in cattle (Furugouri, 1984).

E. Bone Marrow Iron

Cytological examinations of bone marrow are sometimes useful for evaluating disorders of iron metabolism (Fairbanks and Beutler, 1983). A Prussian-blue reaction in which ionic iron reacts with an acid ferrocyanide solution to give a blue color is used commonly. Bone marrow smears are fixed in formalin vapor by placing them in a covered staining jar containing a sponge moistened with formalin for about 3 minutes. Equal volumes of 4% potassium ferrocyanide solution and 4% hydrochloric acid solutions are mixed in a staining jar and heated to 56°C. After slides are immersed in the solution for 30 minutes, they are rinsed with tap water and counterstained for 5 minutes with a dilute basic fuchsin solution, and then rinsed with water, absolute ethanol, and water again.

Except for cats (Harvey, 1981) bone marrow from normal adult domestic animals exhibits stainable iron (hemosiderin) within macrophages. Little or no stainable iron is found in animals with iron deficiency.

Mitochondrial iron of immature erythrocytes occurs as amorphous aggregates called ferruginous micelles. Some nucleated erythrocytes in Prussian-blue stained marrow slides contain one to three, small, blue granules in the cytoplasm. In iron deficiency, macrophages and developing erythrocytes do not contain stainable iron granules. On the other hand, when heme synthesis is impaired, mitochondria accumulate excess amorphous iron aggregates and a ring of large blue siderotic granules may encircle the erythrocyte nucleus (Fairbanks and Beutler, 1983).

F. Erythrocyte Protoporphyrin

In the final step of heme synthesis, iron is inserted into protoporphyrin. When heme synthesis is limited by the availability of iron, protoporphyrin accumulates, and zinc substitutes for iron. The zinc chelate is stable and remains in the erythrocytes throughout their lifespans. Protoporphyrin can be measured by two methods (Trundle, 1984). In one method, protoporphyrin is extracted from erythrocytes with acidified organic solvents. After extraction, the protoporphyrin can be assayed spectrophotometrically or fluorometrically. The method is difficult and time consuming and requires large amounts of blood. The other method uses a front face hematofluorometer to measure zinc protoporphyrin directly on small quantities of unhemolyzed blood samples. Although protoporphyrin and zinc protoporphyrin are the only fluorescent compounds in blood, hemoglobin does absorb some fluorescence. Therefore, it is necessary to correct observed values for hemoglobin concentration.

Erythrocyte protoporphyrin also increases in lead poisoning because ferrocheletase, the enzyme responsible for inserting iron into protoporphyrins, is inhibited by lead. In the anemia of chronic disease, protoporphyrin increases because of internal iron unavailability.

G. Tissue Nonheme Iron

Body iron stores can be determined directly by measuring the iron concentrations in various organs (Torrance and Bothwell, 1980). Although it may be desirable to determine iron in each body tissue, nonheme iron usually is determined in organs containing large quantities of iron that are accessible for biopsy, that is, liver and spleen. Total tissue iron does not reflect stored iron because most tissues contain heme iron compounds such as hemoglobin, myoglobin, or heme enzymes. Only the nonheme iron fraction represents iron stored in ferritin or hemosiderin. Because heme resists acid hydrolysis, nonheme iron can be separated from heme iron by extraction in acid alone or acid combined with sodium pyrophosphate. Sodium pyrophosphate forms a stable complex with ferrous iron and aids extraction.

The acid extraction method (Torrance and Bothwell, 1980) is simple and effective. Small pieces of tissue are incubated in a mixture of 3 *N* hydrochloric acid and 0.61 *M* trichloroacetic acid. After samples are incubated for 20 hours at 65°C nonheme, iron is determined colorimetrically or coulometrically. It is not necessary to homogenize the tissue if small pieces are used, and no centrifugation is required (Torrance and Bothwell, 1980).

H. Ferrokinetics

The term *ferrokinetics* refers to measurements made following the intravenous injection of radioiron-labeled transferrin. The technique was introduced by Huff and coworkers (Huff *et al.*, 1950), and the complexity of the technique varies (Cook and Finch, 1980). In the simplest method, radioiron is injected intravenously without prior incubation with plasma transferrin, and four or five samples are taken over a 2-hour period. In the most complicated experiment, radioiron-labeled transferrin is used, samples are taken for a period of 2 weeks, and the data are analyzed by sophisticated computer programs. A ferrokinetic approach that falls between the two extremes usually is used to study erythropoiesis. That approach will be described here (Cook and Finch, 1980).

1. Iron Turnover

In this method, serum iron and TIBC are determined to establish that the UIBC is greater than the amount of radioactive iron that will be used. The recipient's plasma is incubated with radioactive iron for 30 minutes at room temperature. An aliquot of the labeled plasma is removed to determine radioactivity. The plasma with radioactive iron-labeled transferrin is injected intravenously. The volume injected can be determined by weighing the syringe containing the radioactive iron before and after the injection. At various intervals, plasma radioactivity is determined and plotted on semilogarithmic coordinates.

The plasma volume can be calculated by isotope dilution. A line representing radioactivity for the first few minutes or hours is extrapolated to time zero. The plasma volume can be determined by the following formula:

$$\text{Plasma volume (ml)} = \frac{\text{total activity injected}}{\text{activity at zero time/ml of plasma}}.$$

The simplest approach to ferrokinetics is to use the initial rapid plasma disappearance to calculate the plasma iron turnover (Fig. 9.5), which is the amount of iron flowing through the plasma per unit time. It depends on the rate of iron clearance and the size of the circulating iron pool (plasma iron concentration × total plasma volume). The halftime ($T1_{1/2}$) of iron (or time required for 50% of plasma radioactivity to disappear) is determined from the initial single exponential clearance. The plasma iron turnover (PIT) can be calculated by the following formula:

$$\text{PIT } (\mu\text{g/dl plasma/day}) = \frac{0.693 \times 24 \times 60 \times \text{plasma iron } (\mu\text{g/dl})}{T_{1/2}} = \frac{\text{plasma iron } (\mu\text{g/dl}) \times 1000}{T_{1/2}}$$

If it is assumed that all iron cleared from the plasma is used for hemoglobin synthesis, the erythrocyte synthetic rate can be calculated. When the circulating erythrocytes are constant, the daily erythrocyte production equals the erythrocyte destruction and is calculated as

$$\text{Erythrocytes (ml/day)} = \frac{\text{packed cell volume} \times 100}{\text{erythrocyte lifespan (days)}}.$$

Because 1 ml of erythrocytes contains 1.1 mg of iron, the plasma iron turnover needed to maintain homeostasis can be determined. The observed plasma iron turnover is usually higher, because some iron flows to nonerythroid tissues and some refluxes early from the extravascular circulation.

Plasma iron turnover rate can be expected to increase in ineffective erythropoiesis and to decrease in erythroid hypoplasia (Erslev, 1983). Erroneous results can be obtained if the blood volume is abnormal, such as in severe dehydration or hydremia. Normal values are shown in Table 9.4.

If plasma radioactivity is followed for the next several days, the disappearance curve is extremely complex (Pollycove and Mortimer, 1961). Normally, the curve can be resolved into a three-component, polynomial expression (Fig. 9.6). Each component presumably represents some iron pool that is in equilibrium with the plasma iron.

2. Erythrocyte Utilization Rate

After radioactive iron-labeled transferrin is injected, some iron is incorporated into hemoglobin. If the reactivity in erythrocytes is determined serially for 10–14 days (Fig. 9.5), the percent of the iron used for erythrocyte production can be determined as follows:

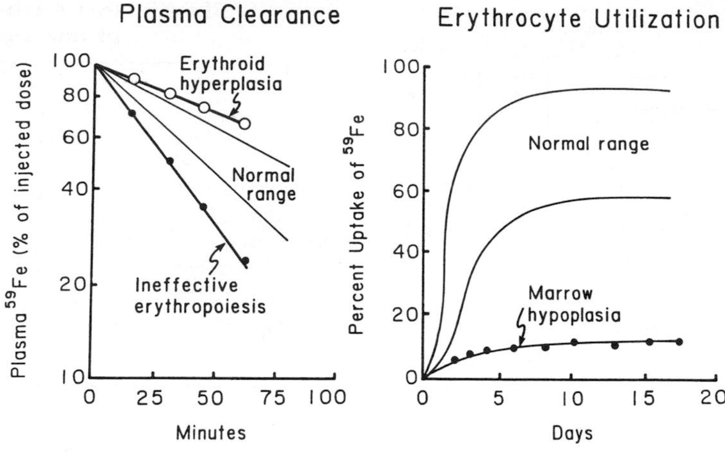

FIGURE 9.5 Iron metabolism in dogs. The left panel shows plasma clearance of injected $_{59}$Fe in normal dogs and those with erythroid hyperplasma and ineffective erythropoiesis. The right panel shows uptake of $_{59}$Fe into erythrocytes as percentage of injected dose and the effect of marrow hypoplasia (Kaneko, 1980; Erslev, 1983).

$$\text{Erythrocyte utilization (\%)} = \frac{\text{radioactivity/ml blood} \times \text{blood volume}}{\text{amount of radioactivity injected}}$$

Normally, erythrocytes incorporate most of the injected radioactive iron (Table 9.4). Decreased utilization can occur in ineffective erythropoiesis and marrow hypoplasia. Those two possibilities can be separated by the plasma iron turnover rate. Ineffective erythropoiesis will shorten the radioactive iron clearance, and marrow hypoplasia will have a prolonged clearance rate. However, increased utilization is difficult to determine, because the normal values are so high (Erslev, 1983).

3. *Marrow Transit Time*

Marrow transit time is the time required for erythrocytes to mature in the bone marrow. Although it is difficult to determine directly, it can be estimated from the appearance of radioactively labeled erythrocytes. It is defined as the time when 50% of the radioiron-labeled erythrocytes have left the marrow and is found by plotting daily erythrocyte iron utilization against

TABLE 9.4 Plasma Iron Turnover (Ferrokinetics) in Domestic Animals[a]

Animal	Half-time $T_{1/2}$ (min)	Fractional transfer rate k (day^{-1})	Transfer rate Tr (mg/day)	Tr or plasma iron turnover rate: Tr/kg body wt (mg/kg day)	Tr/100 ml plasma (mg/100 ml day)	Maximum erythrocyte ^{59}Fe uptake (% dose)	Reference
Horse	75–103		111–153	0.45–0.65	0.77–1.48	74–77	Obara and Nakajima (1961b)
	(88.8)	(10.6)	(132)	(0.55)	(1.18)	(76)	
Cow	187	5.3	106	0.27	0.27	55	Kaneko (1963)
Calf	88–137	8.0–11.3	45–104	0.50–0.64	1.18–1.49	66–82	Kaneko and Mattheeuws (1966)
	(117)	(8.7)	(74)	(0.57)	(1.30)	(73)	
Sheep	85–110	9.9–11.7		0.42–0.65	1.91–2.26	74–87	Baker and Douglas (1957)
	(94)	(10.3)		(0.56)	(1.99)	(78)	
Pig	43–100	2.3–10.0		0.40–1.66	1.30–4.13	72–100	Bush *et al.* (1956)
	(71.4)	(14.0)		(1.11)	(1.57)	(92)	
Dog	39–63	16–40			1.71–2.22)	58–93	Kaneko (1964)
	(56)	(18.2)			(1.96)	(75)	
Cat	(40)	(24.8)			1.75–1.86	19–21	Kaneko (1980)
					(1.80)	(20)	

[a] Values have been recalculated when necessary to maintain consistency. Numbers in parentheses are means.

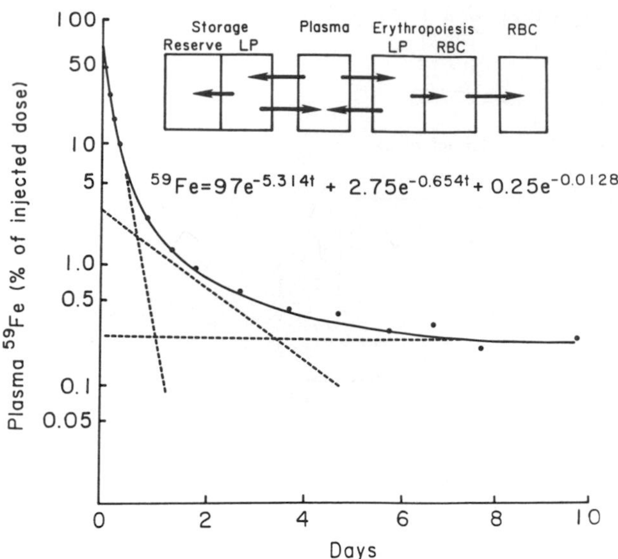

FIGURE 9.6 Plasma ^{59}Fe clearance in a normal steer and the mathematical expression derived from the data, assuming a three-compartment model (Kaneko, 1980). The upper portion shows a possible model of plasma iron disappearance. LP, labile pool (Pollycove and Mortimer, 1961).

time. Thus, it is the time at which 50% of maximum erythrocyte utilization has occurred. A short transit time is found in erythroid hyperactivity and may be caused by early release of reticulocyte. A prolonged marrow transit time indicates erythroid hypoactivity (Erslev, 1983).

V. DISORDERS OF IRON METABOLISM

A. Iron Deficiency

1. General Manifestations of Iron Deficiency

Iron deficiency is probably the most common deficiency in man. Worldwide, it is estimated to occur in hundreds of millions of people. Although iron deficiency is less frequent in domestic animals, it does occur and can adversely affect health and performance.

Three categories of iron deficiency exist in animals: iron deficiency, iron-deficient erythropoiesis, and iron deficiency anemia (Hastka *et al.*, 1994). Various iron analytes can be used to differentiate these categories (Table 9.5). In domestic animals, iron deficiency anemia is the most common form. Transient iron deficiency occurs in the young of most domestic species in association with rapid growth and an all-milk diet. Classical laboratory findings are a microcytic, hypochromic anemia with normal or increased reticulocyte counts. Serum iron is decreased below 50 μg/dl, serum TIBC is increased, serum ferritin is decreased, and erythrocyte

TABLE 9.5 Iron Analytes in Various Stages of Iron Deficiency[a]

Analyte	Iron deficiency	Iron-deficient erythropoiesis	Iron deficiency anemia
Bone marrow iron	↓	↓	↓
Serum ferritin	↓	↓	↓
Serum iron	N	↓	↓
Zinc protoporphyrin	N	↑	↑
Transferrin saturation	N	↓	↓
Hemoglobin	N	N	↓
Erythrocyte size	N	N	↓

[a] Hastka *et al.* (1994).

protoporphyrin is increased. Because the serum iron is decreased while the TIBC is increased, the percent saturation is decreased. That should have an adverse effect on transfer of iron to immature erythrocytes, because transferrin would be mostly monoferric.

Several concepts are important to iron deficiency that occurs in adult animals (Mahaffey, 1986): (1) It results from excessive loss rather than inadequate intake. (2) Iron is usually lost as blood (intestinal malabsorption might be an exception). (3) Blood loss must be prolonged. Normally, storage iron can replace the amount lost in a single bleeding episode, if the animal survives.

The signs of iron deficiency are only partly due to a compromised delivery of oxygen to the tissues from decreased hemoglobin concentration. Iron deficiency depletes important iron-containing compounds in solid tissues. In fact, iron is preferentially shunted from tissue stores and iron-containing enzymes to the erythropoietic tissues. For example, New Hampshire chicks have white feathers instead of a reddish-brown plumage when they are raised on an iron-deficient diet (Davis *et al.*, 1962). Classical signs and symptoms of iron deficiency can occur in human patients phlebotomized for polycythemia vera. The packed cell volumes in those human patients may be greater than 50, yet they may complain of fatigue, weakness, etc.

Biochemical abnormalities have been associated with physiological function in iron deficiency for striated muscle, central nervous system, white blood cells, and the gastrointestinal tract (Dallman *et al.*, 1978).

Iron-deficient rats show decreased work performance that is not reversed by transfusion to correct the associated anemia. Reduced α-glycerophosphate dehydrogenase activity, both during depletion and in response to iron therapy, is temporally related to changes in muscle function. That enzyme is a nonheme iron-containing protein and plays an important role in skeletal muscle metabolism (Finch *et al.*, 1976).

Altered behavior, that is, apathy and irritability, has been associated with iron deficiency. Brain iron decreases rapidly during iron deficiency and the low level persists after iron repletion of other tissues. Aldehyde oxidase, a key enzyme in serotonin degradation, is decreased, and serotonin level is elevated in iron deficiency (Mackler *et al.*, 1978). The accumulation of catecholamines may explain the mental changes associated with the condition.

Iron deficiency often is associated with impaired cell-mediated immunity and ability of polymorphonuclear granulocytes to kill ingested bacteria. Abnormalities result from defective DNA synthesis and decreased activity of ribonucleotide reductase. This enzyme contains iron, but the exact form is uncertain (Hoffbrand *et al.*, 1976).

Neutrophils have many iron-containing compounds. Myeloperoxidase is an iron-containing enzyme that is found in primary granules and contributes to antimicrobial activity. The exact role of the enzyme is unclear, because human patients with a congenital deficiency of myeloperoxidase are asymptomatic. In addition, the neutrophils of some avian species do not normally have myeloperoxidase activity but susceptibility to infections is not increased. Cytochrome is another iron-containing compound that is found in specific granules and is required for the oxidative burst that follows phagocytosis. Cytochrome *c* activity may be preserved preferentially in iron deficiency (Murakawa *et al.*, 1978).

When iron is deficient, acid secretion by the stomach is reduced, and intestinal absorption is impaired. The mucosa of the small intestine can metabolize many foreign compounds. Intestinal transformation plays a major role in the oxidative metabolism of phenacetin and chlorpromazine. When dietary iron is restricted, cytochrome P450 and oxidative enzyme activities fall sharply. Oral iron, but not parenteral iron, will restore cytochrome P450 (Hoensch *et al.*, 1976).

2. *Iron Deficiency in Pigs*

Baby pigs are very susceptible to iron deficiency because they are born with low levels of storage iron (Douglas *et al.*, 1972), they grow rapidly (Braude *et al.*, 1962), and sow's milk is low in iron (Kolb, 1963). The evolution of swine management (to eliminate neonatal pig's access to dirt) has accentuated the problem so that iron compounds are injected parenterally as standard practice. Previously, 100 mg of iron as iron dextran injected within the first 3 days was considered to be adequate, until the pigs were eating enough solid feed to provide needed iron. Although some swine producers continue to inject 100 mg of iron, the recommended level has been increased to 200 mg.

Recently, the growth rate of pigs from birth to weaning at 21 days has increased dramatically. Because weaning weights from 13 to 15 lbs have become increasingly common, an additional 200 mg of iron should be given at 10–14 days (Smith, 1989).

3. *Iron Deficiency in Calves*

Veal calves are purposely fed an iron-deficient diet so meat will be pale. They are anemic, with low erythrocyte numbers, hemoglobin, and packed cell volume. Some neonatal calves may have a congenital iron deficiency, because *in utero* transfer of iron from dam to fetus is impaired (Tennant *et al.*, 1975).

4. *Iron Deficiency in Foals*

Foals have low iron stores during the first 5 months of life as indicated by a decreased serum ferritin concentration, an increased serum TIBC, and a decreased mean erythrocyte volume (Harvey *et al.*, 1987).

5. *Iron Deficiency in Dogs*

Iron deficiency can occur in young animals, but most cases involves chronic blood loss from blood-sucking parasites, gastrointestinal tumors, urinary bladder tumors, or hemorrhagic disorders (Harvey *et al.*, 1982; Dodds and Ward, 1980; Weiser and O'Grady, 1983). Excessive use of dogs as blood donors can deplete iron stores and cause an iron deficiency (Weiser and O'Grady, 1983). Dogs show many of the classical signs and laboratory data of other animals, such as decreased serum iron, lower mean corpuscular volume and mean corpuscular hemoglobin concentration, and variable reticulocytosis, but the serum TIBC does not change significantly. Leptocytosis and erythrocyte fragmentation also occur (Harvey *et al.*, 1982; Weiser and O'Grady, 1983).

6. *Iron Deficiency in Cats*

Transient microcytosis and anemia, which can be attributed to iron deficiency, occur in many 5-week-old kittens. Erythrocyte volume distribution curves are more sensitive than the mean corpuscular volume for detecting microcytosis. Some erythrocyte fragmentation occurs, but hypochromia does not. Serum iron and percent transferrin saturation values decline. The percentage of microcytes (less than 28 fl) can be estimated as follows:

$$\text{Microcytes } (\%) = 1340 \; (\% \text{ transferrin saturation})^{-1.52}.$$

Oral or parenteral iron may be appropriate to decrease neonatal morbidity (Weiser and Kociba, 1983).

B. Iron Overload and Toxicity

Iron overload can be divided into the following categories: (1) genetic, (2) iatrogenic, (3) chronic hemolytic anemia, (4) changes from natural diet, and (5) experimental.

1. *Genetic*

In man, hemochromatosis is an inherited disease that is characterized by increased iron absorption from the intestine, high or total iron saturation of plasma transferrin, and abnormal iron deposition in parenchymal cells. If patients are untreated, excess hepatic iron can cause cirrhosis, and accumulation in heart muscle can result in cardiac failure. Other signs include arthropathy, endocrine abnormalities, and diabetes mellitus. Generally, it is a disease of middle-aged men with Irish-English ancestry. The incidence in the general population is about 1 in 3000.

Primary iron overload or hemochromatosis occurs in the Salers breed of cattle. Affected animals have increased serum iron and transferrin saturation, decreased UIBC, and increased hepatic iron. The probands were the product of a line breeding program and have several common ancestries (House *et al.*, 1994).

Horses with hepatic cirrhosis, excess hepatic liver iron, and liver failure have been diagnosed as having hemochromatosis (Pearson *et al.*, 1994; Lavoie and Teuscher, 1993; Edens *et al.*, 1993). This syndrome probably is not classical hemochromatosis, because it occurs in different breeds of horses and has a normal transferrin saturation. It is unclear whether the iron accumulation occurs secondarily to liver damage, or the hepatic disorders result from the iron accumulation.

2. *Iatrogenic*

Many horse owners and trainers as well as a few veterinarians are convinced that iron deficiency occurs commonly in horses. Thus, they provide iron either in feed or parenteral therapy. One race horse had been treated weekly for 2 years. If the horse did not excrete significant quantities of iron, he would have accumulated about 2.5 pounds of iron (Lewis and Moyer, 1975). When the iron status of 100 clinical horses was evaluated, 10% had serum ferritin values that were greater than 6 standard deviations above the mean (Smith *et al.*, 1986).

Foals and calves can be poisoned acutely by oral administration of iron-containing hematinics (Ruhr, 1983; Mullaney and Brown, 1988). Neonatal calves and foals are particularly susceptible to supplements containing large amounts of iron.

Parenterally injected iron dextran also can cause problems. Anaphylactic shock can occur (Landiges and Garlinghouse, 1981), and if selenium-vitamin E are inadequate, animals can die from acute iron toxicity (Lannek *et al.*, 1962).

3. *Chronic Hemolytic Anemia*

Chronic severe anemias that stimulate erythropoietin production are associated with iron overloading. For example, pyruvate kinase deficiency in Basenji dogs causes a severe hemolytic anemia. Erythroid hyperplasia associated with the anemia results in increased hepatic parenchymal iron deposition and possibly some cirrhosis. When an affected dog's bone marrow is replaced with bone marrow from normal dogs, erythropoiesis becomes normal and hepatic iron overload decreases (Weiden *et al.*, 1981). Similarly, congenital dyserythropoiesis in Polled Hereford cattle is accompanied by a marked increase in stored iron (Steffen *et al.*, 1991).

4. *Changes from Natural Diet*

When exotic animals are translocated from their natural surroundings to captive environments, their diets usually are altered. Iron overload can occur when iron is more bioavailable in the substitute diet. Hemosiderosis occurs in captive black rhinoceroses (Kock *et al.*, 1992), lemurs (Gonzales *et al.*, 1984), African rock hyrax (Frye, 1982), Svakbard reindeer (Borch-Iohnsen and Nilseen, 1987), Afghan pikas (Madarame *et al.*, 1990), and mynahs and other tropical birds (Iancu, 1993).

5. *Experimental*

Iron has been given orally or parenterally to various species of animals to reproduce the signs of human hemochromatosis. Unfortunately, many signs of hemochromatosis, including cirrhosis, cannot be induced in most animal species (Iancu, 1993).

C. Acute Phase Reaction

When bacteria invade an animal, their ability to grow and multiply depends on the availability of iron. The animal responds with several inflammatory processes, collectively called the acute-phase reaction. The major mediator of that response is interleukin-1 released by macrophages. Interleukin-1 causes the following changes within hours or days: decreased serum iron and zinc, increased copper and ceruloplasmin, neutrophilia with increased immature forms, fever, and release of acute-phase proteins from the liver. Those acute-phase proteins include ferritin, haptoglobin, ceruloplasmin, amyloid A protein, C-reactive protein, complement components, and fibrinogen (Dinarello, 1984).

The effect on serum iron is dramatic, with levels decreasing to 10–20% of normal within 15 hours. Serum iron then may return to normal within 2–3 days, if the stimulant is removed. Serum ferritin slowly increases to 1.5–2 times normal and then declines over 2–3 weeks (Smith and Cipriano, 1987). If only serum iron is used to determine the iron status of sick animals, misdiagnosis of iron deficiency is possible.

D. Corticosteriods

Corticosteriods can blunt and interfere with the acute-phase response. They also affect serum iron under normal conditions. Serum iron in horses (Smith *et al.*, 1986) and dogs (Harvey *et al.*, 1987) increases dramatically when corticosteroids are given. Other iron analytes such as serum ferritin and TIBC are not affected. In dogs, endogenous cortisone levels seem to be responsible for the diurnal variation of serum iron. In contrast, cattle respond to dexamethasone by decreasing both serum iron and zinc (Weeks *et al.*, 1988a).

E. Anemia of Chronic Disorders

If an inflammatory response becomes chronic, an anemia develops. Serum iron, TIBC, and percent saturation decrease. Ferrokinetically, an inadequate erythropoietic response is partially due to iron unavailability (Feldman *et al.*, 1981a, 1981b). Normally, iron from senescent erythrocytes is distributed into two pools in macrophages. One pool has a rapid turnover with a half-life of 30 minutes; the other, 7 days (Fillet *et al.*, 1974). During chronic inflammation, iron from the rapid pool is severely impaired, and iron remains sequestered in macrophages. Bone marrow iron and storage iron increase. Because iron is sequestered in the reticuloendothelial system, it is unavailable to the erythropoietic tissue (Freireich *et al.*, 1957).

F. Other Disorders

Several other disorders can affect iron metabolism secondarily. Pigs deficient in copper, pyridoxine, and pteroylglutamic acid have an increased iron turnover and anemia (Bush *et al.*, 1956). Increased iron turnover also occurs in polycythemia vera, erythropoietic porphyria, and familial polycythemia (Kaneko, 1980).

Acknowledgments

This work was supported in part by the Kansas Agricultural Experiment Station and was published as contribution number 95-574-B from the KAES. The author gratefully acknowledges the illustrations by Mallory R. Hoover.

References

Addison, G. M., Beamish, M. R., Hales, C. H., Hodgkins, M., Jacobs, A., and Llewellin, P. (1972). *J. Clin. Pathol.* **25,** 326–329.

Andrews, G. A., Chavey, P. S., and Smith, J. E. (1994). *Vet. Pathol.* **31,** 674–678.

Baker, N. F. and Douglas, J. R. (1957). *Am. J. Vet. Res.* **18,** 142–146.

Basilion, J. P., Rouault, T. A., Massinople, M., Klausner, R. D., and Burgess, W. H. (1994). *Proc. Natl. Acad. Sci. USA* **91,** 574–578.

Beinert, H., and Kennedy, M. C. (1993). *FASEB J.* **7,** 1442–1449.

Borch-Iohnsen, B., and Nilseen, K. J. (1987). *J. Nutr.* **117,** 2072–2078.

Braude, R., Chamberlain, A. G., Kotarbinska, M., and Mitchell, K. G. (1962). *Brit. J. Nutr.* **16,** 427–449.

Bush, J. A., Jensen, W. N., Ashenbrucker, H., Cartwright, G. E., and Wintrobe, M. M. (1956). *J. Exp. Med.* **103,** 161–261.

Chausow, D. G., and Czarnecki-Maulden, G. L. (1987). *J. Nutr.* **117,** 928–932.

Conrad, M. E., and Umbreit, J. N. (1993). *Am. J. Hematol.* **42,** 67–73.

Cook, J. D. and Finch, C. A. (1980). *In* "Iron" (J. D. Cook, I. Chanarin, E. Beutler, and E. B. Brown, eds.), pp. 134–147. Churchill Livingstone, New York.

Crichton, R. R., and Charlotteaux-Wauters, M. (1987). *Eur. J. Biochem.* **164,** 485–506.

Dallman, P. R., Beutler, E. and Finch, C. A. (1978). *Br. J. Haematol.* **40,** 179–183.

Dautry-Varsat, A., and Lodish, H. F. (1984). *Sci. Am.* **251,** 52–58.

Dautry-Varsat, A., Ciechanover, A., and Lodish, H. F. (1983). *Proc. Natl. Acad. Sci. USA* **80,** 2258–2262.

Davis, P. N., Norris, L. C., and Kratzer, F. H. (1962). *J. Nutr.* **78,** 445–453.

Dinarello, C. A. (1984). *N. Engl. J. Med.* **311,** 1413-1418.

Dodds, W. J., and Ward, M. V. (1980). *Modern Vet. Prac.* **61,** 496–497.

Douglas, T. A., Renton, J. P., Watts, C., and Ducker, H. A. (1972). *Comp. Biochem. Physiol.* **43A,** 665–671.

Drabkin, D. L. (1951). *Physiol. Rev.* **31,** 345–453.

Edens, L. M., Robertson, J. L., and Feldman, B. F. (1993). *Equine Vet. Journal.* **25,** 81–84.

Edwards, J., Huebers, H., Kunzler, C., and Finch, C. (1986). *Blood* **67,** 623–628.

Erslev, A. J. (1983). *In* "Hematology" (W. J. Williams, E. Beutler, A. J. Erslev, and M. A. Lichtman, eds.), 3rd ed., pp. 1638–1643. McGraw-Hill, New York.

Fairbanks, V. F., and Beutler, E. (1983). *In* "Hematology" (W. J. Williams, E. Beutler, A. J. Erslev, and M. A. Lichtman, eds.), 3rd ed., pp. 300. McGraw-Hill, New York.

Feldman, B. F., Kaneko, J. J., and Farver, T. B. (1981a). *Am. J. Vet. Res.* **42,** 1109–1113.

Feldman, B. F., Keen, C. L., Kaneko, J. J., and Farver, T. B. (1981b). *Am. J. Vet. Res.* **42,** 1114–1117.

Fillet, G., Cook, J. D., and Finch, C. A. (1974). *J. Clin. Invest.* **53,** 1527–1533.

Finch, C. A. (1980). *In* "Pharmacological Basis of Pharmacology" (A. G. Gilman, L. Goodman, and A. Gilman, eds.), pp. 1315–1330. Macmillan, New York.

Finch, C. A., and Cook, J. D. (1984). *Am. J. Clin. Nutr.* **39,** 471–477.

Finch, C. A., and Huebers, H. A. (1985). *Metal Metabol. Dis.* **4,** 5–10.

Finch, C. A., Miller, L. R., Inamder, A. R., Person, R., Seiler, K., and Mackler, B. (1976). *J. Clin. Invest.* **58,** 447–453.

Freireich, E. J., Miller, A., Emerson, C. P., and Ross, J. F. (1957). *Blood* **12,** 972–983.

Frye, F. L. (1982). *J. Zoo Anim. Med.* **13,** 152–156.

Furugouri, K. (1984). *J. Dairy. Sci.* **46,** 859–865.

Gonzales, J., Bernirschke, K., Saltman, P., Roberts, J., and Robinson, P. T. (1984). *Zoo. Biol.* **3,** 255–265.

Hahn, R. L. (1943). *J. Exp. Med.* **78,** 169–185.
Handy, R. W. (1979). *Clin. Chem.* **25,** 197–198.
Harris, Z. L., Takahashi, Y., Miyajima, H., Serizawa, M., MacGillivray, R. T. A., and Gitlin, J. D. (1995). *Proc. Natl. Acad. Sci. USA* **92,** 2539–2543.
Harrison, P. M. (1977). *Semin. Hematol.* **14,** 55–57.
Harrison, P. M., Treffry, A., and Lilley, T. H. (1986). *J. Inorg. Biochem.* **27,** 287–293.
Harvey, J. W. (1981). *Vet. Clin. N. Am. Small Anim. Pract.* **11,** 349–381.
Harvey, J. W., French, T. W., and Meyer, D. J. (1982). *J. Am. Anim. Hosp. Assoc.* **18,** 946–996.
Harvey, J. W., Asquith, R. L., Sussman, W. A., and Kivipelto, J. (1987). *Am. J. Vet. Res.* **48,** 1348–1352.
Hastka, J., Lasserre, J.-J., Schwarzbeck, A., and Hehlmann, R. (1994). *Clin. Chem.* **40,** 768–773.
Henry, R. J., Cannon, D. C., and Winkelman, J. W. (1974). *In* "Clinical Chemistry: Princples and Techniques," pp. 687–695. Harper & Row, New York.
Hoensch, H., Woo, C. H., Raffin, S. B., and Schmid, R. (1976). *Gastroenterology* **70,** 1063–1070.
Hoffbrand, A. V., Ganeshaguru, K., Hooton, J. W. L., and Tattersall, M. H. N. (1976). *Br. J. Haematol.* **33,** 517–526.
House, J. K., Smith, B. P., Maas, J., Lane, V. M., Anderson, B. C., Graham, T. W., and Pino, M. V. (1994). *J. Vet. Intern. Med.* **8,** 105–111.
Huebers, H., Sciba, E., Huebers, E., and Finch, C. A. (1985). *Proc. Soc. Exp. Biol. Med.* **179,** 222–226.
Huff, R. L., Hennessey, T. G., Austin, R. E., Garcia, J. F., Roberts, B. M., and Lawrence, J.H. (1950). *J. Clin. Invest.* **29,** 1041–1052.
Hunter, J. E. (1978). *J. Nutr.* **108,** 497–595.
Iancu, T. C. (1993). *Adv. Vet. Sci. Comp. Med* **37,** 379–400.
Jacobs, A., Miller, F., Worwood, M., Beamish, M. R., and Wardrop, C. A. (1972). *Br. Med. J.* **4,** 286–290.
Kaneko, J. J. (1963). *N. Y. Acad. Sci.* **104,** 689–700.
Kaneko, J. J. (1964). *Gaines Vet. Symp.* **13,** 2–5.
Kaneko, J. J. (1980). *In* "Clinical Biochemistry of Domestic Animals" (J. J. Kaneko, ed.), pp. 649–669. Academic Press, New York.
Kaneko, J. J., and Mattheeuws, D. R. G. (1966). *Am. J. Vet. Res.* **27,** 923-927.
Klausner, R. D., and Harford, J. B. (1989). *Science* **246,** 870–872.
Kock, N., Foggin, C., Kock, M., and Kock, R. (1992). *J. Zoo Wildl. Med.* **23,** 230–234.
Kolb, E. (1963). *Adv. Vet. Sci.* **8,** 49–114.
Landiges, W. C., and Garlinghouse, L. E., Jr. (1981). *Lab. Anim. Sci.* **31,** 421–422.
Lannek, N., Lindberg, P., and Tollerz, G. (1962). *Nature* **195,** 1006–1007.
Lavoie, J. P., and Teuscher, E. (1993). *Equine. Vet. J.* **25,** 552–551.
Lee, G. R., Nacht, S., Lukens, J. N., and Cartwright, G. E. (1968). *J. Clin. Invest.* **47,** 2058–2069.
Lewis, H. B., and Moyer, W. A. (1975). *In* "Proc First Internat Symp Equine Hematol" (H. Kitchen, and J. D. Krehbiel, eds.), pp. 258–261. Am Assoc Equine Pract, Golden, Company.
Lintzel, W., and Radeff, T. (1931). *Arch. Tiererhr. Tierz.* **6,** 313–358.
Lipschitz, D. A., Cook, J. D., and Finch, C. A. (1974). *N. Engl. J. Med.* **290,** 1213–1216.
Lorier, M. A., Herron, J. L., and Carrell, R. W. (1985). *Clin. Chem.* **31,** 337–338.
Mackler, B., Person, R., Miller, L. R., Inamdar, A. R., and Finch, C. A. (1978). *Pediat. Res.* **12,** 217–222.
Madarame, H., Kumagai, M., Suzuki, J., Watanabe, A., and Konno, S. (1990). *J. Comp. Pathol.* **103,** 351–359.
Maeda, K., McKenzie, H. A., and Shaw, D. C. (1984). *Anim. Blood Grps. Biochem. Genet.* **15,** 299–312.
Mahaffey, E. B. (1986). *In* "Current Veterinary Therapy" (R. W. Kirk, ed.), pp. 521–524. Saunders, Philadelphia, Pennsylvania.
Miyata, Y., Furugouri, K., and Shijimaya, K. (1984). *J. Dairy. Sci.* **67,** 1256–1263.
Mullaney, T. P., and Brown, C. M. (1988). *Equine. Vet. J.* **20,** 119–124.
Murakawa, H., Bland, C. E., Willis, W. T., and Dallman, P. R. (1987). *Blood* **69,** 1464–1468.
Nathanson, M. H., and McLaren, G. D. (1984). *Clin. Res.* **32,** 317.
Newlands, C. E., Houston, D. M., and Vasconcelos, D. Y. (1994). *J. Am. Vet. Med. Assoc.* **205,** 849–851.
NRC. (1975). "Nutrient Requirements of Sheep." National Academy of Sciences, Washington, D.C.
NRC. (1976). "Nutrient Requirements of Beef Cattle." National Academy of Sciences, Washington, D.C.
NRC. (1978a). "Nutrient Requirements of Dairy Cattle." National Academy of Sciences, Washington, D.C.
NRC. (1978b). "Nutrient Requirements of Horses." National Academy of Sciences, Washington, D.C.
NRC. (1979). "Nutrient Requirements of Swine." National Academy of Sciences, Washington, D.C.
NRC. (1981). "Nutrient Requirements of Goats: Angora, Dairy and Meat Goats in Temperate and Tropical Countries." National Academy Press, Washington, D.C.
NRC. (1984). "Nutrient Requirements of Beef Cattle." National Academy Press, Washington, D.C.
NRC. (1985). "Nutrient Requirements of Dogs." National Academy Press, Washington, D.C.
Obara, J., and Nakajima, H. (1961a). *Bull. Natl. Inst. Anim. Health, Tokyo* **42,** 45–55.
Obara, J., and Nakajima, H. (1961b). *Jpn. J. Vet. Sci.* **23,** 247–253.
Pearson, E. G., Hedstrom, O. R., and Poppenga, R. H. (1994). *J. Am. Vet. Med. Assoc.* **204,** 1053–1056.
Pollycove, M., and Mortimer, R. (1961). *J. Clin. Invest.* **40,** 753–853.
Richter, G. W. (1967). *Exp. Mol. Pathol.* **6,** 96–101.
Richter, G. W. (1984). *Lab. Invest.* **50,** 26–35.
Rogers, J., and Munro, H. (1987). *Proc. Natl. Acad. Sci. USA* **84,** 2277–2281.
Rosen, A. C., Rosen, H. R., Huber, K., Bauer, K., Ausch, C., Redlich, K., Klein, M. J., and Moroz, C. (1995).*Gynecol. Obstet. Invest* **39,** 11–14.
Ruhr, L. P. (1983). *J. Am. Vet. Med. Assoc.* **182,** 616–618.
Samaniego, F., Chin, J., Iwai, K., Rouault, T. A., and Klausner, R. D. (1994). *J. Biol. Chem.* **269,** 30904–30910.
Siimes, M. A., Addiego, J. E., Jr., and Dallman, P. R. (1974). *Blood* **43,** 581–659.
Smith, J. E. (1989). *In* "Clinical Biochemistry of Domestic Animals" (J. J. Kaneko, ed.), pp. 256–273. Academic Press, New York.
Smith, J. E., and Cipriano, J. E. (1987). I*Vet. Pathol.* **24,** 354–356.
Smith, J. E., Moore, K., and Schoneweis, D. (1981). *Am. J. Vet. Res.* **42,** 1084–1087.
Smith, J. E., Moore, K., and Boyington, D. (1983). *Biochem. Med.* **29,** 293–297.
Smith, J. E., Moore, K., Boyington, D., Pollman, D. S., and Schoneweis, D. (1984). *Vet. Pathol.* **21,** 597–696.
Smith, J. E., Cipriano, J. E., DeBowes, R., and Moore, K. (1986a). *J. Am. Vet. Med. Assoc.* **188,** 285–2870.
Smith, J. E., DeBowes, R. M., and Cipriano, J. E. (1986b). *J. Am. Vet. Med. Assoc.* **188,** 1296–1298.
Stahl, W. R. (1967). *J. Appl. Physiol.* **22,** 453–546.
Steffen, D. J., Leipold, H. W., Gibb, J., and Smith, J. E. (1991). *Vet. Pathol.* **28,** 234–240.
Stohlman, F., Jr., Howard, D., and Beland, A. (1963). *Proc. Soc. Exp. Biol. Med.* **113,** 986–988.

Stratil, A., and Glasnak, V. (1981). *Anim. Blood Grps. Biochem. Genet.* **12,** 113–212.

Tennant, B., Harrold, D., Reina-Guerra, M., and Kaneko, J. J. (1975). *Cornell Vet.* **65,** 543–556.

Torrance, J. D., and Bothwell, T. H. (1980). *In* "Iron" (J. D. Cook, I. Chanarin, E. Beutler, and E. B. Brown, eds.), pp. 909–115. Churchill Livingstone, New York.

Trundle, D. S. (1984). *Clin. Chem. News* **10,** 8–10.

Underwood, E. J. (1977). *In* "Iron." pp. 13–55. Academic Press, New York.

Walters, G. O., Miller, F. M., and Worwood, M. (1973). *J. Clin. Pathol.* **26,** 770–772.

Ward, C., Saltman, P., Ripley, L., Ostrup, R., Hegenauer, J., Hatlen, L., and Christoper, J. (1977). *Am. J. Clin. Nutr.* **30,** 1054–1063.

Weeks, B. R., Smith, J. E., DeBowes, R. M., and Smith, J. (1988a). *Vet. Pathol.* **26,** 345–346.

Weeks, B. R., Smith, J. E., and Phillips, R. M. (1988b). *Am. J. Vet. Res.* **49,** 1193–1195.

Weeks, B. R., Smith, J. E., and Northrup, J. K. (1989). *Am. J. Vet. Res.* **50,** 198–200.

Weiden, P. L., Hackman, R. C., Deeg, H. J., Graham, T. C., Thomas, E. D., and Storb, R. (1981). *Blood* **57,** 66–70.

Weiser, M. G., and Kociba, G. J. (1983). *Vet. Pathol.* **20,** 1–12.

Weiser, G., and O'Grady, M. (1983). *Vet. Pathol.* **20,** 230–241.

CHAPTER

10

Hemostasis

W. JEAN DODDS

I. INTRODUCTION

Hemostasis, the process by which bleeding is arrested, comprises a complex series of physiological and biochemical events that terminate in the formation of a stable plug that seals the blood vessel. The process involves the following sequence of events: interaction between the blood vessel wall, adhesive glycoproteins, and platelets; blood coagulation; and fibrinolysis (Fig. 10.1). The pathophysiological mechanisms involved with each of these components are summarized below. More detailed descriptions can be found elsewhere (Comp, 1990a; Esmon *et al.*, 1991; Preissner and Jenne, 1991; Bauer *et al.*, 1992; Bornstein, 1992; Breider, 1993; Lüscher and Weber, 1993; Marcus and Safier, 1993; Calvete, 1994; Ruf and Edgington, 1994; Andrew, 1995; Broze, Jr., 1995; Caen and Rosa, 1995; Cerletti *et al.*, 1995; Clemetson, 1995; Davie, 1995; Deng *et al.*, 1995; Dowd *et al.*, 1995; Furie and Furie, 1995a; Ginsberg *et al.*, 1995; Grabowski and Lam, 1995; Hawiger, 1995; Hess *et al.*, 1995; Nemerson, 1995; Rapaport and Rao, 1995; Shattil, 1995; Sixma *et al.*, 1995; Ugarova *et al.*, 1995; Verstraete, 1995; Weiss, 1995; Griendling and Alexander, 1996).

II. MECHANISMS OF HEMOSTASIS

A. Blood Vessels and Adhesion Molecules

When a blood vessel is injured or severed, a brief local reflex vasoconstriction occurs that reduces blood flow at the site. Vascular contraction is maintained by the release of vasoactive compounds from adjacent platelets and surrounding tissues. Simultaneously, platelets in the vicinity adhere to exposed subendothelial collagen fibers. This interaction with collagen causes a "release reaction" whereby platelet constituents, such as adenosine diphosphate (ADP), serotonin, epinephrine, and histamine, are released into the surrounding medium. The role of platelets in sustaining hemostasis has been reviewed by Lüscher and Weber (1993), Marcus and Safier (1993), Caen and Rosa (1993), and Hawiger (1995); the process is summarized in Section II.B.

Other constituents of the blood vessel wall and its endothelial cells play important roles in the formation

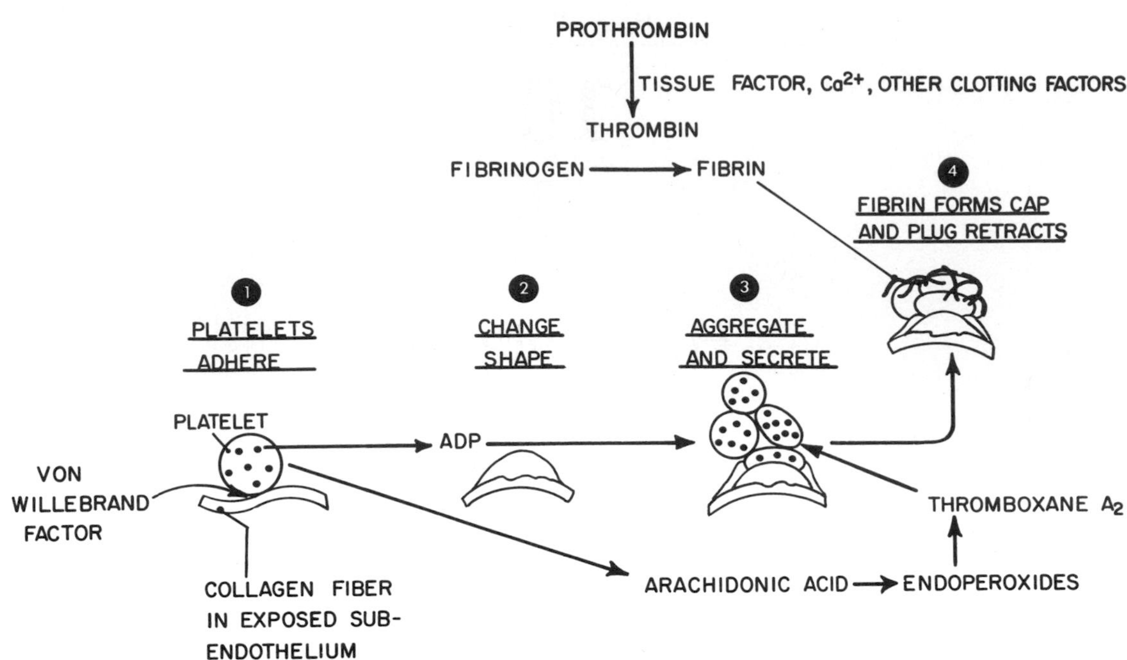

FIGURE 10.1 The hemostatic process (Dodds, 1989a).

and dissolution of hemostatic plugs, thrombi, and atherogenic lesions. The mechanism of interaction among these components has recently been elucidated (Comp, 1990a; Preissner and Jenne, 1991; Breider, 1993; Marcus and Safier, 1993; Caen and Rosa, 1995; Grabowski and Lam, 1995; Hess *et al.*, 1995; Griendling and Alexander, 1996).

One of the basic functions of the intact, normal vascular endothelium is to provide a thromboresistant surface to the flowing blood. Intact endothelium does not promote platelet or leukocyte adherence nor does it activate coagulation. Both active and passive mechanisms apparently play a role in maintaining thromboresistance, and endothelial cells actively contribute to this thromboresistance by regulating a complex balance between the procoagulant and anticoagulant properties of the vasculature (Breider, 1993; Griendling and Alexander, 1996). The procoagulant activities of endothelium include von Willebrand factor, tissue factor, plasminogen activator inhibitor (PAI), and factor V, whereas the anticoagulants of endothelium include its barrier to subendothelium, prostacyclin (prostaglandin I_2 or PGI_2), thrombomodulin, protein S, lipoprotein-associated coagulation inhibitor (LACI), plasminogen activator, and heparan sulfate. Endothelium, particularly in the pulmonary microcirculation, actively removes substances from the circulation that promote platelet aggregation. These include prostaglandin F_1, serotonin, adenine nucleotides, bradykinin, and angiotensin I. In addition, all layers of blood vessels, especially the intima and endothelium, synthesize and release PGI_2, a potent inhibitor of platelet aggregation (Herman *et al.*, 1991; Marcus and Safier, 1993).

The proteoglycan matrix of the blood vessel wall also directly influences thrombogenicity. Three components, heparin, heparan sulfate, and dermatan sulfate, possess anticoagulant activity, whereas other glycosaminoglycans and hyaluronic acid do not. Veins have the highest concentrations of the former compounds and are the most thromboresistant. The antithrombotic proteoglycans differ in potency and mechanism of action. They affect coagulation factors (thrombin, antithrombin III, factor X, and fibrinogen) and platelet functions (inhibit thrombin- and collagen-induced aggregation, and bind to platelet factor 4 and platelet-derived growth factor) (Dodds, 1989a).

Passive mechanisms of thromboresistance also exist but are less contributory. These include protection of the cell surface by its carbohydrate-rich cell coat, the glycocalyx; the negative surface charge of the endothelial lumen, which repels similarly charged cells such as platelets; continuous, nonspecific renewal of cell membranes; and presence at the cell surface of the protease inhibitor α_2-macroglobulin (Breider, 1993; Griendling and Alexander, 1996).

In addition to its role in thromboresistance (Section II.B), the endothelium has important synthetic, metabolic, and pathological functions. Endothelial cells synthesize von Willebrand factor (vWF) (Section II.C.2.g); tissue plasminogen activator (TPA); basement membrane type III and IV collagens and elastin; fibronectin or cold insoluble globulin (CIg), α_2-opsonic protein; numerous metabolic enzymes; and PGI_2. The capacity of endothelium to synthesize large amounts of vWF and fibronectin has permitted these proteins to be used as markers for identifying endothelial cells in tissue specimens and cell cultures.

The endothelium plays an important role in such pathological processes as atherosclerosis, thrombosis, and disseminated intravascular coagulation (DIC), defective hemostasis, inflammation, immune disorders, vascular neoplasia, and metastasis (Comp, 1990a; Esmon *et al.*, 1991; Breider, 1993; Costantini and Zacharski, 1993; Fareed and Callas, 1995; Hemker and Be;aguin, 1995; Griendling and Alexander, 1996). The initiating event in atherogenesis has long been attributed to endothelial injury. After injury, there is increased entrapment of plasma lipoproteins followed by proliferation of arterial smooth muscle cells and/or deposition of free cholesterol and cholesterol esters, *de novo* synthesis of connective tissue, and intracellular and extracellular deposition of lipids in variable amounts. This reaction to injury hypothesis incorporates both the myogenic and lipogenic mechanisms of atherogenesis. Other causes of endothelial injury are infectious agents, especially viruses, gram-negative bacteria, and rickettsia; prolonged hypotension and acidosis or hypoxia; dysproteinemias; primary and secondary immune mechanisms; and acute inflammation. These predispose to either thrombosis and DIC (Joist, 1990), increased vascular permeability (Esmon *et al.*, 1991), or defective hemostasis with purpura or hemorrhage depending on the nature of the incitant (Feldman, 1988; Dodds, 1989a; Mannucci and Giangrande, 1994; Caen and Rosa, 1995).

B. Platelets

Blood platelets are cellular particles produced by megakaryocytes under the influence of a variety of growth regulatory factors (Kaushansky, 1995). The main source is the bone marrow, although the lung and spleen are also sources of platelets. The megakaryocyte is a large (25–50 mm) cell with a polyploid nucleus that is extremely pleomorphic. As the cell matures, pseudopods form and platelets bud off at the extremities (Gewitz, 1995). Megakaryocyte and platelet production are regulated primarily by thrombopoietin, a lineage-specific cytokine that acts as a growth factor for platelet production (Kaushansky, 1995). The normal platelet seen in a Wright's stained blood film is 1–4 mm in diameter and is spherical, oval, or rod shaped. The cytoplasm is pale blue and contains reddish granules. Circulating platelets are disk shaped.

The role of the blood platelet in hemostasis is of equal importance to that of the coagulation mechanism (Lüscher and Weber, 1993; Marcus and Safier, 1993). Platelets are involved with the blood vessel wall and the contact-activated coagulation factors (XII and XI) in the initiation of the hemostatic process. Decrease in the number of circulating platelets (thrombocytopenia) or presence of abnormal, nonfunctional platelets (thrombasthenia, thrombopathia) will impair hemostasis. In some instances of an excess of platelets (thrombocytosis or thrombocythemia), hemostasis may be inadequate, but usually clotting is promoted, which may predispose a patient to thrombosis. The multifaceted role of platelets in biochemical, physiological, and pathological processes has been extensively reviewed (Boudreaux *et al.*, 1989a; Bornstein, 1992; Shukla, 1992; Rao *et al.*, 1993; Calvete, 1994; Gawaz *et al.*, 1994; Caen and Rosa, 1995; Cerletti *et al.*, 1995; Clemetson, 1995; Furie and Furie, 1995a; Ginsberg *et al.*, 1995; Hawiger, 1995; Hess *et al.*, 1995; Nurden, 1995; Shattil, 1995; Sixma *et al.*, 1995; Ugarova *et al.*, 1995; Weiss, 1995).

Platelets promote hemostasis in several ways (Lüscher and Weber, 1993; Marcus and Safier, 1993). When a blood vessel is injured, platelets accumulate at the site of injury. They adhere to the vessel wall, then to each other, and also become involved in the intrinsic coagulation pathway. On exposure of platelets to collagen fibers of the vessel wall, certain active constituents are released from the platelet (serotonin, histamines, ADP) (Sixma *et al.*, 1995). The release of ADP into the ambient fluid induces the adherence and aggregation of platelets to the area. The aggregation process, which requires calcium, fibrinogen, and metabolic energy, forms an initial hemostatic plug of platelets that seals the vessel (Clemetson, 1995). In addition to mediating platelet aggregation, ADP releases a platelet surface phospholipoprotein, platelet factor 3 (PF3), which accelerates coagulation. Platelets also undergo a series of reactions mediated by thrombin and connective tissue that produces marked structural changes (Ugarova *et al.*, 1995). This process has been called viscous metamorphosis. The thrombin associated with *viscous metamorphosis* and evolved from tissue injury rapidly activates the coagulation mechanism, fibrin is formed around the platelet aggregate, and a stable fibrin plug forms.

Platelets are metabolically active in biochemical, physiological, and pathological processes (Table 10.1). In addition to the basic adhesion, aggregation, and release reactions, they agglutinate in the presence of a variety of compounds including bovine factor VIII, ruminant plasma, ristocetin, polylysine, snake venoms, and polybrene (Dodds, 1989a; Smith and Brinkhous, 1991; Brinkhous, 1992; Ermens *et al.*, 1995).

TABLE 10.1 Functions of Blood Platelets[a]

Hemostasis and thrombosis (endothelial cells, coagulation, von Willebrand factor, collagen, adenosine diphosphate, adrenalin, thrombin)
Atherogenesis (platelet-derived growth factor, mitogens, cell-proliferating activity)
Specialized smooth muscle cell (clot retraction)
Inflammation (specialized type of intravascular leukocyte; chemotaxis and phagocytosis)
Prostaglandin metabolism
Immunologic reaction (immune complex disease)
Endotoxin reactivity (Schwartzman reaction)
Monoamines and serotonergic synaptosomes (epinephrine, serotonin)
Interactions with tumor cells (adhesion, aggregation, cell transformation, metastases)
Synthesis of proteins, lipids, carbohydrates, and nucleotides

[a] Modified from Dodds (1989a).

Platelets actively synthesize proteins, lipids, carbohydrates, and nucleotides. Among the proteins synthesized by or contained in platelets are fibrinogen; the components of the contractile system, actomyosin, myosin, actin, tropomyosin, troponin, and sarcoplasmic reticulum relaxing factor; acid hydrolases and other enzymes; and a series of factors including the growth, permeability, bactericidal, and chemotactic activity factors (Dodds, 1989a). The lipids synthesized by platelets include fatty acids, prostaglandins, glycerides, a series of phosphatides, the sphingolipids, sphingomyelin, cerebrosides, and gangliosides; and sterols. The carbohydrates synthesized by or contained within platelets are glycogen, glycosaminoglycans, and glycoproteins, for example, vWF. The adenine nucleotides of platelets include the cyclic AMP and GMP systems, which regulate the production of platelet endoperoxides, thromboxanes, and prostaglandins. Other important components of platelets are serotonin; thrombospondin, the thrombin-sensitive protein contained in platelet alpha granules (Bornstein, 1992); integrins including the platelet glycoprotein IIb/IIIa complex (Calvete, 1994; Ginsberg *et al.*, 1995; Shattil, 1995); vitronectin (Preissner and Jenne, 1991; Hess *et al.*, 1995); P-selectin, a membrane glycoprotein of platelet alpha granules (Furie and Furie, 1995a); calpain, an ubiquitous calcium-activated neutral protease of animal cells (Saido *et al.*, 1994); and metal ions (Ca^{2+}, Mg^{2+}, Zn^{2+}, K^{+}) (Gawaz *et al.*, 1994). Among the group of adhesion proteins that promote hemostasis via platelet–vessel wall interactions are thrombospondin, fibrinogen, and fibronectin (Feldman *et al.*, 1988; Ugarova *et al.*, 1995).

Platelets generate metabolic energy in the form of ATP, via glycolysis, glycogenolysis, and mitochondrial oxidative phosphorylation, which is then utilized to maintain platelet functions (Grauer *et al.*, 1992; Marcus and Safier, 1993). Platelets have many functions, including phagocytosis of viruses, latex, immune complexes and iron; maintenance of vascular integrity by filling gaps that form in the endothelium and by directly supporting endothelial cells; synthesis and release of vWF in humans and some animal species, and fibronectin; participating in surface adhesion and activation processess (Caen and Rosa, 1995; Clemetson, 1995; Nurden, 1995); production and release of potent smooth muscle and endothelial cell proliferating factor(s); and retraction of clots, a process that stabilizes the initial hemostatic plug and activates clot lysis. Clot retraction requires Ca^{2+}, ATP, and thrombasthenin, the platelet contractile protein (Clemetson, 1995).

Additionally, platelets play an important regulatory role via prostaglandin pathways in promoting hemostasis and thrombosis and in maintaining the thromboresistance of intact endothelium (Herman *et al.*, 1991). This involves metabolism of arachidonic acid released from platelet phospholipids. A potent but unstable platelet-aggregating agent and vasoconstrictor, thromboxane A_2 (TxA_2), is produced from cyclic endoperoxides by the action of cyclooxygenase on arachidonic acid (Johnson *et al.*, 1991). The potent aggregating effect of TxA_2 on platelets is inhibited by the production of prostaglandins E_1 and D_2 (PGE_1 and PGD_2) from linolenic and linoleic acids. Thus, platelet phospholipid and lipid metabolism plays a crucial role in the stimulation and inhibition of platelet reactivity. Aspirin is an effective inhibitor of the platelet release reaction because it acetylates cyclooxgenase and inhibits TxA_2 production (Grauer *et al.*, 1992).

A key component of the platelet prostaglandin regulatory mechanism of hemostasis and thrombosis is the production of prostacyclin (PGI_2) by endothelium and other vascular tissues. Prostacyclin is estimated to be about 30 times more potent than PGE_1 and PGD_2 in inhibiting platelet aggregation and also causes relaxation of vascular smooth muscle. Like platelet TxA_2, PGI_2 is formed in blood vessels from cyclic endoperoxides by the action of cyclooxygenase on arachidonic acid. Production of PGI_2 from endogenous or exogenous arachidonic acid is effectively blocked by aspirin and indomethacin, but production from other prostaglandin endoperoxides is not (Herman *et al.*, 1991; Grauer *et al.*, 1992; Goodnight, 1995). The dose of aspirin required to inhibit vascular PGI_2 formation is apparently much greater than that required to prevent platelet synthesis of TxA_2 (Grauer *et al.*, 1992). This observation has considerable significance because the antithrombogenic properties of aspirin, as mediated by inhibition of platelet TxA_2, can be reversed at high aspirin dosage by inhibition of vessel wall PGI_2.

The final points to emphasize about platelets are the age and species differences encountered in their morphological and functional characteristics. Neonatal platelets are less reactive than adult platelets to physiological agonists (Rajasekhar *et al.*, 1994). The comparative aspects of platelet function in animals and humans have been reviewed (Dodds, 1989a).

C. Coagulation

1. General

The process of blood coagulation has been the subject of intensive investigation for more than 50 years (Brinkhous, 1992). It involves a complex series of reactions among a group of coagulation factors, which are designated by the International Committee for the Nomenclature of Blood Clotting Factors (1962) by Roman numerals. The blood coagulation process participates in hemostasis after the initial interactions of the platelets with the vessel wall and with one another (Caen and Rosa, 1995; Davie, 1995). Blood flowing through an injured vessel contacts exposed foreign surfaces, and the injured cells release tissue thromboplastins (Grabowski and Lam, 1995). These surface contacts and released thromboplastins activate coagulation (Ruf and Edgington, 1994).

Coagulation proceeds by an intrinsic or intravascular pathway (Lozier and High, 1990; Bauer *et al.*, 1992; Davie, 1995; Nemerson, 1995) and by an extrinsic or tissue juice pathway, both of which convert prothrombin to thrombin (Hemker and Béguin, 1995; Nemerson, 1995; Rapaport and Rao, 1995). Thrombin converts fibrinogen to soluble fibrin monomer, which, with factor XIII and calcium, is converted to the fully polymerized, insoluble fibrin clot. This final stage of coagulation is also the end point measured in most coagulation tests.

Once initiated, activation of the intrinsic and extrinsic systems continues sequentially and through an interactive reinforcement loop, forming the end products plasma thromboplastin and prothrombinase, respectively (Lozier and High, 1990; Davie, 1995). These prothrombin-converting principles convert factor X to Xa, which initiates the final common pathway of prothrombin conversion. This pathway involves factor V, prothrombin (factor II), fibrinogen (factor I), and factor XIII. The phospholipid necessary for the interaction of factors IX and VIII and for the activation of factors X and V is supplied *in vivo* primarily by platelet factor 3 (PF-3) with some from red cells and tissue juice. *In vitro*, it can be provided by various phospholipid preparations or their substitutes.

The clotting mechanism can be best described as a series of sequential activating steps that produces a plasma thromboplastic and/or a tissue thromboplastic component, both of which can convert prothrombin to thrombin in the presence of calcium ions (Lozier and High, 1990; Davie, 1995). The enzymatically active factors are protein in nature, and, in general, each of the sequential activations is enzymatic hydrolyses of the inactive precursor forms. Note that factor III (tissue thromboplastin) primarily acts in the extrinsic system and that there is no factor VI designated in the nomenclature. Several clotting factors have more recently been identified that participate in the contact phase of intrinsic clotting (factors XII and XI) and are part of the plasma kallikrein–kinin system responsible for capillary permeability and smooth muscle contraction. These, which have not been given numerical designations, include the Fletcher factor (prekallikrein) and the Williams Fitzgerald–Flaujeac factor (high-molecular-weight kininogen, HMWK).

The fibrin produced by blood coagulation forms at the periphery of the initial hemostatic plug of platelets. The platelet–fibrin mass continues to grow and becomes covered with a cap of fibrin. This stabilized hemostatic plug then contracts to seal the vessel. Thrombin generated during the coagulation process also plays an important part in growth and stabilization of the initial platelet plug (Fenton, 1995). Very low concentrations of thrombin, insufficient to produce visible fibrin formation, cause aggregation of platelets, induce the process of viscous metamorphosis and the release reaction, and make the PF-3 on the platelet surface available for coagulation (Guillin *et al.*, 1995). Thus, the processes of platelet aggregation, release, and contraction are directly enhanced by products evolved after activation of the coagulation mechanism.

The coagulation mechanisms of humans, other animals, and less complex forms are remarkably similar (Dodds, 1989a), but only a few studies have systematically compared platelet function, coagulation, and fibrinolysis among species or have provided a reliable comparison using humans as the reference. Recent studies have compared coagulation activities in fetuses, newborn infants, and adults (Hathaway and Corrigan, 1991; Andrew, 1995) and during fetal development in lambs and lamb neonates (Moalic *et al.*, 1989). Tables 10.2 and 10.3 give comparative species data (Dodds, 1989a).

A major consideration in the understanding of coagulation and hemostasis is the effect of physiological stress on coagulation, fibrinolysis, and platelet function. Alterations from normal hemostasis occur during

TABLE 10.2 Comparison of Hemostatic Parameters in Animals with Those in Normal Human Beings[a,b]

Species	Number	Platelet/thrombocyte			Extrinsic clotting	Intrinsic clotting	Fibrinolysis
		Retention	Aggregation	Release			
Nonhuman primate	E	E or D	E	E	E or I	E or I	E or I
Rabbit	I	I	D[c]	E	E	I	E or D
Guinea pig	I	I	D[d]	E or D	D	E	E
Rat	Markedly I	I	D[c]	D	E or I	I	D, E, or I
Mouse	Markedly I		D[c]	D	E or I	E or I	E or I
Hamster	I		D[c]		E or I	I	E
Cat	E or D	E	E	E	E	I	E or I
Dog	I	I	E[d]	E	I	I	I
Pig	I	E or I	D or E[c]	E	E or D	I	E or D
Sheep	I	E or I	D[c]	E	I or E	I	E or D
Goat	I	E	D[c]		E	I	E or D
Cow	I	E	D[c] or E[c]	E	D or E	D, E, or I	E or D
Horse	D	E	E[c]	E	D or E	D	E
Bird and reptile	E		D[e]		E[f] or D	Markedly D	D or Absent

[a] See Dodds (1989a).
[b] D, decreased; E, equivalent; I, increased.
[c] No response to adrenalin.
[d] Occasional response to adrenalin.
[e] Response to thrombin only.
[f] With homologous thromboplastin.

fetal and neonatal life, with age, sex, hormonal changes, pregnancy, ethnicity, exercise, mental stress, physical stress, diurnal variation, obesity, smoking, alimentary lipemia, arteriovenous differences, and blood groups (Johnstone *et al.*, 1989; Blombäck *et al.*, 1992; Crossley *et al.*, 1992; Ivankovic *et al.*, 1992; Conlan *et al.*, 1993; Semrad and Dubielzig, 1993; Henkens *et al.*, 1995). Although the significance of many of these changes is unknown, some affect the interpretation of diagnostic test results or place the patient at risk to bleed or thrombose. For example, neonates have low levels of the prothrombin complex vitamin K-dependent clotting factors (Hathaway and Corrigan, 1991; Andrew, 1995) and plasminogen (Andrew *et al.*, 1992; Schmidt *et al.*, 1993), and individuals who are pregnant, markedly stressed, heavy smokers, or grossly overweight may have enhanced hemostasis and be at risk for thrombotic problems (Conlan *et al.*, 1993, 1994), although the bleeding time has been shown to be prolonged in the last trimester of pregnancy (Ivankovic *et al.*, 1992).

The individual clotting factors are discussed in the following sections; detailed reviews are listed among the references.

2. *The Intrinsic System*

a. Hageman Factor (Factor XII)

Hageman factor is involved in the initial stage of clotting through a surface-mediated or "contact activation" process. Factor XII is a single-chain sialoglycoprotein of 80 kDa that is not adsorbed by $Al(OH)_3$, $BaSO_4$, or $Ca_3(PO_4)$, is adsorbed with kaolin or celite, and is stable at 56°C and at various pHs. It is activated by glass, collagen, skin, stearate, vascular basement membranes, ellagic acid, uric acid, and most foreign surfaces. Factor XII activation is inhibited by cytochrome *c*, lysozyme, ribonuclease, spermine, and Cl-esterase inhibitor. Hageman factor migrates electrophoretically as an α-globulin. It is present in plasma as a precursor to a serine protease. It is split by trypsin or plasma kallikrein into two fragments, a 28-kDa active component and an inactive one of 52 kDa. It contains about 17% carbohydrate and no lipid. Hageman factor is synthesized by the liver and its amino acid sequence is known (Davie, 1995). The half-life of factor XII is 52–60 hours and its plasma concentration is about 15–45 μg/ml (Dodds, 1989a).

TABLE 10.3 Comparison of Coagulation Factor Activities in Animals with Those in Normal Human Beings[a,b]

Species	Factor									
	I (fibrinogen)	II (prothrombin)	V	VII	VIII	IX	X	XI	XII	XIII
Nonhuman primate	E or I	E or I	I or E	I	I or E	E	E	E	I or E	E or D
Rabbit	E or I	Markedly I	Markedly I	D or E	I	I	I	I	E	I
Guinea pig	I or E	D	I	D	I	E	D	E	E	I
Rat	E	E or I	I	E	I	D or E	D	D or E	I	E
Mouse	E	E or D	I	I	E	E	I	E	E	E or I
Hamster	E	E		D	I	E	E	E	E	E or I
Cat	I or E	D or E	Markedly I	E	I	I	Markedly D	I	I	E
Dog	I or E	E or D	Markedly I	I	Markedly I	I	I	I	E or I	I
Pig	I or E	E or D	I or E	E or I	I	I	E	E or I	I	D
Sheep	I	D	I	D	Markedly I	I	D	D	I	D
Goat	I or E	E or D	I	E or D	I	E	D or E	D	I	E
Cow	I	D or E	Markedly I	E or D	E	E	E	E or D	I	Markedly I
Horse	I	E or I	I	E	E or D	D or I	I	D	D	D
Bird	E	D	Markedly D	Markedly D	D or E	Markedly D	E	Markedly D	Absent	E

[a] See Dodds (1989a).
[b] D, decreased; E, equivalent; I, increased.

The functions of activated Hageman factor are numerous and include the conversion of plasma kininogens to kallikrein, the enzyme that elaborates plasma kinins; enhancement of vascular permeability; activation of the fibrinolytic mechanism; initiation of the Schwartzman reaction; activation of the extrinsic coagulation pathway; interaction with the Fletcher (prekallikrein) and Fitzgerald (HMWK) factors in surface-mediated reactions; and mediation of inflammatory and complement (Cl-esterase) reactions (Davie, 1995). The inhibition of the activated first component of complement (Cl-INH) accounts for more than 90% of the plasma inhibitory activity against factor XIIa and its fragments. Most of the properties and functions of Hageman factor apply to plasma thromboplastin antecedent (PTA), which also plays an important role in the initial activation stages of coagulation.

b. Plasma Thromboplastin Antecedent (Factor XI)

Factor XI or PTA is involved in the early stages of coagulation with Hageman factor. It does not require Ca^{2+} for its activity. Factor XI is a fast γ-globulin of about 160 kDa with 5% carbohydrate and a half-life of 30–84 hours. The major plasma protease inhibitor of factor XI is α_1-antitrypsin. Factor XI is present in both serum and plasma and is stable when frozen at $-40°C$ or lower and is stable at room temperature for more than 4 months. Plasma factor XI levels in patients with mild to moderate deficiency tend to increase on storage but not those in the severely deficient patient. The increase in PTA activity of human and canine PTA-deficient plasmas on storage tends to make them unreliable as reagents for factor XI assay. Fortunately, hereditary bovine and several artificial factor XI-deficient plasmas have proved to be excellent substrates (Gentry and Ross, 1993). The adsorbability of factor XI to products such as $Al(OH)_3$, $BaSO_4$, kaolin, and celite can usually be achieved by varying the concentration of adsorbing agent used.

Biochemically, factor XI resembles the Fletcher factor and the precursor of Hageman factor cofactor. It contains two identical polypeptide chains linked by a disulfide bond(s). It is synthesized in the liver and secreted into the plasma as a zymogen that circulates as a complex with HMWK. The cDNA of factor XI has been cloned (Davie, 1995). Normal plasma contains only 2–7 $\mu g/ml$ of PTA.

c. Plasma Prekallikrein (Fletcher Factor)

Plasma prekallikrein is a 79.5-kDa glycoprotein with about 15% carbohydrate. It is synthesized in the liver and secreted into plasma where nearly 75% circulates with factor XI as a complex with HMWK. The conversion of plasma prekallikrein to kallikrein is catalyzed by factor XIIa in a reaction stimulated by HMWK. Plasma contains 35–50 $\mu g/ml$ prekallikrein and it migrates electrophoretically as a fast γ-globulin (Geor *et al.*, 1990; Otto *et al.*, 1991). The structure, kinetics, and function of human and rhesus monkey plasma prekallekreins are similar (Veloso *et al.*, 1992).

d. High-Molecular-Weight Kininogen (HMWK or Williams Fitzgerald–Flaujeac Factor)

HMWK has a molecular weight of about 110 kDa and makes up about 15–20% of the total kininogen in plasma. It is an α-globulin with an isoelectric point of pH 4.3 and a plasma concentration of 70–90 $\mu g/ml$. HMWK serves as the complexing protein to which both prekallikrein and factor XI are bound (Davie, 1995).

e. Christmas Factor (Factor IX)

Christmas factor is a single-chain glycoprotein with about 17% carbohydrate and a molecular weight of about 55 kDa. It is present in both plasma and serum (Lozier and High, 1990; Davie, 1995). Its activity is higher in serum than in plasma and is increased by contact with glass. It is relatively heat labile, has an aminoterminal tyrosine and is absorbed from plasma or serum by $Al(OH)_3$, $BaSO_4$, and $Ca_3(PO4)_2$. Factor IX is a β-globulin, is stable on storage and is synthesized by the liver (Furie and Furie, 1995b). Its plasma concentration is low, about 4 $\mu g/ml$. It is a vitamin K-dependent protein and contains 12 residues of γ-carboxyglutamic acid (Suttie, 1993). Factor IX is activated by the presence of activated factor XI and has an absolute requirement for Ca^{2+}.

Activated factor IX (factor IXa) is a two-chain glycoprotein and acts as a serine protease, which, together with activated factor VIII, phospholipid, and calcium ions, forms a complex that converts factor X to its active form, factor Xa (Bauer *et al.*, 1992). Activated factor VIII acts as the accelerator of this reaction. One of the actions of heparin is to bind to factor IXa and inhibit further activation of the coagulation cascade. The turnover time (18–36 hours) of factor IX in the circulation is longer than that of factor VIII (9–16 hours). The biochemistry of factor IX has been reviewed by Lozier and High (1990) and Furie and Furie (1995b). Factor IX has been cloned and synthesized by recombinant techniques and the genes for human and canine factor IX have been isolated and sequenced (Evans *et al.*, 1989a, 1989b; Lozier and High, 1990; Sommer, 1992; Keith *et al.*, 1994; Davie, 1995; Furie and Furie, 1995b; Sugahara *et al.*, 1996; Walter and High, 1997). This information has led to new developments in gene therapy for hemophilia B in humans and animal models (Palmer *et al.*, 1989; Axelrod *et al.*, 1990). Early experimental data in dogs with hemophilia B have been promising (Kay *et al.*, 1993, 1994; Marx, 1993; Lozier *et al.*, 1994; Walter and High, 1997).

f. Antihemophilic Factor (Factor VIII)

The following summarizes some of the salient points of this key protein that accelerates intrinsic coagulation. Factor VIII circulates with von Willebrand factor to form a very large glycoprotein complex. The factor VIII:coagulant (FVIII:C) moiety expresses procoagulant activity whereas the vWF protein has several hemostatic properties. Factor VIII has a molecular weight of about 330 kDa and is present in trace amounts in plasma, 15–50 ng/ml. The structure of factor VIII has been characterized for the human, bovine, porcine, and canine proteins (Lozier and High, 1990). Most studies before 1980 included vWF contaminants because vWF makes up about 95% of the complex.

Genomic and cDNA clones of the human factor VIII gene have been characterized (Lozier and High, 1990; Antonarakis, 1995; Fallaux *et al.*, 1995; Thompson, 1995). The protein has 2351 amino acids and a molecular weight of 265 kDa when expressed from cDNA clones. Although it is synthesized as a single polypeptide chain, it circulates in plasma primarily as a two-chain molecule arising from cleavage by an unknown protease during either synthesis, secretion, or in the circulation (Lozier and High, 1990; Davie, 1995). There is significant sequence similarity of factor VIII with portions of factor V and ceruloplasmin. This latter observation suggests that factor VIII may bind to metals such as copper.

Factor VIII is the accelerator protein which, together with factor IXa, phospholipid, and Ca^{2+}, forms a complex that activates factor X (Bauer *et al.*, 1992). Thrombin activates factor VIII *in vivo,* apparently by a proteolytic modification with loss of FVIII:C activity unless both factor IXa and phospholipid vesicles are present to maintain stability (Mansell and Parry, 1989). This thrombin-activated factor VIII then accelerates the activation of factor X. Factor VIII also plays a critical role in the activation of factor X via the extrinsic pathway. Plasma factor VIII concentration varies with exercise, epinephrine, age, gender, pregnancy, central nervous stimulation, and use of oral contraceptives (Blombäuck *et al.*, 1992; Conlan *et al.*, 1993).

Factor VIII migrates electrophoretically in the α_2- and β_2-globulin fractions. It is closely associated with fibrinogen so that only ultrapurified factor VIII is uncontaminated with fibrinogen. It is stable for only short periods at room temperature or at 4°C, but stability is enhanced in heparinized rather than citrated plasma, thus the importance of Ca^{2+} in factor VIII structure and function. For maximum stability, it should be quick frozen and stored at −40°C or lower. It is stable at 56–62°C for relatively long periods, but thermostability is quite variable. Currently, commercial, therapeutic plasma factor VIII concentrates are made by recombinant DNA-derived methods, or by heat and/or chemical treatment to destroy contaminant viruses such as hepatitis and AIDS (Aronson, 1990; Bloom, 1991). Factor VIII is not adsorbed by $Al(OH)_3$ or $BaSO_4$, is activated by thrombin, and is inactivated by plasmin. Factor VIII has a short survival time *in vivo* (9–16 hours) and is synthesized primarily in the liver but in several extrahepatic sites as well. The exact cell(s) site(s) of synthesis remains controversial. The FVIII:C component has its biological activity modified during blood coagulation; is deficient or abnormal in hemophilia A; apparently requires complexing with vWF for expression of full biological activity; and does not influence platelet retention, ristocetin-induced platelet aggregation, or the bleeding time. Production of FVIII:C is controlled by the X chromosome, where the gene is located on the tip of its long arm adjacent to the gene for factor IX (Antonarakis, 1995; Davie, 1995; Fallaux *et al.*, 1995; Thompson, 1995).

Current studies of the factor VIII gene and its product have addressed means of disrupting the gene structure to produce an experimental model of hemophilia A (Bi *et al.*, 1995), as well as approaches for gene therapy including *in vitro* production of factor VIII in mice (Dwarki *et al.*, 1995; Walter and High, 1997).

g. von Willebrand Factor

The increasing prevalence of von Willebrand's disease (vWD) and the importance of vWF in hemostasis and thrombosis have only recently been recognized (Ruggeri and Ware, 1993). In contrast to FVIII:C, the vWF component of the factor VIII complex is biologically inactive during clotting, is deficient or abnormal in vWD, is required for platelet retention in glass bead columns and *in vitro* ristocetin-induced platelet agglutination, and is the major component of the factor VIII complex. It is present in human, bovine, feline, and porcine platelets but is very low in canine platelets. vWF is responsible for the immunoprecipitin line produced in agarose by reactions with heterologous antifactor VIII, a property now called vWF antigen (vWF:Ag), and is synthesized and released by endothelial cells in all species studied to date and by megakaryocytes of humans and swine but possibly not of dogs (McCarroll *et al.*, 1988; Waters *et al.*, 1989; Catalfamo *et al.*, 1991; Parker *et al.*, 1991; Mannucci, 1995; Meinkoth and Meyers, 1995). vWF is stable during coagulation, has a plasma concentration of about 10 μg/ml, is present in reduced amounts in serum and several other body fluids and tissues, and is required for the stabilization and full activity of FVIII:C

(Blann, 1990; Rand *et al.*, 1993; Ruggeri and Ware, 1993; Hoogstraten-Miller *et al.*, 1995). Production of vWF is controlled by an autosomal locus or loci.

The human vWF gene has been isolated and sequenced (Ruggeri and Ware, 1993). vWF has an unusual macromolecular structure, being composed of a series of very high molecular weight multimers (estimates are between 1 and 15 million with no defined upper limit). The basic subunit has a molecular weight of about 220 kDa and protomeric dimers or tetramers serve as the basic building blocks of the high-molecular-weight, active multimer. The subunit is a single-chain structure of about 2050 amino acids and is very similar in size to the active subunit of factor VIII (Girma *et al.*, 1995). About 5–15% of the vWF glycoprotein is carbohydrate. The higher the multimeric form, the more active its platelet-adhesive (binding) properties. Thus, the great molecular weight heterogeneity of vWF can affect its clinical and laboratory manifestations. For example, lower molecular weight multimers may be active immunologically but may not support platelet agglutination/adhesion efficiently, whereas larger multimers bind efficiently to factor VIII.

Most of the vWF in plasma is produced and secreted by endothelial cells (Blann, 1990; Meyers *et al.*, 1990b; Ruggeri and Ware, 1993; Meinkoth and Meyers, 1995), with some from megakaryocytes (humans and swine), which explains its presence in platelet α-granules (Mannucci, 1995). Curiously, appreciable amounts of vWF are not found in canine megakaryocytes or platelets although the protein is plentiful in vascular endothelium (McCarroll *et al.*, 1988; Catalfamo *et al.*, 1991; Parker *et al.*, 1991). This may explain why canine serum has much less vWF activity than that of human serum because canine platelets release little or no stored vWF during clotting. Cat platelets contain higher amounts of vWF but less than the levels found in human and porcine platelets (Waters *et al.*, 1989). The species differences in vWF are not surprising given the well-recognized species-specific behavior of the platelet- and plasma-agglutinating and enhancing properties of vWF in the presence of ristocetin, certain snake venoms, and desmopressin (DDAVP) (Feldman, 1988; Dodds, 1989a; Giger and Dodds, 1989; Meyers *et al.*, 1990a; Smith and Brinkous, 1991).

3. The Extrinsic System

a. Tissue Factor (Factor III)

Tissue factor is a species-specific thromboplastin extracted from tissue juice and is a potent activator of the extrinsic coagulation mechanism. The functions of tissue factor in the extrinsic pathway have been reviewed by Ruf and Edgington (1994), Nemerson (1995), and Rapaport and Rao (1995). Tissue factor also possesses factor VIII bypassing thromboplastic activity. The species specificity of tissue extract disappears if it is first incubated with homologous serum. Purified human brain thromboplastin is more effective for use as a universal activator in coagulation tests than either bovine or rabbit brain thromboplastins, but for practical purposes the latter are satisfactory. Tissue factor has a molecular weight of 37 kDa and accelerates the activation of factor X in the presence of factor VII and Ca^{2+}, thus forming factor Xa according to the scheme shown in Fig. 10.2 (Nemerson, 1995). In combination with factor VII, tissue factor amplifies the procoagulant response via components of the intrinsic pathway (factors VIII, IX, and XI), and so plays an important role in initiating and sustaining coagulation (Broze, 1995; Rapaport and Rao, 1995). The structure of tissue factor includes a protein and a lipid component, neither of which is active alone. The most active lipid in restoring activity to the protein component of tissue factor is phosphatidylethanolamine although lecithin has some activity (Nemerson, 1995). Unsaturated fatty acid side chains are required and mixed lipids are as effective as purified compounds in restoring activity.

A potent inhibitor of tissue factor-mediated coagulation, tissue factor pathway inhibitor (TFPI), has recently been purified and characterized (Broze, 1995). Its structure is unique; TFPI directly inhibits factor Xa and independently produces feedback inhibition of the factor VIIa–tissue factor catalytic complex. In plasma, TFPI forms are predominantly 34–41 kDa, but higher molecular weight forms are also present. The plasma concentration is low (about 2 nmol/liter) and much of it is bound to lipoproteins, especially low-density lipoprotein (LDL). TFPI levels fluctuate with LDL levels.

b. Proconvertin (Factor VII)

Factor VII is a single-chain, stable, glycoprotein that functions primarily in the extrinsic pathway of coagulation. It is present in both plasma and serum in trace amounts (5–10 ng/ml) (Howard *et al.*, 1994). Factor VII has the most rapid turnover rate of any coagulation factor, with a half-life of 2–7 hours (Dodds, 1989a). It is interesting that a factor not essential for primary hemostasis is synthesized and metabolized so rapidly (Brinkhous, 1992). Its molecular weight has been estimated to be about 50 kDa (Davie, 1995). Factor VII is synthesized by the liver and secreted into the blood as a zymogen (Suttie, 1993; Davie, 1995). It is the second component, along with tissue factor, that rapidly accelerates coagulation via the extrinsic pathway. Calcium ions are required, and the reaction proceeds by activating factor IX to factor IXa and factor X to factor Xa.

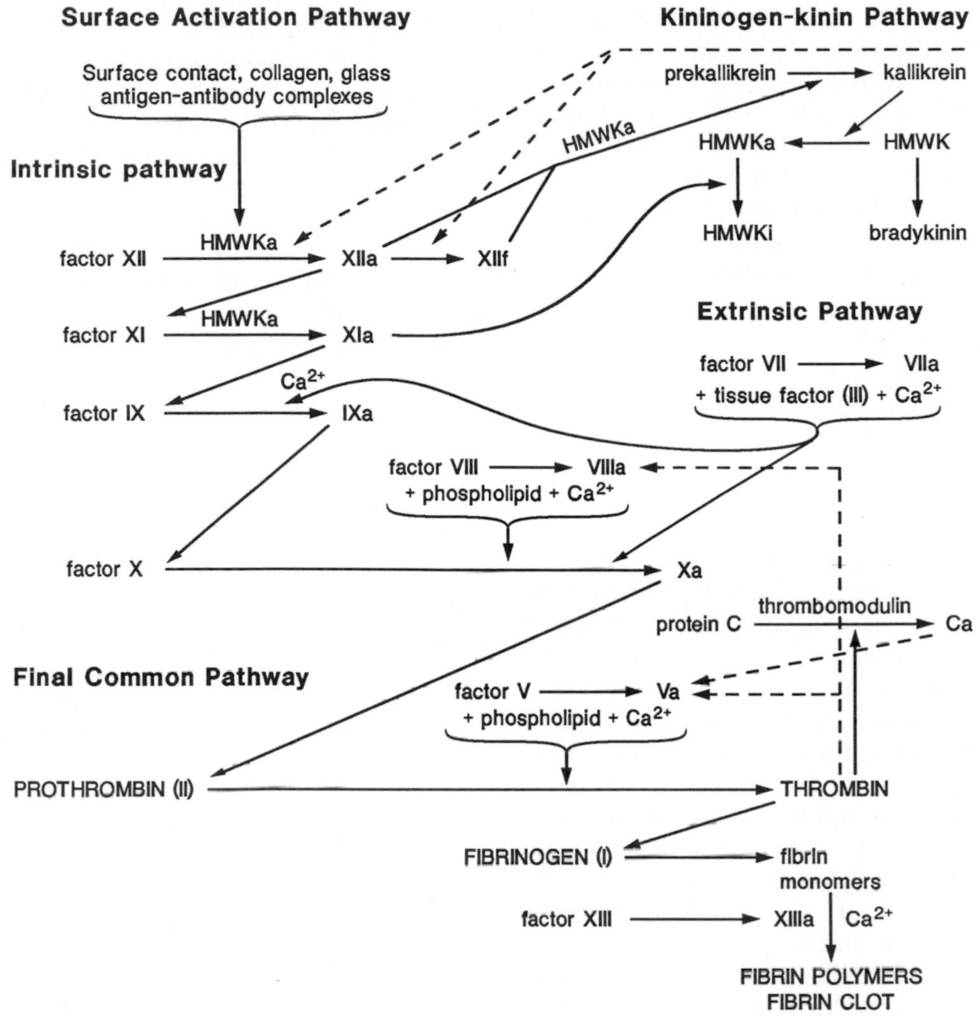

FIGURE 10.2 Instrinsic and extrinsic systems of blood coagulation. HMWK, high molecular weight kininogen.

Thus, many of the early stages of coagulation in the intrinsic pathway can be bypassed, although it is still unclear which of these two activation reactions is more important physiologically (Chabbat *et al.*, 1989). The extrinsic system needs only factors VII, X, V, prothrombin, and fibrinogen to form fibrin (Kitchen and Preston, 1994).

Bovine factor VII consists of 85% protein and 15% carbohydrate and has 406 amino acids. There is marked sequence similarity in the aminotermini of factor VII, prothrombin, factors IX, X, and protein C (Suttie, 1993). Despite structural similarities to the other vitamin K-dependent proteins, factor VII differs significantly in its metabolic behavior. The single chain form of factor VII as found in plasma, after enzymatic cleavage, becomes a two-chain form with a 60- to 80-fold increase in coagulant activity. Factor VII activity can be increased in a variety of ways, including storage at 0–4°C, contact activation with glass, kaolin, or celite, and either intrinsic or extrinsic coagulation. Enhancement of factor VII by contact activation requires factors XII, X, IX, and plasmin (Nemerson, 1995; Rapaport and Rao, 1995). Formation of factor Xa by either pathway causes a 900-fold increase in the rate of factor VII activation in the presence of phospholipid and Ca^{2+}. Thus, factor Xa is a positive feedback regulator enzyme for factor VII activity during coagulation.

4. *The Common System*

a. Stuart Factor (Factor X)

Factor X is activated by the action of factor IXa, phospholipid, and Ca^{2+}. Activation is accelerated by factor VIII of the intrinsic system in combination with phospholipid and Ca^{2+} and is accelerated by tissue factor of the extrinsic system (Lozier and High, 1990).

Factor Xa activates factor V. Thus, the intrinsic and extrinsic pathways converge at the step involving the activation of factor X to proceed via a common pathway to fibrin formation (Davie, 1995). Factor X is an α-globulin of about 59 kDa in plasma and 36 kDa in serum, is synthesized by the liver in the presence of vitamin K and is secreted into the plasma as a precursor to a serine protease. It is present in both plasma and serum at about 10 μg/ml. A number of compounds, such as Russell's viper venom (Stypven), promote the activation of factor X. Its half-life is 20–40 hours. Factor X is destroyed by heating to 56°C and is stable when frozen at −40°C or lower for several months. Its activity disappears from blood that is clotted with an excess of brain extract and can be adsorbed out of plasma by $Al(OH)_3$, $BaSO_2$, and $Ca_3(PO4)_2$.

Because of the pivotal role of factor X in both coagulation pathways, there has been considerable emphasis on the biochemistry and function of factor X in coagulation. The inactive zymogen is converted to its active form during clotting. Factor X has two aminoterminal amino acids and activation involves cleavage of a specific peptide bond to yield factor Xa. The activated form has very potent coagulant and esterolytic activities that are inhibited by plasma antithrombin III (heparin cofactor) and a variety of other serine proteases. The major substrate for factor Xa is prothrombin. In the presence of factor V, a cofactor that accelerates the reaction, phospholipid, and Ca^{2+}, factor Xa readily converts prothrombin to thrombin.

Most (85–90%) of the factor X in plasma is present in its monomeric form and the rest as a dimer. Human and bovine factor X are glycoprotein dimers with 15% carbohydrate and are linked by disulfide bonds. The heavy chain is 38–42 kDa, and the light chain is 16.2–17 kDa. The heavy chain contains the catalytic domain and all the carbohydrate. The light chain is highly acidic and remains intact on activation of the parent molecule. It also contains two potential growth factor domains having sequence similarity with human epidermal growth factor (Davie, 1995). The dimeric structure of factor X is unique since the zymogens of the other serine proteases and the prothrombin complex clotting factors all exist as a single chain. Activation of factor X is stimulated by intrinsic or extrinsic clotting, Stypven, insolubilized trypsin, and 25% sodium citrate. During prolonged activation, the coagulant activity of factor Xa is reduced but the esterase activity remains unchanged. Factor Xa is under positive feedback control via its activation of factor VII. Additional feedback control of factor X activation includes the activation of factor IX by factor Xa and thus the intrinsic pathway primes the extrinsic pathway but does not initiate it (Davie, 1995).

b. Proaccelerin or Accelerator Globulin (Factor V)

Factor V is also required for both the intrinsic and extrinsic coagulation pathways. When activated by thrombin, factor V accelerates the factor Xa-mediated conversion of prothrombin to thrombin up to 278,000-fold (Lozier and High, 1990). Factor V is extremely labile and disappears from plasma as it ages. The factor V of bovine plasma is more stable and found in much higher concentrations than in human plasma. Bovine and human factor V have been isolated and characterized (Davie, 1995). In human plasma, factor V concentrations are 4–14 μg/ml.

Originally, both factors V and VIII were thought to be enzymes converted to their active forms during coagulation. It is now known that factor V is a high-molecular-weight glycoprotein, which, like factor VIII, is nonenzymatic and acts as an accelerator or cofactor of specific coagulation reactions thereby producing enormous biological amplification. In the presence of factor Xa, phospholipid, and Ca^{2+}, factor V converts prothrombin to thrombin. It also serves as a cofactor for the thrombin-catalyzed activation of protein C, and activated protein C rapidly destroys factor Va. Factor V is the receptor on platelets for factor Xa binding to accelerate factor Xa hydrolysis and to inhibit platelet thromboxane A_2 formation. Finally, factor V serves as a receptor on endothelial cells to promote prothrombin conversion and inhibition of prostaglandin synthesis by factor Xa. Thus, with its numerous activities, the factor V–factor Xa interaction affords precise control of the final stages of both intrinsic and extrinsic coagulation (Davie, 1995).

The molecular weight of human and bovine factor V is 330 kDa (Davie, 1995). Calcium is thought to play a significant role in maintaining structural integrity and as a chelating agent to rapidly and irreversibly inactivate factor V. Factor V has very high affinity for hydrophobic surfaces such as phospholipid, and its reactivity is increased about 10-fold with partial proteolysis by thrombin. Factor V also reacts with other proteolytic enzymes, especially Stypven. Factor V is an α-globulin with 10–20% carbohydrate and high sialic acid content. Although the sialic acid is not essential for activity, the penultimate galactose residues are essential. The isoelectric point of factor V is 4.65. It requires lipid for its coagulant properties and avidly binds lipids. In the presence of phospholipid and Ca^{2+} but not prothrombin, factor V increases the esterase activity of factor Xa more than threefold. Thus, factor V directly or indirectly, via binding to phospholipid, regulates factor Xa activity. Procoagulant phospholipids, such as phosphatidylethanolamine, are much more active in complexing with factor V than are other phospholipids (phosphatidylcholine, phosphatidylserine, or cardiolipin). A surprising feature of factor V, in contrast to prothrombin and factor Xa, is its ability to bind to phospholipid micelles in the absence of calcium. Each factor V molecule, however, has a single, tightly bound calcium ion that is important for its interaction with phospholipid.

Factor V is synthesized by endothelial cells, megakaryocytes, and hepatocytes and is tightly bound to platelet membranes, an association regulated by its carbohydrate moiety. Platelet factor V originates from megakaryocytic synthesis as well as binding from plasma. Other coagulation factors, such as fibrinogen, vWF, and factor XIII are present in both platelets and plasma. The half-life of factor V varies between 12–36 hours (Davie, 1995).

c. Protein C

Protein C is a vitamin K-dependent glycoprotein of 67 kDa that acts as a regulator of blood coagulation (Comp, 1990a; Walker and Fay, 1992; Davie, 1995). It is another of the precursors to serine proteases that regulate blood coagulation and inflammation (Welles *et al.*, 1990b; Esmon *et al.*, 1991). It is converted to activated protein C by thrombin in the presence of thrombomodulin, an endothelial cell protein (Fig. 10.2). Activated protein C then inactivates factors Va and VIIIa by a proteolytic process (Davie, 1995). Human protein C is composed of heavy (41-kDa) and light (21-kDa) chains held together by a disulfide bond. The light chain contains γ-carboxyglutamic and β-hydroxyaspartic acids as well as two potential epidermal growth factor domains similar to factor X (Davie, 1995). The heavy chain contains the catalytic domain. The amino acid sequences of human and bovine protein C have been established by sequence analysis and cDNA cloning (Walker and Fay, 1992; Aiach *et al.*, 1995). They express significant structural similarity with factors IX and X. The human protein contains 419 amino acids and has a plasma concentration of about 4 μg/ml.

d. Prothrombin (Factor II) and Thrombin

Prothrombin is a glycoprotein of 71.6 kDa and 8.2% carbohydrate (Davie, 1995). It contains three N-linked carbohydrate chains and is an α-globulin present in plasma at a concentration of 300 Iowa units/ml or 100 μg/ml. Prothrombin is converted to the active enzyme, thrombin, by the action of several factors known as plasma or tissue thromboplastins (Fig. 10.2). These prothrombin activators are a complex mixture. Prothrombin is synthesized by the liver, and its half-life is as short as 10–12 hours and as long as 60 hours, averaging about 36 hours. Both the synthesis and release of prothrombin require vitamin K (Pechlaner *et al.*, 1992). Warfarin (a coumarin derivative), the ingredient of first-generation rat poisons, blocks prothrombin synthesis at an intermediate step, and this can be reversed by the administration of vitamin K_1 (Vermeer and Hamulyák, 1991; Woody *et al.*, 1992). Vitamin K is also a known requirement for four other clotting factors, factors VII, IX, X, and protein C (Mount and Kass, 1989; Miletich, 1990; Soute *et al.*, 1992; Suttie, 1993). The amino acid sequences of human and bovine prothrombin is known (Davie, 1995). The mature polypeptide chain contains 579 amino acids.

Interest has been renewed interest in prothrombin synthesis. The vitamin K-dependent coagulation factors are synthesized from inactive precursor proteins by the action of vitamin K, which converts their glutamic acid groups to active γ-carboxyglutamic acid residues (Suttie, 1993; Dowd *et al.*, 1995). The clotting proteins are then anchored by the dicarboxyl groups on these residues, via Ca^{2+}, to a phospholipid surface during coagulation.

Prothrombin is a metal-binding protein containing two high affinity and several lower affinity metal-binding sites. The presence of 10 γ-carboxyglutamic acid residues near the aminoterminus of the molecule is an integral part of its metal-binding properties. These amino acids have unique properties and participate as metal ligands in prothrombin. Metal ions play a key role in the activation of prothrombin to thrombin, and the interaction of all the vitamin K-dependent clotting proteins is metal dependent.

Thrombin, the active enzyme form of prothrombin, has two chains of unequal size (α and β) held together by a disulfide bond (Fenton, 1995). The active site of thrombin, located on the α chain, includes a serine residue that, like similar serine proteases, can readily be inhibited by diisopropyl fluorophosphate (Guillin *et al.*, 1995). The molecular mass of thrombin is 36 kDa. Heparin acts as an effective anticoagulant primarily by inhibiting the action of thrombin on fibrinogen and, to a lesser extent, by preventing activation of factor IX by factor XIa.

Thrombin has many critical functions in hemostasis, including the formation of fibrin from fibrinogen, regulation of factor XIIIa (fibrinoligase) production, activation of factors V, VII, and VIII, and participation in platelet reactions (Hemker and Béguin, 1995). Thrombin is inhibited by antithrombin (heparin cofactor), fibrin degradation products, α_1-antitrypsin, and α_2-macroglobulin (Section II.E) (Fareed and Callas, 1995). In addition to its role in hemostasis, thrombin is a key regulatory enzyme in many tissue events including inflammation, complement activation, wound healing, macrophage function, and fibroblast function (Fenton,

1995). It functions as a potent mitogen for various cells, a mitogenic conditioner, and an angiogenic substance.

e. Fibrinogen (Factor I) and Fibrin

Fibrinogen is a complex glycoprotein, the substrate for thrombin, and the precursor of fibrin. It occurs in plasma at levels of 2–4 mg/ml. The molecular mass of the human and bovine protein is 340 kDa. They have a dimeric structure composed of three pairs of peptide chains (α, β, γ) joined by disulfide bonds (Davie, 1995). Most vertebrate fibrinogens share a similar structure. Each unit of the dimer is held together by disulfide bridges between the two α chains and between the two γ chains near the aminoterminus of the molecule. The β and γ chains contain carbohydrate but the α chain does not. Following the action of thrombin on fibrinogen, two major fibrinopeptides (A and B) are released from the aminotermini of the α and β chains, respectively. The remaining fibrin monomers are then polymerized to form insoluble, cross-linked fibrin by the interaction of fibrin-stabilizing factor (factor XIII) and Ca^{2+} (Coyne *et al.*, 1992).

Fibrinogen is believed to be synthesized as a heterogeneous group of molecules. This degree of heterogeneity may explain the large number of abnormal or dysfibrinogens known to occur as inherited disorders, in fetal life, or in the presence of underlying disease, especially liver disease (Moalic *et al.*, 1989; Hathaway and Corrigan, 1991; Andrew, 1995). The α chain is especially heterogeneous and has been shown to vary in size during evolutionary development. Except in the orangutan, there is a continuous decrease in the size of the α chains with a slight increase in the size of the β chains during development up the evolutionary scale from closely related higher primates to humans. The mature polypeptide chains contain 610, 461, and 411 amino acid residues for the α, β, and γ chains, respectively (Davie, 1995).

Fibrinogen is an important cofactor in mediating platelet aggregation responses *in vivo* and *in vitro* and platelets contain relatively large amounts of fibrinogen. Fibrinogen is synthesized by hepatic parenchymal cells and has a half-life of about 36 hours. Its synthesis and turnover rate can be markedly accelerated during the acute phase response induced by tissue damage, inflammation, stress, and ACTH (Davie, 1995). Other properties of fibrinogen and fibrin include their interactions with plasmin and with other enzymes such as trypsin, chymotrypsin, brinase, pronase, elastase, and collagenase. They also have high sequence similarity, as demonstrated by the high degree of immunological cross-reactivity among individual vertebrate classes, especially mammals and birds.

f. Fibrin Stabilizing Factor (Factor XIII)

Discovered in 1948, fibrin stabilizing factor (FSF) is the latest one to be given a Roman numeral according to the International Committee classification. It is synthesized by monocytes and hepatic parenchymal cells (Coyne *et al.*, 1992). A very small amount (2–10%) of factor XIII is sufficient for adequate hemostasis. The half-life of FSF is 3–7 days. FSF is a thrombin-labile protein, requires Ca^{2+} for its activity, and is in the α_2-globulin fraction in plasma. It is present in platelets, but occurs in only trace amounts in serum. It is inhibited by silver, lead and zinc and by snake venoms. Its molecular weight is about 320 kDa and it contains two subunits structurally similar to fibrinogen. Its function is to convert unstable, soluble fibrin monomers into insoluble, stable fibrin polymers by forming cross-links between adjacent fibrin strands. Fibrin bonding is strengthened by the FSF-dependent formation of intermolecular γ-glutamyl-ε-lysine bridges.

Factor XIII in plasma is the inactive zymogen for factor XIIIa or fibrinoligase, a transamidase formed by the action of thrombin and Ca^{2+}. Thrombin, therefore, has a dual regulatory role since it controls the rate of fibrin network formation and the generation of fibrinoligase. Factor XIIIa activity can be assayed by clot solubility in 1% monochloroacetic acid or 5 M urea, direct or indirect quantitation of the number of γ-glutamyl-ε-lysine peptides formed, electrophoresis of reduced and nonreduced fibrin in sodium dodecyl sulfate gels or by incorporation of monodansylcadaverine into purified casein substate (Coyne *et al.*, 1992).

The FSF zymogen of human and bovine plasma has a heterologous structure with two subunits differing in amino acid composition and size (*a*, 75 kDa; *b*, 80 kDa). It exists in an equilibrium; $2ab = a_2b_2$. The platelet zymogen, in contrast, is composed of two identical subunits but has no b subunit. The FSF activity of platelets is about 20% of that of whole blood.

D. Fibrinolysis

1. General

The counterpart of the coagulation mechanism is a group of zymogens and enzymes that forms the fibrinolytic system. Fibrinolysis involves a series of events critical to the removal of the hemostatic plug in the course of vessel healing and repair (Fareed *et al.*, 1995; Lijnen and Collen, 1995; Verstraete, 1995). An outline is shown in Fig. 10.3, and a detailed description is given by Verstraete (1995). After direct or indirect activation of the fibrinolytic system, plasminogen is converted to plasmin. Plasmin actively digests fibrin, fibrinogen, and factors V and VIII. This mechanism is

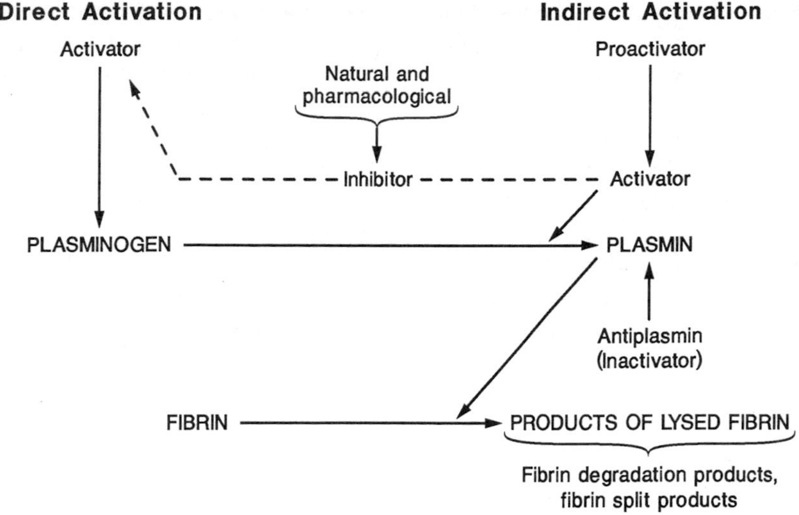

FIGURE 10.3 Scheme of the fibrinolytic mechanism. Solid lines indicate direct activation of precursor zymogen to enzyme; dashed lines show paths of positive and negative feedback.

analogous to that of coagulation in that an active enzyme, plasmin, is formed by activation of its precursor, plasminogen (McKeever *et al.*, 1990). The fibrinolytic system is controlled physiologically by natural fibrinolytic inhibitors. *In vivo*, fibrinolysis is usually limited to the area in and around the fibrin deposits. Plasmin within the hemostatic plug is provided by the action of fibrin in adsorbing plasminogen, which in turn is activated by substances released by the vessel wall and the cells of the blood. Plasmin and activators escaping from the plug are immediately diluted by the blood and neutralized by antiplasmins (Comp, 1990a).

2. *Plasminogen and Plasmin*

Mammalian plasminogens, from plasma or serum, are single-chain monomers with multiple isoelectric forms. The plasminogens of humans, cat, dog, pig, cow, rabbit,and horse have been characterized (Welles *et al.*, 1990a). Plasminogen is a β-globulin of about 85 kDa in either the glutamate or lysine form and is synthesized by the kidney and liver. The human protein contains 790 amino acids. Plasminogen levels are low in newborns, the cause of their impaired fibrin clot lysis (Andrew *et al.*, 1992; Schmidt *et al.*, 1993).

Many naturally occurring substances are plasminogen activators. These include the tissue activators known as cytokinases–urokinases in urine, tears, saliva, milk, bile, and prostatic, cerebrospinal, synovial, and amniotic fluids; streptokinase from β-hemolytic streptococci, niacin, and vitamin B_1; vascular endothelial and circulating activators; erythrocyte erythrokinase; neutrophilic granulocyte activator; and the factor XII-dependent activator which requires factor XII, prekallikrein, and Hageman factor cofactor for its activity. Thus, activators of plasminogen are present in blood, the vessel wall, body fluids, and most tissues.

Plasmin, the active enzyme form of plasminogen, is a serine protease with trypsin-like specificity which has two chains: a heavy (A) chain and a light (B) chain. These chains also have multiple isoelectric forms and molecular weights of about 49 and 26 kDa, respectively. The lysine form of plasmin has a molecular weight of about 77 kDa. Plasmin readily hydrolyzes a variety of proteins, such as factor V, gelatin, casein, arginine esters, and lysine, in addition to fibrinogen and fibrin (Hajjar, 1995).

In contrast to plasminogen, plasmin is normally absent from blood and body fluids because a group of circulating antiplasmins readily inactivates free plasmin. These include α_2-macroglobulin, α_1-antitrypsin, Cl inactivator, and antithrombin III.

3. *Tissue Plasminogen Activator (TPA)*

Tissue plasminogen activator (TPA), a serine protease of 68 kDa, has been isolated from many mammalian tissues (Gerding *et al.*, 1994; Lijnen and Collen, 1995). Its principal site of synthesis is the endothelial cell. The enzyme is present in very small amounts in tissues and is tightly bound to particulate cell constituents, making its isolation very difficult. The cDNA sequence and primary amino acid sequence of TPA are known. TPA is released into the circulation by stimuli such as exercise, venous occlusion, and intravenous injection of vasoactive drugs. Vascular PA isolated from plasma of rigorously exercised subjects appears to be identical in size and properties to TPA. Both TPA and vascular PA are specifically inhibited by a rapid-acting PA inhibitor.

A zymogen form of TPA has not yet been identified. The single-chain human TPA molecule is not a zymogen and is readily converted by plasmin to a double-chain form. In the presence of fibrin, both forms are equally active (Fareed and Callas, 1995; Verstraete, 1995). TPA has four domains: the fibronectin finger, epidermal growth factor, kringle, and serine protease domains. TPA has received great attention in recent years because of its potent thrombolytic ability in lysing clots of patients experiencing thrombotic states (Lijnen and Collen, 1995; Verstraete, 1995; Verstraete and Zoldhelyi, 1995).

E. Inhibitors of Coagulation and Fibrinolysis

1. Coagulation Inhibitors

Much less is known about the naturally occurring inhibitors of coagulation than about the procoagulants of blood. Without an inhibitor mechanism to neutralize the effects of thromboplastins and thrombin, however, widespread intravascular coagulation and thrombosis could result from a simple vascular injury (Fareed *et al.*, 1995). The known types of natural anticoagulants include the circulating antithrombins (designated I through VI), heparin, protein S (a vitamin K-dependent anticoagulant), and anticoagulants to factors IXa, Xa, and XIa (Comp, 1990a, 1990b; Fareed and Callas, 1995; Henkens *et al.*, 1995). The most widely studied antithrombins are I (a reversible inhibitor) and III (a heparin cofactor, progressive irreversible inhibitor and an α_2-globulin). Of the six major protease inhibitors of blood, four have inhibitory action against one or more clotting factors and bind thrombin *in vitro* and *in vivo.* These are antithrombin III, Cl inactivator, α_2-macroglobulin, and α_1-antitrypsin (Exner, 1995).

a. Antithrombin III

Antithrombin III (ATIII) is a relatively heat-stable α_2-globulin of about 54 kDa with a carbohydrate content of about 15%. It is a single-chain glycoprotein synthesized in the liver and secreted into the plasma. This plasma protease inhibitor plays an important role in the regulation of blood coagulation. It is the principal physiological thrombin inhibitor in human plasma and constitutes about 50% of the total progressive antithrombin activity while the remaining activity is shared by α_2-macroglobulin and α_1-antitrypsin. The total progressive antithrombin activity of human plasma is decreased in inherited disorders of ATIII and in certain disease states such as DIC (Menache *et al.*, 1992; Conlan *et al.*, 1994). Similar decreases have been reported in animals with DIC (Boudreaux *et al.*, 1989b; Johnstone *et al.*, 1989). Other functions of ATIII are neutralization of factor Xa, which is accelerated 50- to 100-fold in the presence of heparin, and heparin-potentiated inhibition of factors VII, IXa, XIa, and XIIa and plasma kallikrein. The amino acid sequence of human ATIII is known (Davie, 1995) and the glycoprotein has 432 amino acid residues.

b. Cl Inactivator

Cl inactivator is a neuraminoglycoprotein capable of inhibiting its natural substrate, Cl-esterase, as well as plasma kallikrein, and factors XIa and XIIa, but not factor Xa or plasmin. It has a molecular weight of about 105 kDa and a 35% carbohydrate content.

c. α_1-Antitrypsin

α_1-Antitrypsin, the major trypsin inhibitor of plasma, is a glycoprotein with a molecular weight of about 50 kDa. In addition to inhibiting trypsin, it inhibits other proteases such as chymotrypsin, plasmin, and elastase in an irreversible reaction and is the major inhibitor of XIa (Verstraete, 1995).

d. α_2-Macroglobulin

α_2-Macroglobulin, a large (650-kDa) glycoprotein with subunits of 185 kd, complexes with a number of proteases, including thrombin, kallikrein, and plasmin. The binding of α_2-macroglobulin to thrombin is slow whereas its binding to plasmin is rapid.

e. Factor XIa Inhibitor

Factor XIa inhibitor, one of several known inhibitors of factor XIa, has a molecular weight of 65 kDa and is an α-globulin.

f. Lipoprotein Factor Xa Inhibitor

The lipoprotein factor Xa inhibitor is contained in the LDL fraction of plasma. Its action is facilitated by Ca^{2+}.

g. Protein S

Protein S, a vitamin K-dependent protein, has a molecular weight of 80 kDa, is present in plasma at a concentration of 25 μg/ml, and functions as an anticoagulant by facilitating the action of activated protein C on factors VIIIa and Va (Comp, 1990b; Aiach *et al.*, 1995; Henkens *et al.*, 1995). It is a membrane-binding protein and has a unique carboxyterminal structure that bears no sequence similarity to regions of other proteins. Protein S is synthesized as a single chain and exists in blood in two forms, a free form and one bound to C4b-binding protein.

2. *Fibrinolytic Inhibitors*

The naturally occurring inhibitors of the fibrinolytic system have been summarized by Verstraete (1995). These can be categorized as inhibitors of plasmin, plasminogen activator inhibitors, inhibitors of the factor XII-dependent fibrinolytic pathway, and tissue inhibitors of fibrinolysis.

a. Plasmin Inhibitors

Plasma inhibitors include α_2-macroglobulin, α_1-antitrypsin, Cl inactivator, ATIII, and other uncharacterized antiplasmins of plasma and platelets.

b. Plasminogen Activator Inhibitors

Plasminogen activator inhibitors of various molecular masses have been identified, including those of 70 and 20 kDa. There is a coagulation-dependent inhibitor and another released from platelets. α_2-Macroglobulin also possesses PAI activity, but α_1-antitrypsin and ATIII apparently do not.

c. Factor XII-Dependent Inhibitors

Factor XII-dependent inhibitors are discussed in Section II.C.2.a.

d. Tissue Inhibitors

Inhibitors of fibrinolysis have been isolated from placenta and have molecular weights of 105 and 43 kDa (Broze, 1995).

III. LABORATORY DIAGNOSIS OF BLEEDING DISORDERS

A. General

The following briefly outlines the major points to consider for the diagnosis of bleeding disorders. Further details can be found elsewhere (Dodds, 1989a; Forsythe and Willis, 1989; Mansell and Parry, 1989; Comp, 1990b; Miletich, 1990; Tollefson, 1990; Welles *et al.*, 1990a, 1990b; Benson *et al.*, 1991, 1992; Poggio *et al.*, 1991; Šrámek *et al.*, 1992; Andreasen and Lovelady, 1993; Bath, 1993; Brooks and Catalfamo, 1993; Johnstone, 1993; Howard *et al.*, 1994; Giger, 1994; Kitchen and Preston, 1994; Brandt *et al.*, 1995; Exner, 1995; Brooks *et al.*, 1996b).

1. *Sample Collection and Preparation*

The importance of proper collection and preparation of blood samples for laboratory diagnosis of coagulation disorders cannot be overemphasized (Dodds, 1989a; Johnstone, 1993). Scrupulous cleanliness and avoidance of rough handling or rough surfaces must be observed in order to obtain satisfactory results. Smooth surfaces are a must to prevent activation of factor XII and to inhibit spontaneous platelet clumping. Therefore, plastic or siliconized glassware, syringes, and test tubes should be used in all sample preparations and storage. Blood must be taken by clean venipuncture to avoid contamination with tissue juices that will activate the coagulation system within a few seconds. Collection of samples from intravenous catheters is generally not advisable except in certain clinical or experimental situations where alternative sites are not readily available. A small clot in a blood sample will activate and consume enough clotting activity to invalidate the interpretation of results. Because most animal blood coagulates more rapidly than human blood, unless very rapid blood flow is achieved during collection, it is best to add the anticoagulant to the syringe beforehand and then draw the sample.

After collection, blood is centrifuged immediately for platelet-poor plasma (PPP) or platelet-rich plasma (PRP). Platelet-poor plasma is obtained for routine studies by centrifugation at 800–1000 *g* for 15 minutes and contains about 30,000–80,000 platelets/μl. If relatively platelet-free (<10,000/μl) or hard-spun plasma is desired for specialized or research tests, centrifugation at 3500–5000 *g* for 30 minutes or, preferably, 10,000–29,000 *g* for 15 minutes is required. After centrifugation the PPP is drawn off with plastic or siliconized pipets or with plastic syringes and tubing into plastic or siliconized glass tubes. It is then mixed, dispensed in small aliquots, and tested immediately or frozen at −20°C or preferably at −70°C for assay in the near future. Frozen plasma should be thawed only once and tested immediately; clotting factors are labile, and repeated freezing and thawing denatures proteins and destroys their biological activity.

PRP for platelet function tests (250,000–450,000 platelets/μl) can be prepared at room temperature by centrifugation at either 300 *g* for 2–3 minutes or 150 *g* for 5–10 minutes. Platelets undergo shape changes in the cold and improper freezing disrupts their membranes. After preparation, PRP is drawn off, put into plastic tubes (preferably polycarbonate), allowed to sit undisturbed for 30 minutes for equilibration of intra- and extracellular constituents, and then used within 2 hours.

2. *Anticoagulants*

Trisodium citrate is the anticoagulant of choice for coagulation studies. The preferred technique is to use 1 part 3.8% trisodium citrate to 9 parts blood. Maintaining this fixed citrate-to-blood ratio is important to ensure reliable results (Johnstone, 1993). Sodium oxalate is also used but less frequently. Heparin, which inhibits thrombin and factor IX activation and activates platelets, interferes with the assay of coagulation factors, whereas EDTA inhibits platelet interactions, so that neither of these anticoagulants is appropriate for assessing coagulation with biological activity methods (Lewis and Meyers, 1994; Greinacher, 1995). Of course, if the animal is receiving anticoagulant therapy (e.g., heparin or coumarin), assay results and methods will need to be adjusted accordingly (Dodds, 1989a).

3. *Standard Reference Plasmas and Controls*

Because of the many physiological factors and species specificities affecting hemostatic parameters (Tables 10.2 and 10.3; Dodds, 1989a), it is extremely important that the appropriate reference and control plasmas be used for diagnostic evaluation. Reference and control reagents include fresh-frozen, pooled, citrated plasmas deficient in specific clotting factor activities (either congenitally deficient or artificially depleted), which are used as substrates in specific coagulation assays and as reference blanks; and fresh-frozen, pooled plasma from healthy animals of the species being studied. This homologous, pooled plasma should be obtained from at least eight animals (four of each sex) and should be age-matched whenever possible with the patient sample. In some cases, both age- and sex-matched reference plasma is desirable (Johnstone *et al.*, 1989; Moalic *et al.*, 1989; Hathaway and Corrigan, 1991; Blombäck *et al.*, 1992; Conlan *et al.*, 1993, 1994; Henkens *et al.*, 1995).

The deficient substrate plasmas need not be obtained from the same species as the reference standard or test samples. In fact, heterologous deficient substrates have been used successfully for years in the quantitation of specific coagulation factor activities (Dodds, 1989a). Examples are the use of canine factor VII-, VIII-, and IX-deficient plasmas for the assay of their respective clotting factor activities in all mammalian plasmas including human; bovine factor XI-deficient plasma,an excellent reagent for the assay of factor XI in all species (Gentry and Ross, 1993); and coagulation factor-deficient plasmas from human patients used to assay clotting factor levels in other animal species (MacGregor *et al.*, 1991; Ni and Giles, 1992). Therefore, deficient substrate plasmas are interchangeable, provided that the reference control plasma used to quantitate or to compare with the patient sample is a pool from the same species.

For practical purposes, general screening tests of coagulation and assays of specific coagulation factors are best performed on fresh-frozen patient and reference plasmas. Frozen reference plasmas should be stored in small aliquots in plastic tubes at −40°C or lower. The comparison of fresh patient samples with an individual, fresh control plasma is not advisable, as mild or moderately abnormal results can easily be masked. There is a relatively wide normal range for most coagulation assays and the range for the specific method and laboratory should be established by testing of at least 20 normal, healthy individuals. Results obtained with fresh plasma differ from those obtained with frozen samples so comparison of fresh patient plasma with a frozen standard plasma is misleading. Most laboratories find it impractical to prepare a fresh, homologous plasma pool each time assays are run. It is often advisable, when test samples are shipped frozen to a diagnostic laboratory, to request an individual control sample as well, which will serve as a monitor of sample collection and processing.

B. Evaluation of Platelet Function

1. *Quantitative Disorders*

Thrombocytopenic states and thrombocytoses are diagnosed by direct or indirect platelet counting (see Table 10.8 in Section IV.B, and also Section IV.C).

a. Platelet Count

The first step in assessment of platelet function is a direct or indirect platelet count. The direct method uses diluents, such as 3.8% trisodium citrate, 2% disodium EDTA, 1% ammonium oxalate, or commercial counting fluids (Bath, 1993). The sample is counted by a hemacytometer, preferably using phase-contrast microscopy or by an electronic cell counter. The normal range for most animal species except rodents is 200,000–500,000/μl, and less than 100,000/μl can be considered significant. Rodents usually have platelet counts of at least 900,000/μl. The indirect method of platelet counting is more practical for routine screening purposes and gives a satisfactory estimate. The platelets per oil immersion field (/oil) on a stained blood smear are counted and compared with the number of red or white cells. For example, the number of platelets per 100 white blood cells multiplied by the total white count is an estimate of the platelet count. Another method is to simply count the number of platelets per oil immersion field where one/oil is equivalent to 15,000/μl.

b. Clot Retraction

The phenomenon whereby a whole blood or PRP clot retracts from the sides of a glass container depends on normal platelet quantity and function. Clot retraction time may be reduced in anemia and accelerated in polycythemia because the hematocrit tends to influence the retraction of whole blood clots. A variation of this test, which is more sensitive to differences in the platelet count, uses blood diluted 1 : 10 with cold, buffered saline and clotted with a standardized concentration of thrombin (Boudreaux *et al.*, 1989b; Dodds, 1989a). Specifically, 0.5 ml of fresh whole blood is taken by plastic syringe and added to 4.5 ml of the saline. Two milliliters of this diluted mixture is added in duplicate to two small glass test tubes, each containing 1 unit of bovine thrombin (0.1 ml of a 10 U/ml solution), and mixed by inversion. The tubes are kept in an ice bath or refrigerator for 30 minutes and transferred to a 37°C water bath. They are examined for retraction every 30 minutes for up to 2 hours. For most animal species, normal samples are maximally retracted by 1–2 hours, after which time the clot commences to lyse (Boudreaux *et al.*, 1989). Clot lysis time can also be recorded here by monitoring the sample until lysis is complete.

c. Russell's Viper Venom Time (RVVT)

The Russell's viper venom or Stypven assay measures extrinsic clotting and is described in more detail in Section III.C. It also, however, assesses platelet function since the venom needs a source of phospholipid for its activation. In PPP or PRP, the phospholipid is supplied by an adequate number of functional platelets. Normal animals have a 9–14 second RVVT. The assays of thrombocytopenic or thrombopathic patients are usually prolonged by 3–10 seconds.

d. Bleeding Time

The bleeding time measures hemostasis from a standardized skin or mucosal incision or by transection of a nail or the tail (Feldman, 1988; Forsythe and Willis, 1989; Kraus *et al.*, 1989; Brassard and Meyers, 1991; Cowles *et al.*, 1992; Šrámek *et al.*, 1992; Brooks and Catalfamo, 1993; Bright *et al.*, 1994; Schermerhorn *et al.*, 1994; Panciera and Johnson, 1996). Meaningful results can be obtained only if the test is performed under controlled conditions that regulate the size and depth of the incision or transection. The skin, gingival biopsy, and tail transection bleeding times are performed under anesthesia, whereas the toenail or mucosal methods can be performed on awake, sedated, or anesthetized animals (Forsythe and Willis, 1989; Brooks and Catalfamo, 1993). Choice of skin sites is complicated in animals by their relatively thick, hairy skin, but the tip of the ear, lip, and inner thigh can be used. Normal skin bleeding times vary from 1 to 5 minutes in dogs, and 1 to 3.5 minutes in cats and times are prolonged in platelet defects, vascular lesions, and vWD.

The toenail (transection) method involves severing the apex of the nail bed with a sharp guillotine-type nail cutter. Repeated bleeding times are readily obtained on the same animal by cutting one or more nails. The animal should be in lateral recumbency for the procedure. Normal transection bleeding times are between 2 and 5 minutes in dogs and 1.5 and 3 minutes in cats (Brooks and Catalfamo, 1993). Using the buccal mucosal technique, the mean bleeding time for 34 healthy dogs was 2.62 ± 0.49 m (Forsythe and Willis, 1989; Brooks and Catalfamo, 1993). In healthy horses and cattle, the lip bleeding time is 8–10 minutes.

The transection bleeding time measures all components of hemostasis (vessel wall, platelets, coagulation). As such it is a practical way to assess the bleeding risk of a patient prior to surgery as well as the efficacy of therapeutic response after treatment has been initiated to control or correct a bleeding diathesis (Dodds, 1992; Brooks and Catalfamo, 1993).

e. Whole Blood Clotting Time (WBCT)

The WBCT in plastic or silicone-coated tubes may be prolonged in thrombocytopenic states. The glass tube clotting time is usually normal.

2. *Qualitative Disorders*

a. Platelet Aggregation and Release

The effect of various aggregating agents such as ADP, collagen, adrenalin, and thrombin on PRP or platelet suspensions is used to evaluate platelet function. The assay measures the change in optical density of standardized platelet preparations kept at 37°C and stirred at 1000 rpm after the addition of an aggregating agent. The optimal concentration of these agents varies with the species from which the platelets were prepared. In general, low doses of ADP cause reversible aggregation, whereas higher doses trigger the release reaction and a secondary wave of irreversible aggregation. Platelet aggregation with collagen or connective tissue suspensions results from adherence of platelets to collagen fibers followed by the release of platelet ADP, which in turn causes secondary aggregation and is irreversible. Epinephrine- and thrombin-induced aggregation are also mediated by platelet ADP release. Marked species differences are noted in the responses to aggregating agents (Table II; Dodds, 1989a). The platelet release reaction can be assessed by the degree of secondary ADP-mediated aggregation. It can be quantitated by measuring the amount of PF-3, platelet factor 4, nucleotide, [^{14}C] serotonin, or other materials released by optimal concentrations of agents that induce release (Callan *et al.*, 1995).

Newer techniques have been developed for assessing platelet function in whole blood (Catalfamo *et al.*, 1989; Forsythe *et al.*, 1989; Willis *et al.*, 1989; Grauer *et al.*, 1992; Rajasekhar *et al.*, 1994; Schermerhorn *et al.*, 1994). The latter referenced method is based on adding 1 ml of fresh, citrate anticoagulated whole blood from fasted normal and patient animals to each of paired tubes containing 0.1 ml 0.15 *M* NaCl (control) and 0.1 ml 50 *M* ADP (agonist), respectively. The paired tubes are immediately capped and mixed *simultaneously* at room temperature for 8–12 gentle inversions. This mixing process should last for at least 15 but not longer than 30 seconds. The reaction is then stopped immediately by adding 1 ml of 2% formalin stabilizer, first to the ADP-containing tube, and then to the saline control tube for each of the normal and patient animals. The tubes are recapped and mixed by gentle inversion three or four times. These completed sets of normal and patient tubes are left upright undisturbed overnight (16–18 hours) in a test tube rack at room temperature. The following day, the reaction is complete and the heaviest blood cells have settled to the bottom of the tubes. The supernatant from each tube is carefully aspirated down to about $\frac{1}{16}$ inch above the settled cells (not less or more), and transferred to another set of prelabeled tubes. The extent of platelet reactivity to ADP is measured by comparing the number of free, unaggregated platelets in the ADP-containing sample to that of the nonreactive saline control tube. If aggregability of the patient sample is impaired, a relatively high number of platelets should be left in the supernatant of the ADP tube versus the patient's saline control tube. This same methodology can also be used to assess platelet responses to other agonists including ristocetin.

Methods for platelet function testing have been reviewed (Catalfamo and Dodds, 1988; Boudreaux *et al.*, 1989a,b; Brassard and Meyers, 1991; Johnson *et al.*, 1991; Barr *et al.*, 1992; Bath, 1993; Brooks and Catalfamo, 1993; Boudreaux *et al.*, 1994; Behrend *et al.*, 1996).

b. Platelet Retention (Adhesiveness)

Retention of platelets in a glass bead column or filter of standard size measures their ability to adhere to foreign surfaces (Brassard and Meyers, 1991). When normal animal blood is passed through these columns, more than 75% of the platelets are retained so that the platelet count in the filtrate is low. In animals with platelet function defects or vWD, platelet retention is abnormally low, varying from 0 to 50%. A similar finding occurs with human blood except that platelet retention in normal subjects varies with the assay method and is generally lower (25–80%) than in animals. The most widely used platelet retention assays are the Salzman method, in which blood is collected directly from the patient's vein and passed though a glass bead column by vacuum, and the Bowie technique, in which blood is taken by syringe into heparin and then pumped or drawn by vacuum through a bead column. A two-stage method has been described that depends on fibrinogen for adhesion in the first stage and vWF in the second stage, thus permitting a distinction between these two adhesive proteins (Dodds, 1989a).

In animal species, a modification of the Salzman technique has given the best results for clinical use. Six milliliters of whole blood are drawn as quickly as possible into a plastic syringe. A 1-ml aliquot is immediately added to and mixed in a plastic tube containing 0.1 ml of 3.8% trisodium citrate. The remainder is passed by syringe pump at an even flow rate of 6–8 ml / minute over a standardized l-g glass bead column and collected in four successive l-ml aliquots in a series of plastic tubes containing 0.1-ml citrate anticoagulant. The percentage of platelet retention is calculated as follows:

$$\%\ \text{Retention} = \frac{\text{init. plt. ct.} - \text{av. plt. ct. of last 3 ml}}{\text{initial platelet count}} \times 100.$$

C. Evaluation of Coagulation

1. General Screening Tests

A great variety of screening tests is available for the diagnosis of hemostatic and thrombotic disorders (Table 10.4). These screening tests are not very sensitive to minor abnormalities. This should be kept in mind, especially when a diagnostic workup is performed on patients with only mild bleeding tendencies. If a defect is not found on screening but the history strongly sug-

TABLE 10.4 Practical Screening Tests for Bleeding Disorders

Toenail or buccal mucosal bleeding time
Platelet count
Dilute whole blood clot retraction and lysis test
One-stage prothrombin time (OSPT or PT)
Activated partial thromboplastin time or activated coagulation time (APPT; ACT)
von Willebrand factor (vMF) assay
Thrombin time (TT)
Fibrinogen assay
Assay for fibrin–fibrinogen degradation products (FDP)[a]

[a] Performed for definitive diagnosis of suspected intravascular coagulation.

gests a hemorrhagic problem, more specialized coagulation tests should be performed. Obviously, the screening assays may not demonstrate an abnormality in a mildly affected patient but are often diagnostic in severe deficiencies. The methods for platelet counting, and whole blood clot retraction and lysis are discussed in Sections III.B.1.a and b.

a. One-Stage Prothrombin Time (PT or OSPT)

The PT measures the activity of clotting factors I, II, V, VII, and X in plasma after activation with tissue thromboplastin and recalcification. The most commonly used commercial thromboplastins are made from crude extracts of rabbit brain although research coagulation laboratories may prepare their own homologous brain extracts for optimal activation (Poggio *et al.*, 1991; Kitchen and Preston, 1994). Regardless of the brain extract used, it is essential that homologous, normal plasmas serve as the assay control because erroneous interpretation often occurs when the control and patient plasmas are from different species. Animal plasmas frequently clot very rapidly (<10 seconds) when activated by commercial rabbit brain extract. Therefore, it is difficult to detect minor deficiencies that prolong the assay by only 1 or 2 seconds. This problem can be overcome by diluting the commercial reagent until the clotting time of the control ranges between 10 and 15 seconds. With this adjusted assay system, mild deficiencies usually produce a 2–5 second longer clotting time. A commercial PT assay kit was recently evaluated for use in dogs and cats (Monce *et al.*, 1995).

b. Activated Partial Thromboplastin Time (APTT)

The APTT performed on kaolin- or ellagic acid-activated plasma is a reliable indicator of intrinsic clotting. The test using activated plasma gives shorter clotting times and is more widely used and reliable than tests in which plasma is not exposed to surface activators. The APTT varies with different species, with the source of partial thromboplastin, and with the method and with the type of instrument used to determine the end point. Furthermore, dilution of animal plasmas in order to slow the coagulation process in the APTT has given variable results. Pooled, homologous plasma is therefore essential as a control. With commercial partial thromboplastins of rabbit origin, the APTT of dog and human plasmas are usually 14–18 and 25–45 seconds respectively.

If the APTT is abnormal, differential mixing tests can be used qualitatively to determine which clotting factor(s) is reduced. An equal mixture of test plasma and plasma from a human or animal known to have a specific coagulation abnormality is assayed by the APTT. Failure to correct the APTT to normal or nearly normal identifies the defect.

A simplified variation of the APTT, which measures the intrinsic clotting activity of whole blood, is the activated coagulation time (ACT). The advantage of using this test is that it requires only whole blood, activator, and calcium and depends for activation on the platelets supplied by the sample. Thus, the ACT is sensitive to significant changes in the platelet count. It has been used successfully in small and large animals as a screening test, although the APTT is still preferred by most coagulation laboratories (Dodds, 1989a).

c. Thrombin Time (TT)

The time required for a standardized thrombin solution to clot plasma is a measure of the rate of conversion of fibrinogen to fibrin. This test provides important information on fibrinogen clottability and gives abnormal results in hypofibrinogenemic states, with heparin or heparin-like anticoagulants, in fibrinolytic disease, and in dysfibrinogenemia. The TT is thus an indicator of quantitative and/or qualitative fibrinogen disorders. In DIC with secondary fibrinolysis, the presence of fibrinogen and/or fibrin breakdown products prolongs the TT. Dysfibrinogenemias caused by an abnormal fibrinogen molecule are often detectable only by this test, because fibrinogen concentration is normal.

d. Fibrinogen Assays

Fibrinogen concentrations can be quantitated by several methods, including biological, physical, and immunological assays. Physical methods include an estimation of fibrinogen levels by determining the amount of precipitate formed when plasma is heated to 56°C or treated with 25% ammonium sulfate. A practical method for clinical use in human and veterinary medicine is based on heat precipitability. Plasma samples are placed in microhematocrit tubes and the amount of precipitate formed after heating the tubes for 3–9 minutes at 56°C is quantitated by ocular micrometry. More accurate, biological methods are the quantitative Ratnoff–Menzie assay and its modifications and the measurement of clot weight. The first assay is based on thrombin clottability; the latter method determines the weight of dried recalcified plasma clots. Another common technique is the immunological assay of fibrinogen, which quantitates the immunoprecipitate formed with specific antifibrinogen antisera.

e. Assay for Fibrin–Fibrinogen Degradation Products (FDP)

A commercially available latex agglutination method for measuring FDP in human blood can also be used for several animal species, including dogs. The test, although expensive, is helpful and reliable for confirming selected cases of DIC. Other tests include the ethanol gel test, which depends on the separation of FDP from fibrinogen by ethanol, and the protamine paracoagulation test, which detects fibrin monomers and early FDP by their ability to cause gelation or to form fibrin strands when mixed with a 1% solution of protamine sulfate at pH 6.5. More specialized, quantitative tests are the staphylococcal clumping test, in which FDP induce clumping of most strains of *Staphylococcus aureus* by activating their cell wall coagulase and the very sensitive hemagglutination inhibition assay, which measures FDP in the presence of antifibrinogen serum and tanned, formalinized, fibrinogen-coated homologous erythrocytes.

f. Other Assays

The *RVVT* or *Stypven time* is a measure of extrinsic elotting like the PT, except that it is insensitive to factor VII and requires phospholipid (supplied by PF-3) for its activation. A long PT and normal RVVT are diagnostic of factor VII deficiency. The Stypven assay gives essentially the same clotting time in all species and so eliminates the problems encountered with species specificities of brain thromboplastin. Stypven is quite expensive and is not practical for routine use.

The *WBCT* is measured on a small blood sample (1.5–2 ml) collected by venipuncture into a plastic syringe. The syringe is tilted back and forth until a solid clot forms or the blood is rapidly dispensed into small glass, plastic, or siliconized tubes and the clotting time recorded. The results are more referable to the *in vivo* situation if the tubes are held at 37°C. The clotting time in plastic syringes and tubes or in siliconized tubes is usually about twice that in glass tubes because glass contact activates factor XII and enhances clotting. Canine blood clots in about 2–10 minutes in glass and 9–20 minutes in plastic or silicone. Because of its variability, this test is not recommended for diagnostic use.

The *recalcification* or *plasma clot time* is a more sensitive measure of intrinsic thromboplastin formation than is the WBCT. It times the fibrin end point at 37°C when calcium chloride is added to plasma. Since the recalcification time varies inversely with the platelet count, the speed and time of centrifugation for plasma collection must be standardized for control and test samples. Normal plasma clots in 65–100 seconds; abnormal times occur in thrombocytopenias, deficiencies of the intrinsic system, or in the presence of circulating anticoagulants. Inhibitors can be ruled out by showing correction of the defect in the test plasma after it has been mixed with an equal volume of normal plasma and incubated for 30–60 minutes at 37°C; a double aliquot of normal plasma incubated in parallel serves as the control.

The *prothrombin consumption* (PCT) or *serum clot time* measures the amount of prothrombin remaining after a clot has formed and is actually the prothrombin time of serum. To be reliable, the assay must be performed immediately on the serum of fresh control and test blood incubated at 37°C for 1–2 hours. A source of factor V and fibrinogen is required for the test and is supplied by aluminum hydroxide-adsorbed normal rabbit or bovine plasma. The test serum and adsorbed plasma are activated by rabbit brain tissue extract and recalcified. In normal individuals, coagulation is complete and there is very little prothrombin left in the serum. The PCT is thus long (>40 seconds). Patients with defective intrinsic clotting, thrombocytopenia, or thrombopathia cannot consume prothrombin properly and there is an excess of prothrombin in their serum. The PCT is therefore short in severe defects such as hemophilia, where it may be less than 10 seconds. Thrombocytopenia or thrombopathia can be ruled out by allowing the whole blood for this test to clot in the presence of a source of PF-3 activity such as inosithin.

2. *Specific Assays*

a. Functional Activity Assays

Specific quantitative assays of intrinsic clotting activity are relatively difficult to perform, time consuming, and costly. These assays measure factors VIII:C, IX, XI, and XII and are based on either manual or automated one-stage (PTT) or two-stage (thromboplastin generation) techniques. The clotting factor in question is measured in dilutions of test plasma mixed with standard aliquots of substrate plasma specifically deficient in this factor. The clotting activities of the diluted test plasma are compared by double logarithmic plots (calculated manually or by automated instrumentation) to those obtained for a standard pool of normal plasma from the same species and are reported as a percentage of the normal plasma or reported as units per milliliter. (One unit is defined as the clotting activity present in 1 ml of normal plasma.) As long as the control and test plasmas are homologous, the deficient substrate plasma can be heterologous (animal or human source).

Specific quantitative assays of extrinsic clotting activity are based on the PT and / or RVVT tests. Measurements of factors II, V, VII, and X are usually based on these techniques. Animal or human plasma known to be deficient in one of these factors serves as the substrate and is mixed with dilutions of test plasma (Poggio *et al.*, 1991). Equivalent dilutions of a normal, pooled homologous plasma serve as the control. The results are compared by double logarithmic plot and the activity of the patient's sample is reported as the percentage of normal or as units per milliliter.

Factor X assays usually use either brain tissue extract or Stypven for activation but may use both of these activators. Factor V and VII assays use the PT technique although factor V can also be measured by RVVT and by a more complex, two-stage technique.

Quantitative assays of prothrombin (factor II) are either one-stage methods, based on specific activation of factor II by snake venoms (Smith and Brinkhous, 1991), or two-stage clottability assays. Although the one-stage assay is more easily performed, availability of the Taipan or tiger snake venom used in the reaction is limited by its expense.

Fibrinogen assays are discussed in Section III.C.I.d. The clot solubility time depends on the concentration of factor XIII and is thus a measure of its activity. Recalcified PRP or PPP clots are put into a solution of 5 *M* urea or 1% monochloroacetic acid and held at room temperature for 48 hours. Normal clots do not dissolve, whereas factor XIII-deficient clots usually dissolve within 24 hours.

b. Immunological Assays

The most widely used immunological assay for diagnosis of bleeding disorders measures vWF : Ag [formerly called factor VIII-related antigen (FVIIIR : Ag)]. The functions of this component of the factor VIII complex are discussed in Section II.C.2.g. vWF : Ag is usually quantitated by the very reliable and sensitive ELISA technique, including the commercially available patented test developed by this author and colleagues (Zymtec vWF, Iatric Corp., Tempe, Arizona) (Benson *et al.*, 1991, 1992), and other ELISA-based methods (Meyers *et al.*, 1990a, 1990b; Johnstone and Crane, 1991; Hoogstraten-Miller *et al.*, 1995; Meinkoth and Meyers, 1995). Alternative methods are the quantitative radioimmunoassay of Kraus *et al.*, 1989 (described in Feldman, 1988), and the original Laurell electroimmunoassay used until the ELISA technology was perfected (Dodds, 1989a). Humans and animals affected with the common forms of vWD (types 1 and 3) will typically have low levels of plasma vWF : Ag (Brooks *et al.*, 1991b, 1992, 1996b; Dodds *et al.*, 1993; Johnstone *et al.*, 1993; Ruggeri and Ware, 1993; Ermens *et al.*, 1995; Konkle, 1995), and patients with severe vWD may also have low FVIII : C (Stokol *et al.*, 1995). When used in conjunction with the FVIII : C assay, the ratio of FVIII : C to vWF : Ag is then calculated to confirm a suspected diagnosis of hemophilia A and can also be used to assist in identifying hemophilic carrier females because of their lowered FVIII : C but normal vWF : Ag levels (Murtaugh and Dodds, 1988; Dodds, 1991b, 1992; Joseph *et al.*, 1996). Immunological assays have also been developed for the antigens related to most of the other coagulation proteins (Dodds, 1989a), and include a new ELISA method for thrombin-ATIII complexes (Topper *et al.*, 1996). These assays include neutralizing, precipitating, and radioimmunoassays and have been used to examine whether the plasma of patients with specific functional deficiencies of clotting factors contains normal, reduced, or undetectable amounts of the related antigens (Howard *et al.*, 1994).

c. Other Assays

Quantitative assays of vWF activity include various measurements of ristocetin- and venom-induced platelet agglutination and macroscopic platelet aggregation tests (Ermens *et al.*, 1995; Konkle, 1995). There are many demonstrable species differences in the interaction of mammalian plasmas with ristocetin (Feldman, 1988; Dodds, 1989a). These differences have necessitated modification of existing assays designed for measuring ristocetin cofactor in human plasma to permit its quantitation in animal plasmas. One such modification for canines involves diluting normal and patient plasmas to a final concentration of not more than 2.5% in the reaction mixture (50 μl diluted test plasma; 350 μl washed, formalin-fixed human platelets at 200,000/μl; and 100 μl of ristocetin at 10 mg/ml). The optical density settings of a platelet aggregometer are set at zero baseline for each plasma dilution using a blank tube containing 350 μl Tris buffered saline, 50 μl diluted test plasma and 100 μl ristocetin at 10 mg/ml. It is important to recognize that specific assays for the platelet-related property of human vWF may not be applicable to other species. The snake venom (botrocetin) cofactor assay appears to work well with many species and thus avoids the problems encountered with ristocetin (Smith and Brinkhous, 1991).

Chromogenic substrates are also used to assay several coagulation and kinin components. For example, chromogenic assays are available for prothrombin, thrombin, ATIII, factors VII and X, plasminogen, kininogen, and prekallikrein (Tollefsen, 1990; Welles *et al.*, 1990a; Andreasen and Lovelady, 1993).

D. Evaluation of Fibrinolysis

1. General Screening Tests

Tests of the fibrinolytic system are used with the appropriate coagulation and platelet function assays to distinguish between primary fibrinolysis, a rare entity, and the fibrinolysis commonly seen secondary to DIC and consumption coagulopathy (Feldman, 1988;

Comp, 1990a; McKeever *et al.*, 1990; Hargis and Feldman, 1991; Welch *et al.*, 1992; Joist *et al.*, 1994; Fareed *et al.*, 1995; Lijnen and Collen, 1995; Verstraete, 1995).

a. Clot Lysis Time

The time required for whole blood clots to lyse at 37°C is proportional to the plasmin activity of the blood. It is also dependent on the degree of clot retraction and fibrinogen content of the sample. Poor clot retraction retards clot lysis whereas hypofibrinogenemic clots are friable. A simple, sensitive method for measuring clot retraction and lysis in dilute whole blood is described in Section III.B.l.b. Normal human blood clots take about 16–36 hours to lyse. Most animals have more active fibrinolytic mechanisms and their blood is usually lysed by 8–20 hours. Clot lysis tends to be more active in females than males and to increase with age and stress.

b. Euglobulin Lysis Time

The euglobulins of plasma, such as fibrinogen, plasminogen, plasmin, and plasminogen activator, precipitate on dilution with water. Fibrinolytic inhibitors, such as antiplasmin and antiactivator, do not. The plasma euglobulin lysis test measures the time for redissolved, thrombin-clotted euglobulin precipitates to be lysed by endogenous plasmin. The assay should be performed on fresh PPP prepared at 4°C since platelets contain fibrinolytic inhibitor activity and chilling retards the inactivation of plasminogen activator. The use of a tourniquet or other trauma at the site of venipuncture should be minimal because it enhances fibrinolysis.

c. Fibrin Plate Test

The ability of plasma to lyse standardized fibrin plates is a measure of its lytic capacity and depends on the plasminogen and plasminogen activator concentration. A standard fibrinogen solution (250 mg/ml) is clotted uniformly with bovine thrombin (50 NIH units/ml) in a petri dish. The fresh test plasma or its euglobulin precipitate and a standard urokinase solution are applied to duplicate fibrin plates and incubated for 18 hours at 37°C. The same precautions apply to plasma collection and preparation as described for the euglobulin lysis test. The area of lysis on each fibrin plate is expressed in square millimeters as the product of its two perpendicular diameters.

2. *Specific Assays*

a. Functional Activity Assays

An accurate functional measure of plasma plasminogen content is obtained with a caseinolytic assay based on the ability of plasmin to digest a standard solution of α-casein. The plasma is first acidified to decrease its antiplasmin content and then converted by streptokinase (for human plasma) or urokinase (for animal plasmas) to plasmin. The caseinolytic activity of the resultant plasmin is quantified and compared to that of a known plasmin standard.

b. Other Assays

A variety of other assays are used to measure plasminogen levels by quantifying the plasmin generated. These include use of chromogenic substrates, rate assays of plasmin activity using basic amino acid ester substrates, specific active-site titration, fluorometric immunoassays, radioimmunoassays, protein inhibitor, and affinity chromatographic methods (Fareed *et al.*, 1995; Verstraete, 1995). Assays for TPA are also available (Gerding *et al.*, 1994; Fareed *et al.*, 1995; Lijnen and Collen, 1995).

E. Interpretation of Practical Screening Tests

Table 10.5 outlines the diagnosis of bleeding disorders based on results of screening tests. These can be expanded on the basis of results for the APTT, PT, TT, and platelet count. Similar approaches are used in humans to interpret the significance of hemostatic tests (Hathaway and Corrigan, 1991; Blombäck *et al.*, 1992; Schmidt *et al.*, 1993; Andrew, 1995; Exner, 1995).

1. *Elevated TT, APTT, and PT, Variable Platelet Count*

These cases usually have a variable reduction in platelet count and may have abnormally shaped red blood cells on the blood smear. The prolonged TT indicates a reduced amount of functional fibrinogen most likely caused by the presence of *in vivo* coagulation with production of FDP. The APTT and PT are prolonged whenever fibrinogen-thrombin clottability is impaired. The probable diagnosis is DIC with its requisite underlying cause.

2. *Elevated APTT (Slight), Normal PT and TT, Decreased Platelet Count*

Thrombocytopenia results in less available PF-3 to promote intrinsic coagulation and the APTT is thus slightly prolonged. vWD may present this picture especially if there is concomitant hypothyroidism. A vWD test is needed to confirm the diagnosis.

3. *Elevated APTT (Moderate), Normal PT and TT, Normal or Increased Platelet Count*

A moderate intrinsic system clotting defect is present. If the patient is a young male with an active bleeding history, hemophilia is a primary consideration.

TABLE 10.5 Diagnosis of Bleeding Disorders

Test Results						
Bleeding time	**APTT**	**PT**	**TT**	**Platelet count**	**vWF**	**Probable diagnoses**
Elevated	Elevated	Elevated	Elevated	±Decreased[a]	±Decreased	Disseminated intravascular coagulation (DIC), acute
±Elevated	±Elevated	±Elevated	±Elevated	±Decreased	±Decreased	DIC, chronic
Elevated	Slightly elevated	Normal	Normal	Decreased	Normal	Thrombocytopenia
Elevated	Slightly elevated	Normal	Normal	±Decreased	Decreased	vWD (especially with concomitant hypothyroidism)
±Elevated[b]	Moderately elevated	Normal	Normal	±Elevated	Normal	Hemophilias A or B; Factors X or XI deficiencies
±Elevated[b]	Markedly elevated	Normal	Normal	±Elevated	Normal	Hemophilias A or B; Factors X or XI deficiencies
Normal	Markedly elevated	Normal	Normal	Normal	Normal	Factor XII deficiency (Hageman trait), especially in cats; prekallikrein deficiency (Fletcher trait)
±Elevated	Markedly elevated	Normal	Normal	Decreased	Normal	Factor XII deficiency with concomitant thrombocytopenia (cats)
±Elevated	Slightly to moderately elevated	Moderately to markedly elevated	Normal	±Decreased	Normal	Rodenticide toxicosis, severe liver disease
Normal	Slightly to moderately elevated	Slightly to moderately elevated	Normal	Normal	Normal	Factors II or X deficiencies
Normal	Normal	Moderately to markedly elevated	Normal	Normal	Normal	Factor VII deficiency or early liver disease

[a] ±, variably.
[b] Primary (platelet-dependent) bleeding time is usually normal but secondary (fibrin-dependent) bleeding time is prolonged; rebleeding is also commonly seen.

During the stress caused by bleeding, fibrinogen levels will increase rapidly with resultant shortening of the APTT despite the presence of underlying hemophilia. Alternatively, this profile may indicate another moderate defect of intrinsic coagulation (e.g., factor X or XI deficiency), but these are relatively rare. Hemophilia should be ruled out first. Specific assays of factors VIII:C (hemophilia A) and IX (hemophilia B)are needed for confirmation. The vWF levels should be normal unless both hemophilia and vWD are present (e.g., in Doberman pinschers where vWD is so prevalent).

4. Elevated APTT (Marked), Normal PT and TT, Normal or Increased Platelet Count

A severe, intrinsic system clotting defect is present. In a young male with an active bleeding history, hemophilia is most likely. Both hemophilia and vWD could be present in an affected Doberman pinscher. In other cases of a significant bleeding history, some other intrinsic clotting defect may be present (e.g., factor X or XI deficiency). In the absence of a bleeding history, especially in cats with very long (>60 seconds) APTTs, congenital factor XII deficiency (Hageman trait) should be the first choice. This defect by itself is not of clinical consequence. In animals with a mild to moderate bleeding history and another disease process (e.g., kidney or liver disease), a very long APTT may reflect Fletcher trait (prekallikrein deficiency).

A prolonged PT with normal APTT and TT is most likely caused by factor VII deficiency or early liver disease. If the defect is vitamin K responsive, liver disease is the likely cause.

5. Elevated APTT (Marked), Normal PT and TT, Decreased Platelet Count

When a mild bleeding diathesis is present, especially in cats, the very long APTT is usually caused by an incidental familial factor XII deficiency (Hageman trait), and the bleeding results from concomitant thrombocytopenia.

6. Elevated APTT and PT (Moderate to Marked), Normal TT and Platelet Count

A significantly prolonged PT with a concomitant prolonged APTT and normal TT is most likely caused by rodenticide toxicosis or severe liver disease. The platelet count may also be reduced. The source of the rodenticide should be investigated and treatment with vitamin K_1 instituted for 1–6 weeks depending on the type of rodenticide involved (Mount and Kass, 1989).

7. Elevated APTT, PT (Slight to Moderate), Normal TT and Platelet Count

Mild to moderate prothrombin or factor X deficiencies can prolong both the APTT and PT somewhat but the TT is normal.

8. Normal APTT, TT, and Platelet Count, Elevated PT

A prolonged PT with normal APTT and TT is most likely caused by factor VII deficiency or early liver disease. If the defect is vitamin K responsive, liver disease is the likely cause.

9. Diagnosis of von Willebrand's Disease

Table 10.6 gives a breakdown of the various categories for diagnosing vWD based on the bleeding time and vWF antigen level.

10. Distinguishing Between Rodenticide Toxicosis and Thrombosis

Animals admitted on an emergency basis for acute nontraumatic bleeding pose a difficult diagnostic challenge. While the most likely diagnosis is rodenticide toxicosis, the clinician can be misled if the client insists that there has been no opportunity for exposure. With the newer toxicants available today, a nontarget companion animal can become poisoned secondarily, without the owner's knowledge upon ingesting a poisoned rodent that enters the premises. The presumptive diagnosis of rodenticide poisoning must be distinguished from bleeding associated with DIC because the clinical signs and presentation can be similar. In the case of thrombosis, an underlying cause must be present but may not be immediately obvious.

A complete coagulation profile can readily distinguish between rodenticide poisoning and DIC if it detects the presence of FDP either by direct measurement or with the TT. This test is dependent not only on the quantity of fibrinogen present, but also on its ability to be coagulated by thrombin. In the presence of significant amounts of FDP, the TT is elevated because the fibrinogen molecule has been cleaved by the *in vivo* enzymatic action of thrombin and thereby has been rendered less coagulable.

The typical results of coagulation screening tests for either rodenticide toxicosis or DIC are mild to moderate thrombocytopenia; variable prolongation of the PT and APTT—often exceeding 60 seconds each; and quantitative fibrinogen levels varying from low to elevated. At this point, the critical assay that distinguishes between these two potentially life-threatening problems is the TT. If the TT is normal, the diagnosis is rodenticide toxicosis; if it is elevated, DIC is present.

In mild cases of rodenticide exposure or thrombosis, prolongation of the PT and APTT may be subtle. For example, patients undergoing the stress of a bleeding episode may have elevated fibrinogen and platelet numbers as an acute phase response so that the PT and APTT are shortened below the normal ranges. This also occurs in the early hypercoagulative phase of a thrombotic event. In the case of rodenticide poisoning or acute, generalized liver disease, the PT may appear relatively prolonged in comparison to the APTT, TT, and fibrinogen levels, which are all shortened from increased activity. Thus, the PT may be at the upper limit of normal range or slightly prolonged when the APTT and TT are at the lower range or below the normal range.

This discordance (elevated PT with shortened APTT and TT) is sufficient to diagnose impairment of the vitamin K-dependent prothrombin complex by either rodenticide exposure, liver disease, or even vitamin K-deficiency in neonates or from sterilization of the bowel by prolonged antibiotic usage. This pattern of coagulation test results is explained by two opposing actions: the increased generation of thromboplastic activity following a stress event, which shortens the intrinsic coagulation pathway and elevates fibrinogen, as measured by the APTT and TT, respectively, versus the inhibition of prothrombin complex / vitamin K metabolism by rodenticide, which blocks shortening of the PT in parallel. The difference between the PT and APTT results can be as little as a 1–2 seconds lengthening of the PT and a 1–2 seconds shortening of the APTT, with a resulting gap of 4–5 seconds between the end points of both tests.

IV. BLEEDING DISORDERS

A. General

During the past 40 years, a variety of hemorrhagic diseases have been recognized and studied in animals and humans. Historically, these disorders in animals

TABLE 10.6 Diagnosis of von Willebrand's Disease

Bleeding time[a]	von Willebrand factor antigen (%)[b]	Interpretation
Normal	≥70	Normal range is 70–180%.
Normal[c]	70–79	Lower end of normal range; caution advised for breeding stock. Mates should have higher levels and pups should be checked.
Normal	50–69	Borderline normal (equivocal result) or heterozygous carrier of vWD gene. Recommend retesting and/or breeding only to higher testing mates. Pups should be checked.
Normal or elevated[c]	<50	Type I vWD. This is the most common form of vWD. Heterozygous carrier of vWD.[c] Recommend breeding only to higher testing mates. Pups should be checked.
Elevated[c]	<50	Type I vWD. Has exhibited some bleeding problems (e.g., hematuria, epistaxis, melena, postsurgical bleeding).[c] Clinically affected animals should not be used for breeding. Normal or <50 Type II vWD. Clinically elevated[c] affected animals exhibit severe bleeding symptoms. Multimeric analysis of vWF reveals absence of high-molecular-weight multimers. These animals should not be used for breeding. Has been reported in German shorthair and wirehaired pointers.
Elevated	<0.01	Type III vWD. Exhibits bleeding symptoms and is homozygous for vWD. This animal is the product of two asymptomatic, heterozygous carrier parents and should not be used for breeding.
Normal	>180	Probably reflects stress, an improper sample, or activation from disease. Recommend retesting. Test invalid for prediction of genetic status for vWD trait.

[a] Abnormal bleeding time is not specific for vWD; hemostatic disorders including thrombocytopenia and platelet dysfunction will also cause variable prolongation of bleeding time tests.
[b] Formerly called factor VIII-related antigen; measured by ELISA methodology. When measured by Laurell or RIA methods, normal ranges may vary slightly.
[c] Development of concomitant thyroid dysfunction may aggravate existing vWD or increase the risk of bleeding in some previously asymptomatic carriers.

have been among the most successfully exploited areas of biomedical research in comparative medicine (Dodds, 1988; Brinkous, 1992). Today, reliable and useful animal models exist for nearly all of the inherited and acquired hemostatic and thrombotic disorders (Dodds, 1988, 1989a, 1992). Some deficiencies (e.g., factors V and XIII) have yet to be discovered in domestic or other animals. Numerous reviews describe the clinical, pathophysiological, genetic, management, and treatment aspects of these diseases in animals (Catalfamo and Dodds, 1988; Feldman, 1988; Dodds, 1989a, 1992; McKeever *et al.*, 1990; Prasse *et al.*, 1993; Stefanon *et al.*, 1993; Carr and Johnson, 1994; Mannucci and Giangrande, 1994; Nichols and Hohenhaus, 1994; Sullivan, 1994; Peterson *et al.*, 1995).

1. Medical History

A list of the common causes of hemorrhage in animals is given in Table 10.7. A complete medical history should point to one or more of these causes of bleeding. This should include information about the current bleeding problem, previous bleeding problems, family history, environmental influences, and drugs. Knowledge of the patient's present and past bleeding episodes often provides the key to diagnosis since bleeding problems, especially when internal, can mimic a variety of other disease states. It is therefore important to reevaluate previous illnesses in light of possible clinical signs of bleeding.

2. Physical Examination

A thorough physical examination is essential to determine the location, severity, and nature of the bleeding and to identify the underlying disease that may be present. The type and site of bleeding should be considered carefully. Is the hemorrhage superficial or deep? Is there epistaxis, hematuria, melena, petechiation or ecchymosis on mucosal surfaces or hematoma formation in soft tissues or body cavities? Chronic mucosal surface bleeding suggests a platelet disorder or von Willebrand's disease, whereas large hematomas are more common in clotting factor disorders. Other findings, which suggest acquired rather than inherited causes are splenomegaly, hepatomegaly, and lymphadenopathy.

3. Management and Treatment

Appropriate physiological and physical environments for hemostasis, tissue repair, and prevention of recurrence are important adjuncts to management and treatment of bleeding disorders. Drugs known to interfere with hemostasis are contraindicated for patients

TABLE 10.7 Some Causes of Hemorrhage in Animals

Hereditary defects
- Coagulation disorders
  - X-chromosome-linked recessive traits: hemophilia A, hemophilia B (Christmas disease)
  - Autosomal traits: factor VII deficiency, factor X deficiency, factor XI deficiency, factor XII deficiency, von Willebrand's disease, fibrinogen deficiency (afibrinogenemia, dysfibrinogenemia, and hypofibrinogenemia), prothrombin deficiency (dysprothrombinemia and hypoprothrombinemia)
- Platelet disorders
  - Autosomal traits: thrombasthenia (Glanzmann's disease), thrombopathia (storage-pool disease, thrombocytopathy)

Acquired defects
- Trauma: accidental or from surgical intervention
- Poisoning: warfarin, moldy sweet clover, cottonseed meal, aspirin overdose
- Vitamin deficiency: scurvy, absorptive failure, vitamin K deficiency
- Liver disease: obstructive jaundice, infectious canine hepatitis, tumors, liver failure
- Thrombocytopenia: idiopathic, virus-induced, autoimmune disease, septicemia, splenomegaly, aplastic anemia
- Intravascular coagulation and fibrinolysis syndrome: obstetric complications, sepsis, malignancy, shock, liver disease, heat stroke, incompatible transfusions, heartworm disease
- Platelet function defects: uremia, hyperestrogenism, allergies, drugs, chronic disease, malignancy
- Drug-induced defects: aspirin, steroids, phenylbutazone, live virus vaccines, furacins, sulfonamides, antihistamines, local anesthetics, tranquilizers

with moderate to severe hemostatic defects since they impair platelet function and further compromise the stability of the hemostatic plug (Dodds, 1989a, 1991b, 1992; Dodds *et al.*, 1993). These include aspirin, promazine tranquilizers, phenylbutazone, nitrofurans, potentiated sulfonamides, glycosaminoglycans, penicillins, phenothiazines, antihistamines, local anesthetics, estrogens, antiinflammatory drugs, plasma expanders such as dextran and hydroxyethyl starch, and live virus vaccines (Dodds, 1992; Gentry *et al.*, 1992; Dodds, *et al.*, 1993) (Section IV.C). The latter, like virus infections, affect platelet and/or endothelial function during the viremic phase (3–10 days after vaccination or exposure). Elective surgical procedures, such as ear cropping, spaying, castration, or dentistry, should be performed within 48 hours after vaccination or postponed for 10–14 days (Dodds, 1989a, 1992; Brooks, 1992; Dodds *et al.*, 1993).

Adequate replacement therapy of the correct type is essential for the control of moderate or severe hemostatic defects (Cotter, 1991; Dodds, 1991a; Hohenhaus, 1992; Wagner and Dunlop, 1993; Dodds *et al.*, 1993; Giger, 1994). In general, guidelines for treating human patients with hemostatic disorders are followed for parallel problems in animals. Because platelets and coagulation factors are relatively labile, whole blood, plasma or platelet-enriched products should be processed and stored for use later or transfused as soon as possible after collection and preparation. Although plasma is typically frozen soon after preparation (i.e., fresh-frozen plasma) and then stored at −20°C or lower for long periods, newer techniques for freezing and storing human and animal red blood cells and platelets have not yet been applied to the clinical care of animal patients (Cotter, 1991; Hohenhaus, 1992; Rao *et al.*, 1993; Wagner and Dunlop, 1993; Machin *et al.*, 1995). An efficient method of obtaining maximal utilization of blood products involves removing and freezing the plasma from anticoagulated fresh whole blood and storing the packed red blood cells in nutrient solution (e.g., Adsol, Optisol) at 4°C for 5–6 weeks. The collection, storage, and utilization of whole blood, packed red blood cells, fresh–frozen plasma or other blood components obtained from animal species are best accomplished under conditions used for human blood. Plastic bags are preferable to vacuum bottles for collection and red cells retain viability longer in citrate-phosphate-dextrose (CPD) anticoagulant than in acid-citrate-dextrose (ACD) (Cotter, 1991; Dodds, 1991a; Hohenhaus, 1992).

Until recently, most veterinarians have had to resort to using random (unmatched) blood donors, because typing sera for detemining animal blood groups are generally not commercially available (Dodds, 1993). Blood typing is typically obtained, therefore, by sending samples to one of the few specialized laboratories offering this service for domestic animals. With respect to the dog and cat, a commercial card-typing system for the canine DEA 1.1 antigen has been introduced and a parallel card system for the feline A and B antigens has been developed (Iatric Corp., Tempe, Arizona).

The ideal canine blood donor (a universal donor) is negative at the DEA 1.1, 1.2 and 7 loci; is a healthy, young adult (1–6 years) of good temperament; weighs at least 50 lb; and has never been transfused. A donor dog can safely donate 10 ml of blood/lb of body weight every 3–4 weeks or 5 ml of blood/lb every 2 weeks. Donor dogs should be given an iron-vitamin dietary supplement to enhance hematopoiesis if they are to be bled repeatedly. Today, veterinarins can purchase blood from the handful of private commercial blood banks recently established for dogs and horses. Alternatively, practitioners may keep DEA 1-negative dog donors in their hospital to provide fresh blood for emergency use to treat patients in need of repeated transfusions (Dodds, 1993). These include dogs with such conditions as inherited or acquired bleeding disease, autoimmune hemolytic anemia, other chronic anemias of primary or secondary origin, and animals known to have received prior transfusions of unknown type (e.g., for puppies with severe hookworm anemia, accident cases given blood transfusions for shock or blood loss).

Cross-matching blood from donor and recipient dogs before transfusion is not a substitute for blood typing because, unless the recipient dog has been previously sensitized by an incompatible transfusion, the serum will not contain red cell antibodies that react in the cross-match test. However, if commercial blood products or local compatible donors are unavailable, the best alternative is to cross-match donor and recipient (Dodds, 1993). The recipient's blood must be compatible in the major cross-match (donor red cells and recipient serum) at 37°C, room temperature, and at 4°C and should be compatible in the minor cross-match (recipient red cells and donor serum) at least at 37°C. Incompatibilities by cross-matching indicate prior sensitization of the recipient. Animals in need of repeated transfusions or those with autoimmune hemolytic anemia should always be given blood from typed and compatible (universal) donors.

The major blood types in cats are types A and B with type AB being very rare. In some cat breeds, 30–50% of the animals are blood type B (Hohenhaus, 1992). The ideal feline blood donor has blood type A, which is the most common type found in more than 90% of random-source cats and 50% or more or purebred cats; is a healthy, young adult of good temperament; weighs at least 8 lb; and has never been transfused. Unlike the dog, cats have naturally occurring isoantibodies against the other blood types. Blood type A cats have weak anti-B antibodies, whereas type B cats have more potent anti-A antibodies. This has important clinical significance because transfusion of type A blood to a type B cat, even the first time, can produce a serious adverse transfusion reaction and mating of a type B queen with a type A tom can result in hemolytic disease of newborn kittens of type A (Cotter, 1991; Hohenhaus, 1992). Donor cats can safely donate 5 ml of blood/lb body weight for a maximum of 60 ml every 3–4 weeks and should be supplemented with hematinics. Cat blood is not presently offered by any of the commercial animal blood banks so practitioners must maintain their own feline donors or donor colonies (Hohenhaus, 1992).

Whole blood is not the treatment of choice nor is it desirable for the primary therapy of most veterinary transfusions (Cotter, 1991; Hohenhaus, 1992; Dodds, 1993). Processing freshly collected blood into several clinically useful components is a more cost-effective, efficient, and safer use of blood. The most commonly used blood components in veterinary medicine parallel those in human medicine, namely, packed red blood cells and fresh-frozen plasma. The red blood cells are primarily used to treat acute blood loss anemia from trauma, surgery, or acute hemolytic disease and for chronic anemia caused by internal and external parasites, bone marrow failure, and chronic hemolytic disease. Fresh-frozen plasma is used to treat or control bleeding disorders and to provide passive immunity to alleviate or protect against acute or chronic infectious diseases, especially those of viral origin such as parvovirus or herpes virus disease. Specialty canine blood products such as PRP (for clinically severe platelet defects) and cryoprecipitate (for hemophiliacs and severe cases of vWD) are also available for clinical use at some veterinary schools and from this author's private blood bank (Dodds, 1993; Ching *et al.*, 1994). The high purity and recombinant factor VIII and IX clotting factor concentrates used for treating human hemophiliacs or vWD patients (Aronson, 1990; Bloom, 1991; Hanna *et al.*, 1994) are unlikely to even become available in parallel homologous form for treating animal patients, because of the need for large volumes of starting plasma (low yields) and the unrealistically high preparatory costs (Cotter, 1991; Hohenhaus, 1992). Blood substitutes are also used in veterinary medicine in a manner similar to that of human medicine (Dodds, 1991a).

B. Hereditary Disorders

Table 10.7 lists the inherited coagulation and platelet function defects currently recognized in animals. These disorders have been reviewed in detail elsewhere (Catalfamo and Dodds, 1988; Feldman, 1988; Raymond *et al.*, 1990; Dodds, 1991b, 1992; Brooks *et al.*, 1992; Dodds *et al.*, 1993; Stefanon *et al.*, 1993; Carr and Johnson, 1994; Peterson *et al.*, 1995).

1. Fibrinogen (Factor I) Deficiencies

Hereditary fibrinogen deficiencies have been recognized in goats and dogs. These can be caused by a complete lack of fibrinogen (afibrinogenemia), reduced fibrinogen (hypofibrinogenemia), or an abnormal fibrinogen (dysfibrinogenemia). These defects produce a mild to severe bleeding diathesis and have an autosomal inheritance pattern (Dodds, 1989a, 1992).

Fibrinogen defects, quantitative or qualitative, result in a complete failure of plasma or whole blood to clot in any coagulation test or when thrombin is added and plasma does not form a precipitate when heated to 56°C or treated with 25% ammonium sulfate. Some abnormalities of platelet function, a prolonged bleeding time and abnormal clot retraction also may occur. Although biological assays of fibrinogen are abnormal, the presence of fibrinogen is usually detected immunologically.

Caprine afibrinogenemia was recognized in the Netherlands in a large family of Saanen dairy goats (Dodds, 1989a). It was characterized by a severe hemorrhagic diathesis in newborn and young kids, incompletely dominant inheritance, umbilical bleeding, recurrent hemarthroses, and bleeding into subcutaneous tissues and from mucous membranes. Fibrinogen was undetectable by bio- and immunoassays. Heterozygotes had hypofibrinogenemia.

Canine hypofibrinogenemia was recognized in a family of Saint Bernards from West Germany. Both biological and immunological assays of fibrinogen were decreased. The propositus succumbed to a severe hemorrhagic crisis. Related dogs appeared to be heterozygous for the trait. Inherited dysfibrinogenemia has been observed in families of borzoi and vizslas and in a collie.

2. *Prothrombin (Factor II) Deficiency*

Prothrombin defects prolong the extrinsic system tests (PT and RVVT) while intrinsic coagulation tests are normal. Both hypo- and dysprothrombinemia are recognized as clinical entities in humans. These are autosomal characteristics that produce mild to moderately severe bleeding problems.

Canine hypoprothrombinemia is the only inherited prothrombin abnormality seen in animals (Dodds, 1989a). Two generations of an affected family of boxers from Texas were studied. The defect was characterized by epistaxis and umbilical bleeding in newborn puppies and mild mucosal surface bleeding in young adults. Although the disease appeared initially to resemble dysprothrombinemia with reduced prothrombin activity and normal antigen, additional studies indicated a metabolic or kinetic problem in turnover of all the vitamin K-dependent clotting factors. These dogs were also warfarin sensitive and had a considerably delayed warfarin turnover time. Prothrombin deficiency has also been reported in an English cocker spaniel.

3. *Vitamin K-Dependent Coagulation Factor Deficiency*

Combined deficiency of all vitamin K-dependent coagulation factor (II, VII, IX and X) has occasionally been reported in humans (Pechlaner *et al.*, 1992) and was thoroughly investigated in a family of Devon Rex cats (Maddison *et al.*, 1990; Soute *et al.*, 1992). The feline disorder was vitamin K responsive and was inherited as an autosomal trait (Maddison *et al.*, 1990). A defective vitamin K-dependent carboxylase was identified that had decreased affinity for both vitamin K hydroquinone and propeptide (Soute *et al.*, 1992).

4. *Factor VII Deficiency*

Human factor VII deficiency, although relatively rare, has been thorougly characterized from both genetic and molecular perspectives (Tuddenham *et al.*, 1995). By contrast, this disorder is commonly reported worldwide in beagle dogs from large commercial breeding colonies. The apparent high incidence of the defect in this breed probably reflects the widespread use of beagles in biomedical research (Dodds, 1989a). It has also been described in an Alaskan malamute, bulldogs, and mixed breeds. The disorder is usually discovered fortuitously during routine hematological screening for drug testing, research, or clinical workup.

Canine factor VII deficiency is a mild disease with no overt bleeding tendency except for easy bruisability and a predisposition to systemic demodecosis, presumably because the animals' defective extrinsic clotting provides an ideal, moist environment for mange mites (Dodds, 1989a). Homozygotes are detected by the presence of a long PT, normal RVVT, and reduced factor VII activity. Heterozygotes are more difficult to identify with screening tests only because the PT may fall within the upper end of normal range or be only slightly elevated. Affected dogs have less than 15% and usually below 5% factor VII, whereas heterozygotes have about 15–65% of normal levels. The prothrombin consumption time of factor VII-deficient dogs is usually prolonged because factor VII is absent from their serum.

Factor VII-deficient dogs have served as important models for biomedical research (Dodds, 1989a). Plasma from affected dogs is an excellent standard source of deficient substrate for quantitating factor VII in humans and other species and for quality control of various tissue thromboplastins. The model has been used to investigate the role of factor VII in mediating endotoxin-induced intravascular coagulation and shock. Cross-circulation studies between normal and factor VII-deficient dogs confirmed the rapid turnover and short half-life of this clotting factor as predicted from studies in humans (Dodds, 1989a). A human plasma concentrate containing mostly activated factor VII (FVIIa) and proteins C and S has been given to dogs with hemophilia A to demonstrate its hemostatic benefit in bypassing the intrinsic system requirement for factor VIII (Chabbat *et al.*, 1989). This approach may be useful in treating hemophilic patients with coagulation inhibitors.

5. *Hemophilia A (Factor VIII:C Deficiency, Classic Hemophilia)*

The most common of the severe inherited coagulopathies, hemophilia A occurs in humans, dogs, horses, sheep, cattle, and cats (Feldman, 1988; Dodds, 1992, 1995b; Stefanon *et al.*, 1993). Clinical signs and laboratory studies have shown the disease to be similar if not identical in these species (Feldman, 1988; Carr and Johnson, 1994; Peterson *et al.*, 1995). Mild, moderate, and severe forms of hemophilia have been recognized in humans and dogs, whereas the equine defect is usually severe and the feline defect tends to be mild. The canine disease has been reported in nearly every pure breed of dogs and in mongrels (Dodds, 1992; Carr and Johnson, 1994; Joseph *et al.*, 1996), but is most common in German shepherds worldwide having descended from the obligatory carrier progeny of an influential hemophilic stud (Feldman, 1988; Stefanon *et al.*, 1993; Stokol *et al.*, 1994). Smaller breeds tend to be less severely affected than larger breeds, which appears to be related to weight bearing and is consistent with the severe expression in horses and the mild problem in cats. Thoroughbred, standardbred, and quarter horses have all developed this disease as have several breeds and mixed breeds of cats (Henninger, 1988; Dodds, 1991b; Peterson *et al.*, 1995). The bovine defect arose in Australian Hereford cattle and the ovine in Dutch sheep (Dodds, 1989a).

Hemophilia is an X chromosome-linked recessive trait in humans and other animals. Affected males are hemizygous for the trait, whereas carrier females are heterozygous and affected females (the product of a hemizygous and heterozygous mating) are homozygotes. Affected individuals have markedly prolonged screening tests of intrinsic clotting and very low levels of FVIII:C activity but normal or elevated levels of vWF:Ag. In accord with the Lyon hypothesis whereby heterozygous females have one active and one inactivated X chromosome, hemophilic carriers have about half of the normal amount of FVIII:C. They synthesize a full complement of vWF:Ag, however, so carrier females may be identified by their moderately reduced FVIII:C (40–60% of normal) and normal or elevated vWF:Ag. This finding (Section II.C.2.f) has been used effectively for carrier detection in both humans and animals (Feldman, 1988; Dodds, 1989a, 1991a, 1992, 1995b). More recently, prenatal diagnosis by amniocentesis has been used in humans to predict accurately the genotype of a developing fetus in known or suspected carrier females (Lozier and High,1990; Antonarakis, 1995).

Studies of animals with hemophilia A have contributed significantly to current knowledge of the analogous human disease and to knowledge of the biochemical and physiological roles of factor VIII (Dodds, 1988, 1989a; Feldman, 1988; Patterson *et al.*, 1988; Arai *et al.*, 1989; Chabbat *et al.*, 1989; Ni and Giles, 1992; Tinlin *et al.*, 1993). Hemophilic dogs have been used to study hemostatic plug formation, development of spontaneous canine–anticanine factor VIII inhibitors, *in vivo* half-life, and sites of synthesis of factor VIII (Brinkhous, 1992; Ni and Giles, 1992; Tinlin *et al.*, 1993). They have provided an extremely valuable standard source of deficient substrate for quantitating FVIII:C in all species and helped in the development of the APTT assay method. Affected animals have also been used to investigate the biochemical nature and molecular structure of factor VIII and to test the *in vivo* efficacy of new homologous and heterologous factor VIII and IX concentrates and the factor VIII-bypassing activity of tissue factor, factor VIIa, factor Xa, and phospholipid vesicles (Chabbat *et al.*, 1989; Ni and Giles, 1992). The safety and efficacy of heat-treated factors VIII and IX concentrates were established in hemophilia A and B dogs, respectively.

Infusions of highly purified rDNA-derived human FVIII:C concentrates in dogs with hemophilia have proved the efficacy and safety of this type of product leading to its present clinical use in humans (Aronson, 1990). This recombinant form of factor VIII is now available for routine treatment of hemophilia, thus further reducing the risk of transfusion-transmitted diseases, such as hepatitis B and C, and acquired immune deficiency syndrome (Bloom, 1991; Dodds, 1991a).

Genetic studies with hemophilic dogs have characterized the hemophilic phenotype and genotype, provided the proof of X chromosome-linked recessive inheritance and examined linkage of other closely or distantly located X chromosomal markers (e.g., the production of canine double hemophilia, hemophilia AB, Section IV.B.12.b). The specific gene deletion or substitution in many human and several canine hemophilic families has been identified by cDNA cloning techniques (Lozier and High, 1990; Antonarakis, 1995).

The most challenging and potentially important research with animal models of hemophilia A is directed at molecular genetics and gene therapy (Bi *et al.*, 1995; Dwarki *et al.*, 1995; Fallaux *et al.*, 1995; Thompson, 1995). The canine hemophilia models have and continue to play a key role in this effort (Section IV,B,6) (Walter and High, 1997).

6. *Hemophilia B (Factor IX Deficiency, Christmas Disease)*

Hemophilia B, also an X chromosome-linked recessive trait, is less common than hemophilia A and occurs in humans, about 20 breeds of dogs, and several breeds of cats as well as mixed breeds (Feldman, 1988; Brooks and Dodds, 1989; Dodds, 1989a; Maggio-Price and Dodds, 1993; Feldman *et al.*, 1995). Christmas disease occurs as a mild to moderate disorder of small dog breeds (e.g., Cairn terrier, American cocker spaniel, Shetland sheepdog, Scottish terrier, and French bulldog) and cats and as a severe diathesis in larger breeds (e.g., black and tan coonhounds, Saint Bernards, Alaskan malamutes, Old English sheepdogs, and German shepherds) (Dodds, 1992; Stefanon *et al.*, 1993; Carr and Johnson, 1994; Peterson *et al.*, 1995).

As in classic hemophilia, carrier females can be identified by reduced (40–60%) levels of factor IX (Maggio-Price and Dodds, 1993; Feldman *et al.*, 1995). However, in contrast to hemophilia A, in which mild, moderate, and severe factor VIII:C deficiencies exist, most reported cases of hemophilia B in animals have had less than 5% factor IX. Gene cloning studies with the human and canine disease have identified two types of patients, those with undetectable or very low levels of factor IX antigen and a few with normal amounts of a mutant antigen (Evans *et al.*, 1989a, 1989b; Sommer, 1992). An intriguing hemophilia B mutant was recently described in which affected family members gradually recover under the influence of androgens (Crossley *et al.*, 1992).

It is important to perform accurate diagnostic tests to differentiate between the hemophilias as both disorders have markedly prolonged intrinsic system clotting tests. The propositi in several of these dog and cat families were referred for confirmation of suspected hemophilia A or severe vWD and were found to have hemophilia B (Maggio-Price and Dodds, 1993; Feldman *et al.*, 1995). The plasma defect of hemophilia B is corrected by addition of fresh normal serum, whereas that of hemophilia A is not because serum contains factor IX but not FVIII:C activity. Like hemophilia A dogs, hemophilia B dogs are important models for biomedical science (Dodds, 1988; Evans *et al.*, 1989a, 1989b) and have played a pivotal role in recent gene therapy experiments (Axelrod *et al.*, 1990; Kay *et al.*, 1993, 1994; Marx, 1993; Lozier *et al.*, 1994; Sugahara *et al.*, 1996; Walter and High, 1997). Their plasma is an excellent deficient substrate of standard source for quantitating factor IX activity and they were used in organ transplantation studies that established the liver to be the sole site of factor IX synthesis (Kay *et al.*, 1993, 1994; Furie and Furie, 1995b). Genetic cross-breeding studies with hemophilia A dogs produced double hemophiliacs (hemophilia AB; see Section IV.B.12.b) and infusion studies determined the half-life of factor IX and efficacy of human factor IX concentrates. The occasional thrombogenicity of human prothrombin complex clotting factor concentrates prompted comparison studies with highly purified factor IX in a canine thrombogenic model (MacGregor *et al.*, 1991). As some human patients with mild hemophilia respond to danazol treatment given as a nontransfusional form of therapy, cats with hemophilia B were also treated with danazol but no therapeutic benefit was observed (Boudreaux and Dillon, 1988).

Development of highly purified rDNA-derived factor IX has permitted *in vivo* pharmacokinetic studies in rats and dogs (Keith *et al.*, 1994). Definitive studies of factor IX biosynthesis have elucidated practical approaches for factor IX gene therapy (Palmer *et al.*, 1989; Sommer, 1992; Furie and Furie, 1995b). Initial experiments with skin fibroblasts in factor IX-deficient dogs (Axelrod *et al.*, 1990) began a series of factor IX gene therapy studies including development of a partially hepatectomized hemophilic puppy model (Kay *et al.*, 1993, 1994; Marx, 1993) and other less invasive approaches (Lozier *et al.*, 1994; Fallaux *et al.*, 1995; Thompson, 1995; Sugahara *et al.*, 1996; Walter and High, 1997). Although a cure for hemophilia is still distant, results to date are encouraging. The main stumbling block is the need to ellicit long-term vector expression of functional factor IX (Walter and High, 1996).

7. *Von Willebrand's Disease*

von Willebrand first described the complex, multifaceted syndrome known as vWD in 1926 (Ginsburg and Sadler, 1993; Ruggeri and Ware, 1993). The bleeding diathesis, usually more mild than the hemophilias, primarily involves mucous membranes and skin, with gastrointestinal and urogenital bleeding and epistaxis as common symptoms (Raymond *et al.*, 1990; Brooks, 1992; Mazurier, 1992; Dodds *et al.*, 1993; Lock, 1994; Sullivan *et al.*, 1994). The prolonged bleeding time also results in excessive surgically induced hemorrhage (Dodds *et al.*, 1993; Lock, 1994). vWD is the most common inherited bleeding disorder of humans and has been recognized in swine, more than 60 dog breeds, several breeds of cats, a quarterhorse family, and inbred strains of laboratory rabbits and mice (Dodds, 1988, 1991b, 1992; Feldman, 1988; Sweeney *et al.*, 1990; Brooks *et al.*, 1991b, 1992, 1996a; Huss and Ettinger, 1992; Dodds *et al.*, 1993; Carr and Johnson, 1994). The affected breeds with a high prevalence of the gene are Doberman pinscher (70% prevalence), German shepherd, miniature schnauzer, golden retriever, Shetland sheepdog, basset hound, standard poodle, keeshond, rottweiler, dachshund, Scottish terrier, Manchester terrier, and Pembroke Welsh corgi. In other breeds, the disorder is either less prevalent or the true prevalence is unknown because too few animals have been studied.

By far the most common form of canine vWD is type 1, a situation analogous to human vWD (Feldman, 1988; Dodds *et al.*, 1993; Sadler, 1994; Sadler *et al.*, 1995). Type 1 vWD is inherited as an autosomal, incompletely dominant trait with variable clinical and laboratory expression based on the degree of penetrance of the mutant gene. Homozygosity is lethal in type 1 vWD; affected pups are resorbed *in utero*, born dead, or live only a few days (Dodds, 1988). Heterozygotes are either asymptomatic carriers or express a mild to moderately severe bleeding diathesis, especially involving mucosal surfaces or associated with surgery.

In four dog breeds, Scottish terriers, Chesapeake Bay retrievers, Shetland sheepdogs, and German wirehaired pointers (Dodds, 1988, 1989a, 1991b, 1992; Raymond *et al.*, 1990; Dodds *et al.*, 1993; Johnstone *et al.*, 1993; Brooks *et al.*, 1996a), in Himalayan cats and in Poland-China swine, vWD is analogous to the autosomal recessive type 3 vWD of humans. The homozygote manifests a moderate to severe bleeding tendency and heterozygotes are detected by laboratory tests but are otherwise asymptomatic. The clinical and laboratory spectrum of canine vWD closely parallels that of human vWD (McCarroll *et al.*, 1987, 1988; Dodds, 1988, 1992; Feldman, 1988; Kraus *et al.*, 1989; Waters *et al.*, 1989; Meyers *et al.*, 1990a, 1990b; Ruggeri and Ware, 1993; Konkle, 1995; Meinkoth and Meyers, 1995; Sadler *et al.*, 1995; Brooks, 1996a, 1996b).

Only severely affected individuals will exhibit prolonged screening tests of intrinsic clotting, whereas the bleeding time is consistently prolonged. Individuals with type 1 disease have a variable reduction in FVIII:C but a definite reduction in vWF:Ag and platelet vWF activity (Dodds *et al.*, 1993; Stokol *et al.*, 1995; Zang, 1995). In contrast, the homozygous type 3 vWD patient has low FVIII:C and undetectable vWF:Ag and platelet vWF activity (Raymond *et al.*, 1990; Brooks *et al.*, 1992, 1996a; Johnstone *et al.*, 1993). Platelet retention in glass bead columns is impaired as is ristocetin- or botrocetin-induced platelet agglutination (Feldman, 1988; Smith and Brinkhous, 1991). The other phenomenon observed in classic vWD is a paradoxical increase in FVIII:C, which does not parallel the behavior of vWF:Ag or bleeding time, following transfusion of normal or hemophilic plasma. Correction of the other components of the factor VIII complex is usually more immediate and transient, whereas the increase in FVIII:C is delayed. The vWF:Ag present in the infused material provides the missing vWF component which somehow causes the patient to release or synthesize FVIII:C at some remote site not yet identified (Sections II.C.2.f).

In addition to the more commonly recognized types 1 and 3 vWD, many variants exist in humans (Ginsburg and Sadler, 1993; Sadler, 1994; Sadler *et al.*, 1995), and variants have been described in a families of German shorthair pointer dogs (Feldman, 1988), quarterhorses (Brooks *et al.*, 1991b) and other dog breeds (Brooks *et al.*, 1991a).

The genetics of vWD are complex and have been studied in humans by several groups (Ginsburg and Sadler, 1993; Ruggeri and Ware, 1993; Girma *et al.*, 1995; Sadler *et al.*, 1995; Zang, 1995). In the nationwide genetic screening program for canine vWD begun by our group in 1980 (Dodds, 1995b), about 30,000 Doberman pinschers, 12,000 Shetland sheepdogs, 9000 golden retrievers, 8000 Scottish terriers, 5000 Pembroke Welsh corgis, 5000 standard and miniature poodles, and several thousand miniature schnauzers, basset hounds, Akitas, and rottweilers have been tested for vWD (Raymond *et al.*, 1990; Brooks *et al.*, 1992). For other breeds, smaller numbers have been examined.

Studies of canine vWD have sustained interest in the physiological and disease states that alter factor VIII metabolism. In dogs, vWD is exacerbated by concurrent hypothyroidism (Dodds, 1988, 1992, 1995a; Avgeris *et al.*, 1990; Dodds *et al.*, 1993) so that asymptomatic carriers of vWD may exhibit a bleeding tendency if they develop autoimmune thyroiditis and become hypothyroid, a common situation found in many breeds but especially prevalent in Doberman pinschers (Dodds *et al.*,1993; Dodds 1995b). The concomitant prevalence of vWD and hypothyroidism in these breeds of dogs suggests a link between the synthesis and/or metabolic regulation of thyroid hormones and vWF (Dodds, 1995a). Furthermore, hypothyroid dogs may exhibit thrombocytopenia and associated mucosal surface bleeding. While this connection has come into question from results of experimental and other studies (Panciera and Johnson, 1994, 1995, 1996), the clinical evidence for a regulatory relationship between vWD and hypothyroidism is solid (Avgeris *et al.*, 1990; Dodds, 1991a, 1995a).

In humans, hemostatic abnormalities may be an early manifestation of hypothyroidism (Dodds, 1991a). These include easy bruising and thrombocytopenia, low platelet retention in glass bead columns, which is normalized after treatment with L-thyroxine, low thrombin-induced platelet serotonin release, decreased levels of FVIII:C, and an acquired form of vWD (Avgeris *et al.*, 1990; Dodds *et al.*, 1993; van Genderen, 1994; Dodds, 1995a). Clinical experience with use of thyroid supplement, which nonspecifically shortens the bleeding time in animals with inherited or acquired vWD and other platelet dysfunctions, has supported the efficacy, safety, and low cost of this approach (Dodds, 1991a, 1995a; Dodds *et al.*, 1993).

Thyroid supplementation alone may suffice to control bleeding in mild to moderate vWD, a situation analogous to the use of desmopressin (DDAVP) or danazol to control bleeding in humans and animals (Dodds, 1989b, 1991a; Giger and Dodds, 1989; Kraus *et al.*, 1989; Meyers *et al.*, 1990a; Mansell and Parry, 1991; Lattuada *et al.*, 1992; Pietraszek *et al.*, 1992; Nichols and Hohenhaus, 1994). DDAVP treatment has recently been shown to improve vWF levels in dogs with vWD, although relatively high doses are required and the response is short lived (Giger and Dodds, 1989; Meyers *et al.*, 1990a; Dodds, 1991a). Because of the important role of vWF in sustaining platelet adhesion, animals that are asymptomatic heterozygotes for vWD (as determined by reduced levels of vWF:Ag, Brooks *et al.*, 1996b) are at risk to express a bleeding tendency if some other hemostatic disorder develops (e.g., rodenticide toxicosis, thrombocytopenia, liver disease, hypothyroidism) (Dodds, 1992; Dodds *et al.*, 1993).

Another area of interest is the relationship of vWF to atherogenesis (Hawiger, 1995; Griendling and Alexander, 1996). Studies of normal and vWD swine showed that affected pigs had significantly fewer vascular lesions and less atherosclerosis than did control swine fed the same diet and raised in the same environment. These observations led to the theory that platelet components of the factor VIII complex are involved in the endothelial and vessel wall proliferation of atherogenesis (Meyers *et al.*, 1990b; Rand *et al.*, 1993; Caen and Rosa, 1995; Hawiger, 1995).

8. Factor X Deficiency

A rare coagulation disorder, factor X deficiency was first described in humans in the mid-1950s in the United Kingdom and the United States. In the early 1970s it was recognized in a family of American cocker spaniels (Dodds, 1989a, 1992). Since then, factor X deficiency has been diagnosed occasionally in mongrel dogs and more recently in the Jack Russell terrier (Cook *et al.*, 1993). Homozygotes and strongly penetrant heterozygotes have very low levels of factor X (<6–35%) and a clinically expressed bleeding disease, whereas weakly penetrant heterozygotes (40–70% factor X) are usually asymptomatic. Severely affected humans have <1–10% factor X, and affected dogs with less than 20% factor X usually do not survive neonatal life. The exception was an affected mongrel puppy with 6% factor X that lived for nearly a year before experiencing a fatal bleeding episode. Complete absence of factor X is thought to be a lethal mutation because of its central role in coagulation.

In addition to reduced factor X levels, clinically affected individuals have mild to moderately prolonged APTT, PT, and RVVT. In heterozygotes, these screening tests may be slightly prolonged. Two types of disease have been recognized in humans, one with undetectable antigen related to factor X (classic form) and another with normal or reduced amounts of mutant protein detectable by immunoassay. In affected newborn pups, as in infants, serious bleeding occurs. This disease mimics the "fading puppy syndrome," in which severely affected pups are stillborn or fade and die in the first week or two of life. Necropsies show massive internal bleeding. Signs in adults are mild and referable to mucosal surfaces (Cook *et al.*, 1993).

9. Factor XI (PTA) Deficiency

Factor XI deficiency is a rare disorder of humans in which more than 90% of cases occurs in person of Jewish background. Clinical signs are mild (hematuria, bruising, epistaxis, menorrhagia) unless the patient is subjected to surgical procedures. Bleeding usually starts 12–24 hours after surgery and can be severe and protracted. Lethal bleeding has also been reported after minor procedures such as biopsies and tonsillectomy. The disorder has also been recognized in Holstein cattle, English springer spaniels, great Pyrenees, and Kerry blue terrier dogs (Dodds, 1989a; Gentry and Ross, 1993; Knowler *et al.*, 1994).

In cattle and dogs, PTA deficiency is similar to that of humans—there is protracted bleeding after surgery. Clinically affected animals are homozygotes and have very low factor XI levels, prolonged APTT, recalcification time, and abnormal PCT. Heterozygotes are asymptomatic and have 25–50% factor XI, which is somewhat lower than levels seen in human heterozygotes (Gentry and Ross, 1993). The defect has not been erradicated from the Kerry blue terrier or great Pyrenees breeds. The affected springer spaniel family is now deceased. Affected cattle are still being maintained and they are a valuable source of large quantities of deficient substrate plasma for measuring factor XI levels in humans and other species (Gentry and Ross, 1993). Factor XI activity is also deficient in several early diverging mammalian and nonmammalian species such as cetaceans and fowl. This is a normal phenomenon, however and the significance of the apparent absence of PTA-like activity is unknown. Many of these species also lack Hageman and Fletcher factor activities (Sections IV,B,10 and IV,B,12,c). A cat with an inhibitor of PTA has been described (Feldman, 1988).

10. Factor XII Deficiency (Hageman Trait)

Hageman trait is an asymptomatic coagulation deficiency recognized in humans, dogs, and cats (Otto *et al.*, 1991; Peterson *et al.*, 1995). The first feline case involved one animal discovered fortuitously by prolonged screening tests of intrinsic clotting. It had less than 5% factor XII and died without progeny. The second case, also discovered by screening tests, had less than 1% factor XII and has provided several generations of progeny. Affected individuals have significantly prolonged APTT, recalcification times, and WBCT in glass tubes and usually do not bleed unless hemostasis is otherwise compromised (Peterson *et al.*, 1995). Hageman trait appears to be quite common in cats.

In addition to these defects, the absence of detectable biological or immunological factor XII is a normal phenomenon of a variety of other vertebrates and invertebrates (Dodds, 1989a) such as whales, birds (including the common domestic fowl and waterfowl), reptiles, and possibly fish. Other uncommon mammalian and nonmammaliam species that contain Hageman activity are seals, echidnas, marsupials, and amphibians (Dodds, 1989a). However, the difficulties of interpreting results of clotting factor tests in nonmammalian plasmas measured with human factor XII-deficient substrate should be considered.

11. Platelet Function Defects

a. Thrombasthenia (Glanzmann's Disease)

The term *thrombasthenia* literally means "weak thrombocytes" and was coined by Glanzmann to describe an autosomal recessive bleeding disorder in which there was no clot retraction. There may be low normal platelet counts or mild thrombocytopenia but platelet morphology is essentially normal. Affected patients have long bleeding times, defective platelet aggregation with standard aggregating agents, especially ADP, and abnormal platelet retention. In most cases platelet vWF is normal. Platelet fibrinogen levels, PF-3 availability, and platelet membrane glycoproteins IIb and IIIa are usually significantly reduced. The latter abnormalities are thought to be responsible for the defective aggregation and clot retraction (Catalfamo and Dodds, 1988). The bleeding diathesis is severe and of the purpuric type and epistaxis is common and profuse. In severely affected humans, low-level estrogen treatment has reduced and controlled bleeding.

In 1967, a family of otterhounds was described with an inherited platelet function defect similar to human Glanzmann's disease except for the presence of some bizarre giant platelets (Catalfamo and Dodds, 1988). The morphological abnormality in these giant platelets resembled that of human Bernard Soulier syndrome but the functional disorder was indistinguishable from thrombasthenia, including the typical platelet membrane glycoprotein abnormalities (Nurden, 1995). The disease was therefore termed thrombasthenic–thrombopathia and was inherited as an autosomal dominant trait with variable expression. Both homozygotes and heterozygotes could be identified by specific platelet function tests.

A variant form of this original otterhound platelet defect has surfaced again in several animals related distantly to the original affected family. The present defect may be either a mutated or self-selected variant of the initial problem. Eight affected dogs have been studied, and six were found to have platelet abnormalities detectable by traditional platelet function tests but the other two did not. The new expression of the platelet disorder is characterized by bleeding episodes that are triggered by stress events.

b. Thrombopathia

A group of inherited thrombopathias has been described in humans, many of which have unique characteristics. These include primary or function thrombopathy, Bernard–Soulier syndrome, storage-pool disease or deficit thrombopathy and other platelet release defects. The specific features of each of these disorders are described elsewhere (Catalfamo and Dodds, 1998; Patterson *et al.*, 1989; Boudreaux *et al.*, 1994; Bray, 1994; Callan *et al.*, 1995).

Animal thrombopathias have been recognized in fawn-hooded rats, basset hounds, spitz dogs, American cocker spaniels, Simmental cattle, and cats (Catalfamo and Dodds, 1988; Patterson *et al.*, 1989). The defect in fawn-hooded rats and American cocker spaniels is similar to that of human platelet storage-pool disease (Catalfamo and Dodds, 1988; Callan *et al.*, 1995). Affected rats have reduced platelet serotonin and release. Interesting concomitant findings are a focal glomerulonephritis and proteinuria associated with aging and docile behavior, thought to be related to reduced content of brain serotonin. Hypertension is also a feature of their defect. These rats are thus useful models for hemostatic, immunological, pathological, and behavioral studies. The disorder in American cocker spaniels has been attributed to a deficient platelet δ-granule storage pool of ADP (Callan *et al.*, 1995) and is similar to the platelet defect seen in Chediak–Higashi disease of mice and cats (Pratt *et al.*, 1991; Cowles *et al.*, 1992).

Canine thrombopathia (CTP) is an autosomally inherited bleeding disorder of the basset hound (Catalfamo and Dodds, 1988). The recently described platelet defect of spitz dogs is similar (Boudreaux *et al.*, 1994). These conditions produce a bleeding tendency primarily of mucosal surfaces and intrinsic to the platelet. Thrombopathic platelets fail to respond normally to very high concentrations of most physiologic agents that activate platelets. Of these, only thrombin causes CTP platelets to aggregate at a measurable though reduced rate. The preservation of this key platelet activation pathway probably explains why affected dogs can survive.

Despite the fact that CTP platelets have normal amounts of fibrinogen and surface receptor proteins, they fail to interact normally with fibrinogen (Catalfamo and Dodds, 1988). The platelet dysfunction observed in CTP is associated with abnormally high levels of cyclic adenosine monophosphate (cAMP), which couples the signal generated by a stimulatory agent to platelet activation and ultimately hemostatic plug formation. Metabolism of cAMP and utilization of mobilized ionized calcium are abnormal in CTP platelets (Boudreaux *et al.*, 1994).

In 1990, Searcy and Petrie described a bleeding disorder of eight Simmental cattle attributed to a platelet dysfunction. Subsequent studies by this group (Searcy *et al.*, 1990, 1994) characterized the defect as having abnormal platelet cytoskeletal assembly and reduced platelet aggregation in response to ADP. A similar clinical disorder was described in a newborn Simmental calf (Navarre *et al.*, 1995) and a different explanation for the bleeding diathesis of this cattle breed was reported by Sullivan *et al.* (1994). In this latter case, the bleeding diathesis was attributed to a variant vWF molecule of truncated structure, which apparently contributed to the reduced platelet adhesion and aggregation in response to ADP observed in earlier studies.

12. Other Defects

a. Complement Deficiencies

A variety of inherited complement deficiencies are recognized in humans. In contrast, there are only three known complement deficiencies of other species. These are C4 deficiency in guinea pigs and rats, C5 deficiency in mice, and C6 deficiency in hamsters and rabbits (Dodds, 1989a). Of these, the hemostatic mechanism has been evaluated only in the C4-deficient guinea pig and C6-deficient rabbit. Intrinsic coagulation was found to be reduced in both models relative to species-specific controls. These animals have proved to be useful models for pathophysiological studies of hemostasis and thrombosis and the interrelationships between the complement and hemostatic mechanisms (Esmon *et al.*, 1991).

b. Double Hemophilia (Hemophilia AB)

A planned cross-breeding study between dogs with hemophilia A and hemophilia B was undertaken to study possible linkage between these two X chromosome genes. Dogs with double hemophilia were readily produced, thus demonstrating that the loci for hemophilia A and B are not linked and are located far apart on the X chromosome (Dodds, 1989a).

c. Prekallikrein Deficiency (Fletcher Trait)

Prekallikrein activity has been recognized in apes, swine, guinea pigs, mice, goats, sheep, and horses but is absent or minimal in dogs, cats, cattle, rabbits, whales, ducks, and chickens (Veloso *et al.*, 1992). These results must be interpreted with caution, because human Fletcher factor deficient plasma was used to make these measurements. For example, rabbit plasma without prekallikrein activity actually contains a prekallikrein-like protein. Fletcher trait has recently been described in dogs and in miniature and Belgian horses (Geor *et al.*, 1990; Otto *et al.*, 1991).

d. Coagulopathies Not Recognized in Animals

Other coagulopathies of humans include factor V and XIII (FSF) deficiencies, both rare autosomal traits. The factor V deficiencies have a moderate to severe bleeding diathesis, prolonged APTT, PT, WBCT, short serum PCT, and very low factor V activity. Factor XIII-deficient patients have poor wound healing, bleed at delivery, bruise easily, and bleed after minor surgery, especially dental extractions. FSF-deficient clots dissolve within 24 hours when placed in 5 *M* urea or 1% monochloroacetic acid. A factor XIIIa-like enzyme from equine plasma has recently been evaluated for therapeutic potential (Coyne *et al.*, 1992).

Another defect not recognized in animals is HMWK (Williams Fitzgerald–Flaujeac trait) deficiency. Several nonmammalian species, however, normally lack this activity. HMWK activity is detectable in primate, dog, and whale plasmas, is very low in rabbit and cattle plasmas, and is apparently absent in fowl, reptilian, and amphibian plasmas using conventional assays.

C. Acquired Disorders

Acquired bleeding disorders are numerous and more common than the inherited deficiencies of clotting factors and platelets (Table 10.7). The major causes are briefly discussed in the following subsections, and additional information can be found elsewhere (Johnstone *et al.*, 1991; Dodds, 1992; McClay *et al.*, 1992; Welch *et al.*, 1992; Prasse *et al.*, 1993; Welles *et al.*, 1993a, 1993b, 1994; Carr and Johnson, 1994; Joist *et al.*, 1994; Mannucci and Giangrande, 1994; Goodnight, 1995; Verstraete and Zoldhelyi, 1995).

1. Platelet Function Defects

a. Quantitative (Thrombocytopenias and Thrombocytosis)

Quantitative platelet defects are listed in Table 10.8. Of these, immune-mediated thrombocytopenia accounts for the majority of chronic cases (Grindem *et al.*,

TABLE 10.8 Quantitative Platelet Function Defects[a]

Thrombocytopenia
- Increased platelet destruction, utilization, or sequestration
  - Immunological: primary (ITP), secondary to incompatible blood transfusions, autoimmune diseases (SLE, AIHA, RA), lymphoproliferative disorders, allergies, drugs, toxic agents, live virus vaccines
  - Nonimmunological: hemolytic disease of newborn, severe infections, malignancy, drugs (heparin), DIC, hemolytic uremic syndrome, splenomegaly, hypothermia
- Decrease platelet production
  - Congenital or hereditary: neonatal virus disease, X-chromosome-linked trait (Wiskott–Aldrich syndrome), autosomal traits (various diseases such as Bernard–Soulier syndrome)
  - Acquired: marrow depression (drugs, anemias), marrow infiltration, infections especially viral, drugs (diuretics, estrogens), cyclic thrombocytopenia, paroxysmal nocturnal hemoglobinuria

Thrombocytosis
- Increased platelet production
  - Myeloproliferative disorders: polycythemia vera, chronic leukemia, thrombocythemia
  - Chronic inflammatory disorders: autoimmune diseases (RA), cirrhosis, granulomatosis, tuberculosis, sarcoidosis, chronic pneumonitis, osteomyelitis
  - Malignancy: lymphomas, carcinomas
  - Miscellaneous: acute hemorrhage, iron deficiency, postoperative rebound, osteoporosis
- Release from tissue stores (spleen, lung)
  - Response to exercise
  - Drugs, adrenalin, vincristine

[a] AIHA, autoimmune hemolytic anemia; DIC, disseminated intravascular coagulation; ITP, idiopathic thrombocytopenia purpura; RA, rheumatoid arthritis; SLE, systemic lupus erythematosus.

1991; Northern and Tvedten, 1992). The immunological basis has been examined in humans and in the dog, cat, and horse (Grindem *et al.*, 1991; Jordan *et al.*, 1993). Primary immunological thrombocytopenia, of unknown etiology, has been termed idiopathic thrombocytopenic purpura (Kristensen *et al.*, 1994a; Sullivan, 1994; Kelton, 1995; Lewis *et al.*, 1995b). The majority of cases, however, appear secondary to a variety of underlying conditions such as use of heparin or estrogens, thrombosis, neoplasia, viral diseases, vaccine-associated reactions, or other drugs and chemicals (Grindem *et al.*, 1994; Hart and Nolte, 1994; Kristensen *et al.*, 1994b; Greinacher, 1995; McCann *et al.*, 1995; Moake, 1995; Sullivan *et al.*, 1995).

Nonimmunological thrombocytopenias are less common and have a better prognosis than the immunological cases, if the causative agent for disease can be eliminated. Several diagnostic tests are available for detecting circulating humoral or cell-mediated antibodies directed against autologous or homologous platelets. Methods include a variety of platelet release tests ([^{14}C] serotonin, PF-3 release), platelet migration inhibition, radioimmunoassays, complement fixation, platelet-bound immunoglobulin, latex particle assay, hemagglutination inhibition assays, and flow cytometry (Ramos *et al.*, 1992; Kristensen *et al.*, 1994b; Lewis and Meyers, 1994; Kelton, 1995; Lewis *et al.*, 1995a, 1995b; Sullivan *et al.*, 1995).

b. Qualitative

The major causes of qualitative platelet function defects are listed in Table 10.9. In addition to hereditary defects, a group of diseases and a large number of drugs are now known to produce thrombopathias (Herman *et al.*, 1991; Bright *et al.*, 1994; Coller *et al.*, 1995). Most drugs act by inhibiting the adhesion of platelets to subendothelium (aspirin, which blocks platelet cyclic endoperoxides) and/or the platelet release reaction (phenylbutazone, sulfonamides, antiinflammatory drugs, ticlopidine, promazine tranquilizers) (Boudreaux *et al.*, 1991a, 1991b; Barr *et al.*, 1992; Gentry *et al.*, 1992; Grauer *et al.*, 1992; Semprad and Dubielzig, 1993; Grindem *et al.*, 1994; Caen and Rosa, 1995; Goodnight, 1995; Behrend *et al.*, 1996). Drugs that interfere with platelet function are contraindicated or must be used with caution in moderate or severely affected individuals (Section IV.A.3). Similarly, elective surgery should be avoided during the viremic phase (3–10 days) after live virus vaccination or viral exposure.

The most common diseases manifesting a bleeding tendency attributable to platelet dysfunction are renal failure with uremia (Forsythe *et al.*, 1989; Wardrop *et al.*, 1989; Pietraszek *et al.*, 1992; Escolar *et al.*, 1993; Joist *et al.*, 1994) and liver disease (Willis *et al.*, 1989). Less common causes are the dysproteinemias such as myelomas and macroglobulinemias and estrogen toxicity. The classic clinical case of uremic bleeding is that of

TABLE 10.9 Qualitative Platelet Function Defects

Hereditary
- Thrombasthenia (Glanzmann's disease)
- Thrombopathia: Bernard–Soulier syndrome, storage-pool disease, albinism, Ehlers–Danlos syndrome, Wiskott–Aldrich syndrome, osteogenesis imperfecta, others

Acquired
- Uremia, myeloproliferative disorders, macroglobulinemia, liver disease, fibrinolysis, thrombocytopenia, systemic lupus erythematosus, congenital heart disease, anemias, leukemias
- Drugs: aspirin, phenylbutazone, promazine tranquilizers, estrogens, plasma expanders, nitrofurans, sulfonamides, antiinflammatory drugs, local anesthetics, phenothiazines, live virus vaccines

the old dog with compensated chronic interstitial nephritis, inflamed gums, and chronic periodontal disease. Dentistry on such a patient frequently results in excessive and prolonged gingival bleeding.

2. *Disseminated Intravascular Coagulation with Fibrinolysis*

The combined syndrome of DIC and secondary fibrinolysis is an important acute, subacute, or chronic disease process of humans and other animals. Several reviews discuss the pathophysiology of this syndrome (Comp, 1990a; Joist, 1990; Vermeer and Hamulyák, 1991; Andrew, 1995; Hirsh *et al.*, 1995; Verstraete, 1995; Verstraete and Zoldhelyi, 1995). The major causes of DIC and thrombosis in humans and animals include viral, bacterial, protozoal, and parasitic infections; neoplasia; obstetric complications; and miscellaneous conditions such as trauma, shock, laminitis, heat stroke, colic, burns, drowning, liver disease, and canine heartworm disease (Gerhards and Eberhardt, 1988; Boudreaux *et al.*, 1989a, 1989b, 1991a, 1991b,; Prasse *et al.*, 1990, 1993; Boudreaux and Dillon, 1991; Al-Mondhiry *et al.*, 1992; Welch *et al.*, 1992; Edens *et al.*, 1993; Kitoh *et al.*, 1994; Baty and Harpster, 1995; Laste and Harpster, 1995).

a. Infections

A number of spontaneous and induced viral infections cause DIC in animals and serve as models for research. Among these are hog cholera, African swine fever, infectious canine hepatitis, feline infectious peritonitis, fowl plague, epizootic hemorrhagic disease of deer and equine viral arteritis (Boudreaux *et al.*, 1989b). Bacterial infections include leptospirosis in cattle and guinea pigs, gram-negative sepsis and the endotoxin-mediated Schwartzman reaction of rabbits, horses, rats, guinea pigs, and dogs. Malaria in monkeys and trypanosomiasis and babesiosis of dogs and cattle also produce DIC. Some dogs with the pulmonary or vena caval lesions of complicated or treated heartworm disease and occult heartworm disease succumb to DIC (Boudreaux *et al.*, 1989a; Kitoh *et al.*, 1994).

b. Neoplasia

Neoplasia is a common cause of DIC in animals, perhaps because many neoplasms are not diagnosed early or the owner elects not to treat the animal (Feldman *et al.*, 1988; Costantini and Zacharski, 1993). Naturally occurring adenocarcinomas of the mammary gland, testicles, and thyroid gland, and carcinomas of the lymphatic and hematopoietic systems, liver, and spleen have been the common tumors involved (Hammer *et al.*, 1991; Hargis and Feldman, 1991).

c. Obstetric Complications

A life-threatening complication of pregnancy in women, DIC is also associated with obstetric problems in animals. These include dystocias, eclampsia, and retained fetus(es). Other causes are amniotic fluid embolism and abruptio placentae.

d. Heat Stroke

Spontaneous heat stroke in dogs and pigs has been associated with DIC and is often fatal.

e. Atrial Thrombosis

A high incidence of atrial thrombosis and concomitant DIC has been reported to occur with aging in Syrian hamsters (Dodds, 1989a). The diagnosis of DIC is complicated by the various stages of the syndrome, which include initial activation of hemostatic elements (hypercoagulability) followed by thrombosis and consumption, which depletes hemostatic elements. This results in the "consumption coagulopathy" phase of the syndrome with secondary enhancement of fibrinolysis and bleeding.

The pathophysiological process of DIC is dynamic and proceeds, without treatment, to death of the host or becomes reversible once healing and repair have restored homeostatic balance. The selection of the most appropriate management and treatment is often a crucial decision. Procoagulants given at the wrong stage may promote further thrombosis, whereas antiplatelet and fibrinolytic drugs or anticoagulants may facilitate bleeding (Gerhards and Eberhardt, 1988; Boudreaux *et al.*, 1991a, 1991b; Semrad and Dubielzig, 1993; Coller *et al.*, 1995; Fareed and Callas, 1995; Goodnight, 1995; Hirsh *et al.*, 1995; Laste and Harpster, 1995; Lijnen and Collen, 1995; McCann *et al.*, 1995; Monreal *et al.*, 1995; Verstraete and Zoldhelyi, 1995). In many cases, removal of the primary cause is the only way to manage the patient effectively.

Diagnostic tests for ongoing DIC have been described (Section III.C.1.e). Classically, the consumption phase of DIC shows thrombocytopenia, prolonged APTT, PT, and TT, reduced factor V and VIII : C activities, and enhanced fibrinolysis with elevated FDP. There may also be reduced antithrombin activity or protein C deficiency (Boudreaux *et al.*, 1989b; Boudreaux and Dillon, 1991; Edens *et al.*, 1993).

3. *Liver Disease*

Because the liver is the primary site of clotting factor synthesis, acute or chronic generalized hepatic disease often results in a bleeding tendency (Willis *et al.*, 1989). Hepatitis may also produce DIC and platelet dysfunction. Usually, the most significant findings in liver disease are a prolonged PT with low factor VII early in the disease, followed by reduction of other prothrombin complex clotting factors as the disease progresses and a variable prolongation of APTT and TT.

4. *Vitamin K Deficiency and Rodenticide Toxicity*

The vitamin K-dependent clotting factors (II, VII, IX, X) are reduced in rodenticide toxicity, malabsorption syndromes, and in sterilization of the gut by prolonged use of antibiotics (Mount and Kass, 1989; Maddison *et al.*, 1990; Vermeer and Hamulyák, 1991; Woody *et al.*, 1992; Neer and Savant, 1992; Soute *et al.*, 1992; Brooks, 1996). Diagnosis is confirmed by identifying the cause and finding a moderate to markedly prolonged PT, variable APTT, normal TT, and low normal platelet count or mild to moderate thrombocytopenia. A therapeutic response to whole blood or plasma transfusion and/or vitamin K_1 is diagnostic of vitamin K antagonism or depletion. The newer second-generation rodenticides are 20–50 times more potent and longer lasting (several weeks) than first-generation warfarin compounds. Poisoning with these new toxicants is thus more serious and expensive to treat (Mount and Kass, 1989; Dodds, 1992; Brooks, 1996).

5. *Other Causes*

Monoclonal gammopathies produce a nonspecific bleeding tendency characterized by such problems as epistaxis, oozing from surface abrasions, and mucosal surface bleeding (Ruiz-Gopegui *et al.*, 1994). *Amyloidosis* is also associated with an acquired coagulopathy but this has been restricted in most cases to isolated factor X deficiency although some patients also have factor IX deficiency or platelet defects. *Lupus anticoagulants* are antiphospholipid-protein-based antibodies that appear spontaneously in some patients with systemic lupus erythematosus (Triplett, 1995). The methodology used for detecting these antibodies has been standardized (Brandt *et al.*, 1995; de Groot and Derksen, 1995).

Acknowledgments

The author thanks Ms. Lynette Palma for her excellent secretarial assistance in preparing this chapter.

References

Aiach, M., Gandrille, S., and Emmerich, J. (1995). *Thromb. Haemost.* **74,** 81–89.

Al-Mondhiry, H., Pae, W. E., Miller, C. A., and Pierce, W. S. (1992). *Thromb. Haemost.* **67,** 413–416.

Andreasen, C. B., and Lovelady, S. (1993). *Vet. Clin. Pathol.* **22,** 42–44.

Andrew, M. (1995). *Thromb. Haemost.* **74,** 415–425.

Andrew, M., Brooker, L., Leaker, M., Paes, B., and Weitz, J. (1992). *Thromb. Haemost.* **68,** 325–330.

Antonarakis, S. (1995). *Thromb. Haemost.* **74,** 322–328.

Arai, M., Scandella, D., and Hoyer, L. W. (1989). *J. Clin. Invest.* **83,** 1978–1984.

Aranda, E., Pizarro, M., Pereira, J., and Mezzano, D. (1994). *Thromb. Haemost.* **71,** 488–492.

Aronson, D. L. (1990). *Transfusion* **30,** 748–758.

Avgeris, S., Lothrop, C. D. Jr., and McDonald, T. P. (1990). *J. Am. Vet. Med. Assoc.* **196,** 921–924.

Axelrod, J. H., Read, M. S., Brinkhous, K. M., and Verma, I. M. (1990). *Proc. Natl. Acad. Sci.* **87,** 5173–5177.

Barr, S. C., Ludders, J. W., Looney, A. L., Gleed, R. D., and Erb, H. N. (1992). *Am. J. Vet. Res.* **53,** 2067–2070.

Bath, P. M. W. (1993). *Thromb. Haemost.* **70,** 687–690.

Baty, C. J., and Harpster, N. K. (1995). *In* "Current Veterinary Therapy XII" (J. Bonagura, ed.), pp. 868–873. Saunders, Philadelphia, Pennsylvania.

Bauer, K. A., Mannucci, P. M., Gringeri, A., Tradati, F., Barzegar, S., Kass, B. L.,Ten Cate, H., Kestin, A. S., Brettler, D. B., and Rosenberg, R. D. (1992). *Blood* **79,** 2039–2047.

Behrend, E. N., Grauer, G. F., Greco, D. S., Rose, B. J., and Thrall, M. A. (1996). *J. Am. Anim. Hosp. Assoc.* **32,** 11–18.

Benson, R. E., Catalfamo, J. L., Brooks, M., and Dodds, W. J. (1991). *J. Immunoassay* **12,** 371–390.

Benson, R. E., Catalfamo, J. L., and Dodds, W. J. (1992). *J. Lab. Clin. Med.* **119,** 420–427.

Bi, L., Lawler, A. M., Antonarakis, S. E., High, K. A., Gearhart, J. D., and Kazazian, H. H. (1995). *Nature Genet.* **10,** 119–121.

Blann, A. D. (1990). *Biologicals* **18,** 351–353.

Blombäck, M., Eneroth, P., Landgren, B. M., Lagerström, M., and Anderson, O. (1992). *Thromb. Heamost.* **67,** 70–75.

Bloom, A. L. (1991). *Thromb. Haemost.* **66,** 166–177.

Bornstein, P. (1992). *FASEB J.* **6,** 3290–3299.

Boudreaux, M. K., and Dillon, A. R. (1988). *Vet. Clin. Pathol.* **17,** 84–85.

Boudreaux, M. K., and Dillon, A. R. (1991). *Am. J. Vet. Res.* **52,** 1986–1990.

Boudreaux, M. K., Dillon, A. R., and Spano, J. S. (1989a). *Am. J. Vet. Res.* **50,** 1544–1547.

Boudreaux, M. K., Weiss, R. C., Cox, N., and Spano, J. S. (1989b). *Am. J. Vet. Res.* **50,** 1910–1913.

Boudreaux, M. K., Dillon, A. R., Sartin, E. A., Ravis, W. R., and Spano, J. S. (1991a). *Am. J. Vet. Res.* **52,** 1992–1999.

Boudreaux, M. K., Dillon, A. R., Sartin, E. A., Ravis, W. R., and Spano, J. S. (1991b). *Am. J. Vet. Res.* **52,** 2000–2006.

Boudreaux, M. K., Crager, C., Dillon, A. R., Stanz, K., Toivio-Kinnucan, M. (1994). *J. Vet. Intern. Med.* **8,** 93–98.

Brandt, J. T., Triplett, D. A., Alving, B., and Scharrer, I. (1995). *Thromb. Haemost.* **74,** 1185–1190.

Brassard, J. A., and Meyers, K. M.(1991). *Thromb. Res.* **65,** 191–195.

Bray, P. F. (1994). *Thromb. Haemost.* **72,** 492–502.

Breider, M. A. (1993). *J. Am. Vet. Med. Assoc.* **203,** 300–306.

Bright, J. M., Sullivan, P. S., Melton, S. L., Schneider, J. F., and McDonald, T. P. (1994). *J. Vet. Intern. Med.* **8,** 247–252.

Brinkhous, K. M. (1992). *FASEB J.* **6,** 925–926.

Brooks, M. (1992). *Prob. Vet. Med.* **4,** 636–646.

Brooks, M. (1996). *J. Am. Vet. Med. Assoc.* **208,** 354–355.

Brooks, M., and Catalfamo, J. L. (1993). *Thromb. Haemost.* **70,** 777–780.

Brooks, M. B., and Dodds, W. J. (1989). *J. Am. Anim. Hosp. Assoc.* **25,** 153–155.

Brooks, M., Catalfamo, J. L., and Dodds, W. J. (1991a). *Blood* **78,** suppl. 1, 68a.

Brooks, M., Leith, G. S., Allen, A. K., Woods, P. R., Benson, R. E., and Dodds, W. J. (1991b). *J. Am. Vet. Med. Assoc.* **198,** 114–116.

Brooks, M., Dodds, W. J., and Raymond, S. L. (1992). *J. Am. Vet. Med. Assoc.* **200,** 1123–1127.

Brooks, M. B., Raymond, S. L., and Catalfamo, J. L. (1996a). *J. Am. Vet. Med. Assoc.* **209,** 926–929.

Brooks, M. B, Raymond, S. L., and Catalfamo, J. L. (1996b). *J. Am. Vet. Med. Assoc.* **209,** 930–933.

Broze, G. J. Jr. (1995). *Thromb. Haemost.* **74,** 90–93.

Bussel, J. B. (1995). *Thromb. Haemost.* **74,** 426–428.

Caen, J. P., and Rosa, J. P. (1995). *Thromb. Haemost.* **74,** 18–24.

Callan, M. B., Bennett, J. S., Phillips, D. K., Haskins, M. E., Hayden, J. E., Anderson, J. G., and Giger, U. (1995). *Thromb. Haemost.* **74,** 949–953.

Calvete, J. J. (1994). *Thromb. Haemost.* **72,** 1–15.

Carr, A. P., and Johnson, G. S. (1994). *J. Am. Anim. Hosp. Assoc.* **30,** 475–482.

Catalfamo, J. L., and Dodds, W. J. (1988). *Hemostasis* **18,** 185–193.

Catalfamo, J. L., Raymond, S. L., Taiman, S., and Dodds, W. J. (1989). Unpublished observations.

Catalfamo, J. L., Benson, R. E., Brooks, M., and Dodds, W. J. (1991). *Blood* **78,** suppl. 1, 487a.

Cerletti, C., Evangelista, V., Molino, M., and de Gaetano, G. (1995). *Thromb. Haemost.* **74,** 218–223.

Chabbat, J., Hampikian-Lenin, S., Toully, V., Gaillandre, A., Péjaudier, L., and Steinbuch, M. (1989). *Thromb. Res.* **54,** 603–612.

Ching, Y. N. L. H., Meyers, K. L., Brassard, J. A., and Wardrop, K. J. (1994). *Am. J. Vet. Res.* **55,** 102–110.

Clemetson, K. J. (1995). *Thromb. Haemost.* **74,** 111–116.

Coller, B. S., Anderson, K., and Weisman, H. F. (1995). *Thromb. Haemost.* **74,** 302–308.

Comp, P. C. (1990a). *Semin. Thromb. Hemost.* **16,** 158–161.

Comp, P. C. (1990b). *Semin. Thromb. Hemost.* **16,** 177–181.

Conlan, M. G., Folsom, A. R., Finch, A., Davis, C. E., Sorlie, P., Marcucci, G., and Wu, K. K. (1993). *Thromb. Haemost.* **70,** 380–385.

Conlan, M. G., Folsom, A. R., Finch, A., Davis, C. E., Marcucci, G., Sorlie, P., and Wu, K. K. (1994). *Thromb. Haemost.* **72,** 551–556.

Cook, A. K., Werner, L. L., O'Neill, S. L., Brooks, M., and Feldman, B. F. (1993). *Vet. Clin. Pathol.* **22,** 68–71.

Costantini, V., and Zacharski, L. R. (1993). *Thromb. Haemost.* **69,** 406–414.

Cotter, S. M. (ed.) (1991). *Adv. Vet. Sci. Comp. Med.* **36,** 1–343.

Cowles, B. E., Meyers, K. M., Wardrop, K. J., Menard, M. and Sylvester, D. (1992). *Thromb. Haemost.* **67,** 708–712.

Coyne, C. P., Smith, J. E., and DeBowes, R. M. (1992). *Am. J. Vet. Res.* **53,** 695–705.

Crossley, M., Ludwig, M., Stowell, M., De Vos, P., Olek, K., and Brownlee, G. G. (1992). *Science* **257,** 377–379.

Davie, E. W. (1995). *Thromb. Haemost.* **74,** 1–6.

de Groot, P. G., and Derksen, R. H. W. M. (1995). *Vessels* **1**(2), 22–26.

Deng, G., Royle, G., Seiffert, D., and Loskutoff, D. J. (1995). *Thromb. Haemost.* **74,** 66–70.

Dodds, W. J. (1988). *J. Am. Vet. Med. Assoc.* **193,** 1157–1160.

Dodds, W. J. (1989a). *In* "Clinical Biochemistry of Domestic Animals" (J. J. Kaneko, ed.), 4th ed., pp. 274–315. Academic Press, San Diego.

Dodds, W. J. (1989b). *Adv. Sm. Anim. Med. Surg.* **1**(11), 1–3.

Dodds, W. J. (1991a). *Adv. Vet. Sci. Comp. Med.* **36,** 257–290.

Dodds, W. J. (1991b). *In* "Consultations in Feline Internal Medicine (J.R. August, ed.), pp. 383–388. Saunders, Philadelphia, Pennsylvania.

Dodds, W. J. (1992). *In* "Handbook of Small Animal Practice" (R. V. Morgan, ed.), 2nd ed., pp. 765–777. Churchill Livingstone, New York.

Dodds, W. J.(1993). *Vet. Pract. STAFF* **5**(2), 1,4–7.

Dodds, W. J. (1995a). *J. Am. Vet. Med. Assoc.* **206,** 594–595.

Dodds, W. J. (1995b). *Adv. Vet. Sci. Comp. Med.* **39,** 29–96.

Dodds, W. J., Raymond, S. L., and Brooks, M. B. (1993). *Vet. Pract. STAF* **5**(4 and 5), 14–17, 21–23.

Dowd, P., Hershline, R., Ham, S. W., and Naganathan, S. (1995). *Science* **269,** 1684–1691.

Dwarki, V., Belloni, P., Nijjar, T., Smith, J., Couto, L., Rabier, M., Clift, S., Berns, A., and Cohen, L. K. (1995). *Proc. Natl. Acad. Sci. USA* **92,** 1023–1027.

Edens, L. M., Morris, D. D., Prasse, K. W., and Anver, M. R. (1993). *J. Vet. Intern. Med.* **7,** 190–193.

Ermens, A. A. M., de Wild, P. J., Vader, H. L., and van der Graaf, F. (1995). *Clin. Chem.* **41,** 510–514.

Escolar, G., Dıaz-Ricart, M., Cases, A., Castillo, R., Ordinas, A., and White, J. G. (1993). *Am. J. Pathol.* **143,** 823–831.

Esmon, C. T., Taylor, F. B. Jr., and Snow, T. R. (1991). *Thromb. Haemost.* **66,** 160–165.

Evans, J. P., Watzke, H. H., Ware, J. L., Stafford, D. W., and High K. A., (1989a). *Blood* **74,** 207–212.

Evans, J. P., Brinkhous, K. M., Brayer, G. D., Reisner, H. M., and High, K. A. (1989b). *Proc. Natl. Acad. Sci. USA* **86,** 1445–1449.

Exner, T. (1995). *Thromb. Haemost.* **74,** 338–344.

Fallaux, F. J., Hoeben, R. C., and Briët, E. (1995). *Thromb. Haemost.* **74,** 263–273.

Fareed, J., and Callas, D. D. (1995). *Vessels* **1**(4), 15–24.

Fareed, J., Bick, R. L., Hoppensteadt, D. A., and Bermes, E. (1995). *Clin. Appl. Thromb. Hemost.* **1,** 87–102.

Feldman, B. F., ed. (1988). *Vet. Clin. North Am. Small Anim. Pract.* **18,** 1–282.

Feldman, B. F., Brummerstedt, E., and Larson, L. S. (1988). *Am. J. Vet. Res.* **49,** 1017–1019.

Feldman, D. G., Brooks M. B., and Dodds, W. J. (1995). *J. Am. Vet. Med. Assoc.* **206,** 1901–1905.

Fenton, J. W., II. (1995). *Thromb. Haemost.* **74,** 493–498.

Forsythe, L. T., and Willis, S. E. (1989). *Can. Vet. J.* **30,** 344–345.

Forsythe, L. T., Jackson, M. L., and Meric, S. M. (1989). *Am. J. Vet. Res.* **50,** 1754–1757.

Furie, B., and Furie, B. C. (1995a). *Thromb. Haemost.* **74,** 224–227.

Furie, B. C., and Furie, B. (1995b). *Thromb. Haemost.* **74,** 274–277.

Gawaz, M., Reininger, A. J., and Neumann, F. J. (1994). *Thromb. Haemost.* **72,** 912–918.

Gentry, P. A., and Ross, M. L. (1993). *Can. J. Vet. Res.* **57,** 242–247.

Gentry, P. A., Mansell, P. D., Mason, D. J., and Conlon, P. D. (1992). *Can. J. Vet. Res.* **56,** 365–369.

Geor, R. J., Jackson, M. L., Lewis, K. M., and Fretz, P. B. (1990). *J. Am. Vet. Med. Assoc.* **197,** 741–745.

Gerding, P. A., Jr., Hamor, R. E., Ramsey, D. T., Vasaune, S., and Schaeffer, D. J. (1994). *Am. J. Vet. Res.* **55,** 1368–1370.

Gerhards, H., and Eberhardt, C. (1988). *Am. J. Vet. Res.* **49,** 13–18.

Gewitz, A. M. (1995). *Thromb. Haemost.* **74,** 204–209.

Giger, U. (1994). *Proc. 12th ACVIM Forum,* pp. 155–157.

Giger, U., and Dodds, W. J. (1989). *Vet. Clin. Pathol.* **18,** 39–42.

Ginsberg, M. H., Du, X., O'Toole, T. E., and Loftus, J. C. (1995). Platelet integrins. *Thromb. Haemost.* **74,** 352–359.

Ginsburg, D., and Sadler, J. E. (1993). *Thromb. Haemost.* **69,** 177–184.

Girma, J. P., Ribba, A. S., and Meyer, D. (1995). *Thromb. Haemost.* **74,** 156–160.

Goodnight, S. H.(1995). *Thromb. Haemost.* **74,** 401–405.

Grabowski, E. F., and Lam, F. P. (1995). *Thromb. Haemost.* **74,** 123–128.

Grauer, G. F., Rose, B. J., Toolan, L., and Thrall, M. A. H., and Colgan, S. P. (1992). *Am. J. Vet. Res.* **53,** 1631–1635.

Greinacher, A. (1995).*Vessels* **1**(2), 17–20.

Griendling, K. K., and Alexander, R. W. (1996). *FASEB J.* **10,** 283–292.

Grindem, C. B., Breitschwerdt, E. B., Corbett, W. T., and Jans, H. E. (1991). *Vet. Clin. Pathol.* **20,** 38–43.

Grindem, C. B., Breitschwerdt, E. B., Corbett, W. T., Page, R. L., and Jans, H. E. (1994). *J. Vet. Intern. Med.* **8,** 400–405.

Guillin, M. C., Bezeaud, A., Bouton, M. C., and Jandrot-Perrus, M. (1995). *Thromb. Haemost.* **74,** 129–133.

Hajjar, K. A. (1995). *Thromb. Haemost.* **74,** 294–301.

Hammer, A. S., Couto, C. G., Swardson, C., and Getzy, D. (1991). *J. Vet. Intern. Med.* **5,** 11–14.

Hanna, W. T., Bona, R. D., Zimmerman, C. E., Carta, C. A., Hebert, G. Z., and Rickles, F. R. (1994). *Thromb. Haemost.* **72,** 173–179.

Hargis, A. M., and Feldman, B. F. (1991). *J. Am. Vet. Med. Assoc.* **5,** 891–894.

Hart, S. W., and Nolte, I. (1994). *J. Vet. Intern. Med.* **8,** 355–362.

Hathaway, W., and Corrigan, J. (1991). *Thromb. Haemost.* **65,** 323–325.

Hawiger, J. (1995). *Thromb. Haemost.* **74,** 369–372.

Hemker, H. C., and Béguin, S. (1995). *Thromb. Haemost.* **74,** 134–138.

Henkens, C. M. A., Bom, V. J. J., van der Schaaf, W., Pelsma, P. A., Smit Sibinga, C. Th., de Kam, P. J., and van der Meer, J. (1995). *Thromb. Haemost.* **74,** 1271–1275.

Henninger, R. W. (1988). *J. Am. Vet. Med. Assoc.* **193,** 91–93.

Herman, F., Magyar, K., and Filep, J. G. (1991). *Thromb. Haemost.* **65,** 296–299.

Hess, S., Kanse, S. M., Kost, C., and Preissner, K. T. (1995). *Thromb. Haemost.* **74,** 258–265.

Hirsh, J., Siragusa, S., Cosmi, B., and Ginsberg, J. S. (1995). *Thromb. Haemost.* **74,** 360–363.

Hohenhaus, A. E. (1992). *Prob. Vet. Med.* **4,** 555–670.

Hoogstraten-Miller, S., Bellinger, D., Reddick, R., Read, M., Sigman, J., and Madden, V. (1995). *Lab. Anim. Sci.* **45,** 151–159.

Howard, P. R., Bovill, E. G., Pike, J., Church, W. R., and Tracy, R. P. (1994). Factor VII antigen levels in a healthy blood donor population. *Thromb. Haemost.* **72,** 21–27.

Huss, B. T., and Ettinger, S. J. (1992). *J. Am. Anim. Hosp. Assoc.* **28,** 221–225.

Ivankovic, M., Pereira, J., Germain, A., Opazo, A., and Mezzano, D. (1992). *Thromb. Haemost.* **68,** 375.

Johnson, G. J., Leis, L. A., and King, R. A. (1991). *Thromb. Haemost.* **65,** 578–580.

Johnstone, I. B. (1993). *Can. Vet. J.* **34,** 627–628.

Johnstone, I. B., and Crane, S. (1991). *Can. J. Vet. Res.* **55,** 11–14.

Johnstone, I. B., Physick-Sheard, P., and Crane, S. (1989). *Am. J. Vet. Res.* **50,** 175–173.

Johnstone, I. B., Viel, L., and Whiting, T. (1991). *Can. J. Vet. Res.* **55,** 101–106.

Johnstone, I. B., Norris, A. M., and Hirzer, L. (1993). *Can. Vet. J.* **34,** 679–681.

Joist, J. H. (1990). H *Semin. Thromb. Hemost.* **16,** 151–157.

Joist, J. H., Remuzzi, G., and Mannucci, P. M. (1994). *In* "Hemostasis and Thrombosis" (R. W. Colman, J. Hirsh, V. J. Marder, and E. W. Salzman, eds.), 3rd ed., pp. 921–935. Lippincott, Philadelphia, Pennsylvannia.

Jordan, H. L., Grindem, C. B., and Breitschwerdt, E. B. (1993). *J. Vet. Intern. Med.* **7,** 261–265.

Joseph, S. A., Brooks, M. B., Coccari, P. J., and Riback, S. C. (1996). *J. Am. Anim. Hosp. Assoc.* **32,** 25–28.

Kaushansky, K. (1995). *Thromb. Haemost.* **74,** 521–525.

Kay, M. A., Rothenberg, S., Landen, C. N., Bellinger, D. A., Leland, F., Toman, C., Finegold, M., Thompson, A. R., Read, M. S., Brinkhous, K. M., and Woo, S. L. C. (1993). *Science* **262,** 117–119.

Kay, M. A., Landen, C. N., Rothenberg, S. R., Taylor, L. A., Leland, F., Wiehle, S., Fang, B., Bellinger, D., Finegold, M., Thompson, A. R., Read, M., Brinkhous, K. M., and Woo, S. L. C. (1994). *Proc. Natl. Acad. Sci. USA* **91,** 2353–2357.

Keith, J. C. Jr., Ferranti, T. J., Misra, B., Frederick, T., Rup, B., McCarthy, K., Faulkner, R., Bush, L., and Schaub, R. G. (1994). *Thromb. Haemost.* **73,** 101–105.

Kelton, J. G. (1995). *Thromb. Haemost.* **74,** 228–233.

Kitchen, S., and Preston, F. E. (1994). *Thromb. Haemost.* **71,** 720.

Kitoh, K., Watoh, K., Kitagawa, H., and Sasaki, Y. (1994). *Am. J. Vet. Res.* **55,** 1542–1547.

Knowler, C., Giger, U. Dodds, W. J., and Brooks, M. (1994). *J. Am. Vet. Med. Assoc.* **205,** 1557–1561.

Konkle, B. A. (1995). *Clin. Chem.* **41,** 48–490.

Kraus, K. H., Turrentine, M. A., Jergens, A. E. and Johnson, G. S. (1989). *Vet. Surg.* **18,** 103–109.

Kristensen, A. T., Weiss, D. J., and Klausner, J. S. (1994a).*J. Vet. Intern. Med.* **8,** 323–327.

Kristensen, A. T., Weiss, D. J., Klausner, J. S., Laber, J., and Christie, D.J. (1994b). *Am. J. Vet. Res.* **55,** 1111–1114.

Laste, N. J., and Harpster, N. K. (1995). *J. Am. Anim. Hosp. Assoc.* **31,** 492–500.

Lattuada, A., Varanukulsak, O., Castaman, G. C., and Mannucci, P. M. (1992). *Thromb. Res.* **67,** 467–471.

Lewis, D. C., and Meyers, K. M. (1994). Effect of anticoagulant and blood storage time on platelet-bound antibody concentrations in clinically normal dogs. *Am. J. Vet. Res.* **55,** 602–605.

Lewis, D. C., McVey, D. S., Shuman, W. S., and Muller, W. B. (1995a). *Am. J. Vet. Res.* **56,** 1555–1558.

Lewis, D. C., Meyers, K. M., Callan, M. B., Bücheler, J., and Giger, U. (1995b). *J. Am. Vet. Med. Assoc.* **206,** 47–52.

Lijnen, H. R., and Collen, D. (1995).*Thromb. Haemost.* **74,** 387–390.

Lippert, L. E., Fisher, L. A., and Schook, L. B. (1990). *Thromb. Haemost.* **64,,** 564–568.

Lock, K. (1994). *Aust. Vet. J.* **71,** 263.

Lozier, J. N., and High, K. A. (1990). *Hematol. Pathol.* **4,** 1–26.

Lozier, J. N., Thompson, A. R., Hu, P. C., Read, M., Brinkhous, K. M., High, K., and Curiel, D. T. (1994). *Human Gene Therapy* **5,** 313–322.

Lüscher, E. F., and Weber, S. (1993). *Thromb. Haemost.* **70,** 234–237.

McCann, M. E., Watson, T. D. G., Boudinot, F. D., and Moore, J. N. (1995). *Am. J. Vet. Res.* **56,** 1070–1074.

McCarroll, D. R., Lothrop, S. A., Dolan, M. C., and McDonald, T. P. (1987). *Exp. Hematol.* **15,** 1060–1067.

McCarroll, D. R., Waters, D. C., Steidley, K. R., Clift, R., and McDonald, T. P. (1988). *Exp. Hematol.* **16,** 929–937.

McClay, C. B., Weiss, D. J., Smith, C. M., and Gordon, B. (1992). *Am. J. Vet. Res.* **53,** 1380–1385.

MacGregor, I. R., Ferguson, J. M., McLaughlin, L. F., Burnouf, T., and Prowse, C. V. (1991). *Thromb. Haemost.* **66,** 609–613.

Machin, S. J., Kelsey, H., Seghatchian, M. J., Warwick, R., and Mackie, I. J. (1995). *Thromb. Haemost.* **74,** 246–252.

McKeever, K. H., Hinchcliff, K. W., Kociba, G. J., Reed, S. M., and Muir, W. W. III. (1990). *Am. J. Vet. Res.* **51,** 1335–1339.

Maddison, J. E., Watson, A. D. J., Eade, I. G., and Exner, T. (1990). *J. Am. Vet. Med. Assoc.* **197,** 1495–1497.

Maggio-Price, L., and Dodds, W.J. (1993). *J. Am. Vet. Med. Assoc.* **203,** 1702–1704.

Mannucci, P. M. (1995). *Proc. Natl. Acad. Sci.* **29,** 2428–2432.

Mannucci, P. M., and Giangrande, P. L. F. (1994). *In* "Haemostasis and Thrombosis" (A. L. Bloom, C. D. Forbes, D. P. Thomas, and E. G. D. Tuddenham, eds.), pp. 949–968. Churchill Livingstone, New York.

Mansell, P. D., and Parry, B. W. (1989). *Can. J. Vet. Res.* **53,** 264–267.

Mansell, P. D., and Parry, B. W. (1991). *J. Vet. Int. Med.* **5,** 191–194.

Marcus, A. J., and Safier, L. B. (1993). T*FASEB J.* **7,** 516–522.

Marx, J. (1993). *Science* **262,** 29–30.

Mazurier, C. (1992). *Thromb. Haemost* **67,** 391–396.

Meinkoth, J. H., and Meyers. K. M. (1995). *Am. J. Vet. Res.* **56,** 1577–1585.

Menache, D., Grossman, B. J., and Jackson, C. M. (1992). *Transfusion* **32,** 580–588.

Meyers, K. M., Wardrop, K. J., Dodds, W. J., and Brassard, J. (1990a). *Thromb. Res.* **57,** 97–108.

Meyers, K. M., Wardrop, K. J, Helmick, C., and White, F. (1990b). *Thromb. Res.* **57,** 109–116.

Miletich, J. P. (1990). *Semin. Thromb. Haemost.* **16,** 169–176.

Moake, J. L. (1995). *Thromb. Haemost.* **74,** 240–245.

Moalic, P., Gruel, Y., Foloppe, P., Delahousse, B., Leclerc, M. H., and Leroy, J. (1989). *Am. J. Vet. Res.* **50,** 59–63.

Monce, K. A., Atkins, C. E., and Loughman, C. M. (1995). *J. Am. Vet. Med. Assoc.* **207,** 581–584.

Monreal, L., Villatoro, A. J., Monreal, M., Espada, Y., Anglés, A. M., and Ruiz-Gopegui, R. (1995). *Am. J. Vet. Res.* **56,** 1281–1285.

Mount, M. E., and Kass, P. H. (1989). *Am. J. Vet. Res.* **50,** 1704–1709.

Murtaugh, R. J., and Dodds, W. J. (1988). *J. Am. Vet. Med. Assoc.* **193,** 351–352.

Navarre, C. B., Kasari, T. R., and Green, R. A. (1995). *J. Vet. Intern. Med.* **9,** 283–285.

Neer, T. M., and Savant, R. L. (1992). *J. Am. Vet. Med. Assoc.* **200,** 1344–1345.

Nemerson, Y. (1995). *Thromb. Haemost.* **74,** 180–184.

Ni, H. Y., and Giles, A. R. (1992). *Thromb. Haemost.* **67,** 264–271.

Nichols, R., and Hohenhaus, A. E. (1994). *J. Am. Vet. Med. Assoc.* **205,** 168–173.

Northern, J., Jr., and Tvedten, H. W. (1992). *J. Am. Vet. Med. Assoc.* **200,** 368–376.

Nurden, A. T. (1995). *Thromb. Haemost.* **74,** 345–351.

Otto, C. M., Dodds, W. J., and Greene, C. E. (1991). *J. Am. Vet. Med. Assoc.* **198,** 129–131.

Palareti, G., Biagi, G., Legnani, C., Bianchi, D., Serra, D., Savini, R., and Coccheri, S. (1989). *Thromb. Haemost.* **62,** 1053–1056.

Palmer, T. D., Thompson, A. R., and Miller, A. D. (1989). *Blood* **73,** 438–445.

Panciera, D. L., and Johnson, G. S. (1994). *J. Am. Vet. Med. Assoc.* **205,** 1550–1553.

Panciera, D. L., and Johnson, G. S. (1995). *J. Am. Vet. Med. Assoc.* **206,** 595–596.

Panciera, D. L., and Johnson, G. S. (1996). *J. Vet. Int. Med.* **10,** 62–64.

Parker, M. T., Turrentine, M. A., and Johnson, G. S. (1991). *Am. J. Vet. Res.* **52,** 119–125.

Patterson, D. F., Haskins, M. E., Jezyk, P. F., Giger U., Meyers-Wallen, V. N., Aquirre, G., Fyfe, J. C., and Wolfe, J. H. (1988). Research on genetic diseases: reciprocal benefits to animals and man. *J. Am. Vet. Med. Assoc.* **193,** 1131–1144.

Patterson, W. R., Estry, D. W., Schwartz, K. A., Borchert, R. D., and Bell, T. G. (1989). *Thromb. Haemost.* **62,** 1011–1015.

Pechlaner, C., Vogel, W., Erhart, R., Pümpel, E., and Kunz, F. (1992). *Thromb. Haemost.* **68,** 617.

Peterson, J. L., Couto, C. G., and Wellman, M. L. (1995). *J. Vet. Intern. Med.* **9,** 298–303.

Pietraszek, M., Azzadin, A., Buczko, W., Malyszko, J., and Mysliwiec, M. (1992). *Thromb. Haemost.* **68,** 625.

Poggio, M., Tripodi, A., Mariani, G., and Mannucci, P. M. (1991). *Thromb. Haemost.* **65,** 160–164.

Prasse, K. W., Allen, D., Jr., Moore, J. N., and Duncan, A. (1990). *Am. J. Vet. Res.* **51,** 1950–1955.

Prasse, K. W., Topper, M. J., Moore, J. N., and Welles, E. G. (1993). *J. Am. Vet. Med. Assoc.* **203,** 685–693.

Pratt, H. L, Carroll, R. C, Jones, J. B., and Lothrop, C. D., Jr. (1991). *Am. J. Vet. Res.* **52,** 945–950

Preissner, K. T., and Jenne, D. (1991). *Thromb. Haemost.* **66,** 189–194.

Rajasekhar, D., Kestin, A. S., Bednarek, F. J., Ellis, P. A., Barnard, M. R., and Michelson, A. D. (1994). *Thromb. Haemost.* **72,** 957–963.

Ramos, R. R., Curtis, B. R., and Chaplin, H. (1992). *Transfusion* **32,** 235–238.

Rand, J. H., Wu, X. X., Potter, B. J., Uson, R. R., and Gordon, R. E. (1993). *Am. J. Pathol.* **142,** 843–850.

Rao, G. H. R., Escolar, G., and White, J. G. (1993). *Transfusion* **33,** 766–778.

Rapaport, S. I., and Rao, L. V. M. (1995). *Thromb. Haemost.* **74,** 7–17.

Raymond, S. L., Jones, D. W., Brooks, M. B., and Dodds, W. J. (1990). *J. Am. Vet. Med. Assoc.* **197,** 1342–1346.

Ruf, W., and Edgington, T. S. (1994). *FASEB J.* **8,** 385–390.

Ruggeri, Z. M., and Ware, J. (1993). *FASEB J.* **7,** 308–316.

Ruiz-Gopegui, R., Espada, Y., Vilafranca, M., Cuadradas, C., Fontcuberta, E., Milla, F., Roncales, J., and Ruzafa, A. (1994). *Vet. Clin. Pathol.* **23,** 70–71.

Sadler, J. E. (1994). *Thromb. Haemost.* **71,** 520–525.

Sadler, J. E., Matsushita, T., Dong, Z., Tuley, E. A., and Westfield, L. A. (1995). *Thromb. Haemost.* **74,** 161–166.

Saido, T. C., Sorimachi, H., and Suzuki, K. (1994).*FASEB J.* **8,** 814–822.

Schermerhorn, T., Barr, S. C., Stoffregen, D. A., Koren-Roth, Y., and Erb, H. N. (1994). *Am. J. Vet. Res.* **55,** 1602–1607.

Schmidt, B., Vegh, P., Johnston, M., Andrew, M., and Weitz, J. (1993). *Thromb. Haemost.* **69,** 418–421.

Searcy, G. P., and Petrie, L. (1990). *Can. Vet. J.* **31,** 101–103.

Searcy, G. P., Sheridan, D., and Dobson, K. A. (1990). *Can. J. Vet. Res.* **54,** 394–396.

Searcy, G. P., Frojmovic, M. M., McNicol, A., Robertson, C., Wong, T., and Gerrard, J. M. (1994). *Thromb. Haemost.* **71,** 240–246.

Semrad, S. D., and Dubielzig, R. (1993). *Am. J. Vet. Res.* **54,** 1339–1346.

Shattil, S. J. (1995). *Thromb. Haemost.* **74,** 149–155.

Shukla, S. D. (1992). *FASEB J.* **6,** 2296–2301.

Sixma, J. J., van Zanten, G. H., Saelman, E. U. M., Verkleij, M., Lankhof, H., Nieuwenhuis, H. K., and de Goot, P. G. (1995). *Thromb. Haemost.* **74,** 454–459.

Smith, S. V., and Brinkhous, K. M. (1991). *Thromb. Haemost.* **66,** 259–263.

Sommer, S. S. (1992). *FASEB J.* **6,** 2767–2774.

Soute, B. A. M., Ulrich, M. M. W., Watson, A. D. J., Maddison, J. E., Ebberink, R. H. M., and Vermeer, C. (1992). *Thromb. Haemost.* **68,** 521–525.

Šrámek, R., Šrámek, A., Koster, T., Briët, E., and Rosendaal, F. R. (1992). *Thromb. Haemost.* **67,** 514–518.

Stefanon, G., Stefanon, B., Stefanon, G. G., and Dodds, W. J. (1993). *Can. Pract.* **18**(3), 15–23.

Stokol, T., Parry, B. W., Mansell, P. D., and Richardson, J. L. (1994). *J. Am. Anim. Hosp. Assoc.* **30,** 239–243.

Stokol, T., Parry, B. W., Mansell, P. D. (1995). *Vet. Clin. Pathol.* **24,** 81–90.

Sugahara, Y., Catalfamo, J., Brooks, M., Hitomi, E., Bajaj, S. P., and Kurachi, K. (1996). *Thromb. Haemost.* **75,** 450–455.

Sullivan, P. S. (1994). *Proc. 12th ACVIM Forum,* pp. 149–151.

Sullivan, P. S., Grubbs, S. T., Olchowy, T. W. J., Andrews, F. M., White, J. G., Catalfamo, J. L., Dodd, P. A., and McDonald, T. P. (1994). *J. Am. Vet. Med. Assoc.* **205,** 1763–1766.

Sullivan, P. S., Manning, K. L., and McDonald, T. P. (1995). *J. Am. Vet. Med. Assoc.* **206,** 332–334.

Suttie, J. W. (1993). *FASEB J.* **7,** 445–452.

Sweeney, J. D., Novak, E. K., Reddington, M., Takeuchi, K. H., and Swank, R. T. (1990). *Blood* **76,** 2258–2265.

Thompson, A. R. (1995). *Thromb. Haemost.* **74,** 45–51.

Tinlin, S., Webster, S., and Giles, A. R. (1993). *Thromb. Haemost.* **69,** 21–24.

Tollefsen, D. M. (1990). *Semin. Thromb. Hemost.* **16,** 162–168.

Topper, J. M., Prasse, K. W., Morris, M. J., Duncan, A., and Crowe, N. A. (1996). *Am. J. Vet. Res.* **57,** 427–431.

Triplett, D. A., (1995). *Thromb. Haemost.* **74,** 329–337.

Tuddenham, E. G. D., Pemberton, S., and Cooper, D. N. (1995). *Thromb. Haemost.* **74,** 313–321.

Ugarova, T., Agbanyo, F. R., and Plow, E. F. (1995). *Thromb. Haemost.* **74,** 253–257.

van Genderen, P. J. J. (1994). *Thromb. Haemost.* **73,** 890–895.

Veloso, D., Smith, J. I., Denny, S., and Cosgriff, T. M. (1992). *Thromb. Haemost.* **68,** 526–533.

Vermeer, C., and Hamulyák, K. (1991). *Thromb. Haemost.* **66,** 153–159.

Verstraete, M. (1995). *Thromb. Haemost.* **74,** 25–35.

Verstraete, M., and Zoldhelyi, P. (1995). *Drugs* **49,** 856–884.

Wagner, A. E., and Dunlop, C. I. (1993). *J. Am. Vet. Med. Assoc.* **203,** 40–45.

Walker, F. J., and Fay, P. J. (1992).*FASEB J.* **6,** 2561–2567.

Walter, J., and High, K. A. (1997). *Adv. Vet. Sci. Comp. Med.* **40,** 119–134.

Wardrop, K. J., Dhein, C. R., Frenier, S., and Meyers, K. M. (1989). *J. Am. Anim. Hosp. Assoc.* **25,** 325–329.

Waters, D. C., Eaton, A. H., Steidley, K, R., and McCarroll, D. R. (1989). *Am. J. Vet. Res.* **50,** 201–204.

Weiss, H. J. (1995). *Thromb. Haemost.* **74,** 117–122.

Welch, R. D., Watkins, J. P., Taylor, T. S., Cohen, N. D., and Carter, G. K. (1992). *J. Vet. Intern. Med.* **6,** 29–35.

Welles, E.G ., Prasse, K. W., and Duncan, A. (1990a). *Am. J. Vet. Res.* **51,** 1080–1085.

Welles, E. G., Prasse, K. W., Duncan A., and Morris, M. J. (1990b). *Am. J. Vet. Res.* **51,** 1075–1079.

Welles, E. G., Boudreaux, M. K., and Tyler, J. W. (1993a). *Am. J. Vet. Res.* **54,** 1235–1243.

Welles, E. G., Williams, M. A., Tyler, J. W., and Lin, H. C. (1993b). *Am. J. Vet. Res.* **54,** 1230–1234.

Welles, E. G., Boudreaux, M. K., Crager, C. S., and Tyler, J. W. (1994). *Am. J. Vet. Res.* **55,** 619–627.

Willis, S. E., Jackson, M. L., Meric, S. M., and Rousseaux, C. G. (1989). *Am. J. Vet. Res* **50,** 1893–1897.

Woody, B. J., Murphy, M. J., Ray, A. C., and Green, R. A. (1992). *J. Vet. Intern. Med.* **6,** 23-28.

Zhang, Z. (1995). Doctoral Thesis, Karolinska Institute, Stockholm, Sweden.

C H A P T E R

11

Neutrophil Function

JOSEPH G. ZINKL AND MAHENDRA B. KABBUR

I. INTRODUCTION

The main function of neutrophils is degrading and killing microbes, especially bacteria. Neutrophils are distributed to the tissues by the circulatory system. They adhere to endothelial cells, migrate through vessel walls into tissues, phagocytize microorganisms and foreign subtances, and degrade and kill microorganisms. This chapter provides an overview of the mechanisms of neutrophil adherence, migration (chemotaxis), phagocytosis, and bacteria killing. Defects of neutrophils that prevent them from performing these functions are reviewed. Methods for evaluating the various functions of the phagocytes are briefly described.

II. NEUTROPHIL FUNCTIONS

A. Adherence

In order for neutrophils in the blood to migrate into tissues they must adhere to the endothelial cells. The blood neutrophil population is made up of marginated (adhered) cells and circulating (nonadhered) cells. These populations are in continual flux as neutrophils marginate, demarginate, and remarginate. Eventually, marginated neutrophils strongly adhere to the endothelium before they migrate into tissues under the influence of chemotactic agents. Adherence of neutrophils to endothelial cells involves the interaction of blood dynamics and cellular adhesion molecules and their receptors or ligands. The adhesive actions are initial contact of the neutrophil with the endothelial cell, rolling of the neutrophil along the endothelium, firm adhesion to the endothelial cells, and migration through the endothelium (Jones *et al.*, 1995). This process is complex and is driven by inflammatory mediators derived from pathogens, endothelial cells, leukocytes, platelets, cells of the tissues, and substances in the plasma.

Although there are many factors that influence the interaction of neutrophils and endothelial cells, the action of adhesion molecules with their ligands that results in the attachment of the neutrophil to the endothelial cell is the most important. There are two classes of adhesion molecules that promote the attachment of neutrophils to endothelial cells, the selectins and the integrins (Meuer, 1994).

There are three members of the selectin family of adhesion molecules, L-selectin, P-selectin, and E-selectin. The three selectins are structurally related, differing primarily in the lengths of their complement binding domain (L-selectin is the shortest, P-selectin is the longest, and E-selectin is slightly shorter than P-selectin). They all have a terminal lectin domain. L-selectin is expressed by leukocytes, whereas E-selectin and P-selectin are expressed by endothelial cells. P-selectin is also expressed by platelets (Forrest and Paulson, 1995). In endothelial cells, P-selectin and E-selectin synthesis is induced by several substance, including thrombin and histamine (P-selectin) and interleukin-1 (IL-1) and tumor necrosis factor-α (TNFα) (E-selectin). L-selectin is constitutively expressed on the surface of mature neutrophils (Oppenheimer-Marks and Lipssky, 1994). Selectins are involved early in neutrophil to endothelial cell binding when the cells roll along the endothelium (Jones *et al.*, 1995) . They apparently interact with ligands that are oligosaccharides, particularly those containing sialic acid and fucose moieties of glycoproteins and glycolipids (Forrest and Paulson, 1995). They may also bind to certain carbohydrate moieties containing esterified sulfate groups, the sulfatides (Mulligan *et al.*, 1995). The sialylated and fucosylated glycoproteins and glycolipids to which selectins bind are on the surface of many cells, including endothelial cells and leukocytes (Forrest and Paulson, 1995). The adhesive interactions between the selectins and their ligands are of low affinity, resulting in temporary attachment followed by detachment so that the leukocyte slowly rolls along the endothelium. This event, along with the initial step in inflammation, vascular dilation, slows the neutrophil and places it near the endothelial cell, where integrin activity results in stronger binding to the endothelial cell and flattening of the leukocyte along the endothelium (Shappell and Smith, 1994).

The integrin family consists of interrelated heterodimers, consisting of α- and β-subunits. There are three distinct subfamilies based on the nature of the β-subunit used to form the heterodimer. Neutrophils primarily express the $\beta 2$ subunit CD18 associated with one of three different α-subunits, CD11a, CD11b, and CD11c (Meuer, 1994) . The combination of these subunits results in the development of a group of unique integrins, each of which has primary affinity for certain ligands although there apparently is some overlap in their ligand affinity. The combination of CD18 and CD11a is named leukocyte function-associated antigen-1 (LAF-1) and has particular affinity for intercellular adhesion molecule-1 (ICAM-1). CD18 combined with CD11b is named macrophage antigen-1 (Mac-1) and has particular affinity for inactivated third component of complement (iC3b). CD18 combined with CD11c is named p150,95, but its most specific ligand has not been determined (Meuer, 1994; Anderson, 1995) . Integrins are preformed in cells and are, in part, stored in granules. Activation of the cells by a variety of substances causes the cells to externalize or express the preformed integrins. The binding affinities of integrins for their receptors are stronger than those of selectins. This results in greater affinity of the leukocyte to the endothelial cells, essentially anchoring and flattening the leukocyte prior to its movement through the endothelium into tissues.

Endothelial cells express ICAM-1 and ICAM-2 on their surface as well as VCAM-1. ICAM-2 is constitutively expressed on endothelial cells, whereas ICAM-1 and vascular cell adhesion molecule-1 (VCAM-1) are increased by a variety of factors, including IL-1 and TNF-α (Oppenheimer-Marks and Lipsky, 1994).

Neutrophils undergo four interrelated activities in order to move from the bloodstream into the tissues: initial contact, rolling, firm adhesion, and migration (Jones *et al.*, 1995). The interaction of selectins and integrins with their binding molecules is necessary for these activities to occur. Initial contact and rolling are functions of blood-flow dynamics and binding of selectin to glycoproteins and glycolipids. Firm adhesion, flattening, and migration rely on the actions of the integrins and their receptors. Selectin binding is transient, and it is weaker than that of integrins. Selectin-mediated binding of neutrophils to endothelial cells can occur at the wall shear stress found in postcapillary venules (Jones *et al.*, 1995; Doré *et al.*, 1995). This binding decreases the velocity of the leukocyte enough that integrin interaction with their binding molecules can occur. At wall shear stresses of 1-4 dyn/cm^2, neutrophils adhere to endothelial cells through the action of selectins, whereas integrins acting alone do not adhere until the wall shear stress is less than 1 dyn/cm^2 (Jones *et al.*, 1995; Doré *et al.*, 1995). Not only is selectin binding weaker than intregin binding, it is probably more flexible. In addition, neutrophils contain approximately one-tenth as many selectin molecules as integrin moieties ($2–4 \times 10^4$ sites per cell, vs 30×10^4 Mac-1 sites per cell) (Jones *et al.*, 1995). The rolling action mediated through selectin activity causes the velocity of the cells to slow sufficiently that newly exposed and upregulated integrins can interact with their exposed and upregulated ligands, causing the leukocyte to attach firmly to the endothelial cell (Jones *et al.*, 1995; Doré *et al.*, 1995). At this time the neutrophil changes from a spherical shape, to a flat disk shape, which increases the contact area between it and the endothelial cell and causes the neutrophil to flatten onto endothelial cells. In order for neutrophils to move away from this pavemented zone, it is necessary for the integrin-mediated attachment to weaken slightly. The precise mechanism for this weakening is unknown.

Numerous substances, some of which are also chemoattractants, upregulate integrins, selectins, or their ligands. Among the upregulating factors are complement factor 5a (C5a), formyl methionylleacyl phenyalinine (fMLP), leukotriene B_4 (LTB_4), platelet-activating factor (PAF), granulocyte monocyte-colony stimulating factor (GM-CSF), tumor necrosis factor-α (TNF-α), lipopolysaccharide (LPS), interleukin-1 (IL-1), interleukin-6 (IL-6), interferon-γ (INFγ), histamine, thrombin, and hydrogen peroxide (Shappel and Smith, 1994). Some of these factors can also act as downregulating agents. For example, C5a downregulates L-selectin but upregulates Mac-1 (CD18/CD11a) (Anderson, 1995). Endothelial-derived relaxing factor (EDRF), which is nitric oxide radical (NO·) produced by endothelial cells, is a powerful antiadhesive agent that may inhibit the function of $\beta 2$ integrins (Thom *et al.*, 1994).

The activities that allow the neutrophil to detach from the endothelial cell and move between or through the endothelial cells have not been defined. It is possible that the endothelial cells actively guide or move attached neutrophils to cell junctions, or that the neutrophil migrates along the membrane of the endothelial cell by responding to chemoattractants bound to the endothelial cell surface (a process called heptotaxis) in order to move to and find the locations through which it can migrate. Adhesion molecules, such as CD31 (also known as platelet endothelial cell adhesion molecule—PECAM-1), present at intercellular junctions may also be involved (Argenbright, 1995).

B. Chemotaxis

After neutrophils adhere to the endothelium they migrate through the endothelium into the interstitial spaces. Leukocytes move in the direction of higher concentration of the attractant. The ability of cells to respond to chemotaxins implies that they have sensory mechanisms that are able to detect differences in concentration of chemotaxins in the local environment. Furthermore, the sensory mechanism must be sensitive to concentrations over a very short distance, the diameter of the cell. Cell-surface receptors provide the sensory apparatus. Detection of the chemical gradient is caused by variations in the degree of saturation of the receptors across the diameter of the cell. The cell is stimulated to move in the direction of increasing saturation of the receptors. However, at very high concentrations of a chemotaxin, cell movement stops even though a chemical gradient exists, because the receptors are completely saturated and the cell is unable to detect the gradient (Ramsey, 1974).

Leukocytes move by adhering to and moving along solid surfaces. Locomotion is dependent on the nature of the adhesion of the cell with the surface on which it moves, the substratum (Wilkinson, 1982). Neutrophils change from round, smooth cells to elongated, ruffled cells with pseudopodia. The pseudopodia are broad, thin lamellipodia that extend in the direction of increasing chemoattractant concentration. A uropod is formed behind the cell. Neutrophils extend their lamellipodia in the direction of the gradient and retract their uropodia toward the cell body (Senda *et al.*, 1975; Wilkinson, 1982; McPhail and Harvath, 1993). The protrusion is thin and ruffled, and it rapidly changes shape. It does not contain cytoplasmic organelles, but it is rich in actin and actin-associated proteins (Oliver *et al.*, 1978). A distinct posterior tail or uropod may remain. The process recurs, resulting in a slow, undulatory shape change as the cell moves forward.

Because inflammatory cells crawl during chemotaxis, it is essential for the cells to adhere to surfaces. Adhesion must be strong enough for the cell to pull itself along, but not so strong that the cell becomes immobilized. Integrins bind to extracellular matrix components such as fibronectin, fibrinogen, collagen, and other extracellular matrix components, as well as membrane components of other cells (Edwards, 1994). Reversible adherence to these subtances facilitates movement of phagocytes.

At least three activities are required for cells to move. A pseudopodium forms in the direction of movement, the cell body and the pseudopodium anchor to the substratum, and contraction of the actin–myosin occurs to pull the cell toward the pseudopodium. Actin–myosin contraction revolves around a sol–gel transition that interacts with the cytoskeletal network and is regulated by localized calcium changes (Edwards, 1994).

Energy is required to drive the contractile machinery of the neutrophil. This energy is supplied mainly through ATP generated by anaerobic glycolysis (Klebanoff and Clark, 1978). Glucose may be supplied from the cell's environment or from endogenous glycogen. When neutrophils contact an active chemotactic factor, glucose metabolism is stimulated (Goetzl and Austen, 1974). Some substances that inhibit chemotaxis appear to increase cAMP levels, whereas chemoattractants appear to increase the intracellular levels of cGMP (Smith and Lumsden, 1983). Prostaglandins may be involved in the regulation of the relative concentrations of cAMP and cGMP and thus regulate chemotaxtic ability of neutrophils (DeChatelet, 1979).

There are many substances that are chemoattractants for neutrophils, including *N*-formyl peptides, C5a, LTB_4, interleukin-8 (IL-8), and platelet-activating factor (PAF). Neutrophils have receptors for each of these substances, and they are members of the GTPase-coupled receptor superfamily. These substances have seven transmembrane domains (McPhail and Harvath, 1993). Chemoattractant receptors couple with G proteins (also known as GTPase proteins). These proteins carry out the cycle of guanine nucleotide exchange and GTP hydrolysis that is involved in generating second messengers, which initiate the activity of the contractile proteins of the cell. There are two general classes of G proteins: heterotrimeric, consisting of α-, β-, and γ-subunits, and monomeric or small G proteins.

The α-subunit of the heterotrimeric G proteins activates phospholipases to produce a variety of substances, including inositol phosphate (IP_3). Calcium concentrations within the cells increase markedly through binding of IP_3 to intracellular calciosomes and influx of calcium from the external medium (Edwards, 1994). The increased intracellular calcium activates protein kinase C. Phosphoproteins interact with actin and myosin, resulting in cell motility (Edwards, 1994; Shepro, 1995).

Actin exists in two forms in neutrophils, monomeric (G) actin and polymeric (F) actin. Actin comprises about 5–8% of the total protein in the neutrophil, and in resting cells about 50–70% of the actin is in the monomeric form. In resting cells the large, unpolymerized pool is spread throughout the cytoplasm. This provides the cell with the advantage of rapid polymerization of actin at whatever location it may be required in the cell without the need for depolymerization and redistribution of the monomers first. Polymerized actin is primarily concentrated in the submembrane layer in resting cells. The initial step in actin polymerization is the activation of actin monomers by binding to divalent cations. Monomeric actin binds 1 mol ATP, which is hydrolyzed to ADP during the assembly into F-actin (Edwards, 1994).

There are several proteins that bind to actin and are important for affecting filament growth, cross-linking, anchoring, and function. Profilin, gelsolin, and acumentin affect filament growth by binding to G-actin (profilin) and capping the barbed ends (gelsolin) or pointed ends (acumentin) of growing F-actin filaments. Actin filaments form thick filaments by cross linking through the activities of filamin, caldesmon, and α-actinin. Two filamin subunits bind actin filaments at 40-nm intervals to form perpendicularly cross-linked, branched filaments. Caldesmon also cross-links F-actin chains. Its effects are inhibited by calcium and calmodulin. α-Actinin is a short, rodlike proteins that joins actin filaments side by side as bundles. Vinculin is one of a number of proteins that bind the actin network to the plasma membrane. Vinculin function is calcium dependent, and it is found in close association with α-actinin. Myosin is composed of a pair of heavy chains (200 kDa) and two pairs of light chains (15–20 kDa) that are located near the heads of the heavy chains. Myosin molecules form short, bipolar filaments in which the rodlike tails form the filament and globular heads project from both ends of the shaft of the filament. The heads of these filaments interact with F-actin. Myosin headpieces bind both ADP and ATP and possess magnesium-dependent ATPase activity. In the resting state, myosin binds ADP and inorganic phosphate. Upon stimulation the headpiece binds to actin in a perpendicular manner, the ADP and inorganic phosphate are released, and the head tilts 45° to the actin molecule, causing filament movement. The head then binds ATP and detaches from actin, and the ATP is hydrolyzed to ADP and inorganic phosphate. The kinase is activated by calcium and calmodulin. Increase in calcium concentration stimulate the binding of actin to the myosin–ADP–inorganic phosphate complex (Edwards, 1994).

The cytoskeleton of neutrophils is made up of microtubules and intermediate filaments. The microtubules are composed of equal molar amounts of α- and β-subunits of tubulin, which form a helical complex of interrelated heterodimers in the shape of a tube 25 nm in diameter with a wall width of 5 nm. A large amount of the tubulin is not assembled in the resting neutrophil, but rapid assembly into microtubules occurs upon stimulation. Intermediate filaments of neutrophils are made up of vimentin. Vimentin readily polymerizes to produce filaments 10–12 nm in diameter (Edwards, 1994).

C. Phagocytosis

When a neutrophil reaches the site of bacterial infection, the cell phagocytizes the organism. Before particles can be phagocytized, they must first be opsonized (DeChatelet, 1979). Opsonization causes bacteria to adhere to neutrophil receptors before they are internalized. The most important opsonins are those derived from complement and the immunoglobulins (Klebanoff and Clark, 1978). The most important opsonins of the complement pathway are bound fragments of C3 and C4 generated during complement activation on the particle surface (Morgan, 1990). Activation of the classical or alternative pathway on a target particle causes deposition of C3b. C3b generated by the classical pathway is degraded to C3bi, but C3b generated by the alternative pathway is degraded slowly. C3b has greater affinity for neutrophil receptors, but C3bi is generally in higher density on target particles. Activation of the classical pathway results in deposition of fragments of C4, including the opsinin C4b. Complement receptors on neutrophils include CR1, CR3, and CR4. The primary ligand for CR1 (CD35) is C3b, but it also finds C3bi and C4b with considerably lower affinities. CR1 increases up to 10-fold on the surface of neutrophils and monocytes upon stimulation of the cell. CR3 and CR4 are derived from the CD18/CD11 heterodimer integrins. CR3 (CD18/CD11b or Mac-1) recognizes C3bi. The binding requires high concentrations of calcium and magnesium. CR4 (CD18/CD11c or p150,95) apparently also binds C3bi (Morgan, 1990).

The immunoglobulins serve both to identify organisms and to initiate complement activity. The Fab portion of the antibody binds to the bacteria, whereas the Fc portion binds to the phagocyte (Quie, 1972). The receptor sites for IgG on phagocytes are distinct from the sites for attachment of the opsonically active fragment of C3 (Stossel, 1974). There are three families of receptors for IgG: FcγRI, FcγRII, and FcγRIII. Neutrophils and monocytes also have receptors that specifically bind IgA. The immunoglobulin receptors generally have extracellular, transmembrane, and intracytoplasmic domains, although FcγRIII may only contain an extracellar domain (Edwards, 1994). Interferon γ upregulates the expression of FcγRI on neutrophils (Rosales and Brown, 1993). The functions of the immunoglobulin receptors are mediated through binding of IgG. In general, they bind IgG_3 and IgG_1 better than IgG_2 and IgG_4. These immunoglobulin receptors have numerous functions, including initiation of phagocytosis, activation of the respiratory burst, removal of immune complexes, release of inflammatory mediators, and antibody-dependent cell-mediated cytotoxicity (ADCC) (Rosales and Brown, 1993). Apparently, these receptors act in a cooperative manner, such that one receptor may bind the immunoglobulin ligand while the other actually transduces a signal into the cell (Rosales and Brown, 1993).

Fibronectin, serum amyloid P, and laminin promote phagocytosis. They bind to a large variety of substrates, including fibrin, collagen, heparin, and some bacteria. Plasma fibronectin binds weakly to macrophages, but its binding is enhanced by heparin (Hormann, 1982). Fibronectin increases the attachment of bacteria to neutrophils (Proctor *et al.*, 1982). Fibrinectin receptors on phagocytes recognize the Arg–Gly–Asp sequence of several of the extracellar matrix proteins. Occupancy of the fibrinectin receptors some how activates CR1 and CR3 (Edwards, 1994) or a distinct integrin of the β_3 subfamily (Rosales and Brown, 1993).

Phagocytosis by neutrophils, monocytes, and macrophages is enhanced by a small peptide called tuftsin (Thr–Lys–Pro–Arg) (Najjar, 1983; Nishioka *et al.*, 1994). Tuftsin acts directly on the phagocyte. The spleen acts on the IgG_1 molecule leukokinin in the blood to release leukokinin-S. Then, a protease of the neutrophil membrane cleaves leukokinin-S to tuftsin (Najjar, 1983). Tuftsin markedly increases the rate of phagocytosis (Najjar, 1983), increases the number of organisms phagocytized per cell (Nishioka *et al.*, 1994; Kubo *et al.*, 1994), stimulates respiratory burst (Singh *et al.*, 1992), and increases $NO^{\cdot}$ production (Cillari *et al.*, 1994).

In order to avoid destruction, organisms may evade opsonization, prevent attachment or phagocytosis, or inhibit phagosome and granule fusion. The degree of hydrophobicity of a bacteria relative to that of the neutrophil influences the ability of neutrophils to ingest bacteria (van Oss and Gillman, 1972). Bacteria more hydrophobic than neutrophils are easily phagocytized, whereas those less hydrophobic (or more hydrophilic) are more difficult to phagocytize. Besides their opsonization activity, another action of complement fragments and immunoglobulins may be to increase the hydrophobicity of bacteria, thus making them more susceptible to phagocytosis (van Oss and Gillman, 1973; van Oss *et al.*, 1974). Lipopolysaccharide capsules of gram-negative bacteria may partially increase their virulence by preventing phagocytosis, possibly because of the hydrophilic nature of the capsule (van Oss and Gillman, 1972; Horwitz and Silverstein, 1980; Verhoef and Visser, 1993). Other bacteria cell-wall components such as peptidoglycan, teichoic acids, protein A, M protein, and pili may affect the susceptibility of organisms to phagocytosis (Hendricks *et al.*, 1986). Microorganisms may escape opsonization by varying their antigenic structure (Verhoef and Visser, 1993). Protein A binds to the Fc terminal of IgG, resulting in complement consumption or preventing antibody activity (Verhoef and Visser, 1993). M protein and perhaps protein A of *Streptococcus* organisms interfere with opsonization, probably because they are highly negatively charged. The negative charge may inhibit contact between bacteria and phagocytes. Some organisms such as *Mycobacterium* inhibit the fusion of phagosomes and granules (Verhoef and Vissser, 1993).

After attachment of the bacterium to the neutrophil, the cell extends pseudopodia over and around the particle. Eventually the cytoplasmic processes meet and fuse and enclose the particle to form a phagosome. Glycolysis provides the energy for phagocytosis, and the divalent cations calcium and magnesium are necessary for phagocytosis by neutrophils (Klebanoff and Clark, 1978; DeChatelet, 1979). C3b apparently enhances the action of the cations (DeChatelet, 1979). Actin filaments are preferentially associated with the sites of particle ingestion. Myosin and filamin are also located at these sites. Apparently, assembly of actin filaments occurs at the pointed ends during phagocytosis, but at the barbed ends during locomotion (Edwards, 1994).

D. Bactericidal Mechanisms

The antibacterial mechanisms of phagocytes are classified as either oxygen dependent or oxygen independent (Klebanoff, 1975). The oxygen dependent mechanisms involve a series of reactions that produce highly reactive oxygen containing reactants that interact with other reactants or directly with bacteria. The oxygen independent mechanisms are varied.

1. Oxygen-Dependent Mechanisms

The oxygen-dependent mechanism is initiated by phagocytosis or disturbances of the cell membrane (Rossi *et al.,* 1985). Membrane-bound nicotinamide–adenine dinucleotide phosphate (NADPH) oxidase is activated (Babior, 1978a; Rossi *et al.,* 1985). This reaction depends upon the interaction of membrane associated cytochrome b and other proteins within the cell (Borregaard, 1985; Edwards, 1994). The NADPH oxidase system has at least four components consisting of two closely associated membrane proteins and two cytosolic components. The membrane component, cytochrome b_{558}, is a heterodimer consisting of a 91-kDa glycoprotein (β-subunit) and a 22-kDa protein (α-subunit). The cytosolic compontents are a 47-kDa protein (p47-*phox*) and a 67-kDa component (p67-*phox*) (McPhail and Harvath, 1993). Upon stimulation the cytosolic proteins are translocated to the membrane and the complete NADPH oxidase system is assembled. In addition, a G protein, *rap1A,* is strongly associated with the membrane cytochrome heterodimer, perhaps by guiding the assembly and/or activating NADPH oxidase (McPhail and Harvath, 1993). G proteins may act as receptors for second messengers generated through membrane disturbance (McPhail and Harvath, 1993). Other cytosolic compontents including flavoproteins are also involved with the electron transport system of NADPH oxidase (Cross, 1992; Edwards, 1994).

The NADPH oxidase system catalyzes the reduction of molecular oxygen to superoxide anion ($O_2^{\bar{\cdot}}$). The rapid increase in oxidase activity causes a burst of oxygen consumption (the "respiratory burst"). The respiratory burst reactions are followed by a rapid production of toxic oxygen derivatives in the phagocytic vacuole (Babior, 1978a; Badwey and Karnovsky, 1980). The reaction is dependent upon a continuous reduction of NADP to supply the electron acceptor NADPH for the reduction of O_2 to $O_2^{\bar{\cdot}}$ (Babior, 1978a; Rossi *et al.,* 1985). NADPH is supplied by a marked increase in glucose metabolism through the hexose monophosphate shunt pathway (Fig. 11.1) (Babior, 1978a).

After $O_2^{\bar{\cdot}}$ is produced, it is rapidly dismutated to H_2O_2. $O_2^{\bar{\cdot}}$ and H_2O_2 give rise to $\cdot OH$ and $^1O_2$. Myeloperoxidase catalyzes the formation of hypohalous acids from H_2O_2 and halide ions. Because the concentration of Cl^- is greater than that of Br^- or I^-, hypochlorous acid (HOCl) is formed (Edwards, 1994). The reactive oxygen substances produce secondary oxidative decarboxylation, oxidative deamination, peroxidation, and halogenation reactions that destroy bacteria and other cells (Babior, 1978a; Borregaard, 1985). The formation of hypochlorous acid through the myeloperoxidase reaction is not essential for antimicrobial activity because humans with a deficiency of myeloperoxidase do not have severe bacterial infections. Furthermore, chicken heterophils lack myeloperoxidase, yet they are quite capable of destroying bacteria (Brune and Spitznagel, 1973).

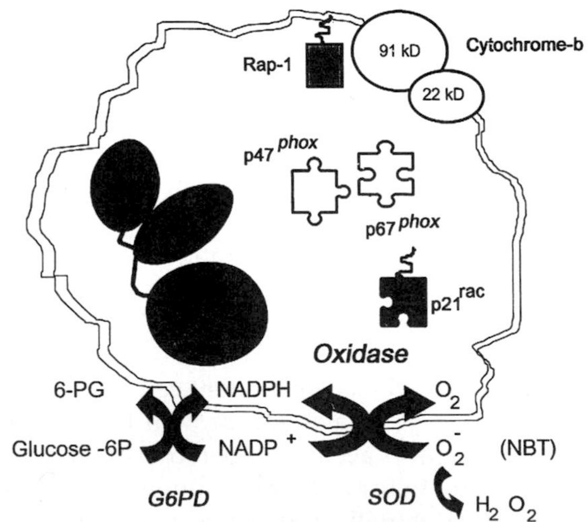

FIGURE 11.1 The neutrophil–NADPH oxidase complex consists of two cytochrome b membrane subunits, α (22 kDa) and β (91 kDa), and two cytosolic cofactors (67 kDa) and 47 (kDa) required of the respiratory burst. RAC p21 and RAP-1, members of the RAS superfamily of GTP-binding proteins, also participate in the stimulation of NADPH oxidase. NADPH serves as a substrate for the reduction of O_2 to superoxide $O_2^{\bar{\cdot}}$, which in turn is rapidly dismutased either spontaneously or by superoxide dismutase (SOD) to H_2O_2, which then gives rise to hydroxyl radical (OH) and singlet oxygen ($^1O_2$) (not shown). NADPH is generated from glucose-6-phosphate through the hexose monophosphate shunt by glucose-6-phosphate dehydrogenase. (From Baehner, 1995, with permission.)

Macrophages generate nitric oxide radicals ($NO^{\cdot}$), which may be bacteriacidal. In addition, $NO^{\cdot}$ reacts with $O_2^{\cdot}$ to form peroxynitrite anion ($ONOO^-$), which can denature nucleic acids, proteins, carbohydrates, and lipids (Ward and Mulligan, 1992). The importance of $NO^{\cdot}$ as an antibacterial material has not yet been determined. Along with other neutrophil-derived oxidants, it can cause tissue damage (Ward and Mulligan, 1992).

The products of the respiratory burst are toxic to the cells (Badwey and Karnovsky, 1980; Gabig and Babior, 1981; Rossi *et al.*, 1985; Ward and Mulligan, 1992). Cytoplasmic superoxide dismutase and glutathione peroxidase, glutathione reductase, and perhaps catalase convert these products to water (Klebanoff and Clark, 1978). The glutathione needed for the reaction is maintained in the reduced state by reduced NADPH supplied by the HMP (Babior, 1978a).

2. *Oxygen-Independent Mechanisms*

The oxygen-independent mechanisms of the granulocyte are heterogenous. Primary granules contain, in addition to the oxygen-dependent systems, cationic proteins including defensins, cathepsin G, azurocidin, and bactericidal/permeability-increasing (BPI) protein (Klebanoff and Clark, 1978; Lehrer and Ganz, 1990). These proteins are rich in arginine or cysteine or both (Ganz *et al.*, 1986). In addition, the primary granules contain hydrolases such as elastase, collagenase, and β-glucuronidase (Lehrer and Ganz, 1990). The secondary granules contain much of the lysozyme, lactoferrin, vitamin B_{12}-binding protein, adhesin receptors, chemoattractant receptors, and gelatinase (Leffell and Spitznagel, 1972; Kane and Peters, 1975; Klebanoff and Clark, 1978; Lehrer and Ganz, 1990). Both primary and secondary granules contain lysozyme. A third granule has been found in neutrophils of several species, including humans, rabbits, cows, sheep, goats, dogs, and horses (Bertram, 1985). In some species gelatinase and some acid hydrolases have been identified in these granules (Klebanoff and Clark, 1978; Baggiolini and DeWald, 1984). The large tertiary granules of bovine neutrophils are probably different from the tertiary granules of rabbits and humans (Bertram, 1985). The tertiary granules of bovine neutrophils have a potent antimicrobial system that is distinct from those of other species (Gennaro *et al.*, 1983).

The pH in phagolysosomes becomes slightly alkaline after activation of NADPH oxidase because the enzyme is a proton pump. Thereafter, the phagolysosome become acidic (Edwards, 1994). An antibacterial action of neutrophils is acidification, probably as the result of increased lactic acid production. A lactate dehydrogenase with affinity for NADPH rather than NADH has been identified in neutrophils (Evans and Karnovsky, 1962). This enzyme has low activity at pH 7.4, but its activity is high at pH 6.0. This lactate dehydrogenase enzyme is probably linked with the hexose monophosphate pathway that provides the NADPH needed for activity and acidification.

Defensins are a family of cysteine- and arginine-rich, low-molecular-weight, antimicrobial peptides. Defensins have been isolated from heterophils or neutrophils of rabbits, rats, guinea pigs, and humans (Ganz *et al.*, 1985; Lehrer and Ganz, 1990; Evans and Harmon, 1995). They contain six invariant cysteines and are arginine rich (Lehrer and Ganz, 1990). The six cysteine residues form disulfide bonds, and the pairing of the first and sixth residues creates a cyclic structure. A similar group of peptides called β-defensins have been isolated from bovine neutrophils (Selsted *et al.*, 1993). They are larger than defensins, having 38–42 amino acid residues, and their cysteine residues vary slightly from those of defensins. Similar peptides have been isolated from heterophils of chickens (Evans and Harmon, 1995). Cysteine-rich antimicrobial proteins have been found in equine neutrophils (Couto *et al.*, 1992a,b). These are also larger than defensins and contain 10 cysteine residues. Small peptides containing 16–18 residues and 4 cysteines have been isolated from pig neutrophils (Evans and Harmon, 1995). The defensin-like peptides have a broad spectrum of antimicrobial activity at micromolar concentrations. They are active against gram-positive, gram-negative, and anaerobic bacteria, several fungi, some enveloped viruses, protozoa, and cells (Ganz *et al.*, 1986; Lehrer *et al.*, 1991; Evans and Harmon, 1995).

Cathepsin G is found in neutrophils and monocytes (Ohlsson *et al.*, 1977; Edwards, 1994). It is rich in arginine but not in cysteine (Ganz *et al.*, 1986). It has microbicidal activity against both gram-positive and gram-negative bacteria, as well as against some fungi (Lehrer *et al.*, 1975; Odeberg and Olsson, 1975). Cathepsin G inhibits bacterial respiration and energy dependent transport systems, as well as protein, DNA, and RNA biosynthesis. (Edwards, 1994). The microbicidal activity decreases with increased ionic strength. Cathepsin G loses its proteolytic activity when heated, but it retains its microbicidal activity (Odeberg and Olsson, 1975; Edwards, 1994).

BPI protein is a large (55–60 kDa), lysine-rich protein. The large amounts of lysine in BPI account for most of its positive charge (Weiss *et al.*, 1978). BPI protein is highly active against *E. coli* and *Salmonella typhimurium*, but it lacks activity against gram-positive bacteria and fungi (Ganz *et al.*, 1986; Lehrer and Ganz, 1990). Strains of bacteria with short-chain polysaccharide envelopes ("rough" strains) are more sensitive than bacteria whose envelopes contain long-chain polysaccharides ("smooth" strains) (Weiss *et al.*, 1978; Lehrer and Ganz, 1990). Apparently, BPI rapidly binds to the outer membrane of susceptible gram-negative bacteria and increases outer membrane permeability to hydrophobic molecules and activation of enzymes that degrade envelope phospholipids and peptidoglycans. BPI also affects inner bacterial membrane functions. In addition, BPI inhibits the respiration of intact susceptible *E. coli* (Lehrer and Ganz, 1990).

Azurocidin, a 29-kDa protein, kills *E. coli*, *S. fecalis*, and *C. albicans in vitro*, showing maximum activity in low-ionic-strength, acid conditions. A related polypeptide, p29b, also has antimicrobial activity *in vitro*. Another protein, 37-kDa CAP (cationic antimicrobial protein), which may be identical to azurocidin, has considerable activity against several gram-negative bacteria (Lehrer and Ganz, 1990).

Granulocytes also contain a variety of neutral proteases with optimal activities at about neutral pH (Spitznagel, 1984). Most neutral proteases are in the primary granules. The major neutral proteases are cathepsin G, elastase, and collagenase. The antimicrobial activity of cathepsin G is related to its action as a cationic protein and has little to do with its enzymic (chymotrypsin-like) activity (Odeberg and Olsson, 1975). Elastase, a highly cationic glycoprotein, degrades bacterial cell-wall protein and potentiates the lytic activity of lysozyme and the microbicidal activity of cathepsin G (Janoff and Scherer, 1968; Blondin and Janoff, 1976; Lehrer and Ganz, 1990). Elastase enhances the activity of cathepsin G and cathepsin B (another protease), probably by exposing proteins that are susceptible to the action of these enzymes (Spitznagel, 1984). By predigesting bacterial cell-wall structures, elastase and other proteases may facilitate penetration of other antimicrobial substances with subsequent disposal of dead microorganisms by proteolysis (Ganz *et al.*, 1986). The microbicidal action of collagenase has not been determined. Perhaps collagenase has a function in maintaining the inflammatory reaction and breaking down extracellular connective-tissue components (Klebanoff and Clark, 1978; Bertram, 1985).

Lysozyme is a very cationic protein of approximately 14,400 Da found in both the primary and the secondary granules of neutrophils, the granules of monocytes, and macrophages and in plasma, tears, saliva, and airway secretions (Ganz *et al.*, 1986). Lysozyme hydrolyzes bacterial cell walls by attacking the *B*1–4 glycosidic linkage that joins *N*-acetyl muramic acid and *N*-acetyl glucosamine of peptidoglycan (Spitznagel, 1984; Ganz *et al.*, 1986). Lysozyme is absent or has very low activity in a number of species (Padgett and Hirsch, 1967; Rausch and Moore, 1975). A wide variety of bacteria are sensitive to the action of lysozyme, but resistance is frequently encountered. Group A streptococci, staphylococci, and nearly all Gram-negative organisms resist the action of lysozyme. However, many resistant bacteria may become susceptible to the action of lysozyme when exposed to synergistic bactericidal activity of antibody and complement or H_2O_2–ascorbic acid (Wilson and Spitznagel, 1968; Miller, 1969). Lysozyme may also be capable of bactericidal activity by nonenzymatic mechanisms (Ganz *et al.*, 1986).

Lactoferrin is a 78-kDa, slightly basic, iron-binding protein related to transferrin. Lactoferrin is found in many secretions, including tears, semen, and milk, as well as in the secondary granules of neutrophils. It has two binding sites for ferric iron. The bacteriostatic activity of lactoferrin is associated with iron-unsaturated lactoferrin. Presumably the inhibitory action of lactoferrin is due to its ability to sequester iron from bacteria that require it for synthesis of multiple components (Bullen *et al.*, 1978). Lactoferrin also exerts bactericidal effects that are independent of its bacteriostatic effects (Lehrer and Ganz, 1990).

Acid hydrolases are present in the primary granules and probably act after the microbe has been killed (Bertram, 1985). These enzymes have broad specificities to a variety of substrates. The acid hydrolases include acid phosphatases, cathepsins, aryl sulfatases, neuraminidase, and nucleases (Klebanoff and Clark, 1978). The acid hydrolases are inhibited by sulfated glycosaminoglycans, heparin, and chondroitin sulfate at acid pH (Avila and Convit, 1976).

III. LEUKOTRIENES AND PROSTAGLANDINS

Prostaglandins and leukotrienes are synthesized by many cells, including neutrophils, macrophages, mast cells, basophils, and eosinophils. Arachidonic acid and other C_{20} polyunsaturated fatty acids containing three, four, and five carbon–carbon double bonds are the immediate precursors to the prostaglandins and leukotrienes. Since arachidonic acid and most of its metabolites contain 20 carbon atoms, they are referred to as eicosanoids. The precursor eicosanoids are released from phospholipids by the action of phospholipase A_2 or phospholipase C and diacylglycerol lipase (Van der Bosch, 1980; Bell *et al.*, 1979; Bates, 1995). Dihomo-γ-linoleic acid (C_{20}-Δ8,11,14) is the precursor for PGE_1 and $PGE_{1\alpha}$; arachidonic acid (C_{20}-Δ5,8,11,14) is the precursor for PGE_2, $PGE_{2\alpha}$, PGG_2, PGH_2, prostacyclin (PGI_2), the thromboxanes (TXA_2 and TXB_2), and the leukotrienes;

and eicosopentaenoic acid (C_{20}-Δ5,8,11,14,17) is the precursor for PGE_3 and $PGF_{3\alpha}$ (Glew, 1992). Among the leukocytes, macrophages appear to be the major producer of prostaglandins (Kurland and Bockman, 1978). Prostaglandin and thromboxane synthesis begins with the action of cyclooxygenase on arachidonic acid to produce endoperoxides PGG_2 and PGH_2. PGH_2 is the common substrate for PGE_2, PGD_2, $PGF_{2\alpha}$, PGI_2 (prostacyclin), and TXA_2 (Roubin and Benveniste, 1985).

In contrast to the broad array of cells that synthesize prostaglandins, the synthesis of leukotrienes is restricted. Neutrophils, monocytes, macrophages, eosinophils, and mast cells are the major producers of leukotrienes (Lewis and Austen, 1984). Leukotriene synthesis begins when 5-lipoxygenase catalyzes the production of 5-monohydroperoxy-eicosatetraenoic acid (5-HPETE) from arachidonic acid. 5-HPETE is spontaneously or catalytically converted to 5-hydroxyeicosatetraenoic acid (5-HETE). In addition, 5-lipoxygenase converts 5-HPETE to LTA_4 (Roubin and Benveniste, 1985). Separate pathways synthesize LTB_4, and the sulfidopeptide leukotrienes (LTC_4, LTD_4, LTE_4 and LTF_4). LTB_4 may be oxidized to less active compounds through oxidation, and the sulfidopeptide leukotrienes may be oxidized to sulfoxides followed by cleavage of the peptide portion to produce 5,12-diHETEs. Further metabolism of the leukotrienes probably involves extensive β-oxidation (Davies *et al.*, 1984).

Prostaglandins and leukotrienes are synthesized and immediately released rather than being stored. The synthesis of these substances is triggered by lymphokines and phagocytosis (Roubin and Benveniste, 1985). Large amounts of prostaglandins and leukotrienes are released by stimulated macrophages. In addition, macrophages are able to synthesize prostaglandins and leukotrienes from exogenous sources of arachidonic acid such as the lipids of necrotic cells. Among the cells of inflammation, there is a remarkable specificity in the generation of the leukotrienes. Neutrophils synthesize mostly LTB_4, eosinophils and mast cells synthesize most LTC_4 (and presumably the other sulfidopeptide leukotrienes), and macrophages and monocytes synthesize large amounts of both leukotriene series (Lewis and Austen, 1984).

The activities of the prostaglandins and leukotrienes are broad. Their activities include enhanced vascular permeability and affecting blood pressure, platelet aggregation, slow-reacting substance activity, and chemotaxis. The actions produced by some of these products may be counteracted by an opposite effect of another product. For example, thromboxane A_2 is a very potent platelet-aggregating agent, whereas prostacyclin antagonizes this effect (Roubin and Benveniste, 1985).

LTB_4 has both a priming and a direct agonist effect on neutrophils. LTB_4 is a potent chemotactic and chemokinetic agent for neutrophils, monocytes, and eosinophils. It stimulates aggregation of neutrophils and enhances lysosomal enzyme releases, superoxide production, expression of CD11a/CD18 and CD11b/CD18, endothelial cell adherence, and complement-dependent killing reactions (Ford-Hutchinson, 1983; Koenig *et al.*, 1983; Czarnetzki and Grabbe, 1983; Lewis and Austen, 1984; Bates, 1995).

LTC_4 and LTD_4 cause contraction of smooth muscle such as that of the small airways and intestinal tract. They also increase neutrophil adherence and microvascular permeability. These slow-reacting substances of anaphylaxis (SRS-As) also affect arteriolar constriction and augment vascular permeability. Apparently, the constrictive effects are immediate and direct effects, and the increased permeability is a subsequent response that may last for several hours (Lewis and Austen, 1984).

Macrophages are the main PGE_2-producing cell of the phagocytic/immune system. PGE_2 production is caused by a variety of stimuli. PGE_2 causes downregulation of the immune activity or immunosuppression by causing T cells to produce lymphokines that are able to inhibit the activity of lymphocytes (Cueppens and Goodwin, 1982). Prostaglandins may inhibit chemotaxis by causing the intracellular production of cAMP; they may inhibit phagocytosis and the release of lysosomal enzymes (DeChatelet, 1979).

IV. CYTOKINES AND HOST RESISTANCE TO DISEASE

Cytokines are a diverse collection of molecules that are involved in activation or priming of macrophages (Mannel *et al.*, 1980). Neutrophils can also be primed by cytokines to have enhanced functional activity (Bajaj *et al.*, 1992; Sample and Czuprynski, 1991). Cytokines have been defined as a group of protein cell regulators produced by a wide variety of cells in the body. They play an important role in many physiological responses, and they are involved in the pathophysiology of many diseases. They also have therapeutic potential (Balkwill and Burke, 1989). The immune system produces cytokines and other humoral factors of biological importance to protect the host when threatened by inflammatory agents or microbial infections.

In general, cytokines are involved in regulating immune and inflammatory responses. They are not found in cells as preformed molecules, and their production is transient. Many individual cytokines are produced by diverse cell types. For example, interferons, tumor necrosis factors, interleukin-1, and colony-stimulating factors (G-CSF, GM-CSF) are individually or collectively produced and released by leukocytes (Bienhoff *et al.*, 1992; Simon and Willoughby, 1981; Scales, 1992; Tiku *et al.*, 1986), epidermal cells (Luger *et al.*, 1985), and other cells in response to biological and physical injury. Research during the last few years has shown that (1) a single cytokine is not only produced from one or multiple cell types, but it also interacts with more than one cell type, (2) a single cytokine has multiple biological activities, (3) a single cell can interact with more than one cytokine, and (4) many cytokines have overlapping biological activities (Mayer *et al.*, 1989; Tsuji and Tori, 1992; Scales, 1992).

Cells that respond to various cytokines must bear receptors for these ligands. The lack of sequence homology among various cytokines suggested that each cytokine would bind a unique receptor type (Rapoport *et al.*, 1992; Steinbeck and Roth, 1989; Tkatch and Tweardy, 1993). For example, two types of receptors have been described for interleukin-1: The type I receptor is found on T-lymphocytes, fibroblasts, endothelial cells, and hepatocytes (Dinarello, 1988; Urdal *et al.* 1988), whereas the type II receptor is found on B-lymphocytes, neutrophils, and bone marrow cells (Dinarello, 1988). Similarly, two distinct receptor types, A and B, specifically bind TNF-α and TNF-β and are found on a wide variety of cells (Dembic *et al.*, 1990; Schall *et al.*, 1990; Loetscher *et al.*, 1990). The receptor for IFN-γ has been identified and its gene localized to chromosome 6 in humans (Pfizenmaier *et al.*, 1988). A single class of high-affinity binding sites for G-CSF (Fukunaga *et al.*, 1990; Larsen *et al.*, 1990) and both high-affinity and low-affinity binding sites for GM-CSF have been detected on various hematopoietic cells, including granulocytes (Goodall *et al.*, 1993; Park *et al.*, 1992; Lopez *et al.*, 1992). Recently, it has been possible to define the structure of some of these cytokine receptors at the molecular level (Schreiber *et al.*, 1992; Lederer and Czuprynski, 1992). The cell surface receptor density may vary from as little as a few hundred to several thousand binding sites per cell (Rapoport *et al.*, 1992; Steinbeck and Roth, 1989; Cairo, 1991; Tkatch and Tweardy, 1993).

By binding to specific receptors on the neutrophil surface, cytokines augment different cell functions. *In vitro* and *in vivo* studies on neutrophils from humans and animals indicate that cytokines can augment their functional activities, including the ability to localize at the site of inflammation, phagocytic activity, production of oxygen metabolites, and release of lysosomal enzymes (Steinbeck and Roth, 1989; Rapoport *et al.*, 1992; Cairo, 1991; Gasson, 1991; Tkatch and Tweardy, 1993). For example, *in vivo* enhancement of neutrophil functions in humans by rhIFN-γ and rhG-CSF has been associated with increased resistance to microbial infections (Morrison *et al.*, 1989; Daifuku *et al.*, 1993; Hengge *et al.*, 1992). Administration of recombinant bovine IFN-γ and rhG-CSF in dairy cows was found to reduce development of experimental *E. coli* (Sordillo *et al.*, 1991) and *S. aureus* mastitis, respectively (Nickerson *et al.*, 1989). The profound influence of these cytokines on phagocytic cells suggests a possible clinical application for prevention of bacterial infections.

V. NEUTROPHIL-MEDIATED TISSUE INJURY

Products of neutrophils cause tissue damage. In numerous experimental models of inflammation, depletion of neutrophils attenuates the intensity of tissue damage. Proteases and reactive oxygen products, including $O_2^{\bar{\cdot}}$, H_2O_2, $^{\cdot}OH$, and HOCl, as well as $NO^{\cdot}$ and its metabolites, contribute to the tissue damage. In addition, neutrophil elastase may convert endothelial cell xanthine dehydrogenase to xanthine oxidase, which generates $O_2^{\bar{\cdot}}$ from xanthine or hypoxanthine. The active oxygen products injure the cell. In addition, $O_2^{\bar{\cdot}}$ may react with $NO^{\cdot}$ to produce peroxynitrite anion ($ONOO^-$), which also damages tissue (Ward and Mulligan, 1992). Tissue damage during inflammation can be reduced by *N*-monomethyl-L-arginine (NMA). NMA competitively inhibits nitric oxide synthetase (Marletta, 1989).

VI. DEFECTS IN NEUTROPHIL FUNCTION

Numerous defects in neutrophil function have been described in humans. Despite this, only a few defects have been recognized in animals. Leukocyte adhesion defects have been described in Holstein–Friesian calves and Irish setter dogs. Chediak–Higashi syndome has been found in several species of animals. A poorly characterized syndrome probably caused by defective neutrophils has been found in Doberman pinscher dogs.

A. Granule Abnormalities

Specific granule deficiency is a very rare condition (Gallin, 1985). Humans with this disorder have depressed inflammatory response and recurrent severe bacterial infections of the skin and deep tissues (Malech and Gallin, 1987). Neutrophils have small, elongated granules that appear late in maturation. The granules associated with defensins are smaller (Parmley *et al.*, 1989). In addition, nuclear morphological abnormalities such as bilobulation and nuclear blebs and clefts are present (Mills and Noya, 1993). Neutrophils of patients with specific granule deficiency lack lactoferrin, vitamin B_{12}-binding protein, and defensins. Tertiary granule gelatinase activity is decreased. Myeloperoxidase activity is normal, and nonstimulated adherence, aggregation, and respiratory-burst activity are normal or enhanced. However, upregulation of receptors does not occur, because these are normally present on the mobilizable pools on the specific granule membranes. (Edwards, 1994). Chemotaxis of these neutrophils is impaired, probably because of the failure to upregulate receptors stored within granules (Mills and Noya, 1993).

Chediak–Higashi (CH) syndrome has been described in children and in beige mice, mink, Hereford cattle, and a killer whale (Jain, 1986). Humans and animals have partial oculocutaneous albinism, frequent infections, and mild bleeding diatheses. Melanocytes, neutrophils, and monocytes contain giant granules (Edwards, 1994). Increased incidence of infections occurs because granule membrane fusion and activation are defective, and the discharge of granule contents into phagocytic vacuoles is delayed (Mills and Noya, 1993). Defective chemotaxis has been described in children, mink, and the beige mouse with CH (Clark, 1978; Smith and Lumsden, 1983). However, this appears to be, at least in part, an artifact caused by performing the test in filters with small pore size. When chemotaxis is evaluated with large-pore-size filters the chemotaxis approachs that of normal cells (Wilkinson, 1982).

B. Adhesion Defects

A genetic disorder termed leukocyte adhesion deficiency (LAD) has been recognized and extensively described in humans (Anderson and Springer, 1987; Kishimoto *et al.*, 1987; Todd and Freyer, 1988). Similar disorders have been described in Irish setter dogs and Hostein–Friesian cattle (Renshaw and Davis, 1979; Giger *et al.*, 1987; Hagemoser *et al.*, 1983; Trowald-Wigh *et al.*, 1992).

This autosomal recessive trait is characterized by recurrent soft-tissue infections, severely impaired pus formation, constant leukocytosis, and profound *in vitro* abnormalities of a wide spectrum of adhesion-dependent functions of granulocytes, monocytes, and lymphocytes. Leukocytes of human patients with LAD have a deficiency or total lack of the family of structurally and functionally related CD11/CD18 glycoproteins, including Mac-1, LFA-1, and p150,95 (Sanchez-Madrid *et al.*, 1983). Each of these molecule contains an α- and a β-subunit, noncovalently associated in an $\alpha_1\beta_1$ structure. They share an identical β-subunit (CD18) and are distinguished immunologically by their α-subunits, designated CD11a, CD11b, and CD11c (LFA-1α, Mac-1α, and p150,95α, respectively). Heterogeneous mutations of the gene encoding the common β subunits (CD18) have been identified as the primary basis for disease in all human cases (Kishimoto *et al.*, 1987; Dana *et al.*, 1987; Kishimoto *et al.*, 1989).

Although the pathogenesis of disease in Irish setters has not been completely studied, the reported findings of deficient expression of Mac-1, LFA-1, and p150,95 on leukocyte surfaces suggests a similar underlying defect of the gene encoding canine CD18 (Giger *et al.*, 1987; Trowald-Wigh *et al.*, 1992).

Bovine leukocyte adhesion deficiency (BLAD), formerly called bovine granulocytopathy syndrome, is a neutrophil function defect of Holstein–Friesian cattle (Hagemoser *et al.*, 1983; Takahashi *et al.*, 1987; Kehrli *et al.*, 1990). Clinical features of this syndrome include recurrent infection of soft tissues and delayed wound healing. Prominent features include persistent and progressive neutrophilia, decreased neutrophil infiltration into infected tissues, and impaired neutrophil chemotaxis, phagocytosis, and oxidative burst (Kehrli *et al.*, 1990; Nagahata *et al.*, 1994). The molecular basis is similar to that of human LAD, the genetic deficiency of CD11/CD18 adhesion molecules. Specifically, a calf with the granulocytopathy syndrome was severely deficiency in Mac-1 (CD11b/CD18) (Kehrli *et al.*, 1990). The genetic lesion in calves was determined to be a homozygous point mutation of adenine to guanine at nucleotide 383 in the gene encoding CD18. This point mutation causes replacement of asparate with glycine at amino acid 128 and resides in the highly conserved extracellular domain, amino acids 96–389 (Shuster *et al.*, 1992a, b). A defect in this region either prevents association with the α-subunit to form LFA-1, MAC-1, and P150,95, or alters adhesion to ICAM-1 (CD54) and ICAM-2 (CD102) (Springer, 1990; Shuster *et al.*, 1992a, b).

A second syndrome of defective neutrophil adhesion has been described in humans. This syndrome involves the selectin system of adherence rather than the β_2 integrins. The molecular mechanism underlying this syndrome is probably a fucose metabolism defect that results in an inability of leukocytes to produce sialylated glycoprotein ligands for the vascular selectins (Jones *et al.*, 1995). As a consequence, the defect involving β_2 integrins has been renamed leukocyte adhesion deficiency type 1 (LAD-1) and the defect involving selectins has been named type 2 (LAD-2) (Jones *et al.*, 1995).

C. Defects in Chemotaxis

Abnormalities of chemotaxis can be divided into cellular defects and chemotactic factor defects. In the former, the patient's leukocytes respond poorly to chemoattractants or there are cell-directed inhibitors of chemotaxis. In the latter, there is a failure of synthesis of chemoattractants, or the serum contains inhibitors of chemoattractants (Clark, 1978; Wilkinson, 1982; Quie, 1983).

Several defects in the chemotactic ability (or locomotor ability) of neutrophils have been described in humans. The "lazy leukocyte syndrome," originally described in two children, is characterized by neutrophils that have decreased random and directional locomotion (Miller *et al.,* 1971). Neutropenia accompanies some of the cases and is ascribed to an inability to mobilize neutrophils from the marrow (Wilkinson, 1982).

Neutrophil actin dysfunction was found in a male infant who had repeated infections from birth and failed to produce pus (Boxer *et al.,* 1974). The neutrophils of the patient were morphologically normal. Serum was active in generating chemotactic factors and opsonins. However, the neutrophils failed to migrate toward chemotactic substances, and they phagocytized at a markedly depressed rate. Actin from the neutrophils failed to polymerize.

Neutrophils of patients with Pelger–Huet anomaly have impaired locomotion through narrow-pore-size filters and less impairment through filters of wider pore size (Park *et al.,* 1977). Neutrophils of dogs with Pelger–Huet anomaly have given variable results in chemotaxis tests (Smith and Lumsden, 1983). If chemotaxis is impaired in the Pelger–Huet anomaly, the impairment is not severe enough to result in severe infections. Furthermore, the impairment of chemotaxis that has been found in patients with left shifts may be related to the small pore size used in the assay system causing the nuclei of band neutrophils, metamyelocytes, and so forth, to detain the cells in the filter pores (Frei *et al.,* 1978).

Neutrophils of humans with hyperimmunoglobulin E-recurrent infection syndrome have defective chemotaxis. These people have depressed inflammatory response. Despite marked local infection, neutrophils are present only in low numbers. Chemotaxis is variable and probably does not account fully for impaired host protection. The basis for the association between excess IgE and impaired chemotaxis may be the IgE-mediated release of histamine causing the activation of H_2 receptors on the neutrophil with subsequent elevation of the cAMP levels in the cell (Hill, 1984). Although the mechanism is unknown, the ratio of cAMP to cGMP is important for normal chemotactic function of neutrophils. The effect of cyclic nucleotides appears to be on the microtubular system (Hill, 1978).

Other conditions in which a cellular defect in chemotaxis has been observed include hyperimmunoglobulin A (IgA), chronic mucocutaneous candidiasis, diabetes mellitus, thermal injury, hypophosphatemia, and juvenile periodontitis. Inhibitors of cellular mechanism of chemotaxis has been suggested in a few people with IgA myeloma (Clark, 1978).

Defective production of the chemoattractant C5a is associated with the total absence of C3 and C5 and results in defective inflammatory responses and severe and recurrent infections. Deficiencies in C1, C2, and C4 are not as important, because C3, the activator of C5, can be activated by either the alternative or the classical pathway. However, the more rapid generation of complement derived chemoattractant occurs through the classical pathway (Clark, 1978). Abnormalities in the formation of the Hageman factor-dependent chemotactic agents, kallikrein and plasminogen activator, have been demonstrated (Clark, 1978). Inhibitors of chemotactic factors have been demonstrated in various conditions, including uremia and cirrhosis (Clark, 1978).

D. Defective Phagocytosis

Defects of phagocytosis often accompany cellular defects of chemotaxis (Hill, 1984). In addition, failure to opsonize bacteria can also result in inability to phagocytize the bacteria. Impaired opsonization of particles by serum occurs with deficiencies of complement components, immunoglobulin, or tuftsin (Klebanoff and Clark, 1978). The ability of some microorganisms to resist phagocytosis probably accounts for their virulence. Perhaps the surface components decrease the hydrophobicity of the organisms and make them less susceptible to phagocytosis (Van Oss *et al.,* 1974). Phagocytosis by neutrophils is better in colostrum-fed calves and lambs than in colostrum-deprived calves and lambs (Lamotte and Eberhart, 1976; Piercy, 1973). Presumably, the decreased ability of the neutrophils of the colostrum-deprived animals is caused by decreased opsonization of organisms rather than by a defect in the neutrophils that could be corrected by a factor in colostrum.

E. Defective Bacteria Killing

In humans, there are several conditions that can lead to chronic infections due to defects in the respiratory burst mechanisms. The most severe of these conditions is chronic granulomatous disease (CGD). In addition, glutathione peroxidase deficiency, severe glucose-6-phosphate deficiency, and myeloperoxidase deficiency may produce defects in bacteria killing by neutrophils (Babior, 1978b; van der Valk and Herman, 1987).

CGD is characterized by recurrent and severe bacterial infections with abscess formation from early childhood. The infections are usually not fatal. The functional defect in CGD is the inability to mount a respiratory burst. The defect is due to an inability to activate the NADPH oxidase system (Segal *et al.*, 1983; Borregaard, 1985). Four different types of CGD have been characterized. These correspond to defects in the proteins integral to assembly of complete NADPH oxidase. A defective β-subunit is the most common form of CGD. It is inherited in an X-linked fashion.The lesion in the gene encoding gp91-*phox* can lead to absence of the protein product, presence of a nonfunctional protein, partial deficiency of gp91-*phox,* or assembly of an abnormal oxidase (Mills and Noya, 1993). A small percentage of people with CGD have abnormal p22-*phox.* Absence of the cytosolic components of NADPH oxidase has also been described, with absence of p46-*phox* being much more common than absence of p67-*phox* (Mills and Noya, 1993). The diagnosis of CGD is based on the demonstration of absent or extemely low respiratory burst in stimulated neutrophils (Mills and Noya, 1993).

Glutathione peroxidase (GPx) deficiency causes a condition similar to CGD (Klebanoff and Clark, 1978; Babior, 1978b). Apparently this is due to an inability to maintain the electron-transport chain necessary for the generation of NADPH. Impaired neutrophil bactericidal ability has been found in selenium-deficient cattle and in selenium- and copper-deficient cattle (Boyne and Arthur, 1979, 1981). The requirement for selenium in glutathione metabolism and the relationship of glutathione to the HMP and respiratory burst suggests that selenium deficiency may result in a defect in the oxygen dependent pathway of bacteria killing (Serfass and Ganther, 1975).

Severe glucose-6-phosphate deficiency (G6P-D) also causes impairment of the respiratory burst in neutrophils. The failure of the neutrophils to mount a respiratory burst is also due to insufficient generation of NADPH (Babior, 1978).

Myeloperoxidase (MPO) deficiency produces few clinical signs (van der Valk and Herman, 1987). Destruction of bacteria by MPO cells is delayed but is usually complete. Oxidizing radicals apparently fulfill the antibacterial role of the peroxide–halide–MPO system in the deficient cells (Babior, 1978b). Myeloperoxidase deficiency is the most common of the inherited neutrophil defects. Flow cytometric methods for determining neutrophil differential counts by peroxidase staining of neutrophil has revealed that the defect is quite common in humans. Yet the number of individuals that develop severe infections is very small. This observation indicates that alternative pathways for killing organisms, which may be slower than the hypochlorous acid mechanism, act to prevent severe, chonic infections in people with MPO deficiency (Mills and Noya, 1993; Edwards, 1994).

In glutathione reductase (GR) deficiency, there is no defect in the killing ability of the cells, but rather there is an inability to detoxify excess H_2O_2. H_2O_2 is toxic to cells and tissues. The oxidant stress precipitates acute hemolysis. People with GR deficiency also develope cataracts early in life. No increase in susceptibility to infections has been noted (Babior, 1978b).

Defective bacteria killing by neutrophils in several closely related Doberman pinschers has been reported. Neutrophils phagocytized bacteria normally, but had impaired bactericidal ability. Stimulated neutrophils had reduced ability to reduce nitroblue tetrazolium (NBT) and generate oxygen radicals. Neutrophils phagocytized opsonized Oil Red O and *Staphylococcus epidermidis* normally. However, killing of *S. .epidermidis* was depressed in neutrophils of the affected dogs. The authors concluded that the disease was similar to CGD or complement receptor deficiency in man (Breitschwerdt *et al.*, 1987) . Because several of the dogs continued to live with no or mild infections, it is unlikely that they had a complement receptor deficiency such as CD18/CD11b deficiency of man. Although decreased NBT reduction was found, it was not as pronounced as found in human CGD neutrophils. It is possible that the dogs had a variation of some other oxygen-dependent bactericidal mechanism such as myeloperoxidase deficiency.

Bovine, ovine, and porcine neutrophils produced less superoxide and consumed less oxygen than human neutrophils when they were challenged with serum-treated zymosan (Young and Beswick, 1986). The iodination reaction, which evaluates the activity of the myeloperoxidase-H_2O_2-halide antibacterial system and the NBT reduction, is decreased in young calves when compared to that in older calves (Hauser *et al.*, 1986).

Neutrophils of humans and dogs with primary ciliary dyskinesia may have defective chemotaxis. Neutrophils of dogs with primary ciliary dyskinesia show increased distance of random migration, but fewer of the neutrophils migrate. Phagocytosis is enhanced and cytochrome c reduction, iodination, and antibody-dependent, cell-mediated cytotoxicity are unaffected in dogs with primary ciliary dyskinesia. The contribution of neutrophil chemotaxis dysfunction to the increased susceptibility of dogs with primary ciliary dyskinesia to respiratory diseases is not known. It is likely that this susceptibility is due to respiratory ciliary dysfunction rather than to neutrophil dysfunction (Morrison *et al.*, 1987).

VII. METHODS FOR TESTING NEUTROPHIL FUNCTION

A. Cell Isolation

Many functions of neutrophils can be tested *in vitro*. Although some functions can be evaluated using whole blood, a preparation of nearly pure neutrophils is usually required. Neutrophils can be isolated from blood through sedimentation, differential centrifugation, or erythrocyte lysis.

A mixed population of leukocytes may be obtained from horse blood by allowing an anticoagulated (EDTA or heparin) sample to stand vertically for 15 to 30 minutes (Zinkl and Brown, 1982). The leukocytes of some other species can be isolated by causing erythrocyte rouleaux with a 3–6% suspension of high-molecular-weight dextran (200,000–500,000) (Metcalf *et al.*, 1986).

Differential centrifugation of whole blood or isolated leukocytes in a density gradient formed by sodium diatrizoate (Hypaque) and ficoll can be used to separate erythrocytes from leukocytes or separate the mononuclear leukocytes from the segmented leukocytes (English and Andersen, 1974). The cells separate based on their densities, with the lighter platelets, the mononuclears (lymphocytes and monocytes), and basophils found in the upper layer. The neutrophils and the eosinophils are intermediate in specific gravity and are above the heavy erythrocytes (Metcalf *et al.*, 1986).

Centrifugation of blood over a discontinuous gradient of Percoll [silica particles coated with poly(vinyl pyrrolidone)] can isolate leukocytes from blood (Riding and Willadsen, 1981; Sedgwick *et al.*, 1986). A discontinuous gradient of Percoll is made by layering a 60–68% Percoll solution over a 70–75% solution. (The concentrations of the Percoll that will successfully isolate the various leukocytes will vary with the species. Initially, several concentrations should be used until a pair of concentrations is found that allows separation of the cells into mononuclear, neutrophil, and erythrocyte layers.) After centrifugation at 200 to 500 *g* for 10 to 15 minutes, the blood will be separated into an upper plasma layer, a mononuclear zone just under the plasma, a polymorphonuclear zone at the interface of the two Percoll concentrations, and an erythrocyte layer at the bottom of the tube.

Techniques that lyse erythrocytes while leaving leukocytes intact can be used to isolate leukocytes from blood or to remove erythrocytes that contaminate leukocytes isolated by another technique. Short (20 to 60 seconds, depending upon the species) hypotonic shock lyses the erythrocytes, and isotonicity is restored with a hypertonic solution before the leukocytes are adversely affected (Carlson and Kaneko, 1973).

B. Adherence

Adherence can be studied by several methods. Most techniques involve sticking the cell to a glass surface, nylon fibers, or plastic. Whole blood or purified neutrophil fractions can be used (Metcalf *et al.*, 1986).

C. Chemotaxis

There are several methods to evaluate neutrophil chemotaxis. The most commonly used techniques are the micropore filter methods (Boyden, 1962; Nelson *et al.*, 1975). Many modifications of Boyden's original method have been used, but the basic technique remains unchanged. A chamber consisting of two compartments separated by a membrane filter with pores of about 2–8 μm is used to separate cells from the chemoattractant. The chemoattractant is placed in the lower compartment, and the suspension of leukocytes is placed in the upper compartment. The chamber is incubated at body temperature for 1 to 4 hours. The leukocytes respond to the chemoattractant by migrating into or through the filter. The number of cells migrating through the filter is determined and compared to the numbers migrating spontaneously. This method has been criticized because it does not distinguish between chemotactic and chemokinetic response. The checkerboard assay modification of the filter technique allows the differentiation of chemotactic and chemokinetic responses. With this modification, chemoattractant at various combinations of concentrations is placed in the cell compartment and the attractant compartment. The distances migrated are compared in order to determine the degree of response due to chemotaxis and that due to chemokinesis (Wilkinson, 1982; Metcalf *et al.*, 1986).

Migration under agarose and direct visulalization with specially designed chambers have also been useful for evaluating chemotaxis (Nelson *et al.*, 1975; Zigmond, 1978). A simple microdroplet assay for chemokinesis has been developed in which the migration of leukocytes out of agarose is measured (Smith and Walker, 1980). This technique has been used to determine that LTB_4 and PGE_2 are chemotactic to equine neutrophils (Lees *et al.*, 1986).

The ability of cells to respond to chemotactic stimuli can be determined by observation of shape change.

This technique is a rapid assay for the study of prechemotactic events in neutrophils. Isolated neutrophils are incubated with potential chemoattractants, the cells are fixed in 2% glutaraldehyde, and wet mounts of the fixed cells are examined by phase-contrast microscopy. If the substance to which the neutrophils are exposed has chemotactic activity, the neutrophils have a bipolar shape. They may or may not have a uropod. Unstimulated neutrophils remain round (Smith *et al.*, 1979).

In vivo techniques for evaluation of chemotaxis include the Rebuck skin window and similar methods in which cells are collected in chambers that have been adhered to abraded skin or implanted in various locations in the body. A chemoattractant is placed in the chamber and migrating cells are collected and counted (Klebanoff and Clark, 1978). A modification that eliminates the chamber is to inject chemoattractant intradermally and collect cells from the intradermal site (Hayashi *et al.*, 1974).

D. Phagocytosis

Phagocytosis can be measured by several methods. A common technique is to allow bacteria, yeast, or other particles and neutrophils to incubate together, and then to determine by light microscopy the number of neutrophils that contain particles, as well as the number of particles phagocytized per neutrophil (Metcalf *et al.*, 1986; Silva *et al.*, 1988; Kabbur *et al.*, 1991).

Ingestion of paraffin oil particles containing Oil Red O and bacterial lipopolysaccharide by neutrophils has been used to evaluate phagocytosis. After incubation is complete, extracellular paraffin is removed by washing and centrifugation, the cells are lysed, and the concentration of Oil Red O is determined by spectrophotometry (Stossel, 1973).

Other methods for determining phagocytosis include using *Candida albicans*, zymosan (cell-wall ghosts of *Saccharomyces cerevisiae*), and ^{51}Cr-labeled sheep erythrocytes (Metcalf *et al.*, 1986).

More recently, flow cytometric technology has been applied to evaluate phagocytosis by neutrophils. Latex particles, zymosan, and bacteria labeled with fluorescent material are incubated with neutrophils. The number of particles in numerous neutrophils can be determined rapidly with flow cytometric analysis (Dunn and Tyrer, 1981; Saad and Hageltorn, 1985; Kabbur *et al.*, 1995).

E. The Respiratory Burst

There are many methods for determining the activity of the oxygen-dependent mechanisms of bacteria killing. These include methods that depend upon the intact pathway for the generation of active oxygen species, as well as methods that determine the activity of the individual enzymes of the system. Among those methods that indicate the complete function of the oxygen-dependent pathways are reduction of nitrobluetetrazolium, chemiluminescence of stimulated cells, reduction of cytochrome c, oxygen consumption, production of H_2O_2, and detection of intracellular oxidized 2′,7-dichlorofluorescein diacetate dye (Babior, 1978a; Bertram, 1985; Rossi *et al.*, 1985; van der Valk and Herman, 1987; Bass *et al.*, 1983).

Oxygen consumption can be measured using an oxygen electrode. The rate of oxygen consumption is determined in resting neutrophils and in cells that have been stimulated (Metcalf *et al.*, 1986).

Production of H_2O_2 can be determined in several ways, including oxidation of leucodiacetyl-2,7-dichlorofluorescein, oxidation of scopletin, and the release of oxygen by catalase (Klebanoff and Clark, 1978; Metcalf *et al.*, 1986).

Ferricytochrome c reduction is useful to determine the amount of $O_2^{\cdot}$ produced. The assay is run with and without the addition of superoxide dismutase. Reduced ferricytochrome c absorbs light at 550 nm. The rate of superoxide dismutase-inhibitable ferricytochrome c reduction is proportional to the production of O_2-(Klebanoff and Clark, 1978).

Nitrobluetetrazolium is reduced to formazan by $O_2^{\cdot}$. NBT is a blue-black insoluble material that precipitates intracellularly. Microscopic visualization of the formazan provides a qualitative means of identifying cells producing $O_2^{\cdot}$. Formazan can also be quantified spectrophotometrically after extraction with *N,N*-dimethylformamide or hot pyridine (Metcalf *et al.*, 1986). The NBT reduction test has been combined with the phagocytosis test to provide information on phagocytic ability and active oxygen production simultaneously (Stossel, 1973; Jabs *et al.*, 1980; Valinoti *et al.*, 1988).

The respiratory burst of neutrophils causes generation of low-intensity light energy known as chemiluminescence (CL). CL is dependent upon the production of singlet oxygen (O^12). Singlet oxygen is produced when one of the unpaired electrons of oxygen is raised to a higher orbit with an inversion of spin. Light is emitted when the electron reverts to the ground state. The light is enhanced by luminol (5-amino-2,3-dihydro-1,4-phthalazinedione) and is measured with equipment designed to detect short luminescent events. The number of luminescent events over a specific time is recorded in stimulated and nonstimulated cells and provides an index of the respiratory burst (Anderson *et al.*, 1980).

Flow cytometric technology has also applied to measure respiratory burst activities of neutrophils. 2′,7-Dichlorofluorescein diacetate dye has been used in flow cytometric studies to measure respiratory burst activity. 2′,7-Dichlorofluorescein diacetate (DCFH-DA) is a stable, nonfluorescent, nonpolar compound that can diffuse through granulocyte cell membranes. Once inside the cell, the acetyl groups are cleaved by cytosolic enzymes. The intracellular 2′,7′-dichlorofluorescein (DCFH) produced is polar and is thus trapped within the cell. DCFH is nonfluorescent, but it is rapidly oxidized to highly fluorescent 2′,7′-dichlorofluorescein (DCF) in the presence of hydrogen peroxide generated during the oxidative burst of neutrophils. With flow cytometry used to measure cellular fluorescence, the DCF assay provides quantitative assessment of the oxidative metabolic burst in individual granulocytes (Bass *et al.*, 1983).

To measure the activity of myeloperoxidase, a homogenate of cells is added to an H_2O_2 and 4-aminoantipyrine mixture. The rate of oxidation of 4-aminoantipyrine is measured by the increase in absorption at 510 nm (Metcalf *et al.*, 1986).

Iodination ability is determined by incubating neutrophils with zymosan and ^{125}I. After a prescribed incubation time, the protein is precipitated with trichloroacetic acid and the ^{125}I activity in the protein is determined (Klebanoff and Clark, 1978).

F. Other Enzymes of the Phagocytes

The activities of other enzymes such as *N*-acetyl-β-Glucosaminidase, β-glucuronidase, elastase, lysozyme, lactoferrin, vitamin B_{12}-binding protein, catalase, and other enzymes can be determined using isolated granules or cell homogenates (Metcalf *et al.*, 1986).

References

Anderson, D. C. (1995). *In* "Physiology and Pathophysiology of Leukocyte Adhesion" (D. N. Granger. and G. W. Schmid-Schönbein, eds.), p 3. Oxford University Press, New York.

Anderson, D. C., and Springer, T. A. (1987). *Annu. Rev. Med.* **38,** 175.

Anderson, D. C., Edwards, M. S., and Baker, C. J. (1980). *J. Infect. Dis.* **141,** 370.

Argenbright, L. W. (1995). In "Physiology and Pathophysiology of Leukocyte Adhesion" (D. N. Granger. and G. W. Schmid-Schönbein, eds.), p. 171. Oxford University Press, New York.

Avila, J. L., and Convit, J. (1976). *Biochem. J.* **160,** 129.

Babior, B. M. (1978a). *N. Engl. J. Med.* **298,** 659.

Babior, B. M. (1978b). *N. Engl. J. Med.* **298,** 721.

Badwey, J. A., and Karnovsky, M. L. (1980). *Ann. Rev. Biochem.* **49,** 695.

Baehner, R. L. (1995). *In* "Blood Diseases of Infancy and Childhood" (D. R. Miller, and R. L. Baehner, eds., p. 593. Mosby-Year Book Inc., St. Louis, Missouri.

Baggiolini, M., and DeWald, B. (1984). *Contemp. Top. Immunobiol.* **14,** 221.

Bajaj, M. S., Kew, R. R., Webster, R. D., and Hyers, T. M. (1992). *Inflammation* **16,** 241.

Balkwill, F. R., Burke, F. (1989). *Immunol Today* **10,** 299.

Bass, D. A., Parce, J. W., Dechatelet, L. R., Szejda, P., Seeds. M. C. and Thomas, M. (1983). *J. Immunol.* **130,** 1910.

Bates, E. J. (1995). *Prostaglandins, Leukotrienes and Essential Fatty Acids* **53,** 75.

Bell, R. L., Kennerly, D. A., Stanford N., and Majerus, P. W. (1979). *Proc. Natl. Acad. Sci. USA* **76,** 3228.

Bertram, T. A. (1985). *Adv. Vet. Sci. Comp. Med.* **30,** 91.

Bevilacqua, M. P. (1993). *Annu. Rev. Immunol.* **11,** 767.

Bienhoff, S. E., Allen, G. K., Berg, J. N.. (1992) *Vet. Immunol. Immunopathol.* **30,** 341.

Blondin, J., and Janoff, A. (1976). *J. Clin. Chem.* **58,** 971.

Borregaard, N. (1985). *Immunol. Let.* **11,** 165.

Boxer, L. A., Hedley-Whyte, E. T., and Stossel, T. P. (1974). *New. Engl. J. Med.* **291,** 1093.

Boyden, S. V. (1962). *J. Exp. Med.* **115,** 453.

Boyne, R., and Arthur, J. R. (1979). *J. Comp. Pathol.* **89,** 151.

Boyne, R., and Arthur, J. R. (1981). *J. Comp. Pathol.* **91,** 271.

Breitschwerdt, E. B., Brown, T. T., and De Buysscher, E. V. (1987). *Am. J. Vet. Res.* **48,** 1054.

Brune, K., and Spitznagel, J. K. (1973). *J. Infect. Dis.* **127,** 84.

Bullen, J. J., Rogers, H. J., and Griffiths, E. (1978). *Curr. Top Microbiol. Immunol.* **80,** 1.

Cairo, M. S. (1991). Clin. Perinatol. **18,** 343.

Carlson, G. P., and Kaneko, J. J. (1973). *Proc. Soc. Exp. Biol. Med.* **142,** 853.

Cillari, E, Arcoleo, F., Dieli, M., D 'Agostino, R., Gromo, G., Leoni, F., and Milano, S. (1994). *Infect. Immunity* **62,** 2649.

Clark, R. A. (1978). *In* "Leukocyte Chemotaxis" (J. I. Gallin, and P. G. Quie, eds.), p. 329. Raven Press, New York.

Couto, M. A. (1992a). *Infect. Immunity* **60,** 3065.

Couto, M. A. (1992b). *Infect. Immunity* **60,** 5042.

Cross, A. R. (1992). *In* "The Molecular Basis of Oxidative Damage by Leukocytes" (A. J. Jesaitis and E. A. Dratz, eds), p. 37. CRC Press, Boca Raton, Florida.

Cueppens, J. L., and Goodwin, J. A. (1982). *Cell Immunol.* **70,** 41.

Curnutte, J. T. (1993). *Clin. Immunol. Immunopathol.* **67,** S2.

Czarnetzki, B. M., and Grabbe, J. (1983). *Agents and Actions Supplement* **12,** 204.

Daifuku, R., Andresen, J., and Morstyn, G. (1993). *J. Antimicrob. Chemother.* **32,** Suppl A, 91.

Dana, N., Clayton, L. K., Tennen, D. G., Pierce, M. W., Lachmann, P. J., Law, S. A., and Arnaout, M. A. (1987). *J. Clin. Invest.* **79,** 1010.

Davies, P., Bailey, P. J., Goldenberg, M. M., and Ford-Hutchinson, A. W. (1984). *Ann. Rev. Immunol.* **2,** 335.

DeChatelet, L. R. (1979). *In* "Phagocytosis and Cellular Immunity" (H. H. Gadebusch, ed.), p. 1. CRC Press, Boca Raton, Florida.

Dembic, Z., Loetscher, H., Gubler, U., Pan, Y. C., Lahm, H. W., Gentz, R., Brockhaus, M., Lesslayer, W. (1990). *Cytokine* **2,** 231.

Dinarello, C. A. (1988). *Blood Purif.* **6,** 164.

Doré, M., Simon, S. I., Hughes, B. J., Entman, M. L., and Smith, C. W. (1995). *Vet. Pathol.* **32,** 258.

Dunn, P. A., and Tyrer, H. W. (1981). *J. Lab. Clin. Med.* **98,** 374.

Edwards, S. W. (1994). "Biochemistry and Physiology of the Neutrophil." Cambridge University Press, New York.

English, D., and Andersen, B. R. (1974). *J. Immunol. Methods* **5,** 249.

Evans, E. W. and Harmon, B. G. (1995). *Vet. Clin. Pathol.* **24,** 109.

Evans, W. H., and Karnovsky, M. L. (1962). *Biochemistry* **1,** 159.

Ford-Hutchinson, A. W. (1983). *Agents and Actions* sup. **12,** 154.

Forrest, M., and Paulson, J. C. (1995). *In* "Physiology and Pathophysiology of Leukocyte Adhesion" (D. N. Granger. and G. W.,

Schmid-Schönbein, eds.), p. 43. Oxford University Press, New York.
Frei, P. C., Hermanovicz, A., and Pecond, A. (1978). *J. Lab. Clin. Med.* **92,** 577.
Fukunaga, R., Ishizaka-Ikeda, E., Seto, Y., and Nagata, S. (1990). *Cell* **61,** 341.
Gabig, T. G., and Babior, B. M. (1981). *Ann. Rev. Med.* **32,** 313.
Gallin, J. I. (1985). *Annu. Rev. Med.* **36,** 263.
Ganz, T., Selsted, M. E., Szklanek, D., Harwig, S. S. L., Daher, K, and Lehrer, R. I. (1985). *J. Clin. Invest.* **76,** 1427.
Ganz, T., Selsted, M. E., and Lehrer, R. I. (1986). *Semin. Respir. Infect.* **1,** 107.
Gasson, J. C. (1991). *Blood* 77, 1131.
Gennaro, R., Odzani, L., and Romeo, D. (1983). *Infect. Immun.* **40,** 684.
Giger, U., Boxer, L. A., Simpson, P. J., Lucchesi, B. R., and Todd, R. F. III (1987). *Blood* **69,** 1622.
Glew, R. H. (1992). *In* "Textbook of Biochemistry," 3rd ed. (T. M. Devlin, ed.), p. 423. Wiley-Liss Inc., New York.
Goetzl, E. J., and Austen, K. F. (1974). *J. Clin. Invest.* **53,** 591.
Goodall, G. J., Bagley, C. J., Vadas, M. A., and Lopez, A. F. (1993). *Growth Factors* **8,** 87.
Hagemoser, W. A., Roth, J. A., Lofstedt, J., and Fagerland, J. A. (1983). *J. Am. Vet. Med. Assoc.* **183,** 1093.
Hauser, M. A., Koob, M. D., and Roth, J. A. (1986). *Am. J. Vet. Res.* **47,** 152.
Hayashi, H., Yoshinaga, M., and Yamamoto, S. (1974). *Antibiotics Chemotherapy* **19,** 296.
Hendricks, P. A. J., Verhoef, J., and Nijkamp, F. P. (1986). *Vet. Res. Commun.* **10,** 165.
Hengge, U. R., Brockmeyer, N. H., and Goos, M. (1992). *Clin. Investig.* **70,** 922.
Hill, H. R. (1978). *In* "Leukocyte Chemotaxis: Methods, Physiology, and Clinical Implications" (J. I. Gallin and P. G. Quie, eds.), p. 179. Raven Press, New York.
Hill, H. R. (1984). *Contemp. Topics Immunobiol.* **14,** 345.
Hormann, H. (1982). *Klin. Wochenschr.* **60,** 1265.
Horwitz, M. A., and Silverstein, S. C. (1980). *J. Clin. Invest.* **65,** 82.
Jabs, A. D., Regen, M., Horita, M., Yokoyama, M. M., and Tseng, H. C. (1980). *Lab. Manage.* **18,** 37.
Jain, N. C. (1986). "Schalm's Veterinary Hematology," 4th ed. Lea and Febiger, Philadelphia.
Janoff, A., and Scherer, J. (1968). *J. Exp. Med.* **128,** 1137.
Jones, D. A., Smith C. W., McIntire, L. V. (1995). In "Physiology and Pathophysiology of Leukocyte Adhesion" (D. N. Granger and G. W. Schmid-Schönbein, eds.), p. 148. Oxford University Press, New York.
Kabbur, M. B., Jain, N. C., Zinki, J. C., and Farver, T. B. (1991). *Am. J. Vet. Res.* **52,** 2023.
Kabbur, M. B., Jain, N.C., and Farver, T. B. (1995). *Comp. Haematol. Intl.* **5,** 47.
Kane, S. P., and Peters, T. J. (1975). *Clin. Sci. Mol. Med.* **49,** 171.
Kehrli, M. E., Jr., Schmalstieg, F. C., Anderson, D. C., Van der Maaten, M. J., Hughes, B. J., Ackermann, M. R., Wilhelmsen, C. L., Brown, G. B., Stevens, M. G., and Whetstone, C. A. (1990). *Am. J. Vet. Res.* **51,** 1826.
Kishimoto, T. K. (1991). *J. NIH Res.* **3**(9), 75.
Kishimoto, T. K., Hollander, N., Roberts, T. M., Anderson, D. C., and Springer, T. A. (1987). *Cell* **50,** 193.
Kishimoto, T. K., O'Conner, K., and Springer, T. A. (1989). *J. Biol. Chem.* **264,** 3588.
Klebanoff, S. J. (1975). *Semin. Hematol.* **12,** 117.
Klebanoff, S. J., and Clark R. A. (1978). "The Neutrophil: Function and Clinical Disorders" North-Holland, New York.
Koenig, W., Bremm, K. D., Mueller, P., Kunau, W. H., Borgeat, P., Spur, B., Crea, A. E. G., and Falsone, G. (1983). *Agents and Actions Supplement* **12,** 167.
Kubo, S., Rodriguez, T., Jr., Roh, M. S., Oyedeji, C., Romsdahl, M. M., and Nishioka, K. (1994). *Hepatology* **19,** 1044.
Kurland, J. I., and Bockman, R. (1978). *J. Exp. Med.* **147,** 952.
Lamotte, G. B., and Eberhart, R. J. (1976). *Am. J. Vet. Res.* **37,** 1189.
Larsen, A., Davis, T., Curtis, B. M., Gimbel, S., Sims, J. E., Cosman, D., Park, L., Soransen, F., March, G. J., and Smith, C. H. (1990). *J. Exp. Med.* **172,** 1559.
Lederer, J. A., and Czuprynski, C. J., (1992). *J. Leukoc. Biol.* **51,** 586.
Lees, P., Dawson, J., and Sedgwick, A.D. (1986). *Equine Vet J.* **18,** 493.
Leffell, M. S., and Spitznagel, J. K. (1972). *Infect. Immun.* **6,** 761.
Lehrer, R.I., and Ganz, T. (1990). *Blood* **76,** 2169.
Lehrer, R. I., Ladra, K. M., and Hake, R. B. (1975). *Infect. Immun.* **11,** 1226.
Lehrer, R.I., Ganz, T., and Selsted, M.E. (1991). *Cell* **64,** 229.
Lewis, R. A., and Austen, K. F. (1984). *J. Clin. Invest.* **73,** 889.
Loetscher, H., Pan, Y. C., Lahm, H. W., Gentz, R., Brockhaus, M., Tabuchi, H., and Lesslauer, W. (1990). *Cell* **61,** 351.
Lopez, A. F., Elliott, M. J., Woodcock, J., Vodas, M. A. (1992). *Immunol. Today* **13,** 495.
Luger, T. A., Kock A., and Danner, M. (1985). *Scand. J. Immunol.* **21,** 455.
McPhail, L. C., and Harvath L. (1993). *In* "The Neutrophil" (J. S. Abramson and J. G. Wheeler, eds.), p. 63. IRL Press, New York.
Malech, H. L., and Gallin, J. I. (1987). *N. Engl. J. Med.* **317,** 687.
Mannel, D. N., Moore R. N., and Mergenhagen, S. E. (1980). *Infect Immun* 30, 523
Marletta, M. A. (1989) *Trends in Biochem. Sci.* **14,** 488.
Metcalf, J. A., Gallin, J. I., Nauseef, W. M., and Root, R. K. (1986). *In* "Laboratory Manual of Neutrophil Function." Raven Press, New York.
Meuer, S. C. (1994). *In* "Adhesion Molecules" (C. D. Wegner, ed), p. 9. Academic Press, New York.
Meyer, T., Regenass, U., Fabbro, D., Alteri, E., Rosel, J., Muller, M., Caravatti, G., and Matter, A. (1989). *Int. J. Cancer* **43,** 851.
Miller, T. E. (1969). *J. Bacteriol.* **98,** 949.
Miller, M. E., Oski, F. A., and Harris, M. B. (1971). *Lancet* **1,** 665.
Mills, E. L., and Noya, F. J. D. (1993). *In* "The Neutrophil" (J. S. Abramson and J. G. Wheeler, eds.), p. 183. IRL Press, New York.
Morgan, P. B. (1990). "Complement: Clinical Aspects and Relevance to Disease." Academic Press, San Diego.
Morrison, W. B., Frank, D. E., Roth, J. A., and Wilsman, N. J. (1987). *J. Am. Vet. Med. Assoc.* **191,** 425.
Morrison, C. J., Brummer, E., and Stevens, D. A. (1989). *Infect. Immun.* **57,** 2953.
Mulligan, M. S., Miyaska, M., Suzuki, M., Kawashima, H., Iizuka, M., Hasegawa, A., Kiso, M., Warner, R. L., Word, P. A., and Suzuki, T. (1995). *Int. Immunol.* **7,** 1107.
Nagahata, H., Kehrli, M. E., Jr., Murata, H., Okada, H., Noda, H., Kociba, G.J. (1994). *Am. J. Vet. Res.* **55,** 40.
Najjar, V. A. (1983). *Ann. NY Acad. Sci.* **419,** 1.
Nelson, R. D., Quie, P. G., and Simmons, R. L. (1975). *J. Immunol.* **115,** 1650.
Nickerson, S. C., Owens, W. E., and Watts, J. L. (1989). *J. Dairy Sci.* **72,**3286.
Nishioka, K., Wagle, J. R., Rodriguez, T. Jr., Maeta, M., Kubo, S., and Dessens, S. E. (1994). *J. Surg. Res.* **56,** 94.
Odeberg, H., and Olsson, I. (1975). *J. Clin. Invest.* **56,** 1118.
Ohlsson, K., Olsson, I., and Spitznagel, J. K. (1977). *Hoppe Seylers Z. Physiol. Chem.* **358,** 361.
Oliver, J. M., Krawiec, J. A., and Becker, E. L. (1978). *J. Reticuloendothel. Soc.* **24,** 697.

Oppenheimer-Marks, N., and Lipsky, P.E. (1994). *In* "Adhesion Molecules" (C. D. Wegner, ed.), p. 141. Academic Press, New York.
Padgett, G. A., and Hirsch, J. G. (1967). *Aust. J. Exp. Biol. Med. Sci.* **45,** 569.
Park, B. H., Dolen, J., and Snyder, B. (1977). *Proc. Soc. Exp. Biol. Med.* **155,** 51.
Park, L. S., Luhr, S., Morrissey, P. J., Cosman, D., and Larsen, A. (1992). *Proc. Natl. Acad. Sci. USA* **89,** 4295.
Parmley, R. T., Gilbert, C. S., and Boxer, L. A. (1989). *Blood* **73,** 838.
Pfizenmaier, K., Wiegmann, K., Scheurich, P., Kronke, M., Merlin, G., Aguet, M., Knowles, B. B., and Ucer, U. (1988). *J. Immunol.* **141,** 856.
Piercy, D. W. T. (1973). *Res. Vet. Sci.* **14,** 350.
Proctor, R. A., Mosher, D. F., and Olbrantz, P. J. (1982). *J. Biol. Chem.* **147,** 88.
Quie, P. G. (1972). *Pediatrics* **50,** 264.
Quie, P. G. (1983). *Agents and Actions,* Sup. 12, 398.
Ramsey, W. S. (1974). *Exp. Cell. Res.* **72,** 489.
Rapoport, A. P., Abboud, C. N., and DiPersio, J. F. (1992). *Blood Rev.* **6,** 43.
Rausch, P. G., and Moore, T. G. (1975). *Blood* **46,** 913.
Renshaw, H. W., and Davis, W. C. (1979). *Am. J. Pathol.* **95,** 731.
Riding, G. A., and Willadsen, P. (1981). *J. Immunol. Methods* **46,** 113.
Rosales, C., and Brown, E. J. (1993). *In* "The Neutrophil" (J. S. Abramson and J. G. Wheeler, eds.), p. 23. IRL Press, New York.
Rossi, F., Della Bianca, V., and de Togni, P. (1985). *Comp. Immunol. Microbiol. Infect. Dis.* **8,** 187.
Roubin, R., and Benveniste, J. (1985). *Comp. Iummunol. Microbiol. Infect. Dis.* **8,** 109.
Saad, A. M., and Hageltorn, M. (1985). *Acta Vet. Scand.* **26,** 289.
Sample, A.K., and Czuprynski, C. J. (1991). *J. Leukoc. Biol.* **49,** 107.
Sanchez-Madrid, F., Nagy, J. A., Robbins, E., Simon, P., and Springer, T. A. (1983). *J. Exp. Med.* **158,** 1785.
Scales, W. E. (1992). *In* "Cytokines in Health and Disease" (S. L. Kunkel and D. G. Remick, eds). p. 15. Marcel Dekker, Inc., New York.
Schall, T. J., Lewis, M., Koller, K. J., Lee, A., Rice, G. C., Wong, G. H., Gatanaga, T., Granger, G. A., Lentz, R., and Raab, H. (1990). *Cell* **61,** 361.
Schreiber, R. D., Farrar, M. A., Hershey, G. K., and Fernandez-Luna, J. (1992). *Int. J. Immunopharmacol.* **14,** 413.
Sedgwick, A. D., Morris, T., Russell, B. A., and Lees, P. (1986). *Vet. Res. Commun.* **10,** 445.
Segal, N. A., Cross, A. R., Garcia, R. C., Borregaard, N., Valerius, N. H., Soothill J. F., and Jones, O. T. G. (1983). *N. Engl. J. Med.* **308,** 245.
Selsted, M. E., Tang, Y. Q., Morris, W. L., McGuire, P. A., Novotny, M. J., Smith, W., Henschen, A. H., and Cullor, J. S. (1993). *J. Biol. Chem.* **268,** 6641.
Senda, N., Tamura, N., Shibata, N., Yoshitake, J., Kondo, K., and Tanaka, K. (1975). *Exp. Cell. Res.* **91,** 393.
Serfass, R. E., and Ganther, H. E. (1975). *Nature* **255,** 640.
Shappell , S. B., and Smith, C.W. (1994). *In* "Adhesion Molecules" (C. D. Wegner, ed.), p. 29. Academic Press, New York.
Shepro, D. (1995). *In* "Physiology and Pathophysiology of Leukocyte Adhesion" (D. N. Granger and T. W. Schmid-Schönbein, eds.), p. 196. Oxford University Press, New York.
Shuster, D. E., Bowworth, B. T., and Kehrli, M. E., Jr. (1992a). *Gene* **114,** 267.
Shuster, D. E., Kehrli, M .E., Jr., Ackermann, M. R., and Gilbert, R. O.(1992b). *Proc. Natl. Acad. Sci. USA* **289,** 9225.
Silva, I. D., Jain, N. C., Farver, T. B., Zinkl, J. G. (1988). *J. Dairy Sci.* **71,** 2513.
Simon, P. L., Willoughby, W. F. (1981). *J. Immunol.* **126,** 1534.
Singh, S. P., Chhabra, R, and Srivastava, V. M. (1992). *Experientia* **48,** 994.
Smith, G. S., and Lumsden, J. H. (1983). *Vet. Immunol. Immunopathol.* **4,** 177.
Smith, M. J. H., and Walker, J. R. (1980). *Brit. J. Pharmacol.* **69,** 473.
Smith, C. W., Hollers, J. C., Patrick, R. A., and Hassett, C. (1979). *J. Clin. Invest.* **63,** 221.
Sordillo, L. M., Peel J., and Babiuk, L. A. (1991). National Mastitis Council, Arlington, p. 50.
Spitznagel, J. K. (1984). *Contemp. Top. Immunol.* **14,** 283.
Steinbeck, M. J., and Roth, J. A. (1989). *Rev. Infect. Dis.* **11,** 549.
Stossel, T. P. (1973). *Blood* **42,** 121.
Stossel, T. P. (1974). *N. Engl. J. Med.* **290,** 717.
Takahashi, K., Miyagawa, K., Abe, S., Kurosawa, T., Sonoda, M., Nakade, T., Nagahata, H., Noda, H., Chihaya, Y., and Isogai, E. (1987). *Jpn. J. Vet. Sci.* **49,** 733.
Thom, S. R., Ohnishi, S. T., and Ischiropoulos, H. (1994). *Toxicol. Applied Pharmacol.* **128,** 105.
Tiku, K., Tiku, M. L., and Skosey, J. L. (1986). *J. Immunol.* **136,** 3677.
Tkatch, L. S., and Tweardy, D. J. (1993). *Lymphokine Cytokine Res.* **12** 477.
Todd, R. F., Freyer, D. R. (1988). *Hematol. Oncol. Clin. North Am.* **2,** 13.
Trowald-Wigh, G., Hakansson, L., Johannisson, A., Norrgren, L., and Hard de Segerstad, C. (1992). *Vet. Immunol. Immunopathol.* **32,** 261.
Tsuji, Y., Tori, M. F. (1992). *In* "Cytokines in Health And Disease" (S. L. Kunkel and D. G. Remick eds), p. 131. Marcel Dekker, Inc, New York.
Urdal, D. L., Call, S. M., Jackson, J. L., and Dower, S. K. (1988). *J. Biol. Chem.* **263,** 2870.
Valinoti, J. M., Zinkl, J. G., and Jain, N. C. (1988). *Vet. Clin. Path* **17,** 99.
Van der Bosch, H. (1980). *Biochim. Biophys. Acta* **614,** 191.
van der Valk, P., and Herman, H. J. (1987). *Lab. Invest.* **57,** 127.
van Oss, C. J., and Gillman, C. F. (1972). *J. Reticuloendothel. Soc.* **12,** 497.
van Oss, C. J., and Gillman, C. F. (1973). *Immunol. Commun.* **2,** 415.
van Oss, C. J., Gillman, C. F., and Neumann, A. W. (1974). *Immunol. Commun.* **3,** 77.
Verhoef, J. and M. R. Visser (1993). *In* "The Neutrophil" (J. S. Abramson and J. G. Wheeler, eds.), p. 109. IRL Press, New York.
Ward, P. A., and M. S. Mulligan (1992). *In* "The Molecular Basis of Oxidative Damage by Leukocytes" (A. J. Jesaitis and E. A. Dratz, eds.), p. 139. CRC Press, Boca Raton, Florida.
Weiss, J., Elsbach, P., Olsson, I., and Odeberg, H. (1978). *J. Biol. Chem.* **253,** 2664.
Wilkinson, P. C. (1982). *In* "Chemotaxis and Inflammation," 2nd ed. p. 1. Churchill Livingston, New York.
Wilson, L. A., and Spitznagel, J. K. (1968). *J. Bacteriol.* **96,** 1339.
Young, S., and Beswick, P. (1986). *J. Comp. Path.* **96,** 193.
Zigmond, S. H. (1978). *In* "Leukocyte Chemotaxis" (J. L. Gallin and P. G. Quie, eds.) pp. 57–66. Raven, New York.
Zinkl, J. G., and Brown, P. D. (1982). *Am. J. Vet. Res.* **43,** 613.

CHAPTER

12

Clinical Enzymology

JOHN W. KRAMER AND WALTER E. HOFFMANN

I. INTRODUCTION

Clinical enzymology has made great advances since the introduction of serum alkaline phosphatase as a diagnostic aid in 1927 by King and Armstrong. Advancement in bioengineering and computer technology linked enzymology, immunology, and molecular genetics to produce new clinical diagnostic procedures. Qualitative and quantitative nonradioisotopic immunodiagnostics with enzyme-linked monoclonal antibodies enhances the specificity and sensitivity of assays for metabolites, hormones, and infectious agents. Immunocytochemistry has advanced the specificity and sensitivity of light and electron microscopic histology and cytology. Endonucleases combined with DNA probes enable the diagnostician to detect the fingerprints of prepatent genetic defects and infectious agents before or without the aid of endogenous antibody development. Although enzymes appear to have only a passive part in these forms of biotechnology, understanding their fundamental action is essential to clinical diagnostics and basic research.

This chapter is intended to be an introduction to enzymes as primary clinical diagnostic aids and their secondary role in other diagnostic techniques. Additional information on clinical enzymology can be located in the following serial publications: *Veterinary Clinical Pathology* and *Clinical Chemistry,* the respective publications of the American Society of Veterinary Clinical Pathology and the American Association of Clinical Chemistry, a division of Animal Clinical Chemistry. *Advances in Clinical Chemistry* and *Methods in Enzymology* are annual serials containing information on clinical enzymology. Organizations that help standardize enzymology are the International Union of Biochemistry's (IUB) Nomenclature Committee (EC) and the Expert Panel of the International Federation of Clinical Chemistry (IFCC).

I. BASIC ENZYMOLOGY

A. Enzyme Nomenclature

Enzyme nomenclature is based on the reaction an enzyme catalyzes. The suffix "-ase" identifies an enzyme. The IUB established a commission on enzyme nomenclature to systematize, categorize, and catalog enzymes (Webb, 1984). The system is numerical and based on the type of reaction catalyzed: a formal name, a common name, initials, and a number for each enzyme. Lactate dehydrogenase (LDH) illustrates this system. Lactate dehydrogenase is an "oxidoreductase" that can catalyze an oxidation or reduction reaction. When the enzyme incorporates a redox cofactor, such as NAD+, it is a dehydrogenase. Therefore, the formal name for LDH is L-lactate: NAD oxidoreductase, its common name is lactate dehydrogenase, and its Enzyme Commission number is EC 1.1.1.27. The abbreviation, common name, and EC number of some enzymes are listed in Table 12.1 and others are obtainable from the IUB listings (Webb, 1984).

Glutamic pyruvate transaminase (GPT) and glutamic oxaloacetate transaminase (GOT) are the former common names and abbreviations of enzymes frequently used in clinical medicine. The IUB commission recommends that these enzymes be classified as alanine aminotransferase (ALT) and aspartate aminotransferase (AST), respectively. The formal names and abbreviations recommended by the IUB for some clinically diagnostic enzymes are listed in Table 12.1.

TABLE 12.1 Enzymes Used in Veterinary Medicine

Abbreviation	Recommended name	EC number
ALT (GPT)	Alanine aminotransferase	2.6.1.2
AP	Alkaline phosphatase	3.1.3.1
Amyl	α-Amylase	3.2.1.1
ARG	Arginase	3.5.3.1
AST (GOT)	Aspartate aminotransferase	2.6.1.1
ChE	Cholinesterase	3.1.1.8
CK (CPK)	Creatine kinase	2.7.3.2
GGT	γ-Glutamyltransferase	2.3.2.2
GPx	Glutathione peroxidase	1.11.1.9
LDH	L-Lactate dehydrogenase	1.1.1.27
LIP	Lipase (triacylglycerol lipase)	3.1.1.3
OCT	Ornithine carbamoyltransferase	2.1.3.3
SDH	Sorbitol dehydrogenase (L-iditol dehydrogenase)	1.1.1.14
PK	Pyruvate kinase	2.7.1.40
TK	Transketolase	2.2.1.1
—	Trypsin	3.4.21.4

B. Units of Enzyme Activity

The concentration of an enzyme can be expressed directly as mass or indirectly as activity. Currently the most accepted expression of enzyme concentration is the rate of the reaction catalyzed, activity, per unit mass or volume. It would be more precise to express the concentration as mole of enzyme per unit volume. The concentrations, however, of most enzymes in serum and tissues are usually very low and immeasurable as mass by current state-of-the-art enzymology. As an example, ALT occurs in plasma at less than 1 mg/liter. With a total plasma protein concentration of 70 g/per liter, the proportion of enzyme protein to total protein would be less than 1 in 7000. Purification, isolation, and determination of enzyme concentration against this background would be an impractical clinical chemistry procedure.

The IUB defines an international unit (U) of enzyme activity as the amount of enzyme that catalyzes the conversion of 1 mol of a substrate or product per minute. The U is expressed per liter. Unit only specifies amount of substrate consumed or product formed per unit time by defined volume of sample. Variables between laboratories are time, temperature, pH, and substrate concentrations. These variables play an essential role in the evaluation of results among laboratories, and for these reasons laboratories must establish their own normal values. To avoid discrepancies, the discussion of enzyme activity changes should include a relative magnitude of change from normal rather than in actual units of enzyme activity.

The IUB and the IUPAC recommended a new unit to express enzyme activity, the katal (1 kat = 1 mol/s), which is consistent with the Système Internationale d'Unites. One U per liter is equal to 16.67 nkat per liter. The U remains the preferred unit of enzyme activity in the United States.

C. Kinetics

Enzymes catalyze reactions by complexing with the substrates to lower the energy of activation of the reaction without changing the equilibrium constant. In the absences of enzyme the reaction will occur, but at a much slower velocity. Enzymes can catalyze reactions in four ways: general acid–base catalysis, covalent catalysis, approximation of reactant catalysis, and induction of strain in the substrate or enzyme. As the product forms, the enzyme–product complex disassociates, freeing the enzyme to associate with another mole of substrate.

Enzymes are characterized on the basis of their affinity with their specific substrate and cofactors and on the effects of inhibitors, activators, pH, and temperature on the reaction.

An enzyme's optimum substrate affinity is expressed as the Michaelis constant (K_m), the substrate concentration at one-half the maximum velocity (V_{max}) of the reaction. Advanced recording systems permit a direct measurement of the enzyme's affinity with its substrate (K_s).

Calculation of the K_m for two substrate reactions, such as for the dehydrogenases, requires a complex approach. The concentration of the primary substrate is varied while the enzyme and secondary substrate, the cofactor, such as NADH, are supplied in sufficient concentration that they do not limit the velocity of the reaction. Once the K_m for the primary substrate is determined, it is held constant at its V_{max} and the concentration of the cofactor is varied.

The affinity of the inhibitor or activator and its mechanism of binding is characterized by plotting (Lineweaver–Burke plot) the effect of the inhibitor on the substrates V_{max} and K_m. The lower the numerical value of the inhibitor constant (K_i) or activator constant (K_a) the greater the affinity with the enzyme.

D. Enzyme Structure

An active enzyme, the holoenzyme, consists of a complex of proteins with posttranslational, nonprotein modifications. The protein portion consists of individual protomers. One or more such protomers may make up the protein, apoenzyme, portion of the holoenzyme. The apoenzyme may undergo posttranslational glycosylation, acylation, phosphorylation, deamidation, sulfhydryl oxidation, or aggregation within the confines of the sinus of the rough endoplastic reticulum. Posttranslation modification of a single protomer results in the formation of a monomer. Monomers can combine to form dimers, tetramers, and oligomers to make up the conformation of the holoenzyme. The three-dimensional configuration of the protein gives it its specific catalytic activity. Modification of the enzyme's environment, or matrix, such as temperature, pH, protein, or urea concentration, alters its configuration. Therefore, the assay matrix determines an enzyme's conformation and, in turn, its catalytic specificity and kinetic characteristics.

No specific amino acid sequence characterizes a protein as an enzyme. Among some enzymes there are similarities in the amino acid sequences of their catalytic site. The serine proteases are a group of enzymes that have serine residue common to their catalytic site. This similarity more closely reflects phylogeny than the essential nature of the enzyme's catalytic site. An enzyme's antigenic site may not be the same as its active site, and if dissimilar, antibody binding may or may not interfere with the enzyme's activity. Enzymes can be bound to the walls of plastic tubes or fibers of filters with little or no loss of their activity.

Enzymes secreted by cells for extracellular function can be synthesized and secreted as inactive proenzymes or zymogens. They are given the suffix "-ogen" or the prefix "pro-" to connote their inactive form. Proteases cleave the proenzyme into its holoenzyme and a remnant peptide. An example is the trypsinogen of trypsin. Coagulation, fibrinolysis, and complement systems contain proenzyme proteases that have a cascading proteolytic effect.

E. Isoenzymes

Isoenzymes are multiple structural forms of enzymes with the same catalytic activity, in the same species, but with apoenzymes derived from different genes. Previously, the IUB definition of the isoenzyme did not require derivation from separate genes, only that they be located in different tissues and have different physical features. The study of posttranslational changes, development of monoclonal antibodies, amino acid sequencing and DNA hybridization has increased the specificity of the isoenzyme definition. Posttranslational modifications of the apoenzyme to form the holoenzyme consist of glycosylation, acylation, phosphorylation, deamidation, sulfhydryl oxidation, or aggregation. An example is the alkaline phosphatase (AP) isoenzymes. When human hepatic, osseus, and renal AP are stripped of their sialic acid residues they have similar catalytic characteristics and electrophoretic mobility, but differ from the AP of intestine and placental. It appears, therefore, that human placental, intestinal and the three multiple forms of tissue-nonspecific APs are derived from three separate genes (Stigbrand *et al.*, 1982). By IUB definition there are only three, not the formerly defined five, AP isoenzymes. The IUB definition for an isoenzyme is not widely accepted in clinical chemistry in the United States. The definition of the isoenzyme as having multiple forms, pretranslational or posttranslational, continues in use.

Allelozymes are isoenzymes derived from the same loci, but with a differing or an allelic genome. When there is a high frequency of different allelozymes for a locus, it is described as being *polymorphic.* Human placental AP has a large amount of allelozyme polymorphism, but other species' placental AP contains very little. Allelozymes are well characterized in only a few nonhuman species. The X-linked phosphoglycerate kinase of mice and the glucose-6-phosphate dehydrogenases (G-6-PD) in the hare are allelozymes, and the artificially produced mosaicism of crossbred domestic cats with the Geoffrey cat X-linked G-6-PD results from alleloenzyme polymorphism. These X-linked allelozymes in the mouse and cat were used to demonstrate the monoclonal origin of neoplasia (Abkowitz *et al.,* 1985).

Hybrid isoenzymes result from the aggregation of two or more protomers. In the LDH isoenzyme, two different protomers aggregate to form the tetramers of the five isoenzymes. Of these five LDH isoenzymes, only LDH-1 and LDH-5 are made up of all of the same protomer. The remaining three isoenzymes, LDH-2, LDH-3, and LDH-4, are hybrids of the various combinations of the two protomers.

Macroenzymes are complexes of an enzyme or enzymes bound to immunoglobulins of the IgG, IgA, and sometimes IgM classes (Remaley and Wilding, 1989). Most diagnostically significant serum enzymes may exist in part as macroenzymes: macroamylase, macro-LDH, macro-AST, macrolipase, and macro-AP. They occur in varying amounts in most species where they have been sought. Although macroenzyme formation may be an autoimmune phenomenon, no association with autoimmune disease process has been established (Corazza *et al.,* 1994; Galasso *et al.,* 1993). Unfortunately they present a diagnostic problem. Macroenzymes may have longer half-lives than the native unbound enzyme and their serum values may change without an accompanying change in the enzyme's tissue of origin (Corazza *et al.,* 1994). The irregular formation of macroenzymes decreases diagnostic specificity and broadens the reference ranges, which in turn decreases their clinical sensitivity.

Characterization of isoenzymes is by zone and column electrophoresis, kinetically with different substrates, inhibitors, and activators, and immunologically with antibodies against the isoenzyme (Weiser and Pardue, 1987). Zone electrophoresis as a function of total enzyme activity has been the most common clinical procedure used to quantify isoenzymes.

Isoenzymes are valuable for phenotyping the genetic pool of both plants and animals (Rattazzi *et al.,* 1982). Quantification of isoenzymes has added to the specificity and sensitivity of clinical enzymology in human medicine, but less so in veterinary medicine. In human medicine LDH and CK isoenzymes are highly efficacious diagnostic and prognostic aids in cardiac infarction.

F. Substrate

Although enzymes are specific for the reaction they catalyze, they may not be specific for the substrate on which the reaction is catalyzed. An enzyme's affinity for one substrate may not be the same for a second. This absence of substrate specificity *in vitro* leads to confusion when comparing data between two or more laboratories using different substrates and reporting their values as U. This confusion is illustrated with AP. Synthetic substrates have been used to assay AP since its inception as a diagnostic aid in the 1920s. Activity has been expressed as an eponym unit, for example, Bodansky and King-Armstrong units. With the advent of the U, investigators calculated a numerical conversion factor to convert the eponym unit to U/liter. They did not standardize the assay and the user of the conversion factor failed to recognize the difference among assays. The most appropriate method of comparing values among reports of various laboratories is on the basis of the magnitude of change, rather than numerical units of activity.

Some enzymes have specificity for groups of substrates, such as hexokinase (HK), which has the same specificity for D-glucose, D-fructose, and D-mannose. For enzymes, such as L(+)-LDH (EC 1.1.1.27), used to measure blood lactate, there is a strict requirement for L(+)-lactate and only poorly oxidizes its mirror image, D(−)-lactate. This specificity of L(+)-LDH is especially important when measuring the plasma lactate in herbivores with grain overload leading to lactate acidosis. The patient absorbs large amounts of D(−)-lactate produced by rumen bacteria from grains. Use of an L(+)-LDH to assay for D(-)-lactate underestimates total plasma lactate concentration (Buttery and Pannall, 1986).

When measuring enzyme activity for diagnostic purposes, the activity of the enzyme must be maximized. This requires knowledge of the equilibrium of the reaction and concentration of substrate. An equilibrium constant (K_{eq}) of 100 indicates that when the reaction has gone to completion there will be 100 times more product than substrate. However, if $K_{eq} = 1$, as in the ALT reaction, only 50% of substrate is converted to product at equilibrium and the reaction appears to stop. If the reaction is reversible, the product may be reconverted to substrate as quickly as product is formed. This reconversion is stopped by converting the product into a second irreversible product. This process is called *trapping*. In the reverse LDH reaction, lactate is converted to pyruvate by LDH. To drive the reaction in the direction of pyruvate, hydrazine is used to "trap" pyruvate as an irreversible pyruvate–hydrazine product. This technique permits a large amount of LDH activity to be determined with a smaller amount of substrate than would have been possible without trapping the pyruvate.

"Substrate exhaustion" is a technical reason for erroneously reporting low activity in samples with a large amount of activity. This is the result of an unusually large amount of enzyme activity consuming the substrate before the first reading is taken for either an end point or kinetic assay. Substrate exhaustion errors can be avoided by the setting of minimum or maximum starting absorption limits. When the limits are exceeded, the sample is diluted with an appropriate diluent and the assay repeated.

When a substrate or product is immeasurable, measurement of a cofactor may suffice. Two convenient and accurately measured cofactors as stoichiometric markers of enzyme activity are NADH and NADPH. Their absorption peaks are at 340 nm and they have absorption coefficients (formerly the extinction coefficient) of 6.22×10^3 *M*/liter. The rate of appearance or disappearance of NADH is stoichiometrically related to substrate consumption and it is commonly used to express the oxidation–reduction of enzyme activity.

No cofactor is required in some reactions and the substrate or the product may be immeasurable. In these enzymes the primary reaction is "coupled" or "linked" to a second reaction utilizing a measurable product in the coupled reaction. An example of a coupled reaction in which three reactions are linked is that of creatine kinase (CK) assay [Equations (12.1)–(12.3)].

$$\text{Creatine phosphate} + \text{ADP} \xrightarrow{\text{CK}} \text{creatine} + \text{ATP}, \tag{12.1}$$

$$\text{ATP} + \text{glucose} \xrightarrow{\text{hexokinase}} \text{glucose-6-phosphate} + \text{ADP}, \tag{12.2}$$

$$\text{Glucose-6-phosphate} + \text{NADP} \xrightarrow{\text{G-6-PD-D}} \text{6-phosphogluconate NADPH} + \text{H}^+. \tag{12.3}$$

In the example the coupled CK assay, the substrate of the first reaction, cofactors, and secondary enzymes are supplied in unlimited quantities. In a coupled reaction, only the primary enzyme's activity can be a variable, in this case CK. The rate of generation of NADPH [Eq. (12.3)] is then stoichiometrically proportional to the ADP consumed.

Synthetic substrates are available for many enzymes having products or substrates not easily quantified. Those of hemostasis, trypsin, plasmin, cathepsin, and kinases, are a few with diagnostic value. Some poorly or undetectable natural substrates can be conjugated to chromogens to form synthetic substrates. When the reaction occurs, chromogen is freed or modified to a form that can be spectrophotometrically measured and its absorption coefficient used to calculate the rate of the reaction.

The development of synthetic substrates for proteolytic enzymes of the coagulation pathway has been revolutionized in the study of hemostasis (Triplett, 1986). Assay of a single coagulation enzyme formerly required complete linkage of a large portion of the pathway to form a fibrin clot. Now a specific enzyme of the coagulation pathway can be assayed independent of the remainder of the pathway's enzyme.

G. Cofactors

Cofactors are small molecules, loosely or closely associated with enzymes as cyclic secondary bisubstrates. As cyclic substrates they are common to a number of enzymes, where they are specific for the type of reaction catalyzed. They are secondary to the enzyme's primary specific substrate. All water-soluble vitamins, except C, are precursors of cofactors that have no stable storage forms. The animal's inability to synthesize and store water-soluble vitamins and its cofactors' "loose" binding with enzymes make them pivotal points in homeostasis. An example of a water-soluble vitamin's importance is niacin, vitamin B^{3}. Niacin is the substrate for the cofactors NAD and NADP, the essential electron transporters of respiration. Twenty-five percent of the body's enzymes have metal cofactors and are called *metalloenzymes*. Some metalloenzymes are complexed with vitamin-derived cofactors, such as the metalloflavoproteins of the oxidation–reduction enzymes.

For optimum enzyme assay reaction rates, loosely bound cofactors must be supplied in optimum amounts. However, when tightly bound, the addition of cofactor makes no difference to the assay unless the patient is deficient in the factor. When a deficiency of a tightly bound cofactor is suspected, such as a mineral, an assay of the appropriate metalloenzyme before and after supplementation can serve to confirm the diagnosis and a prognosis. Selenium and copper deficiencies are detected by measuring their respective metalloenzymes, red blood cell (RBC) glutathione peroxidase, and superoxide dismutase.

Cofactor-activation enzyme assays are used to detect deficiencies of the loosely bound vitamin cofactors (Bayoumi and Rosalki, 1976). They are performed by deleting and adding the suspected deficient cofactor to the appropriate enzyme assay. A difference in the reaction rate between the deleted and added cofactor in a sample from affected versus the reference animals serves as the indicator of a deficiency. An activation assay is used in the study of polioencephalomalacia, an idiopathic disease of cattle and sheep with cerebrocortical necrosis and low RBC transketolase activity. Low transketolase activity is related to the deficiency of thiamin cofactor, thiamin pyrophosphate (TPP) (Loew *et al.*, 1975; Edwin *et al.*, 1979; Jackman, 1985). Transketolase activity is measured before and after adding TPP to the assay. The change in activity that results from the addition of TPP to a patient's sample is compared to that observed in the control sample. The greater the amount of transketolase activation, the greater the thiamin deficiency.

H. Inhibitors and Activators

Inhibitors and activators modify the rate of a reaction in which they are not substrates. They are generally small molecules and ions, and can even be other enzymes. Their action may be reversible or irreversible. If covalently bound, they are irreversible. Reversible inhibition and activation are initiated by electrostatic, hydrogen, and van der Waals' bonds. The action may be competitive with the substrate at the enzyme's catalytic site, or noncompetitive at an allosteric site elsewhere on the enzyme, or, less commonly, uncompetitive.

Feedback inhibition of enzymes by the pathway's end products are well-recognized physiological controls and occur in *in vitro* enzyme assays. Some inhibitors, essential to the homeostasis of the body, occur naturally in plasma. The acute phase reactants of plasma, α_2-antichymotrypsin, α_2-antitrypsin, anti-α-trypsin inhibitor, antithrombin III, and antiplasmin inhibit the proteolytic enzymes of digestion and coagulation.

Inhibitors and activators are therapeutic agents as well as controls of enzymatic assays in the laboratory. Heparin, a commonly used enzyme inhibitor, activates plasma antithrombin, the inhibitor of thrombin and other serine proteases of the coagulation pathway. Without heparin, antithrombin has low inhibitory activity, and without antithrombin, heparin has no anticoagulation activity. Other inhibitors are lead, an inhibitor of the enzymes of the porphyrin synthetic pathway, and the organic phosphate insecticides, inhibitors of esterases, including cholinesterase of nerve endings.

Activators accelerate enzyme reaction rates by promoting the active state of the substrate or the enzyme. Unlike cofactors, activators do not enter into the reaction as bisubstrates. Some activators are proteases, such as those that activate the plasma and pancreatic zymogens. Metal ions that activate enzymes do so by stabilizing the conformation of the active site.

Some divalent metal ions are essential to an enzyme's action. They play an integral role in the structure of the enzyme and cannot be readily exchanged, whereas others can be freely exchanged with the surrounding matrix. The anticoagulant EDTA chelates Ca^{2+} and Mg^{2+} thereby inactivating Ca^{2+}-dependent alkaline phosphatase. Creatine kinases and all other kinases require Mg^{2+} for phosphate transfer.

The substrates of allosteric enzymes serve as self-activators. Allosteric enzymes are polymers with two or more reactive sites. Binding of the first mole of substrate results in a conformational change that enhances the binding of additional substrate. Allosteric activation produces a sigmoidal curve, similar to that of the O_2 saturation curve of hemoglobin, and is inconsistent with the Michaelis–Menten model.

I. Temperature Effects

Change in the temperature of incubation alters the rate of the reaction and rate of denaturation of the enzyme. The higher the temperature, the more rapid the reaction rate, and the faster the denaturation of the enzyme. However, with computerized recording systems it is possible to perform enzyme assays in short periods of time, at higher temperatures with little or no denaturation during the assay period. Enzyme assays are performed at 37°, 30°, 25°, and 22°C. The recommended temperature for reference assays, is 30°C, but most clinical laboratories in the United States prefer to use a 37°C incubation temperature

The rate at which temperature changes the velocity of a reaction is fairly constant for the enzyme. A 0.1° temperature difference is readily detectable as a change in enzyme activity. A reaction conducted at 37°C is about twice as fast as one performed at 25°C (Appendix III). Temperature conversion factors are available for approximate conversion in comparative studies (Hafkenschied and Kholer, 1986; Junge and Grutzman, 1987). When the activity is converted it should be reported as converted, along with the method of conversion.

J. pH Effects

Modification of pH alters the enzyme's configuration and its reaction characteristics. An enzyme assay's optimal pH has a narrow range, and its buffer type and concentration are critical. When the assay's pH is in the range of the substrate's pK, a change in pH changes the concentration of the disassociated and the undissociated substrate forms.

The type of buffer used in an assay should be one that has a pK_a within 1 pH unit of the optimum pH of the assay. Some buffers chelate metal ions and would be inappropriate for an assay of a metal requiring enzyme. Some products are ionized and require buffering. It is essential that the buffer be capable of maintaining the optimum pH for the reaction throughout the assay period for as high a concentration of product as is anticipated.

III. CLINICAL LABORATORY ENZYMOLOGY

A. Natural History of a Clinically Diagnostic Serum Enzyme

A number of variables affect the appearance of an enzyme in plasma. In the steady state, the amount of any single enzyme in serum is a function of its concentration in the cell, the total tissue mass, magnitude of injury to the cell, normal death of the cell, apoptosis, and the degradation rate of the enzyme in plasma or its half-life.

An enzyme's plasma concentration is proportional to its intracellular concentration or specific activity. When the an enzyme's cellular specific activity increases or deceases so will the amount of activity entering plasma. As an example, normal dog liver contains little or no intestinal alkaline phosphatase (IAP) and little is normally found in plasma. When a dog is treated with glucocorticoid, the hepatocellular IAP gene is expressed and a hyperglycosylated form of IAP appears in serum. Without cell death, the newly synthesized hepatic IAP-like enzyme enters plasma to increase the total serum AP activity (Solter and Hoffmann, 1995; Sanecki *et al.*, 1993). The same is true for a decrease in specific activity. In cyclosporin-treated dogs and rats, the hepatocyte ALT's specific activity decrease is accompanied by a decrease in plasma ALT activity (Rhodes *et al.*, 1986).

A change in the total amount of tissue mass results in a change in a serum enzyme's activity without induction or loss of integrity of the cell. The amount of serum CK increases in proportion to muscle mass. A decrease in liver size is accompanied by a decrease in hepatic enzymes in plasma.

Cells need not die to release their enzymes. A short period of hypoxia is sufficient to disrupt the integrity of a cell membrane and potentially allow soluble cytosolic enzymes to escape or "leak" into their surrounding matrix to be drained away in lymph (Mattenheimer and Friedel, 1977; Diederichs *et al.*, 1979; Lindena *et al.*, 1979). Alternatively, hypoxic or toxic injury results in exocytosis or formation of membrane blebs where their cytosolic enzymes are released into the surrounding plasma (Gores *et al.*, 1990). The shedding of blebs and release of cytosolic LDH in the absences of cell necrosis have been demonstrated in the isolated perfused rat liver (Lemasters *et al.*, 1983).

Another consideration is the detectability of an enzyme once it leaves the cell and is diluted by plasma or other body fluids. A small tissue, such as the adrenal glands, has specific enzymes. But, even if the adrenals were to undergo complete acute lysis, the enzymes would be so diluted by plasma they would be clinically undetectable.

Degradation of plasma enzymes' is poorly understood. An enzymes mass is not easily measured, but rather it is the rate of disappearance of its activity, half-life ($T_{1/2}$), that is accepted as the $T_{1/2}$ of the enzyme. An enzyme's $T_{1/2}$ is determined by grinding the tissue of origin, centrifugating and then extracting the enzyme from the supernatant. The extract is then injected into a homologous species and the disappearance rate plotted. It is assumed that the enzyme in the injected preparation is homogenous and the rate of degradation constant. Some enzymes, like GGT and AP, may be soluble but unless extracted with detergents, a variable amount of the activity occurs as a suspended, particulate high-molecular-weight form. These two forms of AP can be cleared at different rates. An example of how an enzyme complex's size makes a difference when calculating $T_{1/2}$ is apparent in the study of macroenzymes, naturally occurring endogenous enzyme–autoantibody complexes. Macroenzymes have greater $T_{1/2}$'s than their native holoenzyme. Their formation and delayed degradation increases total plasma enzyme activity to above the reference range, falsely alerting the clinician to tissue damage (Remaley and Wilding, 1989).

There are many reasons for the loss of enzyme activity, but it is principally a result of the change in functional conformation. Unlike the cytosol, plasma is not conducive to the prevention and reduction of oxidation of peptides. Intracellularly, enzymes are protected from degradation when bound to their substrates and cofactors. In plasma enzymes, substrates and cofactors are dispersed and binding uncommon, leaving the enzyme more susceptible to degradation.

Little is known about the clearance of enzyme peptide from plasma. Small ones are cleared from plasma into urine. Some are assumed to follow the same non-specific proteolysis as the plasma enzymes of coagulation, complement, and the fibrinolytic pathways. Others are desialated, recognized by the asialoglycoprotein receptor on the hepatocytes, endocytosed, and degraded by the phagolysosome. An example of the latter is intestinal AP, which has unsialated terminal galactose and leads to a short half-life (Kuhlenschmidt *et al.* 1991).

B. Development of a Clinical Enzyme Assay

An enzyme, to be valuable as a clinical diagnostic aid, must be economically and readily assayed, and reasonably reflect pathological change in a specific tissue, organ, or group of organs. Even when an enzyme meets these requirements, it may not be commonly accepted in the clinical laboratory. An example is the comparison of the three liver-specific enzymes, arginase (ARG), sorbitol dehydrogenase (SDH), and alanine-amino transferase (ALT). Serum arginase and SDH are specific, sensitive markers of hepatocyte injury in most species. Serum ALT is liver specific in only some species but is used in those species, dog and cat, in preference to SDH and ARG. The reason for this acceptance of ALT and not ARG is that the ALT assay is simpler to perform. In addition, the ALT assay became available first and a large amount of clinical background data for ALT was developed before ARG and SDH assays became commercially available.

An enzyme's clinical diagnostic sensitivity and specificity for a disease or organ injury must be critically evaluated. Diagnostic tests should have reasonably high sensitivity and very high specificity. However, since in most instances serum enzyme determinations are utilized as screening tests to detect organ involvement or disease, high sensitivity and reasonable specificity are most desirable. Although sensitivity and specificity data are not always available for enzymes in current usage, their credibility has been established by years of use. Development of new enzyme assays for clinical use demands evaluation of diagnostic sensitivity and specificity to achieve credibility.

Once cytoplasmic enzymes are free of the cell and in the extracellular fluids, plasma, cerebrospinal fluid (CSF), urine, or milk, their activity can be measured as an index of the cell's integrity. This is less true of the membrane-bound enzymes such as GGT and AP whose plasma activity may be altered without a loss of the cell's integrity. Right heart insufficiency results in portal blood stasis and hepatic hypoxia becomes evident by an increase in serum ALT, but not necessarily by light microscopic histopathology.

To be of diagnostic value, a serum enzyme must have more than high tissue-specific activity. Variables to be considered when determining diagnostic value of an enzyme are range of activity of individuals within the reference population, anatomic location of the cell, location of the enzyme in or on the cell, the total mass of the tissue or the cell type, and plasma $T_{1/2}$. A wide range in the reference population's enzyme activity plasma and serum, such as serum AP in ruminants, makes it difficult to determine when an individual's value becomes abnormal thereby lowering sensitivity. For experimental purposes, however, the same enzyme within an individual is frequently sufficiently constant from day to day to be used to detect sequential changes.

When a serum enzyme lacks high specificity for an organ, a second serum enzyme can be combined with the first to increase its diagnostic value. Serum AP activity increases in bone and liver disorders. To confirm that an increased AP is of hepatic origin, serum ALT or α-glutamyltransferase (GGT) are included in the hepatic enzyme profile.

Only a few of a cell's enzymes are of diagnostic significance in serum. One reason is their dilution in the plasma of the extracellular space. Another reason is the compartment into which the cell discharges its contents. As an example, the respiratory, intestinal, and urinary tracts have high GGT-specific activity, but when damaged there is no clinically significant change in serum GGT activity. Their major cell populations discharge their contents into the lumen of the alveolus, bowel, or tubule. The discharged intracellular enzymes are then diluted by urine, feces, and pulmonary transudates and eliminated from the body, rather than accumulated as they are in plasma.

Urine, gastric, intestinal and pulmonary fluids, blood cells and fixed tissue, and plasma can contain diagnostic enzymes. Clinical enzymology of nonplasma fluids present a problem in quantification not found in plasma. Homeostasis of plasma volume and solutes makes it possible to express enzyme activity per unit of serum volume as the reference point. However, in the nonplasma body samples, water content is not a constant and expressing enzyme activity as unit per volume inappropriate. In tissue, activity is expressed as unit per weight of tissue, protein, DNA, number of cells, or nuclei. Urinary enzyme activity is expressed per mass of creatinine, because urinary water volume is highly variable but creatinine is excreted in an approximate proportion to the body's skeletal muscle mass. Fluids from the lung, stomach, and intestine are of variable volume of water and provide no consistent reference point for expressing enzyme activity. Whichever point of reference is used for expressing enzyme activity, it must be one that will permit consistent valid results within and between assays and among laboratories.

Histochemical enzymology and enzyme-linked immunocytochemistry serve to identify cell types in neoplasms and enzyme deficiencies. Histochemical procedures can only be semiquantifiable by stringent simultaneous staining of patient and control samples of the same tissue thickness.

C. Validation of an Assay

To establish the validity of a new clinical assay, the accuracy of the reference and clinical procedures is compared. The accuracy of an assay refers to how closely the assay can measure the true value of the enzyme activity as determined by the reference method. This is desirable but not always possible with unstable enzymes. A commercially available assayed quality control serum can be used for stable enzymes. The new assay method can then be compared with the reference method at low, high, and medium amounts of activity anticipated in the clinical laboratory. If the method is new only from the standpoint of the species, it can be compared over the anticipated range and sample conditions encountered in-house with the established clinical procedure.

Statements concerning confidence limits and quality assurance are commonly omitted from reports concerning enzyme activity. Precision is of particular importance because of the instability of some enzymes. The precision is a measure of the assay's reproducibility, which is characterized by the standard deviation (SD) and coefficient of variation (CV). The CV should be determined at least two and preferably three amounts of enzyme activity, including the activity at which medical decisions are made. An acceptable CV at high enzyme activity does not ensure precision at the point at which decisions are most important. Within-run and between-run precision must be determined. Research and clinical chemistry laboratories' quality assurance programs should be stated in published reports of enzyme data (Harris, 1993; Henderson, 1993).

D. Specimen Requirements

Many enzymes require metallic ions for maximal activity, and plasma containing the metal chelating anticoagulants, EDTA, citrate, or oxalate is unsatisfactory for their assay. Heparinized plasma can be used in some procedures. When in doubt use serum.

Some serum enzymes are very stable at room temperature. Refrigeration and freezing preserve many enzymes, but others deteriorate even when frozen. Stability of an enzyme's activity in one species does not mean stability in a second (Jones, 1985a, 1985b). Horse serum SDH, but not cattle, sheep, or goat, is unstable at room temperature, refrigerated, and frozen (Horney *et al.*, 1993). Urine GGT is more stable at room temperature than frozen (Van Breda Vriesman, 1968). Freezing results in the concentration of salts and enzymes, and disaggregation of some weakly bonded peptides (Beck and Sammons, 1975; Adams *et al.*, 1985; Stokke, 1974). When thawed, the protomers may randomly reaggregate into inactive or less active configurations. A list of some enzymes and their storage characteristics are found in Table 12.2.

Hemolysis and lipemia can interfere with an enzyme assay, either increasing or decreasing apparent enzyme activity. In some cases hemolysis results in the release of RBC enzymes into plasma. White and red blood cells must be separated from serum as quickly as possible because hypoglycemic blood cells leak LDH and AST into serum even before hemolysis is apparent. Lipemia can cause a false decrease or increase in activity by nature of the light defraction. Lipemic samples are to be avoided, but if there is no other choice, background subtraction with a sample blank is an acceptable method of correcting for the defraction. The ability of various analyzers to eliminate or minimize the effect of hemolysis or lipemia is quite variable (Ryder *et al.*, 1991; Glick *et al.*, 1987). Although some dry chemistry analyzers appear to use a whole blood sample for enzyme analysis, they in fact incorporate a cell separation process such as filtration and the blood must still be free of hemolysis and lipemia.

Glassware or plasticware specimen collection vials must be carefully cleaned. Detergents interfere with many assays, and only a thorough rinsing will reduce the detergent residue to a point that will not infer with the assay. Whenever possible commercially available collection vials are recommended.

A complete clinical history is essential to the interpretations of laboratory values because many drugs are themselves the cause for changes in enzyme activity (Young *et al.*, 1975).

E. Assay of Enzymes in the Clinical Laboratory

Differences in performing clinical assays are inevitable. The constant improvement in technology requires assay procedures to be modified, but the slightest modification may change the stoichiometry of the assay. Modifications of an enzyme assay for whatever reason require revalidation to ensure that zero-order kinetics are not altered (Goren and Davis, 1986). Kits commercially prepared for human medicine are commonly modified for veterinary medicine. A change may appear to be a subtle, but, in fact, all changes require

TABLE 12.2 Enzyme Stability in Serum[a]

Enzyme	Storage	Time (days)	Activity (%)
α-Amylase (EC 3.2.1.1)	Room temperature	8	100
Cholinesterase (EC 3.1.1.8)	Room temperature	8	90
	0–4°C	8	90
	Frozen	8	94
Creatine kinase[b] (EC 2.7.3.2)	Room temperature	1	25
	0–4°C	1	32–65
	Frozen	8	25
Glutamate dehydrogenase (EC 1.4.1.2)	Room temperature	2	60
	0–4°C	2	60–100
	Frozen	2	60
Aspartate aminotransferase (EC 2.6.1.1)	Room temperature	4	90
	0–4°C	8	87
	Frozen	2	90
Alanine aminotranferase (EC 2.6.1.2)	Room temperature	4	75
	0–4°C	8	78
	Frozen	8	31
L-Lactate dehydrogenase (EC 1.1.1.27)	Room temperature	8	74–88
	0–4°C	8	81
	Frozen	8	81
Alkaline phosphatase (EC 3.1.3.1)	Room temperature	8	71
	0–4°C	8	71
	Frozen	8	68
Sorbitol dehydrogenase (EC 1.1.1.14; L-iditol dehydrogenase)	Species dependent		
	0–4°C	8	71

[a] Reproduced in part with permission from Bergmeyer (1974).
[b] An unactivated enzyme assay procedure was used to establish these values.

revalidation of the procedure. Even a subtle change such as using a fresh versus frozen sample requires revalidation. Some species have unusually high amounts of basal activity in serum, such as amylase in dogs and AST in horses, which may exceed limits of linearity for kits prepared for human use. Normal equine and bovine serum have a large amount of background absorption in the 340-nm spectrum that is not observed in people or many other species. This background requires a patient blank or the starting absorption limits must be set higher.

Clinical biochemical enzyme assays are generally either end point or kinetic procedures. In the end point or "fixed-time" procedure, a reaction is started by the addition of sample to a reaction mixture. After a fixed incubation period the reaction is ended by the addition of a reagent that destroys or inhibits the enzyme's activity. The amount of substrate used or product produced for that period of time is determined.

Kinetic or "continuous-monitoring" assay procedures require either manual or automatic sequential recording systems and are the most common methodology in automated analysis. Kinetic assays are more rapid, sensitive, accurate, and easily controlled than end point assays, because the reaction rates can be visualized throughout the time of the assay. A constant recording device permits the graphic display of the linearity of the reaction rate. The primary advantages of the kinetic method versus the end point method are as follows: (1) When high enzyme activities are encountered, the reaction rate can be determined before substrate is exhausted, whereas, in an end point assay, when the substrate is exhausted the assay must be repeated; and (2) if an activator is present, it can be detected on the graph.

When the rate of enzyme activity is very high, approximation of zero-order kinetics becomes limited. If precision quantification is required for an enzyme assay, zero-order kinetics must be maintained. Reagents for an assay generally have only sufficient substrate to maintain zero-order kinetics for the assay's incubation period or for two to three times greater than the upper limit of the enzyme's reference values. When a high level of precision is not required and the rate of the reaction exceed the limits of linearity, results are reported as "greater than" the established upper limit of linearity. Alternatively, the assay can be repeated on a diluted sample, or the size of the sample can be reduced and appropriate corrections for dilution made.

Immunoassays for enzymes are available. Monoclonal antibody technology permits production of specific antienzyme antibodies. These enzyme-specific antibodies are used to isolate pure enzyme for further study and as an assay standard. Combining radiolabeling with immunoassays permits the development of very specific, sensitive assays for inactive, active, and inactivated enzymes such as the zymogens of serum pepsinogen and trypsinogen. High specificity and absence of cross-reactivity in the immune-enzyme assay have disadvantages in veterinary clinical diagnostics. Without species cross-reactivity the veterinary laboratory would have to maintain a number of species-specific immunoassays for the same enzyme. There is a good correlation between the immunoenzyme assays and enzyme activity assays that have been performed. Immunoassays tend to be more expensive and complex than enzyme activity assays and may never have the multiple species advantage of measuring enzyme activity.

F. Enzymuria

Renal diseases and nephrotoxicosis are reflected in changes in enzyme activity of urine (enzymuria), but not serum. The majority of the kidneys cells face the lumen of the nephron. When the cell's integrity is lost, their contents are discharged into the luminal contents, urine. Only a few plasma enzymes escape into urine normally, but when proteinuria of a glomerulopathy occurs more plasm proteins and enzymes appear in urine. Pancreatic amylase enzymuria is diagnostic of active pancreatitis in people, but not in dogs with the same disorder. To date, in veterinary medicine, enzymuria has only been of diagnostic value in renal disease, specifically in nephrotoxicities and renal transplantation rejection (Raab, 1972; Price, 1982; Junge *et al.*, 1986; Prescott, 1982).

The enzyme content of the nephrons proximal and distal tubular epithelium differs. Attempts to localize the damage through ratios and patterns of enzymuria have met with limited success. Subcellular localization of damage in the microvillus versus intracellular appears to be possible by measuring urinary AP and GGT as indicators of nephrotoxicosis and urinary N-acetylglucosamine (NAG) as an indicator of renal hypoxia in transplant rejection. Hyperenzymuria can be a more sensitive indicator of nephrotoxicities than are light microscopic histopathology or renal function tests.

The nephron as flow-through system does not allow for the accumulation of enzymes such as occurs in plasma. If the damage to the renal epithelium is transitory, the enzymes released may clear with the urine before sampling can occur. However, if nephrotoxicity is anticipated, say with specific nephrotoxic drugs, sequential urine enzyme analysis may serve to detect onset of toxicity.

Urinary enzyme assays are more difficult to perform than are serum enzyme assays. Enzyme activity's reference is generally activity per milligram of creatinine. Urine pH, molarity, and variety of solutes differ considerably from those of serum. Some urine solutes may be inhibitors at one concentration but not at another (Beck and Sammons, 1975; Adams *et al.*, 1985; Stokke, 1974). Depending on which enzyme is being measured, inhibitors may have to be removed by dialysis or if the inhibitor is a constant, the specific gravity may have to be standardized before the assay is performed. As an example, urea in high concentration disrupts weak hydrogen bonds. Enzymes dependent on aggregation by weak hydrogen bonds may be rendered inactive.

Few if any valid commercial enzyme assays for urine are available. Before a serum enzyme assay kit is used for urine, it must be revalidated. The validation must include high and low sensitivity, limits of linearity, presence of inhibitors or activators by molecular sieve column chromatography or dialysis, storage stability, reproducibility at various solute dilutions, capability of the buffers to maintain the reaction pH over the anticipated range of salt concentrations, and pH of the urine. Urine is a very poor environment for a molecule that depends on its conformation to function.

G. Quality Assurance

Quality control and quality assurance in clinical enzymology are part of the total quality management plans and procedures of a laboratory and differ little from other areas of the clinical laboratory. Quality control (QC) is that part of a laboratory's quality assurance program which is concerned with evaluation of testing proceedures and includes reagent and equipment checks, linearity checks, and statistical control procedures. Quality assurance is concerned with broader variables ranging from sampling and identification to data reporting and turnaround time (Tiersten, 1986).

Unlike many metabolites measured in clinical chemistry laboratories, some enzymes are insufficiently stable to lend themselves to a QC program. Assayed and unassayed lyophilized QC serum with normal or abnormal activity can be purchased from commercial sources. Unassayed QC serum is less expensive and more appropriate than assayed QC serum in the individual laboratory where the methods differs from the method used by the commercial QC sera source. Commercial programs offer a number of benefits. They generally supply a uniform QC serum over a long period of time, keep track of its rate of deterioration, and permit individual laboratories to compare their results with other laboratories using the same or a different procedure. This comparison is of particular value when evaluating infrequently performed assays.

When the QC serum's value is established, the QC serum is used each time an assay is performed, the results are recorded, and the value is plotted on a graph. The within-run and within-day precisions should be better than the day-to-day precision, because day-to-day values reflect a greater number of variables, such as reagent decay and operator idiosyncrasies. The graph will show trends or progressive changes in the value of the QC. The trends serve to detect the procedural errors before the variable becomes so great that the data are valueless.

IV. ENZYME-LINKED DIAGNOSTICS

A. Enzyme Histochemistry and Cytochemistry

Cytochemical enzyme staining is a common cytological and histochemical diagnostic procedure for cell identification. Linking enzymes to monoclonal antibodies against specific cell epitopes increases the sensitivity and specificity of detecting undifferentiated neoplastic cells (Grindem *et al.*, 1986; Facklam and Kociba, 1986) and permits identification of cell structure and products by light and electron microscopy.

B. Immobilized Enzymes

The immobilization of enzymes adsorbed to solid supports such as membranes, beads, and tubes permits physical separation of the enzyme from its substrate, buffer, and product. This separation allows for the conservation of the enzyme in diagnostics and industrial preparative methodology. Immobilized enzyme techniques applied to the potentiometric technology of ion-selective electrodes facilitates the direct assay of glucose and urea in whole blood. Membrane bound glucose oxidase catalyzes the oxidation of glucose in the sample to form hydrogen peroxide. Upon contact with a platinum electrode, the hydrogen peroxide produces an electrical current that is measured by a voltmeter. The current produced is proportional to the glucose concentration. This procedure has the potential for *in vivo* monitoring of blood metabolites.

C. Enzyme Immunodiagnostics

Conjugation of enzyme with antibody has been a major development in competitive protein binding assays. Enzyme immunoassays (EIA) have are very sensitive, are low cost, eliminate radioisotope use, and aid automation (Kurstak, 1986). Antibody specificity determines the assay's specificity. Monoclonal antibodies are more specific than polyclonal antibodies. Although radioimmunoassay (RIA) is of equal specificity and more sensitive than EIA, EIA is frequently the preferred method. The EIA reagents have relatively long shelf lives, and involve little or no government regulation. The catalytic activity of the assay enzyme amplifies the antigen–antibody complex and has the potential for measuring more than one analyte at a time in the aliquot of sample. With the use of chemiluminescence in the detection phase of the EIA, the assays are equally or more sensitive than RIA and are likely to replace RIA procedures in the future.

Enzymes are the source of sensitivity of the EIA. An EIA for mouse IgG can detect 24,000 molecules of IgG/ml (Shalev *et al.*, 1980). The sensitivity of the EIA is increased by the use of radioactive substrates. By combining (^{3}H) AMP and AP, a detection system measuring 600 molecules/ml is produced (Harris *et al.*, 1979). Two basic EIA procedures are in use to measure soluble antigens bound to solid phases. They differ in the need to separate the enzyme–antibody complex from the enzyme–antibody–antigen complex before measuring the enzyme activity. The homogeneous or nonseparation enzyme-multiplied immunosorbent technique (EMIT, Syva, Palo Alto, California) does not require a separation step. The unique feature of EMIT is that the enzyme is inactive or has limited activity when complexed with antigen (Bastiani and Wilcox-Thole, 1982). It becomes activated when the enzyme–antigen aggregates with antibody. Free antigen of the sample competes with the enzyme-labeled antigen for the antibody to produce an inverse relationship between the amount of enzyme activity and the unlabeled analyte of the sample. The EMIT is limited by the enzyme–antigen–antibody complex kinetics and to low-molecular-weight analities in toxicology, pharmacology, and endocrinology.

The second EIA procedure is a heterogenous, or separation, enzyme-linked immunosorbent assay (ELISA). In contrast to the EMIT, the enzyme in the ELISA is always active. In the ELISA procedure, the unbound enzyme-antibody and bound enzyme-antigen complexes must be separated. This separation step makes the ELISA a more complex procedure and more versatile than the EMIT. Immobilizing the ELISA complex on a solid phase simplifies the separation step to inverting the tube or plate over a sink. ELISA is often used for semiquantifiable systems in which color changes are detected with the unaided eye. They are used to detect parvovirus in canine feces and lymphosarcoma virus in feline blood.

Enzyme-linked immunosorbent histochemistry has provided a very specific diagnostic procedure. By using antibodies against various markers of cells it has become possible to separate cell types whose light and electron microscopic features are the same on the basis of their membrane epitopes.

D. Molecular Genetics

Linking EIA with nucleotide hybridization has resulted in the development of diagnostic procedures designed to detect the nucleotide sequence of aberrant genes and latent infectious diseases. Selective use of enzymes permits the isolation of specific RNA and DNA nucleotide sequences. Given the amino acid sequence of a protein, it or a portion of its gene's nucleotide sequence can be synthesized and labeled with a radioisotope or nonisotopic labels, to provide a probe. This probe can then be hybridized with the native DNA or RNA of a cell under investigation to locate mRNA, abnormal genes, infectious agents, or genetic defects.

Restriction endonucleases of bacteria are enzymes that cleave DNA at specific sites to produce fragments of DNA. The combination of restriction endonucleases and DNA probe hybridization has produced a new technology used to identify bacteria and viruses and to phenotype genetic diseases. In the future this technology has the potential of selecting desirable as well as undesirable genetic traits in pretransplanted ova.

V. ENZYMES OF CLINICAL DIAGNOSTIC IMPORTANCE

A. Alkaline Phosphatase

Alkaline phosphatase (AP; EC 3.1.3.1.) has been utilized as an indicator of hepatic injury since the 1920s and is now the subject of more publications than any other enzyme (Stigbrand *et al.*, 1984). The endogenous substrate for the alkaline phosphatases is unknown but the multiple forms of the enzyme are capable of hydrolyzing many phosphate esters (Bretaudiere and Spillman, 1983). Numerous substrates have been utilized for AP activity determination; however, most automated analyzers utilize *p*-nitrophenylphosphate as the substrate in kinetic assays. The buffer systems utilized also differ, resulting in considerable difference in apparent AP activity in a sample and, in turn, different reference values dependent on the procedure used.

Alkaline phosphatase is a glycoprotein with subunit masses ranging from 40 to 83 kDa. In serum, the APs exist principally as a dimer and are hydrophylic. High-molecular-weight hydrophobic forms of AP have been identified as a substantial portion of serum AP in humans with cholestasis (DeBroe *et al.*, 1985; Kihn *et al.*, 1991) but high-molecular-weight forms of AP make up less than 5% of AP in cholestatic dogs (Solter and Hoffmannn, unpublished data). In tissues the APs may exist as a tetramer (Hawrylak and Stinson, 1988).

The APs are found primarily in intestine, kidney, liver, and bone. Kidney and intestine have by far the greatest activity per gram of tissue. In liver, AP is located primarily on the bile canalicular membrane of hepatocytes of healthy dogs and rats; in man it is also present on the sinusoidal surfaces.

An extensive amount of effort and interest has been directed at the AP isoenzymes in an effort to increase the diagnostic specificity of the determination of serum AP. The isoenzymes of AP are as a result of both expression of different genes and differences in post-translational modifications of the enzyme. In man and higher primates, there are at least three genes coding for AP, but in other mammalian species there exist only two genes (Goldstein *et al.*, 1980). These are referred to as intestinal AP (IAP) and tissue unspecific alkaline phosphatase (UAP). With some exceptions, the intestinal gene is expressed only in intestine and the UAP gene is expressed in bone, liver, kidney, placenta, and in lesser quantities in most other organs. In rabbits both genes are expressed in liver, kidney, and intestine (Noguchi and Yamashita, 1987). In horses, the intestinal gene along with the tissue unspecific gene is expressed in kidney (Hoffmann *et al.*, 1983a). In dogs the intestinal gene is expressed in the liver under the influence of excess glucocorticoids to produce a hyperglycosylated form of intestinal AP (CAP) (Sanecki *et al.*, 1990).

Renal AP is generally not found in serum, whereas osseus and hepatic AP have been identified in the sera of all species studied. Intestinal AP is present in rat serum under normal conditions and increases following a meal or following gavage with corn oil (Eliakim *et al.*, 1991; Hoffmann *et al.*, 1994). Although IAP has been reported in serum of horses with and without gastrointestinal disorders (Blackmore and Elton, 1975), subsequent studies failed to identify IAP in horse serum under any conditions (Hoffmann *et al.*, 1983b;

Hank *et al.*, 1993). Likewise, little or no IAP activity is present in serum of dogs and cats.

The corticosteroid-induced isoenzyme (CAP) is present in serum of glucocorticoid treated dogs, dogs with hyperadrenalcorticism, and secondarily in some dogs with prolonged illness (Sanecki *et al.*, 1990; Solter *et al.*, 1993). Little or no CAP is present in normal dog serum.

Numerous techniques have been described for the separation, identification, and quantification of AP isoenzymes. However, in general, previous assays for AP isoenzymes depended on manual techniques such as electrophoretic separation, isoelectric focusing, heat inhibition, catalytic activity, and immunologic recognition (Hoffmann and Dorner, 1977a, 1977b; Wellman *et al.*, 1982; Saini and Saini, 1978a, 1978b). Because of the cumbersome and labor-intensive nature of these assays, they have been infrequently used. More recently an automated assay that allows the quantitative determination of CAP in dog serum or intestinal AP in rat serum has been developed and utilizes the selective inhibition of osseus and hepatic AP with levamisole (Hoffmann *et al.*, 1988; Hoffmann *et al.*, 1994). In addition, a technique has been described in human AP isoenzyme analysis that involves pretreatment of serum with wheat germ lectin to selectively precipitate osseus AP (Behr and Barnert, 1986; Rosalki and Foo, 1984). Pretreatment of serum with wheat germ lectin retards electrophoretic migration of osseus AP in both dog and rat serum, allowing clear separation of osseus and hepatic AP (Kidney and Jackson, 1988; Unakami *et al.*, 1989). A combination of pretreatment of serum with WGL to precipitate osseus AP and the automated assay for CAP or intestinal IAP, allows a quantitative determination of each of the AP isoenzymes in canine serum, equine serum, and rat serum (Sanecki *et al.*, 1993; Hank *et al.*, 1993; Hoffmann *et al.*, 1994). These latter assays not only allow quantitative analysis with good precision and accuracy, but provide a cost-effective analysis of a large number of samples.

Half-lives of the various AP isoenzymes in circulation are quite variable. In dogs the $T_{1/2}$ for intestinal, renal, and placental AP is less than 6 minutes, whereas for CAP and hepatic AP it is approximately 72 hours (Hoffmann and Dorner, 1977a). Th $T_{1/2}$ for osseus AP has not been determined but its presence in serum suggests it may be of the order of hepatic AP and CAP. The short half-life of intestinal or renal AP explains, in part, their absence in serum. Likewise, the short half-life for cat intestinal AP of approximately 2 minutes and horse intestinal AP of approximately 8 minutes explains the absence of these enzymes in the serum of these two species (Hoffmann and Dorner, 1977b; Hoffmann *et al.*, 1983b). Clearance studies of rat intestinal AP resulted in a biphasic disappearance of the enzyme with the first phase having a half-life of 2.8 minutes, while the second phase was 68.3 minutes (Young *et al.*, 1984). This slower disappearance of intestinal AP in the second phase may explain the appearance of intestinal AP in rat serum.

An increase in serum UAP is well recognized as an indicator of cholestasis in many species including dogs, rats, and man. Numerous studies have shown that this increase in AP activity in liver is a result of increased synthesis regulated either at the level of transcription or translation and is not a result of increased proliferation of cells as described for GGT (Kaplan *et al.*, 1983; Komoda *et al.*, 1984; Schlaeger, 1975; Seetharam 1986). Serum increases in AP activity following bile duct ligation begin at 6–7 hours postligation in rats and reach a maximum of 7–10 times normal serum AP activity at 12–24 hours post ligation (Kaplan *et al.*, 1983; Kaplan and Righetti, 1970; Schlaeger *et al.*, 1982; Schulz and Schlaeger, 1979). In dogs an increase is not seen until approximately 24 hours after ligation with a maximum of 30–40 times normal AP activity at 4–7 days (Guelfi *et al.*, 1982; Noonan and Meyer, 1979). To some extent these differences may be attributed to the absence of a gall bladder in rats. Hepatocellular necrosis in rats treated with CCl_4 results in a five- to sixfold increase in hepatic AP activity, but only a minimal one- to twofold increase in serum AP (Schlaeger *et al.*, 1982).

The mechanism of release of AP from liver and appearance in serum is poorly understood. Although there is evidence of disruptive changes within tight junctions of hepatocytes during cholestasis (Desmet and De Vos, 1982; Boyer, 1983), it is doubtful if these alterations permit passage of macromolecules the size of AP (Jones *et al.*, 1984; DeBroe *et al.*, 1985). Recent studies using a choledochocaval shunt model show that within 12 hours of shunting of bile or taurocholic acid into blood, there is marked induction of AP synthesis and appearance of AP on the basolateral membranes and a parallel increase of serum AP (Ogawa *et al.*, 1990). This occurs in the absence of any increase in biliary pressure and any evidence of alterations in tight junctions (Toyota *et al.*, 1983). Bile acids, therefore, appear to participate in both induction and release of AP into serum. Direct release of AP from the basolateral membrane or release of membrane fragments is a more likely hypothesis for the mechanism of increased AP activity into blood than is biliary regurgitation. Additionally, there is recent evidence that induction of hepatic UAP in short-term treatment of dogs with prednisone (4mg/Kg body weight) results in markedly increased UAP in hepatic tissue and serum with no increase in hepatic tissue bile acids as seen in dogs with bile duct ligation (Solter *et al.*, 1994). This treatment results in UAP on the sinusoidal as well as lateral and bile canalicular surfaces of hepatocytes where it could be be directly released into serum or hepatic lymph in the absence of biliary regurgitation (Solter and Hoffmann, unpublished data). The release of AP into serum by some mechanism other than biliary regurgitation is also consistant with the observation that in the absence of cholestasis in dogs with spontaneous hyperadrenocorticism or chronically treated with glucocorticoids there is a marked increase in the serum CAP activity with CAP located on the sinusoidal membrane and situated for release into lymph or plasma (Solter and Hoffmann, 1995).

Both UAP and CAP are attached to the membrane by a unique glycan-phosphatidylinositol anchor at the C′-terminal end. Serum CAP and serum UAP have the glycan inositol remnant of that anchor (Solter and Hoffmann, 1995). It is likely that UAP and CAP are released from the membrane by a serum phosphatidylinositol-specific phospholipase D known to exist in serum (Low and Prasad, 1988).

Many drugs induce AP in dogs. Glucocorticoids are well-recognized inducers of AP with the initial response to be induction of the UAP or hepatic AP isoenzyme in the first 2 days followed by induction of CAP after 5–7 days (Solter *et al.*, 1994). Glucocorticoids have considerably less effect in other species. Serum CAP is a good screening test for hyperadrenocorticism because of its high sensitivity for detecting increased cortisol secretion. It is not a diagnostic test because of its low specificity for the diagnosis of hyperadrenocorticism (Solter *et al.*, 1993; Teske, *et al.*, 1989; Kidney and Jackson, 1988). Increased serum CAP activity observed in animals with disease processes other than hyperadrenocorticism is consistent with a presumed increase in cortisol secretion as indicated by reported abnormal low dose dexamethasone suppression tests or ACTH stimulation tests in dogs with nonadrenal disease (Chastain *et al.*, 1985: Kaplan *et al.*, 1995).

In young, growing animals osseus AP is the predominant form of serum AP which decreases as maturation progresses and the epiphysis closes. Osseus AP can be as much as 10-fold greater in a 1-month-old puppy than osseus AP in the adult dog (Sanecki *et al.*, 1993). In newborn foals osseus AP can be nearly 100-fold greater than in adult serum (Hank *et al.*, 1993). Serum osseus AP increases have been observed in dogs with hyperparathyroidism, renal disease, and osteosarcoma, with the greatest increases seen in selected cases of osteosarcoma. The diagnostic value of osseus AP in domestic and laboratory animals remains to be determined.

B. Creatine Kinase

Creatine kinase isoenzymes (CK; EC 2.7.3.2) catalyze the reversible phosphorylation of creatine to form creatine phosphate.

The four CK isoenzymes are dimers of two protomers with individual masses of 40 kDa. The first three isoenzymes are numbered CK_1, CK_2, and CK_3. Brain contains the homogenous CK^1 (CK-BB) made up of two B protomers. Skeletal and heart muscle mostly contain the homogenous CK^3 (CK-MM), which is made up of two M protomers and has the longest half-life of the canine CKs (Anderson, 1976; Argiroudis *et al.*, 1982; Boyd, 1983; Fugii *et al.*, 1980; Rapaport, 1975). The CK^2 is the hybrid isoenzyme (CK-MB). A fourth variant, CK-Mt, is located between the mitochondrial membranes and makes up 15% of the total cardiac CK activity.

A number of conditions and compounds inhibit CKs. All CKs are activated by Mg^{2+}, but inhibited by an excess. Due to rapid oxidation of sulfhydryl groups at the active site they are unstable when stored at room, refrigerator, or freezing temperatures. Thiol agents added to the assay reactivate the enzymes to a degree dependent on how long they have been inactive.

Numerous methods are available for serum CK isoenzyme separation. The most common ones are electrophoresis, ion-exchange chromatography, and immunological. The immunological methods require species-specific, monoclonal, anti-CK isoenzyme antibodies. Cross-reactivities between the antihuman CK isoenzymes and animal CKs are not reported. In human medicine the immunological methods are most commonly used. Because of the need for species-specific antibodies for the immunological methods, electrophoresis and ion-exchange chromatography remain the methods in use in animals.

In human medicine serum CK isoenzymes serve as sensitive and specific indicators of cardiac infarction (Moss and Henderson, 1994). Cardiac infarction is infrequently encountered in clinical veterinary medicine where serum CK isoenzymes are of no more diagnostic value than total serum CK determination. In experimental medicine CK isoenzymes are commonly used to quantify cardiac and muscular damage (Akatas *et al.*, 1993). In domestic species, CK isoenzyme analysis has not been shown to be of significant value (Hoffmann, 1990).

The CKs may be too sensitive as indicators of muscle damage for some clinical purposes. Many cell types contain CK, with the highest specific activity in skeletal muscle. Only large increases in serum CK activity are of clinical significance (Lewis, 1978). Small amounts of bruising and intramuscular injections produce detectable rises in serum CK. Moderate transient muscle ischemia following exercise (Querengaesser *et al.*, 1994), prolonged recumbency, convulsions, or shivering results in increased serum CK_2 and CK_3.

As a result of the short serum half-life of CK activity rapidly returns to normal after a muscle damaging incident. This problem can be overcome by complementing muscle-specific CK with the less specific but longer circulating muscle AST activity. A small increase in serum CK and marked rise in AST serves as an indicator of muscle ischemia incident days before the day of sampling.

The magnitude of serum CK increase is reported to be a quantifiable marker of exercise and intramuscular injection muscle destruction (Akatas *et al.*, 1993; Lefebvre *et al.*, 1994; Janssen *et al.*, 1989). Frequent flaws in some of these studies are the assumptions that enzyme release occurs only on cell death, release into plasma is complete, and isoenzyme degradation rates are constants. They are not (Rapaport 1975; Akatas *et al.*, 1995). At best, serum CK elevation is a crude semiquantifiable marker of reversible and irreversible muscle injury. The greater the serum CK rise, the more muscle injured.

Nutritional vitamin E and selenium deficiency myopathies result in muscle membrane instability and release of CK into serum. In the absence of exercise-induced changes in otherwise clinically normal cattle and sheep, elevated total serum CK activity suggests a tentative diagnosis of vitamin E and or selenium deficiency (Smith *et al.*, 1994). Diagnosis is confirmed by the rapid return to normal of serum CK and RBC selenium metalloenzyme, glutathione peroxidase activity per unit of hemoglobin after treatment (Blincoe and Dye, 1958; Erskine *et al.*, 1990; Manktelow, 1963; Noguchi *et al.*, 1973; Orstadius *et al.*, 1959; Owen *et al.*, 1977; Turner and Finch, 1990; Van Vleet, 1975; Van Vleet *et al.*, 1975, 1977).

A large proportion of the brain is myelin, a lipoprotein. If CK activity is expressed per unit of protein or cell rather than per gram of tissue, the brain would have the highest CK-specific activity in the body. Little CK, however, is found in normal cerebrospinal fluid (CSF). Cerebrospinal fluid CK rises in demyelinating diseases and neoplasia in a number of disorders of dogs, cats, cattle, and horses, but the sensitivity of this rise is unclear (Edwin, 1970; Edwin and Jackman, 1974; Evans *et al.*, 1975; Furr and Tyler, 1990; Mayhew *et al.*, 1977; Smith and Healy, 1968; Wilson, 1977). Total serum CK increases in some neurologic disorders not as the result of brain CK entering serum, but as the result of muscle ischemia of convulsions and prolonged immobility (Smith and Healy, 1968).

C. Alanine Aminotransferase

Cytoplasmic alanine aminotransferase (ALT; EC 2.6.1.2) (formerly glutamic pyruvic transaminase; GPT) catalyzes the reversible transamination of L-alanine and 2-oxoglutarate to pyruvate and glutamate. Pyridoxal 5′-phosphate (PP) is the cofactor of ALT and other amino transferases that form the holoenzyme. The ALT apo- and its holoenzyme bound to PP are in serum in varying ratios.

Comparing ALT or AST assay results between laboratories requires knowledge of fundamental assay differences. To obtain maximum transferase activity, both the apo- and holoenzyme must be measured. To convert the inactive apoenzyme to its functional form, PP must be added to the specimen assay and a preincubation time allowed before the assay begins (Christen and Metzler, 1984; Horder and Rej, 1983). This preincubation inconveniently lengthens the assay period and some manufactures omit the addition of pyridoxal phosphate. The low serum ALT and AST activity in hemodialysis patients returns to normal when the patient is treated with pyridoxal phosphate (Chimata *et al.*, 1994). Cephalosporin treatment of rats and dogs results in a decrease in tissue and, consequently, serum ALT and AST activity (Bailey *et al.*, 1974; Dhami *et al.*, 1979; Rhodes *et al.*, 1986). The low activity is the result of a cephalosporin metabolite and is partially corrected by the addition of PP to the assay. The assays for ALT and AST are frequently linked to LDH. When the sample contians a high concentration of lactate, the lag phase of the assay is prolonged, interfering with the transaminase assay (Rogers and Osberg, 1974).

Human and porcine ALT are relatively stable at room temperature and refrigerated, but when frozen as much as 60% of its activity is lost. This freezing instability may be an expression of the ratio of apo- to holoenzyme (Rej, 1990).

In primates, dogs, cats, rabbits, and rats the liver has the greatest ALT-specific activity. In these species an increase in serum ALT is an established sensitive, specific indicator of hepatocyte damage. Muscle also contains ALT and care must be taken to distinguish between serum ALT of muscle versus liver origin (Valentine *et al.*, 1990). Hepatic, ALT-specific activity in pigs, horses, cattle, sheep, or goats is insuffcient to be of diagnostic value (Boyd, 1983).

D. Aspartate Aminotransferase

Aspartate aminotransferase (AST; EC 2.6.1.1) (formerly glutamic oxaloacetic transaminase; GOT), catalyzes the transamination of L-aspartate and 2-oxoglutarate to oxaloacetate and glutamate. It shares many of the features of ALT such as pyridoxal-5′-phosphate cofactor. The two AST isoenzymes, cytosolic and mitochondrial, have masses of about 92 kDa and multiple posttranslational forms. The presence of AST in so many tissues makes the serum enzyme a nonspecfic but sensitive marker of soft tissue damage, but precludes its use as an organ-specific enzyme (Boyd, 1983). Although CK is a more specific marker of muscle damage than AST, AST is frequently used to complement CK changes.

The upper normal limit of equine serum AST activity is considerably greater than that of other species (Rej and Horder, 1983). Therefore, assays established for other species are usually unsatisfactory for the measurement of increased equine serum AST. When equine serum AST is high, the specimen is diluted one-half to one-third, or the sample volume reduced and corrected for. In contrast to ALT, serum AST is stable frozen, at room temperature, and refrigerated. RBC high specfic AST activity can be released into plasma before hemolysis is observed visually and causes a falsely elevated serum value.

E. Sorbitol Dehydrogenase

Sorbitol dehydrogenase (SDH; EC 1.1.1.14; also called iditol dehydrogenase, IDH) reversibly oxidizes D-sorbitol to D-fructose with the cofactor NAD (Gerlich, 1983; Lessing and McGuinness, 1982). Sheep SDH, a metalloenzyme, has a 152-kDa mass whose four identical protomers have catalytic sites containing a single loosely bound Zn^{2+} (Reiersen *et al.*, 1994, Lindstad *et al.*, 1994).

Early reports of *in vitro* instability of human serum SDHs discouraged its use in veterinary medicine. Considerable species variation is reported. Frozen equine and bovine serum lose as much as 25% of their SDH activity in a week (Horney *et al.*, 1993).

A commercial SDH assay kit with a relatively stable, lyophilized control is available. The assay of the reversible reaction is in the direction of sorbitol formation. Because sorbitol inhibition occurs when 10% of the substrate is converted to product, a high fructose substrate concentration is required. The fructose solution's syrup-like texture requires a thorough mixing of assay solution. High endogenous keto acids, such as pyruvate and ketones, are common in the plasma of ruminents. Before recording the change in absorbtion in an SDH assay, a preincubation period is required to permit the reduction of keto acids by the other serum dehydrogenases (Gerlich, 1983). With end point assays a patient blank is necessary because of the high starting absorbance of the SDH assay and the inherently high 340-nm background absorbance of equine and bovine serum. The enzyme is inhibited when its Zn^{2+} is chelated by EDTA and oxalate anticoagulants. Heparinized plasma can be used, but serum is preferred.

Although SDH mRNA is in a number of tissues (Estonius *et al.*, 1993) appreciable amounts of enzyme activity are only found in the testis and hepatocyte (Boyd, 1983). Hepatocytes have the high SDH specific activity and increased plasma SDH is consistent with hepatocyte damage in many species (Asquith *et al.*, 1980; Lechtenberg and Nagaraja, 1991). Although liver-specific in all species, previously established ALT usage in dogs and cats has limited SDH clinical usage as an indicator of hepatocellular damage to horses, cattle, sheep, and goats.

F. Lactate Dehydrogenase

With the cofactor NAD, lactate dehydrogenase (LDH; EC 1.1.1.27) catalyzes the reversible oxidation of pyruvate to L(+)-lactate. Lactate dehydrogenase isoenzymes are tetramers of two protomers with molecular weights of about 35 kDa. The equilibrium favors lactate formation but because pyruvate is an inhibitor, the preferred assay is in the direction of pyruvate formation where it is combined with a trapping agent.

The two homomers are LDH-1 (heart; LDH-HH) and LDH-5 (liver and muscle; LDH-LL or LDH-MM). The hybrids are LDH-2 (LDH-H^1L^3), LDH-3 (LDH-H^2L^2), and LDH-4 (LDH-H^1L^3). Multiple forms of LDH isoenzymes have been reported and a sixth form, LDH-X, has been reported in humans. Electrophoretic and immunoinhibition methods are used to separate the isoenzyme (Henderson, 1983). The instability of homomers LDH-4 and LDH-5 permits refrigeration of specimens for isoenzyme separation, but not freezing. If frozen, the LDH protomers disassociate. When thawed they randomly reassociate.

Tissues contain various amounts of the LDH isoenzymes. Damaged tissue is identified by ratios of electrophoretically seperated serum isoenzyme profiles (Prasse, 1969). There is 150-fold greater LDH activity in RBCs than an equal volumne of plasma. Thus, minimal hemolysis appreciably increases plasma LDH. Anticoagulants such as EDTA and oxalates indirectly inhibit the enzyme. Heparinized plasma or serum is the preferred sample.

Lactate dehydrogenase isoenzyme profiles were the first isoenzyme profiles used in veterinary medicine to detect organ injury (Moore and Feldman, 1974), but they proved to have little specificity and little diagnostic value. More organ-specific procedures such as serum CK have superseded serum LDH isoenzyme profiles.

G. Cholinesterase

Two distinctly different cholinesterases (ChE) are found in the body. Both have the neurotransmitter acetylcholine as their substrate. The acetylcholinesterase (AChE; EC 3.1.1.7), "true" ChE, is located at the myoneural junction where it hydrolyzes acetylcholine to reestablish and prepare the junction for the next signal (Whittaker, 1983). The myoneural junction AChE is also found in RBC, mouse, pig brain, and rat liver. Only a small amount of AChE is in plasma.

The ChE of plasma is "pseudo" cholinesterase or butyrylcholinesterase (ButChE; EC 3.1.1.8), which hydrolyzes butyrylcholine four times faster than acetylcholine. It is located in plasma, white matter of the brain, liver, pancreas, and intestinal mucosa. Both AChE and ButChE have similar inhibitors and activators. Because AChE is low in plasma, plasma ButChE is used as an indicator of the AChE activity at the myoneural junction. People have as many as 12 macroenzymes of ButChE's single monomer in serum, with molecular weights ranging from 80 to 340 kDa. They occur in varying amounts between individuals. While activity in an individual is constant, the population's reference ranges are wide.

There are a number of inhibitors of the ChEs including drugs and naturally occurring substances. The two most clinically important ones are organophosphate insecticides and nerve gas. The phosphoryl group of the organophosphate irreversibly binds to the ChEs, preventing acetylcholine hydrolysis and thereby allowing persistent nerve stimulation.

High environmental background concentrations of organophosphates make it necessary to establish local geographic reference values. Following changes in serum ChE in animals is a strategy for biomonitoring environmental background levels of organophosphate contamination (Halbrook *et al.*, 1992). Decreases in serum ChE activity occur in people with acute infection, pulmonary infection, muscular dystrophy, chronic renal disease, and pregnancy, and insecticide intoxication.

Serum is the specimen of choice. Small amounts of hemolysis do not interfere with the assay. Serum ChEs have variable stability. Refrigerated, equine serum ChE is stable for at least a week (Plumlee *et al.*, 1994).

H. Lipase

Serum pancreatic lipase (EC 3.1.1.3) catalyzes the hydrolysis of triglycerides preferentially at the one and three positions, releasing two fatty acids and a 2'-monoglyceride. Human pancreatic lipase is a glycoprotein with a mass of about 48 kDa (Moss and Henderson, 1994). Bile salts and the cofactor, colipase, form a complex at the lipid surface necessary for lipase's optimum activity. Lipase and colipase are of pancreatic origin. The enzyme is water soluble, but the substrate is not. The reaction takes place at the water–lipid interphase of the lipid micelle where a micelle of bile salt, colipase, enzyme complex forms. Colipase is cleared by the kidney but lipase is not. This disproportionate loss of colipase versus lipase requires colipase addition to the assay for maxium lipase activity. Colipase addition to commercial lipase assays raised the reference ranges (Junge, 1983; Lott *et al.*, 1986).

Calcium enhances lipase activity. It is speculated to do that by trapping fatty acids, shifting the equilibrium to favor product production (Tietz and Shuey, 1993).

An appreciable amount of, but not all, canine serum lipase comes from the pancreas. Total pancreatectomy in the dog results in a depletion of 50–75% of serum lipase activity (Simpson *et al.*, 1991). Gastric mucosa lipase contributes appreciably to intestinal digestion, but not to serum activity.

Clinical serum lipase assays included all lipase forms. Serum lipases are stable at room temperature, refrigerated, and frozen. Lipoprotein lipase is normally low in serum, but following treatment with heparin it contributes a recordable amount of activity to the total serum lipase activity (Greten *et al.*, 1964).

When used alone serum lipase appears to be a more specific indicator of active exocrine pancreatitis than serum amylase (Strombeck *et al.*, 1981). In the dog and cat, pancreatitis results in serum lipase increases of two to three times greater than the upper reference range limit. Serum amylase is more commonly used than lipase to confirm active pancreatitis because the lipase assays are expensive and more difficult to perform than amylase assays. Idiopathic low magnitudes of increased serum lipase activity occur in diseases and treatments other than pancreatitis (Walter *et al.*, 1992, Cook *et al.*, 1993). Prednisone and dexamethasone treatment of dogs results in varying rises in serum lipase activity. These increases have led to the erroneous diagnosis of pancreatitis (Parent, 1982; Fittschen and Bellamy, 1984).

I. α-Amylase

Isoenzymes of α-amylases (EC 3.2.1.1) are Ca^{2+}-dependent metalloenzymes that randomly catalyze the internal hydrolysis of complex carbohydrates. As an example, they hydrolyze glycogen, at α-1-4 linkages to maltose and limited dextrins. They require activator ions such as Cl^+ or Br^+ (Wahlefeld, 1983). Plasma with metal chelating anticoagulants such as EDTA and citrate are inappropriate for an amylase assay. Amylase binds with other amylases and immunoglobulins to form macroamylases.

Many canine tissues, except salivary glands, contain α-amylase activity and its mRNA (Simpson *et al.*, 1989; Mocharla *et al.*, 1990). Canine α-amylase isoenzymes I through IV are electrophoretically classified. The slow migrating isoamylase IV appears to contain the macroamylases. Seventy-seven percent of dogs have between 5 and 62% of their total serum α-amylase activity precipitable as macroamylase (Corazza *et al.*, 1994). Human serum macroamylases have longer half-lives than the α-amylase holoenzyme. This variation in half-lives of α-amylase may be the reason for the wide range in normal canine serum activity.

Pancreas and duodenum have as much as six times the specific α-amylase activity as that found in other organs, but normally make only a small contribution to total serum α-amylases (Jacobs, 1989; Simpson *et al.*, 1991). So little pancreatic amylase is in normal canine serum that it is of no value as a marker of exocrine pancreatic deficiency. Marked increases in total serum α-amylase activity, primarily isoamylase III, are a specific indicator of active exocrine pancreatitis (Jacobs *et al.*, 1982; Cook *et al.*, 1993). The specificity of serum amylase increase as a clinical marker of exocrine pancreatitis is reduced by minor increases in activity that accompany renal diseases and the idiopathic macroamylasemias (Polzin *et al.*, 1983; Corazza *et al.*, 1994). The diagnostic significance, if any, of hypermacroamylasemia in dogs and people is not established.

Although amylaseuria is of diagnostic value in the acute pancreatitis of people, it is not of value in dogs because canine serum α-amylase is not cleared by the kidney (Eto *et al.*, 1969; Johnson *et al.*, 1977; Hudson and Strombeck, 1978; Jacobs, 1988).

Three basic α-amylase assays are saccharogenic, amyloclastic, and chromogenic procedures. Saccharogenic procedures measure the rate of genesis of glucose from starch, but are inappropriate for canine α-amylase. Canine plasma contains α-amylase and a glucoamylase (O'Donnell and McGeeny, 1975; Franzini and Bonini, 1967). The canine glucoamylase hydrolyzes the terminal maltose, produced by α-amylase, to glucose, falsely elevating the apparent α-amylase activity. Canine serum glucoamylase varies with no correlation with α-amylase or pancreatitis (Rapp, 1962).

Amyloclastic α-amylase assays measure the rate of disappearance of the starch and are unaffected by canine glucoamylase activity. A blue dye is bound to the starch. As the starch is hydrolyzed the dye is freed and the rate of release of the blue dye provides the measure of enzyme activity. Amyloclastic starch procedures have given way to synthetic substrates. They are still retained for electrophoresis to measure relative isoamylase activities.

Chromogenic α-amylase assays using dyes bound to synthetic substrates are the current clinical amylase assay. Blocked α-4-nitrophenylmaltoheptaoside and non-blocked β-4-nitrophenylmaltoheptaoside are acceptable synthetic substates for the canine α-amylase with the former the better than the latter. Unblocked α-4-nitrophenymaltoheptaoside is unacceptable substrate for the dog (Braun *et al.*, 1990).

Commercial α-amylase kits are designed for human serum and may contain insufficient substrate for canine serum. Normal canine serum, α-amylase activity is five to six times greater than human serum amylase. Kits designed for human amylase are modified for dog serum by reducing the amount of serum used or by diluting the serum with saline.

J. γ-Glutamyltransferase

γ-glutamyltransferase (GGT; EC 2.3.2.2) is a dimeric, carboxypeptidase that cleaves C-terminal glutamyl groups from synthetic substrates and transfers them to peptides and other suitable acceptors such as glycylglycine (Shaw, 1983; IFCC Expert Panel on Enzymes, 1983). As a result of a large amount of variable post-translational glycosylation (Nemesanszky and Lott, 1985; Tate *et al.*, 1988), the number of GGT isoenzymes is unclear (Milne and Doxey, 1985). Its membrane association makes it difficult to determine its mass and a consistent tissue activity. Estimates of mass range from 90 to 350 kDa. Dilution of the saline suspended kidney homogenates with either urine or serum results in an increase in activity. This increase is thought to be the result of disassociation of aggregates of enzyme and the exposure of additional active sites.

The physiological function of GGT is speculated to be associated with glutathione metabolism. All cells except muscle have some cytosolic and membrane GGT activity (Boyd, 1983; Milne and Doxey, 1985). The greatest amount of cellular GGT is in the brush borders of renal and bile duct epithelium. Serum GGT is possibly derived from the high-molecular-weight fragments of liver that also contain AP, 5′-nucleotidase, L-leucyl-β-naphthylamindase, and nucleotide pyrophosphatase. Basal serum and hepatic GGT is very low in dogs, cats, and rats as compared to ruminants, horses, and guinea pigs.

Equine serum lost 50% of its activity when stored at −30°C for 4 weeks but appreciably less at −20°C (West, 1989). Urinary GGT is relatively stable at room temperature and refrigerated, but is inactivated to varying degrees by cryoconcentration with urea (Beck and Sammons, 1975; Adams *et al.*, 1985). Inactivation of urinary GGT can be prevented by dialysis or the addition of albumin or dimethylsulfoxide to the specimen before freezing (Stokke, 1974).

Cholestatic disorders of all species examined result in increased serum GGT activity (Braun *et al.*, 1987). Prednisolone treatment induces canine hepatic and serum GGT activity with no evidence of cholestasis or bile duct hyperplasia (Solter *et al.*, 1994). The mechanism of GGT increases in serum with hepatic disease has only recently been addressed. Several models of hepatic disease in rats have been utilized to develop an understanding of the conditions and events leading to increased serum GGT. Carbon tetrachloride intoxication results in very little increase in serum GGT, which is consistent with the location of GGT primarily on biliary epithelial cells and its membranous attachment. Bile duct ligation results in a continuing increase of serum GGT activity, which parallels the increase in liver GGT activity (after day 1), which in turn is proportional or parallels the morphometrically determined bile duct volume (Bulle *et al.*, 1990; Leonard *et al.*, 1984) A second model utilizing treatment with alpha naphthyl isothiocyanate (ANIT) to cause biliary cell necrosis results in a peak activity at 1–2 days and then slow return to normal. Persistent exposure to ANIT results in biliary epithelial cell proliferation and continual increases in serum GGT (Leonard *et al.*, 1984). The initial response following both bile duct ligation and ANIT treatment is a very minimal increase of serum GGT activity with a majority of the GGT lost from liver. As opposed to induction of synthesis of GGT as described for hepatic AP in cholestasis, the increase in serum GGT in both models indicates that the increase in serum GGT activity is primarily dependent on the magnitude of biliary hyperplasia. In both bile duct ligation and ANIT treatment, biliary pressure increases either as a direct result of the ligation or blockage of the biliary system with necrotic epithelial cells. A unique model, choledochocaval fistula or shunt (CCS), allows the bile to flow directly from the bile duct into the anterior vena cava resulting in increased liver and blood bile acid concentration and increased bile flow and bile acid excretion rate but no increase in biliary pressure. This model resulted in serum GGT activity nearly equal to the BDL model, and suggests that bile acids or other bile constituents actually mediate the release of GGT into serum (Putzki *et al.*, 1989). Increased biliary pressure and regurgitation of GGT through tight junctions may not be necessary to cause an increase of serum GGT.

Cow, ewe, and doe colostrums, but not mare colostrum, contain a large amount of GGT. This colostrum GGT and the colostral antibodies are transferred across the neonate's intestinal wall into the plasma.

Renal GGT is the source of urinary GGT and with urinary AP, is a marker of active nephrotoxicity (Braun *et al.*, 1987; Greco *et al.*, 1985). On a day-to-day basis, urinary GGT in normal dogs and horses is relatively constant when expressed as enzyme per gram of urinary creatinine (Brobst *et al.*, 1986; Gossett *et al.*, 1987; Uechi, *et al.*, 1994). Reports of instabilty of urinary GGTs reveal the need for strict validation of specimen storage methods.

The induction of GGT provides a means for selecting and breeding of tolerance to toxins (Campbell, 1962; Towers *et al.*, 1983). There is a rapid elevation of serum GGT in sheep after a natural exposure to sporidesmin, the hepatotoxin, of facial eczema. This serum GGT increase is the result of bile duct proliferation. Only some lines of sheep experience this bile duct disease and increase in serum GGT. By selecting the sheep that do not experience the rise, an inherently sporidesmin-resistant line of sheep is obtained.

K. Trypsin

Trypsins (EC 3.4.21.4) are serine proteases that hydrolyze the peptide bonds formed by lysine or arginine with other amino acids. Trypsinogen originates in the pancreas and is converted to trypsin by intestinal enterokinase or trypsin itself. Trypsin isoenzymes I and II are immunologically separable and have differing physical characteristics. They are activated by Ca^{2+} and Mg^{2+}. Natural irreversible inhibitors occur in soy beans, lima beans and egg whites. Both pancreatic trypsinogen and trypsin occur in plasma where trypsin is inhibited by α_1-antitrypsin and α_2-macroglobulin. Trypsins have molecular masses of 23 and 26 kDa. They and their plasma zymogens are stable at room temperature for several days and when frozen for at least 5 years.

Fecal trypsin's proteolytic activity was formally used as a mark of pancreatic sufficiency. The tests used are described in greater detail in the chapter on pancreas. Their values are limited by their destruction in the intestinal track and interference by bacterial proteases.

Plasma trypsin, trypsinogen, and antitrypsin–trypsin complex occur in plasma in proportion to pancreatic mass. Immunoassays for canine serum trypsin-like materials (TLI) measure all three forms of trypsin. Increases in serum TLI occur in canine acute pancreatitis with a great range in activity. Serum TLI decreases in canine pancreatic exocrine insufficiency (Williams and Batt, 1983, 1988; Batt, 1993) and has become the diagnostic aid of choice in the diagnosis of pancreatic insufficiency (Simpson *et al.*, 1991).

L. Glutathione Peroxidase

Glutathione peroxidase (GPx; EC 1.11.1.9) is a metalloenzyme containing four Se^{+} per mole of enzyme. It catalyzes the oxidation of reduced glutathione by peroxide to form water and oxidized glutathione.

In assays, it is coupled to glutathione reductase to produce a detectable chromogen (Faraji *et al.*, 1987). There are two enzymes with glutathione peroxidase activity. One is selino-dependent selino-GPx, and the other a selino-independent glutathione transferase. The seleno-GPx can use either hydrogen peroxide or organic hydroperoxides as substrate. Glutathione transferase uses organic hydroperoxides as its substrates and has low activity with hydrogen peroxide (Lawrence *et al.*, 1978; Lee *et al.*, 1979). It does not have a correlation with Se concentration (Scholz *et al.*, 1981).

There is good direct correlation between RBC activity and tissue selenium concentration (Anderson *et al.*, 1978; Ammerman *et al.,*, 1980). This correlation has made whole blood selino-GPx a diagnostic aid in the detection of selenium deficiency. There are discrepancies concerning selino-GPx stability in frozen hemolysates (Sheppard and Miller, 1981; Jones, 1985a,b). This lack of prolonged stability has made GPx quality control difficult. Expression of the amount of GPx activity has been based on blood volume. GPx is an intracellular enzyme and it is more appropriate to use U/mg hemoglobin or U/dl red cells.

References

Abkowitz, J. L., Ott, R. L., Nakamura, J. E., Steinmann, L., Fialkow, P. J., and Adamson, J. W. (1985). *Clin. Invest.* **75,** 133.

Adams, R., McClure, J. J., Gossett, K. A., Koonce, K. L., and Ezigbo, C. (1985). *Am. J. Vet. Res.* **46,** 47.

Akatas, M., Auguste D., Lefebvre, H. P., Toutain, P. L., and Braun, J. P. (1993). *Vet. Res. Commun.* **17,** 353.

Akatas, M., Lefebvre, H. P., Toutain, P. L., and Braun, J. P. (1995) *J. Vet. Pharmacol. Ther.* **18,** 1.

Ammerman, C. B., Chapman, M. L., Bouwman, G. W., Pontenot, J. P., Blagyey, C. P., and Moxon, A. L. (1980). *J. Animal Sci.* **51,** 1381.

Anderson, M. G. (1976). *Res. Vet. Sci.* **20,** 191.

Anderson, P. H., Berretts, S., and Patterson, D. S. P. (1978). *J. Comp. Pathol.* **88,** 181. Argiroudis, S. A., Kent, J. E., and Blackmore D. J. (1982). *Equine Vet. J.* **14,** 317.

Asquith, R. L., Edds, G. T., Aller, W. W., and Bortell, W. W. (1980). *Am. J. Vet. Res.* **41,** 925.

Bailey, W. C., Weill, H., DeRonen, T. A., Ziskind, M. M., Jackson, H. A., and Greenberg, H. B. (1974). *Ann. Int. Med.* **81,** 200.

Bastiani, R., and Wilcox-Thole, W. L. (1982). *Clin. Lab. Ann.* **1,** 289.

Batt, R. M. (1993). *Vet. Clin. North Am.* **23,** 595.

Bayoumi, R. A., and Rosalki, S. B. (1976). *Clin. Chem.* **22,** 327.

Beck, P. R., and Sammons H. G. (1975). *Clin. Chim. Acta* **65,** 351.

Behr, W., and Barnert, J. (1986). *Clin. Chem.* **32,** 1960.

Bergmeyer, H. U. (1974). *In* "Methods of Enzymatic Analysis" Academic Press, New York.

Blackmore, D. V., and Elton, D. (1975). *Equine Vet. J.* **7,** 34.

Blackmore, D. J., and Plamer, A. (1977) *Res. Vet. Sci.* **23,** 146.

Blincoe, C., and Dye, W. B. (1958). *J. Anim. Sci.* **17,** 224.

Boyd, B. A. (1983). *Vet. Clin. Pathol.* **12,** 9.

Boyd, V. W. (1962). *Res. Vet. Sci.* **3,** 256.

Boyer, J. L. (1983). *Hepatology* **3,** 614.

Braun, J. P., Siest, G., and Rico, A. G. (1987). *Adv. Vet. Sci. Comp. Med.* **31,** 151.

Braun, J. P., Ouedraogo, G., Thorel, B., M'edaille, C., and Rico, A. G. (1990) *J. Clin. Chem. Clin. Biochem.* **28,** 493.

Bretaudiere, J. P., and Spillman, T. (1983). In "Methods of Enzymatic Analysis" (V. Bergmeyer, J. Bergmeyer, and M. Grabl, eds.), 3rd ed., Vol. 4, pp. 75–92. Verlag Chemie, Deerfield Beach, Florida.

Brobst, D. F., Carroll, R. J., and Bayly, W. M. (1986). *Cornell Vet.* **76,** 299.

Bulle, F., Mavier, P., Zafrani E. S., *et al.* (1990) Hepatology **11,** 545,

Buttery, J. E., and Pannall, P. R. (1986). *Clin. Chem.* **32,** 2100.

Campbell, J. R. (1962). *Vet. Rec.* **74,** 1340.

Chastain, C. B., Franklin, R. R., Ganjam, V. K., and Madsen, R. W. (1985). *J. Am. Anim. Hosp. Assoc.* **22,** 435.

Chimata, M., Masoka, H., Fujimaki, M., Hamada, N., Oda, K., Sudo, M., and Nagase, M. (1994). *Nip. Jinzo. Gak. Shi.* **36,** 389.

Christen, P., and Metzler, D., eds. (1984). "Transaminases." Wiley, New York.

Commission on Biochemical Nomenclature, IUPAC-IUB: Nomemclature of multiple forms of enzymes. (1977). *J. Biol. Chem.* **252,** 5939.

Cook, A. K., Breitschwerdt, E. B., Levine, J. F., Bunch, S. E., and Linn, L. O. (1993). *J. Am. Vet. Med. Assoc* **203,** 673.

Corazza, M., Tognetti, R., Guidi, G., and Buonaccorsi, A. (1994). *J. Am. Vet. Med. Assoc.* **205,** 438.

DeBroe, M. E., Roels, F., Nouwen, E. J., Claeys, L., and Wieme, R. J. (1985). *Hepatology.* **5,** 118.

Desmet, V. J., and De Vos, R. (1982). *Prog. Liver Dis.* **7,** 31.

Dhami, M. S., Drangova, R., Frakas, R., Balazs, T., and Feuer, G. (1979). *Clin. Chem.* **25,** 1263.

Diederichs, F., Muhlhaus, K., Trautschould, I., and Friedel, R. (1979). *Enzyme* **24,** 404.

Edwin, E. E. (1970). *Vet. Rec.* **87,** 396.

Edwin, E. E., and Jackman, R. (1974). *J. Sci. Food Agric.* **25,** 357.

Edwin, E. E., Markson, L. M., Shreeve J., Jackman, R., and Carroll, P. J. (1979). *Vet. Rec.* **104,** 4.

Eliakim, R., Mahmood, A., and Alpers, D. H. (1991). *Biochim. Biophys. Acta* **1091,** 1.

Erskine, R. J., Eberhart, R. J., and Scholz, R. W. (1990). *Am. J. Vet. Res.* **51,** 1107.

Estonius, M., Danielsson, O., Hoog, J. O., and Persson, H. (1993). *Advan. Exp. Med. Biol.* **328,** 545.

Eto, K., Pairent, F. W., Appert, H. E., and Howard, J. M. (1969). *Arch. Surg.* **98,** 241.

Evans, W. C., Evans, A. I., Humphreys, D. J., Lewin, B., Davies, W. E. J., and Axford, R. F. E. (1975). *J. Comp. Pathol.* **85,** 253.

Facklam, N. R., and Kociba, G. C. (1986). *Vet. Path.* **23,** 155.

Faraji, B., Kang, H. K., and Valintine, J. L. (1987). *Clin. Chem.* **33,** 539.

Fittschen, C., and Bellamy, J. E. C. (1984). *Can. J. Comp. Med.* **48,** 136.

Franzini, C., and Bonini, P. A. (1967). *Experimentia* **23,** 373.

Fraser, C. G., and Peterson, P. H. (1993). *Clin. Chem.* **39,** 1447.

Fugii, Y., Ikeda, S., and Watanabe, H. (1980). *Bull. Equine Res. Inst.* **17,** 21.

Furr, M. O., and Tyler, R. D. (1990). *J. Am. Vet. Med. Assoc.* **197,** 245.

Galasso, P. J., Litin, S. C., and O'Brain, J. F. (1993). *Mayo. Clin. Proc.* **68,** 349.

Gerlich, U. (1983). In "Methods of Enzymatic Analysis" (V. Bergmeyer, J. Bergmeyer, and M. Grabl, eds.), 3rd ed., Vol. 3, pp. 112–117. Verlag Chemie, Deerfield Beach, Florida.

Glick, M. R., Ryder, K. W., and Glick, S. J. (1987). In "Interferographs," 2nd ed., Science Enterprises, Indianapolis, Indiana.

Goldstein, D. J., Rogers, C. E., and Harris, H. (1980). *Proc. Natl. Acad. Sci. USA 77,* 2857.

Goren P. G., and Davis, J. E. (1986). *Clin. Chem.* **32,** 2021.

Gores G. J., Herman, B., and Lemasters, J. J. (1990). *Hepatology* **11,** 690.

Gossett, K. A., Turnwald, G. H., Kearney, M. T., Greaco, D. S., and Cleghorn, B. (1987). *Am. J. Vet Res.* **48,** 981.
Greco, D. S., Turrwald, G., Adams, R., Gossett, K. A., Kearney, M., and Casey, H. (1985) *Am. J. Vet. Res.* **46,** 2332.
Greten, H., Levy, R. J., and Fredrickson, D. S. (1964). *Biochem. Biophys. Acta.* **164,** 185.
Grindem, C. B., Stevens, J. B., and Perman, V. (1986). *Vet. Pathol.* **23,** 103.
Guelfi, J. F., Braun, J. P., and Genard, P. (1982). *Res. Vet. Sci.* **33,** 309.
Hafkenschied, J. C. M., and Kholer, B. E. M. (1986). *Clin. Chem.* **32,** 1616.
Halbrook, R. S., Shurgart, L. R., Watson, A. P., Munro, N. B., and Linnabary, R. D. (1992). *J. Am. Vet. Med. Assoc.* **201,** 714.
Hank, A. M., Hoffmannn, W. E., Sanecki, R. K., and Schaeffer, D. J. (1993). *J. Vet. Intern. Med.* **7,** 20.
Harris, C. C., Yolkin, R. H., Krokan, H., and Chang Hsu, I. (1979). *Proc. Natl. Acad. Sci. USA* **76,** 5336.
Harris, E. K. (1993). *Clin. Chem.* **39,** 927.
Hawrylak, K., and Stinson, R. A. (1988). *J. Biol. Chem.* **263,** 14368.
Henderson, A. R. (1983). In "Methods of Enzymatic Analysis" (V. Bergmeyer, J. Bergmeyer, and M. Grabl, eds.), 3rd ed., Vol. 3, pp. 138–155. Verlag Chemie, Deerfield Beach, Florida.
Henderson, A. R. (1993) *Clin. Chem.* **39,** 929.
Hoffmann, W. E. (1990). *Proc. IVth Congress of the Int. Soc. Animal Clinical Biochemistry,* p. 179.
Hoffmann, W. E., and Dorner, J. L. (1975). *J. Am. Anim. Hosp. Assoc.* **11,** 283.
Hoffmann, W. E., and Dorner, J. L. (1977a). *Am. J. Vet. Res.* **38,,** 1553.
Hoffmann, W. E., and Dorner, J. L. (1977b). *Am. J. Vet. Res.* **38,** 1637.
Hoffmann, W. E., Dorner, J. L., and Morris, H. (1983a). *Enzyme* **30,** 269.
Hoffmann, W. E., Dorner, J. L., and Morris, H. (1983b). *Vet. Clin. Pathol.* **12,** 33.
Hoffmann, W. E., Sanecki, R. S., and Dorner, J. L. (1988). *Vet. Clin. Pathol.* **22,** 17.
Hoffmann, W. E., Everds, N., Pignatello, M., and Solter, P. F. (1994). *Toxicol. Pathol.* **22,** 633.
Horder, M., and Rej, R. I. (1983). *In* "Methods of Enzymatic Analysis" (V. Bergmeyer, J. Bergmeyer, and M. Grabl, eds.), 3rd ed., Vol. 3, pp. 444–456. Verlag Chemie, Deerfield Beach, Florida.
Horney, B. S., Honor, D. J., MacKenzie, A., and Burton, S. (1993). *Vet. Clin. Pathol.* **22,** 5.
Hudson, E. B., and Strombeck, D. R. (1978). *Am. J. Vet. Res.* **39,** 1316.
Shaw, L. M., Stromme, J. H., London, J. L., and Theodorson, L. (1983). IFCC Expert Panel on Enzymes: IFCC methods for the measurement of the catalytic concentrations of enzymes, Part 9. IFCC method for gamma-glutamyl transferace. *Chem. Clin. Biochem.* **21,** 633.
Jackman, R. (1985). *Vet. Ann.* 71.
Jacobs, R. M. (1989). *Vet. Pathol.* **26,** 525.
Jacobs, R. M. (1988). *Vet. Pathol.* **25,** 443.
Jacobs, R. M., Hall, R. L., and Rogers, W. A. (1982). *Vet. Clin. Pathol.* **11,** 26.
Janssen, G. M. E., Kuipers, H., and Willems, G. M. (1989). *Int. J. Sports. Med.* **10,** S160.
Johnson, S., Ellis, C., and Levitt, M. (1977). *Clin. Res.* **25,** 313A.
Jones, A. L., Hradek, G. T., Schmucker, D. L., and Underdown, B. J. (1984). *Hepatology* **4,** 1173.
Jones, D. G. (1985a). *Res. Vet. Sci.* **38,** 301.
Jones, D. G. (1985b). *Res. Vet. Sci.* **38,** 307.
Junge, K., and Grutzman, K. D. (1987). *Clin. Chem.* **33,** 200.
Junge, K., Diego, J., Strobelt, V., Scholz, D., and Schreber, G. (1986). *Clin Chem.* **32,** 1807.
Junge, W. (1983). *In* "Methods of Enzymatic Analysis" (V. Bergmeyer, J. Bergmeyer, and M. Grabl, eds.), 3rd ed., Vol. 4, pp. 15–34. Verlag Chemie, Deerfield Beach, Florida.
Kaplan, A. J., Peterson, M. E., and Kemppainen, R. J. (1995). *J. Am. Vet. Med. Assoc.* **207,** 445.
Kaplan, M. M., and Righetti, A. (1970). *J. Clin. Invest.* **49,** 508.
Kaplan, M. M., Ohkubo, A., Quaroni, E. G., and Sze-Tu, D. (1983). *Hepatology* **3,** 368.
Kidney, B. A., and Jackson, M. L. (1988). *Can. J. Vet. Res.* **52,** 106.
Kihn, L., Dinwoodie, A., and Stinson, R. A. (1991). *Am. J. Clin. Pathol.* **96,** 476.
Komoda, T., Kumegawa, K., Yajima, T., Tamura, G., and Alpers, D. H. (1984). *Am. J. Physiol.* **246,** G393.
Kuhlenschmidt, M. S., Hoffmann, W. E., and Rippy, M. R. (1991). *Biochem. Med. Metab. Biol.* **46,** 152.
Kurstak, E. (1986). "Enzyme Immunodiagnosis." Academic Press, New York.
Lawrence, R. A., Parkhill, L. K., and Burk, R. F. (1978). *J. Nutr.* **108,** 981.
Lechtenberg, K. F., and Nagaraja, T. G. (1991). *Am. J. Vet. Res.* **52,** 803.
Lee, T. H., Layman, D. K., and Bell, R. R. (1979). *Nutr. Rep. Int.* **20,** 573.
Lefebvure, H. P, Tiutain, P. L., Serthelon, J. P., Lassourd, V., Gardey, L., and Braun, J. P. (1994). *Am. J. Vet. Res.* **55,** 487.
Lemasters, J. J., Stemkowski, C. J., Ji, S., Thurman, R. G. (1983). *J. Cell Biol.* **97,** 778.
Leonard T. B., Neptun D. A., and Popp, J. A. (1984). *Am. J. Pathol.* **116,** 262.
Lessing, N., and McGuinness, E. T. (1982). *In* "Methods in Enzymology" (S. P. Colowick, and N. O. Kaplan, eds.), Vol. 89, pp. 135–140. Academic Press, New York.
Lewis, H. B. (1978). *Vet. Clin. Pathol.* **7,** 6.
Lindena, J., Kupper, W., Friedal, R., and Trautschold, I. (1979). *Enzyme* **24,** 120.
Lindstad, R. I., Hermansen, L. F., and Mckinley-McKee, J. S. (1994). *Euro. J. Biochem.* **221,**847.
Loew, F. M., Bettany, J. M., and Halifax, C. E. (1975). *Can. J. Comp. Med.* **39,** 291.
Lott, J. A., Patal, S. T., Sawheny, A. K., Kazmierczak, S. C., and Love, J. E. (1986). *Clin. Chem.* **32,** 1290.
Low, M. G., and Prasad, A. R. S. (1988). *Proc. Natl. Acad. Sci. USA* **85,** 980.
Manktelow, B. W. (1963). *N. Z. Vet. J.* **11,** 52.
Mattenheimer, H., And Friedel, R. (1977). *Ann. Clin. Lab. Sci.* **7,** 344.
Mayhew, I. G., Whitlock, R. H., and Tasker, J. B. (1977). *Am. J. Vet. Res.* **38,** 1271.
Milne, E. M., and Doxey, D. L. (1985) *Res. Vet. Sci.* **39,** 385.
Mendel, L. B., and Underhill, F. P. (1907). *J. Biol. Chem.* **3,** 135.
Mocharla, H., Mocharla, R., and Hodges, M. E. (1990) *Nucl. Acids Res.* **18,** 1031.
Moore, W., and Feldman, B. (1974). *J. Am. Anim. Hosp. Assoc.* **10,** 420.
Moss, D. W., and Henderson, A. R. (1994). *In* "Tietz Textbook of Clinical Biochemistry" (C. A. Burtis and E. R. Ashwoodl, eds.), 2nd ed., p. 811. Saunders, Philadelphia, Pennsylvania.
Nemesanszky, E., and Lott, J. (1985). *Clin. Chem.* **31,** 797.
Noguchi, T., and Yamashita, Y. (1987). *Biochem. Biohyps. Res. Comm.* **143,** 15.
Noguchi, T., Cantor, A. H., and Scott, M. C. (1973). *J. Nutr.* **103,** 1502.
Noonan, N. E., and Meyer, D. J. (1979). *Am. J. Vet. Res.* 40, 942.
O'Donnell, M. D., and McGeeny, K. F. (1975). *Comp. Biochem. Physiol.* **50,** 269.
Ogawa, H., Mink, J., Hardison, W. G. M., and Miyai, K. (1990). *Lab. Invest.* **62,** 87.
Orstadius, K., Wretlind, B., Lindberg, P., Dordstrom, G., and Lannek, N. (1959). *Zentralbl. Veterinaermed.* **6,** 971.
Owen, R., Moore, J. N., Hopkins, J. B., and Arthur, D. (1977). *J. Am. Vet. Med. Assoc.* **171,** 343.
Parent, J. (1982). *J. Am. Vet. Med. Assoc.* **16,** 831.

Plumlee, K. H., Richardson, E. R., Gardner, I. A., and Galey, E. F. D. (1994) *J. Vet. Diag. Invest.* **6,** 247.
Polzin, D. J., Osborne, C. A., Stevens, J. B., and Hayden, D. W. (1983). *Am. J. Vet. Res.* **44,** 404.
Prasse, K. W. (1969). *Am. J. Vet. Res.* **30,** 2181.
Prescott, L. F. (1982). *Br. J. Clin. Pharmac.* **13,** 303.
Price, R. G. (1982). *Toxicology* **23,** 99.
Putzki, H., Reichert, B., Heymann, H. (1989). *Clin. Chim. Acta* **181,** 81.
Querengaesser, A., Iben, C. and Leibetseder, J. (1994). *J. Nutri.* **124,** 2760S.
Raab, W. P. (1972). *Clin. Chem.* **18,** 5.
Rapaport, E. (1975). *Card. Res.* **9,** 473.
Rapp, J. (1962). *Am. J. Vet. Res.* **23,** 342.
Rattazzi, M. C., Scandalios, J. G., and Whitt, G. S., eds. (1982). *In* "Isoenzymes, Current Topics in Biological and Medical Research", Vol. 1–12. Alan R. Liss, New York.
Reiersen, H., Lindstad, R. I., and Mckinley-McKee, J. S. (1994). *Arch. Biochem. Biophys* **311,** 450.
Rej, R. (1990). *Clin. Chem.* **37,** 131.
Rej, R. I., and Horder, M. (1983). *In* "Methods of Enzymatic Analysis", (V. Bergmeyer, J. Bergmeyer, and M. Grabl, eds.), 3rd ed., vol. 1, pp. 416–443. Verlag Chemie, Deerfield Beach, Florida.
Remaley, A. T., and Wilding P. (1989) *Clin. Chem.* **35,** 2261.
Rhodes, D. C., Dring, H. N., Blackman, and Lewis, H. B. (1986). *Abstr. Ann. Mtg. Am. Soc. Vet. Clin. Pathol.* Denver, Colorado.
Rogers, D. O., and Osberg, I. M. (1974) *Clin. Chem.* **20,** 43.
Rosalki, S. B., and Foo, A. Y., (1984). *Clin. Chem.* **30,** 1182.
Ryder, K. W., Trundle, D. S., Bode, M. A., Cole, R. E., Moorehead, W. R., and Glick M. R. (1991) *Clin. Chem.* **37,** 1134.
Saini, P. K., and Saini, S. K. (1978a). *Am. J. Vet. Res.* **39,** 1506.
Saini, P. K., and Saini, S. K. (1978b). *Am. J. Vet. Res.* **39,** 1510.
Sanecki, R. K., Hoffmannn, W. E., Dorner, J. L., and Kuhlenschmidt, M. S. (1990). *Am. J. Vet. Res.* **51,** 1964.
Sanecki, R. K., Hoffmannn, W. E., Hansen, R., and Schaeffer, D. J. (1993). *Vet. Clin. Pathol.* **22,** 17.
Schlaeger, R. (1975). *Z. Klin. Chem. Klin. Biochem.* **13,** 277.
Schlaeger, R., Harx, P., and Kattermann, R. (1982). *Enzyme* **38,** 3.
Scholz, R. W., Cook, L. S., and Todhunter, D. A. (1981). *Amer. J. Vet. Res.* **42,** 1724.
Schulz, B. O., and Schlaeger, R. (1979). *Enzyme* **24,** 173.
Seetharam, S., Sussman, N. L., Komoda, T., and Alpers, D. H. (1986). *Hepatology.* **6,** 374.
Shalev, A., Greenberg, A. H., and McAlpine, P. J. (1980). *J. Immunol. Meth.* **38,** 135.
Shaw, L. M. (1983). *In* "Methods of Enzymatic Analysis" (V. Bergmeyer, J. Bergmeyer, and M. Grabl, eds.), 3rd ed., Vol. 3, pp. 349–364. Verlag Chemie, Deerfield Beach, Florida.
Sheppard, A. D., and Miller, K. R. (1981). *N. Z. Vet. J.* **29,** 77.
Simpson, K. W., Doxey, D. L. and Keay, G. (1989) *Vet. Res. Comm.* **13,** 441.
Simpson, K. W., Simpson, J. W., Lake, S., Morton, T. B., and Batt, R. M. (1991). *Res. Vet. Sci.* **51,** 78.
Smith, G. M., Fry, J. M., Allen, J. G., and Costa, N. D. (1994). *Aust. Vet. J.* **71,** 12.
Smith, J. B., and Healy, P. J. (1968). *Clin. Chem. Acta* **21,** 295.
Solter, P. F., Hoffmannn, W. E. (1995) *Am. J. Physiol.* **269,** G278.
Solter, P. F., Hoffmannn, W. E., Hungerford, L. L., and Schaeffer, D. J. (1993). *J. Am. Vet. Med. Assoc.* **203,** 534.
Solter, P. F., Hoffmannn, W. E., Chambers, M. D., Schaeffer, D. J., and Kuhlenschmidt, M. S. (1994). *Am. J. Vet. Res.* **55,** 1086.
Stigbrand, T., Millan, J. L., and Fishman, W. H. (1982). *In* "Isoenzymes: Current Topics in Biological and Medical Research" (M. C. Rattazzi, J. G. Scandalios, and G. S. Whitt, eds.), **6,** 93–117.
Stigbrand, T., Millan, J. L., and Fishman, W. H., eds. (1984). *In* "Human Alkaline Phosphatase." Allen R. Liss, New York.
Stokke, O. (1974). *Clin. Chim. Acta* **57,** 143.
Strombeck, D. R., Farver, T., and Kaneko, J. J. (1981). *Am. J. Vet. Res.* **42,** 1966.
Tate, S. S., Khadse, V., and Wellner, D. (1988). *Arch. Biochem. Biophys.* **262,** 397.
Teske, E., Rothuizen, J., de Brijne, J. J., and Rijnberk, A. (1989). *Vet. Rec.* **125,** 12.
Tiersten, D., ed. (1986). *In* "Clinical and Economic Practice of Quality Control, Clinics in Laboratory Medicine." Saunders, Philadelphia, Pennsylvania.
Tietz, N. W., and Shuey, D. F. (1993) *Clin. Chem.* **39,** 746.
Towers, N. R., Meyer, H. H., Campbell, A. G., and Southey, C. A. (1983) *Proc. N. Z. Soc. Anim. Prod.* **43,** 221.
Toyota, N., Miyai, D., and Hardison, W. G. M. (1983). *Lab. Invest.* **50,** 536.
Triplett, D. A. (1986). *Clin. Lab. Ann.* **1,** 243.
Turner, R. J., and Finch, J. M. (1990). *J. Comp. Pathol.* **102,** 99.
Uechi, M., Terui, H., Nakayama, T., Mishina, M., Wakao, Y., and Takahashi, M. (1994). *J. Vet. Med. Sci.* **56,** 849.
Unakami, S., Hirata, M., Ichinohe, K., Tanimoto, Y., and Iizuka, H. (1989). *Exp. Anim.* **38,** 85.
Valentine, B. A., Blue, J. T., Shelley, S. M., and Cooper B. J. (1990). *J. Vet. Intern. Med.* **4,** 140.
Van Breda Vriesman, P. J. C. (1968). *In* "Enzymes in Urine and Kidney, Current Problems in Clinical Biochemistry" (U. C. Dubach, ed.), Vol. 2, p. 117. Hans Huber, Berne.
Van Vleet, J. F. (1975). *J. Am. Vet. Med. Assoc.* **166,** 769.
Van Vleet, J. F., Meyer, K. B., Olander, H. J., and Ruth, G. R. (1975). *Am. J. Vet. Res.* **36,** 387.
Van Vleet, J. F., Crawley, R. R., and Amstutuz, H. E. (1977). *J. Am. Vet. Med. Assoc.* **171,** 443.
Wahlefield, N. (1983). *In* "Methods of Enzymatic Analysis" (V. Bergmeyer, J. Bergmeyer, and M. Grabl, eds.), 3rd ed., Vol. 4, pp. 161–167. Verlag Chemie, Deerfield Beach, Florida.
Walter, G. L., McGraw, P., and Tvedten, H. W. (1992) *Vet. Clin. Pathol.* **21,** 23.
Webb, E. C. (1984). *In* "Enzyme Nomenclature 1984: Recommendations of the Nomenclature Committee of the International Union of Biochemistry on the Nomenclature and Classification of Enzyme-Catalysed Reactions." Academic Press, New York.
Weiser, W. E., and Pardue, H. L. (1987) *Anal. Chem.* **58,** 2523.
Wellman, M. L., Hoffmann, W. E., Dorner, J. L., and Mock, R. E. (1982). *Am. J. Vet. Res.* **43,** 1200.
West, H. J. (1989). *Res. Vet. Sci.* **46,** 301.
Whittaker, M. (1983). *In* "Methods of Enzymatic Analysis" (V. Bergmeyer, J. Bergmeyer, and M. Grabl, eds.), 3rd ed., Vol. 4, pp. 52–74. Verlag Chemie, Deerfield Beach, Florida.
Williams, D. A., and Batt, R. A. (1983). *J. Small Anim. Pract.* **34,** 583.
Williams, D. A., and Batt, R. A. (1988). *J. Am. Vet. Med. Assoc.* **192,** 195.
Wilson, J. W. (1977). *J. Am. Vet. Med Assoc.* **171,** 200.
Young, D. S., Pestaner, L. C., and Gibberman, V. (1975). *Clin. Chem.* **21,** 1D.
Young, G. P., Rose, I. S., Cropper, S., and Alpers, D. H. (1984). *Am. J. Physiol.* **246,** G419.

CHAPTER

13

Hepatic Function

BUD C. TENNANT

I. INTRODUCTION

The liver has an essential role in nutrient metabolism, including the control and maintenance of the blood glucose level, in detoxification and excretion of hydrophobic metabolites and xenobiotics, in the synthesis of most plasma proteins, and in digestion through synthesis, biliary secretion, and conservation of bile acids that are essential both for optimum hydrolysis of dietary fat and for intestinal absorption of fatty acids and other lipids, including fat-soluble vitamins. The clinical manifestations of hepatic disease are directly attributable to alterations in the metabolic, excretory, synthetic, and digestive functions of the liver. The liver has great functional reserve, and signs of hepatic failure often do not develop until 70% or more of the functional capacity of the liver is lost. Importantly, even when a major fraction of the hepatocellular mass has been lost in acute hepatic injury (e.g., acute hepatitis, hepatic necrosis), recovery is possible because of the unique capacity of the liver to regenerate.

Since this chapter was written originally for the first edition of this textbook by Dr. Charles E. Cornelius, remarkable advances have been made in our understanding of the pathophysiological, biochemical, and molecular mechanisms responsible for hepatic disease. Our current knowledge of veterinary hepatology is the result of the collective work of individuals from a variety of disciplines, including practicing veterinarians, biomedical scientists, and, more recently, molecular biologists. No one individual has had a more important or sustained impact than Dr. Cornelius, and no one has been more active in maintaining a comparative perspective on the subject (Cornelius, 1993). In this chapter, the biochemical mechanisms responsible for the cardinal clinical manifestations of hepatic insufficiency are summarized, and the biochemical tests useful in the assessment of hepatic function and in the clinical diagnosis of liver diseases are described. As in previous editions, the goal of this chapter is to provide students of veterinary medicine at all stages of career development with information useful for the diagnosis and treatment of the diseases of animal patients.

II. FUNCTIONAL ANATOMY

The liver develops embryologically as an outgrowth of the primitive gut. When fully developed, the liver is located cranial to the abdominal viscera and all other abdominal organs, and between the splanchnic and systemic circulatory systems. Unlike other mammalian organs, afferent blood to the liver is derived from two sources, the hepatic artery and the hepatic portal vein. Efferent blood leaves the liver by the hepatic vein and enters the systemic circulation via the caudal vena cava. Of the afferent blood, 10–20% comes from the hepatic artery, and the remainder from the hepatic portal vein, which drains the pancreas, spleen, stomach, small intestine, and all but the most terminal portion of the large intestine.

Hepatocytes are the principal cell type of the liver and make up at least 70% of its total volume. Like other cells of epithelial origin, the hepatocyte is functionally polarized, and the plasma membrane contains three morphologically and functionally separate domains. The sinusoidal domain is equivalent to the basolateral domain of other epithelial cells. Contact between hepatocytes is made by the intercellular domain that contains desmosomes and gap junctions. The canalicular domain is responsible for many of the excretory functions of the liver and for the initial phase of bile formation.

Hepatocytes are arranged in single-cell plates separated by sinusoids lined by vascular endothelial cells and through which blood from both the hepatic artery and the hepatic portal vein flows. Hepatic sinusoids differ from other capillaries in two important ways. First, hepatocytes normally do not rest on a conventional basement membrane, but are separated from endothelial cells by the perisinusoidal space of Dissé. Second, fenestrations in the sinusoidal lining cells allow formation of hepatic lymph in the space of Dissé that has a protein content much higher than lymph formed in conventional capillary beds, which is an ultrafiltrate of plasma characteristically low in protein content.

Terminal branches of the hepatic artery and the hepatic portal vein enter the classical liver lobule at its periphery. Blood from both sources mixes and percolates through hepatic sinusoids, then leaves the lobule via the central vein and exits the liver through branches of the hepatic vein. Obstruction of hepatic-vein outflow increases formation of hepatic lymph that is rich in protein. This may occur in congestive right heart failure, in mechanical obstruction of hepatic-vein outflow (Budd–Chiari syndrome), and in early stages of hepatic fibrosis. In advanced cirrhosis, more dense intracellular matrix forms in the space of Dissé, "capillarization" of the sinusoid decreases fenestration of the sinusoidal lining cells, and hepatic lymph is low in protein content, closely resembling the lymph produced by other normal tissues (see discussion of ascites, later).

Bile is secreted initially into the canaliculus and flows toward the periphery of the lobule in channels formed by the canaliculi of adjacent cells. Such channels may involve canalicular membranes of two or as many as four or five hepatocytes. These ultimately converge at the periphery of the lobule, forming the canals of Herring, which drain into the bile ductules of the portal tracts and ultimately into larger bile ducts that form the biliary tree. Hepatic lymph also flows toward the periphery of the lobule in the space of Dissé and exists the lobule via lymphatics located in the portal tracts. Hepatic lymph leaves the liver primarily via the hilar lymphatics, the hilar lymph nodes, and the thoracic duct. A minor fraction of hepatic lymph leaves the liver by lymph vessels associated with the hepatic vein.

The peripheral border of the classical liver lobule is formed by the most peripheral row of hepatocytes (the terminal plate) and by two, three, or more portal tracks that contain branches of the hepatic artery and the hepatic portal vein, and a bile ductule (portal triad). Blood flows from terminal branches of the hepatic artery and hepatic portal vein to the center of the lobule and the central vein. Although this lobular structure remains important in the morphologic descriptions of pathological conditions, most analyses indicate that the functional unit of the liver is the hepatic acinus, in which blood flows from one portal track toward two or more terminal collecting veins (central veins). Significant structural heterogeneity has been demonstrated between periportal hepatocytes of the hepatic acinus (Zone 1), midzonal hepatocytes (Zone 2), and perivenous hepatocytes (Zone 3) (Jungermann and Katz, 1989).

Cells of the periportal zone are more likely to divide than other hepatocytes (Grisham, 1959). Mitochondria are larger and more numerous in periportal hepatocytes than in those of the pericentral zone (Loud, 1968; Uchiyama and Asari, 1984). Fenestrae of periportal sinusoidal endothelial cells are larger than those of the pericentral region, and this may account for selective uptake of large complex molecules such as remnants of chylomicrons in periportal hepatocytes (Wisse *et al.*, 1985).

A significant oxygen gradient has been demonstrated between sinusoids of the periportal and pericentral zones. The concentrations of glucose and amino acids that arrive primarily from the hepatic portal vein are higher in periportal sinusoids during digestion. Such differences in sinusoidal nutrient concentrations are associated with important metabolic differences between zones of the hepatic acinus. The enzymes of glycolysis, gluconeogenesis, and glycogen metabolism have different activities within zones of the acinus.

Glucose-6-phosphatase, phosphoenolpyruvate carboxykinase, and fructose-1,6-diphosphatase activities are higher in periportal hepatocytes, whereas glucokinase and pyruvate kinase activities are higher in pericentral hepatocytes (Zakim, 1996). Glycogen appears to be uniformly distributed within the cells of the acinus during steady-state conditions, but during fasting, glycogen of periportal hepatocytes is utilized more rapidly, and during feeding, it is replaced more rapidly.

Two plasma-membrane transporters for glucose are expressed in the liver. GLUT-2 is the primary glucose transporter of the liver and appears to be expressed in plasma membranes of all hepatocytes. The K_m of GLUT-2 for glucose is 15 to 20 m*M*, a glucose concentration that can be reached or exceeded in the hepatic portal vein during and after feeding. Under these conditions, glucose is transported into hepatocytes for the synthesis of glycogen, amino acids, and triglycerides. Between meals, the glucose concentration in the portal vein decreases to that of the peripheral circulation, approximately 5 m*M*. During the interdigestive period, the concentration of glucose in hepatocytes is high relative to that of sinusoidal blood and GLUT-2, facilitating transport of glucose from the cytoplasm of hepatocytes into the sinusoid and ultimately to the systemic circulation to meet the energy requirements of other tissues. The GLUT-1 transporter is present only in the plasma membranes of the most pericentral hepatocytes (Tal *et al.*, 1990). The affinity of GLUT-1 for glucose is much higher (K_m 1–2 m*M*) than that of GLUT-2. Interestingly, the GLUT-1 gene is transcribed and translated by hepatocytes throughout the acinus; however, by means of a posttranslational control mechanism, it is inserted into the plasma membrane of only pericentral hepatocytes (Bilir *et al.*, 1993).

The liver plays a critical role in the removal of ammonia from the blood. Two separate reactions within the liver acinus are involved. The concentration of ammonia in the periportal sinusoids is high compared to that in the pericentral sinusoids, and most ammonia entering the liver diffuses into the hepatocytes of Zones 1 and 2. Relatively small amounts of ammonia reach hepatocytes of Zone 3. Hepatocytes of Zones 1 and 2 contain carbamoyl phosphate synthase and other enzymes of the urea cycle, and urea is synthesized in these cells. Glutamine synthase actively is confined to hepatocytes located adjacent to the terminal hepatic vein, and in these cells, glutamine synthesis from ammonia occurs. The K_m of carbamoyl phosphate synthase for ammonia is approximately 1.2 m*M*, whereas that of glutamine synthase for ammonia is 0.3 m*M* (Gumucio and Berkowitz, 1992). Synthesis of urea from ammonia in periportal hepatocytes and that of glutamine from ammonia in pericentral hepatocytes represent complementary enzymatic processes. Ammonia is first seen by periportal hepatocytes that synthesize urea utilizing a low-affinity, high-capacity system. At the end of passage through the hepatic sinusoid, the small amount of ammonia remaining is removed for glutamine synthesis, which is a high-affinity, low-capacity mechanism.

Periportal and midzonal hepatocytes are responsible primarily for the bile salt-dependent fraction of bile formation and also may be the primary site of bile salt synthesis. The enzymes necessary for fatty acid synthesis, CoA carboxylase and fatty acid synthase, are located primarily in pericentral hepatocytes. Drug-metabolizing enzymes such as cytochrome P450 are located predominantly in midzonal and perivenous hepatocytes. This may explain why carbon tetrachloride induces hepatocellular damage and fatty metamorphosis preferentially in perivenous hepatocytes (Jungermann and Katz, 1989).

III. CLINICAL MANIFESTATIONS OF HEPATIC INSUFFICIENCY

A. Icterus

1. Bile Pigment, Metabolism, and Excretion

Bilirubin is a yellow pigment produced by enzymatic degradation of heme. Approximately 80% of the bilirubin produced by normal mammals is derived from the removal of aged erythrocytes from the circulation by the reticuloendothelial systems (Robinson *et al.*, 1966; Landau and Winchell, 1970). Degradation of heme from other sources (e.g., myoglobin, the cytochromes, peroxidase, and catalase) accounts for the remaining bilirubin production. The liver contains large amounts of microsomal cytochromes (P450 and b_5) and is the most important source of bilirubin from nonerythroid sources.

The initial step in bilirubin formation is the opening of the heme (ferroprotoporphyrin) ring at the α-methene bridge (Fig. 13.1). This reaction is catalyzed by microsomal heme oxygenase (Tenhunen *et al.*, 1969, 1970a,b), a mixed-function oxidase. Cytochrome P450 serves as the terminal oxidase and requires molecular oxygen and reduced nicotinamide adenine dinucleotide phosphate (NADPH). The products of the heme oxygenase reaction are equimolar amounts of iron, biliverdin, and carbon monoxide. This apparently is the only reaction in mammalian tissues in which carbon monoxide is produced, and quantitation of respiratory carbon monoxide can be used as a measure of heme catabolism and indirectly of bilirubin production.

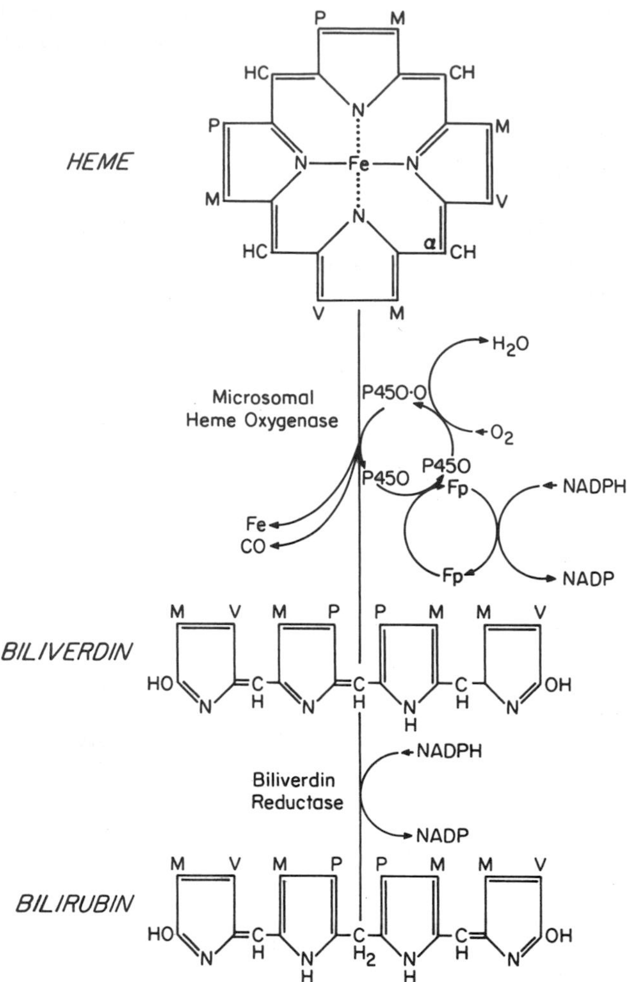

FIGURE 13.1 Enzymatic degradation of heme (ferroprotoporphyrin) and formation of bilirubin.

The second step in bilirubin formation is the reduction of biliverdin by the cytosolic enzyme biliverdin reductase, which, like heme oxygenase, requires NADPH (Tenhunen *et al.*, 1970a). In most mammals, hepatic biliverdin reductase activity is sufficient and is not normally rate-limiting in bilirubin synthesis. Biliverdin reductase activity, however, is almost completely lacking in birds, and in avian species, biliverdin is the major pigment of bile. Biliverdin also is found in significant quantities in the bile of certain mammalian species such as the rabbit, presumably because of decreased biliverdin reductase activity.

Heme oxygenase is most active in the tissues that are sites of erythrocyte removal and heme degradation. The spleen is the most important in this regard, followed by the liver and bone marrow (Tenhunen *et al.*, 1969, 1970b).

The kidney normally plays a minor role in heme degradation, but in hemolytic disorders, in which there is intravascular hemolysis and hemoglobinuria, the kidney has a quantitatively more important role. When intravascular hemolysis occurs, glomerular filtration of hemoglobin is initially prevented by the binding of hemoglobin to the plasma protein haptoglobin. When the haptoglobin binding capacity is exceeded, glomerular filtration of hemoglobin occurs and, depending on the amount filtered, some hemoglobin is reabsorbed by the tubular epithelium. This induces rapid formation of heme oxygenase in the tubule and significant bilirubin formation by the kidney (Pimstone *et al.*, 1971; de Schepper and Van Der Stock, 1972a,b). Renal degradation of hemoglobin may have a homeostatic function intended to conserve iron and minimize renal injury associated with hemoglobinuria (Pimstone, 1972).

2. *Hepatic Bilirubin Excretion*

Unconjugated bilirubin is nonpolar and is almost completely insoluble at physiologic pH. In plasma, unconjugated bilirubin is bound to albumin, which allows transport in stable aqueous solution to the liver. The hepatic mechanisms utilized in excretion of bilirubin, in general, are similar to those utilized by many other organic anions. The initial step is uptake of bilirubin by the hepatocyte. An organic acid binding protein that has been purified from basolateral (sinusoidal) plasma membranes of hepatocytes appears to play a role as both a bromosulfophthalein (BSP) and bilirubin transporter (Wolkoff *et al.*, 1985; Jacquemin *et al.*, 1991, 1994). Bilirubin disassociates from albumin prior to crossing the plasma membrane (Arias, 1974). The rate of hepatic uptake of bilirubin is regulated by the cytosolic protein ligandin (Y-protein), a low-molecular-weight cytosolic protein (4.4 to 4.6 kDa) that makes up approximately 5% of cytosolic liver protein and is identical to glutathione transferase B, the enzyme responsible for conjugation of BSP glutathione (Habig *et al.*, 1974; Kaplowitz *et al.*, 1973). In neonatal liver, the ligandin concentration is low and the neonate has a decreased capacity to excrete bilirubin. Ligandin synthesis can be induced by drugs such as phenobarbital that stimulate bilirubin excretion. Ligandin also binds other organic anions (e.g., porphyrins, BSP, indocyanine green, and certain steroid hormones such as cortisol), and intracellular binding appears to be important in their hepatic uptake.

Following uptake, bilirubin is converted to polar conjugates such as bilirubin diglucuronide. Conjugation with glucuronic acid is catalyzed by glucuronyl transferase, a microsomal enzyme that requires uridine diphosphoglyceronic acid (UDPG). Glucuronic acid esters of bilirubin have been identified in the bile of a variety of species, including the dog (Talafant, 1956), rat (Grodsky and Carbone, 1957), and guinea pig (Schmid, 1956), and also in the bile of the horse, pig, cat, sheep, and cow (Cornelius *et al.*, 1960). Gordon *et al.* (1976) have confirmed that the diglucuronide is the major bilirubin conjugate excreted in canine bile. There is evidence, however, that the bile of dogs (Fevery *et al.*, 1971; Heirwegh *et al.*, 1975; Noir, 1976) and other species (Cornelius *et al.*, 1975a) also contain bilirubin conjugates of other carbohydrates, including glucose and xylose.

The final step in bilirubin excretion by the liver is the transport of conjugated bilirubin across the bile canaliculus into the biliary system. Intravenous infusion of unconjugated bilirubin at rates exceeding the maximal hepatic excretory capacity results in the accumulation of conjugated bilirubin in the plasma. This indicates that in the overall process by which bilirubin is transferred from plasma to bile, the rate-limiting step is canalicular transport of the conjugated pigment rather than hepatic conjugation or uptake (Arias *et al.*, 1961). The concentration gradient of bilirubin glucuronide between the hepatocyte and bile is reported to be as high as 150-fold. An ATP-dependent transport system located in the canalicular membrane is responsible for the specific unidirectional transfer of bilirubin glucuronides and conjugates of glutathione from the cytoplasma into the bile (Chowdhury *et al.*, 1994). This transport system is functionally distinct from the ATP-dependent bile acid canalicular transporter (Muller *et al.*, 1991; Arias *et al.*, 1993). Although mechanistically separate from bile acid excretion (Alpert *et al.*, 1969), bile salt excretion enhances bile flow and in so doing, increases the maximum transport capacity for bilirubin (Goresky *et al.*, 1974).

3. *Extrahepatic Metabolism of Bilirubin*

Following entry into bile, conjugated bilirubin enters the intestine. Conjugated bilirubin, a polar compound, is poorly absorbed in the small intestine and passes to the large intestine, where it is reduced to a series of colorless derivatives collectively called urobilinogens (stercobilinogens). Reduction is catalyzed by dehydrogenases of anaerobic colonic bacteria. In germ-free animals that lack intestinal microorganisms, bilirubin passes unaltered into the feces and urobilinogen is not produced (Gustafsson and Lanke, 1960).

Most of the urobilinogen formed in the colon is passed in the feces. A small fraction is absorbed into the portal circulation, transported to the liver, and excreted in the bile. A fraction (1 to 5%) of the absorbed urobilinogen passes into the general circulation and is excreted by the kidney. In the dog, urobilinogen is excreted by both glomerular filtration and tubular secretion, the latter being enhanced in acid urine (Levy *et al.*, 1968).

Although the liver is the principal site of bilirubin conjugation and excretion, alternative pathways have been demonstrated. In normal animals, these alternative mechanisms are of minor significance but may become quantitatively more important in liver disease. After total hepatectomy, dogs have been shown to develop moderate hyperbilirubinemia and bilirubinuria (Bollman and Mann, 1932). In addition to unconjugated bilirubin, the plasma of hepatectomized dogs contains the monoglucuronide (Hoffman *et al.*, 1960) and, in some studies, the diglucuronide of bilirubin (Royer *et al.*, 1965). The kidney and intestine both have been shown experimentally to be sites of conjugation of bilirubin (Royer *et al.*, 1974). Differences in extrahepatic metabolism of bilirubin may explain remarkable differences between species in the bilirubin levels reached after bile-duct obstruction (see following section).

4. *Icterus*

The clinical sign of *icterus* or *jaundice* develops when the yellow pigment bilirubin accumulates in plasma and other tissues. Yellow discoloration of tissues can first be noted by careful observation when the plasma bilirubin value exceeds 2 to 3 mg/dl and can be appreciated even by an untrained observer when the concentration exceeds 3 to 4 mg/dl. The correlation between the plasma bilirubin concentration and the degree of clinical icterus is not, however, perfect. Elevated plasma bilirubin values are usually present for one or more days before clinical icterus is apparent, and there may be a delay between the time plasma bilirubin returns to normal and the clearance of the yellow discoloration of tissues. Conjugated bilirubin is said to have a greater affinity for connective tissue than the unconjugated pigment, possibly because conjugated pigment is less acidly bound to albumin. The degree of clinical icterus is greater in cholestatic liver disease for any given level of serum bilirubin because of the predominantly conjugated hyperbilirubinemia (With, 1968).

Visible yellow discoloration of tissues is readily recognized in animals in the unpigmented sclera. The normal red color of the visible mucous membranes makes detection of a slight yellow cast more difficult. It is possible to apply pressure to the mucous membranes and temporarily reduce blood flow to the area, so that the underlying color of the tissue can be better assessed.

The color of plasma (icteric index) may be useful clinically in the evaluation of icterus. Normal canine, feline, and ovine plasma is often water-clear and free of yellow color. The finding of yellow plasma in these species is highly suggestive of hyperbilirubinemia. Cattle absorb and transport significant quantities of carotene in plasma, and the icteric index, because it varies with the dietary intake of carotene, has limited use in this species. Equine plasma normally has a high icteric index, which in part is due to a plasma bilirubin concentration that normally is higher than that of other domestic species. There are other as yet uncharacterized non-carotene pigments, however, that also may contribute to the color of equine serum.

Notable species differences occur in the frequency with which icterus is observed in association with liver disease. In sheep and cattle dying of terminal hepatic insufficiency, there usually is a significant biochemical elevation in plasma bilirubin, but the value may not be sufficiently elevated to produce clinical icterus (Hjerpe *et al.*, 1971; Finn and Tennant, 1974). This is due possibly to the capacity of the liver to excrete bilirubin or to extrahepatic mechanisms for bilirubin excretion and/or degradation. Clinical icterus, when present in ruminants, is often associated with hemolytic anemia in which acute overproduction of bilirubin exceeds excretory capacity, for example, anaplasmosis in cattle and copper poisoning in sheep. In profound fatty liver, cattle that are critically ill may exhibit some degree of clinical icterus.

The assessment of clinical icterus in the horse is somewhat more complicated than in other species. The sclera and visible mucous membranes of most normal horses do not appear icteric, but in 10 to 15% of normal horses, a slight but definite (1/4) yellow discoloration of the sclera and/or oral mucous membranes can be detected (Tennant *et al.*, 1975). Scleral icterus of a moderate degree (2/4) may also be observed in horses with a variety of illnesses that do not involve the liver directly, such as pneumonia, impaction of the large intestine, and enteritis. Reduction in food intake is a common factor in such disorders, and fasting in the horse produces a rapid increase in plasma bilirubin concentration. In both hemolytic anemia and hepatic failure in the horse, the degree of icterus is usually remarkably greater than that seen under physiologic conditions or that is associated with reduced food intake. In the horse, severe clinical icterus is almost invariably present in acute hepatic necrosis (Thomsett, 1971; Tennant *et al.*, 1975). However, in chronic hepatic disease, icterus may be a more variable sign. In a series of 34 cases of hepatic cirrhosis in the horse, significant icterus was a presenting sign in 70% (Tennant *et al.*, 1975). Icterus was even less frequent (40%) in another series of horses with cirrhosis (Gibbons *et al.*, 1950).

The dog and cat appear to be intermediate between ruminants and the horse in their propensity to develop clinical icterus. Hemolytic disease, hepatocellular dysfunction, and extrahepatic bile-duct obstruction are characteristically associated with icterus in both dogs and cats. In experimental extrahepatic bile-duct obstruction in the dog, the plasma bilirubin increases at once following obstruction and clinical icterus is observed within I to 3 days. After 2 to 3 weeks, however, the plasma bilirubin of some dogs declines and clinical icterus may disappear. As in sheep and cattle, this may be related to adaptation of extrahepatic mechanisms of bilirubin excretion, particularly the kidney. The kidney of the dog is capable of adapting so that the rate of renal excretion of bilirubin equals the rate of formation. In cats with complete extrahepatic bile-duct obstruction, however, no such decrease is observed and persistent hyperbilirubinemia and deep icterus are characteristic.

B. Hepatic Encephalopathy

Hepatic encephalopathy is the syndrome of disturbances in cerebral function that accompany hepatic insufficiency or hepatic failure. The severity of neurologic signs may vary from subtle and intermittent behavioral changes associated with lethargy or stupor, to bizarre behavior, mania, convulsions, and coma (hepatic coma). Typically, such signs are associated with severe abnormalities in hepatic function.

Hepatic encephalopathy is a prominent clinical feature of hepatic failure in the horse. In one series of cases, 82% of horses presenting with acute hepatitis and 32% with cirrhosis had prominent neurologic abnormalities (Tennant *et al.*, 1973). Varying degrees of central-nervous derangement may be observed. Some horses stand quietly with the feet apart and the head lowered, nodding the head occasionally and appearing somnolent. The pupillary response to light may be normal or only moderately sluggish, but in some cases, vision is lost. Compulsive walking in a circle or in a single direction is often observed, and affected individuals may appear completely oblivious to their surroundings, walking over or through objects in their path (''walking disease''; Rose *et al.*, 1957). In fulminant cases, horses may become delirious with the head pressed forcibly against a wall for long periods of time. Horses may assume a variety of unusual positions or fall suddenly to the ground. Numerous unproductive attempts to rise are followed by periods of violent thrashing. When successful in rising, horses with hepatic encephalopathy may be completely uncontrollable, lunging forward violently and becoming a menace to attending personnel and equipment.

The syndrome of hepatic encephalopathy in cattle may have an abrupt onset but characteristically represents a terminal manifestation of chronic liver disease (Fowler, 1968; Pearson, 1977). Affected calves initially may appear dull, standing apart from other calves, and are anorectic. Other behavioral abnormalities may include violent charging or unusual, unrestrained bawling. Progressive dysmetria and ataxia are followed by recumbency, affected animals being unable to rise or to assume a sternal position. Tenesmus has been reported as a prominent clinical feature associated with dribbling of urine and prolapse of the rectal mucosa. Characteristically, death occurs within 2 days after onset of central nervous system signs (Finn and Tennant, 1974).

Hepatic encephalopathy is observed frequently in the dog with congenital or acquired portosystemic vascular shunts and may be one of the most prominent presenting clinical features of this form of liver disease (Ewing *et al.*, 1974; Barrett *et al.*, 1976; Cornelius *et al.*, 1975a,b; Audell *et al.*, 1974; Maddison, 1992). Neurologic signs associated with portosystemic shunts often are episodic and may be present for some months before recognition of the underlying hepatic disturbance. Depression and stupor with amaurotic blindness are observed in approximately half the cases of congenital portacaval shunts in dogs, with circling, head pressing, and intermittent seizures observed less frequently. Hepatic encephalopathy also has been associated with other primary diseases of the canine liver (Oliver, 1965; Strombeck *et al.*, 1975a; Center, 1996). Two cases of encephalopathy have been reported that were associated with apparent deficiency of argininosuccinate synthetase, a urea-cycle enzyme (Strombeck *et al.*, 1975b).

Hepatic encephalopathy must be differentiated clinically from primary inflammatory, degenerative, or neoplastic diseases of the brain, and this can be accomplished by demonstration of severe underlying hepatic disease. In the horse with acute hepatitis, clinical icterus almost always is present at the time neurologic signs are observed. In the dog and in cattle, frank clinical icterus is observed infrequently in animals with hepatic encephalopathy, so that tests of hepatic function and/or liver biopsy are required for confirmation. In cases of primary hyperammonemia due to deficiency of urea cycle enzymes, conventional hepatic function tests are not expected to be abnormal (Strombeck *et al.*, 1975b).

Factors responsible for the encephalopathy observed in hepatic failure are not completely understood. One of the important functions of the liver is the synthesis of urea from ammonia. Ammonia is present in normal peripheral blood at a concentration of 2 to 5 m*M*/liter. In portal venous blood, the concentration may be five times as high. Most of the ammonia in the hepatic portal vein is removed by the normal liver to form urea, with only a small fraction passing into the systemic circulation. In hepatic failure, synthesis of urea is reduced, and in the horse (Tennant *et al.*, 1975; Cornelius *et al.*, 1975b) and dog (Barrett *et al.*, 1976; Strombeck, 1975a), significant elevations of blood ammonia have been demonstrated. Ammonia has potent neurotoxic effects, and many of the neurologic signs of hepatic encephalopathy can be produced when toxic doses of ammonium salts are administered intravenously (Hooper, 1972).

The reactions of blood ammonia are determined by the physicochemical principles that apply to gases in solution and to the dissociation of weak bases. The ammonia:ammonium ion buffer system of blood can be described by the Henderson–Hasselbalch equation:

$$pH = pK_a - \log\left(\frac{NH_3}{NH_4^+}\right).$$

The pK_a for this system in the dog is approximately 9.1 (Bromberg *et al.*, 1960). This means that at physiologic pH (7.4), almost all of the ammonia of blood is ionized to form NH_4^+. As blood pH increases, the relative amount of free ammonia (NH_3) increases, and as the pH decreases, NH_3 decreases. Cells are almost impermeable to NH_4^+ but are permeable to NH_3, which passes through plasma membranes by nonionic diffusion (Stabenau *et al.*, 1959; Warren and Nathan, 1958; Castell and Moore, 1971). These principles are important in determining the amount of ammonia absorbed from the gastrointestinal tract or the amount that can pass from blood into brain tissue and other tissues (Dimski, 1994).

The potassium status, because of its influence on acid–base parameters, also may be an important determinant in NH_3 toxicity. Potassium deficiency ultimately favors the development of metabolic alkalosis, which in turn causes a shift in the $NH_3 \leftrightarrows NH_4^+$ equilibrium in the direction of the toxic free ammonia.

Blood ammonia ultimately is derived from dietary nitrogen. The gastrointestinal tract is the major source of blood NH_3, but NH_3 also is produced by other tissues, such as muscle and kidney. Renal ammonia is produced from glutamine and, to a lesser extent, from other amino acids. Synthesis of ammonium ion by the kidney represents a normal physiologic mechanism for H^+ excretion, and renal excretion of NH_4^+ is related directly to the pH gradient between blood and urine. The total urinary excretion of NH_4^+ is high in acid urine, and with alkaline urine, NH_4^+ excretion in the urine is low.

The conclusion that the gastrointestinal tract is the major source of blood NH_3 is based on the high concentration of NH_3 found in portal blood compared to that in peripheral venous blood. Part of the ammonia in the hepatic portal vein is derived from the action of bacterial enzymes on dietary amino and amide nitrogen, and part is derived from urea, which is present in the normal secretions of the alimentary canal and is hydrolyzed by the bacterial urease within the lumen of the bowel, primarily in the stomach and large intestine. The question of whether intestinal urease is produced in part by mammalian cells or whether it is entirely of bacterial origin was the subject of controversy for many years. Using germfree rats, Levenson and Tennant (1963) demonstrated that intestinal urease was exclusively of bacteria in origin, and this observation has been confirmed in germfree dogs (Nance *et al.*, 1974).

The relative importance of NH_3 produced by intestinal bacteria and that produced from nonbacterial sources is still not fully known. Nance *et al.* (1971, 1974) and Nance and Kline (1971) have demonstrated that germfree dogs with Eck fistulae develop encephalopathy associated with hyperammonemia. This suggests that at least with vascular shunts, endogenous sources of NH_3 contribute significantly to encephalopathy. In this regard, the intestine can metabolize significant quantities of glutamine independent of intestinal bacteria, and approximately 30% of the glutamine nitrogen that reaches the intestine appears in the portal blood as NH_3/NH_4^+ (Windmueller and Spaeth, 1974).

The liver plays a critical role in maintenance of the blood glucose concentration, and marked hypoglycemia is sometimes associated with liver failure. In fulminant hepatic failure in the horse, the blood glucose has been reported in some cases to be as low as 20 mg/dl or less (Tennant *et al.*, 1975; Hjerpe, 1964). Hypoglycemia also has been reported in dogs with hepatic insufficiency associated with vascular shunts (Ewing *et al.*, 1974; Cornelius *et al.*, 1975a).

Other potentially neurotoxic substances have been suggested to be involved in the pathogenesis of hepatic encephalopathy, and the role of these factors has been reviewed (Maddison, 1992; Center, 1996). Indole and indolyl derivatives that are formed from tryptophan by intestinal bacteria have been suggested as encephalotoxic compounds capable of inducing hepatic coma (Zieve *et al.*, 1974). Other studies have incriminated short-chain fatty acids and, in experimental hepatic failure, total volatile fatty acids (VFAs) increase significantly prior to death (Zieve *et al.*, 1968). Increased plasma VFA concentrations also have been observed in spontaneous hepatic encephalopathy and, when infused intravenously into experimental animals, VFA produces cerebral depression followed by coma (Takahashi *et al.*, 1966). There is increasing evidence that cerebral edema is an important factor in the pathogenesis of hepatic encephalopathy (Butterworth, 1994), possibly related to increased glutamine accumulation within the brain.

C. Hepatic Photosensitivity

Photosensitivity results from hypersensitivity to sunlight induced by the presence of exogenous or endogenously produced photodynamic substances. Clinical signs of photosensitization develop when photosensitive animals are exposed to light and are caused by inflammation and necrosis of unpigmented skin (photodermatitis). A distinction is made between sunburn and photosensitization. In animals with unpigmented skin maintained for long periods of time indoors, abrupt exposure to sunlight may induce sunburn. Sunburn is a direct response of the unpigmented, unprotected skin to ultraviolet radiation (320 μm) in otherwise normal animals. Photosensitization is characteristically a more severe reaction of the skin caused by the interaction of a photodynamic substance and solar radiation. The effective wavelength of light causing photosensitization is determined by the absorption spectrum of the sensitizing substance, which may extend into the visible region of light. Sunburn and photosensitization also differ in that sunburn apparently can develop in the absence of molecular oxygen (Blum *et al.*, 1935), whereas photosensitization occurs only in the presence of molecular oxygen (Cook and Blum, 1959; Schothorst *et al.*, 1970). With sunburn, there is a characteristic delay between exposure to light and the development of erythema of the skin, soreness, or pruritus, whereas in photosensitization, initial clinical signs may be noted within minutes after exposure to sunlight (Clare, 1945).

Three forms of photosensitization are recognized in domestic animals. A primary cause of photosensitivity is ingestion of photodynamic substances not normally present in the diet. Examples are the diseases caused by ingestion of the poisonous plants *Hypericum perforatum* (St. John's wort, Klamath weed) and *Fagopyrum esculentum* (buckwheat). Administration of the parasiticide phenothiazine also causes a form of primary photosensitization. In such a case, the photosensitizing compound is phenothiazine sulfoxide, which penetrates the aqueous humor as well as the skin and may induce keratitis.

A second form of photosensitization results when the photosensitizing compound is produced endogenously. Congenital porphyria (pink tooth) of cattle is the best characterized of this group of diseases, although porphyria associated with photosensitization has also been observed in cats and swine. In cattle, there is a marked increase in production of uroporphyrin I, which is deposited in the teeth and bones and excreted in large quantities in the urine, causing the teeth and urine to fluoresce with ultraviolet light. There is also an increase in the protoporphyrin content of erythrocytes. Photodermatitis and hemolytic anemia that are present in the disease appear to be directly related to the photodynamic effects of these porphyrins (Kaneko *et al.*, 1971; Scott *et al.*, 1979).

A third group of diseases associated with photosensitivity are those that are secondary to hepatic disease (hepatic photosensitization). Photodermatitis associated with liver disease is recognized clinically only in herbivorous animals. Hepatic photosensitivity is observed in both acute and chronic liver diseases. Southdown sheep with congenital photosensitivity have no morphologic abnormalities of the liver in the early stages of the disease, but photodermatitis is always associated with significant biochemical defects in hepatic excretory function (Cornelius and Gronwall, 1968).

The photodynamic agent responsible for hepatic photosensitivity is phylloerythrin, a porphyrin derivative from chlorophyll (Clare, 1945; Rimington and Quin, 1934). Chlorophyll is converted to phylloerythrin by microorganisms of the rumen or large intestine that remove the magnesium atom from the chlorophyll molecule and that hydrolyze the phytyl and carboxymethoxy side chains (Quin *et al.*, 1935), leaving the porphyrin nucleus of chlorophyll intact. The phylloerythrin produced in the alimentary tract is excreted primarily in the feces. A small fraction of the relatively nonpolar porphyrin is absorbed into the portal circulation. In normal animals, phylloerythrin is quantitatively removed by the liver and excreted in the bile. Phylloerythrin is found in the bile and feces of herbivores consuming chlorophyll-containing diets and also may be demonstrated in other species that ingest chlorophyll. The comparatively large amounts excreted by ruminants are attributed to their frequently high chlorophyll intake and to the favorable conditions for microbial production of phylloerythrin within the gastrointestinal tract.

In hepatic insufficiency, phylloerythrin is incompletely cleared from the hepatic portal circulation, enters the systemic circulation, and ultimately accumulates in the skin. In the superficial layers of the skin, phylloerythrin absorbs solar energy, resulting in free radical formation. By causing peroxidation of cellular lipids and other cellular components, free radicals damage cellular organelles, including lysosomes. Inflammation and necrosis of the skin are the result of direct oxidative injury and the secondary action of lysosomal enzymes (Slater and Riley, 1965). The critical range of wavelengths (action spectrum) that results in photodermatitis in hepatic photosensitivity was shown in geeldikkop to be between 380 and 650 nm (Riemerschmid and Quin, 1941) and in facial eczema between 400 and 620 nm (Clare, 1944), ranges consistent with the known absorption spectrum of phylloerythrin.

The types of hepatic disease of ruminants and horses associated with photosensitivity vary considerably, but the effects attributable to the photodynamic action of phylloerythrin are similar. The nature and severity of the cutaneous lesions depend upon the amount of phylloerythrin in the skin, and the intensity and duration of light exposure. The most common site of photodermatitis in the horse is the muzzle, which has a sparse protective covering of hair and often is unpigmented. Unpigmented areas of the distal extremities also are frequently affected (Tennant *et al.*, 1973; Fowler, 1965). In cattle, unpigmented areas of the muzzle, back, and escutcheon and the lateral aspects of teats are especially susceptible.

The areas of skin affected in sheep are those that receive the greatest exposure to light and that lack protection of pigment or wool. These include the ears, eyelids, face, lips, and coronets (Riemerschmid and Quin, 1941). The first clinical signs of photodermatitis in sheep may be apparent restlessness with shaking of the head or rubbing of affected parts. Individual animals may seek relief in the shade. Erythema and edema are the first cutaneous manifestations of photosensitization. Swelling of the lips, ears, and face have led to the descriptive terms "big head" and "facial eczema." Following edema, serum may ooze from damaged skin. Ultimately, second- or third-degree burns may develop, and the morbidity and mortality attributable to lesions of the skin may be more important than any other aspect of the underlying liver disease (Riemerschmid and Quin, 1941).

D. Ascites

The clinical sign of ascites is the result of abnormal accumulation of fluid in the peritoneal cavity. In normal animals, there is significant bidirectional movement of fluid, electrolytes, and, to a lesser degree, protein across the mesenteric capillary bed, through the interstitial space, and across the peritoneal mesothelium into the abdominal cavity. Such movements are determined by osmotic and hydrostatic forces that are described by Starling's equation: Plasma colloidal osmotic pressure minus ascitic fluid colloidal osmotic pressure equals portal capillary pressure minus intra-abdominal hydrostatic pressure. Normal portal capillary pressure on the arterial side of the capillary bed favors formation of an ultrafiltrate of plasma that is nearly protein free. On the venous side of the capillary bed, reabsorption of interstitial fluid occurs because of a decrease in hydrostatic pressure below that of the colloidal osmotic pressure within the capillary bed exerted by plasma proteins, primarily albumin. Under normal conditions, only a small volume of free fluid is present in the peritoneal cavity.

During investigation of the mechanical factors that influence formation of lymph, Starling (1894) observed that obstruction of hepatic venous flow by ligation of the thoracic vena cava cranial to the site of entry of the hepatic veins produced a significant increase in lymph flow through the thoracic duct, and the lymph was high in protein. Obstruction of the hepatic portal vein as it entered the liver also produced an increase in the flow of thoracic duct lymph, but the protein content was low. Starling concluded that the increased flow of thoracic duct lymph following ligation of the thoracic vena cava arose from hepatic lymph, and after ligation of the portal vein, increased thoracic duct lymph was derived from mesenteric capillaries.

Hepatic lymph is produced primarily in the sinusoids and accounts normally for 25 to 50% of lymph flow in the thoracic duct (Brauer, 1963). In the dog (and other species), hepatic lymph has a much higher protein content than lymph from other tissues because of the unique permeability of the sinusoids to plasma proteins, which is due to the absence of a conventional basement membrane associated with hepatocytes and to fenestrations in the endothelial cells lining the sinusoids (Bissel and Maher, 1996; Field *et al.*, 1934). In experimental cirrhosis in the dog and other species, the flow rate of thoracic duct lymph is increased 2 to 5 times and the protein content is increased (Nix *et al.*, 1951a,b).

Ascites caused by cirrhosis of the liver is associated with increased portal vein pressure (portal hypertension). Experimental ligation of the portal vein before it enters the liver, however, results only in minimal and transient ascites or no ascites (Schilling *et al.*, 1952; Volwiler *et al.*, 1950; Berman and Hull, 1952). In the dog, ligation of either the hepatic vein (Orloff and Snyder, 1961a,b; Orloff *et al.*, 1963, 1964a,b, 1966) or the caudal vena cava at a site cranial to entry of the hepatic vein (Schilling *et al.*, 1952; Berman and Hull, 1952; Witte *et al.*, 1968, 1969a,b) produces prompt and intractable ascites. Lymph fluid has been observed to form droplets that "weep" from the surface of the liver following experimental obstruction of hepatic vein outflow (Hyatt *et al.*, 1955). Because of its origin as hepatic lymph, the protein content of such ascitic fluid may be 3.0 to 3.5 g/dl or higher. This observation is consistent with the clinical finding that the protein content of ascitic fluid in the initial stages of cirrhosis may be higher than that of conventional transudates (a modified transudate) because most of the ascitic fluid in cirrhosis is derived from the liver. As fibrosis and capillarization of the hepatic sinusoids becomes progressively more severe (Bissell and Maher, 1996), the protein content of hepatic lymph decreases and the protein content of ascitic fluid is low. It is probable that in many cases of cirrhosis, both hepatic lymph and mesenteric lymph are produced at increased rates and that ascites develops when the return of both sources of lymph to the systemic venous circulation fails to keep pace (Witte *et al.*, 1971a,b). The protein content of ascitic fluid in hepatic disease is influenced not only by the relative proportions of mesenteric and hepatic lymph that contribute to ascites, but also by the protein concentration of plasma. In advanced cirrhosis, when hypoproteinemia is present, the protein content of ascitic fluid can be expected to be proportionately low.

Although experimental portal vein obstruction per se does not result in ascites, only transient portal hypertension is actually produced by this procedure. When persistent portal hypertension is produced experimentally by aortic-portal anastomosis or when such anastomoses occur congenitally, the ascitic fluid has a characteristically low protein content because of its origin in the mesenteric capillary bed.

Serum albumin is produced exclusively in the liver and is the major determinant of plasma and tissue-fluid oncotic pressure. Hypoalbuminemia associated with chronic liver disease has been considered a factor contributing to the development of ascites, but current evidence suggests the role is not primary. For example, the intravascular and total-body albumin pool may not be greatly diminished in cirrhosis, although the concentrations of albumin in plasma may be decreased (Rothschild *et al.*, 1973; Witte *et al.*, 1971a,b). For ascites to develop, there must be an expansion in total body sodium and body water. It is known that excessive sodium chloride intake greatly enhances development of ascites (Berman and Hull, 1952), and ascites is preceded by increased sodium retention by the kidney. Aldosterone levels have been shown to be significantly increased in dogs with ascites caused by hepatic-vein obstruction, but the underlying mechanisms responsible for this steroid response are not resolved. The liver is the primary site of renin degradation, and it has been suggested that decreased degradation of renin by the diseased liver results in increased production of bradykinin, which in turn stimulates aldosterone release from the adrenal cortex (Orloff *et al.*, 1965; Howards *et al.*, 1968).

When caused by liver disease, ascites is indicative of a chronic process and characteristically is associated with cirrhosis. There are important species differences in the occurrence of ascites in chronic liver disease. In dogs and cats with advanced hepatic cirrhosis, ascites is a relatively common sign, but it is almost never observed in horses with cirrhosis (Tennant *et al.*, 1975). Conspicuous ascites also is unusual in cattle with cirrhosis (Finn and Tennant, 1974; Whitlock and Brown, 1969), but has been observed at necropsy (Pearson, 1977). Ascites has been reported in association with liver abscesses in cattle where thrombosis of the hepatic vein adjacent to the abscess causes marked hepatomegaly (Breeze *et al.*, 1976; Selman *et al.*, 1974). In sheep, ascites has been observed in cirrhosis, but is unusual with the sclerosing cholangitis associated with fascioliasis (Hjerpe *et al.*, 1971).

It is important clinically to differentiate ascites caused by liver disease from ascites associated with other primary diseases. The distinction is made based on history and clinical signs, and on the examination of ascitic fluid. Biochemical and cytologic examinations of ascitic fluid may be useful but alone are seldom diagnostic. The protein concentration of ascitic fluid in cirrhosis is characteristically low (Center, 1996) but is variable depending on the stage of disease. In early stages, the protein content of ascitic fluid may exceed 2 to 2.5 g/dl because it is a reflection of the high protein content of lymph from the liver. Later, when serum albumin has decreased and when sinusoidal fibrosis and capillarization have developed, the protein content of hepatic lymph is low and ascitic fluid will be low in protein (1–1.5 g/dl). The protein content of ascitic fluid is generally high in peritonitis and in neoplastic diseases of the abdomen, and protein concentrations below 1.5 to 2.0 g/dl are unusual in ascitic fluid. Total nucleated cell counts in ascitic fluid from dogs with cirrhosis are seldom greater than 1000 to 2000 μl. Bloody or turbid fluid typically results from inflammatory or neoplastic processes (e.g., feline infectious peritonitis, bacterial peritonitis, neoplasia), and nucleated cell counts are elevated. Fluid deeply stained with bile pigment indicates a direct communication between the biliary system and the peritoneal cavity. In domestic animals, so-called "bile ascites" or bile peritonitis is a problem recognized most frequently in the dog and cat, caused by abdominal trauma.

IV. LABORATORY ASSESSMENT OF HEPATIC FUNCTION

The pathogenesis of hepatic disease in domestic animal species is remarkably complex, involving acute and chronic forms of hepatitis, cirrhosis, bile-duct obstruction, intrahepatic forms of cholestasis, and neoplasia. The frequency of these diseases varies with species breed, age, and, in some cases, with environment (diet, geographic location). The differential diagnosis of hepatic disease involves the evaluation of clinical history, physical examination, biochemical tests, hepatic imaging, and histopathologic examination of hepatic biopsies. In the following section, the biochemical tests that are used for the assessment of hepatic disease are described.

There are several diagnostic categories with which the clinician dealing with problems of liver disease must be concerned. In clinical patients with a history and signs suggestive of hepatic disease, laboratory tests are used for confirmation. Laboratory tests are used to assess the severity of liver disease, to establish prognosis, to define treatable causes of the complications of hepatic insufficiency (e.g., ascites, encephalopathy), and to monitor progress of hepatic disease. Finally, biochemical tests of hepatic function may be performed on clinically healthy patients that are known to have a high risk of developing liver disease, e.g., exposure to infectious agents that cause hepatitis, or a familial history of chronic liver disease requiring screening for genetically determined diseases.

A. Hepatic Enzymes

The serum enzymes used routinely in clinical diagnosis are present in high concentration in the liver. In hepatocellular or in cholestatic forms of liver injury, these hepatic enzymes are released into the serum, and increased serum activity becomes useful diagnostically. The duration of elevation in the serum activity of enzymes of hepatic origin is dependent on a variety of factors, including molecular size, intracellular location, rate of plasma clearance, rate of enzyme inactivation, and, in some cases (e.g., alkaline phosphatase [AP] and gamma glutamyl transpeptidase [GGT]), the rate of increased hepatic production.

The serum enzyme activities that are elevated when hepatic necrosis is present are alanine aminotransferase (ALT), aspartate aminotransferase (AST), ornithine carbamoyltransferase (OCT), glutamic dehydrogenase (GD), sorbitol dehydrogenase (SDH), and arginase. Elevated serum activities that suggest cholestasis (intrahepatic or extrahepatic) are AP, GGT, and 5′ nucleotidase (5′-ND).

1. *Serum Alanine and Aspartate Aminotransferases*

In all domestic species, the activity of AST is high in the liver and, in acute and chronic liver injury, the AST activity of serum is increased. AST activity also is high in kidney, pancreas, and erythrocytes and, when cells of these tissues are damaged, the AST activity of serum also can be expected to increase. There is

TABLE 13.1 Serum Alanine Aminotransferase (ALT) Activity of Normal Animals

Species	U/liter	Reference
Dog	15–50	Crawford *et al.* (1985)
	2.5–25	Van Vleet and Alberts (1968)
	0–69	Abdelkader and Hauge (1986)
	20–45	Bunch *et al.* (1985)
	5–80	Johnson *et al.* (1982)
	3–61	Mia and Koger (1979)
Cat	0–36	Center *et al.* (1985c)
	10–80	Peterson *et al.* (1983)
	16 ± 9	Meyer (1983)
	30 (1–59)	Mia and Koger (1979)
Minipig	35 ± 12	Kroker and Romer (1984)
Pig	71 (37–106)	Mia and Koger (1979)

no simple and specific method for determining the origin of increased serum AST activity, and additional tests are required. Two isozymes of AST are present in the liver; one is mitochondrial and the other is from the cytoplasm.

The ALT activity of the liver of dogs and cats also is high. Significant ALT activity is confined to the cytoplasm of the liver. Elevation in the activity of serum ALT (Table 13.1) is considered to be specific for hepatic injury in dogs and cats, and in small animals with hepatocellular injury, the fractional increase in activity of ALT is 4–8 times higher than the corresponding increase in AST (Center, 1996). In the large domestic species, the activity of ALT in the liver is low, and in liver injury the serum ALT is not remarkably elevated.

AST in all domestic species may be increased in skeletal muscle disease, including trauma (e.g., intramuscular injections). To differentiate AST elevations due to liver disease from those due to muscle disease, it is possible to measure serum creatine kinase (CK). In muscle disease, AST and CK both would be expected to be elevated. In acute muscle injury, CK may appear before AST is maximally elevated, and CK activity may decrease before AST activity fully declines. When myopathy coincides with liver disease in small animals, the ALT measurement may be useful, but severe muscle disease may be associated with increased serum ALT activity.

The major value of serum AST and ALT measurements is in detecting hepatocellular injury and in monitoring the progress in acute hepatitis. Both enzymes are increased in many hepatic diseases of small domestic animals and have limited value in differential diagnosis. Aminotransferases are considered useful in differentiating hepatocellular from cholestatic forms of liver injury. In severe liver disease, however, both hepatocellular and cholestatic forms of hepatic disease often coexist. Very high aminotransferase levels suggest acute hepatitis, but more modest increases in aminotransferase activity are seen in many types of liver disease, including chronic hepatocellular disease, cirrhosis, parasitic hepatopathy, and primary or metastatic neoplasia.

2. *Sorbitol Dehydrogenase*

Because of the lack of significant ALT activity in the livers of large domestic animals (Cornelius, 1963; Keller *et al.*, 1985), other liver-specific enzymes have been developed. The serum activity of SDH (Table 13.2) has been shown to be useful in assessment of hepatocellular injury in most domestic species, including dogs, horses, and ruminants, but is used primarily as a liver-specific enzyme in large domestic animals. SDH activity is not stable in serum and declines rapidly. Analysis therefore should be performed as soon after the sample is taken as possible and optimally within 8–12 hours.

3. *Arginase*

Arginase is present in significant concentration only in the liver of ureotelic mammals (Cornelius *et al.*, 1963; Dittrich *et al.*, 1974), and a relatively simple assay is available for its measurements (Mia and Koger, 1978). The plasma arginase activity in acute liver injury returns to normal more rapidly than ALT or AST (Cornelius *et al.*, 1963). In progressive hepatic necrosis, serum arginase activity remains elevated and may suggest an unfavorable prognosis. Serum arginase elevations have been demonstrated in naturally occurring liver disease of horses (Wolf *et al.*, 1967), cattle, sheep (Ross, 1966), goats (Adam *et al.*, 1974), and dogs (Harvey and Hoe, 1971). When serum arginase and

TABLE 13.2 Serum Sorbitol (Iditol) Dehydrogenase (SDH) Activity of Normal Animals

Species	U/liter	Reference
Dog	28 ± 20	Zinkl *et al.* (1971)
	5.5–18	Noonan and Meyer (1979)
	1–9	Abdelkader and Hauge (1986)
	0–6	Keller (1981)
	10 ± 3	Anwer *et al.* (1976)
Pony	14.7 ± 3.6	Anwer *et al.* (1976)
Calf	14.7 ± 1.3	Anwer *et al.* (1976)
Cow	4.3–15.4	Putnam *et al.* (1986)
Sheep	16.5 ± 1.5	Anwer *et al.* (1986)
	7.9 ± 2.3	Alemu *et al.* (1977)
Pig	2.2 (0–6.8)	Osuna (1979)

TABLE 13.3 Serum Glutamic Dehydrogenase (GD) Activity of Normal Animals

Species	U/liter	Reference
Dog	0–9	Abdelkader and Hauge (1986)
	1–6	Keller (1981)
Horse	1.5 ± 1	Ikeda *et al.* (1976)
	0–1.2	Freedland *et al.* (1965)
Cow	3.1 ± 0.5	Boyd (1962)
Sheep	0–9	Alemu *et al.* (1977)
	2.6 ± 1.0	Harvey and Obeid (1974)
Goat	3.4 ± 0.9	Harvey and Obeid (1974)

GGT are measured simultaneously, their respective specificities for necrosis and cholestasis provide valuable diagnostic information (Noonan and Meyer, 1979).

4. *Glutamic Dehydrogenase*

GD has been shown to be of value as an enzyme for assessing hepatic necrosis in sheep, goats, and cattle (Table 13.3). The enzyme is highly concentrated in ovine and bovine liver (Boyd, 1962; Keller, 1971), as well as in the livers of other domestic species (Keller *et al.*, 1985). In ruminants, GD has been reported to be elevated in hepatic necrosis (Fowler, 1971; Boyd, 1962); at parturition (Treacher and Collis, 1977); and associated with bile-duct obstruction (Ford and Gopinath, 1976).

5. *Ornithine Carbamoyltransferase*

OCT is also considered a liver-specific enzyme for the detection of hepatocellular necrosis in domestic species (Treacher and Sansom, 1969). Nearly all OCT activity is confined to the liver of cattle (Treacher and Collis, 1977) and of pigs (Dittrich *et al.*, 1974). OCT and ALT are similar in sensitivity as diagnostic tests for hepatic necrosis in the dog (Litchfield and Garland, 1974). Markiewicz *et al.* (1975) have reported that the serum activity of OCT is correlated with the severity of hepatic fascioliasis in cattle. OCT also has been suggested as a useful marker for hepatocellular injury in swine (Wilson *et al.*, 1972).

6. *Serum Alkaline Phosphatase*

The APs are a group of zinc metalloenzymes that are present in most tissues. High concentrations are found in the intestine, kidney, bone, and liver. Light and electron microscopic studies have demonstrated that alkaline phosphatase activity is highest on the absorptive or secretory surfaces of cells (Kaplan, 1972). Within liver cells, alkaline phosphatase is bound to membranes, and when liver homogenates are subjected to high-speed centrifugation, alkaline phosphatase activity sediments primarily with the microsomal and plasma membrane fractions (Emmelot *et al.*, 1964).

The actual physiologic functions of alkaline phosphatase are not fully understood. Localization of the enzyme to cell surfaces known to be responsible for active secretion or absorption suggests a role in membrane transport. There is circumstantial evidence that the alkaline phosphatase of osteoblasts may be involved in bone calcification. Activity against nucleotides (natural and artificial) may indicate a role in nucleic acid metabolism. It has been suggested that intestinal calcium-stimulated adenosine triphosphatase, which has a role in active calcium transport, and alkaline phosphatase are different activities of the same enzyme (Haussler *et al.*, 1970). A similar relationship has been demonstrated between alkaline phosphatase and other enzymes involved in active transport; for example, sodium, potassium-stimulated adenosine triphosphatase of several tissues, including the liver, has significant hydrolytic activity against substrates of alkaline phosphatase such as *p*-nitrophenyl phosphate (Ahmed and Judah, 1964).

In normal animals, the AP of serum (Table 13.4) originates primarily from liver and bone (Rogers, 1976; Hoffman and Dorner, 1975). Elevations of serum AP are observed in normal growing animals or in adult animals with increased osteoblastic activity. Serum AP activity may be elevated in both acute and chronic liver diseases, but marked elevations are indicative

TABLE 13.4 Total Serum Alkaline Phosphatase (AP) Activity of Normal Adult Animals

Species	U/liter	Reference
Dog	39–222	Abdelkader and Hauge (1986)
	30.6 ± 9.9	Meyer and Noonan (1981)
	10–82	Bunch *et al.* (1982)
Cat	8 ± 0.7	Spano *et al.* (1983)
	10–80	Peterson *et al.* (1983)
	8.4 ± 2.9	Meyer (1983)
	1–39	Center *et al.* (1986a)
Horse	184 ± 57	Gossett and French (1984)
Cow	41 ± 16	Rico *et al.* (1977a)
	7–43	Putnam *et al.* (1986)
	2–809	Allcroft and Folley (1941)
Sheep	91 ± 41	Braun *et al.* (1978)
	63 ± 28	Alemu *et al.* (1977)
	21–1178	Allcroft and Folley (1941)
	35–234	Leaver (1968)
Pig	100 ± 35	Rico *et al.* (1977c)
	<27	Van Leenhoff *et al.* (1974)
Minipig	49 ± 11	Kroker and Romer (1984)

of cholestasis, with the highest plasma concentrations observed in animals with cholangitis, biliary cirrhosis, or extrahepatic bile-duct obstruction.

The alkaline phosphatase isozymes of various tissues may be differentiated on the basis of differences in heat stability, urea denaturation, and inhibition by L-phenylalanine, or by electrophoretic mobility (Nagode *et al.*, 1969a,b; Ruegnitz and Schwartz, 1971). Alternatively, determination of the origin and significance of elevated serum AP can be facilitated by measuring other serum enzymes that are more specific for biliary tract disease. These include leucine aminopeptidase (Everett *et al.*, 1977), 5′-nucleotidase (Righetti and Kaplan, 1972), and GGT. When the serum AP is significantly elevated, overt clinical signs may allow separation of diseases of the liver from those of other tissues such as bone.

In contrast to serum AST and ALT, elevations in AP are not due simply to leakage of enzyme through damaged cells. It was once believed that the high AP level of serum observed in cholestatic liver disease was the result of decreased biliary excretion of the enzyme (Gutman *et al.*, 1940). It now is known that experimental obstruction of bile flow stimulates *de novo* synthesis of hepatic AP (Kaplan and Righetti, 1969, 1970), and the newly synthesized enzyme is refluxed into the circulation. Partial hepatectomy also stimulates increased synthesis of AP in regenerating hepatic tissue (Pekarthy *et al.*, 1972). It seems likely that increased synthesis of AP is involved in clinical extrahepatic bile-duct obstruction, in intrahepatic cholestasis, and in infiltrative diseases of the liver (e.g., lymphoma, metastasis) in which terminal branches of the biliary tree are obstructed, as well as in the regenerative processes that occur in the liver following injury.

There are significant species differences in the magnitude of elevation of serum AP activity in bile-duct obstruction (Fig. 13.2). For many years, it has been recognized that cats differed from dogs in that cats with extrahepatic bile-duct obstruction had "inconsistent" or negligible elevations of AP (Cantarow *et al.*, 1936), and this was thought to be due to urinary excretion of AP (Flood *et al.*, 1937). Other studies have established that cholestatic liver disease in cats can be expected to cause modest but significant elevations in AP (Everett *et al.*, 1977; Dalgaard, 1948), and induction of alkaline phosphatase synthesis following bile duct obstruction occurs in cats as in other species (Sebesta *et al.*, 1964).

In dogs, three AP isoenzymes (intestinal, steroid-induced, and hepatic) have been identified (Wellman *et al.*, 1982a). The steroid-induced isoenzyme is of hepatic origin (Wellman *et al.*, 1982b). Increased activity of the hepatic AP isoenzyme in cholestasis is due to enhanced

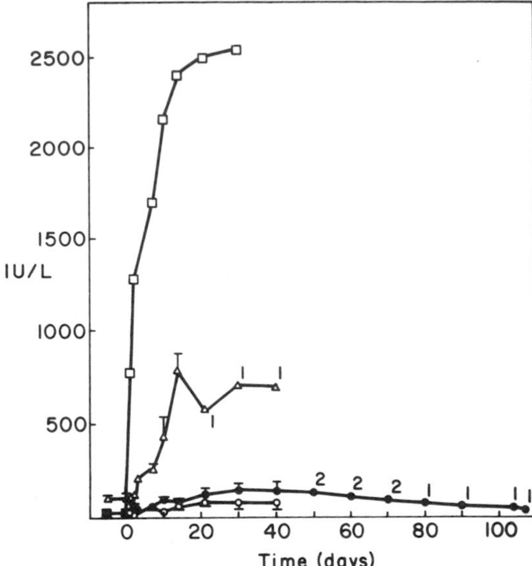

FIGURE 13.2 Comparative changes in serum alkaline phosphatase activity associated with bile duct obstruction in sheep, cats, horses, and dogs (courtesy of Dr. D. Levy). ●, ovine (5); ○, feline (6); △, equine (3); □, canine (2).

RNA translation and not to increased transcription (Seetharam *et al.*, 1986). Hepatic AP enters the serum either from the biliary canaliculus via the paracellular shunt pathway, or directly from plasma membranes. Increased bile-acid concentrations associated with cholestasis are believed to be necessary for the release and transport of solubilized hepatic AP to the serum (Schlaeger *et al.*, 1982).

Although measurement of serum GGT activity has the advantage of specificity, total serum AP activity remains the test most often performed to assess cholestasis in horses, dogs, and cats. Serum AP is less valuable in the evaluation of cholestatic syndromes of cattle and sheep because of wide fluctuations in normal AP activity (Ford, 1958; Harvey and Hoe, 1971).

Modest increases in serum AP do occur in hepatic necrosis (Noonan and Meyer, 1979). Following the experimental production of hepatic necrosis in dogs, the serum activities of arginase, ALT, and AP increase within one day, a time point at which GGT is not elevated. Following bile-duct obstruction, both AP and GGT levels markedly increase along with moderate elevations in ALT and AST, but arginase activity does not increase. This has suggested that arginase (for necrosis) and GGT (for cholestasis) may have the highest specificity in evaluating the type of hepatobiliary disease in the dog (Noonan and Meyer, 1979). Although serum GGT activity may be less affected during hepatocellular necrosis than AP, its activity is not as highly elevated as that of AP in bile-duct obstruction (Guelfi et al., 1982). Increases in serum AP levels have been described in a variety of canine diseases associated with cholestasis (Hoe and Jabara, 1967; Abdelkader and Hauge, 1986; Center *et al.*,1985b).

7. γ-*Glutamyltranspeptidase*

GGT is a membrane-bound enzyme that catalyzes the transfer of γ-glutamyl groups from γ-glutamylpeptides such as glutathione to other peptides and amino acids. It is found primarily in cells with high rates of secretion or absorption. Significant GGT activity is present in the liver, kidney, pancreas, and intestine. GGT is considered a serum marker (Table 13.5) primarily for diseases of the hepatobiliary system associated with cholestasis (Braun *et al.*, 1983) and is now in general use for the diagnosis of liver diseases in animals (Table 13.6). GGT activity is relatively high in the livers of cows, horses, sheep, and goats, but GGT activity is considerably lower in dogs and cats. Although GGT activity is present in many tissues, remarkable elevations in serum activity are observed primarily in diseases of the liver. Urinary excretion of GGT has been measured to assess renal injury (Ford, 1974; Shaw, 1976).

TABLE 13.5 Serum γ-Glutamyltransferase (GGT) Activity of Normal Animals

Species	U/liter	Reference
Dog	11 ± 10	Guelfi *et al.* (1982)
	0–11	Abdelkader and Hauge (1986)
	0–10	Shull and Hornbuckle (1979)
	<5	Bunch *et al.* (1985)
	2 ± 1	Szasz (1974)
	2–4	Bunch *et al.* (1982)
	3 ± 1	Meyer and Noonan (1981)
Cat	0.4 ± 0.3	Center *et al.* (1986a)
	0.3 ± 0.2	Meyer (1983)
	0–2	Krebs (1979)
Horse	4.5–32.5	Yamaoka *et al.* (1978)
	13 ± 6	Rico *et al.* (1977a)
	6–24	Braun *et al.* (1982)
Cow	19 ± 6	Rico *et al.* (1977b)
	6–17	Keller (1978)
	1 ± 5	Unglaub *et al.* (1973)
Calf	15 ± 4	Braun *et al.* (1978a)
	11 ± 2	Weiss (1978)
Sheep	33 ± 7	Braun *et al.* (1978b)
	17–69	Towers and Stratton (1978)
	23 ± 5	Malherbe *et al.* (1977)
Goat	38 ± 13	Orliac (1980)
	27 ± 3	Moursi *et al.* (1979)
Pig	35 ± 21	Rico *et al.* (1977c)
	32 ± 20	Bostedt (1978)
Piglet (8 weeks)	16 ± 8	Enigk *et al.* (1976)

In experimental bile-duct obstruction, serum GGT activity is increased significantly in the dog (Noonan and Meyer, 1979; Shull and Hornbuckle, 1979), sheep (Ford, 1974), and cow (Grunder, 1977). The sensitivity of GGT has been reported to be similar to that of AP as an indicator of cholestasis in the cat (Zawie and Garvey, 1984; Spano *et al.*, 1983). Serum GGT activity within a given species often correlates directly with serum AP activity in cholestatic liver injury. GGT is not increased, however, as markedly as AP in hepatic necrosis (Meyer, 1983).

B. Serum Bilirubin

Bilirubin in serum is measured by the *van den Bergh* or *"diazo"* reaction, in which bilirubin is coupled with diazotized sulfanilic acid. Azo pigments produced by this reaction are dipyrroles. Azo pigments are stable, and this characteristic has been useful in studies of the structure of bilirubin conjugates. Conjugated bilirubin reacts with diazotized sulfanilic acid in aqueous solution (the van den Bergh "direct reaction"), but unconjugated bilirubin reacts very slowly. Only after addition of an accelerator such as methanol or ethanol to the aqueous solution is the diazo reaction with unconjugated bilirubin complete ("the indirect reaction"). It is said that approximately 10% of unconjugated bilirubin in plasma will react with the diazo reagent and give a false "direct" reaction (Arias, 1974).

The requirement of an organic solvent for the diazo reaction to occur with unconjugated bilirubin suggested that the delay was related to water insolubility. There is evidence (Schmid, 1974), however, that intramolecular hydrogen bonding is more important than solubility in preventing the reaction of unconjugated bilirubin with the diazo reagent (Fog and Jellum, 1963; Nichol and Morrell, 1969). The two propionic acid side chains of bilirubin are esterified with glucuronic acid or other carbohydrates, disrupting the intramolecular hydrogen bonding (Fog and Jellum, 1963), allowing the direct diazo reaction to occur. Accelerators of the van den Bergh reaction may have a similar effect on the intramolecular hydrogen bonds of the unconjugated pigment.

Figure 13.3 summarizes the normal production and excretion of bilirubin and other bile pigments. Table 13.7 summarizes the pathophysiologic mechanisms that induce hyperbilirubinemia. Unconjugated hyperbilirubinemia is observed when there is increased production of bilirubin, as in hemolytic anemia (Fig. 13.3) or when either hepatic uptake or conjugation of bilirubin is diminished. Although the unconjugated bilirubin of serum may be significantly increased in such disorders, essentially none is filtered by the glomerulus

TABLE 13.6 Serum γ-Glutamyltransferase (GGT) Activity in Animals with Hepatobiliary

Species	Condition	References
Dog	Bile-duct obstruction; chronic hepatitis	Braun *et al.* (1983)
	Lipidosis; necrosis; cirrhosis; neoplasia	Hauge and Abdelkader (1984)
	Corticoid therapy	De Novo and Prasse (1983)
Cat	Bile-duct obstruction; cholangiohepatitis; lipidosis; cirrhosis; lymphysarcoma; necrosis	Center *et al.* (1986a)
Horse	Toxic hepatic failure	Divers *et al.* (1983)
	Subclinical hepatopathy	Yamaoka *et al.* (1978)
	Hyperlipemia	Wensing *et al.* (1973)
Cow	Ragwort poisoning; fascioliasis lipidosis	Blackshaw (1978)
	Fascioliasis metacercariae migrations and chronicity	Simesen *et al.* (1973)
	Acute fascioliasis	Meissonier and Rousseau (1976)
	Metacercariae migrations	Bulgin and Anderson (1984)
	Senecio poisoning	Johnson and Molyneux (1984)
Sheep	Bile-duct obstruction; sporidesmin toxicity; fascioliasis	Ford and Evans (1985)
	Lupinosis	Malherbe *et al.* (1977)
	Fascioliasis	Ben-Said (1979)
	Cobalt deficiency (white liver disease)	Sutherland *et al.* (1979)
	Ketosis	Meissonier and Rousseau (1976)
Pig	Cysticercus tenuicolis infection	Enigk *et al.* (1976)
	Arsanilic acid toxicity	Ferslew and Edds (1979)

because the pigment is tightly bound to albumin. Consequently, bilirubinuria is not characteristic in animal patients with unconjugated hyperbilirubinemia. In hemolytic disease, the amount of bilirubin excreted by the liver, and therefore the amount that reaches the intestine, may be remarkably increased. This results in increased formation and urinary excretion of urobilinogen (Fig. 13.4).

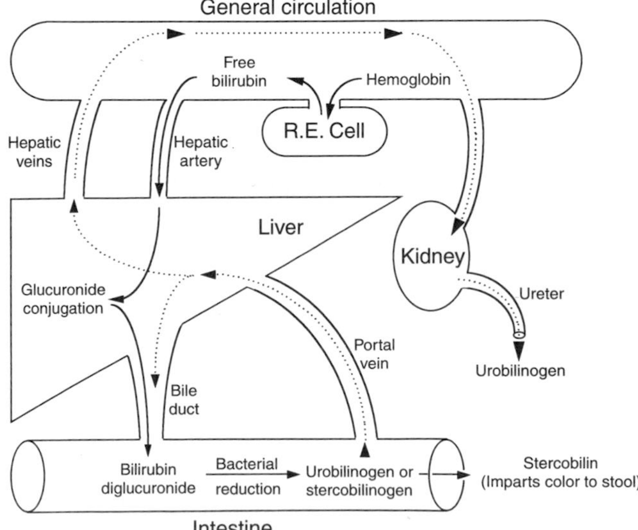

FIGURE 13.3 Normal formation, excretion, and enterohepatic circulation of bilirubin and other bile pigments.

Hyperbilirubinemia of the unconjugated type is caused either by intrahepatic cholestasis (Fig. 13.5) or extrahepatic bile duct obstruction (Fig. 13.6). When the primary defect is impaired excretion of bilirubin, hepatic uptake and conjugation may proceed at relatively normal rates, but conjugated bilirubin is refluxed into the plasma. The plasma concentration of bilirubin increases, and the conjugated pigment, which is water soluble and less firmly bound to albumin, is readily filtered by the glomerulus with resulting bilirubinuria (Fulop *et al.*, 1965; Laks *et al.*, 1963). Because bilirubin excretion into the intestine is either significantly reduced or absent, formation of urobilinogen by intestinal bacteria is remarkably reduced, and the test for urinary urobilinogen is characteristically negative in complete extrahepatic obstruction (Fig. 13.6). The therapeutic use of oral, broad-spectrum antibiotics may diminish the metabolic activity of intestinal bacteria and result in a spuriously negative test for urobilinogen in the urine.

The biochemical differentiation between unconjugated and conjugated hyperbilirubinemia using the van den Bergh reaction can be useful in assessing prehepatic and posthepatic causes of hyperbilirubinemia. In primary, severe hepatitic diseases, however, all major excretory steps (uptake, conjugation, and excretion) may be deranged and result in elevations of both conjugated and unconjugated pigment.

Important species characteristics should be considered when interpreting results of the van den Bergh

TABLE 13.7 Pathophysiologic Mechanisms Responsible for Hyperbilirubinemia[a]

	Plasma bilirubin		Urine	Urine
	Unconjugated	Conjugated	bilirubin	urobilinogen
Increased bilirubin production (hemolytic anemia; resorption from hemorrhage, hematoma)	↑	N	0	↑
Impaired hepatic uptake of unconjugated bilirubin (neonatal hyperbilirubinemia; fasting hyperbilirubinemia; benign unconjugated hyperbilirubinemia of horses; congenital photosensitivity [Southdown sheep])	↑	N	0	↓ or N
Impaired conjugation of bilirubin (glucuronyl transferase deficiency [Gunn rat]; neonatal hyperbilirubinemia)	↑	N	0	↓ or N
Impaired biliary excretion of bilirubin (intrahepatic cholestasis; black liver disease [Corriedale sheep]; congenital photosensitivity [Southdown sheep]; biliary cirrhosis; bile-duct obstruction)	↑	↑	↑	↓ or 0

[a] N, normal; 0, absent; ↑, increased; ↓, decreased.

reaction. In general, the interpretation in dogs and cats is similar. Typically, in cholestatic disease, the conjugated fraction is elevated, representing 50 to 75% of the total serum bilirubin (Fig. 13.7). The normal horse has a much higher total serum bilirubin than any of the other domestic species (Fig. 13.8), and values as high as 4.0 mg/dl or higher have been observed in otherwise healthy individuals. In addition to hepatic and hemolytic diseases, hyperbilirubinemia is observed in horses with intestinal obstruction and a variety of other serious systemic diseases. Food restriction alone causes an abrupt increase in the unconjugated serum bilirubin of the horse (Gronwall and Mia, 1972; Tennant *et al.*, 1975), and decreased bile flow is the probable cause of the hyperbilirubinemia observed in fasting horses (Fig. 13.9).

In cattle and sheep, hyperbilirubinemia of sufficient magnitude to produce clinical icterus (≥3 mg/dl) is caused most frequently by hemolytic disease. Biochemical hyperbilirubinemia (1–2 mg/dl) without clinical icterus may be observed in sheep and cattle with fatty liver associated with ketosis/acetonemia. In such cases, the unconjugated bilirubin pigment predominates. Greater elevations in serum bilirubin and clinical icterus in ruminants associated with ketosis are unusual. Mild to moderate conjugated hyperbilirubinemia has been observed in sheep with sclerosing cholangitis caused by *Fasciola hepatica* infestation (Hjerpe

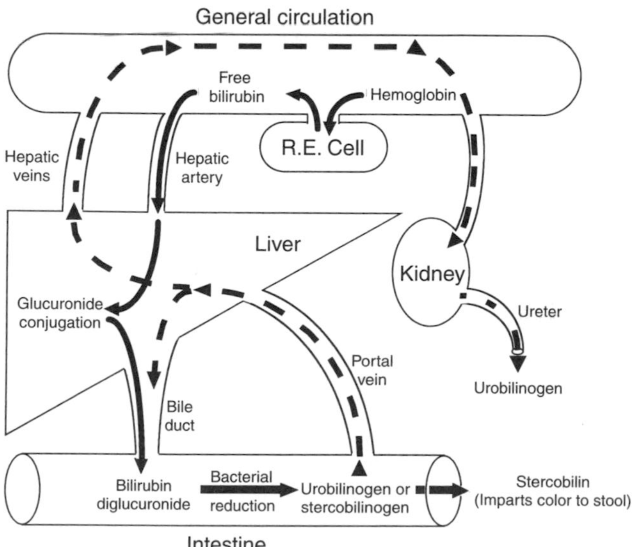

FIGURE 13.4 Formation, excretion, and enterohepatic circulation of bilirubin and other bile pigments associated with overproduction of bilirubin due to hemolysis.

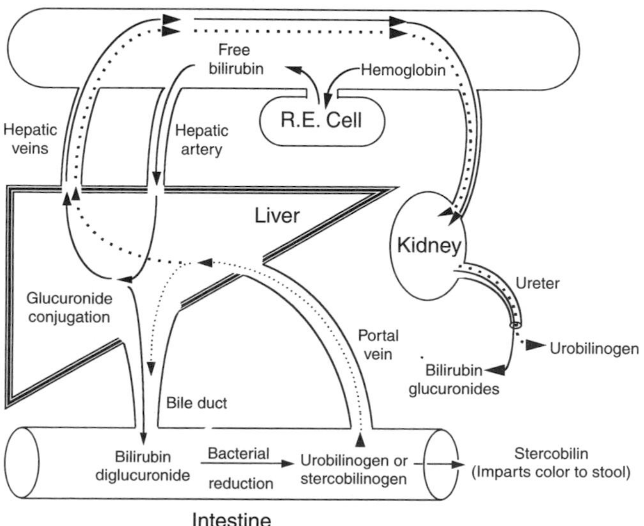

FIGURE 13.5 Formation, excretion, and enterohepatic circulation of bilirubin and other bile pigments associated with hepatocellular injury and intrahepatic cholestasis.

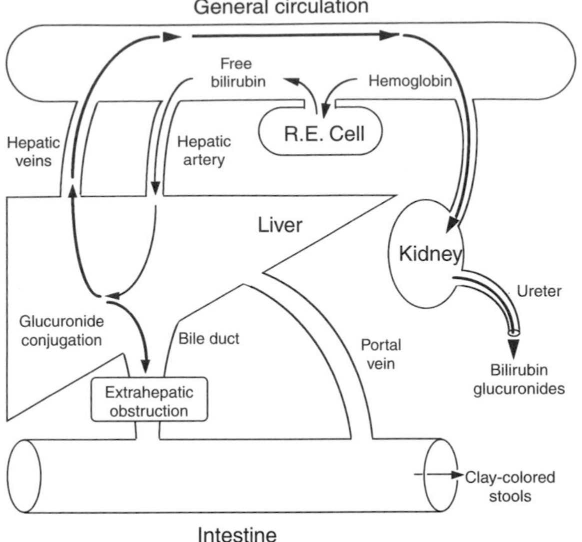

FIGURE 13.6 Formation, excretion, and enterohepatic circulation of bilirubin and other bile pigments in extrahepatic bile-duct obstruction.

et al., 1971), and in cattle with hepatic cirrhosis (Finn and Tennant, 1974).

C. Serum Bile Acids

The primary bile acids are synthesized exclusively within the liver from cholesterol, and this represents the primary pathway of cholesterol catabolism. The

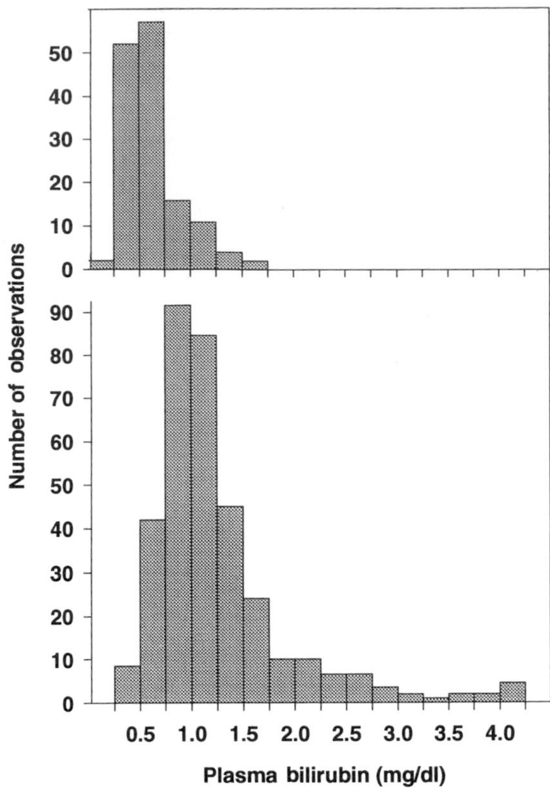

FIGURE 13.8 Total serum bilirubin concentrations of 23 normal Shetland ponies (upper; 143 observations) and 103 normal standard-sized horses (lower; 345 observations). Note the higher median values of standard-sized horses and the wide range of values in clinically healthy adults.

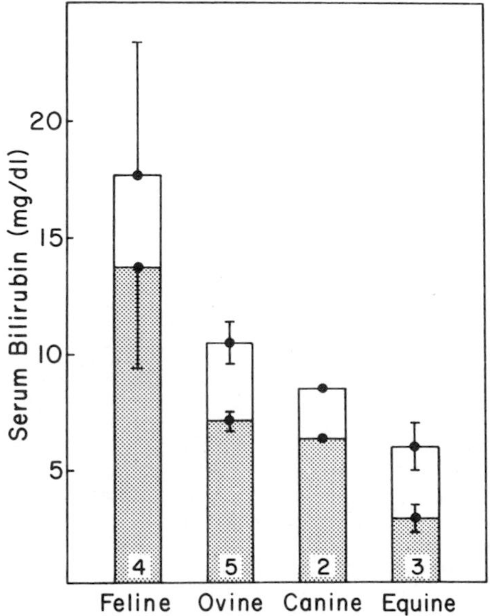

FIGURE 13.7 Comparative changes in total and conjugated ("direct reacting") serum bilirubin 14 days following bile-duct obstruction in cats, sheep, dogs, and horses (courtesy of Dr. D. Levy). Shaded portion of bar: direct reacting; bar height: total.

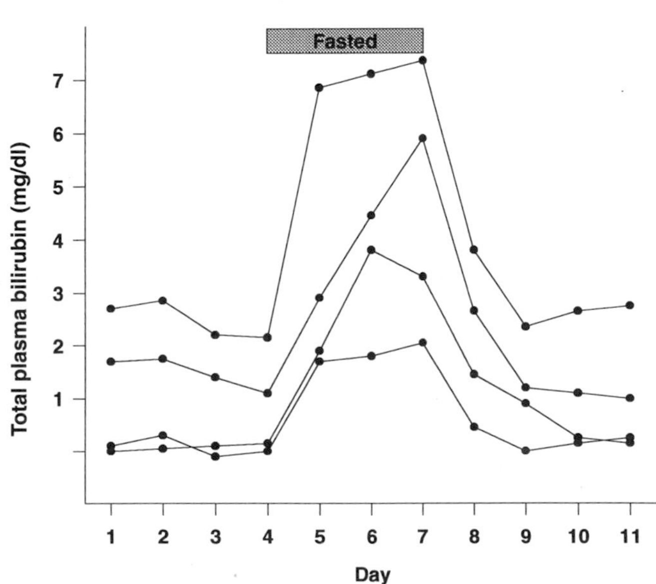

FIGURE 13.9 Serum bilirubin of horses before, during, and following a 3-day period of total fasting.

two primary bile acids of most domestic species are cholic acid, a trihydroxy bile acid, and chenodeoxycholic acid, a dihydroxy bile acid (Fig. 13.10). The major primary bile acid of swine is hyocholic acid, a trihydroxy bile acid closely synthesized from chenodeoxycholic acid (see Chapter 15). Following synthesis, the primary bile acids are secreted into bile as amino acid conjugates of either taurine or glycine. Taurine conjugates predominate in the dog, horse, and sheep. In the cat, bile acids are conjugated exclusively with taurine. Bile acids are the major component of bile, representing 67% of the total solids of bile; phospholipids make up approximately 22%, and protein, cholesterol, and bile pigments account for the remainder of biliary total solids.

Bile acid secretion is the major determinant of bile flow. In the proximal small intestine, bile acids play an important role in the digestion and absorption of fat and fat-soluble compounds (see Chapter 15). In the ileum, bile acids are efficiently absorbed by active transport mechanisms, but a small fraction of intestinal bile acids passes to the large intestine where bacterial deconjugation and dehydroxylation produces secondary bile acids. Dehydroxylation of cholic acid at the 7α position produces deoxycholic acid, and 7α dehydroxylation of chenodeoxycholic acid produces lithocholic acid. The secondary bile acids thus formed in the colon are absorbed into the portal circulation. The normal liver removes the primary and secondary bile acids efficiently from the hepatic portal vein. After reconjugation of the secondary bile acids, both primary and secondary bile acids are secreted again into the bile, completing the enteropathic circulation.

Bile acid metabolism is affected by liver disease in several ways. The synthesis of primary bile acids may be decreased and the proportions of cholic acid and chenodeoxycholic acid may changed. Unusual bile acids also may be produced. Importantly, removal of bile acids by the liver from the hepatic portal vein may be diminished, either by impaired hepatocellular function or by diversion of portal blood through shunts from the liver to the general circulation. This is particularly noticeable after meals in animals with either congenital or acquired hepatoportal shunts. The plasma bile acid concentration is increased continuously in biliary obstruction and, typically, there is increased urinary excretion of bile acids. Increases in the serum bile acid concentrations also are seen in other forms of hepatic disease, and an increased fasting serum bile acid level is specific for a significant deficit in hepatic function.

The measurement of total serum bile acids for assessment of liver disease has been greatly facilitated by development of a simplified spectrophotometric assay that has now been used for several domestic species. The predictive value of the serum bile acid test is remarkably high in the dog (Center *et al.*, 1984). In some dogs with portocaval vascular shunts, the fasting serum bile acid concentration may be within normal limits, but is increased diagnostically 2 hours following a meal (Center *et al.*, 1985a). The predictive value also may be enhanced in cats by comparing the fasting and

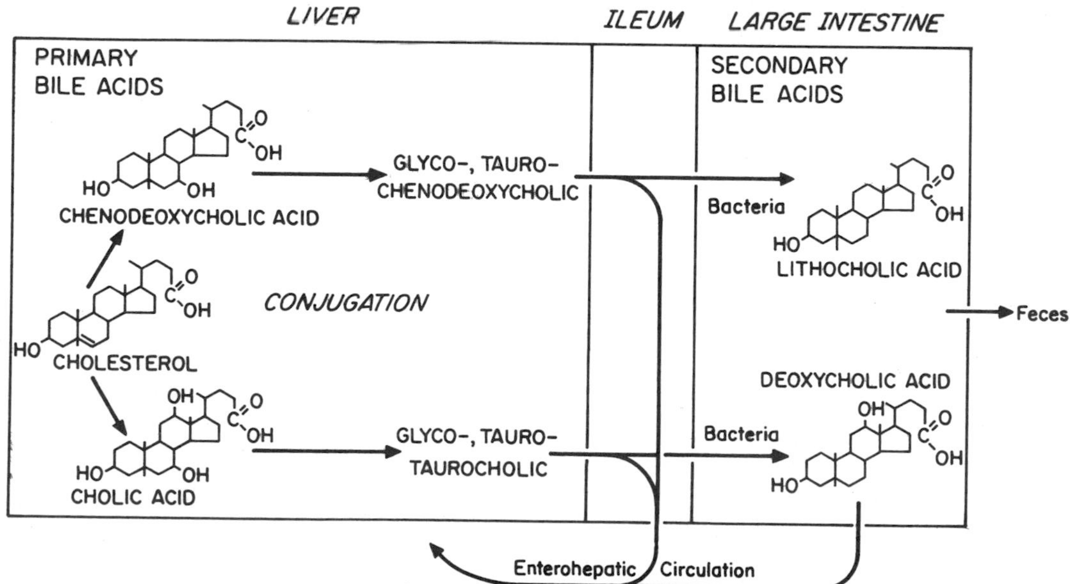

FIGURE 13.10 Metabolism of bile acids in the liver and intestinal tract. The primary bile acids are formed in the liver and the secondary bile acids in the large intestine.

2-hour postprandial bile acid levels (Center *et al.*, 1995). In the horse, fasting alone has been reported to increase the plasma bile acid concentration by decreasing hepatic bile acid clearance (Engelking and Gronwall, 1979), and food intake should be considered when interpreting the serum bile acid test in equine species.

D. Serum Proteins

The liver is the exclusive site of synthesis of albumin, the most abundant of the plasma proteins. Almost all plasma proteins are glycoproteins; in contrast, albumin contains no carbohydrate. Albumin is synthesized exclusively in the liver. Degradation of albumin occurs in the liver, as well as in other tissues, including muscle, kidney, and skin. Degradation of albumin is probably favored in the liver because of the fenestrated endothelial lining cells that allow access of almost all plasma proteins directly to the space of Dissé and to the sinusoidal surface of the hepatocyte. In the general circulation, albumin has two major functions. It is the most important determinant of plasma oncotic pressure (colloid osmotic pressure) and is a major transport protein for many hydrophobic endogenous metabolites and xenobiotics, substances that because of albumin binding remain in stable aqueous solution in the plasma.

The plasma albumin concentration is determined by the hepatic synthetic rate, which normally is in equilibrium with degradation. Hypoalbuminemia may be a clinical sign of severe hepatocellular liver disease because of defective albumin synthesis. Hypoalbuminemia also may be caused by increased albumin loss due to glomerulopathy (protein-losing nephropathy), to severe inflammation of the intestine, or to intestinal lymphangiectasia (protein-losing enteropathy). In severe, chronic hepatopathy, there is a tendency for elevations in IgM, IgG, and IgA, and both decreased albumin and increased globulin cause a decrease in the albumin/globulin ratio.

The liver is the exclusive site of synthesis of all blood coagulation proteins except factor VIII, which is synthesized in many organs. Synthesis of coagulation proteins tends to be diminished in liver disease. Plasma prothrombin (factor II) may be decreased in animals with either acute or chronic liver disease, with a corresponding increase in the prothrombin time. Factors that contribute to the abnormal prothrombin time include diminished hepatic protein synthesis, increased consumption of clotting factors associated with hemorrhage and fibrinolysis, and, in some cases, vitamin K deficiency due to decreased intake and/or decreased absorption. Vitamin K is essential for the hepatic synthesis not only of prothrombin, but also of factors VII, IX, and X. Parenteral administration of vitamin K to individual animal patients may result in improvement in prothrombin time, but coagulation time may remain prolonged. Individuals with obstructive jaundice absorb vitamin K poorly, and defects in their clotting can be improved rapidly by parenteral vitamin K administration. Fibrinogen is an acute-phase reactant and its concentration in plasma may be greatly increased in chronic inflammatory diseases or in neoplasia. Plasma fibrinogen synthesis is generally normal in mild or moderate liver disease, but may be detectably decreased in severe liver disease. Because of the rapid turnover of fibrinogen and prothrombin, the concentrations of the proteins in the plasma decrease rapidly in severe, acute liver injury, whereas albumin, the turnover rate of which is much longer, is diminished primarily in chronic liver disease with loss of significant hepatocellular mass.

The liver also is the site of synthesis of other plasma proteins important in regulation of coagulation. Plasmin, a serine protease synthesized by the liver, is necessary for fibrin degradation. Anti-thrombin 3 has potent protease activity against thrombin, plasmin, and other coagulation factors, and its primary physiological function appears to be to modulate the activity of thrombin. Heparin enhances inhibition of thrombin by antithrombin 3 by at least 1000-fold.

The complement system plays a critical role in the inflammatory response and in host defense mechanisms against infection, and most of the plasma proteins of the complement system are synthesized in the liver. *De novo* synthesis of C2, C3, C4, and factor B have been demonstrated in cultured rat hepatocytes, and syntheses of other complement proteins have been demonstrated in cultured guinea-pig hepatocytes (Ramadori *et al.*, 1986; Anthony *et al.*, 1985). The liver also is the site of synthesis of protease inhibitors, including 1-anti-trypsin, a major plasma protein that inhibits serine proteases, including granulocyte elastase and α2-macroglobulin, an inhibitor of a variety of other proteases.

E. Dye Excretion

The liver can excrete a variety of organic anions. The rate of removal of such substances can be determined and used to assess the functional capacity of the liver and hepatic blood flow. Of these, bromosulfophthalein (BSP, Bromsulphalein, Fig. 13.11) and indocyanine green (ICG, Fig. 13.12) are most frequently used in liver function studies. Following intravenous administration, these dyes are removed rapidly from the plasma, primarily by the liver, and are subsequently excreted in the bile. Delayed plasma clearance

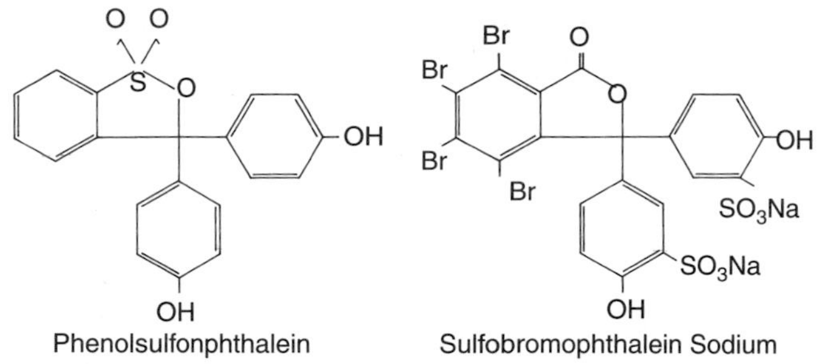

Phenolsulfonphthalein

Sulfobromophthalein Sodium

FIGURE 13.11 Phenolsulfonphthalein (PSP) is a dye that is cleared exclusively by the kidney and used in renal function tests. Sulfobromophthalein (BSP) is closely related structurally, but the modifications result in excretion almost entirely by the liver, allowing its use in the assessment of hepatic function.

indicates abnormal hepatocellular or biliary-tract function.

The overall process of hepatic excretion of BSP is similar but not identical to that of bilirubin. BSP and bilirubin appear to be transported across the plasma membrane by different routes in that membrane binding sites are not shared (Cornelius *et al.*, 1967). Within the cell, however, bilirubin and BSP compete for binding on the cytosolic protein, *ligandin*. Ligandin, by serving as an ion "sink," is thought to be an important

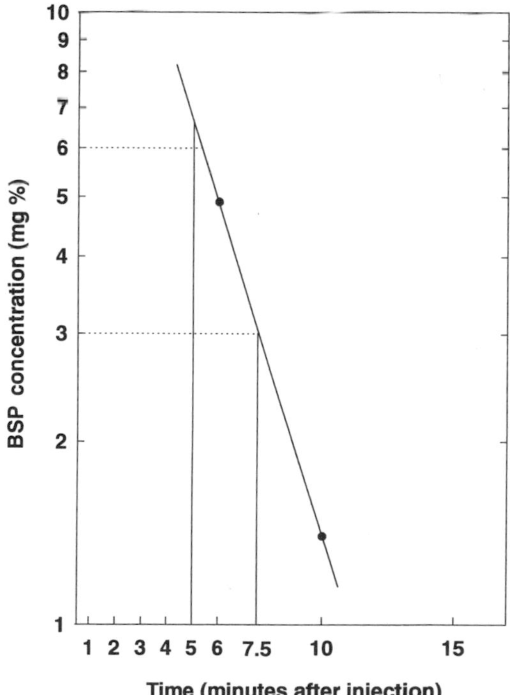

FIGURE 13.12 Calculation of $t_{1/2}$ for bromosulfophthalein (BSP) in the horse. After bolus intravenous injection, serum samples were obtained at 6 and 10 minutes, and $t_{1/2}$ calculated to be 2.5 minutes (normal).

driving force in the overall process of hepatic uptake (Levi *et al.*, 1969). The conjugation mechanisms for bilirubin and BSP are completely separate. BSP is conjugated with glutathione by action of the cytosolic enzyme S-aryl transferase-B. Although conjugation facilitates excretion of BSP, it is not an obligatory step, because approximately 50% of the BSP in bile is unconjugated. Conjugation of bilirubin with glucuronic acid or other sugars requires one of the glycosyl transferases, which are microsomal enzymes. Conjugation appears to be critical in the hepatic excretion of bilirubin. The canalicular transport of BSP is similar to that of conjugated bilirubin, but whether there is competition between the two for a carrier mechanism is unknown.

In the icteric patient, the question arises whether competition between BSP and bilirubin will alter the results of the BSP excretion test. In general, the BSP test is seldom justified in such patients, because hepatic disease is evident and no important additional information is likely to be provided. In general, the BSP test is most useful for situations in which a suggestion of occult liver disease exists but in which the results of other liver-function tests are equivocal. The net competitive effect of bilirubin on BSP excretion is not great. For example, horses starved for 72 hours developed a threefold elevation in total bilirubin, but BSP excretion was prolonged less than 25% (Tennant, unpublished data).

In the dog and cat, a standard dose of 5 mg/kg of BSP is administered as an intravenous bolus. A sample of blood is removed at 30 minutes and the BSP concentration is determined spectrophotometrically. It is assumed that the original dose of BSP (5 mg/kg) is distributed in a plasma volume of 5 ml/kg so that the concentration of BSP in plasma at time zero is (by definition) I mg/ml. The percent retention at 30 minutes is calculated from the ratio of the concentration at time zero and at 30 minutes. Retention of 5% or less is considered to be within normal limits (Cornelius, 1970).

In the large domestic species, it often is difficult or inconvenient to obtain body-weight measurements. Because the initial slope of the disappearance curve is independent of BSP dose, a standard 1-g dosage is administered to normal-sized horses (450 kg) and 0.5 g to smaller horses. Blood samples are obtained at 6, 9, and 12 minutes following injection, the BSP concentration is determined in each, and the fractional clearance rate or plasma half time ($t_{1/2}$) is calculated (Fig. 13.13). In the horse, normal plasma $t_{1/2}$ values vary between 2.5 and 3.5 minutes. In cattle, the BSP excretion rate is similar to that in the horse. Sheep have a more rapid excretion rate, requiring samples to be

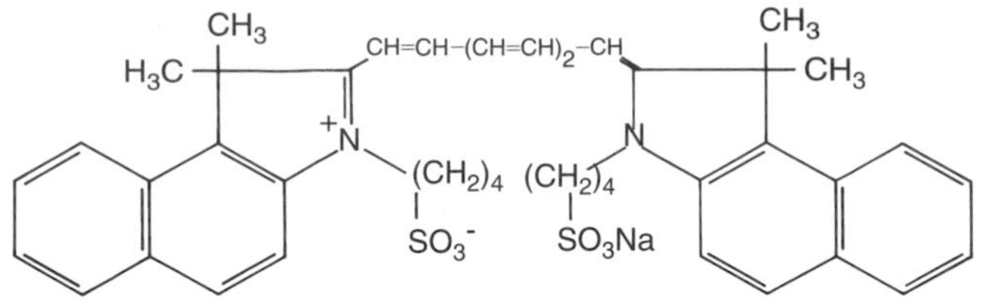

FIGURE 13.13 Indocyanine green used for quantitative assessment of hepatic excretory function.

taken at 3, 5, and 7 minutes following injection. Normal $t_{1/2}$ values range from 1.6 to 2.7 minutes (Cornelius, 1970).

The BSP test is safe, although the dye is very irritating if infiltrated perivascularly. The test should be used only when cardiovascular function is normal. If hypovolemia or hypotension is present, hepatic perfusion will be reduced and erroneously prolonged clearance rates may be observed.

1. *Indocyanine Green*

The clearance rate of indocyanine green also provides a useful estimate of hepatic excretion. The hepatic extraction ratio approaches 1 and the hepatic clearance corresponds to hepatic blood flow. In contrast to BSP, the plasma disappearance of ICG generally follows a single exponential. ICG, because of its commercial availability, is the dye excretion test most readily available for clinical use.

ICG clearance has been used for some time to estimate circulation time and hepatic blood flow (Keiding and Skak, 1985). The ICG clearance rate from plasma is measured after intravenous injection of various bolus doses of dye (Warren *et al.*, 1984). Because the disappearance of dye from plasma follows Michaelis–Menten kinetics, Lineweaver–Burke plot analysis will yield apparent K_m and V_{max} values for ICG removal (Paumgartner *et al.*, 1970). This method can be used to estimate the maximal dye removal capacity from a small number of submaximal clearance values. Although such multiple tests are not useful for routine practice, they provide the investigator with an extremely sensitive index of hepatic dye clearance that is independent of blood flow.

The clinically useful liver function test using ICG measures $t_{1/2}$ and fractional clearance (K) after the intravenous injection of 0.5 mg/kg (0.64 μmol/kg). Three plasma samples are usually taken between 3 and 15 minutes postinjection; ICG is measured at 805 nm and plotted on semilog paper as described earlier for BSP clearance in the horse. The value of K can be calculated from the $t_{1/2}$ value as

$$K = \ln 2/t_{1/2},$$

(where ln 2 equals 0.693). In the dog, one can also estimate the plasma volume and hepatic blood flow by using the average hepatic extraction ratio of 18% (Ketterer *et al.*, 1960).

Normal ICG plasma clearance $t_{1/2}$ and K values for the dog using 0.5 mg/kg ICG (0.64 μmol/kg) are reported to be 8.4 ± 2.3 minutes and 0.089 ± 0.027/minute (Bonasch and Cornelius, 1964), or, using 1.0 mg/kg (1.3 μmol/kg), have been reported to be 9 ± 2 minutes and 0.081 ± 0.017/minute (Center *et al.*, 1983a); for cats at 1.5 mg/kg (1.9 μmol/kg), the values are 3.8 ± 0.9 minutes and 0.19 ± 0.045/minute (Center *et al.*, 1983b). For sheep using an ICG dose of 0.5 mg/kg (0.64 μmol/kg), the $t_{1/2}$ = 4.8 ± 0.5 minutes, and for heifers using a dose of 0.75 mg/kg (0.97 μmol/kg), the $t_{1/2}$ = 3.5 ± 0.8 minutes (Sato, 1984).

V. OVERVIEW AND CONCLUSIONS

Conventional tests of hepatic disease provide information about the integrity of the hepatocytes (ALT, AST, SDH) and the integrity of the biliary system (AP, GGT). Hepatic function can be assessed by testing the excretory capacity of the liver (bilirubin; bile acids) and its synthetic function (NH_3/urea, albumin, fibrinogen, and prothrombin).

For optimal value in interpreting the results of clinical laboratory tests, it is essential that the specific purpose(s) of the test(s) being performed be defined. As described in Section IV, laboratory tests for hepatic disease are performed clinically for a variety of purposes, including confirmation of the presence of liver disease, for assessment of the nature (e.g., hepatocellular injury, cholestasis) and severity of the disease to establish prognosis, for monitoring the clinical course of liver disease, and for screening of individuals considered to be at risk for liver disease to establish the existence of occult liver disease.

Many of the standard tests for liver disease are based on rather simple biochemical procedures that are readily automated for use in multichannel autoanalyzers. One seldom obtains the result of a single test for liver disease, but rather a panel of results or a "liver profile." In some situations, multiple tests results for liver function are received even when there is no specific clinical indication because with autoanalyzers, it may be easier to perform a large series of tests than to be selective.

In a given population of animals, some will have liver disease and some will not. If a test is applied to the whole population, a certain number of those with the disease will have a positive test result (true positives), and some with the disease will have a negative test result (false negatives). Similarly, among those without the disease, some will have a positive test result (false positives) and in some, the test will be negative (true negatives). A test is said to be *sensitive* to the extent that it detects individuals with the disease (true positives divided by the total number with the disease; i.e., true positives plus false negatives). The test is said to be *specific* to the extent that a negative result detects patients that are free of the disease (true negatives divided by the number of subjects without the disease; i.e., true negatives plus false positives).

In the clinical setting, the sensitivity and the specificity of hepatic tests are less important than the *positive predictive value,* defined as the probability that a positive test result indicates the presence of the disease (true positives divided by true positives plus the false positives), and the *negative predictive value,* the probability that a negative test result is indicative of the absence of the disease (true negatives divided by the true negatives plus false negatives). The predictive value of a test, unfortunately, depends on the population being studied and the proportion of individuals in the population with the disease (prevalence). Even when test sensitivity and specificity are high and the number of false positives tests is low, if there are few subjects with the disease in the population, a positive test result will have relatively low positive predictive value, while a negative test result will have proportionately high negative predictive value. If the prevalence of a disease in a population is high, however, a positive result for a test with even low sensitivity and specificity will have high positive predictive value, while a negative result would have proportionately low predictive value for the absence of the disease.

Although the sensitivity of a test is often discussed, the prevalence of the disease and the positive and negative predictive values of tests are often ignored in discussions of interpretation of laboratory results. Importantly, laboratory test results from a selected group of individuals with a high prevalence of liver disease cannot be compared to a population in which the prevalence is low or to one in which the predictive value of a test in one population is similar to that of the other population.

As indicated earlier, it is unusual to perform a single test for hepatic function or disease; rather, a "profile" of hepatic test results is ordinarily obtained. The combined results of a panel of tests often provide increased sensitivity and specificity and improved predictive value in assessing severity and, in some cases, in the differentiation between acute and chronic forms of liver disease. The differential diagnosis of specific problems often is not possible with multiple laboratory tests alone, and hepatic imaging and/or liver biopsy are required. Hepatic tests, however, continue to have an important place in evaluation and monitoring of clinical patients with liver disease and in understanding the underlying pathophysiological mechanisms essential for initiation of successful treatment.

References

Abdelkader, S. V., and Hauge, J. G. (1986). *Acta Vet. Scand.* **27,** 59.

Adam, S. E. I., Oberd, H. M. A., and Artour, G. (1974). *Acta Vet. (Brno)* **43,** 225.

Ahmed, K., and Judah, J. D. (1964). *Biochim. Biophys. Acta* **93,** 603.

Alemu, P., Forsyth, G. W., and Searcy, G. P. (1977). *Can. J. Comp. Med.* **41,** 420.

Allcroft, W. M., and Folley, S. J. (1941). *Biochem. J.* **35,** 254.

Alpert, S., Mosher, M., Shanske, A., and Arias, I. M. (1969). *J. Gen. Physiol.* **53,** 238.

Anthony, R., Morrison, L., MacSween, R. N. M., and Whaley, K. (1985). *Biochem. J.* **232,** 93.

Anwer, M. S., Engelking, L. R., Gronwall, R., and Klentz, R. D. (1976). *Res. Vet. Sci.* **20,** 127.

Arias, I. M. (1974). *In* "The Liver and Its Diseases" (F. Schaffner, S. Sherlock, and C. M. Leevy, eds.), p. 97. Intercontinental Medical Book, New York.

Arias, I. M., Johnson, L., and Wolfson, S. (1961). *Am. J. Physiol.* **200,** 1091.

Arias, I. M., Che, M., Gtmaitan, Z., Leville, C., Nishida, T., and St. Pierre, M. (1993). *Hepatology* **17,** 318.

Audell, L., Jonsson, L., and Lannek, B. (1974). *Zentralbl. Veterinaemmed. {A}* **21,** 797.

Barrett, R. E., deLahunta, A., Roenick, W. J., Hoffer, R. E., and Coons, F. H. (1976). *J. Small Anim. Pract.* **17,** 71.

Ben-Said, N. S. (1979). Thesis. Doct. Med. Vet., University of Tunisia.

Berman, J. K., and Hull, J. E. (1952). *Surgery* **32,** 67.

Bilir, B. M., Gong, L., Kwasiborski, V., Shen, C. S., Fillmore, C. S., Berkowitz, C. M., and Gumucio, J. J. (1993). *J. Biol. Chem.* **268,** 19776.

Bissell, D. M. and Maher, J. J. (1996). *In* "Hepatology, A Textbook of Liver Disease" (D. Zakim and T. D. Boyer, eds.), 3rd Ed., Vol. 1, p. 506. Saunders, Philadelphia, Pennsylvania.

Blackshaw, C. (1978). *New Zealand Vet. J.* **26,** 25.

Blum, H. F., Watrous, W. G., and West, R. J. (1935). *Am. J. Physiol.* **113,** 350.

Bollman, J. L., and Mann, F. C. (1932). *Arch. Surg.* **24,** 675.

Bonasch, H., and Cornelius, C. E. (1964). *Am. J. Vet. Res.* **25,** 254.

Bostedt, H. (1978). *Ber. Muench. Tieraertzl. Wschr.* **87,** 375.

Boyd, J. W. (1962). *Res. Vet. Sci.* **3,** 256.

Brauer, R. W. (1963). *Physiol. Rev.* **43,** 115.

Braun, J. P., Rico, A. G., Benard, P., Thouvenot, J. P., and Bonnefis, M. J. (1978a). *J. Dairy Sci.* **61,** 596.

Braun, J. P., Rico, A. G., Benard, P., Thouvenot, J. P., and Bonnefis, M. J. (1978b). *Res. Vet. Sci.* **25,** 37.

Braun, J. P., Benard, P., Burgat, V., and Rico, A. G. (1983). *Vet. Res. Commun.* **6,** 77.

Breeze, R. G., Pirie, H. M., Selman, I. E., and Wiseman, A. (1976). *J. Pathol.* **119,** 229.

Bromberg, P. A., Robin, E. D., and Forkner, C. E., Jr. (1960). *J. Clin. Invest.* **39,** 332.

Bulgin, M. S., and Anderson, B. C. (1984). *Res. Vet. Sci.* **37,** 167.

Bunch, S. E., Castleman, W. L., Hornbuckle, W. E., and Tennant, B. C. (1982). *J. Am. Vet. Med. Assoc.* **181,** 357.

Bunch, S. E., Polak, D. W., and Hornbuckle, W. E. (1985). *J. Am. Vet. Med. Assoc.* **187,** 1032.

Butterworth, R. O. (1994). *In* "The Liver. Biology and Pathobiology" (I. M. Arias, J. L. Boyer, N. Fausto, W. B. Jacoby, D. Schachter, and D. A. Schafritz, eds.), p. 1193. Raven Press, New York.

Cantarow, A., Stewart, H. L., and McCool, S. G. (1936). *Proc. Soc. Exp. Biol. Med.* **35,** 87.

Castell, D. O., and Moore, E. W. (1971). *Gastroenterology* **60,** 33.

Center, S. A. (1996). *In* "Strombeck's Small Animal Gastroenterology" (W. G. Guilford, S. A. Center, D. R. Strombeck, D. A. William, and D. J. Meyer, eds.), p. 553. Saunders, Philadelphia, Pennsylvania.

Center, S. A., Bunch, S. E., Baldwin, B. H., Hornbuckle, W. E., and Tennant, B. C. (1983a). *Am. J. Vet. Res.* **44,** 722.

Center, S. A., Bunch, S. E., Baldwin, B. H., Hornbuckle, W. E., and Tennant, B. C. (1983b). *Am. J. Vet. Res.* **44,** 727.

Center, S. A., Baldwin, B. H., King, J. M., and Tennant, B. C. (1983c). *Am. J. Vet. Res.* **44,** 1822.

Center, S. A., Leveille, C. R., Baldwin, B. H., and Tennant, B. C. (1984). *Am. J. Vet. Res.* **45,** 2043.

Center, S. A., Baldwin, B. H., deLahunta, A., Dietze, A. E., and Tennant, B. C. (1985a). *J. Am. Vet. Med. Assoc.* **186,** 1090.

Center, S. A., Baldwin, B. H., Erb, H. N., and Tennant, B. C. (1985b). *J. Am. Vet. Med. Assoc.* **187,** 935.

Center, S. A., Baldwin, B. H., Dillingham, S., Erb, H. N., and Tennant, B. C. (1986). *J. Am. Vet. Med. Assoc.* **188,** 507.

Center, S. A., Erb, H. N., and Joseph, S. A. (1995). *J. Am. Vet. Med. Assoc.* **207,** 1048.

Chowdhury, J. R., Chowdhury, N. R., Wolkoff, A. W., and Arias, I. M. (1994). *In* "The Liver: Biology and Pathophysiology" (I. M. Arias, J. L. Boyer, N. Fausto, W. B. Jacoby, D. Schachter, and D. A. Shafritz, eds.), p. 471. Raven Press, New York.

Clare, N. T. (1944). *NZ J. Sci. Tech.* **25A,** 202.

Clare, N. T. (1945). *NZ J. Sci. Tech.* **27A,** 23.

Cook, J. S., and Blum, H. F. (1959). *J. Cell. Comp. Physiol.* **53,** 41.

Cornelius, C. E. (1963). *Nature* **200,** 580.

Cornelius, C. E. (1970). *In* "Clinical Biochemistry of Domestic Animals" (J. J. Kaneko and C. E. Cornelius, eds.), Vol. 1, p. 161. Academic Press, New York.

Cornelius, C. E. (1993). *Adv. Vet. Sci. Comp. Med.* **37,** 1.

Cornelius, C. E., and Gronwall, R. R. (1968). *Am. J. Vet. Res.* **29,** 291.

Cornelius, C. E., Kilgore, W. W., and Wheat, J. D. (1960). *Cornell Vet.* **50,** 47.

Cornelius, C. E., Douglas, G. M., Gronwall, R. R., and Freedland, R. A. (1963). *Cornell Vet.* **53,** 181.

Cornelius, C. E., Arias, I. M., and Osburn, B. I. (1965a). *J. Am. Vet. Med. Assoc.* **146,** 709.

Cornelius, C. E., Gazmuri, G., Gronwall, R., and Rhode, E. A. (1965b). *Cornell Vet.* **55,** 110.

Cornelius, C. E., Ben-Ezzer, J., and Arias, I. M. (1967). *Proc. Exp. Biol. Med.* **124,** 665.

Cornelius, L. M., Thrall, D. E., Halliwell, W. H., Frank, G. M., Kern, A. J., and Woods, C. B. (1975b). *J. Am. Vet. Med. Assoc.* **167,** 220.

Cornelius, C. E., Kelley, K. C., and Himes, J. A. (1975a). *Cornell Vet.* **65,** 90.

Crawford, M. A., Schall, W. D., Jensen, R. K., and Tasker, J. B. (1985). *J. Am. Vet. Med. Assoc.* **187,** 1343.

Dalgaard, J. B. (1948). *Acta Physiol. Scand.* **15,** 290.

De Novo, R. C., and Prasse, K. W. (1983). *Am. J. Vet. Res.* **44,** 1703.

de Schepper, J., and Van Der Stock, J. (1972a). *Arch. Int. Physiol. Biochim.* **80,** 279.

de Schepper, J., and Van Der Stock, J. (1972b). *Arch. Int. Physiol. Biochim.* **80,** 339.

Dimski, D. S. (1994). *J. Vet. Int. Med.* **8,** 73.

Dittrich, C., Stockl, W., and Desser, H. (1974). *Zentralbl. Veterinaermed. Reihe A* **21,** 165.

Divers, T. J., Warner, A., Vaala, W. E., Whitlock, R. H., Acland, H. A., Mansmann, R. A., and Palmer, J. E. (1983). *J. Am. Vet. Med. Assoc.* **183,** 1407.

Emmelot, P., Bos, C. J., Benedetti, E. L., and Rumke, P. H. (1964). *Biochim. Biophys. Acta* **90,** 126.

Engelking, L. R., and Gronwall, R. (1979). *Proc. Soc. Exp. Biol. Med.* **161,** 123.

Enigk, K., Feder, H., and Dey-Hazra, A. (1976). *Zbl. Vet. Med.* **23A,** 255.

Everett, R. M., Duncan, J. R., and Prasse, K. W. (1977). *Am. J. Vet. Res.* **38,** 963.

Ewing, G. O., Suter, P. F., and Bailey, C. S. (1974). *J. Am. Anim. Hosp. Assoc.* **10,** 463.

Ferslew, K. K., and Edds, G. T. (1979). *Am. J. Vet. Res.* **1,** 1365.

Fevery, J., Van Hees, G. P., Leroy, P., Compernolle, F., and Heirwegh, K. P. M. (1971). *Biochem. J.* **125,** 803.

Field, M. E., Leigh, O. C., Heim, J. W., and Drinker, C. K. (1934). *Am. J. Physiol.* **110,** 174.

Finn, J. P., and Tennant, B. (1974). *Cornell Vet.* **64,** 136.

Flood, C. A., Gutman, E. B., and Gutman, A. B. (1937). *Am. J. Physiol.* **120,** 696.

Fog, J., and Jellum, E. (1963). *Nature* **198,** 88.

Ford, E. J. H. (1958). *J. Anat.* **92,** 447.

Ford, E. J. H. (1974). *J. Comp. Pathol.* **84,** 231.

Ford, E. J. H., and Gopinath, C. (1976). *Res. Vet. Sci.* **21,** 12.

Ford, E. J. H., and Evans, J. (1985). *Res. Vet. Sci.* **39,** 103.

Fowler, M. E. (1965). *J. Am. Vet. Med. Assoc.* **147,** 55.

Fowler, M. E. (1968). *J. Am. Vet. Med. Assoc.* **152,** 1131.

Fowler, J. S. (1971). *Br. Vet. J.* **237,** 304.

Freedland, R. A., Hjerpe, C. A., and Cornelius, C. E. (1965). *Res. Vet. Sci.* **6,** 18.

Fulop, M., Sandson, J., and Brazeau, P. (1965). *J. Clin. Invest.* **44,** 666.

Gibbons, W. J., Hokanson, J. F., Wiggins, A. M., and Schmitz, M. B. (1950). *North Am. Vet.* **31,** 229.

Gordon, E. R., Goresky, C. A., Chang, T.-H., and Perlin, A. S. (1976). *Biochem. J.* **155,** 477.

Goresky, C. A., Haddad, H. H., Kluger, W. S., Nadeau, B. E., and Bach, G. G. (1974). *Can. J. Physiol. Pharmacol.* **52,** 389.

Gossett, K. A., and French, D. D. (1984). *Am. J. Vet. Res.* **45,** 354.

Grisham, J. W. (1959). *Fed. Proc.* **18,** 478.

Grodsky, G. M., and Carbone, J. V. (1957). *J. Biol. Chem.* **226,** 449.

Gronwall, R., and Mia, A. S. (1972). *Am. J. Dig. Dis.* **17,** 473.

Grunder, H. D. (1977). *Le Point Veterinaire* **6,** 49.

Guelfi, J. F., Braun, J. P., and Rico, A. G. (1982). *Res. Vet. Sci.* **33,** 309.

Gumucio, J. J., and Berkowitz, C. M. (1992). *In* "Liver and Biliary Disease (N. Kaplowitz, ed.), p. 2. Williams and Wilkins, Baltimore.

Gumucio, J. J., Bilir, B. M., Moseley, R. H., and Berkowitz, C. M. (1994). *In* "The Liver: Biology and Pathology" (I. M. Arias, J. L. Boyer, N. Fausto, W. B. Jacoby, D. Schachter, and D. A. Shafritz, eds.), p. 1143. Raven Press, New York.

Gustafsson, B. E., and Lanke, L. S. (1960). *J. Exp. Med.* **112,** 975.

Gutman, A. B., Hogg, B. M., and Olson, K. B. (1940). *Proc. Soc. Exp. Biol. Med.* **44,** 613.

Habig, W. H., Pabst, M. J., Fleischner, G., Gatmaitan, Z., Arias, I. M., and Jakoby, W. B. (1974). *Proc. Natl. Acad. Sci. USA* **71,** 3879.

Harvey, D. G., and Hoe, C. M. (1971). *Vet. Rec.* **29,** 562.

Harvey, D. G., and Obeid, H. M. (1974). *Br. Vet. J.* **130,** 544.

Hauge, J. G., and Abdelkader, S. V. (1984). *Acta Vet. Scand.* **25,** 495.

Haussler, M. R., Nagode, L. A., and Rasmussen, H. (1970). *Nature (Lond.)* **228,** 1199.

Heirwegh, K. P. M., Fevery, J., Michiels, R., Van Hees, G. P., and Compernolle, F. (1975). *Biochem. J.* **145,** 185.

Hjerpe, C. A. (1964). *J. Am. Vet. Med. Assoc.* **144,** 734.

Hjerpe, C. A., Tennant, B. C., Crenshaw, G. L., and Baker, N. F. (1971). *J. Am. Vet. Med. Assoc.* **159,** 1266.

Hoe, C. M., and Jabara, A. G. (1967). *J. Comp. Pathol.* **77,** 245.

Hoffmann, W. E., and Dorner, J. L. (1975). *J. Am. Anim. Hosp. Assoc.* **11,** 283.

Hoffman, H. N., Whitcomb Jr., F. F., Butt, H. R., and Bollman, J. L. (1960). *J. Clin. Invest.* **39,** 132.

Hooper, P. T. (1972). *Vet. Rec.* **90,** 37.

Howards, S. S., Davis, J. O., Johnston, C. I., and Wright, F. S. (1968). *Am. J. Physiol.* **214,** 990.

Hyatt, R. E., Lawrence, G. H., and Smith, J. R. (1955). *J. Lab. Clin. Med.* **45,** 274.

Ikeda, S., Yamaoka, S., Uratanable, H., Wantanabe, H., and Kameya, T. (1976). *Exp. Rep. Equine Health Lab.* **13,** 1.

Jacquemin, E., Hagenbuch, B., Stieger, B., Wolkoff, A. W., and Meier, P. J. (1991). *J. Clin. Invest.* **88,** 2146.

Jacquemin, E., Hagenbuch, B., Stieger, B., Wolkoff, A. W., and Meier, P. J. (1994). *Proc. Nat. Acad. Sci. USA* **91,** 133.

Johnson, A. E., and Molyneux, R. J. (1984). *Am. J. Vet. Res.* **45,** 26.

Johnson, G. F., Zawie, D. A., Gilbertson, S. R., and Sternlieb, I. (1982). *J. Am. Vet. Med. Assoc.* **180,** 1438.

Jungermann, K., and Katz, N. (1989). *Physiol. Rev.* **69,** 708.

Kaneko, J. J., Zinkl, J. G., and Keeton, K. S. (1971). *Am. J. Vet. Res.* **32,** 1981.

Kaplan, M. M. (1972). *Gastroenterology* **62,** 452.

Kaplan, M. M., and Righetti, A. (1969). *Biochim. Biophys. Acta* **184,** 667.

Kaplan, M. M., and Righetti, A. (1970). *J. Clin. Invest.* **49,** 508.

Kaplowitz, N., Percy-Robb, I. W., and Javitt, N. B. (1973). *J. Exp. Med.* **138,** 483.

Keiding, S., and Skak, C. (1985). *J. Hepatol. Suppl.* **2,** S262.

Keller, P. (1971). *Schweiz. Arch. Tierhk.* **113,** 615.

Keller. P. (1978). *Schweiz. Arch. Tierhk.* **120,** 189.

Keller, P. (1981). *Am. J. Vet. Res.* **42,** 575.

Keller, P., Ruedi, D., and Gutzwiller, A. (1985). *J. Zoo Anim. Med.* **16,** 28.

Ketterer, S. G., Wiegand, B. D., and Rapapon, E. (1960). *Am. J. Physiol.* **199,** 481.

Krebs, C. (1979). *Inaug. Diss. Vet. Med.*, Munchen.

Kroker, R., and Romer, C. (1984). *Zbl. Vet. Med.* **31,** 287.

Laks, M. M., Pincus, I. J., and Goldberg, D. (1963). *Gastroenterology* **44,** 469.

Landau, S. A., and Winchell, H. S. (1970). *Blood* **36,** 642.

Leaver, D. D. (1968). *Res. Vet. Sci.* **9,** 265.

Levenson, S. M., and Tennant, B. (1963). *Fed. Proc.* **22,** 109.

Levi, A. J., Gatmaitan, Z., and Arias, I. M. (1969). *J. Clin. Invest.* **48,** 2156.

Levy, M., Lester, R., and Levinsky, N. G. (1968). *J. Clin. Invest.* **47,** 2117.

Litchfield, M. H., and Garland, C. J. (1974). *Toxicol. Appl. Pharmacol.* **30,** 117.

Loud, A. V. (1968). *J. Cell Biol.* **37,** 27.

Maddison, J. E. (1992). *J. Vet. Int. Med.* **6,** 341.

Malherbe, W. D., Kellerman, T. S., Krick, N. P. J., and Haupt, W. H. (1977). *Onderstepoort J. Vet. Res.* **44,** 29.

Markiewicz, K., Kuleta, Z., and Romanrick, K. (1975). *Acta Parasitol. Pol.* **23,** 177.

Meissonier, E., and Rousseau, P. (1976). *Int. Congr. Maladies Betail IT CF, 9th, Maisse,* p. 633.

Meyer, D. J. (1983). *J. Am. Anim. Hosp. Assoc.* **19,** 1023.

Meyer, D. J., and Noonan, N. E. (1981). *J. Am. Anim. Hosp. Assoc.* **17,** 261.

Mia, A. S., and Koger, H. D. (1978). *Am. J. Vet. Res.* **39,** 1381.

Mia, A. S., and Koger, H. D. (1979). *Vet. Clin. Pathol.* **8,** 9.

Moursi, S. A. H., Atef, M., and Al-Khaygot, A. A. (1979). *Zbl. Vet. Med.* **26A,** 715.

Muller, M., Ishikawa, R., Berger, U., Klunemann, C., Luckas, L., Schreyer, A., Kannich, C., Rutter, W., Kurz, C., and Keppler, K. (1991). *J. Biol. Chem.* **266,** 18920.

Nagode, L. A., Koestner, A., and Steinmeyer, C. L. (1969a). *Clin. Chim. Acta* **26,** 45.

Nagode, L. A., Koestner, A., and Steinmeyer, C. L. (1969b). *Clin. Chim. Acta* **26,** 55.

Nance, F. C., and Kline, D. G. (1971). *Ann. Surg.* **174,** 856.

Nance, F. C., Batson, R. C., and Kline, D. G. (1971). *Surgery* **70,** 169.

Nance, F. C., Kaufman, H. J., and Kline, D. G. (1974). *Gastroenterology* **66,** 108.

Nichol, A. W., and Morrell, D. B. (1969). *Biochim. Biophys. Acta* **177,** 599.

Nix, J. T., Flock, E. V., and Bollman, J. L. (1951a). *Am. J. Physiol.* **164,** 117.

Nix, J. T., Mann, F. C., Bollman, J. L., Grindlay, J. H., and Flock, E. V. (1951b). *Am. J. Physiol.* **164,** 119.

Noir, B. A. (1976). *Biochem. J.* **155,** 365.

Noonan, N. E., and Meyer, D. J. (1979). *Am. J. Vet. Res.* **40,** 942.

Oliver, J. E., Jr. (1965). *V.M./S.A.C.* **60,** 498.

Orliac, D. (1980). Thesis Doct. Vet., Ecole Nationale Veterinaire. Toulouse, France.

Orloff, M. J., and Snyder, G. B. (1961a). *Surgery* **50,** 220.

Orloff, M. J., and Snyder, G. B. (1961b). *Surgery* **50,** 789.

Orloff, M. J., Wall, M. H., Hickman, E. B., and Spitz, B. R. (1963). *Surgery* **54,** 627.

Orloff, M. J., Ross, T. H., Baddeley, R. M., Nutting, R. O., Spitz, B. R., Sloop, R. D., Neesby, T., and Halasz, N. A. (1964a). *Surgery* **56,** 83.

Orloff, M. J., Spitz, B. R., Wall, M. H., Thomas, H. S., and Halasz, N. A. (1964b). *Surgery* **56,** 784.

Orloff, M. J., Lipman, C. A., Noel, S. M., Halasz, N. A., and Neesby, T. (1965). *Surgery* **58,** 225.

Orloff, M. J., Wright, P. W., DeBenedetti, M. J., Halasz, N. A., Annetts, D. L., Musicant, M. E., and Goodhead, B. (1966). *Arch. Surg.* **93,** 119.

Osuna, O. (1979). Ph.D. dissertation. College of Veterinary Medicine, University of Florida, Gainesville, Florida.

Paumgartner, G., Probst, P., Kraines, R., and Levy, C. D. (1970). *NY Acad. Sci.* **170,** 134.

Pearson, E. G. (1977). *Mod. Vet. Pract.* **58,** 421.

Pekarthy, J. M., Short, J., Lansing, A. I., and Lieberman, I. (1972). *J. Biol. Chem.* **247,** 1767.

Peterson, M. E., Kintzer, P. P., Cavanagh, P. G., Fox, P. R., Ferguson, D. C., Johnson, G. F., and Becker, D. V. (1983). *J. Am. Vet. Med. Assoc.* **183,** 103.

Pimstone, N. R. (1972). *In* "Physiology and Disorders of Hemoglobin Degradation" (R. Schmid, E. R. Jaffe, and P. M. Miescher, eds.), p. 3142. Grune and Stratton, New York.

Pimstone, N. R., Engel, P., Tenhunen, R., Seitz, P. T., Marver, H. S., and Schmid, R. (1971). *J. Clin. Invest.* **50,** 2042.

Putnam, M. R., Qualls, C. W., Rice, L. E., Dawson, L. J., and Edwards, W. C. (1986). *J. Am. Vet. Med. Assoc.* **189,** 77.

Quin, J. I., Rimington, C., and Roets, G. C. S. (1935). *Onderstepoort J. Vet. Sci. Anim. Ind.* **4,** 463.

Ramadori, G., Heinz, H.-P., Martur, H., zum Buschenfelde, K.-H. M., and Loos, M. (1986). *Eur. J. Immunol.* **16,** 1137.

Rico, A. G., Braun, J. P., Benard, P., El-Hassan, A. A., and Cazieux, A. (1977a). *Equine Vet. J.* **9,** 100.

Rico, A. G., Braun, J. P., Benard, P. C., and Thouvelot, J. P. (1977b). *J. Dairy Sci.* **60,** 1283.

Rico, A. G., Braun, J. P., Benard, P. C., and Thouvelot, J. P. (1977c). *Res. Vet. Sci.* **23,** 395.
Riemerschmid, G., and Quin, J. I. (1941). *Onderstepoort J. Vet. Sci. Anim. Ind.* **17,** 89.
Righetti, A. B.-B., and Kaplan, M. M. (1972). *Gastroenterology* **62,** 1034.
Rimington, C., and Quin, J. I. (1934). *Onderstepoort J. Vet. Sci. Anim. Ind.* **3,** 137.
Robinson, S. H., Tsong, M., Brown, B. W., and Schmid, R. (1966). *J. Clin. Invest.* **45,** 1569.
Rogers, W. A. (1976). *J. Am. Vet. Med. Assoc.* **168,** 934.
Rose, A. L., Gardner, C. A., McConnell, J. D., and Bull, L. B. (1957). *Aust. Vet. J.* **33,** 49.
Ross, D. B. (1966). *Br. Vet. J.* **122,** 279.
Rothschild, M. A., Oratz, M., and Schreiber, S. S. (1973). *Gastroenterology* **64,** 324.
Royer, M., Noir, B., de Walz, A. T., and Lozzio, B. (1965). *Rev. Int. Hepat.* **15,** 1351.
Royer, M., Noir, B. A., Sfarcich, D., and Nanet, H. (1974). *Digestion* **10,** 423.
Rubarth, S. (1947). *Acta Pathol. Microbiol. Scand.* **69,** suppl., 1.
Ruegnitz, P. C., and Schwartz, E. (1971). *Am. J. Vet. Res.* **32,** 1525.
Sato, T. (1984). *Jpn. J. Vet. Sci.* **46,** 687.
Schilling, J. A., McCoord, A. B., Clausen, S. W., Troup, S. B., and McKee, F. W. (1952). *J. Clin. Invest.* **31,** 702.
Schlaeger, R., Haux, P., and Kattermann, P. (1982). *Enzyme* **28,** 3.
Schmid, R. (1956). *Science* **124,** 76.
Schmid, R. (1974). *In* "The Liver and Its Diseases" (F. Schaffner, S. Sherlock, and C. M. Leevy, eds.), p. 85. Intercontinental Medical Book, New York.
Schothorst, A. A., Van Steveninck, J., Went, L. N., and Suurmond, D. (1970). *Clin. Chim. Acta* **28,** 41.
Scott, D. W., Mort, J. D., and Tennant, B. C. (1979). *Cornell Vet.* **69,** 145.
Sebesta, D. G., Bradshaw, F. J., and Prockop, D. J. (1964). *Gastroenterology* **47,** 166.
Seetharam, S., Sussman, N. L., Komoda, T., and Alpers, D. H. (1986). *Hepatology* **6,** 374.
Selman, I. E., Wiseman, A., Petrie, L., Pirie, H. M., and Breeze, R. G. (1974). *Vet. Rec.* **94,** 459.
Shaw, F. D. (1976). *Res. Vet. Sci.* **20,** 226.
Shull, R. M., and Hornbuckle, W. (1979). *Am. J. Vet. Res.* **40,** 1321.
Simesen, M., Nielsen, K., and Nansen, P. (1973). *Res. Vet. Sci.* **15,** 32.
Slater, T. F., and Riley, P. A. (1965). *Biochem. J.* **96,** 39P.
Spano, J. S., August, J. R., Henderson, R. A., Dumas, M. B., and Groth, A. H. (1983). *Am. J. Vet. Res.* **44,** 2049.
Stabenau, J. R., Warren, K. S., and Rall, D. P. (1959). *J. Clin. Invest.* **38,** 373.
Starling E. H. (1894). *J. Physiol.* **16,** 224.
Strombeck, D. R., Weiser, M. G., and Kaneko, J. J. (1975a). *J. Am. Vet. Med. Assoc.* **166,** 1105.
Strombeck, D. R., Meyer, D. J., and Freedland, R. A. (1975b). *J. Am. Vet. Med. Assoc.* **166,** 1109.
Sutherland, R. J., Condes, D. O., and Carthew, G. C. (1979). *New Zealand Vet. J.* **17,** 227.
Szasz, G. (1974). In "Methods of Enzymatic Analysis," 2nd ed., (H. U. Bergmeyer, ed.), p. 715. Verlag Chemie, New York.
Takahashi, Y., Muto, Y., Nakao, K., and Okinaka, S. (1966). *Proc. 3rd World Congr. Gastroenterol. [Tokyo]* **3,** 510.
Tal, M., Schneider, D. L., Thorens, B., and Lodish, H. F. (1990). *J. Clin. Invest.* **86,** 986.
Talafant, E. (1956). *Nature* **178,** 312.
Tenhunen, R., Marver, H. S., and Schmid, R. (1969) *J. Biol. Chem.* **244,** 6388.
Tenhunen, R., Ross, M. E., Marver, H. S., and Schmid, R. (1970a). *Biochemistry* **9,** 298.
Tenhunen, R., Marver, H. S., and Schmid, R. (1970b). *J. Lab. Clin. Med.* **75,** 410
Tennant, B., Evans, C. D., Schwartz, L. W., Gribble, D. H., and Kaneko, J. J. (1973). *Vet. Clin. North Am.* **3,** 279.
Tennant, B., Baldwin, B., Evans, C. D., and Kaneko, J. J. (1975). *21st Proc. Am. Assoc. Equine Pract., Boston,* p. 410.
Thomsett, L. R. (1971). *Equine Vet. J.* **3,** 15.
Towers, N. R., and Stratton, C. G. (1978). *New Zealand Vet. J.* **26,** 109.
Treacher, R. J., and Collis, K. A. (1977). *Res. Vet. Sci.* **22,** 101.
Treacher, R. J., and Sansom, B. F. (1969). *Res. Vet. Sci.* **10,** 461.
Uchiyama, Y., and Asari, A. (1984). *Cell Tissue Res.* **236,** 305.
Unglaub, W., Afschar, A., and Marx, D. (1973). *Dtsch. Tierarztl. Wschr.* **80,** 13.
Van Leenhoff, J. W., Hickman, R., Saunders, S. J., and Terblanche, J. (1974). *S. Afr. Med. J.* **12,** 1201.
Van Vleet, J. F., and Alberts, J. O. (1968). *Am. J. Vet. Res.* **29,** 2119.
Volwiler, W., Grindlay, J. H., and Bollman, J. L. (1950). *Gastroenterology* **14,** 40.
Warren, K. S., and Nathan D. G. (1958). *J. Clin. Invest.* **37,** 1724.
Warren, D., Cornelius, C., and Ford, B. (1984). *Vet. Hum. Toxicol.* **26,** 295.
Weiss, K. H. (1978). *Tierartzl Umschau.* **33,** 152.
Wellman, M. L., Hoffmann. W. E., Domer. J. L., and Mock, R. E. (1982a). *Am. J. Vet. Res.* **43,** 1200.
Wellman, M. L., Hoffmann, W. E., Domer, J. L., and Mock. R. E. (1982b). *Am. J. Vet. Res.* **43,** 1204.
Wensing, T., Schotman, A. J. H., and Kroneman, J. (1973). *Tidjschr. Diergeneesk.* **14,** 673.
Whitlock, R. H., and Brown, W. R. (1969). *Cornell Vet.* **59,** 515.
Wilson, G. D., Harvey, D. G., and Snook, C. R. (1972). *Br. Vet. J.* **128,** 596.
Windmueller, H. G., and Spaeth, A. E. (1974). *J. Biol. Chem.* **249,** 5070.
Wisse, E., DeZanger, R. B., Charles, K., van der Smissen, P., and McCuskey, R. S. (1985). *Hepatology* **5,** 683.
With, T. K. (1968). *"Bile Pigments. Chemical, Biological and and Clinical Aspects."* Academic Press, New York.
Witte, C. L., Witte, M. H., Dumont, A. E., Frist, J., and Cole, W. R. (1968). *Ann. Surg.* **168,** 567.
Witte, M. H., Dumont, A. E., Cole, W. R., Witte, C. L., and Kintner, K. (1969b). *Ann. Intern. Med.* **70,** 303.
Witte, C. L., Chung, Y. C., Witte, M. H., Sterle, O. F., and Coke, W. R. (1969b). *Ann Surg.* **170,** 1002.
Witte, C. L., Witte, M. H., Kintner, K., Cole, W. R., and Dumont, A. E. (1971a). *Surg. Gynecol. Obstet.* **133,** 65.
Witte, M. H., Witte, C. L., and Dumont, A. E. (1971b). *Gastroenterology* **61,** 742.
Wolf, W. A., Lumb, W. V., and Rarnsay, M. K. (1967). *Am. J. Vet. Res.* **28,** 1363.
Wolkoff, A. W., Sosiak, A., Greenblatt, H. C., Van Renswoude, J., and Stockert, R. J. (1985). *J. Clin. Invest.* **76,** 454.
Yamaoka, S., Ikeda, S., Watanabe, H., Hasegawa, Y., Takizawa, I., and Hasegawa, M. (1978). *Exp. Rep. Equine Health Lab.* **15,** 62.
Zakim, D. (1996). *In* "Hepatology, A Textbook of Liver Disease" (D. Zakim and T. D. Boyer, eds.), 3rd Ed., Vol. 1, p. 58. Saunders, Philadelphia, Pennsylvania.
Zawie, D. A., and Garvey, M. S. (1984). *Vet. Clin. North Am.* **14,** 1201.
Zieve, L., Nicoloff, D. M., and Mahadevan, V. (1968). *Gastroenterology* **54,** 1285.
Zieve, L., Doizaki, W. M., and Zieve, F. J. (1974). *J. Lab. Clin. Med.* **83,** 16.
Zinkl, J. G., Bush, R. M., Cornelius, C. E., and Freedland, R. A. (1971). *Res. Vet. Sci.* **12,** 211.

CHAPTER

14

Pancreatic Function

DUANE F. BROBST

I. INTRODUCTION

The pancreas is composed of two organs, the endocrine and exocrine portions, within one stroma. The functions of the endocrine pancreas are considered with carbohydrate metabolism elsewhere in this volume. Although the functions of the endocrine and exocrine portions of the pancreas may in some way be interrelated, this chapter is concerned with the exocrine pancreas.

The signs of exocrine pancreatic disease are often nonspecific, and physical and radiographic examination of the patient are seldom diagnostic. Because the pancreas is so difficult to evaluate using these approaches, the veterinary clinician has come to rely on biochemical function tests in the diagnosis of pancreatic disease. Still, many aspects of exocrine pancreatic physiology and pathophysiology remain enigmas despite long years of investigation. Neither have satisfactory laboratory procedures been developed that adequately identify ongoing acute pancreatitis and reflect its severity. However, an appreciation for the use and understanding of the fundamentals of the more commonly used function tests, and for tests more recently developed, requires familiarity with the physiology of the exocrine pancreas and mechanisms involved in diseases of this organ. To this end, the chapter is presented.

II. PHYSIOLOGY OF THE PANCREAS

Pancreatic secretion is composed of a mixture of viscid, enzyme-rich fluid and a watery portion containing electrolytes with a high concentration of bicarbonate (HCO_3^-). The enzymes are synthesized by the pancreatic acinar cells, which are arranged in clusters about the terminal pancreatic ductules. The epithelial cells lining the terminal and interlobular ducts, which carry pancreatic juice toward the duodenum, provide the mucin and electrolytes in the pancreatic secretion.

A. Pancreatic Fluid and Electrolytes

Stimulation of the exocrine pancreas with secretin results in production of a large volume of pancreatic fluid with a HCO_3^- concentration that can reach 140–160 mmol/liter. The mechanisms by which ductal epithelial cells secrete HCO_3^- are not well established, but secreted HCO_3^- is probably derived from CO_2 in the blood. Also, duct cells secrete HCO_3^- and H^+ ions from their apical and basolateral plasma membranes, respectively (Raeder, 1992). Secretin-dependent HCO_3^- secretion rate is greater at high pCO_2 values (but normal blood pH) than at normal pCO_2 values. Studies indicate that HCO_3^- is secreted into the duct lumen through the action of a chloride (Cl^-)/HCO_3^- exchange mechanism (Hootman and Ondarza, 1993). The HCO_3^- and Cl. concentrations in pancreatic juice have an inverse relationship with HCO_3^- concentration increasing as secretion rate increases. The involvement of Na^+, K^+-activated ATPase and carbonic anhydrase activity in ductal HCO_3^- secretion is also well established. Both enzymes are at high levels in ductal epithelial cells.

Pure pancreatic juice of dogs has a pH between 8 and 8.3 and is isosmotic with plasma. The volume of secretion by the canine pancreas is greater than by the pancreas of normally fed sheep (Taylor, 1962). In canine pancreatic juice, the HCO_3^- concentration ranges between 60 and 148 mmol/liter; in ovine pancreatic juice the HCO_3^- concentration is 15–30 mmol/liter. There is normally little fluctuation in the rate of pancreatic juice flow when food is eaten by a ruminant. This is because ingesta normally flows almost continuously into the duodenum. When food is withheld for 48 hours, however, ruminant pancreatic secretion decreases by about 50% (Taylor, 1962). The pancreas of the dog secretes two to three times as much fluid per gram of tissue and more enzyme than does the ovine pancreas. The concentration of sodium and potassium in pancreatic secretion tends to parallel plasma concentrations. Calcium concentration in pancreatic juice of the dog, however, tends to be lower than in plasma. Zimmerman *et al.* (1967) demonstrated that under conditions in which pancreatic enzymes were being formed in large amounts calcium output was greatest. They suggested calcium was not secreted with the electrolyte components of pancreatic juice, but rather with the nonelectrolyte portion, and that calcium may be a part of the amylase molecule.

Goblet cells are also present in the pancreatic ducts and they secrete mucus. The pancreatic mucus formed by these cells may produce a protective barrier against back flux of HCO_3^- and degradation of duct epithelium by digestive enzymes (Hootman and Ondarza, 1993).

Pancreatic juice of dogs plays a role in the regulation of the bacterial flora of the proximal region of the small intestine and dogs with exocrine pancreatic insufficiency (EPI) may have a greater than normal number of duodenal bacteria (Williams *et al.*, 1987). Nonenzymic and enzymic components of pancreatic secretion can influence the number and type of bacteria in the small bowel. In the absence of alkaline pancreatic juice, the increased acidity favors growth of acid-tolerant organisms such as *Lactobacillus* and *Streptococcus* species (Simpson *et al.*, 1990). Inadequate digestion of dietary constituents could also provide substrate for an altered bacterial flora. When the bacterial overgrowth in the small intestine includes obligate anaerobes, proteases from these bacteria may be responsible for reduction of mucosal enzyme activity and partial villous atrophy (Williams *et al.*, 1987).

Overgrowth of folate-producing bacteria in the proximal small intestine may result in an increase in serum folate levels and a reduction in serum cobalamin (vitamin B_{12}) levels. The reduction of vitamin B_{12} is the result of binding of the vitamin B_{12} intrinsic factor complex by intestinal bacteria (Batt and Morgan, 1982). The pancreatic secretion of dogs does have an important function in the absorption of cobalamin. Dogs with induced EPI had decreased cobalamin absorption, which was not restored by oral administration of pancreatic enzymes, despite improvement of steatorrhea (Simpson *et al.*, 1989b). In contrast, malabsorption of cobalamin was reversed by oral administration of canine pancreatic juice. Hence, both pancreatic and gastric intrinsic factors have a role in the normal absorption of cobalamin in dogs (Batt and Horadagoda, 1989).

B. Pancreatic Enzymes

The acinar cell of the exocrine pancreas secretes a mixture of enzymes involved with digestion of proteins, fats, and carbohydrates. These exocrine-derived proteins are synthesized in the area of the cell occupied by the rough endoplasmic reticulum and are packaged in condensing vacuoles near the Golgi complex. Later these proteins are in the apical portion of the cell, in zymogen granules, and are stored here until discharged from the cell by hormone-responsive exocytosis. During transport of these secretory proteins, they are at all times segregated within membrane-bound compartments and are excluded from the cytosolic space.

The type of food ingested by some animals may, with time, alter the enzymatic composition of the pancreas. Lavau *et al.* (1974) found that rats on a diet high in starch or glucose and low in fat had increased concentrations of amylase and decreased lipase in pancreatic tissue. Protein malnutrition may also be associated with alteration in the composition of pancreatic enzymes. Gyr *et al.* (1975) determined that moderate protein deficiency in monkeys caused a significant decrease in pancreatic secretion of amylase, lipase, and chymotrypsin.

Malaisse-Lagae *et al.* (1975) showed that in the rat, the concentration of enzymes in acinar cells surrounding the islets of Langerhans (peri-insular portion) differed from the rest, the teleinsular part of the pancreas. The concentration of amylase, relative to lipase, was invariably higher in the teleinsular than in the peri-insular exocrine pancreas. Their findings indicate that the secretory partition of the exocrine pancreas could be responsible for modulation in the composition of pancreatic juice under different conditions. The peri-insular acini may differ from the teleinsular acini in the level of their secretory activity, their response to various secretagogues, and their relative content of different hydrolases. Functionally, the endocrine and exocrine cells of the pancreas may exert a mutual and direct control of their secretory activity. That is, insular hormones modify exocrine function, and exocrine factors may alter islet cell responsiveness to certain secretagogues. The close anatomic relationship between endocrine islets and exocrine acini, together with the fact that islet hormones affect exocrine pancreatic function, led to the concept of an islet–acinar axis with islet cell hormones being delivered to the exocrine interstitial space via an islet–acinar portal system or islet–acinar microcirculation. Nakagawa *et al.* (1993) established the presence of islet–acinar axis activity in the rat pancreas. They proposed that the islet–acinar axis may be responsible for increasing exocrine cell growth and enzyme synthesis, via insulin, over the long term, whereas the islet hormones somatostatin and pancreatic polypeptide provide a means to regulate rapidly the suppression of enzyme release from acinar cells. The histological differences between acinar cells surrounding islets and those located in the periphery are considered to be the result of more synthesis and less secretion of enzymes in peri-insular regions. The increase in insulin following a meal could stimulate digestive enzyme synthesis in peri-insular cells, whereas somatostatin and/or pancreatic polypeptide may inhibit immediate exocrine release of stored enzymes. Matsushita *et al.* (1994) observed that insulin exerted a direct effect on rat pancreatic acinar cells and potentiated exocrine secretion of amylase evoked by secretin in combination with cholecystokinin (CCK), in part by increasing Na^+, K^+-ATPase activity.

1. *Proteolytic Enzymes*

The major proteolytic enzymes, which are secreted as inactive proenzymes, are trypsinogen, chymotrypsinogen, and procarboxypeptidase. A concept for the activation of these pancreatic proenzymes is that enterokinase in the intestine activates trypsinogen to trypsin and the other proenzymes are then activated by trypsin; premature activation of trypsin is thought to be prevented by pancreatic-specific trypsin inhibitor. Borgström *et al.* (1991) studied the activation of proenzymes using a lecithin assay and synthetic substrates. It was observed that the first protein to be activated was procolipase followed by the activation of prophospholipase, followed by chymotrypsin and, lastly, the activation of trypsinogen. The addition of increasing amounts of pancreatic-specific trypsin inhibitor could delay the activation of the proenzymes and therefore it was believed the activation was mediated by trypsin and not enterokinase.

2. *Lipases*

Pancreatic lipase is secreted in an active form, but its activity is enhanced by bile salts. Bile salts enhance the efficiency of lipolysis by increasing the surface area of oil–water interfaces at which water-soluble lipase is effective. Colipase, a small protein synthesized in the pancreas, allows pancreatic lipase to function in spite of micellar concentrations of conjugated bile salts. Bile salts by themselves hinder lipase absorption onto triglycerides by covering the whole water–substrate interface. Colipase tends to prevent this and acts as an anchor for lipase adsorption, thus allowing lipase to hydrolyze substrate (Vandermeers-Piret *et al.*, 1977). Calcium may also be required for activation of lipase (Janowitz and Banks, 1976).

Pancreatic lipase exhibits optimal activity under alkaline conditions and hydrolyzes triglycerides to fatty acids and glycerol, but mono- and diglycerides are also end products. Pancreatic lipase has greater activity against short-chain than long-chain triglycerides (Cohen *et al.*, 1971). Phospholipases A and B also are present in pancreatic juice. Phospholipase A splits off a fatty acid from lecithin or cephalin to form lysolecithin and lysocephalin. Phospholipase B is capable of splitting off a fatty acid and lysolecithin to form glycerophosphorylcholine.

Carriere *et al.* (1993) demonstrated that, in man, gastric and pancreatic lipases act as complementary enzymes in the digestion of triglyceride although pancreatic lipase was the main enzyme responsible for lipid digestion. These investigators confirmed that gastric lipolysis resulted only from gastric lipase activity, and that bile salts, at physiological concentrations, had no effect on gastric lipolysis. Also they determined that the majority of gastric lipase secreted in the stomach remained active in the duodenum. This may partially explain why human beings with complete absence of pancreatic lipase continue to absorb a significant amount of their ingested dietary fat (Muller *et al.*, 1975).

Studies of survival of pancreatic enzymes during small intestinal transit demonstrated that the most dramatic loss of enzymatic activity occurred for lipase. In the midjejunum of man, only about 10% of duodenal lipase activity was present, whereas in the ileum, only about 1% of residual lipase activity was measured. About 74% of amylase activity survived transit from duodenum to terminal ileum and about 20% of the trypsin activity reached the ileum (Layer *et al.*, 1986). Proteases are responsible for the loss of lipase activity in duodenal juice with pancreatic lipase being more inactivated by chymotrypsin than by trypsin (Thiruvengadam and Di Magno, 1988). Inactivation of proteases, therefore, improves the intraluminal activity of lipase (Layer and Gröger, 1993). Lipase is also the most susceptible pancreatic enzyme to acid denaturation. At a pH of 4, lipase is almost completely inactivated, whereas amylase and trypsin are still active (Adler *et al.*, 1993). This explains why fat malabsorption develops earlier compared with protein and starch malabsorption (Layer and Gröger, 1993). The digestive products of fat in the duodenum regulate the rate of gastric emptying. Therefore, only with sufficient lipase activity in the duodenum would the rate of lipid hydrolysis create an adequate luminal concentration of fatty acids to slow gastric emptying (Malfertheiner and Domínguez-Muñoz, 1993).

3. *Amylase*

Pancreatic amylase is an α-amylase that catalyzes the hydrolysis of starch and glycogen to form maltoses and residual glucose. Amylase, as found in pancreatic or other body tissues and fluids, is a metalloenzyme with an absolute requirement for calcium ions (Janowitz and Banks, 1976). Optimal activity is obtained also only in the presence of a variety of inorganic anions, the most effective being chloride.

A considerable number of organs, other than pancreas, contain amylases, and differentiating these amylases from pancreatic amylase may present a problem when attempting to make a diagnosis of active pancreatitis. The maintenance of serum amylase levels following removal of the pancreas in some species suggests that nonpancreatic sources of amylase are important in the regulation of normal circulating amylase levels. Jacobs *et al.* (1982) observed four peaks of isoamylase activity in the sera and tissue of dogs. The majority of activity was present in peak 4 and the pancreas, duodenum, kidney, lung, spleen, testes, and uterus–ovaries contained peak 4 isoamylase. The pancreas contained peak 3 in addition to peak 4 isoamylase. An increase in peak 3 isoamylase was present in dogs with pancreatitis and the relative increase in peak 3 was greater than that seen for total amylase or the other isoamylases. In chickens and in mammals properties of amylase from the intestine and pancreas were found to differ. Chicken and mammalian pancreatic amylase were more dependent on chloride for activation than were intestinal amylases, and pancreatic amylase also functioned at a pH optimum greater than intestinal amylase (Osman, 1982).

C. Regulation of Pancreatic Secretion

The digestive enzymes in pancreatic juice enter the intestinal lumen at a rate that is regulated by a complex interplay of nervous and hormonal stimuli. A basal rate of pancreatic secretion exists in most animals, but with the ingestion of food a series of hormonal and neural mechanisms is set into action that activates and controls the pancreatic secretion of water, electrolytes, and digestive enzymes. The dominant hormones are secretin and CCK. Secretin is released from the upper small intestinal mucosa by the entry of H^+ into the duodenum and this stimulates the centroacinar and ductal cells of the pancreas to secrete the electrolyte-containing fluid. The HCO_3^- secreted by the pancreas neutralizes the HCl secreted by the stomach. Plasma secretin levels were measured in dogs by radioimmunoassay (Llanos *et al.*, 1977). When the pH of the stomach content of dogs decreased to less than 3, the plasma secretin level increased to 43% above basal values, and the pancreatic secretion of HCO_3^- was near maximum. The threshold for release of secretin and HCO_3^- was pH 4. Cholecystokinin is released from the duodenal mucosal cells in response to products of protein digestion, and this hormone causes some secretion of pancreatic enzymes and contraction of the gallbladder. Li and Owyang (1994) demonstrated that in rats endogenous CCK under physiological conditions acts via stimulation of a vagal afferent pathway and their studies support a neural physiological role for CCK on pancreatic enzyme secretion. Doses of exogenous CCK that produce physiological plasma CCK levels also act via stimulation of the vagal pathway. In contrast, doses of CCK that produce supraphysiological plasma CCK levels act on intrapancreatic neurons and pancreatic acini. Exocrine pancreatic secretion in conscious dogs was also stimulated by the intravenous administration of gastrin-releasing peptide (GRP) (Nustede *et al.*, 1993). CCK-related mechanisms are believed to be responsible for this stimulation since administration of a CCK receptor antagonist causes a marked reduction of GRP-induced exocrine pancreatic secretion.

Mechanisms of inhibitory control of pancreatic secretion appear little studied in comparison to stimulation of secretion. Exocrine pancreatic secretion in man is inhibited by intraduodenal infusion of either trypsin or reinfusion of pancreatic–biliary juice. Therefore, this protease-mediated negative feedback of pancreatic secretion appears to be controlled by duodenal mechanisms (Malfertheiner and Domínguez-Muñoz, 1993). Somatostatin may inhibit exocrine pancreatic secretion by inhibiting release of hormones such as CCK and secretin. Somatostatin may also alter the pancreatic response to a meal by inhibition of gastric secretion or gastric and gallbladder emptying (Heintges *et al.*, 1994).

III. PANCREATIC DISEASE

Pancreatic disease, in its various forms, is not uncommon in dogs and cats and may occur in other animals as well. The exocrine pancreas may be affected by acute or chronic disease processes that may lead to the digestive problems associated with pancreatic insufficiency.

A. Acute Pancreatitis

1. *Etiology*

Acute pancreatitis in any species can be a disease of variable intensity, ranging from mild edema to generalized pancreatic necrosis. The etiology of inflammatory pancreatic disease is poorly understood but some of the etiologic factors that may be responsible for acute pancreatic disease can be classified as follows.

a. Mechanical Factors

The reflux of bile from the bile duct into the pancreatic duct has been considered a possible means of pancreatitis initiation. Duodenal reflux of fatty acids following ingestion of a high-fat meal may also result in pancreatic inflammation (Attix *et al.*, 1981). Traumatic hemorrhagic pancreatitis has been reported in cats that have survived falls from buildings (Suter and Olsson, 1969). Pancreatolithiasis occurs more frequently in cattle than in any other animal, and Verine (1973) reported a frequency of 1/2700 in slaughtered cattle. The associated pancreatic lesions, however, were chronic in nature.

b. Infectious Disease

Sterile bile perfused into the main pancreatic duct of cats at basal pressure and at duct pressure in the upper physiologic range produced no inflammation. The perfusion of bile infected with *Escherichia coli* at basal pressure caused a severe acute edematous pancreatitis and at upper physiologic pressure acinar necrosis resulted. Infected bile was found to raise basal pancreatic ductal pressure by 30% (Arendt, 1993). High physiologic duct pressure may result in the conversion of mild inflammation into acinar necrosis.

c. Ischemia

When the isolated canine pancreas was subjected to reduced blood flow with low arterial pO_2, pancreatic ischemia was severe and significant injury occurred in the pancreas (Broe *et al.*, 1982). It has also been postulated that ischemia may play a role in the conversion of edematous pancreatitis to acute hemorrhagic pancreatitis.

d. Nutritional Factors

Goodhead (1971) found dogs on a poor plane of nutrition failed to develop as severe a degree of induced pancreatitis as well-nourished dogs. Orally induced zinc toxicosis in veal calves produced lesions that included necrosis of acinar tissue (Graham *et al.*, 1988).

e. Miscellaneous

Induced hypercalcemia in cats has a deleterious effect on the pancreas that results in acinar and ductal cell necrosis and eventually pancreatitis (Frick *et al.*, 1990). Following the administration of excessive amounts of calcium to a dog with hypocalcemia, pancreatic hemorrhage and necrosis also ensued (Neuman, 1975). Hormonal factors in mice (Rao *et al.*, 1982) and corticosteroids given at high levels to dogs (Attix *et al.*, 1981) may also cause pancreatitis.

2. *Pathophysiology*

The pancreas is very susceptible to ischemia and ischemia of this organ has been considered an initiating or promoting factor of active pancreatitis (Yotsumoto *et al.*, 1993). The reduction of pancreatic blood flow may be caused by various vasoactive mediators such as histamine, myocardial depressant factor, prostaglandins, free radicals, platelet-activating factor, and kinin. Since the pancreas contains kallikrein and is able to release kinin, the kallikrein-kinin system was investigated in rabbits to determine the effect of this system on induction of hypotension in pancreatic blood vessels and the effect of bradykinin on the aggravation of acute pancreatitis (Yotsumoto *et al.*, 1993). In rabbits with induced acute edematous pancreatitis that were administered bradykinin, blood flow to the pancreas was diminished and serum amylase and lipase levels rose significantly. Pancreatic necrosis and inflammatory reaction were greater in these rabbits than in controls. The results suggest that bradykinin has an aggravating or promoting effect on edematous pancreatitis. Although activated trypsin may have initiated the edematous pancreatitis, bradykinin appeared to have a direct effect on pancreatic acinar cells, and was a key enzyme in the activation of phospholipase A_2 and the generation of prostaglandins together with platelet-activating factor (PAF) from biomembrane phospholipids (Fig. 14.1). Induced prostaglandins and

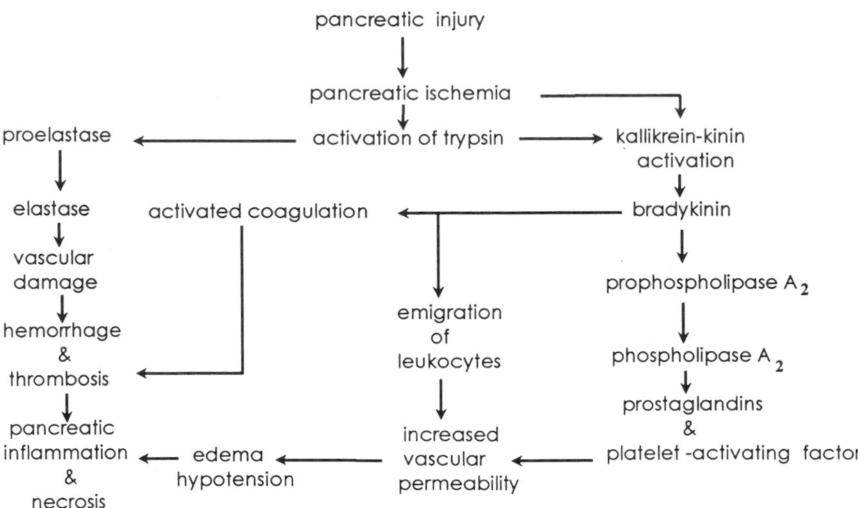

FIGURE 14.1 Etiologic and pathogenic factors in acute pancreatitis.

leukotriene activate leukocytes, cause vasodilation, and increase vascular permeability. The increased vascular permeability may enhance edema and stimulate emigration of leukocytes. It is possible that free radicals and PAF are generated in the activated leukocytes and injured vascular endothelial cells causing pancreatic injury. The activation of the blood coagulation cascade by bradykinin and vascular endothelial cell damage caused by free radicals or PAF may produce microthrombi in microcapillaries that can induce pancreatic necrosis (Yotsumoto *et al.*, 1993). The effects of PAF that may be released endogenously during the evolution of acute pancreatitis can be suppressed by a selective PAF antagonist. In rats with induced acute pancreatitis, treatment with PAF antagonist 30 minutes after induction of pancreatitis significantly inhibited the elevation of serum amylase levels and altered the progression of the disease (Formela *et al.*, 1993). The activation of elastase, probably by trypsin, may also cause vascular damage associated with hemorrhage and thrombosis (Fig. 14.1).

A point in understanding the pathogenesis of acute pancreatitis is determining where the initiating event occurs. It has been believed that pancreatic juice from small pancreatic ducts escapes into the interstitial space of the pancreas and with activation of digestive enzymes, the inflammatory process begins in the interstitial space. Steer and Meldolesi (1987) suggest the triggering action of acute pancreatitis occurs not within the interstitial space of the pancreas but within acinar cells and results from deranged transport or secretion of enzymes. These investigators have observed, in the acinar cells of both mice with pancreatitis induced by a choline-deficit diet and of animals with pancreatitis caused by excessive stimulation of pancreatic secretion, large vacuoles containing a mixture of digestive enzymes and lysosomal hydrolases. Lysosomal hydrolases are capable of activating trypsinogen and theoretically the mixture of digestive enzymes and hydrolases, and the fragility of the large vacuoles containing them were considered responsible for the release of activated enzymes into the acinar cell cytoplasm. Thus, pancreatic autodigestion may begin within the acinar cells themselves.

Studies of the pathophysiology of acute pancreatitis using a rat model have shown that the synthesis of some pancreatic enzymes was repressed during the acute phase, which might explain why the serum levels of these enzymes decrease after their initial burst (Iovanna *et al.*, 1991). The decrease in the synthesis of the trypsinogen I, chymotrypsinogen B, procarboxypeptidase A, proelastase I, and amylase was believed to be due to repressed pancreatic gene expression encoding for these exocrine enzymes. The suppression of hydrolase synthesis might be part of a defense mechanism to limit acute pancreatitis, because hydrolase enzyme activity may be involved in causing pancreatitis. In contrast to the decrease in synthesis of hydrolase enzymes in acute pancreatitis, the nonenzymatic secretory proteins actin, lithostathine, and pancreatitis-associated protein (PAP) were overexpressed by the pancreas during the acute phase. Increases in actin and lithostathine expression may be associated with pancreatic regeneration, which follows pancreatitis (Iovanna *et al.*, 1991).

3. Laboratory Diagnostic Aids

In addition to the hemogram, tests measuring the amount of pancreatic enzyme activity in the blood or other body fluids, and other laboratory procedures, can be useful in establishing a diagnosis of acute pancreatitis. Some of these laboratory procedures, both well established and experimental, follow.

a. Serum Amylase

i. Source Experimentally induced pancreatitis in dogs produces hyperamylasemia that peaks within 12–48 hours postinduction and subsides to normal within 8–14 days (Brobst *et al.*, 1970). Hyperamylasemia may be one indication of acinar cell damage and pancreatic duct obstruction.

Serum amylase concentrations in dogs may, however, also increase secondary to extrapancreatic disorders, which may have clinical signs similar to those of pancreatitis. Several tissues of dogs, including the intestine, kidney, and uterus, have been found to have amylase activity (Stickle *et al.*, 1980). Amylase activity greater than serum values was observed in tissue extracts from the pancreas, duodenum, ileum, ovary, and testicles.

Experimentally induced acute pancreatitis in the cat, unlike the dog, is not associated with hyperamylasemia. Kitchell *et al.* (1986) have shown that in induced acute pancreatitis in cats the serum amylase level decreased during most of the experimental period. This hypoamylasemia may be due to decreased release of amylase from acinar cells, increased metabolism, or the presence of circulatory amylase inhibitory substances. Naturally occurring pancreatitis in four cats, likewise, was not associated with hyperamylasemia (Simpson *et al.*, 1994). Cats with pancreatic lesions induced by experimental hypercalcemia also had normal serum amylase activity but had a marked increase in urinary amylase (Frick *et al.*, 1990). Therefore, urine amylase activity may reflect pancreatitis in this species.

Both acute and chronic pancreatitis have been described in horses (McClure, 1987). In documented cases of pancreatitis in horses, both serum amylase and lipase activities have been elevated. The activities of these enzymes may be elevated in other gastrointestinal diseases but will generally be less than in pancreatitis.

ii. Significance It is believed that increases in serum amylase activity in the dog are generally indicative of acute pancreatic cell damage and ductal obstruction. However, because increased serum amylase activity is not specific for acute pancreatitis, some investigators have reasoned that serum amylase activity should be increased by three- to fourfold greater than the reference value to be diagnostic for pancreatitis in dogs (Strombeck *et al.*, 1981). It was concluded that serum amylase activity, without knowledge of serum lipase activity, was of little value in diagnosing pancreatitis. High serum amylase activity sometimes accompanied problems that did not involve the pancreas, and low serum amylase activity did not always rule out pancreatitis.

The evaluation of serum isoamylases, as done by electrophoresis, may provide a means of specifically identifying the acutely diseased pancreas. Simpson *et al.* (1991), however, determined that in dogs pancreatectomy did not significantly alter the activities of serum isoamylase peaks 1, 2, and 3 (the purported pancreatic-specific isoamylase) or of lipase and amylase. These findings indicated that the pancreas is not the sole source of circulating amylase, isoamylase, and lipase activities.

Serum amylase activity should also be interpreted in terms of renal function. Dogs with spontaneously occurring primary renal failure were found to have mean serum amylase activity 2.5-fold greater than normal dogs (Polzin *et al.*, 1983). When azotemia is present in dogs, additional laboratory data are required to determine whether the hyperamylasemia results from acute pancreatitis, primary renal failure, or both.

For dogs with primary renal failure, a satisfactory explanation for the hyperamylasemia is not apparent. Humans, guineas pigs, and baboons with acute pancreatitis have an increased renal clearance of amylase, relative to creatinine, and this increased clearance, in these species, has been considered useful in the diagnosis of acute pancreatitis. The dog, however, excretes little α-amylase in the urine (Brobst *et al.*, 1970). Jacobs (1989) observed that canine serum amylase, because of its polymerization with various serum proteins, has a large and heterogenous molecular size that does not permit it to be readily filtered by glomeruli. A linear relationship between urinary amylase activity and urinary protein loss supported the conclusion that urinary amylase activity is dependent on glomerular lesions. In a study of dogs with proteinuria Corazza *et al.* (1994) observed that urinary amylase activity was greater in dogs with renal insufficiency. Total serum amylase activity and macroamylasemia were also higher in dogs with renal insufficiency. Their data indicate that the mechanism responsible for high serum amylase activities in dogs with renal failure may be due to the macroamylasemia. In this study, high circulating immunocomplex concentrations in 60% of the dogs suggested a cause-and-effect relationship for the formation of macroamylasemia. The reason all dogs with proteinuria did not develop macroamylasemia was not investigated in this study.

iii. Methodology Amylase is a rather stable protein at 5°C. The enzyme requires calcium ion for activity, and thus plasma from blood samples with calcium-chelating agents present is not acceptable for amylase determination (Hardy and Stevens, 1975).

The principle of serum amylase determination is the enzymatic hydrolysis of starch into maltose and limit dextrins. The rate of this hydrolysis is measured either by the rate of disappearance of starch (amyloclastic) or by the rate of appearance of reducing sugars (saccharogenic), that is, glucose and maltose, in the incubation mixture. The saccharogenic method for determining canine serum or plasma amylase activity gave results 70% higher than those of amyloclastic methods (Rapp, 1962; O'Donnell and McGeeney, 1975). The excess of saccharogenic activity was shown to correlate with glucoamylase activity and glucoamylase is a different molecular entity than α-amylase (O'Donnell and McGeeney, 1975). Therefore, serum amylase activity in the dog should be determined by amyloclastic methods specific for α-amylase. Glucoamylase activity is also present in the serum and plasma of cats and can interfere with the measurement of amylase activity (Ouedraogo *et al.,* 1992). As with the dog, methods specific for α-amylase such as the iodine method, starch dye, or "blocked" chromogenic substrate were recommended.

Turbidimetric and nephelometric methods are also employed to determine serum amylase activity. When a starch solution is hydrolyzed by amylase, the molecular size of the polysaccharides decreases owing to fragmentation by amylase action. This results in diminished turbidity and light scatter of the original solution. The decrease in turbidity can be measured absorptiometrically, and the decrease in light scatter can be measured nephelometrically and related to amylase activity. Turbidimetric and nephelometric methods can be simple, but the major disadvantages of these procedures are the lack of proper standards and poor substrate stability (Wu and Kao, 1984).

b. Serum Lipase

i. Source and Significance The pancreas is the primary source of serum lipase. Only pancreatic lipase is effective in liberating C_{14}, C_{16}, and C_{18} fatty acids from their triglycerides at appreciable rates; lipolytic enzymes from the duodenum and liver are much less effective (Lott *et al.,* 1986). Some investigators are of the opinion that in acute pancreatitis serum lipase concentration becomes elevated for a longer period of time than amylase. Brobst *et al.* (1970), however, determined in dogs with experimentally induced pancreatitis that serum amylase and lipase activities tend to parallel each other. Kasahara *et al.* (1975) also observed that serum activities of lipase, amylase, and elastase in dogs with experimentally induced pancreatitis parallel each other. The activities of these enzymes were higher in the pancreatic vein than in the femoral vein, indicating the enzymes did originate in the pancreas.

Hyperlipasemia, like hyperamylasemia, may, however, be associated with conditions other than acute pancreatitis. Spontaneously occurring and surgically induced renal failure in dogs caused serum lipase activity to increase three- to fivefold (Polzin *et al.,* 1983). Strombeck *et al.* (1981) also observed that serum lipase activity increased not only with pancreatitis in dogs but also with renal and hepatic diseases. On the other hand, the results of their study indicted that serum lipase activity was useful to identify pancreatitis, and low serum lipase activity almost always ruled out the possibility of pancreatitis. In their evaluation of the relationships of serum amylase and lipase activity in dogs with pancreatitis, they concluded little change in mean amylase concentration occurred until serum lipase values exceeded a set point (800 U/liter). Bellah and Bell (1989) demonstrated that exploratory laparotomy in dogs, with examination of all viscera but without any surgical procedure being done, caused serum lipase activity to rise threefold or greater although serum amylase activity did not increase. The increased lipase activity decreased to normal range by 48 hours following the laparotomy. The increase in lipase activity may have resulted from manipulation of the pancreas, stomach, liver, or intestines, however, gross and histological examination of the pancreas did reveal evidence of trauma or pancreatic disease. Likewise, in dogs treated with dexamethasone, serum lipase activity increased significantly although serum amylase activity decreased (Parent, 1982). Serum lipase activity had increased by the eighth day of administration, for dogs treated with low and high doses of dexamethasone, although evidence of pancreatic disease was not observed. It was considered possible that the increase in serum lipase activity resulted from dexamethasone-induced release from tissues other than the pancreas.

Although cats with acute experimentally induced pancreatitis had decreased serum amylase activity, their serum lipase activity increased (Kitchell *et al.,* 1986). Increased serum lipase activity appeared to be the earliest and most consistent indicator of acute pancreatitis in cats. Naturally occurring acute necrotizing pancreatitis in four cats was accompanied by hyperlipasemia in only one animal and all cats had normal serum amylase activity (Simpson *et al.,* 1994). The cat with hyperlipasemia was azotemic and decreased renal clearance of lipase may have been responsible for the increased serum lipase activity. Pancreatic lesions in the four cats were detected by abdominal ultrasonography.

Two major forms of lipase have been identified in human plasma. Using gel-filtration techniques, a single peak of lipase activity was identified in plasma from healthy subjects, whereas a separate peak of activity was present in patients with pancreatitis (Arzoglou *et al.*, 1986). The two forms of lipase were distinguished primarily by their relative molecular masses and their avidity for human pancreatic lipase antibodies.

ii. Methodology Serum lipase determination has traditionally been performed by variations of the Cherry–Crandall titrimetric procedure. This procedure is based on the hydrolysis of a lipid substrate into its constituent fatty acids. The quantity of NaOH required to neutralize the fatty acids provides a measure of lipase activity. The Cherry–Crandall method is time consuming and therefore several modifications of the procedure have evolved (Brobst and Brester, 1967). An enzymatic, kinetic procedure for determining serum lipase activity in dogs has been found suitable for automated methods and it correlated well with the Cherry–Crandall procedure (Walter *et al.*, 1992). Using the kinetic method it was determined that diseases of dogs associated with greater than twofold elevation in serum lipase activity included pancreatitis, gastritis with liver disease, and oliguric renal failure with metabolic acidosis. The serum lipase activity remained within the reference range or increased less than twofold with decreases such as gastritis, cholestasis, and colitis.

In a comparative study of current serum lipase procedures (with a titrimetric procedure serving as the standard), it was determined that those procedures utilizing colipase were in better agreement with the standard (Lott *et al.*, 1986). Colipase also increased analytical sensitivity and specificity. Colipase is present in the plasma of patients with pancreatitis, but it is usually inadequate to activate pancreatic lipase fully (Lott *et al.*, 1986). Serum and peritoneal lipase activity can also be determined by turbidimetric methods that assay triolein degradation (Robert *et al.*, 1986).

c. Miscellaneous Procedures

i. Pancreatitis-Associated Protein Acute pancreatic disease in dogs and cats is a disease of variable severity ranging from mild edema to severe necrosis. The increased activities of serum amylase and lipase are considered to be crude indicators of early pancreatitis but these enzymes do not reflect the clinical severity of the disease. It has been determined in rats with induced pancreatitis that the synthesis of pancreatic enzymes was repressed in the acute phase but that a nonenzymatic secretory protein, pancreatitis-associated protein (PAP) was overexpressed by the pancreas (Iovanna *et al.*, 1991, 1994). In human patients with acute pancreatitis, PAP values were much greater in those with pancreatic necrosis than those with mild pancreatitis. Continuous elevation of serum PAP concentration informed the clinician that pancreatitis was still in progress and stabilization of the PAP value reflected the climax of the crisis. Sustained decrease in PAP concentration reflected an improvement in the patient's condition. Thus, serum PAP values may provide information on the development and severity of acute pancreatitis.

ii. Interleukin-6 Serum concentrations of interleukin-6 (IL-6) were greater in human patients with severe acute pancreatitis than with mild pancreatitis (Heath *et al.*, 1993). The investigators believed serum concentrations of interleukin-6 could be useful as an indicator of severity of acute pancreatitis. Interleukin-6 appears to be the principal mediator of the acute-phase protein response of which C-reactive protein is an important component. C-reactive protein is a nonspecific indicator of inflammation and its serum concentration, as that of IL-6 rises with acute pancreatitis in human patients. The serum concentration of IL-6, however, peaked earlier than that of C-reactive protein (Heath *et al.*, 1993), and therefore was considered the better indicator of acute pancreatitis.

iii. Pancreatic Polypeptide Serum pancreatic polypeptide (PP) is a peptide hormone produced by cells dispersed throughout both the exocrine and endocrine pancreas. The mechanisms of action for PP may involve modulation of digestive processes by decreasing pancreatic and biliary secretion. Serum immunoreactive PP concentrations in dogs with induced pancreatitis peaked between 4 and 12 hours after induction of pancreatitis and were normal at 96 hours after induction of pancreatitis (Murtaugh *et al.*, 1985). This short period in which immunoreactive PP is elevated following the onset of pancreatitis may limit its usefulness as a diagnostic procedure; however, it may serve as an indicator of peracute inflammation.

iv. Plasma and Ascitic Fluid Trypsin and Trypsinogen Sensitive and specific radioimmunoassay techniques have been developed to characterize the molecular forms of immunoreactive pancreatic trypsin in experimental acute pancreatitis in dogs. In dogs with more mild pancreatitis trypsinogen predominated in the plasma, while in the more severe cases of acute pancreatitis significant amounts of trypsin were present bound to the protease inhibitors α_2-macroglobulin and α_1-protease. Dogs dying of acute pancreatitis had greater amounts of trypsin-like immunoreactivity (TLI) in the ascitic fluid than in the plasma, although the plasma levels increased during the course of the disease. The survivor dogs had little activation of trypsinogen and lower concentrations of TLI in ascitic fluid and plasma (Geokas *et al.*, 1981). This work would suggest that the activation of trypsinogen resulting in inhibitor-bound forms of trypsin in ascitic fluid and plasma is important in the pathogenesis of acute pancreatitis.

The measurement of serum TLI in dogs may prove to be a specific test for acute pancreatitis. This procedure measures both serum trypsinogen and trypsin by radioimmunoassay. The immunoreactive material present in the serum of normal dogs is most likely trypsinogen and may result from leakage into the pancreatic lymphatic and venous vessels. Inflammatory or obstructive processes in residual pancreatic tissue could result in elevated serum TLI activity. Following the ligation of pancreatic ducts in dogs (Simpson *et al.*, 1989a), plasma TLI concentration was observed to increase within 24 hours and tended to peak before and to decrease more rapidly than activities of lipase and amylase. Because TLI in dogs is specifically pancreatic in origin, the assay for plasma or serum TLI concentration may be able to provide an earlier indication of acute pancreatitis than serum amylase or lipase activity. Plasma lipase activity remained elevated for longer periods of time than did TLI and thus a normal TLI concentration would not rule out pancreatitis. Increased serum TLI concentration, like increased lipase and amylase activities, may possibly also result from renal insufficiency or disease of other organs that appear to influence the half-lives of these enzymes (Simpson *et al.*, 1991). In preliminary studies using an assay specific for feline TLI, Medinger *et al.* (1993) determined that cats with acute pancreatitis also had serum TLI concentrations greater than reference range.

v. Trypsinogen-Activation Peptides (TAP) Intrapancreatic activation of trypsinogen is believed to occur either as a cause or a consequence of acute pancreatitis. In experimental acute pancreatitis in rats, extraintestinal trypsinogen activation causes a direct release of TAP into the circulation. The concentration of TAP in the plasma and urine of the rats was measured by radioimmunoassay and provided an accurate and early prediction of mortality in rats with pancreatitis ranging in severity from minimal to lethal (Schmidt *et al.*, 1992). The assay for TAP has some advantages over the TLI assay. The TAP assay has no cross-reactivity with trypsin or trypsinogen and reflects the amount of extraintestinal trypsinogen activation. Unlike TLI, TAP does not bind to α_1-protease inhibitor or α_2-macroglobulin and, thus, is entirely accessible for antibody binding.

vi. Hyperlipemia Pathologic hyperlipemia has been observed in some dogs with acute pancreatitis (Anderson and Straufuss, 1971). The lipemia is usually transient and may occur with exacerbations of acute pancreatitis. The most striking feature is an elevation of plasma triglyceride levels, which imparts a milky turbidity to the plasma. Kessler *et al.* (1962) experimentally induced pancreatitis in rabbits and found an elevation of plasma triglycerides with lipemia being present. Lipemic plasma from animals with pancreatitis inhibited the "lipemia-clearing factor," lipoprotein lipase, of normal rabbits. Correlation was close between the degree of inhibition of lipoprotein lipase and the level of plasma triglyceride. It was considered possible that inhibitors of lipoprotein lipase were released from necrotic pancreatic tissue.

B. Pancreatic Insufficiency

1. Etiology

Exocrine pancreatic insufficiency in the dog is a disorder that develops because of inadequate pancreatic digestive enzymes. The disorder is characterized by weight loss and polyphagia, and steatorrhea may be present at times. The main causes are recognized as pancreatic acinar atrophy and chronic pancreatitis. The causes of canine pancreatic atrophy are unknown. The abnormality is more common in large breeds of dogs and occurs generally in their first year of life (Anderson and Low, 1965). The acinar cells appear to undergo progressive loss of cytoplasm and zymogen granules.

Chronic pancreatis reflects repeated attacks of acute pancreatitis with progressive destruction of acinar tissue and replacement with fibrous connective tissue. The early clinical signs of chronic pancreatitis may resemble those of acute pancreatitis, with the necrotizing process smoldering asymptomatically. Chronic pancreatitis is commonly seen in the canine species. Duffell (1975) observed that chronic and subacute pancreatitis are common in the cat.

The oral administration of zinc compounds, which have been used to prevent sporidesmin poisoning in ruminants, may cause chronic pancreatitis. Sheep that were administered zinc oxide experimentally developed various degrees of pancreatic fibrosis with decreased output of pancreatic amylase and protein (Smith and Embling, 1984). Induced zinc toxicosis in veal calves produced lesions that included pancreatic atrophy, necrosis of acinar tissue, and multifocal fibrosis of pancreatic acini. Pancreatic lesions were most severe and frequent in acinar tissue compared to islet tissue. The calves developed severe diarrhea believed to be caused by EPI (Graham *et al.*, 1988). The importance of pancreatic exocrine secretion in ruminants is not well understood, but ewes and lambs deprived of pancreatic juice have been shown to have much less lipid content in intestinal lymph than did normal sheep (Gooden and Lascelles, 1973).

Chronic fibrosing pancreatitis also occurs in horses. An affected horse has a generalized, progressive dermatitis that is part of a multisystemic, chronic eosinophilic disease of unknown etiology (Nimmo Wilkie *et al.*, 1985). The pancreas of an affected horse contains eosinophilic granulomas and has loss of exocrine tissue.

2. *Pathophysiology*

With almost total destruction of the acinar tissue of the pancreas, EPI occurs (Anderson, 1972). Without trypsin, the feces contain undigested meat fiber, and there is nitrogen loss. In the absence of lipase, neutral fat is in the feces. Without pancreatic amylase, starch is lost in the feces; however, this is of less importance than the resulting disturbed protein and fat digestion. The reserve capacity of the pancreas is great, and, in humans, steatorrhea was not observed until lipase output was less than 10% of normal. Creatorrhea was observed only when trypsin output was less than 10% of normal (Di Magno *et al.*, 1973).

After total resection of the pancreas of dogs, the digestion and absorption of fat and protein is markedly disturbed, although this is not true of carbohydrates. Yoshizawa *et al.* (1976) demonstrated that in total resection of the pancreas of dogs, the absorption of olive oil was a severely impaired by the duodenal mucosa, but absorption of oleic acid and medium-chain triglycerides (MCT) was less affected. The hydrolysis of MCT by pancreatic lipase is more rapid than that of long-chain triglycerides. The degree of steatorrhea in pancreatic insufficiency thus may be reduced by feeding MCT as the source of fat.

Valdivieso (1972) demonstrated that rats with surgically induced pancreatic insufficiency absorb a substantial amount of MCT. By the use of isotope-labeled lipid, it was shown that MCTs were not hydrolyzed in the jejunum but diffused intact into the gut epithelial cells. In the cell, MCTs were partially hydrolyzed by intestinal lipase and then transferred to the portal blood partly as intact triglyceride and partly as fatty acid. In the cecum, MCTs were hydrolyzed in the lumen by a hydrolytic system independent of the pancreas. The liberated fatty acids diffused through the cecal mucosa to the portal circulation. Hypertrophy of the cecum occurred in the rats after induction of pancreatic insufficiency.

Pigeon (1982) observed that dogs with experimentally induced pancreatic insufficiency that were fed a dry dietary food, formulated for use in gastrointestinal disorders, excreted approximately 60% less fat and dry matter than when fed a popular dry dog food. Improved dry matter digestibility was found to be associated with improved fat digestion and absorption. Conditions that increased the fecal water content lead to reduced assimilation of fat.

3. *Laboratory Diagnostic Aids*

a. Serum Amylase and Lipase

Serum amylase and lipase levels may be elevated if acute exacerbations of pancreatitis occur. However, with complete destruction of acinar tissue and healing by fibrosis, serum enzyme elevations would not be expected, and enzyme activity may even be decreased.

b. Fecal Examination

Microscopic examination of feces to detect excess fecal fat has employed the use of Sudan stains to stain lipid droplets (Brobst, 1989). These screening tests, however, can be difficult to interpret and are not precise. Many dogs with established EPI may not always have excess lipid in their feces and a negative test does not rule out steatorrhea.

Fecal proteolytic activity estimates, using either the x-ray film gelatin digestion test or tubes of gelatin, have also been used as an indirect indicator of pancreatic enzyme activity. These gelatin digestion tests unfortunately are an unreliable assay of fecal proteolytic activity and give many false-negative and false-positive results. Fecal proteolytic activity assayed using an azocasein or casein substrate, however, is a reliable procedure and may be used in the diagnosis of EPI when the serum TLI test is not available. Reference values for fecal proteolytic activity in dogs have been established for a colorimetric procedure using an azocasein substrate and for a radial enzyme diffusion into agar gels containing casein substrate (Williams and Reed, 1990). These assays permit evaluation of pancreatic function provided several fecal samples are assayed. Proteolytic activity remaining after treatment of fecal samples with specific trypsin inhibitor indicated that trypsin accounted for 0–71% of the proteolytic activity. This may signify that trypsin is degraded as it passes through the intestinal tract or that there is a variation in the proportion of trypsin to total protease activity in pancreatic juice.

c. Fat Absorption

The fat absorption test is based on the principle that dietary fats must be hydrolyzed to fatty acids and glycerol prior to absorption. With reduced ability to assimilate fats, the defect may lie in either a deficiency of pancreatic lipase or an inability of the small bowel to absorb properly digested fats. In the test, a heparinized blood sample is drawn from the fasted animal and centrifuged. A meal rich in digestible oil, such as corn or peanut oil, is fed to the animal and a second heparinized blood sample is taken 3 hours later. The turbidities of the pre- and postfeeding sample are compared. Normally, the prefeeding sample has clear plasma, whereas that postfeeding has a turbid appearance (lipemia). If the plasma samples are equally clear, one can assume either that pancreatic exocrine function is deficient or that the intestine is incapable of proper absorption (Brobst and Funk, 1972). The two conditions can be differentiated by repeating the fat meal at a later date, this time supplemented with pancreatic extract as a source of lipase. A cloudy postfeeding sample indicates that absorption from the intestine was normal and the pancreas was deficient in secretion of lipase. The degree of lipemia can also be measured spectrophotometrically or the increase in plasma triglyceride concentration over basal levels can be measured following the lipid-rich meal. A problem with any oral tolerance test, such as this, may arise when the gastric emptying time is delayed. The simultaneous ingestion of carbohydrate and protein may also reduce alimentary lipemia.

Pilsworth and Lehner (1986) demonstrated that the use of cimetidine (an inhibitor of gastric acid secretions) was a useful adjunct to oral pancreatic extract therapy in dogs with pancreatic insufficiency that failed to respond to pancreatic enzyme replacement alone. It was estimated that 90% of administered pancreatic extract was destroyed in the stomach because of its high acidity. The use of cimetidine elevates the duodenal pH, and it is suggested that solubilization of fats is increased because of bile acid actions that are inhibited at acid pH.

d. Pancreatic Chymotrypsin Activity

The duodenal juice of normal dogs contains 487 ± 129 mU/ml of chymotrypsin, and dogs with EPI have been found to contain only 1.3 ± 0.2 mU/ml (Batt and Mann, 1981). The ability of the pancreas to secrete chymotrypsin, as a measure of exocrine pancreatic function, is based on the oral administration of a synthetic peptide, *N*-benzoyl-L-tyrosyl-*p*-aminobenzoic acid (BT-PABA), and the cleaving of BT-PABA by chymotrypsin to release *p*-aminobenzoic acid (PABA). After absorption, PABA appears in the blood and urine. In normal dogs, plasma PABA concentration peaks at 60–90 minutes and then gradually declines (Batt and Mann, 1981; Strombeck and Harrold, 1982). In normal dogs, the 6-hour PABA urinary excretion has been suggested to be any value greater than 46–50% of the administered dose (Strombeck and Harrold, 1982). The measurement of plasma PABA in a 60-minute test did not identify dogs with pancreatic insufficiency as consistently as the measurement of 6-hour urinary PABA excretion. Dogs with EPI and with 6-hour PABA urinary excretion of less than 20% required pancreatic enzyme replacement therapy for remission of diarrhea (Strombeck, 1978). The combined BT-PABA: xylose absorption test has been used for simultaneous evaluation of exocrine pancreatic function and intestinal absorptive function in dogs (Rogers *et al.*, 1980). Using similar techniques for the BT-PABA:xylose test in healthy cats, the mean peak concentration of PABA was approximately half that for dogs (Sherding *et al.*, 1982). It was concluded that the large variations among cats may limit the usefulness of this test in this species. It has been observed that an appreciable but subnormal increase in plasma PABA concentration may occur in some dogs with EPI (Williams and Batt, 1988). Overgrowth of duodenal microflora is common in dogs with EPI and these investigators were of the opinion that some bacteria, particularly *Clostridium* species, may produce peptidase with a chymotrypsin-like substrate specificity. In general, procedures such as the BT-PABA test, which involve gastric emptying and intestinal factors as well as pancreatic function, may lack some specificity in evaluating pancreatic function.

Burrows and Orfely (1989) observed that when BT-PABA was administered orally to dogs as a solution in propylene glycol, the resulting plasma PABA concentration was higher than when an equal dose of BT-PABA suspended in water was administered. Therefore, published reports on the use of the BT-PABA test should state whether dissolved or suspended BT-PABA was administered. Using the solution form of BT-PABA and interpreting the results by comparison with normal absorption curves obtained after administration of the suspension form of BT-PABA could contribute to failure to diagnose EPI.

e. Serum Trypsin-Like Immunoreactivity

The determination of serum TLI concentration may be the most reliable procedure for the diagnosis of EPI in the dog. This assay can be done on a single 12-hour fasting sample. The reference range for normal dogs is 5.2–34 μg/liter and for dogs with EPI is <2.5 μg/liter (Williams and Batt, 1988). Pancreatomy in the dog was followed by a decrease in the TLI concentration to <2.5 μg/liter (Simpson *et al.*, 1991). Clinical signs of EPI are usually not observed until serum TLI concentration is <3.0 μg/liter (Williams, 1988). Dogs fed diets with serum protein content ranging from 6.8 to 39% had a serum TLI reference range of 5.7–20.1 μg/liter

(Carro and Williams, 1989). For every 1% increase in dietary protein content, the calculated serum TLI concentration increased 0.16 μg/liter. Thus, feeding of a very low protein diet may lead to decreased synthesis of pancreatic proteases. Observation of a family of dogs in which EPI had been reported suggested that the serum TLI test could reveal the presence of EPI in young puppies before clinical signs were seen and when fecal proteolytic activity was still normal (Boari *et al.*, 1994). Preliminary investigation of the serum TLI assay in cats indicated the procedure would likely be of value in the diagnosis of EPI (Medinger *et al.*, 1993).

References

Adler, G., Mundlos, S., Kühnelt, P., Dreyer, E. (1993). *Digestion* **54,** suppl. 2, 3–9.

Anderson, N. V. (1972). *Vet. Clin. North Am.* **2,** 79–97.

Anderson, N. V., and Low, D. G. (1965). *Anim. Hosp.* **1,** 101–109.

Anderson, N. V., and Straufuss, A. C. (1971). *J. Am. Vet. Med. Assoc.* **159,** 885–891.

Arendt, T. (1993). *Dig. Dis. Sci.* **38,** 39–44.

Arzoglou, P. L., Lessinger, J. M., and Ferard, G. (1986). *Clin. Chem.* **32,** 50–52.

Attix, E., Strombeck, D. R., Wheeldon, E. B., and Stern, J. S. (1981). *Am. J. Vet. Res.* **42,** 1668–1674.

Batt, R. M., and Horadagoda, N. U. (1989). *Am. J. Physiol.* **257,** G344–349.

Batt, R. M., and Mann, L. C. (1981). *Vet. Rec.* **108,** 303–307.

Batt, R. M., and Morgan, J. O. (1982). *Rec. Vet. Sci.* **32,** 17–22.

Bellah, J. R., and Bell, G. (1989). *Am. J. Vet Res.* **50,** 1638–1641.

Boari, A., Williams, D. A., and Famigli-Bergamini, P. (1994). *J. Small Anim. Pract.* **35,** 247–250.

Borgström, A., Erlanson-Albertsson, C., and Borgström, B. (1991). *Digestion* **49,** A20, p. 10.

Brobst, D. F. (1989). *In* "Clinical Biochemistry of Domestic Animals" (J. J. Kaneko, ed.), 4th ed., pp. 398–416. Academic Press, San Diego.

Brobst, D., and Brester, J. E. (1967). *J. Am. Vet. Med. Assoc.* **150,** 767–771.

Brobst, D. F., and Funk, A. (1972). *J. Am. Vet. Med. Assoc.* **161,** 1412–1417.

Brobst, D., Ferguson, A. B., and Carter, J. M. (1970). *J. Am. Vet. Med. Assoc.* **157,** 1697–1702.

Broe, P. J., Zuidema, G. D., and Cameron, J. L. (1982). *Surgery* **91,** 377–382.

Burrows, C. F., and Orfely, C. (1989). *Res. Vet. Sci.* **46,** 125–126.

Carriere, F., Barrowman, J. A., Verger, R., and Laugier, R. (1993). *Gastroenterology* **105,** 876–888.

Carro, T., and Williams, D. A. (1989). *Am. J. Vet. Res.* **50,** 2105–2107.

Cohen, M., Morgan, R. G. H., and Hofmann, A. F. (1971). *Gastroenterology* **60,** 1–15.

Corazza, M., Tognetti, R., Guidi, G., and Buonaccorsi, A. (1994). *J. Am. Vet. Med. Assoc.* **205,** 438–440.

DiMagno, E. P., Go, V. L. W., and Summerskill, W. H. J. (1973). *New Engl. J. Med.* **288,** 813–815.

Duffell, S. J. (1975). *J. Small Anim. Pract.* **16,** 365–374.

Formela, L. J., Whittaker, M., and Kingsnorth, A. N. (1993). *Digestion* **54,** A38, p. 277.

Frick, T. W., Hailemariam, S., Heitz, P. U., Largiader, F., and Goodale, R. L. (1990). *Gastroenterology* **98,** 1675–1681.

Geokas, M. C., Largman, C., Durie, P. R., Brodrick, J. W., Ray, S. B., O'Rourke, M., and Vollmer, J. (1981). *Am. J. Pathol.* **105,** 31–39.

Gooden, J. M., and Lascelles, A. K. (1973). *Aust. J. Biol. Sci.* **26,** 625–633.

Goodhead, B. (1971). *Arch. Surg.* **103,** 724–727.

Graham, T. W., Holmberg, C. A., Keen, C. L., Thurmond, M. C., and Clegg, M. S. (1988). *Vet Pathol.* **25,** 484–491.

Gyr, K., Wolf, R. H., Imondi, A., and Felsenfeld, O. (1975). *Gastroenterology* **68,** 488–494.

Hardy, R. M., and Stevens, J. B. (1975). *In* "Textbook of Veterinary Internal Medicine" (S. J. Ettinger, ed.), Vol. 2, pp. 1247–1269. Philadelphia, Pennsylvania.

Heath, D. I., Cruickshank, A., Gudgeon, M., Jehanli, A., Shenkin, A., and Imrie, C. W. (1993). *Gut* **34,** 41–45.

Heintges, T., Lüthen, R., and Niederau, C. (1994). *Digestion* **55,** suppl. 1, 1–9.

Hootman, S. R., and Ondarza, J. (1993). *Digestion* **54,** 323–330.

Iovanna, J. L., Keim, V., Michel, R., and Dagorn, J. C. (1991). *Am. J. Physiol.* **261,** G485–489.

Iovanna, J. L., Keim, V., Nordback, I., Montalto, G., Camarena, J., Letoublon, C., Levy, P., Berthezene, P., and Dagorn, J. C. (1994). *Gastroenterology* **106,** 728–734.

Jacobs, R. M. (1989). *Vet. Pathol.* **26,** 349–350.

Jacobs, R. M., Hall, R. L., and Rogers, W. A. (1982). *Vet. Clin. Pathol.* **11,** 26–32.

Janowitz, H. D., and Banks, P. A. (1976). *In* "Disorders of the Gastrointestinal Tract, Disorders of the Liver, Nutritional Disorders" (J. M. Dietschy, ed.), pp. 193–195. Grune and Stratton, New York.

Kasahara, K., Carballo, J. R., Takada, Y., Appert, H. E., and Howard, J. M. (1975). *Surg. Gynecol. Obstet.* **141,** 347–351.

Kessler, J., Finkel, M., Ho, P., and Janowitz, H. D. (1962). *Proc. Soc. Exp. Biol. Med.* **110,** 24–26.

Kitchell, B. E., Strombeck, D. R., Cullen, J., and Harrold, D. (1986). *Am. J. Vet. Res.* **47,** 1170–1173.

Lavau, M., Bazin, R., and Herzog, J. (1974). *J. Nutr.* **104,** 1432–1437.

Layer, P., and Gröger, G. (1993). *Digestion* **54,** suppl. 2, 10–14.

Layer, P., Go, V. L. W., and DiMagno, E. P. (1986). *Am. J. Physiol.* **251,** G475–G480.

Li, Y., and Owyang, C. (1994). *Gastroenterology* **107,** 525–531.

Llanos, O. L., Konturek, S. J., Rayford, P. L. and Thompson, J. C. (1977). *Am. J. Physiol.* **233,** E41–E46.

Lott, J. A., Patel, S. T., Sawhney, A. K., Kazmierczak, S. C., and Love, J. E. (1986). *Clin. Chem.* **32,** 1290–1302.

Malaisse-Lagae, F., Ravazzola, M., Robberecht, P., Vandermeers, A., Malaisse, W. J., and Orci, L. (1975). *Science 190,* 795–797.

Malfertheiner, P., and Domínguez-Muñoz, J. E. (1993). *Digestion* **54,** suppl. 2, 15–20.

McClure, J. J. (1987). *In* "Current therapy in equine medicine-2" (N. E. Robinson, ed.) pp. 46–47. Saunders, Philadelphia, Pennsylvania.

Matsushita, K., Okabayashi, Y., Koide, M., Hasegawa, H., Otsuki, M., and Kasuga, M. (1994). *Gastroenterology* **106,** 200–206.

Medinger, T. L., Burchfield, T., and Williams, D. A. (1993). *Proc. 11th Am. College Vet. Intern. Med. Forum, Washington, D.C.,* A87, p. 946.

Muller, D. P. R., McCollum, J. P. K., Trompter, R. S., and Harries, J. T. (1975). *Gut* **16,** 838.

Murtaugh, R. J., Jacobs, R. M., Sherding, R. G., and O'Dorisio, T. M. (1985). *Am. J. Vet. Res.* **46,** 654–565.

Nakagawa, A., Stagner, J. I., and Samols, E. (1993). *Gastroenterology* **105,** 868–875.

Neuman, N. B. (1975). *J. Am. Vet. Med. Assoc.* **166,** 381–383.

Nimmo Wilkie, J. S., Yager, J. A., Nation, P. N., Clark, E. G., Townsend, H. G. G., and Baird, J. D. (1985). *Vet. Pathol,* **22,** 297–305.

Nustede, R., Schmidt, W. E., Lohmann, C., Köhler, H., Schlemminger, R., and Schafmayer, A. (1993). *Digestion* **54,** 79–83.
O'Donnell, M. D., and McGeeney (1975). *Comp. Biochem. Physiol.* **50B,** 269–274.
Osman, A. M. (1982). *Comp. Biochem. Physiol.* **73B,** 571–574.
Ouedraogo, G., Rico, A. G., and Braun, J. P. (1992).*Vet. Clin. Pathol.* **21,** 47–49.
Parent, J. (1982). *J. Am. Vet. Med. Assoc.* **180,** 743–746.
Pigeon, G. (1982). *J. Am. Vet. Med. Assoc.* **181,** 232–235.
Pilsworth, R. C., and Lehner, R. P. (1986).*Vet Rec.* **119,** 240.
Polzin, D. J., Osborne, C. A., Stevens, J. B., and Hayden, D. W. (1983). *Am. J. Vet. Res.* **44,** 404–410.
Raeder, M. (1992). *Gastroenterology* **103,** 1674–1684.
Rao, K. N., Eagon, P. K., Okamura, K., Van Thiel, D. H., Gavaler, J. S., Kelly, R. H., and Lombardi, B. (1982). *Am. J. Pathol.* **109,** 8–14.
Rapp, J. P. (1962). *Am. J. Vet. Res.* **23,** 343–350.
Robert, J. H., Meyer, P., and Rohner, A. (1986). *Ann. Surg.* **203,** 163–168.
Rogers, W. A., Stradley, R. P., Sherding, R. G., Powers, J., and Cole, C. R. (1980). *J. Am. Vet. Med. Assoc.* **177,** 1128–1131.
Schmidt, J., Fernandez-Del Castillo, C., Rattner, D. W., Lewandrowski, K., Compton, C. C., and Warshaw, A. L. (1992). *Gastroenterology* **103,** 1009–1016.
Sherding, R. G., Stradley, R. P., Rogers, W. A., and Johnson, S. E. (1982). *Am. J. Vet. Res.* **43,** 2272–2273.
Simpson, K. W., Batt, R. M., McLean, L., and Morton, D. B. (1989a). *Am. J. Vet. Res.* **50,** 629–632.
Simpson, K. W., Morton, D. B., and Batt, R. M. (1989b). *Am. J. Vet. Res.* **50,** 1233–1236.
Simpson, K. W., Batt, R. M., Jones, D., and Morton, D. B. (1990). *Am. J. Vet. Res.* **51,** 203–206.
Simpson, K. W., Simpson, J. W., Lake, S., Morton, D. B., and Batt, R. M. (1991). *Res. Vet. Sci.* **51,** 78–82.
Simpson, K. W., Shiroma, J. T., Biller, D. S., Wicks, J., Johnson, S. E., Dimski, D., and Chew, D. (1994). *J. Small Anim. Pract.* **35,** 93–99.
Smith, B. L., and Embling, P. P. (1984). *Res. Vet. Sci.* **37,** 58–62.
Steer, M. L., and Meldolesi, J. (1987). *Engl. J. Med.* **316,** 144–150.
Stickle, J. E., Carlton, W. W., and Boon, G. D. (1980). *Am. J. Vet. Res.* **41,** 506–509.
Strombeck, D. R. (1978). *J. Am. Vet. Med. Assoc.* **173,** 1319–1323.
Strombeck, D. R., and Harrold, D. (1982). *J. Am. Vet. Med. Assoc.* **180,** 419–421.
Strombeck, D. R., Farver, T., and Kaneko, J. J. (1981). *Am. J. Vet. Res.* **42,** 1966–1970.
Suter, P. F., and Olsson, S. (1969). *J. Am. Vet. Radiol. Soc.* **10,** 4–11.
Taylor, R. B. (1962). *Res. Vet. Sci.* **3,** 63–71.
Thiruvengadam, R., and DiMagno, E. P. (1988). *Am. J. Physiol.* **255,** G476–G481.
Valdivieso, V. (1972). *Am. J. Dig. Dis.* **17,** 129–137.
Vandermeers-Piret, M. C., Vandermeers, A., Wijns, W., Rathe, J., and Christophe, J. (1977). *Am. J. Physiol.* **232,** E131–E135.
Verine, H. J. (1973). *Comp. Pathol. Bull.* **5,** 3–4.
Walter, G. L., McGraw, P., and Tvedten, H. W. (1992). *Vet. Clin. Path.* **21,** 23–27.
Williams, D. (1988). *Proc. 6th Am. Coll. Vet. Intern. Med. Forum,* pp. 452–454.
Williams, D. A., and Batt, R. M. (1988). *J. Am. Vet. Med. Assoc.* **192,** 195–201.
Williams, D. A., and Reed, S. D. (1990). *Vet. Clin. Pathol.* **19,** 20–24.
Williams, D. A., Batt, R. M., and McLean, L. (1987). *J. Am. Vet. Med. Assoc.* **191,** 201–206.
Wu, W. T., and Kao, Y. S. (1984). *In* "Clinical Diagnosis and Management by Laboratory Methods" (J. B. Henry ed.), 17th ed., pp. 537–549. Saunders, Philadelphia, Pennsylvania.
Yoshizawa, Y., Mizumoto, R., Nakase, A., and Hono, I. (1976). Histochemical studies on fat absorption in the jejunal mucosa after total pancreatectomy in the dog. *Am. J. Surg.* **131,** 689–696.
Yotsumoto, F., Manabe, T., and Ohshio, G. (1993).*Digestion* **54,** 224–230.
Zimmerman, M. J., Dreiling, D. A., Rosenberg, I. R., and Janowitz, H. D. (1967). *Gastroenterology* **52,** 865–870.

CHAPTER

15

Gastrointestinal Function

WILLIAM E. HORNBUCKLE AND BUD C. TENNANT

I. INTRODUCTION

The digestive system is composed of the gastrointestinal (GI) tract or alimentary canal, the salivary glands, the liver, and the exocrine pancreas. The principal functions of the gastrointestinal tract are the digestion and absorption of ingested nutrients and the excretion of waste products of digestion. Most nutrients are ingested in a form that is too complex for absorption, is insoluble and therefore indigestible, or is incapable of being digested. Within the GI tract, many of these substances are solubilized and further degraded enzymatically to simple molecules whose form and small size permits their absorption across the mucosal epithelium. This chapter describes the normal biochemical processes of intestinal secretion, digestion, and absorption and, with these in perspective, the pathogenesis of the important gastrointestinal diseases of domestic animals and the biochemical basis for their diagnosis and treatment.

II. SALIVARY SECRETIONS

A. Mechanisms of Secretion

Saliva is produced by three major pairs of salivary glands and by small glands distributed throughout the buccal mucosa and submucosa. Two types of secretory cells are found in the acinar portions of the salivary glands: (1) the mucous cells, which contain droplets of mucus, and (2) the serous cells, that contain multiple secretory granules. In those species that produce salivary amylase, the secretory granules are the zymogen precursors of this enzyme. A third cell type is found lining the striated ducts. The striations along the basal borders of these cells are caused by vertical infoldings of the cell membrane, a characteristic of epithelial cells involved in rapid movement of water and electrolytes. The primary secretion of the acinar cells is modified by active transport processes of the ductal epithelium.

The distribution of the different types of secretory cells in the salivary glands varies among species. The parotid glands of most animals are serous glands that produce a secretion of low specific gravity and osmolality containing electrolytes and proteins, including certain hydrolytic enzymes. The mandibular (submaxillary) and sublingual glands are mixed salivary glands containing both mucous and serous types of cells and producing a more viscous secretion that contains large amounts of mucus (Dukes, 1955).

B. Composition of Saliva

1. *Mucus*

Mucus is an aqueous mixture of proteoglycans and glycoproteins. One of the most completely studied glycoproteins is mucin. Salivary mucins are *O*-glycosylated, consisting of peptides with many oligosaccharides linked covalently to the hydroxyamino acids serine or threonine. The carbohydrate portion of submaxillary mucin from sheep is a disaccharide of *N*-acetylneuraminic acid (sialic acid) and *N*-acetylgalactosamine (Carlson *et al.*, 1973). The enzymes that link protein with hexosamine have been purified from sheep (Carlson *et al.*, 1973) and swine (Schachter *et al.*, 1971) mandibular glands.

The physiological functions of mucin are related to its high viscosity. *N*-Acetylneuraminic acid is the component responsible for the formation of viscous aqueous solutions and, at physiological pH, causes expansion and stiffening of the mucin molecule. The resistance of mucin to enzymatic breakdown is also due to the presence of disaccharide residues. Removal of terminal *N*-acetylneuraminic acid residues by action of neuraminidase significantly increases the susceptibility of peptide bonds to trypsin.

2. *Electrolytes*

The principal inorganic constituents of saliva are sodium, potassium, chloride, and bicarbonate that, with the exception of bicarbonate, originate directly from the plasma. Rates of salivary flow vary depending on stimulation and there are wide variations in electrolyte concentration. Saliva is formed by a process that initially requires uptake of sodium and other electrolytes from the interstitium of the terminal structural unit of the salivary gland, the acinus or end piece. Water flows passively. This primary or precursor fluid has a sodium concentration similar to that of plasma, and the potassium concentration is similar to or slightly higher than that of plasma. As the primary fluid passes from the acinus along the duct system, the concentration of sodium, potassium, and other electrolytes is changed. In most species, there is net sodium absorption and potassium secretion. Wide variations in electrolyte composition may occur depending on flow rate (Young and Schneyer, 1981), the salivary gland of origin, and the species (Table 15.1).

3. *Amylase*

The saliva of rodents contains the α-amylase ptyalin, but this enzyme activity is absent in the saliva of dogs, cats, horses, cattle, and sheep (Dukes, 1955; Young and Schneyer, 1981). Salivary amylase splits the α-1,4-glucosidic bonds of various polysaccharides and is similar in major respects to pancreatic α-amylase, which is described in Section V.B. Salivary amylase initiates digestion of starch and glycogen in the mouth in those species that secrete the enzyme. The optimal pH for amylase activity is approximately 7, so this activity ceases when the enzyme mixes with acidic gastric contents.

4. *Lipase*

Lingual lipase is secreted by Von Ebner's gland of the tongue and is important in the digestive processes of the newborn (Cook *et al.*, 1994).

C. Functions of Saliva

Saliva continuously bathes the oral cavity, which serves to protect the surface epithelium. Ingested food is moistened and lubricated by saliva, thereby facilitat-

TABLE 15.1 Electrolyte Concentration of Mandibular Gland and Parotid Gland Saliva Observed during Maximum Rates of Secretion (mmol/liter)

	Mandibular gland			Parotid gland		
	Na^+	K^+	HCO_3^-	Na^+	K^+	HCO_3^-
Sheep	20	7	23	160–175	9–10	113–140
Dog	70–100	12–15	10–30	80–110	6–14	50
Cat	40–51	9–10	26	—	—	—
Rabbit	50–100	10–40	25	110–140	10	12–30

ing mastication and swallowing. The teeth also are protected from decay by saliva, which washes food particles from the surfaces of the teeth and, because of its buffering capacity, neutralizes the organic acids produced by bacteria normally present in the mouth. Saliva is necessary for vocalization and, in some species that groom themselves, promotes cooling as it evaporates (Cook *et al.*, 1994). Additionally, it may be a source of pheromones. Salivary glands contain large number of growth factors, vasoactive serine proteases, and regulatory peptides (Cook *et al.*, 1994). There is reason to believe that these glandular constituents affect a wide range of biologic functions not necessarily limited to the alimentary system.

Ruminants produce much greater quantities of saliva than do simple-stomached animals, and their saliva has a higher pH and bicarbonate-ion concentration. In ruminants, saliva serves several unique functions (Phillipson, 1977). It is required for maintenance of the fluid composition of the contents of the rumen. The great buffering capacity of ruminant saliva is necessary to neutralize the large amounts of organic acids that are end products of rumen fermentation.

The urea in saliva can be utilized by rumen bacteria for protein synthesis. Protein synthesized in the rumen is then used to meet dietary protein requirements. In this way, urea nitrogen can be "recycled" through the amino acid pool of the body and, in ruminants, need not be considered an end stage in protein catabolism. The ability to reutilize urea has also been demonstrated in the horse; this may be of particular benefit during periods of protein deficiency (Houpt and Houpt, 1971).

III. GASTRIC SECRETIONS

The stomach is divided into two main regions on the basis of secretory function. The oxyntic-gland area corresponds approximately to the body of the stomach in most species of domestic animals and also to the fundus in the dog and cat. The oxyntic glands contain (1) oxyntic or parietal cells that produce hydrochloric acid (HCl), (2) peptic (zymogenic, chief) cells that produce pepsinogen, and (3) mucous cells. The pyloric gland area contains mucus-producing pyloric glands whose secretion is slightly alkaline. This area also contains the G cells that produce the polypeptide hormone gastrin.

A. Composition of Gastric Secretion

1. Basal vs Stimulated Secretion

There are two components of gastric secretion. The basal component is secreted continuously by the surface epithelial cells and other mucus-producing cells. This component is neutral or slightly alkaline pH. The electrolyte composition is similar to that of an ultrafiltrate of plasma (Table 15.2). The basal secretion contains large amounts of mucus, which has a cytoprotective effect on the epithelium. The secretory component produced by the oxyntic-gland cells in response to stimulation contains free HCl and pepsinogen, the principal enzyme of gastric digestion.

The composition of gastric juice depends on the relative amounts of the basal and secretory components in the juice. These amounts, in turn, are functions of the flow rate of each. In the dog, gastric juice is produced in the resting state at a rate of approximately 5 ml/hour (Gray and Bucher, 1941). The composition is similar to that of the basal component, containing practically no peptic activity or HCl. When the flow of gastric juice is stimulated maximally, the dog may produce 80 ml or more per hour of a secretion containing large amounts of peptic activity and HCl. Na^+, the principal cation in the basal secretion, is replaced to a large extent by H^+ ion. The concentration of K^+ is similar in both basal and stimulated secretions and therefore remains relatively constant at the various rates of flow.

HCl and pepsinogen are secreted by separate mechanisms, but their production appears closely linked under physiological conditions. Stimulation of the vagus nerve or intravenous injection of gastrin increases pepsinogen and HCl levels together. Other stimuli may affect the two processes differently, for example, in the dog histamine infusion stimulates HCl production maximally but appears to inhibit pepsinogen secretion (Emas and Grossman, 1967).

TABLE 15.2 Composition of Parietal and Nonparietal Secretions of Canine Gastric Mucosa

Component	Parietal secretion[a] (mmol/liter)	Nonparietal secretion[a] (mmol/liter)	Nonparietal secretion[b] (mmol/liter)
Na^+	—	155.0	138.0
H^+	159.0	—	—
K^+	7.4	7.4	4.0
Ca^{2+}	—	3.7	5.0
Cl^-	166.0	133.0	117.0
pH	<1.0	7.54[c]	7.42

[a] Determined *in vivo* using dogs with gastric fistulas (Gray and Bucher, 1941).

[b] Determined *in vitro* with isolated gastric mucosa (Altamirano, 1963).

[c] Calculated from bicarbonate concentration assuming pCO_2 of 40 torr.

2. *Pepsin*

Pepsinogen is the zymogen, or inactive precursor, of pepsin, the principal proteolytic enzyme of gastric juice. Pepsinogen was first crystallized from the gastric mucosa of swine (Herriott, 1938), and several pepsinogens have now been separated. The porcine pepsinogen has a molecular weight of approximately 43 kDa and is composed of the pepsin molecule and several smaller peptides. One of these peptides has a molecular weight of 3.2 kDa and is an inhibitor of peptic activity. Activation of pepsin from pepsinogen occurs by selective cleavage of this small basic peptide from the parent pepsinogen (Neurath and Walsh, 1976). Autocatalytic conversion begins below pH 6.0. At pH 5.4, the inhibitor peptide dissociates from the parent molecule, and at pH 3.5–4.0 the inhibitor is completely digested by pepsin.

Pepsin has a very acidic isoelectric point, being stable in acidic solution below pH 6.0 but irreversibly denatured at pH 7.0 or above. In contrast, pepsinogen is stable in neutral or slightly alkaline solution. The optimal pH for peptic activity is generally between 1.6 and 2.5, but the effect of pH may vary with the substrate. Pepsin is capable of hydrolyzing peptide bonds of most proteins, mucin being one important exception. Pepsin splits bonds involving phenylalanine, tyrosine, and leucine most readily but can hydrolyze almost all other peptide bonds.

3. *Rennin*

Rennin is another proteolytic enzyme produced by the gastric mucosa and has some characteristics that are similar to those of pepsin. It has been separated from pepsin in preparations from the stomachs of newborn calves. Rennin splits a mucopeptide from casein to form paracasein, which then reacts with calcium ion to form an insoluble coagulum. The coagulated milk protein probably delays gastric emptying and increases the efficiency of protein digestion in young calves.

4. *HCl*

HCl is produced by the oxyntic cells. When the normal mucosa is stimulated, both Cl^- and H^+ are secreted together, but current evidence suggests that H^+ and Cl^- are secreted by separate, closely coupled mechanisms. Small amounts of Cl^- are secreted continuously by unstimulated oxyntic cells in the absence of H+ secretion, and this mechanism is responsible for the negative charge of the resting mucosal surface of the stomach relative to that of the serosa. For every H^+ secreted, an electron is removed that ultimately is accepted by oxygen to form OH^-, which is neutralized within the cell by H+ from H_2CO_3. The HCO_3^- then enters the venous blood by means of a Cl^-/HCO_3^- exchange and during HCl secretion , the pH of gastric venous blood frequently is greater than that of arterial blood (Davenport, 1966).

The membrane-bound enzyme responsible for transport of H^+ by the oxyntic cell is a K^+-stimulated ATPase (Sachs *et al.*, 1976; Wallmark *et al.*, 1980) that serves as H^+/K^+ exchange pump. At the time of oxyntic cell stimulation, the secretory membrane is altered to provide augmented K^+ and Cl^- conductances (Wolosin, 1985). KCl leaves the apical cell membrane passively, and net production of HCl results from the electroneutral exchange of K^+ for H^+ (Fig. 15.1).

B. Control of Gastric Secretion

1. *General*

A variety of stimuli can initiate gastric secretion. The sight or smell of food or the presence of food within the mouth causes gastric secretion by a reflex mechanism involving the vagus nerve. The presence of certain foods within the stomach or distension of the stomach alone can also initiate both intrinsic and vagal nerve reflexes which cause secretion of gastric fluid. In addition to neural reflexes, these stimuli also cause release of gastrin from the pyloric-gland area that enters the bloodstream, stimulating gastric secretion. The release of gastrin from G cells is inhibited by excess H^+; this negative-feedback mechanism is important in the control of HCl production.

2. *Gastrin*

Gastrin has been isolated in pure form from the antral mucosa of swine (Gregory *et al.*, 1964). When administered intravenously, the purified hormone causes the secretion of HCl and pepsin and stimulates gastrointestinal motility and pancreatic secretion. Two separate peptides have been obtained from porcine gastric mucosa and have been designated gastrin I and gastrin II. Gastrin is a heptadecapeptide amide, with a pyroglutamyl N-terminal residue and with the amide of phenylalanine as the C-terminal residue (Fig. 15.2). In the center of the molecule is a sequence of five glutamyl residues that give the molecule its acidic properties. Gastrin II differs from gastrin I only in the presence of a sulfate ester group linked to the single tyrosyl residue. The C-terminal tetrapeptide amide, Trp–Met–Asp–Phe–NH_2, is identical in all species so far studied (Gregory, 1967). The tetrapeptide has all of the activities of the natural hormone. It is not as potent

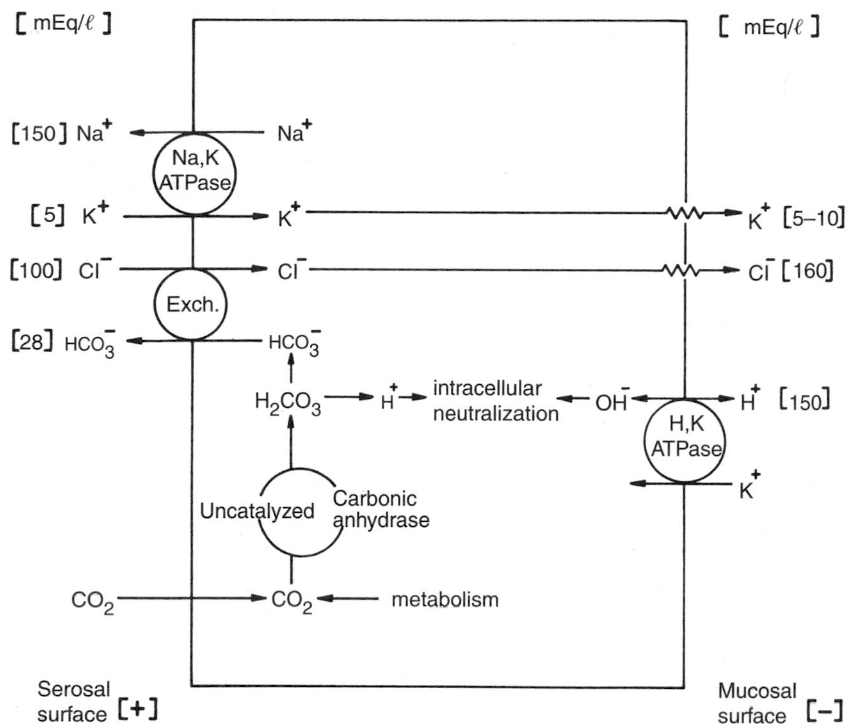

FIGURE 15.1 Movement of ions across the mucosal (apical) and serosal (basal) cell membranes of the parietal during HCl secretion.

as the parent molecule, but activity can be increased by lengthening the peptide chain.

Gastrin is released in response to vagal stimulation by distension of the pyloric antrum and by direct luminal contact with food, particularly partially hydrolyzed protein (Walsh and Grossman, 1975). Gastrin is the only hormone known to stimulate HCl secretion (Walsh and Grossman, 1975). The exact mechanism of action is not yet known.

3. *Histamine*

Histamine secreted locally within the mucosa has a major effect on the function of oxyntic cells (Soll and Grossman, 1978). Histamine has been recognized as a potent stimulant of HCl production for many years, but this effect was not inhibited by traditional antihistaminic drugs (H-1 antagonists) and, until the demonstration of H-2 receptors in the stomach (the atrium and uterus) by Black *et al.* (1972), the physiological role of histamine in HCl secretion was controversial. Specific H-2 antagonists (cimetidine) now have been shown to inhibit the secretory response not only to histamine, but to other secretory stimuli as well (Grossman and Konturek, 1974).

The complex of oxyntic cell receptors involved in the control of oxyntic cell function is shown in Fig. 15.3. When the H-2 receptor of the oxyntic cell is occupied by histamine, basal lateral adenylate cyclase is activated, resulting in increased cellular cyclic AMP (cAMP) and in a sustained secretory response. The secretagogue action of cAMP is mediated by the activity of cAMP-dependent protein kinases (Chew, 1985).

Cholinergic stimulation of the oxyntic cell involves type 1 muscarinic receptors and a calcium activation pathway. Calmodulin inhibitors such as trifluoroperazine inhibit H^+ secretion (Raphael and Machen, 1984). The Ca–calmodulin system may influence the rate of cAMP synthesis, and a more distal site of action has been suggested by the identification of a Ca-dependent protein kinase activity in a membrane fraction prepared from oxyntic cells that was rich in H^+–K^+-

FIGURE 15.2 Amino acid sequence of porcine gastrin I (Gregory, 1966). Gastrin II differs from gastrin I by the presence of a sulfate ester group on the single tyrosyl residue.

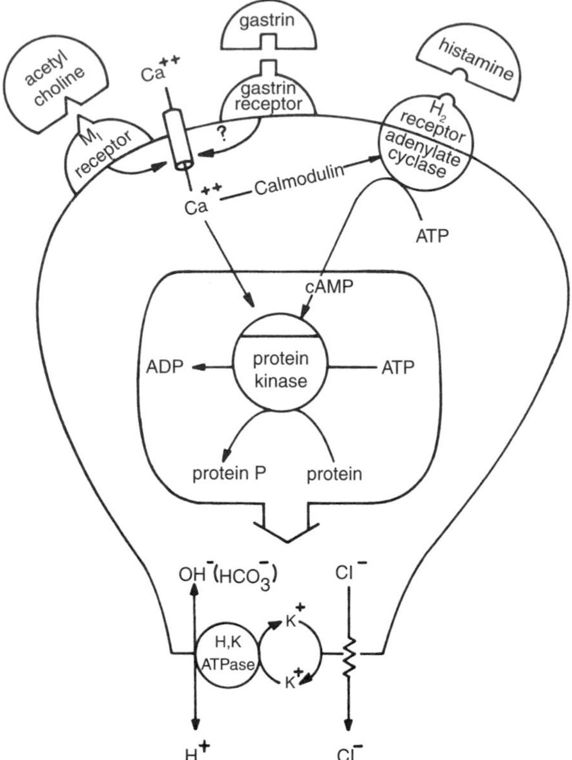

FIGURE 15.3 Pathways of secretagogue action on the parietal cell. Stimulation by gastrin and acetylcholine are mediated by entry of Ca^{2+} into the cell. Histamine activates adenylate cyclase with production of cAMP, the action of which is mediated by protein kinase.

ATPase (Schlatz *et al.*, 1981). A specific receptor for gastrin has been demonstrated on oxyntic cells, and a specific gastrin antagonist, proglutamide, inhibits H^+ production. Gastrin appears to act synergistically with histamine and acetylcholine (AcCh) to regulate H^+ production, but the actual mechanism of action of gastrin is unknown.

4. *Prostaglandins*

In addition to inhibiting HCl secretion, prostaglandins also act on a mucosal cell population that is distinct from oxyntic cells that secret cytoprotective substances (mucin, glycosaminoglycans). The ulcerogenic effects of inhibitors of prostaglandin synthesis (indomethacin, aspirin) apparently are the result of inhibition of the protective effect of endogenous prostaglandins.

Knowledge of the molecular aspects of receptor function of HCl secretion by oxyntic cells now provides the opportunity for specific pharmacological intervention for control and treatment of ulcerative diseases of the upper gastrointestinal tract that appear to be the

TABLE 15.3 Inhibitors of Oxyntic Cell Function

A. Inhibitors of H^+–K^+-ATPase: omeprazole, verapmil, vanadate
B. Inhibitors of carbonic anhydrase: acetazolamide
C. Inhibitors of cell activation or response
   1. Calcium channel antagonists: verapamil, lanthanum
   2. Prostaglandin E_2
D. Receptor antagonists
   1. H_2-receptor antagonists: cimetidine, ranitidine
   2. Gastrin antagonists: proglumide, benzotript
   3. Anticholinergic agents: atropine
E. Inhibitors of calmodulin: trifluoroperazine

result of HCl-induced mucosal injury. Gastric ulcers represent a serious problem in foals (Aclund *et al.*, 1983; Becht and Byars, 1986) and the H-2 receptor antagonist cimetidine is now used for treatment of foals with clinical signs of gastric ulceration. Ranitidine, also an inhibitor of the H-2 receptor and more potent and longer-acting than cimetidine, has been shown experimentally to inhibit HCl production in the horse (Campbell-Thompson and Merritt, 1987). Other potential therapeutic target sites have been demonstrated. Of particular importance are drugs that specifically inhibit the H^+–K^+-ATPase (Table 15.3).

IV. Bile

A. Composition of Bile

Bile is secreted continuously by the hepatocytes into the bile canaliculi and is transported through a system of ducts to the gallbladder, where it is modified, concentrated, and stored. During digestion, bile is discharged into the lumen of the duodenum, where it aids in emulsification, hydrolysis, and solubilization of dietary lipids. The digestive functions of bile are accomplished almost exclusively by the detergent action of its major components, the bile salts and phospholipids.

B. Properties of Bile

The carboxyl group of the bile acid (BA) is completely ionized at the pH of bile and is neutralized by Na^+, resulting in the formation of bile salts. These bile salts are effective detergents. They are amphipathic molecules that have both hydrophobic and hydrophilic regions. In low concentrations, bile salts form molecular or ideal solutions, but when their concentration increases above a certain critical level, they form polymolecular aggregates known as micelles. The concentration at which these molecules aggregate is called the critical micellar concentration (CMC).

Bile salt micelles are spherical and consist of a central nonpolar core and an external polar region. Fatty acids, monoglycerides, and other lipids are solubilized when they enter the central core of the micelle and are covered by the outside polar coat. Solubilization occurs only when the CMC is reached. For the bile salt–monoglyceride–fatty-acid–water system present during normal fat digestion, the CMC is approximately 2 m*M*, which normally is exceeded both in bile and in the contents of the upper small intestine (Hofmann, 1963). Phospholipids, principally lecithin, are also major components of bile. In the lumen of the small intestine, pancreatic phospholipase catalyzes the hydrolysis of lecithin, forming free fatty acid and lysolecithin. The latter compound also is a potent detergent that acts with the bile salts to disperse and solubilize lipids in the aqueous micellar phase of the intestinal contents.

C. Synthesis of Bile Acids

The primary bile acids are C-24 carboxylic acids synthesized by the liver from cholesterol. BA synthesis is the major end stage pathway for cholesterol metabolism (Danielsson, 1963). Cholic acid ($3\alpha,7\alpha,12\alpha$-trihydroxy-5β-cholanoic acid) (CA) and chenodeoxycholic acid ($3\alpha,7\alpha$-dihydroxy-5β-cholanoic acid) (CDCA) are the primary BA sunthesized by most species of domestic animals. In swine, CDCA is hydroxylated at the 6α position by the liver to yield hyocholic acid (HCA), which is a major primary BA in this species (Haslewood, 1964).

BAs are secreted as amino acid conjugates of either glycine or taurine. Taurine conjugates predominate in the dog, cat, and rat. In the rabbit, the conjugating enzyme system appears to be almost completely specific for glycine (Bremer, 1956). Both taurine and glycine conjugates are present in ruminants. In the newborn lamb, 90% of the bile acids are conjugated with taurine. As the lamb matures, glycine conjugates increase to reach one-third of the total BA in mature sheep (Peric-Golia and Socic, 1968).

Under normal conditions, only conjugated BAs are present in the bile and in the contents of the proximal small intestine. In the large intestine, the conjugated BAs are hydrolyzed rapidly by bacterial enzymes so that in the contents of the large intestine and in the feces, free or unconjugated BAs predominate. Several genera of intestinal bacteria, including *Clostridium, Enterococcus, Bacteroides,* and *Lactobacillus* (Midtvedt and Norman, 1967), are capable of splitting the amide bonds of conjugated BAs.

Intestinal bacteria also modify the basic structure of the BAs. One such reaction is the removal of the α-hydroxyl group at the 7 position of CA or CDCA. These bacterial reactions yield the secondary BAs, deoxycholic acid (DCA) and lithocholic acid (LCA) (Gustafsson *et al.,* 1957). LCA is relatively insoluble and is not reabsorbed to any great extent (Gustafsson and Norman, 1962). DCA is reabsorbed from the large intestine in significant quantities and is either rehydroxylated by the liver to CA and secreted (Lindstedt and Samuelsson, 1959) or secreted as the conjugated DCA. The extent to which bacteria transform the primary BAs depends on the nature of the diet, the composition of the intestinal microflora, and the influences of these and other factors on intestinal motility (Gustafsson *et al.,* 1966; Gustafsson and Norman, 1969a,b).

D. Enterohepatic Circulation of Bile Acids

The enterohepatic circulation begins as conjugated BAs enter the duodenum and mix with the intestinal contents, forming emulsions and micellar solutions. The BAs are not absorbed in significant amounts from the lumen of the proximal small intestine. Absorption occurs primarily in the ileum (Lack and Weiner, 1961, 1966; Weiner and Lack, 1962), where an active-transport process has been demonstrated (Dietschy *et al.,* 1966). The absorbed conjugated BAs pass unaltered into the portal circulation (Playoust and Isselbacher, 1964) and return to the liver, where the cycle begins again. This arrangement provides for optimal concentrations of BAs in the proximal small intestine, where fat digestion occurs, and then for efficient absorption after these functions have been accomplished. Absorption of unconjugated BAs from the large intestine accounts for 3–15% of the total enterohepatic circulation (Weiner and Lack, 1968).

In dogs, the total BA pool was estimated to be 1.1–1.2 g. The half-life of the bile acids in the pool ranged between 1.3 and 2.3 days and the rate of hepatic synthesis was 0.3–0.7 g/day (Wollenweber *et al.,* 1965). Because the daily requirement for bile acids greatly exceeds the normal synthetic rate, the repeated reutilization of the BA is facilitated by the enterohepatic circulation. Under steady-state conditions, the total BA pool passes through the enterohepatic circulation approximately 10 times each day (Hofmann, 1966).

The size of the BA pool depends upon the diet, the rate of hepatic synthesis, and the efficiency of the enterohepatic circulation. Surgical removal of the ileum in dogs interrupts the enterohepatic circulation, resulting in an increase in BA turnover and a reduction in the size of the BA pool (Playoust *et al.,* 1965). In diseases of the ileum, there may be defective BA reabsorption and a bile-salt deficiency. If the deficiency is severe, impaired utilization of dietary fat may occur, resulting in steatorrhea and impaired absorption of the fat-soluble vitamins.

V. SECRETIONS OF THE EXOCRINE PANCREAS

The exocrine pancreas is an acinous gland with a general structure similar to that of the salivary glands. The cytoplasm of the secretory cells contains numerous zymogen granules, which vary in size and number depending on the activity of the gland. These granules contain the precursors of the hydrolytic enzymes responsible for digestion of the major components of the diet. Cells of the terminal ducts appear to secrete the HCO_3^- responsible for neutralizing the HCl that enters the duodenum from the stomach.

A. Composition of Pancreatic Juice

1. *Electrolyte Composition*

The cation content of pancreatic secretion is similar to that of plasma, Na^+ being the predominant cation and the concentrations of K^+ and Ca^{2+} being much lower. A unique characteristic of pancreatic fluid is its high HCO_3^- concentration and alkaline pH. In the dog, the pH ranges from 7.4 to 8.3, depending on HCO_3^- content. The volume of pancreatic secretion is directly related to its HCO_3^- content, and the pH increases and Cl^- concentration decreases as the rate of flow is increased. The Na^+ and K^+ concentrations and osmolality appear to be independent of secretory rate (Fig. 15.4).

2. *α-Amylase*

The amylase produced by the pancreas catalyzes the specific hydrolysis of α-1,4-glucosidic bonds that are present in starch and glycogen (α-1,4-glycan-4-glycan hydrolase). Pancreatic amylase appears to be essentially identical to the amylase of saliva. It is a calcium-containing metalloenzyme. Removal of calcium by dialysis inactivates the enzyme and markedly reduces the stability of the apoenzyme. Pancreatic amylase has an optimal pH for activity of 6.7–7.2 and is activated by Cl^-.

After synthesis of pancreatic α-amylase in the ribosomes, the enzyme is transferred from the endoplasmic reticulum to cytoplasmic zymogen granules for storage. It is secreted in active form upon stimulation of the acinar cells. Newborn calves (Huber *et al.*, 1961) and pigs (Walker, 1959) secrete amylase at a significantly lower rate than do mature animals. The rate of synthesis is also influenced by diet. Animals fed a high-carbohydrate diet synthesize amylase at several times the rate of animals on a high-protein diet (Ben Abdeljlil and Desnuelle, 1964).

Unbranched α-1,4-glucosidic chains, such as those found in starch, are hydrolyzed in two steps. The first is rapid and results in formation of the maltose and maltotriose. The second step is slower and involves hydrolysis of maltotriose into glucose and maltose. Polysaccharides such as amylopectin and glycogen contain branched chains with both α-1,4- and α-1,6-glucosidic linkages. When α-amylase attacks these compounds, the principal products are maltose (α-1,4-glycosidic bond), isomaltose (α-1,6-glucosidic bond), and small amounts of glucose. Final hydrolysis of the maltose and isomaltose occurs at the surface of the mucosal cell, where the enzymes maltase and isomaltase are integral parts of the microvillous membrane.

3. *Proteolytic Enzymes*

The proteolytic enzymes of the pancreas are responsible for the major portion of protein hydrolysis that occurs within the lumen of the gastrointestinal tract. Two types of peptidases are secreted by the pancreas. Trypsin, chymotrypsin, and elastase are endopeptidases that attack peptide bonds along the polypeptide chain to produce smaller peptides. The exopeptidases attack either the carboxy-terminal or the amino-terminal peptide bonds, releasing single amino acids. The principal exopeptidases secreted by the pancreas are carboxypeptidases A and B. The endopeptidases and exopeptidases act in complementary fashion (Table 15.4), ultimately producing free amino acids or very small peptides. The free amino acids are absorbed directly and the small peptides are further hydrolyzed by the aminopeptidases of the intestinal mucosa.

The pancreatic peptidases are secreted as the inactive proenzymes (zymogens), trypsinogen, chymotrypsinogen, and the procarboxypeptidases A and B. Trypsinogen is converted to active trypsin in two ways. At alkaline pH, trypsinogen can be converted autocatalytically to trypsin. The activated enzyme is then capable of converting more zymogen to active enzyme. Trypsinogen also can be activated by the enzyme enterokinase, which is produced by the duodenal mucosa. The latter reaction is highly specific in that enterokinase will activate trypsinogen but not chymotrypsinogen. Chymotrypsinogen, proelastase, and the procarboxypeptidases A and B are converted to active enzymes by the action of trypsin.

The amino acid sequences and other structural characteristics of bovine trypsinogen and chymotrypsino-

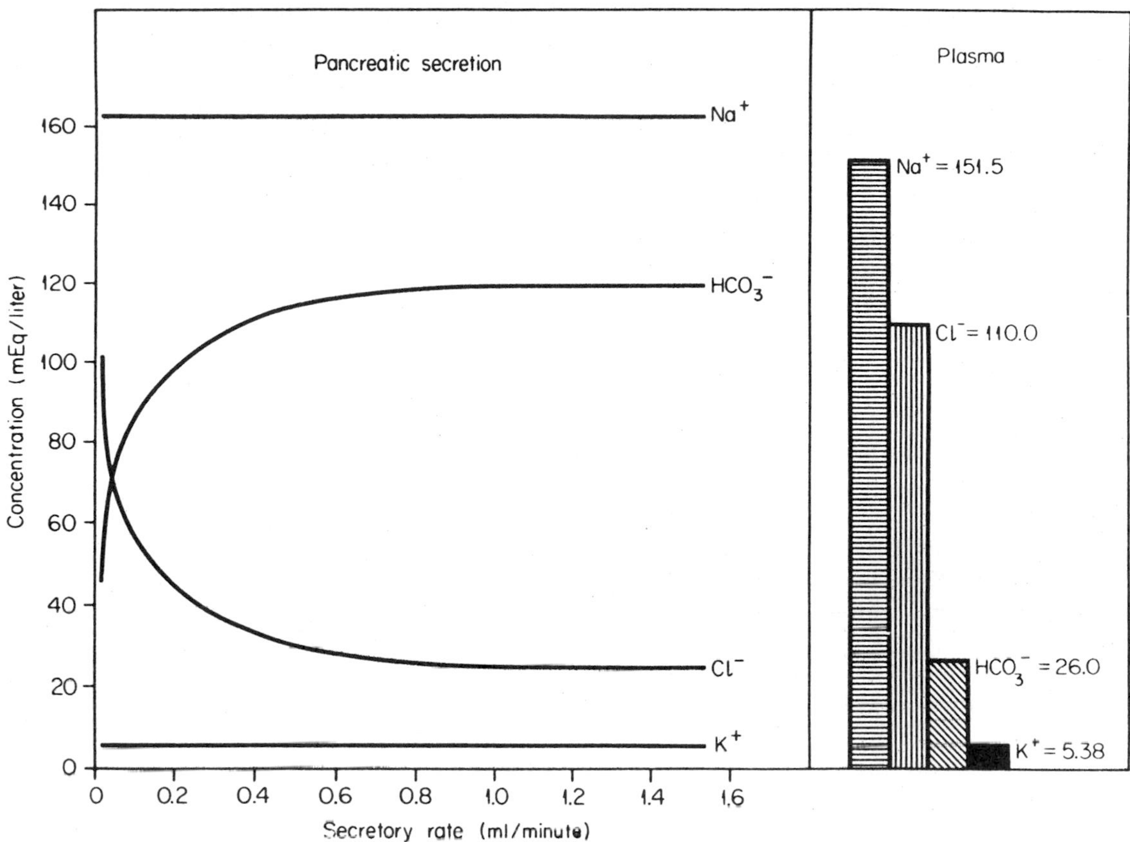

FIGURE 15.4 Influence of secretory rate on the electrolyte composition of canine pancreatic juice. From Bro-Rasmussen *et al.*, (1956).

gen have been determined (Hartley *et al.*, 1965; Hartley and Kauffman, 1966; Brown and Hartley, 1966). The polypeptide chain of trypsinogen contains 229 amino acid residues. Activation of trypsinogen occurs with hydrolysis of a single peptide bond located in the 6 position between lysine and isoleucine. As the C-terminal hexapeptide is released, enzyme activity appears along with a helical structure of the parent molecule. Chymotrypsinogen A is composed of 245 amino acid residues and has numerous structural similarities to trypsinogen. Activation of the chymotrypsinogen also occurs with cleavage of a single peptide bond.

TABLE 15.4 Relationship among the Activities of Pancreatic Endopeptidase and Exopeptidases

Enzyme	Type	Activity
Trypsin	Endopeptidase	Produces peptides with C-terminal basic amino acids
Carboxypeptidase B	Exopeptidase	Removes C-terminal basic amino acids
Chymotrypsin	Endopeptidase	Produces peptides with C-terminal aromatic amino acids
Elastase	Endopeptidase	Produces peptides with C-terminal nonpolar amino acids
Carboxypeptidase A	Exopeptidase	Removes C-terminal aromatic and nonpolar amino acids

4. *Lipase*

The pancreas produces several lipolytic enzymes with different substrate specificities. The most important of these from a nutritional viewpoint is the lipase responsible for hydrolysis of dietary triglyceride. This enzyme has the unique property of requiring an oil–water interface for activity so that only emulsions can be effectively attacked. The principal products of lipolysis are glycerol, monoglycerides, and fatty acids. The monoglycerides and fatty acids accumulate at the oil–water interface and can inhibit lipase activity. Transfer of these products from the interface to the aqueous phase is favored by HCO_3^- secreted by the pancreas and by the bile salts.

Two other carboxylic ester hydrolases have been characterized in pancreatic secretion. Both enzymes have an absolute requirement for bile salts, in contrast to glycerol ester hydrolase, which is actually inhibited by bile salts at pH 8. One of the enzymes requiring bile salts is a sterol ester hydrolase responsible for hydrolysis of cholesterol esters and the other enzyme hydrolyzes various water-soluble esters. The pancreas also secretes phospholipase A, which in the presence of bile converts lecithin to lysolecithin, an effective detergent that contributes to the emulsification of dietary fat.

B. Control of Pancreatic Secretions

1. *Hormonal Control*

Pancreatic secretion is controlled and coordinated by both neural and endocrine mechanisms. When ingesta or HCl enters the duodenum, the hormone secretin, which is produced by the duodenal mucosa, is released into the circulation. Secretin increases the volume, pH, and HCO_3^- concentration of the pancreatic secretion.

Secretin is a polypeptide hormone containing 27 amino acid residues, and all 27 amino acids are required to maintain the helical structure of the molecule and its activity (Bodanszky *et al.*, 1969). The C-terminal amide of secretin is a property shared with other polypeptide hormones such as gastrin and vasopressin that act on the flow of water in biological systems (Mutt and Jorpes, 1967). In addition to its effects on the pancreas, secretin also increases the rate of bile formation.

The secretin-stimulated pancreatic juice has a large volume and a high HCO_3^- concentration, but a low enzyme activity. Stimulation of the vagus nerve causes a significant rise in pancreatic-enzyme concentration. This type of response also is produced by cholecystokinin (pancreozymin), another polypeptide hormone produced by the duodenal mucosa that also causes contraction of the gallbladder. The C-terminal pentapeptide of cholecystokinin–pancreozymin is exactly the same as that of gastrin. This fascinating relationship suggests that gastrin and cholecystokinin–pancreozymin participate in some integrated yet poorly understood system of digestive control.

2. *Other Gastrointestinal Polypeptide Hormones*

In recent years, a large number of polypeptides have been isolated from the gastrointestinal mucosa and have been classified as gut hormones (Table 15.5). Some of these substances have not yet met all the rigid physiological requirements of true hormones. Some may have paracrine rather than endocrine activities, that is, they act on cells and tissues in the immediate vicinity of the cells of origin rather than being released into the vascular system.

TABLE 15.5 Gastrointestinal Peptide Hormones

Hormone	Source	Action
Gastrin	G cells of pyloric antrum	Gastric acid secretion
Secretin	S cells of duodenum and jejunum	Pancreatic fluid and HCO_3^- secretion, bile secretion
Cholecystokinin (pancreozymin)	Duodenal and jejunal mucosa; myenteric plexus	Pancreatic enzyme secretion, gall bladder contraction, and sphincter of Oddi relaxation
Somatostatin	D cells of pancreas, CNS, gastric and intestinal mucosa	Inhibits effect of gastrin on gastric secretion, inhibits pancreatic enzyme secretion, stimulates ileal water and NaCl
Enteroglucagon	L cells of small intestine, canine stomach	Control of intestinal cell growth
Gastric inhibitory polypeptide	Duodenal and jejunal mucosa	Inhibits gastric secretion and stimulates intestinal secretion
Motilin	Upper small-intestinal mucosa	Stimulates gastrointestinal motility

a. Motilin

Motilin is a polypeptide containing 22 amino acids that was originally isolated from porcine duodenal mucosa (Brown *et al.*, 1971). The amino acid composition and sequence have been described (Brown *et al.*, 1972, 1973). Immunoreactive motilin has been found in the enterochromaffin cells of the duodenum and jejunum of several species (Polak *et al.*, 1975) and, by means of radioimmunoassay, motilin has been identified in the plasma of dogs (Dryburgh and Brown, 1975). Motilin has been shown to stimulate pepsin output and motor activity of the stomach (Brown *et al.*, 1971) and to induce lower esophageal sphincter contractions (Jennewein *et al.*, 1975). Studies by Itoh *et al.* (1978) suggest that motilin plays an important role in initiating interdigestive gastrointestinal contractions.

b. Somatostatin

Somatostatin, which is named for its activity of inhibitory release of growth hormone from the pituitary gland, has been purified from ovine and bovine hypothalamus. The hypothalamic hormone is composed of 14 amino acids. Somatostatin also has been demonstrated in the stomach, pancreas, and intestinal mucosa in concentrations higher than in the brain (Pearse *et al.*, 1977). Somatostatin from porcine intestine has been isolated and sequenced and contains 28 amino acids and apparently is a prohormone (Pradayrol *et al.*, 1980). Somatostatin is a potent inhibitor of insulin and glucagon release. It also inhibits gastrin release and gastric acid secretion (Barros D'Sa *et al.*, 1975; Bloom *et al.*, 1974), apparently acting independently on parietal cells and on G cells. These and a variety of other physiological effects suggest that somatostatin has important gastrointestinal regulatory functions.

c. Enteroglucagon

Enteroglucagon is the hyperglycemic, glycogenolytic factor isolated from the intestinal mucosa. It occurs in two forms, one a 3.5-kDa form and another somewhat larger (Valverde *et al.*, 1970). Enteroglucagon differs from pancreatic glucagon biochemically, immunologically, and in its mode of release. The physiological function of enteroglucagon is not known, but its release from the mucosa following a meal and the associated increase in circulating blood levels have suggested a regulatory role on bowel function (Pearse *et al.*, 1977). Enteroglucagon also differs significantly from the glucagon produced by the A cells of the gastric mucosa of the dog (Sasaki *et al.*, 1975). Canine gastric glucagon is biologically and immunochemically identical to pancreatic glucagon. Gastric glucagon appears to be unique to the dog, similar activity not being observed in the stomach of the pig or the abomasum of cattle and sheep (Sutherland and de Duve, 1948).

VI. DIGESTION AND ABSORPTION

A. Water and Electrolyte Absorption

1. Mechanisms of Mucosal Transport

The microvillous membrane of the intestinal mucosa, because of its lipid composition, acts as a barrier to water and water-soluble substances. Water and polar solutes penetrate the mucosa in one of three ways. They may pass through aqueous pores or channels that connect the luminal surface of the cell with the apical cytoplasm, they may attach to membrane carriers that facilitate passage through the lipid phase of the mucosal cell membrane, or they may pass paracellularly through tight junctions (shunt pathway). Transport of water and water-soluble compounds is influenced by the permeability characteristics of the limiting membrane and by the nature of the driving forces that provide energy for transport. Passive movement occurs either by simple diffusion or as a result of pH or concentration gradients (activity), osmotic pressure, or electrical potential that may exist across the membrane. The movement of an ion in the direction of an electrochemical gradient is considered passive. Active transport is said to occur when a substance moves in a direction opposite to that of an established electrochemical gradient.

Most water-soluble compounds, such as monosaccharides and amino acids, cannot diffuse across the intestinal mucosal membrane at rates that are adequate to meet nutritional requirements. The transport of these nutrients requires membrane carriers that are integral parts of the membrane and whose binding is highly specific. Carrier-mediated transport systems can be saturated and competitively inhibited by related compounds.

Three types of carrier transport mechanisms are recognized (Curran and Schultz, 1968).

1. *Active transport,* as stated previously, involves movement of electrolytes against an electrochemical gradient. In the case of nonelectrolytes such as glucose, active transport is defined as movement against a concentration gradient. Active transport requires metabolic energy and is inhibited by various metabolic blocking agents or by low temperature.
2. *Facilitated diffusion* occurs when the passive movement of a substance is more rapid than can be accounted for by simple diffusion. Facilitated-diffusion systems may increase the rate of movement across the membrane by two or three orders of magnitude. The responsible carrier mechanism is similar to that involved in active transport in that it displays saturation kinetics, may be inhibited competitively, and is temperature dependent. However, transport does not occur against concentration or electrochemical gradients, and direct expenditure of energy is not required.
3. *Exchange diffusion* is a transfer mechanism similar to facilitated diffusion and was postulated originally by Ussing (1947) to explain the rapid transfer of radioactive Na^+ across epithelial cell membranes *in vitro.* The mechanism involves the exchange of one ion for another of like charge (e.g., Na^+ and H^+ or Cl^- and HCO_3^-), not giving rise to net transport but contributing in a major way to unidirectional flux rate.

In the intestine, net water absorption is the result of bulk flow through pores. Diffusion in the usual sense plays no important role in water movement. When bulk flow of water occurs, it is possible for solutes to move across the membrane in the direction of flow by a phenomenon called *solvent drag*. The effect of solvent drag on the transport of a given solute depends on the rate of volume flow and upon the reflection coefficient, an expression of the relationship between the radius of membrane pores and the radius of the solute molecule being transported. By means of solvent drag, it is possible for a solute such as urea to be transported by the intestine against a concentration gradient (Hakim and Lifson, 1964).

2. *Sodium and Chloride Absorption*

Na^+ and Cl^- are the major ions in the fluid that are transported by the intestine during absorption or secretion; under most conditions, transport of these two ions is coupled. The transport of water and electrolytes by the intestinal mucosa is a dynamic process, with rapid unidirectional fluxes of both occurring continuously. Net absorption occurs when the flow from lumen to plasma exceeds that from plasma to lumen. Active transport of Na^+ can occur along the entire length of the intestine, but the rate and net absorption is greatest in the ileum and colon. Na^+ transport is by an energy-requiring "sodium pump" mechanism that is intimately associated with the Na^+–K^+-ATPase located within the basolateral cell membrane of the absorptive epithelial cell. Three mechanisms exist for the entry of Na^+ at the brush border: (1) electrodiffusion down a concentration gradient (2) cotransport of electrolytes that either enter (Cl^-) or exit (H^+) the cell as Na^+ enters, and (3) Na^+ entry coupled with organic nonelectrolytes (glucose, amino acids). Current evidence suggests that in the absence of the absorption of nonelectrolytes, electroneutral uptake accounts for most NaCl absorption. At the brush border, Na^+ enters down a concentration gradient but exits at the basolateral cell surface against a substantial gradient. Maintenance of the transmembrane Na^+ gradient by the Na pump requires continual metabolism and generation of ATP. The Na^+–K^+-ATPase can be inhibited by cardiac glycosides such as ouabain, which are effective inhibitors of Na^+ transport. The Na^+ gradient ultimately serves as an energy source for transport of other solutes (Schultz and Curran, 1970).

In the jejunum, net absorption of sodium occurs slowly unless nonelectrolytes, such as glucose or amino acids, are absorbed simultaneously. In the ileum, Na^+ absorption is independent of glucose absorption. Net water absorption in the jejunum is almost entirely dependent upon the absorption of glucose and other nonelectrolytes, whereas absorption from the ileum is unaffected by glucose. The differential effect of glucose on absorption from the jejunum and ileum is the result of fundamental differences in electrolyte transport mechanisms in these two regions of the intestine.

As Na^+ is transported across the mucosa, an equivalent amount of anion must be transported to maintain electrical neutrality. A major fraction of Cl^- absorption can be accounted for by passive cotransport with Na^+. Under certain circumstances, Cl^- enters the cell in exchange for HCO_3^-.

3. *Potassium Absorption*

Dietary K^+ is absorbed almost entirely in the proximal small intestine. Absorption across the intestinal mucosa occurs down a concentration gradient (high luminal concentration to low plasma concentration). The intestinal fluid reaching the ileum from the jejunum has a K^+ concentration and a Na^+/K^+ ratio that are similar to those of plasma. In the ileum and colon, the rate of Na^+ absorption is much greater than that of K^+ so that, under normal conditions, the Na^+/K^+ ratio in the feces is much lower than that in plasma, approaching a ratio of 1.

4. *Water Absorption*

The absorption of water has been one of the most extensively studied aspects of intestinal transport.Water movement is the result of bulk flow through membranous pores, and simple diffusion plays only a minor role. The question of whether water is actively or passively transported has been the subject of considerable controversy and the controversy itself points to the fundamental difficulties that arise in trying to establish a definition of active transport. Hypertonic saline solutions can be absorbed from canine intestine *in vivo* and from canine and rat intestine *in vitro*. These observations indicate that water absorption can occur against an activity gradient and that the process is dependent upon metabolic energy. This suggests that an active transport process is involved, but Curran (1965) presents an alternative interpretation that is now generally accepted. This view is that water transport occurs secondarily to active solute transport and is the result of local gradients established within the mucosal membrane. Water transport is then coupled to the energy-dependent processes responsible for solute transport, but is one step removed from it.

In the dog and probably other carnivores, the ileum is the main site of net Na^+ and water absorption. In the dog, the colon accounts for no more than perhaps 20% of the total. In herbivorous animals that have a well-developed large intestine, there may actually be net secretion of water within the small intestine during digestion. For example, in the guinea pig (Powell *et al.*, 1968) and horse (Argenzio, 1975) all net absorption of water takes place in the cecum and colon.

B. Carbohydrate Digestion and Absorption

1. *Polysaccharide Digestion*

a. Starch and Glycogen

Carbohydrate is present in the diet primarily in the form of polysaccharides. The most common polysaccharides are starch, glycogen, and cellulose. Starch and glycogen are composed of long chains of glucose molecules linked together by repeating α-1,4-glucosidic bonds. Branch points of the chains are linked by α-1,6-glucosidic bonds. In those species that secrete salivary amylase, digestion of starch and glycogen begins in the mouth when this enzyme mixes with food. The action of salivary amylase is interrupted in the stomach, however, because of the low pH of the gastric secretion.

Starch digestion begins again in the proximal small intestine with the highly specific action of pancreatic amylase on α-1,4-glucosidic bonds. This enzyme catalyzes a series of stepwise hydrolytic reactions, resulting in formation of the principal end products of starch digestion, the disaccharides maltose and isomaltose and small amounts of glucose. Glucose is absorbed directly by the intestinal mucosa and transported to the portal vein. The disaccharides are hydrolyzed further by enzymes of the intestinal cell brush border.

b. Cellulose

Cellulose, like starch, is a polysaccharide of glucose, but it differs from starch in that the glucose molecules are linked by β-1,4-glucosidic bonds. Starch can be utilized by all species, but cellulose is utilized indirectly as a significant source of energy only by animals that have extensive bacterial fermentation within the gastrointestinal tract. Ruminant species digest cellulose most efficiently, but other animals in which the large intestine is well developed, such as the horse, also utilize cellulose as an important energy source.

In ruminants, hydrolysis of cellulose is accomplished by cellulytic bacteria that are part of the complex rumen microflora. The primary end products of cellulose fermentation are short-chain fatty acids—acetic, propionic, and butyric acids. These are absorbed directly from the rumen and serve as the major source of energy for ruminants. Propionic acid is the major precursor for carbohydrate synthesis in mature ruminants.

2. *Disaccharide Digestion*

Maltose and isomaltose are the disaccharide (glucose–glucose) end products of starch digestion. The diet also may contain lactose (galactose–glucose) and sucrose (fructose–glucose). Disaccharide digestion is completed at the surface of the cell by disaccharidases (Gray, 1975), which are components of the brush border (Table 15.6).

The disaccharidases have been solubilized from the brush border and partially purified. Sucrase and isomaltase have been purified together as a two-enzyme complex (Kolinska and Semenza, 1967; Gray *et al.*, 1979) and this enzyme complex accounts for the total hydrolysis of the products of amylase digestion (Gray *et al.*, 1979; Rodriguez *et al.*, 1984). The mutual mucosa contains two enzymes with lactase activity. One of these is a nonspecific β-galactosidase that hydrolyzes synthetic β-galactosides effectively but that hydrolyzes lactose at a slow rate. This enzyme has an optimal pH of 3 and is associated with the lysosomal fraction of the cell. The other lactase hydrolyzes lactose readily, is associated with the brush-border fraction of the cell, and is the enzyme of primary importance in the digestive process (Alpers, 1969).

Maltase, isomaltase, and sucrase are almost completely absent from the intestine in newborn pigs (Hartman *et al.*, 1961; Dahlqvist, 1961) and calves (Huber *et al.*, 1961). The activity of these disaccharidases increases after birth and reaches adult levels during the first months of life. Lactase activity is highest at birth and decreases gradually during the neonatal period. The relatively high lactase activity may be an advantage to the newborn in utilizing the large quantities of lactose present in their diets. Bywater and Penhale (1969) demonstrated lactase deficiency following acute enteric infections and suggested that lactose utilization may be decreased in such cases.

3. *Monosaccharide Transport*

a. Specificity of Monosaccharide Transport

Regardless of whether monosaccharides originate in the lumen of the intestine or are formed at the surface of the mucosal cell, transport across the mucosa involves processes that have a high degree of chemical specificity. Glucose and galactose are absorbed from the intestine more rapidly than other monosaccharides. Fructose is absorbed at approximately one-half of the rate of glucose, and mannose is absorbed at less than one-tenth the rate of glucose (Kohn *et al.*, 1965).

Glucose and galactose can be absorbed against a concentration gradient. The monosaccharides which

TABLE 15.6 Enzymes of the Intestinal Brush Border

Enzyme	Substrate	Product	Reference
Lactase	Lactose	Glucose, galactose	Alpers (1969), Forstner *et al.* (1968)
Sucrase	Sucrose; 1,4α-dextrins	Glucose, fructose; residual 1,6-oligosaccharides	Gray *et al.* (1979)
Isomaltase	1,6α-dextrins	Glucose	Gray *et al.* (1979), Rodriguez *et al.* (1984)
α-Limit dextrinase	1,6α-dextrins	Glucose	Taraval *et al.* (1983)
Trehalase	Trehalose	Glucose	Eichholtz (1967), Nakano *et al.* (1977)
Enterokinase	Trypsinogen	Trypsin	Grant and Herman-Taylor (1976)
Aminopeptidase A	Acidic amino-terminal amino acids	Acidic amino acids	Benajaiba and Maroux (1980)
Aminopeptidase N	Neutral amino-terminal amino acids	Neutral amino acids	Kim and Brophy (1976), Erickson *et al.* (1983)
γ-Glutamyl transferase	Peptides with γ-glutamyl bonds	γ-Glutamyl amino acids	Benajaiba and Maroux (1980), Hughey and Curthoys (1976)
Alkaline phosphatase	Phosphate esters	Inorganic phosphate	Eichholz (1967), Forstner *et al.* (1968)

are transported most efficiently against gradients have common structural characteristics: (1) the presence of a pyranose ring, (2) a carbon atom attached to C-5, and (3) a hydroxyl group at C-2 with the same stereoconfiguration as D-glucose. However, these features are not absolute requirements. Both D-xylose, which has no substituted carbon atom at C-5, and D-mannose, which lacks the appropriate hydroxyl configuration at C-2, can be transported against concentration gradients under certain specific experimental conditions (Csaky and Lassen, 1964; Csaky and Ho, 1966; Alvarado, 1966b).

Glucose transport is competitively inhibited by galactose (Cori, 1925; Fisher and Parsons, 1953) and by a variety of substituted hexoses that compete with glucose for carrier binding sites. The glucoside phlorizin is a potent inhibitor (Parsons *et al.*, 1958; Alvarado and Crane, 1962). Phlorizin also competes for binding sites, but has a much higher affinity for these sites than does glucose.

The absorptive surface of the mucosal cell is the microvillous membrane, or brush border. It is through this part of the plasma membrane that glucose must pass during the initial phase of mucosal transport. Techniques have been developed for isolating highly purified preparations of microvillous membranes from mucosal homogenates (Eichholz and Crane, 1965; Forstner *et al.*, 1968). Faust *et al.* (1967) studied the binding of various sugars to these isolated membrane fractions. They found that D-glucose was bound by the membrane preferentially to L-glucose or to D-mannose, and that glucose binding was completely inhibited by 0.1 m*M* phlorizin. The specificity of their observations suggested that binding represented an initial step in glucose transport, namely, attachment to a membrane carrier.

b. Sodium Requirement

The absorption of glucose and other monosaccharides is influenced significantly by Na^+ (Schultz and Curran, 1970; Kimmich, 1973). When Na^+ is present in the solution bathing the intestinal mucosa, glucose is absorbed rapidly, but when Na^+ is removed and replaced by equimolar amounts of other cations, glucose absorption virtually stops (Riklis and Quastel, 1958; Csaky, 1961; Bihler and Crane, 1962; Bihler *et al.*, 1962). Glucose absorption is inhibited by ouabain, digitalis, and other cardiac glycosides that are also inhibitors of Na^+–K^+-ATPase activity and Na^+ transport (Csaky and Hara, 1965; Schultz and Zalusky, 1964). These observations demonstrate the close relationship between the transport of glucose and that of Na^+.

c. Characteristics of the Na^+–Glucose Transporter (Carrier)

The concentrative step in the active transport of glucose occurs at the brush-border membrane and energy for this process is derived from an electrochemical Na^+ gradient (Schultz and Curran, 1970; Schultz, 1977). Under conditions of net influx, Na^+ and glucose enter in a ratio of 1 : 1 (Goldner et al., 1969; Hopfer and Groseclose, 1980). Cotransport of glucose and Na^+ involves a membrane transporter or carrier that is believed to be a 75-kDa polypeptide (Wright and Peerce, 1985). Na^+ activates glucose transport primarily by increasing the affinity of the carrier for glucose. A model showing two hypothetical forms of the glucose carrier is shown in Fig. 15.5. A channel or pore mechanism has been proposed in which the glucose binding site is located within the membrane. The translocation of glucose in this model is believed to be the result of a Na^+-induced conformational change in the transporter (Semenza *et al.*, 1984).

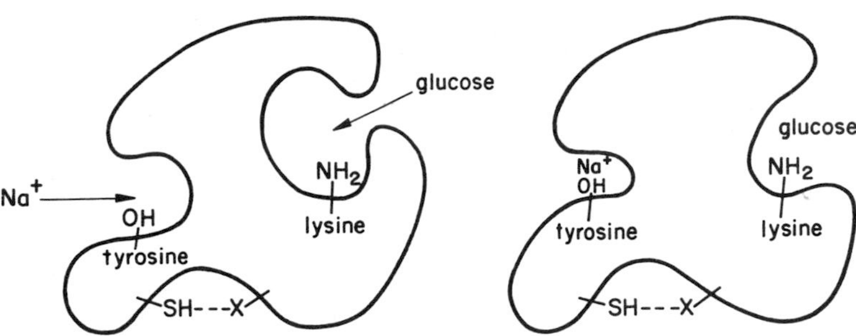

FIGURE 15.5 Model of a Na^+-activated glucose carrier of the intestinal brush border (after Wright and Peerce, 1985).

C. Protein Digestion and Absorption

1. Enzymatic Hydrolysis

The initial step in protein digestion is the enzymatic hydrolysis of peptide bonds by proteases with formation of smaller peptides and amino acids. The endopeptidases hydrolyze peptide bonds within the protein molecule and also hydrolyze certain model peptides. Exopeptidases hydrolyze either the carboxy-terminal (carboxypeptidase) or the amino-terminal (aminopeptidase) amino acids of peptides and certain proteins. Thus, a mixture of exopeptidases and endopeptidases cleave long-chain polypeptides from the ends as well as within the length of the chain, resulting in sequentially shorter and shorter polypeptide chains and amino acids.

Dietary proteins first come in contact with proteolytic enzymes in the stomach. The best known of the gastric proteases is the family of pepsins (Samloff, 1971), which hydroyze most proteins with the exception of keratins, protamines, and mucins. Pepsins are relatively nonselective and hydrolyze peptide bonds involving many amino acids, the most readily hydrolyzed of which involve leucine, phenylalanine, tyrosine, and glutamic acid (Meyer and Kelly, 1977).

The extent of proteolysis in the stomach depends on the nature of the dietary protein and the length of time the protein remains in the stomach. The food bolus mixed with saliva has a neutral or slightly alkaline pH as it enters the stomach and a period of time is required for it to mix with gastric secretions and become acidified. Proteolytic digestion begins when the pH of the gastric contents approaches 4 and occurs optimally in two pH ranges, 1.6–2.4 and 3.3–4.0 (Taylor, 1959a,b). Because of the relative lack of specificity of the pepsins, some peptide bonds of almost all dietary proteins are split during passage through the stomach. The gastric phase of protein digestion may have a minor and possibly dispensable role in overall protein assimilation (Freeman and Kim, 1978), but the reservoir function of the stomach contributes to the gradual release of nutrients, ensuring more efficient utilization in the small intestine.

Partially digested peptides pass from the stomach to the duodenum, where the acidic contents are neutralized by sodium bicarbonate present in bile and pancreatic juice. Peptic activity persists in the duodenum only during the period required to raise the pH above 4.0. The major peptidases that are active within the lumen of the small intestine are the pancreatic enzymes trypsin, chymotrypsin, elastase, and carboxypeptidases A and B. The action of these enzymes is integrated so that the endopeptidases produce peptides with C-terminal amino acids that then become substrates for the exopeptidases. Trypsin produces peptides with basic C-terminal amino acids that are particularly suited for the action of carboxypeptidase B. Chymotrypsin produces peptides with aromatic amino acids in the C-terminal position, and elastase produces peptides with C-terminal amino acids that are nonpolar. Carboxypeptidase A hydrolyzes both types of C-terminal peptide bonds (Table 15.4).

The final steps in peptide digestion are associated with mucosal epithelial cells. Almost all of the aminopeptidase activity is associated with the mucosa and very little activity is present in luminal contents. Mucosal aminopeptidase activity is located both in the cytosol and in the brush-border membrane fractions of the epithelial cell (Heizer and Laster, 1969; Kim *et al.*, 1972). These physically separate enzymes have remarkably different substrate specificities (Kim *et al.*, 1974). The brush-border enzyme has more than 50% of the activity for tripeptides, yet less than 10% of the total activity for dipeptides relative to the cytosolic enzyme(s) (Peters, 1970; Kim *et al.*, 1972). Almost all activity for tetrapeptides is present in the brush border (Freeman and Kim, 1978). Proline-containing peptides are hydrolyzed almost exclusively by cytosolic peptidases, whereas leucine aminopeptidase activity is located primarily in the brush border. The brush border peptidases appear to have digestive functions similar to those of the disaccharidases and oligosaccharidases of the brush border.

Endopeptidase activity of the intestinal mucosa is associated primarily within the lysosomal fraction of the cell.

2. *Absorption of Proteolytic Products*

Although the subject of this section has been studied for many years, the relative amounts of the various protein digestion products (i.e., peptides vs amino acids) that are actually absorbed by intestinal mucosal cells during normal digestion remains problematic. It is a difficult process to investigate because the products of proteolysis are absorbed rapidly after they are formed, and therefore studies of luminal contents give only an estimate of the overall rate of protein digestion. Equally important, dietary protein is continually mixed with endogenous protein in the form of digestive secretions and extruded mucosal cells. Most endogenous proteins are hydrolyzed and the amino acids absorbed in a manner similar to that of dietary protein, and the two processes occur simultaneously. Endogenous protein accounts for a significant fraction of the amino acids of the intestinal contents. Even when dietary protein is labeled with a radioactive tracer, there is such rapid utilization that the tracer soon reenters the lumen in the form of endogenous protein secretion.

In adult mammals, protein is not absorbed from the intestine in quantities of nutritional significance without previous hydrolysis. Most neonatal animals absorb significant amounts of immunoglobulin and other colostral proteins, but this capacity is lost soon after birth. The intestinal mucosa, however, is not totally impermeable to large polypeptide molecules. The absorption of insulin (MW 5700, Laskowski *et al.*, 1958; Danforth and Moore, 1959), ribonuclease (MW 13,700, Alpers and Isselbacher, 1967), ferritin (Bockman and Winborn, 1966), and horseradish peroxidase (Cornell *et al.*, 1971) have all been demonstrated.

During the digestion of protein, the amino acid content of portal blood increases rapidly, but attempts to demonstrate parallel increases in peptides in the portal blood have not been uniformly successful. This has been regarded as evidence that only amino acids can be absorbed by the intestinal mucosa and that peptides are not absorbed. Although it seems clear that most dietary protein is absorbed by the mucosal epithelium in the form of free amino acids, peptides also may be taken up by the mucosal cell in quantitatively significant amounts. Peptides thus absorbed may be hydrolyzed either at the cell surface or intracellularly, and individual amino acids finally enter the portal circulation via the basolateral cell membrane.

Small peptides, under certain circumstances, may cross the intestinal epithelium intact and enter the portal circulation. Webb (1986) suggested that intact peptide absorption accounted for more than half of luminal amino acid nitrogen in the calf. The amount of peptide nitrogen entering the portal circulation in other species characteristically has been lower and variable depending upon the source of protein and on the digestibility of the peptide being investigated (Gardner, 1982, 1984).

3. *Transport of Amino Acids*

Amino acids, like glucose and certain other monosaccharides, are absorbed and transferred to the portal circulation by active transport processes. The same type of saturation kinetics observed in studies of monosaccharide absorption are observed with amino acids, which suggests the presence of carrier transport mechanisms. Certain monosaccharides inhibit amino acid transport (Saunders and Isselbacher, 1965; Newey and Smyth, 1964) and although inhibition generally has been of the noncompetitive type, competitive inhibition between galactose and cycloleucine has been demonstrated (Alvarado, 1966a), which suggests that a common carrier may be involved.

Most amino acids are transported against concentration and electrochemical gradients, and the overall transport process requires metabolic energy. The chemical specificity of these transport mechanisms is shown by the fact that the natural L-forms of various amino acids are absorbed more rapidly than the corresponding D-forms, and that only the L-amino acids appear to be actively transported. For most transport systems, Na^+ is necessary for absorption of amino acids as it is for a variety of other nonelectrolyte substances (Schultz and Curran, 1970; Gray and Cooper, 1971).

Separate transport systems exist for different groups of amino acids. Each member of a group inhibits the transport of other members competitively, suggesting that they share the same carrier. There is demonstrable overlap between groups, indicating that the overall transport process is complex (Christensen, 1984, 1985; Stevens *et al.*, 1984). The following is a summary of the designations and substrates of the recognized amino acid transport systems of the intestinal brush border (Stevens *et al.*, 1984):

1. The neutral brush-border (NBB) pathway is responsible for monoaminomonocarboxylic (neutral) amino acids and histidine. Na^+ is required and these amino acids show mutual competition for transport.
2. The monoaminodicarboxylic acids (aspartic and glutamic acid) pathway (XGA) requires Na^+. Aspartic and glutamic acids are not transported against concentration gradients. Following uptake, they are transami-

nated by the intestinal mucosa and under physiological conditions enter the portal vein as alanine.

3. The imino pathway transports imino acids, proline, hydroxyproline, methylaminoisobutyric acid, and *N*-substituted glycine derivatives sarcosine (*N*-methylglycine) and betaine (*N*-dimethylglycine). This pathway also has a Na^+ requirement.

4. Dibasic amino acids, including lysine, arginine, ornithine, and the neutral amino acid cystine, use the Y^+ pathway.

5. Phenylalanine and methionine share the PHE amino acid transport system.

The τ-glutamyl cycle has been proposed as a possible transport system for amino acids (Meister and Tate, 1976). τ-Glutamyltransferase (GGT) is a membrane-bound enzyme that is present in a number of mammalian tissues and catalyzes the initial step in glutathione degradation. The τ-glutamyl moiety of glutathione is transferred to amino acid (or peptide) receptors with the production of cysteinylglycine:

$$\text{Glutathione} + \text{amino acid} \xrightarrow{\text{GGT}} \tau\text{-glutamyl-amino acid} + \text{Cys–Gly}.$$

The highest GGT activity is present in tissues that are known to transport amino acids actively, such as the jejunal villus, the proximal convoluted tubule of the kidney, and the liver. Meister and Tate (1976) have suggested that GGT may function in translocation by interaction with extracellular amino acids and with intracellular glutathione. The hypothetical mechanism involves the noncovalent binding of extracellular amino acids to the plasma membrane, whereas intracellular glutathione interacts with GGT to yield a τ-glutamyl–enzyme complex. When the τ-glutamyl moiety is transferred to the membrane-bound amino acid, a τ-glutamyl–amino acid complex is formed that, when released from the membrane binding site, moves into the cell. The τ-glutamyl-amino acid complex is split by the action of τ-glutamyl cyclotransferase, an enzyme appropriately located in the cytosol. Glutathione is regenerated by means of the τ-glutamyl cycle, which provides good substrates for GGT (Thompson and Meister, 1975).

The τ-glutamyl cycle does not require sodium, and the previously demonstrated sodium dependence for amino acid transport would not be explained by the cycle. The cycle is not considered to be the only amino acid transport system, and its quantitative significance in individual tissues is unknown.

4. Neonatal Absorption of Immunoglobulin

At birth, most domestic species, including the calf, foal, lamb, pig, kitten, and pup, absorb significant quantities of colostral protein from the small intestine (Brambell, 1958). Immune globulin (Ig) either is absent in the serum of domestic species at birth or is present at low levels. Within a few hours after ingestion of colostrum, the serum Ig levels rise. This represents the principal mechanism by which the young of most domestic animal species acquire maternal immunity. Under normal environmental conditions, ingestion of colostrum is an absolute requirement for health during the neonatal period (Fig. 15.6). The rabbit is the exception in that maternal Ig is received primarily *in utero* by transplacental transfer.

Protein enters the neonatal absorptive cell by pinocytosis and passes through the cell to the lymphatics. The process is not selective, because many proteins other than Ig can be absorbed (Payne and Marsh, 1962a,b). The ability to absorb intact protein is lost by domestic species soon after birth. In the piglet, "closure" occurs within 1 to 2 days (Leary and Lecce, 1978; Westrom *et al.*, 1984) beginning in the duodenum and occurring last in the ileum (Murata and Namioka,

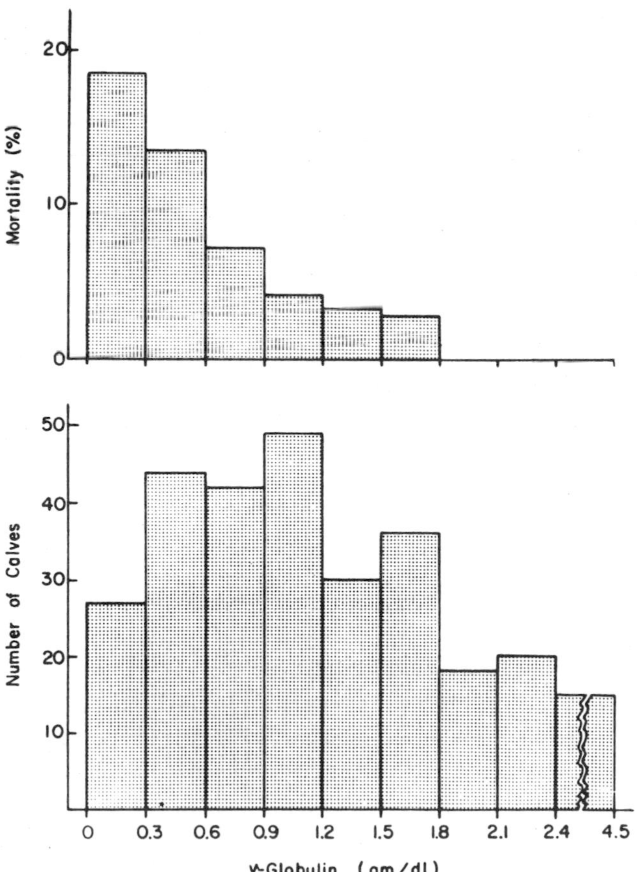

FIGURE 15.6 Histogram of the distribution of serum γ-globulin in market calves and the relation of hypogammaglobulinemia to mortality. From Braun and Tennant (1979; unpublished observations).

1977). In rodents, protein absorption normally continues for approximately 3 weeks. The mechanism of intestinal "closure" was studied by Lecce and coworkers (1964; Lecce, 1966; Lecce and Morgan, 1962), who found that complete starvation of pigs lengthened the period of protein absorption to 4–5 days, whereas early feeding shortened the period. Feeding different fractions of colostrum including lactose and galactose resulted in loss of protein absorptive capacity. The route of feeding may not be the critical factor, however. Calves that are prevented from eating but that receive nutrients parenterally lose the ability to absorb protein at the same time as control calves (Deutsch and Smith, 1957).

In the neonatal calf, Ig deficiency due to failure of colostral Ig absorption plays a role in the pathogenesis of gram-negative septicemia (Smith, 1962; Gay, 1965). Most calves deprived of colostrum develop septicemia early in life and may develop acute diarrhea before death (Smith, 1962; Wood, 1955; Tennant *et al.*, 1975). Hypogammaglobulinemia is almost always demonstrable in calves dying of gram-negative septicemia (Fey, 1971) and is the result either of insufficient Ig intake or of insufficient intestinal absorption. The Ig fraction is the essential factor in colostrum that protects against systemic infections (Penhale *et al.*, 1971).

Serum immunoglobulin values of neonatal calves vary, and a 10% incidence of hypogammaglobulinemia may occur in clinically normal calves (Tennant *et al.*, 1969a,b; House and Baker, 1968; Smith *et al.*, 1967; Thornton *et al.*, 1972; Braun *et al.*, 1973). Most such individuals probably have insufficient colostrum intake. Even when calves were given the opportunity to ingest colostrum, however, a surprising number were hypogammaglobulinemic. Some of the reasons for varying gammaglobulinemia values are recognized, but the relative importance of each is not known. The concentration of lactoglobulin, the volume consumed (Bush *et al.*, 1971; Selman *et al.*, 1971), the time elapsed from birth to ingestion of colostrum (Selman *et al.*, 1971), and the method of ingestion (natural suckling vs bucket feeding) may have an important influence on the serum IgG (Smith *et al.*, 1967; McBeath *et al.*, 1971). Calves that suckle their dams usually attain serum IgG concentrations that are higher than those attained by calves given colostrum from a bucket. The frequency of hypogammaglobulinemia may be influenced by season (Gay *et al.*, 1965b; McEwan *et al.*, 1970a), although this relationship is not consistent (Smith *et al.*, 1967; Thornton *et al.*, 1972). Familial factors also may influence development of hypogammaglobulinemia (Tennant *et al.*, 1969a).

Regardless of cause, the mortality of hypogammaglobulinemic calves is higher than that of calves with normal serum IgG levels (Gay, 1965; House and Baker, 1968; Thornton *et al.*, 1972; McEwan *et al.*, 1970a; Boyd, 1972; Naylor *et al.*, 1977). In addition to having more septicemic infections (Smith, 1962: Gay *et al.*, 1965a; Roberts *et al.*, 1954; Wood, 1955; Fey, 1971; McEwan *et al.*, 1970a), hypogammaglobulinemic calves have a greater prevalence of acute diarrheal disease (Boyd, 1972; Naylor *et al.*, 1977; Penhale *et al.*, 1970; Gay *et al.*, 1965b), which indicates that the local protective effects of Ig in the intestine are important (Fisher *et al.*, 1975; Logan and Penhale, 1971).

The prevalence of hypogammaglobulinemia and the high mortality associated with it has led to the development of several rapid tests for identification of hypogammaglobulinemic calves (McBeath *et al.*, 1971; Aschaffenburg, 1949; Fisher and McEwan, 1967a; Patterson, 1967; Stone and Gitter, 1969). The zinc sulfate turbidity test (Kunkel, 1947) was the first to be used for determination of serum immunoglobulin concentrations of neonatal calves (McEwan *et al.*, 1970b). A close correlation has been established between test results and the amount of serum IgG and IgM (Fisher and McEwan, 1967a,b; McEwan *et al.*, 1970b; Penhale *et al.*, 1967).

The sodium sulfite turbidity test is similar to the zinc sulfate test and also has been used to identify hypogammaglobulinemic calves (Stone and Gitter, 1969; Pfeiffer and McGuire, 1977). Failure of turbidity to develop when serum is added to a saturated solution of sodium sulfite indicates immunoglobulin deficiency. A semiquantitative assessment of the Ig concentration is made by grading the degree of turbidity (Stone and Gitter, 1969).

The refractometer is used as a rapid test for Ig deficiency (McBeath *et al.*, 1971: Boyd, 1972). There is a close relationship between the concentration of IgG and total serum protein (TSP) in neonatal calves (Tennant *et al.*, 1969a) and the wide variations in TSP were due to variations in IgG. Direct linear correlation between the refractive index (RI) and the Ig concentration has also been observed (McBeath *et al.*, 1971). The regression line for this relationship was independently confirmed (Tennant *et al.*, 1978). The Y intercepts in these studies were identical (4 g/dl). The refractometer has a value as a rapid field instrument for the assessment of Ig status, but in cases of hemoconcentration it has limitations (Boyd, 1972).

The glutaraldehyde coagulation test was used originally in cattle to detect hypergammaglobulinemia in samples of whole blood (Sandholm, 1974). Glutaraldehyde has also been used in a semiquantitative test to evaluate IgG in canine (Sandholm and Kivisto, 1975) and human serum (Sandholm, 1976). This procedure has been modified to detect hypogammaglobulinemic

TABLE 15.7 **Relationship between Results of the Glutaraldehyde Coagulation Test, Serum γ-Globulin Concentration, and Death Rate**

Source of calves	No.	Glutaraldehyde reaction	Serum γ-globulin (g/dl) Mean (±SD)	Extremes	Death rate (%)
Calves before ingestion of colostrum	10	Negative	0.18 (± 0.06)	0.1–0.25	—[a]
Calves from production unit	60	Negative	0.35 (± 0.13)	0.11–0.63	16.7[b]
	13	Incomplete	0.60 (± 0.13)	0.42–0.85	7.7
	208	Positive	1.46 (± 0.63)	0.42–4.4	3.4

[a] Samples of serum were obtained at birth, but no follow-up of calves was made.

[b] The death rate of calves that were test negative was significantly ($p < 0.1$) greater than that of test-positive calves, using *t*-test for significance of differences between two percentages.

calves. Calves with a negative test result (serum IgG ≤0.4 g/dl) had markedly higher mortality than calves with positive results (Table 15.7) (Tennant *et al.*, 1979), which is similar to results obtained by using the zinc sulfate turbidity test (Gay *et al.*, 1965a; McEwan *et al.*, 1970a) or other estimates of circulating IgG. Many tests can be initiated quickly using the glutaraldehyde coagulation test, and results can be evaluated rapidly without instrumentation.

D. Lipid Digestion and Absorption

1. *Absorption of Fats*

a. Luminal Phase

The fat in the diet is primarily in the form of triglycerides or long-chain fatty acids. The initial step in utilization of triglycerides occurs in the lumen of the proximal small intestine, where hydrolysis is catalyzed by pancreatic lipase. Lipase is secreted by the pancreas in active form. Because the enzyme requires an oil–water interface for activity, only emulsions of fat can be hydrolyzed. Enzyme activity is directly related to the surface area of the emulsion so the smaller the emulsion particle, the greater the total surface area of a given quantity of triglyceride and the greater the rate of hydrolysis (Benzonana and Desnuelle, 1965). Bile salts are not an absolute requirement, but favor hydrolysis by their detergent action, which causes formation of emulsions with small particle sizes, and by stimulating lipase activity within the physiological pH range of the duodenum. A colipase is present in the pancreatic secretion that facilitates the interaction of lipase with its triglyceride substrate and protects lipase from inactivation (Borgstrom and Erlanson, 1971).

Pancreatic lipase splits the ester bonds of triglycerides preferentially at the 1 and 3 positions so that the major end products of hydrolysis are 2-monoglycerides and free fatty acids. Both compounds are relatively insoluble in water but are brought rapidly into micellar solution by the detergent action of bile salts. The mixed micelles so formed have a diameter of approximately 2.0 nm and are believed to be the form in which the products of fat digestion are actually taken up by the mucosal cell (Hofmann and Small, 1967). The intraluminal events that occur in fat absorption are schematically summarized in Fig. 15.7.

b. Mucosal Phase

The initial step in intestinal transport of fat is the uptake of fatty acids and monoglycerides by the mucosal cell from micellar solution. The precise mechanism is yet unclear, but present evidence suggests that the lipid contents of the micelle are somehow discharged at the cell surface and enter the mucosal cell in molecular rather than micellar form (Isselbacher, 1967). The net effect is the absorption of the end products of lipolysis

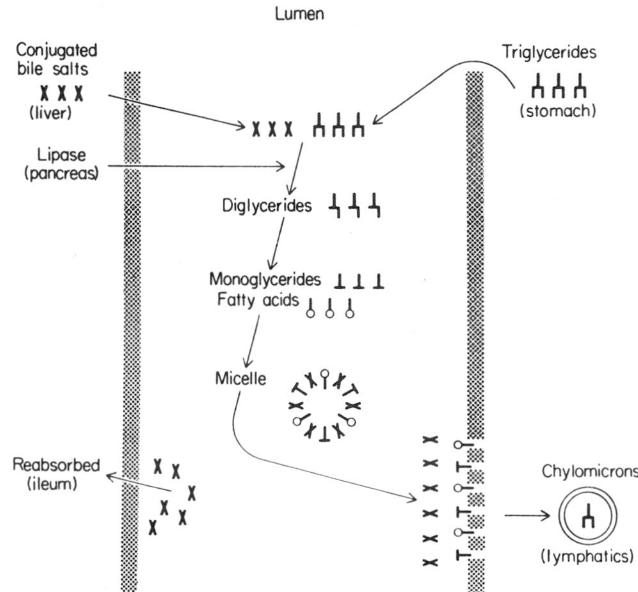

FIGURE 15.7 Intraluminal events during fat absorption. From Isselbacher (1967).

and the exclusion of bile salts, which are absorbed further down the intestine, primarily in the ileum. Uptake of fatty acids appears to be a passive process having no requirement for metabolic energy.

Within the mucosal cell, the fatty acids are transported by a soluble binding protein to the endoplasmic reticulum, where the fatty acids and monoglycerides are rapidly reesterified to triglyceride (Ockner and Manning, 1974; Ockner and Isselbacher, 1974). The two biochemical pathways for triglyceride biosynthesis in the intestine are summarized in Fig. 15.8. Direct acylation of monoglyceride occurs in the intestine and is the major pathway for lipogenesis in the intestine during normal fat absorption. The initial step in this series of reactions involves activation of fatty acids by acyl-CoA synthetase, a reaction that requires Mg^{2+}, ATP, and CoA and that has a marked specificity for long-chain fatty acids. This specificity explains the observation by Bloom *et al.* (1951) that medium- and short-chain fatty acids are not incorporated into triglycerides during intestinal transport, but enter the portal circulation as nonesterified fatty acids. The activated fatty acids then react sequentially with mono- and diglycerides to form triglycerides in steps catalyzed by mono- and diglyceride transacylases. The enzymes responsible for this series of reactions are present in the microsomal fraction of the cell (Rao and Johnston, 1966). These enzymes occur together in the endoplasmic reticulum as a "triglyceride–synthetase" complex.

An alternative route that is available for fatty acid esterification involves L-α-glycerophosphate derived either from glucose or from dietary glycerol by the action of intestinal glycerokinase. Activated fatty acid CoA derivatives react with L-α-glycerophosphate to form lysophosphatidic acid (monoglyceride phosphate), which by a second acylation forms phosphatidic acid (diglyceride phosphate). Phosphatidic acid phosphatase then hydrolyzes the phosphate ester bond, forming diglyceride, and by means of a transacylase step similar to that described earlier, triglyceride is formed. Although this pathway appears to be of minor importance for triglyceride synthesis in the intestine, intermediates in this sequence of reactions are important in the synthesis of phospholipids that are essential for stabilization of the chylomicron.

The next step in fat transport is formation of chylomicrons within the endoplasmic reticulum. The chylomicron is composed primarily of triglyceride and has an outer membranous coating of cholesterol, phospholipid, and protein (Zilversmit, 1965). The β-lipoprotein component of the chylomicron is synthesized by the intestinal mucosal cell. Inhibition of protein synthesis by puromycin or acetoxycycloheximide interferes with chylomicron formation and significantly reduces fat transport (Sabesin and Isselbacher, 1965).

The final step in fat absorption is extrusion of the chylomicra into the intercellular space opposite the basal lateral portion of the absorptive cell by reverse pinocytosis. From the intercellular space, the chylomicra pass through the basement membrane and enter the lacteal. The chylomicra then pass from the lacteals into lymph ducts and into the general circulation, thereby completely bypassing the liver during the initial phase of absorption.

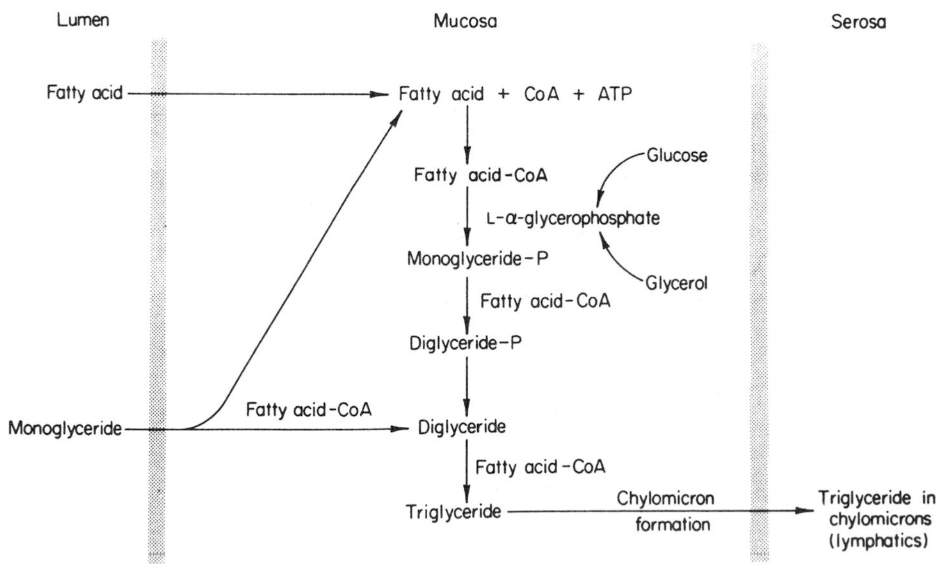

FIGURE 15.8 Biochemical reactions involved in intestinal transport of long-chain fatty acids and monoglycerides. From Isselbacher (1966).

2. *Absorption of Other Lipids*

a. Cholesterol

Dietary cholesterol is present in both free and esterified forms, but only nonesterified cholesterol is absorbed. Cholesterol esters are hydrolyzed within the lumen of the intestine by sterol esterases secreted by the pancreas. Bile salts are required both for the action of this enzyme and for the absorption of nonesterified cholesterol. In the mucosal cell, cholesterol is reesterified and transferred by way of the lymph to the general circulation. The type of triglyceride present in the diet significantly affects the absorption of cholesterol and its distribution in lymph lipids (Ockner *et al.*, 1969).

b. Vitamin A

The diet contains vitamin A activity in two principal forms: (1) as esters of preformed vitamin A alcohol (retinol) and fatty acids and (2) as provitamin A, primarily β-carotene. Vitamin A ester is hydrolyzed by a pancreatic esterase within the lumen (Murthy and Ganguly, 1962), and the free alcohol is absorbed in the upper small intestine. Vitamin A alcohol is reesterified in the mucosa primarily with palmitic acid. The vitamin A ester is absorbed by way of the lymph and after reaching the general circulation, it is rapidly cleared from the plasma and stored in the liver. In the postabsorptive state, vitamin A circulates as the free alcohol, the form released as needed from the liver by the action of hepatic retinylpalmitase esterase. The blood level of vitamin A is independent of liver reserves, and as long as a small amount of vitamin A is present in the liver, the blood level remains normal (Dowling and Wald, 1958).

In diets that lack animal fat, the carotenes, primarily β-carotene, serve as the major precursor of vitamin A. The intestinal mucosa has a primary role in conversion of provitamin A to the active vitamin, although conversion can occur to a limited degree in other tissues. The mechanism involves central cleavage of β-carotene into two active vitamin A alcohol molecules that are subsequently esterified and absorbed by the lymphatics as with preformed vitamin A.

Bile salts are required for the mucosal uptake of β-carotene and for the conversion of β-carotene to vitamin A. Uptake of carotene and release of vitamin A ester into the lymph are rate-limiting steps. Cattle absorb substantial amounts of β-carotene without prior conversion to vitamin A, and these pigments are responsible for much of the yellow color of the plasma. Most other species have no β-carotene in the plasma and it is thought that extraintestinal conversion is more efficient in these species than in cattle (Ganguly and Murthy, 1967).

c. Vitamin D

Vitamin D, like cholesterol, is a sterol that is absorbed by the intestine and transported via the lymph (Schachter *et al.*, 1964). Intestinal absorption differs, however, in that vitamin D is transported to the lymph in nonesterified form. The uptake of vitamin D by the mucosal cell is favored by the presence of bile salts. Simultaneous absorption of fat from micellar solutions increases transport of vitamin D out of the cell into the lymph, the limiting step.

One of the major actions of vitamin D is to enhance the intestinal absorption of calcium. The mechanism of action of vitamin D has been described by Wasserman and co-workers (1968; Wasserman and Taylor, 1966, 1968). They have shown that vitamin D causes synthesis of a calcium-binding protein that plays a central role in the transport of calcium.

VII. DISTURBANCES OF GASTROINTESTINAL FUNCTION

A. Vomition

Vomiting is a coordinated reflex act which results in rapid, forceful expulsion of gastric contents through the mouth. The reflex may be initiated by local gastric irritation caused by a variety of toxic irritants, infectious agents, foreign bodies, gastric tumors, or obstructions of the pyloric canal or the small intestine, or by drugs such as apomorphine or other toxic substances that act centrally on the "vomiting center" of the medulla.

Severe vomiting produces loss of large quantities of water and of H^+ and $C1^-$ ions. These losses cause dehydration, metabolic alkalosis with increased plasma HCO_3^-, and hypochloremia. Chronic vomiting may also be associated with the loss of significant tissue K^+ and with hypokalemia. The K^+ deficit is caused primarily by increased urinary excretion resulting from alkalosis (Leaf and Santos, 1961). Gastric secretions contain significant quantities of K^+ and losses in the vomitus also contribute to the K^+ deficiency. K^+ deficiency, which develops initially because of the alkalosis, perpetuates the alkalotic state by interfering with the ability of the kidney to conserve H^+ (Koch *et al.*, 1956; Darrow, 1964). Both K^+ and the hypovolemia caused by dehydration may result in renal tubular damage and in renal failure.

Vomiting occurs frequently in the dog, cat, and pig, but is an unusual sign in the horse, which has anatomical restrictions of the esophagus that interfere with expulsion of gastric contents. In cattle, sheep, and goats, the physiological process of rumination utilizes neuromuscular mechanisms similar to those involved in vomiting. Uncontrolled expulsion of ruminal contents is an uncommon sign, most frequently occurring after ingestion of toxic materials or associated with traumatic reticulitis and resulting "vagal indigestion." The contents of the abomasum are not expelled directly even when the pyloric canal is obstructed. Pyloric outflow obstruction does occur in cattle that is similar metabolically to that observed in nonruminants. This obstruction may be observed in right-sided displacement of the abomasum with or without torsion, occasionally with left-sided displacement of the abomasum, in cows with functional pyloric obstruction as a result of reticuloperitonitis, and from "vagal indigestion." When the pylorus is obstructed, abomasal contents are retained, causing distension of the abomasum, which in turn stimulates further secretion and retention. Retained abomasal contents may be regurgitated into the large reservoir of the rumen and sequestered there from other fluid compartments of the body. The net result is loss of H^+ and Cl^- ions and development of metabolic alkalosis, hypochloremia, and hypokalemia (Espersen and Simesen, 1961; Svendsen, 1969). This metabolic syndrome often is associated with fluid distension of the rumen related to pyloric outflow obstruction. Similar distention of the rumen in the absence of hypochloremic, hypokalemic metabolic alkalosis suggests more proximal obstruction of rumen outflow, namely the omasum.

Chronic hypertrophic gastritis, which resembles Menetrier's disease in humans, has been demonstrated in the dog (van der Gagg *et al.*, 1976; Van Kruiningen, 1977; Happe and van der Gagg, 1977; Kipnis, 1978). Van Kruiningen's series of cases were Basenjis that had concomitant lymphocytic–plasmocytic enteritis. The primary disease, however, has been observed in other breeds without intestinal lesions. Signs of illness usually involve chronic vomiting, weight loss, and occasionally diarrhea. Hypoalbuminemia occurs in most cases. In man, hyperchlorhydria or achlorhydria can occur. The morphological changes in the stomach wall (hypertrophic rugae) and some of the clinical features help to differentiate this disease from gastric neoplasia.

Zollinger–Ellison syndrome has been reported in the dog (Straus *et al.*, 1977; van der Gagg *et al.*, 1978; Breitschwerdt *et al.*, 1986; English *et al.*, 1988). Vomiting, diarrhea, inappetence and weight loss are frequently observed. The disease is caused by pancreatic non-β islet-cell tumors that result in hypergastrinemia, hyperchlorhydria, hypertrophic gastritis, peptic esophagitis, and duodenal ulcers.

B. Gastric Dilatation–Volvulus

Gastric dilatation–volvulus (GDV) is an acute gastrointestinal disorder associated with high mortality (Wingfield *et al.*, 1976; Todoroff, 1979; Morgan, 1982; Leib and Blass, 1984). It typically occurs in large deep-chested dogs but has been reported in smaller dogs, the cat, and other species (Van Kruiningen *et al.*, 1974). Gastric dilatation precedes development of volvulus and is the result of the accumulation of gas (Caywood *et al.*, 1977; Rogolsky and Van Kruiningen, 1978; Warner and Van Kruiningen, 1978) and fluid in the stomach as a result of either mechanical or functional disturbances in pyloric outflow. As the stomach distends and rotates about the distal esophagus, displacement and occlusion of the pylorus and duodenum occur. Necrosis and perforation of the stomach wall and peritonitis are common causes of death.

Distension and displacement of the stomach causes obstruction of the caudal vena cava and portal vein resulting in venous stasis and sequestration of blood in splanchnic, renal, and posterior muscular capillary beds (Wingfield *et al.*, 1976). This decrease in circulating blood volume (venous return) and subsequent decrease in cardiac output, arterial blood pressure, and tissue perfusion culminates in hypovolemic shock. Endotoxemia, a consequence of portal-vein occlusion, contributes to the shock syndrome. The release of myocardial depressant factors from ischemic pancreatic tissue impairs the clearance of endotoxins by the reticuloendothelial system as well as causing direct cardiodepressant effects. Altered microvascular perfusion with hypoxemia and endotoxemia favors development of disseminated intravascular coagulopathy (DIC) (Lees *et al.*, 1977).

Increased plasma gastrin immunoreactivity has been reported in dogs with GDV (Leib *et al.*, 1984). Preexisting conditions of relative hypergastrinemia may predispose to GDV. Gastrin can increase caudal esophageal sphincter pressure, delay gastric emptying, and predispose to pyloric outflow obstruction by causing gastric mucosal and pyloric muscular hypertrophy.

Experimental gastric dilatation and dilatation with torsion has been studied in the dog (Wingfield *et al.*, 1974; Merkley *et al.*, 1976a, 1976b). Hyperkalemia and hyperphosphatemia were consistent findings in dogs with gastric dilatation and torsion (Wingfield *et al.*, 1974; Merkley *et al.*, 1976b). This was the result of hypovolemia, decreased renal perfusion, and renal insufficiency on the one hand and the loss of intracellular K^+ from damaged tissue on the other. Increased blood urea nitrogen (BUN) and serum creatinine (Cr) levels persisted after decompression of the stomach. Hemoconcentration and increased TSP were attributed to fluid shifts from the vascular compartment into the lumen of the alimentary tract, wall of the stomach, and peritoneal cavity. Increased alanine aminotransferase (ALT) and aspartate aminotransferase (AST) were most apparent following decompression of the stomach and were attributed to alteration of hepatocytes and smooth muscle of the stomach and spleen. Increased of creatine kinase (CK) resulted from the effects of tissue hypoxia on striated muscle. Metabolic acidosis was attributed in part to increased production of lactic acid caused by tissue hypoxia.

A wide range of acid–base and electrolyte disturbances have been reported in clinical patients with GDV (Muir, 1982; Wingfield *et al.*, 1982; Kagan and Schaer, 1983). Dogs presenting with GDV may have normal acid–base status. Metabolic acidosis and hypokalemia commonly occur. Metabolic alkalemia and respiratory alkalosis also have been observed. Hyperkalemia is unusual. The absence of an increase in anion gap in one study indicated that the production of volatile fatty acids and lactic acid was not excessive (Wingfield *et al.*, 1982).

C. Ischemia–Reperfusion Injury

Ischemia–reperfusion injury is a contributing cause of death in horses with strangulating intestinal obstruction (Moore *et al.*, 1995) and in dogs with GDV. Together with luminal occlusion of the alimentary tract, functional constriction or mechanical obstruction of intestinal vasculature occurs. Depending on the duration and severity of ischemia, oxygenation of tissue is compromised and there is a subsequent attenuation of oxidative phosphorylation and a decrease in ATP. Anaerobic glycolysis ensues, leading to intracellular acidosis and increased intracellular concentrations of Ca^{2+}. Unless timely restoration of blood flow and oxygenation occurs, these metabolic derangements eventually contribute to cellular edema, lysosomal release of degradative enzymes, autolytic destruction of cellular organelles, and cell death.

When intestinal obstruction is relieved and tissue perfusion is reestablished, reoxygenation of tissue can result in a cascade of biochemical events that can aggravate ischemic-induced tissue injury. The resulting reperfusion injury is caused in part by oxygen free radicals (OFR), particularly superoxide (O_2^-), and hydroxyl free radicals ($OH^·$), and is characterized by increased microvascular and mucosal permeability and mucosal necrosis (Moore *et al.*, 1995). The formation of OFRs is preceded by accumulation of hypoxanthine in endothelial cells and intestinal mucosal cells during ischemia. The conversion of xanthine dehydrogenase to xanthine oxidase also occurs during ischemia, a reaction that is facilitated by high intracellular levels of calcium ions and the protease calpain. When reperfusion occurs, xanthine oxidase converts hypoxanthine to uric acid and superoxide radicals. O_2^- and hydrogen peroxide (H_2O_2), a product of superoxide dismutase (SOD) reduction, are converted to highly reactive $OH^·$ in the presence of an iron catalyst. $OH^·$ initiates structural and functional cellular membrane damage via lipid peroxidation. The release of inflammatory mediators attending lipoperoxidation contributes to tissue injury.

Malondialdehyde (MDA) is a stable by-product of lipoperoxidation, and its detection can be utilized by investigators as an indicator of ischemia–reperfusion injury (Moore *et al.*, 1995).

Neutrophils are recruited into ischemic and reperfused tissue by xanthine oxidase-derived OFRs and chemoattractants released from cellular membranes during lipid peroxidation. Increased cytosolic calcium concentrations during ischemia and subsequent lipoperoxidation activate phospholipase A2, which in turn causes the release of platelet-activating factor (PAF), metabolites of arachidonic acid (leukotrienes and prostaglandins), and lysophosphatidylcholine. Leukotriene B4, thromboxane A2, and PAF are the primary products of phospholipid metabolism that promote infiltration and degranulation of neutrophils in affected tissue.

When neutrophils attach to endothelium, they release elastase and lactoferrin, which promotes extravasation (Moore *et al.*, 1995). The conversion of oxygen to O_2^- within neutrophils is facilitated by the NADPH oxidase system. This O_2^- is metabolized to H_2O_2, and the latter reacts with Cl^- to form hypochlorous acid. Myeloperoxidase (MPO), an enzyme contained in neutrophils, catalyzes this reaction. MPO activity in intestinal mucosal correlates well with the degree of neutrophil infiltration and mucosal injury.

Serine proteases are believed to play a contributing role in ischemia–reperfusion injury (Moore *et al.*, 1995). The pancreas is an important source of endoproteases (trypsin, chymotrypsin, and elastase) that can cause mucosal injury, particularly in the small intestine. Proteases produced by granulocytes, as well as lysosomes, are more important in mucosal injury of the large bowel. Elastase, neutral proteases, and cathepsin G are released from granulocytes during phagocytosis. Cathepsin B is a lysosomal protease that has trypsin-like activity.

Several pharmacological agents have been used in experimental and clinical studies of ischemia–

reperfusion injury (Moore *et al.*, 1995). The mechanistic rationale for many of these agents is comparable to the role of endogenous antioxidants. Examples of commonly used agents include xanthine oxidase inhibitors (allopurinol), deferroxamine, 21-aminosteroids, inhibitors of PLA2, cyclooxygenase, and lipoxygenase. Superoxide dismutase, catalase, and glutathione peroxidase (Gpx) are free-radical scavenging enzymes. Mannitol, albumin, dimethyl sulfoxide (DMSO), dimethyl thiourea, and manganese chloride represent nonenzymatic free-radical scavengers. Other agents that have been studied include nitric acid, protease inhibitors, hydroxyethylstarch, and neutrophil directed agents. Although the aforementioned agents have demonstrated efficacy in some studies, there are many inconsistencies. From a clinical perspective, success has been limited with single agents and there is more interest in combination or multimodal therapy.

D. Acute Diarrheas

The term "diarrhea" is used generically to describe the passage of abnormally fluid feces with increased frequency, increased volume, or both. The significance of diarrhea depends primarily on the underlying cause and on the secondary nutritional and metabolic disturbances that are caused by excessive fecal losses.

There are theoretically three factors that can act independently or in combination to produce diarrhea. An increase in the rate of intestinal transit is one factor believed important in functional disorders of the gastrointestinal tract in which "hypermotility" has been considered to be the primary cause. Although increased intestinal motility may be a factor in certain types of diarrheal disease, when motility patterns have been investigated, diarrheal disease has actually been associated with decreased motility (Christensen *et al.*, 1972). A second factor in the pathogenesis of diarrhea is decreased intestinal assimilation of nutrients that may result either from decreased intraluminal hydrolysis of nutrients, such as *maldigestion* due to pancreatic exocrine insufficiency or bile-salt deficiency, or from defective mucosal transport of nutrients, that is, *malabsorption,* that results from various types of inflammatory bowel disease, villus atrophy, intestinal lymphoma, or intrinsic biochemical defects in the mucosal cell that interfere with digestion and/or absorption. Finally, increased intestinal secretion of water and electrolytes is a major factor in the pathogenesis of certain types of acute diarrhea.

Enteropathogenic strains of *Escherichia coli* produce soluble enterotoxins (Smith and Halls, 1967; Kohler, 1968; Moon, 1978), which alter bidirectional Na^+ and water flux (Fig. 15.9). The most extensively studied enterotoxin is that produced by *Vibrio cholerae.* This bacterium produces a high-molecular-weight, heat-labile toxin (CT), one subunit of which has properties similar to those of the heat-labile (LT) enterotoxin produced by certain strains of *E. coli* (Richards and Douglas, 1978). The mechanism of action of CT is believed to involve the activation of adenylate cyclase. This membrane-bound enzyme converts ATP to cAMP, which through the action of protein kinase is responsible for the greatly increased secretion of water and electrolytes by the intestinal mucosa. Although species differences have been observed (Hamilton *et al.*, 1978a,b; Forsyth *et al.*, 1978), this mechanism appears to be important in the mode of action of LT of *E. coli* as well (Richards and Douglas, 1978).

Additional extensive studies have centered on the molecular mechanism of action of CT. Under physiological conditions, adenylate cyclase is activated by the binding of guanosine triphosphate to the inactive enzyme. An associated GTPase inactivates the enzyme by converting enzyme-bound GTP to GDP and inorganic phosphate. This GTP–GDP system plays a critical role in the physiological regulation of adenylate cyclase. Cholera toxin is believed to bind to the adenyl cyclase in a way that inhibits hydrolysis of GTP, thereby maintaining the enzyme in an activated state (Levinson and Blume, 1977; Johnson *et al.*, 1978; Cassel and Pfeuffer, 1978) (Fig. 15.10).

Certain enteropathogenic strains of *E. coli* produce a low-molecular-weight, heat-stable toxin (ST) alone or in addition to LT (Richards and Douglas, 1978; Moon, 1978; Hamilton *et al.*, 1978a). In epidemiological studies of neonatal diarrheal diseases of calves, most isolated strains of *E. coli* produce only ST (Moon *et al.*, 1976; Braaten and Myers, 1977; Lariviere *et al.*, 1979). In contrast to LT and CT, which induce intestinal Na^+ and water secretion only after a lag phase of several hours, ST induces intestinal secretion immediately. ST induces intestinal secretion by activating guanylate cyclase, and the mediator of intestinal secretion induced by ST is cyclic 3′,5′-guanosine monophosphate (Hughes *et al.*, 1978; Field *et al.*, 1978).

Enterotoxin-induced intestinal secretion may be blocked by cycloheximide, an inhibitor of protein synthesis (Serebro *et al.*, 1969). The lack of specificity and the toxicity of cycloheximide precluded its clinical use, but acetazolamide has been shown to inhibit intestinal fluid secretion (Norris *et al.*, 1969; Moore *et al.*, 1971). Ethacrynic acid, another potent diuretic, has been shown to inhibit enterotoxin-induced fluid secretion (Carpenter *et al.*, 1969). Unfortunately, the diuretic effects of these drugs preclude their clinical use, but similar drugs with "intestinal specificity" would have

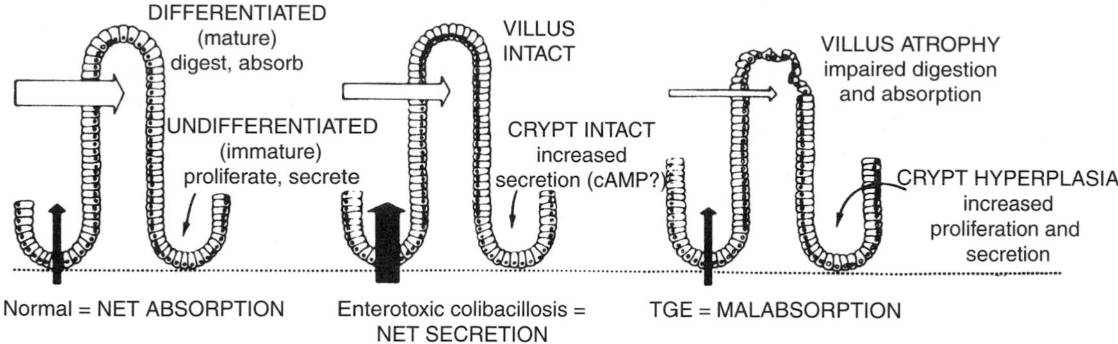

FIGURE 15.9 Pathogenesis of diarrhea caused by *E. coli* enterotoxin and by coronavirus. After Moon (1978).

significant therapeutic potential. Adenosine analogs also have been shown to inhibit cholera toxin-stimulated intestinal adenylate cyclase.

Prostaglandin E_1 (PgE_1) and CT have similar effects on electrolyte transport in rabbit ileum. Application of either to the mucosa inhibits NaCl absorption and stimulates Cl^+ secretion. Both indomethacin (Gots *et al.*, 1974) and aspirin (Farris *et al.*, 1976) inhibit enterotoxin-induced intestinal secretion in laboratory animal models, and the prostaglandins do not function as mediators in the pathogenesis of cholera (Schwartz *et al.*, 1975). However, Jones *et al.* (1977) demonstrated a positive therapeutic response to a new prostaglandin inhibitor in calves with acute enteritis.

The effects of the *E. coli* ST can be inhibited *in vitro* by the calcium channel blockers diltiazem and lodoxamide tromethamine, and the prostaglandin synthesis inhibitors indomethacin and quinacrine (Knoop and Abbey, 1981; Thomas and Knoop, 1982). Neither class

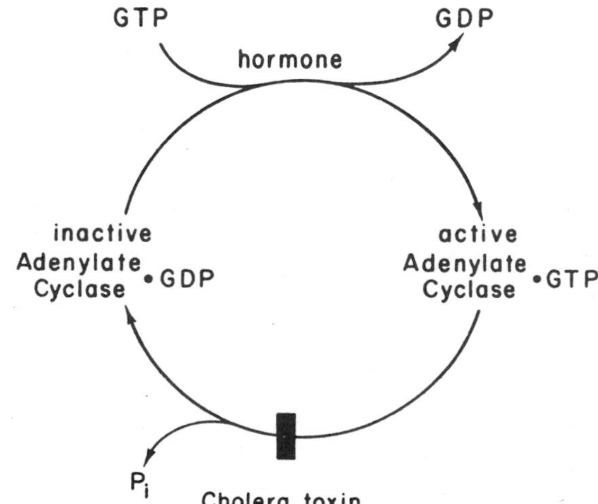

FIGURE 15.10 Mechanism of action of cholera toxin, which inhibits hydrolysis of GTP, thereby increasing adenylate cyclase activity. After Cassell and Selinger (1978).

of drug blocks the effect of cGMP, suggesting that calcium and prostaglandin influence the earliest step(s) in ST response: either its brush-border binding or the activation of guanylate cyclase.

The autonomic nervous system has important effects on intestinal ion transport and water absorption (Tapper *et al.*, 1978). Catecholamines stimulate formation of cAMP in a variety of mammalian cells (Sutherland and Rall, 1960; Schultz *et al.*, 1975), apparently by activating the GTP–GDP system described earlier (Cassel and Selinger, 1978; Ciment and deVellis, 1978). Adrenergic blocking agents, such as chlorpromazine (Holmgren *et al.*, 1978) and propranolol (Donowitz *et al.*, 1979), have significant inhibitory effects on enterotoxin-induced intestinal secretion. Although the mechanism of action of these two adrenergic blockers is not known, they represent still another class of drugs that may be of therapeutic benefit.

The intestinal "adsorbent" drug Pepto-Bismol, containing bismuth subsalicylate, and Attapulgite, a heat-treated silicate, have antienterotoxic effects (Drucker *et al.*, 1977; Ericsson *et al.*, 1977; Gyles and Zigler, 1978). Therapeutic trials with bismuth subsalicylate show significant therapeutic benefit in certain large-volume diarrheal diseases of humans that are enterotoxigenic in origin (Portnoy *et al.*, 1976; DuPont *et al.*, 1977; DuPont, 1978). The mechanism of the intestinal secretion inhibition is not known, but the chemical relation of bismuth subsalicylate to other known prostaglandin inhibitors is known. It is possible that such drugs, by decreasing endogenous production of prostaglandin, decrease the basal level of cyclic nucleotides, which in turn causes an increase in the threshold of response to enterotoxin. Salicylates also may stimulate sodium chloride absorption (Powell *et al.*, 1979). Collectively, these observations suggest that new, innovative methods for therapy and control of acute clinical diarrheal disease may be developed.

Acute diarrhea represents the leading cause of morbidity and mortality in neonatal calves and pigs. The pathogenesis of the neonatal enteric infection is complex, often involving nutritional or environmental factors as well as infectious agents, such as enteropathogenic strains of *E. coli*, the transmissible gastroenteritis virus (TGE), rotaviruses, and other bacterial and viral pathogens. The severe clinical signs and frequently fatal outcome of acute diarrheal disease are often directly related to dehydration and to associated H^+ and electrolyte disturbances (Dalton *et al.*, 1965; Fisher and McEwan, 1967b; Tennant *et al.*, 1972, 1978).

In acute diarrhea with large-volume, watery stools, the fecal fluid originates primarily from the small intestine. The electrolyte composition of the stool in such cases is similar to that of the fluid found normally in the lumen of the small intestine, which in turn is similar to that of an ultrafiltrate of the plasma. The rapid dehydration that accompanies acute enteritis in the newborn soon produces hemoconcentration and leads to hypovolemic shock. These cases are characterized by metabolic acidosis (Dalton *et al.*, 1965; Phillips and Knox, 1969) caused by (1) decreased excretion of H^+ due to decreased renal perfusion and (2) increased production of organic acids, the result of decreased tissue oxygenation, which leads to excessive anaerobic glycolysis. Hyperkalemia also is observed characteristically in young, severely dehydrated animals. Hyperkalemia in such cases is the result of increased movement of cellular K^+ into the extracellular fluid and to decreased renal excretion. Cardiac irregularities caused by hyperkalemia can be demonstrated with the electrocardiogram, and cardiac arrest related to hyperkalemia is a direct cause of death in calves with acute diarrhea (Fisher, 1965; Fisher and McEwan, 1967b). Marked hypoglycemia also has been observed occasionally prior to death in calves with acute enteric infections. Hypoglycemia is believed to be due to decreased gluconeogenesis and increased anaerobic glycolysis, the result of hypovolemic shock (Tennant *et al.*, 1968). The sequence of metabolic changes that occur during acute neonatal diarrhea is summarized in Fig. 15.11.

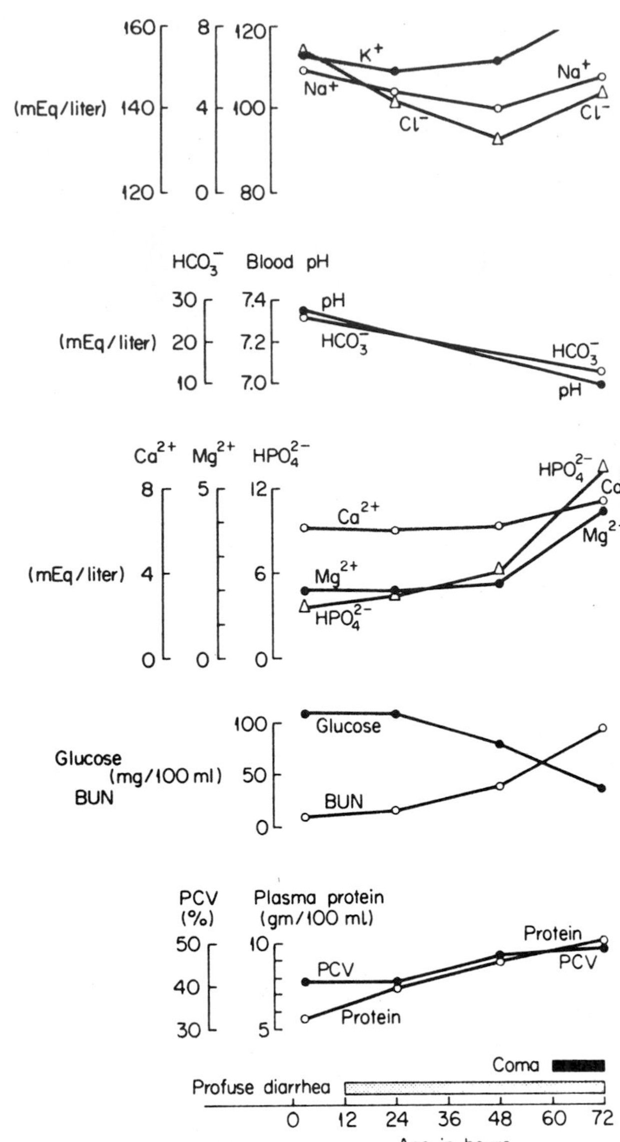

FIGURE 15.11 Metabolic alterations during the course of fatal enteric infection in a neonatal calf. From Tennant *et al.* (1972).

In chronic forms of diarrheal disease, excessive fecal losses of electrolyte and fluid are compensated in part by renal conservation mechanisms and in part by ingestion. If water is consumed without adequate ingestion of electrolytes, hyponatremia and hypokalemia may develop (Tasker, 1967; Patterson *et al.*, 1968). In such cases, the osmolality of the plasma is significantly decreased and hypotonic dehydration occurs. In longer-standing cases of chronic diarrhea, the plasma K^+ concentration may become dangerously low. It is imperative, in this case, that intravenous fluids contain sufficient K^+ to prevent further reduction in concentration and to avoid additional cardiac irregularities or cardiac arrest.

E. Intestinal Malabsorption

Decreased absorption of nutrients may occur either as a result of defective intraluminal digestion (maldigestion) associated with pancreatic insufficiency (juvenile pancreatic atrophy, chronic pancreatitis) (Anderson and Low, 1965a,b) or because of defects in mucosal transport (malabsorption). Intestinal malabsorption is associated with several types of intestinal disease including chronic inflammatory diseases (lymphocytic–plasmacytic enteropathy, eosinophilic enteritis), granulomatous diseases (Johne's disease, intestinal parasitism), and lymphoma. The cardinal clinical signs of malabsorption include persistent or recurrent diarrhea, steatorrhea, and weight loss. In the horse, small-intestinal malabsorption such as that associated with granulomatous enteritis may be associated with weight loss, but diarrhea may not be present because of the compensatory capacity of the uninvolved cecum and colon.

The initial reports of primary or idiopathic intestinal malabsorption in dogs (Miller, 1960; Vernon, 1962; Kaneko *et al.*, 1965) were compared to nontropical sprue (adult celiac disease, gluten-induced enteropathy) of humans, but association with gluten sensitivity was not demonstrated. Wheat-sensitive enteropathy has been described in the Irish setter breed (Batt *et al.*, 1984d). Most of the dogs were seen between 7 months and 2 years of age and had poor weight gain, weight loss, inappetence, or hyperphagia. Diarrhea was not a consistent observation. The most consistent morphologic abnormality in peroral jejunal biopsies was partial villus atrophy. Enzymatic changes included decreased mucosal alkaline phosphatase and peptidases, whereas disaccharidases and GGT activities were unaffected. Recovery of morphological and biochemical abnormalities occurred in affected dogs that received cereal-free diets, but recurred when wheat flour was added to the ration. A variety of other causes of intestinal malabsorption have been reported in the dog (Van Kruiningen, 1968; Ewing, 1971; Van Kruiningen and Hayden, 1972; Hill, 1972; Hill and Kelly, 1974; Schall, 1974; Anderson, 1975, 1977; Burrows *et al.*, 1979). An enteropathy said to resemble tropical sprue in humans has been described in German shepherds by Batt *et al.* (1983a). Affected dogs were five years of age or older and had diarrhea and weight loss for at least four months before the diagnosis was made. Peroral jejunal biopsies revealed partial villus atrophy and variable infiltrations of lymphocytes and plasma cells in the lamina propria. Subcellar biochemical studies of jejunal enterocytes revealed decreased activity of many brush-border enzymes and increased lysosomal enzymes.

Enteropathy associated with bacterial overgrowth of the small intestine also has been observed in German shepherds (Batt *et al.*, 1984). The dogs were 2 years of age or less, and all had chronic histories of intermittent diarrhea with or without weight loss. Bacterial counts of greater than 10^6 colonies per milliliter were observed in duodenal fluid. Enterococci, *E. coli,* and *Clostridium* spp. were identified in cultures. Peroral jejunal biopsies revealed no characteristic histopathologic changes. A deficiency of the immunoglobulin, IgA, may explain the vulnerability of these German shepherds to intestinal bacterial overgrowth (Whitbread *et al.*, 1984).

Intestinal malabsorption is reported less frequently in the cat than in the dog (Theran and Carpenter, 1968; Wilkinson, 1969). Malabsorption syndromes similar to those recognized in dogs are being recognized with increased frequency in farm animals (Blood *et al.*, 1979). Meuten *et al.* (1978), Cimprich (1974), and Merritt *et al.* (1976) have reported malabsorption in the horse secondary to chronic granulomatous enteritis, and specific amino acid malabsorption has been reported in Johne's disease (Patterson and Berrett, 1969).

Steatorrhea, the presence of excessive amounts of fat in the feces, is a prominent sign of intestinal malabsorption in dogs. The stools are bulky, gray or tan, and, grossly, may have an oily appearance. The normal dog excretes 3–5 g fat in the stool each day. This level of fecal fat is quite constant and is independent of dietary fat intake over a wide range of 15 to 48 g/day (Heersma and Annegers, 1948). In intestinal malabsorption, the ability to absorb fat is decreased and fecal fat excretion increases significantly. Under these conditions, the amount of fecal fat excreted becomes proportional to dietary intake.

Merritt *et al.*, (1979) reported that body weight is an important factor in fat output. In small dogs (i.e., less than 10–15 kg body weight) with intestinal malabsorption, the abnormality in fecal fat output was quantitatively less severe than in larger dogs. Fecal fat excretion for normal dogs was 0.24 $\pm$ 0.01 g/kg body weight per day.

Steatorrhea can be documented qualitatively by staining the fresh stool with a lipophilic stain, such as Sudan III, and observing increased numbers of oil droplets under the light microscope. In experienced hands, this method is a reliable diagnostic procedure (Drummey *et al.*, 1961). The following methods can be used to demonstrate neutral and split fats. For neutral fat, two drops of water are added to a stool sample on a glass slide and mixed. Two drops of 95% ethanol are then added and mixed, followed by several drops of a saturated solution of Sudan III in 95% ethanol. A coverslip is applied to the mixture, which is then examined for yellow or pale-orange refractile globules of fat, particularly at the edges of the coverslip. Normally, two or three fat droplets per high-power field are present. A large number of neutral fat droplets suggests a lack of pancreatic lipase activity, that is, exocrine pancreatic insufficiency.

For free fatty acids (split fats), several drops of 36% acetic acid are added to a stool sample on a glass slide and mixed. Several drops of Sudan III solution are then added and mixed. A coverslip is applied, and the slide gently heated over an alcohol burner until it begins to boil. The slide is air cooled and then quickly heated again. This procedure is repeated two or three times. The warm slide is examined for stained free fatty acid droplets, which, when warm, appear as deep-orange fat droplets from which spicules and soaps, resembling the pinna of the ear, form as the preparation cools. Normal stools may contain many tiny droplets of fatty acids (up to 100 per high-power field). With increasing amounts of split fats, the droplets become larger and more numerous, which suggests an abnormality in fat absorption.

Quantitation of fecal fat is the most accurate method of assessing steatorrhea (Burrows *et al.*, 1979) with dietary fat balance being determined for a period of 48–72 hours. Fecal fat is analyzed using a modification of the technique of van de Kamer *et al.* (1949), which employs ether extraction of fecal lipid and titration of fatty acids. The results are expressed as grams of neutral fat excreted per 24 hours. Merritt *et al.* (1979) have suggested that dogs be fed 50 g fat per kilogram per day for two to three days prior to fecal collection. Analysis of a 24-hour collection of stool when this is done is believed to be as accurate as a 72-hour stool collection. Results are expressed as fat excretion in grams per kilogram body weight.

In addition to malabsorption of fat, the canine malabsorption syndrome is associated with decreased absorption of other nutrients. These defects in absorption are responsible for the progressive malnutrition that is a cardinal feature of the disease. There may be malabsorption of vitamin D and/or calcium that results in osteomalacia. Anemia may result from malabsorption of iron or of the B vitamins required for normal erythropoiesis. Malabsorption of vitamin K can result in hypoprothrombinemia and delayed clotting of blood. Glucose malabsorption has been documented by Kaneko *et al.* (1965), and it is likely that amino acids are similarly malabsorbed at the small-intestinal level. Carbohydrate and fat malabsorption unquestionably contribute to the energy deficit that results in weight loss. Amino acid malabsorption may contribute to development of hypoproteinemia, although increased intestinal loss of plasma proteins is believed to be more important.

The diagnosis of idiopathic canine malabsorption is made only after other primary inflammatory, neoplastic, or parasitic diseases of the intestine and diseases of the pancreas, liver, or stomach that result in defective intraluminal digestion are ruled out. The presence of parasitic infection is established by examining the feces for parasite cysts or ova. Other inflammatory or neoplastic diseases of the intestine may be suggested on the basis of clinical or radiological examination, but a definitive diagnosis usually depends on histopathological examination of an intestinal biopsy specimen.

Both idiopathic and secondary intestinal malabsorption must be differentiated from those diseases in which there is decreased intraluminal hydrolysis of nutrients. The latter are due most frequently to pancreatic exocrine insufficiency as a result of chronic pancreatitis or juvenile atrophy. In these diseases, hydrolysis of the major dietary constituents is reduced because of the lack of pancreatic enzymes. Intraluminal hydrolysis of fat may also be decreased because of a deficiency of bile salts caused either by decreased hepatic secretion or by bile-duct obstruction. Experimentally, however, complete diversion of bile flow in the dog actually has a quantitatively small effect on fat absorption (Wells *et al.*, 1955; Hill and Kidder, 1972a).

The problems of pancreatic exocrine deficiency are discussed in detail elsewhere, in the chapter on pancreatic function. The simplest and perhaps most widely used test to differentiate intestinal malabsorption from pancreatic exocrine insufficiency is that described by Jasper (1954). The test is employed to detect reduction in trypsin-like activity in the feces of dogs with decreased pancreatic exocrine secretion (Grossman, 1962). There is wide variation in normal activity, making interpretation of the test difficult (Frankland, 1969; Hill and Kidder, 1970; Burrows *et al.*, 1979).

The test reveals only the presence or absence of hydrolysis of gelatin and does not differentiate gelatinase activity produced by intestinal bacteria from that secreted by the pancreas. There is evidence in some species that trypsin is almost completely destroyed by bacteria during its passage through the intestine, and that the proteinase activity of the feces is primarily of bacterial origin (Borgstrom *et al.*, 1959). Despite these deficiencies, the test is clinically useful. Fecal gelatinase activity is detected consistently in cases of intestinal malabsorption and is virtually always absent in severe pancreatic exocrine insufficiency. Burrows *et al.* (1979) observed that in comparison to clinically normal dogs, the 24-hour trypsin output in dogs with pancreatic insufficiency was significantly lower, whereas in dogs with malabsorption the trypsin output was significantly higher.

An indirect method to detect chymotrypsin activity has been described as a means of differentiating dogs with pancreatic exocrine insufficiency from those with intestinal malabsorption (Imondin *et al.*, 1972; Freudiger and Bigler, 1977; Strombeck, 1978; Strombeck and Harrold, 1982; Batt *et al.*, 1979; Batt and Mann, 1981; Zimmer and Todd, 1985). The synthetic peptide *n*-benzoyltyrosine-*p*-aminobenzoic acid (bentiromide) is orally administered to dogs. If chymotrypsin is present in the duodenum, hydrolysis of the bentiromide occurs, and *p*-aminobenzoic acid (PABA) is released, which is subsequently absorbed and then excreted in the urine within 6 hours. The urine and/or plasma are analyzed for PABA. Less than 43% PABA excretion identifies dogs with suspected pancreatic exocrine insufficiency (Strombeck, 1978). Thirty- or 60-minute blood levels of PABA are used to detect dogs with pancreatic exocrine disease (Zimmer and Todd, 1985), but this method did not identify dogs with exocrine pancreatic insufficiency as consistently as the 6-hour urinary excretion (Strombeck and Harrold, 1982). Factors that may influence results of the bentiromide test include the rate of gastric emptying, intestinal absorption of the PABA, and peptide cleavage by other peptidases (Batt *et al.*, 1979).

A radioimmunoassay for trypsin-like immunoreactivity (TLI) is currently widely used to identify dogs with pancreatic exocrine insufficiency (Williams and Batt, 1983; Williams, *et al.*, 1987) and is useful in differentiating maldigestion from primary malabsorption. The TLI in normal dog serum is trypsinogen. The route of entry of trypsinogen into the systemic circulation is believed to be the pancreatic venous or lymphatic vessels. Trypsinogen release in inflammatory pancreatic disease (i.e., acute or chronic pancreatitis) may increase TLI values. Increased TLI values have been reported in a dog with confirmed pancreatic exocrine insufficiency that had normal PABA values and fecal proteolytic activity (Williams and Batt, 1986). In most cases of pancreatic exocrine insufficiency, TLI is remarkably reduced compared to that of normal dogs or dogs with intestinal malabsorption.

F. Tests of Intestinal Absorption

1. *Oleic Acid and Triolein Absorption*

Several tests have been developed for the clinical evaluation of intestinal absorptive capacity. The absorption of ^{131}I-labeled oleic acid and ^{131}I-labeled triolein has been studied extensively in normal dogs (Turner, 1958; Michaelson *et al.*, 1960), and Kaneko *et al.* (1965) used this test to study dogs with intestinal malabsorption. The day before administration of the ^{131}I-labeled compound, a small amount of Lugol's iodine solution is administered to block thyroidal uptake of the isotope. Tracer amounts of the test substances are mixed with nonradioactive carrier and are administered orally. Absorption is determined by measuring the radioactivity of the plasma at intervals following administration and calculating the percentage of the dose absorbed on the basis of plasma volume.

The ^{131}I-oleic acid and ^{131}I-triolein tests performed in sequence are used to differentiate steatorrhea caused by pancreatic enzyme deficiency from that caused by a primary defect in absorption (Kallfelz *et al.*, 1968). If steatorrhea is caused by a lack of pancreatic lipase, oleic acid absorption will be normal, whereas that of triolein, which requires lipolysis for absorption, will be significantly reduced. The absorption of both compounds is reduced in intestinal malabsorption.

2. *Vitamin A Absorption*

The vitamin A absorption test measures intestinal lipid absorption (Hayden and Van Kruiningen, 1976). Normal absorption of vitamin A requires secretion of bile and pancreatic enzymes. After oral administration of 200,000 units of vitamin A in normal dogs, serum vitamin A concentrations reach their peak at 6–8 hours, with values ranging between three and five times fasting serum levels. There are small differences in vitamin A absorption between breeds, and delayed gastric emptying will also alter results.

3. *Glucose Absorption*

The absorption of glucose can be evaluated by the oral glucose tolerance test (OGTT) where a oral test dose of glucose is given and the blood glucose levels are measured at half-hour intervals for 3 to 4 hours. In canine malabsorption, the OGTT curve of blood glucose is diminished or flat (Kaneko *et al.*, 1965). The test also has been used in the horse for evaluation of small intestinal malabsorption (Roberts and Hill, 1973). Dogs with pancreatic exocrine deficiency may have "diabetic" or high OGTT curves (Hill and Kidder, 1972b). The major disadvantage of this test is that it does not differentiate between decreased intestinal absorption and increased tissue uptake after absorption. This problem can be alleviated by comparing results of the OGTT with those of the intravenous glucose tolerance test (IVGTT). Hill and Kidder (1972a) reported that normal dogs on low-carbohydrate/high protein diets can have "diabetic" tolerance curves; thus, test dogs should be on a high-carbohydrate diet for 3 to 5 days before testing.

4. *D-Xylose Absorption*

D-Xylose is used clinically to evaluate intestinal absorption (Craig and Atkinson, 1988). D-Xylose is not metabolized by the body to any significant degree and the problems of evaluating tissue utilization that occur with glucose are avoided. Because large amounts of D-xylose must be used, the rate of absorption is proportional to luminal concentration and independent of active transport processes.

A D-xylose absorption test for dogs has been described by Van Kruiningen (1968). In this procedure, a standard 25-g dose of D-xylose is administered by stomach tube. During the 5-hour period after administration, the patient is confined in a metabolism cage and urine is collected. At the end of the 5-hour test period, the urine remaining in the bladder is removed by catheter and the total quantity excreted in 5 hours is determined. Normal dogs excreted an average of 12.2 g of the 25-g dose during the test period, with a range of 9.1–16.5 g. Because this test is dependent on the rate of intestinal absorption as well as the rate of renal excretion, it is necessary to establish that kidney function is normal.

A modified D-xylose tolerance test is now most widely used clinically (Hill *et al.*, 1970; Hayden and Van Kruiningen, 1973). Dogs are fasted overnight, a baseline blood sample is taken, and D-xylose is administered by stomach tube at the rate of 0.5 g/kg. A control test is performed on a normal dog simultaneously. Blood samples are taken at $\frac{1}{2}$, 1, 2, 3, 4, and 5 hours after administration. The D-xylose concentration in the blood is determined by the method of Roe and Rice (1948) or by the phloroglucinol microassay (Merritt and Duelly, 1983). The phloroglucinol procedure is more economical, requires less plasma, and is technically easier than the orcinol–ferric chloride procedure of Roe and Rice (Merritt and Duelly, 1983). Maximal blood levels of D-xylose are normally reached at 1 hour after administration. A D-xylose level of at least 45 mg/dl within 60–90 minutes is expected in normal dogs (Hill *et al.*, 1970).

The D-xylose absorption test is also used for differential diagnosis of equine diarrheal diseases (Roberts, 1974). Bolton *et al.* (1976) reported that a dosage of 0.5 g D-xylose per kilogram body weight was useful in detecting horses with intestinal malabsorption. The peak plasma concentration in normal horses is less than one-third of that seen in normal dogs given D-xylose at comparable doses.

5. *Vitamin B_{12} and Folic Acid Absorption*

Serum folate and vitamin B_{12} levels (Waters and Mollin, 1961) have been recommended as screening tests for dogs suspected of having intestinal malabsorption (Batt and Morgan, 1982b). The primary site of folate absorption is the jejunum and vitamin B_{12} is absorbed primarily in the ileum. Structural and functional disturbances of the small intestinal mucosa are associated with reduced serum levels of folate, vitamin B_{12}, or both depending on the segment(s) of the small intestine involved. Although serum folate levels can decrease markedly within several days, the folate concentration within erythrocytes decreases much more slowly, so low erythrocyte folate values may be a more accurate indicator of a chronic disorder. Overgrowth of folate-producing bacteria in the proximal small intestine can result in a spurious increase in serum folate levels and reduced serum levels of vitamin B_{12}. The reduction of vitamin B_{12} is the result of binding of the vitamin B_{12}–intrinsic factor complex by intestinal bacteria.

6. *Other Tests for Assessment of Intestinal Function*

Simultaneous evaluation of pancreatic exocrine function and intestinal absorptive function is used in dogs (Stradley *et al.*, 1979; Rogers *et al.*, 1980) and cats (Sherding *et al.*, 1982; Hawkins *et al.*, 1986). The combined bentiromide and D-xylose absorption tests has proven to be useful diagnostically in dogs. Blood is normally taken at 0, $\frac{1}{2}$, 1, $1\frac{1}{2}$, 2, $2\frac{1}{2}$, and 3 hours after oral administration of the test solution, but a single blood sample taken at $1\frac{1}{2}$ hours was adequate for differential diagnostic purposes (Stradley *et al.*, 1979). The combined bentiromide/D-xylose absorption test was of limited usefulness in cats because of marked individual variations. Peak blood PABA levels (60–120 minutes) and peak blood D-xylose levels (30–120 minutes) in healthy cats were less than those of normal dogs, and blood D-xylose levels in cats with infiltrative small bowel disease were not abnormal (Hawkins *et al.*, 1986).

The content of exhaled hydrogen gas has been evaluated as an indicator of carbohydrate malassimilation in dogs (Washabau *et al.*, 1986), cats (Muir *et al.*, 1991), calves (Holland *et al.*, 1986), and humans (Perman, 1991). Unabsorbed carbohydrate is fermented by bacteria in the colon to H_2 and organic acids. Of the H_2, 10–14% is absorbed and excreted by the lungs (Washabau *et al.*, 1986). Increases in pulmonary H_2 excretion can occur in normal dogs fed rations containing wheat or corn flour. Increased H_2 excretion normally occurs in most species receiving lactulose. Mild increases in H_2 excretion occur in normal humans and dogs receiving xylose, but not in cats.

Breath H_2 excretion has diagnostic value in determining mouth-to-cecum transit time and for identifying small-intestinal bacterial overgrowth (Muir *et al.*, 1991). False-negative H_2 breath tests have been seen in humans receiving antibiotics. Diet as well as variations in bacterial flora can also cause false-positive test results.

The nitrosonapthol test qualitatively measures urinary excretion of 4-hydroxyphenylacetic acid and related compounds that are intestinal bacterial degradation products of tyrosine. The test has been used to differentiate pancreatic or small-intestinal diarrheal diseases from those associated primarily with large-bowel disease (Burrows and Jezyk, 1983). The test was positive in 77% of the dogs with pancreatic and small-intestinal disease and in only 9.5% of those dogs with large-bowel disease. Positive tests were associated with bacterial overgrowth of the small intestine and became negative during antibiotic treatment that resulted in clinical improvement. The test may be useful in dogs to select patients with small-intestinal bacterial overgrowth that might respond to antibiotic therapy.

G. Small-Intestinal Bacterial Overgrowth

Small-intestinal bacterial overgrowth (SIBO) is defined as an increase in the number of bacterial organisms cultured from the duodenum or proximal jejunum during the interdigestive phase (Burrows *et al.*, 1994). In the dog, total bacterial counts exceeding 10^5 colony-forming units per milliliter (cfu/ml) of proximal jejunal or duodenal fluid and anaerobic bacterial counts exceeding 10^4 cfu/ml are seen. Healthy cats have higher numbers of bacterial flora in their small intestine than do other species (Johnston *et al.*, 1993), in numbers that approximate those for SIBO in dogs and man. In most clinical situations, SIBO occurs secondary to partial intestinal obstructions or other forms of primary mucosal injury, defective or deficient immunologic responses, and intestinal motility disorders. SIBO has been reported in dogs with chronic intestinal diseases (Rutgers *et al.*, 1988, 1993, 1995) and pancreatic exocrine insufficiency (Williams *et al.*, 1987; Simpson *et al.*, 1990). Much of the literature pertains to German shepherds with subnormal levels of IgA (Westermarck *et al.*, 1993; Delles *et al.*, 1993, 1994; Willard *et al.*, 1994). SIBO has also been reported in beagle dogs with normal IgA levels (Batt *et al.*, 1992).

In the absence of primary disease, dogs with SIBO may be subclinical or have chronic manifestations of diarrhea and/or weight loss. Chronic SIBO can cause inflammatory bowel disease, but histopathologic findings usually are mild to negligible. The lesions consist of villlus atrophy and infiltrations of lymphocytes and plasmacytes in the lamina propria. There is substantial but reversible biochemical injury to enterocytes of the brush-border membrane (Batt and McLean, 1987). Aerobic bacteria, such as enterococci and *Escherichia coli,* cause a selective loss of brush-border alkaline phosphatase activity and peroxisomal catalase, as well as changes that are consistent with mitochondrial disruption. There are exceptions but aerobic overgrowth is typical for the dog in contrast to anerobic overgrowth in humans. The high floral counts in the intestines of cats are thought to predispose them to certain nutritional deficiencies, such as taurine deficiency, and intestinal disturbances attributed to deconjugated bile salts (Johnston *et al.*, 1993).

All of the aforementioned histopathologic and biochemical changes can be improved by antibiotic administration (Batt *et al.*, 1993). Dietary supplementation with fructo-oligosaccharide in IgA-deficient German shepherds resulted in decreased bacterial counts in luminal fluid and intestinal mucosa tissue (Willard *et al.*, 1995). Plasma-cell infiltrations in jejunal villi were decreased by feeding different protein sources (Edwards *et al.*, 1995).

The culturing and quantitation of intestinal bacterial flora (Simpson *et al.*, 1990) is the definitive means of diagnosing SIBO. Less invasive diagnostic methods include determination of serum folate and cobalamin levels and the nitrosonaphthal test on urine (Burrows *et al.*, 1995). Deconjugation of bile acids by intestinal flora, with a subsequent disproportionate increase in unconjugated bile acids in the circulation, is seen in humans with bacterial overgrowth (Einarsson *et al.*, 1992).

H. Gastric Helicobacteriosis

Gastric helicobacteriosis has been associated with chronic gastritis, atrophic gastritis, peptic ulcer disease, adenocarcinoma, and lymphoma in humans (Handt *et al.*, 1994; Isaacson 1994; Parsonnett *et al.*, 1991). Therapeutic modalities in human medicine have been directed predominately at *Helicobacter pylori* (Labenz *et al.*, 1993; Cutler and Schubert, 1993). Spontaneous and experimental disease caused by *H. pylori* is also seen in nonhuman primates, dogs, cats, and pigs. Helicobacter infections in animals, based on studies of *H. pylori* and *H. felis*, has serious public health implications (Handt *et al.*, 1994). *H. felis, H. heilmannii* (*Gastrospirillum hominis*), and other large "gastrospirillum-like" bacteria have also been associated with clinical signs of vomiting and/or chronic gastritis in dogs, cats, and cheetahs (Geyer *et al.*, 1993; Eaton *et al.*, 1991). *H. mustelae* causes similar problems in the ferret (Gottfried *et al.*, 1990). Experimental disease induced by *H. pylori* and *H. felis* has been produced in gnotobiotic dogs, casts, pigs, rats, and mice (Lee *et al.*, 1992; Radin *et al.*, 1990).

The ulcerogenic and neoplastic manifestations of helicobacter-induced disease observed in humans has not been seen in the dog and cat. Lymphoplasmacytic cellular infiltration of the lamina propria of the stomach, accompanied by eosinophils and some neutrophils, is characteristic of *H. pylori* or *H. felis* infections. Lymphoid follicle formation, particularly in the antrum of the stomach, is a prominent histologic lesion in some cases. Helicobacter organisms can be demonstrated by impression smears, contrast-phase microscopy, urease mapping, and histopathology of gastric mucosa. Serologic activity of helicobacter infection can be monitored. Unique culture characteristics, fatty-acid composition, and DNA typing are utilized to identify subspecies of the organism.

The majority of dogs and cats with helicobacter infections are asymptomatic. Transmission of infection is thought to be by oral–oral or anal–oral mechanisms. The importance of this infection is its zoonotic potential to infect people, in whom much more serious health problems are encountered.

I. Intestinal Permeability

Changes in intestinal mucosal permeability can be a factor in the pathogenesis of mucosal injury and subsequent gastrointestinal disease (Burrows *et al.*, 1995, Sanderson and Walker, 1993). Whether as a primary or secondary disorder, increased permeability predisposes to the passage of intraluminal macromolecules across the intestinal mucosa. Depending on the noxious or antigenic characteristics of these macromolecules, pathologic features of toxic or immune-mediated injury may occur. A primary mucosal permeability defect is suspected in humans and Irish setters with gluten-induced enteropathy (Batt *et al.*, 1984; Hall and Batt, 1991a,b). Enhanced mucosal permeability due to small intestinal bacterial overgrowth has been reported in clinically healthy beagles (Batt *et al.*, 1992). Seondary permeability disorders have been resolved by appropriate treatment of giardiasis and bacterial overgrowth in dogs (Batt *et al.*, 1988; Hall and Batt, 1990).

Clinicopathologic evaluation of intestinal permeability is based on the oral administration of simple, nondigestible molecules (probes) and their recovery in urine (Papasouliatis *et al.*, 1993; Elwood *et al.*, 1993). Inappropriate levels of these probes in urine indicates abnormal macromolecular permeation through transcellular and/or paracellular pathways. Polyethylene glycols, ^{51}CR-labeled ethylenediaminetetraacetate (^{51}EDTA) (Hall *et al.*, 1989, 1990, Hall and Batt, 1991; Batt *et al.*, 1992) and nonhydrolyzable sugars have been used in permeability tests. The disaccharides cellobiose and lactulose and the monosaccharides mannitol and L-rhamnose are unable to penetrate healthy enterocytes (Papasouliatis *et al.*, 1993). In the presence of abnormal mucosal permeability, the disaccharides passively diffuse through the mucosa via paracellular pathways and the monosaccharides diffuse passively transcellularly.

Differential sugar absorption and calculated disaccharide-to-monosaccharide excretion ratio is preferred over single sugar measurements. The use of lactulose and mannitol in the evaluation of intestinal permeability has been reported in healthy cats (Papasouliotis *et al.*, 1993; Bulsma *et al.*, 1995). The cellobiose-to-mannitol urinary excretion ratio was increased in Irish setters with gluten-sensitive enteropathy (Hall and Batt, 1991). Simultaneous quantification of rhamnose, lactulose, 3-*O*-methyl-D-glucose, and xylose in urine by a unique chromatographic technique has been reported to assess both intestinal function and permeability (Sorensen *et al.*, 1993).

J. Protein-Losing Enteropathy

Albumin, IgG, and other plasma proteins are present in low concentration in normal gastrointestinal secretions. Because protein usually undergoes complete degradation within the intestinal lumen, it has been suggested that the gastrointestinal tract must have a physiological role in the catabolism of plasma proteins. The relative significance of this pathway, however, has been the subject of considerable controversy. Some investigators have concluded, for example, that as much as 50% or more of the normal catabolism of albumin (Glenert *et al.*, 1961, 1962; Campbell *et al.*, 1961; Wetterfors, 1964, 1965; Wetterfors *et al.*, 1965) and τ-globulin (Andersen *et al.*, 1963) may occur in the gastrointestinal tract. Others believe that the physiological role of the intestine in plasma protein catabolism is far less significant, accounting for about 10% of the total catabolism (Waldmann *et al.*, 1967, 1969; Katz *et al.*, 1960; Franks *et al.*, 1963a,b).

Regardless of questions concerning the physiological significance of the gastrointestinal tract in plasma protein catabolism, it is well established that normal intestinal losses are substantially increased in a variety of gastrointestinal diseases, collectively referred to as the protein-losing enteropathies (PLEs). The increased loss causes hypoproteinemia (especially hypoalbuminemia), which may be observed in various types of chronic enteric diseases. The excessive losses are the result of ulcerations or other mucosal changes that alter permeability or obstruct lymphatic drainage from the intestine. If severe, hypoalbuminemia may result in retention of fluid with development of ascites and subcutaneous edema of pendant areas.

Excessive plasma protein loss has been seen in swine with chronic ileitis (Nielsen, 1966), in calves with acute enteric infections (Marsh *et al.*, 1969), in cattle with parasitic or other inflammatory abomasal disease (Nielsen and Nansen, 1967; Halliday *et al.*, 1968; Murry, 1969) and in Johne's disease (Patterson *et al.*, 1967; Nielsen and Andersen, 1967; Patterson and Berrett, 1969). In addition to the classic mucosal and submucosal lesions of Johne's disease, Nielsen and Andersen (1967) demonstrated the presence of secondary intestinal lymphangiectasia. Meuten *et al.* (1978) observed PLE associated with granulomatous enteritis in two horses.

PLE is seen with some frequency in the dog (Campbell *et al.*, 1968; Farrow and Penny, 1969; Hill, 1972; Finco *et al.*, 1973; Hayden and Van Kruiningen, 1973; Mattheeuws *et al.*, 1974; Hill and Kelly, 1974; Milstein and Sanford, 1977; Barton *et al.*, 1978; Olson and Zimmer, 1978). The most common cause appears to be lymphocytic–plasmacytic enteropathy (Tams and Twedt, 1981; Tams, 1987). Intestinal lymphangiectasia also has been reported as a cause of increased intestinal protein loss. Increased plasma protein loss from the stomach has been seen in dogs with hypertrophic gastritis.

Increased intestinal protein loss is the most likely cause of the hypoalbuminemia associated with certain other enteric diseases, including lymphoma and malabsorptive syndromes. Munro (1974) demonstrated that protein loss in dogs with experimentally induced protein-losing gastropathy occurs by an intercellular route. Isotope-labeled polyvinylpyrrolidine (^{131}I-PVP), ^{51}Cr-labeled ceruloplasmin, and ^{51}Cr-labeled albumin have been used to evaluate enteric protein loss in the dog (Finco *et al.*, 1973; Barton *et al.*, 1978; Hill and Kelly, 1974; van der Gagg *et al.*, 1976; Olson and Zimmer, 1978). Fecal α-1-protease inhibitor (α1-PI) is minimally degraded as it passes down the gastrointestinal tract. In conditions where there is excessive loss of plasma protein into the gut, there is an increase in fecal α1-PI (Williams, 1991).

K. Canine Ulcerative Colitis

Canine ulcerative colitis was first described by Cello (1964). Since then, canine ulcerative colitis, including granulomatous colitis of boxer dogs, has been reported by others (Van Kruiningen *et al.*, 1965; Kennedy and Cello, 1966; Koch and Skelley, 1967; Sander and Langham, 1968; Ewing and Gomez, 1973, Gomez *et al.*, 1977; Russell *et al.*, 1971). In boxer dogs, the disease is characterized by intractable diarrhea that is often hemorrhagic. Histopathologically, there is a granulomatous or histiocytic submucosal infiltrate and the macrophages are laden with periodic-acid-Schiff-positive material. Electron photomicrography demonstrated bacteria in the macrophages (Russell *et al.*, 1977). Cases of ulcerative colitis in dogs have been attributed to trichuriasis, balantidiasis, protothecosis, histoplasmosis, eosinophilic ulcerative colitis, or neoplasia (Lorenz, 1975). Severe ulcerative colitis also has been reported in the cat, and in some the feline leukemia virus (FELV) is demonstrated. Shindel *et al.* (1978) have described colonic lesions in cats caused by feline panleukopenia.

Biochemical manifestations of ulcerative colitis depend upon the duration and severity of illness, the degree of colorectal involvement, and the presence of systemic complications. In severe cases of long duration with extensive colorectal involvement, hypoalbuminemia and hypergammaglobulinemia are often observed. Hypoalbuminemia is attributed to increased loss of plasma through the denuded and inflamed colorectal mucosa and hypergammaglobulinemia is the response to the continuing chronic inflammation.

VIII. DISTURBANCES OF RUMEN FUNCTION

The digestive process of ruminants differs from that of other animals because rumen microbial digestion occurs prior to other normal digestive processes. The short-chain fatty acids (acetic, propionic, and butyric acids) are the primary end products of rumen fermentation and are the chief sources of energy available to ruminants from the diet (Hungate *et al.*, 1961). Cellulose, which undergoes only limited digestion in most simple-stomached animals, is readily digested because of the cellulytic bacteria in the rumen. Significant quantities of nonprotein nitrogen (NPN) can also be used by ruminal bacteria for protein synthesis, and this bacterial protein subsequently can be utilized to meet the protein requirements of the animal. Under experimental conditions, ruminants may grow and reproduce while receiving diets containing only NPN, for example, urea, as sources of nitrogen. Bacterial production of vitamins can also meet essentially all the requirements of ruminants.

Although it is nutritionally essential, bacterial fermentation within the rumen presents certain unusual hazards for ruminants. For example, when rapid changes in diet occur, the products of fermentation can be released more rapidly than they can be removed or utilized. Acute rumen tympany, acute indigestion, or D-lactic acidosis and urea poisoning are diseases that result from such abrupt changes in diet (Hungate, 1966, 1968).

A. Acute Rumen Indigestion (Rumen Overload, Lactic Acidosis)

Acute rumen indigestion occurs in sheep or cattle consuming high-roughage diets when they inadvertently are allowed access to large amounts of readily fermentable carbohydrate, such as grain or apples (Dunlop, 1972). *Streptococcus bovis* is the rumen microorganism believed to be chiefly responsible for rapid fermentation and for production of large quantities of lactic acid (Hungate *et al.*, 1952; Krogh, 1963a,b).

When lactic acid accumulates more rapidly than it is absorbed, rumen pH falls and rumen atony develops. Rumen bacteria produce a racemic mixture of lactic acid. Some L-lactate is absorbed and metabolized by the liver and other tissues, but D-lactate cannot be utilized and contributes significantly to the acid load of the body. The excessive lactic acid production results in metabolic acidosis, characterized by reduced blood pH and HCO_3^- concentration and by a fall in urine pH from a normal alkaline value to as low as pH 5.0. Fluid accumulates in the rumen because of the increased osmolality of the rumen fluid. This accumulation of fluid into the rumen causes hemoconcentration, which in turn may lead to hypovolemic shock and death (Hyldgaard-Jensen and Simesen, 1966). If affected animals survive the initial period of rapid fermentation, chemical rumenitis induced by the hyperosmolality of the rumen fluid and by the excess lactic acid may develop. Secondary mycotic rumenitis may then follow, which in severe cases can be fatal. In surviving cattle, metastatic hepatic abscesses may also occur.

B. Acute Rumen Tympany (Bloat)

The rumen of mature cattle can produce 1.2–2.0 liters of gas per minute (Hungate *et al.*, 1965). The gas is the product of rumen fermentation and is composed primarily of carbon dioxide (CO_2) and methane. CO_2 is also released when salivary HCO_3^- comes in contact with the organic acids in the rumen. Under normal conditions, these large amounts of gas are continually removed by eructation.

Any factor that interferes with eructation can produce acute tympany of the rumen (bloat), leading to rapid death. Interruption of the normal eructation reflex or mechanical obstruction of the esophagus typically results in free-gas bloat. The most important form of bloat, however, is seen in cattle consuming large quantities of legumes or in feedlot cattle on high-concentrate diets. The primary factor in these more common forms of bloat is a change in the ruminal contents to a foamy or frothy character because of altered surface tension. Gas becomes trapped in small bubbles within the rumen and cannot be eliminated by eructation (Clarke and Reid, 1974).

The chemical changes that cause foam to form within the rumen are not fully understood. Some reports (Nichols, 1966; Nichols and Deese, 1966) suggest that plant pectin and pectin methyl esterase, a plant enzyme, are critical factors. The enzyme acts on pectin to release pectic and galacturonic acids, which greatly increase the viscosity of the rumen fluid, resulting in formation of a highly stable foam. A soluble legume protein fraction with ribulose diphosphate carboxylase activity has been suggested as another important dietary factor in the pathogenesis of bloat (Howarth, 1975). Slime-producing bacteria also have been incriminated in the pathogenesis of frothy bloat. These microorganisms produce an extracellular polysaccharide that results in stable foam formation.

Effective medical treatment and control are directed toward decreasing or preventing foam formation. This has been accomplished with certain nonionic detergents with surfactant properties that break up or prevent formation of foam within the rumen (Bartley, 1965). Another approach has been the prophylactic administration of sodium alkyl sulfonate, which inhibits pectin methyl esterase activity and prevents foam formation by eliminating the products of this enzyme reaction (Nichols, 1963). Antifoaming agents such as poloxalene administered before ingestion of bloat-producing diets have been shown to be effective prophylactically (Howarth, 1975; Stiles *et al.*, 1971). Silicone antifoaming agents also have been used for this purpose (Clark and Reid, 1974). Genetic selection of cattle that are less susceptible to rumen tympany has also been pursued (Howarth, 1975).

C. Urea Poisoning

Unlike monogastric animals, ruminants, via their microbial flora, can effectively use nonprotein nitrogen to meet some of their dietary protein requirements. Urea, biuret (Oltjen *et al.*, 1969) and ammonium salts (Webb *et al.*, 1973) can serve as dietary NPN sources. Urea, which is the most frequently used, is hydrolyzed by ruminal bacterial urease into CO_2 and NH_3. The free NH_3 is incorporated into amino acids and protein by the rumen microorganisms. The bacterial protein is digested and absorbed in the abomasum and small intestine along with dietary protein.

Signs of urea poisoning typically develop within minutes after consumption of food containing toxic amounts of urea. Clinical manifestations reflect the encephalotoxic effects of excess absorbed NH_3 (Word *et al.*, 1969; Elmer and Barclay, 1971). Tolerance to urea may be significantly increased by increasing the amount of urea in the diet gradually or by adding readily fermentable carbohydrate to the diet. Ruminants can actually adapt and thrive on a diet in which urea is the sole source of dietary nitrogen. However, if urea is fed at more than 3% in the diet in unadapted animals, toxic effects are very likely to occur.

Urea poisoning may occur accidentally when animals gorge on large amounts of urea-containing dietary supplement, when there has been an error in formulation of bulk feed, or when the urea-containing additive is incompletely mixed. Oral administration of acetic acid has been shown to reduce acute urea toxicity, apparently by decreasing absorption of free NH_3 from the rumen. Normally, NH_3 is in equilibrium,

$$NH_3 + H^+ \longrightarrow NH_4^+$$

with only 1% in the free form. Acidification shifts the equilibrium further to the right, thereby reducing the amount of the NH_3. Because only the free form crosses cell membranes, the net effect is a reduction of absorption of NH_3 by the cell. Acetic acid is used as a treatment for urea poisoning, but it is of more value as a prophylactic agent (Word *et al.*, 1969).

References

Aclund, H. M., Gunson, D. E., and Gillette, D. M. (1983). *Vet. Path.* **20,** 653.

Alpers, D. H. (1969). *J. Biol. Chem.* **244,** 1238.

Alpers, D. H., and Isselbacher, K. J. (1967). *J. Biol. Chem.* **242,** 5617.

Altamirano, M. (1963). *J. Physiol. (London)* **168,** 787.

Alvarado, F. (1966a). *Science* **151,** 1010.

Alvarado, F. (1966b). *Biochim. Biophys. Acta* **112,** 292.

Alvarado, F., and Crane, R. K. (1962). *Biochim. Biophys. Acta* **56,** 170.

Andersen, S. B., Glenert, J., and Wallevik, K. (1963). *J. Clin. Invest.* **42,** 1873.

Anderson, N. V. (1975). *In* "Textbook of Veterinary Internal Medicine—Diseases of the Dog and Cat" (S. J. Ettinger, ed.), Vol. 2, p. 1150. Saunders, Philadelphia, Pennsylvania.

Anderson, N. V. (1977). *In* "Current Veterinary Therapy" (R. W. Kirk, ed.), 6th ed., p. 912. Saunders, Philadelphia, Pennsylvania.

Anderson, N. V., and Low, D. G. (1965a). *J. Am. Anim. Hosp. Assoc.* **1,** 101.

Anderson, N. V., and Low, D. G. (1965b). *J. Am. Anim. Hosp. Assoc.* **1,** 189.

Argenzio, R. (1975). *Cornell Vet.* **65,** 303.

Aschaffenburg, R. (1949). *Br. J. Nutr.* **3,** 200.

Barros D'Sa, A. A. J., Bloom, S. R., and Baron, J. H. (1975). *Lancet I*, 886.

Bartley, E. E. (1965). *J. Am. Vet. Med. Assoc.* **147,** 1397.

Barton, C. L., Smith, C., Troy, G., Hightower, D., and Hood, D. (1978). *J. Am. Anim. Hosp. Assoc.* **14,** 85.

Batt, R. M., and McLean, L. (1987). *Gastroenterology* **93,** 986.

Batt, R. M., and Mann, L. C. (1981). *Vet. Rec.* **108,** 303.

Batt, R. M., and Morgan, J. O. (1982). *Res. Vet. Sci.* **32,** 17.

Batt, R. M., Bush, B. M., and Peters, T. J. (1979). *J. Small Anim. Pract.* **20,** 185.

Batt, R. M., Bush, B. M., and Peters, T. J. (1983). *Am. J. Vet. Res.* **44,** 1492.

Batt, R. M., Carter, M. W., and McLean, L. (1984a). *Clin. Sci.* **66,** 38.

Batt, R. M., Carter, M. W., and McLean, L. (1984b). *Res. Vet. Sci.* **37,** 339.

Batt, R. M., McLean, L., and Loughran, M. (1984d). *Gastroenterology* **86,** 1021.

Batt, R. M., Carter, M. W., and McLean, L. (1985). *Res. Vet. Sci.* **39,** 80.

Batt, R. M., McLean, L., and Riley, J. E. (1988). *Gut* **29,** 473.

Batt, R. M., Hall, E. J., McLean, L., and Simpson, K. W. (1992). *Am. J. Vet. Res.* **53,** 1935.

Becht, J. L., and Byars, T. D. (1986). *Equine Vet. J.* **18,** 307.

Ben Abdeljlil, A., and Desnuelle, P. (1964). *Biochim. Biophys. Acta* **81,** 136.

Benajaiba, A., and Maroux, S. (1980). *Eur. J. Biochem.* **107,** 381.

Benzonana, G., and Desnuelle, P. (1965). *Biochim. Biophys. Acta* **105,** 121.

Bihler, I., and Crane, R. K. (1962). *Biochim. Biophys. Acta,* **59,** 78.

Bihler, I., Hawkins, K. A., and Crane, R. K. (1962). *Biochim. Biophys. Acta* **59,** 94.

Black, J. W., Duncan, W. A., Durant, C. J., Ganellin, C. R., and Parsons, E. M. (1972). *Nature (London)* **236,** 385.

Blood, D. C., Henderson, J. A., and Radostits (1979). "Veterinary Medicine," pp. 101, 132, 535. Lea & Febiger, Philadelphia, Pennsylvania.

Bloom, B., Chaikoff, I. L., and Reinhardt, W. O. (1951). *Am. J. Physiol.* **166,** 451.

Bloom, S. R., Mortimer, C. H., and Thornen, M. G. (1974). *Lancet ii,* 1106.

Bockman, D. E., and Winborn, W. B. (1966). *Anat. Rec.* **155,** 603.

Bodanszky, A., Ondetti, M. A., Mutt, V., and Bodanszky, M. (1969). *J. Am. Chem. Soc.* **91,** 944.

Bolton, J. R., Merritt, A. M., Cimprich, R. E., Ramberg, C. F., and Street, W. (1976). *Cornell Vet.* **66,** 183.

Borgstrom, B., and Erlanson, C. (1971). *Biochim. Biophys. Acta* **242,** 509.

Borgstrom, B., Dahlqvist, A., Gustafsson, B. E., Lundh, G., and Malmquist, J. (1959). *Proc. Soc. Exp. Biol. Med.* **102,** 154.

Boyd, J. W. (1972). *Vet. Rec.* **90,** 645.

Braaten, B. A., and Myers, L. L. (1977). *Am. J. Vet. Res.* **38,** 1989.

Brambell, F. W. R. (1958). *Biol. Rev. Cambridge Philos. Soc.* **33,** 488.

Braun, R. K., Norcross, M. L., and Campbell, S. G. (1973). *J. Am. Vet. Med. Assoc.* **163,** 1188.

Breitschwerdt *et al.* (1986). p. 55.

Bremer, J. (1956). *Biochem. J.* **63,** 507.

Bro-Rasmussen, F., Killmann, S.-A., and Thaysen, J. H. (1956). *Acta Physiol. Scand.* **37,** 97.

Brown, J. R., and Hartley, B. S. (1966). *Biochem. J.* **101,** 214.

Brown, J. C., Mutt, V., and Dryburgh, J. R. (1971). *Can. J. Physiol. Pharmacol.* **49,** 399.

Brown, J. C., Cook, M. A., and Dryburgh, J. R. (1972). *Gastroenterology* **62,** 401.

Brown, J. C., Cook, M. A., and Dryburgh, J. R. (1973). *Can. J. Biochem.* **51,** 533.

Burrows, C. F. and Jezyk, P. F. (1983). *J. Am. Vet. Med. Assoc.* **183,** 318.

Burrows, C. F., Merritt, A. M., and Chiapella, A. M. (1979). *J. Am. Vet. Med. Assoc.* **174,** 62.

Burrows, C. F., Hall, E. J., Willard, M. D., and Williams, D. A. (1994). *Veterinary Exchange,* supplement to *Compendium,* December.

Burrows, C. F., Batt, R. M., and Sherding, R. G. (1995). *In* "Textbook of Veterinary Medicine" (S. J. Ettinger and E. C. Feldman, eds.), 4th ed., p. 1169. Saunders, Philadelphia, Pennsylvania.

Bush, L. J., Aguilera, M. A., Adams, G. D., and Jones, E. W. (1971). *J. Dairy Sci.* **54,** 1547.

Bywater, R. J., and Penhale, W. J. (1969). *Res. Vet. Sci.* **10,** 591.

Campbell, R. M., Cuthbertson, D. P., Mackie, W., McFarlane, A. S., Phillipson, A. T., and Sudsaneh, S. (1961). *J. Physiol. (London)* **158,** 113.

Campbell, R. S. F., Brobst, D. F., and Bisgard, G. (1968). *J. Am. Vet. Med. Assoc.* **153,** 1050.

Campbell-Thompson, M. L., and Merritt, A. M. (1987). *Am. J. Vet. Res.* **48,** 1511.

Carlson, D. M., McGuire, E. J., Jourdian, G. W., and Roseman, S. (1973). *J. Biol. Chem.* **248,** 5763.

Carpenter, C. C. J., Curlin, G. T., and Greenough, W. B. (1969). *J. Infect. Dis.* **120,** 332.

Cassel, D., and Pfeuffer, T. (1978). *Proc. Natl. Acad. Sci. USA* **75,** 2669.

Cassel, D., and Selinger, Z. (1978). *Proc. Natl. Acad. Sci. USA* **75,** 4155.

Caywood, D., Teague, H. D., Jackson, D. A., Levitt, M. D., and Bond, J. H. (1977). *J. Am. Anim. Hosp. Assoc.* **13,** 459.

Chew, C. S. (1985). *J. Biol. Chem.* **260,** 7540.

Christensen, H. N. (1984). *Biochim. Biophys. Acta* **779,** 255.

Christensen, H. N. (1985). *J. Membr. Biol.* **84,** 97.

Christensen, J., Weisbrodt, N. W., and Hauser, R. L. (1972). *Gastroenterology* **62,** 1167.
Ciment, G., and deVellis, J. (1978). *Science* **202,** 765.
Cimprich, R. E. (1974). *Vet. Pathol.* **11,** 535.
Clarke, R. T. J., and Reid, C. S. W. (1974). *J. Dairy Sci.* **57,** 753.
Cook, D. I., Van Lennep, E. W., Roberts, M. L., and Young, J. A. (1994). *In* "Physiology of the Gastrointestinal Tract," 3rd ed. (Leonard R. Johnson, ed.). Vol. 2, p. 1061, Raven Press, New York.
Cori, C. F. (1925). *J. Biol. Chem.* **66,** 691.
Cornell, R., Walker, W. A., and Isselbacher, K. J. (1971). *Lab. Invest.* **25,** 42.
Craig, R. M., and Atkinson, A. J. (1988). *Gastroenterology* **95,** 223.
Csaky, T. Z. (1961). *Am. J. Physiol.* **201,** 999.
Csaky, T. Z., and Hara, Y. (1965). *Am. J. Physiol.* **209,** 467.
Csaky, T. Z., and Ho, P. M. (1966). *Life Sci.* **5,** 1025.
Csaky, T. Z., and Lassen, U. V. (1964). *Biochim. Biophys. Acta* **82,** 215.
Curran, P. F. (1965). *Fed. Proc., Fed. Am. Soc. Exp. Biol.* **24,** 993.
Curran, P. F., and Schultz, S. G. (1968). *In* "Handbook of Physiology" (J. Field, ed.), Am. Physiol. Soc., Sect. 6, Vol. III, p. 1217. Williams & Wilkins, Baltimore.
Cutler, A. F. and Schubert, T. T. (1993). *Gastroenterology* **88,** 505.
Dahlqvist, A. (1961). *Nature (London)* **190,** 31.
Dalton, R. G., Fisher, E. W., and McIntyre, W. I. M. (1965). *Br. Vet. J.* **121,** 34.
Danforth, E., and Moore, R. O. (1959). *Endocrinology* **65,** 118.
Danielsson, H. (1963). *Adv. Lipid Res.* **1,** 335.
Darrow, D. C. (1964). "A Guide to Learning Fluid Therapy." Thomas, Springfield, Illinois.
Davenport, H. W. (1966). "Physiology of the Digestive Tract," 2nd ed. Yearbook Publ., Chicago.
Delles, E. K., Willard, M. D., Simpson, R. B., Fossum, T. W., Slater, M. R., and Reinhart, G. A. (1993). *J. Vet. Int. Med.* **3,** 132 (abstract).
Delles, E. K., Willard, M. D., Simpson, R. B., Fossum, T. W., Slater, M. R., Kelp, D., Lees, G. E., Helman, R., and Reinhart, G. A. (1994). *Am. J. Vet. Res.* **55,** 957.
Deutsch, H. F., and Smith, V. R. (1957). *Am. J. Physiol.* **191,** 271.
Dietschy, J. M., Salomon, H. S., and Siperstein, M. D. (1966). *J. Clin. Invest.* **45,** 832.
Donowitz, M., Charney, A. N., and Hynes, R. (1979). *Gastroenterology* **76,** 482.
Dougherty, R. W., Riley, J. L., and Cook, H. M. (1975a). *Am. J. Vet. Res.* **36,** 827.
Dougherty, R. W., Riley, J. L., Baetz, A. L., and Cook, H. M. (1975b). *Am. J. Vet. Res.* **36,** 833.
Dowling, J. E., and Wald, G. (1958). *Proc. Natl. Acad. Sci. USA* **44,** 648.
Drucker, M. M., Ogra, P. L., Goldhar, J., and Neter, E. (1977). *Infection* **5,** 211.
Drummey, G. D., Benson, J. A., and Jones, C. M. (1961). *N. Engl. J. Med.* **264,** 85.
Dryburgh, J. R., and Brown, J. C. (1975). *Gastroenterology* **68,** 1169.
Dukes, H. H. (1955). "The Physiology of Domestic Animals," 7th ed. Cornell Univ. Press (Comstock), Ithaca, New York.
Dunlop, R. H. (1972). *Adv. Vet. Sci. Comp. Med.* **16,** 259.
DuPont, H. L. (1978). *J. Am. Vet. Med. Assoc.* **173,** 649.
DuPont, H. L., Sullivan, P., Pickering, L. K., Haynes, G., and Ackerman, P. B. (1977). *Gastroenterology* **73,** 713.
Eaton, K. A., Radin, M. J., Kramer, L., Wack, R., Sherding, R., Krakowka, S., and Morgan, D. R. (1991). *Scand. J. Gastroenterol.* **26,** 38.
Edwards, J. F., Fossum, T. W., Willard, M. D., Cohen, N. D., Patterson, W. B., and Carey, D. P. (1995). *Am. J. Vet. Res.* **56,** 340.
Eichholz, A. (1967). *Biochim. Biophys. Acta* **135,** 475.
Eichholz, A., and Crane, R. K. (1965). *J. Cell Biol.* **26,** 687.
Einarsson, K., Bergstrom, M., Eklof, R., Nord, C. E., and Bjorkhem, I. (1992). *Scand. J. Clin. Lab. Invest.* **52,** 425.
Elwood, C. M., Rutgers, H. C., Sorensen, S. H., Proud, F. J., and Batt, R. M. (1993). *J. Vet. Int. Med.* **7,** 131 (abstract).
Emas, S., and Grossman, M. I. (1967). *Gastroenterology* **52,** 29.
Erickson, R. H., Bella, A. M., Brophy, E. J., Kobata, A., and Kim, Y. S. (1983). *Biochim. Biophys. Acta* **751,** 258.
Ericsson, C. D., Evans, D. G., DuPont, H. L., Evans, D. J., and Pickering, L. K. (1977). *J. Infect. Dis.* **136,** 693.
Espersen, G., and Simesen, M. G. (1961). *Nord. Veterinaermed.* **13,** 147.
Ewing, G. O. (1971). *In* "Current Veterinary Therapy" (R. W. Kirk, ed.), 4th ed., p. 551. Saunders, Philadelphia, Pennsylvania.
Ewing, G. O., and Gomez, J. A. (1973). *J. Am. Anim. Hosp. Assoc.* **9,** 395.
Farris, R. K., Tapper, E. J., Powell, D. W., and Morris, S. M. (1976). *J. Clin. Invest.* **57,** 916.
Farrow, B. R. H., and Penny, R. (1969). *J. Small Anim. Pract.* **10,** 513.
Faust, R. G., Wu, S. L. and Faggard, M. L. (1967). *Science* **155,** 1261.
Fey, H. (1971). *Ann. NY Acad. Sci.* **176,** 49.
Field, M., Graf, L. H., Laird, W. J., and Smith, P. L. (1978). *Proc. Natl. Acad. Sci.* **75,** 2800.
Finco, D. R., Schall, W. B., Hooper, B. E., Chandler, F. W., and Keating, K. A. (1973). *J. Am. Vet. Med. Assoc.* **163,** 262.
Fisher, E. W. (1965). *Br. Vet. J.* **121,** 132.
Fisher, E. W., and McEwan, A. D. (1967a). *Vet. Rec.* **80,** 290.
Fisher, E. W., and McEwan, A. D. (1967b) *Br. Vet. J.* **123,** 4.
Fisher, R. B., and Parsons, D. S. (1953). *J. Physiol. (London)* **119,** 224.
Fisher, E. W., Martinez, A. A., Trainin, Z., and Meirom, R. (1975). *Br. Vet. J.* **131,** 402.
Forstner, C. G., Sabesin, S. M., and Isselbacher, K. J. (1968). *Biochem. J.* **106,** 381.
Forsyth, G. W., Hamilton, D. L., Goertz, K. E., and Johnson, M. R. (1978). *Infect. Immun.* **21,** 373.
Frankland, A. L. (1969). *J. Small Anim. Pract.* **10,** 531.
Franks, J. J., Mosser, E. L., and Anstadt, G. L. (1963a). *J. Gen. Physiol.* **46,** 415.
Franks, J. J., Edwards, K. W., Lackey, W. W., and Fitzgerald, J. B. (1963b). *J. Gen. Physiol.* **46,** 427.
Freeman, H. J., and Kim, Y. S. (1978). *Annu. Rev. Med.* **29,** 99.
Freudiger, U. and Bigler, B. (1977). *Kleintier-Praxis* **22,** 73.
Gardner, M. G. (1982). *Quart. J. Exp. Physiol.* **67,** 629.
Gardner, M. G. (1984). *Biol. Rev.* **59,** 289.
Gay, C. C. (1965). *Bacteriol. Rev.* **29,** 75.
Gay, C. C., Anderson, N., Fisher, F. W., and McEwan, A. D. (1965a). *Vet. Rec.* **77,** 148.
Gay, C. C., Fisher, E. W., and McEwan, A. D. (1965b). *Vet. Rec.* **77,** 994.
Geyer, C., Colbatzky, F., Lechner, J., and Hermanns, W. (1993). *Vet. Rec.*, July 3, p. 18.
Glenert, J., Jarnum, S., and Riemer, S. (1961). *Acta Chir. Scand.* **121,** 242.
Glenert, J., Jarnum, S., and Riemer, S. (1962). *Acta Chir. Scand.* **124,** 63.
Goldner, A. M., Schultz, S. G., and Curran, P. E. (1969). *J. Gen. Physiol.* **53,** 362.
Gomez, J. A., Russell, S. W., Trowbridge, J., and Lee, J. (1977). *Am J. Dig. Dis.* **22,** 485.
Gots, R. E., Formal, S. B., and Giannella, R. A. (1974). *J. Infect. Dis.* **130,** 280.
Gottfried, M. R., Washington, K., and Harrell, L. J. (1990). *Gastroenterology* **85,** 813.
Grant, D. A. W. and Herman-Taylor, J. (1976). *Biochem. J.* **155,** 243.
Gray, G. M. (1975). *N. Engl. J. Med.* **292,** 1225.
Gray, J. S., and Bucher, G. R. (1941). *Am. J. Physiol.* **133,** 542.
Gray, G. M., and Cooper, H. L. (1971). *Gastroenterology* **61,** 535.
Gray, G. M., Lally, B. C., and Conklin, K. A. (1979). *J. Biol. Chem.* **254,** 6038.
Gregory, R. A. (1966). *Gastroenterology* **51,** 953.

Gregory, R. A. (1967). *In* "Handbook of Physiology" (J. Field, ed.), Am. Physiol. Soc., Sect. 6, Vol. II, p. 827. Williams & Wilkins, Baltimore.

Gregory, R. A., Hardy, P. M., Jones, D. S., Kenner, G. W., and Sheppard, R. C. (1964). *Nature (London)* **204,** 931.

Grossman, M. I. (1962). *Proc. Soc. Exp. Biol. Med.* **110,** 41.

Grossman, M. I., and Konturek, S. J. (1974). *Gastroenterology* **66,** 517.

Gustafsson, B. E., and Norman, A. (1962). *Proc. Soc. Exp. Biol. Med.* **110,** 387.

Gustafsson, B. E., and Norman, A. (1969a). *Br. J. Nutr.* **23,** 429.

Gustafsson, B. E., and Norman, A. (1969b). *Br. J. Nutr.* **23,** 627.

Gustafsson, B. E., Bergstrom, S., Lindstedt, S., and Norman, A. (1957). *Proc. Soc. Exp. Biol. Med.* **94,** 467.

Gustafsson, B. E., Midtvedt, T., and Norman, A. (1966). *J. Exp. Med.* **123,** 413.

Gyles, C. L., and Zigler, M. (1978). *Can. J. Comp. Med. Vet. Sci.* **42,** 260.

Hakim, A. A., and Lifson, N. (1964). *Am. J. Physiol.* **206,** 1315.

Hall, E. J., and Batt, R. M. (1990). *J. Am. Med. Assoc.* **196,** 91.

Hall, E. J., and Batt, R. M. (1991a). *Gut* **32,** 749.

Hall, E. J., and Batt, R. M. (1991b). *Res. Vet. Sci.* **51,** 83.

Halliday, G. J., Mulligan, W., and Dalton, R. G. (1968). *Res. Vet. Sci.* **9,** 224.

Hamilton, D. L., Forsyth, G. W., Roe, W. E., and Nielsen, N. O. (1978a). *Can. J. Comp. Med. Vet. Sci* **42,** 316.

Hamilton, D. L., Johnson, M. R., Forsyth, G. W., Roe, W. E., and Nielsen, N. O. (1978b). *Can. J. Comp. Med. Vet. Sci.* **42,** 327.

Handt, L. K., Fox, J. G., Dewhirst, F. E., Fraser, G. J., Paster, B. J., Yan, L. L., Rozmirek, H., Rufo, R., and Stalis, I. H. (1994). *Infect. Immun.* **62,** 2367.

Happe, R. P., and van der Gagg, I. (1977). *J. Small Anim. Pract.* **18,** 179.

Hartley, B. S., and Kauffman, D. L. (1966). *Biochem. J.* **101,** 229.

Hartley, B. S., Brown, J. R., Kauffman, D. L., and Smillie, L. B. (1965). *Nature (London)* **207,** 1157.

Hartman, P. A., Hays, V. W., Baker, R. O., Neagle, L. H., and Catron, D. V. (1961). *J. Anim. Sci.* **20,** 114.

Haslewood, G. A. D. (1964). *Biol. Rev. Cambridge Philos. Soc.* **39,** 537.

Hawkins, E. C., Meric, S. M., Washabau, R. J., Feldman, E. C., and Turrel, J. M. (1986). *Am. J. Vet. Res.* **47,** 567.

Hayden, D. W., and Van Kruiningen, H. J. (1973). *J. Am. Vet. Med. Assoc.* **162,** 379.

Hayden, D. W., and Van Kruiningen, H. J. (1976). *J. Am. Anim. Hosp. Assoc.* **12,** 31.

Heersma, J. R., and Annegers, J. H. (1948). *Am. J. Physiol.* **153,** 143.

Heizer, W. D., and Laster, L. (1969). *Biochim. Biophys. Acta* **185,** 409.

Herriott, R. M. (1938). *J. Gen. Physiol.* **21,** 501.

Hill, F. W. G. (1972). *J. Small Anim. Pract.* **13,** 575.

Hill, F. W. G., and Kelly, D. F. (1974). *Am. J. Dig. Dis.* **19,** 649.

Hill, F. W. G., and Kidder, D. E. (1970). *J. Small Anim. Pract.* **11,** 191.

Hill, F. W. G., and Kidder, D. E. (1972a). *J. Small Anim. Pract.* **13,** 23.

Hill, F. W. G., and Kidder, D. E. (1972b). *Br. Vet. J.* **128,** 207.

Hill, F. W. G., Kidder, D. E., and Frew, J. (1970). *Vet. Rec.* **87,** 250.

Hofmann, A. F. (1963). *Biochem. J.* **89,** 57.

Hofmann, A. F. (1966). *Gastroenterology* **50,** 56.

Hofmann, A. F., and Small, D. M. (1967). *Annu. Rev. Med.* **18,** 333.

Holland, R. E., Herdt, T. H., and Refsal, K. R. (1986). *Am. J. Vet. Res.* **47,** 2020.

Holmgren, J., Lange, S., and Lonnroth, I. (1978). *Gastroenterology* **75,** 1103.

Hopfer, U., and Groseclose, R. (1980). *J. Biol. Chem.* **255,** 4453.

Houpt, T. R., and Houpt, K. A. (1971). *Am. J. Vet. Res.* **32,** 579.

House, J. A., and Baker, J. A. (1968). *J. Am. Vet. Med. Assoc.* **152,** 893.

Howarth, R. E. (1975). *Can. Vet. J.* **16,** 281.

Huber, J. T., Jacobson, N. L., Allen, R. S., and Hartman, P. A. (1961). *J. Dairy Sci.* **44,** 1494.

Hughes, J. M., Murad, F., Chang, B., and Guerrant, R. L. (1978). *Nature (London)* **271,** 755.

Hughey, R. P., and Curthoys, N. P. (1976). *J. Biol. Chem.* **251,** 7863.

Hungate, R. E. (1966). "The Rumen and Its Microbes." Academic Press, New York.

Hungate, R. E. (1968). *In* "Handbook of Physiology" (J. Field, ed.), Am. Physiol. Soc., Sect. 6, Vol. V, p. 2725. Williams & Wilkins, Baltimore.

Hungate, R. E., Dougherty, R. W., Bryant, M. P., and Cello, R. M. (1952). *Cornell Vet.* **42,** 423.

Hungate, R. E., Mah, R. A., and Simesen, M. (1961). *Appl. Microbiol.* **9,** 554.

Hungate, R. E., Fletcher, D. W., Dougherty, R. W., and Barrentine, B. F. (1965). *Appl. Microbiol.* **13,** 161.

Hyldgaard-Jensen, J., and Simesen, M. G. (1966). *Nord. Veterinaermed.* **18,** 73.

Imondi, A. R., Stradley, R. P., and Wolgemuth, R. (1972). *Gut* **13,** 726.

Isaacson, P. G. (1994). *N. Engl. J. Med.* **330,** 1310.

Isselbacher, K. J. (1966). *Gastroenterology* **50,** 78.

Isselbacher, K. J. (1967). *Fed. Proc., Fed. Am. Soc. Exp. Biol.* **26,** 1420.

Itoh, Z., Takeuchi, S., Aizawa, I., Mori, K., Taminato, T., Seino, Y., Imura, H., and Yanaihara, N. (1978). *Am J. Dig. Dis.* **23,** 929.

Jasper, D. E. (1954). *North Am. Vet.* **35,** 523.

Jennewein, H. M., Hummelt, H., Siewert, R., and Waldeck, F. (1975). *Digestion* **13,** 246.

Johnson, G. L., Kaslow, H. R., and Bourne, H. R. (1978). *Proc. Natl. Acad. Sci. USA* **75,** 3113.

Johnston, K., Lamport, A., and Batt, R. M. (1993). *Vet. Rec.* **132,** 362.

Jones, E. W., Hamm, D., and Bush, L. (1977). *Bovine Pract.* **10,** 48.

Kagan, K. G., and Schaer, M. (1983). *J. Am. Vet. Med. Assoc.* **182,** 703.

Kallfelz, F. A., Norrdin, R. W., and Neal, T. M. (1968). *J. Am. Vet. Med. Assoc.* **153,** 43.

Kaneko, J. J., Moulton, J. E., Brodey, R. S., and Perryman, V. D. (1965). *J. Am. Vet. Med. Assoc.* **146,** 463.

Katz, J., Rosenfeld, S., and Sellers, A. L. (1960). *Am. J. Physiol.* **200,** 1301.

Kennedy, P. C., and Cello, R. M. (1966). *Gastroenterology* **51,** 926.

Kim, Y. S., and Brophy, S. J. (1976). *J. Biol. Chem.* **251,** 3199.

Kim, Y. S., Birtwhistle, W., and Kim, Y. W. (1972). *J. Clin. Invest.* **51,** 1419.

Kim, Y. S., Kim, Y. W., and Sleisenger, M. H. (1974). *Biochim. Biophys. Acta* **370,** 283.

Kimmich, G. A. (1973). *Biochim. Biophys. Acta* **300,** 31.

Kipnis, R. M. (1978). *J. Am. Vet. Med. Assoc.* **173,** 182.

Knoop, F. C., and Abbey, D. M. (1981). *Can. J. Micro.* **27,** 754.

Koch, S. A., and Skelley, J. F. (1967). *J. Am. Vet. Med. Assoc.* **150,** 22.

Koch, A. R., Brazeau, P., and Gilman, A. (1956). *Am. J. Physiol.* **186,** 350.

Kohler, E. M. (1968). *Am. J. Vet. Res.* **29,** 2263.

Kohn, P., Dawes, E. D., and Duke, J. W. (1965). *Biochim. Biophys. Acta* **107,** 358.

Kolinska, J., and Semenza, G. (1967). *Biochim. Biophys. Acta* **146,** 181.

Krogh, N. (1963a). *Acta Vet. Scand.* **4,** 27.

Krogh, N. (1963b). *Acta Vet. Scand.* **4,** 41.

Kunkel, H. G. (1947). *Proc. Soc. Exp. Biol. Med.* **66,** 217.

Labenz, J., Gyenes, E., Ruhl, G. H., and Borsch, G. (1993). *Gastroenterology* **88,** 491.

Lack, L., and Weiner, I. M. (1961). *Am. J. Physiol.* **200,** 313.

Lack, L., and Weiner, I. M. (1966). *Am. J. Physiol.* **210,** 1142.

Lariviere, S., Lallier, R., and Moria, M. (1979). *Am. J. Vet. Res.* **40,** 130.

Laskowski, M., Jr., Haessler, H. A., Miech, R. P., Peanasky, R. J., and Laskowski, M. (1958). *Science* **127,** 1115.

Leaf, A., and Santos, R. F. (1961). *N. Engl. J. Med.* **264,** 335.

Leary, H. L., and Lecce, J. G. (1978). *Biol. Neonat.* **34,** 174.

Lecce, J. G. (1966). *Biol. Neonat.* **9,** 50.
Lecce, J. G., and Morgan, D. O. (1962). *J. Nutr.* **78,** 263.
Lecce, J. G., Morgan, D. O., and Matrone, G. (1964). *J. Nutr.* **84,** 43.
Lee, A., Krakowka, S., Fox, J. G., Otto, G., Eaton, K. A., and Murphy, J. C. (1992). *Vet Pathol.* **29,** 487.
Lees, G. E., Leighton, R. L., and Hart, R. (1977). *J. Am. Anim. Hosp. Assoc.* **13,** 463.
Leib, M. S., and Blass, C. E. (1984). *Compend. Cont. Educ. Prac. Vet.* **6,** 961.
Leib, M. S., Wingfield, W. E., Twedt, D. C., and Bottoms, G. D. (1984). *J. Am. Vet. Med. Assoc.* **185,** 205.
Levinson, S. L., and Blume, A. J. (1977). *J. Biol. Chem.* **252,** 3766.
Lindstedt, S., and Samuelsson, B. (1959). *J. Biol. Chem.* **234,** 2026.
Logan, E. F., and Penhale, W. J. (1971). *Vet. Rec.* **89,** 628.
Lorenz, M. D. (1975). *In* "Textbook of Veterinary Internal Medicine—Diseases of the Dog and Cat" (S. J. Ettinger, ed.), Vol. 2, p. 1203. Saunders, Philadelphia, Pennsylvania.
McBeath, D. G., Penhale, W. J., and Logan, E. F. (1971). *Vet. Rec.* **88,** 266.
McEwan, A. D., Fisher, E. W., and Selman, I. E. (1970a). *J. Comp. Pathol.* **80,** 259.
McEwan, A. D., Fisher, E. W., Selman, I. E., and Penhale, W. J. (1970b). *Clin. Chim. Acta* **27,** 155.
Marata and Nameska (1977). p. 43.
Marsh, C. L., Mebus, C. A., and Underdahl, N. R. (1969). *Am. J. Vet. Res.* **30,** 163.
Mattheeuws, A., DeRick, H., Thoonen, H., and Van der Stock, J. (1974). *J. Small Anim. Pract.* **15,** 757.
Meister, A., and Tate, S. S. (1976). *Annu. Rev. Biochem.* **45,** 559.
Merkley, D. F., Howard, D. R., Eyster, G. E., Krahwinkel, D. J., Sawyer, D. C., and Krehbiel, J. D. (1976a). *J. Am. Anim. Hosp. Assoc.* **12,** 143.
Merkley, D. F., Howard, D. R., Krehbiel, J. D., Eyster, G. E., Krahwinkel, D. J., and Sawyer, D. C. (1976b). *J. Am. Anim. Hosp. Assoc.* **12,** 149.
Merritt, A. M., and Duelly, P. (1983). *Am. J. Vet. Res.* **44,** 2184.
Merritt, A. M., Cimprich, R. E., and Beech, J. (1976). *J. Am. Vet. Med. Assoc.* **169,** 603.
Merritt, A. M., Burrows, C. F., and Cowgill, L. (1979). *J. Am. Vet. Med. Assoc.* **174,** 59.
Meuten, D. J., Butler, D. G., Thomsen, G. W., and Lumsden, J. H. (1978). *J. Am. Vet. Med. Assoc.* **172,** 326.
Meyer, J. H., and Kelly, G. A. (1977). *Am. J. Physiol.* **231,** 682.
Michaelson, S. M., El-Tamami, M. Y., Thomson, R. A. E., and Howland, J. W. (1960). *Am. J. Vet. Res.* **21,** 364.
Midtvedt, T., and Norman, A. (1967). *Acta Pathol. Microbiol. Scand.* **71,** 629.
Miller, R. M. (1960). *Mod. Vet. Pract.* **41,** 34.
Milstein, M., and Sanford, S. E. (1977). *Can. Vet. J.* **18,** 127.
Moon, H. W. (1978). *J. Am. Vet. Med. Assoc.* **172,** 443.
Moon, H. W., Whipp, S. C., and Skartvedt, S. M. (1976). *Am. J. Vet. Res.* **37,** 1025.
Moore, W. L., Jr., Bieberdorf, F. A., Morawski, S. G., Finkelstein, R. A., and Fordtran, J. S. (1971). *J. Clin. Invest.* **50,** 312.
Moore, R. M., Muir, W. W., and Granger, D. N. (1995). *J. Vet. Intern. Med.* **9,** 115.
Morgan, R. V. (1982). *Compend. Cont. Educ. Prac. Vet.* **4,** 677.
Muir, W. W. (1982). *J. Am. Vet. Med. Assoc.* **181,** 229.
Muir, P., Papasouliotis, K., Gruffydd-Jones, T. J., Cripps, P. J., and Harbour, D. A. (1991). *J. Am. Vet. Res.* **52,** 1104.
Munro, D. R. (1974). *Gastroenterology* **66,** 960.
Murata, H., and Namioka, S. (1977). *J. Comp. Path.* **87,** 431.
Murry, M. (1969). *Gastroenterology* **56,** 763.
Murthy, S. K., and Ganguly, J. (1962). *Biochem. J.* **83,** 460.
Mutt, V., and Jorpes, J. E. (1967). *Recent Prog. Horm. Res.* **23,** 483.
Nakano, M., Sumi, Y., and Miyakawa, M. (1977). *J. Biochem.* **81,** 1041.
Naylor, J. M., Kronfeld, D. S., Bech-Nielsen, S., and Bartholomew, R. C. (1977). *J. Am. Vet. Med. Assoc.* **171,** 635.
Neurath, H., and Walsh, K. A. (1976). *Proc. Natl. Acad. Sci. USA* **73,** 3825.
Newey, H., and Smyth, D. H. (1964). *Nature (London)* **202,** 400.
Nichols, R. E. (1963). *J. Am. Vet. Med. Assoc.* **143,** 998.
Nichols, R. E. (1966). *Am. J. Vet. Res.* **27,** 369.
Nichols, R. E., and Deese, D. (1966). *Am. J. Vet. Res.* **27,** 623.
Nielsen, K., and Andersen, S. (1967). *Nord. Veterinaermed.* **19,** 31.
Nielsen, K., and Nansen, P. (1967). *Can. J. Comp. Med. Vet. Sci.* **31,** 106.
Norris, H. T., Curran, P. F., and Schultz, S. G. (1969). *J. Infect. Dis.* **119,** 117.
Ockner, R. K., and Isselbacher, K. J. (1974). *Rev. Physiol. Biochem. Pharmacol.* **71,** 107.
Ockner, R. K., and Manning, J. (1974). *J. Clin. Invest.* **54,** 326.
Ockner, R. K., Hughes, F. B., and Isselbacher, K. J. (1969). *J. Clin. Invest.* **48,** 2367.
Olson, N. C., and Zimmer, J. F. (1978). *J. Am. Vet. Med. Assoc.* **173,** 271.
Oltjen, R. R., Williams, E. E., Slyter, L. L., and Richardson, G. V. (1969). *J. Animal Sci.* **29,** 816.
Papasouliotis, K., Gruffydd-Jones, T. J., Sparkes, A. H., Cripps, P. J., and Millard, W. J. (1993). *Am. J. Vet. Res.* **54,** 840.
Parsonnet, J., Friedman, M. D., Vandersteen, D. P., Chagn, Y., Vogelman, J. H., Orentreich, N., and Sibley, R. K. (1991). *N. Engl. J. Med.* **325,** 1127.
Parsons, B. J., Smyth, D. H., and Taylor, C. B. (1958). *J. Physiol. (London)* **144,** 387.
Patterson, D. S. P. (1967). *Vet. Rec.* **80,** 260.
Patterson, D. S. P., and Berrett, S. (1969). *J. Med. Microbiol.* **2,** 327.
Patterson, D. S. P., Allen, W. M., and Lloyd, M. K. (1967). *Vet. Rec.* **81,** 717.
Patterson, D. S. P., Allen, W. M., Berret, S., Ivins, L. N., and Sweasey, D. (1968). *Res. Vet. Sci.* **9,** 117.
Payne, L. C., and Marsh, C. L. (1962a). *J. Nutr.* **76,** 151.
Payne, L. C., and Marsh, C. L. (1962b). *Fed. Proc., Fed. Am. Soc. Exp. Biol.* **21,** 909.
Pearse, A. G. E., Polak, J. M., and Bloom, S. R. (1977). *Gastroenterology* **72,** 746.
Penhale, W. J., Christie, G., McEwan, A. D., Selman, I. E., and Fisher, E. W. (1967). *Vet. Rec.* **81,** 416.
Penhale, W. J., Christie, G., McEwan, A. D., Fisher, E. W., and Selman, I. E. (1970). *Br. Vet. J.* **126,** 30.
Penhale, W. J., Logan, E. F., and Stenhouse, A. (1971). *Vet. Rec.* **89,** 623.
Peric-Golia, L., and Socic, H. (1968). *Am. J. Physiol.* **215,** 1284.
Perman, J. A. (1991). *Can. J. Physiol. Pharmacol.* **69,** 111.
Peters, T. J. (1970). *Biochem. J.* **120,** 195.
Pfeiffer, N. E., and McGuire, T. C. (1977). *J. Am. Vet. Med. Assoc.* **170,** 809.
Phillips, R. W., and Knox, K. L. (1969). *J. Comp. Lab. Med.* **3,** 1.
Phillipson, A. T., (1977). *In* "Dukes Physiology of Domestic Animals" (M. J. Swenson, ed.), p. 250. Cornell Univ. Press, Ithaca, New York.
Playoust, M. R., and Isselbacher, K. J. (1964). *J. Clin. Invest.* **43,** 878.
Playoust, M. R., Lack, L., and Weiner, I. M. (1965). *Am. J. Physiol.* **208,** 363.
Polak, J. M., Pearse, A. G. E., and Heath, C. M. (1975). *Gut* **16,** 225.
Portnoy, B. L., DuPont, H. L., Pruitt, D., Abdo, J. A., and Rodriguez, J. T. (1976). *J. Am Vet. Med. Assoc.* **236,** 844.
Powell, D. W., Malawer, S. J., and Plotkin, G. R. (1968). *Am. J. Physiol.* **215,** 1226.
Powell, D. W., Tapper, E. J., and Morris, S. M. (1979). *Gastroenterology* **76,** 1429.

Pradayrol, L., Jornvall, H., Mutt, V., and Ribet, A. (1980). *FEBS Lett.* **109,** 55.
Radin, M. J., Eaton, K. A., Krakowa, S., Morgan, D. R., Lee, A., Otto, G., and Fox, J. (1990). *Infect. Immun.* **58,** 2606.
Rao, G. A., and Johnston, J. M. (1966). *Biochim. Biophys. Acta* **125,** 465.
Raphael, N., and Machen, T. E. (1984). *Am. J. Physiol.* **247,** G366.
Richards, K. L., and Douglas, S. D. (1978). *Microbiol. Rev.* **42,** 592.
Riklis, E., and Quastel, J. H. (1958). *Can. J. Biochem. Physiol.* **36,** 347.
Roberts, M. C. (1974). *Equine Vet. J.* **6,** 28.
Roberts, M. C., and Hill, F. W. G. (1973). *Equine Vet. J.* **5,** 171.
Rodriquez *et al.* (1984). p. 32, Table 15.6.
Roe, J. H., and Rice, E. W. (1948). *J. Biol. Chem.* **173,** 507.
Rogers, W. A., Stradley, R. P., Sherding, R. G., Powers, J., and Cole, C. R. (1980). *J. Am. Vet. Med. Assoc.* **177,** 1128.
Rogolsky, B., and Van Kruiningen, H. J. (1978). *J. Am. Anim. Hosp. Assoc.* **14,** 504.
Russell, S. W., Gomez, J. A., and Trowbridge, J. (1971). *Lab. Invest.* **25,** 509.
Rutgers, H. C., Batt, R. M., and Kelly, D. F. (1988). *J. Am Vet. Med. Assoc.* **192,** 1739.
Rutgers, H. C., Lamport, A., and Simpson, K. W. (1993). *J. Vet. Intern. Med.* **7,** 1993.
Rutgers, H. C., Batt, R. M., Elwood, E. M., and Lamport, A. (1995). *J. Am. Vet. Med. Assoc.* **206,** 187.
Sabesin, S. M., and Isselbacher, K. J. (1965). *Science* **147,** 1149.
Sachs, G., Chang, H. H., Rabon, E., Schackmann, R., Lewin, M., and Saccomani, G. (1976). *J. Biol. Chem.* **251,** 7690.
Samloff, I. M. (1971). *Gastroenterology* **60,** 586.
Sander, C. H., and Langham, R. F. (1968). *Arch. Pathol.* **85,** 94.
Sanderson, I. R., and Walker, W.,A. (1993). *Gastroenterology* **104,** 622.
Sandholm, M. (1974). *Res. Vet. Sci.* **17,** 32.
Sandholm, M. (1976). *Clin. Biochem.* **9,** 39.
Sandholm, J., and Kivisto, A. K. (1975). *J. Small Anim. Pract.* **16,** 201.
Sasaki, H., Rubacalva, B., Baetes, D., Blazquez, E., Srikant, C. B., Orci, L., and Unger, R. H. (1975). *J. Clin. Invest.* **56,** 135.
Saunders, S. J., and Isselbacher, K. J. (1965). *Biochim. Biophys. Acta* **102,** 397.
Schachter, D., Finkelstein, J. D., and Kowarski, S. (1964). *J. Clin. Invest.* **43,** 787.
Schachter, H., McGuire, E. J., and Roseman, S. (1971). *J. Biol. Chem.* **246,** 5321.
Schall, W. D. (1974). *In* "Current Veterinary Therapy" (R. W. Kirk, ed.), 5th ed., p. 742. Saunders, Philadelphia, Pennsylvania.
Schlatz, L. J., Bools, C., and Reimann, E. M. (1981). *Biochim. Biophys. Acta* **637,** 539.
Schultz, S. G. (1977). *Am. J. Physiol.* **233,** E249.
Schultz, S. G., and Curran, P. F. (1970). *Physiol. Rev.* **50,** 637.
Schultz, S. G., and Zalusky, R. (1964). *J. Gen. Physiol.* **47,** 1043.
Schultz, G., Schultz, K., and Hardman, J. G. (1975). *Metabolism* **24,** 429.
Schwartz, C. J., Kimberg, D. V., and Ware, P. (1975). *Gastroenterology* **68,** 94.
Selman, I. E., McEwan, A. D., and Fisher, E. W. (1971). *Res. Vet. Sci.* **12,** 1.
Semenza, G., Kessler, M., Hosang, M., Weber, J., and Schmidt, U. (1984). *Biochim. Biophys. Acta* **779,** 343.
Serebro, H. A., Iber, F. L., Yardley, J. H., and Hendrix, T. R. (1969). *Gastroenterology* **56,** 506.
Sherding, R. G., Stradley, R. P., Rogers, W. A., and Johnson, S. E. (1982). *Am. J. Vet. Res.* **43,** 2272.
Shindel, N. M., Van Kruiningen, H. J., and Scott, F. W. (1978). *J. Am. Anim. Hosp. Assoc.* **14,** 738.
Simpson, K. W., Batt, R. M., Jones, D., and Morton, D. B. (1990). *Am. J. Vet. Res.* **51,** 203.
Smith, H. W. (1962). *J. Pathol. Bacteriol.* **84,** 147.
Smith, H. W., and Halls, S. (1967). *J. Pathol. Bacteriol.* **93,** 531.
Smith, H. W., O'Neil, J. A., and Simmons, E. J. (1967). *Vet. Rec.* **80,** 664.
Soll, A. H., and Grossman, M. I. (1978). *Am. Rev. Med.* **29,** 495.
Sorensen, S. H., Proud, F. J., Adam, A., Rutgers, H. C., and Batt, R. M. (1993). *Clinica Chimica Acta* **221,** 115.
Stevens, B. R., Kaunitz, J. D., and Wright, E. M. (1984). *Ann. Rev. Physiol.* **46,** 417.
Stiles, D. A., Bartley, E. E., Kilgore, G. L., Boren, F. W., and Perry, H. B. (1971). *J. Dairy Sci.* **54,** 65.
Stone, S. S., and Gitter, M. (1969). *Br. Vet. J.* **125,** 68.
Stradley, R. P., Stern, R. J., and Heinhold, N. B. (1979). *Am. J. Vet. Res.* **40,** 1201.
Straus, E., Johnson, G. F., and Yalow, R. S. (1977). *Gastroenterology* **72,** 380.
Strombeck, D. R. (1978). *J. Am. Vet. Med. Assoc.* **173,** 1319.
Strombeck, D. R., and Harrold, D. (1982). *J. Am. Vet. Med. Assoc.* **180,** 419.
Sutherland, E. W., and de Duve, C. (1948). *J. Biol. Chem.* **175,** 663.
Sutherland, E. W., and Rall, T. W. (1960). *Pharmacol. Rev.* **12,** 265.
Svendsen, P. (1969). *Nord. Veterinaermed.* **21,** Suppl. I.
Tams, T. R. (1987). *Compend. Cont. Educ. Pract. Vet.* **9,** 1192.
Tams, T. R., and Twedt, D. C. (1981). *Compend. Cont. Educ. Pract. Vet.* **3,** 105.
Tapper, E. J., Powell, D. W., and Morris, S. M. (1978). *Am. J. Physiol.* **235,** E402.
Taraval, F. R., Datsura, R., Woloszczuk, W., Marshall, J. J., and Whelan, W. J. (1983). *Eur. J. Biochem.* **130,** 147.
Tasker, J. B. (1967). *Cornell Vet.* **57,** 668.
Taylor, W. H. (1959a). *Biochem. J.* **71,** 73.
Taylor, W. H. (1959b). *Biochem. J.* **71,** 373.
Tennant, B., Harrold, D., and Reina-Guerra, M. (1968). *Cornell Vet.* **58,** 136.
Tennant, B., Harrold, D., Reina-Guerra, M., and Laben, R. C. (1969a). *Am. J. Vet. Res.* **30,** 345.
Tennant, B., Reina-Guerra, M., Harrold, D., and Goldman, M. (1969b). *J. Nutr.* **97,** 65.
Tennant, B., Harrold, D., and Reina-Guerra, M. (1972). *J. Am. Vet. Med. Assoc.* **161,** 993.
Tennant, B., Harrold, D., and Reina-Guerra, M. (1975). *Cornell Vet.* **65,** 457.
Tennant, B., Ward, D. E., Braun, R. K., Hunt, E. L., and Baldwin, B. H. (1978). *J. Am. Vet. Med. Assoc.* **173,** 654.
Tennant, B., Baldwin, B. H., Braun, R. K., Norcross, N. L., and Sandholm, M. (1979). *J. Am. Vet. Med. Assoc.* **174,** 848.
Theran, P., and Carpenter, J. (1968). *J. Am. Vet. Med. Assoc.* **152,** 65.
Thomas, D. D., and Knoop, F. C. (1982). *J. Infect. Dis.* **147,** 450.
Thompson, G. A., and Meister, A. (1975). *Proc. Natl. Acad. Sci. USA* **72,** 1985.
Thornton, J. R., Willoughby, R. A., and McSherry, B. J. (1972). *Can. J. Comp. Med.* **36,** 17.
Todoroff, R. J. (1979). *Compend. Cont. Educ. Prac. Vet.* **1,** 42.
Turner, D. A. (1958). *Am. J. Dig. Dis.* **3,** 594.
Ussing, H. H. (1947). *Nature (London)* **160,** 262.
Valverde, I., Rigopoulou, D., Marco, J., Faloona, G. R., and Unger, R. H. (1970). *Diabetes* **19,** 614.
van de Kamer, J. H., ten Bokkel Huinink, H., and Weyers, H. A. (1949). *J. Biol. Chem.* **177,** 347.
van der Gagg, I., Happe, R. P., and Wolvehomp, W. T. C. (1976). *Vet. Pathol.* **13,** 172.
van der Gagg, I., Happe, R. P., and Lamers, C. B. H. W. (1978). *Vet. Pathol.* **15,** 573.
Van Kruiningen, H. J. (1971). *In* "Current Veterinary Therapy" (R. W. Kirk, ed.), 4th ed., p. 544, Saunders, Philadelphia, Pennsylvania.
Van Kruiningen, H. J. (1977). *Vet. Pathol.* **14,** 19.

Van Kruiningen, H. J., and Hayden, D. W. (1972). *Vet. Clin. North Am.* **2,** 29.
Van Kruiningen, H. J., Montali, R. J., Strandberg, J. D., and Kirk, R. W. (1965). *Pathol. Vet.* **2,** 521.
Van Kruiningen, H. J., Gregorie, K., and Meuten, D. J. (1974). *J. Am. Anim. Hosp. Assoc.* **10,** 294.
Vellenga, L., Egberts, H. J. A., Wensing, T., van Dijk, J. E., Mouwen, J. M. V. M., and Brevkink, H. J. (1992). *Am. J. Vet. Res.* **53,** 1180.
Vernon, D. F. (1962). *J. Am. Vet. Med. Assoc.* **140,** 1062.
Waldmann, T. A., Morell, A. G., Wochner, R. D., Strober, W., and Sternlieb, I. (1967). *J. Clin. Invest.* **46,** 10.
Waldmann, T. A., Wochner, R. D., and Strober, W. (1969). *Am. J. Med.* **46,** 275.
Walker, D. M. (1959). *J. Agric. Sci.* **52,** 357.
Wallmark, B., Stewart, H. B., Rabon, E., Saccomani, G., and Sachs, G. (1980). *J. Biol. Chem.* **255,** 5313.
Walsh, J. H., and Grossman, M. I. (1975). *N. Engl. J. Med.* **292,** 1324.
Warner, N. S., and Van Kruiningen, H. J. (1978). *J. Am. Anim. Hosp. Assoc.* **14,** 618.
Washabau, R. J., Strombeck, D. R., Buffinton, C. A., and Harrold, D. (1986). *J. Am. Vet. Res.* **47,** 1402.
Wasserman, R. H., and Taylor, A. N. (1966). *Science* **152,** 791.
Wasserman, R. H., and Taylor, A. N. (1968). *J. Biol. Chem.* **243,** 3987.
Wasserman, R. H., Corradino, R. A., and Taylor, A. N. (1968). *J. Biol. Chem.* **243,** 3978.
Waters, A. H., and Mollin, D. L. (1961). *J. Clin. Path.* **14,** 325.
Webb, K. E. (1986). *Fed. Proc.* **45,** 2268.
Webb, D. W., Bartley, E. E., and Meyer, R. M. (1973). *J. Anim. Sci.* **35,** 1263.
Weiner, I. M., and Lack, L. (1962). *Am. J. Physiol.* **202,** 155.
Weiner, I. M., and Lack, L. (1968). *In* "Handbook of Physiology" (J. Field, ed.), Am. Physiol. Soc., Sect. 6, Vol. III, p. 1439. Williams & Wilkins, Baltimore.
Wells, M. H., Shingleton, W. W., and Saunders, A. P. (1955). *Proc. Soc. Exp. Biol. Med.* **90,** 717.
Westermarck, E., Batt, R. M., Vaillant, C., and Wilberg, M. (1993). *Am. J. Vet. Res.* **51,** 203.
Westrom, B. R., Svendsen, J., Ohlsson, B. G., Tagesson, C., and Karlsson, B. W. (1984). *Biol. Neonat.* **46,** 20.
Wetterfors, J. (1964). *Acta Med. Scand.* **176,** 787.
Wetterfors, J. (1965). *Acta Med. Scand.* **177,** 243.
Wetterfors, J., Liljedahl, S.-O., Plantin, L.-O., and Birke, G. (1965). *Acta Med. Scand.* **177,** 227.
Whitbread, T. J., Batt, R. M., and Garthwaite, G. (1984). *Res. Vet. Sci.* **37,** 350.
Wilkinson, G. T. (1969). *J. Small Anim. Pract.* **10,** 87.
Willard, M. D., Simpson, R. B., Fossum, T. W., Cohen, N. D., Delles, E. K., Kolp, D. L., Carey, D. P., and Reinhart, G. (1994). *J. Am. Vet. Med. Assoc.* **204,** 1201.
Willard, M. D., Simpson, R. B., Delles, E. K., Cohen, N. D., Fossum, T. W., Kolp, D., and Reinhart, G. A. (1995). *Am. J. Vet. Res.* **55,** 654.
Williams, D. A., and Batt, R. M. (1983). *J. Small Anim. Pract.* **24,** 583.
Williams, D. A., and Batt, R. M. (1986). *J. Am. Anim. Hosp. Assoc.* **22,** 672.
Williams, D. A., Batt, R. M., and McLean, L. (1987). *J. Am. Vet. Med. Assoc.* **191,** 201.
Wingfield, W. E., Cornelius, L. M., and DeYoung, D. W. (1974). *J. Small Anim. Pract.* **15,** 41.
Wingfield, W. E., Betts, C. W., and Rawlings, C. A. (1976). *J. Am. Anim. Hosp. Assoc.* **12,** 136.
Wingfield, W. E., Twedt, D. C., Moore, R. W., Leib, M. S., and Wright, M. (1982). *J. Am. Vet. Med. Assoc.* **180,** 1070.
Wollenweber, J., Kottke, B. A., and Owen, C. A., Jr. (1965). *Clin. Res.* **13,** 410.
Wolosin, J. M. (1985). *Am. J. Physiol.* **248,** 6596.
Wood, P. C. (1955). *J. Pathol. Bacteriol.* **70,** 179.
Word, J. D., Martin, L. C., Williams, D. L., Williams, E. I., Panciera, R. J., Nelson, T. E., and Tillman, A. D. (1969). *J. Anim. Sci* **27,** 786.
Wright, E. M., and Peerce, B. E. (1985). *Ann. NY Acad. Sci.* **456,** 108.
Young, J. A., and Schneyer, C. A. (1981). *Austr. J. Exp. Biol. Med. Sci.* **59,** 1.
Zilversmit, D. B. (1965). *J. Clin. Invest.* **44,** 1610.
Zimmer, J. F., and Todd, S. E. (1985). *Cornell Vet.* **75,** 426.

C H A P T E R

16

Skeletal Muscle Function

GEORGE H. CARDINET III

I. INTRODUCTION

Skeletal muscles comprise approximately 50% of the body's weight; hence, skeletal muscle cells (myofibers) constitute the largest mass of cells in the body that have similar morphological and physiological properties. The protoplasmic properties of contractility and conductility that characterize myofibers are an expression of specialized sarcoplasmic organelles, principally membranes and filaments, which are highly ordered and compartmentalized to serve those functions which in turn impart a distinctive morphology to the myofibers. Studies of muscles have provided excellent examples of the integration of form and function whereby molecular and organelle structure and function can be directly related to cellular morphology and function, which in turn can be related to the morphology and function of the organs and organisms they serve.

Skeletal muscles of the body are appropriately considered to be organs. The forces developed during contraction differ qualitatively and quantitatively among different muscles of the body. In addition, skeletal muscles differ in their vascular and nerve supplies and constituent myofiber populations, all of which confer differences in their morphology and function. Whereas skeletal muscles participate in the functioning of virtually all organ systems of the body, skeletal muscle function must be considered when assessing other organ systems in healthy and diseased states.

An attempt is made in this chapter to outline the great diversity that exists in skeletal muscles. By so doing, it is hoped that the knowledge and techniques developed over the past three decades will find their rightful introduction and application to increase our recognition and understanding of neuromuscular disorders in clinical veterinary medicine.

There are, no doubt, numerous neuromuscular disorders in domestic animals that remain unrecognized today, and the observations made by Innes (1951) more than 45 years ago are worthy of citing here in hopes of stimulating the interest of young veterinary pathologists and clinicians to be diligent in their pursuit of disorders with muscular weakness:

> I have drawn attention before to the fact that no diseases are recognized until they are found by looking with great care. One need only scan the veterinary literature of the last twenty years for this to be confirmed, and a long list of conditions could be compiled which might have been discovered years previously if more meticulous methods of pathological examinations had been used. Veterinary Pathology has still a long road to travel before it achieves the profundity of human pathology, and this is not said in any deprecatory sense of past achievements. There are many reasons why this is so, but lack of volume of material studies and inattention to the lessons to be learned from medical work have played a part. In veterinary work, unfortunately, also the question of economics always intrudes, and in the end we are constantly influenced by a consideration of what is, or is not, important.
>
> In conducting autopsy examinations of animals there is usually an immense amount of bias in selection of tissues for histological work; in the case of skeletal muscles (from my own experience) no doubt they are mostly ignored unless macroscopical changes are very obvious.

Since 1951, significant advances have been made. Most notable have been the comparative biomedical investigations (Harris, 1979) that resulted from the original recognition and descriptions of hereditary muscular dystrophies in the mouse (Michelson *et al.*, 1955), chicken (Asmundson and Julian, 1956), hamster (Homburger *et al.*, 1962), turkey (Harper and Parker, 1964, 1967), sheep (McGavin and Baynes, 1969), mink (Hegreberg *et al.*, 1974; Hamilton *et al.*, 1974), dog (Cooper *et al.*, 1988; Kornegay *et al.*, 1988), and cat (Carpenter *et al.*, 1989; Gaschen *et al.*, 1992).

In domestic animals, which are more commonly considered of clinical importance, significant advances have occurred in recent years. The methods of biotechnology and molecular genetics have revealed insights into the etiology and pathogenesis of neuromuscular disorders; however, there is a continuing need for vast improvements in our knowledge and understanding of neuromuscular diseases in animals. In compiling the revisions for this chapter, the author is pleased to find and summarize significant advances with regard to newer descriptions of hereditary, metabolic, and immune disorders of muscle in domestic animals.

II. SPECIALIZATION OF THE SARCOLEMMA AND SARCOPLASM FOR MUSCULAR CONTRACTION

A. Neuromuscular Transmission: Excitation–Conduction

The plasmalemma of the skeletal myofiber (sarcolemma) is specialized for the protoplasmic properties of excitation and conduction. These properties are largely due to the presence of ion channels for sodium, potassium, calcium, and chloride that regulate the selective and nonselective conductance of these ions across the sarcolemma, either into or out of the myofiber. Recent advances in our knowledge have revealed that abnormal functioning of these ion channels accounts for muscle weakness through altered excitability of the sarcolemma.

The neuromuscular junction or motor end plate is the synaptic site for chemical transmission of excitation from the presynaptic axon terminal of a motoneuron to the postsynaptic skeletal myofiber (Fig. 16.1). The axon terminal rests within a primary depression of sarcolemma, the primary cleft, and contains numerous small clear vesicles that contain acetylcholine (ACh), the neurotransmitter for excitation of skeletal myofibers. Each vesicle contains a quantum of neurotransmitter, consisting of 6000 to 8000 molecules of ACh. Arising from the primary cleft underlying the axon terminal are numerous smaller secondary clefts and complementary folds. The space within the primary and secondary clefts, located between the axon terminal and the postsynaptic sarcolemma, comprises the synaptic cleft. This space is filled with basal lamina containing acetylcholinesterase (AChE).

Arrival of a nerve action potential at the axon terminal results in activation of voltage-gated calcium ion channels associated with the active zone of the presynaptic membrane. The calcium influx initiates a calcium-dependent fusion of approximately 200 vesicles with the active zone of the presynaptic membrane and the subsequent release of ACh by exocytosis. The ACh released diffuses across the synaptic cleft to bind with acetylcholine receptors (AChRs), which are concentrated on the crests of the secondary folds of the postsynaptic sarcolemma. Somewhat deeper within the troughs of the secondary folds are voltage-gated so-

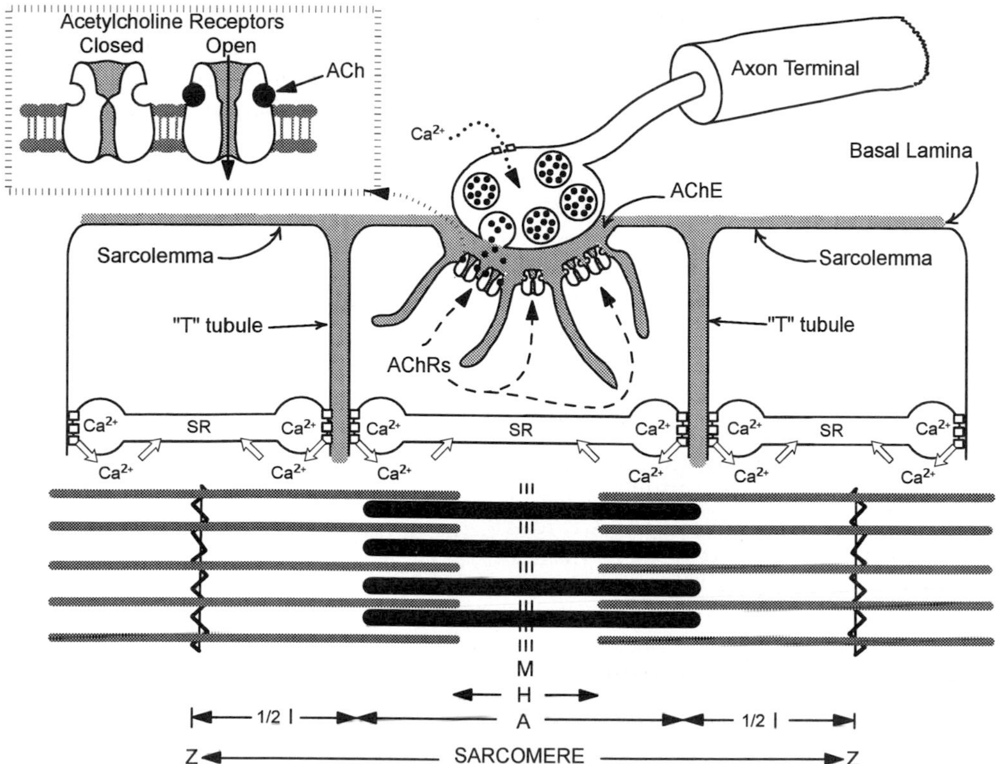

FIGURE 16.1 Excitation of myofibers to contract involves neuromuscular transmission and the subsequent release of calcium ions into the sarcoplasm. Arrival of an impulse at the axon terminal activates voltage-gated calcium ion channels, resulting in the influx of calcium ions that initiate the calcium-dependent release of the neurotransmitter acetylcholine (ACh) by exocytosis. Liberated ACh diffuses across the synaptic cleft to bind with ACh receptors (two molecules of ACh per receptor) on the postsynaptic sarcolemma. Binding of ACh with AChRs increases the conductance of sodium and potassium ions across the postsynaptic membrane to produce a local end-plate potential at the neuromuscular junction. The end-plate potential generates a muscle action potential that spreads away from the neuromuscular junction in all directions over the surface of the myofiber and into its depths via the transverse (T) tubules. Within the depths of the myofibers, excitation is coupled to contraction through the release of calcium ions from terminal cisternae of the sarcoplasmic reticulum (SR) through calcium release channels of the terminal cisternae. The calcium release channels form small "feet" that extend from the terminal cisternae to the T tubules. The liberated calcium ions bind to the regulatory protein troponin and release the inhibitory action of the regulatory proteins on the contractile events that lead to sliding of the thin (actin) and thick (myosin) filaments. The liberated ACh is subsequently hydrolyzed by AChE (acetylcholinesterase) within the basal lamina of the synaptic cleft.

dium ion channels, which are also present within the sarcolemma throughout nonjunctional regions of the myofiber.

Excitation of the myofiber is initiated by the reversible binding of ACh with AChRs. The structure of the AChR is very similar among animal species. The AChR molecule is an integral transmembrane protein that forms a nonselective cation channel composed of five subunits consisting of two α-subunits and single β-, γ-, and δ-subunits. Each α-subunit possesses a binding site for ACh. The binding of ACh with AChR (two ACh molecules / receptor) results in a local depolarization of the postsynaptic membrane caused by the transient increased conductance of the AChR cation ion channels to sodium and potassium ions. The amplitude of the end-plate potential (depolarization) is proportional to the number of ACh–AChR complexes formed.

In addition to the quantal release of ACh in response to a nerve action potential (impulse) at the axon terminal, release of single quanta of ACh occurs spontaneously, in the absence of an impulse. In both instances, the quantal release of ACh is dependent on the activation of voltage-gated calcium ion channels and calcium conductance into the axon terminal; hence, low calcium ion concentrations at the axon terminal will reduce the quantal release of ACh.

At rest, individual quanta of ACh are spontaneously released at a slow rate and cause transient, low-amplitude depolarizations at the end plate. These are referred to as miniature end-plate potentials (MEEPs). With the arrival of a nerve action potential, approximately 200 quanta are released and with the increased number of ACh–AChR combinations, there is a greater conductance of sodium and potassium ions that form a large amplitude depolarization, the end-plate potential (EEP). When the amplitude of the EEP exceeds threshold, a wave of depolarization (muscle action potential, MAP) is generated over the sarcolemma, away from the end plate in all directions. The MAP is propagated by voltage-gated sodium channels over the surface of the myofiber and into its depths via transverse (T) tubules. The T-tubules are invaginations of the sarcolemma that transverse the long axis of the myofiber and their lumina openly communicate with the extracellular fluid space. The transient binding of ACh to AChRs is abolished by diffusion of ACh away from the receptors and its hydrolysis by AChE.

B. Coupling Excitation to Contraction

Excitation–contraction coupling involves the transformation of depolarizing events in the sarcolemma into the initiation of mechanical shortening of the myofibrils within the myofiber by calcium ions released from the terminal cisternae of the sarcoplasmic reticulum (SR). These events occur within the depths of the myofiber at "triads" where the T-tubules form junctional complexes with adjacent terminal cisternae of the SR. The sarcoplasmic reticulum functions in the uptake, storage, and release of calcium ions to regulate the concentration of calcium ions in the aqueous sarcoplasm bathing the myofilaments and other organelles. The concentration of calcium in the SR is aided by the presence of calsequestrin, a calcium-binding protein present within the lumen of the cisternae.

At the T–SR junctional complex of triads, the sarcolemma contains voltage-sensitive dihydropyridine receptors (DHPRs) and the terminal cisternae of the SR possess ryanodine-sensitive calcium ion channels (ryanodine receptors) that form "feet" that fill the gap between the terminal cisternae and T-tubules. With depolarization of the sarcolemma within the T-tubules, DHPRs interact with ryanodine receptors to mediate the voltage-dependent release of calcium ions from the SR into the sarcoplasm, elevating the calcium ion concentration from $\leq 10^{-7}$ to $\leq 10^{-6}$ M. This elevation in calcium ion concentrations initiates contraction through its interaction with the regulatory proteins that serve to inhibit shortening at low calcium ion concentrations. Relaxation is initiated by a reduction in the sarcoplasmic calcium ion concentration by active transport of calcium ions into the lumen of the SR by a calcium-ATPase pump within the SR membrane. For further details concerning the structures and functions involved in neuromuscular transmission and coupling of excitation to contraction refer to Engel (1994a), Horowitz (1994), Lindstrom (1994), and Magleby (1994).

C. Muscular Contraction

Muscular contraction results from the transformation of chemical energy into mechanical energy. The energy for contraction is derived from the hydrolysis of adenosine triphosphate (ATP) into adenosine diphosphate (ADP) and inorganic phosphate (Fig. 16.2). That hydrolysis is catalyzed by the actin-activated adenosine triphosphatase (ATPase) activity present in the S_1 fraction of the myosin molecule.

Chemically, the transformation of energy is associated with the cyclical association and disassociation of the contractile proteins actin and myosin, whereas mechanically the transformation is associated with shortening of sarcomeres, which is achieved by conformational changes of the myosin molecules that result in sliding of the overlapping arrays of thick (myosin) and thin (actin) myofilaments (Huxley, 1983).

The ATP required for contraction is not stored in significant quantities. Therefore, ATP must be readily produced through the metabolism of fats, carbohydrates, and creatine phosphate stores to support the energy requirements for contraction. Aerobically, ATP is produced in muscle mitochondria by oxidative phosphorylation coupled to electron transport through the oxidation of (1) fatty acids mobilized from triglyceride fat stores in muscle and fat depots and (2) glucose from liver and muscle glycogen stores. The oxidation of glucose through acetyl-CoA also produces ATP by substrate phosphorylation of ADP in the sarcoplasm. Anaerobically, ATP is produced in the aqueous sarcoplasm through substrate phosphorylation of ADP by (1) creatine kinase, utilizing creatine phosphate stores; (2) adenylate kinase, utilizing ADP produced by the ATP hydrolysis; and (3) glyco(geno)lysis, utilizing glucose derived from both liver and muscle glycogen stores.

1. Myofilaments and Contractile Proteins

The contractile proteins are myosin, the principal component of thick myofilaments, and actin, the principal component of thin myofilaments. Lateral projections of the thick myofilaments (myosin cross-bridges) are the constituents of myosin which form the reactive sites of cyclical association and disassociation that occurs between myosin and actin during contraction. The

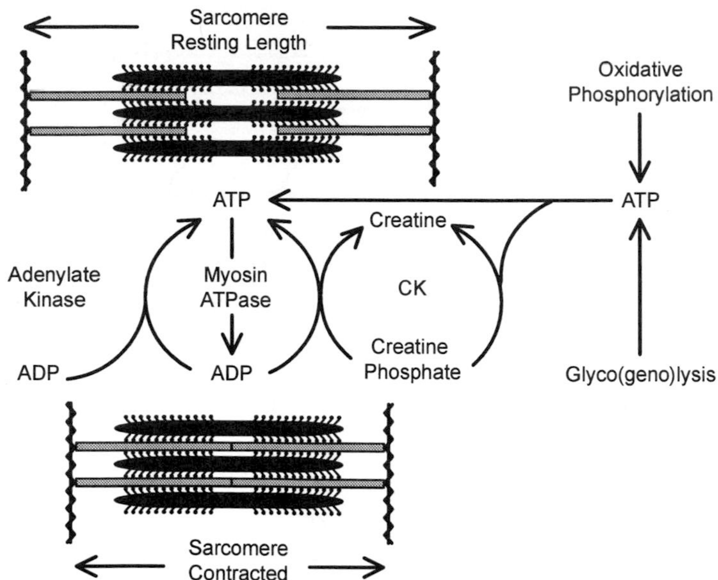

FIGURE 16.2 Muscular contraction involves the shortening of sarcomeres by sliding of the overlapping arrays of thick (myosin) and thin (actin) myofilaments. The energy for contraction is derived from the hydrolysis of ATP in the presence of an actin-activated ATPase present within the head regions of the myosin thick filament cross-bridges. The ATP is generated by the energy metabolism of the myofiber, principally by anaerobic glycolysis or oxidative phosphorylation. The utilization of ATP may be direct from those sources or indirect from the phosphorylation of ADP from creatine phosphate by creatine kinase (CK).

force-generating step for sliding of the filaments past each other results from changes in the angle of the cross-bridge attachments (Fig. 16.3).

a. Thick Myofilaments and Myosin

To gain an understanding of the physicochemical changes that occur at the cross-bridges, the composition and properties of myosin need to be considered. Myosin is an asymmetric protein with both structural and enzymatic protein properties. It is composed of two myosin heavy chains (polypeptide chains with an approximate molecular mass of 200,000 Da; Gershman *et al.*, 1969; Gazith *et al.*, 1970), and four myosin light chains (polypeptide chains with molecular masses ranging from 16,000 to 27,000 Da; Lowey and Risby, 1971; Sarkar *et al.*, 1971; Weeds and Lowey, 1971; Lowey, 1994).

The two myosin heavy chains are arranged in a double helix to form a long fibrous portion and at one end each heavy chain is folded to form a globular head domain (one head/heavy chain; two heads/myosin molecule). The four myosin light chains are contained within the globular heads (two per head) near the junction of the head and neck domains. There are two classes of light chains: (1) two identical DTNB light chains, which disassociate from their globular heads with the thiol reagent 5, 5′-dithiobis(2-nitrobenzoic acid) (DTNB), and (2) two related but different species of "alkali light chains," (A1 and A2) which disassociate at a high pH (Weeds, 1969; Gazith *et al.*, 1970; Weeds and Lowey, 1971; Lowey, 1994).

The principal biochemical properties of myosin are its ATP-binding capacity, actin-binding capacity, and actin-activated ATPase activity. The actin-activated ATPase activity of myosin appears to reside primarily in the heavy chains; however, disassociation of the alkali light chains from the globular head regions results in some loss of activity, whereas disassociation of the DTNB light chains does not (Wagner and Weeds, 1977). Myosin isoforms exist based on their alkaline light chain composition; that is, A1 myosin, A2 myosin, and A1,A2 myosin (Lowey *et al.*, 1979; Lowey, 1994).

The myosin molecule is composed of head, neck, and tail domains. Proteolytic digestion of myosin with trypsin results in the formation of two fragments: (1) heavy meromyosin (HMM), which is composed of the two globular heads of myosin and a short neck domain composed of the initial segment of the fibrous portion, and (2) light meromyosin (LMM), the tail domain composed of the remaining long fibrous portion of the molecule. The HMM fragments correspond

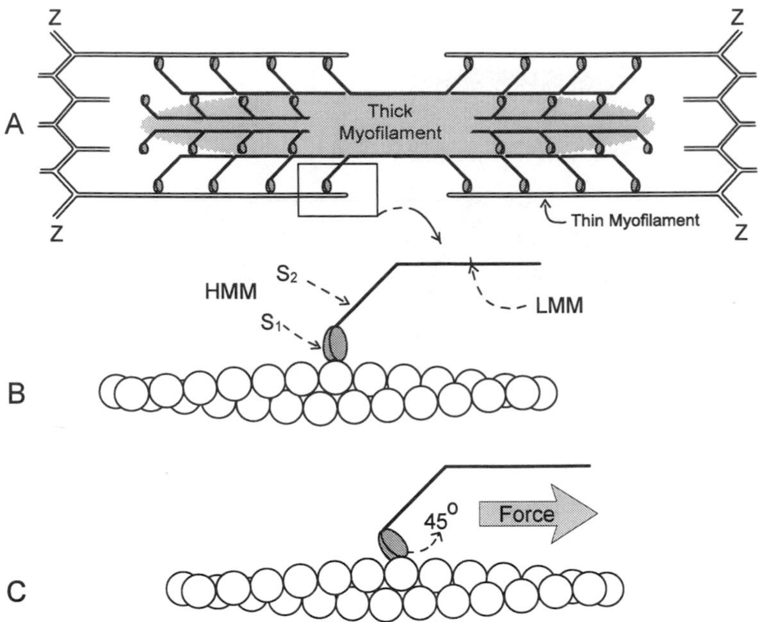

FIGURE 16.3 Schematic presentation of myosin cross-bridges on thick myofilaments. (A) Portions of the myosin molecules (cross-bridges) project from the thick myofilaments and make contact with the thin myofilaments. (B) The light meromyosin (LMM) portion of myosin molecules form the major structural component (backbone) of the thick myofilament, and the heavy meromyosin (HMM) component forms the cross-bridge connections between the thick and thin myofilaments. The cross-bridges of myosin are composed of two fractions: the S_1 fraction, a globular protein fraction composed of two heads, each possessing binding capacities for ATP and actin and the actin-activated ATPase activity of myosin, and the S_2 fraction, a fibrous protein fraction that forms the flexible linkage between the S_1 fraction and the LMM portion of myosin. (C) The force for sliding of the myofilaments results from a change in the angle of attachment (i.e., a change from 90 to 45°) between the S_1 globular head and actin filament.

structurally to the cross-bridges, whereas the LMM fragments comprise the bulk of the thick myofilaments (Huxley, 1983; Fig. 16.3). Papain digestion results in two subfragments of HMM: subfragment 1 (S_1), the globular head domain, and subfragment 2 (S_2), the short fibrous neck domain. The actin-binding capacity, ATP-binding capacity, and actin-activated ATPase activity of myosin reside in the S_1 fragment, whereas the S_2 fragment comprises the flexible linkage between the globular head of the cross-bridge and the LMM fragment. Thus, the S^1 portion of the myosin molecule that possesses the characteristic chemical properties of myosin is structurally located on the terminal portion of the cross-bridges at the site of interaction with the actin of the thin myofilaments.

b. Thin Myofilaments and Actin

The thin filaments are composed of two F-actin strands arranged in a double helical configuration. The F-actin strands are polymers of the globular protein G-actin and each G-actin monomer possesses a complementary binding site for the myosin S_1 fragment. Upon combining with myosin, actin activates the ATPase activity of the myosin S_1 fragment (Huxley, 1983).

c. Myosin–Actin Cross-Bridge Cycle

Contraction is a physiochemical cycle in which the cyclic interaction of actin and myosin (association, change in angulation of cross-bridge attachments, and disassociation) and hydrolysis of ATP are interdependent and complex (Lymn and Taylor, 1971; Eisenberg and Kielley, 1974; Fig. 16.4). In the noncontracting state, actin and myosin are combined at the cross-bridges (step 1, Fig. 16.4), and the angle of attachment between the cross-bridge heads and the actin filaments is 45°. With the addition of ATP, ATP is bound to each globular head (two molecules ATP/myosin molecule) and there is a rapid disassociation of actin and myosin (step 2, Fig. 16.4). The ATP is cleaved to form a myosin–products complex, and the globular head moves to a

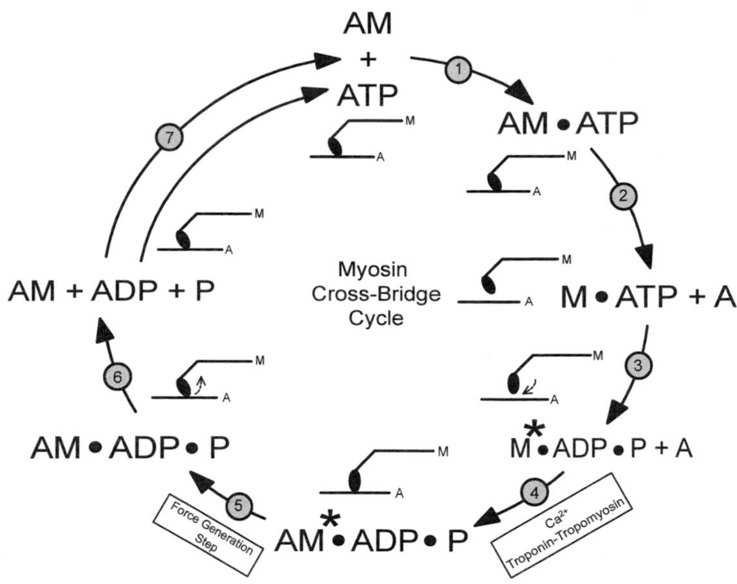

FIGURE 16.4 Muscular contraction results from the cyclical association and disassociation of actin (A) and myosin (M) in which conformational changes occur in the cross-bridge linkages between the thick and thin myofilaments, associated with the hydrolysis of ATP.

new location on the thin filament (step 3, Fig. 16.4), which permits the angle of attachment to become 90° when the myosin–products complex recombines with the actin filament (Step 4, Fig. 16.4). This recombination step between the myosin–products complex and actin is controlled by the regulatory proteins troponin and tropomyosin in response to calcium ion concentrations (discussed later). The force for contraction is generated by movement of the cross-bridge head to a 45° angle of attachment (step 5, Fig. 16.4), and the cycle is completed with the detachment of the hydrolytic products of ATP from the head (step 6, Fig. 16.4). With the formation of ATP through rephosphorylation (step 7, Fig. 16.4), the cycle may be repeated. Measurements indicate that each cycle (stroke) shortens a sarcomere by 12 nm (Barden and Mason, 1978).

Rigor mortis, the rigid and stiff condition of skeletal muscles that develops following death, involves cessation of the cross-bridge cycle in the post–force-generating step (step 6, Fig. 16.4). After death, when all ATP stores have been utilized, disassociation of actin and myosin will not occur and the contraction cycle is terminated with a large number of actin–myosin complexes formed with the myosin heads set at 45°.

2. Thin Myofilaments and Regulatory Proteins

Two proteins (tropomyosin and troponin) working in concert with calcium regulate muscle contraction (Ebashi *et al.,* 1969). Tropomyosin, a fibrous protein, is arranged along the length of the thin filaments, within the grooves of the two F-actin strands. Troponin is a globular protein complex composed of three subunits: TN-I (troponin inhibitory component), TN-T (tropomyosin-binding component), and TN-C (calcium-binding component). The TN-T component attaches the complex to tropomyosin at intervals along the thin myofilaments. With low sarcoplasmic calcium concentrations ($\leq 10^{-7}$ *M*), tropomyosin molecules block the myosin-binding sites on actin, which prevents the interaction of actin and myosin. At higher concentrations ($\leq 10^{-6}$ *M*), calcium ions combine with the TN-C component to initiate a conformational change in the TN-I component, which results in the movement of tropomyosin to free the myosin-binding sites on actin. With the myosin-binding site on actin exposed, actin and myosin combine and initiate the cyclical changes associated with that interaction. When calcium ion concentrations are reduced through uptake of calcium by the SR calcium–ATPase pump, the process is reversed and the interaction of actin and myosin is inhibited.

3. Cytoskeletal Filaments and Accessory Proteins

The organization of myofilaments within sarcomeres and the organization of myofibrils is supported by a complex cytoskeletal network of intermediate filaments (Wang and Ramirez-Mitchell, 1983). Intermediate filaments and a number of accessory proteins that form fine filaments function to (1) maintain the alignment of myofilaments and sarcomeres, (2) attach and maintain alignment of adjacent myofibrils, (3) attach the sarcomeres of peripheral myofibrils to the sarcolemma, and (4) connect terminal sarcomeres to the sarcolemma at myotendinous junctions. Collectively, the cytoskeletal filaments function to maintain the structural and functional relationships of the myofilaments and transfer the forces developed by the myofilaments to the sarcolemma.

a. Alignment of Myofilaments, Sarcomeres, and Myofibrils

Thick myofilaments are attached to Z-lines by small filaments composed of the protein titan (Maruyama, 1986; Pierbon-Bormoli *et al.*, 1989). The titan filaments arise near the M-line within the axes of the thick filaments and span the length of the thick filament as well as the I-band region to attach to the Z-line. Within the I-band region, the titan filaments provide an elastic attachment to the Z-line, which imparts a passive elasticity to sarcomeres. An additional protein, nebulin, forms small filaments that run the length of thin myofilaments and may regulate the length of thin myofilaments.

At the periphery of myofibrils, adjacent Z-lines within the same sarcomere are connected by intermediate filaments of desmin. Also, intermediate filaments of desmin encircle the circumference of Z-lines and appear to form linkages with Z-lines of adjacent myofibrils to aid in the alignment of sarcomeres in register with adjacent myofibrils (Tokuyasu *et al.*, 1983).

b. Attachment of Myofilaments to the Sarcolemma

At the periphery of myofibrils adjacent to the sarcolemma there are riblike attachments (costomeres) which are present on either side of Z-lines (Pardo *et al.*, 1983). Desmin filaments appear to be anchored to the sarcolemma by a number of adhesion proteins such as vinculin.

At myotendinous junctions, the thin myofilaments of the last sarcomere attach to the sarcolemma, which is thrown into numerous villous projections. The thin myofilaments are anchored by the proteins α-actinin and vinculin, among others. Growth in the length of muscle fibers occurs at the myotendinous junctions by the addition of new sarcomeres (William and Goldspink, 1971).

III. HETEROGENEITY OF SKELETAL MUSCLE

A. Gross Muscle Coloration

Differences in gross coloration of muscles have long been recognized and reviewed by Needham (1926). A variation in coloration is noted not only among species of animals, but also among individual muscles within the same individual. The coloration of some muscles ranges from deep red to pale, whereas other muscles exhibit intermediate shades of coloration. Early workers believed that the differences in coloration were due to differences in the content of blood within the muscles; however, it was later shown that the red coloration was associated with the presence of myoglobin, and other cytochromes within the myofibers.

As a result of these differences in coloration, the terms "red" and "white" were introduced to distinguish between muscles of different gross coloration. Subsequently, numerous biochemical, histochemical, and physiological studies have been conducted with selected muscles from a variety of species. As a result, the terms "red" and "white" have come to imply more specific meanings, such as the physiological properties and populations of myofiber types within a muscle.

B. Physiological Properties

Muscle coloration was observed to be associated with significant differences in the physiological properties of muscular contraction. It was demonstrated that the speed of contraction of red muscles was most often slower than that of white muscles in a variety of animals (Lee *et al.*, 1916; Denny-Brown, 1929). In addition, redness of a muscle was associated with the development of tetanus at lower frequencies of stimulation, the development of smaller twitch tensions, and a greater resistance to fatigue. Conversely, white muscles required greater frequencies of stimulation for the development of tetanus, developed larger twitch tensions, and tended to fatigue quickly.

Whereas muscles were found to differ in their physiological properties of contraction, the terminology of slow-contracting or slow-twitch and fast-contracting or fast-twitch muscles evolved. Moreover, since speed of contraction was closely associated with gross muscle coloration, the terms "red" and "white" came to be used interchangeably with "slow" and "fast," respectively. However, there are numerous exceptions to this association of gross coloration with physiological properties of contraction. Therefore, direct associations must not be assumed.

C. Motor Units

The morphological and functional unit of skeletal muscles is the motor unit. The motor unit is composed of (1) the motoneuron, consisting of its cell body located within the central nervous system (selected cranial nerve nuclei or the ventral horn of the spinal cord), and its axon, which extends along the ventral root and peripheral nerve; (2) the neuromuscular junctions, and (3) the myofibers innervated by the motoneuron.

Motoneurons may differ based on their rates of discharge: (1) phasic motoneurons with a fast discharge rate and (2) tonic motoneurons with a slow discharge rate (Granit *et al.*, 1957). In addition, the phasic motoneurons are characterized by shorter after-hyperpolarization potentials, faster conduction velocities, and larger axons than the tonic motoneurons (Eccles *et al.*, 1958). Investigations of these parameters in motoneurons of slow-contracting and fast-contracting muscles indicate that tonic motoneurons, which discharge at rates of 10–20 per second, innervate slow-contracting muscles, and phasic motoneurons, which discharge at rates of 30–60 per second, innervate fast-contracting muscle (Eccles *et al.*, 1958). Thus, there are at least two types of motor units that differ in their physiological properties and type of motoneuron innervation.

Physiological measurements performed on isolated motor units in the cat have revealed two types of fast-twitch motor units and one type of slow-twitch unit (Burke *et al.*, 1973; Burke and Tsairis, 1973; Burke, 1994). Some fast-twitch motor units are resistant to fatigue and designated FR units (i.e., fast twitch, resistant to fatigue); others fatigue rapidly and are designated FF units (i.e., fast twitch, fatigable). All of the slow-twitch units were resistant to fatigue, and were merely designated S units (i.e., slow twitch). The average number of muscle fibers per motor unit ranged from 550 to 650.

D. Quantitative Biochemistry

1. Metabolic Pathways That Generate Energy (ATP) for Contraction

Quantitative differences in enzyme activities and various substrate concentrations have been reported between red and white muscles. Those biochemical differences between red and white muscles reflect differences in their principal metabolic pathways active in the generation of energy (ATP) for muscular contraction (Fig. 16.5).

a. Aerobic and Anaerobic Energy Metabolism

In general, white muscles are biochemically suited to derive energy for contraction by substrate phosphorylation via anaerobic glyco(geno)lysis. White muscles tend to have higher concentrations of glycogen and creatine phosphate as well as higher activities for enzymes associated with glycogenolysis and glycolysis, with the exception of hexokinase. Red muscles, on the other hand, generally have higher concentrations of triglycerides and myoglobin and are better suited to derive their energy by oxidative phosphorylation via the electron transport system following the oxidation of fatty acids and glucose via the Krebs cycle (Green, 1951; Lawrie, 1952, 1953; Ogata, 1960; Domonkos, 1961; Domonkos and Latzkovits, 1961a, 1961b; George and Talesara, 1961; Beatty *et al.*, 1963, 1966b, 1967; Blanchaer, 1964; Dawson and Romanul, 1964; George and Bokdawala, 1964; George and Iype, 1964; Dawson and Kaplan, 1965; Cosmos, 1966; Cosmos and Butler, 1967; Beecher *et al.*, 1968; Bass *et al.*, 1969; Kubista *et al.*, 1971; Crabtree and Newsholme, 1972; Staudte and Pette, 1972; Baldwin *et al.*, 1973b; Pette, 1975).

Whereas anaerobic glycolysis and glycogenolysis are greater in white muscles, red muscles have (1) greater glycogen synthetase activities (Stubbs and Blanchaer, 1965; Bocek and Beatty, 1966; Beatty and Bocek, 1970), (2) higher glycogenesis (Bocek *et al.*, 1966a, 1966b), and (3) higher hexokinase activities (Peter *et al.*, 1968; Bass *et al.*, 1969; Cardinet *et al.*, 1972; Crabtree and Newsholme, 1972; Pette, 1975). Explanations for those differences are not self-evident because similar relative activities for glycogenesis and glycogenolysis as well as hexokinase and glycolysis might be anticipated. Rather than having a relationship with other glycolytic enzymes, hexokinase activities demonstrate a constant relationship with mitochondrial citrate synthase activity.

Aerobic glycolysis via the pentose cycle does not appear to be a principal metabolic pathway in skeletal muscle (Glock and McLean, 1954; Green and Landau, 1965; Beatty *et al.*, 1966a, 1966c; Beatty and Bocek, 1970). The role of amino acids in the energy metabolism of muscle is uncertain. Transaminase activities are higher in red muscle of the chicken (Cardinet *et al.*, 1972). Also, amino acid uptake, incorporation, and turnover in skeletal muscle protein are greater in red muscle (Goldberg, 1967). Studies with 3H-labeled amino acids have shown that, with the exception of alanine, all amino acids tested are more rapidly incorporated into red muscle protein (Citoler *et al.*, 1966). Alanine produced in muscle may function in the transport of amino groups to the liver, particularly during exercise (Felig and Wahren, 1971; Felig, 1977); however, that role is disputed by some. In the liver, alanine may be reconverted to glucose and transported back to muscle as part of the glucose-alanine cycle between muscle and liver.

Oxidation of fatty acids is an important energy source for muscle and is dependent on several factors, including carbohydrate availability, state of physical conditioning, and the intensity and duration of exercise (Issekutz *et al.*, 1965, 1966; Paul and Issekutz, 1967; Mole *et al.*, 1971; Holloszy *et al.*, 1971; Pande and Blanchaer, 1971; Therriault *et al.*, 1973; Issekutz *et al.*, 1975). Red muscle has been shown to contain larger quantities of lipid than white muscle (Beatty and Bo-

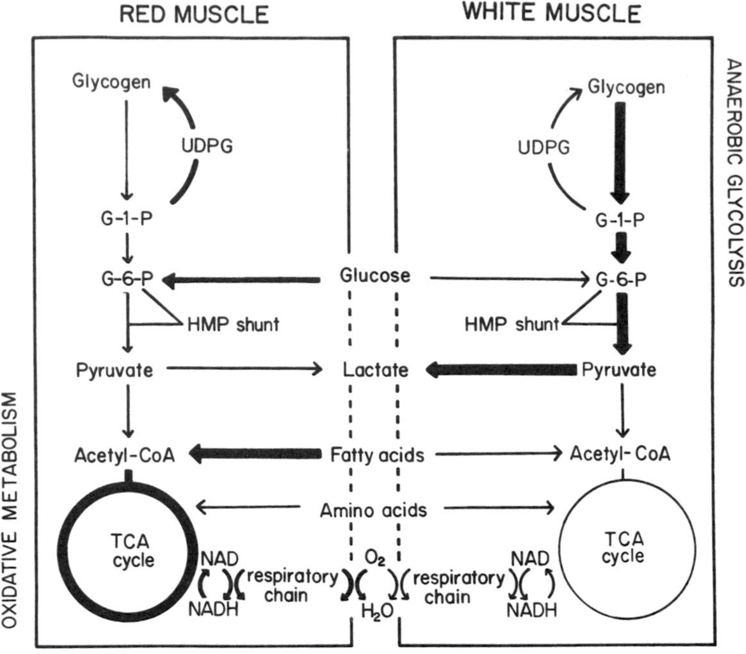

FIGURE 16.5 Schematic representation of some differences in energy-yielding metabolic pathways of red and white muscles. In red muscles, energy for contraction is derived primarily by oxidative phosphorylation resulting from the oxidation of fatty acids, carbohydrates, and perhaps amino acids via the tricarboxylic acid (TCA) cycle. White muscles derive their energy primarily via anaerobic glycogenolysis and glycolysis through the degradation of glycogen and glucose to lactate. Aerobic glycolysis via the HMP (hexose monophosphate) shunt is a minor pathway in both types of muscle.

cek, 1970). The activities of palmitate-activating enzyme and carnitine acyltransferases were twice as high in red as in white muscle mitochondria. Studies indicate that red muscle is potentially capable of deriving more usable energy from free fatty acids than is white muscle (Pande and Blanchaer, 1971).

b. Purine Nucleotide Cycle

During muscular contraction ammonia is produced through the deamination of AMP to inosine monophosphate (IMP) by AMP deaminase as part of the purine nucleotide cycle (Lowenstein, 1972). The cycle involves (1) the deamination of AMP to form IMP and ammonia, (2) the deamination of aspartate and the hydrolysis of guanosine triphosphate to form adenylsuccinate, and (3) the cleavage of adenylsuccinate to form fumarate and AMP. The activity of AMP deaminase was found to be greater in white muscles than red muscles.

The suggested functions of the purine nucleotide cycle subserve the energy requirements for muscular contraction through (1) maintenance of a high ATP/ADP ratio by regulating the relative AMP, ADP, and ATP levels through the removal of AMP, (2) regulation of phosphofructokinase (PFK) activity by elevations in ammonia, (3) regulation of phosphorylase activity by accumulation of IMP, (4) replenishment of citric acid cycle intermediates by the production of fumarate, and (5) deamination of amino acids for oxidative metabolism through the formation of aspartate (Terjung *et al.*, 1986).

From the foregoing, it is evident that the deamination of AMP can serve to promote ATP synthesis by (1) stimulating anaerobic glyco(geno)lysis through the activation of phosphorylase *b* by IMP, and the activation of PFK by ammonia, and (2) supporting oxidative metabolism through the production of intermediates into the TCA cycle.

The degradation of adenine nucleotides in equine muscle appears to occur mainly through deamination of AMP. Reported AMP deaminase activities are greatest for equine middle gluteal muscles, which were approximately double reported values for muscles of the rat and rabbit (Cutmore *et al.*, 1986). In the dog, the degradation of adenine nucleotides via AMP deaminase appears to be limited and occurs principally through the formation of adenosine by the action of 5′-nucleotidase (Brockman and McKenzie, 1983).

2. *Metabolic Pathway for Utilization of Energy (ATP) for Contraction: Myosin-ATPase*

As previously discussed, the actin-activated myosin ATPase activity catalyzes the cyclical physiochemical interactions of actin and myosin during contraction (Fig. 16.4). Furthermore, the intrinsic speed of contraction (sarcomere shortening) has been demonstrated to be proportional to the activity of actin-activated myosin-ATPase (Barany, 1967; Mommaerts, 1969) and differences exist between the myosin-ATPase activities of fast-contracting and slow-contracting muscles (Close, 1972). From those observations it is postulated that the rate-limiting step of sarcomere shortening during contraction is the hydrolysis of ATP. Concomitant with the different actin-activated ATPase activities of myosins from fast-contracting and slow-contracting muscles, their alkali light-chain compositions also differ (Lowey and Risby, 1971; Sarkar *et al.*, 1971). Slow- and fast-contracting muscles contain similar amounts of myosin; however, the ATPase activities of slow-contracting muscles are lower than in fast-contracting muscles and their pH dependency and liability in acid and alkaline conditions also differ (Barany *et al.*, 1965; Seidel, 1967).

E. Histology and Histochemistry

1. *Morphologic Variations of Myofibers*

Different types of myofibers were recognized by early microscopists. By examination of unstained sections or in combination with various lipid stains, two basic types of myofibers were described. One type of myofiber was dark or opaque and contained numerous granules and lipid droplets between the myofibrils. The other type of myofiber was light or translucent and contained few granules and lipid droplets. These myofibers were referred to as dark and light myofibers, respectively. In addition, myofibers intermediate to dark and light myofibers were described (Bullard, 1912). The most common finding was that mammalian muscles contained variable percentages of dark and light myofibers and therefore were mixed with respect to their myofiber-type composition. However, in certain instances, such as in the rabbit, guinea pig, and chicken, some muscles were found to contain predominantly dark myofibers and their gross coloration was red, whereas other muscles were found to contain predominantly light myofibers and their gross coloration was white. Although there was no unanimous agreement by early investigators, the tendency for myofibers of red muscles to be darker than the myofibers of white muscles led to the designation of dark and light myofibers as red and white myofibers, respectively.

2. *Histochemical Properties of Myofibers*

Histochemical and biochemical studies of red and white muscles tended to support the concept that gross muscle coloration is associated with the myofiber-type composition of a muscle (Nachmias and Padykula, 1958; Dubowitz and Pearse, 1960; Stein and Padykula, 1962; Romanul, 1964; Beatty *et al.*, 1966b; Bocek and Beatty, 1966; Cosmos and Butler, 1967; Gauthier, 1969). With the development and introduction of histochemical techniques to studies of muscle, the heterogeneity of myofibers within a muscle became more obvious. Since these techniques localized enzyme systems at the cellular level, their application involved implications of biochemical and functional heterogeneity of myofibers that had been suggested by the early microscopists.

a. Histoenzymic Properties Associated with Aerobic and Anaerobic Energy Metabolism

Histochemical studies of succinate dehydrogenase (SDH), reduced nicotinamide adenine dinucleotide–tetrazolium reductase (NADH-TR), and reduced nicotinamide adenine dinucleotide phosphate–tetrazolium reductase (NADPH-TR) demonstrated that the red or dark granular myofibers had higher activities for these enzymes than the white or less granular light myofibers (Padykula, 1952; Wachstein and Meisel, 1955; Nachmias and Padykula, 1958; Dubowitz and Pearse, 1960). In conjunction with electron microscopic observations, the activities of these enzymes were localized in mitochondria. The granules and higher activities of these enzymes in the dark or red myofibers correspond to greater numbers of mitochondria. Studies of red and white myofibers of the rat diaphragm and semitendinosus have revealed that the mitochondria of red myofibers are large with abundant cristae, and the white myofibers have fewer mitochondria, which are smaller and have fewer cristae. Associated with the large mitochondria of the red myofibers are lipid inclusions. Therefore, the granules and lipid droplets described by early microscopists correspond to mitochondria and lipid inclusions, respectively (Padykula and Gauthier, 1963, 1967; Gauthier and Padykula, 1966; Gauthier, 1969).

A reciprocal histochemical profile for red and white myofibers was observed between mitochondrial enzymes and phosphorylase, an indicator of anaerobic glyco(geno)lysis (Dubowitz and Pearse, 1960). Red myofibers with high mitochondrial enzyme activities had low phosphorylase activities, whereas the reverse was true of white myofibers and a classification of myofibers was introduced in which red and white myofibers were designated as type I and type II myofibers, respectively (Dubowitz and Pearse, 1960). Owing to the fact that various intermediate histochemical reactions of myofibers exist between the extreme or classical characteristics described for red or type I myofibers and white or type II myofibers, other classifications were proposed. Stein and Padykula (1962) proposed a classification of A, B, and C myofibers based on their SDH reactions that roughly corresponds with the classification of white, intermediate, and red muscle myofibers, respectively (Padykula and Gauthier, 1967). As many as eight myofiber types were described in rat muscle, where esterase activity was found to be more precisely reciprocal with phosphorylase activity, suggesting a reciprocity between anaerobic glycolysis and lipid metabolism (Romanul, 1964).

Histochemical localization of glycogen synthetase is varied. In human muscle its activity was usually parallel with that of phosphorylase in type II or white fibers (Engel, 1962), but reciprocal and higher in type I or red fibers of rat and monkey muscle (Hess and Pearse, 1961; Bocek and Beatty, 1966), a finding that is in agreement with quantitative results in the rat and monkey.

b. *Histoenzymic Properties Associated with Myosin-ATPase*

In addition to employing histochemical stains for enzymes of ATP-generating metabolic pathways, myofibers have been differentiated histochemically into type 1 and type 2 myofibers based on their staining reaction for myofibrillar ATPase (Engel, 1962, 1965). Furthermore, type 2 myofibers, classified by the myofibrillar ATPase staining reaction, may be further subdivided into type 2A, type 2B, and type 2C myofibers based on the lability of their ATPase activity following preincubation in acid media (Brooke and Kaiser, 1970). This classification scheme is generally applicable to all mammalian species except for canine muscles, which do not contain classical type 2B myofibers (Braund *et al.*, 1978; Orvis and Cardinet, 1981; Snow *et al.*, 1982).

Of all the histochemical techniques applied to date, the myofibrillar ATPase reaction best differentiates myofiber types and it is the preferred method for classification of myofiber types. If the actin-activated myosin ATPase activity is rate limiting in the speed of contraction, it follows that the histochemical method for myofibrillar ATPase would be anticipated to be a specific method for the differentiation of myofiber types in association with their speed of contraction. Further, it has been found that fast type myosin is required for the histochemical demonstration of type 2 fibers (Staron and Pette, 1986).

Most limb muscles are "mixed," and contain variable proportions of type 1, type 2A, and type 2B myofibers. Type 2C myofibers are normally rare in mature muscle. These fibers presumably represent a transition stage between type 1 and type 2A or 2B myofibers. Neonatal myofibers are initially type 2C–like, which differentiate into type 1, type 2A, and/or type 2B myofibers (Brooke and Kaiser, 1970; Braund and Lincoln, 1981; Cardinet *et al.*, 1982).

Beyond the conventional type 1, 2A, 2B, and 2C classification of myofibers in limb muscles, an additional subset of the type 2 myofiber has been identified in the masticatory muscles of carnivores and primates (Maxwell *et al.*, 1980; Orvis and Cardinet, 1981; Rowlerson *et al.*, 1981, 1983; Shelton *et al.*, 1985). The histochemical myosin ATPase reaction of masticatory type 2 myofibers in those species is stable following acid preincubation at pH 4.2–4.3, a property which is similar to that of type 2C fibers. These fibers have been designated type 2M.

c. *Relationships with Functional Properties*

Because gross muscle coloration was often associated with the myofiber-type composition of a muscle and its contraction speed, myofiber types became equated with the speed of contraction. Therefore, red or type 1 myofibers were referred to as slow or tonic and white or type 2 myofibers as fast or phasic. That concept was based primarily on indirect evidence and while most evidence tended to support that concept, there were many exceptions. Some red muscles that were slow contracting contained myofibers with intermediate staining intensities for mitochondrial enzymes, whereas some fast-contracting muscles contained significant numbers of red myofibers. It was suggested that fast-twitch myofibers were both red and white, and slow-twitch myofibers were intermediate (Barnard *et al.*, 1970). Also, studies of diaphragm muscles in small mammals (Gauthier and Padykula, 1966) and thyroarytenoid muscles in rabbits (Hall-Craggs, 1968) showed that some red muscles were fast contracting rather than slow. Because those muscles contained myofibers with many characteristics of red myofibers based on mitochondrial enzyme histochemistry, it was concluded that not all red myofiber types were slow contracting.

More direct evidence that supports and clarifies the interrelationships of physiological properties and histochemical properties of myofibers has come from combined studies of single motor units. In the rat, fatigability of motor units was correlated with the histochemical properties of the myofibers comprising the units (Edstrom and Kugelberg, 1968). Fatigable motor units were composed of type A myofibers, whereas fatigue-resistant motor units were composed of type C myofibers according to the myofiber typing classification of Stein and Padykula (1962).

In the cat, three physiological types of motor units were classified in the medial gastrocnemius muscle on the basis of their contraction times and fatigability (Burke *et al.*, 1973; Burke and Tsairis, 1973; Burke, 1994). Two types of motor units were fast contracting, but differed in their sensitivity to fatigue. One type of motor unit, designated FR, was fast twitch and resistant to fatigue, and the second type of motor unit, designated FF, was fast twitch but fatigued rapidly. The third type of motor unit, designated S, was slow twitch and resistant to fatigue.

The myofibers present in each motor unit type within the muscle were identified in sections by their glycogen depletion patterns following tetanic stimulation. By staining adjacent sections for the histochemical demonstration of their myofibrillar ATPase and oxidative enzyme activities, it was found that the slow-twitch motor units were composed of type 1 myofibers (light staining for myofibrillar ATPase) and the fast-twitch motor units were composed of type 2 myofibers (dark staining for myofibrillar ATPase). The myofibers comprising the FR and FF units were further characterized by the lability of their myofibrillar ATPase activity to acid preincubation and it was found that FR units were composed of type 2A myofibers and FF units were composed of type 2B myofibers. The fatigue resistance of the S and FR units was found to be related to the high oxidative enzyme capacities of the type 1 and type 2A myofibers. The functional characteristics of type 2C and type 2M motor units have not been established. Based on their histochemical features, they are presumed to be fast twitch and fatigue resistant.

The physiological and histochemical properties and classifications of myofibers are summarized in (Table 16.1) and illustrated in Fig. 16.6. Assessment of the histochemical profiles of myofibers in each physiological type of unit revealed that each motor unit was homogeneous with respect to its myofiber-type composition (Fig. 16.7).

d. Myofiber Types and Myosin Isoforms

Each myofiber type is composed of a different myosin isoform (reviewed by Pette and Vrbova, 1985). With regard to myosin heavy chains (HC) and light chains (LC), slow-twitch, type 1 myofibers possess heavy chains (HC-s), alkali light chains (LC-s1a, LC-s1b), and DTNB light chains (LC-s2), which are distinct from fast-twitch, type 2 myofiber heavy chains (HC-f), alkali light chains (LC-f1, LC-f3), and DTNB light chains (LC-f2).

Type 2A and type 2B myofibers have identical light-chain compositions but different heavy chains (HC-fA and HC-fB, respectively). Type 2M fibers also have a distinct heavy-chain (HC-m) and light-chain (LC-f1M and LC-f2M) composition (Shelton *et al.*, 1985a). Type 2C fibers contain a mixture of fast and slow heavy and light chains (Billeter *et al.*, 1980; Snow *et al.*, 1981). From the foregoing and in consideration of the nine HC and LC combinations possible for myosin isoforms, more than the three defined myofiber (motor unit) types are likely to be present in limb muscles. Isoform variations are also present for the regulatory proteins troponin and tropomyosin in association with the fiber type specific myosin isoforms.

IV. NEURONAL TROPHIC INFLUENCES ON MUSCLE

Trophic influences of nerve on muscle may be defined as those functions of the nerve that affect or regulate the metabolism of the muscle (Guth, 1968). This definition should encompass structural and physiological properties as well, since the structure of the myofiber and its organelles, physiological properties, and biochemical properties are all interdependent.

The implication of neuronal influences on muscle integrity has long been recognized by studies of denervation whereby numerous morphological, physiological, and biochemical changes result when a myofiber is separated from its axon (Gutmann, 1962). Precisely how motoneurons exert their trophic influence on myofibers has been a subject of considerable interest. Two general postulates have been considered: (1) the influence of motoneurons is due to the frequency (activity) of their impulse discharges (Vrbova, 1963; Salmons and Vrbova, 1969; Pette *et al.*, 1973; Sreter *et al.*, 1973; Lomo *et al.*, 1974) and/or (2) the influence of motoneurons is mediated by specific trophic substances liberated by the motoneuron at the neuromuscular junction (Buller *et al.*, 1960).

A. Cross-Innervation Studies

More precise implications concerning the influence of the motoneuron on speed of contraction and myofiber types were first demonstrated in a series of experiments in which nerves to fast and slow muscles were cross-united (Buller *et al.*, 1960; Buller and Lewis, 1965; Close, 1965; Romanul and Van Der Meulen, 1966, 1967; Dubowitz, 1967; Robbins *et al.*, 1969; Barany and Close,

TABLE 16.1 Physiological and Histochemical Properties and Classifications of Principal Myofiber Types in Limb Muscles

	Myofiber types		
Physiological properties			
Twitch contraction	Slow twitch	Fast twitch	Fast twitch
Fatiguability	Resistant	Resistant	Fatiguable
Histochemical properties			
Myofibrillar ATPase stain (pH 9.4)	Light	Dark	Dark
Preincubation at pH 4.5	Dark	Light	Dark
Preincubation at pH 4.2	Dark	Light	Light
NADH-TR stain	Intermediate to dark	Dark	Light
Phosphorylase stain	Light	Dark	Dark
Classifications			
Engel (1962, 1965)	I	II	II
Stein and Padykula (1962)	B	C	A
Padykula and Gauthier (1967)	Intermediate	Red	White
Brooke and Kaiser (1970)	1	1A	2B
Ashmore and Doerr (1971)	β-Red	α-Red	α-White
Peter *et al.* (1972)	Slow twitch–oxidative	Fast twitch–oxidative glycolytic	Fast twitch–glycolytic
Burke *et al.* (1973)	S (Slow twitch–fatigue resistant)	FR (Fast twitch–fatigue resistant)	FF (Fast twitch–fatiguable)

1971; Weeds *et al.*, 1974; Sreter *et al.*, 1975). In this experimental design, motoneurons that normally innervate slow muscles come to innervate muscles that are normally fast, and motoneurons that normally innervate fast muscles come to innervate muscles that are normally slow. These experiments resulted in a reversal of contractile properties; that is, fast-twitch muscles become slow, and slow-twitch muscles become fast. Accompanying these changes in the speed of contraction was a corresponding change in the enzyme histochemical profiles of the myofibers. Hence, slow muscles that have myofiber populations with histochemical characteristics of high oxidative, low glycolytic, and low myofibrillar ATPase activities were changed to myofiber populations that tended to have low oxidative, high glycolytic, and high myofibrillar ATPase activities when innervated by a nerve that normally innervates a fast-twitch muscle. Converse changes occur by cross-union of a fast-twitch muscle with the nerve of a slow-contracting muscle (Romanul and Van Der Meulen, 1966, 1967; Dubowitz, 1967; Robbins *et al.*, 1969). Therefore, the motoneuron influences (1) the type of energy metabolism employed by a myofiber and all the structural changes in fiber organelles that this implies and (2) the myofiber's physiological properties of contraction. The changes in contraction speed brought about by neuronal influences are presumably due to a direct effect on the type of myosin-ATPase produced within a myofiber (Barany and Close, 1971; Weeds *et al.*, 1974; Sreter *et al.*, 1975).

The evidence derived from these experiments suggests that the motoneuron directs the phenotypic expression of myofibers and that the myofiber maintains the potential for differentiation and redifferentiation of its phenotype at the direction of the nervous system.

B. Chronic Stimulation Studies

With the abolition of low-frequency (tonic) impulse stimulation through deafferentation or tenotomy, the slow-twitch soleus muscle becomes faster (Buller *et al.*, 1960; Buller and Lewis, 1965). Also, high-frequency (phasic) impulse stimulation of the nerve to the slow-twitch soleus muscle resulted in its conversion to a fast-twitch muscle (Salmons and Vrbova, 1969; Pette *et al.*, 1973; Sreter *et al.*, 1973; Hennig and Lomo, 1987).

Conversely, low-frequency stimulation or merely increased impulse stimulation of the nerves to fast-twitch muscles results in their transformation to slow twitch (reviewed by Pette and Vrbova, 1985). These findings provided evidence that it was the activity of motoneurons that determined the morpho-physiochemic properties of myofibers. In studies employing stimulation of the nerve supply, it is difficult to separate the effects of evoked muscle activity and putative neurotrophic substances because changes in the nerve impulse activity may also change the amount and type of trophic substances (Lomo, 1986). However, direct stimulation of denervated muscles has effected the

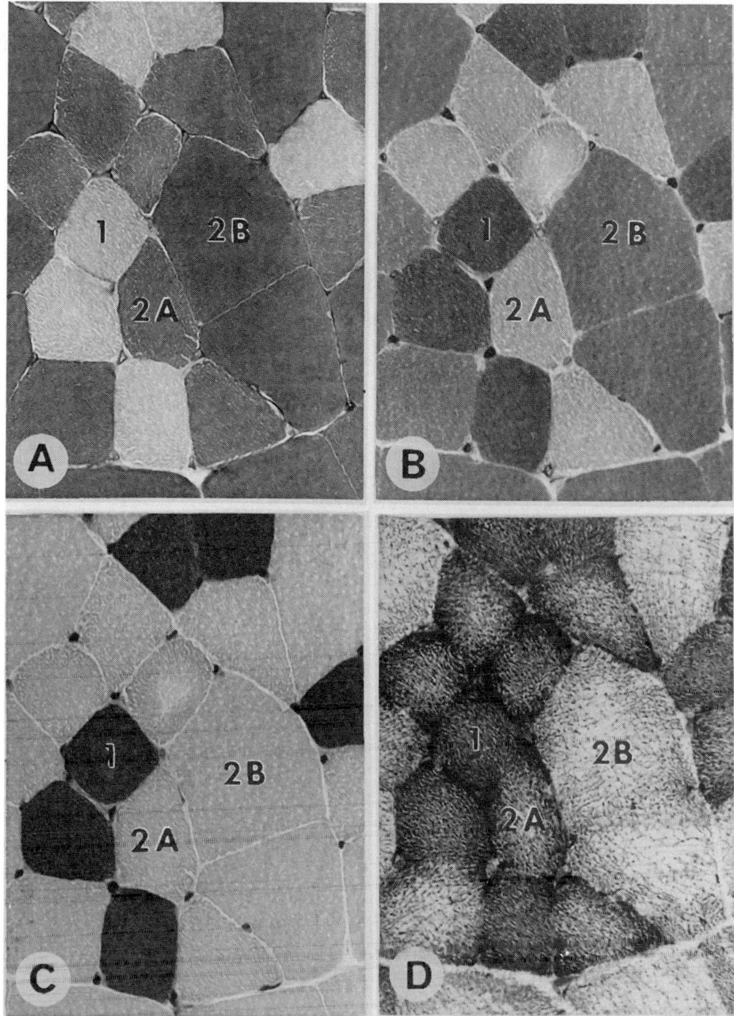

FIGURE 16.6 Serial sections of cat medial gastrocnemius muscle incubated for the histochemical demonstration of (A) myofibrillar ATPase, incubated at pH 9.4; (B) myofibrillar ATPase, preincubated at pH 4.5; (C) myofibrillar ATPase, preincubated at pH 4.2; and (D) NADH-TR. Type 1 myofibers comprise slow-twitch motor units that are resistant to fatigue. Type 2A myofibers comprise fast-twitch motor units that are resistant to fatigue, and type 2B myofibers comprise fast-twitch motor units that fatigue rapidly.

same transformations (Lomo *et al.*, 1974; Hennig and Lomo, 1987), a finding that further supports the concept that "activity" per se is the trophic influence. It is concluded that "activity" controls gene expression in mammalian myofibers (Pette and Vrbova, 1985).

V. EXERCISE, ADAPTATIONS TO TRAINING, AND PERFORMANCE

Exercise induces major biochemical adaptations in skeletal muscle. The nutritional state, intensity and duration of exercise, and degree of physical fitness are all factors that qualitatively and quantitatively affect the metabolic pathways used in the generation of energy for muscular contraction.

A. Exercise Intensity and Sources of Energy

The rate of energy utilization during intense exercise can be as much as 200 times greater than at rest and the rate of ATP utilization is closely associated with the rate of ATP synthesis (Holloszy, 1982). Hence, the availability of ATP is a central requirement for sustaining the rate and duration of exercise. Since the stores of creatine phosphate and ATP available for immediate use within myofibers are small, the metabolic path-

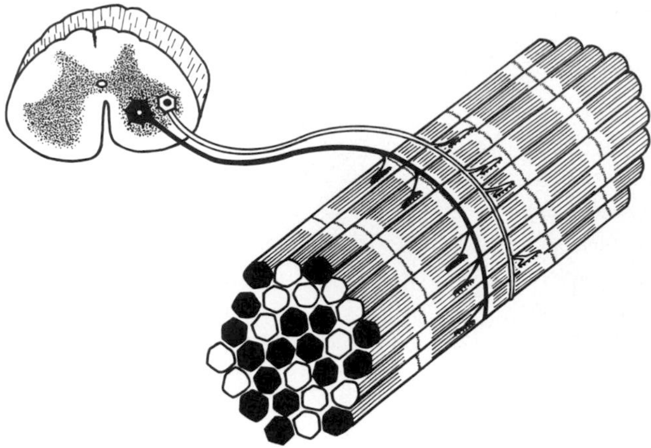

FIGURE 16.7 Schematic representation of the homogeneity of skeletal muscle motor units. Motor units have a homogeneous myofiber-type composition whereby slow-twitch motor units are composed of only type 1 myofibers (light staining for myosin ATPase). Fast-twitch units are composed of only type 2A (fast twitch, fatigue resistant) myofibers or only type 2B (fast twitch, fatigable) myofibers (dark staining for myosin ATPase).

ways for ATP synthesis serve a vital function in the maintenance of exercise.

The main fuels for muscular contraction are fatty acids and glucose, which are supplied by intramuscular and extramuscular depots during exercise. Glucose or glycogen may be metabolized to synthesize ATP through either aerobic or anaerobic pathways, whereas fat metabolism is only aerobic. It is estimated that 65% or more of the oxygen utilization during moderate to heavy exercise is accounted for by the oxidation of carbohydrates and that exhaustion coincides with glycogen depletion (Holloszy, 1982; Sahlin, 1986). Intramuscular depots are the triglyceride and glycogen inclusions of the sarcoplasm, and the extramuscular depots are adipose tissues and the liver (Issekutz *et al.*, 1966; Paul and Issekutz, 1967; Holloszy *et al.*, 1971; Paul, 1971; Terjung *et al.*, 1972; Baldwin *et al.*, 1973a; Therriault *et al.*, 1973; Lindholm *et al.*, 1974a). The rate-limiting factor in the extramuscular supply of glucose to working muscle is glucose uptake by the myofiber. The rate-limiting factor in the plasma free fatty acid (FFA) supply appears to be the rate of FFA release from adipose tissues (Issekutz *et al.*, 1965, 1966).

At rest, oxidation of FFA and stored triglycerides contributes the bulk of the fuel used for maintaining muscle tone, whereas the oxidation of glucose accounts for only 10–20% of the CO_2 produced (Havel, 1971). At the onset of exercise, energy is initially derived from creatine phosphate and anaerobic glyco(geno)lysis, which is accompanied by an increased production of lactate. However, as the duration of exercise is increased and blood flow increases, there is a shift to aerobic metabolism in which glucose, FFA, and triglycerides are oxidized and lactate production decreases (Havel, 1971; Rowell, 1971; Wahren *et al.*, 1971; Baldwin *et al.*, 1973b). At low to moderate exercise intensities, the oxidation of fatty acids provides the major source of energy. At moderate to high exercise intensities, the oxidation of fatty acids decreases and carbohydrates account for 50% or more of the amount of substrate utilized (Hagenfeld and Wahren, 1971; Saltin and Karlsson, 1971; Therriault *et al.*, 1973; Wahren, 1977). At moderately high exercise intensities, muscle glycogen and triglyceride depots decrease and muscle glycogen accounts for approximately 70% of the glucose oxidized, and 70% of the fatty acids oxidized are derived from muscle triglyceride depots. At maximal workloads, glycogen utilization is high and the major source of fuel (Hultman and Nilsson, 1971; Froberg *et al.*, 1971).

Evidence exists that not all fibers are activated at once but that motor units are recruited with respect to their oxidative capacities (Baldwin *et al.*, 1973b). Measurements of rates of glycogen and triglyceride depletion in rats indicate that, in prolonged work of moderate intensity, the red and intermediate fibers are the first to become activated. Once the workload exceeds the capacity of the red and intermediate fibers, the white motor units are then activated by the central nervous system (Baldwin *et al.*, 1973a, 1973b). Similarly, glycogen depletion studies of fiber types in the horse suggest that the order of myofiber recruitment is type 1, type 2A, and type 2B (Lindholm *et al.*, 1974; Snow *et al.*, 1981, 1982; Essen-Gustavsson *et al.*, 1984; Hodgson *et al.*, 1984a; Valberg, 1986). With moderate intensity, type 1 and type 2A myofibers are preferentially recruited, whereas moderate intensity of long duration or maximal exercise intensity is required for recruitment of type 2B myofibers. With maximal intensity (galloping) exercise in thoroughbreds, glycogen depletion is greatest in type 2A and type 2B fibers, and type 1 fibers are either not recruited or there is a glycogen sparing effect during high-intensity, short-duration exercise (Hodgson *et al.*, 1984b). On the other hand, maximal intensity (trotting) exercise in standardbreds results in glycogen depletion of all type 1, most type 2A, and many type 2B myofibers (Valberg, 1986).

B. Adaptations to Exercise Training

The principal metabolic adaptation of skeletal muscles to training is an increase in oxidative capacity to utilize fat, carbohydrate, and ketones (Holloszy, 1982; Holloszy *et al.*, 1986. Oxygen consumption is similar in trained and untrained subjects for a given workload, but there is a lower production of CO_2 and hence a lower respiratory quotient (RQ) in trained subjects (Holloszy *et al.*, 1971, 1977; Saltin and Karlsson, 1971). Although several factors may affect the RQ, a major factor would be the source of energy for contraction, and fat utilization yields lower RQ values than carbohydrate utilization (Paul, 1971). The capacity for oxidation of palmitate, oleate, linoleate, palmityl-CoA, and palmitylcarnitine, as well as the activities for carnitine palmityltransferase, palmityl-CoA, dehydrogenase and mitochondrial ATP-dependent palmityl-CoA synthetase were doubled in rats subjected to a training program (Mole *et al.*, 1971). Also, the turnover of FFA is significantly greater in trained than in untrained dogs (Issekutz *et al.*, 1966).

Myoglobin content is greater in trained rats (Holloszy *et al.*, 1971) and pigs (Jorgensen and Hyldgaard-Jensen, 1975) in which the myoglobin functions to facilitate O_2 utilization by enhancing the intrafiber O_2 transport between the sarcolemma and mitochondria. The mitochondrial protein content increases approximately 60% (Holloszy, 1967; Mole *et al.*, 1971), which is associated with increased size and numbers of mitochondria in trained rats (Gollnick and King, 1969). An increase in mitochondrial size but not number, has been reported in trained men (Morgan *et al.*, 1972). Cristae double their density within muscle mitochondria of trained rats (Holloszy *et al.*, 1971). Mitochondrial coupling factor 1, mitochondrial ATPase, and cytochrome *c* are increased in trained subjects, whereas mitochondrial creatine kinase and adenylate kinase activities were not increased (Oscai and Holloszy, 1971). Increased SDH activities following training have been reported in humans (Gollnick *et al.*, 1972; Saltin *et al.*, 1977), rats (Holloszy, 1967), and pigs (Jorgensen and Hyldgaard-Jensen, 1975) as have levels of NADH-cytochrome *c* reductase, succinate oxidase, cytochrome oxidase, citrate synthase, and NADH in trained rats (Holloszy, 1967; Kowalski *et al.*, 1969; Holloszy *et al.*, 1971) and guinea pigs (Barnard *et al.*, 1970). In horses, training results in increased activities for citrate synthase, hexokinase, and 3-OH-acyl-CoA dehydrogenase (Cutmore *et al.*, 1985; Hodgson *et al.*, 1985).

Muscles of trained horses have a higher proportion of type 2A to type 2B fibers than untrained horses (Essen-Gustavsson and Lindholm, 1985) and training increases the oxidative capacity of equine type 2B fibers.

With training, muscle glycogen is increased (Gollnick *et al.*, 1972). The rate of lactate production is closely related to the intensity of exercise (Hermansen, 1971; Wahren, 1977) and lactate production is lower in trained than in untrained subjects performing the same exercise. The increased plasma lactate concentrations are directly correlated with decreases in plasma FFA due to a reduction in FFA release from adipose tissues (Edwards *et al.*, 1971; Hermansen, 1971; Jorfeldt, 1971; Karlsson, 1971; Saltin and Essen, 1971; Saltin and Karlsson, 1971, Lindholm and Saltin, 1974; Issekutz *et al.*, 1975; Snow and Mackenzie, 1977). Glycogenolytic and glycolytic enzyme activities are unchanged or only slightly decreased in both red and white muscles of trained rats with the exception of hexokinase. Hexokinase activities are markedly increased in trained rats and guinea pigs (Holloszy *et al.*, 1971; Baldwin *et al.*, 1973b; Peter *et al.*, 1968; Lamb *et al.*, 1969). Although the increase occurred in both red and white muscles, the percent increase was greatest in red muscles (Baldwin *et al.*, 1973b). In the horse, training increases AMP deaminase activity, which would serve to facilitate glyco(geno)lysis through activation of PFK.

In summary, the major metabolic consequences of the adaptations of muscle to endurance exercise are a slower utilization of muscle glycogen and blood glucose, a greater reliance on fat oxidation, and less lactate production during a given intensity (Holloszy and Coyle, 1984). At workloads below maximal O_2 utilization, aerobic pathways are the principal sources of energy through the oxidation of fatty acids and glucose. This correlates with the evidence that type 1 and type 2A (red) myofibers are activated first and their metabolic orientation is toward aerobic pathways. With workloads approaching maximal O_2 utilization, the sources of energy are derived principally from anaerobic pathways through glyco(geno)lysis. Type 2B (white) myofibers are recruited only with increased workload intensities and their metabolic energy derivation is mainly by anaerobic pathways. All of the complexities of energy derivation and fiber-type activities are not totally understood for all combinations of workload intensities and durations but one can conceive a general scheme of alterations in enzyme activities correlating to specific myofiber types in response to the type of exercise and the energy demands to support that exercise.

C. Performance and Athletic Potential

Various forms of exercise usually emphasize speed or endurance, or some combination of the two. Athletic ability reflects the efficiency of an individual to attain the desired speed and/or endurance required through the combined effects of genetic and environmental factors such as training. Obviously, the ability to manipulate myofiber properties to attain the appropriate balance of speed and endurance would be desirable for developing the best athletes for selected activities.

Studies of myofiber types in dogs and horses reveal that genetic selection of breeds for their speed (e.g.

greyhounds and thoroughbreds) has resulted in more fast-twitch, type 2 myofibers within their muscles of propulsion compared to breeds selected for sustained, low-intensity activities (Gunn, 1978; Snow and Guy, 1980; Guy and Snow, 1981).

Whereas cross-innervation and chronic stimulation studies suggest that "activity" controls gene expression in muscle, "activity" in the form of exercise training holds the potential for improving speed and/or endurance. As we have seen, training activity improves endurance performance through improved oxidation of fats and carbohydrates. For example, in standardbred horses training results in greater type 2A/2B ratios, and the best performers had the highest ratios (Essen-Gustavsson and Lindholm, 1985). Beyond the effects of training on ATP synthetic pathways, evidence also exists that high-intensity interval training can result in transformations of fiber types in the order of type 2B-to-type 2A-to-type 1 in both rats and humans (Green *et al.*, 1984; Luginbuhl *et al.*, 1984; Simoneau *et al.*, 1985). For now, there is little evidence for transformations from slow, type 1 to fast, type 2 through exercise training.

VI. DIAGNOSTIC LABORATORY METHODS FOR THE EVALUATION OF NEUROMUSCULAR DISORDERS

Muscular weakness is the principal clinical sign of neuromuscular disorders. Manifestations of muscular weakness may be functional (e.g., paresis, paralysis, gait abnormalities, exercise-related weakness, dysphagia, regurgitation, dyspnea, and dysphonia) and/or physical (e.g., gross atrophy, hypotrophy, hypertrophy, and skeletal deformities).

Instituting measures of prevention and/or therapy of neuromuscular disorders depends on an accurate definition of the functional and physical manifestations of muscle weakness and the identification of the specific pathoanatomic motor unit component(s) involved (i.e., neurons in neuropathies, neuromuscular junctions in disorders of neuromuscular transmission, or myofibers in myopathies) and, when possible, identification of the specific cellular dysfunctions underyling the muscular weakness.

The evaluation of neuromuscular disorders requires a coordinated approach and special examinations, some of which fall outside the scope of this chapter. These involve the neurologic examination and include the signalment (e.g., species, breed, age, sex), history (e.g., congenital or acquired, course of the disease, exposures, and responses to treatment), findings (e.g., presence and distribution of signs, neurologic deficits, and abnormal reflexes) and electrodiagnostic tests that involve electromyography (EMG), and the evaluation of sensory and motor nerve conduction velocity measurements and evoked MAPs.

Standard hematologic and clinical chemistry panels are indicated to provide general screening that would suggest possible infectious, immune, or metabolic abnormalities.

In addition to these evaluations, there are some more specific tests that provide insight into the pathoanatomic involvement and in some instances specific identification of the etiology and pathogenesis of the muscular weakness.

A. Muscle-Specific Serum Enzyme Determinations Used in the Diagnosis of Neuromuscular Disorders

A valuable adjunct to the clinical diagnosis of neuromuscular diseases has been the utilization of serum enzyme determinations. This involves the detection of enzymes in serum or plasma. The activities or concentrations of the enzymes are usually low in serum or plasma because they are normally located within healthy myofibers. Necrosis of myofibers is a primary example of a process by which serum activities of intracellular enzymes are elevated and the elevations are roughly proportional to the mass of tissue involved. Elevations in serum enzyme activities may also occur in association with increased cell permeability (leakage), increased enzyme production by the parenchymal cells, obstructions to normal enzyme excretory routes, increased amount of enzyme-forming tissue, delayed removal or inactivation of enzyme (Cornelius, 1967), and perhaps cell secretion.

The initial examination of a patient with muscle weakness should always include the measurement of muscle-specific enzyme activities. This provides immediate information concerning the possible presence of muscle necrosis and provides a course-grain analysis for distinguishing between myopathies and neuropathies. Despite the ready availability of such tests, they are often underutilized or employed only after the fact and their usefulness has passed.

1. *Creatine Kinase*

The most widely used serum enzyme determination in neuromuscular diseases of domestic animals is that for creatine kinase (CK), previously designated creatine phosphokinase (CPK). In muscle, this enzyme functions in making ATP available for contraction by the phosphorylation of ADP from creatine phosphate (Fig. 16.2).

The use of serum CK has offered greater promise of organ specificity in diseases of muscle than most other enzyme determinations employed to date. Analysis of tissues from humans indicates that significant CK activities are present in skeletal muscle, myocardium, and brain, with lesser amounts in the gastrointestinal tract, uterus, urinary bladder, kidney, and thyroid (Dawson and Fine, 1967). The diversity of organs tested in other animals is not as broad but those tested correlate well with these findings. The liver has negligible amounts of CK (Oliver, 1955; Tanzer and Gilvarg, 1959; Colombo *et al.*, 1962; Eppenberger *et al.*, 1962; Cardinet *et al.*, 1967; Dawson and Fine, 1967).

Some normal values for CK activity have been determined in domestic animals (Table 16.2); these vary with physical activity, restraint, biopsy (Anderson, 1975; Blackmore and Elton, 1975; Rose *et al.*, 1977; Tarrant and McVeigh, 1979), age, and sex (Heffron *et al.*, 1976). Intramuscular injections may also increase CK activities due to local areas of muscle necrosis (Nevins *et al.*, 1973; Steiness *et al.*, 1978). The amount of CK liberated following intramuscular injection of a drug depends on the properties of the injected solution and on such muscle factors as species differences in muscle CK activity, local blood flow, susceptibility of the muscles and local muscle binding of the drug (Steiness *et al.*, 1978). Therefore, an accurate history is important in evaluating CK activities.

TABLE 16.2 Normal Values of Serum Creatine Kinase Activity in Some Domestic Animals

			CK activity (U/liter)		
Species	**Age**	**Sex**	**Mean**	**Std. dev.**	**Range**
Canine[a]	0–4 months	Male	4.7	±0.3	1.2–8.2
		Female	3.0	±0.2	1.2–5.5
	4–6 months	Male	2.4	±0.2	0.1–6.6
		Female	1.8	±0.2	0.4–6.2
	6–12 months	Male	1.8	±0.2	0.4–5.6
		Female	0.8	±0.2	0.0–7.5
	>12 months	Male	1.2	±0.3	0.2–2.6
		Female	0.8	±0.1	0.0–1.8
Canine[b]	>12 months	Both sexes	22.7	±7.3	9.9–28.8
Caprine[a]		Female	0.9	±0.9	0.0–2.5
Equine[a,c]	3–12 years		1.3	±0.9	0.0–3.6
Feline[a]		Male	1.7	±1.1	0.4–3.4
		Female	1.9	±1.3	0.0–4.5
Ovine[a]		Male	0.8	±0.6	0.0–2.9
		Female	0.5	±0.3	0.0–0.9
Gallus domesticus[d]	27–134 days		—	—	38.1–59.1
	1 year	Male	109	—	—
		Female	50	—	—

[a] Enzyme activities determined without reducing agents. Data from Cardinet (1969).

[b] Enzyme activities determined in the presence of glutathione. Data from Heffron *et al.* (1976).

[c] Unexercised; not in training. Enzyme activities determined without reducing agents. Data from Cardinet (1969).

[d] Enzyme activities determined in the presence of cysteine. Data from Holliday *et al.* (1965).

There are three principal isoenzymic forms of CK. Creatine kinase has a dimeric structure consisting of M (muscle) subunits and B (brain) subunits, which combine to form the three heterogeneous MM (or CK3), MB (or CK2), and BB (or CK1) isoenzymes (Dawson *et al.*, 1965, 1967). A fourth variant form, CK-Mt, is found in mitochondrial membranes and may account for up to 15% of the total cardiac CK activity. The isoenzymes can be separated by three different methods: (1) electrophoresis, (2) immunological techniques, and (3) ion-exchange chromatography (Fiolet *et al.*, 1977).

The pattern of isoenzyme distribution varies among the organs of different species. Thus, identification of the isoenzymes present can be used to help determine the tissue source of elevations in CK (Eppenberger *et al.*, 1964a; Dawson *et al.*, 1965; Dawson and Fine, 1967; Van der Veen and Willebrands, 1966; Sherwin *et al.*, 1969; Klein *et al.*, 1973). Ontogenic studies in the rat revealed that all organs investigated contained only BB-CK in early stages of fetal development. In skeletal muscle, BB-CK forms slowly disappear and are initially replaced by MB-CK forms followed by CK-MM forms. Mixtures of isoenzymes occur during the transition. In the adult pattern, MB-CK forms have been variously reported to be present or absent. The inconsistency in noting the presence of MB-CK forms in skeletal muscle of animals may be due to the source of skeletal muscle sampled since all skeletal muscles may not contain the MB-CK isoenzyme (Sherwin *et al.*, 1967; Thorstensson *et al.*, 1976). However, other studies indicate that, although there is more MM-CK in white muscle of the rat than in red, there is no MB-CK fraction in either (Dawson and Fine, 1967). Reasons for the discrepancies in the detection of MB-CK forms in mammalian skeletal muscles are not evident. The adult isoenzyme pattern in muscles of the rat appears at 90 days after birth, whereas in cardiac muscle, the shift occurs earlier and the adult pattern has both MB-CK and MM-CK forms. In the brain, BB-CK is the major isoenzyme throughout life (Eppenberger *et al.*, 1964a). An unusual sex linkage of muscle CK isoforms has been reported in Harris' hawks (Morizot *et al.*, 1987). The total CK activity of breast muscle was similar in males and females; however, the females had three isoenzymic forms, but males had only one, which was judged to be BB-CK.

Determinations of serum isoenzyme patterns have found clinical application in human medicine. The determination of MB forms is generally advocated as the best biochemical diagnostic tool for acute myocardial infarction (Fiolet *et al.*, 1977). However, studies of the isoenzyme composition in cardiac muscle of the horse reveals that less than 1.5 to 3.9% of the total CK activity is attributable to the MB-CK form. Hence, its determination cannot be used for detection of myocardial disorders in the horse (Fujii *et al.*, 1980; Argiroudis *et al.*, 1982).

Changes in serum isoenzyme patterns have been observed in various neuromuscular disorders and Duchenne muscular dystrophy (Goto *et al.*, 1969; Somer *et al.*, 1976). In the horse, one study found that horses with a previous history of rhabdomyolysis had higher serum MM-CK and BB-CK activities, but lower MB-CK activities than control horses (Johnson and Perce, 1981). With modest exercise the affected horse had a significant decrease in serum MM-CK and BB-CK, and increase in MB-CK. The significance of those findings has not been elucidated.

Elevations in total CK activities have been reported in the hereditary muscular dystrophies of chickens (Holliday *et al.*, 1965; Wagner *et al.*, 1971) and Syrian hamsters (Eppenberger *et al.*, 1964b); in selenium-vitamin E deficiencies of cattle (Allen *et al.*, 1975; Van Vleet *et al.*, 1977), sheep (Whanger *et al.*, 1970), and swine (Ruth and Van Vleet, 1974); in myodegeneration due to ingestion of toxic plants in cattle (Henson *et al.*, 1965); in the arthrogryposis-hydranencephaly syndrome in calves (Hamada, 1974); in capture myopathies involving a moose (Haigh *et al.*, 1977) and pronghorns (Chalmers and Barrett, 1977); and in paralytic myoglobinuria involving horses (Gerber, 1964; Cardinet *et al.*, 1967; Lindholm *et al.*, 1974b) and a double-muscled heifer (Holmes *et al.*, 1972). Elevations in CK activities have also been reported in polioencephalomalacia and focal symmetric encephalomalacia of sheep and it has been suggested that CK determinations may be of value in diagnosing diseases of the central nervous system (Smith and Healy, 1968). However, in diseases of the central nervous system that involve motor function, elevations of CK may be from skeletal muscle rather than from the central nervous system. This appears to be the case in central nervous system disorders in man (Cao *et al.*, 1969).

Diseases of muscle are classified, whenever possible, as to the origin or site of the primary lesion. Myopathies are those diseases in which the primary defect or disease process is considered to be limited to the myofibers, and neuropathies are those diseases of muscle which are secondary changes due to defects or diseases of the neuron (e.g., denervation). Although CK determinations may be specific for diseases of muscle, they do not provide information relative to the origin of the disease process. However, elevations of CK are generally higher in myopathies than neuropathies because myonecrosis is much more prevalent in myopathies than neuropathies. More precise information regarding the origin of muscle diseases can be obtained by the use of histological and histochemical examination of muscle biopsies.

In a comparative study, the behavior of CK was found to be distinctly different from that of serum aspartate aminotransferase (AST or GOT) during the course of paralytic myoglobinuria in horses (Fig. 16.8). The latter enzyme is discussed in the next section. Elevations in AST activities were present for weeks after the onset of clinical disease, whereas CK activities remained elevated for only a few days. The course of elevations of these enzymes in this disease can be directly attributed to different disappearance rates of their activity in the plasma (Fig. 16.9).

Although CK is more specific for myonecrosis than AST, the simultaneous determinations of AST and CK in the horse are potentially valuable diagnostic and prognostic aids owing to the different disappearance rates of their serum or plasma activities. (1) Elevated CK activities indicate that myonecrosis is active or has recently occurred; (2) persistent elevations of CK indicate that myonecrosis continues to be active; and (3) elevated AST due to myonecrosis accompanied by decreasing or normal CK activities indicates that myonecrosis is no longer active. It has not been established whether there are similar differences in the disappearance rates of AST and CK activities in the plasma of other animal species. Therefore, it is not known whether the same assessment of myonecrosis through

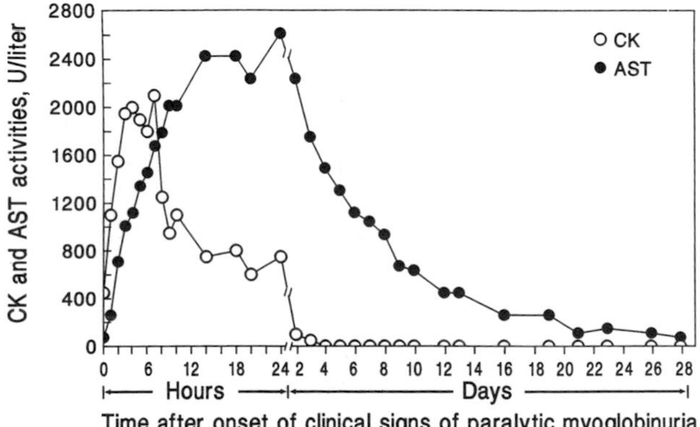

FIGURE 16.8 The difference in the time course of elevations in AST and CK activities due to muscle necrosis (equine paralytic myoglobinuria). The AST activity remained elevated for much longer periods than CK activity. Adapted from Cardinet *et al.* (1967).

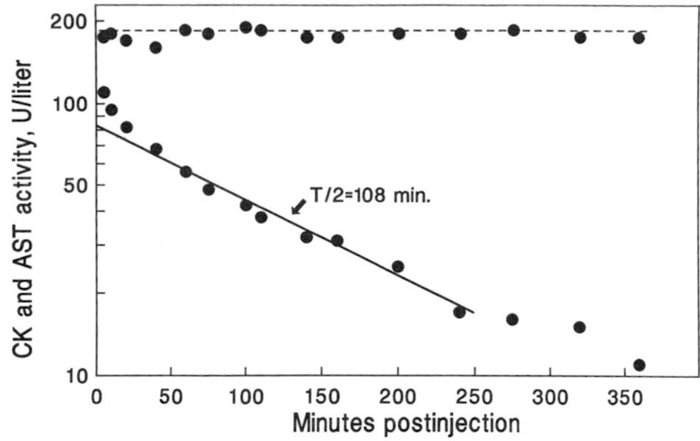

FIGURE 16.9 Disappearance of AST (dotted-dashed line) and CK (dotted-solid line) activities in the serum of a horse following the intravenous injection of AST and CK. The differences in the course of serum elevations of these two enzymes due to necrosis in the horse are the result of differences in their disappearance rates in the serum. Adapted from Cardinet *et al.* (1967).

the simultaneous determination of AST and CK activities can be applied to species other than the horse.

The normal CK values presented in Table 16.2 were determined by the method of Tanzer and Gilvarg (1959). Storage of serum at 0 to −5°C results in a loss of activity; however, the addition of reducing agents such as cysteine or glutathione to the incubating medium increases the serum activities, and the losses due to storage are minimized (Okinaka *et al.*, 1964; Weismann *et al.*, 1966; Hess *et al.*, 1968). The cysteine or glutathione reactivates the CK activity lost during storage, therefore, the addition of those reducing agents to the incubating medium is the method of choice and necessitates the establishment of normal "activated" values.

2. *Aspartate Aminotransferase*

Another enzyme that has been used as a diagnostic aid in neuromuscular disorders of domestic animals is serum aspartate aminotransferase (AST) formerly called glutamic oxaloacetic transaminase (GOT). Some normal values of AST activity reported for domestic animals are summarized in Table 16.3. Normal values do not appear to differ greatly between sexes, although reported values for cows (Cornelius *et al.*, 1959a) are somewhat higher than values for bulls (Roussel and Stallcup, 1966). Differences associated with age have been reported in sheep (Lagace *et al.*, 1961) and there are seasonal differences in bulls (Roussel and Stallcup, 1966). Also, physical activity is associated with higher values in horses (Cornelius and Burnham, 1963; Cardinet *et al.*, 1963, 1967; Blackmore and Elton, 1975).

Elevations of AST activities have been reported in white muscle disease of lambs, cows, and swine (Blincoe and Dye, 1958; Kuttler and Marble, 1958; Swingle *et al.*, 1959; Blincoe and Marble, 1960; Whanger *et al.*, 1970; Ruth and Van Vleet, 1974), tying up and paralytic myoglobinuria in horses (Cornelius and Burnham, 1963; Cardinet *et al.*, 1963, 1967; Lindholm *et al.*, 1974b), azoturia in a double-muscled heifer (Holmes *et al.*, 1972), hereditary muscular dystrophy in chickens (Cornelius *et al.*, 1959b), myodegeneration due to ingestion of toxic plants in cattle (Henson *et al.*, 1965), and capture myopathy in a moose (Haigh *et al.*, 1977) and in pronghorns (Chalmers and Barrett, 1977). Although the use of AST determinations has proven valuable as a diagnostic aid, the enzyme lacks organ specificity since in addition to high concentrations in skeletal and cardiac muscle, AST activities are also high in the liver as well as other organs and tissues, including the red blood cells (RBCs) (Cornelius *et al.*, 1959a; Nagode *et al.*, 1966; Cardinet *et al.*, 1967).

3. *Lactate Dehydrogenase*

Lactate dehydrogenase (LDH) activities are high in various tissues of the body. Therefore, measurements of LDH are not organ specific. Molecules of LDH are tetrameric, made up of four subunits of the two parent molecules, M (muscle) and H (heart). Various combinations of those subunits result in five isoenzymes of LDH, which can be separated by electrophoresis. The M monomer is found in purest form in skeletal muscle as the isoenzyme M4 (or LDH5), whereas the H monomer is found predominantly in the heart muscle as the isoenzyme H4 (or LDH1). The other three forms are molecular hybrids forming the isoenzymes M3H (or LDH4), M2H2 (or LDH3), and MH3 (or LDH2), and they are found in various amounts in different organs. The H4 isoenzyme is maximally active at low concentrations of pyruvate and strongly inhibited by excess pyruvate (Dawson *et al.*, 1964), which favors the oxidation of lactate (Dawson *et al.*, 1964; Sund, 1969). The M form, on the other hand, maintains activity at relatively high pyruvate concentrations (Dawson *et al.*, 1964), which favors anaerobic reduction of pyruvate (Dawson *et al.*, 1964; Sund, 1969). Thus, tissues with essentially aerobic metabolism, such as heart muscle, contain mostly heart-specific isoenzymes, whereas tissues with more viable or flexible metabolic properties, such as skeletal muscle, contain predominantly the muscle-specific isoenzyme (Dawson *et al.*, 1964).

The LDH isoenzyme pattern within skeletal muscle seems to be under genetic control (Dawson *et al.*, 1964), but is influenced by environmental factors (Sjodin, 1976). Total LDH activity and muscle-specific LDH activity are higher in fast-twitch myofibers than in slow-twitch myofibers. Positive correlations have been

TABLE 16.3 Normal Values of Serum Aspartate Aminotransaminase Activity in Some Animals

Species	AST activity (U/liter)				
	Comment	Mean	Std. dev.	Range	Reference[a]
Bovine	Bull, 1–97 weeks	11.4	±8.3	—	1
	Cows, 2–10 years	21.0	±2.7	—	2
	Calves, 7–27 days	11.3	±1.8	—	2
Canine	>9 months	10.9	±2.6	—	2
	<2 years	12.8	±0.8	—	3
	4–12 months	10.6	±2.6	6–23	4
	1–5 years	12.4	±3.7	6–25	4
	>5 years	10.4	±2.5	7–14	4
Equine	Unexercised, not in training	165	±33.8	—	2
	>1 year	79.2	±16.2	58–161	5
		89.3	±8.6	53–91	6
		72.5	±25.0	48–170	7
	Exercised, in training >1 year	166	±75.4	48–456	5
		120	±14.9	88–154	6
Feline	1 month	9.1	±2.3	6–13	8
Ovine	Lambs, 7–35 days	56.5	±14.9	—	9
	0–4 weeks	—	—	21–25	10
	5–8 weeks	—	—	<60	10
	>10 weeks	—	—	<39	10
Wild bighorn sheep	At capture, day 1	65.8	±7.2	—	11
	After handling (day 3)	244	±48.5	—	11
	After captivity (day 15)	65.3	±52.8	—	11
	Grand mean	125	±94.1	—	11
Domestic sheep		61.4	—	—	11
Caprine	Adult	107	±23.1	—	12
Porcine	1–3 years	1.5	±6.8	—	2
Gallus domesticus	6 months	178	±89.3	—	13

[a] 1, Roussel and Stallcup (1966); 2, Cornelius *et al.* (1959a); 3, Hibbs and Coles (1965); 4, Crawley and Swenson (1965); 5, Cornelius and Burnham (1963); 6, Cardinet *et al.* (1963); 7, Cardinet *et al.* (1963); 8, Cornelius and Kaneko (1960); 9; Blincoe and Marble (1960); 10, Lagace *et al.* (1961); 11, Franzmann and Thorne (1970); 12, Lewis (1976); 13, Cornelius (1960).

found between both the individual percentage of fast-twitch myofibers and muscle lactate concentration, and between lactate concentration and total LDH activity and muscle-specific LDH activity, respectively (Tesch *et al.*, 1978). Because the fast-twitch myofibers derive more of their energy for contraction via anaerobic metabolism than do slow-twitch myofibers, it seems reasonable that fast-twitch myofibers should contain more of the muscle-specific LDH isoenzymes. Skeletal muscle LDH activity was found to increase with training in men (Suominen and Heikkinen, 1975; Sjodin *et al.*, 1976) and swine (Jorgensen and Hyldgaard-Jensen, 1975), whereas it has been reported to decrease with training in horses (Guy and Snow, 1977). The reasons for this difference are not obvious.

Elevated LDH activities have been reported in selenium-vitamin E deficiency of cattle (Allen *et al.*, 1975), sheep (Whanger *et al.*, 1970), and swine (Ruth and Van Vleet, 1974) and in myoglobinuria in a double-muscled heifer (Holmes *et al.*, 1972) and in horses (Rose *et al.*, 1977). Elevations in LDH activity have also been reported in a variety of hepatic disorders. Therefore, unless isoenzyme analysis is utilized, the measurements of LDH elevations are not organ specific.

4. *Serum Pyruvate Kinase*

Measurements of serum pyruvate kinase (PK) activities have been found to be more discriminatory than creatine kinase in distinguishing between stress-susceptible pigs (Duthie and Arthur, 1987; Duthie *et al.*, 1988). Employing PK measurements, homozygous, halothane-reacting (malignant hyperthermic) pigs could be discriminated from homozygous nonreacting pigs. Isoenzymes of PK exist and their distinction may be important in the application of this enzymic assay.

B. Muscle-Specific Serum Proteins and Antibodies

With the advent of immunochemical procedures such as enzyme-linked immunosorbent assay (ELISA), radial immunodiffusion assays, radioimmunoassays, and immunocytochemical assay, sensitive tests for the detection of tissue-specific proteins and antibodies in serum are being applied to the diagnosis of neuromuscular disorders. The potential sensitivity and specificity of these procedures will no doubt in the future replace a large number of the clinical enzymology assays currently employed.

1. *Myoglobin*

Myoglobin is a 17,500-Da heme protein that stores and transports oxygen in myofibers. Elevated levels of myoglobin have been found in myopathies of humans (Penn, 1986). The specificity of myoglobin for skeletal and cardiac muscle and its plasma clearance make myoglobin determinations a potentially effective method for monitoring myonecrosis (Holmgren and Valberg, 1992). Plasma myoglobin concentrations were compared to aspartate aminotransferase (AST) and creatine kinase (CK) activities following exercise in horses susceptible to rhabdomyolysis (Valberg *et al.*, 1993). The authors report that the detection of myoglobin was the most sensitive indicator of myonecrosis.

2. *Myosin Light Chains and Troponin*

Application of myosin light-chain assays are evolving for use in the diagnosis of inflammatory muscle disease (Mader *et al.*, 1994) and myocardial infarctions (Nicol *et al.*, 1993), whereas troponin I assays are being promoted for use in the diagnosis of acute myocardial infarctions in humans (Edleman, 1995). These assays offer the potential for improved specificity and sensitivity and are likely to find applications in veterinary diagnostic clinical chemistry.

3. *Acetylcholine Receptor Antibodies*

Detection of circulating autoantibodies to acetylcholine receptors (AChR) by an immunoprecipitation assay (Lindstrom *et al.*, 1981) is a valuable adjunct to the diagnosis of immune-mediated myasthenia gravis (MG) in humans (Engel, 1994c) and dogs (Lennon *et al.*, 1978; Shelton *et al.*, 1988). It has been estimated that approximately 80% of human patients with MG have detectable AChR antibodies (Engel, 1994c).

A valuable immunocytochemical screening test for circulating AChR antibodies in canine MG employs staphylococcal protein A conjugated to horseradish peroxidase (SPA-HRP), a reagent that localizes IgG. Control sections of muscle containing neuromuscular junctions when incubated with canine MG patient sera and subsequently with SPA-HRP localizes IgG at neuromuscular junctions (Cardinet and Bandman, 1985). This immunoreagent has also served to detect antinuclear antibodies, antistrial antibodies, and sarcolemmal associated antibodies in immune-mediated myasthenia gravis and inflammatory muscle disorders in the dog (Shelton and Cardinet, 1987).

VII. MUSCLE BIOPSY AND HISTOCHEMISTRY IN THE DIAGNOSIS OF NEUROMUSCULAR DISORDERS

The use of muscle biopsies in the evaluation of neuromuscular disorders (motor unit diseases) allows examination of myofibers, neuromuscular junctions, intramuscular nerve branches, connective tissues, and blood vessels. As discussed previously, histochemical examination of skeletal muscle provides information relative to the morphological, biochemical, and presumed physiological properties of myofibers. Therefore, the application of histochemical techniques in conjunction with routine light and electron microscopic examination of muscle biopsies offers the potential to evaluate and integrate the pathoanatomical, biochemical, and physiological manifestations of neuromuscular disorders. Further, the advent of immunocytochemistry has extended our ability to recognize immunopathologic mechanisms and disorders as well.

The application of histochemical techniques has become an essential diagnostic procedure for the evaluation of neuromuscular disorders in humans (Engel, 1962, 1965, 1967, 1970; Dubowitz, 1968; Dubowitz *et al.*, 1973). Their application has been most helpful in determining which portion of the motor unit (neuron, myofiber, or both) is involved in the disease process. Their use has been particularly successful in distinguishing between neuropathies and myopathies and/or providing profiles specific for selected neuromuscular disorders (Engel, 1970; Dubowitz *et al.*, 1973). Routine histochemical procedures have been less informative in the evaluation of junctionopathies (disorders of neuromuscular transmission). However, immunocytochemical methods add to the evaluation of immune-mediated diseases of the neuromuscular junction in which deposits of immune complexes are localized along the end plate in myasthenia gravis (Engel *et al.*, 1977; Pflugfelder *et al.*, 1981).

Histochemical techniques have been employed in the evaluation of neuromuscular disorders in animals of comparative biomedical research interest such as in the hereditary muscular dystrophies of animals. However, only in recent years have such techniques begun to be routinely applied to the evaluation of muscle biopsies in clinical cases of neuromuscular diseases in animals. A detailed consideration of muscle biopsy techniques in the evaluation of neuromuscular diseases is beyond the scope of this chapter. For details, the reader is referred to Dubowitz *et al.*, (1973), Engel (1994b), and Banker and Engel (1994).

During the past two decades, there has been an evolving recognition of the broad spectrum of neuromuscular disorders that occur in animals. Neuromuscular disorders in animals are associated with spontaneous and inherited endocrine, immune-mediated, infectious, metabolic, and neoplastic diseases. With continued application of those procedures and newer technologies, undoubtedly many heretofore unrecognized neuromuscular disorders in animals will be recognized (Cardinet and Holliday, 1979). To date, disorders of carbohydrate, lipid, electron transport and electrolyte metabolism have been recognized in skeletal muscles of domestic animals common to clinical veterinary medicine.

VIII. SELECTED NEUROMUSCULAR DISORDERS OF DOMESTIC ANIMALS

A. Ion Channelopathies

1. Acetylcholine Receptor Ion Channels and Myasthenia Gravis

Myasthenia gravis is a disorder of neuromuscular transmission in which there is a reduction in the number of ligand-gated AChR, ion channels on the postsynaptic sarcolemmal membrane (PSM). This condition results in weakness due to the reduced sensitivity of the PSM to the transmitter, ACh. Two basic forms of MG exist: (1) acquired autoimmune MG and (2) congenital MG. Different mechanisms are responsible for the reduction of AChRs in these two disorders.

a. Acquired Autoimmune Myasthenia Gravis

Acquired MG is an immune-mediated disorder of humans (Engel, 1994c), dogs (Lennon *et al.*, 1978, 1981), and cats (Indrieri *et al.*, 1983) in which autoantibodies are produced against AChRs. The density of AChRs is reduced by the complement-mediated destruction, accelerated internalization, and degradation of AChRs by cross-linking of the receptors by antibody. In humans (Tzartos *et al.*, 1982) and dogs (Shelton *et al.*, 1988), the autoantibody response is heterogeneous. Most antibodies are IgG and directed against the main immunogenic region (MIR), a specific external portion of the α-subunit that is distinct from the ACh-binding site; however, autoantibodies are also produced against all of the other subunits. Only a small percentage of antibodies is directed against the ACh-binding sites on the α-subunits.

Dogs with MG usually exhibit some form of muscular weakness; however, this can be quite variable and may include the following presentations: acute quadriplegia, degrees of exercise-related weakness, gait abnormalities, or no apparent limb muscle weakness with dysphasia or regurgitation associated with a megaesophagus. There appears to be a bimodal distribution in the onset of this disease in dogs (early and late onset). Dogs are rarely affected before one year of age and peak frequencies were found at 3 and 10 years of age, respectively; the prevalence did not appear to be gender related (Shelton *et al.*, 1988). Frequently, overt signs may be limited to esophageal dysfunction. In a study of 152 dogs afflicted with idiopathic megaesophagus, 40–57 dogs (26% to 38%) had MG (Shelton *et al.*, 1990).

A definitive diagnosis of MG is provided by detection of circulating antibodies to the AChR. Additional diagnostic tests that provide for a presumptive diagnosis include clinical, pharmacologic, electrodiagnostic, and immunocytochemical methods of evaluation. When clinical signs permit the objective assessment of strength, pharmacologic testing can be employed through the intravenous administration of 1–10 mg of edrophonium chloride, an ultra-short-acting anticholinesterase agent. Improved strength with edrophonium provides a presumptive diagnosis of MG. A presumptive diagnosis of MG is also suggested when the application of low-frequency (2–10 Hz), repetitive nerve stimulation results in the reduced amplitude of the first few evoked compound MAPs (decrementing response).

Two immunocytochemical procedures provide sensitive presumptive tests for MG. In muscle biopsies of human and canine MG patients that contain neuromuscular junctions, it is possible to localize the IgG bound to the PSM using the immunoreagent SPA-HRP (Engel *et al.*, 1977; Pflugfelder *et al.*, 1981). Similarly, patient sera applied to control sections and subsequently treated with SPA-HRP will detect circulating antibodies that bind to the PSM of neuromuscular junctions (Cardinet and Bandman, 1985; Cain *et al.*, 1986; Shelton and Cardinet, 1987).

b. Congenital Myasthenia Gravis

Congenital MG is a developmental disorder of humans (Engel, 1994d) and dogs (Lennon *et al.*, 1981; Miller *et al.*, 1983; Oda *et al.*, 1984a). In the dog, the synthesis of AChRs appears to be normal and degradation does not appear to be accelerated. The reduced

AChR density is believed due to a low insertion rate of AChRs into the PSM (Oda *et al.*, 1984b). In humans, several congenital myasthenic syndromes are recognized and include putative failure of packaging or resynthesis of ACh, the absence of AChE from the basal lamina, and a deficiency of AChRs and abnormal channel function. Because congenital MG is not immune mediated, the immunodiagnostic tests used in acquired MG are of no value in establishing the diagnosis of congenital MG.

2. *Sodium Ion Channels and Periodic Paralysis*

Equine hyperkalemic periodic paralysis (HYPP) is a dominantly inherited disorder of muscle that causes episodes of stiffness (myotonia), weakness, tremors, and/or paralysis in association with elevated serum potassium (Spier *et al.*, 1990, 1993). Weakness or paralysis can be induced by the ingestion of potassium.

Linkage studies have revealed that HYPP cosegregates with the equine adult skeletal muscle sodium channel α-subunit gene (Rudolph *et al.*, 1992a) involving a Phe to Leu mutation in the transmembrane domain IVS3 (Rudolph *et al.*, 1992b). Patients are heterozygous and express both normal and mutant α-subunits. The primary physiological defect in mutant sodium channels is impaired inactivation (Cannon *et al.*, 1995). With incomplete inactivation, sodium ion conductance into myofibers is sustained, which in turn sustains depolarization of the sarcolemma. It appears likely that the clinical variability and severity of signs is associated with the ratio of mutant to normal sodium ion channels expressed in the skeletal muscles of affected horses (Zhou *et al.*, 1994).

This disorder in horses is similar to primary HYPP in humans in which there is also a mutation involving the gene encoding the sodium ion channel α-subunit. A number of mutations involving the gene encoding the α-subunit of the skeletal muscle sodium ion channel have been identified in humans that are responsible for the HYPP disorder and paramyotonia congenita, an exercise and cold-induced stiffness and weakness (Lehmann-Horn *et al.*, 1994).

Diagnosis of equine HYPP is aided by knowledge of the breed and pedigree, clinical signs, and the assessment of serum potassium ion concentrations during an episode of weakness or stiffness. Oral administration of potassium chloride can be used as a test to induce weakness in susceptible horses. A definitive diagnosis is possible by base-pair analysis of the DNA sequence responsible for encoding of the α-subunit (Rudolph *et al.*, 1992b).

3. *Chloride Ion Channels and Myotonia*

Myotonia is a clinical sign in which the uncontrolled, prolonged, and painless contraction of skeletal muscles occurs. The condition is due to hyperexcitability of the sarcolemma and the abnormal production of repetitive depolarizations of the sarcolemma followed by delayed repolarization and relaxation. Affected patients exhibit varying degrees of muscle stiffness with the onset of exercise. The stiffness will often subside with continued exercise or repeated movements and is not aggravated by cold. Muscles may be grossly hypertrophied with well-defined muscle groups. Percussion of muscles results in local contractions that create dimpling of the surface overlying the contracting muscles.

Myotonia congenita (MC) occurs in goats in which there is impaired conductance of chloride ions across the sarcolemma of T-tubules (Bryant and Morales-Aguilera, 1971; Adrian and Bryant, 1974). Unlike, HYPP, MC is a nonprogressive disorder that improves with exercise ("warm-up") and is not induced by cold. Histopathologic changes in skeletal muscle are usually minimal and nonspecific.

Two principal myotonic disorders occur in humans: (1) myotonia congenita (Rudel *et al.*, 1994) and (2) myotonic dystrophy (MD; Harper and Rudel, 1994). Myotonia congenita is a nonprogressive childhood disorder in which there is a diminished chloride conductance across the sarcolemma caused by mutations of the skeletal muscle chloride ion channel. Two forms of MC are recognized in which the mutations of the chloride ion channel involve autosomal dominant (Thomsen's disease) and recessive (Becker's myotonia) modes of inheritance (Koch *et al.*, 1992).

Myotonic dystrophy is an autosomal dominant myotonic disorder in humans that differs from MC in that it is progressive and variably involves a variety of other systems (e.g., smooth muscle of hollow organs, heart, brain and peripheral nerves, endocrine glands, eyes, skeletal system, and integument). The underlying cause of this disorder has not been established even though the gene has been localized. Onset of signs may be recognized in childhood or later in adults, hence MD can also have a congenital presentation. The systemic features are most helpful in differentiating between MC and MD in addition to histopathologic features, which include increased central nuclei, ringed fibers, sarcoplasmic masses, and type 1 fiber atrophy.

In addition to goats, myotonic disorders have been described in horses (Steinberg and Botelho, 1962; Jamison *et al.*, 1987; Reed *et al.*, 1988) and dogs (Kortz, 1989). In attempting to draw comparisons between these disorders and the human disorders, some confusion exists since most reports describe a congenital onset compatible with MC but physical and histopathologic features that are similar to MD. In the horse, two reports provide clinical and histopathologic features in foals that appear comparable to MD (Jamison *et al.*, 1987; Reed *et al.*, 1988). In dogs, congenital presentations are described in chow chows (Griffiths and Duncan, 1973; Wentink *et al.*, 1974; Amann *et al.*, 1985) among a number of other breeds. The histopathologic features observed in these dogs have been variable. More detailed studies of membrane function and molecular genetics will be required to elucidate fully the pathogenesis and comparative features of these disorders.

4. *Voltage-Sensitive Calcium Ion Channels of the Sarcoplasmic Reticulum and Malignant Hyperthermia*

Malignant hyperthermia (MH) is an inherited, autosomal dominant, pharmacogenetic disorder of humans and swine, and possibly dogs and horses, in which there is a rapid increase in body temperature and muscle rigidity associated with the uncontrolled contraction and metabolism of muscles leading to rhabdomyolysis, and death (Gronert, 1994). The condition is triggered by the administration of volatile anesthetic agents and exacerbated by succinylcholine. In swine, stresses such as fighting, transport, and exercise also trigger its onset.

In swine, MH is due to an alteration in the calcium ion release channel (ryanodine receptor) of the SR due to a mutation in the gene for encoding the ryanodine receptor (Fujii *et al.*, 1991). The alteration increases the sensitivity of the calcium release channels and results in their prolonged open state and release of calcium into the sarcoplasm. In humans, the same mutation has been observed in some families; however, other mutations are likely and some cases do not appear to be linked to the ryanodine receptor gene (MacLennan and Phillips, 1992).

Diagnostic testing for susceptibility to MH involves a standardized *in vitro* contracture test employing muscle biopsy samples in which contractures induced by caffeine or halothane are measured (Larach, 1989). During an attack, serum enzyme activities are markedly elevated due to extensive myonecrosis. No specific histopathologic features are present in susceptible individuals. With the identification of the mutation in the gene for encoding the ryanodine receptor, molecular genetic testing will provide a definitive test in swine (Houde *et al.*, 1993; O'Brien *et al.*, 1993).

There are reports of MH-like presentations in horses associated with halothane anesthesia (Hildebrand and Howitt, 1983; Manley, *et al.*, 1983) and in dogs associated with exercise (Rand and O'Brien, 1987). Contracture testing (Nelson, 1991) and calcium homeostasis defects (O'Brien *et al.*, 1990) suggest the stress / exercise-induced syndrome in dogs may be MH.

B. Cytoskeletal Dystrophin Deficiency and Muscular Dystrophy

Duchenne muscular dystrophy is an X-linked recessive disorder of skeletal muscle in humans (Engel *et al.*, 1994), dogs (Kornegay *et al.*, 1988; Cooper *et al.*, 1988), cats (Carpenter *et al.*, 1989; Gaschen *et al.*, 1992) and mice (Bulfield *et al.*, 1984).

The disorder is due to a deficiency of dystrophin (Hoffman *et al.*, 1987), a subsarcolemmal cytoskeletal protein that participates in the attachment of myofibrils to the sarcolemma. Dystrophin in concert with a transmembrane protein complex (dystrophin-associated protein) is believed to provide stability to the sarcolemma. Presumably, the deficiency of dystrophin creates instability of the sarcolemma; however, the precise mechanisms for initiating myofiber necrosis are not known.

A number of mutations of the dystrophin gene have been identified. In the dog, dystrophin deficiency is due to a splice-site mutation in which there is an RNA processing error (Sharp *et al.*, 1992). The mutation in the cat has not been identified.

The expression of the disease varies among species but may be characterized as a progressive degenerative disorder of muscle in which there is a marked increase in serum enzyme activities, gross hypertrophy of some muscle groups, and a stiff, "bunny hopping" gait in dogs and cats. Affected individuals may also have a cardiomyopathy since the dystrophin deficiency also involves cardiac myofibers (Moise *et al.*, 1991). In dogs, the onset of clinical signs is usually evident by 2–4 months of age and somewhat later in cats. In dogs, this disorder was first observed in golden retrievers and has subsequently been identified in other breeds, including Irish terriers, rottweilers, and samoyeds.

Histologic sections reveal focal lesions consisting of myofiber clusters undergoing the spectrum of change from myonecrosis through macrophage infiltration and phagocytosis to regeneration. Individual fibers may be atrophic or hypertrophic, calcified, and hypercontracted and possess central nuclei. The focal clustering of necrotic fibers may be due to the local effects of mast cell accumulations and degranulation (Gorospe *et al.*, 1994).

Beyond clinical signs and biopsy features, immunoblotting and immunocytochemical staining for dystrophin are valuable diagnostic methods for detecting deficiencies of dystrophin within muscle biopsies (Cooper *et al.*, 1990; Gaschen *et al.*, 1992).

C. Immune-Mediated Canine Masticatory Muscle Myositis

The muscles of mastication are selectively affected in two inflammatory muscle disorders in dogs known as *eosinophilic* and *atrophic myositis.* Limb muscles are essentially spared. These disorders may be distinct and separate entities or merely variations of the same disorder that distinguish between acute and chronic onsets, respectively.

The muscles of mastication in the dog are principally composed of type 2M myofibers, fast-twitch fibers that possess a unique myosin isoform, heavy chain, and light chains (Shelton *et al.* 1985a). Recent evidence suggests this disorder is an autoimmune disease (Shelton *et al.*, 1985b, 1987). Dogs afflicted with this disorder produce autoantibodies directed against type 2M fibers and there is little cross-reaction of the antibodies against type 2A fibers of limb muscles. The specificity of these antibodies for 2M fibers provides presumptive evidence for its immune pathogenesis and the limited distribution of lesions to the muscles of mastication.

Diagnosis of the disorder includes muscle biopsy demonstration of an inflammatory disorder, localization of immunoglobulins fixed to type 2M fibers within the biopsy, and demonstration of circulating antibodies against type 2M fibers employing SPA-HRP.

D. Disorders of Glyco(geno)lysis Affecting Skeletal Muscle

Disorders of glyco(geno)lysis affecting skeletal muscles variably involve some excess storage of glycogen within affected myofibers, resulting in the presence of glycogen-containing vacuoles. In humans, glycogen storage diseases (GSD) affecting muscle include deficiencies of α-1,4-glucosidase (GSD-II), debranching enzyme (GSD-III), branching enzyme (GSD-IV), myophosphorylase (GSD-V), phosphofructokinase (GSD-VII), phosphoglycerate kinase (GSD-IX), phosphoglycerate mutase (GSD-X), and lactate dehydrogenase (GSD-XI; Engel and Hirschhorn, 1994; DiMauro and Tsujino, 1994). Though variably documented, some of these disorders also occur in domestic animals (Walvoort, 1983).

1. α-1,4-Glucosidase Deficiency (GSD II)

Lysosomal α-1,4-glucosidase deficiency, also known as Pompe's disease, generalized type II glycogenosis, and acid maltase deficiency, occurs in humans with childhood (infantile and juvenile) and adult forms, which are variations of the same disorder based on the age of onset and tissue and organ involvement. This disorder is inherited as an autosomal recessive trait (Engel and Hirschhorn, 1994).

This disorder has also been reported in shorthorn and Brahman cattle (Jolly *et al.*, 1977; Richards *et al.*, 1977; O'Sullivan *et al.*, 1981). Both infantile and late onset equivalent variations have been described (Howell *et al.*, 1981). Clinical signs in Brahman calves become evident at 2–3 months of age with a loss of condition, poor growth, and lethargy followed by incoordination and muscle tremors with death by 9 months of age. Although the onset of clinical signs may also occur within the first 2–3 months of age, some affected shorthorn calves appear clinically normal until 5–9 months of age when weight gains are not maintained and progressive muscular weakness develops with death by 12–16 months of age. Excessive accumulation of glycogen occurs in skeletal and cardiac muscle, brain, and spinal cord. The disorder in cattle is also inherited as an autosomal recessive trait (Howell *et al.*, 1981).

In other species, single cases of generalized glycogenoses in which α-glucosidase deficiency were demonstrated in the Lapland dog (Walvoort *et al.*, 1982) and Japanese quail (Murakami *et al.*, 1980). In addition, generalized glycogenoses have also been described in sheep (Manktelow and Hartley, 1975), cats (Sandstrom *et al.*, 1969) and dogs (Mastofa, 1970; Walvoort *et al.*, 1981); however, those conditions were not characterized biochemically.

2. Debranching Enzyme Deficiency (GSD III)

Debranching enzyme possesses two activities, α-1,4-glucan transferase and α-1,6-glucosidase. In the hydrolysis of glycogen, myophosphorylase acts on the α-1,4 linkages of the terminal glucose residues up to the last four glucose residues preceding the α-1,4 linkages. The α-1,4-glucan transferase transfers the last three residues to another branch and the α-1,6-glucosidase hydrolyses the α-1,6 branch point.

Several presumed cases of debranching enzyme deficiency have been described in the dog (Rafiquzzaman *et al.*, 1976; Ceh *et al.*, 1976; Otani and Mochizuki, 1977). Biochemical demonstration of debranching enzyme deficiency is limited to a single case (Ceh *et al.*, 1976). There is diffuse organ involvement with glycogen storage. Onset of signs appear at 2 months of age and the disorder is progressive with death by 10–15 months.

3. Branching Enzyme Deficiency (GSD IV)

Glycogen storage associated with branching enzyme deficiency has been reported in a family of Norwegian forest cats (Fyfe *et al.*, 1992). Clinical signs and involved organs include skeletal muscle, heart, and the central nervous system. Abnormal glycogen was evident in tissues at birth. Hepatic involvement and function was reportedly normal.

4. Myophosphorylase Deficiency (GSD V)

Myophosphorylase deficiency (McArdle's disease) is an inherited, autosomal recessive, glycogenosis in humans (DiMauro and Tsujino, 1994) and Charolais cattle (Angelos *et al.*, 1995).

In cattle, clinical signs include recumbency and fatigue with forced exercise and elevated serum enzymes. Muscle glycogen concentrations were elevated 1.6 times higher than controls. Histopathologic changes are modest with some vacuolated myofibers; however, the vacuoles did not contain glycogen. A rapid diagnosis is possible by employing the histochemical staining method for demonstrating myophosphorylase activity in frozen sections of biopsy specimens, which may be confirmed by quantitative analyses.

5. Phosphofructokinase Deficiency (GSD VII)

Phosphofructokinase (PFK) is a key enzyme of the Embden–Meyerhof pathway in all tissues and inherited deficiencies of this enzyme in humans are expressed primarily as a myopathy in which it is designated as type VII glycogen storage disease (GSD-VII). There is a partial expression of hemolysis and the deficiency is otherwise quite heterogeneous (Rowland *et al.*, 1986; Kaneko, 1987).

Deficiency of PFK was reported in springer spaniel dogs that presented a clinical picture remarkably similar to that seen in humans. The dogs presented with a history of intermittent severe hemolytic episodes, and total erythrocyte PFK activity was 10% of normal controls (Giger *et al.*, 1985). The disorder is inherited as an autosomal recessive trait (Giger *et al.*, 1986).

Mammalian PFK is present in different tissues as tetramers of three subunits, PFK-M (muscle), PFK-L (liver), and PFK-P (platelets). Human muscle and liver contain homogenous tetrameric PFK-M4 and PFK-L4, respectively. The erythrocytes contain an admixture of PFK-M and PFK-L tetramers. In normal dogs, a similar isoenzymic distribution pattern exists except in the erythrocytes where the PFK tetramers consist of an admixture of PFK-M and PFK-P subunits (Vora *et al.*, 1985). As in humans, PFK deficiency was found to be a deficiency of the PFK-M subunits and the erythrocytic hybrids consisted of the PFK-L and PFK-P subunits. Giger *et al.* (1985) speculate that exertional stresses inducing hyperventilation and respiratory alkalosis in turn directly or indirectly enhance the alkaline fragility of their erythrocytes and subsequent hemolysis.

Early reports suggested that affected dogs did not manifest severe muscle-related signs, which were possibly masked by signs referable to hemolysis. However, biochemical studies have revealed reduced glycolysis in muscle (Giger *et al.*, 1988a, 1988b), and pathologic studies have established the presence of a myopathy that included the presence of PAS positively stained polysaccharide storage vacuoles in up to 10% of the myofibers (Harvey *et al.*, 1990).

6. Polysaccharide Storage Myopathy

A polysaccharide storage disorder has been described in quarterhorses, quarterhorse crossbreds, american paints, and appaloosa horses (Valberg *et al.*, 1992). The horses had recurrent episodes of exertional rhabdomyolysis and myoglobinuria. Up to 5% of the type 2 muscle fibers had vacuoles that contained acid mucopolysaccharide inclusions that were brilliantly stained with the periodic-acid Schiff's (PAS) stain. These inclusions consisted of β-glycogen particles and fibrillar material. When treated with amylase, the PAS staining intensity was reduced only slightly. Glycogen concentrations were approximately 45% greater in the muscles of horses with these polysaccharide inclusions.

An *in vitro* biochemical method was employed for screening deficiencies of enzymes involved in anaerobic glyco(geno)lysis. Affected muscles were capable of utilizing glycogen, glucose-1-phosphate, glucose-6-phosphate, fructose-6-phosphate, and fructose-1,6-phosphate as substrates to produce lactate; however, the amount of lactate produced was significantly lower than the controls. This study strongly points to an underlying metabolic abnormality that remains to be precisely defined.

E. Mitochondrial Myopathies

A large number of complex mitochondrial myopathies have been described in humans (Morgan-Hughes, 1994) and, recently, there have been reports of mitochondrial muscle disorders in animals. When the utilization of oxygen is reduced, lactate formation is favored and results in elevated blood lactate even at low-intensity levels of exercise. Disorders of the respiratory chain often result in increased numbers of mitochondria and aggregates of mitochondria under the sarcolemma. In sections, these mitochondrial aggregates impart a "ragged-red" appearance to the periphery of myofibers when stained with the modified trichrome stain. The presence of ragged red myofibers and elevated lactate concentrations with exercise warrant investigation of metabolic abnormalities involving mitochondria.

1. *Respiratory Chain Complex I (NADH Ubiquinone Oxioreductase) Deficiency*

A deficiency of Complex I was observed in an Arabian mare with severe limitations to even mild exercise in which there was an elevated venous pO_2, low oxygen consumption, and lactic acidosis (Valberg *et al.*, 1994). Complex I of the respiratory chain is one of four complexes involved in oxidative phosphorylation and transfers electrons from NADH to CoQ in the conversion of oxygen to water.

2. *Cytochrome* c *Oxidase Deficiency*

An episodic weakness reported in Old English sheepdog littermates is accompanied by elevated serum enzymes, lactic acidosis, and increased pO_2 (Breitschwerdt *et al.*, 1992). This disorder may involve a reduction in cytochrome *c* oxidase (Vijayasarathy *et al.*, 1994).

F. Endocrine Myopathies

Signs referable to muscle weakness are frequently observed as part of the clinical presentations of endocrine disorders.

1. *Corticosteroid Myopathy*

Hyperadrencorticism causes muscle wasting (atrophy) and weakness in dogs with Cushing's disease and following corticosteroid administration (Braund *et al.*, 1980a, 1980b; Duncan and Griffiths, 1977). The muscle wasting is due to a rather selective anguloid-to-angular atrophy of type 2 myofibers; however, quantitative studies reveal atrophy of type 1 fibers as well. Myotonia is a variable accompanying sign of this disorder (Duncan and Griffiths, 1977). Muscle wasting appears to be due to catabolism with decreased protein synthesis and increased protein degradation mediated by altered transcription in protein metabolism (Kaminski and Ruff, 1994).

2. *Hypothyroid Myopathy*

Thyroid status has a profound affect on skeletal muscle and hypothyroid states are often accompanied by manifestations of neuromuscular disease. However, descriptions of muscle disorders in clinical veterinary medicine are limited (Braund *et al.*, 1981). Selective type 2 myofiber atrophy and type 1 myofiber predominance (or type 2 myofiber paucity) are common findings in canine hypothyroidism. Experimental studies reveal that the proportion of myofiber types is influenced by thyroid status in which thyroidectomy results in type 1 myofiber predominance, and thyroid excess results in type 2 myofiber predominance (Ianuzzo *et al.*, 1980). These changes may be mediated via neural influences since denervation abolishes these effects of thyroidectomy (Johnson *et al.*, 1980).

References

Adrian, R. H., and Bryant, S. H. (1974). *J. Physiol.* **240,** 505.

Allen, W. M., Bradley, R., Berrett, S., Parr, W. H., Swannack, K., Barton, C. R. Q., and Macphee, A. (1975). *Br. Vet. J.* **131,** 292.

Amann, J. F., Tomilson, J., and Hakinson, J. K. (1985). *JAVMA* **187,** 415.

Anderson, M. G. (1975). *Equine Vet J.* **7,** 160.

Angelos, S., Valberg, S. J., Smith, B. P., McQuarrie, P. S., Shanske, S., Tsujino, S., DiMauro, S., and Cardinet III, G. H. (1995). *Muscle Nerve* **18,** 736.

Argiroudis, S. A., Kent, J. E., and Blackmore, D. J. (1982). *Equine Vet. J.* **14,** 317.

Ashmore, C. R., and Doerr, L. (1971). *Exp. Neurol.* **31,** 408.

Asmundson, V. S., and Julian, L. M. (1956). *J. Hered.* **47,** 248.

Baldwin, K. M., Reitman, J. S., Terjung, R. L., Winder, W. W., and Holloszy, J. O. (1973a). *Am. J. Physiol.* **225,** 1045.

Baldwin, K. M., Winder, W. W., Terjung, R. L., and Holloszy, J. O. (1973b). *Am. J. Physiol.* **225,** 962.

Banker, B. Q., and Engel, A. G. (1994). *In* "Myology" (A. G. Engel and C. Franzini-Armstrong, eds.), Vol. 1, pp. 832–888. McGraw-Hill, New York.

Barany, M. (1967). *J. Gen. Physiol.* **50,** 197.

Barany, M., and Close, R. I. (1971). *J. Physiol.* **213,** 455.

Barany, M., Barany, K., Reckard, T., and Volpe, A. (1965). *Arch. Biochem. Biophys.* **109,** 185.

Barden, J. A., and Mason, P. (1978). *Science* **199,** 1212.

Barnard, R. J., Edgerton, V. R., and Peter, J. B. (1970). *J. Appl. Physiol.* **28,** 762.

Bass, A., Brdiczka, D., Eyer, P., Hofer, S., and Pette, D. (1969). *Eur. J. Biochem.* **10,** 198.

Beatty, C. H., and Bocek, R. M. (1970). *In* "Physiology and Biochemistry of Muscle as a Food" (E. J. Briskey, R. G. Cassens, and B. B. Marsh, eds.), Vol. 2, p 155. Univ. of Wisconsin Press, Madison, Wisconsin.

Beatty, C. H., Peterson, R. D., and Bocek, R. M. (1963). *Am. J. Physiol.* **204,** 939.

Beatty, C. H., Basinger, G. M., and Bocek, R. M. (1966a). *Arch. Biochem. Biophys.* **117,** 275.

Beatty, C. H., Basinger, G. M., Dully, C. C., and Bocek, R. M. (1966b). *J. Histochem. Cytochem.* **14,** 590.

Beatty, C. H., Peterson, R. D., Basinger, G. M., and Bocek, R. M. (1966c). *Am. J. Physiol.* **210,** 404.

Beatty, C. H., Basinger, G. M., and Bocek, R. M. (1967). *J. Histochem. Cytochem.* 15, 93.

Beecher, G. R., Kastenschmidt, L. L., Cassens, R. G., Hoekstra, W. G., and Briskey, E. J. (1968). *J. Food Sci.* **33,** 84.

Billeter, R., Weber, H., Lutz, H., Howard, H., Eppenberger, H. M., and Jenny, E. (1980). *Histochem.* **65,** 249.

Blackmore, D. J., and Elton, D. (1975). *Equine Vet. J.* **7,** 34.

Blanchaer, M. C. (1964). *Am. J. Physiol.* **206,** 1015.

Blincoe, C., and Dye, W. B. (1958). *J. Anim. Sci.* **17,** 224.

Blincoe, C., and Marble, D. W. (1960). *Am. J. Vet. Res.* **21,** 866.

Bocek, R. M., and Beatty, C. H. (1966). *J. Histochem. Cytochem.* **14,** 549.
Bocek, R. M., Basinger, G. M., and Beatty, C. H. (1966a). *Am. J. Physiol.* **210,** 1108.
Bocek, R. M., Peterson, R. D., and Beatty, C. H. (1966b). *Am. J. Physiol.* **210,** 1101.
Braund, K. G., and Lincoln, C. E. (1981). *Am. J. Vet. Res.* **42,** 407.
Braund, K. G., Edwin, J. H., and Richardson, K. E. Y. (1978). *Am. J. Vet. Res.* **39,** 561.
Braund, K. G., Dillon, A. R., and Mikeal, R. I. (1980a). *Exp. Neurol.* **68,** 50.
Braund, K. G., Dillon, A. R., Mikeal, R. I., and August, J. R. (1980b). *Vet. Pathol.* **17,** 134.
Braund, K. G., Dillon, A. R., August, J. R., and Ganjam, V. K. (1981). *Vet. Pathol.* **18,** 589.
Breitschwerdt, E. B., Kornegay, J. N., Wheeler, S. J., Stevens, J. B., and Baty, C. J. (1992). *JAVMA* **201,** 731.
Brockman, K. G., and McKenzie, K. K. (1983). *Am. J. Physiol.* **244,** H552.
Brooke, M. H., and Kaiser, K. K. (1970). *Arch. Neurol.* **23,** 369.
Brooke, M. H., Williamson, E., and Kaiser, K. K. (1971). *Arch. Neurol.* **25,** 360.
Bryant, S. H., and Morales-Aguilera, A. (1971). *J. Physiol.* **219,** 367.
Bulfield, G., Siller, W. G., and Wright, P. A. L. (1984) *PNAS* **81,** 1189.
Bullard, H. H. (1912). *Am. J. Anat.* **14,** 1.
Buller, A. J., and Lewis, D. M. (1965). *J. Physiol.* **178,** 343.
Buller, A. J., Eccles, J. C., and Eccles, R. M. (1960). *J. Physiol.* **150,** 417.
Burke, R. E. (1994). In "Myology" (A. G. Engel and C. Franzini-Armstrong, eds.), Vol. 1, pp. 464–483. McGraw-Hill, New York.
Burke, R. E., and Tsairis, P. (1973). *J. Physiol.* **234,** 749.
Burke, R. E., Levine, D. N., Tsairis, P., and Zajac, F. E. (1973). *J. Physiol.* **234,** 723.
Cain, G. R., Cardinet, III, G. H., Cuddon, P.A., Gale, R. P., and Champlin, R. (1986). *Transplantation* **41,** 21.
Cannon, S. C., Hayward, L. J., Beech, J., and Brown, R. H. (1995). *J. Neurophys.* **73,** 1892.
Cao, A., De Virgilis, S., Lippi, C., and Trabalza, N. (1969). *Clin. Chim. Acta.* **23,** 475.
Cardinet, G. H., III (1969). Unpublished observations.
Cardinet, G. H., III, and Bandman, E. (1985). *J. Vet. Med, Ser. C, Anat. Histol. Embrylogia.* **14**(1), 81.
Cardinet, G. H., III, and Holliday, T. A. (1979). *Ann. N.Y. Acad. Sci.* **317,** 290.
Cardinet, G. H., III, Fowler, M. E., and Tyler, W. S. (1963). *Am. J. Vet. Res.* **24,** 980.
Cardinet, G. H., III, Litterell, J. F., and Freedland, R. A. (1967). *Res. Vet. Sci.* **8,** 219.
Cardinet, G. H., III, Freedland, R. A., Tyler, W. S., and Julian, L. M. (1972). *Am. J. Vet. Res.* **33,** 1671.
Cardinet, G. H., III, Leong, C. L., and Means, P. S. (1982). *Muscle Nerve* **5,** 665.
Cardinet, G. H., III, Shelton, G. D., and Willis, S. E. (1986). *J. Vet. Med, Ser. C, Anat. Histol. Embryologia.* **15,** 168.
Carpenter, J. L., Hoffman, E. P., Romanul, F. C., Kunkel, L. M., Rosales, R. K., Ma, N. S., Dasbach, J. J., Rae, J. F., Moore, F. M., and McAfee, M. B. (1989). *Am. J. Pathol.* **135,** 909.
Ceh, L., Hauge, J. G., Svenkerud, R., and Strande, A. (1976). *Acta Vet. Scand.* **17,** 210.
Chalmers, G. A., and Barrett, M. W. (1977). *J. Am. Vet. Med. Assoc.* **171,** 918.
Citoler, P., Benitez, L., and Maurer, W. (1966). *Exp. Cell. Res.* **45,** 195.
Close, R. (1965). *Nature (London)* **206,** 831.
Close, R. I. (1972). *Physiol. Rev.* **52,** 129.
Colombo, J. P., Richterich, R., and Rossi, E. (1962). *Klin. Wochenschr.* **40,** 37.
Cooper, B. J., Winand, N. J., Stedman, H., Valentine, B. A., Hoffman, E. P., Kunkel, L. M., Scott, M. O., Fischbeck, K. H., Kornegay, J. N., and Avery, R. J. (1988). *Nature* **334,** 154.
Cooper, B. J., Gallagher, E. A., Smith, C. A., Valentine, B. A., and Winand, N. J. (1990). *Lab. Invest.* **62,** 171.
Cornelius, C. E. (1960). *Calif. Vet.* **13,** 22.
Cornelius, C. E. (1967). *Proc. Am. Anim. Hosp. Assoc. 34th Ann. Meeting,* p. 82.
Cornelius, C. E., and Burnham, L. G. (1963). *J. Am. Vet. Med. Assoc.* **142,** 639.
Cornelius, C. E., and Kaneko, J. J. (1960). *J. Am. Vet. Med. Assoc.* **137,** 62.
Cornelius, C. E., Bishop, J., Switzer, J., and Rhode, E. A. (1959a). *Cornell Vet.* **49,** 116.
Cornelius, C. E., Law, G. R. J., Julian, L. M., and Asmundson, V. S. (1959b). *Proc. Soc. Exp. Biol. Med.* **101,** 41.
Cosmos, E. (1966). *Dev. Biol.* **13,** 163.
Cosmos, E., and Butler, J. (1967). *In* "Exploratory Concepts in Muscular Dystrophy and Related Disorders" (A. T. Milhorat, ed.), pp. 197–204. *Excerpta Med. Found.,* Amsterdam.
Crabtree, B., and Newsholme, E. A. (1972). *Biochem. J.* **126,** 49.
Crawley, G. J., and Swenson, M. J. (1965). *Am. J. Vet. Res.* **26,** 1468.
Cutmore, C. M. M., Snow, D. H., and Newsholme, E. A. (1985). *Equine Vet. J.* **17,** 354.
Cutmore, C. M. M., Snow, D. H., and Newsholme, E. A. (1986). *Equine Vet. J.* **18,** 72.
Dawson, D. M., and Fine, J. H. (1967). *Arch. Neurol.* **16,** 175.
Dawson, D. M., and Kaplan, N. O. (1965). *J. Biol. Chem.* **240,** 3215.
Dawson, D. M., and Romanul, F. C. A. (1964). *Arch. Neurol.* **11,** 369.
Dawson, D. M., Goodfriend, T. L., and Kaplan, N. O. (1964). *Science* **143,** 929.
Dawson, D. M., Eppenberger, H. M., and Kaplan, N. O. (1965). Biochem. *Biophys. Res. Commun.* **21,** 346.
Dawson, D. M., Eppenberger, H. M., and Kaplan, N. O. (1967). *J. Biol. Chem.* **242,** 210.
Denny-Brown, D. E. (1929). *Proc. R. Soc. London, Ser. B* **104,** 371.
DiMauro, S., and Tsujino, S. (1994). *In* "Myology" (A. G. Engel and C. Franzini-Armstrong, eds.), Vol. 2, p. 1554–1576. McGraw-Hill, New York.
Domonkos, J. (1961). *Arch. Biochem. Biophys.* **95,** 138.
Domonkos, J., and Latzkovits, L. (1961a). *Arch. Biochem. Biophys.* **95,** 144.
Domonkos, J., and Latzkovits, L. (1961b). *Arch. Biochem. Biophys.* **95,** 147.
Dubowitz, V. (1967). *J. Physiol.* **193,** 481.
Dubowitz, V. (1968). "Developing and Diseased Muscle." Spastics Int. Med. Publ., London.
Dubowitz, V., and Pearse, A. G. E. (1960). *Histochemie* **2,** 105.
Dubowitz, V., Brooke, M. H., and Neville, H. E. (1973). "Muscle Biopsy: A Modern Approach." Saunders, Philadelphia, Pennsylvania.
Duncan, I. D., and Griffiths, I. R. (1977). *Vet. Rec.* **100,** 30.
Duthie, G. G., and Arthur, J. R. (1987). *Am. J. Vet. Res.* **48,** 309.
Duthie, G. G., Arthur, J. R., Simpson, S. P., and Nicol, F. (1988). *Am. J. Vet. Res.* **49,** 508.
Ebashi, S., Endo, M., and Ohtsuki, I. (1969). *Quart. Rev. Biophys.* **2,** 351.
Eccles, J. C., Eccles, R. M., and Lundberg, A. (1958). *J. Physiol.* **142,** 275.
Edleman, R. (1995). *Clin. Lab. News* **21,** 1.
Edstrom, L., and Kugelberg, E. (1968). J. Neurol. Neurosurg. Psychiat. 31, 424.
Edwards, R. H. T., Melcher, A., Hesser, C. M., and Wigertz, O. (1971). *Adv. Exp. Med. Biol.* **11,** 425.
Eisenberg, E., and Kielley, W. W. (1974). *J. Biol. Chem.* **249,** 4742.

Engel, A. G. (1994a). *In* "Myology" (A. G. Engel and C. Franzini-Armstrong, eds.), Vol. 1, p. 261. McGraw-Hill, New York.

Engel, A. G. (1994b). *In* "Myology" (A. G. Engel and C. Franzini-Armstrong, eds.), Vol. 1, p. 822. McGraw-Hill, New York.

Engel, A. G. (1994c). *In* "Myology" (A. G. Engel and C. Franzini-Armstrong, eds.), Vol. 2, p. 1769. McGraw-Hill, New York.

Engel, A. G. (1994d). *In* "Myology" (A. G. Engel and C. Franzini-Armstrong, eds.), Vol. 2, p. 1798. McGraw-Hill, New York.

Engel, A. G., and Hirschhorn, R. (1994). *In* "Myology" (A. G. Engel and C. Franzini-Armstrong, eds.), Vol. 2, p. 1533. McGraw-Hill, New York.

Engel, A. G., Lambert, E. H., and Howard, F. M. (1977). *Mayo Clin. Proc.* **52,** 267.

Engel, A. G., Yamamoto, M., and Fischbeck, K. H. (1994). *In* "Myology" (A. G. Engel, and C. Franzini-Armstrong, eds.), Vol. 2, p. 1133. McGraw-Hill, New York.

Engel, W. K. (1962). *Neurology* **12,** 778.

Engel, W. K. (1965). *In* "Neurohistochemistry" (C. W. M. Adams, ed.), p. 622. Elsevier, Amsterdam.

Engel, W. K. (1967). *Pediat. Clin. North Am.* **14,** 963.

Engel, W. K. (1970). Arch. Neurol. 22, 97.

Eppenberger, H. M., von Fellenberg, R., Richterich, R., and Aebi, H. (1962). *Enzymol. Biol. Clin.* **2,** 139.

Eppenberger, H. M., Eppenberger, M., Richterich, R., and Aebi, H. (1964a). *Dev. Biol.* **10,** 1.

Eppenberger, M., Nixon, C. W., Baker, J. R., and Homburger, F. (1964b). Proc. Soc. *Exp. Biol. Med.* **117,** 465.

Essen-Gustavsson, B., and Lindholm, A. (1985). *Equine Vet. J.* **17,** 434.

Essen-Gustavsson, B., Karlstrom, K., and Lindholm, A. (1984). *Equine Vet. J.* **16,** 197.

Felig, P. (1977). *Ann. N.Y. Acad. Sci.* **301,** 57.

Felig, P., and Wahren, J. (1971). *Adv. Exp. Med. Biol.* **11,** 205.

Fiolet, J. W. T., Willebrands, A. F., Lie, K. I., and Ter Welle, H. F. (1977). *Clin. Chim. Acta* **80,** 23.

Franzmann, A. W., and Thorne, E. T. (1970). *J. Am. Vet. Med. Assoc.* **157,** 647.

Froberg, S. O., Carlson, L. A., and Ekelund, L. G. (1971). *Adv. Exp. Med. Biol.* **11,** 307.

Fujii, J., Otsu, K., Zorzato, F., de Leon, S., Khanna, V. K., Weiler, J. E., O'Brien, P. J., and MacLennan, D. H. (1991). *Science* **253,** 448.

Fujii, Y., Ikeda, S., and Watanabe, H. (1980). *Bull. Equine Res. Inst.* **17,** 21.

Fyfe, J. C., Giger, U., Van Winkle, T. J., Haskins, M. E., Steinberg, S. A., Wang, P., and Patterson, D. F. (1992). *Ped. Res.* **32,** 719.

Gaschen, F. P., Hoffman, E. P., Gorospe, J. R., Uhl, E. W., Senior, D. F., Cardinet, G. H., and Pearce, L. K. (1992). *J. Neurol. Sci.* **110,** 149.

Gauthier, G. F. (1969). *Z. Zellforsch. Mikrosk. Anat.* **95,** 462.

Gauthier, G. F., and Padykula, H. A. (1966). *J. Cell. Biol.* **28,** 333.

Gazith, J., Himmelfarb, S., and Harrington, W. F. (1970). *J. Biol. Chem.* **245,** 15.

George, J. C., and Bokdawala, F. D. (1964). *J. Anim. Morphol. Physiol.* **11,** 124.

George, J. C., and Iype, P. T. (1964). *Ind. J. Ornithol.* **2,** 84.

George, J. C., and Talesara, C. L. (1961). *J. Cell. Comp. Physiol.* **58,** 253.

Gerber, H. (1964). *Zentralbl. Veterinaermed.* **11,** 135.

Gershman, L. C., Stracher, A., and Dreizen, P. (1969). *J. Biol. Chem.* **244,** 2726.

Giger, U., Harvey, J. W., Yamaguchi, R. A., McNutty, P. K., Chiapella, A., and Buetler, E. (1985). *Blood* **65,** 345.

Giger, U., Reilly, M. P., Asakura, T., Baldwin, C. J., and Harvey, J. W. (1986). *Anim. Genetics* **17,** 15.

Giger, U., Argov, Z., Schnall, M., Bank, W. J., and Chance, B. (1988a). *Muscle Nerve* **11,** 1260.

Giger, U. Kelly, A. M., and Teno, P. S. (1988b). *Enzyme* **40,** 25.

Glock, G. E., and McLean, P. (1954). *Biochem. J.* **56,** 171.

Goldberg, A. L. (1967). *Nature* (*London*) **216,** 1219.

Gollnick, P. D., and King, D. W. (1969). *Am. J. Physiol.* **216,** 1502.

Gollnick, P. D., Armstrong, R. B., Saubert, C. W., Piehl, K., and Saltin, B. (1972). *J. Appl. Physiol.* **33,** 312.

Gorospe, J. R., Tharp, M. D., Hinckley, J., Kornegay, J. N., and Hoffman, E. P. (1994). *J. Neurol. Sci.* **122,** 44.

Goto, I., Nagamine, M., and Katsuki, S. (1969). *Arch. Neurol.* **20,** 422.

Granit, R., Phillips, C. G., Skoglund, S., and Steg, G. (1957). *J. Neurophysiol.* **20,** 470.

Green, D. E. (1951). *In* "Enzymes and enzyme systems" (J. T. Edsall, ed.), p. 15. Harvard Univ. Press, Cambridge, Massachusetts.

Green, H. J., Klug, G. A., Reichmann, H., Seedorf, U., Wiehrer, W., and Pette, D. (1984). *Pflugers Arch.* **400,** 432.

Green, M. R., and Landau, B. R. (1965). *Arch. Biochem. Biophys.* **111,** 569.

Griffiths, I. R., and Duncan, I. D. (1973). *Vet. Rec.* **93,** 184.

Gronert, G. A. (1994). *In* "Myology" (A. G. Engel and C. Franzini-Armstrong, eds.), Vol. 2, p. 1661. McGraw-Hill, New York.

Gunn, H. M. (1978). *J. Anat.* **127,** 615.

Guth, L. (1968). *Pathol. Rev.* **48,** 645.

Gutmann, E. (1962). "The Denervated Muscle." Publ. House Czech. Acad. Sci., Prague.

Guy, P. S., and Snow, D. H. (1977). *Biochem. Biophys. Res. Commun.* **75,** 863.

Guy, P. S., and Snow, D. H. (1981). *Res. Vet. Sci.* **31,** 244.

Hagenfeldt, L., and Wahren, J. (1971). *Adv. Exp. Med. Biol.* **11,** 153.

Haigh, J. C., Stewart, R. R., Wobeser, G., and MacWilliams, P. S. (1977). *J. Am. Vet. Med. Assoc.* **171,** 924.

Hall-Craggs, E. C. B. (1968). *J. Anat.* **102,** 241.

Hamada, T. (1974). *Vet. Record* **95,** 441.

Hamilton, M. J., Hegreberg, G. A., and Gorham, J. R. (1974). *Am. J. Vet. Res.* **35,** 1321.

Harper, J. A., and Parker, J. E. (1964). *Poult. Sci.* **43,** 1326.

Harper, J. A., and Parker, J. E. (1967). *J. Hered.* **58,** 189.

Harper, P. S., and Rudel, R. (1994). *In* "Myology" (A. G. Engel and C. Franzini--Armstrong, eds.), Vol. 2, p. 1192. McGraw-Hill, New York.

Harris, J. B. (1979). *Ann. N.Y. Acad. Sci.* **317,** 1.

Harvey, J. W., Calderwood Mays, M. B., Gropp, K. E., and Denaro, F. J. (1990). *Vet. Pathol.* **27,** 1.

Havel, R. J. (1971). *Adv. Exp. Med. Biol.* **11,** 315.

Heffron, J. J. A., Bomzon, L., and Pattinson, R. A. (1976). *Vet. Record* **98,** 338.

Hegreberg, G. A., Camacho, Z., and Gorham, J. R. (1974). *Arch. Pathol.* **97,** 225.

Hennig, R., and Lomo, T. (1987). *Acta Physiol. Scand.* **130,** 115.

Henson, J. B., Dollahite, J. W., Bridges, C. H., and Rao, R. R. (1965). *J. Am. Vet. Med. Assoc.* **147,** 142.

Hermansen, L. (1971). *Adv. Exp. Med. Biol.* **11,** 401.

Hess, J. W., Murdock, K. J., and Natho, G. J. W. (1968). *Am. J. Clin. Pathol.* **50,** 89.

Hess, R., and Pearse, A. G. E. (1961). *Proc. Soc. Exp. Biol. Med.* **107,** 569.

Hibbs, C. M., and Coles, E. H. (1965). *Proc. Soc. Exp. Biol. Med.* **118,** 1059.

Hildebrand, S. V., and Howitt, G. A. (1983). *Am. J. Vet. Res.* **44,** 2280.

Hodgson, D. R., Rose, R. J., Allen, J. R., and DiMauro, J. (1984a). *Cornell Vet.* **75,** 366.

Hodgson, D. R., Rose, R. J., Allen, J. R., and DiMauro, J. (1984b). *Res. Vet. Sci.* **36,** 169.

Hodgson, D. R., Rose, R. J., DiMauro, J., and Allen, J. R. (1985). *Equine Vet. J.* **17,** 300.

Hoffman, E. P., Brown, R. H., and Kunkel, L. M. (1987). *Cell* **51,** 919.

Holliday, T. A., Asmundson, V. S., and Julian, L. M. (1965). *Enzymol. Biol. Clin.* **5,** 209.
Holloszy, J. O. (1967). *J. Biol. Chem.* **242,** 2278.
Holloszy, J. O. (1982). *Arch. Phys. Med. Rehabil.* **63,** 231.
Holloszy, J. O., and Coyle, E. F. (1984). *J. Appl. Physiol: Respirat. Environ. Exer. Physiol.* **56,** 831.
Holloszy, J. O., Oscai, L. B., Mole, P. A., and Don, I. J. (1971). *Adv. Exp. Med.* **11,** 51.
Holloszy, J. O., Rennie, M. J., Hickson, R. C., Conlee, R. K., and Hagberg, J.M. (1977). *Ann. N.Y. Acad. Sci.* **301,** 440.
Holloszy, J. O., Dalsky, G. P., Nemeth, P. M., Hurley, B. F., Martin, W. H. III, and Hagberg, J. M. (1986). *In* "International Series on Sport Sciences" (B. Saltin, ed.), Vol. 16, p. 183. Human Kinetics, Champaign, Illinois.
Holmes, H. G., Ashmore, C. R., Robinson, D. W., Finn, J. P., and O'Dell, J. (1972). *Vet. Record* **90,** 625.
Holmgren, N., and Valberg, S. (1992). *Am. J. Vet. Res.* **53,** 957.
Homburger, F., Baker, J. R., Nixon, C. W., and Wilgram, G. (1962). *Arch. Intern. Med.* **110,** 660.
Horowitz, P. (1994). Excitation-contraction coupling in skeletal muscle. *In* "Myology" (A. G. Engel and C. Franzini-Armstrong, eds.), Vol. 1, p. 423. McGraw-Hill, New York.
Houde, A., Pommier, S. A., and Roy, R. (1993). *J. Anim. Sci.* **71,** 1414.
Howell, J. M. C., Dorling, P. R., Cook, R. D., Robinson, W. F., Bradley, S., and Gawthorne, J. M. (1981). *J. Pathol.* **134,** 266.
Hultman, E., and Nilsson, L. H. (1971). *Adv. Exp. Med. Biol.* **11,** 143.
Huxley, H. E. (1983). *In* "Muscle and Nonmuscle Motility" (A. Stracher, ed.), Vol. 1, p. 1. Academic Press, New York.
Ianuzzo, C. D., Patel, P., Chen, V., and O'Brien, P. (1980). *In* "Plasticity of Muscle" (D. Pette, ed.), p. 593. Walter de Gruyter, Berlin.
Indrieri, R. J., Creighton, S. R., Lambert, E. H., and Lennon, V. A. (1983). *JAVMA* **182,** 57.
Innes, J. R. M. (1951). *Br. Vet. J.* **107,** 131.
Issekutz, B., Jr., Miller, H. I., Paul, P., and Rodahl, K. (1965). *J. Appl. Physiol.* **20,** 293.
Issekutz, B., Jr., Miller, H. I., and Rodahl, K. (1966). *Fed. Proc., Fed. Am. Soc. Exp. Biol.* **25,** 1415.
Issekutz, B., Jr., Shaw, W. A. S., and Issekutz, T. B. (1975). *J. Appl. Physiol.* **39,** 349.
Jamison, J. M., Baird, J. D., Smith-Maxie, L. L., and Hulland, T. J. (1987). *Equine Vet. J.* **19,** 353.
Johnson, B. D., and Perce, R. B. (1981). *Equine Pract.* **3,** 4.
Johnson, M. A., Mastaglia, F. L., Montgomery, A., Pope, B., and Weeds, A. G. (1980). *In* "Plasticity of Muscle" (D. Pette, ed.), p. 607. Walter de Gruyter, Berlin.
Jolly, R. D., Van-De-Water, N. S., Richards, R. B., and Dorling, P. R. (1977). *Aust. J. Exp. Biol. Med. Sci.* **55,** 141.
Jorfeldt, L. (1971). *Adv. Exp. Med. Biol.* **11,** 407.
Jorgensen, P. F., and Hyldgaard-Jensen, J. F. (1975). *Acta Vet. Scand.* **16,** 368.
Kaneko, J. J. (1987). *Clin. Chim. Acta* **165,** 1.
Kaminski, H. J., and Ruff, R. L. (1994). *In* "Myology" (A. G. Engel and C. Franzini-Armstrong, eds.), Vol. 2, p. 1726. McGraw-Hill, New York.
Karlsson, J. (1971). *Adv. Exp. Med. Biol.* **11,** 383.
Klein, M. S., Shell, W. E., and Sobel, B. E. (1973). *Cardiovasc. Res.* **7,** 412.
Koch, M. C., Steinmeyer, K., Lorenz, C., Ricker, K., Wolf, F., Otto, M., Zoll, B., Lehmann-Horn, F., Grzeschik, K. H., and Jentsch, T. J. (1992). *Science* **257,** 797.
Kornegay, J. N., Tuler, S. M., Miller, D. M., and Levesque, D. C. (1988). *Muscle Nerve* **11,** 1056.
Kortz, G. (1989). *Semin. Vet. Med. and Surg.* (*Small Animal*) **4,** 141.
Kowalski, K., Gordon, E. E., Marinez, A., and Adamek, J. (1969). *J. Histochem. Cytochem.* **17,** 601.
Kubista, V., Kubistova, J., and Pette, D. (1971). *Eur. J. Biochem.* **18,** 553.
Kuttler, K. L., and Marble, D. W. (1958). *Am. J. Vet. Res.* **19,** 632.
Lagace, A., Bell, D. S., Moxon, A. L., and Pounden, W. D. (1961). *Am. J. Vet. Res.* **22,** 686.
Lamb, D. R., Peter, J. B., Jeffress, R. N., and Wallace, H. A. (1969). *Am. J. Physiol.* **217,** 1628.
Larach, M. G. (1989). Anesth. Analg. 69, 511.
Lawrie, R. A. (1952). Nature (London) 170, 122.
Lawrie, R. A. (1953). Biochem. J. 55, 298.
Lee, F. S., Guenther, A. E., and Meleney, H. E. (1916). *Am. J. Physiol.* **40,** 446.
Lehmann-Horn, F., Engel, A. G., Ricker, K., and Rudel, R. (1994). *In* "Myology" (A. G. Engel and C. Franzini-Armstrong, eds.), Vol. 2, p. 1303. McGraw-Hill, New York.
Lehmann-Horn, F., Sipos, I. Jurkat-Rott, K., Heine, R., Brinkmeier, H., Fontaine, B., Kovacs, L., and Melzer, W. (1995). *Soc. Gen. Physiol. Ser.* **50,** 101.
Lennon, V. A., Palmer, A. C., Pflugfelder, C., and Indrieri, R. J. (1978). *In* "Genetic Control of Autoimmune Disease" (N. R. Rose, P. E. Bigazzi, and N. L. Warner, eds.), Vol. 1. p. 295. Elsevier North Holland, New York.
Lennon, V. A., Lambert, E. H., Palmer, A. C., Cunningham, J. G., and Christie, T. R. (1981). *In* "Myasthenia Gravis: Pathogenesis and Treatment" (E. Satoyoshi, ed.). p. 41. Univ. of Tokyo Press, Tokyo.
Lewis, J. H. (1976). *Am. J. Vet. Res.* **37,** 601.
Lindholm, A., and Saltin, B. (1974). *Acta Vet. Scand.* **15,** 310.
Lindholm, A., Bjerneld, H., and Saltin, B. (1974a). *Acta Physiol. Scand.* **90,** 475.
Lindholm, A., Johansson, H. E., and Kjaersgaard, P. (1974b). *Acta Vet. Scand.* **15,** 325.
Lindstrom, J. (1994). *In* "Myology" (A. G. Engel and C. Franzini-Armstrong, eds.), Vol. 1, p. 585. McGraw-Hill, New York.
Lindstrom, J., Einarson, B. and Tzartos, S. (1981). *Methods Enzymol.* **74,** 432.
Lomo, T. (1986). *In* "International Series on Sport Sciences: Biochemistry of Exercise VI" (B. Saltin, ed.), Vol. 16, p. 27. Human Kinetics, Champaign, Illinois.
Lomo, T., Westgaard, R. H., and Dahl, H. A. (1974). *Proc. R. Soc. London Ser. B* **187,** 99.
Lowenstein, J. M. (1972). *Physiol. Rev.* **52,** 382.
Lowey, S. (1994). *In "Myology" (A. G. Engel and C. Franzini-Armstrong, eds.), Vol. 1, p. 485. McGraw-Hill, New York.*
Lowey, S., and Risby, D. (1971). Nature (London) 234, 81.
Lowey, S., Benfield, P. A., Silberstein, L., and Lang, L. M. (1979). *Nature* (*London*) **282,** 522.
Luginbuhl, A. J., Dudley, G. A., and Staron, R. S. (1984). *Histochemistry* **81,** 55.
Lymn, R. W., and Taylor, E. W. (1971). *Biochemistry* **10,** 4617.
McGavin, M. D., and Baynes, I. D. (1969). *Path. Vet.* **6,** 513.
MacLennan, D. H., and Phillips, M. S. (1992). *Science* **256,** 789.
Mader, R., Nicol, P. D., Turley, J. J., Billbao, J., and Keystone, E. C. (1994). *J. Med. Sci.* **30,** 902.
Magleby, K. L. (1994). *In* "Myology" (A. G. Engel and C. Franzini-Armstrong, eds.), Vol. 1, p.442. McGraw-Hill, New York.
Manktelow, B. W., and Hartley, W. J. (1975). *J. Comp. Pathol.* **85,** 139.
Manley, S. V., Kelly, A. B., and Hodgson, D. (1983). *JAVMA* **183,** 85.
Maruyama, K. (1986). *Int. Rev. Cytol.* **104,** 81.
Mastofa, I. E. (1970). Acta Vet. Scand. 11, 197.
Maxwell, L. C., Carlson, D. S., and Brangwin, C. E. (1980). *Histochem. J.* **12,** 209.
Michelson, A. M., Russell, E. S., and Harman, P. J. (1955). *Proc. Natl. Acad. Sci. USA* **41,** 1079.

Miller, L. M., Lennon, V. A., Lambert, E. H., Reed, S. M., Hegreberg, G. A., Miller, J. B., and Ott, R. L. (1983). *JAVMA* **182,** 694.

Moise, N. S., Valentine, B. A., Brown, C. A., Erb, H. N., Beck, K. A., Cooper, B. J., and Gilmour, R. F. (1991). *J. Am. College Cardiol.* **17,** 812.

Mole, P., Oscal, L. B., and Holloszy, J. O. (1971). *J. Clin. Invest.* **50,** 2323.

Mommaerts, W. F. H. M. (1969). *Physiol. Rev.* **49,** 427.

Morgan, T. E., Cobb, L. A., Short, F. A., Ross, R., and Gunn, D. R. (1972). *Adv. Exp. Med. Biol.* **11,** 87.

Morgan-Hughes, J. A. (1994). *In* "Myology" (A. G. Engel and C. Franzini-Armstrong, eds.), Vol. 2, p. 1610. McGraw-Hill, New York.

Morizot, D. C., Bednarz, J. C., and Ferrell, R. E. (1987). *Cytogenet. Cell Genet.* **44,** 89.

Murakami, H., Takagi, A., Nonaka, S., Ishiura, S. Sugita, H., and Mizutani, M. (1980). *Exp. Anim. (Tokyo)* **29,** 475.

Nachmias, V. T., and Padykula, H. A. (1958). *J. Biophysic. Biochem. Cytol.* **4,** 47.

Nagode, L. A., Frajola, W. J., and Loeb, W. F. (1966). *Am. J. Vet. Res.* **27,** 1385.

Needham, D. M. (1926). *Physiol. Rev.* **6,** 1.

Nelson, T. E. (1991). *JAVMA* **198,** 989.

Nevins, M. A., Saran, M., Bright, M., and Lyon, L. J. (1973). *J. Am. Vet. Med. Assoc.* **224,** 1382.

Nicol, P. D., Matsueda, G. R., Haber, E., and Khaw, B. A. (1993). *J. Nucl. Med.* **34,** 2144.

O'Brien, P. J., Pook, H. A., Klip, A., Britt, B. A., Kalow, B. I., McLaughlin, R. N., Scott, E., and Elliott, M. E. (1990). *Res. Vet. Sci.* **48,** 124.

O'Brien, P. J., Shen, H., Cory, C. R., and Zhang, X. (1993). *JAVMA* **203,** 842.

Oda, K., Lambert, E. H., Lennon, V. A., and Palmer, A. C. (1984a). *Muscle Nerve* **7,** 705.

Oda, K., Lennon, V. A., Lambert, E. H., and Palmer, A. C. (1984b). *Muscle Nerve* **7,** 717.

Ogata, T. (1960). *J. Biochem.* **47,** 726.

Okinaka, S., Sugita, H., Momoi, H., Toyokura, Y., Watanabe, T., Ebashi, F., and Ebashi, S. (1964). *J. Lab. Clin. Med.* **64,** 299.

Oliver, I. T. (1955). *Biochem. J.* **61,** 116.

Orvis, J. A., and Cardinet, G. H., III (1981). *Muscle Nerve* **4,** 354.

Oscai, L. B., and Holloszy, J. O. (1971). *J. Biol. Chem.* **246,** 6968.

O'Sullivan, B. M., Healy, P. J., Fraser, I. R., Nieper, R. E., Whittle, R. J., and Sewell, C. A. (1981). *Aust. Vet. J.* **57,** 227.

Otani, T., and Mochizuki, H. (1977). *Exp. Anim. (Tokyo)* **26,** 172.

Padykula, H. A. (1952). *Am. J. Anat.* **91,** 107.

Padykula, H. A., and Gauthier, G. F. (1963). *J. Cell. Biol.* **18,** 87.

Padykula, H. A., and Gauthier, G. F. (1967). *In* "Exploratory Concepts in Muscular Dystrophy and Related Disorders" (A. T. Milhorat, ed.), p. 117. Excerpta Med. Found., Amsterdam.

Pande, S. V., and Blanchaer, M. C. (1971). *Am. J. Physiol.* **220,** 549.

Pardo, J. V., Siciliano, J. D. F., and Craig, S. W. (1983). *PNAS* **80,** 1008.

Paul, P. (1971). *Adv. Exp. Med. Biol.* **11,** 225.

Paul, P., and Issekutz, B. (1967). *J. Appl. Physiol.* **22,** 615.

Penn, A. S. (1986). *In* "Myology" (A. G. Engel and B. Q. Banker, eds.), Vol. 2, p. 1785. McGraw-Hill, New York.

Peter, J. B., Jeffress, R. N., and Lamb, D. R. (1968). *Science* **160,** 200.

Peter, J. B., Barnard, R. J., Edgerton, V. R., Gillespie, C. A., and Stempel, K. E. (1972). *Biochemistry* **11,** 2627.

Pette, D. (1975). *Acta Histochem.* **14,** 47.

Pette, D., and Vrbova, G. (1985). Muscle Nerve **8,** 676.

Pette, D., Smith, M. E., Staudte, H. W., and Vrbova, G. (1973). *Pflugers Arch.* **338,** 257.

PPflugfelder, C. M., Cardinet, G. H., III, Lutz, H., Holliday, T. A., and Hansen, R. J. (1981). *Muscle Nerve* **4,** 289.

Pierbon-Bormoli, S., Betto, R., and Salviati, G. (1989). *J. Muscle Res. Cell Motil.* **10,** 446.

Rafiquzzaman, M., Svenkerud, R., Strande, A., and Hauge, J. G. (1976). *Acta Vet. Scand.* **17,** 196.

Rand, J. S., and O'Brien, P. J. (1987). *JAVMA* **190,** 1013.

Reed, S. M., Hegreberg, G. A., Bayly, W. M., Brown, C. M., Paradis, M. R., and Clemmons, R. M. (1988). *Muscle Nerve* **11,** 291.

Richards, R. B., Edwards, J. R., Cook, R. D., and White, R. R. (1977). Neuropathol. *Appl. Neurobiol.* **3,** 45.

Robbins, N., Karpati, G., and Engel, W. K. (1969). *Arch. Neurol.* **20,** 318.

Romanul, F. C. A. (1964). *Arch. Neurol.* **11,** 355.

Romanul, F. C. A., and Van Der Meulen, J. P. (1966). *Nature (London)* **212,** 1369.

Romanul, F. C., and Van Der Meulen, J. P. (1967). *Arch. Neurol.* **17,** 387.

Rose, R. J., Purdue, R. A., and Hensley, W. (1977). *Equine Vet. J.* **9,** 122.

Roussel, J. D., and Stallcup, O. T. (1966). *Am. J. Vet. Res.* **27,** 1527.

Rowell, L. B. (1971). *Adv. Exp. Med. Biol.* **11,** 127.

Rowland, L. P. , DiMauro, S., and Layzer, R. B. (1986). *In* "Myology" (A. G. Engel and B. Q. Banker, eds.), Vol. 2, p. 1603. McGraw-Hill, New York.

Rowlerson, A., Pope, B., Murray, J., Whalen, R. B., and Weeds, A. G. (1981). *J. Muscle Res. Cell Motil.* **2,** 415.

Rowlerson, A., Mascarello, F., and Veggetti, A. (1983). *J. Muscle Res. Cell Motil.* **4,** 443.

Rudel, R., Lehmann-Horn, F., and Ricker, K. (1994). *In* "Myology" (A. G. Engel and C. Franzini-Armstrong, eds.), Vol. 2, p. 1291. McGraw-Hill, New York.

Rudolph, J. A., Spier, S. J., Byrns, and Hoffman, E. P. (1992a). *Anim. Genet.* **23,** 241.

Rudolph, J. A., Spier, S. J., Byrns, G., Rojas, C. V., Bernoco, D., and Hoffman, E. P. (1992b). *Nature Genet.* **2,** 144.

Ruth, G. R., and Van Vleet, J. F. (1974). *Am. J. Vet. Res.* **35,** 237.

Sahlin, K. (1986). *In* "International Series on Sports Sciences: Biochemistry of Exercise VI" (B. Saltin, ed.), Vol. 16, p. 323. *Human Kinetics Publishers, Champaign, Illinois.*

Salmons, S., and Vrbova, G. (1969). J. Physiol. **201,** 535.

Saltin, B., and Essen, B. (1971). *Adv. Exp. Med. Biol.* **11,** 419.

Saltin, B., and Karlsson, J. (1971). Adv. Exp. Med. Biol. 11, 289.

Saltin, B., Henriksson, J., Nygaard, E., Andersen, P., and Jansson, E. (1977). *Ann. N.Y. Acad. Sci.* **301,** 3.

Sandstrom, B. Westman, J., and Ockerman, P. A. (1969). *Acta. Neuropathol.* **14,** 194.

Sarkar, S., Sreter, F. A., and Gergely, J. (1971). *Proc. Natl. Acad. Sci. USA* **68,** 946.

Seidel, J. C. (1967). *J. Biol. Chem.* **242,** 5623.

Sharp, N. J., Kornegay, J. N., Van Camp, S. D., Herbstreith, M. H., Secore, S. L., Kettle, S., Hung, W. Y., Constantinou, C. D., Dykstra, M. J., and Roses, A. D. (1992). *Genomics* **13,** 115.

Shelton, G. D., and Cardinet, G. H., III (1987). *J. Vet. Intern. Med.* **1,** 36.

Shelton, G. D., Bandman, E., and Cardinet, G. H., III (1985a). *Am. J. Vet. Res.* **46,** 493.

Shelton, G. D., Cardinet, G. H., III, and Bandman, E. (1985b). *Muscle Nerve* **8,** 783.

Shelton, G. D., Cardinet, G. H., III, and Bandman, E. (1987). *Muscle Nerve* **10,** 753.

Shelton, G. D., Cardinet, G. H., III, and Lindstrom, J. M. (1988). *Neurology* **38,** 1417.

Shelton, G. D., Willard, M. D., Cardinet, III, G. H., and Lindstrom, J. (1990). *J. Vet. Intern. Med.* **4,** 281.

Sherwin, A. L., Siber, G. R., and Elhilali, M. M. (1967). *Clin. Chim. Acta.* **17,** 245.

Sherwin, A. L., Norris, J. W., and Bulcke, J. A. (1969). *Neurology* **19,** 993.

Simoneau, J. A., Lortie, G., Boulay, M. R., Thibault, M. C., and Bouchard, C. (1985). *Eur. J. Appl. Physiol.* **54,** 250.
Sjodin, B. (1976). Acta Physiol. Scand. 436, 1.
Sjodin, B., Thorstensson, A., Frith, K., and Karlsson, J. (1976). *Acta Physiol. Scand.* **97,** 150.
Smith, J. B., and Healy, P. L. (1968). *Clin. Chim. Acta.* **21,** 295.
Snow, D. H., and Guy, P. S. (1980). *Res. Vet. Sci.* **28,** 137.
Snow, D. H., and Mackenzie, G. (1977). *Equine Vet. J.* **9,** 134.
Snow, D. H., Billeter, R., and Jenny, E. (1981). *Res. Vet. Sci.* **30,** 381.
Snow, D. H., Billeter, R., Mascarello, F., Carpene, E., Rowlerson, A., and Jenny, E. (1982a). *Histochemistry.* **75,** 53.
Snow, D. H., Kerr, M. G., Nimmo, M. A., and Abbott, E. M. (1982b). *Vet. Record.* **110,** 377.
Somer, H., Dubowitz, V., and Donner, M. (1976). *J. Neurol. Sci.* **29,** 129.
Spier, S. J., Carlson, G. P., Holliday, T. A., Cardinet, III, G. H., and Pickar, J. G. (1990). *JAVMA* **197,** 1009.
Spier, S. J., Carlson, G. P., Harrold, D., Bowling, A., Byrns, G., and Bernoco, D. (1993). *JAVMA* **202,** 933.
Sreter, F. A., Gergely, J., Salmons, S., and Romanul, F. (1973). *Nature (London), New Biol.* **241,** 17.
Sreter, F. A., Luff, A. R., and Gergely, J. (1975). *J. Gen. Physiol.* **66,** 811.
Staron, R. S., and Pette, D. (1986). *Histochemistry* **86,** 19.
Staudte, H. W., and Pette, D. (1972). *Comp. Biochem. Physiol.* **31,** 533.
Stein, J. M., and Padykula, H. A. (1962). *Am. J. Anat.* **110,** 103.
Steinberg, S., and Botelho, S. (1962). *Science* **137,** 979.
Steiness, E., Rasmussen, F., Svendsen, O., and Nielsen, P. (1978). *Acta Pharmacol. Toxicol.* **42,** 357.
Stubbs, S. S., and Blanchaer, M. C. (1965). *Can. J. Biochem.* **43,** 463.
Sund, H. (1969). *In* "Biological Oxidations" (T. P. Singer, ed.), p. 603. Wiley (Interscience), New York.
Suominen, H., and Hiekkinen, E. (1975). *Eur. J. Appl. Physiol.* **34,** 249.
Swingle, K. F., Young, S., and Dang, H. C. (1959). *Am. J. Vet. Res.* **20,** 75.
Tanzer, M. L., and Gilvarg, C. (1959). *J. Biol. Chem.* **234,** 3201.
Tarrant, P. V., and McVeigh, J. M. (1979). *Res. Vet. Sci.* **27,** 325.
Terjung, R. L., Baldwin, K. M., Mole, P. A., Klinkerfuss, G. H., and Holloszy, J. O. (1972). *Am. J. Physiol.* **223,** 549.
Terjung, R. L., Dudley, G. A., Meyer, R. A., Hood, D. A., and Gorski, J. (1986). *In* "International Series on Sport Sciences: Biochemistry of Exercise VI" (B. Saltin, ed.), Vol. 16, p. 131. Human Kinetics, Champaign, Illinois.
Tesch, P., Sjodin, B., and Karlsson, J. (1978). *Acta Physiol. Scand.* **103,** 40.
Therriault, D. G., Beller, G. A., Smoake, J. A., and Hartley, L. H. (1973). *J. Lipid Res.* **14,** 54.
Thorstensson, A., Elwin, K., Sjodin, B., and Karlsson, J. (1976). *Scand. J. Clin. Lab.* **36,** 821.
Tokuyasu, K. T., Dutton, A. H., and Singer, S. J. (1983). *J. Mol. Biol.* **96,** 1727.
Tzartos, S., Seybold, M., and Lindstrom, J. M. (1982). *Proc. Natl. Acad. Sci. USA* **79,** 188.
Valberg, S. (1986). *Equine Vet. J.* **18,** 479.
Valberg, S. J., Cardinet, III, G. H., Carlson, G. P., and DiMauro, S. (1992). *Neuromusc. Disord.* **2,** 351.
Valberg, S., Jonsson, L., Lindholm, A., and Holmgren, N. (1993). *Equine Vet. J.* **25,** 11.
Valberg, S. J., Carlson, G. P., Cardinet, III, G. H., Birks, E. K., Jones, J. H., Chomyn, A., and DiMauro, S. (1994). *Muscle Nerve* **17,** 305.
Van Der Veen, K. J., and Willebrands, A. F. (1966). *Clin. Chim. Acta.* **13,** 312.
Van Vleet, J. F., Crawley, R. R., and Amstutz, H. E. (1977). *J. Am. Vet. Med. Assoc.* **171,** 443.
Vijayasarathy, C., Giger, U., Procuiuk, U., Patterson, D. F., Breitschwerdt, E. B., and Avadhani, N. G. (1994). *Comp. Biochem. Physiol. A Physiol* **109,** 887.
Vora, S., Giger, U., Turchen, S., and Harvey, J. W. (1985). *Proc. Natl. Acad. Sci. USA* **82,** 8109.
Vrbova, G. (1963). *J. Physiol.* **169,** 513.
Wachstein, M., and Meisel, E. (1955). *J. Biophys. Biochem. Cytol.* **1,** 483.
Wagner, P. D., and Weeds, A. G. (1977). *J. Mol. Biol.* **109,** 455.
Wagner, W. D., Peterson, R. A., and Anido, V. (1971). *Am. J. Vet. Res.* **32,** 2091.
Wahren, J. (1977). Ann. N.Y. Acad. Sci. 301, 45.
Wahren, J., Ahlborg, G., Felig, P., and Jorfeldt, L. (1971). *Adv. Exp. Med. Biol.* **11,** 189.
Walvoort, H. C. (1983). *J. Inher. Metab. Dis.* **6,** 3.
Walvoort, H. C., Van der Ingh, T. S. G. M., and Van Nes, J. J. (1981). *Berl. Munch. Tierarztl. Wochenschr.* **94,** 39.
Walvoort, H. C., Slee, R. G., and Koster, J. F. (1982). *Biochim. Biophys. Acta* **715,** 63.
Wang, K., and Ramirez-Mitchell, R. (1983). *J. Cell Biol.* **96,** 562.
Weeds, A. G. (1969). *Nature (London)* **223,** 1362.
Weeds, A. G., and Lowey, S. (1971). J. Mol. Biol. 61, 701.
Weeds, A. G., Trentham, D. R., Kean, C. J. C., and Buller, A. J. (1974). *Nature (London)* **247,** 135.
Wentink, G. H., Hartman, W., and Koeman, J. P. (1974). *Tijdschr. Diergeneesk.* **99,** 729.
Whanger, P. D., Weswig, P. H., Muth, O. H., and Oldfield, J. E. (1970). *Am. J. Vet. Res.* **31,** 965.
Wiesmann, U., Colombo, J. P., Adam, A., and Richterich, R. (1966). *Enzym. Biol. Clin.* **7,** 266.
William, P. E., and Goldspink, G. (1971). *J. Cell Sci.* **9,** 751.
Zhou, J., Spier, S. J., Beech, J., and Hoffman, E. P. (1994). *Human Mol. Genet.* **3,** 1599.

CHAPTER

17

Kidney Function

DELMAR R. FINCO

I. INTRODUCTION

In mammals, the kidneys function as a major excretory organ for elimination of metabolic wastes from the body. In most species, death occurs within a week after total cessation of renal function. Partial loss of renal function results in variable deviations from normal, depending on the quantity of functional tissue remaining. The term *azotemia* refers to accumulation of nitrogenous wastes in the blood. Blood concentrations of creatinine and urea are measured as indices of azotemia, although neither imparts significant toxicity because of its accumulation. Animals with moderate to severe azotemia may have a constellation of clinical signs, including lethargy, anorexia, mucosal ulcers, vomiting, diarrhea, weight loss, anemia, and altered urine output. These signs are referred to as *renal failure, uremia,* or uremic syndrome, and reflect the development of abnormalities in a multitude of tissues secondary to subnormal renal functions.

The role of the kidneys in maintaining life is a composite of several functions. Water and many electrolytes are conserved by the kidneys in times of negative body balance and are excreted in times of positive balance. During the processing of body fluids, the kidneys avidly conserve nutrients such as glucose and amino acids; consequently, urine is almost devoid of such materials. Hydrogen ions are excreted or conserved so that blood pH is kept within narrow limits. End products of nitrogen metabolism, such as urea, creatinine, and allantoin, are removed from the body via the urine so that blood levels remain low and relatively constant. The kidneys produce several hormones, including renin, erythropoietin, and prostaglandins, and perform a vital hydroxylation of vitamin D that is required for its activation. The kidneys also respond to several hormones, including antidiuretic hormone (ADH), parathyroid hormone (PTH), aldosterone, thyroid hormone, and others.

The story of the evolution of the kidneys to fullfill the changing needs of the organism has been told eloquently by Homer Smith (1959), a pioneer in the study of the kidney. Single-celled and primitive organisms had little need for excretory specialization because simple diffusion of materials to and from their iso-osmotic aqueous environment was adequate both for intake and for excretory processes. As the complexity of organisms grew, an extracellular space developed that mimicked the osmolality and electrolyte makeup of the original aqueous environment. Simple excretory organs sufficed because extracellular water and wastes could be extruded, and needs could be replaced easily by ingestion. However, the threat of osmotic imbalance occurred as the seas increased in salinity. Some animals such as the shark solved this problem by retaining urea as a compound to balance the osmolality of the seas. Others depended on renal excretion of excess salt via renal tubular systems.

The animals that eventually moved to freshwater from the ocean faced a different problem. The osmolality of their environment being nil, there was an influx of water from the environment into their tissues. A high glomerular filtration rate evolved to facilitate the removal of this water. But then there was a movement of some animals from freshwater to land and a need to conserve water rather than to excrete it. The high-filtration kidney was then modified so that most of the filtered water was reabsorbed while solute wastes were still excreted.

An understanding of the evolutionary processes previously outlined provides insight into an apparent paradox involving the mammalian kidney. In mammals, there is a continuous potential loss of sodium-rich fluid equal to 4–6% of cardiac output via glomerular filtration. Energy is then expended by the tubules to reabsorb more than 99% of the filtered sodium, together with accompanying water. Certainly an engineer asked to design an excretory organ would be unlikely to choose this devious approach to the end result.

The evolutionary development of the kidney has clinical relevance. In renal failure, one must deal not only with problems of inadequate excretion because of decreased glomerular filtration, but also with problems of excess losses arising from alterations in function of the tubular system reclaiming filtrate. Polyuria of renal failure is an example of such a problem.

II. NORMAL RENAL FUNCTIONS

A. The Nephron as a Structural and Functional Unit

Most renal functions are sums of the activities of thousands of nephrons. Each nephron consists of several anatomical divisions (Fig. 17.1). Glomeruli are composed of a tuft of capillaries supplied by an afferent arteriole and drained by an efferent arteriole. Bowman's capsule surrounds the glomerular capillaries

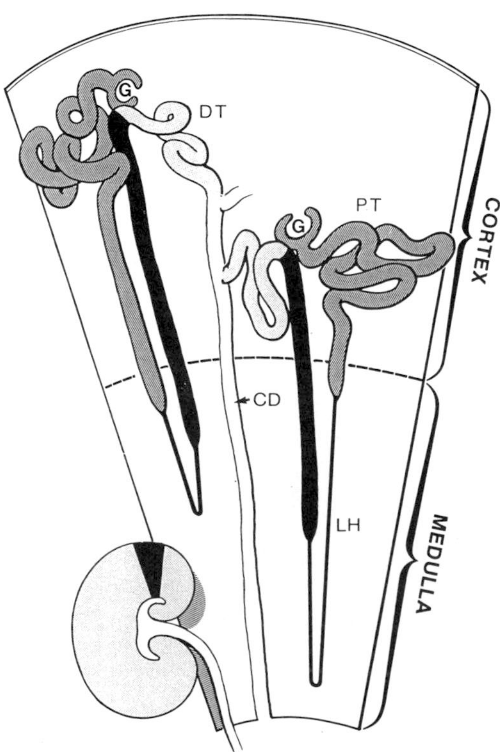

FIGURE 17.1 Nephrons as functional units in the kidney. Outer cortical and juxtamedullary nephrons have some differences in structure and in function. Subdivisions of nephrons include the glomerulus (G), proximal tubule (PT), loop of Henle (LH), distal tubule (DT), and the collecting duct (CD).

and channels filtrate into the proximal tubule. The convoluted portion of the proximal tubule assumes a meandering path within the cortex before the straight portion (pars recta) descends toward the medulla. A marked thinning of the straight portion occurs abruptly as it descends; this constitutes the beginning of the loop of Henle. This structure courses into the medulla, makes a hairpin turn, and ascends into the cortex. At some point in its ascent, the epithelium becomes thicker, which persists to the point of apposition of the loop of Henle with the glomerulus from which the tubule originates. The afferent and efferent arterioles of the glomerulus are adjacent to the ascending tubule. These structures, at the site of their confluence, are referred to as the *juxtaglomerular apparatus.* Modified tubular cells at this site are called the *macula densa.* The distal tubule is considered to be that segment of the nephron between the macula densa and the first union of two distal tubules into a collecting duct. Collecting ducts empty into the renal pelvis.

The number of nephrons varies among species, with larger species having more of them. For example, the cat has about 190,000, the dog about 430,000, and the cow about 4 million per kidney (Smith, 1951). Within the canine species, where there are large body size differences between breeds, nephron size rather than nephron number varies with body size (Finco and Duncan, 1972; Zhu *et al.*, 1992).

A simple view of nephron function can be summarized as follows. Perfusion of glomerular capillaries with blood under pressure results in the passage of material (filtrate) through the capillary walls into Bowman's space. Filtrate then passes through the tubular system of each nephron and is modified in composition during its passage. Modification occurs primarily by reabsorption of lumen contents by tubular cells for return to the body. In addition, important secretory functions are performed by tubule cells, whereby materials delivered to the renal interstitium by the blood are transported from the interstitium to the tubule lumen for excretion. The end result of these tubular activities is a marked reduction in the volume of filtrate and a modification of its character.

B. Nephron Heterogeneity

Common clinical methods for measuring renal function determine the sum of operation of all functional nephrons. However, both anatomical and physiological data indicate that all nephrons are not the same. Internephron heterogeneity has been described in a variety of species (Lameire *et al.*, 1977; Valtin, 1977; Sands *et al.*, 1992). Anatomic differences between outer cortical (OC) and juxtamedullary (JM) nephrons are most apparent. Physiologic differences appear to exist also (Walker and Valtin, 1982). Shunting of blood from the OC nephrons to the JM nephrons could alter renal function by decreasing function of the OC nephrons and increasing function of the JM nephrons. However, studies that have investigated the role of blood shunting in pathologic processes have not led to unanimity of opinion on its relevance (Lameire *et al.*, 1977; Brenner *et al.*, 1986; Steinhausen *et al.*, 1990).

C. Renal Perfusion

1. *Renal Vasculature*

Although some variations occur, vascular patterns of renal perfusion are similar in the mammalian species that have been studied. The renal artery divides into several interlobar arteries that are of about equal size. Further vascular subdivisions occur (arcuate, interlobular arteries) until afferent arterioles supply the glomeruli. Efferent arterioles drain the glomeruli and divide into another capillary network. The site perfused by this network is dependent on the anatomical location of the glomeruli from which the efferent arteriole originates (Beeuwkes, 1971). In the superficial cortex, efferent arterioles supply only the tubule of the parent glomerulus. In the midcortex, efferent arterioles supply either superficial or deep tubules of different glomerular origin. In the juxtamedullary area, efferent arterioles provide capillary networks for the juxtamedullary zone and straight vessels (vasa recta) that course into the medulla.

Venous drainage of renal blood is similar in most mammals, but the cat has a system of subcapsular veins that drains the outer cortex exclusively (Nissen and Galskov, 1972).

Owing to lack of collateral circulation, abrupt blockage of renal arterial vasculature results in infarction of the renal tissue normally supplied (Siegel and Levinsky, 1977). However, slow occlusion of the renal artery divisions may result in survival of some renal parenchyma because of communications between extrarenal arteries and interlobar or arcuate arteries. Such communications apparently do not carry significant quantities of blood under usual conditions, but do dilate when normal renal vasculature is slowly compromised (Christie, 1980).

2. *Renal Blood Flow*

The kidneys receive roughly 15–25% of cardiac output (Kaikara *et al.*, 1969). This rate of perfusion is very high compared to that of most other tissues. Clearance procedures (discussed in Section VI) and electromagnetic flowmeters provide reliable methods for measuring total renal blood flow (RBF) but give no information on intrarenal distribution of blood. Other methods (inert gas washout, microsphere injection) have been used for intrarenal blood flow measurements but all have been criticized (Dworkin and Brenner, 1991). However, renal medullary blood flow rate is known to be comparable to that of other tissues of the body and much lower than the high cortical flow rate.

3. *Autoregulation*

Both RBF and glomerular filtration rate (GFR) remain fairly constant during variations in systemic arterial pressure from 10.7 to 24 kPa (80 to 180 mm Hg) (Pitts, 1974). This phenomenon is known as *autoregulation,* and it occurs secondary to modulation of afferent arteriolar resistance. It occurs in the isolated perfused kidney and thus is not dependent on renal innervation or on extrarenal humoral agents. Two major theories are provided to explain autoregulation (Dworkin and Brenner, 1991). The myogenic theory postulates a direct vascular response to changes in systemic arterial pressure. The tubuloglomerular feedback theory postulates that some intratubular signal is transmitted via the macula densa to the adjacent arterioles, causing their response to changing arterial pressure.

Autoregulation of RBF and GFR operates under physiologic conditions, but it may be abolished with disease, which has significance in clinical medicine.

4. *Role of Neural Stimuli in Renal Perfusion*

The kidney receives sympathetic fibers, with evidence for direct innervation of renal tubular cells (Barajas and Mueller, 1973), which may play some role in renal function (Gottschalk *et al.,* 1985). Current thought suggests that the kidneys are under minimal basal neural tone so that denervation produces minimal changes in renal function. However, neural stimulation beyond the basal tone may alter renal function, particularly by vasoconstriction and decreased renal blood flow. An example is given in Section III.B.2.

D. Glomerular Filtration

1. *Nature of Filtrate*

Material that passes from the capillary lumen into Bowman's space has been collected by micropuncture techniques and compared to the composition of blood. The composition of glomerular filtrate is nearly identical to that of blood from which the cells and protein have been removed. A small difference in ion concentration exists as a consequence of Gibbs–Donnan factors generated secondary to protein restriction to the capillary lumen (Pitts, 1974). Another difference is that glomerular filtrate is not totally devoid of protein. Filtrate harvested from proximal tubules of dogs by micropuncture techniques had albumin concentrations of 20 to 30 mg/liter (2.0 to 3.0 mg/dl), and contained lower molecular weight proteins as well (Dirks *et al.,* 1965; Maack *et al.,* 1992). The relevance of this protein is discussed in Section V.C.

2. *Anatomic Considerations for Filtration*

In order for material to pass from the glomerular capillary lumen into Bowman's space, it must traverse three anatomic barriers: the capillary endothelium, the basement membrane, and the capillary epithelium. Glomerular capillaries are different from other capillaries of the body in several ways. The endothelial fenestrae are larger and account for proportionately more surface area, the basement membranes are thicker, adventitial connective tissue is absent, and the epithelial cells have an elaborate arrangement of interdigitating foot processes (Farquhar, 1975). An intercapillary cell, the mesangial cell, acts as a scavenger cell for particles lodged in the capillary wall. The mesangial cell also has contractile properties and may influence the filtration area of the glomerulus (Raij and Keane, 1985).

It is generally acknowledged that the endothelium can function only to restrain cellular elements since the fenestrae are 50 to 100 nm in diameter. Athough the gaps between the epithelial foot processes are very wide (25 nm), slit membranes which cover the gaps could act as filters. The uninterrupted nature of the basement membrane suggests that it is likely to be the major filtration barrier (Farquhar, 1975).

3. *Glomerular Filtration Dynamics*

Forces relevant to glomerular filtration are shown in Fig. 17.2. Hydraulic (hydrostatic) pressure of cardiac origin provides the energy for glomerular filtration. The major force opposing filtration is the oncotic pressure of plasma proteins. An additional force opposing filtration is hydrostatic force in Bowman's space. One potential force, oncotic pressure in Bowman's space, is usually ignored because protein concentration in filtrate is so low compared to that of plasma.

The permeability and the surface area of the capillary filtering area are other factors that influence the rate of filtration (Maddox and Brenner, 1991). As fluid traverses the glomerular capillary bed, plasma proteins are concentrated because filtrate leaves the capillary. This concentration results in a progressive rise in oncotic pressure as the efferent arteriole is approached. In rats, oncotic pressure in the capillary matches net effective hydraulic pressure before the end of the capil-

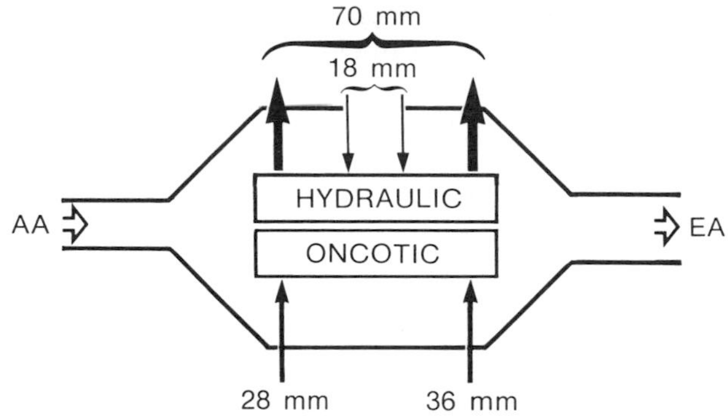

FIGURE 17.2 Schematic representation of forces significant in glomerular filtration. Hydraulic (blood) pressure is the major factor forcing fluid from the capillary lumen into Bowman's space. This pressure is opposed by plasma oncotic pressure and hydraulic pressure in Bowman's space.

lary, causing filtration to cease prematurely. This is called *filtration pressure equilibrium.* With these conditions, an increase in plasma flow rate increases GFR as follows. An increase in flow rate would decrease the proportion of plasma that became filtrate. Intraluminal oncotic pressure would not rise as rapidly as it would at a slower flow rate, so a positive filtration pressure would exist over a greater span of the capillary loop. Thus, with filtration pressure equilibrium, plasma flow rate markedly influences GFR (Tucker and Blantz, 1977). However, filtration pressure equilibrium does not exist in dogs and probably does not exist in most other species (Fried and Stein, 1983).

4. *Filtration of Macromolecules*

Common methods of detecting proteinuria that are used as a part of urinalysis are negative in normal animals because the glomerular filter does not allow passage of large quantities of protein into filtrate. The idea that pores in the filtration barrier prevent macromolecules and cells from entering filtrate led to the concept that the size and shape of molecules influenced their filtration (Pitts, 1974). In addition, filtration is influenced by electrical charge (Rennke and Venkatachalam, 1977; Brenner *et al.*, 1978). Negatively charged sialoproteins are components of all three layers of the capillary wall. These negative charges facilitate passage of cationic macromolecules, and impede passage of anionic macromolecules through the filter. Albumin has a size and shape that marginally prevents its passage through the glomerular capillary wall. However, its negative charge at blood pH further impedes its loss into filtrate. If the negatively charged sialoproteins lining the filtration barrier are removed experimentally or with disease, then albuminuria occurs (Brenner *et al.*, 1978).

For quantitative comparisons, the passage of macromolecules through the glomerular barrier can be compared to water movement through the barrier (Table 17.1).

E. Tubular Modification of Filtrate

Death from hypovolemia and electrolyte depletion would occur in less than 1 hour if all the glomerular filtrate formed were lost from the body. The renal tubules modify glomerular filtrate to achieve body homeostasis for water and most electrolytes. Large quantities of filtrate (65–80%) are reabsorbed in the proximal tubules. As the reduced volume of filtrate traverses the more distal portions of the nephron, further reabsorption takes place. Secretion of certain endogenous materials (i.e., protons, potassium, organic anions) as well as exogenous materials (certain drugs, toxins) from tubular cells into the tubular lumen also is important in renal excretory functions.

The concept of tubular maxima (T_m) for reabsorption and secretion applies to many materials handled by the tubules. This concept originated before cell receptors and transport kinetics were studied in other systems. Whole kidney T_m values, however, represent the composite of all segments of the nephron and the action of both the luminal and the basolateral cell membranes. The T_m value for maximal tubular reabsorption or secretion by the kidney is expressed as milligrams transported per 100 ml of filtrate, or milligrams transported per unit time.

Another measurement that evaluates the role of the tubules in body homeostasis is the determination of fractional excretion (FE). The FE is defined as that fraction of filtered substance (S) that is excreted in the urine (i.e., not reabsorbed by tubules). The FE of a substance S can be calculated from values for its filtered load and its urinary excretion. For an S that passes freely through the glomerular barrier, the filtered load is calculated as the product of GFR × plasma concen-

TABLE 17.1 Permeability of the Glomerular Capillary Barrier

Substance	MW	Permeability coefficient[a]
Urea	60	1
Creatinine	113	1
Glucose	180	1
Inulin	5,500	0.98
Myoglobin	17,000	0.75
Hemoglobin	65,000	0.03
Albumin	69,000	<0.001

[a] Coefficient = permeability compared to water.

tration. The FE value is calculated as urinary excretion divided by filtered load,

$$\text{FE} = \frac{S\ (U\text{: mmol/liter}) \times \text{volume}\ (U\text{: ml})}{\text{GFR (ml/m)} \times S\ (P\text{: mmol/liter})},$$

where S is substance, for example, sodium; U, urine; and P, plasma concentration. All units will cancel when the timed urine collection period is expressed in minutes.

In health, values for FE are not fixed, but vary depending on requirements placed on the kidney for maintaining homeostasis. For example, if intestinal absorption of inorganic phosphate increases and body needs for phosphate are constant, then FE_P would increase. Clinical use of FE measurements is discussed in Section VI.

F. Conservation of Nutrients

1. *Glucose*

Glucose (MW = 180 Da) passes freely through the glomerular filtration barrier and attains the same concentration in filtrate as in plasma. The normal mammalian renal tubule has evolved into a highly efficient structure for reclaiming filtered glucose. At normal blood levels of glucose, reabsorption of filtered glucose is nearly complete within the first 20% of the proximal tubule. At the luminal membrane, glucose is reabsorbed by secondary active transport coupled to sodium reabsorption. Glucose attains a high concentration in the tubule cell and is transported into the renal interstitium through the basolateral membrane by a sodium-independent mechanism. Although nearly all glucose is normally reabsorbed in the proximal tubule, more distal sites for reabsorption also have been identified (Wen, 1976). Despite this system, urine from a normal animal still contains a small quantity of glucose that is undetectable by the urinalysis methods of glucose assay.

The kidney has definite limits for glucose reabsorption. Blood and thus filtrate concentration of glucose may exceed the T_m for glucose. Rate of fluid flow through the tubule also influences whether glucosuria occurs (Pitts, 1974). The interpretation of glucosuria is discussed in Section VE.

2. *Amino Acids*

Several hundred millomoles of amino acids is potentially lost daily by glomerular filtration. In nearly all species, more than 99% is reabsorbed by specific mechanisms by the tubules (Silbernagl, 1985). An exception is the domestic cat, which normally has 1 to 1.8 g/liter of a sulfur-containing amino acid (felinine) in its urine. The role of this amino acid in feline metabolism and its relevance in urine are unclear (Westall, 1953; Tallan *et al.*, 1954; Trippet, 1957; Avizonis and Wriston, 1959; Greaves and Scott, 1960).

The bulk of amino acid reabsorption occurs in the early portions of the proximal tubule. If blood levels of amino acids are increased so that greater amounts are filtered, the late portions of the proximal tubule are recruited for reabsorption. However, the more distal portions of outer cortical nephrons, evaluated by micropuncture, are incapable of amino acid reabsorption.

Several distinct transport mechanisms exist for amino acid reabsorption. These mechanisms are carrier mediated and saturable, and they obey standard kinetic principles. At least one mechanism exists for the acidic amino acids, at least one for the basic amino acids, and several for the neutral amino acids. The transport mechanisms are ion dependent (sodium, potassium, chloride) for many of the amino acids (Silbernagl, 1985).

3. *Other Metabolites*

Metabolic intermediates may be present in the blood in low concentrations and could be lost in the urine because of their low molecular weight. Tubular reabsorption of these compounds occurs, but with high blood levels some may appear in the urine when tubular maxima are exceeded (Pitts, 1974). Acid–base status also may influence urinary loss of citrate, α-ketoglutarate, and succinate (Packer *et al.*, 1995). Blood forms of lipids apparently do not pass the glomerular filter, and thus the urine of most species is devoid of fat. However, fat is a normal constituent of the urine of dogs and cats, presumably because of exocytosis from the tubular cells into the lumen.

G. Electrolyte Homeostasis

The kidneys play a significant role in maintaining the concentration of many plasma electrolytes within narrow limits. These electrolytes include sodium, potassium, inorganic phosphate, chloride, and magnesium. The kidneys play a prominent role in electrolyte homeostasis because of the lack of discrimination by the intestinal tract in absorption of electrolytes. In carnivores, nearly all of the sodium, potassium, and chloride and much of the inorganic phosphate ingested is absorbed from the gut without regard for body need. The kidneys serve as a monitor of the blood to excrete these electrolytes when body excess occurs.

1. *Sodium Homeostasis*

A large quantity of sodium is filtered daily and most is reabsorbed in the proximal tubules. At this site, sodium moves from the tubular lumen into the cell passively because of both concentration and electrical gradients. Sodium is actively extruded from the cell across the basolateral membrane by operation of the sodium–potassium ATPase system. Water passively follows the sodium according to osmotic gradients, so that fluid leaving the proximal tubule is isotonic. Additional sodium is removed from luminal fluid in the ascending limb of the loop of Henle. Transport is passive in the thin portion of the loop, but active in the thick portion. As in the proximal tubule, sodium reabsorption occurs in the loop without regard for body need. The diuretics furosemide and ethacrynic acid act on the luminal border of the cells of the thick portion of the loop of Henle to prevent electrolyte reabsorption.

Further reabsorption of sodium occurs in the more distal portions (distal tubule, collecting duct) of the nephron. Control of body sodium balance rests primarily with homeostatic mechanisms in these areas. However, perturbations in proximal tubule or loop of Henle function by disease or drugs may result in flooding of the control areas so that their capacity is exceeded.

Body sodium balance is closely related to water balance; sodium retention usually results in water retention. Baroreceptors, osmoreceptors, and intrarenal receptors detect abnormalities in plasma volume and osmolality and cause appropriate renal responses. Several mechanisms act in concert to direct the kidneys to achieve homeostasis, including intrarenal angiotensin II effects on sodium reabsorption, the renin–angiotensin–aldosterone system, pressure natriuresis, renal sympathetic nerve activity, and atrial natriuretic peptide (Gonzales-Campoy and Knox, 1992).

2. *Potassium Homeostasis*

Both extrarenal and renal mechanisms exist for potassium homeostasis. After potassium is ingested and absorbed from the gut, it is quickly transported into cells to prevent hyperkalemia. Cell uptake is facilitated by insulin and epinephrine. Only later are appropriate adjustments made to body potassium content by renal excretion of excess (Rosa *et al.*, 1992).

In the kidneys, potassium freely passes through glomerular capillary walls and appears in filtrate in the same concentration as in blood. Nearly 70% of filtered potassium is reabsorbed in the first two-thirds of the proximal tubule. Both passive influx into the descending limb and efflux out of the thick ascending limb of the loop of Henle occur.

The major nephron sites for potassium homeostasis are the terminal distal tubule and the cortical collecting duct. Potassium is secreted from cells into the tubular lumen at these sites. This secretion occurs passively via a favorable electrochemical gradient, but the amount excreted is controlled to maintain body potassium balance. Several factors influence tubular potassium secretion including aldosterone, tubular fluid flow rate, tubular sodium delivery rate, acid–base state, ADH, and the presence of unreabsorbable anion in the distal nephron (Wright and Giebisch, 1992).

The normal kidney is able to undergo adaptive changes when dietary intake of potassium increases. Aldosterone secretion is increased by several stimuli, including hyperkalemia. Aldosterone enhances renal potassium excretion in association with an increase in activity of the sodium–potassium ATPase system in the basolateral membranes of the renal tubular cells (Sebastian *et al.*, 1986). However, aldosterone-dependent changes may not be the exclusive mechanism for renal adaptation to potassium excess. Enhanced potassium excretion can be demonstrated in isolated perfused kidneys (Silva *et al.*, 1977), and the sodium–potassium ATPase system is stimulated in animals in renal failure by aldosterone-independent mechanisms (Schon *et al.*, 1974).

Cattle normally excrete a much larger percentage of the filtered load of potassium than do carnivores because of higher dietary potassium intake (Pickering, 1965). It appears that this excretory pattern is an acquired one, since dogs develop the same increase in excretory capacity if placed on a high potassium diet (Pickering, 1965).

3. *Inorganic Phosphorus (Phosphate)*

The inorganic portion of blood phosphorus exists in both the mono and dibasic salts; a ratio of four HPO_4^{2-} to one $H_2PO_4^-$ exists at normal blood pH levels (Knochel and Jacobson, 1986). The kidney is the major organ for control of plasma phosphate concentration. The concentration of phosphate in glomerular filtrate is nearly the same as plasma, and nearly all phosphate reabsorption occurs in the proximal tubule by a sodium-dependent mechanism.

It is well established that parathyroid hormone (PTH) inhibits tubular reabsorption of phosphate. Cellular mechanisms of PTH action include the generation of cyclic AMP (Knox, 1977). In states of hyperphosphatemia, increased plasma levels of PTH result in an increase in the FE of phosphate. Earlier studies considered PTH to be the exclusive mechanism for control of renal phosphate excretion, but subsequent investigations indicate that thyroparathyroidectomized animals can adequately respond to an excess or deficit of phosphate with appropriate modification of urinary excretion of phosphate (Trohler *et al.*, 1976; Haramati *et al.*, 1983). Many factors are now known to alter renal tubular phosphate reabsorption, including vitamin D metabolites, insulin, thyroid hormone, growth hormone, glucocorticoids, acid–base balance, extracellular volume state, and the diuretic furosemide (Berndt and Knox, 1992).

4. *Chloride*

Chloride is freely filtered into Bowman's space and is reabsorbed by the renal tubular system according to body needs. As the most prevalent anion in filtrate, chloride often is reabsorbed in concert with sodium by sodium-dependent mechanisms. However, distinct sodium-independent mechanisms also exist, which explains the clinical observation that plasma sodium and chloride concentrations may vary independently. Specific cellular mechanisms of chloride transport are less defined than for sodium (Berry and Rector, 1991).

5. *Calcium*

In most carnivores and omnivores, intestinal absorption of calcium is the major controlling factor in calcium homeostasis, and the role of the kidney is minor.

Regarding renal handling of calcium, plasma-ionized and chelated calcium readily pass the glomerular filtration barrier, but protein-bound calcium is retained in the blood. In the tubule, calcium is reabsorbed roughly in parallel to sodium in the proximal tubule, but differences exist between handling of calcium and sodium more distally (Costanzo and Windhager, 1992).

Hypercalcemia due to a variety of causes may exist in animals with normal renal function. This fact emphasizes the limited role that the kidney plays in calcium homeostasis in many species. However, dietary intake of calcium can modestly influence urinary calcium excretion, and may be a factor in formation of calcium-containing uroliths. Horses normally excrete large amounts of calcium in the urine; the kidneys may play a more significant role in calcium homeostasis in this species.

Factors known to modify renal excretion of calcium include PTH, vitamin D metabolites, extracellular fluid volume, acid–base balance, phosphate depletion, hypercalcemia, and diuretics (Costanzo and Windhager, 1992).

6. *Magnesium*

The kidney is responsible for control of plasma magnesium concentration. From 70 to 80% of plasma magnesium is filterable through the glomerular capillary, implying less protein binding of magnesium than of calcium (Quamme, 1992). Tubular handling of magnesium seems unique compared to other electrolytes. Proximal tubule reabsorption is only 20–30% of the amount filtered. The major site of reabsorption under normal conditions is the thick ascending limb of the loop of Henle (Quamme, 1992). About 10% of filtered magnesium is normally reabsorbed more distally. Major renal control of magnesium homeostasis rests in tubular detection of peritubular magnesium concentration (Quamme, 1992).

H. Acid–Base Balance

The lungs and the kidneys are the primary organs concerned with acid–base regulation. Although the lungs may cause sudden and dramatic changes in blood pH by altering pCO_2, the maneuver is not associated with loss of hydrogen ions (H^+) from the body. When pCO_2 is reestablished at normal levels, the H^+ reappears due to dissociation of H_2CO_3 to H^+ and HCO_3^- according to the law of mass action. Thus, the kidney is ultimately responsible for H^+ excretion.

1. *Conservation and Generation of Bicarbonate*

Bicarbonate is an important buffer and is available to neutralize H^+ ingested or produced during metabolism. Since bicarbonate passes freely through the glomerular filter, it is necessary that it be reabsorbed. Most bicarbonate is reabsorbed in the proximal tubules by a process that requires H^+ secretion and action of carbonic anhydrase. Filtered bicarbonate is converted to CO_2 by the reaction

$$HCO_3^- + H^+ \rightarrow H_2CO_3 \rightarrow H_2O + CO_2.$$

After CO_2 is formed, it diffuses into tubular cells and is reconverted to bicarbonate. Since H^+ are recycled in the process, no net secretion occurs in association with bicarbonate reabsorption.

If a tendency for decreased plasma bicarbonate exists, all filtered bicarbonate is reclaimed, and new bicarbonate is synthesized utilizing CO_2 derived from renal cell metabolism. The synthesis of new bicarbonate in renal tubular cells requires net secretion of H^+ into the tubular lumen.

If a tendency for an increased plasma bicarbonate exists, then bicarbonate reabsorption is incomplete and the excess bicarbonate is excreted in the urine.

A T_m for bicarbonate has been found in intact animals. However, this T_m is apparently a regulatory maximum rather than an absolute one since isolated perfused tubules do not demonstrate a T_m phenomenon for bicarbonate. In the intact animal, the amount of bicarbonate reabsorbed can be modified by several factors including extracellular volume, plasma concentrations of certain blood constituents (systemic pH and pCO_2, aldosterone, PTH, glucocorticoids, thyroid hormone, ANP), and body stores of potassium and chloride (Gennari and Maddox, 1992).

2. *Secretion of H^+ (Protons)*

In addition to preservation of bicarbonate as a buffer, the kidney secretes H^+ as a direct defense against acidosis. Parts of the nephron differ in their ability to secrete H^+. Large quantities of H^+ are cycled by the proximal tubules in association with bicarbonate reabsorption, but no net proton secretion occurs with this process. Although proximal tubules have a high capacity to secrete H^+, the gradient against which they can secrete H^+ is modest. Thus, fluid leaving the proximal tubule has a pH no lower than 6.0 even in conditions of acidosis. In the distal tubule and collecting duct, the capacity for H^+ secretion is less than the proximal tubule, but the cell-to-lumen concentration gradient against which H^+ can occur is much greater (1:1000). When such gradients are attained, for instance, during periods of metabolic acidosis, urine pH values of about 4.5 occur.

3. *Facilitation of H^+ Excretion by Urinary Buffers*

Buffers that accept H^+ effectively remove protons from solution and negate their pH effect. Appropriate urinary buffers allow secretion of more H^+ since the secretory process for H^+ is gradient limited. The urinary excretion of phosphate provides a buffering action to facilitate further H^+ secretion. The HPO_4^{2-} that is filtered can accept a H^+ to be excreted as $H_2PO_4^-$. Since urinary excretion of phosphate is an ongoing part of phosphate homeostasis, this mechanism operates without induction. By contrast, the ammonia buffer system is induced in response to needs of the animal to combat an increased acid load. Ammonia is synthesized in the nephron, predominantly from the amino acid glutamine (Halperin *et al.*, 1985). This synthesis can be increased 5 to 10-fold during acidosis (Pitts, 1974). Ammonia that is generated diffuses from renal cells into the tubular lumen. In the distal tubule where luminal pH is low, H^+ secretion is enhanced because H^+ are removed by the reaction

$$NH_3 + H^+ \rightarrow NH_4^+.$$

The pK of this reaction is about 9.15, so in the cell the ratio of NH_4^+ to NH_3 is less than 1000 to 1, while in the tubule lumen (pH less than 5) it is greater than 10,000 to 1.

An alternate explanation for the beneficial effects of the ammonia system has been proposed (Halperin and Jungas, 1983). This explanation states that an anion that is metabolized to neutral products consumes H^+ in the process. Deamination of glutamine yields α-ketoglutarate$^-$, which can be metabolized either to water and CO_2 or to glucose. Both metabolic pathways for α-ketoglutarate degradation consume H^+. Ammonium from deamination, if converted to urea in the liver, consumes bicarbonate during the urea cycle, leaving the overall effects of glutamine metabolism neutral. Excretion of ammonium in urine results in body alkalinization, because only the H^+-consuming portion of glutamine metabolism occurs.

The functions of the kidney in acid–base balance may be simplistically defined as both conservation of bicarbonate and excretion of H^+ in times of acidosis, and excretion of bicarbonate and retention of H^+ in times of alkalosis. However, the interrelationships between acid–base functions and homeostatic functions for other electrolytes sometimes result in compromises of function. For example, metabolic alkalosis can occur under certain circumstances because of complex relationships between extracellular volume, chloride balance, and acid–base control mechanisms (Jacobson and Seldin, 1983). Interaction between potassium and acid–base balances also have been described (Gennari and Cohen, 1975).

I. Water Homeostasis

1. *Overview*

All mammals and birds have some ability to concentrate urine above the osmolality of plasma. This capability provides a mechanism for conserving water when intake is sporadic. Animals also have the ability to excrete hypotonic urine so that excess body water can be excreted without solute depletion.

The massive amount of water potentially lost because of the high GFR of mammals dictates that most filtrate be reabsorbed even in instances of overhydration. Thus, water reabsorption in the proximal tubule, descending limb of the loop of Henle, and early distal tubule occurs without regard for the state of water balance.

In the proximal tubule, 60–80% of filtrate is reabsorbed secondary to solute reabsorption. In the descending limb, loop of Henle, and in the proximal portion of the distal tubule, further rebsorption occurs so that only about 9% of filtrate remains as fluid enters the terminal portion of the distal tubule and the collecting duct. Urine volume is determined by the amount of this 9% that is reabsorbed. In times of body water deficit, nearly all of the 9% is reabsorbed and the urine voided is very low in volume and high in solute concentration. In times of body water excess, little if any of the 9% is reabsorbed, so that urine volume is high and solute concentration is low.

The proximal tubules and loop of Henle normally are not major sites of water homeostasis, but impairment of their function by disease or drugs may have profound effects on urine volume. If the more proximal portions of the nephron do not function to reabsorb water at the normal rate, then the distal portions are flooded with large volumes of fluid that they are unable to handle, and polyuria ensues. Osmotic diuretics such as mannitol cause polyuria in this way (Gennari and Kassirer, 1974).

2. *Mechanism of Water Conservation*

Reabsorption of water throughout the nephron occurs secondary to osmotic gradients. In the proximal tubule, the reabsorption of solute (sodium, chloride, etc.) is associated with water reabsorption because proximal tubule cells are water permeable. The ascending limb of the loop of Henle is impermeable to water but solute moves into the interstitium because of both active and passive transport. This solute movement, coupled with an anatomic arrangement of blood vessels (vasa recta) that minimizes removal of solute, contributes to the development of medullary interstitial hypertonicity. This hypertonicity provides an osmotic gradient for water reabsorption from the lumen of the distal tubule and collecting duct into the interstitium when cells at these sites are made water permeable by the action of ADH.

3. *Events in the Tubule*

Proportionate reabsorption of solute and water in the proximal tubule results in an isotonic fluid entering the descending limb of the loop of Henle (Fig. 17.3). As this fluid descends into the medulla, water is absorbed into the hypertonic interstitium, causing a reduction in fluid volume and an increase in sodium concentration. The high concentration of sodium achieved at the loop exceeds that in the interstitium, allowing passive movement of sodium from the lumen into the interstitium in the thin ascending limb.

The entire ascending limb is impermeable to water. Consequently, as the fluid column ascends into the cortex, volume does not change. Active transport of sodium chloride from the thick ascending limb into the interstitium increases interstitial osmolality and makes intraluminal fluid hypo-osmolar to plasma.

Both solute and water are removed from early distal tubule fluid, in proportions that vary with species. Urine remains hypo-osmolar in the dog, chinchilla, some strains of rats, and monkeys. It becomes iso-

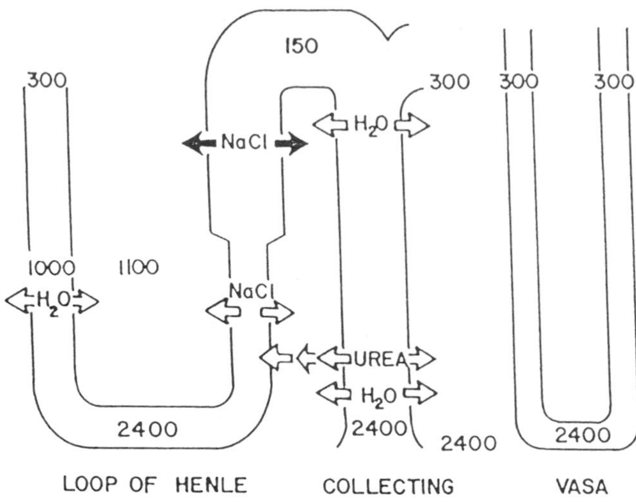

FIGURE 17.3 Schematic representation of a popular theory of the urine-concentrating mechanism in the renal medulla. Numbers represent milliosmoles per liter. Solute in the descending loop of Henle is concentrated by passive movement of water into the hyperosmolar interstitium. The ascending limb is impermeable to water, but NaCl moves into the interstitium passively (open arrows) in the inner medulla and by active transport (solid arrow) in the outer medulla, creating dilute urine in the distal tubule (150 mOsm/liter). Water may be reabsorbed in both the outer and inner medulla if collecting duct cells are rendered water permeable by ADH. Urea passively reabsorbed in the inner medulla in the presence of ADH contributes to interstitial hyperosmolarity. The vasa recta provide nutrition to the medulla and remove quantities of solute and solvent beyond that necessary to maintain normal interstitial volume and osmolality.

osmolar in the cat and in some strains of rats (Jamison, 1976; Suzuki, 1971).

Fluid has now reached the segment of the tubule in which urine volume is controlled. In the terminal distal tubule and cortical collecting duct, ADH affects the permeability of luminal cell membranes to water. With ADH present, water moves from the lumen through the cell into the hypertonic interstitium according to osmotic gradients. In the absence of ADH, very little water is absorbed. The inner medullary portion of the collecting duct is affected by ADH in the same manner. In addition, urea permeability of cells of this segment is facilitated by ADH, allowing urea to enter the interstitium.

4. *Cellular Role of Antidiuretic Hormone in Urine Concentration*

ADH is released in response to several stimuli, including increased plasma osmolality. Circulating ADH is bound to receptors on the basolateral membrane of collecting duct cells. Binding initiates generation of cAMP, which mediates intracellular events that culminate in a change in the luminal membrane from water impermeable to water permeable. Water channels studied in more detail in other tissues probably explain the change in membrane permeability (Agre *et al.*, 1993).

A large number of agents have been identified that modify the intracellular events initiated by ADH. These include prostaglandin E_2, calcium, protein kinase C, aldosterone, and certain amino acids (Hays *et al.*, 1987). Compounds that modify events initiated by ADH are probably important in the pathogenesis of polyuria in many diseases of animals which are unresponsive to ADH.

5. *Role of Urea in Urine Concentration*

In 1931 Jollife and Smith observed that urine-concentrating ability improved when dogs were changed from a low protein diet to a high protein diet. The role of urea in the concentration of urine has been elucidated more recently, and it explains the effect of protein previously observed. As filtered urea flows through the nephron, it is concentrated as water is reabsorbed. Accordingly, fluid reaching the inner medullary collecting duct has a high urea concentration. Cells of the medullary collecting duct are permeable to urea when ADH is present, and the high intraluminal urea concentration favors diffusion from tubule lumen into the interstitium (Fig. 17.3). The high interstitial concentration of urea achieved in this manner results in some urea passing into the lumen of the thin ascending limb of the loop of Henle, further increasing the concentration of urea in the medullary collecting duct and creating a cycle to concentrate urea. The addition of urea to the sodium chloride in the medullary interstitium increases the osmotic gradient for water reabsorption beyond that achieved with sodium chloride alone and thus increases maximum concentrating ability by the animal.

6. *Renal Medullary Factors*

The efferent arterioles of juxtamedullary glomeruli give rise to vessels (vasa recta) which course straight down into the medulla, make a hairpin turn deep in the medullary area, and then return to the cortex (Fig. 17.3). These vessels function to provide nutrients to medullary structures and to facilitate removal of materials from the medulla. The path of these vessels and the slow rate of blood flow through them are factors instrumental in the preservation of medullary hypertonicity. Equilibrium is achieved between interstitium and blood at all levels of the medulla, so that influx of salt and urea occurring as the vasa recta descends into the medulla is reversed by efflux as fluid ascends toward the cortex. Rapid blood flow through the medulla would not allow for equilibration and "washout" of solute from the medulla would occur (Jamison, 1976).

Renal medullary cells are subjected to an environment with a very high osmolality. To balance high interstitial osmolality, cells of medullary structures generate and maintain high intracellular levels of organic osmolytes. Polyols, methylamines, and amino acids are used; mechanisms exist for rapid changes in intracellular osmolality to accommodate interstitial fluctuations (Handler and Kwon, 1993).

7. *Factors Impairing Urine Concentration*

Several factors may impair reabsorption of water from tubule lumens. Such factors can result in the production of volumes of urine larger than are appropriate for maintaining water balance. Polyuria is a common sign detected in a variety of diseases of domestic animals. Based on existing knowledge concerning the mechanisms of urine concentration, all concentrating defects resulting in polyuria fall into three categories (Table 17.2).

J. Renal Maturation and Senescence

Renal function in the neonatal animal has relevance in clinical practice in all species. Unfortunately, our knowledge about renal maturation is limited. Structur-

TABLE 17.2 Mechanisms for Development of Polyuria in Animals

Defect	Examples of diseases or conditions causing this defect
1. Retention of particles in tubule lumens	
a. High blood concentration of filtered particles	Hyperglycemia (diabetes, mellitus, iatrogenic) Mannitol administration
b. Lack of particle reabsorption	Fanconi syndrome (impaired absorption of solute in the proximal tubules)
2. Decrease in interstitial osmolality	"Washout" secondary to increased urine flow rate of any cause "Washout" due to increased medullary blood flow Treatment with loop diuretics (furosemide, ethacrynic acid)
3. Impermeability of collecting duct to water	Diabetes insipidus (deficiency of ADH) Nephrogenic diabetes insipidus (kidneys normal except for response to ADH)

ally, species vary at birth in the degree of renal maturation.

In newborn dogs, nephrogenesis is incomplete and immature glomeruli are visible in the subcapsular area until about 3 weeks of age. Most growth of kidney that occurs with maturation is due to lengthening of tubules. Functionally, the GFR of individual superficial nephrons of pups increased sevenfold between 2 and 77 days of age (Horster and Valtin, 1971). This increase was due not only to increased glomerular volume but also to increased glomerular permeability. Tubular functions appear to lag behind glomerular functions. Canine fetuses 10 days or less from whelping had urine specific gravity readings of 1.008–1.025 (Rahill and Subramanian, 1973), indicating some concentrating ability. Random urine samples in pups at 2 days of age had an osmolality about twice that of plasma, and the ratio increased to seven times that of plasma at 77 days of age (Horster and Valtin, 1971). Thus, pups have a rapid increase in concentrating ability with maturation. Studies on 2-day-old pups demonstrated that they could excrete an oral water load, but their ability to do so was inferior to that of adults (McCance and Widdowson, 1955). Newborn pups have less capacity for excretion of a potassium load than adult dogs, but the same distal tubule site of potassium excretion exists for both (Lorenz *et al.*, 1986).

The ability of young calves to concentrate urine seems limited, because, with diarrhea, mean urine osmolality values were only 406 mOsm/liter (Thornton and English, 1976). With indiscriminant water drinking, young calves may develop water intoxication because of limited abililty for renal excretion of a water load (Kirkbride and Frey, 1967).

Few data are available for domestic animals concerning the effects of aging on renal function. In human beings, slight decreases in GFR and concentrating ability were observed with age (Dontas *et al.*, 1972; Slack and Wilson, 1976; Lindeman, 1986), but the effects were minor when compared to the total quantity of renal reserve. Morphological changes observed in association with aging in human beings include shrinkage in nephron size after the fourth decade, as determined by nephron dissection, and a gradual increase in thickness of basement membranes of both glomeruli and the tubules (Darmady *et al.*, 1973). Differences in renal clearance were found in dogs of different ages by Asheim *et al.* (1961), but the range of ages studied was not reported. In a more comprehensive study, dogs were uninephrectomized at 7 to 8 years of age, and the effects of high-protein and low-protein diets were compared. The GFR values increased to more than 75% of normal following uninephrectomy, and no subsequent decrement in GFR occurred over 4 years. These results indicated that reduction of renal mass to 50% of normal did not place the remaining kidney at risk for functional deterioration, nor did the high-protein diet have adverse effects (Finco *et al.*, 1994). Thus, deterioration of renal functions is not inevitable in aging dogs, and routine dietary protein restriction seems unneeded.

K. Renal Metabolism

A high metabolic rate exists in the kidneys. Although both kidneys make up about 0.5% of body weight, they account for about 10% of total oxygen consumption by the body under basal conditions (Klahr and Hammerman, 1985). Anaerobic metabolism also occurs in some regions of the kidney.

Uptake of substrate from the blood by the kidney is somewhat selective, and not necessarily related to the concentration of substrate in the blood. The kidney uses primarily free fatty acids (mostly palmitic), lactate, glutamine, glucose, and citrate. Other substrates used by the kidney, but probably accounting for an insignificant amount of the total renal metabolism, are inositol, α-ketoglutarate, pyruvate, fructose, glycerol, acetoacetate, β-hydroxybutyrate, low-molecular-weight proteins, polypeptides, and certain amino acids (Klahr and Hammerman, 1985).

Products of certain metabolic pathways are released by the kidney into the blood. During starvation the kidney produces glucose via gluconeogenic pathways to provide up to 50% of the body glucose synthesized (Owen *et al.*, 1969). Alanine is released by the kidney after its synthesis from pyruvate, and serine is produced from glycine. Both amino acids are used by the liver for gluconeogenesis (Klahr and Hammerman, 1985).

Earlier studies on renal metabolism suggested that renal requirements for O_2 were attributable to the energy requirements for reabsorption of the filtered load of sodium. However, this belief has been questioned and at present it appears that endergonic metabolic syntheses in the kidney (glucose, serine, creatine) explain energy usage as well (Cohen, 1986).

Areas of the kidney differ in their metabolic characteristics. Oxidative metabolism appears to be highest in the ascending loop of Henle and in the proximal tubule, based on the plethora of mitochondria and enzymes of the citric acid cycle at these sites. Gluconeogenesis is a function of the renal cortex and probably occurs mostly in the proximal tubules. Anaerobic glycolysis appears to be the major metabolic pathway in the medulla and papilla, anatomic sites at which thin loops of Henle and medullary collecting ducts dominate.

III. ALTERATIONS IN RENAL FUNCTION DUE TO EXTRARENAL FACTORS

The kidneys perform many of their functions in synchrony with activities of other organs. Under certain circumstances the kidneys may be anatomically normal but may function abnormally because of extrarenal factors. When azotemia occurs as a consequence of this situation, it is called *extrarenal azotemia*. Extrarenal azotemia may be the result of several factors. When it is due to increased generation of nitrogenous wastes or to faulty delivery of such wastes to the kidney, it is called *prerenal azotemia*. Extrarenal azotemia also may be due to impairment of flow of urine from the body; this is called *postrenal azotemia*. Some of the extrarenal factors affecting renal function are subsequently discussed. It is important in clinical medicine to be aware of such factors so that extrarenal causes of renal dysfunction are not confused with renal disease.

A. Diet and Protein Catabolism

Blood urea nitrogen (BUN) values of some species are modestly increased when protein intake of a normal animal is increased, apparently because of the generation of more urea (Watson *et al.*, 1981; Epstein *et al.*, 1984). The increase in BUN occurs despite the fact that a high-protein meal may increase GFR in dogs by up to 100% for several hours (Smith, 1951). Dietary protein intake also affects GFR in sheep, although the increase is not as large as in dogs (Rabinowitz *et al.*, 1973). The effects of diet on BUN and its relevance to clinical interpretation are discussed in Section VI. The effects of diet on the progression of preexisting renal failure are discussed in Section IV.

Tissue protein catabolism also has been incriminated as a cause of extrarenal azotemia. Some of these factors are discussed in Section VI.

B. Renal Perfusion

Because the kidney depends on hydraulic force generated by the cardiovascular system for filtration, it is not surprising that factors influencing glomerular perfusion pressure may alter renal functions. Because blood flow to the kidneys far exceeds the quantity required for sustaining renal tissues in a viable state, perfusion defects (ischemia) can influence renal excretory functions without causing renal damage. The degree and duration of renal ischemia determines whether renal dysfunction is accompanied by renal damage. Several diseases and conditions occur in which renal parenchymal damage does not occur, but prerenal azotemia develops because of a decrease in GFR. At the other extreme, complete occlusion of the renal artery for 4 hours under normothermic conditions results in ischemic necrosis of the kidneys that is irreversible (Hamilton *et al.*, 1948).

1. *Systemic Circulatory Abnormalities*

a. Dehydration

Dehydration can cause hypovolemia, which leads to impaired excretion of urea and creatinine secondary to reduced renal blood flow (RBF) and GFR. Dehydration azotemia was experimentally induced in dogs by controlled occlusion of the gastric pylorus. The dehydration that occurred secondary to vomiting caused reduced cardiac output, decreased RBF, and markedly reduced GFR. The pronounced effect on filtration occurred as a consequence of increased oncotic pressure of plasma proteins. The autoregulatory capacity of the kidney also was lost (Balint *et al.*, 1975; Balint and Sturcz, 1959; Balint and Fekete, 1960; Balint and Forgacs, 1965; Balint and Visy, 1965). Dehydration in dogs also may cause changes in intrarenal patterns of perfusion. Outer cortical flow may be preserved longer during the period of dehydration than inner cortical flow. In addition, sudden and intermittent periods of ischemia of 1 to 60 minutes in duration may occur in localized areas of the cortex during dehydration (Kirkebo and Tyssebotn, 1977).

b. Decreased Cardiac Output

Any cardiac disease causing decreased cardiac output has the potential to produce prerenal azotemia by affecting renal perfusion. Cardiac tamponade in dogs caused a fall in GFR, apparently because of decreased cardiac output (Mandin and Davidman, 1978).

In some instances, several factors may interact to cause a renal perfusion defect. For example, dogs with induced right atrioventricular valve insufficiency, pulmonic stenosis, and renal denervation had normal RBF during exercise. However, dogs with the same vascular defects but with renal nerves intact had RBF reduced to 30% of normal during exercise (Millard *et al.*, 1972).

2. *Selective Renal Hypoperfusion*

Dehydration and cardiac dysfunction sufficient to cause prerenal azotemia usually are easily detected clinically. By contrast, selective renal hypoperfusion cause by increased resistance in renal afferent arterioles may be clinically undetectible.

Hepatorenal syndrome is a term used to describe a condition in human beings in which hepatic failure leads to azotemia. In many instances, the azotemia is prerenal in cause, renal lesions are absent, and kidneys function normally when transplanted to people without hepatic failure (Conn, 1973; Kew, 1972; Papper, 1975). The pathogenesis of the renal failure remains speculative, but roles for renal nerves (DiBona, 1984), cardiovascular dysfunction (Better, 1986), and a humoral agent (Levy, 1992) have been supported.

Because selective renal hypoperfusion is undetectible by routine clinical procedures, it is difficult to differentiate from acute renal failure.

C. Renal Outflow Impairment

Obstruction of urine outflow leads to azotemia because kidney function is affected by the obstruction. As is the case with prerenal factors, urinary obstruction may or may not cause injury to renal parenchyma, depending on the duration and severity of the obstruction.

1. *Acute Effects—Unilateral Obstruction*

Complete ureteral occlusion results in rapid changes in determinants of glomerular filtration. Due to occlusion, pressure within the proximal tubule increases rapidly to twice normal values. Within 5 hours pressure begins to decline, however, and by 24 hours it is normal or subnormal. The GFR decreases as intratubular pressure increases but it does not return to normal as intratubular pressure falls because afferent arteriolar constriction occurs, resulting in a decreased capillary hydraulic pressure. The afferent arteriolar constriction may be due to production of thromboxane by infiltrating cells and mesangial cells (Klahr, 1991). Despite complete ureteral occlusion, glomerular filtration does not completely stop because tubular reabsorption of filtrate accommodates formation of new filtrate (Klahr *et al.*, 1986).

2. *Acute Effects—Bilateral Obstruction*

With bilateral ureteral obstruction, the increase in intratubular pressure is greater and more sustained than with unilateral obstruction. The decrease in GFR that occurs with bilateral obstruction is due solely to the increase in intratubular pressure because normal values for intracapillary pressure are found (Klahr *et al.*, 1986). Another difference between unilateral and bilateral obstruction is that a diuresis occurs following release of bilateral obstruction, but not after release of unilateral obstruction.

3. *Events after Release of Acute Ureteral Obstruction*

Following release of unilateral obstruction, GFR remains decreased because of persistent vasoconstriction. Effects vary from nephron to nephron; some are filtering at a subnormal rate whereas others are nonfiltering. The GFR increases with time. One hour (3600 seconds) after release of unilateral ureteral obstruction of 18,000 seconds (5 hours) in duration in the dog, GFR of the affected kidney was 70% of normal (Vaughan *et al.*, 1971).

Tubular pressure declines after release of bilateral obstruction, but not to normal levels. Glomerular capillary pressures are higher than with unilateral obstruction and more nephrons are filtering, the net result being a higher GFR. Recovery of function depends on the duration of obstruction. Rats with 30 hours of bilateral ureteral obstruction had a return to nearly normal function after 5 days (McDougal and Wright, 1972).

Some events after release of acute bilateral obstruction may represent homeostatic responses of the kidney to alterations in body composition that occurred as a consequence of obstruction. Postobstructive diuresis is attributed to positive water and electrolyte balance and retention of nitrogenous wastes that accumulated during obstruction. However, renal dysfunction also may influence postobstructive events. Dehydrated cats that should be conserving water have impaired concentrating ability after relief of acute obstruction (Finco and Cornelius, 1977). Hypotheses to explain postobstructive diuresis include washout of the medullary hypertonic gradient, impaired response of the collecting ducts to ADH, and accumulation of atrial natriuretic factor (Klahr, 1991). An acidifying defect occurs in dogs following removal of acute, complete ureteral obstruction (Thirakomen *et al.*, 1976), which was not present in dogs studied after recovery (Kerr, 1956). After release of obstruction, cats with hyperkalemia may develop hypokalemia. The pathogenesis of the hypokalemia has not been studied, but it is presumed to be a consequence of transient renal dysfunction.

4. *Function after Release of Chronic Obstruction*

Postrenal azotemia may lead to primary renal azotemia because damage to the kidney during obstruction may be irreversible. The duration and degree of obstruction are two important factors governing the amount of renal damage. In dogs, 2 weeks of total unilateral ureteral obstruction resulted in 25% of original function immediately after release, and stabilization at 50% of function of the contralateral kidney at 2 months (Kerr, 1956). In another study in dogs, the duration of total obstruction was varied and maximum recovery of function was subsequently measured. Full recovery followed 7 days of obstruction, 70% recovery followed 2 weeks of obstruction, and 30% function followed 4 weeks of obstruction. Almost no function was recovered after 42 days of obstruction (Kerr, 1954).

In dogs, partial ureteral occlusion for 14 to 60 days resulted in 100% recovery of GFR with 14 days of occlusion, 31% with 28 days, and 8% with 60 days; recovery was measured 28 days after relief of obstruction (Leahy *et al.*, 1989).

IV. PRIMARY RENAL DYSFUNCTION

A. Acute Renal Failure

Acute renal failure can be defined as primary renal dysfunction of sudden onset. Clinically, it is important to differentiate acute renal failure from chronic renal failure because the acute disease is potentially reversible but the chronic form is not.

1. Causes

a. Ischemia

As previously indicated, perfusion defects that initially result in prerenal azotemia may progress to acute primary renal failure. Species differences may exist regarding vulnerability to ischemic renal damage. Experimental results suggest that the dog may not tolerate hypovolemia as well as the human being; dogs may succumb to shock rather than surviving to develop acute ischemic renal failure (Phillips *et al.*, 1946).

b. Drugs and Other Nephrotoxins

The kidney is particularly vulnerable to effects of noxious agents because of its high perfusion rate, its ability to concentrate many substances in the tubule lumen, and its high rate of metabolism. Many materials are nephrotoxic to domestic animals but nephrotoxic antibiotics may be the most common cause of acute renal failure. In addition, heavy metals, organic chemicals (ethylene glycol, carbon tetrachloride, chloroform, herbicides, pesticides), acorn poisoning, hemoglobin and myoglobin, and snake venom nephrotoxicity are described in animals (Kosek *et al.*, 1972; Churchill *et al.*, 1974).

c. Infectious Agents

Some infectious agents, such as *Leptospira* spp. may cause acute nephritis and acute renal failure (Low *et al.*, 1956). In domestic animals this cause is uncommon compared to nephrotoxic agents, but may occur occasionally. Infection of dogs with *L. grippotyphosa* has been reported, presumably because existing vaccines may not protect against this serotype (Rentko *et al.*, 1992).

2. Pathogenesis

Acute renal failure has been studied extensively in laboratory animals. Models of nephrotoxic acute failure have been produced with a variety of agents including glycol, uranium salts, mercury, or hemoglobin. Ischemic renal failure has been produced by clamping the renal artery or injecting epinephrine. Several theories have been proposed to explain the pathogenesis of acute renal failure. One theory is that impaired GFR is due to renal vasoconstriction. Another theory is that tubular blockage occurs as a consequence of intraluminal debris and interstitial edema. Another theory surmises that glomerular filtrate forms but is totally reabsorbed (passive backflow) because of tubular necrosis. Changes in the glomerular ultrafiltration barrier also have been postulated (Harrington and Cohen, 1975; Oken, 1975; Levinsky, 1977; Brezis *et al.*, 1986). Many paradoxes and contradictions can be found about the role of these factors when individual studies are compared, and it is apparent that no theory of pathogenesis has been proved to the exclusion of others. It is likely that different causes of acute renal failure have different pathogenetic mechanisms and that a unified theory of pathogenesis of acute renal failure for all causes is unrealistic.

B. Chronic Renal Failure

Chronic renal failure is often the consequence of slow, insidious destruction of renal parenchyma. The reserve capacity of the kidneys and ongoing compensatory processes that occur coincident with renal damage result in the absence of clinical signs until very little viable renal tissue remains. Thus, chronic damage to renal tissue is irreversible, and clinical improvement can be achieved only by removal of extrarenal factors which are superimposed on the renal disease, or by replacement of renal function by dialysis or renal transplantation.

1. Etiology

Causes of chronic renal failure in domestic animals remain poorly defined. Some causes (i.e., chronic pyelonephritis, chronic glomerulonephritis) have been identified in individual animals, but it is likely that other presently unidentified factors are responsible for most cases. An old theory derived from studies in rats blamed high dietary protein intake as a cause of renal failure, but this idea has been disproved (see next section).

2. Progression of Renal Failure

Progressive destruction of renal tissue has been presumed to occur because of persistence of the initiating cause. Because the cause is usually unknown, the hypothesis can rarely be tested. More recently it has been proposed that renal damage becomes a self-perpetuating phenomenon once functional renal mass has been reduced to some critical level (Hostetter *et al.*, 1981). The self-perpetuation hypothesis implies that renal damage will be progressive despite disappearance or removal of the initial cause. In young dogs, reduction of renal mass by three-fourths was not associated with functional impairment or development of significant renal lesions in residual tissue after periods of up to 4 years (Bovee *et al.*, 1979; Robertson *et al.*, 1986). However, 11/12 or greater reduction of renal mass in dogs is associated with development of renal lesions in as little time as 3 months (Finco *et al.*, 1985; Polzin *et al.*, 1988). Lesions are similar to those observed in dogs with naturally occurring chronic renal failure. Some dogs with 15/16 reduction of renal mass have progressive decreases in GFR and develop uremia (Finco *et al.*, 1992). All of these observations support the hypothesis that renal disease is a self-perpetuating phenomenon in dogs, but only when renal mass has already been markedly reduced.

Several dietary factors have been hypothesized to influence the rate of progression of renal failure. Dietary protein has received the most attention as a potentially harmful nutrient. As previously indicated, increased protein intake causes an increase in GFR and RBF by causing afferent arteriolar dilation and an increase in hydraulic pressure within glomerular capillaries. With time, renal hypertrophy accompanies the increase in RBF and GFR and either the pressure or the hypertrophy have been incriminated as causes of renal damage. Although dogs with induced renal failure have increased glomerular capillary pressures (Brown *et al.*, 1990), several studies have failed to verify that high-protein diets cause decrements in renal function or structural damage to the kidneys (Finco *et al.*, 1992; Polzin *et al.*, 1993; Finco *et al.*, 1994). Caloric restriction has been documented as a protective factor in progression of renal failure in rats (Tapp *et al.*, 1989). A combination of protein and calorie deprivation was considered beneficial in cats with induced chronic renal disease but the role of protein versus calorie deprivation was not established (Adams *et al.*, 1992).

Intake of dietary phosphate also has been blamed for progression of chronic renal failure. Studies in dogs have demonstrated a benefit of phosphate restriction on survival and decline in GFR, but benefits on renal morphology and renal mineralization were not observed (Finco *et al.*, 1992). Theories explaining toxic effects of phosphate include tissue calcification, increased metabolic rate by renal tubule cells (Lau, 1989), and cell damage by uptake of calcium (Harris *et al.*, 1986).

3. *Pathophysiology of Chronic Renal Failure*

In contrast to the abrupt sequence of events in acute renal failure, most changes take place gradually during chronic failure. Insidious reduction of renal function goes undetected during the months or years prior to development of clinical and biochemical abnormalities. Many abnormalities are only apparent when GFR is reduced to less than one-third of normal.

a. Functional Adaptations

A compensatory response occurs in concert with tissue destruction so that remaining viable tissue maximizes its function. The damaged kidneys retain remarkable ability to attend to the functions which are required for water, electrolyte, and acid–base homeostasis. Individual nephrons respond to stimuli for homeostasis in an appropriate fashion, even though some morphologic abnormalities may be present. Thus, patients with chronic renal failure have no clinical and few detectable biochemical changes until most tissue is destroyed. These orderly and appropriate (i.e., "normal") responses by residual nephrons have been referred to as the *intact nephron hypothesis* (Bricker *et al.*, 1960; Bricker, 1969).

Several examples of adaptation and successful homeostasis by the diseased kidney are available. Single-nephron GFR (SNGFR) increases as renal mass is reduced. For example, SNGFR increased from 71 nl/minute in normal dogs to to 132.5 nl/minute in dogs with three-fourths nephrectomy and to 161.8 in dogs with seven-eights nephrectomy (Brown *et al.*, 1990). Despite the increased filtration by individual nephrons, whole body GFR is decreased because there are fewer functional nephrons. The decrease in GFR results in a decrease in filtration of blood components. If tubular adaptation to this decrease did not occur, blood components would be retained to accumulate in the body. A progressive increase in FE occurs for sodium, chloride, phosphate, and magnesium as GFR declines, resulting in clinically normal plasma concentrations until GFR is very low.

Another example of adaptation is an increased tubular secretion of potassium. Reduced GFR results in less potassium elimination via filtration; increased tubular secretion by residual functional nephrons prevents hyperkalemia until GFR is reduced drastically.

b. Morphologic Adaptations

Mature animals are incapable of generating new nephrons and thus adaptive mechanisms are restricted to altering existing ones. Hypertrophy is the major mechanism for structural adaptation, but hyperplasia also occurs in immature animals and in adults with a greater loss of renal mass (Kaufman *et al.*, 1976; Larsson *et al.*, 1980; Van Urk *et al.*, 1978).

The events involved in compensatory change in existing nephrons begin very soon after reduction of renal mass. Increased incorporation of phospholipid into renal cell membranes is detectable 5 minutes after contralateral nephrectomy (Toback *et al.*, 1974). An increase in the ratio of RNA to DNA in tubular cells (hypertrophy) was noted 6 hours after contralateral nephrectomy of rats, and an increase in kidney weight occurred at 12 hours (Van Urk *et al.*, 1978). A marked increase in kidney weight exists 10 days after unilateral nephrectomy of dogs (Carriere and Gagnan-Brunette, 1977), but maximal increase in size does not occur until 8 weeks (Rous and Wakim, 1967). Dissection of nephrons of dogs with chronic renal disease indicates that hypertrophy involves a marked increase in glomerular size and in proximal tubule diameter and length (Oliver *et al.*, 1941).

Tubular cells are capable of producing and responding to several growth factors; these factors may be important in the process of hypertrophy (Fine *et al.*, 1992).

C. Consequences of Renal Failure

1. *Uremia*

The clinical signs associated with renal failure (uremia) may have multiple causes. An early concept was that the signs and effects of renal failure could be explained by the retention of a compound normally excreted by the kidneys. The success of hemodialysis in alleviating most signs of renal failure supported this idea. The culprits accused of being responsible for uremia include ammonia, uric acid, hippuric acid, leucine, tyrosine, sulfates, phosphates, chloride, potassium, acidosis, organic anions, indicans, guanidine and its derivatives (i.e., methyl guanidine, guanidinosuccinic acid), magnesium, phenols, aromatic oxy acids, urinary alkaloids, and hyperosmolality (Schreiner and Maher, 1961) as well as parathyroid hormone (Massry, 1977). However, no concensus has been reached that any of these factors alone is the uremic toxin. Urea and creatinine, although commonly used as indices of renal function, have little or no toxicity until blood concentration exceed levels commonly detected with renal failure (Giovannetti and Barsotti, 1975).

An alternate theory to explain signs of uremia is that a multitude of biochemical abnormalities that exist in extracellular fluid during renal failure result in abnormal cell metabolism, which is manifested as uremic signs (England and Mitch, 1993).

2. *Clinical Signs of Uremia*

Almost all body systems are affected by uremia. The nervous system is affected both centrally and peripherally. Cerebral dysfunction (depression) occurs in many species, and cattle grind their teeth. Uremic dogs had EEG alterations that correlated with increases in brain calcium levels after induction of uremia (Arieff *et al.*, 1975). In brain synaptosomes from rats with chronic renal failure, derangements in acetylcholine metabolism (Smogorzewki and Massry, 1993) and in calcium ATPase were found, but both were preventable by parathyroidectomy. Peripheral neuropathy also occurs with chronic uremia. Motor nerve conduction velocity is decreased in human beings in association with primary axonal degeneration and segmental demyelination (Raskin and Fishman, 1976).

The gastrointestinal tract is severely affected by uremia. Anorexia is common, and both acute and chronic renal failure may cause vomiting in carnivores. The vomiting may be partially inhibited by ablation of the chemoreceptor trigger zone in the medulla, suggesting a circulating factor as one cause (Borison and Herbertson, 1959). Chronic renal failure also may cause oral and gastrointestinal ulcers. The ulcers have been attributed to ammonia released by bacterial action on urea (Black, 1970). The gastrointestinal lesions may contribute to vomiting by initiating afferent impulses to the vomiting center. It is suggested that uremia causes abomasal atony in cattle (Divers *et al.*, 1982). Diarrhea or constipation also has been associated with renal failure. A typical "uremic breath" may be noted in patients with advanced renal failure. Studies in human beings indicate that dimethylamine and trimethylamine are responsible for this odor (Simenhoff *et al.*, 1977).

Endocrine, muscular, skeletal, cardiovascular, respiratory, ocular, and dermatologic signs also are described in uremia (Mujais *et al.*, 1986). Many signs attributable to these systems have been described in domestic animals, but they appear less common than the neurologic and gastrointestinal problems previously discussed.

3. *Derangements in Water Homeostasis*

Changes in urine volume have long been associated with renal failure. A very marked decrease in GFR is associated with oliguria or anuria. This may occur secondary to massive, sudden injury (acute failure) or as a terminal event with chronic failure. The effects of oliguria or anuria on water balance are dependent on water intake and extrarenal losses; fluid retention and edema are potential complications of oliguric or anuric renal failure.

Polyuria is a common sign observed with renal failure. The pathogenesis of the polyuria has been investigated to determine whether it was a consequence of renal damage or secondary to homeostatic changes in renal function dictated by reduced functional mass. Studies of dogs with unilateral renal disease produced by infection or aminonucleoside nephrosis indicated that the diseased kidney could concentrate urine as well as the normal kidney (Bricker *et al.*, 1959). Likewise, dogs with remnant kidneys could concentrate urine as well as normal dogs when the solute load was adjusted with urea to reflect GFR (Coburn *et al.*, 1965). Remnant kidneys also had the same ability to concentrate urine in the uremic as in the nonuremic environment (Morrin *et al.*, 1970). Thus, several lines of evidence indicate that the major cause of polyuria during renal failure is an adaptation occurring for purposes of homeostasis. However, a decrease in medullary interstitial hypertonicity occurs with renal failure (Eknoyan, 1977), which could be secondary to the diuresis and thus pathologic. Another factor favoring a role of renal damage is a diminished response of collecting ducts to ADH (Fine *et al.*, 1978).

Polyuria, although studied in models of chronic renal failure, may occur in acute renal failure as well (Anderson *et al.*, 1977). Some data from human beings indicate that more severe acute renal failure results in oliguria, but that less severe acute failure may result in polyuria (Brezis *et al.*, 1986).

Failure to dilute urine adequately also occurs in dogs with reduced renal mass (Coburn *et al.*, 1965). Nevertheless, some dogs with chronic renal disease may have hyposthenuria, presumably due to some retention of function by the ascending limb of the loop of Henle (Finco, 1979).

Inability to regulate water excretion predisposes the patient with polyuric renal failure to water imbalance. Dehydration is a common finding when water losses from polyuria are not replaced. Severe dehydration can cause prerenal uremia, which is then superimposed on existing polyuric renal failure.

4. Hematologic Abnormalities

A nonregenerative, normocytic, normochromic anemia often is observed in association with chronic renal failure. Anemia is frequently but not invariably present in dogs and cats when chronic azotemia is discovered.

Erythropoietin (EPO) is a glycoprotein probably produced by renal interstitial cells, which circulates to the bone marrow and stimulates erythrocyte progeners at several stages of maturation. EPO production is decreased during renal failure, and this deficit is primarily responsible for the anemia of chronic renal failure. The human gene controlling EPO synthesis has been cloned (Jacobs *et al.*, 1985) and EPO is commercially available. Use of human EPO in uremic dogs and cats has resulted in a significant increase in hematocrit, but in dogs some adverse reactions have occurred (Cowgill, 1992).

Although EPO deficiency is the major factor responsible for anemia during renal failure, serum from uremic animals contains factors that are directly toxic to bone marrow cell cultures (Lamperi *et al.*, 1974) or inhibitory to EPO (Wallner *et al.*, 1977). In addition, red cell survival time is shortened in uremic human beings (Naets, 1975). Some studies in dogs suggest that parathyroid hormone may shorten red cell survival (Akmal *et al.*, 1985), but other studies do not support this theory (Anagnostou and Kurtzman, 1986).

Abnormalities in leukocyte numbers and in platelet function also have been described in the uremic state (Anagnostou and Kurtzman, 1986). Platelet defects may cause clotting problems in uremic people; hemostatic abnormalities related to abnormal platelet adhesion also have been reported in dogs (Brassard *et al.*, 1994). Clinically, bleeding abnormalities are uncommon in uremic dogs, but are fairly frequent in cattle (Divers *et al.*, 1982).

5. Blood Chemical Effects

a. Plasma Electrolyte Concentrations

The homeostatic mechanisms previously described maintain plasma electrolyte concentrations in the normal range until GFR reaches very low levels. Control of some electrolytes may be better than others, and species variation may exist.

In dogs, plasma sodium concentration remains normal until terminal stages of chronic renal failure. Hyponatremia has been described in acute renal failure in cattle (Brobst *et al.*, 1978; Divers *et al.*, 1982), and in chronic renal failure in horses (Koterba and Coffman, 1981; Tennant *et al.*, 1982). The pathogenesis of the hyponatremia has not been well studied; possibilities include lack of sodium intake, sequestering of sodium, or dilution of sodium because of water retention.

In most species, plasma potassium concentration remains normal during the polyuric phase of renal failure, but hyperkalemia develops with oliguria or anuria. The life of the animal with anuria is threatened because of the cardiotoxic effects of hyperkalemia. In cattle with acute renal failure and cattle with bilateral nephrectomy, hypokalemia was common (Divers *et al.*, 1982). This finding was attributed to anorexia, increased salivary excretion of potassium, and impaired intestinal absorption (Watts and Campbell, 1970; Divers *et al.*, 1982).

In cats fed a commercial food with 0.35% potassium, renal failure, increased FE of potassium, acidosis, and hypokalemia occurred (Dow *et al.*, 1987). Polymyopathy also was observed in cats fed this diet. Subsequent studies incriminated ingestion of a low-potassium, acidifying diet in the renal dysfunction (Dow *et al.*, 1990).

Hypermagnesemia occurs in advanced renal failure in monogastric species, but it has received little attention clinically.

Calcium and phosphate metabolism are markedly altered during uremia, and are involved in the development of renal secondary hyperparathyroidism in monogastrics. Although all details are not known, it seems that early in the course of renal failure there is an analytically imperceptible hyperphosphatemia and hypocalcemia. The hyperphosphatemia is due to decreased GFR. The hypocalcemia may be a direct physicochemical effect of hyperphosphatemia (calcium × phosphate = constant), a deficiency of the active form(s) of vitamin D, or both. The hypocalcemia stimulates PTH release, which in turn induces phosphaturia by inhibiting renal tubular reabsorption of filtered phosphate. The enhanced phosphaturia results in normalization of plasma phosphate and calcium values, but PTH secretion must be maintained at the increased rate to sustain the effect. As renal function deteriorates, increased PTH secretion is required to maintain normophosphatemia (Slatopolsky *et al.*, 1971).

Although PTH affects renal excretion of phosphate in the manner just described, renal adaptation to hyperphosphatemia also occurs with PTH absent (Cramer and McMillan, 1980; Caverzasio *et al.*, 1982). The renal mechanisms for phosphate homeostasis, whether PTH dependent or independent, are effective in maintaining normophosphatemia unless renal dysfunction is severe. Hyperphosphatemia occurs at about the same level of renal dysfunction as does azotemia. When renal failure advances further, hypocalcemia may be present. The hypocalcemia has several causes, including impaired intestinal calcium absorption because of lack of renal hydroxylation of 25-hydroxycholecalciferol, precipitation of calcium from blood because of hyperphosphatemia, and skeletal resistance to the effects of PTH in mobilizing bone calcium.

Occasionally, hypercalcemia occurs during renal failure in both dogs (Finco and Rowland, 1978) and in horses (Tennant *et al.*, 1974, 1982). Caution must be used in interpreting total calcium values, since only the ionized portion of calcium is biologically active. During renal failure, accumulation of agents that chelate calcium may lead to increased plasma concentration of total calcium while ionized calcium is normal. However, there is evidence for increased ionized calcium as well. Studies indicate that the setpoint for PTH release is raised in dispersed cells from hyperplastic parathyroid glands (Brown *et al.*, 1982). Studies in dogs indicate that hypocalcemia may not be required for stimulation of PTH secretion (Lopez-Hilker *et al.*, 1986). These findings suggest that moderate hypercalcemia may be perceived to be appropriate by the hyperplastic parathyroid glands of some dogs in chronic renal failure.

Hypercalcemia in uremic horses may have a different cause. Normally the horse avidly absorbs ingested calcium from the gut and excretes the excess in the urine. With renal failure, renal excretion is impaired and hypercalcemia may ensue if dietary intake of calcium persists (Tennant *et al.*, 1982; Divers, 1983).

Plasma phosphate concentrations seem variable during renal failure in both cattle and horses, apparently because of the interplay between ingestion, renal excretion, and extrarenal excretion (Watts and Campbell, 1970; Koterba and Coffman, 1981; Tennant *et al.*, 1982; Divers *et al.*, 1982). The status of PTH secretion during renal failure in these species does not seem well established.

In species in which renal secondary hyperparathyroidism occurs, its effect on the skeleton may be dramatic. During uremia in the growing animal, bone growth is impaired and calcification is faulty. Pups may be "dwarfed" in size compared to littermates with normal renal function. Decalcification also occurs in the adult; "rubber jaw" syndrome may occur in some dogs and on rare occasions exostosis occurs (Norrdin, 1975).

b. Acid–Base Alterations

The prominent role of the kidneys in acid–base balance suggests that perturbations would be likely with renal dysfunction. Metabolism of the usual foodstuffs of carnivores and omnivores results in net production of protons that must be neutralized or excreted. A reserve capacity exists for these functions in health; compensatory hypertrophy that follows renal injury increases the capacity of individual nephrons to respond during renal disease. Effects of compensatory hypertrophy include enhanced proximal tubule bicarbonate reabsorption and ammonia generation, and enhanced proton secretion by the distal nephron. A consequence of these factors is that with generalized renal failure, acidosis usually does not develop until GFR is less than one-fourth of normal (Harris *et al.*, 1986).

The acid–base status of uremic ruminants may be different from that of carnivores. Although data are limited, it appears that a metabolic alkalosis occurs fairly frequently in cattle, possibly in association with abomasal stasis (Brobst *et al.*, 1978; Divers *et al.*, 1982).

6. Metabolic Alterations Associated with Uremia

Several alterations in body metabolism have been noted as a result of renal failure. Glucose intolerance has been reported in dogs with induced renal failure (Swenson *et al.*, 1973; Akmal *et al.*, 1985) but this may be a terminal event (Finco *et al.*, 1994).

Alterations in serum lipids and lipid metabolism have been found in several species during renal failure. Hypercholesterolemia occurs in association with the nephrotic syndrome in several species. Increases in blood levels of very low density lipoproteins (VLDL), which are mainly triglycerides, have been found in non-nephrotic human beings and rats with chronic renal failure. The elevations are apparently due to increased liver production and decreased peripheral uptake of VLDL. A decrease in high-density lipoproteins in chronic uremia of man also occurs (Mujais *et al.*, 1986). Other than for hypercholesterolemia, the presence of lipid alterations during uremia in domestic animals has not been well examined.

Protein metabolism in renal failure is altered but the mechanisms are not all known. Animals with chronic renal failure often suffer from a caloric deficit and weight loss due to anorexia. Alterations that occur in protein metabolism may be due either to malnutrition itself or to the deleterious effects of uremia on metabolism. Plasma and intracellular profiles for free amino acids change with uremia (Mujais *et al.*, 1986). Both dialysis and dietary manipulation in man seem to normalize the amino acid profiles (Bergstrom *et al.*, 1975; Furst *et al.*, 1980).

Dietary manipulation of protein, amino acids, and amino acid analogs has been conducted in patients with renal failure for several reasons. These reasons include normalization of plasma and tissue amino acid profiles on the assumption that such normalization would be beneficial to the patient, protein restriction to avoid catabolism of amino acids and accumulation of nitrogenous wastes, reversal of the negative nitrogen balance observed with both acute and chronic uremia, and a halt to the progression of renal damage. Clinical observation supports restriction of dietary protein intake for alleviation of some signs of uremia, but other beneficial effects of dietary manipulations are not as well documented. Many conflicting studies are reported (Mitch and Walser, 1986).

7. Effects of Renal Failure on Drug Metabolism

Compromised renal function affects the excretion and, therefore, the blood levels of drugs that are normally excreted by this route. Plasma binding of drugs also may be affected by renal failure. In uremia, anionic drugs have decreased plasma protein binding, whereas cationic drugs have normal or increased binding (Reidenberg, 1976, 1977a). Plasma binding affects both distribution and renal excretion of drugs. In addition, biotransformation of drugs is altered during the uremic state (Reidenberg, 1977b). During uremia, drug oxidations by microsomal oxidative systems are normal or accelerated while reductions are slowed. These observations probably apply to domestic species, although made in experimental animals.

8. Effects of Uremia on the Immune System

Uremia causes suppression of the immune system in a variety of species. Both humoral and cellular factors are impaired (LoGrippo *et al.*, 1970; Souhami, 1973; Touraine *et al.*, 1975; Dobbelstein, 1976; Raskova and Morrison, 1978). Abnormalities are found in acute as well as chronic uremia but seem more severe in the latter. Several explanations have been given for the suppressed immune response, but none has been satisfactory. The degree of suppression of immunity seems to vary from patient to patient, but is generally mild.

V. URINALYSIS

Urinalysis usually is performed as a screening method for acquiring a large amount of information about several body systems, with minimal expenditure of resources. It also is performed specifically for evaluation of the kidneys. Its value in studying the kidneys is related to the detection of abnormal components in the urine that may be of renal origin (casts, protein) and to the relation of urine-concentrating ability to kidney function. In the subsequent discussion, emphasis is placed on the underlying principles and interpretations rather than on methodology.

Analysis of urine is conducted routinely on samples collected at random. Such collections offer convenience in obtaining the sample but have the disadvantage of sampling an unknown and possibly unrepresentative portion of total daily urine production. For example, a sample of urine obtained from a dog or cat in the postprandial period may be alkaline because of the alkaline tide that follows food ingestion, but a composite of daily production may be acidic. Likewise, analysis of 10 ml of urine of a daily output of 100 ml represents a 10% sample of daily urine components; a 10-ml sample of 1 liter represents a 1% sample.

Quantitative urinalysis has been used in people and advocated for use in dogs and cats (DiBartola *et al.*,

1980; Russo *et al.*, 1986). Although cumbersome, such measurements provide much more reliable information about daily urine composition. Because of their infrequent use, the subsequent discussion deals with urinalysis of random samples.

The specific gravity of random samples of urine should be considered when all other urinalysis tests are interpreted. Urine volume usually is inversely related to its specific gravity, and therefore urinary components are concentrated in urine of high specific gravity and diluted in urine of low specific gravity.

The method of collection of urine is of utmost importance in interpreting a urinalysis, and should be part of the urinalysis report. In that way, the urinalysis results can be properly interpreted beyond the time period of mental recall of the method of urine procurement. Spontaneously voided urine may contain bacteria and white blood cells (WBCs) that are normal constituents of the genitalia. In the normal animal, samples obtained by cystocentesis are devoid of bacteria, contain but few WBCs, but may contain abnormal numbers of red blood cells (RBCs) due to trauma of cystocentesis. Samples of urine obtained by catheterization are devoid of lower tract contamination if collected properly but may have increased numbers of RBCs due to iatrogenic trauma.

Urine samples should be obtained in clean containers, or sterile ones if microbial studies are desired. The sample should be analyzed within 30 minutes; if urinalysis cannot be performed within that time, the sample should be refrigerated to minimize changes in urine composition. Refrigerated specimens should probably be examined within 43,200 seconds (12 hours) (Osborne and Stevens, 1981). Refrigerated samples should be warmed to room temperature prior to analysis, particularly to avoid errors in specific gravity measurement if a urinometer is used.

A. Urine Color, Odor, and Turbidity

Urine color is normally yellow in all domestic species. The cause of coloration is reported to be urochrome, which is a combination of urobilin and urobilinogen with a peptide (Anonymous, 1965). The intensity of the color usually varies with the urine concentration; thus, concentrated urine is dark yellow, and dilute urine is pale yellow.

Abnormal constituents may alter the color of urine. Bilirubin or its degradation products may give urine a darker yellow color than normal. Hemoglobin or hematuria may impart a red or brown color, and myoglobin a reddish brown appearance. Some drugs or diagnostic reagents also may alter the color of urine.

Urine odor varies somewhat from species to species, and the specific reasons for the differences have not been well studied. Sex differences in urine odor within a species may also exist. Infection with urea-splitting organisms may result in a highly ammoniacal urine odor, and certain drugs may impart an odor to urine.

Turbidity of urine varies among species and among samples in the same individual. Horse urine is normally markedly turbid because of calcium carbonate content and from mucus originating from glands in the renal pelvis. Cow urine becomes turbid on standing because of calcium carbonate crystals. Dog and cat urine is often clear but may be turbid in the absence of abnormalities. Some abnormal components (cells, mucus, bacteria) or normal components present in excess amounts (crystals) also may cause turbidity.

B. Specific Gravity

As previously discussed, the amount of water in urine is determined by passive movement of water from tubules to the renal interstitium secondary to osmotic gradients. Osmotic forces are determined by numbers of particles per unit of solvent. Specific gravity is used to estimate the number of particles per unit of solvent. Making this estimate has clinical value since urine specific gravity may be "abnormal" under a variety of circumstances. However, caution must be exercised in interpreting urine specific gravity readings because of the homeostatic role played by the kidney in water balance. In times of ingestion of large amounts of water, urine specific gravity as low as 1.001 is normal because of the need to excrete water for homeostasis. In times of dehydration, a marked conservation of water is expected, and urine specific gravity is increased to the maximum possible for that animal. Thus, the "normal" range of specific gravity readings encompasses the extremes between maximal diluting and maximal concentrating capabilities and includes specific gravity readings that are considered abnormal.

Apart from the normal range, it is appropriate to consider the range of specific gravity readings appropriate for a species under existing environmental conditions and husbandry practices. Because carnivores and most herbivores do not normally overdrink, some concentration of urine is usually anticipated. Interpretation of urine specific gravity as normal or abnormal is facilitated by considering the state of water balance during the time the urine was excreted. A urine of low specific gravity produced by a dehydrated animal indicates abnormal water conservation by the kidneys. Several causes of low urine specific gravity have been identified in domestic animals. States in which dilute urine may be produced in the dog are given in Table 17.3.

Some error exists in using specific gravity measurements made with a urinometer for estimating osmolality of urine. Specific gravity readings are a reflection not only of the number of particles but of their size and attraction for one another. Thus, protein may falsely elevate specific gravity by 0.001 per 400 mg/dl; glucose falsely elevates readings by 0.001 per 270 mg/dl (Wolf, 1969).

Refractometry has the advantages over urinometry of accuracy, simplicity, and the requirement of but a few drops of urine. With some refractometers, urine with specific gravity readings above 1.035 must be diluted (1:1) with distilled water to obtain readings. The actual specific gravity is obtained by doubling results to the right of the decimal point. Refractometry is the preferred method of measuring urine concentration because it is more accurate than urinometer readings and it is simpler to perform than osmometry. Osmometry is considered the definitive method of measuring solute–solvent ratios in urine since osmometers measure one of the colligative properties of urine (freezing point depression, vapor pressure). Whereas specific gravity readings are expressions of the relative density of urine compared to water, osmotic pressure readings are expressed as milliosmoles per kilogram of water. Normal serum osmolality in domestic animals is about 280–300 mOsm/kg. Urine values are discussed in Section VI.B.

TABLE 17.3 Causes of Dilute Urine in the Dog

Condition	Postulated or known pathogenesis
Overdrinking	Compensatory polyuria
Generalized renal failure	Osmotic diuresis, medullary washout, ADH refractoriness
Hepatic failure	Decreased urea production; toxic effects of hyperammonemia on tubules; primary polydipsia
Hyperadrenocorticism	Interference with ADH activity
Diabetes mellitus	Glucose diuresis
Diabetes insipidus	ADH deficiency
Nephrogenic diabetes insipidus	ADH-refractory tubules
Pyometra	Endotoxin effects on tubules
Hypercalcemia	Interference with ADH activity; later, renal failure
Pyelonephritis	Medullary washout
Postobstruction (urinary)	Osmotic diuresis; natriuretic hormone; renal damage
Hypoadrenocorticism	Natriuresis, medullary washout, ADH resistance
Potassium deficiency	Unknown
Hyperreninemia	Primary polydipsia, secondary polyuria
Fanconi syndrome	Osmotic diuresis

C. Urine Protein Determination

As previously discussed, a small amount of protein is present in the urine of all species. The amount found in normal urine is determined by the net result of normal leakage through glomerular capillaries, protein reabsorption by the proximal tubules, and addition of proteins to the filtrate as it passes through the tubules, pelvis, ureters, bladder, urethra, and genital tract. It is important to establish whether the degree of proteinuria is normal or abnormal. In the clinical laboratory, an effort is made to distinguish normal levels of protein in urine from pathologic proteinuria. A "negative" protein reaction is anticipated when amounts in urine are not large enough to be indicative of disease. A study of common methods used to detect proteinuria revealed that false-positive and false-negative results were obtained with all procedures (Thysell, 1969). This suggests that a degree of skepticism should exist in interpreting protein results.

Dipstick methods are in common use for the assessment of proteinuria in domestic animals. The procedure is dependent on color changes in tetrabromphenol blue at various protein levels. However, pH may also cause color changes. To keep pH constant, the reagent square is impregnated with buffer. In domestic animals, it appears that false-positive values occur frequently. This may be related to the pH dependence of the sticks and leaching of buffer from the strip by prolonged contact with alkaline urine. The sticks are also more sensitive to albumin than to globulins (Bowie *et al.*, 1977), and severe proteinuria due to globulins may give a false-negative reading. Difficulty in interpretation of the color change also appears to be a problem since, in one study, different readings were obtained by different technicians on the same urine (James *et al.*, 1978). In women with hypertension and pregnancy, dipsticks were considered to give an unacceptable number of both false-positive and false-negative results for proteinuria (Meyer *et al.*, 1994).

All urine samples should be evaluated by an alternate procedure if they are positive for protein by the dipstick procedure. The sulfosalicylic acid test and the nitric acid ring test are two semiquantitative methods that may be used. The sulfosalicylic acid test may give false-positive values with X-ray contrast media and large doses of penicillin, cephaloridin, and sulfisoxazole.

Semiquantitative methods of urine protein determination that are a part of routine urinalysis on random samples may be inadequate in cases of equivocal proteinuria, or in cases where the measurement of daily urinary loss of protein is required. Quantitative methods of protein measurement must be combined with provisions for consideration of 24-hour urine output in order to increase the reliability of the assessment. For quantitative measurement of protein concentration, the Coomassie blue method (Bradford, 1976) is a simple spectrophotometric procedure that can be used. Quantitative measurement of urine protein excretion can be determined on 86,400-second (24-hour) urine collection periods but such collections usually require a metabolism cage and urinary catheterization to be assured of accurate collections. The 24-hour urinary excretion of protein by normal dogs has been measured (Table 17.4). The 24-hour urinary excretion of protein by normal cats was 17.4 ± 9.1 mg/kg/day (Russo *et al.*, 1986), 12.7 ± 1.6 (Monroe *et al.*, 1989), and 4.93 ± 1.34 mg/kg/day (Adams *et al.*, 1992); in 4- to 12-week-old kittens protein excretion was 2.54 to 11.39 mg/kg (Hoskins *et al.*, 1991).

An alternative to 24-hour collections is the determination of the urine protein-to-creatinine ratio (UP/C) on a random sample of urine, based on the following logic. Creatinine excretion can be expected to be fairly constant between 86,400-second (24-hour) periods except in animals with acute renal dysfunction or severe renal failure. The concentration of creatinine in urine depends on urine volume just as protein concentration does. With daily creatinine excretion being constant, proteinuria can be related to creatinine concentration, and the urine volume factor cancels,

$$\text{Urine P/C} = \frac{\text{U protein (mg/dl)} \times \text{U volume (ml)}}{\text{U creatinine (mg/dl)} \times \text{U volume (ml)}} = \frac{\text{Protein}}{\text{Creatinine}} = \text{P/C ratio,}$$

where U is urine; P, protein; and C, creatinine. The use of urine P/C ratio allows more precise evaluation of proteinuria than determination of protein concentration on a random sample, and a good correlation between values obtained between 10 A.M. and 2 P.M. versus 24-hour collections has been found in dogs (White *et al.*, 1984). Subsequent studies revealed excellent correlation between samples taken randomly throughout the day (Grauer *et al.*, 1985) as well as day versus night (McCaw *et al.*, 1985). Significant differences were not found between urine samples collected by voiding (midstream), cystocentesis, or catheterization (Barsanti and Finco, 1979). However, in both dogs and cats, voided urine from males had a higher protein concentration than that from females (Barsanti and Finco, 1979; Monroe *et al.*, 1989). Increased dietary protein intake significantly increased UP/C in both normal and remnant kidney cats (Adams *et al.*, 1992), but effects of dietary protein in dogs is equivocal (Jergens *et al.*, 1987). Because dietary protein may affect GFR, and thus creatinine clearance, confounding factors (creatinine excretion, protein leak) may affect overall results. As expected, hemorrhage and inflammatory disease of the urinary tract increased the UP/C values (Bagley *et al.*, 1991). Values for UP/C in normal dogs and cats are <0.5 when both protein and creatinine are expressed as mg/dl; values above 1.0 are clearly abnormal (Lulich and Osborne, 1990).

Once it has been established that the magnitude of proteinuria is abnormal, its source and cause should be considered. Proteinuria in the absence of RBCs or inflammatory cells is usually of renal origin and may be due to a glomerular leak or to a lack of proximal tubular reabsorption, or both. Proteinuria in the presence of RBCs and WBCs in the urine is difficult to interpret. Hemorrhage is associated with direct leakage of RBCs and plasma proteins; the presence of WBCs implies inflammatory responses with potential leakage of plasma proteins. When RBCs or WBCs are present and the protein reaction is positive, the possibility of concomitant disease at separate sites or of differing causes exists. It may be necessary in such instances to eliminate the cause of the WBC or RBC and establish whether proteinuria disappears or persists.

In human beings, proteinuria is reported in such conditions as fever, passive renal congestion, and the biped postural stance in the absence of renal or urinary tract disease (Rubin and Balish, 1971). In beagle dogs, neither running nor swimming caused an increase in proteinuria comparable to that observed in pathologic processes (Huisman *et al.*, 1982; Joles *et al.*, 1984).

Proteinuria has been classified into types in human beings (Hardwicke, 1975), and the same types exist in domestic animals. Overflow proteinuria occurs as a consequence of increased levels of low-molecular-

TABLE 17.4 Twenty-Four-Hour Protein Excretion by Normal Dogs

Author	Analytical method	No dogs	Protein excretion (mg/kg/24 hours) Mean	Range
Barsanti and Finco, 1979	Coomassie blue	10	7.0	2.4–19.7
Barsanti and Finco, 1979	Ponceau S	10[a]	3.8	0.8–15.1
Biewenga *et al.*, 1982	Ponceau S	29	6.6	2.7–23.2
White *et al.*, 1984	Ponceau S	8	4.8	1.9–11.1
McCaw *et al.*, 1985	Coomassie blue	14	7.7	1.8–22.4
Grauer *et al.*, 1985	Coomassie blue	16	2.3	0.6–5.1
Center *et al.*, 1985	Turbidometric	19	2.5	0.2–7.7

[a] Same dogs as Barsanti; Coomassie blue method.

weight proteins in the blood. Such proteins pass through the normal glomerular filter and appear in the urine when tubular resorptive mechanisms are saturated. Newborn calves may have proteinuria as a consequence of absorption of proteins of colostrum from the gut into the blood. Low-molecular-weight components are subsequently lost in the urine. The proteinuria disappears when gut absorption of protein ceases after the first few days of life (Pierce, 1959, 1961). The appearance of Bence Jones light-chain proteins in the urine of animals with multiple myeloma is another example of overflow proteinuria. These proteins may not be detectable by conventional urine protein methods but are detectable by urine electrophoresis.

Glomerular proteinuria occurs as a consequence of injury to glomeruli. Amyloidosis and glomerulonephritis are two broad catagories of disease in domestic animals which cause glomerular proteinuria. The defect in the filter may vary in magnitude among diseases or in the same animal at different stages of disease. Glomerular proteinurias are often relatively selective so that albumin is the main component of the urinary protein. Severe proteinuria in the absence of RBCs and WBCs is usually of glomerular origin. Tubular proteinuria occurs as a consequence of normal passage of low-molecular-weight proteins through the glomerular filter combined with defective tubular reabsorption. Proteins in the urine of these patients are predominantly low-molecular-weight globulins, and amino acids may also be present (Fanconi syndrome). Basenji dogs with glucosuria, aminoaciduria, and proteinuria probably have this type of abnormality. In general, tubular proteinurias result in mild to moderate protein loss (Rubin and Balish, 1971).

A high-molecular-weight mucoprotein (Tamm–Horsfall mucoprotein) has been isolated from the normal urine of several species, including dogs and cats (Lewis *et al.*, 1970; Schenk *et al.*, 1971). This mucoprotein is apparently secreted by cells of the ascending loop of Henle and the macula densa portion of the distal tubule. It is believed to be a component of some urinary casts. Its concentration in the urine is normally too low to be detected by semiquantitative screening methods.

D. Occult Blood Reactions

The dipstick method used for the detection of blood is highly sensitive and fairly specific. It detects both myoglobin and hemoglobin. The positive reaction is due to the peroxidase activity of the heme molecule. Hydrogen peroxide in the strip is converted to oxygen, which then oxidizes *o*-tolidine to cause the formation of a blue color. Microbial peroxidase or oxidizing contaminants such as H_2O_2 or sodium hypochlorite may cause false-positive reactions.

A positive occult blood reaction on urine indicates either leakage of free hemoglobin through glomeruli (hemoglobinuria), of myoglobin through glomeruli (myoglobinuria), extravasation of RBCs directly into the urogenital tract (hematuria), or of contaminants previously listed. The size and shape of the hemoglobin molecule are such that a small percentage of hemoglobin in the plasma that is unbound to haptoglobin passes the glomerular filter and appears in urine. This occurs in diseases in which intravascular hemolysis is occurring. Hemorrhage anywhere from glomerular capillaries to the tip of the genitalia may cause hematuria; thus, its presence should not be considered a localizing sign. Attempts have been made in humans to localize the source of hematuria by examining red cell morphology (Offringa and Benbasset, 1992) and by the coincident excretion of red cells and specific urinary proteins (Guder and Hofmann, 1993) but such methods have not been applied to domestic animals.

Differentiating hemoglobin from myoglobin in urine requires a special analysis since sequestered blood may have the same color as myoglobin (Glauser *et al.*, 1972). Differentiating hemoglobinuria from hematuria may be difficult, but the following observations are helpful. With hemoglobinuria, plasma may also be observed to have hemolysis. The urine may be clear red with hemoglobinuria but cloudy red with hematuria. The urine sediment may have no RBCs or a disproportionately low number of RBCs with hemoglobinuria. Care must be exercised to distinguish intravascular hemolysis from hemolysis of RBCs once they are in urine. Red cells may hemolyze in urine of low osmolality (Vaughan and Wyker, 1971).

E. Glucosuria

Dipstick procedures use the glucose oxidase method of glucose measurement. This enzyme converts glucose to gluconic acid. In the presence of hydrogen peroxide and peroxidase incorporated into the strip, an indicator in the strip (o-tolidine) is oxidized to cause a color change. Although the method is specific for glucose compared to other sugars, contamination of urine with H_2O_2 or sodium hypochlorite can cause false-positive reactions. Conversely, ascorbic acid and formaldehyde in urine may prevent positive samples from giving a positive reaction (Mayson *et al.*, 1972). Because of the enzymatic basis for the test, temperature affects the speed of the color change, and it is important that the test be done with components at room temperature.

The strip is not sufficiently sensitive to detect the small quantity of glucose found in the urine of normal animals but gives positive results when glucose concentration is 0.56 mmol/liter (10 mg/dl) or more (Mayson *et al.*, 1972).

Glucosuria occurs when the renal tubular maximum for reabsorption from filtrate is exceeded. This may occur with any cause of hyperglycemia or with decreased tubular glucose reabsorptive capacity. Generally, hyperglycemia greater than 10 mmol/liter (180 mg/dl) results in glucosuria, but the flow rate of fluid through the tubules and the nephron heterogeneity with regard to resorptive capacity are also factors. Common causes of hyperglycemia sufficient to cause glucosuria are parenteral therapy with dextrose and diabetes mellitus. Any cause of the Fanconi syndrome may result in glycosuria during normoglycemia because faulty renal tubular reabsorption occurs with this syndrome.

F. Urine pH

The pH of urine usually is determined by renal regulation of blood bicarbonate and H^+ levels. Since pH is a measure of free ions, the actual amount of acid being excreted by the kidney depends not only on pH, but on the amount of undissociated hydrogen present in urine buffers. The titratable acidity of urine is defined as the amount of base required to titrate urine back to the pH of blood. Measurement of titratable acidity and H^+ excreted as NH_4^+ gives an estimate of hydrogen ion excretion, while urine pH does not. Nevertheless, measurement of urine pH is useful since unusual results provide one with insight concerning many abnormalities. The urine pH of domestic animals usually is determined by dietary generation of acid. Animals on diets of high animal protein content usually have acidic urine, whereas those on cereal diets or mostly forages have neutral or slightly alkaline urine. Anorexia usually results in acid urine in all species. The extremes of urine pH values in domestic mammals are 4.5 to 8.5.

Some generalizations can be made about the acid–base status of an animal by measuring urine pH, but the generalizations must be qualified by many exceptions. A very low (acidic) urine pH implies a large excretion of H^+ by the kidneys. Whether this excretion has adequately maintained blood pH and bicarbonate values cannot be ascertained by urine pH measurement alone. Likewise, a high (alkaline) urine pH may indicate a need to conserve H^+ and excrete bicarbonate in order to maintain an acid–base balance. Whether the kidneys were successful in this endeavor can only be determined by blood pH and plasma bicarbonate determinations.

Generalizations about acid–base status from urine pH must take the state of renal function into consideration. Impaired ability of the kidneys to manipulate bicarbonate or excrete H^+ can lead to abnormal acid–base states which are not reflected in urine pH values. With chronic generalized renal failure of carnivores and omnivores, urine pH is usually moderately to markedly acidic when acidosis is present. Proximal renal tubular acidosis and distal renal tubular acidosis are two groups of abnormalities of human beings that probably exist in several species of domestic animals. In proximal tubular acidosis, renal reabsorption of bicarbonate is faulty. This may initially cause alkaline urine, but it soon becomes acidic as the plasma bicarbonate level and the filtered load of bicarbonate fall while normal distal H^+ secretion persists. With distal tubular acidosis, proximal bicarbonate reabsorption is normal, but distal acidification is faulty, so that urine pH is acidic but greater than 5.3 (Thier *et al.*, 1977).

Urine pH may be altered by bacterial contamination after urine is procured. Acidification or alkalinization may occur depending on the end products of metabolism of the contaminating organism. Urine pH does not change in sterile urine stored at room temperature for several hours (Finco, 1979).

Aciduria during alkalemia occurs in the vomiting dog; this paradox may be associated with anion reabsorption, as previously mentioned.

G. Acetonuria

Ketone bodies refer to acetoacetate, β-OH-butyrate, and acetone. These compounds are products of fat degradation; the quantity produced is markedly increased under conditions in which oxidation of fats provides the bulk of the energy. Ketone bodies appear in filtrate and are reabsorbed by the tubules until the T_m is exceeded. Under normal conditions, very low plasma levels of ketone bodies exist, and they are absent from the urine. Species differences exist regarding circumstances under which ketone bodies appear in the urine. They are commonly found in the ruminant during starvation, but rarely in the dog and cat. Diabetes mellitus is the disease classically associated with ketonuria.

The urine ketone reaction is based on the reaction of acetoacetic acid or acetone with nitroprusside. No reaction occurs with β-OH-butyrate. False-positive reactions may occur with BSP, phenyl ketones, or L-dopa metabolites.

H. Bile Pigments in Urine

The interpretation of bilirubinuria or urobilinogenuria is discussed in another chapter.

I. Urine Sediment

1. General Considerations

The examination of urine sediment is the single most important part of urinalysis. Because this process is slightly more time consuming than the dipstick analysis, there may be a tendency among veterinarians to perform only dipstick tests and specific gravity determinations. This approach is inexcusable considering the value of sediment examination and the false sense of security that is left when urinalysis is reported to be done but sediment is not examined.

Quantitative overinterpretation of the sediment is another problem. Although the analyzer provides numerical values for many microscopic findings (i.e., number of cells, casts, etc., per low-power or high-power field) the many variables that influence these counts must be kept in mind. Early studies of urine sediment in human beings were made to determine the 86,400-second (24-hour) urinary content of various elements. Few studies have been reported in domestic animals, and so our concept of normal versus abnormal is based on the examination of random samples of urine and the interpretation of normal on semiquantitative bases. Some of the variables that influence the concentration of elements in urinary sediment include the volume of the urine used, how that volume relates to daily urine production, the period of time between formation of urine and its examination (some components disintegrate crystals may form), the speed and time of centrifugation, the volume used to resuspend the sediment for viewing, the volume of suspension used to prepare the slide, and the thickness of the slide preparation. Each laboratory must standardize its procedure and define the limits for normal with those procedures in order to obtain meaningful results.

An acceptable procedure is to centrifuge 5 ml of urine in a conical tube for 5 minutes at 700 *g*. Although the speed of centrifugation for urine sediment examination is often stated as revolutions per minute, this value is meaningless for determining force unless the length of the centrifuge arm is known. The supernatant fluid is removed except for 0.5 ml, which is used to gently resuspend the sediment. One drop is placed on a clean slide, and a 22-mm^2 coverslip is placed on the drop. After 1 minute to allow for sedimentation, the slide is scanned under low power (×100) with reduced light. Final identifications are usually made at ×440 under reduced light. At least 10 fields should be counted and numbers of elements averaged (Duncan and Prasse, 1976).

Several stains for urinary sediment are available commercially. The experienced technologist generally finds these stains unnecessary, but they may be helpful to the novice. New methylene blue stain may be used by mixing a drop directly with the sediment. However, cellular elements do not always take up the stain. Heat-drying of the slide with subsequent application of stain gives a better stain. Neither type of stained slide should be used for semiquantitative measurements since the concentration of components in the sediment has been altered by the manipulations.

Components of urinary sediment decompose with time, depending on conditions. For example, casts are likely to solubilize in alkaline urine. Urine sediment should be examined as soon as possible after urine is obtained to minimize errors in interpretation of urinalysis.

2. Epithelial Cells

These cells are normal components of urine sediment. Their number may vary depending on the presence of inflammation or degeneration, neoplasia, and the method of urine collection. The concept that the site of origin of epithelial cells can be deduced from their morphological character appears to have been borrowed from human medicine. A study in dogs and cats revealed that neither the presence of, nor specific morphologic characteristics of transitional cells, caudate cells, or squamous epithelial cells could be used to localize their source in the urogenital tracts (Batamuzi and Kristensen, 1995).

Epithelial cells in sediment have little diagnostic significance in themselves. It is necessary to correlate their presence with other findings in the urinalysis or with other studies in the animal.

3. Casts

Casts are cylindrical structures that are formed from cells, cell fragments, and macromolecules that are present in the lumens of renal tubules. They conform to the shape of the lumen from which they originate. When relatively intact RBCs, WBCs, or epithelial cells make up the bulk of the cast, they are identified by reference to these cells. Hyaline casts are clear, structureless, colorless objects barely discernible from the background. Immunofluorescence studies have revealed that hyaline casts from human patients are made up of Tamm–Horsfall mucoprotein but do not contain serum proteins (Orita *et al.*, 1977). One study suggested that adherence of mucoprotein was essential for cast formation (Lindner and Haber, 1983). Granular casts are homogenous in appearance except for the presence of small (fine) to large (coarse) granules. Some immunofluorescence studies revealed that the granules are composed of fractions of various serum proteins (Rutecki *et al.*, 1971; Orita *et al.*, 1977). These findings contradict those of others, who concluded that granules in casts were remnants of tubular epithelial cells (Relman and Levinsky, 1971). In another study it was concluded that subtypes of granular casts exist; type I were from leukocytes; type II were from renal tubular epithelial cells; type III were from unidentified mucopolysaccharides; and type IV were from bacteria (Lindner *et al.*, 1983). Waxy casts are similar to hyaline casts except that they are distinctive because of a hard, refractile outline. They are believed to be degenerative stages of cellular casts.

The size of casts appears to be related to their site of formation. Small-diameter casts are probably of more proximal origin, whereas broad casts are likely to be of collecting duct origin.

Some of the conditions conducive to cast formation have been identified. Tamm–Horsfall mucoprotein is readily precipitated by highly acidic urine pH and by high electrolyte concentrations (McQueen and Engel, 1966). Thus, casts are often formed in acidic urine and may be dissolved *in vitro* by alkaline urine. Rate of urine flow also seems to be a factor, with slow flow in the distal segments of the nephron favoring cast formation.

The presence of casts in the urine is interpreted as an indication of abnormal concentrations of their precursors in the tubular lumens. Thus, proteinuria, pyuria, hematuria, and cell desquamation of renal origin may provide the material for cast formation. Urine flow rate and pH are factors that interact with these components.

A few casts per low-power field are considered normal. Data for domestic animals are not available, but in normal human beings 5000–10,000 casts are excreted per day (Relman and Levinsky, 1971). Most of the casts normally found are of the hyaline type.

4. *Crystals*

Several types of crystals may appear in the urine of domestic animals. These may be categorized as one group which is formed from normal components of urine and another group which originates from exogenous compounds, abnormal metabolism, or abnormal excretion.

The formation of crystals from compounds normally in the urine occurs as a consequence of physiochemical factors, such as changes in urine concentration or pH. Many components of urine are present in a supersaturated state; disruption of this state results in precipitation of its components. Phosphate crystals in dogs and cats and carbonate crystals of cattle and horses fall in this category. They reflect conditions of urine formation and their presence should not be interpreted as abnormal. Since dalmatian dogs normally excrete urates as well as allantoin as end products of purine metabolism, urate crystals also are normal for this species. Bilirubin crystals may also be normal in concentrated urine specimens.

Abnormal crystalluria usually is indicated by the presence of the remaining crystals listed in Table 17.5. Ammonium biurate and tyrosine crystals are observed in canine urine in association with hepatic dysfunction, oxalate and hippuric acid crystals may be observed in some cases of ethylene glycol toxicity (Thrall *et al.*,

TABLE 17.5 Components of Urinary Sediments

- Cells
  - Epithelial cells (normal, neoplastic)
  - RBC
  - WBC
- Crystals
  - Magnesium ammonium phosphate
  - Urate
  - Ammonium biurate
  - Oxalate
  - Hippuric acid
  - Cystine
  - Bilirubin
  - Tyrosine
  - Drugs (sulfonamides)
- Casts
  - Granular
  - Cellular (epithelial, RBC, WBC)
  - Hyaline
  - Fatty
  - Waxy
- Miscellaneous
  - Lipid droplets
  - Spermatazoa
  - Mucous threads
  - Amorphous material
- Infectious agents
  - Bacteria
  - Mycotic agents
  - Parasites, parasite ova
    - *Capillaria*
    - *Stephanurus*
    - *Dioctophyme*
    - *Dirofilaria microfilaria*
- Physical contaminants
  - Pollen
  - Hair
  - *Alternaria*
  - Other

1984; Kramer *et al.*, 1984), and sulfonamide crystals may be observed after therapy with these drugs.

5. *Miscellaneous Components*

Other constituents of urine sediment that are considered normal include lipid droplets, which are particularly common in dog and cat urine. Spermatozoa are normal in urine from the male and may be present in urine obtained from dogs by cystocentesis (Hubbert, 1972). Mucous threads and amorphous debris vary considerably from one urine sample to another within a species. No specific pathologic significance seems to be attached to their presence.

6. *Infectious Agents*

Infectious agents may appear in the urine as a consequence of contamination of normal urine during collection, infection localized to the genitourinary tract, or occasionally polysystemic infection, which includes the urinary tract.

The presence of bacteria in urinary sediment must be interpreted with caution because of possible contamination of urine during or after procurement. Conversely, cocci are difficult to visualize by examination of urinary sediment and their presence may become apparent only with culture of the urine. Mycotic agents in the urine are usually contaminants or evidence of polysystemic infection.

Parasite ova in the urine usually present no diagnostic problem. Occasionally, ova of enteric origin may contaminate voided urine and must be distinguishable from the parasites listed in Table 17.5.

VI. TESTS OF RENAL FUNCTION

A. Tests for Azotemia

Blood, serum, or plasma urea (expressed as urea, or urea nitrogen) and serum or plasma creatinine concentrations (SC) are used as indices of retention of nitrogenous wastes by the kidneys. The quantity of urea in blood may be expressed in terms of the entire urea molecule (blood urea) or in terms of only the nitrogen portion (BUN). Blood urea and BUN levels are related on the basis of molecular weight (60 versus 28, respectively). Thus, BUN values are only 0.47 of blood urea levels.

An understanding of the genesis, metabolism, and excretion of urea and creatinine should aid in their assessment as renal function tests. The subsequent discussion summarizes information obtained in domestic animals and humans.

1. *Urea Metabolism*

Small quantities of urea are ingested in animal foodstuffs because urea is present in all tissues of mammals. Urea also enters the gut in pancreatic secretions (Altman and Dittmer, 1971) and secondary to water movement into the intestines. Little or no urea is found in the stool of azotemic human beings (Brown *et al.*, 1971), so urea is either well absorbed or it is metabolized as it passes through the gastrointestinal tract.

Urea synthesis provides a mechanism for excretion of ammonium. The urea cycle incorporates two molecules of ammonium into each urea molecule. Formation of urea is an energy-requiring reaction, which occurs almost exclusively in the liver (White *et al.*, 1973). Although apparently rare in domestic animals, biochemical defects in the urea cycle were found in a dog with signs of hyperammonemia (Strombeck *et al.*, 1975).

Protein is the major source of ammonium for urea synthesis, and thus the rate of urea formation depends on the rate of protein (i.e., amino acid) catabolism. An increase in BUN may reflect an accelerated rate of protein catabolism rather than decreased urinary excretion of urea. In a study involving 14 clinically normal dogs fed diets of known protein content for 3 days, blood urea decreased from 5.71 mmol/liter (BUN = 16.0 mg/dl) with an 8.5% protein diet to 4.89 mmol/liter (BUN = 13.7 mg/dl) with a 5% protein diet. Preprandial and 4-hour postprandial studies indicated that blood urea increased from 4.89 to 5.71 mmol/liter (BUN = 13.7 to 16.0 mg/dl) with the 5% protein diet and from 5.71 to 9.57 mmol/liter (BUN = 16.0 to 26.8 mg/dl) with the 8.5% protein diet (Anderson and Edney, 1969). In another study, blood urea was elevated for 10–18 hours after food ingestion (Street *et al.*, 1968). The degree of elevation of blood urea in these studies was probably not clinically important since the normal canine range of 3.57–10.71 mmol/liter (BUN = 10–30 mg/dl) was not exceeded. However, higher protein diets could feasibly cause clinically important increases, and thus an 18-hour fast has been recommended to eliminate the influence of diet on BUN (Street *et al.*, 1968). Patients with marginal renal function and impaired urea clearance may have a greater increase in BUN as a result of diet than normal animals. For this reason, patients with an elevation of BUN that decreases into the normal range with fasting should have their renal function evaluated with more sensitive tests.

The rate of urea formation may be more rapid in cattle than in dogs or people. Nephrectomized cattle had a more rapid increase in BUN than was reported for other species (Watts and Campbell, 1970). The role that rumen metabolism of urea played in the BUN values has not been established.

Any process inducing protein catabolism can result in an increased BUN. In man, specific causes include (1) hemorrhage into the small bowel with digestion, absorption, and catabolism of amino acids; (2) fever; (3) burns; (4) corticosteroid administration; (5) starvation; (6) infection; and (7) tetracycline administration (Dossetor, 1966; Kopple and Coburn, 1974; Shils, 1963). The magnitude of increase in BUN associated with enteric hemorrhage is apparently related to the amount of hemorrhage. Hypovolemia associated with hemorrhage or dehydration may lead to concomitant prerenal retention of nitrogenous wastes, as previously discussed (Section III). Administration of large amounts of whole blood by gavage to dogs with reduced renal function caused a detectible but not massive increase in BUN (Finco, 1995), suggesting a minor role for intestinal hemmorhage as a cause of prerenal azotemia in dogs. Fever apparently results in increased BUN because of increased tissue catabolism. The magnitude of increase in BUN with burns is infrequently stated, but one human being with 50% third-degree burns had blood urea values of 21.4–28.6 mmol/liter (BUN of 60–80 mg/dl) coincident with a normal SC (Dossetor, 1966).

The effect of corticosteroids on BUN was determined in a study comparing azotemic human renal transplant recipients given prednisolone with chronic renal failure patients of comparable dysfunction who had not been given corticosteroids. The SC was from 177 to more than 442 mmol/liter (2–5 mg/dl) in both groups. Patients treated with an average of 70 mg prednisolone per day had BUN values from 1.5 to 2 times that of azotemic patients not given corticosteroids (Kopple and Coburn, 1974). Dogs with a transplanted kidney treated with prednisone and azathioprine had larger BUN-to-creatinine ratios than dogs with induced renal failure, suggesting that catabolic drugs or catabolic events affect BUN in dogs as well (Finco *et al.*, 1984).

Starvation was reported in a man in which SC was 62 mmol/liter (0.7 mg/dl) and blood urea was 89.3 mmol/liter (BUN 250 mg/dl). The BUN decreased to normal with an adequate diet (Kumar *et al.*, 1972).

Administration of tetracycline to azotemic human beings was followed by acidosis, anorexia, vomiting, and weight loss. The BUN and serum phosphorous concentration increased, but the SC remained unchanged. Patients without renal compromise given tetracycline at the same dosage did not develop these symptoms. Because SC was not altered by tetracycline in the azotemic group, it was concluded that the increase in BUN was due to protein catabolism rather then nephrotoxicosis (Shils, 1963). Six dogs with impaired renal function given tetracycline at a dose of 50 mg/kg body weight TID had no significant change in BUN, although two dogs had moderate increases in both BUN and creatinine while remaining clinically normal. These results were interpreted as reason for caution when using tetracycline in dogs with azotemia, because of potential renal effects (Finco, 1995).

Nonrenal factors reported to cause decreased BUN values are (1) anabolic steroids, (2) diminished protein intake, and (3) severe hepatic insufficiency. Anabolic steroids apparently prevent the increase in BUN associated with the use of tetracycline in man (Shils, 1963).

The ingestion of decreased quantities of high-quality protein results in decreased BUN because nearly all ingested protein is used for protein synthesis. However, inadequate caloric intake coincident with low protein intake may not result in decreased BUN because the protein ingested may be catabolized for energy. Conversely, intravenous administration of large quantities of amino acids and adequate amounts of nonprotein sources of energy to azotemic human beings and dogs results in anabolism and decreased BUN (Abel *et al.*, 1973; Van Buren *et al.*, 1972). Thus, the fate of amino acids (i.e., protein anabolism versus catabolism) as well as their quantity must be considered in relating diet to BUN.

Acid–base status may influence the rate of BUN increase in acute renal failure. Rats made azotemic by ureteral ligation that had acidosis prevented (blood pH 7.32) by treatment with sodium bicarbonate had blood urea values of 79.6 mmol/liter (BUN 223 mg/dl). Untreated rats and physiological saline-treated rats had blood pH values of 7.04 and blood urea values of 106.7 and 108.2 mmol/liter, respectively (BUN 299 and 300 mg/dl, respectively) (Simon and Luke, 1971). Other studies in rats documented that acidosis stimulated protein catabolism *in vivo* as well as in isolated muscle preparations (Mitch *et al.*, 1994).

Vigorous prolonged exercise (cross-country ski racing) in human beings caused about a 60% increase in BUN. During racing, production rates of urea increased 60–80% over resting values, and urinary excretion of urea was reduced (Refsum and Stromme, 1974).

Once formed, urea is carried from the liver via the vascular system and passively diffuses throughout total body water. Urea administered to nephrectomized dogs reached equilibrium with body fluids within 5400 seconds (1.5 hours) (Schloerb, 1960). Brains of dogs with induced acute uremia had equal blood and brain concentrations of urea (Arieff *et al.*, 1975), indicating that urea passes the blood-brain barrier in the azotemic patient.

2. *Urea Excretion*

a. Extrarenal

Most urea is excreted by the kidneys, but lesser routes of excretion have been identified. In clinically normal human beings, studies with isotope-labeled urea indicated that at least 25% of the urea synthesized is subsequently degraded by enteric organisms (Walser and Bodenlos, 1959). Bacterial ureases catalyze the breakdown of urea to ammonia and carbon dioxide. When enteric bacteria were killed by oral administration of neomycin, nearly all ^{14}C-labeled urea given orally was recovered in the urine. In azotemic human beings, the number of strains of anaerobic bacteria that produce urease increases as BUN increases (Brown *et al.*, 1971). The hydrolysis of urea, and presumably hydrogen ion trapping by ammonia, results in a pH of the colon that is higher in acidotic, uremic human beings than in clinically normal human beings (Brown *et al.*, 1971).

Although these findings would imply that the gastrointestinal tract serves as a significant route of exit of urea from the body, such is not the case. Nearly all ammonia from bacterial degradation is used to resynthesize urea in the liver. Thus, a futile cycle occurs, and enteric loss of urea because of bacterial degradation is minimal (Mitch *et al.*, 1975).

Such a futile cycle does not occur during ruminant digestion. In ruminants, ammonia derived from urea may be utilized to form amino acids from keto and hydroxy acid precursors that are normal products of digestion (Houpt, 1959).

In man, small quantities of urea may be lost via perspiration (Altman and Dittmer, 1971). Prior to the use of hemodialysis, induction of profuse sweating was considered an adjuvant treatment of uremia (Sobel, 1964). Because of poorly developed sweat glands in the dog and cat, urea loss by perspiration is probably negligible.

Severe vomiting and diarrhea have been suggested as mechanisms by which urea may be lost from the body (Dossetor, 1966). Quantitative data on loss by these routes are not available.

b. Renal

Urea appears in glomerular filtrate in the same concentration as in plasma as a result of simple filtration through the glomerular basement membrane (Smith, 1951). However, not all filtered urea is excreted in the urine, owing to passive reabsorption in the collecting ducts. It is important clinically to note that the amount of filtered urea that is reabsorbed is dependent on the rate of water movement through the collecting ducts. More rapid passage of fluid allows less reabsorption of urea. At the highest urine flow rate attainable in the dog, the fractional excretion of urea is about 60%, but it decreases as urine flow rate decreases (Shannon, 1936). Thus, changing urine flow rate can influence the FE for urea and BUN concentration independent of changes in GFR. Dehydrated patients with renal failure that are treated with large quantities of fluids may experience a large decrease in BUN unrelated to changes in GFR, merely because the FE for urea has increased.

Urea reabsorbed from filtrate enters the interstitium and the general circulation via the renal vasculature. Its role in the interstitium as a urine concentrating agent is discussed in Section II.I.

3. *Creatinine Metabolism*

Small quantities of creatinine are ingested by animals consuming diets containing animal tissues, since creatinine is distributed throughout body water. Muscle tissue contains phosphocreatine, which undergoes spontaneous cyclization with loss of inorganic phosphate to form creatinine. Creatinine is apparently well absorbed from the intestinal tract (Goldman, 1954).

Most of the creatinine eventually excreted originates from endogenous creatine. The amino acids arginine and glycine combine to form guanidinoacetate in the pancreas, kidneys, and small intestine. In the liver, methionine provides a methyl group for conversion of guanidinoacetate to creatine (Tyler, 1972). Creatine circulates in the plasma and is taken up by muscle, where it stores energy in the form of phosphocreatine. The serum concentration of creatine in man is about 34 mol/liter (0.45 mg/dl), but values for domestic animals are not available (Altman and Dittmer, 1971).

Creatine undergoes no catabolic reactions other than decomposition to creatinine (Bloch *et al.*, 1941). Creatine is excreted in small quantities in the urine of man (Bleiler and Schell, 1962; Bloch *et al.*, 1941; Hoberman *et al.*, 1948), but older literature indicates that large quantities of creatine are present in the urine of the dog (Hawk, 1910). Another study indicates that creatine clearance in the dog approaches zero at low plasma concentrations (Pitts, 1934). These conflicting data indicate that conclusions concerning urinary excretion of creatine by the dog cannot be made. This is important since creatine is the sole precursor of creatinine; an understanding of creatine excretion is needed to evaluate its effect on creatinine metabolism.

Creatine conversion to creatinine is a nonenzymatic, irreversible process that occurs at a rate of 1.6–2% daily (Bloch *et al.*, 1941; Hoberman *et al.*, 1948). Creatinine is distributed throughout body water, but it diffuses more slowly than urea and requires about 14,400 seconds (4 hours) for equilibration in the dog (Schloerb, 1960).

The quantity of creatinine formed each day depends on the total body content of creatine, which in turn depends on dietary intake, rate of synthesis of creatine, and muscle mass. Withdrawal of meat from the diet of clinically normal human beings for a period of 6 weeks resulted in a decrease in SC of about 25% (Bleiler and Schell, 1962). However, dietary intake of creatine may inhibit its endogenous synthesis (Walker, 1960, 1961). Severe, prolonged exercise of man caused an increase in SC and BUN of about 60% (Refsum and Stromme, 1974). Because urine creatinine excretion during the exercise period was normal, it was concluded that the increased SC was due to increased creatinine production.

Factors influencing muscle mass such as disease of muscle, tissue wasting, and character of muscle as influenced by physical training, may also affect the size of the creatine pool and thus the daily production of creatinine. Horses with myoglobinuria were reported to have higher SC values than anticipated from the degree of renal dysfunction, because of rhabdomyolysis (Koterba and Coffman, 1981).

Even in patients with normal renal function, a direct relationship seems to exist between muscle mass and SC. The SC level is 30% higher in clinically normal human males than in females (Doolan *et al.*, 1962). A SC concentration of 133 mol/liter (1.5 mg/dl) may be normal in a muscular working animal but may indicate impairment of renal function in a sedentary one. The decrease in SC associated with loss of muscle mass was termed "significant" (Doolan *et al.*, 1962; Newman, 1971), but specific values were not given.

Some diurnal variation in SC occurs in human beings. In one study, afternoon samples were about 10% higher than morning values (Doolan *et al.*, 1962).

4. Creatinine Excretion

a. Extrarenal

The concentration of creatinine in sweat ranges from 9 to 115 mol/liter (0.1 to 1.3 mg/dl) (Altman and Dittmer, 1971). Lowered ratios of BUN to SC in uremic human beings who are vomiting profusely (Dossetor, 1966) suggest proportionately less creatinine than urea in vomit, but quantitative studies have not been reported. Creatinine is absent from the feces of uremic human beings, probably because of intestinal reabsorption or bacterial metabolism or both. Bacterial metabolism of creatinine appears to be a major method of removal from the body. Colonic flora of the rat can be induced to catabolize creatinine (Jones and Burnett, 1972). Isotope studies in azotemic human beings revealed that 15.9–65.7% of the creatinine formed was metabolized or excreted via nonrenal routes (Jones and Burnett, 1974). Metabolic products identified were CO_2 and methylamine. More recently, recycling of some creatinine breakdown products back to creatinine was found in uremic human beings (Mitch *et al.*, 1980), but the amount of recycling appears to be much less than what occurs with urea.

b. Renal

In all mammalian species, creatinine is freely filtered through the glomerulus and appears in glomerular filtrate in the same concentration as plasma (Smith, 1951). Species differences exist with regard to tubular action on creatinine. In man, creatinine is secreted by renal tubules. In the male dog, an extremely weak proximal tubular secretory mechanism exists for creatinine (O'Connell *et al.*, 1962; Swanson and Hakim, 1962). The mechanism is absent in the female dog. Studies in cats (Finco and Barsanti, 1982) and ponies (Finco and Groves, 1985) indicated that no tubular secretion of creatinine occurred in these species, whereas considerable tubular secretion of creatinine occurred in the goat (Brown and Finco, 1987). In species secreting creatinine, the process is by active transport in the proximal tubule; thus tubular secretion of creatinine is not influenced by urine flow rate.

5. Precision of BUN and SC Measurement

Two general techniques are used for urea analysis (Davidsohn and Henry, 1969). With enzymatic methods, urea is converted to ammonia by urease. The ammonia concentration is measured by a color reaction. These methods are specific because urease acts only on urea. Free blood ammonium concentration does not interfere since it is present in such low concentration compared to urea. The other general method of urea analysis is based on the formation of colored products from direct combination of urea with compounds such as diacetylmonoxamine. This method is less specific, and dibasic amino acids may also cause a color reaction. With spectrophotometric methods of urea analysis, the two methods give comparable results when blood urea is <36 mmol/liter (BUN <100 mg/dl). Above this value, results with the diacetylmonoxamine method are 5–15% higher. A dipstick method of estimating blood urea by the urease method is available, but the accuracy is poor. It has been recommended that dipsticks be used only for emergency analyses and that samples reading greater than 7.1 mmol/liter for blood urea (20 mg/dl for BUN) be analyzed by more reliable methods (Bold *et al.*, 1970).

Clinical measurement of SC involves a reaction between alkaline picrate solution and creatinine. Several products of the creatinine–picric acid reaction have been proposed (Blass *et al.*, 1974). Unfortunately, the reaction is not specific; more than 50 compounds interfere, including glucose, fructose, ascorbic acid, hippuric acid, protein, urea, ketones, histidine, arginine, glycocyamidine, and pyruvic acid. Interfering compounds are collectively called *noncreatinine chromogens* (Henry, 1964; Newman, 1971; Narayanan and Appleton, 1972; Blass *et al.*, 1974).

More specific methods of measuring SC have been devised. A clay (Lloyd's reagent) can be used to absorb creatinine selectively prior to reaction with alkaline picrate, but the method is cumbersome (Henry, 1964). Enzyme-specific methods for creatinine analysis also are available and have been applied to the serum of domestic species (Jacobs *et al.*, 1991). The enzymatic methods are not in widespread usage presumably because of cost.

Noncreatinine chromogens were reported to be absent from the urine of the dog (Balint and Visy, 1965). However, values for urine creatinine in dogs were about 6% lower with an enzymatic creatinine assay than with the Jaffe method (Finco *et al.*, 1993).

The relative absence of urinary excretion of noncreatinine chromogens is important for two reasons. First, creatinine clearance in animals is erroneous if both urine and serum are analyzed by the alkaline picrate method. Urine creatinine concentration U_c is accurately measured, but plasma creatinine concentration P_c is erroneously high. Since the U_c/P_c ratio is calculated in making clearance determinations, apparent creatinine clearance will be low compared to actual clearance. Second, it indicates that extrarenal mechanisms exist for the elimination of noncreatinine chromogens. This implies that noncreatinine chromogen concentration in blood does not increase in parallel with creatinine during the progression of renal dysfunction. Consequently, SC determined by the alkaline picrate method becomes a progressively more accurate estimate of actual SC concentration as the degree of renal function decreases.

6. *Interpretation of BUN and SC Values*

Both BUN and SC are relatively insensitive in detecting renal dysfunction. The GFR must be reduced to 25% of normal or less before BUN values outside the normal range are obtained (Bernstein, 1965). There are no data to indicate that either compound is more sensitive than the other in detecting renal dysfunction.

In most domestic species, serum concentrations of both urea and creatinine are crude estimates of GFR.

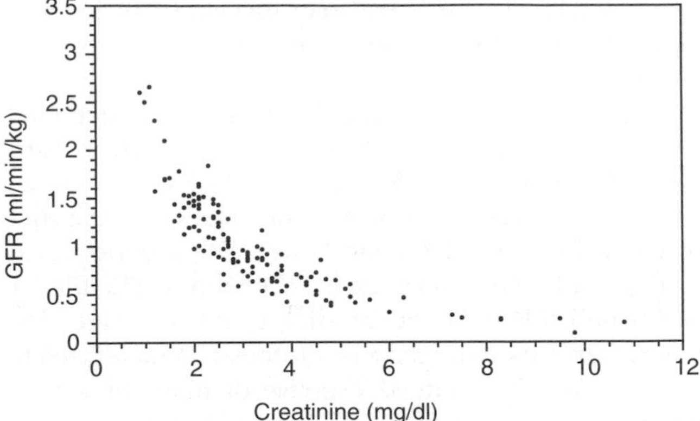

FIGURE 17.4 Relationship between serum creatinine concentration and GFR in 129 clinically normal dogs with reduced renal function. Comparison with Fig. 17.5 (BUN versus GFR in the same dogs) indicates that SC is a better predictor of GFR than BUN.

On the basis of the mechanism of renal excretion of the two compounds, creatinine is a better indicator of GFR (Figs. 17.4 and 17.5). However, studies in humans indicate that SC does not increase in proportion to the decrease in creatinine clearance (Doolan *et al.*, 1962; Enger and Blegen, 1964; Goldman, 1954). This discrepancy is attributable to the extrarenal losses of creatinine previously discussed and suggests that extrarenal losses increase progressively as the degree of azotemia increases. Studies in dogs also indicate that SC values do not increase in proportion to the decrease in GFR, but the deviation in SC from values predicted by GFR is not large (Finco *et al.*, 1995). The practical application of recognition of extrarenal creatinine loss is that a patient whose creatinine has doubled may have considerably less than half the previous GFR. It may also mean that two patients of the same size with the same SC may have different GFR values, if the two differ in extrarenal creatinine disposition. The same statements

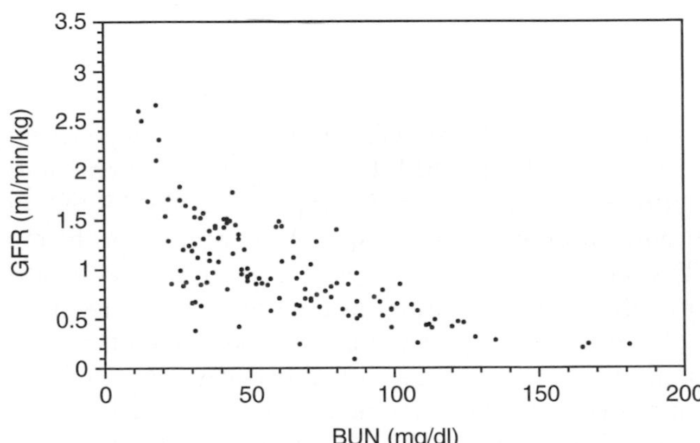

FIGURE 17.5 Relationship between GFR and BUN in the 129 clinically normal dogs with renal dysfunction.

may apply to BUN values and emphasize that measurement of serum levels alone is not a precise method of assessing renal function.

In man it is reported that BUN correlates more directly with symptoms of uremia than SC. Patients with a high BUN are more likely to have uremic symptoms than patients with a low BUN, even when the two groups have similar SC values (Kassirer, 1971).

Levels of BUN and SC should be interpreted while considering two types of errors. Concluding that renal disease does not exist when it actually does constitutes one error. Because of factors previously discussed, it should be apparent that animals with GFR values as low as 25% of normal may not have elevated BUN and SC values. Concluding that renal disease exists when it actually does not constitutes the second type of error. All causes of prerenal azotemia fall in this category. Human patients referred to a nephrology service were studied to compare GFR with BUN and SC values. Normal BUN and SC values were found in patients proved abnormal by GFR measurements in 26% and 22% of the evaluations, respectively. Abnormally high BUN and SC values were found in patients with normal renal function in 9% and 4% of the patients, respectively (Rickers *et al.*, 1978). Although similar statistics have not been generated for domestic animals, it is likely that both types of errors are made. The error of false-positive values occurs with readings that are slightly above the normal range. When such values are obtained, prerenal factors should be considered, and more sensitive tests of renal function should be considered.

Some effort has been made to use BUN-to-SC ratios to evaluate the pathogenesis of azotemia in human beings. However, studies in the dog revealed that BUN-to-SC ratios were of no value in differentiating prerenal, renal, or postrenal failure or in differentiating acute renal failure from chronic renal failure (Finco and Duncan, 1976). Because ruminant microflora may metabolize a greater percentage of urea than do enteric flora in nonruminants, BUN values may not increase proportionately to SC levels in renal failure in ruminants until anorexia occurs. This phenomenon should be considered when evaluating BUN values.

Progressive deterioration of renal function can be monitored with measurement of either BUN or SC, subject to the variables previously enumerated for each test. Observations of SC values from people in renal failure revealed that the reciprocal of SC declined linearly with time in many but not all patients as disease progressed (Mitch, 1985). The plot was used to predict when hemodialysis or transplantation was required or to monitor effects of diet on the progression of renal failure. Plotting of the reciprocal of SC against time in dogs with renal failure has resulted in linear relationships in some cases (Allen *et al.*, 1987) but not in others (Barsanti, 1987), suggesting that either extrarenal creatinine disposition or progression of renal failure is not invariably related to time in a linear fashion.

B. Urine Concentration Tests

Polydipsia and polyuria are related signs in which one factor is primary and the other is secondary (i.e., a response to the primary factor). Primary polydipsia with compensatory polyuria occurs in domestic animals. When water is restricted in such animals, urine volume decreases and urine concentration increases.

Primary polyuria is normally accompanied by secondary polydipsia. When water is restricted in such animals, neither urine volume nor urine concentration changes markedly.

Unfortunately, both renal and nonrenal causes of primary polyuria exist (Table 17.3). Nevertheless, urine concentration tests are valuable for distinguishing primary polydipsia from primary polyuria, and for aiding in differentiating causes of primary polyuria from one another.

1. *Indication for Urine Concentration Tests*

When polydipsia and polyuria exist, a urine concentration test may distinguish the primary factor from the compensatory one. For the diagnosis of generalized renal disease, the test is indicated when polyuria but not azotemia is present. As indicated in Section IV, polyuria precedes azotemia during the progression of chronic renal failure in most species, the cat being a notable exception (Ross and Finco, 1981). Urine concentration tests serve no purpose in animals with unequivocal azotemia since renal dysfunction has already been established.

2. *Types of Urine Concentration Tests*

Stimulus for the production of a concentrated urine may be elicited by inducing endogenous secretion of ADH by withholding water. ADH secretion occurs in dogs with a body water deficit of 2–3% (Pitts, 1974). Both abrupt and gradual water deprivation tests have been described. The gradual test has been advocated because of concern that any cause of polyuria may secondarily affect the concentrating mechanism by causing medullary washout of solute. Under such circumstances, production of concentrated urine would be impossible until the medullary hypertonicity was reestablished.

The amount of time required for the normal kidney to reestablish medullary hypertonicity is not well defined in domestic animals. One study in dogs found that repletion following water diuresis had not occured after 10 hours (Boylan and Asshauer, 1962). In dogs, washout reported as a consequence of furosemide occurred within 5 minutes of intravenous administration of the drug. On the other hand, not all causes of polyuria are associated with medullary washout. Dogs with induced diabetes insipidus maintain a degree of medullary hyperosmolality (Appelbloom *et al.*, 1965), and dogs with spontaneous diabetes insipidus respond to Pitressin tannate within 24 hours (Finco, 1979). Various other diseases in domestic animals have not been examined for the status of the renal medullary gradients, nor have comparative trials of abrupt and gradual water deprivation tests been made. Because the cost of hospitalization makes the gradual deprivation test expensive to perform and because its superiority has not been documented, the abrupt deprivation test is a logical choice over the gradual test for initial study. If the animal fails to concentrate urine with the abrupt test, a gradual deprivation test may be considered.

An alternative form of urine concentration test is performed by supplying an exogenous source of ADH. Regardless of the animal's state of hydration, concentrated urine should be produced in the presence of ADH if the kidney is normal. Aqueous pitressin or DDAVP (1-deamino-8-D-arginine-vasopressin) may be used for the test. Since the half-life of aqueous Pitressin is extremely short, constant infusion is required to maintain adequate blood levels for the test. This complicates the test compared to the use of DDAVP. The question of secondary washout of medullary gradient is more germane to a concentration test performed with aqueous Pitressin during minutes compared to conducting a DDAVP test over a period of many hours. The DDAVP has activity in the dog for about 12 hours after mucosal application (Greene *et al.*, 1979).

The combined use of water deprivation and Pitressin in a single procedure has been advocated for the diagnosis of polyuria problems in both the human being and the dog (Mulnix *et al.*, 1976).

3. Methods of Conducting Urine Concentration Tests

The various tests for urine concentration probably all have merit if conducted carefully and interpreted properly. The tests subsequently described are those the author has found to be satisfactory in dogs and cats.

a. Abrupt Water Deprivation Test

This test is conducted so that a definite stimulus for ADH release is elicited prior to the conclusion of the test. In the past, the procedure involved withholding water for a specific number of hours and then determining the degree of urine concentration. This technique is inaccurate because of variation among patients in rate of dehydration following water deprivation. With this technique some patients may be dehydrated to dangerous degrees and others may not be challenged. The preferable end point is a specific degree of dehydration rather than a specified time period of water deprivation.

Prior to conducting the test, it is necessary to evaluate the animal's state of hydration. Performing a water deprivation test in a dehydrated animal is meaningless and dangerous. It is meaningless because clinically detectable dehydration is usually 5% or greater (Finco, 1972), and such an animal would already be secreting ADH and concentrating urine if capable. It is dangerous because imposing water deprivation on an animal that is already dehydrated may lead to complications of hypovolemia. Performing a water deprivation test in an azotemic animal is also contraindicated. The test gives no added information in cases of primary renal azotemia, and dehydration aggravates the azotemia by superimposing a prerenal factor on the underlying disease. Patients with prerenal azotemia are usually producing a concentrated urine spontaneously.

The stepwise procedure for conducting the abrupt water deprivation test in the dog or cat is the following:

1. While still providing water, withhold food for 4 hours.
2. After the fast, obtain a blood sample for measurement of packed cell volume (PCV), total plasma solids, and plasma osmolality if feasible. Obtain urine for refractometry and measurement of osmolality if feasible. Osmolality determinations should be deferred until the end of the test, when all samples can be analyzed at one time.
3. Weigh the animal with an accurate balance scale with the urinary bladder empty.
4. Remove all sources of drinking water. Fasting is preferable, but dry food may be given if the period of deprivation is prolonged.
5. Evaluate the patient at appropriate intervals for evidence of dehydration. Body weight measurements are the primary guide. Plasma total solids, PCV determinations, and clinical signs are less accurate and should be used only as secondary indices. Appropriate time intervals between observations vary considerably among patients; times can be chosen based on weight changes that occur during the early stages of the test. The first observation after initiating the test should be made at 14,400 seconds (4 hours). All body weight measurements should be made with the bladder empty, and consideration should be given to the weight changes that occur from eating, defecation, or other losses.

6. Terminate the test when 5–7% of body weight is lost or if undesirable signs develop during the test. Since urine produced during maximal dehydration should give the best indication of concentrating ability, it is desirable to empty the animal's bladder once sufficient weight loss has occurred and to wait for newly formed urine for analysis. Both plasma and urine are obtained for analysis (osmolality, refractometry) at the conclusion of the procedure. The interpretation of results of the test is discussed subsequently.

b. The 1-Deamino-8-D-Arginine Vasopressin (Desmopressin) Test

DDAVP is a synthetic vasopressin and it may be used in instances when it may be risky to deprive the patient of water. The test is conducted as follows:

1. Water need not be withheld for the test, but overnight fasting is preferable.
2. The same preinjection measurements listed for the abrupt water deprivation test are made.
3. The DDAVP is administered slowly intravenously at a dose of 3–10 μg, depending on the size of the dog or cat.
4. The animal is allowed water ad libitum during the study. Although not essential, comparison of water intake and urine volume for 24-hour periods before and after injection may aid in the interpretation of results.
5. Urine is obtained at 1, 2, 4, 6, 12, and 24 hours.

4. *Interpretation of Results of Urine Concentration Tests*

The same results would be anticipated whether exogenous or endogenous ADH were utilized, but some studies indicate that urine-concentrating ability is better after water deprivation than after injection of a long-acting Pitressin preparation (West *et al.*, 1955). Results with DDAVP would be expected to be comparable to the Pitressin preparation. Although not confirmed by controlled studies, it is the author's impression that the water deprivation test is a more reliable test than the DDAVP test. In the past, a urine specific gravity of 1.025 has been considered to be an adequate indication of concentrating ability for the dog and cat (Osborne *et al.*, 1972). This value was transposed from studies of human beings, whose maximal concentrating ability is about 1.040. Studies on normal beagle dogs indicated that 57,600 seconds (16 hours) or more of water deprivation resulted in specific gravity readings of 1.058 ± 0.008 (Balazs *et al.*, 1971). In another study of 20 dogs, mean concentrating ability was 1.062 ± 0.007 for specific gravity and 2289 ± 251 mOsm/kg for osmolality after a mean of 141,480 seconds (39.3 hours) of deprivation that resulted in 4–16% weight loss (Hardy and Osborne, 1979).

In studies of normal cats fed commercial dry catfood, water deprivation causing 5–8% loss of body weight resulted in urine osmolality of 2196 ± 533 mOsm/kg and urine specific gravity of 1.064 ± 0.015 (Ross and Finco, 1981). However, whether values obtained from abrupt deprivation of normal animals would be the same as values from normal animals that had polyuria prior to deprivation has not been established. Another problem is the influence of diet on concentrating ability, as mentioned in Section II.

Any guidelines to be used for differentiating normal from abnormal concentrating ability must be tentative until more information is collected. At present, for the dog and the cat with the use of the abrupt water deprivation test, urine specific gravity readings of 1.040 or above are considered evidence of normal concentrating ability. Readings between 1.030 and 1.040 are considered questionable and worthy of further investigation. Readings below 1.030 are considered abnormal. For the DDAVP test, values slightly lower (0.005) may be used.

If osmolality readings are made, pretest and posttest plasma samples may be compared for the water deprivation test in order to document the adequacy of water deprivation. Plasma osmolality should be higher in the post-test sample. Food ingestion seems to markedly alter plasma osmolality results, and thus at least a 4-hour fast is recommended prior to sampling. Posttest urine and plasma osmolality ratios may also be examined to establish the effectiveness of the renal concentrating mechanism.

C. Clearance Methods

The development of the clearance concept represented a major advancement in the measurement of renal function (Smith, 1951). Not only could specific renal functions be measured quantitatively, but a marked increase in sensitivity in detecting renal dysfunction was achieved. When performed properly, clearance procedures should detect a 20% decrement in renal function, compared to a 75% decrement required before BUN and SC change.

The clearance formula is expressed as follows

$$C = \frac{U_v U_c}{P_c}$$

where C is clearance; U_v, urine volume; U_c, urine concentration; and P_c, plasma concentration.

Physiology textbooks can be consulted for a more detailed description of the clearance concept and its application. Data have accumulated to indicate that inulin can be used to measure GFR, and *p*-

aminohippuric acid (PAHA) at low blood levels can be used to measure renal plasma flow. Although the clearance formula and inulin and PAHA clearances are standards by which other measures of renal function are judged, they are too time consuming for routine clinical use in veterinary medicine. Since both inulin and PAHA are foreign to the body, it is necessary to induce and maintain constant blood levels by giving priming doses followed by constant infusion of solutions. In research studies on dogs and cats, the determination of inulin and creatinine clearances requires several hours of human effort per animal.

1. *Importance of Hydration State*

Studies have demonstrated a significant effect of state of hydration on renal functions. In one study, creatinine clearance in clinically normal euhydrated dogs was increased by 14% when dogs received water (equal to 3% body weight) by gavage 40 minutes prior to clearance procedures. In the same dogs, clearance was decreased by 23% from the euhydrated state when dogs had dehydration induced (Tabaru *et al.*, 1993). Such changes in renal function according to the state of hydration probably affect scintigraphic and single-injection techniques of renal function evaluation as well as clearance procedures. Since clinical assessment of hydration state is insensitive, it is possible that many patients with marginal hydration yield spuriously low results from renal function tests. For short-term clearance determinations it is appropriate to administer water routinely by gavage (3% body weight) to avoid volume depletion during the procedure. Increased urine flow rate achieved by water administration also increases the reliability of the urine collections.

2. *Endogenous Creatinine Clearance*

In contrast to clearance methods previously mentioned, the endogenous creatinine clearance is feasible with clinic cases. Blood levels of creatinine are relatively constant over short intervals of time unless some factor acutely affects renal function. Thus, the blood level of creatinine can be measured once while an accurately timed collection of urine is made. Both 20-minute and 24-hour endogenous creatinine clearance values have been determined for the dog (Finco, 1971; Bovee and Joyce, 1979). Twenty-minute clearance determinations performed on 27 normal dogs of both sexes weighing from 5.6 to 31.3 kg gave values of 2.8 ± 0.96 ml/minute per kilogram of body weight or 60.0 ± 21.9 ml/minute per square meter of body surface area (Finco, 1971). Comparable values are obtained on 20-minute and 24-hour collections (Finco *et al.*, 1993). Discrepancies between laboratories have been noted and may be due to several factors. One important factor is the chromogen problem with creatinine measurement discussed in Section VI.A.5.

Because noncreatinine chromogens are present in plasma but absent in urine, endogenous creatinine clearance measurements underestimate the actual clearance of the compound. In one study, when endogenous and exogenous creatinine clearances were compared with inulin clearance in 10 healthy dogs, exogenous creatinine clearance was 1.03 times inulin clearance, but endogenous creatinine clearance was only 0.70 times inulin clearance (Finco *et al.*, 1981). This difference occurred despite the use of a kinetic method for creatinine analysis presumed to be specific for creatinine. Thus it is likely that values for endogenous creatinine clearance underestimate GFR in the dog, and probably other species that have insignificant tubular secretion of creatinine such as the cat and the pony.

Although endogenous creatinine clearance may not accurately measure GFR, it still has utility as a renal function test because it is a more sensitive indicator of renal function than BUN or SC measurements. However, meaningful interpretation depends on a knowledge of normal values for clearance with the method of creatinine analysis employed.

More recently, an enzymatic method of creatinine analysis was employed during clearance tests to determine if the error inherent in the Jaffe method could be eliminated (Finco *et al.*, 1993). With this method, values for inulin clearance and endogenous creatinine clearance in dogs were nearly identical (Fig 17.6). This study indicates that at least in dogs and possibly in other species that do not have tubular secretion of creatinine (cats, horses), endogenous creatinine clearance accu-

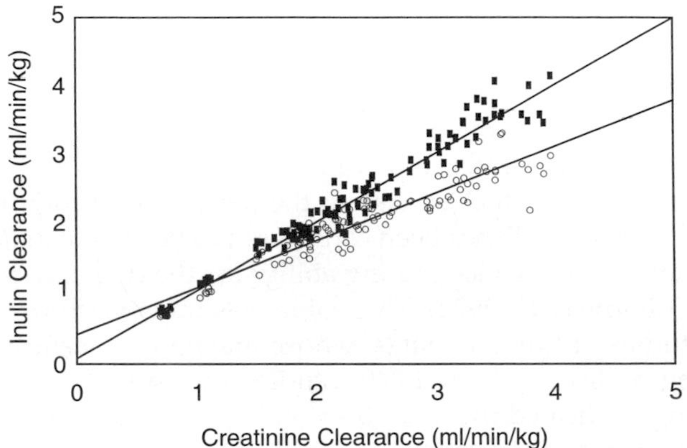

FIGURE 17.6 Relationship between inulin clearance and endogenous creatinine clearance in dogs with normal and impaired renal function. Analysis of creatinine by the Jaffe method (open circles) gave clearance values less than inulin, whereas analysis by a specific, enzymatic method for creatinine (solid squares) gave nearly identical results for creatinine and inulin.

rately measures GFR when specific (enzymatic) methods are used for creatinine analysis.

The technique for performing endogenous creatinine clearance is simple, but certain precautions are necessary. The test is dependent on total collection of urine during the time interval that is chosen. If urine is erroneously left in the bladder at the beginning of the test, creatinine will be measured that was not cleared from the blood during the test time chosen. Clearance values will be erroneously high. If all urine is not removed from the bladder at the termination of the test, not all creatinine cleared from the blood during the test period will be measured, and clearance values will be erroneously low. The precision required in emptying the bladder increases progressively as the time period for collection is decreased. When the 20-minute clearance test is performed, it is necessary to rinse the bladder several times with sterile saline both prior to initiating the test and at its conclusion. The rinses prior to the start are discarded; the rinses made prior to the conclusion of the test are added to the urine that is collected. The amount of creatinine excreted during the test period is the product of creatinine concentration of the rinse–urine mixture times its volume. Clearance is calculated by dividing urine creatinine excretion by plasma creatinine concentration.

3. *Exogenous Creatinine Clearance*

In dogs, cats, and horses, administration of creatinine to obtain constant, elevated blood levels is a proven way of accurately measuring GFR with the Jaffe method of creatinine analysis in normal animals (Smith, 1951) and in animals with renal dysfunction (Finco *et al.*, 1991, Fig. 17.7). By elevating plasma concentration of creatinine, the error due to noncreatinine chromogens becomes negligible. Since constant intravenous infusions make clearance procedures complicated, subcutaneous administration of creatinine has been employed (Finco *et al.*, 1981). Measurement of plasma creatinine concentration after subcutaneous injection revealed that levels were reasonably constant, and comparison of clearance values with those of inulin infused intravenously indicated that the creatinine clearance values were valid measures of GFR. Directions for application of this method to clinical practice have been published (Finco *et al.*, 1982).

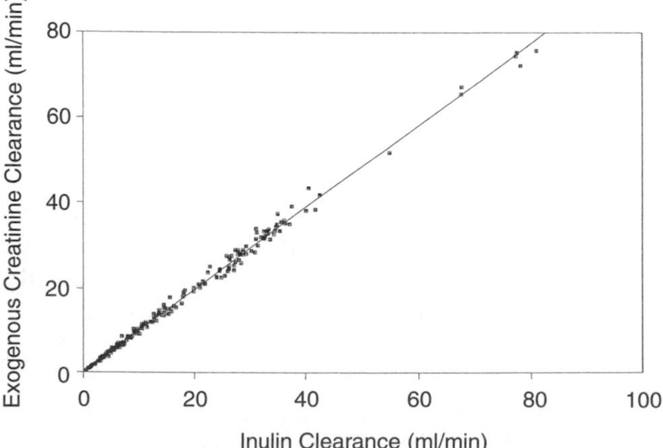

FIGURE 17.7 Relationship between inulin clearance and exogenous creatinine clearance in male and female dogs with subnormal to normal renal function. Exogenous creatinine clearance was the same as inulin clearance in both male and female dogs (Finco *et al.*, 1991).

D. Quantitative Renal Scintigraphy

Quantitative renal scintigraphy utilizes a strongly γ-emitting radioisotope, tagged to a compound handled by the kidney in a known manner. The compound is injected intravenously and the accumulation of the isotope in the kidneys is monitored with a detector placed over the kidneys. The detector simultaneously monitors plasma concentration of the isotope. The compound ^{99m}Tc-diethylenetriaminepentaacetic acid (DTPA) often is used to measure renal function by this technique. It is excreted by the kidney exclusively by glomerular filtration, and thus can be used to measure GFR. During the early phase of renal scintigraphy, accumulation of the isotope in the kidney is a function of glomerular filtration and plasma concentration of the isotope. Because both are being measured at many instants in time, values for GFR for each kidney can be generated, using classic clearance measurements of inulin for validation and calibration (Cowgill and Hornof, 1986 Krawiec *et al.*, 1986 Uribe *et al.*, 1992). In addition, information about kidney size can be obtained. A computer algorithm was developed for dogs to define kidney borders accurately (Cowgill and Hornof, 1986). This method of evaluating the kidneys has the advantages of rapid, noninvasive measurements, and assessment of both size and function of each kidney separately. In addition to use of DTPA for estimation of GFR in both the dog and cat, ^{99m}Tc-mercaptoacetyltriglycine has been used to estimate renal plasma flow (RPF) in the dog (Itkin *et al.*, 1994). Unfortunately, in some studies the correlation between these methods and classic clearance methods has not been impressive. In addition, cost of the equipment and isotope restriction are disadvantages.

E. Single-Injection Techniques

Following a single injection of a compound into the body, blood concentration is dictated by its distribution, metabolism, and excretion. For nonmetabolizable materials excreted only by the kidneys, rate of disappearance of the material from the blood is related to renal function. This is the basis for single-injection techniques devised to measure renal function.

Several methods have been employed to analyze data from single-injection techniques. One method is to determine plasma half-life of the injected material, once its distribution in the body has occurred. Results from this method provide a value for half-life that cannot be converted to a numerical value for classical renal functions (GFR, RPF) but which may have clinical utility when compared to half-life values from normal animals. Compounds utilized in using this approach include sulfanilate and phenolsulfonphthalein (Rosenfeld, 1955 Mixner and Anderson, 1958 Brobst *et al.*, 1967 Carlson and Kaneko, 1971 Osbaldiston and Fuhrman, 1970).

Another approach to measuring renal function with a single injection of a test substance is to collect urine to determine what percent of the injected material has been excreted in a given amount of time. The urinary excretion of phenolsulfonphthalein in the 20-minute period after injection has been used in dogs to evaluate renal function (Finco, 1971). Such urine collection methods have the advantage over plasma decay methods of measuring only the renal contribution to removal of material from the blood, but have the disadvantage of requiring urinary catheterization and careful urine collection.

From data available, neither plasma half-life nor urine measurements following single injections of test substances seem as sensitive as classic clearance procedures for evaluating renal function.

In addition, pharmacokinetic modeling of plasma decay curves has been used to measure renal functions. Several models have been used and advocated, usually based on correlation between classic clearance values and clearances derived from the model computations. One study (Hall *et al.*, 1977) found that two methods advocated for calculation of GFR from single-injection decay curves (slope–intercept, two-compartment decay) gave spurious data, particularly when the volume of extracellular fluid was changed. In Hall's study, a third method (area under the curve) was devised using iodothalamate as the test substance. This method gave results that compared well with inulin clearance in 11 normal dogs and in 3 dogs with volume expansion. Similar methods have been used by others for both dogs and cats (Rogers *et al.*, 1991; Brown, 1994).

Some studies have used pharmacokinetic modeling to measure GFR, but failed to test the method by direct comparison with classic clearance techniques in normal animals, and in animals with abnormal extracellular fluid volumes (Powers *et al.*, 1977 Fettman *et al.*, 1985). A study in cats demonstrated a poor correlation between exogenous creatinine clearance and two pharmacokinetic models of plasma clearance of inulin (Rogers *et al.*, 1991). Validation of pharmacokinetic methods by comparison with classic clearance methods is required before such techniques can be considered acceptable measures of GFR.

Single-injection methods for measuring renal function have not had widespread use in veterinary practice because neither the test substances used nor their analysis for plasma concentration is routinely available. Iohexol is a nonmetabolizable radiocontrast medium that is excreted exclusively by glomerular filtration. It is readily available and has been used to determine GFR following a single injection into humans (Almen, 1994 Frennby *et al.*, 1994 Lindblad and Berg, 1994). An apparatus is commercially available which measures iohexol concentration and computes GFR from the measurements. Comparison of GFR measurements made by classic clearance methods with the single-injection method for iohexol gave good agreement (Brown *et al.*, 1995).

F. Urinary Enzymes

Increases in serum enzyme activity commonly are used to detect organ abnormalities. Such enzyme measurements are not true tests of function, but they are sensitive indicators of organ damage. Unfortunately, increases in serum enzyme activity that are specific for the kidney have not been documented secondary to renal damage.

The appearance of enzymes of renal origin in urine provides a unique opportunity for detection of renal injury. Studies on enzyme assays of urine for detection of renal injury have been reviewed (Mattenheimer, 1971, 1977 Raab, 1972 Dubach and Schmidt, 1978). Attention has been focused on enzymes present in the brush border of the proximal tubules and on enzymes in lysosomes. Factors complicating analysis are the presence of enzyme inhibitors and accelerators in urine that often must be removed prior to analysis. Of the domestic species, some results have been reported from the dog (Conzelman *et al.*, 1970 Ellis *et al.*, 1973 Greco *et al.*, 1985 Gossett *et al.*, 1987). However, results have not been encouraging for clinical application.

G. Single-Sample ("Spot") Fractional Excretion Determinations

As indicated in Section II.E., FE refers to the fraction of material filtered that appears in urine. The FE of any material reflects either the efforts of the kidney to maintain homeostasis or defects in its ability to do so. Several examples demonstrate the clinical utility of FE measurements. In the normal animal, the FE of an ingested element that is well absorbed from the intestines (i.e., sodium) is expected to increase if dietary intake increases, in order for body balance to be maintained. If GFR is decreased but intestinal absorption is constant, the FE of sodium and phosphorus must be increased if body balance is to be maintained. The increase in FE is required because the filtered load (GFR × plasma concentration) is decreased, and thus a smaller percent of the filtered material need be reabsorbed to maintain body balance. An animal with a tubular absorption defect (i.e., Fanconi syndrome) is not able to respond to needs for homeostasis, and FE values in such animals will be higher than is appropriate for the diet being consumed.

Values for FE are determined by measuring GFR, plasma concentration, and urinary excretion of the substance in question. Values published as "normal" may be extremely misleading unless dietary intake and intestinal absorption of the material for which FE is being determined are defined.

Because determination of GFR is laborious, the "spot" method used for quantifying proteinuria (Section V.C.) has been applied to FE determinations of other materials (Traver *et al.*, 1977). This procedure may have merit in determining tubular homeostatic activity, but several qualifications must be imposed. The SC is not a valid marker of GFR because of plasma noncreatinine chromogens (Section VI.A.5) and because in some species it is secreted by the tubules. Consequently, values obtained are not equal to true FE values. Random samples of urine taken without regard to the time of food intake may not provide samples representative of 24-hour urinary excretion. In dogs, urinary sodium excretion is greatest during the first 6 hours after eating and diminishes substantially subsequently. Urinary potassium and phosphorus excretion are low soon after eating, but increase between 6 and 18 hours after eating (Groves *et al.*, 1987). Random samples of urine represent an accumulation over an unknown period of time, further complicating intake-related fluctuations in urinary excretion. Plasma concentrations of some electrolytes also are influenced by time of eating (Groves *et al.*, 1987). The presence of circadian rhythms in excretion of urinary components has been poorly studied in domestic animals but if present could complicate "spot" clearances as well as short-term FE measurements by classical techniques. The presence of renal dysfunction dictates an increase in FE to maintain homeostasis failure to account for level of renal function may lead to misinterpretation of results (Adams *et al.*, 1991).

The utility of "spot" FE measurements will be determined when the influence of these factors is determined in each species. It is possible that results will be found to be more reliable in herbivores than in carnivores because food intake is more constant in herbivores.

H. A Perspective on Renal Function Tests and Their Use

In veterinary medicine, the question of whether renal function is normal or abnormal is of frequent concern. The question should be answered reasonably quickly and at the least cost to the owner of the animal. This implies that the evaluation should be simple and quick to do.

The relative simplicity of performing BUN or SC measurements and determination of the specific gravity of a random sample of urine makes them logical choices for initial assessment of renal function. Normal values for BUN or SC and production of a concentrated urine indicate that at least one-third of renal mass is functional. This quantity of tissue is adequate for existence without signs of renal failure, and probably is sufficient for maintaining an animal through most surgical procedures. Clinical investigation is often discontinued at this point unless there is a specific reason to question the state of the kidneys. However, it must be recognized that normal values for these measurements could still mean loss of up to two-thirds of renal function.

Elevated levels for BUN or SC indicate that prerenal, renal, or postrenal azotemia exists. When azotemia is found, there is little if any additional information attained from more sensitive renal function tests. Renal function tests do not differentiate these categories, and other means must be used to distinguish one from the other.

Normal values for BUN or SC accompanied by a low specific gravity on a random sample of urine rule out uremia but introduce the question of the cause of low urine specific gravity. In euhydrated animals, water deprivation or ADH administration is helpful in determining whether concentrating ability is present. However, lack of concentrating ability is not specific for generalized renal dysfunction, and other means must be used to differentiate nonrenal causes from generalized renal failure.

Tests other than BUN, SC, and urine concentration are indicated under a variety of circumstances. One instance is when a urine-concentrating defect exists, but values for BUN or SC are normal or equivocal. Another example may be when a potentially nephrotoxic drug is to be administered to an animal in which renal disease is suspected (glomerular proteinuria or urinary casts). The choice of the more sensitive test to be used in such instances depends on the use to be made of the data. If relative rather than absolute values for renal function are adequate, such tests as endogenous creatinine clearance or plasma decay curve half-life data may be adequate, as long as normal values are available for the hospital and laboratory in which they are to be used. When a specific measure of GFR is sought, creatinine clearance may be performed in the species in which it is valid (dog, cat, horse). Endogenous creatinine can be used if enzymatic methods of creatinine analysis are used; otherwise, exogenous creatinine clearance is required to overcome the error of the Jaffe method. Alternatives for GFR measurement are validated pharmacokinetic models or reliable renal scintigraphy.

References

Abel, R. M., Beck, C. H., Abbott, W. M., Ryan, J. A., Barnett, G. O., and Fischer, J. F. (1973). *N. Engl. J. Med.* **288,** 696–699.

Adams, L. G., Polzin, D. J., Osborne, C. A., and O''Brien, T. D. (1991). *Am. J. Vet. Res.,* **52,** 718–722.

Adams, L. G., Polzin, D. J., Osborne, C. A., and O'Brien, T. D. (1992). *J. Vet. Int. Med.* **6,** 36–40.

Agre, P., Preston, G. M., Smith, B. L., Jung, J. S., Raina, S., Moon, C., Guggino, W. B., and Nielsen, S. (1993). *Am. J. Physiol.* **265,** F463–F476.

Akmal, M., Telfer, N., Ansari, A. N., and Massry, S. G. (1985). *J. Clin. Invest.* **76,** 1695–1698.

Allen, T., Jaenke, R., and Fettman, M. (1987). *J. Am. Vet. Med. Assoc.* **190,** 866–868.

Almen, T. (1994). *Invest. Radiol.* **29,** S2–S6.

Altman, P. L., and Dittmer, D. S. (1971). *Fed. Am. Soc. Exp. Biol.,* Washington, D.C.

Anagnostou, A., and Kurtzman, N. (1986). *In* "The Kidney" (B. M. Brenner and F. C. Rector, eds.), pp. 1631–1656. Saunders, Philadelphia, Pennsylvania.

Anderson, R. J., Linas, S. L., Berns, A. S., Henrich, W. L., Miller, T. R., Gabow, P. A., and Schrier, R. W. (1977). *N. Engl. J. Med.* **296,** 1134–1138.

Anderson, R. S., and Edney, A. T. (1969). *Vet. Rec.* **84,** 348–349.

Anonymous (1965). *In* "Hawk's Physiological Chemistry" (B. L. Oser, ed.), pp. 1153–1205. McGraw-Hill, New York.

Appelbloom, J. W., Brodsky, W. A., and Scott, W. M. (1965). *Am. J. Physiol.* **208,** 38–45.

Arieff, A. E., Guisado, R., and Massry, S. G. (1975). *Kidney Int.* **7,** S194–S200.

Asheim, A., Person, F., and Person, S. (1961). *Acta Physiol. Scand.* **51,** 150–162.

Avizonis, P. V., and Wriston, J. C. (1959). *Biochem. Biophys. Acta* **34,** 279–281.

Bagley, R. S., Center, S. A., Lewis, R. M., Shin, S., Dougherty, S. A., Randolph, J. F., and Erb, H. (1991). *J. Vet. Int. Med.* **5,** 66–70.

Balazs, T., Sekella, R., and Pauls, J. F. (1971). *Lab. Anim. Sci.* **21,** 546–548.

Balint, P., and Fekete, A. (1960). *Acta Physiol. Acad. Sci. Hung.* **17,** 277–286.

Balint, P., and Forgacs, I. (1965). *Acta Physiol. Acad. Sci. Hung.* **27,** 47–58.

Balint, P., and Sturcz, J. (1959). *Acta Med. Acad. Sci. Hung.* **13,** 9–17.

Balint, P., and Visy, M. (1965). *Acta Physiol. Acad. Sci. Hung.* **28,** 265–272.

Balint, P., Laszlo, K., Szocs, E., and Tarjan, E. (1975). *Acta Med. Acad. Sci. Hung.* **32,** 193–205.

Barajas, L., and Mueller, J. (1973). *J. Ultrastruct. Res.* **43,** 107–132.

Barsanti, J. A. (1987). Personal communication.

Barsanti, J. A., and Finco, D. R. (1979). *Am. J. Vet. Res.* **40,** 1583–1589.

Batamuzi, E. K., and Kristensen, F. (1995). *J. Sm. Anim. Pract.* **36,** 17–21.

Beeuwkes, R. (1971). *Am. J. Physiol.* **221,** 1361–1374.

Bergstrom, J., Furst, P., and Noree, L. O. (1975). *Clin. Nephrol.* **3,** 187–192.

Berndt, T. J., and Knox, F. G. (1992). *In* "The Kidney: Physiology and Pathophysiology" (D. W. Seldin and G. Giebisch, eds.), pp. 2511–2532. Raven Press, New York.

Bernstein, L. M. (1965). "Renal Function and Renal Failure." Williams & Wilkins, Baltimore.

Berry, C. A., and Rector, F. C. (1991). *In* "The Kidney" (B. M. Brenner and F. C. Rector, eds.), pp. 245–283. Saunders, Philadelphia, Pennsylvania.

Better, O. (1986). *Kidney Int.* **25,** 841–853.

Biewenga, W. J., Gruys, E., and Hendricks, H. J. (1982). *Res. Vet. Sci.* **33,** 366–374.

Black, D. A. (1970). *Arch. Intern. Med.* **126,** 906–909.

Blass, K. G., Thibert, R. J., and Lam, L. K. (1974). *Z. Klin. Chem. Klin. Biochem.* **12,** 336–343.

Bleiler, R. A., and Schell, H. P. (1962). *J. Lab. Clin. Med.* **59,** 945–955.

Bloch, K., Schoenheimer, P., and Rittenberg, D. (1941). *J. Biol. Chem.* **138,** 155–166.

Bold, A. M., Menzies, I. S., and Walker, G. (1970). *J. Clin. Pathol.* **23,** 85–89.

Borison, H. L., and Herbertson, L. M. (1959). *Am. J. Physiol.* **197,** 850–852.

Bovee, K. C., and Joyce, T. (1979). *J. Am. Vet. Med. Assoc.* **174,** 488–491.

Bovee, K. C., Kronfeld, D. S., Ramburg, C., and Goldschmidt, M. (1979). *Invest. Urol.* **16,** 378–384.

Bowie, L., Smith, S., and Gochman, N. (1977). *Clin. Chem.* **23,** 128–130.

Boylan, J. W., and Asshauer, E. (1962). *Pluegers Arch.* **276,** 99– 116.

Bradford, M. M. (1976). *Anal. Biochem.* **72,** 248–254.

Brassard, J. A., Meyers, K. M., Person, M., and Dhein, C. R. (1994). *J. Lab. Clin. Med.* **124,** 48–54.

Brenner, B. M., Hostetter, T. H., and Humes, H. D. (1978). *N. Engl. J. Med.* **298,** 826–833.

Brenner, B. M., Zatz, R., and Ichikawa, I. (1986). *In* "The Kidney" (B. M. Brenner and F. C. Rector, eds.), pp. 93–123. Saunders, Philadelphia, Pennsylvania.

Brezis, M., Rosen, S., and Epstein, F. H. (1986). *In* "The Kidney" (B. M. Brenner and F. C. Rector, eds.), pp. 735–799. Saunders, Philadelphia, Pennsylvania.

Bricker, N. S. (1969). *Am. J. Med.* **46,** 1–11.

Bricker, N. S., Dewey, R. R., Lubowitz, H., Stokes, J., and Kirkensgaard, T. (1959). *J. Clin. Invest.* **38,** 516–523.

Bricker, N. S., Morrin, P. A., and Kime, S. W., Jr. (1960). *Am. J. Med.* **28,** 77–98.

Brobst, D. F., Carter, J. M., and Herron, M. (1967). *Gaines Vet. Symp., Proc. Gaines Res. Cent.,* White Plains, New York.

Brobst, D. F., Parish, S. M., Torbeck, R. L., Frost, O. L., and Bracken, F. K. (1978). *J. Am. Vet. Med. Assoc.* **173,** 481–485.

Brown, C. L., Hill, M. J., and Richards, P. (1971). *Lancet* **ii,** 406–408.

Brown, E., Wilkson, R., Eastman, R., Pallata, J., and Marynick, S. (1982). *Clin. Endocrinol. Metab.* **54,** 172–179.

Brown, S. A. (1994). *Am. J. Vet. Res.* **55,** 1470–1473.

Brown, S. A., and Finco, D. R. (1987). Unpublished data.

Brown, S. A., Finco, D. R., Crowell, W. A., Choat, D. C., and Navar, L. G. (1990). *Am J Physiol.* **258,** F495–F503.

Brown, S. A., Finco, D. R., Boudinot, F. D., Wright, J., Tarver, S. L., and Cooper, T. (1995). *Am. J. Vet. Res.* **57,** 105–110.

Carlson, G. P., and Kaneko, J. J. (1971). *J. Am. Vet. Med. Assoc.* **158,** 1229–1234.

Carriere, S., and Gagnan-Brunette, M. (1977). *Can. J. Physiol. Pharmacol.* **55,** 105–110.

Caverzasio, J., Gloor, H., Fleisch, H., and Bonjour, J. (1982). *Kidney Int.* **21,** 471–476.

Center, S. A., Wilkinson, E., Smith, C. A., Erb, H., and Lewis, R. M. (1985). *JAVMA* **187,** 820–824.

Christie, B. A. (1980). *Am. J. Vet. Res.* **41,** 1519–1525.

Churchill, D., Knaack, J., Chirito, E., Barre, P., Cole, C., Muehrcke, R., and Gault, M. H. (1974). *Am. J. Med.* **56,** 575–582.

Coburn, J. W., Gonick, H. C., Rubini, M. E., and Kleeman, C. R. (1965). *J. Clin. Invest.* **44,** 603–614.

Cohen, J. J. (1986). *Kidney Int.* **29,** 32–40.

Conn, H. O. (1973). *Gastroenterology* **65,** 321–340.

Conzelman, G. M., Flanders, L. E., Springer, K., and Crout, D. W. (1970). *Lab. Anim. Care* **20,** 192–197.

Costanzo, L. S., and Windhager, E. E., (1992). *In* "The Kidney: Physiology and Pathophysiology" (D. W. Seldin and G. Giebisch, eds.), pp. 2375–2393. Raven Press, New York.

Cowgill, L. D. (1992). *Semin. Vet. Med. Surg.* **7,** 175–182.

Cowgill, L., and Hornof, W. (1986). *In* "Current Veterinary Therapy IX," (R. W. Kirk, ed.), pp. 1108–1111. Saunders, Philadelphia, Pennsylvania.

Cramer, C. F., and McMillan, J. (1980). *Am. J. Physiol.* **239,** G261–G265.

Darmady, E. M., Offer, J., and Woodhouse, M. A. (1973). *J. Pathol.* **109,** 195–207.

Davidsohn, I., and Henry, J. B. (1969). "Clinical Diagnosis by Laboratory Methods." Saunders, Philadelphia, Pennsylvania.

DiBartola, S. P., Chew, D. J., and Jacobs, G. (1980). *J. Am. Anim. Hosp. Assoc.* **16,** 537–546.

DiBona, G. (1984). *Kidney Int.* **25,** 841–853.

Dirks, J. H., Clapp, J. R., and Berliner, R. W. (1965). *J Clin Invest.* **45,** 916–921.

Divers, T. J. (1983). *Comp. Cont. Ed.* **5,** S310–S317.

Divers, T. J., Crowell, W. A., Duncan, J. R., and Whitlock, R. H. (1982). *J. Am. Vet. Med. Assoc.* **181,** 694–699.

Dobbelstein, H. (1976). *Nephron* **15,** 409–414.

Dontas, A. S., Marketos, S. G., and Papanayiotou, P. (1972). *Postgrad. Med. J.* **48,** 295–303.

Doolan, P. D., Olsen, E. L., and Meil, G. B. (1962). *Am. J. Med.* **32,** 65–79.

Dossetor, J. B. (1966). *Ann. Intern. Med.* **65,** 1287-1299.

Dow, S. W., Fettman, M. J., LeCouter, R. A., and Hamar, D. W. (1987). *J. Am. Vet. Med. Assoc.* **191,** 1569–1574.

Dow, S. W., Fettman, M. J., Smith, K. R., Hamar, D. W., Nagode, L. A., Refsel, K. R., and Wilke, W. L. (1990). *J. Nutr.* **120,** 569–578.

Dubach, U. C., and Schmidt, U. (1978). *J. Clin. Chem. Clin. Biochem.* **16,** 197–211.

Duncan, J. R., and Prasse, K. W. (1976). *Vet. Clin. North Am.* **6,** 647–661.

Dworkin, L. D., and Brenner, B. M. (1991). The renal circulations. *In* "The Kidney" (B. Brenner, F. Rector, eds.), pp. 164–204. Saunders, Philadelphia, Pennsylvania.

Eknoyan, G. (1977). *In* "Pathophysiology of the Kidney" (N. A. Kurtzman and M. Martinez-Maldonado, eds.), pp. 842–881. Thomas, Springfield, Illinois.

Ellis, B. G., Price, R. G., and Topham J. C. (1973). *Chem.-Biol. Interact.* **7,** 101–113.

Enger, E., and Blegen, E. M. (1964). *Scand. J. Clin. Lab. Invest.* **16,** 273–280.

England, M. D., and Mitch, W. E. (1993). *Contrib. Nephrol.* **102,** 48–58.

Epstein, M. E., Barsanti, J. A., Finco, D. R., and Cowgill, W. A. (1984). *J. Am. Anim. Hosp. Assoc.* **20,** 779–782.

Farquhar, M. G. (1975). *Kidney Int.* **8,** 197–211.

Fettman, M. J., Allen, T. A., Wilke, W. L., Radin, M. J., and Eubank, M. C. (1985). *Am. J. Vet. Res.* **46,** 482–485.

Field, M. J., and Giebisch, G. J. (1985). *Kidney Int.* **27,** 379–387.

Finco, D. R. (1971). *JAVMA* **159,** 336–340.

Finco, D. R. (1972). *J. Am. Anim. Hosp. Assoc.* **8,** 155–165.

Finco, D. R. (1979). Unpublished observations.

Finco, D. R. (1995). Unpublished observations.

Finco, D. R., and Barsanti, J. A. (1982). *Am. J. Vet. Res.* **43,** 2207–2209.

Finco, D. R., and Cornelius, L. (1977). *Am. J. Vet. Res.* **38,** 941–947.

Finco, D. R., and Duncan, J. R. (1972). *Am. J. Vet. Res.* **33,** 2447–2450.

Finco, D. R., and Duncan, J. R. (1976). *J. Am. Vet Med. Assoc.* **168,** 593–601.

Finco, D. R., and Groves, C. (1985). *Am. J. Vet. Res.* **46,** 1625–1628.

Finco, D. R., and Rowland, G. N. (1978). *J. Am. Vet. Med. Assoc.* **173,** 990–994.

Finco, D. R., Coulter, D. B., and Barsanti, J. A. (1981). *Am. J. Vet. Res.* **42,** 1874–1877.

Finco, D. R., Coulter, D. B., and Barsanti, J. A. (1982). *J. Am. Anim. Hosp. Assoc.* **18,** 804–806.

Finco, D. R., Crowell, W. A., and Barsanti, J. A. (1985). *Am. J. Vet. Res.* **46,** 646–653.

Finco, D. R., Barsanti, J. A., and Rawlings, C. A. (1984). *JAVMA* **185,** 664–665.

Finco, D. R., Brown, S. A., Crowell, W. A., and Barsanti, J. A. (1991). *Am. J. Vet. Res.* **52,** 1029–1032.

Finco, D. R., Brown, S. A., Crowell, W. A., Groves, C. A., Duncan, J. R., and Barsanti, J. A. (1992). *Am. J. Vet. Res.* **53,** 157–163.

Finco, D. R., Tabaru, H., Brown, S. A., and Barsanti, J. A. (1993). *Am. J. Vet. Res.* **54,** 1575-1578.

Finco, D. R., Brown, S. A., Crowell, W. A., Brown, C. A., Barsanti, J. A., Carey, D. P., and Hirakawa, D. A. (1994). *Am. J. Vet. Res.* **55,** 1282–1290.

Finco, D. R., Brown, S. A., Vaden, S. L., and Ferguson, D. C. (1995). *J. Vet. Pharm. Therap.* **18,** 418–421.

Fine, L. G., Schlondorff, D., and Trizna, W. (1978). *J. Clin. Invest.* **61,** 1519–1526.

Fine, L. G., Hammerman, M. R., and Abboud, H. E. (1992). *J. Am. Soc. Nephrol.* **2,** 1163–1170.

Frennby, B., Sterner, G., Almen, T., Hagstam, K-E., and Jacobsson, L. (1994). *Invest. Radiol.* **29,** S234–S235.

Fried, T. A., and Stein, J. H. (1983). *Arch. Int. Med.* **143,** 787–791.

Furst, P., Alvesstrand, A., and Bergstrom, J. (1980). *Am. J. Clin. Nutr.* **33,** 1387–1395.

Gennari, F., and Cohen, J. (1975). *Kidney Int.* **8,** 1–5.

Gennari, F. J., and Kassirer, J. P. (1974). *N. Engl. J. Med.* **294,** 714–720.

Gennari, F. J., and Maddox, D. A. (1992). *In* "The Kidney: Physiology and Pathophysiology" (D. W. Seldin and G. Giebisch, eds.), pp. 2695–2732. Raven Press, New York.

Giovannetti, S., and Barsotti, G. (1975). *Nephron.* **14,** 123–133.

Glauser, S. C., Wagner, H., and Glauser, E. M. (1972). *Am. J. Med. Sci.* **264,** 135–139.

Goldman, R. (1954). *Proc. Soc. Exp. Biol. Med.* **85,** 446–448.

Gonzales-Campoy, J. M., and Knox, F. G. (1992). *In* "The Kidney: Physiology and Pathophysiology" (D. W. Seldin and G. Giebisch, eds.), pp. 2041–2097. Raven Press, New York.

Gossett, K. A., Turnwald, G. H., Kearney, M. T., Greco, D. S., and Cleghorn, B. (1987). *Am. J. Vet. Res.* **48,** 455–457.

Gottschalk, C. W., Moss, N. G., and Colindres, R. (1985). *In* "The Kidney: Physiology and Pathophysiology" (D. W. Seldin and G. Giebisch, eds.), pp. 581–611. Raven Press, New York.

Grauer, G. F., Thomas, C. B., and Eicker, S. W. (1985). *Am. J. Vet. Res.* **46,** 2116–2119.

Greaves, J. P., and Scott, P. P. (1960). *Nature (London)* **187,** 242.

Greco, D., Turnwall, G., and Adams, R. (1985). *Am. J. Vet. Res.* **46,** 2332–2335.

Greene, C. E., Wong, P., and Finco, D. R. (1979). *J. Am. Anim. Hosp. Assoc.* **15,** 371–377.

Groves, C., Finco, D., White, J., and Brown, S. (1987). Unpublished data.

Guder, W. G., and Hofmann, W. (1993). *Clin. Biochem.* **26,** 277–282.

Hall, J. E., Guyton, A. C., and Fair, B. M. (1977). *Am. J. Physiol.* **232,** F72–F76.

Halperin, M., Goldstein, M. B., Stinebaugh, B. J., and Jungas, R. L. (1985). *In* "The Kidney: Physiology and Pathophysiology" (D. W. Seldin and G. Giebisch, eds.), pp. 1471–1490. Raven Press, New York.

Halperin, M. L., and Jungas, R. L. (1983). *Kidney Int.* **24,** 709–713.

Hamilton, P. B., Phillips, R. A., and Hiller, A. (1948). *Am. J. Physiol.* **152,** 517–522.

Handler, J. S., and Kwon, H. M. (1993). *Am. J. Physiol.* **265,** C1449–C1455.

Haramati, A., Haas, J. A., and Knox, F. G. (1983). *Am. J. Physiol.* **244,** F265–F269.

Hardwicke, J. (1975). *Clin. Nephrol.* **3,** 37–41.

Hardy, R. M., and Osborne, C. A. (1979). *J. Am. Vet. Med. Assoc.* **174,** 479–484.

Harrington, J. T., and Cohen, J. J. (1975). *N. Engl. J. Med.* **292,** 89–91.

Harris, R., Meyer, T., and Brenner, B. (1986). *In* "The Kidney" (B. M. Brenner and F. C. Rector, eds.), pp. 1553–1585. Saunders, Philadelphia, Pennsylvania.

Hawk, P. B. (1910). *J. Biol. Chem.* **8,** 465-477.

Hays, R., Franki, N., and Ding, G. (1987). *Kidney Int.* **31,** 530–537.

Henry, R. J. (1964). "Clinical Chemistry." Harper, New York.

Hoberman, H. D., Sims, E. A. H., and Peters, J. H. (1948). *J. Biol. Chem.* **172,** 45–58.

Horster, M., and Valtin, H. (1971). *J. Clin. Invest.* **50,** 779–795.

Hoskins, J. D., Turnwald, G. H., Kearney, M. T., Gossett, K. A., and Fakier, N. (1991). *Am. J. Vet. Res.* **52,** 1295–1299.

Hostetter, T. H., Olson, J. L., and Rennke, H. G. (1981). *Am. J. Physiol.* **241,** F85–F93.

Houpt, T. R. (1959). *Am. J. Physiol.* **197,** 115–120.

Hubbert, W. T. (1972). *Cornell Vet.* **62,** 13–20.

Huisman, G. H., Joles, J. A., Kraan, W. J., Visschedijk, A. H. J., Velthuizen, J., and Charbon, G. J. (1982). *Eur. J. Appl. Physiol.* **49,** 231–242.

Itkin, R. J., Krawiec, D. R., Twardock, A. R., and Gelberg, H. B. (1994). *Am. J. Vet. Res.* **55,** 1660–1665.

Jacobs, K., Shoemaker, C., Rudersdorf, R., Neill, S. D., Kaufman, R. J., Mufson, A., Seehra, J., Jones, S. S., Hewick, R., Fritsch, E. F., Kawakita, M., Shimizu, T., and Miyake, T. (1985). *Nature* **313,** 806–810.

Jacobs, R. M., Lumsden, J. H., Taylor, J. A., and Grift, E. (1991). *Can. J. Vet. Res.* **55,** 150–154.

Jacobson, H., and Seldin, D. (1983). *Am. J. Physiol.* **14,** F425–F432.

James, G. P., Bee, D. E., and Fuller, J. B. (1978). *Clin. Chem.* **24,** 1934–1939.

Jamison, R. L. (1976). *In* "The Kidney" (B. M. Brenner and F. C. Rector, eds.), pp. 391–441. Saunders, Philadelphia, Pennsylvania.

Jergens, A. E., McCaw, D. L., and Hewett, J. E. (1987). *Am. J. Vet. Res.* **48,** 1106–1109.

Joles, J. A., Sanders, M., Velthuizen, J., Hertog, J. M. D., and Van Dijk, C. (1984). *Int. J. Sports Med.* **5,** 311–316.

Jolliffe, N., and Smith, H. W. (1931). *Am. J. Physiol.* **98,** 572–581.

Jones, J. D., and Burnett, P. C. (1972). *Clin. Chem.* **18,** 280–284.

Jones, J. D., and Burnett, P. C. (1974). *Clin. Chem.* **20,** 1204–1212.

Kaikara, S., Rutherford, R. B., Schwentker, E. P., and Wagner, H. N., Jr. (1969). *J. Appl. Physiol.* **27,** 218–222.

Kassirer, J. P. (1971). *N. Engl. J. Med.* **285,** 385–389. Kaufman, J. M., Siegel, N., Lytton, B., and Hayslett, J. P. (1976). *Invest. Urol.* **13,** 441–444.

Kerr, W. S. (1954). *J. Appl. Physiol.* **6,** 762–772.

Kerr, W. S. (1956). *Am. J. Phys.* **184,** 521–526.

Kew, M. (1972). *Gut* **13,** 748–753.

Kirkbride, C. A., and Frey, R. A. (1967). *J. Am. Vet. Med. Assoc.* **151,** 742–746.

Kirkebo, A., and Tyssebotn, I. (1977). *Acta Physiol. Scand.* **101,** 257–263.

Klahr, S. (1991) *Sem. Nephrol.* **18,** 689–699.

Klahr, S., and Hammerman, M. (1985). *In* "The Kidney: Physiology and Pathophysiology" (D. W. Seldin and G. Giebisch, eds.), pp. 699–718. Raven Press, New York.

Klahr, S., Buerkert, J., and Morrison, A. (1986). *In* "The Kidney" (B. M. Brenner and F. C. Rector, eds.), pp. 1443–1490. Saunders, Philadelphia, Pennsylvania.

Knochel, J. P., and Jacobson, H. R. (1986). *In* "The Kidney" (B. M. Brenner and F. C. Rector, eds.), pp. 619–662. Saunders, Philadelphia, Pennsylvania.

Knox, F. G. (1977). Physiologist **20,** 25–31.

Kopple, J. D., and Coburn, J. W. (1974). *J. Am. Med. Assoc.* **227,** 41–44.

Kosek, J. C., Mazze, R. I., and Cousins, M. J. (1972). *Lab. Invest.* **27,** 575–580.

Koterba, A. M., and Coffman, J. R. (1981). *J. Eq. Med. Surg.* **3,** S461–S469.

Kramer, J. W., Bistline, D., Sheridan, P., and Emerson, C. (1984). *J. Am. Vet. Med. Assoc.* **184,** 584.

Krawiec, D. R., Badertscher, R. R., II., Twardock, A. R., Rubin, S. I., and Gelberg, H. B. (1986). *Am. J. Vet. Res.* **47,** 2175–2179.

Kumar, R., Steen, P., and McGeown, M. G. (1972). *Lancet* **ii,** 1005.

Lameire, N. H., Lifschitz, M. D., and Stein, J. H. (1977). *Annu. Rev. Physiol.* **39,** 159–184.

Lamperi, S., Bandiani, G., Fiorio, P., Muttini, P., and Scaringi, G. (1974). *Nephron.* **13,** 278–287.

Larsson, L., Aperis, A., and Wilton, P. (1980). *Kidney Int.* **18,** 29–35.

Lau, K. (1989). *Kidney Int.* **36,** 918–937.

Leahy, A. L., Ryan, P. C., McEntree, G. M., Nelson, A. C., and Fitzpatrick, J. M. (1989). *J Urol.* **142,** 199–203.

Levinsky, N. G. (1977). *N. Engl. J. Med.* **296,** 1453–1458.

Levy, M. (1992). *In* "The Kidney: Physiology and Pathophysiology" (D. W. Seldin and G. Giebisch, eds.), pp. 3305–3326. Raven Press, New York.

Lewis, R. A., Schwartz, R. H., and Schenk, E. A. (1970). *Fed. Proc., Fed. Am. Soc. Exp. Biol.* **29,** 574.

Lindblad, H. G., and Berg, U. B. (1994). *Acta. Pediatr.* **83,** 418–422.

Lindeman, R. D. (1986). *Exper. Gerontol.* **21,** 379–406.

Lindner, L. E., and Haber, M., H. (1983). *Am. J. Clin. Pathol.* **80,** 347–352.

Lindner, L. E., Vacca, D., and Haber, M. H. (1983). *Am. J. Clin. Pathol.* **80,** 353–358.

LoGrippo, G. A., Rupe, C. E., Hayaski, H., and LeSher, D. (1970). *Henry Ford Hosp. Med. J.* **18,** 83–90.

Lopez-Hilker, S., Galceran, T., Chan, Y., Rapp, N., Martin, K. J., and Slatopolsky, E. (1986). *J. Clin. Invest.* **78,** 1097–1102.

Lorenz, J. M., Kleinman, L. I., and Disney, T. A. (1986). *Am. Physiol. Soc.* **251,** F513–F519.

Low, D. G., Hiatt, C. W., Gleiser, C. A., and Bergman, E. N. (1956). *J. Infect. Dis.* **98,** 249–259.

Lulich, J. P., and Osborne, C. A. (1990). *The Compendium.* **12,** 59–73.

Maack, T., Park, C., and Camargo, M. (1992). *In* "The Kidney: Physiology and Pathophysiology" (D. W. Seldin and G. Giebisch, eds.), pp. 3005–3038. Raven Press, New York.

McCance, R. A., and Widdowson, E. M. (1955). *J. Physiol. (London)* **129,** 628–635.

McCaw, D. L., Knapp, D. W., and Hewett, J. E. (1985). *Am. J. Vet. Res.* **46,** 1665–1669.

McDougal, W. S., and Wright, F. S. (1972). *Kidney Int.* **2,** 304–317.

McQueen, E. G., and Engel, G. B. (1966). *J. Clin. Pathol.* **19,** 392–396.

Maddox, D. A., and Brenner, B. M. (1991). The renal circulations. *In* "The Kidney" (B. Brenner, F. Rector, eds.), pp. 205–244. Saunders, Philadelphia, Pennsylvania.

Mandin, H., and Davidman, M. (1978). *Am. J. Physiol.* **234,** F117–F122.

Massry, S. (1977). *Nephron* **19,** 125–130.

Mattenheimer, H. (1971). *Med. Clin. North Am.* **55,** 1493–1508.

Mattenheimer, H. (1977). *Ann. Clin. Lab. Sci.* **7,** 422–432.

Mayson, J. S., Schumaker, O., and Nakamura, R. M. (1972). *Am. J. Clin. Pathol.* **52,** 297–299.

Meyer, N. L., Mercer, B. M., Friedman, S. A., and Sibia, B. M. (1994). *Am. J. Obstet. Gynecol.* **170,** 137–141.

Millard, R. W., Higgins, C. B., Franklin, D., and Vatner, S. F. (1972). *Circ. Res.* **31,** 881–888.

Mitch, W. E. (1985). *In* "The Kidney: Physiology and Pathophysiology" (D. W. Seldin and G. Giebisch, eds.), pp. 2059–2081. Raven Press, New York.

Mitch, W. E., and Walser, M. (1986). *In* "The Kidney" (B. M. Brenner and F. C. Rector, eds.), pp. 1759–1790. Saunders, Philadelphia, Pennsylvania.

Mitch, W. E., Lietman, P. S., and Walser, M. (1975). *Proc. Am. Soc. Nephrol. Anni. Meet., 8th,* Washington, D.C. Abstr. 65.

Mitch, W. E., Collier, V. U., and Walser, M. (1980). *Cln. Sci.* **58,** 327–335.

Mitch, W. E., Price, S. R., May, R. C., Jurkovitz, C., and England, B. K. (1994). *Am. J. Kidn. Dis.* **23,** 224-228.

Mixner, J. P., and Anderson, R. R. (1958). *J. Dairy Sci.* **41,** 306-313.

Monroe, W. E., Davenport, D. J., and Saunders, G. K. (1989). *Am. J. Vet. Res.* **50,** 1906–1909.

Morrin, P. A., Joint, M. S., Handa, S. P., and Frame, J. (1970). *Invest. Urol.* **8,** 273–283.

Mujais, S., Sabatini, S., and Kurtzman, N. (1986). *In* "The Kidney" (B. M. Brenner and F. C. Rector, eds.), pp. 1587–1630. Saunders, Philadelphia, Pennsylvania.

Mulnix, J. A., Rijnberk, A., and Hendricks, H. J. (1976). *J. Am. Vet. Med. Assoc.* **168,** 1327–1330.

Naets, J. P. (1975). *Nephron.* **14,** 181–194.

Narayanan, S., and Appleton, H. D. (1972). *Clin. Chem.* **18,** 270–274.

Newman, G. H. (1971). *Postgrad. Med.* **50,** 236–239.

Nissen, O. I., and Galskov, A. (1972). *Circ. Res.* **30,** 82–96.

Norrdin, R. W. (1975). *Cornell Vet.* **65,** 173–186.

O"Connell, J. B., Romeo, J. A., and Mudge, G. H. (1962). *Am. J. Physiol.* **203,** 985–990.

Offringa, M., and Benbasset, J. (1992). *Postgrad. Med. J.* **68,** 648–654.

Oken, D. E. (1975). *Annu. Rev. Med.* **26,** 307–319.

Oliver, J., Bloom, F., and MacDowell, M. (1941). *J. Exp. Med.* **73,** 141–159.

Orita, Y., Nobuko, I., Naohiko, U., Keiko, S., Ando, A., Fujiwara, Y., Hirano, S., and Abe, H. (1977). *Nephron.* **19,** 19–25.

Osbaldiston, G. W., and Fuhrman, W. (1970). *Can. J. Comp. Med.* **34,** 138–141.

Osborne, C., and Stevens, J. (1981). "Handbook of Canine and Feline Urinalysis." Ralston Purina Co., St. Louis, Missouri.

Osborne, C. A., Low, D. G., and Finco, D. R. (1972). "Canine and Feline Urology." Saunders, Philadelphia, Pennsylvania.

Owen, O. E., Felig, P., Morgan, A. P., Wahren, J., and Cahill, G. F. (1969). *J. Clin. Invest.* **48,** 574–583.

Packer, R. K., Curry, C. A., and Brown, K. M. (1995). *J. Am. Soc. Nephrol.* **5,** 1624–1629.

Papper, S. (1975). *Clin. Nephrol.* **4,** 41–44.

Phillips, R. A., Dole, V. P., Hamilton, P. B., Kendall, E., Archibald, R. M., and Van Slyke, D. D. (1946). *Am. J. Physiol.* **145,** 314–336.

Pickering, E. C. (1965). *Proc. Nutr. Soc.* **24,** 73–80.

Pierce, A. E. (1959). *J. Physiol. (London)* **148,** 469–488.

Pierce, A. E. (1961). *J. Physiol. (London)* **156,** 136–149.

Pitts, R. F. (1934). *Am. J. Physiol.* **109,** 532–541.

Pitts, R. F. (1974). "Physiology of the Kidney and Body Fluids." Yearbook Publ., Chicago, Illinois.

Polzin, D. J., Leininger, J. R., and Osborne, C. A. (1988). *Lab. Invest.* **58,** 172–183.

Polzin, D. J., Osborne, C. A., O"Brien, T. D., and Hostetter, T. H. (1993). *Proc. 11th Annual Vet. Med. Forum* **938.**

Powers, T. E., Powers, J. D., and Garg, R. C. (1977). *Am. J. Vet. Res.* **38,** 1933–1936.

Quamme, G. A. (1992). *In* "The Kidney: Physiology and Pathophysiology" (D. W. Seldin and G. Giebisch, eds.), pp. 2339–2356. Raven Press, New York.

Raab, W. P. (1972). *Clin. Chem.* **18,** 5–25.

Rabinowitz, L., Gunther, R. A., Shoji, E. S., Freedland, R. A., and Avery, E. H. (1973). *Kidney Int.* **4,** 188–207.

Rahill, W. J., and Subramanian, S. (1973). *Lab. Anim. Sci.* **23,** 92–96.

Raij, L., and Keane, W. F. (1985). *Am. J. Med.* **79,** 24–30.

Raskin, N. H., and Fishman, R. A. (1976). *N. Engl. J. Med.* **294,** 204–210.

Raskova, J., and Morrison, A. B. (1978). *Lab. Invest.* **38,** 103–109.

Refsum, H. E., and Stromme, S. B. (1974). *Scand. J. Lab. Invest.* **33,** 247–254.

Reidenberg, M. M. (1976). *Clin. Pharmacokinet.* **1,** 121–125.

Reidenberg, M. M. (1977a). *Am. J. Med.* **62,** 466–470.

Reidenberg, M. M. (1977b). *Am. J. Med.* **62,** 482–485.

Relman, A. S., and Levinsky, N. G. (1971). *In* "Diseases of the Kidney" (M. B. Strauss and L. G. Welt, eds.), pp. 87–139. Little, Brown, Boston, Massachusetts.

Rennke, H. G., and Venkatachalam, M. A. (1977). *Fed. Proc., Fed. Am. Soc. Exp. Biol.* **36,** 2619–2626.

Rentko V. T., Clark, N., Ross, L. A., and Schelling, S. H. (1992). *J Vet. Int. Med.* **6,** 235–244.

Rickers, H., Brochner-Mortensen, J., and Rodbro, P. (1978). *Scand. J. Urol. Nephrol.* **12,** 39–44.

Robertson, J. L., Goldschmidt, M., Kronfeld, D. S., and Bovee, K. C. (1986). *Kidney Int.* **29,** 511–519.

Rogers, K. S., Komkov, A., Brown, S. A., Lees, G. E., Hightower, D., and Russo, E. A. (1991). *Am. J. Vet. Res.* **52,** 961–964.

Rosa, R. M., Williams, M. E., and Epstein, F. H. (1992). *In* "The Kidney: Physiology and Pathophysiology" (D. W. Seldin and G. Giebisch, eds.), pp. 2165–2191. Raven Press, New York.

Rosenfeld, S. (1955). M.S. Thesis, Univ. of California, Davis.

Ross, L., and Finco, D. (1981). *Am. J. Vet. Res.* **42,** 1704–1710.

Rous, S. N., and Wakim, K. G. (1967). *J. Urol.* **98,** 30–35.

Rubin, M. I., and Balish, T. (1971). *Pediatr. Clin. North Am.* **18,** 245–264.

Russo, E. A., Lees, G. E., and Hightower, D. (1986). *Am. J. Vet. Res.* **47,** 1308–1312.

Rutecki, G. J., Goldsmith, C., and Schreiner, G. E. (1971). *N. Engl. J. Med.* **284,** 1049–1052.

Sands, J. M., Kokko, J. P., and Jacobson, H. R. (1992). *In* "The Kidney: Physiology and Pathophysiology" (D. Seldin and G. Geibisch, eds.), pp. 1087–1155. Raven Press, New York.

Schenk, E. A., Schwartz, R. H., and Lewis, R. A. (1971). *Lab. Invest.* **25,** 92–95.

Schloerb, P. R. (1960). *Am. J. Physiol.* **199,** 661–665.

Schon, D. A., Silva, P., and Hatslett, J. P. (1974). *Am. J. Physiol.* **227,** 1323–1330.

Schreiner, G. E., and Maher, J. F. (1961). "Uremia: Biochemistry, Pathogenesis, and Treatment." Thomas, Springfield, Illinois.

Sebastian, A., Hernandez, R. E., and Schambelan, M. (1986). *In* "The Kidney" (B. M. Brenner and F. C. Rector, eds.), pp. 519–549. Saunders, Philadelphia, Pennsylvania.

Shannon, J. A. (1936). *Am. J. Physiol.* **117,** 206–225.

Shils, M. E. (1963). *Ann. Intern. Med.* **58,** 389–408.

Siegel, M. B., and Levinsky, N. G. (1977). *Circ. Res.* **41,** 227–231.

Silbernagl, S. (1985). *In* "The Kidney: Physiology and Pathophysiology" (D. W. Seldin and G. Giebisch, eds.), pp. 1677–1701. Raven Press, New York.

Silva, P., Brown, R. S., and Epstein, F. H. (1977). *Kidney Int.* **11,** 466–475.

Simenhoff, M. L., Burke, J. F., Saukkonen, J. J., Ordinario, A. T., and Doty, R. (1977). *N. Engl. J. Med.* **297,** 132–135.

Simon, D., and Luke, R. G. (1971). *Proc. Soc. Exp. Biol. Med.* **137,** 1073–1074.

Slack, T., and Wilson, D. M. (1976). *Mayo Clin. Proc.* **51,** 296–300.

Slatopolsky, E., Cagler, S., Pennell, J. P., Taggart, D. D., Canterbury, J. M., Reiss, E., and Bricker, N. S. (1971). *J. Clin. Invest.* **50,** 492–499.

Smith, H. W. (1951). "The Kidney Structure and Function in Health and Disease." Oxford Univ. Press, London and New York.

Smith, H. W. (1959). "From Fish to Philosopher. The Story of Our Internal Environment." Ciba, Summit, New Jersey.

Smogorzewski, M., and Massry, S. G. (1993). *Kidney Int.* **44,** 630–637.

Sobel, J. (1964). *Lancet* **ii,** 760–761.

Souhami, R. L. (1973). *Br. J. Exp. Pathol.* **54,** 380–387.

Steinhausen, M., Endlich, K., and Wiegman, D. L., (1990). *Kidney Int.* **38,** 769–784.

Street, A. E., Chesterman, H., Smith, G. K., and Quinton, R. M. (1968). *Toxicol. Appl. Pharmacol.* **13,** 363–371.

Strombeck, D. R., Meyer, D. J., and Freedland, R. A. (1975). *J. Am. Vet. Med. Assoc.* **166,** 1109–1111.

Suzuki, A. (1971). *Jpn. J. Pharmacol.* **21,** 75–85.

Swanson, R. E., and Hakim, A. A. (1962). *Am. J. Physiol.* **203,** 980–984.

Swenson, R. S., Peterson, D. T., Eshleman, M., and Reaven, G. M. (1973). *Kidney Int.* **4,** 267–272.

Tabaru, H., Finco, D. R., Brown, S. A., and Cooper, T. (1993). *Am. J. Vet. Res.* **54,** 1758–1764.

Tallan, H. H., Moore, S., and Stein, W. H. (1954). *J. Biol. Chem.* **211,** 927–938.

Tapp, D. C., Kobayashi, S., Fernandes, G., and Venbatachalam, M. A. (1989). *Seminars in Nephrol.* **9,** 343–353.

Tennant, B., Bettleheim, P., and Kaneko, J. J. (1982). *J. Am. Vet. Med. Assoc.* **180,** 630–634.

Tennant, B. C., Lowe, J. E., and Tasker, J. B. (1974). *Fed. Proc., Fed. Am. Soc. Exp. Biol.* **33,** 670.

Thier, S. O., McCurdy, D. K., and Rastegar, A. (1977). *In* "Pathophysiology of the Kidney" (N. A. Kurtzman and M. Martinez-Maldonado, eds.), pp. 335–355. Thomas, Springfield, Illinois.

Thirakomen, K., Koslov, N., Arruda, J. A., and Kurtzman, N. A. (1976). *Am. J. Physiol.* **231,** 1233–1239.

Thornton, J. R., and English, P. B. (1976). *Aust. Vet. J.* **52,** 335–337.

Thrall, M., Grauer, G., and Mero, K. (1984). *J. Am. Vet. Med. Assoc.* **184,** 37–41.

Thysell, H. (1969). *Acta Med. Scand.* **185,** 401–407.

Toback, F. G., Smith, P. D., and Lowenstein, L. M. (1974). *J. Clin. Invest.* **56,** 91–97.

Touraine, J. L., Touraine, F., Revillard, J. P., Brochier, J., and Traeger, J. (1975). *Nephron.* **14,** 195–208.

Traver, D. S., Salem, C., Coffman, J. R., Garner, H. E., Moore, J. N., Johnson, J. H., Tritschler, L. G., and Amend, J. F. (1977). *J. Eq. Med. Surg.* **1,** 378–382.

Tripett, S. (1957). *J. Chem. Soc.* 1929–1930.

Trohler, U., Bonjour, J. P., and Fleisch, H. (1976). *J. Clin. Invest.* **57,** 264–273.

Tucker, B. J., and Blantz, R. C. (1977). *Am. J. Physiol.* **232,** F477–F483.

Tyler, F. H. (1972). *In* "Metabolic Basis of Inherited Disease" (J. B. Stanbury, ed.), pp. 1204–1218. McGraw-Hill, New York.

Uribe, D., Krawiec, D. R., Twardock, A. R., and Gelberg, H. B. (1992). *Am. J. Vet. Res.* **53,** 1101–1107.

Valtin, H. (1977). *Am. J. Physiol.* **233,** F491–F501.

Van Buren, C. T., Dudrick, S. J., and Baumbauer, E. (1972). *Surg. Forum.* **23,** 83–84.

Van Urk, H., Malamud, D., Soler-Montesinos, L., and Malt, R. A. (1978). *Lab. Invest.* **38,** 674–676.

Vaughan, E. D., Jr., and Wyker, A. W. (1971). *J. Urol.* **105,** 709–711.

Vaughan, E. D., Sorenson, E. J., and Gillenwater, J. Y. (1971). *Invest. Urol.* **8,** 450–461.

Walker, J. B. (1960). *J. Biol. Chem.* **235,** 2357–2361.

Walker, J. B. (1961). *J. Biol. Chem.* **236,** 493–498.

Walker, L. A., and Valtin, H. (1982). *Ann. Rev. Physiol.* **44,** 203–219.

Wallner, S. F., Kurnick, J. E., Vautrin, R., and Ward, H. P. (1977). *Am. J. Hematol.* **3,** 45–55.

Walser, M., and Bodenlos, L. J. (1959). *J. Clin. Invest.* **38,** 1617–1626.

Watson, A. D. J., Church, D. B., and Fairburn, A. J. (1981). *Am. J. Vet. Res.* **42,** 1878–1880.

Watts, C., and Campbell, J. R. (1970). *Res. Vet. Sci.* **11,** 508–514.

Wen, S. F. (1976). *Am. J. Physiol.* **231,** 468–475.

West, C. D., Traeger, J., and Kaplan, S. A. (1955). *J. Clin. Invest.* **34,** 887–898.

Westall, R. G. (1953). *Biochem. J.* **55,** 244–248.

White, A., Handler, P., and Smith, E. L. (1973). "Principles of Biochemistry." McGraw-Hill, New York.

White, J. V., Olivier, N. B., Reimann, K., and Johnson, C. (1984). *JAVMA* **185,** 882–885.

Wolf, A. V. (1969). *Am. J. Med.* **46,** 837–843.

Wright, F. S., and Giebisch, G. (1992). *In* "The Kidney: Physiology and Pathophysiology" (D. W. Seldin and G. Giebisch, eds.), pp. 2209–2247. Raven Press, New York.

Zhu, M. Q., Vaneerdeweg, W., Buyssens, N., and DeBroe, M. E. (1992). *Nephron.* **62,** 187–191.

CHAPTER

18

Fluid, Electrolyte, and Acid–Base Balance

GARY P. CARLSON

I. INTRODUCTION

The body fluids are arranged in dynamic but orderly functional compartments. Maintenance of these compartments in terms of volume and composition is essential for maintenance of normal physiologic and biochemical events. The electrolytes dissolved in body fluids fulfill vital roles in virtually all of life's processes. Transmembrane movements of electrolytes are responsible for the electrical events that result in nerve conduction and muscular contraction and also serve as essential cofactors in many enzymatically mediated metabolic reactions. The pH of body fluids is maintained within narrow limits. This is necessary for the maintenance of protein structure and function, which is an essential condition for normal progression of metabolic events. Virtually every organ system participates in the maintenance of fluid balance and/or is adversely affected by imbalances. In many disease states, impaired fluid intake, excessive fluid losses, or organ damage and dysfunction results in a state of altered fluid and electrolyte balance. Regardless of whether fluid and electrolyte alterations are the primary problem or simply represent secondary manifestations of some other disease process, successful patient management depends on correct evaluation and appropriate therapy. To do this one must have a clear understanding of the anatomy and physiology of the body fluids, of the pathologic mechanisms by which normal processes become deranged, of the means by which these disturbances can be identified accurately, and finally the procedures that can be used to correct such disturbances in a prompt, safe, and effective manner (Tasker, 1980).

II. PHYSIOLOGY OF FLUID AND ELECTROLYTE BALANCE

A variety of different units have been used in the quantitative evaluation of biologic specimens. To avoid confusion, the units of measure that apply directly to the body fluids and electrolytes are described. An international standard for clinical chemistry units, the "Système Internationale d'Unites" (SI units), was developed to provide consistent terminology and usage. Although SI units are the international standard, they may not be familiar to all students or clinicians. All solute concentrations are expressed in moles or millimoles per liter, blood gas partial pressure in kilopascals, and osmolality in millikelvins of freezing point depression. See Appendix I for conversion tables.

Electrolytes are substances which exist as positive or negative charged particles in aqueous solution. The positively charged particles are *cations* and the negatively charged particles are *anions*. For univalent ions such as sodium, potassium, chloride, and bicarbonate, 1 mole equals 1 equivalent. For multivalent ions, 1 equivalent is equal to the molecular weight in grams (i.e., 1 mole) divided by the charge on the particle. The maintenance of electrical neutrality in biologic fluids requires that there be an equal number of equivalents or milliequivalents of anions and cations in solution. Electrolytes in solution combine equivalent for equivalent, not on a gram for gram or mole for mole basis.

The osmotic properties of a solute in solution are related to the number of particles in solution and not to its weight or its charge. One osmole of a nondissociable substance is equal to its molecular weight in grams. One osmole of any substance which dissociates in solution into two or more particles is equal to the molecular weight in grams divided by the number of particles into which each molecule dissociates. *Osmolarity* is defined as the number of osmoles per liter of final solution, whereas *osmolality* is the number of osmoles per kilogram of water. Although the expressions are very similar, osmolality more correctly describes the osmotic properties as measured in the clinical laboratory.

Most solutes in biologic fluids are present in relatively dilute concentrations and it is more convenient to express these concentrations as millimoles, milliequivalents, or milliosmoles. These simply represent one-thousandth of the standard unit. Conventional terms are milligrams per deciliter (mg/dl), millimole per liter (mmol/liter), milliequivalent/liter (mEq/liter) and milliosomole per kilogram of water (mOsm/kg). The concentrations of the principal anions and cations in plasma are presented in Table 18.1 as expressed in these conventional terms.

III. BODY FLUID COMPARTMENTS

Before proceeding with a discussion of the assessment of fluid deficits or imbalances, we need to consider the organization and composition of the fluid compartments from which these losses occur. An understanding of the forces which govern the relative volume and composition of the body fluid compartments is central to understanding both the clinical and clinicopathologic manifestations of altered fluid balance.

A. Total Body Water

Water is the most abundant compound in the body and all of life's essential processes take place in this aqueous environment. Although there is substantial variation, the total body water (TBW) of most domestic animals is approximately 60% of body weight (0.60 liter/kg). In a 500-kg horse this amounts to approximately 300 liters (Carlson, 1983b); in a 20-kg dog it amounts to just over 12 liters (Kohn and DiBartola, 1992).

Adipose tissue contains very little water and the amount of body fat has a major impact on the relative TBW. The average body water content of women is 0.45 to 0.50 liter/kg as compared to 0.55 to 0.60 liter/kg for men (Elkinton and Danowski, 1955; Edelman *et al.*, 1958). This is largely the result of the larger fat deposits in the adult woman and the larger muscle mass of the adult man (Elkinton and Danowski, 1955). Clear sex-associated differences in body fat are not appreciated in domestic animals. However, certain species of domestic animals such as fattened swine or sheep have a large amount of body fat. Although lighter sheep had a TBW of near 0.65 liter/kg (Wade and Sasser, 1970), the TBW of these fattened animals may be less than 0.50 liter/kg (Hansard, 1964; English,

TABLE 18.1 Plasma Electrolyte Concentrations Expressed in Different Units[a]

Plasma electrolyte	mg/dl	Normal concentration mmol/liter	mEq/liter	mOsm/kg
Cations				
Sodium (Na^+)	326.6	142.0	142.0	142.0
Potassium (K^+)	16.8	4.3	4.3	4.3
Calcium (Ca^{2+})[b]	5.0	1.25	2.5	1.25
Magnesium (Mg^{3+})[b]	1.3	0.55	1.1	0.55
Total cations	349.7	148.1	149.9	148.1
Anions				
Chloride (Cl^-)	369.2	104.0	104.0	104.0
Bicarbonate (HCO_3^-)	146.4	24.0	24.0	24.0
Phosphate (H2PO_4^-, HPO_4^{2-})[c]	3.4	1.1	2.0	1.1
Proteins[c]	7000.0	2.5	14.0	2.5
Others (sulfate lactate, etc.)	—	5.5	5.9	5.5
Total anions	7519.0	135.5	149.9	137.0
Total milliosmoles from electrolytes				285.1

[a] From Tasker (1980).
[b] Only ionized calcium and magnesium have been considered here.
[c] Phosphate concentration is that of organic phosphorus. Since a variable equilibrium exists between $H_2PO_4^-$ and HPO_4^{2-}, the actual valence and milliequivalents must be estimated. The same variability is true of protein anions as well.

1966b), whereas the TBW of the athletic horse is generally greater than 0.65 liter/kg (Robb *et al.*, 1972; Dieterich and Holleman, 1973; Judson and Mooney, 1983). The relative water content of newborn animals is much higher than adults. Data in human infants, calves, foals, and lambs suggest a water content in excess of 75% of body weight at birth (Edelman and Leibman, 1959; Dalton, 1964; Phillips *et al.*, 1971; Pownall and Dalton, 1973; Bennett, 1975). The large TBW is primarily due to the very large extracellular fluid (ECF) volume, which exceeds 0.40 liter/kg at birth in most species (Tollertz, 1964; Bennett, 1975; Kami *et al.*, 1984; Spensley *et al.*, 1987). An initial rapid decline occurs during the first few days to weeks of life with TBW and ECF volumes approaching adult levels by 6 months of age (Spensley *et al.*, 1987).

The TBW consists of two major compartments, the intracellular fluid (ICF) volume and the ECF volume. The distribution of body water is illustrated in Fig. 18.1 indicating the normal fluid balance of a 450-kg horse. The ICF accounts for approximately one-half to two-thirds of the TBW and the ECF accounts for the remainder. Although these two compartments differ markedly in electrolyte composition, they are in osmotic equilibrium and water is freely diffusible between them. The relative volume distribution of water between these two compartments is largely governed by the number of osmotically active particles in each compartment. The ECF volume is determined by the ECF sodium content, and ICF volume is a function of ICF potassium content. The relationship between the exchangeable cation content (sodium in the ECF, potassium in the ICF) and the total body water was defined by Edelman *et al.*, (1958):

$$\text{Serum sodium mEq/liter } H_2O \approx \frac{(\text{Exchangeable sodium + potassium})}{\text{Total body water}} \quad (18.1)$$

Since there are no major osmotic gradients between the ECF and the ICF, serum sodium concentration and osmolality reflect the osmolality of the ICF compartment as well as that of the ECF compartment (Edelman *et al.*, 1958; Scribner, 1969; Saxton and Seldin, 1986).

B. Extracellular Fluid Volume

The ECF should be viewed as a physiologic rather than a strictly definable anatomic space (Carlson *et al.*, 1979a). The ECF volume of adult animals ranges from 0.15 to 0.30 liter/kg body weight (Hix *et al.*, 1953; English, 1966b; Evans, 1972; Hankes *et al.*, 1973; Zweens *et al.*, 1975; Thornton and English, 1977; Carlson *et al.*, 1979a; Kohn, 1979; Spurlock *et al.*, 1985) depending on the species and the volume dilution procedure used. Regulation of ECF volume is a complex process in which a variety of factors interact. The ECF consists of all the fluids located outside the cells and includes the plasma (0.05 liter/kg), interstitial fluid and lymph (0.15 liter/kg), and the transcellular fluids (Edelman and Leibman, 1959; Rose, 1984; Saxton and Seldin, 1986). The transcellular fluids, which include the fluid content of the gastrointestinal tract, are generally con-

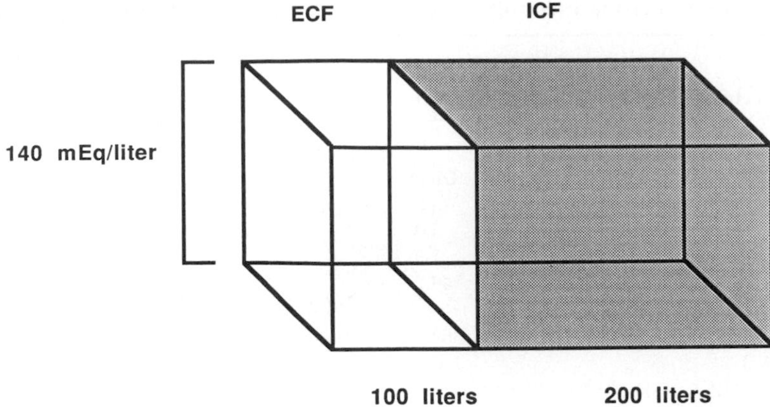

FIGURE 18.1 Body fluid compartments in a normal 450-kg horse. Serum sodium concentration is 140 mEq/liter (140 mmol/liter). Extracellular fluid (ECF) volume is 100 liters and intracellular fluid (ICF) volume is 200 liters.

sidered a subcomponent of the ECF. In small animal species the fluid content of the gastrointestinal tract is relatively small (Strombeck, 1979). In the large animal herbivore species, a substantial volume of fluid is normally present within the gastrointestinal tract. In the horse this may amount to 30–45 liters (Carlson, 1979b) and in cattle the forestomach may contain as much as 30–60 liters of fluid (Phillipson, 1977). During periods of water restriction and certain other forms of dehydration, this gastrointestinal fluid reservoir can be called on to help maintain effective circulating volume (McDougall *et al.*, 1974). All of the fluids of the ECF contain sodium in approximate concentrations of 130–150 mEq/liter H_2O. Sodium provides the osmotic skeleton for the ECF and the sodium content is the single most important determinant of ECF volume (Rose, 1984). Sodium deficits result in decreases in ECF volume, whereas sodium excess is most often associated with water retention and results in edema (McKeown, 1984; Rose, 1984).

C. Intracellular Fluid Volume

The ICF volume represents the fluid content within the body's cells. This volume cannot be measured directly, but is calculated as the difference between the measured TBW and the measured ECF volume. Potassium provides the osmotic skeleton for the ICF in much the same way that sodium provides the osmotic skeleton for the ECF. Because water is freely diffusible into and out of the cell, changes in the tonicity of the ECF are rapidly reflected by similar changes in ICF tonicity (Saxton and Seldin, 1986). This is largely the result of the movement of water across the cell membrane with resultant changes of ICF volume. Thus, although plasma sodium concentration decreases in response to water retention, ICF volume increases (Humes, 1984). On the other hand, with water depletion resulting in hypernatremia ICF volume decreases (Humes, 1984). Relatively little is known about the organization of intracellular water into the various subcellular compartments and organelles.

IV. REGULATION OF BODY FLUIDS AND ELECTROLYTES

A. Effective Circulating Volume

The effective circulating volume refers to that part of the ECF that is within the vascular space and is effectively perfusing the tissues (Rose, 1984). Effective circulating volume tends to vary with ECF volume and both parameters vary with the total body sodium stores (Rose, 1984). Sodium loading produces volume expansion, whereas sodium depletion leads to volume depletion.

Effective circulating volume is not a quantitatively measurable entity but refers to the rate of perfusion of the capillary circulation. Effective circulating volume is maintained by varying vascular resistance, cardiac output, and renal sodium and water excretion (Rose, 1984). Decreases in effective circulating volume result in decreased venous return, decreased cardiac output, and decreased blood pressure. Decreased volume and pressure are recognized by special volume receptors in the cardiopulmonary circulation and kidney which trigger increased sympathetic tone, resulting in increased arterial and venous constriction as well as increased cardiac contractility and heart rate. These responses tend to correct for the volume deficit by increasing cardiac output and systemic blood pressure.

Volume and pressure changes associated with decreases in effective circulating volume also result in activation of the renin–angiotensin system with subsequent enhancement of aldosterone secretion by the adrenal cortex (Brobst, 1984). Aldosterone acts to enhance renal sodium resorption, which is a critical factor in the restoration of effective circulating volume. Additional factors that influence sodium resorption in response to changes in fluid volume, include alterations in glomerular filtration rate, renal hemodynamics, atrial natriuretic factor, and plasma sodium concentration.

B. Antidiuretic Hormone

Antidiuretic hormone (ADH) plays a primary role in the regulation of the osmolality of the body fluids. Antidiuretic hormone is synthesized in the hypothalamus and released in response to changes in plasma osmolality. Because sodium concentration is the primary determinant of plasma osmolality, ADH release is closely correlated to plasma sodium concentration. Increases in plasma osmolality are recognized by special sensors in the hypothalamus and the normal response is increased thirst to enhance water intake and the release of ADH, which increases water reabsorption by the renal collecting tubules. Antidiuretic hormone exerts its activity on the collecting tubules by activating adenyl cyclase, which results in the generation of cyclic adenosine monophosphate (AMP) and protein kinases, which in turn alter the permeability of the tubules to water (Rose, 1984). Antidiuretic hormone also is released in response to decreases in effective circulating fluid volume although the renin–angiotensin system exerts primary control over volume changes. Antidiuretic hormone acts extrarenally as an arterial vasoconstrictor, thus increasing blood pressure. Plasma osmolality decreases in response to a water load and ADH release is inhibited. The resultant reduction in ADH-mediated reabsorption of water in the collecting tubules allows for appropriate renal excretion of the water load and a return of plasma osmolality toward normal. This highly sensitive system responds rapidly to small changes in osmolality and, as a result, plasma osmolality is normally maintained within a relatively narrow range.

C. Renin–Angiotensin

The renin–angiotensin system plays an important role in the maintenance of effective circulating fluid volume. Renin is a proteolytic enzyme produced by special juxtaglomerular cells of the glomerular afferent arteriole. Renin is released in response to reduced renal perfusion produced by hypotension, volume depletion, or increased sympathetic activity. Renin converts the circulating globulin angiotensinogen to angiotensin I, which is subsequently converted by an enzyme in the lung and vascular endothelial cells to the biologically active form, angiotensin II. Angiotensin II exerts a variety of systemic effects which tend to correct hypovolemia and hypotension. Angiotensin II increases renal retention of sodium and water by enhancing secretion of aldosterone from the adrenal cortex; it also has direct effects on the renal tubule. Angiotensin II exerts hemodynamic effects, which tend to increase blood pressure by inducing arteriolar vasoconstriction.

D. Aldosterone

Aldosterone plays a central role in the maintenance of effective circulating fluid volume and potassium balance largely through its effects on renal resorption of sodium in exchange for potassium and hydrogen ion. Aldosterone is produced in the adrenal cortex and exerts its effects on sensitive cells such as the renal collecting tubules by interacting with specific cytoplasmic receptors. The aldosterone receptor complexes subsequently enhance RNA-mediated production of specific proteins, which actually mediate the physiologic effects of the hormone. There is evidence for aldosterone-mediated effects on gastrointestinal sodium and potassium absorption as well as effects on sweat glands to alter the electrolyte composition of sweat in response to sodium depletion (Michell, 1974). Aldosterone secretion is enhanced by the renin–angiotensin system in response to changes in effective circulating fluid volume.

E. Atrial Natriuretic Factor

Atrial natriuretic factor (ANF), also known as atrial natriuretic peptide (ANP), exists as a group of diverse peptide hormones produced in the heart and released into the circulation in response to processes which increase central venous pressure and result in stretching of the atrial wall (Inagami, 1994). The actions of ANF, which tend to reduce cardiac output and systemic blood pressure, are mediated by transmembrane receptors which result in the production of cyclic guanidine monophosphate generated by quanylyl cyclase (Inagami, 1994). ANF results in natriuresis and diuresis by the kidneys. ANF also causes vasodilatation and fluid volume reduction by direct action on vascular smooth muscle and inhibition of the release of aldosterone from the adrenal cortex and norepinephrine from peripheral adrenergic neurons. Additionally, ANF has been found in the brain and centrally medicated effects on fluid volume regulation may be important. Elevated plasma ANF has been noted in humans with a variety of diseases ranging from congestive heart failure to obstructive lung disease to chronic renal failure. However, the physiologic importance of ANF in these disease processes is not completely resolved. In humans, Bartter's syndrome and Gordon's syndrome are thought to be due to an excess or deficiency, respectively of ANF (Christensen, 1993). In horses, elevation of plasma ANF has been associated with treadmill exercise (McKeever *et al.*, 1991).

V. PHYSIOLOGY OF ACID–BASE BALANCE

The hydrogen ion concentration of the ECF is maintained within remarkably narrow limits, and is normally approximately 40 nmol/liter. This is 40×10^{-6} mmol/liter or roughly one-millionth the concentration of the other common electrolytes. Even at these extremely low concentrations, hydrogen ions have profound effects on metabolic events largely through interaction with cellular proteins. These interactions result in altered protein configuration and thus altered protein function. Most enzymatic reactions have a narrowly defined range of pH optimum, and changes in hydrogen ion concentration have direct effects on the rates of reaction and, thus, many basic biologic processes.

A. Definition of pH

Although the hydrogen ion concentration can be expressed in nanomoles per liter, a more useful expression is that of pH. The pH of a solution is equal to the negative logarithm of the hydrogen ion concentration.

$$\mathrm{pH} = -\log[\mathrm{H^+}] \tag{18.2}$$

It is important to remember that pH varies inversely with hydrogen ion concentration. When hydrogen concentration in the blood increases, pH decreases and the animal develops an acidosis. When the hydrogen ion concentration in the blood decreases, pH rises and the animal develops an alkalosis. The traditional view of acid–base balance involves the following:

1. Extracellular and intracellular buffering
2. Regulation of the rate of alveolar ventilation to control carbon dioxide concentration
3. Regulation of renal hydrogen excretion

B. Buffers

A buffer system consists of a weakly dissociated acid and the salt of that acid. The body buffers are able to take up or release hydrogen ion so that changes in hydrogen ion concentration are minimized. According to the law of mass action the dissociation of a weak acid, HA, into H^+ and A^- can be written

$$K_a = \frac{[\mathrm{H^+}]\,[\mathrm{A^-}]}{[\mathrm{HA}]}, \tag{18.3}$$

where K_a is the dissociation constant for the reaction. The equation can be rewritten

$$[\mathrm{H^+}] = K_a \frac{[\mathrm{HA}]}{[\mathrm{A^-}]} \quad \text{or} \quad [\mathrm{H^+}] = K_a \frac{[\mathrm{Acid}]}{[\mathrm{Salt}]}. \tag{18.4}$$

Taking the negative logarithm of both sides of the equation yields

$$-\log[\mathrm{H^+}] = -\log K_a - \log \frac{[\mathrm{HA}]}{[\mathrm{A^-}]}. \tag{18.5}$$

Substituting pH for $-\log[\mathrm{H^+}]$, $+\log[\mathrm{A^-/HA}]$ for $-\log[\mathrm{HA/A^-}]$, and defining pK_a as $-\log K_a$, the formula now reads

$$\mathrm{pH} = pK_a + \log \frac{[\mathrm{A^-}]}{[\mathrm{HA}]}. \tag{18.6}$$

This is the familiar Henderson–Hasselbalch equation for the dissociation of a weak acid. The buffering capacity of the body includes the extracellular buffers, intracellular buffers, and bone. The extracellular buffers include the bicarbonate (HCO_3^-/H_2CO_3) and phosphate ($HPO_4{=}/H_2PO_4^-$) buffer pairs as well as the plasma proteins. The intracellular buffers include protein, organic and inorganic phosphates, and in the red cell hemoglobin. Cation exchange involving the intracellular movement of hydrogen ion in exchange for potassium and, to a much lesser extent, sodium is an additional and important means whereby cellular mechanisms buffer an acid load. The carbonate in bone provides a large and often overlooked buffer store. Although difficult to measure accurately, it has been estimated that bone carbonate may contribute up to 40% of the buffering capacity of an acute acid load. It is important to remember that a change in hydrogen ion concentration will affect all the body's buffer pairs. Thus, evaluation of any one buffer pair reflects the changes which occur in all of the buffer pairs.

1. *Bicarbonate/Carbonic Acid Buffer System*

Large amounts of carbon dioxide are produced by oxidative metabolism each day. Although carbon dioxide is not an acid, it combines with water as it is added to the bloodstream, resulting in the formation of carbonic acid. The enzyme carbonic anhydrase in the red cells facilitates this reaction:

$$CO_2 + H_2O \rightarrow H_2CO_3 \rightleftarrows H^+ + HCO_3^-. \tag{18.7}$$

The ensuing elevation in the hydrogen concentration is

minimized because most of the excess H^+ ions combine with the intracellular buffers, particularly hemoglobin (Hb):

$$H_2CO_3 + Hb^- \rightleftarrows HHb + HCO_3^-. \quad (18.8)$$

The bicarbonate generated by this reaction then leaves the erythrocyte and enters the extracellular fluid in exchange for extracellular chloride ion. The net effect is that CO_2 is primarily carried in the venous circulation as HCO_3^- with little change in the extracellular pH. These processes are reversed in the alveoli. As HHb is oxygenated, H^+ is released and the H^+ combines with HCO_3^- to form H_2CO_3, which dissociates to CO_2 and H_2O. Carbon dioxide then is excreted by alveolar ventilation.

From a clinical and physiologic standpoint the bicarbonate–carbonic acid buffer pair is clearly the most important. Bicarbonate is present in relatively high concentrations, it is relatively easy to measure, and it is the buffer system over which the body has the greatest control. The Henderson–Hasselbalch equation applied to this buffer pair becomes

$$pH = 6.1 + \log \frac{[HCO_3^-]}{[H_2CO_3]}, \quad (18.9)$$

where 6.1 = pK for the HCO_3^-/H_2CO_3 buffer pair.

Plasma pH is determined by the concentration of bicarbonate and carbonic acid or, more importantly, the ratio between bicarbonate and carbonic acid. An equilibrium exists between the partial pressure of CO_2 in the alveolar air, the partial pressure of gaseous CO_2 dissolved in the blood, and the carbonic acid concentration of the blood. The conventional method of evaluating carbonic acid is the determination of the partial pressure of carbon dioxide (pCO_2) in the blood. The carbonic acid concentration can be calculated by multiplying pCO_2 by 0.03 (0.03 is the solubility constant for carbon dioxide in plasma). The Henderson–Hasselbalch equation for this buffer pair then becomes

$$pH = 6.10 + \log \frac{[HCO_3^-]}{0.03(pCO_2)}. \quad (18.10)$$

The four primary acid–base imbalances and their compensating responses are presented in Table 18.2.

TABLE 18.2 Acid–Base Imbalances and Compensating Responses

Disorder	pH	[H+]	Primary imbalance	Compensating response
Metabolic acidosis	↓	↑	↓ $[HCO_3^-]$	↓ pCO_2
Metabolic alkalosis	↑	↓	↑ $[HCO_3^-]$	↑ pCO_2
Respiratory acidosis	↓	↑	↑ pCO_2	↑ $[HCO_3^-]$
Respiratory alkalosis	↑	↓	↓ $[pCO_2$	↓ $[HCO_3^-]$

Acidosis is associated with a decrease in pH due to an increase in hydrogen ion concentration. Alkalosis is due to a decrease in hydrogen ion concentration, which is reflected by an increase in pH. In the metabolic disorders, the primary imbalance is due to changes in bicarbonate concentration. The compensating response is mediated by the respiratory system which, within limits, alters the pCO_2 so as to counterbalance the primary imbalance and to partially restore the pH toward normal. As the name implies, the primary imbalances of the respiratory disorders are related to alterations in alveolar ventilation, which result in increases in pCO_2 in respiratory acidosis and decreases in pCO_2 with respiratory alkalosis. The compensating responses for these primary respiratory imbalances are mediated by the kidney through alterations in the excretion or retention of hydrogen ions or bicarbonate.

Blood samples drawn for acid–base evaluation must be drawn anaerobically and sealed so as to avoid alteration in the blood gas tension. Heparin is the anticoagulant of choice. The rectal temperature of the patient should be taken so that appropriate temperature corrections can be made. For evaluation of blood gases during intense exercise, central blood temperature should be used for this correction because rectal temperature may not accurately reflect blood temperature under these non-steady-state conditions. The temperature correction usually has a greater impact on the pO_2 and pCO_2 than bicarbonate or base balance. Arterial blood samples are generally preferred and are essential for the evaluation of primary respiratory disorders or the patient's status under general anesthesia. Venous blood samples provide reliable data on metabolic acid–base abnormalities and because they are easier to obtain are routinely used. As a general rule blood gas determination should be made as soon after collection as possible. However, appropriately collected blood samples can be held in ice water for as long as 4 hours and still yield reliable results.

The effects of various sampling sites (arterial, venous, and capillary blood) on blood gas determination have been evaluated in dogs (van Sluijs *et al.*, 1983; Ilkiw *et al.*, 1991), horses (Littlejohn and Mitchell, 1969; Speirs, 1980), and swine (Hannon *et al.*, 1990). Consistent differences were demonstrated between arterial and venous blood. Arterial blood samples yield higher values for pH and lower values for pCO_2 and bicarbonate than venous blood, but the calculated base balance tends to be very similar for both arterial and venous blood samples. In horses, venous–arterial differences in bicarbonate can exceed 10 mEq/liter during intense exercise. The higher venous bicarbonate value reflects the important role of bicarbonate as a means of CO_2 transport in the venous circulation (Carlson, 1995). At rest and during exercise, more than 70% of the CO_2 produced in the tissues is transported in the venous circulation to the lungs as bicarbonate. This process is greatly facilitated by erythrocyte carbonic anhydrase and the mechanism of the *chloride shift*. The effects of temperature on oxygen content of dog blood have been studied (Hedley-Whyte and Laver, 1964) as has the oxygen affinity and Bohr coefficient of dog blood (Reeves *et al.*, 1982). Recumbency in neonatal patients and body position during general anesthesia may have a significant impact on both arterial and mixed venous blood gas data (Steffey *et al.*, 1977; Mason *et al.*, 1987; Madigan *et al.*, 1992). Normograms have been developed relating the effects of temperature, CO_2 content, and hemoglobin saturation for dog blood (Rossing and Cain, 1966). The use of blood gas data for the evaluation of acid–base imbalances for clinical problems has been reviewed (DiBartola, 1992a, 1992b, 1992c; DiBartola and De Morais, 1992; De Morais, 1992a; George, 1994).

C. Acidosis

1. *Metabolic*

Metabolic acidosis is characterized by a decrease in pH and bicarbonate. Metabolic acidosis, as traditionally viewed, can be produced by the addition of hydrogen ions or a loss of bicarbonate ions. The initial buffering of an acid load is by the ECF buffers, primarily the bicarbonate–carbonic acid buffer pair (Rose, 1984). Intracellular buffers, particularly protein and phosphate, assist in the buffering process. The intracellular movement of hydrogen in exchange for potassium helps to prevent an excessive increase of the ECF hydrogen ion concentration in the face of an acid load. This exchange is called the *cation shift* and can result in hyperkalemia even though the total body potassium stores have been depleted due to renal or gastrointestinal losses.

a. Causes of Metabolic Acidosis

The most common causes include lactic acidosis, ketoacidosis, gastrointestinal loss of bicarbonate due to diarrhea, and renal failure, which may result in a decreased ability to excrete hydrogen and thus to retain bicarbonate (Emmett and Nairns, 1977). A profound metabolic acidosis without dehydration leading to depression, recumbency, and death has been described in goat kids (Tremblay *et al.*, 1991) and appears to be similar to reports in calves (Kasari and Naylor, 1984, 1986). The cause was undetermined, but the acidosis was usually associated with an increased anion gap. Sodium bicarbonate therapy, if initiated early in the course of the disease, was often curative. Additional causes of a metabolic acidosis include ingestion of certain medications or toxic compounds such as salicylate, methanol, ethylene glycol, or paraldehyde, which results in the accumulation of exogenous anions (DiBartola 1992b).

b. Compensation

A metabolic acidosis is recognized very quickly and the compensating respiratory response of increased ventilation will begin reduction of the pCO_2 within minutes. In dogs the anticipated respiratory response is a reduction of pCO_2 by 0.7 mm Hg for each mEq/liter decrease in bicarbonate (DeMorais, 1992a). This minimizes the fall in pH but the protective effects of the respiratory response are relatively short lived, lasting only a few days. Long-term correction of a metabolic acidosis requires renal bicarbonate retention and enhanced renal acid excretion primarily as ammonium ion since there is little ability to increase the titratible acidity, which consists primarily of phosphate buffers (Rose, 1984). Complete correction of a metabolic acidosis may be difficult in patients with intrinsic renal disease or diseases that would impair the kidney's ability to excrete acid or retain bicarbonate such as renal tubular acidosis.

2. *Respiratory*

A respiratory acidosis is characterized by a decrease in pH and an increase in pCO_2. Respiratory acidosis develops because of decreased effective alveolar ventilation. The initial buffering of the acid load produced by a respiratory acidosis is almost exclusively by the intracellular buffers. The principal ECF buffer, the bicarbonate–carbonic acid buffer pair, cannot buffer a respiratory acidosis. Carbon dioxide diffuses through the lung much more readily than O_2, thus, diseases that compromise ventilation normally result in decreases in pO_2 before significant increases in pCO_2 develop. The respiratory center is extremely sensitive to minor changes in pCO_2 and increased pCO_2 normally provides the major stimulus to ventilation (Rose, 1984). In contrast, hypoxemia does not begin to promote enhanced ventilation until the arterial pCO_2 is substantially decreased. If, however, the arterial pCO_2 is held at normal values or is elevated because of intrinsic lung disease, then ventilation begins to be enhanced as the arterial pO_2 falls below 70–80 mm Hg (Rose, 1984).

a. Causes of Respiratory Acidosis

Any disorder that interferes with normal effective ventilation may produce a respiratory acidosis. The most common causes are primary pulmonary diseases ranging from acute upper respiratory obstruction, to pneumonia, to pneumothorax, and chronic obstructive lung disease. Diseases or drugs that affect the central nervous system may inhibit the medullary respiratory center and can produce a profound respiratory acidosis. An additional cause of special importance in veterinary medicine is general anesthesia with volatile agents using a closed system. Under these conditions ventilation may be seriously reduced without producing hypoxia. The high oxygen content of the gas mixture maintains high pO_2 in the blood but depression of the respiratory center may result in insufficient alveolar ventilation so that CO_2 accumulates. This problem can be overcome through the use of a positive pressure ventilatory apparatus and careful monitoring of arterial blood gases during general anesthesia.

b. Compensation

The compensating response for a respiratory acidosis is renal retention of bicarbonate. This response requires several days and, thus, the response is only seen in a chronic respiratory acidosis. In dogs with chronic respiratory acidosis, a compensating increase of 0.35 mEq/liter of bicarbonate is anticipated for each mm Hg increase in pCO_2 (De Morais, 1992a). The extent of the rise in the plasma bicarbonate concentration in chronic respiratory acidosis is determined by increased renal hydrogen secretion (Rose, 1984). Exogenous bicarbonate is unnecessary and should bicarbonate be administered to patients with a respiratory acidosis it would be excreted without affecting the final plasma bicarbonate concentration.

D. Alkalosis

1. Metabolic Alkalosis

Metabolic alkalosis is characterized by an increase in pH and bicarbonate. Metabolic alkalosis occurs with some frequency in domestic animals and is commonly seen in association with digestive disturbances in ruminants. The development of a metabolic alkalosis requires an initiating process capable of generating an alkalosis and then additional factors which are necessary for the continued maintenance of the alkalosis (Rose, 1984). Generation of a metabolic alkalosis can be due to excessive hydrogen loss, bicarbonate retention, or as a contraction alkalosis. A contraction alkalosis occurs with reduction of ECF fluid volume due to a loss or sequestration of sodium and chloride containing fluid without commensurate loss of bicarbonate (Garella *et al.*, 1975). Excessive hydrogen ion losses can result in a metabolic alkalosis.

a. Causes of Metabolic Alkalosis

The most common causes of increased hydrogen loss are gastrointestinal losses due to vomiting in small animals (Strombeck, 1979) or sequestration of chloride-rich fluid in the abomasum and forestomach of ruminants (Gingerich and Murdick, 1975b; McGuirk and Butler, 1980). Excessive renal hydrogen loss associated with mineralocorticoid excess, diuretic usage (particularly the loop diuretics such as furosemide), and low chloride intake may cause or contribute to the generation of a metabolic alkalosis (Rose, 1984). Most of these disorders are also associated with the development of significant sodium and chloride deficits and resultant decreases in effective circulating volume. These deficits and the responses that decreased effective circulating volume induce are central features of the processes which serve to maintain and perpetuate a metabolic alkalosis. Hydrogen loss from the ECF can also occur with hydrogen movement into the cells in response to potassium depletion (Irvine and Dow, 1968). Excessive bicarbonate administration is an additional potential cause of metabolic alkalosis. Most normal animals can tolerate large doses of bicarbonate and excesses are rapidly eliminated by renal excretion (Rumbaugh *et al.*, 1981). However, patients with decreases in effective circulating blood volume or with potassium or chloride deficits may not tolerate a bicarbonate load because renal clearance of excess bicarbonate is likely to be impaired.

The factors responsible for the maintenance of a metabolic alkalosis all impair renal bicarbonate excretion. These factors might include decreased glomerular filtration of bicarbonate seen in some types of renal failure. However, the most common factor is increased renal tubular bicarbonate resorption, which is associated with the renal response to decreases in the effective circulating fluid volume, potassium depletion, and/or chloride depletion (Rose, 1984). Sodium resorption is enhanced in response to hypovolemia in order to help restore normal effective circulating fluid volume. The maintenance of electroneutrality requires that sodium resorption in the proximal tubule be accompanied by a resorbable anion such as chloride, whereas in the distal tubule sodium resorption is associated with the secretion of a cation, usually hydrogen or, to a lesser extent, potassium. The only resorbable anion normally present in appreciable quantities in the proximal tubular fluid is chloride. In a metabolic alkalosis, plasma bicarbonate is increased and chloride concentration is generally decreased as the result of disproportionately high chloride losses due to vomiting, sequestration of gastric fluid (Whitlock *et al.*, 1975b), diuretic usage, or heavy sweat losses in exercising horses (Carlson, 1975, 1979a). The relative lack of the resorbable anion, chloride, in the proximal tubule thus allows a larger amount of sodium to reach the distal tubule where the action of aldosterone enhances hydrogen loss into the tubular lumen in exchange for sodium. The maintenance of effective circulating volume is so critical that the body chooses to maintain circulating volume by enhanced sodium resorption by whatever means necessary, even at the expense of extracellular pH. Renal hydrogen excretion is directly linked with bicarbonate resorption and thus it is not possible to eliminate the excess bicarbonate, and the metabolic alkalosis is maintained (Rose, 1984). This is the reason for the paradoxic acid urine seen in some patients with metabolic alkalosis (Gingerich and Murdick, 1975a, 1975b; McGuirk and Butler, 1980). Hypokalemia is another factor that contributes to the maintenance of a metabolic alkalosis. Hypokalemia is associated with an increase in intracellular hydrogen ion concentration. Increased renal tubular cell hydrogen ion concentration may enhance hydrogen secretion and thus bicarbonate reabsorption by the tubular cells.

b. Compensation

Chemoreceptors in the respiratory center sense the alkalosis and the respiratory response to a metabolic alkalosis is hypoventilation resulting in an increase in pCO_2. In dogs the expected compensating response is an increase of pCO_2 of 0.7 mm Hg for each mEq/liter increase in bicarbonate.

2. *Respiratory Alkalosis*

Respiratory alkalosis is associated with an increase in pH and a decrease in pCO_2.

a. Causes of Respiratory Alkalosis

Respiratory alkalosis is due to hyperventilation which may be stimulated by hypoxemia associated with pulmonary disease, congestive heart failure, or severe anemia. Hyperventilation may also be associated with psychogenic disturbances or neurologic disorders which stimulate the medullary respiratory center such as salicylate intoxication, or gram-negative sepsis. Respiratory alkalosis may be seen in animals in pain or under psychological stress. Hyperventilation may occur in dogs and other nonsweating animals because they employ respiratory evaporative processes for heat loss to prevent overheating (Tasker, 1980).

b. Compensation

The initial compensating response to an acute respiratory alkalosis is a modest decline in ECF bicarbonate concentration as the result of cellular buffering. Subsequent renal responses result in decreased ECF bicarbonate concentration through reduced renal bicarbonate reabsorption. These responses require 2 to 3 days for completion. The decline in bicarbonate is partially offset by chloride retention in order to retain electroneutrality. Thus, hyperchloremia and decreased pCO_2 may be associated with compensated respiratory alkalosis as well as compensated metabolic acidosis. Compensating responses for chronic respiratory alkalosis lasting several weeks may actually be sufficient to return pH to normal. In dogs anticipated renal compensation for a chronic respiratory alkalosis results in a decrease of bicarbonate of 0.55 mEq/liter for each mm Hg decrease in pCO_2 (DeMorais, 1992a).

E. Mixed Acid–Base Imbalances

Mixed acid–base disorders occur when several primary acid–base imbalances coexist (De Morais, 1992a). Metabolic acidosis and alkalosis can coexist and either or sometimes both of these metabolic abnormalities may occur with either respiratory acidosis or alkalosis (Nairns and Emmett, 1980; Wilson and Green, 1985). Evaluation of mixed acid–base abnormalities requires an understanding of the anion gap, the relationship between the change in serum sodium and chloride concentration, and the limits of compensation for the primary acid–base imbalances (Wilson and Green, 1985; Saxton and Seldin, 1986). Clinical findings and history are also necessary in order to define the factors that may contribute to the development of mixed acid–base disorders. The following are important considerations in evaluation of possible mixed acid–base disorders:

1. Compensating responses to primary acid–base disturbances do not result in overcompensation.
2. With the possible exception of chronic respiratory acidosis, compensating responses for primary acid–base disturbances rarely correct pH to normal. In patients with acid–base imbalances, a normal pH indicates a mixed acid–base disturbance.
3. A change in pH in the opposite direction to that predicted for a known primary disorder indicates a mixed disturbance.
4. With primary acid–base disturbances, bicarbonate and pCO_2 always deviate in the same direction. If these parameters deviate in opposite directions a mixed abnormality exists.

Although mixed acid–base abnormalities undoubtedly occur in animals and have been documented in the veterinary literature, they are often overlooked (Wilson and Green, 1985). An appreciation of the potential for the development of mixed abnormalities is essential for the correct interpretation of clinical and clinicopathologic data which would otherwise be confusing. Care should be taken when evaluating suspected mixed acid–base abnormalities that sufficient time has elapsed such that anticipated compensating responses could have occurred (DeMorais, 1992a).

F. Anion Gap

The anion gap can be calculated as the difference between the major cation (sodium) and the measured anions (chloride + bicarbonate) (Emmett and Nairns, 1977). Some investigators prefer to use the formula

$$\text{Anion gap} = (\text{Sodium} + \text{Potassium}) - (\text{Chloride} + \text{Bicarbonate}). \quad (18.11)$$

The addition of potassium to the equation, however, adds little to the diagnostic utility of this calculation (Emmett and Nairns, 1977; Man and Carroll, 1977; Epstein, 1984); the anion gap calculated with the inclusion of potassium concentration will be about 4 mEq/liter higher. Because most of the published data on the anion gap in animal species are the result of calculations using the second equation, this form is used in Table 18.4. Provided the component determinations are valid, the calculated anion gap provides an approximation of the so-called "unmeasured anions." Normally these unmeasured anions consist primarily of negatively charged plasma proteins, because the charges of the unmeasured cations (potassium, calcium, and magnesium) tend to balance out the charges of the unmeasured anions (phosphate, sulfate, and organic ions). The anion gap for most species of domestic animals appears to be similar to that defined for human subjects, i.e., approximately 10–20 mEq/liter (10 to 20 mmol/liter). However, there do appear to be significant differences in the normal range of the anion gap of different species as indicated in Table 18.4 (Adrogue *et al.*, 1978; Shull, 1978, 1981; Feldman and Rosenberg, 1981; Bristol, 1982; Polzin *et al.*, 1982; Gossett and French, 1983). Age-related changes in anion gap have been reported in horses (Gossett and French, 1983), with young foals having a significantly larger anion gap than adults. Further experimental data will be necessary to establish more clearly the normal range for the anion gap of animals under varying conditions.

TABLE 18.3 Causes of Alterations in Anion Gap

- Decreased anion gap
  - Increased cationic protein
    - Polyclonal gammopathy (IgG)
  - Hypoalbuminemia
  - Hyperchloremic acidosis
    - Altered protein anionic equivalents
  - Laboratory error
- Increased anion gap
  - Metabolic acidosis
    - Organic acids (lactic, keto acids)
      - Hypovolemic shock
      - Anaerobic exercise
      - Diabetes
      - Grain overload
      - Ketosis
    - Nonmetabolizable acids
      - Inorganic acids (sulfate, phosphate)
        - Uremic acidosis
      - Intoxication or poisoning
        - Salicylate
        - Paraldehyde
        - Metaldehyde
        - Methanol
        - Ethylene glycol
  - Laboratory error

This simple calculation (Eq. 18.11) can be extremely helpful in the categorization of acid–base disorders with regard to potential causal factors (Table 18.3) and may serve as a prognostic guide in a variety of circumstances (Shull, 1978; Bristol, 1982; Garry and Rings, 1987). Decreases in the anion gap can be seen with increases in cationic proteins associated with polyclonal gammopathy or multiple myeloma. Decreases in anion gap due to decreases in unmeasured anions occur most commonly with hypoalbuminemia and hyperchloremic metabolic acidosis, but also may be noted with overhydration. The causal factors associated with a hyperchloremic metabolic acidosis with a normal to low anion gap can often be differentiated based on the serum potassium concentration. Hyperchloremic metabolic acidosis associated with gastrointestinal fluid losses from diarrhea or renal causes such as renal tubular acidosis most often manifest a hypokalemia (Saxton and Seldin, 1986; Ziemer *et al.*, 1987a, 1987b). Hyperchloremic metabolic acidosis associated with decreased mineralocorticoid secretion or activity, such as that seen in Addison's disease or renal failure, generally presents with a hyperkalemia (Saxton and Seldin, 1986). There are indications that changes in hydrogen ion concentration may alter protein equivalency and thus alter the anion gap in either an acidosis or alkalosis (Adrogue *et al.*, 1978; Madias *et al.*, 1979).

Dehydration and alkalosis are potential but minor causes of increased anion gap. Most commonly, elevations of the anion gap are associated with the development of a metabolic acidosis in which there is an increase in anions which are not routinely measured in the clinical laboratory. This is called a *high anion gap acidosis* and may be associated with an accumulation of metabolizable acids as in a lactic acidosis associated with anaerobic exercise, grain overload, or hypovolemic shock or ketoacidosis due to diabetes or ketosis or with the accumulation of nonmetabolizable acids as in uremic acidosis or various intoxications (see Table 18.3). The presence of a metabolic acidosis with a high anion gap thus provides grounds to undertake a thorough investigation of disease processes capable of pro-

TABLE 18.4 Reference Values for Blood Gas and Electrolyte Determinations[a]

Electrolyte or gas	Dog	Cat	Horse	Ox	Sheep	Goat	Pig
Sodium (mEq/liter)	140–155	147–156	132–146	132–152	139–152	142–155	135–150
Potassium (mEq/liter)	3.7–5.8	4.0–5.3	2.6–5.0	3.9–5.8	3.9–5.4	3.5–6.7	4.4–6.7
Chloride (mEq/liter)	105–120	115–123	99–109	97–111	95–103	99–110	94–106
Calcium (mg/dl)	9.0–11.3	6.2–10.2	11.2–13.6	9.7–12.4	11.5–12.9	8.9–11.7	7.1–11.6
Magnesium (mg/dl)	1.8–2.4	2.2	2.2–2.8	1.8–2.3	2.2–2.8	2.8–3.6	2.7–3.7
Phosphorus (mg/dl)	2.6–6.2	4.5–8.1	3.1–5.6	5.6–6.5	5.0–7.3	6.5	5.3–9.6
Anion gap (mEq/liter)	15–25		6.6–14.7	13.9–20.2			
pH	7.31–7.42	7.24–7.40	7.32–7.44	7.35–7.50	7.32–7.54		7.39
pCO_2 (mm Hg)	29–42	29–42	38–46	35–44	37–46		44.3
Bicarbonate (mEq/liter)	17–24	17–24	24–30	20–30	20–25		25.6
Osmolality (mOsm/kg H_2O)	280–305	280–305	270–300	270–300			

[a] In part from Tasker (1980).

ducing an accumulation of these unmeasured anions. The anion gap also may be useful in the identification of mixed acid–base imbalances. When the change in the anion gap does not approximate the change in bicarbonate, a mixed metabolic acid–base imbalance should be suspected.

G. Bicarbonate and Total CO_2

If respiratory disturbances can be eliminated, the metabolic component of acid–base balance is indicated by the bicarbonate concentration. Bicarbonate is usually estimated by determination of the "CO_2 content" or "total CO_2" of plasma or serum samples.

Bicarbonate actually accounts for approximately 95% of the measured total CO_2 and thus the total CO_2 provides a measure of metabolic changes in acid–base balance. The bicarbonate determined in this fashion will be decreased in a metabolic acidosis and increased in a metabolic alkalosis. Estimates of bicarbonate are often provided in automated chemistry profiles. These determinations may provide an indication of the metabolic acid–base status. However, if acid–base abnormalities are suspected, a proper blood gas evaluation should be undertaken.

H. Buffer Base, Standard Bicarbonate, and Base Excess or Deficit

These values are mathematically derived from the measurements of blood pH and pCO_2 and provide an indication of the metabolic component of acid–base balance. Note that the metabolic changes indicated by these parameters do not always reflect the primary acid–base imbalances, but may represent compensating responses for primary respiratory disorders.

The *buffer base* provides an indication of the sum of all the buffer anions in blood under standardized conditions. The *standard bicarbonate* is the plasma bicarbonate concentration that would be found under specific conditions which eliminate respiratory influences on the values obtained. The *base excess,* which is sometimes considered to be the *base deficit* when the value is negative, indicates the deviation of the buffer base from normal. This derived value is often supplied in routine assessment of acid–base balance and is generally taken as an indication of the deviation of bicarbonate from normal. In an animal with a metabolic acidosis, the calculated base deficit provides a means of estimating the amount of bicarbonate required to correct the acid–base balance to normal. This estimate is calculated by multiplying the base deficit by the probable bicarbonate space (which is variably estimated from 0.25 to 0.55 liter/kg body weight). In newborn animals the bicarbonate may be even higher, 0.40–0.65 liter/kg body weight. The usual figure used is 0.3–0.4 liter/kg. For a 20-kg animal with a base deficit of 10 mmEq/liter (10 mmol/liter), the bicarbonate required would be calculated as

$$\begin{aligned}&\text{Bicarbonate required}\\&\quad = 20\ \text{kg} \times 0.3\ \text{liter/kg} \times 10\ \text{mEq/liter}\\&\quad = 60\ \text{mEq.}\end{aligned} \tag{18.12}$$

It should be appreciated that this calculation provides only a crude guide to bicarbonate requirements, but can be a useful step in the quantitative approach to correction of a serious primary metabolic acidosis.

I. Nontraditional Approach to Acid–Base Balance

Some years ago Peter Stewart developed a quantitative physiochemical approach to acid–base balance (Stewart, 1981, 1983). He recognized that the acid–base status of the aqueous solutions of the body was determined not only by the Henderson–Hasselbalch equation but also by a number of other relationships which could be represented by a series of simple equations, all of which must be satisfied simultaneously. In this approach, acid–base balance is determined by three independent variables: (1) strong ion difference, [SID], which for our purposes can be considered as (sodium + potassium) − (chloride); (2) the partial pressure of CO_2; and (3) the total concentration of nonvolatile weak acids, [Atot], the principal component of which is the plasma proteins but also includes inorganic phosphate. Bicarbonate and hydrogen ion concentration and, thus, pH are dependent variables determined solely by the independent variables listed. The appeal of Stewart's approach is the focus on factors which are causally related to acid–base balance, the independent variables. However, many of these concepts are not intuitively obvious. The interested reader is referred to Stewart's original work and to a number of papers that have attempted to adapt this approach for practical application (Fencil and Rossing, 1989, Jones, 1990; Fencil and Leith, 1993; De Morais, 1992b; Kowalchuck and Scheuermann, 1993; Gilfix *et al.*, 1993; Frischmeyer and Moon, 1994; Whitehair *et al.*, 1995).

It is useful to consider Stewart's three independent variables and how they relate to the more traditional view of acid–base balance. Strong electrolytes are completely dissociated in aqueous solution and chemically nonreactive. The [SID] is simply the difference between the total concentration of strong cations (sodium, potassium, and magnesium) and the total concentration of strong anions (chloride, sulphate, lactate, acedoacetate, and β-hydroxybutyrate). Since they are present in higher concentrations in the body fluids sodium, potassium and chloride are normally the principal determinants of [SID]. The [SID] is synonymous with buffer base as described by Singer and Hastings (1948), and as such can be considered to be roughly equivalent to the metabolic component of the traditional approach to acid–base balance. Abnormalities in pCO_2 are viewed in essentially the same manner in both the traditional and nontraditional approach to acid–base balance. Primary increases in pCO_2 result in a respiratory acidosis and decreases in pCO_2 result in a respiratory alkalosis. The contribution of plasma proteins to acid–base balance is not considered in the traditional approach to acid–base balance. The plasma proteins, or perhaps more correctly plasma albumen, make up the majority of [Atot], whereas inorganic phosphate normally accounts for less than 5% of [Atot]. The [Atot] in body fluids exist in both dissociated [A^-] and undissociated [HA] forms. A decrease in [Atot] due to hypoalbuminemia causes an alkalosis with an increase in bicarbonate; hyperalbuminemia has the opposite effect. Hypoalbuminemia is one of the most common causes of alkalosis in older human patients (McAuliffe *et al.*, 1986). Changes in A^- associated with changes in albumen concentration also have a direct, but often unrecognized, effect on anion gap. Increases in A^- result in an increase in anion gap, whereas decreases in A^- cause a decrease in anion gap (McAuliffe *et al.*, 1986). In some circumstances change in protein concentration will potentate or ameliorate the effects of alterations in [SID] on acid–base balance. As an example, in a vomiting dog the elevated protein associated with dehydration may reduce the bicarbonate increase anticipated for a given change in [SID].

Protein and inorganic phosphate often remain constant in the face of an acid–base disturbance and thus acid–base balance is largely controlled by changes in pCO_2 mediated by the respiratory system, whereas changes in [SID] are largely under the control of the kidneys. Compensating changes for increases in [SID] are brought about by alterations in pCO_2 via the respiratory tract; the reverse is true of primary respiratory disorders. Thus renal compensation for primary respiratory disorders and respiratory compensation for primary metabolic acid–base disturbances are similar in both the traditional and nontraditional approach. Precise quantitation of the anticipated compensating responses to primary acid–base disturbances based on change in [SID] has not yet been determined.

One might consider why such diverse conditions as vomiting in a dog, heavy sweat loss in an endurance horse, displaced abomasum in a cow, and the administration of the loop diuretic furosemide in a cat result in a similar acid–base disturbance. In each circumstance, we see a disproportionate loss of chloride relative to sodium, which results in a hypochloremia and an increase in [SID]. Correction of the alkalosis is not brought about by the administration of hydrogen ions, but by the provision of chloride, generally as sodium chloride or potassium chloride, which results in a decrease in [SID] and thus a return of the dependent variables, bicarbonate and hydrogen ion, toward normal. The data presented in Fig. 18.2 illustrate the close interrelationship between the relative balance of cations and anions in the normal animal and animals with metabolic acid–base disturbances. The horse with the lactic acidosis had just completed a race and in the traditional view the high anion gap metabolic acidosis was due to the high plasma lactate and a markedly decreased bicarbonate. From a nontraditional point of view, the principal reason for the acidosis was the marked increase in the strong anion, lactate, which resulted in decreased [SID]. Intense exercise caused profound but transient compartmental fluid shifts, which increase the concentration of sodium and potassium relative to chloride in the venous circulation.

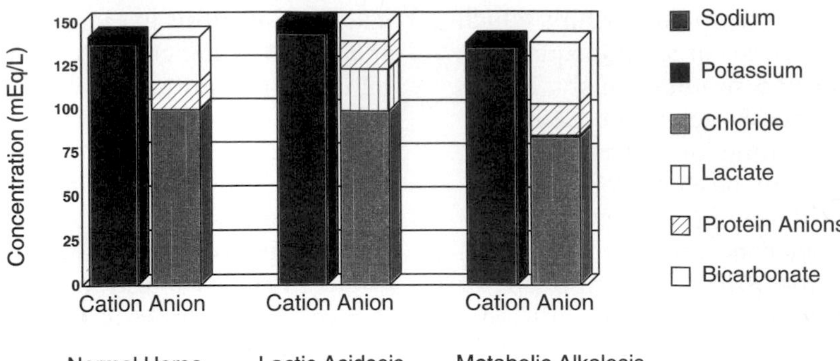

FIGURE 18.2 The cation–anion balance of three horses with differing acid–base status. The anion gap (AG) is calculated as (sodium + potassium) − (chloride + bicarbonate). In this figure AG is represented by the sum (protein anions + lactate). The [SID] is calculated as (sodium + potassium) - (chloride + lactate) and [SID] is equal to the sum (bicarbonate + protein anions).

This tends to ameliorate in part the decrease in [SID] brought about by the accumulation of the strong anion lactate (Carlson, 1995). The horse with the metabolic alkalosis had just completed an endurance race and was moderately dehydrated from heavy sweat losses. The disproportionately high loss of chloride in equine sweat was the principal reason for the hypochloremia with a resultant increase in [SID] and the development of a moderate metabolic alkalosis. The increase in [Atot] and A^- associated with elevated plasma proteins in this horse contributed to a modest increase in anion gap and blunted the increase in bicarbonate associated with the increased [SID]. In almost all circumstances there is a close correlation between bicarbonate and [SID]. In the cerebrospinal fluid where there is no appreciable protein, [SID] and bicarbonate concentration are essentially the same.

The nontraditional approach to acid–base balance has not replaced the traditional approach in routine medical matters. However, calculation of [SID] is simple and provides useful insight in patients with metabolic acid–base disturbances. Factors that influence [SID] range from changes in free water, to sodium–chloride imbalances resulting from excessive losses or disproportionate retention of sodium or chloride, to the accumulation of strong organic anions. Organic acidosis can be produced by the accumulation of exogenous as well as endogenous organic anions. Examples of exogenous anions include salicylate, glycolate, and formate associated with the ingestion of aspirin, ethylene glycol, and methanol, respectively. Many of these endogenous and exogenous organic anions are not routinely monitored in the diagnostic laboratory. This can create problems when calculating the [SID] because the presence of these unmeasured strong anions may not be appreciated. Although the anion gap can be helpful, it does not always accurately predict the presence of these compounds. More sophisticated mathematical methods have been suggested as a means for the detection of unmeasured anions for use in this calculation (Stewart, 1981). In animals with major changes in protein or albumen concentration, the primary concern must be a thorough investigation of the cause of the increase or decrease in protein. The acid–base consequences of change in protein and albumen concentration tend to be modest, but are a potential source of confusion when evaluating acid–base data.

With our present state of knowledge, the nontraditional approach to acid–base balance has had limited acceptance by the veterinary medical community. This limited acceptance is largely the result of our inability to translate this information effectively into the practical decision-making process. However, as we learn to harness its power, this approach has the potential to be extremely helpful for the evaluation of the interrelationship between fluid, electrolyte, and acid–base balance.

J. Dietary Factors in Acid–Base Balance

Dietary factors, particularly the dietary cation–anion balance (DCAB), have been extensively studied in cattle, swine, poultry, and horses. The calculation of the DCAB of a dry feed ration is remarkably similar to the calculation of [SID] for body fluids. The DCAB in mEq is generally represented as (sodium + potassium) − (chloride + sulfate) per kilogram of dry matter of the diet. Diets with a high DCAB, such as alfalfa hay, have an alkalinizing effect and are an important factor in the alkaline urine of most herbivores. High-grain rations tend to have a lower DCAB.

Manipulation of the DCAB has been employed to enhance milk yield in dairy cattle, to reduce the incidence and severity of gastric ulceration in swine, to decrease the incidence of milk fever in cattle, and to alter the urine pH and calcium balance in horses. The addition of sodium bicarbonate to the ration of dairy cattle to raise the DCAB from −100 to +200 mEq/kg diet DM resulted in increased milk production of more than 8%, which was due in part to more effective ruminal digestion (Block, 1984). On the other hand, supplementation of the diet of cattle with calcium chloride or ammonium chloride so as to lower the DCAB has an acidifying effect and has been shown to reduce the incidence of milk fever by enhancing the mobilization of calcium from the bone (Billig, 1984). As the application of these dietary practices become more widespread, we need to appreciate the implications of dietary factors and electrolyte supplementation on mineral metabolism and acid–base balance (Fredeen *et al.*, 1988).

Sodium bicarbonate supplementation has been used as a prerace ergogenic aid in race horses. Relatively large doses, 500 g or more of sodium bicarbonate, often mixed with sugar and water and referred to as "milk shakes," have been given via nasogastric tube and result in a marked metabolic alkalosis. Although experimental studies have often failed to detect a measurable performance benefit from sodium bicarbonate supplementation, practical experience suggests that some horses, particularly Standardbreds, show marked improvement in race times. Administration of any substance with intent to alter the performance is illegal in most racing jurisdictions. In many racing states, horses must meet specific guidelines for venous blood pH and bicarbonate or risk disqualification.

VI. EVALUATION OF IMBALANCES

It is important to understand the difference between volume regulation and osmoregulation. Osmoregulation is governed by osmoreceptors influencing ADH and thirst; volume disturbances are sensed by multiple volume receptors which activate effectors such as aldosterone. Antidiuretic hormone increases water resorption (and therefore urine osmolality), but does not affect sodium transport. Aldosterone enhances sodium reabsorption but not directly that of water. Thus, osmoregulation is achieved by changes in water balance and volume regulation primarily through changes in sodium balance.

Water balance is achieved when water intake from all sources is equal to water output by all routes. Water is available as drinking water, as water content of feedstuffs, and as metabolic water derived by oxidative metabolism. Oxidation of 1 g of fat, carbohydrate, or protein results in the production of 1.07, 0.06, or 0.41 g of water, respectively. Water is lost from the body by four basic routes: urine, feces, insensible respiration and cutaneous water loss, and as sensible perspiration or sweat in some animal species. Water intake and output may vary considerably from day to day, but normal animals are able to maintain water balance within remarkably narrow limits and at the same time maintain the critical interrelationship between water balance and electrolyte balance.

For human subjects, the normal values for water intake and output via various routes are well established. Although there is a substantial amount of data on water balance for many domestic animals (Leitch and Thomson, 1944; English, 1966a; Tasker, 1967a; Yoshida *et al.*, 1967; Fonnesbeck, 1968; Kamal *et al.*, 1972; Hinton, 1978; Sufit *et al.*, 1985), these data vary markedly from species to species and are only valid for the specific experimental conditions under which they were collected. Animals eat to meet their caloric requirements. The nursing or grazing animal may have a feed intake that is greater than 90% water as compared to animals on dry hay or dried prepared pet food, which may contain less than 10% water. Some desert rodents are so well adapted that they are able to maintain water balance without water intake and rely on the water content of feedstuffs and metabolic water derived from oxidative metabolism. The koala in its native state in Australia obtains virtually all of its water from the leaves of specific species of eucalyptus trees, which constitute its entire diet. Dehydration due to water restriction with and without heat stress has been studied widely in a variety of animal species (Elkinton and Taffel, 1942; Hix *et al.*, 1953; Bianca *et al.*, 1965; Tasker, 1967b; Kamal *et al.*, 1972; Schultze *et al.*, 1972; Rumsey and Bond, 1976; Carlson *et al.*, 1979a; Brobst and Bayly, 1982; Rumbaugh *et al.*, 1982; Genetzky *et al.*, 1987).

A. Water

1. Depletion—Dehydration

Dehydration is a relatively common problem in domestic animals. Dehydration results from inadequate fluid intake in the face of normal to increased fluid losses. When water losses occur with little or no electrolyte losses (i.e., panting or feed and water restriction), serum sodium concentration and osmolality increase. This is called *hypertonic dehydration* and occurs when water losses exceed losses of the exchangeable cations sodium and potassium (Carlson, 1987). This imbalance between total body water and exchangeable cations is best characterized as a *relative water deficit* (Scribner, 1969). The effects of a pure water loss of 30 liters in a 450-kg horse are illustrated in Fig. 18.3. In this theoretical example, there has been a 10% loss of body water, but no change in electrolyte balance. Plasma sodium concentration has increased from a normal of 140 mEq/liter (140 mmol/liter) to 155 mEq/liter (155 mmol/liter). Fluid losses are shared proportionately by the ICF and ECF and few other clinical or clinicopathologic abnormalities will be noted until the fluid losses become more severe. Hypernatremia is associated with contraction of the ICF volume and shrinkage of the cells.

When water losses are associated with proportionate losses of exchangeable cations [i.e., 130–150 mEq (130–150 mmol) of sodium plus potassium per liter of water lost] an isotonic fluid volume contraction develops. The effects of an isotonic fluid loss of 30 liters of water, 1400 mEq (1400 mmol) of sodium, and 2800 mEq (2800 mmol) of potassium in a 450-kg horse are illustrated in Fig. 18.4. In this theoretical example there has been an approximate 10% loss of body water as well as a 10% loss of the exchangeable sodium and potassium. Despite these losses, the plasma sodium concentration and osmolality remain unchanged. The ECF and ICF share the fluid losses proportionally because the losses of sodium and potassium were proportional to the initial content in their respective fluid compartments.

In many instances isotonic dehydration may occur in which sodium loss exceeds potassium loss. This type of dehydration is seen with heavy sweat loss in horses, with acute diarrhea in most species, and with inappropriate diuretic administration. The effects of an isotonic fluid loss of 30 liters of water, 3800 mEq (3800 mmol) of sodium, and 400 mEq (400 mmol) of potassium in a 450-kg horse are illustrated in Fig. 18.5. With this type of isotonic fluid loss, plasma sodium concentration remains within normal limits despite the development of large sodium deficits. These animals manifest clinical signs of inadequate circulating fluid volume reflecting the sodium deficit and the associated decrease in plasma and ECF volume as has been shown in human subjects (McCance, 1937, 1938). When a substantial portion of the water deficit is replaced by water consumption or free water administration in these animals, the serum sodium concentration and osmolality decline and a hypotonic, hypovolemic dehydration can develop (Sufit *et al.*, 1985). The hyponatremia noted in this circumstance is best considered as an indication of a *relative water excess* (Scribner, 1969).

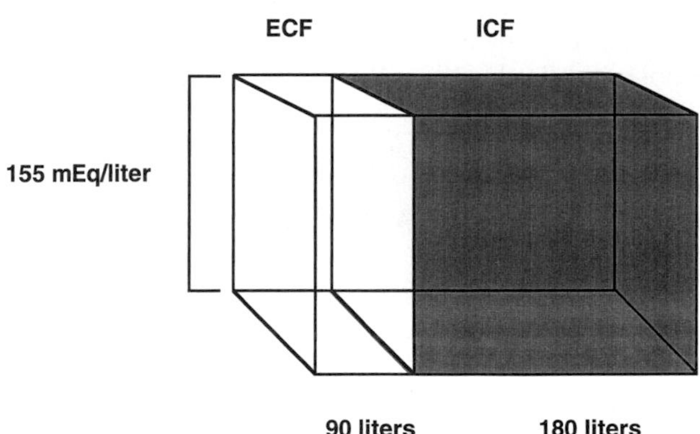

FIGURE 18.3 Body fluid compartments in a 450-kg horse with a pure water loss of 30 liters. Hypertonic fluid volume contraction is indicated by the increase in serum sodium concentration from 140 mEq/liter (140 mmol/liter) to 155 mEq/liter (155 mmol/liter). Fluid losses are shared by the extracellular fluid (ECF) volume and the intracellular fluid (ICF) volume.

2. *Water Excess — Overhydration*

The effects of the administration and retention of 30 liters of water in a 450-kg horse are illustrated in Fig. 18.6. In this example, total body water has been increased by 30 liters and there is an absolute as well as relative water excess. The primary effect of this water load is to dilute the electrolytes in the body fluids, producing a substantial decline in plasma sodium concentration and osmolality. These changes occur despite the fact that there has been no change in sodium or potassium balance. In this example the ECF and ICF share the water excess proportionately. The hyponatremia is associated with expansion of ICF and thus swelling of the cells.

Overhydration rarely occurs in normal individuals. The large water load described in Fig. 18.5, if administered to a normal animal, would produce only transient changes and the excess water would be eliminated by renal excretion. Even animals with psychogenic polydipsia ordinarily are able to maintain normal water balance through appropriate renal water excretion unless sodium depletion and renal medullary washout occur. However, overhydration can occur iatrogenically as the result of excessive fluid administration to patients with compromised renal function. If these fluids provide free water, as with 5% dextrose, plasma sodium concentration will decrease, reflecting the change in relative water balance. If these fluids consist of isotonic sodium-containing replacement fluids such as saline or lactated Ringer's solution, there will be no change in plasma sodium concentration (Cornelius *et al.*, 1978; Carlson and Rumbaugh, 1983), but there will be an increase in plasma and ECF volume with the potential for cardiovascular overload, pulmonary

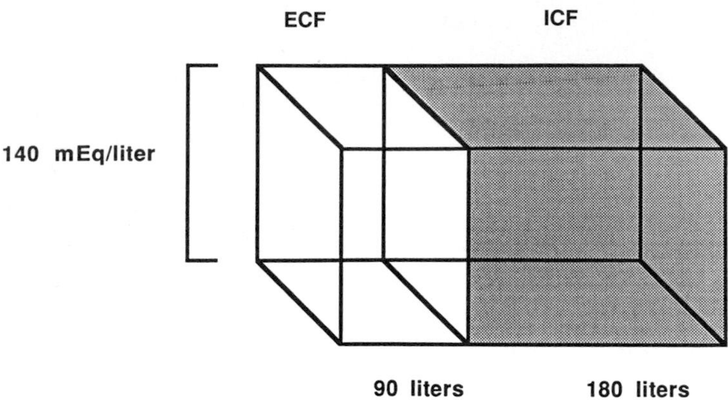

FIGURE 18.4 Isotonic fluid volume contraction. Body fluid compartments in a 450-kg horse with an isotonic fluid loss of 30 liters of water, 1400 mEq (1400 mmol) sodium, and 2800 mEq (2800 mmol) potassium. Serum sodium concentration remains unchanged at 140 mEq/liter (140 mmol/liter) and fluid losses are shared proportionately by the extracellular fluid (ECF) volume and intracellular fluid (ICF) volume.

edema, or generalized edema formation. In this instance, the primary problem is sodium retention and the changes in water balance are secondary.

B. Sodium

The ECF volume contains approximately one-half to two-thirds of the body's sodium. Most of the remaining sodium is bound in skeletal bone, relatively little of which is rapidly exchangeable (McKeown, 1984). The ECF volume thus contains essentially all of the body's readily available and exchangeable sodium. The exchangeable sodium content is the principal determinant of ECF volume, and sodium deficits are the principal causes of decreased ECF volume (Saxton and Seldin, 1986). Increases in sodium content result in expansion of ECF volume, which may lead to the development of hypertension or edema formation (McKeown, 1984; Dow *et al.*, 1987a). In either instance, the observed plasma sodium concentration will depend primarily on relative water balance. Because daily monitoring of electrolyte balance is difficult in most animal species, urinary fractional excretion or creatinine clearance ratios have been useful to provide and index of daily intake or potential deficits of sodium, potassium, chloride, and other electrolytes. Nor-

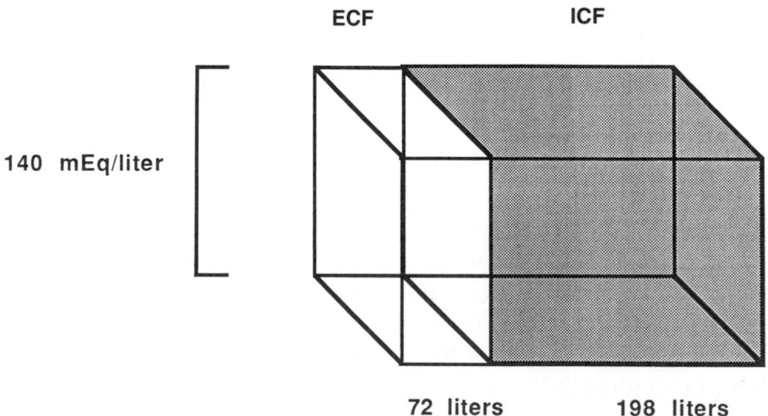

FIGURE 18.5 Isotonic fluid volume contraction sodium depletion. Body fluid compartments in a 450-kg horse with an isotonic fluid loss of 30 liters of water, 3800 mEq (3800 mmol) sodium, and 400 mEq (400 mmol) potassium. Serum sodium remains unchanged at 140 mEq/liter (140 mmol/liter) despite the development of the substantial sodium deficit. Fluid losses are borne primarily by the extracellular fluid (ECF) volume.

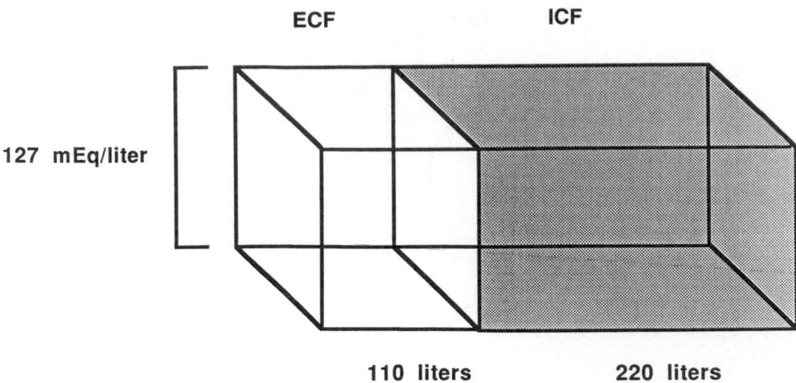

FIGURE 18.6 Hypotonic fluid volume expansion (relative water excess). Body fluid compartments in a 450-kg horse after administration and retention of 30 liters of water. Although no sodium or potassium losses have occurred, serum sodium concentration has declined from 140 mEq/liter (140 mmol/liter) to 127 mEq/liter (127 mmol/liter) reflecting a relative water excess. Water retention results in proportionate expansion of the extracellular fluid (ECF) volume and the intracellular fluid (ICF) volume.

mal values have been established for dogs and cats, horses (Morris *et al.*, 1984), and cattle (Fleming *et al.*, 1992).

1. *Sodium Depletion*

Sodium depletion is rarely the result of dietary sodium deficiency (Aitken, 1976). This is true even for herbivores whose feedstuffs are normally very low in sodium. Chronic sodium depletion has been reported in lactating dairy cows on a low-salt diet (Whitlock *et al.*, 1975a), but the sodium deficit was most probably the result of sodium losses in milk (Michell, 1985). Mastitis markedly enhances sodium loss in milk and could play a role in sodium depletion in lactating cows maintained on a low-salt diet (Michell, 1985). Sodium depletion is almost invariably associated with excessive losses of sodium-containing fluid (McKeown, 1984; Rose, 1984), most often occurring as the result of gastrointestinal losses through vomiting or diarrhea (Fisher and Martinez, 1976; Lakritz *et al.*, 1992). Excessive renal sodium losses can occur with intrinsic renal disease or as the result of diuretic therapy (Rose and Carter, 1979; Rose, 1984; Rose *et al.*, 1986). Cutaneous losses via sweating are very important in the exercising horse and may occur in any animal with extensive exfoliative dermatitis or burns. Salivary losses as the result of esophagostomy have been reported as a cause of sodium depletion in horses (Stick *et al.*, 1981b).

A history suggestive of excessive loss of sodium-containing fluid is thus an extremely important criteria for the diagnosis of sodium depletion in domestic animals. The foregoing discussion applies indirectly to the so-called "third space" problems associated with a sequestration or compartmentalization of a portion of the ECF volume (Rose, 1984). This can occur with obstructive bowel disease or with the sudden accumulation of fluid within the abdomen or thorax resulting from peritonitis, ruptured bladder, ascites, or pleural effusion. These fluids have an electrolyte composition similar to the ECF and initially are drawn from the plasma volume and interstitial fluids. The resultant changes in plasma volume and effective circulating fluid volume produce the same clinical and clinicopathologic changes as those observed with excessive external losses of sodium-containing fluid (Billig and Jordan, 1969). In these cases fluids have not been lost from the body, but there has been an internal sequestration of fluid which can be mobilized if treatment is effective.

2. *Sodium Excess*

Sodium excess occurs most often in association with an increase in body water leading to an isotonic expansion of ECF volume and the development of hypertension or generalized edema (Saxton and Seldin, 1986). Congestive heart failure, hypoalbuminemia, and hepatic fibrosis may lead to a failure to maintain effective circulating volume, which in turn results in compensating renal sodium retention (Rose, 1984). In these cases expanded ECF volume represents an attempt to restore effective circulating fluid volume, and, at least initially, plasma volume may be decreased while ECF volume expands. An expansion of ECF volume of 20–30% may be necessary before edema is first evident (Scribner, 1969). Sodium excess and edema can develop iatrogenically as the result of the excessive administration of sodium-containing fluid to patients with severely compromised renal function.

Most domestic animals can tolerate a very large sodium intake provided they have adequate drinking water (Pierce, 1957; Aitken, 1976; Buck *et al.*, 1976). Excessive salt intake may occur when animals that had been on salt-restricted diets are first allowed free access to salt (Mitchell, 1985). Salt intoxication can occur in cattle that are feeding in reclaimed saltwater marshes or on pastures contaminated by oil field wastes when they are deprived of freshwater (Pierce, 1957; Aitken, 1976; Sandals, 1978; McCoy and Edwards, 1979; Michell, 1985). Salt poisoning in swine also occurs in association with water restriction followed by access to water (Aitken, 1976; Buck *et al.*, 1976). Salt intoxication is generally associated with increases in plasma and/or cerebrospinal fluid sodium concentrations. Neurologic signs associated with excessive salt consumption coupled with water restriction are related to development of cortical edema and in swine a characteristic eosinophilic meningoencephalitis (Aitken, 1976).

C. Potassium

Potassium is largely an intracellular ion with more than 98% of the exchangeable potassium located intracellularly (Strombeck, 1979; Brobst, 1986). This distribution of potassium is coupled with the active extrusion of sodium from the cells which is maintained by an energy-dependent sodium : potassium pump at the cell membrane (Tannen, 1984; Brobst, 1986). Potassium distribution across the cell membrane plays a critical role in the maintenance of cardiac and neuromuscular excitability. Changes in potassium concentration which alter the ratio of intracellular to extracellular potassium alter membrane potential (Tannen, 1984).

In general hypokalemia increases membrane potential, producing a hyperpolarization block resulting in weakness or paralysis. Hyperkalemia decreases membrane potential, causing hyperexcitability (Patrick, 1977). These features depend on the state of total body potassium content but are also dependent on the speed with which hypokalemia or hyperkalemia develops (Saxton and Seldin, 1986). As an example, acute potassium depletion may result in hypokalemia prior to the development of a marked change in intracellular potassium, producing a substantial alteration in the ratio of intracellular to extracellular potassium. Acute hypokalemia can produce a much greater alteration in membrane potential and thus more marked clinical signs than a gradually developing hypokalemia of the same magnitude (Saxton and Seldin, 1986). Potassium homeostasis involves regulation of internal balance (i.e., the distribution of potassium between the ECF and ICF) as well as external balance (i.e., the relation of potassium input to output) (Rose, 1984; Brobst, 1986). Internal potassium balance is influenced by changes in acid–base status, glucose and insulin administration, exercise, and catecholamine release and is discussed in greater detail in the section on factors which influence plasma potassium concentration.

1. *Potassium Depletion*

Potassium depletion is the result of altered external balance in which potassium losses by all routes exceed intake. Potassium is present in relatively high concentrations in most animal feeds. Therefore, dietary deficiency alone is not a common cause for potassium depletion (Tasker, 1980). However, dietary factors were associated with hypokalemia hospitalized cats, particularly when associated with diseases linked to increased potassium loss (Dow *et al.*, 1987b). Herbivores such as the horse fed an all-hay diet may have a daily potassium intake of 3000–4000 mEq (3000–4000 mmol) per day (Tasker, 1967a; Hintz and Schryver, 1976). Most of this potassium is absorbed in the small intestine and colon with subsequent renal excretion of more than 90% of the daily potassium intake (Hintz and Schryver, 1976). These animals are thus highly adept at excretion of a large daily potassium intake. However, renal compensation for deficient potassium intake is not very efficient and renal conservation of potassium may be delayed for several days when animals normally fed a high potassium diet are suddenly taken off feed or develop anorexia (Tasker, 1967b).

Potassium depletion most commonly develops as the result of gastrointestinal losses from vomiting or diarrhea. Excessive renal losses can occur as the result of diuretic usage, mineralocorticoid excess, renal tubular acidosis, and in the diuresis state which follows relief of urinary obstruction (Saxton and Seldin, 1986). Potassium depletion may result in decreased ICF volume, altered membrane potential, altered intracellular pH, and alterations of potassium-dependent enzymatically mediated reactions (Elkinton and Winkler, 1944). Clinical features include muscle weakness, ileus, cardiac arrhythmias, rhabdomyolysis, and renal dysfunction (Earley and Daugharty, 1969; Tannen, 1984; Dow *et al.*, 1987a, 1987b).

2. *Potassium Excess*

Potassium excess occurs relatively rarely and is generally a consequence of some alteration of renal excretion of potassium. Potassium excess may be associated with Addison's disease, some forms of renal disease, and clinical situations associated with hypovolemia and renal shutdown (Rose, 1984; Weldon *et al.*, 1992). Care should be taken to avoid the rapid administration of large amounts of potassium salts orally or intravenously to patients with severely compromised renal function. Even normal animals can develop significant electrocardiographic abnormalities with potassium (Glazier *et al.*, 1982; Epstein, 1984; Dhein and Wardrop, 1995) or calcium infusions (Glazier *et al.*, 1979).

D. Chloride

Modest changes in hydration tend to produce roughly proportional changes in plasma sodium and chloride relative to sodium concentration. Acid–base alterations are associated with disproportionate changes in plasma chloride concentration (Saxton and Seldin, 1986). A disproportionate hyperchloremia is seen in association with a low to normal anion gap metabolic acidosis. Chloride concentration increases in this type of acidosis as the result of proportionately smaller losses of chloride than bicarbonate and enhanced renal chloride resorption in response to decreased bicarbonate (Saxton and Seldin, 1986). Disproportionate hypochloremia is a consistent feature of metabolic alkalosis (Rose, 1984). Chloride depletion develops in these animals due to excessive loss or sequestration of fluids with high chloride content. Changes in water balance can result in modest alterations in the relative concentrations of plasma sodium and chloride. A pure water deficit produces an increased sodium concentration which exceeds the increases in chloride. This contributes to the development of a contraction alkalosis with an increase in bicarbonate. A pure water excess has the opposite effect and can result in an expansion acidosis.

VII. CLINICAL FEATURES OF FLUID AND ELECTROLYTE BALANCE

It is convenient to discuss the theoretical ramifications of pure water loss or specific electrolyte deficits. However, in practical clinical situations the issue is almost never this clearly defined and most often there is a combination of fluid, electrolyte, and acid–base alterations. Many medical problems result in a consistent pattern of fluid and electrolyte loss with predictable changes in fluid volume, electrolyte concentration, and acid–base balance. A clear understanding of the interrelationships between specific deficits and their clinicopathologic consequences is essential if an appropriate initial diagnosis is to be made. Once treatment has been initiated, all clinical and clinicopathologic data will be influenced not only by the primary medical problem and compensating responses, but also by the effects of chemotherapeutic agents and fluid therapy as well. Therapeutic intervention may not be always appropriate, organ function may be impaired and, thus, the anticipated clinicopathologic responses become less predictable. These situations represent the bulk of clinical case material and laboratory data evaluated by clinicians and clinical pathologists. Under these circumstances, it is essential to understand the basic mechanisms which underlie changes in clinicopathologic data and how these changes relate to specific imbalances.

Rational fluid therapy depends on accurate evaluation of the fluid and electrolyte deficits, the associated acid–base alterations, and the primary disease processes which underlie these imbalances. Evaluation must include an accurate history, a complete physical examination, and, of course, laboratory evaluation of appropriate parameters.

A. History

An accurate history is absolutely essential for the evaluation and management of the patient with fluid and electrolyte imbalances. Basic signalment: age, sex, breed, pregnancy, and stage of lactation are important because these factors influence the incidence and severity of many disorders. The presence of a preexisting or coexisting disease process as well as an accurate drug history can be exceedingly important not only in the evaluation of fluid and electrolyte disorders but also in fluid selection and patient management. Of particular importance is the history of prior renal disease, diuretic usage, or exposure to potentially nephrotoxic drugs. Status of feed and water intake is exceedingly useful. Most animals that continue to eat and drink normally are able to maintain fluid balance even in the face of excessive fluid losses. However, reduced or restricted fluid intake in the face of normal to enhanced fluid losses can quickly result in dehydration. Inadequate fluid intake may result from neurologic disorders or traumatic injuries to the head or neck, and painful or obstructive lesions in the mouth, pharynx, or gastrointestinal tract may restrict feed and water intake. Inadequate water intake is often the result of management errors, broken or frozen water lines, etc. A history of polydipsia suggests that excessive fluid losses have occurred. Vomiting and diarrhea are obvious causes of fluid and electrolyte loss but these findings also reflect gastrointestinal disorders which may contribute to inadequate fluid and electrolyte intake or absorption.

Excessive fluid losses may be associated with vomiting, diarrhea, polyuria, excessive salivation, copious drainage from cutaneous wounds or burns, and as the result of heavy sweat losses in exercising horses. The water losses that occur in these situations are generally associated with significant sodium depletion and subsequent decreases in the effective circulating fluid volume. Vomiting in small animals (Clark, 1980), gastrointestinal stasis in ruminants (Gingerich and Murdick, 1975b), and excessive sweat losses in endurance horses (Carlson, 1983c) are associated with large losses or compartmentalization of chloride-rich fluids which contribute to the metabolic alkalosis that frequently accompanies these disorders.

B. Clinical Signs

Dehydration is defined as a loss of body water. Clinical signs of dehydration are said to be first apparent with fluid losses equivalent to 4–6% of body weight. Moderate dehydration is said to be present with fluid losses of 8–10% and severe dehydration when fluid losses are greater than 12% of body weight. It is important to realize that, although these guidelines have been clinically useful, there is relatively little documentation of this precise quantitative relationship in most animal species. Accurate measurement of fluid intake from all sources and output by all routes is not possible in most clinical situations. In acute situations, changes in body weight provide the most accurate guide to change in net water balance. Repeated measurement of body weight is a key component of the monitoring of patients on fluid therapy. The clinical signs of dehydration include weight loss, altered skin turgor, sunken eyes, and dry mucous membranes. If control of renal function in normal, urine volume is generally markedly reduced. The clinical consequences of dehydration depend much more on the pattern of electrolyte loss than the absolute water deficit.

Clinical signs associated with acute sodium deficits are largely related to hypovolemia and decreases in the effective circulating volume. These signs include increased pulse rate, decreased pulse pressure, delayed jugular distensibility, increased capillary refill time, and decreased blood pressure. Urine output is generally decreased and urine sodium and chloride concentrations are normally reduced. Decreases in ECF volume are always reflected by decreases in plasma volume, but the reverse is not always true. In the absence of blood or protein loss, the packed cell volume and total plasma protein concentration increase, reflecting the decrease in plasma and ECF volume, and are discussed more fully in section VIII.A.

VIII. CLINICOPATHOLOGIC INDICATORS OF FLUID AND ELECTROLYTE IMBALANCE

A. Packed Cell Volume and Total Plasma Protein

The packed cell volume (PCV) and total plasma protein (TPP) concentration are simple, convenient, and useful tools for the evaluation of acute fluid and electrolyte alterations. If we assume that there has been no gain or loss of erythrocytes or protein from the vascular compartment, changes in PCV and TPP reflect changes in plasma volume (Boyd, 1981). For the estimation of alterations in plasma volume in human subjects, the change in PCV or hemoglobin concentration is preferred over change in protein concentration (van Beaumont *et al.*, 1973). This is due to fluxes of protein both into and out of the plasma volume which may occur in certain rapidly changing clinical circumstances (Landis and Pappenheimer, 1963; Senay, 1970). There is a substantial extravascular protein pool and the volume distribution of both albumin and globulins is approximately two to three times that of the plasma volume (Landis and Pappenheimer, 1963; Mattheeuws *et al.*, 1966). This is one of the reasons why plasma transfusions have limited effects on the plasma protein concentration of hypoproteinemic patients.

In most animal species, there is a much wider range of normal for PCV than for TPP. In some of these species, most notably the horse, mobilization of erythrocytes from splenic reserves in response to excitement, pain, or catecholamine release can result in marked variability in PCV, erythrocyte count, or hemoglobin concentration (Persson, 1967; Persson, *et al.*, 1973; Carlson, 1987). For these reasons it has proven most useful to utilize the change in both PCV and TPP as a crude index of change in plasma volume in domestic animals. For these estimations to be valid, initial values for PCV and TPP must be known and there must have been no loss of blood or protein. The most common causes for decreases in plasma volume are the sequestration or loss of sodium-containing fluid such as occurs in obstructive bowel disease, diarrhea, vomiting, renal disease, or in heavily sweating horses. In these circumstances both PCV and TPP increase in relation to the change in plasma volume, but not necessarily in the same proportion (Boyd, 1981). The following example illustrates this point.

If we assume that the quantity of plasma proteins within the plasma volume remains constant (although this assumption is not always true), it is possible to calculate the percentage change in plasma volume based on protein concentration (Boyd, 1981):

$$\%PV = [(PP_1/PP_2) - 1]\ 100, \qquad (18.13)$$

where PV indicates plasma volume, PP_1 indicates initial plasma protein concentration, and PP_2 indicates final plasma protein concentration. It is also possible to calculate the change in plasma volume based on the change in PCV provided it is assumed that the number of erythrocytes within the circulating blood volume and the mean corpuscular volume remain constant (Boyd, 1981)

$$\%PV = \left(\frac{PCV_1\ (1 - PCV_2)}{PCV_2\ (1 - PCV_1)} - 1\right)100, \qquad (18.14)$$

where PV again indicates plasma volume, PCV_1 indicates the initial PCV, and PCV_2 indicates the final PCV. Using this approach, the calculated increase of PCV and plasma protein concentration associated with a progressive reduction in plasma volume in a 20-kg dog is presented in Fig. 18.7. In this example the dog was assumed to have had an initial plasma volume of 5% of body weight (0.05 liter/kg), an initial PCV of 40% (0.4 liter/liter), and an initial TPP concentration of 6.5 gm/dl (65 g/liter). A 50% reduction in plasma volume results in a 100% increase in TPP from 6.5 to 13.0 gm/dl, whereas PCV only increased from 40 to 55% (Fig. 18.8). It will be noted that as plasma volume declines the percentage change in PCV is always less than the percentage change in plasma protein concentration. This is true whether the changes are due to an increase or decrease in plasma volume.

In most clinical situations the initial PCV and plasma protein concentration are not known. To make these estimates, we must assume that these values were within the normal range prior to dehydration. A large disparity in the change in PCV and TPP concentration in the face of a history of loss of sodium-containing fluid and clinical evidence of reduced effective circulating volume suggests that either erythrocytes or protein have been lost from the circulation. Marked increases in PCV with normal to low plasma proteins are frequently encountered in animals with acute toxic enteritis due to losses of protein into the gastrointestinal tract (Merritt *et al.*, 1977). This finding suggests erosive bowel disease but also could be seen in animals with low protein due to preexisting renal or hepatic disease. Increases in plasma protein with little or no change in PCV are seldom seen in dehydrated animals, but could occur with a preexisting anemia or hyperproteinemia. Blood loss generally results in a decrease in both PCV and TPP concentration.

Sodium deficits result in decreases in ECF volume, which are usually reflected by changes in plasma volume (Hopper *et al.*, 1959; Saxton and Seldin, 1986). As plasma volume decreases, the PCV and TPP concentration increase. Changes in PCV and TPP are, thus, particularly useful in evaluating the need for and the response to sodium-containing replacement fluid (Carlson, 1983a). When sodium-containing fluids are administered to dehydrated volume-depleted subjects, plasma volume reexpansion is reflected by the return of PCV and TPP toward the normal range (Hayter *et al.*, 1962). Declining PCV and TPP in response to fluid therapy are the two most useful laboratory indicators of return of effective circulating fluid volume and

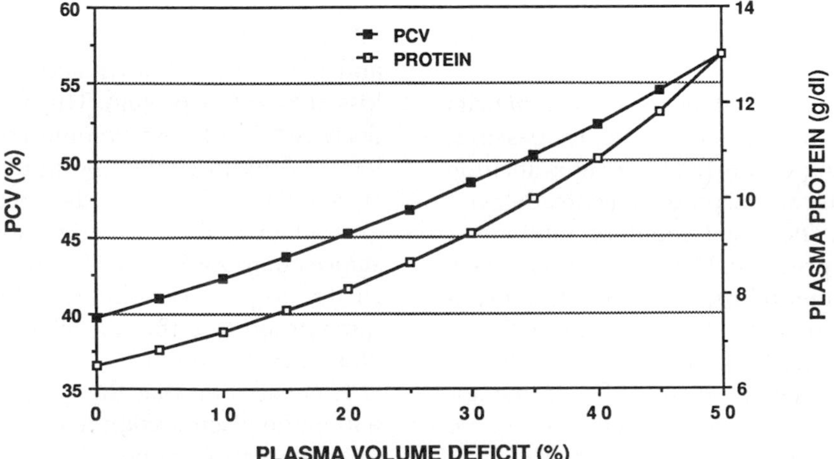

FIGURE 18.7 Effects of decreasing plasma volume on PCV and plasma protein concentration as calculated using Eqs. (18.13) and (18.14).

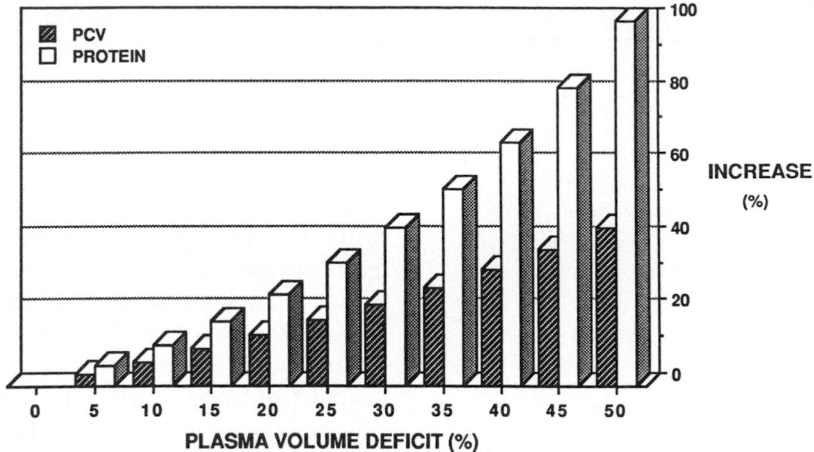

FIGURE 18.8 Calculated effects of plasma volume depletion. Percentage increase in PCV and plasma protein concentration with progressive plasma volume deficits.

should be correlated with clinical evidence such as return of normal pulse pressure, capillary refill time, and jugular distensibility.

B. Serum Sodium

Serum sodium concentration varies within relatively narrow limits in the normal individual but there is substantial interspecies variation in the normal range of sodium, chloride, and osmolality as indicated in Table 18.4. A serum sodium concentration of 134 mEq/liter (134 mmol/liter), while quite normal for a horse or cow, represents a significant hyponatremia in a dog or cat.

Before proceeding with a discussion of the significance of alterations of sodium concentration, some comment on the methods used for electrolyte determination is appropriate. Flame photometry for many years had been the standard method for the determination of both sodium and potassium. Recently, methods employing ion-specific electrodes have achieved wider use. Although both procedures yield accurate and reproducible results, there may be consistent differences between the two procedures. Flame photometry yields results in mEq or mmol/liter of plasma or serum, whereas the ion-specific electrodes are generally considered to measure the electrolyte concentration only in the aqueous phase of the sample. The water content of serum or plasma samples is normally between 93 and 94%, with most of the remaining volume occupied by protein. For this reason, electrolyte determinations performed by flame photometry tend to be 6–7% lower than those obtained using ion-specific electrodes. Interestingly, ion-specific electrode instruments that dilute samples tend to yield values which are very similar to values reported from the flame photometer. This may be a function of the mathematical algorithm employed in these devices to convert changes in electric potential to electrolyte concentration. There may be interfering substances in the urine of some animal species which render urine potassium determinations inaccurate when assessed by ion-specific potentiometry (Brooks *et al.*, 1988) The relationship between protein concentration and serum or plasma water content has been determined for man and the horse (Eiseman *et al.*, 1936; Carlson and Harrold, 1977; Harris *et al.*, 1987). This relationship can be used to correct serum or plasma electrolyte concentration determined by flame photometry for the effects of variation in protein concentration and thus water content. Electrolyte concentration then can be expressed per liter of serum or plasma water.

$$\text{Sodium mEq/liter } H_2O = \frac{\text{Measured sodium (mEq/liter)}}{\text{\% Water}} \times 100 \qquad (18.15)$$

1. Hyponatremia

The common causes of hyponatremia are listed in Table 18.5. A falsely low sodium concentration may be noted when there is marked hyperlipemia or hyperproteinemia. Large quantities of lipid or protein occupy a significant volume in a serum or plasma sample and because electrolytes are dissolved only in the aqueous phase the measured concentrations will be falsely low. The presence of obvious lipemia or markedly elevated serum protein concentration should alert the clinician to the probable cause of an accompanying hypo-

TABLE 18.5 Causes of Hyponatremia

False hyponatremia
- Hyperlipidemia
- Hyperproteinemia
- Hyperglycemia

Hyponatremia (relative water excess)
- Decreased effective circulating volume
  - Vomiting
  - Diarrhea
  - Excessive sweating
  - Cutaneous loss, burns
  - Blood loss
  - Repeated pleural drainage
    - Pleuritis, chylothorax
  - Adrenal insufficiency
    - "Third space problems"
  - Sequestration of fluid
    - Peritonitis
    - Ascites
    - Ruptured bladder
- Excess circulating volume
  - Congestive heart failure
  - Chronic liver failure
  - Nephrotic syndrome

Normal effective circulating volume
- Physchogenic water drinking
- Renal disease
- Inappropriate ADH secretion

natremia. Should this information not be available, the presence of a false hyponatremia can be determined by comparison of the measured serum osmolality and the calculated osmolality based on sodium, glucose, and urea concentration as explained in Section VIII.E. This potential cause for confusion in interpretation of hyponatremia can be avoided if ion-specific electrodes are used for electrolyte determination.

Marked hyperglycemia associated with diabetes mellitus or the administration of glucose at an excessive rate generally produces a hyponatremia. As glucose concentration increases in the ECF, osmotic forces are generated that result in the movement of cellular water into the ECF, diluting serum sodium concentration. The actual mechanics of this process are complicated by the fact that cells and tissues are variably permeable to glucose and, thus, the glucose space clearly exceeds the ECF volume. For practical purposes, we can anticipate that serum sodium concentration will decline 1.6 mEq/liter (1.6 mmol/liter) for each 100 mg/dl (5.55 mmol/liter) increase in glucose concentration (Saxton and Seldin, 1986). Serum osmolality may be increased by hyperglycemia but this should not cause a large disparity between the measured and calculated serum osmolality.

Changes in water balance are principally responsible for changes in serum sodium concentration (Leaf, 1962). Hyponatremia should be considered an indication of a *relative water excess* (Scribner, 1969). Hyponatremia is often but not invariably associated with conditions which cause sodium depletion and resultant decreases in effective circulating volume. These conditions include vomiting, diarrhea, excessive sweat losses, and adrenal insufficiency. Dehydration and volume depletion induce neurohormonal responses that result in increased water consumption via increased thirst and enhanced renal conservation of water as well as sodium (Rose, 1984). Fluid losses in these forms of dehydration are most often hypotonic or isotonic and initial fluid and electrolyte deficits do not result in hyponatremia until water intake and/or renal water retention disturb the balance between the remaining exchangeable cations and the total body water. Thus, while substantial sodium and potassium deficits are associated with these conditions, plasma sodium concentration does not always reflect these deficits and diagnosis of sodium depletion should be based on other grounds (Scribner, 1969).

The accumulation of sodium-containing fluid within body cavities as a result of ascites, peritonitis, or a ruptured bladder is referred to as a *third space problem* (Rose, 1984). The fluid that accumulates in the "third space" has a composition similar to the ECF. When this accumulation of fluid occurs rapidly, plasma volume is reduced and serum sodium concentration then may decrease as the compensating responses result in water retention. A classic example of this situation is the marked hyponatremia associated with ruptured bladder in neonatal foals. As the dilute urine of the neonate accumulates in the abdomen, osmotic equilibrium is established first with the ECF and then with all of the body fluid compartments. Sodium and chloride as well as other ions are drawn from the rest of the ECF into this progressively expanding fluid compartment and hyponatremia develops despite the fact that there has been no appreciable loss of sodium from the body. The severe neurologic signs associated with ruptured bladder in foals are related in large part to the effects of the sudden and marked hypotonic hyponatremia on the central nervous system. Hyponatremia associated with excessive retention of water also can occur without the development of significant sodium depletion or decreases in effective circulating fluid volume. Hyponatremia may be observed with psychogenic polydipsia if the rate of water consumption exceeds renal capacity for free water clearance due to intrinsic renal disease or renal medullary washout (Tyler *et al.*, 1987). This also may occur in patients with impaired free water clearance due to renal disease or when under the influence of inappropriate release of ADH (McKeown, 1984). Hyponatremia and associated neurologic disturbances can develop with naturally occurring disease (Lakritz *et al.*, 1992) or iatrogenically if excessive amounts of free water are administered to patients with altered renal function (Arieff, 1986; Sterns *et al.*, 1986).

Urine sodium concentration can be useful in the differentiation of the causes of hyponatremia as indicated in Fig. 18.9. Renal adaptive responses normally result in sodium retention and production of urine with very low sodium concentration in patients with sodium depletion resulting from vomiting, diarrhea, excessive sweat loss, or third space problems (Rose, 1984). Hyponatremia and hypokalemia of adrenal insufficiency are generally associated with a relatively high urine sodium concentration (Rose, 1984). With the syndrome of inappropriate secretion of antidiuretic hormone (SIADH), urine sodium tends to be high in the presence of hyponatremia, whereas urine sodium concentration tends to be low in animals with psychogenic polydipsia. Urine sodium concentration in animals with renal failure can be quite variable.

A form of moderate to pronounced hyponatremia associated with elevated levels of ADH which develops in the absence of an appropriate osmotic or volume stimuli has been well recognized in human subjects. This syndrome has been associated with several systemic disorders including malignant neoplasms which produce an ADH-like material, with a variety of brain diseases which apparently stimulate synthesis and release of ADH, or in association with certain pulmonary diseases which may result in abnormal neural inputs from the lung that trigger the inappropriate ADH release from the pituitary.

The diagnosis of SIADH in the absence of readily available procedures for ADH determination is dependent on ruling out a number of other potential causes for persistent hyponatremia. The diagnostic criterion for SIADH for human subjects include the following (McKeown, 1984):

1. Demonstration of a hypotonic hyponatremia without hypovolemia or edema
2. Normal renal, adrenal, and thyroid function
3. Inappropriately elevated urine osmolality relative to plasma osmolality
4. Relatively high urine sodium concentration
5. Correction of the hyponatremia by strict fluid restriction

SIADH has been reported in the dog (Breitschwerdt and Root, 1979; Giger and Gorman, 1984; Houston *et al.*, 1989; Crow and Stockham, 1985), but the incidence and importance of this problem in other animal species is uncertain. A variant of SIADH has been described in certain chronically ill and malnourished human sub-

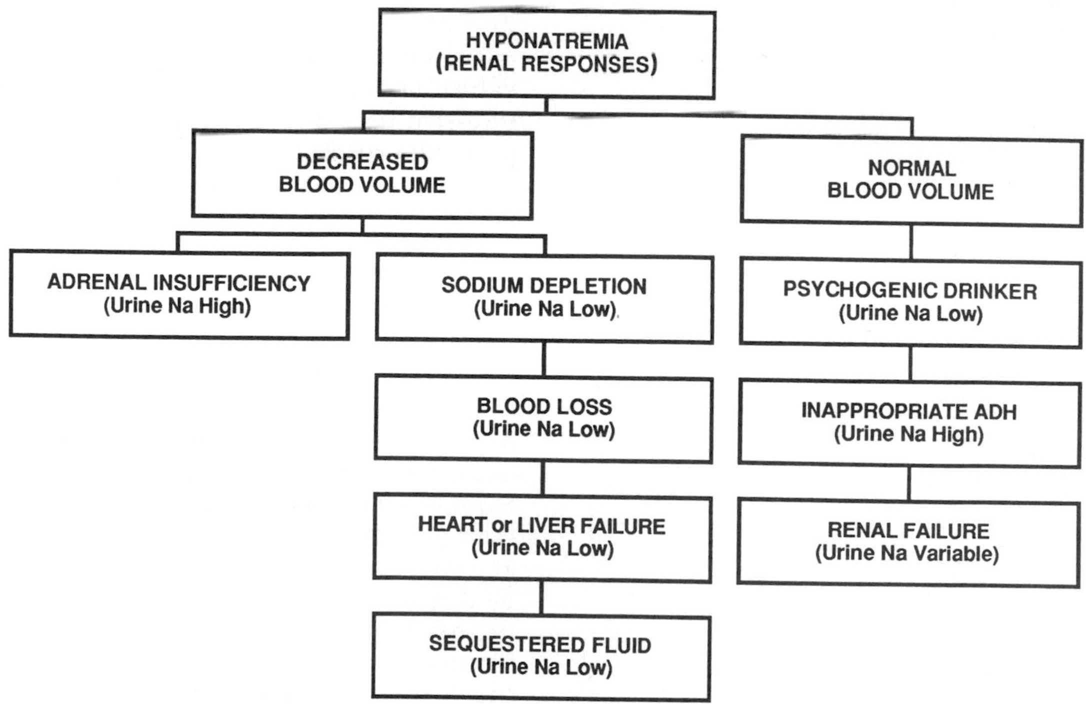

FIGURE 18.9 The urinary sodium concentration as an aid in the differentiation of possible causes of hyponatremia. Urinary electrolyte concentration is also influenced by dietary intake.

jects that maintain a mild but persistent hyponatremia. In these individuals ADH secretion remains under osmotic control but it would appear that the osmoreceptor threshold functions at a lower value than normal (McKeown, 1984). Thus, it is the resetting of the osmostat which results in persistent chronic hyponatremia. Resolution of the underlying medical problem usually results in correction of this variant form of SIADH.

2. *Hypernatremia*

Hypernatremia almost always is associated with elevation of serum osmolality. Hypernatremia occurs in dehydrated subjects when water losses exceed losses of sodium and potassium and should be considered as an indication of a *relative water deficit* (Scribner, 1969). This can occur in the initial stages of diarrhea, vomiting, or renal disease if losses of water exceed the electrolytes lost (Saxton and Seldin, 1986). Common causes of hypernatremia are listed in Table 18.6. As water losses are replaced by increased water consumption and / or enhanced renal water retention, serum sodium concentration tends to decline into or below the normal range. Hypernatremia also develops as the result of an essentially pure water loss, such as the evaporative respiratory water loss in panting animals (Tasker, 1980). Hypernatremia is associated with excessive renal free water loss with either central or nephrogenic diabetes insipidus if water intake is restricted (Breukink *et al.*, 1983). Food and water deprivation in normal individuals is associated with substantial reduction of renal and fecal output (Tasker, 1967b; Carlson *et al.*, 1979b; Brobst and Bayly, 1982; Genetzky *et al.*, 1987). However, continued cutaneous and respiratory insensible water loss may result in hypernatremia (Elkinton and Taffel, 1942; Rumsey and Bond, 1976; Carlson *et al.*, 1979b; Genetzky *et al.*, 1987).

TABLE 18.6 Causes of Hypernatremia

Water losses in excess of electrolyte loss
Digestive
Vomiting
Diarrhea
Cutaneous
Burns
Renal
Diuretics
Diabetes mellitus
Manitol
Intrinsic renal disease
Pure water losses
Insensible
Panting
Diabetes insipidus
Central
Nephrogenic
Inadequate water intake
Water deprivation
Abnormal thirst mechanism
Sodium excess (water restriction)
Hypertonic saline or sodium bicarbonate
Salt poisoning
Mineralocorticoid excess

Abnormal thirst mechanisms with resultant hypernatremia have been reported in young dogs (Crawford *et al.*, 1984; Hoskins and Rothschmitt, 1984). Hypernatremia may occur transiently following administration of hypertonic saline or sodium bicarbonate if water intake is restricted or impaired. The hypernatremia observed with salt poisoning in cattle and swine is, in fact, triggered by water restriction in animals with a high salt intake (Padovan, 1980; Pearson and Kallfelz, 1982). As long as adequate water is available, salt poisoning does not occur. In human subjects, hypernatremia is reported with mineralocorticoid excess (McKeown, 1984).

C. Serum Potassium

Serum potassium concentration does not always reflect potassium balance but is influenced by factors which alter internal balance, that is, the distribution of potassium across the cell membrane between the ECF and ICF as well as factors which change external balance, that is, potassium intake and output (Patrick, 1977; Rose, 1984; Brobst, 1986). The effective adjustment of external and internal balance in normal individuals in response to either large potassium loads or excessive potassium losses usually maintains serum potassium concentration within normal limits. However, changes in potassium concentration occur in a wide variety of clinical circumstances and have profound neuromuscular effects largely due to changes in cell membrane potential (Brobst, 1986). The compensating responses to change in circulating fluid volume distribution and acid–base balance can result in confusing and contradictory findings. As an example, calves with acute diarrhea may develop significant depletion of body potassium stores due to excessive losses and inadequate intake (Phillips and Knox, 1969). However, the serum potassium concentration in these calves may actually be normal to increased as the result of renal shutdown and the metabolic acidosis induced by dehydration and sodium depletion with resultant decreases in the effective circulating fluid volume. Electrolyte replacement fluids given to these calves should include potassium (Fettman *et al.*, 1986). Correct interpretation of serum potassium concentration thus requires a knowledge of probable intake and sources of excessive loss as well as the status of renal function and acid–base balance. Measurement of erythrocyte potassium concentration has been suggested as an aid in the assessment of the potassium status of horses with exercise-related myopathy (Muylle *et al.*, 1984a, 1984b; Bain and Merritt, 1990). However, experimental studies have failed to demonstrate a close correlation between erythrocyte potassium concentration and potassium depletion.

1. *Hypokalemia*

Hypokalemia occurs relatively frequently in domestic animals and may result from depletion of the body's potassium stores or from a redistribution of potassium from the ECF into the ICF space (Brobst, 1986), as indicated in Table 18.7. Hypokalemia most often is associated with excessive potassium losses from the gastrointestinal tract as the result of vomiting or diarrhea (Tasker, 1967c). Excessive renal loss of potassium results from the action of mineralocorticoid excess, the action of certain diuretics, and as the result of altered renal tubular function in animals with renal tubular acidosis or postobstructive states. Chronic dietary potassium deficiency eventually can lead to modest hypokalemia even in normal individuals (Aitken, 1976; Dow *et al.*, 1987b). A rapidly developing and profound hypokalemia can develop in animals with reduced dietary intake due to anorexia when coupled with other causes of excessive potassium loss (Tasker, 1980).

Hypokalemia may develop without potassium depletion as the result of intracellular movement of potassium from the ECF space. This occurs with an acute alkalosis (Burnell *et al.*, 1966) and in patients treated with insulin and glucose infusions (Tannen, 1984). In fact, medical management of severe life-threatening hyperkalemia often involves the administration of glucose, insulin, and in some circumstances bicarbonate in an effort to shift potassium intracellularly. Catecholamines exert a biphasic effect on potassium concentration (Tannen, 1984). The initial response to catecholamine administration is a modest transient increase in potassium as the result of α-adrenergic stimulation followed by hypokalemia as the result of β-adrenergic receptor responses (Clausen *et al.*, 1980; Bendheim *et al.*, 1985; Williams *et al.*, 1985). Beta-adrenergic agents have been used for the treatment of hyperkalemic periodic paralysis (Bendheim *et al.*, 1985). The relationship between catecholamine release, receptor stimulation, and potassium balance may play a significant role in the development of exertional fatigue in human and equine athletes (Sjogaard *et al.*, 1985; Carlson, 1987).

2. *Hyperkalemia*

Hyperkalemia can be produced iatrogenically by the excessive administration of potassium salts but does not usually result from high dietary intake in individuals with normal renal function (Aitken, 1976). The three major causes of hyperkalemia are indicated in Table 18.8. Hyperkalemia may develop *in vitro* in blood sample containers due to hemolysis or prolonged storage of blood samples prior to the separation of serum or plasma from erythrocytes in the sample. This leakage of erythrocyte potassium can result in significant errors in those species with a high potassium content in their erythrocytes; the horse, pig, and most cattle (Tasker, 1980). The erythrocytes of cats and most dogs have a high sodium content and relatively low potassium. Slight hemolysis will have little effect on serum or plasma potassium concentration in these

TABLE 18.7 Causes of Hypokalemia

Hypokalemia due to altered external balance
Gastrointestinal losses
Vomiting
Diarrhea
Renal losses
Mineralocorticoid excess
Diuretics
Renal tubular acidosis
Postobstruction diuresis
Hypokalemic nephropathy in cats
Dietary deficiency
Hypokalemia due to altered internal balance
Excessively rapid bicarbonate administration
Insulin with glucose administration
Catecholamine release
Hypokalemic periodic paralysis

TABLE 18.8 Causes of Hyperkalemia

False hyperkalemia
Hemolysis
Markedly elevated leukocyte or platelet count
Hyperkalemia due to altered external balance
Addison's disease
Renal disease
Ruptured bladder
Urethral obstruction
Hypovolemia with renal shutdown
Hyperkalemia due to altered internal balance
Metabolic acidosis
Diabetes
Tissue necrosis
Hyperkalemic periodic paralysis
Vigorous exercise

species. Polymorphism however, occurs in intracellular cation content of certain breeds of sheep, cattle, and dogs which relate to sodium–potassium ATPase activity of mature erythrocytes. Release of potassium from leukocytes or platelets into the serum following clot formation in the sample collection vial is a potential cause of hyperkalemia in subjects with a marked leukocytosis or thrombocytosis (Mandell *et. al.*, 1988; Degan, 1986).

Hyperkalemia may be associated with excessive renal potassium retention in conditions such as Addison's disease, acute renal failure, and renal shutdown secondary to inadequate effective circulatory fluid volume. Hyperkalemia results, in a number of circumstances, from the movement of intracellular potassium into the ECF without change in external potassium balance. Hyperkalemia develops in association with a metabolic acidosis, particularly when the acidosis results from volume depletion complicated by renal shutdown. Hyperkalemia may be noted in patients with diabetes or transiently in animals with massive muscle necrosis. Interestingly, most horses with exertional rhabdomyolysis do not develop hyperkalemia or a metabolic acidosis (Koterba and Carlson, 1982). Vigorous short-term exercise at high intensity results in a profound hyperkalemia in horses. Serum potassium as high as 9–10 mEq/liter (9–10 mmol/liter) has been observed transiently in horses exercising at maximal intensity on a high-performance treadmill (Harris and Snow, 1986). Muscular exhaustion in these horses appeared to be related to the hyperkalemia and the profound lactic acidosis seen in the immediate postexercise state (Kryzwanek, 1974; Kryzwanek *et al.*, 1976; Milne *et al.*, 1976; Carlson, 1987). An interesting and unusual condition of episodic hyperkalemia and muscular weakness has been recognized in the dog (Jezyk, 1982) and certain quarterhorses (Cox, 1985; Steiss and Naylor, 1986; Spier *et al.*, 1990, 1993; Pickar *et al.*, 1991). In the horse the condition closely resembles the heritable disease hyperkalemic periodic paralysis, which has been reported in human subjects, and is due to an alteration in the voltage regulated sodium channel (Rudolph *et al.*, 1992a, 1992b). Sudden marked increases in serum potassium concentration result from the transcellular movement of potassium. Serum potassium concentration can reach 8–9 mEq/liter (8–9 mmol/liter) and is associated with profound electrocardiographic abnormalities and fluid shifts which result in marked increases in PCV and protein concentration. The disease is inherited as an autosomal dominant and all affected horses can be traced back to a common ancestor. A DNA test which can detect the single base pair substitution responsible for this disease has been developed (Rudolph *et al.*, 1992a). Using this procedure it is possible to identify individuals which are heterozygous or homozygous for this trait.

D. Serum Chloride

It long has been assumed that the anion chloride which combines with sodium to form common salt simply follows sodium in the physiologic processes which regulate body fluid and electrolyte balance. It is becoming increasingly apparent that this may not always be true and that some of the problems ascribed to sodium retention do not occur unless chloride is present in excess as well (Kurtz *et al.*, 1987). Causes of alterations in chloride concentration are given in Table 18.9. The hyperchloremia and hypochloremia which are normally seen in association with roughly proportional changes in sodium concentration are due to changes in body water balance.

Changes in chloride concentration which are not associated with a similar change in sodium concentration are usually associated with acid–base imbalances (Divers *et al.*, 1986). Chloride concentration tends to vary inversely with bicarbonate concentration. A disproportionate increase in chloride most commonly is associated with a normal to low anion gap hyperchloremic metabolic acidosis, and may be seen as a compensating response for a primary respiratory alkalosis (Saxton and Seldin, 1986). Disproportionate decreases in chloride characteristically are seen in a metabolic alkalosis but also may be seen as part of the compensating response for a chronic primary respiratory acidosis (Saxton and Seldin, 1986).

E. Osmolality

It has been demonstrated that the concentration of sodium in serum water is closely correlated with the serum osmolality over an extremely wide range

TABLE 18.9 Causes of Alterations in Chloride Concentration

Hyperchloremia
- With proportional increase in sodium
  - Dehydration (relative water deficit)
- Without proportional increase in sodium
  - Hyperchloremic metabolic acidosis
  - Compensation for respiratory alkalosis

Hypochloremia
- With proportional decrease in sodium
  - Overhydration (relative water excess)
- Without proportional decrease in sodium
  - Hypochloremic metabolic alkalosis
  - Compensation for respiratory acidosis

of physiologic and pathologic states provided appropriate corrections are made for the contributions made by variations in glucose and urea concentrations (Edelman *et al.*, 1958). The measurement of serum osmolality has two specific and very useful purposes (Gennari, 1984). First, to determine whether serum water content deviates widely from normal, and second to screen for the presence of foreign low-molecular-weight substances in the blood. Interpretation of serum osmolality for these purposes requires simultaneous comparison of the measured osmolality and the calculated osmolality as determined from the measured concentrations of the major solutes in serum (sodium, glucose, and urea). The difference between the measured and the calculated osmolality is sometimes referred to as the *osmolal gap* (Feldman and Rosenberg, 1981; Shull, 1978, 1981).

Sodium is the principal cation in serum and sodium is balanced by a number of different anions (chloride, bicarbonate, protein, sulphate, and phosphate). Sodium concentration thus provides a reasonable estimate of the total electrolyte concentration (anions and cations) in the sample, and in this calculation is usually multiplied by a factor of 2. As has been mentioned earlier the water content of serum samples is approximately 94%. Correction for the water content of serum is not necessary since, fortuitously, it is counterbalanced by the fact that sodium chloride does not dissociate completely and has an osmotic coefficient of 0.93 in serum (Wolf, 1966; Dahms *et al.*, 1968; Rose, 1984). The osmolality can thus be calculated using the measured sodium concentration and the concentration of the two nonelectrolyte components of serum which are normally present in amounts sufficient to influence osmolality:

$$\text{mOsm/kg } H_2O = 2 \times \text{Sodium} + \text{Glucose} + \text{Urea}. \quad (18.16)$$

This calculation is valid if concentrations of sodium, glucose, and urea are expressed in mmol/liter. Conversion of glucose concentration from mg/dl to mmol/liter requires division by 18, whereas urea concentration in mg/dl can be converted to mmol/liter by dividing by 2.8.

Decreases in osmolar gap indicate laboratory error. Increases in osmolar gap (>10 mOsm/kg) could also represent laboratory error but generally result from one of two circumstances: either a decrease in serum water content or the addition of low-molecular-weight substances in serum. Decreases in serum water content occur with marked hyperlipidemia or hyperproteinemia and the calculated osmolality will exceed the measured osmolality. This is the cause of the false hyponatremia discussed in Section VI.B. The presence of a similar disparity between the measured and calculated osmolality in the absence of hyperlipidemia or hyperproteinemia should prompt an alert for the presence of exogenous substances in abnormally high concentrations in the serum. These substances could include a variety of exogenous and potentially toxic compounds such as mannitol, ethanol, methanol, ethylene glycol, isopropanol, ethyl ether, acetone, trichlorethane, and paraldehyde (Saxton and Seldin, 1986).

References

Adrogue, H. J., Brensilver, J., and Madias, N. E. (1978). *Am. J. Physiol.* **235,** F291–F297.

Aitken, F.C. (1976). *In* "Sodium and Potassium in Nutrition of Mammals," Technical Communication No. 26, Commonwealth Bureau of Nutrition, Commonwealth Agricultural Bureaux, Farnham Royal, England.

Arieff, A. I. (1986). *N. Engl. J. Med.* **314,** 1529–1535.

Bain, F. T., and Merritt, A. M. (1990) *J. Am. Vet. Med. Assoc.* **196,** 1259–1261.

Bendheim, P. E., Reale, E. O., and Berg, B. O. (1985). *Neurology* **35,** 746–749.

Bennett, E. J. (1975). *Anesthesiology* **43,** 210–224.

Bianca, W., Findlay, J. D., and McLean, J. A. (1965). *Res. Vet. Sci.* **6,** 38–55.

Billig, D. M., and Jordan, P. H. (1969). *Surg. Gynecol. Obstet.* **129,** 1274–1282.

Billig, E. (1984). *J. Dairy Sci.* **67,** 2939–2948.

Block, E. (1984). *J. Dairy Sci.* **67,** 2939–2948.

Brooks, C. L., Garry, F., and Swartout, M. S. (1988). *Am. J. Vet. Res.* **49,** 710–714.

Boyd, J. W. (1981). *Br. Vet. J.* **137,** 166–172.

Breitschwerdt, E. B., and Root, C. R. (1979). *J. Am. Vet. Med. Assoc.* **175,** 181–186.

Breukink, H. J., Van Wegen, P., and Schotman, J. H. (1983). *Equine Vet. J.* **15,** 284–287.

Bristol, D. G. (1982). *J. Am. Vet. Med. Assoc.* **181,** 63–65.

Brobst, D. (1984). *Calif. Vet.* **38(9),** 20–23, 48.

Brobst, D. (1986). *J. Am. Vet. Med. Assoc.* **188,** 1019–1025.

Brobst, D. F., and Bayly, W. M. (1982). *J. Equine Vet. Sci.* **2,** 51–56.

Buck, W. B., Osweiler, G. D., and Van Gelder, G. A. (1976). *In* "Clinical and Diagnostic Veterinary Toxicology" (G. A. Van Gelder, ed.), 2nd ed., pp. 83–86. Kendall/Hunt Publishing Co., Dubuque, Iowa.

Burnell, J. M., Villamil, M. F., Uyeno, B. T., and Scribner, B. H. (1966). *J. Clin. Invest.* **35,** 935–939.

Carlson, G. P. (1975). *Proc. Int. Symp. Equine Hematol., 1st,* East Lansing, Michigan, pp. 473–480. Am. Assoc. Equine Pract., Golden, Colorado.

Carlson, G. P. (1979a). *Proc. Annu. Meet. Am. Assoc. Equine Pract., 25th,* Miami, Florida pp. 459–468.

Carlson, G.P. (1979b). *Vet. Clin. N. Am. (Lg. Anim. Pract.)* **1,** 313–329.

Carlson, G. P. (1983a). *In* "Current Therapy in Equine Medicine" (N. E. Robinson, ed.), pp. 311–318. Saunders, Philadelphia, Pennsylvania.

Carlson, G. P. (1983b). *In* "Equine Exercise Physiology" (D. H. Snow, S. G. B. Persson, and R. J. Rose, eds.), pp. 291–309. Granta Editions, Cambridge, England.

Carlson, G .P. (1983c). *In* "Current Therapy in Equine Medicine" (N. E. Robinson, ed.), pp. 318–321. Saunders Co., Philadelphia, Pennsylvania.

Carlson, G. P. (1987). *In* "Equine Exercise Physiology 2" (J. R. Gillespie and N. E. Robinson, eds.), pp. 393–426. ICEEP Publications, Davis, California.

Carlson, G. P. (1995). *Equine Vet J. Suppl.* **18,** 261–265.

Carlson, G. P., and Harrold, D. R. (1977). *Vet. Clin. Pathol.* **6,** 18–20.

Carlson, G. P., and Rumbaugh, G. E. (1983). *Am. J. Vet. Res.* **44,** 964–968.

Carlson, G. P., Harrold, D., and Rumbaugh, G. E. (1979a). *Am. J. Vet. Res.* **40,** 587–589.

Carlson, G. P., Rumbaugh, G. E., and Harrold, D. R. (1979b). *Am. J. Vet. Res.* **40,** 982–985.

Chinard, F. P. (1951). *In* "Methods in Medical Research" (M. B. Vessehers, ed.), Vol. 4, pp. 38–48. Year Book Medical Publishers, Chicago.

Christensen, G. (1993). *Scand. J. Lab. Invest.* **53,** 203–209.

Clark, A. M. (1980). *Vet. Rec.* **106,** 146–149.

Clausen, T., Wang, P., Orskov, H., and Kristensen, O. (1980). *Scand. J. Clin. Lab. Invest.* **40,** 211–220.

Cornelius, L. M., Finco, D. R., and Culver, D. H. (1978). *Am. J. Vet. Res.* **39,** 1185–1190.

Cox, J. H. (1985). *Proc. Annu. Meet. Am. Assoc. Equine Pract., 31st,* Toronto, pp. 383–391.

Crawford, M. A., Kittleson, M. D. and Fink, G. D. (1984). *J. Am. Vet. Med. Assoc.* **184,** 818–821.

Crow, S. E. and Stockham, S. L. (1985). *J. Am. Anim. Hosp. Assoc,* **21,**393–400.

Dahms, H. Rock, R., and Seligson, D. (1968). *Clin. Chem.* **14,** 859–870.

Dalton, R. G. (1964). *Br. Vet. J.* **120,** 378–384.

Degan, M. A. (1986). *Vet. Clin. Path.* **15,** 20–22.

De Morais, H. S. A. (1992a). *In* "Fluid Therapy in Small Animal Practice" (S. P. DiBartola ed), pp. 276–296. Saunders, Philadelphia, Pennsylvania.

De Morais, H. S. A. (1992b). *In* "Fluid Therapy in Small Animal Practice" (S. P. DiBartola ed), pp 297–320. Saunders, Philadelphia, Pennsylvania.

Dhein, C. R. and Wardrop, K. J. (1995). *J. Am. Vet. Med. Assoc.* **206,** 1565–1566.

DiBartola, S. P. (1992a). *In* "Fluid Therapy in Small Animal Practice" (S. P. DiBartola ed), pp. 193–215. Saunders, Philadelphia, Pennsylvania.

DiBartola, S. P. (1992b). *In* "Fluid Therapy in Small Animal Practice" (S. P. DiBartola ed), pp. 216–243. Saunders, Philadelphia, Pennsylvania.

DiBartola, S. P. (1992c). *In* "Fluid Therapy in Small Animal Practice" (S. P. DiBartola ed), pp. 244–257. Saunders, Philadelphia, Pennsylvania.

DiBartola, S. P., and De Morais, H. S. A. (1992). *In* "Fluid Therapy in Small Animal Practice" (S. P. DiBaratola, ed.), pp. 258–275. Saunders, Philadelphia, Pennsylvania.

Dieterich, R. A., and Holleman, D. F. (1973). *Can J. Zool.* **51,** 867–873.

Divers, T. J., Freeman, D. E., Ziemer, E. L., and Becht, J. L. (1986). *Proc. Annu. Meet. Am. Assoc. Equine Pract., 32nd,* Nashville, Tennessee, pp. 69–80.

Dow, S. W., LeCouteur, R. A., Fettman, M. J., and Spurgeon, T. L. (1987a). *J. Am. Vet. Med. Assoc.* **191,** 1563–1568.

Dow, S. W., Fettman, M. J., LeCouteur, R. A., and Hamar, D. W. J. (1987b). *J. Am. Vet. Med. Assoc.* **191,** 1569–1575.

Earley, L. E., and Daugharty, T. M. (1969). *N. Engl. J. Med.* **281,** 72–86.

Edelman, I. S., and Leibman, J. (1959). *Am. J. Med.* **27,** 256–277.

Edelman, I. S., Leibman, J., O'Meara, P., and Birkenfeld, L. W. (1958). *J. Clin. Invest.* **37,** 1236–1256.

Eiseman, A. J., MacKenzie, L. B., and Peters, J. P. (1936). *J. Biol. Chem.* **116,** 35–45.

Elkinton, J. R., and Danowski, T. S. (1955). "The Body Fluids. Basic Physiology and Practical Therapeutics." Williams and Wilkins, Baltimore. pp. 68–111.

Elkinton, J. R., and Taffel, M. (1942). *J. Clin. Invest.* **21,** 787–794.

Elkinton, J. R., and Winkler, A. W. (1944). *J. Clin. Invest.* **23,** 93–101.

Emmett, M. and Nairns, R. G. (1977). *Medicine* **56,** 38–54.

English, P. B. (1966a). *Res. Vet. Sci.* **7,** 233–257.

English, P. B. (1966b). *Res. Vet. Sci.* **7,** 258–275.

Epstein, V. (1984). *Equine Vet. J.* **16,** 453–456.

Evans, J. W. (1972). *J. Anim. Sci.* **33,** 1001–1004.

Feldman, B. F., and Rosenberg, D. P. (1981). *J. Am. Vet. Med. Assoc.* **178,** 396–398.

Fencil, V. and Leith D. H. (1993) *Resp. Physiol.* **91,** 1–16.

Fencil, V., and Rossing, T. H. (1989). *Ann. Rev. Med.* **40,** 17–29.

Fettman, M. J., Brooks, P. A., Burrows, K. P., and Phillips, R. W. (1986). *J. Am. Vet. Med. Assoc.* **188,** 397–401.

Fisher, E. W., and Martinez, A. A. (1976). *Res. Vet. Sci.* **20,** 302–305.

Fleming, S. A., Hunt, E. L., Brownie, C., Rakes, A., and McDaniel, B. (1992). *Am. J. Vet. Res.* **53,** 222–224.

Fonnesbeck, P. V. (1968). *J. Anim. Sci.* **27,** 1350–1356.

Fredeen, A. H., DePeters, E. J., and Baldwin, R. L. (1988). *J. Anim. Sci.* **66,** 174–84.

Frischmeyer, K. J., and Moon, P.F. (1994). *Am. J. Vet. Res.* **55,** 1153–1157.

Gamble, J. L., and Robertson, J. S. (1959). *Am. J. Physiol.* **179,** 659–667.

Garella, S., Chang, B. S., and Kahn, S. I. (1975). *Kidney Int.* **8,** 279–283.

Garry, F., and Rings, D. M. (1987). *Proc. Annu. Vet. Med. Forum,* 5th, Washington, D.C., pp. 823–826.

Genetzky, R. M., Loparco, F. V., and Ledet, A. E. (1987). *Am. J. Vet. Res.* **48,** 1007–1011.

Gennari, F. J. (1984). *N. Engl. J. Med.* **310,** 102–105.

George, J. W. (1994). *In* "Veterinary Laboratory Medicine Clinical Pathology" (J. R. Duncan, K. W. Prasse, and E. A. Mahaffey, eds.), 3rd ed., pp. 94–111. Iowa State University Press, Ames.

Giger, U. and Gorman, N.T. (1984). *Compend. Cont. Ed. Pract. Vet.* **6,** 805–812.

Gilfix B. M., Bique, M., and Magder, S. (1993), *J. Crit. Care* **8,** 187–197.

Gingerich, D. A., and Murdick, P. W. (1975a). *Am. J. Vet. Res.* **36,** 663–668.

Gingerich, D. A., and Murdick, P. W. (1975b). *J. Am. Vet. Med. Assoc.* **166,** 227–230.

Glazier, D. B., Littledike, E. T., and Evans, R. D. (1979). *J. Equine Med. Surg.* **3,** 489–494.

Glazier, D. B., Littledike, E. T., and Evans, R. D. (1982). *Am. J. Vet. Res.* **43,** 1934–1937.

Gossett, K. A., and French, D. D. (1983). *Am. J. Vet. Res.* **44,** 1744–1745.

Hankes, G. H., Nelson, A. W., and Swan, H. (1973). *Am. J. Vet. Res.* **34,** 221–229.

Hannon J. P., Bossone, C. A., and Wade, C. E. (1990). *Lab. Anim. Sci.* **40,** 293–299.

Hansard, S. L. (1964). *Am. J. Physiol.* **206,** 1369–1372.

Harris, P., and Snow, D. H. (1986). *J. Physiol.* **376,** 46P.

Harris, R. C., Harman, J. C., Marlin, D. J. and Snow, D. H. (1987). *In* "Equine Exercise Physiology 2" (J.R. Gillespie and N.R. Robinson, eds.),pp. 464–475. ICEEP Publications, Davis, California.

Hayter, C. J., Clapham, W. F., Mills, I. H., and Wardener, H. E. (1962). *Clin. Sci.* **23,** 229–236.

Hedley-Whyte, J., and Laver, M. B. (1964). *J. Appl. Physiol.* **19,** 901.

Hinton, M. (1978). *Equine Vet. J.* **10,** 27-31.

Hintz, H. F., and Schryver, H. F. (1976). *J. Anim. Sci.* **3,** 637–643.

Hix, E. L., Evans, L. E., and Underbjerg, G. K. L. (1953). *J. Anim. Sci.* **12,** 459–473.

Hix, E. L., Underbjerg, G. K. L., and Hughes, J. S. (1959). *Am. J. Vet. Res.* **20,** 184–191.

Hopper, J., Elkinton, J. R., and Winkler, A. W. (1959). *J. Clin. Invest.* **23,** 111–117.
Hoskins, J. D., and Rothschmitt, J. (1984). *Vet. Med.* **79,** 489–491.
Houston, D. M., Allen, D. G., and Kruth, S. A.(1989). *Can Vet. J.* **30,** 423–425.
Humes, H. D. (1984). *In* "Fluids and Electrolytes" (J. P. Kokko and R. L. Tannen, eds.), pp. 119–149. Saunders, Philadelphia, Pennsylvania.
Ilkiw J. E., Rose, R. J., and Martin I. C. A. (1991). *J. Vet. Int. Med.* **5,** 294–298.
Inagami, T. (1994). *J Clin. Pharmacol.* **34,** 424–426.
Irvine, R. O. H., and Dow, J. W. (1968). *Aust. Ann. Med.* **17,** 206–213.
Jezyk, P. F. (1982). *J. Am. Anim. Hosp. Assoc.* **18,** 977–980.
Jones, N. L. (1990). *Clin. Biochem.* **23,** 189–195.
Judson, G. J., and Mooney, G. J. (1983). *In* "Equine Exercise Physiology" (D. H. Snow, S. G. B. Persson, and R. J. Rose, eds.), Granta Editions, Cambridge, England.
Kamal, T. H., Shehata, O., and Elbanna, I. M. (1972). *In* "Isotope Studies on the Physiology of Domestic Animals." *Proc. Symp. Int. Atomic Energy Agency,* Vienna, pp. 95–102.
Kami, G., Merritt, A. M., and Duelly, P. (1984). *Equine Vet. J.* **16,** 356–358.
Kasari, T. R., and Naylor, J. M. (1984). *Can. Vet. J.* **25,** 394–399.
Kasari, T.R., and Naylor, J. M. (1986). *Can. J. Vet. Res.* **50,** 502–508.
Kohn, C. W. (1979). *Vet. Clin. N. Am. (Large Anim. Pract.)* **1,** 289–311.
Kohn, C. W. and DiBartola, S.P. (1992). *In* "Fluid Therapy in Small Animal Practice" (S. P. DiBaratola, ed.), pp. 1–34. Saunders, Philadelphia, Pennsylvania.
Koterba, A., and Carlson, G. P. (1982). *J. Am. Vet. Med. Assoc.* **180,** 303–306.
Kowalchuk, J. M., and Scheuermann, B. W. (1993). *Can. J. Physiol. Pharmacol.* **72,** 818–826.
Kryzwanek, H. (1974). *J.S. Afr. Vet. Assoc.* **45,** 355–360.
Kryzwanek, H., Milne, D. W., Gabel, A. A., and Smith, L. G. (1976). *Am. J. Vet. Res.* **37,** 291–294.
Kurtz, T. W., Al-Bander, H. A., and Morris, R. C. (1987). *New Engl. J. Med.* **317,** 1043–1048.
Lakritz, J., Madigan, J., and Carlson, G. P. (1992). *J. Am. Vet. Med. Assoc.* **200,** 1114–1116.
Landis, E. M., and Pappenheimer, J. P. (1963). *In* "Handbook of Physiology" (W. F. Hamilton, ed.), Sect. 2, Vol. II, pp. 961–1034, Am. Physiol. Soc. Williams and Wilkins, Baltimore.
Leaf, A. (1962). *N. Engl. J. Med.* **267,** 77–83.
Leitch, I., and Thomson, M. A. (1944). *Nutr. Abstr. Rev.* **14,** 197–223.
Littlejohn, A., and Mitchell, B. (1969). *Res. Vet. Sci.* **10,** 260–262.
McAuliffe, J. J., Lind, L. J., Leith, D. E., and Fencl, V. (1986). *Am. J. Med.* **81,** 86–90.
McCance, R. A. (1937). *Biochem. J.* **31,** 1278–1284.
McCance, R. A. (1938). *J. Physiol.* **92,** 208–218.
McCoy, C. P., and Edwards, W. C. (1979). *Oklahoma Vet.* **31,** 12–14.
McDougall, J. G., Coghlan, J. P., Scoggins, B. A., and Wright, R. D. (1974). *Am. J. Vet. Res.* **35,** 923–929.
McGuirk, S. M., and Butler, D. G. (1980). *J. Am. Vet. Med. Assoc.* **177,** 551–554.
McKeever, K. H., Hinchcliff, K. W., Schmall, L. M., Lamb, D. R. and Muir, W. W. (1991). In "Equine Exercise Physiology 3" (S. G. B. Persson, A. Lindholm, and L. B. Jeffcott, eds.), pp. 368–373. ICEEP Publications, Davis, California.
McKeown, J. W. (1984). *In* "Fluids and Electrolytes" (J. P. Kokko and R .L. Tannen, eds.), pp. 63–117. Saunders, Philadelphia, Pennsylvania.
Madias, N. E., Ayus, J. C., and Adrogue, H.J. (1979). *N. Engl. J. Med.* **300,** 1421–1423.
Madigan, J. E., Thomas, W.P., Backus, K.Q., and Powell, W.E. (1992). *Equine Vet. J.* **24,** 399–401.
Man, S. O., and Carroll, H. J. (1977). *N. Engl. J. Med.* **297,** 814–817.
Mandell, C. P., Degen, M. A., Hopper, P. E., and Zinkl, J. G. (1988). *Vet. Clin. Path.* **17,** 32–33.
Mason, D. E., Muir, W. W., and Wade, A. (1987). *J. Am. Vet. Med. Assoc.* **190,** 989–994.
Mattheeuws, D. R. G., Kaneko, J. J., Loy, R. G., Cornelius, C. E., and Wheat, J. D. (1966). *Am. J. Vet. Res.* **27,** 699–705.
Merritt, A. M., Kohn, C. W., Ramberg, C. F., Cimprich, R. E., Reid, C. F., and Bolton, J. R. (1977). *Am. J. Vet. Res.* **38,** 1769–1774.
Michell, A. R. (1974). *Br. Vet. J.* **130,** vi–viii.
Michell, A. R. (1985). *Vet. Rec.* **116,** 653–657.
Milne, D. W., Skarda, R. T., Gabel, A. A., Smith, L. G., and Ault, B. S. (1976). *Am. J. Vet. Res.* **37,** 285–290.
Morris, D. D., Divers, T. J., and Whitlock, R. H. (1984). *Am. J. Vet. Res.* **45,** 2431–435.
Muylle, E., Van den Hende, C., Nuytten, J., Deprez, P., Vlaminck, K., and Oyaert, W. (1984a). *Equine Vet. J.* **16,** 447–449.
Muylle, E., Nuytten, J., Van den Hende, C., Vlaminck, K., and Oyaert, W. (1984b). *Equine Vet. J.* **16,** 450–451.
Nairns, R. G., and Emmett, M. (1980). *Medicine* **59,** 161–187.
Padovan, D. (1980). *Cornell Vet.* **70,** 153–159.
Patrick, J. (1977). *Kidney Int.* **11,** 476–490.
Pearson, E. G., and Kallfelz, F. A. (1982). *Cornell Vet.* **72,** 142–149.
Persson, S. (1967). *Acta Vet. Scand. Suppl.* **19,** 1–189.
Persson, S. G. B., Ekman, L., Lydin, G., and Tufvesson, G. (1973). *Zentralbl. Veterinaermed. Reihe A.* **20,** 441–455.
Phillips, R. W., and Knox, K. L. (1969). *J. Comp. Lab. Med.* **3,** 1–3.
Phillips, R. W., Lewis, L. D., and Knox, K. L. (1971). *Ann. N.Y. Acad. Sci.* **176,** 231–243.
Phillipson, A. T. (1977). *In* "Duke's Physiology of Domestic Animals" (M. J. Swenson, ed.), 9th ed., pp. 250–286. Comstock Publishing Co., Ithaca, New York.
Pickar, J. G., Spier, S. J., Snyder, J. R., and Carlsen, R. C. (1991). *Am. J. Physiol.* **260,** C926–C933.
Pierce, A. W. (1957). *Vet. Rev. Annot.* **3,** 37–43.
Polzin, D. J., Stevens, J. B., and Osborne, C. A. (1982). *Comp. Cont. Ed. Pract. Vet.* **4,** 1021–1036.
Pownall, R., and Dalton, R. G. (1973). *Br. Vet. J.* **129,** 583–588.
Reeves *et al.*, (1982). *J. Appl. Physiol.* **53,** 87–95.
Robb, J., Harper, H. F., Hintz, H. F., Reid, J. T., Lowe, J. E., Schryver, H. F., and Rhee, M. S. S. (1972). *Anim. Prod.* **14,** 25–34.
Rose, B. D. (1984). *In* "Clinical Physiology of Acid–Base and Electrolyte Disorders," 2nd ed. McGraw-Hill, New York.
Rose, R. J., and Carter, J. (1979). *J. Vet. Pharmacol. Ther.* **2,** 215–221.
Rose, R. J., Carter, R. J., and Ilkiw, J. E. (1980). *J. Vet. Pharmacol. Ther.* **3,** 111–119.
Rose, R. J., Gibson, K. T., and Suann, C. J. (1986). *Vet. Rec.* **19,** 522–525.
Rossing, R. G., and Cain, S. M. (1966). *J. Appl. Physiol.* **21,** 195–201.
Rudolph, J. A., Spier, S. J., Byrns, G., Rojas, C. V., Bernoco, D. and Hoffman, E. P. (1992a) *Nat. Genet.* **2,** 144–147.
Rudolph, J. A., Spier, S. J., Byrns, G. and Hoffman, E. P. (1992b) *Anim. Genet.* **23,** 241–250.
Rumbaugh, G. E., Carlson, G. P., and Harrold, D. R. (1981). *J. Am. Vet. Med. Assoc.* **178,** 267–271.
Rumbaugh, G. E., Carlson, G. P., and Harrold, D. (1982). *Am. J. Vet. Res.* **43,** 735–737.
Rumsey, T. S., and Bond, J. (1976). *J. Amin. Sci.* **42,** 1227–1238.
Sandals, W. C. D. (1978). *Can. Vet. J.* **19,** 136–137.
Saxton, D. R., and Seldin, D. W. (1986). *In* "Fluids and Electrolytes" (J. P. Kokko and R. L. Tannen, eds.), pp. 3–62. Saunders, Philadelphia, Pennsylvania.

Schultze, G., Kirsch, K., and Rocker, L. (1972). *Pflügers Arch.* **337,** 351–366.

Scribner, B. H. (1969). *In* "Teaching Syllabus for the Course on Fluid and Electrolyte Balance." School of Medicine, University of Washington, Seattle.

Senay, L. C. (1970). *J. Physiol.* (*London*) **210,** 617–635.

Shull, R. (1978). *Vet. Clin. Pathol.* **7,** 7–14.

Shull, R. (1981). *J. Am. Vet. Med. Assoc.* **178,** 1119–1123.

Singer, R. B., and Hastings, A. B. (1948). *Medicine* **27,** 223–243.

Sjogaard, G., Adams, R. P., and Saltin, B. (1985). *Am. J. Physiol.* **248,** R190–R196.

Speirs, V. C. (1980). *Am. J. Vet. Res.* **41,** 199–203.

Spensley, M. S., Carlson, G. P., and Harrold, D. (1987). *Am. J. Vet. Res.* **48,** 1703–1707.

Spier, S. J., Carlson, G. P., Holliday, T. A., Cardinet, G. H. D., and Pickar, J. G. (1990). *J. Am. Vet. Med. Assoc.* **197,** 1009–1017.

Spier, S. J., Carlson, G. P., Harrold, D., Bowling, A., Byrns, G., and Bernoco, D. (1993). *J. Am. Vet. Med. Assoc.* **202,** 933–937.

Spurlock, G. H., Landry, S. L., Sams, R., McGuirk, S., and Muir, W. W. (1985). *Am. J. Vet. Res.* **46,** 1117–1120.

Steffey, E. P., Wheat, J. D., Meagher, D. M., Norrie, R. D., Mckee, J. Brown, M., and Arnold, J. (1977). *Am. J. Vet. Res.* **38,** 379–382.

Steiss, J. E., and Naylor, J. M. (1986). *Can. Vet. J.* **27,** 332–335.

Sterns, R. H., Riggs, J. E., and Schochet, S. S. (1986). *N. Engl. J. Med.* **314,** 1535–1542.

Stewart, P. A. (1981). *In* "How to Understand Acid–Base: A Quantitative Acid–Base Primer for Biology and Medicine." Elsevier North Holland, New York.

Stewart, P. A. (1983). *Can. J. Physiol. Pharmacol.* **61,** 1444–1461.

Stick, J. A., Robinson, N. E., and Krehbiel, J. D. (1981). *Am. J. Vet. Res.* **42,** 733–737.

Strombeck, D. R. (1979). *In* "Small Animal Gastroenterology." Stonegate Publishing, Davis, California.

Sufit, E., Houpt, K. A., and Sweeting, M. (1985). *Equine Vet. J.* **17,** 12–16.

Tannen, R. L. (1984). *In* "Fluids and Electrolytes." (J. P. Kokko and R. L. Tannen, eds.), pp. 150–228. Saunders, Philadelphia, Pennsylvania.

Tasker, J. B. (1967a). *Cornell Vet.* **57,** 649–657.

Tasker, J. B. (1967b). *Cornell Vet.* **57,** 658–667.

Tasker, J. B. (1967c). *Cornell Vet.* **57,** 668–677.

Tasker, J. B. (1980). *In* "Clinical Biochemistry of Domestic Animals" (J. J. Kaneko, ed.), 2nd ed., pp. 402–447. Academic Press, New York.

Thornton, J. R., and English, P. B. (1977). *Res. Vet. Sci.* **22,** 298–302.

Tollertz, G. (1964). *Acta. Vet. Scand.* **5,** 24–34.

Tremblay, R. R. M., Butler, D. G., Allen, J. W., and Hoffman, A. M. (1991). Can. Vet. J. **32,** 308–310.

Tyler, R. D., Qualls, C. W., Heald, R. D., Cowell, R. L., and Clinkenbeard, K. D. (1987). *J. Am. Vet. Med. Assoc.* **191,** 1095–1097.

van Beaumont, W., Strand, J. C., Petrofsky, J. S., Hipskind, S. G., and Greenleaf, J. E. (1973). *J. Appl. Physiol.* **34,** 102–106.

van Sluijs, F. J., de Vries, H. W., De Bruijne, J. J., and van den Brom, W. E. (1983). *Am. J. Vet. Res.* **44,** 459–462.

Wade, L., and Sasser, L. B. (1970). *Am. J. Vet. Res.* **31,** 1375–1378.

Weldon, A. D., Moise, N. S. and Rebhun, W. C. (1992) *J. Vet. Int. Med.* **6,** 294–297.

Whitehair, K. J., Haskins, S. C., Whitehair, J. G., and Pascoe, P. J. (1995) *J. Vet. Int. Med.* **9,** 1–11.

Whitlock, R. H., Kessler, M. J., and Tasker, J. B. (1975a). *Cornell Vet.* **65,** 512–526.

Whitlock, R. H., Tasker, J. B., and Tennant, B. (1975b). *Dig. Dis. Sci.* **20,** 595–596.

Williams, M. E., Gervino, E. V., Rosa, R. M., Landsberg, L., Young, J. B., Silva, P., and Epstein, F. H. (1985). *N. Engl. J. Med.* **312,** 823–827.

Wilson, E. A., and Green, R. A. (1985). *Comp. Cont. Ed. Pract. Vet.* **7,** S364–S371.

Winkler, A. W., Elkington, J. R., Hopper, J., and Hoff, H. (1944). *J. Clin. Invest.* **23,** 103–109.

Wolf, A. V. (1966). *In* "Aqueous Solutions and Body Fluids: Their Concentrative Properties and Conversion Tables." Harper & Row, New York.

Yoshida, S., Sudo, H., Noro, K., and Nokaya, C. (1967). *Exp. Rep. Equine Hlth. Lab.* **4,** 30–36.

Ziemer, E. L., Parker, H. R., Carlson, G. P., Smith, B. P., and Ishizaki, G. (1987a). *J. Am. Vet. Med. Assoc.* **190,** 289–293.

Ziemer, E. L., Parker, H. R., Carlson, G. P., and Smith, B. P. (1987b). *J. Am. Vet. Med. Assoc.* **190,** 294–296.

Zweens, J., Frankena, H., Rispens, P., and Zijlstra, W. G. (1975). *Pflügers Arch.* **357,** 275–290.

CHAPTER

19

Pituitary Function

JAN A. MOL AND AD RIJNBERK

I. HYPOTHALAMUS–PITUITARY SYSTEM

The hypothalamus–pituitary system is a preeminent example of integration of neural and endocrine control. It consists of three major systems: (1) a neuroendocrine system connected to an endocrine system by a portal circulation, (2) a neurosecretory pathway, and (3) a direct neural regulation of endocrine secretion (Fig. 19.1).

The neuroendocrine system involves clusters of peptide- and monoamine-secreting cells in the anterior and midportion of the ventral hypothalamus. Their products reach the median eminence by axonal transport. From there they are released into the capillary vessels of the hypothalamus–pituitary portal system and transported to the pituitary to regulate the secretion of hormones from the anterior lobe (AL) of the adenohypophysis.

The neurosecretory pathway runs from the anterior hypothalamus, traverses the floor of the ventral hypothalamus, and terminates in the neural lobe (NL) of the neurohypophysis on fenestrated blood vessels (Page, 1986). The system is involved in osmoregulation through the production and release of vasopressin, and in parturition and nursing through the secretion of oxytocin.

In the intermediate lobe (IL) the secretory activity is regulated via direct neuronal inhibitory and stimulatory influences (Tilders *et al.*, 1985). In amphibians (Verburg-van Kemenade *et al.*, 1986) and reptiles (Dores *et al.*, 1987) the IL plays an important role in adaptation to background color. The function of IL cells in mammals has not been fully established, but they may play a role in opioid-regulated functions.

A. Anatomical Considerations

The hypothalamus and pituitary control vital functions such as growth, reproduction, lactation, basal metabolism, stress response, parameters of immune function, and the state of hydration. Understanding of the complicated functional relationship of the hypothalamus to the pituitary requires an appreciation of the anatomical relationships.

1. Hypothalamus

Hypophysiotropic neurohormones are produced in several areas of the hypothalamus. For example, in an immunofluorescence study of hypothalami of dogs, the majority of cell bodies with immunoreactivity to corticotropin-releasing hormone (CRH) were found in

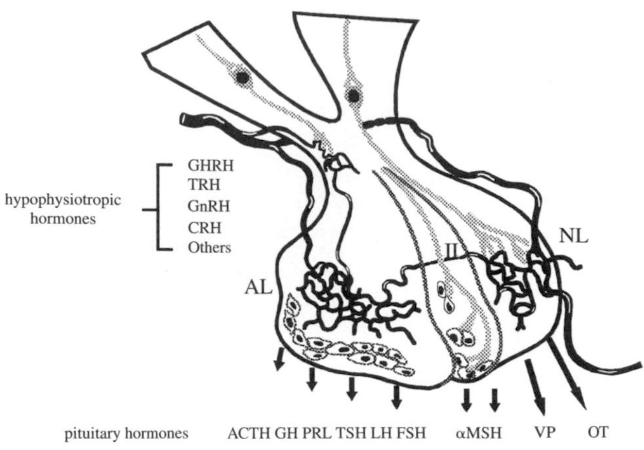

FIGURE 19.1 Schematic representation of the relationship between the hypothalamus and the pituitary in a generalized mammal. The hypothalamus exerts control over the anterior lobe (AL) by releasing and inhibiting factors. These hormones reach the AL cells via capillaries of the pituitary portal system. The neural lobe (NL) of the pituitary is a downward projection of the hypothalamus. The intermediate lobe (IL) is under direct neurotransmitter control.

the region of the periventricular and paraventricular nuclei, but they were also found in the supraoptic and suprachiasmatic area as well as craniodorsal to the mamillary bodies (Stolp *et al.*, 1987).

The cell bodies of the neurohormone-producing neurons that project to the median eminence are in part intermingled with cell bodies that also synthesize these neurohormones but project to other brain areas. The majority of the neurons that project to the median eminence are found in the preoptic and suprachiasmatic region of the hypothalamus. Axons containing the same neurohormone may have synaptic contacts that enable regulation of cellular function between these neurons.

The neurohormone-producing cells receive a complex neural input from a variety of chemical messengers, such as neurotransmitters and other neurohormones. Not only the neurohormones CRH and arginine vasopressin (AVP), but in general the combination of a neurohormone and another chemical messenger, may colocalize within a single neuron.

The neurons of the neurohypophyseal system represent a more anatomically distinct entity, with cell bodies located in the paraventricular and supraoptic hypothalamic nuclei (Sawchenko and Swanson, 1983). However, within these areas there are also neurons producing a variety of other neuropeptides.

2. *Neurotransmitter Systems*

The major neurotransmitter systems for intercellular communication within the central nervous system consist of monoamines and peptides. These chemical messengers regulate the biosynthesis and release of the hypophysiotropic neurohormones. Through a network of axodendritic and axoaxonic contacts, these neurons are connected to the neurohormone-producing cells. In addition, many monoamines and peptides are found within the hypophysiotropic hormone-producing cells, where they are released together with the neurohormones into the portal system and modify the effect of hypothalamic hormones on the pituitary.

The biogenic amine neurotransmitters, known to play a regulating/modulating role in the hypothalamus–pituitary system, include catecholamines (dopamine, noradrenalin, adrenalin), indolamines (serotonin, melatonin), acetylcholine, γ-aminobutyric acid (GABA), and histamine. Neuropharmacologic agents can be used to alter neurotransmitter effects and, as a consequence, hypothalamic and pituitary hormone release.

Many of the peptides with potential effects on hypophysiotropic hormone release are widely distributed in hypothalamic and extrahypothalamic areas of the brain. They include, among many others, peptides common to the gastrointestinal tract, such as gastrin, cholecystokinin, and pancreatic polypeptide, as well as bombesin, angiotensin II, galanin, substance P, neurotensin, enkephalins, neuropeptide Y, natriuretic peptide, and, from a common precursor, the vasoactive intestinal peptide (VIP) and the peptide histidine isoleucine (PHI). The last three peptides may, through vasoconstriction and vasodilatation activities, play an important role in the control of the portal blood flow.

3. *Vascular System*

The releasing and inhibiting hormones are stored in nerve terminals in the median eminence, where their concentrations are 10 to 100 times as great as elsewhere in the hypothalamus. The uniquely organized capillary plexus (see Page, 1986) of the median eminence is in close proximity to nerve terminals of the hypophysiotropic neurons. In contrast to other brain regions, the blood–brain barrier in the area of the median eminence is incomplete, permitting protein and peptide hormones as well as other charged particles to move to the intercapillary spaces and the nerve terminals contained therein. These terminals respond to humoral and neuronal stimuli by secreting releasing and inhibiting factors into the portal system.

The portal capillaries coalesce into a series of vessels that descend through the pituitary stalk and form a second capillary plexus that surrounds the AL cells (Fig. 19.1).

Inferior hypophyseal arteries supply the neurohypophysis. From the primary plexus of the neural lobe

(NL), blood flows not only to the systemic circulation but also to the AL and the hypothalamus. There is evidence for some degree of circulatory flow within the pituitary, that is, from the AL to the NL, from there to the infundibulum, and then back to the AL. The primary capillary plexus of the NL appears to be well positioned in the mini-circulatory system, controlling all of the afferent vascular events and many of the efferent vascular events in the pituitary (Page, 1986).

The vascularization of the intermediate lobe is closely linked to that of the neurohypophysis. However, in spite of the rich blood supply of the NL, the IL is a poorly vascularized structure.

4. *Pituitary*

During embryogenesis the adenohypophysis develops from Rathke's pouch, which arises from the primitive roof of the mouth in contact with the base of the brain. Rathke's pouch subsequently separates by constriction from the oral cavity. The anterior wall thickens and forms the anterior lobe of the adenohypophysis. This largest portion of the adenohypophysis remains separated from the intermediate lobe by the hypophyseal cleft, which is the residual lumen of Rathke's pouch. In several species (Fig. 19.2) the adenohypophysis also extends into a pars tuberalis that forms a cuff or collar around the proximal neurohypophysis and may even envelop part of the median eminence (Hullinger, 1979; Batten and Ingleton, 1987).

The posterior wall of Rathke's pouch is closely opposed to the neural tissue of the NL, thereby forming the intermediate lobe (IL), which is well developed in most mammals, but not in man and birds (Fig. 19.2). In man only during fetal life is a distinct IL found (Visser and Swaab, 1979), whereas there is some debate about the view that in adults "invading" cells of the posterior lobe are homologous with the IL cells of lower vertebrates (Coates *et al.*, 1986; McNicol, 1986; Lamberts, 1987).

The pituitary stalk (infundibulum) and the neurohypophysis (posterior lobe, neural lobe) develop from the basal outgrowth of the diencephalon in connection with the development of Rathke's pouch. The cells of the diencephalic outgrowth later develop into glial cells (pituicytes), whereas nerve fibers from the supraoptic and paraventricular nuclei grow into the NL.

5. *Cells of the Anterior Lobe*

The peptide hormones secreted by the AL can be divided into three categories: (1) the somatomammotropic hormones growth hormone (GH) and prolactin (PRL), (2) the glycoprotein hormones thyrotropin (TSH), follicle-stimulating hormone (FSH), and luteinizing hormone (LH), and (3) the corticomelanotropins, which include α-melanotropin (MSH), adrenocorticotropin (ACTH), β-endorphin (β-END), and β-lipotropin (β-LPH).

Over the years identification of the adenohypophyseal cells has been developed by histochemical staining techniques, immunocytological methods, and ultrastructural studies. Staining and immune reactions de-

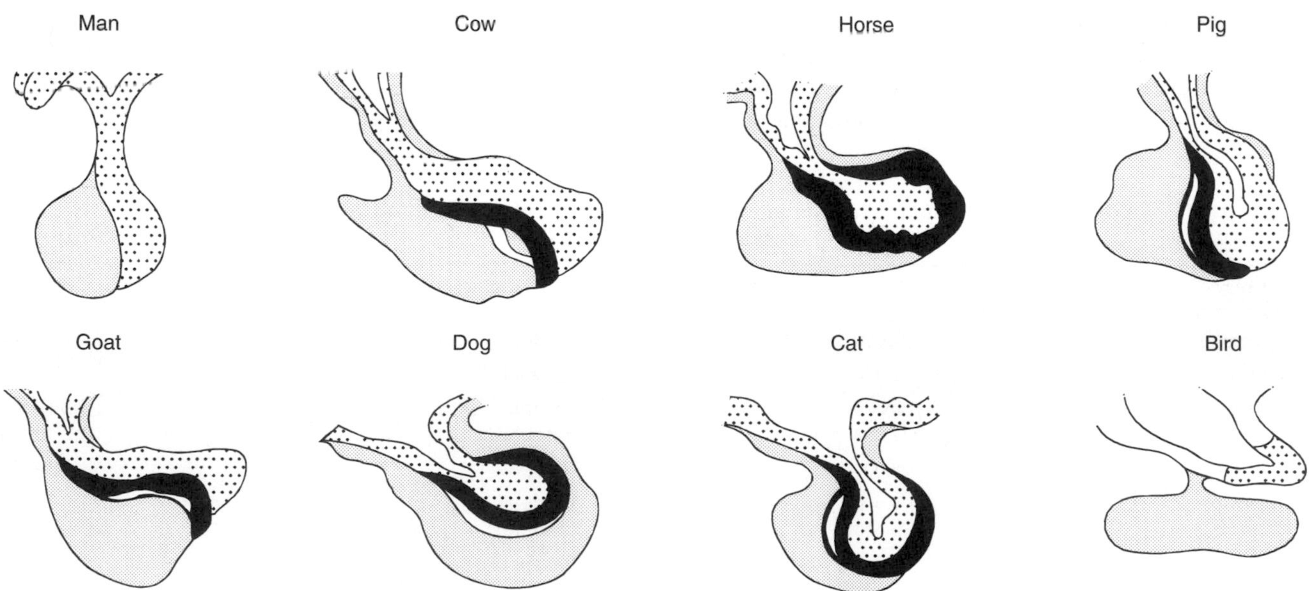

FIGURE 19.2 Schematic illustration of median sections through mammalian and avian pituitaries. Key ▨, anterior lobe; ■, intermediate lobe; ⁙ neural lobe. Redrawn from Seiferle (1975) and Batten and Ingleton (1987).

pend on the chemical nature of the hormones that are stored in granules within the cytoplasm. Special stains allow histochemical identification of particular cell types. A general identification of adenohypophyseal cells, as derived from Baker (1974), Bentley (1976), and Batten and Ingleton (1987), is given in Table 19.1, although it must be realized that the various cell types do not always give exactly the same staining reaction in different animal species.

As far as ultrastructural features are concerned, the adenohypohyseal cells have the typical cytoplasmic organelles for peptide synthesis and release. Characteristics such as the form and location of the various organelles and the size and shape of the cells allow identification of adenohypophyseal cells at the ultrastructural level (Mikami, 1986; Batten and Ingleton, 1987).

In the embryonic development of the cells of the AL, two main lineages arise from a progenitor cell, the acidophilic (A) cells that result in the mammasomatotropic cells and the basophilic (B) cells that differentiate into gonadotrope, thyrotrope, and corticotrope. During the fetal development of the mouse AL the expression of the α-subunit of the glycoproteins is among the first that becomes apparent. Then hormone biosynthesis is found in the corticotropes and then in the gonadotropes. The differentiation of the thyrotropes and somatomammotropes depends on the expression of a specific transcription factor called Pit-1 or GHF-1. Mutations in the Pit-1 gene may result in the complete absence of TSH-, GH-, and PRL-producing cells in the pituitary.

The concept of one type of pituitary cell producing one hormone now appears to be an oversimplification. Not only are the majority of gonadotropes and somatomammotropes multihormonal cells that contain LH and FSH or GH and PRL, respectively, but other combinations such as gonadotrope hormones and ACTH or TSH and ACTH have been described. As already has been discussed for the hypothalamus, the hormone-producing AL cells, or the so-called "folliculostellate" cells, also may synthesize a variety of chemical messengers, intrapituitary growth factors, and cytokines that exert local paracrine effects on cell function and proliferation (Denef, 1994).

6. Cells of the Intermediate Lobe

The predominant IL cell in mammals is the melanotrope, a cell with immunoreactivity for α-MSH, which is very sparse in the AL (Halmi and Krieger, 1983). In some species, including the dog (Halmi *et al.*, 1981) and the horse (Amann *et al.*, 1987), the IL is cytologically heterogeneous. ACTH-containing cells (B cells) have been found to be dispersed among the predominant melanotropes (A cells). In the dog the immunoreactive ACTH content of the IL even exceeds that in the AL (Halmi *et al.*, 1981). In agreement with the previously mentioned direct neural regulation of IL secretory activity, the presence of neural elements was demonstrated in the bovine IL (Boyd, 1987). Evidence is accumulating that the cells of the intermediate lobe are also involved in the biosynthesis and release of a yet-unknown prolactin-releasing factor. The blood supply of the IL is very poor, and therefore the peptides released from the IL are thought to act mainly by diffusion in a paracrine manner.

7. Cells of the Neural Lobe

The NL contains axonal nerve fibers, often swollen by being packed with neurosecretory granules. These nerve fibers and the glial cells (pituicytes) have synap-

TABLE 19.1 Molecular Weights of Adenohypophyseal Hormones and Staining Properties of Adenohypophyseal Cells

Cell type	Hormone	MW[a]	Subunits	Staining	Orange G	PAS[b]	Color[c]
Somatotrope	GH	22,000	—	Acidophil	+	−	Yellow
Lactotrope	PRL	22,500	—	Acidophil	+	−	Yellow
Gonadotrope	LH	30,000	α/β	Basophil	−	+	Blue
	FSH	32,000	α/β	Basophil	−	+	Blue
Thyrotrope	TSH	28,000	α/β	Basophil	−	+	Blue
Corticotrope	ACTH	4,500	—	Basophil	−	+	Red
	β-END	3,500	—	Basophil	−	+	Red
Melanotrope	α-MSH	1,700	—	Basophil	−	+	Red

[a] Molecular weight.
[b] Periodic acid Schiff reaction.
[c] Color obtained after co-staining with PAS-Orange G and performic acid–Alcian blue.

toid contacts. The pituicytes appear to play an intermediary role in the regulation of the release of vasopressin and oxytocin (Van Leeuwen and De Vries, 1983).

B. Regulation of Pituitary Functions

For the regulation of each of the five major adenohypophyseal hormone systems (ACTH, LH and FSH, TSH, GH, and PRL) there is a feedback (closed-loop) system. AL hormone and hypophysiotropic hormone secretion are suppressed by the products of target endocrine glands such as the thyroid, adrenal, and gonad. Apart from this long-loop feedback, some hormones (e.g., PRL) regulate their own secretion directly by acting on the hypothalamus (short-loop feedback). On this powerful feedback control with primarily bloodborne signals, other signals are superimposed. These may originate within the central nervous system (open loop) and can be mediated through neurotransmitters and hypophysiotropic hormones. Thus, influences are exerted that represent the environment (temperature, light–dark), stress (pain, fear), and intrinsic rhythmicity.

These regulatory factors influence peptide synthesis and/or release in adenohypophyseal cells, where each of the steps in hormone synthesis and ultimate secretion represents a potential control point in the regulation of circulating hormone levels.

1. Hypophysiotropic Hormones

The main hypothalamic neurohormones may stimulate or inhibit the release of a single hormone, or affect several hormone-producing cells (Fig. 19.3). The predominant influence of the hypothalamic hormones on the pituitary is stimulatory. Interference with the integrity of the hypothalamus–pituitary connections results in decreased secretion of pituitary hormones. The exception is PRL, the secretion of which is increased when hypothalamic influence is removed.

Hypothalamus

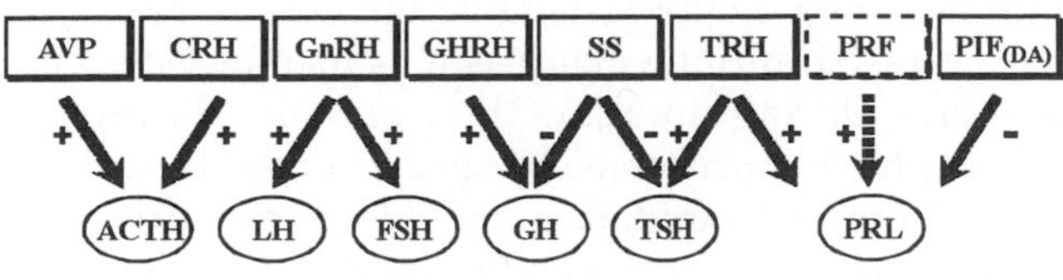

Adenohypophysis

FIGURE 19.3 Hypophysiotropic regulation of the secretion of pituitary hormones. Solid lines denote hormones whose structures have been determined. Dashed lines indicate a factor whose identity is still unknown.

As the complexity of the peptide structure of the hypophysiotropic hormones increases, species variation in sequence may occur. Whereas the structures of TRH, GnRH, and SS (3, 10, and 14 amino acids, respectively) are identical in all mammals, those of GHRH and CRH (44 and 41 amino acids) exhibit species specificity. The one nonpeptide hypophysiotropic hormone is dopamine. In addition to its major role as a neurotransmitter, dopamine is the most important inhibitor of PRL.

2. Regulation of Gene Expression

The developments in recombinant DNA technology have provided better understanding of the genes encoding pituitary hormones and the regulatory elements involved in gene transcription. The main elements regulating the gene transcription will be mentioned briefly (Fig. 19.4). Eukaryotic genes encoding peptide hormones consist of a promoter (regulatory) unit and a transcription unit encoding the primary transcript that after appropriate processing will form the messenger RNA (mRNA).

The promoter unit is the upstream (5′) part of the gene. The promoter has specific DNA sequences (response elements) permitting the binding of transcription factors that enhance or inhibit gene expression by changes in the stability of the RNA polymerase–TATA box factors complex at the constitutive promoter. At a greater distance from the gene, specific enhancer elements may also regulate gene transcription. The highly specialized hormone production is also regulated by tissue-specific silencers and enhancers that are mandatory for gene expression.

Specific binding of hypothalamic neurohormones to membrane receptors of individual AL cells will result in changes in second messenger concentrations of, for instance, cyclic AMP or inositol triphosphate within the cell, a process called signal transduction. These changes will result in differences in phosphorylation of transcription factors that bind to a cAMP response element (CRE) or tetradecanoylphorbol acetate response element (TRE) in the promotor unit. Complexes of steroid hormones, thyroid hormone, or the retinoids with their respective receptors may also bind to specific areas of the promoter. For example, a glucocorticoid response element (GRE) in the promoter of the gene encoding pro-opiomelanocortin binds the glucocorticoid–receptor complex, resulting in inhibition of gene transcription (negative GRE). Transcription is thus regulated by responses to extracellular signals, but tissue-specific intracellular transcription factors also play a crucial role in determining gene expression.

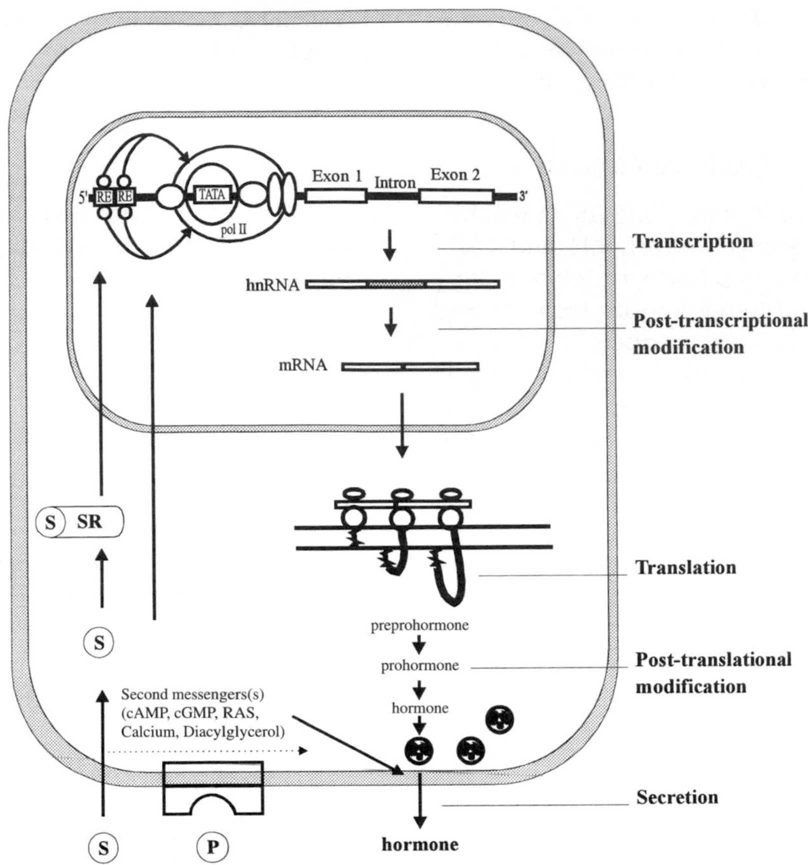

FIGURE 19.4 Schematic representation of potential control points in pituitary hormone synthesis and secretion.

After initiation of transcription a primary transcript is made, which contains an RNA copy of the entire transcription unit, the heteronuclear RNA (hnRNA). After excision of the intron areas, a process called splicing, the formation of a 7-methylguanine cap at the 5′ end, and the addition of a poly(A) tail at the 3′ end, a mature mRNA is formed. Through alternative splicing reactions, length variants of the mRNA may ultimately result in variation of the coding sequence (see section on growth hormone) or changes in mRNA stability as found for the insulin-like growth factors (IGFs). Through differences in exon coupling, even completely different peptides can be obtained from a single gene, as in the case of the gene encoding calcitonin.

3. *Prohormone Biosynthesis, Processing, and Release*

The process of peptide synthesis occurs principally on the rough endoplasmatic reticulum (rER). mRNA, encoded by nuclear DNA, passes to cytosolic ribosomes, and by sequential processing of transfer RNAs (tRNA) with their attached amino acids, the translation process of mRNA to a peptide starts. The beginning of the growing peptide forms a specific signal peptide that facilitates the attachment of the translation complex to the rER and enables the passage into the lumen of the rER. The signal sequence of the pre-prohormone is cleaved and the remaining prohormone undergoes several modifications such as disulfide formation and glycosylation. The peptide passes along the rER lumen into the Golgi complex, where peptides are packaged and released into the cytoplasm as membrane-bound granules. During storage of these granules the prohormone is further processed by specific proteolytic cleavage, C-terminal amidation, or N-terminal carboxylation. Characteristic proteolytic cleavage sites are pairs of the basic amino acids arginine and lysine. The granules are stored until the hormone is released by exocytosis. This process involves fusion of the granule membrane with the cell membrane.

II. ANTERIOR LOBE AND INTERMEDIATE LOBE

A. Proopiomelanocortin-Derived Peptides

The corticotropic cells of the AL and the melanotropic cells of the IL are both able to synthesize proopiomelanocortin (POMC), the common prohormone for ACTH, α-MSH, β-LPH, and a family of β-END-related peptides (Fig. 19.5).

1. *ACTH/α-MSH*

a. Gene Expression

The gene encoding POMC contains three exon areas. Exon 1 encodes a 5′ untranslated region (5′-UTR) and exon 2 encodes the signal peptide and the N-terminal part of the POMC prohormone. The majority of the prohormone, including γ-MSH, is encoded by exon 3. After the splicing process a mRNA of approximately 1100 nucleotides is formed in the pituitary of most species, including the dog (Mol *et al.*, 1991). By alternative splicing of the intron A sequence a longer transcript is found in the hypothalamus. Ectopic human ACTH-producing tumors contain shorter POMC mRNA sequences (Jacobson and Drouin, 1994).

The expression of the POMC gene in corticotropes and melanotropes is regulated by the same promoter elements, as shown by experiments with transgenic mice. Synergistic stimulation of gene expression requires binding of yet partially characterized transcription factors in the distal and central part of the POMC promoter. Binding of corticotropin-releasing hormone mediates the activation of the corticotrope in the AL, resulting in increased cAMP production within the cell (Jacobson and Drouin, 1994). The sequence of CRH appears to be very homologous among species. The peptide sequence CRH in the dog is identical to that in the human, rat, and horse (Mol *et al.*, 1994), and there are only small differences in CRH in cattle and sheep. Glucocorticoids inhibit POMC gene expression via type II glucocorticoid receptors in the AL corticotrope.

POMC expression in the IL melanotrope is also stimulated by cAMP. *In vitro* experiments in various species point to a possibility that CRH activates this signal transduction system. However, chronic infusion of CRH stimulates the POMC mRNA concentration in the AL, but inhibits mRNA concentration in the IL (Höllt and Haarmann, 1984). The POMC gene expression in the IL is inhibited by dopaminergic and GABAergic compounds (Chronwall *et al.*, 1987; Loeffler *et al.*, 1986), but not by glucocorticoids.

b. (Pro)hormone

POMC prohormone sequences are known for a variety of species and share a common structure (Fig. 19.5). Among species a high sequence homology is found between the N-terminal site (N-POC) of the prohormone (including γ-MSH), the ACTH region, and the β-endorphin region. In contrast, the regions between γ-MSH and ACTH and the first part of β-lipotropin are rather heterogeneous (Numa and Imura, 1985; Mol *et al.*, 1991). Species differences in the sequence of ACTH occur only in the N-terminal part (Fig. 19.6), whereas the ACTH(1-24) sequence, necessary for full biological activity, is identical among mammals.

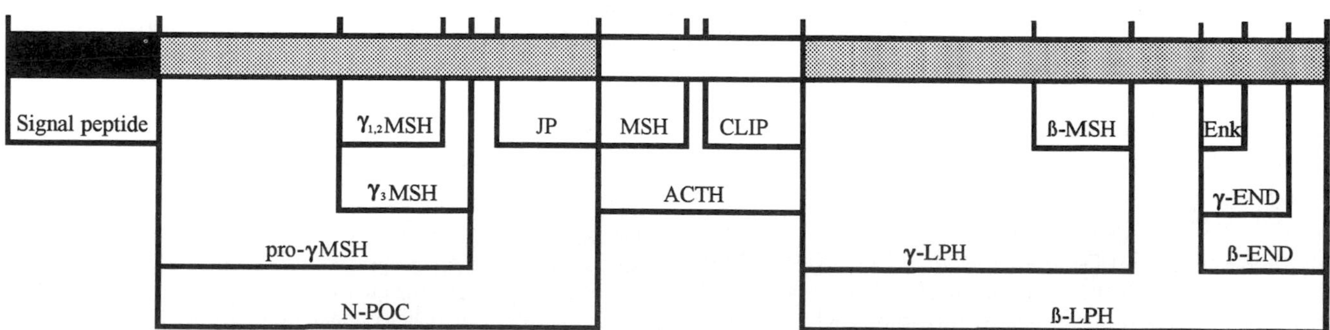

FIGURE 19.5 Schematic representation of preproopiomelanocortin (horizontal bar), with four main domains: the signal peptide, which is cleaved after entrance to the lumen of the RER; the N-terminal peptide (N-POC), containing the (pro-)γMSH sequences and the joining peptide (JP); the ACTH domain, from which MSH and corticotropin-like intermediate lobe peptide (CLIP) can be generated; and the β-lipotropin (β-LPH) domain, including the endorphin (END) family of peptides and a metenkephalin sequence (Enk). Pairs of basic amino acid residues are indicated with vertical lines, representing potential sites of proteolytic cleavage. In the anterior lobe, major cleavage products are N-POC, ACTH, and β-LPH. In the intermediate lobe, the major products are N-POC, γ-MSH, α-MSH, and β-endorphin.

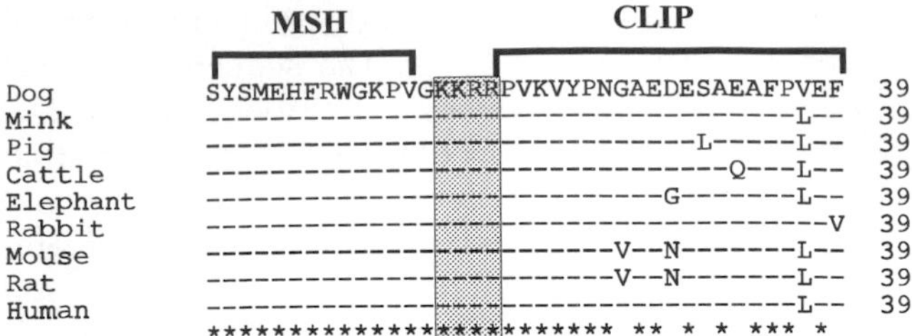

FIGURE 19.6 Comparison of the ACTH(1–39) sequence from various species using the one-letter code for amino acids. Identical amino acids are given by a single line (-). The asterisk (*) means an identical sequence for all species. The shaded box indicates the two pairs of basic amino acids prone to proteolytic cleavage.

The proteolytic enzymes PC1/PC3 and PC2 are involved in the processing of prohormones in neuroendocrine cells. They cleave the precursor at pairs of basic amino acids, resulting in the formation of biologically active hormones. The PC1/PC3 enzyme present in the AL cleaves POMC to ACTH and β-LPH. PC2 generates the ACTH(1–13) fragment in both AL and IL. The combination of both enzymes is present in the IL.

C-Terminal amidation and N-terminal acetylation of ACTH(1–13) results in α-MSH. Various degrees of acetylation of MSH result in the storage of desacetyl-, monoacetyl-, and diacetyl-α-MSH in secretory granules (Eipper and Mains, 1980; Krieger, 1980). The desacetyl form is predominant in the AL. In the rat the IL contains diacetyl-α-MSH, whereas the canine IL contains predominantly the monoacetyl-α-MSH (Young *et al.*, 1992).

c. Secretion by the AL

ACTH is released in frequent pulses, as demonstrated in the pituitary venous effluent of the horse (Redekopp *et al.*, 1986a,b). By measurements in peripheral blood, Kemppainen and Sartin (1984) documented episodic secretion of ACTH in dogs, with an average of nine peaks per 24-hour period.

The release of ACTH by the AL is influenced by many factors (Table 19.2). A number of these factors modulate the release of CRH and AVP, which are considered to be the predominant stimulating neurohormones *in vivo* (Antoni, 1986; Keller-Wood and Dallman, 1984). The relative contribution of CRH and AVP to ACTH release varies among species and circumstances. In dogs (Van Wijk *et al.*, 1994) and pigs (Minton and Parsons, 1993) both exogenously administered CRH and LVP induce comparably high plasma ACTH concentrations. In sheep, AVP is the predominant ACTH-releasing factor. In the rat the stimulation of ACTH release by CRH and AVP is synergistic, meaning that the response to CRH + AVP is greater than the sum of the reactions to CRH and AVP separately (Buckingham, 1987).

The basal release of ACTH is regulated by the occupancy of type I or mineralocorticoid receptors (MR) in the hippocampus (Jacobson and Sapolsky, 1991). Decreases in brain MR binding capacity of the dog during aging results in enhanced basal activity of the hypothalamus–pituitary–adrenal axis (Rothuizen *et al.*, 1993). In sheep, basal ACTH release is inhibited by active immunization against CRH (Guillaume *et al.*, 1992a) but not by immunization against AVP (Guillaume *et al.*, 1992b).

Exercise and insulin-induced hypoglycemia stimulate ACTH release. In the horse, exercise-induced ACTH release is accompanied by increased AVP concentration in pituitary venous blood without changes in CRH concentrations (Alexander *et al.*, 1991). In sheep, mild hypoglycemia is accompanied by increases in AVP and CRH concentrations in portal blood, but in severe hypoglycemia the AVP secretion is relatively much higher than the CRH release (Caraty *et al.*, 1990). AVP also regulates the ACTH response on hypoglycemia in the neonatal rat (Muret *et al.*, 1992).

Hypotension induced by nitroprusside (Kemppainen and Sartin, 1987) or by hemorrhage (Lilly *et al.*, 1983) causes ACTH release in the dog, together with large increases in plasma AVP derived from the NL (Raff *et al.*, 1988). Selective neurohypophysectomy results in greatly attenuated ACTH and AVP responses to hypotension and angiotensin II, whereas the ACTH response to CRH injection remains unchanged. By substitution with adequate AVP infusion, the ACTH response to hypotension can be restored (Raff *et al.*, 1992). In sheep, chronic absence of ovarian hormones after ovariectomy reduces the ACTH response to hypotension also, but not to CRH, AVP, or hypoglycemia (Pecins-Thompson and Keller-Wood, 1994).

TABLE 19.2 Factors Modulating ACTH Release from the Adenohypophysis

	Stimulating	Inhibiting
Hormones	CRH	Glucocorticoids
	Vasopressin	Enkephalin
	Angiotensin II	
	Cholecystokinin-8	
	TRH, GnRH (in the dog)	
Biogenic amines	(Nor)adrenalin	Dopamine
	Serotonin	
	GABA	
Others	Cytokines	
	Hypoglycemia	
	Hypoxia	
	Stress (physical, emotional)	

TABLE 19.3 Basal Concentrations of POMC-Derived Peptides in Plasma of Healthy Animals[a]

Species (*n*)	Hormone	Time of sampling (hours)	Range (pmol/liter) (mean ± SEM)	Reference
Dog (31)	ACTH		4.4–22	Feldman *et al.* (1977)
Dog (15)	ACTH	08.00–14.00	(7.0 ± 0.7)	Kemppainen *et al.* (1986)
Dog (160)	ACTH	08.00–10.00	2.2–19.8	Peterson *et al.* (1986)
Dog (19)	ACTH	08.00	(17.6 ± 2.2)	Rijnberk *et al.* (1987)
Dog (11)	α-MSH	08.00–14.00	(7.2 ± 0.6)	Kemppainen and Sartin (1986)
Dog (160)	α-MSH	08.00–10.00	1.5–15	Peterson *et al.* (1986)
Dog (14)	α-MSH	08.00	(12.6 ± 1.9)	Rijnberk *et al.* (1987)
Dog (160)	β-END	08.00–10.00	1.5–17.4	Peterson *et al.* (1986)
Cat (6)	ACTH	08.00–13.00	(16 ± 3)	Willemse *et al.* (1993)
Cat (100)	ACTH	09.00	1.1–22	Peterson *et al.* (1994)
Cat (6)	α-MSH	08.00–13.00	(77 ± 7)	Willemse *et al.* (1993)
Cat (100)	α-MSH	09.00	3.6–200	Peterson *et al.* (1994)
Cat (100)	β-END/LPH	09.00	3.8–130	Peterson *et al.* (1994)
Horse	ACTH		(7 ± 1)	Orth *et al.* (1982)
Horse	α-MSH		(14.4 ± 1.2)	Orth *et al.* (1982)
Horse	β-END		(18.3 ± 4.4)	Orth *et al.* (1982)
Horse	ACTH		3.5–15.0	Hodson *et al.* (1986)
Sheep	ACTH	10.00	(116 ± 19)	Clarke *et al.* (1986)
Sheep	α-MSH	10.00	0–4.2	Clarke *et al.* (1986)

[a] Most of the values have been converted to SI units. 1 ng ACTH/liter = 0.22 pmol/liter; 1 ng α-MSH/liter = 0.60 pmol/liter; 1 ng β-END/liter = 0.29 pmol/liter.

In the dog the release of ACTH is stimulated by β-adrenergic agonists (isoproteronol), dopaminergic antagonists (haloperidol), and serotoninergic agonists (quipazine maleate). TRH and GnRH stimulate the release of cortisol, probably by stimulating ACTH release (Stolp *et al.*, 1982). Although direct stimulatory effects of catecholamines on the *in vitro* release of ACTH from ovine pituitary cells has been found, central stimulation of predominantly noradrenergic, but also adrenergic, pathways evokes the highest ACTH response (Liu *et al.*, 1991).

Endogenous opiates (metenkephalin, dynorphin, and β-endorphin) inhibit the release of ACTH in man (Besser *et al.*, 1987). Conflicting results have been reported on the effect of metenkephalin in the rat, but β-endorphin and dynorphin may exert tonic inhibition of CRH release (Plotsky, 1986). The metenkephalin agonist DAMME stimulates the release of ACTH in the dog (Meij *et al.*, 1990).

Activation of the immune system by infections results in the enhanced production of the cytokine interleukin-1 (IL-1β), which has the ability to stimulate CRH secretion from the hypothalamus and thus activates the hypothalamus–pituitary–adrenal axis. The biosynthesis and release of IL-6 has been found in the folliculostellate cells of the AL. IL-6 also stimulates the HPA-axis, at both the hypothalamic and the pituitary level (Sweep *et al.*, 1991).

Endogenous corticosteroids inhibit ACTH release predominantly at hypothalamic sites. Synthetic steroids such as dexamethasone may act primarily at the pituitary level (De Kloet *et al.*, 1974). In a review on corticosteroid-mediated feedback, Keller-Wood and Dallman (1984) suggested three different time schedules: a fast feedback that acts on the corticotropic cell but may not be related to nuclear receptor binding, an intermediate feedback that probably acts by inhibition of CRH release, and a slow feedback that acts by a decrease in mRNA encoding POMC in the pituitary gland. The delay in inhibiting ACTH production may be caused by the high stability of the mRNA encoding POMC and not by the absence of direct inhibition of the transcription. Dexamethasone inhibits gene transcription *in vivo* within 30 minutes in the rat (Fremeau *et al.*, 1986). In the dog the intermediate-delayed feedback is determined by the mean change in corticosteroid concentration over time (Keller-Wood, 1989). Acute lowering of plasma cortisol in the horse by inhibition of the synthesis in the adrenal gland resulted first in an increased ratio of ACTH : CRH in pituitary venous blood before CRH concentrations started to rise (Alexander *et al.*, 1993), indicating that the first effect is the opposite of the fast feedback and is mediated by increased sensitivity of the corticotrope.

d. Secretion by the IL

The release of POMC-derived peptides by the IL is under direct neural control. The rat and the mouse have been the mammals in which most studies on the regulation and processing of POMC in the IL have been carried out thus far. In the rat the release of POMC-derived peptides is regulated predominantly via tonic dopaminergic inhibition and β-adrenergic stimulation (Berkenbosch *et al.*, 1981; Tilders *et al.*, 1985), although GABAergic innervation of the IL has also been demonstrated (Oertel *et al.*, 1982). In addition, Proulx-Ferland *et al.* (1982) demonstrated that CRH is a potent stimulator of α-MSH secretion by the IL. In line with the absence of a glucocorticoid receptor in the IL (Antakly *et al.*, 1985), the α-MSH response to CRH could not be suppressed by dexamethasone administration. The expression of the glucocorticoid receptor in the IL is suppressed by dopamine (Antakly *et al.*, 1987), whereas the CRH receptor content of the rat IL is stimulated by dopamine (Shiver *et al.*, 1992).

In the dog, *in vitro* studies (Mol *et al.*, 1987) and *in vivo* and immunocytochemical observations (Middleton *et al.*, 1987a) have revealed the IL to be resistant to glucocorticoid suppression. There is also evidence from *in vivo* studies that dopaminergic pathways play a regulatory role in canine IL function (Kemppainen and Sartin, 1986). However, in other respects the situation in the dog is different from that in the rat with regard not only to the heterogeneous cytology (see Section I.A.5), but also to some of the regulation characteristics. Despite the fact that CRH-immunoreactive fibers have been identified in the canine neurointermediate lobe (Stolp *et al.*, 1987) and although *in vitro* CRH stimulates ACTH release from the neurointermediate lobe (Mol *et al.*, 1987), there is no convincing evidence that CRH can stimulate release of ACTH from the IL *in vivo* (Kemppainen and Sartin, 1986, 1987; Middleton *et al.*, 1987b), whereas no (Kemppainen and Sartin, 1986) or a very small (Rijnberk *et al.*, 1987) α-MSH response to CRH stimulation has been observed. In contrast with a significant increase of plasma MSH concentrations after administration of the dopamine antagonist haloperidol, even after administration of dexamethasone to inhibit the contribution of the AL, no release of IL ACTH is observed after haloperidol.

In the cat as well, no stimulation of MSH occurs after CRH administration (Willemse and Mol, 1994). However, cats undergoing handling and skin testing without anesthesia show significant increases in plasma MSH concentrations (Willemse *et al.*, 1993). *In vitro* experiments revealed the sensitivity of feline IL MSH release to dopaminergic inhibition (Willemse and Mol, 1994).

In fetal and newborn lambs and in adult sheep (Newman *et al.*, 1987) the administration of a dopamine-receptor antagonist results in α-MSH release. Elimination of the inhibitory hypothalamic control in sheep by hypothalamus–pituitary disconnection results in increased α-MSH release (Clarke *et al.*, 1986). The dopamine inhibition of ACTH secretion in the hyperadrenocorticoid horse (Wilson *et al.*, 1982) suggests that in the normal horse the IL is under dopaminergic control. In the rabbit the dopaminergic control of the IL is absent (Schimchowitsch *et al.*, 1986).

e. Action

The predominant action of ACTH is stimulation of steroidogenesis and corticosteroid release from the adrenals (see the chapter on adrenal function). ACTH also exerts a growth-stimulating effect on the adrenal cortex. Moreover, non-ACTH portions of POMC, that is, N-terminal POMC peptides, are involved in adrenocortical growth (Lowry *et al.*, 1987). In pharmacological dosages ACTH may promote lipolysis in fat cells and amino acid uptake in muscle. The role of ACTH produced in hypothalamic neurons projecting to higher brain centers remains to be elucidated. Distribution patterns have been reported for the cat (Rao *et al.*, 1986).

The biological effect of α-MSH in mammals remains uncertain, although the presence of specific α-MSH receptors in many tissues suggests that this peptide may affect the function of several organs (Tatro and Reichlin, 1987). Stimulation of nerve regeneration by α-MSH has been reported (Verhaagen *et al.*, 1987; De Koning and Gispen, 1987). The C-terminal fragment of ACTH(18–39), called corticotropin-like intermediate lobe peptide (CLIP), has been found to be a potent stimulator of *in vitro* adrenal DNA synthesis (Wulffraat *et al.*, 1987). α-MSH is present in the fetal pituitary of all species studied, and there is evidence that it may play a role in the regulation of intrauterine growth (Mulchahey *et al.*, 1987).

f. Disease

Lesions at the hypothalamic and pituitary level may result in altered synthesis and release of POMC-derived peptides. There have been no reports of the occurrence of isolated ACTH deficiency in domestic animals. There are a few reports of dogs with tumorous (supra)hypophyseal lesions, with indirect evidence for multiple adenohypophyseal and neurohypophyseal deficits (Rijnberk, 1971; Eigenmann *et al.*, 1983a), and one dog reported to have secondary hypoadrenocorticism without information about other pituitary functions (Peterson *et al.*, 1992).

In contrast to the few descriptions of ACTH deficiency, pituitary-dependent hyperadrenocorticism is a common disorder in the dog (Peterson, 1987a), in which the adenomas producing the ACTH excess may originate in the AL or the IL (Capen *et al.*, 1967; Peterson *et al.*, 1986). From the significantly lower CRH concentrations in cerebrospinal fluid of dogs with pituitary-dependent hyperadrenocorticism as compared to control dogs, it is concluded that the excessive ACTH secretion is not caused by chronic hyperstimulation with CRH (Van Wijk *et al.*, 1992). The ACTH secretion appeared also to be less sensitive to stimulation with CRH than with LVP (Van Wijk *et al.*, 1994). Evidence for a genetic involvement in tumorigenesis was found in a family of Dandie Dinmont terriers (Scholten-Sloof *et al.*, 1992). In the horse the disease originates primarily in the IL (Wilson *et al.*, 1982; Orth *et al.*, 1982). In agreement with the characteristics described earlier for the secretion of POMC-derived peptides by the IL, ACTH release by tumors of IL origin in both the dog and the horse tends to be strongly resistant to suppression by dexamethasone (Orth *et al.*, 1982; Peterson *et al.*, 1986). In dogs the highest plasma α-MSH concentrations are found in individuals with dexamethasone-resistant ACTH secretion. This suggests an IL origin of the disease, although there is evidence that the pituitary lesions do not always maintain the characteristics of the lobe of origin (Rijnberk *et al.*, 1988b). A dog with diabetes insipidus has been described, in which the pituitary tumor released primarily biologically inactive POMC-derived peptides (Goossens *et al.*, 1995). In contrast with the equine IL tumors, these tumors in dogs respond poorly to administration of dopamine agonists in terms of diminished ACTH secretion (Rijnberk *et al.*, 1988b).

For details on clinical manifestations, laboratory findings, diagnostics, and treatment of these diseases, including iatrogenic hypoadrenocorticism due to corticosteroid therapy, the reader is referred to textbooks (Feldman and Nelson, 1996; Rijnberk, 1996) and the chapter on adrenocortical function in this volume.

g. Tests

Basal levels of circulating POMC-derived peptides are measured for diagnostic purposes in situations of suspected hypo- as well as hypersecretion. ACTH values below or just within the reference range (Table 19.3) may be found in cases of hypothalamus–pituitary disease, as well as in situations in which endogenous or exogenous glucocorticoid excess suppresses hormone synthesis in the corticotropic cells. This makes the measurement of basal ACTH levels a useful tool in the differentiation between pituitary-dependent hyperadrenocorticism and hyperadrenocorticism due to adrenocortical tumor (Feldman, 1983; Peterson, 1986). In pituitary-dependent hyperadrenocorticism, ACTH values exceeding the reference range may be found, but there is considerable overlap. Basal MSH concentrations have been found to be elevated in horses (Orth *et al.*, 1982) and dogs (Peterson *et al.*, 1986; Rijnberk *et al.*, 1987) with pituitary-dependent hyperadrenocorticism of IL origin.

Of the dynamic tests, the dexamethasone suppression tests (see chapter on adrenocortical function) are still the best for the diagnosis and differential diagnosis of excessive ACTH and glucocorticoid secretion. In animals suspected of having pituitary–adrenocortical insufficiency, the secretory capacity for POMC-derived peptides can be measured by provocative testing.

CRH and AVP are also used in the differentiation between pituitary-dependent hyperadrenocorticism and hyperadrenocorticism due to adrenocortical tumor. The chronically suppressed corticotropic cells are presumed to be unresponsive to these stimuli. However, Meijer *et al.* (1978) found considerable overlap in plasma cortisol values in the AVP test, whereas much less is claimed to occur in the CRH test (Peterson, 1986). It has been demonstrated that LVP stimulates cortisol release by adrenal tumors in a direct way (Van Wijk *et al.*, 1994).

It has been shown that dogs with pituitary-dependent hyperadrenocorticsm usually do not respond supranormally to CRH administration (Rijnberk *et al.*, 1987). The cells of these pituitary lesions were found to be less responsive to CRH *in vitro* than were normal corticotropes (Mol *et al.*, 1987).

As far as the possibility for manipulation of canine IL function is concerned, of the substances tested only the dopamine agonist haloperidol (Kemppainen and Sartin, 1986; Rijnberk *et al.*, 1987) caused significant increases in circulating α-MSH. More detailed information pertaining to some of these tests is presented in Section IV of this chapter and in Table 19.4.

2. *β-Endorphin/β-Lipotropin*

a. Gene Expression and Biosynthesis

β-Lipotropin (β-LPH) consists of the 91 C-terminal amino acids of the POMC precursor. It is synthesized in the corticotropic cells of both the AL and IL. The C-terminal sequence (36–91) is remarkably similar in human, porcine, and ovine pituitaries, whereas the N-terminal sequence (1–36) is rather heterogeneous. β-

TABLE 19.4 Maximal Plasma Concentrations of POMC-Derived Peptides Following Intravenous Administration of a Variety of Stimulants

Species (n)	Substance	Dose	Range (pmol/liter) (mean ± SEM)	Reference
ACTH				
Dog (16)	CRH	1 μg/kg	(48.6 ± 11.7)[a]	van Wijk *et al.* (1994)
Dog (6)	AVP	0.6 μg/kg	(50.4 ± 15.0)	Kemppainen and Sartin (1987)
Dog (16)	LVP	0.2 U/kg	(48.6 ± 4.8)	van Wijk *et al.* (1994)
Dog (19)	Haloperidol	0.06 mg/kg	(61.6 ± 6.6)	Rijnberk *et al.* (1987)
Dog (6)	Insulin	0.5 U/kg	(84.9 ± 19.6)	Kemppainen and Sartin (1987)
Cat (6)	CRH	1 μg/kg	(29.0 ± 3.2)	Willemse and Mol (1994)
Cat (4)	Haloperidol	2 mg/kg	(81 ± 54)	Peterson *et al.* (1994)
Sheep (5)	CRH	1 μg/kg	(125 ± 52)	Pradier *et al.* (1986)
Sheep (5)	AVP	1 μg/kg	(202 ± 77)	Pradier *et al.* (1986)
Pig (4)	CRH	0.5 μg/kg	109 ± 2%[b]	Minton and Parsons (1993)
Pig (3)	LVP	1 μg/kg	452 ± 5%[b]	Minton and Parsons (1993)
α-MSH				
Dog (6)	Haloperidol	0.2 mg/kg	94.2[c]	Kemppainen and Sartin (1987)
Dog (19)	Haloperidol	0.06 mg/kg	(39.6 ± 3.6)	Rijnberk *et al.* (1987)
Cat (6)	CRH	1 μg/kg	no effect	Willemse and Mol (1994)
Cat (4)	Haloperidol	2 mg/kg	(125 ± 36)	Peterson *et al.* (1994)
β-END				
Dog (6)	Haloperidol	0.2 mg/kg	179[c]	Kemppainen and Sartin (1987)

[a] Increment.
[b] Value relative to basal concentration.
[c] Mean.

MSH (sequence 37–58) appears to be an extraction artifact and plays no physiological role.

By proteolytic cleavage at amino acid 61, γ-LPH (β-LPH(1-61)) and a family of endorphins is formed. Through specific deletion of C-terminal amino acids and N-terminal acetylations, various β-END-related peptides are formed. The three main compounds are β-END(61–91), γ-END(61–77), and α-END(61–76). The corticotropic cells of the AL synthesize approximately equal or higher concentrations of β-LPH than of β-END(1–31) in most species. The equine AL, however, contains primarily β-END(1–31) and N-acetyl-β-END(1–27) (Millington *et al.*, 1992). The corticotropic/melanotropic cells of the IL produce relatively more β-END and related peptides. In the IL of the normal dog the ratio β-LPH/β-END is less than 0.1 (Krieger, 1983). β-END(1–27) is the most abundant form in the canine IL, followed in decreasing order of concentration by the nonacetylated β-END(1–31) and (1–26) forms, and the N-acetyl (1–27), (1–26), and (1–31) forms (Young and Kemppainen, 1994).

b. Secretion

The stimuli that induce ACTH or α-MSH release from either AL or IL cells also cause a release of the β-LPH-related peptides from the same cell. The cellular subset of specific proteolytic, amidating, and acetylating enzymes present in the secretory vesicles define the ultimate composition of POMC-derived peptides that are secreted into the blood.

c. Action

The main action of β-LPH is to mobilize fat from adipose tissues, as demonstrated in the rabbit. Its biological function in man and other species has not been fully elucidated. Brain tissue can break down β-LPH to form β-END-related peptides. However, it is questionable whether these pituitary peptides are involved in brain function; in conscious sheep, hemorrhagic stress elevates β-END concentrations in plasma but not in cerebrospinal fluid (Smith *et al.*, 1986).

β-END is an endogenous opiate with a potent morphinomimetic action. It is also produced in the hypothalamus where the entire POMC precursor is present. Brain and pituitary endorphin are probably part of separate systems. The role of β-END and other opioid peptides in the secretion of pituitary hormones is has been studied with long-acting analogues, opiates, and opioid receptor antagonists (Besser, 1987; Van Wimersma Greidanus, 1987). It appears that there are opioid mechanisms in the hypothalamus and that stimulation, for example, causes secretion of PRL (Grossman *et al.*, 1981; Delitala *et al.*, 1982) and inhibition of the release of oxytocin, whereas the release of vasopressin is differentially affected (van Wimersma Greidanus, 1987; Hellebrekers *et al.*, 1987). Cross-species studies have revealed marked differences in the amount of hypothalamic opiate receptors of the various subtypes of opioid peptides.

d. Disease/Tests

Hypersecretion of β-END is associated with the ACTH hypersecretion of IL tumors in both equine and canine pituitary-dependent hyperadrenocorticism (Orth *et al.*, 1982; Peterson *et al.*, 1986; Rijnberk *et al.*, 1987). As the regulation of the secretion of β-END is similar to that of other POMC-derived peptides from the AL and IL, testing procedures can be applied as mentioned in the section on ACTH and MSH. Elevation of plasma β-END has been documented in horses following running and shipping (Li and Chen, 1987) and in sheep during electroimmobilization and shearing procedures (Jephcott *et al.*, 1987). Treadmill exercise of Thoroughbred horses induces a variable response in plasma β-END concentration (Art *et al.*, 1994). Elevated plasma β-END concentrations are found in dogs with congestive heart failure (Himura *et al.*, 1994) and are further elevated after naloxone treatment.

B. Glycoprotein Hormones

The gonadotropes and thyrotropes synthesize the hormones LH, FSH, and TSH, each of which consists of two different peptide chains, termed the α- and β-subunits. The amino acid sequence of the α-subunit is identical for LH, FSH, and TSH within a species, as it is for placental chorionic gonadotropin. The β-subunits have unique structures and determine the hormone specificity. Carbohydrate substituents account for 10–20% of the molecular weights of these hormones.

1. α-Subunit

a. Gene Expression

The gene encoding the α-subunit varies between 8 and 16.5 kb among the species and contains four exons. The α-subunit comes to expression in a variety of cells such as thyrotropes, gonadotropes, and syncytiotrophoblasts. Its promotor sequence contains a cAMP response element (CRE), a negative thyroid hormone response element (nTRE), a negative glucocorticoid response element (nGRE), and thyrotrope-specific and GnRH/TRH response sites.

b. (Pro)hormone

Translation of the mRNA results in the formation of a precursor peptide with a molecular weight of approximately 13,500. The α-subunit has a high degree of homology among species (Fig. 19.7). The human sequence contains 92 amino acids, rather than 96 as found in all other species studied thus far. The shorter human sequence is due to a deletion of 12 nucleotides at the beginning of exon 3. After cleavage of the signal peptide, five disulfide bridges are formed. Two asparagine residues are prone to N-glycosylation, but there is also a putative O-glycosylation site (Fig. 19.7). The α-subunit is produced in excess of the β-subunit that determines the hormone specificity. The β-subunit formation is rate limiting in the formation of the hetero $\alpha\beta$ dimer.

1. TSH

The thyrotropic cells of the anterior pituitary produce thyroid-stimulating hormone (TSH), which stimulates both the synthesis and secretion of thyroid hormone.

a. Gene Expression

The rat and human TSH gene consists of three exons, whereas the mouse gene has five exons. The most prominent stimulator of gene transcription is the hypothalamic thyrotropin-releasing hormone (TRH). TRH activates the thyrotrope via a signal transduction process that is mediated by activation of the phospholipase C pathway. At least two ultimately activated transcription factors are known. The first is Pit-1/GHF-1, which is essential for cells producing TSH, GH, and PRL. Secondly, heterodimers of *c-fos*/*c-jun* forming an AP-1 complex activate the TSH-β gene. Both α-subunit and TSH-β-subunit expression are strongly inhibited by a negative thyroid hormone response element (nTRE) in the pituitary thyrotrope. The α and β types of the thyroid hormone receptor may form homodimers, and heterodimers, or even form heterodimers with the retinoic acid receptors RAR and RXR. Which forms are

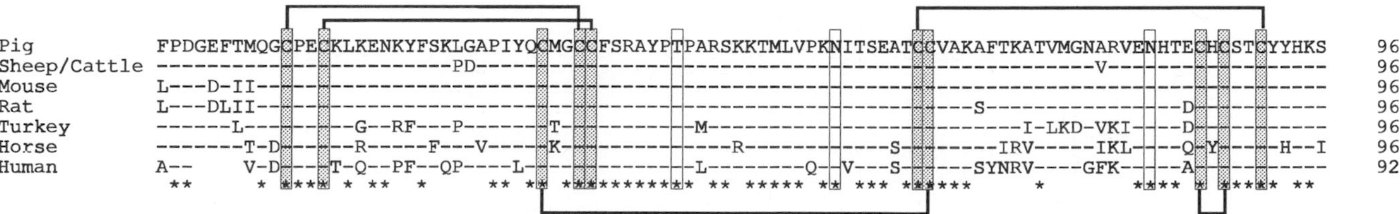

FIGURE 19.7 Sequence comparison of the α-subunit for glycoprotein hormones (LH, FSH, and TSH). Lines between shaded boxes represent intrachain disulfide bridges. The open boxes indicate potential glycosylation sites. See legend to Fig. 19.6.

important for negative feedback is the subject of current research.

b. (Pro)hormone

The TSH-β chain consists of 118 amino acids and forms six intrachain disulfide bonds (Fig. 19.8). The TSH-β of the species in this figure share 74% identical amino acids. There are no free cysteine residues, consistent with the fact that subunits form heterodimers noncovalently. The β-chain of TSH has one putative N-glycosylation site.

c. Secretion

The release of TSH is stimulated by the hypothalamic TRH. Somatostatin (SS) has an inhibitory effect on TSH release. Thyroid hormone exerts a long-loop negative feedback. Thus, secretion of TSH is regulated by the interplay of TRH, SS, and thyroid hormone feedback (Reichlin, 1982). Although this feedback depends on 3,3′,5-triiodothyronine (T_3) binding to the nuclear receptor of the thyrotrope, plasma TSH correlates better with the thyroxine (T_4) plasma concentration (Larsen, 1982). A specific 5′-deiodinating enzyme is responsible for T_3 generation in the thyrotropic cell.

d. Action

TSH stimulates both synthesis and secretion of thyroid hormones. After receptor binding, TSH stimulates the production of cAMP, which acts as a second messenger. In addition, intracellular Ca^{2+} may modulate the biological effect of TSH via the phosphoinositol pathway. As a result of receptor activation, T_4, and to a much lesser extent T_3, are secreted into the blood. Prolonged stimulation of the thyroid with TSH results not only in hypersecretion of thyroid hormone, but also in enlargement of the thyroid gland.

e. Disease/Tests

Sensitive TSH assays became available in 1995 (Klingler *et al.*, 1995; Nachreiner *et al.*, 1995). At the time this chapter was written, the first clinical evaluations were encouraging for the usefulness of one of these assays for the diagnosis of primary hypothyroidism (Williams *et al.*, 1996).

Long-term glucocorticoid excess reduces T_4 levels in dogs (Belshaw, 1983; Peterson *et al.*, 1984). As in man, this may be the result of diminished pituitary sensitivity to TRH (Visser and Lamberts, 1981) or to inhibited TRH release. However, in dogs there is some indirect evidence for intact TSH release, whereas it has been postulated that glucocorticoids interfere with basal T_4 secretion by inhibiting lysosomal hydrolysis of colloid in the thyroid follicular cell (Kemppainen *et al.*, 1983). Low plasma thyroxine values may also be due to decreased plasma TBG concentrations. Also, the use of certain drugs, such as antiepileptics, results in low plasma total thyroxine concentrations in the dog.

In the dog hypothyroidism is most often due to a primary defect in the thyroid gland. Secondary hypothyroidism may be caused by a pituitary tumor (Rijnberk, 1971; Chastain *et al.*, 1979; Dunbar and Ward, 1982) or panhypopituitarism caused by a suprasellar tumor (Eigenmann *et al.*, 1983a). Low plasma thyroxine values in combination with low TSH values may thus be caused by either secondary hypothyroidism or hyperadrenocorticism and certain drugs. In the first circumstance a true hypothyroidism exists, whereas in the latter cases the dogs should be essentially euthyroid. Unfortunately, free thyroxine measurements in the dog have not been found to discriminate between these circumstances, and thus true secondary hypothyroidism due to impaired TSH secretion should be docu-

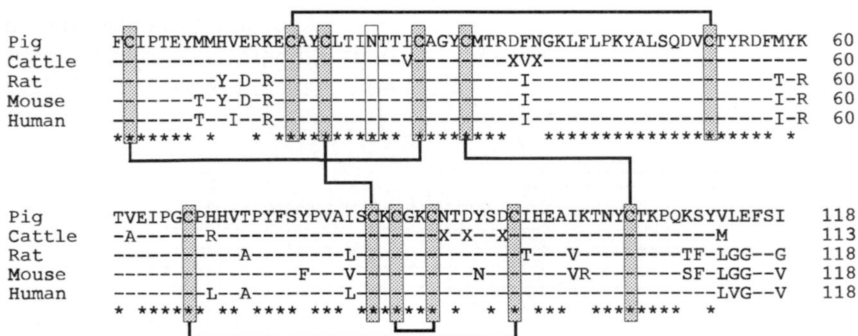

FIGURE 19.8 Sequence comparison of the β-subunit of thyroid-stimulating hormone (TSH). Lines between shaded boxes represent intrachain disulfide bridges. The open box indicates a potential glycosylation site. See legend to Fig. 19.6.

mented by performing a TRH test and measurement of TSH secretion. For further details the reader is referred to the chapter on thyroid function.

2. *LH and FSH*

Luteinizing hormone (LH), which is identical to the interstitial-cell stimulating hormone, and follicle-stimulating hormone (FSH) are produced by the gonadotropic cells of the AL.

a. Gene Expression

In primates multiple copies of LHβ-related genes have been identified, the chorionic gonadotropins (CG), which are expressed in the placenta. The horse is the only known nonprimate in which a CGβ protein is formed in the placenta for the maintenance of early pregnancy. In the horse this CGβ in the placenta is derived from the same gene that encodes pituitary LHβ. Pulsatile GnRH and castration both stimulate the LHβ mRNA levels in the pituitary. Pulsatile treatment with GnRH also increases the length of the polyadenylation of the LHβ mRNA. The consequences for mRNA stability or translation efficiency are unknown. In the horse three transcription start sites have been found (Sherman *et al.*, 1992).

The gene encoding FSHβ contains three exons and two introns. Exon 1 encodes only untranslated mRNA sequences, exon 2 encodes the signal peptide and the first 18 amino acids of the prohormone, and exon 3 encodes the remaining amino acids of the prohormone and a 3′-untranslated region (3′UTR) that is long relative to other β-subunits. The bovine gene is transcribed into a single mRNA, whereas in humans there are several transcripts of different lengths as the result of alternative splicing in exon 1 or variations in the polyadenylation length.

Little is known about the response elements in the FSHβ-promoter. However, it has been demonstrated that GnRH, activin, inhibin, and gonadal steroids regulate the gene expression.

b. (Pro)hormone

After cleavage of the signal peptide, LHβ proteins are formed with a variation in amino acid content among the species of 115 to 121 amino acids (Fig. 19.9). The mature FSHβ chain contains 111 amino acids in most species, except in the horse, in which it contains 118 amino acids (Fig. 19.10). Both subunits contain six intrachain disulfide bridges. LHβ has a single N-glycosylation site, whereas FSHβ has two N-glycosylation sites. The carbohydrate chain formed by the glycosylation of LHβ is sulfated, in contrast to FSHβ, in which a sialic acid residue is added to the carbohydrate chain. This difference may play a role in the hormone sorting within the gonadotrope and certainly results in prolonged plasma half-life of FSH.

c. Secretion and Action

As with other pituitary hormones, the gonadotropins are released in a pulsatile fashion, stimulated by the hypothalamic gonadotropin-releasing hormone GnRH. The pulsatile secretion of GnRH has been demonstrated in sheep (Clarke and Cummins, 1982) and the horse (Irvine and Alexander, 1986). Continuous infusion of GnRH in rhesus monkeys switched off gonadotropin secretion (Belchetz *et al.*, 1978) because of down regulation of GnRH receptors (Naor *et al.*, 1984). In the female, FSH stimulates follicle development and

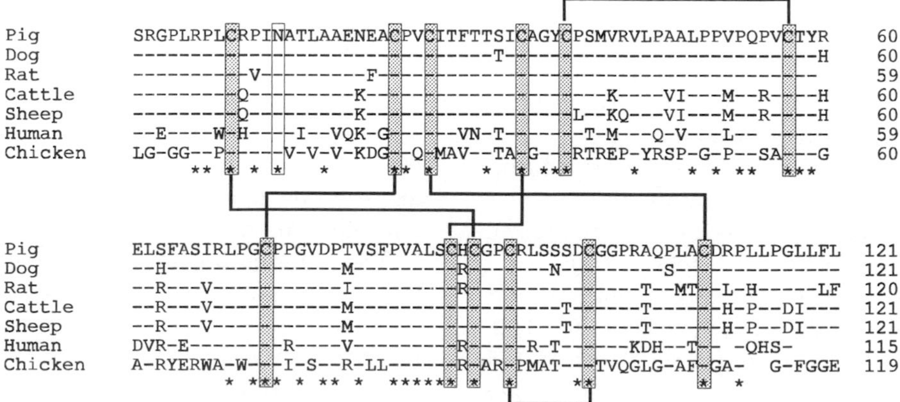

FIGURE 19.9 Sequence comparison of the β-subunit for luteinizing hormones (LH). Lines between shaded boxes represent intrachain disulfide bridges. The open box indicates a potential glycosylation site. See legend to Fig. 19.6.

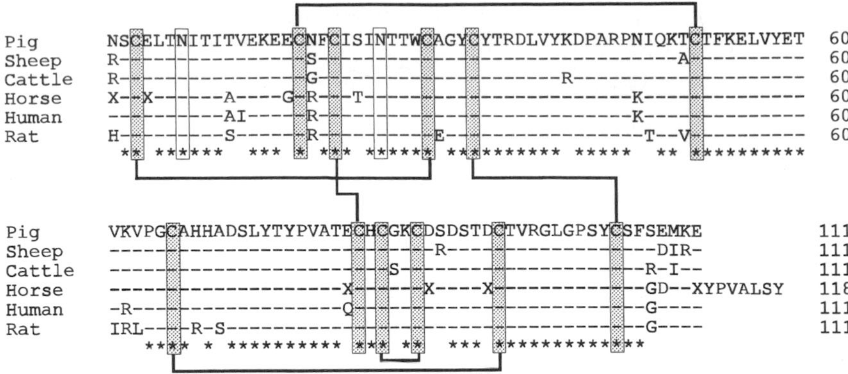

FIGURE 19.10 Sequence comparison of the β-subunit for follicle-stimulating hormone (FSH). Lines between shaded boxes represent intrachain disulfide bridges. The open boxes indicate potential glycosylation sites. See legend to Fig. 19.6.

ripening in the ovaries. Within the theca interna, LH stimulates the production of androstenedione that is converted, by FSH stimulation, into estradiol-17β. Estradiol stimulates the pulse frequency of the GnRH release. Above a threshold level of estradiol, a massive LH surge is stimulated by a hypothalamic GnRH surge, resulting in initiation of ovulation. Administration of PMSG in cows results in increased follicular growth. The increased number of preovulatory follicles severely suppresses endogenous FSH secretion (Bevers *et al.*, 1989).

After ovulation, LH stimulates the formation of a corpus luteum in the ovary, and this secretes high concentrations of progesterone into the plasma. This progesterone inhibits the release of gonadotropins by decreasing the pulse frequency of GnRH release, or by decreasing the sensitivity of the gonadotrope to stimulation with GnRH. During the course of early to late anestrus, the sensitivity of pituitary LH release of the female dog to stimulation with GnRH increases (Van Haaften *et al.*, 1994).

The ovaries also synthesize a nonsteroidal factor that attenuates or inhibits the midcycle gonadotropin surge. After partial purification by which all detectable inhibin and follistatin activity (see later discussion) was removed, the so-called "gonadotropin surge inhibiting factor" (GnSIF) appeared to inhibit the GnRH-stimulated LH secretion *in vitro* without affecting the basal FSH release (Danforth and Cheng, 1993).

In the male animal, LH stimulates the synthesis and release of testosterone in the Leydig cells of the testis, which may in turn exert a negative feedback on LH secretion. This feedback may depend upon aromatization of testosterone to estradiol in the brain. In the male dog both testosterone and estradiol are the major inhibitors of LH and FSH release (Winter *et al.*, 1982). The opioid agonist bremazocine exerts a stimulatory effect on the LH peak level, but not on the peak frequency during the follicular phase of the estrous cycle of heifers. Both peak height and frequency are decreased after administration of an opioid antagonist. In pigs and sheep there are also reports on the role of opioid control in various reproductive states (Short *et al.*, 1987).

FSH secretion is also stimulated by GnRH. The LH surge is accompanied by an FSH surge. The release of FSH can be selectively inhibited *in vitro* by inhibin, a glycoprotein hormone produced in the Sertoli cells of the testis and the granulosa cells of the ovary (De Jong, 1979). A simple bioassay has been developed for inhibin-like activity using ovariectomized ewes. Injection of follicular fluid, which is rich in inhibin activity, resulted in a selective lowering of the plasma FSH (Miller and Bolt, 1987). FSH in turn has been found to increase inhibin secretion by cultured Sertoli cells and granulosa cells (Ultee-Van Gessel and De Jong, 1987). The amino acid sequences of bovine and porcine inhibin have been elucidated (Forage *et al.*, 1986). In dogs with Sertoli-cell tumors, increased inhibin bioactivity and mRNA for the α- and βB-subunits is found in tumor tissue. The increased inhibin concentrations in plasma are likely to be the cause of the suppressed FSH plasma concentrations, although plasma LH and testosterone concentrations were also suppressed (Grootenhuis *et al.*, 1990).

The activins are formed in the pituitary gonadotrope. Originally these proteins, consisting of homo- and heterodimers of activin-A and activin-B, were purified from ovarian follicular fluid. Within the pituitary the activin-B homodimer stimulates FSH synthesis without affecting LH synthesis. The action of the activins is antagonized by the glycoprotein follistatin, which is also produced both in folliculostellate cells and in gonadotropes of the pituitary.

d. Tests

Plasma LH can be detected by heterologous radioimmunoassays. Using a β-unit-specific monoclonal antibody against equine LH, parallel displacement curves have been reported for LH preparations from the horse, sheep, pig, cow, dog, cat, rabbit, rat, and kangaroo (Matteri *et al.*, 1987). Cross reactivity ranged from 45 to 265% for LH of these species, but there was only 15% cross reactivity for human LH. For details readers are referred to the chapter on reproduction.

C. Somatomammotropic Hormones

Growth hormone (GH), prolactin (PRL), and placental lactogen show homology in amino acid composition and some biological activities, and thus may be grouped together as a family of somatolactotropic hormones. There is increasing evidence that these hormones evolved from a single ancestral gene (Wallis, 1984; Seo, 1985).

1. GH

The somatotropic cells, producing GH, are the most abundant cells of the anterior lobe.

a. Gene Expression

In primates multiplication of the GH gene has resulted in a cluster of five GH-related genes. The GH-N gene is expressed in the pituitary and the other four, consisting of the chorionic somatomammotropins (so-called "placental lactogens") and a variant gene (GH-V), are expressed in the placenta during pregnancy. In nonprimates a single gene encodes GH, but a family of PRL-like genes (also including placental lactogens) are present.

The Pit-1 (also called GHF-1) transcription factor is essential for the development of pituitary cells producing GH, PRL, and TSH. Within the somatotrope the Pit-1 gene expression is stimulated by the hypothalamic growth-hormone-releasing hormone (GHRH) and inhibited by somatostatin (SS). Both neurohormones modulate GH gene transcription by stimulation or inhibition, respectively, of the adenylate cyclase activity. The activins as formed by the gonatropes may also inhibit the expression of the GH gene. This effect is additive to the inhibition by SS.

A synergistic stimulation of GH gene transcription is mediated by retinoic acid and by both thyroid hormone and glucocorticoids. The retinoic acid receptor (RAR) and the thyroid hormone receptor (TRα or TRβ) may form heterodimers with the 9-*cis* retinoic acid receptor (RXR) and thus strongly promote GH gene transcription. The glucocorticoid-mediated gene transcription does not result in enhanced plasma GH concentrations *in vivo*, because glucocorticoids also stimulate the release of somatostatin from the hypothalamus and thus inhibit GH release from the pituitary.

The gene encoding GH-N contains five exon areas. By alternative splicing, the 5′ part of exon 3 is missing in a minority of pituitary transcripts, and this results in the formation of a GH protein missing 15 amino acids in the N-terminal part of the molecule (Baumann, 1991).

b. (Pro)hormone

GH is a single-chain polypeptide. It contains two intrachain disulfide bridges and has a molecular weight of approximately 22,000. The amino acid sequence of GH belongs to the best-known sequences of pituitary hormones among species (Fig. 19.11). Using DNA technology the cDNAs for human and bovine GH have been cloned in bacteria and brought to expression, leading to the production of recombinant GH (Olson *et al.*, 1981), which is used to stimulate human growth and to enhance bovine milk production (Bauman *et al.*, 1985).

In cattle a polymorphism in the GH gene exists resulting in four GH variants that can arise from two possible N-terminal amino acids (phenylalanine or alanine) and two amino acids in position 127 (leucine or valine). The transmission of the trait of high milk production was greater for homozygous leucine-127 in Holstein cows and valine-127 in Jersey cows (Lucy *et al.*, 1993).

c. Secretion

A variety of factors modulate the release of GH (Table 19.5). The integration of all of these factors results in pulsatile release of GH with pulse intervals of 4–6 hours, as shown in the dog (Watson *et al.*, 1987; French *et al.*, 1987). The hypothalamic neurohormones GHRH and SS play a central role in the pulsatile GH release, where bursts of GHRH result in the episodic release of GH and the intervening troughs in GH secretion are maintained by SS (Fig. 19.12).

GHRH was first isolated from human pancreatic islet cell tumors (Rivier *et al.*, 1982; Guillemin *et al.*, 1982). Using this knowledge, the amino acid sequence of hypothalamic GHRH of various species was elucidated. A GHRH(1–44)NH_2 and a GHRH(1–40)OH form is found in the hypothalamus; both have GH-releasing capacity because the biologic activity resides in the 1–31 sequence. Outside the central nervous system, GHRH mRNA or immunoreactivity is found in leukocytes, testis, ovary, and placenta. Continuous infusion of a high dose of GHRH results in elevated GH levels, followed by a decline to low values due to

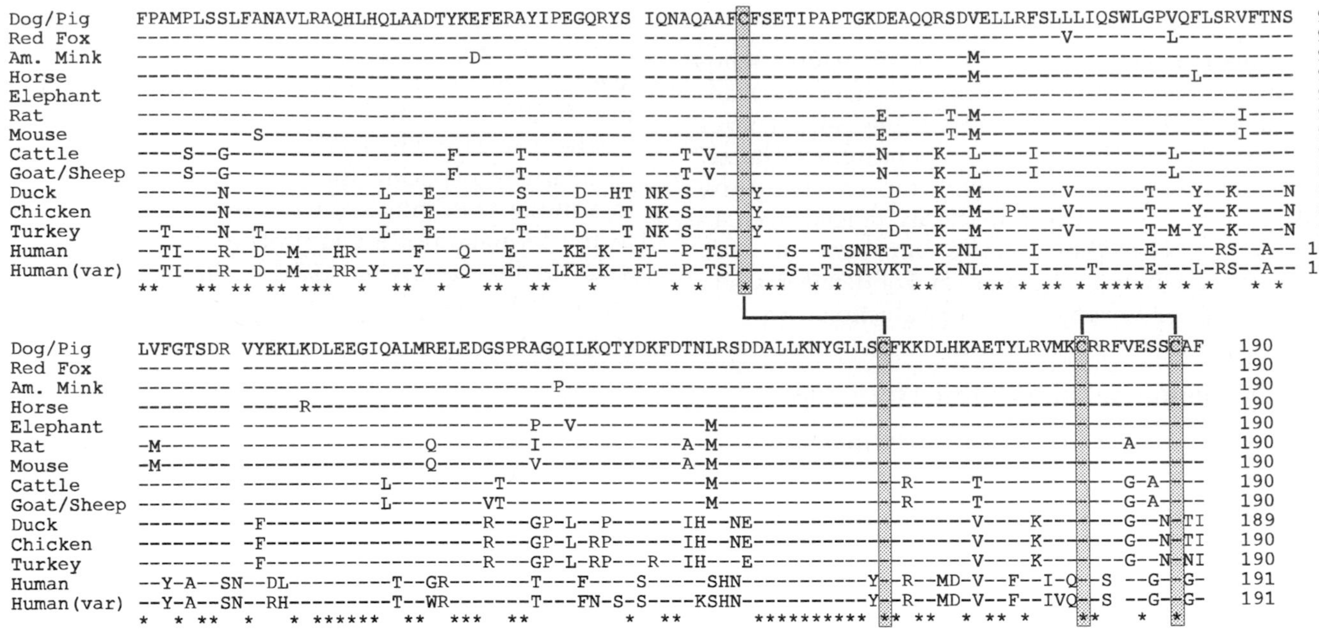

FIGURE 19.11 Sequence comparison of growth hormone (GH). Lines between shaded boxes represent intrachain disulfide bridges. See legend to Fig. 19.6.

desensitization of the somatotropic cell as a result of down regulation of GHRH binding sites. This phenomenon is a time- and concentration-dependent process (Bilezikjian *et al.*, 1986). The GHRH-induced GH release is potentiated by protein kinase C activation, indicating that the phosphoinositol system also plays a role in the GH release. A candidate for this activation is the yet unknown endogenous counterpart of the synthetic GH-releasing peptide (GHRP-6) that stimulates GH release *in vivo* in a variety of species, including the dog (Muruais *et al.*, 1993). TheTRH effect on GH release is more complex. In healthy mammals, including the dog (Rutteman *et al.*, 1987a), TRH does not stimulate GH release (Scanes *et al.*, 1986). In cattle, TRH induces some GH release, but when used together with GHRH these two releasing hormones act in strong synergy (Lapierre *et al.*, 1987) (see also Table 19.7).

Somatostatin (SS) denotes a family of peptides including the 14-amino-acid-containing S-14 and the N-terminal-extended S-28 (Reichlin, 1982). Passive immunization of chickens with SS antiserum increases the basal plasma GH concentrations and augments TRH-induced GH secretion (Harvey *et al.*, 1986). In compari-

TABLE 19.5 Putative Factors Modulating GH Release

	Stimulating	Inhibiting
Hormones	GHRH	Somatostatin (SS)
	Glucagon	IGF-1, IGF-II
	Pentagastrin	Corticosteroid excess
	Enkephalin	Hypothyroidism
Biogenic amines	α-Adrenergic agonists	α-Adrenergic antagonists
	β-Adrenergic antagonists	β-Adrenergic agonists
	Dopamine	Serotonin antagonists (cyproheptadine)
Others	Hypoglycemia	Hyperglycemia
	Fall in free fatty acids	Rise in free fatty acids
	Amino acids (arginine)	
	Sleep	
	Stress (emotional)	
	Exercise	

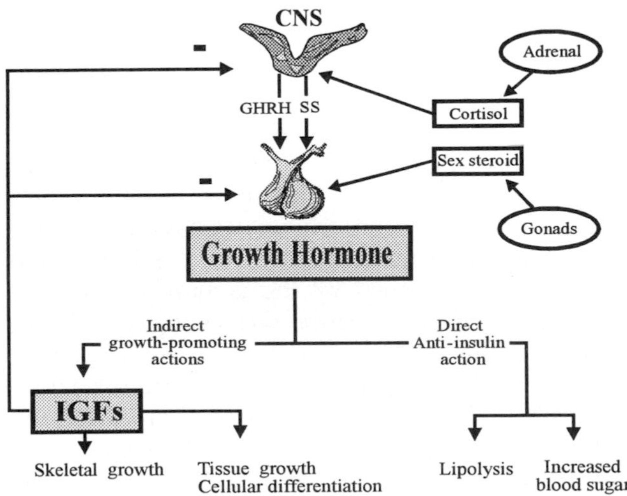

FIGURE 19.12 Hypothalamic regulation of pituitary GH release, peripheral actions of GH, and the effects of IGFs (feedback), glucocorticoids, and sex steroids on the hypothalamic–pituitary system.

son with that of S-14, the suppressive effect of S-28 on GH release is much longer. SS appears mainly to inhibit GH release and to have little effect on the synthesis of GH. The apparent ineffectiveness of SS in some studies may be due to its extremely short half-life (Harvey and Scanes, 1987), which was found to be 1.82 minutes in the dog (Schusdziarra *et al.*, 1979). However, in dogs, GH pulsatile secretion could also not be inhibited by the administration of the long-acting SS analogue SMS 201–995 in a dose of 1 μg/kg (Watson *et al.*, 1987). Continuous infusion of SS in a dose of 0.15 μg/kg/min abolished the secretory bursts, but no effect was seen on basal secretion rates (Cowan *et al.*, 1984). In the chicken, bolus injections of SS also failed to inhibit basal GH release (Harvey *et al.*, 1986).

The pulsatile GH secretion is age and gender related. Although there are conflicting reports on the effect of aging on basal GH release, it is generally assumed that the pulse height of GH decreases with age in man. This age-related decline has been confirmed in domestic fowl (Harvey *et al.*, 1987) and cattle (Verde and Trenkle, 1987). In the rat there is a gender-related pattern of GH secretion. The male rat has high pulses and a very low trough, whereas the female rat has lower bursts but higher basal levels (see also Section II.C.1.e.).

There are also indirect neuronal influences that modulate GH secretion (see also Table 19.5) (Hall *et al.*, 1986). The main physiological stimuli for GH secretion are sleep, physical exercise, stress, fasting, catecholamines, hypoglycemia, and certain amino acids, but not all apply to all species. For example, GH secretion in dogs is not related to sleep or day–night cycles, although GH peaks may occur after forced wakefulness at the onset of sleep (Takahashi *et al.*, 1981). Insulin-induced hypoglycemia and arginine administration do not consistently result in GH release in the dog (Eigenmann and Eigenmann, 1981a). Of the neurotransmitter systems involved, adrenergic systems seem to play a major role. α_2-Adrenergic agonists promote GH secretion, whereas β-adrenergic agonists are inhibitory. Thus, clonidine, a central α_2-adrenergic agonist, is also an effective stimulator of GH secretion in the dog (Eigenmann and Eigenmann, 1981a).

There is increasing evidence that the GH-stimulated release of insulin-like growth factor I (IGF-I) exerts a negative long-loop feedback on pituitary GH release. IGF-I stimulates the release of somatostatin by the hypothalamus and also inhibits pituitary GH release directly (Ceda *et al.*, 1987). In the domestic fowl, injection of IGF-I results in decreased GH responses to stimulation with TRH and GHRH (Buonomo *et al.*, 1987). The elevated GH levels in fasting dogs have also been explained as the result of impaired feedback inhibition due to low IGF-I levels in plasma (Eigenmann *et al.*, 1985). A low level of feeding also results in reduced plasma IGF-I concentrations and significantly increased plasma GH concentrations in steers (Breier *et al.*, 1986). GH responsiveness of the liver appears to change under different nutritional conditions through changes in receptor number, thereby having a dominant influence on the regulation of GH release.

After secretion, GH may circulate in the blood partially bound to specific GH-binding proteins. One of these binding proteins appears to be identical to the amino acid sequence of the extracellular domain of the GH receptor, whereas the other seems to be unrelated to the GH receptor (Bingham *et al.*, 1994). There was no difference in the binding of ^{125}I-labeled hGH to serum proteins of poodles differing considerably in body size, but binding to serum of dwarf pigs was reduced in comparison to binding to serum of normal pigs (Lauteriq *et al.*, 1988).

d. Action

The effects of growth hormone can be divided into two main categories: rapid or metabolic actions and slow or hypertrophic actions. The (acute) metabolic responses are due to direct interaction of growth hormone with the target cell, whereas the slow hypertrophic effects, or those on cartilage, bone, and other tissues, are indirect. The direct effects of GH result in enhanced lipolysis and restricted glucose transport due to insulin resistance (Eigenmann, 1984). Administration of recombinant-DNA-derived human GH induces glucose intolerance in the dog (Shaar *et al.*, 1986). Treatment of pigs for 7 days with purified pituitary GH or recombinant-DNA GH causes insulin resistance together with decreases in insulin- and IGF-I-stimulated lipogenesis of adipose tissue (Walton *et al.*, 1987). The GH responsiveness of adipose tissue is less dependent on nutrition than that of the liver, which may explain the reduced adiposity in the absence of growth enhancement in GH-treated ruminants (Breier *et al.*, 1986).

In contrast to these direct catabolic effects, the indirect actions are anabolic and appear to be mediated by the insulin-like growth factors, IGF-I and IGF-II (Daughaday *et al.*, 1987). In their chemical structure the IGFs have approximately 50% homology with insulin/proinsulin, suggesting they have evolved from a common ancestral molecule. The IGFs are bound to a family of six different IGF binding proteins (IGFBP). The main IGFBP in plasma is IGFBP-3, which is formed and released by the liver upon stimulation with GH. The IGF–IGFBP complex binds in plasma to an acid-labile subunit, resulting in a 150-kDa complex. As a consequence of this binding the IGFs have a very long plasma half-life, which is consistent with their long-term growth-promoting action. Insulin and IGF seem to complement each other, insulin being the acute and IGF the long-term regulator of anabolic processes (for a review, see Eigenmann, 1987). The other IGFBPs play a more local role in the modulation of the IGF effects in specific tissues. These effects may be inhibitory as well as stimulatory.

Competitive binding experiments have revealed that IGF receptors are distinct from insulin receptors and that there are two subtypes of IGF receptors (type I and type II), which differ in their affinity for IGF-I and IGF-II. These two subtypes of IGF receptors have been characterized in terms of structure, phosphorylation, and regulation (Rechler and Nissley, 1985).

There is increasing evidence that the action of GH cannot be categorized by two opposing biological activities in such absolute terms as described earlier. GH exerts its growth-promoting effect not only via IGF-I produced in the liver, but also by a direct effect on cells in the growth plate. It has been proposed that GH stimulates cell differentiation directly and clonal expansion indirectly through the local production of insulin-like growth factors (Nilsson *et al.*, 1986).

Circulating IGF-I concentrations appear to be associated with body size. In a study in different breeds of dogs the IGF levels were found to correlate with body size (Eigenmann *et al.*, 1984a). By studying genetic subgroups within one breed, i.e., standard, miniature, and toy poodles, Eigenmann *et al.* (1984b) found IGF-I levels (and not GH) to parallel body size. Dwarf chickens also have relatively low circulating levels of this growth factor (Huybrechts *et al.*, 1987).

e. Extrapituitary GH Synthesis

In dogs, endogenous progesterone and exogenous progestins may induce considerable rises in plasma growth hormone levels, resulting in acromegalic changes and insulin resistance with the possibility of development of frank diabetes mellitus (Eigenmann and Rijnberk, 1981; Eigenmann *et al.*, 1983b; Selman *et al.*, 1994b). GH excess has only been found to occur in intact female dogs during diestrus (progesterone phase) or in dogs treated with progestins (Concannon *et al.*, 1980; Rijnberk *et al.*, 1980; Eigenmann and Rijnberk, 1981). Estradiol priming may enhance the effect of exogenous progestins in inducing GH overproduction (Eigenmann and Eigenmann, 1981b). Progestin withdrawal and/or ovariohysterectomy results in a reduction of the elevated plasma GH and IGF-I concentrations (Eigenmann *et al.*, 1984a) and reversal of the soft-tissue changes and the insulin resistance (Eigenmann *et al.*, 1983b). A decrease in the plasma GH levels has also been achieved by administration of a synthetic antiprogestin (Watson *et al.*, 1987).

Selman *et al.* (1991) made progress in the characterization of this phenomenon. Progestin-induced GH excess was insensitive to stimulation with GHRH or clonidine and also insensitive to inhibition with the SS analogue SMS 201-995, whereas the dosage SMS used inhibited the GHRH- and clonidine-stimulated GH release in healthy male dogs (Selman *et al.*, 1991). Taken together these observations strongly supported the autonomous character of the progestin-induced GH secretion. Hypophysectomy of progestin-treated ovariohysterectomized female dogs did not result in decreases in plasma GH concentrations (Selman *et al.*, 1994a), supporting an autonomous, nonpituitary production. Analyses of various tissues revealed the highest immunoreactive GH concentrations in the mammary gland. The mammary origin of elevated plasma GH was then confirmed by an arteriovenous gradient across the mammary gland, a rapid decrease and normalization of plasma GH concentrations after complete mammectomy (Selman *et al.*, 1994a), and the presence of GH mRNA in the canine mammary gland by reverse transcriptase PCR (Mol *et al.*, 1995a). From the sequence identity between the pituitary and the mammary PCR product it was concluded that a single gene encodes pituitary and mammary GH in the dog.

Until recently, the progestin-induced GH synthesis was only found in the dog. In cats Peterson (1987b) found no significant rise in plasma GH concentrations during 12 months of treatment with megestrol acetate. However, it was demonstrated recently that in cats with progestin-induced fibroadenomatous changes of the mammary gland the GH mRNA is also present in these tissues (Mol *et al.*, 1995a). Moreover, GH mRNA has now been demonstrated in the normal and neoplastic human mammary gland as well (Mol *et al.*, 1995b, 1996).

In normal cyclic female dogs, significantly higher basal plasma concentration of GH and IGF-I are found during the progesterone-dominated metestrus than during anestrus. This increased basal concentration is accompanied by a decreased responsiveness to stimulation with GHRH or clonidine (Rutteman *et al.*, 1987b; Selman *et al.*, 1991). In the light of the progesterone-induced synthesis in mammary tissue, it is possible that these changes during metestrus are caused by a contribution of the mammary gland to the circulating plasma GH concentrations. Whether extrapituitary GH synthesis is also the cause of gender-related differences in GH secretion in other species remains to be elucidated.

f. Disease

Inadequate GH secretion at an early age causes retardation of growth. Apart from occasional reports on dwarfism in dogs and cats, GH-deficiency dwarfism seems to occur primarily as a genetically transmitted condition in the German shepherd (Andresen *et al.*,

1974) and in the Carelian bear dog (Andresen and Willeberg, 1976). The abnormality has been documented as a primary GH deficiency (Scott *et al.*, 1978; Eigenmann, 1981) with low levels of circulating IGF-I (Eigenmann *et al.*, 1984c) and seems to be due to pressure atrophy of the adenohypophysis by cysts of Rathke's pouch (Müller-Peddinghaus *et al.*, 1980).

In adults, GH deficiency causes much less impressive changes. The clinical features remain confined to the skin and coat. Affected dogs present with alopecia and hyperpigmentation of the skin (Parker and Scott, 1980; Eigenmann and Patterson, 1984). The skin problems may improve by treatment with heterologous GH, although the administration of hGH to the dog is not effective and gives rise to GH antibody formation (Van Herpen *et al.*, 1993). The condition is not well defined and may be an isolated GH deficiency. It has been suggested that in some cases of alopecia the low basal GH levels and the absence of response to stimulation by clonidine or GHRH were due to enhanced somatostatin release as a result of mild and fluctuating hyperadrenocorticism (Rijnberk *et al.*, 1993).

In cats, as in man (Melmed *et al.*, 1983), excessive amounts of growth hormone may be secreted by a pituitary tumor. Since the publication of Gembardt and Loppnow (1976) on adenohypophyseal adenomas in diabetic cats, there has been growing awareness that GH hypersecretion may be involved in the pathogenesis of diabetes mellitus (Peterson et al., 1990).

Hypothyroidism leads to a markedly reduced growth-hormone secretion, with lack of response to provocative stimuli (Chernausek *et al.*, 1983). Apart from low serum GH levels, hypothyroidism is associated with low GHBP, IGF-I, and IGFBP-3 concentrations (Rodriguez-Arnao *et al.*, 1993).

In hyperadrenocorticism, GH secretion is decreased and the response to provocative stimuli is impaired. In hyperadrenocorticoid dogs this proved to be reversible when normocorticism was achieved, suggesting that the impaired GH secretion is the result of hypercorticism per se (Peterson and Altszuler, 1981) and mediated by glucocorticoid-induced increase of the hypothalamic SS tone (Takahashi *et al.*, 1992).

g. Tests

As GH secretion is pulsatile, single values are of no great diagnostic value. In healthy individuals, resting plasma GH concentrations (Table 19.6) may be very low, even below the detection limit of the assay. Hence when GH deficiency is suspected, a stimulation test is needed. In the dog, stimulation with clonidine, xylazine, or GHRH has proven to be reliable for this purpose,

TABLE 19.6 Basal GH and IGF-I Concentrations in Plasma of Healthy Animals[a]

Species (*n*)	Age, breed (condition)	Range (μg/liter) (mean ± SEM)	Reference
GH			
Dog (63)	Adult	(1.9 ± 0.1)	Eigenmann and Eigenmann (1981a)
Dog (5)	7 weeks, Great Dane	(13.7 ± 2.2)	Nap *et al.* (1993)
Dog (5)	27 weeks, Great Dane	(0.7 ± 0.4)	Nap *et al.* (1993)
Dog (6)	Beagle (anestrous)	(0.5 ± 0.1)	Selman *et al.* (1991)
Dog (6)	Beagle (metestrous)	(2.2 ± 0.4)	Selman *et al.* (1991)
Cat (25)	Adult	(3.2 ± 0.7)	Eigenmann *et al.* (1984d)
Mare (12)	7–11 year, Selle Français	(2.8 ± 0.7)	Davicco *et al.* (1994)
Calf (6)	3 days	(11.0 ± 1.8)	Coxam *et al.* (1987)
Steer (6)	267–303 days	15.3–32.6	Wheaton *et al.* (1986)
Pig (4)	5 days	(9.0 ± 1.4)	Lee *et al.* (1993a)
Pig (4)	170 days	(1.5 ± 0.1)	Lee *et al.* (1993a)
Sheep (6)		(3.0 ± 0.2)	Krysl *et al.* (1987)
IGF-I			
Dog (8)	Cocker spaniel	36 ± 27	Eigenmann *et al.* (1984a)
Dog (10)	Beagle	87 ± 33	Eigenmann *et al.* (1984a)
Dog (10)	Keeshond	117 ± 34	Eigenmann *et al.* (1984a)
Dog (13)	German shepherd	280 ± 23	Eigenmann *et al.* (1984a)
Mare (12)	7–11 year, Selle Français	154–318	Davicco *et al.* (1994)
Pig (4)	170 days	(182 ± 30)	Lee *et al.* (1993a)
IGF-II			
Dog (8)	7 ± 0.5 year, beagle	(390 ± 55)	Mol *et al.* (1997)
Pig (4)	170 days	(326 ± 19)	Lee *et al.* (1993a)

[a] The values can be converted to SI units: 1 μg GH/liter = 0.0455 nmol/liter; 1 μg IGF/liter = 0.133 nmol/liter.
[b] Means of determinations in samples collected at 15-min intervals for 24 hours.

TABLE 19.7 Maximal Plasma Concentrations of GH Following Intravenous Administration of a Variety of Stimulants[a]

Species (*n*)	Substance	Dose	Range (μg/liter) (mean ± SEM)	Reference
Dog (6)	Clonidine	10 μg/kg	(19.2 ± 6.2)	Selman *et al.* (1991)
Dog (6)	GHRH	1 μg/kg	(10.0 ± 5.1)	Selman *et al.* (1991)
Cow (4)	GHRH	0.4 μg/kg	16.5	Enright *et al.* (1987)
Calf (4)	GHRH	0.4 μg/kg	(107 ± 55)	Enright *et al.* (1987)
Cow (6)	GHRH	3.3 μg/kg	20.4	Lapierre *et al.* (1987)
Cow (6)	TRH	1.1 μg/kg	4.0	Lapierre *et al.* (1987)
Cow (6)	TRH + GHRH	1.1 μg/kg each	71.4	Lapierre *et al.* (1987)
Lamb	GHRH	0.5 μg/kg	(50.1 ± 19.1)	Barenton *et al.* (1987)

[a] See also footnote to Table 19.6.

whereas in calves and lambs GHRH stimulation has been used (Table 19.7).

Elevated GH values are not definitive proof of acromegaly, not only because of the pulsatile nature of the GH secretion but also because environmental factors may cause sharp increases. Although hypersecretion should be tested by an inhibition test, in the case of progestin-induced GH secretion a blunted stimulation of high basal GH concentrations may support the diagnosis, together with elevated plasma IGF-I concentrations (Selman *et al.*, 1991). It should be kept in mind that the reference values for the IGF-I concentrations in the dog are breed (= body size) dependent.

2. *Prolactin*

a. Gene Expression

Prolactin (PRL) is encoded by a single gene in the human genome. In nonprimates such as cattle, however, a family of PRL-related genes is found, and only one gene encodes GH. The PRL-related genes of the rat can be divided into the lactogens (placental lactogen I, II, I-variant, and the decidual luteotropin) and the nonlactogens (PRL-like protein (PLP) A, PLP-B, PLP-C, and the proliferin-1 and -2 proteins).

The appearance of specific PRL-producing lactotropes is one of the latest events in pituitary development. It has been proposed that most, if not all, lactotropes are derived from presomatotropes under the influence of the lactotrope-specific transcription factor (LSF-1). In humans and rats the majority of PRL-producing cells are also capable of synthesizing GH and are called mammosomatotropes. Even in adulthood the balance between somatotropes and lactotropes may change in the direction of PRL-producing cells under the influence of chronic stimulation with GnRH (Cooke, 1995). It is not clear whether this is due to conversion of somatotropes to lactotropes or whether new lactotropes are generated from pituitary stem cells.

The Pit-1 transcription factor is a prerequisite for the expression of PRL, as it is for expression of GH and TSH. The Pit-1 factor appears to interact with the estrogen response element (ERE) located on a more distal enhancer site, resulting in augmented transcription. More specific for the PRL promoter is the binding of the LSF-1 transcription factor. Glucocorticoids may inhibit PRL gene expression by a negative GRE.

The PRL gene, like the GH gene, has five exon areas. In the extrapituitary expression of PRL in decidua and lymphocytes, a different promoter and exon 1 are used, located upstream to the pituitary promotor.

b. (Pro)hormone

Prolactin (PRL) is synthesized as a single polypeptide that, after cleavage of the signal peptide, has a molecular weight of approximately 22,500 and three intrachain disulfide bridges. There is a good deal of variability in the sequences of PRLs of different species (Fig. 19.13). A minority of prolactin molecules may be present in the pituitary in a glycosylated form. This form has a decreased potency in receptor binding and biological assays.

Expression of the PRL gene within the hypothalamus–neurohypophyseal system results in full-length PRL mRNA. After translation the hormone is processed differently from the AL by specific proteolytic enzymes that generate a 16- or 14-kDa N-terminal fragment. This fragment is able to inhibit angiogenesis by a specific receptor (Clapp *et al.*, 1994).

c. Secretion

For two reasons PRL has a unique place among the AL hormones. First, it is the only hormone that is under tonic inhibition by the hypothalamus. After transplan-

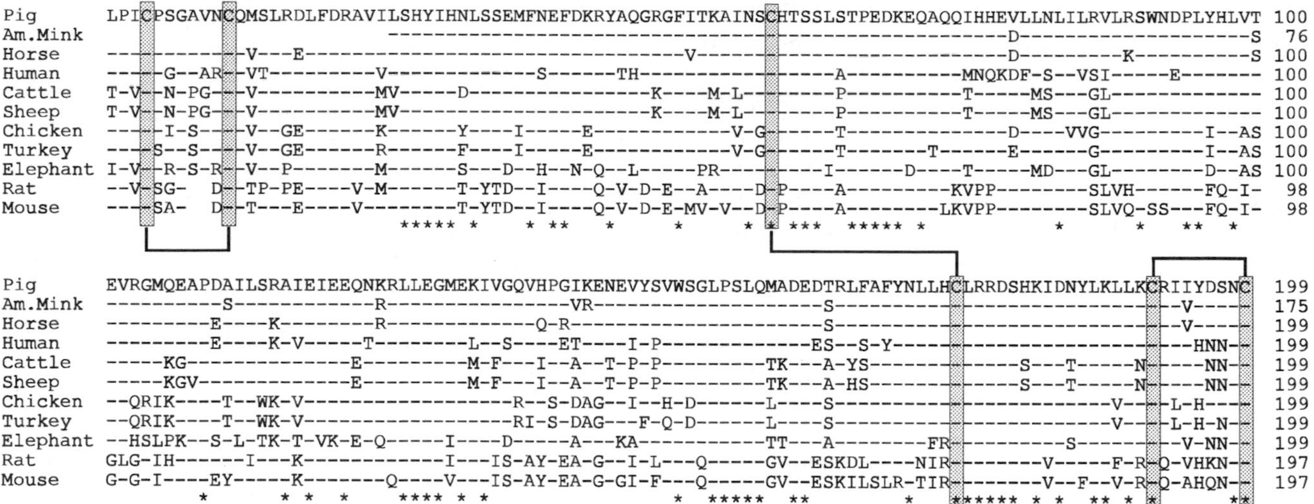

FIGURE 19.13 Sequence comparison of prolactin (PRL). Lines between shaded boxes represent intrachain disulfide bridges. See legend to Fig. 19.6.

tation of the pituitary under the kidney capsule, PRL synthesis is remarkably enhanced. Secondly, PRL lacks a specific target organ that produces factors exerting negative feedback.

The main hypothalamic prolactin-inhibiting factor (PIF) is dopamine. After the release in the median eminence, it may reach the lactotrope via the portal vasculature or by diffusion from the synaptic contact of dopaminergic terminals that end on the melanotropes of the IL. The relative contribution of these systems is not clear. Several isoforms of the endothelins, known for their vasoconstrictor properties, are potent inhibitors of PRL release *in vitro*, but no effect has been demonstrated *in vivo*. Both basal and TRH-induced PRL secretion is also inhibited by calcitonin *in vitro*.

The suckling-, stress-, or estrogen-induced PRL surge cannot be explained by changes in dopaminergic inhibition alone. Therefore, the presence of a prolactin-releasing factor (PRF) has been suggested. Several candidates for PRF have been proposed. One of the components of hypothalamic extracts known to have a stimulatory effect on PRL release is TRH. This was demonstrated in several mammalian species, including the pig (Bevers and Willemse, 1982). However, TRH is probably neither the sole nor the major physiological PRF. Another strong candidate for PRF is vasoactive intestinal peptide (VIP) (Shimatsu *et al.*, 1985).

The rat posterior pituitary was found to contain a potent PRF (Hyde *et al.*, 1987). Further analysis demonstrated that two compounds, oxytocin from the NL and an unidentified peptide from the IL, could function as PRF. The low potency of oxytocin made it a less likely candidate for the physiological regulation of PRL release. The PRF from the IL appeared to be a small peptide that is present in the posterior pituitary of many species. Its chemical nature remains to be determined.

Like other pituitary hormones, PRL is released in a pulsatile manner, with fluctuations during different stages of the reproductive cycle. Apart from an increase around the time of ovulation (McNeilly *et al.*, 1982), PRL concentrations in plasma tend to increase in the luteal phase of the sexual cycle in dogs (DeCoster *et al.*, 1983; Okkens *et al.*, 1985) and cows (Dieleman *et al.*, 1986), but not in cats (Banks *et al.*, 1983). During lactation very high PRL concentrations have been found in the sow by Bevers *et al.* (1978) and in the dog by Concannon *et al.* (1978). In addition, distinct increases in PRL concentration have been found in relation to parturition (Taverne *et al.*, 1978/1979; Concannon *et al.*, 1978). The effect of the lighting regime on PRL secretion in rams is thought to be a direct effect of melatonin on the pituitary gland (Lincoln and Clarke, 1994).

Important modulating factors in the control of PRL secretion are estrogens, especially 17-β-estradiol. Estrogens modulate the TRH receptor levels in the pituitary (DeLean *et al.*, 1977) and cause a biphasic increase in transcription of the prolactin gene (Gorski *et al.*, 1985). From experiments in rats, Deis and Alonso (1985) concluded that progesterone may also stimulate the release of PRL. Rutteman *et al.* (1987a) found that estradiol rapidly induces an enhanced PRL response to TRH in dogs, without changing basal PRL levels. Subsequent administration of medroxyprogesterone acetate did not further affect these findings.

Neurogenic factors also influence PRL secretion. Milking and suckling are almost immediately followed by PRL release. Removal of litters from their dams, for example, piglets from sows (Bevers *et al.*, 1978), results in a rapid decline in PRL levels in plasma. Following return of the litters, PRL concentrations rise again.

d. Action

The most familiar role of PRL in mammals is stimulation of mammary gland growth and lactation. PRL increases mitosis of mammary gland epithelial cells not only during development, but also during pregnancy and lactation. It has been relatively difficult to demonstrate *in vitro* effects of PRL on cell proliferation in mammary glands (Friesen *et al.*, 1985). This can be explained by the increasing evidence that the growth-promoting effect of PRL has much in common with that of GH, i.e., intermediate factors comparable to IGF are required for effects on cell proliferation (Nicoll *et al.*, 1985).

PRL has a wide variety of other physiological actions among vertebrates. It may affect water and electrolyte balance, metabolism, gonadal function, and behavior. Of these, the effect on the ovary has received much attention. PRL has a luteotrophic effect in some animals, such as rodents, sheep, and ferrets (see the review by McNeilly *et al.*, 1982), but not in the cow (Bevers and Dieleman, 1987). Bovine follicles do not bind (ovine) PRL (Bevers *et al.*, 1987).

The reciprocal relationship between PRL and LH has been well demonstrated in several species, including the sow (Bevers *et al.*, 1983) and the cow (Dieleman *et al.*, 1986). This can explain the reduced fertility during lactation that is known in many species. It has been suggested that PRL may also interfere directly at the ovarian level (McNeilly, 1982). These effects of PRL also appear to play a role in the maintenance of the long interestrous interval in the bitch. Treatment of bitches with the dopamine agonist bromocriptine results in considerable shortening of the interestrous interval (Okkens *et al.*, 1985).

Mammalian maternal behavior in several species consists of nest building and caring for offspring. Prolactin-induced maternal behavior has been demonstrated in some animals. However, the primary role of PRL in inducing mammalian maternal behavior has been questioned (Scapagnini *et al.*, 1985). There appears to be diversity in the hormonal basis of maternal behavior, and in some species estrogens and progesterone play a crucial role (Rosenblatt, 1984).

e. Disease

The prolactinoma is the most frequently diagnosed functioning pituitary tumor in man. Mixed GH–PRL adenomas are known to occur in a substantial number of patients with acromegaly, and the concomitant production and secretion of PRL by corticotrope adenomas is not unusual (Ishibashi and Yamaji, 1985). There have been no well-documented reports on the occurrence of prolactinomas in animals, although in dogs with pituitary-dependent hyperadrenocorticism (PDH) Stolp *et al.* (1986) found PRL concentrations in plasma to be higher than those in healthy control dogs and in dogs with hyperadrenocorticism due to adrenocortical tumor. Moreover, the PRL concentrations in dogs with PDH responded supranormally to stimulation with TRH. This suggests that the pituitary lesions of dogs with PDH contain prolactin cells.

Thus, there do not seem to be pathological hyperprolactinemic states in domestic animals that require treatment with dopaminergic drugs. These drugs are nevertheless often used, the main application being in a physiological condition in the bitch called pseudopregnancy (Arbeiter and Windig, 1977; Janssens, 1981; Mialot *et al.*, 1981). Administration of the dopamine agonist bromocriptine results in rapid disappearance of the signs suggesting impending parturition. As treatment with bromocriptine lowers PRL concentrations in plasma (Okkens *et al.*, 1985; Stolp *et al.*, 1986), it is likely that the lactation and maternal behavior of the diestrual bitch are PRL dependent.

f. Tests

Despite the variations in amino acid sequences of PRLs from different species, plasma concentrations can be measured in heterologous assays, for example, employing labeled porcine PRL with an antibody against ovine PRL in an assay for canine PRL (Stolp *et al.*, 1986). PRL levels in the blood may vary during the reproductive cycle, but the basal levels in both male and anestrous female dogs are within similar narrow limits (Stolp *et al.*, 1986). Reference values are presented in Table 19.8.

When there is a suspicion of prolactin deficiency, the secretory capacity can be tested with TRH. In healthy dogs intravenous administration of 10 μg TRH per kg body weight resulted in PRL increments of 13.1–42.5 μg/liter (Rutteman *et al.*, 1987a). In cats, 75 μg TRH caused PRL elevation to seven times the basal levels (Banks *et al.*, 1983). Prolactin concentrations can be suppressed with the dopamine agonist bromocriptine. Both 20 μg/kg intravenously (Stolp *et al.*, 1986) and 10 μg/kg orally (Rijnberk *et al.*, 1987) resulted in protracted decreases lasting at least 8 hours after administration.

TABLE 19.8 Concentrations of PRL in Plasma of Healthy Animals[a]

Species (*n*)	Condition	Range (ug/liter) (mean ± SEM)	Reference
Dog	Anestrous female	(9.1 ± 1.2)	Knight *et al.* (1977)
Dog (6)	1.5 weeks of lactation	(86 ± 19)	Concannon *et al.* (1978)
Dog (6)	8–32 hours prepartum	(117 ± 24)	Concannon *et al.* (1978)
Dog (23)	Males + anestrous females	(0.9 ± 10.5)[b]	Stolp *et al.* (1986)
Dog (5)	Ovariectomized females	7.9–11.5	Rutteman *et al.* (1987a)
Cat (8)	Early gestation	(7.0 ± 0.3)	Banks *et al.* (1983)
Cat (8)	End of gestation	(43.5 ± 4.5)	Banks *et al.* (1983)
Cat (8)	4 weeks post partum	(40.6 ± 7.2)	Banks *et al.* (1983)
Cow (5)	Luteal phase	(23.3 ± 4.8)	Dieleman *et al.* (1986)
Cow (5)	Follicular phase	(15.8 ± 2.7)	Dieleman *et al.* (1986)
Sow (3)	2nd week of lactation	(9.1 ± 1.4–26.1 ± 5.0)[c]	Bevers *et al.* (1978)
Sow (3)	Piglets removed	(1.4 ± 0.3–1.9 ± 0.1)[c]	Bevers *et al.* (1978)
Sheep (8)	Males	68.6 ± 15.3	Matthews and Parrott (1992)

[a] When necessary PRL concentrations are converted to SI units: 1 μg PRL/liter = 0.0444 nmol/liter.
[b] Means of 6 measurements throughout the day.
[c] Means of 8 measurements at 15-min intervals.

III. NEUROHYPOPHYSIS

A. Vasopressin

1. *Gene Expression*

Although there is considerable homology between the genes for vasopressin (VP) and oxytocin (OT), each of the relevant hypothalamic neurons produces only one hormone, either VP or OT. The VP gene consists of three exons. Exon 1 encodes the signal peptide, the nonapeptide VP, and the N-terminal part of neurophysin II (NP-II). Exon 2 encodes the central part of NP-II, and exon 3 encodes the C-terminal part of NP-II and a glycoprotein (GP). The major factor that regulates VP synthesis is osmotic stimulation. Rats given hypertonic saline chronically have increased VP and OT mRNA concentrations in the magnocellular neurons in the hypothalamus. Similar increases in VP mRNA are observed in hypovolemic conditions. Acute increases in plasma osmolality or hypovolemia result in a rapid lengthening of the poly(A) tail of the mRNA, followed by a slow increase in mRNA abundance.

2. *(Pro)hormone*

After cleavage of the signal peptide, the GP moiety of the VP prohormone is glycosylated, disulfide bonds are generated, and then the prohormone is packaged into a secretory vesicle. The correct sorting of the prohormone into the secretory pathway requires the formation of aggregates. Specific association of VP and NP-II results in dimeric and tetrameric units, which is essential for sorting (De Bree and Burbach, 1994). The secretory vesicles are axonally transported to the axon endings in the posterior pituitary. Studies in the dog suggest that approximately 1.5 hours are required from the time of synthesis to possible release of the nonapeptide (Richter and Ivell, 1985). During this transport the neurophysin II, VP, and glyprotein are liberated by proteolytic cleavage. The C-terminal glycine residue of the VP molecule is finally amidated. The mature VP nonapeptide has a disulfide bridge between cysteine residues at positions 1 and 6 (Fig. 19.14).

In most mammals VP contains an arginine residue at position 8 (arginine vasopressin, AVP). In the members of the pig family, VP contains a lysine residue at position 8. This lysine vasopressin (LVP) is the only form present in the domestic pig. Heterozygotic peccaries, warthogs, and hippopotami may possess both AVP and LVP. Substitution of the phenylalanine residue at position 3 by isoleucine leads to arginine vasotocin (AVT). This peptide has been identified as a uniquely nonmammalian substance and is probably the most primitive and certainly the most commonly occurring neurohypophyseal peptide. Neurohypophysis of some strains of chickens (but not ducks or turkeys) contain not only AVT but also AVP (Choy and Watkins, 1986).

In the rat and the calf the cDNAs encoding the neurophysins of AVP and oxytocin have a high degree of homology. Amino acid analysis of the AVP- and oxytocin-related neurophysins also shows a strong homology for bovine, porcine, and equine neurophysin.

	1	2	3	4	5	6	7	8	9
Arginine vasopressin (AVP)	Cys	Tyr	Phe	Gln	Asn	Cys	Pro	Arg	Gly-NH_2
Lysine vasopressin (LVP)	-	-	-	-	-	-	-	Lys	-
Arginine vasotocin (AVT)	-	-	Ile	-	-	-	-	-	-
Conopressin S (snail venom)	-	Phe	Ile	Arg	-	-	-	-	-
Mesotocin (MT)	-	-	Ile	-	-	-	-	Ile	-
Oxytocin (OT)	-	-	Ile	-	-	-	-	Leu	-
	*				*	*	*		*

FIGURE 19.14 Sequence comparison of vasopressin (VP) and oxytocin (OT). Lines between shaded boxes represent intrachain disulfide bridges. See legend to Fig. 19.6.

The glycoprotein sequence appears to be highly conserved among various species, but its physiological role remains to be clarified (Majzoub, 1985).

3. *Secretion*

The major determinant in the release of VP is plasma osmolality. Below a certain plasma osmolality threshold, which may vary considerably between individuals, plasma VP concentration is suppressed to levels that allow maximal free water clearance (Wade *et al.*, 1982). Once plasma osmolality rises to the threshold, the pituitary secretes VP. Although the concentration of plasma VP causing maximal antidiuresis is about 5–10 pmol/liter, the release of VP is related to plasma osmolality over a much broader range (Robertson, 1983). Biewenga *et al.* (1987) developed a nomogram for this relationship in the dog, which allows analysis of the osmoregulation of VP secretion in terms of sensitivity and threshold.

Apart from the influence of osmolality, a significant decrease in circulating blood volume may also cause enhanced VP secretion. The decrease in left atrial pressure triggers receptors. Denervation of these receptors prevents the hypovolemia-mediated VP release. The functional properties of this regulatory system differ from those of the osmoregulatory system. Although the relationship between plasma osmolality and VP appears to be linear, the relationship between blood volume and plasma VP is best described by an exponential function (Vokes and Robertson, 1985a).

Under ordinary circumstances, changes in VP levels are not observed until blood volume decreases by 5–10%. With further decreases, plasma VP rises exponentially. The relative insensitivity of VP to modest changes in blood volume means that this regulatory mechanism plays little or no role in the physiological control of water balance. Under ordinary circumstances, total body water rarely changes by more than 1–2%, an amount far too small to affect VP secretion through hemodynamic influences (Wang and Goetz, 1985). In dogs, 24 hours of fluid deprivation increases plasma VP because of changes in both extracellular volume and tonicity, but the increase in tonicity plays a greater role than the reduction in volume (Wade *et al.*, 1983).

It has been proposed that hemodynamic influences modulate VP secretion by raising or lowering the osmostat (Robertson, 1978). According to this concept, a decrease in plasma volume and/or pressure will permit normal osmoregulation of VP, but the threshold for release will be lowered by an amount proportional to the degree of hypovolemia or hypotension, associated with increased sensitivity of the VP response to rising plasma osmolality. Conversely, hypervolemia or hypertension increases the threshold and decreases the sensitivity of release (Quillen and Cowley, 1983).

There are numerous other factors that may influence VP secretion. Of these, nausea/vomiting (Raschler *et al.*, 1986) and insulin-induced hypoglycemia (Vokes and Robertson, 1985b) are the most potent. The effects of opiates and opioid peptides on VP secretion have been studied for many years and have been the subject of some controversy. Both stimulatory and inhibitory effects have been reported. Apart from the substances, doses, and routes of administration used, the animal species is an important factor (Van Wimersma Greidanus, 1987). Intracerebroventricular administration of β-END in conscious rats decreases basal and stimulated VP release (ten Haaf *et al.*, 1986). An opioid inhibition of dehydration-induced VP release has been reported in dogs (Wade, 1985). However, Hellebrekers *et al.* (1988) could not confirm this endogenous opioid modulation of osmolality-regulated VP release in conscious dogs subjected to hypertonic saline infusion. In contrast, sharp increases in plasma VP have been found in conscious dogs after intravenous administration of the μ-type opiate receptor agonist methadone (Hellebrekers *et al.*, 1987).

4. *Action*

The major role of VP is to regulate body fluid homeostasis by affecting water resorption. An increase in plasma VP results in increased water retention, which maintains plasma osmolality between narrow limits. The antidiuretic effect is achieved by promoting the reabsorption of solute-free water in the distal and collecting tubules of the kidney. In the absence of VP, this portion of the nephron is not permeable to water and the hypotonic filtrate of the ascending limb of Henle's loop passes unmodified through the distal tubule and collecting duct. In this condition urine osmolality decreases to around 80 mOsm/kg. The action of VP on the tubular cells is mediated by V2 receptors that, after VP binding, activate a signal transduction system leading to an intracellular increase in cAMP (Jard, 1983).

Besides this well-known role in the regulation of fluid homeostasis, VP exerts a large variety of effects. Among these is the vasopressor effect, which is mediated by V1 receptors (Liard, 1986). In addition, VP increases glucogenolysis by liver cells, increases ACTH release by adenohypophysis (Lowry *et al.*, 1987), and has several effects on animal behavior (De Wied and Versteeg, 1979). The extrarenal effects on hepatocytes and vascular smooth muscle are exerted through cAMP-independent, calcium-dependent V1 receptors (Jard, 1983).

There is some evidence that in birds AVT, apart from its established antidiuretic action, also has oxytocic properties and participates in normal oviposition (Shimada *et al.*, 1986, 1987). MT does not seem to function in oviposition, and its release was found to be negatively correlated with plasma osmolality (Koike *et al.*, 1986).

5. *Disease*

Disorders of the hypothalamic–neurohypophyseal system resulting in deficiency or excess of VP are known to occur in domestic animals. Deficient VP release causes the syndrome of diabetes insipidus, which is characterized by persistent inappropriately dilute urine in the presence of strong osmotic stimuli for VP release. The disease is not uncommon in the dog and cat (see Feldman and Nelson, 1996; Rijnberk, 1996) and rare in other species.

VP excess is known as the syndrome of inappropriate ADH (SIADH) secretion or the Schwartz–Bartter syndrome. The VP release is inappropriately high in relation to the plasma osmolality. The resulting defect in water excretion causes hyponatremia, which is the hallmark of SIADH. This condition has been documented as a disease of the dog, and at the same time studies of the threshold and sensitivity of VP secretion revealed that it occurred in two forms (Rijnberk *et al.*, 1988; Houston *et al.*, 1989).

The glucocorticoid excess in Cushing's syndrome is accompanied by (mild) polyuria in the dog. A marked impairment of the osmolality-regulated AVP release at the pituitary level may, in concert with a partial AVP resistance at the kidney level, be the cause of this corticosteroid-induced polyuria (Biewenga *et al.*, 1991). Chronic liver disease is accompanied by both enhanced activity of the pituitary–adrenocortical axis and disturbances in sodium and water homeostasis. Also in these cases a profoundly impaired osmoregulation of AVP release was found (Rothuizen *et al.*, 1995). In dogs with spontaneous pericardial effusion, moderately elevated plasma AVP levels declined rapidly after pericardiocentesis (Stokhof *et al.*, 1994).

6. *Tests*

The diagnosis of VP deficiency requires that it be differentiated from other causes of water diuresis. Following exclusion of conditions such as hyperadrenocorticism, hyperthyroidism, and hypercalcemia, both nephrogenic diabetes insipidus and primary polydipsia remain as the main differential diagnoses. Slightly elevated plasma osmolality may suggest neurogenic or nephrogenic diabetes insipidus, whereas low plasma osmolality may be observed in primary polydipsia. However, in many cases this parameter is not conclusive.

The procedure that is widely used to differentiate these disorders is the modified water deprivation test, as introduced for the dog by Mulnix *et al.* (1976). In this test maximal urine concentration is induced by several hours of dehydration. Once a plateau in urine osmolality is reached, the effect of an injection with VP is investigated. A further increase in urine osmolality by 50% or more is regarded as diagnostic for VP deficiency.

Although these indirect criteria for VP secretion can usually differentiate between complete neurogenic and complete nephrogenic diabetes insipidus, they cannot differentiate among partial neurogenic, partial nephrogenic, and dipsogenic polyurias. In these situations, direct measurements of plasma VP during hypertonic saline infusion are required (Biewenga *et al.*, 1989).

Hypersecretion of VP can only be diagnosed by measurements of the circulating concentrations of the hormone (Table 19.9). The basic criterion is the presence of "inappropriately" high VP concentrations in relation to the hypoosmolality of the extracellular fluid. The type of osmoregulatory defect can best be judged by repeatedly measuring plasma osmolality and VP during administration of hypertonic saline (Rijnberk *et al.*, 1988). Details of this test are given in Section IV of this chapter.

B. Oxytocin

1. *Gene Expression*

The oxytocin (OT) gene expression is predominantly found in the hypothalamic supraoptic nucleus and paraventricular nucleus (Ivell, 1986). The neurons proj-

TABLE 19.9 Concentrations of Neurohypophyseal Hormones in Plasma of Healthy Animals[a]

Species (*n*)	Peptide	Condition	Range (pmol/liter) (mean ± SEM)	Reference
Dog (5)	AVP	Basal	0.5–2.9	Dogterom *et al.* (1978)
Dog (12)	AVP	Basal	(3.3 ± 0.4)	Simon-Oppermann *et al.* (1983)
Dog (9)	AVP	24 hours dehydration	(7.7 ± 1.5)	Szczepanska-Sadowska *et al.* (1983)
Dog (25)	AVP	Basal	0.9–6.0	Biewenga *et al.* (1987)
Dog (10)	OT	First week of lactation	15–66	Eriksson *et al.* (1987)
Cow (334)	OT	Before milking	(1.6 ± 0.6)	Schams (1983)
Sow (9)	OT	6–12 days postpartum	<10	Porter *et al.* (1992)
Sow (9)	OT	idem, suckling	24.9–74.6	Porter *et al.* (1992)
Sheep (12)	OT	Spontaneous pulse height	(18.3 ± 9.5)	Payne and Cooke (1993)
Goat	AVP	Basal	(1.3 ± 0.2)	Thornton *et al.* (1986)
Goat	OT	Basal	(4.5 ± 1.0)	Thornton *et al.* (1986)
Calf (16)	AVP	Basal	(1.3 ± 0.2)[b]	Doris and Bell (1984)
Calf (16)	AVP	4 days dehydration	(16.9 ± 1.9)[b]	Doris and Bell (1984)
Hen	AVT	Basal	(3.3 ± 0.3)[b]	Shimada *et al.* (1986)
Hen	AVT	Oviposition	(14.8 ± 0.6)[b]	Shimada *et al.* (1986)

[a] When necessary the values were converted to SI units: 1 ng/liter AVP = 0.932 pmol/liter; 1 ng/liter AVT = 0.953 pmol/liter; 1 ng/liter OT = 0.994 pmol/liter.
[b] In μU/ml.

ect mainly to the posterior pituitary, where OT is secreted for endocrine stimulation of parturition and lactation, whereas other neurons project to higher brain centers, where OT may be converted by brain peptidases and exert an effect as a neurotransmitter or neuromodulator (Burbach, 1986). Central-nervous OT may attenuate learning and memory processes, and it may function in maternal behavior (Kovács, 1986). OT has also been demonstrated in the placenta, ovary, corpus luteum, testis, and adrenal gland. At least in the bovine corpus luteum, OT is synthetized locally (Ivell *et al.*, 1985) and acts as a paracrine compound by influencing the prostaglandin-mediated regression of the corpus luteum (Flint and Sheldrick, 1982).

The organization of the OT gene closely resembles that of the VP gene. The exon–intron organization is similar, with the exception that exon 3 lacks a glycoprotein domain. In the cow, among 194 bases of the 3′ part of exon 2, no single base is different from the corresponding one in the VP exon 2. Exon 3 has the least homology between the VP and OT genes (Ivell and Richter, 1985).

Apart from an identical increase in OT and VP mRNA after hyperosmotic or hypovolemic stimulation, differences exist between the expression of both genes. The OT gene expression increases in estrus, whereas VP gene expression is not cycle dependent. In the suprachiasmatic nucleus there is a circadian rhythm in VP expression, but not in OT mRNA. During pregnancy and lactation an increased poly(A) tail of the OT mRNA is observed. The length of the poly(A) tail is considered to be related to stability of mRNA or may stimulate the translation efficiency.

2. *(Pro)hormone*

After translation of the mRNA, OT is cleaved from neurophysin I by proteolytic enzymes and a C-terminal amidation of OT occurs. This posttranslational processing is time dependent, as for the biotransformation of the VP prohormone (Ivell, 1986).

3. *Secretion*

The release of OT is most abundant during parturition and the lactation period. The main stimuli for the release of OT from the posterior pituitary therefore arise either from the nipples during milking (Wachs *et al.*, 1984) or suckling, or from vaginal and cervical distension. The suckling-related release of OT has been documented in the dog (Eriksson *et al.*, 1987). The neural regulation of OT release is complex. Both dopamine and opioid peptides (enkephalin and β-endorphin) modulate OT release at the nerve terminals in the NL (Forsling, 1986). This interaction may be mediated by opioid effects on posterior pituitary pituicytes. There is some evidence that OT is released in response to osmotic challenge in the rat and some carnivores, but not in primates or ungulates (Thornton *et al.*, 1987).

Plasma OT levels may also vary during the reproductive cycle. Generally, a similar secretion pattern is found among domestic animals. Plasma levels increase after the preovulatory LH surge and decrease at the time of luteal regression, in the cow, goat, and sheep (Schams, 1983). In the mare, plasma OT levels are lower in the midluteal phase than on day 2 of estrus and day 5 after ovulation (Burns *et al.*, 1981). The latter is closer to the pattern of OT release in man. The high OT content of the corpus luteum of the sheep and the cow may mean that the majority of the circulating OT is derived from the ovary in these species.

Ovarian estrogen may enhance pituitary OT release, whereas progesterone inhibits it. In male animals, mating may be a stimulatory factor. Nonreproductive factors such as stress and plasma osmolality may also influence OT release (Forsling, 1986).

4. *Action*

OT causes contraction of the myoepithelial elements of the excretory ducts of the mammary glands, resulting in milk ejection. In addition, it exerts a constricting effect on the uterine muscle. Estrogens increase the response of the uterus to OT, whereas progesterone inhibits it. During pregnancy, the sensitivity of the uterus to OT is greatly reduced. However, during delivery and immediately thereafter, the uterus is particularly sensitive to OT. In birds, AVT is far more active in stimulating the contraction of the oviduct than is OT (Bentley, 1971).

5. *Disease/Tests*

To our knowledge no pituitary dysfunction concerning OT release in domestic animals has been reported. Basal plasma levels of OT are in approximately the same range as plasma VP concentrations (Table 19.9). In man and the goat (Thornton *et al.*, 1986), OT release cannot be stimulated by hypertonic saline infusion, but does increase in response to insulin-induced hypoglycemia (Nussey *et al.*, 1986). In the rat, nausea-producing chemical agents and cholecystokinin cause a dose-dependent increase in OT (Verbalis *et al.*, 1986).

IV. ASSESSMENT OF PITUITARY FUNCTION

Endocrine studies in animals with suspected hypothalamus–pituitary disease may be carried out for three reasons: (1) documentation of specific endocrine deficits, (2) insight into the type, size, and progression of the lesion, and (3) characterization of hyperfunction. Specific diagnostic tests for several of the pituitary hormones have been described in the previous sections. Here the discussion will be concentrated on a more general approach, in an attempt to determine the extent of involvement of AL, IL, and/or NL.

A. Adenohypophysis

1. *Anterior Lobe*

The species specificity of some of the AL hormones precludes their measurement in several animal species with the assays now available. For many other hormones, however, homologous or heterologous assays are available, allowing assessment of AL function.

Although baseline hormone levels can provide much information, the pulsatile nature of AL hormone secretion often makes interpretation difficult. It is particularly difficult to distinguish between normal and low levels of these hormones. For this reason provocative tests should be used to examine hypothalamus–pituitary adequacy. Now that hypothalamic-releasing hormones have become available and are applicable in most species (Tables 19.4–19.8), appropriate testing of AL function is possible.

In all instances blood samples are collected at −15, 0, 10, 20, 30, and 60 minutes for measurement of the pituitary hormones. At time zero the test substance is injected intravenously. It may be one of the following:

1. 1 μg GHRH/kg or 10 μg clonidine/kg to test GH secretory capacity (see also Table 19.7).
2. 1 μg CRH/kg or 0.6 μg lysine vasopressin/kg to test ACTH secretory capacity (see also Table 19.5).
3. 10 μg TRH/kg to test PRL secretion (see Section II.C.2 of this chapter)

Because of the rapid proteolytic degradation of some pituitary hormones, blood samples should always be chilled on ice as soon as possible and stored at ≤20°C (see also Section V.E of the chapter on adrenocortical function).

Combined function tests are coming into use in man. The anterior pituitary is stimulated with four hypothalamic releasing hormones in a single procedure. Despite the fact that there is some interaction of the different releasing hormones, combined administration of releasing hormones seems to be a useful test for the assessment of pituitary function (Schopohl *et al.*, 1986). Such a combined function test has been introduced for the dog. In this test the anterior pituitary is stimulated with four hypothalamic releasing hormones (CRH, GHRH, TRH and GnRH) with measurements of ACTH (+cortisol), GH, PRL and LH (Meij *et al.*, 1996). Compared with the single administration of these secretagogues, there is little interference in this test, except for the LH response, which is lower in the combined test then following single GnRH administration.

2. *Intermediate Lobe*

α-MSH is secreted specifically by the IL. Its release can be stimulated with dopamine agonists such as haloperidol (Table 19.5). The times of sampling can be as for the tests of adenohypophyseal function (see Section IV.A.1 of this chapter).

B. Neurophypohysis

As explained in Section III.B of this chapter, VP secretion can be examined indirectly by means of a modified water-deprivation test and directly by measuring plasma VP during hypertonic saline infusion. Both procedures are described here in detail, as they have been developed for use in the dog (Mulnix *et al.*, 1976; Biewenga *et al.*, 1987). It should be noted that because of interference of plasma proteins, the radioimmunoassays for AVP and OT require prior extraction of the peptide. Antisera against AVP may possess 100% cross reactivity with AVT (Choy and Watkins, 1986; Hellebrekers *et al.*, 1987).

1. *Modified Water Deprivation Test*

a. Indication

Differentiation between neurogenic and nephrogenic diabetes insipidus, when the administration of DDAVP (Minrin) has not resolved the diagnosis.

b. Procedure

Following 12 hours of fasting, water is withheld and plasma and urine are collected every hour or every 2 hours, depending on the severity of the polyuria. Osmolality is measured in both samples. At each collection the animal is weighed. When the weight loss approaches 5% of initial body weight, the test should be stopped.

When, in the presence of an adequate osmotic stimulus (Posm > 305 mOsm/kg), urine concentration is maximal (less than 5% increase in Uosm between consecutive collections), 2 IU of lysine vasopressin are administered subcutaneously. Uosm is measured again 1 hour later.

c. Interpretation

In both nephrogenic diabetes insipidus and neurogenic diabetes insipidus, Uosm will remain low during water deprivation. In neurogenic diabetes insipidus Uosm will rise by 50% or more following the injection of vasopressin, whereas in nephrogenic diabetes insipidus there will be little or no rise in Uosm. However, as mentioned earlier, in some cases the outcome may not be conclusive and erroneous results may also occur.

2. *AVP Measurements during Hypertonic Saline Infusion*

a. Indications

Differential diagnosis of polyuric conditions and suspicion of inappropriate VP secretion.

b. Procedure

The euhydrated animal is infused for 2 hours via the jugular vein with 20% NaCl solution at a rate of 0.03 ml/kg body weight per minute. Samples for plasma AVP and Posm are obtained at 20-minute intervals. Especially in the severely polyuric animal, there is a risk of inducing critical hypertonicity. In these cases the samples collected at 0, 40, and 80 minutes should be checked for Posm immediately.

c. Interpretation

The slope of the regression line for Posm and PAVP is used as a measure of the sensitivity of the osmoregulatory system. In the nomogram developed by Biewenga *et al.* (1987), the 90% range for sensitivity was 0.24–2.47 (pg/ml)/(mOsm/kg). The 90% range for the threshold of the system was 276–309 mOsm/kg.

Acknowledgments

The authors gratefully acknowledge the contribution to the manuscript of Dr. B. E. Belshaw.

References

Alexander, S. L., Irvine, C. H. G., Ellis, M. J., and Donald, R. A. (1991). *Endocrinology* **128,** 65.

Alexander, S. L., Irvine, C. H. G., Livesey, J. H., and Donald, R. A. (1993). *Endocrinology* **133,** 860.

Amann, J. F., Smith, R. M., Ganjam, V. K., Paull, W. K., McClure, R. C., Green, E. M., and Garner, H. E. (1987). *Amer. J. Vet. Res.* **48,** 323.

Andresen, P., and Willeberg, P. (1976). *Hereditas* **83,** 232.

Andresen, P., Willeberg, P., and Rasmussen, P. G. (1974). *Nord. Vet. Med.* **26,** 692.

Antakly, T., Sasaki, A., Liotta, A. S., Palkovits, M., and Krieger, D. T. (1985). *Science* **229,** 277.

Antakly, T., Mercille, S., and Côté, J. P. (1987). *Endocrinology* **120,** 1558.

Antoni, F. A. (1986). *Endocr. Rev.* **7,** 351.

Arbeiter, K., and Windig, W. (1977). *Kleintier Praxis* **22,** 271.

Art, T., Franchimont, P., and Lekeux, P. (1994). *Vet. Rec.* **135,** 499.

Baker, B. L. (1974). *In* "The Pituitary Gland and Its Neuroendocrine Control, Part 1" (E. Knobel and W. H. Sawyer, eds.), p. 45. Williams and Wilkins, Baltimore.

Banks, D. R., Paape, S. R., and Stabenfeldt, G. H. (1983). *Biol. Reprod.* **28,** 923.

Barenton, B., Chabanet, C., and Pelletier, J. (1987). *Proc. Soc. Exp. Biol. Med.* **185,** 312.

Batten, T. F. C., and Ingleton, P. M. (1987). *In* "Fundamentals of Comparative Vertebrate Endocrinology" (I. Chester-Jones, P. M. Ingleton, and J. G. Phillips, eds.), p. 285. Plenum Press, New York.

Bauman, D. E., Eppard, P. J., DeGeeter, M. J., and Lanza, G. M. (1985). *J. Dairy Sci.* **86,** 1352.

Baumann, G. (1991). *Endocr. Rev.* **12,** 224.

Belchetz, P. E., Plant, T. M., and Nakai, Y. (1978). *Science* **202,** 631.

Belshaw, B. E. (1983). *In* "Textbook of Veterinary Internal Medicine II" (S. J. Ettinger, ed.), p. 1592. W.B. Saunders & Co, Philadelphia.

Bentley, P. J. (1971). *In* "Endocrines and Osmoregulation," p. 111. Springer Verlag, Berlin.

Bentley, P. J. (1976). *In* "Comparative Vertebrate Endocrinology," p. 61. Cambridge University Press, Cambridge.

Berkenbosch, F., Vermes, I., Binnenkade, R., and Tilders, F. J. H. (1981). *Life Sci.* **29,** 2249.

Besser, G. M. (1987). *In* "Highlights on Endocrinology" (C. Christiansen and B. J. Riis, eds.), p. 233. Proceedings of the First European Congress on Endocrinology, June 21–25, 1987.

Bevers, M. M., and Dieleman, S. J. (1987). *Anim. Reprod. Sci.* **14,** 95.

Bevers, M. M., and Willemse, A. H. (1982). *Theriogenology* **18,** 303.

Bevers, M. M., Willemse, A. H., and Kruip, Th. A. M. (1978). *Biol. Reprod.* **19,** 628.

Bevers, M. M., Willemse, A. H., and Kruip, Th. A. M. (1983). *Acta Endocrinol.* **104,** 261.

Bevers, M. M., Dieleman, S. J., Kruip, Th. A. M., and Willemse, A. H. (1987). *In* "Follicular Growth and Ovulation Rate in Farm Animals" (J. F. Rocke and D. O'Callaghan, eds.), p. 45. Martinus Nijhoff Publishers, Dordrecht, Boston.

Bevers, M. M., Dieleman, S. J., Van Tol, H. T. M., Blankestein, D. M., and Van den Broek, J. (1989). *J. Reprod. Fert.* **87,** 745.

Biewenga, W. J., Van den Brom, W. E., and Mol, J. A. (1987). *Front. Horm. Res.* **17,** 139.

Biewenga, W. J., Rijnberk, A., and Mol, J. A. (1989). *Vet. Q.* **11,** 193.

Biewenga, W. J., Rijnberk, A., and Mol, J. A. (1991). *Acta Endocrinol (Copenh.)* **124,** 583.

Bilezikjian, L. M., and Vale, W. W. (1992). *Trends Endocrinol. Metab.* **3,** 218.

Bilezikjian, L. M., Seifert, H., and Vale, W. (1986). *Endocrinology* **118,** 2045.

Bingham, B., Oldham, E. R., and Baumbach, W. R. (1994). *Proc. Soc. Exp. Biol. Med.* **206,** 195.

Boyd, W. H. (1987). *Anat. Anzeiger, Jena* **164,** 117.

Breier, B. H., Bass, J. J., Butler, J. H., and Gluckman, P. D. (1986). *J. Endocrinol.* **111,** 209.

Buckingham, J. C. (1987). *J. Endocrinol.* **113,** 389.

Buonomo, F. C., Lauterio, T. J., Baile, C. A., and Daughaday, W. H. (1987). *Gen. Comp. Endocrinol.* **66,** 274.

Burbach, J. P. H. (1986). *Curr. Top. Neurobiol.* **6,** 55.

Burns, P. J., Kumaresan, P., and Douglas, R. H. (1981). *Theriogenology* **16,** 531.

Capen, C. C., Martin, S. L., and Koestner, A. (1967). *Path. Vet.* **4,** 301.

Caraty, A., Grino, M., Locatelli, A., Guillaume, V., Boudouresque, F., Conte-Devolx, B., and Oliver, C. (1990). *J. Clin. Invest.* **85,** 1716.

Ceda, G. P., Davis, R. G., Rosenfeld, R. G., and Hoffman, A. R. (1987). *Endocrinology* **120,** 1658.

Chastain, C. B., Riedesel, D. H., and Graham, C. L. (1979). *Canine Pract.* **6,** 59.

Chernausek, D. R., Umderwood, L. E., Utiger, R. D., and Van Wijk, J. J. (1983). *Clin. Endocrinol.* **19,** 337.

Choy, V. J., and Watkins, W. B. (1986). *Neuropeptides* **8,** 183.

Chronwall, B. M., Millington, W. R., Griffin, W. S. T., Unnerstall, J. R., and O'Donohue, T. L. (1987). *Endocrinology* **120,** 1201.

Clapp, C., Torner, L., Gutierrez-Ospina, G., *et al.* (1994). *Proc. Natl. Acad. Sci. USA* **91,** 10384.

Clarke, I. J., and Cummins, J. T. (1982). *Endocrinology* **111,** 1737.

Clarke, I. J., Clements, J. A., Cummins, J. T., Dench, F. Smith, A. I., Robinson, P. M., and Funder, J. W. (1986). *Neuroendocrinology* **44,** 508.

Coates, P. J., Doniach, I., Hale, A. C., and Rees, L. H. (1986). *J. Endocrinol.* **111,** 335.

Concannon, P. W., Butler, W. R., Hansel, W., Knight, P. J., and Hamilton, J. M. (1978). *Biol. Reprod.* **19,** 1113.

Concannon, P. W., Altszuler, N., Hampshire, J., Butler, W. R., and Hansel, W. (1980). *Endocrinology* **106,** 1173.

Cooke, N. E. (1995). *In* "Endocrinology" (L. J. DeGroot, ed.), p. 368. Saunders, Philadelphia, Pennsylvania.

Cowan, J. S., Gaul, P., Moor, B. C., and Kraicer, J. (1984). *Can. J. Physiol. Pharmacol.* **62,** 199.

Coxam, J. S., Davicco, M. J., Opmeer, F. A., Ravault, J. P., and Barlet, J. P. (1987). *Reprod. Nutr. Dev.* **27,** 533.

Danforth, D. R., and Cheng, C. Y. (1993). *Hum. Reprod.* **8,** 117.

Daughaday, W. H., Hall, K., Salmon, W. D., Van den Brande, J. L., and Van Wyk, J. J. (1987). *J. Clin. Endocrinol. Metab.* **65,** 1075.

Davicco, M. J., Faulconnier, Y., Coxam, V., Dubroeucq, H., Martin-Rosset, W., and Barlet, J. P. (1994). *Arch. Int. Physiol. Biochim. Biophys.* **102,** 115.

DeBree, F. M., and Burbach, J. P. (1994). *Biochimie* **76,** 315.

DeCoster, R., Beckers, J. F., Beerens, D., and De Mey, J. (1983). *Acta Endocrinol.* **103,** 473.

Deis, R. P., and Alonso, N. (1985). *In* "Prolactin. Basic and Clinical Correlates" (R. M. MacLeod, M. O. Thorner, and U. Scapagini, eds.), p. 307. Fidia Research Series, Vol. 1. Liviana Press, Padova.

De Jong, F. H. (1979). *Mol. Cell. Endocrinol.* **13,** 1.

De Kloet, E. R., Van der Vies, J. and De Wied, D. (1974). *Endocrinology* **94,** 61.

De Koning, P., and Gispen, W. H. (1987). *Peptides* **8,** 415.

DeLean, A., Ferland, L., Drouin, J., Kelly, P. A., and Labrie, F. (1977). *Endocrinology* **100,** 1496.

Delitala, G., Giusta, M., and Rodriguez, E. (1982). *Acta Endocrinol.* **100,** 321.

Denef, C. (1994). *In* "The Pituitary Gland" (H. Imura, ed.), p.351. Raven, New York.

De Wied, D., and Versteeg, D. H. G. (1979). *Fed. Proc.* **38,** 2348.

Dieleman, S. J., Bevers, M. M., Van Tol, H. T. M., and Willemse, A. H. (1986). *Anim. Reprod. Sci.* **10,** 275.

Dogterom, J., Van Wimersma Greidanus, Tj. B., and De Wied, D. (1978). *Amer. J. Physiol.* **234,** E463.

Dores, R. M., Wilhelm, M. W., and Sandoval, D. M. (1987). *Gen. Comp. Endocrinol.* **68,** 153.

Doris, P. A., and Bell, F. R. (1984). *Neuroendocrinology* **38,** 290.

Dunbar, M., and Ward, B. C. (1982). *J. Amer. Anim. Hosp. Assoc.* **18,** 737.

Eigenmann, J. E. (1981). *J. Amer. Anim. Hosp. Assoc.* **17,** 798.

Eigenmann, J. E. (1984). *Vet. Clin. North Amer. Small Anim. Pract.* **14,** 827.

Eigenmann, J. E. (1987). *Front. Horm. Res.* **17,** 161.

Eigenmann, J. E., and Eigenmann, R. Y. (1981a). *Acta Endocrinol.* **98,** 514.

Eigenmann, J. E., and Eigenmann, R. Y. (1981b). *Acta Endocrinol.* **98,** 603.

Eigenmann, J. E., and Patterson, D. F. (1984). *J. Amer. Anim. Hosp. Assoc.* **20,** 741.

Eigenmann, J. E., and Rijnberk, A. (1981). *Acta Endocrinol.* **98,** 599.

Eigenmann, J. E., Lubberink, A. A. M. E., and Koeman, J. P. (1983a). *J. Amer. Anim. Hosp. Assoc.* **19,** 377.

Eigenmann, J. E., Eigenmann, R. Y., Rijnberk, A., Van der Gaag, I., Zapf, J., and Froesch, E. R. (1983b). *Acta Endocrinol.* **104,** 167.
Eigenmann, J. E., Patterson, D. F., Zapf, J., and Froesch, E. R. (1984a). *Acta Endocrinol.* **105,** 294.
Eigenmann, J. E., Patterson, D. F., and Froesch, E. R. (1984b). *Acta Endocrinol.* **106,** 448.
Eigenmann, J. E., Zanesco, S., Arnold, U., and Froesch, E. R. (1984c). *Acta Endocrinol.* **105,** 289.
Eigenmann, J. E., De Bruijne, J. J., and Froesch, E. R. (1985). *Acta Endocrinol. (Copenh.)* **108,** 161.
Eipper, B. A., and Mains, R. E. (1980). *Endocr. Rev.* **1,** 1.
Enright, W. J., Zinn, S. A., Chapin, L. T., and Tucker, H. A. (1987). *Proc. Soc. Exp. Biol. Med.* **184,** 483.
Eriksson, M., Lindén, A., Stock, S., and Uvnäs-Moberg, K. (1987). *Peptides* **8,** 411.
Feldman, E. C. (1983). *J. Am. Vet. Med. Assoc.* **183,** 195.
Feldman, E. C., and Nelson, R. W. (1996). *In* "Canine and Feline Endocrinology and Reproduction." W. B. Saunders Co., Philadelphia.
Feldman, E. C., Bohannon, N. V., and Tyrrell, J. B. (1977). *Am. J. Vet. Res.* **38,** 1643.
Flint, A. P. F., and Sheldrick, E. L. (1982). *Nature* **297,** 587.
Forage, R. G., Ring, J. M., Brown, R. W., McInerney, B. V., Cobon, G. S., Gregson, R. P., Robertson, D. M., Morgan, F. J., Hearn, M. T. W., Findlay, J. K., Wettenhall, R. E. H., Burger, H. G., and de Kretser, D. M. (1986). *Proc. Natl. Acad. Sci. USA* **83,** 3091.
Forsling, M. L. (1986). *Curr. Top. Neurobiol.* **6,** 19.
Fremeau, R. T., Lundblad, J. R., Pritchett, D. B., Wilcox, J. N., and Roberts, J. L. (1986). *Science* **234,** 1265.
French, M. B., Vaitkus, P., Cukerman, E., Sirek, A., and Sirek, O. V. (1987). *Am. J. Physiol.* **252,** E268.
Friesen, H. G., Gertler, A., Walker, A., and Elsholtz, H. (1985). *In* "Prolactin. Basic and Clinical Correlates" (R. M. MacLeod, M. O. Thorner, and U. Scapagnini, eds.), Fidia Research Series, No. 1. Liviana Press, Padova.
Gembardt, C., and Loppnow, H. (1976). *Berl. Münch. Tierärtl. Wschr.* **89,** 336.
Goossens, M. M. C., Rijnberk, A., Mol, J. A., Wolfswinkel, J., and Voorhout, J. (1995). *J. Vet. Int. Med.* **9,** 361.
Gorski, J., Shull, J., Weber, J., and Durrin, L. (1985). *In* "Prolactin. Basic and Clinical Correlates" (R. M. MacLeod, M. O. Thorner, and U. Scapagini, eds.), Fidia Research Series, Vol. 1, p. 259. Liviana Press, Padova.
Grossman, A., Stubbs, W. A., and Gaillard, R. C. (1981). *Clin. Endocrinol.* **14,** 381.
Grootenhuis, A. J., Van Sluijs, F. J., Klaij, I. A., Steenbergen, J., Timmerman, M. A., Bevers, M. M., Dieleman, S. J., and De Jong, F. H. (1990). *J. Endocrinol.* **127,** 235.
Guillaume, V., Conte-Devolx, B., Magnan, E., Boudouresque, F., Grino, M., Cataldi, M., Muret, L., Priou, A., Deprez, P., Figaroli, J. C., and Oliver, C. (1992). *Endocrinology* **130,** 2291.
Guillemin, R., Brazeau, P., Böhlen, P., Esch, F., Ling, N., and Wehrenberg, W. B. (1982). *Science* **218,** 585.
Hall, T. R., Harvey, S., and Scanes, C. G. (1986). *Comp. Biochem. Physiol.* **84A,** 231.
Halmi, N. S., and Krieger, D. (1983). *In* "The Anterior Pituitary Gland" (A. S. Bhatnagar, ed.). Raven, New York.
Halmi, N. S., Peterson, M. E., Colurso, G. J., Liotta, A. S., and Krieger, D. T. (1981). *Science* **211,** 72.
Harvey, S., and Scanes, C. G. (1987). *Comp. Biochem. Physiol.* **87A,** 315.
Harvey, S., Lam, S.-K., and Hall, T. R. (1986). *J. Endocrinol.* **111,** 91.
Harvey, S., Hoshino, S., and Suzuki, M. (1987). *Gen. Comp. Endocrinol.* **65,** 92.
Hellebrekers, L. J., Mol, J. A., Van den Brom, W. E., and Van Wimersma Greidanus, Tj. B. (1987). *Eur. J. Pharm.* **136,** 279.
Hellebrekers, L. J., Lagerweij, E., De Vries, H. W., and Van Wimersma Greidanus, Tj. B. (1988). *Acta Endocrinol.* **119,** 106.
Himura, Y., Liang, C. S., Imai, N., Delehanty, J. M., Woolf, P. D., and Hood, W. B. (1994). *J. Am. Coll. Cardiol.* **23,** 194.
Hodson, N. P. Wright, J. A., and Hunt, J. (1986). *Vet. Record* **118,** 148.
Houston, D. M., Allen, D. G., Kruth, S. A., Pook, H., Spinato, M. T., and Keough, H. (1989). *Can. Vet. J.* **30,** 423.
Höllt, V., and Haarmann, I. (1984). *Biochem. Biophys. Res. Comm.* **124,** 407.
Hullinger, R. L. (1979). *In* "Miller's Anatomy of the Dog," 2nd Ed. (H. E. Evans and G. C. Christensen, eds.), p. 602. Saunders, Philadelphia, Pennsylvania.
Huybrechts, L. M., Kühn, E. R., Decuypere, E., Mérat, P., and Scanes, C. G. (1987). *Reprod. Nutr. Dévelop.* **27,** 547.
Hyde, J. E., Murai, J., and Ben-Jonathan, N. (1987). *Endocrinology* **121,** 1531.
Irvine, C. H. G., and Alexander, S. L. (1986). *J. Endocrinol.* **113,** 183.
Ishibashi, M., and Yamaji, T. (1985). *In* "Prolactin. Basic and Clinical Correlates" (R. M. MacLeod, M. O. Thorner, and U. Scapagnini, eds.), Fidia Research Series, Vol. 1, p. 693. Liviana Press, Padova.
Ivell, R. (1986). *Curr. Top. Neuroendocrinol.* **6,** 1.
Ivell, R., and Richter, D. (1985). *In* "Oxytocin, Clinical and Laboratory Studies" (J. A. Amico and A.G. Robinson, eds.), p. 115. Elsevier, Amsterdam.
Ivell, R., Brackett, K., Fields, M. J., and Richter, D. (1985). *FEBS Lett.* **190,** 263.
Jacobson, L., and Drouin, J. (1994). *In* "The Pituitary Gland" (H. Imura, ed.), p. 117. Raven, New York.
Jacobson, L., and Sapolsky, R. (1991). *Endocr. Rev.* **12,** 118.
Janssens, L. A. (1981). *Tijdschr. Diergeneeskd.* **106,** 767.
Jard, S. (1983). *Progr. Brain Res.* **60,** 383.
Jephcott, E. H., McMillen, I. C., Rushen, J. P., and Thorburn, G. D. (1987). *Res. Vet. Sci.* **43,** 97.
Keller-Wood, M. (1989). *Am. J. Physiol.* **257,** R427.
Keller-Wood, M. E., and Dallman, M. F. (1984). *Endocr. Rev.* **5,** 1.
Kemppainen, R. J., and Sartin, J. L. (1984). *J. Endocrinol.* **103,** 219.
Kemppainen, R. J., and Sartin, J. L. (1986). *Acta Endocrinol. (Copenh.)* **113,** 471.
Kemppainen, R. J., and Sartin, J. L. (1987). *Front. Horm. Res.* **17,** 18.
Kemppainen, R. J., Thompson, F. N., Lorenz, M. D., Munnell, J. F., and Chakraborty, P. K. (1983). *J. Endocrinol.* **96,** 293.
Kemppainen, R. J., Filer, D. V., Sartin, J. L., and Reed, R. B. (1986). *Acta Endocrinol. (Copenh.)* **112,** 12.
Klingler, W., Sustarsic, D., ElShami, A. S., Williams, D. A., and Hoppen, H. O. (1995). *Proceedings WSAVA,* 696.
Knight, P. J., Hamilton, J. M., and Hiddleston, W. A. (1977). *Vet. Res.* **101,** 202.
Koike, T. I., Neldon, H. L., McKay, D. W., and Rayford, P. L. (1986). *Gen. Comp. Endocrinol.* **63,** 93.
Kovács, G. L. (1986). *Curr. Top. Neurobiol.* **6,** 91.
Krieger, D. T. (1980). *Recent Prog. Horm. Res.* **36,** 277.
Krieger, D. T. (1983). *Endocr. Rev.* **4,** 22.
Krysl, L. J., Branine, M. E., Galyean, M. L., Estell, R. E., and Hoefler, W. C. (1987). *J. Anim. Sci.* **64,** 1178.
Lamberts, S. W. J. (1987). *Front. Horm. Res.* **17,** 93.
Lapierre, H., Petitclerc, D., Dubreuil, P., Pelletier, G., Gaudreau, P., Morisset, J., Couture, Y., and Brazeau, P. (1987). *Reprod. Nutr. Dévelop.* **27,** 60.
Larsen, P. R. (1982). *N. Engl. J. Med.* **306,** 23.
Lauteriq, T. J., Trivedi, B., Kapadia, M., and Daughaday, W. H. (1988). *Comp. Biochem. Physiol.* **91a,** 15.

Lee, C. Y., Chung, C. S., and Simmen, F. A. (1993a). *Mol. Cell. Endocrinol.* **93,** 71.

Lee, C. Y., Bazer, F. W., and Simmen, F. A. (1993b). *J. Endocrinol.* **137,** 473.

Li, W. I., and Chen, C. L. (1987). *Life Sci.* **40,** 1411.

Liard, J.-F. (1986). *Clin. Sci.* **71,** 713.

Lilly, M. P., Engeland, W. C., and Gann, D. S. (1983). *Endocrinology* **112,** 681.

Lincoln, G. A., and Clarke, I. J. (1994). *J. Neuroendocrinol.* **6,** 251.

Liu *et al.* (1991). p. 10.

Liu, J., Clarke, I. J., Funder, J. W., and Engler, D. (1994). *J. Clin. Invest.* **93,** 1439.

Loeffler, J. Ph., Demeneix, B. A., Pittius, C. W., Kley, N., Haegele, K. D., and Höllt, V. (1986). *Peptides* **7,** 253.

Lowry, P. J., Linton, E. A., and Jackson, S. (1987). *Front. Horm. Res.* **17,** 1.

Lucy, M. C., Hauser, S. D., Eppard, P. J., Krivi, G. G., Clark, J. H., Bauman, D. E., and Collier, R. J. (1993). *Domest. Anim. Endocrinol.* **10,** 325.

McNeilly, A. S., Glasier, A., Jonassen, J., and Howie, P. W. (1982). *J. Reprod. Fertil.* **65,** 55.

McNicol, A. M. (1986). *J. Pathol.* **150,** 169.

Matteri, R. L., Roser, J. F., Baldwin, D. M., Lipovetsky, V., and Papkoff, H. (1987). *Domest. Anim. Endocrinol.* **2,** 157.

Matthews, S. G., and Parrott, R. F. (1992). *Exp. Physiol.* **77,** 357.

Majzoub, J. A. (1985). *In* "Vasopressin" (R. W. Schrier, ed.), p. 465. Raven Press, New York.

Meij, B. P., Rijnberk, A., and Mol, J. A. (1990). *J. Endocrinol.* **127,** 265.

Meij, B. P., Mol, J. A., Hazewinkel, H. A. W., Bevers, M. M., and Rijnberk, A. (1996). *Domest. Anim. Endocr.* **13,** 61.

Meijer, J. C., De Bruijne, J. C., Rijnberk, A., and Croughs, R. J. M. (1978). *J. Endocrinol.* **77,** 111.

Melmed, S., Braunstein, G. D., Horvath, E., Ezrin, C., and Kovacs, K. (1983). *Endocr. Rev.* **4,** 271.

Mialot, J. P., Lagneau, F., Chaffaux, St., and Badinand, F. (1981). *Rec. Méd. vét.* **157,** 351.

Middleton, D. J., Rijnberk, A., Bevers, M. M., Goos, H. J. Th., Beeftink, E. A., Thijssen, J. H. H., and Croughs, R. J. M. (1987a). *Front. Horm. Res.* **17,** 10.

Middleton, D. J., Rijnberk, A., Bevers, M. M., Goos, H. J. Th., Beeftink, E. A., Thijssen, J. H. H., and Croughs, R. J. M. (1987b). *Horm. Metab. Res.* **19,** 632.

Mikami, S.-J. (1986). *Internat. Rev. Cytology* **103,** 189.

Miller, K. F., and Bolt, D. J. (1987). *Domest. Anim. Endocrinol.* **4,** 1.

Millington, W. R., Dybdal, N. O., Mueller, G. P., and Chrinwall, B. M. (1992). *Gen. Comp. Endocrinol.* **85,** 297.

Minton, J. E., and Parsons, K. M. (1993). *J. Anim. Sci.* **71,** 724.

Mol, J. A., Slob, A., Middleton, D. J., and Rijnberk, A. (1987). *Front. Horm. Res.* **17,** 61.

Mol, J. A., van Mansfeld, A. D., Kwant, M. M., van Wolferen, M., and Rothuizen, J. (1991). *Acta Endocrinol. (Copenh.)* **125,** 77.

Mol, J. A., van Wolferen, M., Kwant, M. M., and Meloen, R. (1994). *Neuropeptides* **27,** 7.

Mol, J. A., van Garderen, E., Selman, P. J., Wolfswinkel, J., Rijnberk, A., and Rutteman, G. R. (1995a). *J. Clin. Invest.* **95,** 2028.

Mol, J. A., Henzen-Logmans, S. C., Hageman, Ph., Misdorp, W., Blankenstein, M. R., and Rijnberk, A. (1995b). *J. Clin. Endocr. Metab.* **80,** 3094.

Mol, J. A., van Garderen, E., Rutteman, G. R., and Rijnberk, A. (1996). *J. Steroid Biochem. Mol. Biol.* **57,** 67.

Mol, J. A., Selman, P. J., Sprang, E. A. P., Oosterlaken-Dijksterhuis, M., and Van Neck, H. (1997). Submitted.

Mulchahey, J. J., DiBlasio, A. M., Martin, M. C., Blumenfeld, Z., and Jaffe, R. B. (1987). *Endocr. Rev.* **8,** 406.

Müller-Peddinghaus, R., El Etreby, M. F., Siefert, J., and Ranke, M. (1980). *Vet. Pathol.* **17,** 406.

Mulnix, J. A., Rijnberk, A., and Hendriks, H. J. (1976). *J. Amer. Vet. Med. Assoc.* **169,** 1327.

Muret, L., Priou, A., Oliver, C., and Grino, M. (1992). *Endocrinology* **130,** 2725.

Muruais, J., Penalva, A., Dieguez, C., and Casanueva, F. F. (1993). *J. Endocr.* **138,** 211.

Nachreiner, R. E., Forsberg, M., Johnson, C. A., and Refsal, K. R. (1995). *Am. Coll. Vet. Internal Med.,* 184.

Naor, Z., Amsterdam, A., and Catt, K. J. (1984). *In* "Hormone Receptors in Growth and Reproduction" (B. B. Saxena, ed.), p. 1. Raven, New York.

Nap, R. C., Mol, J. A., and Hazewinkel, H. A. W. (1993). *Domest. Anim. Endocrinol.* **10,** 237.

Newman, C. B., Wardlaw, S. L., Stark, R. I., Daniel, S. S., and Frantz, A. G. (1987). *Endocrinology* **120,** 962.

Nicoll, C. S., Anderson, T. R., Hebert, N. J., and Russell, S. M. (1985). *In* "Prolactin. Basic and Clinical Correlates" (R. M. MacLeod, M. O. Thorner, and U. Scapagnini, eds.), Fidia Research Series, Vol. 1. Liviana Press, Padova.

Nilsson, A., Isgaard, J., Lindahl, A., Dahlström, A. Skottner, A., and Isaksson, O. G. P. (1986). *Science* **233,** 571.

Numa, S. and Imura, H. (1985). *In* "The Pituitary Gland" (H. Imura, ed.), p. 83. Raven Press, New York.

Nussey, S. S, Ang, V. T. Y., Finer, N., and Jenkins, J. S. (1986). *Clin. Endocrinol.* **24,** 97.

Oertel, W. H., Mugaini, E., Tappaz, M. L., Weise, V. K., Dahl, A., Schmechel, D. E., and Kopin, I. J. (1982). *Proc. Natl. Acad. Sci. USA* **79,** 675.

Okkens, A. C., Bevers, M. M., Dieleman, S. J., and Willemse, A. H. (1985). *Vet. Quart.* **7,** 173.

Olson, K. C., Fenno, J., and Lin, N. (1981). *Nature* **293,** 408.

Orth, D. N., Holscher, M. A., Wilson, M. G., Nicholson, W. E., Plue, R. E., and Mount, C. D. (1982). *Endocrinology* **110,** 1430.

Page, R. B. (1986). *In* "Morphology of Hypothalamus and Its Connections" (D. Ganten and D. Pfaff, eds.), p. 1. Springer Verlag, Berlin.

Payne, J. H., and Cooke, R. G. (1993). *Res. Vet. Sci.* **55,** 38.

Parker, W. M., and Scott, D. W. (1980). *J. Am. Anim. Hosp. Assoc.* **16,** 824.

Pecins-Thompson, M., and Keller-Wood, M. (1994). *Endocrinology* **134,** 678.

Peterson, M. E. (1984). *Vet. Clin. North Am.* **14,** 731.

Peterson, M. E. (1986). *In* "Current Veterinary Therapy IX" (R. W. Kirk, ed.), p. 963. Saunders, Philadelphia, Pennsylvania.

Peterson, M. E. (1987a). *Front. Horm. Res.* **17,** 37.

Peterson, M. E. (1987b). *Res. Vet. Sci.* **42,** 354.

Peterson, M. E., and Altszuler, N. (1981). *Am. J. Vet. Res.* **42,** 1881.

Peterson *et al.* (1984). p. 17.

Peterson, M. E., Orth, D. N., Halmi, N. S., Zielinski, A. C., Davis, D. R., Chavez, F. T., and Drucker, W. D. (1986). *Endocrinology* **119,** 720.

Peterson, M. E., Taylor, R. S., Greco, D. S., Nelson, R. W., Randolph, J. F., Foodman, M. S., Moroff, S. D., Morrison, S. A., and Lothrop, C. D. (1990). *J. Vet. Int. Med.* **4,** 192.

Peterson, M. E., Kemppainen, R. J., and Orth, D. N. (1992). *Am. J. Vet. Res.* **53,** 421.

Peterson, M. E., Kemppainen, R. J., and Orth, D. N. (1994). *Am. J. Vet. Res.* **55,** 295.

Plotsky, P. M. (1986). *Regul. Peptides* **16,** 235.

Porter, D. G., Ryan, P. L., and Norman, L. (1992). *J. Reprod. Fert.* **96,** 251.

Pradier, P., Davicco, M. J., Safwate, A., Tournaire, C., Dalle, M., Barlet, J. P., and Delost, P. (1986). *Acta Endocrinol.* **111,** 93.

Proulx-Ferland, L., Labrie, F., Dumont, D., Côté, J., Coy, D. H., and Sveiraf, J. (1982). *Science* **217,** 62.

Quillen, E. W., and Cowley, A. W. (1983). *Am. J. Physiol.* **244,** H73.

Raff, H., Skelton, M. M., and Cowley, A. W. (1988). *J. Endocrinol.* **122,** 41.

Raff, H., Papanek, P. E., and Cowley, A. W. (1992). *Am. J. Physiol.* **263,** R382.

Rao, J. K., Hu, H., Prasad, A., and Jayaraman, A. (1986). *Neurosci. Lett.* **71,** 48.

Raschler, W., Rauh, W., Brandeis, W. E., Huber, K.-H., and Schärer, K. (1986). *Acta Paediatr. Scand.* **75,** 111.

Rechler, M. M., and Nissley, S. P. (1985). *Ann. Rev. Physiol.* **47,** 425.

Redekopp, C., Irvine, C. H. G., Donald, R. A., Livesey, J. H., Sadler, W., Nicholls, M. G., Alexander, S. L., and Evans, M. J. (1986a). *Endocrinology* **118,** 1410.

Redekopp, C., Livesey, J. H., Sadler, W., and Donald, R. A. (1986b). *J. Endocrinol.* **108,** 309.

Reichlin, S. (1982). *N. Engl. J. Med.* **309,** 1495.

Richter, D., and Ivell, R. (1985). *In* "The Pituitary Gland" (H. Imura ed.). Raven, New York.

Rijnberk, A. (1971). Thesis, University of Utrecht, Elinkwijk, Utrecht.

Rijnberk, A. (1996). *In* "Clinical Endocrinology of Dogs and Cats." Kluwer Academic Publishers, Boston.

Rijnberk, A., Eigenmann, J. E., Belshaw, B. E., Hampshire, J., and Altszuler, N. (1980). *J. Am. Vet. Med. Assoc.* **177,** 534.

Rijnberk, A., Mol, J. A., Rothuizen, J., Bevers, M. M., and Middleton, D. J. (1987). *Front. Horm. Res.* **17,** 48.

Rijnberk, A., Biewenga, W. J., and Mol, J. A. (1988a). *Acta Endocrinol. (Copenh.),* **117,** 59.

Rijnberk, A., Mol, J. A., Kwant, M. M., and Croughs, R. J. (1988b). *J. Endocrinol.* **118,** 271.

Rijnberk, A., van Herpen, H., Mol, J. A., and Rutteman, G. R. (1993). *Vet. Rec.* **133,** 542.

Rivier, J., Spiess, J., Thorner, M., and Vale, W. (1982). *Nature* **300,** 276.

Robertson, G. L. (1978). *In* "The Year Book in Endocrinology" (S. H. Ingbar, ed.), p. 205. Plenum, New York.

Robertson, G. L. (1983). *J. Lab. Clin. Med.* **101,** 351.

Rodriguez-Arnao, J., Miell, J. P., and Ross, R. J. M. (1993). *Trends Endocrinol. Metab.* **4,** 169.

Rosenblatt, J. S. (1984). *In* "Prolactin Secretion: A Multidisciplinary Approach" (F. Mena and C. M. Valverde, eds.). Academic Press, New York.

Rothuizen, J., Reul, J. M., van Sluijs, F. J., Mol, J. A., Rijnberk, A., and de Kloet, E. R. (1993). *Endocrinology* **132,** 161.

Rothuizen, J., Biewenga, W. J., and Mol, J. A. (1995). *Domest. Anim. Endocr.* **12,** 13.

Rutteman, G. R., Stolp. R., Rijnberk, A., Loeffler, S., Bakker, J. A., Bevers, M. M., Meulenberg, P. M. M., and Eigenmann, J. E. (1987a). *Acta Endocrinol.* **114,** 275.

Rutteman, G. H., Rijnberk, A., and Misdorp, W. (1987b). *Front. Horm. Res.* **17,** 173.

Sawchenko, P. E., and Swanson, L. W. (1983). *Progr. Brain Res.* **60,** 19.

Scanes, C. G., Denver, R. J., and Bowen, S. J. (1986). *Poultry Sci.* **65,** 384.

Scapagnini, U., Drago, F., Continella, G., Spadaro, F., Pennisi, G., and Gerenday, I. (1985). *In* "Prolactin. Basic and Clinical Correlates" (R. M. MacLeod, M. O. Thorner, and U. Scapagnini, eds.), Fidia Research Series, Vol. 1. Liviana Press, Padova.

Schams, D. (1983). *Acta Endocrinol.* **103,** 180.

Schimchowitsch, S., Palacios, J. M., Stoeckel, M. E., Schmitt, G., and Porte, A. (1986). *Neuroendocrinology* **42,** 71.

Scholten-Sloof, B. E., Knol, B. W., Rijnberk, A., Mol, J. A., Middleton, D. J., and Ubbink, G. J. (1992). *J. Endocrinol.* **135,** 535.

Schopohl, J., Losa, M., König, A., Müller, O. A., Stalla, G. K., and Von Werder, K. (1986). *Klin. Wochenschr.* **64,** 314.

Schusdziarra, V., Harris, V., and Unger, R. H. (1979). *Endocrinology* **104,** 109.

Scott, D. W., Kirk, R. W., Hampshire, J., and Altszuler, N. (1978). *J. Am. Anim. Hosp. Assoc.* **14,** 183.

Seiferle (1975). Fig. 19.2.

Selman, P. J., Mol, J. A., Rutteman, G. R., and Rijnberk, A. (1991). *Acta Endocrinol. (Copenh.)* **125,** 42.

Selman, P. J., Mol, J. A., Rutteman, G. R., van Garderen, E., and Rijnberk, A. (1994a). *Endocrinology* **134,** 287.

Selman, P. J., Mol, J. A., Rutteman, G. R., and Rijnberk, A. (1994b). *Eur. J. Endocrinol.* **131,** 413.

Seo, H. (1985). *In* "The Pituitary Gland" (Imura, H. ed.), p. 57. Raven Press, New York.

Shaar, C. J., Grinnan, E. L., Short, W. G., Powell, J. G., Bryan, N., Bemis, K. G., and Tinsley, F. C. (1986). *Endocrinol. Res.* **12,** 21.

Sherman, G. B., Wolfe, M. W., and Farmerie, T. A. (1992). *Mol. Endocrinol.* **6,** 951.

Shimada, K., Neldon, H. L., and Koike, T. I. (1986). *Gen. Comp. Endocrinol.* **64,** 362.

Shimada, K., Saito, N., Itogawa, K., and Koike, T. I. (1987). *J. Reprod. Fert.* **80,** 143.

Shimatsu, A., Kato, Y., Ohta, H., Tojo, K., Kabayama, Y., Inoue, T., and Imura, H. (1985). *In* "Prolactin. Basic and Clinical Correlates" (R. M. MacLeod, M. O. Thorner, and U. Scapagini, eds.), Fidia Research Series, Vol. 1, p. 73. Liviana Press, Padova.

Shiver, T., Familari, M., and Aguilera, G. (1992). *Endocrinology* **130,** 2299.

Short, R. E., Brooks, A. N., Peters, A. R., and Lamming, G. E. (1987). *J. Reprod. Fert.* **80,** 213.

Simon-Oppermann, C. S., Gray, D., Szczepanska-Sadowska, E., and Simon, E. (1983). *Am. J. Physiol.* **245,** R541.

Smith, R., Owens, P. C., Lovelock, M., Cheng Chan, E., and Falconer, J. (1986). *Endocrinology* **118,** 2572.

Stokhof, A. A., Overduin, L. M., Mol, J. A., and Rijnberk, A. (1994). *Eur. J. Endocrinol.* **130,** 357.

Stolp, R., Croughs, R. J. M., Meijer, J. C., and Rijnberk, A. (1982). *J. Endocrinol.* **93,** 365.

Stolp, R., Bevers, M. M., Rijnberk, A., Croughs, R. J. M., and Rutteman, G. R. (1986). *Horm. Metab. Res.* **18,** 595.

Stolp, R., Steinbusch, H. W. M., Rijnberk, A., and Croughs, R. J. M. (1987). *Neurosci. Lett.* **74,** 337.

Sweep *et al.* (1991). p. 11.

Szczepanska-Sadowska, E., Gray, D., and Simon-Opperman, C. (1983). *Am. J. Physiol.* **245,** R549.

Takahashi, Y., Ebihara, S., Nakamura, Y., and Takahashi, K. (1981). *Endocrinology* **109,** 262.

Takahashi, H., Bando, H., Zhang, C., Yamasaki, R., and Saito, S. (1992). *Acta Endocrinol.* **127,** 13.

Tatro, J. B., and Reichlin, S. (1987). *Endocrinology* **121,** 1900.

Taverne, M., Willemse, A. H., Dieleman, S. J., and Bevers, M. (1978/1979). *Anim. Reprod. Sci.* **1,** 257.

Ten Haaf, J. A., Van Wimersma Greidanus, Tj. B., Maigret, C., and De Wied, D. (1986). *Neuroendocrinol.* **44,** 102.

Thornton, S. N., Leng, G., Bicknell, R. J., Chapman, C., and Purdew, T. (1986). *J. Endocrinol.* **110,** 335.

Thornton, S. N., Parrott, R. F., and Delaney, C. E. (1987). *Acta Endocrinol. (Copenh.)* **114,** 519.

Tilders, F. J. H., Berkenbosch, F., and Smelik, P. G. (1985). *Front. Horm. Res.* **14,** 161.

Ultee-van Gessel, A. M., and De Jong, F. H. (1987). *J. Endocrinol.* **113,** 103.

Van Haaften, B., Bevers, M. M., Van den Brom, W. E., Okkens, A. C., Van Sluijs, F. J., Willemse, A. H., and Dieleman, S. J. (1994). *J. Reprod. Fert.* **101,** 221.

Van Herpen, H., Rijnberk, A., and Mol, J. A. (1994). *Vet. Rec.* **134,** 171.
Van Leeuwen, F. W., and De Vries, G. J. (1983). *Progr. Brain Res.* **60,** 343.
Van Wijk, P. A., Rijnberk, A., Croughs, R. J., Voorhout, G., Sprang, E. P., Mol, J. A. (1992). *Endocrinology* **131,** 2659.
Van Wijk, P. A., Rijnberk, A., Croughs, R. J., Wolfswinkel, J., Selman, P. J., and Mol, J. A. (1994). *Eur. J. Endocrinol.* **130,** 410.
Van Wimersma Greidanus, Tj. B. (1987). *Front. Horm. Res.* **17,** 61.
Verbalis, J. G., McCann, M. J., McHale, C. M., and Stricker, E. M. (1986). *Science* **232,** 1417.
Verburg-van Kemenade, B. M. L., Tonon, M. C., Jenks, B. G., and Vaudry, H. (1986). *Neuroendocrinol.* **44,** 446.
Verde, L. S., and Trenkle, A. (1987). *J. Anim. Sci.* **64,** 426.
Verhaagen, J., Edwards, P. M., Jennekens, F. G. I., Schotman, P., and Gispen, W. H. (1987). *Brain Res.* **404,** 142.
Visser, T. J., and Lamberts, S. W. J. (1981). *Acta Endocrinol.* **96,** 480.
Visser, M., and Swaab, D. F. (1979). *J. Dev. Physiol.* **1,** 161.
Vokes, T., and Robertson, G. L. (1985a). *In* "Neuroendocrine Perspectives," Vol. 4 (E. E. Müller, R. M. MacLeod, and L. A. Frohman, eds.), p. 1. Elsevier, Amsterdam.
Vokes, T., and Robertson, G. L. (1985b). *In* "Vasopressin" (R. W. Schrier, ed.), p. 271. Raven, New York.
Wachs, E. A., Gorewit, R. C., and Currie, W. B. (1984). *Domest. Anim. Endocrinol.* **1,** 141.
Wade, C. E. (1985). *Am. J. Physiol.* **249,** E634.
Wade, C. E., Bie, P., Keil, L. C., and Ramsay, D. J. (1982). *Am. J. Physiol.* **243,** E287.
Wade, C. E., Keil, L. C., and Ramsay, D. J. (1983). *Neuroendocrinology* **37,** 349.
Wallis, M. (1984). *In* "Prolactin Secretion, A Multidisciplinary Approach" (F. Mena and C. Valverde, eds.), p. 1. Academic Press, New York.
Walton, P. E., Etherton, T. D., and Chung, C. S. (1987). *Domest. Anim. Endocrinol.* **2,** 183.
Wang, B. C., and Goetz, K. L. (1985). *In* "Vasopressin" (R. W. Schrier, ed.), p. 221. Raven, New York.
Watson, A. D. J., Rutteman, G. R., Rijnberk, A., and Mol, J. A. (1987). *Front. Horm. Res.* **17,** 145.
Wheaton, J. E., Al-Raheem, S. N., and Marcek, J. M. (1986). *J. Anim. Sci.* **62,** 1267.
Willemse, T., and Mol, J. A. (1994). *Am. J. Vet. Res.* **55,** 1677.
Willemse, T., Vroom, M. W., Mol, J. A., and Rijnberk, A. (1993). *Am. J. Vet. Res.* **54,** 69.
Wilson, M. G., Nicholson, W. E., Holscher, M. A., Sherrel, B. J., Mount, C. D., and Orth, D. N. (1982). *Endocrinology* **11,** 941.
Winter, M., Pirmann, J., Falvo, R. E., Schanbacher, B. D., and Miller, J. (1982). *J. Reprod. Fert.* **64,** 449.
Wulffraat, N. M., Drexhage, H. A., Jeucken, P., Van der Gaag, R. D., and Wiersinga, W. M. (1987). *J. Endocrinol.* **115,** 505.
Young, D. W., and Kemppainen, R. J. (1994). *Am. J. Vet. Res.* **55,** 567.
Young, D. W., Zerbe, C. A., and Kemppainen, R. J. (1992). *Peptides* **13,** 1061.

CHAPTER

20

Adrenocortical Function

AD RIJNBERK AND JAN A. MOL

I. INTRODUCTION

In their daily activities the authors of this chapter are not involved with livestock or with experimental animals. They have accepted the invitation of the editor to contribute to this book on clinical biochemistry under the condition that they could confine themselves largely to companion animals. This seems justified, as most of the clinical problems of the adrenal cortex are seen in these species. Other species are covered in terms of general comparative aspects of mammals and birds and with the presentation of some additional species-specific data.

II. ANATOMICAL CONSIDERATIONS

In mammals the adrenal glands are bilateral structures located craniomedial to the kidneys. The adrenal is, in fact, made up of two glands of different embryological origin. The center of the gland, the medulla, comprises coalesced chromaffin cells of neuroectodermal origin that secrete epinephrine and/or norepinephrine. The surrounding adrenal cortex arises from mesoderm and can be divided histologically into three zones: (1) zona glomerulosa (or arcuata), (2) zona fasciculata, and (3) zona reticularis. The cells of the zona glomerulosa are arranged in ovoid groups immediately inside the connective tissue capsule. The region may be particularly obvious (e.g., sheep) or difficult to discern (e.g., small rodents). The zona fasciculata accounts for most of the adrenal cortex (>60%) and appears histologically in radial lines. The inner zona reticularis comprises networks of cell cords surrounding the large blood sinuses.

In birds the adrenal glands are also separate, encapsulated organs. The glands lie cranial to the kidney, often wholly or partly covered by the gonads. The adrenocortical and chromaffin tissues are intermingled to varying degrees, depending on the species. The basic unit of avian adrenocortical tissue is a cord of a double row of parenchymal cells, with their long axes in the transverse plane of the strand. The cords radiate from the center of the gland, branch, and anastomose frequently. Groups of chromaffin cells lie between the cords and also ensheathed within the connective tissue reticulum (Chester-Jones, 1987).

Blood flow of the adrenals is centripetal. In species with separated medulla and cortex, this results in high medullary corticoid concentrations that induce an enzyme in the chromaffin cells, converting norepinephrine to epinephrine. Therefore, species with anatomically separated medulla and cortex predominantly secrete epinephrine, and species with intermingled chromaffin and adrenocortical tissue secrete predominantly norepinephrine (Marks, 1983).

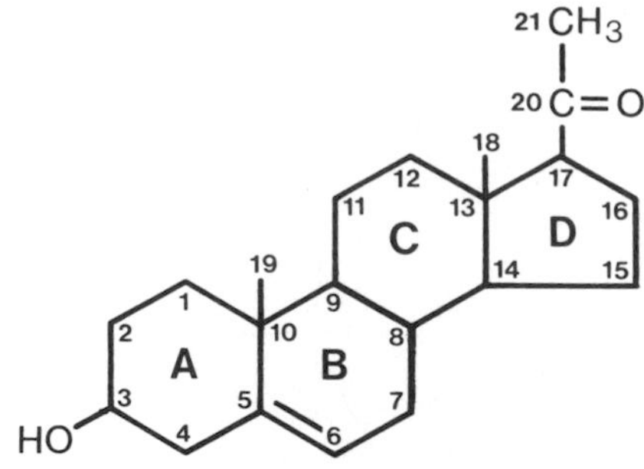

FIGURE 20.1 The numbers of the carbon atoms and the letters designating the rings of the pregnenolone molecule [recommendations of the International Union of Pure and Applied Chemistry; IUPAC-IUB 1967 (1968)].

III. PHYSIOLOGY OF ADRENOCORTICAL HORMONES

The secretion of the mammalian adrenal cortex comprises three main categories of hormones, which can be related to some extent to the anatomical zonation just described. The zona glomerulosa produces mineralocorticoids (aldosterone and deoxycorticosterone), which maintain salt balance. The cells of the zona fasciculata secrete glucocorticoids (cortisol and corticosterone), which are primarily involved in carbohydrate metabolism. The third category of adrenocorticol hormones, the androgens (e.g., androstenedione), are produced in the zona reticularis. This zone to a minor degree also secretes glucocorticoids and other hormones such as progesterone and estrogens.

In birds aldosterone and corticosterone are the main corticosteroids secreted. Zonation of the avian adrenal is less clear than in the mammalian adrenal. However, the outer subcapsular cells, looping in a manner similar to the zona glomerulosa, appear to be the predominant aldosterone secretors. The cells reaching toward the central part of the gland form corticosterone (Kime, 1987).

A. Steroid Nomenclature

The adrenal steroids contain as their basic structure a cyclopentanoperhydrophenanthrene nucleus consisting of three six-carbon rings (A, B, and C) and a single five-carbon ring (D). The letter designations for the carbon rings and the numbers of the carbon atoms are shown for pregnenolone (Fig. 20.1), a key biosynthetic intermediate. The Greek letter Δ indicates a double bond, as does the suffix -ene. The position of a substituent below or above the plane of the steroid ring is indicated by α and β, respectively. The α-substituent is drawn with a broken line (e.g., --OH), and the β-substituent with a solid line (e.g., —OH). The C_{18} steroids, which are devoid of a side chain at C-17 and have a substituent at C-18, are estrogens. The C_{19} steroids, which have substituent methyl groups at positions C-18 and C-19, are androgens (see also Fig. 20.2).

Steroids that have a ketone group at C-17 are termed 17-ketosteroids. The C_{21} steroids, the corticosteroids and progestagens, are those that have a two-carbon side chain (C-20 and C-21) attached at C-17 and, in addition, have substituent methyl groups at C-18 and C-19. The C_{21} steroids that also possess a hydroxyl group at position 17 are termed 17-hydroxycorticosteroids and may have predominantly glucocorticoid properties.

B. Biosynthesis

Cholesterol, derived from food and from endogenous synthesis via acetate (Fig. 20.2), is the principal starting compound in steroidogenesis. The adrenal gland is enriched in receptors that internalize low- and high-density lipoproteins. This uptake mechanism is increased when the adrenal is stimulated and provides the major cholesterol source. Subsequent steps occur in the mitochondrion or endoplasmic reticulum.

The zonal difference in adrenocortical hormone production is due to two steroidogenic enzyme systems. The mithochondrial cytochrome $P450_{CMO}$ (corticosterone methyloxidase), which converts corticosterone to aldosterone, is found only in the zona glomerulosa. The characteristic enzyme of the inner zones is the microsomal cytochrome $P450_{17\alpha}$ (17-hydroxylase/lyase), which catalyzes the 17α-hydroxylation of pregnenolone and progesterone as well as the side-chain fission of 17α-hydroxy C_{21} steroids. The other steroidogenic enzymes occur in all the three zones of the adrenal cortex (Müller, 1986).

The characteristic 11β- and 21-hydroxylations appear to have developed at a very early stage of evolu-

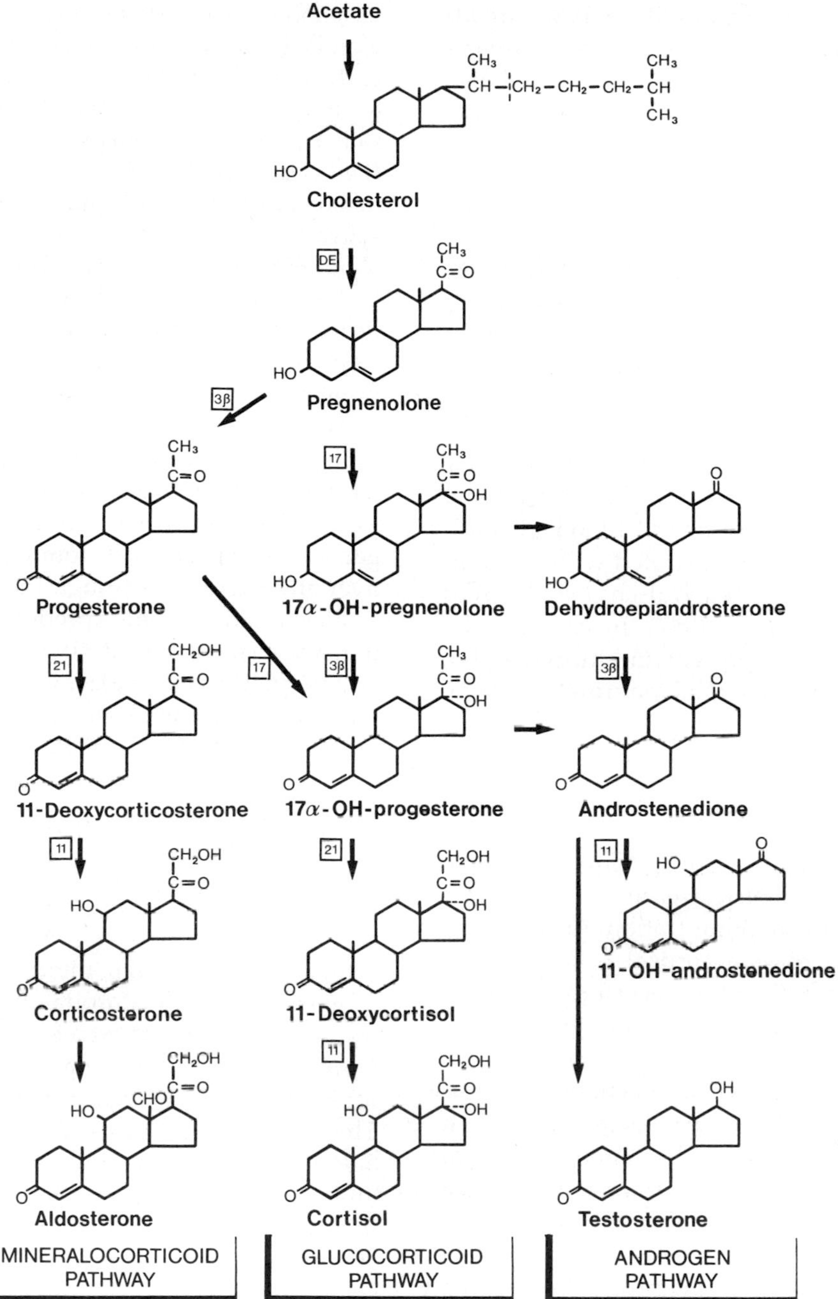

FIGURE 20.2 Biosynthetic pathways for adrenal steroid production. Letters and numbers by arrows denote specific enzymes: DE, debranching enzyme; 3*β*, 3*β*-ol-dehydrogenase with △4,5-isomerase; 11, C-11 hydroxylase; 17, C-17 hydroxylase; 21, C-21 hydroxylase.

tion and are present in all vertebrates. The 18-oxygenated corticosteroids also retain their importance as mineralocorticoids in all vertebrates. For the 17*α*-hydroxylase potential, the situation is different. Most mammals secrete cortisol as the predominant glucocorticoid. However, rodents and also birds secrete predominantly 17-deoxycorticosteroids such as corticosterone.

In line with this, steroid determinations in adrenal venous blood of dogs (Hirose *et al.*, 1977) have revealed cortisol/corticosterone ratios that range from about 3 to 7. Dor *et al.* (1973) have found ratios of secretion rates ranging from 1.2 to 2.7 and corticosterone/aldosterone secretion ratios ranging from 7 to 25. With HPLC analysis in peripheral blood of ACTH-stimulated dogs, the cortisol/corticosterone ratios ranged from 2.4 to 9.7

with a mean of 5.0 (Lothrop and Oliver, 1984). In addition to glucocorticoids and aldosterone, adrenal-vein blood of dogs and pigs contains androstenedione, 11-hydroxyandrostenedione, androsterone, pregnenolone, progesterone, and 11-hydroxyprogesterone (Heap *et al.*, 1966; Holzbauer and Newport, 1969) as well as very small amounts of estradiol and estrone (Dor *et al.*, 1973).

In cats the cortisol / corticosterone ratio has been estimated to range from 1.6 to 12.4, whereas in kittens the ratio is less than unity (Ilett and Lockett, 1969). In peripheral blood of horses, higher cortisol / corticosterone ratios (16:0.5 and 7:0.5) have been reported (Zolovick *et al.*, 1966; James *et al.*, 1970).

In birds, corticosterone, a glucocorticoid with minor mineralocorticoid properties, is by far the main adrenal secretory product. In the peripheral blood of adult birds no cortisol could be demonstrated with radioimmunoassay (Zenoble *et al.*, 1985; Walsch *et al.*, 1985) or HPLC analysis (Lumeij *et al.*, 1987). In chickens the 17α-hydrolyase activity is present in embryonic life but decreases after hatching (Nakamura *et al.*, 1978; Carsia *et al.*, 1987).

C. Transport

At normal concentrations, only about 10% of the total blood cortisol and corticosterone is in the free form, that is, susceptible to ultrafiltration. At body temperature, 70% of the plasma cortisol is bound to a globulin called transcortin or corticosteroid-binding globulin (CBG). Transcortin has a high affinity for cortisol and corticosterone, but its binding capacity is limited. Another 20% of plasma cortisol is bound to albumin, although its affinity for cortisol is much less than that of transcortin. In line with these percentages, in the dog the free fraction has been estimated to range from 5 to 12% (Kemppainen *et al.*, 1991; Meyer and Rothuizen, 1993).

Transcortin is ubiquitous in mammals, but plasma concentrations vary considerably, resulting in species differences in total cortisol concentration. Most domestic animals have little corticosteroid-binding activity compared to that of humans (Rosner, 1969). As the free rather than the protein-bound steroid is biologically active, methods have been developed to measure free cortisol. By employing the combination of ultrafiltration and equilibrium dialysis, Meyer and Rothuizen (1994) demonstrated that in dogs with portosystemic encephalopathy the associated hyperadrenocorticism is characterized not only by an increased total cortisol concentration in plasma, but also by an increase in the free fraction of plasma cortisol.

Unbound steroids readily diffuse into the salivary glands. Because of the close relationship between free cortisol in blood and saliva (Riad-Fahmy *et al.*, 1982), techniques have been developed for the collection of saliva from cattle (Murphy and Connell, 1970), sheep (Fell *et al.*, 1985), and dogs (Phillips *et al.*, 1983). Detailed background information on steroid hormones in saliva is available in Ferguson (1984). In cow's milk, about 60% of the cortisol is present in the ultafiltrate. Following parturition, the percentage of unbound cortisol (in colostrum) decreases to 40% (Shutt and Fell, 1985) owing to the higher concentrations of CBG-like protein.

The physiological significance of protein binding probably lies in a buffering effect, which prevents rapid variations of the plasma cortisol level. Transcortin restrains the active cortisol from reaching the target organ and also protects it from rapid inactivation by the liver and excretion through the kidneys.

Plasma aldosterone is predominantly bound to albumin, which has a low affinity. The relatively low degree of protein binding of plasma aldosterone partially explains the very low plasma concentration and the short biological half-life of this hormone.

D. Metabolic Breakdown and Excretion

Only unbound cortisol and its metabolites are filterable at the glomerulus. Most of this filtered cortisol is reabsorbed, whereby a tubular maximum is only achieved at very high filtered loads of free cortisol (Boonayathap and Marotta, 1974). Less than 20% of the filtered cortisol is excreted unchanged in the urine. Nevertheless, in most mammals the kidneys account for 50–80% of the excretion of the metabolized steroids. The remainder is lost via the gut. To render them suitable for renal elimination, the steroids are inactivated and made more water soluble through enzymatic modifications. The liver is the major organ responsible for steroid inactivation and conjugation to form water-soluble compounds, although in the dog—in contrast to the situation in humans—the kidney and the gastrointestinal tract also contribute to the metabolic clearance of cortisol (McCormick *et al.*, 1974). In the canine kidney, cortisol glucuronide is both secreted and reabsorbed, without a tubular maximum or a plasma threshold (Boonayathrap and Marotta, 1974).

Cortisol is cleared from the plasma with a half-life of 60 minutes or less. For pigs, the metabolic clearance rate of cortisol was calculated to be about 11 $hour^{-1}$ kg^{-1} (Hennesy *et al.*, 1986). In dogs about 60% of infused cortisol is eliminated within 24 hours in the urine (Rijnberk *et al.*, 1968). The 11β-hydroxyl group of cortisol can be oxidized to the ketone, forming cortisone (Fig. 20.3). the reaction is reversible, and in general the equilibrium is shifted to favor the 11β-hydroxyl group. However, because the adrenal cortex produces much more cortisol than cortisone (if any), there is substantial cortisol-to-cortisone conversion. These two steroids have similar subsequent metabolic fates.

Apart from this 11β-hydroxylation, cortisol metabolism in the dog involves the following: (i) reduction of ring A to tetrahydro derivatives, (ii) reduction of the 20-keto group to a hydroxyl, and (iii) conjugation with glucuronic acid to form glucuronidates (Gold, 1961). In addition, unconjugated 20-hydroxycortisol/cortisone has been found in canine urine.

In the total glucuronide fraction of urinary corticoids in dogs, at least 60% is represented by steroids reduced at C-20. This is of prime importance when it comes to assessment of adrenocortical function by measuring urinary cortisol metabolites. Measurements directed at steroids containing a 17α, 21-dihydroxy, 20-keto arrangement, i.e., the Porter–Silber reaction, will detect only a small part of the cortisol metabolites and therefore have limited value (Siegel, 1965). Instead, preferably, measurements are preformed involving reduction of the urine metabolites at C-20, followed by oxidation to 17-ketosteroids, which are then quantitated. Further details on this measurement of "total 17-hydroxycorticosteroids" are given in Section V.

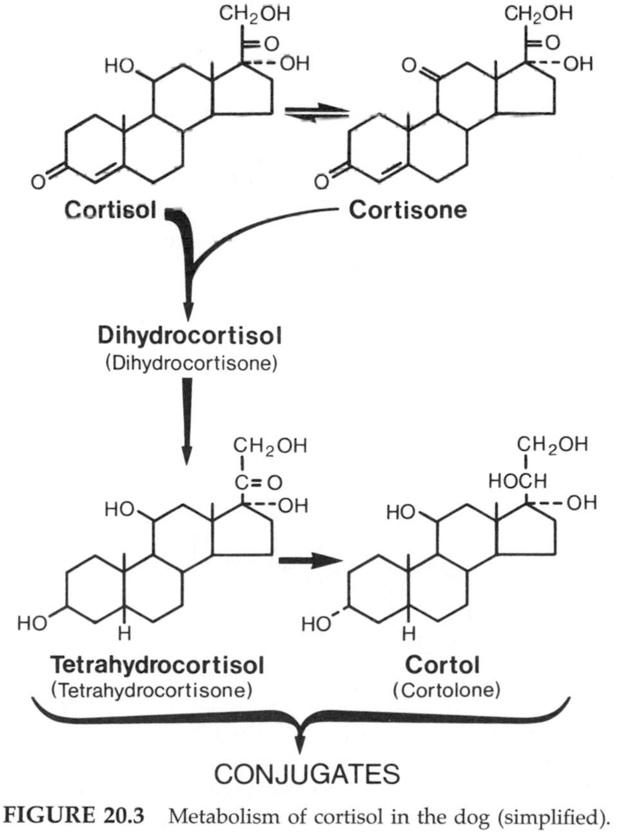

Aldosterone is converted not only to tetrahydroaldosterone-3-glucuronide, but also to aldosterne-18-glucuronide. Most aldosterone metabolism takes place in the liver and kidney, but the intestine and spleen might also contribute to a minor degree (Balikian, 1971).

In domestic animals, the catabolism of androgens has not been studied in any detail. For the dog it is known that very little of the secreted androgens can be measured as 17-ketosteroids in urine (Siegel, 1967; Rijnberk *et al.*, 1968a). Probably most of the excretion occurs via the bile (J. Van der Vies, personal communication, 1968).

In the cat glucocorticoid excretion is also largely biliary (Taylor, 1971; Rivas and Borrell, 1971). Feline liver function has several specific features, including a relatively low glucuronyl-transferase activity. Hence, formation of glucuronide conjugates is low and the conjugates are mainly sulfates, which are mostly excreted via the biliary route. Despite this low urinary excretion of metabolites and conjugates, it was recently shown that renal excretion of free cortisol is sufficient for diagnostic purposes (Goossens *et al.*, 1995) (see also Section V.B of this chapter).

CH2OH
=O
--OH
HO
O
Cortisol
CH2OH
=O
--OH
O
O
Cortisone
Dihydrocortisol
(Dihydrocortisone)
CH2OH
C=O
--OH
HO
HO
H
Tetrahydrocortisol
(Tetrahydrocortisone)
CH2OH
HOCH
--OH
HO
HO
H
Cortol
(Cortolone)
CONJUGATES

FIGURE 20.3 Metabolism of cortisol in the dog (simplified).

E. Regulation of Secretion

The steroidogenic activity of the two inner zones of the adrenal cortex is almost exclusively controlled by the pituitary hormone ACTH. The production of aldosterone in the zona glomerulosa is adapted to the sodium and potassium status of the organism by a complex, multifactorial, and mainly extrapituitary control system. For details on the secretion of ACTH by the pituitary and its regulation, readers are referred to Chapter 19, this volume. In puppies the feedback control of the hypothalamic-pituitary-adrenal axis is already operative at the time of birth (Muelheims *et al.*, 1969), although up to the age of 8 weeks, basal plasma cortisol concentrations are lower than in mature dogs. There is indirect evidence that this is related to low binding to transport proteins (Randolph *et al.*, 1995).

The action of ACTH on the adrenal gland is rapid; within minutes of its release, there is an increased concentration of steroids in the adrenal venous blood. The most likely mechanism by which ACTH stimulates steroidogenesis is via activation of membrane-bound adenylate cyclase. This increases the level of cyclic adenosine 3′,5′-monophosphate (cAMP), which activates adrenocortical protein kinases. This results in the phosphorylation of enzymes that enhance the rate of conversion of cholesterol to pregnenolone.

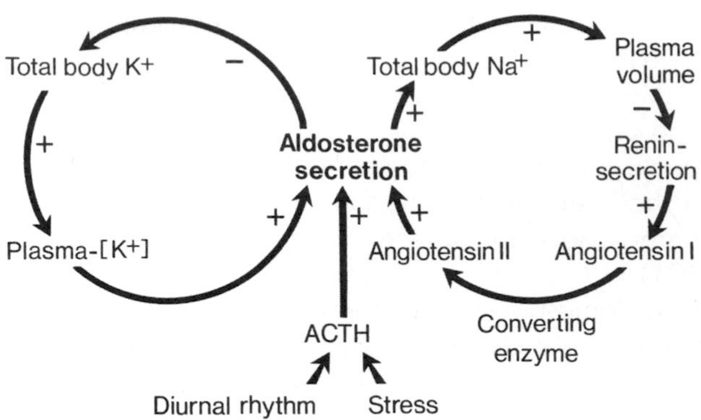

FIGURE 20.4 Physiological control of aldosterone seretion. After Müller and Labhart (1986).

In the regulation of aldosterone secretion, the two most important effectors are the peptide hormone angiotensin and the extracellular potassium concentrations. Regulation of aldosterone secretion by the potassium status is direct and rather simple (Fig. 20.4). The second loop, which adapts aldosterone secretion to the sodium balance, is much more complex. Apart from this renin–angiotensin system and potassium, a number of known, unknown, or poorly defined stimulators and inhibitors may contribute to the regulation of aldosterone biosynthesis and secretion. ACTH also belongs to a group of substances with short-term stimulatory effects on aldosterone biosynthesis.

F. Actions

1. *Glucocorticoids*

Of the naturally occurring glucocorticoids (cortisol, cortisone, and corticosterone), cortisol is the most potent. Several of the synthetic analogs of cortisol are more potent than cortisol itself (Table 20.1). The structural features that determine glucocorticoid potency are illustrated in Fig. 20.5.

TABLE 20.1 Relative Potencies of Corticosteroids

Corticosteroid	Mineralocorticoid activity	Glucocorticoid activity
Cortisol	1	1
Cortisone	0.7	0.7
Corticosterone	0.2	2
11-Deoxycorticosterone (DOC)	0.0	20
Aldosterone	0.1	400
9α-Fluorocortisone	10	400
Prednisone	4	0.7
Prednisolone	4	0.7
Dexamethasone	30	2
Triamcinolone	3	0.0
6α-Methylprednisolone	5	0.5

In their target organ cells, glucocorticoids act in the same manner as other steroid hormones: diffusion through the cell membrane; binding to the cytoplasmic, glucocorticoid (GR), and mineralocorticoid (MR) receptors; translocation of the hormonereceptor complex to the nucleus; and subsequent stimulation of the synthesis of specific RNAs, leading to synthesis of specific enzymes. The latter include key enzymes in gluconeogenesis such as fructose-1,6-disphosphatase, glucose-6-phosphatase, and pyruvate carboxylase. In the dog, corticosteroid excess also results in induction of an isoenzyme of alkaline phosphatase. This corticosteroid-induced form of alkaline phosphatase has not yet been found in other species. The marked stability at 65°C allows easy quantitation in plasma as a diagnostic screening tool for cases of suspected hyperadrenocorticism (Teske *et al.*, 1986, 1989).

Glucocorticoids may also act in a different manner mediated via a glucocorticoid receptor in the plasma membranes. The receptor seems to play a role in the cellular response in lymphoid tissues, exerting immunosuppressive effects. Ultimately, this may result in cytolysis, a property that has important therapeutic implications (Gametchu, 1987).

The overall effect of glucocorticoids on metabolism is to supply glucose to the organism by the transformation of proteins. This occurs via the just-mentioned induction of gluconeogenic enzymes in the liver. Thus, glucocorticoids divert metabolism from a phase of growth and storage toward increased physical activity and energy consumption, whereas chronic excess leads to catabolic effects such as muscle wasting, skin atrophy, and osteoporosis. The tendency to hyperglycemia is opposed by increased secretion of insulin, which in turn tends to enchance fat synthesis. This along with the increased food intake owing to central appetite stimulation (Debons *et al.*, 1986) explains the (centripe-

FIGURE 20.5 Structure–function relationships of corticosteroids. The bold lines and letters indicate the structure common to all glucocorticoids. Substituents that may enhance glucocorticoid activity are represented by light lines and letters.

tal) fat deposition, manifested by abdominal enlargement.

In situations of glucocorticoid deficiency, water excretion is impaired, whereas glucocorticoid excess may result in polyuria, being most pronounced in the dog. This is in part due to antagonism of cortisol to the action of vasopressin. In addition, glucocorticoid excess decreases the sensitivity of the osmoregulation of vasopressin release (Biewenga *et al.*, 1991). Even physiological increases in cortisol may inhibit basal vasopressin release in dogs (Papanek and Raff, 1994).

Glucocorticoids have long been known to have effects on blood cells, including a reduction in the numbers of eosinophils and lymphocytes and an increase in the number of neutrophils and hence in the total number of leukocytes. Glucocorticoid deficiency leads to normochromic, normocytic anemia.

As far as effects on other endocrine glands are concerned, Peterson and Altszuler (1981) demonstrated that canine hyperadrenocorticism results in reversible suppression of growth-hormone secretion. The frequently observed lowering of circulating thyroxine concentrations has been ascribed to changes in the thyroid hormone binding capacity of the plasma and to inhibition of lysosomal hydrolysis of colloid in the thyroid follicular cell (Kemppainen *et al.*, 1983; Woltz *et al.*, 1983). Glucocorticoids have multiple effects on peripheral transfer, distribution, and metabolism (Kaptein *et al.*, 1992), but thyroid function does not seem to be affected (Rijnberk, 1995). The glucocorticoid prednisone was found to inhibit LH secretion in male dogs, leading to reduced testosterone concentrations (Kemppainen *et al.*, 1983).

2. *Adrenal Androgens*

In health, adrenocortical production of androgens is trivial in comparison with the production of these hormones by the gonads. However, a pathological excess of adrenal androgens might induce virilization, that is, the development of masculine secondary sex characteristics, which would be most noticeable in the female or immature male. So far, no virilization as a consequence of enzyme deficiency has been reported in domestic animals, but in equine hyperadrenocorticism, hirsutism is a common feature (Pauli *et al.*, 1974; Van der Kolk *et al.*, 1993).

3. *Mineralocorticoids*

The widespread MR have equal affinity for aldosterone and the glucocorticoids cortisol and corticosterone, whereas the latter two hormones circulate at much higher concentrations than does aldosterone. However, in the classical aldosterone targets (kidney, colon, salivary gland) the enzyme 11β-hydroxysteroid dehydrogenase converts cortisol and corticosterone (but not aldosterone!) to the 11-keto analogs. These analogs cannot bind to MR, thereby enabling aldosterone to occupy this receptor (Funder *et al.*, 1988). Thus, not the receptor but a prereceptor modification provides the tissue specificity. The synthetic steroid 9α-fluorocortisone (Table 20.1) binds tightly to mineralocorticoid receptors and is used for mineralocorticoid replacement therapy, because it is more stable than aldosterone after oral administration. Mineralocorticoid antagonists such as spironolactone bind to these receptors and in this way block aldosterone action. Aldosterone controls the volume and the cationic composition of the extracellular fluid by regulating sodium and potassium balance. Its main action is on the tubular apparatus of the kidneys, but aldosterone receptors are also found in the gut, the salivary glands, and the sweat glands.

In the kidney the effect of aldosterone on the distal tubule consists almost entirely of an exchange of sodium with potassium and hydrogen ions. Aldosterone also promotes the excretion of magnesium and ammonium ions.

G. Physiological Variation, Stress, and the Immune System

In the absence of extraordinary stress, the plasma cortisol concentration of healthy animals varies within certain limits, although adrenocortical secretion does not occur evenly throughout the day, but rather in bursts. In humans, most of the secretory bursts occur between midnight and early morning, leading to a diurnal rhythm of circulating cortisol levels. The temporal coincidence between secretory bursts and plasma ACTH concentrations indicates neuroendocrine control.

In domestic animals and certainly in the dog, initially there has been some controversy as to the occurrence of circadian variation in cortisol concentrations. Later, Kemppainen and Sartin (1984b) presented definite evidence for episodic but not circadian fluctuations in plasma cortisol concentrations in dogs. In the horse (Dybdal *et al.*, 1994) and the pig (Bottoms *et al.*, 1972; Larsson *et al.*, 1979; Benson *et al.*, 1986; Janssens *et al.*, 1995c), as well as in sheep (Fulkerson and Tang, 1979), bulls (Thun *et al.*, 1981), pigeons (Joseph and Meier, 1973; Westerhof *et al.*, 1994a), and chickens (Lauber *et al.*, 1987), the occurrence of circadian variations is generally acknowledged. Also, in the cat there have been reports of the occurence of an (opposite) diurnal rhythm. However, in later reports this was not confirmed (Johnston and Mather, 1979; Leyva *et al.*, 1984).

As in the dog, the secretion is episodic without evidence for a diurnal rhythm (Peterson and Randolph, 1989). Apart from relatively short-term changes such as episodic secretion and diurnal variation, plasma cortisol concentration may change with age and during the estrus cycle, during pregnancy, and around the time of parturition. For example—as mentioned in Section III.E of this chapter—basal cortisol concentrations in puppies 8 weeks of age or younger are lower than in mature dogs, most probably a result of reduced binding to plasma proteins rather than of an altered secretion pattern (Randolph *et al.*, 1995). In gilts a discrete cortisol peak occurs during the early follicular phase of the sexual cycle, coinciding with the decline in plasma progesterone levels, that is, at 4 days prior to the plasma LH surge (Janssens *et al.*, 1995a). Plasma cortisol concentrations of pregnant cows decrease significantly during the fourth month of pregnancy. Thereafter, they remain fairly constant until the sixth and fifth day before parturition, when a sharp rise is seen. Also, at around 24 hours before parturition, plasma corticoids increase again (Eissa and El-Belely, 1990).

Independent of these physiological variations, it is clear that stress may activate the pituitary–adrenocortical system. Factors such as housing (Koelkebeck and Cain, 1984; Carlstead *et al.*, 1993; Cockram *et al.*, 1994), lactation (Gwazdauskas *et al.*, 1986), exercise (Foss *et al.*, 1971; Dybdal *et al.*, 1980; Rossdale *et al.*, 1982), surgery (Robertson *et al.*, 1994), anesthesia, heat (Gould and Siegel, 1985), emotional strain (James *et al.*, 1970; Kirkpatrick *et al.*, 1977; Bobek *et al.*, 1986), food deprivation (Messer *et al.*, 1995), and anticipation of feeding (Murayama *et al.*, 1986) all lead to increased corticosteroid secretion. Guthrie *et al.* (1980, 1982) demonstrated in the horse that during exercise increased plasma renin activity results in a considerable rise in aldosterone secretion. From the lack of any increase in plasma corticosterone in homing pigeons during flight, Viswanathan *et al.* (1987) concluded that the animals were not under serious stress.

With chronic stress, such as during prolonged tethered housing of pigs, an increased responsiveness of the adrenal cortex to ACTH is induced, whereas the sensitivity of the pituitary to stimulation with CRH and/or vasopressin remains unaltered and the challenge (nose sling)-induced ACTH response is lower than in loosely housed pigs. This indicates that during chronic stress, mitigating mechanisms become operational, most probably mediated by endogenous opioids (Janssens, 1995b).

According to Selye's theory of general adaptation, corticosteroids are needed for the (metabolic and circulatory) defense reaction. Administration of cortisol to improve nonspecific resistance is indicated in established pituitary or adrenal insufficiency. From an immunological point of view, stress suppresses defense reactions. Stress-induced glucocorticoid secretion prevents the overshooting of the animal's reactions to stress. The glucocorticoids modulate the mediators of the immune system such as the various lymphokines and mediators of the inflammatory reactions: prostaglandins, leukotrienes, kinins, serotonin, and histamine (Munck *et al.*, 1984).

It is now well established that the interaction of the neuroendocrine system and the immune system is not a one-way but a bidirectional communication. Tissue injury or inflammation elicits production of immunoregulatory cytokines (lymphokines and monokines) by macrophages and monocytes. These cytokines also activate the pituitary–adrenal axis and increase glucocorticoid concentrations, whereas the production and action of these immune mediators are inhibited by glucocorticoids. Thus, there is strong evidence for the existence of a feedback circuit, in which immunoregulatory cytokines act as afferent and ACTH and glucocorticoids as efferent hormonal signals (Woloski *et al.*, 1985; Besedovsky *et al.*, 1986). The regulatory actions of the cytokines are exerted at the level of the hypothalamus, where CRH is the major mediator of the response (Berkenbosch *et al.*, 1987), although cytokines also exert an influence at the level of the pituitary and the adrenals (Gaillard, 1994).

IV. ADRENOCORTICAL DISEASES

As this chapter is not meant to cope with clinical and pathological details, discussion of adrenocortical diseases is confined to definitions, short statements on occurrence in species, and referral to key references. In textbooks on endocrine diseases of companion animals (Drazner, 1994; Rijnberk, 1995; Feldman and Nelson, 1996) the adrenocortical diseases of the dog and the cat are described in detail.

A. Hypoadrenocorticism

Adrenocortical hypofunction includes all conditions in which the secretion of adrenal steroid hormones falls below the requirements of the animal. It may be divided into two categories:

1. Primary hypoadrenocorticism or Addison's disease is the result of deficiency of both glucocorticoid and mineralocorticoid secretion from the adrenal cortices (primary adrenocortical failure). "Idiopathic" or immune-mediated

atrophy (Schaer *et al.*, 1986) of all zones is the most frequently observed histopathologic lesion, although the lymphocytic adrenalitis may be confined to the zona fasciculata and zona reticularis (Kooistra *et al.*, 1995). It is well known in the dog and rare in the cat (Peterson and Randolph, 1989).

2. Secondary hypoadrenocorticism is the result of pituitary ACTH deficiency, causing decreased glucocorticoid secretion. Because of the almost unaltered aldosterone secretion, it will usually remain unnoticed. It is observed occasionally in dogs as a result of a destructive (neoplastic) pituitary lesion.

Animals receiving long-term corticosteroid treatment, despite physical and biochemical hyperadrenocorticoid–changes, develop secondary adrenocortical insufficiency because of prolonged hypothalamo–pituitary suppression. Atrophy of the two inner zones of the adrenal cortex results from the loss of endogenous ACTH stimulation. This is observed in the species in which corticosteroid therapy is common practice, that is, the dog (Greco and Behrend, 1995; Rijnberk, 1996) and the horse (Hoffsis and Murdock, 1970; Toutain and Brandon, 1983). It has also been observed in dogs and cats treated with progestagens, owing to the glucocorticoid activity that is intrinsic to these drugs (Chastain *et al.*, 1981; Mansfield *et al.*, 1986; Middleton *et al.*, 1987; Selman *et al.*, 1994).

B. Hyperadrenocorticism

In principle the adrenal cortex of animals might, as in humans, give rise to three distinct clinical syndromes of hyperfunction, that is, mineralocorticoid excess (Conn's syndrome), glucocorticoid excess (Cushing's syndrome), and androgen excess (adrenogenital syndrome). In animals, thus far only the occurrence of glucocorticoid excess has been reported. This form of hyperadrenocorticism is common in the dog while rare in the cat and the horse.

In about 15% of hyperadrenocorticoid dogs, the disease is due to a primary adrenocortical tumor. In the other 85% of cases, the disease is associated with a pituitary lesion, producing excess ACTH. The pituitary hyperplasias and neoplasias may reside in either the anterior lobe or the intermediate lobe (Peterson *et al.*, 1982; Kemppainen and Peterson, 1994). In cats the disease may also be either primary adrenocortical or pituitary dependent (Meijer *et al.*, 1978b; Peterson and Randolph, 1989). In the horse the disease originates almost invariably in the intermediate lobe of the pituitary (Orth *et al.*, 1982). In domestic ferrets hyperadrenocorticism is quite common, and it is primarily associated with adrenocortical tumor or nodular adrenocortical hyperplasia (Rosenthal *et al.*, 1993a).

V. ASSESSMENT OF ADRENOCORTICAL FUNCTION

A. Routine Laboratory Diagnostics

In both hypoadrenocorticism and hyperadrenocorticism, a number of abnormal laboratory findings are more common than in other diseases. Many of these abnormalities can be derived from the action of glucocorticoids and mineralocorticoids (described in Section III.F). The essentials are summarized here.

In primary hypoadrenocorticism the decreased aldosterone production results in hyponatremia and hyperkalemia. The associated hypovolemia leads to prerenal uremia, which is suggested when the ratio of urea to creatinine in plasma (both expressed in micromoles per liter) exeeds 150. In hyperadrenocorticism the classic hematologic abnormality is eosinopenia, which may be associated with lymphopenia and occasionally with leukocytosis and erythrocytosis. A very common biochemical abnormality is the elevated plasma concentration of alkaline phosphatase (Eckersall and Nash, 1983) that may occur in conjunction with mild elevation of alanine aminotransferase. In addition, hyperglycemia, hyperlipidemia, and low thyroxine concentrations may be observed. Urine specific gravity is usually low, and in 5–10% of cases glucosuria is found.

Of these abnormalities the elevated alkaline phosphatase (AP) concentration is the most common laboratory abnormality in dogs with corticosteroid excess (either exogenous or endogenous). This increase is due to the induction of a specific isoenzyme, which has greater heat stability at 65°C than other AP isoenzymes (Teske *et al.*, 1986) and is therefore easily measured by a routine laboratory procedure. An abnormally elevated AP-65°C value may point to hyperadrenocorticism, but it is unsuitable as a diagnostic test because of its low specificity (Teske *et al.*, 1989). The low thyroxine (T_4) concentrations in plasma that may be observed in hyperadrenocorticism seem to be a consequence of altered transport, distribution, and metabolism of T_4, rather than of hyposecretion (Rijnberk, 1996).

There may be other conditions causing these abnormalities, however; in addition, some cases do not present with typical routine laboratory characteristics. Therefore, a detailed understanding of the specific laboratory tests commonly used to investigate adrenocortical function is essential. In addition, the spontaneous forms of hyperadrenocorticism may arise from pituitary ACTH overproduction or from autonomously hypersecreting adrenocortical tumors. Each of these requires a different mode of treatment, and therefore insight into the tests used in the differential diagnosis is of prime importance.

B. Tests of Basal Adrenocortical Function

The availability of highly specific radioimmunoassays has greatly enhanced and simplified assessment of adrenocortical function. Some of the older (chemical) methods, which are now seldom used, are mentioned briefly. Methods in common use are described in more detail, which applies especially to the dynamic tests, part of which is used in the differential diagnosis (Section V.C).

1. Cortisol Production Rate

In the radionuclide dilution method, the rate of cortisol production is determined through administration of radiolabeled cortisol and isolation of cortisol metabolites from urine that is collected over at least 24 hours. The specific activities of labeled cortisol and tetrahydrocortisol (or tetrahydrocortisone) are compared. Some preliminary work (Rijnberk *et al.*, 1968a) indicated that this test worked for the dog and might potentially be very useful. However, the laborious character of the procedure rules out general availability.

2. Plasma Corticoids

Plasma corticoid levels are subject to considerable variation owing to the pulsatile nature of the secretion, physiological variation (see Section III.G of this chapter), and alterations in transport proteins (Meyer and Rothuizen, 1994). Therefore, single determinations are of little diagnostic value in assessing hypo- or hyperadrenocorticism. Nevertheless, in a considerable percentage of cases resting values outside the reference ranges may be found. There are few data on sex and breed differences, and they are not dealt with in Table 20.2. Nevertheless there is evidence that they should be taken into account. In dogs circulating cortisol concentrations do not differ between males and females, but in small dogs higher values are found than in large-breed dogs (Reimers *et al.*, 1990).

3. Urinary Corticoids

With measurements of urinary corticoids and/or their metabolites, an integrated reflection of corticoid production over a period of time is obtained, thereby adjusting for the fluctuations in the plasma levels. Indeed, in dogs measurements of 17-hydroxycorticosteroids in 24-hour urine samples were found to have a high discriminatory power (Table 20.3; Rijnberk *et al.*, 1968b). However, the method has been abandoned because of the complicated and time-consuming chemical analysis, in combination with the difficulty of accurate collection of 24-hour urine samples.

Meanwhile, for the dog, another measurement reflecting cortisol secretion over a period of time has been introduced by Stolp *et al.* (1983). In this approach the 24-hour urine collection has been replaced by morning samples. The urinary corticoids (largely cortisol) are measured by radioimmunoassay and related to the creatinine concentration. In an assessment (Rijnberk *et al.*, 1988), the corticoid/creatinine ratios were found to have a higher accuracy in the diagnosis of hyperadrenocorticism than the commonly used dexamethasone screening test (see Section V.C.3).

The method has also been introduced for diagnostic use in the cat (Goossens *et al.*, 1995) and the ferret (Gould *et al.*, 1995). Also, for these species the determination of the urinary corticoid/creatinine ratio appears to be valuable for the diagnosis of hyperadrenocorticism.

4. Salivary Cortisol

In saliva, the "free" (biologically active) fraction rather than the "total" cortisol (Section III.C) is measured. Thus far in animals, salivary collection devices have been used primarily for research purposes. In the past a disadvantage has been the anesthesia involved, although it does not inhibit salivary flow (Phillips *et al.*, 1983).

In recent years there has been renewed interest in the approach, especially now that a technique has been introduced for saliva collection in unrestrained animals. Cortisol measurements in saliva may serve as a noninvasive indicator of stress in animals. When dogs are allowed to chew on large wads of cotton for up to 30 seconds, adequate saliva samples can be obtained (Vincent and Michell, 1992). The same method has been used succesfully in pigs (Parrott and Misson, 1989) and sheep (Cooper *et al.*, 1989). It does not seem to be advisable to increase saliva flow by pilocarpine (Blackshaw and Blackshaw, 1989). It has also been reported that in calves (Fell and Shutt, 1986) and goats (Greenwood and Shutt, 1992) saliva can be aspirated from the side of the mouth with minimum restraint.

5. Cortisol in Milk

Monitoring adrenocortical activity by measurements in milk may be of value in searching for stressors in the livestock industry (Stephens, 1980). Some of the results of cortisol measurements in milk have been incorporated in Table 20.2.

TABLE 20.2 Basal Corticosteroid Concentrations in Healthy Animals[a]

Species (n)	Hormone (fluid)[b]	Time of sampling	Range (nmol/l) (mean ± SEM)	Reference
Dog (6)	Cortisol (p)	8 samples/24 hr	<8–214	Johnston and Mather (1978)
Dog (21)	Cortisol (p)	8.00 hr	27–188	Meijer *et al.* (1979)
Dog	Aldosterone (p)		0.08–0.94	Willard *et al.* (1987)
Dog (24)	Cortisol (p)	9.30 hr	42–205	Hansen *et al.* (1994)
Cat (12)	Cortisol (p)	8.00 hr	9–71	Kemppainen *et al.* (1984)
Cat (6)	Cortisol (p)	8.00 hr	(87 ± 16)	Willemse *et al.* (1993)
Cat (100)	Cortisol (p)	9.00 hr	5–140	Peterson *et al.* (1994)
Cat (6)	Cortisol (p)		(30 ± 14)	Peterson and Kemppainen (1993)
Horse (11)	Cortisol (p)	11.00 hr	36–81	Hodson *et al.* (1986)
Horse (6)	Cortisol (p)	8.00 hr	(153 ± 18)	Messer *et al.* (1995)
Horse (7)	Cortisol (p)	09.00 hr	103–217	Van der Kolk *et al.* (1993)
Cow (23)	Cortisol (p)	5.00 hr	(17 ± 2)[c]	Gwazdauskas *et al.* (1986)
Cow (11)	Cortisol (s)	Milking time	(12 ± 2)	Shutt and Fell (1985)
Cow (11)	Cortisol (m)	Milking time	(1 ± 0.2)	Shutt and Fell (1985)
Cow (8)	Cortisol (c)	Day 1 postpartum	(46 ± 4)	Shutt and Fell (1985)
Calf (8)	Cortisol (sa)	Afternoon	(3.3 ± .5)	Cooper *et al.* (1989)
Calf (19)	Cortisol (sa)		0.3–3.4	Fell and Shutt (1986)
Pig (8)	Cortisol (s)	8.00 hr	(82 ± 3)	Becker *et al.* (1985)
Pig (8)	Free cortisol	Hourly	(5 ± 0.5)	Hemsworth *et al.* (1986)
Pig (6)	Cortisol (p)	10.00 hr	(49 ± 8)	Janssens *et al.* (1995b)
Pig (6)	Cortisol (p)	18.00 hr	(20 ± 4)	Janssens *et al.* (1995b)
Sow (20)	Cortisol (p)	10.00 hr	(29 ± 5)	Von Borrell and Hurnik (1991)
Sheep (11)	Cortisol		(62 ± 10)	Jephcott *et al.* (1986)
Sheep (6)	Cortisol (sa)	07.00 hr	(3.1 ± 1.6)	Cooper *et al.* (1989)
Goat (5)	Cortisol (p)		(65 ± 8)	Frandsen (1987)
Goat (6)	Cortisol (sa)		(3.0 ± 1.3)	Greenwood and Shutt (1992)
Pigeon (30)	Corticosterone (p)	11.30 hr	<5–34	Lumeij *et al.* (1987)
Pigeon (14)	Corticosterone (p)	04.00 hr	(13 ± 3)	Westerhof *et al.* (1994a)
White leghorn (9)	Corticosterone (p)	19.30–22.0 hr	(5 ± 2)	Craig and Craig (1985)
White plymouth (20)	Aldosterone		0.06–0.07	Rice *et al.* (1985)
Ferret (8)	Cortisol (p)		26–235	Rosenthal *et al.* (1993b)
Ferret (8)	Corticosterone (p)		2–47	Rosenthal *et al.* (1993b)

[a] Values have been converted to SI units when necessary.
[b] p, plasma; s, serum; m, milk; c, colostrum; sa, saliva.
[c] Two days after drying off.

6. *Plasma ACTH*

Determinations of resting plasma ACTH concentrations are not very useful in the diagnosis of hyperadrenocorticism because of the overlapping values for normal dogs and dogs with hyperadrenocorticism. Their value lies mainly in the differential diagnosis of hyperadrenocorticism (see later discussion). In animals with primary adrenocortical failure, plasma ACTH concentrations are usually extremely high. However, in most cases such a determination is primarily of academic interest, as the diagnosis can be made with the more practical ACTH stimulation test (see Section V.C.1). Plasma ACTH measurements are useful to resolve doubt about a primary or secondary adrenocortical insufficiency (Kooistra *et al.*, 1995).

C. Dynamic Tests

Apart from the just-described measurements of reflections of basal adrenocortical secretions, various maneuvers have been introduced to test the physiology

TABLE 20.3 Urinary Corticoids

Method	Species (n)	Range (mean ± SD)	Reference
17-Hydroxycorticoids			
Appleby *et al.* (1955)	Dog (21)	0.06–0.32 mg kg^{-1}/24 hr	Rijnberk *et al.* (1968a)
Siegel (1965)	Dog	2.13–6.98 mg/24 hr	Siegel (1968)
Metcalf (1963)	Dog	(2.3 ± 0.7) mg/24 hr	Bell *et al.* (1971)
Corticoids (radioimmunoassay)			
Stolp *et al.* (1983)	Dog (20)	0.3–3.6 nmol kg^{-1}/24 hr	Stolp *et al.* (1983)
Corticoid/creatinine ratio			
	Dog (28)	1.2–6.9 × 10^{-6}	Stolp *et al.* (1983)
	Dog (12)	(9 ± 4 × 10^{-6})	Jones *et al.* (1990)
	Dog (20)	0.5–17.7 × 10^{-6}	Feldman and Mack (1992)
	Dog (31)	0.1–31.2 × 10^{-6}	Smiley and Peterson (1993)
	Cat (42)	2–36 × 10^{-6}	Goossens *et al.* (1995)
	Ferret (51)	0.04–1.66 × 10^{-6}	Gould *et al.* (1995)

of the hypothalamic–pituitary–adrenocortical axis. As in other areas of endocrinology, stimulation tests are used when hypofunction is suspected, and suppression tests are used when hyperfunction is suspected.

1. *Tests of Adrenocortical Reserve*

There are a number of modifications of the test for adrenal reserve, including intramuscular (IM) injection of ACTH rather than intravenous (IV) (Table 20.4). The critical feature of the test, however, is that a substantial increase in plasma (and/or urinary) corticoids must occur in response to ACTH if one is to prove the existence of adrenal reserve. One method is described in detail in Section V.F.

In veterinary medicine, ACTH stimulation has also been used for confirmation of adrenocortical hyperfunction. This is based on the assumption that hyperplastic adrenal cortices have abnormally large cortisol reserves. Indeed, hyperresponsiveness to ACTH or lysine–vasopressin (Meijer *et al.*, 1978a) stimulation is found in pituitary-dependent hyperadrenocorticism, but there is considerable overlap with normal dogs. Over the years, several modifications/combinations of the ACTH stimulation test (Eiler and Oliver, 1980; Eiler *et al.*, 1984; Feldman, 1985, 1986) for the diagnosis of hyperadrenocorticism have been used, but it still appears to be a dynamic test with less diagnostic accuracy than the low-dose dexamethasone suppression test (LDDST) (Feldman, 1986; Peterson, 1986). This seems to hold true for the horse as well (Dybdal *et al.*, 1994), although the ACTH stimulation test is recommended for the diagnosis of equine hyperadrenocorticism (Van der Kolk *et al.*, 1993). The value of the ACTH stimulation test lies in the diagnosis of hypoadrenocorticism (see Section V.F). When the test is used for the diagnosis of hyperadrenocorticism, combining it with a dexamethasone suppression test is not at all desirable, as doses of ≥0.1 mg of dexamethasone /kg can alter the results of the ACTH-stimulation test for at least three days (Kemppainen *et al.*, 1989).

2. *Tests of Pituitary (ACTH) Reserve*

In animals with adrenals responsive to ACTH, administration of the 11β-hydroxylase inhibitor metyrapone decreases cortisol secretion and consequently increases ACTH secretion. As a result, cortisol precursor synthesis increases, which can be measured as urinary 17-hydroxycorticosteroids. Although effective in the dog (Siegel, 1968), the procedure is cumbersome, which has prohibited its wider use in veterinary medicine. At present, pituitary ACTH reserve is tested by use of physiological suprapituitary stimuli, that is, vasopressin or CRH (see Chapter 19). Such tests are mainly useful in differential diagnosis, whereby it should be noted that vasopressin is the least useful for this purpose, as it can exert a direct stimulatory effect on cortisol release by adrenocortical tumors (Van Wijk *et al.*, 1994).

3. *Tests of Pituitary–Adrenocortical Suppressibility*

The integrity of the feedback system can be tested by giving a potent glucocorticoid and judging suppression of ACTH secretion by measuring either steroids excreted in urine or plasma cortisol levels. A potent glucocorticoid such as dexamethasone is used, so that the administered compound can be given in small amounts that will not contribute significantly to the steroids to be analyzed. In pituitary-dependent hyperadrenocorticism, there is loss of normal sensitivity to suppression, and as a result higher values are found than in normal individuals following dexamethasone administration (Table 20.5).

TABLE 20.4 Plasma Corticoids (nmol/liter) Following ACTH Stimulation

Species (*n*)	ACTH preparation	Dose and route	Time of sampling (min)	Range (mean ± SEM)	Reference
Cortisol					
Dog (6)	ACTH gel	2.2 IU/kg IM	60–120	188–308	Johnston and Mather (1978)
Dog (21)	ACTH (1–24)	250 μg IV	90	120–620	Meijer *et al.* (1978c)
Dog (8)	ACTH (1–24)	250 μg IV	90	(291 ± 38)	Hansen *et al.* (1994)
Dog (8)	ACTH (1–24)	250 μg IM	90	(280 ± 39)	Hansen *et al.* (1994)
Cat (4)	ACTH (1–24)	125 μg IM	30	132–210	Kemppainen *et al.* (1984)
Cat (5)	ACTH (1–24)	125 μg IM	30	(277 ± 41)	Smith and Feldman (1987)
Cat (6)	ACTH (1–24)	125 μg IV	60	(274 ± 19)	Peterson and Kemppainen (1993)
Cow (5)	ACTH gel	200 IU	Peak	(134 ± 3)[a]	Gwazdauskas *et al.* (1986)
Cow (5)	ACTH (1–24)	200 IU IM	480	(185 ± 17)	Shutt and Fell (1985)
Horse (7)	ACTH (1–24)	250 μg IV	Peak	213–384	van der Kolk *et al.* (1993)
Horse (9)	ACTH gel	1 U/kg IV	4 hr	(455 ± 110)	Dybdal *et al.* (1994)
Goat (5)	pACTH	0.35 IU/kg	120	(96 ± 19)	Frandsen (1987)
Ferret (8)	ACTH (1–24)	1 μg/kg IV	60	143–308	Rosenthal *et al.* (1993b)
Corticosterone					
Ferret (8)	ACTH (1–24)	1 μg/kg IV	60	30–72	Rosenthal *et al.* (1993b)
Pigeon	ACTH (1–24)	125 μg IM	30–120	80–419	Lumeij *et al.* (1987)
Psittac (35)	ACTH gel	16 IU IM	60	(73 ± 26)	Lothrop *et al.* (1985)
Eagle (12)	ACTH (1–24)	125 μg IM	60	298–541	Zenoble *et al.* (1985)
Condor (6)	ACTH (1–24)	125 μg IM	60	243–365	Zenoble *et al.* (1985)
White leghorn (9)	pACTH	50 IU IM	30	(40 ± 5)	Craig and Craig (1985)
Broiler chick (11)	ACTH (1–24)	4 IU/kg	30	(7 ± 1)	Van Mourik *et al.* (1986)
Aldosterone					
Dog (9)	ACTH gel	2.2 IU/kg IM	60	0.48–1.59	Willard *et al.* (1987)

[a] Two days in dry period.

In the dog the most frequently used test is still the low-dose dexamethasone suppression test (LDDST) introduced by Meijer *et al.* (1978a). In this test, 0.01 mg of dexamethasone/kg of body weight is administered intravenously and a plasma sample for cortisol determination is taken after 8 hours. In healthy dogs at this time the plasma cortisol concentrations are still depressed (Kemppainen and Sartin, 1984a). This test is described in detail in Section V.F. Others recommend intramuscular administration of 0.015 mg of dexamethasone/kg and the collection of plasma samples for cortisol determination 2, 4, 6, and 8 hours after injection (Peterson, 1984). Although not necessary for the diagnosis, the results of a 3- or 4-hour sample may be informative for differential diagnosis (see Sections V.D and V.F).

TABLE 20.5 Upper Reference Values for Post-dexamethasone Cortisol Concentrations in the Low-Dose Dexamethasone Suppression Test

Species	Dose and route	Time of sampling	Plasma cortisol (nmol/liter)	Reference
Dog	10 μg/kg iv	8 h	<40	Meijer *et al.* (1979)
Dog	15 μg/kg im	2, 4, 6, 8 h	<28	Peterson (1986)
Cat	10 μg/kg iv	8 h	<36	Smith and Feldman (1987)
Cat	15 μg/kg iv	8 h	<14	Peterson and Graves (1988)
Horse	40 μg/kg iv	15 h	<30	Dybdal *et al.* (1994)
Pigeon	0.5 μg/kg iv	4 h	<5	Westerhof *et al.* (1994b)

D. Tests for Differential Diagnosis

In animals with primary hypoadrenocorticism, the physical and biochemical features are very much determined by the electrolyte disturbances caused by the insufficient mineralocorticoid secretion. In secondary hypoadrenocorticism, mineralocorticoid secretion is practically unchanged, and the presenting signs are usually completely different. Thus only very rarely is there a need for additional tests to distinguish between these two forms of hypoadrenocorticism. However, when doubt exists about the background of the hyporesponsiveness to exogenous ACTH, plasma ACTH concentrations should be measured (see also Section V.B.6) and/or the pituitary should be tested for ACTH reserve (see Section V.C).

For hyperadrenocorticism, the situation is completely different. Once the diagnosis of hyperadrenocorticism has been made by either a basal test or a dynamic test, it is necessary to distinguish between pituitary-dependent hyperadrenocorticism and hyperadrenocorticism arising from an adrenocortical tumor. In principle these forms of the disease require different modes of treatment. Probably owing to the additional secretion of steroids with mineralocorticoid activity by adrenocortical tumors, a higher incidence of hypokalemia is found than in animals with pituitary-dependent hyperadrenocorticism (Meijer, 1980). However, this sign is not specific enough to allow differentiation between the two forms. As there are no further signs that may be helpful to distinguish the two entities, specific tests are needed. The two most helpful procedures are the measurement of basal plasma ACTH levels an the high-dose dexamethasone suppression test (HDDST). A test of pituitary ACTH reserve may also be helpful.

Normal to high plasma ACTH concentrations (≥40 pg/ml) in hyperadrenocorticoid dogs usually indicate ACTH excess of pituitary origin. Low to undetectable concentrations of ACTH are found in dogs with hyperadrenocorticism arising from an adrenocortical tumor in which the feedback control of pituitary ACTH secretion is undisturbed. ACTH concentration may not be low when both disease entities occur simultaneously (Nothelfer and Weinhold, 1992; Van Sluijs *et al.*, 1995). However useful this measurement may be, in practice it is not the first choice when it comes to the differential diagnosis of hyperadrenocorticism because of the requirements for collecting and handling the samples and the technically difficult assay (see also Section V.E).

In contrast to the determination of endogenous ACTH, the HDDST is readily available. Despite a decreased sensitivity to the suppression by glucocorticoids, the ACTH secretion of most animals with pituitary-dependent hyperadrenocorticism can be suppressed with a 10-fold dose of dexamethasone, resulting in a decrease of the adrenocortical secretion. The autonomous hypersecretion by adrenocortical tumors will not be influenced by the high dose of dexamethasone.

Two procedures are used in the dog, one employing plasma cortisol as a reflection of adrenocortical secretion (Meijer *et al.*, 1979) and the other using urinary corticoid/creatinine ratios (Rijnberk *et al.*, 1987, 1988). In both, a greater than 50% decline from baseline values is regarded as diagnostic for pituitary-dependent hyperadrenocorticism (see Section V.F). When suppression is less than 50%, the hyperadrenocorticism may be due to either an adrenocortical tumor or a pituitary ACTH excess that is extremely resistant to dexamethasone suppression. For the differentiation between these two forms, the measurement of endogenous ACTH may be necessary, or a CRH test may be performed. In dogs with pituitary-dependent hyperadrenocorticism, CRH administration (1 μg/kg IV) results in elevations of both plasma ACTH and cortisol, whereas no such rise is seen in dogs with hyperadrenocorticism caused by adrenocortical tumors (Van Wijk *et al.*, 1994). The high cost of the commercially available powder preparation and the necessary pharmaceutical preparation prohibit general use of CRH.

E. Collection and Handling of Samples

It has become common practice for clinicians to perform endocrine function studies and to mail samples to a laboratory for measurement of cortisol. It is therefore important to rule out nonpathological factors that can alter hormone concentrations. Apart from stress (see Section V.B), the nutritional state of the animal may also play a role. Reimers *et al.* (1986) found that mean serum concentrations of cortisol in dogs fasted 12 or 24 hours were lower than those in dogs that were not fasted. The concentrations were not further affected by continued fasting for 36 hours.

Olson *et al.* (1981) studied several aspects of sample handling, such as uncentrifuged storage and storage time. They concluded that either serum or plasma of dogs is suitable for radioimmunoassay of cortisol and that samples (with and without added anticoagulants) may be left uncentrifuged at 4°C for up to 40 hours without cortisol degradation. However, prolonged storage of serum at room temperature is detrimental, particularly for samples having large concentrations of cortisol. Samples allowed to defrost and sit for more than 3 days en route to a laboratory may have a lower cortisol concentration than at the time of collection.

The peptide ACTH is readily inactivated at room temperature. Therefore, blood for ACTH measurements should be collected in EDTA-coated tubes placed in ice, and the blood should be centrifuged at 4°C. The plasma should then be placed in polypropylene tubes and kept frozen until assayed. These requirements are most easily fulfilled when the samples are collected in a clinic with the necessary facilities. If the samples have to be transported, it is imperative that they be kept frozen with dry ice in a Styrofoam container.

It has become clear that these strict rules can be alleviated to some extent by adding a protease inhibitor (aprotinin) to the collection tubes. Kemppainen *et al.* (1994) have found that aprotinin-containing plasma samples shipped (second-day delivery) unfrozen in an insulated container with frozen refrigerant packs are not affected by transport.

F. Protocols

The protocols presented in this section are applicable to the dog. With the data presented in some of the tables, extrapolation to other species may be possible.

1. The ACTH Stimulation Test

a. Indications

The ACTH stimulation test is performed when there is suspicion of decreased adrenocortical reserve capacity: (1) primary adrenocortical insufficiency (Addison's disease) and (2) (iatrogenic) secondary adrenocortical insufficiency.

b. Performance

Blood for cortisol measurements is collected immediately before and 90 minutes after intravenous administration of 0.25 mg synthetic ACTH (Cortrosyn [Organon]). In cases in which treatment for adrenocortical insufficiency was already started, on the morning of the test the cortisone administration is postponed until after completion of the test.

c. Interpretation

In healthy dogs, the cortisol concentrations rise to 270–690 nmol/liter. In Addison's disease, the control value is usually low and does not increase following ACTH administration. In animals with secondary adrenocortical insufficiency the basal cortisol values may be low as well, and, depending on the severity (duration) of the ACTH deficiency, the cortisol rise is subnormal or absent.

2. The Low-Dose Dexamethasone Suppression Test

a. Indication

The low-dose dexamethasone suppression test is used when hyperadrenocorticism (Cushing's syndrome) is suspected.

b. Performance

In the morning 0.01 mg dexamethasone per kg body weight is administered intravenously. Blood for cortisol measurements is collected immediately before and at 3 and 8 hours after dexamethasone administration.

c. Interpretation

A plasma cortisol concentration exceeding 40 nmol/liter at 8 hours after dexamethasone administration can be regarded as diagnostic for hyperadrenocorticism with a diagnostic accuracy of 0.83 (95% confidence limits 0.76–0.88) (Rijnberk *et al.*, 1988). The measurements at 0 and 3 hours are not needed for the diagnosis of hyperadrenocorticism, but may be informative in the differential diagnosis. Quite commonly the high value at 8 hours is preceded by a lower value at 3 hours. Thus, at 8 hours the pituitary–adrenocortical system escapes from suppression by dexamethasone. If the value at 3 hours is at least 50% lower than the basal value, the desease may be regarded as pituitary dependent.

3. The High-Dose Dexamethasone Suppression Test

a. Indication

Differentiation between pituitary-dependent hyperadrenocorticism and hypercorticism arising from adrenocortical tumors is possible with the high-dose dexamethasone suppression test.

b. Performance

Blood for cortisol determination is collected immediately before and 3 hours after administration of 0.1 mg dexamethasone per kg body weight.

c. Interpretation

If the plasma cortisol declines by more than 50%, the diagnosis of pituitary-dependent hyperadrenocorticism is justified. A decrease of less than 50% can be due either to an adrenocortical tumor or to dexamethasone-resistant pituitary-dependent hyperadrenocorticism. Differentiation between these two forms requires additional tests (see Section V.D).

4. The Corticoid/Creatinine Ratios (with High-Dose Dexamethasone Test)

a. Indication

Determination of corticoid/creatinine ratios can be performed when hyperadrenocorticism is suspected.

b. Performance

The owner is asked to collect morning urine samples at set times (e.g., 7 A.M.) on three consecutive days. On the preceding evenings the animal should have its last walk at identical times (e.g., 11 P.M.). After collection of the second urine sample oral administration of dexamethasone is begun. At 8-hour intervals the owner administers 0.1 mg dexamethasone per kg body weight.

c. Interpretation

Corticoid/creatinine ratios exceeding 10×10^{-6} can be regarded as compatible with hyperadrenocorticism with a diagnostic accuracy of 0.91 (95% confidence limits 0.85–0.95) (Rijnberk *et al.*, 1988). When the ratio of the third urine sample is 50% lower than the mean of the first two ratios, the diagnosis of a pituitary-dependent hyperadrenocorticism is justified. A lesser decrease may be due to either adrenocortical tumors or dexamethasone-resistant pituitary-dependent hyperadrenocorticism (for differentiation, see Section V.D).

d. Comment

This test is now generally regarded as a sensitive screening test for the diagnosis of hyperadrenocorticism (Feldman, 1995). There has been some debate on its specificity (Rijnberk and Teske, 1992; Feldman, 1992), and it has been stated that the test lacks specificity (Feldman, 1995). Indeed, as with other screening tests the specificity is not 100% (Smiley and Peterson, 1993; Kaplan *et al.*, 1995) and in conditions that are associated with increased adrenocortical function, such as hepatic encephalopathy (Rothuizen *et al.*, 1995), elevated ratios can be found.

In this discussion two points remain somewhat underexposed. First, a high sensitivity and not a high specificity is essential for a screening test. The predictive value of a positive test result is dependent on the sensitivity of the test and also on the prevalence of the disease in the population studied. When the testing is limited to dogs suspected of the disease, the population under study will have a high prevalence of disease and as a result the predictive value of a positive test result will be high. Secondly, as always, the diagnostic accuracy of a test also depends on the quality of the test procedure. In this respect it should be mentioned that easily elevated C/C ratios can be obtained when the urines are collected under stressful conditions, such as in the hospital. This will increase the number of false positive test results, i.e., lower the specificity.

When C/C ratios are measured in a population suspected of hyperadrenocorticism, with urine collections at home, not only is the sensitivity of the test high, but also the specificity (0.77) is comparable to that of the low-dose dexamethasone suppression test (0.73) (Rijnberk *et al.*, 1988).

References

Appleby, J. T., Gibson, G., Norymberski, J. K., and Stubbs, R. D. (1955). *Biochem. J.* **60,** 453.

Balikian, H. M. (1971). *Endocrinology* **89,** 1309.

Becker, B. A., Nienaber, J. A. Christenson, R. K., Manak, R. C., DeShazer, J. A., and Hahn, G. L. (1985). *Am. J. Vet. Res.* **46,** 1034.

Bell, E. T., Christie, D. W., and Parkes, M. F. (1971). *Acta Endocrinol.* **68,** 387.

Benson, G. J., Langner, P. H., Thurmon, J. C., Nelson, D. R., Neff-Davis, C. Davis, L. E., Tranquilli, J. C., and Gustafsson, B. K. (1986). *Am. J. Vet. Res.* **47,** 1071.

Berkenbosch, F., van Oers, J., Del Rey, A., Filders, F., and Besedovsky, H. (1987). *Science* **238,** 524.

Besedovsky, H., Del Rey, A., Sorkin, E., and Dinarello, C. A. (1986). *Science* **233,** 652.

Biewenga, W. J., Rijnberk, A., and Mol, J. A. (1991). *Acta Endocrinol.* **124,** 583.

Blackshaw, J. K., and Blackshaw, A. W. (1989). *Vet. Res. Comm.* **13,** 265.

Bobek, S., Niezgoda, J., Pierzchala, K., Litynski, P., and Sechman, A. (1986). *J. Vet. Med. Assoc.* **33,** 698.

Boonayathap, U., and Marotta, S. F. (1974). *Horm. Metab. Res.* **6,** 74.

Bottoms, G. D., Roesel, O. F., Rausch, F. D., and Akins, E. L. (1972). *Am. J. Vet. Res.* **33,** 785.

Carlstead, K., Brown, J. L., and Strawn, W. (1993). *Appl. Anim. Behav.* **38,**143.

Carsia, R. V., Morin, M. E. Rosen, H. D., and Weber, H. (1987). *Proc. Soc. Exp. Biol. Med.* **184,** 436.

Chastain, C. B., Graham, C. L., and Nichols, C. E. (1981). *Am. J. Vet. Res.* **42,** 2029.

Chester-Jones, I. (1987). *In* "Fundamentals of Comparative Vertebrate Endocrinology" (I. Chester-Jones, P. M. Ingleton, and J. G. Philips, eds.), p. 95. Plenum, New York.

Cockram, M. S., Ranson, M., Imlak, P., Goddard, P. J., Burrels, C., and Harkiss, G. D. (1994). *Anim. Prod.* **58,** 389.

Cooper, T. R., Trunkfield, H. R., Zanella, A. J. and Booth, W. D. (1989). *J. Endocrinol.* **123,** R13.

Craig, J. V., and Craig, J. A. (1985). *Poultry Sci.* **64,** 809.

Debons, A. F., Zurek, L. D., Tse, C. S., and Abrahamsen, S. (1986). *Endocrinology* **118,** 1678.

Dor, P., Keymolen, V., De Roods, M., Levin, S. and Borkowski, A. (1973). *Eur. J. Cancer* **9,** 687.

Drazner, F. H. (1994). "Small Animal Endocrinology." Saunders, Philadelphia.

Dybdal, N. O., Gribble, D., Madigan, J. E., and Stabenfeldt, G. H. (1980). *Equine Vet. J.* **12,** 137.

Dybdal, N. O., Hargreaves, K. M., Madigan, J. E., Gribble, D. H., Kennedy, P. C., Stabenfeldt, G. H. (1994). *J. Am. Vet. Med. Assoc.* **204,** 627.

Eckersall, P. D., and Nash, A. S. (1983). *Res. Vet. Sci.* **34,** 310.

Eiler, H., and Oliver, J. W. (1980). *Am. J. Vet. Res.* **41,** 1243.

Eiler, H., Oliver J. W., and Legendre, A. M. (1984). *J. Am. Vet. Med. Assoc.* **185,** 289.

Eissa, H. M., and El-Belely, M. S. (1990). *Brit. Vet. J.* **146,** 24.

Feldman, E. C. (1985). *J. Am. Vet. Med. Assoc.* **187,** 49.

Feldman, E. C. (1986). *J. Am. Vet. Med. Assoc.* **189,** 1562.

Feldman, E. C. (1992). *J. Am. Vet. Med. Ass.* **201,** 1140.

Feldman, E. C. (1995). *In* "Textbook of Veterinary Internal Medicine" (S. J. Ettinger and E. C. Feldman, eds.), p. 1557. Saunders, Philadelphia, Pennsylvania.

Feldman, E. C., and Mack, R. E. (1992). *J. Am. Vet. Med. Ass.* **200,** 1637.

Feldman, E. C. and Nelson, R. W. (1996). "Canine and Feline Endocrinology and Reproduction." Saunders, Philadelphia, Pennsylvania.

Fell, L. R. and Shutt, D. A. (1986). *Can. J. Anim. Sci.* **66,** 637.

Fell, L. R., Shutt, D. A., and Bentley, C. J. (1985). *Aust. Vet. J.* **62,** 403.

Ferguson, D. B. (1984). "Steroid Hormones in Saliva," Frontiers in Oral Physiology, Vol.5. Karger, Basel.

Foss, M. L., Barnard, R. J., and Tipton, C. M. (1971). *Endocrinology* **89,** 96.

Frandsen, J. C. (1987). *Vet. Parasitol.* **23,** 43.

Fulkerson, W. J., and Tang, B. Y. (1979). *J. Endocrinol.* **81,** 135.

Funder, J. W., Pearce, P. T., Smith, R., and Smith, A. T. (1988). *Science* **242,** 583.

Gaillard, R. C. (1994). *Trends Endocr. Metab.* **5,** 303.

Gametchu, B. (1987). *Science* **236,** 456.

Gold, N. I. (1961). *J. Biol. Chem.* **236,** 1924

Goossens, M. M. C., Meyer, H. P., Voorhout, G., and Sprang, E. P. M., (1995). *Domest. Anim. Endocrinol.* **12,** 355.

Gould, N. R., and Siegel, H. S. (1985). *Poultry Sci.* **64,** 144.

Gould, W. J., Reimers, T. J., Bell, J. A., Laurence, H. J., Randolph, J. F., Rowland, P. H., and Scarlett, J. M. (1995). *J. Am. Vet. Med. Ass.* **206,** 42.

Greco, D. S., and Behrend, E. N. (1995). *In* "Kirk's Current Veterinary Therapy XII" (J.D. Bonagura, ed.), p. 413. Saunders, Philadelphia.

Greenwood, P. L. and Shutt, D. A. (1992). *Aust. Vet. J.* **69,** 161.

Guthrie, G. P., Cecil, S. G., and Kotchen, T. A. (1980). *J. Endocrinol.* **85,** 49.

Guthrie, G. P., Cecil, S. G., Darden, E. D., and Kotchen, T. A. (1982). *Gen. Comp. Endocrinol.* **48,** 296.

Gwazdauskas, F. C., Keys, J. E., and McGilliard, M. L. (1986). *J. Dairy Sci.* **69,** 2134.

Hansen, B. L., Kemppainen, R. J., and MacDonald, J. M. (1994). *J. Amer. Anim. Hosp. Ass.* **30,** 38.

Heap, R. B., Holzbauer, M., and Newport, H. M. (1966). *J. Endocrinol.* **36,** 159.

Hemsworth, P. H., Barnett, J. L., and Hansen, C. (1986). *Appl. Anim. Behav. Sci.* **15,** 303.

Hennesy, D. P., Conn, R. J., and Wan, S. S. (1986). *Res. Vet. Sci.* **41,** 361.

Hirose, T., Matsumoto, J. Aikawa, T., and Suzuki, T. (1977). *J. Endocrinol.* **73,** 539.

Hodson, N. P., Wright, J. A., and Hunt, J. (1986). *Vet. Rec.* **118,** 148.

Hoffsis, G. F., and Murdock, P. W. (1970). *J. Am. Vet. Med. Assoc.* **157,** 1590.

Holzbauer, M., and Newport, H. M. (1969). *J. Physiol.* **200,** 821.

Ilett, K. F., and Lockett, M. F. (1969). *J. Endocrinol.* **43,** 313.

IUPAC-IUB 1967. (1968). *Biochim. Biophys. Acta* **164,** 453.

James, V. H. T., Horner, M. W., Moss, M. S., and Rippon A. E. (1970). *J. Endocrinol.* **48,** 319.

Janssens, C. J. J. G., Houwing, H., Helmond, F. A., Schrama, J. W., and Wiegant, V. M. (1995a). *In* "Chronic Stress and Pituitary–Adrenal Function in Female Pigs" (Thesis, C. J. J. G. Janssens, Wageningen, The Netherlands).

Janssens, C. J. J. G., Helmond, F. A., Loyens, L. W. S., Schouten, W. G. P., and Wiegant, V. M. (1995b). *Endocrinology* **136,** 1468.

Janssens, C. J. J. G., Helmond, F. A., and Wiegant, V. M. (1995c). *Domest. Anim. Endocrinol.* **12,** 167.

Jephcott, E. H., McMillen, I. C. Rushen, J., Hargreaves, A., and Thorburn, G. D. (1986). *Res. Vet. Sci.* **41,** 371.

Johnston, S. D., and Mather, E. C. (1978). *Am. J. Vet. Res.* **39,** 1766.

Johnston, S. D., and Mather, E. C. (1979). *Am. J. Vet. Res.* **40,** 190.

Jones, C. A., Refsal, K. R., Lippert, A. C., Nachreiner, R. F., and Schwacha, M. M. (1990). *Domest. Anim. Endocrinol.* **7,** 559.

Joseph, M. M., and Meier, A. H. (1973). *Gen. Comp. Endocrinol.* **20,** 326.

Kaplan, A. J., Peterson, M. E., and Kemppainen, R. J. (1995). *J. Am. Vet. Med. Ass.* **207,** 445.

Kaptein, E. M, Moore, G. E., Ferguson, D. C., and Hoenig, M. (1992). *Endocrinology* **130,** 1669.

Kemppainen, R. J., and Peterson, M. E. (1994). *Trends Endocrinol. Metab.* **5,** 21.

Kemppainen, R. J., and Sartin, J. L. (1984a). *Am. J. Vet. Res.* **45,** 472.

Kemppainen, R. J., and Sartin, J. L. (1984b). *J. Endocrinol.* **103,** 219.

Kemppainen, R. J., Thompson, F. N., Lorenz, M. D., Munnell, J. F., and Chakraborty, P. K. (1983). *J. Endocrinol.* **96,** 293.

Kemppainen, R. J., Mansfield, P. D., and Sartin, J. L. (1984). *J. Am. Anim. Hosp. Assoc.* **20,** 737.

Kemppainen, R. J., Sartin, J. L., and Peterson, M. E. (1989). *Am. J. Vet. Res.* **50,** 1914.

Kemppainen, R. J., Peterson, M. E., and Sartin, J. L. (1991). *Domest. Anim. Endocrinol.* **10,** 45.

Kemppainen, R. J., Clark, T. P., and Peterson, M. E. (1994). *Domest. Anim. Endocrinol.* **11,** 355.

Kime, D. E. (1987). *In* "Fundamentals of Comparative Vertebrate Endocrinology" (I. Chester-Jones, P. M. Ingleton, and J. G. Philips, eds.), p. 38. Plenum, New York.

Kirkpatrick, J. F., Baker, C. B., Wiesner, L., and Angle, M. (1977). *Comp. Biochem. Physiol.* **57A,** 179.

Koelkebeck, K. W., and Cain, J. R. (1984). *Poultry Sci.* **63,** 2123.

Kooistra, H. S., Rijnberk, A., and Van den Ingh, T. S. G. A. M. (1995). *Vet. Quart.* **17,** 59.

Larsson, M., Edqvist, L. E., Ekman, L., and Persson, S. (1979). *Acta Vet. Scand.* **20,** 16.

Lauber, J. K., Vriend, J., and Oishi, T. (1987). *Comp. Biochem. Physiol.* **86A,** 73.

Leyva, H., Addiego, L., Stabenfeldt, G. (1984). *Endocrinology* **115,** 1729.

Lothrop, C. D., and Oliver, J. W. (1984). *Am. J. Vet. Res.* **45,** 2304.

Lothrop, C. D., Olsen, J. H., Loomis, M. R., Jensen, J. M., and Lenhard, A. (1985). *J. Am. Vet. Med. Assoc.* **187,** 1113.

Lumeij, J. T., Boschma, Y., Mol, J. A., De Kloet, E. R., and Van den Brom, W. E. (1987). *Avian Pathol.* **16,** 199.

McCormick, J. R., Hermann, A. H., Lien, W. M., and Egdahl, R. H. (1974). *Endocrinology* **94,** 17.

Mansfield, P. D., Kemppainen, R. J., and Sartin, J. L. (1986). *J. Am. Anim. Hosp. Assoc.* **22,** 515.

Marks, V. (1983). *In* "Biochemistry in Clinical Practices" (D. L. Williams and V. Marks, eds.), p. 617. William Heinemann Medical Books Ltd., London.

Meijer, J. C. (1980). In "Current Veterinary Therapy VII" (R. W. Kirk, ed.), p. 975. Saunders, Philadelphia, Pennsylvania.

Meijer, J. C., De Bruijne, J. J. Rijnberk, A., and Croughs, R. J. M. (1978a). *J. Endocrinol.* **77,** 111.

Meijer, J. C., Lubberink, A. A. M. E., and Gruys, E. (1978b). *Tijdschr. Diergeneesk.* **103,** 1048.

Meijer, J. C., Mulder, G. H., Rijnberk, A., and Croughs, R. J. M. (1978c). *J. Endocrinol.* **76,** 209.

Meijer, J. C., Lubberink, A. A. M. E., Rijnberk, A., and Croughs, R. J. M. (1979). *J. Endocrinol.* **80,** 315.

Messer, N. T., Johnson, P. J., Refsal, K. R., Nachreiner, R. F., Ganjam, V. K., and Krause, G. F. (1995). *Am. J. Vet. Res.* **56,** 116.

Metcalf, M. G. (1963). *J. Endocrinol.* **26,** 415.

Meyer, H. P., and Rothuizen, J. (1993). *Domest. Anim. Endocrinol.* **10,** 45.

Meyer, H. P., and Rothuizen, J. (1994). *Domest. Anim. Endocrinol.* **11,** 317.

Middleton, D. J., Watson, A. D. J., Howe, C. J., and Caterson, I. D. (1987). *Can. J. Vet. Res.* **51,** 60.

Muelheims, G. H., Francis, F. E., and Kinsella, R. A. (1969). *Endocrinology* **85,** 365.

Müller, J. (1986). *In* "Clinical Endocrinology. Theory and Practice" (A. Labhart, ed.), p. 351. Springer Verlag, Berlin.

Müller, J., and Labbart, A. (1986). *In* "Clinical Endocrinology. Theory and Practice" (A. Labhart, ed.), p. 361. Springer Verlag, Berlin.

Munck, A., Guyre, P. M., and Holbrook, N. J. (1984). *Endocr. Rev.* **5,** 25.

Murayama, S., Moriya, K., and Sasaki, Y. (1986). *Jpn. J. Zootech. Sci.* **57,** 317.

Murphy, G. M., and Connell, J. A. (1970). *Aust. Vet. J.* **46,** 595.

Nakamura, T., Tanabe, Y., and Hirano, H. (1978). *Gen. Comp. Endocrinol.* **35,** 302.

Nothelfer, H. B., and Weinhold, K. (1992). *Berl. Münch. Tierärztl. Wschr.* **105,** 305.

Olson, P. N., Bowen, R. A., Husted, P. W., and Nett, T. M. (1981). *Am. J. Vet. Res.* **42,** 1618.

Orth, D. N., Holscher, M. A., Wilson, M. G., Nicholson, W. E., Plue, R. E., and Mount, C. D. (1982). *Endocrinology* **110,** 1430.

Papanek, P. E., and Raff, H. (1994). *Am. J. Physiol.* **266,** R1744.

Parrott, R. F., and Misson, B. H. (1989). *Brit. Vet. J.* **145,** 501.

Pauli, B. U., Rossi, G. L., and Straub, R. (1974). *Vet. Pathol.* **11,** 417.

Peterson, M. E. (1984). *Vet. Clin. N. Am.* **14,** 731.

Peterson, M. E. (1986). *In* "Current Veterinary Therapy IX" (R. W. Kirk, ed.), p. 963. Saunders, Philadelphia, Pennsylvania.

Peterson, M. E., and Altszuler, N. (1981). *Am. J. Vet. Res.* **42,** 1881.

Peterson, M. E., and Graves, T. K. (1988). *Res. Vet. Sci.* **44,** 38.

Peterson, M. E., and Kemppainen, J. R. (1993). *Am. J. Vet. Res.* **54,** 300.

Peterson, M. E., and Randolph, J. F. (1989). *In* "The Cat. Diseases and Clinical Management" (R. G. Sherding, ed.), p. 1135. Churchill Livingstone, New York.

Peterson, M. E., Krieger, D. T., Drucker, W. D., and Halmi, N. S. (1982). *Acta Endocrinol.* **101,** 15.

Peterson, M. E., Winkler, B., Kintzer, P. P., and Altszuler, N. (1986). *Domest. Anim. Endocrinol.* **3,** 117.

Peterson, M. E., Kemppainen, R. J., and Orth, D. N. (1994). *Am. J. Vet. Res.* **55,** 295.

Phillips, P. A., Newcomer, C. E., and Schutz, D. S. (1983). *Lab. Anim. Sci.* **33,** 465.

Randolph, J. F., Center, S. A., Reimers, T. J., Scarlett, J. M., Corbett, J. R. (1995). *Am. J. Vet. Res.* **56,** 511.

Reimers, T. J., McGarrity, M. S., and Strickland, D. (1986). *Am. J. Vet. Res.* **47,** 2485.

Reimers, T. J., Lawler, D. F., Sutaria, P. M., Correa, M. T. and Erb, H. N. (1990). *Am. J. Vet. Res.* **51,** 454.

Riad-Fahmy, D., Read, G. F., Walker, R. F., and Griffiths, K. (1982). *Endoc. Rev.* **3,** 367.

Rice, G. E., Arnason, S. S., Arad, Z., and Skadhauge, E. (1985). *Comp. Biochem. Physiol.* **81A,** 769.

Rijnberk, A. (1996). *In* "Clinical Endocrinology of Dogs and Cats" (A. Rijnberk, ed.), p. 47. Kluwer, Dordrecht.

Rijnberk, A., and Teske, E. (1992). *J. Am. Vet. Med. Ass.* **201,** 1139.

Rijnberk, A., der Kinderen, P. J., and Thijssen, J. H. H. (1968a). *J. Endocrinol.* **41,** 387.

Rijnberk, A., der Kinderen, P. J., and Thijssen, J. H. H. (1968b). *J. Endocrinol.* **41,** 396.

Rijnberk, A., Mol, J. A., Rothuizen, J., Bevers, M. M., and Middleton, D. J. (1987). *Front. Horm. Res.* **17,** 48.

Rijnberk, A., van Wees, A., and Mol, J. A. (1988). *Vet. Rec.* **122,** 178.

Rivas, C., and Borrell, S. (1971). *J. Endocrinol.* **51,** 283.

Robertson, I. S., Kent, J. E., and Molony, V. (1994). *Res. Vet. Sci.* **56,** 8.

Rosenthal, K. L., Peterson, M. E., Quesenberry, K. E., Hillyer, E. V., Beeber, N. L., Moroff, S. D., and Lothrop, C. D. (1993a). *J. Am Vet. Med. Ass.* **203,** 271.

Rosenthal, K. L., Peterson, M. E., Quesenberry, K. E., and Lothrop, C. D. (1993b). *Am. J. Vet. Res.* **54,** 29.

Rosner, W. (1969). *New Engl. J. Med.* **281,** 658.

Rossdale, P. D., Burguez, P. N., and Cash, R. S. G. (1982). *Equine Vet. J.* **14,** 293.

Rothuizen, J., Biewenga, W. J., and Mol, J. A. (1995). *Domest. Anim. Endocrinol.* **12,** 13.

Schaer, M., Riley, W. J., Buergelt, C. D., Bowen, D. J., Senior, D. F., Burrows, C. F., and Campbell, G. A. (1986). *J. Am. Anim. Hosp. Assoc.* **22,** 789.

Selman, P. J., Mol, J. A., Rutteman, G. R., and Rijnberk, A. (1994). *Eur. J. Endocrinol.* **131,** 422.

Selye, H. (1971). *In* "Hormones and Resistance." Springer Verlag, Berlin.

Shutt, D. A., and Fell, L. R. (1985). *J. Dairy Sci.* **68,** 1832.

Siegel, E. T. (1965). *Am. J. Vet. Res.* **26,** 1152.

Siegel, E. T. (1967). *Am. J. Vet. Res.* **28,** 287.

Siegel, E. T. (1968). *Am. J. Vet. Res.* **29,** 173.

Smiley, L. E., and Peterson, M. E. (1993). *J. Vet. Int. Med.* **7,** 163.

Smith, M. C., and Feldman, E. C. (1987). *Am. J. Vet. Res.* **48,** 1719.

Stephens, D. B. (1980). *Adv. Vet. Sci. Comp. Med.* **24,** 179.

Stolp, R., Rijnberk, A., Meijer, J. C., and Croughs, R. J. M. (1983). *Res. Vet. Sci.* **34,** 141.

Taylor, W. (1971). *Vitam. Horm.* **29,** 201.

Teske, E., Rothuizen, J., de Bruijne, J. J., and Mol, J. A. (1986). *J. Chromatogr.* **369,** 349.

Teske, E., Rothuizen, J., de Bruijne, J. J., and Rijnberk, A. (1989). *Vet. Rec.* **125,** 12.

Thun, R., Eggenberger, E., Zerobin, K., Lüscher, T., and Vetter, W. (1981). *Endocrinology* **109,** 2208.

Toutain, P. L., and Brandon, R. A. (1983). *In* "Veterinary Pharmacology and Toxicology" (Y. Ruckebush, P. L. Toutain, and G. D. Kovitz, eds.). MIP Press, Boston.

Van der Kolk, J. H., Kalsbeek, H. C., van Garderen, E., Wensing, T., and Breukink, H. J. (1993). *Vet. Rec.* **133,** 594.

Van Mourik, S., Outch, K. H., Cumming, R. B., and Stelmasiak, T. (1986). *Comp. Biochem. Physiol.* **84A,** 397.

Van Sluijs, F. J., Sjollema, B. E., Voorhout, G., Van den Ingh, T. S. G. A. M., and Rijnberk, A. (1995). *Vet. Quart.* **17,** 113.

Van Wijk, P. A., Rijnberk, A., Croughs, R. J. M., Wolfswinkel, J., Selman, P. J., and Mol, J. A. (1994). *Eur. J. Endocrinol.* **130,** 410.

Vincent, I. C., and Michell, A. R. (1992). *Res. Vet. Sci.* **53,** 342.

Viswanathan, M., John, T. M., George, J. C., and Etcher, R. J. (1987). *Horm. Metab. Res.* **19,** 400.

Von Borrell, E., and Hurnik, J. F. (1991). *Physiol. Behav.* **49,** 709.

Walsh, M. T., Beldegreen, R. A., Clubb, S. L., and Chen, C. L. (1985). *Am. J. Vet. Res.* **46,** 1584.

Westerhof, I., Mol, J. A., Van den Brom, W. E., Lumeij, J. T., and Rijnberk, A. (1994a). *Avian Dis.* **38,** 428.

Westerhof, I., Van den Brom, W. E., Mol, J. A., Lumeij, J. T., and Rijnberk, A. (1994b). *Avian Dis.* **38,** 435.

Willard, M. D., Refsal, K., and Thacker, E. (1987). *Am. J. Vet. Res.* **48,** 1713.

Willemse, T., Vroom, M. W., Mol, J. A., and Rijnberk A. (1993). *Am. J. Vet. Res.* **54,** 69.

Woloski, B. M. R. N. J., Smith, E. M., Meyer, W. J., Fuller, G. M., and Blalock, J. E. (1985). *Science* **230,** 1035.

Woltz, H. H., Thompson, F. N., Kemppainen, R. J., Munnell, J. F., and Lorenz, M. D. (1983). *Am. J. Vet. Res.* **44,** 2000.

Zenoble, R. D., Kemppainen, R. J., Young, D. W., and Carpenter, J. W. (1985). *J. Am. Vet. Med. Assoc.* **187,** 1119.

Zolovick, A., Upson, D. W., and Eleftherion, B. E. (1966). *J. Endocrinol.* **35,** 249.

C H A P T E R

21

Thyroid Function

J. JERRY KANEKO

I. INTRODUCTION

Disorders of the thyroid gland are the most common endocrine disorders in humans, and an extensive historical and scientific literature is available. Among the domestic animals, thyroid function and its diseases are well known in companion animals but less so in livestock. In livestock, nutritional iodine deficiencies have been of greater importance than thyroid-gland diseases, particularly in the iodine-deficient areas of the world. The importance of thyroid function and its diseases has also become progressively more important as the longevity of companion animals such as the dog and cat has increased.

Advances in thyroid physiology, the pathogenesis of its diseases, and the continued development and refinement of methods of testing thyroid function have added impetus to the study of thyroid disease in all animals. This chapter reviews the anatomy and physiology of the thyroid gland and its diseases as a corollary to understanding the pathophysiology of the thyroid gland in disease. The physiological bases of a variety of thyroid function tests, most of which are now readily available to the veterinary clinician, are emphasized.

II. ANATOMICAL CONSIDERATIONS

The thyroid gland of animals is a bilobed structure that overlays the trachea at a point just below the larynx. Anatomical variations of the gland are quite marked between species and, to some extent, within a given species. The isthmus connecting the two lobes of the thyroid is the region that varies most markedly between species. Humans and the pig have a large discrete isthmus that forms a pyramidal lobe connecting the two lateral lobes. The cow has a fairly wide band of glandular tissue that forms the connecting isthmus. In the horse, sheep, goat, cat, and dog, the isthmus is a narrow remnant of tissue and may be nonexistent. The size of the gland relative to body weight is extremely small in all animals, approximating 0.20% of body weight. The size of the gland varies and may be enlarged in iodine deficiency, ingestion of goitrogenic toxins, tumors, or hyperactivity. In primary hypothyroidism, the gland may be reduced to fibrotic and inactive remnants of thyroid tissue.

Accessory or extrathyroidal tissue is quite commonly seen in the dog, particularly near the thoracic inlet, though it may be found anywhere along the esophagus. This tissue is fully functional physiologically, synthesizes hormone, and can be located by its uptake of radionuclides.

The thyroid gland is a highly vascularized tissue with a large blood flow. The functional unit of the thyroid gland is the thyroid follicle, a spherical structure composed of an outer monolayer of follicular cells surrounding an inner core of colloid, the thyroglobulin–hormone complex, which is the storage reservoir of thyroid hormone. The colloid stored in the lumen is a clear, viscous fluid. The individual follicular cells vary from 5 to 10 μm in height and the entire follicle may vary from 25 to 250 μm in diameter. The size of the follicles and the height of their cells vary according to the functional state of the gland. The cells may vary from an inactive squamous cell to the highly active, tall columnar cell. Interspersed between the follicles are the thyroid C cells, the source of calcitonin, the hypocalcemic hormone associated with calcium metabolism. A third type of hormonal tissue, the parathyroid, is embedded within the thyroid or located in close proximity. The parathyroids are the source of parathormone, the hypercalcemic hormone. Removal of the parathyroids is virtually unavoidable during surgical thyroidectomies, so that postsurgical hypocalcemias are important consequences to be considered.

III. NUTRITIONAL REQUIREMENTS

The thyroid gland is unique among the endocrine glands in that an integral part of its hormone, L-thyroxine (T_4), is a trace mineral, iodine, that is available to the animal in only limited amounts. Marine plants are known to be good sources of iodine, but on land, iodine is limited and many regions of the world are known to be iodine deficient. The minimum daily iodine requirement of the adult beagle dog is 140 μg (1103 nmol) (Belshaw *et al.*, 1975), and the recommended daily requirement is 35 μg/kg bw (276 nmol/kg bw) for the adult dog and 70 μg/kg bw (551.6 nmol/kg bw) for the growing puppy. Milk is a very poor source of hormonal iodide, contributing only about 4–7% of the maintenance requirements for hormone (Akasha and Anderson, 1984).

The very small requirements are compensated for by a very efficient intestinal absorption mechanism and by conservation and recycling of internal iodine. Very little iodine is lost from the body by the various excretory routes such as urine, saliva, tears, milk, sweat, and feces. Also, although most endocrine glands store little of their hormones, the thyroid manages to store large quantities of hormone, sufficient for 1 to 3 weeks depending upon the species. The thyroid gland contains about 20% of the total body iodine. Its iodine content and size varies with iodine intake and the state of thyroid function but it usually contains 10–40 mg (78.8–315.2 μmol) iodine/100 g tissue or 4–16 mg (31.5–126.1 μmol) iodine in a 20-kg dog.

IV. IODINE METABOLISM

Iodine can be absorbed in any of its soluble chemical forms, but in the intestines it is usually absorbed in the form of iodides (I^-) or iodates (IO_4^-), or as the hormonal forms. Iodides may be absorbed from any moist body surface, including the mucous membranes, and is absorbed easily through broken epithelia. Normally, the chief route of entry into the general circulation is by absorption through the mucosal cells of the small intestine. The I^- in the circulation is trapped almost exclusively by the thyroid gland; small amounts are trapped by the salivary gland and minimal amounts by the gastric mucosa, placenta, and mammary gland. In the ruminant, 70–80% of an oral dose is absorbed in the rumen and 10% in the omasum (Barua *et al.*, 1964).

The main route of excretion of I^- is by the kidneys, through which almost all the I^- that was not trapped by the thyroid is lost in the urine. A small but significant amount is lost in the saliva, and minimal amounts are lost in tears, feces, sweat, and milk. In ruminants, a significant amount may be lost in the feces as well (Bustad *et al.*, 1957). A minute amount of free hormone, that fraction not bound to serum proteins, is also lost in the urine. These routes of excretion are especially important considerations when patients are being treated with radioiodine.

V. FUNCTIONS OF THE THYROID GLAND

The main functions of the thyroid gland are the trapping of I^- and the synthesis, storage and release of thyroid hormones (Fig. 21.1), and these activities are under the control of thyrotrophic or thyroid-stimulating hormone (TSH). Although all steps in hormonogenesis are stimulated by TSH, the trapping of I^- and the release of hormone are the two major sites of its action.

A. The Thyroid Hormones

The principal thyroid hormones elaborated by the thyroid are the two active hormones, 3,5,3′,5′-tetraiodothyronine (T_4) and 3,5,3′-triiodothyronine (T_3), and the inactive hormone, 3,3′,5′-triiodothyronine (reverse T_3 or rT_3). rT_3 is the inner deiodination product of T_4. The structures of the individual hormones are given in Fig. 21.2. The T_4 molecule contains 65.3% iodine and the T_3 molecule contains 58.5% iodine. T_3 is the active hormone in the target cell. T_4 functions as the transport form and as the feedback regulator of the thyroid gland.

B. Hormonogenesis and Release

1. *Trapping of Iodide*

The I^- in the general circulation is taken up by the thyroid follicular cells by a highly efficient trapping and concentrating mechanism. It does this against a large concentration gradient, which can be from 1 : 20 to 1 : 500 across the thyroid cell membrane and is stimulated by TSH. The trapping process is catalyzed by a trapping enzyme, requires oxygen, and is an active transport or "pump" mechanism catalyzed by a Na^+–K^+-ATPase and dependent upon ATP. It is the high efficiency of this trapping system that concentrates virtually all of the body iodine in the thyroid gland. It also accounts for the microgram nutrient requirement for iodine. In addition, the thyroid follicular cells have a high capability for compensatory hypertrophy when there is a scarcity of iodine; hence the development of iodine-deficiency goiters. The efficiency of I^- trapping is also the basis for the radioactive iodine uptake (^{131}I uptake) test of thyroid function, ^{131}I thyroid imaging, and ^{131}I thyroid therapy. The trapping of I^- is stimulated by TSH and blocked by goitrogenic agents such as thiocyanate (SCN^-) and perchlorate (ClO_4^-) and by large amounts of I^-. The sites of these and other blocks in the thyroxine biosynthetic pathway are also shown in Fig. 21.1. Other compounds are trapped by the thyroid gland; a clinically useful one is ^{99m}Tc-pertechnetate, which is used for thyroid imaging.

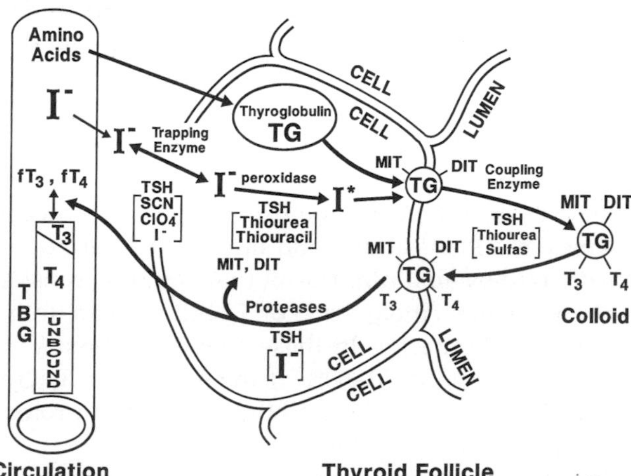

FIGURE 21.1 Pathways of iodine metabolism and thyroid hormone synthesis. Goitrogenic blocking agents are shown in brackets.

2. *Synthesis of Thyroid Hormones*

After trapping, there is an oxidation of I^- catalyzed by a peroxidase, and the product is a highly active form of iodine, most likely a free radical, I*. This reaction is inhibited by thyrotoxic agents such as the thiouracils or thioureas and stimulated by TSH. Propylthiouracil is commonly used in the treatment of hyperthyroidism. The I* almost instantaneously binds to the phenyl groups of the tyrosine moieties of thyroglobulin at the 3 and/or 5 position to form a monoiodotyrosine (MIT) or a diiodotyrosine (DIT). This iodination occurs while the tyrosines remain in polypeptide linkage within the thyroglobulin molecule. Next, the iodinated phenyl groups of the tyrosines are coupled by the oxidative condensation of an iodinated phenyl group of one DIT to another DIT to form T_4, or of an MIT group to a DIT to form T_3. These iodination reactions occur mainly at the follicular cell membrane–colloid interface. These iodination reactions are energy requiring and are sensitive to blocking by sulfa drugs, thioureas, and *para*-aminobenzoic acid (PABA).

3. *Storage of Hormone*

Thyroglobulin is the thyroidal glycoprotein of high molecular weight (660,000) synthesized exclusively by the thyroidal follicular cells. After synthesis, it moves to the cell membrane, where the iodinations occur and the iodinated thyroglobulin, or colloid, is released into the lumen of the follicle where it is stored. The amount stored can be quite large, as evidenced by the mass of protein contained within a normal follicle.

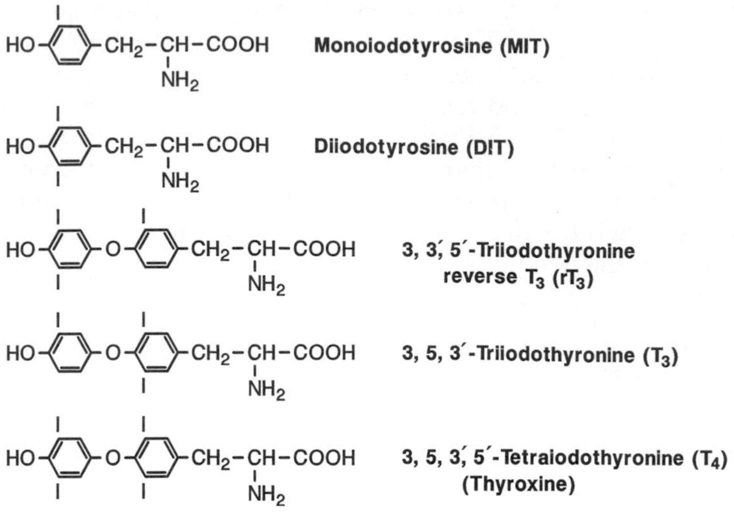

Monoiodotyrosine (MIT)

Diiodotyrosine (DIT)

3, 3′, 5′-Triiodothyronine reverse T_3 (rT_3)

3, 5, 3′-Triiodothyronine (T_3)

3, 5, 3′, 5′-Tetraiodothyronine (T_4) (Thyroxine)

FIGURE 21.2 Chemical structures of the major iodinated compounds of the thyroid gland.

4. *Release of Hormone*

Although all steps in the hormonogenesis and release process are stimulated by TSH, TSH stimulus of the release of hormones is the second of its two principal sites of action. Within a few minutes after giving TSH, small packets of colloid are taken up by the follicular cell membrane, and vesicles are formed and taken into the cell by endocytosis. The vesicles merge with lysosomes in the cell, and the lysosomal proteases hydrolyze the colloid and release their MIT, DIT, T_3, and T_4. The released MIT and DIT are enzymatically degraded by microsomal tyrosine deiodinases and their iodine is recycled within the gland. The T_4 and T_3 are released into the circulation by a simple diffusion process. Of the total hormone released, about 90% is T_4 and 10% is T_3. Within the gland, there is also some deiodination of the T_4 of the inner phenyl group to form rT_3 but most of this deiodonation occurs in the peripheral tissues. This rT_3 is the inactive form of thyroid hormone and is on a degradation pathway.

VI. TRANSPORT OF HORMONE: PROTEIN BINDING

The thyroid hormones in the circulation are T_4, T_3, and rT_3. Immediately on entering the circulation, these hormones are bound to transport proteins, mainly to thyroxine binding globulin (TBG) and in lesser amounts to thyroxine binding prealbumin (TBPA) and to albumin. There is a wide spectrum of species variation in hormone binding by serum proteins (Table 21.1). TBPA is present in all species, in contrast to earlier reports indicating that TBPA was present only in humans and in rhesus monkey, horse, cat, rabbit, pigeon and chicken. The early work was based on electrophoretic migration (Tanabe *et al.*, 1969; Refetoff *et al.*, 1970), but later work was based on chemical properties that identified TBPA or its analogs at different migration sites (Larsson *et al.*, 1985). The differences in migration are probably due to the amino acid composition of these proteins among the various species. By chemical criteria, all species have TBPA.

TBG is the major binding protein for hormone, but not all species have TBG. In those species without TBG, albumin is the major binding protein. The binding constant of TBG for T_4 is about 10^{10} liter/mol so that some 99.97% of the plasma T_4 is bound to TBG and only 0.03% is free or unbound (fT_4). The binding constant for T_3 is about 10^9 liter/mol so that about 99.7% of the plasma T_3 is bound to TBG and 0.3% is free. Therefore, TBG or albumin transport most of the hormones.

In the cat, rabbit, rat, mouse, Guinea pig, pigeon, or chicken, TBG is absent and most of the hormone is transported by albumin. The albumin binding constant is about 10^5 liter/mole for T_4 or T_3, but with an unlimited binding capacity. In these species without TBG, albumin transports 50–80% of the hormones. T_3 (and likely rT_3) appears to bind to these transport proteins in parallel with T_4 binding.

Protein binding has several functions. It serves to solubilize these lipid-soluble hormones for transport in the aqueous plasma. The bound forms also do not readily pass through the renal glomerular membrane,

TABLE 21.1 Thyroxine-Binding Proteins of Various Species, Their Electrophoretic Migration, and Their Relative Thyroxine Binding[a]

Species	Electrophoretic migration position		
	α-Globulin	Albumin	Prealbumin
Human	TBG (73)	ALB (10)	TBPA (17)
Rhesus monkey	TBG (59)	ALB (8)	TBPA (33)
Cattle[b]	TBG (60) TBPA (20)	ALB (20)	
Sheep	TBG (86)	TBPA (14)	
Goat	TBG (63)	TBPA (37)	
Water buffalo	TBG (78)	TBPA (22)	
Horse	TBG (61)	ALB (17) TBPA (22)	
Pig	TBPA (0) TBG (93)	ALB (7)	
Dog	TBG (60) TBPA (17) α-G (11)	ALB (12)	
Cat[b]	TBPA (39)	ALB (61)	
Rabbit[b]	TBPA (73)	ALB (27)	
Rat[b]		ALB (80) TBPA (20)	
Mouse[b]		ALB (80) TBPA (20)	
Guinea pig[b]		ALB (81) TBPA (19)	
Chicken	TBPA (10)	ALB (75) TBPA (15)	
Pigeon[b]		ALB (50)	TBPA (50)

[a] TBG, thyroxine-binding globulin; ALB, albumin; TBPA, thyroxine-binding prealbumin; α-G, α-globulin. Positioned as shown relative to the albumin position in each species. Numbers in parentheses are % thyroxine binding. Constructed from the data of Tanabe *et al.* (1969), Refetoff *et al.* (1970), and Larsson *et al.* (1985).

[b] Approximation.

so they minimize urinary loss of hormones; the free hormones pass and are lost. The bound forms also serve as a large and readily accessible reservoir of the active hormones for delivery to the target organs and cells. Finally, the protein binding equilibrium is the fundamental basis for protein- or immunoprotein-binding assays of hormone, indirect assay of TBG, and the free thyroxine index (FTI) tests of thyroid function.

VII. MECHANISM OF THYROID HORMONE ACTION

A. General Effects of Thyroid Hormones

After the administration of thyroid extracts or T_4, the first physiological effects are noted in 24–28 hours and maximal effects are noted in 7–10 days (Table 21.2). The T_4 requirements for these effects varies. The requirement for T_3 also varies; less is required for equivalent activity and it acts more quickly. T_3 is now recognized as the active form of the thyroid hormone within the target cell. The T_4 that is transported into the target cell is rapidly deiodinated to the active T_3. However, rT_3, which is also produced in the cell, is an inactive form of thyroid hormone.

B. Molecular Basis of Thyroid-Hormone Action

The molecular basis of thyroid-hormone action at the cellular level has been frequently reviewed and a multifaceted concept of its action has evolved (Stanbury and Dumont, 1983). For many years, the mitochondrion was considered to be the site of thyroid-hormone action. Uncoupling of oxidative phosphorylation (ox-phos) in the mitochondria was a viable hypothesis for many years. Under normal conditions, 3 mol of ATP (P) are synthesized per atom of oxygen ($\frac{1}{2}O_2 = O$) used in the cytochrome oxidase system; hence, the P:O ratio equals 3. If less than 3 mol ATP are formed per unit O in a system in the presence of a compound such as the thyroid hormone, the system is said to be uncoupled, that is, the P:O ratio is less than 3. In this event, more O_2 would need to be utilized to generate an equivalent amount of ATP, and O_2 consumption would increase. T_4 has repeatedly been shown to uncouple oxidative phosphorylation in

TABLE 21.2 Effects of Thyroid Hormone

Category	Effect
Clinical	Tremors, nervousness, exophthalmos, hyperactivity, weight loss
Physiological	Increase temperature, heart function
Calorigenic	Increase in basal metabolic rate (O_2 consumption)
Carbohydrate metabolism	Increased glucose turnover, absorption
Protein metabolism	Anabolic, positive N balance
Lipid metabolism	Decrease in blood cholesterol
Development	Stimulation of growth and maturation
Reproductive	Fertility, pregnancy, ovulation
Hematological	Erythropoiesis

in vitro systems, but it does so only in large, unphysiological amounts. These findings were extended to the whole animal to explain the increased oxygen consumption by T_4. The large amounts of oxidative energy not incorporated into ATP increased body temperature and was dissipated as heat. Thus, T_4 action was theorized to be the result of the uncoupling of oxidative phosphorylation.

Another now well-known effect of thyroid hormone is the stimulation of cellular protein synthesis (Tata *et al.*, 1963; Tapley, 1964). This occurs during the latent period when the calorigenic effect of thyroid hormone occurs. T_3 is now known to stimulate messenger RNA (mRNA) transcription, increasing translation and protein synthesis and accounting for the anabolic effects of thyroid hormones. This also means that the site of action is at the cell nucleus.

Another action of thyroid hormone is to stimulate the "sodium pump" (Na–K-ATPase) at the cell membrane, an action that would increase O_2 consumption (Edelman, 1974). Oubain, an inhibitor of Na–K-ATPase, also inhibits the increased O_2 consumption induced by T_4 or T_3. Thus, stimulation of the sodium pump is an important way in which thyroid hormones stimulate increased oxygen consumption and accounts for almost half of the increase.

A direct effect of thyroid hormones on cell membrane transport has also been suggested from studies on 2-deoxyglucose uptake (Segal *et al.*, 1977). Pliam and Goldfine (1977) proposed that T_3 binds to receptors on the cell membrane and is transported by carriers into the cell.

A combination of the preceding mechanisms is the most likely explanation of the mechanism of action of thyroid hormone. Stanbury and Dumont (1983) describe a model patterned after the generalized steroid hormone model in which (1) T_3 is bound to a receptor site on the plasma membrane; (2) T_3 is transported across the membrane into the cell by carrier protein; (3) T_3 in the cell is again bound to a carrier protein; (4) T3 rapidly exchanges between receptor sites on the mitochondrion or nucleus; and (5) T3 interacts in metabolic processes and protein synthesis.

VIII. CATABOLISM AND EXCRETION OF THYROID HORMONES

The thyroid hormones undergo deiodination, conjugation, or oxidative reactions. The deiodination reaction occurs widely in peripheral tissues and is catalyzed by deiodinases specific for each iodine position, and the released iodine is returned to the iodine pool. The details of the specific reactions, however, are yet not known. The deiodinase that catalyzes the deiodination of T_4 to T_3 in target cells is actually participating in an activation reaction. Deiodination at the 3 position of T_4 gives rT_3, the inactive form on the degradation pathway. Within a cell, therefore, some of the T_4 is converted to its active form and some is converted to its inactive form for degradation. Some of the thyroid hormones are conjugated in the liver and excreted in the bile either as glucuronides or as sulfates. Also, the thyroid hormones, being amino acids, are deaminated and decarboxylated in the liver to form corresponding tetraiodoacetic acid and triiodoacetic acid, which are excreted. The small amounts of free hormones in the blood are also excreted by the kidney.

IX. REGULATION OF THE THYROID GLAND

A. Production and Regulation of Thyroid-Stimulating Hormone

The thyroid gland is under the control of thyroid-stimulating hormone, secreted by the anterior pituitary gland, which in turn is mediated through the hypothalamus and its thyrotrophin-releasing factor (TRF). TRF is synthesized in the anterior hypothalamus by neurosecretory cells and then transported down axonic processes to the anterior pituitary. There, TRF stimulates the thyrotrophic cells to synthesize TSH. The secretion of TSH is regulated by a classic negative-feedback control system based on the product of the target gland, thyroxine. A high plasma concentration of free hormone, fT_4, depresses TSH secretion by the pituitary and to some extent, the TRF from the hypothalamus.

The pituitary, with its direct inhibition by increased fT_4, is the faster-responding of the two tissues. At low concentrations of fT_4, however, the hypothalamus responds quickly by increasing its synthesis and release of TRF, which in turn increases the synthesis and release of pituitary TSH. In consequence, this regulatory mechanism, which is sensitive to either high or low concentrations of fT_4, maintains the circulatory hormone concentrations at the normal homeostatic level for that species.

B. Action of Thyroid-Stimulating Hormone on the Thyroid Gland

TSH has a number of direct effects on the thyroid gland: The gland increases in size, the height of the follicular cell increases, and there is a loss of colloid. The response of the thyroid gland to TSH is modified by the intake of stable iodine. When iodine intake is high, the action of TSH is suppressed and the height, size, and activity of the follicular cells are decreased. When iodine intake is low, there is a compensatory hypertrophy of the gland with an increase in number, height, and size of the follicular cells. This is characteristic of iodine-deficiency goiters or nontoxic goiters.

C. Action of Thyroid-Stimulating Hormone on Hormonogenesis

The effects of TSH upon hormone synthesis by the thyroid follicle are outlined in Fig. 21.1 and discussed in Section V. TSH affects every reaction in the hormone synthetic pathway, beginning with the uptake of iodine, its activation, tyrosine iodination, coupling of the iodinated phenyls, and hydrolysis of colloid. The two most important sites of TSH action are the initial uptake of iodine and the final hydrolysis of colloid to release hormones into the circulation.

D. Long-Acting Thyroid Stimulators

A second thyroid stimulatory factor has been identified in the serum of human patients with Graves' disease. This factor appears to be closely involved in the mechanism of thyrotoxicosis (Adams, 1958; McKenzie, 1958). It differs from TSH in that it is cleared from the blood more slowly and produces its thyroid stimulating effects many hours (8–24) after that of TSH. It is therefore referred to as long-acting thyroid stimulator, or LATS. Studies of LATS have been largely in humans, where it has been closely correlated with hyperthyroidism (Lipman, 1967). Since the initial observations, many other forms of delayed-acting thyroid stimulators have been found, and all, including LATS, are now known to be immunoglobulins of the IgG class. These thyroid-stimulating immunoglobulins (TSI) or thyroid-stimulating antibodies (TSAb) do not appear to have a role in normal thyroid physiology, nor have corresponding stimulators been found in animals.

X. THYROID FUNCTION TESTS

The diagnosis of thyroid disease is usually obscured by the very nonspecific nature and variety of clinical signs. A thorough physical examination is essential for the detection of a potential thyroid disease and as a rationale for laboratory tests of thyroid function. Initial routine laboratory tests such as hematology or urinalysis are of limited value. The initial parallel biochemical screen is also limited except for hypercholesterolemia. In consequence, the specific tests of thyroid function are of great importance in the diagnosis of thyroid disease. Fortunately, improvements of standard tests and the development of new tests have made several specific and direct tests of thyroid function readily available to the veterinary clinician.

A. Indirect Tests of Thyroid Function

1. Hematology

A moderate normocytic normochromic anemia is sometimes associated with clinical hypothyroidism in the dog. This anemia has also been observed in human hypothyroidism and in experimental animals and is known to be of a depression type or the anemia of chronic disease (Cline and Berlin, 1963). The stained blood smear characteristically has little or no evidence of active erythrogenesis such as anisocytosis, polychromasia, or nucleated red cells. Leptocytosis may be prominent in some cases. The hemogram, therefore, is characteristic of the nonresponsive anemia of chronic diseases such as neoplasia or chronic infection. This anemia is not diagnostic for hypothyroidism, but conversely, in cases of unexplained hypoplastic or nonresponsive anemia, hypothyroidism is an important differential diagnosis.

2. The Basal Metabolic Rate

The basal metabolic rate (BMR) measures the O_2 consumption of an animal under rigidly standardized basal conditions and is the classical test of thyroid function in humans. The test is based upon the decreased O_2 consumption in hypothyroidism and the increased O_2 consumption found in hyperthyroidism. The BMR has found little application in clinical veterinary medicine because of the difficulty of maintaining animals under basal conditions. Even under anesthesia, basal conditions are seldom maintained.

3. *Cholesterol*

The serum cholesterol generally varies inversely with thyroid activity. The net effect of thyroid hormone on cholesterol metabolism is to increase the rate of its catabolism by the liver (Koppers and Palumbo, 1972), thereby lowering the cholesterol. In hypothyroidism, the net effect is a decrease in cholesterol catabolism and an increase in cholesterol.

The cholesterol is carried in dog plasma equally by the low-density lipoproteins (LDL-Chol) and the high-density lipoproteins (HDL-Chol) (Mahley and Weisgraber, 1974). In humans, cholesterol is carried mainly by the LDL-Chol and only about 20% by the HDL-Chol. This partitioning of cholesterol is important because increased LDL-Chol is associated with atherosclerosis, whereas increased HDL-Chol is associated with a reduced risk for heart disease. A detailed discussion of cholesterol metabolism and its association in various diseases is given in a separate chapter. Table 21.3 gives the normal cholesterol reference ranges of a number of domestic animals.

Early on, increased cholesterol was often the only index of thyroid function used because the protein-bound iodine (PBI) was unreliable. Cholesterol alone is also of limited value because hypercholesterolemia can result from a variety of causes unrelated to thyroid activity. These include the diet, nephrotic syndrome, hepatic function, biliary obstruction, and diabetes mellitus. The diagnostic accuracy of serum cholesterol for hypothyroidism in the dog is about 66%. However, when the concentrations are very high, >500 mg/dl (>12.9 mmol/liter), and diabetes mellitus is eliminated, serum cholesterol's diagnostic accuracy is greatly increased. Therefore, increased cholesterol is a signal for further investigation of thyroid disease. Recently, cholesterol used in conjunction with the fT_4 was found to have the highest predictive value among a battery of thyroid function tests (Larsson, 1988).

Similarly, hypocholesterolemia has little value as an index of hyperthyroidism. On the other hand, cholesterol decreases consistently in response to thyroxine replacement therapy, so it has value as a guide to

TABLE 21.3 Serum Cholesterol Concentration in Various Species

Species	Total (mg/dl)	Free (mg/dl)	Ester (mg/dl)	Reference
Cat	95–130			Kritchevsky (1958)
	93 ± 24	30 ± 10	63 ± 23	Boyd (1942)
	98 ± 7.3			Morris and Courtice (1955)
Cow	80–120			Kritchevsky (1958)
Pregnant, nonlactating	110 ± 32	37 ± 15	73 ± 15	Boyd (1942)
Lactating, nonpregnant	241.9			Lennon and Mixner (1957)
Heifer, 15–18 months	96.1			Lennon and Mixner (1957)
Calves	105.6			Lennon and Mixner (1957)
	123			Lennon and Mixner (1957)
Dog	125–250			Kaneko (1980)
	110–135			Kritchevsky (1958)
	110 ± 28	51 ± 20	59 ± 19	Boyd (1944)
	194 ± 35			Morris and Courtice (1955)
Low-fat diet	140			Boyd and Oliver (1958)
High-fat diet	280			Boyd and Oliver (1958)
Fowl (chicken, ducks, geese)	100–200			Kritchevsky (1958)
Cockerel	100 ± 23	34 ± 9	66 ± 19	Boyd (1942)
Nonlaying hens	116 ± 2–152 ± 2			Leveille *et al.* (1957)
Laying hens	208 ± 15–285 ± 24			Leveille *et al.* (1957)
Goat	80–130			Kritchevsky (1958)
Horse	83–140			Kritchevsky (1958)
	96.8 ± 2.8	15.7	81.1	Norcia and Furman (1959)
	128 ± 12			Morris and Courtice (1955)
Rabbit	30–70			Kritchevsky (1958)
	45 ± 18	22 ± 13	23 ± 12	Boyd (1942)
Sheep	70			Boyd and Oliver (1958)
	64 ± 12			Morris and Courtice (1955)
Swine (low-fat diet)	35.6 ± 53.7	5.7–10.9	28–48	Reiser *et al.* (1959)

therapeutic response. In thyroidectomized horses with clinical evidence of hypothyroidism, Lowe *et al.* (1974) demonstrated a 50% decrease in serum cholesterol shortly after feeding iodinated casein.

B. Direct Tests of Thyroid Function

A direct approach to thyroid evaluation is to measure the amount of the hormones in the blood. Because there are thyroid inhibitory effects among a wide variety of iodine-containing compounds, it is critical that any form of iodine-containing medication, including thyroid hormones, be uncovered in the history of the patient. As a general rule, any iodine-containing medication or thyroid hormones being given to the patient should be withdrawn for at least 2 weeks before any thyroid function tests are undertaken.

1. *Serum Protein-Bound Iodine*

The classic protein-bound iodine test measured all the precipitable protein-bound hormonal forms (T_4, T_3, rT_3) plus all coprecipitable iodine. Clearly, this method overestimated the amounts of hormone, but it was a useful index of thyroidal activity because the PBI is proportional to the amount of thyroid hormones. Because most of the iodine bound to plasma protein is hormonal iodine, the PBI mainly represents hormone concentration. The partitioning between hormonal and nonhormonal forms is quite constant, so that the PBI is directly proportional to the amount of thyroid hormone.

The PBI method is fraught with logistical and technical difficulties and, though useful in its time, is no longer used. The sample and the chemical procedure are readily contaminated by any source of iodine: antiseptics, internal or external iodine-containing medicaments, iodinated radiographic contrast media, and thyroid medications (Table 21.4). Reliability is improved by prior extraction of the thyroid hormones from the serum with butanol (butanol-extractable iodine, BEI). A further refinement was to separate the hormones by column chromatography (Pilleggi *et al.*, 1961; Fisher *et al.*, 1965). Automation of the iodine assay increased its reliability, but immunoassays for thyroid hormones are now the tests of choice. Reference ranges for animals are given in Table 21.5; note that these ranges are considerably lower than in humans.

2. *Competitive Protein-Binding Assays*

The elucidation of the protein-binding equilibrium and the competition between labeled and nonlabeled hormones for binding sites on the proteins provided the basis for the determination of T_4, and indirectly, the TBG and fT_4:

$$\text{TBG–T}_4 \xrightarrow{k} \text{TBG} + \text{fT}_4. \tag{21.1}$$

Since the TBG–T_4 is the total T_4, one can substitute and reformulate Eq. (21.1):

$$\text{fT}_4 = k\frac{\text{T}_4}{\text{TBG}}. \tag{21.2}$$

It can be seen that the amount of fT_4 is directly proportional to the amount of T_4 and inversely proportional to the amount of TBG, and that knowing these two factors permits the calculation of fT_4. The calculated fT_4 is expressed as the free thyroxine index (FTI), but it is not valid for animals. T_4 is now measured directly by competitive protein binding (CPB) assay, by radioimmunoassay (RIA), or by enzyme-labeled immunoassay (EIA). The TBG is measured indirectly by competitive binding of labeled T_3 to a secondary binding agent such as red cells, resins, charcoal, or hemoglobin. TBG can also be assayed directly by immunoassay.

a. Thyroxine by Competitive Protein Binding

The competitive protein binding (T_4-CPB) method was the forerunner of all the competitive protein binding methods for T_4 and does not depend on the colorimetric determination of iodine (Murphy and Pattee, 1964). The method is based on the competitive binding of TBG for patient T_4 and ^{131}I-labeled T_4. The labeled T_4 and patient T_4 bind to the TBG in proportion to their concentrations, so that labeled T_4 binding is inversely proportional to patient T_4 concentration. The use of TBG has been replaced by immunoglobulins as the binding proteins, so CPB is now seldom used. Reference values in animals are given in Table 21.6.

There is minimal interference from diphenylhydantoin and salicylates when given in large amounts (Sparagana *et al.*, 1969), so T_4-CPB is a reliable method (Lucis *et al.*, 1969). Comparisons with T_4-RIA indicated no significant difference (Kaneko *et al.*, 1975). Also, in horses, T4-CPB decreased dramatically to almost zero at 2 days after thyroidectomy (Lowe *et al.*, 1974). Feeding of iodinated casein to these horses resulted in a rapid rise of T_4 to above normal. Iodinated casein contains about 2.5 μg (3.85 nmol) T_4 and 1.25 μg (1.61 nmol) T_3 per gram (Kaneko, 1979).

b. Thyroxine by Radioimmunoassay

Thyroxine by radioimmunoassay (T_4-RIA) is a typical immunoassay method in which an antibody is used as the binding protein. Owing to their high specificities, antibodies are useful for the assay of proteins, polypeptides, and haptens, including most of the hormones.

TABLE 21.4 Some Drugs Affecting Thyroid Function Tests

Compound	Effect on ^{131}I uptake	Effect on PBI	Effect on T_3 uptake	Duration of effect (average period)
Iodides				
Lugol's, cough syrup, vitamin preparations	Decrease	Increase	None	10–30 days
Iodine antiseptics	Decrease	Increase	None	10–30 days
Iodine-containing drugs	Decrease	Increase	None	10–30 days
X-Ray contrast media				
Iodoalphionic acid (pheniodal)	Decrease	Increase	None	3–12 months
Iodopyracet (Diodrast)	Decrease	Increase	None	2–7 days
Iodized oil (oleum iodatum)	Decrease	Increase	None	0.5–3 years
Most iv contrast media	Decrease	Increase	None	2–6 weeks
Most gallbladder media	Decrease	Increase	None	3 weeks–3 mo
Hormones				
Thyroid extract	Decrease	Increase	Increase	4–6 weeks
Triiodothyronine	Decrease	Decrease	Decrease	2–4 weeks
ACTH	Decrease	None	Increase	2 weeks
Estrogens	None	Increase	None	—
Oral contraceptives	None	Increase	None	—
Androgens	None	None	Increase	—
Thiocarbamide compounds				
Thiouracil	Decrease	Decrease	None	5–7 days
Propylthiouracil	Decrease	Decrease	None	5–7 days
Thiocyanate	Decrease	Decrease	None	14–21 days
Phenylbutazone	Decrease	None	Increase	14 days
Bromides	Decrease	None	—	10–30 days
Antihistamines	Decrease	None	—	7 days
Diphenylhydantoin (dilantin)	None	Decrease	Increase	7–10 days
Salicylates	None	None	Increase[a]	—

[a] Effect depends on the method.

The principles and details of radioimmunoassay are described in the chapter on clinical enzymology. Polyclonal antibodies give accurate results and are usually used for these hormone assays. Reference values in animals are given in Table 21.6. Belshaw and Rijnberk (1979) report that other than in hypothyroidism of dogs, only diphenylhydantoin, corticosteroids, or Cushing's disease result in very low concentrations of either T_4-RIA or T_3-RIA. This further emphasizes the need for a careful history of the patient prior to assay for thyroid hormones. The mean normal T_4-RIA in dogs is 2.3 ± 0.8 μg/dl (29.6 ± 10.3 nmol/liter) with an observed range of 0.6–3.6 μg/dl (7.7–46.3 nmol/liter) (Kaneko *et al.*, 1978), which is comparable to the 1.5–3.6 μg/dl (19.3–46.3 nmol/liter) of Sims *et al.* (1977) and Belshaw and Rijnberk (1979). T_4-RIA is now widely used in cats because of the high prevalence of hyperthyroidism in this species. The reference range for cats is 0.1–2.5 μg/dl (1.3–32.3 nmol/liter). In the horse, T_4-RIA is also quite low , 0.9–2.8 μg/dl (11.6–36.0 nmol/liter). Messer *et al.* (1995) report a similar finding, a mean T_4-RIA of 21.42 ± 3.46 nmol/liter in 12 adult horses.

c. *Thyroxine by Enzyme-Labeled Immunoassay*

An important advance in hormone assay is the development of an enzyme-labeled immunoassay (EIA) test comparable in every way to radioimmunoassay except that the labeling is by an enzyme rather than radioiodine. This method has some obvious advantages in that there is no need to use radioactivity and the enzyme can be assayed in any laboratory. One system labels T_4 with malate dehydrogenase (MD), which, in competition with unlabeled T_4, binds to antibody. This method has the acronym EMIT, for enzyme multiplied immunosorbent test.[1] This method has another advantage in that the labeled and unlabeled fractions need not be separated. The MD is inactive when bound to T_4; when it binds to the immunoglobulin, it is activated. The activity of malate dehydrogenase is assayed by standard enzyme methodology, and the T_4 is read from a standard curve as in a radioimmunoassay.

Another system uses two separate recombinant fragments of the enzyme β-galactosidase (Horn *et al.*,

[1] BioRad Laboratories, Richmond, CA 94804; ICL Scientific, Fountain Valley, CA 92708.

TABLE 21.5 Concentration of Serum Protein-Bound Iodine in Domestic Animals

Species	PBI (μg/dl serum)[a]	Reference
Dog	2.3 (1.5–3.5)	Siegel and Belshaw (1968)
	3.4 ± 1.0 (1.5–5.1)	Mallo and Harris (1967)
	2.3 ± 0.8 (1.1–4.3)	Quinlan and Michaelson (1967)
	1.99 ± 0.24	Theran and Thornton (1966)
	2.6 ± 0.18	O'Neal and Heinbecker (1953)
	2.7 (1.8–4.5)	Kaneko (1980)
Cat	3.5 (2.5–6.0)	Kaneko (1980)
Horse, Thoroughbred	1.86 ± 0.29 (1.2–2.5)	Irvine (1967b)
	2.2 ± 0.6 (1.5–3.5)	Kaneko (1964)
	(1.67–2.7)	Trum and Wasserman (1956)
Pig		
Landrace	2.7 ± 0.1	Sorenson (1962)
Large white	4.4 ± 0.2	Sorenson (1962)
Dairy Cattle	(2.73–4.11)	Long *et al.* (1951)
Lactating	3.7 ± 0.3	Sorenson (1962)
Pregnant heifers	5.0 ± 0.7	Sorenson (1962)
Nonpregnant heifers	3.3 ± 0.1	Sorenson (1962)
Beef cattle	2.19	Long *et al.* (1951)
Steer	2.5 ± 0.3	Sorenson (1962)
Sheep	3.8 ± 1.0 (3.6–4.0)	Hackett *et al.* (1957)
Lamb	(4–13)	Falconer (1987)
Goat, miniature	4 (2–5)	Ragan *et al.* (1966)

[a] Values are means with their standard deviations, if available. Ranges are given in parentheses.

1991). The individual fragments, enzyme donor (ED) and enzyme acceptor (EA), are inactive, but when they recombine, they form the active enzyme. Thyroxine is bound to ED. In the presence of thyroxine antibody, the thyroxine–ED–antibody complex inhibits recombination with EA and no active β-galactosidase is formed. When sample thyroxine and a standard amount of thyroxine–ED are mixed, they compete for a standard amount of antibodies. Thus, unbound thyroxine–ED accumulates in direct proportion to the amount of sample thyroxine. The unbound thyroxine–ED is free to bind with EA to form the active β-galactosidase. In this case, enzyme activity is directly proportional to the amount of thyroxine. This method had a very high correlation coefficient compared to that of an RIA (r = 0.969), an EIA (r = 0.966), and a fluorescence polarization immunoassay (FPIA, r = 0.939).

Another variation of the nonisotopic immunoassay is the fluorescence immunoassay, in which a fluorochrome is tagged to the T_4. The T_4 concentration is inversely proportional to the fluorescence, as in the RIA. This is also a very sensitive test, but it requires a sensitive spectrofluorometer for the assay.

Enzyme immunoassays have not replaced radioimmunoassays for hormones, but the principle is now widely used for antigen or antibody assays using horseradish peroxidase (HRP) as the enzyme label. The procedure is popularly known as the enzyme-labeled immunoadsorbent assay, or by its acronym, ELISA. This procedure has been adapted for T_4 by labeling T_4 with HRP, then coupling the HRP to a dye that indicates enzyme activity and hence T_4 concentration.

d. *Triiodothyronine by Radioimmunoassay*

Triiodothyronine is assayed by radioimmunoassay (T_3-RIA), and reference values in animals are given in Table 21.6. In the dog, the mean normal T_3-RIA is 107 ± 18 ng/dl (1.6 ± 0.3 nmol/liter) with an observed range of 82–138 ng/dl (1.26–2.12 nmol/liter) (Kaneko, 1980). It closely parallels T_4-RIA in the dog, so that the simultaneous determination of T_4-RIA and T_3-RIA will increase the diagnostic accuracy of either one alone. In cats, T_3-RIA is less widely used than T_4-RIA. The reference range for cats is 15–50 ng/dl (0.23–0.77 nmol/liter). In the horse, Messer *et al.* (1995) report a mean T_3-RIA of 0.85 ± 0.52 nmol/liter.

3. *"Free" Thyroxine and "Free" Triiodothyronine*

"Free" thyroxine (fT_4) is the unbound fraction of the total circulating T_4, and its concentration is controlled by the equilibrium between TBG and TBG-T_4 (Section III). Equilibrium dialysis or ultrafiltration are the best methods for determining the free hormones, but they are too labor intensive for use in most clinical

TABLE 21.6 Serum Thyroxine and Triiodothyronine Concentrations in Animals[a]

Species	T_4 CPB (μg/dl)	T_4 RIA (μg/dl)	T_3 RIA (ng/dl)	Reference
Dog	0.3–2.3 (1.3 ± 0.5)			Kaneko (1980)
		0.6–3.6 (2.3 ± 0.8)		Kaneko *et al.* (1978)
		0.74–4.1 (2.48 ± 0.52)	45–175 (94 ± 24)	Belshaw and Rijnberk (1979)
			82–138 (107 ± 18)	Kaneko (1980)
	1.13–4.71 (3.1 ± 1.1)			Kallfelz (1973)
		1.4–3.6	45.4–117.5	Larsson (1988)
Cat	0.1–2.5 (0.1 ± 0.5)			Kaneko (1980)
Male	1.0–3.6 (1.7 ± 0.5)			Bigler (1976)
Female	0.8–3.7 (2.1 ± 0.6)			Bigler (1976)
	(0.95 ± 0.5)			Ling *et al.* (1974)
		0.8–3.8 (2.1 ± 0.1)	15–104 (52.3 ± 1.8)	Peterson *et al.* (1983)
Horse	0.9–2.8 (1.9)			Kaneko (1980)
	1.46–3.38			Kallfelz and Lowe (1970)
	(2.57 ± 0.71)			Thomas and Adams (1978)
	(2.70)			Irvine and Evans (1975)
		(1.55 ± 0.27)	(55.34 ± 33.85)	Messer *et al.* (1995)
Foal		2.0–3.3 (2.67 ± 0.50)	110–130 (117 ± 12)	Osame and Ichijo (1994)
Weanling		2.1–2.8	48–62	Glade and Luba (1987)
Cow	4.2–8.6 (6.3 ± 1.0)			McCrady *et al.* (1973)
	(5.10 ± 1.30)			Kallfelz and Erali (1973)
Pig	(2.10 ± 0.42)			Kallfelz and Erali (1973)
Goat	(5.25 ± 2.08)			Kallfelz and Erali (1973)
Sheep	(6.05 ± 1.64)			Kallfelz and Erali (1973)
Monkey (*Macaca mulata*)		(4.1 ± 0.6)	(160 ± 34)	Belchetz *et al.* (1978)
Baboon		(9.9 ± 2.2)	(121 ± 18)	Maul *et al.* (1982)

[a] Observed ranges; means ± SD in parentheses.

laboratories. An equilibrium dialysis method for free thyroxine has recently become commercially available. Other methods for free thyroid hormone assay have been reviewed by Wilke (1986). Labeled thyroxine-analog methods of determining free T_4 are inaccurate and should not be used (Alexander, 1986). Free T_4 as determined by RIA (fT_4-RIA), however, is accurate and ought to be more widely used in conjunction with the T_4-RIA and T_3-RIA.

The fT_4 concentration for dogs is 0.52–2.7 ng/dl (6.7–34.7 pmol/liter) (Larsson, 1988) and for the beagle dog is 3.53 ± 0.34 ng/dl (45.4 ± 4.4 pmol/liter) (Michaelson, 1969). Eckersall and Williams (1983) found the total T_4-RIA by a commercial kit for humans to be inaccurate, but the fT_4-RIA was highly accurate and readily discriminated hypothyroid dogs. Larsson (1988) found that the fT_4 and the serum cholesterol were the best combination of initial screening tests to detect hypothyroid dogs. On the other hand, in the hyperthyroid cat, Hays *et al.* (1988) found no differences in dialyzable T_4 from the normal. They infer that the total T_4 is sufficient for diagnosis and that the free hormone is not needed in the cat. In the horse, Messer *et al.* (1995) report a mean fT_4-RIA of 14.0 ± 1.16 pmol/liter.

The fT_3-RIA parallels fT_4-RIA in its binding characteristics. Since fT_3 is the physiologically active form of the hormone, it is potentially the single most reliable test of thyroid function. In the horse, Messer *et al.* (1995) report a mean fT_3-RIA of 0.89 ± 0.53 pmol/liter.

4. Triiodothyronine Uptake

The uptake of radiolabeled triiodothyronine by red cells (T_3 uptake) *in vitro* as a test of thyroid function was reported by Hamolsky *et al.* (1959). This test measured the partitioning of radioiodine-labeled T_3 between TBG and a secondary binding agent, in this case, RBC. Serum TBG is the primary binding agent, and it binds T_4 more firmly than it does T_3 (Section VI). Therefore when ^{131}I-T_3 is added to the serum, it will bind first to any unbound receptor sites on the TBG. When RBC or another secondary binding agent such as resin or charcoal is added, excess unbound labeled T_3 will bind to the secondary agent. The radioactivity of the RBC (or resin) will thus be inversely proportional to the amount of unbound TBG. The T_3-uptake test is therefore an indirect measure of the amount of unbound TBG. The uptake is low in hypothyroidism where the low T_4 leaves excess unbound binding sites on the TBG and high in hyperthyroidism where there are few binding sites on the TBG. The T_3 uptake should not be confused with the T_3-RIA.

There are many modifications to the test, basically in the choice of secondary binding agents. Pain and Oldfield (1969) surveyed six T3-uptake methods including the red-cell uptake, resin uptake, resin-sponge uptake, Sephadex, charcoal–hemoglobin, and the thyroid binding index. They concluded that the original RBC T_3 uptake gave overall satisfactory results, but only two of the newer methods, Sephadex and charcoal, gave satisfactory results for both hyper- and hypothyroidism. On the basis of ease of performance, the T_3 uptake by charcoal–hemoglobin (Irvine and Standeven, 1968) is the test of choice.

The T_3-uptake test is widely used in humans for the indirect assay of unbound TBG, from which the free thyroxine index (FTI) is calculated. In animals, however, the results are so poorly correlated with thyroid activity as to be virtually useless. Wide variations in the hormone binding proteins and their degree of binding account for these inaccuracies (Table 21.1).

5. Thyroxine Uptake

This assay is a modification of the β-galactosidase EIA described in Section VIII.B. (Horn *et al.*, 1991). In this case, no antibodies are added to the system, and the thyroxine enzyme donor fragment is free to bind with TBG in the serum. The remaining thyroxine enzyme donor fragment binds to the thyroxine acceptor fragment to generate the active enzyme. Thus, the activity of β-galactosidase is inversely proportional to the amount of TBG in the serum.

6. Free Thyroxine Index

The total T_4 represents the thyroxine bound to TBG (T_4–TBG) and is readily measured as the T_4-RIA. The T_3 uptake or the T_4 uptake as described earlier represents the unbound TBG and, as indicated previously, is also easily measured. Referring to the equilibrium equation (21.2), it can be readily seen that

$$fT_4 = k\frac{T_4\text{–TBG}}{\text{TBG}} = T_4 \times T_3 \text{ uptake (or } T_4 \text{ uptake)} = \text{FTI}.$$

The product, $T_4 \times T_3$ uptake, is the free thyroxine index (FTI) and is the calculated index of the amount of fTT_4. The FTI is still widely used in humans, where it correlates well with thyroid disease (Wilke and Eastment, 1986), but is invalid for use in animals. The free hormones should be determined in animals using direct fT_3 and fT_4 immunoassays.

7. Thyroid-Binding Globulin, Thyroglobulin, and Thyroid Autoantibodies

Thyroid-binding globulin (TBG) and thyroglobulin (Tg) or colloid are measured by RIA. The standard technique for thyroglobulin antibodies (TgAb) is hemagglutination, but it is being supplanted by the ELISA (Voller *et al.*, 1980) as a superior method (Roman *et al.*, 1984). Haines *et al.* (1984), using ELISA, detected TgAb in a high percentage of dogs with hypothyroidism, dogs with other endocrine diseases, and dogs that were closely related to the TgAb-positive dogs, but found a low percentage in healthy unrelated dogs. They conclude that thyroid autoimmunity is strongly genetically influenced in the dog.

8. Thyroid-Stimulating Hormone

Serum TSH is measured by radioimmunoassay (TSH-RIA) or by the two-site immunoradiometric assay (IRMA) in humans. The cross reactivity of the anti-human-TSH antibody for dog TSH is insufficient for it to be used as an accurate diagnostic test in the dog. Using the human TSH-RIA assay method for the dog, the normal values were 5.9 ± 4.1 μU/ml (Kaneko *et al.*, 1975), and 3.50 ± 1.67 μU/ml (Larsson, 1981). The human assay method used in monkeys (*M. mulatta*) gave 0.2–2.6 μU/ml (1.53 μU/ml) (Belchetz *et al.*, 1978). A TSH-RIA using an anti-canine TSH has become commercially available specifically for use in dogs.

9. *Thyrotrophin-Releasing Factor*

As with TSH, no reliable thyrotrophin-releasing factor (TRF) assay has been developed for use in domestic animals. Purified TRH, however, is available and is used in the TRF response test.

C. Radionuclide Uptake Tests of Thyroid Function

1. *Radioiodine Uptake*

Radioiodine is taken up by the thyroid gland in exactly the same manner as the nonradioactive isotope, and its uptake remains as one of the most definitive tests for thyroid function. With the development of direct hormone assays, radioiodine uptake tests are now largely used in nuclear medicine for imaging and localizing of active thyroid nodules and "hot spots" and for estimating therapeutic doses of radioiodine. Radioactive ^{99m}Tc-pertechnetate is also taken up and is now widely used for imaging.

The normal uptake at 72 hours is 10–40% in the dog (Kaneko *et al.*, 1959), compared to a range of 7–37% reported by Lombardi *et al.* (1962). The normal uptake at 24 hours is 7–35% and at 48 hours is 8–38%. The correlation of the 72-hour uptake with thyroid function has made it a reliable diagnostic test of thyroid disease in dogs. The normal uptake in cats at 24 hours is 9.2% (Peterson *et al.*, 1983) whereas the maximal uptake is about 33% at 3–5 days postinjection (Broome *et al.*, 1988). Iodine-containing compounds, including exogenous thyroid hormone, also interfere with thyroid uptake (Table 21.4), so the uptake test should be deferred for at least 2 weeks if iodine compounds have been administered.

2. *The Conversion Ratio*

The conversion ratio (CR) gives the fraction of the injected radioiodine that has been converted to thyroxine during the radioiodine uptake test. The serum proteins with their bound radioactive hormones are separated from non-protein-bound serum iodides using a resin and then counted for radioactivity (Scott and Reilly, 1954; Zieve *et al.*, 1956). The result, when expressed as the ratio of protein-bound radioactivity to total plasma radioactivity, is the conversion ratio. Therefore, it reflects the amount of labeled thyroid hormone that was formed by the gland from the injected radioiodine. A CR of 2–6% has been reported (Lombardi *et al.*, 1962) for normal dogs. This test is useful in the diagnosis of hyperthyroidism in humans, but its value in animals is unknown.

3. *Thyroxine Secretion Rate*

The output of thyroxine by the thyroid gland is also a direct indicator of thyroid function and can be determined by several methods. Most data on thyroxine secretion rate (TSR) in domestic animals have been obtained by a technique based on the amount of exogenous L-thyroxine necessary to inhibit the release of radioactivity by the thyroid, that is, by thyroxine suppression. Another method has been to calculate the TSR from the fractional turnover rate of injected radiothyroxine and the T_4 or PBI. The TSR has been determined in many domestic animals, including the cow (Pipes *et al.*, 1963), sheep (Henneman *et al.*, 1955), goat (Flamboe and Reineke, 1959), pig (Sorenson, 1962), horse (Irvine, 1967), and dog (Kallfelz, 1973). A wide variation in the reported values is probably the result of differences in technique and conditions of study. The TSR is also affected by age, lactation, diet, season, and training in the case of horses. In animals, the TSR appears to vary from a low of about 0.108 mg (0.139 μmol)/100 kg/day in the horse to a high of about 0.46 mg (0.59 μmol)/kg/day in the cow or 0.49 mg (0.63 μmol)/kg/day for the dog.

D. Trophic Hormone Response Tests

1. *Thyroid-Stimulating Hormone Response*

The response of the thyroid to TSH injection is a means of evaluating thyroid activity, as well as of differentiating a primary hypothyroidism due to a thyroid lesion from a hypothyroidism secondary to a pituitary lesion. The responsiveness of the thyroid to the TSH injection is evaluated by an increase (or failure to increase) as evidence of thyroid activity (or lack of activity). A variety of dosages of TSH have been used, as well as a variety of procedures, to detect this increase in activity, among them the radioiodine uptake test (Kaneko *et al.*, 1959), the radioiodine uptake curve (Siegel and Belshaw, 1968), the PBI (Siegel and Belshaw, 1968), the T_4-CPB (Hoge *et al.*, 1974), the T_4-RIA (Kaneko *et al.*, 1978), and, more recently, imaging. In the general procedure, thyroidal activity (or lack of activity) is first established by the measurement of serum hormones, and this is followed by the TSH response test. In a primary hypothyroidism where the lesion is localized in the thyroid, there is no response to the exogenous TSH. If the hypothyroidism is due to a pituitary hypofunction with a deficiency of TSH, or a hypothalamic lesion with a lack of TRH, the thyroid will respond to the exogenous TSH, as shown by a significant increase in serum hormone concentrations. Glucocorticoids and phenylbutazone are also well known to depress thyroid activity, so the TSH response test is useful in detecting low hormone concentrations due to drugs or in Cushing's disease.

A useful procedure for the TSH response test is first to obtain a serum sample for baseline T_4 or T_3, then to inject 10 IU of bovine TSH intravenously. Oliver and Waldrop (1983) and Held and Oliver (1984) recommend a minimum of 5 IU for the dog and horse, respectively. After 4 hours, a second serum sample is taken and hormone again measured. The normal response in dogs is a doubling or more of the hormone above baseline level. In the primary hypothyroid individual, there a virtual absence of a response. In the drug-induced or Cushing's patient with low hormone there will be a response to well within the normal hormone concentrations. The secondary (TSH) or tertiary (TRH) hypothyroid patient will have a response similar to the drug-induced or Cushing's patient. In cats, Peterson *et al.* (1983) found T_4 to increase almost threefold above the baseline at 4 hours post-TSH. In the baboon, the TSH response test had peaks of more than double the baseline at 8 and 12 hours for T_3 and T_4, respectively (Maul *et al.*, 1982). In horses, Messer *et al.* (1995), at 6 hours after TSH administration, found significant increases in T_4 , T_3, and fT_4, but not significantly in fT_3. They suggest that the TSH response test may not be as valuable for thyroid disease diagnostics in the horse as it is in the dog and cat.

2. *Thyrotrophin-Releasing Factor Response*

The response to thyrotrophin-releasing factor (TRF) as developed for humans has been used in dogs and cats. In the dog, there was no T_4-RIA, T_3-RIA, or T_3 uptake response to exogenous human TRF (Kraft and Gerbig, 1977). TSH could not be measured, but presumably the human TRF did not stimulate TSH release by the pituitary. Larsson (1981) measured TSH after giving TRH, and, although a significant difference was found, the overlap was too great to be of diagnostic value. Lothrop *et al.* (1984), however, found a doubling of T_4-RIA at 6 hours postinjection in dogs and cats. Therefore, the TRF response test must be further evaluated in animals. Some (Jones, 1993) recommend the TRH response test in cats in preference to the TSH test, particularly in hyperthyroid cats. At 4 hours after the IV administration of TRH, normal cats have a greater than 50% increase in their serum T_4, whereas hyperthyroid cats have a minimal or no increase.

3. *Triiodothyronine Suppression Test*

In cats, T_3 has an inhibitory effect on the secretion of TSH and injection of T_3 is followed by a fall in serum T_4 levels. In hyperthyroid cats, because of their autonomous production of T_4, administered T_3 will have no suppressing effect on their high levels of serum T_4. For the test, after a pretest serum sample is taken, 25 μg T_3 is given t.i.d. for 2 days. Four hours after the last dose, a serum sample for T_4 is taken. In hyperthyroid cats, there is little or no fall in serum T_4, in contrast to a greater than 50% suppression in normal cats (Jones, 1993).

XI. DISEASES OF THE THYROID

Thyroid disease has been most extensively studied and reviewed in the dog (Belshaw, 1983). In the cat, the incidence of hyperthyroidism has increased greatly in the last decade and is the most frequently encountered endocrinopathy in this species (Peterson *et al.*, 1994). In the horse, Lowe *et al.* (1974) described the clinical effects of experimental thyroidectomy in mares and stallions, and hypothyroidism remains as a frequent consideration in breeding problems. In ruminants, a congenital goiter has been described in Merino sheep and in Afrikander cattle. Local enlargements or nodules are also seen in all animals; these are usually benign tumors.

A. Goiter

Goiter may be defined as an enlargement of the thyroid gland that is not due to inflammation or malignancy. There are two general types of goiters: (1) nontoxic goiters, which produce either normal amounts of hormone (simple goiter) or below-normal amounts of hormone (hypothyroid), and (2) toxic goiters, which produce excess amounts of hormone (hypertrophy). Furthermore, a defect or deficiency at any trophic step can also result in thyroid disease. Iodine deficiency (endemic goiter) is well known in iodine-deficient areas of the world. Goitrogenic materials, either natural substances or drugs, induce goiters by their blocking effects on steps in the hormonogenic pathways. There are also rare types of familial goiters associated with defects in hormone synthesis (dyshormonogenesis) in humans (Stanbury and Dumont, 1983) and that find their counterparts in Merino sheep (Rac *et al.*, 1968) and in Afrikander cattle (Van Zyl *et al.*, 1965). Falconer (1987) reviewed the congenital goiter in the Merino sheep in which the fundamental defect is a failure of thyroglobulin synthesis. The goiter is inherited as an autosomal recessive and is frequently seen in Australia. The similar congenital goiter in the Afrikander cattle (Ricketts *et al.*, 1985) is also inherited as an autosomal recessive. These cattle have a thyroglobulin synthesis defect involving defective gene splicing of the thyroglobulin gene transcript. In Bongo antelopes, goiter seen in a group of adults was associated with synthesis of an abnormal 19S thyroglobulin (Schiller *et al.*, 1995). The goitrous antelopes were hormonally euthyroid, but had other manifestations of hypothyroidism such as reproductive difficulties. Iodine-deficiency goiter observed in seven thoroughbred foals in the northern island of Hokkaido, Japan (Osame and Ichijo, 1994), was attributed to iodine deficiency in the soil of the region. All foals had readily detectable thyroid enlargements, and four of the seven had clinical signs of thyroid deficiency as well. The goiters receded after iodine supplementation of the feed.

Simple goiter is a compensatory increase in thyroid glandular mass (hyperplasia and hypertrophy) so that the gland maximizes iodine uptake and is able to synthesize and release a normal amount of T_4. At this time, the patient is physiologically normal, but the gland can become quite large. Ultimately, in iodine deficiency, the goitrous gland fails to synthesize sufficient T_4 and hypothyroidism occurs.

B. Hypothyroidism

Hypothyroidism may be the result of a variety of causative factors. Thyroiditis, with similarities to Hashimoto's thyroiditis in humans, has been reported in about 12% of beagle dogs (Beierwaltes and Nishiyama, 1968). Antithyroglobulin antibodies were found in these dogs. In the adult dog, follicular atrophy is probably the most common cause of hypothyroidism (Clark and Meier, 1958). Finally, hypothyroidism may be secondary to a pituitary insufficiency.

The hypothyroid dog is typically obese, lethargic, has myxedema, a dry skin, and a sparse hair coat. Hypothyroidism is therefore an important differential in the diagnoses of dermatoses. The requirement of T_4 for normal reproduction, growth, and development is well known, so hypothyroidism is an important differential in reproductive failures. Experimentally thyroidectomized mares and stallions failed to grow, were lethargic, had coarse, dull hair coats and increased serum cholesterol (Lowe *et al.*, 1974). Hypothyroid horses tend toward obesity and crestiness.

In the initial screen, an increased cholesterol is often the first clue to hypothyroidism. Larsson (1988) concluded that fT_4 and cholesterol are the best indicators of canine hypothyroidism. Definitive laboratory findings in hypothyroidism of animals are a low T_4 and/or T_3 with little or no response to the TSH response test. Therefore, the recommended algorithm is to first obtain the total T_4 and T_3 (and the fT_4 and fT_3 if available). If the results are equivocal, this is followed by the TSH response test. Other definitive studies such as the ^{131}I uptake and the ^{99m}Tc-pertechnetate scans can be used in specialized hospital settings for the identification of isolated thyroid nodules.

In human thyroid disease diagnostics, TSH is now considered to be the single best test of thyroid status (Beckett, 1994) and to be more cost-effective than T_4. TSH can now be readily assayed with a functional sensitivity of 0.01–0.02 mU/liter. Because of this sensitivity, human thyroidologists now recommend that TSH be used as the initial test and that T_4 or preferably fT_4 be used only on a selected basis. In the event that similar degrees of accuracy and sensitivity evolve for dog TSH, this may well become the definitive test for hypothyroidism in the dog.

C. Hyperthyroidism

Hyperthyroidism or toxic goiter is characterized by weight loss, hyperactivity, a voracious appetite, and increased thyroid hormones. Hyperthyroidism is rarely seen in dogs (Meier and Clark, 1958), but in the cat, in the past decade, the high incidence of hyperthyroidism has been recognized as a common endocrinopathy (Holzworth *et al.*, 1980; Peterson *et al.*, 1983). The most common form of hyperthyroidism in the cat is a functional thyroid adenoma.

Increases in T_4 and/or T_3 are virtually pathognomonic signs of hyperthyroidism in the cat. In 131 cases, T_4 was increased to between 4.0–54.1 μg/dl (51.5–696.3 nmol/liter) in all cats, and T_3 was between 54–1000 ng/dl (0.83–15.36 nmol/liter) in 97% of the cats (Peterson *et al.*, 1983). Hays *et al.* (1988) suggest that T_4 and T_3 are sufficient for diagnosis and that the free hormones are not needed for the diagnosis. In the 131 cases, the mean 24-hour ^{131}I uptake was 39.1%, compared to 9.2% in normal cats. There was no increment of response above the baseline value to the TSH response test in these cats, as would be expected for a tumor. ^{99m}Tc-pertechnetate scans demonstrated increased uptake and size in one or both lobes of the thyroid.

The TRH response test is equivocal for use in the hyperthyroid cat, but Jones (1993) found little or no increase in serum T_4 in these cases. The T_3-suppression test similarly had no decrease in serum T_4 in hyperthyroid cats and is currently considered to be the most useful of the function tests in hyperthyroid cats.

D. Tumors of the Thyroid Gland

Except for the dog and cat, tumors of the thyroid are rare in animals (Lucke, 1964). In the dog, Brodey and Kelly (1968) found no evidence of clinical thyroid disease associated with thyroid tumors. Loar (1986), however, found that about 20% of thyroid tumors were functional. Belshaw (1983), using scintigraphic imaging in dogs, identified functional thyroid tumors and their metastases. Interestingly, a majority of thyroid carcinomas in dogs were associated with high serum thyroglobulin (Tg), but there was no direct correlation with T_4 (Verschueren and Gosling, 1992).

In feline hyperthyroidism, a functional thyroid adenoma is the most common finding.

References

Adams, D. D. (1958). *J. Clin. Endocrinol. Metab.* **18,** 699.

Akasha, M., and Anderson, R. R. (1984). *Proc. Soc. Exp. Biol. Med.* **177,** 360.

Alexander, N. M. (1986). *Clin. Chem.* **32,** 417.

Barua, J., Cragle, R. G., and Miller, J. K. (1964). *J. Dairy Sci.* **47,** 539.

Beckett, G. J. (1994). *JIFCC* **6,** 368.

Beierwaltes, W. H., and Nishiyama, R. H. (1968). *Endocrinology* **83,** 501.

Belchetz, P. E., Gredley, G., and Himsworth, R. (1978). *J. Endocrinol.* **76,** 427.

Belshaw, B. E. (1983). *In* "Textbook of Veterinary Internal Medicine," Vol. II (S. J. Ettinger, ed.), 2nd ed., p. 1592. Saunders, Philadelphia, Pennsylvania.

Belshaw, B. E., and Rijnberk, A. (1979). *J. Am. Anim. Hosp. Assoc.* **15,** 17.

Belshaw, B. E., Cooper, T. B., and Beeker, D. V. (1975). *Endocrinology* **96,** 1280.

Bigler, B. (1976). *Schweiz. Arch. Tierheilk.* **118,** 417.

Boyd, E. M. (1942). *J. Biol. Chem.* **143,** 131.

Boyd, E. M. (1944). *Canad. J. Res.* **22,** 39.

Boyd, G. S., and Oliver, M. F. (1958). *In* "Cholesterol" (R. P. Cook, ed.), p. 187. Academic Press, New York.

Brodey, R. S., and Kelly, D. (1968). *Cancer* **22,** 406.

Broome, M. R., Turrel, J. M., and Hays, M. T. (1988). *Am. J. Vet. Res.* **49,** 193.

Bustad, L. K., George, L. A., Jr., Marks, S., Warner, D. E., Barnes, C. M., Herde, K. E., and Kornberg, H. A. (1957). *Radiat. Res.* **6,** 380.

Clark, S. T., and Meier, H. (1958). *Zentralbl. Veterinaermed.* **5,** 17.

Cline, M. J., and Berlin, N. I. (1963). *Am. J. Physiol.* **204,** 415.

Eckersall, P. D., and Williams, M. E. (1983). *J. Small Anim. Pract.* **24,** 525.

Edelman, I. S. (1974). *N. Engl. J. Med.* **290,** 1303.

Falconer, I. R. (1987). *Comp. Path. Bull.* **19**(3), 2.

Fisher, D. A., Oddie, T. H., and Epperson, J. (1965). *J. Clin. Endocrinol. Metab.* **25,** 1580.

Flamboe, E. E., and Reineke, E. P. (1959). *J. Animal Sci.* **18,** 1135.

Freinkel, N., and Ingbar, S. H. (1955). *J. Clin. Endocrinol.* **15,** 598.

Glade, M. J., and Luba, N. K. (1987). *Am. J. Vet. Res.* **48,** 578.

Hackett, P. L., Gaylor, D. W., and Bustad, L. K. (1957). *Am. J. Vet. Res.* **18,** 338.

Haines, D. M., Lording, P. M., and Penhale, W. J. (1984). *Am. J. Vet. Res.* **45,** 1493.

Hamolsky, M. W., Golodetz, A., and Freedberg, A. S. (1959). *J. Clin. Endocrinol. Metab.* **19,** 103.

Hays, M. T., Turrel, J. M. and Broome, M. R. (1988). *J. Am. Vet. Med. Assoc.* **192,** 1.

Henneman, H. A., Reineke, E. P., and Griffin, S. A. (1955). *J. Anim. Sci.* **14,** 419.

Hoge, W. R., Lund, J. E., and Blakemore, J. C. (1974). *J. Am. Anim. Hosp. Assoc.* **10,** 167.

Holzworth, J., Theran, P., Carpenter, J. L., Harpster, N. K. and Todoroff, R. J. (1980). *J. Am. Vet. Med. Assoc.* **176,** 345.

Horn, K., Castineiras, M. J., Ortola, J., Kock, R., Perriard, F. C., Bittner, S., Pairet, J. V., Ers, P., Boulanger, J., Zeidner, S., Maier, R., Boege, F., Dubois, H., McGovern, M., and Opitz, I. (1991). *Eur. J. Clin. Chem. Clin. Biochem.* **29,** 697.

Irvine, C. H. G. (1967). *J. Endocrinol.* **39,** 313.

Irvine, C. H. G., and Evans, M. J. (1975). *J. Reprod. Fertil. Suppl.* **23,** 709.

Irvine, W. J., and Standeven, R. M. (1968). *J. Endocrinol.* **41,** 31.

Jones, B.R. (1993). *Vet. Focus* **3,** 3.

Kallfelz, F. A. (1973). *Am. J. Vet. Res.* **34,** 535.

Kallfelz, F. A., and Erali, R. P. (1973). *Am. J. Vet. Res.* **34,** 1449.

Kallfelz, F. A., and Lowe, J. E. (1970). *J. Am. Vet. Med. Assoc.* **156,** 1888.

Kaneko, J. J. (1964). *Proc. 10th Annu. Conv. Am. Assoc. Equine Practitioners, Denver,* p. 125.

Kaneko, J. J. (1979). Unpublished.

Kaneko, J. J. (1980). *In* "Clinical Biochemistry of Domestic Animals" (J. J. Kaneko, ed.), 3rd ed., p. 491. Academic Press, New York.

Kaneko, J. J., Tyler, W. S., Wind, A. P., and Cornelius, C. E. (1959). *J. Am. Vet. Med. Assoc.* **135,** 10.

Kaneko, J. J., Baker, B., and Mills, R. (1975). *Bull. Am. Soc. Vet. Clin. Path.* **4,** 30.

Kaneko, J. J., Comer, K. M., and Ling, G. V. (1978). *Calif. Vet.* **32,** 9.

Koppers, L. E., and Palumbo, P. J. (1972). *Med. Clin. North Am.* **56,** 1013.

Kraft, W., and Gerbig, T. (1977). *Dtsch. Tierarztl. Wochenschr.* **84,** 185.

Kritchevsky, D. (1958). "Cholesterol," p. 279. John Wiley & Son, New York.

Larsson, M. (1981). *Acta Vet. Scand.* **22,** 589.

Larsson, M. (1988). *J. Am. Anim. Hosp. Assoc.* **24,** 209.

Larsson, M., Pettersson, T. and Carlstrom, A. (1985). *Gen. Comp. Endocrinol.* **58,** 360.

Lennon, H. D., and Mixner, J. P. (1957). *J. Dairy Sci.* **40,** 1424.

Leveille, G., Fisher, H., and Weiss, H. S. (1957). *Proc. Soc. Exp. Biol. Med.* **94,** 383.

Ling, G. V., Lowenstine, L. W., and Kaneko, J. J. (1974). *Am. J. Vet. Res.* **35,** 1247.

Lipman, L. M. (1967). *Am. J. Med.* **43,** 486.

Loar, A. S. (1986). *In* "Kirk's Current Veterinary Therapy" (R. W. Kirk, ed.), 9th ed., p. 1033. W. B. Saunders, Philadelphia.

Lombardi, M. H., Comar, C. L., and Kirk, R. W. (1962). *Am. J. Vet. Res.* **23,** 412.

Long, J. F., Gilmore, L. O., Curtis, G. M., and Rife, D. C. (1951). *J. Anim. Sci.* **10,** 1027.

Lothrop, C. D., Jr., Tamas, P. M., and Fadok, V. A. (1984). *Am. J. Vet. Res.* **45,** 2310.

Lowe, J. E., Baldwin, B. H., Foote, R. H., Hillman, R. B., and Kallfelz, F. A. (1974). *Cornell Vet.* **64,** 276.

Lucis, O. M., Cummings, G. T., Matthews, S., and Burry, C. (1969). *J. Nucl. Med.* **10,** 160.

Lucke, V. M. (1964). *J. Small Anim. Pract.* **5,** 351.

McCrady, J. D., Hightower, D., Williams, G. D., Kyzar, J. R., and Bridges, C. H. (1973). *Cardiovasc. Res. Centr. Bull.* **12,** 21.

McKenzie, J. M. (1958). *Endocrinology* **63,** 372.

Mahley, R. W., and Weisgraber, K. H. (1974). *Circ. Res.* **35,** 713.

Mallo, G. L., and Harris, A. L. (1967). *Vet. Med. Small Anim. Clin.* **62,** 533.

Maul, D. H., Rosenberg, D. P., Henrickson, R. V. and Kaneko, J. J. (1982). *Lab. Anim. Sci.* **32,** 267.

Messer, M. T., Ganjam, V. K., Nachreiner, R. F., and Krause, G. F. (1995). *J. Am. Vet. Med. Assoc.* **206,** 63.

Michaelson, S. M. (1969). *Mod. Vet. Practice* **50,** 43.

Morris, B., and Courtice, F. C. (1955). *Q. J. Exp. Physiol. Cog. Med. Sci.* **40,** 127.

Murphy, B. E. P., and Pattee, C. J. (1964). *J. Clin. Endocrinol. Metab.* **24,** 187.
Norcia, L. M., and Furman, R. H. (1959). *Proc. Soc. Exp. Biol. Med.* **100,** 759.
Oliver, J. W. and Waldrop, V. (1983). *J. Am. Vet. Med. Assoc.* **182,** 486.
O'Neal, L. W., and Heinbecker, P. (1953). *Endocrinology* **53,** 60.
Osame, S., and Ichijo, S. (1994). *J. Vet. Med. Sci.* **56,** 771.
Pain, R. W., and Oldfield, R. K. (1969). *Tech. Bull. Regist. Med. Technol.* **39,** 139.
Peterson, M. E., Kintzer, P. P., Cavanagh, P. G., Fox, P. R., Ferguson, D. C., Johnson, G. F., and Becker, D. V. (1983). *J. Am. Vet. Med. Assoc.* **183,** 103.
Peterson, M. E., Randolph, J. F., and Mooney, C. T. (1994). *In* "The Cat: Diseases and Clinical Management" (R. G. Sherding, ed.), p. 1403. Churchill Livingstone, New York.
Pilleggi, V. J., Lee, N. D., Golub, O. J., and Henry, R. J. (1961). *J. Clin. Endrocinol. Metab.* **21,** 1272.
Pipes, G. W., Bauman, T. R., Brooks, J. R., Comfort, J. E., and Turner, C. W. (1963). *J. Anim. Sci.* **22,** 476.
Pliam, N. B., and Goldfine, I. D. (1977). *Biochem. Biophys. Res. Commun.* **79,** 166.
Quinlan, W., and Michaelson, S. M. (1967). *Am. J. Vet. Res.* **28,** 179.
Rac, R., Hill, G. N., Pain, R. W., and Mulhearn, C. J. (1968). *Res. Vet. Sci.* **9,** 209.
Ragan, H. A., Horstman, V. G., McClellan, R. D., and Bustad, L. K. (1966). *Am. J. Vet. Res.* **116,** 161.
Refetoff, S., Robin, N. I. and Fang, V. S. (1970). *Endocrinology* **86,** 793.
Reiser, R., Sorrels, M. F., and Williams, M. C. (1959). *Circ. Res.* **7,** 833.
Ricketts, M. H., Pohl, V., de Martynoff, G., Boyd, C. D., Bester, A. J., Van Jaarsveld, P. P., and Vassart, G. (1985). *EMBO J.* **4,** 731.
Roman, S. H., Korn, F., and Davies, T. F. (1984). *Clin. Chem.* **30,** 246.
Schiller, C. A., Montali, R. J., and Grollman, E. F. (1995). *Vet. Pathol.* **32,** 242.
Scott, K. G., and Reilly, W. A. (1954). *Metab. Clin. Exp.* **3,** 506.
Segal, J., Schwartz, H., and Gordon, A. (1977). *Endocrinology* **101,** 143.
Siegel, E. T., and Belshaw, B. E. (1968). *In* "Current Veterinary Therapy" (R. W. Kirk, ed.), 3rd ed., p. 545. Saunders, Philadelphia, Pennsylvania.
Sims, M. H., Redding, R. W., and Nachreiner, R. F. (1977). *J. Am. Vet. Med. Assoc.* **171,** 178.
Sorenson, P. H. (1962). *In* "Use of Radioisotopes in Animal Biology and the Medical Sciences," Vol. I, p. 455. Academic Press, New York.
Sparagana, M., Phillips, G., and Kucera, L. (1969). *J. Clin. Endocrinol. Metab.* **29,** 191.
Stanbury, J. B. and Dumont, J. (1983). *In* "The Metabolic Basis of Inherited Disease" (J. B. Stanbury, J. B. Wyngaarden, D. S. Fredrickson, J. L. Goldstein, and M. S. Brown, eds.), 5th ed., p. 231. McGraw-Hill, New York.
Tanabe, Y., Ishii, T., and Tamaki, Y. (1969). *Gen. Comp. Endocrinol.* **13,** 14.
Tapley, D. F. (1964). *Mayo Clinic Proc.* **39,** 626.
Tata, J. R., Ernster, L., Lindberg, O., Arrhenius, E., Pederson, S., and Hedman, R. (1963). *Biochem. J.* **86,** 408.
Theran, P., and Thornton, G. W. (1966). *J. Am. Vet. Med. Assoc.* **148,** 562.
Thomas, C. L., and Adams, J. C. (1978). *Am J. Vet. Res.* **39,** 239.
Trum, B. F., and Wasserman, R. H. (1956). *Am. J. Vet. Res.* **17,** 271.
Van Zyl, A., Schulz, K., Wilson, B., and Pansegrouw, D. (1965). *Endocrinology* **76,** 353.
Verschueren, C. P., and Goslings, B. M. (1992). *In* "Clinico-pathological and Endocrine Aspects of Canine Thyroid Cancer" (C. P. Verschueren, ed.), p. 95. C. P. Vershueren, Brussels.
Voller, A., Bidwell, D. E., and Burek, C. L. (1980). *Proc. Soc. Exp. Biol. Med.* **163,** 402.
Wilke, T. J. (1986). *Clin. Chem.* **32,** 585.
Wilke, T. J., and Eastment, H. T. (1986). *Clin. Chem.* **32,** 1746.
Zieve, L., Vogel, W. C., and Schultz, A. L. (1956). *J. Lab. Clin. Med.* **47,** 663.

CHAPTER

22

Clinical Reproductive Endocrinology

LARS-ERIK EDQVIST AND MATS FORSBERG

I. INTRODUCTION

Clinical reproductive endocrinology includes the study of diseases of the endocrine glands involved in reproduction and their secretory products, the reproductive hormones. To obtain a satisfactory understanding of the complex endocrinological events that occur during normal and abnormal reproductive function, quantification of specific hormones is necessary. Initially, biological assay systems that measured the effect of a hormone on its target tissue were used; weight-gain change was mainly used as the measure of hormone concentration. These assays were relatively imprecise, time consuming, and expensive. Later they were replaced by chemical determinations for steroid hormones. These assay systems usually required large volumes (often 1 liter) of plasma or serum, which made serial blood sampling of individual animals impossible. Some steroid hormone patterns were studied via urine analysis in 24-hour urine aliquot. Collection of such urine aliquots from domestic species was difficult and not practical under field conditions.

Major progress in hormone analytical techniques occurred as the result of the development of immunoassay and related systems. The first of the assays, radioimmunoassay, was developed in 1959 (Berson and Yalow, 1959) with the competitive protein binding assay following a few years later (Murphy, 1964). Nonradiometric assays, such as enzyme immunoassay, were developed in the 1970s (Engvall and Perlmann, 1971; van Weemen and Schuurs, 1971). These assay systems are sensitive, specific, and relatively inexpensive, and they require small amounts of assay material. They have been of special value for studying endocrinological reproductive function in domestic animals, in that they have made possible the study of dynamic endocrine changes through the assay of serial blood samples from the same animal. The immunoassay systems have also been useful as diagnostic aids for the identification and elucidation of clinical reproductive problems. In clinical practice, these methods are important from both a diagnostic and a therapy-monitoring point of view.

Because this book deals with clinical biochemistry in domestic animals, the main emphasis of the chapter is on the determination of hormones and the use of the data as diagnostic aids. General reproductive endocrinology in domestic species is broadly covered. Readers specifically interested in this subject are referred to specialized books dealing with this matter.

A. Definition of Hormones

The best-understood humoral control system in the body is the endocrine system. This system uses specific messengers, termed hormones, to regulate important body functions. By the classical definition, hormones are chemical substances that are synthesized and secreted by ductless endocrine glands in minute quantities directly into the blood vascular system and are transported to a remote target organ where they regulate the rates of specific biochemical processes. The classic endocrine glands include the pituitary, thyroid, parathyroid, adrenal, pancreas, ovary, testis, placenta, and pineal glands.

In the case of reproduction, the pituitary and pineal glands, gonads, and placenta play a primary role in controlling the system. Other endocrine glands such as the adrenal and thyroid glands also have some influence on reproductive function. Other organs such as the uterus and the hypothalamus, although they may not fulfill the strict definition of endocrine glands, can synthesize and secrete hormones that have a profound influence on reproductive function.

B. Chemical Classes of Reproductive Hormones

1. Peptide and Protein Hormones

a. Releasing Hormones

Several types of hormones are involved in the regulation of reproduction. Releasing hormones are peptide hormones that are produced within the hypothalamus and transferred via the hypothalamo–hypophyseal portal veins to the adenohypophysis, where they regulate the synthesis and/or release of adenohypophyseal hormones. Gonadotropin-releasing hormone (GnRH), a decapeptide with the sequence pGlu–His–Trp–Ser–Tyr–Gly–Leu–Arg–Pro–Gly–NH_2, regulates the release of two important reproductive hormones, namely luteinizing hormone (LH) and follicle-stimulating hormone (FSH). Thyrotropin-releasing hormone, a tripeptide (pGlu–His–Pro–NH_2) that regulates the synthesis and release of thyroid-stimulating hormone (TSH), also causes the release of prolactin in several species.

b. Hypophyseal Hormones

Luteinizing hormone and FSH are glycoproteins containing 13–25% carbohydrate. The molecular weight of LH in domestic animals (bovine, ovine, porcine, equine) is about 30,000. Ovine and equine FSH have molecular weights of about 32,000. Prolactin is a protein with a molecular weight of approximately 23,000 (bovine, ovine, porcine).

The cells of the adenohypophysis can be divided into basophils (affinity for basic stains) and acidophils (affinity for acid stains). Luteinizing hormone and FSH are produced within basophilic cells; it has been demonstrated that LH and FSH can be present within the same cell. Prolactin, on the other hand, has been localized in acidophilic cells. As pointed out previously, the releasing hormones are important regulators of the synthesis and/or release of these hormones from the adenohypophysis.

c. Neurohypophyseal Hormones

The posterior pituitary is responsible for storage and release of oxytocin, an important reproductive hormone, and antidiuretic hormone (vasopressin). These two hormones are synthesized primarily in the regions of the paraventricular and supraoptic nuclei of the hypothalamus. The hormones are transported to the posterior pituitary by axoplasmic fluid. Release of these hormones occurs as a result of stimulation of the nerve cell bodies in the nuclei. Oxytocin is an octapeptide with a molecular weight of 1000.

d. Placental Gonadotropins

A gonadotropin called pregnant mare serum gonadotropin (PMSG) is produced by mares during early pregnancy (days 40–140) by fetal trophoblastic cells of the chorionic girdle, which attach to, invade, and phagocytose the maternal epithelium and become embedded within the uterus as specialized endometrial cups. This process begins on day 36 of pregnancy. The hormone has recently been renamed equine chorionic gonadotropin (eCG) due to its close relationship with human chorionic gonadotropin (hCG) (Farmer and Papkoff, 1979). No specific placental gonadotropins have been demonstrated in other domestic species.

e. Subunits

Luteinizing hormone and FSH, as well as TSH, consist of two nonidentical subunits designated α and β. Chemical and biological studies have indicated that the

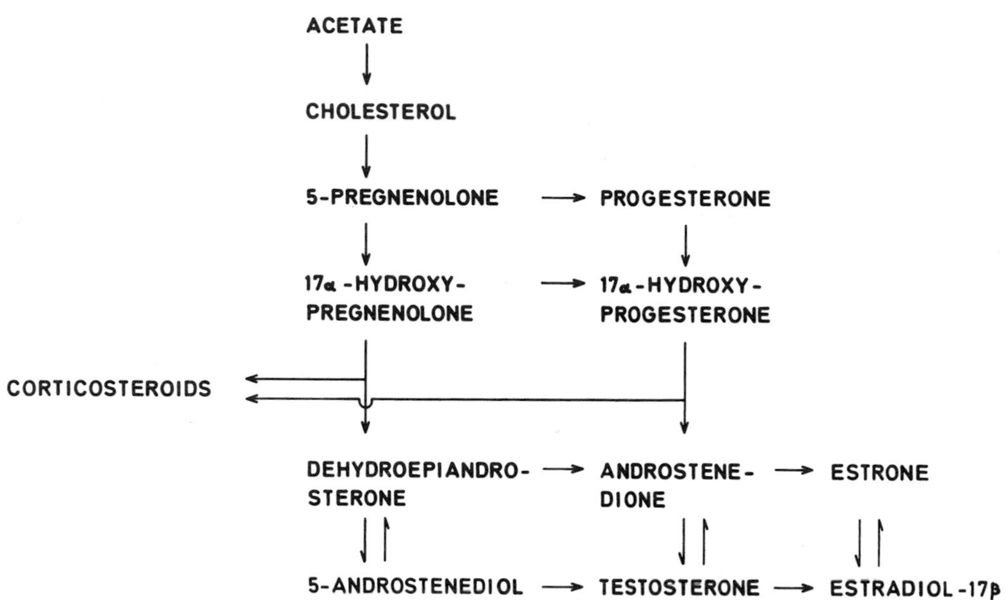

FIGURE 22.1 Pathway for the synthesis of biologically active steroids from acetate. The steroids secreted from the gonads and the adrenal are formed from acetate and cholesterol.

α-subunit is identical for these glycoprotein hormones within a species, whereas the β-subunit, unique for each hormone within a species, determines the biological activity. Subunits, by themselves, possess little or no biological activity. Individual subunits are probably not released into the circulatory system under normal physiological conditions.

2. *Steroid Hormones*

Steroid hormones are derived from a common precursor molecule, cholesterol, via the metabolic pathway schematically outlined in Fig. 22.1. More than 1,500 biologically active steroids have been isolated from biological material or have been produced synthetically. The molecular weight of steroid hormones is low, usually below 500 (Table 22.1). Examples of steroids that play an important role in reproductive processes are estrogens, androgens, and progestogens, with the main source being the gonads. The structures of the most important sex steroids are given in Fig. 22.2. The most common steroid hormones are usually designated by a trivial name, such as estradiol, testosterone, or progesterone. The International Union of Pure and Applied Chemistry (IUPAC) has recommended systematic names for steroid hormones. These

TABLE 22.1 Nomenclature[a] and Molecular Weights of Some Biologically Important Steroids and Prostaglandins

Trivial name	Systematic name	Molecular weight
Androstenedione	4-Androstene-3,17-dione	286
Cortisol	11β,17α,21-Trihydroxy-4-pregnene-3,20-dione	363
17β-Estradiol	1,3,5(10)-Estratriene-3,17β-diol	272
Estrone	3-Hydroxy-1,3,5(10)-estratrien-17-one	270
17α-Hydroxyprogesterone	17α-Hydroxy-4-pregnene-3,20-dione	331
Pregnenolone	3β-Hydroxy-5-pregnen-20-one	317
Progesterone	4-Pregnene-3,20-dione	315
Testosterone	17β-Hydroxy-4-androsten-3-one	288
$PGF_{2\alpha}$	9α,11α,15-Trihydroxyprosta-5,13-dienoic acid	354
15-Keto-13,14-dihydro-$PGF_{2\alpha}$	9α,11α-Dihydroxy-15-ketoprost-5-enoic acid	354

[a] Revised Tentative Rules for Nomenclature of Steroids (1968). *Biochim. Biophys. Acta* **164,** 453.

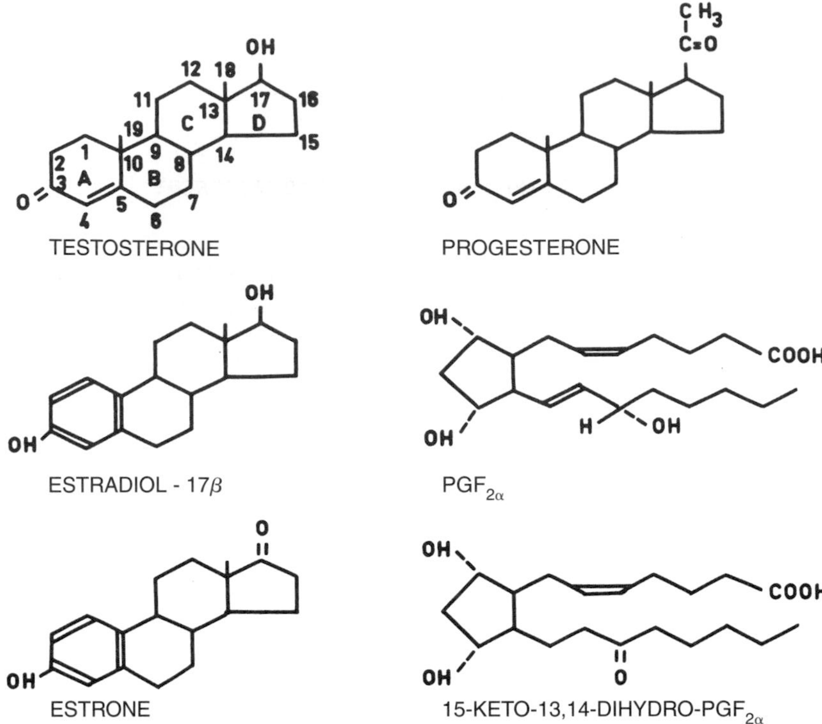

FIGURE 22.2 The number sequence for the carbon atoms of the steroid skeleton and lettering sequence for the four rings are shown for testosterone. The structures of three other important sex steroid hormones, estrone, estradiol-17β, and progesterone, as well as the structures of prostaglandin $F_{2\alpha}$ and its blood plasma metabolite 15-keto-13,14-dihydroprostaglandin $F_{2\alpha}$, are also depicted.

systematic names describe the chemical and steroisometric characteristics of the particular steroid hormone (Table 22.1).

3. *Prostaglandins*

Prostaglandins constitute a group of 20-carbon unsaturated fatty acids with molecular weights usually between 300 and 400 (Table 22.1). Prostaglandins are not hormones in the strictest sense, and the expressions "parahormones" or "local hormones" have been used to describe these substances. This is because prostaglandins are not secreted from any particular gland, and the biological half-life of prostaglandins is usually extremely short, allowing, in most cases, only a local action. Several different prostaglandins are found in a number of types of mammalian tissues. One prostaglandin released from the uterus, prostaglandin $F_{2\alpha}$ ($PGF_{2\alpha}$), plays an important role in regulating reproductive cycles in domestic species through the control of luteal activity in nonpregnant animals and the initiation of delivery in pregnant animals. The structures of prostaglandin $F_{2\alpha}$ and its main metabolite, 15-keto-13,14-dihydroprostaglandin $F_{2\alpha}$, are given in Fig. 22.2.

C. Hormone Receptors

Steroid hormones are fat soluble, so they are able to enter all cells of the body because the lipid cellular boundaries present no barrier to the steroid. The question thus arises as to how specific actions occur in this physiological situation. Steroid hormone concentrations in plasma are very low compared to those of many compounds, including their important precursor, cholesterol. For example, plasma estrogens in nonpregnant domestic animals range from as low as 10 pmol/liter to as high as 150 pmol/liter. In this situation, most cells within the body have very low concentrations of estrogen. The specificity of tissue response occurs because cells of tissues that have need of estrogen stimulation should have receptors that enable those particular cells to concentrate the hormone within the cell and, more importantly, to elicit particular cellular responses. Hence the important generalization that specific tissue responses require specific receptors to be present within the cell, in this case, for a particular steroid hormone.

Binding of a hormone to a receptor in a target cell can be considered to be the primary event in the action of that hormone. Such a hormone–receptor interaction will cause a measurable and distinct biological response for each hormone. Receptors are defined by the criteria of having a limited binding capacity (receptors are saturable, which limits the number of hormone molecules that can enter a target cell), of binding specific hormones (e.g., estrogen receptors are specific for estrogenic compounds), and of creating a biological response upon binding.

Steroid hormones enter cells by passive diffusion and bind to receptors inside the cell. It is assumed that only non-protein-bound or free hormone can enter target cells. The protein-bound steroid is thus virtually biologically inactive, and only the minute quantity of free (non-protein-bound) steroid can enter the cell. The protein binding is, however, reversible, and as free hormone leaves the vascular system into target cells, protein-bound hormone will be released to replace the deficit. On entry, the steroid interacts with its receptor in the target cell. When the steroid interacts with its receptor, a steroid–receptor complex is formed. The hormone–receptor complex is then "activated" and alters gene expression. The target cell responds by increased RNA synthesis with the transcription of specific mRNA that enters cytoplasm and stimulates protein synthesis. The specific effect of steroid hormones on target cells is altered cell function related to a change in the pattern of protein synthesis.

Protein hormones such as LH and FSH do not enter the target cell to exert their effects, but interact with their receptors, which are located on the plasma membranes of the cell. The binding of the hormone to the cell surface receptor activates one or more second messengers, such as cyclic AMP (3′,5′-AMP). The second messenger is the intracellular mediator of many actions of LH and FSH in the ovary and the testis. Second messenger is thought to activate another intracellular enzyme, protein kinase, which will influence the transport of cholesterol into the mitochondrion and the conversion of cholesterol to pregnenolone, which is the rate-limiting step in the biosynthetic pathway for the steroids that play a significant role in reproductive processes.

D. Interconversion of Steroids in Target Tissues

The effects of steroid hormones on cells can be accentuated or modulated by the conversion of the entering hormone to another form. For example, many of the tissues that are particularly responsive to androgens have the enzyme 5α-reductase, which converts testosterone to 5α-dihydrotestosterone (5α-DHT). 5α-DHT has a much higher affinity for the androgen receptor within the target cell, which makes 5α-DHT more biologically active than testosterone in terms of its androgenic effects. The androgenic potency of 5α-DHT is twice that of testosterone.

Another important interconversion of steroid hormones is the one resulting from the increase in circulating cortisol concentrations that occurs in the fetal lamb prior to parturition. The elevated fetal cortisol concentrations stimulates 17α-hydroxylase, C17–C20 lyase activity, and probably aromatase activity in the placenta. These enzymes make it possible for progesterone to be converted to estrogens in the placenta. Estrogens then affect the synthesis of prostaglandin $F_{2\alpha}$, which precipitates delivery. The interconversion of progesterone to estrogen is well documented in the sheep and probably occurs in a similar way in the goat and the cow.

E. Synthesis and Clearance of Hormones

The determination of concentrations of hormones in biological fluids, including plasma, urine, saliva, and feces, has been useful in determining the reproductive status of animals. Although there are a number of factors that can influence hormone concentrations, the overriding factors are synthesis and clearance. Of greatest interest is the rate of synthesis of a hormone from a particular endocrine gland, because factors that govern clearance are usually stable; therefore, the concentration of a hormone usually reflects its rate of synthesis or secretion.

The synthesis of steroid hormones of the reproductive system is under the control of gonadotropins that are released in pulsatile fashion. This mechanism has a profound influence on the secretion of testosterone in the male, in that changes in pulsatile rate can occur a number of times a day with increases in pulse rate resulting in greatly increased concentrations of testosterone. For example, in males of many domestic species, testosterone values can vary from 3.5 to 20 nmol/liter within a period of a few hours, with the extremes still representing normal production of testosterone by the testes. The usual judgement as to normalcy is based on an animal having at least the minimal or basal concentration. In the female, estrogen and progesterone synthesis by the ovary is also under the control of a pulsatile mode of gonadotropin secretion. The pulse rate usually remains relatively stable over limited periods of time so that fluctuations in concentration of these hormones are not as acute as for androgens. Thus, gonadotropin pulsatility is not a consideration in interpreting the significance of progesterone and estrogen concentrations in biological fluids.

In the female, synthesis rates for ovarian steroid hormones are obviously related to ovarian function.

Progesterone concentrations, relatively stable during the luteal phase of the estrous cycle, decline rapidly over a 24- to 36-hour period during luteolysis. Estrogen values continually increase during the follicular phase of the cycle, declining with the onset of the gonadotropin preovulatory surge as the granulosa is converted from estrogen to progesterone production. Even though secretion rates can change for both progesterone and estrogen, analysis of these hormones usually provides useful information as to luteal or follicular activity, respectively. If one wishes to use hormone values (in blood, for example) as an indication for secretory activity of an endocrine organ, one other factor must be considered: the conversion of steroid hormones by peripheral tissues. For example, in primates, estrone concentrations are derived mainly from the conversion of ovarian estradiol-17β and adrenal androstenedione by tissues such as the liver.

Steroids are eliminated via conjugation with glucuronic acid and/or sulfates to form inactive mono- or diglucuronides or sulfates. These conjugates are all water soluble, with excretion occurring via urine or bile (feces). Conjugation occurs mainly in the liver, and the conjugates lack steroidal activity. Steroids are also rendered inactive by their metabolism to compounds that have greatly reduced biological activity. In this way, steroids are rapidly cleared from the bloodstream. Clearance is defined as the volume of blood that would be totally cleared of a particular steroid per unit time. Clearance can thus be expressed as liters/minute, and the clearance for most steroid hormones is around 1 liter/minute. In most situations, the clearance rate of steroids is relatively constant, so that blood concentrations are a relatively good measure of fluctuations in production rates.

Placental gonadotropic hormones such as hCG and eCG are produced in high concentrations and have much longer half-lives than the pituitary gonadotropins and prolactin. The latter have half-lives around 10–30 minutes, whereas the corresponding figures for the placental hormones are from 1.5 days for hCG to 6 days for eCG. One exception is equine LH, which has structural similarities to eCG and also has a much longer half-life (days) than LH from other species. The half-lives are increased because the molecules contain a larger carbohydrate moiety than do FSH and LH of most other species.

In the blood, prostaglandins are rapidly metabolized to their respective 15-keto-13,14-dihydro compounds (Fig. 22.2). Primary prostaglandins such as $PGF_{2\alpha}$ have a half-life in the peripheral circulation that is less than 20 seconds, whereas 15-keto-13,14-dihydro-$PGF_{2\alpha}$ has a somewhat longer half-life of about 8 minutes. Ninety percent or more of $PGF_{2\alpha}$ is metabolized during one passage through the lungs. The 15-keto metabolites are biologically inactive and are degraded into short dicarboxylic acids before being excreted into urine. An example of the rapid metabolism of prostaglandins is the fact that more than 90% of a 25-mg intramuscular injection of $PGF_{2\alpha}$ in the cow is excreted in the urine and feces (2:1 ratio) over a 48-hour period (Neff *et al.*, 1981).

II. ASSAY METHODS

The radioimmunoassay technique was originally introduced for the measurement of plasma insulin (Berson and Yalow, 1959), and the enzyme immunoassay techniques for the quantitative determination of immunoglobulin G (Engvall and Perlmann, 1971). The techniques are competitive and utilize the same basic principle, which is based on the ability of nonlabeled hormone to compete with a fixed amount of isotopically or enzymatically labeled hormone for the binding sites on a fixed amount of protein. The nonlabeled hormone reduces the number of free binding sites on the protein, thus decreasing the availability of the binding sites to the labeled hormone. At equilibrium, the free hormone is separated from the protein-bound hormone, and the reaction is quantified by the determination of the amount of labeled hormone that is antibody-bound or free (Fig. 22.3). The degree of inhibition of binding of the labeled hormone to the binding protein is a function of the concentration of nonlabeled hormone present in the solution. As a basis for the quantification, a standard curve is developed with fixed amounts of labeled hormone and binding protein incubated together in the presence of a known and graded concentration of unlabeled hormone (Fig. 22.3).

Certain disadvantages exist to the use of radioisotopes as labels in immunoassays, including the limited shelf life and stability of radiolabeled compounds, the need for relatively expensive counting equipment (especially for tritium-labeled compounds), the need for well-trained personnel and specialized laboratory equipment, and problems in disposal of radioactive waste. Consequently, attempts have been made to develop assays that use nonisotopic labels such as enzymes, fluorogens, and chemiluminescent precursors. Enzyme immunoassays can be as sensitive, accurate, and precise as radioimmunoassays, and specificity depends on the quality of the antibody, as is true for radioimmunoassay systems (Munro and Stabenfeldt, 1984). Of particular interest in domestic animals has been the use of these techniques in the determination of progesterone in blood (Munro and Stabenfeldt, 1984; Meyers *et al.*, 1988; Lopate and Threlfall, 1991) and in milk (Arnstadt and Cleere, 1981; Sauer *et al.*, 1981; Allen and Foote, 1988; Etherington *et al.*, 1991). Enzyme

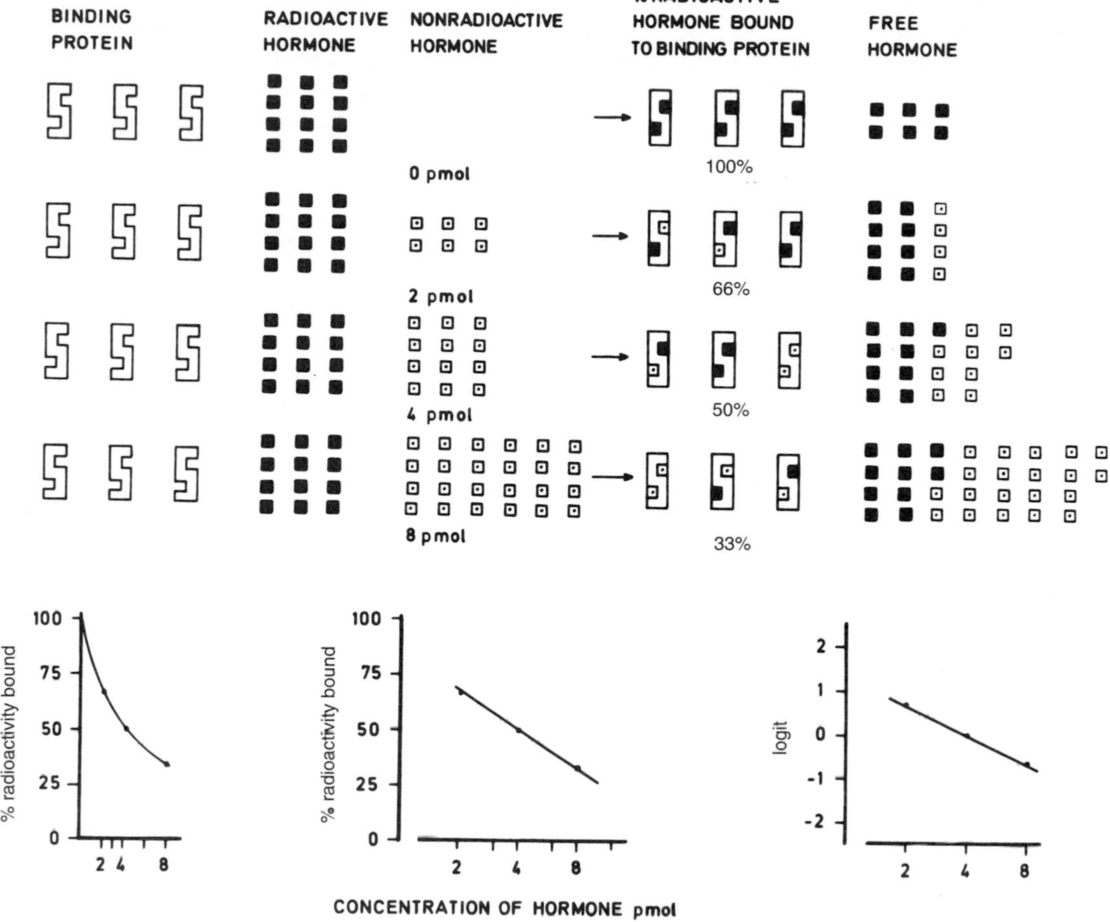

FIGURE 22.3 The principle of an immunoassay technique is based upon the ability of the binding protein to bind the labeled hormone. Excess labeled hormone is added to ensure saturation of the hormone-binding sites on the binding protein. The addition of increasing amounts of nonlabeled hormone (2, 4, and 8 pmol) results in a proportional decrease in the quantity of labeled hormone bound to the binding protein. Separation of labeled hormone bound to the binding protein from free hormone must be achieved before quantification can be done. In the lower part of the figure, this reaction is depicted in three different ways. In the panel to the left and in the middle, percent labeled hormone bound to the binding protein is on the ordinate, and the amount of hormone on a linear scale (left) and log scale (middle) is on the abscissa. In the panel to the right, the logit of the response variable is on the ordinate, and the amount of hormone is plotted on a log scale. The method depicted in the middle and to the right can be used for determination of parallelism. The logit/log transformation (right) is frequently used when immunoassay data are analyzed by computer.

immunoassay in the laboratory setting is very efficient, particularly in the time required for the assay reaction (2 hours or less), in ease of separation of bound from free hormone when using the microtiter plate system (30 seconds to wash 96 wells on a microtiter plate), and in speed of end-point analysis (optical densities can be determined in 1 minute for a 96-well microtiter plate).

Among the various approaches to enzyme immunoassay, the double antibody sandwich method for determination of hormones is frequently used. In this assay system, the plastic wells are coated with antibody. The sample to be processed is then added, and its hormone binds to the antibody-coated plastic well. A second enzyme-labeled antibody directed against another epitope of the hormone is then added. The amount of enzyme-labeled antibody bound is directly proportional to the amount of hormone in the sample. There are two advantages with this methodology: (1) the hormone does not need to be isolated for labeling, and (2) the same general technique can be used to label different antibodies. The method can only be used to assay hormones with at least two binding sites; thus, it is unsuitable for measurement of low-molecular-weight hormones such as steroid hormones.

The analytical equipment used for enzyme immunoassay can be used for a variety of determinations, including those involved in disease surveillance and drug analysis. This flexibility allows for the sharing of one specialized spectrophotometer among several disciplines at a great reduction in cost. Another benefit is that the analytical equipment needs little maintenance. The elimination of the problems engendered by use of radioisotopes is a major advantage for enzyme immunoassay. This advantage and the fact that color change is fundamental to enzyme immunoassay means that the assay can be used visually to determine the presence or absence of a corpus luteum (CL) in domestic and other species. Today, it is possible to use enzyme-immunoassay analytical systems for progesterone in blood or milk on the farm beside the animal to get a direct assessment of the functional ovarian status of an animal (Nebel *et al.*, 1989; Herrier *et al.*, 1990; Matsas *et al.*, 1992; Romagnolo and Nebel, 1993).

Chemiluminescent-labeled immunoassays are commercially available. All cases of chemiluminescence assay systems are based on a reaction producing light in the visible spectrum. Such an analytical system is used in the authors' laboratory for the determination of several different hormones in domestic species (Forsberg *et al.*, 1993a). A microtiter plate with 96 wells can be read and processed in less than 2 minutes.

It is interesting to note from a historical basis that radioimmunoassay originally developed by endocrinologists was later utilized by immunologists. Conversely, enzyme immunoassay developed by immunologists for studying mechanisms of disease was later adopted by endocrinologists (Dray *et al.*, 1975). For the foreseeable future, immunoassay techniques, especially enzyme immunoassays, will be the laboratory methods of choice in reproductive diagnostic endocrinology of domestic species.

A. Immunoassay

1. Production of Antibodies

Immunoassay techniques utilize antibodies as binding protein. Hormones such as LH and FSH, which are glycoproteins with molecular weights of around 30,000, are antigenic because of their size and chemical composition. In general, a lower level of purity is required of the polypeptide hormones for antibody production as compared to the hormone used in the labeling procedure.

If an assay for bovine LH utilizes an antibody to bovine LH or radio- or enzyme-labeled bovine LH as tracer, and bovine LH as the standard, the assay system is completely species specific and is said to be of the homologous type. Such a system represents the ideal immunoassay system for measuring a polypeptide hormone. Because of limited availability and / or lack of suitable purity of polypeptide hormone preparations, heterologous assay systems have been developed. In these cases, an antiserum to a polypeptide hormone of one species has been used for the determination of the same polypeptide hormone in another species. The standard hormone used for quantification of the assay should, however, originate from the same species for which the measurements are performed.

There are polypeptide hormone antisera available that show a high degree of cross reactivity. One such antiserum of special interest in the field of reproductive hormones in domestic species is a polyclonal LH antiserum raised against ovine LH (Niswender *et al.*, 1969). This antiserum reacts specifically with LH from other species and has been used for the determination of LH in approximately 45 species including the cow, sheep, pig, cat, and dog (Millar and Aehnelt, 1977; Madej and Linde-Forsberg, 1991). A monoclonal antibody generated against bovine LH has been reported to have high cross reactivity among species (Matteri *et al.*, 1987; Bravo *et al.*, 1992; Forsberg *et al.*, 1993b).

Steroid hormones and prostaglandins have considerably lower molecular weights and thus are not immunogenic per se. However, these structures can be rendered immunogenic if covalently linked to large carrier molecules such as bovine serum albumin, and specific antibodies can be elicited in this fashion (Landsteiner and Van der Scherr, 1936). In order for such a hormone–protein conjugate to be immunogenic, approximately 10–20 hormone molecules should be present per molecule of protein. In the case of bovine serum albumin, about 30% of the sites available for conjugation should be occupied.

Most naturally occurring steroid hormones and prostaglandins contain hydroxyl or ketone groups that are used to prepare derivatives containing active groups such as carboxyl or amino groups. These groups are then activated so that they react with amino or carboxyl groups of the protein molecule. The specificity of the antisera obtained by immunization with a steroid–protein conjugate is dependent on the site used for conjugating the steroid to the protein. More specific antisera are obtained if the hapten (steroid) is attached to the protein at a site remote from the characteristic functional groups of the hormone (Lindner *et al.*, 1970).

The most frequent species used for production of polyclonal antibodies are sheep and rabbits. One of the most popular and efficient schedules for immunization involves multiple injections in the back and neck of the animal of the antigen emulsified in complete Freund's adjuvant (Vaitakaitus *et al.*, 1971). During immunization, the developing antibody titer is monitored, and a relative large number of milliliters of serum can be obtained when a suitable titer has been achieved. A few milliliters of a high-titer antiserum are usually sufficient for millions of immunoassay determinations. Antisera seem to be quite stable when stored at −20° C, although the usual preferred temperature is −70° C.

A new development in the technology of antibody production was the discovery that hybridomas could be used to produce an endless supply of antibodies with certain specificity (Köhler and Milstein, 1975). The procedure involves the fusion of two cell lines: B lymphocytes selected for the production of a specific antibody and myeloma cells that have the capacity for permanent growth. Antibody production occurs by injection of the cell lines into mice; permanency is assured by maintaining a supply of cells in the frozen state. Monoclonal antibodies can be used for the quantitative immunoassay of hormones. They have the advantages of specificity, unlimited supply over time, and the possibility of standardizing assay methods between laboratories. Disadvantages are that they have lower affinities than do polyclonal antibodies, and they do not always form precipitates with antigens. Mixing of monoclonal antibodies may result in an increase of affinity (Ehrlich *et al.*, 1982).

2. Labeled Hormone

In radioimmunoassay techniques for polypeptide hormones, the antigen (hormone) is most commonly used for preparing the radioactive tracer. Usually, radioactive iodine, ^{125}I, is used for radioiodination of the antigen. The two most frequently used techniques for iodination are free or immobilized chloramine-T (Iodo-beads) (Hunter and Greenwood, 1962; Markwell, 1982) and the lactoperoxidase procedure (Thorell and Johansson, 1971). Peptide hormones containing tyrosyl or histidyl residues can also be iodinated with these techniques.

Many RIA systems for steroid hormones and prostaglandins utilize tritiated forms of these molecules that are available commercially. Because tritium has a considerably longer half-life than iodine, tritium tracers can be used in many cases over several years, whereas the iodinated tracers often have to be prepared monthly. There are, however, certain advantages in using iodinated tracers for steroid hormones and prostaglandins: Simpler and cheaper counting systems can be used, that is, gamma counting as opposed to liquid scintillation counting. Another advantage of radioiodine over tritium is its higher specific activity, which increases the sensitivity of the assay. Direct incorporation of iodine in the skeleton of steroid hormones results in a loss of the immunoreactivity. Thus, the approach taken for radioiodination of steroid hormones has been to link a tyrosyl or histidyl molecule to the steroid molecule, making direct radioiodination possible (Niswender, 1973), or to iodinate a compound such as tyramine and conjugate the iodinated compound to the steroid molecule (Lindberg and Edqvist, 1974).

Most EIA systems for steroid hormones use horseradish peroxidase, alkaline phosphatase, or β-galactosidase as labels. The enzyme-labeled steroid is produced the same way as has been described for the synthesis of steroid–protein conjugates for the production of antibodies.

3. Separation of Antibody-Bound and Free Hormone

An essential part of any immunoassay system is an efficient procedure for the separation of antibody-bound and free hormone. Several different approaches have been taken to achieve a rapid and efficient separation. Currently, the most frequent separation procedures in use are based on (1) antibodies coupled to an insoluble polymer, (2) precipitation of antibody-bound hormone, or (3) adsorption of free hormone.

Antibodies coupled to an insoluble polymer have been used for separating antibody-bound and free hormone in RIA procedures for both protein hormones (Wide and Porath, 1966) and steroid hormones (Abraham, 1969). One procedure involves decanting from polystyrene tubes in which antibodies have been adsorbed to the surface of the tube, which is followed by determination of radioactivity in the antibody-bound (contained in the tube) or free (contained in the eluent) form. In most EIA systems designed for low-molecular-weight hormones, antibody-bound hormone is measured after free hormone has been removed by washing the wells of the microtiter plate that have been previously coated with antibody.

Another common procedure for RIA utilizes antibodies covalently coupled to an insoluble polymer granule (Wide and Porath, 1966). In this case, free and antibody-bound hormone are separated through centrifugation. After removal of the supernatant containing the free hormone, the antibody-bound radioactivity can be determined. Antibody-coated glass beads have been used in both RIA and EIA (Schmidt *et al.*, 1993). Separation is achieved by washing the bead, and then the radioactivity or enzyme activity is determined.

Precipitation of antibody-bound hormone has been achieved through the addition of ammonium sulfate (Mayes and Nugent, 1970) or polyethylene glycol (Desbuquois and Aurbach, 1971), leaving the free hormone in the solution. The latter precipitation procedure has been found advantageous for the precipitation of prostaglandin–antibody complexes (Van Orden and Farley, 1973). Precipitation of the antibody-bound hormone complexes can also be achieved through the addition of a second antibody prepared against the first antibody. Thus, in an RIA technique for bovine LH that utilizes an antiserum to bovine LH raised in a rabbit (first antibody), the second antibody will be an antibody prepared against rabbit gamma globulin. The addition of the second antibody will result in a precipitate containing the antibody-bound LH, which can be separated from the supernatant by centrifugation. The time required for the separation can be decreased by the addition of polyethylene glycol (Eisenman and Chew, 1983). Systems using a second antibody coupled to insoluble particles are efficient and commonly used (Dericks-Tan and Taubert, 1975; Forsberg *et al.*, 1993b).

For steroid hormones, a traditional separation procedure is the adsorption of free hormone to dextran-coated charcoal. After the addition of the charcoal, the separation of free and antibody-bound steroid is achieved through centrifugation. This method is rapid and efficient in separating free and bound steroid hormones. However, the charcoal can also adsorb some of the antibody-bound steroid, which is called "stripping." To control this, timing of the reaction is important. Carrying out the reaction at 4°C has been found to limit this dissociation (Abraham, 1974).

4. Reliability Criteria

The reliability of immunoassay analyses depends on specificity, sensitivity, accuracy, and precision.

a. Specificity

The specificity of an immunoassay, or its freedom from interference by substances other than the one to be measured, is dependent on several different factors, the most important being the specificity of the antiserum used. Demonstration of specificity for the immunoassay of large protein hormones such as LH and FSH is relatively difficult and relies upon indirect criteria. Because it is not possible to synthesize these hormones, they must be isolated and purified from biological material. The purity of such preparations is variable, and the most common cause of nonspecificity for these hormones is impurity of the immunizing material.

A relatively common finding is the cross reaction of TSH with antibodies to LH, and vice versa. Antibody specificity is usually demonstrated by testing the binding of hormones other than the one intended to be measured to the antibody. If, for example, bovine TSH significantly inhibits the binding of bovine LH to an antibody to bovine LH, this indicates that the antiserum used is nonspecific or that the TSH preparation contains LH. If the inhibition curves are parallel, the latter explanation is likely because the parallelism indicates the same binding kinetics. It should be noted that parallelism, in itself, is not adequate proof of specificity. As indicated previously, both LH and TSH are composed of two subunits, an α-subunit that is identical for the two hormones and a β-subunit that is unique for each hormone. It is possible that an antiserum could contain binding sites that will react only with the α-subunit. In such a case, the dose–response curve of LH and TSH utilizing such an antisera will be parallel, the assay system will not be hormone specific, and thus the system will be invalid for the measurement of LH (Niswender and Nett, 1977). Double antibody sandwich methods utilizing monoclonal and/or polyclonal antisera can partly reduce the problem.

In the case of immunoassay techniques for steroid hormones and prostaglandins, the same proof of specificity has to be undertaken. Here the situation is simpler, because lower-molecular-weight hormones can easily be purified and, in most cases, produced synthetically. Furthermore, comparison of immunoassay results with results obtained by mass fragmentography gives very valuable information, because the latter technique can be considered as an absolute proof of structure and thus specificity.

Some idea as to the specificity of an antiserum to a steroid hormone can be gained from the position of the steroid molecule that is used as the anchoring point to the protein (Fig. 22.2). If an antiserum to estradiol-17β is produced through the use of an antigen conjugated via the hydroxyl group at carbon 17 of the steroid, the resulting antiserum will react almost equally well with estrone and thus will have relatively poor specificity. This is because the only structural difference between the estradiol-17β and estrone molecules is the configuration at position 17 (Fig. 22.2). In general, steroid antibodies are more specific for the portion of the steroid molecule that protrudes from the carrier protein and less specific for the portion of the steroid used for linkage to the protein. Thus, in the case of estradiol-17β, highly specific antibodies have been developed after immunization with conjugates when carbon 6 of the B ring has been used as the site of attachment to the protein (Exley *et al.*, 1971). By use of the same fundamental principle concerning the selection of appropriate sites for conjugation, specific antibodies have been produced against progesterone conjugated to the protein through carbon 11 and to testosterone through carbons 1 and 3 (Niswender and Midgley, 1970).

Protein hormone determination by immunoassay is often performed on blood serum or plasma. The influence of serum or plasma on the binding of the tracer to the antibody must be investigated. The assay of different amounts of serum or plasma should result in curves parallel to those obtained with protein hormone standard. In some steroid hormone immunoassays, the hormone is extracted by an organic solvent from a plasma sample. Organic solvents can also be used as a means to selectively remove steroids from biological fluids. For example, most immunoassay procedures designed to measure progesterone utilize antisera developed against a progesterone-11-protein conjugate, which results in a minor cross reaction with corticosteroids (Thorneycroft and Stone, 1972). If a nonpolar solvent such as petroleum ether is used for the extraction of progesterone from serum or plasma samples, about 80–90% of progesterone is extracted, leaving the more polar corticosteroids in the plasma (Johansson, 1969). The use of such a selective extraction system increases the overall assay specificity. Direct analytical systems for steroid hormones omitting the extraction step are also employed. Because steroid hormones in a blood sample to be analyzed are bound to carrier proteins, direct assay systems have to secure that all steroid molecules, both free and protein-bound, in the sample are given equal opportunities to interact with the antibody used in the assay. The synthetic steroid danazol (17α,2,4-pregnanedien-20-yno(2,3-d)isoxazol-17-ol) can be used to displace progesterone from the binding proteins (Carrière and Lee, 1994). Likewise, 8-ANS (8-anilino-1-naphthalene-sulfonic acid) and levonorgestrel (D(−)-norgestrel) can be used to enhance the displacement of testosterone from its binding proteins (Hoyle and Ebert, 1990). Certain immunoassay systems may require purification of the plasma extract to achieve an acceptable specificity.

The main problem in the immunoassay of primary prostaglandins is that they can continue to be formed in large amounts by platelets after the blood sample has been obtained (Samuelsson *et al.*, 1975). Therefore, the concentrations of $PGF_{2\alpha}$ reported in blood serum or plasma appear, in most cases, to be 100–1000 times higher than the actual values (Granström and Kindahl, 1976). The primary prostaglandins have a very short half-life in the circulation (Hamberg and Samuelsson, 1971) and are rapidly converted to their corresponding 15-keto-13,14-dihydro derivatives. The latter have considerably longer half-lives and occur in higher concentrations than the parent compounds (Beguin *et al.*, 1972). Analysis of metabolites of prostaglandin $F_{2\alpha}$ avoids the problem of the overestimation observed for the parent compound in that the metabolites are formed only within the body and values thus remain stable once a blood sample has been obtained. Radioimmunoassay systems utilizing antibodies to 9α,11α-dihydroxy-15-ketoprost-5-enoic acid and 5α,7α-dihydroxy-11-ketotetranorprosta-1,16-dioic acid have been developed (Granström and Samuelsson, 1972). Most problems involved in the determination of the primary prostaglandins are avoided if their main metabolites, the 15-keto-13,14-dihydro compounds, are measured.

b. Sensitivity

The sensitivity of an immunoassay is defined as the smallest quantity of hormone that the assay can detect reliably. Usually, two kinds of sensitivity are evaluated. The sensitivity of the standard curve is defined as the smallest amount of hormone that is significantly different from zero at the 95% confidence limit. However, the most meaningful sensitivity to establish is the smallest amount of hormone that can be measured per unit of biological fluid, for example, per milliliter of plasma.

c. Accuracy

The accuracy of an assay is defined as the extent to which the measurement of a hormone agrees with the exact amount of the hormone. Accuracy is often determined by comparing immunoassay data with values determined by other procedures such as gravimetry, gas–liquid chromatography, and mass spectrometry. For steroid hormones and prostaglandins, accuracy is also often determined by recovery experiments in which different amounts of hormones are added to a biological fluid, such as plasma, that contains low concentrations of the hormone. The amount of hormone measured in the assay is then compared with the amount of hormone added.

d. Precision

Two types of precision are usually evaluated. The within-assay precision is determined from duplicate measurements of the same sample within the same assay. The between-assay precision is determined from replicate analyses of the same sample in different assays. Usually the between-assay variance is greater than the within-assay variance. Assay variance should be checked continuously with each assay of a certain hormone by use of plasma pools containing set amounts of the hormone. Usually three different plasma sets containing low, medium, and high hormone concentrations are used. Within- and between-assay variations in immunoassay procedures are usually greater than for most other routine procedures used in clinical chemistry.

B. Summary

The use of immunoassay techniques has dramatically increased our knowledge and understanding of reproductive endocrinology in domestic species. These procedures are relatively simple to perform, inexpensive, sensitive, and specific. Data gained from these techniques have resulted in characterization of endocrine changes throughout the estrous cycle and pregnancy in domestic animals.

III. PHYSIOLOGY OF REPRODUCTIVE HORMONES IN THE FEMALE

This section is relatively brief in its coverage of endocrinological events during the reproductive cycle. Readers interested in a more complete presentation of hormonal events involved in reproduction are referred to texts dealing specifically with the subject. Also, some hormones with some influence on reproductive processes, such as those of the thyroid and adrenal glands, are not covered in this section.

A. Estrous Cycle

The major endocrine events that precede ovulation have been well documented in the cow (Chenault *et al.*, 1975), ewe (Nett *et al.*, 1974), sow (Shearer *et al.*, 1972), mare (Evans and Irvine, 1975; Palmer and Jousset, 1975; Stabenfeldt *et al.*, 1975), dog (Concannon *et al.*, 1975), and cat (Shille *et al.*, 1979b). In large domestic animals (cattle, horse, pig, sheep, and goat), follicle growth occurs during the luteal phase in spite of the inhibitory nature of progesterone, the main secretory product of the CL. Although follicles are usually not ovulated during the luteal phase in most species, the mare occasionally ovulates during the luteal phase (Hughes *et al.*, 1973).

With regression of the CL, follicles grow rapidly prior to ovulation because of gonadotropin stimulation. The follicles secrete increasing amounts of estrogen during development, which is important for onset of sexual receptivity as well as for the initiation of the surge release of gonadotropins that is essential for the ovulatory process (Fig. 22.4). Estrogens initiate the release of LH and FSH through the release of GnRH (Moenter *et al.*, 1990). In most species, the preovulatory surge of gonadotropins begins approximately 24 hours before ovulation and is usually of short duration, for example, 8–10 hours in the cow. The mare is an exception in that large amounts of LH are released during an 8- to 9-day period with ovulation occurring on the third day (Geschwind *et al.*, 1975). Another important function of GnRH is to elicit sexual receptivity. Thus, the onset of the preovulatory LH surge and sexual receptivity are coordinated via GnRH synthesis and release.

Following ovulation, a CL is formed under the influence of pituitary gonadotropins. In most species, LH is the major luteotropin; prolactin is thought to play a role in sheep and in rodents and may be the predominant luteotropin in these species. If pregnancy does not ensue, the CL regresses, which permits the estrous cycle to be repeated. This well-timed sequence occurs repetitively at set intervals if not interrupted by pregnancy.

A summary of estrous-cycle activity in common domestic animal species is given in Fig. 22.5. Seasonal breeders, such as the mare, ewe, doe, and queen, undergo cyclic ovarian activity only during the breeding season, whereas the cow, sow, and bitch are affected little, if any, by photoperiod and can have cyclic ovarian activity throughout the year. Estrous-cycle length is approximately 21 days in the cow, doe, mare, and sow and 17 days in the ewe. The bitch has a much longer estrous cycle, the luteal phase often being between 50 and 80 days in duration. The interval between cyclic ovarian activity in the bitch is extended even

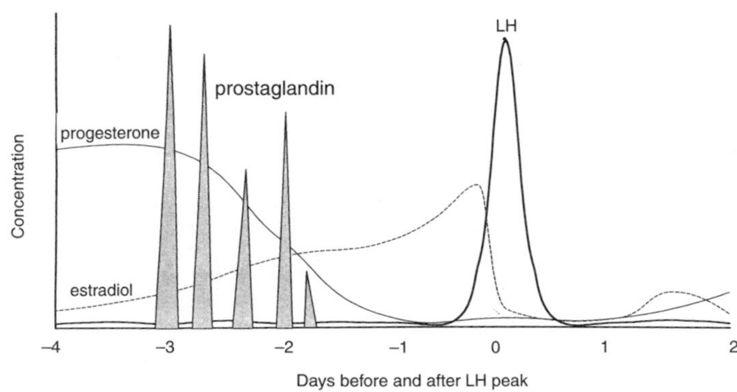

FIGURE 22.4 Schematic presentation of the preovulatory events in general applicable to the cow, doe, ewe, and sow. These endocrine events also occurs in the mare, but in this species, the release of LH occurs over a considerably longer time period.

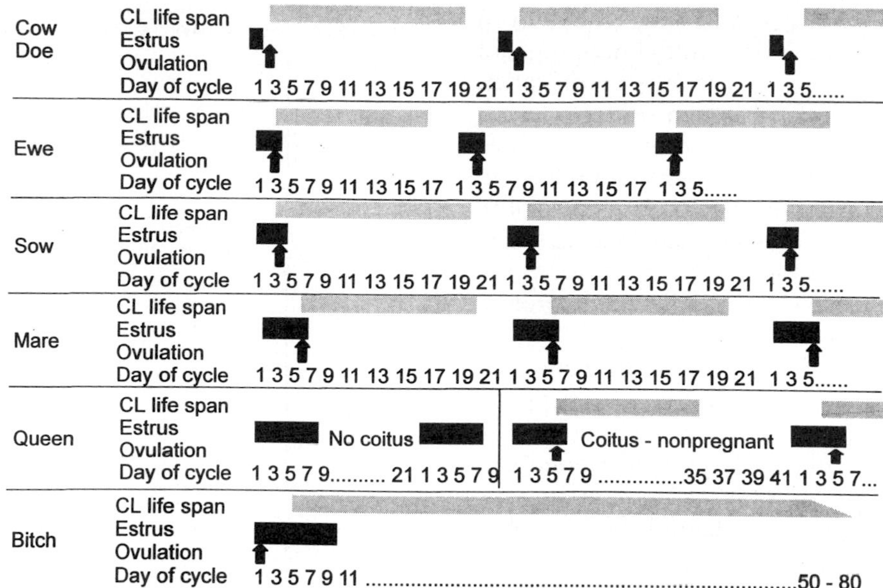

FIGURE 22.5 Comparative data on the duration of estrus, time of ovulation, and duration of CL function in the cow, ewe, sow, mare, queen, and bitch. (Modified from Stabenfeldt, 1974; Edqvist and Stabenfeldt, 1989.)

further by the occurrence of a 4 to 5-month anestrum period following regression of the CL. The cat is an induced ovulator, requiring coital stimulation for ovulation and thus for CL formation. In the absence of coitus, follicles develop every 15–20 days, with follicular growth and regression occupying 5–7 days of each period. An ovulatory, nonfertile mating results in the formation of CL that persist for approximately 35 days and a syndrome termed pseudopregnancy in the cat.

B. Control of the Corpus Luteum

The regression of the CL (luteolysis) is a key event that is responsible for the well-timed estrous cyclicity seen in most domestic species. The importance of the uterus in the control of the lifespan of the CL has been documented through hysterectomy in the cow, ewe, sow, and mare (du Mesnil du Buisson and Dauzier, 1959; Wiltbank and Casida, 1956; Stabenfeldt *et al.*, 1974a). Removal of the uterus from these species during the luteal phase results in prolongation of luteal activity. It is now well established that the uterus in these species synthesizes and releases $PGF_{2\alpha}$, which causes the CL to regress (McCracken *et al.*, 1972).

The temporal release patterns of $PGF_{2\alpha}$ (Fig. 22.4), usually in a pulsatile mode lasting a few hours, have been described in the ewe (Harrison *et al.*, 1972; Barcikowski *et al.*, 1974), sow (Gleeson *et al.*, 1974), doe (Fredriksson *et al.*, 1984), and cow (Nancarrow *et al.*, 1973). Some of the problems involved in determining $PGF_{2\alpha}$, such as its short half-life and formation by platelets at collection, can be avoided if the main plasma metabolite, 15-keto-13,14-dihydro-$PGF_{2\alpha}$, is determined. Data are available on the patterns of the metabolite during luteolysis in the cow (Kindahl *et al.*, 1976a,b), ewe (Peterson *et al.*, 1976), mare (Neely *et al.*, 1979), doe (Fredriksson *et al.*, 1984), and sow (Shille *et al.*, 1979a).

Regression of CL is usually accomplished within 48 hours following the onset of the prostaglandin release. It is likely that estrogens, presumably of ovarian follicle origin, initiate $PGF_{2\alpha}$ release. Estrogen also initiates the formation of endometrial oxytocin receptors (McCracken *et al.*, 1984), which in sheep are important for pulsatile synthesis and release of $PGF_{2\alpha}$ in that oxytocin can initiate the release of $PGF_{2\alpha}$ (Sharma and Fitzpatrick, 1974). In ruminants, oxytocin is synthesized and released from the CL in response to $PGF_{2\alpha}$, which in turn initiates the synthesis and release of $PGF_{2\alpha}$ by the uterus. This mechanism is the basis for the pulsatile secretion of $PGF_{2\alpha}$ (for reviews, see Flint *et al.*, 1992; Whates and Denning-Kendall, 1992). In the dog and cat, $PGF_{2\alpha}$ does not appear to be involved in luteolysis, although its precise role is still uncertain.

C. Early Pregnancy

Modification of $PGF_{2\alpha}$ release is essential for the establishment of pregnancy in the species (cow, ewe, mare, and sow) in which this compound serves as the luteolysin (Kindahl *et al.*, 1976a; Nett *et al.*, 1976; Shille *et al.*, 1979a). The rapid elongation of fetal membranes, which precedes the critical time of the initiation of luteal regression by about 3 days in the nonpregnant animal, appears to be important for the modification of prostaglandin release. Fredriksson *et al.* (1984) for goats and Zarco *et al.* (1988) for sheep have shown that the main change in $PGF_{2\alpha}$ synthesis and release in nonpregnant vs pregnant animals involves a continuous, not pulsatile, mode of secretion. In fact, $PGF_{2\alpha}$ concentrations increase at the onset of pregnancy, but the release is continuous, which allows for the prolongation of luteal activity.

Maternal recognition of pregnancy in the cow, ewe, doe, and sow involves mechanisms that alter the prostaglandin release to protect the CL from luteolysis. In the cow, ewe, and doe, conceptuses secrete trophoblast proteins between days 10 and 21–24 of pregnancy. These substances are produced before the blastocyst attaches to the endometrium, are secreted into the uterine lumen, and bind to uterine oxytocin receptors, blocking the episodic uterine $PGF_{2\alpha}$ secretion (for a review, see Flint *et al.*, 1992). In the sow, in which the maternal recognition of pregnancy is controlled by conceptus-derived estrogens (Perry *et al.*, 1976; Fig. 22.6), similar trophoblast proteins have been isolated from preimplantation blastocysts (La Bonnardière, 1993).

Significant increases in intraluminal $PGF_{2\alpha}$ have been observed in pigs (Frank *et al.*, 1978) and cattle (Thatcher *et al.*, 1979), but not in horses (Zavy, 1979), beginning about two weeks postconception. Possible uterine antiluteolytic factors include PGE (Hoyer *et al.*, 1978), and PGE_2 (Magness *et al.*, 1978), with increased production of PGE_2 by endometrial tissue having been demonstrated in ewes during early pregnancy (Ellinwood *et al.*, 1979).

Modification of the release pattern of $PGF_{2\alpha}$ (pulsatile to continuous) by the luteotropic products from the conceptus and uterus is probably the most important factor that allows luteal activity to continue. The net result is that luteal activity is extended in the cow, ewe, mare, and sow beginning at about 14 days following ovulation. Modification of $PGF_{2\alpha}$ release does not appear to be important for the establishment of pregnancy in the dog and cat.

D. Pregnancy and Parturition

The presence of a CL is necessary for the maintenance of pregnancy in the vast majority of cows (Estergreen *et al.*, 1967). The pig also requires luteal support throughout gestation (du Mesnil du Bussion and Dauzier, 1957). In the ewe, the presence of a CL is required for the first 50–60 days of gestation (Linzell and Heap, 1968). After this time period, the fetoplacental unit

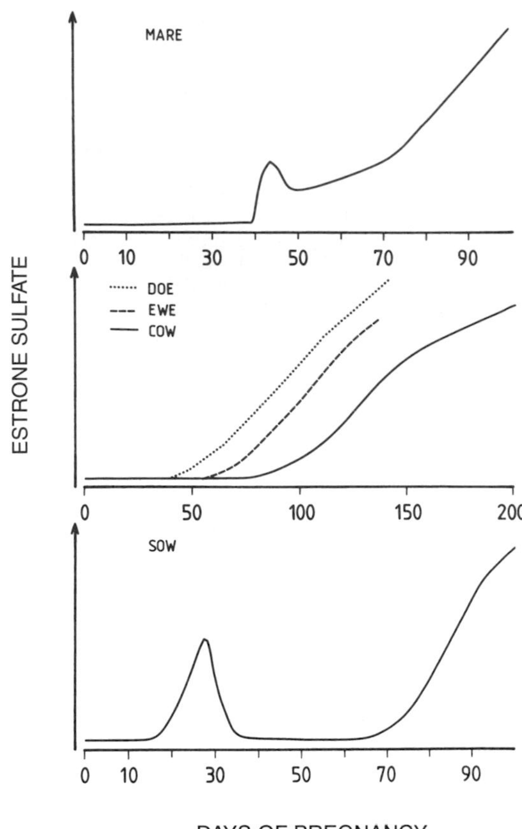

FIGURE 22.6 Blood levels of conjugated estrogens (mainly estrone sulfate) during early pregnancy in the mare, doe, ewe, cow and sow. Data derived from Kindahl *et al.*, 1982 (mare); Chaplin and Holdsworth (1982) (doe); Tsang (1978) (ewe); Gaiani *et al.*, (1982) (cow); and Robertson and King (1974) (sow).

secretes significant amounts of progesterone. The dog requires the presence of a CL for all of gestation. Sokolowski (1971) found that ovariectomy even as late as day 56 postbreeding resulted in premature delivery.

The necessity for secondary CL, formed in the mare between days 40 and 60 of gestation, has been discussed over the years. The secondary CL are the result of eCG secretion by the endometrial cups, which are formed from a circular band of cells of placental origin (chorionic girdle cells) that invade the endometrium and form isolated endocrine organs of temporary function (Allen, 1969). Although secondary CL are not essential for the maintenance of pregnancy in the mare in that the primary CL continues to function for up to 150 days, they do add extra progestational support for the pregnancy during the time placental production of progesterone is being established. Progesterone support of pregnancy in the mare begins to be assumed by the placenta as early as day 50 of gestation but is not complete for all mares until approximately 100 days or later, a time that coincides with the beginning demise of both primary and secondary CL (Holtan *et al.*, 1979).

It has been demonstrated that eCG has a close immunological relationship with equine LH (Farmer and Papkoff, 1979); further, a luteotropic effect of eCG has been shown by incubation studies of CL (Squires *et al.*, 1979). If fetal loss occurs after the formation of endometrial cups, continuing eCG production supports luteal activity, even to the point of making lysis of these CL difficult in conjunction with the pharmacological administration of prostaglandin $F_{2\alpha}$.

The endocrine activity of the fetoplacental unit can be monitored through the measurement of conjugated estrogens, especially estrone sulfate, in peripheral plasma or urine (Fig. 22.6). In the pregnant sow, concentrations of estrone sulfate become detectable at day 17 of pregnancy, increase until about day 28, then decline to low or undetectable values, and then increase again beginning around days 75–80 of pregnancy, remaining high until parturition (Robertson and King, 1974). In the mare, estrone sulfate values begin to increase around days 35–40 of pregnancy with accentuated production at 80–90 days and with the highest values obtained from day 150 to parturition (Terqui and Palmer, 1979; Kindahl *et al.*, 1982). In the cow (Gaiani *et al.*, 1982) and ewe (Tsang, 1978), estrone sulfate levels can be detected from around day 70–80 after conception, and in the goat, concentrations start to increase around day 50 of pregnancy (Chaplin and Holdsworth, 1982). In the pregnant cow, estrone sulphate concentrations in milk start to increase between days 100 and 120 (Hatzidakis *et al.*, 1993; Henderson *et al.*, 1994). In all species, high values of estrone sulfate in the peripheral blood, milk, or urine are strong evidence for the presence of a viable fetus.

The mare produces two estrogens during pregnancy that are unique to equids, equilin and equilenin, both of which have unsaturated B rings (Girard *et al.*, 1932). By fetal gonadectomy in the horse, Pashen and Allen (1979) have shown the importance of fetal gonads in the production of estrogen in cooperation with the placenta. Little evidence of estrogen production during pregnancy has been reported for the dog (Hadley, 1975). Estrogen production by the fetoplacental unit in the cat also appears to be minimal during gestation (Verhage *et al.*, 1976).

An important endocrine change that occurs prior to parturition in the cow, ewe, and sow involves an increase in the synthesis and release of unconjugated estrogens by the fetoplacental unit. This increased estrogen synthesis is reflected in elevated plasma estrone concentrations in the pregnant cow beginning 20–30 days prepartum (Edqvist *et al.*, 1973), in the ewe about 2 days before parturition (Challis *et al.*, 1971), and in the pig about 1 week before delivery (Robertson and King, 1974). In the cow (Stabenfeldt *et al.*, 1970; Edqvist *et al.*, 1973) and bitch (Smith and McDonald, 1974), parturition is preceded by an abrupt fall in progesterone concentrations 24–48 hours prior to delivery. In the ewe (Stabenfeldt *et al.*, 1972), mare (Noden *et al.*, 1978), and sow (Baldwin and Stabenfeldt, 1975), partial withdrawal of progesterone occurs prior to delivery. In the cow (Fairclough *et al.*, 1975; Edqvist *et al.*, 1976, 1978), ewe (Liggins *et al.*, 1972), and bitch (Concannon *et al.*, 1988), it has been demonstrated that prostaglandin release initiates regression of the CL and thus is responsible for the withdrawal of progesterone.

High estrogen and prostaglandin concentrations combined with low progesterone concentrations increase the contractile state of the uterus. Prostaglandins may also initiate cellular changes within the cervix, in addition to the effect of relaxin, which result in cervical softening and dilation. Cervical stimulation, the result of the initial entry of the fetus into the pelvic canal, causes the reflex release of oxytocin from the posterior pituitary. This increases the intensity of uterine contractions and thus aids the final delivery process.

IV. CLINICAL ASPECTS OF REPRODUCTIVE ENDOCRINOLOGY

The development of immunoassay techniques for hormone determinations in domestic species has created laboratory procedures that are useful as diagnostic aids in clinical work. For some reproductive hormones, relatively well defined indications for their clinical use are known at present and more are likely to be identified in the future. Important differences exist between humans and animals with respect to endocrine analysis. For example, steroid hormone concentrations are much lower (10-fold less in the case of estrogens) in animals, which produces a requirement for more rigorous assay systems. In animals, the most useful information comes from assessing gonadal or fetoplacental activity (vs pituitary activity), and thus, the determination of steroid hormones is emphasized. Besides the fact that gonadotropin values are less useful for the assessment of clinical situations, variations in amino acid composition of specific hormones among species mean that multiple systems have to be developed to determine the content of one protein hormone across species lines. Antibodies such as the one developed for LH by Niswender *et al.* (1969), which is able to detect LH in many species, are the exception.

For hormone analyses to be useful as a diagnostic tool, certain criterion have to be fulfilled. The concentration of the hormone at the sampling site (usually a peripheral vein) should closely correlate with the amount of the hormone being released from the endocrine gland. It is preferable that the release pattern of the hormone be steady, which allows valid information to be obtained on the secretory status of the endocrine gland from one sample. Several reproductive hormones do not fulfill the latter criterion, and thus their determination is not useful from a routine diagnostic view. For example, the duration of the LH surge observed in conjunction with ovulation is short in most domestic species except the horse. In the cow, the preovulatory LH peak has a duration of 8–10 hours, requiring samples to be obtained every 4 hours to detect the peak (Karg *et al.*, 1976). In the mare, the duration of the peak is considerably longer, 8–9 days, which allows a less frequent sampling interval (Geschwind *et al.*, 1975). However, the long duration of LH peak in the mare prevents the determination from being useful in predicting ovulation.

One reproductive hormone, progesterone, has been found to be of significant clinical value in females of most domestic species, and overall, its analysis gives the most useful information on the reproductive status of animals. Other hormones with established or potential clinical use will be discussed further for each species. As a percentage of total number of clinical assays, the highest number have been utilized as pregnancy tests. It is worth keeping in mind that the analysis of hormones as a diagnostic aid in solving clinical problems supplements, but does not replace, the information that should be gained by a careful clinical examination.

Although blood has been the usual medium for hormone analysis, milk, urine, saliva, and even feces are also sources for gaining useful endocrine information. The latter substances have the advantage in certain situations of being easier to collect and, at the very least, they avoid the use of venipuncture. It has been shown that the determination of estrogen conjugates in urine is much more effective in revealing ovarian follicle production of estrogens than the analysis of either free or conjugated estrogen in plasma.

A. Cattle

1. Progesterone

Several reports are available on progesterone concentrations in blood during early pregnancy in cattle (Pope *et al.*, 1969). A finding of importance is that progesterone can be measured in the milk of lactating cows and, further, that its concentration accurately reflects the concurrent plasma concentration of progesterone (Laing and Heap, 1971).

The difference that exists in both plasma and milk progesterone concentrations 19–24 days after a fertile breeding as compared to a nonfertile breeding has been used as an early pregnancy test (Shemesh *et al.*, 1968; Robertson and Sarda, 1971; Fig. 22.7). The plasma progesterone concentration in blood of pregnant cows at 21 days postbreeding is almost always at least 2 ng/ml (6 nmol/liter) and usually 4–8 ng/ml (13-26 nmol/liter), as compared to less than 0.5 ng/ml (1.6 nmol/liter) in the nonpregnant animal at the same time. Elevated progesterone concentrations, however, only reflect the presence of luteal tissue and are not directly indicative of the presence of a fetus *in utero*. Furthermore, a slight prolongation of luteal activity in a nonpregnant animal can occur that results in elevated progesterone concentrations at day 21, a situation in which the analytical progesterone result would be positive and the animal would be falsely considered pregnant. The accuracy of the forecast for pregnancy (positive forecast) thus is often lower than desirable, in most cases ranging between 75 and 90%. The negative forecast, however, is more accurate; cows having low progesterone concentrations in milk or blood 21 days postbreeding will almost always not be pregnant. The accuracy of the positive forecast can be increased if progesterone analyses are also carried out on samples obtained at the time of insemination to eliminate animals inseminated during the luteal phase of the estrous cycle. Cows inseminated during the luteal phase will be in the luteal phase 21 days later and the animal will be falsely considered pregnant.

It should be recognized that the main focus of early pregnancy diagnosis is for more efficient management of breeding. In most cases involving early pregnancy diagnosis, pregnancy status needs to be verified again at about 40 days postbreeding. In spite of its limitations, progesterone analysis for pregnancy diagnosis is the most common clinical use of any of the reproductive hormones. Progesterone analysis can also be used for the retrospective confirmation of the absence (or presence) of a CL at the time of insemination in cattle. Since inadequate detection of estrus is the most common cause of low fertility in herds utilizing artificial insemination, the determination of progesterone concentrations at the time of insemination can be a useful tool when herds with fertility problems are encountered. Karg *et al.* (1976) indicated that improper timing of insemination, as judged from milk progesterone determinations, occurred in 15% of the cases in a controlled field test; the figure rose to 26% under practical field conditions. Lower figures have been reported by others, namely 0 to 3% in the Netherlands (Van de Wiel *et al.*, 1978), and 4% in Sweden (Oltner and Edqvist, 1981). The use of progesterone determinations in milk obtained at the time of insemination in

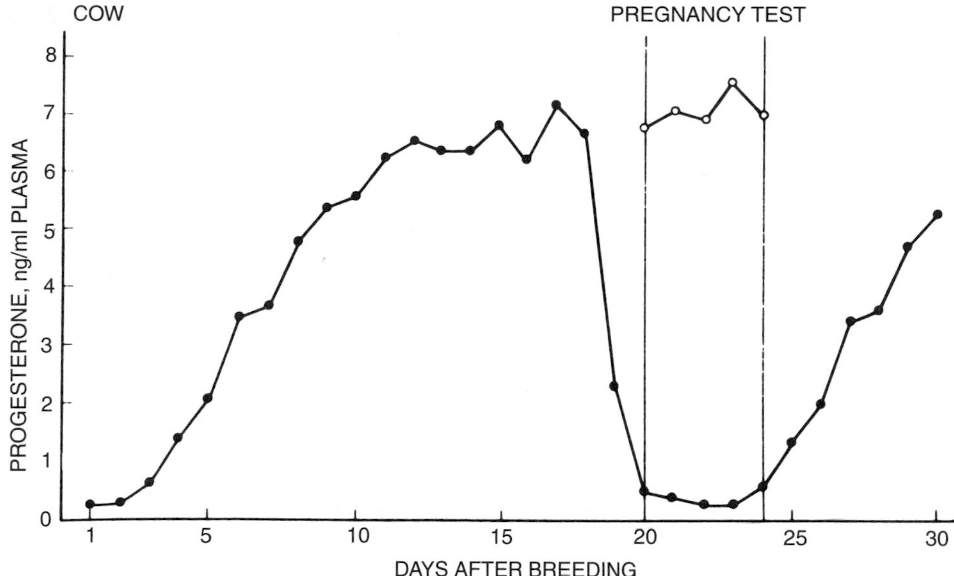

FIGURE 22.7 Time after breeding for utilizing progesterone analysis as a means of pregnancy diagnosis. Progesterone content in nonpregnant (●) cows compared with progesterone content of pregnant animals (○) (Stabenfeldt *et al.*, 1969b).

cows with questionable heat signs and inconclusive genital-tract findings can serve as a valuable tool for educating the staff responsible for insemination (Garcia and Edqvist, 1990).

A potential important area for the diagnostic use of progesterone analysis is the elucidation of clinical syndromes in the postpartum period, in which cows fail to show sexual receptivity for extended periods of time. Animals that have reestablished ovarian activity can often be distinguished from those that have not. Elevated progesterone values indicate that significant ovarian activity is present. Cows that have luteal activity can be manipulated through $PGF_{2\alpha}$ treatment with a reasonable expectation of initiation of a new cycle. Progesterone determinations can also be used to verify the presence of luteal tissue in conjunction with endometritis/pyometra (Pepper and Dobson, 1987) and ovarian cysts (Booth, 1988; Sprecher *et al.*, 1990). Both conditions should respond to prostaglandin therapy. Differentiation of luteal vs follicular ovarian cysts in cattle is not usually done because it is technically difficult by palpation *per rectum* and because the same gonadotropin treatment can be used in both situations. Luteinization of the structure with a luteotropin is followed in 10–14 days with prostaglandin treatment. In both cases though, the use of progesterone analysis is useful in establishing the therapeutic response.

2. *Estrone Sulfate*

High concentrations of estrone sulfate are found in blood beginning at about day 80 of pregnancy and in milk from about day 100 (Fig. 22.6). In studies using estrone sulfate concentrations in milk samples collected at 120 days or more of gestation as an indicator for pregnancy status, an overall accuracy rate of 96% was found (Power *et al.*, 1985). A similar figure has also been reported by McCaughey *et al.* (1982). These authors suggested that estrus and mastitis may influence the accuracy of the test. When comparing peripheral blood concentrations of estrone sulfate among three breeds of cattle, it was found that the breed giving birth to the lightest calves had lower estrone sulfate concentrations in the interval 101–200 days of gestation (Abdo *et al.*, 1991).

3. *Other Substances*

A protein of placental origin (bovine pregnancy-specific protein B; bPSBP) has been observed in the blood of pregnant cows beginning between days 16 and 21 of gestation (Sasser *et al.*, 1983; Sasser and Ruder, 1987). This discovery of a protein that is observed only if an embryo is present opens the way for an early and definitive pregnancy diagnosis in cattle by blood analysis. Humblot *et al.* (1988) compared diagnosis of pregnancy in Friesian cattle by determination of progesterone and bPSBP in blood. The study revealed the accuracy (ratio of cows positive and pregnant to total cows with positive) of the positive forecast to be 67.2% (82/122) for progesterone on day 24 after insemination. The accuracy of the negative forecast for progesterone was 98% (52/53). The accuracy of the positive forecast for bPSPB increased with gestation age from 86.2% (50/58) on day 24 to 98.8% (83/84) on day 70. The accuracy of the negative diagnoses by bPSPB increased from 71.8% (84/117) on day 24 to 100% (83/83) on days 30 to 35. The authors concluded measurement of bPSPB 30 days after insemination is an efficient test both for positive and negative pregnancy diagnoses. A cardinal principle of pregnancy detection is that the test should be effective by the time the estrous cycle would end if the animal was not pregnant. This reemphasizes the main point of early pregnancy detection: It is done so the animal can be rebred if not pregnant. The test fulfilling this criterion has yet to be found.

B. Sheep

1. Progesterone

Progesterone analysis as an early test for pregnancy has been used in sheep (Robertson and Sarda, 1971). In the ewe, the progesterone analysis has to be carried out on blood samples, because most breeds of sheep are not lactating at the time of breeding. Maximal luteal phase progesterone concentrations in the ewe are approximately 2–4 ng/ml (6–13 nmol/liter), whereas the concentrations at estrus range from 0.15 to 0.25 ng/ml (0.5–0.8 nmol/liter) (Stabenfeldt *et al.*, 1969c; Dickie and Holzmann, 1992). Using an amplified enzyme immunoassay technique for plasma progesterone, an accuracy of 100% was reported for the diagnosis of pregnancy using samples taken between days 15 and 16 from a flock of 130 ewes (106 diagnosed pregnant and 24 diagnosed nonpregnant) (McPhee and Tiberghien, 1987).

A relatively marked increase in progesterone values from 2–4 ng/ml (6–13 nmol/liter) to 12–20 ng/ml (38–64 nmol/liter) occurs between days 60 and 125 of pregnancy (Stabenfeldt *et al.*, 1972). This increase is due to increased progesterone production from the fetoplacental unit. The contribution of the CL to progesterone concentrations remains constant throughout pregnancy. Thus, the relatively large difference in progesterone concentrations between nonpregnant ewes with a CL present and ewes with fetus(es) present could be used to minimize the relatively high false-positive forecasts observed for sheep at 17 days postbreeding. In a trial that determined plasma progesterone concentrations in 46 ewes on days 18 and 70 after mating, correct pregnancy diagnoses were made for 80% and 93% of ewes, respectively (Weigl *et al.*, 1975). This approach might be useful for intensive sheep operations for the identification of nonpregnant animals for early culling. The limitations of progesterone analysis as a pregnancy test discussed for cattle apply to sheep as well.

2. Estrone Sulfate

In the pregnant ewe, estrone sulfate produced from the fetoplacental unit is detected in elevated concentrations beginning at around day 70 after conception (Tsang, 1978; Fig. 22.6). Worsfold *et al.* (1986) determined estrone sulfate concentrations in blood from ewes bred 85 days previously and found the accuracy of the nonpregnant vs pregnant interpretations was 44% and 88%, respectively. A considerable overlap in estrone sulfate values between ewes with single and multiple fetuses was found, and even at day 116 of pregnancy, estrone sulfate concentrations overlapped between groups of ewes with single and multiple fetuses. Because of this finding, the authors concluded that estrone sulfate concentrations could not be used as a predictive tool for litter size (Fletcher and Worsfold, 1988).

3. Other Substances

Circulating pregnancy-associated proteins have been reported in sheep. A heat-labile protein with a molecular weight of around 8000 was found in sera of sheep as early as day 6 of gestation (Cerini *et al.*, 1976). It has been demonstrated that the bovine pregnancy-specific protein B also can be used for early detection of pregnancy in the ewe because pregnant ewes have a blood antigen that cross reacts with antibodies to bPSPB (Ruder *et al.*, 1988). In a study comparing the accuracy of detecting pregnancy with an ultrasonic device, a real-time scanning instrument, and RIA for bPSPB in ewes, it was concluded that the RIA for bPSPB detected pregnancy earlier and more accurately than the ultrasonic device and was as accurate as the real-time scanning instrument (Ruder *et al.*, 1988).

Ovine placental lactogen (oPL) has been demonstrated in the peripheral blood of pregnant ewes (Kelly *et al.*, 1974). Increased maternal oPL concentrations occur between 40 and 50 days of pregnancy, reach maximum concentrations between days 120 and 140, and decline as parturition approaches (Chan *et al.*, 1978). The determination of oPL concentrations might be used as a basis for a specific pregnancy test in the ewe.

C. Pig

1. Progesterone

Luteal-phase progesterone concentrations in the pig are considerably higher than in cattle and sheep, namely, 20–50 ng/ml (64–159 nmol/liter), whereas concentrations at estrus are below 0.5 ng/ml (1.6 nmol/liter) (Stabenfeldt *et al.*, 1969a). The difference in progesterone values between nonpregnant and pregnant animals 19–24 days after service has been used as an early pregnancy test (Robertson and Sarda, 1971; Edqvist *et al.*, 1972; Saiz *et al.*, 1988). In the pig, progesterone determination is usually performed on blood. Because of the sensitivity of the assay systems and the concentration of progesterone in blood in pigs during the luteal phase, analyses can be performed on a small volume of blood (about 10 drops), allowing the sample to be obtained through a small incision in an ear vein. The limitations associated with using progesterone determinations as a pregnancy test in the pig are similar to those previously discussed for cattle. Progesterone analyses have also been used to determine ovarian activity in clinically anestrous gilts, as well as to establish the stage of the estrous cycle in gilts and subsequent response to treatment (King *et al.*, 1985). It is also possible to monitor luteal-phase activity in the sow through measurement of fecal gestagens (Hultén *et al.*, 1994).

2. *Estrone Sulfate*

Previous studies of estrogen concentrations in urine revealed a marked increase in estrogen between day 20 and 30 of pregnancy in pigs (Velle, 1958). Studies of blood concentrations of estrone sulfate in pigs during pregnancy showed patterns similar to those determined in urine (Robertson and King, 1974; Fig. 22.6). In early pregnancy in the pig, it has been hypothesized that estrogen synthesized by the early preimplantation embryo may be the messenger for the maternal recognition of pregnancy (Perry *et al.*, 1976) and that the elevated estrone sulfate concentrations in the maternal circulation during early pregnancy reflect fetal synthesis. The determination of estrone sulfate in early pregnancy in the pig is thus a specific pregnancy test (Edqvist *et al.*, 1980, Cunningham *et al.*, 1983). The index of discrimination between estrone sulfate concentrations in blood of pregnant vs nonpregnant pigs around 25 days after breeding is very high. As a pregnancy test, Sugiyama *et al.* (1985) reported an accuracy rate of 98% in pigs between days 20 and 26 of pregnancy. Horne *et al.* (1983) found litter size on days 20–26 of pregnancy to be positively correlated with estrone sulfate values, but such a relationship was not found later in pregnancy. During the latter part of pregnancy, the maternal blood concentrations of estrogens (estrone sulfate, estrone, and estradiol-17β) are very high and can thus be used to confirm pregnancy at this stage of gestation (Fig. 22.6).

In the pregnant sow, the concentration of estrogen in urine and feces follows the same pattern as in blood (Choi *et al.*, 1987). Szenci *et al.* (1993) compared ultrasonography and the determination of unconjugated estrogen in feces at 25 to 30 days after insemination for the diagnosis of pregnancy in pigs. Based on farrowing data, the positive and negative predictive values for the ultrasound were 93.2 and 100%, respectively. For the determination of unconjugated estrogens, the corresponding predictive values were 93.8 and 55.1%, respectively.

D. Horse

1. *eCG (PMSG)*

Although pregnancy diagnosis has been done in mares through the measurement of equine chorionic gonadotropin (eCG), formerly named pregnant mare serum gonadotropin (PMSG), the main drawback to its use is that the presence of eCG does not guarantee the presence of a fetus, but indicates that a viable fetus was present at the time of endometrial cup formation. This situation exists because endometrial cups have autonomy of function and continue to secrete PMSG for a period of time in spite of loss of the fetus (Allen, 1969). This means that both mares with normal pregnancies and mares that experience embryonic mortality after day 40 of gestation will have elevated eCG concentrations in blood. The use of eCG determinations as a positive pregnancy diagnosis test will cause some mares to be diagnosed as pregnant that will not deliver a foal (Mitchell, 1971; Jeffcott *et al.*, 1987; Fig. 22.8).

2. *Progesterone*

Progesterone analysis is useful for establishing the presence or absence of ovarian activity in animals with puzzling behavioral patterns. Agitated or aggressive behavior, often interpreted as sexual in orientation, occurs without regard to luteal status. Progesterone analysis can be helpful because elevated values directly indicate the presence of a CL and, additionally, are evidence that folliculogenesis and ovulation are normal. Relatively low luteal-phase values for progesterone (1–3 ng/ml; 3–10 nmol/liter) vs normal luteal-phase values (>3 ng/ml; >10 nmol/liter) are often associated with the presence of a persistent CL (Stabenfeldt *et al.*, 1974b). Progesterone values of mares with persistent luteal activity are low because some $PGF_{2\alpha}$ synthesis and release often occurs about 14 days postovulation, albeit insufficient to cause complete luteolysis (Neely *et al.*, 1979).

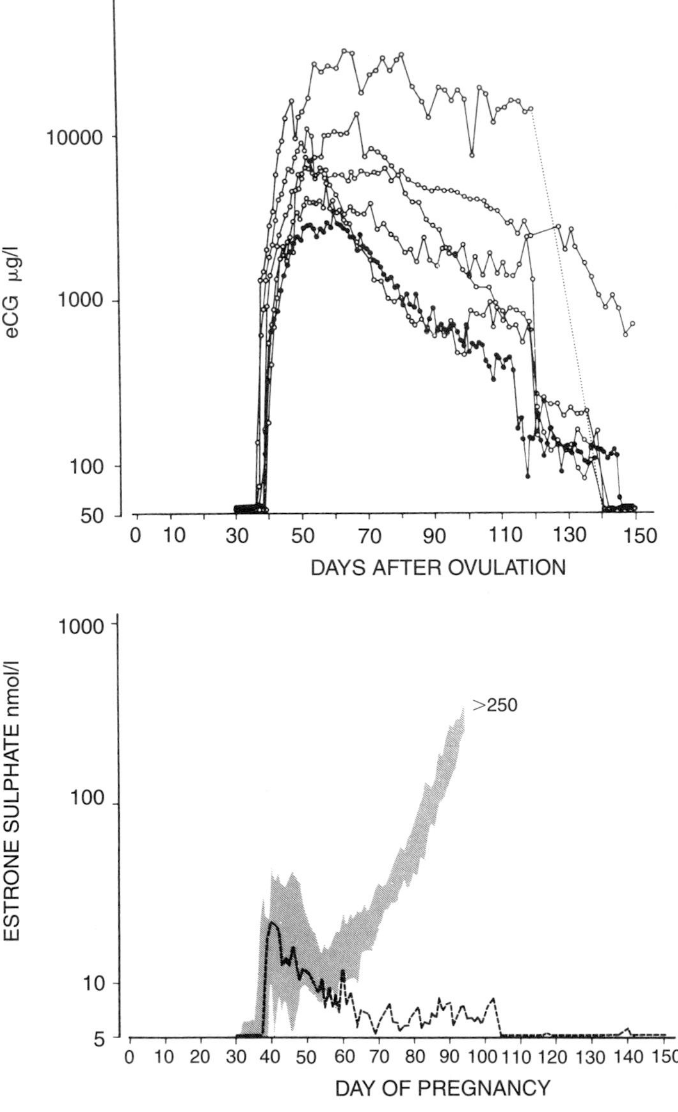

FIGURE 22.8 (Upper) Blood plasma concentrations of eCG in five normal pregnant mares (○) and in one mare in which the fetus died around day 50 of pregnancy (●). (Lower) Geometric means of blood plasma concentrations of estrone sulfate in the five normal mares (shaded area) and the estrone sulfate concentration in the mare in which the fetus died around day 50 and was aborted on day 113 of pregnancy (Darenius *et al.*, 1988).

Progesterone analyses can also be useful in mares that fail to manifest sexual receptivity, yet have cyclic ovarian activity (Hughes *et al.*, 1973). Progesterone analysis at 5-day intervals over 20 days (approximately one estrous cycle length) can verify the presence or absence of cyclic ovarian activity. The time of ovulation can also be predicted within a 2- to 3-day interval, a prediction that can be helpful to the veterinary practitioner anticipating the next time of ovulation. Breeding may have to occur by artificial means in these situations. Hinrichs *et al.* (1988) compared the accuracy of determining day of ovulation ±1 day using three different methods: (1) an immediate, qualitative ELISA for progesterone in blood, (2) a quantitative ELISA for progesterone in blood, and (3) daily teasing to detect estrus. Ovulation was detected by ultrasound examination per rectum. The accuracy in determining day of ovulation ±1 day using the three methods was 72% for the qualitative progesterone assay, 88% for the quantitative progesterone assay, and 86% for teasing.

3. *Conjugated Estrogens*

Estrone in its unconjugated form (free estrogen) reflects important physiological events in the mare beginning at about day 75 of gestation when the fetoplacental unit begins to produce estrone in rapidly increasing amounts (Nett *et al.*, 1975). More importantly, it has been shown that estrone is rapidly conjugated after secretion to water-soluble estrogen conjugates and the ratio between free and conjugated forms is 1 : 100 (Terqui and Palmer, 1979). Terqui and Palmer (1979) and Kindahl *et al.* (1982) have shown that significant increases in estrone conjugate concentrations occur between days 35 and 40 of gestation (Fig. 22.8). Kindahl *et al.* (1982) showed the increase to be 10- to 20-fold between days 35 and 40. Concentrations then decline slightly, with a further increase noted at the same time that free estrone concentrations begin to increase. The initial source of increased estrone production during gestation days 20 and 70 is the ovaries (Daels *et al.*, 1990). Later on, its occurrence is likely driven by the attachment of the embryo and the production of eCG (PMSG). Sist *et al.* (1987) found a majority of pregnant mares to have significantly higher serum estrone sulfate concentrations by day 60 after breeding. They also determined estrone sulfate in milk from pregnant mares and reported concentrations to be lower than in serum but following the same pattern. In the authors' experience, mares should be pregnant for more than 90 days before analyses of estrone sulfate in plasma can serve as a reliable indicator of pregnancy (Fig. 22.8).

Estrone conjugate concentrations in urine can be used to document pregnancy in the mare and, in fact, may be more accurate than plasma analysis because of the concentrating aspects associated with urine formation (Daels *et al.*, 1991). Increased amounts of fecal estrogens and gestagens have also been reported in the pregnant mare (Schwarzenberger *et al.*, 1991).

In addition to being a means of confirmation of pregnancy, estrone conjugate analysis also allows the soundness of the pregnancy to be assessed because estrogen concentrations reflect the dynamics of a growing fetus. Thus, it is possible not only to indicate that pregnancy is in progress by estrone conjugate analysis, but also to indicate whether the pregnancy is proceeding well or is in some state of compromise. Analysis of estrone conjugates is an important aid for the verification of pregnancy in the mare and does not suffer the drawback of false positive diagnosis as is the case for eCG (Fig. 22.8). Estrone conjugate analysis also has been used to assess follicle growth patterns in nonpregnant mares, a test that was difficult to do by determining free estradiol-17β (Makawiti *et al.*, 1983; Daels *et al.*, 1991).

4. *Testosterone*

Testosterone values vary in the mare according to the reproductive state. Values, usually less than 15 pg/ml (52 pmol/liter) during anestrum, range between 20 and 40 pg/ml (69–139 pmol/liter) during cyclic ovarian activity, with the higher values being observed during the follicular phase of the cycle immediately before ovulation. Testosterone determinations have been used to aid the diagnosis of granulosa-theca cell tumors in the mare (Stabenfeldt *et al.*, 1979) and to differentiate granulosa-theca cell tumors from ovarian teratoma (Panciera *et al.*, 1991). Leydig-like cells in the theca appear to be the source of testosterone. In cases of granulosa-theca cell tumors, testosterone values vary with values ranging from 40 pg to over 100 pg/ml (139–347 pmol/liter). These tumors generally develop slowly, and it is not known whether this slow development also reflects a slowly developing capacity for testosterone production or whether there is variability in the number of testosterone-secreting cells among tumors. Aggressive stallion-like behavior is associated with values less than 150 pg/ml (520 pmol/liter) (Meinecke and Gips, 1987). In the authors' experience, this is the only clinical situation involving abnormal behavior in the mare in which a direct relationship to gonadal steroids has been established.

Testosterone determinations in the male horse have been used as an aid in the diagnosis of cryptorchidism. Cox (1975) reported that horses with less than 40 pg/ml (139 pmol/liter) plasma should be considered castrated, whereas animals with concentrations less than 100 pg/ml (347 pmol/llter) should be considered as having testicular tissue present. Although some cryptorchid animals have testosterone concentrations less than 100 pg/ml (347 pmol/liter), most have values range from 200 to 1000 pg/ml (693–3467 pmol/liter) (Cox, 1975). Testosterone concentrations in intact males usually range from 1000 to 2000 pg/ml (3467–6934 pmol/liter) (Cox *et al.*, 1973; Berndtson *et al.*, 1974). hCG administration for the purpose of stimulating testosterone production by the testes has been suggested as a means of resolving cases in which values are between 40 and 100 pg/ml (139–347 pmol/liter). The dosage of injected hCG has been 6000–12,000 I.U. (Cox *et al.*, 1986; Arighi and Bosu, 1989; Silberzahn *et al.*, 1989), and the second sample for analysis of testosterone was obtained after 30–120 minutes (Cox *et al.*, 1986), after 24 hours (Arighi and Bosu, 1989), or after 3 days (Silberzahn *et al.*, 1989). For routine diagnosis, the authors do not recommend the use of hCG because most single baseline determinations are adequate for diagnosis. The main value of a second determination is for substantiation of the first values, and thus the time and expense required for the second sample is not commensurate with the information to be gained (Stabenfeldt and Hughes, 1980).

Analysis for estrone sulfate conjugates for the diagnosis of cryptorchidism may be preferred over that for testosterone because it requires only one analysis and the accuracy is slightly improved (Cox *et al.*, 1986). Arighi and Bosu (1989) recommended both resting testosterone and estrone sulfate to diagnose accurately the presence of testicular tissue. It should be noted, however, that estrone sulfate analysis for cryptorchidism cannot be used in horses less than 3 years of age or in donkeys because little estrone sulfate is produced in either of these situations.

It is the authors' experience that the presence of sexual behavior in suspected cases of equine cryptorchidism is not necessarily dependent upon elevated testosterone concentrations in that many patients (about one-third) referred because of behavioral problems are, in fact, castrated. In essence, some animals can maintain normal libido with low circulating concentrations of testosterone. In this situation, testosterone analysis is very helpful either to eliminate unnecessary surgery or to indicate to the surgeon that testicular tissue is present and should be found on surgical entry.

E. Dog

1. *Progesterone*

In the bitch, there is little difference between the progesterone patterns of the pregnant and nonpregnant luteal phase. Plasma concentrations of progesterone are elevated throughout the luteal phase. However, although mean concentrations of progesterone are higher during the latter part of the pregnant luteal phase than during the nonpregnant luteal phase, these differences are not significant enough to allow progesterone determinations to be used as a pregnancy diagnosis test.

Bitches usually ovulate at the onset of sexual receptivity, remain sexually receptive for about the first week of luteal activity, and, in fact, are fertile during this period of time (Holst and Phemister, 1974). Progesterone analysis can be used to confirm the occurrence of ovulation. Concentrations usually increase slightly above baseline in the periovulatory period, reaching about 15 nmol/liter at ovulation, followed by a sustained increase beginning 24 hours following ovulation (Concannon *et al.*, 1975; Fig. 22.9). This approach may be used in a retrospective analysis of a breeding cycle in an attempt to correlate the time of ovulation with other criteria such as vaginal cytological changes. The determination of progesterone in daily samples of blood obtained during the periovulatory period gives a more precise timing of ovulation than can be obtained by vaginal cytology (Linde and Karlsson, 1984).

In situations wherein a particular pairing of animals does not result in a mutual sexual attraction and where artificial insemination must be used, progesterone analysis is a useful tool to verify ovulation. Artificial insemination with frozen–thawed semen usually results in lower pregnancy rates than artificial insemination using fresh semen. Using progesterone determinations to pinpoint ovulation, it was found that the pregnancy rate following insemination with frozen–thawed semen could be improved when the insemination was performed at progesterone concentrations less than 30 nmol/liter (Linde-Forsberg and Forsberg, 1989, 1993; Fig. 22.9).

Progesterone analysis can also be useful in cases of short estrous-cycle intervals to determine if ovulatory failure has occurred, a situation in which progesterone concentrations are low following the termination of estrus. Progesterone analysis is also useful in cases when there is some question as to the completeness of removal of ovarian tissue during an ovariohysterectomy. An important distinction concerning the dog is that, unlike other domestic species, the bitch is keyed into sexual receptivity at the end of the follicular phase by progesterone. Thus, dogs may be sexually attractive because of odors arising from the vagina that are due to factors such as infection. However, bitches will only accept males in the presence of increased progesterone concentrations, a situation which is almost always associated with luteal tissue as part of a remaining ovarian remnant.

2. *Testosterone*

The most common endocrine test in male dogs is testosterone for the purpose of checking the secretory status of the Leydig cells. Testosterone values in normal dogs range from about 1 ng/ml (3.5 nmol/liter) to about 10 ng/ml (35 nmol/liter) because of the pulsatile release pattern of testosterone. Testosterone concentrations in castrated dogs are less than 0.5 nmol/liter.

Assay of basal testosterone concentration may allow the diagnosis of the absence of testicular tissue in the same manner as described previously for the male horse. Sometimes the positive confirmation of presence of testicular tissue may require the use of hCG or GnRH to stimulate testosterone production. In this situation, a resting plasma sample is collected immediately before the administration of the stimulating hormone, and a second blood sample is collected 1 hour later. A significant increase in plasma testosterone concentration is diagnostic of the presence of testicular tissue.

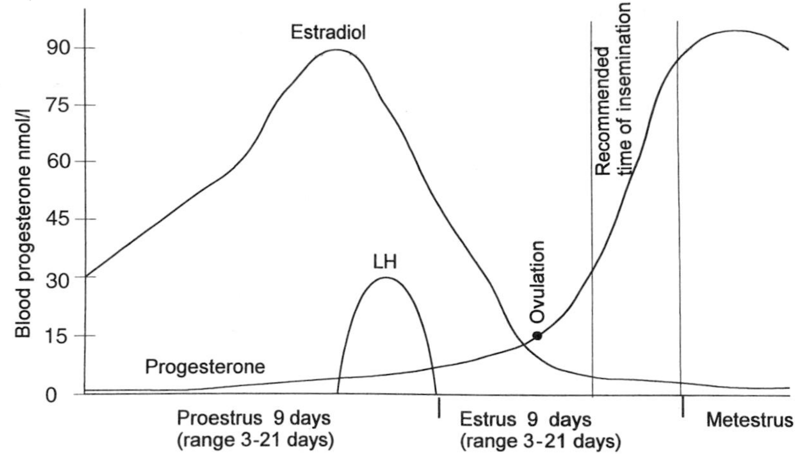

FIGURE 22.9 Schematic representation of the temporal relationships among estradiol, progesterone, peak of luteinzing hormone (LH), and ovulation in the bitch. The recommended time of mating or artificial insemination is based on the progesterone concentration.

Testosterone analysis is done in conjunction with fertility examinations, often in stud animals that are presented as infertility cases following a prolonged show tour. Most of these animals have testosterone concentrations that are compatible with normal spermatogenesis, although sperm counts are often very low. The management of this syndrome is still uncertain.

The authors have observed feminizing syndromes in intact dogs in which concentrations of testosterone are greatly decreased below normal (to 100 pg/ml or less; ≤350 pmol/liter), and estradiol values are 2–3 times normal (30–45 pg/ml vs the normal 15 pg/ml; 110–165 pmol/liter and 55 pmol/liter). It is thought that a majority of these cases involve Sertoli cell tumors. Dogs with confirmed Sertoli cell tumors have high peripheral plasma levels of inhibin and suppressed levels of LH and testosterone (Grootenhuis *et al.*, 1990). These authors were, however, unable to detect differences in blood concentrations of estradiol between dogs with Sertoli cell tumors and control dogs.

3. *Relaxin*

Relaxin is produced in the pregnant dog beginning between days 20 and 25 of gestation but is undetectable during anestrus, throughout nonpregnant ovarian cycles, and in male dogs (Steinetz *et al.*, 1987). Maximal concentrations are attained by days 40–50 of pregnancy and are followed by slight declines before parturition. Relaxin is produced predominantly by the placenta, and it is thus possible to use relaxin analysis to confirm pregnancy in the dog.

F. Cat

1. *Progesterone*

Progesterone analysis can be used to verify the occurrence of ovulation in the cat following coitus. Ovulation usually occurs 24–36 hours after coital contact with a male at the appropriate time of the follicular phase of the estrous cycle (Shille *et al.*, 1983). If the breeding schedule is very limited in time, it is possible for a cat to be bred too early in the follicular phase with coitus failing to elicit LH release. More rarely, females may allow copulation at times other than the follicular phase. In both these situations, ovulatory failure would be documented by the finding of low progesterone values 10 days postbreeding. In one study, progesterone concentrations greater than 1.87 ng/ml (6 nmol/liter) were consistently associated with luteal-phase ovaries, and values less than 0.15 ng/ml (0.5 nmol/liter) were associated with follicular-phase ovaries (Lawler *et al.*, 1991).

2. *Estrogen*

As the cat is an induced ovulator (requires coitus), estradiol analysis can be used to assess the presence of ovarian follicle activity with values ranging from 10 pg/ml (37 pmol/liter) in the interfollicular phase to 60 pg/ml (220 pmol/liter) during folliculogenesis. As indicated previously, queens have ovarian follicle growth patterns that last 5–7 days, followed by a slightly longer interval prior to the next growth phase (Shille *et al.*, 1979b). Estrogen analysis could document the presence or absence of ovarian follicular activity in animals that fail to manifest sexual activity. It also could be used to assess the completeness of an ovario-hysterectomy in cats that are spayed, but that present signs suggestive of sexual receptivity. The usual finding is that of low estrogen concentrations, which indicates that the behavior is not sexual in orientation. In the domestic cat, estrogen metabolites are primarily excreted in the feces (Shille *et al.*, 1990; Möstl *et al.*, 1993). Analyses of fecal steroids is a useful noninvasive approach for monitoring ovarian function in exotic felines (Graham *et al.*, 1993).

3. *Testosterone*

Testosterone analyses can be used to evaluate Leydig cell function in the testis of the male cat. The range of values is usually between 1 and 10 ng/ml (3.5–35 nmol/liter).

4. *Relaxin*

Relaxin is produced by the fetoplacental unit beginning about day 20 of gestation. Maximal concentrations are achieved by day 30–35 (Addiego *et al.*, 1987). Relaxin concentrations thus can be used to assess pregnancy status in the cat and even to assess its normalcy, based on the fact that fetoplacental units in jeopardy produce less relaxin.

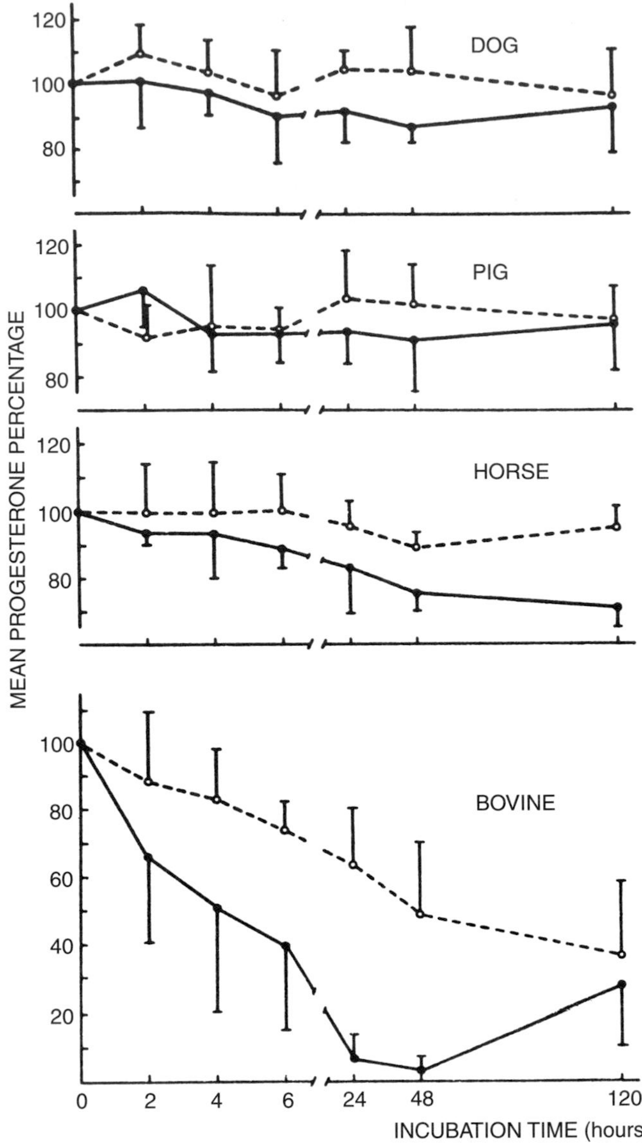

FIGURE 22.10 Mean (and SD) plasma progesterone percentage after storage of heparinized whole blood from 4 dogs, 4 pigs, 4, horses, and 8 cows at 20°C (●) and at 4°C (○). The intial progesterone value at time 0 has arbitrarily been set at 100% (Oltner and Edqvist, 1982).

V. GENERAL COMMENTS

A. Hormone Concentrations

In the foregoing presentation, concentrations of hormones have not been emphasized. This is because there is still some variability in the values reported by various laboratories. It is important that clinical endocrinology laboratories understand and have experience with their assay systems in relation to particular clinical syndromes. For example, the actual concentration of progesterone during the follicular phase of the estrous cycle of domestic animals is approximately 100 pg/ml (318 pmol/liter), certainly no greater than 200 pg/ml (636 pmol/liter) plasma. Some laboratories, however, report basal values of 1–2 ng/ml (3.2–6.4 nmol/liter) for progesterone. The authors have used 1 ng/ml (3.2 nmol/liter) as the lowest concentration of progesterone that is compatible with an actively secreting CL, especially for the cow, ewe, mare, and sow, and pay particular attention to the accuracy of values in this range. Actually, even values between 0.5 and 1.0 ng/ml (1.6–3.2 nmol/liter) are viewed as indicating luteal activity. It is possible to produce useful information with higher basal concentrations; however, the line that divides the presence and absence of CL activity must be precisely known. The reference values developed by laboratories are dependent upon the type of assay used.

It is possible to have a wide range of values for a particular hormone and still have normal physiological conditions. Luteal-phase progesterone concentrations in the cow, for example, range between 2 and 12 ng/ml (6–38 nmol/liter) because of pulsatile gonadotropin secretion (Walters *et al.*, 1984) with no adverse effect on the preparation of the animal for the nurture of a potential embryo. Hormone values can also depend upon the type of material used in the assay. Luteal-phase progesterone values are approximately 4 ng/ml (13 nmol/liter) in fat-free milk, between 5 and 35 ng/ml (16–111 nmol/liter) in whole milk, and approximately 250 ng/ml (795 nmol/liter) in milk fat.

B. Material for Analysis and Storage Effects

1. Blood

The treatment and storage of samples prior to analysis can influence the hormone value obtained for a blood sample. Red blood cells from cows have the capacity to metabolize progesterone to other steroids (Short, 1958; Fig. 22.10). The rapid decrease, about 10–20% reduction per hour, in the progesterone content in heparinized bovine blood samples appears to be due to the conversion of progesterone to 20β-hydroxylated gestagens. Also, other 20-keto-gestagens such as pregnenolone and 17α-hydroxyprogesterone undergo a similar conversion to 20β-hydroxylated gestagens, and their blood concentrations are reduced during storage (Choi *et al.*, 1989). The drop in progesterone concentration in heparinized whole blood follows glycolysis, and when glucose concentrations are low, the enzymatic degradation of progesterone in heparinized cow blood is to some degree inhibited (Oltner and Edqvist, 1982; Fig. 22.10). The use of anticoagulants, such as sodium fluoride, that block glycolysis greatly reduces the rate of decline in whole-blood progesterone concentrations (Vahdat *et al.*, 1979; Pulido *et al.*, 1991). Temperature also plays a significant role in regulating the rate of decline in progesterone; the lower the temperature of storage, the slower the rate of progesterone decline (Fig. 22.10).

The enzymes responsible for progesterone conversion are present in the red blood cells, and harvesting serum or plasma as soon as possible after collection of the sample prevents the metabolism of progesterone (Vahdat *et al.*, 1979). If serum or plasma cannot be separated within 30 minutes, the sample should be put on ice to retard enzymatic processes.

A decline in progesterone concentration also occurs in stored whole blood from the ewe (Van der Molen and Groen, 1968; Wiseman *et al.*, 1982/83), although the decline is not as rapid and pronounced as in the cow. On the other hand, no significant decrease in progesterone of whole blood of female goats was found during storage at room temperature for 24 hours (Navarro *et al.*, 1990). Some decline in progesterone also seems to occur during storage of equine whole blood (Fig. 22.10). For other domestic animals discussed in this chapter, no major effects of storage on progesterone concentrations have been reported.

Storage of bovine whole blood also results in significant decreases of 17-keto-androgens (e.g., androstenedione, dehydroepiandrosterone) and 17-keto-estrogens (e.g., estrone) because 17α-hydroxysteroid dehydrogenase present in red blood cells converts the steroids to epitestosterone and estradiol-17α, respectively (Choi *et al.*, 1989). Concentrations of testosterone and estradiol-17β are not influenced when whole bovine blood is stored for 24 hours at 20°C (Choi *et al.*, 1989).

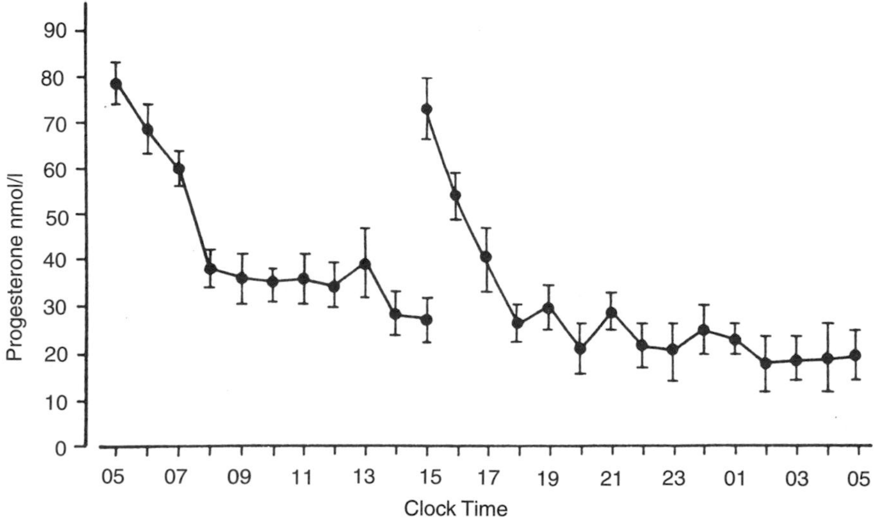

FIGURE 22.11 Hourly changes during 24 hours in whole-milk concentrations of progesterone, mean and SE of 12 cows (Garcia and Edqvist, 1990).

Plasma or serum samples can be stored frozen for long periods of time without significant loss in reproductive hormone concentration. Sample-handling procedures from blood collection to the freezing of serum or plasma should be standardized for each species and hormone determined.

2. *Milk*

Milk samples for progesterone analysis can be preserved with sodium azide, potassium dichromate, or thimerosal for storage at room temperature for about 3 weeks and for several months if the milk is stored at 4–5°C (Nachreiner *et al.*, 1986, 1992). There is a difference in progesterone content among colostrum, composite milk, and strippings because of variation in milk-fat content (Garcia and Edqvist, 1990; Fig. 22.11). This difference is less obvious when skim milk is used for the immunoassay of progesterone, but the temperature must be controlled at centrifugation since warm butterfat will absorb progesterone more readily than cold butterfat (Nachreiner *et al.*, 1986, 1992). A common recommendation for sampling of milk for analyses of progesterone is to obtain strippings after milking.

References

Abdo, G. A., Njuguna, O. M., Fredriksson, G., and Madej, A. (1991). *Acta Vet. Scand.* **32,** 183.

Abraham, G. E. (1969). *J. Clin. Endocrinol. Metab.* **29,** 866.

Abraham, G. E. (1974). *Acta Endocrinol. (Copenhagen), Suppl.*, **75,** 183.

Addiego, L. A., Tsutsui, T., Sewart, D. R., and Stabenfeldt, G. H. (1987). *Biol. Reprod.* **34,** 1165.

Allen, W. R. (1969). *J. Endocrinol.* **43,** 593

Allen, S. E., and Foote, R. H. (1988). *Theriogenology* **29,** 893.

Arighi, M., and Bosu, W. T. K. (1989). *J. Equine Vet. Sci.* **9,** 20.

Arnstadt, K. I., and Cleere, W. F. (1981). *J. Reprod. Fert.* **62,** 173.

Baldwin, D. M., and Stabenfeldt, G. H. (1975). *Biol. Reprod.* **12,** 508.

Barcikowski, B., Carlson, J. C., Wilson, L., and McCracken, J. A. (1974). *Endocrinology* **95,** 1340.

Beguin, F., Bygdeman, M., Green, K., Samuelsson, B., Topozada, M., and Wiqvist, N. (1972). *Acta Physiol. Scand.* **86,** 430.

Berndtson, W. E., Pickett, B. W., and Nett, T. M. (1974). *J. Reprod. Fert.* **39,** 115.

Berson, S. A., and Yalow, R. S. (1959). *Nature (London)* **184,** 1648.

Booth, J. M. (1988). *Vet. Rec.* **123,** 437.

Bravo, P. W., Stabenfeldt, G. H., Fowler, M. E., and Lasley, B. L. (1992). *Biol. Reprod.* **47,** 884.

Carrière, P. D., and Lee, B. (1994). *Can. J. Vet. Res.* **58,** 230.

Cerini, M., Findley, J. K., and Lawson, R. A. S. (1976). *J. Reprod. Fert.* **46,** 65.

Challis, J. R. G., Harrison, F. A., and Heap, R. B. (1971). *J. Reprod. Fert.* **25,** 306.

Chan, J. S. D., Robertson, H. A., and Friesen, H. G. (1978). *Endocrinology* **102,** 1606.

Chaplin, V. M., and Holdsworth, R. J. (1982). *Vet. Rec.* **111,** 224.

Chenault, J. R., Thatcher, W. W., Kalra, P. S., Abrams, R. M., and Wilcox C. J. (1975). *J. Dairy Sci.* **58,** 709.

Choi, H. S., Keisenhofer, E., Gantner, H., Hois, J., and Bamberg. E. (1987). *Anim. Reprod. Sci.* **15,** 209.

Choi, H. S., Möstl, E., and Bamberg, E. (1989). *Theriogenology* **31,** 571.

Concannon, P. W., Hansel, W., and Visek, W. J. (1975). *Biol. Reprod.* **13,** 112.

Concannon, P. W., Isaman, L., Frank, D. A., Michel, F. J., and Currie, W. B. (1988). *J. Reprod. Fert.* **84,** 71.

Cox, J. E. (1975). *Equine Vet. J.* **7,** 179.

Cox, J. E., Williams, J. H., Rowe, P. H., and Smith, J. A. (1973). *Equine Vet. J.* **5,** 85.

Cox, J. E., Redhead, P. H., and Dawson, I. E. (1986). *Equine Vet. J.* **18,** 179.

Cunningham, N. F., Hattersley, J. J .P., and Wrathall, A. E. (1983). *Vet. Rec.* **113,** 229.

Daels, P. F., Shideler, S., Lasley, B. L., Hughes, J. P., and Stabenfeldt, G. H. (1990). *J. Reprod. Fert.* **90,** 55.

Daels, P. F., Ammon, D. C., Stabenfeldt, G. H., Liu, I. K. M., Hughes, J. P., and Lasley, B. L. (1991). *Theriogenology* **35,** 1001.

Darenius, K., Kindahl, H., and Madej, A., (1988). *Theriogenology* **29,** 1215.

Dericks-Tan, J. S. E., and Taubert, H. D. (1975). *Acta Endocrinol. (Copenhagen)* **78,** 451.

Desbuquois, B., and Aurbach, G. D. (1971). *J. Clin. Endocrinol. Metab.* **33,** 732.

Dickie, M. B., and Holzmann, A. (1992). *J. Vet. Med. A* **39,** 525.

Dray, F., Andrieu, J.-H., and Renaud, F. (1975). *Biochim. Biophys. Acta* **403,** 131.

du Mesnil du Bussion, F., and Dauzier, L. (1957). *C. R. Soc. Biol.* **151,** 311.

du Mesnil du Buisson, F., and Dauzier, L. (1959). *Ann. Zootech., Suppl.* **8,** 147.

Edqvist, L.-E., and Stabenfeldt, G. H. (1989). *In* "Clinical Biochemistry of Domestic Animals", (J. J. Kaneko, ed.), Ed. 4, p. 661. Academic Press, New York.

Edqvist, L.-E., Einarsson, S., Ekman, L., Gustafsson, B., and Lamm, A. M. (1972). *Proc. Int. Congr. Anim. Reprod., 7th. Munich* **1,** 635.

Edqvist, L.-E., Ekman, L., Gustafsson, B., and Johansson, E. D. B. (1973). *Acta Endocrinol. (Copenhagen)* **72,** 81.

Edqvist, L.-E., Kindahl, H., and Stabenfeldt, G. H. (1976). *Proc. Int. Congr. Anim. Reprod., 8th. Cracow* **3,** 357.

Edqvist, L.-E., Kindahl, H., and Stabenfeldt, G. H. (1978). *Prostaglandins* 16, 111.

Edqvist, L.-E., Einarsson, S., and Larsson, K. (1980). *Proc. Int. Pig Vet. Soc., Copenhagen,* 27.

Ehrlich, P. H. Moyle, W. R., Moustafa, Z. A., and Canfield, R. E. (1982). *J. Immunol.* **128,** 2709.

Eisenman, J. R., and Chew, B. P. (1983). *J. Dairy Sci.* **66,** 1174.

Ellinwood, W. E.. Nett, T. M., and Niswender, G. D. (1979). *Biol. Reprod.* **21,** 845.

Engvall, E., and Perlmann, P. (1971). *Immunochem.* **8,** 871.

Estergreen, V. L., Frost, O. L., Gomes, W. R., Erb, R. E., and Bullard, J. F. (1967). *J. Dairy Sci.* **50,** 1293.

Etherington, W. G., Christie, K. A., Walton, J. S., Leslie, K. E., Wickstrom, S., and Johnson, W. H. (1991). *Theriogenology* **35,** 731.

Evans, M., and Irvine, C. (1975). *J. Reprod. Fert., Suppl.* **23,** 195.

Exley, D., Johnson, M. W., and Dean, P. D. G. (1971). *Steroids* **18,** 605.

Fairclough, R. J., Hunter, J. T., and Welch, R. A. S. (1975). *Prostagladins* **9,** 901.

Farmer, S. W., and Papkoff, H. (1979). *Biol. Reprod.* **21,** 425.

Fletcher, N. A., and Worsfold, A. I. (1988). *Br. Vet. J.* **144,** 269.

Flint, A. P. F., Stewart, H. J., Lamming, G. E., and Payne, J. H. (1992). *J. Reprod. Fert., Suppl.* **45,** 53.

Forsberg, M., Linde-Forsberg, C., Karlsson, Å., and Carlsson, M.-A. (1993a). *J. Reprod. Fert., Suppl.* **47,** 127.

Forsberg, M., Tagle, R., Madej, A., Molina, J. R., and Carlsson, M.-A. (1993b). *Acta Vet. Scand.* **34,** 1.

Frank, M., Bazer, F. W., Thatcher, W. W., and Wilcox, C. J. (1978). *Prostagladins* **15,** 151.

Fredriksson, G., Kindahl, H., and Edqvist, L.-E. (1984). *Anim. Reprod. Sci.* **7,** 537.

Gaiani, R., Mattioli, M., Galeati, G., and Chiesa, F. (1982). *Arch. Vet. Ital.* **33,** 86.

Garcia, M., and Edqvist, L.-E. (1990). *Theriogenology* **33,** 1091.

Geschwind, I. I., Dewey, R., Hughes, J. P., Evans, J. W., and Stabenfeldt, G. H. (1975). *J. Reprod Fert., Suppl.* **23,** 207.

Girard, A., Sandulesco, G., Fridenson, A., and Rutgers, J. T. (1932). *C. R. Acad. Sci.* **194,** 909.

Gleeson, A. R., Thorburn, G. D., and Cox, R. I. (1974). *Prostaglandins* **5,** 521.

Graham, L. H., Raeside, J. I., Goodrowe, K. L., and Liptrap, R. M. (1993). *J. Reprod. Fert., Suppl.* **47,** 119.

Granström, E., and Kindahl, H. (1976). *In* "Advances in Prostaglandin and Thromboxane Research" (B. Samuelsson and R. Paoletti, eds.), Vol. 1, p. 81. Raven, New York.

Granström, E., and Samuelsson, B. (1972). *FEBS Lett.* **26,** 211.

Grootenhuis, A. J., van Sluijs, F. J., Klaij, I. A., Steenbergen, J., Timmerman, M. A., Bevers, M. M., Deleman, S. J., and de Jong, F. H. (1990). *J. Endocr.* **127,** 235.

Hadley, J. C. (1975). *J. Reprod. Fert.* **44,** 453.

Hamberg, M., and Samuelsson, B. (1971). *J. Biol. Chem.* **246,** 6713.

Harrison, F. A., Heap, R. B., Horton, E. W., and Poyser, N. L. (1972). *J. Endocr.* **53,** 215.

Hatzidakis, G., Katrakili, K., and Krambovits, E. (1993). *J. Reprod. Fert.* **98,** 235.

Henderson, K. M., Camberis, M., Simmons, M. H., Starrs, W. J., and Hardie, A. H. M. (1994). *J. Steroid Biochem. Molec. Biol.* **50,** 189.

Herrier, A., Elsaesses, F., and Niemann, H. (1990). *Theriogenology* **33,** 415.

Hinrichs, K., Sertich, P. L., Solorzano, N. M., and Caldwell, L. A. (1988). *Theriogeneology* **29,** 1123.

Holst, P. A., and Phemister, R. D. (1974). *Am. J. Vet. Res.* **35,** 401.

Holtan, D. W., Squires, E. L., Lapin, D. R., and Ginther, O. J. (1979). *J. Reprod. Fertil., Suppl.* **27,** 457.

Horne, C., Chew, B. P., Wiseman, B. S., and Dziuk, P. J. (1983). *Biol. Reprod.* **29,** 56.

Hoyer, G. L., Colcow, M. L., and Weems, C. W. (1978). *Proc. Ann. Meet. Am. Soc. Anim. Sci. Abstr.* **47,** 367.

Hoyle, N. R., and Ebert, C. (1990). *Joint Meeting AACC/IFCC,* San Francisco.

Hughes, J. P., Stabenfeldt, G. H., and Evans, J. W. (1973). *Proc. Ann. Meet. Am. Assoc. Equine Practitioners.* 18th, 1972, San Francisco, p. 119.

Hultén, F., Dalin, A.-M., Forsberg, M., and Einarsson, S. (1994). *Proc. XVII Nordic Vet. Congr., Rekjavik* **1,** 152.

Humblot, P., Camous, S., Martal, J., Charlerly, J., Jeanguyot, N., Thibier, M., and Sasser, G. (1988). *Theriogeneology* **30,** 257.

Hunter, W. M., and Greenwood, F. C. (1962). *Nature (London)* **194,** 495.

Jeffcott, L. B., Hyland, J. H., MacLean, A. A., Dyke, T., and Robertson-Smith, G. (1987). *J. Reprod. Fert. Suppl.* **35,** 461.

Johansson, E. D. B. (1969). *Acta Endocrinol. (Copenhagen)* **61,** 592.

Karg, H., Claus, R., Hoffmann, B., Schallenberg, E., and Schams, D. (1976). *In* "Nuclear Techniques in Animal Production and Health," p. 487. IAEA, Vienna.

Kelly, P. A., Robertson, H. A., and Friesen, H. G. (1974). *Nature (London)* **248,** 435.

Kindahl, H., Edqvist, L.-E., Bane, A., and Granström, E. (1976a). *Acta Endocrinol. (Copenhagen)* **82,** 134.

Kindahl, H., Edqvist, L.-E., Granström, E., and Bane, A. (1976b). *Prostaglandins* **11,** 871.

Kindahl, H., Knudsen, O., Madej, A., and Edqvist, L.-E. (1982). *J. Reprod. Fert., Suppl.* **32,** 353.

King, G. J., Walton, J. S., and Bellissimo, D. J. (1985). *Canad. Vet. J.* **26,** 2.

Köhler, G., and Milstein, C. (1975). *Nature (London)* **256,** 495.

La Bonnardière, C. (1993). *J. Reprod. Fert., Suppl.* **48,** 157.

Laing, J. A., and Heap, R. B. (1971). *Br. Vet. J.* **127,** xix.

Landsteiner, K., and Van der Scherr, J. (1936). *J. Exp. Med.* **63,** 325.

Lawler, D. F., Evans, R. H., Reimers, T. J., Colby, E. D., and Monti, K. L. (1991). *Am. J. Vet. Res.* **52,** 1747.

Liggins, G. C., Grieves, S. A., Kendall, J. Z., and Knox, B. S. (1972). *J. Reprod. Fert., Suppl.* **16,** 85.

Lindberg, P., and Edqvist, L.-E. (1974). *Clin. Chim. Acta* **53,** 169.

Linde, C., and Karlsson, I. (1984). *J. Small Anim. Pract.* **25,** 77.

Linde-Forsberg, C., and Forsberg, M. (1989). *J. Reprod. Fert., Suppl.* **39,** 299.

Linde-Forsberg, C., and Forsberg, M. (1993). *J. Reprod. Fert., Suppl.* **47,** 313.

Lindner, H. R., Perel, E., and Friedlander, A. (1970). *In* "Research on Steroids IV" (M. Finkelstein, C. Conti, A. Klopper, and C. Cassano, eds.), p. 197. North-Holland, Amsterdam.

Linzell, J. L., and Heap, R. B. (1968). *J. Endocr.* **41,** 433.

Lopate, C., and Threlfall, W. R. (1991). *Theriogenology* **35,** 583.

McCaughey, W. J., Kerr, D. M., Greer, D., and Elliott, C. (1982). *Irish Vet. J.* **36,** 145.

McCracken, J. A., Carlson, J. C., Glew, M. E., Goding, J. R., Baird, D. T., Green, K., and Samuelsson, B. (1972). *Nature (London)* **238,** 129.

McCracken, J. A., Schramm, W., and Okulicz, W. C. (1984). *Anim. Reprod. Sci.* **7,** 31.

McPhee, I. M., and Tiberghien, M. P. (1987). *Vet. Rec.* **121,** 63.

Madej, A., and Linde-Forsberg, C. (1991). *J. Reprod. Fert.* **91,** 463.

Magness, R. R., Huie, J. M., and Weems, C. W. (1978). *J. Anim. Sci., Suppl.* **46,** 376.

Makawiti, D. W., Allen, W. E., and Kilpatrick, H. J. (1983). *J. Reprod. Fert.* **68,** 481.

Markwell, M. A. K. (1982). *Anal. Biochem.* **125,** 427.

Matsas, D. J., Nebel, R. L., and Pelzer, K. D. (1992). *Theriogenology* **37,** 859.

Matteri, R. L., Roser, J. F., Baldwin, D. M., Lipovetsky, V., and Papkoff, H. (1987). *Domest. Anim. Endocr.* **4,** 157.

Mayes, D., and Nugent, C. A. (1970). *Steroids* **15,** 389.

Meinecke, B., and Gips, H. (1987). *J. Vet. Med. A* **34,** 545.

Millar, R. P., and Aehnelt, C. (1977). *Endocrinology* **101,** 760.

Mitchell, D. (1971). *Can. Vet. J.* **12,** 41.

Moenter, S. M., Caraty, A., and Karsch, F. J. (1990). *Endocrinology* **127,** 1375.

Möstl, E., Lehmann, H., and Wenzel, U. (1993). *J. Reprod. Fert., Suppl.* **47,** 540.

Munro, C., and Stabenfeldt, G. (1984). *J. Endocrinol.* **10,** 41.

Murphy, B. E. P. (1964). *Nature (London)* **201,** 679.

Nachreiner, R. F., Oschmann, S. J., Edqvist, L.-E., and Richards, J. I. (1986). *In* "Nuclear and Related Techniques in Animal Production and Health," p. 653. IAEA, Vienna.

Nachreiner, R. F., Oschmann, S. J., Edqvist, L.-E., and Richards, J. I. (1992). *Am. J. Vet. Res.* **53,** 1085.

Nancarrow, C. W., Buckmaster, J., Chamley, W., Cox, R. I., Cumming, I. A., Cummins, L., Drinan, J. P., Findlay, J. K., Goding, J. R., Restall, B. J., Schneider, B. J., and Thorburn, G. D. (1973). *J. Reprod. Fert.* **32,** 320.

Navarro, H., Zarco, L., Ducoing, A., Flores, G., and Valencia, J. (1990). *Theriogenology* **33,** 749.

Nebel, R. L., Altemose, D. L., Munkittrick, T. W., Sprecher, D. J., and McGilliard, M. L. (1989). *Theriogenology* **31**, 753.

Neely, D. P., Kindahl, H., Stabenfeldt, G. H., Edqvist, L.-E., and Hughes, J. P. (1979). *J. Reprod. Fert., Suppl.* **27**, 181.

Neff, A. W., Ruppel, C. C., Gosline, R. E., Jaglan, P. S., and McGrath, J. P. (1981). *Acta Vet. Scand., Suppl.* **77**, 11.

Nett, T. M,. Akbar, A. M., and Niswender, G. D. (1974). *Endocrinology* **94**, 713.

Nett, T. M., Akbar, A. M., Phemister, R. D., Holst, P. A., Reichert, L. E., and Niswender, G. D. (1975). *Proc. Soc. Exp. Biol. Med.* **148**, 134.

Nett, T. M., Staigmiller, R. B., Akbar, A. M., Diekman, M. A., Ellinwood, W. E., and Niswender, G. D. (1976). *J. Anim. Sci.* **42**, 876.

Niswender, G. D. (1973). *Steroids* **22**, 413.

Niswender, G. D., and Midgley, A. R. (1970). *In* "Immunologic Methods in Steroid Determination" (F. G. Peron and B. V. Caldwell, eds.). p. 149. Appleton, New York.

Niswender, G. D., and Nett, T. M. (1977). *In* "Reproduction in Domestic Animals" (H. H. Cole and P. T. Cupps, eds.), p. 119. Academic Press, New York.

Niswender, G. D., Reichert, L. E., Midgley, A. R., and Nalbandov, A. V. (1969). *Endocrinology* **84**, 1166.

Noden, P. A., Oxender, W. D., and Hafs, H. D. (1978). *Am. J. Vet. Res.* **39**, 1965.

Oltner, R., and Edqvist, L.-E. (1981). *Br. Vet. J.* **137**, 78.

Oltner, R., and Edqvist, L.-E. (1982). *Acta Vet. Scand.* **23**, 1.

Palmer, E., and Jousset, B. (1975). *J. Reprod. Fert., Suppl.* **23**, 213.

Panciera, R. J., Slusher, S. A., and Hayes, K. E. N. (1991). *Cornell Vet.* **81**, 43.

Pashen, R. L., and Allen, W. R. (1979). *J. Reprod. Fert., Suppl.* **27**, 499.

Pepper, R. T., and Dobson, H. (1987). *Vet. Rec.* **120**, 53.

Perry, J. S., Heap, R. B., Burton, R. D., and Gadsby, J. E. (1976). *J. Reprod. Fert., Suppl.* **25**, 85.

Peterson, A. J., Tervit, H. R., Fairclough, R. J., Havik, P. G., and Smith, J. F. (1976). *Prostaglandins* **12**, 551.

Pope, G. S., Gupta, S. K., and Munro, I. B. (1969). *J. Reprod. Fert.* **20**, 369.

Power, M. J., Cleeve, W. F., Gosling, J. P., Fottrell, P. F., Langely, O. H., and Sreenan, J. M. (1985). *Irish Vet. J.* **39**, 18.

Pulido, A., Zarco, L., Galina, C. S., Murcia, C., Flores, G., and Posadas, E. (1991). *Theriogenology* **35**, 965.

Robertson, H. A., and King, G. J. (1974). *J. Reprod. Fert.* **40**, 133.

Robertson, H. A., and Sarda, I. R. (1971). *J. Endocrinol.* **49**, 407.

Romagnolo, D., and Nebel, R. L. (1993). *Theriogenology,* **39**, 1128.

Ruder, C. A., Stelflug, J. N., Dahmen, J. J., and Sasser, R. G. (1988). *Theriogenology* **29**, 905.

Saiz, C. F., Cruz, L. M., Martin, R. S., and Bourque, I. C. (1988). *Proc. Int. Congr. Anim. Reprod., 11th, Dublin* **2**, 67.

Samuelsson, B., Granström, E., Green, K., Hamberg, M., and Hammarström, S. (1975). *Ann. Rev. Biochem.* **44**, 669.

Sasser, R. G., and Ruder, C. A. (1987). *J. Reprod. Fert., Suppl.* **34**, 261.

Sasser, R. G., Ivani, K. A., and Ruder, C. A. (1983). *Proc. West. Sect. Soc. Anim. Sci.* **34**, 267.

Sauer, M. J., Foulkes, J. A., and Cookson, A. D. (1981). *Steroids* **38**, 45.

Schmidt, D. E., Brooks, T. L., Mhatre, S., Junghans, R. P., and Khazaeli, M. B. (1993). *BioTechniques* **14**, 1020.

Schwarzenberger, F., Möstl, E., Bamberg, E., Pammer, J., and Schmehlik, O. (1991). *J. Reprod. Fert., Suppl.* **44**, 489.

Sharma, S. C., and Fitzpatrick, R. J. (1974). *Prostaglandins* **6**, 97.

Shearer, I. J., Purvis, K., Jenkin, G., and Haynes, N. B. (1972). *J. Reprod. Fert.* **30**, 347.

Shemesh, M., Ayalon, N., and Lindner, H. R. (1968). *J. Reprod. Fert.* **15**, 161.

Shille, V., Karlbom, I., Einarsson, S., Larsson, K., Kindahl, H., and Edqvist, L.-E. (1979a). *Zentralbl. Veterinaermed. A* **26**, 169.

Shille, V. M., Lundström, K. E., and Stabenfeldt, G. H. (1979b). *Biol. Reprod.* **21**, 953.

Shille, V. M., Munro, C., Farmer, S. W., and Papkoff, H. (1983). *J. Reprod. Fert.* **69**, 29.

Shille, V. M., Haggerty, M. A., Shackleton, C., and Lasley, B. L. (1990). *Theriogenology* **34**, 779.

Short, R. V. (1958). *J. Endocr.* **16**, 415.

Silberzahn, P., Pouret, E. J. M., and Zwain, I. (1989). *Equine Vet. J.* **21**, 126.

Sist, M. D., Williams, J. F., and Geary, A. M. (1987). *J. Equine Vet. Sci.* **7**, 20.

Smith, M. S., and McDonald, L. E. (1974). *Endocrinology* **94**, 404.

Sokolowski, J. H. (1971). *Lab. Anim. Sci.* **21**, 696.

Sprecher, D. J., Strelow, L. W., and Nebel, R. L. (1990). *Theriogenology* **34**, 1149.

Squires, E. L., Stevens, W. B., Pickett, B. W., and Nett, T. M. (1979). *Am. J. Vet. Res.* **40**, 889.

Stabenfeldt, G. H. (1974). *J. Am. Vet. Med. Assoc.* **164**, 311.

Stabenfeldt, G. H., and Hughes, J. P. (1980). *Vet. Clin. North Am: Large Anim. Pract.* **2**, 253.

Stabenfeldt, G. H., Akins, E., Ewing, L. L., and Morrisette, M. C. (1969a). *J. Reprod. Fert.* **20**, 443.

Stabenfeldt, G. H., Ewing, L. L., and McDonald, L. E. (1969b). *J. Reprod. Fert.* **19**, 433.

Stabenfeldt, G. H., Holt, J. A., and Ewing, L. L. (1969c). *Endocrinology* **85**, 11.

Stabenfeldt, G. H., Osburn, B. I., and Ewing, L. L. (1970). *Am. J. Physiol.* **218**, 571.

Stabenfeldt, G. H., Drost, M., and Franti, C. E. (1972). *Endocrinology* **90**, 144.

Stabenfeldt, G. H., Hughes, J. P., Wheat, J. D., Evans, J. W., Kennedy, P. C., and Cupps, P. T. (1974a). *J. Reprod. Fert.* **37**, 343.

Stabenfeldt, G. H., Hughes J. P., Evans, J. W., and Neely, D. P. (1974b). *Equine Vet. J.* **6**, 158.

Stabenfeldt, G. H., Hughes, J. P., Evans, J. W., and Geschwind, I. I. (1975). *J. Reprod. Fert., Suppl.* **23**, 155.

Stabenfeldt, G. H., Hughes, J. P., Kennedy, P. C., Meagher, D. M., and Neely, D. P. (1979). *J. Reprod. Fert., Suppl.* **27**, 277.

Steinetz, B., Goldsmith, L., and Lust, G. (1987). *Biol. Reprod.* **37**, 719.

Sugiyama, S., Nakao, T., Tsunoda, N., and Kawata, K. (1985). *Br. Vet. J.* **141**, 60.

Szenci, O., Taverne, M. A. M., Palme, R., Bertoti, B., and Merics. I. (1993). *Vet. Rec.* **132**, 510.

Terqui, M., and Palmer, E. (1979). *J. Reprod. Fert., Suppl.* **27**, 441.

Thatcher, W. W., Wilcox, C. J., Bazer, F. W., Collier, R. J., Eley, R. M., Stover, D. G., and Bartol, F. F. (1979). *Anim. Reprod.* **3**, 259.

Thorell, J. I., and Johansson, B. G. (1971). *Biochim. Biophys. Acta* **251**, 363.

Thorneycroft, I. H., and Stone, S. C. (1972). *Contraception* **5**, 129.

Tsang, C. P. W. (1978). *Theriogenology* **10**, 97.

Vahdat, F., Hurtgen, J. P., Whitmore, H. L., Johnston, S. D., and Ketelson, C. L. (1979). *Theriogenology* **12**, 371.

Vaitakaitus, J., Robbins, J. B., Nieschlag, E., and Ross, G. T. (1971). *J. Clin. Endocrinol. Metab.* **33**, 988.

Van der Molen, H. J., and Groen, D. (1968). *Acta Endocrinol. (Copenhagen)* **58**, 419.

Van de Wiel, D. F. M., Van Eldik, J., Koops, W., Postma, A., and Oldenbroek, J. K. (1978). *Tijdschr. Diergeneeskd.* **103**, 91.

Van Orden, D. E., and Farley, D. B. (1973). *Prostaglandins* **4**, 215.

Van Weemen, B. K., and Schuurs, A. H. W. H. (1971). *FEBS Lett.* **15**, 232.

Velle, W. (1958). *Am. J. Vet. Res.* **19**, 405.

Verhage, H. G., Beamer, N. B., and Brenner, R. M. (1976). *Biol. Reprod.* **14,** 579.

Vermeulen, A., and Verdonck, L. (1968). *Steroids* **11,** 609.

Walters, D. L., Schams, D., and Schallenberger, E. (1984). *J. Reprod. Fert.* **71,** 479.

Weigl, R. M., Tilton, J. E., Haugse, C. N., Light, M. R., and Buchanan, M. L. (1975). *N.D. Farm Res.* **33,** 11.

Whates, D. C., and Denning-Kendall, P. A. (1992). *J. Reprod. Fert., Suppl.* **45,** 39.

Wide, L., and Porath, J. (1966). *Biochim. Biophys. Acta* **130,** 257.

Wiltbank, J. N., and Casida, L. E. (1956). *J. Anim. Sci.* **15,** 134.

Wiseman, B. S., Vincent, D. L., Thomford, P. J., Sceffrahn, N. S., Sargent, G. F., and Kesler, D. J. (1982/83). *Anim. Reprod. Sci.* **5,** 157.

Worsfold, A. I., Chamings, R. J., and Booth, J. H. (1986). *Br. Vet. J.* **142,** 195.

Zarco, L., Stabenfeldt, G. H., Basu, S., Bradford, G. E., and Kindahl, H. (1988). *J. Reprod. Fert.* **83,** 527.

Zavy, M. T. (1979). Ph.D. Thesis, University of Florida, Gainesville, Florida.

CHAPTER

23

Calcium-Regulating Hormones and Diseases of Abnormal Mineral (Calcium, Phosphorus, Magnesium) Metabolism

THOMAS J. ROSOL AND CHARLES C. CAPEN

I. CALCIUM METABOLISM

A. Introduction

Calcium (Ca) is a mineral that plays a central role in maintaining the homeostasis of vertebrate animals, including muscle contraction, blood coagulation, enzyme activity, neural excitability, hormone secretion, and cell adhesion, in addition to being an essential structural component of the skeleton (Capen and Rosol, 1993b). The precise control of calcium ion in extracellular fluids is vital, therefore, to the health of man and animals. To maintain a constant concentration of calcium, despite variations in intake and excretion, endocrine control mechanisms have evolved that primarily consist of the interactions of three major hormones. Although the direct roles of parathyroid hormone (PTH), calcitonin (CT), and vitamin D frequently are emphasized in the control of blood calcium, other hormones, such as adrenal corticosteroids, estrogens, thyroxine, somatotropin, glucagon, and parathyroid hormone-related protein, may contribute to the maintenance of calcium homeostasis under certain conditions (Capen, 1989; Capen and Rosol, 1993b). Calcium is also involved in the pathogenesis of metabolic diseases that disrupt the normal regulation of Ca balance and may result in hypercalcemia or hypocalcemia (Chew *et al.*, 1992).

B. Calcium in the Body

The majority of the Ca of the body (99%) is present in the inorganic matrix of bone as hydroxyapatite (Fig. 23.1). Most of the remaining Ca (0.9%) is sequestered in the plasma membrane and endoplasmic reticulum of cells. Extracellular fluid contains 0.1% of the body's Ca mass, with a total Ca concentration of about 2.5 mmol/liter (Brown, 1994). Approximately 50% of the extracellular Ca (1.2 mmol/liter) is in the ionized form (Ca^{2+}), which is the biologically active form of Ca. Neonatal animals have slightly greater extracellular Ca concentrations compared to those of adults (Hazewinkel, 1991; Chew *et al.*, 1992; Szenci *et al.*, 1994b). There is very little Ca in the cytosol of cells (approximately 100 n*M*), which is predominantly in the ionized form.

Vertebrate animals, such as the marine fishes, originally evolved in an environment with a high concentration of Ca^{2+} (Copp, 1969b). Seawater contains approximately 10 mmol/liter Ca^{2+}, but the extracellular fluids of marine fishes contain less than 2 mmol/liter Ca^{2+}. Therefore, fishes have to limit and regulate Ca^{2+} absorption from the intestinal tract, skin, or gills and to develop mechanisms to excrete Ca^{2+} efficiently. This resulted in the development of hormones that reduce serum Ca^{2+} concentration and include calcitonin and stanniocalcin. Fishes lack parathyroid glands, which have the primary role of reducing the loss of body Ca^{2+} and maintaining serum Ca^{2+}. Parathyroid glands first appear phylogenetically in amphibians that spend all or part of their life cycle on land, an environment low in Ca^{2+}. In terrestrial vertebrates, both the parathyroid glands and the kidneys are important in preserving total body Ca. Because there is less need for promoting the excretion of Ca or lowering serum Ca^{2+} in land animals, hormones such as calcitonin are less critical in maintaining Ca homeostasis.

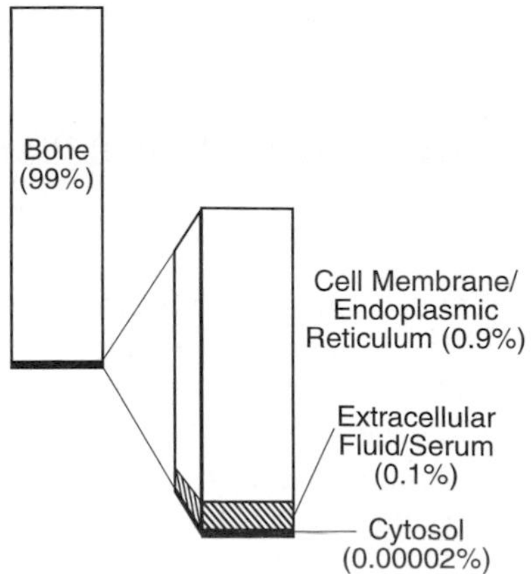

FIGURE 23.1 Distribution of calcium (Ca) in the body. (Rosol *et al.*, 1995a.)

C. Function of Calcium

Calcium serves two primary functions in the body: (1) to maintain the structural integrity of bones and teeth, and (2) as a messenger or regulatory ion (Ebashi, 1985). The 10,000-fold concentration gradient of Ca^{2+} between the extracellular fluid (1.2 mmol/liter) and the cytoplasm (100 n*M*) permits Ca^{2+} to function as a signaling ion to activate intracellular processes. The lipid bilayer of the cell membrane has a low permeability to Ca^{2+}; therefore, influx of Ca^{2+} into the cytoplasm is controlled by a heterogeneous group of Ca channels regulated by membrane potential, cell membrane receptors, or intracellular secondary messengers (Miller, 1992; Brown, 1994). Influx of Ca^{2+} into cells can: (1) regulate cellular function by interactions with Ca-binding proteins (e.g., calmodulin) and Ca-sensitive protein kinases, and (2) stimulate biologic responses such as neurotransmitter release, contraction, and secretion. Ionized calcium also plays an important role in cell adhesion and blood coagulation. In addition, Ca^{2+} may regulate cellular function by binding to a G-protein-linked Ca^{2+}-sensing receptor in the cell membrane, such as in parathyroid chief cells or renal epithelial cells (Brown *et al.*, 1995).

It is critical for cells to maintain the normal low level of intracellular Ca^{2+}. If cellular Ca homeostasis fails as a result of anoxia, an energy-deprived state, or perturbed membrane integrity, cell viability is threatened due to uncontrolled entry of Ca^{2+} through the plasma membrane or from intracellular stores (Siesjö, 1989).

D. Forms of Calcium

Extracellular and serum Ca exists in three forms: (1) ionized Ca^{2+}; (2) complexed Ca^{2+} (5% of total Ca) to anions such as citrate, bicarbonate, phosphate, or lactate; and (3) protein-bound Ca^{2+} (45% of total Ca) (Fig. 23.2) (Chew *et al.*, 1992; Capen and Rosol, 1993b). The protein-bound fraction of Ca^{2+} is principally bound to negatively charged sites on albumin, with smaller amounts bound to globulins. The protein-bound form of Ca^{2+} is dependent on serum pH, and alterations in serum pH will change the [Ca^{2+}]. As the pH of serum becomes more acidic, the [Ca^{2+}] will increase because of competition of H^+ for binding to the negatively charged sites on serum proteins. The ionized and complexed Ca^{2+} compose the ultrafilterable fraction of Ca^{2+} and represent the fraction that is present in the glomerular filtrate. The concentration of ionized Ca^{2+} in the serum is approximately 1.25–1.6 mmol/liter (5.0–6.4 mg/dl) in most domestic animals.

E. Measurement of Calcium

Total serum Ca concentration in serum, heparinized plasma, or urine can be measured directly using atomic absorption spectrophotometry or a colorimetric assay using *o*-cresolphthalein complexone (Table 23.1) (Fraser *et al.*, 1987). Total serum Ca concentration in dogs is partially dependent on serum albumin or total protein content, but can be standardized using the following formulas: (1) Adjusted Ca (mg/dl) = Ca (mg/dl) − albumin (g/dl) + 3.5, or (2) Adjusted Ca (mg/dl) = Ca (mg/dl) − 0.4[total serum protein (g/dl)] + 3.3 (Meuten *et al.*, 1982). This correction procedure has not been validated with ionized Ca to determine whether it has true biological significance. In cats, horses, and cattle, there is a significant linear relationship between serum Ca and albumin concentrations (Bienzle *et al.*, 1993). Variability in the serum concentrations of Ca and albumin precluded determination of a dependable correction formula in cats (Flanders *et al.*, 1989). Serum or urine, but not plasma, can be stored at 4°C for weeks or months. Urine should be acidified with 6 *M* HCl to a pH of 3–4 during or immediately after collection to prevent Ca precipitation with anions or proteins. Urine Ca concentration may be low (<4 mg/dl, 1.6 mmol/liter) in normal animals of most species and below the sensitivity of colorimetric assays for Ca. Therefore, atomic absorption spectrophotometry often is necessary for measurement of Ca in urine. Calcium content of feces, tissue, or food that has been ashed also can be measured by atomic absorption spectrophotometry.

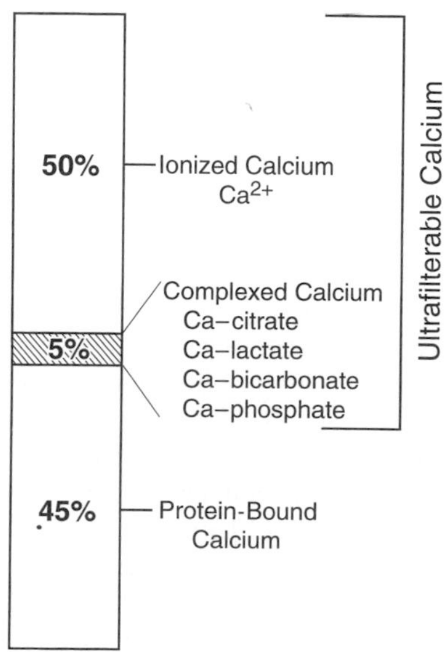

FIGURE 23.2 Fractions of extracellular calcium (Ca). (Rosol *et al.*, 1995a.)

Ionized blood or plasma [Ca^{2+}] can be measured using an instrument equipped with a Ca^{2+} ion-selective electrode (Table 23.1) (Fraser *et al.*, 1987). It is most informative to measure the biologically active form of Ca^{2+} (ionized Ca^{2+}), but this is not always practical because it requires special instrumentation and proper sample handling. In most cases, changes in serum total Ca parallel changes in serum Ca^{2+} (Lincoln and Lane, 1990; Chew *et al.*, 1992). Ionized [Ca^{2+}] can be measured in heparinized whole blood, plasma, or serum. Heparin can bind Ca^{2+} and variably lower the serum Ca^{2+} concentration artifactually; therefore, the amount of heparin should be standardized for all samples (Forman and Lorenzo, 1991). Serum or plasma can be stored 48 hour or longer at 4°C (Szenci *et al.*, 1991, 1994a; Schenck *et al.*, 1995). Anaerobic collection is critical to prevent changes in the pH of the sample. It is important to measure [Ca^{2+}] at the same pH as the blood at time of collection for the most biologically relevant value, because correction formulas to adjust Ca^{2+} to pH 7.4 have not been validated in animals (Kohn and Brooks, 1990).

F. Daily Calcium Balance and Movement

Calcium is absorbed from food in the gastrointestinal tract, with the greatest absorption occurring in the duodenum (McDowell, 1992a). In normal animals an equivalent amount of Ca is excreted primarily in the urine, with small losses in sweat and intestinal secretions. The Ca released from bone by osteoclastic bone resorption and deposited in bone by bone formation are balanced in normal adult animals. Therefore, intestinal Ca absorption is the major determinant of the amount of Ca excreted in the urine in adult animals.

Calcium can move across epithelial barriers by two routes: (1) transport between cells, and (2) transport through cells (Yanagawa and Lee, 1992; Brown, 1994). Transport of Ca between cells occurs by two mecha-

TABLE 23.1 Measurement of Calcium in Blood and Urine

	Test	Sample requirements/ handling	Methods	Comments	Reference ranges (OSU values)[a]
Blood calcium	Total calcium	Blood, serum, plasma	Atomic absorption, colorimetry		9.4–12.0 mg/dl (dog)[b] 8.4–11.0 mg/dl (cat) 11.7–13.6 mg/dl (horse) 8.7–11.4 mg/dl (cow)
	Ionized calcium	Anaerobic collection of blood, serum, or plasma	Calcium ion-specific electrode	Obtain blood pH.	5.2–6.0 mg/dl (dog) 4.3–5.9 mg/dl (cat) 6.0–7.2 mg/dl (horse)
Urine calcium handling	Total calcium	Acidify urine to prevent calcium precipitation	Atomic absorption, colorimetry	Colorimetry may be too insensitive in some species. Total Ca value is of no use.	NA[c]
	Ca/Cr ratio			Good measure of Ca excretion at one time point. Collection from fasting animals is an indirect measure of bone resorption.	$0.5–2.3 \times 10^{-2}$ (fasted dogs)
	Fractional calcium excretion	Collect urine from fasted animal		Measure of renal calcium reabsorption; decreased by parathyroid hormone.	$0.05–0.14 \times 10^{-2}$ (fasted dogs)
	24-hour calcium excretion			Good measure of daily calcium loss in urine.	0.17–0.70 mg/kg/d (fasted dogs)

[a] Reference values at the Ohio State University, Veterinary Medical Teaching Hospital and Laboratories for normal adult animals.
[b] To convert mg/dl to mmol/liter, divide by 4.
[c] NA = not appropriate.

nisms, convection and diffusion. Convection (solvent drag) is the movement of Ca with the flow of water. A change in Ca concentration on either side of the epithelium is not necessary for Ca movement between the cells. Convection of Ca occurs in the process of glomerular filtration and with water reabsorption in the proximal convoluted tubules. Diffusion is the passive movement of Ca down an electrical or chemical gradient. For example, Ca diffuses from the lumen of the proximal convoluted tubules to the renal interstitium because there is a higher [Ca^{2+}] in the lumen.

In order for Ca to move through cells, mechanisms have evolved for the regulation of Ca transport into the cytosol, buffering in the cytoplasm, and release from the cell (Yanagawa and Lee, 1992; Brown, 1994). Large concentration and electrical gradients are present for Ca^{2+} to enter cells. There is a 10,000-fold concentration gradient from extracellular to intracellular fluids and the interiors of cells are negatively charged (−50 to −100 mV). The regulation of Ca absorption by cells is accomplished by Ca channels in the cell membrane (Miller, 1992; Brown, 1994). After the Ca^{2+} channels are opened, there is rapid diffusion of Ca^{2+} into the cell. The rise in intracellular Ca^{2+} can stimulate biological processes by interaction with Ca^{2+}-binding enzymes and regulatory proteins (Rasmussen, 1986a, 1986b). It is also possible for an increase in the transcellular flux of Ca^{2+} to stimulate biological processes without a net increase in intracellular [Ca^{2+}]. In some cells or tissues, the intracellular Ca^{2+} signal oscillates (e.g., hepatocytes responding to norepinephrine) (Berridge, 1990) or can be transmitted intercellularly by gap junctions (Sanderson *et al.*, 1994). The Ca^{2+} that enters the cytosol is rapidly buffered by Ca-binding proteins (e.g., calbindin and calmodulin) and by transport into mitochondria and endoplasmic reticulum. It is important to buffer intracellular Ca^{2+}, because high concentrations of intracellular Ca^{2+} can be toxic to cells.

Calcium can be transported from the cytoplasm to intracellular vesicles (e.g., endoplasmic reticulum) or the extracellular space by membrane-associated Ca^{2+}-ATPases (Grover and Khan, 1992; Raeymaekers and Wuytack, 1996) and a cell membrane Na^+/Ca^{2+} exchanger (Dominguez and Juhaszova, 1992). The Na^+/Ca^{2+} exchanger permits the transport of one Ca^{2+} ion outside of the cell with the entry of three Na^+ ions. The Na^+ enters the cell down a chemical and electrical gradient. The Ca^{2+}-ATPases are dependent on intracellular [ATP] and the Na^+/Ca^{2+} exchanger is indirectly dependent on [ATP], because ATP is necessary for the cell membrane Na^+/K^+ ATPase to maintain a steep gradient of Na^+ across the cell membrane.

A final mechanism by which Ca^{2+} is transported through cells is vesicular transport (Yanagawa and Lee, 1992). In this form of Ca^{2+} movement, Ca^{2+} is absorbed as a constituent of a vesicle (e.g., pinocytosis) and is transported across the cell in the vesicle. There is no need for Ca^{2+} buffering, because it does not directly enter the cytoplasm. This form of Ca^{2+} transport currently is not well understood; however, the mammary gland secretes most of its Ca by secretory vesicles. Intracellular Ca is pumped by a Ca^{2+} ATPase into the Golgi apparatus of mammary epithelial cells, the Ca^{2+} binds to citrate and casein, and the complexed Ca^{2+} is secreted in the form of secretory vesicules (Atkinson *et al.*, 1995).

G. Renal Handling of Calcium

The kidney normally reabsorbs 98% or more of the filtered Ca. This high degree of reabsorption is important to maintain balance of Ca in the body, but permits the kidneys, if necessary, to excrete large amounts of Ca in the urine (Bronner, 1989; Rouse and Suki, 1990; Bindels, 1993). Most filterable Ca (ionized and complexed Ca) enters the glomerular filtrate by convection and is reabsorbed by the renal tubules. The kidneys reabsorb approximately 40-fold more Ca than is absorbed by the intestinal tract because of the high degree of blood flow and ultrafiltration in the glomerulus. Reduction of glomerular filtration impairs the ability of the kidneys to excrete Ca.

Most filtered Ca (70%) is reabsorbed in an unregulated manner in the proximal convoluted tubules by diffusion and convection with water uptake between the epithelial cells (Fig. 23.3). The thick ascending loop of Henle absorbs about 20% of the filtered Ca, but the precise mechanism is unclear. Much of the Ca reabsorption appears to be passive, but an active component may also be present (Rouse and Suki, 1990). The site of active regulation of Ca reabsorption is the distal convoluted tubule that reabsorbs approximately 10% of the filtered Ca. The principal stimulator of Ca reabsorption in the distal convoluted tubule is parathyroid hormone (Bindels, 1993). Ca reabsorption in the distal convoluted tubule is an active transcellular process that requires the presence of Ca channels in the luminal cell membrane (Miller, 1992), intracellular Ca-binding proteins (calbindins) (Bouhtiauy *et al.*, 1994a, 1994b), and a Ca^{2+}-ATPase (Grover and Khan, 1992) and Na^+/Ca^{2+} exchanger (Dominguez and Juhaszova, 1992) in the basolateral cell membranes (Bronner, 1989; Rouse and Suki, 1990). Factors associated with increased or decreased Ca excretion are presented in Table 23.2 (Canzanello *et al.*, 1990; Chew *et al.*, 1992; Bindels, 1993). Renal epithelial cells express the Ca^{2+}-sensing receptor on their cell membranes, and the distribution of the receptor overlaps with the localization of PTH receptors, so the kidneys may partially autoregulate the renal reabsorption of calcium based on the blood Ca^{2+} concentration (Brown and Hebert, 1996).

Measurements of Ca excretion in animals include urine [Ca], Ca/Cr ratio in the urine, fractional Ca excretion, and 24-hour Ca excretion (Table 23.1) (Nordin *et al.*, 1967; DiBartola *et al.*, 1980; Morris *et al.*, 1984; Fleming *et al.*, 1991; Lulich *et al.*, 1991; Chew *et al.*, 1992;

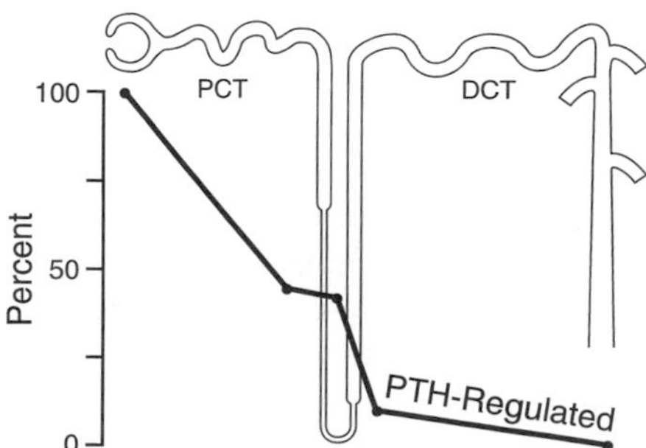

FIGURE 23.3 Sites of calcium (Ca) reabsorption in the nephron. The greatest amount of filtered Ca is reabsorbed passively in the proximal convoluted tubule (PCT). The site of active regulation of Ca reabsorption by parathyroid hormone (PTH) is in the distal convoluted tubule (DCT) (Rosol *et al.*, 1995a.)

TABLE 23.2 Factors Affecting Urinary Calcium Excretion

Factors that increase urinary Ca excretion
Saline infusion or high dietary Na^+ content
High dietary protein
Hypercalcemia (due to increased filtered Ca)
Vitamin D deficiency or excess
Acidosis
Mineralocorticoid/glucocorticoid excess
Osmotic diuretics
Loop diuretics (e.g., furosemide)
Factors that decrease urinary Ca excretion
Parathyroid hormone
Calcitonin
Dehydration
Alkalosis
Hypercalcemia (due to reduction of GFR)
Thiazide diuretics

Fleming *et al.*, 1992). Urinary Ca excretion correlates with dietary absorbance in normal adult animals. The concentration of Ca in the urine is of little value because of large fluctuations in normal urine and is usually <4 mg/dl (1.6 mmol/liter). Horses excrete larger amounts of Ca in the urine, and urine [Ca] is typically much higher than 4 mg/dl (1.6 mmol/liter). The Ca/Cr ratio of the urine is a better indicator of Ca excretion because it corrects for errors in timing of urine collections, urine concentration or dilution, and differences in lean body mass (Weaver, 1990). Calculation of fractional Ca excretion requires measurement of urine and serum Ca and Cr concentrations and is an indication of renal calcium reabsorption at the time of analysis as well as the degree of excretory renal function (DiBartola *et al.*, 1980). Fractional Ca excretion is best measured in a fasting animal to eliminate the effect of dietary Ca on renal Ca excretion. Measurement of Ca excretion in a fasting animal is an indirect measurement of bone resorption, because Ca released from bone and the obligate renal Ca loss are the major sources of urinary Ca when there is little gastrointestinal Ca absorption. The 24-hour Ca excretion is a good measurement of daily Ca loss and may be used to investigate daily Ca balance.

H. Intestinal Calcium Absorption

There are two components of Ca absorption from the intestinal tract, namely, saturable or transcellular transport and nonsaturable or intercellular (paracellular) transport (Favus, 1992). Percent intestinal absorption is proportional to dietary intake (Hazewinkel, 1991). Low-Ca diets are associated with high absorption rates (up to 95%) and high-Ca diets have low absorption rates (about 40%). Although high-Ca diets have low absorption rates, they can increase serum Ca concentrations because of the presence of nonsaturable intestinal absorption. In contrast, diets deficient in Ca are associated with normal serum Ca concentrations because of compensation by parathyroid hormone-stimulated bone resorption, renal Ca reabsorption, and increased 1,25-dihydroxyvitamin D synthesis.

Saturable transport is a carrier-mediated, vitamin D–dependent process and occurs predominantly in the duodenal segment of the small intestine, but also occurs in the cecum and colon (Favus, 1992; McDowell, 1992a; Karbach and Feldmeier, 1993). Saturable transport requires influx of Ca into intestinal epithelial cells via Ca^{2+} channels, movement and buffering in the cytoplasm, and basolateral exit by a Ca^{2+}-ATPase. The active form of vitamin D (1,25-dihydroxyvitamin D) stimulates transcellular transport of Ca^{2+}. One function of 1,25-dihydroxyvitamin D in the intestinal epithelial cell is to increase the expression of calbindin, an intracellular Ca-binding protein. In contrast, nonsaturable Ca transport occurs throughout the small intestine and is the main mechanism of Ca absorption in animals with vitamin D deficiency. Nonsaturable Ca^{2+} transport is dependent on the luminal [Ca^{2+}]. As dietary intake of Ca increases, much of the Ca in the intestinal lumen is unavailable for nonsaturable absorption because of precipitation of Ca salts or complexes formed with anions (McDowell, 1992a).

Fractional intestinal Ca absorption is approximately 20–40% in adult animals and can exceed 60% during increased demand for Ca (Hazewinkel, 1991). Fractional absorption is increased during pregnancy, lactation, and growth, and when animals are fed low-Ca diets (Favus, 1992). The primary adaptive influence on Ca absorption is the circulating 1,25-dihydroxyvitamin D concentration. Factors that increase intestinal Ca absorption directly or indirectly by stimulation of 1,25-dihydroxyvitamin D synthesis include parathyroid hormone, growth hormone, testosterone, estrogen, and furosemide (Favus, 1992). Factors that reduce intestinal absorption of Ca include glucocorticoids, thyroid hormones, chronic acidosis, and lumenal conditions that induce complexation of Ca^{2+} (high concentrations of phosphate, phytates, oxalate, fatty acids, pH > 6.1, and other anions) (Favus, 1992).

I. Bone and Calcium Balance

There are two sources of Ca^{2+} in bone that can enter the circulation: (1) readily mobilizeable Ca salts in the extracellular fluid, and (2) hydroxyapatite crystals that require digestion by osteoclasts before Ca^{2+} can be released from bone. The nature and regulation of the readily mobilizable Ca in bone is poorly understood; however, it is present in small amounts and is likely to play a role in the fine regulation of serum Ca concentration (Parfitt, 1987). If there is a significant need for Ca from bone it must come from osteoclastic resorption of hydroxyapatite crystals. In adult animals there is a stable balance between Ca deposition associated with bone formation and Ca release associated with osteoclastic bone resorption. In young animals, bone has a positive Ca balance because of the relative excess of bone formation. Conditions that result in excessive bone resorption (e.g., humoral hypercalcemia of malignancy, osteolytic bone metastases, or primary hyperparathyroidism) can release large amounts of Ca from bone and contribute to the development of hypercalcemia (Rosol and Capen, 1992).

J. Cell Membrane Calcium Ion–Sensing Receptor

The concentration of ionized calcium in serum and extracellular fluid can regulate cellular function by interacting with a recently identified Ca^{2+}-sensing receptor in the plasma membrane of various cells (Brown *et al.*, 1993; Chattopadhyay *et al.*, 1996). The cell membrane Ca^{2+} receptor belongs to the family of 7-transmembrane domain G-protein-linked receptors and is unique because its ligand is an ion. The Ca^{2+} receptor plays an important role in the regulation of extracellular Ca^{2+} homeostasis and is present on parathyroid chief cells, thyroid C-cells, renal epithelial cells, brain, and placenta, among other tissues. The Ca^{2+} receptor is responsible for sensing serum Ca^{2+} concentration and modifying parathyroid hormone secretion, calcitonin secretion, and calcium transport by renal epithelial cells (Pollak *et al.*, 1993; Brown and Hebert, 1996; Chattopadhyay *et al.*, 1996). Mutations in one or both of the Ca^{2+}-sensing receptor genes in humans result in familial hypocalciuric hypercalcemia or neonatal severe hypercalcemia, respectively, because of an inadequate ability to sense the extracellular Ca^{2+} concentration and coordinate the appropriate cellular response (Pollak *et al.*, 1993). Familial hypocalciuric hypercalcemia is a benign condition characterized by mild hypercalcemia, decreased urinary excretion of Ca, and inappropriately normal parathyroid hormone (PTH) levels. In contrast, neonatal severe hypercalcemia is a lethal condition unless total parathyroidectomy is performed early in life to reduce markedly elevated PTH levels.

II. CALCIUM-REGULATING HORMONES

A. Parathyroid Hormone

1. *Development and Anatomy of the Parathyroid Gland*

Parathyroid glands are present in all air-breathing vertebrates. Phylogenetically, parathyroids first appear in amphibians coincidentally with the transition from an aquatic to a terrestrial life. Parathyroid glands in tetrapods have arisen from the need to protect against the development of hypocalcemia and the necessity to conserve calcium to maintain skeletal integrity in terrestrial animals, which often are in a relatively low calcium–high phosphorus environment (Copp *et al.*, 1962; Coleman, 1969).

Embryologically, parathyroids are of entodermal origin, being derived from the III and IV pharyngeal pouches in close association with primordia of the thymus. The entodermal bud that forms the thyroid gland arises on the midline at the level of the first pharyngeal pouch. Calcitonin-secreting C cells of neural crest origin reach the postnatal thyroid gland by migrating into the ultimobranchial body. This lateral pharyngeal pouch moves caudally in mammals to fuse with the primordia of the thyroid gland and distributes C cells into each thyroid lobe.

In the dog and cat both the external and internal parathyroids are close to the thyroid gland. The external parathyroid (III) in the dog is from 2 to 5 mm in length and is found in the loose connective tissue cranial and slightly lateral to the anterior pole of the thyroid. The internal parathyroid (IV) is smaller, flatter, and situated on the medial surface of the thyroid beneath the fibrous capsule. The blood supply of the two glands in the dog is separate, with the external parathyroid being supplied by a branch from the cranial thyroid artery and the internal parathyroid by minute ramifications of the arterial supply to the thyroid (Smithcors, 1964).

In other species such as cattle and sheep the larger external parathyroid gland is located a considerable distance cranial to the thyroid in the loose connective tissue along the common carotid artery. The smaller internal parathyroid is situated on the dorsal and medial surface of the thyroid gland. The larger lower parathyroid gland in horses is located a considerable distance from the thyroid in the caudal cervical region near the bifurcation of the bicarotid trunk at the level of the first rib. Pigs have only a single pair of parathyroids found cranial to the thyroid, embedded either in thymus (young animals) or in adipose connective tissue (adult pigs). Rats also have a single pair of parathyroid glands that are located close to the thyroid.

2. *Functional Cytology*

The parathyroid glands contain a single basic type of secretory cell whose principal function is secretion of parathyroid hormone (Capen and Roth, 1973). The parathyroids of man and animals are composed of chief cells in different stages of secretory activity (Capen, 1975). Experimental and pathological evidence has accumulated to suggest that certain fine structural characteristics of chief cells are associated with different stages of synthetic and secretory activity (Munger and Roth, 1963; Capen, 1971).

Chief cells in an inactive or resting stage predominate in the parathyroid glands of man and most animal species. Inactive chief cells are cuboidal, have uncomplicated interdigitations between contiguous cells, and contain poorly developed organelles and infrequent secretory granules. The cytoplasm has either numerous lipid bodies and lipofuscin droplets (e.g., bovine glands) or aggregations of glycogen granules (e.g., human and feline glands). The Golgi apparatus is small, composed of straight or curved stacks of agranular membranes, and associated with few prosecretory granules and vesicles in the process of formation. Individual profiles of granular endoplasmic reticulum, ribosomes, and small mitochondria are dispersed throughout the cytoplasm. Chief cells in the active stage of secretion occur less frequently in the parathyroid glands of most species. The cytoplasm of active chief cells has an increased density of organelles, secretory granules, and loss of glycogen particles and lipid bodies (Roth and Capen, 1974).

3. Biosynthesis

A larger biosynthetic precursor (preproparathyroid hormone, preproPTH) is the initial translation product synthesized on ribosomes of the rough endoplasmic reticulum in chief cells (Arnaud *et al.*, 1971; Cohn and MacGregor, 1981). It is composed of 115 amino acids and contains a hydrophobic signal or leader sequence of 25 amino acid residues that facilitates the penetration and subsequent vectorial discharge of the nascent peptide into the cisternal space of the rough endoplasmic reticulum (Fig. 23.4) (Habener and Potts, 1978). PreproPTH is rapidly converted within 1 minute or less of its synthesis to proparathyroid hormone (proPTH) by the proteolytic cleavage of the N-terminal sequence of 25 amino acids (Fig. 23.4) (Habener, 1981).

The intermediate precursor, proPTH, is composed of 90 amino acids and moves within membranous channels of the endoplasmic reticulum to the Golgi apparatus (Fig. 23.4). Enzymes with trypsin-like and carboxypeptidase-B-like activity within membranes of the Golgi apparatus cleave a hexapeptide from the N-terminal (biologically active) end of the molecule forming active parathyroid hormone (PTH) (Habener *et al.*, 1977; MacGregor *et al.*, 1978). Active PTH is packaged into membrane-limited, macromolecular aggregates in the Golgi apparatus for subsequent storage in chief cells (Fig. 23.5). Under conditions of increased demand, PTH may be released directly from chief cells without being packaged into secretion granules. Biologically active PTH secreted by chief cells is a straight chain polypeptide consisting of 84 amino acid residues with a molecular weight of approximately 9500 (Rosol *et al.*, 1996).

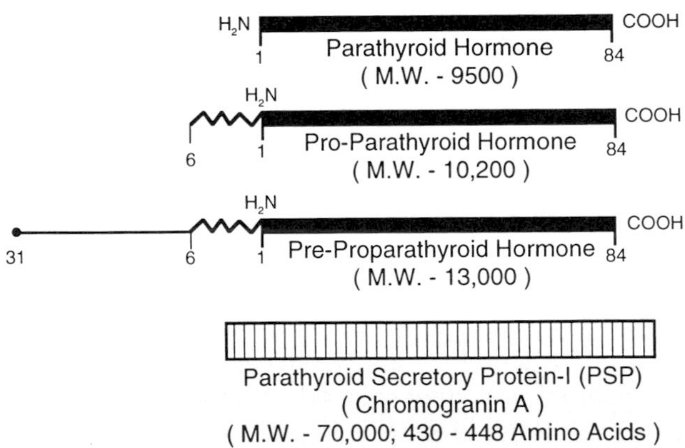

FIGURE 23.4 Chemistry of parathyroid hormone and related peptides synthesized by chief cells. Active parathyroid hormone is first synthesized as part of a large biosynthetic precursor molecule. Pre-proparathyroid hormone (115 amino acids) is the initial translational product from ribosomes and is rapidly converted to proparathyroid hormone in the rough endoplasmic reticulum. Proparathyroid hormone (90 amino acids) is converted enzymatically to active parathyroid hormone (84 amino acids) in the Golgi apparatus as the hormone is packaged into secretory (storage) granules. Parathyroid secretory protein (chromogranin A) is a high-molecular-weight molecule synthesized by chief cells, incorporated into storage granules with active parathyroid hormone, and secreted in parallel with active hormone in response to changes in blood calcium. It is necessary for normal secretory granule function and maturation. (Capen, 1989.)

In the early phases of secretory activity the endoplasmic reticulum of chief cells aggregates into large lamellar arrays, and free ribosomes group into clusters (Roth and Raisz, 1966) . It is at this stage that preproPTH is synthesized by the chief cells (Fig. 23.5). This is followed by packaging of the PTH, in which the Golgi apparatus enlarges and often appears as multiple complexes in several parts of the chief cell associated with many prosecretory granules. As this occurs, the granular endoplasmic reticulum involutes and disperses.

4. Storage and Secretion

a. Secretory Granules

The secretory granules develop by sequential accumulation and condensation of finely granular material within cisternae of the Golgi apparatus. Prosecretory granules are concentrated in the vicinity of the Golgi apparatus and occasionally are observed in the process of becoming detached from the membranes of the Golgi complex. Secretory ("storage") granules have been demonstrated readily at the level of ultrastructure within chief cells of the parathyroid glands in man and all animal species examined (Capen and Roth, 1973; Roth and Capen, 1974). The paucity of secretory granules in certain species (e.g., rat) initially led some investigators to erroneously suggest that chief cells do not have a mechanism for the storage of preformed hormone.

The secretory granules in chief cells usually are small, ranging from 100 to 300 nm in their greatest diameter, and are composed of fine, dense particles. They are electron dense, are surrounded by a delicate, closely applied limiting membrane, and are round to oval. The number of granules within chief cells varies considerably between species, with bovine parathyroids having consistently more secretory granules than man and other animals (Capen *et al.*, 1968). Chief cells

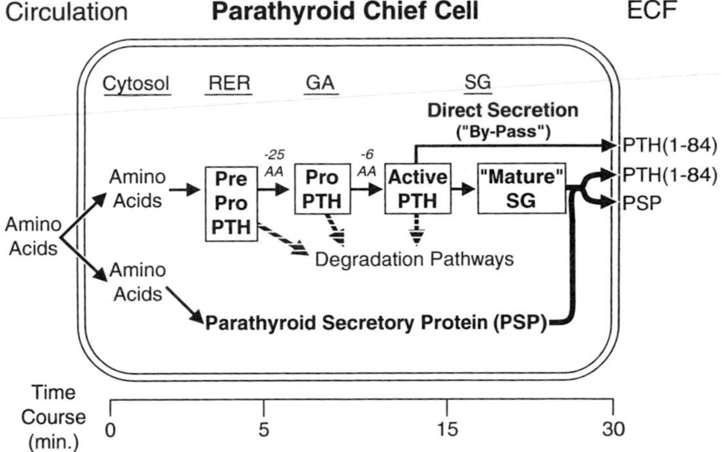

FIGURE 23.5 Subcellular compartmentalization, transport, and cleavage of precursors of parathyroid hormone (PTH). Pre-proparathyroid hormone (pre-pro-PTH) is the initial translation product from ribosomes of the rough endoplasmic reticulum (RER), which is rapidly converted to proparathyroid hormone (pro-PTH). The hydrophobic sequence at the amino-terminal end of prepro-PTH molecule facilitates penetration of the leading portion of the nascent peptide into the lumen of the endoplasmic reticulum. Pro-PTH is transported to the Golgi apparatus (GA), where it is converted enzymatically by a carboxypeptidase to biologically active PTH. A major portion of the biosynthetic precursors and active PTH is degraded by lysosomal enzymes and is not secreted by chief cells under normal conditions.

have fewer storage granules than many other endocrine cells concerned with the biosynthesis of polypeptide hormones, e.g., adenohypophysis and calcitonin-secreting C cells of the thyroid (Fig. 23.6) (Capen and Young, 1967).

The secretory granules migrate peripherally in chief cells, and their limiting membrane fuses with the plasma membrane of the cell. An internal cytoskeleton composed of microtubules and contractile filaments is important in the control of peripheral movement of

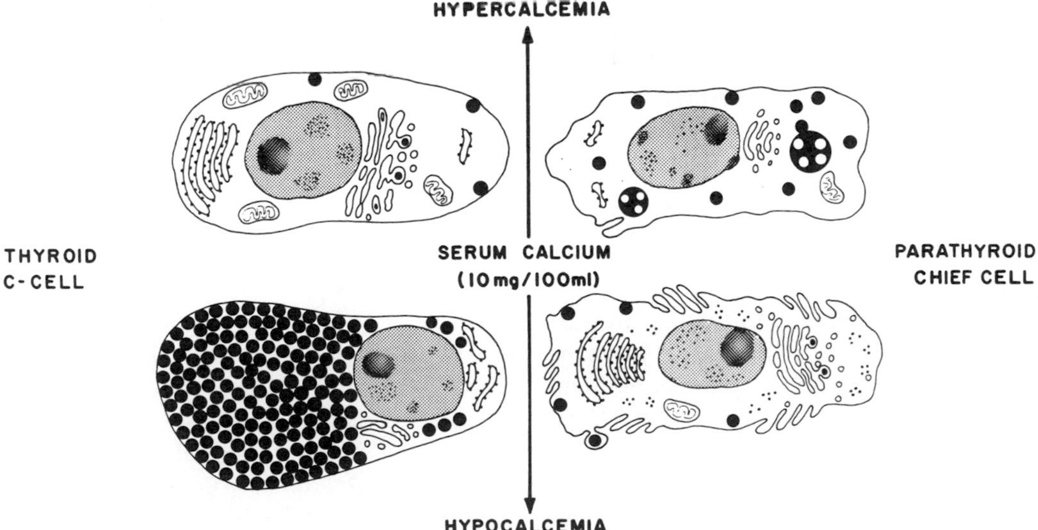

FIGURE 23.6 Response of parathyroid chief cells and thyroid C cells to hypercalcemia and hypocalcemia. C cells accumulate secretory granules in response to hypocalcemia, whereas chief cells are nearly degranulated but have an increased development of synthetic and secretory organelles. In response to hypercalcemia, C cells are degranulated and chief cells are predominantly inactive with small numbers of secretory granules. (Capen, 1989.)

secretory granules and liberation of secretory products from chief cells (Youshak and Capen, 1970). Secretory granules are extruded from chief cells by exocytosis (emiocytosis) into perivascular spaces (Fetter and Capen, 1970; Capen, 1971).

b. Chromogranin A

In addition to parathyroid hormone, secretory granules in chief cells (Fig. 23.5) also contain chromogranin A (parathyroid secretory protein-I). Chromogranin A (CGA) was first isolated from secretory granules of the bovine adrenal medulla (Winkler and Fischer-Colbrie, 1992). Chromogranin A, a 49-kDa peptide, is a major constituent of secretory granules of the adrenal medulla, pituitary, parathyroid, thyroid C cells, pancreatic islets, endocrine cells of the gastrointestinal tract, and sympathetic nerves and composes up to 50% of the total protein secreted by the parathyroid. Chromogranin A shares considerable homology between species. Immunologic cross reactivity to mammalian proteins has been observed in reptiles, amphibians, fish, and *Drosophila* tissues. Chromogranin A is synthesized as a pre-protein and is directed to the internal cavity of the rough endoplasmic reticulum (rER) by the N-terminal pre-region of the peptide. Once inside the rER, the pre-region is cleaved by a signal peptidase.

Although the functions of CGA are still under investigation, several roles have been postulated. Chromogranin A is suspected to play an important role in the maturation of secretory granules. Inside the Golgi apparatus, CGA is involved in the packaging of contents into newly formed vesicles. Chromogranin A precipitates as it diffuses into the trans-Golgi network. Other secretory products such as parathyroid hormone become entrapped in the growing CGA conglomerate and subsequently are packaged into a secretory granule (Cohn *et al.*, 1994). Chromogranin A has a large calcium-binding capacity that may enhance vesicle stability. As granules mature, they accumulate large amounts of calcium (up to 40 m*M*), which also may serve as a route of Ca^{2+} secretion. However, free (ionized) calcium concentrations remain in the micromolar range because most Ca^{2+} is bound to CGA. Chromogranin A–calcium complexes are important in maintaining the integrity of the secretory granule, because the absence of calcium will cause dissociation of protein complexes and result in osmotic lysis of the vesicle. Therefore, the intragranular functions of CGA include hormone packaging, stabilization of the granule against osmotic gradients, and excretion of intracellular calcium.

During the process of secretion the contents of secretory granules are extruded into the pericapillary spaces. The pH and calcium concentration of the extracellular fluid promote dissociation of CGA complexes and solubilization of its bound calcium and other contents of the granule. Once solubilized, extracellular peptidases cleave CGA into biologically active peptides that act as paracrine or autocrine regulators of secretion (Deftos, 1991). Most of the CGA-derived peptides have been reported to decrease hormone secretion.

5. Control of Secretion

Secretory cells in the parathyroid gland store small amounts of preformed hormone, but are capable of responding to minor fluctuations in calcium concentration by rapidly altering the rate of hormonal secretion and more slowly by altering the rate of hormonal synthesis. The parathyroids have a unique feedback control by the concentration of calcium (and to a lesser extent magnesium) ion in serum.

Interaction of serum ionized calcium with the Ca^{2+}-sensing receptor on chief cells results in the formation of an inverse sigmoidal relationship between serum Ca^{2+} and PTH concentrations (Fig. 23.7) (Silver, 1992; Cloutier *et al.*, 1993; Brown, 1994). The serum [Ca^{2+}] that results in half-maximal PTH secretion is defined as the serum calcium "set point" and is stable for an individual animal. The sigmoidal relationship between serum [Ca^{2+}] and PTH secretion permits the chief cells to respond rapidly to a reduction in serum Ca^{2+}. Binding of Ca^{2+} to the Ca^{2+}-sensing receptor results in an increase in the intracellular Ca^{2+}concentration of chief cells and a reduction in PTH secretion. This makes the parathyroid chief cells unique, because increased intracellular Ca^{2+} concentrations are a stimulus for secretion in most cell types. The major inhibitors of PTH synthesis and secretion are increased serum [Ca^{2+}] and

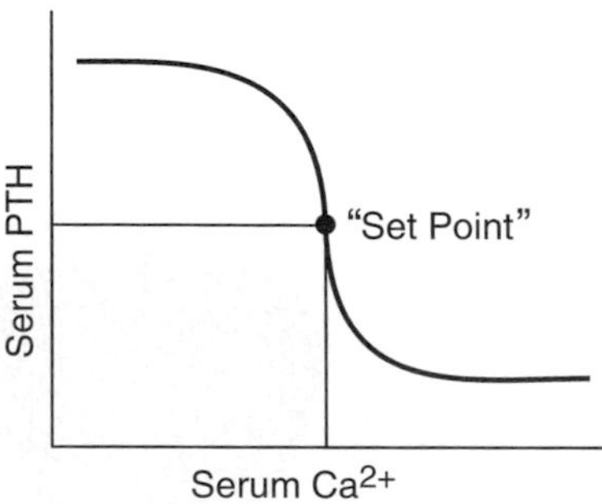

FIGURE 23.7 The serum Ca "set point" is defined as the concentration of serum Ca at which the concentration of serum parathyroid hormone (PTH) is half-maximal. The sigmoidal relationship between serum Ca and PTH enables the parathyroid glands to rapidly respond to minor reductions in serum Ca and increase PTH secretion and serum PTH concentration to return serum Ca to normal levels. The concentration of serum Ca in most normal animals is maintained at a level less than the "set-point" concentration. (Rosol *et al.*, 1995a.)

1,25-dihydroxyvitamin D. Inhibition of PTH synthesis by 1,25-dihydroxyvitamin D completes an important endocrine feedback loop between the parathyroid chief cells and the renal epithelial cells, because PTH stimulates renal production of 1,25-dihydroxyvitamin D (Fig. 23.8).

If the blood calcium is elevated by the intravenous infusion of calcium, there is a rapid and pronounced reduction in circulating levels of immunoreactive parathyroid hormone (iPTH); however, a small percentage of PTH secretion is nonsuppressible (Fig. 23.9). Therefore, there are always low concentrations of circulating PTH in the blood even under conditions of hypercalcemia. Conversely, if the blood calcium is lowered by infusion by EDTA (ethylenediaminetetraacetic acid), there is a brisk and substantial increase in iPTH levels (Fig. 23.9). The concentration of blood phosphate has no direct regulatory influence on the synthesis and secretion of PTH; however, certain disease conditions with hyperphosphatemia in both animals and man are associated clinically with hyperparathyroidism. An elevated blood phosphorus level may lead indirectly to parathyroid stimulation by virtue of its ability to lower blood calcium (Krook and Lowe, 1964). If the blood phosphorus is elevated significantly by an infusion of phosphate and calcium administered simultaneously in amounts to prevent the accompanying reduction of blood calcium, plasma iPTH levels remain within the normal range (Fig. 23.9).

Magnesium ion has an effect on parathyroid secretory rate similar to that of calcium, but its effect is not equipotent to that of calcium (Mayer and Hurst, 1978). The reduced potency of Mg^{2+} compared to that of Ca^{2+} may be due to reducing binding affinity of Mg^{2+} for the Ca^{2+}-sensing membrane receptor on chief cells. The more potent effects of Ca^{2+} in the control of PTH secretion, together with its preponderance over Mg^{2+} in the extracellular fluid, suggest a secondary role for magnesium in parathyroid control.

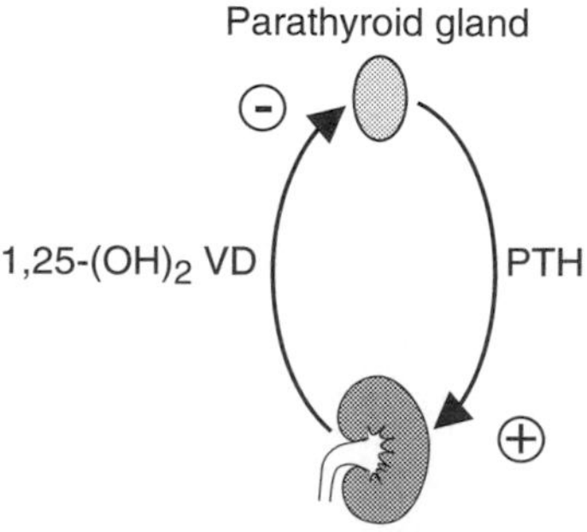

FIGURE 23.8 Endocrine negative-feedback loop between the parathyroid gland and the kidney involving serum concentrations of parathyroid hormone (PTH) and 1,25-dihydroxyvitamin D (1,25-$(OH)_2$VD). (Rosol *et al.*, 1995a.)

Calcium ion not only controls the rate of biosynthesis and secretion of parathyroid hormone, but also other metabolic and intracellular degradative processes within chief cells (Chu *et al.*, 1973). An increase of calcium ions in extracellular fluids rapidly inhibits the uptake of amino acids by chief cells, synthesis of proPTH, and conversion to PTH, and secretion of stored PTH and increases the intracellular degradation of PTH (Cohn *et al.*, 1974). The shifting of the percentage of flow of proparathyroid hormone from the degradative pathways to the secretory route represents a key adaptive response of the parathyroid gland to a low-calcium diet. During periods of long-term calcium restriction the enhanced synthesis and secretion of PTH would be accomplished by an increased capacity of the entire pathway in individual hypertrophied chief cells and through hyperplasia of active chief cells.

Recently synthesized and processed PTH may be released directly in response to increased demand and bypass the chief cell's storage pool of mature secretory granules in the cytoplasm. Bypass secretion of calcium can be stimulated only by a low circulating concentration of calcium ion and not by other secretagogues for PTH (Fig. 23.10). Degradation of PTH in secretory granules by lysosomal enzymes occurs after prolonged exposure of chief cells to a high-calcium environment.

6. *Biological Action*

Parathyroid hormone is the principal hormone involved in the minute-to-minute fine regulation of blood calcium in mammals. It exerts its biologic actions by directly influencing the function of target cells primarily in bone and kidney and indirectly in the intestine to maintain plasma calcium at a level sufficient to ensure the optimal functioning of a wide variety of body cells (Fig. 23.11).

In general the most important biological effects of PTH are to (1) elevate the blood calcium concentration, (2) decrease the blood phosphorus concentration, (3) increase the urinary excretion of phosphate by decreased tubular reabsorption, (4) increase tubular reabsorption of calcium, resulting in diminished calcium loss into the urine, (5) increase the rate of skeletal remodeling and the net rate of bone resorption, (6) increase osteolysis and the numbers of osteoclasts on bone surfaces, (7) increase the urinary excretion of Type I collagen metabolites, such as hydroxyproline and collagen telopeptides, (8) activate adenylate cyclase or stimulate increased intracellular Ca^{2+} concentrations in target cells, and (9) accelerate the formation of the principal active vitamin D metabolite (1,25-dihydroxycholecalciferol) (1,25$[OH]_2$D) by the kidney through a trophic effect on the 1-α-hydroxylase in mi-

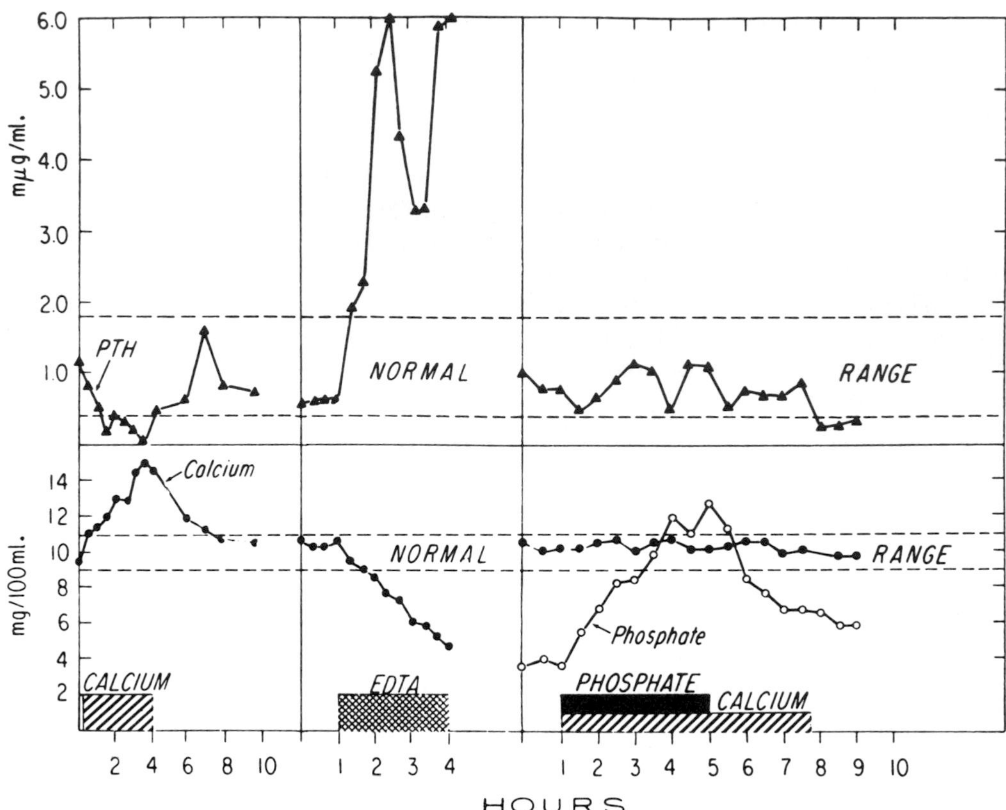

FIGURE 23.9 Changes of plasma immunoreactive parathyroid hormone in response to hypercalcemia induced by calcium infusion, hypocalcemia produced by EDTA infusion, and hyperphosphatemia with normocalcemia in a cow. From Aurbach and Potts (1964).

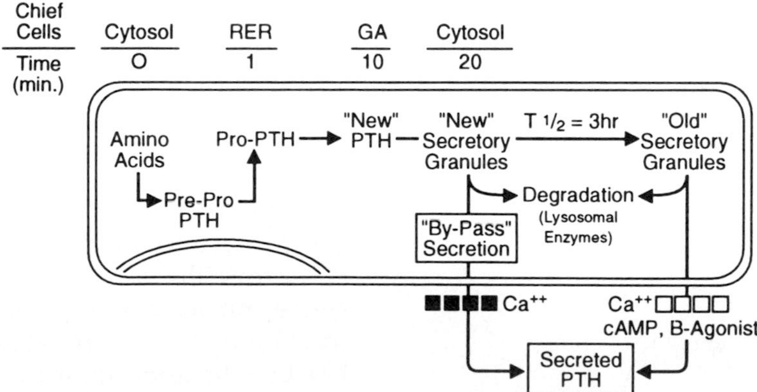

FIGURE 23.10 Bypass secretion of parathyroid hormone in response to increased demand signaled by decreased blood calcium ion concentration. Recently synthesized and progressed active PTH(1–84) may be released directly and not enter the storage pool of mature ("old") secretory granules in the cytoplasm of chief cells. PTH from the storage pool can be mobilized by cyclic adenosine monophosphate (cAMP) and β (B)-agonists (such as epinephrine, norepinephrine, and isoproterenol) as well as by lowered blood calcium ion, whereas secretion from the pool of recently synthesized PTH can be stimulated only by a decreased calcium ion concentration. RER, Rough endoplasmic reticulum; GA, Golgi apparatus. Redrawn from Cohn and MacGregor (1981).

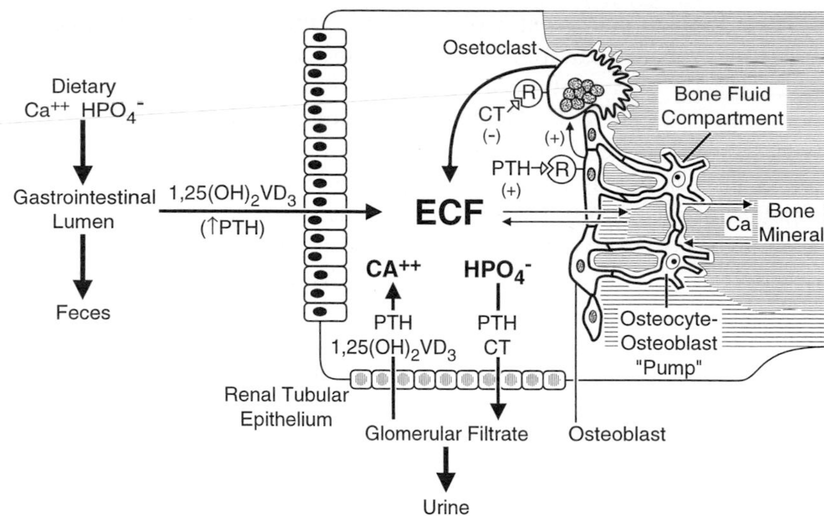

FIGURE 23.11 Interrelations of parathyroid hormone (PTH), calcitonin (CT), and 1,25-dihydroxycholecalciferol [1,25-$(OH)_2VD_3$] in the hormonal regulation of calcium and phosphorus in extracellular fluids (ECF).

tochondria of renal epithelial cells in the proximal convoluted tubules.

An important action of PTH on bone is to mobilize calcium from skeletal reserves into extracellular fluids (Fig. 23.11) (Canalis *et al.,* 1994). The administration of PTH causes an initial decline followed by a sustained increase in circulating levels of calcium. This transitory decrease in blood calcium is considered to be the result of a sequestration of calcium phosphate in bone and soft tissues (Parsons and Robinson, 1971). The subsequent increase in blood calcium results from an interaction of parathyroid hormone with receptors on osteoblasts that stimulate increased calcium release from bone and direct an increase in osteoclastic bone resorption (High *et al.,* 1981a). The mechanisms by which osteoblasts mediate increased osteoclastic bone resorption are currently incompletely understood.

The response of bone to parathyroid hormone is biphasic. The immediate effects are the result of increasing the activity of existing bone cells. This rapid effect of parathyroid hormone depends upon the continuous presence of hormone and results in an increased flow of calcium from deep in bone to bone surfaces through the action of an osteocyte–osteoblast "pump" in order to make fine adjustments in the blood calcium concentration (Fig. 23.11) (Parfitt, 1977).

The later effects of parathyroid hormone on bone are potentially of a greater magnitude of response and not dependent upon the continuous presence of hormone. Osteoclasts are primarily responsible for the long-term action of PTH on increasing bone resorption and overall bone remodeling (Canalis *et al.,* 1994). This is interesting in light of recent findings that have failed to demonstrate receptors for PTH on osteoclasts, but receptors were present on osteoblasts (Segre, 1994). These cells are flat and cover bone surfaces under normal conditions. The initial binding of PTH to osteoblasts lining bone surfaces appears to cause the cells to contract, thereby exposing the underlying mineral to osteoclasts (Rodan and Martin, 1981). The change in shape of osteoblasts associated with PTH may be related to calcium entry into the cell and alteration in microtubule and microfilament function, and appears to be critical to mediation of osteoclastic bone resorption stimulated by the hormone. Bone resorption products, particularly osteocalcin, can attract osteoclast precursors and enhance the resorption process. In addition, osteoblasts elaborate undefined chemical mediators to stimulate osteoclasts directly (McSheehy and Chambers, 1986).

If the increase in PTH is sustained, the size of the active osteoclast pool in bone is increased by activation of osteoclast progenitor cells in the endosteal bone-cell envelope (Heersche *et al.,* 1994). The plasma membrane of osteoclasts in intimate contact with the resorbing bone surface is modified to form a series of membranous projections referred to as the brush "ruffled" border. This area of active bone resorption is isolated from the extracellular fluids by adjacent transitional ("sealing") zones, thereby localizing the lysosomal enzymes and acidic environment to the immediate area undergoing dissolution. The mineral and organic components (e.g., hydroxyproline) released from bone are phagocytized by osteoclasts and transported across the cell in transport vesicles to be released into the extracellular fluid compartment (Baron, 1996; Roodman, 1996).

Parathyroid hormone also has the potential to serve as an anabolic agent in bone and stimulate osteoblastic bone formation (Canalis *et al.*, 1994). The physiologic role of the anabolic action of PTH *in vivo* is uncertain, but intermittent administration of exogenous PTH has been reported to increase bone mass in humans and animals.

The major effect of PTH under physiologic conditions is exerted on bone cells on endosteal surfaces of cancellous bone and in Haversian canals of cortical bone. In disease states of hyperparathyroidism, the activation of osteoclast progenitor cells occurs in the periosteum as well, leading to the formation of active sites of bone resorption and formation on the periosteal surface. This process results in the characteristic subperiosteal areas of bone resorption seen radiographically in long-standing hyperparathyroidism.

Parathyroid hormone has a rapid (within 5 to 10 minutes) and direct effect on renal tubular function leading to decreased reabsorption of phosphorus and phosphaturia (see Section III on phosphorus metabolism). The site of action of PTH in blocking tubular reabsorption of phosphorus has been localized by micropuncture methods to the proximal tubules of the nephron (Fig. 23.12). In addition, PTH leads to an increased urinary excretion of potassium, bicarbonate, sodium, cyclic adenosine monophosphate, and amino acids.

Although the effect of PTH on the tubular reabsorption of phosphorus has been considered to be of major importance, the ability of PTH to enhance the renal reabsorption of calcium is of considerable importance in the maintenance of calcium homeostasis. This effect of PTH upon tubular reabsorption of calcium is due to a direct action on the distal convoluted tubule (Yanagawa and Lee, 1992). The urinary excretion of magnesium, ammonia, and titratable acidity also are decreased by PTH. The other important effect of PTH on the kidney is in the regulation of the conversion of 25-hydroxycholecalciferol to 1,25-$(OH)_2D$ and other metabolites of vitamin D. The role of PTH as a trophic hormone in the metabolic activation of cholecalciferol will be discussed further in the section of this chapter on vitamin D.

Parathyroid hormone has been shown to promote the absorption of calcium from the gastrointestinal tract in animals under a variety of experimental conditions (Nemere and Norman, 1986; Favus, 1992). This effect is not as rapid as the action on the kidney and is not observed in vitamin D-deficient animals. The increased intestinal calcium transport is due principally to an indirect effect of PTH by its action of stimulating the renal synthesis of the biologically active metabolite of vitamin D ($1,25[OH]_2D$); however, PTH also

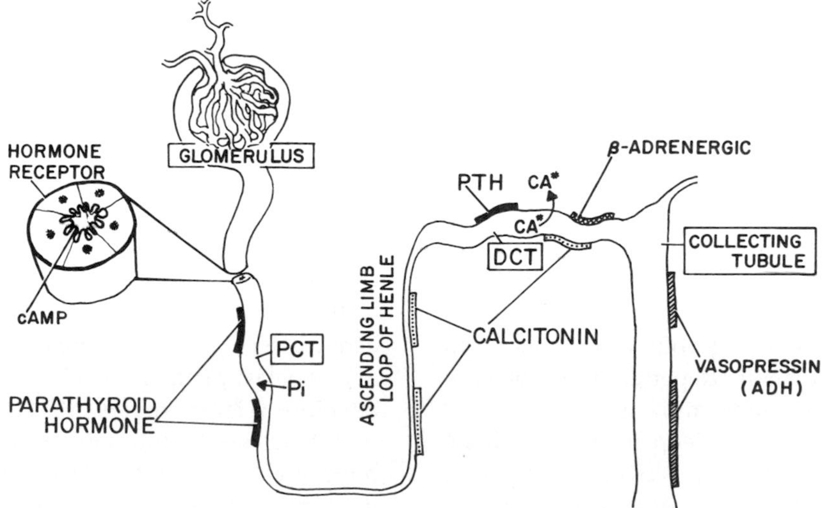

FIGURE 23.12 Distribution of adenylate cyclase-linked hormone receptors (target cells for parathyroid hormone and calcitonin) in the nephron. Parathyroid hormone diminishes tubular reabsorption of phosphorus in the proximal convoluted tubules (PCT), whereas increased calcium reabsorption occurs in cells located in the distal convoluted tubules (DCT). Cells with receptors for calcitonin that are situated in the ascending limb of the loop of Henle and distal convoluted tubules also diminish tubular reabsorption of phosphorus and cause phosphaturia. (Capen, 1989.)

may play a minor role by directly stimulating Ca absorption by intestinal epithelial cells.

7. *Parathyroid Hormone Receptor*

The receptor for N-terminal PTH (1–34), the region important in calcium regulation, has been cloned and sequenced (Abou-Samra *et al.*, 1992; Segre, 1994). It is a 7-transmembrane domain receptor that is expressed in renal epithelial cells and osteoblasts, among other cells. The N-terminal regions of PTH and PTHrP bind this receptor with equal affinity. Binding of PTH or PTHrP to the receptor results in increased levels of both cytoplasmic cAMP and Ca^{2+} by stimulation of adenylate cyclase and the phosphatidyl inositol pathways (Abou-Samra *et al.*, 1992; Coleman *et al.*, 1994). The PTH receptor is also located on many cell types, such as dermal fibroblasts, that are not associated with the action of PTH. It is assumed that the receptor functions as the binding protein for PTHrP in these tissues. Therefore, this receptor may be best termed the PTH/PTHrP receptor. It is not known whether there are distinct receptors for the N-terminal region of PTHrP; however, there are is some experimental data to suggest that alternate receptors may exist for the midregion and C-terminal regions of PTH and PTHrP (Murray *et al.*, 1994).

8. *Subcellular Mechanism of Action*

The calcium-mobilizing and phosphaturic activities of parathyroid hormone are mediated through the intracellular accumulation of 3′,5′-adenosine monophosphate (cAMP) or Ca^{2+} in target cells (Fig. 23.13) (Coleman *et al.*, 1994). Binding of PTH to PTH/PTHrP receptors on target cells results in activation of the receptor, binding of the receptor to stimulatory or inhibitory G proteins, and stimulation of adenylyl cyclase or phosphatidylinositol hydrolysis in the plasma membrane. Stimulation of adenylyl cyclase stimulates the conversion of ATP to cAMP in target cells. The accumulation of cAMP in target cells functions as an intracellular mediator or second messenger of parathyroid hormone action to increase permeability for calcium ion. Cytosolic Ca^{2+} concentration also may be increased by the actions of inositol triphosphate to release Ca^{2+} from intracytoplasmic stores or by stimulation of Ca^{2+} transport through transmembrane calcium channels. The resultant increase in cytosol calcium content in combination with cAMP accumulation initiates biochemical reactions in bone cells and renal epithelial cells to conduct the intracellular functions of PTH.

In addition, PTH contributes to the regulation of the rate of formation of the active form of vitamin D by

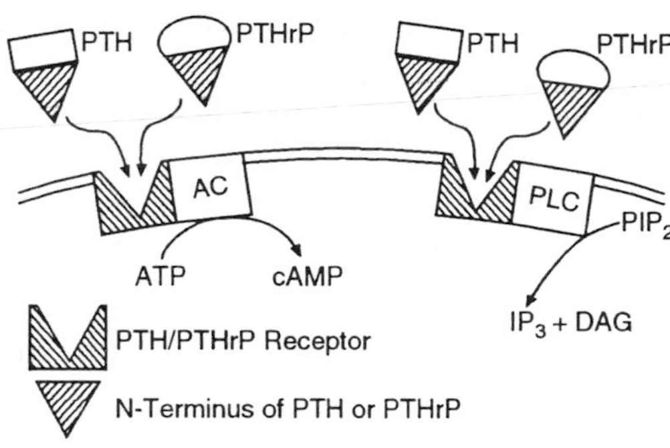

FIGURE 23.13 Mechanism of parathyroid hormone and parathyroid hormone-related protein action. The biologically active N-terminal ends of the hormones PTH and PTHrP bind to specific PTH/PTHrP receptors on the surface of target cells. The receptor–hormone complex is coupled to the catalytic subunit of adenylyl cyclase (AC) and phospholipase C (PLC) in the cell membrane by a nucleotide regulatory protein (G protein). This results in the conversion of ATP to cyclic adenosine monophosphate (cAMP) and phosphatidylinositol to inositol triphosphate (IP_3) and diacylglycerol (DAG). IP_3 stimulates release of Ca^{2+} from intracellular stores. Both cAMP and Ca^{2+} serve as second messengers for polypeptide hormones such as PTH and PTHrP in target cells and result in expression of the biological response of the hormones. (Rosol and Capen, 1992.)

mitochondria in renal tubular epithelial cells. The active metabolites of vitamin D make bone cells more sensitive to the direct effect of PTH ("permissive effect") and greatly enhance the gastrointestinal absorption of calcium, thereby amplifying the effect of PTH upon plasma calcium concentration.

9. *Assays for Parathyroid Hormone*

The metabolism of PTH and the formation of multiple circulating forms of PTH has made development of clinically useful immunoassays challenging (Kronenberg *et al.*, 1994). The principal circulating forms of PTH include intact PTH (1–84) and C-terminal peptides (e.g., PTH 35–84). Parathyroid hormone has two functional domains. N-Terminal PTH (PTH 1–34) is the region of major biological activity in relation to calcium regulation. N-Terminal PTH does not exist in the circulation in a biologically relevant concentration, but administration of exogenous PTH (1–34) will induce the typical actions of PTH on bone and kidney cells (Everhart-Caye *et al.*, 1996). The function of C-terminal PTH peptides is less well understood (Mallette, 1994).

PTH is secreted in two forms: intact PTH (1–84) and C-terminal peptides. PTH (1–84) is the circulating form of biologically active PTH. PTH (1–84) is rapidly removed from the circulation by endopeptidases on Kupffer cells or glomerular filtration, and its half-life is less than 5 minutes (Arnaud and Pun, 1992). Some of the C-terminal peptides are released from the Kupffer cells back into the circulation and are excreted by glomerular filtration. Therefore, C-terminal PTH has a greater serum half-life than PTH (1–84) and is present in the serum at greater concentrations (50–90% of total PTH) than PTH (1–84), especially in states of hyperparathyroidism. End organs (bone and kidney) also are responsible for degradation of a small fraction of serum PTH (1–84). This involves internalization of PTH–receptor complexes and intracellular degradation by lysosomal enzymes. Rapid removal of PTH (1–84) from the circulation allows the *in vivo* actions of PTH to be controlled by the rate of PTH secretion from the parathyroid chief cells.

Early immunoassays for PTH were single-site radioimmunoassays (RIA) for C-terminal peptides. These assays were suboptimal, because the biologically active form of PTH was not measured. However, they were clinically useful in states of hyperparathyroidism in patients with normal renal function. Midregion and C-terminal RIAs measure both intact PTH and C-terminal PTH, which makes them less clinically relevant (Jorch *et al.*, 1982; Mallette and Tuma, 1984; Flanders and Reimers, 1991). In addition, conditions that reduce glomerular filtration rate, such as renal failure, will result in a large increase in the serum concentration of C-terminal PTH.

Intact serum PTH concentrations are best measured by two-site immunoradiometric assay (IRMA) or N-terminal RIA (Nussbaum and Potts, 1994). Serum intact PTH can be measured in dogs, cats, and horses with assays developed for human PTH (Nichols Institute Diagnostics, San Juan Capistrano, CA; INCSTAR Corp., Stillwater, MN; Diagnostic Products Corp., Los Angeles, CA) because of the cross reactivity of the antisera used in the assays (Torrance and Nachreiner, 1989; Nagode and Chew, 1991; Barber *et al.*, 1993; Cloutier *et al.*, 1993) (Table 23.3). Intact PTH has been successfully measured in dogs or cats with either a human N-terminal RIA or two-site IRMA (Nagode and Chew, 1991; Barber *et al.*, 1993). The correlation between the two assays in dogs is good (Nagode and Chew, 1991). High concentrations of circulating C-terminal PTH have been reported to decrease the measurement of intact PTH in cats, but this can be overcome when the sample is diluted (Barber *et al.*, 1993). Currently, intact PTH is best measured in horses using an N-terminal RIA (L. A. Nagode, personal communication, October 1996).

a. Canine Parathyroid Hormone

The complementary DNA that encodes for canine PTH was cloned and sequenced (Rosol *et al.*, 1995b, 1996). The predicted mature protein consists of 84 amino acids (aa) and has a high degree of homology to bovine, porcine, human, and rat PTH (Fig. 23.14). There are only two substitutions in the first 40 aa of canine PTH compared to those of human PTH at positions 7 and 16. These substitutions are not unique to canine PTH because they also occur in bovine (position 7 and 16) and porcine PTH (position 16) (Kronenberg *et al.*, 1994). Knowledge of the sequence of cPTH will permit synthesis of peptides and development of immunoassays specific for the dog.

B. Parathyroid Hormone-Related Protein

1. Introduction

Parathyroid hormone-related protein (PTHrP) is not strictly a calcium-regulating hormone; however, it was identified in 1982 as an important PTH-like factor that plays a central role in the pathogenesis of humoral hypercalcemia of malignancy (Moseley and Gillespie, 1995). Since its discovery, it has become known that PTHrP is widely produced in the body and has numerous actions in the normal fetus and adult animal independent of its role in cancer-associated hypercalcemia (Gröne *et al.*, 1994; Philbrick *et al.*, 1996). This is in contrast to PTH, which is produced by the parathyroid glands and functions principally in regulation of calcium balance. Some of the actions of PTHrP involve normal regulation of calcium metabolism (Rosol and Capen, 1992; Rosol *et al.*, 1995a). For example, PTHrP functions as a calcium-regulating hormone in the fetus and is produced by the fetal parathyroid gland and placenta (MacIsaac *et al.*, 1991b). In the adult, PTHrP circulates in the blood in very low concentrations (<1 p*M*), but is produced by many different tissues and functions principally as a paracrine cellular regulator. PTHrP may play a role in the transport of Ca into milk during lactation. PTHrP acts as an abnormal systemic calcium-regulating hormone and mimics the actions of PTH in patients with humoral hypercalcemia of malignancy (see Section IV.B on hypercalcemia).

Parathyroid hormone-related protein (PTHrP) is a 139–173 amino acid peptide originally isolated from human and animal tumors associated with humoral hypercalcemia of malignancy (Rosol and Capen, 1992). PTHrP shares 70% sequence homology with PTH in its first 13 amino acids. The N-terminal region of PTHrP (amino acids 1–34) binds and stimulates PTH receptors in bone and kidney cells with affinity equal to that of PTH and results in PTHrP functioning similarly to PTH in patients with humoral hypercalcemia of malignancy (Orloff *et al.*, 1994).

PTHrP not only plays a major role in most forms of humoral hypercalcemia of malignancy (HHM), but

TABLE 23.3 Measurement of Serum Hormone Concentrations

Test	Sample requirements/ handling	Methods	Comments	Reference ranges (OSU values)[a]
PTH, intact	Serum	N-Terminal RIA (Nichols Inst.)	Human assays work well in animals. Commercial assays are no longer available.	23 ± 5 pg/ml (dog) 13 ± 2 pg/ml (cat)
		Two-site IRMA (Nichols Inst.) (Diagnostic Products Corp.)	Human assays work well in animals.	18–122 pg/ml (2–13 p*M*) (dog)[b] 0–38 pg/ml (0–4 p*M*) (cat)[b] 2–19 pg/ml (0.2–2.0 p*M*) (horse)[b]
PTH, midregion, or C-terminal	Serum	RIA	Less clinically relevant; may be increased in animals with renal failure.	Human assays are variable and unreliable in animals.
PTHrP	Frozen plasma collected with protease inhibitors (for INCSTAR RIA and Nichols IRMA)	N-terminal RIA for PTHrP (1–36) (INCSTAR Corp.)	Human assays work well in dogs.	<2 pM (dogs)
		Two-site IRMA for PTHrP (1–86) (Nichols Inst., INCSTAR Corp.)	INCSTAR IRMA requires EDTA–plasma. No other protease inhibitors are necessary.	<2 pM (dogs)
Calcitonin		Canine-specific RIA	Low species cross-reactivity.	ND[c]
1,25-dihydroxyvitamin D	Serum	Radioreceptor assay or RIA	Good species cross-reactivity.	35 ± 7 pg/ml (adult dogs) 88 ± 13 pg/ml (12-wk-old dogs) 28 ± 3 pg/ml (adult cats) 49 ± 14 pg/ml (10-wk-old cats) <6 pg/ml (horses)
25-hydroxyvitamin D	Serum	Radioreceptor assay	Good species cross-reactivity.	82–285 nmol/Liter (dogs)[b] 50–165 nmol/Liter (bovine)[b]

[a] Reference values at Ohio State University, Veterinary Medical Teaching Hospital and Laboratories for normal adult animals unless otherwise indicated.

[b] Endocrine Diagnostic Section, Animal Health Diagnostic Laboratory, Lansing, MI.

[c] ND = not determined.

```
                   .         .         .         .         .
Canine    SVSEIQFMHNLGKHLSSMERVEWLRKKLQDVHNFVALGAPIAHRDGSSQR   (50)
Bovine    A......................................S..Y.......   (50)
Porcine   ......L..........L.....................S.V....G...   (50)
Human     ......L........N........................L.P..AG...   (50)
Rat       A.....L........A.V..MQ.............S..VQM.A.E..Y..   (50)

Canine    PLKKEDNVLVESYQKSLGEADKADVDVLTKAKSQ                   (84)
Bovine    .R..........H...............I...P.                   (84)
Porcine   .R..........H..........A....I...P.                   (84)
Human     .R..........HE...........N........                   (84)
Rat       .T...E...DGNS.....G........V.....                    (84)
```

FIGURE 23.14 Predicted amino acid sequences of canine, bovine, porcine, human, and rat parathyroid hormone (1–84). Identical amino acids are represented by dots.

has been demonstrated in many normal tissues, including epithelial cells of the skin and other organs; endocrine glands; smooth, skeletal, and cardiac muscle; lactating mammary gland; placenta; fetal parathyroid glands; bone; brain; and lymphocytes (Moseley *et al.*, 1991; Gröne *et al.*, 1994; Wysolmerski and Broadus, 1994; Philbrick *et al.*, 1996). Therefore, PTHrP functions as (1) a hormone in an endocrine manner in the fetus, (2) a paracrine factor in many fetal and adult tissues, and (3) an abnormal hormone in an endocrine manner in adults with HHM (Fig. 23.15). It is not known if there is a specific receptor for PTHrP in tissues other than bone and kidney; however, the mid- and C-terminal regions of PTHrP have functions that are independent of its PTH-like effects. Therefore, PTHrP is a polyhormone with multiple biologically active regions whose function depends on enzymatic processing of PTHrP in the serum or tissue of origin (Mallette, 1994). The midregion of PTHrP is responsible for stimulating Ca uptake by the fetal placenta (MacIsaac *et al.*, 1991b) and the C-terminal region can inhibit osteoclastic bone resorption (Fenton *et al.*, 1991b).

2. *Parathyroid Hormone-Related Protein in the Fetus*

a. Fetal Calcium Balance

Fetuses maintain higher concentrations of serum Ca compared to those of their dams. Because fetal parathyroid glands produce low levels of PTH, the mechanism of maintaining increased serum Ca concentrations in fetuses was unknown until recently (Care, 1991). Investigations have demonstrated that PTHrP functions to maintain Ca balance in the fetus (MacIsaac *et al.*, 1991a, 1991b; Abbas *et al.*, 1994). It is the major hormone secreted by fetal parathyroid chief cells and is produced by the placenta to stimulate Ca uptake by the fetus. Thyroparathyroidectomy of sheep fetuses results in a reduction in fetal serum Ca concentration and reduced placental transport of Ca (MacIsaac *et al.*, 1991b). The midregion of PTHrP is the most active portion that stimulates Ca and Mg transport by the placenta.

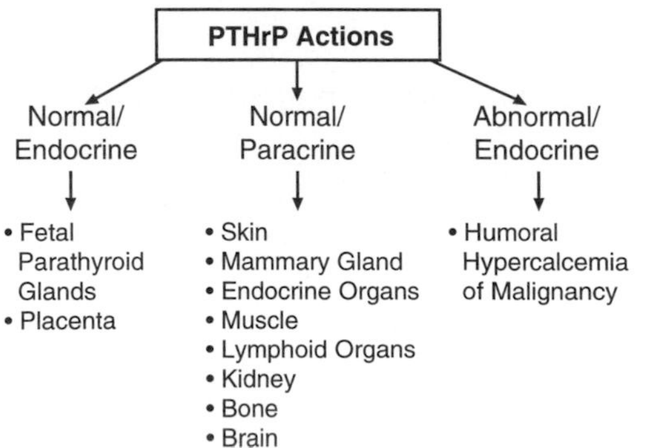

FIGURE 23.15 Actions of parathyroid hormone-related protein (PTHrP). (Rosol and Capen, 1992.)

b. Fetal Bone Development

PTHrP plays a role in differentiation of many tissues during gestation. It is widely expressed in numerous fetal tissues (Campos *et al.*, 1991; Moseley *et al.*, 1991) and is especially important in bone growth and development. Mice with "knock-out" or disruption of the PTHrP genes have decreased cartilage growth, shortened bones, and prematurely ossified bones, and die *in utero* or at birth (Karaplis *et al.*, 1994). Death at the time of birth is due to an inability to respire because the entire rib cage is ossified and rigid, the face is shortened, and the tongue protrudes from the mouth.

PTHrP regulates growth of cartilage at the epiphyseal plate by increasing chondrocyte proliferation and inhibiting maturation of chondrocytes from the proliferative zone to the hypertrophic zone (Vortkamp *et al.*, 1996). PTHrP gene "knockout" mice have short ossified bones at birth due to a lack of cartilage proliferation and early maturation of chondrocytes (Karaplis *et al.*, 1994). PTHrP increases chondrocyte proliferation and thickness of the proliferative chondrocyte zone at the growth plate by inhibiting apoptosis of chondrocytes. Inhibition of apoptosis is dependent on translocation of PTHrP to the nucleus and nucleolus (Henderson *et al.*, 1995). PTHrP contains a nucleolar localization sequence between amino acids 87 and 107. It is unknown whether PTHrP is translocated to the nucleus after secretion from the cell, binding to cell membrane receptors and receptor internalization, or whether translocation represents movement from the point of synthesis in the cytoplasm to the nucleus without exit from the cell (intracrine effect). A rare activating mutation of the PTH/PTHrP receptor in humans (Jansen's chondrodystrophy) results in increased thickness of the proliferating zone of chondrocytes at the growth plate (Schipani *et al.*, 1996).

3. *Parathyroid Hormone-Related Protein in the Normal Adult*

a. Introduction

PTHrP is produced by many tissues in adult animals, including endocrine glands, muscle (smooth, skeletal, and cardiac), brain, lymphocytes, lactating mammary gland, kidney, prostate gland, lung, skin and other simple epithelia, and bone (Capen and Rosol, 1993a; Gröne *et al.*, 1994; Philbrick *et al.*, 1996). The function of PTHrP in most of these tissues is poorly understood, but it probably functions as an autocrine or paracrine regulatory factor, because circulating concentrations of PTHrP in normal humans and animals are low (<1 p*M*) (Burtis, 1992; Rosol *et al.*, 1992b). The PTH/PTHrP receptor is often expressed on the same or adjacent cells in tissues that synthesize PTHrP.

b. PTHrP in the Skin

Epidermal keratinocytes produce PTHrP, which plays a role in their proliferation or differentiation (Werkmeister *et al.*, 1993). Keratinocytes do not contain the classic PTH/PTHrP receptor, but PTHrP induces an increase in intracellular Ca^{2+}, which suggests that keratinocytes have an alternative PTHrP receptor (Orloff *et al.*, 1992). The classic PTH/PTHrP receptor is present on dermal fibroblasts (Hanafin *et al.*, 1995). PTHrP production by keratinocytes may act in an autocrine manner in the epidermis and a paracrine manner in dermis. Inhibition of PTHrP action in the skin with PTH/PTHrP receptor antagonists results in increased hair formation and prolonged anagen phase of the hair cycle (Holick *et al.*, 1994).

c. PTHrP in the Mammary Gland

The greatest concentration of PTHrP is found in milk (10–100 nmol/L) and is 10,000- to 100,000-fold greater than that in the serum (Ratcliffe, 1992; Riond *et al.*, 1995a). The function of PTHrP in the mammary gland and in milk is poorly understood at present. Overexpression of PTHrP in the mammary gland during gland development prior to lactation results in glandular hypoplasia due to a reduction in branching morphogenesis of the mammary ducts (Wysolmerski *et al.*, 1995).

Abundant and biologically active PTHrP is produced by alveolar epithelial cells during lactation, which results in the high concentration of PTHrP in milk. Synthesis of PTHrP by the mammary gland abruptly ceases when suckling stops and the gland undergoes involution (Thiede, 1994). PTHrP is enzymatically cleaved in milk, but the N-terminal PTHrP retains biological activity. The mammary gland actively transports calcium from serum into milk, and PTHrP may play a role in stimulating the transport of calcium by alveolar epithelial cells; however, this has not been confirmed in all species (Barlet *et al.*, 1992; Philbrick *et al.*, 1996).

Circulating concentrations of PTHrP may be minimally increased in lactating dams, but PTHrP from the mammary gland likely plays a minor role in systemic calcium balance of lactating animals (Melton *et al.*, 1990; Barlet *et al.*, 1993). No relationship has been demonstrated between PTHrP and the pathogenesis of parturient hypocalcemia and paresis in lactating dairy cattle (Riond *et al.*, 1996). Alternatively, milk PTHrP could have a physiologic function in suckling neonates, such as regulation of growth or differentiation of the gastrointestinal tract (Thiede, 1994). Infusion of PTHrP (1–34) into the mammary artery of lactating sheep resulted in increased blood flow to the mammary gland (Thompson, 1993). This was likely due to relaxation of the smooth muscle in the mammary arteries.

d. PTHrP and Smooth Muscle

PTHrP is widely produced by smooth muscle in the body, including blood vessels, uterus, oviduct of the hen, urinary bladder, and gastrointestinal tract (Philbrick *et al.*, 1996). In general, PTHrP expression is increased when smooth muscle is mechanically stretched; PTHrP induces relaxation of smooth muscle and attenuation of contraction. With progressive distension of the uterus during pregnancy or descent of the ovum in the hen's oviduct, PTHrP expression by the myometrium and myosalpinx increases. PTHrP expression in the urinary bladder smooth muscle is proportional to distension with urine. Both PTHrP and PTH are potent vasodilators and inducers of hypotension due to relaxation of vascular smooth muscle when administered intravenously. PTHrP probably functions as a paracrine regulator of vascular tone, causing vasodilatation and modulating vasoconstriction of other vasoactive compounds.

4. *Assays for Parathyroid Hormone-Related Protein*

Two-site immunoradiometric (IRMA) and N-terminal radioimmunoassays are available for the measurement of human PTHrP (Table 23.3) (Bilezikian, 1992; Kremer and Goltzman, 1994). These assays are useful to measure PTHrP in the dog (see Section IV.B) (Rosol *et al.*, 1992b; Weir, 1992) because of the high degree of sequence homology in PTHrP between species, especially in the first 111 amino acids (Burtis, 1992). An N-terminal RIA for human PTHrP has not proven useful to measure circulating PTHrP in a small number of horses (Rosol *et al.*, 1994). PTHrP concentrations are best measured in fresh or frozen plasma using EDTA as an anticoagulant and with the addition of protease inhibitors, such as aprotinin and leupeptin (Pandian *et al.*, 1992).

The circulating forms of PTHrP are not completely understood, because it rapidly undergoes proteolysis intracellularly and extracellularly after secretion into blood (Pandian *et al.*, 1992). The natural proteolytic cleavage sites of PTHrP (1–141) have not been completely identified (Burtis, 1992; Bowden *et al.*, 1993; Orloff *et al.*, 1994). The forms of PTHrP that are present *in vivo* include an N-terminal peptide, a combined N-terminal and midregion peptide, a midregion peptide, and a C-terminal peptide (Burtis *et al.*, 1994; Yang *et al.*, 1994). The N-terminal peptide extends from amino acid (aa) 1 to 36. The cleavage site may be at the Arg^{37}. The N-terminal peptide also contains three Arg residues at aa 19–21 that could serve as a proteolytic cleavage site. Proteolysis at this site would destroy the PTH-like bioactivity of PTHrP. The midregion of PTHrP extends from the end of the N-terminal peptide to approximately aa 86. There are many Arg and Lys residues between aa 86 and aa 111 that could serve as proteolytic cleavage sites.

Proteolysis of PTHrP makes sample collection and storage important criteria for accurate measurement of circulating PTHrP. The fragments that would be expected to have PTH-like biologic activity *in vivo* include N-terminal PTHrP (1–36), PTHrP (1–86), and intact PTHrP (1–141). The two-site IRMAs measure intact PTHrP (1–141) and PTHrP (1–86) because one antibody binds the N-terminus and one binds the midregion. N-Terminal RIAs measure intact PTHrP (1–141), PTHrP (1–86), and N-terminal PTHrP (1–36). C-Terminal PTHrP accumulates in the serum of human patients with renal failure, which suggests that C-terminal PTHrP peptides are excreted by the kidney as occurs with PTH (Burtis *et al.*, 1990).

5. *Canine Parathyroid Hormone-Related Protein*

The complementary DNA for canine PTHrP has been cloned and sequenced (Rosol *et al.*, 1995b). The PTHrP gene has been reported to be complex with multiple promoters, up to nine exons, and alternate splicing of exons that encode the C-terminal peptides (Gillespie and Martin, 1994). The sequence of the canine PTHrP cDNA indicated that the dog PTHrP gene is more closely related to the human PTHrP gene than are the PTHrP genes in rats, mice and chickens.

The coding region of canine PTHrP predicts a 177-amino acid protein with four regions: signal peptide; N-terminal or parathyroid hormone-like region; midregion; and C-terminal region (Fig. 23.16) (Burtis, 1992; Orloff *et al.*, 1994). The deduced amino acid sequence of the N-terminal region of the mature protein (aa 1–36) is identical in the four mammalian species (dog, human, rat, and mouse) and there is a very high degree of homology in these species of the midregion of PTHrP (Fig. 23.17) (Suva *et al.*, 1987; Yasuda *et al.*, 1989; Mangin *et al.*, 1990; Rosol *et al.*, 1995b). The high degree of interspecies homology indicates the importance of the N-terminus and midregion to the function of PTHrP.

There is less homology of the C-terminal region of canine PTHrP with that region of PTHrP from other species. The function of the C-terminal region is unknown. It has been reported that PTHrP (107–111) and PTHrP (107–139) inhibit osteoclastic bone resorption (Fenton *et al.*, 1991a). Increased urine concentrations of C-terminal PTHrP have been demonstrated in humans and mice with cancer-associated hypercalcemia (Imamura *et al.*, 1991; Kasahara *et al.*, 1992). The circulating concentrations of the C-terminal peptide are increased in patients with renal failure (Burtis *et al.*, 1990). This mimics the finding of increased C-terminal PTH in the serum of patients with renal failure and indicates that the kidney is an important site of excretion of C-terminal PTHrP and that it may have a longer serum half-life than N-terminal or midregion PTHrP.

C. Calcitonin (Thyrocalcitonin)

1. *Development and Structure of Thyroid C Cells*

Calcitonin (thyrocalcitonin, CT) is a more recently discovered hormone than PTH (Copp, 1994). Phylogenetically, the early appearance of calcitonin in primitive elasmobranch fish precedes the first appearance of PTH in amphibians. Calcitonin is secreted by a second endocrine cell population in the mammalian thyroid gland. C cells are distinct from follicular cells in the thyroid that secrete thyroxine and triiodothyronine. They are situated either within the follicular wall be-

FIGURE 23.16 Different regions of canine PTHrP protein: signal peptide (aa −36 to −1); N-terminal parathyroid hormone (PTH)-like region (aa 1–13), which shares primary sequence homology with PTH; N-terminal PTH-like region (aa 1–34), which binds to the PTH receptor and is identical in dogs, humans, and rodents; midregion with high interspecies homology; and C-terminal region of low interspecies homology.

```
                    .         .         .         .         .
Canine     AVSEHQLLHDKGKSIQDLRRRFFLHHLIAEIHTAEIRATSEVSPNSKPAP   ( 50)
Human      ................................................S.   ( 50)
Rat        ..................................................   ( 50)
Mouse      ..................................................   ( 50)
Chicken    .....................I..QN..EGVN.............P...T   ( 50)

Canine     NTKNHPVRFGSDDEGRYLTQETNKVETYKEQPLKTPGKKKKGKPGKRKEQ   (100)
Human      ..................................................   (100)
Rat        ...............................................R..   (100)
Mouse      ...............................................R..   (100)
Chicken    ....Y......E............SQ........VS.....A........   (100)

Canine     EKKKRRTRSAWLNSGVAESGLEGDHPYDISATS--LELNLRRH          (141)
Human      ............D...TG.......LS.T.T..--R..DS...          (141)
Rat        ...........--P.TTG...LE.PQPHT.P..TS..PSS.T.          (141)
Mouse      ...........--PST.A...LE.PLPHT.R..--..PS..T.          (139)
Chicken    ......A........MYG.NVTESPVL.N.V.T--HNHI..--          (139)
```

FIGURE 23.17 Predicted amino acid sequences of the mature form of canine parathyroid hormone-related protein (1–141) and of human, rat, mouse, and chicken PTHrP. Note the high degree of homology in the N-terminal 111 amino acids and complete homology of PTHrP 1–36 (the PTH-like region that binds to PTH/PTHrP receptors) in dogs, humans and rodents. Identical amino acids are represented by dots.

tween follicular cells or as small groups of cells between thyroid follicles. C cells do not border the follicular colloid directly, and their secretory polarity is oriented toward the interfollicular capillaries. The distinctive feature of C cells is the presence of numerous small membrane-limited secretory granules in the cytoplasm (Fig. 23.6). Immunocytochemical techniques have localized calcitonin of C cells to these secretory granules.

Calcitonin-secreting C cells have been shown to be derived embryologically from cells of the neural crest. Primordial cells from the neural crest migrate ventrally and become incorporated within the last (ultimobranchial) pharyngeal pouch. They move caudally with the ultimobranchial body to the point of fusion with the midline primordia that give rise to the thyroid gland. The ultimobranchial body is incorporated into the thyroid near the hilus in mammals, and C cells subsequently are distributed throughout the gland. Although C cells are present throughout the thyroid gland in postnatal life of man and most other mammals, they often remain more numerous near the hilus and point of fusion with the ultimobranchial body. In submammalian species C cells and calcitonin activity remain segregated in the ultimobranchial gland, which is anatomically distinct from both the thyroid and the parathyroid glands.

2. Chemistry

Calcitonin is a polypeptide hormone composed of 32 amino acid residues arranged in a straight chain with a 1–7 disulfide linkage (Copp, 1970). It is a smaller molecule than parathyroid hormone (84 amino acids) and is synthesized as part of a larger biosynthetic precursor molecule. Pre-procalcitonin is the initial translational product formed in the rough endoplasmic reticulum (Fig. 23.18). Procalcitonin is transported to the Golgi apparatus where it is converted to calcitonin prior to packaging in membrane-limited secretory granules. Depending upon the need for calcitonin, a proportion of the precursors and active hormone undergo degradation prior to release from C cells. Under certain pathologic conditions (e.g., neoplasia) these neural crest-derived C cells may secrete other humoral factors, including serotonin, bradykinin, ACTH, and prostaglandins (Fig. 23.18).

The calcitonin gene is expressed differently in thyroid (C cells) and neural tissues (Jacobs, 1985). In thyroid C cells the mRNA encodes primarily for pre-procalcitonin (molecular weight: 17,400 Da), whereas in neural tissues there is alternative RNA processing and encoding for pre-procalcitonin gene-related peptide (CGRP) (15,900 Da). CGRP is a neuropeptide composed of 37 amino acids and is concerned with nociception, ingestive behavior, and modulation of the nervous and endocrine systems.

The complete amino acid sequence of calcitonin from porcine, bovine, ovine, canine, salmon, and other animal species, as well as human calcitonin, has been determined (Mol *et al.*, 1991). Synthetic human calcitonin has been prepared and shown to be biologically active. The structure of calcitonin differs considerably between species. The molecules in eight selected species share only eight of the 32 amino acid residues (Fig. 23.19). However, the amino terminal of the calcitonin molecule is similar in all species. It consists of a seven-

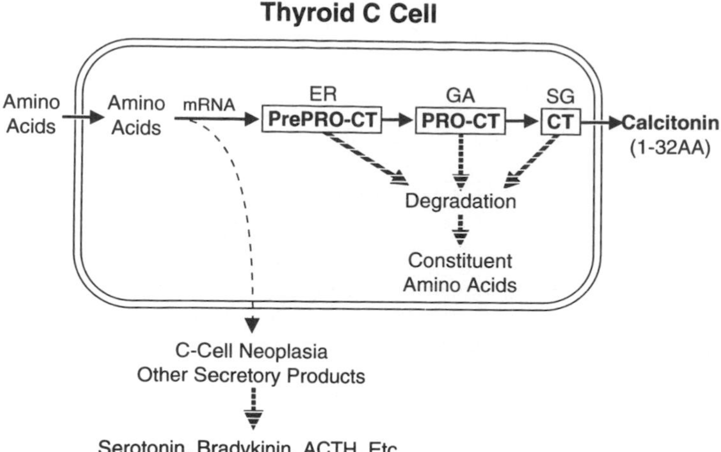

FIGURE 23.18 Biosynthesis of calcitonin (CT) by thyroid C cells. ER, endoplasmic reticulum; GA, Golgi apparatus; SG, secretory granule.

membered ring enclosed by an intrachain disulfide bridge. The complete sequence of 32 amino acids and the disulfide bond are essential for full biologic activity. Salmon calcitonin is more potent in lowering blood calcium on a weight basis than any of the other calcitonins when administered to mammals, including man. The reason for the greater biologic potency of salmon calcitonin in mammals is uncertain, but probably is related to an increased resistance to metabolic degradation or a greater affinity for receptor sites in bone and other target tissues.

3. Regulation of Secretion

The concentration of calcium ion in plasma and extracellular fluids is the principal physiologic stimulus for the secretion of calcitonin by C cells (Care, 1992). Calcitonin is secreted continuously under conditions of normocalcemia, but the rate of secretion of calcitonin increases greatly in response to an elevation in blood calcium. Magnesium ion has an effect on calcitonin secretion similar to that of calcium, but these effects are observed only under experimental conditions with nonphysiologic levels of magnesium. When blood Ca^{2+} concentration increases, the intracellular Ca^{2+} concentration in C cells increases, resulting in enhanced calcitonin secretion. C cells express the same Ca^{2+}-sensing receptor as parathyroid chief cells; the receptor is responsible for sensing the extracellular Ca^{2+} concentration and likely contributes to the regulation of calcitonin secretion along with a voltage-sensitive Ca^{2+} channel (Chattopadhyay *et al.*, 1996).

C cells store substantial amounts of calcitonin in their cytoplasm in the form of membrane-limited secretory granules (Fig. 23.6). In response to hypercalcemia, stored hormone from C cells is rapidly discharged into interfollicular capillaries (Capen and Young, 1969). If a hypercalcemic stimulus is sustained, this is followed by an increased development of cytoplasmic organelles concerned with the synthesis and secretion of calcitonin. The endoplasmic reticulum with attached ribosomes is hypertrophied, and the Golgi apparatus is enlarged and associated with prosecretory granules in the process of formation. Hyperplasia of C cells occurs in response to long-term hypercalcemia (Capen and Young, 1969; Collins *et al.*, 1977). When the blood

```
Canine    C S N L S T C V L G T Y S K D L N N F H T F S G I G F G A E T P
Bovine    . . . . . . . . . S A . W . . . . . Y . R . . . M . . . P . . .
Ovine     . . . . . . . . . S A . W . . . . . Y . R Y . . M . . . P . . .
Porcine   . . . . . . . . . S A . W R B . . . . . R . . . M . . . P . . .
Human     . G . . . . . M . . . . T Q . F . K . . . . P Q T A I . V G A .
Rat       . G . . . . . M . . . . T Q . . . K . . . . P Q T S I . V G A .
Chicken   . A S . . . . . . . K L . Q E . H K L Q . Y P R T B V . S G . .
Salmon    . . . . . . . . . . K L . Q E . H K L Q . Y P R T N T . S G . .
```

FIGURE 23.19 Amino acid sequences of calcitonin in different species. Amino acids 1, 4–7, 9, 28, and 32 are shared by all species. The cysteines at positions 1 and 7 form the N-terminal disulfide bridge. Notice the similarities between ovine, bovine and porcine calcitonin, human and rat calcitonin, and chicken and salmon calcitonin. Identical amino acids are represented by dots. Modified from Mol *et al.* (1991).

calcium is lowered, the stimulus for calcitonin secretion is diminished and numerous secretory granules accumulate in the cytoplasm of C cells (Fig. 23.6). The storage of large amounts of preformed hormone in C cells and rapid release in response to moderate elevations in blood calcium probably are a reflection of the physiologic role of calcitonin as an emergency hormone to protect against the development of hypercalcemia.

Calcitonin secretion is increased in response to a high-calcium meal often before a significant rise in plasma calcium can be detected. The cause of this increase in calcitonin secretion could be either a small undetectable rise in plasma ionized calcium or a direct stimulation of certain gastrointestinal hormones by the oral calcium load; the hormones in turn act as secretagogues for calcitonin release from the thyroid gland (Cooper *et al.*, 1972). Gastrin, pancreozymin, and glucagon all have been demonstrated to stimulate calcitonin release under experimental conditions in animals (Care *et al.*, 1970). These findings suggest that gastrointestinal hormones are important in triggering the early release of calcitonin to prevent the development of hypercalcemia following the ingestion of a high-calcium meal (Fig. 23.20) (Care, 1992).

4. *Biological Action*

The administration of calcitonin or stimulation of endogenous secretion results in the development of varying degrees of hypocalcemia and hypophosphatemia. These effects of calcitonin on plasma calcium and phosphorus are most evident in young animals or older

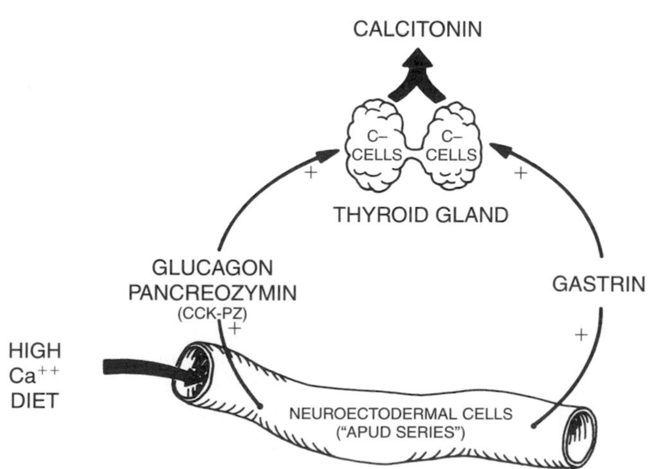

FIGURE 23.20 The gastrointestinal hormone–thyroid C-cell axis provides a mechanism for rapid release of calcitonin from the thyroid in response to a high-calcium diet before a significant elevation in blood calcium occurs. (Capen, 1989.)

animals with increased rates of skeletal turnover. Calcitonin exerts its function by interacting with target cells, primarily in bone and kidney and to a lesser extent in the intestine. The action of PTH and CT is antagonistic on bone resorption (Fig. 23.11) but synergistic on decreasing the renal tubular reabsorption of phosphorus (Fig. 23.12). The hypocalcemic effects of calcitonin are primarily the result of decreased entry of calcium from the skeleton into plasma due to a temporary inhibition of PTH-stimulated bone resorption (Heersche, 1992). The hypophosphatemia develops from a direct action of calcitonin on increasing the rate of movement of phosphate out of plasma into soft tissue and bone (Talmage *et al.*, 1972), as well as from the inhibition of bone resorption. The action of CT is not dependent on vitamin D, because it acts both in vitamin D-deficient animals and after the administration of large doses of vitamin D.

The action of calcitonin on inhibiting bone resorption stimulated by PTH and other factors is from blockage of osteoclastic osteolysis (Fig. 23.11). Specific structural alterations are produced in osteoclasts by calcitonin (Chambers and Moore, 1983). Osteoclasts withdraw from resorptive surfaces and the brush border and transitional zone become atrophic (Weisbrode and Capen, 1974; Matthews, 1992). In addition, the rate of activation of osteoprogenitor cells to preosteoclasts and osteoclasts decreases, resulting in decreased osteoclast number. Although CT can block bone resorption completely, the inhibition is a transitory effect. The continuous administration of CT *in vivo* and *in vitro* in the presence of PTH leads to an "escape phenomenon" whereby the effects of PTH on increasing bone resorption become manifest in the presence of CT (Heersche, 1992).

The effects of calcitonin on bone formation are less dramatic. Initially, there appears to be an increase in the rate of bone formation, but the long-term administration of calcitonin appears to lead to a reduction in both bone resorption and formation. Calcitonin-secreting C-cell neoplasms in both bulls and human beings are associated with a low rate of skeletal turnover and densely mineralized bone (Black *et al.*, 1973b). Unfortunately, calcitonin has not proven to be an effective therapeutic agent to induce a long-term increase in bone formation in human patients with postmenopausal osteoporosis. Calcitonin can partially ameliorate the increase in bone resorption that occurs in certain forms of osteoporosis (Lyles *et al.*, 1993; Copp, 1994) or can be used to treat hypercalcemia for a short duration (Dougherty *et al.*, 1990).

Calcitonin and PTH both decrease renal tubular reabsorption of phosphate to cause phosphaturia; however, the adenylate cyclase-linked receptors for calcitonin are found in the ascending limb of Henle and the distal convoluted tubule (Fig. 23.12). In addition, calcitonin results in diuresis of sodium, chloride, and calcium. The physiologic significance of calcitonin on renal electrolyte excretion in mammals is uncertain, and this function as a regulator of ionic balance may be more important in primitive vertebrates such as fish. Likewise, the physiologic importance of calcitonin on gastrointestinal absorption of calcium is equivocal.

5. *Physiological Significance*

Calcitonin and parathyroid hormone acting in concert provide a dual negative-feedback control mechanism to maintain the concentration of calcium in extracellular fluids within narrow limits. Present evidence suggests that parathyroid hormone is the major factor concerned with the minute-to-minute regulation of blood calcium under normal conditions. This probably is related to the fact that protection against the development of hypocalcemia by PTH in most higher mammals living in a relatively low calcium–high phosphorus environment is a life-sustaining function. Calcitonin functions more as an emergency hormone to (1) prevent the development of "physiological" hypercalcemia during the rapid postprandial absorption of calcium, and (2) protect against excessive loss of calcium and phosphorus from the maternal skeleton during pregnancy. For example, thyroidectomized cows without a source of calcitonin develop significant hypercalcemia following a high-calcium meal (40 g/day) compared to intact control cows fed a similar diet (Barlet, 1972).

6. *Calcitonin Assays*

Serum calcitonin is best measured by radioimmunoassay. Because of the low degree of homology of calcitonin between species, there is poor cross reactivity of RIAs for calcitonin. There is relatively little need to measure calcitonin in clinical veterinary medicine because of the scarcity of calcitonin-secreting neoplasms or clinical disorders resulting from abnormal blood levels of calcitonin. In humans, RIA for calcitonin is useful to identify functional C-cell (medullary) tumors of the thyroid gland (Ziegler *et al.*, 1996). RIAs for calcitonin are necessary also for experimental investigations of regulation of C cells *in vivo* or *in vitro*. Since the sequencing of canine calcitonin, a canine-specific RIA has been developed and has been used to measure serum calcitonin in dogs (Hazewinkel, 1991).

D. Cholecalciferol (Vitamin D)

1. *Introduction*

The third major hormone involved in the regulation of calcium metabolism and skeletal remodeling is cholecalciferol (vitamin D_3) or irradiated ergosterol (vitamin D_2). Although these compounds have been considered to be vitamins for a long time, they can equally be considered hormones (Bell, 1985). Cholecalciferol is ingested in small amounts in the diet and can be synthesized in the epidermis from precursor molecules (e.g., 7-dehydrocholesterol) through a previtamin D_3 intermediate form (Fig. 23.21). This reaction is catalyzed by ultraviolet irradiation (wavelength 290 to 320 μm) from the sun (Holick, 1981). A high-affinity vitamin D-binding protein (DBP) transports cholecalciferol from its site of synthesis in the skin by the bloodstream to the liver (Holick and Clark, 1978). Carnivores, such as the dog and cat, may be less able to synthesize vitamin D_3 in the skin and are more dependent on dietary sources of vitamin D compared to herbivores and omnivores (How *et al.*, 1994).

In response to prolonged exposure to sunlight, previtamin D_3 is converted to lumisterol and tachysterol (Fig. 23.21). Because the DBP has no affinity for lumisterol and minimal affinity for tachysterol, the translocation of these photoisomers into the circulation is negligible and they are sloughed off with the natural turnover of the skin.

2. *Metabolic Activation*

Vitamin D must be metabolically activated before it can produce its known physiological functions in target cells. Vitamin D_3 from dietary sources is absorbed by facilitated diffusion and bound to an α-2-globulin in the blood for transport. Cholecalciferol from endogenous sources is synthesized in the skin from 7-dehydrocholesterol by a photochemical reaction in the presence of ultraviolet irradiation (Fig. 23.22) (Holick and Clark, 1978). It is bound to a vitamin D-binding protein in the blood and transported to the liver.

The first step in the metabolic activation of vitamin D is the conversion of cholecalciferol to 25-hydroxycholecalciferol (25-OH-CC) in the liver. The enzyme responsible for controlling this reaction is a hepatic microsomal enzyme, referred to as calciferol-25-hydroxylase, associated with the endoplasmic reticulum (Armbrecht *et al.*, 1992). Although 25-OH-CC can exert biologic effects when substantial amounts are present, it primarily serves as a precursor for the formation of more active vitamin D metabolites. Consid-

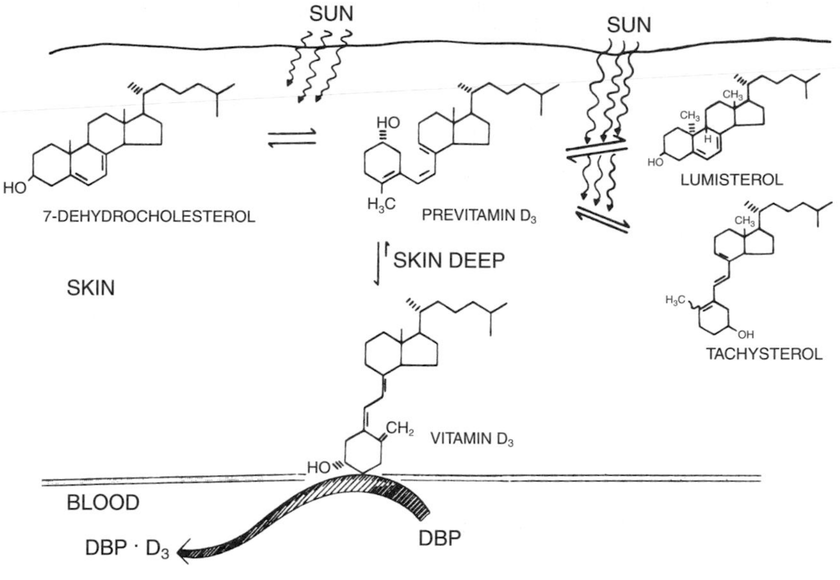

FIGURE 23.21 Prolonged exposure to sunlight results in the photochemical conversion of previtamin D_3 to lumisterol and tachysterol. These photoisomers remain primarily in the epidermis and are lost with the natural turnover of the skin because the vitamin D_3-binding protein has a relatively low affinity for them. From Holick (1981).

erably larger amounts of protein-bound 25-OH-CC circulate than with the more hydroxylated metabolites, such as 1,25-$(OH)_2D$, which are present in extremely low levels in the blood.

This first metabolite of cholecalciferol (25-OH-CC) is transported to the kidney and undergoes further transformation to a more polar and active metabolite (Fig. 23.22) (DeLuca, 1973; Armbrecht *et al.*, 1992). The principal active metabolite of 25-OH-CC formed in the kidney is 1,25-$(OH)_2D$, but other metabolites are formed, such as 24,25$(OH)_2D$. The

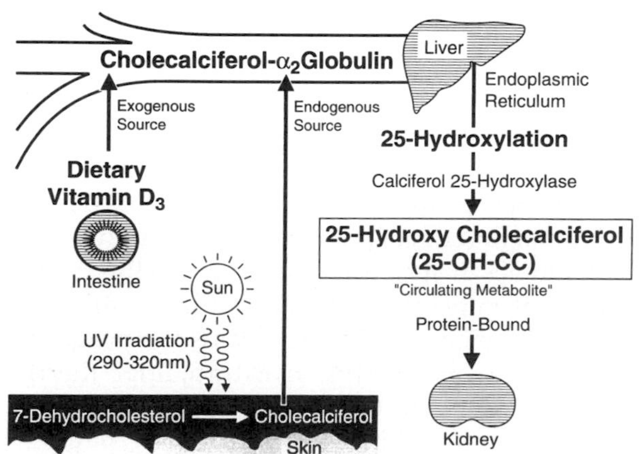

FIGURE 23.22 Metabolism of vitamin D. The initial step of metabolic activation of vitamin D_3 from endogenous and dietary sources is in the liver to form 25-hydroxycholecalciferol.

rate of formation of 1,25-$(OH)_2D$ is catalyzed by 25-dihydroxycholecalciferol-1α-hydroxylase in mitochondria of renal epithelial cells in the proximal convoluted tubules. It has been shown that the 1α-hydroxylase enzyme system is a mixed-function steroid hydroxylase similar to those in the adrenal cortex, contains cytochrome P450, and requires molecular oxygen. 1,25-Dihydroxycholecalciferol exerts strong feedback inhibition on the 1α-hydroxylase in the kidney. The conversion of 25-OH-CC to 1,25-$(OH)_2D$ is the rate-limiting step in vitamin D metabolism and is primarily the reason for the delay between administration of vitamin D and expression of its biological effects.

The control of this final step in the metabolic activation of vitamin D is complex and is regulated by the plasma calcium concentration and its influence on the rates of secretion of PTH and possibly CT (Hulter *et al.*, 1985). The two other major hormones controlling calcium metabolism appear to have opposite effects on this 1α-hydroxylase and formation of 1,25-$(OH)_2D$ (Armbrecht *et al.*, 1992). Parathyroid hormone and conditions that stimulate PTH secretion increase the transformation of 25-OH-CC to 1,25-$(OH)_2D$, whereas calcitonin under certain conditions inhibits the conversion. In addition, a low concentration of phosphorus in the blood increases the formation of 1,25-$(OH)_2D$, whereas a high concentration suppresses the activity of the 1α-hydroxylase. Minute-to-minute changes in 1α-hydroxylase activity may result from changes in the ionic environment of renal mitochondria caused by the accumulation or release of calcium or inorganic phosphate.

The rates of synthesis of 24,25-$(OH)_2D$ and 1,25-$(OH)_2D$ are reciprocally related and controlled by similar factors (Tanaka and DeLuca, 1974). When 1,25-$(OH)_2D$ synthesis increases, the synthesis of 24,25-$(OH)_2D$ declines, and vice versa. For example, feeding a high-calcium diet decreases the formation of 1,25-$(OH)_2D$, and 25-OH-CC is diverted to form primarily 24,25-$(OH)_2D$, a metabolite that is much less active in stimulating intestinal calcium transport. 24,25-$(OH)_2D$ may play a role in bone formation (Ornoy *et al.*, 1978) and egg hatchability (Henry and Norman, 1978), and, with 1,25-$(OH)_2D$, may exert negative feedback control on the parathyroid gland (Fig. 23.23).

Other hormones may increase the activity of renal 1α-hydroxylase and the formation of 1,25-$(OH)_2D$. Prolactin, estradiol, placental lactogen, and growth hormone enhance 1α-hydroxylase activity (Reichel *et al.*, 1989). Increased secretion of these hormones, either alone or in combination, is important in efficient adaptation to the major calcium demands during life. The physiological adjustments in calcium homeostasis imposed by pregnancy, lactation, and growth may be mediated primarily by an increased intestinal absorption of calcium stimulated by 1,25-$(OH)_2D$.

1,25-Dihydroxycholecalciferol is the major biologically active metabolite of cholecalciferol. It interacts with target cells in the intestine and bone under physiological conditions to enhance the rates of existing reactions and increase calcium mobilization. Its onset of action is more rapid and the degree of potency is much greater than those of either cholecalciferol or 25-OH-CC. A similar two-step process of metabolic activation also occurs with irradiated ergosterol (vitamin D_2).

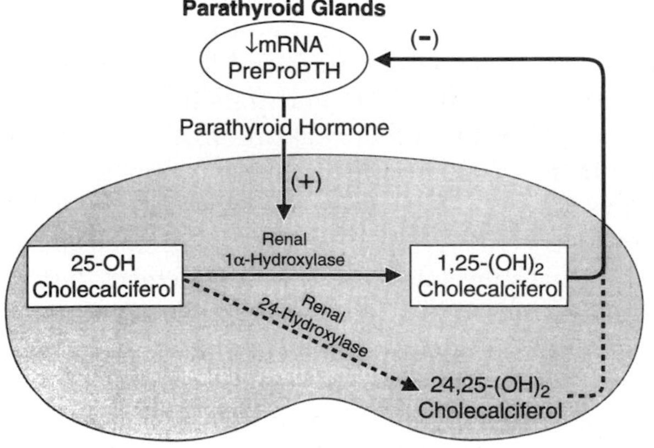

FIGURE 23.23 Negative feedback exerted by vitamin D metabolites on parathyroid chief cells to decrease the rate of parathyroid hormone secretion, which in turn diminishes the rate of formation of additional 1,25-dihydroxycholecalciferol.

3. *Chemistry*

The chemical structure of cholecalciferol (vitamin D_3) resembles that of other steroid hormones. It is a seco-steroid in which the B ring of the basic steroid nucleus has been opened by breakage of the 9-carbon–10-carbon bond. Photoactivation by ultraviolet irradiation (290 to 320 μm) of 7-dehydrocholesterol in epidermis of the skin results in a cleavage between the 9- and 10-carbons and unfolding of the B ring of the basic steroid nucleus. During metabolic activation of cholecalciferol (or irradiated ergosterol), hydroxyl groups are successively attached to the steroid nucleus by specific hydroxylases at positions 1 and 25 in the liver and kidney to form the hormonal (biologically active) form of vitamin D. In other natural metabolites of cholecalciferol, the two hydroxyl groups are attached at either 24 or 25 and 26 positions.

There are a number of other sterols closely related to cholecalciferol. Vitamin D_2 is formed by the irradiation of the plant sterol referred to as ergosterol. When irradiated ergosterol is ingested and absorbed from the intestine, it undergoes a series of steps of metabolic activation similar to those described for cholecalciferol (vitamin D_3). However, the more active metabolites of ergosterol are 25-hydroxyergosterol and 1,25-dihydroxyergosterol, synthesized by the liver and kidney, respectively. Another related sterol of considerable therapeutic interest is dihydrotachysterol. The A ring of the steroid nucleus in this compound is rotated so that the hydroxyl in position 3 occupies a position sterically equivalent to the hydroxyl position 1 of 1,25-$(OH)_2D$. Dihydroxytachysterol undergoes metabolic transformation to 25-dihydrotachysterol, but subsequent hydroxylation of position 1 does not occur.

4. *Biological Action*

Vitamin D and its active metabolites function to increase the absorption of calcium and phosphorus from the intestine, thereby maintaining adequate levels of these electrolytes in the extracellular fluids in order to permit the appropriate mineralization of bone matrix. From a functional point of view vitamin D can be thought to act in such a way as to cause the retention of sufficient mineral ions to ensure mineralization of bone matrix, whereas PTH maintains the proper ratio of calcium to phosphorus in extracellular fluids. In addition, a small amount of vitamin D is needed to permit PTH to exert its action on bone ("permissive effect").

The major target tissue for 1,25-dihydroxycholecalciferol is the mucosa of the small intestine. In the proximal part it increases the active transcellular transport of calcium, and in the distal part the transport of phosphorus. Following synthesis in the kidney, 1,25-$(OH)_2D$ is transported in a protein-bound form to specific target cells in the intestine and bone (Fig. 23.24). Circulating levels of this hormonal form of cholecalciferol are extremely low. Free 1,25-$(OH)_2D$ penetrates the plasma membrane of target cells and initially binds to a cytoplasmic receptor in cells of the intestine (Fig. 23.24) (Walters, 1992). Subsequently, the hormone–receptor complex is transferred to the nucleus and 1,25-$(OH)_2D$ binds to specific receptors in the nuclear chromatin, where it stimulates gene expression leading to increased synthesis of vitamin D-dependent proteins such as calcium-binding protein (CaBP) (calbindin) by intestinal cells (Christakos *et al.*, 1989) (Fig. 23.24). Calbindin has been isolated from several tissues (e.g., small intestine, kidney, placenta, bone, uterus, nervous system, and the shell gland of laying hens) across which significant amounts of calcium are transported.

The absorptive capacity of the intestine for calcium is a direct function of the amount of CaBP present (Wasserman and Fullmer, 1983). Administering vitamin D or feeding low-calcium and low-phosphorus diets has been shown to stimulate the synthesis of CaBP, which contributes to the increased intestinal absorption of calcium. The physiological functions of CaBP are related to the transcellular transport of calcium from the luminal to the basilar border of intestinal absorptive cells and the regulation of intracellular calcium concentration (Fig. 23.24). At the basilar aspect of intestinal absorptive cells, calcium is exchanged for sodium and enters the extracellular fluids.

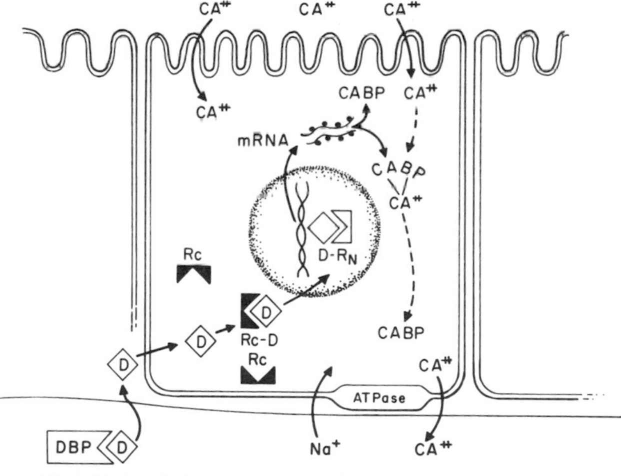

FIGURE 23.24 Molecular mechanisms of action of 1,25-dihydroxycholecalciferol in the intestine. The active metabolite of vitamin D is transported to the intestine by a vitamin D-binding protein (DBP). The hydrophilic steroid penetrates the plasma membrane, binds to a cytoplasmic receptor, and is transported to the nucleus, where it interacts with the nuclear chromatin to increase the formation of mRNA. The mRNA becomes associated with ribosomes on the endoplasmic reticulum and directs the synthesis of new proteins such as calcium-binding protein (CABP, calbindin). The CABP is involved in the transcellular transport of calcium to the basilar aspects of the intestinal absorptive cells, where calcium is exchanged for sodium and enters the extracellular fluid compartment. (Capen, 1989.)

The active metabolites of cholecalciferol also act on bone (High *et al.*, 1981b; Finkelman and Butler, 1985). In addition to its indirect effect on mineralization of bone matrix, vitamin D is necessary for osteoclastic resorption and calcium mobilization from bone. Small amounts of vitamin D or its active metabolite are necessary to permit osteolytic cells to respond to PTH ("permissive effect") under physiologic conditions. 1,25-$(OH)_2D$, 25-OH-CC, and cholecalciferol in pharmacologic doses will stimulate osteoclastic proliferation and the resorption of bone *in vitro* and *in vivo*. 1,25-Dihydroxycholecalciferol is about 100 times more potent on a weight basis in stimulating bone resorption *in vitro* than is 25-OH-CC (Reynolds *et al.*, 1973).

Less is known regarding the action of cholecalciferol and its active metabolites on the kidney. Present evidence suggests that active metabolites of vitamin D stimulate the retention of calcium and phosphorus by increasing proximal tubular reabsorption.

The active metabolites of vitamin D also have a direct effect on the parathyroid gland, in addition to their well-characterized action on intestine and bone. The parathyroids selectively localize 1,25-$(OH)_2D$ and contain specific cytoplasmic and nuclear receptors for the active metabolite of vitamin D (Hughes and Haussler, 1978). Therefore, a negative feedback loop exists whereby vitamin D metabolites (either alone or in combination) directly interact with parathyroid cells to diminish the secretion of PTH, which in turn diminishes the formation of 1,25-$(OH)_2D$ (Fig. 23.23).

It has been recognized that the active metabolites of vitamin D, in particular 1,25-dihydroxycholecalciferol, have many roles in health and disease beyond the regulation of calcium metabolism (Reichel *et al.*, 1989; Walters, 1992). The vitamin D receptor is widespread in most tissues (including hematopoietic cells, muscle, reproductive system, nervous system, endocrine glands, gastrointestinal tract, urinary tract, lungs, and skin), and vitamin D functions to regulate cell growth and differentiation. In general, vitamin D inhibits cell growth and stimulates cell differentiation, but there are exceptions to this paradigm. For these reasons, 1,25-dihydroxycholecalciferol has potential therapeutic value for treatment of psoriasis and certain forms of cancer, among other diseases. Hypercalcemia due to increased intestinal Ca absorption and increased osteoclastic bone resorption are significant side effects of therapy using the active forms of vitamin D. Therefore, there is considerable interest in developing vitamin D analogs that minimize the calcemic actions of vitamin D, but retain the ability of vitamin D to alter cell growth or differentiation (Bikle, 1992).

5. *Toxicosis*

A disease of cattle known as "enteque seco" in the Argentine and "espichamento" in Brazil is characterized by widespread mineralization of the cardiovascular system, lung, kidney, and other soft tissues (Worker and Carrillo, 1967; Done *et al.*, 1976). A similar disease affects cattle and horses in many other countries worldwide, including the United States. Affected cattle develop a chronic debilitating disease characterized by weight loss, painful gait, kyphosis, polyuria, and dyspnea. The lesions of soft tissue mineralization are remarkably similar to those produced experimentally with large doses of vitamin D (Capen *et al.*, 1966). Studies have shown that ingestion of small amounts of dried leaves from an indigenous plant (*Solanum malacoxylon*) greatly increased the rate of intestinal calcium and phosphorus absorption to produce hypercalcemia and hyperphosphatemia, osteosclerosis, parathyroid atrophy, and hyperplasia of thyroid C cells. Extracts of the leaves of *S. malacoxylon* contain an extremely potent, water-soluble active principle that stimulates the intestinal calcium transport system (de Boland *et al.*, 1976). The active principle has been shown to be 1,25-$(OH)_2$D conjugated to a glycoside. *Cestrum diurnum* ("day-blooming jessamine") also contains substantial amounts of 1,25-$(OH)_2$D and has caused a similar debilitating disease in cattle and horses in the southeastern part of the United States (Wasserman *et al.*, 1975). See also the section on hypercalcemia due to vitamin D or vitamin D metabolites.

6. *Assays for Vitamin D Metabolites*

Most assays for vitamin D and its metabolites employ competitive protein or radioreceptor techniques using vitamin D binding proteins isolated from tissues and tritium-labeled tracers; however, newer ^{125}I-based radioimmunoassays have been developed (Hollis, 1986; Iqbal, 1994; Hollis *et al.*, 1996). 25-Hydroxyvitamin D and 1,25-dihydroxyvitamin D are the most frequently measured metabolites. Because the structures of the common vitamin D metabolites do not vary between species, the same assays can be used to measure serum metabolites in animals and humans. 1,25-Dihydroxyvitamin D is the most potent metabolite in regulating calcium balance; therefore, its measurement is often necessary to understand abnormal calcium regulation. The circulating concentration of 1,25-dihydroxyvitamin D is approximately 25 pg/ml in adult animals and is greater in younger animals (Table 23.3) (Nagode and Chew, 1991). Measurement of serum 25-hydroxyvitamin D is useful for evaluating the potential for dietary vitamin D deficiency or excess, because serum levels are much greater and less tightly controlled compared to those of 1,25-dihydroxyvitamin D. Normal serum concentrations of 25-hydroxyvitamin D are 10–50 ng/ml (Dougherty *et al.*, 1990). Animals that have ingested excess vitamin D may have serum 25-hydroxyvitamin D concentrations 4–20-fold greater than normal (Carothers *et al.*, 1994).

III. PHOSPHATE METABOLISM

A. Introduction

Phosphate in the mammalian body is present predominantly (90%) as hydroxyapatite $[Ca_{10}(PO_4)_6(OH)_2]$ in the mineralized matrix of bone, with most of the remaining 10% occurring intracellularly in soft tissues. Phosphate is the major intracellular anion existing in organic (e.g., phospholipids, nucleic acids, phosphoproteins, ATP) or inorganic forms and plays an integral role in many metabolic processes (e.g., energy metabolism, delivery of O_2 to tissues, muscle contraction, and skeletal integrity) (Dennis, 1996). Adequate fluorometric or ion-sensitive methods to measure intracellular phosphate are lacking. Rapid translocations can occur between intracellular and serum phosphate pools, which can dramatically change serum phosphate concentrations. In nonruminant animals on normal diets with adequate amounts of vitamin D, the kidneys are the major regulators of serum phosphate concentration (Yanagawa and Lee, 1992).

B. Serum Phosphate

Serum phosphate is measured as inorganic orthophosphate because the organic forms are not routinely evaluated. Methods for colorimetric determination of phosphate rely on the formation of a complex of phosphate ion with molybdate (Fiske and Subbarow, 1925; Fraser *et al.*, 1987). Although inorganic phosphate is measured, it is often expressed as elemental phosphorus (P_i). The atomic weight of phosphorus is 31, so 3.1 mg P_i/dl serum is equivalent to 1 mmol/liter phosphorus or 1 mmol/liter phosphate. Serum P_i ranges from 2.5 to 6.0 mg/dl (0.8–1.9 mmol/liter) in adult animals; however, serum P_i concentrations are greater in young animals (especially giant-breed dogs) and may be outside the normal range for adult animals. Most serum inorganic phosphate (80%) is in the dibasic form, HPO_4^{2-}, and the remaining 20% is primarily in the monobasic form, $H_2PO_4^-$. This results in an average valence of serum inorganic phosphate as −1.8, and the milliequivalence of serum phosphate can be estimated by the following calculation: 1 mmol/liter phosphate = 1.8 meq/liter phosphate.

It is important to prevent hemolysis of blood samples, which will artificially increase the measurement of serum phosphate because of the release of intracellular stores. Phosphate circulates as a free anion bound to Na^+, Mg^{2+}, or Ca^{2+}, or bound to protein (10–20% of total serum phosphate). Serum phosphate is an unreliable indicator of body stores and may be higher in younger animals because growth hormone increases renal phosphate reabsorption. Feeding of high-carbohydrate diets or glucose infusions will decrease serum phosphate because of a shift of phosphate intracellularly in response to increased glycolysis and the need for phosphorylated intermediates. High-meat diets may increase serum phosphate as a result of their high phosphate content.

C. Intestinal Phosphate Absorption

Absorption of dietary phosphate is approximately 60–70% and occurs by active transport using a Na/Phosphate cotransporter and by passive diffusion (Favus, 1992; Civitelli and Avioli, 1994). In ruminants the transporter may be coupled to H^+ rather than Na^+ (Shirazi-Beechey *et al.*, 1996). Absorption takes place principally in the forestomachs (ruminants) and duodenum and jejunum (monogastric animals and ruminants) (Care, 1994). Horses may absorb some phosphate from the large intestine (Barlet *et al.*, 1995). The active form of vitamin D, 1,25-dihydroxyvitamin D, increases intestinal phosphate absorption. Low dietary levels of phosphate result in adaptive changes in the intestine that result in increased net phosphate absorption (Schröder *et al.*, 1995). In addition, increased renal production of 1,25-dihydroxyvitamin D and adaption of the kidney to increase renal phosphate reabsorption can compensate for low levels of dietary phosphate.

The source of dietary phosphorus affects its availability (Soares, 1995). Inorganic sources of phosphate and bone and meat meals have high bioavailability (95%). Some diets contain substances or nutrients that can antagonize phosphate absorption, including aluminum and magnesium. Diets high in calcium and fat raise the requirement for phosphorus (Gerloff and Swenson, 1996). Most phosphorus in concentrate sources is in the organic form of phytate, which is poorly utilized in nonruminants. Ruminants have phytase in their rumens to release the phosphate from the sugar skeleton, but nonruminants do not. However, addition of microbial phytase to concentrate diets increases phosphorus bioavailablity for nonruminants. Phytates from different foodstuffs will vary in their digestibility for ruminants.

In ruminants, large amounts of endogenous phosphate enter the gastrointestinal tract from salivary secretions (Care, 1994; Barlet *et al.*, 1995; Shirazi-Beechey *et al.*, 1996). Parotid saliva in ruminants is 16–40 mmol/liter phosphate, and total phosphate secretion by the salivary glands is significantly greater than the dietary supply of phosphate. The endogenous salivary secretion of phosphate is complemented by intestinal phosphate absorption and massive recycling of phosphate, resulting in greater phosphate absorption in the gastrointestinal tracts of ruminants than in those of monogastric animals. Phosphate plays a role as a buffer for volatile fatty acids and as a nutrient for microorganisms in the rumen. High-fiber diets that increase saliva production also increase the total salivary secretion of phosphate. Increased endogenous salivary secretion of phosphate will lead to increased intestinal absorption, but also to increased endogenous fecal loss, resulting in a net loss of phosphate. Ruminants fed a high-roughage diet use endogenous fecal loss of phosphate as the principal mechanism to regulate phosphate excretion. In contrast, ruminants on concentrate diets excrete more phosphate in the urine because of reductions in both saliva flow and endogenous fecal loss. Therefore, the quantity of saliva and regulation of saliva's phosphate concentration are important determinants of phosphate excretion in ruminants.

D. Phosphate Excretion

Renal excretion of phosphate is determined by the glomerular filtration rate (GFR) and the maximum tubular reabsorption rate (TmP) (Yanagawa and Lee, 1992). The majority of renal phosphate reabsorption occurs in the proximal convoluted tubule with small amounts in the distal nephron. Reabsorption is sodium dependent because phosphate transport is performed by a brush-border Na–phosphate cotransporter (Biber *et al.*, 1996). The sodium gradient is maintained by a Na/K ATPase, and therefore phosphate reabsorption is indirectly energy dependent. The regulation of renal phosphate reabsorption can adapt to the body's need for phosphate. Reabsorption is increased with growth, lactation, pregnancy, and low-phosphate diets and is decreased during slow growth, renal failure, or excess dietary phosphorus (Lötscher *et al.*, 1996; Silverstein *et al.*, 1996). The major hormonal regulator is parathyroid hormone, which decreases the TmP and increases renal phosphate excretion (Murer *et al.*, 1996). Other hormones that inhibit Na/P_i cotransport include calcitonin, atrial natriuretic peptide, epidermal growth factor, transforming growth factors α and β, and parathyroid hormone-related protein. In contrast, insulin, growth hormone, and insulin-like growth factor I stimulate Na/P_i cotransport by renal epithelial cells (Kempson, 1996).

E. Hypophosphatemia

Hypophosphatemia may impair red and white blood-cell function and lead to muscle weakness, recumbency, intestinal ileus, anorexia, or vomition (DiBartola, 1992; Hodgson and Hurley, 1993; Gerloff and Swenson, 1996). Red blood-cell fragility and hemolysis can be increased by the lack of ATP in the erythrocyte due to the necessity of phosphate for the erythrocytic glycolytic pathway. Intravascular hemolysis is uncommon, but the occurrence of hypophosphatemia in cattle with postparturient hemoglobinuria may predispose erythrocytes to hemolysis when the animals are exposed to hemolytic agents in plants. The production of 2,3-diphosphoglycerate (2,3-DPG) is dependent on phosphate. Erythrocytes with low 2,3-DPG concentrations have increased binding affinity for O_2 and reduced delivery of O_2 to peripheral tissues. This may be responsible for muscle weakness in patients with hypophosphatemia. White blood cells may have decreased chemotaxis and phagocytosis also due to a lack of ATP during hypophosphatemia. The causes of hypophosphatemia include maldistribution secondary to a large carbohydrate load, respiratory alkalosis, metabolic acidosis, catecholamine release, or insulin treatment of an animal with diabetes mellitus. Hypophosphatemia also can be caused by reduced renal reabsorption in animals with primary hyperparathyroidism or renal tubular defects, reduced intestinal absorption during vitamin D deficiency, or oncogenic osteomalacia (see metabolic disease section).

F. Hyperphosphatemia

Hyperphosphatemia leads to a reciprocal reduction in serum ionized calcium concentration because of the mass law interactions between phosphate and calcium ions and because of decreased renal 1,25-hydroxyvitamin D synthesis by the kidney. The clinical signs of acute hyperphosphatemia may be due to hypocalcemia and consist of tetany and soft-tissue mineralization (especially when the serum calcium $\times$ phosphorus product is greater than 70 mg/dl). The principal causes of hyperphosphatemia include massive cellular lysis (e.g., during chemotherapy, rhabdomyolysis, hemolysis), vitamin D intoxication, chronic renal failure (see section on secondary renal hyperparathyroidism), hypoparathyroidism, hypersomatotropism, and hyperthyroidism (Chew and Meuten, 1982; DiBartola, 1992; Yanagawa and Lee, 1992; Brooks, 1995).

IV. METABOLIC DISEASES OF ABNORMAL CALCIUM/PHOSPHORUS METABOLISM

A. Hyperparathyroidism

1. Secondary Renal Hyperparathyroidism

Secondary hyperparathyroidism as a complication of chronic renal failure (CRF) is a metabolic disease characterized by an excessive, but not autonomous, rate of PTH secretion. This disorder is encountered most frequently in dogs and cats but also occurs in other animal species. The secretion of hormone by the hyperplastic parathyroid glands in this disorder usually remains responsive to fluctuations in blood calcium.

The primary etiologic mechanism in this form of hyperparathyroidism is a long-standing, progressive renal disease resulting in severely impaired function. Chronic renal insufficiency in older animals results from interstitial nephritis, glomerulonephritis, nephrosclerosis, or amyloidosis. Several congenital anomalies such as cortical hypoplasia, polycystic kidneys, and bilateral hydronephrosis may result in renal insufficiency in younger animals. When the renal disease progresses to the point at which there is significant reduction in glomerular filtration rate, phosphorus is retained and progressive hyperphosphatemia develops (Fig. 23.25).

The pathogenesis of secondary renal hyperparathyroidism is complex, but it has been recently recognized that 1,25-dihydroxyvitamin D (calcitriol) plays an important role in this disease (Fig. 23.26) (Nagode and Chew, 1992; Felsenfeld and Llach, 1993). Animals with advanced CRF have decreased circulating concentrations of 1,25-dihydroxyvitamin D caused by decreased renal synthesis. Initially, the elevated blood phosphorus depresses the activity of the 1α-hydroxylase in the kidney. In the later stage of chronic renal disease there are decreased numbers of tubular epithelial cells with 1α-hydroxylase activity to produce the active form of vitamin D. Early stages of CRF are often associated with normal concentrations of circulating 1,25-dihydroxyvitamin D (even with the loss of nephrons) because of the correcting effects of increased PTH on renal 1α-hydroxylase that enhance renal tubular synthesis of 1,25-dihydroxyvitamin D.

1,25-Dihydroxyvitamin D is an important regulator of parathyroid chief cell function and acts to decrease PTH mRNA expression, increase expression of the vita-

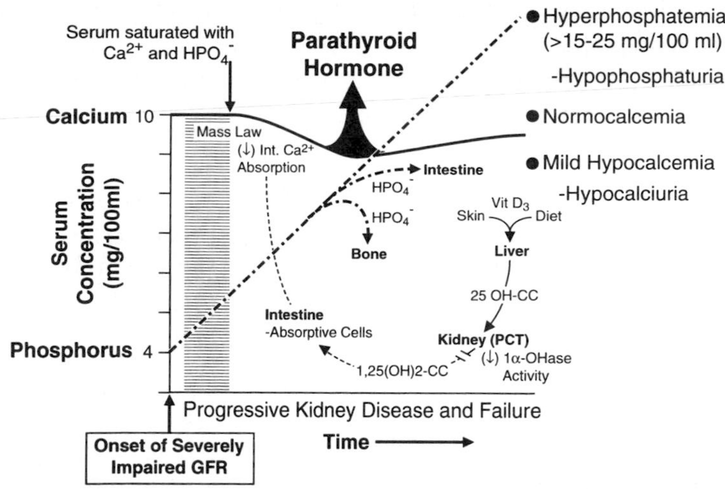

FIGURE 23.25 Alterations in serum calcium and phosphorus during the pathogenesis of secondary hyperparathyroidism associated with progressive renal failure.

min D receptor, and control the set point of chief cell responsiveness to negative feedback by the serum Ca concentration. The function of 1,25-dihydroxyvitamin D serves to complete the endocrine feedback loop with the parathyroid gland to control serum Ca concentration (Fig. 23.8). Decreased circulating levels of 1,25-dihydroxyvitamin D in animals with CRF results in chief cell hyperplasia and increased secretion of intact PTH. The increased circulating PTH correlates with the degree of uremia and has been suggested to play a role in both the clinical signs and progression of CRF by acting as an uremic toxin (Fig. 23.27) (Nagode and Chew, 1992).

The use of dietary phosphorus restriction and intestinal phosphate binders in animals with mild uremia is capable of reducing serum phosphate. Reduced serum phosphate results in decreased serum PTH concentrations because of the removal of phosphate inhibition of 1,25-dihydroxyvitamin D production by the kidneys, but may not return PTH levels to normal (Nagode and Chew, 1992). The oral administration of low doses of 1,25-dihydroxyvitamin D (1.7–3.4 ng/kg) has been shown to normalize serum PTH in dogs and cats with naturally occurring CRF and may reduce the progression of disease (Nagode *et al.*, 1996). Removal of the parathyroid glands did not reduce the progression of

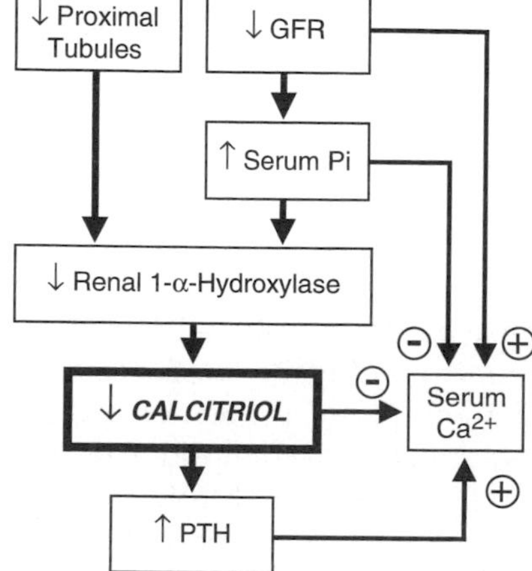

FIGURE 23.26 Hyperparathyroidism and chronic renal failure. Role of 1,25-dihydroxyvitamin D (calcitriol) in the pathogenesis of renal secondary hyperparathyroidism. GFR, glomerular filtration rate, P_i, serum phosphorus, PTH, serum parathyroid hormone. From Rosol *et al.* (1995a).

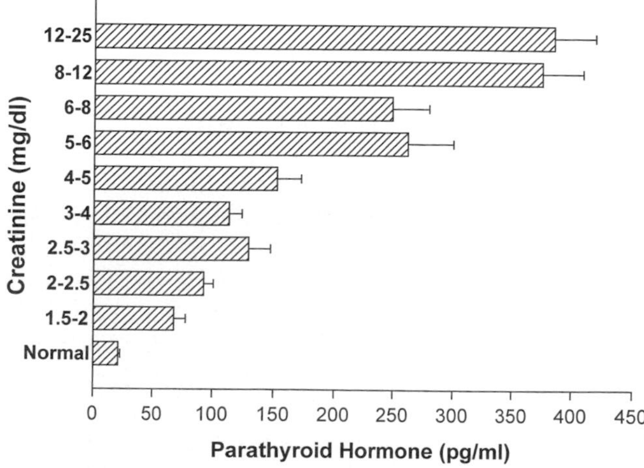

FIGURE 23.27 Relationship of serum N-terminal parathyroid hormone to levels of serum creatinine in 35 normal dogs and 333 canine patients with uremia (mean ± SEM). From Nagode and Chew (1992).

experimentally induced renal failure in dogs; however, serum PTH levels were not as high in the experimental animals as in dogs with spontaneous, advanced renal disease (Finco *et al.*, 1994).

Parathyroid stimulation in patients with chronic renal disease also may be attributed to hypocalcemia that develops in the pathogenesis of the disease. As the phosphorus concentration increases, blood calcium decreases reciprocally (Krook and Lowe, 1964). Impaired intestinal absorption of calcium due to an acquired defect in vitamin D metabolism also plays a role in the development of hypocalcemia in chronic renal insufficiency and uremia. Increased PTH secretion compensates for the mild hypocalcemia, and most animals with CRF are normocalcemic. For a further discussion of serum calcium in animals with CRF, see the sections on hypercalcemia and hypocalcemia.

All four parathyroid glands are enlarged because of organellar hypertrophy initially and cellular hyperplasia later, as compensatory mechanisms to increase hormonal synthesis and secretion in response to the hypocalcemic stimulus. Because the parathyroid gland stores comparatively little preformed hormone, the rate of peptide synthesis appears to be rate limiting for hormonal secretion. Although the parathyroids are not autonomous, the concentration of PTH in the peripheral blood in patients with chronic renal failure may exceed that of primary hyperparathyroidism. A greatly increased number and size of chief cells is required, therefore, to sustain the increased rates of hormonal secretion in patients with long-standing renal disease.

Parathyroid hormone accelerates osteoclastic resorption, results in release of stored calcium from bone, and returns the blood calcium toward normal (Fig. 23.25). The long-standing increase in bone resorption eventually results in the metabolic bone disease associated with chronic renal insufficiency. Progressive glomerular and tubular dysfunction with loss of target cells interferes with expression of the phosphaturic response by the increased circulating PTH in renal disease. Phosphorus is retained and the blood concentration continues to rise in spite of the secondary hyperparathyroidism (Fig. 23.25).

Although skeletal involvement is generalized with hyperparathyroidism, it does not affect all parts uniformly. Bone lesions become apparent earlier and reach a more advanced degree in certain areas. Resorption of alveolar socket bone and loss of lamina dura dentes occur early in the course of the disease. This results in loose teeth that may be dislodged easily and interfere with mastication. Cancellous bones of the maxilla and mandible also are sites of predilection in hyperparathyroidism. Because of the accelerated resorption the bones become softened and readily pliable (i.e., "rubber-jaw disease"), and the jaws fail to close properly. Long bones of the abaxial skeleton are less dramatically affected. Lameness, stiff gait, and fractures after relatively minor trauma may result from the increased bone resorption. Areas of subperiosteal resorption by numerous osteoclasts may disrupt the osseous attachment of tendons, leading to elevation and stretching of the periosteum, bone pain, and an inability to support the body's weight.

2. *Nutritional Secondary Hyperparathyroidism*

The increased secretion of parathyroid hormone with this metabolic disorder is a compensatory mechanism directed against a disturbance in mineral homeostasis induced by nutritional imbalances. The disease occurs in cats (Rowland *et al.*, 1968), dogs, certain nonhuman primates (Hunt *et al.*, 1967), and laboratory animals, as well as in many farm-animal species (Krook and Lowe, 1964; McKenzie *et al.*, 1981; Capen and Martin, 1982; Gilka and Sugden, 1984).

Dietary mineral imbalances of etiologic importance in the pathogenesis of nutritional secondary hyperparathyroidism are (1) a low content of calcium, (2) excessive phosphorus with normal or low calcium, and (3) inadequate amounts of vitamin D. The significant end result is hypocalcemia, which results in parathyroid stimulation. A diet low in calcium fails to supply the daily requirement even though a greater proportion of ingested calcium is absorbed, and hypocalcemia develops (Fig. 23.28). Ingestion of excessive phosphorus results in increased intestinal absorption and elevation of blood phosphorus and can reduce intestinal calcium absorption by forming complexes with calcium in the intestinal lumen. Hyperphosphatemia does not stimulate the parathyroid gland directly, but does so indirectly by virtue of its ability to lower blood calcium when the serum becomes saturated with respect to these two ions. Diets containing inadequate amounts of vitamin D_3 (even with normal levels of vitamin D_2) cause diminished intestinal calcium absorption and hypocalcemia in certain New World monkeys.

In response to the nutritionally induced hypocalcemia, all parathyroid glands undergo cellular hypertrophy and hyperplasia (Capen *et al.*, 1968). Active chief cells stimulated by diet-induced hypocalcemia become larger and more tightly arranged compared to those in parathyroids of normal animals. Because kidney function is normal, the increased levels of PTH result in diminished renal tubular resorption of phosphorus and increased reabsorption of calcium, and blood levels return toward normal (Fig. 23.28). In addition, bone resorption is accelerated and release of calcium elevates blood calcium levels to the low-normal

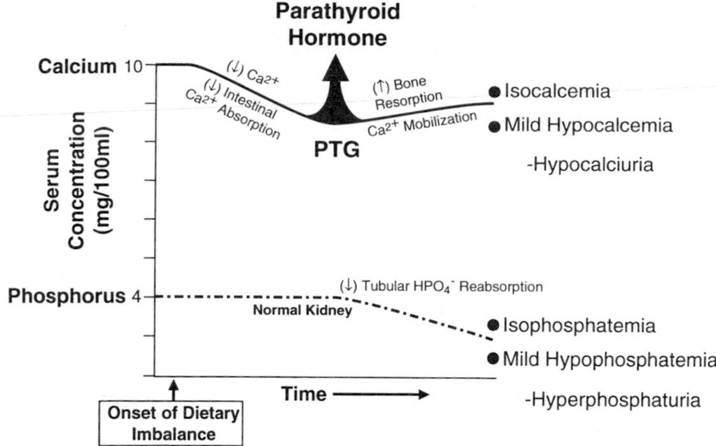

FIGURE 23.28 Nutritional secondary hyperparathyroidism. Alterations in serum calcium and phosphorus in the pathogenesis of nutritional secondary hyperparathyroidism caused by feeding a diet low in calcium or deficient in cholecalciferol but with a normal amount of phosphorus.

range. Continued ingestion of the imbalanced diet sustains the state of compensatory hyperparathyroidism, which leads to progressive development of metabolic bone disease.

Nutritional secondary hyperparathyroidism develops in young cats fed a predominantly meat diet (Pedersen, 1983). For example, beef heart or liver contains minimal amounts of calcium (7 to 9 mg/100 g) and has a markedly imbalanced calcium to phosphorus (Ca : P) ratio (1 : 20 to 1 : 50). An adequate diet for kittens up to 6 months of age should supply 200 to 400 mg calcium. Kittens fed beef heart or liver develop functional disturbances within approximately 4 weeks. Clinical signs are dominated by disturbances in locomotion manifested by a reluctance to move, posterior lameness, and an uncoordinated gait (Rowland *et al.*, 1968). The cortex of long bones is progressively thinned because of increased resorption, and the medullary cavity is widened. Affected kittens become quiet, are reluctant to play, and assume a sitting position or are in sternal recumbency. Normal activities may result in sudden onset of severe lameness due to incomplete or folding fractures of one or more bones. The high content of digestible protein (over 50% on a wet basis) and fat promotes rapid growth in kittens fed beef heart. They appear well nourished and their hair coat maintains a good luster.

In general, kittens are more susceptible and develop more severe skeletal lesions than adult cats fed a similar diet. The disease develops rapidly because the dietary imbalance is wide and the skeletal metabolic rate of kittens is high. Vertebral fractures with compression of spinal cord and paralysis are common complications in kittens but are infrequent in adult cats.

The feeding of a monotonous meat diet to dogs of any age results in secondary hyperparathyroidism with the development of skeletal disease of varying severity by mechanisms similar to those described previously (Saville and Krook, 1969; Krook *et al.*, 1971a). The low calcium content and unfavorable Ca : P of nonsupplemented all-meat diets are unable to fulfill the daily requirements either for growing pups (240 mg calcium and 200 mg phosphorus/lb body weight/day) or adult dogs (120 mg calcium and 100 mg phosphorus/lb body weight/day).

Lameness is the initial functional disturbance in growing dogs and may vary from a slight limp to complete inability to walk. The bones are painful on palpation, and folding fractures of long bones and vertebrae are not uncommon (Goddard *et al.*, 1970; Morris *et al.*, 1971). Clinical signs often are related to resorption of jaw bones in adult dogs. Parathyroid hormone-stimulated resorption of alveolar socket bone results in loss of lamina dura dentes, loosening and subsequent loss of teeth from their sockets, and recession of gingiva with partial root exposure in advanced cases (Krook *et al.*, 1971a).

Secondary hyperparathyroidism of nutritional origin also occurs in collections of aviculturists (domestic and captive birds) (Wallach and Flieg, 1969; Arnold *et al.*, 1974; Long *et al.*, 1983), zoological parks (caged lions and tigers, green iguanas, crocodiles, etc.) (Wallach and Hoessle, 1968; Wallach, 1971; Anderson and Capen, 1976a, 1976b), and laboratory animals used for research (ground squirrels, nonhuman primates, etc.) (Krook and Barrett, 1962; Hunt *et al.*, 1967; Lehner *et al.*, 1976).

In horses the most frequent nutritional imbalance involves the ingestion of excessive phosphorus. This results in increased intestinal absorption of phosphorus, elevation of the blood phosphorus concentration, and inhibition of intestinal calcium absorption. Hyperphosphatemia does not stimulate the parathyroid gland directly, but does so indirectly by virtue of its ability to lower blood calcium. Horses that develop the disease usually have been fed high-grain diets with below-average-quality roughage. Evidence of high phosphorus intake may be difficult to establish inasmuch as the excess phosphorus may be fed by the owner in the form of a bran supplement added to a grain diet in order to improve the health of the horse. The diet usually is palatable and nutritious except for the unbalanced phosphorus (excessive amounts) and calcium (marginal or deficient) content. A diet deficient in calcium fails to supply the daily requirement even though a greater proportion of ingested calcium is absorbed, and hypocalcemia develops. Occasionally, horses may develop nutritional hyperparathyroidism after pasturing on grasses with a high oxalate content. This results in intestinal malabsorption of calcium (Walthall and McKenzie, 1976). The oxalates form insoluble complexes with calcium in the intestine, resulting in an elevated fecal calcium : phosphorus ratio (2.3 : 1) compared with horses on a similar calcium and phosphorus intake but without the oxalate-rich plants (fecal calcium:phosphorus ratio 1.2 : 1). The interference in intestinal calcium absorption results in the development of progressive hypocalcemia that leads to parathyroid stimulation and development of the metabolic bone disease.

Initial clinical signs in horses with nutritional secondary hyperparathyroidism usually include a transitory shifting lameness in one or more limbs, generalized tenderness of joints, and a stilted gait (Krook and Lowe, 1966). The lameness develops as a result of increased osteoclastic resorption of outer circumferential lamellae with disruption of tendinous insertions and bone trabeculae supporting the articular cartilage, resulting in disruption of joint cartilage on weight bearing. Resorption of alveolar socket bone and loss of lamina dura dentes occur early and may result in loose teeth. Later in the course of the disease, severe lesions develop in bones of the skull, especially the maxilla and mandible, resulting in bilateral firm enlargements of the facial bones immediately above and anterior to the facial crests. The horizontal rami of the mandibles are irregularly thickened by a progressive hyperostotic fibrous osteodystrophy ("big head") that develops in horses with nutritional secondary hyperparathyroidism. The hyperostosis of skull bones results from osteoid and fibrous connective tissue deposition in excess of the volume of bone resorbed. The chronic excess intake of phosphorus and increased secretion of PTH result in stimulation of osteoblasts to form osteoid in excess of the amount of bone resorbed by osteoclasts and progressive enlargement of skeletal bones. The hyperostotic fibrous osteodystrophy may impinge upon the nasal cavity, resulting in dyspnea, especially after exertion.

Changes in urine calcium and phosphorus are more consistent and useful in the clinical diagnosis of nutritional secondary hyperparathyroidism in horses. The increased secretion of PTH acts on the normal kidneys to markedly increase urinary phosphorus excretion but decrease calcium loss in the urine compared to normal horses (Joyce *et al.*, 1971). Blood urea nitrogen, serum creatinine, and other parameters used to assess renal function are within normal limits in horses with nutritional hyperparathyroidism. Serum alkaline phosphatase levels often are in the high-normal range or elevated in horses with overt bone disease, reflecting the increased osteoblastic activity in hyperparathyroidism. Under experimental conditions, horses fed a high-phosphorus diet (phosphorus 1.4%; calcium 0.7%) are able to normalize their blood calcium more rapidly following an EDTA-induced hypocalcemia than are controls fed a balanced diet (phosphorus 0.6%; calcium 0.7%), owing to the increased parathyroid activity (Argenzio *et al.*, 1974).

A metabolic bone disorder has been recognized in primates for many years and has received numerous appellations, including cage paralysis, simian bone disease, and osteomalacia. It is now understood that hypocalcemia resulting from either inadequate dietary vitamin D_3 intake in New World laboratory primates housed indoors or excessive phosphorus in the diet of pet monkeys leads to long-term stimulation of the parathyroid glands. The monkeys become inactive, offer less resistance to handling, and have difficulty masticating their food. In the more advanced stages there is maxillary hyperostosis due to osteoid deposition and proliferation of fibrous connective tissue, joint pain, and distortion of limbs by palpable fractures without mineralized calluses. There is radiographic evidence of generalized skeletal demineralization, loss of lamina dura dentes, subperiosteal cortical bone resorption, bowing deformities, and multiple folding fractures of long bones. Cortical bone is thinned because of the activity of increased numbers of osteoclasts. The resorbed bone is partially replaced by the proliferation of immature fibroblasts and neocapillaries.

3. *Primary Hyperparathyroidism*

Parathyroid hormone is produced in excess of normal in primary hyperparathyroidism by a functional lesion in the parathyroid gland for no apparent useful purpose. This disease is encountered less frequently in older dogs and cats than is the relatively common secondary hyperparathyroidism.

The normal control mechanisms for PTH secretion by the concentration of blood calcium ion are lost in primary hyperparathyroidism. Hormone secretion is autonomous, and the parathyroid gland produces excessive amounts of hormone in spite of a sustained increase in blood calcium. Cells of the renal tubules are particularly sensitive to alterations in the amount of circulating PTH. The hormone acts on these cells initially to promote the excretion of phosphorus and the retention of calcium (Fig. 23.29). A prolonged increased secretion of PTH results in accelerated osteoclastic bone resorption. Mineral is removed from the skeleton and replaced by immature fibrous connective tissue. The bone lesion of fibrous osteodystrophy is generalized throughout the skeleton but is accentuated in local areas.

The lesion responsible for the excessive secretion of PTH usually is an adenoma composed of active chief cells with interspersed oxyphil and water-clear cells (Krook, 1957; Leav *et al.*, 1976; Capen and Martin, 1977; Sherding *et al.*, 1980; Kallet *et al.*, 1991). Chief-cell adenomas usually are single and result in enlargement of the parathyroid gland. They can be located either in the cervical region near the thyroid or, infrequently, within the thoracic cavity near the base of the heart (Cheville, 1972). Other lesions occasionally associated with primary hyperparathyroidism are primary chief-cell hyperplasia and chief-cell carcinoma (Capen and Roth, 1973; DeVries *et al.*, 1993). C cells in the thyroid gland are usually hyperplastic but are unable to return the blood calcium to normal.

The functional disturbances observed clinically are the result of hypercalcemia and weakening of bones by excessive resorption in long-standing cases. Hypercalcemia results in anorexia, vomiting, constipation, depression, and generalized muscular weakness due to decreased neuromuscular excitability. Lameness due to fractures of long bones may occur after physical trauma. Compression fractures of vertebral bodies can exert pressure on the spinal cord and nerves, resulting in motor and/or sensory dysfunction. Excessive resorption of cancellous bone of the skull resulting in a loosening or loss of teeth from alveolar sockets and hyperostosis of the mandible and maxilla due to proliferation of woven bone have been observed in dogs with primary hyperparathyroidism. Radiographic evaluation reveals areas of subperiosteal cortical resorption, loss of lamina dura around the teeth, soft-tissue mineralization, bone cysts, and a generalized decrease in bone density with multiple fractures. Mineralization of renal tubules and formation of calculi may occur in advanced cases of primary hyperparathyroidism in the dog with substantial elevations of blood calcium (Klausner *et al.*, 1986).

The most practical laboratory tests to aid in establishing the diagnosis of primary hyperparathyroidism are quantitation of total and ionized blood calcium and intact PTH immunoassay. Hypercalcemia is a consistent finding and results from accelerated release of calcium from bone and increased renal calcium reabsorption (although total urine calcium excretion is increased because of increased filtered calcium) (Fig. 23.29). The blood phosphorus is low (4 mg/dl [1.3 mmol/liter] or less) or in the low-normal range because of inhibition of renal tubular reabsorption of phosphorus by the excess PTH. Activity of alkaline phosphatase (an enzyme involved in bone apposition)

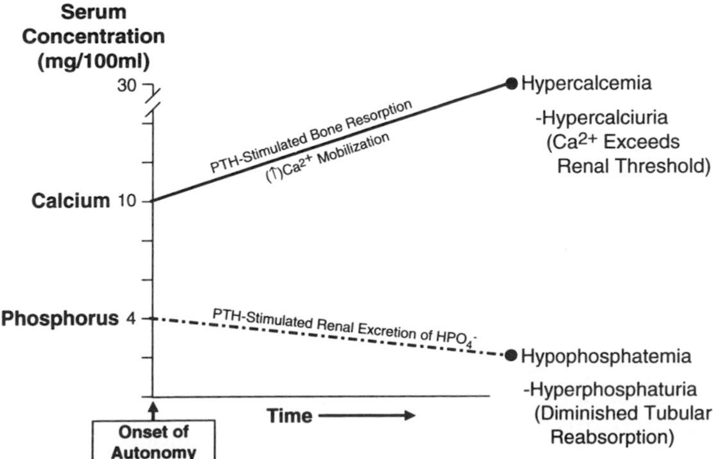

FIGURE 23.29 Primary hyperparathyroidism. Alterations in serum calcium and phosphorus in response to an autonomous secretion of parathyroid hormone in primary hyperparathyroidism.

may be elevated in the serum of animals with overt bone disease. The increased activity of this enzyme results from direct stimulation by PTH and a compensatory increase in osteoblastic activity along trabeculae as a response to mechanical stress in bone weakened by excessive resorption. Serum intact PTH may be increased or in the high normal range (Table 23.3). Normal serum PTH concentration with concurrent hypercalcemia is considered an inappropriate response of the parathyroid gland(s) and is consistent with mild primary hyperparathyroidism.

Successful removal of the functional parathyroid lesion results in a rapid decrease in circulating PTH levels because the half-life of PTH in plasma is less than 10 minutes. It should be emphasized that plasma calcium levels in patients with overt bone disease may decrease rapidly and be subnormal within 12 to 24 hours after operation, resulting in hypocalcemic tetany. Serum calcium levels should be monitored frequently following surgical removal of a parathyroid neoplasm. Postoperative hypocalcemia (5 mg/dl [1.25 mmol/liter] and lower) can be the result of (1) depressed secretory activity of chief cells due to long-term suppression by the chronic hypercalcemia or injury to the remaining parathyroid tissue during surgery, (2) abruptly decreased bone resorption due to lowered PTH levels, and (3) accelerated mineralization of osteoid matrix formed by the hyperplastic osteoblasts but previously prevented from undergoing mineralization by the elevated PTH levels. In some dogs with hypercalcemia and parathyroid adenoma, acute hypocalcemia may occur as a result of spontaneous infarction and necrosis of the adenoma (Rosol *et al.*, 1988b). Infusions of calcium gluconate to maintain the serum calcium between 7.5 (1.88) and 9.0 (2.25) mg/dl (mmol/liter), feeding high-calcium diets, and supplemental vitamin D therapy will correct this serious postoperative complication.

Primary hyperparathyroidism also has been described in German shepherds associated with autonomous chief-cell hyperplasia. The pups develop hypercalcemia, hypophosphatemia, increased iPTH, and increased fractional clearance of inorganic phosphate in the urine (Thompson *et al.*, 1984). Clinical signs include stunted growth, muscular weakness, polyuria, polydipsia, and a diffuse reduction in bone density. Intravenous infusion of calcium failed to suppress the autonomous secretion of parathyroid hormone by the diffuse hyperplasia of chief cells in all parathyroids. Other lesions include nodular hyperplasia of thyroid C cells and widespread mineralization of lungs, kidney, and gastric mucosa. The disease is inherited and may represent an inactivating mutation in the Ca^{2+}-sensing receptor, as occurs in humans with neonatal primary hyperparathyroidism (see Section I.J) (Brown *et al.*, 1995).

B. Hypercalcemia

Hypercalcemia is a common disorder that affects animals. Hypercalcemia has many causes, but is most frequently associated with neoplastic diseases (Chew *et al.*, 1992). Hypercalcemia results from an imbalance of calcium released from bones, calcium excretion by the kidney, and/or calcium absorption from the intestinal tract (Rosol and Capen, 1992). Pathogenic mechanisms of hypercalcemia include (1) excessive secretion of parathyroid hormone (PTH) or parathyroid hormone-related protein (PTHrP), which induce hypercalcemia by stimulating PTH/PTHrP receptors in bone and kidney, (2) excessive absorption of calcium from the intestines due to increased vitamin D or its metabolites, (3) excessive bone resorption due to local bone lesions or abnormal stimulation of osteoclastic bone resorption, (4) reduced renal excretion of calcium, and (5) increased total serum calcium due to increased protein-bound and/or complexed calcium with normal ionized calcium levels (Table 23.4).

The clinical signs of hypercalcemia are similar regardless of its underlying cause and depend on the rapidity of onset of increased serum ionized calcium levels (Rosol *et al.*, 1995a). Animals with serum calcium

TABLE 23.4 Causes of Hypercalcemia

Stimulation of parathyroid hormone receptors
Primary hyperparathyroidism
Secondary renal hyperparathyroidism (uncommon)
Humoral hypercalcemia of malignancy
Excessive intestinal absorption of calcium
Vitamin D intoxication
1,25-Dihydroxyvitamin D intoxication
Granulomatous diseases associated with excessive 1,25-dihydroxyvitamin D production
Hematologic malignancies associated with excessive 1,25-dihydroxyvitamin D production
Excessive bone resorption (independent of PTH receptors)
Immobilization
Hyperthyroidism
Osteomyelitis
Solid-tumor metastases to bone
Hematologic malignancies in bone marrow
Vitamin A toxicity
Reduced renal excretion of calcium
Thiazide diuretics
Hypocalciuric hypercalcemia (familial disease of humans)
Increased protein-bound or complexed serum calcium
Hypoadrenocorticism
Renal failure
Hemoconcentration

values in excess of 16.0 mg/dl (4.0 mmol/liter) generally have the most severe clinical signs. Exceptions to this rule occur, and occasionally animals with severe hypercalcemia have mild clinical signs. Horses (Table 23.1) and rabbits (Hong *et al.*, 1995) have normal total serum calcium concentrations greater than those of other domestic animals, which should be considered before hypercalcemia is diagnosed in these species. Metabolic acidosis will enhance the severity of clinical signs because it will result in an increase in the ionized fraction of serum calcium.

Increased serum Ca^{2+} will induce clinical signs relating to the gastrointestinal, neuromuscular, cardiovascular, and renal systems (Rosol *et al.*, 1995a). Decreased contractility of the gastrointestinal smooth muscle may be associated with anorexia, vomition, or constipation. There may be generalized locomotive weakness due to decreased neuromuscular excitability. Behavioral changes, depression, stupor, coma, seizures, and muscle twitching have been observed in dogs with hypercalcemia. Lameness and bone pain from demineralization of bone or pathologic fractures may be clinical signs with long-standing hypercalcemia. Hypercalcemia results in increased myocardial excitability and diminished ventricular systole, which may result in weakness and syncope associated with cardiac dysrhythmia. There is shortening of the Q–T interval and prolongation of the P–R interval (first-degree heart block). Ventricular fibrillation may develop in severe hypercalcemia (Chew and Capen, 1980). Hypercalcemia may predispose some animals to developing pancreatitis. The pathogenesis of pancreatitis associated with hypercalcemia is unknown, but may be related to degeneration of pancreatic acinar cells and leakage of cytoplasmic enzymes (Cates *et al.*, 1988; Frick *et al.*, 1995).

Polyuria and polydipsia are commonly encountered and may be the reason for an animal owner to seek medical attention. Initially, polyuria and polydipsia are due to impaired renal concentrating ability (Chew and Capen, 1980). The mechanism of this defect is not completely understood, but it appears that hypercalcemia inhibits the antidiuretic hormone-dependent resorption of NaCl in the diluting segment of the nephron (Sejersted *et al.*, 1984). Urine specific gravity is often low (<1.020) and may be hyposthenuric (1.001–1.007). Sodium excretion usually remains unchanged due to the vasoconstrictor effect of hypercalcemia, which results in a reduction of the glomerular filtration rate.

Hypercalcemia also has a toxic effect on renal tubules, either directly or from ischemia induced by vasoconstriction (Chew and Capen, 1980). Renal failure is an important consequence of severe or long-standing hypercalcemia (Kruger *et al.*, 1996). Tubular epithelial cells undergo degeneration, with the collecting system most severely affected. There is mineralization of epithelial cells and basement membranes of tubules. There may be glycosuria due to failure of tubular reabsorption and granular cast formation from degenerate tubular epithelial cells. Azotemia will occur when renal injury is severe, and polyuria and polydipsia will be secondary to renal failure. The magnitude of mineralization and tubular damage may be reduced by phosphate restriction (Harris *et al.*, 1986).

1. *Cancer-Associated Hypercalcemia*

The most common cause of hypercalcemia in animals and human beings is cancer-associated hypercalcemia (Rosol and Capen, 1992). There are three mechanisms of increased serum calcium induced by neoplasms: (1) humoral hypercalcemia of malignancy, (2) hypercalcemia induced by metastases of solid tumors to bone, and (3) hematologic malignancies (Fig. 23.30).

a. Humoral Hypercalcemia of Malignancy

Humoral hypercalcemia of malignancy (HHM) is a syndrome associated with diverse animal and human malignant neoplasms (Rosol and Capen, 1992). Characteristic clinical findings in patients with HHM include hypercalcemia, hypophosphatemia, hypercalciuria (often with decreased fractional calcium excretion), increased fractional excretion of phosphorus, increased nephrogenous cAMP, and increased osteoclastic bone resorption. Hypercalcemia is induced by humoral effects on bone, kidney, and possibly the intestine (Fig. 23.31) (Rosol and Capen, 1988). Increased osteoclastic bone resorption is a consistent finding in HHM and increases calcium release from bone. The kidney plays a critical role in the pathogenesis of hypercalcemia,

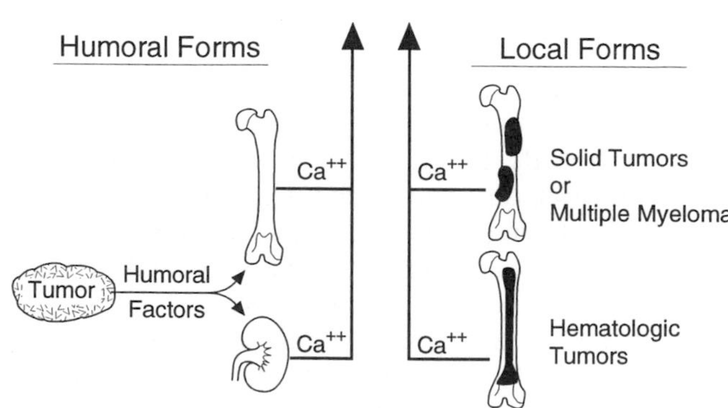

FIGURE 23.30 Pathogenesis of cancer-associated hypercalcemia. Humoral and local forms of cancer-associated hypercalcemia increase circulating concentrations of calcium by stimulating osteoclastic bone resorption or increased tubular reabsorption of calcium. (Rosol and Capen, 1992.)

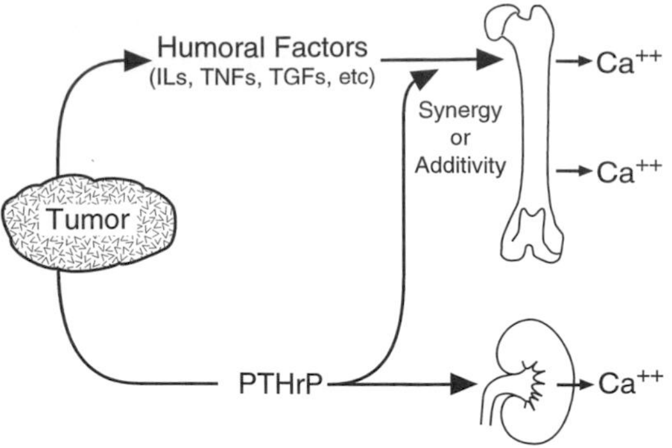

FIGURE 23.31 Humoral factors such as parathyroid hormone-related protein (PTHrP), interleukin-1 (IL-1), tumor necrosis factors (TNF), or transforming growth factors (TGF) produced by tumors induce humoral hypercalcemia of malignancy (HHM) by acting as systemic hormones and stimulating osteoclastic bone resorption or increasing tubular reabsorption of calcium. (Rosol and Capen, 1992.)

because calcium reabsorption is stimulated by PTHrP which binds and activates renal PTH/PTHrP receptors. In some forms of HHM, there are increased serum 1,25-dihydroxyvitamin D levels, which may increase Ca absorption from the intestine (Rosol *et al.*, 1992b).

Malignancies that are commonly associated with HHM in animals include the adenocarcinoma derived from apocrine glands of the anal sac in dogs, some T-cell lymphomas of dogs, and miscellaneous neoplasms that induce HHM in various species, such as cats and horses, in a sporadic pattern (Rosol and Capen, 1988, 1992; Rosol *et al.*, 1994). Excessive secretion of biologically active PTHrP plays a central role in the pathogenesis of hypercalcemia in most forms of HHM; however, cytokines such as interleukin-1 (IL-1), tumor necrosis factor α, or transforming growth factors α and β or 1,25-dihydroxyvitamin D may have synergistic or cooperative actions with PTHrP (Fig. 23.31). Before PTHrP was identified it was well understood that tumors associated with humoral hypercalcemia of malignancy induced a syndrome that mimicked primary hyperparathyroidism due to secretion of a PTH-like factor that was antigenically unrelated to PTH. Purification of the PTH-like activity from the canine anal sac adenocarcinoma and multiple human tumors associated with HHM resulted in the discovery of PTHrP (see section on parathyroid hormone-related protein) (Moseley *et al.*, 1987; Weir *et al.*, 1988a).

PTHrP binds to the N-terminal PTH/PTHrP receptor in bone and kidney, but does not cross-react immunologically with native PTH (Fig. 23.13). PTHrP stimulates adenylyl cyclase and increases intracellular Ca^{2+} in bone and kidney cells by binding to and activating the cell membrane PTH/PTHrP receptors. This results in a stimulation of osteoclastic bone resorption, and in increased renal tubular calcium reabsorption and decreased renal tubular phosphate reabsorption. Interleukin-1 stimulates bone resorption *in vivo* and *in vitro* and is synergistic with PTHrP (McCauley *et al.*, 1991; Rosol and Capen, 1992). Transforming growth factors α and β can stimulate bone resorption *in vitro* and have been identified in tumors associated with HHM, including adenocarcinomas derived from apocrine glands of the anal sac in dogs (Merryman *et al.*, 1989).

i. Canine Adenocarcinoma Derived from Apocrine Glands of the Anal Sac The adenocarcinoma derived from apocrine glands of the anal sac of dogs has been described as consistently fulfilling the criteria for HHM (Meuten *et al.*, 1981, 1983b). This unique tumor appears primarily in middle-aged (mean of 10 years) female dogs with no breed predilection and rarely metastasizes to bone. Clinical signs are referable to hypercalcemia and a mass in the perineum (60% of the cases). Apocrine adenocarcinomas may require rectal palpation to confirm their presence, because their size ranges from 7 mm to 6 × 8 cm. Canine patients with this tumor were reported to have hypercalcemia, decreased levels of immunoreactive PTH, normal levels of prostaglandin E_2 metabolite (13,14-dihydro-15-keto-prostaglandin E_2), and increased urinary excretion of calcium, phosphorus, and cAMP. Bone histomorphometric evaluation revealed a significantly decreased trabecular bone volume and increased osteoclastic bone resorption in lumbar vertebrae from dogs with the adenocarcinoma and hypercalcemia (Meuten *et al.*, 1983b). Parathyroid glands from dogs with the apocrine adenocarcinoma were atrophic or inactive, and there were areas of nodular hyperplasia of C cells in the thyroid glands.

Most dogs with HHM have increased circulating concentrations of PTHrP (Fig. 23.32). Plasma concentrations of PTHrP are greatest in dogs with apocrine adenocarcinomas of the anal sac and sporadic carcinomas associated with HHM (10–100 p*M*) (Rosol *et al.*, 1992b). The serum calcium concentrations in these dogs correlate well with circulating PTHrP concentrations and are consistent with the concept that PTHrP plays a primary role in the pathogenesis of HHM in these dogs (Fig. 23.33).

Some dogs with apocrine adenocarcinomas have inappropriate levels of 1,25-dihydroxyvitamin D (maintenance of normal range or increased) for the degree of hypercalcemia (Fig. 23.34) (Rosol *et al.*, 1992b). This suggests that the humoral factors produced by the neoplastic cells are capable of stimulating renal 1α-hydroxylase and increasing the formation of 1,25-dihydroxyvitamin D even in the presence of increased blood calcium. Plasma immunoreactive PTH was not

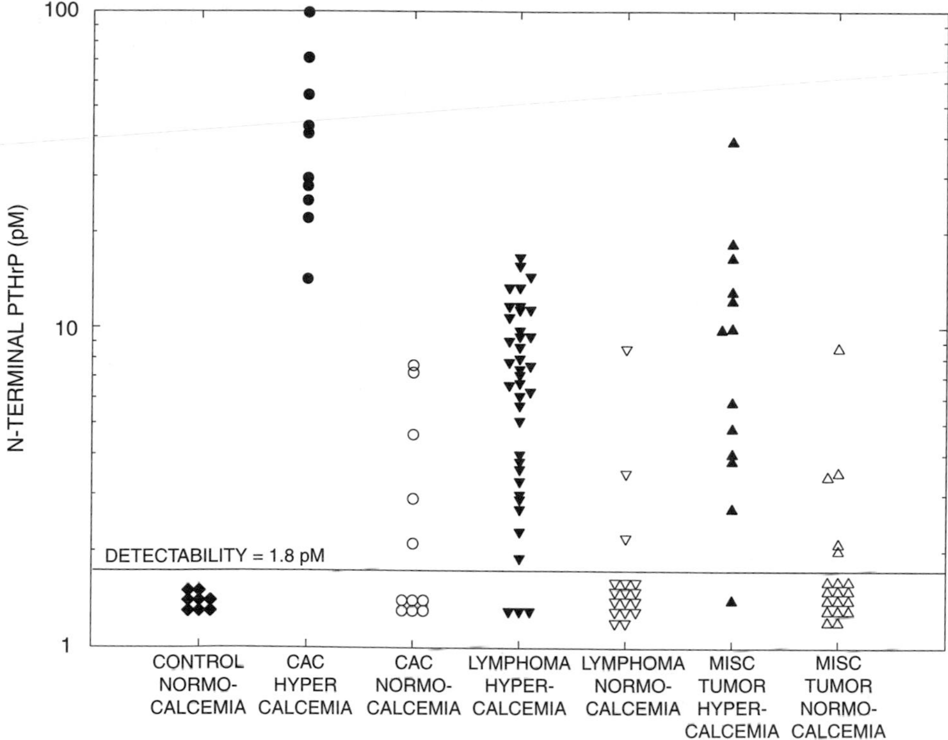

FIGURE 23.32 Circulating N-terminal parathyroid hormone-related protein (PTHrP) concentrations in normal dogs (CONTROL); dogs with hypercalcemia (>12 mg/dl) and anal-sac adenocarcinomas (CAC), lymphoma, or miscellaneous tumors (MISC TUMOR); and dogs with normocalcemia (<12 mg/dl) and anal sac adenocarcinomas, lymphoma, or miscellaneous tumors. From Rosol *et al.* (1992b).

increased in hypercalcemic dogs and was significantly less than that in dogs with primary hyperparathyroidism (Fig. 23.35). Surgical removal or radiation therapy of the adenocarcinoma results in a rapid return to normal of serum calcium and phosphorus, increased serum PTH, and decreased 1,25-dihydroxyvitamin D (Rosol *et al.*, 1992b). Postsurgical survival in dogs with adenocarcinoma and hypercalcemia ranged from 2 to 21 months, with a mean of 8.8 months. Sublumbar metastases occurred in a high percentage (94%) of the dogs and were associated with a recrudescence of the biochemical alterations in serum and urine.

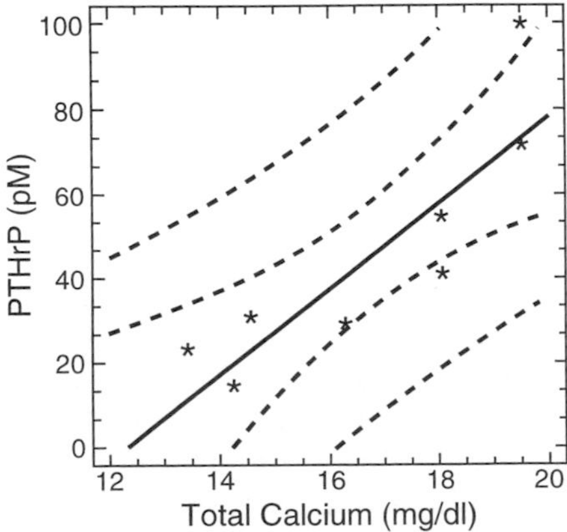

FIGURE 23.33 Regression of PTHrP and total calcium in dogs with anal sac Ca. There was a significant linear correlation (0.87, $P < 0.01$) between serum calcium and N-terminal PTHrP concentrations in dogs with humoral hypercalcemia of malignancy and adenocarcinomas derived from apocrine glands of the anal sac.

ii. Nude-Mouse Model of the Canine Anal Sac Adenocarcinoma A model of HHM has been developed in nude mice utilizing a serially transplantable anal sac adenocarcinoma (CAC-8) derived from a hypercalcemic dog (Rosol *et al.*, 1986). Nude mice with transplanted CAC-8 developed severe hypercalcemia and hypophosphatemia. Serum calcium concentration returned to the normal range in two days after surgical removal of the tumors. Serum 1,25-dihydroxyvitamin D levels were significantly increased and were correlated with serum calcium levels in CAC-8-bearing nude mice. Histomorphometric analysis of lumbar vertebrae revealed increased bone formation and resorption. The adenocarcinoma (CAC-8) produces PTHrP,

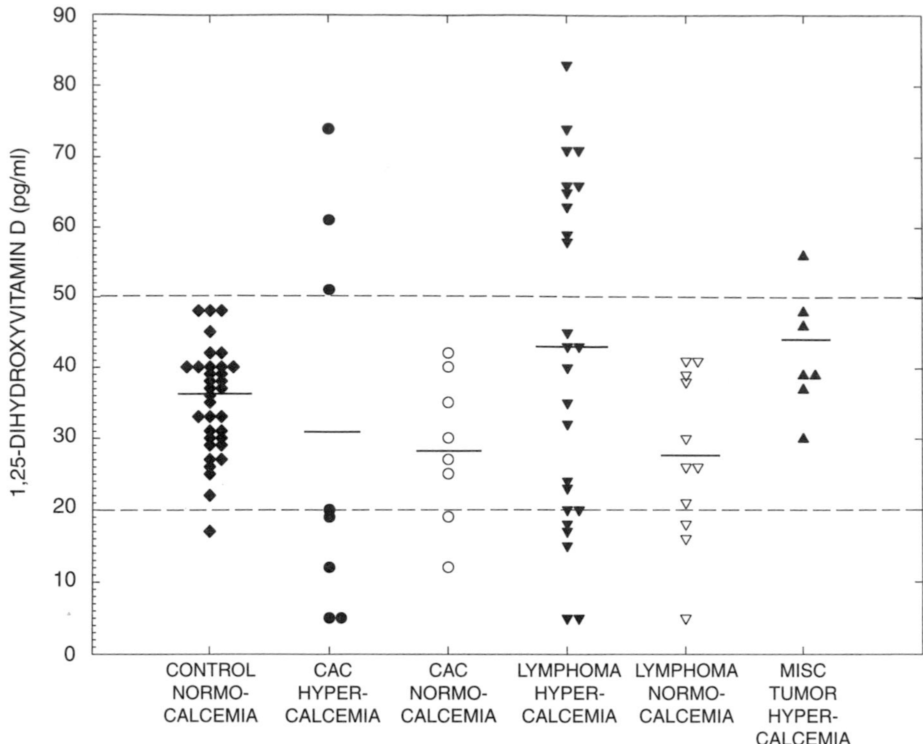

FIGURE 23.34 Serum 1,25-dihydroxyvitamin D concentrations in normal dogs (CONTROL); dogs with hypercalcemia (>12 mg/dl) and anal-sac adenocarcinomas (CAC), lymphoma, or miscellaneous tumors (MISC TUMOR); and dogs with normocalcemia (<12 mg/dl) and anal-sac adenocarcinomas or lymphoma. The normal range was 20–50 pg/ml (dashed lines). Bar = mean. From Rosol *et al.* (1992b).

TGFβ, and TGFα, which may work cooperatively to induce bone resorption and hypercalcemia, but PTHrP is likely to be the central humoral factor in the pathogenesis of hypercalcemia (Rosol *et al.*, 1988a; Weir *et al.*, 1988b; Merryman *et al.*, 1989; Rosol *et al.*, 1990).

Hypercalcemic nude mice with transplanted CAC-8 fed a low-calcium diet (<0.01%) had a significant reduction in serum calcium; however, the serum calcium concentration was not returned to the normal range (Rosol and Capen, 1987). The urine calcium excretion was reduced to control values in the hypercalcemic tumor-bearing nude mice fed the low-calcium diet. This suggested that calcium absorption from the intestine contributes to the hypercalcemia induced by CAC-8. Nude mice with CAC-8 have increased levels of serum 1,25-dihydroxycholecalciferol, which may mediate the effect of dietary calcium intake on serum calcium concentration. However, calcium absorption from the intestine cannot completely account for the hypercalcemia induced by the adenocarcinoma, because nude mice fed the low-calcium diet still had significant hypercalcemia.

Dichlorodimethylene bisphosphonate, a bisphosphonate known to inhibit osteoclastic bone resorption, significantly reduced serum Ca in CAC-8-bearing nude mice, but the serum calcium levels did not reach the normal range for non-tumor-bearing nude mice (Rosol and Capen, 1987). This emphasizes the role of the kidney in the pathogenesis of hypercalcemia in HHM, because complete inhibition of bone resorption was not capable of normalizing serum calcium levels.

iii. Lymphoma Canine lymphoma is associated with hypercalcemia in 20–40% of the cases. Some dogs with lymphoma and hypercalcemia have HHM. Hypercalcemic lymphomas associated with HHM were usually of the T-cell subset (Weir *et al.*, 1988c). The affected dogs have increased fasting and 24-hour Ca excretion, increased fractional P excretion, and increased nephrogenous cAMP. Increased osteoclastic resorption was present in bones without evidence of tumor metastasis. Dogs with HHM and lymphoma may have a pathogenesis of hypercalcemia similar to that occurring in humans with HTLV-I virus-induced lymphoma or leukemia. Neoplastic cells from humans with HTLV-I-induced lymphoma have increased PTHrP production due to stimulation of PTHrP transcription by the virally encoded Tax transcription factor (Prager *et al.*, 1994).

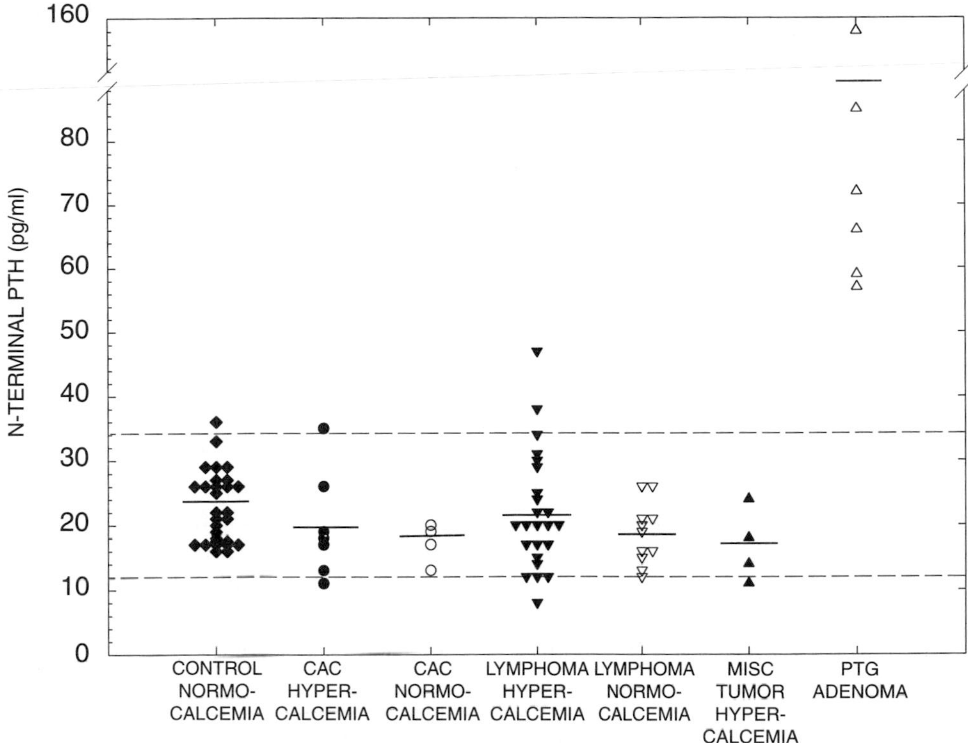

FIGURE 23.35 Serum N-terminal PTH concentrations in normal dogs (CONTROL); dogs with hypercalcemia (>12 mg/dl) and anal-sac adenocarcinomas (CAC), lymphoma, miscellaneous tumors (MISC TUMOR), or parathyroid adenomas (PTG ADENOMA); and dogs with normocalcemia (<12 mg/dl) and anal-sac adenocarcinomas or lymphoma. The normal range was 12–34 pg/ml (dashed lines). Bar = mean. From Rosol *et al.* (1992b).

Most dogs with lymphoma and HHM have significantly increased circulating PTHrP concentrations, but levels are lower (2–15 p*M*) than in dogs with carcinomas and HHM, and there is no correlation with serum calcium concentration (Figs. 23.32 and 23.36) (Rosol *et al.*, 1992b). This indicates that PTHrP is an important marker of dogs with HHM and lymphoma, but is not the sole humoral factor responsible for the stimulation of osteoclasts and development of hypercalcemia. It is likely that cytokines, such as interleukin-1 or tumor necrosis factor, may function synergistically with PTHrP to induce HHM in dogs with lymphoma (Fig. 23.31) (Rosol and Capen, 1992). Some dogs and human patients with lymphoma and hypercalcemia have increased serum 1,25-dihydroxyvitamin D levels which may be responsible for or contribute to the induction of hypercalcemia (Fig. 23.34) (Rosol *et al.*, 1992b; Seymour and Gagel, 1993).

b. Hematologic Malignancies

Some forms of hematologic malignancies present in the bone marrow induce hypercalcemia by the local induction of bone resorption (Rosol and Capen, 1992). This occurs most commonly with multiple myeloma and lymphoma. A number of paracrine factors or cytokines may be responsible for the stimulation of bone resorption. The cytokines most often implicated in the pathogenesis of local bone resorption include interleukin-1, tumor necrosis factor α, and tumor necrosis fac-

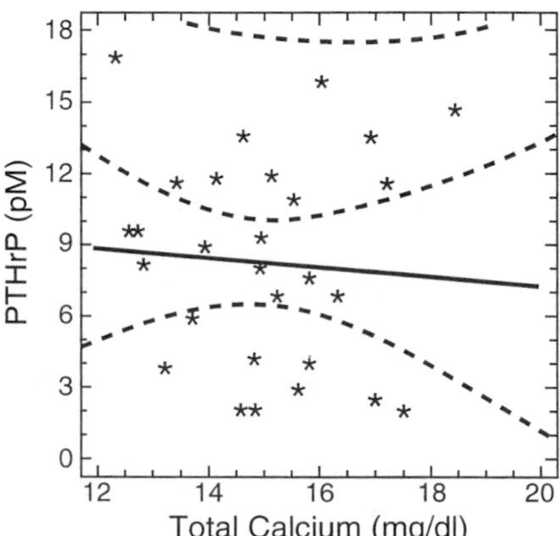

FIGURE 23.36 Regression of PTHrP and total calcium in dogs with lymphoma. There was no correlation between serum calcium and N-terminal PTHrP concentrations in dogs with humoral hypercalcemia of malignancy and lymphoma, which suggests that PTHrP probably functions cooperatively with other cytokines to induce hypercalcemia.

tor β (lymphotoxin) (Martin and Grill, 1992). Other cytokines or factors that may play a role include interleukin-6, transforming growth factors α and β, and PTHrP (Black and Mundy, 1994). Production of low levels of PTHrP by a tumor in bone may stimulate local bone resorption without inducing a systemic response because of increased circulating concentrations of PTHrP. Prostaglandins (especially prostaglandin E_2) also may be responsible for the local stimulation of bone resorption.

Some dogs with lymphosarcoma and hypercalcemia have localized bone resorption associated with metastases to medullary cavities without evidence of increased bone resorption at sites distant from the tumor metastases (Meuten *et al.*, 1983a). Hypercalcemic dogs with lymphosarcoma and bone metastases had decreased serum PTH and 1,25-dihydroxyvitamin D levels; increased excretion of Ca, P, and hydroxyproline; and increased serum prostaglandin E_2 metabolite, 13,14-dihydro-15-keto-prostaglandin E_2. The mediator of local bone resorption has not been identified; however, prostaglandin E_2 may be an important primary or secondary local mediator of bone resorption in these dogs. Other potential mediators include the cytokines, interleukin-1, or tumor necrosis factors.

c. Tumor Metastatic to Bone

Solid tumors that metastasize widely to bone can produce hypercalcemia by the induction of local bone resorption associated with tumor growth. This is not common in animals, but is an important cause of cancer-associated hypercalcemia in human beings (Rosol and Capen, 1992). Tumors that often metastasize to bone and induce hypercalcemia in human patients include breast and lung carcinomas.

The pathogenesis of enhanced bone resorption is not well understood, but the two primary mechanisms are (1) secretion of cytokines or factors that stimulate local bone resorption, and (2) indirect stimulation of bone resorption by tumor-induced cytokine secretion from local immune or bone cells (Garrett, 1993). Cytokines or factors that may be secreted by tumor cells and stimulate local bone resorption include PTHrP (Powell *et al.*, 1991), transforming growth factors α and β, or prostaglandins (especially prostaglandin E_2). In some cases, bone-resorbing activity can be inhibited by indomethacin, which suggests that prostaglandins are either directly or indirectly associated with the stimulation of bone resorption. The cytokines most often implicated in indirect stimulation of bone resorption by local immune cells include interleukin-1 and tumor necrosis factor.

Malignant neoplasms with osseous metastases may cause moderate to severe hypercalcemia and hypercalciuria, but the serum alkaline phosphatase activity and phosphorus are usually normal or moderately elevated. These changes are believed to be due to release of calcium and phosphorus into the blood from areas of bone destruction at rates greater than can be cleared by the kidney and intestine. Bone involvement can be multifocal but usually is sharply demarcated and localized to the area of metastasis.

2. *Miscellaneous Causes*

a. Hypoadrenocorticism (Addison's Disease)

Hypercalcemia (up to 15 mg/dl [3.7 mmol/liter]) occurs in approximately 30% of dogs with hypoadrenocorticism (Peterson and Feinman, 1982; Chew *et al.*, 1992). The elevated serum calcium concentration returns to normal when the disease is treated with adequate corticosteroid replacement therapy. The mechanism of hypercalcemia is not known, but the ionized fraction of serum calcium remains normal, whereas the nonionized fraction increases, and there is inappropriate hypocalciuria for the degree of hypercalcemia. Because the serum ionized calcium concentration is not increased, the hypercalcemia often is not deleterious to the canine patient; however, some of the clinical signs of hypoadrenocorticism (vomiting, anorexia, polyuria, and polydipsia) are similar to those of hypercalcemia.

b. Renal Failure in Horses

The kidneys normally excrete large amounts of calcium into the urine in horses. Horses with renal failure have been reported to be at greater risk for development of hypercalcemia than other species, but this has not been consistently documented (Tennant *et al.*, 1982; Kohn and Chew, 1987). Decreased glomerular filtration of Ca (decreased Ca clearance) may be important in horses that continue to absorb Ca across the intestinal tract. Horses with experimental bilateral nephrectomy develop a striking magnitude of hypercalcemia in association with hypophosphatemia (Tennant *et al.*, 1981), a situation quite different from that in dogs and cats, which develop mild hypercalcemia (or hypocalcemia) and moderate to severe hyperphosphatemia with advancing renal disease.

c. Renal Failure in Dogs and Cats

Hypercalcemia may occur in some dogs in the polyuric phase of acute renal failure (Chew *et al.*, 1992). The mechanism of mild hypercalcemia in the diuretic phase of acute renal disease is not known; however, there may be rapid lowering of serum phosphorus and transient secondary hypercalcemia. The hypercalcemia of acute renal failure resolves without specific treatment.

Mild to moderate hypercalcemia occurs in 10–15% of dogs and cats with chronic renal failure (CRF) (Kruger *et al.*, 1996). The mechanisms underlying the development of hypercalcemia during CRF are less well understood than those for the development of hypocalcemia. Hypercalcemia in dogs and cats with CRF may be due to an altered set point of PTH secretion in parathyroid chief cells (secondary to low 1,25-dihydroxyvitamin D levels) and to excessive secretion of biologically active PTH despite high serum Ca concentrations. Increased secretion of PTH by parathyroid chief cells leads to a stimulation of osteoclastic bone resorption (Ritz *et al.*, 1992). This occurs most often in animals in the late stages of CRF with very high serum intact PTH levels and low serum 1,25-dihydroxyvitamin D levels. Ionized calcium is usually normal at the time of increased total Ca concentration in dogs and cats with CRF, implying increased concentrations of either the protein-bound or the complexed fraction of Ca (Chew *et al.*, 1992). Chronic renal failure can be associated with increased serum concentrations of substances that complex Ca, such as phosphate, citrate, and sulfate. The possible relationship of increased citrate and sulfate concentrations to the magnitude of hypercalcemia has not been determined. Although hyperphosphatemia is common in animals with CRF, its magnitude does not determine whether or not hypercalcemia will develop.

d. Hypercalcemia Due to Vitamin D or Vitamin D Metabolites

Excessive vitamin D, vitamin D-like compounds, or active metabolites of vitamin D will induce hypercalcemia because of increased absorption of calcium from the intestines and stimulation of osteoclastic bone resorption at high dosages. The possible causes of hypercalcemia include (1) intoxication with vitamin D or one of its active metabolites, e.g., excessive dietary supplementation (Harrington and Page, 1983) or ingestion of vermicides, (2) consumption of plants that contain active vitamin D-like substances (Capen, 1992), (3) granulomatous diseases associated with excessive production of 1,25-dihydroxyvitamin D by macrophages, e.g., some dogs with blastomycosis (Dow *et al.*, 1986), and (4) hematologic malignancies associated with excessive production of 1,25-dihydroxyvitamin D, e.g., some lymphomas (see Section IV.B.1.a.iii on hypercalcemia and lymphoma).

The incidence of hypercalcemia due to vitamin D intoxication in small animals has increased because of renewed use of vitamin D (cholecalciferol) as a pesticide and rodenticide (Carothers *et al.*, 1994). The ingested vitamin D is converted to 25-hydroxyvitamin D in the liver and is the major toxic metabolite (Vieth, 1990). 25-Hydroxyvitamin D has a much reduced binding affinity for the vitamin D receptor (500- to 1000-fold) than does 1,25-dihydroxyvitamin D (active form of vitamin D), but is present at such high concentrations that it stimulates hypercalcemia by increasing intestinal absorption of Ca. Only small amounts of the 25-hydroxyvitamin D are converted to the active form of vitamin D, 1,25-dihydroxyvitamin D, which often is present in the circulation at normal to subnormal levels. This is due to inhibition of renal 1-α-hydroxylase by low serum PTH levels, hypercalcemia, hyperphosphatemia, and negative feedback from 25-hydroxyvitamin D that binds to the vitamin D receptor.

The hypercalcemia of hypervitaminosis D is usually accompanied by hyperphosphatemia (because vitamin D also increases intestinal absorption of phosphate in addition to calcium) and normal serum alkaline phosphatase activity. Skeletal disease is not a consistent feature, because the increased concentrations of blood calcium and phosphorus are derived principally from augmented intestinal absorption.

e. Hemoconcentration/Laboratory Error/Artifact

Hypercalcemia may be detected occasionally in dehydrated animals. The magnitude of elevation in blood calcium is mild and is attributed to fluid volume contraction that results in hyperproteinemia and an increased relative concentration of ionized and nonionized calcium. The hypercalcemia rapidly resolves following fluid therapy. The majority of dehydrated animals do not develop hypercalcemia. Small fluctuations in serum calcium are possible after eating.

Technical error is unlikely in the measurement of serum calcium concentrations; however, it is wise to substantiate that hypercalcemia is a repeatable finding by confirmation with a second test.

C. Hypocalcemia

1. Parturition/Lactation-Associated Hypocalcemic Syndromes

a. Parturient Hypocalcemia ("Milk Fever")

Parturient hypocalcemia is a metabolic disease of high-producing dairy cows characterized by the development of severe hypocalcemia, hypophosphatemia, and muscle paresis near the time of parturition. The pathogenic mechanisms responsible for the rapid and precipitous decrease in calcium and phosphorus levels in the blood are complex and involve several interrelated factors (Capen, 1972; Horst *et al.*, 1994). Total and ionized calcium levels decrease progressively beginning several days before parturition (Lincoln and Lane, 1990). Serum magnesium may increase reciprocally as calcium declines (Riond *et al.*, 1995b). The blood glucose concentration often is increased in response to hypocalcemia because of an interference with the secretion of insulin from beta cells. An adequate amount of calcium ion in extracellular fluids is required for insulin secretion in response to glucose and other secretagogues for insulin (Witzel and Littledike, 1973). The uptake of calcium by beta cells appears to stimulate the contraction of microfilaments, which triggers the peripheral migration of storage granules and release of insulin.

The development of parturient hypocalcemia in dairy cows was considered to be the result of an inadequate response of the parathyroid glands to the substantial demands for calcium imposed by the mineralization of fetal bones and the initiation of lactation. However, ultrastructural studies indicated that parathyroid chief cells in cows with parturient hypocalcemia were capable of responding to the increased demands for calcium by the secretion of stored hormone and hypertrophy of secretory organelles concerned with the synthesis of new hormone (Capen and Young, 1967). Either chief cells were depleted of storage granules, or the secretory granules had migrated peripherally and were fused with the plasma membrane. These structural findings suggest an active secretory response by parathyroid glands in cows with parturient hypocalcemia and are in agreement with biochemical studies (Mayer *et al.*, 1969). An immunoassay was used to measure plasma levels of parathyroid hormone and detected equal or greater levels of hormone in cows with parturient hypocalcemia equal to or greater than those in normal parturient cows (Fig. 23.37). Therefore, the ability of parathyroid glands to respond to the challenge for extra calcium mobilization with increased hormonal synthesis and secretion does not appear to be defective in cows that develop parturient hypocalcemia (Mayer, 1970).

Target cells in bone and skeletal calcium reserves of cows with parturient hypocalcemia are temporarily refractory to the action of the elevated levels of parathyroid hormone. Investigations have shown that the elevation of serum calcium in response to exogenous parathyroid extract is less when the extract is administered prepartum than when it is given postpartum (Martig and Mayer, 1973). Bone turnover, particularly resorption, is low in cows with parturient hypocalcemia, and only a few osteoclasts are present on smooth, inactive trabecular bone surfaces (Rowland *et al.*, 1972). The urinary excretion of hydroxyproline, derived from the breakdown of bone matrix, does not increase significantly during late gestation in cows that develop the disease, as occurs in cows that maintain their serum calcium near normal through parturition

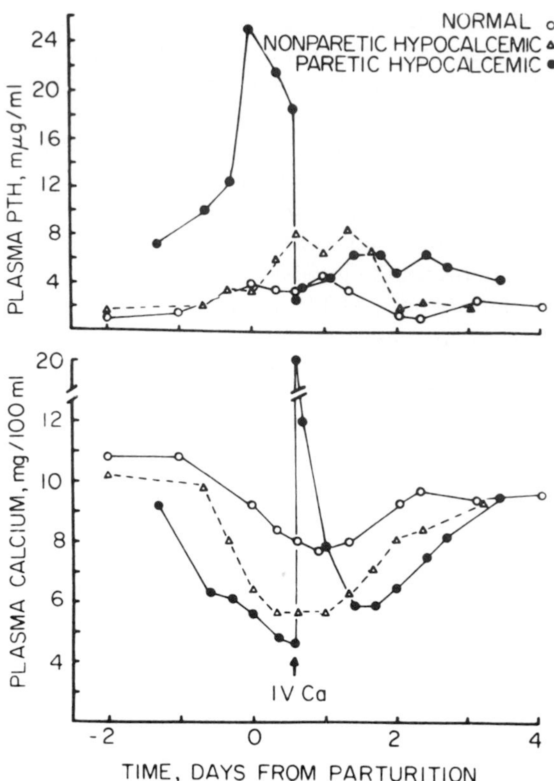

FIGURE 23.37 Development of varying degrees of hypocalcemia in cows near parturition with corresponding increases in plasma PTH levels. The cow developing severe hypocalcemia [below 5 mg/100 ml (1.25 mmol/liter)] had a considerably greater increase in plasma PTH than the moderate rise detected in nonparetic hypocalcemic and normal cows. Note that PTH levels decline rapidly following treatment of the paretic cow with intravenous calcium. From Mayer (1970).

and early lactation (Black and Capen, 1971). A secretion of calcitonin prepartum in certain cows, especially those fed high-calcium diets, may be one factor that contributes to the inability of increased parathyroid hormone levels to mobilize calcium rapidly from skeletal reserves and maintain blood levels during the critical period near parturition. The thyroid content of calcitonin is reduced (14% of control cows) and many C (parafollicular) cells are degranulated (Young and Capen, 1970). Elevated plasma levels of calcitonin have been reported in cows prior to the development of profound hypocalcemia by some investigators (Black and Capen, 1973) but not by others (Mayer, 1971). A syndrome of hypocalcemia analogous to parturient paresis has been induced with long-term intravenous infusion of calcitonin in lactating cows (Barlet, 1968).

The decline in blood calcium may be rapid in certain high-producing dairy cows near parturition and the initiation of lactation. Because bone resorption during the prepartal period often is relatively low owing to a substantial intake of dietary calcium, there is a relatively small pool of active bone-resorbing cells capable of responding rapidly to PTH. Therefore, when PTH secretion increases because of the rapidly developing hypocalcemia, the increase in activity of the few active osteoclasts present on bone surfaces is insufficient to restore the plasma calcium concentration to normal. The activation of osteoprogenitor cells and the subsequent conversion of preosteoclasts to osteoclasts under the influence of increased PTH with expansion of the pool of active bone-resorbing cells requires time (as long as 48 to 72 hours in an adult cow) and an adequate concentration of calcium in extracellular fluids. If the extracellular fluid calcium falls below a critical level, the increased circulating concentration of PTH may be ineffective in causing an elevation of cytosol calcium in target cells to activate new bone-resorbing cells. Neither an increased endogenous secretion of PTH nor the exogenous administration of parathyroid extract to cows will restore homeostasis once the hypocalcemia is profound. Only the administration of calcium and elevation of calcium in extracellular fluids will restore the responsiveness to PTH, trigger bone resorption, and correct the hypocalcemia.

The composition of the prepartal diet fed to dairy cows is known to be a significant factor in the pathogenesis of parturient hypocalcemia. High-calcium diets have been incriminated in significantly increasing the incidence of the disease (Ender *et al.*, 1962). Conversely, low-calcium diets or prepartal diets supplemented with pharmacologic doses of vitamin D have been reported to reduce the incidence of the disease (Hibbs and Conrad, 1960; Muir *et al.*, 1968).

Although cows fed a high-calcium diet (Ca 150 g:P 25 g/day) have higher blood calcium levels prepartum, they are less able to maintain serum calcium near the critical time of parturition (Black *et al.*, 1973a). Plasma immunoreactive parathyroid hormone levels are lower prepartum than in cows fed a balanced diet (Ca 25 g:P 25 g/day) and decline further at 48 hours postpartum. Inactive chief cells predominate in the parathyroid glands of cows fed high-calcium diets, whereas actively synthesizing chief cells are most numerous in parathyroids of cows fed balanced prepartal diets (Black *et al.*, 1973a). In response to the elevated blood calcium in cows fed high-calcium prepartal diets, thyroid stores of calcitonin are diminished, and C cells are partially degranulated and appear to actively synthesize more hormone (Black *et al.*, 1973a). This stimulation of C cells is accompanied by a decrease in bone turnover near parturition. Trabecular bone surfaces are inactive, and there are few osteoclasts resorbing bone.

A test of the immediately available calcium reserves by EDTA-induced hypocalcemia illustrates the long-term effects of feeding high-calcium and balanced prepartal diets on the function of parathyroid glands and bone. Cows fed a balanced diet respond to the experimental hypocalcemic challenge with a more rapid and greater increase in plasma PTH levels (Fig. 23.38) and the return of blood calcium to preinfusion levels is faster than in cows fed a high-calcium diet. This is accompanied by a marked increase in urinary hydroxyproline excretion, suggesting increased bone-matrix catabolism in response to the PTH secretion. These findings suggest that the long-term feeding of a high calcium diet prepartum will partially suppress chief cells in the parathyroid glands, so that they are less able to respond rapidly by increased PTH synthesis and secretion to a hypocalcemic challenge either induced by EDTA infusion or associated with parturition and the initiation of lactation (Fig. 23.39).

Calcium homeostasis in pregnant cows fed a high-calcium diet appears to be maintained principally by

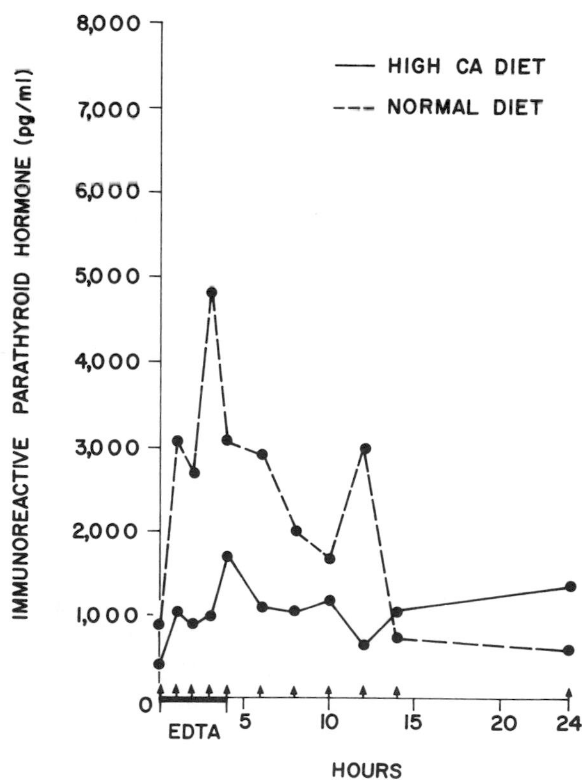

FIGURE 23.38 Parathyroid gland response to experimental hypocalcemia induced by EDTA in pregnant cows fed high-calcium and normal balanced diets. Cows fed the normal prepartal diet responded to the hypocalcemia with a more rapid and greater increase in immunoreactive PTH than cows fed a high-calcium diet. From Black *et al.* (1973a).

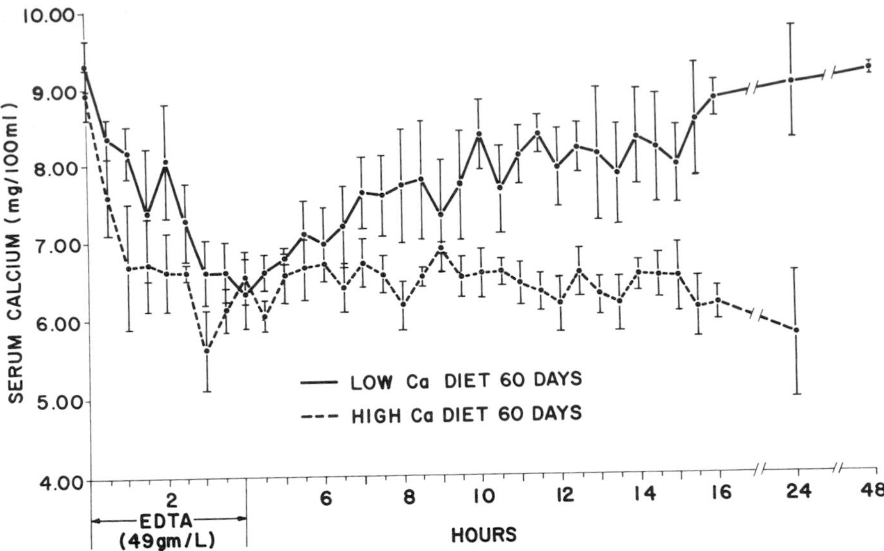

FIGURE 23.39 Immediately available calcium reserves in pregnant cows fed a low-calcium diet compared to those in pregnant cows fed a high-calcium diet prior to parturition. In response to a hypocalcemic challenge provided by EDTA infusion, cows fed the low-calcium diet are able to mobilize calcium from skeletal reserves and return the blood calcium level to the normal range more rapidly than cows fed a high-calcium diet, thereby preventing the development of hypocalcemia. In cows fed the high-calcium diet, chief cells in the parathyroid gland are inactive, and there are few osteoclasts on bone surfaces. Cows fed the low-calcium diet have predominantly actively secreting chief cells in the parathyroids and frequent osteoclasts on bone surfaces. (Capen, 1989.)

intestinal calcium absorption (Fig. 23.40). This greater reliance on intestinal absorption than on PTH-stimulated bone resorption probably is a significant factor in the more frequent development of profound hypocalcemia near parturition in cows fed high-calcium prepartal diets. These cows would be more susceptible to the decreased calcium available for absorption as a result of the anorexia often associated with the high blood estrogen levels at parturition (Muir *et al.*, 1972).

Calcium homeostasis in cows fed balanced or relatively low-calcium diets prepartum appears to be more

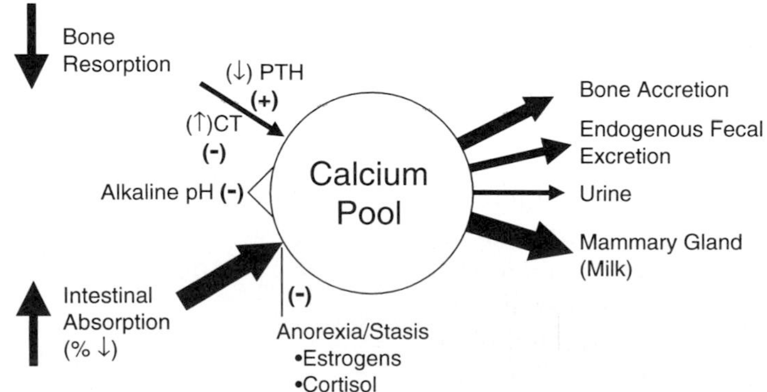

FIGURE 23.40 Calcium homeostasis in cows fed a high-calcium prepartal diet is primarily dependent on intestinal calcium absorption. The rate of bone resorption is low, and parathyroid glands are inactive. Anorexia and gastrointestinal stasis that can often occur near parturition interrupt the major inflow into the extracellular fluid calcium pool. Outflow of calcium with the onset of lactation exceeds the rate of inflow into the calcium pool, and the cows develop a progressive hypocalcemia and paresis. Modified from Mayer (1971).

under the fine control of PTH secretion with the approach of parturition (Yarrington *et al.*, 1977). The higher levels of PTH secreted during the prepartal period by an expanded population of actively synthesizing chief cells results in a larger pool of active bone-resorbing cells to fulfill the increased needs for calcium mobilization at the critical time near parturition and the initiation of lactation (Fig. 23.40). These cows would be less susceptible to the influence of decreased calcium absorption and flow into the extracellular calcium pool resulting from the anorexia and intestinal stasis associated with parturition (Fig. 23.41).

A syndrome biochemically and clinically resembling naturally occurring parturient paresis has been produced experimentally in cows by the prepartal administration of bisphosphonates (Yarrington *et al.*, 1976). These compounds are synthetic analogues of pyrophosphate and are potent inhibitors of bone resorption. When cows with no known history of parturient paresis were fed a low-calcium diet and administered dichlorodimethylene bisphosphonate (CL_2MDP) postpartum, they consistently developed hypocalcemia and hypophosphatemia with muscular weakness, incoordination, and eventually sternal or lateral recumbency (Yarrington *et al.*, 1976). These studies suggest an important role for the skeleton in the maintenance of calcium homeostasis at parturition in dairy cows.

Pharmacologic doses of vitamin D_2 have been administered to dairy cows prepartum to develop an effective method for the prevention of parturient hypocalcemia and paresis. High levels of the parent vitamin D compound are known to increase the rate and quantity of calcium and phosphorus absorbed from the intestinal tract in cattle, to elevate the blood calcium and phosphorus levels progressively after 3 to 5 days, and to increase the net calcium deposition and retention in areas of new bone growth. Because of the inherent difficulty in accurately predicting the date of parturition, however, clinical and pathological evidence of toxicity have been observed in cattle given vitamin D in large doses for an extended period (Capen *et al.*, 1966). These included gastrointestinal stasis, diuresis and anorexia, abnormal cardiac function, reduced rumination, weight loss, and extensive mineralization of the cardiovascular system. The 3- to 5-day delay preceding a significant elevation in blood calcium and phosphorus and the cardiovascular toxicity following administration for extended periods have limited the usefulness of the parent vitamin D compound in preventing the development of hypocalcemia in susceptible cows.

Active vitamin D metabolites appear to be promising in the prevention of parturient hypocalcemia and paresis. The more rapid onset of action (6 to 8 hours), greater potency (up to several hundredfold), and ability to regulate more precisely the magnitude of elevation in blood calcium, thereby minimizing potential toxic effects, are distinct advantages of the principal active metabolite of vitamin D_3 (1,25-$[OH]_2$D) over the parent vitamin D compound in the development of an effective prophylactic regimen for parturient paresis-

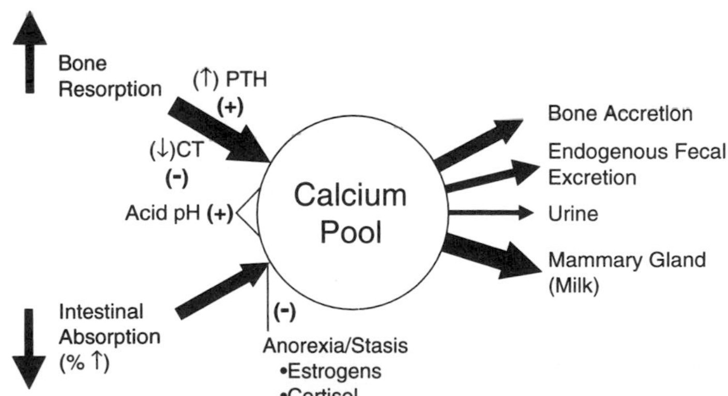

FIGURE 23.41 Calcium homeostasis in cows fed a low-calcium prepartal diet. Bone resorption and intestinal absorption both contribute substantially to the inflow of calcium to the pool in extracellular fluids. The anorexia and gastrointestinal stasis that often occur near parturition may temporarily interrupt one inflow pathway. However, there is more likely to be an adequate pool of active bone-resorbing cells capable of responding to the increased PTH secretion under these dietary conditions to maintain an approximate balance between calcium inflow and outflow, thereby preventing the development of progressive hypocalcemia. Total inflow–total outflow. Modified from Mayer (1971).

susceptible dairy cows. The administration of 1,25-$(OH)_2D$ by the intramuscular route results in a dose-related increase in serum calcium. Approximately 20% of cows with parturient paresis relapse after intravenous calcium therapy. These cows may fail to produce adequate levels of 1,25-dihydroxyvitamin D because of temporary refractoriness of the renal 25-hydroxyvitamin D 1-α-hydroxylase to the rise in serum PTH that occurs in hypocalcemic cows (Goff *et al.*, 1989).

Studies in parturient paresis-susceptible pregnant cows indicate that the intramuscular administration of 600 μg of 1,25-$(OH)_2D$ between 24 and 48 hours prepartum, before a decline in serum calcium and phosphorus had occurred, was effective in maintaining blood levels near the time of parturition. Blood calcium and phosphorus were stable or increasing at parturition and during initiation of lactation (Capen *et al.*, 1979). Blood calcium and phosphorus were stabilized by the injection of a smaller (270 μg) dose of steroid at 48- and 96-hour intervals until the actual time of parturition. Figure 23.42 illustrates the effects on serum calcium of a 600-μg dose of 1,25-$(OH)_2D$, which was administered at 5.5 days prepartum in a Jersey cow, followed after 72 hours by 270 μg of 1,25-$(OH)_2D$. The serum calcium increased progressively for 60 hours following the 600-μg injection, was 10.9 mg/dl (2.7 mmol/liter) at parturition, and remained at 9.3 mg/dl (2.3 mmol/liter) 42 hours postpartum. The serum phosphorus was 6.2 mg/dl (2.0 mmol/L) at parturition and had increased to 7.2 mg/dl (2.3 mmol/liter) at 42 hours postpartum.

b. Puerperal Tetany (Eclampsia in Dogs)

Less is known about the development of hypocalcemic syndromes in animal species other than the cow. Puerperal tetany is most frequently encountered in the small, hyperexcitable breeds of dogs. The clinical course is rapid, and the bitch may proceed from premonitory signs of restlessness, panting, and nervousness to ataxia, trembling, muscular tetany, and convulsive seizures in 8 to 12 hours (Resnick, 1972; Martin and Capen, 1980).

Functional disturbances associated with hypocalcemia in the bitch are primarily the result of neuromuscular tetany, whereas in cows the clinical signs are related to paresis. The occurrence of either tetany or paresis in response to hypocalcemia appears to be the result of basic physiologic differences in the function of the neuromuscular junction of the cow and the bitch (Bowen *et al.*, 1970a). The release of acetylcholine and transmission of nerve impulses across neuromuscular junctions are blocked by hypocalcemia in the cow (but not in the dog), leading to muscle paresis. The dog appears to have a higher margin of safety in neuromuscular transmission, in that the degree to which the endplate potential exceeds the firing threshold is greater in the dog than in the cow. Excitation–secretion coupling is maintained at the neuromuscular junction in the bitch with hypocalcemia. Tetany occurs in the

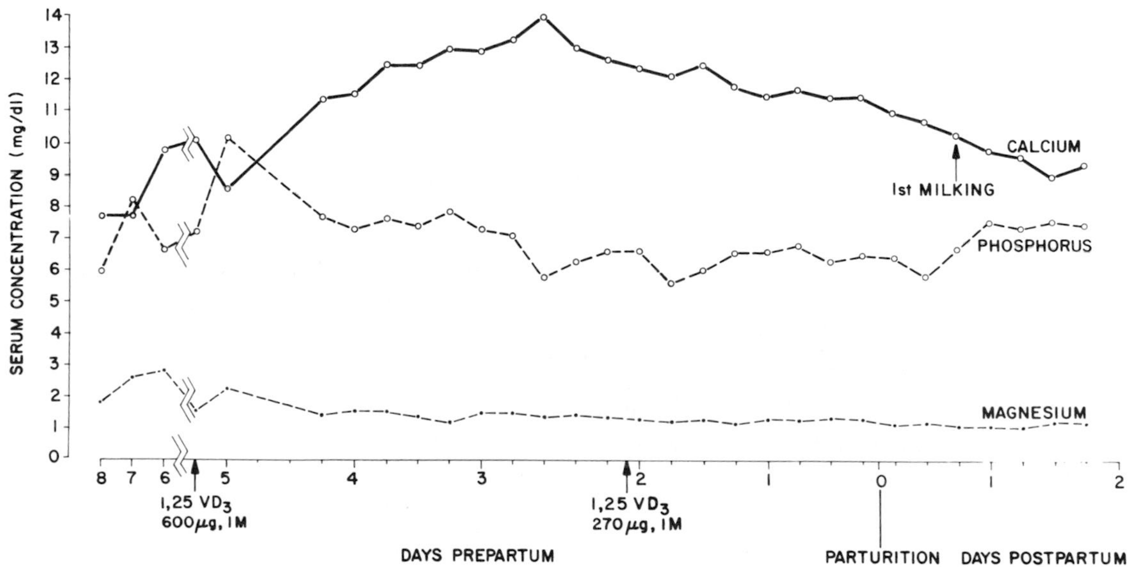

FIGURE 23.42 Effect of two doses (600 and 270 μg) of 1,25-dihydroxyvitamin D separated by a 3-day interval on serum calcium, phosphorus, and magnesium in a parturient paresis-susceptible Jersey cow. From Capen *et al.* (1979).

dog as a result of spontaneous repetitive firing of motor nerve fibers. Because of the loss of stabilizing membrane-bound calcium, nerve membranes become more permeable to ions and require a stimulus of lesser magnitude to depolarize.

There is no evidence to suggest that puerperal tetany (eclampsia) in heavily lactating bitches is the result of an interference in parathyroid hormone secretion. Severe hypocalcemia and often hypophosphatemia develop near the time of peak lactation (approximately 1 to 3 weeks postpartum), probably the result of temporary failure in homeostasis resulting in an imbalance between the rates of inflow to and outflow from the extracellular fluid calcium pool. The administration of intravenous calcium to stop the tetany, combined with temporarily decreasing the lactational drain of calcium by removing the pups, usually corrects the disruption of calcium homeostasis in the majority of bitches. Supplemental dietary calcium and vitamin D have proven useful in preventing relapses in certain bitches with puerperal tetany (Martin and Capen, 1980).

c. *Other Species*

A number of metabolic disorders characterized primarily by the development of hypocalcemia and associated manifestations occur in several different animal species (including the queen, ewe, goat, mare, sow, and chinchilla). Many of these hypocalcemic syndromes develop near the time of the increased calcium demand associated with parturition and probably are a reflection of a temporary failure of calcium homeostatic mechanisms. A comparable hypocalcemic syndrome associated with parturition does not occur in human beings. The eclampsia that develops in women is related to a toxemia of pregnancy with degenerative changes in the liver, kidney, and placenta, rather than to disturbances in mineral homeostasis.

2. *Primary Hypoparathyroidism*

In hypoparathyroidism, either subnormal amounts of parathyroid hormone are secreted by pathologic parathyroid glands or the hormone secreted is unable to interact normally with target cells. Hypoparathyroidism has been recognized occasionally in dogs, particularly in the smaller breeds such as schnauzers and terriers (Sherding *et al.*, 1980; Hulter *et al.*, 1981; Peterson, 1982; Jones and Alley, 1985). However, the incidence is much less than that of hyperparathyroidism in both dogs and cats.

Several pathogenic mechanisms can result in an inadequate secretion of parathyroid hormone. The parathyroid glands may be damaged or inadvertently removed during the course of surgery on the thyroid gland. If the parathyroid glands or their vascular supply have only been damaged but not removed, functional parenchyma often regenerates and clinical manifestations subsequently disappear.

Agenesis of both pairs of parathyroids is a rare cause of congenital hypoparathyroidism in pups. Idiopathic hypoparathyroidism in adult dogs usually is associated with diffuse lymphocytic parathyroiditis resulting in extensive degeneration of chief cells with partial replacement by fibrous connective tissue. In the early stages of lymphocytic parathyroiditis, there is extensive infiltration of the gland with lymphocytes and plasma cells and nodular regenerative hyperplasia of the remaining chief cells. Later, the parathyroid gland is completely replaced by lymphocytes, fibroblasts, and neocapillaries with only an occasional viable chief cell. The lymphocytic parathyroiditis may develop by an immune-mediated mechanism, because a similar destruction of secretory parenchyma and lymphocytic infiltration has been produced experimentally in dogs by repeated injections of parathyroid tissue emulsions (Lupulescu *et al.*, 1968). Other possible causes of hypoparathyroidism include invasion and destruction of parathyroids by primary (thyroid, salivary, etc.) or metastatic neoplasms in the anterior cervical area. Trophic atrophy of parathyroids occurs with severe hypercalcemia resulting from vitamin D intoxication, ectopic production of parathyroid hormone-related protein by nonendocrine neoplasms, and multifocal osteolytic lesions with release of calcium associated with tumor metastases.

In human beings a variant of the syndrome of hypoparathyroidism (pseudohypoparathyroidism) has been reported in which target cells in kidney and bone are unable to respond to the secretion of parathyroid hormone (Levine *et al.*, 1994). This is due to inactivating mutations in the PTH receptor, the G proteins that couple the PTH receptor to adenylyl cyclase, or the catalytic subunit of adenylyl cyclase in the plasma membrane, resulting in an inability to form cAMP in target cells (Fig. 23.13). Severe hypocalcemia develops in patients with pseudohypoparathyroidism even though parathyroid glands are hyperplastic (Roth and Capen, 1974) and immunoreactive parathyroid hormone levels are elevated.

The functional disturbances and clinical manifestations of hypoparathyroidism primarily are the result of increased neuromuscular excitability and tetany. Bone resorption is decreased and blood calcium levels diminish progressively (4 to 6 mg/dl) (1.0 to 1.5 mmol/liter) because of the lack of parathyroid hormone (Fig. 23.43). Affected dogs are restless, nervous, and ataxic

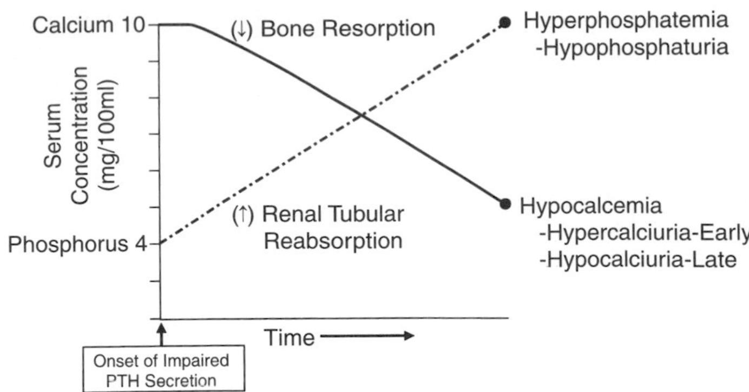

FIGURE 23.43 Hypoparathyroidism. Alterations in serum calcium and phosphorus in response to an inadequate secretion of parathyroid hormone. Progressive increases in serum phosphorus levels and a marked decline in serum calcium levels result in a neuromuscular tetany.

with intermittent tremors of individual muscle groups that progress to generalized tetany and convulsive seizures. Concurrently, blood phosphorus levels are substantially elevated because of increased renal tubular reabsorption.

3. *Hypocalcemia and Renal Secondary Hyperparathyroidism*

Disturbances in the circulating levels of Ca-regulating hormones occur frequently in animals with chronic renal failure, but most animals remain normocalcemic because of the continued coordination of PTH and 1,25-dihydroxyvitamin D on Ca metabolism. Mild to moderate hypercalcemia occurs in 10–15% of dogs and cats with chronic renal failure, whereas hypocalcemia occur more frequently (Chew *et al.*, 1992). When hypocalcemia occurs, it is largely the result of decreased intestinal absorption of Ca due to decreased serum 1,25-dihydroxyvitamin D. An additional effect can occur because of mass-law interactions of Ca with very high serum concentrations of phosphate as calcium phosphate salts are deposited in soft tissues. Consequently, hypocalcemia usually reflects advanced loss of nephron mass.

4. *Hypocalcemia and Exercise*

Blood levels of ionized calcium may decrease during intense or prolonged exercise for multiple reasons, including binding to serum proteins, complexation with lactate, changes in calcium-regulating hormones, and losses in the urine, gastrointestinal tract, or sweat. Hypocalcemia may result in increased muscular excitability or tetany and fatigue, and it may contribute to the development of synchronous diaphragmatic flutter in horses. Horses in a 3-day cross-county competition had decreased serum Ca^{2+} concentrations associated with increased serum albumin, total protein, phosphorus, and lactate concentrations (Geiser *et al.*, 1995).

D. Hypercalcitonism and Hypocalcitoninism

Clinical syndromes associated with abnormalities in secretion of calcitonin are recognized much less frequently than disorders of parathyroid hormone secretion in both animal and man (Copp, 1994). The syndromes identified so far have been primarily the result of excess secretion of calcitonin rather than a lack of secretion. A hypersecretion of calcitonin has been reported in human beings and bulls (Black *et al.*, 1973b; Capen and Black, 1974) with medullary (ultimobranchial) thyroid neoplasms derived from C cells. In human beings the syndrome often is familial and affects many individuals in a kindred.

Calcitonin-secreting C-cell neoplasms occur in populations of adult to aged bulls. The chronic stimulation of ultimobranchial derivatives in the thyroid by the high-calcium diets fed to bulls may be related to the pathogenesis of the neoplasm (Krook *et al.*, 1969, 1971b). Cows do not develop proliferative C-cell lesions under similar dietary conditions, probably because of the high physiologic requirements for calcium imposed by pregnancy and lactation. The higher plasma immunoreactive calcitonin levels in bulls than in cows may be related to their greater intake of dietary calcium relative to physiologic requirements (Deftos *et al.*, 1972) (Fig. 23.44).

C-cell (medullary) tumors of the thyroid also occur sporadically in dogs and horses as a firm mass in the anterior cervical region (Leav *et al.*, 1976). Calcitonin activity can be localized to the cytoplasm of tumor cells by immunohistochemical techniques (Leblanc *et al.*, 1991). In addition to calcitonin, medullary C-cell

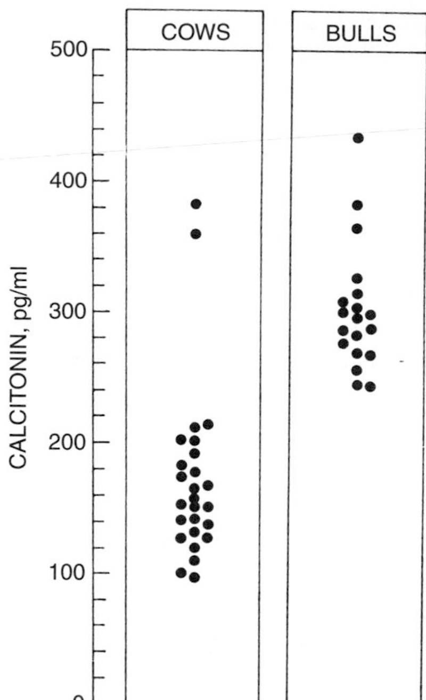

FIGURE 23.44 The higher plasma immunoreactive calcitonin levels in bulls (303 ± 13 pg/ml) than in cows (165 ± 12 pg/ml) may be related to the high dietary intake of calcium relative to physiologic requirements. This prolonged stimulation of calcitonin-secreting C cells by the high calcium intake in the diet has been suggested to be important in the pathogenesis of the more frequent incidence of C-cell (ultimobranchial) neoplasms in older bulls than in cows. From Deftos *et al.* (1972).

tumors may secrete other humoral substances, such as prostaglandins, serotonin, and bradykinin. The incidence of occurrence of C-cell tumors in dogs is uncertain, but it was reported that 7 of 200 thyroid tumors in dogs were derived from C-cells (Zarrin, 1977). They often are firm on palpation because of the presence of large amounts of amyloid in the stroma. Thyroid neoplasms of C-cell origin can be readily differentiated ultrastructurally by the presence of numerous membrane-limited secretory granules in the cytoplasm. Small granules of this type are not present in thyroid tumors derived from follicular cells.

C-cell tumors in both human beings and bulls often are associated with the simultaneous occurrence of pheochromocytomas in the adrenal medulla and neoplasms in other endocrine organs of neural-crest origin (Capen and Black, 1974). Serum calcium and phosphorus levels in adult animals with a chronic excessive secretion of calcitonin usually remain in the low-normal range because of the relatively slow turnover rate of bone. Osteosclerotic changes have been reported in bulls with this syndrome of long-term and excessive calcitonin excretion.

Specific disease syndromes resulting from a lack of calcitonin secretion have not been recognized in either man or animals. However, experimentally thyroidectomized animals are less able than normal to manage a calcium load and may develop postprandial hypercalcemia (Copp, 1969a).

E. Rickets and Osteomalacia

Rickets is a disease of bone that occurs in young animals because of phosphorus or vitamin D deficiency, whereas osteomalacia is a disease of the adult skeleton due to the same causes (Palmer, 1993). Rickets and osteomalacia are characterized by a failure of mineralization and an accumulation or excess of nonmineralized osteoid in areas of bone formation. In rickets there is, in addition, failure of mineralization at the physeal zones of endochondral ossification. This results in thickening of the physis by a zone of proliferating cartilage cells that deforms and bows because it cannot support the body weight. Animals with osteomalacia do not develop widened physes if the deficiency of P or vitamin D occurs after closure of the physis. Osteomalacia is characterized by an accumulation of unmineralized osteoid on trabecular endosteal surfaces and in Haversian systems.

Serum alkaline phosphatase is increased in most forms of rickets and osteomalacia because of an increase in bone-specific alkaline phosphatase. The elevation in serum alkaline phosphatase reflects the increased osteoblastic activity in bone in these diseases. Serum alkaline phosphatase also may be increased in other metabolic bone diseases such as primary and secondary hyperparathyroidism (Delmas, 1995). Bone-specific alkaline phosphatase constitutes the major fraction of total serum activity in young animals, but represents only about 50% of the total activity in adults. There are three genes for placental, intestinal, and bone–liver–kidney alkaline phosphatase isoenzymes. The bone, liver, and kidney forms differ in electrophoretic mobility (even though they are translated from the same mRNA), probably because of different post-translational modifications of carbohydrate side chains. The role of alkaline phosphatase in bone homeostasis is not completely understood. Hypophosphatasia is an inherited autosomal recessive defect in the bone–liver–kidney alkaline phosphatase gene that results in severe rickets in humans. This suggests a function of alkaline phosphatase in the mineralization process that may include (1) the release of local inorganic phosphate and (2) hydrolysis of inorganic pyrophosphate, which is an inhibitor of hydroxyapatite crystal formation.

1. *Rickets Due to Vitamin D or Phosphorus Deficiency*

In young animals, vitamin D is required for the orderly growth and mineralization of cartilage in the physis (growth plate) (Capen, 1985). Young animals fed diets deficient in vitamin D and housed indoors without exposure to ultraviolet radiation develop rickets. In the absence of vitamin D, mineral granules do not accumulate within hypertrophied chondrocytes in the physes of long bones, and there may be a decrease in serum Ca and P levels due to impaired intestinal absorption. Mineralization of the cartilaginous matrix fails to occur, and the physis becomes irregularly thickened as the proliferating chondrocytes and cartilaginous matrix persist because they are not removed by endochondral ossification. Endochondral ossification (vascular invasion and resorption of cartilage with deposition of osseous tissue on residual cartilage septae) requires mineralization of the growth plate. The administration of cholecalciferol, 25-hydroxyvitamin D, or 1,25-dihydroxyvitamin D leads to reestablishment of a normal calcification front in the osteoid on bone surfaces and the physes, often before there is a change in the mineral-ion concentration of extracellular fluids. Secondary hyperparathyroidism may occur as a late manifestation of rickets associated with vitamin D deficiency due to the decrease in serum Ca concentration.

Phosphorus deficiency also will result in rickets because of the failure to maintain an adequate ion product of serum Ca and P at the zones of mineralization in bone. Phosphorus deficiency will not result in hypocalcemia, and serum 1,25-dihydroxyvitamin D levels will be normal or increased.

Rickets seldom occurs in suckling animals unless the dam provides too little milk. Rickets is rare in foals, has been reported in calves and sheep, and occurs in puppies and kittens because of vitamin D or phosphorus deficiency. Rickets may be particularly severe in pigs owing to their rapid growth rate.

2. *Rickets Due to Hereditary Defects in Vitamin D Action*

Several forms of rickets occur in children, pigs, and laboratory animals because of abnormal metabolism of vitamin D (Liberman, 1996). Vitamin D-refractory syndromes occur in individuals with the morphologic lesions of rickets when fed normal diets in the absence of malabsorption or when exposed to adequate sunlight.

a. Vitamin D-Dependent Rickets, Type I (Hereditary Deficiency in 1,25-Dihydroxyvitamin D Production)

Vitamin D-dependent rickets, Type I, in both pigs and human beings is a familial disease inherited by an autosomal recessive gene (Wilke *et al.*, 1979). Newborn pigs appear healthy and have normal blood calcium and phosphorus concentrations. At 4–6 weeks of age, blood Ca and P decrease, serum alkaline phosphatase activity increases, and clinically detectable rickets develops during the following 3–4 weeks. The pigs develop deformities of bone in the axial and abaxial skeleton, severe pain, and classic lesions of rickets.

In response to the hypocalcemia, plasma immunoreactive PTH levels in pigs with vitamin D-dependent rickets are elevated and serum 25-hydroxycholecalciferol levels are markedly increased (75 ng/ml) compared to those of control pigs (8 ng/ml) (Wilke *et al.*, 1979). Serum 1,25-dihydroxyvitamin D levels are depressed in rachitic pigs (17 pg/ml) compared with levels in normal pigs (47 pg/ml). Homozygous, rachitic animals have no renal 25-hydroxycholecalciferol-1-hydroxylase in kidney homogenates or mitochondrial preparations, which is responsible for the insufficient renal production of 1,25-dihydroxyvitamin D (Winkler *et al.*, 1986). Repeated administration of 25-hydroxycholecalciferol will not significantly change the plasma concentrations of Ca, P, and immunoreactive PTH; however, administration of 1,25-dihydroxyvitamin D daily for 3–4 weeks results in increased plasma Ca, decreased P, and return of immunoreactive PTH and alkaline phosphatase levels to near the normal range.

b. Vitamin D-Dependent Rickets, Type II (Hereditary Defects in the Vitamin D Receptor–Effector System)

Vitamin D-dependent rickets, Type II, is a group of disorders characterized by target organ resistance to 1,25-dihydroxyvitamin D (Liberman, 1996). The disease is characterized by increased serum 1,25-dihydroxyvitamin D levels, hypocalcemia, and secondary hyperparathyroidism, and clinically by rickets or osteomalacia. The disease in humans responds variably to high doses of 1,25-dihydroxyvitamin D. New World primates have a relative end-organ resistance to 1,25-dihydroxyvitamin D and require high levels of vitamin D_3 in their diet. The common marmoset has been shown to represent a model of vitamin D-dependent rickets, Type II (Yamaguchi *et al.*, 1986). Marmosets have increased serum 1,25-dihydroxyvitamin D concentrations compared to those of rhesus monkeys, and some marmosets develop osteomalacia even when on high-vitamin D_3 diets. The resistance to 1,25-dihydroxyvitamin D is likely due to low affinity binding of 1,25-dihydroxyvitamin D to its nuclear receptor (Gacad and Adams, 1993).

3. *X-linked Hypophosphatemic Rickets*

X-linked hypophosphatemia is a disorder in mice and humans characterized by hypophosphatemia, normocalcemia, decreased tubular reabsorption of phosphorus, and defective mineralization of cartilage and bone (Drezner, 1996). The disease is due to impaired phosphate reabsorption in the proximal convoluted tubules of the kidney. There may be decreased circulating levels of 1,25-dihydroxyvitamin D for the degree of hypophosphatemia. Treatment with phosphorus and pharmacologic doses of vitamin D or one of its metabolites is required.

4. *Osteomalacia Due to Phosphorus or Vitamin D Deficiency*

Osteomalacia literally means "softening of the bones" and is a generalized disease characterized by defective mineralization of osteoid in mature bone (Palmer, 1993). Osteomalacia due to P or vitamin D deficiency results from an interference with mineralization of normal bone matrix (osteoid). In the early stages there is a greater extent of osteoid-covered bone surfaces, and later there is an increase in osteoid seam width. There usually is much less inhibition of bone matrix formation than impairment of matrix mineralization.

Adequate local concentrations of Ca and P are necessary for proper deposition of mineral at the mineralization front. Hypophosphatemia plays a major role in the pathogenesis of osteomalacia. Phosphorus deficiency inhibits normal osteoblast function and reduces the Ca $\times$ P ion product. Vitamin D deficiency will also result in osteomalacia that resolves upon adequate supplementation or ultraviolet light exposure. The mechanisms of vitamin D deficiency-induced osteomalacia are not completely understood. The active form of vitamin D, 1,25-dihydroxyvitamin D, stimulates Ca and P absorption from the intestine. In addition, there are receptors for 1,25-dihydroxyvitamin D in osteoblasts, and 1,25-dihydroxyvitamin D stimulates increased expression of alkaline phosphatase and other proteins in osteoblasts. It is likely that proper levels of vitamin D are necessary for adequate Ca and P absorption and normal osteoblast function. The vitamin D metabolite 24,25-dihydroxycholecalciferol may also play a role in the normal mineralization of osteoid (Ornoy *et al.*, 1978).

The presence of osteomalacia is difficult to assess clinically. The diagnosis usually requires an interpretation of bone pain, serum and urine Ca and P concentrations and excretion rates, serum concentrations of vitamin D metabolites, radiography, and most conclusively, histomorphometric analysis of a mineralized bone biopsy.

Osteomalacia also may occur during fat malabsorption syndromes due to inhibition of absorption of Ca and P, and the loss of the fraction (20–30%) of 25-hydroxyvitamin D that undergoes enterohepatic recirculation (Marel *et al.*, 1986). Adults with vitamin D-refractory syndromes, X-linked hypophosphatemia, or Fanconi syndrome will also develop osteomalacia. One of the manifestations of a Fanconi-like syndrome in dogs is a reduction in the tubular reabsorption of P in the kidney (Bovee *et al.*, 1979). Metabolic acidosis is associated with hypercalciuria, loss of skeletal Ca, and renal P wasting, which may induce rickets and osteomalacia (Marel *et al.*, 1986).

5. *Tumor-Induced Osteomalacia*

Benign or occasionally malignant mesenchymal tumors have been associated with a paraneoplastic syndrome in human beings characterized by hypophosphatemic osteomalacia (Lyles, 1996). Tumors reported include hemangioma, fibroma, giant-cell tumor, and hemangiopericytoma. The syndrome is characterized by hypophosphatemia associated with increased fractional P excretion and low circulating levels of 1,25-dihydroxyvitamin D due to inhibition of renal 1α-hydroxylase. It appears that the tumors release a humoral factor that causes phosphaturia and abnormal vitamin D metabolism, because removal of the tumors results in amelioration of clinical signs and of the osteomalacia.

6. *Osteomalacia and Chronic Renal Failure*

Most chronic renal-failure patients develop fibrous osteodystrophy to some degree because of renal secondary hyperparathyroidism; however, a minority of dialyzed human patients develop severe osteomalacia (Marel *et al.*, 1986). Aluminum (Al) accumulation in bone may play a central role in the development of renal osteomalacia. The incidence of renal osteomalacia correlates well with the presence of Al in bone, the content of aluminum in dialysate water, and oral administration of aluminum hydroxide to control hyperphosphatemia by binding intestinal phosphorus. Aluminum is normally excreted by the kidney and accumulates in bone at sites of mineralization. Aluminum may also suppress PTH secretion, which accentuates the osteomalacia; however, administration of Al to dogs did not impair the PTH response to hypocalcemia (Henry *et al.*, 1984).

F. Osteoporosis

Osteoporosis is defined as a reduction in bone density that predisposes bone to fractures due to an imbalance in bone formation and bone resorption. It may occur as a result of malnutrition (starvation or calcium or phosphorus deficiency), during protein-wasting diseases such as parasitism, as a result of local or generalized skeletal disuse, or as an aging change (Palmer, 1993). Osteopenia is defined as a reduction in bone mass without an interpretation on its quality. Many forms of osteopenia occur secondary to endocrinopathies or from excessive or prolonged use of drugs. Age-related osteoporosis is an important disease of humans, but is not a major disease problem in domestic animals (Rodan *et al.*, 1996). There are two types of age-related osteoporosis in humans: Type I or postmenopausal osteoporosis occurs predominantly in women (51–75 years of age) and is associated with loss of trabecular bone, and Type II or senile osteoporosis occurs only slightly more often in women (2:1), is seen in older individuals (>70 years old), and is associated with loss of cortical and trabecular bone. There is a mild, transient increase in bone remodeling in dogs after ovariohysterectomy, but the degree of bone loss associated with ovarian insufficiency is not clinically significant (Boyce *et al.*, 1990).

Bone remodeling in the adult skeleton takes place in discrete foci termed bone remodeling units (BRU). At each BRU bone is first resorbed by osteoclasts and then formed by osteoblasts. The BRU is represented by the formation of osteons in cortical bone and bone structural units in trabecular bone. With advancing age, bone resorption by osteoclasts exceeds bone formation by osteoblasts, resulting in an imbalance in bone remodeling. If bone remodeling is increased, i.e., there are more BRUs, the imbalance will be enhanced. Bone density in humans is at its maximum at about 35 years of age. There is a reduction in bone density of 50 and 35% in trabecular and cortical bone, respectively, over the human lifespan. Bone loss is enhanced after menopause because of the loss of estrogens, or during androgen deficiency in males. The mechanism of action of the sex hormones is not known, but the principal effect of estrogen lack or androgen withdrawal is increased bone resorption. Estrogen deficiency leads to increased activity of local factors in bone that may be responsible for increasing bone resorption. These factors include interleukin-1, interleukin-6, tumor necrosis factor α, and prostaglandins.

The development of osteoporosis also will be enhanced by chronic liver or kidney disease, adrenal glucocorticoid excess, thyroid hormone excess, immobilization, and low calcium intake. Glucocorticoids decrease calcium absorption from the intestine, which results in secondary hyperparathyroidism, and are potent inhibitors of bone formation (Norrdin *et al.*, 1988). Thyroid hormones increase bone remodeling, but may stimulate bone resorption to a greater extent than bone formation (Stern, 1996).

G. Biochemical Markers of Bone Metabolism

One of the most accurate methods to assess bone remodeling, including both formation by osteoblasts and resorption by osteoclasts, is the cancellous bone biopsy (Rosol *et al.*, 1992a; Recker, 1996). This is a technically demanding procedure and requires special facilities for processing and special expertise for evaluation. Recent advances have been made in measurement of circulating biochemical markers of bone turnover (Table 23.5) (Delmas, 1995; Eriksen *et al.*, 1995; Calvo *et al.*, 1996). Some of these markers provide useful indirect measures of bone resorption and bone formation *in vivo*. Bone resorption can be estimated by measuring the serum or urine concentrations of osteoclast-derived enzymes or collagen degradation products. Bone formation can be indirectly evaluated by measuring serum concentrations of osteoblast proteins, enzymes, or precursor peptides released during collagen synthesis.

Because of wide normal variation in the serum markers or lack of bone specificity, individual samples may have limited diagnostic value. However, samples collected over time to monitor changes in concentrations or samples from groups of animals are more valuable diagnostically or in experimental investigations. Most of the serum markers are higher in young animals than in adults because of the rapid turnover of bone during growth. Bone turnover *in vivo* follows a pronounced circadian pattern with increased bone resorption after food intake or during the night (Mühl-

TABLE 23.5 Biochemical Markers of Bone Turnover

Bone resorption: serum markers
Tartrate-resistant acid phosphatase
Bone resportion: urine markers
Pyridinoline and deoxypyridinoline (collagen cross-links) and cross-linked collagen telopeptides
Fasting calcium concentration
Hydroxyproline
Hydroxylysine glycosides
Bone formation: serum markers
Alkaline phosphatase (total or bone-specific isozyme)
Osteocalcin (bone Gla protein)
Procollagen I carboxy-terminal extension peptide

bauer and Fleisch, 1995). Therefore, sample collection should always be at the same time of day.

1. *Bone Resorption*

a. *Tartrate-Resistant Acid Phosphatase*

Serum acid phosphatase activity is derived from lysosomes from cells in bone, prostate, platelets, erythrocytes, and the spleen. There are at least six isoenzymes. In bone, acid phosphatase is produced by osteoclasts. Acid phosphatase produced by osteoclasts in bone and macrophages throughout the body is resistant to inhibition by L(+)-tartrate. Serum tartrate-resistant acid phosphatase activity is an insensitive measure of bone resorption because of the lack of osteoclast specificity and the potential for enzyme inhibitors in serum. Future development of immunoassays specific for tartrate-resistant acid phosphatase may be more useful (Price *et al.*, 1995).

b. *Pyridinoline and Deoxypyridinoline (Collagen Cross-Links)*

Pyridinoline (Pyr) and deoxypyridinoline (D-Pyr) are covalent cross-links between collagen molecules generated from lysine and hydroxylysine and are potentially useful markers of bone resorption (Delmas, 1995). Pyr and D-Pyr are absent from skin, which is an abundant source of collagen. Most Pyr and D-Pyr in serum and urine is likely derived from bone because of the large amount of bone collagen in the body and its constant rate of turnover. Pyr and D-Pyr are not metabolized *in vivo* and are excreted in urine in free form (30–40%) or bound to N-terminal and C-terminal collagen telopeptides (60–70%). Total urine Pyr and D-Pyr can be measured by fluorometry after reverse-phase HPLC purification of hydrolyzed urine (Pratt *et al.*, 1992). Widespread measurement of urinary Pyr and D-Pyr will require development of more convenient assays, such as immunoassays. Immunoassays have been developed for Pyr and D-Pyr-crosslinked collagen telopeptides present in urine and serum, but the clinical utility of these assays is currently under investigation (Delmas, 1995).

c. *Fasting Urine Calcium*

The concentration of calcium in the urine (corrected for creatinine) of a fasted animal is an indirect measure of bone resorption because of the lack of contribution of calcium absorption from the gastrointestinal tract. This is an insensitive measure and will only detect large increases in bone resorption. Calcium-regulating hormones, such as PTH, also will affect the resorption rate of calcium by the kidneys.

d. *Urine Hydroxyproline*

Hydroxyproline is present mainly in collagen (representing 13% of the amino acids), but is also present in the C1q fraction of complement. Hydroxyproline is released from collagen during degradation and is not reutilized. Because bone collagen represents 50% of the body's collagen, urine hydroxyproline is an indirect marker of bone resorption. However, it is an insensitive measure of such resorption because 90% of the filtered hydroxyproline is reabsorbed by the kidney and oxidized in the liver. Hydroxyproline can be measured by colorimetric assay of a hydrolyzed urine sample (Kivirikko *et al.*, 1967) or HPLC after hydrolysis and derivatization (Paroni *et al.*, 1992).

e. *Urine Hydroxylysine*

Hydroxylysine is a unique amino acid present in collagen that is not reutilized after collagen degradation, but is less abundant than hydroxyproline (Delmas, 1995). Hydroxylysine exists in two forms in collagen (galactosyl hydroxylysine and glucosyl hydroxylysine), and the forms differ in proportion in soft tissues and bone. Therefore, assay of urinary excretion of galactosyl hydroxylysine and glucosyl hydroxylysine may be a more sensitive measure of bone resorption than a similar assay for hydroxyproline. The usefulness of measurement of these two compounds is limited by the HPLC technique used in the assay.

2. *Bone Formation*

a. *Alkaline Phosphatase*

Total serum alkaline phosphatase activity is a combination of isoenzymes produced in the liver, bone, kidney, placenta, and intestine. The liver and bone isoenzymes represent the greatest fractions. Bone alkaline phosphatase is present in osteoblast cell membranes and is released by an unknown mechanism during bone formation. Measurement of total alkaline-phosphatase activity spectrophotometrically using *p*-nitrophenyl phosphate is an insensitive measure of bone formation because of the contribution of multiple isoenzymes. Bone and liver alkaline phosphatase may be separated by electrophoresis, but this is cumbersome and leads to only minor improvements in the specificity of the assay. Both the liver and bone isoenzymes are encoded by the same gene, but differ in posttranslational glycosylation. This has enabled development of an immunoassay to human bone-specific alkaline phosphatase with a low degree of cross reactivity to other isoenzymes and has increased the usefulness of alkaline phosphatase measurement as a marker of bone formation (Delmas, 1995).

b. Osteocalcin

Osteocalcin (γ-carboxyglutamic acid-containing protein, bone Gla protein) is a vitamin K-dependent protein that is the most abundant noncollagenous protein in bone and is produced by osteoblasts and odontoblasts (Ducy and Karsenty, 1996). It has 49 amino acids and represents 1–2% of total bone protein. The active form of vitamin D, 1,25-dihydroxyvitamin D, is a potent inducer of osteocalcin production by osteoblasts. Measurement of serum osteocalcin by radioimmunoassay is one of the most useful markers of bone formation (Delmas, 1995).

Osteocalcin has a short half-life and is rapidly cleared by the kidney, so serum levels are dependent on renal function. Osteocalcin is quickly degraded in serum *in vivo* and *in vitro* at room temperature. Circulating forms of osteocalcin include the intact peptide, an N-terminal/midregion fragment, and a C-terminal fragment. These forms may interfere with certain radioimmunoassays. Antibodies to osteocalcin cross react between species. Antiserum to bovine osteocalcin can be used to measure human and equine osteocalcin by RIA (Hope *et al.*, 1993). Canine osteocalcin protein has been isolated from bone and sequenced, and a sensitive RIA has been developed (Colombo *et al.*, 1993; Fanti *et al.*, 1993)

c. Procollagen I Carboxy-Terminal Extension Peptide

Collagen Type I constitutes 97% of the collagen in bone. Collagen type I is synthesized as procollagen by osteoblasts during bone formation; it contains peptide extensions at the C- and N-terminal ends that are cleaved during synthesis of collagen fibrils. The procollagen peptides that are released into the circulation have a short half-life and are cleared by the liver. Serum concentrations of C-terminal extension peptides can be measured by radioimmunoassay; however, peptide concentrations are insensitive measures of bone formation and may have a low correlation with bone formation *in vivo* (Delmas, 1995).

V. MAGNESIUM METABOLISM

Magnesium (Mg) is an essential dietary element for animals (Speich and Bousquet, 1991; McDowell, 1992b). Green plants are an excellent dietary source of Mg for animals because of the presence of Mg^{2+} in chlorophyll (Wilkinson *et al.*, 1990). Magnesium metabolism has been studied most extensively in cattle and sheep, because clinical disorders related to magnesium deficiency occur most commonly in those species (Birch, 1990). However, disorders related to magnesium metabolism also are recognized in other animals, including cats, dogs, goats, and horses.

Magnesium ions (Mg^{2+}) and Ca^{2+} bind to negatively charged molecules (e.g., ATP, DNA, RNA), enzymes, and other binding proteins. Mg^{2+} and Ca^{2+} differ only in their radius and polarizability. Mg^{2+} is smaller than Ca^{2+} and binds to many ligands with greater affinity than does Ca^{2+} (Martin, 1990). Interactions between the two ions occur both intra- and extracellularly (Vormann and Günther, 1993). However, different types of interactions occur depending on the binding protein. Such interactions cannot be easily predicted without empirical testing. Depending on the physiological system or binding molecule, Mg^{2+} and Ca^{2+} can interact in the following ways: (1) Mg^{2+} and Ca^{2+} can have similar biological effects, (2) Mg^{2+} and Ca^{2+} can bind to molecules with different affinities resulting in partial synergy or antagonism, (3) the cations can be direct antagonists to each other with no cross-reactivity, or (4) effects can be dependent on Mg^{2+} or Ca^{2+} with no interaction between cations (Vormann and Günther, 1993). The pathophysiology of Mg-related disorders is often complicated by the Ca^{2+} status of the animal, as occurs in cattle with grass tetany. Conditions of hypomagnesemia will lead to exaggerated effects of Ca^{2+} because the antagonistic effects of Mg^{2+} will not be present. In contrast, conditions of hypermagnesemia may produce antagonistic effects on Ca^{2+} metabolism.

A. Distribution of Magnesium in the Body

Magnesium (atomic mass 24.32) is the fourth most prevalent cation in the body, being surpassed by calcium, sodium, and potassium, but ranks second to potassium as a soft-tissue cation and resembles K^+ in distribution (Altura *et al.*, 1987). The body of domestic animals contains 0.05% magnesium by weight, of which 60% is in the skeleton, 38% in the soft tissue, and 1–2% in the extracellular component (Martens and Rayssiguier, 1980). Magnesium content of bone ranges from 4 to 8 gm/kg with a Ca : Mg ratio of 50–100 : 1 (Meyer and Zentek, 1990). About 70% of bone Mg is removable by treatment with dilute acid and represents Mg bound to the surface of bone crystals, whereas the remaining 30% is an integral part of the bone crystal lattice (Taylor, 1959). Most of the soft-tissue component resides in skeletal muscle (320–420 mg/kg wet tissue) and liver (Al-Khamis, 1995).

Some of the Mg in bone is available for mobilization to soft tissues in young animals when dietary intake is inadequate. Rib Mg content decreased 30% in calves fed Mg-deficient diets (McKim *et al.*, 1984). The adult is less able to mobilize Mg from the skeleton or muscle during dietary deficiency, and animals may die in tetany with little depletion of bone Mg. Because there is little body reserve available to offset Mg deficiency, a continuous intake of Mg is necessary for health (Wallach, 1987). However, Mg mobilization from bone in adult sheep on Mg-deficient diets may partially protect them from developing clinical signs of hypomagnesemia (Matsui *et al.*, 1994).

Soft-tissue Mg occurs intracellularly at approximately 360 mg/kg compared with the plasma concentration of about 2.4 mg/dl (1.0 mmol/liter). Intracellular magnesium exists in a free ionized form (1%) that is readily exchangeable and a bound form (99%) that is chelated, typically with ATP and apoenzyme proteins. A simple diffusion equilibrium probably exists between the labile, exchangeable form and the Mg^{2+} in the extracellular fluid, whereas a more complex relation occurs between the two intracellular forms.

The blood–brain barrier in neonatal animals permits Mg^{2+} transport, but forms a functional barrier in the postnatal period (Rivera *et al.*, 1991). However, Mg^{2+} concentration in the cerebrospinal fluid (CSF) is usually greater than in the serum, which indicates that Mg^{2+} is actively transported by the choroid plexus (Vormann and Günther, 1993). This is in contrast to the Ca^{2+} concentration in CSF, which is 50% less than serum. The CSF [Mg^{2+}] decreased in proportion to serum [Mg^{2+}] in calves with experimental hypomagnesemia (Sarode *et al.*, 1992). The concentration of Mg in CSF is relatively stable (approximately 2.2 mg/dl [0.91 mmol/liter]) when serum Mg concentration is greater than 1.0 mg/dl (0.41 mmol/liter), but can fall sharply when serum Mg decreases below this level (Meyer and Zentek, 1990).

It is difficult to assess the Mg status of an animal, because plasma Mg (the easiest measured source of Mg) represents less than 1% of total body Mg (Elin, 1987b; Reinhart, 1988; Nowitzki-Grimm *et al.*, 1991; Elin, 1994). Total Mg content of body fluids or tissues can be measured by colorimetric assay, kinetic assay, and atomic or atomic-emission spectrophotometry, and ionized Mg can be measured using an ion-selective electrode (Elin, 1987b; Seiler, 1990; Blanchflower and Kennedy, 1991; Altura *et al.*, 1994; Anderson and Talcott, 1994; Ising *et al.*, 1995). Erythrocyte Mg content can be used to estimate Mg status of the body, but it may not accurately reflect the Mg content of all tissues (Nowitzki-Grimm *et al.*, 1991; Nozue *et al.*, 1991). Mg depletion will lead to a sudden decrease in plasma Mg concentration and a progressive decline in erythrocyte Mg concentration (Mulei and Daniel, 1989; Geven *et al.*, 1990). Magnesium content of mononuclear white blood cells may be a better estimate of Mg status compared to erythrocyte Mg concentration (Elin, 1987a). Measurement of Mg retention by quantifying total urine Mg after parenteral administration is a potentially useful method to assess body Mg status (Martin, 1994). Magnesium balance is principally controlled by absorption from the gastrointestinal tract and excretion by the kidney. There appears to be no primary hormonal control of plasma Mg concentration.

The physiological functions of Mg may be divided into intracellular and extracellular categories.

1. *Intracellular Function of Magnesium*

Magnesium is essentially an intracellular cation and functions as an activator or a catalyst for more than 300 enzymes in the body, including phosphatases and enzymes that involve ATP (Heaton, 1990). All reactions that utilize ATP require Mg^{2+} because the intracellular substrate is the complex, MgATP. The action of Mg extends to all major anabolic and catabolic processes. Thus, Mg plays a pivotal role in muscle contraction, protein, fat, and carbohydrate metabolism, methyl-group transfer, oxidative phosphorylation, functional properties and stabilization of membranes, cell division, and immune responses. Magnesium regulates ribosomal RNA and DNA structure, thereby affecting cell growth and membrane structure (Günther, 1990). Magnesium is required for maintenance of normal cellular potassium (Ryan, 1993), and Mg deficiency can lead to intracellular K depletion and excessive K excretion (Abbott and Rude, 1993). Magnesium regulates mitochondrial membrane permeability, and one of the earliest ultrastructural changes observed in Mg deficiency is swelling and disruption of mitochondrial cristae (Altura *et al.*, 1987); however, intracellular Mg^{2+} is usually constant in different physiologic or pathophysiologic states.

Total intracellular magnesium can be measured with atomic absorption spectrophotometry, and free intracellular Mg^{2+}, $[Mg^{2+}]_i$, can be measured using a specific Mg^{2+} microelectrode, ^{31}P nuclear magnetic resonance, or Mg^{2+}-specific fluorescent dyes (Fujise *et al.*, 1991; London, 1991; Murphy, 1993). Intracellular free concentrations of Mg^{2+} in nucleated mammalian cells usually range from 0.3 to 3.0 mmol/liter, with the highest levels in myocardial myocytes (2.5–3.1 mmol/liter) and lower concentrations in vascular smooth muscle cells (0.1–1.0 mmol/liter) and erythrocytes (0.3 mmol/liter) (Altura *et al.*, 1987; Fujise *et al.*, 1991; Fenwick and Daniel, 1991). The $[Mg^{2+}]_i$ represents a balance among influx, efflux, and intracellular buffering or chelation (Günther, 1993). In general, Mg^{2+} enters the cell passively and is actively extruded from the cytoplasm (Beyenbach, 1990). Much of the intracellular Mg^{2+} is bound to ATP in the form of $MgATP^{2-}$. Intracellular

Mg^{2+} concentration may be relatively stable, vary between cell regions, or be regulated by various hormones depending on the cell type, but the mechanisms that regulate $[Mg^{2+}]_i$ are poorly understood (Günther, 1986; Romani *et al.*, 1993). Alterations in $[Mg^{2+}]_i$ are slower and last longer than changes in intracellular Ca^{2+}. Therefore, $[Mg^{2+}]_i$ may play a role in the sensitization or desensitization of cellular response systems rather than an acute "on–off" signal as occurs with changes in $[Ca^{2+}]_i$ (Maguire, 1990).

Magnesium deficiency is associated with increased insulin and epinephrine secretion. Insulin increases cellular Mg, and epinephrine decreases cellular Mg; therefore, there is near-normal cellular Mg equilibrium during Mg deficiency. However, insulin and epinephrine can cause cell-membrane depolarization, increased cell Ca and P_i (which may result in cellular calcinosis), and increased cGMP. These effects can be antagonized by the Mg-like effects of taurine, which is mobilized during Mg deficiency. Taurine metabolism may be one of the factors that account for the occurrence of latent clinical disease in states of Mg deficiency (Durlach and Durlach, 1984; Durlach *et al.*, 1987).

2. *Extracellular Function of Magnesium*

Magnesium interacts with extracellular Ca^{2+} by binding to many shared ligands. In general, Mg^{2+} acts as a Ca^{2+} antagonist, and hypomagnesemia is associated with accentuated actions of Ca^{2+}. Mg^{2+} binds to chief cells in the parathyroid glands and inhibits secretion of PTH, but the effect is not as great as that of Ca^{2+}.

Magnesium is important in the production and decomposition of acetylcholine. Low concentrations of Mg or a low Mg^{2+}/Ca^{2+} ratio potentiates the release of acetylcholine at neuromuscular junctions. Therefore, a low concentration of Mg in the extracellular fluid may lead to tetany due to the increased release of acetylcholine at motor end plates. Cholinesterase activity is unaffected by Mg deficiency.

B. Absorption and Excretion

Dietary Mg has a low availability compared to that of dietary Ca. The apparent digestibility of Mg (the percentage of dietary Mg that is not excreted in the feces) ranges up to 40% in ruminants and up to 60–65% in dogs, cats, and horses (Hintz and Schryver, 1972; Meyer and Zentek, 1990). Digestibility of dietary Mg is greater in suckling animals than in adults. The availability of dietary Mg varies considerably depending on type of feed and individual animal variation.

1. *Endogenous Fecal Magnesium (Fecal Mg from Digestive Secretions)*

Magnesium in the feces is not solely unabsorbed dietary Mg, because digestive secretions may contain a considerable amount of endogenous Mg. The apparent digestibility of Mg (food Mg − fecal Mg/food Mg) is a valid estimate of true digestibility only when endogenous Mg in the feces is a small part of total fecal Mg. Endogenous Mg has been estimated in sheep, calves, cattle, and horses, but the estimates varied because of differences in experimental methods (Simesen, 1980). Mean values of endogenous Mg ranged from 0.5 to 5.1 mg/kg body weight/day. Endogenous fecal excretion of Mg is of substantial magnitude in ruminants and may be increased by a greater flow of saliva stimulated by diets high in roughage (Care, 1980). Endogenous Mg excretion may relate to the greater susceptibility of ruminants to hypomagnesemia.

2. *Absorption*

Magnesium is absorbed as a freely diffusing ion, as is calcium, by three mechanisms: (1) paracellular diffusion, (2) paracellular solvent drag, and (3) transcellular active transport (Leonhard *et al.*, 1990; Flatman, 1991; Stojevic *et al.*, 1993). In general, Mg absorption occurs principally in the distal small intestine and colon of monogastric animals and forestomachs of ruminants (Hintz and Schryver, 1972; Meyer and Zentek, 1990; Kayne and Lee, 1993; Stojevic *et al.*, 1993; Rémond *et al.*, 1996). The absorptive efficiency of Mg decreases with increasing Mg concentration of the diet when plasma Mg concentration is normal. Increased dietary Mg will lead to increased serum, erythrocyte and tissue Mg concentrations, especially bone (Chester-Jones *et al.*, 1989, 1990). The concentration of ionized Mg in the small intestine is low because of binding of Mg^{2+} to ingesta. Calves absorb Mg from the large and small intestine until 1 month of age, but once the forestomachs become functional, Mg will be absorbed from this part of the digestive tract (Kolb, 1985). The concentration gradient between the rumenal lumen and blood is too small to account for net Mg absorption, and an active transport process utilizing a Na/K ATPase may be involved (Martens *et al.*, 1978; Martens, 1983; Martens and Käsebieter, 1983), along with passive diffusion down an electrochemical gradient (Care, 1980). Rumen Mg^{2+} absorption may be increased by lower pH (which increases Mg^{2+} solubility) and decreased by microbial sequestration (Kennedy and Bunting, 1991).

The fetus has a plasma Mg concentration that is greater than that of the mother because of active transport of Mg across the placental barrier (Barri *et al.*,

1990). The gradient is dependent on the fetal parathyroid glands in sheep. The midmolecule portion of parathyroid hormone-related protein may be one humoral factor that is involved in stimulating the transport of Mg across the placenta (MacIsaac *et al.*, 1991b). Parathyroid hormone-related protein is produced by both the fetal parathyroid glands and placenta.

The absorption of Mg is increased by the active form of vitamin D (1,25-dihydroxyvitamin D), but the mechanism of the transport is unknown, and Ca and Mg differ in the degree to which absorption is stimulated (Ebel, 1990). However, 1,25-dihydroxyvitamin D administration results in a decreased plasma Mg concentration (Boling and Evans, 1979; Care, 1980). This may be a result of enhanced Mg excretion due to a direct effect of 1,25-dihydroxyvitamin D on tubular reabsorption of Mg, or a secondary effect due to reduced serum PTH concentrations (Fontenot *et al.*, 1989). A common intestinal transport system for Ca and Mg has been hypothesized (Ebel, 1990). Calcium and magnesium interfere with each other's absorption, and high dietary supplementation with Ca decreases the apparent absorption and retention of Mg (Shiga, 1988). In addition, feeding the chelating agent EDTA (which binds Ca in preference to Mg) inhibits the absorption of Ca from the small intestine, but promotes the absorption of Mg in the proximal small intestine in sheep (van't Klooster and Care, 1966). Diets high in phosphate reduce intestinal Mg absorption (Hintz and Schryver, 1972; Ebel, 1990). Increased rumenal concentrations of potassium lead to an increased potential difference across the lumenal epithelial cells and result in reduced Mg absorption (Care, 1988; Martens *et al.*, 1988; Fisher *et al.*, 1994).

3. *Excretion*

The three principal routes of Mg excretion are the gastrointestinal tract, the kidney, and the mammary gland during lactation. Magnesium loss can also occur in animals that sweat. Pregnant animals lose large amounts of Mg during fetal bone development in the latter parts of gestation.

a. Gastrointestinal Tract

Most of the Mg that is excreted is present in the feces. Fecal Mg includes endogenous fecal Mg, which is higher for ruminants than for monogastric animals, and unabsorbed Mg from the diet. Malabsorption syndromes, enteropathies, and cholestatic liver disease can lead to increased intestinal loss of Mg.

b. Kidney

The kidneys play a major role in regulating Mg balance and serum concentration by controlling tubular reabsorption principally in the ascending limb of the loop of Henle (De Rouffignac and Quamme, 1994; Quamme, 1993). Magnesium absorbed in the gastrointestinal tract in excess of body requirements is excreted by the kidney. The tubular reabsorption rate is regulated by dietary availability of Mg, certain hormones, serum concentrations of calcium and magnesium, and urinary calcium excretion (Ryan, 1990). Dietary deficiency of Mg is associated with an increase in renal reabsorption. Fractional Mg clearance is approximately 5–10%, but can vary from less than 1 to 100% depending on the degree of dietary Mg absorption and serum Mg concentration (Fleming *et al.*, 1991; Sutton and Domrongkitchaiporn, 1993). Magnesium appears in the urine when the filtered load exceeds the maximal tubular reabsorptive capacity (Littledike and Cox, 1979). Maximum tubular reabsorption of Mg and filtration rate determine the threshold concentration of Mg in the serum. The renal threshold of serum Mg is 1.8–1.9 mg/dl (0.7–0.8 mmol/liter) in the bovine (Rook and Storry, 1962).

The majority (50–60%) of ultrafilterable Mg is reabsorbed in the thick ascending limb of the loop of Henle, and regulation of renal Mg excretion occurs at this site (Quamme, 1993). Mg is reabsorbed in this segment through the paracellular pathway. Movement of Mg can be controlled by regulation of paracellular permeability and transepithelial voltage. Hypercalcemia and hypermagnesemia inhibit renal reabsorption and increase excretion of Mg. Multiple hormones can increase renal Mg reabsorption and include parathyroid hormone, parathyroid hormone-related protein, calcitonin, antidiuretic hormone, glucagon, β-adrenergic agonists, and insulin (Rizzoli *et al.*, 1989; Ryan, 1990; De Rouffignac *et al.*, 1993). Impaired Mg reabsorption and excess excretion occur with osmotic diuresis (including diabetes mellitus), loop diuretics, hypercalciuria, tubular acidosis, and various toxicities (including aminoglycosides and *cis*-platinum) (Reinhart, 1988; Kes and Reiner, 1990; Sutton and Domrongkitchaiporn, 1993). Excess urinary excretion of Mg could lead to hypomagnesemia, especially if dietary sources are marginal or deficient. Hypermagnesemia may occur secondary to reduced glomerular filtration or chronic renal failure, especially if there is excess dietary intake of Mg.

Urinary Mg concentration can be used as an indicator of dietary supply, but is not representative of current plasma Mg concentration. Magnesium status or adequacy of dietary Mg intake in cattle can be determined more accurately with urinary Mg than with serum Mg. The minimum range of urine Mg indicating adequate herd status is 3.7–9.8 mg/dl (1.5–4.0 mmol/liter) (Horber *et al.*, 1979; Alexander, 1985). The higher range of urine Mg values is present in high-producing dairy cows fed adequate levels of dietary Mg. If urinary Mg fractional clearance is less than 10% or creatinine-corrected urinary Mg concentration is less than 2.4 mg/dl (1.0 mmol/liter) in dairy cows, then a positive response to Mg supplementation is likely (Sutherland *et al.*, 1986). A serum Mg level of 1.5 mg/dl (0.63 mmol/liter) is regarded as the minimum safe figure for cattle.

Renal antagonism between Ca and Mg may be one of the causes of increased plasma Mg associated with hypocalcemia (Halse, 1984). This may partially explain why (1) some cows with postparturient paresis have plasma Mg levels greater than 2.4 mg/dl (1.0 mmol/liter), (2) plasma Mg decreases when milk fever is treated (Marr *et al.*, 1955), and (3) there is a delay in clinical hypomagnesemia seen in cows with fasting hypocalcemia. Plasma Mg may also be increased in cows during subclinical postparturient hypocalcemia (Dishington, 1975). Phosphate depletion leads to urinary Mg wasting and may result in hypomagnesemia (Ryan, 1990).

c. *Mammary Gland*

The mammary gland actively secretes Mg into milk, and the ratio of milk Mg : serum Mg is approximately 5 : 1. The output of Mg in milk is 3–6 mg/kg body weight/day, and the requirements of lactating animals is 20–50 mg/kg body weight/day (Meyer and Zentek, 1990). Milk Mg concentration is steady during lactation in cattle and dogs, but increases in sheep and rabbits. Heavily lactating cows can lose from 2 to 4 g of Mg per day from the mammary gland, which represents a large proportion of dietary Mg absorbed from the gut. Magnesium concentration of milk is stable under conditions of reduced feed or Mg intake or during hypomagnesemia (Meyer and Zentek, 1990). Production increases can be expected from correcting hypomagnesemia in dairy cows (Wilson, 1980; Alexander, 1985). Milk yield significantly decreased in cows 1 week before and after clinical hypomagnesemia (Lucey *et al.*, 1986).

VI. SERUM MAGNESIUM

The method of choice for determination of total Mg in biologic materials is atomic absorption spectrophotometry (Anderson and Talcott, 1994). Colorimetric and kinetic methods also are available for measurement of total serum Mg (Seiler, 1990; Blanchflower and Kennedy, 1991). Serum ionized Mg can be measured with a Mg^{2+}-selective electrode (McGuigan and Blatter, 1989; Altura *et al.*, 1994; Riond *et al.*, 1995b). If a Mg^{2+}-selective electrode is not available, measurement of ultrafilterable Mg may be a satisfactory alternative for evaluating physiologically significant Mg concentrations (Córdova and Pérez-Gallardo, 1993). Changes in serum total Mg often parallel changes in serum ionized Mg.

Serum Mg concentration ranges from 1.5 to 5 mg/dl (0.8–2.1 mmol/liter) depending on species (Table 23.6) About 70–80% of serum Mg is ultrafilterable (free ion or complexed to phosphate, citrate, or other compounds), and the ultrafilterable fraction is independent of the total serum content (Evans and Parsons, 1988). The protein-bound fraction of serum Mg (20%) is bound predominantly to albumin and also to globulins, but serum total Mg concentration is independent of albumin concentration (Evans and Parsons, 1988). Serum ionized Mg concentration represents approximately 70% of total serum Mg and parallels changes in total serum Mg concentration (Riond *et al.*, 1995b). In animals other than cattle, Mg occurs at a higher concentration in erythrocytes than in plasma, and there is no apparent exchange of Mg between plasma, and erythrocytes. Serum Mg varies between breeds of cattle, with lower values in Herefords (1.59 mg/dl, 0.65 mmol/liter) and higher values in Angus (1.91 mg/dl, 0.79 mmol/liter) (Littledike *et al.*, 1995). Plasma Mg may decrease during lipolysis associated with stress, cold, or starvation (Rayssiguier, 1984). Leakage of Mg from tissues into blood after death precludes the use of postmortem blood samples for evaluation of serum Mg concentration.

The validity of using Mg concentration of eye fluids after death to diagnose hypomagnesemia is questionable. Magnesium concentration of aqueous humor after death was reported to have no relationship to serum Mg before death and may actually increase postmortem (Whitaker *et al.*, 1986). It was demonstrated that Mg concentration in bovine vitreous humor (1.8–2.7 mg/dl [0.7–1.1 mmol/liter]) paralleled serum Mg for 48 hours post mortem at temperatures of 4 and 23°C, but decreased at 30°C (Lincoln and Lane, 1985). In contrast, low correlations were demonstrated between Mg concentrations in serum and aqueous or vitreous humor in cows (McCoy and Kennedy, 1994). However, most cows in these investigations were not hypomagnesemic, so the efficacy of using vitreous humor Mg concentrations for evaluation of hypomagnesemia has not been established.

A. Regulation of Serum Magnesium

Serum Mg concentration is less well controlled than that of Ca, and less is known concerning the regulation of serum Mg. There is a reciprocal relationship between Mg and Ca in the serum. Insufficient dietary Mg will lead to hypomagnesemia. Magnesium homeostasis is

TABLE 23.6 Concentration of Magnesium in Animals

Species	Fraction[a]	No. of animals	Concentration (mg/dl) (mean ± 2 SD)	Concentration (mmol/liter) (mean ± 2 SD)	Reference
Dog	P	78	1.99 ± 0.24	0.82 ± 0.10	(Evans and Parsons, 1988)
	U	78	1.39 ± 0.19	0.57 ± 0.08	(Evans and Parsons, 1988)
	P	10	1.82 ± 0.34	0.75 ± 0.14	(Geven *et al.*, 1990)
	U	10	1.17 ± 0.39	0.48 ± 0.16	(Geven *et al.*, 1990)
	R	10	7.03 ± 1.02	2.89 ± 0.42	(Geven *et al.*, 1990)
Cattle	I	17	1.75 ± 0.34	0.72 ± 0.14	(Riond *et al.*, 1995b)
	S	17	2.55 ± 0.49	1.05 ± 0.20	(Riond *et al.*, 1995b)
Steers	S	24	1.92 ± 0.39	0.79 ± 0.16	(Chester-Jones *et al.*, 1990)
Steers	R	24	1.77 ± 0.78	0.73 ± 0.32	(Chester-Jones *et al.*, 1990)
Cows	R	64	2.41 ± 0.58	0.99 ± 0.24	(Fenwick and Daniel, 1991)
Sheep	S	6	2.00 ± 0.64	0.82 ± 0.26	(Lucas *et al.*, 1982)
Ewes[b]	P	27	2.50	1.03	(Sansom *et al.*, 1982a)
Ewes[c]	P	12	2.14	0.88	(Sansom *et al.*, 1982a)
Goats	S	5	3.28 ± 0.60	1.35 ± 0.25	(Hazarika and Pandey, 1993)
Pigs					
Sows	P	12	1.60	0.66	(Thielscher *et al.*, 1994)
Sows	P	12	1.46	0.60	(Thielscher *et al.*, 1994)
Horses					
Pony	S	147	1.82 ± 0.73	0.75 ± 0.30	(Unkel, 1984)
Camel					
Dromedary	S	100	2.60 ± 0.94	1.07 ± 0.39	(Biagi and Salutini, 1983)
Rat					
Dam	S	3	2.06 ± 0.24	0.85 ± 0.10	(Günther *et al.*, 1988)
Dam[d]	S	4	2.14 ± 0.19	0.88 ± 0.08	(Günther *et al.*, 1988)
Fetus	S	4 (pooled)	4.37 ± 1.17	1.80 ± 0.48	(Günther *et al.*, 1988)

[a] P = plasma, S = serum, U = ultrafilterable fraction, I = ionized, R = erythrocyte.
[b] Three weeks prior to lambing.
[c] Three weeks after lambing.
[d] Pregnant animals.

a result of balance between intestinal absorption and renal excretion with additional regulation by the adrenals, thyroids, and parathyroid glands. However, no endocrine gland exerts a primary regulatory role on plasma Mg. Renal failure may result in an increased serum Mg concentration. In an immature animal the skeleton is a partially labile source of Mg, whereas in the adult the skeleton is largely inert in relation to Mg mobilization.

1. *Adrenal Glands*

Hyperaldosteronism is associated with a negative Mg balance, hypomagnesemia, and increased excretion of Mg in urine and feces (Simesen, 1980; Charlton and Armstrong, 1989). In adrenal insufficiency, the opposite effect occurs and serum Mg concentration is increased. The effect of mineralocorticoids on Mg homeostasis is secondary to their role in Na and K homeostasis. There is no effect of Mg concentration on aldosterone secretion. Aldosterone has no direct effect on Mg absorption by the rumen (Martens and Hammer, 1981); however, it reduces salivary Na concentration and results in compensatory increases in salivary K concentration that could enhance the inhibition of rumen Mg absorption by diets high in K (Care, 1988; Dua and Care, 1995). Although aldosterone may play a role in the pathogenesis of hypomagnesemia, its presence is not necessary, because hypomagnesemia occurred in adrenalectomized sheep that were changed from dry food to young grass (Care and Ross, 1963).

2. *Thyroid Glands*

Increased thyroid activity tends to depress the plasma Mg concentration (Simesen, 1980). However, thyroxine is not an important factor in the pathogenesis of hypomagnesemic tetany. Calves fed a milk diet that led to hypomagnesemia developed decreased thyroid activity and increased plasma glucose concentrations (Mulei and Daniel, 1988). Calcitonin has little or no effect on plasma Mg concentration.

3. *Parathyroid Glands*

Elevation of plasma Mg and severe hypomagnesemia inhibit secretion of parathyroid hormone (PTH), whereas moderate hypomagnesemia stimulates secretion of PTH (Mayer and Hurst, 1978; Care, 1980; Shiga *et al.*, 1980; Fiore *et al.*, 1990). However, plasma Ca is the major physiologic control of parathyroid gland function. Parathyroid hormone predominantly increases plasma Ca relative to plasma Mg, although PTH can increase plasma Mg by increasing intestinal absorption, increasing tubular reabsorption, and increasing bone resorption (Care, 1980; Littledike and Goff, 1987). Hypermagnesemia that occurs during postparturient paresis in cattle may be a result of increased PTH levels during hypocalcemia. However, hyperparathyroidism is not associated with increased serum concentrations of Mg (Claeyssens *et al.*, 1990), possibly because of the antagonistic effects of PTH and hypercalcemia on Mg excretion.

The mechanism by which hypomagnesemia results in hypocalcemia is not known; however, hypomagnesemia can impair PTH secretion and induce end-organ resistance to PTH (Abbott and Rude, 1993). Parathyroid hormone activates adenylyl cyclase in bone and kidney, which requires Mg ions. Hypomagnesemia may reduce the target-organ response to PTH (Care, 1980; Rude and Oldham, 1987). Plasma PTH remains normal, decreases, or increases in cows with hypomagnesemia and hypocalcemia, but the PTH present may have reduced physiologic effectiveness. Calves fed synthetic Mg-deficient milk developed decreased plasma Mg, Ca, and P concentrations and no change in plasma PTH or calcitonin concentrations (Rayssiguier *et al.*, 1977). Therefore, hypocalcemia associated with Mg deficiency may be attributed to the lack of PTH secretion associated with decreased serum Ca and lack of responsiveness of target organs to PTH.

VII. DISTURBANCES OF MAGNESIUM METABOLISM

Cattle and sheep are subject to clinical disorders due to Mg deficiency; however, few animals with hypomagnesemia develop clinical signs (Smith and Edwards, 1988; Hoffsis *et al.*, 1989). The disease is often a herd problem and can be classified into a rapidly developing or a slowly developing type (Allcroft, 1954). In calves and beef cattle the disease is generally of the slow type, whereas in milking cattle and sheep the onset is usually rapid. The occurrence usually coincides with the first flush grass in spring. Pregnant and lactating cows are more susceptible to hypomagnesemia because of their increased requirements for Mg.

The cause of multiple clinical signs associated with hypomagnesemic tetany is unknown, but hypomagnesemia is usually asymptomatic. Magnesium is an essential regulator of excitability in the central nervous system and blocks chemical transmission in the peripheral synapses by competitively inhibiting Ca^{2+}-dependent, presynaptic excitation–secretion coupling (Chutkow, 1990). Small decreases in [Mg^{2+}] of the cerebrospinal fluid can lead to generalized convulsions. The depletion of Mg^{2+} in the extracellular fluid may be responsible for the tetanic syndrome in animals with grass tetany because of increased release of acetylcholine at neuromuscular junctions. The major clinical signs of hypomagnesemia are restlessness, muscular twitching, excitement, staggering, salivation, involuntary eye movements, increased sensitivity to strange noises, paresis and tetany, and finally convulsions and death (Chausow *et al.*, 1985; Dua and Nauriyal, 1988).

The onset of symptoms associated with hypomagnesemia cannot be directly correlated with serum Mg concentration. It is likely that serum Ca plays a concomitant role in many cases of hypomagnesemic tetany, because animals with hypomagnesemia are often hypocalcemic (Baker *et al.*, 1979). Cows and steers mobilized Ca less readily during experimental hypocalcemic stress when they were hypomagnesemic (Sansom *et al.*, 1982b). Hypocalcemia associated with hypomagnesemia may be due to reduced feed consumption, but could not be explained by decreased absorption or increased excretion of Ca in hypomagnesemic sheep (Khan *et al*, 1982). Impaired PTH secretion, failure of activation of parathyroid hormone-sensitive adenylate cyclase in bone and kidney, and decreased Ca^{2+} release from bone may also contribute to the development of hypocalcemia (Levi *et al.*, 1974; Rayssiguier *et al.*, 1977; Care, 1980; Reinhart, 1988).

In animals with Mg deficiency, intracellular Mg^{2+} is normal or only slightly reduced. However, secondary effects on cell composition occur and include a loss of K and accumulation of Ca and Na (Heaton and Rayssiguier, 1987). Decreased competition of Mg^{2+} with Ca^{2+} can lead to increased intracellular Na^{+} and Ca^{2+} concentrations, which may be important in the pathogenesis of the clinical signs of hypomagnesemia. Mineralization of blood vessels or soft tissues is a characteristic finding of Mg deficiency in laboratory animals or ruminants (Haggard *et al.*, 1978; Plannells *et al.*, 1995).

A. Hypomagnesemia in Calves

This syndrome has been intimately associated with the feeding of whole milk to calves for an extended period of time and is due to a simple dietary deficiency of Mg (Caple and West, 1992). The onset of hypomagnesemia is related to the reduction in availability of dietary Mg that occurs with increasing age. Hypomagnesemia may be exacerbated by diarrhea (Groutides and Michell, 1990). Calves suckling beef cows are usually less susceptible to hypomagnesemia because they consume solid food with a higher Mg content (Sansom, 1981). The Mg requirement of growing calves is approximately 16–18 mg/dl (6.6-7.4 mmol/liter of milk (Simesen, 1980), which is greater than the Mg content [12 mg/dl (5.0 mmol/liter)] of cow's milk. The fastest-growing calves show the greatest depression of serum Mg, and characteristic symptoms of hypomagnesemia include excitability, muscle fasciculation, constant movement, stiff gait, twitching of the ears, and intermittent convulsions (Hoffsis *et al.*, 1989; Caple and West, 1992). Tetany may develop when the concentration of serum magnesium decreases to 0.7 mg/dl (0.3 mmol/liter) and death may occur at 0.5 mg/dl (0.2 mmol/liter). However, serum Mg concentration may be increased during the final stages of convulsions, presumably because of the release of Mg by rapidly contracting muscles.

Tetany and sudden death associated with Mg deficiency have been reported in 4-month-old suckling beef calves (Haggard *et al.*, 1978). The cattle were consuming hay and silage with a low Mg content. Deposition of calcium salts occurred in the elastic fibers of arteries in the heart, lungs, and spleen. Treatment with intravenous Mg salts and free-choice Mg-containing supplement was effective in preventing clinical signs and death.

1. Bone–Plasma Relationship

Bone Mg may serve as a source of plasma Mg under conditions of deficiency in calves. Normal bone ash Mg in cattle 6–18 months old is 1.2%, which decreases to 0.66% in cattle greater than three years old (Beighle *et al.*, 1994). Mg content of bone ash is greater in steers (1.0%) than in female cattle (0.65%) (Beighle *et al.*, 1994). Mg in bone ash will be reduced to 0.40–0.48% when plasma levels have declined to 0.7 mg/dl (0.3 mmol/liter) (Smith, 1959). Differences in bone-ash Mg may exist between bones during moderate Mg depletion (Simesen, 1980). Several weeks of oral Mg supplementation or repeated subcutaneous injections of magnesium sulfate (1 g Mg/day) will restore bone ash Mg to normal levels.

Magnesium depletion of bone apparently results from the exchange of Mg by Ca on the bone-crystal surface. Therefore, the Ca/Mg ratio of bone ash is a more sensitive indicator of Mg deficiency than bone Mg content under field conditions (Blaxter and Sharman, 1955).

2. Vitamin D and Serum Calcium and Magnesium

Addition of vitamin D (5–10 IU/kg body weight/day) to milk fed to calves does not prevent hypomagnesemia, but it does prevent hypocalcemia (Simesen, 1980). Calves with moderate hypomagnesemia (1.4–1.5 mg/dl [0.58–0.62 mmol/liter]) and hypocalcemia due to a lack of vitamin D suddenly collapse and remain almost motionless with their necks and legs stiff for periods of a few minutes to half an hour. Calves with only hypomagnesemia demonstrate nervousness, twitching ears, staring eyes, staggering, and inability to stand when serum magnesium concentrations drop below 1 mg/dl (0.4 mmol/liter). Intrarumenal administration of potassium chloride and citric acid to calves can be performed to experimentally induce severe hypomagnesemia and death (Dua *et al.*, 1988; Dua and Nauriyal, 1988).

B. Hypomagnesemia in Adult Cattle

1. Slow Type

Subclinical hypomagnesemia can occur in some animals for a long time, often for months. When serum Mg approaches critically low levels, it is accompanied by mild clinical symptoms such as nervousness and increased excitability. Certain conditions, such as fasting or reduced feed intake, may result in tetany or paresis. Often a combination of several adverse factors is found, but in most cases low dietary Mg and a low plane of nutrition are essential. The incidence of the disease seems to increase with age, and there is a tendency for recurrence in some animals. Plasma Mg and Ca levels are lower in mature than in young cows (McAdam and O'Dell, 1982). Clinical hypomagnesemia occurred in 1% of dairy cows in a survey of England and Wales, and subclinical hypomagnesemia (serum Mg less than 1.9 mg/dl [0.8 mmol/liter]) in 7% of milking cows and 15% of dry cows (Whitaker and Kelly, 1982). Subclinical hypomagnesemia increases the susceptibility to milk fever by decreasing the ability of cows to mobilize Ca in response to hypocalcemia (Sansom, 1983; Kelly, 1988; van Mosel *et al.*, 1991). Serum Mg levels less than 1.9 mg/dl (0.8 mmol/liter) were associated with decreased blood urea N, suggesting that marginal Mg deficiency may cause decreased feed intake (Scott *et al.*, 1980).

a. Seasonal Hypomagnesemia and Winter Tetany

Seasonal hypomagnesemia occurs in the bovine during the winter months and is associated with poor feed quality, cold, wet, or windy periods, and conditions when heat loss from the body would be the greatest. Supplementary feeding will increase serum Mg, but usually not to normal levels. Clinical hypomagnesemia occurs in animals with very low serum Mg concentrations, or various degrees of hypocalcemia, and coincides with the time of minimal serum Mg concentration.

Persistent low levels of Mg may be found in beef cattle during winter months, but hypomagnesemia is accompanied by few or no clinical symptoms. Some cows may develop moderate incoordination, paresis, or tetany (winter tetany) often associated with recent parturition. Hyperirritability is common. Hypomagnesemia is associated with various degrees of hypocalcemia, and variation in clinical signs may be associated with fluctuations in the Ca/Mg ratio.

b. Effect of Feed Intake

Underfeeding or fasting is effective in lowering serum Mg concentration because of its dependence on dietary Mg and absorption from the intestinal tract. Therefore, any condition associated with a reduction in feed intake will exacerbate a preexisting hypomagnesemic state. Such conditions may include recent parturition, underfeeding, a sudden change in the quality or quantity of the ration, estrus, and weather conditions that reduce grazing or rumination time.

Serum Ca and Mg concentrations decrease in fasting, lactating cows, and starved cows may develop a syndrome similar to milk fever with serum calcium levels as low as 7.0 mg/dl (1.7 mmol/liter) and serum Mg less than 1.0 mg/dl (0.4 mmol/liter). During refeeding, serum calcium concentration quickly recovers to normal values, whereas serum Mg levels remain low during the early stages of refeeding (Fig. 23.45). Energy deficiency will reduce rumen formation of volatile fatty acids and CO_2, which are necessary for normal Mg absorption from the rumen (Martens and Rayssiguier, 1980).

c. Effect of Artificially Low Magnesium Intakes

Lactating cows develop hypomagnesemia and may develop tetany and die, if fed diets with extremely low Mg (about 2 g Mg/day) that are adequate in other nutrients. The occurrence of clinical symptoms is variable between animals, but appears to be more common in animals if serum Mg falls rapidly.

2. Rapid Type: Grass Tetany (Hypomagnesemic Tetany)

Grass tetany is the result of hypomagnesemia often associated with pasturing cattle on spring grass following winter feeding. It occurs most frequently in the first weeks of the grazing season. Grass tetany may also occur sporadically in autumn, especially during wet conditions with relatively low temperatures. After the first frost, the number of cases declines rapidly.

Winter feedstuffs that are deficient in Ca and Mg will lead to increased secretion of PTH and depletion of the mobilizeable bone Mg. Spring pastures that have increased Ca compared to winter feeds will result in increased Ca absorption, reduced secretion of PTH, and reduced bone resorption. This will lead to increased Mg uptake by bone and could result in a rapid reduction of serum Mg concentration that would predispose cattle to the development of grass tetany (Langley and Mann, 1991).

a. Clinical Biochemical Findings

Grass tetany is associated with pronounced hypomagnesemia that is in most cases combined with moderate hypocalcemia. Tetany is not dependent on the concurrent hypocalcemia (Meyer, 1977); however, cows with grass tetany may have lower plasma Ca than asymptomatic, hypomagnesemic cows (Kappel *et al.*, 1983). Low magnesium levels are usually found in clinically normal animals from the same herd. Levels below 1.8 mg/dl (0.7 mmol/liter) are classed as hypomagnesemic, and those below 1.1 mg/dl (0.5 mmol/liter) as severely hypomagnesemic. When blood Mg declines below 1.0 mg/dl (0.4 mmol/liter), appetite decreases. There is a pronounced individual susceptibility to hypomagnesemia. Serum phosphate levels may be normal or low, and there are no consistent alterations in serum Na, K, or Cl levels or in pH of the blood.

Magnesium concentration in the urine can be used as a differential-diagnostic criterion between milk fever and grass tetany. Mg content of the urine is about 50 mg/dl (21 mmol/liter) during milk fever (Sjollema, 1930), whereas during hypomagnesemia the urine has a low Mg concentration [(1.0–3.0 mg/dl) (0.4–1.2 mmol/liter]) (Simesen, 1980). Semiquantitative determinations of urine Mg may be used in the clinical setting, but it is very important to avoid fecal contamination of the urine sample. Postmortem blood samples for diagnostic purposes are of little value, because the Mg level rises rapidly after death. In such cases, analysis of a urine sample carefully removed from the bladder, in connection with analysis of several blood samples from susceptible animals in the herd, may aid the diagnosis. Criteria for Mg in the urine are as follows: >10 mg/dl (4.1 mmol/liter) is adequate; 2.0–10 mg/dl (0.8–4.1 mmol/liter) is marginal; and <2.0 mg/dl (0.8 mmol/liter) is a severe deficiency and an indicator of danger of tetany (McDowell, 1992b).

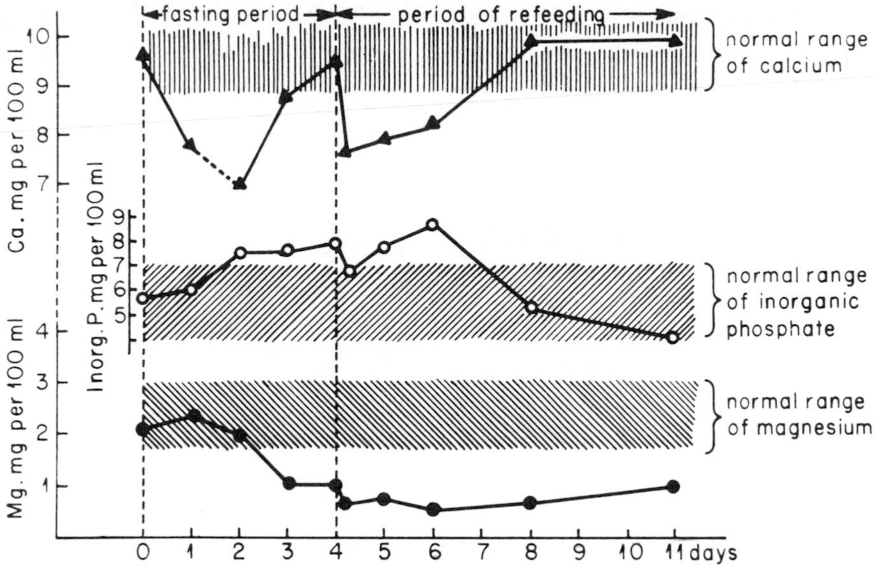

FIGURE 23.45 Changes in serum calcium, magnesium, and inorganic phosphorus during a 4-day period of fasting. From Simesen (1980).

b. Etiological Considerations

The etiology of hypomagnesemic tetany is not completely elucidated, and there are multiple opinions on its pathogenesis. Hypomagnesemia may not be the sole cause of grass tetany. Hypomagnesemic tetany is associated with (1) serum Mg concentration that falls rapidly, within a few days; (2) no depletion of bone Mg in adults; and (3) feeding grass with normal Mg content (Simesen, 1980). A major factor in the pathogenesis of hypomagnesemia is reduced absorption of Mg from the gastrointestinal tract.

The primary change resulting in tetany does not appear to be in the muscle fibers. In hypomagnesemic sheep there were no changes in motor unit potentials evoked from voluntary movements, muscle evoked potentials, nerve conduction velocity, or muscle excitability (Sims *et al.*, 1980). Clinical signs may be more closely associated with the Mg content of cerebrospinal fluid (CSF) than with blood Mg levels. A decrease in blood Mg is followed by a delayed reduction in CSF Mg concentration, which may explain why animals with very low blood Mg levels are symptom free for extended periods of time (Meyer, 1977).

c. Chemical and Mineral Composition of Pasture

The greatest single etiological factor in hypomagnesemic tetany is the decrease in net absorption of Mg from the alimentary tract. The factors responsible for reduced dietary availability of Mg are not clear; however, they may be important only during conditions of marginal supply of Mg (Simesen, 1980). Mineral indices or individual mineral constituents of grass have not demonstrated consistent relationships to hypomagnesemia in grazing cattle. Grass tetany may result either from inadequate intestinal absorption of Mg in the grass, or from substances in the grass completing with Mg absorption. Low serum Mg levels occur when Mg content of the herbage is less than 0.20%, and an increase in serum Mg can be induced by application of Mg-containing fertilizer to the pasture. Hypomagnesemia rarely occurs in cattle grazing clover-rich pastures that are high in Ca and Mg content. Acid rain has the potential to leach Mg from plants and soil (Wilkinson *et al.*, 1990).

Change from winter stall feeds to spring herbage usually involves a decrease in dietary Mg intake and a decrease in availability of dietary Mg. In the spring when young grass is growing rapidly, grass content of K, N, organic acids, and the ratio of K/(Ca + Mg) increase and the percent dry matter decreases (Grunes and Welch, 1989). These conditions predispose animals to the development of hypomagnesemia, especially in lactating cows with increased demand for Mg.

A high N and K content of grass promotes the incidence of hypomagnesemia by inhibiting Mg absorption (Kolb, 1985; Martens and Gäbel, 1986). It is unclear whether high dietary K can directly reduce plasma Mg and result in hypomagnesemic tetany, but dietary K^+ supplementation will reduce Mg absorption in ruminants (Ammerman and Henry, 1987; Beardsworth *et al.*, 1987; Martens *et al.*, 1988). A high-K diet decreased absorption of Mg from the intestinal tract in cows, but the effect was important in the etiology of hypomagnesemia only when dietary intake of Mg was low (Field and Suttle, 1979). It was reported that a high-K (4%) and low-Mg (0.01%) diet synergistically lowered plasma Mg in sheep (Miller *et al.*, 1982). The additional hypomagnesemic effect of K must have resulted from causes other than reduced absorption of Mg.

The Na/K ratio of rumen contents may be more important than Na and K concentrations (Martens, 1980). The greater the Na/K ratio, the greater the absorption of Mg. Hypomagnesemia was reported in cattle fed a diet with normal Mg content (Luthman and Björkman, 1981), and it was suggested that the Na/K ratio of the fodder or uptake of Mg by body fat may have been important in the pathogenesis of hypomagnesemia. High K concentration of forage (due to fertilization) will reduce Mg absorption in the rumen. Low Na content of spring pastures will increase aldosterone secretion and result in increased K secretion by the rumen and salivary glands. This will further exacerbate the effect of diets high in K content (Dua and Care, 1995).

Grazing of *Phalaris tuberosa* pastures in Argentina is associated with a tendency to induce tetany in cattle. Gradual decreases in serum Mg in cattle were associated with a high K/(Ca + Mg) ratio in the pasture (Cseh *et al.*, 1984). A forage K/(Ca + Mg) ratio greater than 2.2 was a more sensitive indicator of tetanic pasture conditions than Mg content alone (Grunes and Welch, 1989).

Plasma insulin levels were increased in cows on a fescue–bluegrass pasture and were negatively correlated with plasma Mg. In addition, injection of insulin into two cows reduced plasma K and Mg by 20% (Miller *et al.*, 1980). Elevated plasma insulin levels and hypomagnesemia were also present in 10 cows overwintered on pasture with low Mg content and moved to lush pasture with high K content (Ramsey *et al.*, 1979b). In contrast, there was no change in plasma insulin, Ca, or K in hypomagnesemic cows fed fescue pasture fertilized with N and K or hay from a pasture that produced grass tetany in grazing cows (Ramsey *et al.*, 1979a).

d. Rumen Ammonia

Ammonia concentration increases in the rumen of cows during the first few days of grass feeding, but there are conflicting results on the effect of dietary crude protein and rumen ammonia concentration on Mg absorption. It has been suggested that only an acute rise in rumen ammonia may reduce Mg absorption, and that this effect is diminished as the rumen adapts to elevated NH_3 concentrations (Martens and Rayssiguier, 1980). Physiologic concentrations of ammonium ion decreased Mg absorption in an isolated rumen model (Martens *et al.*, 1988). Energy deficiency will also lead to elevated rumen ammonia concentrations.

e. Rumen Aluminum

High concentrations of Al in forage may be associated with an increased incidence of grass tetany, but this is not due to decreased intestinal Mg absorption (Fontenot *et al.*, 1989). Aluminum does not affect Mg absorption, excretion, or retention, but can decrease Mg concentration in serum, bone, and kidney and interfere with Mg-dependent ATP activity (Allen, 1987). Interactions of Al with Ca and parathyroid hormone metabolism may secondarily result in hypomagnesemia (Fontenot *et al.*, 1989). In a series of 32 cows, rumen Al concentration was greater in cows with grass tetany than in subclinical hypomagnesemic and control cows (Kappel *et al.*, 1983). Addition of $AlSO_4$ to the rumen decreased plasma Mg 34% in cows with limited amounts of Mg in the feed. Administration of Al (4 and 8 g/Kg feed) reduced serum Mg 32% in steers (Allen *et al.*, 1980).

f. Trigger Effect

The clinical symptoms of hypomagnesemia may be precipitated by (1) sudden reduction in feed intake, (2) adverse weather, (3) estrus, (4) sudden reductions in dietary Mg, (5) sudden change in diet, (6) transportation, or (7) any stressful environmental stimulus (Simesen, 1980). Muscular fibrillations may be present a few days prior to tetany. Epinephrine-induced lipolysis may result in the sequestration of Mg in fat-cell membranes (Rayssiguier, 1977). Infusion of norepinephrine reduced serum Mg and Ca in sheep, and energy deficiency prolonged these effects (Terashima *et al.*, 1983; Wittwer *et al.*, 1995b).

Cold (2–5°C) and deficient dietary intake of Mg synergistically reduced plasma Mg in sheep (Terashima *et al.*, 1982). Low environmental temperature further reduced plasma Mg in hypomagnesemic sheep on a Mg-deficient diet, but increased plasma Mg in sheep fed a diet with adequate Mg (Hilgers, 1979). Urinary excretion of Mg was increased and serum Mg concentration was decreased in four ewes at an environmental temperature of 8.8°C (Shiga *et al.*, 1979).

In beef cows in Ontario, hypomagnesemia developed 24 hours before introduction to pasture, possibly because of the stress of prepasture handling (Hidiroglou *et al.*, 1981). No grass tetany was observed in these cows, and Mg in mineral feeds did not prevent the hypomagnesemia.

3. Methods of Prevention

The essential aim of prevention is to increase daily intake of Mg. Extra dietary Mg may be provided in multiple forms and by a number of routes (Allen and

Davies, 1981; Mayland and Wilkinson, 1989; Caple and West, 1992; Wittwer *et al.*, 1995a). A practical and cost-effective method of supplementing dietary Mg is to provide free-choice Mg supplements (Robinson *et al.*, 1989). Magnesium alloy bullets have not been useful in preventing hypomagnesemia because of variable rates or too-slow decomposition in the reticulorumen (Stuedemann *et al.*, 1984). Absorption of Mg from the oxide form is more efficient than from sulfate, carbonate, or dolomite forms. Intestinal Mg absorption is greater in Brahman than in Jersey, Holstein, or Hereford cows (Greene *et al.*, 1989). Therefore, selective breeding has the potential to reduce the incidence of hypomagnesemia-related disorders in cattle. In addition, selective plant breeding has the ability to increase grass Mg content (Sleper *et al.*, 1989).

C. Miscellaneous Conditions

1. *Hypomagnesemia in Sheep*

Clinical symptoms of hypomagnesemia occur in sheep, and the pathogenic factors in the development of hypomagnesemia are similar to those for cows (Kelly, 1979; Caple and West, 1992). There is a marked variation in susceptibility between animals. Ewes with twins are more prone to hypomagnesemia than ewes with single lambs. Plasma Mg increases during the last 3 weeks of pregnancy, but decreases at lambing and for 3 weeks after lambing (Sansom *et al.*, 1982a). This is in contrast to plasma Ca, which is lowest 3–4 weeks prior to lambing due to formation of the fetal skeleton. Sheep are most susceptible to hypomagnesemia at about 2–6 weeks after lambing, which is exacerbated by lush spring pastures (Kelly, 1979).

Magnesium supplementation of ewes with twin lambs did not increase serum Mg concentration from a mean of 1.8 mg/dl (0.7 mmol/liter) and did not alter milk consumption or weight gain of lambs, even though 36% of the ewes had serum Mg concentrations less than 1.7 mg/dl (0.7 mmol/liter) (Bray *et al.*, 1987). Sheep with poor dental health have lower serum Mg concentrations than do flocks with normal dental health (Spence *et al.*, 1985). Old ewes with twins and worn teeth developed postpartum hypomagnesemia and ketosis that resulted in anorexia, blindness, and incoordination (Jopp and Quinlivan, 1981). Treatment was not effective, but feeding good-quality hay served as a preventative.

Transport tetany of lambs occurs within 10 days of arrival at the feedlot (Lucas *et al.*, 1982). Affected lambs develop hypomagnesemia, hypocalcemia, and hyperphosphatemia as a result of fasting and stress.

2. *Wheat Pasture Poisoning*

Wheat pasture poisoning is a form of hypomagnesemia that occurs during grazing of lush wheat pastures that contain high K and protein. Such rations may induce tetany by interfering with absorption or retention of Mg. Mixed wheat and rye pastures that were associated with tetany had marked increases in K, protein, digestibility, and a peak K/(Ca + Mg) ratio of 3.2 (Bohman *et al.*, 1983b). Plasma Mg concentrations in affected cattle were low (1.7 mg/dl [0.7 mmol/liter]), but were lower 15 days before and 6 days after tetany (1.2 mg/dl [0.5 mmol/liter]) (Bohman *et al.*, 1983a). At the time of tetany, plasma Ca was 4.1 mg/dl (1.0 mmol/liter) and was associated with increased plasma parathyroid hormone and 1,25-dihydroxyvitamin D. Tetany was thus due to hypocalcemia and moderate hypomagnesemia. The hypomagnesemia did not alter PTH secretion or 1,25-dihydroxyvitamin D production in response to hypocalcemia.

3. *Effect of Magnesium Administration*

The anesthetic level of serum Mg in the ruminant and dog was approximately 14 mg/dl (5.8 mmol/liter) and the lethal level about 20 mg/dl (8.2 mmol/liter) (Bowen *et al.*, 1970b). High serum concentrations of Mg can induce coma, respiratory arrest, and diastolic cardiac arrest. The level of Mg necessary to cause death is much higher when Ca is administered concurrently. The anesthetic or lethal effect of Mg is due to blockade of neuromuscular transmission without depression of the central nervous system (Bowen *et al.*, 1970b). Therefore, infusion of magnesium sulfate should not be used as a method of euthanasia. Clinical administration of magnesium sulfate may be useful therapy for muscular seizures associated with eclampsia (Lucas *et al.*, 1995).

Excessive dietary supplementation with Mg will lead to hypermagnesemia, reduced serum calcium concentration, and potential toxicosis, including diarrhea, weight loss, drowsiness, and death (Thiemann *et al.*, 1987). Sheep fed high levels of Mg (up to 2.4% of the diet) developed diarrhea with no other clinical signs (Chester-Jones *et al.*, 1989), whereas cattle fed up to 4.7% Mg in the diet developed diarrhea and lethargy (Chester-Jones *et al.*, 1990). Administration of $Mg(OH)_2$ to cattle for gastrointestinal disorders has the potential to increase serum Mg concentrations and induce metabolic acidosis (Kasari *et al.*, 1990).

4. *Reproduction*

Some investigations have provided data to support a role for Mg in fertility and the incidence of retained placenta in cattle. Magnesium supplementation has reduced the incidence of retained placenta and puerperal endometritis and has increased the conception rate in cows with hypomagnesemia (Krupnik, 1985). Supplementation of cows with Ca and Mg increased conception rates of first inseminations, and the interval from calving to conception was correlated with plasma Mg (Ingraham *et al.*, 1987a). In one study no association was found between hypomagnesemia and retained placenta in cattle (Gibasiewicz, 1984), whereas a reduced incidence of retained placenta in cattle with subclinical hypomagnesemia and supplementation of dietary Mg was reported (Krupnik and Marcinkowski, 1983). Cows supplemented with Mg and Cu had improved fertility, but cows supplemented with Mg or Cu alone did not (Ingraham *et al.*, 1987b). Erythrocyte Mg concentration was correlated with fertility in dairy cows (Mulei *et al.*, 1988).

5. *Hypomagnesemia in Goats*

Goats grazing fertilized grass pastures with high levels of K and N and low Na developed decreased plasma Mg concentration that was associated with poor growth and decreased milk yield (Rayssiguier, 1984). Experimental hypomagnesemia in goats resulted in tetany, convulsions, and death with hepatic and renal degeneration (Hazarika *et al.*, 1991; Hazarika and Pandey, 1993).

6. *Hypomagnesemia and Critical-Care Patients*

Critically ill humans have a high incidence of hypomagnesemia, and recent evidence indicates that the same is true for animal patients (Michell, 1991; Martins *et al.*, 1995). The causes include decreased Mg intake in anorectic animals, lack of Mg supplementation to parenteral fluids in animals on long-term fluid therapy or dialysis, excess gastrointestinal loss of Mg, redistribution of Mg intracellularly, or tissue sequestration of Mg. Some drugs can also increase Mg renal loss, including diuretics (e.g., furosemide, thiazides, and mannitol), cardiac glycosides, and aminoglycosides. Magnesium can be shifted intracellularly by administration of glucose, insulin, or amino acids. Renal excretion of Mg is increased in patients with diabetes mellitus, hyperthyroidism, and primary hyperparathyroidism. Hypomagnesemia also may result from excessive catecholamine secretion in critically ill or stressed animals and bacteremia during sepsis (Elisaf and Siamopoulos, 1995).

Magnesium is a vital ion, required for normal myocardial function. Therefore, it is important to prevent hypomagnesemia in cardiac critical-care patients (de Morais and Hansen, 1995; Stepien, 1995). Magnesium is a necessary cofactor for the Na/K ATPase in myocytes. Mg deficiency may be associated with decreased Na/K ATPase activity, which would lead to increased intracellular K^+ concentration, increased Purkinje-fiber excitability, and potential arrhythmia generation. In addition, Mg deficiency is associated with increased circulating levels of macrophage-derived cytokines (interleukin-1, interleukin-6, and tumor necrosis factor α) that potentiate cardiovascular pathology (Weglicki *et al.*, 1992).

Mg concentration is not commonly measured in animal patients, but certain electrolyte abnormalities may occur simultaneously with hypomagnesemia and serve as predictors of hypomagnesemia. These include hypokalemia, hypophosphatemia, hyponatremia, and hypocalcemia (Whang *et al.*, 1984). Hypokalemia is the best predictor of the four electrolyte disturbances.

7. *Magnesium-Associated Urolithiasis and Enteroliths*

a. Ruminants

Calves fed a diet with excess Mg (1.4% as MgO) had decreased feed intake and growth rate, increased serum creatinine and Mg, and decreased serum Ca (Kallfelz *et al.*, 1987). Calves fed excess Mg or Mg and P (1.6%) developed urinary-tract obstructions due to uroliths composed of calcium apatite, whereas calves fed excess Mg and Ca (1.8%) did not develop urolithiasis. Excess dietary Ca had a protective effect on Mg-associated urolithiasis. Diets with Mg contents greater than 0.35% also may predispose male lambs to urolithiasis (Poole, 1989), but the effect of high dietary Mg is not observed in all investigations (Cuddeford, 1987). Lowering the phosphorus content of diets that are high in Mg and that induce urolithiasis reduces the incidence of urolithiasis (Poole, 1989).

b. Feline

Struvite ($MgNH_4PO_4 \cdot 6H_2O$) urolithiasis predisposes cats to urinary-tract blockage, especially in male cats or cats with lower urinary-tract disease (Buffington, 1994; Buffington *et al.*, 1996). The incidence of struvite urolithiasis in cats is related to the diet (ash and Mg content) and the acid–base status of the animal. Urinary Mg concentration and excretion are highly correlated with Mg intake (Pastoor *et al.*, 1995b). Diets rich in fat and energy decrease dry matter and Mg intake, increase urine volume, and decrease fecal water excretion (Sauer *et al.*, 1985). Increased calcium or phosphate intake will decrease Mg absorption and urinary Mg concentration in cats (Pastoor *et al.*, 1994b, 1995a). Diets high in Mg (especially the alkaline forms, e.g., MgO) and alkaline urine are associated with struvite urolithiasis. The epidemiology of urinary calculi in cats has changed, with a decrease in struvite calculi and an increase in calcium oxalate calculi (Buffington, 1994). This is likely due to feeding cats diets with low Mg content and to the use of urine acidifiers (methionine and ammonium chloride). Acidifying diets may induce depletion of body stores of potassium and increase fractional excretion of calcium.

Obstructive urinary disease can be induced in cats fed high levels of MgO in dry (0.45% Mg) and moist diets (0.75–1.0% Mg) (Finco *et al.*, 1985). Magnesium homeostasis is maintained by increased urinary excretion of Mg. However, urinary Mg did not correlate well with dietary Mg, possibly because of differences in intestinal absorption of Mg between diets or because of precipitation of struvite in the urinary tract. Urethral obstruction may result from blockage by struvite crystals or uroliths. Not all cats with high urine Mg concentrations developed obstruction, indicating that factors other than Mg ingestion play a role in the development of obstruction. The effect of Mg on struvite urolithiasis depends on urinary pH and the form of Mg in the diet (Buffington *et al.*, 1985, 1990). In cats fed purified diets containing 0.5% Mg, MgO, or $MgCl_2$, the urinary pH was the lowest in the $MgCl_2$-fed group. Urinary struvite crystals were found in all control (0.05% Mg diet) and MgO-fed cats, but not in $MgCl_2$-fed cats. Feeding of $CaCl_2$ to cats will also reduce urinary pH (Pastoor *et al.*, 1994a). Meal-fed cats had higher urinary pH than *ad libitum*-fed cats. Alkaline urine was present in obstructed cats fed MgO and dry diets; however, acid urine was present in obstructed cats fed MgO and moist diets (Finco *et al.*, 1985).

Idiopathic lower urinary-tract disease (feline idiopathic cystitis, FIC) in cats is a common cause of increased frequency and urgency of urination and is associated with sterile urine. The pathogenesis of this disease may be related to abnormal regulation of neuropeptides in the bladder and the induction of neurogenic inflammation (Buffington, 1994; Buffington *et al.*, 1996). Struvite urolithiasis is no longer a frequent symptom in cats with idiopathic lower urinary-tract disease. Experimental or natural Mg-induced urinary obstruction due to struvite urolithiasis differs from idiopathic urinary-tract disease. However, struvite urolithiasis can serve as one predisposing factor in a multitude of factors associated with the pathogenesis of urinary-tract blockage in certain cats with idiopathic urinary-tract disease.

c. Horses

Horses fed high levels of Mg in combination with high protein and phosphate are predisposed to develop ammonium–magnesium–phosphate enteroliths in the colon. In the horse, Mg may be excreted in the colon and can precipitate with high concentrations of phosphate that enter the colon from the small intestine (Blue, 1979; Meyer and Zentek, 1990).

8. Hypermagnesemia

Hypermagnesemia may occur in animals with chronic renal failure, hypocalcemia, cows with parturient paresis, and sheep administered epidermal growth factor (Halse, 1984; Brautbar *et al.*, 1990; Gow *et al.*, 1992b; Riond *et al.*, 1995b). The mechanism of hypermagnesemia in cows with parturient paresis is unknown, but increased renal reabsorption of Mg induced by PTH could play a significant role (Riond *et al.*, 1995b). Hypermagnesemia in sheep administered epidermal growth factor is associated with decreased urinary Mg excretion (Gow *et al.*, 1992a).

References

Abbas, S. K., Ratcliffe, W. A., Moniz, C., Dixit, M., Caple, I. W., Silver, M., Fowden, A., and Care, A. D. (1994). The role of parathyroid hormone-related protein in calcium homeostasis in the fetal pig. *Exp. Physiol.* **79,** 527–536.

Abbott, L. G., and Rude, R. K. (1993). Clinical manifestations of magnesium deficiency. *Miner. Electrolyte Metab.* **19,** 314–322.

Abou-Samra, A.-B., Jüppner, H., Force, T., Freeman, M. W., Kong, X.-F., Schipani, E., Urena, P., Richards, J., Bonventre, J. V., Potts Jr, J. T., Kronenberg, H. M., and Segre, G. V. (1992). Expression cloning of a common receptor for parathyroid hormone and parathyroid hormone-related peptide from rat osteoblast-like cells: A single receptor stimulates intracellular accumulation of both cAMP and inositol triphosphates and increases intracellular calcium. *Proc. Natl. Acad. Sci. (USA)* **89,** 2732–2736.

Alexander, A. M. (1985). Magnesium status of dairy cows. *N. Z. Vet. J.* **33,** 171–172.

Al-Khamis, K. I., Al-Hadiyah, B. M., Bawazir, S. A., Ibrahim, O. M., and Al-Yamani, M. J. (1995). Quantification of muscle tissue magnesium and potassium using atomic absorption spectrometry. *Anal. Lett.* **28,** 1033–1053.

Allcroft, R. (1954). Hypomagnesaemia in cattle. *Vet. Rec.* **66,** 517–522.

Allen, V. G. (1987). Influence of aluminum on magnesium metabolism. *In* "Magnesium in Cellular Processes and Medicine" (B. M. Altura, J. Durlach, and M. S. Seelig, eds.), pp. 50–66. Karger, Basel.

Allen, W. M., and Davies, D. C. (1981). Milk fever, hypomagnesaemia and the "downer cow" syndrome. *Br. Vet. J.* **137,** 435–441.

Allen, V. G., Robinson, D. L., and Hembry, F. G. (1980). Aluminum in the etiology of grass tetany. *J. Anim. Sci.* **51 (Suppl. 1),** 44.

Altura, B. M., Durlach, J., and Seelig, M. S. (1987). "Magnesium in Cellular Processes and Medicine," pp. 1–4. Karger, Bern.

Altura, B. T., Shirey, T. L., Young, C. C., Dell'orfano, K., Hiti, J., Welsh, R., Yeh, Q., Barbour, R. L., and Altura, B. M. (1994). Characterization of a new ion selective electrode for ionized magnesium in whole blood, plasma, serum, and aqueous samples. *Scand. J. Clin. Lab. Invest.* **54,** 21–36.

Ammerman, C. B., and Henry, P. R. (1987). Magnesium requirements of animals and man. *In* "Magnesium in Cellular Processes and Medicine" (B. M. Altura, J. Durlach, and M. S. Seelig, eds.), pp. 5–26. Karger, Basel.

Anderson, M. P., and Capen, C. C. (1976a). Nutritional osteodystrophy in captive green iguanas (*Iguana iguana*). *Virchows Arch. [Cell Pathol.]* **21,** 229–247.

Anderson, M. P., and Capen, C. C. (1976b). Ultrastructural evaluation of parathyroid and ultimobronchial glands in iguanas with experimental nutritional osteodystrophy. *Gen. Comp. Endocrinol.* **30,** 209–222.

Anderson, K. A., and Talcott, P. A. (1994). Magnesium. *In* "Handbook on Metals in Clinical and Analytical Chemistry" (H. G. Seiler, A. Sigel, and H. Sigel, eds.), pp. 453–466. Marcel Dekker, New York.

Argenzio, R. A., Lowe, J. E., Hintz, H. F., and Schryver, H. F. (1974). Calcium and phosphorus homeostasis in horses. *J. Nutr.* **104,** 18–27.

Armbrecht, H. J., Okuda, K., Wongsurawat, N., Nemani, R. K., Chen, M. L., and Boltz, M. A. (1992). Characterization and regulation of the vitamin D hydroxylases. *J. Steroid Biochem. Molec. Biol.* **43,** 1073–1081.

Arnaud, C. D., and Pun, K. (1992). Metabolism and assay of parathyroid hormone. *In* "Disorders of Bone and Mineral Metabolism" (F. L. Coe and M. J. Favus, eds.), pp. 107–122. Raven Press, New York.

Arnaud, C. D., Sizemore, G. W., Oldham, S. B., Fischer, J. A., Tsao, H. S., and Littledike, E. T. (1971). Human parathyroid hormone: Glandular and secreted molecular species. *Am. J. Med.* **50,** 630–638.

Arnold, S. A., Kram, M. A., Hintz, H. F., Evans, H., and Krook, L. (1974). Nutritional secondary hyperparathyroidism in the parakeet. *Cornell Vet.* **64,** 37–46.

Atkinson, S., Alston-Mills, B., Lönnerdal, B., and Neville, M. C. (1995). Major minerals and ionic constituents of human and bovine milks. *In* "Handbook of Milk Composition" pp. 593–622. Academic Press, San Diego.

Aurbach, G. D., and Potts Jr., J. T. (1964). The parathyroids. *In* "Advances in Metabolic Diseases" (R. Levine and R. Luft, eds.), pp. 45–93. Academic Press, New York.

Baker, R. M., Boston, R. C., Boyes, T. E., and Leaver, D. D. (1979). Variations in the response of sheep to experimental magnesium deficiency. *Res. Vet. Sci.* **26,** 129–133.

Barber, P. J., Elliott, J., and Torrance, A. G. (1993). Measurement of feline intact parathyroid hormone: Assay validation and sample handling studies. *J. Sm. Anim. Pract.* **34,** 614–620.

Barlet, J. P. (1968). [Experimental induction of a syndrome analogous to vitullary fever by administration of thyrocalcitonin to cows during lactation]. [French]. *Comptes Rendus Hebdomadaires des Seances de l Academie des Sciences—D: Sciences Naturelles* **267,** 2010–2013.

Barlet, J. P. (1972). Calcium homeostasis in the normal and thyroidectomized bovine. *Horm. Metab. Res.* **4,** 300–303.

Barlet, J., Champredon, C., Coxam, V., Davicco, M. J., and Tressol, J. C. (1992). Parathyroid hormone-related peptide might stimulate calcium secretion into the milk of goats. *J. Endocrinol.* **132,** 353–359.

Barlet, J., Abbas, S. K., Care, A. D., Davicco, M., and Rouffet, J. (1993). Parathyroid hormone-related peptide and milking-induced phosphaturia in dairy cows. *Acta Endocrinol.* **129,** 332–336.

Barlet, J. P., Davicco, M. J., and Coxam, V. (1995). Physiologie de l'absorption intestinale du phosphore chez l'animal. *Reprod. Nutr. Dev.* **35,** 475–489.

Baron, R. (1996). Molecular mechanisms of bone resorption: An update. *Acta Orthop. Scand.* **66 (Suppl. 266),** 66–70.

Barri, M., Abbas, S. K., Pickard, D. W., Hammonds, R. G., Wood, W. I., Caple, I. W., Martin, T. J., and Care, A. D. (1990). Fetal magnesium homeostasis in the sheep. *Exp. Physiol.* **75,** 681–688.

Beardsworth, L. J., Beardsworth, P. M., and Care, A. D. (1987). The effect of increased K^+ concentration on the absorption of magnesium from the reticulo-rumen of conscious sheep. *J. Physiol.* **386,** 87P.

Beighle, D. E., Boyazoglu, P. A., Hemken, R. W., and Serumaga-Zake, P. A. (1994). Determination of calcium, phosphorus, and magnesium values in rib bones from clinically normal cattle. *Am. J. Vet. Res.* **55,** 85–89.

Bell, N. H. (1985). Vitamin D-endocrine system. *J. Clin. Invest.* **76,** 1–6.

Berridge, M. J. (1990). Calcium oscillations. *J. Biol. Chem.* **265,** 9583–9586.

Beyenbach, K. W. (1990). Transport of magnesium across biological membranes. *Magnesium Trace Elem.* **9,** 233–254.

Biagi, G., and Salutini, E. (1983). Blood biochemistry of the Somalian dromedary: III. Calcium, phosphorus, and magnesium of both sexes at different ages. *Annals della Facolta di Medicinia Veterinaria de Pisa.* **35,** 201–216.

Biber, J., Custer, M., Magagnin, S., Hayes, G., Werner, A., Lötscher, M., Kaissling, B., and Murer, H. (1996). Renal Na/Pi-cotransporters. *Kidney Int.* **49,** 981–985.

Bienzle, D., Jacobs, R. M., and Lumsden, J. H. (1993). Relationship of serum total calcium to serum albumin in dogs, cats, horses, and cattle. *Can. Vet. J.* **34,** 360–364.

Bikle, D. D. (1992). Clinical counterpoint: Vitamin D: New actions, new analogs, new therapeutic potential. *Endocr. Rev.* **13,** 765–784.

Bilezikian, J. P. (1992). Clinical utility of assays for parathyroid hormone-related protein. *Clin. Chem.* **38,** 179–181.

Bindels, R. J. M. (1993). Calcium handling by the mammalian kidney. *J. Exp. Biol.* **184,** 89–104.

Birch, N.J. (1990). Magnesium in biology and medicine: an overview. *In* "Metal Ions In Biological Systems, Vol. 26, Magnesium and Its Role in Biology, Nutrition, and Physiology" (H. Sigel, and A. Sigel, eds.), pp. 105–117. Marcel Dekker, New York.

Black, H. E., and Capen, C. C. (1971). Urinary and plasma hydroxyproline during pregnancy, parturition and lactation in cows with parturient hypocalcemia. *Metabolism* **20,** 337–343.

Black, H. E., and Capen, C. C. (1973). Plasma calcitonin-like activity and urinary cyclic adenosine monophosphate during pregnancy, parturition, and lactation in cows with parturient hypocalcemia. *Horm. Metab. Res.* **5,** 297–302.

Black, K. S., and Mundy, G. R. (1994). Other causes of hypercalcemia: Local and ectopic secretion syndromes. *In* "The Parathyroids" (J. P. Bilezikian, R. Marcus, and M. A. Levine, eds.), pp. 341–358. Raven Press, New York.

Black, H. E., Capen, C. C., and Arnaud, C. D. (1973a). Ultrastructure of parathyroid glands and plasma immunoreactive parathyroid hormone in pregnant cows fed normal and high calcium diets. *Lab. Invest.* **29,** 173–185.

Black, H. E., Capen, C. C., and Young, D. M. (1973b). Ultimobranchial thyroid neoplasms in bulls. A syndrome resembling medullary thyroid carcinoma in man. *Cancer* **32,** 865–878.

Blanchflower, W. J., and Kennedy, D. G. (1991). Evaluation of a kinetic test kit for the determination of magnesium in bovine serum or plasma. *Vet. Rec.* **128,** 254–255.

Blaxter, K. L., and Sharman, G. A. M. (1955). Hypomagnaesmic tetany in beef cattle. *Vet. Rec.* **67,** 108–115.

Blue, M. G. (1979). Enteroliths in horses—a retrospective of 30 cases. *Equine Vet. J.* **11,** 76–84.

Bohman, V. R., Horn, F. P., Littledike, E. T., Hurst, J. G., and Griffin, D. (1983a). Wheat pasture poisoning. II. Tissue [plasma] components of cattle grazing cereal forages and related to tetany. *J. Anim. Sci.* **57,** 1364.

Bohman, V. R., Horn, F. P., Stewart, B. A., Mathers, A. C., and Grunes, D. L. (1983b). Wheat pasture poisoning. I. An evaluation of cereal pastures as related to tetany in beef cows. *J. Anim. Sci.* **57,** 1352.

Boling, J. A., and Evans, J. M. (1979). Influence of intramuscular or intravenous injection of cholecalciferol on serum calcium, phosphorus and magnesium levels in the bovine. *Int. J. Vit. Nutr. Res.* **49,** 29–34.

Bouhtiauy, I., Lajeunesse, D., Christakos, S., and Brunette, M. G. (1994a). Two vitamin D_3-dependent calcium binding proteins increase calcium reabsorption by different mechanisms I. Effect of CaBP 28K. *Kidney Int.* **45,** 461–468.

Bouhtiauy, I., Lajeunesse, D., Christakos, S., and Brunette, M. G. (1994b). Two vitamin D_3-dependent calcium binding proteins increase calcium absorption by different mechanisms II. Effect of CaBP 9K. *Kidney Int.* **45,** 469–474.

Bovee, K. C., Joyce, T., Blazer-Yost, B., Goldschmidt, M. S., and Segal, S. (1979). Characterization of renal defects in dogs with a syndrome similar to the Fanconi syndrome in man. *J. Am. Vet. Med. Assoc.* **174,** 1094–1099.

Bowden, S. J., Hughes, S. V., and Ratcliffe, W. A. (1993). Molecular forms of parathyroid hormone-related protein in tumours and biological fluids. *Clin. Endocrinol.* **38,** 287–294.

Bowen, J. M., Blackmon, D. M., and Heavner, J. E. (1970a). Neuromuscular transmission and hypocalcemic paresis in the cow. *Am. J. Vet. Res.* **31,** 831–839.

Bowen, J. M., Blackmon, D. M., and Heavner, J. E. (1970b). Effect of magnesium ions on neuromuscular transmission in the horse, steer, and dog. *J. Am. Vet. Med. Assoc.* **157,** 164–173.

Boyce, R. W., Franks, A. F., Jankowsky, M. L., Orcutt, C. M., Piacquadio, A. M., White, J. M., and Bevan, J. A. (1990). Sequential histomorphometric changes in cancellous bone from ovariectomized dogs. *J. Bone Mineral Res.* **5,** 947–953.

Brautbar, N., Roy, A. T., and Hom, P. (1990). Hypomagnesemia and hypermagnesemia. *In* "Metal Ions In Biological Systems, Vol. 26, Compendium on Magnesium and Its Role in Biology, Nutrition, and Physiology" (H. Sigel and A. Sigel, eds.), pp. 285–320. Marcel Dekker, New York.

Bray, A. R., Burton, R. N., and Cox, B. T. (1987). Magnesium supplementation of lactating ewes and lamb growth. *N. Z. Vet. J.* **35,** 55–57.

Bronner, F. (1989). Renal calcium transport: Mechanisms and regulation—an overview. *Am. J. Physiol.* **257,** F707–F711.

Brooks, D. G. (1995). Acute tumor lysis syndrome in dogs. *Small Animal Oncology* **17,** 1103–1106.

Brown, E. M. (1994). Homeostatic mechanisms regulating extracellular and intracellular calcium metabolism. *In* "The Parathyroids" (J. P. Bilezikian, R. Marcus, and M. A. Levine, eds.), pp. 15–54. Raven Press, New York.

Brown, E. M., and Hebert, S. C. (1996). A cloned extracellular Ca^{2+}-sensing receptor: Molecular mediator of the actions of extracellular Ca^{2+} on parathyroid and kidney cells? *Kidney Int.* **49,** 1042–1046.

Brown, E. M., Gamba, G., Riccardi, D., Lombardi, M., Butters, R., Kifor, O., Sun, A., Hediger, M. A., Lytton, J., and Hebert, S. C. (1993). Cloning and characterization of an extracellular Ca^{2+}-sensing receptor from bovine parathyroid. *Nature* **366,** 575–580.

Brown, E. M., Pollak, M., Seidman, C. E., Seidman, J. G., Chou, Y.-H. W., Riccardi, D., and Hebert, S. C. (1995). Calcium-ion-sensing cell-surface receptors. *N. Engl. J. Med.* **333,** 234–240.

Buffington, C. A. (1994). Lower urinary tract disease in cats—new problems, new paradigms. *J. Nutr.* **124,** 2643S–2651S.

Buffington, C. A., Rogers, Q. R., Morris, J. G., and Cook, N. E. (1985). Feline struvite urolithiasis: Magnesium effect depends on urinary pH. *Feline Pract.* **15,** 29–33.

Buffington, C. A., Rogers, Q. R., and Morris, J. G. (1990). Effect of diet on struvite activity product in feline urine. *Am. J. Vet. Res.* **51,** 2025–2030.

Buffington, C. A., Chew, D. J., and DiBartola, S. P. (1996). Interstitial cystitis in cats. *Vet. Clin. N. Am. Sm. An. Pract.* **26,** 317–326.

Burtis, W. J. (1992). Parathyroid hormone-related protein: Structure, function, and measurement. *Clin. Chem.* **38,** 2171–2183.

Burtis, W. J., Brady, T. G., Orloff, J. J., Ersbak, J. B., Warrell, R. P. J., Olson, B. R., Wu, T. L., Mitnick, M. E., Broadus, A. E., and Stewart, A. F. (1990). Immunochemical characterization of circulating parathyroid hormone-related protein in patients with humoral hypercalcemia of malignancy. *N. Engl. J. Med.* **322,** 1106–1112.

Burtis, W. J., Dann, P., Gaich, G. A., and Soifer, N. E. (1994). A high abundance midregion species of parathyroid hormone-related protein: Immunological and chromatographic characterization in plasma. *J. Clin. Endocrinol. Metab.* **78,** 317–322.

Caddell, J. L. (1987). Magnesium requirements of animals and man. *In* "Magnesium in Cellular Processes and Medicine" (B. M. Altura, J. Durlach, and M. S. Seelig, eds.), pp. 77–88. Karger, Basel.

Calvo, M. S., Eyre, D. R., and Gundberg, C. M. (1996). Molecular basis and clinical application of biological markers of bone turnover. *Endocr. Rev.* **17,** 333–368.

Campos, R. V., Asa, S. L., and Drucker, D. J. (1991). Immunocytochemical localization of parathyroid hormone-like peptide in the rat fetus. *Cancer Res.* **51,** 6351–6357.

Canalis, E., Hock, J. M., and Raisz, L. G. (1994). Anabolic and catabolic effects of parathyroid hormone on bone and interactions with growth factors. *In* "The Parathyroids" (J. P. Bilezikian, R. Marcus, M. A. Levine, eds.), pp. 65–82. Raven Press, New York.

Canzanello, V. J., Bodvarsson, M., Kraut, J. A., Johns, C. A., Slatopolsky, E., and Madias, N. E. (1990). Effect of chronic respiratory acidosis on urinary calcium excretion in the dog. *Kidney Int.* **38,** 409–416.

Capen, C. C. (1971). Fine structural alterations of parathyroid glands in response to experimental and spontaneous changes of calcium in extracellular fluids. *Am. J. Med.* **50,** 598–611.

Capen, C. C. (1972). Endocrine control of calcium metabolism and parturient hypocalcemia in dairy cattle. *In* "Proc. 4th Ann. Conv. Am. Assoc. Bovine Pract." (E. I. Williams, ed.), pp. 189–198. Heritage Press, Stillwater.

Capen, C. C. (1975). Functional and fine structural relationships of parathyroid glands. *In* "Advances In Veterinary Sciences And Comparative Medicine," pp. 249–286. Academic Press, New York.

Capen, C. C. (1985). Calcium-regulating hormones and metabolic bone disease. *In* "Textbook of Small Animal Orthopaedics" (C. D. Newton, and D. M. Nunamaker, eds.), pp. 673–722. J. B. Lippincott Co., Philadelphia, Pennsylvania.

Capen, C. C. (1989). The calcium regulating hormones: Parathyroid hormone, calcitonin, and cholecalciferol. *In* "Veterinary Endocrinology and Reproduction" (L. E. McDonald and M. H. Pineda, eds.), pp. 92–185. Lea & Febiger, Philadelphia, Pennsylvania.

Capen, C. C. (1992). The endocrine glands. *In* "Pathology of Domestic Animals" (K. V. F. Jubb, P. C. Kennedy, and N. Palmer, eds.), pp. 267–348. Academic Press, San Diego.

Capen, C. C., and Black, H. E. (1974). Calcitonin-secreting ultimobranchial neoplasms of the thyroid gland in bulls: An animal model for medullary thyroid carcinoma in man (Sipple's syndrome). *Am. J. Pathol.* **74,** 377–380.

Capen, C. C., and Martin, S. L. (1977). Calcium metabolism and disorders of parathyroid glands. *Vet. Clin. North Am.* **7,** 513–555.

Capen, C. C., and Martin, S. L. (1982). Calcium regulating hormones and diseases of the parathyroid glands. *In* "Textbook Of Veterinary Internal Medicine, Second Edition" (S. J. Ettinger, ed.), pp. 1550–1592. Saunders, Philadelphia, Pennsylvania.

Capen, C. C., and Rosol, T. J. (1993a). Pathobiology of parathyroid hormone and parathyroid hormone-related protein: Introduction and evolving concepts. *In* "Pathobiology of the Parathyroid and Thyroid Glands" (V. D. LiVolsi, and R. A. DeLellis, eds.), pp. 1–33. Williams & Wilkins, Baltimore.

Capen, C. C., and Rosol, T. J. (1993b). Hormonal control of mineral metabolism. *In* "Disease Mechanisms in Small Animal Surgery" (M. J. Bojrab, ed.), pp. 841–857. Lea & Febiger, Philadelphia, Pennsylvania.

Capen, C. C., and Roth, S. I. (1973). Ultrastructural and functional relationships of normal and pathologic parathyroid cells. *In* "Pathobiology Annual" (H. L. Ioachim, ed.), pp. 129–175. Appleton, New York.

Capen, C. C., and Young, D. M. (1967). The ultrastructure of the parathyroid glands and thyroid parafollicular cells of cows with parturient paresis and hypocalcemia. *Lab. Invest.* **17,** 717–737.

Capen, C. C., and Young, D. M. (1969). Fine structural alterations in thyroid parafollicular cells of cows in response to experimental hypercalcemia induced by vitamin D. *Am. J. Pathol.* **57,** 365–382.

Capen, C. C., Cole, C. R., and Hibbs, J. W. (1966). The pathology of hypervitaminosis D in cattle. *Pathol. Vet.* **3,** 350–378.

Capen, C. C., Cole, C. R., and Hibbs, J. W. (1968). The influence of vitamin D on calcium metabolism and the parathyroid glands of cattle. *Fed. Proc.* **27,** 142–152.

Capen, C. C., Hoffsis, G. F., and Norman, A. W. (1979). 1,25-Dihydroxycholecalciferol and the prevention of parturient hypocalcemia and paresis ("milk fever") in pregnant dairy cows. *In* "Proceedings of the 4th Workshop on Vitamin D" (A. W. Norman, ed.), p. 176. Walter de Gruyter, Berlin.

Caple, I. W., and West, D. M. (1992). Ruminant hypomagnesemic tetanies. *In* "Current Veterinary Therapy 3, Food Animal Practice" (J. L. Howard, ed.), pp. 318–321. Saunders, Philadelphia, Pennsylvania.

Care, A. D. (1980). Magnesium homeostasis in relation to grass tetany. *In* "Scientific Foundations of Veterinary Medicine" (A. T. Phillipson, L. W. Hall, W. R. Prichard, eds.), pp. 104–108. William Heinemann Medical Books Limited, London.

Care, A. D. (1988). A fresh look at hypomagnesaemia. *Br. Vet. J.* **144,** 3–4.

Care, A. D. (1991). Placental transfer of calcium. *J. Dev. Physiol.* **15,** 253–257.

Care, A. D. (1992). The regulation of the secretion of calcitonin. *Bone Miner.* **16,** 182–185.

Care, A. D. (1994). The absorption of phosphate from the digestive tract of ruminant animals. *Br. Vet. J.* **150,** 197–205.

Care, A. D., and Ross, D. B. (1963). The role of the adrenal cortex in magnesium homeostasis and in the aetiology of hypomagnesaemia. *Res. Vet. Sci.* **4,** 24–38.

Care, A. D., Bates, R. F. L., Phillippo, M., Lequin, R. M., Hackeng, W. H. L., Barlet, J., and Larvor, P. (1970). Stimulation of calcitonin release from bovine thyroid by calcium and glucagon. *J. Endocrinol.* **48,** 667–668.

Carothers, M. A., Chew, D. J., and Nagode, L. A. (1994). 25-OH-Cholecalciferol intoxication in dogs. *Proc. Amer. Coll. Vet. Int. Med. Forum.* **12,** 822–825.

Cates, M. C., Singh, S. M., Peick, A. L., Harvey, M. H., and Reber, H. A. (1988). Acute hypercalcemia, pancreatic duct permeability, and pancreatitis in cats. *Surgery* **104,** 137–141.

Chambers, T. J., and Moore, A. (1983). The sensitivity of isolated osteoclasts to morphological transformation by calcitonin. *J. Clin. Endocrinol. Metab.* **57,** 819–824.

Charlton, J. A., and Armstrong, D. G. (1989). The effect of an intravenous infusion of aldosterone upon magnesium metabolism in the sheep. *Q. J. Exp. Physiol.* **74,** 329–337.

Chattopadhyay, N., Mithal, A., and Brown, E. M. (1996). The calcium-sensing receptor: A window into the physiology and pathophysiology of mineral ion metabolism. *Endocr. Rev.* **17,** 289–307.

Chausow, D. G., Forbes, R. M., Czarnecki, G. L., and Corbin, J. E. (1985). Experimentally-induced magnesium deficiency in growing kittens. *J. Anim. Sci.* **61 (Suppl. 1),** 295.

Chester-Jones, H., Fontenot, J. P., Veit, H. P., and Webb Jr, K. E. (1989). Physiologic effects of feeding high levels of magnesium to sheep. *J. Anim. Sci.* **67,** 1070–1081.

Chester-Jones, H., Fontenot, J. P., and Veit, H. P. (1990). Physiological and pathological effects of feeding high levels of magnesium to steers. *J. Anim. Sci.* **68,** 4400–4413.

Cheville, N. F. (1972). Ultrastructure of canine carotid body and aortic body tumors. Comparison with tissues of thyroid and parathyroid origin. *Vet. Pathol.* **9,** 166–189.

Chew, D. J., and Capen, C. C. (1980). Hypercalcemic nephropathy and associated disorders. *In* "Current Veterinary Therapy VII" (R. W. Kirk, ed.), pp. 1067–1072. Saunders, Philadelphia, Pennsylvania.

Chew, D. J., and Meuten, D. J. (1982). Disorders of calcium and phosphorus metabolism. *Vet. Clin. N. Am. Sm. An. Pract.* **12,** 411–438.

Chew, D. J., Nagode, L. A., and Carothers, M. (1992). Disorders of calcium: Hypercalcemia and hypocalcemia. *In* "Fluid Therapy in Small Animal Practice" (S. P. DiBartola, ed.), pp. 116–176. Saunders, Philadelphia, Pennsylvania.

Christakos, S., Gabreilides, C., and Rhoten, W. B. (1989). Vitamin D-dependent calcium binding proteins: Chemistry, distribution, functional considerations, and molecular biology. *Endocr. Rev.* **10,** 3–26.

Chu, L. L. H., MacGregor, R. R., Anast, C. S., Hamilton, J. W., and Cohn, D. V. (1973). Studies on the biosynthesis of rat parathyroid hormone and proparathyroid hormone: Adaptation of the parathyroid gland to dietary restriction of calcium. *Endocrinology* **93,** 915–924.

Chutkow, J. G. (1990). Magnesium and the peripheral (extradural) nervous system: Metabolism, neurophysiological functions, and clinical disorders. *In* "Metal Ions in Biological Systems: Vol. 26, Compendium on Magnesium and Its Role in Biology, Nutrition, and Physiology" (H. Sigel, and A. Sigel, eds.), pp. 417–440. Marcel Dekker, Inc. New York.

Civitelli, R., and Avioli, L. V. (1994). Calcium, phosphate, and magnesium absorption. *In* "Physiology of the Gastrointestinal Tract" (L. R. Johnson, ed.), pp. 2173–2181. Raven Press, New York.

Claeyssens, S., Lavoinne, A., Daragon, A., Josse, S., Godin, M., Matray, F., and Kuhn, J. M. (1990). Total and ultrafiltrable plasma magnesium in hyper- and hypoparathyroidism, and in calcium-related metabolic disorders. *Clin. Chim. Acta.* **195,** 107–114.

Cloutier, M., Rousseau, L., Gascon-Barré, M., and D'Amour, P. (1993). Immunological evidences for post-translational control of the parathyroid function by ionized calcium in dogs. *Bone Miner.* **22,** 197–207.

Cohn, D. V., and MacGregor, R. R. (1981). The biosynthesis, intracellular processing, and secretion of parathormone. *Endocr. Rev.* **2,** 1–26.

Cohn, D. V., MacGregor, R. R., Chu, L. L. H., Huang, D. W. Y., Anast, C. S., and Hamilton, J. W. (1974). Biosynthesis of proparathyroid hormone and parathyroid hormone. Chemistry, physiology and role of calcium in regulation. *Am. J. Med.* **56,** 767–773.

Cohn, D. V., Fasciotto, B. H., Zhang, J., and Gorr, S.-U. (1994). Chemistry and biology of chromogranin A (secretory protein I) of the parathyroid and other endocrine glands. *In* "The Parathyroids" (J. P. Bilezikian, M. A. Levine, and R. Marcus, eds.), pp. 107–119. Raven Press, New York.

Coleman, R. (1969). Ultrastructural observations on the parathyroid glands of *Xenopus laevis* Daudin. *Z. Zellforsch. Mikrosk. Anat.* **100,** 201–214.

Coleman, D. T., Fitzpatrick, L. A., and Bilezikian, J. P. (1994). Biochemical mechanisms of parathyroid hormone action. *In* "The Parathyroids" (J. P. Bilezikian, R. Marcus, and M. A. Levine, eds.), pp. 239–258. Raven Press, New York.

Collins, W. T., Jr., Capen, C. C., Dobereiner, J., and Takarnia, C. H. (1977). Ultrastructural evaluation of parathyroid glands and thyroid C cells of cattle fed *Solanum malacoxylon. Am. J. Pathol.* **87,** 603–614.

Colombo, G., Fanti, P., Yao, C., and Malluche, H. H. (1993). Isolation and complete amino acid sequence of osteocalcin from canine bone. *J. Bone Mineral Res.* **8,** 733–743.

Cooper, C. W., Schwesinger, W. H., Ontjes, D. A., Mahgoub, A. M., and Munson, P. L. (1972). Stimulation of secretion of pig thyrocalcitonin by gastrin and related hormonal peptides. *Endocrinology* **91,** 1079–1089.

Copp, D. H. (1969a). Endocrine control of calcium homeostasis. *J. Endocrinol.* **43,** 137–161.

Copp, D. H. (1969b). The ultimobranchial glands and calcium regulation. *In* "Fish physiology. Volume II: Endocrine System" (W. S. Hoar, and D. J. Randall, eds.), pp. 377–398. Academic Press, New York.

Copp, D. H. (1970). Endocrine regulation of calcium metabolism. *Ann. Rev. Physiol.* **32,** 61–86.

Copp, D. H. (1994). Calcitonin: Discovery, development, and clinical application. *Clin. Invest. Med.* **17,** 268–277.

Copp, D. H., Cameron, E. C., and Cheney, B. A. (1962). Evidence for calcitonin—a new hormone from the parathyroid that lowers blood calcium. *Endocrinology* **70,** 638–649.

Córdova, L., and Pérez-Gallardo, L. (1993). Ultrafiltrable magnesium: Simple quantitative measurement by two rapid methods. *Magnesium-Bulletin* **15,** 47–49.

Cseh, S. B., Fay, J. P., and Casaro, A. (1984). Changes in blood composition of pregnant cows during the onset of hypomagnesaemia. *Vet. Rec.* **115,** 567–570.

Cuddeford, D. (1987). Role of magnesium in the aetiology of ovine urolithiasis in fattening store lambs and intensively fattened lambs. *Vet. Rec.* **121,** 194–197.

de Boland, A. R., Skliar, M. I., Boland, R. L., Carrillo, B. J., and Ruksan, B. (1976). A method for the isolation of the active principle of *Solanum malacoxylon. Anal. Biochem.* **75,** 308–313.

de Morais, H. A., and Hansen, B.D. (1995). Chloride and magnesium: The forgotten ions. *Proceedings of the 13th American College of Veterinary Internal Medicine Forum,* pp. 628–631.

De Rouffignac, C., and Quamme, G. (1994). Renal magnesium handling and its hormonal control. *Physiological Reviews.* **74,** 305–322.

Deftos, L. (1991). Chromogranin A: Its role in endocrine function and as an endocrine and neuroendocrine tumor marker. *Endocr. Rev.* **12,** 181–187.

Deftos, L. J., Murray, T. M., Powell, D., Habener, J. F., Singer, F. R., Mayer, G. P., and Potts Jr., J. T. (1972). *In* "Calcium, Parathyroid Hormone, and the Calcitonins" (R. V. Talmage, and P. L. Munson, eds.), pp. 140–151. Excerpta Medica, Amsterdam.

Delmas, P. D. (1995). Biochemical markers of bone turnover. *Acta Orthop. Scand.* **66,** 176–182.

DeLuca, H.F. (1973). The kidney as an endocrine organ for the production of 1,25-dihydroxyvitamin D3, a calcium-mobilizing hormone. *N. Engl. J. Med.* **289,** 359–365.

Dennis, V. W. (1996). Phosphate metabolism: Contribution of different cellular compartments. *Kidney Int.* **49,** 938–942.

DeVries, S. E., Feldman, E. C., Nelson, R. W., and Kennedy, P. C. (1993). Primary parathyroid hyperplasia in dogs: Six cases (1982–1991). *J. Am. Vet. Med. Assoc.* **202,** 1132–1136.

DiBartola, S. P. (1992). Disorders of phosphorus: Hypophosphatemia and hyperphosphatemia. *In* "Fluid Therapy in Small Animal Practice" (S. P. DiBartola, ed.), pp. 177–192. Saunders, Philadelphia, Pennsylvania.

DiBartola, S. P., Chew, D. J., and Jacobs, G. (1980). Quantitative urinalysis including 24-hour protein excretion in the dog. *J. Am. An. Hosp. Assoc.* **16,** 537–546.

Dishington, I. W. (1975). Prevention of milk fever (hypocalcemic paresis puerpalis) by dietary salt supplements. *Acta Vet. Scand.* **16,** 503–512.

Dominguez, J. H., and Juhaszova, M. (1992). The renal sodium-calcium exchanger. *J. Lab. Clin. Med.* **1028,** 2298.

Done, S. H., Dobereiner, J., and Tokarnia, C. H. (1976). Systemic connective tissue calcification in cattle poisoned by *Solanum malacoxylon:* A histologic study. *Br. Vet. J.* **132,** 28–38.

Dougherty, S. A., Center, S. A., and Dzanis, D. A. (1990). Salmon calcitonin as adjunct treatment for vitamin D toxicosis in a dog. *J. Am. Vet. Med. Assoc.* **196,** 1269–1272.

Dow, S. W., Legendre, A. M., Stiff, M., and Greene, C. (1986). Hypercalcemia associated with blastomycosis in dogs. *J. Am. Vet. Med. Assoc.* **188,** 706–709.

Drezner, M. K. (1996). Phosphorus homeostasis and related disorders. *In* "Principles of Bone Biology" (J. P. Bilezikian, L. G. Raisz, G. A. Rodan, eds.), pp. 263–276. Academic Press, San Diego.

Dua, K., and Care, A. D. (1995). Impaired absorption of magnesium in the aetiology of grass tetany. *Br. Vet. J.* **151,** 413–426.

Dua, K., and Nauriyal, D. C. (1988). Clinico-biochemical aspects of experimentally induced hypomagnesemia in buffalo calves. *Indian J. An. Sci.* **58,** 38–44.

Dua, K., Gupta, P. P., and Nauriyal, D. C. (1988). Pathological changes in experimental hypomagnesemia in buffalo calves. *Indian J. Anim. Sci.* **58,** 656–662.

Ducy, P., and Karsenty, G. (1996). Skeletal Gla proteins: Gene structure, regulation of expression, and function. *In* "Principles of Bone Biology" (J. P. Bilezikian, L. G. Raisz, and G. A. Rodan, eds.), pp. 183–195. Academic Press, San Diego.

Durlach, J., and Durlach, V. (1984). Speculations on hormonal controls of magnesium homeostasis: A hypothesis. *Magnesium* **3,** 109–131.

Durlach, J., Bara, M., Guiet-Bara, A., and Rinjard, P. (1987). Taurine and magnesium homeostasis: New data and recent advances. *In* "Magnesium in Cellular Processes and Medicine" (B. M. Altura, J. Durlach, and M. S. Seelig, eds.), pp. 219–238. Karger, Basel.

Ebashi, S. (1985). Ca^{2+} in biological systems. *Experientia* **41,** 978–981.

Ebel, H. (1990). *Intestinal magnesium absorption. In* "Metal Ions In Biological Systems, Vol. 26, Compendium on Magnesium and Its Role in Biology, Nutrition, and Physiology" (H. Sigel and A. Sigel, eds.), pp. 227–248. Marcel Dekker, New York.

Elin, R. J. (1987a). Overview of problems in the assessment of magnesium status. *In* "Magnesium in Cellular Processes and Medicine" (B. M. Altura, J. Durlach, and M. S. Seelig, eds.), pp. 67–76. Karger, Basel.

Elin, R. J. (1987b). Assessment of magnesium status. *Clin. Chem.* **33,** 1965–1970.

Elin, R. J. (1994). Magnesium: The fifth but forgotten electrolyte. *Am. J. Clin. Pathol.* **102,** 616–622.

Elisaf, M., and Siamopoulos, K. C. (1995). Bacteremia-induced hypomagnesemia. *Magnesium-Bulletin* **17,** 70–73.

Ender, F., Dishington, I. W., and Helgebostad, A. (1962). Parturient paresis and related forms of hypocalcemic disorders induced experimentally in dairy cows. *Acta Vet. Scand.* **3(Suppl.1),** 1–52.

Eriksen, A. F., Brixen, K., and Charles, P. (1995). New markers of bone metabolism: Clinical use in metabolic bone disease. *Eur. J. Endocrinol.* **132,** 251–263.

Evans, G., and Parsons, C. E. (1988). Inter-relationships of plasma calcium, magnesium, zinc and albumin in healthy Beagle dogs. *In* "Animal Clinical Biochemistry—The Future" (D. J. Blackmore, ed.), pp. 302–306. Cambridge University Press, Cambridge.

Everhart-Caye, M., Inzucchi, S. E., Guinness-Henry, J., Mitnick, M. A., and Stewart, A. F. (1996). Parathyroid hormone (PTH)-related protein (1–36) is equipotent to PTH(1–34) in humans. *J. Clin. Endocrinol. Metab.* **81,** 199–208.

Fanti, P., Colombo, G., Yao, C., Brown, S. A., Vernon, M. W., and Malluche, H. H. (1993). Development and characterization of a polyclonal antiserum-based radioimmunoassay for dog osteocalcin. *J. Bone Mineral Res.* **8,** 745–752.

Favus, M. J. (1992). Intestinal absorption of calcium, magnesium, and phosphorus. *In* "Disorders of Bone and Mineral Metabolism" (F. L. Coe and M. J. Favus, eds.), pp. 57–81. Raven Press, New York.

Felsenfeld, A. J., and Llach, F. (1993). Parathyroid gland function in chronic renal failure. *Kidney Int.* **43,** 771–789.

Fenton, A. J., Kemp, B. E., Hammonds, R. G. J., Michelhill, K., Moseley, J. M., Martin, T. J., and Nicholson, G. C. (1991a). A potent inhibitor of osteoclastic bone resorption within a highly conserved peptapeptide region of parathyroid hormone-related protein: PTHrP[107–111]. *Endocrinology* **129,** 3424–3426.

Fenton, A. J., Kemp, B. E., Kent, G. N., Moseley, J. M., Zheng, M.-H., Rowe, D. J., Britto, J. M., Martin, T. J., and Nicholson, G. C. (1991b). A carboxy-terminal peptide from the parathyroid hormone-related protein inhibits bone resorption by osteoclasts. *Endocrinology* **129,** 1762–1768.

Fenwick, D. C., and Daniel, R. C. W. (1991). Monthly variation and distribution of erythrocyte Na, K, and Mg concentrations in normal dairy cattle. *J. Vet. Med. A.* **38,** 485–493.

Fetter, A. W., and Capen, C. C. (1970). The ultrastructure of the parathyroid glands of young pigs. *Acta Anat.* **75,** 359–372.

Field, A. C., and Suttle, N. F. (1979). Effect of high potassium and low magnesium intakes on the mineral metabolism of monozygotic twin cows. *J. Comp. Path.* **89,** 431–439.

Finco, D. R., Barsanti, J. A., and Crowell, W. A. (1985). Characterization of magnesium-induced urinary disease in the cat and comparison with feline urologic syndrome. *Am. J. Vet. Res.* **46,** 391–400.

Finco, D. R., Brown, S. A., Cooper, T., Crowell, W. A., Hoenig, M., and Barsanti, J. A. (1994). Effects of parathyroid hormone depletion in dogs with induced renal failure. *Am. J. Vet. Res.* **55,** 867–873.

Finkelman, R. D., and Butler, W. T. (1985). Vitamin D and skeletal tissues. *J. Oral Pathol.* **14,** 191–215.

Fiore, C. E., Clementi, G., Prato, A., and Tamburino, C. (1990). Influence of magnesium supplementation on parathyroid hormone and bone Gla protein concentration in normal rats. *Magnesium Trace Elem.* **9,** 289–293.

Fisher, L. J., Dinn, N., Tait, R. M., and Shelford, J. A. (1994). Effect of level of dietary potassium on the absorption and excretion of calcium and magnesium by lactating cows. *Can. J. Anim. Sci.* **74,** 503-509.

Fiske, C. H., and Subbarow, Y. (1925). The colorimetric determination of phosphorus. *J. Biol. Chem.* **66,** 375–400.

Flanders, J. A., and Reimers, T. J. (1991). Radioimmunoassay for parathyroid hormone in cats. *Am. J. Vet. Res.* **52,** 422–425.

Flanders, J. A., Scarlett, J. M., Blue, J. T., and Neth, S. (1989). Adjustment of total serum calcium concentration for binding to albumin and protein in cats: 291 cases (1986–1987). *J. Am. Vet. Med. Assoc.* **194,** 1609–1611.

Flatman, P. W. (1991). Mechanisms of magnesium transport. *Annu. Rev. Physiol.* **53,** 259-271.

Fleming, S. A., Hunt, E. L., Riviere, J. E., and Anderson, K. L. (1991). Renal clearance and fractional excretion of electrolytes over four 6-hour periods in cattle. *Am. J. Vet. Res.* **52,** 5–8.

Fleming, S. A., Hunt, E. L., Brownie, C., Rakes, A., and McDaniel, B. (1992). Fractional excretion of electrolytes in lactating dairy cows. *Am. J. Vet. Res.* **53,** 222–224.

Fontenot, J. P., Allen, V. G., Bunce, G. E., and Goff, J. P. (1989). Factors influencing magnesium absorption and metabolism in ruminants. *J. Anim. Sci.* **67,** 3445–3455.

Forman, D. T., and Lorenzo, L. (1991). Ionized calcium: Its significance and clinical usefulness. *Ann. Clin. Lab. Sci.* **21,** 297–304.

Fraser, D., Jones, G., Kooh, S. W., and Radde, I. C. (1987). Calcium and phosphate metabolism. *In* "Fundamentals of Clinical Chemistry" (N. W. Tietz, ed.), pp. 705–728. Saunders, Philadelphia, Pennsylvania.

Frick, T. W., Mithöfer, K., Fernandez-del Castillo, C., Rattner, D. W., and Warshaw, A. L. (1995). Hypercalcemia causes acute pancreatitis by pancreatic secretory block, intracellular zymogen accumulation, and acinar cell injury. *Amer. J. Surg.* **169,** 167–172.

Fujise, H., Cruz, P., Reo, N. V., and Lauf, P. K. (1991). Relationship between total magnesium concentration and free intracellular magnesium in sheep red blood cells. *Biochim. Biophys. Acta* **1094,** 51–54.

Gacad, M. A., and Adams, D. A. (1993). Identification of a competitive binding component in vitamin-D resistant New World primate cells with a low affinity but high capacity for 1,25-dihydroxyvitamin D_3. *J. Bone Mineral Res.* **8,** 27–35.

Garrett, I. R. (1993). Bone destruction in cancer. *Semin. Oncol.* **20,** 4–9.

Geiser, D. R., Andrews, F. M., Rohrbach, B. W., White, S. L., Maykuth, P. L., Green, E. M., and Provenza, M. K. (1995). Blood ionized calcium concentrations in horses before and after the cross-country phase of three-day event competition. *Am. J. Vet. Res.* **56,** 1502–505.

Gerloff, B. J., and Swenson, E. P. (1996). Acute recumbency and marginal phosphorus deficiency in dairy cattle. *J. Am. Vet. Med. Assoc.* **208,** 716–720.

Geven, W. B., Vogels-Mentink, G. M., Willems, J. L., Joordens, J. J. M., Hilbers, C. W., and Monnens, L. A. H. (1990). Experimental magnesium depletion in the dog. *Magnesium-Bulletin* **12,** 166–169.

Gibasiewicz, W. A. (1984). Relationship between the frequency of placental retention and serum magnesium values in cows. *Medycyna Weterynaryjna* **40,** 622–623.

Gilka, F., and Sugden, E. A. (1984). Ectopic mineralization and nutritional hyperparathyroidism in boars. *Canad. J. Comp. Med.* **48,** 102–107.

Gillespie, M. T., and Martin, T. J. (1994). The parathyroid hormone-related protein gene and its expression. *Mol. Cell. Endocrinol.* **100,** 143–147.

Goddard, K. M., Williams, G. D., Newberne, P. M., and Wilson, R. B. (1970). A comparison of all-meat, semi-moist, and dry-type dog foods as diets for growing beagles. *J. Am. Vet. Med. Assoc.* **157,** 1233–1236.

Goff, J. P., Reinhardt, T. A., and Horst, R. L. (1989). Recurring hypocalcemia of bovine parturient paresis is associated with failure to produce 1,25-dihydroxyvitamin D. *Endocrinology* **125,** 49–53.

Gow, C. B., Silvapulle, M. J., and Moore, G. P. (1992a). Epidermal growth factor alters the electrolyte profile of lactating ewes (*Ovis aries*). *Comp. Biochem. Physiol. A.* **103,** 687–693.

Gow, C. B., Wilkinson, M., Silvapulle, M. J., and Moore, G. P. (1992b). Fluid balance, electrolyte profiles and plasma parathyroid hormone concentrations in ewes treated with epidermal growth factor. *J. Endocrinol.* **135,** 91–101.

Greene, L. W., Baker, J. F., and Hardt, P. F. (1989). Use of animal breeds and breeding to overcome the incidence of grass tetany: a review. *J. Anim. Sci.* **67,** 3463-3469.

Gröne, A., Werkmeister, J. R., Steinmeyer, C. L., Capen, C. C., and Rosol, T. J. (1994). Parathyroid hormone-related protein in normal and neoplastic tissues: immunohistochemical localization and biochemical extraction. *Vet. Pathol.* **31,** 308–315.

Groutides, C. P., and Michell, A. R. (1990). Changes in plasma composition in calves surviving or dying from diarrhoea. *Br. Vet. J.* **146,** 205–210.

Grover, A. K., and Khan, I. (1992). Calcium pump isoforms: Diversity, selectivity and plasticity. *Cell Calcium.* **13,** 9–17.

Grunes, D. L., and Welch, R. M. (1989). Plant contents of magnesium, calcium and potassium in relation to ruminant nutrition. *J. Anim. Sci.* **67,** 3485–3494.

Günther, T. (1986). Functional compartmentation of intracellular magnesium. *Magnesium* **5,** 53–59.

Günther, T. (1990). Functional compartmentation of intracellular magnesium. *In* "Metal Ions In Biological Systems, Vol. 26, Magnesium and Its Role in Biology, Nutrition, and Physiology" (H. Sigel, and A. Sigel, eds.), pp. 193–213. Marcel Dekker, New York.

Günther, T. (1993). Mechanisms and regulation of Mg^{2+} efflux and Mg^{2+} influx. *Miner. Electrolyte Metab.* **19,** 259–265.

Habener, J. F. (1981). Recent advances in parathyroid hormone research. *Clin. Biochem.* **14,** 223–229.

Habener, J. F., and Potts, J. T., Jr. (1978). Biosynthesis of parathyroid hormone. *N. Engl. J. Med.* **299,** 580–644.

Habener, J. F., Chang, H. T., and Potts, J. T., Jr. (1977). Enzymatic processing of proparathyroid hormone by cell-free extracts of parathyroid glands. *Biochem.* **16,** 3910–3917.

Haggard, D. L., Whitehair, C. K., and Langham, R. F. (1978). Tetan deficiency in suckling beef calves. *J. Am. Vet. Med. Assoc.* **172,**

Halse, K. (1984). Calcium effects on renal conservation of magnesium in cows. *Acta Vet. Scand.* **25,** 213–228.

Hanafin, N. M., Chen, T. C., Heinrich, G., Segre, G. V., and Holick, M. F. (1995). Cultured human fibroblasts and not cultured human keratinocytes express a PTH/PTHrP receptor mRNA. *J. Invest. Dermatol.* **105,** 133–137.

Harrington, D. D., and Page, E. H. (1983). Acute vitamin D_3 toxicosis in horses: Case reports and experimental studies of the comparative toxicity of vitamins D_2 and D_3. *J. Am. Vet. Med. Assoc.* **182,** 1358–1369.

Harris, D. C. H., Gabow, P. A., Linas, S. L., Rosendale, D. E., Guggenheim, S. P., and Schrier, R. W. (1986). Prevention of hypercalcemia-induced renal concentrating defect and tissue calcium accumulation. *Am. J. Physiol.* **251,** F642–F646.

Hazarika, G. C., and Pandey, N. N. (1993). Clinical and biochemical changes during experimental chronic hypomagnesemia in goats and its treatment. *Indian Vet. J.* **70,** 247–250.

Hazarika, G. C., Pandey, N. N., and Prasad, M. C. (1991). Pathological changes in experimentally induced hypocalcemia/hypomagnesemia in goats. *Indian J. An. Sci.* **61,** 798–802.

Hazewinkel, H. A. (1991). Dietary influences on calcium homeostasis and the skeleton. *In* "Proceedings of the Purina International Nutrition Symposium," pp. 51–59. Eastern States Veterinary Conference, Orlando, Florida.

Heaton, F. W. (1990). Role of magnesium in enzyme systems. *In* "Metal Ions In Biological Systems, Vol. 26, Compendium on Magnesium and Its Role in Biology, Nutrition, and Physiology" (H. Sigel, and A. Sigel, eds.), pp. 119–133. Marcel Dekker, Inc. New York.

Heaton, F. W., and Rayssiguier, Y. (1987). Magnesium deficiency and membrane properties. *In* "Magnesium in Cellular Processes and Medicine" (B. M. Altura, J. Durlach, and M. S. Seelig, eds.), pp. 121–130. Karger, Basel.

Heersche, J. N. M. (1992). Calcitonin effects on osteoclastic resorption: The "escape phenomenon" revisited. *Bone Miner.* **16,** 174–177.

Heersche, J. N. M., Bellows, C. G., and Aubin, J. E. (1994). Cellular actions of parathyroid hormone on osteoblast and osteoclast differentiation. *In* "The Parathyroids" (J. P. Bilezikian, R. Marcus, and M. A. Levine, eds.), pp. 83–92. Raven Press, New York.

Henderson, J. E., Amizuka, N., Warshawsky, H., Biasotto, D., Lanske, B. M. K., Goltzman, D., and Karaplis, A. C. (1995). Nucleolar localization of parathyroid hormone-related peptide enhances survival of chondrocytes under conditions that promote apoptotic cell death. *Mol. Cell Biol.* **15,** 4064–4075.

Henry, H. L., and Norman, A. W. (1978). Vitamin D: Two dihydroxylated metabolites are required for normal chicken egg hatchability. *Science* **201,** 835–837.

Henry, D. A., Goodman, W. G., Nudelman, R. K., DiDomenico, N. C., Alfrey, A. C., Slatopolsky, E., Stanley, T. M., and Coburn, J. W. (1984). Parenteral aluminum administration in the dog: I. Plasma kinetics, tissue levels, calcium metabolism, and parathyroid hormone. *Kidney Int.* **25,** 362–369.

Hibbs, J. W., and Conrad, H. R. (1960). Studies of milk fever in dairy cows. VI. Effect of three prepartal dosage levels of vitamin D on milk fever incidence. *J. Dairy Sci.* **43,** 1124–1129.

Hidiroglou, M., Thompson, B. K., Ho, S. K., and Proulx, J. G. (1981). Hypomagnesaemia in beef cows wintered in Ontario. *Can. J. Comp. Med.* **45,** 124–129.

High, W. B., Black, H. E., and Capen, C. C. (1981a). Histomorphometric evaluation of the effects of low dose parathyroid hormone administration on cortical bone remodeling in adult dogs. *Lab. Invest.* **44,** 449–454.

High, W. B., Capen, C. C., and Black, H. E. (1981b). Effects of 1,25-dihydroxycholecalciferol, parathyroid hormone and thyroxine on trabecular bone remodeling in adult dogs: A histomorphometric study. *Am. J. Pathol.* **105,** 279–287.

Hilgers, H. P. (1979). Influence of low environmental temperature on the mineral (Mg^{2+}, Na^+, K^+) and water balance of sheep fed different amounts of magnesium. *Inaugural Dissertation, Tierärzliche Hochschule, Hannover.*

Hintz, H. F., and Schryver, H. F. (1972). Magnesium metabolism in the horse. *J. An. Sci.* **35,** 755–759.

Hodgson, S. F., and Hurley, D. L. (1993). Acquired hypophosphatemia. *Endocrinol. Metab. Clin. North. Am.* **22,** 397–409.

Hoffsis, G. F., Saint-Jean, G., and Rings, M. D. (1989). Hypomagnesemia in ruminants. *Comp. Cont. Ed.* **11,** 519–526.

Holick, M. F. (1981). The cutaneous photosynthesis of previtamin D_3: A unique photoendocrine system. *J. Invest. Dermatol.* **77,** 51–58.

Holick, M. F., and Clark, M. B. (1978). The photobiogenesis and metabolism of vitamin D. *Fed. Proc.* **37,** 2567–2574.

Holick, M. F., Ray, S., Chen, T. C., Tian, X., and Persons, K. S. (1994). A parathyroid hormone antagonist stimulates epidermal proliferation and hair growth in mice. *Proc. Natl. Acad. Sci. (USA)* **91,** 8014–8016.

Hollis, B. W. (1986). Assay of circulating 1,25-dihydroxyvitamin D involving a novel single-cartridge extraction and purification procedure. *Clin. Chem.* **32,** 2060–2063.

Hollis, B. W., Kamerud, J. Q., Kurkowski, A., Beaulieu, J., and Napoli, J. L. (1996). Quantification of circulating 1,25-dihydroxyvitamin D by radioimmunoassay with an ^{125}I-labeled tracer. *Clin. Chem.* **42,** 586–592.

Hong, M. K., Vossoughi, J., Haudenschild, C. C., Wong, S. C., Zuckerman, B. D., and Leon, M. B. (1995). Vascular effects of diet-induced hypercalcemia after balloon artery injury in giant Flemish rabbits. *Am. Heart J.* **130,** 758–764.

Hope, E., Johnston, S. D., Hegstad, R. L., Geor, R. J., and Murphy, M. J. (1993). Effects of sample collection and handling on concentration of osteocalcin in equine serum. *Am. J. Vet. Res.* **54,** 1017–1020.

Horber, H., Eigenmann, U., Jucker, H., and Leemann, W. (1979). Magnesium status of dairy cows at the start of the grazing season and its assessment with a new rapid urine test. *Schweizer Archiv für Tierheilkunde* **121,** 187–193.

Horst, R. L., Goff, J. P., and Reinhardt, T. A. (1994). Calcium and vitamin D metabolism in the dairy cow. *J. Dairy Sci.* **77,** 1936–1951.

How, K. L., Hazewinkel, H. A., and Mol, J. A. (1994). Dietary vitamin D dependence of cat and dog due to inadequate cutaneous synthesis of vitamin D. *Gen. Comp. Endocrinol.* **96,** 12–18.

Hughes, M. R., and Haussler, M. R. (1978). 1,25-Dihydroxyvitamin D_3 receptors in parathyroid glands. *J. Biol. Chem.* **253,** 1065–1073.

Hulter, H. N., Toto, R. D., Bonner Jr., E. L., Ilnicki, L., and Sebastian, A. (1981). Renal and systemic acid-base effects of chronic hypoparathyroidism in dogs. *Am. J. Physiol.* **241,** 495–501.

Hulter, H. N., Halloran, B. P., Toto, R. D., and Peterson, J. C. (1985). Long term control of plasma calcitriol concentration in dogs and humans dominant role of plasma calcium concentration in experimental hyperparathyroidism. *J. Clin. Invest.* **76,** 695–702.

Hunt, R. D., Garcia, F. G., and Hegsted, D. M. (1967). A comparison of vitamin D_2 and D_3 in new world primates. I. Production and regression of osteodystrophia fibrosa. *Lab. Anim. Care.* **17,** 222–234.

Imamura, H., Sato, K., Shizume, K., Satoh, T., Kasono, K., Ozawa, M., Ohmura, E., Tsushima, T., and Demura, H. (1991). Urinary excretion of parathyroid hormone-related protein fragments in patients with humoral hypercalcemia of malignancy and hypercalcemia tumor-bearing nude mice. *J. Bone Mineral Res.* **6,** 77–84.

Ingraham, R. H., Kappel, L. C., Morgan, E. B., and Srikandakumar, A. (1987a). Correction of subnormal fertility with Ca and Mg supplementation. *J. Dairy Sci.* **70,** 167–180.

Ingraham, R. H., Kappel, L. C., Morgan, E. G., and Srikandakumar, A. (1987b). Correction of subnormal fertility with copper and magnesium supplementation. *J. Dairy Sci.* **70,** 167–180.

Iqbal, S. J. (1994). Vitamin D metabolism and the clinical aspects of measuring metabolites. *Ann. Clin. Biochem.* **31,** 109–124.

Ising, H., Bertschat, F., Günther, T., Jeremias, E., and Jeremias, A. (1995). Measurement of free magnesium in blood, serum and plasma with an ion-sensitive electrode. *Eur. J. Clin. Chem. Clin. Biochem.* **33,** 365–371.

Jacobs, J. W. (1985). Calcitonin gene expression. *Bone Mineral Res.* **3,** 151–179.

Jones, B. R., and Alley, M. R. (1985). Primary idiopathic hypoparathyroidism in St. Bernard dogs. *New Zealand Vet. J.* **33,** 94–97.

Jopp, A. J., and Quinlivan, T. D. (1981). Ovine post-parturient hypomagnesemic ketosis. *New Zealand Vet. J.* **29,** 37–38.

Jorch, U. M., Anderson, C., Delaquerriere-Richardson, L. F. O., and Cook, M. A. (1982). Concentrations of plasma C-terminal immunoreactive parathyroid hormone in the standardized research Beagle. *Am. J. Vet. Res.* **43,** 350–351.

Joyce, J. R., Pierce, K. R., Romane, W. M., and Baker, J. M. (1971). Clinical study of nutritional secondary hyperparathyroidism in horses. *J. Am. Vet. Med. Assoc.* **158,** 2033–2042.

Kallet, A. J., Richter, K. P., Feldman, E. C., and Brum, D. E. (1991). Primary hyperparathyroidism in cats: Seven cases (1984–1989). *J. Am. Vet. Med. Assoc.* **199,** 1767–1771.

Kallfelz, F. A., Ahmed, A. S., Wallace, R. J., Sasangka, B. H., and Warner, R. G. (1987). Dietary magnesium and urolithiasis in growing calves. *Cornell Vet.* **77,** 33–45.

Kappel, L. C., Youngberg, H., Ingraham, R. H., Hembry, F. G., Robinson, D. L., and Cherney, J. H. (1983). Effect of dietary aluminum on magnesium status of cows. *Amer. J. Vet. Res.* **44,** 770–773.

Karaplis, A. C., Luz, A., Glowacki, J., Bronson, R. T., Tybulewicz, V. L. J., Kronenberg, H. M., and Mulligan, R. C. (1994). Lethal skeletal dysplasia from targeted disruption of the parathyroid hormone-related peptide gene. *Genes & Devel.* **8,** 277–289.

Karbach, U., and Feldmeier, H. (1993). The cecum is the site with the highest calcium absorption in rat intestine. *Digest. Dis. Sci.* **38,** 1815–1824.

Kasahara, H., Tsuchiya, M., Adachi, R., Horikawa, S., Tanaka, S., and Tachibana, S. (1992). Development of a C-terminal-region-specific radioimmunoassay of parathyroid hormone-related protein. *Biomed. Res.* **13,** 155-161.

Kasari, T. R., Woodbury, A. H., and Morcom-Kasari, E. (1990). Adverse effect of orally administered magnesium hydroxide on serum magnesium concentration and systemic acid–base balance in adult cattle. *J. Am. Vet. Med. Assoc.* **196,** 735–742.

Kayne, L. H., and Lee, D. B. N. (1993). Intestinal magnesium absorption. *Miner. Electrolyte Metab.* **19,** 210–217.

Kelly, J. (1988). Magnesium and milk fever. *In Practice* **10,** 168–170.

Kelly, P. (1979). Hypomagnesaemia in sheep: A review. *A. D. A. S. Quarterly Rev.* **34,** 151–166.

Kempson, S. A. (1996). Peptide hormone action on renal phosphate handling. *Kidney Int.* **49,** 1005–1009.

Kennedy, D. W., and Bunting, L. D. (1991). Alterations in ruminal utilization of magnesium and zinc in lambs fed different ratios of concentrate: Forage. *Int. J. Vit. Nutr. Res.* **61,** 67–71.

Kes, P., and Reiner, Z. (1990). Symptomatic hypomagnesemia associated with gentamicin therapy. *Magnesium Trace Elem.* **9,** 54–60.

Khan, S. A., Miller, J. K., Schneider, M. D., White, P. K., Ramsey, N., and Mraz, F. R. (1982). Calcium metabolism in hypomagnesemic sheep. *J. Dairy Sci.* **65 (Suppl. 1),** 218.

Kivirikko, K. I., Laitinen, O., and Prockop, D. J. (1967). Modifications of a specific assay for hydroxyproline in urine. *Anal. Biochem.* **19,** 249–255.

Klausner, J. S., Fernandez, F. R., O'Leary, T. P., Johnson, S. D., and Osborne, C. A. (1986). Canine primary hyperparathyroidism and its association with urolithiasis. *Vet. Clin. N. Am. Sm. An. Pract.* **16,** 227–239.

Kohn, C. W., and Brooks, C. L. (1990). Failure of pH to predict ionized calcium percentage in healthy horses. *Am. J. Vet. Res.* **51,** 1206–1210.

Kohn, C. W., and Chew, D. J. (1987). Laboratory diagnosis and characterization of renal disease in horses. *Vet. Clin. N. Am. Eq. Pract.* **3,** 585–615.

Kolb, E. (1985). Review of recent biochemical findings on magnesium metabolism in ruminants and the origin and treatment of hypomagnesaemia with reference to the use of Tetamag. *Monatshefte für Veterinärmedizin.* **40,** 615–619.

Kremer, R., and Goltzman, D. (1994). Assays for parathyroid hormone-related protein. *In* "The Parathyroids" (J. P. Bilezikian, R. Marcus, M. A. Levine, eds.), pp. 321–340. Raven Press, New York.

Kronenberg, H. M., Bringhurst, F. R., Segre, G. V., and Potts, J. T., Jr. (1994). Parathyroid hormone biosynthesis and metabolism. *In* "The Parathyroids" (J. P. Bilezikian, R. Marcus, and M. A. Levine, eds.), pp. 125–138. Raven Press, New York.

Krook, L. (1957). Spontaneous hyperparathyroidism in the dog. *Acta Pathol. Microbiol. Scand* **41(Suppl. 122),** 1–81.

Krook, L., and Barrett, R. B. (1962). Simian bone disease—a secondary hyperparathyroidism. *Cornell Vet.* **52,** 459

Krook, L., and Lowe, J. E. (1964). Nutritional secondary hyperparathyroidism in the horse. *Pathol. Vet.* **1(Suppl. 1),** 1–93.

Krook, L., Lutwak, L., and McEntee, K. (1969). Dietary calcium, ultimobranchial tumors and osteopetrosis in the bull. A syndrome of calcitonin excess? *Am. J. Clin. Nutr.* **22,** 115–118.

Krook, L., Lutwak, L., Henrickson, P. A., Kallfelz, F., Hirsch, C., Romanus, B., Belanger, L. F., Marier, J. R., and Sheffy, B. E. (1971a). Reversibility of nutritional osteoporosis: Physicochemical data on bones from an experimental study in dogs. *J. Nutr.* **101,** 233–246.

Krook, L., Lutwak, L., McEntee, K., Henrikson, P. A., Braun, K., and Roberts, S. (1971b). Nutritional hypercalcitoninism in bulls. *Cornell Vet.* **61,** 625–639.

Kruger, J. M., Osborne, C. A., Nachreiner, R. F., and Refsal, K. R. (1996). Hypercalcemia and renal failure: Etiology, pathophysiology, diagnosis, and treatment. *Vet. Clin. N. Am. Sm. An. Pract.* **26,** 1417–1445.

Krupnik, A. (1985b). Effect of magnesium supplementation on the frequency of placental retention and puerperal endometritis in cows. *Zeszyty Naukowe Akademii Rolniczej we Wroclawiu, Weterynaria* **42,** 91-110.

Krupnik, A., and Marcinkowski, K. (1983). Chronic magnesium deficiency and placental retention in cows. *Medycyna Weterynaryjna* **39,** 421–423.

Langley, W. F., and Mann, D. (1991). Central nervous system magnesium deficiency. *Arch. Intern. Med.* **151,** 593–596.

Leav, I., Schiller, A. L., Rijnberk, A., Legg, M. A., and der Kinderen, P. J. (1976). Adenomas and carcinomas of the canine and feline thyroid. *Am. J. Pathol.* **83,** 61–122.

Leblanc, B., Parodi, A. L., Lagadic, M., Hurtrel, M., and Jobit, C. (1991). Immunocytochemistry of canine thyroid tumors. *Vet. Pathol.* **28,** 370–380.

Lehner, N. D. M., Bullock, B. C., and Clarkson, T. B. (1976). Biological activities of vitamin D_2 and D_3 for growing squirrel monkeys. *Lab. Anim. Care.* **17,** 483–493.

Leonhard, S., Smith, E., Martens, H., Gäbel, G., and Ganzoni, E. (1990). Transport of magnesium across an isolated preparation of sheep rumen: a comparison of $MgCl_2$, Mg aspartate, Mg pidolate, and Mg-EDTA. *Magnesium Trace Elem.* **9,** 265–271.

Levi, J., Massry, S. G., Coburn, J. W., Llach, F., and Kleeman, C. R. (1974). Hypocalcemia in magnesium-depleted dogs: Evidence for reduced responsiveness to parathyroid hormone and relative failure of parathyroid gland function. *Metabolism* **23,** 323–335.

Levine, M. A., Schwindinger, W. F., Downs Jr., R. W., and Moses, A. M. (1994). Psuedohypoparathyroidism: Clinical, biochemical, and molecular features. *In* "The Parathyroids" (J. P. Bilezikian, R. Marcus, and M. A. Levine, eds.), pp. 781–800. Raven Press, New York.

Liberman, U. A. (1996). Hereditary deficiencies in vitamin D action. *In* "Principles of Bone Biology" (J. P. Bilezikian, L. G. Raisz, and G. A. Rodan, eds.), pp. 903–916. Academic Press, San Diego.

Lincoln, S. D., and Lane, V. M. (1985). Postmortem magnesium concentration in bovine vitreous humor: comparison with antemortem serum magnesium concentration. *Am. J. Vet. Res.* **46,** 160–162.

Lincoln, S. D., and Lane, V. M. (1990). Serum ionized calcium concentration in clinically normal dairy cattle, and changes associated with calcium abnormalities. *J. Am. Vet. Med. Assoc.* **197,** 1471–1474.

Littledike, E. T., and Cox, P. S. (1979). Clinical, mineral, and endocrine interrelationships in hypomagnesemic tetany. *In* "Grass Tetany ASA Special Publication Number 35" (V. V. Rendig and D. L. Grunes, eds.), pp. 1–51. American Society of Agronomy, Madison, Wisconsin.

Littledike, E. T., and Goff, J. (1987). Hormonal control of magnesium metabolism parathyroid hormone and vitamin D metabolites. *In* "Magnesium in Cellular Processes and Medicine" (B. M. Altura, J. Durlach, and M. S. Seelig, eds.), pp. 164–182. Karger, Basel.

Littledike, E. T., Wittum, T. E., and Jenkins, T. G. (1995). Effect of breed, intake, and carcass composition on the status of several macro and trace minerals of adult beef cattle. *J. Anim. Sci.* **73,** 2113–2119.

London, R. E. (1991). Methods for measurement of intracellular magnesium: NMR and fluorescence. *Annu. Rev. Physiol.* **53,** 241–258.

Long, P., Choi, G., and Rehmel, R. (1983). Oxyphil cells in a red-tailed hawk (*Buteo jamaicensis*) with nutritional secondary hyperparathyroidism. *Avian Dis.* **27,** 839–843.

Lötscher, M., Wilson, P., Nguyen, S., Kaissling, B., Biber, J., Murer, H., and Levi, M. (1996). New aspects of adaptation of rat renal Na–Pi contransporter to altertions in dietary phosphate. *Kidney Int.* **49,** 1012–1018.

Lucas, M. J., Huffman, E. M., and Johnson, L. W. (1982). Clinical and clinicopathologic features of transport tetany of feedlot lambs. *J. Am. Vet. Med. Assoc.* **181,** 381–383.

Lucas, M. J., Leveno, K. J., and Cunningham, F. G. (1995). A comparison of magnesium sulfate with phenytoin for the prevention of eclampsia. *N. Engl. J. Med.* **333,** 201–205.

Lucey, S., Rowlands, G. J., and Russell, A. M. (1986). Short term associations between disease and milk yield of dairy cows. *J. Dairy Sci.* **53,** 7–15.

Lulich, J. P., Osborne, C. A., Polzin, D. J., Johnston, S. D., and Parker, M. L. (1991). Urine metabolite values in fed and nonfed clinically normal beagles. *Am. J. Vet. Res.* **52,** 1573–1578.

Lupulescu, A., Potorac, E., Pop, A., Heitmanek, C., Merculiev, E., Chisiu, N., Oprisan, R., and Neacsu, C. (1968). Experimental investigation on immunology of the parathyroid gland. *Immunology* **14,** 475–482.

Luthman, J., and Björkman, C. (1981). Magnesium content of fodder from herds with cows affected with hypomagnesemia. *Svensk Veterinärtidning* **33,** 105–107.

Lyles, K. M. (1996). Oncogenic osteomalacia. *In* "Principles of Bone Biology" (J. P. Bilezikian, L. G. Raisz, and G. A. Rodan, eds.), pp. 935–940. Academic Press, San Diego.

Lyles, K. W., Jackson, T. W., Nesbitt, T., and Quarles, L. D. (1993). Salmon calcitonin reduces vertebral bone loss in glucocorticoid-treated beagles. *Amer. J. Physiol.* **264,** E938–E942.

McAdam, P. A., and O'Dell, G. D. (1982). Mineral profile of blood plasma of lactating dairy cows. *J. Dairy. Sci.* **65,** 1219–1226.

McCauley, L. K., Rosol, T. J., Stromberg, P. C., and Capen, C. C. (1991). *In vivo* and *in vitro* effects of interleukin-1α and cyclosporin A on bone and lymphoid tissues in mice. *Toxicol. Pathol.* **19,** 1–10.

McCoy, M. A., and Kennedy, D. G. (1994). Evaluation of post mortem magnesium concentration in bovine eye fluids as a diagnostic aid for hypomagnesaemic tetany. *Vet. Rec.* **135,** 188–189.

McDowell, L. R. (1992a). Calcium and phosphorus. *In* "Minerals in Animal and Human Nutrition," pp. 26–77. Academic Press, San Diego.

McDowell, L. R. (1992b). Magnesium. *In* "Minerals in Animal Nutrition," pp. 115–136. Academic Press, San Diego.

MacGregor, R. R., Hamilton, J. W., and Cohn, D. V. (1978). The mode of conversion of proparathormone to parathormone by a particulate converting enzymic activity of the parathyroid gland. *J. Biol. Chem.* **253,** 2012–2017.

McGuigan, J. A. S., and Blatter, L. A. (1989). Measurement of free magnesium using magnesium selective electrodes. *Magnesium-Bulletin* **11,** 139–142.

MacIsaac, R. J., Caple, I. W., Danks, J. A., Diefenbach-Jagger, H., Grill, V., Moseley, J. M., Southby, J., and Martin, T. J. (1991a). Ontogeny of parathyroid hormone-related protein in the ovine parathyroid gland. *Endocrinology* **129,** 757–764.

MacIsaac, R. J., Heath, J. A., Rodda, C. P., Moseley, J. M., Care, A. D., Martin, T. J., and Caple, I. W. (1991b). Role of the fetal parathyroid glands and parathyroid hormone-related protein in the regulation of placental transport of calcium, magnesium, and inorganic phosphate. *Reprod. Fertil. Dev.* **3,** 447–457.

McKenzie, R. A., Gartner, R. J. W., Blaney, B. J., and Glanville, R. J. (1981). Control of nutritional secondary hyperparathyroidism in grazing horses with calcium plus phosphorus supplementation. *Aust. Vet. J.* **57,** 554–557.

McKim, K. A., Bell, M. C., Miller, J. K., Sims, M. H., Reynolds, C. K., and Ramsey, N. (1984). Hypomagnesemia in calves and effects of magnesium repletion on plasma and cerebrospinal fluid minerals. *J. Dairy Sci.* **67 (Suppl. 1),** 216.

McSheehy, P. M., and Chambers, T. J. (1986). Osteoblastic cells mediate osteoclastic responsiveness to parathyroid hormone. *Endocrinology* **118,** 824–828.

Maguire, M. E. (1990). Magnesium: a regulated and regulatory cation. *In* "Metal Ions In Biological Systems, Vol. 26, Compendium on Magnesium and Its Role in Biology, Nutrition, and Physiology" (H. Sigel and A. Sigel, eds.), pp. 135–153. Marcel Dekker, New York.

Mallette, L. E. (1994). Parathyroid hormone and parathyroid hormone-related protein as polyhormones. *In* "The Parathyroids" (J. P. Bilezikian, R. Marcus, and M. A. Levine, eds.), pp. 171–184. Raven Press, New York.

Mallette, L. E., and Tuma, S. N. (1984). A new radioimmunoassay for the midregion of canine parathyroid hormone. *Mineral Electrolyte Metab.* **10,** 43–47.

Mangin, M., Ikeda, K., and Broadus, A. E. (1990). Structure of the mouse gene encoding parathyroid hormone-related protein. *Gene* **95,** 195–202.

Marel, G. M., McKenna, M. J., and Frame, B. (1986). Osteomalacia. *In* "Bone and Mineral Research/4" (W. A. Peck, ed.), pp. 335–412. Elsevier Science Publishers, Amsterdam.

Marr, A., Moodie, E. W., and Robertson, A. (1955). Some biochemical and clinical aspects of milk fever. *J. Comp. Path.* **65,** 347–365.

Martens, H. (1980). Recent findings on the pathogenesis of hypomagnesemia in ruminants. *Fortschritte der Veterinärmedizin* **30,** 261–269.

Martens, H. (1983). Saturation kinetics of magnesium efflux across the rumen wall in heifers. *Brit. J. Nutr.* **49,** 153–158.

Martens, H., and Gäbel, G. (1986). Physiologic aspects of the pathogenesis and prophylaxis of grass tetany. *Dtsch. Tierärztl. Wschr.* **93,** 170–177.

Martens, H., and Hammer, U. (1981). Magnesium and sodium absorption from the isolated sheep rumen during intravenous aldosterone infusion. *Dtsch. Tierärztl. Wschr.* **88,** 404–407.

Martens, H., and Käsebieter, H. (1983). *In vitro* studies on the influence of sodium and postassium ions on magnesium transport across the isolated rumen mucosa of sheep. *Zentralblatt für Veterinärmedizin.* **30,** 1–14.

Martens, H., and Rayssiguier, Y. (1980). *In* "Digestive Physiology and Metabolism in Ruminants" (Y. Ruckebusch, and P. Thivend, eds.), pp. 447–466. AVI Publishing Co., Westport, CT.

Martens, H., Harmeyer, J., and Michael, H. (1978). Magnesium transport by isolated rumen epithelium of sheep. *Res. Vet. Sci.* **24,** 161–168.

Martens, H., Heggemann, G., and Regier, K. (1988). Studies on the effect of K, Na, $NH4^+$, VFA and CO_2 on the net absorption of magnesium from the temporarily isolated rumen of heifers. *J. Vet. Med. A.* **35,** 73–80.

Martig, J., and Mayer, G. P. (1973). Diminished hypercalcemic response to parathyroid extract in prepartum cows. *J. Dairy Sci.* **56,** 1042–1046.

Martin, L. G. (1994). Intravenous magnesium loading test as a method of evaluating magnesium status in the dog. *Proceedings Intern. Vet. Emerg. & Critical Care Symp.* 1–3.

Martin, R. B. (1990). Bioinorganic chemistry of magnesium. *In* "Metal Ions In Biological Systems, Vol. 26, Compendium on Magnesium and Its Role in Biology, Nutrition, and Physiology" (H. Sigel and A. Sigel, eds.), pp. 1–13. Marcel Dekker, New York.

Martin, S. L., and Capen, C. C. (1980). Puerperal tetany. *In* "Current Veterinary Therapy VII," pp. 1027–1029. Saunders, Philadelphia, Pennsylvania.

Martin, T. J., and Grill, V. (1992). Hypercalcemia and cancer. *J. Steroid Biochem. Molec. Biol.* **43,** 123–129.

Martins, L. G., Van Pelt, D. R., and Wingfield, W. E. (1995). Magnesium and the critically ill patient. *In* "Kirk's Current Veterinary Therapy XII Small Animal Practice" (J. D. Bonagura, ed.), pp. 128–131. Saunders, Philadelphia, Pennsylvania.

Matsui, T., Yano, H., Kawabata, T., and Harumoto, T. (1994). The effect of suppressing bone resorption on Mg metabolism in sheep (*Ovis aries*). *Comp. Biochem. Physiol.* **107A,** 233–236.

Matthews, J. L. (1992). Effect of calcitonin on bone cell ultrastructure. *Bone Miner.* **16,** 178–181.

Mayer, G. P. (1970). The roles of parathyroid hormone and thyrocalcitonin in parturient paresis. *In* "Parturient Hypocalcemia" (J. J. B. Anderson, ed.), pp. 177–193. Academic Press, New York.

Mayer, G. P. (1971). A rational basis for the prevention of parturient paresis. *Bovine Pract.* **6,** 2–8.

Mayer, G. P., and Hurst, J. G. (1978). Comparison of the effects of calcium and magnesium on parathyroid hormone secretion rate in calves. *Endocrinology* **102,** 1803–1807.

Mayer, G. P., Ramberg, C. F., Jr., Kronfeld, D. S., Buckle, R. M., Sherwood, L. M., Aurbach, G. D., and Potts, J. T., Jr. (1969). Plasma parathyroid hormone concentration in hypocalcemic parturient cows. *Am. J. Vet. Res.* **30,** 1587–1597.

Mayland, H. F., and Wilkinson, S. R. (1989). Soil factors affecting magnesium availability in plant–animal systems: A review. *J. Anim. Sci.* **67,** 3437–3444.

Melton, M. E., D'Anza, J. J., Wimbicus, S. A., Grill, V., Martin, T. J., and Kukreja, S. C. (1990). Parathyroid hormone-related protein and calcium homeostasis in lactating mice. *Am. J. Physiol.* **259,** E792–E796.

Merryman, J. I., Rosol, T. J., Brooks, C. L., and Capen, C. C. (1989). Separation of parathyroid hormone-like activity from transforming growth factor-α and -β in the canine adenocarcinoma (CAC-8) model of humoral hypercalcemia of malignancy. *Endocrinology* **124,** 2456–2563.

Meuten, D. J., Cooper, B. J., Capen, C. C., Chew, D. J., and Kociba, G. J. (1981). Hypercalcemia associated with an adenocarcinoma derived from the apocrine glands of the anal sac. *Vet. Pathol.* **18,** 454–471.

Meuten, D. J., Chew, D. J., Capen, C. C., and Kociba, G. J. (1982). Relationship of serum total calcium to albumin and total protein in dogs. *J. Am. Vet. Med. Assoc.* **18,** 63–67.

Meuten, D. J., Kociba, G. J., Capen, C. C., Chew, D. J., Segre, G. V., Levine, L., Tashjian, A. H. J., Voelkel, E. F., and Nagode, L. A. (1983a). Hypercalcemia in dogs with lymphosarcoma: Biochemical, ultrastructural, and histomorphometric investigations. *Lab. Invest.* **49,** 553–562.

Meuten, D. J., Segre, G. V., Capen, C. C., Kociba, G. J., Voelkel, E. F., Levine, L., Tashjian, A. H., Chew, D. J., and Nagode, L. A.

(1983b). Hypercalcemia in dogs with adenocarcinoma derived from apocrine glands of the anal sac: Biochemical and histomorphometric investigations. *Lab. Invest.* **48,** 428–435.

Meyer, H. (1977). Pathogenesis of the clinical symptoms of hypomagnesaemia in ruminants. *Vet. Sci. Comm.* **1,** 43–50.

Meyer, H., and Zentek, J. (1990). Magnesium in animal nutrition. *In* "Metal Ions In Biological Systems, Vol. 26, Compendium on Magnesium and Its Role in Biology, Nutrition, and Physiology" (H. Sigel and A. Sigel, eds.), pp. 57–83. Marcel Dekker, New York.

Michell, A. R. (1991). Intracellular cations: Out of sight, out of mind? *J. Nutr.* **121,** S97–S98.

Miller, R. J. (1992). Voltage sensitive Ca^{2+} channels. *J. Biol. Chem.* **267,** 1403–1406.

Miller, J. K., Madsen, F. C., Lentz, D. E., Wong, W., Ramsey, N., Tysinger, C. E., and Hansard, S. L. (1980). Blood plasma magnesium, potassium, glucose, and immunoreactive insulin changes in cows moved abruptly from barn feeding to early spring pasture. *J. Dairy Sci.* **63,** 1073–1079.

Miller, J. K., White, P. K., Khan, S. A., and Schneider, M. D. (1982). Hypomagnesemia in sheep changed abruptly to low magnesium, high potassium diets. *J. Dairy Sci.* **65 (Suppl. 1),** 219.

Mol, J. A., Kwant, M. M., Arnold, I. C. J., and Hazewinkel, H. A. W. (1991). Elucidation of the sequence of canine (pro)-calcitonin. A molecular biological and protein chemical approach. *Regulatory Peptides* **35,** 189–195.

Morris, D. D., Divers, T. J., and Whitlock, R. H. (1984). Renal clearance and fractional excretion of electrolytes over a 24-hour period in horses. *Am. J. Vet. Res.* **45,** 2431–2435.

Morris, M. L., Teeter, S. M., and Collins, D. R. (1971). The effects of the exclusive feeding of an all-meat dog food. *J. Am. Vet. Med. Assoc.* **158,** 477–488.

Moseley, J. M., and Gillespie, M. T. (1995). Parathyroid hormone-related protein. *Crit. Rev. Clin. Lab. Sci.* **32,** 299–343.

Moseley, J. M., Kubota, M., Diefenbach-Jagger, H., Wettenhall, R. E. H., Kemp, B. E., Suva, L. J., Rodda, C. P., Ebeling, P. R., Hudson, P. J., Zajac, J. D., and Martin, T. J. (1987). Parathyroid hormone-related protein purified from a human lung cancer cell line. *Proc. Natl. Acad. Sci. USA* **84,** 5048–5052.

Moseley, J. M., Hayman, J. A., Danks, J. A., Alcorn, D., Grill, V., Southby, J., and Horton, M. A. (1991). Immunohistochemical detection of parathyroid hormone-related protein in human fetal epithelia. *J. Clin. Endocrinol. Metab.* **73,** 478–484.

Mühlbauer, R. C., and Fleisch, H. (1995). The diurnal rhythm of bone resorption in the rat. Effect of feeding habits and pharmacological inhibitors. *J. Clin. Invest.* **95,** 1933–1940.

Muir, L. A., Hibbs, J. W., and Conrad, H. R. (1968). Effect of vitamin D on the ability of cows to mobilize blood calcium. *J. Dairy Sci.* **51,** 1046–1050.

Muir, L. A., Hibbs, J. W., Conrad, H. R., and Smith, K. L. (1972). Effect of estrogen and progesterone on feed intake and hydroxyproline excretion following induced hypocalcemia in dairy cows. *J. Dairy Sci.* **55,** 1613–1620.

Mulei, C. M., and Daniel, R. C. W. (1988). The association of plasma magnesium concentration and thyroid gland activity in young calves. *J. Vet. Med. A* **35,** 516–521.

Mulei, C. M., and Daniel, R. C. W. (1989). Changes in concentrations of certain plasma and erythrocyte minerals associated with hypomagnesaemia in young calves. *J. Vet. Med. A* **36,** 783–788.

Mulei, C. M., Daniel, R. C. W., and Green, D. (1988). Changes in erythrocyte Mg, Na and K concentrations in late pregnancy and early lactation and their relationship with subsequent fertility and milk production in dairy cows. *J. Vet. Med. A.* **35,** 522–528.

Munger, B. L., and Roth, S. I. (1963). The cytology of the normal parathyroid glands of man and Virginia deer. *J. Cell Biol.* **16,** 379–400.

Murer, H., Lötscher, M., Kaissling, B., Levi, M., Kempson, S. A., and Biber, J. (1996). Renal brush border membrane Na/Pi-cotransport: Molecular aspects in PTH-dependent and dietary regulation. *Kidney Int.* **49,** 1769–1773.

Murphy, E. (1993). Measurement of intracellular ionized magnesium. *Miner. Electrolyte Metab.* **19,** 250–258.

Murray, T. M., Rao, L. G., and Rizzoli, R. E. (1994). Interactions of parathyroid hormone, parathyroid hormone-related protein, and their fragments with conventional and nonconventional receptor sites. *In* "The Parathyroids" (J. P. Bilezikian, M. A. Levine, and R. Marcus, eds.), pp. 185–212. Raven Press, New York.

Nagode, L. A., and Chew, D. J. (1991). The use of calcitriol in treatment of renal disease of the dog and cat. *In* "Proceedings of the Purina International Nutrition Symposium," pp. 39–49. Eastern States Veterinary Conference, Orlando, Florida.

Nagode, L. A., and Chew, D. J. (1992). Nephrocalcinosis caused by hyperparathyroidism in progression of renal failure: Treatment with calcitriol. *Semin. Vet. Med. Surg. Sm. Anim.* **7,** 202–220.

Nagode, L. A., Chew, D. J., and Podell, M. (1996). Benefits of calcitriol therapy and serum phosphorus control in dogs and cats with chronic renal failure. *Vet. Clin. N. Am. Sm. An. Pract.* **26,** 1293–1330.

Nemere, I., and Norman, A. W. (1986). Parathyroid hormone stimulates calcium transport in perfused duodena from normal chicks: Comparison with the rapid (transcaltachic) effect of 1,25-dihydroxyvitamin D_3. *Endocrinology* **119,** 1406–1408.

Nordin, B. E. C., Hodgkinson, A., and Peacock, M. (1967). The measurement and meaning of urinary calcium. *Clin. Orthop.* **52,** 293–322.

Norrdin, R. W., Carpenter, T. R., Hamilton, B. F., and Brewster, R. D. (1988). Trabecular bone morphometry in beagles with hyperadrenocorticism and adrenal adenomas. *Vet. Pathol.* **25,** 256–264.

Nowitzki-Grimm, S., Grimm, P., Thöni, H., and Classen, H. G. (1991). Diagnosis of magnesium status. *Magnesium-Bulletin* **13,** 107–115.

Nozue, T., Kodama, T., and Kobayashi, A. (1991). Improved method for the determination of the magnesium concentration of mononuclear blood cells. *Clin. Chim. Acta.* **203,** 411–412.

Nussbaum, S. R., and Potts, J. T. Jr. (1994). Advances in immunoassays for parathyroid hormone. *In* "The Parathyroids" (J. P. Bilezikian, R. Marcus, and M. A. Levine, eds.), pp. 157–170. Raven Press, New York.

Orloff, J. J., Ganz, M. B., Ribaudo, A. E., Burtis, W. J., Reiss, M., Milstone, L. M., and Stewart, A. F. (1992). Analysis of PTHRP binding and signal transduction mechanisms in benign and malignant squamous cells. *Am. J. Physiol.* **262,** E599–E607.

Orloff, J. J., Reddy, D., dePapp, A. E., Yang, K. H., Soifer, N. E., and Stewart, A. F. (1994). Parathyroid hormone–related protein as a prohormone: Posttranslational processing and receptor interactions. *Endocr. Rev.* **15,** 40–60.

Ornoy, A., Goodwin, D., Noff, D., and Edelstein, S. (1978). 24,25-Dihydroxyvitamin D is a metabolite of vitamin D essential for bone formation. *Nature* **276,** 517–519.

Palmer, N. (1993). Bones and joints. *In* "Pathology of Domestic Animals" (K. V. F. Jubb, P. C. Kennedy, and N. Palmer, eds.), pp. 1–182. Academic Press, San Diego.

Pandian, M. R., Morgan, C. H., Carlton, E., and Segre, G. V. (1992). Modified immunoradiometric assay of parathyroid hormone-related protein: Clinical application in the differential diagnosis of hypercalcemia. *Clin. Chem.* **38,** 282–288.

Parfitt, A. M. (1977). The cellular basis of bone turnover and bone loss. *Clin. Orthop.* **127,** 236–247.

Parfitt, A. M. (1987). Bone and plasma calcium homeostasis. *Bone* **8 (Suppl. 1),** S1–S8.

Paroni, R., De Vecchi, E., Fermo, I., Arcelloni, C., Diomede, L., Magni, F., and Bonini, P. A. (1992). Total urinary hydroxyproline determined with rapid and simple high-performance liquid chromatography. *Clin. Chem.* **38,** 407–411.

Parsons, J. A., and Robinson, C. J. (1971). Calcium shift into bone causing transient hypocalcemia after injection of parathyroid hormone. *Nature (London)* **230,** 581–582.

Pastoor, F. J. H., Opitz, R., van 't Klooster, A. T., and Beynen, A. C. (1994a). Dietary calcium chloride vs calcium carbonate reduces urinary pH and phosphorus concentrations, improves bone mineralization and depresses kidney calcium level in cats. *J. Nutr.* **124,** 2213–2222.

Pastoor, F. J. H., Van 'T Klooster, A. T., Mathot, J. N. J. J., and Beynen, A. C. (1994b). Increasing calcium intakes lower urinary concentrations of phosphorus and magnesium in adult ovariectomized cats. *J. Nutr.* **124,** 299–304.

Pastoor, F. J. H., Van 'T Klooster, A. T., Mathot, J. N. J. J., and Beynen, A. C. (1995a). Increasing phosphorus intake reduces urinary concentrations of magnesium and calcium in adult ovariectomized cats fed purified diets. *J. Nutr.* **125,** 1334–1341.

Pastoor, F. J. H., van't Klooster, A. T., Opitz, R., and Beynen, A. C. (1995b). Effect of dietary magnesium level on urinary and faecal excretion of calcium, magnesium and phosphorus in adult, ovariectomized cats. *Br. J. Nutr.* **74,** 77–84.

Pedersen, N. C. (1983). Nutritional secondary hyperparathyroidism in a cattery associated with the feeding of a fad diet containing horsemeat. *Feline Pract.* **13,** 19–26.

Peterson, M. E. (1982). Treatment of canine and feline hypoparathyroidism. *J. Am. Vet. Med. Assoc.* **181,** 1434–1436.

Peterson, M. E., and Feinman, J. M. (1982). Hypercalcemia associated with hypoadrenocorticism in sixteen dogs. *J. Am. Vet. Med. Assoc.* **181,** 802–804.

Philbrick, W. M., Wysolmerski, J. J., Galbrath, S., Holt, E., Orloff, J. J., Yang, K. H., Vasavada, R. C., Weir, E. C., Broadus, A. E., and Stewart, A. F. (1996). Defining the roles of parathyroid hormone-related protein in normal physiology. *Physiol. Rev.* **76,** 127–173.

Planells, E., Llopis, J., Peran, F., and Aranda, P. (1995). Changes in tissue calcium and phosphorus content and plasma concentrations of parathyroid hormone and calcitonin after long-term magnesium deficiency in rats. *J. Amer. Coll. Nutr.* **14,** 292–298.

Pollak, M. R., Brown, E. M., Chou, Y. W., Hebert, S. C., Marx, S. J., Steinmann, B., Levi, T., Seldman, C. E., and Seldman, J. G. (1993). Mutations in the human Ca^{2+}-sensing receptor gene cause familial hypocalciuric hypercalcemia and neonatal severe hypercalcemia. *Cell* **75,** 1297–1303.

Poole, D. B. R. (1989). Observations on the role of magnesium and phosphorus in the aetiology of urolithiasis in male sheep. *Irish Vet. J.* **42,** 60–63.

Powell, G. J., Southby, J., Danks, J. A., Stillwell, R. G., Hayman, J. A., Henderson, M. A., Bennett, R. C., and Martin, T. J. (1991). Localization of parathyroid hormone-related protein in breast cancer metastases: Increased incidence in bone compared to other sites. *Cancer Res.* **51,** 3059–3061.

Prager, D., Rosenblatt, J. D., and Ejima, E. (1994). Hypercalcemia, parathyroid hormone-related protein expression and human T-cell leukemia virus infection. *Leukemia and Lymphoma* **14,** 395–400.

Pratt, D. A., Daniloff, Y., Duncan, A., and Robins, S. P. (1992). Automated analysis of the pyridinium crosslinks of collagen in tissue and urine using solid phase extraction and reversed-phase high-performance liquid chromatography. *Anal. Biochem.* **207,** 168–175.

Price, C. P., Kirwan, A., and Vader, C. (1995). Tartrate-resistant acid phosphatase as a marker of bone resorption. *Clin. Chem.* **41,** 641–643.

Quamme, G. A. (1993). Magnesium homeostasis and renal magnesium handling. *Miner. Electrolyte Metab.* **19,** 218–225.

Raeymaekers, L., and Wuytack, F. (1996). Calcium pumps. *In* "Biochemistry of Smooth Muscle Contraction," pp. 241–253. Academic Press, San Diego.

Ramsey, N., Miller, J. K., and Bell, M. C. (1979a). Effects of dietary imbalance of magnesium, calcium, and potassium on plasma insulin in sheep and cattle. *J. Anim. Sci.* **49,** 13–14.

Ramsey, N., Miller, J. K., and Bell, M. C. (1979b). Hypomagnesemia and insulin in cows overwintered on stockpiled fescue pasture or changed abruptly from stored feed to lush spring pasture. *Tennessee Farm & Home Science Progress Report.* **No. 112,** 16, 109–120.

Rasmussen, H. (1986a). The calcium messenger system (part I). *N. Engl. J. Med.* **314,** 1094–1101.

Rasmussen, H. (1986b). The calcium messenger system (part II). *N. Engl. J. Med.* **314,** 1164–1170.

Ratcliffe, W. A. (1992). Role of parathyroid hormone-related protein in lactation. *Clin. Endocrinol.* **37,** 402–404.

Rayssiguier, Y. (1977). Hypomagnesemia resulting from adrenaline infusions in ewes: Its relation to lipolysis. *Horm. Metab. Res.* **9,** 309–314.

Rayssiguier, Y. (1984). Magnesium and disorders associated with magnesium deficiency (in goats). *Les maladies de la chevre, colloque international, Niort, France* **28,** 411–414.

Rayssiguier, Y., Garel, J. M., Davicco, M.-J., and Barlet, J. P. (1977). Magnesium deficiency induced in calves with magnesium-deficient milk (synthetic). *Horm. Metab. Res.* **9,** 438–439.

Recker, R. R. (1996). Biopsies. *In* "Principles of Bone Biology" (J. P. Bilezikian, L. G. Raisz, and G. A. Rodan, eds.), pp. 1333–1342. Academic Press, San Diego.

Reichel, H., Koeffler, P., and Norman, A. W. (1989). The role of the vitamin D endocrine system in health and disease. *N. Engl. J. Med.* **320,** 980–991.

Reinhart, R. A. (1988). Magnesium metabolism: A review with special reference to the relationship between intracellular content and serum levels. *Arch. Intern. Med.* **148,** 2415–2420.

Rémond, D., Meschy, F., and Boivin, R. (1996). Metabolites, water and mineral exchanges across the rumen wall: Mechanisms and regulation. *Ann. Zootech.* **45,** 97–119.

Resnick, S. (1972). Hypocalcemia and tetany in the dog. *Vet. Med. Small Anim. Clin.* **67,** 637–641.

Reynolds, J. J., Holick, M. F., and DeLuca, H. F. (1973). The role of vitamin D metabolites in bone resorption. *Calcif. Tissue Res.* **12,** 295–301.

Riond, J.-L., Kocabagli, N., Forrer, R., and Wanner, M. (1995a). Repeated daytime measurements of the concentrations of PTHrP and other components of bovine milk. *J. Anim. Physiol. a. Anim. Nutr.* **74,** 194–199.

Riond, J.-L., Kocabagli, N., Spichiger, U. E., and Wanner, M. (1995b). The concentration of ionized magnesium in serum during the periparturient period of non-paretic dairy cows. *Vet. Res. Comm.* **19,** 195–203.

Riond, J.-L., Kocabagli, N., Cloux, F., and Wanner, M. (1996). Parathyroid hormone-related protein in the colostrum of paretic post parturient dairy cows. *Vet. Rec.* **138,** 333–334.

Ritz, E., Matthias, S., Seidel, A., Reichel, H., Szabo, A., and Hörl, W. H. (1992). Disturbed calcium metabolism in renal failure—pathogenesis and therapeutic strategies. *Kidney Int.* **42, Suppl. 38,** S37–S42.

Rivera, L. I., Gootman, P. M., Lin, R., and Gootman, N. (1991). Effects of elevated plasma magnesium concentration on cerebrospinal fluid levels of magnesium in neonatal swine. *Proc. Soc. Exp. Biol. Med.* **197,** 98–101.

Rizzoli, R., Caverzasio, J., Chapuy, M. C., Martin, T. J., and Bonjour, J. P. (1989). Role of bone and kidney in parathyroid hormone-related peptide-induced hypercalcemia in rats. *J. Bone Mineral Res.* **4,** 759–765.

Robinson, D. L., Kappel, L. C., and Boling, J. A. (1989). Management practices to overcome the incidence of grass tetany. *J. Anim. Sci.* **67,** 3470–3484.

Rodan, G. A., and Martin, T. J. (1981). Role of osteoblasts in hormonal control of bone resorption—a hypothesis. *Calcif. Tissue Int.* **33,** 349–351.

Rodan, G. A., Raisz, L. G., and Bilezikian, J. P. (1996). Pathophysiology of osteoporosis. *In* "Principles of Bone Biology" (J. P. Bilezikian, L. G. Raisz, and G. A. Rodan, eds.), pp. 979–990. Academic Press, San Diego.

Romani, A., Marfella, C., and Scarpa, A. (1993). Cell magnesium transport and homeostasis: Role of intracellular compartments. *Miner. Electrolyte Metab.* **19,** 282–289.

Roodman, G. D. (1996). Advances in bone biology: The osteoclast. *Endocr. Rev.* **17,** 308–332.

Rook, J. A. F., and Storry, J. E. (1962). Magnesium in the nutrition of farm animals. *Nutr. Abstr. Rev.* **32,** 1055–1071.

Rosol, T. J., and Capen, C. C. (1987). The effect of low calcium diet, mithramycin, and dichlorodimethylene bisphosphonate on humoral hypercalcemia of malignancy in nude mice transplanted with the canine adenocarcinoma tumor line (CAC-8). *J. Bone Mineral Res.* **2,** 395–405.

Rosol, T. J., and Capen, C. C. (1988). Pathogenesis of humoral hypercalcemia of malignancy. *Dom. Anim. Endocrinol.* **5,** 1–21.

Rosol, T. J., and Capen, C. C. (1992). Biology of disease: Mechanisms of cancer-induced hypercalcemia. *Lab. Invest.* **67**(6), 680–702.

Rosol, T. J., Capen, C. C., Weisbrode, S. E., and Horst, R. L. (1986). Humoral hypercalcemia of malignancy in nude mouse model of a canine adenocarcinoma derived from apocrine glands of the anal sac. *Lab. Invest.* **54,** 679–688.

Rosol, T. J., Capen, C. C., and Horst, R. L. (1988a). Effects of infusion of human parathyroid hormone-related protein (1–40) in nude mice: Histomorphometric and biochemical investigations. *J. Bone Mineral Res.* **3,** 699–706.

Rosol, T. J., Chew, D. J., Capen, C. C., and Sherding, R. G. (1988b). Acute hypocalcemia associated with infarction of parathyroid gland adenomas in two dogs. *J. Am. Vet. Med. Assoc.* **192,** 212–214.

Rosol, T. J., Capen, C. C., Danks, J. A., Suva, L. J., Steinmeyer, C. L., Hayman, J., Ebeling, P. R., and Martin, T. J. (1990). Identification of parathyroid hormone-related protein in canine apocrine adenocarcinoma of the anal sac. *Vet. Pathol.* **27,** 89–95.

Rosol, T. J., Chew, D. J., Couto, C. G., Ayl, R., Nagode, L. A., and Capen, C. C. (1992a). Effects of mithramycin on calcium metabolism and bone in dogs. *Vet. Pathol.* **29,** 223–229.

Rosol, T. J., Nagode, L. A., Couto, C. G., Hammer, A. S., Chew, D. J., Peterson, J. L., Ayl, R. D., Steinmeyer, C. L., and Capen, C. C. (1992b). Parathyroid hormone (PTH)-related protein, PTH, and 1,25-dihydroxyvitamin D in dogs with cancer-associated hypercalcemia. *Endocrinology* **131**(3), 1157–1164, © The Endocrine Society.

Rosol, T. J., Nagode, L. A., Robertson, J. T., Leeth, B. D., Steinmeyer, C. L., and Allen, C. M. (1994). Humoral hypercalcemia of malignancy associated with ameloblastoma in a horse. *J. Am. Vet. Med. Assoc.* **204,** 1930–1933.

Rosol, T. J., Chew, D. J., Nagode, L. A., and Capen, C. C. (1995a). Pathophysiology of calcium metabolism. *Vet. Clin. Pathol.* **24**(2), 49–63. Published by Veterinary Practice Publishing Company, P.O. Box 4457, Santa Barbara, CA 93140, phone: 805-965-1028; fax: 805-965-0722.

Rosol, T. J., Steinmeyer, C. L., McCauley, L. K., Gröne, A., DeWille, J. W., and Capen, C. C. (1995b). Sequences of the cDNAs encoding canine parathyroid hormone-related protein and parathyroid hormone. *Gene* **160,** 241–243.

Rosol, T. J., McCauley, L. K., Steinmeyer, C. L., and Capen, C. C. (1996). Nucleotide sequence of canine preproparathyroid hormone. *In* "The Comparative Endocrinology of Calcium Regulation" (C. Dacke, J. Danks, I. Caple, G. Flik, eds.), pp. 201–203. Journal of Endocrinology, Ltd., Bristol.

Roth, S. I., and Capen, C. C. (1974). Ultrastructural and functional correlations of the parathyroid glands. *Int. Rev. Ex. Pathol.* **13,** 162–221.

Roth, S. I., and Raisz, L. G. (1966). The course and reversibility of the calcium effect on the ultrastructure of the rat parathyroid gland in organ culture. *Lab. Invest.* **15,** 1187–1211.

Rouse, D., and Suki, W. N. (1990). Renal control of extracellular calcium. *Kidney Int.* **38,** 700–708.

Rowland, G. N., Capen, C. C., and Nagode, L. A. (1968). Experimental hyperparathyroidism in young cats. *Pathol. Vet.* **5,** 504–519.

Rowland, G. N., Capen, C. C., Young, D. M., and Black, H. E. (1972). Microradiographic evaluation of bone from cows with experimental hypervitaminosis D: Diet-induced hypocalcemia and naturally occurring parturient paresis. *Calcif. Tissue Res.* **9,** 179–193.

Rude, R. K., and Oldham, S. B. (1987). Hypocalcemia of Mg deficiency: Altered modulation of adenylate cyclase by Mg^{++} and Ca^{++} may result in impaired PTH secretion and PTH end-organ resistance. *In* "Magnesium in Cellular Processes and Medicine" (B. M. Altura, J. Durlach, and M. S. Seelig, eds.), pp. 183–195. Karger, Basel.

Ryan, M. P. (1990). The renal handling of magnesium. *In* "Metal Ions In Biological Systems, Vol. 26, Compendium on Magnesium and Its Role in Biology, Nutrition, and Physiology" (H. Sigel, and A. Sigel, eds.), pp. 249–269. Marcel Dekker, New York.

Ryan, M. P. (1993). Interrelationships of magnesium and potassium homeostasis. *Miner. Electrolyte Metab.* **19,** 290–295.

Sanderson, M. J., Charles, A. C., Boitano, S., and Dirksen, E. R. (1994). Mechanisms and function of intercellular calcium signalling. *Mol. Cell. Endocrinol.* **98,** 173–187.

Sansom, B. F. (1981). Hypomagnesaemia in calves: Treatment and control. *Vet. Annual.* **21,** 74–79.

Sansom, B. F. (1983). Magnesium and milk fever. *Vet. Rec.* **112,** 447–449.

Sansom, B. F., Bunch, K. L., and Dew, S. M. (1982a). Changes in plasma calcium, magnesium, phosphorus and hydroxyproline concentrations in ewes from twelve weeks before until three weeks after lambing. *Brit. Vet. J.* **138,** 393–401.

Sansom, B. F., Manston, R., Vagg, M. J., Mallinson, C. B., and Contreras, P. A. (1982b). Calcium mobilization in hypomagnesemic steers and cows. *In* "Proceedings of the XIIth World Congress on Diseases of Cattle, The Netherlands, Volume 1." A.R.C. Institute of Research on Animal Disease, Compton, United Kingdom.

Sarode, D. B., Nisal, M. B., and Sapre, V. A. (1992). Alteration in composition of cerebrospinal fluid and blood of experimentally induced hypomagnesemia in crossbred calves. *Indian Vet. J.* **69,** 1001–1004.

Sauer, L. S., Hamar, D., and Lewis, L. D. (1985). I. Effect of dietary mineral composition on urinary mineral concentration and excretion by the cat. II. Effect of diet composition on water intake and excretion by the cat. *Feline Pract.* **15,** 10–21.

Saville, P. D., and Krook, L. (1969). Gravimetric and isotopic studies in nutritional hyperparathyroidism in beagles. *Clin. Orthop.* **62,** 15–24.

Schenck, P. A., Chew, D. J., and Brooks, C. L. (1995). Effects of storage on normal canine serum ionized calcium and pH. *Am. J. Vet. Res.* **56,** 304–307.

Schipani, E., Langman, C. B., Parfitt, A. M., Jensen, G. S., Kikuchi, S., Kooh, S. W., Cole, W. G., and Jüppner, H. (1996). Constitutively activated receptors for parathyroid hormone and parathyroid hormone-related peptide in Jansen's metaphyseal chondrodysplasia. *N. Engl. J. Med.* **335,** 708–714.

Schröder, B., Käppner, H., Failing, K., Pfeffer, E., and Breves, G. (1995). Mechanisms of intestinal phosphate transport in small ruminants. *Br. J. Nutr.* **74,** 635–648.

Scott, P. R., Kelly, J. M., Whitaker, D. A., and Cameron, N. D. (1980). Marginal magnesium deficiency as a possible cause of reduced voluntary feed intake in commercially managed dairy cows. *Vet. Res. Comm.* **4,** 225–229.

Segre, G. V. (1994). Receptors for parathyroid hormone and parathyroid hormone-related protein. *In* "The Parathyroids" (J. P. Bilezikian, R. Marcus,and M. A. Levine, eds.), pp. 213–230. Raven Press, New York.

Seiler, H. G. (1990). Determination of magnesium in biological material. *In* "Metal Ions In Biological Systems, Vol. 26, Compendium on Magnesium and Its Role in Biology, Nutrition, and Physiology" (H. Sigel and A. Sigel, eds.), pp. 611–624. Marcel Dekker, New York.

Sejersted, O. M., Steen, P. A., and Kiil, F. (1984). Inhibition of transcellular NaCl reabsorption in dog kidneys during hypercalcemia. *Acta Physiol. Scand.* **120,** 543–549.

Seymour, J. F., and Gagel, R. F. (1993). Calcitriol: The major humoral mediator of hypercalcemia in Hodgkin's and non-Hodgkin's lymphomas. *Blood* **82,** 1383–1394.

Sherding, R. G., Meuten, D. J., Chew, D. J., Knaack, K. E., and Haupt, K. H. (1980). Primary hypoparathyroidism in the dog. *J. Am. Vet. Med. Assoc.* **176,** 439–444.

Shiga, A. (1988). Effects of calcium carbonate supplementation on Ca, Mg and P metabolism in ewes. *Jpn. J. Vet. Sci.* **50,** 175–182.

Shiga, A., Miyachi, T., Sato, M., Shinozaki, K., Saeki, T., and Hayashi, M. (1979). Experimental studies on hypomagnesaemia of ruminants. IV. Effect of cold stress on the mineral metabolism of sheep fed on herbage. *Journal of the Faculty of Agriculture, Iwate University* **14,** 173–183.

Shiga, A., Kominato, A., and Shinozaki, K. (1980). Experimental studies on hypomagnesemia in ruminants. V. Metabolism of phosphate and plasma parathyroid hormone in sheep fed diets of varying magnesium and calcium concentrations. *Jap. J. Vet. Sci.* **42,** 221–230.

Shirazi-Beechey, S. P., Penny, J. I., Dyer, J., Wood, I. S., Tarpey, P. S., Scott, D., and Buchan, W. (1996). Epithelial phosphate transport in ruminants, mechanisms and regulation. *Kidney Int.* **49,** 992–996.

Siesjö, B.K. (1989). Calcium and cell death. *Magnesium* **8,** 223–237.

Silver, J. (1992). Regulation of parathyroid hormone synthesis and secretion. *In* "Disorders of Bone and Mineral Metabolism" (F. L. Coe and M. J. Favus, eds.), pp. 83–106. Raven Press, New York.

Silverstein, D., Barac-Nieto, M., and Spitzer, A. (1996). Mechanism of renal phosphate retention during growth. *Kidney Int.* **49,** 1023–1026.

Simesen, M. G. (1980). Calcium, phosphorus, and magnesium metabolism. *In* "Clinical biochemistry of domestic animals" (J. J. Kaneko, ed.), pp. 576–648. Academic Press, New York.

Sims, M. H., Bell, M. C., and Ramsey, N. (1980). Electrodiagnosis evaluation of hypomagnesemia in sheep. *J. Anim. Sci.* **50,** 539–546.

Sjollema, B. (1930). On the nature and therapy of grass staggers. *Vet. Rec.* **10,** 425–430.

Sleper, D. A., Vogel, K. P., Asay, K. H., and Mayland, H. F. (1989). Using plant breeding and genetics to overcome the incidence of grass tetany. *J. Anim. Sci.* **67,** 3456–3462.

Smith, R. H. (1959). Calcium and magnesium metabolism in calves. *Biochem. J.* **71,** 609–614.

Smith, R. A., and Edwards, W. C. (1988). Hypomagnesemic tetany of ruminants. *Vet. Clin. N. Am. Food An. Pract.* **4,** 365–377.

Smithcors, J. F. (1964). The endocrine system. *In* "Anatomy of the Dog" (M. E. Miller, G. C. Christensen, and H. E. Evans, eds.), pp. 822–826. Saunders, Philadelphia, Pennsylvania.

Soares, J. H., Jr. (1995). Phosphorus bioavailability. *In* "Bioavailability of Nutrients for Animals: Amino Acids, Minerals, and Vitamins," pp. 257–294. Academic Press, San Diego.

Speich, M., and Bousquet, B. (1991). Magnesium: Recent data on metabolism, exploration, pathology and therapeutics. *Magnesium-Bulletin* **13,** 116–121.

Spence, J. A., Sykes, A. R., Atkinson, P. J., and Aitchison, G. U. (1985). Skeletal and blood biochemical characteristics of sheep during growth and breeding: A comparison of flocks with and without broken mouth. *J. Comp. Pathol.* **95,** 505–524.

Stepien, R. L. (1995). Common complications of therapy for chronic heart failure. *Proceedings of the 13th American College of Veterinary Medicine Internal Medicine Forum,* pp. 245–249.

Stern, P. H. (1996). Thyroid hormone and bone. *In* "Principles of Bone Biology" (J. P. Bilezikian, L. G. Raisz, and G. A. Rodan, eds.), pp. 521–531. Academic Press, San Diego.

Stojevic, Z., Emanovic, D., Milinkovic-Tur, S., and Duzel, S. (1993). Magnesium absorption from the swine caecum. 1. Effect of different magnesium and ammonium concentrations in the contents on the intensity of magnesium absorption. *Vet. Arhiv.* **63,** 265–272.

Stuedemann, J. A., Wilkinson, S. R., and Lowrey, R. S. (1984). Efficacy of a large magnesium alloy rumen bolus in the prevention of hypomagnesemic tetany in cows. *Amer. J. Vet. Res.* **45,** 698–702.

Sutherland, R. J., Bell, K. C., McSporran, K. D., and Carthew, G. W. (1986). A comparative study of diagnostic tests for the assessment of herd magnesium status in cattle. *N. Z. Vet. J.* **34,** 133–135.

Sutton, R. A. L., and Domrongkitchaiporn, S. (1993). Abnormal renal magnesium handling. *Miner. Electrolyte Metab.* **19,** 232–240.

Suva, L. J., Winslow, G. A., Wettenhall, R. E. H., Hammonds, R. G., Moseley, J. M., Diefenbach-Jagger, H., Rodda, C. P., Kemp, B. E., Rodriguez, H., Chen, E. Y., Hudson, P. J., Martin, T. J., and Wood, W. I. (1987). A parathyroid hormone-related protein implicated in malignant hypercalcemia: Cloning and expression. *Science* **237,** 893–896.

Szenci, O., Brydl, E., and Bajcsy, C. A. (1991). Effect of storage on measurement of ionized calcium and acid–base variables in equine, bovine, ovine, and canine venous blood. *J. Am. Vet. Med. Assoc.* **199,** 1167–1169.

Szenci, O., Besser, T. E., Stollar, Z., and Brydl, E. (1994a). Effect of storage time and temperature on canine ionized calcium concentration in blood, plasma, and serum. *J. Am. An. Hosp. Assoc.* **30,** 495–499.

Szenci, O., Chew, B. P., Bajcsy, A. C., Szabo, P., and Brydl, E. (1994b). Total and ionized calcium in parturient dairy cows and their calves. *J. Dairy Sci.* **77,** 1100–1105.

Talmage, R. V., Anderson, J. J. B., and Cooper, C. W. (1972). The influence of calcitonins on the disappearance of radiocalcium and radiophosphorus from plasma. *Endocrinology* **90,** 1185–1191.

Tanaka, Y., and DeLuca, H. F. (1974). Stimulation of 24,25-dihydroxyvitamin D_3 production by 1,25-dihydroxyvitamin D_3. *Science* **183,** 1198–1200.

Taylor, T. G. (1959). The magnesium of bone mineral. *J. Agric. Sci.* **52,** 207–216.

Tennant, B., Lowe, J. E., and Tasker, J. B. (1981). Hypercalcemia and hypophosphatemia in ponies following bilateral nephrectomy. *Proc. Soc. Exp. Biol. Med.* **167,** 365–368.

Tennant, B., Bettleheim, P., and Kaneko, J. J. (1982). Paradoxic hypercalcemia and hypophosphatemia associated with chronic renal failure in horses. *J. Am. Vet. Med. Assoc.* **180,** 630–634.

Terashima, Y., Tucker, R. E., Deetz, L. E., Degregorio, R. M., Muntifering, R. B., and Mitchell, G. E. (1982). Plasma magnesium levels as influenced by cold exposure in fed or fasted sheep. *J. Nutr.* **112,** 1914–1920.

Terashima, Y., Sudoh, S., and Itoh, H. (1983). Hypocalcemia and hypomagnesemia induced by IV infusion of noradrenaline in sheep under various feeding regimens. *Jap. J. Zootechnical Sci.* **54,** 336–341.

Thiede, M. A. (1994). Parathyroid hormone-related protein: A regulated calcium-mobilizing product of the mammary gland. *J. Dairy Sci.* **77,** 1952–1963.

Thielscher, H. H., Kretschmer, M., and Thielscher, M. (1994). Content of magnesium and calcium in the blood plasma of pigs of two breeds differing in fitness. *Monatshefte für Veterinärmedizin.* **49,** 281–284.

Thiemann, K. G., Uhl, W., and Knauer, G. (1987). Acute magnesium poisoning in dairy cows. *Monatshefte für Veterinärmedizin* **42,** 701–702.

Thompson, G. E. (1993). Parathyroid hormone-related protein and mammary blood flow in the sheep. *Exp. Physiol.* **78,** 499–501.

Thompson, K. G., Jones, L. P., Smylie, W. A., Quick, C. B., Segre, G. V., Meuten, D. J., and Petrites-Murphy, M. B. (1984). Primary hyperparathyroidism in German shepherd dogs: A disorder of probable genetic origin. *Vet. Pathol.* **21,** 370–376.

Torrance, A. G., and Nachreiner, R. (1989). Human-parathormone assay for use in dogs: validation, sample handling techniques, and parathyroid function testing. *Am. J. Vet. Res.* **50,** 1123–1127.

Unkel, M. (1984). Concentrations of K^+, Ca^{2+}, Mg^{2+}, Na^+ and phosphate in blood serum from Icelandic ponies. *Tierärztliche Umschau.* **39,** 989–994.

van Mosel, M., Van 'T Klooster, A. T., and Wouterse, H. S. (1991). Effects of a deficient magnesium supply during the dry period on bone turnover of dairy cows at parturition. *Vet. Quarterly* **13,** 199–208.

van't Klooster, A. T., and Care, A. D. (1966). Inhibition by chelation of intestinal calcium absorption in sheep. *Biochem. J.* **99,** 2p.

Vieth, R. (1990). The mechanisms of vitamin D toxicity. *Bone Miner.* **11,** 267–272.

Vormann, J., and Günther, T. (1993). Extracellular and intracellular magnesium–calcium interactions. *Magnesium-Bulletin* **15,** 111–120.

Vortkamp, A., Lee, K., Lanske, B., Segre, G. V., Kronenberg, H. M., and Tabin, C. J. (1996). Regulation of rate of cartilage differentiation by Indian hedgehog and PTH-related protein. *Science* **273,** 613–622.

Wallach, J. D. (1971). Environmental and nutritional diseases of captive reptiles. *J. Am. Vet. Med. Assoc.* **159,** 1632–1643.

Wallach, S. (1987). Magnesium exchangeability and bioavailability in magnesium deficiency. *In* "Magnesium in Cellular Processes and Medicine" (B. M. Altura, J. Durlach, and M. S. Seelig, eds.), pp. 27–49. Karger, Basel.

Wallach, J. D., and Flieg, G. M. (1969). Nutritional secondary hyperparathyroidism in captive birds. *J. Am. Vet. Med. Assoc.* **155,** 1046–1051.

Wallach, J. D., and Hoessle, C. (1968). Fibrous osteodystrophy in green iguanas. *J. Am. Vet. Med. Assoc.* **153,** 863–865.

Walters, M. R. (1992). Newly identified actions of the vitamin D endocrine system. *Endocr. Rev.* **13,** 719–764.

Walthall, J. C., and McKenzie, R. A. (1976). Osteodystrophia fibrosa in horses at pasture in Queensland: Field and laboratory observations. *Aust. Vet. J.* **52,** 11–16.

Wasserman, R. H., and Fullmer, C. S. (1983). Calcium transport proteins, calcium absorption, and vitamin D. *Ann. Rev. Physiol.* **45,** 375–390.

Wasserman, R. H., Corradino, R. A., and Krook, L. (1975). *Cestrum diurnum.* A domestic plant with 1,25-dihydroxycholecalciferol-like activity. *Biochem. Biophys. Res. Commun.* **62,** 85–91.

Weaver, C. M. (1990). Assessing calcium status and metabolism. *J. Nutr.* **120,** 1470–1473.

Weglicki, W. B., Phillips, T. M., Freedman, A. M., Cassidy, M. M., and Dickens, B. F. (1992). Magnesium-deficiency elevates circulating levels of inflammatory cytokines and endothelin. *Mol. Cell. Biochem.* **110,** 169–173.

Weir, E. C. (1992). Hypercalcemia and malignancy. *Proceedings of the 10th American College of Veterinary Internal Medicine Forum,* pp. 640–612.

Weir, E. C., Burtis, W. J., Morris, C. A., Brady, T. G., and Insogna, K. L. (1988a). Isolation of 16,000-dalton parathyroid hormone-like proteins from two animal tumors causing humoral hypercalcemia of malignancy. *Endocrinology* **123,** 2744–2751.

Weir, E. C., Centrella, M., Matus, R. E., Brooks, M .L., Wu, T., and Insogna, K. L. (1988b). Adenylate cyclase-stimulating, bone-resorbing and B TGF-like activities in canine apocrine cell adenocarcinoma of the anal sac. *Calcif. Tissue Int.* **43,** 359–365.

Weir, E. C., Norrdin, R. W., Matus, R. E., Brooks, M. B., Broadus, A. E., Mitnick, M., Johnston, S. D., and Insogna, K. L. (1988c). Humoral hypercalcemia of malignancy in canine lymphosarcoma. *Endocrinology* **122,** 602–608.

Weisbrode, S. E., and Capen, C. C. (1974). Ultrastructural evaluation of the effects of calcitonin on bone in thyroparathyroidectomized rats administered vitamin D. *Am. J. Pathol.* **77,** 445–464.

Werkmeister, J. R., Rosol, T. J., McCauley, L. K., Merryman, J. I., Horton, J. E., and Capen, C.C. (1993). Parathyroid hormone-related protein production by normal human keratinocytes in vitro. *Exp. Cell Res.* **208,** 68–74.

Whang, R., Oei, T. O., Aikawa, J. K., Watanabe, A., Vannatta, J., Fryer, A., and Markanich, M. (1984). Predictors of clinical hypomagnesemia: Hypokalemia, hypophosphatemia, hyponatremia, and hypocalcemia. *Arch. Intern. Med.* **144,** 1794–1796.

Whitaker, D. A., and Kelly, J. M. (1982). Incidence of clinical and subclinical hypomagnesaemia in dairy cows in England and Wales. *Vet. Rec.* **110,** 450–451.

Whitaker, D. A., Kelly, J. M., and Ritchie, J. S. (1986). Magnesium in aqueous humour. *Vet. Rec.* **118,** 520

Wians, F. H., Strickland, D. M., Hankins, G. D. V., and Snyder, R. R. (1990). The effect of hypermagnesemia on serum levels of osteocalcin in an animal model. *Magnesium Trace Elem.* **9,** 28–35.

Wilke, R., Harmeyer, J., von Grabe, C., Hehrmann, R., and Hesch, R. D. (1979). Regulatory hyperparathyroidism in a pig breed with vitamin D dependency rickets. *Acta Endocrinol.* **92,** 295–308.

Wilkinson, S. R., Welch, R. M., Mayland, H. F., and Grunes, D. L. (1990). Magnesium in plants: Uptake, distribution, function, and utilization by man and animals. *In* "Metal Ions In Biological Systems, Vol. 26, Compendium on Magnesium and Its Role in Biology, Nutrition, and Physiology" (H. Sigel and A. Sigel, eds.), pp. 33–56. Marcel Dekker, New York.

Wilson, G. F. (1980). Effects of magnesium supplements on the digestion of forages and milk production of cows with hypomagnesaemia. *Anim. Prod.* **31,** 153–157.

Winkler, H., and Fischer-Colbrie, R. (1992). Chromogranins A and B: The first 25 years and future perspectives. *Neuroscience* **49,** 497–528.

Winkler, I., Schreiner, F., and Harmeyer, J. (1986). Absence of renal 25-hydroxycholecalciferol-1-hydroxylase activity in a pig strain with vitamin D-dependent rickets. *Calcified Tissue International* **38,** 87–94.

Wittwer, F., Contreras, P. A., Klein, R., and Böhmwald, H. (1995a). Efecto de la administración oral de oxido de magnesio y sulfato de magnesio en vacas con hipomagnesemia inducida. *Arch. Med. Vet.* **27,** 19–27.

Wittwer, F. M., Moreira, M. E., Klein, R. W., and Böhmwald, H. L. (1995b). Efectos de la administración de adrenalina en las concentraciones sanguíneas y urinarias de minerales de ovejas. *Vet. Mex.* **26,** 209–213.

Witzel, D. A., Littledike, E. T. (1973). Suppression of insulin secretion during induced hypocalcemia. *Endocrinology* **93,** 761–766.

Worker, N. A., and Carrillo, B. J. (1967). "Enteque seco," calcification and wasting in grazing animals in the Argentine. *Nature* **215,** 72–74.

Wysolmerski, J. J., and Broadus, A. E. (1994). Hypercalcemia of malignancy: The central role of parathyroid hormone-related protein. *Annu. Rev. Med.* **45,** 189–200.

Wysolmerski, J. J., McCaughern-Carucci, J. F., Daifotis, A. G., Broadus, A. E., and Philbrick, W. M. (1995). Overexpression of parathyroid hormone-related protein or parathyroid hormone in transgenic mice impairs branching morphogenesis during mammary gland development. *Development* **121,** 3539–3547.

Yamaguchi, A., Kohno, Y., Yamazaki, T., Takahashi, N., Shinki, T., Horiuchi, N., Suda, T., Koizumi, H., Tanioka, Y., and Yoshiki, S. (1986). Bone in the marmoset: A resemblance to vitamin D-dependent rickets, type II. *Calcified Tissue International* **39,** 22–27.

Yanagawa, N., and Lee, D. B. N. (1992). Renal handling of calcium and phosphorus. *In* "Disorders of Bone and Mineral Metabolism" (F. L. Coe and M. J. Favus, eds.), pp. 3–40. Raven Press, New York.

Yang, K. H., dePapp, A. E., Soifer, N. E., Dreyer, B. E., Wu, T. L., Porter, S. E., Bellantoni, M., Burtis, W. J., Insogna, K. L., Broadus, A. E., Philbrick, W. M., and Stewart, A. F. (1994). Parathyroid hormone-related protein: Evidence for isoform- and tissue-specific posttranslational processing. *Biochem.* **33,** 7460–7469.

Yarrington, J. T., Capen, C. C., Black, H. E., Re, R., Potts, J. T., Jr., and Geho, W. B. (1976). Experimental parturient hypocalcemia in cows following prepartal chemical inhibition of bone resorption. *Am. J. Pathol.* **83,** 569–587.

Yarrington, J. T., Capen, C. C., Black, H. E., and Re, R. (1977). Effects of low calcium prepartal diet on calcium homeostatic mechanisms in the cow: Morphologic and biochemical studies. *J. Nutr.* **107,** 2244–2256.

Yasuda, T., Banville, D., Rabbani, S., Hendy, G., and Goltzman, D. (1989). Rat parathyroid hormone-like peptide: Comparison with the human homologue and expression in malignant and normal tissue. *Mol. Endocrinol.* **3,** 518–525.

Young, D. M., and Capen, C. C. (1970). Thyrocalcitonin content in the thyroid glands of cows with vitamin D-induced hypercalcemia. *Endocrinology* **86,** 1463–1466.

Youshak, M. S., and Capen, C. C. (1970). Fine structural alterations in parathyroid glands of chickens with osteopetrosis. *Am. J. Pathol.* **60,** 257–274.

Zarrin, K. (1977). Naturally occurring parafollicular cell carcinoma of the thyroids in dogs. *Vet. Pathol.* **14,** 556–566.

Ziegler, R., Deutschle, U., and Raue, F. (1996). Calcitonin in human pathophysiology. *Hormone Res.* **20,** 65–73.

CHAPTER

24

The Vitamins

ROBERT B. RUCKER AND JAMES G. MORRIS

I. INTRODUCTION AND BRIEF HISTORY

The concept that food components are linked to tissue growth and repair was evident in the writings of Greek philosophers as early as the fifth century B.C. Nutrition was one of the topics in the Hippocratic collection. The first disease to be recognized as nutrition-related may have been night blindness. An ancient Egyptian medical text, the *Papyrusebers* (written about 1550–1570 B.C.), prescribed "beef liver, roasted, squeezed, placed against the eye" for various eye diseases, including night blindness. In China, liver applied topically to the eye was also used as a treatment for night blindness (Guggenheim, 1981). By the mid-1700s, the curative effects of certain foods had been linked to a number of diseases. James Lynn, a physician in England, assembled his classic treatise that fresh fruits and vegetables seemed effective in curing scurvy. By the 1800s, the association of corn with pellagra was made, and by the 1900s, Eijkman, a Dutch physician working in Java, reported that consumption of polished rice was related to polyneuritis, that is, beriberi. However, the concept that specific diseases could be caused by the lack of a specific compound in the diet did not evolve until the beginning of the twentieth century. Following the success of Pasteur and the "germ theory," many diseases now recognized as nutritional in origin were initially attributed to infectious agents. It was widely held that only the gross constituents of the diet, that is, carbohydrates, protein, fat, and minerals, were needed for complete nourishment. As F. G. Hopkins, one of the founders of nutrition as a science, noted in his 1929 Nobel lecture, "the quantitative data obtained and the attractive circumstance that such data appeared to supply a real measure of nutritional needs . . . caused a feeling that knowledge concerning nutrition was adequate and complete." Nevertheless, the view that small amounts of certain factors seemed necessary for optimal growth and development soon evolved. A foundation was provided from these early studies for the formulation of the vitamin theory. The term vitamin was suggested by Casimir Funk (Funk, 1911) to indicate "vital amines," a class of substances whose lack he proposed as being responsible for diseases such as scurvy, rickets, pellagra, and beriberi (Guggenheim, 1981; Bender, 1922).

The effort to define the nature of vitamins was first directed at lipid substances that were demonstrated to be essential in the diet of animals. McCollum and Davis at Wisconsin showed that butter or egg yolk, but not lard, supplied a lipid-soluble factor that was necessary for growth in rats. As a consequence of their studies, the first fat-soluble substance with growth-promoting properties (designated as Vitamin A) was reported in the early 1900s, a time considered as the beginning of the "age" of vitamin exploration (Guggenheim, 1981; Darby and Jukes, 1992). Now, there is constant awareness and sensitivity to the possibility of dietary vitamin deficiencies (and excesses). Nutritional deficiencies are not uncommon in animals, particularly animals fed monotonous diets for therapeutic or other reasons. A number of subsidiary and contributory factors may also lead to vitamin-related diseases, in addition to a low concentration of the vitamin in the diet. These factors include interference with normal food intake, loss of appetite (anorexia), impaired absorption or utilization, increased excretion, and the presence of antagonists. Stressful physiological states that increase nutrient demands, such as lactation, may also perturb the vitamin status of animals.

II. DEFINITION, GENERAL PROPERTIES, AND OVERVIEW OF FUNCTIONS

No definition for vitamins is totally satisfactory. As a group of compounds, vitamins have been defined as organic substances present in minute amounts in natural foodstuffs that are essential to normal metabolism, the lack of which in the diet causes deficiency diseases. This definition, however, is not specific and could apply to a number of compounds derived from the secondary metabolism of amino acids, simple sugars, and fatty acids. Table 24.1 lists the vitamins. The compounds listed are categorized as vitamins because in most mammals they represent essential organic compounds not easily classified with the macronutrients. Some may be synthesized, but in insufficient amounts to meet normal needs during critical developmental periods.

Vitamins can also be classified according to chemical and physical properties, such as whether they are soluble in aqueous solution or lipid solvents (Friedrich, 1988). Those vitamins that are soluble in lipid solvents (vitamins A, D, E, and K) are absorbed and transported by conventional lipid transport processes. For water-soluble vitamins, respective solubility coefficients are major factors that dictate ease of absorption. Within physiological ranges of intake, water-soluble vitamins are usually absorbed by active processes. At high concentrations (5–10 times or more the typical requirements), passive processes may also be involved. Further, most vitamins are absorbed in the duodenum. Exceptions are folic acid and vitamin B_{12}. Folic acid is absorbed throughout the upper and mid-intestine; B_{12} is absorbed only in the ileum of some species. The details of these processes will be outlined later. The diversity and complexity of vitamin metabolism and processing, however, should be appreciated at the outset. Vitamins in foods are often present as cofactors or in highly modified forms. Pancreatic and intestinal cell-derived enzymes are required to initiate normal uptake in absorption. Nucleosidases, phosphatases, and peptidases are key factors in processing cofactors to vitamins (Fig. 24.1, Rose, 1996).

Vitamins serve a broad range of functions. For example, some of the actions of vitamin A and vitamin D are best described as consistent with the actions of steroid hormones; derivatives of vitamin A are also important signal transduction mediators. In contrast, vitamin K acts principally as an enzymatic cofactor. Vitamin E is novel in that it acts as a chemical agent scavenging free-radical containing lipids and oxidants, independent of any direct association with an enzyme. In contrast , most of the water-soluble vitamins act as cofactors or serve coenzymes.

The varied functions of vitamins have also complicated the development of a simple system of classification or nomenclature. When the vitamins were originally discovered they were isolated as fractions from selected foods, and as their exact chemical composition was seldom known, a system of letter designations was developed. However, this system became complicated when it was discovered that some of the functions originally ascribed to vitamins were due to other substances, such as one of the essential amino acids. Consequently, the designation of vitamins by letters was not systematically pursued. Similarly, the lack of chemical composition data resulted in a complex system of expressing dosages as arbitrarily defined units, in which a unit was defined in relationship to a biological phenomenon in a given animal model. With advancements in analytical methodology, the chemical structure of vitamins was defined, their functional roles clarified, and assays developed to define an adequate status.

Indeed, the problem of assessing the dietary requirement and adequacy for various vitamins has long plagued nutritionists. Measurements of the concentration in the diet do not account for availability of the vitamin. Four approaches are used to assess the vitamin status of animals: 1) measuring vitamin concentra-

TABLE 24.1 The Vitamins

Vitamin	Function	Relative need per 1000 Kcal or 4180 J	Common sources
Fat-soluble			
Vitamin A	Gene expression, vision, signal transduction	0.3–0.6 μg as retinal equivalents	Plants (carotenes); most organ meats, particularly liver, high in retinyl esters
Vitamin D	Gene expression, signal transduction, calcium metabolism	1–3 μg cholecalciferol, if Ca and P intakes are adequate	Marine fish, eggs, liver oils, meat, UV-irradiated plant and animal sterols
Vitamin E	Antioxidant	2–5 mg	Vegetable and seed oils, green leaves
Vitamin K	Enzyme cofactor for γ-carboxyglutamic acid formation, blood clotting, osteocalcin formation	50–150 μg	Green plants, egg yolk, milk and cheese, bacteria
Water-soluble			
Ascorbic acid	Redox cofactor, antioxidant, antiscurvy	10–40 mg	Citrus fruits, fruits and vegetables in general
Niacin	Redox cofactor, mono- and polyadenylations, antipellagra factor	2–10 mg	Tryptophan-rich foods, milk, meat
Riboflavin (B_2)	Redox cofactor	300–600 μg	Milk, organ meats, eggs, nuts, seeds, fungi, bacteria
Thiamin (B_1)	Carbohydrate metabolism, decarboxylation, transketolation antiberiberi factor	300–600 μg	Seeds, nuts, meat, yeast, bacteria
Pyridoxine (B_6)	Decarboxylation, aldol condensations, transamination	400–600 μg	Yeast, organ meats, wheat germ, vegetables
Pantothenic acid	Acyl and acetyl transport	2–4 mg	Yeast, plants, grains, eggs, organ meats
Biotin	Carboxylation and transcarboxylation	20–40 μg	Yeast, egg yolks, organ meats, bacteria, grains
Folic acid	Single-carbon transfers, antianemia factor, DNA and purine synthesis	75–150 μg	Green plants, organ meats, eggs, cheese
Cobalamin (B_{12})	Single-carbon transfers, mutase reactions, antianemia factor	0.5 μg	Animal products, milk, fish, eggs, meat
Choline	Lipid transport, neural signal transduction	100–150 mg	Meat, cereal, legumes, lecithin, egg yolk
Inositol	Signal transduction	50–100 mg	Meat, milk, grains (oil), legumes

tions in biological fluids, such as blood plasma; 2) administering the vitamin to the animal and measuring urinary excretion of the vitamin or metabolized products, as an index of the saturation of body stores; 3) loading tests, in which the animal is given a compound that requires the vitamin for its metabolism, and 4) urinary excretion of a product is measured; and enzyme stimulation tests, in which rate of a reaction is measured before and after the addition of the vitamin. Examples that are highlighted in subsequent sections are the histidine loading test for the assessment of folate and vitamin B_{12}, and enzyme stimulation tests such as the erythrocyte glutathione reductase and transketolase tests for riboflavin and thiamin, respectively, and the aminotransferase stimulation tests for vitamin B_6.

In the sections that follow, each of the known vitamins and substances that act as "conditional vitamins" will be described. A goal of each section is to provide an overview pertaining to both physiological and biochemical functions of given vitamins. Issues important to nutritional assessment, deficiencies, and effects of excesses will be addressed. For additional references and information, one or two seminal reviews or monographs are cited for each section.

I. FAT-SOLUBLE VITAMINS

A. Vitamin A

1. *Introduction*

Vitamin A-related deficiency diseases and their cures have been described throughout written history (Guggenheim, 1981). From a modern perspective, the observations by Hopkins, Stepp, and others that a growth-stimulating factor could be extracted from milk into alcohol was a seminal step that eventually led to the identification of vitamin A. The growth-promoting factor was described as being present in egg yolk, but-

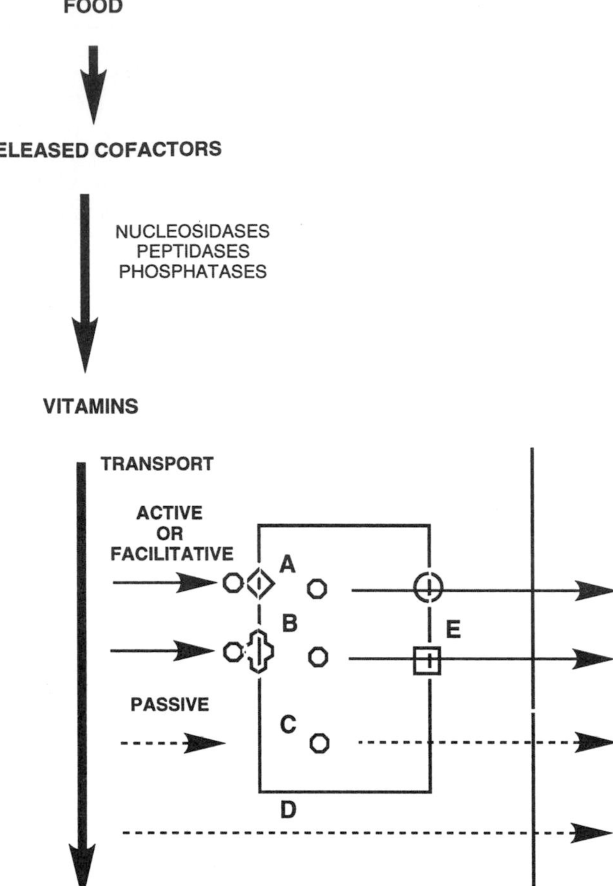

FIGURE 24.1 Steps in vitamin absorption, digestion, and processing. The processing of vitamins requires a number of steps: mastication, hydration, acidification, and the action of numerous enzymes. The vitamins usually enter the body by active or facilitative processes (A or B). At high concentrations the absorption of many vitamins is passive (C) or can occur by pericellular diffusion (D). The uptake of vitamins from the luminal surface maybe facilitative, whereas active transport is often involved for delivery of the vitamin into circulation from the serosal membrane (E). In some cases the transport across the intestinal cell involves packaging into a transport particle, such as chylomicra.

ter, and cod liver oil. In nature, vitamin A is largely present as lipid esters in animal tissues and as provitamin forms of vitamin A in plants. These provitamins belong to the carotenoid family of compounds, for example, β-carotene.

The structures and recommended names of naturally occurring and commercial forms of vitamin A and carotenoids are shown in Fig. 24.2. Once chemical features for the carotenoids and retinoids were resolved in the 1940s and 1950s, studies of their biological function were undertaken and commercial synthesis of vitamin A and vitamin A-like molecules proceeded rapidly.

2. *Proforms of Vitamin A: The Carotenoids*

More than 600 carotenoids have been isolated; however, only about 50 appear to have some degree of provitamin A activity. The proforms of vitamin A, the carotenoids, represent an usual class of biological pigments. In plants and prokaryotes, carotenoids serve as mediators of photo-energy-related processes by capturing energy from light. Carotenoids can also quench singlet oxygen and may act as both antioxidants and pro-oxidants. Carotenoids are rich in conjugated double bonds, and besides being designed to interact with light, they are also readily destroyed by intense light, particularly UV light (Sporn *et al.*, 1994). In order to act as a provitamin A, a carotenoid must contain a β-ionone structure (Fig. 24.2).

In plants, carotenoids occur in association with chloroplasts, complexed with protein and other lipids, and provide the main source of provitamin A for animals. The provitamin A potency of plants is largely retained when forage crops are made into hays, but declines with the uptake of oxygen on storage. Grains, with some exceptions (e.g., yellow corn), are minor sources of provitamin A. Among the legume grains, chickpeas and green and black gams are the best sources of provitamin A. The richest source of carotenoid is red palm oil, which contains 500 μg of mixed α- and β-carotene per milliliter. About 7 ml/day of red palm oil can easily meet the nutritional needs of a child or small animal.

3. *Metabolism*

Following ingestion, retinoyl esters in animal products are hydrolyzed to retinol by pancreatic hydrolases (esterases) or lipid hydrolases localized on the surface of the brush border of intestinal cells. Bile and dietary lipid facilitate the absorption process, as retinoyl esters must be part of a lipid micelle to be absorbed. The micellar structures enhance fusion into the microvillus of intestinal cells. Similarly, lipid micelles enhance the uptake of carotenoids into intestinal cells. The bioavailability and digestion of vitamin A and carotenoids are affected by the overall nutritional status and the integrity of the intestinal microvillus. Absorption of physiological doses of vitamin A in most animals is 70 to 90%, but the efficiency of absorption for carotenoids added to diets is 40 to 60%, depending on the type of carotenoid. Carotenoids contained in plant chloroplasts, however, are often very poorly absorbed (less than 10%) because of the low digestibility of chloroplasts and release of carotenoids.

In the intestinal mucosal cell, some carotenoids are oxidized to both carotinals or retinals (Fig. 24.3). Most of the retinal is next reduced by alcohol dehydrogenases to retinol and is reesterified. Retinol and associ-

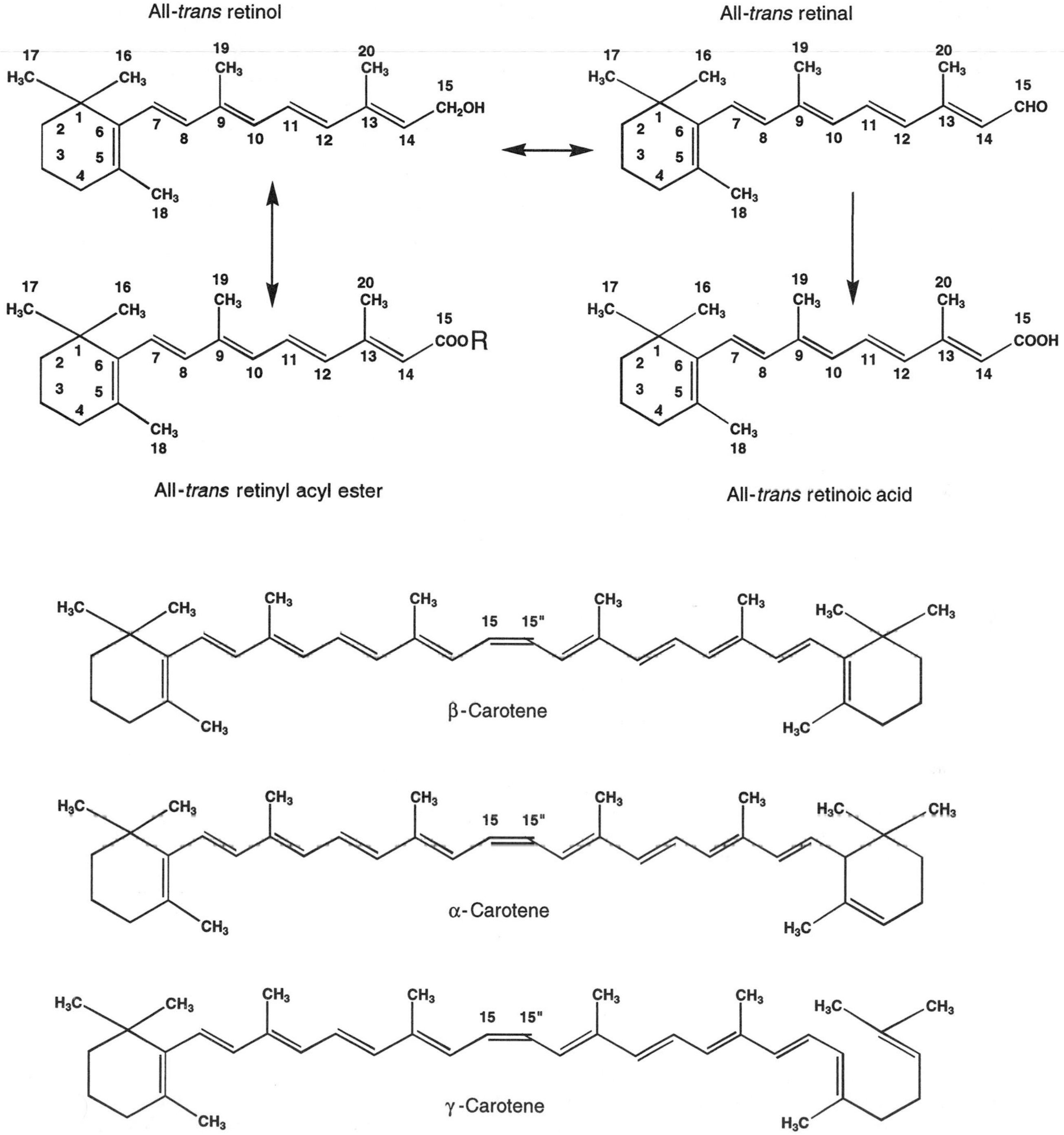

FIGURE 24.2 Structures of common retinoids and carotenoids. The structures for all-*trans* retinol, all-*trans* retinal, all-*trans* retinyl acyl (designated R) ester, and all-*trans* retinoic acid are shown. Note that the oxidation of retinal to retinoic acid is not reversible. β-Carotene is the most potent vitamin A precursor. For vitamin A activity the presence of a β-ionone is essential. β-carotene contains two β-ionone rings. If β-carotene is cleaved at the 15,15′ position, theoretically two retinals are produced. In contrast, α-carotene and γ-carotene result in only one active retinoid when cleaved.

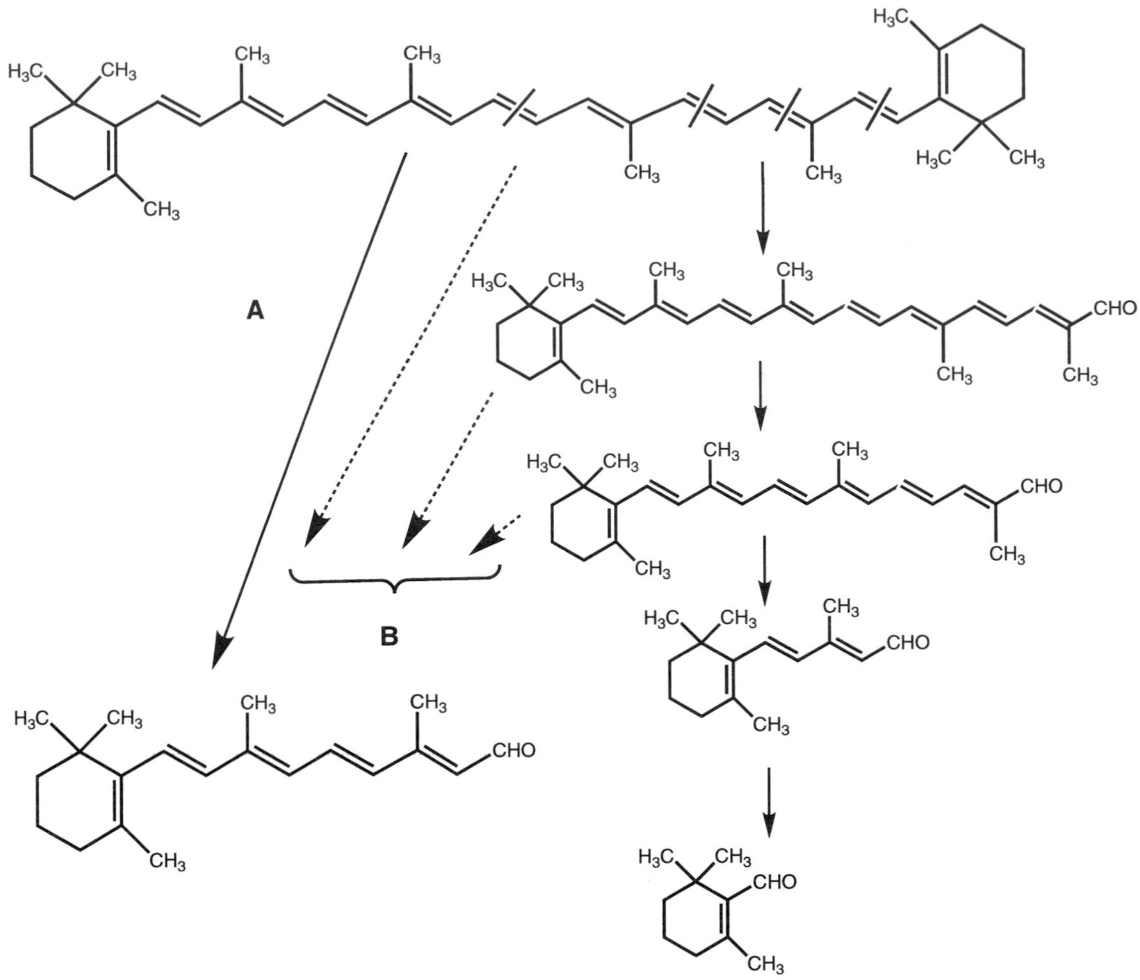

FIGURE 24.3 Cleavage of carotenoids occurs by two distinct routes. Route A represents enzymatic or intrinsic cleavage at the 15,15′ position. However, chemical or extrinsic cleavage (routes designated B) can result at several sites to generate complex mixtures of carotenals.

ated esters are then incorporated into chylomicra, which are released into the lymph. Both intrinsic and extrinsic pathways are proposed for the oxidative cleavage of carotenoids with provitamin A activity to retinal (Sporn *et al.*, 1994; Packer, 1990). Evidence for the intrinsic pathway is based on the direct production of retinal from β-carotene, presumably by the action of a 15,15′-β-carotene dioxygenase. The extrinsic pathway is based on evidence of nonenzymatic chemical cleavage of carotenoids to carotinals.

Chylomicra particles in lymph are carried to the liver, where about 75% of the retinol-derived products are cleared in most animals. In liver, the processing of vitamin A occurs primarily in liver stellate cells, where vitamin A is stored as retinoyl ester, passed on to adjacent parenchymal cells, directed to retinol-binding protein (RBP), a unique retinol-transport protein, or oxidized. Figure 24.4 describes the major feature of vitamin A metabolism in liver cells.

As vitamin A is needed by the body, retinoyl ester in the liver is hydrolyzed and released as retinol bound to RBP, which binds one molecule of vitamin A as retinol per molecule of RBP. When released into circulation, RBP exists as a protein complex with transthyretin, which binds thyroxin. The RBP–transthyretin complex transports vitamin A and thyroxins to their respective target cells. The primary target cells for vitamin A are the epithelial cells.

The association constant (K_a) between retinol and the transthyretin–RBP complex is relatively low, approximately 10^6 M/liter. This association constant is about the same as that for the binding of retinol to other proteins, such as albumin, which does not imply a high degree of specificity. In cell cultures, RBP is not

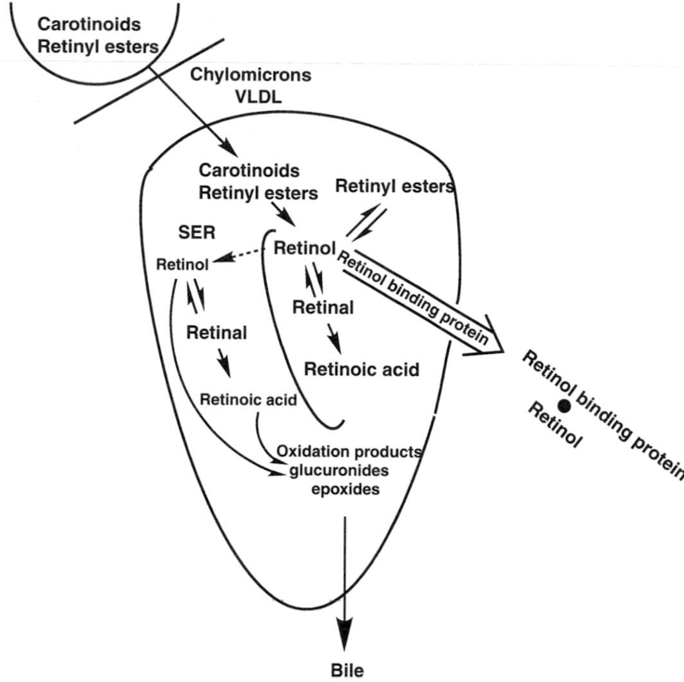

FIGURE 24.4 Major steps in the processing of vitamin A. Carotenoids, retinoids, and retinyl esters associated with chylomicra and VLDL particles are brought to the liver sininoids via the lymphatic system or portal systems in avians and reptiles. Vitamin A is next taken up by liver stellate cells. There is excellent communication between stellate cells and adjacent liver parenchymal cells. Each of these cells produce retinol-binding protein (RBP), which facilitates its transport into secretory pathways that allow delivery of retinol into blood. Excess retinol in cells is stored as retinyl ester (RE). Excess retinol is also oxidized by dehydrogenases in the cytosol, as well as by microsomal oxidases associated with the smooth endoplasmic reticulum (SER). The SER oxidase system results in a variety of oxidation products as well as retinyl glucuronides. Some of these products are delivered via the bile into the intestine. Drugs such as alcohol and barbiturates stimulate microsomal oxidase activity and can cause depletion of retinol.

essential for retinol uptake. However, RBP is designed for binding retinol in a hydrophobic pocket within the protein that most likely protects retinol from oxidation and destruction during transport (Packer, 1990).

Inside the targeted cells, vitamin A, as retinol, interacts with cellular binding proteins that function to control its subsequent metabolism, for example, oxidation to retinal and to retinoic acid. Some of these cellular binding proteins are also a part of the superfamily of glucocorticoid–retinoid–thyroxine transcriptional factors. It is the role of such proteins in transcription and regulation that makes vitamin A important to so many facets of cellular regulation.

Oxidation of retinoids in liver cells proceeds by two pathways, the principal one being cytosolic and involving one of the alcohol dehydrogenase isozymes. The other pathway involves so-called "microsomal oxidases," which may be induced by dietary excess, including alcohol, and many drugs. High doses of phenobarbital or ethanol cause depletion of liver retinol by induction of a microsomal oxidase system that promotes retinoid oxidation. Some of the oxidized products are sufficiently polar to be excreted in the urine. However, the major route of disposal is bile.

4. *Functions*

The major roles of Vitamin A are in cellular differentiation, tissue growth, and vision (Sherman, 1986). In quantitative terms, only a small fraction of the total vitamin A requirement is involved in the visual process because of extensive recycling of retinal. In vision, vitamin A, as a component of rhodopsin, facilitates the efficient transfer of energy from photons of light to electrochemical signals (Fig. 24.5). The events leading

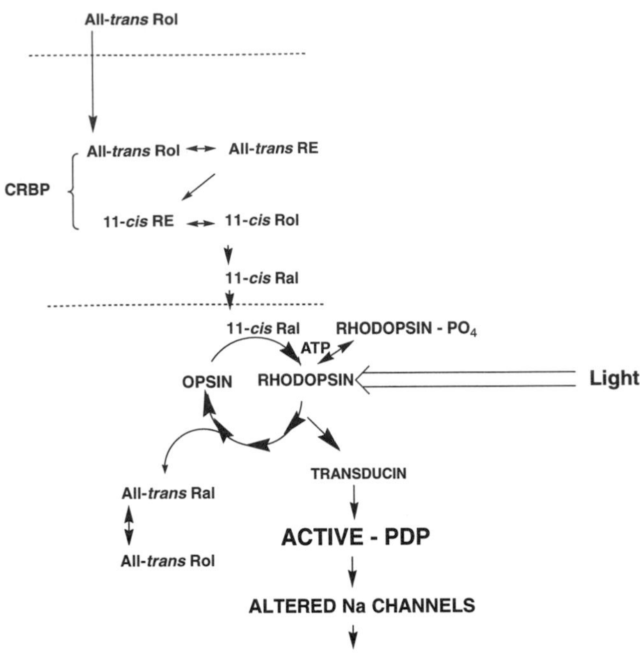

FIGURE 24.5 Vitamin A and vision. When all-*trans* retinol enters the rod cell, it first combines with cytosol retinoid binding protein (CRBP). Retinal and the all-*trans* retinal ester are capable of isomerization. Reactions involving all-*trans* retinal and retinyl esters occur by pathways that utilize CRBP–retinoid complexes as substrates. *cis*-Retinal reacts with opsin to form rhodopsin. The multiple arrows from rhodopsin back to opsin represent conformational changes in rhodopsin upon stimulation by light. The activated rhodopsin interacts with transducin, which in turn activates phosphodiesterase, which eventually causes a decreases in cGMP. This sequence of reactions causes an alteration in sodium channels and alters membrane polarization to activate optic nerve signal propagation. A number of additional events aid in controlling the process, such as inactivation of rhodopsin by phosphorylation.

up to the propagation of this signal are as follows: Vitamin A, as 11-*cis* retinal, forms a protonated Schiff base by binding to a lysine residue at position 296 of the membrane-bound protein opsin to yield rhodopsin. When a photon of light strikes rhodopsin, *cis–trans* isomerization occurs and the process results in a highly strained form of rhodopsin (bathorhodopsin), which is next converted to metarhodopsin with subsequent deprotonation. The deprotonated metarhodopsin interacts with transducin (one of the proteins in the transmembrane G-protein family). This interaction causes a subunit of transducin to bind GTP, stimulating cGMP phosphodiesterase activity. This results in a decease in cGMP, which constitutes a significant amplification of the initiating event, the conversion of light-derived energy through 11-*cis* to *trans* isomerization of retinal and specific changes in protein conformation.

The local decrease in cGMP results in changes in cation flux (Na and Ca ions) across rod cell membranes, resulting in more ionic calcium in the cytosol. The increase in calcium blocks the sodium channel and initiates depolarization, the firing of cells of the optic nerve. Further, metarhodopsin is phosphorylated during these final steps and interacts with a protein designated as arrestin. The metarhodopsin–arrestin complex inhibits the transducin response and causes the release of all-*trans* retinal and the return to rhodopsin (opsin), thus completing the cycle.

In dietary vitamin A deficiency there is an inability to appropriately saturate opsin with 11-*cis* retinal to form rhodopsin and its subsequent complexes, which in turn decreases the sensitivity of the visual apparatus so that light of low intensity is not perceived, leading to the clinical condition of nyctalopia or night blindness.

5. *Growth and Cell Differentiation*

As work on the role of vitamin A in growth progressed, it was soon recognized that in addition to retinal, retinol and its esters, retinoic acid would support growth and many features of normal development, but not vision, for which retinol (retinal) is essential. It is now appreciated that much of the growth-related response of deficient animals to vitamin A is related to maintenance of epithelial cells. With vitamin A deficiency, mucus-secreting cells are replaced by squamous, keratin-enriched cells (Sucov and Evans, 1995; Sherman, 1986).

Within cells, all-*trans* retinol associates with cytosolic retinol-specific binding proteins (CRBPs), and the resulting complexes become vehicles for subsequent processing . For example, all-*trans* retinol may be oxidized and isomerized to 13-*cis* retinoic acid, which subsequently binds to retinoic-acid-specific cytosolic binding proteins, which act as transcription factors in protein regulation and cellar differentiation.

In response to very low doses of retinoids, epithelial cells undergo a "terminal differentiation." Vitamin A derivatives control the expression of various proteins important to mucus formation and cytoskeletal integrity, such as keratin and transglutaminase, and the rate of cell cycling. In response to an abnormally low level of retinoids, cells lose their normal columnar shape, become flattened or squamous, and increase their cytosolic content of keratin (stabilized by transglutaminase-catalyzed cross-links). In dermis, this process results in a protective outer layer, scales, and other specialized surfaces. However, where the primary function of the epithelial cell is provision of a moist surface or absorption, squamous overkeratinization leads to loss of functional integrity. Lack of protective mucous secretions facilitates the establishment of infections of the lungs and other tissues that depend on a mucus barrier. In the intestine, keratinization induces premature sloughing of enterocytes and malabsorption.

A number of hypotheses have been proposed for a gradient delivery of retinoids to epithelial cells, which help explain why some epithelial cells undergo terminal differentiation, whereas others undergo cell cycling and periodic turnover. However, much is still required to clearly define the process. Other cells that are responsive to retinoids include phagocytic cells and cells associated with the immune response; for example, the normal proliferation of the B cells and T cells requires vitamin A.

6. *Requirements*

The appearance, signs, and symptoms of vitamin A deficiency in a number of animal species have been described in detail. For any given animal, the requirement for vitamin A depends upon age, sex, rate of growth, and reproductive status. For optimal maintenance, the allowance for many animals ranges from 100 to 200 I.U. per kilogram of body weight per day (one I.U. is equal to 0.3 μg of retinol). In young animals, however, a more precise method of expressing the vitamin A requirement is on an energetic basis. In animal feeds, 4000 to 10,000 I.U. per kilogram of feed is considered adequate in the United States to provide vitamin A requirements for most animals.

Pathological conditions that influence vitamin A status include malabsorption, including pancreatic insufficiency and cholestatic disease; cystic fibrosis; liver disease; and kidney disease. Many forms of liver disease interfere with the production or release of RBP, which results in a lower plasma level of vitamin A. Renal failure can result in loss of RBP in urine. Factors that impair lipid absorption and transport also influence vitamin A status.

7. *Evaluation of Vitamin A Status of Animals*

The vitamin A status of animals may be evaluated on the basis of physiological, clinical, and biochemical procedures. Clinical testing for night blindness and elevation of CSF pressure have been used to indicate vitamin A insufficiency. The concentrations of retinol and its esters are readily measured in biological samples by HPLC using various detectors and provides an indication of the vitamin A status. As the concentration of retinol in plasma is well maintained until liver reserves are grossly depleted, plasma retinol is not a reliable index of vitamin A reserves. Analysis of biopsy samples of liver provides a useful index of reserves. In many carnivores , the plasma contains, in addition to retinol, equal or greater concentrations of retinyl palmitate and retinyl stearate. Plasma retinol concentrations in excess of 30 μg/dl generally indicate that vitamin A is not limiting. In most species, liver concentrations of 100 μg retinol/g liver indicate adequate status.

8. *Pharmacology and Toxicity*

Vitamin A and various retinoids are used increasingly to treat skin disorders (acne and psoriasis) and certain forms of cancer (Sporn *et al.*, 1994; Gillian, 1992). Retinoyl-β-glucuronide and hydroxyethyl retinamide are commercial preparations of retinoids that have such activity, but are less toxic than retinoic acid. The mechanisms by which these agents function most probably relate to the complex pathways involved in epithelial and epidermal cell differentiation.

Vitamin toxicities may be classified under three broad categories: acute, chronic, and teratogenic. When a single dose of vitamin A (greater than 100 milligrams) is injected into animals in the 20–50 kg weight range, symptoms such as nausea, vomiting, increased cerebral spinal fluid pressure, and impaired muscular coordination result. A lethal dose of vitamin A (100 mg) given to young monkeys has been reported to cause coma, convulsions, and eventual respiratory failure.

Chronic toxicity may be induced by intakes of vitamin A in amounts 10 times the normal requirements. Intakes of vitamin A in this range can lead to alopecia, ataxia, bone and muscle pain, and pruritis. Although cats have a high tolerance to excessive intakes of vitamin A, naturally occurring hypervitaminosis A occurs in cats that are given a diet based largely on liver. Affected cats exhibit skeletal deformations, particularly exostoses of the cervical vertebra, which precludes effective grooming. Vitamin A is also a powerful teratogen. A single large dose during pregnancy (in the 50- to 100-mg range) for an animal weighing 20–50 kg can result in fetal malformations. Chronic intakes (exceeding 10 times the requirements for given animals) can also be teratogenic. Carotenoids, unlike retinoids, are generally nontoxic, and many animals routinely ingest gram amounts of carotenoids on a daily basis with no deleterious effects.

B. Vitamin D

1. *Introduction*

Sir Edward Mellanby in 1921 reported the induction of rickets in dogs through dietary manipulation. He discovered that the disease could be corrected with cod liver oil. McCollum in 1922 reported that the curative factor in cod liver oil was not vitamin A , but appeared to be another fat-soluble substance. This substance was later identified as vitamin D, based on the ability to inactivate the vitamin A factor in cod liver by mild oxidation, but in the process retain the antirachitic activity (Bender, 1992).

2. *Sources, Functions, and Metabolism of Vitamin D*

The D vitamins are a family of 9,10-secosteroids that differ only in the structure of the side chain attached to carbon-17. The two forms of vitamin D significant in veterinary medicine are ergocalciferol (vitamin D_2) and cholecalciferol (vitamin D_3). The differences in the side chain result in the vitamins having disparate potencies with some species of animal and differing in toxicity when consumed in large amounts. These two forms of vitamin D are produced in a two-step reaction when their respective sterols, ergocalciferol and 7-dehydrocholesterol, absorb ultraviolet radiation and undergo photolysis, which is then followed by thermal isomerization. Excessive ultraviolet radiation of the sterols produces inactive compounds (Fig. 24.6). Under most circumstances, animals can synthesize sufficient quantities of cholecalciferol if they receive adequate exposure to ultraviolet light of wavelength 280–320 nm, particularly when the calcium and phosphorus requirements of the animal are met. As vitamin D_3 is produced at one site (skin) and acts at other sites including bone and intestine, it fulfills the definition of a prohormone (Diplock, 1985).

In most animals, 7-dehydrocholesterol is abundant in skin, being the ultimate precursor for cholesterol, which is synthesized *de novo* from acetate. However, the skin of cats and dogs and possibly other carnivores contains only small quantities of 7-dehydrocholesterol, which does not permit adequate synthesis of vitamin D, and these animals are solely dependent on the diet for this vitamin.

With the exception of animal products, most natural foods contain low vitamin D activity. Fish, in particular saltwater fish such as sardines, salmon, and herring, and fish liver oils contain significant to large quantities of vitamin D. Many plants also contain hydroxylated ergosterol derivatives, some of which have potent vitamin D activities (Collins and Norman, 1991).

Initially, vitamin D was assumed to be a cofactor for reactions that served to maintain calcium and phosphorus (as phosphate). When isotopes of calcium became available, it was soon appreciated that there was considerable time lag between the administration of vitamin D, and its effect on calcium-related metabolism. This lag was shown to be due to the conversion of vitamin D to an active form. Investigations throughout the 1960s and 1970s led to the sequence of events outlined in Fig. 25.6 (Wasserman and Fullmer, 1995). For example, the kidneys were identified as the site of 1, 25-dihydroxycholecalciferol (calcitriol, 1,25-$(OH)_2D_3$) production. This discovery, together with the finding that 1,25-$(OH)_2D_3$ is found in the nuclei of intestinal cells, suggested that vitamin D was functioning in a manner analogous to that of steroid hormones. The production of calcitriol is normally closely regulated through feedback control by 1,25-$(OH)_2D$ and parathyroid hormone (PTH) of the activities of the 25-OH-D 1α- and 25-OH-D 24-hydroxylases. A fall in plasma calcium triggers the release of PTH from the parathyroid gland, which stimulates 1α-hydroxylase production and leads to increased output of calcitriol from the kidney. A separate hydroxylase, which catalyzes 24,25-$(OH)_2D_3$ production, is activated under eucalcemic and hypercalcemic states. Whether 24,25-$(OH)_2D_3$ has unique activity is controversial. There is evidence that 24,25-$(OH)_2$-D_3 is required for some of the biological responses attributed to vitamin D. Norman and co-workers have shown that hatchablity in chickens is markedly improved if both 1,25-$(OH)_2D_3$ and 24,25-$(OH)_2D_3$ are administered into eggs containing viable embryos from hens rendered rachitic (vitamin D-deficient) prior to egg production. The two major sites of action of calcitriol in relation to calcium homeostasis are bone, where it acts rapidly in concert with PTH in response to hypocalcemia, and intestine, where the response time is longer.

Calbindin, a calcium-binding protein, is a major product synthesized in intestinal cells in response to calcitriol (Balmain, 1991; Norman *et al.*, 1987). Calbindin influences the movement of calcium across the intestinal cell. Binding of calcium to this protein allows the intracellular concentration of calcium to be elevated. The hormone forms of cholecalciferol also stimulate the production of the calcium- and sodium-dependent ATPases, which reside on the basal-lateral membrane surface of the intestinal cell, thus facilitating the vectorial movement of calcium out of the cell into circulation. In addition, recent evidence also indicates that 1,25-$(OH)_2D_3$ can stimulate secondary messenger systems, such as protein-kinase- and adenyl-cylase-controlled dependent messenger systems.

In addition to 1,25-$(OH)_2D_3$ and 24,25 $(OH)_2D_3$, more than 20 other hydroxylated intermediates and end products have been identified. Clearly defined functional roles for these products have not been elucidated.

In addition to intestinal cells, the osteoblasts of bone are a key target of vitamin D metabolites and play a major role in short-term calcium homeostasis. In addi-

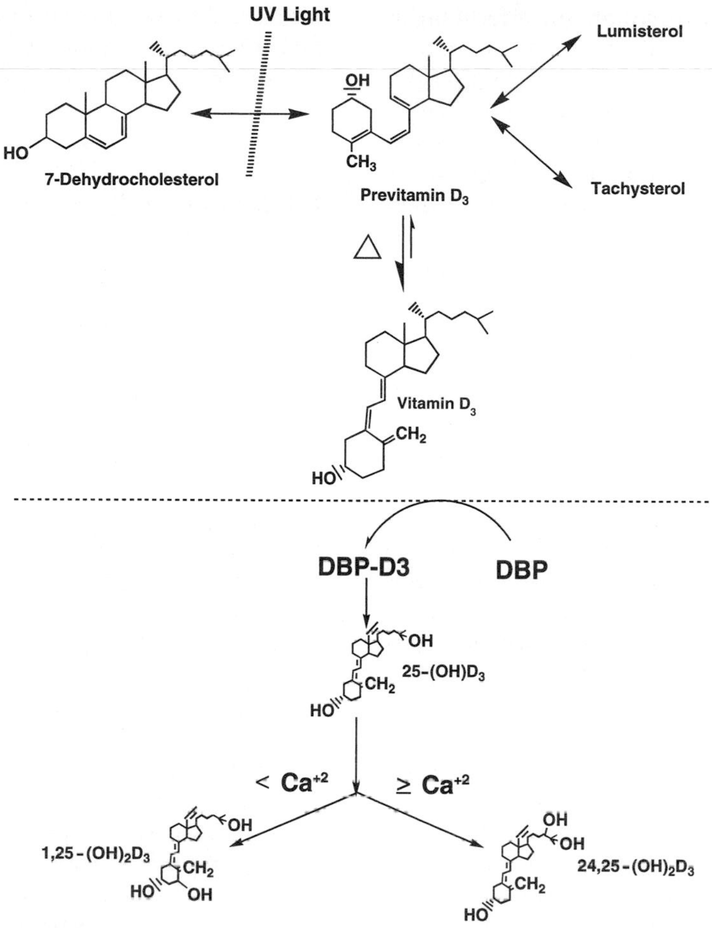

FIGURE 24.6 Vitamin D metabolism. 7-Dehydrocholesterol in skin can be converted to pre-vitamin D by the action of UV light. At body temperature pre-vitamin D spontaneously is converted to vitamin D_3. Related derivatives, such as lumisterol and tachysterol, are also formed. Vitamin D-binding protein (DBP) aids in transport of vitamin D_3 from skin to the liver, where vitamin D_3 is converted to 25-hydroxy-vitamin D_3. In the kidneys, in response to a hypocalcemic state, 1α,25-dihydroxycholecalciferol production occurs. Under a eucalcemic or hypercalcemic state, 24,25-dihydroxycholecalciferol production occurs.

tion, 1,25-$(OH)_2D$ is indirectly required for normal bone mineralization during skeletal growth and remodeling of bone. Vitamin D receptors (VDR) in bone are located in osteoblasts and progenitor cells of bone; they control synthesis and secretion of a number of bone-specific proteins in osteoblasts, such as osteocalcin, osteopontin, collagen, and alkaline phosphatase. Although osteocalcin and osteopontin synthesis have been shown to be regulated at the transcriptional level of their respective genes, for the most part vitamin D metabolites attenuate the action of polypeptide hormones such as PTH or calcitonin, which stimulate bone resorption and accretion respectively. Of these two processes, maintaining bone resorption is the most important, because under normal conditions the serum calcium and phosphate ion concentrations are at levels that favor bone apposition or accretion.

3. *Other Functions of Vitamin D*

Vitamin D receptors (VDRs) have been found in a large number of cell types, ranging from skeletal muscle cells to cells important to immune and phagocytic functions, such as macrophages. In pancreatic β cells, 1,25-$(OH)_2D_3$ has also been observed to be important to normal insulin secretion. Vitamin D increases insulin release from isolated perfused pancreatic cells. Moreover, vitamin D metabolites can suppress immunoglobulin production by activated B-lymphocytes. T cells are also affected by vitamin D metabolites. 1,25-

$(OH)_2D_3$ exhibits permissive or enhancing effects on T-cell suppressor activity (Norman, 1995).

A specific transport protein delivers 1,25-$(OH)_2D_3$ and other active forms of vitamin D to targeted cells. The active form of vitamin D then interacts with receptor proteins, which in turn signal enhanced expression of selected proteins. Even in a tissue such as the intestine, VDR represents only about 0.001% of the total protein. Recently, the total size of the deduced amino acid sequence of the VDR protein has been determined for rats as 423, and for humans as 427. The cysteine-rich DNA-binding domain of the amino terminus is 100% conserved between these species, and the steroid-binding domain at the carboxyl terminus is 93% conserved.

4. *Requirements and Toxicity*

Most animals require 5 μg or less of vitamin D per 1000 Kcal of diet. When intake exceeds 5 to 10 times this amount, there is a risk of toxicity, characterized by hypercalcemia and calcification of soft tissues, particularly in the blood vessels of the lung, kidney, and heart. Acute doses of vitamin D (100 times the requirement) can eventually result in a negative calcium balance, because bone resorption is accelerated. Some plants (*Solanum malacoxylon, Cestrum diurnun,* and *Trisetum flavescens*) contain compounds with vitamin D activity (mostly glycosylated forms of 1,25-dihydroxyvitamin D), and consumption of these plants by grazing animals causes an induced vitamin D toxicity and calcinosis manifested by deposition of calcium in soft tissues. Rodenticides containing cholecalciferol as the active ingredient have resulted in toxicity in companion animals that ingest the bait directly, or ingest carcasses of rodents that have ingested the bait (Collins and Norman, 1991).

5. *Assessment of Vitamin D Status*

Reliable assays for the measurement of vitamin D, calcidiol, and calcitriol in plasma are available. Calcitriol occurs in picomolar concentrations (normal values 40 to 150 pmol/liter or 17 to 63 pg/ml) and has a half life of about 4–6 hours in humans. Concentrations of vitamin D in plasma after oral administration are in the nanomolar range ("normal" values range from 0 to 310 nmol/liter or 0 to 120 ng/ml). Vitamin D has a half-life of 24 hours, so the plasma concentration reflects immediate intake rather than overall status. In contrast, 25-OH vitamin D has a half-life of about 3 weeks. It provides the useful index of vitamin D status and is the measurement of choice. Plasma concentrations of 25-OH vitamin D of 20–150 nmol/liter or 8 to 60 ng/ml cover the normal range for most animals. Much higher levels than these have been observed in cats given diets containing high levels of cholecalciferol without apparent deleterious effects.

C. Vitamin E

1. *Introduction*

In the early 1920s, Herbert Evans and Kathryn Bishop observed that rats failed to reproduce when fed diets containing rancid lard, unless they were supplemented with lettuce or whole wheat (Machlin, 1980). Later it was found that germ oil, particularly wheat germ oil, contained an active principle that seemed responsible for improving reproductive performance. These early studies provided yet another function for a fat-soluble substance. By the early 1930s, it was recognized that this substance was a factor that differed from vitamin A or vitamin D. The compound was designated as vitamin E by Sure, and later as α-tocopherol, from the Greek word "tokos" meaning childbirth or reproduction. By 1940, a number of compounds in the tocopherol family had been identified and purified. With elucidation of tocopherol structures and eventual chemical synthesis (Fig. 24.7), studies demonstrating that embryonic failure resulted from vitamin E deficiency quickly followed. Pappenheimer, Olcott, and others observed that muscle degeneration was also a common deficiency symptom. Next, other signs and symptoms were identified, including oxidative diathesis and encephalomalcia in chickens and liver necrosis and hemolytic anemias in other vitamin E-deficient animals (Guggenheim, 1981).

2. *Chemistry, Metabolism, and Sources*

The principal dietary sources of tocopherols are plant oils. Tocopherols are unique among the vitamins as they act primarily at a chemical level as antioxidants, that is, they do not serve as cofactors or appear to be involved directly as a specific factor in cellular regulation. Primarily, vitamin E protects unsaturated fatty acids in the phospholipids of cell membranes. The quinone moiety of tocopherols is capable of quenching free radicals, such as free radical of hydrogen ($H^\cdot$), superoxide radicals ($O_2^\cdot$), hydroxyl radicals ($OH^\cdot$), and other lipid-derived radical species ($LOO^\cdot$). Vitamin E is very reactive, and in the course of its action as a free-radical scavenger it is sacrificed; however, in this process it inhibits the formation of lipid-derived oxidation products (Marks, 1977).

Cell membranes contain vitamin E at a concentration of approximately 1 mg per 5–10 g lipid membrane, a concentration sufficient to retard membrane-lipid oxidation. Membrane lipids are constantly engaged in

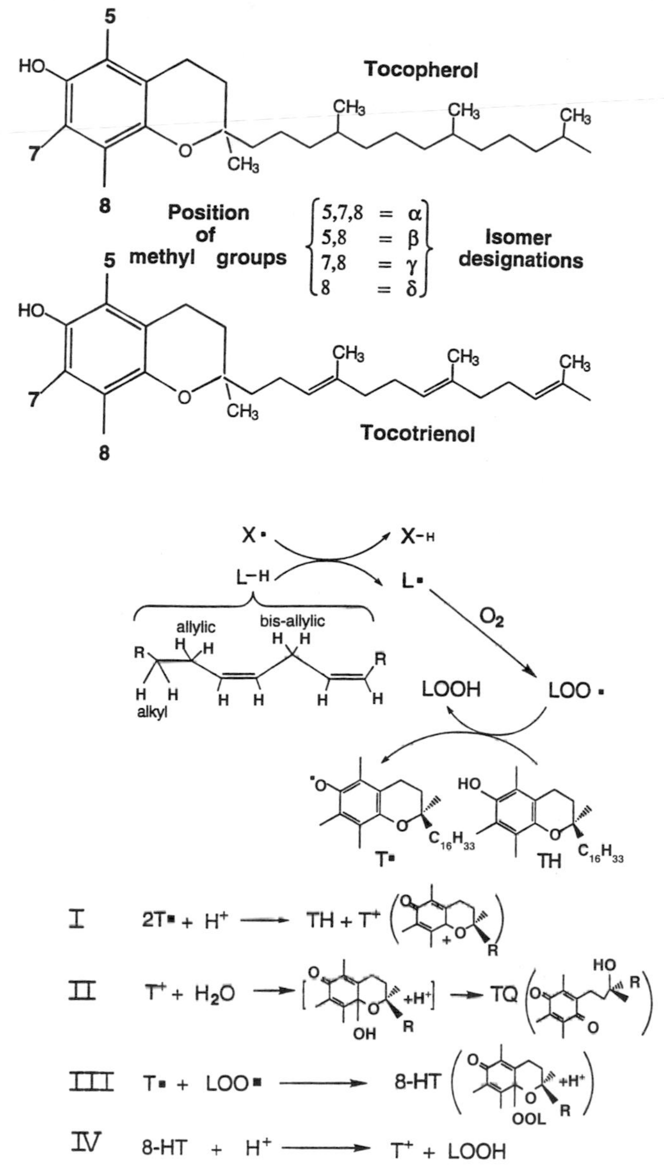

FIGURE 24.7 Vitamin E. The two principal forms of vitamin E, tocopherol and tocotrienol, are shown. Methyl groups are found at the 5, 7, or 8 positions. Tocopherol is the most potent of the various forms of vitamin E in biological systems. Some of the basic mechanisms involving free-radical quenching are also shown. Vitamin E is particularly important in quenching free radicals that generate from the allelic and bis-allelic unconjugated bonds found in membrane polyunsaturated lipids. Resolution of the vitamin E and various known intermediates in the process are given in steps I and IV.

the process of turnover and repair. By prolonging the initiation time before a free-radical chain reactions occurs, vitamin E gives cells time to replace damaged membrane lipids through the process of normal cell turnover.

With regard to absorption and transport, tocopherols first partition into the intestinal micelles prior to absorption (Kayden and Traber, 1995). Following absorption, vitamin E is transferred into the lymph associated with chylomicrons and intestinally derived VLDL particles, similar to other fat-soluble vitamins. Vitamin E is cleared from chylomicrons and VLDL by the lung and the liver. From the liver, most of the vitamin E is found in association with VLDL and LDL particles. Although no specific plasma or serum carrier protein for vitamin E has been unequivocally elucidated, phospholipid transfer protein, which accelerates exchange and transfer of vitamin E between lipoproteins, has been shown to have high specificity for the α-tocopherol form of vitamin E. Phospholipid transfer protein may thus determine the types of tocopherols that are returned to the liver from plasma via HDL- and LDL-derived lipoproteins. For example, the α-tocopherol form of vitamin E is preferred, and the LDL particles contain the highest concentration of vitamin E. From the perspective of human nutrition, this is important in that high concentrations of vitamin E protect the LDL particle from oxidation. It has been proposed that oxidized LDL particles are important mediators of vascular disease (Traber and Packer, 1995).

Vitamin E enters cells by processes similar to those for LDL uptake. LDL membrane receptors, through receptor-mediated endocytosis, appear responsible for vitamin E uptake. Once in cells, vitamin E is incorporated into lipid membranes. About 40% of vitamin E is found in nuclear membranes; the remaining 60% is divided among lysosomal, mitochondrial, and outer cell-wall membranes. From a pharmacological perspective, vitamin E has been associated with improved immune response, improved meat quality, and improved wound healing (Bendich, 1993; Liu *et al.*, 1995).

3. *Requirements*

The nutritional status of vitamin E is often difficult to assess. A number of factors can influence the concentration of tocopherols in cells. Vitamin E acts as the last line of defense for lipid oxidation, primarily residing in lipid membranes. Consequently, enzymes such as superoxide dismutases, catalase, glutathione peroxidase, and related systems for oxidant defense can moderate the absolute need for vitamin E. Further, high dietary intakes of polyunsaturated dietary fats increase the vitamin E requirement because of their eventual deposition in cell membranes and higher susceptibility to oxidation.

4. *Evaluation of Vitamin E Status*

Tocopherols in biological tissues can be measured by HPLC. Although α-tocopherol can be separated readily from other tocopherols, the separation of the β and γ isomers is difficult. For nutritional assessment of vitamin E, the current indices are based on changes in total tocopherol concentrations in plasma and serum. Measurement of tocopherol concentration in erythrocytes may be an even better indicator for tissue vitamin E than plasma or serum levels. The platelet concentration of vitamin E is also a sensitive measure of vitamin E intake. Moreover, in the experimental setting the measurement of adipose levels of tocopherols seems to be a reliable index for assessing long-term vitamin E status. As in other cells, vitamin E partitions primarily into the membrane lipid compartments of adipose cells. Thus, the concentration of vitamin E per adipose tissue mass may even increase when there is loss of non-membrane stored triglycerides. Because plasma tocopherol concentration, is affected by lipid concentration, an α-tocopherol/total lipid ratio of 0.6 to 0.8 mg/g total lipids has been suggested as indicating adequate nutritional status. Functional tests such as the erythrocyte hemolysis in the presence of 2% peroxide have also been used to indicate status.

D. Vitamin K

1. Introduction

In 1929, Henrik Dan reported what was first thought to be an essential role for cholesterol in the diet of chickens. He noted that chicks fed diets that had been extracted with nonpolar solvents to remove sterols developed subdural and muscular hemorrhages, and that blood seemed to clot at a slower rate. These studies were extended by MacFarland, who observed clotting defects in chicks fed diets based on lipid-extracted fish or meat meals. MacFarland showed that this defect could not be reversed by providing any of the other known vitamins. As a consequence, hemorrhagic disease in chicks was associated with a new factor, designated as vitamin K. Later it was demonstrated that hemorrhagic disease in chicks could be reversed by extracts of alfalfa. In the 1940s it became clear that substances synthesized by bacteria also contained vitamin K activity (Friedrich, 1988; Bender,1992). As this work progressed, information also became available regarding a compound in spoiled clover and grasses that seemed to cause hemorrhagic disorders in cattle and serve as antagonist to vitamin K (Fig. 24.8).

2. Function and Metabolism

With the isolation and identification of vitamin K, an understanding of the mechanism of its action evolved, though not without controversy. At first there was the problem of reconciling how compounds present in the spoiled sweet clover acted as vitamin K antagonists. A number of questions were also raised regarding the structural requirements for vitamin K activity. Now it is appreciated that a number of compounds in the 1,4-naphthoquinone series—even relatively simple compounds such as menadione—possess vitamin K activity. Menadione combines with isoprenoids arising from the cholesterol synthesis pathway, resulting in an active phylloquinone. Dietary phylloquinones are transported to liver by chylomicrons and intestinal VLDL particles and from liver by VLDL and LDL. Studies of vitamin K clearance show that the total pool of vitamin K in the body is replaced rapidly; within hours to days, in contrast to the slower turnover of the other fat-soluble vitamins (weeks to months).

The mechanism of action for vitamin K became clear after it was demonstrated that the formation of γ-carboxyglutamic acid residues (GLA) in thrombin and other proteinases associated with the blood clotting cascade was vitamin K dependent (Suttie, 1995). The formation of GLA residues is a key, as they serve as calcium-binding sites in the proforms of proteinases associated with blood coagulation. Calcium binding is a requisite for their eventual activation. In this regard, vitamin K serves as cofactor for microsomal carboxylases that are responsible for GLA formation. The vitamin K-dependent carboxylase utilizes oxygen and bicarbonate as substrates. The reaction only occurs if glutamic acid is a part of a polypeptide with the correct sequence for specificity. Only the reduced form of vitamin K serves as a cofactor (Fig. 25.8), so a reductase system is necessary for vitamin K regeneration, and one of the intermediate forms is a vitamin K epoxide. As this pathway was resolved, it became apparent that many of the vitamin K antagonists functioned as inhibitors of reductases important for vitamin K regeneration.

GLA residues are also found in bone proteins. The GLA-containing proteins in bone (osteocalcins) appear to be involved in the regulation of new bone growth and formation. The presence of GLA protein in bone helps to explain why administration of the vitamin K antagonist at levels that cause hemorrhagic diseases also may result in bone defects, particularly in neonates. The mineralization disorders are characterized by complete fusion of the proximal tibia growth plate and cessation of longitudinal bone growth (Suttie, 1995).

3. Nutritional Requirements

The establishment of the dietary requirement for many animals has been difficult, in part due to (1) the short half-life of vitamin K, (2) the fact that large amounts of vitamin K may be synthesized by intestinal bacteria, and (3) the extent to which different

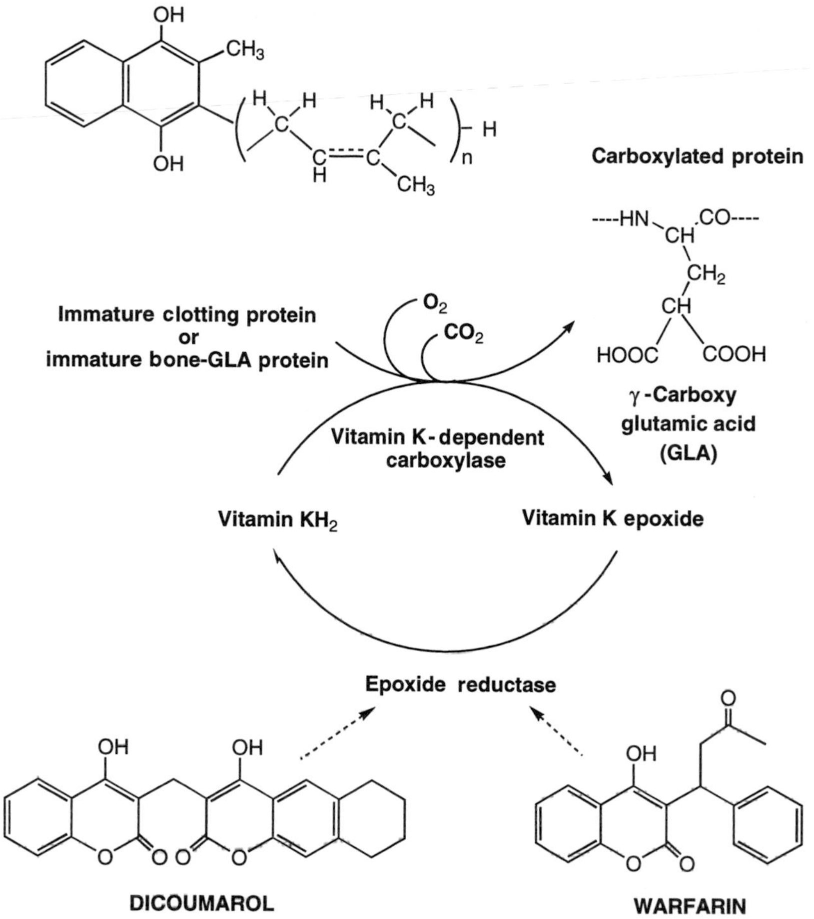

FIGURE 24.8 Vitamin K and its functions. Vitamin K consists of compounds with the general structure shown. Forms of vitamin K in the phylloquinone series (K_1) have a saturated side chain of isoprene units. In the menaquinone series, the isoprene side chain is unsaturated. The isoprene side chain can be as long as 20 carbons in length. Vitamin K functions as a cofactor for microsomal carboxylases, which act on immature clotting proteins in the liver or immature proteins in developing and metabolic active bone to cause the formation of γ-carboxyglutamic acid. During the course of carboxylation, vitamin K is converted to vitamin K epoxide. The epoxides may be converted back to vitamin K. The site of action of vitamin K antagonists such as dicoumarol and warfarin is the reductase regeneration system.

animal species practice coprophagy. Birds tend to have relative high requirements for vitamin K; thus, chickens are often used as experimental animals in vitamin K studies. Recent work suggests that the vitamin K requirement is dependent upon the relative content of vitamin K epoxide reductase activity. A low level of epoxide reductase activity can result in an increased requirement for vitamin K (Suttie, 1995). Ruminal microorganisms synthesize large amounts of vitamin K; thus, ruminants do not need an external source.

Recent assessments of nutritional requirements suggest that small animals should obtain approximately 500–1000 μg as phylloquinone per kg diet. Oxidized squalene and high intakes of vitamin E may act as vitamin K antagonists. Insufficient vitamin K can also occur with antibiotic therapy, treatment with coccidiostatic drugs, or long-term parenteral hyperalimentation without vitamin K supplements. Poultry and swine diets are regularly supplemented with menadione, but the need to supplement the diet of other species is questionable. Few hazards have been attributed to long-term ingestion of vitamin K in amounts of 1–10 mg per kg diet as phylloquinone. However, menadione in amounts corresponding to 10–100 mg per kg diet may act as a pro-oxidant, and high dietary concentrations produce hemolysis. Phylloquinone (vitamin K_1) rather than menadione should be used parenterally to treat animals that have ingested warfarin or other anticoagulants. Menadione, being water-soluble, produces hemolysis.

IV. WATER-SOLUBLE VITAMINS

We have chosen to organize the discussion of water-soluble vitamins based on physiological function. Most vitamins serve eventually as enzymatic cofactors. For example, niacin, riboflavin, and ascorbic acid serve primarily as redox cofactors (Friedrich, 1988; Combs, 1991). The roles of thiamin, pyridoxine (vitamin B_6), and pantothenic acid (as a component of coenzyme A) are distinguished because of their unique importance to carbohydrate, protein and amino acid, and acyl and acetyl transport, respectively. Biotin, folic acid, and vitamin B_{12} (cobalamin) will be discussed in relationship to their roles in single-carbon metabolism.

A. Ascorbic Acid

1. Introduction

Ascorbic acid functions primarily as a cofactor for microsomal mono-oxygenases (hydroxylases) and oxidases. In most animals, ascorbic acid is synthesized from glucose in the liver or kidney (Fig. 24.9). In some animals, however, a deficiency of gulonolactone oxidase, the last step in ascorbic acid synthesis, results in the need for a dietary source . Because of its importance in humans, vitamin C deficiency (scurvy) often determined the course of history. Outbreaks of scurvy influenced the outcome of military campaigns, explorations, and the ability to survive long ocean voyages. It was the search for the anti-scurvy factor that eventually lead to the isolation of ascorbic acid.

2. Chemistry

Ascorbic acid is the most powerful reducing agent available to cells and is of general importance as an antioxidant, because of its high reducing potential (Friedrich, 1988). However, under some conditions ascorbic acid can also act as an pro-oxidant. Ascorbic acid is 2,3-enediol-L-gulonic acid. Both of the hydrogens of the enediol group can dissociate, which results in the strong acidity of ascorbic acid ($pK_1 = 4.2$). Enediols are also excellent reducing agents; the reaction usually occurs in a stepwise fashion, with monodehydroascorbic acid as a semiquinone intermediate. This intermediate then disproportionates to ascorbic acid and dehydroascorbic acid. Dehydroascorbic acid is not as hydrophilic as ascorbic acid, because it exists in a deproteinated form. As such, this form of ascorbic acid can move easily across cell membranes. The dehydro form, however, is easily cleaved by alkali—for example, to oxalic acid and threonic acid.

3. Absorption, Tissue Distribution, and Metabolic Functions

Dietary ascorbic acid is absorbed from the duodenum and proximal jejunum (Friedrich, 1988). Measurable amounts can also cross the membranes of the mouth and gastric mucosa. Although some controversy exists regarding the relationship between ascorbic acid intake and the intestinal absorption of ascorbic acid, most careful studies indicate that within the physiological ranges of intake (20–400 mg per kg dry food), 80–90% of the vitamin may be absorbed.

In tissues, the highest concentration of ascorbic acid is found in the adrenal and pituitary glands, followed the liver, thymus, brain, and pancreas. Cellular uptake of ascorbic acid occurs by both active and simple diffusion processes. In diabetic animals, the ascorbic acid content of tissue is often depressed, which suggests that factors responding to hypergylcemic states can compromise ascorbic acid status.

Ascorbic acid is maintained in cells by several mechanisms. Ascorbate reductases maintain L-ascorbic acid in the reduced form, which prevents leakage from the cell as dehydroascorbic acid. Significant amounts of ascorbic acid, particularly in fish, may also exist as the 2-sulfate derivative. In rats, about 5% of a labeled dose of ascorbic acid is recovered in urine as 2-*O*-methyl ascorbic acid. The ability to modify ascorbic acid as the 2-sulfate or 2-*O*-methyl derivative as well as to oxidize ascorbic acid has considerable impact on the ability of cells to compartmentalize or modulate functional ascorbic acid levels. In the neonate, glutathione is important to ascorbate recycling and regeneration (Fig. 24.10). An argument can be made for a dietary need for ascorbic acid in the neonates of some species in which the adult does not normally have a dietary requirement. For example, the levels of glutathione are relatively low in neonate rat and mouse tissue. Ascorbate is oxidized to dehydroascorbic acid, which is further catabolized; hence the need for continual replacement.

As a cellular reducing agent, ascorbic acid plays a number of very important roles. It serves as a cofactor for mixed-function oxidations that result in the incorporation of molecular oxygen into various substrates (England and Seifer, 1986). Examples include the hydroxylation of proline in collagen, elastin, C1q complement, and acetylcholine esterase. Hydroxylases (monooxygenases) and some P450-dependent hydroxylases

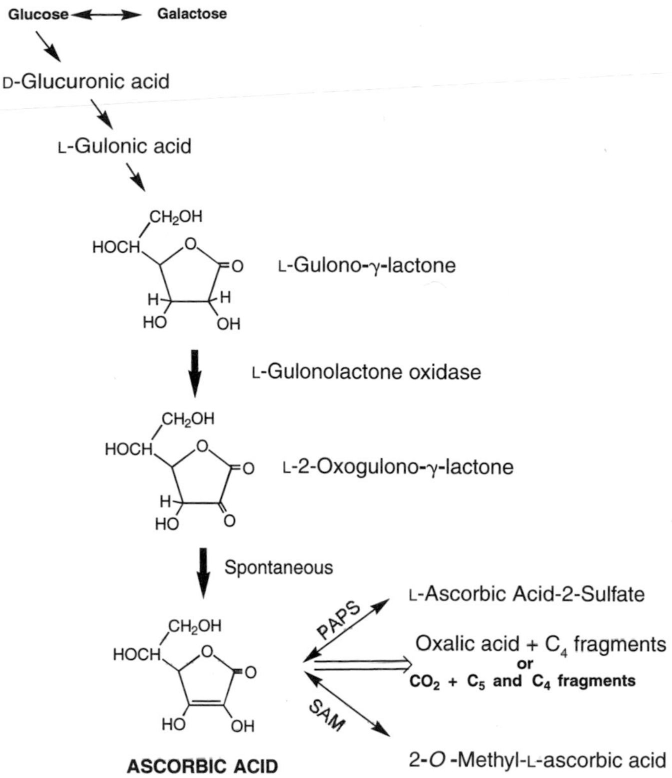

FIGURE 24.9 Ascorbic acid. Ascorbic acid is synthesized in most animals by conversion of glucose through the direct oxidative pathway to L-gulono-γ-lactone and eventually to L-2-oxogulonyl-γ-lactone, which spontaneously oxidizes to ascorbic acid. Also shown are important steps in the metabolism of ascorbic acid. Many cells are capable of converting ascorbic acid to L-ascorbic acid 2-sulfate or to 2-0-methyl-L-ascorbic acid. Phosphoadenosyl phosphasulfate (PAPS) is a requisite cofactor for L-ascorbic acid 2-sulfate synthesis. *S*-Adenosyl methionine (SAM) is the cofactor for 2-methyl-L-ascorbic acid formation. When ascorbic acid is in excess in cells, catabolic enzymes that decarboxylate and cleave ascorbic acid are induced.

that carry out the hydroxylation of steroids, drugs, and other xenobiotics also utilize ascorbic acid as a reductant. Moreover, the hydroxylation steps in the biosynthesis of carnitine and the hydroxylation of tyrosine in the formation of catecholamines represent other important catalytic functions of ascorbic acid. Most of the enzymes involved in these processes are metal-requiring enzymes, in which the role of ascorbic acid is to maintain the metal (usually Cu or Fe) in a reduced state (Fig. 24.11).

Ascorbic acid also been suggested to play a number of regulatory roles. Ascorbic acid influences histamine metabolism in some animals, particularly humans. There is an inverse correlation between ascorbic acid levels and serum histamine levels. Further, steps in the transcriptional regulation of certain proteins, such as the fibrillar collagens, also appear to be influenced by the presence or absence of strong reducing agents such as ascorbic acid (England and Seifer, 1986).

4. *Requirements and Toxicity*

Ascorbate is synthesized by most animals, with the exception of primates, guinea pigs, some snakes, fruit-eating bats, birds such as passerines, and salmonid fish. For these animals, impaired collagen synthesis is a principal feature of ascorbate deficiency, which is evidenced by capillary fragility, bleeding gums, delayed wound healing, and impaired bone formation. In fish kyphosis, scoliosis and gill pathology also occurs. Connective-tissue lesions are primarily a result of underhydroxylated collagen (at specific prolyl and lysyl

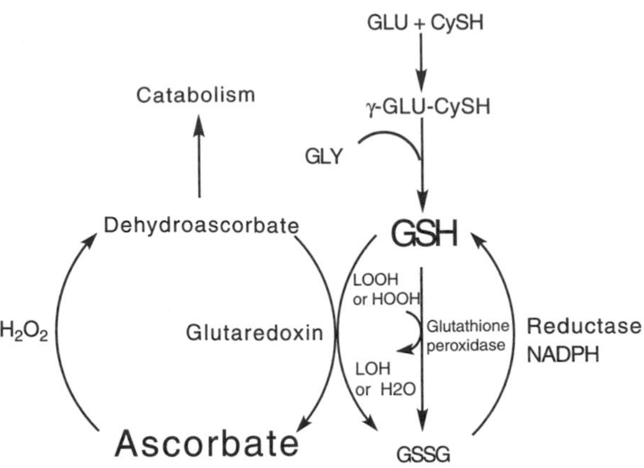

FIGURE 24.10 The relationship between glutathione and ascorbic acid. Glutathione is the most important cosubstrate for reactions catalyzed by glutathione peroxidase. Glutathione (GSH) is also important in maintaining ascorbic acid in its reduced form by serving as a cosubstrate for glutaredoxin.

residues) being abnormally susceptible to degradation. In addition, the inability to deal with metabolic stress requiring normal adrenal gland function and the reduced ability to metabolize fatty acids (carnitine synthesis) contribute to signs of scurvy.

To maintain the functions requiring ascorbate, most animals generate 10–60 mg of ascorbic acid per 1000 Kcal utilized in the course of normal metabolism. Similarly, requirements for ascorbic acid, when needed in the diet, range from 50 to 250 mg per kg diet. When ascorbic acid is consumed in gram amounts (per kilogram of diet), gastrointestinal distress and bleeding may occur. Also, when fed in excess of metabolic need, tissue levels of ascorbic acid are homoeostatically maintained. Homeostasis occurs through the induction of ascorbic acid decarboxylases and cleavage enzymatic activity, which results in enhanced degradation of ascorbate to CO_2 plus ribulose or oxalic acid plus threonic acid. These conversions are probably to protect cells against nonspecific and Fenton reactions.

B. The B Vitamins: Niacin

1. Introduction

Throughout out the 18th and 19th centuries, the disease pellagra was prevalent in western Europe and the southern region of the United States. This disease would eventually become associated with the consumption of corn (maize). Through the elegant work of Goldberger and others, pellagra was then linked to a dietary deficiency of niacin. Dogs played an important role as models for pellagra, as they exhibit a condition designated "black tongue" when given a diet similar to one that produced pellagra in humans. Black tongue is characterized by initial reddening of the mucosa of the lips and mouth, which progresses to necrosis of the mucosa accompanied by ropy saliva, a fetid odor, and diarrhea (Henderson, 1983).

Niacin deficiency is a consequence of consumption of foodstuffs, such as corn, that are low in bioavailable niacin and the amino acid tryptophan. Tryptophan is important to niacin status, because niacin can be generated upon tryptophan degradation (Fig. 24.12). Although niacin deficiency is observed infrequently in free-ranging animals, it nevertheless is a good example of a vitamin-related disease that occurs from consuming a monotonous diet. In corn, the bioavailability of niacin is poor unless the corn (maize) grain is finely ground and processed under alkaline conditions, for example, ground in the presence of limestone. This was not the practice in the human populations of western Europe and the southern United States that developed pellagra, although it was the practice in Central and South America. Normally, niacin is derived from food by hydrolysis of NAD (nicotinamide adenosyl dinucleotide) and NADP (nicotinamide adenosyl phosphodinucleotide) to niacin by the action of pancreatic or intestinal nucleosides and phosphatases (Fig. 24.1). As most animals consume diets that contain adequate tryptophan and available NAD and NADP, niacin deficiency is usually not a problem. An exception to this generalization is cats. In this species the degradation of tryptophan does not proceed along a pathway that leads to nicotinic acid, even though all the enzymes for the pathway are present. High activity of the enzyme picolinic carboxylase at a branch point in the pathway results in diversion from eventual NAD production. For cats and probably all other felids, but not dogs, available niacin is an obligatory dietary factor.

2. Functions

NAD and its phosphorylated form, NADP, are two coenzymes derived from niacin. Both contain an unsubstituted pyridine 3-carboxamide that is essential to function in redox reactions with a chemical potential near −0.32 V. Virtually all cells are capable of converting niacin to NAD. Most enzymes that require NAD are oxidoreductases (dehydrogenases) that aid the catalysis of a diverse array of reactions, such as the conversion of alcohols and polyols to aldehydes or ketones (Fig. 24.13). The most common mechanisms involve the stereospecific abstraction of a hydride ion (H:) from the substrate with its subsequent transfer. Cells generally delegate NAD to enzymes in catabolic path-

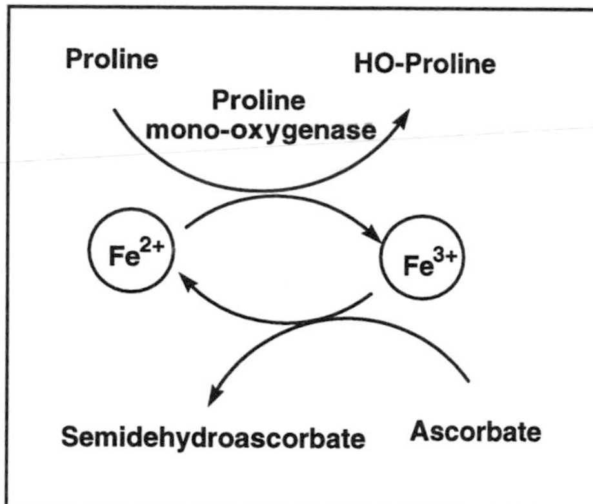

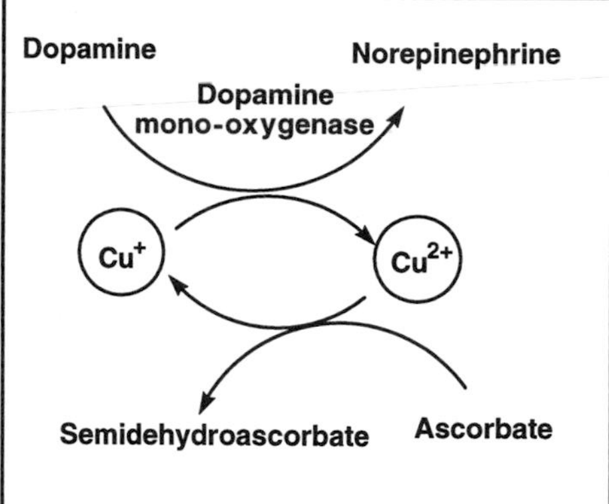

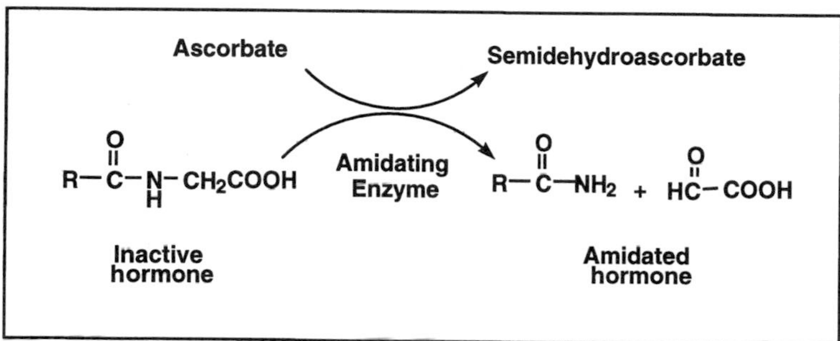

1

2
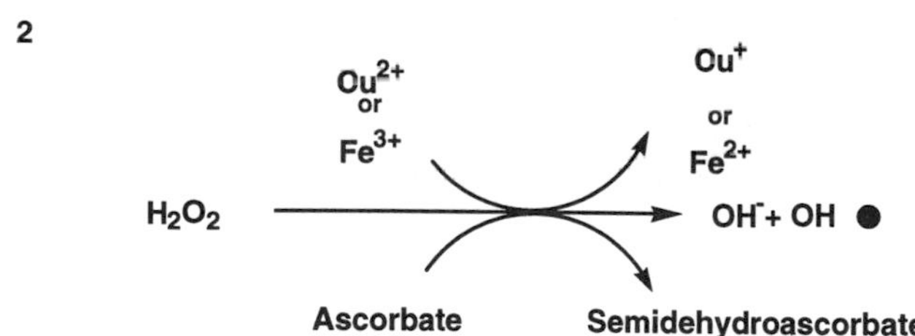

FIGURE 24.11 Selected functions of ascorbic acid. Ascorbic acid is a common mono-oxygenase cofactor. Shown are reactions typical of those for prolyl hydroxynase (mono-oxygenase), dopamine dehydrogenase, and amidating enzyme. Further, ascorbic acid can act as a reductant (reaction 1) or pro-oxidant (reaction 2).

ways, whereas NADP is utilized in synthetic pathways. An additional and equally important function of NAD is its role as a substrate in poly- and monoribosylation reactions. Mono- and polyribosylations are important to many cellular regulatory functions. Enzymes that undergo monoribosylation can become activated or deactivated upon addition of ADP–ribose. Somewhat analogous to phosphorylation, ribosylation represents another example of covalent modification as a regulatory control. In the nuclei of cells, polyribosylation of histones precedes the normal process of DNA repair. This latter phenomenon is important in that pellagra-related lesions of skin following exposure to UV light. UV damage of epidermal cell DNA is an underlying mechanism for the dark-pigmented lesions associated with pellagra. Lack of niacin and therefore NAD is thought to be contributing factor to the skin lesions because of the inability of cells to carry out polyribosy-

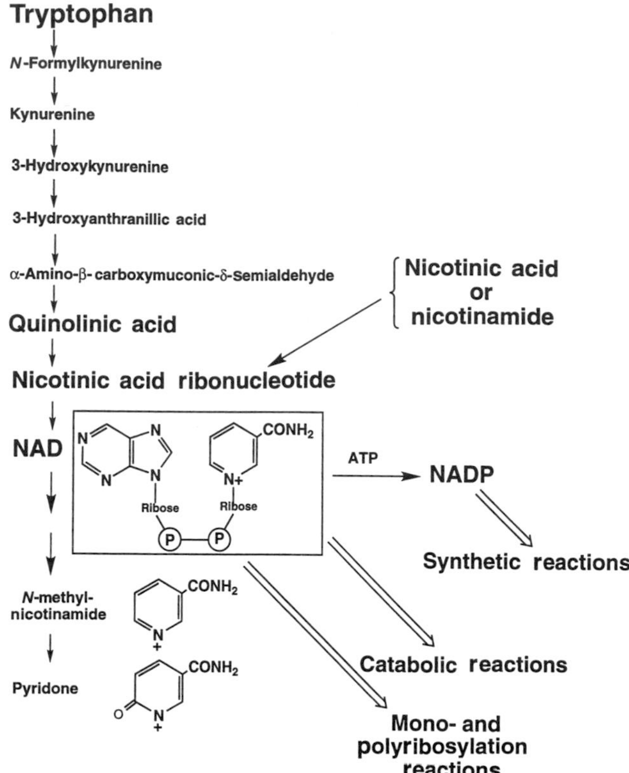

FIGURE 24.12 Metabolism of tryptophan to niacin and NAD. NAD may be derived from tryptophan through the quinolinate pathway. Tryptophan, however, is not a precursor for NAD in those animals that divert α-amino-β-carboxymuconic-δ-semialdehyde to alternative products, e.g., picolinic acid, rather than toward quinolinic acid formation. For example, in cats, nicotinic acid or nicotinamide is needed for NAD production. Note that the phosphorylated form of NAD (NADPH) is used primarily for synthetic reactions. NAD serves as a cofactor for catabolic reactions and as a substrate in mono- and polyadenylation reactions. When NAD is degraded, the nicotinamide component can be methylated or oxidized to pyridone. In birds, more complex conjugated forms of nicotinamide are observed (not shown).

lation reactions. It is this nonredox function of NAD that accounts for the rapid turnover of NAD in cells. Some estimates suggest that as much as 40 to 60% of the NAD in cells is involved in mono- or polyribosylation reactions (Weiner and Eys, 1983; Poirier and Moreau, 1992).

3. *Requirements and Pharmacology*

Niacin is needed in amounts corresponding to 10–25 mg/kg of diet. Depending on species, the conversion of tryptophan to niacin produces about 1 mg of niacin for every 60 mg of tryptophan degraded. In some animals, e.g., cats, high levels of the enzyme picolinic carboxylase causes conversion of α-amino-β-carboxymuconic-δ-semialdehyde to picolinic acid rather than quinolinic acid; cats therefore have an obligatory requirement for niacin (*see* Fig. 24.12). Niacin (nicotinamide) is relatively nontoxic, although nicotinic acid can cause vasodilation when consumed in excess of 100 mg per kg diet. Consequently, there are a number of therapeutic uses for pharmacologic doses of niacin-derived compounds, when increased blood flow is desirable.

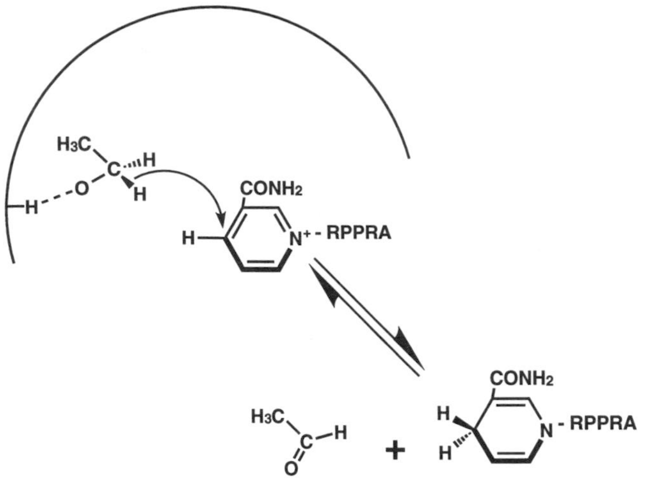

FIGURE 24.13 Role of NAD in the oxidation of ethanol to acetaldehyde.

4. *Determination of Niacin Status*

A nicotinamide loading test has been used to determine niacin status of patients. The patient is given an oral dose of nicotinamide and the urinary excretion products, which are species dependent, are measured. For humans, monkeys, dogs, rats, and swine, largely methylated products are produced.

C. The B Vitamins: Riboflavin

1. *Introduction*

Riboflavin was one of the first of the B vitamins identified (Fig. 24.14). Originally, it was thought to be the heat-stable factor responsible for the prevention of pellagra. Riboflavin is present in tissue and cells as FAD (flavin adenine dinucleotide) and FMN (flavin adenine mononucleotide). FAD and FMN are cofactors in aerobic processes, usually as cofactors for oxidases, although FAD also can function in anaerobic environments as a dehydrogenase cofactor (Cooperman and Lopez, 1991). Many flavin-containing proteins are found in the smooth endoplasmic reticulum of cells, for example, as microsomal enzymes.

2. *Functions*

Enzymes containing flavins are distinguished because they are capable of transferring hydrogen directly to molecular oxygen with the formation of hy-

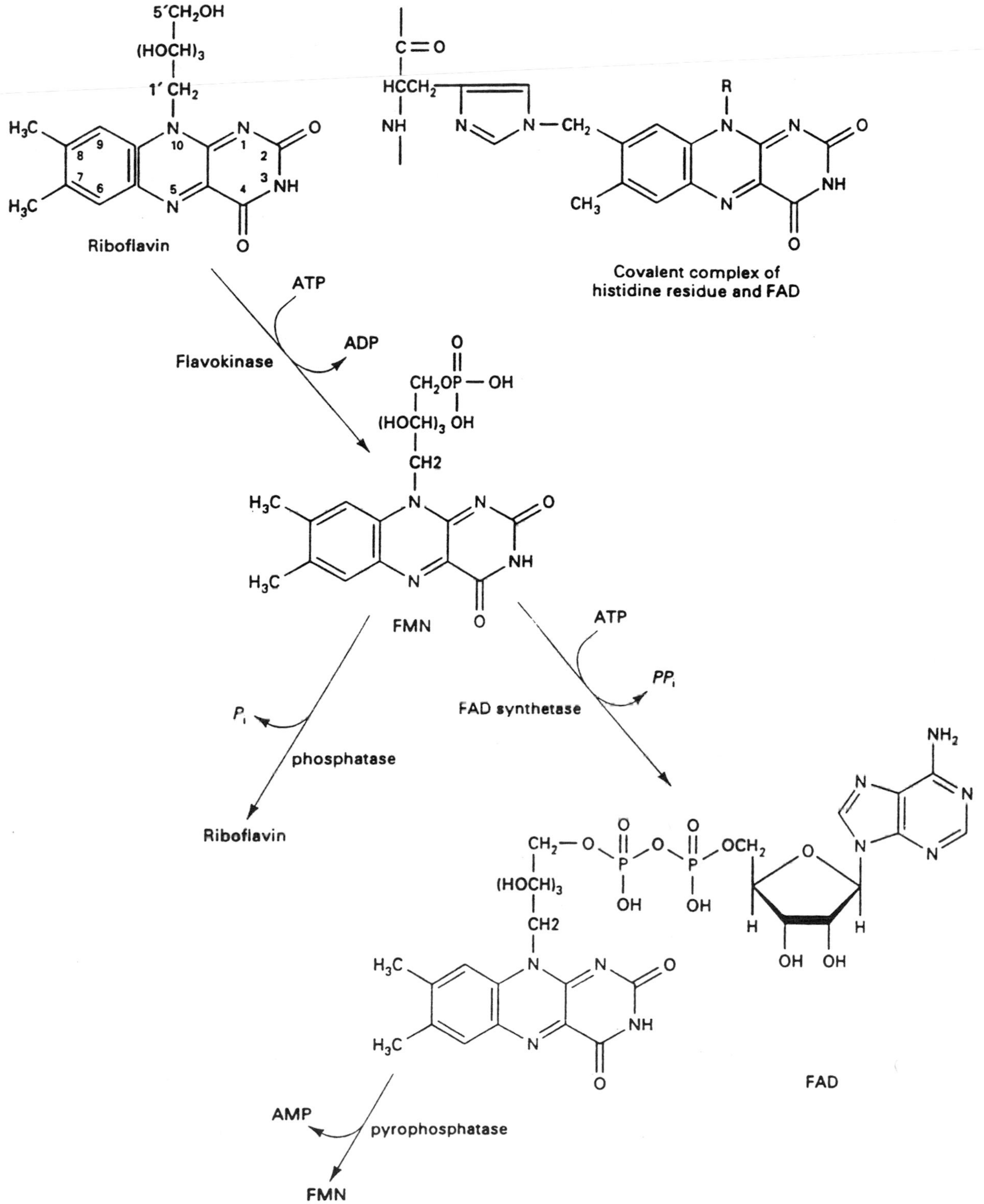

FIGURE 24.14 Riboflavin metabolism. The major forms of riboflavin found in cells are flavin adenine mononucleotide (FMN) and flavin adenine dinucleotide (FAD). In some enzymes, riboflavin may also be covalently bound to the enzymes that it serves, for example, succinic dehydrogenase.

drogen peroxide as a product. Oxygen preferentially participates in reactions involving one electron, one hydrogen transfer proceeding in a stepwise manner. The chemical characteristics of riboflavin are ideally suited for such reactions. Riboflavin-containing cofactors permit biological systems to carry out a range of redox reactions utilizing mechanisms that involve ion hydride transfers (via NAD or NADP), radical hydrogen ion transfers (via FMN or FAD) or ascorbic acid, and one-electron plus one-proton transfers (via FMN or FAD).

3. *Metabolism and Requirements*

FMN and FAD in foods is hydrolyzed in the upper gut to free riboflavin. Riboflavin is absorbed by active processes and is transported in blood to target tissues in association with albumin. Once in cells, riboflavin is phosphorylated to FMN. FMN is also released from cells and may bind to albumin for reutilization by other cells. Active-transfer mechanisms are responsible for the uptake of FMN. In this regard certain drugs, such as penicillin and theophylline, can displace riboflavin from binding proteins that are important to its transport. Urine is the major route of excretion for riboflavin, although some FAD is excreted in bile.

Requirements for riboflavin are lower than those for niacin or ascorbic acid, primarily because riboflavin is tightly associated with the oxidases and dehydrogenases it serves as cofactor. Riboflavin turnover is dependent on the turnover of the proteins to which it is associated. In some cases, FMN is covalently bound, such as in succinic dehydrogenase. Therefore, in most animals the half-life of riboflavin is several weeks to several months, in contrast to days to several weeks for niacin. Most small animals require 2–6 mg riboflavin per kg diet.

Riboflavin deficiency signs include lesions of the oral cavity, the lips, and the angle of the mouth (cheilosis), inflammation of the tongue (glossitis), and accompanying seborrheic dermatitis. In severe cases of riboflavin deficiency, the filiform papillae of the tongue are lost and the tongue changes color from its usual pink to magenta. Anemia and increased vascularization of the eye are present in some animals with riboflavin, deficiency. As grains are a poor source of riboflavin, deficiencies frequently occur in animals given diets predominantly based on cereal grains. One of the more striking and specific signs of riboflavin deficiency in birds is "curled toe syndrome." Curled toe paralysis has been of economic significance to the broiler industry. In young animals, growth failure, loss of feathers, or alopecia has also been observed.

4. *Determination of Riboflavin Status*

The erythrocyte glutathione reductase activity coefficient (EGRAC) is the preferred clinical test of riboflavin adequacy. This is an enzyme stimulation test and measures the reduction of oxidized glutathione by the enzyme glutathione reductase with and without the addition of exogenous FAD. In humans a ratio of 1.0 to 1.3 is normal and a ratio between 1.3 and 1.8 indicates deficiency. In dogs a ratio of greater than 1.3 has been taken to indicate deficiency.

D. B Vitamins Directed at Specific Features of Carbohydrate, Protein, or Lipid Metabolism

1. *Thiamin*

a. Introduction

Studies of thiamin played an important role in the development of early concepts of the function and importance of vitamins. The demonstration by the Dutch medical officer Christian Eijkman that polyneuritis, which was associated with human beriberi, could also be produced in an experimental animal by dietary manipulation presaged the use of animal models in the study of vitamins (Friedrich, 1988). Eijkman and his colleagues fed a diet of polished rice, presumably low in thiamin, to chickens and observed the characteristic feature of head retraction. The focus on rice and the observation that there appeared to be a curative principle in rice bran led to the eventual isolation of thiamin. This sequence of discoveries formed the basis of the discovery of vitamins as precursors to cofactors and their roles as regulators (Sable and Gubler, 1982).

b. General Functions

Thiamin in cells occurs either as the pyrophosphate (TPP) or the triphosphate, TPPP (Fig. 24.15). There are two general types of reactions wherein TPP functions as a magnesium-coordinated coenzyme for active aldehyde transfer reactions (Fig. 24.16). One example is the decarboxylation of α-keto acids, designated as a transketolase reaction. Decarboxylation of α-keto acids occurs twice in the tricarboxylic acid cycle (TCA): in the conversion of pyruvic acid to acetyl-CoA and in the conversion of α-ketoglutarate to succinyl-CoA. The other reaction of TPP is the facilitation of the transformation of "ketols" (ketose phosphates) in the pentose phosphates pathway. In this pathway, NADP is reduced to NADPH, an essential reducing agent for synthetic reactions (see the discussion of niacin). A deficiency of thiamin impairs metabolism of carbohydrates because of defective TCA cycle regulation. Further,

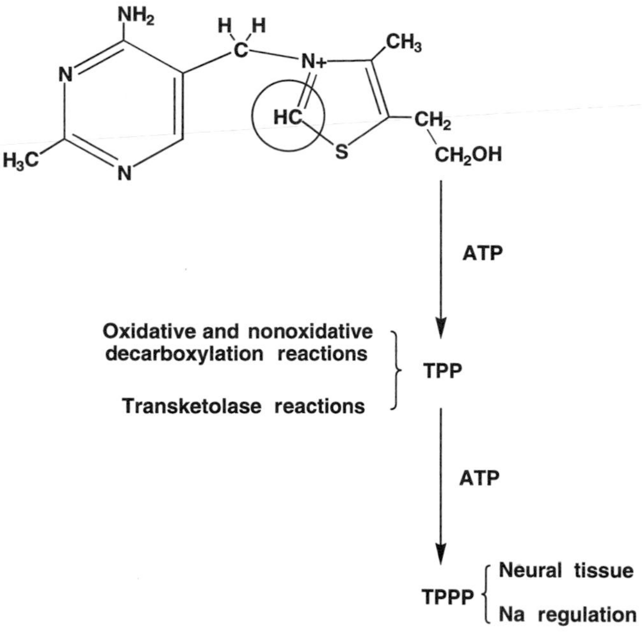

FIGURE 24.15 Thiamin. The five-membered (thiazole) ring of thiamin contains the (-N = CH − S-) arrangement of atoms called an ylid. The ylid arrangement results in a carbon with considerable negative charge, that is a carbanion. The circled carbon essentially acts as an electron-rich center for reactions that are commonly characterized as decarboxylations and transketolations. The cofactor form of thiamin is either as the diphosphate (TPP) or the triphosphate (TPPP). TPP predominates in neural tissue.

perturbations in the pentose phosphate-related carbohydrates pathway lead to decreased production of NADPH, which affects synthetic processes such as fatty acid biosynthesis (Bender, 1992).

TPPP predominates in neural tissue and in brain. In the brain, TPPP is proposed to be involved in sodium gating processes, that is, the flux of sodium ions across neuronal cell membranes. This aspect of thiamin metabolism may be related to the psychosis and impairment of neuromuscular control that is observed in humans with thiamin deficiency.

c. Requirements

Thiamin status should be routinely considered in disease assessment, because a number of factors influence thiamin availability and may induce a deficiency. Because thiamin is heat and alkali labile, extensive destruction of thiamin can occur in the various steps of food processing and preservation. Canning, which involves prolonged elevated temperatures and often alkaline conditions, can result in very low recoveries of thiamin in the final product. Thiamin can also be destroyed enzymatically by thiaminases, which are abundant in the flesh of some fish, particularly spoiled fish, and bacteria associated with fermentation processes. Thiamin deficiency has been observed in fish-eating birds, when spoiled or uncooked fish has been fed. Thiaminase activity is strikingly high in tuna and sardines, catalyzing the inactivation of milligram quantities of thiamin per hour per gram of fish muscle. Thiamin deficiency induced by thiaminases has also been reported in foxes fed uncooked fish products and in cats both given fresh fish, as well as in cats given canned cat food that has suffered excessive processing losses.

In herbivores, thiamin deficiency has occurred following ingestion of bracken fern (*Pteridium aquilinum*), horsetail fern (*Equisetum arvense*), or nardoo (*Marsilea drumen*), which contain thiaminases. In both herbivores and simple-stomached animals, predominant characteristic of thiamin deficiency is polioencephalomalacia. Primarily, lesions of the brain are hemorrhages of the periventricular gray matter and pathology of the inferior colliculi, medial vestibular, and lateral geniculate nucleii. A relationship between excessive production of hydrogen sulfide in the rumen of cattle and sheep and polioencephalomalacia has been demonstrated. The bioavailability of thiamin in foodstuffs is reduced by high levels of tannins. As a general requirement, animals should receive from 4 to 10 mg of thiamin per kg dry food (Combs, 1991).

d. Determination of Thiamin Status

Traditionally, the erythrocyte transketolase saturation test, which is a measure of the stimulation of the transketolase reaction, has been used to assess thiamin status. A stimulation of greater than 16% has been taken as deficiency. A more sensitive test, however, is the measurement of thiamin phosphorylated esters in plasma: The level of these esters declines in plasma before any change occurs in erythrocyte transketolase values. Thiamin loading tests that measure the urinary excretion of thiamin following an oral dose of thiamin have also been used, but lack the sensitivity of the measurement of esters in blood (Sable and Gubler, 1982).

2. *Pyridoxine*

a. Introduction

Vitamin B_6 is a collective term for pyridoxine, pyridoxal, and pyridoxine (Fig. 24.17). Pyridoxine is most abundant in plants, and pyridoxal and pyridoxine are most abundant in animal tissues (Combs, 1991; Dakshinamurti, 1990). Each of these compounds can be interconverted. The active form pyridoxal (B_6 PLP) is phos-

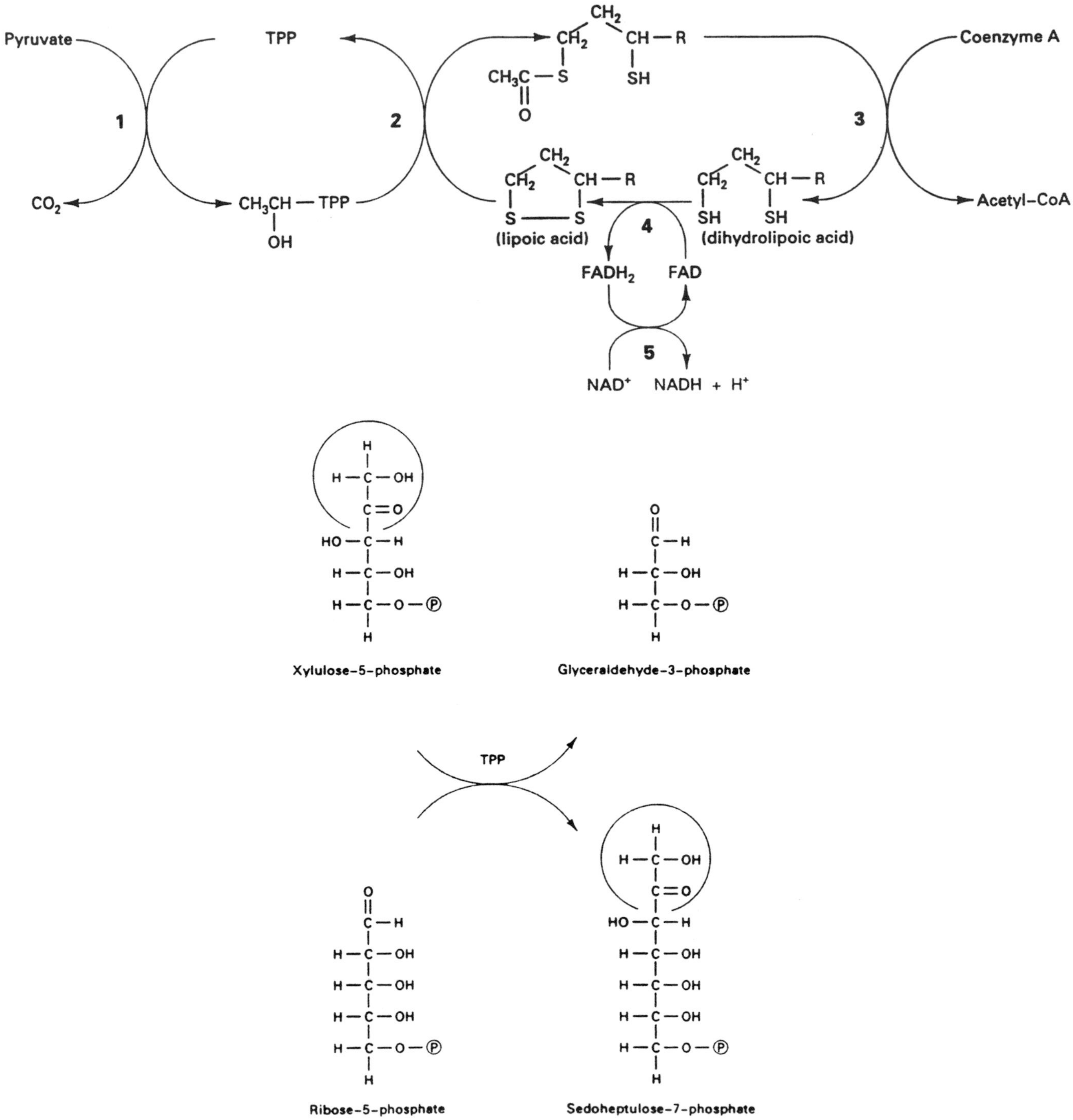

FIGURE 24.16 Reactions involving thiamin. The oxidative decarboxylation of pyruvic acid, and the transketolation reaction, such as the formation of sedoheptulose-7 phosphate and glyceraldehyde-3-phosphate from xyulose-5-phosphate and ribose-5-phosphate, are classic examples of reactions involving thiamin.

phorylated. When pyridoxal-5-phosphate is in excess, it can be converted to pyridoxic acid, which is excreted.

The reactions carried out by vitamin B_6 fall into three general categories related to the metabolism and interconversion of amino acids (Bender, 1992; Coburn *et al.*, 1984; Coburn, 1994). The most common of these reactions is the transaminase reaction. Transaminations are essential to the interconversion of amino acids

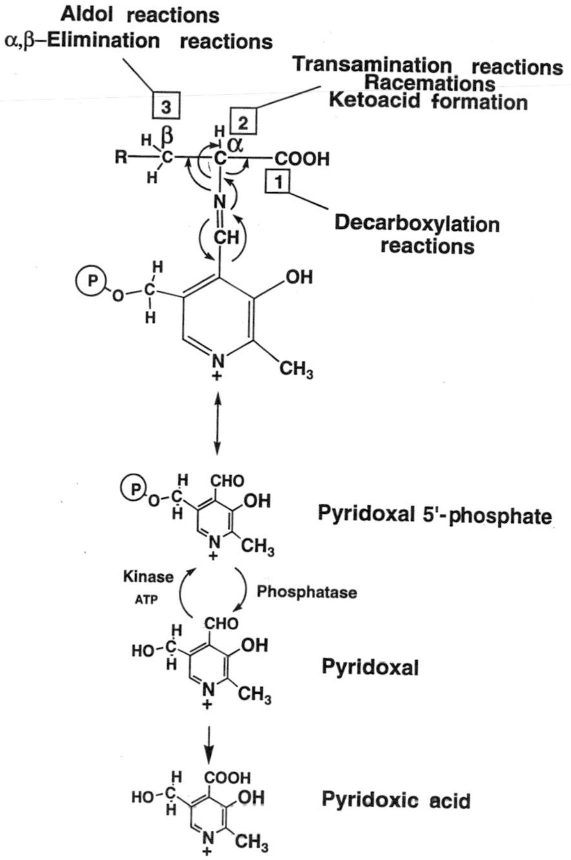

FIGURE 24.17 Vitamin B_6. Vitamin B_6 is a collective term for pyridoxal and the amidated and reduced derivatives pyridoxamine and pyridoxine (structures not shown). The active form of pyridoxal is phosphorylated and carries out decarboxylation reactions, transamination reactions (including racemations and keto acid formations and aldol reactions (including α–β elimination reactions).

and corresponding α-keto acids. The transamination mechanism also applies for reactions important to producing racemic amino acid mixtures, for example, the conversion of L-alanine to D-alanine, and α,β-addition or elimination reactions. Examples of α, β-elimination reactions are the conversion of serine to pyruvic acid and the conversion of homocysteine plus serine to cystathionine. The basic feature of the transamination-type mechanism involves electron withdrawal from the α-carbon, resulting in a proton liberation that sets the stage for substitution and addition reactions (Fig. 24.17).

The second most common reaction involves electron withdrawal from the α-carbon and carboxylic acid group carbon, which facilitates decarboxylation. Decarboxylation reactions include the conversions of tyrosine to tyramine, 5-hydroxytryptophan to serotonin, histidine to histamine, and glutamate to γ-aminobutyric acid (GABA). The convulsions associated with vitamin B_6 deficiency are attributed to insufficient activity of PLP-dependent L-glutamate decarboxylase, leading to a deficit of the inhibitory neurotransmitter GABA.

A third type of reactions involves electron withdrawal from the α- and β-carbons of amino acids. This sets the stage for hydride condensations or aldol reactions. A good example of an aldol reaction is the conversion of serine to glycine with the transfer of the β-carbon (as formaldehyde) to another vitamin cofactor, tetrahydrofolic acid. An excellent example of an hydride condensation is the formation of α-aminolevulinic acid, the first step in heme biosynthesis.

Another important function of vitamin B_6 (as pyridoxal 5′-phosphate), independent of amino acid metabolism, is its role in glycogen phosphorylase (Fig. 24.18). Glycogen phosphorylase catalyzes the hydrolysis of ether (α-1,4) bonds in glycogen to form glucose-6-phosphate. Ether bonds are best catalyzed through acid-mediated mechanisms. The acid proton in this instance is derived from the phosphate group of pyridoxal 5′-phosphate. This important mechanism was elucidated relatively recently. Previously, it was speculated that the association of vitamin B_6 with glycogen phosphorylase was primarily some type of storage mechanism (Palm *et al.*, 1990; Coburn, 1994).

b. Metabolism and Requirements

The requirement of vitamin B_6 by animals is positively related to their intake of protein and amino acids (Coburn,1994). Vitamin B_6 deficiency is rarely seen in animals, as most diets provide adequate amounts. Normally, vitamin B_6 is needed in amounts that range from 2 to 6 mg/kg diet. Ruminants and many herbivores meet a substantial part of their vitamin B_6 requirement from intestinal microbial synthesis. Drug-induced vitamin B_6 deficiency can occur following administration of the tuberculostatic drug isoniazid (isonicotinic acid hydrazide). This drug forms a hydrazone derivative with pyridoxal or pyridoxal phosphate that inhibits the pyridoxal-containing enzymes. Patients receiving long-term isoniazid therapy respond to the administration of supplemental vitamin B_6. Penicillamine (β,β-dimethylcysteine) used in the treatment of Wilson's disease also induces a vitamin B_6 deficiency because of the formation of an inactive thiazole derivative. A naturally occurring antagonist to vitamin B_6, linatine (1-amino-D-proline), is present in flax seed and forms a stable product with pyridoxal phosphate. The most important signs of B_6 deficiency relate to defects in amino acid metabolism. Neurological signs occur as

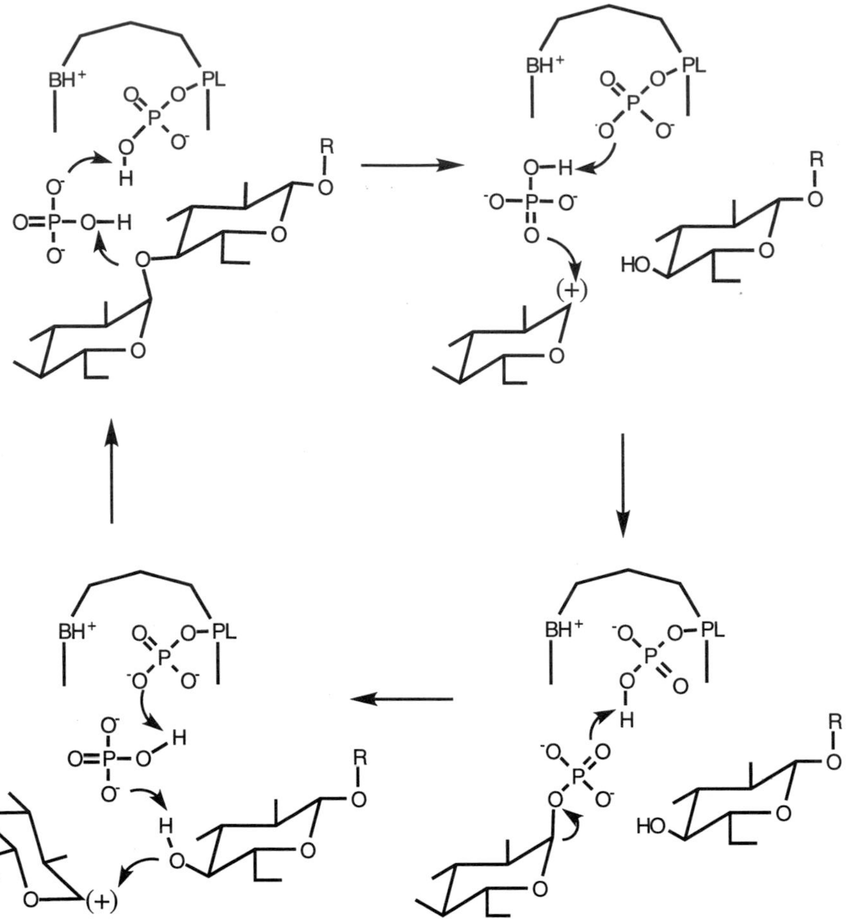

FIGURE 24.18 The role of vitamin B_6 in glycogen phosphorylase. Glycogen phosphorylase catalyzes the hydrolyses of ether bonds in glycogen to form glucose-6-phosphate. Ether bond cleavage is best catalyzed by acid-mediated mechanisms. In this case the acid proton comes from the phosphate group of pyridoxal-5-phosphate.

result of reduced synthesis of important biogenic amines from amino acid precursors, and anemia is a consequence of decreased heme synthesis. Some animals, such as cats, have a marked oxaluria when deficient in vitamin B_6.

c. Determination of Vitamin B_6 Status

A number of tests have been used as an index of vitamin B_6 status. These include measurement of the activities of enzymes that require pyridoxal phosphate as a coenzyme, such as kynureninase and aminotransferases. However, the most sensitive method involves the direct measurement of pyridoxal and/or pyridoxal phosphate in plasma. The relative ratio of these two forms and their response to dietary intake of pyridoxine depends on the species of animal.

3. Pantothenic Acid

a. Introduction

Pantothenic acid was discovered in 1933 by Roger William, who observed that it was an essential growth factor for yeast and lactic acid bacteria. Later, Elvehjen, Jukes, and others demonstrated pantothenic acid to be essential for animals. Pantothenic acid is a component of coenzyme A (Fig. 24.19). Pantothenic acid (as a part of phosphopantothene) is also present at the active site of acyl carrier protein (ACP), a component of the fatty acid synthesis complex, as phosphopantoteine. Both forms are present in foods. Consequently, absorbed pantothenic acid must first be released from coenzyme A and ACP, steps that involve the actions of peptidases and nucleosidases (Combs, 1991; Fox, 1991).

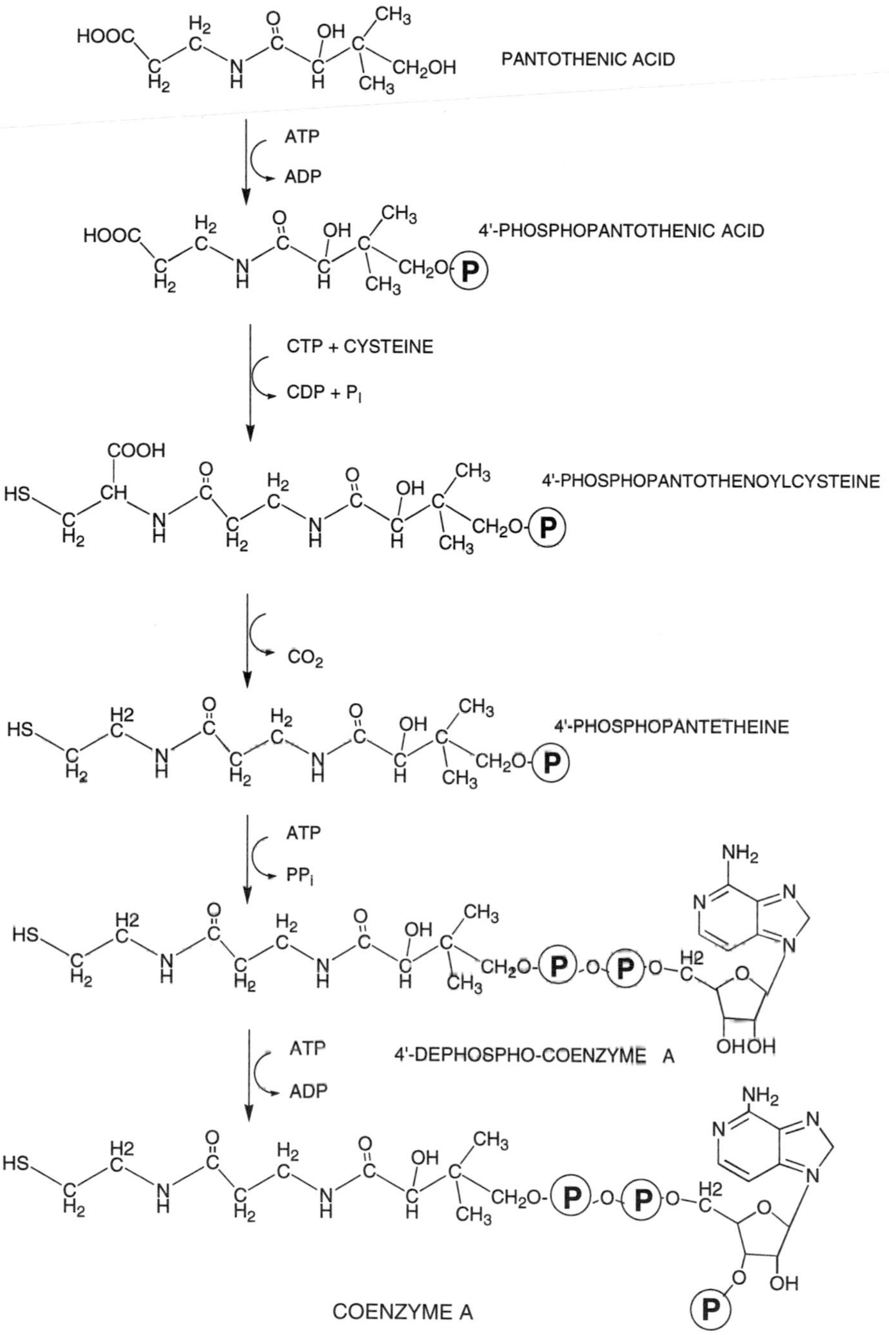

FIGURE 24.19 Pantothenic acid and coenzyme A.

b. Metabolism Functions and Requirements

CoA is the principal moiety for the vectoral transport of acyl and acetyl groups in synthetic and catabolic reactions, and a deficiency is characterized by impaired acetyl and acyl metabolism. The ability to utilize fatty acids as fuels is compromised. There is also an increased production of short-chain fatty acids and ketone bodies, which can lead to severe metabolic acidosis. In addition, dermal lesions occur because of impaired fatty acid metabolism. For most animals, the requirement for pantothenic acid is 10–20 mg per kg diet.

c. Assessment of Pantothenic Acid Status

Because of the widespread occurrence of the vitamin in foods, deficiencies are extremely rare. Urinary output of pantothenate is directly proportional to dietary input and has been used as an index of adequacy. For humans, mean excretion is 4 mg/day, and <1 mg/day is abnormally low.

E. Vitamins Involved in Single-Carbon Metabolism

1. Biotin

a. Introduction

Biotin functions in enzymatic carboxylation reactions as a cofactor for the CO_2-fixing enzymes: acetyl-CoA carboxylase, which is essential for fatty acid synthesis; propionyl-CoA carboxylase, which participates in odd-chain fatty acid metabolism; and pyruvate carboxylase (Bonjour, 1991), which is involved in the formation of oxaloacetate, an important obligatory step in reverse glycolysis and gluconeogenesis. Biotin is also involved as the coenzyme in the carboxylation of β-methylcrotonyl-CoA (a product of leucine metabolism).

b. Metabolism and Requirements

Biotin is found in highest concentrations in the liver. In food, biotin is present in relatively high concentrations in cereals, including soybeans, rice , barley, oats, corn, and wheat. Biotin is covalently bound to the enzymes that it serves as cofactor; the chemical linkage is to a peptide bond between the carboxylic acid moiety on biotin and the ε-amino function of peptidyl lysine in the enzyme. The biotin–enzyme peptide bond requires an ATP-dependent step (Fig. 24.20).

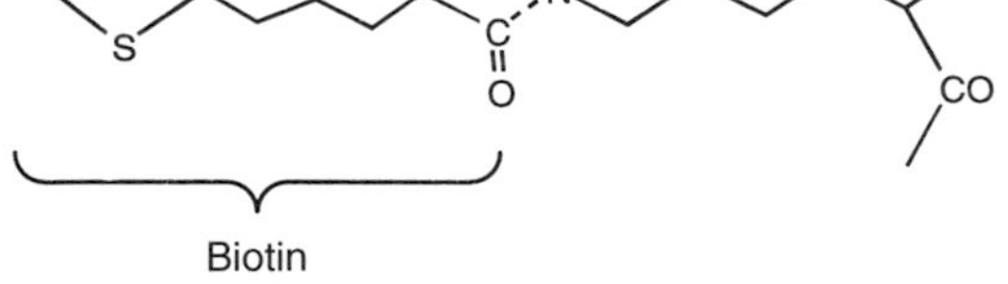

FIGURE 24.20 Biotin. Biotin is covalently bound in carboxylases and transcarboxylases by peptidyl linkage between the carboxylic acid moiety of biotin and the ε-amine function of peptidyl lysine. The biotin–lysine adduct is called biocytin. Biocytin can be cleaved by the enzyme biocytinase to generate free biotin.

When biotin-containing carboxylases are degraded, biotin is released as biocytin. Biocytinase is an important liver enzyme, that catalyzes the cleavage of the peptide linkage between biotin and lysine to release free biotin for reutilization. In the absence of the enzyme, biotin deficiency occurs in the newborn. The biotin requirement of animals is relatively low (in the microgram per kilogram range), which is due to at least three considerations. Biotin is covalently attached to a limited number of enzymes that it serves as a cofactor, it is extensively reutilized, and some biotin is produced by the gut microflora. Nevertheless, nutritional problems associated with biotin status can occur. Biotin and biocytin have a high affinity for certain proteins, particularly avidin in egg white. The consumption of raw egg albumin can induce biotin deficiency because the strong association of biotin with avidin renders dietary biotin unavailable. Ingestion of significant quantities of raw egg white by fur-bearing animals and pigs has lead to the condition described as "egg-white injury."

Biotin deficiency leads to impairment of gluconeogenesis and fat metabolism. Biotin deficiencies can also induce severe metabolic acidosis. The inability to carry out fat metabolism also markedly affects the dermis with biotin deficiency. Alopecia and dermatitis are characteristics of biotin deficiency in most animals and birds.

c. Assessment of Biotin Status

Deficiencies of biotin in the absence of consumption of raw egg white (avidin) are rare. Biotin has been measured in biological samples by microbiological assay methods following proteolytic digestion. Functional measurements of carboxylase activity (e.g., propionyl-CoA carboxylase) in liver samples have confirmed deficiencies in experimental animals.

2. Folic Acid and Vitamin B_{12}

a. Introduction

Knowledge of folic acid and B_{12} evolved from efforts to better understand macrocytic anemias and certain degenerative neurologic disorders (Brody *et al.*, 1991; Friedrich, 1988). The Scottish physician Combe recognized in the early 1800s that a certain form of macrocytic anemia appears related to a disorder of the digestive organs. In classic studies by Minot and Murphy, Castle, and others, it became clearer that the disorder was associated with gastric secretions and in some cases could be reversed by consuming raw or lightly cooked liver. Through careful clinical investigations and inferences, Castles postulated the existence of an intrinsic factor in gastric juice that appeared to combine with a dietary extrinsic factor to modulate the severity of the anemia (Bender, 1992; Combs, 1991).

In parallel studies, folic acid was also associated with macrocytic anemia. Large-scale efforts by a number of pharmaceutical companies throughout the 1940s and 1950s and careful clinical and basic studies at academic institutions eventually led to the isolation of folic acid and vitamin B_{12} and our current understanding of the novel interactions between these two vitamins, as well as the complexity of their chemistry and unique features associated with absorption and transport.

b. Chemistry and Functions

The structures of folic acid and vitamin B_{12} are given in Fig. 24.21 and Fig. 24.22. Folic acid is a member of a family of compounds with a pteridine moiety. In the case of folic acid, the pteridine moiety is conjugated by a methylene bridge to *para*-aminobenzoic acid, which in turn is joined to glutamyl residues by a peptide linkage. An overview of one-carbon transfers involving tetrahydrofolate (THF) coenzymes and their metabolic origins is given in Fig. 24.21. The reactions include the generation and utilization of formaldehyde, formimino, and methyl groups. For these conversions to occur, folic acid must be in its completely reduced state. The reductions occur at positions 5, 6, 7, and 8 to form a tetrahydrofolic acid (THFA). Reduction brings the nitrogens at positions 5 and 10 closer together and changes the electrochemical properties of both, which facilitates the formation of the various THFA single-carbon derivatives that are involved in the metabolic conversions shown in Fig. 24.18. The formyl, methanyl, and methylene forms are utilized for purine synthesis and important steps in thymidy-

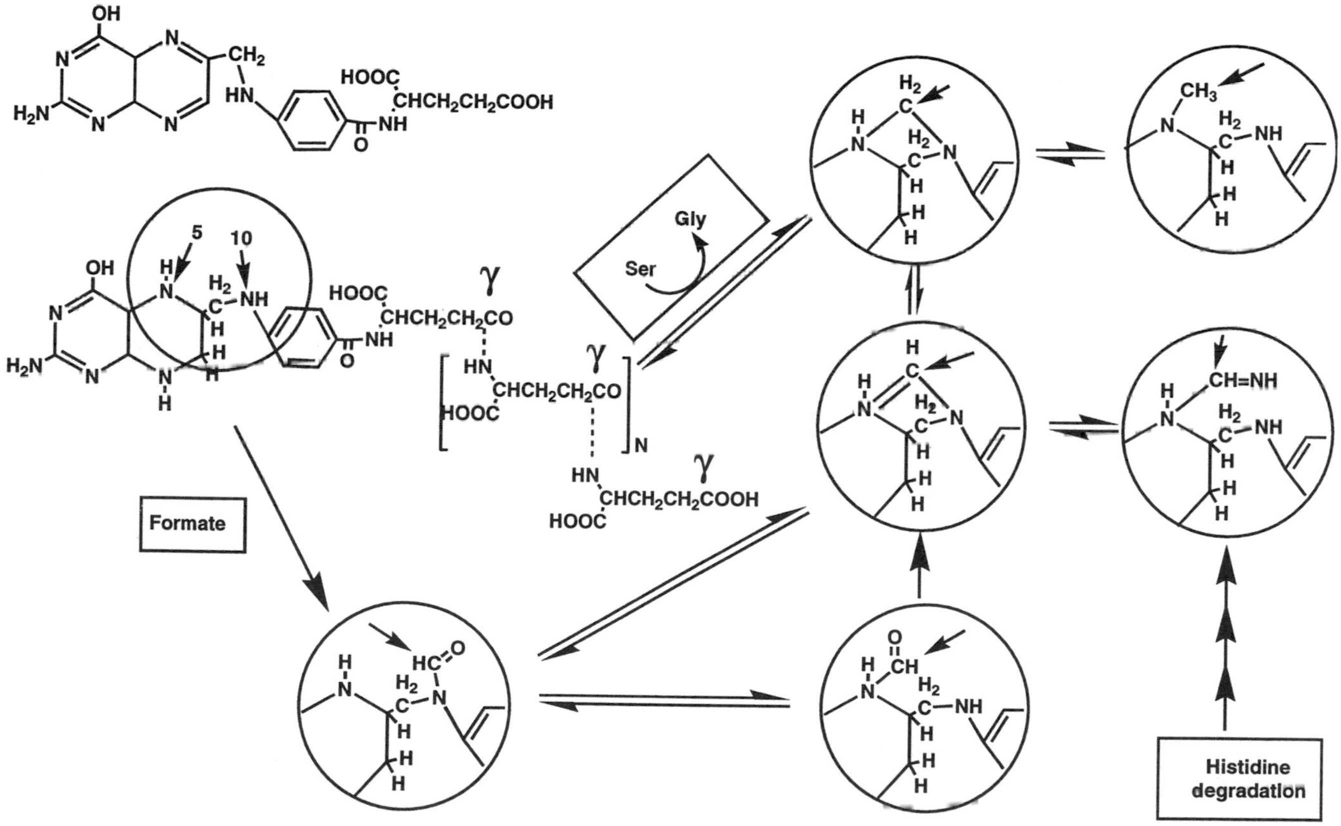

FIGURE 24.21 Folic acid. Structure of oxidized and reduced tetrahydrofolic acid (THFA) are shown. The 5 and 10 positions of the molecule are highlighted because these sites are important to single-carbon transfer. Folic acid is found in foods as γ-linked polyglutamyl folic acid (*n* ranges from 3 to 5 units). In the intestine, γ-peptidases (also referred to as conjugases) cleave the polyglutamyl residues to the single glutamate (as shown for the oxidized structure of folic acid). Single carbon units can enter into THFA by a number of routes. Formimino groups can arise from glycine and histidine degradation. Cylization results in N^5,N^{10}-methenyl THFA. Reduction of this product results in N^5,N^{10}-methylene THFA. This product can also be derived directly from an aldol-type condensation reaction arising from the conversion of serine to glycine (see Fig. 24.17). This step represents a major source of single carbon units. Subsequently, all of the forms of methylated THFA may be ultimately reduced to N^5-methyl THFA.

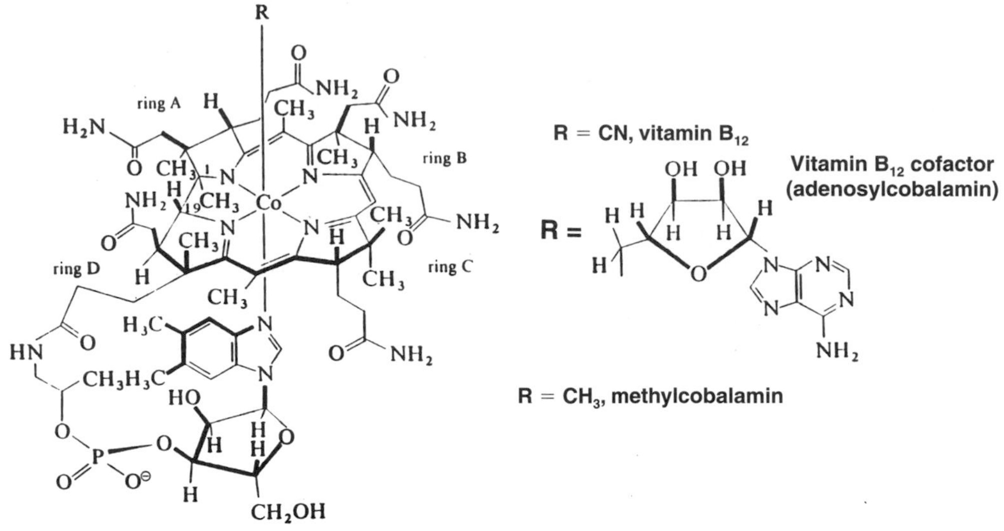

FIGURE 24.22 Structure of vitamin B_{12}. At the position designated as R, vitamin B_{12} can be methylated or coordinated with H_2O or a deoxyadenosyl moiety (see text). Commercial preparations of vitamin B_{12} are complexed with cyanide, for example, cyanocobalamine.

late (i.e., DNA-related) synthesis. These reactions are essential to cell division and proliferation and so are of obvious importance to animals. Folate also participates in reactions involved in the interconversion and catabolism of amino acids. Much of the folate in the body is in the form of methyl-THF, but for the transfer of methylene groups required for purine synthesis, folate has to be to in the form of THF. Removal of the methyl group from methyl-THF, and its eventual transfer to homocysteine, is effected by vitamin B_{12} and represents an important interaction between these two vitamins (Fig. 24.23).

Two vitamin B_{12} coenzymes participate in mammalian metabolism (Bender, 1992): vitamin B_{12} in the methylated form in the THFA–homocysteine transmethylase system, in which a methyl group is transferred from methyl THFA to homocysteine; and in the adenosylated form as 5-deoxyadenosylcobalamin, as a coenzyme of methylmalonyl-CoA mutase. In this reaction, methylmalonyl-CoA is converted to succinyl-CoA for ultimate use as a metabolic fuel (Fig. 24.24). This reaction is essential for the delivery of carbon from odd-chained fatty acids into the TCA cycle. In the absence of vitamin B_{12}, methylmalonic acid accumulates in blood and is excreted in the urine. This reaction of is particular significance to some herbivores and ruminants, which depend heavily on propionate, an odd-chain fatty acid, as a major gluconeogenic precursor.

c. Metabolism

Steps in the absorption, transport, and utilization of folic acid and vitamin B_{12} are more complex than for other water-soluble vitamins. In the case of folic acid, the conjugated glutamyl residues must be removed for effective absorption, for example, to monoglutamyl tetrahydrofolate. The enzymes in the intestinal cells that carry out the hydrolysis of conjugated glutamyl residues are commonly referred to as conjugases. Spe-

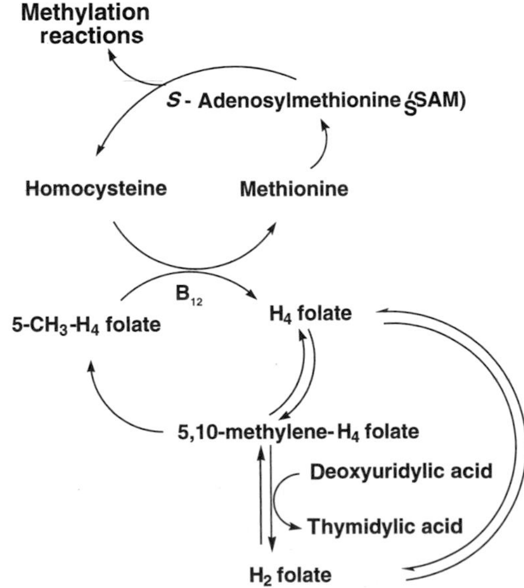

FIGURE 24.23 Interactive reactions involving THFA and vitamin B_{12}. N^5, N^{10}-methylene THFA is important for thymidylic acid formation from deoxyuridylic acid. The single carbon unit of other forms of THFA, such as the formyl and N^5,N^{10}-methenyl derivatives, are ultimately transferred through vitamin B_{12} to form methionine from homocysteine. Methionine can be activated by ATP to produce S-adenosylmethionine. This intermediate is important to methylation of phospholipids and DNA, and to production of methylated forms of various amino acids and carbohydrates.

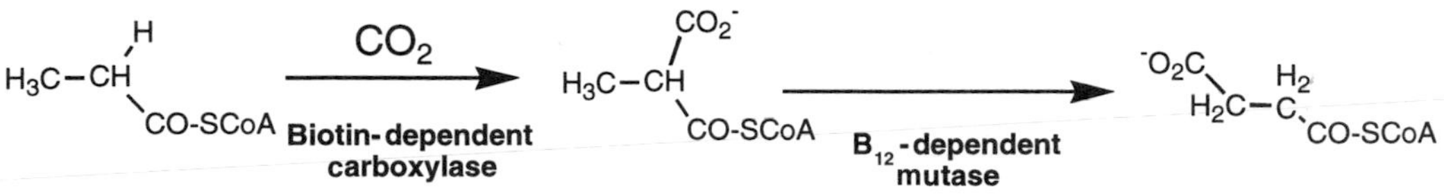

$H_3C-CH(H)-CO\text{-}SCoA \xrightarrow[\text{Biotin-dependent carboxylase}]{CO_2} H_3C-CH(CO_2^-)-CO\text{-}SCoA \xrightarrow[\text{B}_{12}\text{-dependent mutase}]{} {}^-O_2C-H_2C-C(H_2)-CO\text{-}SCoA$

FIGURE 24.24 A vitamin B_{12}-dependent mutase reaction.

cific serum transport proteins exist for folic acid, and cellular uptake is by active processes . As folic acid is found in circulation primarily as methyl tetrahydrofolate, reduction and methylation are important steps in the eventual transfer of folic acid across intestinal cells into circulation.

For B_{12}, the steps important to processing involve first release of B_{12} from foods under acidic conditions (Ellenbogen and Cooper, 1991). Vitamin B_{12} then binds to proteins produced by cells of the gastric fundus (and also by the pancreas and salivary gland in some species). Two proteins have been identified, which are usually designated as R protein and intrinsic factor. Vitamin B_{12} first binds to R protein and is apparently released in the intestinal lumen by the action of pancreatic and intestinal proteinases and peptidases, to allow vitamin B_{12} to bind to intrinsic factor. The vitamin B_{12}–intrinsic factor complex then interacts with receptors on the intestinal brush border localized in the midgut, that is, the ileum.

Interference with R-protein intrinsic-factor production, inflammatory disease affecting the ileum, or overproduction of intestinal microflora can adversely affect the availability of vitamin B_{12}. With bacterial overproduction, there is competition between the host and bacteria for vitamin B_{12} and production of bacterial proteins that bind B_{12} and interact with its uptake. Gut bacteria can also be a source of B_{12}. Many animals obtain vitamin B_{12} through coprophagy. In ruminants, vitamin B_{12} is synthesized in large quantities by ruminal bacteria provided an adequate concentration of cobalt is present in the diet.

Once vitamin B_{12} is taken up by luminal cells, it is transported into the lysosomes where the vitamin B_{12}–intrinsic factor complex is degraded and the vitamin B_{12} is released and vectorally directed for release into plasma. Vitamin B_{12} is transported in plasma by one of at least three specific transport proteins, transcobalamins I, II, or III. The transcobalamins transport vitamin B_{12} to cells where it is again transferred into targeted cells by endocytotic mechanisms.

d. Requirements and Deficiency

The requirements for folic acid range from 1 to 10 mg per kg diet for most animals. There are some conditions in which the folic acid requirements are conditionally high, such as when either natural or pharmacological folic acid agonists are present in the diet. With the discovery that THFA is required for DNA synthesis, a number of antimetabolites that function as inhibitors of folic acid reductase were developed in the 1950s and 1960s. The best example is methotrexate, which ultimately inhibits the proliferation and regeneration of rapidly replicating cells. Cell division is blocked in the S phase because of impaired DNA synthesis. As a consequence, drugs such as methotrexate are widely used in cancer chemotherapy, particularly for tumors of the lymphoreticular system.

The requirement for vitamin B_{12} for most animals is in the range of 2–15 μg per kg diet. Although deficiencies of folic acid and vitamin B_{12} are uncommon in free-ranging nonruminant animals, diseases of the stomach, proximal duodenum, or ileum and pancreatic insufficiency can affect folic acid and vitamin B_{12} absorption, respectively.

Deficiencies of both vitamin B_{12} and folic acid produce clinical signs of macrocytic anemia and dissynchronies in growth and development owing to the importance of folic acid to purine and DNA synthesis. Chronic deficiencies of either folic acid or B_{12} can also promote fatty liver disease and indirectly influence extracellular matrix maturation stability by causing abnormal elevations in homocysteine. Such signs and symptoms are attributable to both THFA and B_{12} deficiencies because of the integral relationship between vitamin B_{12} and THFA regeneration. Dietary intakes of folic acid sufficient to maintain functional THFA levels can mask the initial signs of vitamin B_{12} deficiency, such as macrocytic and megaloblastic anemia. Prolonged vitamin B_{12} deficiency in humans results in serious neurological disorders, such as degeneration of the myelin sheath.

e. Assessment of Vitamin B_{12} and Folate Status of Animals

The independent roles of vitamin B_{12} in propionate metabolism and folate in histidine metabolism provide the basis for methods of assessment of their adequacy, independent of their mutual role in methyl transfer. Vitamin B_{12} in a component of the coenzyme of methylmalonyl-CoA mutase that catalyzes the conversion of L-methylmalonyl-CoA to succinyl-CoA. Administration of a loading dose of valine (1 g/kg body weight), which is a precursor for methylmalonyl-CoA, results in excretion of methylmalonic acid in urine in the vitamin B_{12}-deficient animal. Similarly, in histidine metabolism THF is required for the removal of the formimino group from formiminoglutamic acid. When a folate-deficient animal is given a loading dose of histidine (0.2 g/kg body weight) there is an enhanced urinary excretion of formiminoglutamic acid.

V. VITAMIN-LIKE COMPOUNDS

Vitamin-like compounds are products derived from carbohydrate, amino acid, or fatty acid metabolic pathways. These compounds primarily perform specialized transport functions or are associated with signal transduction mediators in cells. A nutritional case can be made that in some animal species, these compounds have important "conditional" requirements, that is, developmental periods may be identified in which a dietary source is required to maintain balance.

A. Lipotropic Factors

Nutritional requirements exist for a number of compounds at specific periods in development, particularly neonatal development, and periods of rapid growth. These compounds typically perform specialized transport functions particularly in relation to fatty acids. Apart from specific amino acids such as methionine, examples include choline, inositol, and carnitine.

1. Choline

Choline plays a key role in methyl group metabolism, carcinogenesis, and lipid transport and is a component of lecithin (Chan,1991). Choline is generally the major source of methyl groups in the diet, but can also be synthesized *de novo* from ethanolamine when methionine, dimethylcysteine, or betaine are in adequate supply. Formation of betaine from choline provides an important sources of labile methyl groups for transmethylation reactions. Sufficient amounts of choline are normally produced if the diet contains adequate amounts of precursors; however, in young growing animals a positive growth response can occur upon addition of choline, commercially available as trimethyl hydroxyethylammonium chloride or as the bitartrate (Fig. 24.25).

Choline is one of the precursors of acetylcholine. Choline is also a component of sphingomyelin and lecithin (see Fig. 24.25). The most abundant source of choline in the diet is lecithin. The primary sign of a choline deficiency is fatty liver. In monkeys, dogs, and rats it has also been shown that prolonged choline

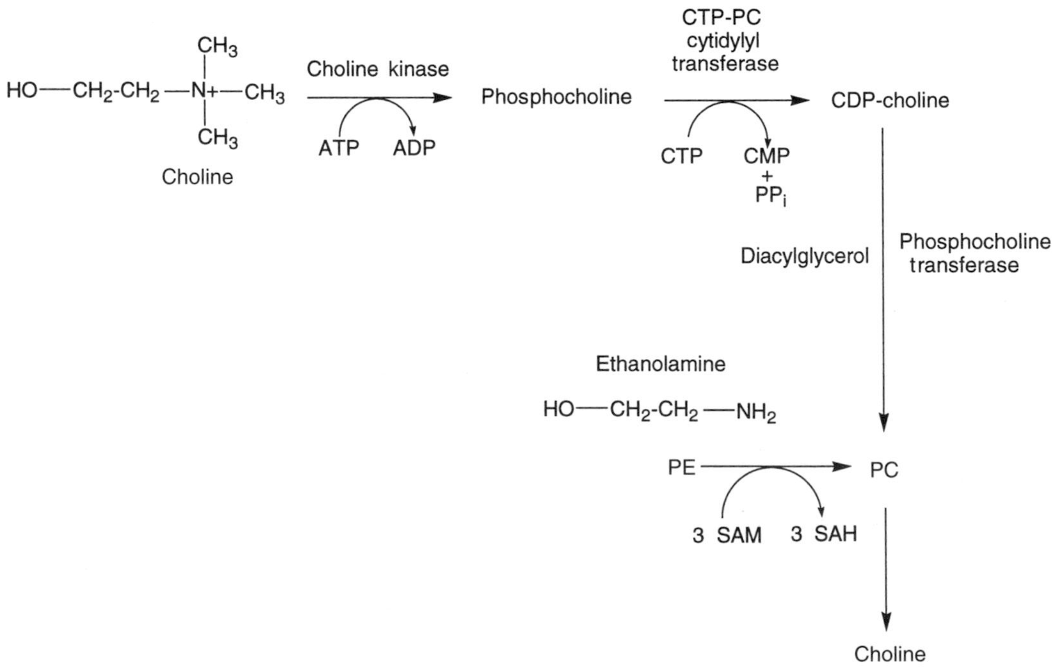

FIGURE 24.25 Choline and phosphatidylcholine (PC). Choline is essential for phosphatidylcholine (PC) synthesis. PC is one of the major phospholipids found in membranes and in circulating lipid particles. PC may be synthesized by several pathways. Choline as phosphocholine may be transferred directly to diacylglycerol following activation. PC may also be derived from diacylglycerol ethanolamine. In this pathway diacylglycerol ethanolamine is methylated by SAM to form PC.

deficiency results in cirrhosis. In mice and rats, prolonged deficiency ultimately results in hepatocellular cancer, a unique example of nutrition deficiency resulting in neoplasm. The dietary requirement for choline is dependent on the total sum of methyl donors in the diet, particularly methionine. Often, 500–1000 mg of choline are added per kilogram of diet.

2. *Inositol*

Inositol is also a component of phospholipids, and as with to choline, a dietary deficiency results in a fatty liver if the compound is limiting (Holub, 1987). Inositol is synthesized from glucose-6-phosphate after cyclization. In some animals, particularly gerbils and hamsters, a nutritional need for inositol can be demonstrated when they are given diets containing coconut oil. Myoinositol is plentiful in most foodstuffs. The estimated daily intake for large animals can be as high as 1 or 2 grams per day. Inositol is particularly important in signal transduction and phospholipid assembly. Plasma levels of inositol are increased during renal disease and nephrectomy. Otherwise, there is constant turnover due to incorporation and release from phospholipids.

3. *Carnitine*

Oxidation of fatty acids requires their transportation from the cytosol into the mitochondrial matrix where they undergo β-oxidation (Borum,1986). Carnitine plays a major role in this transport process by accepting activated fatty acids at the outer mitochondrial membrane Carnitine comes from both the diet and synthesis from peptidyl lysine (Fig. 24.26). After release of trimethyl lysine, the next steps involve its ascorbic acid-dependent hydroxylation. This step sets the stage for cleavage followed by a second hydroxylation step. These steps are not carried out efficiently in some neonates. Given the importance of carnitine to β-oxidation of long-chain fatty acids, carnitine deficiency can have profound effects on lipid utilization. An inherited carnitine deficiency has been recognized in some breeds of dogs.

Meats and dairy products, in contrast to plant foods, are well supplied with carnitine. Cereal grains, besides being low in carnitine, are also generally low in the precursors of carnitine: lysine and methionine.

4. *Taurine*

Although taurine (Fig. 24.27) is not generally considered a vitamin, the requirement for taurine by some animals is of the similar order as that for choline (Sturman, 1992). Taurine, or 2-aminoethanesulfonic acid, is present in all animal tissues and is one of the principal free amino acids in cells. Some tissues such as the retina, olfactory bulb, and granulocytes have particularly high concentrations of taurine. Taurine and/or glycine are used by many animals as conjugates for the bile acids. Some of the taurine excreted in the bile

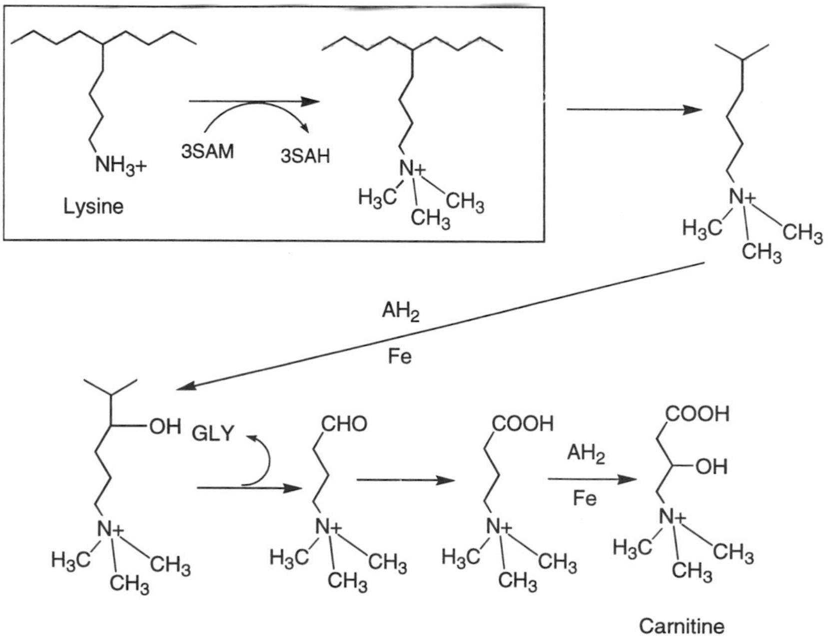

FIGURE 24.26 Synthesis of carnitine from peptidyl lysine. Peptidyl lysine is first methylated to trimethyl lysine. Subsequent hydroxylation and cleavage steps ultimately result in carnitine.

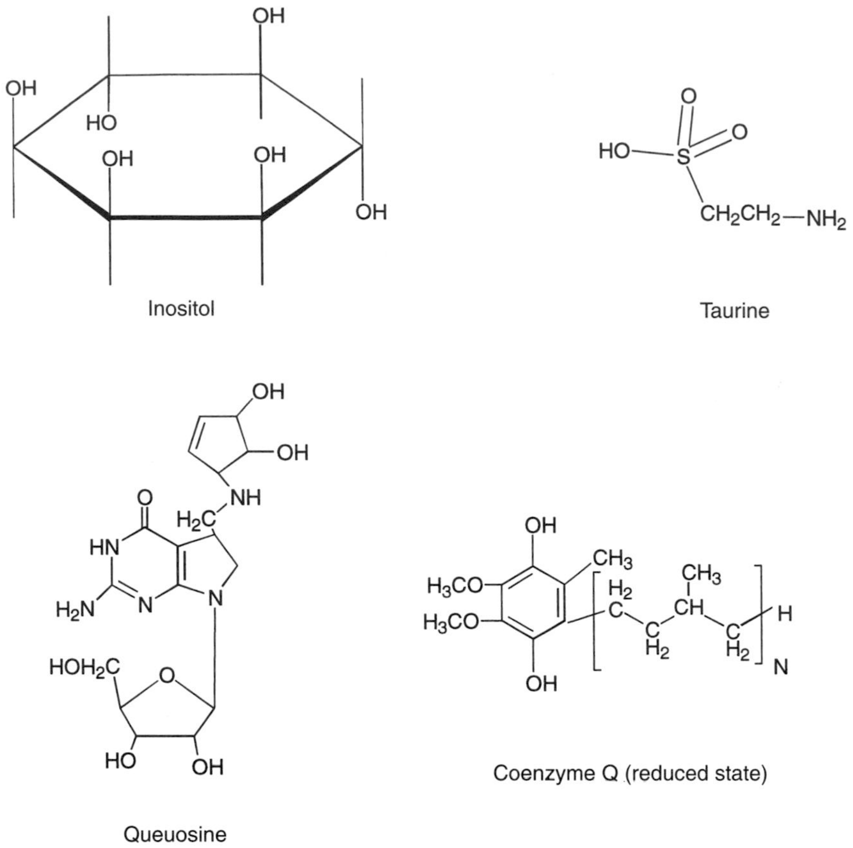

Inositol

Taurine

Queuosine

Coenzyme Q (reduced state)

FIGURE 24.27 Inositol, taurine, queuosine, and coenzyme Q.

is returned to the liver in the enterohepatic circulation. Most animals can synthesize adequate amounts of taurine from the oxidation of cysteine, to provide for the daily loss in the enterohepatic circulation. However, some animals, particularly domesticated and wild felids, and human infants do not synthesize adequate amounts of taurine. Dogs normally synthesize sufficient taurine to be independent of a dietary need for taurine; however, when dogs are given diets low in sulfur amino acids, taurine may become limiting. Cats are particularly susceptible to taurine deficiency. Defective synthesis in cats is a result of low activity of two enzymes in the synthetic pathway, cysteine dioxygenase and cysteinesulfinic acid decarboxylase, and an obligatory requirement for taurine to conjugate bile acids. A wide array of clinical signs have been described in taurine-deficient cats, including central retinal degeneration, reversible dilated cardiomyopathy, reproductive failure in queens, teratogenic defects, and abnormal brain development in kittens.

Dietary concentrations of taurine required to maintain adequate levels in plasma and whole blood are a function of type of diet, which affects the extent of microbial degradation that occurs in the enterohepatic circulation. For most expanded diets, 1 g taurine/kg dry matter is adequate, but canned diets may require concentration up to 2.5 g taurine/kg dry matter.

5. Assessment of Taurine Status

Plasma and whole blood concentrations of 40 and 300 μM of taurine appear to be adequate in cats for reproduction, which is the most demanding physiological state for taurine.

B. Other Cofactors and Potential Vitamins

The following compounds are highlighted because of their known roles as coenzymes in prokaryotes and potential roles as probiotics (growth-promoting substances) in higher animals. These compounds include queuosine, coenzyme Q, pteridines (other than folic acid) such as biopterin and the pteridine cofactor for the Mo–Fe flavoproteins, lipoic acid, and pyrroloquinoline quinone.

1. *Queuosine*

Queuosine is included because it represents a novel product arising from a microbe–host interaction. Queuine is the nucleoside base that is modified to queuosine (Fig. 24.27). Queuosine resembles guanidine and is preferably utilized in some tRNAs. The importance of this interaction has yet to be fully understood. Germ-free animals survive without a source of queuine or queuosine (Farkas, 1980).

2. *Coenzyme Q*

Although claims have been made for a dietary requirement for coenzyme Q, more work is needed to fully clarify a true nutritional role for this compound. Ubiquinone or coenzyme Q is found in mitochondria. Coenzyme Q (Fig. 24.27) is structurally similar to vitamins E and K. Quinines are ideally suited to interact with cytochromes to effect the flow of electrons in the mitochondrial respiratory chain. Coenzyme Q can be synthesized and is easily absorbed from the intestine by the same route as other fat soluble vitamins. However, there is no known requirement for coenzyme Q in higher animals.

Of the lipophilic substances with redox cycling capacity, the ubiquinones (coenzyme Q) are a group of ubiquitous 2,3-dimethoxy-5-methyl benzolquinones substituted at position 6 with terpenoid chains of varying lengths. In mitochondria, coenzyme Q affects two-electron processes and helps initiate two single-electron transfers through semiquinone intermediates. Coenzyme Q is found mainly in the mitochondrial intermembrane. Although there is no apparent dietary requirement for coenzyme Q, it is present in food and promoted for various putative health benefits. After absorption from the intestine, coenzyme Q is transported by the same transport system as vitamin E and vitamin K.

3. *Pteridines*

In animals, tetrahydrobiopterin is an important redox cofactor, best known for its role at the catalytic site for phenylalanine hydoxylase (Kaufman, 1993). Tetrahyhydrobiopterin is presumably made in sufficient quantities from pathways important to guanine synthesis. A related pteridine cofactor is the molybdenum pteridine cofactor found in xanthine and other sulfur oxidases (Fig. 24.28). The importance of molybdenum to the synthesis of the molybdenum pteridine cofactor is the basis for molybdenum being an essential element.

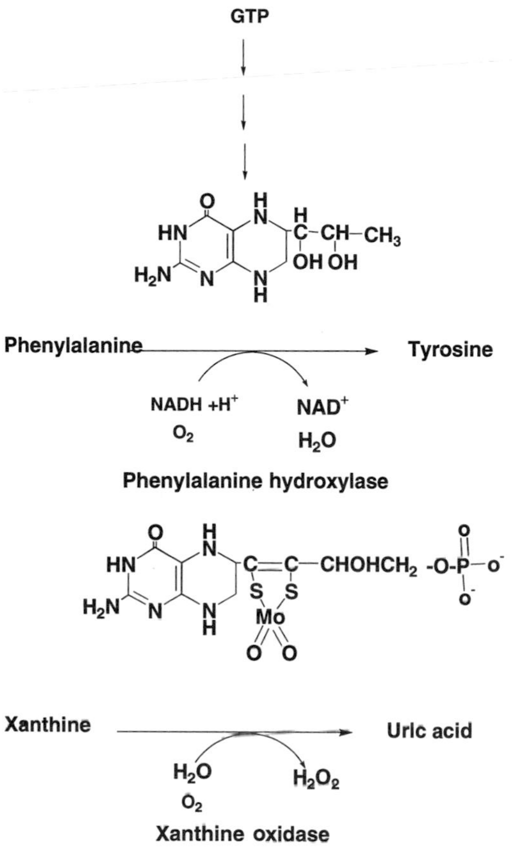

GTP

Phenylalanine → Tyrosine

NADH +H^+ O_2 → NAD^+ H_2O

Phenylalanine hydroxylase

Xanthine → Uric acid

H_2O O_2 → H_2O_2

Xanthine oxidase

FIGURE 24.28 Biopterin (found in phenylalanine hydroxylase) and the molybdenum cofactor (found in xanthine oxidase).

4. *Lipoic Acid*

Lipoic acid is presumably made in the livers of most animals (Bender, 1992). This coenzyme is linked by amide linkage to lysyl residues within transacetylases. Lipoyl moieties functions in the transfer of electrons and activated acyl groups from the thiazole moiety of thiamin pyrophosphate to CoASH (Fig. 24.16). In this process the disulfide bond is broken and dihydrolipoyl transiently generated. Reoxidation is required to reinitiate this cycle. Although most reactions in biological systems may be described as nucleophilic in nature, reactions involving oxidized lipoic acid involve electrophilic mechanisms owing to the oxidized state of the two sulfur atoms in lipoic acids.

5. *Pyrroloquinoline Quinone*

Pyrroloquinoline quinone (PQQ) (Fig. 24.29) is a recently identified cofactor that was originally isolated from methylotrophic bacteria (Smidt *et al.*, 1991). PQQ is utilized in bacteria as a redox cycling cofactor. It has been shown to be present in mammalian tissue; however, its primary function is still not clear. PQQ

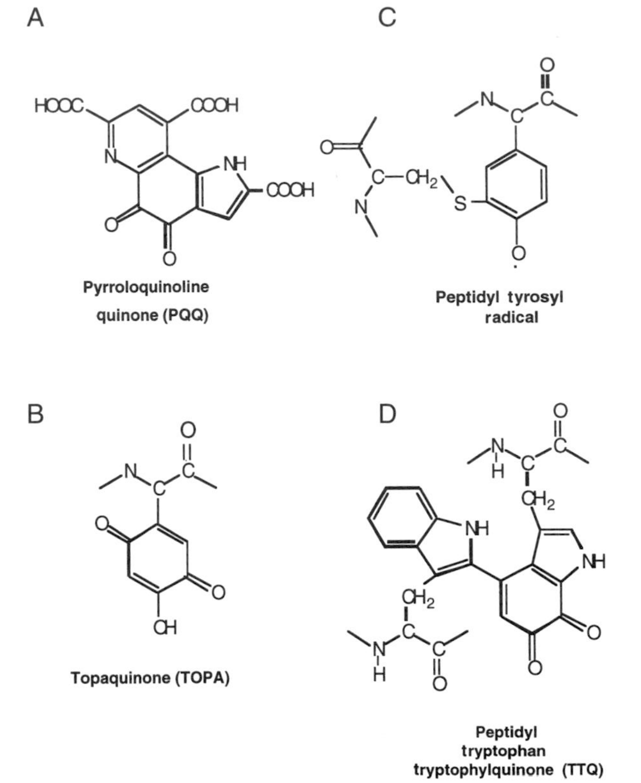

A

Pyrroloquinoline quinone (PQQ)

C

Peptidyl tyrosyl radical

B

Topaquinone (TOPA)

D

Peptidyl tryptophan tryptophylquinone (TTQ)

FIGURE 24.29 Quinone cofactors. Pyrolloquinoline quinone (PQQ) is a dissociable redox cofactor important to the metabolism of methylotrophic bacteria. It has growth-promoting properties in animals (a). TOPA is found in mammalian amine oxidases (b). The tyrosyl radical is at the active site of ribonucleotide reductase (c). TTQ, derived from tryptophan, is found in methylamine dehydrogenase in prokaryotes (d).

in amounts corresponding to 300 or more micrograms per kilogram diet is a potent growth stimulant in mice fed chemically defined diets. If there is a more universal nutritional need for PQQ, however, it has yet to be established. The growth response in mice only occurs in offspring from dams that have been nutritionally deprived of PQQ throughout their adult lives. Nevertheless, knowledge of PQQ's functions may prove to be important. It has the potential of serving as a cofactor. PQQ may also be viewed as similar to some of the bioflavonoid compounds, such as guerncin, rutin, and catechin, that appear to have health promoting benefits because of their ability to also act as redox cycling agents and/or antioxidants.

VI. CONCLUSIONS

The goals of this chapter were to highlight selected functions of vitamins and to develop concepts and principles relative to vitamins that may aid in clinical assessment. Vitamins represent a diverse array of molecules, which have equally diverse functional roles in an array of organisms and species. However, there are many common themes and similarities that have practical significance. For example, when expressed on an energy basis, vitamin requirements are often of the same order across species. Differences in requirements between species are usually due to unique pathways for production, degradation, or disposal. Ascorbic acid and niacin are examples of vitamins that cannot be synthesized by some animals and therefore are true vitamins. Similarly, there is a dietary requirement for vitamin D in cats and dogs, but the requirement is conditional for most other species. Taurine is an example of a nutrient where continual enteric degradation and loss results in a nutritional need.

Vitamin deficiencies are often the result of animals consuming monotonous diets, that is, diets based on a limited number of food sources. In nonruminant animals, the consumption of diets composed of a restricted array of foods clearly increases the probability of vitamin deficiency. Further, young and growing animals may have relatively higher need for some nutrients. Many species have a requirement for vitamins during the neonatal period that later in life may be produced in adequate amounts, for example, choline, carnitine, or inositol. There are also numerous possibilities for interactions that can have deleterious physiological consequences. The processing of foods can result in substantial losses of natural and added vitamins such as thiamin, and the presence of thiaminase (as in raw fish) is another potential source of loss. Although single vitamin deficiencies of the fat-soluble vitamins occur naturally, frequently multiple vitamin deficiencies of the B complex are encountered. Foods that are poor sources of one of the B vitamins tend to be poor sources of several B vitamins.

With respect to functions, vitamins evolved to serve unique and complex roles: as cofactors, as signaling agents in cells, as regulators of gene expression, and as redox and free-radical quenching agents. As is the case for any substance that is essential to a given function, all vitamins at some point in development can be viewed as limiting nutrients, the absence of which results in specific deficiency signs and symptoms.

References

Balmain, N. (1991). Calbindin, a vitamin D-dependent, calcium-binding protein in mineralized tissues. *Clinical Orthopaedics and Related Research* **265**, 265–276.

Bender, D. A. (1992). "Nutritional Biochemistry of the Vitamins." Cambridge University Press, Cambridge, U.K.

Bendich, A. (1993). Physiological role of antioxidants in the immune system. *J. Dairy Sci* **76**, 2789–2794.

Bonjour, J.-P. (1991). *In* "Handbook of Vitamins" (L. J. Machlin, ed.). Marcel Dekker, New York.

Borum, P. R. (1986). "The Clinical Aspects of Human Carnitine Deficiency." Pergamon Press, New York.

Brody, T., Shane, B., and Stokstad, E. L. R. (1991). *In* "Handbook of Vitamins: Aspects" (L. J. Macklin, ed.). Marcel Dekker, New York.

Chan, M. M. (1991). Choline. *In* "Handbook of Vitamins" (L. Machlin, ed.). Marcel Dekker, New York.

Coburn, S. P. (1994). A critical review of minimal vitamin B6 requirments for growth in various species with a proposed method of calculation. *Vitamins and Hormones* **48,** 259–300.

Coburn, S. P., Mahuren, J. D., and Gularte, T. R. (1984). *J. Nutr* **114,** 2269.

Collins, E. D., and Norman, A. W. (1991). Vitamin D. *In* "Handbook of Vitamins" (L. J. Machlin, ed.). Marcel Dekker, New York.

Combs, G. F. (1991). "The Vitamins." Academic Press, San Diego.

Cooperman, J. M., and Lopez, R. (1991). *In* "Handbook of Vitamins" (L. J. Macklin, ed.). Marcel Dekker, New York.

Dakshinamurti, K., ed. (1990). "Vitamin B-6." Annals New York Acad. Sci., Vol. 585, New York.

Darby, W. J., and Jukes, T. H. (1992). "Founders of Nutrition Science," Vol. 1 and 2. American Institute of Nutrition, Bethesda, Maryland.

Diplock, A. T., ed. (1985). "Fat Soluble Vitamins." William Heinemann Medical Books, London.

Ellenbogen, L., and Cooper, B. A. (1991). Vitamin B-12. *In* "Handbook of Vitamins" (L. Machlin, ed.). Marcel Dekker, New York.

England, S., and Seifer, S. (1986). The biochemical functions of ascorbic acid. *Ann. Rev. Nutr.* **6,** 365–406.

Farkas, W. R. (1980). Effect of diet on the queuosine family of tRNAs of germ-free mice. *J. Biol Chem* **255,** 6832–6835.

Fox, H. M. (1991). Pantothenic acid. *In* "Handbook of Vitamins: Nutritional, Biochemical and Clinical Aspects" (L. J. Macklin, ed.). Marcel Dekker, New York.

Friedrich, Wilhelm (1988). "Vitamins." Walter de Grader Publisher, New York.

Funk, C. (1911). *J. Physiol.* **43,** 395.

Gillian, Morris-Kay, ed. (1992). "Retinoid and Teratogenesis." Oxford University Press, New York.

Guggenheim, Acaule Karl (1981). "Nutrition and Nutritional Diseases: The Evolution of Concepts." Collator Press, New York.

Henderson, L. M. (1983). Niacin. *Ann. Rev. Nutr.* **3,** 289–297.

Holub, B. J. (1987). The cellular forms and functions of inositol phospholipids and their metabolic derivatives. *Nutrition Rev.* **45,** 65–71.

Kaufman, S. (1993). New tetrahydrobiopterin-dependent systems. *Ann. Rev. Nutr.* **13,** 261–286.

Kayden, H. J., and Traber, M. G. (1995). Absorption, lipoprotein transport, and regulation of plasma concentrations of vitamin E in humans. *J. Lipid Res.* **34,** 343–358.

Liu, O. Lanai, M. C., and Schaefer, D. M. (1995). A review of dietary vitamin E supplementation for improvement of beef quality. *J. Anim. Sci.* **73,** 3131–1340.

Machlin, L. J. (1980). "Vitamin E—A Comprehensive Treatise." Marcel Dekker, New York.

Marks, J. (1977). *In* "A Guide to the Vitamins." Medical and Technical Publications, New York.

Norman, A. W. (1980). *In* "Vitamin D, Molecular Biology and Clinical Nutrition." Marcel Dekker, New York.

Norman, A. W. (1995). Pleiotropic actions of 1-α, 25-dihydroxyvitamin D_3: An overview. *J. Nutr.* **125**(6 Suppl), 1687S–1689S.

Norman, A. T., Vanaman, T. C., and Means, A. R., eds. (1987). "Calcium-Binding Proteins in Health and Disease." Academic Press, San Diego.

Packer, L., ed. (1990). Retinoid. *In* "Methods in Enzymology," Vol. 189–190. Academic Press, San Diego.

Palm, D., Klein, H. W., Schinzel, R., Buehner, M., and Helmreich, E. J. (1990). The role of pyridoxal 5′-phosphate in glycogen phosphorylase catalysis. *Biochemistry* **29,** 1099–1107.

Poirier, Guy G., and Moreau, P., eds. (1992). "ADP-Ribosylation Reactions." Springer-Verlag, New York.

Rose, R. C. (1996). Intestinal absorption of water-soluble vitamins. *Proceedings of the Society for Experimental Biology and Medicine* **212,** 191–198.

Sable, H. Z., and Gubler, C. J. (1982). Thiamin. *Annals New York Acad. Sci.* **378.**

Sherman, M. I., Ed. (1986). "Retinoid and Cell Differentiation." CRC Press, Boca Raton, Florida.

Smidt, C. R., Steinberg, F., and Rucker, R. B. (1991). Physiologic importance of pyrroloquinoline quinone. *Proc. Soc. Expl. Biol. Med.* **197,** 19–26.

Sporn, M. B., Roberts, A. B., and Goodman, D. S., eds. (1994). "The Retinoid: Biology, Chemistry, and Medicine," 2nd ed. Raven Press, New York.

Sturman, J. (1992). Review: Taurine deficiency and the cat. *Adv. Exp. Med. Biol.* **315,** 1–5.

Sucov, H. M., and Evans, R. M. (1995). Retinoic acid and retinoic acid receptors in development. *Molecular Neurobiology* **10,** 169–184.

Suttle, J. (1995). The importance of menaquinone in human nutrition. *Annu. Rev. Nutr.* **15,** 399–417.

Traber, M. G., and Packer, L. (1995). Vitamin E: Beyond antioxidant function. *Amer. J. Clin. Nutr.* **62**(6 Suppl), 1501S–1509S.

Wasserman, R. H., and Fullmer, C. S. (1995). Vitamin D and intestinal calcium transport: Facts, speculations and hypotheses. *J. Nutr.* **125,** 1971S–1979S.

Weiner, M., and Eys, J. V. (1983). "Nicotinic Acid." Dekker Press, New York.

CHAPTER

25

Lysosomal Storage Diseases

MARK HASKINS AND URS GIGER

I. LYSOSOMAL BIOLOGY

In 1955, de Duve *et al.* named the cytoplasmic particles that contain a series of hydrolytic enzymes lytic bodies, or "lysosomes." These organelles have a single lipoprotein membrane and contain several dozen different acid hydrolase enzymes (Holtzman, 1989), which typically catalyze catabolic reactions $A-B + H_2O \rightarrow A-H + B-OH$. Lysosomes and their "housekeeping" enzymes function in degrading many substrates that are found in all nucleated mammalian cells. Deficiencies of these enzymes lead to lysosomal accumulation of these substrates, thereby causing disease.

In normal cells, most lysosomal hydrolases are synthesized as preproenzymes on rough endoplasmic reticulum (ER) ribosomes. Through a signal-recognition particle complex, the enzymes are translocated into the lumen of the ER, where high-mannose oligosaccharides are added (Fig. 25.1; reviewed in Kornfeld and Sly, 1995). These oligosaccharides are trimmed and the glycoprotein moves to the Golgi apparatus, where further shortening occurs. Further posttranslational modification results from the action of two enzymes that add a mannose-6-phosphate (Man-6-P) marker. The Man-6-P moiety can be recognized by two similar integral membrane glycoprotein receptors that transfer the enzyme to the lysosome. These two receptors are (1) small and cation-dependent for binding, and (2) large and cation-independent, which in some species also bind insulin-like growth factor II. Both receptors appear responsible for the transport of the enzymes from the Golgi apparatus via clathrin-coated vesicles to the prelysosomal/endosomal compartment. Once the lysosomal enzymes dissociate, the receptors recycle to the Golgi apparatus.

A proportion of the Man-6-P modified enzyme in the Golgi may also leave the cell via secretory granules (Fig. 25.1). Different enzymes appear to be secreted from cells in varied amounts (Dobrenis *et al.*, 1994). Thus, the level of activity in serum of any particular enzyme is related to how much is secreted and its stability at plasma pH. Secreted enzyme can also ultimately reach the lysosome because the cation-independent receptor is present in the plasma membrane on some cells such as fibroblasts (Kaplan *et al.*, 1977; Distler *et al.*, 1979; Natowicz *et al.*, 1979). Thus, enzymes that connect with this receptor can be internalized and transferred to lysosomes. This pathway provides the mechanism for therapy for lysosomal storage diseases discussed later.

Although posttranslational glycosylation is common to most lysosomal enzymes, other modifications or activator proteins are necessary for the function of a subset of the hydrolases. For example, the eight lysosomal sulfatases undergo an additional co- or posttranslational modification, which converts a cysteine

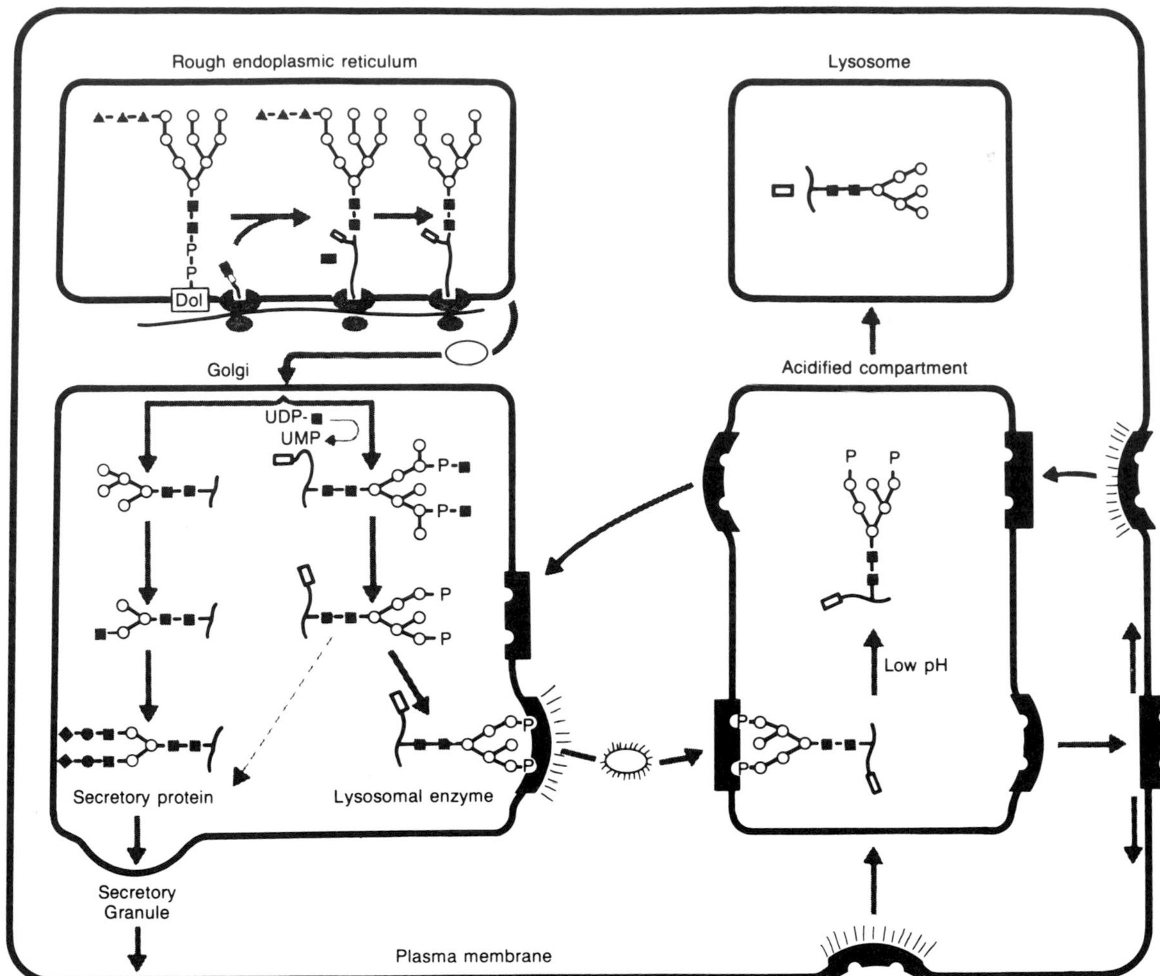

FIGURE 25.1 Schematic diagram of how lysosomal enzymes are processed and delivered to the lysosome. (From Kornfeld, S., 1987. Trafficking of lysosomal enzymes. *FASEB J.* **1**, 463, with permission.)

residue common to all known eukaryotic sulfatases to 2-amino-3-oxopropionic acid (Schmidt *et al.*, 1995). The absence of this conversion results in multiple sulfatase deficiency. The degradation of sphingolipids with short hydrophilic head groups requires sphingolipid activator proteins (SAPs), which are small, nonenzymatic glycoproteins (reviewed in Sandhoff *et al.*, 1995). Deficiency in activity of SAPs is also known to cause lysosomal storage diseases.

Lysosomes function in degrading large complex substrates that have been taken into a cell by endocytosis, or autophagy (the degradation/turnover of a cell's own molecules). The endosome containing the substrates fuses with a primary lysosome, producing a secondary lysosome, which contains the mixture of hydrolases and substrates. Degradation of most substrates occurs by the activity of a series of hydrolases in a stepwise fashion, each step requiring the action of the previous hydrolase to modify the substrate, thereby permitting catabolism to proceed to the next enzyme step in the pathway. If one step in the process fails, further degradation ceases. For example, the glycosaminoglycans, formerly known as mucopolysaccharides, are long molecules of repeating subunits and are, as part of proteoglycans, a component of the ground substance of the extracellular matrix. The series of hydrolases that are responsible for the stepwise degradation of one of the glycosaminoglycans (GAGs), dermatan sulfate, are illustrated in Fig. 25.2. Each enzyme works sequentially. Each of the enzymes in this pathway has been described as deficient in a domestic animal (Table 25.1).

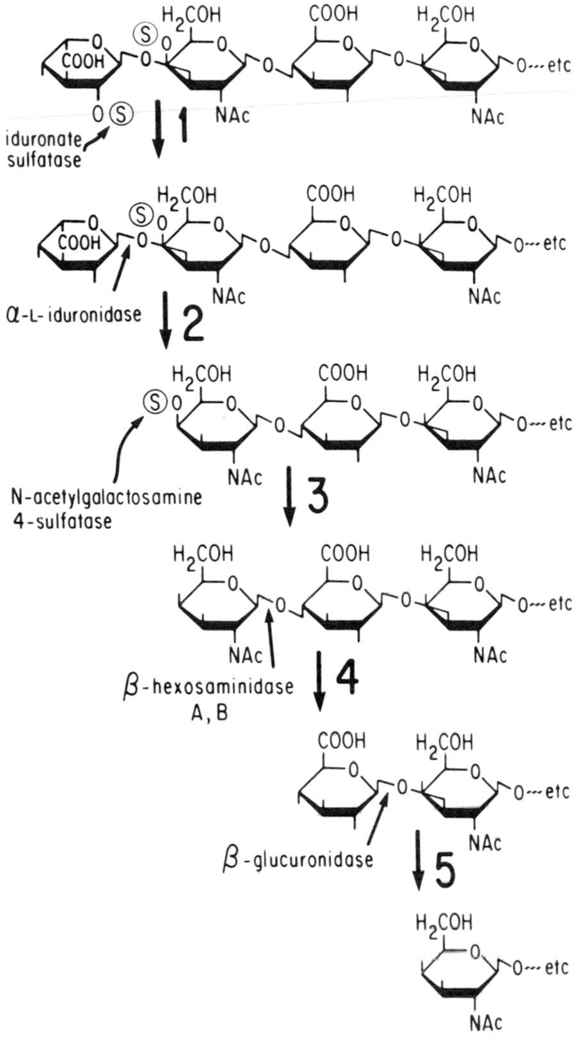

FIGURE 25.2 The stepwise degradation of the glycosaminoglycan dermatan sulfate by a series of lysosomal enzymes, all of which have been determined to be deficient in activity in domestic animals. [From Neufeld, E. F., and Muenzer, J. (1995). The mucopolysaccharidoses. *In* "The Metabolic and Molecular Bases of Inherited Disease, Seventh Edition, Vol. II. (C. R. Scriver, A. L. Beaudet, W. S. Sly, and D. Valle, eds.), p. 2467. McGraw Hill, Inc., New York, with permission.]

II. LYSOSOMAL STORAGE DISEASES

The lysosomal storage diseases (LSDs) are defined as a group of individually rare genetic disorders of cellular catabolism. The earliest detailed clinical reports of a LSD were in humans by Tay (1881) and Sachs (1887). Eighty one years later the stored material in this disease (Tay–Sachs disease) was defined as G_{M2} ganglioside (Svennerholm, 1962); seven years later the enzyme that is deficient in activity (β-hexosaminidase A) was defined (Okada and O'Brien, 1969; Sandhoff, 1969). Isolation of the cDNA coding for the alpha subunit of β-hexosaminidase A was reported 15 years later (Myerowitz *et al.*, 1984, 1985; Korneluk *et al.*, 1985), which was quickly followed by the identification of the first of more than 50 mutations responsible for Tay–Sachs disease (Myerowitz and Hogikyan, 1986, 1987) and the cloning of the gene (Proia and Soravia, 1987).

Most LSDs are inherited as autosomal recessive traits and result from mutations in the coding sequence of one of the acid hydrolases located in the lysosome. Point mutations, deletions, and other alterations in sequence may occur anywhere along the length of DNA coding the enzyme protein. Each individual alteration will produce a unique change in the protein.

The consequence of such a genetic abnormality is the reduction or elimination of the catalytic activity of the particular enzyme. This reduction in enzyme activity results in the accumulation within the lysosome of the substrate of that enzyme (Fig. 25.3), hence the name lysosomal storage disease. The reduction in the amount of product of the metabolic pathway does not appear to produce disease. LSDs are classified by the primary substrates that accumulate and are defined by the individual enzyme that is deficient in activity. For example, the mucopolysaccharidoses (MPSs) are a group of diseases resulting from defective catabolism of glycosaminoglycans. Each of the MPSs is caused by impaired function of one of 10 enzymes required for normal glycosaminoglycan degradation. In humans, these disorders were initially defined by clinical phenotype, then by the particular glycosaminoglycan(s) (heparan, dermatan, chondroitin, and keratan sulfates) present in the urine of patients. Now, in addition to being defined by the enzyme that is deficient in activity, many of the diseases are subdivided by the particular mutation in the coding sequence of the gene responsible for the defect.

Different mutations of the same gene may produce (1) very similar diseases, as seen in two families of cats with MPS VI (McGovern *et al.*, 1985), or (2) somewhat different diseases, as is seen in humans with MPS IH and MPS IS, with and without CNS disease, respectively (Neufeld and Muenzer, 1995). Affected individuals either may be homozygous for the same mutation in both alleles, or may be heteroallelic (having one mutation in the allele on one chromosome and a different mutation in the allele on the other chromosome). In addition, if the substrates being stored in different diseases have similar pathologic effects, mutations of different lysosomal enzyme genes may produce very similar diseases. This has been described in humans with MPS III A–D, each caused by a different enzyme deficiency, yet all have a very similar phenotype (Neu-

TABLE 25.1 Lysosomal Storage Diseases Described in Animals

Disease	Deficient enzyme	Species and selected references
Fucosidosis	Alpha-fucosidase	Springer spaniel dog (Kelly *et al.*, 1983, Healy *et al.*, 1984; Friend *et al.*, 1985; Smith *et al.*, 1996)
Galactosylceramide lipidosis (globoid cell leukodystrophy; Krabbe disease)	Galactosylceramidase (galactocerebroside beta-galactosidase)	Cairn terrier dog (Fankhauser *et al.*, 1963; Hirth and Nielsen, 1967; Austin *et al.*, 1968a,b; McGrath *et al.*, 1969; Fletcher *et al.*, 1966, 1971, 1977; Fletcher and Kurtz, 1972; Suzuki *et al.*, 1970, 1974; Howell and Palmer, 1971) West highland white terrier dog (Fankhauser *et al.*, 1963; Jotner and Jonas, 1968; Fletcher *et al.*, 1971, 1972) Dorset sheep (Pritchard *et al.*, 1980) Twitcher mouse (Duchen *et al.*, 1980; Kobayashi *et al.*, 1980) Domestic cat (Johnson, 1970) Miniature poodle dog (Suzuki *et al.*, 1974) Beagle dog (Johnson *et al.*, 1975) Blue tick hound dog (Boysen *et al.*, 1974) Rhesus monkey (Luzi *et al.*, 1997)
Glucocerebrosidosis (Gaucher disease)	Acid beta-glucosidase (glucocerebrosidase)	Sydney silky terrier dog (Hartley and Blakemore, 1973; Van de Water *et al.*, 1979; Farrow *et al.*, 1982) Sheep (Laws and Saal, 1968) Pig (Sandison and Anderson, 1970)
Glycogen storage disease II (Pompe disease)	Acid alpha-glucosidase	Lapland dog (Mostafa, 1970; Walvoort *et al.*, 1982, 1985; Walvoort, 1983) Cat (Sandstrom *et al.*, 1969) Corriedale sheep (Manktelow and Hartley, 1975) Shorthorn cattle (Richards *et al.*, 1977; Howell *et al.*, 1981) Brahman cattle (O'Sullivan *et al.*, 1981; Wisselaar *et al.*, 1993) Japanese quail (Nunoya *et al.*, 1983; Higuchi *et al.*, 1987; Suhara *et al.*, 1989; Fujita *et al.*, 1991)
G_{M1} gangliosidosis	Beta-galactosidase	Siamese cat (Baker *et al.*, 1971; Handa and Yamakawa, 1971; Farrell *et al.*, 1973; Holmes and O'Brien, 1978a,b) Domestic cat (Blakemore, 1972; Purpura *et al.*, 1978a,b; Purpura and Baker, 1977) Korat cat (Baker *et al.*, 1976) Beagle mix dog (Read *et al.*, 1976; Rittman *et al.*, 1980; Rodriguez *et al.*, 1982; Alroy *et al.*, 1985) Springer spaniel (Alroy *et al.*, 1985, 1992; Kaye *et al.*, 1992) Portuguese water dog (Saunders *et al.*, 1988; Shell *et al.*, 1989; Alroy *et al.*, 1992) Friesian cattle (Donnelly *et al.*, 1973a,b) Suffolk sheep (Murnane *et al.*, 1989, 1991a,b, 1994) Sheep (Ahern-Rindell *et al.*, 1988a,b, 1989)
G_{M2} gangliosidosis (Sanhoff disease)	Beta-hexosaminidase A and B	Domestic cat (Cork *et al.*, 1977, 1978; Walkley *et al.*, 1990; Barker *et al.*, 1986) Korat cat (Neuwelt *et al.*, 1985; Muldoon *et al.*, 1994) Japanese Spaniel dog (Cummings *et al.*, 1985; Ishikawa *et al.*, 1987) Yorkshire pig (Read and Bridges, 1968; Pierce *et al.*, 1976; Kosanke *et al.*, 1978, 1979) German short-haired pointer dog (Karbe and Schiefer, 1967; McGrath *et al.*, 1968; Gambetti *et al.*, 1970; Bernheimer and Karbe, 1970; Karbe, 1973; Singer and Cork, 1989)
Mucolipidosis II (I-cell disease)	*N*-Acetylglucosamine-1-phosphotransferase	Cat (Bosshard *et al.*, 1995; Hubler *et al.*, 1995)
Alpha mannosidosis	Alpha-mannosidase	Persian cat (Burditt *et al.*, 1980; Vandevelde *et al.*, 1982; Jezyk *et al.*, 1986; Warren *et al.*, 1986; Raghavan *et al.*, 1988; Cummings *et al.*, 1988; Maenhout *et al.*, 1988; Castagnar, 1990) Angus and Murray gray cattle (Jolly, 1971, 1974, 1975, 1978; Hocking *et al.*, 1972; Phillips *et al.*, 1974; Jolly *et al.*, 1973, 1974, 1975, 1980a) Galloway cattle (Embury and Jarrett, 1985)
Beta mannosidosis	Beta-mannosidase	Anglo-Nubian goat (Jones and Dawson, 1981; Jones *et al.*, 1983; Lovell and Jones, 1983; Kumar *et al.*, 1986) Saler cattle (Abbitt *et al.*, 1991; Patterson *et al.*, 1991; Healy *et al.*, 1992; Bryan *et al.*, 1993)
Mucopolysaccharidosis I (Hurler, Scheie, and Hurler/Scheie syndromes)	Alpha-L-iduronidase	Domestic cat (Haskins *et al.*, 1979a,b, 1983) Plott hound dog (Shull *et al.*, 1982, 1984; Spellacy *et al.*, 1983; Shull and Hastings, 1985; Menon *et al.*, 1992; Stoltzfus *et al.*, 1992)

(*continues*)

TABLE 25.1 *(Continued)*

Disease	Deficient enzyme	Species and selected references
Mucopolysaccharidosis II (Hunter syndrome)	Iduronate sulfatase	Labrador retriever dog (Prieur *et al.*, 1995)
Mucopolysaccharidosis III A (Sanfilippo A syndrome)	Heparan *N*-sulfatase	Wirehaired dachshund dog (Fischer *et al.*, (1996)
Mucopolysaccharidosis III D (Sanfilippo D syndrome)	*N*-Acetylglucosamine 6-sulfatase	Nubian goat (Thompson *et al.*, 1992; Friderici *et al.*, 1995)
Mucopolysaccharidosis VI (Maroteaux–Lamy syndrome)	*N*-Acetylglucosamine 4-sulfatase (arylsulfatase B)	Siamese cat (Cowell *et al.*, 1976; Jezyk *et al.*, 1977; Haskins *et al.*, 1979c, 1980; McGovern *et al.*, 1985; DiNatale *et al.*, 1992) Domestic short-haired cat (Giger *et al.*, unpublished) Miniature pinscher dog (Neer *et al.*, 1992, 1995) Miniature schnauzer dog (Thrall, personal communication) Welsh corgi dog (Giger *et al.*, unpublished) Rat (Yoshida *et al.*, 1993a,b, 1994)
Mucopolysaccharidosis VII (Sly disease)	Beta-glucuronidase	Mix-breed dog (Haskins *et al.*, 1984; Schuchman *et al.*, 1989) Gus mouse (Birkenmeier *et al.*, 1989; Vogler *et al.*, 1990; Sands and Birkenmeier, 1993) Cat (Gitzelmann *et al.*, 1994)
Sphingomyelinosis A and B (Niemann–Pick A and B diseases)	Acid spingomyelinase	Siamese cat (Chrisp *et al.*, 1970; Wenger *et al.*, 1980; Snyder *et al.*, 1982; Yamagami *et al.*, 1989) Miniature poodle dog (Bundza *et al.*, 1979) Mouse[a]

[a] Mutation analysis (Horinouchi *et al.*, 1993) has shown that the two mouse models previously described (Pentchev *et al.*, 1980, 1984; Miyawaki *et al.*, 1982) lack a molecular defect in the acid sphingomyelinase gene. Consequently, they represent sphingomyelinosis C or D.

feld and Muenzer, 1995). Animals with LSDs very closely resemble their human counterparts.

Several animal species were diagnosed clinically and pathologically as having a lysosomal storage disorder (included in Table 25.1) before this group of diseases was defined by deficiencies in hydrolase activity. The first definitive discovery of an enzyme deficiency in a nonhuman animal species was G_{M1} gangliosidosis in a Siamese cat by Baker *et al.* in 1971. Since then, naturally occurring LSDs defined by a deficiency in lysosomal enzyme activity have been recognized in cats, cattle, dogs, goats, mice, pigs, rats, and sheep (Table 25.1). Notably, although there are estimated to be more than 5 million horses in the United States and although the horse is a closely observed species, an LSD has not been documented in horses.

It is important to recognize that there is a group of diseases that apparently involve the lysosome, but which at present have not been defined as being caused by a defect in hydrolase activity or transport: ceroid lipofucinosis (Batten disease), sphingomyelinosis C (Niemann–Pick disease C), and Chediak–Higashi syndrome (Table 25.2). Two of the subcategories of ceroid lipofucinosis have recently had an enzyme (palmitoyl-protein thioesterase: Vesa *et al.*, 1995) and gene (CLN3 gene: Lerner *et al.*, 1995) associated with them in humans. The animals listed in Table 25.2 have not yet been characterized with respect to the basic defect.

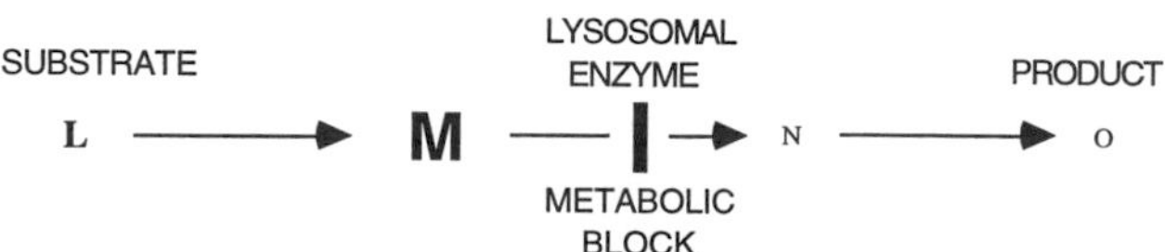

FIGURE 25.3 Metabolic consequences of an enzyme deficiency. The reduction in catalytic activity of the enzyme that converts M to N results in the accumulation of M within the lysosome.

An additional group of storage diseases result from defects in glycogen metabolism and do not involve lysosomal enzymes (Table 25.3). Although they are not LSDs, they are presented here for completeness and to avoid confusion. Note that glycogenosis type II is an LSD and is included in Table 25.1. Several mouse models of LSDs have now been created by gene knockout technology (Table 25.4). In creating a knockout, the phenotype has ranged from essentially no disease to being fatal soon after birth. New knockout models of LSDs will continue to be created in mice to provide more information about the pathogenesis of these debilitating disorders, and for evaluating therapy.

III. PATHOGENESIS

In LSDs, the continued presentation of substrates to the cell for degradation results in storage and the production of large lysosomes. By electron microscopy, ly-

TABLE 25.2 Diseases That Appear to Involve the Lysosome, but Are Not Presently Known to Involve a Defective Lysosomal Hydrolase

Disease	Deficient enzyme	Species and selected references
Ceroid lipofuscinosis	Palmitoyl-protein thioesterase; CLN3 gene	Australian cattle dog "blue heelers" (Cho *et al.*, 1986; Wood *et al.*, 1987) Border collie dog (Taylor and Farrow, 1988) Chihuahua dogs (Rac and Giesecke, 1975) Cocker spaniel dogs (Fankhauser, 1965; Beck, 1974; Nimmo Wilkie and Hudson, 1982) Dachshund dogs (Cummings and de Lahunta, 1977; Vandevelde and Fatzer, 1980) Dalmatian dog (Malkusch, 1982; Goebel *et al.*, 1988) English setter dog (Koppang, 1969, 1973, 1979, 1988) Saluki dogs (Appleby *et al.*, 1982) Tibetan terrier dog (Riis *et al.*, 1992) Yugoslavian shepherd dog (Bichsel and Vandevelde, 1982) Siamese cat (Green and Little, 1974) Devon cattle (Harper *et al.*, 1988) South Hampshire sheep (Jolly *et al.*, 1980b, 1988, 1989, 1992) Swedish sheep (Järplid and Haltia, 1993)
Sphingomyelinosis C (Niemann–Pick C disease)	Unknown defect in cholesterol transport.	Cat (Lowenthal *et al.*, 1990; Munana *et al.*, 1994) Mouse previously classified as type A (Pentchev *et al.*, 1980; Miyawaki *et al.*, 1982, 1986) Mouse (Adachi *et al.*, 1976; Tanaka *et al.*, 1988; Goldin *et al.*, 1992; Ohno *et al.*, 1992; Butler *et al.*, 1993; Higashi *et al.*, 1991, 1993) Boxer dog (Kuwamura *et al.*, 1993)
Chediak–Higashi syndrome	Unknown	Cats (Kramer *et al.*, 1977) Cattle (Padgett *et al.*, 1964) Foxes (Nes *et al.*, 1983) Killer whales (Taylor and Farrell, 1973) Mice (Lutzner *et al.*, 1967) Mink (Leader *et al.*, 1963)

sosomes within the cytoplasm can be seen as membrane-bound inclusions containing the stored substrate (Fig. 25.4). However, in some LSDs, the substrates that accumulate may be lost during tissue processing. The swollen lysosomes become large enough to be seen by light microscopy (Fig. 25.5). The accumulation of the primary substrate for a particular enzyme pathway may then interfere with other lysosomal hydrolases necessary for different degradative pathways (Kint *et al.*, 1973), thus leading to the secondary accumulation of these additional substrates. As substrates accumulate, the lysosomes swell and occupy more and more of the cytoplasm (Fig. 25.6). This increase in the number and size of lysosomes may obscure the other cellular organelles, and may deform the nuclear outline. As the process continues, the affected cells swell, leading to organomegaly. With the exception of the central nervous system (CNS), and cartilage and bone, the pathophysiology of these disorders is predominantly related to the increase in the cell, tissue, or organ size. For example, in the mucopolysaccharidoses, storage of glycosaminoglycans within the cells of the mitral heart valve causes these normally fusiform cells to become rounded (Fig. 25.6). This, in turn, causes the valve leaflet and cordae tendinea to become thick (Fig. 25.7), interfering with normal cardiac function and producing a murmur. Stor-

TABLE 25.3 Storage Disorders of Glycogen Metabolism That Are Not Lysosomal Storage Diseases

Disease	Deficient enzyme	Species and selected references
Glycogen storage disease Ia	Glucose-6-phosphatase	Maltese dog (Brix *et al.*, 1995, Kishnani *et al.*, 1995)
Glycogen storage disease III	Glycogen debranching	Dog (Ceh *et al.*, 1976)
Glycogen storage disease IV	Glycogen branching	Cat (Fyfe *et al.*, 1992, 1995)
Glycogen storage disease VI	Phosphorylase kinase	Mouse (Bender and Lalley, 1989; Barnard *et al.*, 1990) Rat (Malthus *et al.*, 1980; Clark *et al.*, 1986)
Glycogen storage disease VII	Muscle phosphofructokinase	English springer spaniel dog (Giger *et al.*, 1985, 1988, Smith *et al.*, 1995)

TABLE 25.4 Knockout Mouse Models of Lysosomal Storage Diseases

Disease	Deficient enzyme	Species and selected references
Glucocerebrosidosis (Gaucher disease)	Acid beta-glucosidase (glucocerebrosidase)	Tybulewicz *et al.*, 1992; McKinney *et al.*, 1995
Mucolipidosis-like without clinical signs	Cation-dependent Man-6-P receptor	Ludwig *et al.*, 1993
G_{M2} gangliosidosis (Tay–Sachs disease; Sandhoff disease)	Beta-hexosaminidase A; beta-hexosaminidase B	Yamanaka *et al.*, 1994; Sango *et al.*, 1995
Mucopolysaccharidosis I (Hurler; Scheie; Hurler-Scheie syndrome)	Alpha-L-iduronidase	Clarke *et al.*, 1996
Mucopolysaccharidosis VI (Maroteaux–Lamy Syndrome)	*N*-Acetylglucosamine 4-sulfatase (arylsulfatase B)	Peters *et al.*, 1994
Sphingomyelinosis A and B (Niemann–Pick A and B diseases)	Acid sphingomyelinase	Horinouchi *et al.*, 1995
Shindler disease	Alpha-*N*-acetylgalactosaminidase	Desnick *et al.*, (1995, personal communication)
No spontaneous disease described	Cathepsin D	Saftig *et al.*, 1995
Gangliosidosis + mucopolysaccharidosis	Alpha + beta subunits of beta-hexosaminidase	Sango *et al.*, 1996

age within the cells of the cornea (Fig. 25.8) results in reflection and refraction of light, producing the cloudiness observed grossly and by ophthalmoscopy (Fig. 25.9).

In many LSDs, the central nervous system contains swollen neurons (Fig. 25.10) and lysosomes which contain lamellar substrate (Fig. 25.11). The pathogenesis of the CNS lesions includes the development of meganeurites and neurite sprouting, which appear correlated with alterations in ganglioside metabolism (Purpura and Baker, 1977, 1978; Purpura *et al.*, 1978; Walkley, 1988; Walkley *et al.*, 1988, 1990, 1991; Siegel and Walkley, 1994). Gangliosides, whether stored as a primary substrate (in G_{M1} and G_{M2} gangliosidosis) or secondarily (in MPS I) appear to stimulate neurite sprout development. The presence of new neurites and their synapses apparently plays a role in the CNS dysfunction of these diseases.

I-cell disease (named for the inclusions seen in cultured fibroblasts; Tondeur *et al.*, 1971) is an exception to the usual pathogenesis of LSDs (reviewed in Kornfeld and Sly, 1995). As is frequently observed, exceptions can provide new information about normal processes. Studies of fibroblasts from patients with this disease were

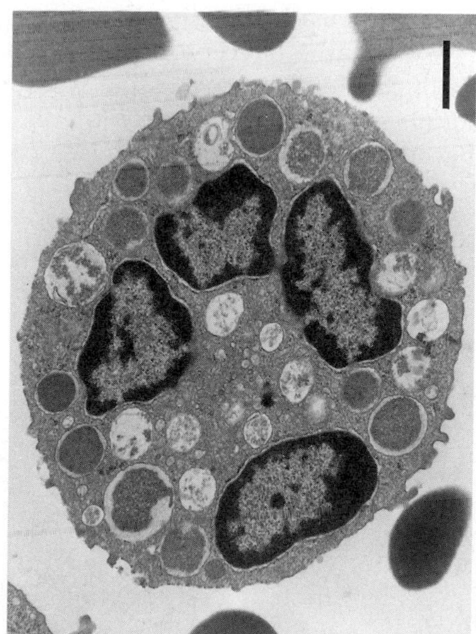

FIGURE 25.4 An electron micrograph of a polymorphonuclear leukocyte from a cat with MPS VI showing the enlarged lysosomes containing granular material (dermatan sulfate). Bar = 1 μm.

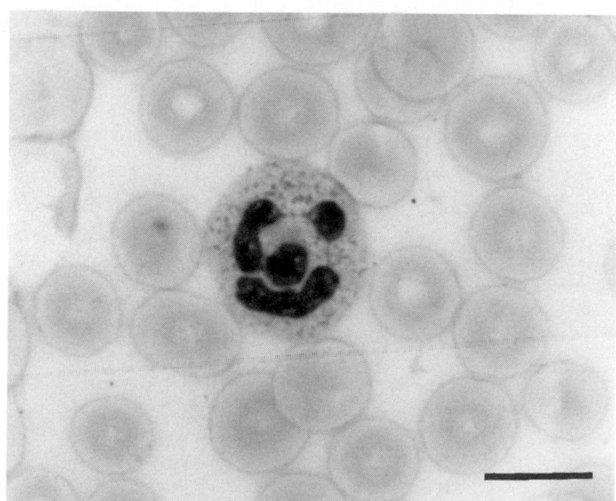

FIGURE 25.5 A light micrograph of a polymorphonuclear leukocyte from a dog with MPS VII, showing the cytoplasmic granules that represent the lysosomes containing GAG, which stain metachromatically with toluidine blue. Bar = 10 μm.

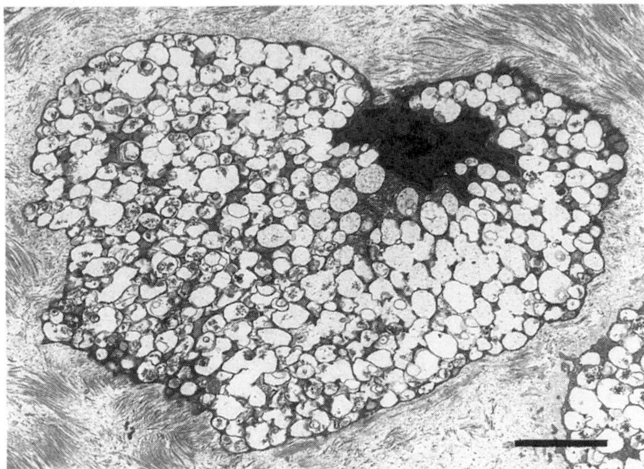

FIGURE 25.6 An electron micrograph of a cell from the mitral heart valve from a cat with MPS I. Note the extreme number of cytoplasmic vacuoles, the loss of recognition of other organelles, and the deformed nuclear outline. Bar = 3 μm.

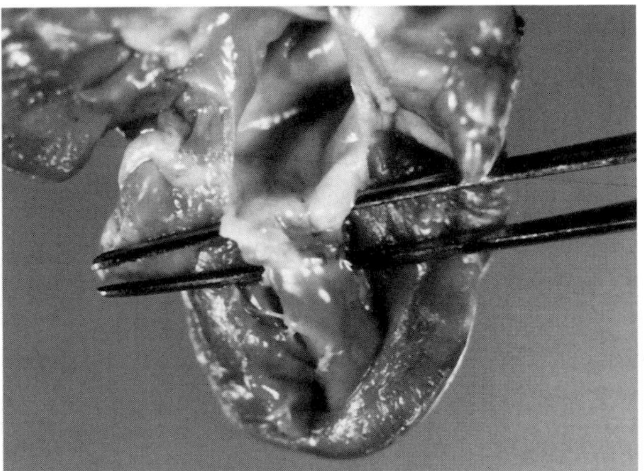

FIGURE 25.7 The mitral valve from a cat with MPS I, illustrating the thickened valve leaflets and cordae tendinae.

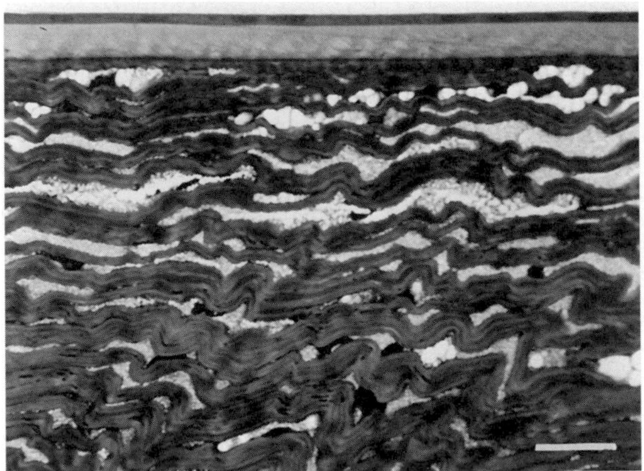

FIGURE 25.8 A light micrograph of the posterior cornea from a cat with MPS VI illustrating the highly vacuolated keratocytes. Bar = 25 μm.

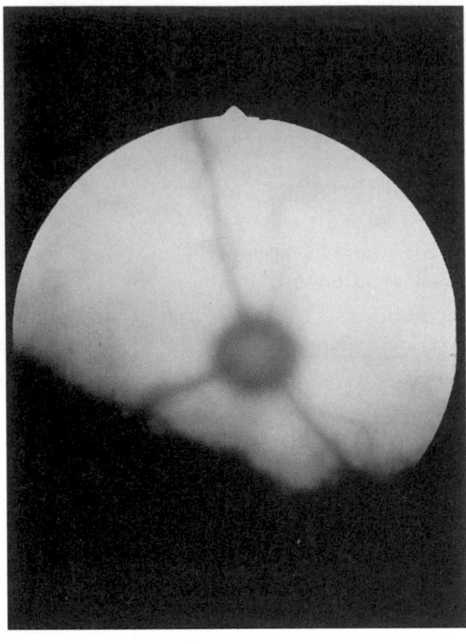

FIGURE 25.9 The appearence of the retina with indistinct optic disc and vessels of an MPS I cat as seen through the cloudy cornea.

seminal in providing insight into the Man-6-P transport system (Hickman and Neufeld, 1972). This disorder results from a failure in the first enzyme in the pathway responsible for the posttranslational phosphorylation of the mannose moiety of most lysosomal hydrolases (Reitman *et al.*, 1981; Hasilik *et al.*, 1981). The consequence of a defect in this phosphotransferase enzyme is production of lysosomal enzymes that lack the signal responsible for efficiently directing the enzymes to the lysosome by the Man-6-P receptor-mediated pathway. Thus, little of the enzyme reaches the lysosome, and large amounts are secreted extracellularly into the plasma. Because the phosphotransferase activity has

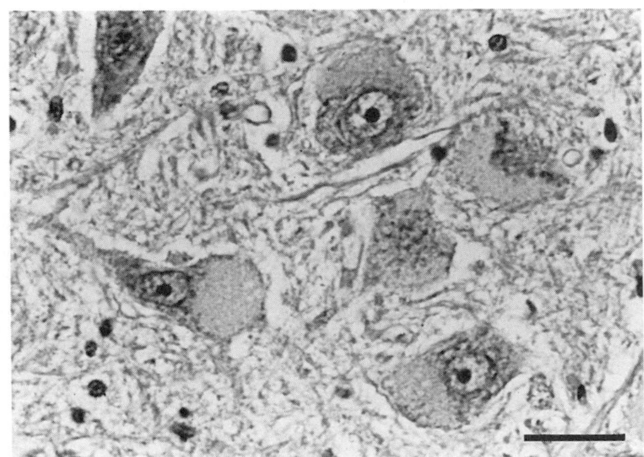

FIGURE 25.10 A light micrograph of swollen neurons in the facial nucleus in the brain of a cat with MPS I. Bar = 25 μm.

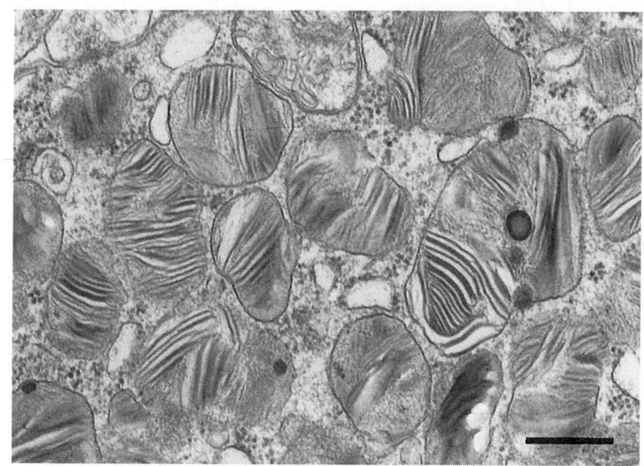

FIGURE 25.11 A electron micrograph of the lysosomes in a neuron from a cat with MPS I showing the lamellar inclusions. These inclusions are not typical of glycosaminoglycans, but rather represent glycolipids that accumulate secondary to the primary substrate storage. Bar = 0.5 μm.

been difficult to measure, the diagnosis of I-cell disease has usually been reached by demonstrating the low intracellular activity of most lysosomal enzymes, and consequent high enzyme activity in serum. Although a clinical and pathologic phenotype that combines all of the lysosomal storage diseases would be expected in I-cell disease, this does not occur. I-cell is a severe disease in children and cats, most of the pathology is found in mesenchymally derived cells; Kupffer cells and hepatocytes are essentially normal (Martin *et al.*, 1975, 1984). Mental retardation is present in children, and death is common before 8 years of age; however, there is relatively little CNS pathology (Nagashima *et al.*, 1977; Martin *et al.*, 1984). All cell types examined to date have been deficient in phosphotransferase activity, yet many organs (including liver, spleen, kidney, and brain) have normal intracellular lysosomal enzyme activities. This observation indicates that there is either an intracellular Man-6-P-independent pathway to lysosomes, or that secreted enzymes are internalized by cell surface receptors that recognize carbohydrates on enzymes, such as mannose, which is not phosphorylated (Waheed *et al.*, 1982). A Man-6-P-independent pathway to the lysosome has been demonstrated for β-glucocerebrosidase and acid phosphatase (Williams and Fukuda, 1990; Peters *et al.*, 1990).

IV. CLINICAL SIGNS

As a group, these disorders are usually chronic and progressive and have a wide spectrum of clinical severity, even within a particular disease. The predominant clinical signs are related to the CNS, skeleton, eye, cardiovascular system, and organomegaly. Most LSDs can be divided clinically into those with and without

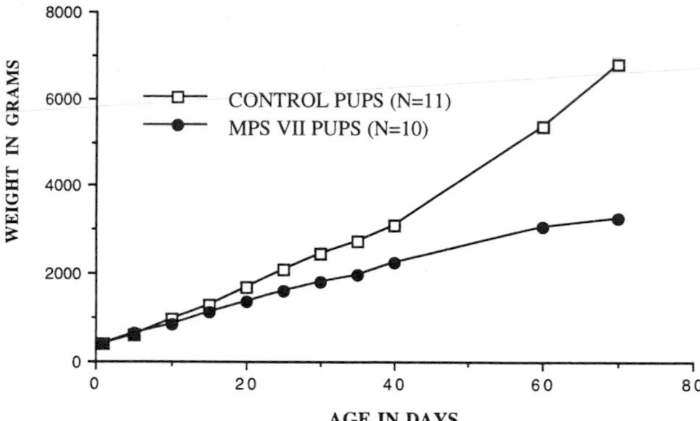

FIGURE 25.12 Growth retardation often seen in animals with lysosomal storage disorders. ●, pups with MPS VII (N = 10); □, control pups (normal littermates; N = 11).

CNS involvement. Head and limb tremors that progress to gait abnormalities, spastic quadriplegia, seizures, and death are commonly observed. The disorders in animals with marked clinical signs of CNS disease include fucosidosis, galactosylceramide lipidosis, G_{M1} and G_{M2} gangliosidosis, α- and β-mannosidosis, MPS III A and D, and sphingomyelinosis A and B.

Other clinical signs associated with lysosomal storage disorders are associated with failure to thrive, growth retardation (Fig. 25.12), umbilical hernia, corneal clouding, hepatosplenomegaly, cardiac murmurs, and skeletal abnormalities including facial dysmorphia and vertebral, rib, and long-bone deformities (Fig. 25.13). The MPS disorders, in general, affect more organ systems than the other diseases. The age of onset and severity of clinical signs is usually relatively consistent within a disease in animals; however, some variability can exist even in a family having the same

FIGURE 25.13 A normal Siamese cat next to a littermate with MPS VI. Note the outward manifestations of the skeletal abnormalities: flattened facies, small size, low posture associated with fusion of the cervical and lumbar spine.

TABLE 25.5 Diseases in Animals in Which a Mutation in the Coding Sequence of the Relevant Enzyme Has Been Determined

Disease	Deficient enzyme	Species and selected references
Fucosidosis	Alpha-fucosidase	Springer spaniel dog (Skelly *et al.*, 1996)
Galactosylceramide lipidosis	Galactosylceramidase (galactocerebroside beta-galactosidase)	Cairn terrier dog, West highland white terrier dog (Wenger, 1995, personal communication)
G_{M2} gangliosidosis	Beta-hexosaminidase A and B	Korat cat (Muldoon *et al.*, 1994)
Alpha mannosidosis	Alpha-mannosidase	Angus and Gallway cattle (Nilssen, 1995)
Beta mannosidosis	Beta-mannosidase	Anglo-Nubian goat (Haithcock *et al.*, 1995; Jones, 1995, personal communication) Saler cattle (Chen *et al.*, 1994)
Mucopolysaccharidosis I	Alpha-L-iduronidase	Domestic cat (Schuchman, 1995, personal communication) Plott hound dog (Menon *et al.*, 1992)
Mucopolysaccharidosis III D	*N*-Acetylglucosamine 6-sulfatase	Nubian goat (Cavanagh *et al.*, 1995)
Mucopolysaccharidosis VI	Arylsulfatase B	Rat (Kunieda *et al.*, 1995) Siamese cat (Yogalingam *et al.*, 1996)
Mucopolysaccharidosis VII	Beta-glucuronidase	GUS mouse (Sands and Birkenmeier, 1993) DSH cat (Fyfe *et al.*, 1996)

mutation. In research colonies of dogs and cats with LSDs kept in a relatively consistent environment, the explanation for varied clinical signs rests with the variable genetic background upon which the mutation is expressed.

Most LSDs are manifest within four months after birth, with some evident before weaning and fewer with adult onset (canine MPS IIIA). In severely affected animals, death often occurs before weaning. In research colonies, heterozygous matings are expected to produce, on average, 25% affected offspring. However, because of neonatal death, significantly fewer affected offspring reach weaning age.

In humans with LSDs, although not always consistently, specific mutations in some diseases have been associated with a particular pattern of clinical severity and progression. Null mutations that produce little RNA or unstable RNA resulting in no enzyme protein synthesis usually have a severe phenotype. Although specific mutations have recently been identified for several LSDs in animals (Table 25.5), there is still not enough information to be useful in prognosis.

V. DIAGNOSIS

The approach to a diagnosis of an LSD includes a complete history and physical examination with evaluation of the chest, abdomen, CNS, skeleton, and eyes. Laboratory tests should include a complete blood count with evaluation of granulocyte and lymphocyte morphology, skeletal radiographs, and urine screening for abnormal metabolites, particularly glycoaminoglycans. It may be helpful to establish a fibroblast culture from a skin biopsy, and to perform a liver biopsy. The disease may progress quickly, and the diagnosis may follow a complete postmortem examination.

A pedigree analysis should be performed as part of the history to determine information about the inbreeding of the parents and the presence of other family members with similar clinical signs or that died early. As most LSDs are inherited as autosomal recessive traits, parents are often related and are carriers (heterozygotes), but are clinically (phenotypically) normal. On average, one-fourth of the offspring of heterozygous parents are affected, two-thirds of unaffected offspring are carriers, and other relatives may also be affected (Fig. 25.14).

Abnormal metabolites may be found in urine, and their presence points toward specific metabolic pathways that warrant further evaluation. A metabolic screen of urine (Fig. 25.15; Jezyk *et al.*, 1982; Giger and Jezyk, 1992) for GAGs is a relatively simple and inexpensive approach to identify the mucopolysaccharidoses and some cases of gangliosidosis (toluidine blue spot test; Fig. 25.16). Thin-layer chromatography of urinary oligosaccharides is helpful in identifying mannosidosis. Urine samples to be evaluated should be kept refrigerated or frozen and sent on ice to an appropriate laboratory.[1]

A final diagnosis for LSDs requires the demonstration of a particular enzyme deficiency. Enzyme assays

[1] One such laboratory is the Metabolic Screening Laboratory, Section of Medical Genetics, Veterinary Hospital of the University of Pennsylvania, 3900 Delancy Street, Philadelphia, PA 19104-6010. A complete history including signalment of the animal should be included with the samples.

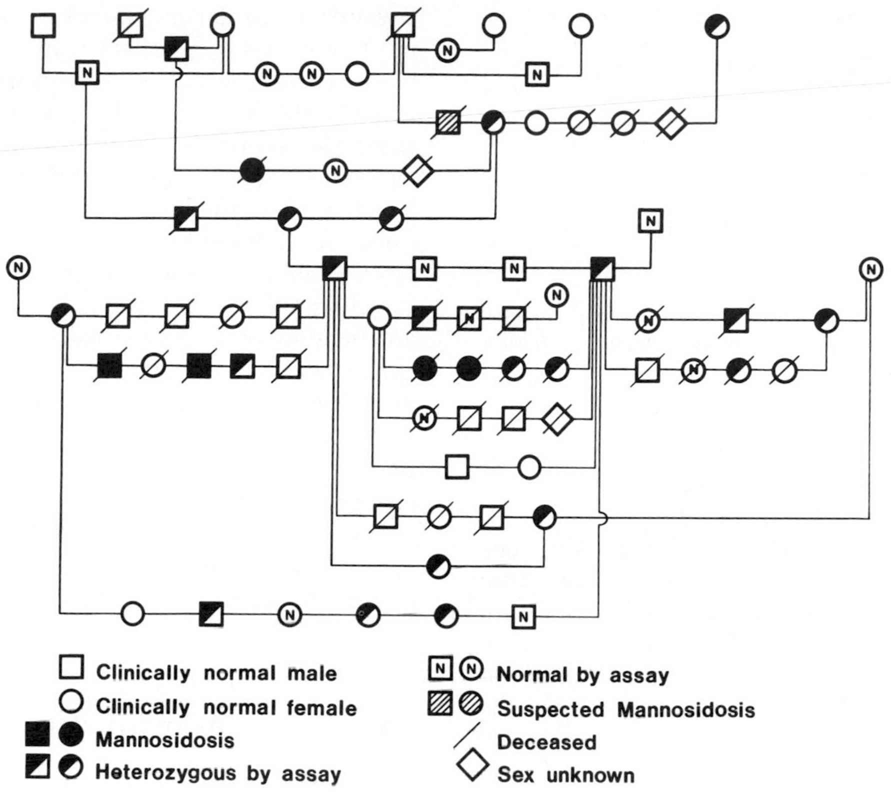

FIGURE 25.14 The pedigree of a family of cats with alpha-mannosidosis.

using artificial substrates can usually be performed on serum, white blood cells, cultured fibroblasts, or liver. Generally, there is a profound deficiency in activity of the enzyme, making the diagnosis straightforward. In addition, the activities of other lysosomal enzymes in the cells or tissues are frequently higher than normal. The biochemical status of the clinically normal parents should be evaluated when possible. In an autosomal recessive disease, heterozygous parents are expected to have half-normal activity of the enzyme in question because each parent carries one normal and one mutant allele. Although in a population heterozygotes (carriers) have on average half-normal activity, there is a marked overlap between the ranges for enzyme values from normal and obligate heterozygous animals (Fig. 25.17). Thus, accurate detection of an individual as normal or a carrier may be difficult with enzyme assays, and can best be achieved by molecular tests for the specific mutation in those diseases and families where the abnormality is known at the gene level (Table 25.5).

An animal suspected of having an LSD that dies or requires euthanasia should have a complete postmortem examination, including the central nervous, skele-

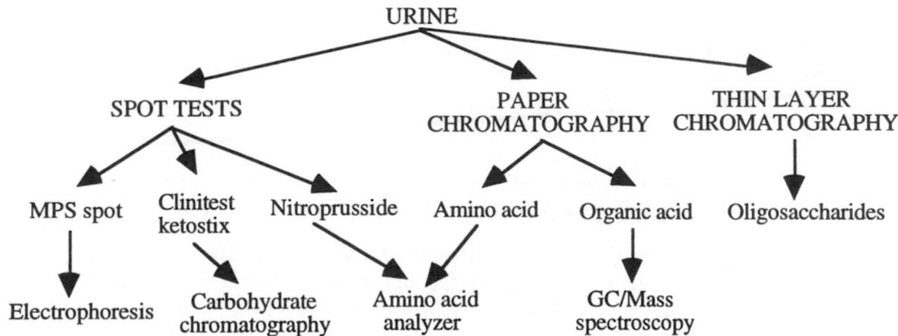

FIGURE 25.15 The scheme used to detect metabolic diseases, including some lysosomal storage disorders, by examining compounds present in urine.

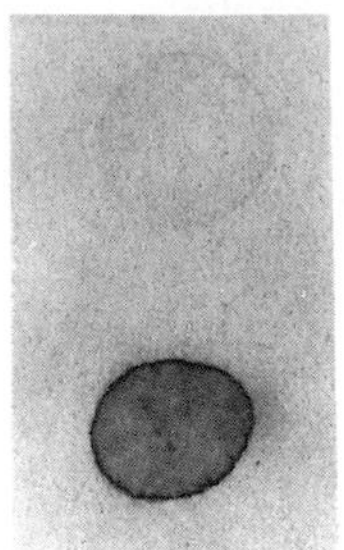

FIGURE 25.16 The results of a urine spot test detecting abnormal amounts of glycosaminoglycans in the urine of animals with MPS.

tal, and ocular systems. The liver should be frozen as quickly as possible for subsequent accumulated substrate and enzyme determinations, and DNA analysis. Fibroblast cultures can be established from skin, linea alba, or pericardium (using sterile technique) for future studies that may require living cells. Liver, brain, and other tissues should be preserved in formalin, and samples should be taken for thin section and electron microscopy (in gluteraldehyde), particularly from the liver and the CNS.

VI. THERAPY

The combination of secretion of lysosomal enzymes by cells and uptake of enzymes by diseased cells via the Man-6-P receptor system forms the basis for the present approaches to therapy for the LSDs. Providing a source of normal enzyme to abnormal cells will permit that enzyme to be taken up by the plasma membrane receptor, resulting in delivery of the normal enzyme to the lysosome where it can catabolize stored substrate. At present, the approaches to deliver normal enzyme to abnormal cells include (1) the parenteral injection of purified recombinant normal enzyme produced in Chinese hamster ovary cells (Barton *et al.*, 1990; Shull *et al.*, 1994), (2) heterologous bone-marrow transplantation, which provides normal bone marrow and bone-marrow-derived cells that release normal enzymes continuously (Haskins *et al.*, 1991), and (3) approaches to gene therapy where a viral vector is used to transfer a copy of the normal enzyme cDNA to a patient's own cells, which then act as a source of normal enzyme (Wolfe *et al.*, 1990, 1992). The amount of enzyme secreted by normal cells will affect the degree of correction achieved; the more enzyme is secreted, the more is available for uptake by diseased cells.

Acknowledgments

The authors wish to thank to Dr. Donald Patterson for his vision and for establishing the first Section of Medical Genetics in a veterinary school, Dr. Peter Jezyk for developing a metabolic screening laboratory for animals, Dr. Gus Aguirre for his contributions to the understanding of ocular disease, Mr. James Hayden of Bio-Graphics for the illustrations, Ms. Ping Wang for enzyme analyses, and Ms. Cathy Just, Ms. Margie Weil, and a cadre of veterinary students for animal care. Supported by NIH grants RR 02512, DK 25795, EY 07705, and NS 33526.

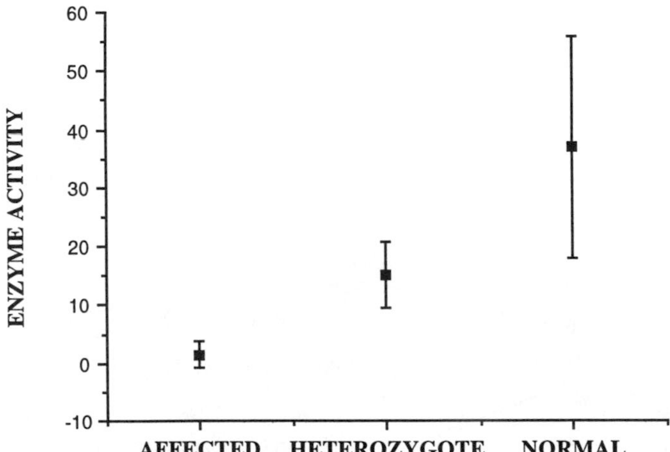

FIGURE 25.17 Mean serum alpha-mannosidase activity of a colony of related cats, illustrating the overlap that exists between normal and heterozygous animals. Although heterozygote detection is possible in a population, detection is more difficult for an individual. Molecular techniques will overcome these difficulties in carrier detection. (Error bars: ±2 SD.)

References

Abbitt, B., Jones, M. Z., Kasari, T. R., Storts, R. W., Templeton, J. W., Holland, P. S. and Castenson, P. E. (1991). Beta-mannosidosis in twelve Salers calves. *J. Vet. Med. Assoc.* **198,** 109–113.

Adachi, M., Volk, B. W., and Schneck, L. (1976). Animal model of human disease: Niemann–Pick Disease type C. *Am. J. Pathol.* **85,** 229–231.

Ahern-Rindell, A. J., Prieur, D. J., Murnane, R. D., Raghavan, S. S., Daniel, P. F., McCluer, R. H., Walkley, S. U., and Parish, S. M. (1988a). Inherited lysosomal storage disease associated with deficiencies of beta-galactosidase and alpha-neuraminidase in sheep. *Am. J. Med. Genet.* **31,** 39–56.

Ahern-Rindell, A. J., Murnane, R. D., and Prieur, D. J. (1988b). Beta-Galactosidase activity in fibroblasts and tissues from sheep with a lysosomal storage disease. *Biochem. Genet.* **26,** 733–746.

Ahern-Rindell, A. J., Murnane, R. D., and Prieur, D. J. (1989). Interspecific genetic complementation analysis of human and sheep fibroblasts with beta-galactosidase deficiency. *Somat. Cell Mol. Genet.* **15,** 525–533.

Alroy, J., Orgad, U., Ucci, A. A., Schelling, S. H., Schunk, K. L., Warren, C. D., Raghavan, S. S., and Kolodny, E. H. (1985). Neurovisceral and skeletal GM1-gangliosidosis in dogs with beta-galactosidase deficiency. *Science* **229,** 470–472.

Alroy, J., Orgad, U., DeGasperi, R., Richard, R., Warren, C. D., Knowles, K., Thalhammer, J. G., and Raghavan, S. S. (1992). Canine GM1-gangliosidosis. A clinical, morphologic, histochemical, and biochemical comparison of two different models. *Am. J. Pathol.* **140,** 675–689.

Appleby, E. C., Longstaffe, J. A., and Bell, F. R. (1982). Ceroid-lipofuscinosis in two saluki dogs. *J. Comp. Pathol.* **92,** 374.

Austin, J., Armstrong, D., and Margolis, G. (1968a). Studies of globoid leukodystrophy in dogs. *Neurology* **18,** 300.

Austin, J., Armstrong, D., and Margolis, G. (1968b). Canine globoid leukodystrophy: A model demyelinating disorder. *Trans. Am. Neurol. Assoc.* **93,** 181–182.

Baker, H. J., Jr., Lindsey, J. R., McKhann, G. M., and Farrell, D. F. (1971). Neuronal GM1 gangliosidosis in a Siamese cat with beta-galactosidase deficiency. *Science* **174,** 838–839.

Baker, H. J., Mole, J. A., Lindsey, J. R., and Creel, R. M. (1976). Animal models of human ganglioside storage diseases. *Fed. Proc.* **35,** 1193–1201.

Barker, C. G., Blakemore, W. F., Dell, A., Palmer, A. C., Tiller, P. R., and Winchester, B. G. (1986). GM1 gangliosidosis (type 1) in a cat. *Biochem. J.* **235,** 151–158.

Barnard, P. J., Derry, J. M., Ryder-Cook, A. S., Zander, N. F., and Kilimann, M. W. (1990). Mapping of the phosphorylase kinase alpha subunit gene on the mouse X chromosome. *Cytogenet. Cell Genet.* **53,** 91-94.

Barton, N. W., Furbish, F. S., Murray, G. J., Garfield, M., and Brady, R. O. (1990) Therapeutic response to intravenous infusions of glucocerebrosidase in a patient with Gaucher disease. *Proc. Natl. Acad. Sci. USA* **87,** 1913–1916.

Beck, E. (1974). Die Lipofuscinose doch keine Einzelerkrankung. *Kleintierpraxis* **19,** 60.

Bender, P. K., and Lalley, P. A. (1989). I/Lyn mouse phosphorylase kinase deficiency: Mutation disrupts expression of the α/α'-subunit mRNAs. *Proc. Natl. Acad. Sci. USA* **86,** 9996–10000.

Bernheimer, H., and Karbe, E. (1970). Morphologische und neurochemische Untersuchungen von 2 Formen der amaurotischen Idiotie des Hundes: Nachweis einer GM2-Gangliosidose. *Acta Neuropathol. (Berl),* **16,** 243–261.

Bichsel, P., and Vandevelde, M. (1982). Un cas de lipofuscinose cerodie chez un berger de Yougoslave. *Schweiz. Arch. Tierheilk.* **124,** 413.

Birkenmeier, E. H., Davisson, M. T., Beamer, W. G., Ganschow, R. E., Vogler, C. A., Gwynn, B., Lyford, K. A., Maltais, L. M., and Wawrzyniak, C. J. (1989). Murine mucopolysaccharidosis type VII. Characterization of a mouse with beta-glucuronidase deficiency. *J. Clin. Invest.* **83,** 1258–1256.

Blakemore, W. F. (1972). GM-1 gangliosidosis in a cat. *J. Comp. Path.* **82,** 179–185.

Bosshard, N. U., Hubler, M., Arnold, S., Briner, J., Spycher, M. A., Sommerlade, H-J., von Figura, K., and Gitzelmann, R. (1996). Feline mucolipidosis, the counterpart of human I-cell disease. *Vet. Pathol.* **33,** 1–13.

Boysen, B. G., Tryphonas, L., and Harries, N. W. (1974). Globoid cell leukodystrophy in the bluetick hound dog. I. Clinical manifestations. *Can. Vet. J.* **15,** 303–308.

Brix, A. E., Howerth, E. W., Mconkie-Rosell, A., Peterson, D., Egnor, D., Well, M. R., and Chen, Y-T. (1995). Glycogen storage disease type Ia in two littermate Maltese puppies. *Vet. Pathol.* **32,** 460–465.

Bryan, L., Schmutz, S., Hodges, S. D., and Snyder, F. F. (1993). Bovine beta-mannosidosis: Pathologic and genetic findings in Salers calves. *Vet. Pathol.* **30,** 130–139.

Bundza, A., Lowden, J. A., and Charlton, K. M. (1979). Niemann–Pick disease in a poodle dog. *Vet. Pathol.* **16,** 530–538.

Burditt, L. J., Chotai, K., Hirani, S., Nugent, P. G., Winchester, B. G., and Blakemore, W. F. (1980). Biochemical studies on a case of feline mannosidosis. *Biochem. J.* 189, 467–473.

Butler, J. D., Vanier, M. T., and Pentchev, P. G. (1993). Niemann–Pick C disease: cystine and lipids accumulate in the murine model of this lysosomal cholesterol lipidosis. *Biochem. Biophys. Res. Commun.* **196,** 154–159.

Castagnaro, M. (1990). Lectin histochemistry of the central nervous system in a case of feline alpha-mannosidosis. *Res. Vet. Sci.* **49,** 375-377.

Cavanagh, K. T., Leipprandt, J. R., Jones, M. Z., and Friderici, K. (1995). Molecular defect of caprine *N*-acetylglucosamine-6-sulfatase deficiency. A single base substitution creates a stop codon in the 5′-region of the coding sequence. *J. Inher. Metab. Dis.* **18,** 96.

Ceh, L., Hauge, J. G., Svenkerud, R., and Strande, A. (1976). Glycogenosis type III in the dog. *Acta Vet. Scand.* **17,** 210–222.

Chen, H., Leipprandt, J. R., Traviss, C. E., Sopher, B., Jones, M. Z., Cavanagh, K. T., and Friderici, K. H. (1994). Lysosomal beta-mannosidase: cDNA cloning and characterization. *Am. J. Human. Genet.* **55,** A130.

Cho, D. Y., Leipold, H. W., and Rudolph, R. (1986). Neuronal ceroidosis (ceroid-lipofuscinosis) in a blue heeler dog. *Acta Neuropathol. (Berl).* **69,** 161.

Chrisp, C. E., Ringler, D. H., Abrams, G. D., Radin, N. S., and Brenkert, A. (1970). Lipid storage disease in a Siamese cat. *J. Am. Vet. Med. Assoc.* **156,** 616–622.

Clark, D. G., Neville, S. D., Brinkman, M., Nelson, P. V., Illman, R. J., Guthberlet, A., and Haynes, W. D. (1986). Age-related augmentation of phosphorylase b kinase in hepatic tissue from the glycogen-storage-disease (GSD/GSD) rat. *Biochem. J.* **238,** 811–816.

Clarke, L. A., Russell, C., Warrington, C., Pownall, S., Borowsik, A., and Jirik, F. R. (1996). A murine model for MPS I. *Proc 4th International Symposium on Mucopolysaccharide and Related Diseases, Wollongong, Australia.*

Cork, L. C., Munnell, J. F., Lorenz, M. D., Murphy, J. V., Baker, H. J., and Rattazzi, M. C. (1977). GM2 ganglioside lysosomal storage disease in cats with beta-hexosaminidase deficiency. *Science* **196,** 1014–1017.

Cork, L. C., Munnell, J. F., and Lorenz, M. D. (1978). The pathology of feline GM2 gangliosidosis. *Am. J. Pathol.* **90,** 723–734.

Cowell, K. R., Jezyk, P. F., Haskins, M. E., and Patterson, D. F. (1976). Mucopolysaccharidosis in a cat. *J. Am. Vet. Med. Assoc.* **169,** 334–339.

Cummings, J. F., and de Lahunta, A. (1977). An adult case of canine neuronal ceroid-lipofuscinosis. *Acta Neuropathol. (Berl).* **39,** 43.

Cummings, J. F., Wood, P. A., Walkley, S. U., de Lahunta, A., and DeForest, M. E. (1985). GM2 gangliosidosis in a Japanese spaniel. *Acta Neuropathol. (Berl).* **67,** 247–253.

Cummings, J. F., Wood, P. A., de Lahunta, A., Walkley, S. U., and Le Boeuf, L. (1988). The clinical and pathologic heterogeneity of feline alpha-mannosidosis. *J. Vet. Intern. Med.* **2,** 163–170.

de Duve, C., Pressman, B. C., Ginnetto, R., Wattiaus, R., and Apelmans, F. (1955). Tissue fractionation studies. VI. Intracellular distribution patterns of enzymes in rat-liver tissue. *Biochem. J.* **60,** 604–617.

Di Natale, P., Annella, T., Daniele, A., Spagnuolo, G., Cerundolo, R., de Caprariis, D., and Gravino, A. E. (1992). Animal models for lysosomal storage diseases: A new case of feline mucopolysaccharidosis VI. *J. Inherit. Metab. Dis.* **15,** 17–24.

Distler, J., Hieber, V., Sahagian, G., Schmickel, R., and Jourdian, G. W. (1979). Identification of mannose 6-phosphate in glycoproteins that inhibit the assimilation of beta-galactosidase by fibroblasts. *Proc. Natl. Acad. Sci. USA* **76,** 4235–4239.

Dobrenis, K., Wenger, D. A., and Walkley, S. U. (1994). Extracellular release of lysosomal glycosidases in cultures of cat microglia. *Molec. Biol. Cell* **5,** 113a.

Donnelly, W. J., Sheahan, B. J., and Kelly, M. (1973a). Beta-Galactosidase deficiency in GM1 gangliosidosis of Friesian calves. *Res. Vet. Sci.* **15,** 139–41.

Donnelly, W. J., Sheahan, B. J., and Rogers, T. A. (1973b). GM1 gangliosidosis in Friesian calves. *J. Pathol.* **111,** 173–179.

Duchen, L. W., Eicher, E. M., Jacobs, J. M., Scaravilli, F., and Teixeira, F. (1980). Hereditary leucodystrophy in the mouse: The new mutant twitcher. *Brain* **103,** 695–710.

Embury, D. H., and Jerrett, I. V. (1985). Mannosidosis in Galloway calves. *Vet. Pathol.* **22,** 548–551.

Fankhauser, R. (1965). Degenerative lipidotische Erkrankung des Zentralnervensystems bei zwei Hunden. *Schweiz. Arch. Tierheilk.* **107,** 73.

Fankhauser, R., Luginbuhl, H., and Hartley, W. J. (1963). Leukodystrophie vom Typus Krabbe beim Hund. *Schweiz. Arch. Tierheilk.* **105,** 198–207.

Farrell, D. F., Baker, H. J., Herndon, R. M., Lindsey, J. R., and McKhann, G. M. (1973). Feline GM 1 gangliosidosis: Biochemical and ultrastructural comparisons with the disease in man. *J. Neuropathol. Exp. Neurol.* **32,** 1–18.

Farrow, B. R., Hartley, W. J., Pollard, A. C., Fabbro, D., Grabowski, G. A., and Desnick, R. J. (1982). Gaucher disease in the dog. *Prog. Clin. Biol. Res.* **95,** 645–653.

Fischer, A., Carmichael, K.P., Munnell, J.F., Matalon, R., and Giger, U. (1996). Mucopolysaccharidosis IIIA (Sanfilippo A) in dachshunds. *Proc. 14th ACVIM Forum,* 678–679.

Fletcher, T. F., and Kurtz, H. J. (1972). Animal model: Globoid cell leukodystrophy in the dog. *Am. J. Pathol.* **66,** 375–378.

Fletcher, T. F., Kurtz, H. J., and Low, D. G. (1966). Globoid cell leukodystrophy (Krabbe type) in the dog. *J. Am. Vet. Med. Assoc.* **149,** 165–172.

Fletcher, T. F., Lee, D. G., and Hammer, R. F. (1971). Ultrastructural features of globoid-cell leukodystrophy in the dog. *Am. J. Vet. Res.* **32,** 177–181.

Fletcher, T. F., Suzuki, K., and Martin, F. B. (1977). Galactocerebrosidase activity in canine globoid leukodystrophy. *Neurology* **27,** 758–766.

Friderici, K., Cavanagh, E. T., Leipprandt, J. R., Traviss, C. E., Anson, D. S., Hopwood, J. J., Jones, M. Z. (1995). Cloning and sequence analysis of caprine *N*-acetylglucosamine 6-sulfatase cDNA. *Biochim. Biophys. Acta.—Molec. Basis of Dis.* **1271,** 369–373.

Friend, S. C., Barr, S. C., and Embury, D. (1985). Fucosidosis in an English springer spaniel presenting as a malabsorption syndrome. *Aust. Vet. J.* **62,** 415–420.

Fujita, T., Nonaka, I., and Sugita, H. (1991). Japanese quail and human acid maltase deficiency: A comparative study. *Brain Dev.* **13,** 247–255.

Fyfe, J. C., Giger, U., Van Winkle, T. J., Haskins, M. E., Steinberg, S. A., Wang, P., and Patterson, D. F. (1992). Glycogen storage disease type IV: Inherited deficiency of branching enzyme activity in cats. *Pediatr. Res.* **32,** 719–725.

Fyfe, J. C., Hawkins, M. G., and Henthorn, P. (1995). Molecular characterization of feline glycogen storage disease type IV. *Am. J. Hum. Genet.* **57,** A212.

Fyfe, J. C, Kurzhals, R. L., Giger, U., Haskins, M. E., Patterson, D. F., Wang, P., Wolfe, J. H., Yunhki, N., and Henthorn, P. S. (1996). A missense mutation causes beta-glucuronidase deficiency in feline MPS VII. *Am. J. Hum. Genet.* **59,** A197.

Gambetti, L. A., Kelly, A. M., and McGrath, J. T. (1970). Biochemical studies in a canine gangliosidosis. *J. Neuropath. Exp. Neurol.* **29,** 137.

Giger, U., and Jezyk, P. F. (1992). Diagnosis of inherited diseases in small animals. *In* "Current Veterinary Therapy XI. Small Animal Practice" (R. W. Kirk and J. D. Bonagura, eds.), pp. 18–22. Saunders, Philadelphia.

Giger, U., Harvey, J. W., Yamagucchi, R. A., McNulty, P. K., Chiapella, A., and Beutler, E. (1985). Inherited phosphofructokinase deficiency in dogs with hyperventilation-induced hemolysis: Increased *in vitro* and *in vivo* alkaline fragility of erythrocytes. *Blood* **65,** 234–351.

Giger, U., Kelly, A. M., and Teno, P. S. (1988). Biochemical studies of canine muscle phosphofructokinase deficiency. *Enzym* **40,** 25–29.

Gitzelmann, R., Bosshard, N. U., Superti-Furga, A., Spycher, M. A., Briner, J., Wiesmann, U., Lutz, H., and Litschi, B. (1994). Feline mucopolysaccharidosis VII due to beta-glucuronidase deficiency. *Vet. Pathol.* **31,** 435–443.

Goebel, H. H., Bilzer, T., Dahme, E., and Dalkusch , F. (1988). Morphological studies in canine (Dalmatian) neuronal ceroid-lipofuscinosis. *Am. J. Med. Genet.* **5,** 127.

Goldin, E., Roff, C. F., Miller, S. P., Rodriguez-Lafrasse, C., Vanier, M. T., Brady, R. O., and Pentchev, P. G. (1992). Type C Niemann–Pick disease: A murine model of the lysosomal cholesterol lipidosis accumulates sphingosine and sphinganine in liver. *Biochim. Biophys. Acta* **19,** 303–311.

Green, P. D., and Little, P. B. (1974). Neuronal ceroid-lipofuscin storage in Siamese cats. *Can. J. Comp. Med.* **38,** 207–212.

Haithcock, B. E., Leipprandt, J. R., Kraemer, S., Cavanagh, K. T., Friderici, K., Dyme, J. L., Chen, H., Oiao, X. T., and Jones, M. Z. (1995). Molecular sequencing of caprine beta-mannosidase cDNA. *Am. J. Hum. Genet.* **57,** A144.

Handa, S., and Yamakawa, T. (1971). Biochemical studies in cat and human gangliosidosis. *J. Neurochem.* **18,** 1275–1280.

Harper, P. A., Walker, K. H., Healy, P. J., Hartley, W. J., Gibson, A. J., and Smith, J. S. (1988). Neurovisceral ceroid-lipofuscinosis in blind Devon cattle. *Acta Neuropathol.* **75,** 632–636.

Hartley, W. J., and Blakemore, W. F. (1973). Neurovisceral glucocerebroside storage (Gaucher's disease) in a dog. *Vet. Pathol.* **10,** 191–201.

Hasilik, A., Waheed, A., and von Figura, K. (1981). Enzymatic phosphorylation of lysosomal enzymes in the presence of UDP-N-acetylglucosamine. Absence of the activity in I-cell fibroblasts. *Biochem. Biophys. Res. Commun.* **98,** 761–767.

Haskins, M. E., Jezyk, P. F., Desnick, R. J., McDonough, S. K., and Patterson, D. F. (1979a). Alpha-L-iduronidase deficiency in a cat: A model of mucopolysaccharidosis I. *Pediatr. Res.* **13,** 1294–1297.

Haskins, M. E., Jezyk, P. F., Desnick, R. J., McDonough, S. K., and Patterson, D. F. (1979b). Mucopolysaccharidosis in a domestic shorthaired cat—a disease distinct from that seen in the Siamese cat. *J. Am. Vet. Med. Assoc.* **175,** 384–387.

Haskins, M. E., Jezyk, P. F., and Patterson, D. F. (1979c). Mucopolysaccharide storage disease in three families of cats with arylsulfatase B deficiency: Leukocyte studies and carrier identification. *Pediatr. Res.* **13,** 1203–1210.

Haskins, M. E., Aguirre, G. D., Jezyk, P. F., and Patterson, D. F. (1980). The pathology of the feline model of mucopolysaccharidosis VI. *Am. J. Pathol.* **101,** 657–674.

Haskins, M. E., Aguirre, G. D., Jezyk, P. F., Desnick, R. J., and Patterson, D. F. (1983). The pathology of the feline model of mucopolysaccharidosis I. *Am. J. Pathol.* **112,** 27–36.

Haskins, M. E., Desnick, R. J., DiFerrante, N., Jezyk, P. F., and Patterson, D. F. (1984). Beta-glucuronidase deficiency in a dog: A model of human mucopolysaccharidosis VII. *Pediatr. Res.* **18,** 980–984.

Haskins, M. E., Baker, H. J., Birkenmeier, E., Hoogerbrugge, P. M., Poorthuis, B. J. H. M., Sakiyama, T., Shull, R. M., Taylor, R. M., Thrall, M. A., and Walkley, S. U. (1991). Transplantation in animal model systems. *In* "Treatment of Genetic Diseases" (R. Desnick, ed.), pp. 183–201 Churchill Livingstone, New York.

Healy, P. J., Farrow, B. R., Nicholas, F. W., Hedberg, K., and Ratcliffe, R. (1984). Canine fucosidosis: A biochemical and genetic investigation. *Res. Vet. Sci.* **36,** 354–359.

Healy, P. J., Kidd, G. N., Reuter, R. E., Bunce, C., Hosie, I., and Stapleton, T. (1992). Beta-mannosidosis in Salers calves in Australia. *Aust. Vet. J.* **69,** 145.

Hickman, S., and Neufeld, E. F. (1972). A hypothesis for I-cell disease: Defective hydrolases that do not enter lysosomes. *Biochem. Biophys. Res. Commun.* **49,** 992–999.

Higashi, Y., Pentchev, P. G., Murayama, S., and Suzuki, K. (1991). Pathology of Niemann–Pick type C: Studies of murine mutants. *In* "Neuropathology in Brain Research" (F. Ikkuta, ed.), p. 85. Elsevier Science, Amsterdam.

Higashi, Y., Murayama, S., Pentchev, P. G., and Suzuki, K. (1993). Cerebellar degeneration in the Niemann-Pick type C mouse. *Acta Neuropathol. (Berl).* **85,** 175–184.

Higuchi, I., Nonaka, I., Usuki, F., Ishiura, S., and Sugita H. (1987). Acid maltase deficiency in the Japanese quail; early morphological event in skeletal muscle. *Acta. Neuropathol. (Berl).* **73,** 32–37.

Hirth, R. S., and Nielsen, S. W. (1967). A familial canine globoid cell leukodystrophy ("Krabbe Type"). *J. Small. Anim. Pract.* **8,** 569–575.

Hocking, J. D., Jolly, R. D., and Batt, R. D. (1972). Deficiency of alpha-mannosidase in Angus cattle. An inherited lysosomal storage disease. *Biochem. J.* **128,** 69–78.

Holmes, E. W., and O'Brien, J. S. (1978a). Hepatic storage of oligosaccharides and glycolipids in a cat affected with GM1 gangliosidosis. *Biochem. J.* **175,** 945–953.

Holmes, E. W., and O'Brien, J. S. (1978b). Feline GM1 gangliosidosis: Characterization of the residual liver acid beta-galactosidase. *Am. J. Hum. Genet.* **30,** 505–515.

Holtzman, E. (1989). Historical fragments; methods; source terminology. *In* "Lysosomes," (P. Siekevitz, ed.), pp. 1–24. Plenum Press, New York.

Horinouchi, K., Sakiyama, T., Pereira, L., Lalley, P. A., and Schuchman, E. H. (1993). Mouse models of Niemann–Pick disease: mutation analysis and chromosomal mapping rule out the type A and B forms. *Genomic* **18,** 450–451.

Horinouchi, K., Erlich, S., Perl, D. P., Ferlinz, K., Bisgaier, C. L., Sandhoff, K., Desnick, R. J., Stewart, C. L., and Schuchman E. H. (1995). Acid sphingomyelinase deficient mice: a model of types A and B Niemann–Pick disease. *Nature Genet.* **10,** 288–293.

Howell, J. M., and Palmer, A. C. (1971). Globoid cell leucodystrophy in two dogs. *J. Small Anim. Pract.* **12,** 633–642.

Howell, J. M., Dorling, P. R., Cook, R. D., Robinson, W. F., Bradley, S., and Gawthorne, J. M. (1981). Infantile and late onset form of generalised glycogenosis type II in cattle. *J. Pathol.* **134,** 267-277.

Hubler, M., Haskins, M. E., Arnold, S., Kaser-Hotz, B., Bosshard, N. U., Briner, J., Spycher, M. A., Gitzelmann, R., Sommerlade, H.-J., von Figura, K., and Rüsch, P. (1995). Mucolipidosis Type II in a domestic shorthair cat. *J. Sm. Anim. Pract.* **37,** 435–441.

Ishikawa, Y., Li, S. C., Wood, P. A., and Li, Y. T. (1987). Biochemical basis of type AB GM2 gangliosidosis in a Japanese spaniel. *J. Neurochem.* **48,** 860-864.

Järplid, B., and Haltia, M. (1993). An animal model of the infantile type of neuronal ceroid-lipofuscinosis. *J Inher. Metab. Dis.* **16,** 274–277.

Jezyk, P. F., Haskins, M. E., Patterson, D. F., Mellman, W. J., and Greenstein, M. (1977). Mucopolysaccharidosis in a cat with arylsulfatase B deficiency: A model of Maroteaux–Lamy syndrome. *Science* **198,** 834–836.

Jezyk, P. F., Haskins, M. E., and Patterson, D. F. (1982). Screening for inborn errors of metabolism in dogs and cats. *Prog. Clin. Biol. Res.* **94,** 93–116.

Jezyk, P. F., Haskins, M. E., and Newman, L. R. (1986). Alpha-mannosidosis in a Persian cat. *J. Am. Vet. Med. Assoc.* **189,** 1483–1485.

Johnson, K. H. (1970). Globoid leukodystrophy in the cat. *J. Am. Vet. Med. Assoc.* **157,** 2057–2064.

Johnson, G. R., Oliver, J. E., Jr., and Selcer, R. (1975). Globoid cell leukodystrophy in a Beagle. *J. Am. Vet. Med. Assoc.* **167,** 380–384.

Jolly, R. D. (1971). The pathology of the central nervous system in pseudolipidosis of Angus calves. *J. Pathol.* **103,** 113–121.

Jolly, R. D. (1974). Animal model of human disease: Mannosidosis of children, other inherited lysosomal storage diseases. *Am. J. Pathol.* **74,** 211–214.

Jolly, R. D. (1975). Mannosidosis of Angus cattle: A prototype control programme for some genetic diseases. *In* "Advances in Veterinary Science and Comparative Medicine" (C. E. Cornelius and C. A. Brandley, eds.), Vol. 19, p. 1-21. Academic Press, New York.

Jolly, R. D. (1978). Mannosidosis and its control in Angus and Murray Grey cattle. *NZ Vet. J.* **26,** 194–198.

Jolly, R. D., Tse, C. A., and Greenway, R. M. (1973). Plasma mannosidase activity as a means of detecting mannosidosis heterozygotes. *NZ Vet. J.* **21,** 64–69.

Jolly, R. D., Digby, J. G. and Rammell, C. G. (1974). A mass screening programme of Angus cattle for the mannosidosis genotype. A prototype programme for control of inherited diseases in animals. *NZ Vet. J.* **2,** 218–222.

Jolly, R. D., Thompson. K. G., and Winchester, B. G. (1975). Bovine mannosidosis—a model lysosomal storage disease. *Birth Defects* **11,** 273–278.

Jolly, R. D., Slack, P. M., Winter, P. J., and Murphy, C. E. (1980a). Mannosidosis: Patterns of storage and urinary excretion of oligosaccharides in the bovine model. *Aust. J. Exp. Biol. Med. Sci.* **58,** 421–428.

Jolly, R. D., Janmaat, A., West, D. M., and Morrison, I. (1980b). Ovine ceroid-lipofuscinosis: A model for Batten's Disease. *Neuropathol. Appl. Neurobiol.* **6,** 195–205.

Jolly, R. D., Shimada, A., Craig, A. S., Kirkland, K. B., and Palmer, D. N. (1988). Ovine ceroid-lipofuscinosis. II: Pathologic changes interpreted in light of biochemical observations. *Am. J. Med. Genet.* **5,** 159–170.

Jolly, R. D., Shimada, A., Dopfmer, I., Slack, P. M., Birtles, M. J., and Palmer, D. N. (1989). Ceroid-lipofuscinosis (Batten's disease): Pathogenesis and sequential neuropathological changes in the ovine model. *Neuropathol. Appl. Neurobiol.* **15,** 371–383.

Jolly, R. D., Shimada, A., Dopfmer, I., Slack, P. M., Birtles, M. J., and Palmer, D. N. (1992). Ceroid-lipofuscinosis (Batten's disease): Pathogenesis and sequential neuropathological changes in the ovine model. *Neuropathol. Appl. Neurobiol.* **15,** 371–383.

Jones, M. Z., and Dawson, G. (1981). Caprine beta-mannosidosis. Inherited deficiency of beta-D-mannosidase. *J. Biol. Chem.* **256,** 5185–5188.

Jones, M. Z., Cunningham, J. G., Dade, A. W., Alessi, D. M., Mostosky, U. V., Vorro, J. R., Benitez, J. T., and Lovell, K. L. (1983). Caprine beta-mannosidosis: Clinical and pathological features. *J. Neuropathol. Exp. Neurol.* **42,** 268–285.

Jortner, B. S., and Jonas, A. M. (1968). The neuropathology of globoid-cell leucodystrophy in the dog. A report of two cases. *Acta Neuropathol. (Berl).* **10,** 171–182.

Kaplan A., Achord, D. T., and Sly, W. S. (1977). Phosphohexosyl components of a lysosomal enzyme are recognized by pinocytosis receptors on human fibroblasts. *Proc. Natl. Acad. Sci. USA* **74,** 2026–2030.

Karbe, E. (1973). Animal model of human disease GM2-gangliosidoses (amaurotic idiocies) type I, II, and 3. *Am. J. Path.* **71,** 151–154.

Karbe, E., and Schiefer, B. (1967). Familial amaurotic idiocy in male German shorthair pointers. *Pathol. Vet.* **4,** 223–232.

Kaye, E. M., Alroy, J., Raghavan, S. S., Schwarting, G. A., Adelman, L. S., Runge, V., Gelblum, D., Thalhammer, J. G., and Zuniga, G. (1992). Dysmyelinogenesis in animal model of GM1 gangliosidosis. *Pediatr. Neurol.* **8,** 255–261.

Kelly, W. R., Clague, A. E., Barns, R. J., Bate, M. J., and MacKay, B. M. (1983). Canine alpha-L-fucosidosis: A storage disease of Springer spaniels. *Acta Neuropathol. (Berl).* **60,** 9–13.

Kint, J. A., Dacremont, G., Carton, D., Orye, E., and Hooft, C. (1973). Mucopolysaccharidosis: Secondarily induced abnormal distribution of lysosomal isoenzymes. *Science* **181,** 352–354.

Kishnani, P. S., Bao, Y., Wu, J. Y., Brix, A. E., and Chen, Y. T. (1995). Mutation in canine glucose-6-phosphatase deficiency (glycogen storage disease type Ia, GSD-Ia). *Am. J. Hum. Genet.* **57,** A216.

Kobayashi, T., Yamanaka, T., Jacobs, J. M., Teixeira, F., and Suzuki, K. (1980). The Twitcher mouse: An enzymatically authentic model of human globoid cell leukodystrophy (Krabbe disease). *Brain Res.* **202,** 479–483.

Koppang, N. (1969). Neuronal ceroid-lipofuscinosis in English Setters. *J. Small Anim. Pract.* **10,** 639–644.

Koppang, N. (1973). Canine ceroid-lipofuscinosis—a model for human neuronal ceroid-lipofuscinosis and aging. *Mech. Aging Dev.* **2,** 421–445.

Koppang, N. (1979). Neuronal ceroid-lipofuscinosis (Batten's disease). *In* "Spontaneous Animal Models of Human Disease" (E. J. Andrews, B. C. Ward, and N. H. Altman, eds.), Vol. 2, p. 136. Academic Press, New York.

Koppang, N. (1988). The English setter with ceroid-lipofuscinosis: A suitable model for the juvenile type of ceroid-lipofuscinosis in humans. *Am. J. Med. Genet. Suppl.* **5,** 117–125.

Korneluk, R. G., Mahuran, D. J., Neote, K., Klavins, M. H., O'Dowd, B. F., Tropak, M., Willard, H. F., Anderson, M. J., Lowden, J. A., and Gravel, R. A. (1986). Isolation of cDNA clones coding for the alpha-subunit of human beta-hexosaminidase. Extensive homology between the alpha- and beta-subunits and studies on Tay–Sachs disease. *J. Biol. Chem.* **261,** 8407–8413.

Kornfeld, S. (1987). Trafficking of lysosomal enzymes. *FASEB J.* **1,** 463.

Kornfeld, S., and Sly, W. S. (1995). I-cell Disease and pseudo-Hurler polydystrophy: Disorders of lysosomal enzyme phosphorylation and localization. *In* "The Metabolic and Molecular Bases of Inherited Disease," Seventh Edition, Vol. II (C. R. Scriver, A. L. Beaudet, W. S. Sly, and D. Valle, eds.), pp. 2495–2508. McGraw Hill, Inc., New York.

Kosanke, S. D., Pierce, K. R., and Bay, W. W. (1978). Clinical and biochemical abnormalities in porcine GM2-gangliosidosis. *Vet. Pathol.* **15,** 685–699.

Kosanke, S. D., Pierce, K. R., and Read, W. K. (1979). Morphogenesis of light and electron microscopic lesions in porcine GM2-gangliosidosis. *Vet. Pathol.* **16,** 6–17.

Kramer, J. W., Davis, W. C., and Prieur, D. J. (1977). The Chediak–Higashi syndrome of cats. *Lab. Invest.* **36,** 554–562.

Kumar, K., Jones, M. Z., Cunningham, J. G., Kelley, J. A. and Lovell, K. L. (1986). Caprine beta-mannosidosis: Phenotypic features. *Vet. Rec.* **118,** 325–327.

Kunieda, T., Simonaro, C. M., Yoshida, M., Ikadai, H., Levan, G., Desnick, R. J., and Schuchman, E. H. (1995). Mucopolysaccharidosis type VI in rats: Isolation of cDNAs encoding arylsulfatase B, chromosomal localization of the gene, and identification of the mutation. *Genomics* **29,** 582–587.

Kuwamura, M., Awakura, T., Shimada, A., Umemura, T., Kagota, K., Kawamura, N., and Naiki, M. (1993). Type C Niemann–Pick disease in a boxer dog. *Acta Neuropathol. (Berl).* **85,** 345–348.

Laws, L., and Saal, J. R. (1968). Lipidosis of the hepatic reticuloendothelial cells in a sheep. *Aust. Vet. J.* **44,** 416–417.

Leader, R. W., Padgett, G. A., and Gorham, J. R. (1963). Studies of abnormal leukocyte bodies in the mink. *Blood* **22,** 477–484.

Lerner, T. J., and the International Batten Disease Consortium. (1995). Isolation of a novel gene underlying Batten disease, CLN3. *Cell* **82,** 949–957.

Lovell, K. L., and Jones, M. Z. (1983). Distribution of central nervous system lesions in beta-mannosidosis. *Acta Neuropathol. (Berl).* **62,** 121–126.

Lowenthal, A. C., Cummings, J. F., Wenger, D. A., Thrall, M. A., Wood, P. A., and de Lahunta, A. (1990). Feline sphingolipidosis resembling Niemann–Pick disease type C. *Acta Neuropathol. (Berl).* **81,** 189–197.

Ludwig, T., Ovitt, C. E., Bauer, U., Hollinshead, M., Remmler, J., Lobel, P., Ruther, U., and Hoflack, B. (1993). Targeted disruption of the mouse cation-dependent mannose 6-phosphate receptor results in partial missorting of multiple lysosomal enzymes. *EMBO J.* **12,** 5225–5235.

Lutzner, M. A., Lowrie, C. T., and Jordan, H. W. (1967). Giant granules in leukocytes of the beige mouse. *J. Hered.* **58,** 299–300.

Luzi, P., Mohammad, A. Rafi, Victoria, T., Baskin, G. B., and Wenger, D. A. (1997). Characterization of the Rhesus monkey galactocerebrosidase (GALC) cDNA and gene and identification of the mutation causing globoid cell leukodystrophy (Krabbe disease) in the primate. *Genomics* (in press).

McGovern, M. M., Mandell, N., Haskins, M., and Desnick, R. J. (1985). Animal model studies of allelism: Characterization of arylsulfatase B mutations in homoallelic and heteroallelic (genetic compound) homozygotes with feline mucopolysaccharidosis VI. *Genetics* **110,** 733–749.

McGrath, J. T., Kelly, A. M., and Steinberg, S. A. (1968). Cerebral lipidosis in the dog. *J. Neuropathol. Exp. Neurol.* **27,** 141.

McGrath, J. T., Schutta, H. Yashen, A., and Steinberg, A. (1969). A morphology and biochemical study of canine globoid leukodystrophy. *J. Neuropathol. Exp. Neurol.* **28,** 171.

McKinney, C. E., LaMarcia, M. E., Kondo, R., and Ginns, E. I. (1995). Murine models of Gaucher disease: Introduction of the N370S point mutation by tag and exchange. *Am. J. Hum. Genet.* **57,** A181.

Maenhout, T., Kint, J. A., Dacremont, G., Ducatelle, R., Leroy, J. G., and Hoorens, J. K. (1988). Mannosidosis in a litter of Persian cats. *Vet Rec,* 122, 351–354.

Malkusch, F. (1984). "Beitrag zur bioptischen Diagnose der neuronalen Ceroid-Lipofuscinose (NCL) des Hundes." Dissertation, Institut für Tierpathologie der Universität München.

Malthus, R., Clark, D. G., Watts, C., and Sneyd, J. G. (1980). Glycogen-storage disease in rats, a genetically determined deficiency of liver phosphorylase kinase. *Biochem. J.* **188,** 99–106.

Manktelow, B. W., and Hartley, W. J. (1975). Generalized glycogen storage disease in sheep. *J. Comp. Path.* **85,** 139–145.

Martin, J. J., Leroy, J. G., Farriaux, J. P., Fontaine, G., Desnick, R. J., and Cabello, A. (1975). I-cell disease (mucolipidosis II): A report on its pathology. *Acta Neuropathol. (Berl).* **33,** 285–305.

Martin, J. J., Leroy, J. G., van Eygen, M., and Ceuterick, C. (1984). I-cell disease. A further report on its pathology. *Acta Neuropathol. (Berl).* **64,** 234–242.

Menon, K. P., Tieu, P. T., and Neufeld, E. F. (1992). Architecture of the canine IDUA gene and mutation underlying canine mucopolysaccharidosis I. *Genomics* **14,** 763–768.

Miyawaki, S., Mitsuoka, S., Sakiyama, T., and Kitagawa, T. (1982). Sphingomyelinosis, a new mutation in the mouse: A model of Niemann–Pick disease in humans. *J. Hered.* **73,** 257–263.

Miyawaki, S., Yoshida, H., Mitsuoka, S., Enomoto, H., and Ikehara, S. (1986). A mouse model for Niemann–Pick disease. Influence of genetic background on disease expression in spm/spm mice. *J. Hered.* **77,** 379–384.

Mostafa, I. E. (1970). A case of glycogenic cardiomegaly in a dog. *Acta Vet. Scand.* **11,** 197–208.

Muldoon, L. L., Neuwelt, E. A., Pagel, M. A., and Weiss, D. L. (1994). Characterization of the molecular defect in a feline model for type II GM2-gangliosidosis (Sandhoff disease). *Am. J. Pathol.* **144,** 1109–1118.

Munana, K. R., Luttgen, P. J., Thrall, M. A., Mitchell, T. W., and Wenger, D. A. (1994). Neurological manifestations of Niemann–Pick disease type C in cats. *J. Vet. Intern. Med.* **8,** 117–121.

Murnane, R. D., Prieur, D. J. Ahern-Rindell, A. J., Parish, S. M., and Collier, L. L. (1989). The lesions of an ovine lysosomal storage disease. Initial characterization. *Am. J. Pathol.* **134,** 263–270.

Murnane, R. D., Hartley, W. J., and Prieur, D. J. (1991a). Similarity of lectin histochemistry of a lysosomal storage disease in a New Zealand lamb to that of ovine GM1 gangliosidosis. *Vet. Pathol.* **28,** 332–335.

Murnane, R. D., Wright, R. W., Jr., Ahern-Rindell, A. J., and Prieur, D. J. (1991b). Prenatal lesions in an ovine fetus with GM1 gangliosidosis. *Am. J. Med. Genet.* **39,** 106–111.

Murnane, R. D., Prieur, D. J., Ahern-Rindell, A. J., Holler, L. D., and Parish, S. M. (1994). Clinical and clinicopathologic characteristics of ovine GM-1 gangliosidosis. *J. Vet. Intern. Med.* **8,** 221–223.

Myerowitz, R., and Hogikyan, N. D. (1986). Different mutations in Ashkenazi Jewish and non-Jewish French Canadians with Tay–Sachs disease. *Science* **232,** 1646–1648.

Myerowitz, R., and Hogikyan, N. D. (1987). A deletion involving Alu sequences in the beta-hexosaminidase alpha-chain gene of French Canadians with Tay–Sachs disease. *J. Biol. Chem.* **262,** 15396–15399.

Myerowitz, R., and Proia, R. L. (1984). cDNA clone for the alpha-chain of human beta-hexosaminidase: Deficiency of alpha-chain mRNA in Ashkenazi Tay–Sachs fibroblasts. *Proc. Natl. Acad. Sci. USA* **81,** 5394–5398.

Myerowitz, R., Piekarz, R., Neufeld, E. F., Shows, T. B., and Suzuki, K. (1985). Human beta-hexosaminidase alpha-chain: Coding sequence and homology with the beta chain. *Proc. Natl. Acad. Sci. USA* **82,** 7830–7834.

Nagashima, K., Sakakibara, K., Endo, H., Konishi, Y., Nakamura, N., Suzuki, Y., and Abe, T. (1977). I-cell disease (mucolipidosis 11). Pathological and biochemical studies of an autopsy case. *Acta Pathol. Jpn.* **27,** 251–264.

Natowicz, M. R., Chi, M. M., Lowry, O. H., and Sly, W. S. (1979). Enzymatic identification of mannose 6-phosphate on the recognition marker for receptor-mediated pinocytosis of beta-glucuronidase by human fibroblasts. *Proc. Natl. Acad. Sci. USA* **76,** 4322–4326.

Neer, T. M., Dial, S. M., Pechman, R., Wang, P., and Giger, U. (1992). Mucopolysaccharidosis VI (Maroteaux–Lamy Syndrome) in a miniature pinscher. *J. Vet. Internal Med.* **6,** 124.

Neer, T. M., Dial, S. M., Pechman, R., Wang, P., and Giger, U. (1995). Mucopolysaccharidosis VI (Maroteaux–Lamy Syndrome) in a miniature pinscher. *J. Vet. Internal Med.* **9,** 429–433.

Nes, N., Lium, B., Braend, M., and Sjaastad, O. (1983). A Chediak–Higashi-like syndrome in Arctic blue foxes. *Finsk. Veterinaertidsskrift* **89,** 313.

Neufeld, E. F., and Muenzer, J. (1995). The mucopolysaccharidoses. *In* "The Metabolic and Molecular Bases of Inherited Disease," Seventh Edition, Vol. II (C. R. Scriver, A. L. Beaudet, W. S. Sly, and D. Valle, eds.), pp. 2465–2508. McGraw Hill, Inc., New York.

Neuwelt, E. A., Johnson, W. G., Blank, N. K., Pagel, M. A., Maslen-McClure, C., McClure, M. J., and Wu, P. M. (1985). Characterization of a new model of GM2-gangliosidosis (Sandhoff's disease) in Korat cats. *J Clin Invest,* **76,** 482–490.

Nilssen, O. (Presented at the American Society of Human Genetics in conjunction with Nilssen, O., Berg, T., Rubenthiran, U., Hansen, G. M., Riise, H. M., Tranejaerg, L. Malm, D., and Tollersrud, O. K.) (1995). α-Mannosidosis: Functional cloning of the α-mannosidase gene and identification of a mutation in affected siblings. *Am. J. Hum. Genet.* **57,** A39.

Nimmo Wilkie, J. S., and Hudson, E. B. (1982). Neuronal and generalized ceroid-lipofuscinosis in a cocker spaniel. *Vet. Pathol.* **19,** 623–628.

Nunoya, T., Tajima, M., and Mizutani, M. (1983). A new mutant of Japanese quail (*Coturnix coturnix japonica*) characterized by generalized glycogenosis. *Lab. Anim.* **17,** 138–142.

Ohno, K., Nanba, E., Miyawaki, S., Sakiyama, T., Kitagawa, T., and Takeshita, K. (1992). A cell line derived from sphingomyelinosis mouse shows alterations in intracellular cholesterol metabolism similar to those in type C Niemann–Pick disease. *Cell Struct. Funct.* **17,** 229–235.

Okada, S. and O'Brien, J. S. (1969). Tay–Sachs disease: Generalized absence of a beta-D-*N*-acetylhexosaminidase component. *Science* **165,** 698–700.

O'Sullivan, B. M., Healy, P. J., Fraser, I. R., Nieper, R. E., Whittle, R. J., and Sewell, C. A. (1981). Generalised glycogenosis in Brahman cattle. *Aust. Vet. J.* **57,** 227–229.

Padgett, G. A., Leader, R. W., Gorham, J. R., and O'Mary, C. C. (1964). The familial occurrence of the Chediak–Higashi syndrome in mink and cattle. *Genetics* **49,** 505–512.

Patterson, J. S., Jones, M. Z., Lovell, K. L., and Abbitt, B. (1991). Neuropathology of bovine beta-mannosidosis. *J. Neuropathol. Exp. Neurol.* **50,** 538–546.

Pentchev, P. G., Gal, A. E., Boothe, A. D., Fouks, J., Omodeo-Sale, F., and Brady, R. O. (1980). A lysosomal storage disorder in mice characterized by the accumulation of several sphingolipids. *Birth Defects* **16,** 225–230.

Pentchev, P. G., Boothe, A. D., Kruth, H. S., Weintroub, H., Stivers, J., and Brady R. O. (1984). A genetic storage disorder in BALB/C mice with a metabolic block in esterification of exogenous cholesterol. *J. Biol. Chem.* 259, 5784–5791.

Peters, C., Braun, M., Weber, B., Wendland, M., Schmidt, B., Pohlmann, R., Waheed, A., and von Figura, K. (1990). Targeting of a lysosomal membrane protein: A tyrosine-containing endocytosis signal in the cytoplasmic tail of lysosomal acid phosphatase is necessary and sufficient for targeting to lysosomes. *EMBO J.* **9,** 3497–3506.

Peters, C., Evers, M., Saftig, P., Hess, B., and von Figura, K. (1994). Mucopolysaccharidosis VI (Maroteaux–Lamy Syndrome): Generation and characterization of a mouse model. *Gene Therapy,* **4,** S18.

Phillips, N. C., Robinson, D., Winchester, B. G. and Jolly, R. D. (1974). Mannosidosis in Angus cattle. The enzymic defect. *Biochem. J.* **137,** 363–371.

Pierce, K. R., Kosanke, S. D., Bay, W. W., and Bridges, C. H. (1976). Animal model of human disease: GM2 gangliosidosis. *Am. J. Pathol.* **83,** 419–422.

Prieur, D. J., Wilkerson, M. J., Lewis, D. C., Kennaway, N. G., Toone, J. R., Applegarth, D. A., Vallance, H., Marks, S. L., and Wood, R. K. (1995). Iduronate-2-sulfatase deficiency in a dog: Canine Hunter syndrome. *Am. J. Hum. Genet.* **57,** A182

Pritchard, D. H., Napthine, D. V., and Sinclair, A. J. (1980). Globoid cell leucodystrophy in polled Dorset sheep. *Vet. Pathol.* **17,** 399–405.

Proia, R. L., and Soravia, E. (1987). Organization of the gene encoding the human beta-hexosaminidase alpha-chain. *J. Biol. Chem.* **262,** 5677 (published erratum appears in *J. Biol. Chem.* **262,** 15322).

Purpura, K. P., and Baker, H. J. (1977). Neurite induction in mature cortical neurones in feline GM1-ganglioside storage disease. *Nature* **266,** 553–554.

Purpura, D. P., Pappas, G. D., and Baker, H. J. (1978). Fine structure of meganeurites and secondary growth processes in feline GM1-gangliosidosis. *Brain Res.* **143,** 1–12.

Purpura, D. P., and Baker, H. J. (1978). Meganeurites and other aberrant processes of neurons in feline GM1-gangliosidosis: A Golgi study. *Brain Res.* **143,** 13–26.

Rac, R., and Giesecke, P. R. (1975). Lysosomal storage disease in chihuahuas. *Aust. Vet. J.* **51,** 403–404.

Raghavan, S., Stuer, G., Riviere, L., Alroy, J., and Kolodny, E. H. (1988). Characterization of alpha-mannosidase in feline mannosidosis. *J. Inherit. Metab. Dis.* **11,** 3–16.

Read, W. K., and Bridges C. H. (1968). Cerebrospinal lipodystrophy in swine. A new disease model in comparative pathology. *Pathol. Vet.* **5,** 67–74.

Read, D. H., Harrington, D. D., Keenana, T. W., and Hinsman, E. J. (1976). Neuronal–visceral GM1 gangliosidosis in a dog with beta-galactosidase deficiency. *Science* **194,** 442–445.

Reitman, M. L., Varki, A., and Kornfeld, S. (1981). Fibroblasts from patients with I-cell disease and pseudo-Hurler polydystrophy are deficient in uridine 5′-diphosphate-*N*-acetylglucosamine: glycoprotein N-acetylglucosaminylphosphotransferase activity. *J. Clin. Invest.* **67,** 1574–1579.

Richards, R. B., Edwards, J. R., Cook, R. D., and White, R. R. (1977). Bovine generalized glycogenosis. *J. Neuropath. Appl. Neurobiol.* **3,** 46–56.

Riis, R. C., Cummings, J. F., Loew, E. R., and de Lahunta, A. (1992). Tibetan terrier model of canine ceroid lipofuscinosis. *Am. J. Med. Genet.* **42,** 615–621.

Rittmann, L. S., Tennant, L. L., and O'Brien, J. S. (1980). Dog GM1 gangliosidosis: Characterization of the residual liver acid beta-galactosidase. *Am. J. Hum. Genet.* **32,** 880–889.

Rodriguez, M., O'Brien, J. S., Garrett, R. S., and Powell, H. C. (1982). Canine GM1 gangliosidosis. An ultrastructural and biochemical study. *J. Neuropathol. Exp. Neurol.* **41,** 618–629.

Sachs, B. (1887). On arrested cerebral development with special reference to its cortical pathology. *J. Nerv. Ment. Dis.* **14,** 541–553.

Saftig, P., Hetman, M., Wolfgang, S., Weber, K., Heine, L., Mossmann, H. Köster, Hess, B., Evers, M., von Figura, K., and Peters, C. (1995). Mice deficient for the lysosomal proteinase cathepsin D exhibit progressive atrophy of the intestinal mucosa and profound destruction of lymphoid cells. *EMBO J.* **14,** 3599–3608.

Sandhoff, K. (1969). Variation of beta-*N*-acetylhexosaminidase pattern in Tay–Sachs disease. *FEBS Lett.* **4,** 351–354.

Sandhoff, K., Harzer, K., and Fürst, W. (1995). Sphingolipid activator proteins. *In* "The Metabolic and Molecular Bases of Inherited Disease" Seventh Edition, Vol. II (C. R. Scriver, A. L. Beaudet, W. S. Sly, and D. Valle, eds.), pp. 2427–2441. McGraw Hill, Inc., New York.

Sandison, A. T., and Anderson, L. J. (1970). Histiocytosis in two pigs and a cow: conditions resembling lipid storage disorders in man. *J. Pathol.* **100,** 207–210.

Sandoff, K., Harzer, K., and Fürst, W. (1995). Sphingolipid activator proteins. *In* "The Metabolic and Molecular Bases of Inherited Disease," Seventh Edition, Vol. II (C. R. Scriver, A. L. Beaudet, W. S. Sly, and D. Valley eds.), pp. 247–241. McGraw Hill, Inc., New York.

Sands, M. S., and Birkenmeier, E. H. (1993). A single-base-pair deletion in the beta-glucuronidase gene accounts for the phenotype of murine mucopolysaccharidosis type VII. *Proc. Natl. Acad. Sci. USA* **90,** 6567–6571.

Sandstrom, B., Westman, J., and Ockerman, P. A. (1969). Glycogenosis of the central nervous system in the cat. *Acta Neuropathol.* **14,** 194–200.

Sango, K., Yamanaka, S., Hoffmann, A., Okuda, Y., Grinberg, A., Westphal, H., McDonald, M. P., Crawley, J. N., Sandhoff, K., Suzuki, K., and Proia, R. L. (1995). Mouse models of Tay–Sachs and Sandhoff diseases differ in neurologic phenotype and ganglioside metabolism. *Nature Genet.* **11,** 170–176.

Saunders, G. K., Wood, P. A., Myers, R. K., Shell, L. G., and Carithers, R. (1988). GM1 gangliosidosis in Portuguese water dogs: pathologic and biochemical findings. *Vet. Pathol.* **25,** 265–269.

Schmidt, B., Selmer, T., Ingendoh, A., and von Figura, K. (1995). A novel amino acid modification in sulfatases that is defective in multiple sulfatase deficiency. *Cell* **82,** 271–278.

Schuchman, E. H., Toroyan, T. K., Haskins, M. E., and Desnick, R. J. (1989). Characterization of the defective beta-glucuronidase activity in canine mucopolysaccharidosis type VII. *Enzyme* **42,** 174–180.

Shell, L. G., Potthoff, A. I., Carithers, R., Katherman, A., Saunders, G. K., Wood, P. A., and Giger, U. (1989). Neuronal–visceral GM1 gangliosidosis in Portuguese water dogs. *J. Vet. Intern. Med.* **3,** 1–7.

Shull, R. M., and Hastings, N. E. (1985). Fluorometric assay of alpha-L-iduronidase in serum for detection of affected and carrier animals in a canine model of mucopolysaccharidosis I. *Clin. Chem.* **31,** 826–827.

Shull, R. M., Munger, R. J., Spellacy, E., Hall, C. W., Constantopoulos, G., and Neufeld, E. F. (1982). Canine alpha-L-iduronidase deficiency. A model of mucopolysaccharidosis I. *Am. J. Pathol.* **109,** 244–248.

Shull, R. M., Helman, R. G., Spellacy, E., Constantopoulos, G., Munger, R. J., and Neufeld, E. F. (1984). Morphologic and biochemical studies of canine mucopolysaccharidosis I. *Am. J. Pathol.* **114,** 487–495.

Shull, R. M., Kakkis, E. D., McEntee, M. F., Kania, S. A., Jonas, A. J., and Neufeld, E. F. (1994). Enzyme replacement in a canine model of Hurler syndrome. *Proc. Natl. Acad. Sci. USA* **91,** 12937–12941.

Siegel, D. A., and Walkley, S. U. (1994). Growth of ectopic dendrites on cortical pyramidal neurons in neuronal storage diseases correlates with abnormal accumulation of GM2 ganglioside. *J. Neurochem.* **62,** 1852–1862.

Singer, H. S., and Cork, L. C. (1989). Canine GM2 gangliosidosis: morphological and biochemical analysis. *Vet. Pathol.* **26,** 114–120.

Skelly, B. J., Sargan, D. R., Herrtage, M. E., and Winchester, B. G. (1996). The molecular defect underlying canine fucosidosis. *J. Med. Genet.* 33, 284–288.

Smith, B. F., Stedman, H., Rajpurohit, Y., Henthorn, P., Wolfe, J. H., Patterson, D. F., and Giger, U. (1995). The molecular basis of canine muscle-type phosphofructokinase deficiency. *J. Biol. Chem.* **271,** 20070–20074.

Smith, M. O., Wenger, D. A., Hill, S. L., Matthews, J. (1996). Fucosidosis in a family of American-bred English springer spaniels. *J. Am. Vet. Med. Assoc.* **209,** 2088–2090.

Snyder, S. P., Kingston, R. S., and Wenger, D. A. (1982). Niemann–Pick disease. Sphingomyelinosis of Siamese cats. *Am. J. Pathol.* **108,** 252–254.

Spellacy, E., Shull, R. M., Constantopoulos, G., and Neufeld, E. F. (1983). A canine model of human alpha-L-iduronidase deficiency. *Proc. Natl. Acad. Sci. USA* **80,** 6091–6095.

Stoltzfus, L. J., Sosa-Pineda, B., Moskowitz, S. M., Menon, K. P., Dlott, B., Hooper, L., Teplow, D. B., Shull, R. M., and Neufeld, E. F. (1992). Cloning and characterization of cDNA encoding canine alpha-L-iduronidase. mRNA deficiency in mucopolysaccharidosis I dog. *J. Biol. Chem.* **267,** 6570–6575.

Suhara, Y., Ishiura, S., Tsukahara, T., and Sugita, H. (1989). Mature 98,000-dalton acid alpha-glucosidase is deficient in Japanese quails with acid maltase deficiency. *Muscle Nerve* **12,** 670–678.

Suzuki, Y., Austin, J., Armstrong, D., Suzuki, K., Schlenker, J., and Fletcher, T. (1970). Studies in globoid leukodystrophy: Enzymatic and lipid findings in the canine form. *Exp. Neurol.* **29,** 65–75.

Suzuki, Y., Miyatake, T., Fletcher, T. F., and Suzuki, K. (1974). Glycosphingolipid beta-galactosidases. 3. Canine form of globoid cell leukodystrophy; comparison with the human disease. *J. Biol. Chem.* **249,** 2109–2112.

Svennerholm, L. (1962). The chemical structure of normal human brain and Tay–Sachs gangliosides. *Biochem. Biophys. Res. Commun.* **9,** 436–446.

Tanaka, J., Nakamura, H., and Miyawaki, S. (1988). Cerebellar involvement in murine phingomyelinosis: A new model of Niemann–Pick disease. *J. Neuropathol. Exp. Neurol.* **47,** 291–300.

Tay, W. (1881). Symmetrical changes in the region of the yellow spot in each eye of an infant. *Trans. Ophthalmol. Soc. UK* **1,** 155–157.

Taylor, R. F., and Farrell, R. K. (1973). Light and electron microscopy of peripheral blood neutrophils in a killer whale affected with Chediak–Higashi syndrome. *Fed. Proc.* **32,** 822.

Taylor, R. M., and Farrow, B. R. (1988). Ceroid-lipofuscinosis in border collie dogs. *Acta Neuropathol. (Berl).* **75,** 627–631.

Thompson, J. N., Jones, M. Z., Dawson, G., and Huffman, P. S. (1992). *N*-Acetylglucosamine 6-sulfatase deficiency in a Nubian goat: A model of Sanfilippo syndrome type D (mucopolysaccharidosis IIID). *J. Inherit. Metab. Dis.* **15,** 760–768.

Tondeur, M., Vamos-Hurwitz, E., Mockel-Pohl, S., Dereume, J. P., Cremer, N., and Loeb, H. (1971). Clinical, biochemical, and ultrastructural studies in a case of chondrodystrophy presenting the I-cell phenotype in tissue culture. *J. Pediatr.* **79,** 366–378.

Tybulewicz, V. L., Tremblay, M. L., LaMarca, M. E., Willemsen, R., Stubblefield, B. K., Winfield, S., Zablocka, B., Sidransky, E., Martin, B. M., Huang, S. P., Mintzer, K. A., Westphal, H., Mulligan, R. C., and Ginns, E. I. (1992). Animal model of Gaucher's disease from targeted disruption of the mouse glucocerebrosidase gene. *Nature* **357,** 407–410.

Van De Water, N. S., Jolly, R. D., and Farrow, B. R. (1979). Canine Gaucher disease–the enzymic defect. *Aust. J. Exp. Biol. Med. Sci.* **57,** 551–554.

Vandevelde, M., and Fatzer, R. (1980). Neuronal ceroid-lipofuscinosis in older dachshunds. *Vet. Pathol.* **17,** 686–92.

Vandevelde, M., Fankhauser, R., Bichsel, P., Wiesmann, U., and Herschkowitz, N. (1982). Hereditary neurovisceral mannosidosis associated with alpha-mannosidase deficiency in a family of Persian cats. *Acta Neuropathol. (Berl).* **58,** 64–68.

Vesa, J., Hellsten, E., Verkruyse, L. A., Camp, L. A., Rapola, J., Santavuori, P., Hofmann, S. L., and Peltonen, L. (1995). Mutations in the palmitoyl protein thioesterase gene causing infantile neuronal ceroid lipofuscinosis. *Nature* **376,** 584–587.

Vogler, C., Birkenmeier, E. H., Sly, W. S., Levy, B., Pegors, C., Kyle, J. W., and Beamer, W. G. (1990). A murine model of mucopolysaccharidosis VII. Gross and microscopic findings in beta-glucuronidase-deficient mice. *Am. J. Pathol.* **136,** 207–217.

Waheed, A., Pohlmann, R., Hasilik, A., von Figura, K., van Elsen, A., and Leroy, J. G. (1982). Deficiency of UDP-*N*-acetylglucosamine: Lysosomal enzyme *N*-acetylglucosamine-1-phosphotransferase in organs of I-cell patients. *Biochem. Biophys. Res. Commun.* **105,** 1052–1058.

Walkley, S. U. (1988). Pathobiology of neuronal storage disease. *International Review Neurobiol.* **29,** 191–244.

Walkley, S. U., Haskins, M. E., and Shull, R. (1988). Alterations in neuron morphology in mucopolysaccharidosis I: A Golgi study. *Acta Neuropathol.* **75,** 611–620.

Walkley, S. U., Wurzelmann, S., Rattazzi, M. C., and Baker, H. J. (1990). Distribution of ectopic neurite growth and other geometrical distortions of CNS neurons in feline GM2 gangliosidosis. *Brain Res.* **510,** 63–73.

Walkley, S. U., Baker, H. J., Rattazzi, M. C., Haskins, M. E., and Wu, J.-Y. (1991). Neuroaxonal dystrophy in neuronal storage disorders: Evidence for major GABAergic neuron involvement. *J. Neurol. Sci.* **104,** 1–8.

Walvoort., H. C. (1983). Glycogen storage diseases in animals and their potential value as models of human disease [review]. *J. Inherit. Metab. Dis.* **6,** 3–16.

Walvoort, H. C., Slee, R. G., and Koster, J. F. (1982). Canine glycogen storage disease type II. A biochemical study of an acid alpha-glucosidase-deficient Lapland dog. *Biochim. Biophys. Acta* **715,** 63–69.

Walvoort, H. C., Dormans, J. A., and van den Ingh, T. S. (1985). Comparative pathology of the canine model of glycogen storage disease type II (Pompe's disease). *J. Inherit. Metab. Dis.* **8,** 38–46.

Warren, C. D., Alroy, J., Bugge, B., Daniel, P. F., Raghavan, S. S., Kolodny, E. H., Lamar, J. J., and Jeanloz, R. W. (1986). Oligosaccharides from placenta: Early diagnosis of feline mannosidosis. *FEBS Lett.* **195,** 247–252.

Wenger, D. A., Sattler, M., Kudoh, T., Snyder, S. P., and Kingston, R. S. (1980). Niemann–Pick disease: A genetic model in Siamese cats. *Science* **208,** 1471–1473.

Williams, M. A., and Fukuda, M. (1990). Accumulation of membrane glycoproteins in lysosomes requires a tyrosine residue at a particular position in the cytoplasmic tail. *J. Cell Biol.* **111,** 955–66.

Wisselaar, H. A., Hermans, M. M., Visser, W. J., Kroos, M. A., Oostra, B. A., Aspden, W., Harrison, B., Hetzel, D. J., Reuser, A. J., and Drinkwater, R. D. (1993). Biochemical genetics of glycogenosis type II in Brahman cattle. *Biochem. Biophys. Res. Commun.* **190,** 941–947.

Wolfe, J. H., Schuchman, E. H., Stramm, L. E., Concaugh, E. A., Haskins, M. E., Aguirre, G. D., Patterson, D. F., Desnick, R. J., and Gilboa, E. (1990). Restoration of normal function in mucopolysaccharidosis type VII cells by retroviral vector-mediated gene transfer. *Proc. Natl. Acad. Sci.* **87,** 2877–2881.

Wolfe, J. H., Sands, M. S., Barker, J. E., Gwynn, B., Rowe, L. B., Vogler, C. A., and Birkenmeier, E. H. (1992). Reversal of pathology in murine mucopolysaccharidosis type VII by somatic cell gene transfer. *Nature* **360,** 749–753.

Wood, P. A., Sisk, D. B., Styer, E., and Baker, H. J. (1987). Animal model: Ceroidosis (ceroid-lipofuscinosis) in Australian cattle dogs. *Am. J. Med. Genet.* **26,** 891–898.

Yamagami, T., Umeda, M., Kamiya, S., and Sugiyama, K. (1989). Neurovisceral sphingomyelinosis in a Siamese cat. *Acta Neuropathol. (Berl).* **79,** 330–332.

Yamanaka, S., Johnson, M. D., Grinberg, A., Westphal, H., Crawley, J. N., Taniike, M., Suzuki, K., and Proia, R. L. (1994). Targeted disruption of the Hexa gene results in mice with biochemical and pathologic features of Tay–Sachs disease. *Proc. Natl. Acad. Sci. USA* **91,** 9975–9979.

Yogalingam, G., Litjens, T., Bielicki, J., Crawley, A. C., Muller, V., Anson, D. S., and Hopwood, J. J. (1996). Feline mucopolysaccharidosis type VI. Characterization of recombinant *N*-acetylgalactosamine 4-sulfatase and identification of a mutation causing the disease. *J. Biol. Chem.* **271,** 27259–27265.

Yoshida, M., Noguchi, J., Ikadai, H., Takahashi, M., and Nagase, S. (1993a). Arylsulfatase B-deficient mucopolysaccharidosis in rats. *J. Clin. Invest.* **91,** 1099–1104.

Yoshida, M., Ikadai, H., Maekawa, A., Takahashi, M., and Nagase, S. (1993b). Pathological characteristics of mucopolysaccharidosis VI in the rat. *J. Comp. Pathol.* **109,** 141–153.

Yoshida, M., Tachibana, M., Kobayashi, E., Ikadai, H. and Kunieda, T. (1994). The locus responsible for mucopolysaccharidosis VI (Maroteaux–Lamy syndrome) is located on rat chromosome 2. *Genomics* **20,** 145–146.

CHAPTER

26

Tumor Markers

BRUCE R. MADEWELL

I. INTRODUCTION

A tumor marker, in the broadest sense, might be considered a laboratory measurement of a substance or a process that provides clinically useful information with regard to tumor diagnosis and patient management. Tumor markers might be unique attributes of neoplasms, or may reflect the neoplastic process by a high level of expression relative to that of normal cells. Ideally, a marker of cancer should be sensitive, specific, and noninvasive; allow identification of affected patients and/or those at risk for the disease; reflect tumor burden; have value as a prognostic indicator; be predictive of tumor recurrence; and serve as an aid for the selection of an effective treatment strategy. Cancer markers should also accurately reflect the disease process, rather than any side effects of medication or treatment; be well understood with regard to any possible influences of the patient's age and gender; be reproducible in measurement and have a clear normal range; and be rapidly responsive to changes, thus reflecting current disease load.

Very few tumor markers have been described that fulfill these idealistic criteria in human medicine, and fewer have undergone rigorous clinical appraisal; few putative tumor markers have been described in veterinary medicine. However, the explosion in new information based on data from immunology and cell and molecular biology has now provided methods for critically examining biological processes associated with cancer, and to contrast those cellular perturbations with those of homeostasis. Those tumor markers that are convincingly demonstrated to distinguish normal from neoplastic cells will then enter clinical appraisal, and perhaps a few will emerge as ideal markers of cancer.

A partial list of *serum tumor markers* is provided in Table 26.1. Categorically, these putative markers include oncofetal proteins, hormones and ectopic hormones, enzymes, immunoglobulins, tumor-associated antigens, and miscellaneous markers. In this chapter, we will also examine the application of new technologies to neoplasms of importance to animals, using the broadest definition of tumor "marker" as an alteration in a cellular attribute that has been shown to be associated with a given tumor type providing a profile for risk assessment, clinical diagnosis, clinical staging, or therapeutic approaches. In particular, we will examine the advances made in *immunohistochemistry* for the evaluation of poorly differentiated neoplasms. *Flow cytometry* will be examined for measurement of DNA ploidy and cell proliferation in clinical samples of tumors, or, using specific immunologic probes, as a method for identification of subpopulations of cells or the detection of tumor cells in heterogenous clinical specimens. The *cytogenetic* changes that have been described in companion-animal neoplasms will be reviewed. New *DNA diagnostic technology,* based on isolation of RNA, DNA, or protein from tumor cells and the analysis of these components for structural or quantitative abnormalities, will also be reviewed with respect to its contemporary application in veterinary medicine.

II. SERUM TUMOR MARKERS

A. Oncofetal Proteins

Carcinoembryonic antigen (CEA) and *alpha-fetoprotein (AFP)* are the classic oncofetal proteins used as tumor markers. Oncofetal proteins are substances that appear to originate within tumor cells and enter the circulation as a result of secretion by the tumor or as breakdown products of tumor cells. These proteins are normally present during variable periods of embryonic or fetal life, do not disappear entirely during adult life, and reappear with certain neoplasms. In human patients, the oncofetal protein CEA has been the most widely studied marker. It was discovered in 1965 by raising antisera against a colon carcinoma, and it is now known to belong to a family of cell surface glycoproteins that are found in various malignancies (Bates, 1991). CEA is used most often as a marker of neoplasms of the lung, colon–rectum, breast, pancreas, ovary, and prostate gland. CEA has had little application as a serum tumor marker for cancer in veterinary practice.

AFP is synthesized by the yolk sac and fetal liver of almost all mammals (Yamada *et al.,* 1995). AFP serves, in human cancer patients, as a marker of nonseminomatous testicular cancer and hepatocellular carcinoma. In animals, AFP was detected in cattle with hepatocellular carcinoma and in fetuses (Kithier *et al.,* 1974). A more recent evaluation of AFP as a serum marker in dogs using reagents specific for human AFP revealed only rare elevations in association with various naturally occurring malignancies (Hahn and Richardson, 1995), but serum AFP concentrations were greatly elevated in dogs with experimentally induced hepatic tumors (Lowseth *et al.,* 1991). Canine AFP was recently purified from amniotic fluid, and an ELISA developed for canine AFP; that study also revealed that AFP levels in newborn puppies are higher than those measured in newborn cattle, pigs, and human infants (Yamada *et al.,* 1995).

TABLE 26.1 Categories of Tumor Markers

Oncofetal proteins: Carcinoembryonic antigen, alpha-fetoprotein
Hormones: Hormones and ectopic hormones, human chorionic gonadotropin (hCG)
Enzymes: Prostatic acid phosphatase, alkaline phosphatase, lactate dehydrogenase
Immunoglobulins
Tumor-associated antigens: CA-125, prostate-specific antigen (PSA)
Others: Fibronectin, sialic acid, acute-phase proteins, tissue polypeptide antigens

B. Hormones and Ectopic Hormones

1. *Human Chorionic Gonadotropin*

Human chorionic gonadotropin (hCG) is used commonly in human medicine as a marker of gestational trophoblastic neoplasia; β-hCG serves as a nearly ideal tumor marker and is elevated in virtually all cases of trophoblastic neoplasia. Recent data provides optimism that β-hCG may be a definitive biomarker for cancer. Complete β-hCG is present in cell membranes of cultured fetal and cancer cells. Its expression in cancer defines the metastatic aggressiveness of the tumors in which it is found. Of fundamental importance is that cells from nonembryonic tissue or benign neoplasms, unlike cancer or fetal cells, do not express β-hCG. Therefore, the presence of β-hCG becomes a defining phenotypic expression of malignant transformation, not only in trophoblastic tumors or tumors of germ cell origin, but also in a wide range of solid tumors (Regelson, 1995). I am not aware of applications of hCG for assessment of animal tumors, perhaps a reflection of the fact that gestational trophoblastic neoplasms are rarely diagnosed in domestic animals.

2. *Serum Parathyroid Hormone*

Measurements of parathyroid hormone are useful for study of clinical disorders of calcium metabolism. Elevation of serum parathyroid hormone concentration at the time of hypercalcemia is strong evidence for the presence of primary hyperparathyroidism (Chew *et al.*, 1995). Although primary hyperparathyroidism is relatively uncommon in dogs and cats, descriptions of both adenomas and carcinomas of the parathyroid glands are recorded in the literature and should be considered in the differential-diagnostic evaluation of any dog or cat with persistent hypercalcemia.

3. *Parathyroid Hormone-Related Protein*

The hypercalcemia of malignancy is one of the most frequently recognized paraneoplastic syndromes in animals. Hypercalcemia associated with malignancy results either from the stimulation of local bone resorption associated with neoplasms metastatic to the skeleton, or via the release of humoral factors that act systemically to increase osteoclastic bone resorption, and to increase calcium resorption from the kidneys or calcium absorption from the GI tract (Rosol and Capen, 1992). The ectopic production of parathyroid hormone per se by tumors is rare. Instead, there is now substantial evidence that inappropriate production of parathyroid hormone-related protein (PTHrP) by tumors can explain most of the clinical and pathologic findings in patients, both animal and human, with the humoral hypercalcemia of malignancy. Parathyroid hormone-related protein was named for its amino acid homology and pathophysiologic mimicry of the actions of PTH (de Papp and Stewart, 1993). However, although data show clear evidence for an association between increased activity of PTHrP and hypercalcemia in dogs with adenocarcinomas derived from the apocrine glands of the anal sac, malignant lymphoma, and miscellaneous other carcinomas, there appears to be a variety of systemically acting humoral factors, such as cytokines, released by tumors that alter calcium metabolism (Rosol and Capen, 1992).

4. *Thyroxine/Thyroglobulin*

Measurement of serum thyroxine concentration, or use of the thyroid-stimulating hormone response test, is used to establish a diagnosis of a functional thryoid neoplasm. It has been reported that approximately 20% of thyroid tumors in dogs are functional (Loar, 1986). In a more recent publication, it was reported that up to 70% of dogs with thyroid carcinomas had high serum concentrations of thyroglobulin, although there was no clear relationship between the thyroglobulin and thyroxine concentrations of those dogs (Verschueren and Goslings, 1992).

Elevation of serum thyroxine is diagnostic of hyperthyroidism in the cat, although single serum throxine measurements may give equivocal results. The serum triiodothyronine (T3) suppression test can be used to confirm hyperthyroidism in these cases (Peterson *et al.*, 1990). Most cats with clinical evidence of hyperthyroidism will not have an underlying carcinoma, however, but rather an adenoma or adenomatous hyperplasia of the thyroid gland.

5. *Adrenocorticotropin/Cortisol*

Plasma cortisol concentration, or alteration of that concentration using adrenocorticotropin (ACTH) stimulation and/or dexamethasone inhibition, is used in a variety of procedures for diagnosis and monitoring of dogs and cats with hyperfunctioning adrenocortical tumors. Similarly, the measurement of ACTH may facilitate the diagnosis of spontaneous adrenal disorders, both primary and secondary to pituitary dysfunction. The specific mechanics of conducting these tests and the interpretation of their results are reviewed elsewhere (Kempainen and Clark, 1995). Some of these same diagnostic methods have been applied to the diagnosis of hyperadrenocorticism associated with adrenal gland tumors in ferrets (Gould *et al.*, 1995).

6. *Sex Steroid Hormones/Hormone Receptors*

In veterinary practice, measurements of sex steroid hormones in plasma have been utilized most frequently for characterization of equine granulosa cell tumors (Stabenfeldt *et al.*, 1979). The granulosa cell tumor is the most commonly recognized ovarian tumor in the mare and is frequently associated with clinical signs resulting from high plasma testosterone or estrogen concentrations (Meinecke and Gips, 1987). Less frequently, plasma progestins are elevated. However, there are considerable differences in the steroid hormone profiles among affected horses reflective of dynamic changes within the neoplastic granulosa and ovarian stroma cells.

Measurements of plasma estradiol have also been used recently to characterize the functional activity of adrenocortical tumors in ferrets (Lipman *et al.*, 1993).

There have been a multitude of manuscripts in the veterinary literature over the past 20 years regarding estrogen and/or progesterone receptor concentrations in dogs and cats with mammary tumors; those studies are reviewed elsewhere (Donnay *et al.*, 1995). The measurements have generally been made from cytosol fractions prepared from fresh mammary tumors, and more recently using immunohistochemistry. Estrogen and progesterone receptors can be detected in normal mammary tissues, and in the dog, it has been determined that ER concentrations were decreased in malignant mammary tumors compared to those in normal mammary tissues, whereas PR concentrations tended to be higher than those detected in the normal gland (Donnay *et al.*, 1995). Poorly differentiated tumors tended to have lower receptor concentrations than those found in benign or well-differentiated tumors; and this was particularly true in those neoplasms that were of large size and associated with a high rate of growth.

C. Enzymes

1. Lactate Dehydrogenase, Serum Alkaline Phosphatase, Prostatic Acid Phosphatase

Patients with malignant disease often have increased lactate dehydrogenase (LDH) activity in serum, although it is usually a nonspecific increase in animals. In human patients, the isoenzyme LDH-1 is elevated in germ cell tumors (teratomas, testicular seminomas, ovarian dysgerminomas). Similarly, alkaline phosphatase measurements assist nonspecifically in the diagnosis of hepatobiliary disease and bone disease associated with increased osteoblastic activity. Prostatic acid phosphatase (PAP) is an enzyme secreted by the normal prostate gland. PAP levels increase in the serum of human patients with advanced prostate cancer, and abnormal levels indicate that the tumor has spread beyond the prostatic capsule. In general, the serum level increases with progressive stages of disease. Although PAP can be localized in the canine prostate gland using immunohistochemical procedures (Aumuller *et al.*, 1987), serologic detection of this secretory antigen has not been described in dogs in association with prostatic disease.

D. Immunoglobulins

Measurement of serum immunoglobulin concentrations in the plasma hyperviscosity syndromes probably represents the oldest and most consistent use of serum tumor markers in clinical veterinary medicine. In monoclonal disorders such as multiple myeloma or Waldenstrom's macroglobulinemia, recognition of a characteristic electrophoretic pattern of serum on cellulose acetate membrane provides evidence for a monoclonal protein (M-protein, monoclonal, myeloma), although it must be remembered that hypergammaglobulinemia is not seen in all cases of myeloma, and indeed, there may be hypogammaglobulinemia and a normal-appearing electrophoretic pattern. Further, cellulose acetate electrophoresis may not detect small monoclonal proteins. The monoclonal spike detected on serum or urine protein electrophoresis can then be assayed qualitatively by immunoelectrophoresis to determine the specific antibody class of the paraprotein, and the protein can then be quantitated by single radial immunodiffusion or other quantitative methods. These markers, therefore, are used to facilitate diagnosis as well as providing a convenient measure for monitoring patients undergoing therapy.

E. Tumor-Associated Antigens

1. CA-125

The tumor-associated antigen CA-125 has been used clinically in human medicine to monitor patients with epithelial ovarian carcinomas; CA-125 is not specific for ovarian cancer, however, and CA-125 levels may also be elevated in patients with nongynecological tumors and in certain other conditions such as the first trimester of preganancy and endometriosis (Helzlsouer *et al.*, 1993). I am unaware of application of CA-125 as a marker for animal neoplasms.

2. Prostate-Specific Antigen

Prostate-specific antigen (PSA) is a tumor-associated antigen and is raised in the serum of human patients with several conditions, including cancer of the prostate, benign prostatic hyperplasia, and prostatitis. It is used along with the digital rectal examination for screening patients for prostatic cancer, and it is also used to monitor patients with prostatic cancer after therapy. It is often used in conjunction with serum measurement of prostatic acid phosphatase. PSA can be detected by immunohistochemistry in canine prostate tissues using anti-human PSA antibodies, but it is not detected in the serum of dogs with prostatic disorders (McEntee *et al.*, 1987).

F. Miscellaneous Serum Tumor Markers

1. Fibronectin

Fibronectin is one of the extracellular matrix proteins. It is a large glycoprotein found in an insoluble form in connective tissue stroma and basement membranes. It is also present in soluble form in plasma. One study of plasma fibronectin concentrations from dogs with a variety of neoplasms revealed dogs with both high and low plasma concentrations (Feldman *et al.*, 1988a). Fibronectin concentrations were also examined in peritoneal effusions from dogs with a variety of intra-abdominal tumors; those data suggested that decreased fibronectin concentrations in peritoneal effusions were associated with a hypercoagulable state (Feldman *et al.*, 1988b).

2. *Sialic Acid*

Serum lipid-associated sialic acid levels have been used as tumor markers because of a rather consistent alteration of glycolipids, glycoproteins, and glycosyltransferases associated with their synthesis in tumor tissues. In one study, serum lipid-associated sialic acid levels were elevated in 22 of 24 dogs with cancer (Kloppel *et al.*, 1978), whereas in another study increased levels were also observed in dogs with malignant tumors compared to levels in healthy dogs, those with nonneoplastic conditions, and dogs with benign tumors (Poli *et al.*, 1986).

3. *Acute-Phase Proteins*

The acute-phase proteins (APP) are those proteins whose plasma concentration rises 25% or more following tissue injury or infection These proteins are elevated in a wide variety of acute and chronic diseases, including cancer. The best-studied human APP include ceruloplasmin, complement components C3 and C4, α_1-acid glycoprotein (AG), α_1-proteinase inhibitor (α_1-anti-trypsin), α_1-anti-chymotrypsin, haptoglobulin (Hp), fibrinogen (Fbn), C-reactive protein (CRP), and serum amyloid A (SAA). There has been some application of these APPs in veterinary medicine as markers for animals with cancer. α_1-Acid glycoprotein (AGP, oromuscoid) is categorized as an APP. Serum concentrations of AGP are increased in human patients with a variety of neoplasms, but like other APPs, its concentration in serum is also increased by acute and chronic inflammatory disorders. In the dog, serum AGP concentrations were shown to be increased in dogs with hematopoietic and nonhematopoietic tumors at the time of diagnosis (Ogilvie *et al.*, 1993). In that study, high serum concentrations of AGP in dogs with malignant lymphoma decreased significantly following induction of clinical remission with anticancer chemotherapy, suggesting a role for these measurements in monitoring the course of malignant lymphoma in dogs receiving treatment.

Tumor necrosis factor-α (TNF-α) is a cytokine produced primarily by macrophages. There are data to show that TNF-α may be an important inflammatory cytokine associated with hematopoietic suppression due to feline leukemia virus infection (Khan *et al.*, 1992).

III. FLOW CYTOMETRY

A. DNA Ploidy

The most frequent application of the flow cytometer for animal cancer patients has been for the evaluation of DNA content; these measurements have been made in a variety of canine and feline neoplasms, including those derived from the mammary glands (Hellmen *et al.*, 1988, 1989, 1995; Minke *et al.*, 1990; Rutteman *et al.*, 1988; Alenza *et al.*, 1995), bone (Fox *et al.*, 1990), hematopoietic system (Frazier *et al.*, 1993), urinary bladder (Clemo *et al.*, 1994), thyroid glands (Verschueren *et al.*, 1991), prostate gland (Madewell *et al.*, 1991), and skin (Ayl *et al.*, 1992). These measurements are based on the higher or lower DNA content of tumor tissues, due either to abnormal DNA (aneuploidy) or increased DNA content and cell division, compared to that of nonmalignant tissues. Because many neoplasms are believed to arise from a single abnormal cell, the resultant chromosome constitution is unique and tends to remain stable. The ratio of the abnormal DNA content of tumor tissue to that of normal cells is called the DNA index. The abnormal DNA is therefore useful for tumor diagnosis as well as the recognition of rare neoplastic cells in early or residual disease states.

The widespread application of flow cytometry for analysis of ploidy in human and animal cancer patients has revealed, in general, that normal tissues, reactive tissues, and tissues of benign tumors have diploid (2N) DNA values. Preneoplastic lesions may contain aneuploid DNA values, whereas in established tumors, the percentage of aneuploidy varies with tumor type. For example, in human patients with leukemias, aneuploid DNA histograms are more likely to be associated with acute leukemias than with chronic leukemias. In solid tumors, the percentage of aneuploidy varies considerably with the tumor type, from as little as 45% aneuploidy in patients with ovarian tumors to as high as 98% in patients with sarcomas (Koss *et al.*, 1989).

Another application of the flow cytometer is for estimation of the frequency of dividing cells. Cells synthesizing DNA have a higher DNA content than those not synthesizing DNA; the portion of cells with the higher level is the S-phase fraction. Many tumors will not possess DNA aneuploid populations, and an assessment of the proliferation rate will be the basis for interpreting the histogram as abnormal. Many studies have shown that proliferation indexes obtained by flow cytometric analysis are comparable to those obtained by a labeling index. In one study, the S-phase fraction of canine mammary tumors was determined by flow cytometry; the S-phase fraction was significantly higher in dogs with malignant tumors than in dogs with benign tumors, and it correlated positively with the histologic grade of malignancy (Alenza *et al.*, 1995).

In flow cytometry, therefore, a specimen can be categorized as having an abnormal (aneuploid) histogram if the DNA content of the population of cells is clearly separated from the normal diploid cell population; if the proliferation index is elevated; and if increased numbers of cells possess DNA in excess of that found in cells in the G_2 phase of the cell cycle (tetraploid).

B. Cell Surface Determinants

Through the development of specific monoclonal antibodies, the flow cytometer also allows characterization of cell populations by cell surface determinants. These reagents allow for the differentiation of both lymphocyte classes and their precursors, as well as many of the lymphocyte subclasses. In addition, monocytes, granulocytes, platelets, and their precursors can be identified. Recently, several laboratories have developed a wide range of well-characterized monoclonal antibodies against canine leukocyte surface antigens (Doveren *et al.*, 1985; Gebhard and Carter, 1992; Moore *et al.*, 1992; Mazza *et al.*, 1993; Cobbold and Metcalfe, 1994; Voss *et al.*, 1994; Watson *et al.*, 1994). The specificities of many of these antibodies were defined in a workshop and a CD classification for canine leukocyte antigens was proposed (Cobbold and Metcalfe, 1994). These monoclonal antibodies developed to normal hematopoietic elements are now being used to characterize the leukemias and lymphomas, thus providing new phenotypic categories that have diagnostic and prognostic relevance. The recent development of monoclonal antibodies to feline myeloid, erythroid, and lymphoid cell lineages should also expedite immunophenotyping of feline hematopoietic neoplasms (Groshek *et al.*, 1994). Further, newly developed antibodies are now being used to characterize solid tumors. For example, hormone receptors can be quantitated on breast tumor cells to assist in treatment planning.

IV. CYTOGENETIC CHANGES

Most advanced neoplasms lose monoclonality and acquire heterogenous genetic constitutions. Cytogenetic analysis examines the gross morphologic features of chromosomes and is best used to detect consistent, nonrandom chromosomal abnormalities such as translocations and deletions. Such abnormalities are informative markers of clonality if they are specific for the neoplasm and are found in sufficient numbers of cells (Nabers, 1994). In human medicine, chromosome studies of tumors have provided important clues to the location of the relevant genes and the mechanisms by which their growth regulatory functions have been altered.

Cytogenetics has been somewhat limited in veterinary medicine because it requires cell culture and the preparation of chromosomes in the metaphase of mitosis, which further serves to make their cellular origin uncertain. Most cytogenetic studies reported in the dog have been compilations of single or few cases; many of these studies are referenced elsewhere (Hahn *et al.*, 1994). Both structural and numeric chromosome alterations have been described in canine neoplasms. Very complex structural changes, including centric fusions and resultant aberrant metacentric chromosomes, have been described in dogs with both solid and hematopoietic neoplasms. These changes have generally been inconsistent and have not yet revealed information of pathogenic significance. It is presumed that increasing complexity of the karyotype is probably correlated in some way with tumor characteristics such as rate of growth or degree of invasiveness, but the present data do not provide unequivocal support for that supposition. The presence of double minute chromosomes may reflect gene amplification, and numeric chromosome aberrations may also lead to gene amplification. Probably the best-documented cytogenetic abnormality in animal neoplasms is that associated with the canine transmissible venereal tumor, in which the normal 78 chromosomes are reduced to 58 or 59, with many metacentrics (Makino, 1963; Murray *et al.*, 1969). There have been few descriptions of cytogenetic abnormalities associated with solid tumors (Mayr *et al.*, 1994a, 1993a), including the recent demonstration of trisomy of chromosome 9 in a canine hemangiopericytoma and a partial deletion in chromosome 1 of several other canine hemangiopericytomas (Mayr *et al.*, 1995a).

G-banded cytogenetic analysis was recently done on canine malignant lymphoma specimens in an effort to correlate cytogenetic findings with other variables that might influence treatment responses. In that study, numerical chromosomal abnormalities were detected more often than structural abnormalities, and the results of the cytogenetic studies did not correlate with other prognostic variables. The cytogenetic findings did, however, predict the length of first clinical remission and survival time (Hahn *et al.*, 1994).

V. PROLIFERATION MARKERS/APOPTOSIS

Measurement of cell proliferation may provide useful information concerning tumor prognosis and diagnosis. Methods directed at different parts of the cell cycle are available.

A. Mitotic Counts

Counting mitoses is the most traditional morphological method for assessing cell proliferation. The mitotic phase of the cell cycle is easily recognized on routinely prepared histologic sections. Mitotic counts are expressed as the number of mitoses per certain number of high dry microscopic fields, or as the number of mitoses present in a certain number of cells (mitotic index). Mitotic counts give prognostic information for some tumors such as fibrosarcomas in cats. The major flaw of the mitotic count is its lack of reproducibility. Mitotic counts are easy to perform, but errors in interpretation can occur because of delays in fixation, variations in section thickness and in the size of the microscope field used, and difficulties in the recognition of mitoses.

In clinical practice, high mitotic index was demonstrated to affect treatment responses adversely for dogs with synovial sarcoma (Vail *et al.*, 1994), sarcomas of the canine spleen (Spangler *et al.*, 1994), and canine and feline fibrosarcomas (Bostock, 1986).

B. Thymidine Labeling Index and BrdU Incorporation

The synthesis of DNA, or the S phase of the cell cycle, can be measured by labeled DNA precursor incorporation. Tritiated thymidine, or a halogenated thymidine analog, bromodeoxyuridine (BrdU), is used for direct estimation of DNA synthesis; thymidine in tissue sections is revealed with photographic emulsion, whereas the development of specific monoclonal antibodies to BrdU has allowed its immunohistochemical detection in paraffin-embedded sections, allowing simultaneous examination of the morphology of the proliferating cells. The thymidine labeling index is defined as the ratio of the number of positively stained cells to the total number of cells; because this method directly assesses DNA synthesis, it is considered the standard for cell cycle analysis and proliferation studies (Sahin *et al.*, 1993). For BrdU, the S-phase fraction of the cell cycle can be determined by counting nuclei labeled with BrdU, thus providing an accurate estimate of the proliferative potential of a tumor. One problem with assessment of tumor proliferation using the BrdU labeling index is the necessity of infusing patients with BrdU prior to surgery, or the necessity of incubating fresh tissue with BrdU in order to permit its incorporation into tumor DNA.

C. Nucleolar Organizing Regions

Silver staining of proteins associated with nucleolar organizing regions (NORs) of interphase chromosomes may be used on conventional histological sections or cytologic preparations for assessment of proliferation rates (Rose *et al.*, 1994). Nucleolar organizing regions are segments of chromosomal DNA that encode for ribosomal RNA. In normal cells, NORs are usually tightly aggregated in one or more nucleoli; in malignant cells, they desegregate from the nucleoli and tend to disperse throughout the nucleus. Because NORs are related to DNA synthesis and metabolic activities, it is generally agreed that the size and number of NORs reflect nuclear activity. The amount of NOR protein is related to the cell cycle, increasing progressively from Go to S-phase. The NORs can be demonstrated histologically by a simple silver-staining method because of the argyrophilia of their nonhistone acidic proteins. These argyrophilic nucleolar organizing regions (AgNORs) indicate actual or potential transcriptional activity of ribosomal DNA and therefore serve as apparent cytologic markers of cell proliferation; increased numbers of AgNORs may reflect increased tumor-cell turnover or cell ploidy and are generally associated with poor cellular differentiation (Schned, 1993). AgNOR counts are expressed as the number of AgNORs per nuclei, or as the percentage of tumor cells with >5 AgNORs per nucleus. The mean AgNOR count is believed to be a reflection of DNA ploidy, whereas the percentage of cells with >5 AgNORs per nucleus is thought to be a reflection of proliferative activity.

In the dog, mean nucleolar size has been shown to correlate with mitotic index in squamous-cell carcinoma of the skin. In that study, a close correlation was found between mean nucleolar area/cell and mitotic activity, and the mean nucleolar area/cell was significantly higher in canine squamous-cell carcinoma than in canine basal-cell carcinoma (DeVico *et al.*, 1995). In another study, the rate of proliferation of canine perianal tumors determined by quantitative analysis of AgNOR proteins revealed a progressive increase in the rate of proliferation parallel with the histologic features of malignancy (Preziosi *et al.*, 1995a). For dogs with mast-cell tumors treated surgically, AgNOR staining was demonstrated to be a better prognostic indicator than either tumor histologic grade or mitotic index (Bostock *et al.*, 1989). The staining of AgNORs in canine transmissible venereal tumors was also used as a measure predictive of response to vincristine chemotherapy (Harmelin *et al.*, 1995). AgNOR staining of canine mammary tumors revealed high counts in tubular and solid carcinomas, particularly in those tumors showing infiltration of surrounding connective tissues; an unfavorable prognosis following surgery was determined for most dogs that had high AgNOR counts (Bostock *et al.*, 1992). AgNOR staining in feline mammary carcinomas was not shown, however, to correlate with other tumor attributes, and therefore was not useful for diagnostic or prognostic purposes (DeVico *et al.*, 1995).

D. Proliferation Markers

Methods for detection of cell cycle related antigens are described in Section VI.E.

E. Image Analysis

Another laboratory measure of ploidy, S-phase fraction, and nuclear area is image analysis; this can be accomplished on cytocentrifuge preparations of Fuelgen-stained nuclei extracted from traditionally prepared tissue specimens (formalin fixation, paraffin embedding). Image analysis allows evaluation of individual nuclei, providing both morphometric and densitometric nuclear values. In one study, 90 canine mammary tumors were examined by image analysis (Destexhe *et al.*, 1995). The mean S-phase fraction of malignant lesions was twice that of benign lesions; the mean nuclear surface was smaller in malignant lesions than in benign lesions; and 37% of the malignant tumors were aneuploid. Tumors shown histologically to invade lymphatic vessels were significantly associated with aneuploidy and a high S-phase fraction.

F. Apoptosis

Apoptosis is a "programmed" form of cell death that leads to the clearance of unwanted cells without disruption of tissue structure or function (Savell, 1994). There are several laboratory methods that allow characterization or recognition of apoptosis. One unique attribute is that in apoptosis, there is activation of endogenous endonucleases. These enzymes lead to a characteristic ladder of oligonucleosomal fragments upon electrophoresis, which is in contrast to the smearing of DNA that has been degraded during necrosis because of uncontrolled access of nucleases to DNA. Another feature of apoptosis is the characteristic cellular morphology. The principle morphologic feature in apoptosis is the condensation of nuclear heterochromatin, resulting in the formation of a crescent apposed to the nuclear membrane. Other changes include cell shrinkage, cytoplasmic condensation, and the formation of buds at the cell membrane that later condense into "apoptotic bodies." Another feature of apoptosis is that the process of cell death is not accompanied by inflammation or necrosis because lysosomal enzymes and granule contents are not released by the dying cells. These features are best detected by the electron microscope, but apoptotic bodies can be visualized in histologic sections using hematoxylin-stained cells or fluorescence microscopy and dyes such as propidium iodide or acridine orange.

One method for visualization of apoptotic cells is *in situ* 3′ labeling, which allows detection of DNA double-strand breaks (Kressel and Groscurth, 1995). Another method for detection of internucleosomal fragmentation of DNA is agarose gel electrophoresis followed by ethidium bromide staining, which allows recognition of the highly ordered DNA fragmentation characteristic of apoptotic cells. This apoptotic DNA is characterized as a "DNA ladder" reflective of the regular fragmentation (180–200 bp multiples) of the nuclear genome. Flow cytometry using propidium iodide staining can also be used to detect the reduced DNA, or subdiploid cells, characteristic of apoptotic cells compared to diploid (normal) cells.

VI. IMMUNOHISTOCHEMISTRY

The introduction of immunohistochemistry as a routine procedure in many veterinary diagnostic laboratories has improved the objective diagnosis of neoplasms. Immunohistochemistry has emerged as an important tool for tumor typing and has led to an improved understanding of cellular differentiation. Immunohistochemistry allows precise determination of the histogenesis of many neoplasms that would be characterized as undifferentiated on the basis of morphologic examination of the hematoxylin- and eosin-stained tissue. Although most of the antibodies used routinely in the immunohistochemistry laboratory do not recognize unique attributes of neoplasms per se, accurate determination of histogenesis may allow appropriate treatment methods to be used and more reliable estimates of prognosis to be established. Some of the immunohistochemical determinations that are used in veterinary medicine are outlined (Table 26.2) and the following is a brief summary of their applications.

A. Histopathology, Intermediate Filaments, Cytokeratins

The distinction of carcinoma from other undifferentiated or poorly differentiated neoplasms may be difficult. The immunohistochemical markers that are used to characterize tumors derived from epithelium include cytokeratin intermediate filaments (IF), epithelial membrane antigen or the milk-fat globule proteins, and carcinoembryonic antigen. The cytokeratin intermediate filament proteins are a component of the cytoskeleton of mammalian cells. This cytoskeleton contains the thin (4–6 nm) actin-containing microfilaments, the thick (20–30 nm) microtubules, and the

TABLE 26.2 Immunohistochemistry: Utility in Veterinary Medicine

Hormones and hormone receptors:
- Somatostatin
- Insulin
- Glucagon
- Thyroglobulin
- Gastrin
- Calcitonin
- Calcitonin gene-related peptide
- Parathyroid hormone
- Estrogen receptor
- Progesterone receptor

Oncoproteins: c-*erb*B-2 oncoprotein, c-*myc* oncoprotein, p53 protein, RB protein

Proliferation markers: Proliferating-cell nuclear antigen (PCNA), Ki-67

Enzymes involved in cytotoxic drug metabolism: P-glycoprotein

Histopathology tumor markers:
- Prostatic acid phosphatase
- Prostate-specific antigen
- von Willebrand factor
- Carcinoembryonic antigen
- Chromogranin A
- Neuron-specific enolase
- S-100 protein
- Lysozyme
- Collagen IV
- Smooth-muscle actin
- Skeletal myosin
- Myoglobin
- Synaptophysin
- Laminins
- Type IV collagen

Histopathology, intermediate filaments: Cytokeratins (pan-cytokeratin, specific molecular weight cytokeratins)

Histopathology, intermediate filaments: Mesenchyme (vimentin, desmin, neurofilaments, glial fibrillary acidic protein)

Histopathology, lymphoid markers:
- Leukocyte common antigen
- Pan T cell
- Pan B cell
- Macrophage
- IgG, IgM, IgA

Virology:
- FeLV
- FIV
- Papillomavirus

intermediate size (10 nm) intermediate filaments. The intermediate filaments are the most complex in terms of protein structure, and at present there are six distinct classes of intermediate filaments, including the acidic (type I, pH 4.5–5.5) and neutral–basic (type II, pH 6.5–7.5) keratins of epithelia; type III proteins, including vimentin of many mesenchymal cells; the type IV neurofilaments; the type V lamins of eukaryotic cells; and nestin, the type VI protein of neuroectodermal stem cells (Steinert, 1993).

The cytokeratin proteins are unique among the IFs, presenting a high degree of diversity of polypeptide units: 20 in human tissues and 22 in bovine and murine tissues. The cytokeratin unit molecule consists of a disulfide-bonded heterodimer between type I and type II cytokeratin polypeptides. These cytoskeletal components form elaborate cagelike networks around the nuclei of mammalian cells (Steinert, 1993). Cytokeratin IFs are highly correlated with the degree of differentiation of the tissue; therefore, specific cytokeratin identification is useful not only for determining the epithelial histogenesis of tumor specimens, but also for distinguishing glandular from squamous differentiation independent of other morphologic features. For epithelial neoplasms undergoing progression from epithelial dysplasia to carcinoma *in situ* to invasive squamous-cell carcinoma, there is concurrent decreased expression of high- and low-molecular-weight cytokeratins characteristic of normal differentiated epithelial cells. Well-differentiated squamous cell carcinomas express high-molecular-weight cytokeratins (de las Mulas *et al.*, 1995). The great majority of canine and feline carcinomas will show reactivity of at least a portion of the tumor cells for cytokeratins (Andreasen *et al.*, 1988a; Cardona *et al.*, 1989; Hellmen *et al.*, 1989; Magnol *et al.*, 1985; Rabanal *et al.*, 1989). Panels of cytokeratin antibodies, including antibodies to both high- and low-molecular-weight keratins, however, are often needed in animal specimens to increase the sensitivity of immunohistochemical recognition of epithelial tumors (Sandusky *et al.*, 1991).

Most studies demonstrating expression of cytokeratin intermediate filaments in domestic animal tissues rely on cross reactivities of those proteins with polyclonal or monoclonal antibodies developed against human IFs. Although cross reactivity is often discerned and subsequently used clinically for demonstration of IF proteins in animal tissues, some caution is advised in interpreting the results of these studies: Clear differences in the immunoreactivities of animal and human tissues for many of these diagnostic antibodies have been observed (Vos *et al.*, 1992).

Specific cytokeratins have been proposed as histodiagnostic markers, for some tumors, usually as a consequence of the change in cytokeratin expression during tumorigenesis. A type I (acidic) 57-kDa cytokeratin was proposed as a molecular marker for malignant transformation in canine mammary tumorigenesis, reacting with the basal monolayer of epidermis and ductal carcinomas (Arai *et al.*, 1994).

Using human-derived monoclonal antibodies as well as feline-derived antibodies to cytokeratin proteins, feline cytokeratins in mammary tissues were determined to be highly homologous to patterns of cytokeratin expression in human tissues (Ivanyi *et al.*, 1992). Luminal epithelial cells reacted with monoclonal antibodies specific for cytokeratin IFs consistent with simple epithelia, whereas myoepithelial cells reacted with antibodies specific for cytokeratin proteins in basal cells of stratified epithelia (Ivanyi *et al.*, 1993). Further, cytokeratins typical of stratified epithelia were detected in areas of squamous metaplasia of adenocarcinoma cells, suggesting that these IFs serve as a marker for squamous metaplasia before it can be detected by the light microscope.

B. Histopathology, Intermediate Filaments, Mesenchyme

1. *Vimentin*

Vimentin intermediate filaments are used to mark nonmuscle sarcomas (Andreasen *et al.*, 1988a). As anticipated, the vaccine-associated sarcomas of cats, predominately fibrosarcomas, are immunoreactive for vimentin IFs, but within this spectrum of soft-tissue sarcomas are tumors that are also reactive for one or more smooth-muscle markers (desmin, smooth muscle-specific actin), suggesting a fibroblastic or myofibroblastic phenotype (Hendrick and Brooks, 1994; Pace *et al.*, 1994). Vimentin expression is lost during the process of differentiation of skeletal muscle and is absent in mature myofibrils of skeletal muscle. Vimentin IFs may be expressed, however, in poorly differentiated rhabdomyosarcomas (Vos *et al.*, 1993).

2. *Desmin*

Desmin is a cytoskeletal component that holds myofibrils in place; it has been used as a marker for canine leiomyomas and leiomyosarcomas (Andreasen and Mahaffey, 1987). During different stages of development, skeletal muscles express desmin. Skeletal muscle tumors are reactive to desmin antibodies in the dog and cat (Andreasen *et al.*, 1988a; Madewell *et al.*, 1988; Sarnelli *et al.*, 1994), and the botryoid variant of canine rhabdomyosarcoma has also been demonstrated to contain desmin (Andreasen *et al.*, 1988b).

3. *Neurofilaments*

Mammalian neurofilaments contain at least three distinct subunits. Antibodies to individual or triplet neurofilaments have been shown to react with canine axons and neurons (Vos *et al.*, 1989); neurofilaments are used to demonstrate neural tumors, as well as carcinoid tumors, pheochromocytomas, islet-cell tumors of the pancreas, and others (Mukai *et al.*, 1986).

4. *Fibrillary Acidic Protein*

Glial fibrillary acidic protein (GFAP), the major subunit of glial filaments, is found not only in glial cells in the central nervous system, but also in some Schwann cells and schwannomas and in some other tissues (Dolman, 1989). In the dog, GFAP reactivity is found in glial cells, predominantly fibrous astrocytes, Schwann cells, axons, and cell bodies of some peripheral ganglia (Vos *et al.*, 1989). Some cells in canine ganglioneuroblastomas will stain positively for GFAP (Schulz *et al.*, 1994).

C. Histopathology—Other Markers

1. *Carcinoembryonic Embryonic Antigen*

Carcinoembryonic embryonic antigen (CEA) is a glycoprotein present in a wide variety of normal and neoplastic tissues. In human beings, adenocarcinomas are frequently positive for CEA, and this feature is used routinely to identify adenocarcinomas derived from the colon, stomach, pancreas, breast, endometrium, thyroid, ovary, and urinary bladder (Ordonez, 1989). In animal specimens, curiously, CEA reactivity was detected in a small number of choroid plexus tumors from the dog (Ribas *et al.*, 1989). CEA reactivity has also been demonstrated in the normal canine exocrine pancreas and in some neoplasms of the exocrine canine pancreas (Rabanal and Fondevila, 1992). CEA reactivity was also detected in apocrine gland-derived tumors in the dog (Ferrer *et al.*, 1990).

2. *Alpha-Fetoprotein*

Using immunohistochemistry, alpha-fetoprotein was detected in only 3 of 11 hepatocellular carcinomas examined in the dog (de las Mulas *et al.*, 1995).

3. *Epithelial Membrane Antigen*

Epithelial membrane antigen (EMA) or milk-fat globule proteins have not been studied systematically in domestic animals. EMA is a mucin produced in large amounts by epithelial cells of the lactating breast and other glandular epithelia. Antibodies to EMA have been used for detection of acinar cells and carcinomas in human patients. EMA reactivity was demonstrated on the surface of canine salivary glandular epithelium *in vitro* (Strandstrom *et al.*, 1989).

4. Von Willebrand's Factor (Factor VIII-Related Antigen)

Von Willebrand's factor, a part of the factor VIII complex, is restricted to endothelial cells, megakaryocytes, and platelets. In the human being, antibodies to factor VIII-rag (F8RA / vWF) are established as markers of endothelial cells, megakaryocytes, and platelets, and immunohistochemical demonstration of factor VIII-rag is considered a useful adjuvant to conventional stains for the diagnosis of equine, feline, and canine vascular neoplasms (Moore *et al.*, 1986; van Buest *et al.*, 1988; Miller *et al.*, 1992). More recently, canine endothelial cells in normal organs and vascular neoplasms were also shown to express CD31, an adhesion molecule also referred to as platelet endothelial cell adhesion molecule (Ferrer *et al.*, 1995).

Another emerging and very useful application of immunohistochemistry using anti-F8RA/vWF or other vascular markers is to highlight the microvessels within tumors. Some pathology laboratories are now measuring intratumoral vessel density as a prognostic indicator. In some tumors, it has been determined that intratumoral vessel density is a better predictor of disease-free survival than tumor stage, grade, or type (Weidner, 1995). This is based on the knowledge that tumor growth is angiogenesis dependent. For any tumor to grow to macroscopic dimensions, it must induce the surrounding tissue either to share its nutrient supply or to provide an alternative supply. Indeed, the mark of successful malignant transformation involves the ability to induce and expand the vascular network to provide the nutrients needed for cellular transformation.

5. Myoglobin

Myoglobin antibody, although skeletal muscle-specific, is generally of limited value in the diagnostic laboratory because myoglobin is identified only in the most differentiated tumors, which usually do not pose diagnostic difficulty (deLellis and Dayal, 1987), and smooth-muscle tumors are characteristically myoglobin negative (Donner *et al.*, 1983).

6. Actin Microfilaments

Actins are a complex group of proteins; vertebrates possess at least six distinct forms of actin that are tissue specific. Four muscle actins have been identified: cardiac, vascular smooth muscle, enteric smooth muscle, and gamma-actin. Skeletal-muscle tumors are reactive for skeletal-muscle actin (Erlandsen, 1989; Sarnelli *et al.*, 1994), and smooth-muscle tumors stain for smooth-muscle actin (Hanzaike *et al.*, 1995).

7. Skeletal Myosin

Skeletal and cardiac-muscle myosin are referred to as sarcomeric myosin; antibodies that discriminate between smooth and sarcomeric myosins are used for diagnosis of smooth- and skeletal-muscle-derived tumors.

8. Prostate-Specific Antigen and Prostate-Specific Acid Phosphatase

PSA and PSAP are useful markers of genitourinary tumors in the dog; demonstration of PSA and PSAP allows diagnosis of otherwise undifferentiated prostatic carcinomas from carcinomas derived from the urinary bladder or colon (Aumuller *et al.*, 1987; McEntee *et al.*, 1987). A canine prostatic secretory protease has also been described (Aumuller *et al.*, 1987).

9. S-100 Proteins

S-100 is an acidic, calcium-binding protein of uncertain function that is composed of two subunits, designated α and β. Originally isolated from brain and thought to be specific for glial cells, S-100 proteins have subsequently been detected in a variety of nonneuroectodermal tissues (Schmitt and Bacchu, 1989). Reactivity for S-100 has been demonstrated in canine malignant melanoma, including amelanotic melanoma (Sandusky *et al.*, 1985), but immunoreactivity for S-100 proteins has also been detected in canine hemangiosarcoma, liposarcoma, leiomyosarcoma, malignant fibrous histiocytoma, and schwannoma (Rabanal *et al.*, 1989; Thoolen *et al.*, 1992; Tanimoto and Ohtsuki, 1992). S-100 reactivity has also been demonstrated in canine myoepithelial cells (Ferrer *et al.*, 1990), but its presence in epithelial cells of the mammary gland in the dog precludes its use as a specific marker for myoepithelial cells (Moller *et al.*, 1994). Although S-100 protein is not a specific marker for melanoma, it is not generally expressed in non-neural-derived epithelium, and therefore it is useful for distinguishing between carcinoma and melanoma. In veterinary specimens, the diagnosis of malignant melanoma is often a challenge. Immunohistochemical demonstration of S-100 proteins, vimentin intermediate filaments, and neuron-specific enolase, coupled with negative reaction for cytokeratin, may allow a tentative diagnosis of malignant melanoma (Sandusky *et al.*, 1985). Several monoclonal antibodies have been developed that are specific for human melanoma-associated antigens, but inconsistent results have often been obtained when these antibodies are applied to animal specimens. However, two commercially available monoclonal antibodies to human melanoma-associated antigens have been shown to react with the majority of canine melanomas that were formalin-fixed and paraffin-embedded (Berrington *et al.*, 1994).

10. Neuron-Specific Enolase

Enolase is an enzyme involved in the glycolytic pathway of metabolism. Five forms that differ in their unit composition and tissue distribution have been identified (Ordonez, 1993). The enolase γ-subunit, characteristic of neural cells, has been designated neuron-specific enolase (NSE). This enzyme was first described in the brain and later in other neuroendocrine cells and their tumors. For example, medullary thyroid tumors derived from the neural crest thyroidal C-cell system may be immunoreactive for NSE (LeBlanc *et al.*, 1991). Antibodies to NSE also stain melanomas, but lack specificity for their diagnosis (Vinores *et al.*, 1984). Chordomas typically stain for NSE (Dunn *et al.*, 1991). Ganglioneuroblastomas in the dog may also stain for neuron-specific enolase (Mattix *et al.*, 1994).

11. Chromogranin A

The chromogranins are a group of acidic peptides present in the secretory granules of endocrine cells; the major chromogranins are A, B, and C. Chromogranin A is the most abundant of the three, and it has proven useful as a marker for endocrine differentiation (Ordonez, 1993) . Antibodies to chromogranin are used to diagnose tumors derived from the adrenal medulla, islet cells of the pancreas, C-cells of the thyroid gland, parathyroid glands, and anterior pituitary (Wilson and Lloyd, 1984).

12. Type IV Collagen and Laminin

Type IV collagen and laminin are the principal components of basement membranes. Although these membrane components are visualized in tissue sections with stains such as the periodic acid Schiff reaction, immunohistochemical methods allow specific and sensitive identification of these basement membrane components. Many studies have revealed that the basement membrane is either interrupted or absent in invasive carcinomas, and this observation may be used as a prognostic finding. Continuous basement membranes are usually found associated with benign tumors, and in one study of canine mammary neoplasms, most of the malignant mammary tumors examined lacked continuous basement membranes, especially the invasive carcinomas (Pena *et al.*, 1995). An interrupted basement membrane suggests that the basement membrane was formed but then degraded by enzymes such as type IV collagenase associated with tumor invasion. In that same study, it was determined that myoepithelial cells were the main source of basement membrane proteins in canine mammary tumors. That observation provides foundation for the lower metastatic rate associated with tumors with myoepithelial proliferation (Pena *et al.*, 1995). In another study, laminin expression was examined in testicular tumors derived from dogs; fragmentation, loss, or absence of the basement membrane was associated with tumor invasion and strongly correlated with histologic classification and proliferative activity (Benazzi *et al.*, 1995).

13. Integrins

The name *integrin* was first used to signify the presumed role of these proteins in integrating the intracellular cytoskeleton with the extracellular matrix (Schwartz and Ingber, 1994). The $\beta 1$ integrins are members of the VLA (very late activation) family of transmembrane glycoproteins. The integrins mediate cell-to-cell adhesions and interactions between cells and extracellular matrix proteins such a fibronectin, collagens, and laminin. There are data in human medicine to demonstrate that alterations of the expression of these adhesive proteins may be related to tumor spread. Immunohistochemistry was used to demonstrate $\beta 1$ integrin expression on the surface of canine mammary tumors (Pena *et al.*, 1994). In one study of normal, dysplastic, and neoplastic canine mammary tumors, evidence suggested that $\beta 1$ expression was decreased in malignant tumors, and that the decreased expression noted in malignancy was correlated to the metastatic potential of the tumors, thus potentially serving as a marker of malignancy (Restucci *et al.*, 1995).

The $\beta 2$-integrin subfamily contains molecules mediating the adhesion of migrating lymphocytes to endothelial cells, the extracellular matrix, and epithelial cells. Canine keratinocytes may express $\beta 2$ integrins such as intercellular adhesion molecules (ICAM-2), which can be detected immunohistochemically. With antibodies and procedures applicable to studies of lymphocyte–keratinocyte interactions, the pathogenesis of epidermotropism of lymphocytes in canine mycosis fungoides was examined (Olivry *et al.*, 1995).

14. Amyloid

Antibodies directed at amyloid of immunoglobulin lambda-light chain origin as well as amyloid-A have been used to identify amyloid in various animal species using formalin-fixed, paraffin-embedded tissues. These reactivities have been used to confirm that the amyloid associated with some plasma-cell tumors is of immunoglobulin origin, and have also served as useful markers for plasmacytoid differentiation in some tumors (Linke *et al.*, 1984; Rowland *et al.*, 1991; Rowland and Linke, 1994).

D. Enzymes Involved in Cytotoxic Drug Metabolism

1. P-Glycoprotein

One of the most important mechanisms of cellular resistance to anticancer drugs involves the expression of the plasma membrane P-glycoprotein (Pgp), which is thought to function as an ATP-dependent efflux pump that reduces the cellular concentration of a wide range of chemotherapeutic agents. Pgp can be detected in tumors of every histologic type, and although some studies have been contradictory, several studies on human leukemias, lymphomas, and some childhood solid tumors have demonstrated a strong association between the Pgp expression in tumor samples and clinical outcome (Benchimol and Ling, 1994). Using Western blotting of membrane preparations made from canine lymphoma cells and an antibody to human Pgp, its expression was demonstrated in tumor cells derived from dogs that were resistant to chemotherapy (Moore *et al.*, 1995).

E. Proliferation Markers

Methods for detection of cell cycle related antigens are now available.

1. Ki-67

The monoclonal Ki67 is the prototype, recognizing a nonhistone nuclear protein expressed in proliferating cells in Gl, S, G2, and M, but not in quiescent (G0) cells. Ki67 labeling correlates well with cell proliferation as assessed by tritiated thymidine uptake, and provides a measure of the number of proliferating cells (proliferative fraction). Immunostaining with antibodies to the Ki67 antigen is now well established as a quick and efficient method for evaluating the growth fractions of various tumor types because of its distinctive reaction patterns that exclusively involve proliferating cells (Ross and Hall, 1995). Proliferative activity assessed by Ki-67 monoclonal antibodies in canine testicular tumors provided good correlation with other proliferation markers (Sarli *et al.*, 1994).

2. Proliferating-Cell Nuclear Antigen

Proliferating-cell nuclear antigen (PCNA), also known as cyclin, is a nonhistone nuclear protein associated with DNA polymerase delta, which is present throughout the cell cycle in proliferating cells. The PCNA gene is transcribed in both proliferating and quiescent cells; PCNA mRNA accumulates only in proliferating cells, reaching its maximum during S-phase (Sahin *et al.*, 1993; Rose *et al.*, 1994). The use of monoclonal antibodies to identify PCNA has been validated in the horse, cat, and dog (Theon *et al.*, 1994; Sarli *et al.*, 1994; Preziosi *et al.*, 1995b). In an applied clinical study, it was demostrated that the proliferation rate in squamous-cell carcinoma of the facial skin affected the prognosis of animals treated with radiation therapy (Theon *et al.*, 1995). A significant difference in the values of PCNA indices was reported between benign and malignant growths in both the dog and cat (Sarli *et al.*, 1995).

F. Hormones and Hormone Receptors

Antibodies to insulin, glucagon, somatostatin, pancreatic polypeptide, gastrin, and adrenocorticotropic hormones have proven valuable for the diagnosis of canine and feline pancreatic endocrine tumors and their metastases (O'Brien *et al.*, 1987; Boosinger *et al.*, 1988; Hawkins *et al.*, 1987; van der Gaag *et al.*, 1988). Canine pancreatic endocrine tumors may also be immunoreactive for islet amyloid polypeptide and calcitonin gene-related peptide (O'Brien *et al.*, 1990). Immunoreactivity for insulin, somatostatin, glucagon, and pancreatic polypeptide has also been demonstrated in the equine endocrine pancreas (Furuoka *et al.*, 1989). The normal canine exocrine pancreas contains cells immunoreactive for amylase, carboxypeptidase A, and α-antitrypsin (Rabanal and Fondevila, 1992). These same antibodies are also useful for detection of pancreatic exocrine carcinomas, with most neoplasms reacting for amylase and carboxypeptidase, and fewer immunoreactive for carcinoembryonic antigen and α-antitrypsin (Rabanal and Fondevila, 1992). These markers, therefore, are useful in the dog for distinguishing metastatic carcinomas of the pancreas from tumors derived from other histologic sites.

Immunohistochemical demonstration of thyroglobulin is useful for determining the histogenesis of primary thyroid carcinomas, those tumors arising in ectopic sites such as the heart base, and metastases of presumed thyroid origin (Moore *et al.*, 1984). Thyroid carcinomas with follicular differentiation consistently stain positively for thyroglobulin, whereas those with compact phenotype are less consistently reactive.

Many thyroid neoplasms will react immunohistochemically with neuron-specific enolase as well (Leblanc *et al.*, 1991). Medullary thyroid tumors derived from the neural crest thyroidal C-cell system may be immunoreactive for calcitonin, calcitonin gene-related peptide, and neuron-specific enolase (Leblanc *et al.*, 1991). These parafollicular cells can be distinguished from compact follicular cells of thyroid tumors on the basis of their immunoreactivity for calcitonin (Carver *et al.*, 1995). Medullary thyroid carcinomas in the horse and sheep may also be immunoreactive for calcitonin (van der Velden and Meulenaar, 1986; Renzoni *et al.*, 1995).

Parathyroid hormone immunoreactivity may be used to localize parathyroid glands in tissue sections. Further, using antibodies to parathyroid hormone-related protein, widespread localization of PTHrP was reported in both normal and neoplastic canine tissues; it is a presumed mediator of hypercalcemia of malignancy in dogs, but its widespread localization in normal canine tissues also demonstrates a physiologic role for the protein (Grone *et al.*, 1994).

1. *Estrogen Receptor; Progesterone Receptor*

The cytosol receptor assays provided evidence for steroid receptor-positive and steroid receptor-negative tumors; using these assays, the prevalence of estrogen and progesterone receptors has been reported to exceed 50% in canine mammary carcinomas, and data suggested that dogs with tumors expressing steroid hormone receptors survived longer than dogs without estrogen receptors or progesterone receptors (Sartin *et al.*, 1992). At the cellular level, however, it is often found that mammary tumors are heterogenous and composed of individual receptor-positive and receptor-negative cancer cells. The availability of specific, monoclonal antibodies for estrogen and progesterone receptors and the development of immunohistochemical methods for the localization of steroid hormone receptors in mammary tumors has resulted in a plethora of articles in human patients that detail close correlation of results with features of tumor grade and histologic types, as well as treatment responses. Although immunohistochemistry has been used to assess steroid receptors in canine mammary tumors, these new biological findings have not yet been extensively correlated with the clinical behavior of the tumors.

2. *Epidermal Growth-Factor Receptor*

Using immunohistochemistry, epidermal growth-factor receptor (EGF-R) was detected in almost one-half of plutonium-239-induced canine pulmonary neoplasms, but it was also expressed in a smaller percentage of nonneoplastic proliferative lung lesions (Gillett *et al.*, 1992).

3. *Oncoproteins*

Immunohistochemical staining with specific monoclonal antibodies for proteins of oncogenes or tumor suppressor genes may serve as biomarkers for cancer. Cytologic specimens, exfoliated cells, needle-biopsy specimens, and fixed tissues, including those from hospital archives, can be suitable for examination and could be particularly informative for some tumors. Examples of proteins to which monoclonal antibodies have been produced for immunohistochemistry include p53 and Rb, Bcl-2, BCRA1, c-*myc*, c-*ras*, c-*erb*B-2, and others. Most of these monoclonal antibodies have been raised to human or rodent antigens, although preliminary data in the veterinary literature demonstrate cross reactivities of at least some of these antibodies with specimens derived from domestic animals. These markers can be tested for their ability to function as tumor markers in animals on the basis of *a priori* knowledge of their functions.

G. Histopathology—Leukocyte Markers

Until recently, research on the canine immune system was done using antibodies against human or murine leukocyte antigens, but monoclonal antibodies with specific reactivities to canine and feline leukocyte antigens have been developed in recent years. Leukocyte antigens are currently defined by cluster-differentiation (CD) numbers assigned in international workshops; the specificities of many of the canine antibodies were established at the First Canine Leukocyte Antigen Workshop, and the canine homologs of the human CD antigens were summarized (Cobbold and Metcalfe, 1994). For example, antibodies with specificities toCD45 recognize B and T cells; antibodies to CD3 and CD5 recognize T cells; CD4 reactive cells include T helper-cell subset (MHC-II restricted); CD8 reactive cells include a cytotoxic T-cell subset (MHC-I restricted T cells); Thy-1 reactive cells include peripheral T cells, thymocytes; MHC-II reactive cells include antigen-presenting B and T cells; and monoclonal antibodies to CD21 and mb-1 recognize B cells. Plasma cell tumors have often been distinguished in the immunohistochemistry laboratory by their reactivity for antibodies to canine IgG $F(ab)_2$ and vimentin (Baer *et al.*, 1989).

In the cat, monoclonal antibodies specific for the feline homologs of CD4 (Ackley *et al.*, 1990; Tompkins *et al.*, 1990), CD8 (Klotz and Cooper, 1986; Tompkins *et al.*, 1990), major histocompatibility class II (Rideout *et al.*, 1990), and a leukocyte common antigen (CD45) have been described (Hunt *et al.*, 1995).

In the dog, the majority of malignant lymphomas are classified as B cell in origin (Teske *et al.*, 1994b). The proliferative lymphocytes in canine mycosis fungoides have been identified as T lymphocytes, although these lymphocytes expressed predominately CD3 and CD8 antigens, in contrast to the predominant CD4 expression of T lymphocytes in human mycosis fungoides (Moore *et al.*, 1994). Other cutaneous lymphoid neoplasms in the dog have been chararacterized by immunohistochemistry; these cutaneous lymphoid tumors predominately involve T lymphocytes, whereas epitheliotrophic lymphomas are exclusively of T-lymphocyte origin (DeBoer *et al.*, 1990; Ferrer *et al.*, 1993; Fivenson *et al.*, 1994; Moore *et al.*, 1990, 1994; Tobey *et al.*, 1994). An equine cutaneous T-cell lymphoma has also been identified (Littlewood *et al.*, 1995).

Several studies demonstrate the value of immunophenotyping for dogs with malignant lymphoma as an aid to predicting response to chemotherapy. Using cytofluorography and monoclonal antibodies to canine or human cell-surface determinants, it was determined that dogs with T-cell tumors had shortened remission and survival times when compared to dogs with B cell-derived tumors (Greenlee *et al.*, 1990). More recently, using immunohistochemistry, an avidin–biotin complex immunoperoxidase technique, and an expanded repertoire of monoclonal antibodies, canine lymphoma cells were categorized as B-cell type, T-cell type, or non-B/non-T cell type (Teske *et al.*, 1994a). In that study, dogs with B cell-derived tumors had significantly longer durations of remission and overall survival times following onset of chemotherapy than those dogs with T cell-derived tumors. Some data are now also evolving in the literature regarding the immunophenotyping of equine lymphomas (Asahina *et al.*, 1994), and the neoplastic cells of sporadic bovine leukosis develop from both B- and T-cell lineages (Asahina *et al.*, 1995).

H. Virology

1. *Feline Leukemia Virus*

Immunohistochemistry allows detection of specific localization of feline leukemia virus in tissues. Using immunofluorescence, "sequestered" feline leukemia virus (FELV) antigen was detected in localized tissues such as the spleen, intestinal crypt epithelium, or lymphomatous tissues in aviremic cats (Hayes *et al.*, 1989). Similarly, we detected an apparent sequestered FELV infection in a cat with a malignant lymphoma arising in the setting of immunosuppression following renal transplantation (Gregory *et al.*, 1991). In another report, retroviral DNA was detected in both blood and tumor tissue from a cat with an olfactory neuroblastoma, but not in nonneoplastic tissues, including the brain of the affected cat (Schrenzel *et al.*, 1990).

2. *Bovine Papillomavirus*

Immunohistochemistry is a rapid and reliable screening method for the detection of papillomaviruses in tissues using antisera to papillomavirus group-specific antigen (Sundberg *et al.*, 1984). This method has been used to identify papillomaviruses in a variety of avian and mammalian neoplasms (Sundberg, 1987); for example, using anti-papillomavirus group-specific structural antigens (bovine papillomavirus subgroup A), papillomavirus was detected in cutaneous papillomatosis from a dog (Shimada *et al.*, 1993). Other specific clinical applications of this detection method are referenced (Campbell *et al.*, 1988; Elzein *et al.*, 1991). Detection of papillomavirus group-specific antigen requires that highly differentiated cells contain large numbers of viral particles; in many papillomavirus infections, however, the resultant lesions differ in the frequency with which mature virus particles are being produced, which alters their immunoreactivity. A negative result, therefore, does not necessarily indicate the absence of papillomavirus infection. Other methods, principally based on nucleic acid hybridization and/or the polymerase chain reaction (PCR), are now being used in many research laboratories to investigate further the molecular pathogenesis of the various avian and mammalian papillomaviruses.

VII. CYTOCHEMISTRY

Hematologic diagnosis has traditionally been based on light microscopy supplemented by cytochemistry. New techniques of immunophenotyping, cytogenetics, and molecular genetics are providing new classification systems; many of these investigative strategies are now moving into the diagnostic laboratory as new sources of information that have influence on patient management. Cytologic assessment remains of major importance, however, in the diagnosis of acute leukemia; cytochemistry facilitates the diagnosis of the majority of cases of acute myelogenous leukemia and allows a provisional diagnosis of acute lymphoblastic leukemia. The most important cytochemical reactions for neutrophilic differentiation are the myeloperoxidase reaction and the Sudan black C stain. Chloroacetate esterase is also used as a marker for neutrophil differentiation. In the dog and cat, alkaline phosphatase reactivity may indicate acute myeloid leukemia because normal neutrophils in these species lack alkaline phosphatase reactivity (Jain, 1993). The most important stains for monocytic differentiation are the "nonspecific" esterase (NSE) reactions, such as α-naphthyl acetate esterase. The diffuse cytoplasmic reactivity of monocytes for nonspecific esterases can be inhibited by sodium fluoride; some T-lymphocyte subsets will also show localized cytoplasmic reactivity for NSE, but the reactivity is resistant to fluoride inhibition. The lipase stain is also used as a marker of monocytic differentiation. Myelomonocytic leukemias are identified by simulataneous cytochemical reactivities, typically alkaline phosphatase and non-specific esterase.

VIII. MOLECULAR ONCOLOGY

A. Methods

Increasing data regarding genetic changes that are often specifically associated with particular types of tumors provides the basis for new and effective strategies for diagnostics. These tests can be based directly on the changes in DNA that occur in tumor cells, or on the RNA that is made from that altered DNA. Cytogenetics was the tool used first to examine gross morphologic features of chromosomes, and to characterize the consistent, nonrandom chromosomal abnormalities such as translocations and deletions. These precedent studies provided important clues to the location of relevant genes and the mechanisms by which their growth regulatory functions have been altered. New advances in molecular technology, however, now allow analysis of genes at the nucleotide level; these methods are beginning to be applied in veterinary medicine. Techniques for identifying genetic changes in tumor cells are discussed next.

1. *Nucleic Acid Hybridization*

Hybridization refers to the base pairing of complementary strands of nucleic acid that leads to the double-stranded molecule. *In situ* hybridization allows identification of nucleic acid sequences within cells that can be components of genes that code for particular proteins. The advantage of *in situ* hybridization is that it allows examination of gene content at the site of expression in the cell, and it can provide evidence of whether or not a specific gene is transcribed within the cell. Reviews of methodologies and applications of *in situ* hybridization are available elsewhere (Mitchell *et al.*, 1992). For cancer diagnosis, *in situ* hybridization provides an opportunity to use complementary nucleic acid sequences to cancer-causing viral genes such as leukemia viruses, immunodeficiency viruses, or the papillomaviruses. *In situ* hybridization has been used extensively in diagnostic pathology and oncology by providing proof that cells synthesize proteins. Whereas immunohisto(cyto)chemistry is used to demonstrate specific tumor-associated proteins by virtue of their gene expression, *in situ* hybridization provides unequivocal evidence of their synthesis by the neoplastic cells. This application is perhaps best illustrated by the demonstration of expression of specific oncogenes in preneoplasia and overt neoplasms. For example, using *in situ* hybridization, *erb*-B RNA was detected in a small number of plutonium-239-induced canine pulmonary neoplasms, demonstrating activation of that oncogene (Gillett *et al.*, 1992).

2. *Restriction Enzymes and Restriction Fragment Length Polymorphisms*

Restriction enzymes (or endonucleases) are a group of endonucleases that cleave double-stranded DNA at a specific recognition site (restriction site) determined by the exact DNA sequence. Hundreds of restriction enzymes with related and different specificities are now available. Restriction enzymes cut DNA into reproducible sizes; however, alterations of that restriction site (i.e., mutation, deletion) might abolish that DNA sequence so that the enzyme does not cut that DNA. Therefore, the uncut DNA fragment will be larger than normal upon electrophoresis. Mutations at sites recognized by restriction enzymes, or polymorphisms at restriction enzyme sites, are extremely useful for genetic analysis, because mutations at restriction sites are inherited like any other genetic trait and can be used for linkage analysis. *Restriction fragment length polymorphism* (RFLP) analysis involves application of restriction enzymes to detect polymorphisms linked to cancer-related genes, and it has been used extensively to examine regions of the genome likely to contain tumor suppressor genes (Rowley *et al.*, 1993).

3. *Southern Blots*

The Southern blot is a standard technique used for the identification of specific DNA sequences. Typically, a DNA sequence (i.e., a PCR product, or chromosomal DNA that has been digested with a restriction enzyme) is denatured (to single-strand form) and transferred or blotted to a special membrane, usually nitrocellulose. A labeled probe (radioactive or nonradioactive) will hybridize to a complementary sequence on the membrane, which can then be detected, for example, by autoradiography or some other imaging procedure. Southern blotting was used to detect c-*myc*, c-*erb*B-2, c-*ros*-1, c-*yes*-1, v-*myc*, v-Ki-*ras*, and v-Ha-*ras* proto-oncogenes in genomic DNA from several domestic animal species, including dogs, cats, horses, and cattle (Miyoshi *et al.*, 1991). In another study, using Southern blotting, c-*erb*B-2 and c-*myc* overexpression was detected in canine melanoma cell lines (Ahern *et al.*, 1993). *Northern* blot analysis is a variation of Southern blotting, whereby steady-state levels of a particular RNA transcript (mRNA), instead of cleaved DNA fragments, are separated by electrophoresis without prior digestion and then probed in a hybridization reaction with complementary sequences. Northern analysis can be used to detect both qualitative and quantitative information on mRNA transcripts; for example, it can be used to identify amplified gene expression.

4. DNA Sequencing

Direct sequencing is the primary method for characterizing genes. Although sequencing methods have been simplified in recent years, they are still laborious and time consuming, and therefore of limited clinical use. Direct DNA sequencing is expected to be a routine procedure in specialized clinical laboratories within 10 years (Kant, 1995). Many genes and their mutations associated with cancer are now being elucidated, and a number of autosomal genes, such as the major ones associated with familial breast cancer (*BCRA1, BCRA2*), cumulatively affect large numbers of individuals in the population. The nucleotide sequences of genes relevant to animals with cancer are now being elucidated.

5. Polymerase Chain Reaction

Another approach to examining genetic structure is the polymerase chain reaction. PCR is an *in vitro* method that uses enzyme synthesis to exponentially amplify specific DNA (or cDNA) sequences. Amplification of a gene or gene fragment of interest from virtually any tumor preparation by means of PCR, coupled with improved detection methodologies, has enhanced diagnostic capabilities, allowing characterization of genes responsible for neoplastic transformation and tumor progression.

Applications of PCR for viral agents that may be associated with, or causative of, cancer in animals include its use in the detection of infections with bovine papillomavirus, bovine immunodeficiency virus, feline leukemia virus, and feline immunodeficiency virus (Belak and Ballagi-Pordany, 1993). For the detection of the bovine leukemia virus (BLV), PCR has been used adjunct to indirect (serologic) methods for diagnosis of BLV infection in circumstances where serology may be unreliable, such as during early stages of infection, in animals persistently infected with bovine viral diarrhea virus, or in young calves that have received colostrum from infected dams (Belak and Ballagi-Pordany, 1993). Most of the PCR strategies for detection of BLV infection are based on demonstration of proviral DNA in blood lymphocytes.

B. Genetic Changes in Cancer Cells

The key molecular events associated with carcinogenesis involve oncogenes and tumor suppressor genes. These genes may be altered in premalignant as well as malignant processes; methods are now available to study these genes directly in clinical specimens. For example, mutated p53 genes have been detected in the urine of human patients with cancer of the urinary bladder; mutant *ras* genes in stools derived from patients with pancreatic and colorectal cancer; and both mutant p53 and *ras* genes in sputum from patients with lung cancer (Duffy, 1995). These findings demonstrate that the genetic alterations associated with cancer may be detected in fluids or secretions that had contact with the malignant tissues. Alterations involve changes in both gene structure and gene expression. Oncogenes contribute to tumor formation either because constitutional activation prevents them from following regulatory signals or because structural changes incapacitate their normal signaling function. Proto-oncogenes may be activated by a number of mechanisms.

1. Amplification

Proto-oncogenes may be inappropriately amplified or overexpressed in neoplasms, and these phenomena have become frequent findings in the molecular analyses of various tumors. Amplified oncogenes were first discovered in cells carrying two cytogenetic abnormalities, double minute chromosomes and homogenously staining chromosome regions. The first reports of somatic ampification of a cellular oncogene involved the c-*myc* oncogene in a promyelocytic leukemia cell line and a colonic carcinoma cell line. *Myc* proteins serve as transcription factors. Overexpression of *myc* promotes cell proliferation, and it appears that dysregulation of *myc* confers a growth advantage to affected cells. In human patients, from a clinical perspective, L-*myc* is amplified in patients with non-small-cell lung cancer, and N-*myc* in patients with neuroblastoma. Amplification correlates with a more rapid disease progression and advanced disease state and is an independent prognostic factor for decreased survival time. In human breast cancer, the c-*erb*B-2 (also known as HER-2/*neu*) is amplified consistently in patients with advanced neoplasms, and those amplifications are associated with poor prognosis. There are preliminary data to show that c-*erb*B is also overexpressed in canine malignant melanoma (Ahern *et al.*, 1993).

2. *Mutation*

Mutation is another mechanism of oncogene activation resulting in altered protein product. Single nucleotide substitutions leading to amino acid changes (missense mutations) are responsible for most disease-causing alterations of genes. Point mutations in proto-oncogenes and tumor suppressor genes are relatively common abnormalities in human tumors, and preliminary data are now emerging regarding the occurrence of similar mutations in some animal tumors. In human medicine, many of these mutations have been shown to have considerable significance in disease processes; for some genes, specific mutations have been used to detect minimum tumor burdens either early in the disease course or following cytoreductive treatment. Specific mutations have been used to categorize individuals predisposed to cancer in high-risk families; other mutations have been shown to correlate with patient response to therapy, thus serving as a measure of prognosis. Still others have served to provide evidence for the role of specific environmental carcinogens in tumorigenesis by virtue of the characteristic mutations induced.

Mutations in the *RAS* family of oncogenes occur with high frequency and predictability at certain neoplasms at codons 12, 13, and 61, conferring transforming properties to *ras* p21 proteins. Mutations of *ras* genes have been identified in a wide variety of neoplasms, occurring most frequently in tumors of the pancreas, colon, thyroid gland, and lung, and in myeloid leukemia. For example, Ki-*ras* activation occurs in approximately 30% of human patients with non-small-cell lung cancer; we have determined similar activation of Ki-*ras* in dogs with lung tumors (Kraegel *et al.*, 1992).

3. *Translocation*

Proto-oncogenes may be activated by translocations and inversions. There are two principal consequences of translocations or inversions. In one situation, either the T-cell receptor or an immunoglobulin protein comes to lie near a proto-oncogene, thereby activating it. Alternatively, the break occurs within a gene on each chromosome involved, creating a fusion gene encoding a chimeric protein. The genes involved in activating translocations often encode transcription factors, indicating that altered transcription plays a role in tumorigenesis. Translocation is a common mechanism in human hematopoietic neoplasms, whereby proto-oncogenes located near the break are activated as a result of the translocation. For example, in Burkitt's lymphoma, c-*myc* is placed by translocation under the influence of immunoglobulin enhancers and promoters. Immunoglobulin and T-cell receptor genes are frequently involved in chromosomal aberrations because they are naturally rearranged to generate active antigen-receptor genes. This process leads, on occasion, to interchromosomal translocations or inversions, as in Burkitt's lymphoma.

One study of 15 dogs with canine lymphoma or leukemia revealed 10 dogs with rearrangements of the immunoglobulin heavy chain gene (IgH), and 4 dogs with TCR*B*-chain gene rearrangements. Most of the IgH chain rearrangements were associated with tumors expressing cell-surface immunoglobulin, whereas 3 of 4 dogs with TCR*B*-gene rearrangements were Thy-1 positive (Momoi *et al.*, 1993).

At present, there are still limited data on the nucleotide sequences of the (proto-)oncogenes and tumor suppressor genes in domestic animals, their chromosomal localizations, their expression in tissues, and their roles in cellular growth, differentiation, and carcinogenesis. c-*myc* expression was described in canine mammary tumors (Engstrom *et al.*, 1987), and overexpression of c-*myc* was reported in explants of canine malignant melanomas (Ahern *et al.*, 1993). A canine genomic oncogene related to the human c-*yes*-1 oncogene was detected in a canine mammary neoplasm, as well as the genomic DNA from the same dog, suggesting a role for that oncogene in the early stages of carcinogenesis (Miyoshi *et al.*, 1991). In another study, c-*yes*-1 protooncogene was amplified in the DNA derived from canine lymphoid leukemia cells (Mina *et al.*, 1994). The canine c-*yes* proto-oncogene has been cloned and sequenced (Zhao *et al.*, 1995). Expression of c-Ha-*ras*, c-Ki-*ras*, and c-N-*ras* was discerned in normal and neoplastic tissues from the dog and cow (Madewell *et al.*, 1989). The nucleotide sequences of the clinically relevant regions of the c-Ki-*ras* and c-N-*ras* genes were subsequently reported (Saunders *et al.*, 1992; Kraegel *et al.*, 1992), as well as preliminary data on the activation of c-Ki-*ras* in canine spontaneous lung tumors (Kraegel *et al.*, 1992). Some data are also now emerging regarding the canine, feline, and equine tumor suppressor genes p53. Genetic sequences of the canine p53 gene, encompassing regions of exons 3–8, were reported (Kraegel *et al.*, 1995). A 284th codon mutation was described in a canine papilloma (Mayr *et al.*, 1994b) and in several thyroid carcinomas (Devilee *et al.*, 1994). Mutated p53 species have also been described in feline hematopoietic tumors and a feline mammary carcinoma (Okuda *et al.*, 1993, 1994; Mayr *et al.*, 1993b; Mayr *et al.*, 1995b). Wild-type equine p53 cDNA sequences between exons 2 and 9 have been described, and two mutations were detected in equine cutaneous squamous-cell carcinomas at sites of known activation by ultraviolet light in human cutaneous carcinomas (Pazzi *et al.*, 1995).

References

Ackley, C. D., Hoover, E. A., and Cooper, M. D. (1990). Identification of a CD4 homologue in the cat. *Tiss. Antig.* **35,** 92–98.

Ahern, T. E., Bird, R. C., Bird, A. E., and Wolfe, L. G. (1993). Overexpression of c-*erbB*-2 and c-*myc*, but not c-*ras*, in canine melanoma cell lines, is associated with metastatic potential in nude mice. *Anticancer Res.* **13,** 1365–1372.

Alenza, M. D. P., Rutteman, G. R., Kuipers-Dijkshoorn, N. J., Pena, L., Montoya, A., Misdorp, W., and Cornelisse, C. J. (1995). DNA flow cytometry of canine mammary tumours: The relationship of DNA ploidy and S-phase fraction to clinical and histological features. *Res. Vet. Sci.* **58,** 238–243.

Andreasen, C. B., and Mahaffey, E. A. (1987). Immunohistochemical demonstration of desmin in canine smooth muscle tumors. *Vet. Pathol.* **24,** 211–215.

Andreasen, C. B., Mahaffey, E. A., and Duncan, J. R. (1988a). Intermediate filament staining in the cytologic and histologic diagnosis of canine soft tissue tumors. *Vet. Pathol.* **25,** 341–349.

Andreasen, C. B., White, M. R., and Swayne, D. E. (1988b). Desmin as a marker for canine botyroid rhabdomyosarcomas. *J. Comp. Path.* **98,** 23–29.

Arai, K., Kaneko, S., Naoi, M., Suzuki, K., Maruo, F., Uehara, K., (1994). Expression of stratified squamous epithelia-type cytokeratin by canine mammary epithelial cells during tumorigenesis: Type I (acidic) 57 kilodalton cytokeratin could be a molecular marker for malignant transformation of mammary epithelial cells. *Jpn. J. Vet. Sci.* **56,** 51–58.

Asahina, M., Murakami, K., Ajito, T., Goryo, M., and Okada, K. (1994). An immunohistochemical study of an equine B-cell lymphoma. *J. Comp. Pathol.* **111,** 445–451.

Asahina, M., Kimura, K., Murakami, K., Ajito, T., Wu, D., Goryo, M., Aida, Y., Davis, W. C., and Okada, K. (1995). Phenotypic analysis of neoplastic cells from calf, thymic, and intermediate forms of bovine leukosis. *Vet. Pathol.* **32,** 683–691.

Aumuller, G., Vedder, H., Enderle-Schmidt, M., Seitz, J. (1987). Cytochemistry and biochemistry of acid phosphatases VII: immunohistochemistry of canine prostatic acid phosphatase. *Prostate* **11,** 1–15.

Ayl, R. D., Couto, C. G., Hammer, A. S., Weisbrode, S., Ericson, J. G., and Mathes, L. (1992). Correlation of DNA ploidy to tumor histologic grade, clinical variables, and survival in dogs with mast cell tumors. *Vet. Pathol.* **29,** 386–390.

Baer, K. E., Patnaik, A. K., Gilbertson, S. R., and Hurvitz, A. I. (1989). Cutaneous plasmacytomas in dogs: A morphologic and immunohistochemical study. *Vet. Pathol.* **26,** 216–221.

Bates, S. E. (1991). Clinical applications of serum tumor markers. *Annal. Intern. Med.* **115,** 623–634.

Belak, S., and Ballagi-Pordany, A. (1993). Applications of polymerase chain reaction (PCR) in veterinary diagnostic virology. *Vet. Res. Commun.* **17,** 55–72.

Benazzi, C., Sarli, G., Preziosi, R., and Marcato, P. S. (1995). Laminin expression in testicular tumours of the dog. *J. Comp. Pathol.* **112,** 141–150.

Benchimol, S., and Ling, V. (1994) P-glycoprotein and tumor progression. *J. Natl. Cancer Inst.* **86,** 814.

Berrington, A. J., Jimbow, K., and Haines, D. M. (1994). Immunohistochemical detection of melanoma associated antigens on formalin-fixed, paraffin-embedded canine tumors. *Vet. Pathol.* **31,** 455–461.

Boosinger, T. R., Zerbe, C. A., Grabau, J. H., and Pletcher, J. M. (1988). Multihormonal pancreatic endocrine tumor in a dog with duodenal ulcer and hypertrophic gastropathy. *Vet. Pathol.* **25,** 237–239.

Bostock, D. E. (1986). Neoplasms of the skin and subcutaneous tissues in dogs and cats. *Br. Vet. J.* **142,** 1–19.

Bostock, D. E., Crocker, J. Harris, K., and Smith, P. (1989). Nucleolar organiser regions as indicators of post-surgical prognosis in canine spontaneous mast cell tumours. *Br. J. Cancer* **59,** 915–918.

Bostock, D. E., Moriarty, J., and Crocker, J. (1992). Correlation between histologic diagnosis mean nucleolar organizer region count and prognosis in canine mammary tumors. *Vet. Pathol.* **29,** 381–385.

Campbell, K. L., Sundberg, J. P., Goldschmidt, M. H., Knupp, C., and Reichmann, M. E. (1988). Cutaneous inverted papillomas in dogs. *Vet. Pathol.* **25,** 67–71.

Cardona, A., Madewell, B. R., Naydan, D. K., and Lund, J. K. (1989). A comparison of six monoclonal antibodies for detection of cytokeratins in normal and neoplastic canine tissues. *J. Vet. Diag. Invest.* **1,** 316–323.

Carver, J. R., Kapatkin, A., and Patnaik, A. K. (1995). A comparison of medullary thyroid carcinoma and thyroid adenocarcinoma in dogs: A retrospective study of 38 cases. *Vet. Surg.* **24,** 315–319.

Chew, D. J., Nagode, L. A., Rosol, T. J., Carothers, M. A., and Schenck, P. (1995). Utility of diagnostic assays in the evaluation of hypercalcemia and hypocalcemia: Parathyroid hormone, vitamin D metabolites, parathyroid hormone-related peptide, and ionized calcium. *In* "Kirk's Current Veterinary Therapy XII—Small Animal Practice" (J. D. Bonagura and R. W. Kirk, eds.), pp. 378–383. Saunders, Philadelphia.

Clemo, F. A. S., DeNicola, D. B., Carlton, W. W., Morrison, W. B., and Walker, E. (1994). Flow cytometric DNA ploidy analysis in transitional cell carcinomas of urinary bladders. *Vet. Pathol.* **31,** 207–215.

Cobbold, S. P., and Metcalfe, S. (1994). Monoclonal antibodies that define canine homologues of human CD antigens: Summary of the First International Canine Leukocyte Antigen Workshop (CLAW). *Tiss. Antig.* **43,** 137–154.

DeBoer, D. J., Turrel, J. M., and Moore, P. F. (1990). Mycosis fungoides in a dog: Demonstration of T-cell specificity and response to radiotherapy. *J. Am. Anim. Hosp. Assn.* **26,** 566–572.

de las Mulas, J. M., de los Monteros, A. E., Bautista, M. J., Gomez-Villamondos, J. C., and Morales, C. (1994). Immunohistochemical distribution patterns of intermediate filament proteins and muscle actin in feline and human mammary carcinomas. *J. Comp. Pathol.* **111,** 365–381.

de las Mulas, J. M., de los Monteros, A., Carrasco, L., van Niel, M., and Fernandez, A. (1995a). Immunohistochemical distribution pattern of intermediate filament proteins in 50 feline neoplasms. *Vet. Pathol.* **32,** 692–701.

de las Mulas, J. M. de, Gomez-Villamondos, J. C., Perez, J., Mozos, E., Estrada, M., and Mendez, A. (1995b). Immunohistochemical evaluation of canine primary liver carcinomas: Distribution of alpha-fetoprotein, carcinoembryonic antigen, keratins and vimentin. *Res. Vet. Sci.* **59,** 124–127.

de Papp, A. E., and Stewart, A. F. (1993). Parathyroid hormone-related protein. A peptide with diverse physiologic functions. *TEM* **4,** 181–186.

deLellis, R. A., and Dayal, Y. (1987). The role of immunohistochemistry in the diagnosis of poorly differentiated malignant neoplasms. *Sem. Oncol.* **14,** 173–192.

Destexhe, E., Bicker, E., and Coignoul, F. (1995). Image analysis evaluation of ploidy, S-phase fraction and nuclear area in canine mammary tumours. *J. Comp. Pathol.* **113,** 205–216.

DeVico, G., Maiolino, P., and Restucci, B. (1995). Silver-stained nucleolar (Ag-NOR) cluster size in feline mammary carcinomas: Lack of correlation with histological appearance, mitotic activity, tumour stage, and degree of nuclear atypia. *J. Comp. Pathol.* **113,** 69–73.

Devilee, P., van Leeuwen, S., Voesten, A., Rutteman, G. R., Vos, J. H., and Cornelisse, C. J. (1994). The canine p53 gene is subject to somatic mutations in thyroid carcinoma. *Anticancer Res.* **14,** 2039–2046.

Dolman, C. L. (1989). Glial fibrillary acidic protein and cartilage. *Acta Neuropathol.* **79,** 101–103.

Donnay, I., Rauis, J., Devleeschouwer, N., Wouters-Ballman, P., Leclercq, G., and Verstegen, J. (1995). Comparison of estrogen and progesterone receptor expression in normal and tumor mammary tissues from dogs. *Am. J. Vet. Res.* **56,** 188–1194.

Donner, L., deLanerolle, P., and Costa, J. (1983). Immunoreactivity of paraffin-embedded normal tissues and mesenchymal tumors of smooth muscle origin. *Am. J. Clin. Pathol.* **80,** 667–681.

Doveren, R. F. C., Buurman, W. A., Schutte, B., Groenewegen, G., and Van Der Linden, C. J. (1985). Class II antigens on canine T lymphocytes. *Tiss. Antig.* **25,** 255–265.

Duffy, M. D. (1995). Can molecular markers now be used for early diagnosis of malignancy? *Clin. Chem.* **41,** 1410–1413.

Dunn, D. G., Harris, R. K., Meis, J. M., and Sweet, D. E. (1991). A histomorphologic and immunohistochemical study of chordoma in twenty ferrets (*Mustela putorius furo*). *Vet. Pathol.* **28,** 467–473.

Elzein, E. T. E., Sundberg, J. P., Housawi, F. M., Gameel, A. A., Ramadan, R. O., and Hassanein, M. M. (1991). Genital bovine papillomavirus infection in Saudi Arabia. *J. Vet. Diag. Invest.* **3,** 36–38.

Engstrom, W., Barrios, C., Azavedo, E., Mollermark, G., Kangstrom, L. E., Eliason, I., and Larsson, O. (1987). Expression of c-*myc* in canine mammary tumours. *Anticancer Res.* **7,** 1235–1238

Erlandsen, R. A. (1989). Cytoskeletal proteins including myofilaments in human tumors. *Ultrastruc. Pathol.* **13,** 155–186.

Feldman, B. F., Brummerstedt, E., Larsen, L. S., and Larsen, S. (1988a). Plasma fibronectin concentration associated with varius types of canine neoplasia. *Am. J. Vet. Res.* **49,** 1017–1019.

Feldman, B. F., Brammerstedt, E., Larsen, L. S., and Larsen, S. (1988b). Fibronectin in neoplastic peritoneal effusions in dogs. *Acta Vet. Scand.* **29,** 273–279.

Ferrer, L., Rabanal, R. M., Fondevila, D., and Prats, N. (1990). Immunocytochemical demonstration of intermediate filament proteins, S-100 protein and CEA in apocrine sweat glands and apocrine gland derived lesions of the dog. *J. Vet. Med. Assn.* **37,** 569–576.

Ferrer, L., Fondevila, D., Rabanal, R., Tarres, J., and Ramis, A. (1993). Immunohistochemical detection of CD3 antigen (pan T marker) in canine lymphomas. *J. Vet. Diag. Invest.* **5,** 616–620.

Ferrer, L., Fondevila, D., Rabanal, R. M., and Vilafranca, M. (1995). Immunohistochemical expression of CD31 antigen in normal and neoplastic canine endothelial cells. *J. Comp. Pathol.* **112,** 319–326.

Fivenson, D. P., Saed, G. M., Beck, E. R., Dunstan, R., and Moore, P. F. (1994). T-cell receptor gene rearrangement in canine mycoisis fungoides; further support for a canine model of cutaneous T-cell lymphoma. *J. Invest. Dermatol.* **102,** 227–230.

Fox, M. H., Armstrong, L. W., Withrow, S. J., Powers, B. E., LaRue, S. M., Straw, R. C., and Gillette, E. L. (1990). Comparison of DNA aneuploidy of primary and metastatic spontaneous canine osteosarcomas. *Cancer Res.* **50,** 6176–6178.

Frazier, K. S., Hines, M. E., Hurvitz, A. I., Robinson, P. G., and Herron, A. J. (1993). Analysis of DNA aneuploidy and c-myc oncoprotein content of canine plasma cell tumors using flow cytometry. *Vet. Pathol.* **30,** 503–511.

Furuoka, H., Ito, H., Hamada, M., Suwa, T., Satoh, H., and Itakura, C. (1989). Immunocytochemical component of endocrine cells in pancreatic islets of horses. *Jpn. J. Vet. Sci.* **51,** 35–43.

Gebhard, D. H., and Carter, P. B. (1992). Identification of T-lymphocyte subsets with monoclonal antibodies. *Vet. Immuno. Immunopathol.* **33,** 187–199.

Gillett, N. A., Stegelmeier, B. L., Kelly, G., Haley, P. J., and Hahn, F. F. (1992). Expression of epidermal growth factor receptor in plutonium-239-induced lung neoplasms in dogs. *Vet. Pathol.* **29,** 46–52.

Gould, W. J., Reimers, T. J., Bell, J. A., Lawrence, H. J., Randolph, J. T., Rowland, P. H., and Scarlett, J. M. (1995). Evaluation of urinary cortisol:creatinine ratios for the diagnosis of hyperadrenocorticism associated with adrenal gland tumors in ferrets. *J. Am. Vet. Med. Assn.* **206,** 2–46.

Greenlee, P. G., Filippaa, D. A., Quimby, F. W., Patnaik, A. K., Calvano, S. E., Matus, R. E., Kimmel, M., Hurvitz, A. I., and Lieberman P. H. (1990). Lymphomas in dogs. A morphologic, immunologic and clinical study. *Cancer* **66,** 480–490.

Gregory, C. R., Madewell, B. R., Griffey, S. M., and Torten, M. (1991). Feline leukemia virus-associated lymphosarcoma following renal transplantation in a cat. *Transplantation* **52,** 1097–1099.

Grone, A., Werkmeister, J. R., Steinmeyer, C. L., Capen, C. C., and Rosol, T. J. (1994). Parathyroid hormone-related protein in normal and neoplastic canine tissues: Immunohistochemical localization and biochemical extraction. *Vet. Pathol.* **31,** 308–315.

Groshek, P. M., Dean, G. A., and Hoover, E. A. (1994). Monoclonal antibodies identifying feline haematopoietic cell lineages. *Comp. Haematol. Intern.* **4,** 181–191.

Hahn, K. A., and Richardson, R. C. (1995). Detection of serum alpha-fetoprotein in dogs with naturally occurring malignant neoplasia. *Vet. Clin. Pathol.* **24,** 18–21.

Hahn, K. A., Richardson, R. C., Hahn, E. A., and Chrisman, C. L. (1994). Diagnostic and prognostic importance of chromosomal aberrations identified in 61 dogs with lymphosarcoma. *Vet. Pathol.* **31,** 528–540.

Hanzaike, T., Ito, I., Ishikawa, T., Ishikawa, Y., and Kadota, K. (1995). Leiomyosarcoma of soft tissue in a cow. *J. Comp. Path.* **112,** 237–242.

Harmelin, A., Zuckerman, A., and Nyska, A. (1995). Correlation of Ag-NOR protein measurements with prognosis in canine transmissible venereal tumour. *J. Comp. Path.* **112,** 429–433.

Hawkins, K. L., Summers, B. A., Kuhajda, F. P., and Smith, C. A. (1987). Immunocytochemistry of normal pancreatic islets and spontaneous islet cell tumors of the dog. *Vet. Pathol.* **24,** 170–179.

Hayes, K. A., Rojko, J. L., Tarr, M. J., Polas, P. J., Olsen, R. G., and Mathes, L. E. (1989). Atypical localised viral expression in a cat with feline leukaemia. *Vet. Rec.* **124,** 334–336.

Hellmen, E. (1994). S100 protein is not specific for myoepithelial cells in the canine mammary gland. *J. Comp. Pathol.* **110,** 49–55.

Hellmen, E., and Lindgren, A. (1989). The accuracy of cytology in diagnosis and DNA analysis of canine mammary tumours. *J. Comp. Pathol.* **101,** 443–450.

Hellmen, E., and Lindgren, A. (1989). The expression of intermediate filaments in canine mammary glands and their tumors. *Vet. Pathol.* **26,** 420–428.

Hellmen, E., and Svensson, S. (1995). Progression of canine mammary tumours as reflected by DNA ploidy in primary tumours and their metastases. *J. Comp. Pathol.* **113,** 327–342.

Hellmen, E., Lindgren, A., Linell, F., Matsson, P., and Nilsson, A. (1988). Comparison of histology and clinical variables to DNA ploidy in canine mammary tumors. *Vet. Pathol.* **25,** 219–226.

Helzlsouer, K. J., Bush, T. L., Alberg, A. J., Bass, K. M., Zacur, H., and Comstock, G. W. (1993). Prospective study of serum CA-125 levels as markers of ovarian cancer. *J. Am. Med. Assn.* **269,** 1123–1126.

Hendrick, M. J., and Brooks, J. J. (1994). Postvaccinal sarcomas in the cat: Histology and immunohistochemistry. *Vet. Pathol.* **31,** 126–129.

Hunt, P., Else, R. W., McConnell, I., and Hopkins, J. (1995). Identification of CD45 (leukocyte common antigen) in the domestic cat. *Res. Vet. Sci.* **59,** 201–204.

Ivanyi, D., Minke, J. M. H. M., Hageman, C., Groeneveld, E., and van Doornewaard, G. (1992). Patterns of expression of feline cytokeratins in healthy epithelia and mammary carcinoma cells. *Am. J. Vet. Res.* **53,** 304–314.

Ivanyi, D., Minke, J. M. H. M., Hageman, C., Groenveld, E., van Doornewaard, G., and Misdorp, W. (1993). Cytokeratins as markers of initial stages of squamous metaplasia in feline mammary carcinomas. *Am. J. Vet. Res.* **54,** 1095–1102.

Jain, N. C. (1993). Interpretation of leukocyte parameters. *In* "Essentials of Veterinary Hematology" (N. C. Jain, ed.), pp. 295–306. Lea and Febiger, Philadelphia, Pennsylvania.

Kant, J. A. (1995). Direct DNA sequencing in the clinical laboratory. *Clin. Chem.* **41,** 1407–1409.

Kempainen, R. J., and Clark, T. P. (1995). CVT update: Sample collection and testing protocols in endocrinology. *In "Kirk's Current Veterinary Therapy—XII, Small Animal Practice"* (J. D. Bonagura and R. W. Kirk, eds.), pp. 335–339. Saunders, Philadelphia.

Khan, K. N. W., Kociba, G. J., Wellman, M. L., and Reiter, J. A. (1992). Tumor necrosis factor levels and the effects of pentoxifylline in cats with feline leukaemia virus infection. *Comp. Haematol. Inter.* **2,** 133–139.

Kithier, K., Al-Sarraf, M., Belamaric, J., Radl, J., Valenta, Z., Ziskovsky, V., and Masopust, J. (1974). Alpha-fetoprotein in bovine hepatocellular carcinoma. *J. Comp. Path.* **84,** 133–141.

Kloppel, T. M., Franz, C. P., Morre, D. J., and Richardson, R. C. (1978). Serum sialic acid levels increased in tumor-bearing dogs. *Am. J. Vet. Res.* **39,** 1377–1380.

Klotz, F. W., and Cooper, M. D. (1986). A feline thymocyte antigen defined by a monoclonal antibody (FT2) identifies a subpopulation of non-helper cells capable of specific cytotoxicity. *J. Immunol.* **136,** 2510–2514.

Koss, L. G., Czerniak, B., Herz, F., and Wersto, R. P. (1989). Flow-cytometric measurements of DNA and other cell components in human tumors: A critical appraisal. *Human Pathol.* **20,** 528–548.

Kraegel, S. A., Gumerlock, P. H., Dungworth, D. H., Oreffo, V. I. C., and Madewell, B. R. (1992). K-*ras* activation in non-small cell lung cancer in the dog. *Cancer Res.* **52,** 4724–4727.

Kraegel, S. A., Pazzi, K. A., and Madewell, B. R. (1995). Sequence analysis of canine p53 in the region of exons 3–8. *Cancer Letters* **92,** 181–186.

Kressel, M., and Groscurth, P. (1995). Distinction of apoptotic and necrotic cell death by *in situ* labelling of fragmented DNA. *Cell Tiss. Res.* **278,** 549–556.

LeBlanc, B., Parodi, A. L., Lagadic, M., Hurtrel, M., and Jobit, C. (1991). Immunocytochemistry of canine thyroid tumors. *Vet. Pathol.* **28,** 370–380.

Linke, R. P., Hol, P. R., Gruys, E., Geisel, O., Nathrath, W. B. J., and Trautwein, G. (1984). Immunohistochemical identification and cross reactions of amyloid-A fibril protein in man and eleven other species. *J. Comp. Pathol.* **94,** 339–356.

Lipman, N. S., Marini, R. P., Murphy, J. C., Zhibo, Z., and Fox, J. G. (1993). Estradiol-17B-secreting adrenocortical tumor in a ferret. *J. Am. Vet. Med. Assn.* **203,** 1552–1555.

Littlewood, J. D., Whitwell, K. E., and Day, M. J. (1995). Equine cutaneous lymphoma: a case report. *Vet. Dermatol.* **6,** 105–111.

Loar, A. S. (1986). Canine thyroid tumors. *In* "Kirk's Current Veterinary Therapy IX, Small Animal Practice" (R. W. Kirk, ed.), pp. 1033–1039. Saunders, Philadelphia.

Lowseth, L. A., Gillet, N. A., Chang, I.-Y., Muggenburg, B. A., and Boecker, B. B. (1991). Detection of serum alpha-fetoprotein in dogs with hepatic tumors. *J. Am. Vet. Med. Assn.* **199,** 735–741.

McEntee, M., Isaacs, W., and Smith, C. (1987). Adenocarcinoma of the canine prostate: immunohistochemical examination for secretory antigens. *The Prostate* **22,** 163–170.

Madewell, B. R., Lund, J., Munn, R., and Pino, M. (1988). Canine laryngeal rhabdomyosarcoma: Am immunohistochemical and light microscopic study. *Jpn. J. Vet. Sci.* **50,** 1079–1084.

Madewell, B. R., Gumerlock, P. H., Saunders, K. A., Myers, F. J., Kraegel, S. A., and Kawasaki, E. S. (1989). *Anticancer Res.* **9,** 1743–1750.

Madewell, B. R., Deitch, A. D., Higgins, R. J., Marks, S. L., and deVere White, R. W. (1991). DNA flow cytometric study of the hyperplastic and neoplastic canine prostate. *The Prostate* **18,** 173–179.

Magnol, J. P., Al Saati, T., and Delsol, G. (1985). Marquage immunocytochimique des cytokeratines des carcinomes canins. *Revue Med. Vet.* **136,** 357–362.

Makino, S. (1963). Some epidemiologic aspects of venereal tumors of dogs as revealed by chromosome DNA studies. *Ann. NY Acad. Sci.* **101,** 1106–1122.

Mattix, M. E., Mattix, R. J., Williams, B. H., Ribas, J. L., and Wilhelmsen, C. L. (1994). Olfactory ganglioneuroblastoma in a dog: A light, ultrastructural, and immunohistochemical study. *Vet. Pathol.* **31,** 262–265.

Mayr, B., Eschborn, U., Loupal, G., and Schleger, W. (1993a). Trisomy 1 in a canine mammary tubular adenocarcinoma, complex type. *Vet. Pathol.* **30,** 311–313.

Mayr, B., Heczka, U., Schellander, K., Schleger, W., and Reifinger, M. (1993b). Sequence of an exon of the feline p53 gene-mutation in a lymphosarcoma. *Br. Vet. J.* **38,** 387–390.

Mayr, B., Reifinger, M., Weissenbock, H., Schleger, W., and Eisenmenger, E. (1994a). Cytogenetic analyses of four solid tumours in dogs. *Res. Vet. Sci.* **57,** 88–95.

Mayr, B., Schellander, L., Schleger, W., and Reifinger, M. (1994b). Sequence of an exon of the canine p53 gene mutation in a papilloma. *Br. Vet. J.* **150,** 81–84.

Mayr, B., Scheller, M., Reifinger, M., and Loupal, G. (1995a). Cytogenetic characterization of a fibroma and three hemangiopericytomas in domestic dogs. *Br. Vet. J.* **151,** 433–441.

Mayr, B., Schaffner, G., Kurzbauer, R., Reifinger, M., and Schellander, K. (1995b). Sequence of an exon of tumor suppressor p53 gene—a comparative study in domestic animals: mutation in a feline solid mammary carcinoma. *Br. Vet. J.* **151,** 325–329.

Mazza, G., Duffus, W. P. H., Elson, C. J., Stokes, C. R., Wilson, A. D., and Whiting, A. H. (1993). The separation and identification by monoclonal antibodies of dog IgG fractions. *J. Immunol. Meth.* **161,** 192–203.

Meinecke, B., and Gips, H. (1987), Steroid hormone secretory patterns in mares with granulosa cell tumours. *J. Vet. Med. A* **34,** 545–560.

Miller, M. A., Ramos, J. A., and Kreeger, J. M. (1992). Cutaneous vascular neoplasia in 15 cats: Clinical, morphologic, and immunohistochemical studies. *Vet. Pathol.* **29,** 329–336.

Mina, R. B., Tateyama, S., Miyoshi, N., Uchida, K., Yamaguchi, R., and Ohtsuka, H. (1994). Amplification of a c-*yes*-1-related oncogene in canine lymphoid leukemia. *J. Vet. Med. Sci.* **56,** 773–774.

Minke, J. M. H. M., Corenlisse, C. J., Stolwijk, J. A. M., Kuipers-Dijkshoorn, N. J., Rutteman, G. R., and Misdorp, W. (1990). Flow cytometric DNA ploidy analysis of feline mammary tumors. *Cancer* **50,** 4003–4007.

Mitchell, B. S., Dhami, D., and Schumacher, U. (1992). *In situ* hybridization: A review of methodologies and applications in the biomedical sciences. *Medical Lab. Sci.* **49,** 107–118.

Miyoshi, N., Tateyama, S., Ogawa, K., Yamaguchi, R., Kuroda, H., Yasuda, N., and Shimizu, T. (1991). Abnormal structure of the canine oncogene, related to the human c-*yes*-1 oncogene, in canine mammary tumor tissue. *Am. J. Vet. Res.* **52,** 2046–2048.

Moller, A. C., and Hellmen, E. (1994). S100 protein is not specific for myoepithelial cells in the canine mammary gland. *J. Comp. Pathol.* **110,** 49–55.

Momoi, Y., Nagase, M., Okamoto, Y., Okuda, M., Sasaki, N., Watari, T., Goitsuka, R., Tsujimoto, H., and Hasegawa, A. (1993). Rearrangements of immunoglobulin and T-cell receptor genes in canine lymphoma/leukemia cells. *Jpn. J. Med. Sci.* **55,** 775–780.

Moore, F. M., Kledzik, G. S., Wolfe, H. J., and DeLellis, R. A. (1984). Thyroglobulin and calcitonin immunoreactivity in canine thyroid carcinomas. *Vet. Pathol.* **21,** 168–173.

Moore, P. F., Hacker, D. V., and Buyukmichi, N. (1986). Ocular angiosarcoma in the horse: Morphological and immunohistochemical studies. *Vet. Pathol.* **23,** 240–244.

Moore, P. F., Rossitto, P. V., and Danilenko, D. M. (1990). Canine leukocyte integrins: Characterization of a 18 homologue. *Tiss. Antig.* **36,** 211–220.

Moore, P. F., Rossito, P. V., Danilenko, D. M., Wielenga, J. J., Raff, R. F., and Stevens, E. (1992). Monoclonal antibodies specific for canine CD4 and CD8 define functional T-lymphocyte subsets and high-density expression of CD4 by canine neutrophils. *Tiss. Antig.* **40,** 75–85.

Moore, P. F., Olivry, T., and Naydan, D. (1994). Canine cutaneous epitheliotropic lymphoma (mycosis fungoides) is a proliferative disorder of CD8+ cells. *Am. J. Pathol.* **144,** 421–429.

Moore, A. S., Leveille, C. R., Reimann, K. A., Shu, H., and Arias, I. M. (1995). The expression of P-glycoprotein in canine lymphoma and its association with multidrug resistance. *Cancer Investig.* **13,** 475–479.

Mukai, M., Torikata, C., Iri, H., Morikawa, Y., Shimizu, L., Shimada, T., Nukina, N., Ihara, Y., and Kageyama, K. (1986). Expression of neurofilament triplet proteins in human neural tumors: An immunohistochemical study of paraganalioma, ganglioneuroma, ganglioneuroblastoma, and neuroblastoma. *Am. J. Pathol.* **122,** 28–35.

Murray, M., James, Z. H., and Martin, W. B. (1969). A study of the cytology and karyotype of the canine transmissible venereal tumour. *Res. Vet. Sci.* **10,** 565–568.

Nabers, S. P. (1994). Molecular pathology—detection of neoplasia. *NEJM* **331,** 1508–1510.

O'Brien, T. D., Hayden, D. W., O'Leary, T. P., Caywood, D. D., and Johnson, K. H. (1987). Canine pancreatic endocrine tumors: Immunohistochemical analysis of hormone content and amyloid. *Vet. Pathol.* **24,** 308–314.

O'Brien, T. D., Westermark, P., and Johnson, K. H. (1990). Islet amyloid polypeptide and calcitonin gene-related peptide immunoreactivity in amyloid tumor cells of canine pancreatic endocrine tumors. *Vet. Pathol.* **27,** 194–198.

Ogilvie, G. K., Walters, L. W., Greeley, S. G., Henkel, S. E., and Salman, M. D. (1993). Concentration of $\alpha 1$-acid glycoprotein in dogs with malignant neoplasia. *J. Am. Vet. Med. Assn.* **203,** 1144–1146.

Okuda, M., Umeda, A., Matsumoto, Y., Momoi, Y., Watari, T., Goitsuka, R., O'Brien, S. J., Tsujimoto, H., and Hasegawa, A. (1993). Molecular cloning and chromosomal mapping of feline p53 tumor suppressor gene. *J. Vet. Med. Sci.* **55,** 801–805.

Okuda, M., Umeda, A., Sakai, T., Ohashi, T., Hwa-Young, Y., Watari, T., Goitsuka, R., Tsujimoto, H., and Hasegawa, A. (1994). Cloning of feline p53 tumor suppressor gene and its aberration in hematopoietic tumors. *Int. J. Cancer* **58,** 602–607.

Olivry, T., Moore, P. F., Naydan, D. K., Danilenko, D. M., and Affolter, V. K. (1995). Investigation of epidermotropism in canine mycosis fungoides: Expression of intercellular adhesion molecule-1 (ICAM-1) and beta-2 integrins. *Arch. Dermatol. Res.* **287,** 186–192.

Ordonez, N. G. (1989). Application of immunohistochemistry in the diagnosis of poorly differentiated neoplasms and tumors of unknown origin. *Cancer Bull.* **41,** 142–141.

Ordonez, N. G. (1993). Immunocytochemistry in the diagnosis of soft-tissue sarcomas. *Cancer Bull.* **45,** 13–23.

Pace, L. W., Kraeger, J. M., Miller, M. A., Turk, J. R., and Fischer, J. R. (1994). Immunohistochemical staining of feline malignant fibrous histocytomas. *Vet. Pathol.* **31,** 168–172.

Pazzi, K. A., Kraegel, S. A., Griffey, S. M., Theon, A. P., and Madewell, B. R. (1996). Analysis of the equine tumor suppressor gene p53 in the normal horse and in eight squamous cell carcinomas. *Canc. Lett.* **107,** 125–130.

Pena, L., Nieto, A., Alenza, M. D. A., Rodriquez, A., Sanchez, M. A., and Castano, M. (1994). Expression of fibronectin and its integrin receptor *A5B1* in canine mammary tumours. *Res. Vet. Sci.* **57,** 358–364.

Pena, L., Castano, M., Sanchez, M. A., Rodriquez, A., and Flores, J. M. (1995). Immunocytochemical study of type IV collagen and laminin in canine mammary tumours. *J. Vet. Med. A* **42,** 50–61.

Peterson, M. E., Graves, T. K., and Gamble, D. A. (1990). Triiodothyronine (T3) suppression test. An aid in the diagnosis of mild hyperthyroidism in cats. *J. Vet. Int. Med.* **4,** 233–238.

Poli, A., Arispici, M., Camillo, F., and Carazzo, M. (1986). Increase of serum lipid-associated sialic acid concentrations in dogs with neoplasms. *Am. J. Vet. Res.* **47,** 607–609.

Preziosi, R., Della Salda, L., Ricci, A., Simoni, P., and Marcato, P. S. (1995a). Quantification of nucleolar organiser regions in canine perianal tumours. *Res. Vet. Sci.* **58,** 277–281.

Preziosi, R., Sarli, G., Benazzi, C., and Marcato, P. S. (1995b). Detection of proliferating cell nuclear antigen (PCNA) in canine and feline mammary tumours. *J. Comp. Pathol.* **113,** 310–313.

Rabanal, R., and Fondevila, D. (1992). Immunocytochemical detection of amylase, carboxypeptidase A, carcinoembryonic antigen and αl-antitrypsin in carcinomas of the exocrine pancreas of the dog. *Res. Vet. Sci.* **52,** 317–223.

Rabanal, R. H., Fondevila, D. M., Montane, V., Domingo, M., and Ferrer, L. (1989). Immunocytochemical diagnosis of skin tumours of the dog with special reference to undifferentiated types. *Res. Vet. Sci.* **47,** 129–133.

Regelson, W. (1995). Have we found the "definitive cancer biomarker"? *Cancer* **15,** 1299–1301.

Renzoni, G., Taccini, E., Lossi, L., Rossi, G., and Braca, G. (1995). Thyroid C-cell carcinoma in a sheep: Histopathological and immunocytochemical study. *Vet. Pathol.* **32,** 727–730.

Restucci, B., deVico, G., and Maiolino, P. (1995). Expression of B1 integrin in normal, dysplastic and neoplastic canine ammary gland. *J. Comp. Pathol.* **113,** 165–173.

Ribas, J. L., Mena, H., and Braund, K. G., and Sesterhenn, I. A., Toivio-Kinnucan, M. (1989). A histologic and immunocytochemical study of choroid plexus tumors of the dog. *Vet. Pathol.* **26,** 55–64.

Rideout, B. A., Moore, P. F., and Pedersen, N. C. (1990). Distribution of MHC class II antigens in feline tissues and peripheral blood. *Tiss. Antig.* **36,** 221–227.

Rose, D. S. C., Maddox, P. H., and Brown, D. C. (1994). Which proliferation markers for routine histopathology? A comparison of five antibodies. *J. Clin. Pathol.* **47,** 1010–1014.

Rosol, T. J., and Capen, C. C. (1992). Mechanisms of cancer-induced hypercalcemia. *Lab. Invest.* **67,** 680–702.

Ross, W., and Hall, P. A. (1995). Ki67: From antibody to molecule to understanding? *J. Clin. Pathol. Mol. Path.* **48,** M113–M117.

Rowland, P. H., and Linke, R. P. (1994). Immunohistochemical characterization of lambda light-chain-derived amyloid in one feline and five canine plasma cell tumors. *Vet. Pathol.* **31,** 390–393.

Rowland, P. H., Valentine, B. A., Stebbins, K. E., and Smith, C. A. (1991). Cutaneous plasmacytomas with amyloid in six dogs. *Vet. Pathol.* **28,** 125–130.

Rowley, J. D., Asler, J. C., and Sklar, J. (1993). The impact of new DNA diagnostic technology on the management of cancer patients. Survey of diagnostic techniques. *Arch. Pathol. Lab. Med.* **117,** 1104–1109.

Rutteman, G. R., Cornelisse, C. J., Dijkshoorn, N. J., Poortman, J., and Misdorp, W. (1988). Flow cytometric analysis of DNA ploidy in canine mammary tumors. *Cancer Res.* **48,** 3411–3417.

Sahin, A. A., Ro, J. Y., and Bruner, J. M. (1993). Morphologic accessment of tumor proliferative fraction. *Cancer Bull.* **45,** 37–41.

Sandusky, G. E., Carlton, W. W., and Wightman, K. A. (1985). Immunohistochemical staining for S100 protein in the diagnosis of canine amelanotic melanoma. *Vet. Pathol.* **22,** 577–581.

Sandusky, G. E., Wightman, K. A., and Carlson, W. W. (1991). Immunocytochemical study of tissues from clinically normal dogs and of neoplasms, using keratin monoclonal antibodies. *Am. J. Vet. Res.* **52,** 613.

Sarli, G., Banazzi, C., Preziosi, R., and Marcato, P. S. (1994). Proliferative activity assessed by anti-PCNA and Ki-67 monoclonal antibodies in canine testicular tumours. *J. Comp. Pathol.* **110,** 357–368.

Sarli, G., Benazzi, C., Preziosi, R., and Marcato, P. S. (1995). Assessment of proliferative activity by anti-PCNA monoclonal antibodies in formalin-fixed, paraffin-embedded samples and correlation with mitotic index. *Vet. Pathol.* **32,** 93–96.

Sarnelli, R., Grassi, F., and Romagnoli, S. (1994). Alveolar rhabdomyosarcoma of the greater omentum in a dog. *Vet. Pathol.* **31,** 473–475.

Sartin, E. A., Barnes, S., Kwapien, R. P., and Wolfe, L. G. (1992). Estrogen and progesterone receptor status of mammary carcinomas and correlation with clinical outcomes in dogs. *Am. J. Vet. Res.* **53,** 2196–2200.

Saunders, K. A., Madewell, B. R., Oreffo, V. I. C., Kraegel, S. A., and Gumerlock, P. H. (1992). Nucleotide sequence of canine c-N-*ras*; codons 1–71. *Am. J. Vet. Res.* **53,** 600–603.

Savell, J. (1994). Apoptosis in disease. *Eur. J. Clin. Invest.* **24,** 715.

Schmitt, F. C., and Bacchu, C. E. (1989). S-100 protein: Is it a tumour marker in diagnostic immunocytochemistry? *Histopathol.* **15,** 281–288.

Schned, A. R. (1993). Nucleolar organizer regions as discriminators for the diagnosis of well-differentiated adenocarcinoma of the prostate. *Arch. Pathol. Lab. Med.* **117,** 1000.

Schrenzel, M. D., Higgins, R. H., Hinrichs, S. H., Smith, M. O., and Torten, M. (1990). Type C viral retroviral expression in spontaneous olfactory neuroblastoma. *Acta Neuropathol.* **80,** 547–549.

Schulz, K. S., Steele, K. E., Saunders, G. K., Smith, M. M., and Moon, M. L. (1994). Thoracic ganglioneuroblastoma in a dog. *Vet. Pathol.* **31,** 716–718

Schwartz, M. A., and Ingber, D. E. (1994). Integrating with integrins. *Mol. Biol. Cell* **5,** 389–393.

Shimada, A., Shinya, K., Awakura, T., Narama, I., Maeda, H., and Umemura, T. (1993). Cutaneous papillomatosis associated with papillomavirus infection in a dog. *J. Comp. Pathol.* **108,** 103–107.

Spangler, W. L., Culbertson, M. R., and Kass, P. H. (1994). Primary mesenchymal (nonangiomatous/nonlymphomatous) neoplasms occurring in the canine spleen: anatomic classification, immunohistochemistry, and mitotic activity correlated with patient survival. *Vet. Pathol.* **31,** 37–47.

Stabenfeldt, G. H., Hughes, J. P., Kennedy, P. C., Meagher, D. M., and Neely, D. P. (1979). Clinical findings, pathological changes and endocrinological secretory patterns in mares with ovarian tumours. *J. Reprod. Fert. Suppl.* **27,** 277–285.

Steinert, P. M. (1993). Structure, function, and dynamics of keratin intermediate filaments. *J. Investig. Dermatol.* **100,** 729–734.

Strandstrom, H.., Madewell, B. R., Deitch, A. D., and Gumerlock, P. H. (1989). Canine 563 cells: An established canine salivary cell line. *J. Vet. Med.* **36,** 361–366.

Sundberg, J. P. (1987). Papillomavirus infections in animals. *In* "Papillomaviruses and human diseases" (K. Srrjanen, L. Gissmann, and L. G. Koss, eds.), pp. 40–103. Springer-Verlag, Heidelberg.

Sundberg, J. P., Junge, R. E., and Lancaster, W. D. (1984). Immunoperoxidase localization of papillomaviruses in hyperplastic and neoplastic epithelial lesions in animals. *Am. J. Vet. Res.* **45,** 1441–1446.

Tanimoto, T., and Ohtsuki, Y. (1992). A solitary Schwannoma in the cecum of a cow. *Vet. Pathol.* **29,** 81–83.

Teske, E., van Heerde, P., Rutteman, G. R., Kurzman, I. D., Moore, P. F., and MacEwen, E. G. (1994a). Prognostic factors for treatment of malignant lymphomas in dogs. *J. Am. Vet. Med. Assn.* **205,** 1722–1738.

Teske, E., Wisman, P., Moore, P. F., and van Heerde, P. (1994b). Histologic classification and immunophenotyping of canine non-Hodgkin's lymphomas: Unexpected high frequency of T cell lymphomas with B cell morphology. *Exp. Hematol.* **22,** 1179–1187.

Theon, A. P., Metzger, L., and Griffey, S. (1994). *In situ* analysis of cellular proliferation in canine, feline and equine tumors by immunohistochemistry: A comparison of bromodeoxyuridine, proliferating cell nuclear antigen and interchromatin-associated antigen immunostaining techniques. *J. Vet. Diag. Invest.* **6,** 453–457.

Theon, A. P., Madewell, B. R., Shearn, V., and Moulton, J. E. (1995) Prognostic factors associated with radiotherapy of the nasal plane in cats. *J. Am. Vet. Med. Assn.* **206,** 991–996.

Thoolen, R. J. M. M., Vos, J. H., van der Linde-Sipman, J. S., deWeger, R. A., van Unnik, J. A., Misdorp, W., and Dijk, J. E. (1992). Malignant fibrous histocytomas in dogs and cats: An immunohistochemical study. *Res. Vet. Sci.* **53,** 198–204.

Tobey, J. C., Houston, D. M., Breur, G. J., Jackson, M. L., and Stubbington, D. A. (1994). Cutaneous T-cell lymphoma in a cat. *J. Am. Vet. Med. Assn.* **204,** 606–609.

Tompkins, M. B., Gebhard, D. H., Bingham, H. R., Hamilton, M. J., Davis, W. C., and Tompkins, W. A. (1990). Characterization of monoclonal antibodies to feline T lymphocytes and their use in the analysis of lymphocyte tissue distribution in the cat. *Vet. Immunol. Immunopathol.* **26,** 305–317.

Vail, D., Powers, B. E., Getzy, D. M., Morrison, W. B., McEntee, M. C., O'Keefe, D. A., Norris, A. M., and Withrow, S. J. (1994). Evaluation of prognostic factors for dogs with synovial sarcomas: 36 cases (1986–1991). *J. Am. Vet. Med. Assn.* **205,** 1300–1307.

van Buest B. R., Suter, M. M., and Summers, B. A. (1988). Factor VIII-related antigen in canine endothelial neoplasms: An immunohistochemical study. *Vet. Pathol.* **25,** 251–255.

van der Gaag, I., van den Ingh, Th. S. G. A. M., and Lamers, C. B. H. W. (1988). Zollinger-Ellison syndrome in a cat. *Vet. Quart.* **10,** 151–153.

van der Velden, M. A., and Meulenaar, H. (1986). Medullary thyroid carcinoma in a horse. *Vet. Pathol.* **23,** 622–624.

Verschueren, C. P., and Goslings, B. M. (1992). Comparative aspects of thyroid cancer, *In* "Clinico-pathological and Endocrine Aspects of Canine Thyroid Cancer" (C. P. Verschueren, ed.), pp. 95–117. C. P. Verschueren, Brussels.

Verschueren, C. P., Rutteman, G. R., Kuipers-Dijkshoorn, N. J., Siollema, B. E., Vos, J. H., Van Dijk, J. E., and Cornelisse, C. J. (1991). Flow-cytometric analysis in primary and metastatic canine thyroic carcinomas. *Anticancer Res.* **11,** 1755–1762.

Vinores, B. A., Bonnin, J. M., Rubinstein, L. C., and Marangos, P. J. (1984). Immunohistochemical demonstration of neuron-specific enolase in neoplasms of the CNS and other tissues. *Arch. Pathol. Lab. Med.* **108,** 536–540.

Vos, J. H., van den Ingh, T. S. G. A. M., Misdorp, W., Ramaekers, F. C. S., van Mil, F. N., and de Neijs, M. (1989). An immunohistochemical study of canine tissues with vimentin, desmin, glial fibriallary acidic protein and neurofilament antisera. *Vet. Med. A* **36,** 561–575.

Vos, J. H., van den Ingh, T. S. G. A. M., de Neijs, M., van Mil, F. N., Ivanyi, D., and Ramaekers, F. C. S. (1992). Immunohistochemistry with keratin monoclonal antibodies in canine tissues: urogenital tract, respiratory tract, (neuro-)endocrine tissues, choroid plexus and spinal cord. *Vet. Med. A* **39,** 721–740.

Vos, J. H., Borst, G. H. A., de las Mulas, J., Ramaekers, F. C. S., van Mil, F. N., Molenbeck, R. F., Ivanyi, D., and van den Ingh, T. S. G. A. M. (1993). Rhabdomyosarcomas in young pigs in a swine breeding farm: A morphologic and immunohistochemical study. *Vet. Pathol.* **30,** 271–279.

Voss, C., Kremmer, E., Schumm, M., Gunther, W., Hoffman-Fezer, G., Kolb, H. J., and Thierfelder, S. (1994). Rat monoclonal antibodies against three different epitopes of canine Thy-1 and their depletion capacity *in vivo. Tiss. Antig.* **43,** 193–197.

Watson, C. J. E., Cobbold, S. P., Davis, H. ff. S., Rebello, P. R. U. B., Thiru, S., McNair, R., Rasmussen, A., Waldmann, H., Caine, R. Y., and Metcalfe, S. M. (1994). Immunosuppression of canine renal allograft recipients by CD4 and CD8 monoclonal antibodies. *Tissue Antigens* **43,** 155–162.

Weidner, N. (1995). Intratumoral vessel density as a prognostic factor in cancer. *Am. J. Pathol.* **147,** 9–15.

Wilson, L., and Lloyd, R. V. (1984). Detection of chromogranin in neuroendocrine cells with a monoclonal antibody. *Am. J. Pathol.* **115,** 458–468.

Yamada, T., Kakinoki, M., Totsuka, K., Ashida, Y., Nishinozo, K., Tsuchiya, R., and Kobayashi, K. (1995). Purification of canine alpha-fetoprotein values in dogs. *Vet. Immunol. Immunopathol.* **47,** 25–33.

Zhao, D. M., Miyoshi, N., Uchida, K., Yamaguchi, R., Yamagami, T., and Hayashi, T. (1995). Sequence of the canine c-*yes* oncogene. *Res. Vet. Sci.* **59,** 230–233.

CHAPTER

27

Cerebrospinal Fluid

CLETA SUE BAILEY AND WILLIAM VERNAU

The analysis of cerebrospinal fluid (CSF) has been described as the central nervous system equivalent of the complete blood count (Jamison and Lumsden, 1988), and the analogy is a good one. A CSF analysis provides a general index of neurologic health and often provides evidence of the presence of disease. Similar to a complete blood count, CSF analysis has reasonable sensitivity but low specificity. The possible alterations of CSF are relatively limited compared to the varieties of neurologic diseases that exist (particularly if the analysis is restricted to total cell counts and total protein determination). Additionally, the type and degree of CSF abnormality seems to be related as much to the location of disease as to the cause or the severity of the lesion; meningeal and paraventricular diseases generally produce greater abnormalities than deep parenchymal diseases. Previous therapy may affect the type, degree, and duration of CSF abnormalities as well. The CSF abnormalities identified upon analysis are also dependent on the CSF collection site with respect to lesion location (Thomson *et al.*, 1990, 1989). Lastly, the CSF of animals with neurological disease is not always abnormal (Tipold, 1995). Only occasionally does CSF analysis provide a specific diagnosis (Kjeldsberg and Knight, 1993)—for example, if infectious agents (bacteria or fungi) or neoplastic cells are observed. In most situations, the chief utility of CSF analysis is to assist in the diagnostic process by excluding the likelihood of certain disease processes being present. As is the case with all tests of relatively low specificity, examination of CSF is most useful when the results are correlated with the history, clinical findings, imaging studies, and ancillary laboratory studies. As stated by Fankhauser (1962), "It is futile to make a diagnosis based solely on the CSF findings and particularly on single alterations of it. Only the entire picture of all findings linked with the other clinical symptoms is of value in reaching a diagnosis."

I. FUNCTIONS OF CEREBROSPINAL FLUID

Cerebrospinal fluid has four major functions: (1) physical support of neural structures, (2) excretion and "sink" action, (3) intracerebral transport, and (4) control of the chemical environment of the central nervous system. Cerebrospinal fluid provides a "water jacket" physical support and buoyancy. When suspended in CSF, a 1500-g brain weighs only about 50 g. The CSF is also protective because its volume changes reciprocally with changes in the volume of intracranial contents, particularly blood. Thus, the CSF protects the brain from changes in arterial and central venous pressure associated with posture, respiration, and exertion. Acute or chronic pathological changes in intracranial contents can also be accommodated, to a point, by changes in the CSF volume (Fishman, 1992; Rosenberg, 1990; Milhorat, 1987).

Excretory function is provided by the direct transfer of brain metabolites into the CSF. This capacity is particularly important because the brain lacks a lymphatic system. The lymphatic function of the CSF is also manifested in the removal of large proteins, and even cells such as bacteria or blood cells, by bulk CSF absorption (see Section II.D, CSF Absorption). The "sink" action of the CSF arises from the restricted access of water-soluble substances to the CSF and the low concentration of these solutes in the CSF. Therefore, solutes entering the brain, as well as those synthesized by the brain, diffuse freely from the brain interstitial fluid into the CSF. Removal may then occur by bulk CSF absorption or, in some cases, by transport across the choroid plexus into the capillaries (Fishman, 1992; Rosenberg, 1990; Milhorat, 1987; Davson and Segal, 1996).

Because CSF bathes and irrigates the brain, including those regions known to participate in endocrine functions, CSF may serve as a vehicle for intracerebral transport of biologically active substances. For example, hormone-releasing factors, formed in the hypothalamus and discharged into the CSF of the third ventricle, may be carried in the CSF to their effective sites in the median eminence. The CSF may also be the vehicle for intracerebral transport of opiates and other neuroactive substances (Fishman, 1992; Milhorat, 1987).

An essential function of CSF is the provision and maintenance of an appropriate chemical environment for neural tissue. Anatomically, the interstitial fluid of the central nervous system and the CSF are in continuity (see Section II.A, Anatomy of Brain–Fluid Interfaces); therefore, the chemical composition of the CSF reflects and affects the cellular environment. The composition of the CSF (and the interstitial fluid) is controlled by cells forming the interfaces, or barriers, between the "body" and the neural tissue. These semipermeable interfaces, the blood–brain barrier, the blood–CSF barrier, and the CSF–brain barrier, control the production and absorption of CSF and provide a fluid environment that is relatively stable despite changes in the composition of blood (Fishman, 1992; Milhorat, 1987; Davson and Segal, 1996).

II. CSF FORMATION, CIRCULATION, AND ABSORPTION

The brain (and the spinal cord) as an organ is isolated in many ways from the body and the systemic circulation. This isolation is accomplished anatomically by several interfaces between brain tissue and systemic fluids (Table 27.1). At these interfaces, selec-

TABLE 27.1 Composition of the Brain–Fluid Interfaces[a]

Interface	Cell type	Junction type
Blood–brain	Brain capillary endothelium	Tight junction
Blood–CSF		
Blood–CSF	Choroid plexus epithelium	Apical tight junction
CSF–blood	Arachnoid cells	Tight junction
	Arachnoid villi	Valve
CSF–brain	Ependyma	Gap junction
	Pia mater	Gap junction

[a] Modified from Rosenberg (1990).

tive carriers and ion pumps transport electrolytes and essential nutrients and thereby control the brain's microenvironment. A substantial portion of this control is achieved through the formation, circulation, and absorption of CSF at these brain–fluid interfaces (Rosenberg, 1990; Davson and Segal, 1996).

A. Anatomy of Brain–Fluid Interfaces

1. The Blood–Brain Barrier

The important blood–brain (and blood–spinal cord) interface is formed by the endothelial cells of the intraparenchymal capillaries. In most areas of the brain and spinal cord, the capillary endothelium differs from that of other body tissues in the following ways: (1) the absence of fenestrae, (2) the presence of tight junctions between adjacent cells, (3) a lower number of pinocytotic pits and vesicles, (4) a higher number of mitochondria, and (5) closely applied, perivascular, astrocytic foot processes. These features result in the capillary endothelium being a selective barrier—the blood–brain barrier—that regulates the entry, and probably the exit, of biologically important substances and aids in the maintenance of a precise, stable environment for the neural tissues (Rosenberg, 1990; Milhorat, 1987; Davson and Segal, 1996).

2. The Blood–CSF Barrier

One part of the blood–CSF interface is formed by the epithelial cells of the circumventricular organs. The circumventricular organs, which include the four choroid plexuses, the median eminence, the neural lobe of the hypophysis, and other specialized areas, border the brain ventricles and are involved with specific secretory activities that appear to require a direct contact with plasma. The capillaries within these organs are fenestrated, similar to capillaries in other organs of the body. Overlying each of the organs are specialized epithelial cells joined by intercellular tight junctions at their apical (ventricular) borders. These epithelial cells also are characterized by an abundance of intracellular organelles and lysosomes. These organelles are probably an important aspect of the barrier and secretory functions of these cells (Rosenberg, 1990; Milhorat, 1987). The choroid plexuses are the major source of CSF. They are formed by evaginations of the ependyma and the pial blood vessels into the ventricles, and they consist of a single row of cuboidal, specialized epithelial cells thrown into villi around a core of blood vessels and connective tissue. The apical (ventricular) surface of the epithelial cells has a brush border of microvilli with occasional cilia. The basal and lateral cell surfaces have numerous infoldings. Overall, the structure of these cells resembles that of other epithelia specialized for fluid transport, such as proximal renal tubular epithelium (Rosenberg, 1990; Milhorat, 1987; Davson and Segal, 1996). Autonomic nerve terminals have also been identified in the choroid plexus, but their function is unclear (Fishman, 1992; Nilsson *et al.*, 1992).

The second part of the blood–CSF interface is formed by the arachnoid membrane at the arachnoid villi. These villi are microscopic evaginations of the arachnoid membrane into the lumen of the dural sinuses. The barrier function of these arachnoid cells is demonstrated by their tight junctions. Their transport function is indicated by giant intracellular vacuoles, some of which have both basal and apical openings, and pinocytotic vesicles. The sinus surface of a villus is covered by sinus endothelium (Rosenberg, 1990; Milhorat, 1987). Endothelium-lined channels may link directly with the subarachnoid space (Davson and Segal, 1996; Bell, 1995). Arachnoid villi are not limited to intracranial venous sinuses, but also are present at the spinal nerve roots penetrating into the spinal veins (Milhorat, 1987; Bell, 1995).

3. The CSF–Brain Interface

The extensive CSF–brain (and spinal cord) interface consists of the ependyma within the cavities of the central nervous system and the pia mater covering the central nervous system. These two layers are each composed of a single layer of cells joined by gap junctions. The ependyma and the pia mater are not important permeability barriers; the CSF (ventricular and subarachnoid) and the brain interstitial fluid are directly continuous (Milhorat, 1987; Davson and Segal, 1996).

B. CSF Formation

1. Choroidal and Extrachoroidal Formation

Cerebrospinal fluid is formed principally by the choroid plexus, with a smaller amount being formed extrachoroidally (Milhorat, 1987; Davson and Segal, 1996). Choroidal formation involves two processes that occur in series: first, filtration across the choroidal capillary wall, and second, secretion by the choroidal epithelium. Within the choroid plexus, hydrostatic pressure of the choroidal capillaries initiates the transfer of water and ions to the interstitial fluid and then to the choroidal epithelium. Water and ions are then transferred into the ventricles by either (1) intracellular movement across the epithelial membranes or (2) intercellular movement across the apical tight junctions between epithelial cells. Both of these processes are probably dependent upon ion pumps. Secretion of CSF results from the active transport of sodium, which is dependent on the membrane-bound, sodium–

potassium activated ATPase present at the apical (ventricular) surface of the choroidal epithelium (Rosenberg, 1990; Davson and Segal, 1996). The presence of autonomic nerve terminals in the choroid plexus suggests a neural control of CSF secretion. However, the functional role of this innervation in normal and pathologic conditions is unknown (Fishman, 1992; Nilsson *et al.*, 1992).

Spurred primarily by clinical evidence that excision of the choroid plexus did not benefit human patients with hydrocephalus, experimental evidence now supports the existence of an extrachoroidal source of CSF. The diffusion of brain interstitial fluid across the ependyma or pia mater is the apparent source of this extrachoroidal CSF component. Formation of the interstitial fluid is thought to occur by active transport processes (secretion) at the cerebral capillaries, but an alternative theory proposes passive permeability of the capillary endothelium and active transport by the surrounding astrocytes (Rosenberg, 1990; Milhorat, 1987). The relative contributions of choroidal and extrachoroidal sources to CSF in normal and pathologic conditions are not certain. Some investigators report the choroid plexus to be the major if not the sole source of CSF, whereas others conclude that at least one-third of newly formed CSF is extrachoroidal (Milhorat, 1987; Davson and Segal, 1996).

2. *Rate of CSF Formation*

Regardless of the amount of extrachoroidal formation, the rate of CSF formation is closely correlated to the weight of the choroid plexus and varies among species (Table 27.2) (Cserr, 1971; Welch, 1975). Increases and decreases in formation rate have been achieved experimentally, but the general tendency is for the formation rate to remain relatively constant. The formation rate directly parallels the rate of sodium exchange, which is linked to the bicarbonate ion. The enzyme carbonic anhydrase plays an important role because it provides the bicarbonate. Inhibition of carbonic anhydrase slows (but does not abolish) sodium, bicarbonate, and chloride flow, resulting in a reduction of CSF secretion (Maren, 1992). Several drugs and conditions inhibit CSF production (Table 27.3), but their clinical utility is limited either by their time frame of action or by their toxicity (Rosenberg, 1990; Pollay, 1992; Davson and Segal, 1996).

Moderate variations in intracranial pressure probably do not affect CSF formation. However, studies of chronically hydrocephalic animals have shown a reduction of CSF formation with increasing intraventricular pressure. The secretion process may also be affected by chronically increased intracranial pressure (Fishman, 1992).

C. CSF Circulation

Cerebrospinal fluid flows in bulk from sites of production to sites of absorption. Fluid formed in the lateral ventricles flows through the paired interventricular foramina (foramina of Monro) into the third ventricle, then through the mesencephalic aqueduct

TABLE 27.2 Rate of CSF Formation in Various Species[a]

Species	Rate (μl/min)
Mouse	0.325
Rat	2.1–5.4
Guinea pig	3.5
Rabbit	10
Cat	20–22
Dog	47–66
Sheep	118
Goat	164
Calf[b]	290
Monkey	28.6–41
Human	350–370

[a] Estimated by ventriculo-cisternal perfusion. Modified from Davson and Segal (1996).
[b] Calhoun *et al.* (1967).

TABLE 27.3 Factors Influencing CSF Formation[a]

Effect	Substance or condition	Site of action
Increase	Cholera toxin	cAMP
	Phenylephrine[b]	Cholinergic pathways
Decrease	Acetazolamide, furosemide	Carbonic anhydrase
	Atrial natriuretic hormone	cGMP
	Diazepam analog[c]	Choroidal benzodiazepine receptor
	Dopamine D_1 receptor agonist	Choroidal dopamine receptor
	Hyperosmolarity	Choroidal capillaries
	Hypothermia	Metabolism (decreased)
	Noradrenaline[b]	cAMP/Choroidal Na^+–K^+-ATPase
	Omeprazole[c]	H^+–K^+-ATPase?
	Ouabain	Na^+–K^+-ATPase
	Serotonin receptor agonist	Choroidal serotonin receptor
	Steroids[c]	Choroidal Na^-–K^+-ATPase
	Vasopressin	Choroidal vasopressin (V_1) receptor

[a] Modified from Fishman (1992).
[b] Nilsson *et al.* (1992).
[c] Davson and Segal (1996).

(aqueduct of Sylvius) into the fourth ventricle. The majority of CSF exits from the fourth ventricle into the subarachnoid space; a small amount may enter the central canal of the spinal cord. In people, CSF enters the subarachnoid space through the lateral apertures (foramina of Luschka) and the median aperture (foramen of Magendie) of the fourth ventricle. Animals below the anthropoid apes do not have a median aperture (Fletcher, 1993; Fankhauser, 1962). Cerebrospinal fluid has also been shown to flow from the spinal subarachnoid space into the spinal perivascular spaces, across the interstitial space, then into the central canal (Stoodley *et al.*, 1996). Mechanisms for propelling the CSF along its route probably include (1) the continuous outpouring of newly formed ventricular fluid, (2) the ciliary action of the ventricular ependyma, (3) respiratory and vascular pulsations, and (4) the pressure gradient across the arachnoid villi (Milhorat, 1987).

D. CSF Absorption

Absorption of CSF occurs by bulk absorption of the fluid and by absorption or exchange of individual constituents of the fluid, that is, ions, proteins, and drugs. Bulk absorption occurs directly into the venous system and depends primarily upon the CSF hydrostatic pressure; as the pressure rises, the absorption rate increases (Davson and Segal, 1996). If intracranial pressure falls below a critical point, bulk absorption decreases, a homeostatic response to stabilize the intracranial pressure and the CSF volume. The primary site of bulk absorption, at least in people, is the arachnoid villi that project into the dural sinuses. Two other routes are through lymphatic channels in the dura and through the perineural sheaths of cranial nerves (particularly the olfactory nerve) and spinal nerves. Perineural absorption may be through arachnoid villi projecting into perineural veins, lymphatics, or connective tissue (Milhorat, 1987; Davson and Segal, 1996). The importance of these various absorption routes varies with the species (Bell, 1995).

Absorption through the arachnoid villi occurs transcellularly through micropinocytotic vesicles and giant intracellular vesicles, but may also occur through endothelium-lined, intercellular clefts. The mechanisms appear to vary among species (Bell, 1995). Absorption is unidirectional from the CSF into the venous blood—the villi act like one-way valves. The basis for the valve-like mechanism appears to be transport by the giant vesicles (see Section II.A.2). Particles ranging in size from colloidal gold (0.2 μm) to erythrocytes (7.5 μm) can be transported across the villi. In disease conditions, accumulations of larger particles (e.g., protein molecules, erythrocytes, leukocytes) within the villi may impair absorption, leading to hydrocephalus (Fishman, 1992; Milhorat, 1987). The choroid plexus also has an absorptive function, acting on specific substances in the CSF rather than by bulk fluid absorption. A variety of compounds are actively transported from the CSF, in a fashion reminiscent of the proximal renal tubule. Solutes may also be cleared from the CSF by diffusion into adjacent brain cells or capillaries (Fishman, 1992; Milhorat, 1987).

III. CELLULAR COMPOSITION OF NORMAL CSF

A. Total Erythrocyte and Nucleated Cell Count

Cerebrospinal fluid normally does not contain erythrocytes (Chrisman, 1992; Rand *et al.*, 1990b; Cook and DeNicola, 1988; Wilson and Stevens, 1977). Most commonly, the presence of erythrocytes in a CSF sample is iatrogenic, due to inadvertant trauma associated with the needle placement. However, CSF erythrocytes may also originate from pathologic hemorrhage. The normal nucleated-cell counts of CSF in domestic animals are given in Table 27.4. The most widely accepted reference ranges for the numbers of leukocytes in the CSF of dogs and cats are 0–5 cells/μl (deLahunta, 1983; Oliver and Lorenz, 1993) to 0–8 cells/μl (Duncan *et al.*, 1994). However, these ranges are a little too broad in our experience and the most comprehensive series in the literature confirm this. Jamison and Lumsden

TABLE 27.4 Total White Blood Cell Count of Normal CSF in Domestic Animals

Species	Collection site[a]	N[b]	Cells/μl[c]	Reference
Dog	C	50	0–2	Jamison (1988)
Dog	C, L	31	0–4	Bailey (1985)
Cat	C	33	0–2	Rand (1990)
Horse	Pooled C and L	44	0–6	Mayhew (1977)
Horse	C	14	0–5	Furr (1994)
Cow	L	16	0.85–3.52[d]	Welles (1992)
Llama	L	17	0–3	Welles (1994)
Sheep	L	NS[e]	0–5	Fankhauser (1962)
Goat	NS	NS	0–4	Brewer (1983)
Pig	NS	NS	0–7	Fankhauser (1962)

[a] C = cerebellomedullary cistern. L = lumbar subarachnoid space.
[b] N = number of animals.
[c] Range.
[d] 95% confidence interval.
[e] NS, not stated.

(1988) examined 50 clinically and histopathologically normal dogs and derived cerebellomedullary CSF reference limits of 0–2 cells/μl. In fact, all except one of these dogs had counts of 0–1 cells/μl (personal communication). Bailey and Higgins (1985) examined 31 dogs that were clinically and histopathologically normal. For cerebellomedullary CSF, the mean was 1.45 cells/μl, with the 95% confidence intervals being 1.04–1.86 and the observed range being 0–4 cells/μl. Of 31 dogs, 26 had counts between 0 and 2 cells/μl. They also found that lumbar CSF had significantly lower counts with a mean of 0.55 cells/μl, a 95% confidence interval of 0.22–0.88 and an observed range of 0–4 cells/μl, although 30/31 dogs had counts of 0–2 cells/μl. We currently feel that a normal count for cerebellomedullary CSF in dogs is 0–2 cells/μl, with 3 cells/μl being in the gray zone and 4 cells/μL being abnormal. Rand *et al.* (1990b), derived reference limits for cerebellomedullary CSF from 33 cats that were clinically and histopathologically normal. The samples did not have blood contamination. The mean $\pm$ 1 SD for the white blood cell count was 0.1 $\pm$ 0.4 with an observed range of 0–2 cells/μl. Of 33 cats, 30 had counts of 0 cells/μl. Three cells or more per microliter is therefore considered abnormal in cerebellomedullary fluid of cats.

B. Differential Cell Count

1. *Leukocytes*

Excellent morphologic descriptions of the cell types normally found in the CSF of domestic animals can be found elsewhere (Rand *et al.*, 1990b; Cook and DeNicola, 1988; Jamison and Lumsden, 1988). Normal CSF consists of varying proportions of small lymphocytes and monocytes. The proportions are species and age dependent (Kjeldsberg and Knight, 1993). In dogs, monocytic type cells predominate (Jamison and Lumsden, 1988), although there is significant individual variation. In cats (Rand *et al.*, 1990b), monocytoid cells also predominate, with a mean of 87%, whereas small lymphocytes have a mean of 9%. This same trend is observed in horses, with 73.6% monocytes and 26.2% lymphocytes (Furr and Bender, 1994). However, small lymphocytes predominate in cattle (Welles *et al.*, 1992) and llamas (Welles *et al.*, 1994). In the human literature prior to 1975, any neutrophils present at all in the CSF were thought to be indicative of disease (Kjeldsberg and Knight, 1993). However, with the advent of techniques for concentrating CSF specimens, such as the cytocentrifuge, it became clear that a very small number of neutrophils may be found in normal human CSF (Fishman, 1992; Kjeldsberg and Knight, 1993). Similar observations have been made in many veterinary species, and rare neutrophils are a common and normal finding in the CSF of all domestic species. Eosinophils are not present in normal CSF, although a single cell is occasionally seen on cytocentrifuge slides of fluids with normal total cell counts. Large foamy activated macrophages or phagocytes are not seen in normal CSF (Fishman, 1992; Christopher *et al.*, 1988) and their presence is nonspecific evidence of an inflammatory disorder. Plasma cells are not seen in normal CSF (Fishman, 1992; Pelc *et al.*, 1981; Kjeldsberg and Knight, 1993) . Their presence is indicative of underlying inflammatory disease. In people, plasma cells are seen particularly in acute viral disease and various chronic inflammatory conditions, including tuberculous meningitis, syphilis, multiple sclerosis, and the Guillain–Barré syndrome (Pelc *et al.*, 1981; Kjeldsberg and Knight, 1993). In animals, plasma cells have been observed in various conditions, including distemper (Vandevelde and Spano, 1977), other viral meningitis (Vandevelde and Spano, 1977), rabies (Green *et al.*, 1992), granulomatous meningoencephalomyelitis (Bailey and Higgins, 1986a; Vandevelde and Spano, 1977), neoplasia, and abscessation (personal observations). Therefore, although they are abnormal, no specificity is associated with their presence in CSF. Similarly, reactive lymphocytes are not found in normal CSF, but their presence has no specificity. They can be seen in active or resolving infectious disease, immune-mediated diseases, and neoplasia (Cook and DeNicola, 1988).

2. *Other Cells*

Cells other than leukocytes can be seen in both normal and abnormal CSF. Cells lining the leptomeninges, choroid plexus cells, and ependymal cells can be seen as single cells, or, more often, as small papillary clusters or sheets. Cytologically, choroid plexus cells and ependymal cells are indistinguishable (Cook and DeNicola, 1988; Kjeldsberg and Knight, 1993). The majority of lining cells seen in normal CSF are thought to be choroid plexus cells (Kjeldsberg and Knight, 1993). Excellent descriptions and photographs of these cell types can be found elsewhere (Rand *et al.*, 1990b; Cook and DeNicola, 1988; Kjeldsberg and Knight, 1993). Chondrocytes are occasionally seen in CSF sampled by lumbar puncture, likely due to the spinal needle nicking the intervertebral disc (Bigner and Johnston, 1981). Squamous cells can be observed in CSF and may be due to skin contamination or an underlying pathological process such as epidermoid cysts (Kornegay and Gorgacz, 1982) or metastatic carcinomas. Bone-marrow cells (immature hematopoietic precursors) have been described in the CSF of people (Kjeldsberg and Knight, 1993) and dogs (Christopher, 1992). In people, bone-marrow cells in the CSF are usually associated with lumbar puncture, usually in infants or in patients with vertebral bone abnormalities that create difficulties during the sampling process. The cells are present because of sampling from the vertebral body or articular process. Christopher (1992) observed hematopoietic cells in the lumbar CSF of two dogs and speculated that it was due either to marrow penetration or to dural extramedullary hematopoiesis (Christopher, 1992). Extramedullary hematopoiesis has been observed in the choroid plexus of five dogs that did not have underlying hematologic abnormalities or the presence of extramedullary hematopoiesis elsewhere (Bienzle *et al.*, 1995). Although the CSF was normal in these dogs and hematopoietic cells were not observed, this site could provide another potential source for the presence of these cells in CSF. Metastatic myeloid leukemia could conceivably produce similar findings, but peripheral blood and marrow examination would clarify the origin of the cells in question. Neurons, astrocytes, glial cells, and neutral tissue have been observed in the CSF of people (Bigner and Johnston, 1981) and also in cerebellomedullary cisternal samples associated with traumatic taps in animals (personal observation).

IV. BIOCHEMICAL CONSTITUENTS OF NORMAL CSF

Because CSF is a product of plasma filtration and membrane secretion, its composition is different from that of plasma. In general, CSF is a clear, colorless, nearly acellular, low-protein fluid. Various ions, enzymes, and other substances are also found in normal CSF. In health, the CSF composition is maintained relatively constant by the various membrane interfaces, although some fluctuations occur with fluctuations in plasma composition. The chemical composition of the CSF of various animal species is summarized in Tables 27.5 through 27.8. These values should serve only as a guide; normal values must be established for individual laboratories.

A. Ontogeny of CSF

In people and animals, differences in CSF appearance and composition exist between neonates and adults. Human neonatal CSF is usually xanthochromic, probably because of a greater protein and bilirubin content than adult CSF. Glucose content is also increased, more closely approximating the blood glucose level. Many of these differences, such as protein content, are attributed to immaturity of the blood–brain barrier. Immaturity of the blood–brain barrier may be due to an increased number of fenestrae in the brain capillaries or inadequate closure of their endothelial tight junctions. Other factors that may contribute to age differences in CSF composition are the integrity of the blood–CSF barrier, the rate of CSF secretion and efficiency of absorption, the volume of the extracellular space of the brain, and the lipid solubility of the substances (Fishman, 1992; Davson and Segal, 1996). Protein also decreases with age in foals and puppies (Furr and Bender, 1994; Rossdale *et al.*, 1982; Meeks *et al.*, 1994). In contrast, two studies of calves found that CSF protein increased with age (St. Jean *et al.*, 1995; Binkhorst, 1982). Foals also had xanthochromia and a higher CSF glucose and creatine kinase levels than adults (Furr and Bender, 1994; Rossdale *et al.*, 1982). The WBC count decreased with age in puppies and calves (Meeks *et al.*, 1994; Binkhorst, 1982).

Studies done in prenatal, neonatal, and adult laboratory animals (including rats, rabbits, pigs, sheep, cats, dogs, and monkeys) and people have shown that, in general, the CSF/plasma concentration ratios (R_{CSF}) of Na^+, Mg^{2+}, and Cl^- increase with age. The R_{CSF} of K^+, HCO_3^-, and urea decrease. In some instances, however (e.g., Cl^- and K^+), changing plasma levels of these substances contribute to the change in the R_{CSF}. The R_{CSF} of total protein, as well as those of the individual proteins, decreases with age. The decreasing concentration of proteins in the CSF compared to plasma protein is an indication of the maturation of the blood–brain and blood–CSF barriers. In rats, the R_{CSF} of amino acids also falls quickly with age, although large individual variations exist. Taurine, for example, has a higher level in the adult than the newborn. This fact, as well as the species-specific transport of some proteins (e.g., albumin) into the CSF indicates a special mechanism of transport based on factors other than molecular weight (Davson and Segal, 1996).

B. Proteins in the CSF

Proteins identified in the CSF are given in Tables 27.9 and 27.10. In general, the concentration of a CSF protein is inversely related to its molecular weight. If the blood–brain barrier is normal, serum proteins with a molecular weight greater than 160,000 Da are largely excluded. However, Felgenhauer (1974) reported CSF : serum protein distribution ratios to be better correlated with the hydrodynamic radii than with the molecular weight of the proteins. Almost all the proteins normally present in CSF are derived from the serum. The exceptions are transthyretin (prealbumin)

TABLE 27.5 Biochemical Constituents of CSF of the Dog[a]

Constitute	Tipold *et al.* (1994)	Sorjonen (1987)	Bailey and Higgins (1985)	Bichsel *et al.* (1984b)	Sorjonen *et al.* (1981)	Krakowka *et al.* (1981)	Coles (1980)	Bleich *et al.* (1964)	Fankhauser (1962)
Methods									
#RBC/μl	NS	≤10	<1500	0	≤10	0	NS	NS	NS
Necropsy	NS	Yes	Yes	NS	NS	Yes	NS	NS	NS
Total protein (mg/dl)									
Cerebellomedullary		27 ± 4.2 (23–35)	13.97 ± 4.54 (3–23)		29.9 ± 1.57 (23–38.5)	27.6 ± 1.1SE (15.5–42)			27.5 (11–55)
Lumbar			28.68 ± 5.52 (18–44)						
Method		Coomassie brilliant blue	Coomassie brilliant blue		Micro-Lowry	Coomassie brilliant blue			NS
Albumin (mg/dl)		37 ± 4.29% (31–44%)		17.1 ± 6.7 (7.5–27.6)	12.43 ± 0.96[b] (10.5–17.4) 11.27 ± 1.0[c] (7.8–19.0)	10.28 ± 0.8SE (5.8–18.9)			27 (16.5–37.5)
Albumin quotient		0.22 ± 0.05 (0.17–0.3)							
Globulin (mg/dl)					17.45 ± 0.83 (14.0–21.1)				9.0 (5.5–16.5)
IgG (mg/dl)				0.85 ± 0.14 (0.71–1.09)	4.68 ± 0.68 (2.5–8.5)	1.16 ± 0.1SE			
IgG index	0.7 ± 0.3 (0.2–1.3)			0.38 ± 0.24 (0.15–0.9)					
IgM (μg/ml)	1.7 (0–5.8)				0	0			
IgA (μg/ml)	0.08 (0–0.2)				0	0			

Alanine transferase (Reitman–Frankel units)	13.7 ± 1.35SE (2–32)		
Aspartate tranferase (Reitman–Frankel units)	20.1 ± 1.64SE (9–46)		
Creatine kinase (SU)	<1		
Bicarbonate (meq/liter)		23.5 ± 0.19SE	
Calcium (mg/dl)			6.56 (5.13–7.40)
Chloride (meq/liter)		130 ± 0.5SE	808 (761–883) mg/dl 667 (602–783) mg/dl
Magnesium (mg/dl)			3.09 (2.58–3.81)
Phosphorus (mg/dl)			3.09 (2.82–3.47)
Potassium (meq/liter)		3.3 ± 0.04SE	
Sodium (meq/liter)		153 ± 0.5SE	
Glucose (mg/dl)			74 (61–116)
pH	10–11	7.36	
Urea (mg/dl)			1.005
Specific gravity			(1.003–1.012)

[a] Mean ± 1 SD, observed range in parentheses, unless otherwise noted. Values are for cerebellomedullary fluid unless otherwise noted. NS, not stated.
[b] By electrophoresis.
[c] By radial immunodiffusion.

TABLE 27.6 Biochemical Constituents of CSF of the Cat[a]

Constituent	Rand *et al.* (1990a)	Hochwald *et al.* (1969)	Ames (1964)	Fankhauser (1962)
Methods				
# RBC/μl	<30	NS		NS
Necropsy	Yes	No		NS
Total protein (mg/dl)				
Cerebellomedullary	18 ± 7[b]	27.0 ± 8.8		<20
Reference range	6–36			
Lumbar		44.0 ± 1.7		
Method	Ponceau S	Biuret		NS
Albumin (mg/dl)				
Cerebellomedullary		6.5 ± 2.1		
Lumbar		10.1 ± 2.9		
Gamma globulin (mg/dl ± SD)				
Cerebellomedullary		1.2 ± 0.27		
Lumbar		1.6 ± 0.30		
IgG (mg/dl)	1.4 ± 1.7			
Reference range	0–5.3			
IgG–total protein index	0.321 ± 0.210 (0.086–1.297)			
Aspartate transferase (U/liter)	17 ± 7			
Reference range	0–34			
Creatine kinase (U/liter)	47 ± 51[b]			
Reference range	2–236			
Lactate dehydrogenase (U/liter)	12 ± 5[b]			
Reference range	0–24			
Calcium (meq/kg H_2O ± SE)			1.50 ± 0.06	5.2 mg/dl
Chloride (meq/kg H_2O ± SE)			144 ± 2	900 mg/dl
Magnesium (meq/kg H_2O ± SE)			1.33 ± 0.02	
Potassium (meq/kg H_2O ± SE)			2.69 ± 0.09	
Sodium (meq/kg H_2O ± SE)			158 ± 4	
Glucose (mg/dl)	4.1 ± 1.3 mmol/l			85
Reference range	1.0–7.2 mmol/l			
pH				Slightly alkaline

[a] Mean ± 1 SD, observed range in parentheses, unless otherwise noted. Values are for cerebellomedullary fluid unless otherwise noted. NS; not stated.

[b] Significantly correlated with CSF RBC count.

and transferrin, which are also synthesized by the choroid plexus, and beta and gamma trace proteins, tau protein (tau fraction, modified transferrin), glial fibrillary acidic protein, and myelin basic protein, which appear to be synthesized intrathecally (Thompson, 1988).

1. *Albumin*

With electrophoretic techniques, protein in the CSF can be separated into prealbumin, albumin, and alpha, beta, and gamma globulins. The major protein in CSF is albumin, which is synthesized only in the liver. The limited entry of albumin into the CSF is dependent upon the blood–brain/CSF barrier to macromolecules. When total CSF protein is increased, the albumin concentration is increased disproportionately. This phenomenon illustrates the role of molecular size in determining the distribution of serum proteins into the CSF (Felgenhauer, 1974).

2. *Alpha and Beta Globulins*

Immunoelectrophoresis can separate the alpha and beta globulins into several proteins (Table 27.9). The origin of tau protein (beta$_2$ transferrin) is uncertain. This protein may be modified serum transferrin (beta$_1$ transferrin) or it may be a unique protein, "tau protein," in the CSF (Fishman, 1992). In veterinary and human medicine, no correlation has been made between changes in the concentrations of these globulins and specific neurologic disease (Fishman, 1992; Sorjonen *et al.*, 1991). Thus, their measurement has limited clinical use at this time.

3. *Gamma Globulins*

Because of the changes found in association with multiple sclerosis and other inflammatory diseases, the gamma globulins have received a great deal of attention. Electrophoretic techniques define the gamma globulins as a heterogeneous group of proteins with migrations at similar rates (see Table 27.9). The gamma globulin fraction contains immunoglobulins. Immunologic assays identify three major immunoglobulins in normal CSF—IgG, IgA, and IgM. Minute amounts of other immunoglobulins have also been detected in normal CSF (Fishman, 1992; Kjeldsberg and Knight, 1993).

a. IgG

The major immunoglobulin in normal CSF is IgG, which normally originates from the serum. An increased level of CSF gamma globulin has been reported in a number of inflammatory central nervous system disorders. In disease conditions, gamma globulin may enter the CSF through dysfunctional blood–brain/CSF barriers, or may be synthesized intrathecally by cells that have migrated into the brain or CSF and are participating in the disease process (Fishman, 1992; Kjeldsberg and Knight, 1993).

b. IgM and IgA

Cerebrospinal fluid IgM and IgA also originate normally from the serum. However, in certain diseases, particularly inflammatory diseases, these immunoglobulins are produced within the central nervous system as well (Fishman, 1992; Kjeldsberg and Knight, 1993). IgM is ontogenetically and phylogenetically the most primitive immunoglobulin and is therefore detected at an earlier stage of the general immune response of the body. IgM is also the first immunoglobulin to return to normal when the offending antigen disappears. The characteristics of IgM and IgA participation in the intrathecal immune response still need to be resolved, however (Tipold *et al.*, 1994; Felgenhauer, 1982).

4. *Other Proteins*

Many other proteins have been identified in CSF, including myelin basic protein, S-100 protein, C-reactive protein, interferon, embryonic proteins, fibronectin, and glial fibrillary acidic protein. In general, the CSF concentrations of these proteins may be altered by a number of neurologic disease processes. The utility of assay of these proteins in clinical veterinary or human medicine has yet to be established (Fishman, 1992; Kjeldsberg and Knight, 1993).

C. Glucose

The CSF glucose is derived solely from the plasma by facilitated diffusion. The concentration of CSF glucose depends upon the blood glucose concentration, the rate of glucose transport into the CSF, and the metabolic rate of the central nervous system. The normal CSF glucose level is about 60–80% of the blood glucose concentration, reflecting at least in part the high metabolic rate of the central nervous system. Equilibration with plasma glucose requires about 1 to 2 hours; thus, ideally, plasma glucose should be determined about 1 hour prior to CSF aspiration and analysis. In people, a glucose gradient exists along the neuraxis; the concentration decreases from ventricular to lumbar fluid (Fishman, 1992; Kjeldsberg and Knight, 1993).

D. Enzymes

Numerous enzymes have been assayed in the CSF of animals(see Tables 27.5 through 27.8) (Rand et al., 1990a; Wilson, 1977; Jackson *et al.*, 1996) and people (Banik and Hogan, 1983). These enzymes have three possible sources: (1) blood, (2) neural tissue or neural tumors, or (3) cells within the CSF (Fishman, 1992; Kjeldsberg and Knight, 1993). In probably every instance, the blood enzyme levels are higher than the CSF levels. Unfortunately, many studies of CSF levels in disease fail to report the concurrent blood level and a measure of blood–brain/CSF barrier integrity. However, studies of CSF creatine kinase (CK) in dogs and horses did not find a relationship between WBC counts, serum CK, or CSF total protein and CSF CK (Furr and Tyler, 1990; Jackson *et al.*, 1996). Regarding correlation of CSF RBC with CSF CK, one study reported a significant correlation (Indrieri *et al.*, 1980), whereas another study did not find a statistical association between the two parameters (Jackson *et al.*, 1996). To date, none of the enzyme assays has been shown to be sufficiently sensitive or specific to warrant its routine use in clinical practice (Rand *et al.*, 1994b; Fishman, 1992; Kjeldsberg and Knight, 1993; Jackson *et al.*, 1996; Indrieri *et al.*, 1980).

E. Neurotransmitters

Because they are produced by neurons, neurotransmitters and their metabolites have been extensively studied in people for their potential use as markers

TABLE 27.7 Biochemical Constituents of CSF of Foals and Horses[a]

Constituent	Rossdale *et al.* (1982)	Andrews *et al.* (1994)	Andrews *et al.* (1990a)	Andrews *et al.* (1990b)	Rossdale *et al.* (1982)	Mayhew *et al.* (1977)[b]	Fankhauser (1962)
Age	<40 hr[c]	≤10 days	4–9 years	NS	Adult	0.75–15 years	Adult?
Methods							
# RBC/μl	NS	<2000	<600	NS	NS	195.15 ± 511.96	NS
Necropsy	NS	No	2 of 12	NS	NS	No	NS
Total protein (mg/dl)							
Cerebellomedullary	138 ± 50 (70–210)	82.8 ± 19.2 (56.7–115)	87.0 ± 17.0 (59–118)		105 ± 38 (40–170)	37.23 ± 28.4[d] 5–100[e]	47.58 (28.75–71.75)
Lumbosacral		83.6 ± 16.1 (60.5–116)	93 ± 16 (65–124)			−0.46 ± 13.7 (LS–CM difference)	
Method	Biuret	Coomassie brilliant blue	Coomassie brilliant blue		Biuret	TCA	NS
Albumin (mg/dl ± SD)							
Cerebellomedullary		52.0 ± 8.6 (34–64)	35.8 ± 9.7 (24–51)				38.64 (22.62–67.94)
Lumbosacral		53.8 ± 15.7 (30–92)	37.8 ± 11.2 (24–56)				
Albumin quotient (±SD)							
Cerebellomedullary		1.86 ± 0.29 (1.55–2.33)	1.4 ± 0.4 (1–2.1)				
Lumbosacral		1.85 ± 0.51 (1.07 ± 2.88)	1.5 ± 0.4 (1–2.4)				
Globulin (mg/dl)							9.34 (3.37–18.37)
IgG (mg/dl ± SD)							
Cerebellomedullary		10.2 ± 5.5 (3–22)	5.6 ± 1.4 (3–8)				
Lumbosacral		9.9 ± 5.7 (3–22.5)	6.0 ± 2.1 (3–10)				
IgG index (±SD)							
Cerebellomedullary		0.519 ± 0.284 (0.0095–0.942)	0.19 ± 0.046 (0.12–0.27)				
Lumbosacral		0.482 ± 0.27 (0.091–2.089)	0.194 ± 0.05 (0.12–0.26)				
Alkaline phosphatase (IU)						0.83 ± 0.95 0–8[e]	
Aspartate transferase (IU)						30.74 ± 6.31 SFU	
Cerebellomedullary	16.6 ± 7.6 (6–26)			4–16[e]	18.27 ± 10.8 (7.5–30)	15–50[e]	
Lumbosacral				0–16[e]			

Creatine kinase (IU)				1.08 ± 3.13	
Cerebellomedullary	15.2 ± 9.2 (4–33)	0–8[e]	5.78 ± 3.7 (3.2–11)	0–8[e]	
Lumbosacral		0–8[e]			
γ-Glutamyl transferase (IU)	1.5 ± 1.5 (0.9–2.3)		2.45 ± 1.9 (0.8–4.2)		
Lactate dehydrogenase (IU)	23.2 ± 10.7 (10–40)	0–8[e]	27.7 ± 8.0 (12–34)	1.54 ± 1.75 0–8[e]	
Calcium (mg/dl)				4.18 ± 0.87 2.5–6.0[e]	6.26 (5.55–6.98)
Chloride (meq/liter)	109 ± 3.4 (104–113)		103.3 ± 13.5 (92–116)	109.22 ± 6.90 95–123[e]	737 mg/dl (690–792)
Magnesium (mg/dl)					1.98 (1.06–2.95)
Phosphorus (mg/dl)				0.83–0.20 0.5–1.5[e]	1.44 (0.87–2.20)
Potassium (meq/liter)	3.6 ± 2.1 (1.3 ± 4.6)		2.9 ± 0.6 (1.9 ± 3.9)	2.95 ± 0.05 2.5–3.5[e]	12.66 mg/dl (10.65–14.20)
Sodium (meq/liter)	142.6 ± 2.8 (139–147)		143.9 ± 2.6 (139–147)	144.58 ± 1.86 140–150[e]	
Cholesterol (mg/dl)				4.76 ± 5.72 0–20[e]	0.36–0.55
Glucose (mg/dl)					57.2 (40–78)
Cerebellomedullary		35–70% of blood glucose[e]		48.0 ± 9.92 30–70[e]	
Lumbosacral		35–70% of blood glucose[e]		55.13 ± 8.22 40–75[e]	
Lactic acid (mg/dl)					
Cerebellomedullary		1.92 ± 0.12			
Lumbosacral		2.3 ± 0.21			
pH					7.13–7.36
Specific gravity		1.003–1.005[e]			1.004–1.008
Urea nitrogen (mg/dl)				11.82 ± 3.26 0–20[e]	

[a] Mean ± 1 SD, observed range in parentheses, unless otherwise noted. Values are for cerebellomedullary fluid unless otherwise noted. NS, not stated; LS, lumbar subarachnoid space CSF; CM, cerebellomedullary cistern CSF.
[b] Except where noted, values are for pooled cerebellomedullary and lumbosacral fluid.
[c] Spontaneously delivered.
[d] Total protein for ponies—60.48 ± 20.45, reference range 20–105 (significantly different from that for horses).
[e] Reference range.

TABLE 27.8 Biochemical Constituents of CSF of the Cow, Sheep, Goat, Pig, and Llama[a]

Constituent	Cow (Welles *et al.*, 1992)[b]	Sheep (Altman and Dittmer, 1974)[c]	Goat (Altman and Dittmer, 1974)[c]	Pig (Altman and Dittmer, 1974)[c]	Llama (Welles *et al.*, 1994)[d]
Methods					
# RBC/μl	(5–1930)	NS	NS	NS	(0–1360)
Necropsy	No	NS	NS	NS	No
Total protein (mg/dl)					
Cerebellomedullary		(8–70)	12	(24–29)	
Lumbosacral	39.16 ± 3.39 (23.4–66.3)				43.1 ± 9.0 (31.2–66.8)
Method	Coomassie brilliant blue	NS	NS	NS	Coomassie brilliant blue
Albumin (mg/dl)					
Cerebellomedullary				(17–24)	
Lumbosacral	15.75 ± 1.53% (8.21–28.71)				17.9 ± 4.45 (11.8–27.1)
Albumin quotient					0523 ± 0.114
Globulin (mg/dl)				(5–10)	(0.38–0.75)
Gamma globulin (mg/dl)	4.84 ± 0.44% (2.46–8.85)				6.4 ± 2.50 (3.4–13.8)
IgG (mg/dl)	9.49 ± 1.03 (4.88–16.57)				
Creatine kinase (U/liter)	11.44 ± 3.43 (2–48)				4.6 ± 4.69 (0.0–15.0)
Lactate dehydrogenase (U/liter)	13.94 ± 1.318 (2–25)				13 ± 5.6 (7–24)
Calcium (mg/dl)		5.6 ± 0.3			
Chloride (meq/liter)		832 mg/dl (750–868)	681 mg/dl		134 ± 6.5 (116–143)
Magnesium (mg/dl)	1.99 ± 0.03 meq/liter (1.8–2.1)	2.88			
Potassium (meq/liter)	2.96 ± 0.03 (2.7–3.2)				3.19 ± 0.10 (2.9–3.3)
Sodium (meq/liter)	140 ± 0.78 (132–142)				154 ± 5.8 (134–160)
Glucose (mg/dl)	42.88 ± 0.99 (37–51)	(48–109)	71	(45–87)	69.3 ± 7.35 (59–86)
pH		7.35 (7.3–7.4)			

[a] Mean ± 1 SD, observed range in parentheses, unless otherwise noted. NS, not stated.
[b] Lumbosacral fluid. Mean ± SEM.
[c] Cerebellomedullary fluid.
[d] Lumbosacral fluid.

of neuronal activity and neurologic and psychiatric disease (Davis, 1990). The concentrations of several neurotransmitters (e.g., γ-aminobutyric acid, dopamine) and their metabolites (e.g., 5-hydroxyindolacetic acid, homovanillic acid, and dihydroxyphenylacetic acid) have been measured in the CSF from various sites in dogs, sheep, goats, cattle, and horses (Vaughn *et al.*, 1989b, 1988a; Sisk *et al.*, 1990; Loscher and Schwartz-Porsche, 1986; Ruckebusch and Costes, 1988; Bardon and Ruckebusch, 1984; Ruckebusch and Sutra, 1984; Faull *et al.*, 1982). Some metabolite concentrations have a gradient along the neuraxis (Vaughn and Smyth,

TABLE 27.9 Cerebrospinal Fluid Proteins Identified by Electrophoresis (Top Row) and Immunoelectrophoresis (Underlying Columns)[a]

Transthyretin (prealbumin)	Albumin	Alpha$_1$ globulin	Alpha$_2$ globulin	Beta globulin	Gamma globulin
Transthyretin	Albumin	Alpha$_1$ antitrypsin	Alpha$_2$ macroglobulin	Beta lipoprotein	IgG
		Alpha$_1$ lipoprotein	Alpha$_2$ lipoprotein	Transferrin	IgA
		Alpha$_1$ glycoprotein	Haptoglobin	Tau protein (modifed transferrin)	IgM
			Ceruloplasmin	Plasminogen	
			Erythropoietin	Complement	IgD
				Hemopexin	IgF
				Beta-trace	Gamma-trace

[a] Modified from Fishman (1992).

1989; Vaughn *et al.*, 1988b; Ruckebusch and Costes, 1988; Ruckebusch and Sutra, 1984), and some are age-related (Smyth *et al.*, 1994; Vaughn *et al.*, 1989b; Ruckebusch and Costes, 1988). Despite intense interest, more research is needed to verify the clinical utility of assay of these substances in the CSF (Fishman, 1992; Kjeldsberg and Knight, 1993).

F. Other CSF Constituents

Many other substances have been measured in CSF in experimental and clinical situations. These include electrolytes, gases, organic and amino acids, ammonia, urea, creatinine, prostaglandins, cytokines, and hormones. Assay of these substances is not particularly helpful in the diagnosis of neurologic disease in people because the substances are not generally associated with specific disease (Fishman, 1992; Kjeldsberg and Knight, 1993). Their usefulness in veterinary medicine has yet to be established.

G. Concentration Gradient along the Neuraxis

In cats (Hochwald *et al.*, 1969), dogs (Bailey and Higgins, 1985; Vaughn *et al.*, 1988b), horses (Andrews *et al.*, 1990a; Vaughn and Smyth, 1989), rhesus macaques (Smith and Lackner, 1993) and people (Fishman, 1992; Davson and Segal, 1996), the total protein concentration increases along the neuraxis from rostral to caudal. For example, in people the total protein concentration of ventricular, cerebellomedullary cistern, and lumbar subarachnoid fluid is about 26, 32, and 42 mg/dl, respectively (Weisner and Bernhardt, 1978). (Total protein, albumin, and globulin contents of cerebellomedullary cistern and lumbar subarachnoid

TABLE 27.10 Protein Content of Cerebellomedullary Cisternal CSF of Healthy Dogs, Cats, and Horses as Identified by Electrophoresis[a]

Protein fraction	Dog (Sorjonen, 1987)	Cat (Rand *et al.*, 1990a)	Horse (Kristensen and Firth, 1977)
Prealbumin			2.0 ± 0.9
Albumin	37 ± 4.29 (31–44)	11 ± 15 (1–53)	43.4 ± 6.8
Alpha globulin	28 ± 5.27 (24–31)	21 ± 11 (0–48)	
Alpha$_1$			5.3 ± 1.3
Alpha$_{2a}$			3.3 ± 0.8
Alpha$_{2bc}$			6.4 ± 1.8
Beta globulin	25 ± 5.31 (19–30)	57 ± 15 (37–91)	
Beta$_1$			17.0 ± 3.2
Beta$_2$			7.8 ± 2.3
Gamma globulin	7.75 ± 1.84 (6–9)	12 ± 7 (0–29)	14.8 ± 3.3

[a] Mean ± SD percentage of total CSF protein; range in parentheses.

CSF for dogs, cats, and horses are given in Tables 27.5, 27.6, and 27.7, respectively.) The concentration of the albumin and globulin fractions also increases from ventricular to lumbar fluid. The increased protein content may be the result of a greater permeability of the spinal blood–CSF barrier than the ventricular barrier to albumin (Fishman, 1992), additions of protein from adjacent nervous tissue (e.g., IgG from lymphocytes located in or near the CSF pathway (Weisner and Bernhardt, 1978)), progressive equilibration of CSF with plasma through the capillary walls (Weisner and Bernhardt, 1978), and low flow rates of lumbar CSF (Davson and Segal, 1996).

A study of healthy dogs also identified a small but significant gradient for the CSF WBCs; lumbar fluid contained significantly fewer cells than cerebellomedullary fluid (Bailey and Higgins, 1985). Another study did not find a difference in WBC counts between fluids from the two sites (Vaughn *et al.*, 1988b). However, 4 of the 10 dogs in this study had CSF total WBC counts $>3/\mu l$, and none of the dogs were necropsied to verify their health. Therefore, some of these dogs may have had subclinical neurologic disease, disguising a small cellular gradient. The small number of WBCs in normal fluid may make a cellular gradient more of a theoretical issue than a practical issue, however. If a cellular gradient exists, it may be due to fewer cells entering the lumbar CSF than the cerebellomedullary CSF, a greater rate of cell lysis in the lumbar CSF, a greater migration rate of WBCs from lumbar CSF back into the blood, or loss of WBCs that entered the CSF rostrally and lysed as CSF circulated to the caudal subarachnoid space.

A gradient has also been reported for CSF neurotransmitter metabolites in the dog (Vaughn *et al.*, 1988b) and the horse (Vaughn and Smyth, 1989). In each species, the neurotransmitter metabolite content of cerebellomedullary CSF was greater than that of lumbar subarachnoid CSF. This gradient probably reflects the major source of the neurotransmitter (brain) and transport of the metabolite from the CSF into the blood along the spinal axis (Vaughn *et al.*, 1988b).

V. CSF COLLECTION AND ANALYTICAL TECHNIQUES

A. Collection

1. General Techniques

Specific details for the collection of CSF from the various species are covered in many excellent articles and textbooks (Holbrook and White, 1992; Mayhew, 1989; deLahunta, 1983; Kornegay, 1981; Brewer, 1983; Fowler, 1989; Brewer, 1987; Boogerd and Peters, 1986) and will not be covered here except for the authors' preferred technique of collection from the cerebellomedullary cistern of dogs and cats (see later discussion). Considerations that apply regardless of species are sterility, use of a spinal needle, and collection from animals with increased intracranial pressure. In order to prevent iatrogenic central nervous system infection, sterility during the collection procedure is vital. A generous area around the puncture site should be clipped and surgically prepared. Preparation of too small an area can lead to contamination if any difficulty in palpating landmarks or entering the subarachnoid site is encountered. Additionally, the use of a fenestrated drape is highly recommended. Spinal puncture is contraindicated in an area of severe pyoderma / furunculosis, cellulitis, or epidural abscess. A needle with a stylet (spinal needle) should be used to prevent implantation of a plug of epidermis in the subarachnoid space, which not only could lead to infection but could also seed an epidermoid tumor. Replacement of the stylet upon withdrawal is controversial, either preventing or causing entrapment and severance or dislocation of nerve root filaments (Fishman, 1992). Aspiration of CSF from animals with increased intracranial pressure may result in brain herniation. Appropriate anesthetic agents, hyperventilation, and mannitol may decrease the probability of herniation. Use of the smallest gauge needle possible may also help prevent herniation by decreasing CSF leakage through the puncture hole in the meninges. Only the minimal amount of CSF necessary to perform the desired tests should be withdrawn. Brain herniation can occur following lumbar taps as well as cerebellomedullary cistern taps.

2. Collection Site

The choice of collection site is influenced by the species and breed of animal, the location of the neurologic lesion, and anesthetic considerations. The size of some animals may make lumbar subarachnoid puncture difficult if not impossible. However, cerebellomedullary puncture usually can be accomplished even in very large or obese animals. Because of differences in anatomy, the type or breed influences the exact site for lumbar puncture in the dog; L3–4 or L4–5 is recommended for large-breed, nonchondrodystrophic dogs (e.g., German shepherd dogs), whereas L4–5 or L5–6 is recommended for small, chondrodystrophic dogs (e.g., dachshunds) (Morgan *et al.*, 1987). The puncture site chosen should be as close to the lesion as possible without penetrating the lesion, or the site should be caudal to the lesion. In animals with spinal disease, cerebellomedullary fluid is abnormal more frequently with cervical disease than it is with thoracolumbar disease, but overall lumbar fluid is abnormal more often than cerebellomedullary fluid. With intracranial disease, CSF from both sites is usually abnormal, perhaps because both sites are caudal to the lesion (Thomson *et al.*, 1990; Scott, 1992; Thomson *et al.*, 1989). Occasionally, CSF is collected from both sites. Although the order of collection, cerebellomedullary or lumbar CSF collected first, appears not to influence the analytical results significantly (Bailey and Higgins, 1985), aspiration from the relatively small lumbar subarachnoid space is easier if the CSF pressure has not just been lowered by cerebellomedullary CSF collection. Cerebellomedullary puncture should be done under general anesthesia. In most instances, lumbar puncture can be done with sedation and local anesthesia. Therefore, if general anesthesia is contraindicated, a lumbar puncture should be done.

3. CSF Collection from the Cerebellomedullary Cistern

The authors' preferred technique for CSF collection from the cerebellomedullary cistern is to utilize the palpable bony landmarks that are the closest to the puncture site. These structures are the vertebral arch of C1 and the external occipital protuberance. After anesthetic induction and intubation, the animal is placed in lateral recumbancy and padding is placed under the neck to align the dorsal cervical and cranial midline parallel to the tabletop. The assistant is instructed to ''tuck in the animal's chin'' and push the external occipital protuberance toward the operator. This procedure flexes the atlanto-occipital joint and maximizes the space between the occipital bone and C1. Asking the assistant to simply flex the neck seems to produce flexion of the midcervical area more than the atlanto-occipital area. The external occipital protuberance, the C2 spinous process, and the C1 vertebral arch are palpated. The last structure is located by rolling a fingertip off the cranial edge of the C2 spinous process and palpating firmly, feeling for a transverse bony ridge (the C1 vertebral arch). The C1 vertebral arch can usually be palpated, and if so, the puncture is made on the midline just in front of the fingertip palpating the vertebral arch. If C1 is not palpable, the distance between the cranial edge of the C2 spinous process and the occipital protuberance is noted, and the puncture is made on the midline about one-third of that distance cranial to the cranial edge of the C2 spinous process. In rare cases, neither C1 nor C2 can be palpated. In this situation, the lateral edge of each C1 transverse process is palpated and a triangle from each edge to the occipital protuberance is constructed visually. The puncture is made on the midline in the center of that triangle.

B. Physical Examination: Clarity, Color, and Viscosity

After collection, the CSF is examined visually and the clarity, color, and viscosity are recorded. Normal CSF is clear and colorless and has essentially the same viscosity as water. For accurate assessment, the CSF can be compared to the same amount of distilled water in the same type of container. The containers can be held against a white, typewritten page to judge color and clarity, and gently shaken to assess viscosity. If the CSF appears abnormal, the color and clarity of the supernatant after centrifugation should be noted.

C. Cytological Analysis

1. General Techniques

Collection of CSF in a plastic or silicon-coated glass tube is preferred because monocytes will adhere to glass and can activate in the process (Fishman, 1992). This can result in erroneous cell counts and also alter morphology. In practical terms, this is of little consequence in those specimens that are rapidly processed, but becomes more significant as the delay between collection and processing increases. A complete cytologic examination includes both a total cell count and a differential, as well as thorough morphologic assessment. A differential and thorough morphologic assessment should be performed routinely, even on those samples that have cell counts within normal limits. In our experience, very low cell counts alone cannot be used as an indicator of normality. In one study that utilized cytocentrifugation, nearly 25% of canine CSF samples with cell counts in the normal range had abnormalities in cell type or morphology (Christopher *et al.*, 1988). Abnormalities included the presence of phagocytic macrophages, increased percentage of neutrophils in the differential, and the presence of reactive lymphocytes and plasma cells. Malignant cells have been observed in samples with normal counts (Bichsel *et al.*, 1984b; Grevel and Machus, 1990). This certainly emphasizes the importance of cytologic assessment, especially in specimens with normal cell counts. In general terms, CSF samples should be processed as soon as possible after collection, as cells degenerate relatively quickly in CSF (Chrisman, 1992; Fishman, 1992; Steele *et al.*, 1986; Kjeldsberg and Knight, 1993). This is because of the marginal hypotonicity of the fluid and very low protein content (in nonpathologic specimens). Proteins and lipids tend to have a membrane-stabilizing effect (Steele *et al.*, 1986). A multitude of veterinary references state that processing must be performed within 30 minutes of collection (Thomson *et al.*, 1990; Oliver and Lorenz, 1993; Cook and DeNicola, 1988; Chrisman, 1983) and this necessitates analysis in the individual veterinary hospital or practice setting. However, there is a complete lack of controlled veterinary studies to confirm this assertion, although, as stated previously, the sooner the better, in general terms. There have been several human studies performed. Steele found that there were significant differences in rates of lysis that were dependent on cell type. At room temperature, neutrophil counts decreased most rapidly and were significantly reduced at 1 hour and 2 hours, 68 ± 10% (SEM) and 50 ± 12% of original values, respectively. Values were not significantly reduced at half an hour. Lymphocyte and monocyte numbers were not significantly altered until 3 hours (69 ± 7% and 66 ± 7%, respectively). However, refrigeration at 4°C markedly reduces the rate of lysis, with only 15% decrease in the total white blood cell numbers at 2 hours (Kjeldsberg and Knight, 1993). Another study (Stokes *et al.*, 1975) reported the following rates of disappearance for white blood cells at room temperature: 4 hours—10% lysis, 18 hours—70% lysis, 36 hours—100%. Because there is differential sensitivity of cell types, both cell counts and the differential are altered with time. The authors have observed this phenomenon. When differential counts are compared between slides made immediately and those made via cytocentrifugation 3–6 hours later, there is a marked decrease in the percentage of neutrophils (personal observation). This would tend to confirm the preceding data. Therefore, the recommendation that analysis be performed within 30 minutes is reasonable, but it is predicated on the conditions that the sample is exposed to. Refrigeration obviously slows lysis, likely long enough for transport to reference laboratories in some instances. Addition of protein to the sample helps preserve cells and also expedites transport of samples to more remote facilities. There was excellent correlation between the total numbers of cells on the slides and the differential between sediment slides processed immediately and those preserved with fetal bovine serum (200 μl of CSF and 200 μl of fetal bovine serum) and cytocentrifuged 2–4 hours later (Rand *et al.*, 1990b). We have observed excellent morphology on slides made from CSF as long as 12 hours postcollection, when the sample was immediately mixed with approximately 20% by volume of autologous, well-centrifuged serum and transported on an ice pack. However, the effects on cell count and differential are unknown under these conditions. This is perhaps an option for practitioners in remote locations without the capability or expertise to process CSF. The duration of cell preservation under different conditions requires further study.

2. *Total Leukocyte and Erythrocyte Counts*

Electronic cell or particle counters are not sensitive enough to be used for enumeration of cells in CSF. The level of background counts with these counters is frequently in excess of the counts present in the majority of CSF samples that are analyzed. Therefore, counts are performed with the use of a standard hemacytometer chamber with a Neubauer ruling (Cook and DeNicola, 1988; Jamison and Lumsden, 1988; Brobst and Bryan, 1989). The chamber is charged with undiluted fluid. Ideally, the cells should be allowed to settle for 10 minutes in a humidified environment. This allows all the cells to be visible in the same plane of focus. The cells in the 9 largest squares on both sides of the chamber are counted (18 squares in total) and the result multiplied by 0.55 to obtain the number of cells per microliter. Numerous references recommend counting the cells in 9 large squares and multiplying by 1.1 to determine the count per microliter. To the untrained observer, unstained leukocytes and erythrocytes may be difficult to differentiate. Leukocytes are larger and the presence of nuclei gives them a more granular appearance than erythrocytes. With experience, nuclear morphology can often be appreciated (Cook and DeNicola, 1988). The cytoplasmic border is usually slightly irregular. In contrast, erythrocytes are usually smaller, smooth, and refractile, although they may become crenated upon standing (Jamison and Lumsden, 1988). A number of techniques have been developed to lyse erythrocytes and stain leukocytes in an effort to expedite counting. Technical details can be found in numerous sources (Chrisman, 1992; Cook, 1985; Brobst and Bryan, 1989), but there are several major disadvantages associated with these methods (Jamison and Lumsden, 1988), and we do not recommend them.

3. *Cytologic Examination*

A wide variety of methods to facilitate cytologic examination of CSF have been reported and compared (Grevel, 1991; Roszel, 1972; Kölmel, 1977; Sörnäs, 1967; Barrett and King, 1976; Steinberg and Vandevelde, 1974; Ducos *et al.*, 1979; Hansen *et al.*, 1974; Woodruff, 1973; Jamison and Lumsden, 1988). There is controversy as to which method is optimal, and all have their strengths and weaknesses. Methods include simple centrifugation, sedimentation and variations thereof, membrane filtration, and cytocentrifugation (Jamison and Lumsden, 1988). The reader is referred to these references for specific methodological details. In the great majority of cases, simple centrifugation produces slides that are completely unsatisfactory for cytologic examination. Membrane filtration techniques have the chief advantage of excellent cellular recovery with yields approaching 90–100% (Barrett and King, 1976). However, the methodology is laborious and time consuming, the morphology relatively poor, many cells are partly hidden in the filter substance, which itself stains variably, and the technique requires specialized, nonroutine staining techniques that the great majority of veterinary cytologists have no experience or expertise in interpreting. For these reasons, they are not recommended. Cytocentrifugation (Hansen *et al.*, 1974; Woodruff, 1973) has become the method of choice in both human (Fishman, 1992; Kjeldsberg and Knight, 1993) and veterinary medicine (Christopher *et al.*, 1988; Jamison and Lumsden, 1988). The technique is rapid and simple and provides slides with good cytologic detail. The technique is enhanced by the addition of protein to the CSF sample prior to centrifugation. This helps preserve cell morphology. Conditions of cytocentrifugation vary from laboratory to laboratory. We personally prefer the method described by Rand (Rand *et al.*, 1990b). The disadvantages of cytocentrifugation include the expense of the instrument and the relatively low cell yield. In one comparative study (Barrett and King, 1976), the following cell yields were determined: Millipore filtration 81 ± 3 % (SEM), Nucleopore filtration 69 ± 3%, and cytocentrifugation 11 ± 1%. The Sornas method of centrifugation results in a cell yield, after staining, of approximately 20% (Sörnäs, 1967). The sedimentation technique of Sayk, modified by Kölmel (Grevel, 1991; Grevel and Machus, 1990; Kölmel, 1977) results in a yield of approximately 30% (Kölmel, 1977), although this can be increased to almost 90% if a membrane filter is substituted for direct sedimentation onto a slide. We have some experience with the Kölmel apparatus and technique and have found the cell morphology to be at least as good as that in cytocentrifugation, with an apparently superior cell yield, although this would need to be confirmed with controlled comparative studies. Therefore, most studies would suggest that sedimentation techniques result in greater cell yields than cytocentrifugation, although there is a least one study that found the yield of cytocentrifugation to be marginally higher than that of sedimentation (Ducos *et al.*, 1979). Standard Romanowsky stains are recommended for staining of slides. They provide good cellular detail on air-dried CSF preparations and are familiar to most observers. These stains include the Wright's and Wright–Giemsa staining methods, as well as a variety of rapid staining methods including Diff-quik and Camco-quik (Jamison and Lumsden, 1988).

4. Immunocytochemistry

There are very few reports in the veterinary literature documenting the use of immunocytochemistry for the assessment of CSF. In people, the value of cytologic diagnosis of CSF can be improved if morphologic studies are appropriately supplemented by immunocytochemistry (Kjeldsberg and Knight, 1993). Immunophenotypic studies of cytocentrifuge slides are useful in the differential diagnosis of leukemia, maligant lymphoma, primary brain tumors, and metastatic tumors (Bigner and Johnston, 1981; Bigner, 1992; Kjeldsberg and Knight, 1993). Panels of monoclonal antibodies are often used, with the greatest limitation being the volume and the cellularity of the specimen that is available for the marker studies. Undifferentiated tumor panels frequently include leukocyte common antigen and cytokeratin antibodies. These can be very helpful in distinguishing single carcinoma cells from lymphocytes or monocytes (Bigner and Johnston, 1981; Bigner, 1992; Kjeldsberg and Knight, 1993). Glial fibrillary acidic protein has proven to be helpful in distinguishing a glial origin, but there are currently no specific markers to distinguish primary brain tumors. Immunocytochemistry can be used also to characterize the lymphocyte subpopulations present in CSF. Of the lymphocytes found in normal human CSF, 75–95% are T cells, with a mean of approximately 85% (Kjeldsberg and Knight, 1993). Within the population of T cells, T helper cells predominate and account for up to 88% of T cells. Alterations of these percentages have been shown to have significant associations with disease in people (Kjeldsberg and Knight, 1993). Similar studies are lacking in domestic animals but would potentially be quite useful. In people, most cental nervous system lymphomas are B cell in origin, and immunocytochemistry can be used to document monoclonality (Bigner and Johnston, 1981; Bigner, 1992). This is strong evidence (but not definitive proof) of malignancy. The B-cell origin can also be confirmed. For patients with T-cell lymphomas, marker studies can be more difficult to interpret as T cells predominate in normal and inflammatory CSF (Kjeldsberg and Knight, 1993). If there is systemic involvement, then comparison with the peripheral phenotype is useful to confirm presence in the CSF. Immunocytochemistry has also been utilized to detect infectious agents such as cytomegalovirus in human patients (Kjeldsberg and Knight, 1993).

5. Polymerase Chain Reaction

Polymerase chain reaction (PCR) technology represents a powerful adjunct to routine cytologic assessment and has the potential to revolutionize CSF analysis by dramatically increasing both the sensitivity and specificity of diagnosis. Because PCR exponentially increases *in vitro* the number of original DNA copies to a final number dependent on the number of cycles programmed, it is uniquely suited to the low volumes and small cell numbers frequently found in CSF samples. In people, one of the most powerful and useful applications of PCR methodology has been its use to confirm malignancy and detect minimal residual disease in lymphomatous meningitis (Rhodes *et al.*, 1996). This is accomplished via detection of clonal immunoglobulin or T-cell receptor gene rearrangements and the detection of clone-specific rearrangements, respectively. However, the exquisite sensitivity may result in false positive results due either to contamination or to very low initial numbers of cells producing an artifactual clonal band. Other applications include detection of a wide variety of infectious agents, including toxoplasma, borrelia, TB, and herpes simplex (Christen *et al.*, 1995; Lin *et al.*, 1995; Novati *et al.*, 1994; Guffond *et al.*, 1994). In the great majority of these studies, PCR has resulted in a more rapid diagnosis with superior sensitivity and specificity when compared to those of standard culture and serologic diagnostic techniques. In veterinary medicine, PCR has been used to detect several infectious agents in CSF samples, including sarcocystis neurona, toxoplasma, and listeria (Fenger, 1994; Peters *et al.*, 1995; Stiles *et al.*, 1996). The same advantages appear to apply in these instances, although some agents such as listeria may not gain access to the meningoventricular system, resulting in negative results when applied to CSF of confirmed positive cases (Peters *et al.*, 1995). Further developmental work in conjunction with prospective studies will be required before the true utility of PCR-based CSF diagnostics can be accurately assessed in domestic animals.

D. Protein Analysis

1. *Measurement of CSF Total Protein*

An increase in the concentration of CSF total protein was recognized as an indicator of neurologic disease soon after the introduction of lumbar puncture in human medicine. A number of tests were developed to assess qualitative changes in CSF protein, such as Lange's colloidal gold test, the Nonne–Appelt test, the Pandy test, and others. These qualitative tests largely have been replaced by quantitative methods. Several techniques have been developed for the quantitative measurement of CSF total protein, including turbidometric methods, biuret procedures, and Lowry's method. The accuracy of these methods in many clinical laboratories is no better than ±5% (Fishman, 1992). Total CSF protein values are reported in numerous articles and vary noticeably with the technique and the laboratory performing the assay. Therefore, clinicians must learn the normal values for their own laboratories.

2. *CSF Protein Fractionation*

A number of techniques for fractionation of CSF proteins have been developed. These include electrophoresis using paper or cellulose acetate, agar, agarose, polyacrylamide, and starch gels. Immunoelectrophoresis, electroimmunodiffusion, radioimmunoassay, and isoelectric focusing are more recent techniques (Fishman, 1992; Kjeldsberg and Knight, 1993). Because of the normally low protein content, most of these methods require concentration of the CSF, which can create technical artifacts in the measured protein content. Techniques that do not require CSF concentration, such as electroimmunodiffusion, are therefore advantageous (Fishman, 1992).

3. *Albumin and the CSF/Serum Albumin Index*

Because albumin is synthesized only extrathecally, increased CSF albumin indicates damage to the blood–brain/CSF barriers, intrathecal hemorrhage, or a traumatic CSF tap. In these conditions, albumin will leak into the CSF in general proportion to its serum concentration. Therefore, in the absence of intrathecal hemorrhage (pathologic or iatrogenic), the ratio of CSF albumin to serum albumin can be used as an indicator of barrier dysfunction (Tibbling *et al.*, 1977; Link and Tibbling, 1977). This ratio is called the albumin index (also known as albumin quota or albumin quotient) and is calculated as follows (Kjeldsberg and Knight, 1993):

$$\text{Albumin index} = \frac{\text{CSF albumin (mg/dl)}}{\text{serum albumin (g/dl)}}.$$

Values above the normal range indicate increased barrier permeability. The use of this index is potentially limited, however, because the large variability of CSF albumin in normal animals (at least in dogs and horses; Andrews *et al.*, 1994, 1990a; Bichsel *et al.*, 1984b; Krakowka *et al.*, 1981) yields a large variability in the values for this index (Davson and Segal, 1996). In people, the albumin index is age dependent, being highest in newborns and adults over 40 years of age (Kjeldsberg and Knight, 1993).

4. *Quantitative Measurement of Immunoglobulins*

a. IgG and the IgG/Albumin Index

The identification of intrathecal production of immunoglobulin is helpful in the diagnosis of neurologic disease. Immunoglobulin G is the dominant CSF immunoglobulin. However, the IgG content of CSF is not a particularly useful measurement by itself, because the IgG present in CSF may be of serum origin (via a dysfunctional blood–brain/CSF barrier, intrathecal hemorrhage, or traumatic puncture) or intrathecally produced (as in various neural diseases). Varied opinions exist regarding the best way to calculate the contribution of IgG from each source (Trotter and Rust, 1989; Thompson, 1988). To determine the probable origin of CSF IgG, it can be related mathematically to a protein of purely extrathecal origin. Because albumin is synthesized entirely extrathecally, it is the preferred comparison protein and is the most widely used (Fishman, 1992). Transferrin and alpha$_2$ macroglobulin have also been recommended because of their extrathecal origin (Schliep and Felgenhauer, 1974).

The simplest formula for correction of the CSF IgG level for extrathecal "contamination" and thereby demonstration of intrathecal IgG synthesis is the IgG/albumin index (Tibbling *et al.*, 1977; Link and Tibbling, 1977). This index is calculated using the CSF and serum concentrations of albumin and IgG as follows (Kjeldsberg and Knight, 1993):

$$\text{IgG index} = \frac{\dfrac{\text{CSF IgG (mg/dl)}}{\text{serum IgG (g/dl)}}}{\dfrac{\text{CSF albumin (mg/dl)}}{\text{serum albumin (g/dl)}}}.$$

The denominator of this index (CSF albumin/serum albumin) is the albumin index . Because albumin is synthesized only extrathecally, the albumin index assesses the amount of albumin crossing the blood–brain/CSF barriers, and therefore is a measure of barrier integrity. Blood contamination of the CSF with as little as 0.2% serum (equivalent to about 5000–10000 RBC/μl) by a traumatic puncture falsely elevates the IgG index in people (Peter and Tourtellotte, 1986). Also, the IgG index loses reliability when CSF protein levels are less than 25 mg/dl or greater than 150 mg/dl (Boerman *et al.*, 1991).

An additional problem with the IgG index is its basic premise that the selectivity of the protein transfer at the blood–CSF barrier is independent of the actual permeability condition. This concept has been shown to be incorrect, and the IgG index, as well as the IgA and IgM indices, vary in a nonlinear fashion with progressive impairment of the barrier (Reiber and Felgenhauer, 1987). Therefore, Reiber and Felgenhauer (1987) developed a formula to calculate the intrathecally synthesized fractions of IgG, IgM, and IgA in the CSF.

b. IgM and IgA Indices

As with IgG, CSF IgM and IgA may be of serum origin or intrathecally produced. Indices for IgM and IgA can be calculated in the same fashion as those for IgG (Fryden *et al.*, 1978). However, because of high variability in normal IgM and IgA levels and the biological variation of these large molecules, the application of the same formula for IgM and IgA indices as used for the IgG index may only provide rough estimates (Tipold *et al.*, 1994; Reiber and Felgenhauer, 1987).

5. *Qualitative Immunoglobulin Assays*

Qualitative assays of CSF immunoglobulins include agarose-gel electrophoresis, acrylamide immunoelectrophoresis, isoelectric focusing, and immunofixation. These tests separate the proteins into "bands" and provide information regarding the CSF protein composition. Although abnormal band patterns are not specific for a particular disease, they do indicate pathology and may indicate a type of disease. Abnormal band patterns may be detected even in patients with a normal IgG index. Thus, both quantitative and qualitative immunoglobulin assays are useful in the assessment of central nervous system disorders, particularly immunologic or inflammatory diseases (Fishman, 1992; Kjeldsberg and Knight, 1993).

E. Antibody/Antigen Tests

A variety of CSF antibody and antigen tests are now available for viruses, fungi, rickettsia, protozoa, and other organisms (Greene, 1990). For antibody titers, two samples taken 2 weeks apart should be assayed. Because of interrun variability, the samples should be assayed at the same time in the same analytical run. Interpretation of CSF antibody titers must take into account the possibility of transudation of serum antibodies through a defective blood–brain/CSF barrier. Serum antibodies could be present because of disease, previous exposure to antigen, or vaccination. Ideally, the CSF/serum albumin index and IgG index are also determined (see Sections V.D.3 and 4) to identify blood–CSF barrier dysfunction and intrathecal production of immunoglobulin. Intrathecal production of antigen-specific antibody (specific Ig) can be determined with an antibody index in the same way as intrathecal IgG production is detected with the IgG index. The formula is (Reiber and Lange, 1991)

$$\text{Antibody index} = \frac{\dfrac{\text{CSF specific Ig}}{\text{serum specific Ig}}}{\dfrac{\text{CSF total Ig}}{\text{serum total Ig}}}.$$

A modification of this formula accounting for large local synthesis of polyclonal IgG in the central nervous system may be necessary (Reiber and Lange, 1991). An antibody index > 1 suggests intrathecal production of the specific antibody (Reiber and Lange, 1991; Munana *et al.*, 1995). Antibody indices have been calculated in human patients with a variety of diseases (Reiber and Lange, 1991). The diagnostic reliability of these indices and application to clinical veterinary medicine needs further study.

Antigen detection tests include immunoelectrophoretic techniques, agglutination tests, and ELISA for bacterial antigens and latex agglutination for cryptococcal antigens. The recently developed PCR procedures detect specific antigen DNA in CSF and are highly sensitive, specific, and rapid (see Section V.C.5). The ELISA and PCR procedures have much promise for the diagnosis of neural infections (Kjeldsberg and Knight, 1993).

F. Microbial Tests

The Gram stain, the Ziehl–Neelson acid-fast stain, and both aerobic and anaerobic cultures of CSF are time-honored methods for diagnosis of bacterial central nervous system infections. Bacteriological tests must be performed as soon as possible after CSF acquisition because some bacteria undergo rapid autolysis in the test tube. Additional tests such as the acridine orange stain for bacteria and tests for microbial antigens by counterimmunoelectrophoresis or agglutination techniques may also be useful (Fishman, 1992).

G. Blood Contamination

Erythrocytes may be present in CSF samples because of subarachnoid hemorrhage or, more commonly, because of traumatic puncture. Blood contamination due to traumatic puncture is a common problem during CSF collection and, depending on its degree, can interfere with cytologic interpretation. Blood contamination is more likely to occur with lumbar puncture as opposed to cerebellomedullary cisternal puncture (Thomson *et al.*, 1990; Bailey and Higgins, 1985; Oliver and Lorenz, 1993). Blood contamination is a source of leukocytes and hence can affect both the count and the differential. In one study of CSF in cats (Rand *et al.*, 1990b), the total leukocyte count, the neutrophil percentage, and the eosinophil percentage were positively correlated with the CSF erythrocyte count once this count exceeded 500 erythrocytes per microliter. However, there was no significant increase in total white blood cell count or alteration in the differential percentages with up to 500 RBC/μl of CSF. Numerous correction factors have been used to correct leukocyte counts for the effect of blood contamination and include the following: In people, 1 white blood cell per 700 red blood cells is subtracted from the total white blood cell count (Fishman, 1992); in dogs, 1 white blood cell per 500 red blood cells is subtracted from the total count (Bailey and Higgins, 1985), and in cats, a maximum of 1 white blood cell per 100 red blood cells is subtracted (Rand *et al.*, 1990b). A more accurate formula takes into account the actual white blood cell and red blood cells counts of the patient and hence compensates for any significant alterations in these counts (Fishman, 1992),

$$W = \mathrm{WBC_F} - \frac{\mathrm{WBC_B} \times \mathrm{RBC_F}}{\mathrm{RBC_B}},$$

where W is the white blood cell count of the fluid before blood was added, that is, the corrected count; $\mathrm{WBC_F}$ is the total white blood cell count in the bloody fluid; $\mathrm{WBC_B}$ is the white blood cell count in the peripheral blood per microliter; and $\mathrm{RBC_F}$ and $\mathrm{RBC_B}$ are the numbers of red blood cells per microliter in the CSF and blood, respectively. Despite all of these elaborate corrections, our own experience is that many thousands of red blood cells in contaminated samples of CSF will frequently be observed without any accompanying white blood cells, suggesting that these correction factors may not be valid. This empirical observation has been made by others (deLahunta, 1983). This lack of validity has been proven by several studies (Wilson and Stevens, 1977; Novak, 1984). Wilson and Stevens (1977) concluded that blood contamination appeared to have little effect on white blood cell numbers and that the preceding correction formula was unreliable. They found 91 samples from both normal and diseased animals where there were numerous red blood cells but no white blood cells. Some of the red blood cell counts exceeded 15,000 RBC/μl, but white blood cells were still absent. Novak (1984) concluded that the standard computations frequently overcorrect white blood cell counts in blood-contaminated CSF, and the magnitude of the overcorrection may obscure disease in some instances—in eight infants with significant blood contamination but proven bacterial meningitis, correction computations normalized or overcorrected the white blood cell counts. The mechanism of this overcorrection could not be defined, but it is clear that that the presence of low numbers of neutrophils should not be immediately discounted when red cells are concurrently found (Christopher *et al.*, 1988).

A study of feline CSF (Rand *et al.*, 1990a) also found that values for CSF total protein, lactate dehydrogenase, creatine kinase, IgG ratio, and γ-globulin percentage were affected by blood contamination. The CSF total protein value of blood-contaminated CSF may be corrected using the formula for white blood cell correction given earlier but substituting the total protein levels of the bloody CSF and the serum for the corresponding white blood cell counts (Kjeldsberg and Knight, 1993). In people, bloody contamination of CSF with as little as 0.2% serum (equivalent to about 5000–10000 RBC / ml) elevates the IgG index (Fishman, 1992).

VI. GENERAL CHARACTERISTICS OF CSF IN DISEASE

A. Physical Characteristics: Clarity, Color, and Viscosity

Normal CSF is clear and colorless and has the consistency of water. In pathological conditions the clarity, color, and / or consistency may change.

1. *Clarity*

Cloudy or turbid CSF is usually due to pleocytosis; about 200 WBC / μl or 400 RBC / μl will produce a visible change. With these low levels of cellularity the CSF may appear opalescent or slightly hazy. Microorganisms, epidural fat, or myelographic contrast agent may also produce hazy or turbid CSF.

2. *Color*

Although the term xanthochromia means yellow color, it has often been used to describe pink CSF as well. The color of CSF is most usefully described as (1) pink or orange, (2) yellow, or (3) brown. These colors correspond to the major pigments derived from red cells: oxyhemoglobin, bilirubin, and methemoglobin. Oxyhemoglobin is red in color, but after dilution in the CSF it appears pink or orange. Oxyhemoglobin is released from lysed red cells and may be detected in the CSF supernatant about 2 hours after red cells enter the CSF. The level of oxyhemoglobin reaches its peak about 36 hours later and disappears over the next 4–10 days. Bilirubin is yellow in color. Bilirubin is the derivative of hemoglobin and is considered to formed by macrophages and other leptomeningeal cells that degrade the hemoglobin from lysed red cells. Bilirubin is detected about 10 hours after red cells enter the CSF, reaches a maximum at about 48 hours, and may persist for 2 to 4 weeks. Bilirubin is also the major pigment responsible for the abnormal color of CSF with a high protein content. Methemoglobin in CSF is dark yellow-brown. Methemoglobin is an oxidation product of hemoglobin characteristically found in encapsulated subdural hematomas and in old, loculated intracerebral hemorrhages (Fishman, 1992; Kjeldsberg and Knight, 1993). Occasionally blackish CSF is encountered in animals with melanin-producing tumors in the nervous system.

Causes of a CSF color change other than red-cell contamination include icterus due to liver disease or hemolytic disease, markedly increased CSF total protein level, and drug effects. Both free and conjugated bilirubin may be present in the CSF, although the amount of bilirubin in the CSF does not correlate well with the degree of hyperbilirubinemia. If the CSF protein level is increased, the color change will be greater because of increased amounts of the albumin-bound bilirubin. High CSF protein content alone can impart a yellow color to the CSF (Fishman, 1992; Kjeldsberg and Knight, 1993). The drug rifampin is well know to impart an orange-red color to body fluids. Rifampin is 90% bound to protein, and hypoproteinemia may result in the staining of CSF (Fishman, 1992).

3. *Viscosity*

Increased viscosity is usually due to a very high CSF protein content, particularly fibrinogen. If pleocytosis is present, a surface pellicle or a clot may form. In this situation, collection of the CSF in a heparinized or EDTA tube may be necessary to obtain an accurate cell count. Cryptococcosis may increase CSF viscosity because of the polysaccharide capsule of the yeast. Epidural fat or nucleus pulposus in the CSF may also increase viscosity, or it may result in globules within the fluid (Fishman, 1992; Kjeldsberg and Knight, 1993).

B. Cytology

An increase in the cellularity of CSF is termed pleocytosis. In general terms, the degree of pleocytosis is dependent upon several factors, including the nature of the inciting cause and the severity and location of the lesion with respect to the subarachnoid space or ventricular system (Cook and DeNicola, 1988). It should also be emphasized that normal CSF analysis does not exclude the presence of disease (Fishman, 1992; Kjeldsberg and Knight, 1993). This is especially true with deep parenchymal lesions that do not communicate with the leptomeninges, and hence the subarachnoid space or the ependymal surfaces. In these cases, despite the presence of neurologic disease that is often severe, the lesion may not affect the CSF cellularity (Cook and DeNicola, 1988). Abnormal CSF findings always indicate the presence of pathologic abnormality.

1. *Neutrophilia*

A marked pleocytosis with neutrophil predominance generally suggests either bacterial meningitis (Kornegay *et al.*, 1996; Kjeldsberg and Knight, 1993) or suppurative, nonseptic (corticosteroid-responsive) meningitis (Tipold and Jaggy, 1994; Meric, 1988, 1992a). Total leukocyte counts in excess of 2000 cells per microliter are frequently encountered in these diseases and may even exceed 10,000 cells per microliter (Meric, 1992a). Observation of bacteria or a positive culture confirms septic meningitis. In our experience, bacteria are more commonly observed in the CSF of large animals afflicted with septic meningitis than in dogs or cats with septic meningitis. Neutrophil nuclear morphology is often used as a criterion for determining the likelihood of sepsis, with nuclear degenerative changes or karyolysis being interpreted as evidence of bacterial disease. However, the neutrophils in confirmed cases of septic meningitis in dogs and cats are frequently well preserved, especially if there has been prior therapy. Therefore, absence of bacteria or degenerative nuclear changes in neutrophils cannot be used to exclude unequivocally a diagnosis of septic meningitis, although it does make it less likely. In people, acute viral meningoencephalitis can initially produce a neutrophilic pleocytosis (Fishman, 1992; Converse *et al.*, 1973; Kjeldsberg and Knight, 1993) that may persist from a few hours to several days prior to the development of the more typical mononuclear reaction. A similar phenomenon has been documented in animals (Green, 1993). Occasionally, distemper virus infection can cause massive encephalomalacia (Vandevelde and Spano, 1977), resulting in a neutrophilic pleocytosis in contrast to the moderate mononuclear pleocytosis that is more typical. Central nervous system neoplasia may result in a neutrophil predominance in the CSF, especially if there is significant necrosis and inflammation associated with the tumor. Moderate to marked pleocytosis with neutrophil predominance is a common finding in canine meningioma (Bailey and Higgins, 1986b). Canine intervertebral disc disease is predominantly characterized by a mild mononuclear pleocytosis (Thomson *et al.*, 1989). However, acute severe disease may result in counts greater than 50 cells per microliter with more than 50% neutrophils (Thomson *et al.*, 1989). This finding is a reflection of acute inflammation that may be exacerbated by myelomalacia in some instances. The authors have seen a similar phenomenon associated with fibrocartilaginous thromboembolism in dogs. A neutrophilic pleocytosis of varying severity often occurs following myelography with iodinated contrast agents (Widmer *et al.*, 1992; Johnson *et al.*, 1985; Carakostas *et al.*, 1983). These changes usually peak at 24 hours post myelogram (see Section VII for further details). Similarly, a neutrophilic pleocytosis has been observed postictally in people. We have occasionally observed similar findings in dogs (see Section VII).

2. *Lymphocytosis*

Alterations in both numbers and morphology of lymphocytes (see Section III.B) in the CSF occur in a variety of diseases. Central nervous system viral infections often result in a predominantly lymphocytic pleocytosis, which has been documented in dogs (Vandevelde and Spano, 1977), cats (Rand *et al.*, 1994a; Dow *et al.*, 1990), horses (Green *et al.*, 1992; Hamir *et al.*, 1992), sheep and goats (Brewer, 1983) and numerous other species. In people, CSF lymphocytosis has been observed in bacterial meningitis following antibiotic therapy (Fishman, 1992; Cargill, 1975; Converse *et al.*, 1973; Kjeldsberg and Knight, 1993). This is a good example of how therapy and chronicity can alter the CSF findings. A similar finding has been reported in the dog (Tipold and Jaggy, 1994; Sarfaty *et al.*, 1986) and the calf (Green and Smith, 1992).

3. *Eosinophilia*

Eosinophils are not present in normal, uncontaminated (by blood) CSF, although single eosinophils are occasionally seen on cytocentrifuge slides from animals with CSF that has normal nucleated counts and protein content. Although the presence of eosinophils in CSF is abnormal and is evidence of underlying disease, no diagnostic specificity is associated with their presence, as they can be found in a variety of diseases (Bosch and Oehmichen, 1978). Additionally, CSF eosinophilia and peripheral blood eosinophilia do not necessarily occur together, and if they do, no positive correlation exists between the magnitude of peripheral blood eosinophilia and the severity of the CSF eosinophilia (Smith-Maxie *et al.*, 1989; Bosch and Oehmichen, 1978). In one case series of eight dogs with eosinophilic meningoencephalitis, five of eight had concurrent peripheral eosinophilia, but no correlation was present between the peripheral and CSF eosinophil counts. The two dogs with the highest CSF eosinophil counts had peripheral eosinophil counts within normal reference limits. In people, central nervous system invasion by parasites, especially *Angiostrongylus cantonensis,* is the most frequent cause of eosinophilic pleocytosis, and in many of these cases eosinophils predominate in the differential cell count (Bosch and Oehmichen, 1978; Kuberski, 1979). CSF eosinophilia can also occur in association with bacterial, fungal, and viral infections and hence can be seen concurrently with suppurative, granulomatous, and lymphocytic inflammatory processes of the central nervous system (Smith-Maxie *et al.*, 1989; Jamison and Lumsden, 1988). However, in many of these cases, eosinophils represent less than 5% of the total cell count in CSF (Smith-Maxie *et al.*, 1989; Bosch and Oehmichen, 1978). Other documented causes in people include neurosyphilis, tuberculosis, rickettsial disease, foreign-body reactions to shunt tubes, intrathecal penicillin or contrast agents, hypereosinophilic syndrome, multiple sclerosis, lymphoma, Hodgkin's disease, leukemia, melanoma, disseminated glioblastoma, idiopathic disorders, and systemic allergic reactions (Fishman, 1992; Smith-Maxie *et al.*, 1989; Kuberski, 1979; Kjeldsberg and Knight, 1993). In animals, CSF pleocytosis that consists predominantly or almost exclusively of eosinophils is rare. We have personally seen CSF eosinophilia in association with idiopathic or steroid responsive eosinophilic meningoencephalitis (Smith-Maxie *et al.*, 1989) and canine neural angiostrongylosis (Mason, 1989). Pleocytosis with eosinophil predominance has also been described in central nervous system cryptococcosis (Vandevelde and Spano, 1977), although this finding is not common in our experience. Other documented causes of CSF eosinophilia (though not predominance) in animals include bacterial encephalitis, distemper, rabies, toxoplasmosis, neosporosis, cuterebral encephalitis, central nervous system nematodiasis, protothecosis, granulomatous meningoencephalomyelitis, lymphoma, astrocytoma, and cerebral infarction (Chrisman, 1992; Lester, 1992; Darien *et al.*, 1988; Vandevelde and Spano, 1977; Jamison and Lumsden, 1988).

4. Neoplastic Cells

The observation of neoplastic cells in CSF samples from animals with central nervous system neoplasia is rare in our experience. Very few veterinary studies have investigated the incidence of positive CSF cytology in confirmed cases of central nervous system neoplasia. In one study involving 77 histopathologically confirmed cases of primary central nervous system neoplasia in dogs, neoplastic cells were not observed in a single case (Bailey and Higgins, 1986b). However, in this study, cytologic assessment was performed only on those cases with an elevated cell count and these only accounted for 41.3% of cases. As stated previously, cases have been reported in the veterinary literature in which tumor cells were observed in the CSF but the CSF cell counts were within normal limits (Grevel *et al.*, 1992; Bichsel *et al.*, 1984b; Grevel and Machus, 1990). Additionally, in the study assessing primary brain tumors in 77 dogs (Bailey and Higgins, 1986b), CSF differentials and cytology were performed on cytospin slides. The cell yield with cytospin smears is very low, approximating 10% (Barrett and King, 1976) in some studies, and this may partly explain the failure to observe neoplastic cells in this study. Other veterinary studies utilizing different techniques report a higher incidence of neoplastic cell observation in the CSF from confirmed cases of central nervous system neoplasia. In two studies utilizing a Kölmel sedimentation apparatus, tumor cells were seen in the CSF of five of eight cases (Grevel and Machus, 1990) and four of nine cases (Grevel *et al.*, 1992). In the former study, two of the five cytologically positive cases had normal cell counts. In another study utilizing a membrane filter technique (Bichsel *et al.*, 1984b), three of nine cases of dogs with central nervous system neoplasia had CSF samples in which neoplastic cells were observed. Both of the techniques utilized in these studies result in a higher cell yield than cytocentrifugation, which may be partly responsible for the increased incidence of neoplastic cell observation in the CSF in these studies.

A much larger number of studies assess the incidence of CSF neoplastic cells in people with central nervous system neoplasia. Overall sensitivities that are frequently quoted are 70% for central nervous system leukemia, 20–60% for metastatic meningeal carcinoma, and approximately 30% for primary central nervous system tumors (Kjeldsberg and Knight, 1993), regardless of the technique utilized. The detection rate of malignant cells in the CSF is significantly improved by the collection of multiple samples (Olson *et al.*, 1974). These figures are supported by one study utilizing cytocentrifugation in 117 cases of histopathologically confirmed central nervous system neoplasia (Glass *et al.*, 1979). Overall, 26% (31/117) were positive. However, if only those cases with leptomeningeal involvement were considered, the incidence increased to 59%. Conversely, of 66 cases in which the tumor did not reach the leptomeninges, only a single case was positive. In another study, only 13.9% of all gliomas had a positive CSF cytology (Balhuizen *et al.*, 1978). This low incidence is likely due to the fact that the great majority of gliomas do not extend into the subarachnoid space (Balhuizen *et al.*, 1978). As a result of these studies, the following generalizations are frequently made in human medicine: (1) A positive CSF cytology is a reliable indicator of central nervous system malignancy and almost always reflects a leptomeningeal tumor (or one involving the ventricular system), and (2) a negative cytology does not exclude the presence of an intracerebral tumor, particularly a deep parenchymal mass that does not breach the pia or the ventricular system. Controlled studies are required in veterinary medicine to determine the incidence of positive CSF cytology in confirmed cases of the different types of central nervous system neoplasia and also to compare the sensitivities of different preparative methods. These studies may be hampered by the general lack of experience at identifying cells derived from central nervous system neoplasms. Tumor cells can be erroneously identified as normal ependymal or choroid plexus cells. Solitary tumor cells from metastatic carcinomas can be mistaken for lymphocytes or monocytes (Kjeldsberg and Knight, 1993). The need for the preceding types of studies has been somewhat decreased by the advent of routine access to advanced imaging techniques.

C. Protein

1. *Changes in CSF Total Protein Content*

An increase in the total protein content of CSF is the single most useful alteration in the chemical composition of the fluid (Fishman, 1992). However, this alteration accompanies many diseases and is therefore nonspecific. Increased total protein may be caused by (1) increased permeability of the blood–brain/spinal cord/CSF barriers allowing passage of serum proteins into the CSF, (2) intrathecal globulin production, and (3) interruption of CSF flow and/or absorption. Particular emphasis has been put on CSF flow rate as a major factor in CSF protein content (Reiber, 1994). In many diseases, two or all three of these mechanisms are at work. In complete spinal subarachnoid space blockage (e.g., by a compressive lesion or arachnoiditis), CSF withdrawn caudal to the block may clot when aspirated. This phenomenon is called Froin's syndrome and results from very high CSF protein levels caused by the defective flow and absorption and blood–spinal cord barrier breakdown (Fishman, 1992; Kjeldsberg and Knight, 1993).

Decreased total protein is much less common. Theoretically, low levels of CSF protein could result from decreased entry of protein into the CSF or increased removal. No evidence exists to support the first mechanism. Increased removal can occur, however, if intracranial pressure is increased while the barriers to serum protein remain normal. In this situation, bulk flow absorption of CSF is increased, while entrance of protein into the CSF remains normal. Protein content of fluid collected from the lumbosacral site could be decreased if large volumes are removed, or if ongoing leakage of CSF from the lumbar area is occurring. In these situations, lumbosacral CSF is replaced more quickly than normal by ventricular CSF, which has a lower protein content than lumbosacral CSF (Fishman, 1992; Kjeldsberg and Knight, 1993). Low CSF protein has also occurred in people with hyperthyroidism, leukemia, or water intoxication (Fishman, 1992; Kjeldsberg and Knight, 1993).

2. *Albuminocytologic Dissociation*

In many disease processes, the CSF cell count and CSF total protein increase in rough parallel. In some disorders, though, the cell count remains normal, or nearly so, whereas the total protein is notably increased, a phenomenon termed albuminocytologic dissociation. Some degenerative disorders, ischemia/infarction, immune-mediated diseases (e.g., polyradiculoneuritis), tumors, and neural compression produce albuminocytologic dissociation (Laterre, 1996).

3. *Increased CSF Albumin and Albumin Index*

Elevation of CSF albumin (which originates in the serum), and consequently an increased albumin index, is indicative of dysfunction of the blood–brain/spinal cord/CSF barriers or contamination of the CSF by blood (from intrathecal hemorrhage or traumatic spinal tap). Barrier damage occurs in most types of neurologic disorders, including inflammatory diseases, neoplasia, trauma, compression, and occasionally metabolic diseases (Sorjonen, 1987; Bichsel *et al.*, 1984b; Sorjonen *et al.*,1991; Krakowka *et al.*, 1981).

4. *Increased CSF IgG and IgG Index*

The CSF IgG can be increased by transudation of protein across damaged blood–brain/CSF barriers, by intrathecal hemorrhage (pathologic or iatrogenic), or by intrathecal IgG synthesis. An elevated CSF IgG content and increased IgG index, indicating intrathecal IgG synthesis, is typical for infectious inflammatory diseases (Tipold *et al.*, 1994, 1993b). In contrast, animals with noninflammatory diseases usually have normal IgG indices (Tipold *et al.*, 1993b). In a few animals with noninfectious disorders, mild intrathecal IgG synthesis occurs, reflecting the presence of inflammatory infiltrates around the lesion (Tipold *et al.*, 1993b). Therefore, the IgG index is often useful for distinguishing between inflammatory and noninflammatory lesions, which is not always possible on the basis of CSF cell counts alone (Bichsel *et al.*, 1984b). In one study (Tipold *et al.*, 1993b), 7 of 66 dogs with noninflammatory lesions had no pleocytosis but had an elevated IgG index; in contrast, 17 of 32 dogs with noninflammatory disease had pleocytotic CSF and a normal IgG index. The authors of this study consider an IgG index ≥ 2.8 as proof of intrathecal synthesis allowing a diagnosis of meningoencephalomyelitis, and an IgG index between 1.3 and 2.8 as suggestive of inflammatory disease. In a few dogs with intense inflammatory lesions and intrathecal IgG production, the IgG index may not be elevated because of marked IgG transudation against which the local IgG synthesis is undetectable (Bichsel *et al.*, 1984b). Traumatic puncture and red blood cell contamination of the CSF can artifactually increase the IgG index. Additionally, the normal IgG index of cerebellomedullary fluid and that of lumbar fluid are likely to be different because of the different protein concentrations of these fluids.

5. Classification of Disease Based on Albumin Index and IgG Index

Alterations of the albumin index and the IgG index can be grouped into three pathogenetic categories: (1) blood–brain/CSF barrier disturbance (increased albumin index), (2) intrathecal IgG synthesis (increased IgG index), and (3) barrier disturbance combined with intrathecal IgG production (both indices increased). These categories correlate somewhat with certain types of disease. Barrier disturbance may be seen in degenerative, inflammatory, metabolic, space-occupying, vascular, and traumatic conditions (Sorjonen, 1987; Bichsel *et al.*, 1984b; Sorjonen *et al.*, 1991). Intrathecal IgG synthesis is typical of inflammatory conditions (Tipold *et al.*, 1994), but also occurs in noninfectious disorders that have secondary inflammation, such as some tumors (Tipold *et al.*, 1993b; Bichsel *et al.*, 1984b). Barrier disturbance coupled with intrathecal IgG production is typical of infectious–inflammatory diseases (Bichsel *et al.*, 1984b).

6. Increased CSF IgM, IgA, and IgM and IgA Indices

The immunoglobulins IgM and IgA may be increased in the CSF of animals with inflammatory neurologic disease. A study of 69 dogs with inflammatory disease detected IgM elevations in 16 dogs and IgA elevations in 40 (Tipold *et al.*, 1994). An increased CSF IgM index is considered by some investigators to be a good indication of recent or persistent immunological stimulation in people (Sharief and Thompson, 1989). In contrast, Felgenhauer (1982) reported IgM was present through all stages of human herpes and bacterial meningitis, and increased and decreased with IgG. Perhaps a transition from IgM to IgG production does not occur in the central nervous system (Tipold *et al.*, 1994); or perhaps, in the presence of a normal or near-normal blood-CSF barrier, IgM accumulates in the CSF (Felgenhauer, 1982). In people with *Borrelia* infection, CSF IgM is persistently produced and the IgM index is a better indicator of this disease than is the IgG index (Fishman, 1992). Further studies need to be done in animals to determine the specificity of the various immunoglobulin alterations.

7. Electrophoretic Patterns of CSF Protein in Disease

Abnormalities in the CSF electrophoretic pattern can suggest categories of disease (Sorjonen, 1987; Sorjonen *et al.*, 1991). In one study, dogs with inflammatory diseases had one of three patterns: (1) little or no blood–brain barrier disturbance (as determined by CSF albumin concentration and the albumin quota) with decreased gamma globulin, (2) mild blood-brain barrier disturbance with markedly increased gamma globulin, and (3) moderate or marked blood–brain barrier disturbance with increased gamma globulin. Dogs with intervertebral disc protrusion or cervical spondylomyelopathy had a pattern of normal barrier function or severe barrier disturbance with decreased alpha globulin. Dogs with brain neoplasia had marked barrier disturbance and normal or mildly increased alpha and beta globulins (Sorjonen *et al.*, 1991).

In the gamma globulin region, three patterns of protein bands can occur—monoclonal, oligoclonal, and polyclonal. Oligoclonal bands are associated with disease and are seen in a high percentage of people with multiple sclerosis or encephalitis. These bands, readily identifiable against the low background of normal polyclonal IgG in the CSF, are thought to represent the products of a limited number of plasma-cell clones. Oligoclonal bands unique to CSF (i.e., not present in serum) indicate intrathecal synthesis of immunoglobulin and may be more sensitive than the IgG index in detecting this synthesis. People with multiple sclerosis may have a normal IgG index yet have CSF oligoclonal banding; thus, the demonstration of these bands is considered by some to be the single most useful test in the diagnosis of multiple sclerosis (Kjeldsberg and Knight, 1993). Oligoclonal bands are also seen in patients with inflammatory diseases and in some patients with neoplasia (Fishman, 1992). Occasionally, a single (monoclonal) band is identified in the CSF electrophoretic pattern of people. Monoclonal bands have been seen in neurologically normal people as well as patients with neurologic disease (Kjeldsberg and Knight, 1993).

8. Other CSF Proteins

Numerous attempts have been made to correlate specific CSF proteins, particularly "brain-specific" proteins, with specific diseases. Proteins such as C-reactive protein, interferon, myelin basic protein, and S-100 are increased in the CSF associated with neurologic disease, but these increases are found in many heterogeneous conditions. This nonspecificity limits the clinical usefulness of these protein assays. However, the measurement of some of these proteins is thought to be useful as a screening procedure for neurologic disease or as an indication of prognosis (Fishman, 1992; Kjeldsberg and Knight, 1993; Lowenthal *et al.*, 1984). The clinical usefulness of these assays needs further investigation.

9. *Plasma Proteins in the CSF*

Alterations in plasma proteins may be reflected in the CSF. For example, the serum protein monoclonal gammopathy of multiple myeloma may be evident in the CSF. Bence–Jones proteins are also readily seen in the CSF. The high-molecular-weight paraproteins do not cross the normal blood–brain barrier, however. Serum protein electrophoresis is indicated in patients with elevated CSF globulins to clarify the source of the globulins (Fishman, 1992).

D. Antibody Titers

The CSF antibody titer can be measured for a number of diseases (Green *et al.*, 1993; Greene, 1990). Interpretation of the results is confounded by the need to differentiate among titers caused by vaccination, exposure to the antigen without development of the disease, and actual disease. Interpretation of CSF antibody titers could be aided by an accurate vaccination history, comparison of CSF and serum titers, assessment of blood–brain/CSF barrier function and intrathecal immunoglobulin production by determination of albumin and immunoglobulin indices, determination of CSF IgM levels, and analysis of acute and convalescent samples (Chrisman, 1992; Green *et al.*, 1993).

E. Glucose

Increased CSF glucose usually reflects hyperglycemia. Decreased CSF glucose occurs with several disorders of the nervous system, particularly acute, bacterial, fungal, amebic, or tuberculous meningitis. In people, low CSF glucose is also characteristic of diffuse carcinomatous meningitis, meningeal cysticercosis or trichinosis, and syphilitic meningitis. The major factors responsible for low CSF glucose levels are inhibition of the entry of glucose due to alteration of membrane glucose transport and increased anaerobic glycolysis by neural tissue. As noted earlier, hyperglycemia elevates the CSF glucose, which may mask a decreased CSF level. Therefore, calculation of a CSF/serum glucose ratio has been recommended to identify pathologically low CSF glucose level (Kjeldsberg and Knight, 1993). A low CSF glucose level in the absence of hypoglycemia indicates a diffuse, meningeal disorder, rather than focal disease (Fishman, 1992). Decreased CSF glucose classically has been associated with bacterial meningitis, but many human patients with bacterial meningitis have normal CSF glucose levels. Therefore, the recommendation has been made that CSF glucose need be measured only if the opening CSF pressure, cell count, cytospin differential, and protein are inconclusive (Hayward *et al.*, 1987).

F. Enzymes

Numerous enzymes have been assayed in the CSF of animals (Rand *et al.*, 1994b; Furr and Tyler, 1990; Wilson, 1977; Jackson *et al.*, 1996). Of these, creatine kinase has received the most attention and opinions of its usefulness are somewhat conflicting. Although Furr and Tyler (1990) confirmed previous observations that CSF creatine kinase activity was elevated in several neurologic diseases, they concluded that the greater frequency of elevation in the CSF of horses with protozoal myelitis vs horses with cervical compressive myelopathy indicated this enzyme assay was useful in differentiating these two diseases. This conclusion was disputed by Jackson *et al.* (1996), who did not find the sensitivity or specificity of creatine kinase measurement sufficient for diagnosis of a specific disease. Jackson *et al.* (1996) also concluded that contamination of the CSF sample with epidural fat or dura mater may contribute to previously unexplained elevations in CSF creatine kinase activity. Their conclusion regarding this enzyme's lack of sensitivity and specificity reflects the current situation with all of the enzymes in CSF studied to date—none have sufficient specificity to warrant their routine use as diagnostic tests (Rand *et al.*, 1994b; Fishman, 1992; Kjeldsberg and Knight, 1993; Jackson *et al.*, 1996; Indrieri *et al.*, 1980). The site of CSF collection with respect to the location of the lesion may be responsible for some of the lack of diagnostic significance in CSF enzyme analysis. Cerebellomedullary fluid may be less affected than lumbar fluid in animals with spinal disease (Indrieri *et al.*, 1980). Measurement of enzyme isomers may increase the specificity (Kjeldsberg and Knight, 1993).

G. Other Constituents

1. *Interferon*

Interferon is increased in the CSF in a large percentage of people with viral encephalitis–meningitis. This finding is not specific, however, as increases are also found in patients with bacterial meningitis (Glimaker *et al.*, 1994) or multiple sclerosis, and occasionally in patients with noninflammatory neurologic disease (Brooks *et al.*, 1983). In an experimental study of canine distemper, interferon appeared to be a valid marker for persistence of the virus in the central nervous system (Tsai *et al.*, 1982). Studies on the clinical application of CSF interferon assay would be interesting.

2. *Neurotransmitters*

γ-Aminobutyric acid (GABA) is a major inhibitory neurotransmitter, whose dysfunction has been suggested to play a role in experimental (Griffith *et al.*, 1991) and clinical seizure disorders. A study of epileptic dogs found the average CSF concentration of GABA to be significantly reduced, a situation similar to that in people (Loscher and Schwartz-Porsche, 1986). Increased CSF levels of the biogenic amine neurotransmitter metabolites homovanillic acid and 5-hydroxyindoleacetic acid were found in 2 of 10 collies experimentally given invermectin (Vaughn *et al.*, 1989a). Both of these collies had severe neurologic deficits. Neurotransmitter metabolite concentrations were also elevated in the CSF of goats demonstrating neurologic abnormalities after experimental boron toxicosis (Sisk *et al.*, 1990). Significant differences in neurotransmitter concentrations were also found between the CSF of normal dogs and that of narcoleptic dogs (Faull *et al.*, 1982). The use of neurotransmitter assays in clinical diagnosis needs more investigation.

3. *Quinolinic Acid*

Quinolinic acid is a neuroexcitotoxic metabolite of L-tryptophan and an agonist of *N*-methyl-D-aspartate receptors. Increased levels have been found in people with a varitey of neurological diseases, including AIDS (Heyes *et al.*, 1992), and in macaques infected with simian immunodeficiency virus (Smith, 1995). Quinolinic acid levels may be elevated in CSF of animals with inflammatory nervous system disease. Therefore, it may be useful as a marker of inflammation and perhaps also as an indicator of prognosis (Smith, 1995).

4. *Lactic Acid*

In human medicine, the measurement of CSF lactic acid has been advocated in differentiating bacterial from viral meningitis. The concentration of CSF lactic acid is elevated in conditions causing severe or global brain ischemia and anaerobic glycolysis. Therefore, many diseases may elevate CSF lactic acid, and the overlapping CSF lactate levels limit the value of CSF lactate assay (Fishman, 1992; Kjeldsberg and Knight, 1993). A study of CSF lactic acid levels in horses with neurologic disease found elevated lactate levels in several types of central nervous system diseases (Green and Green, 1990). Therefore, as with people, increased CSF lactic acid in the horse appears to be a nonspecific indicator of central nervous system disease. Interestingly, in the horses studied, elevated lactic acid was the only CSF abnormality associated with brain abscess (Green and Green, 1990).

5. *3-OH Butyrate*

The measurement of serum 3-OH butyrate concentration is useful in the feeding management of pregnant ewes and in the diagnosis of pregnancy toxemia. Following death, however, rapid autolytic change renders serum biochemical analysis useless. Scott *et al.* (1995) compared the 3-OH butyrate concentrations of serum collected antemortem and aqueous humor and CSF collected within 6 hours of death. Their results indicated that either fluid was suitable for postmortem determination of 3-OH butyrate levels, and that such data could be extrapolated to indicate antemortem serum 3-OH butyrate concentration and the possibility that pregnancy toxemia contributed to the death of the animal.

VII. CHARACTERISTICS OF CSF ASSOCIATED WITH SPECIFIC DISEASES

A. Degenerative Disorders

This group of disorders includes a variety of diseases such as the inherited, breed-specific polyradiculoneuropathies, myelopathies, and encephalopathies; motor neuron diseases; and cerebellar abiotrophies. The storage diseases can also be included in this group. The inclusion of canine degenerative myelopathy is arguable, awaiting further clarification of its pathogenesis. The CSF in animals with degenerative disorders is characteristically normal, reflecting the lack of inflammation in the disease process (Braund, 1994; Oliver and Lorenz, 1993). A mild to moderate increase of CSF total protein may occur in several of these disorders, however. Increased total protein is also found in people with motor neuron disease, Parkinson's disease, and various hereditary neuropathies and myelopathies. The mechanism of the protein increase is unknown. Electrophoretic studies of CSF associated with some human neurodegenerative disorders have shown a transudative pattern. Intrathecal immunoglobulin production has also been found in people with motor neuron disease (Fishman, 1992). In storage diseases such as globoid-cell leukodystrophy, accumulations of metabolic material may be seen in CSF cells (Roszel, 1972).

1. *Canine Degenerative Myelopathy*

Although the CSF of dogs with degenerative myelopathy may be normal, a mild elevation of the white blood cell count is present occasionally (Bichsel *et al.*, 1984b). More common is a normal cell count coupled with a mild to moderate elevation of total protein (approximately 40–70 mg/dl). This albuminocytologic dissociation may support the theory that this disorder is an immune-mediated disease (Waxman *et al.*, 1980). However, the common existence of concurrent, chronic, spinal cord compression by Type II disc protrusion in these dogs complicates the situation, because chronic cord compression may also produce an increase in total protein. The elevated total protein in canine degenerative myelopathy is probably due to increased CSF albumin (Bichsel *et al.*, 1984b). The CSF IgG index is usually normal (Tipold *et al.*, 1993b; Bichsel *et al.*, 1984b), indicating a lack of intrathecal IgG production.

2. *Degenerative Myeloencephalopathy of Llamas*

A degenerative myeloencephalopathy has been identified in two adult llamas. Lesions consist of bilateral white-matter degeneration in all spinal-cord segments, and degenerate neurons in the brain-stem nuclei or degeneration of brain-stem white-matter tracts. Inflammation is not evident. Lumbosacral CSF from both animals was normal (Morin *et al.*, 1994).

3. *Equine Motor Neuron Disease*

The CSF of horses with this disorder is either normal or has albuminocytologic dissociation (Divers *et al.*, 1994; Cummings *et al.*, 1990). In a study of 28 cases (Divers *et al.*, 1994), 9 of 26 horses had elevated CSF protein. The albumin quotient was abnormal in only 2 of 19 horses. The IgG index was increased in 8 of 16 horses. The abnormalities in total protein and IgG index did not appear to be associated with the duration or severity of clinical signs. The increased protein and IgG index in these horses suggest that intrathecal production occurs. Blood–brain barrier damage and intrathecal IgG production also occur in people with motor neuron disease (Apostolski *et al.*, 1991).

B. Idiopathic Disorders

1. *Granulomatous Meningoencephalomyelitis*

The CSF associated with granulomatous meningoencephalomyelitis (GME) is usually abnormal. The fluid may be clear or hazy and is generally colorless. The total white blood cell count is moderately to markedly elevated, as is the total protein. The white blood cell differential is variable, but typically lymphocytes predominate, with monocytes / macrophages and neutrophils making up the remainder in about equal percentages (Bailey and Higgins, 1986a; Braund, 1994; Tipold, 1995; Thomas and Eger, 1989; Sarfaty *et al.*, 1986). A 15–30% neutrophilic component is suggestive of GME, but the white blood cell differential can range from 95% neutrophils (Sorjonen, 1990) to 100% mononuclear cells. Plasma cells, cells undergoing mitosis, and large, mononuclear cells with abundant foamy cytoplasm are occasionally present (Bailey and Higgins, 1986a; Braund, 1994). Lumbar fluid is also abnormal, although it generally has fewer cells and less protein than cerebellomedullary fluid (Bailey and Higgins, 1986a). Electrophoresis of CSF suggests that blood–brain barrier dysfunction is present in the acute stage of disease; intrathecal IgG production with resolution of the barrier dysfunction occurs in chronic disease (Sorjonen, 1990). The albumin quota is elevated (Sorjonen, 1987), and the IgG index is usually elevated (Tipold *et al.*, 1994, 1993b; Bichsel *et al.*, 1984b). If barrier dysfunction is severe, with marked transudation of protein, the IgG index may be normal because the amount of intrathecally produced IgG is small in comparison to the amount of transudated serum IgG (Fishman, 1992; Bichsel *et al.*, 1984b).

2. *Necrotizing Encephalitis of Pug Dogs, Maltese Dogs, and Yorkshire Terriers*

A necrotizing encephalitis of unknown cause has been identified in Pug Dogs (deLahunta, 1983; Cordy and Holliday, 1989), Maltese dogs (Stalis *et al.*, 1995), and Yorkshire terriers (Tipold *et al.*, 1993a). The lesions are similar in each breed, although the distribution of lesions in the Pug and Maltese dogs (large, diffuse, cerebral) is different than in the Yorkshire terriers (well-defined multifocal, brain stem). The CSF associated with the Pug and Maltese-dog diseases has a moderate to marked, predominantly lymphocytic white blood cell count (although one Maltese had 62% neutrophils) and moderate to marked elevation in total protein (Stalis *et al.*, 1995; deLahunta, 1983; Cordy and Holliday, 1989; Bradley, 1991). The CSF of the Yorkshire terriers has mild to moderate increases of white blood cell and protein, with a predominantly mononuclear differential count (Tipold, 1995; Tipold *et al.*, 1993a). Seizures are a consistent clinical sign for the Pugs and the Maltese dogs, but not the Yorkshire terriers. The relationship of the seizures to the more severe CSF abnormalities of the Pug and Maltese dogs compared to those of the Yorkshire terriers is interesting to consider.

C. Immune-Mediated Diseases

1. *Acute Idiopathic Polyradiculoneuritis/ Coonhound Paralysis*

Acute idiopathic polyradiculoneuritis is one of the most common canine polyneuropathies, and coonhound paralysis is the most common form. The disorder resembles Guillain–Barré syndrome of people. In affected dogs, the classical CSF abnormality is albuminocytologic dissociation without pleocytosis. The abnormality is more obvious in lumbosacral CSF than in cerebellomedullary CSF (Cuddon, 1990; Cummings *et al.*, 1982). The CSF IgG level and IgG index may also be increased, indicating intrathecal immunoglobulin production (Tipold *et al.*, 1993b; Cuddon, 1990).

2. *Equine Cauda Equina Neuritis*

This disease is thought to be an autoimmune polyneuritis. The CSF of affected horses may be xanthochromic and typically has a prominent, usually lymphocytic pleocytosis (at least in the chronic stage) and moderately elevated protein. The CSF can also be normal (Mayhew, 1989; Yvorchuk, 1992).

3. *Steroid-Responsive Meningitis/Arteritis*

Steroid-responsive meningitis/arteritis is a common, suppurative meningitis of dogs. The CSF has a marked, often extreme, neutrophilic pleocytosis and moderately to markedly increased protein. Occasionally, a single sample collected early in the disease is normal (deLahunta, 1983; Tipold, 1995; Meric, 1988). The IgG index is typically elevated (Tipold *et al.*, 1994, 1993b), and IgM and IgA levels are often elevated as well (Tipold *et al.*, 1995, 1994). Microbial cultures are negative. In protracted or inadequately treated cases, the pleocytosis is mild to moderate with a mixed population or even a mononuclear cell predominance; the protein level may be normal or slightly elevated. The CSF may even be normal (Tipold and Jaggy, 1994). A polyarteritis/vasculitis reported in beagles, Bernese mountain dogs, German short-haired pointers, and sporadically in other breeds (Meric, 1988) has similar CSF abnormalities and pathologic changes and may be the same disease as steroid-responsive meningitis/arteritis (Tipold and Jaggy, 1994).

D. Infectious Diseases

The variety of CSF abnormalities associated with infectious disease reflects the variety of infectious diseases affecting the central nervous system. If the infection causes inflammation, the total white blood cell count and protein usually will be elevated, but the degree and type of abnormality depend on the infectious agent, the immune status of the animal, the location of the infectious process (e.g., surface-related vs parenchymal), the duration of the infection, and previous treatment. The general rules of inflammation due to infection apply, that is, bacterial infections result in suppurative inflammation, whereas viral infections result in nonsuppurative inflammation. Several important exceptions exist, however.

1. *Bacterial Diseases*

The CSF associated with aerobic or anaerobic bacterial infections of the central nervous system may be clear, hazy or turbid (depending on the cell count), and colorless or amber with moderate to marked elevations of total white blood cell count and total protein. Because of the elevated protein, the CSF may clot or foam when shaken. The white blood cell differential count has a characteristically high percentage of neutrophils (>75%), which may be degenerate (Rand *et al.*, 1994a; Green and Smith, 1992; Scott, 1995; Foreman and Santschi, 1989; Santschi and Foreman, 1989; Dow *et al.*, 1988; Tipold, 1995; Kornegay, 1981; Meric, 1988; Baum, 1994). The protein is composed of albumin that has crossed the diseased blood–brain/CSF barrier and immunoglobulin produced intrathecally; therefore, the IgG index is usually elevated (Tipold *et al.*, 1994, 1993b). The IgM and IgA levels may be normal or increased (Tipold *et al.*, 1994). The CSF of animals with chronic or treated bacterial infections may be nonsuppurative with mild to moderate elevations of total white blood cell count and total protein (Green and Smith, 1992). Occasionally, extracellular or intracellular bacteria may be seen on a Gram stain (Green and Smith, 1992; Foreman and Santschi, 1989; Kornegay, 1981). Because some bacteria undergo rapid autolysis in the test tube, bacterial culture of these infections is often unrewarding. Nonetheless, culture should be attempted. Polymerase chain reaction techniques may be used to detect the presence of bacterial DNA (Peters *et al.*, 1995).

a. Listeriosis

The CSF of cattle with meningoencephalitis caused by *Listeria monocytogenes* typically has mild to moderate elevations in total white blood cell count and total protein, with the white cells being mostly mononuclear cells, in spite of the fact that it is a bacterial infection (Rebhun and deLahunta, 1982). These mild changes probably reflect the characteristic lesions of this disease, which are mononuclear vascular cuffing and parenchymal microabscesses. The disease in sheep may produce a CSF similar to that of infected cattle (Scarratt, 1987). However, two studies reported ovine CSF to have moderate to marked elevations in white blood cell count and protein, with a neutrophilic pleocytosis (53 to 100% neutrophils) (Scott, 1993, 1992). Perhaps the mononuclear CSF reported in cattle reflects a more chronic stage or resolution of the disease (Green and Smith, 1992; Kjeldsberg and Knight, 1993). A study of bacterial culture and polymerase chain reaction (PCR) for the detection of *L. monocytogenes* in the CSF of 14 infected ruminants yielded no positive cultures and only one positive PCR. The authors concluded that *L. monocytogenes* only occasionally gains access to the meningoventricular system in the course of the disease, and that reliable, *in vivo* diagnosis of listeric encephalitis generally cannot be based on the detection of the organism in the CSF (Peters *et al.*, 1995).

b. Neuroborreliosis (Lyme Disease)

Although neuroborreliosis caused by the Lyme disease spirochete, *Borrelia burgdorferi*, has been suspected in dogs (Mandel *et al.*, 1993; Feder *et al.*, 1991) and horses (Hahn *et al.*, 1996; Burgess and Mattison, 1987), the actual incidence in animals is unknown, largely because of diagnostic difficulties. These difficulties arise from a delay or repression of seroconversion after infection; the high number of seropositive, clinically normal animals; the persistence of infection and seropositivity despite resolution of clinical disease; antibody cross reactivity; and difficulty in culturing the organism from tissue or fluid samples (Appel *et al.*, 1993; Madigan, 1993; Parker and White, 1992; Levy *et al.*, 1993). The CSF associated with neuroborreliosis in animals has not been characterized. In people, CSF abnormalities are related to the stage of the disease. When present, typical abnormalities are a mononuclear pleocytosis (T lymphocytes, plasma cells, and IgM-positive B cells (Sindern and Malin, 1995)) with a moderately elevated total protein and normal or decreased CSF glucose (Fishman, 1992). Persistent CSF oligoclonal bands and intrathecal synthesis of IgG, IgM, and IgA occur (Henriksson *et al.*, 1986). Diagnosis is enhanced by the determination of intrathecal synthesis of specific *B. burgdorferi* antibodies (Kaiser and Lucking, 1993), but cross reactivity is a problem (Fishman, 1992). *Borrelia burgdorferi* antibodies have also been detected in the CSF of dogs (Mandel *et al.*, 1993; Feder *et al.*, 1991). PCR techniques for CSF have been recently developed, but the diagnostic success rate is variable (Lebech, 1994). The CSF of a horse was reported PCR positive for *B. burgdorferi* (Hahn *et al.*, 1996).

c. Rickettsial Diseases

Rickettsial diseases such as ehrlichiosis, usually caused by *Ehrlichia canis*, and rocky mountain spotted fever, caused by *Rickettsia rickettsii*, sporadically involve the central nervous system of animals. In dogs with neural ehrlichiosis, the CSF resembles that of viral diseases, that is, the white blood cell count and protein may be normal or slightly to moderately elevated with a predominantly mononuclear pleocytosis (Maretzki *et al.*,1994; Greene *et al.*, 1985; Buoro *et al.*, 1990; Meinkoth *et al.*, 1989; Firneisz *et al.*, 1990). The albumin quotient is reported to be elevated (Sorjonen *et al.*, 1991). Occasionally, *Ehrlichia* morulae may be observed in CSF mononuclear cells or neutrophils (Maretzki *et al.*, 1994). The few reports of CSF associated with Rocky Mountain spotted fever suggest a difference from ehrlichiosis in that the CSF pleocytosis of Rocky Mountain spotted fever may be predominantly neutrophilic, particularly early in the disease (Breitschwerdt, 1995; Breitschwerdt *et al.*, 1985; Greene *et al.*, 1985; Rutgers *et al.*, 1985). A predominantly neutrophilic pleocytosis has also been reported in dogs experimentally infected with *R. rickettsii* (Breitschwerdt *et al.*, 1990). In this same study, IgG or IgM antibodies were not detected in the CSF of experimentally infected dogs, but were detected in the CSF of one naturally infected dog that also had a high serum titer (Breitschwerdt *et al.*, 1990).

d. Thromboembolic Meningoencephalitis

In cattle, *Hemophilus somnus* causes bacteremia and thromboembolism, with some preference for neural tissue. The vascular lesion results in multifocal hemorrhages. Consequently, the CSF is characteristically yellow with a high red blood cell count (not iatrogenic in origin), and moderately to markedly increased white blood cell count (predominantly neutrophils) and protein (George, 1996; Mayhew, 1989; Ames, 1987; Little and Sorensen, 1969). The bacterium can be cultured only occasionally from CSF, and more easily from septicemic animals (Little, 1986; Nayer *et al.*, 1977).

2. *Viral Diseases*

The CSF associated with viral diseases is characterized by nonsuppurative inflammatory changes. The total white blood cell count and total protein are generally mildly to moderately elevated. The white cell population may be mixed with a majority of mononuclear cells or may be entirely mononuclear cells. Occasionally, neutrophils predominate, particularly in the early stages of disease or in certain diseases (see later discussion). The IgG index is commonly elevated (Tipold *et al.*, 1993b; Bichsel *et al.*, 1984b). The IgA and IgM levels may also be elevated. The CSF of viral infections may also be normal, particularly if the meninges or ependyma are not involved (Rand *et al.*, 1994a; Tipold *et al.*, 1994; Fishman, 1992; Fankhauser, 1962; Tipold, 1995).

a. Canine Distemper

The CSF abnormalities associated with canine distemper vary strikingly with the stage of the disease. Dogs with acute, demyelinating, noninflammatory distemper encephalitis may have normal or near-normal CSF (mild elevations of total cell count and total protein) (Tipold, 1995; Johnson *et al.*, 1988). Protein elevation is most likely due to blood–brain barrier dysfunction (Bichsel *et al.*, 1984b). The IgG index may also be normal or occasionally mildly elevated, which correlates with the histological findings of multifocal demyelination with very little or no infiltration of inflammatory cells (Tipold *et al.*, 1994, 1993b; Bichsel *et al.*, 1984b; Vandevelde *et al.*, 1986; Johnson *et al.*, 1988). The acute form of nervous canine distemper is an exception to the usual association of an elevated IgG index with infectious neurologic diseases because infiltration with inflammatory cells occurs only in the chronic stage of distemper encephalitis (Vandevelde *et al.*, 1986). The CSF IgM and IgA content is also usually normal (Tipold *et al.*, 1994; Johnson *et al.*, 1988). The CSF of subacute/chronic, inflammatory distemper usually has a moderately elevated total white blood cell count, primarily mononuclear, and moderately elevated protein (Bichsel *et al.*, 1984b; Tipold, 1995). The IgG index is typically elevated (Bichsel *et al.*, 1984b; Vandevelde *et al.*, 1986) and IgA levels are commonly increased. Interestingly, IgM levels are increased more often in the chronic stage than in the dogs with acute, noninflammatory distemper (Tipold *et al.*,1994, 1993b). The IgM and IgA are presumably of intrathecal origin (Tipold *et al.*, 1994), although blood–brain barrier dysfunction is also present in some dogs and therefore protein could be of serum origin (Sorjonen, 1987; Bichsel *et al.*, 1984b; Sorjonen *et al.*, 1991). Occasionally the CSF is normal or has only mild changes in cell count or total protein content (Tipold et al., 1994, 1993b; Bichsel *et al.*, 1984b; Sorjonen *et al.*, 1991; Tipold, 1995; Vandevelde *et al.*,1986). Antimyelin antibody and antiviral antibody have also been identified in the CSF of inflammatory distemper (Vandevelde *et al.*, 1986). Canine distemper virus antibody is normally absent from CSF; when present it is diagnostic of infection. False-positive results can occur, however, if the CSF is contaminated by serum distemper virus antibody by either iatrogenic or pathologic blood–brain barrier disturbance. The CSF of delayed-onset canine distemper (also known as old-dog encephalitis) has an elevated protein and nonsuppurative, inflammatory cytology. The IgG index is elevated, and much of the CSF IgG is virus specific, suggesting an intrathecal anti-viral immune response. The IgM and IgA concentrations are normal (Johnson *et al.*,1988).

b. Equine Herpesvirus Myeloencephalitis

With its predilection for endothelial cells, the equid herpesvirus 1 (EHV-1) may cause vasculitis and perivascular hemorrhage in the brain and spinal cord. As a result, the CSF is often xanthochromic. The total white blood cell count may be normal while the total protein level is moderately to markedly elevated (albuminocytologic dissociation). The CSF/serum albumin ratio is increased (Klingeborn *et al.*, 1983). In some cases the total protein is normal, perhaps because the CSF is analyzed very early in the course of the disease before the protein level has risen, or late in the disease after the level has subsided (Kohn and Fenner, 1987). Antibodies to the virus may be identified in the CSF (Blythe *et al.*, 1985; Klingeborn *et al.*, 1983; Jackson *et al.*, 1977). Antiviral CSF antibodies are not present routinely in neurologically normal horses, horses vaccinated with modified live EHV-1, or horses with other neurologic diseases (Blythe *et al.*, 1985). However, because of destruction of the blood–brain barrier, serum antiviral antibodies may pass into the CSF and confound the interpretation of the CSF titers (Blythe *et al.*, 1985; Klingeborn *et al.*, 1983; Kohn and Fenner, 1987; Jackson *et al.*, 1977). Determining the CSF IgG index may help to assess the relevance of a positive CSF EHV-1 titer.

c. Feline Infectious Peritonitis

The feline infectious peritonitis (FIP) coronavirus may cause a multifocal, pyogranulomatous meningitis, choroid plexitis, and ependymitis characterized by perivascular granulomas around small blood vessels. The CSF associated with these lesions consistently has a moderate to marked elevation of white blood cells and protein. In one study, the CSF of cats with FIP was distinctive compared to that of cats with other inflammatory central nervous system diseases in having greater than 200 mg/dl total protein (Rand *et al.*, 1994a). Despite the viral nature of the disease, the white cell population is dominated by neutrophils, commonly greater than 70% (Rand *et al.*, 1994a; Kline *et al.*, 1994; Baroni and Heinold, 1995). Prolonged glucocorticosteroid therapy may result in a normal CSF in rare instances; the authors have observed this on at least one occasion.

d. Feline Immunodeficiency Virus

The CSF associated with feline immunodeficiency virus (FIV) neurologic disease typically has a mild, primarily lymphocytic, pleocytosis (Dow *et al.*, 1990; Phillips *et al.*, 1994). In experimentally infected cats, the pleocytosis appears related to the duration, and perhaps route, of infection, as well as to the age of the cat. In one study, pleocytosis appeared within 2–8 weeks of inoculation of adult cats, then disappeared by 20 weeks (Dow *et al.*, 1990). In a study of kittens, the total and differential cell counts were normal at 3 and 12–16 months postinoculation (Podell *et al.*, 1993). The total protein content is typically normal, although the albumin quotient and IgG index may be elevated (Dow *et al.*, 1990; Podell *et al.*, 1993). Antibodies to the virus may be detected in the CSF, and their presence in CSF that has not been contaminated by peripheral blood is presumptive evidence of FIV neural infection

(Dow *et al.*, 1990; Phillips *et al.*, 1994). In experimentally inoculated cats, FIV antibodies developed in the CSF 4–8 weeks after the appearance of CSF pleocytosis (Dow *et al.*, 1990). The virus can be recovered from the CSF of most cats that have intrathecal antibodies (Dow *et al.*, 1990; Phillips *et al.*, 1994). In the immunodeficient, chronic stage of FIV infection, the effect of possible opportunistic neural infections on CSF must be considered.

e. Rabies

Because rabies is an overwhelmingly fatal, zoonotic disease, there is a paucity of information regarding its CSF abnormalities. In people, the total white blood cell count is normal or has a mild, lymphocytic pleocytosis, and total protein is mildly increased. Occasionally, the pleocytosis is marked (Fishman, 1992). The CSF of animals with rabies may be normal or abnormal. Typical abnormalities include a mild to moderate mononuclear pleocytosis and mild to moderate elevations in total protein. The white cells may be predominantly lymphocytes, with macrophages, neutrophils, and occasionally plasma cells and eosinophils (Braund, 1994; Green *et al.*, 1992; Hanlon *et al.*, 1989; Coles, 1980; Hamir *et al.*, 1992). A neutrophilic pleocytosis reported for one horse was thought to reflect an early stage of the disease (Green, 1993). Xanthochromia was detected in the CSF of 3 of 5 horses in one study (Green *et al.*, 1992), perhaps due to antemortem head trauma. The CSF IgM titer increases in 2–3 weeks or more after the onset of clinical rabies (Murphy *et al.*, 1980). Because of this delay, a negative titer result does not eliminate rabies infection as a possibility (Greene and Dreesen, 1990). Infective virus may be isolated from the CSF before clinical signs of the disease appear, and neutralizing antibodies in the CSF may not be identified until after clinical signs develop (Fekadu and Shaddock, 1984). Because of the human health hazard, CSF collection should be avoided if rabies is suspected.

3. Fungal Diseases

Fungal infection of nervous tissue is relatively uncommon, although *Cryptococcus neoformans* has a predilection for the central nervous system. The CSF associated with neural cryptococcosis is quite variable. The total white blood cell count can be near normal or markedly increased. The white blood cell differential count is typically mixed with a majority of neutrophils (Berthelin *et al.*, 1994; Steckel *et al.*, 1982). However, mononuclear CSF has been reported (deLahunta, 1983; Berthelin *et al.*, 1994), as has eosinophilic fluid (deLahunta, 1983; Vandevelde and Spano, 1977). The total protein is typically elevated, although sometimes only marginally so. The albumin quotient and IgG index are mildly to markedly elevated (Sorjonen *et al.*, 1991). Cryptococcal organisms are commonly seen in the CSF (93% in one report/review) (Berthelin *et al.*, 1994) and cultures are often, but not invariably, positive. Latex agglutination for cryptococcal antigen in the CSF may also be positive, but experience with this test with CSF is limited (Berthelin *et al.*, 1994).

The literature contains only a few reports of the CSF abnormalities associated with neural aspergillosis, blastomycosis, coccidioidomycosis, or histoplasmosis. Again, the CSF abnormalities are variable; a mixed pleocytosis and elevated protein is typical (Vandevelde and Spano, 1977; Mullaney *et al.*, 1983; Coates, 1995; Gelatt *et al.*, 1991; Nafe *et al.*, 1983; Schaer *et al.*, 1983; Kornegay, 1981). In a reported case of aspergillosis of the brain of a dog, the CSF had a normal total nucleated cell count (differential count was not done) and a normal protein (Parker and Cunningham, 1971).

4. Prion Disorders

The transmissible spongiform encephalopathies are a group of neurodegenerative diseases of people and animals caused by prions (proteinaceous infectious particles). The diseases in this group include kuru and Creutzfeldt–Jakob disease of people, bovine spongiform encephalopathy, scrapie of sheep and goats, transmissible mink encephalopathy, and spongiform encephalopathies in deer, captive ungulates, and domestic cats (Schreuder, 1994a,b). The CSF associated with the spongiform encephalopathies in animals has normal cytology, protein content, and electrophoretic pattern. Thus, these diseases, although apparently infectious, do not appear to damage the blood–brain barrier or elicit an immune response in the central nervous system (Scott et al., 1990; Lowenthal and Karcher, 1994; Millson *et al.*, 1960; Strain *et al.*, 1984).

E. Ischemic Disorders

In general, neural ischemia causes blood–brain/CSF barrier dysfunction resulting in increased CSF protein. If infarction occurs, the tissue destruction and cellular response may result in CSF pleocytosis. With extensive, particularly acute, infarction the pleocytosis may be substantially neutrophilic (Fishman, 1992). In animals, CSF abnormalities have been reported for fibrocartilaginous embolism and cerebral ischemia/infarction.

1. Fibrocartilaginous Embolism

The characteristics of CSF associated with fibrocartilaginous embolism have been reported for dogs and horses. About one-third of the reported canine cases have normal CSF, about one-third have pleocytosis, usually mononuclear, and increased protein, and about one-third have albuminocytologic dissociation (Cauzinille and Kornegay, 1996; Bichsel *et al.*, 1984a). Similar findings have been reported for two horses (Jackson *et al.*, 1995; Taylor *et al.*, 1977). The type of pleocytosis, neutrophilic or mononuclear, probably depends on the size, location, and age of the infarct. The CSF albumin is reported to be normal, and the IgG index either normal or slightly elevated, the latter reflecting the inflammation around the lesion (Tipold *et al.*, 1993b; Bichsel *et al.*, 1984b). Normal CSF may be associated with a better prognosis for recovery (Cauzinille and Kornegay, 1996).

2. *Cerebral Ischemia/Infarction*

Cerebral infarction (ischemic encephalopathy) has been reported primarily in cats, but also in a few dogs. In cats within the first week of onset, the white blood cell count is normal or mildly elevated with a mixed, predominantly mononuclear, pleocytosis, and protein is mildly to markedly elevated (Rand *et al.*, 1994b; de-Lahunta, 1983). Dogs with cerebral infarction have similar CSF characteristics (Bichsel *et al.*, 1984b; deLahunta, 1983; Vandevelde and Spano, 1977; Joseph *et al.*, 1988), although two dogs in one report had a mixed, but predominantly neutrophilic, pleocytosis and normal protein (Vandevelde and Spano, 1977). The neutrophilic pleocytosis reflected the acute encephalomalacia noted on histopathologic examination (Vandevelde and Spano, 1977). In one dog of another report, the CSF albumin and IgG index were normal (Bichsel *et al.*, 1984b). Another dog with a deep, parenchymal, hemorrhagic infarct in the basal nuclear region had normal CSF (Norton, 1992).

F. Malformations of Neural Structures

Although reports with CSF analysis are relatively few, the CSF of animals with neural malformations is generally normal (Rand *et al.*, 1994b; Vandevelde and Spano, 1977; Milner *et al.*, 1996; Meric, 1992b; Shell *et al.*, 1988; Greene *et al.*, 1976; Wilson *et al.*, 1979). However, if the malformation interferes with CSF circulation or absorption, abnormalities in protein and even cell count may be present. The CSF may also be altered by secondary or additional unrelated processes (Rishniw *et al.*, 1994). For example, intraventricular hemorrhage can occur in hydrocephalic animals, producing xanthochromic CSF with an increased white blood cell count and protein content.

1. *Intracranial Intra-arachnoid Cysts*

Intracranial intra-arachnoid cysts have been identified in six dogs: two Lhasa apsos, two pugs, one shih tzu, and one German short-haired pointer (Vernau *et al.*, 1996). Analysis of cerebellomedullary CSF was performed in three of the dogs. In two, the CSF was normal (clear, colorless, <2 WBC/μl , <25 mg/dl total protein). The CSF of the third dog had a moderate, lymphocytic pleocytosis and moderately increased total protein, suggesting a nonsuppurative, inflammatory process. Necropsy of this dog identified a nonsuppurative, necrotizing meningoencephalitis unrelated to its cyst (Vernau *et al.*, 1996).

G. Metabolic/Nutritional Disorders

Cerebrospinal fluid analysis is not done commonly in animals with metabolic or nutritional neurologic disorders because most of these disorders are diagnosed from historical and physical findings and laboratory tests of blood and urine. When other procedures are nondiagnostic, or when therapy does not eliminate, or perhaps worsens, the neurologic dysfunction, CSF analysis is indicated to investigate other causes of the neurologic signs. In most cases, routine analysis of CSF associated with metabolic or nutritional disorders does not detect abnormalities (Fishman, 1992; Scott, 1995; Bichsel *et al.*, 1984b; Vandevelde and Spano, 1977). Although brain edema is relatively common with some of these disorders (e.g., hypoxia, hyponatremia, or the osmotic dysequilibrium syndromes of hemodialysis or diabetic ketoacidosis), the edema is usually cytotoxic rather than vasogenic. Therefore, the blood–brain/CSF barrier is usually intact and CSF protein is normal. If edema is severe enough to result in brain ischemia, infarction, or herniation, the blood–brain/CSF barrier becomes dysfunctional, vasogenic edema occurs, and CSF protein rises. If neural necrosis ensues, the white blood cell count may also increase. Even in the (apparent) absence of vasogenic edema, blood–brain barrier leakage may occur, perhaps because of the biochemical effects of the disorder on the barrier cells. Animals and people with uremic or hepatic encephalopathies or hypothyroidism may have increased total protein with a normal IgG index (Fishman, 1992; Bichsel *et al.*, 1984b, 1988). People with diabetic neuropathy may also have increased CSF protein (Fishman, 1992). Animals with severe metabolic encephalopathies often have seizures, and the effect of seizures on the CSF must also be considered (see Section VII.H.2). Disorders in which neural necrosis is a primary feature, such as the polioencephalomalacia of thiamine deficiency, typically have a pleocytosis and increased total protein (Bichsel *et al.*, 1984b; George, 1996; deLahunta, 1983). Specific biochemical analysis of CSF may show abnormalities, such as abnormalities in osmolality or electrolyte content with salt or water intoxication (Mayhew, 1989; Kopcha, 1987), abnormalities in amino acid levels (such as glutamine) with hepatic encephalopathy (Schaeffer *et al.*, 1991; Grabner *et al.*, 1991), and elevated citrulline in bovine citrullinemia (Healy *et al.*, 1990).

H. Miscellaneous Conditions

1. *Alterations in CSF Following Myelography*

Changes in the composition of CSF following myelography have been reported in people (Fishman, 1992) and animals (Widmer and Blevins, 1991; Burbidge *et al.*, 1989). Most contrast agents are low-grade leptomeningeal irritants, resulting in leptomeningeal inflammation that is reflected in the CSF. By 90 minutes after myelography, the total white blood cell count and total protein can be elevated and the white blood cell differential count altered. The pleocytosis is typically a mixed mononuclear/neutrophilic response, with the proportion of mononuclear cells to neutrophils varying with the contrast agent used and the time interval after myelography. The pleocytosis may resolve within 10 days (Johnson *et al.*, 1985), although individual animals may have a slightly increased total white blood cell count up to 14 days following contrast injection (Spencer *et al.*, 1982; Wood *et al.*, 1985). In contrast, one study of the contrast agents iohexol and iotrolan did not detect any alteration of total white blood cell count in CSF taken between 1 and 14 days following myelography (van Bree *et al.*, 1991). The CSF specific gravity and Pandy test score can also be elevated, presumably partly because of the presence of the contrast media (Widmer *et al.*, 1992). Increased CSF albumin and immunoglobulin levels may be due predominantly to blood–brain/CSF barrier leakage, and may return to normal levels within 5 days (Johnson *et al.*, 1985). In summary, any alteration of CSF within the first week or two following myelography must be assessed cautiously.

2. *Seizures: Interictal and Postictal CSF Characteristics*

Patients with seizures due to progressive intracranial or some extracranial disorders typically have CSF changes reflecting the disorder. In contrast, the interictal CSF of patients with nonprogressive, intracranial disease should be normal. Postictal CSF is often abnormal, however. Pleocytosis of postictal CSF has been well documented in people (Fishman, 1992; Rider *et al.*, 1995; Barry and Hauser, 1994). The white blood cell counts may be up to 80/μl with a neutrophilic component from 5 to 92%. The cell counts are highest at about 24 hours after the seizure. The mechanism of the pleocytosis is obscure (Fishman, 1992). Convulsive seizures, regardless of cause, may also induce a reversible increase in blood–brain/CSF barrier permeability, resulting in a transient elevation of CSF protein. Brain metabolism is also stimulated during the seizure, resulting in an increase in brain lactate production and a decrease in brain pH (Fishman, 1992). However, differentiating the effects of the local (brain) phenomena from the effects of systemic phenomena that occur during seizures (hypertension, acidosis, hypoxia, etc.) is difficult. For example, severe, experimental hyperthermia in dogs (core body temperature >41.2°C) results in increased CSF enzymes, calcium, and chloride, probably due to increased blood–brain/CSF barrier permeability (Deswal and Chohan, 1981). Interpretation of postictal CSF must be done cautiously because of the potential confusion of a postictal, "idiopathic epileptic" condition with a progressive disease that alters the CSF primarily. For children with seizures, the recommendation has been made that CSF with >20 white blood cell/μl or >10 polymorphonuclear cells/μl not be attributed to the seizures (Rider *et al.*, 1995).

I. Neoplasia

The CSF associated with neoplastic conditions affecting the central nervous is quite variable, reflecting the variety of tumors, locations, and tissue reactions to the disease. The CSF is usually clear and colorless, although xanthochromia may be present if hemorrhage has occurred. The total white blood cell count is often normal, but pleocytosis may occur, particularly with meningiomas and choroid plexus tumors (and occasionally other tumors) (Bailey and Higgins, 1986b; Carrillo *et al.*, 1986). Pleocytosis is usually mononuclear, although meningiomas may have >50% neutrophils (Bailey and Higgins, 1986b). Neural lymphosarcoma often has a lymphocytic/lymphoblastic pleocytosis (Lane *et al.*, 1994; Williams *et al.*, 1992; Couto *et al.*, 1984), except in cattle, in which the tumor is usually extradural (Sherman, 1987). One report indicates that pleocytotic CSF is associated with a significantly shorter survival time than is normal or albuminocytologic CSF (Heidner *et al.*, 1991). The commonest CSF abnormality is increased total protein, with choroid plexus tumors producing the most marked elevations (Rand et al., 1994b; Bailey and Higgins, 1986b; Moore *et al.*, 1994; Sarfaty *et al.*, 1988; Mayhew, 1989; Brehm *et al.*, 1992; Waters and Hayden, 1990; Roeder *et al.*, 1990; Kornegay, 1991). Dogs with neural neoplasia, particularly of the meninges or choroid plexus, commonly have blood–brain/spinal cord barrier disturbance and subsequently an increased albumin quotient (Sorjonen, 1987; Bichsel *et al.*, 1984b; Sorjonen *et al.*, 1991; Moore *et al.*, 1994). In one study, this abnormality was most common with choroid plexus tumors and least common with astrocytomas (Moore *et al.*, 1994).

Alpha and beta globulin levels are usually normal; gamma globulins are normal or mildly increased (Sorjonen, 1987; Sorjonen *et al.*, 1991; Moore *et al.*, 1994). The IgG index may be elevated, reflecting the presence of inflammatory infiltrates around the lesion (Tipold *et al.*, 1993b; Bichsel *et al.*, 1984b).

The CSF associated with spinal neoplasia is reported to be normal more often than is the CSF of brain tumors (Fingeroth *et al.*, 1987; Luttgen *et al.*, 1980; Schott *et al.*, 1990). This finding may reflect the fact that most spinal neoplasia is extradural or that most spinal tumors are relatively small at the time of diagnosis. It may also reflect the site of CSF collection—that most of the samples are cerebellomedullary rather than lumbar, although many reports do not state the puncture site. Cerebrospinal fluid collected caudal to the lesion is abnormal more often than is CSF collected cranial to the lesion (Thomson *et al.*, 1990).

J. Parasitic Diseases

1. *Equine Protozoal Encephalomyelitis*

Equine protozoal encephalomyelitis (EPM), caused by *Sarcocystis neurona* (*S. falcatula* (Dame *et al.*, 1995)), is characterized by multifocal areas of mononuclear, perivascular inflammation and necrosis; severe lesions may be hemorrhagic and have neutrophilic infiltration (Mayhew and deLahunta, 1978; Madigan and Higgins, 1987). The CSF may be normal or have mild to moderate mononuclear pleocytosis and increase in total protein (Mayhew, 1989; Reed *et al.*, 1994). Xanthochromia is occasionally present, as well as neutrophils and eosinophils (Mayhew, 1989). The CSF albumin concentration and albumin quotient are reported to be normal and the IgG index elevated, indicating intrathecal IgG production (Andrews and Provenza, 1995). Antibodies to *S. neurona* can be identified in the CSF by immunoblot analysis (Granstrom, 1993). The test is very sensitive and specific for the diagnosis of EPM; however, its accuracy depends on an intact blood–brain/CSF barrier because many horses have serum antibodies to *S. neurona* but do not have clinical disease (Andrews and Provenza, 1995; Fenger, 1995). Detection of *S. neurona* in the CSF by polymerase chain reaction provides definitive evidence of the presence of the parasite in the central nervous system. The results of polymerase chain reaction assay are independent of serum leakage across the blood–brain/CSF barrier (Fenger, 1994).

2. *Neosporosis, Toxoplasmosis*

Both *Neospora* and *Toxoplasma* can invade the central nervous system, causing necrosis, vasculitis and a multifocal, granulomatous meningoencephalomyelitis. *Neospora* seems to have more of a predilection for the central nervous system than *Toxoplasma*, particularly in young dogs (Dubey *et al.*, 1989, 1988). The CSF associated with neural protozoal infections generally has a mild to moderate increase in white blood cell count and total protein. Typically, the white blood cell differential count shows a mixed pleocytosis with monocytes, lymphocytes, neutrophils, and eosinophils in order of decreasing percentage (Rand *et al.*, 1994a; Cuddon *et al.*, 1992; Tipold, 1995; Vandevelde and Spano, 1977; Kornegay, 1981; Dubey *et al.*, 1990; Averill and deLahunta, 1971; Hass *et al.*, 1989). Occasionally, the white blood cell count and protein are normal (Tipold, 1995; Parish *et al.*, 1987). The CSF IgG index was elevated in three of three dogs studied; in two of two dogs, the IgM was normal and the IgA was elevated (Tipold *et al.*, 1994, 1993b). In a study of experimentally infected cats, *T. gondii*-specific IgG was intrathecally produced, but *T. gondii*-specific IgM was not detected (Munana *et al.*, 1995). Antiprotozoal antibodies in the CSF may be detected by a variety of methods (Ruehlmann *et al.*, 1995; Cole *et al.*, 1993; Patton *et al.*, 1991). However, the presence of antibodies does not necessarily indicate clinical disease (Munana *et al.*, 1995; Dubey and Lindsay, 1993). Recently, polymerase chain reaction techniques have been developed to identify the protozoa in tissue and fluids, including CSF (Stiles *et al.*, 1996; Parmley *et al.*, 1992; Novati *et al.*, 1994). Occasionally the organisms themselves may be seen in CSF cells (Dubey, 1990; McGlennon *et al.*, 1990).

In considering the CSF abnormalities of toxoplasmosis and neosporosis, two issues must be kept in mind. First, reports of toxoplasmosis prior to 1988 (when *Neospora* was identified) must be carefully scrutinized because many of these cases were actually neosporosis. Second, because *T. gondii* is not a primary pathogen, clinical toxoplasmosis is relatively rare and is seen mostly in conjunction with a second disease, particularly canine distemper, which may itself alter the CSF (Dubey *et al.*, 1989).

3. *Migratory Parasites*

Neural invasion by migratory parasites is relatively common in large and exotic animals, yet rare in dogs and cats. The CSF may reflect the physical trauma and consequent inflammatory response, and in some cases an immune reaction to the parasite tissue. The CSF abnormalities are variable and probably depend to some degree on the specific parasite, as well as on its location and the type of incited response. For example, *H. bovis* larvae in the cow normally lodge in the lumbar epidural space and their effect on the spinal cord may be primarily compression. The CSF in such a case could be normal or have only mildly to moderately elevated protein. Parasites that actually invade neural tissue may leave the CSF unchanged or produce CSF with mild to marked pleocytosis and protein elevation, as well as xanthochromia. An eosinophilic pleocytosis is suggestive of parasitism, and typical of some parasites such *Parelaphostrongylus* (Mason, 1989; George, 1996; Pugh *et al.*, 1995; Baum, 1994). However, eosinophilic pleocytosis is not pathognomonic for parasitism, nor does a lack of eosinophils in the CSF rule out neural parasitism (Lester, 1992; Braund, 1994; George, 1996; deLahunta, 1983). The development of an ELISA technique using *Parelaphostrongylus*-specific immunoglobulin to detect parasite antigen in the CSF might be of diagnostic value (Dew *et al.*, 1992).

K. Toxicity

Even though neurologic signs may occur, the CSF associated with toxicity is usually normal (e.g., cows with lead poisoning, tetanus, or botulism) (Fankhauser, 1962; Feldman, 1989). Mild elevations of the white blood cell count and protein may occur if the toxin causes breakdown of blood-brain/CSF barrier or neural degeneration or necrosis, such as in some cases of lead poisoning (Fankhauser, 1962; Swarup and Maiti, 1991; Dorman *et al.*, 1990; Dow *et al.*, 1989; George, 1996; Mayhew, 1989; Little and Sorensen, 1969). Lead has been shown to poison capillary endothelial cells selectively (Goldstein *et al.*, 1977), as well as to cause cerebral cortical necrosis (Christian and Tryphonas, 1970). If necrosis is severe, the white blood cell count and the total protein can be markedly increased with a predominance of neutrophils, as with leukoencephalomalacia caused by moldy corn poisoning in horses. Xanthochromia is also a characteristic of moldy corn poisoning, reflecting the perivascular hemorrhages in the central nervous system (McCue, 1989; Masri *et al.*, 1987). With toxicities, biochemical alterations of the CSF may occur more commonly than alterations of CSF cell counts or protein. At the onset of fatal signs of lead poisoning, CSF glucose, urea, creatinine, and creatine kinase levels are increased (Swarup and Maiti, 1991). Neurostimulatory toxins may result in elevated monoamine metabolites in the CSF (Sisk *et al.*, 1990). Ivermectin toxicity producing recumbency in dogs elevates the CSF concentrations of homovanillic acid and 5-hydroxyindoleacetic acid (Vaughn *et al.*, 1989a). Interestingly, copper poisoning in sheep does not produce significant increases in CSF copper, zinc, or iron levels (Gooneratne and Howell, 1979).

L. Trauma/Compression of Neural Tissue

The CSF abnormalities associated with trauma and/or compression are variable depending on the rate at which the neural insult developed, the degree of neural damage, the location of the lesion (particularly with respect to the CSF collection site), the elapsed time since the onset of the neural insult, and the maintenance or progression of the insult. With acute trauma, the CSF may be pink and hazy or turbid, or actually bloody. After centrifugation, the supernatant can be clear. If hemorrhage occurred more than 48 hours prior to CSF collection, the supernatant may be yellow because of bilirubin. The total red blood cell count may be markedly elevated. The white blood cell count may be mildly to moderately elevated, reflecting either hemorrhage into the subarachnoid space or inflammation instigated by the trauma. Erythrophagocytosis may be present. The pleocytosis is usually a mixed cell population, and a substantial proportion of neutrophils (40–50%) is possible. Total protein may be moderately to markedly elevated because of disruption of blood vessels, interruption of CSF flow and absorption, and necrosis (Thomson *et al.*, 1989; Green *et al.*, 1993). Thus, the CSF of acute trauma may have a distinct, inflammatory character. With spinal cord trauma/compression, lumbar CSF is more consistently abnormal than cerebellomedullary CSF (Thomson *et al.*, 1990). The CSF abnormalities of chronic trauma or sustained compression tend to be milder than the abnormalities of acute damage. The white blood cell count may be normal or mildly elevated with generally a mixed or mononuclear pleocytosis. The cerebellomedullary CSF of horses with cervical stenotic myelopathy is reported to be hypocellular with a reduced number of lymphocytes (Grant *et al.*, 1993). The CSF protein associated with chronic trauma or sustained neural compression may be normal to moderately elevated (Thomson *et al.*, 1989; Mayhew, 1989). The albumin content and the albumin quotient of CSF associated with trauma/compression may be normal or increased, the latter reflecting the vascular damage and edema (Sorjonen, 1987; Sorjonen *et al.*, 1991; Bichsel *et al.*, 1984b; Andrews and Provenza, 1995). The gamma globulin percentage and the IgG index are usually normal. Occasional elevations probably reflect the presence of inflammatory cells in the lesion (Tipold *et al.*, 1993b; Bichsel *et al.*, 1984b; Andrews and Provenza, 1995).

References

Altman, P. L., and Dittmer, D. S. (1974). "Biology Data Book," Vol. III, 2nd Ed., p. 1978. Federation of American Societies for Experimental Biology, Bethesda, Maryland.

Ames, A. (1964). *J. Neurophysiol.* **27,** 672.

Ames, T. R. (1987). *Vet. Clin. North Am. Food Anim. Pract.* **3,** 61.

Andrews, F. M., and Provenza, M. (1995). *Proc. ACVIM* **13,** 600.

Andrews, F. M., Maddux, J. M., and Faulk, D. (1990a). *Prog. Vet. Neurol.* **1,** 197.

Andrews, F. M., Matthews, H. K., and Reed, S. M. (1990b). *Veterinary Medicine* **85,** 1325.

Andrews, F. M., Geiser, D. R., Sommardahl, C. S., Green, E. M., and Provenza, M. (1994). *Am. J. Vet. Res.* **55,** 741.

Apostolski, S., Nikolic.J., Bugarski-Prokopljevic, C., Miletic, V., Pavlovic, S., and Filipovic, S. (1991). *Acta Neurol. Scand.* **83,** 96.

Appel, M. J. G., Allan, S., Jacobson, R. H., Lauderdale, T. L., Chang, Y. F., Shin, S. J., Thomford, J. W., Todhunter, R. J., and Summers, B. A. (1993). *J. Infect. Dis.* **167,** 651.

Averill, D. R., and deLahunta, A. (1971). *J. Am. Vet. Med. Assoc.* **159,** 1134.

Bailey, C. S., and Higgin•, R.J. (1985). *Am. J. Vet. Res.* **46,** 1162.

Bailey, C. S., and Higgins, R.J. (1986a). *J. Am. Vet. Med. Assoc.* **188,** 418.

Bailey, C. S., and Higgins, R.J. (1986b). *J. Am. Vet. Med. Assoc.* **188,** 414.

Balhuizen, J.C., Bots, G.T.A., Schaberg, A., and Bosman, F.T. (1978). *J. Neurosurg.* **48,** 747.

Banik, N. L., and Hogan, E. L. (1983), *In* "Neurobiology of Cerebrospinal Fluid," Vol. 2 (J. H. Wood, ed.), pp. 205. Plenum Press, New York.

Bardon, T., and Ruckebusch, M. (1984). *Neuroscience Letters* **49,** 147.

Baroni, M. and Heinold, Y. (1995). *Prog. Vet. Neurol.* **6,** 88.

Barrett, D. L., and King, E. B. (1976). *Acta Cytol.* **20,** 174.

Barry, E., and Hauser, W. A. (1994). *Arch. Neurol.* **51,** 190.

Baum, K. H. (1994). *Vet. Clin. North Am. Food Anim. Pract.* **10,** 383.

Bell, W. O. (1995). *Pediatr. Neurosurg.* **23,** 42.

Berthelin, C. F., Legendre, A. M., Bailey, C. S., Kass, P. H., and Wolf, A. M. (1994). *Prog. Vet. Neurol.* **5,** 136.

Bichsel, P., Vandevelde, M., and Lang, J. (1984a). *Schweiz. Arch. Tierheilk.* **387,** 397.

Bichsel, P., Vandevelde, M., Vandevelde, E., Affolter, U., and Pfister, H. (1984b). *Res. Vet. Sci.* **37,** 101.

Bichsel, P., Jacobs, G., and Oliver, J. E. (1988). *J. Am. Vet. Med. Assoc.* **192,** 1745.

Bienzle, D., Kwiecien, J. M., and Parent, J. M. (1995). *Vet. Pathol.* **32,** 437.

Bigner, S. H. (1992). *J. Neuropath. and Exp. Neur.* **51,** 235.

Bigner, S. H., and Johnston, W. W. (1981). *Acta Cytol.* **25,** 335.

Binkhorst, G. J. (1982), *In* "Proceedings. XIIth Congress on Diseases of Cattle, The Netherlands," Vol. 2, p. 864. Internationaal Congrescentrum RAI, Amsterdam.

Bleich, H. L., Berkman, P. M., and Schwartz, W. B. (1964). *J. Clin. Invest.* **43,** 11.

Blythe, L. L., Mattson, D. E., and Lassen, E. D. (1985). *Can. Vet. J.* **26,** 218.

Boerman, R. H., Arnoldus, E. P., Peters, A. C., Bloem, B. R., Raap, A. K., and van der Ploeg, M. (1991). *J. Med. Virol.* **33,** 83.

Boogerd, W., and Peters, A. C. B. (1986). *Lab. Anim. Sci.* **36,** 386.

Bosch, I., and Oehmichen, M. (1978). *J. Neurol.* **219,** 93.

Bradley, G. A. (1991). *Vet. Pathol.* **28,** 91.

Braund, K. G. (1994). "Clinical Syndromes in Veterinary Neurology," 2nd Ed., St. Louis, Mosby.

Brehm, D. M., Vite, C. H., Steinberg, H. S., Havilan, F., and Van Winkle, T. (1995). *J. Am. Anim. Hosp. Assoc.* **31,** 349.

Breitschwerdt, E. B. (1995), *In* "Textbook of Veterinary Internal Medicine" 4th ed. (S. J. Ettinger and E. C. Feldman, eds.), p. 376. Saunders, Philadelphia, Pennsylvania.

Breitschwerdt, E. B., Meuten, D. J., Walker, D. H., Levy, M., Kennedy, K., King, M., and Curtis, B. (1985). *Am. J. Vet. Res.* **46,** 2124.

Breitschwerdt, E. B., Levy, M. G., Davidson, M. G., Walker, D. H., Burgdorfer, W., Curtis, B. C., and Babineau, C. A. (1990). *Am. J. Vet. Res.* **51,** 1312.

Brewer, B. D. (1983). *Vet. Clin. North Am. Large Anim. Pract.* **5,** 680.

Brewer, B. D. (1987). *Vet. Clin. North Am. Food Anim. Pract.* **3,** 13.

Brobst, D., and Bryan, G. (1989). *In* "Diagnostic Cytology of the Dog and Cat" (R. L. Cowell and R. D. Tyler, eds.), p. 141. American Veterinary Publications, Goleta, California.

Brooks, B. R., Hirsch, R. L., and Coyle, P. K. (1983). *In* "Neurobiology of Cerebrospinal Fluid" (J. H. Wood, ed.), p. 263. Plenum Press, New York.

Buoro, I. B. J., Kanui, T. I., Atwell, R. B., Njenga, K. M., and Gathumbi, P. K. (1990). *J. Small Anim. Pract.* **31,** 624.

Burbidge, H. M., Kamnegieter, N., and Dickson, L.R. (1989). *Eq. Vet. J.* **21,** 347.

Burgess, E. C., and Mattison, M. (1987). *J. Am. Vet. Med. Assoc.* **191,** 1457.

Calhoun, M. C., Hurt, H. D., Eaton, H. D., Rousseau, J. E., and Hall, R. C. (1967). "Rates of Formation and Absorption of Cerebrospinal Fluid in Holstein Male Calves, Bull. 401." Storrs Agricultural Experiment Station, The University of Connecticut.

Carakostas, M. C., Gossett, K. A., Watters, J. W., and MacWilliams, P. S. (1983). *Vet. Radiol.* **24,** 267.

Cargill, J. S. (1975). *Lancet* Oct. 4, p. 665.

Carrillo, J. M., Sarfaty, D., and Greenlee, P. (1986). *J. Am. Anim. Hosp. Assoc.* **22,** 367.

Cauzinille, L., and Kornegay, J. N. (1996). *J. Vet. Intern. Med.* **10,** 241.

Chrisman, C. L. (1983). *In* "Current Veterinary Therapy VIII—Small Animal Practice" (R. W. Kirk, ed.), p. 676. Saunders, Philadelphia, Pennsylvania.

Chrisman, C. L. (1992). *Vet. Clin. North Am. Small Anim. Pract.* **22,** 781.

Christen, H. J., Eiffert, H., Ohlenbusch, A., and Hanefeld, F. (1995). *Eur. J. Pediatr.* **154,** 374.

Christian, R. G., and Tryphonas, L. (1970). *Am. J. Vet. Res.* **32,** 203.

Christopher, M. M. (1992). *Vet. Clin. Pathol.* **21,** 95.

Christopher, M. M., Perman, V., and Hardy, R. M. (1988). *J. Am. Vet. Med. Assoc.* **192,** 1726.

Coates, J.R. (1995). *J. Am. Vet. Med. Assoc.* **206,** 1333.

Cole, R. A., Lindsay, D. S., Dubey, J. P., and Blagburn, B. L. (1993). *J. Vet. Diagn. Invest.* **5,** 579.

Coles, E. H. (1980). *In* "Clinical Biochemistry of Domestic Animals," 3rd ed. (J. J. Kaneko, ed.), p . 719. Academic Press, New York.

Converse, G. M., Gwaltney, J. M., Strassburg, D. A., and Hendley, J. O. (1973). *J. Pediatr.* **83,** 220.

Cook, J. R., Jr., and DeNicola, D. B. (1988). *Vet. Clin. North Am. Small Anim. Pract.* **18,** 75.

Cordy, D. R., and Holliday, T. A. (1989). *Vet. Pathol.* **26,** 191.

Couto, C. G., Cullen, J., Pedroia, V., and Turrel, J. M. (1984). *J. Am. Vet. Med. Assoc.* **184,** 809.

Cserr, H. F. (1971). *Physiol. Rev.* **51,** 273.

Cuddon, P. A. (1990). *Proc. ACVIM* **8,** 1009.

Cuddon, P., Lin, D. S., Bowman, D. D., Lindsay, D. S., Miller, T. K., Duncan, I. D., deLahunta, A., Cummings, J., Suter, M., and Cooper, B. (1992). *J. Vet. Intern. Med.* **6,** 325.

Cummings, J. F., deLahunta, A., Holmes, D. F., and Schultz, R. D. (1982). *Acta Neuropath. (Berl)* **56,** 167.

Cummings, J. F., deLahunta, A., George, L., Fuhrer, L., Valentine, B. A., Cooper, B. J., Summers, B. A., Huxtable, C. R., and Mohammed, H. O. (1990). *Cornell Vet.* **80,** 357.

Dame, J. B., MacKay, R. J., Yowell, C. A., Cutler, T. J., Marsh, A., and Greiner, E. C. (1995). *J. Parasitol.* **81,** 930.

Darien, B. J., Belknap, J., and Nietfeld, J. (1988). *J. Vet. Intern. Med.* **2,** 201.

Davis, B. A. (1990). "Biogenic Monoamines and Their Metabolites in the Urine, Plasma, and Cerebrospinal Fluid of Normal, Psychiatric, and Neurological Subjects." CRC Press, Boca Raton, Florida.

Davson, H., and Segal, M. B. (1996). "Physiology of the CSF and Blood–Brain Barriers." CRC Press, Boca Raton, Florida.

deLahunta, A. (1983). "Veterinary Neuroanatomy and Clinical Neurology," 2nd ed., Saunders, Philadelphia, Pennsylvania.

Deswal, K., and Chohan, I. S. (1981). *Int. J. Biometeorol.* **25,** 227.

Dew, T. L., Bowman, D. D., and Grieve, R. B. (1992). *J. Zoo Wildl. Med.* **23,** 281.

Divers, T. J., Mohammed, H. O., Cummings, J. F., Valentine, B. A., deLahunta, A., Jackson, C. A., Summers, B. A., and Lahunta, A. (1994). *Eq. Vet. J.* **26,** 409.

Dorman, D. C., Parker, A. J., and Buck, W. B. (1990). *J. Am. Anim. Hosp. Assoc.* **26,** 589.
Dow, S. W., LeCouteur, R. A., Henik, R. A., Jones, R. L., and Poss, M. L. (1988). *J. Vet. Intern. Med.* **2,** 171.
Dow, S. W., LeCouteur, R. A., Poss, M. L., and Beadleston, D. (1989). *J. Am. Vet. Med. Assoc.* **195,** 365.
Dow, S. W., Poss, M. L., and Hoover, E. A. (1990). *J. Acquir. Immune. Defic. Syndr.* **3,** 658.
Dubey, J. P. (1990). *Compend. Contin. Educ. Pract. Vet.* **12,** 653.
Dubey, J. P., and Lindsay, D. S. (1993). *Parasitology Today* **9,** 452.
Dubey, J. P., Carpenter, J. L., Speer, C. A., Topper, M. J., and Uggla, A. (1988). *J. Am. Vet. Med. Assoc.* **192,** 1269.
Dubey, J. P., Carpenter, J. L., Topper, M. J., and Uggla, A. (1989). *J. Am. Anim. Hosp. Assoc.* **25,** 659.
Dubey, J. P., Greene, C. E., and Lappin, M. R. (1990), *In* "Infectious Diseases of the Dog and Cat" (C. E. Greene, ed.), p. 818. Saunders, Philadelphia, Pennsylvania.
Ducos, R., Donoso, J., Weickhardt, U., and Vietti, T. J. (1979). *Cancer* **43,** 1479.
Duncan, J. R., Prasse, K. W., and Mahaffey, E. A. (1994). "Veterinary Laboratory Medicine. Clinical Pathology," 3rd ed. Iowa State University Press, Ames, Iowa.
Fankhauser, R. (1962), *In* "Comparative Neuropathology" (J. R. M. Innes and L. Z. Saunders, eds.),p. 21. Academic Press, New York.
Faull, K. F., Barchas, J. D., Foutz, A. S., Dement, W. C., and Holman, R. B. (1982). *Brain Res.* **242,** 137.
Feder, B. M., Joseph, R. J., Moroff, S. D., Schneider, E. M., and Bosler, E. M. (1991). *J. Vet. Intern. Med.* **5,** 137.
Fekadu, M., and Shaddock, J. H. (1984). *Am. J. Vet. Res.* **4,** 724.
Feldman, B. F. (1989), *In* "Clinical Biochemistry of Domestic Animals" 4th ed. (J. J. Kaneko, ed.), p. 835. Academic Press, New York.
Felgenhauer, K. (1974). *Klin. Wochenschr.* **52,** 1158.
Felgenhauer, K. (1982). *J. Neurol.* **228,** 223.
Fenger, C. K. (1994). *Proc. ACVIM* **12,** 550.
Fenger, C. K. (1995). *Proc. ACVIM* **13,** 597.
Fingeroth, J. M., Prata, R. G., and Patnaik, A. K. (1987). *J. Am. Vet. Med. Assoc.* **191,** 720.
Firneisz, G. D., Cochrane, S. M., Parent, J., and Houston, D. M. (1990). *Can. Vet. J.* **31,** 652.
Fishman, R. A. (1992). "Cerebrospinal Fluid in Diseases of the Nervous System," 2nd ed. Saunders, Philadelphia, Pennsylvania.
Fletcher, T. F. (1993), *In* "Miller's Anatomy of the Dog" 3rd ed. (H. E. Evans, ed.), p. 800. Saunders, Philadelphia, Pennsylvania.
Foreman, J. H., and Santschi, E. M. (1989). *Compend. Contin. Educ. Pract. Vet.* **11,** 640.
Fowler, M. E. (1989). "Medicine and Surgery of South American Camelids." Iowa State University Press, Ames, Iowa.
Fryden, A., Link, H., and Norrby, E. (1978). *Infect. Immun.* **21,** 852.
Furr, M. O., and Bender, H. (1994). *Am. J. Vet. Res.* **55,** 781.
Furr, M. O., and Tyler, R. D. (1990). *J. Am. Vet. Med. Assoc.* **197,** 245.
Gelatt, K. N., Chrisman, C. L., Samuelson, D. A., and Buergelt, C. D. (1991). *J. Am. Anim. Hosp. Assoc.* **27,** 427.
George, L. W. (1996), *In* "Large Animal Internal Medicine," 2nd ed. (B. P. Smith, ed.), p. 1001. Mosby, St. Louis.
Glass, J. P., Melamed, M., Chernik, N. L., and Posner, J. B. (1979). *Neurology* **29,** 1369.
Glimaker, M., Olcen, P., and Andersson, B. (1994). *Scand. J. Infect. Dis.* **26,** 141.
Goldstein, G. W., Wolinsky, J. S., and Csejtey, J. (1977). *Ann. Neurol.* **1,** 235.
Gooneratne, S. R., and Howell, J. M. (1979). *Res. Vet. Sci.* **27,** 384.
Grabner, A., and Goldberg, M. (1991). *Tierarztl. Prax.* **19,** 271.
Granstrom, D. E. (1993). *Proc. ACVIM* **11,** 587.
Grant, B., Roszel, J., Peterson, R., and Perris, E. (1993). *Proceedings of the Annual Convention of the American Association of Equine Pract* **38,** 737.
Green, S. L. (1993). *Vet. Clin. North Am. Equine Pract.* **9,** 337.
Green, E. M., and Green, S. (1990). *Proc. ACVIM* **8,** 495.
Green, S. L., and Smith, L. L. (1992). *J. Am. Vet. Med. Assoc.* **201,** 125.
Green, S. L., Smith, L. L., Vernau, W., and Beacock, S. M. (1992). *J. Am. Vet. Med. Assoc.* **200,** 1133.
Green, E. M., Constantinescu, G. M., and Kroll, R. A. (1993). *Compend. Contin. Educ. Pract. Vet.* **15,** 288.
Greene, C. E. (1990), *In* "Infectious Diseases of the Dog and Cat" (C. E. Greene, ed.), p. 891. Saunders, Philadelphia, Pennsylvania.
Greene, C. E., and Dreesen, D. W. (1990), *In* "Infectious Diseases of the Dog and Cat" (C. E. Greene, ed.), p. 365. Saunders, Philadelphia, Pennsylvania.
Greene, C. E., Vandevelde, M., and Braund, K. (1976). *J. Am. Vet. Med. Assoc.* **169,** 405.
Greene, C.E., Burgdorfer, W., Cavagnolo, R., Philip, R.N., and Peacock, M.G. (1985). *J. Am. Vet. Med. Assoc.* **186,** 465.
Grevel, V. (1991). *Tierarztl. Prax.* **19,** 553.
Grevel, V., and Machus, B. (1990). *Vet. Med. Report* **2,** 403.
Grevel, V., Machus, B., and Steeb, C. (1992). *Tierarztl. Prax.* **20,** 419.
Griffith, N. C., Cunningham, A. M., Goldsmith, R., and Bandler, R. (1991). *Epilepsia* **32,** 767.
Guffond, T., Dewilde, A., Lobert, P.-E., Caparros-Lefebvre, D., Hober, D., and Wattre, P. (1994). *Clin. Infect. Dis.* **18,** 744.
Hahn, C. N., Mayhew, I. G., Whitewell, K. E., Smith, K. C., Carey, D., Carter, S. D., and Read, R. A. (1996). *Eq. Vet. J.* **28,** 84.
Hamir, A. N., Moser, G., and Rupprecht, C. E. (1992). *J. Comp. Pathol.* **106,** 411.
Hanlon, C. A., Zeimer, E. L., Hamir, A. N., and Rupprecht, C. E. (1989). *Am. J. Vet. Res.* **50,** 364.
Hansen, H. H., Bender, R. A., and Shelton, B. J. (1974). *Acta Cytol.* **18,** 259.
Hass, J. A., Shell, L., and Saunders, G. (1989). *J. Am. Anim. Hosp. Assoc.* **25,** 253.
Hayward, R. A., Shapiro, M. F., and Oye, R. K. (1987). *Lancet* **1,** 813.
Healy, P. J., Harper, P., and Dennis, J. A. (1990). *Aust. Vet. J.* **67,** 255.
Heidner, G. L., Kornegay, J. N., Page, R. L., Dodge, R. K., and Thrall, D. E. (1991). *J. Vet. Intern. Med.* **5,** 219.
Henriksson, A., Link, H., and Crus, M. (1986). *Ann. Neurol.* **20,** 337.
Heyes, M. P., Saito, K., Crowley, J. S., Davis, L. E., Demitrack, M. A., Der, M., Dilling, L. A., Elia, J., Kruesi, M. J., and Lackner, A. (1992). *Brain* **115,** 1249.
Hochwald, G. M., Wallenstein, M. C., and Matthews, E. S. (1969). *Am. J. Physiol.* **217,** 348.
Holbrook, T. C., and White, S. L. (1992). *Vet. Clin. North Am. Food Anim. Pract.* **8,** 305.
Indrieri, R. J., Holliday, T. A., and Keen, C. L. (1980). *Am. J. Vet. Res.* **41,** 1299.
Jackson, T. A., Osburn, B. I., Cordy, D. R., and Kendrick, J. W. (1977). *Am. J. Vet. Res.* **38,** 709.
Jackson, W., deLahunta, A., Adaska, A., and Divers, T. J. (1995). *Prog. Vet. Neurol.* **6,** 16.
Jackson, C., deLahunta, A., Divers, T., and Ainsworth, D. (1996). *J. Vet. Intern. Med.* **10,** 246.
Jamison, E. M., and Lumsden, J. H. (1988). *Sem. Vet. Med. Surg. Sm. Anim.* **3,** 122.
Johnson, G. C., Fuciu, D. M., Fenner, W. R., and Krakowka, S. (1985). *Am. J. Vet. Res.* **46,** 1303.
Johnson, G. C., Fenner, W. R., and Krakowka, S. (1988). *J. Neuroimmunol.* **17,** 237.
Joseph, R. J., Greenlee, P. G., Carrilo, J. M., and Kay, W. J. (1988). *J. Am. Anim. Hosp. Assoc.* **24,** 569.

Kaiser, R., and Lucking, C. H. (1993). *J. Neurol. Sci.* **118,** 64.

Kjeldsberg, C. R., and Knight, J. A. (1993). *In* "Body Fluids: Laboratory Examination of Amniotic, Cerebrospinal, Seminal, Serous and Synovial Fluids," 3rd ed. (C. R. Kjeldsberg and J. A. Knight, eds.), p. 65. (p. 65. American Society of Clinical Pathologists, Chicago.

Kline, K. L., Joseph, R. J., and Averill, D. R. (1994). *J. Am. Anim. Hosp. Assoc.* **30,** 111.

Klingeborn, B., Dinter, Z., and Hughes, R. A. C. (1983). *Zentralbl. Veterinarmed.* [*B*] **30,** 137.

Kohn, C. W., and Fenner, W. F. (1987). *Vet. Clin. North Am. Equine Pract.* **3,** 405.

Kolmel, H. W. (1977). *Acta Cytol.* **21,** 154.

Kopcha, M. (1987). *Vet. Clin. North Am. Food Anim. Pract.* **3,** 119.

Kornegay, J. N. (1981). *Compend. Contin. Educ. Pract. Vet.* **3,** 85.

Kornegay, J. N. (1991). *Prob. Vet. Med.* **3,** 309.

Kornegay, J. N., and Gorgacz, E. J. (1982). *Vet. Pathol.* **19,** 646.

Kornegay, J. N., Lorenz, M. D., and Zenoble, R. D. (1996). *J. Am. Vet. Med. Assoc.* **173,** 1334.

Krakowka, S., Fenner, W., and Miele, M. S. (1981). *Am. J. Vet. Res.* **42,** 1975.

Kristensen, F., and Firth, E. C. (1977). *Am. J. Vet. Res.* **38,** 1089.

Kuberski, T. (1979). *Ann. Intern. Med.* **91,** 70.

Lane, S. B., Kornegay, J. N., Duncan, J. R., and Oliver, J. E. J. (1994). *J. Vet. Intern. Med.* **8,** 99.

Laterre, D. C. (1996). *In* "Handbook of Clinical Neurology. Tumors of the Spine and Spinal Cord. Part I" (P. J. Vinken and G. W. Bruyn, eds.), p. 125. American Elsevier Publishing Co., New York.

Lebech, A.-M. (1994). *In* " Lyme Borreliosis" (J. S. Axford and D. E. H. Rees, eds.), p. 303–305. Plenum Press, New York.

Lester, G. (1992). *Compend. Contin. Educ. Pract. Vet.* **14,** 1624.

Levy, S. A., Dombach, D. M., Barthold, S. W., and Wasmoen, T. L. (1993). *Compend. Contin. Educ. Pract. Vet.* **15,** 833.

Lin, J. J., Harn, H. J., Hsu, Y. D., Tsao, W. L., Lee, H. S., and Lee, W. H. (1995). *J. Neurol.* **242,** 147.

Link, H., and Tibbling, G. (1977). *Scand. J. Clin. Lab. Invest.* **37,** 391.

Little, P. B. (1986). *Can. Vet. J.* **27,** 94.

Little, P. B., and Sorensen, D. K. (1969). *J. Am. Vet. Med. Assoc.* **155,** 1892.

Loscher, W. and Schwartz-Porsche, D. (1986). *J. Neurochem.* **46,** 1322.

Lowenthal, A., and Karcher, D. (1994). *Ann. NY Acad. Sci.* **724,** 68.

Lowenthal, A., Crols, R., DeSchutter, E., Gheuens, J., Karcher, D., Noppe, M., and Tasnier, A. (1984). *Int. Rev. Neurobiol.* **25,** 95.

Luttgen, P. J., Braund, K. G., Brawner, W. R., and Vandevelde, M. (1980). *J. Small Anim. Pract.* **21,** 213.

McCue, P. M. (1989). *Compend. Contin. Educ. Pract. Vet.* **11,** 646.

McGlennon, N. J., Jeffries, A. R., and Casas, A. (1990). *J. Small Anim. Pract.* **31,** 102.

Madigan, J. E. (1993). *Vet. Clin. North Am. Large Anim. Pract.* **9,** 429.

Madigan, J. E., and Higgins, R. J. (1987). *Vet. Clin. North Am. Equine Pract.* **3,** 397.

Mandel, N. S., Senker, E. G., Bosler, E. M., and Schneider, E. M. (1993). *Compend. Contin. Educ. Pract. Vet.* **15,** 581.

Maren, T. H. (1992). *In* "Barriers and Fluids of the Eye and Brain" (M.B. Segal, ed.), p. 37. CRC Press, Boca Raton, Florida.

Maretzki, G. H., Fisher, D. J., and Greene, C. G. (1994). *J. Am. Vet. Med. Assoc.* **205,** 1554.

Mason, K. V. (1989). *Aust. Vet. J.* **66,** 152.

Masri, M. D., Olcott, B. M., Nicholson, S. S., McClure, J. J., Schmidt, S. P., Freestone, J. F., and Kornegay, W. R. (1987). *Am. Assoc. Eq. Pract. Proc.* **33,** 367.

Mayhew, I. G. (1989). "Large Animal Neurology." Lea & Febiger, Philadelphia, Pennsylvania.

Mayhew, I. G., and deLahunta, A. (1978). *Cornell Vet.* **68,** Supp. 6, 106.

Mayhew, I. G., Whitlock, R. H., and Tasker, J. B. (1977). *Am. J. Vet. Res.* **38,** 1271.

Meeks, J. C., Christopher, M. L., Chrisman, C. L., Hopkins, A. L., and Homer, B. H. (1994). *Proc. ACVIM* **12,** 1008. (abstract).

Meinkoth, J. H., Hoover, J. P., Cowell, R. L., Tyler, R. D., and Link, J. (1989). *J. Am. Vet. Med. Assoc.* **195,** 1754.

Meric, S. M. (1988). *J. Vet. Intern. Med.* **2,** 26.

Meric, S. M. (1992a). *In* "Current Veterinary Therapy XI—Small Animal Practice" (R. W. Kirk and J. D. Bonagura, eds.), p. 1007. 1009. Saunders, Philadelphia, Pennsylvania.

Meric, S. M. (1992b). *In* "Essentials of Small Animal Internal Medicine" (R. W. Nelson and C. G. Couto, eds.), p. 732. Mosby, St. Louis.

Milhorat, T. H. (1987). "Cerebrospinal Fluid and the Brain Edemas." Neuroscience Society of New York.

Millson, G. C., West, L. C., and Dew, S. M. (1960). *J. Comp. Pathol. Ther.* **70,** 194.

Milner, R. J., Engela, J., and Kirberger, R. M. (1996). *Veterinary Radiology & Ultrasound* **37,** 34.

Moore, M. P., Gavin, P. R., Bagley, R. S., and Harrington, M. L. (1994). *Proc. ACVIM* **12,** 917.

Morgan, J. P., Atilola, M., and Bailey, C. S. (1987). *J. Am. Vet. Med. Assoc.* **191,** 951.

Morin, D. E., Toenniessen, J. G., French, R. A., Knight, B. L., and Zachary, J. F. (1994). *J. Am. Vet. Med. Assoc.* **204,** 938.

Mullaney, T. P., Levin, S., and Indrieri, R. J. (1983). *J. Am. Vet. Med. Assoc.* **182,** 516.

Munana, K. R., Lappin, M. R., Powell, C. C., Cooper, C. M., and Chavkin, M. J. (1995). *Prog. Vet. Neurol.* **6,** 27.

Murphy, F. A., Bell, J. F., Bauer, S. P., Gardner, J. J., Moore, G. J., Harrison, A. K., and Coe, J. E. (1980). *Lab. Invest.* **43,** 231.

Nafe, L.A., Turk, J. R., and Carter, J .D. (1983). *J. Am. Anim. Hosp. Assoc.* **19,** 933.

Nayer, P. S. G., Ward, G. E., Saunders, J. R., and MacWilliams, P. (1977). *Can. Vet. J.* **18,** 159.

Nilsson, C., Lindvall-Axelsson, M., and Owman, C. (1992). *Brain Res. Rev.* **17,** 109.

Norton, F. (1992). *Prog. Vet. Neurol.* **3,** 120.

Novak, R. W. (1984). *Am. J. Clin. Pathol.* **82,** 95.

Novati, R., Castagna, A., Morsica, G., Vago, L., Tambussi, G., Ghezzi, S., Gervasoni, C., Bisson, C., d'Arminio Monforte, A., and Lazzarin, A. (1994). *AIDS* **8,** 1691.

Oliver, J. E. and Lorenz, M. D. (1993). "Handbook of Veterinary Neurology" 2nd ed. Saunders, Philadelphia, Pennsylvania.

Olson, M. E., Chernik, N. L., and Posner, J. B. (1974). *Arch. Neurol.* **30,** 122.

Parish, S. M., Maag-Miller, L., Besser, T. E., Weidner, J. P., McElwain, T., Knowles, D. P., and Leathers, C. W. (1987). *J. Am. Vet. Med. Assoc.* **191,** 1599.

Parker, A. J. and Cunningham, J. G. (1971). *J. Small Anim. Pract.* **12,** 513.

Parker, J. L. and White, K. K. (1992). *Cornell Vet.* **82,** 253.

Parmley, S. F., Goebel, F. D., and Remington, J. S. (1992). *J. Clin. Microbiol.* **30,** 3000.

Patton, S., Legendre, A. M., and Pelletier, D. (1991). *J. Vet. Intern. Med.* **5,** 199.

Pelc, S., De Maertelaere, E., and Denolin-Reubens, R. (1981). *Eur. Neurol.* **20,** 95.

Peter, J. B., and Tourtellotte, W. W. (1986). *Ann. Neurol.* **20,** 167.

Peters, M., Pohlenz, J., Jaton, K., Ninet, B., and Bille, J. (1995). *J. Vet. Med.* **42,** 84.

Phillips, T. R., Prospero-Garcia, O., Puaoi, D. L., Lerner, D. L., Fox, H. S., Olmsted, R. A., Bloom, F. E., Henriksen, S. J., and Elder, J. H. (1994). *J. Gen. Virol.* **75,** 979.

Podell, M., Oglesbee, M., Mathes, L., Krakowka, S., Olmstead, R., and Lafrado, L. (1993). *J. Acquir. Immune. Defic. Syndr.* **6,** 758.

Pollay, M. (1992), *In* "Barriers and Fluids of the Eye and Brain" (M. B. Segal, ed.), p. 49 CRC Press, Boca Raton, Florida.

Pugh, D. G., Causey, M. K., Blagburn, B. L., and Wolfe, D. F. (1995). *Compend. Contin. Educ. Pract. Vet.* 600.

Rand, J. S., Parent, J., Jacobs, R., and Johnson, R. (1990a). *Am. J. Vet. Res.* **51,** 1049.

Rand, J. S., Parent, J., Jacobs, R., and Percy, D. (1990b). *Am. J. Vet. Res.* **51,** 1044.

Rand, J. S., Parent, J., Percy, D., and Jacobs, R. (1994a). *Can. Vet. J.* **35,** 103.

Rand, J. S., Parent, J., Percy, D., and Jacobs, R. (1994b). *Can. Vet. J.* **35,** 174.

Rebhun, W. C., and deLahunta, A. (1982). *J. Am. Vet. Med. Assoc.* **180,** 395.

Reed, S. M., Granstrom, D., Rivas, L. J., Saville, W. A., Moore, B. R., and Mitten, L. A. (1994). *Am. Assoc. Eq. Pract. Proc.* **40,** 199.

Reiber, H. (1994). *J. Neurol. Sci.* **122,** 189.

Reiber, H., and Felgenhauer, K. (1987). *Clin. Chim. Acta* **163,** 319.

Reiber, H., and Lange, P. (1991). *Clin. Chem.* **37,** 1153.

Rhodes, C. H., Glantz, M. J., Glantz, L., Lekos, A., Sorenson, G. D., Honsinger, C., and Levy, N. B. (1996). *Cancer* **77,** 543.

Rider, L. G., Thapa, P. B., Del Beccaro, M. A., Gale, J. L., Foy, H. M., Farwell, J. R., and Mendelman, P. M. (1995). *Pediatr. Emerg. Care* **11,** 226.

Rishniw, M., Wilkerson, M. J., and deLahunta, A. (1994). *Prog. Vet. Neurol.* **5,** 35.

Roeder, B. L., Johnson, J. W., and Cash, W. C. (1990). *Compend. Contin. Educ. Pract. Vet.* **12,** 1175.

Rosenberg, G. A. (1990). "Brain Fluids and Metabolism." Oxford University Press, New York.

Rossdale, P. D., Cash, R. S. G., and Leadon, D. P. (1982). *Eq. Vet. J.* **14,** 134.

Roszel, J. F. (1972). *J. Am. Vet. Med. Assoc.* **160,** 720.

Ruckebusch, M., and Costes, G. (1988). *Revue Med. Vet.* **139,** 1125.

Ruckebusch, M., and Sutra, J. F. (1984). *J. Physiol.* **348,** 457.

Ruehlmann, D., Podell, M., Oglesbee, M., and Dubey, J. P. (1995). *J. Am. Anim. Hosp. Assoc.* **31,** 174.

Rutgers, C., Kowalski, J., Cole, C. R., Sherding, R. G., Chew, D. J., Davenport, D., O'Grady, M., and Murtaugh, R. J. (1985). *J. Am. Anim. Hosp. Assoc.* **21,** 361.

Santschi, E. M., and Foreman, J. H. (1989). *Compend. Contin. Educ. Pract. Vet.* **11,** 479.

Sarfaty, D., Carrillo, J. M., and Greenlee, P. G. (1986). *J. Am. Vet. Med. Assoc.* **188,** 387.

Sarfaty, D., Carrillo, J. M., and Peterson, M. E. (1988). *J. Am. Vet. Med. Assoc.* **193,** 854.

Scarratt, W. K. (1987). *Compend. Contin. Educ. Pract. Vet.* **9,** F28.

Schaeffer, M. C., Rogers, Q. R., Leung, P. M., Wolfe, B. M., and Strombeck, D. R. (1991). *Life Sci.* **48,** 2215.

Schaer, M., Johnson, K. E., and Nicholson, A. C. (1983). *J. Am. Anim. Hosp. Assoc.* **19,** 311.

Schliep, G., and Felgenhauer, K. (1974). *J. Neurol.* **207,** 171.

Schott, H. C., Major, M. D., Grant, B. D., and Bayly, W. M. (1990). *J. Am. Vet. Med. Assoc.* **196,** 1820.

Schreuder, B. E. C. (1994a). *Vet. Q.* **3,** 182.

Schreuder, B. E. C. (1994b). *Vet. Q.* **3,** 174.

Scott, P. R. (1992). *Br. Vet. J.* **148,** 15.

Scott, P. R. (1993). *Br. Vet. J.* **149,** 165.

Scott, P. R. (1995). *Br. Vet. J.* **151,** 603.

Scott, P. R., Aldridge, B. M., Clarke, M., and Will, R. G. (1990). *Br. Vet. J.* **146,** 88.

Scott, P.R., Sargison, N. D., Penny, C. D., and Strachan, W. D. (1995). *Br. Vet. J.* **151,** 459.

Sharief, M. K., and Thompson, E. J. (1989). *J. Neurol. Neurosurg. Psychiatry* **52,** 949.

Shell, L. G., Carrig, C. B., Sponenberg, D. P., and Jortner, B. S. (1988). *J. Am. Anim. Hosp. Assoc.* **24,** 341.

Sherman, D. M. (1987). *Vet. Clin. North Am. Food Anim. Pract.* **3,** 179.

Sindern, E., and Malin, J. P. (1995). *Acta Cytol.* **39,** 73.

Sisk, D. B., Colvin, B. M., and Merrill, A. (1990). *Vet. Hum. Toxicol.* **32,** 205.

Smith, M. O. (1995). *Proc. ACVIM* **13,** 929.

Smith, M. O., and Lackner, A. A. (1993). *Am. J. Vet. Res.* **54,** 1845.

Smith-Maxie, L. L., Parent, J. P., Rand, J., Wilcock, B. P., and Norris, A. M. (1989). *J. Vet. Intern. Med.* **3,** 167.

Smyth, G. B., Vaughn, D. M., and Frischmeyer, K. J. (1994). *Prog. Vet. Neurol.* **5,** 13.

Sorjonen, D. C. (1987). *Am. J. Vet. Res.* **48,** 301.

Sorjonen, D. C. (1990). *J. Am. Anim. Hosp. Assoc.* **26,** 141.

Sorjonen, D. C., Warren, J. N., and Schultz, R. D. (1981). *J. Am. Anim. Hosp. Assoc.* **17,** 833.

Sorjonen, D. C., Golden, D. L., Levesque, D. C., Shores, A., and Moore, M. P. (1991). *Prog. Vet. Neurol.* **2,** 261.

Sornas, R. (1967). *J. Neurol. Neurosurg. Psychiatry* **30,** 568.

Spencer, C. P., Chrisman, C. L., Mayhew, I. G., and Kaude, J. V. (1982). *Am. J. Vet. Res.* **43,** 1958.

St. Jean, G., Yvorchuk-St.Jean, K., Anderson, D. E., and Moore, W. E. (1995). *Proc. ACVIM* **13,** 1008.

Stalis, I. H., Chadwick, B., Dayrell-Hart, B., Summers, B. A., and Van Winkle, T. J. (1995). *Vet. Pathol.* **32,** 230.

Steckel, R. R., Adams, S. B., Long, G. G., and Rebar, A. H. (1982). *J. Am. Vet. Med. Assoc.* **180,** 1085.

Steele, R. W., Marmer, D. J., O'Brien, M. D., Tyson, S. T., and Steele, C. R. (1986). *J. Clin. Microbiol.* **23,** 965.

Steinberg, S. A., and Vandevelde, M. (1974). *Folia Vet. Lat.* **4,** 235.

Stiles, J., Prade, R., and Greene, C. E. (1996). *Am. J. Vet. Res.* **57,** 264.

Stokes, H. B., O'Hara, C. M., Buchanan, R. D., and Olson, W. H. (1975). *Neurology* **25,** 901.

Stoodley, M. A., Jones, N. R., and Brown, C. J. (1996). *Brain Res.* **707,** 155.

Strain, G. M., Barta, O., Olcott, B. M., and Braun, W. F. J. (1984). *Am. J. Vet. Res.* **45,** 1812.

Swarup, D., and Maiti, S. K. (1991). *Indian J. Anim. Sci.* **61,** 942.

Taylor, H.W., Vandevelde, M., and Firth, E. C. (1977). *Vet. Pathol.* **14,** 479.

Thomas, J. B., and Eger, C. (1989). *J. Small Anim. Pract.* **30,** 287.

Thompson, E. J. (1988). "The CSF Proteins: A Biochemical Approach." Elsevier Science Publishing Co., New York.

Thomson, C. E., Kornegay, J. N., and Stevens, J. B. (1989). *J. Small Anim. Pract.* **30,** 685.

Thomson, C. E., Kornegay, J. N., and Stevens, J. B. (1990). *J. Am. Vet. Med. Assoc.* **196,** 1841.

Tibbling, G., Link, H., and Ohman, S. (1977). *Scand. J. Clin. Lab. Invest.* **37,** 385.

Tipold, A. (1995). *J. Vet. Intern. Med.* **9,** 304.

Tipold, A., and Jaggy, A. (1994). *J. Small Anim. Pract.* **35,** 311.

Tipold, A., Fatzer, R., Jaggy, A., Zurbriggen, A., and Vandevelde, M. (1993a). *J. Small Anim. Pract.* **34,** 623.

Tipold, A., Pfister, H., and Vandevelde, M. (1993b). *Res. Vet. Sci.* **54,** 40.

Tipold, A., Pfister, H., Zurbriggen, A., and Vandevelde, M. (1994). *Vet. Immunol. Immunopathol.* **42,** 149.

Tipold, A., Vandevelde, M., and Zurbriggen, A. (1995). *Res. Vet. Sci.* **58,** 103.

Trotter, J. L., and Rust, R. S. (1989), *In* "The Cerebrospinal Fluid" (R. M. Herndon and R. A. Brumback, eds.), p. 179. Kluwer Academic Publishers, Boston.

Tsai, S. C., Summers, B. A., and Appel, M. J. G. (1982). *Arch. Virol.* **72,** 257.

van Bree, H., Van Rijssen, B., and Van Ham, L. (1991). *Am. J. Vet. Res.* **52,** 926.

Vandevelde, M., and Spano, J. S. (1977). *Am. J. Vet. Res.* **38,** 1827.

Vandevelde, M., Zurbriggen, A., Steck, A., and Bichsel, P. (1986). *J. Neuroimmunol.* **11,** 41.

Vaughn, D. M., Simpson, S. T., Blagburn, B. L., Whitmer, W. L., Heddens-Mysinger, R., and Hendrix, C. M. (1989a). *Vet. Res. Commun.* **13,** 47.

Vaughn, D. M., Coleman, E., Simpson, S. T., and Satjawatcharaphong, C. (1988a). *Am. J. Vet. Res.* **49,** 1302.

Vaughn, D. M., Coleman, E., Simpson, S. T., Whitmer, B., and Satjawatcharaphong, C. (1988b). *Am. J. Vet. Res.* **49,** 2134.

Vaughn, D. M. and Smyth, G. B. (1989). *Vet. Res. Commun.* **13,** 413.

Vaughn, D. M., Smyth, G. B., Whitmer, W. L., and Satjawatcharaphong, C. (1989b). *Vet. Res. Commun.* **13,** 237.

Vernau, K. M., Kortz, G. D., Koblik, P. D., LeCouteur, R. A., Bailey, C. S., and Pedroia, V. (1996). *Veterinary Radiology & Ultrasound,* in press.

Waters, D. J., and Hayden, D. W. (1990). *J. Vet. Intern. Med.* **4,** 207.

Waxman, F. J., Clemmons, R. M., and Hinrichs, D. J. (1980). *J. Immunol.* **124,** 1216.

Weisner, B., and Bernhardt, W. (1978). *J. Neurol. Sci.* **37,** 205.

Welch, K. (1975), *In* "Advances in Neurology: Current Reviews" (W. J. Friedlander, ed.), p. 247. Raven Press, New York.

Welles, E. G., Tyler, J. W., Sorjonen, D. C., and Whatley, E. M. (1992). *Am. J. Vet. Res.* **53,** 2050.

Welles, E. G., Pugh, D. G., Wenzel, J. G., and Sorjonen, D. C. (1994). *Am. J. Vet. Res.* **55,** 1075.

Widmer, W. R., and Blevins, W. E. (1991). *J. Am. Anim. Hosp. Assoc.* **27,** 163.

Widmer, W. R., DeNicola, D. B., Blevins, W. E., Cook, J. R., Cantwell, H. D., and Teclaw, R. F. (1992). *Am. J. Vet. Res.* **53,** 396.

Williams, M. A., Welles, E. G., Gailor, R. J., Ewart, S. L., Humburg, J. M., Mullaney, T. P., Stickle. J, , Chang, C. D., and Walter, G. L. (1992). *Prog. Vet. Neurol.* **3,** 51.

Wilson, J. W. (1977). *J. Am. Vet. Med. Assoc.* **171,** 200.

Wilson, J. W., and Stevens, J. B. (1977). *J. Am. Vet. Med. Assoc.* **171,** 256.

Wilson, J. W., Kurtz, H. J., Leipold, H. W., and Lees, G. E. (1979). *Vet. Pathol.* **16,** 165.

Wood, A. K., Farrow, B. R. H., and Fairburn, A. J. (1985). *Acta Radiol.* **26,** 767.

Woodruff, K. H. (1973). *Am. J. Clin. Pathol.* **60,** 621.

Yvorchuk, K. (1992), *In* "Current Therapy in Equine Medicine," 3rd ed. (N. E. Robinson, ed.), p. 569. Saunders, Philadelphia, Pennsylvania.

C H A P T E R

28

Clinical Biochemistry in Toxicology

JAMES R. TURK AND STAN W. CASTEEL

The availability of accurate historical information, including a list of the animal species affected, clinical signs, toxins suspected, potential route of exposure, vehicle, relative amount, and timing of exposure, is often a limiting factor in the diagnosis of toxic disease. This information permits the diagnostician to make a rational selection of samples and tests to be performed by considering the known target organs of the toxins suspected. Unfortunately, this information is often unavailable during the initial stages of an intoxication. In the absence of such detailed history, the identification of target organs using clinical biochemistry may be helpful in the retrospective creation of a list of potential toxins.

There are species differences in the most useful biochemical markers of organ-specific cellular injury and in the susceptibility to various toxins (Kramer, 1989). For this reason, the following discussion is organized on an organ system and species basis. Neither bacterial toxins nor hereditary disease predisposing to intoxication will be discussed.

I. HEPATOTOXICITY

Many toxins induce hepatic injury (Table 28.1). The susceptibility of the liver to toxic insult is in part a consequence of its location between the digestive tract and the rest of the body and the central role it plays in biotransformation and disposition of xenobiotics (Miyai, 1991; Snyder, 1979). Extrahepatic metabolism of toxins by mixed-function oxidases may affect the target organ and potential hepatotoxicity of a given xenobiotic (Gram *et al.*, 1986). A variety of factors, including the induction of these enzyme systems by drugs (Snyder, 1979) and suppression of enzyme activity by infectious agents and cytokines (Monshouwer *et al.*, 1995), may modify the response to a given toxin. Lipophilic compounds tend to be more hepatotoxic than hydrophilic ones because of elimination of the latter by the kidney (Kelly, 1993). Many toxins are hepatotoxic and nephrotoxic, however, and most toxins have multiple target organs.

The cytosolic enzyme alanine aminotransferase (ALT) is nearly specific for hepatocellular injury in the dog and cat. The plasma half-life of this enzyme in these species is approximately 60 hours. Increased serum levels parallel the magnitude of hepatocellular

TABLE 28.1 Hepatotoxins[a]

Hepatotoxin	Disease onset	Geography	Species affected
Aflatoxin B_1 (*Aspergillus flavus*)	Acute to chronic	Worldwide	All
Agave lecheguilla	Acute	US, MEX	Cp, O
Blue-green algae (cyanobacteria)	Acute	Worldwide	All
Chlorinated hydrocarbons	Acute to chronic	Worldwide	All
Copper	Acute to chronic	Worldwide	O, All
Compositae	Acute	AFR, AUS	B, O, P
Corticosteroids	Chronic	Worldwide	Cn
Cresols (pitch)	Acute	Worldwide	P
Cycadales	Acute to chronic	AUS, FL, PR, DOM REP	B, O
Cylcopiazonic acid (*Penicillium cyclopium*)	Acute to chronic	UK	B, E, O
Dimethylnitrosamine	Acute to chronic	Worldwide	B, Cn, F, O
Ethanol	Acute to chronic	Worldwide	All
Fumonisin (*Fusarium moniliforme*)	Chronic	Worldwide	B, E, O, P
Gossypol (cottonseed)	Acute to chronic	Worldwide	B, Cp, E, O, P
Hymenoyxs odorata (bitterweed)	Acute to chronic	US	O
Indospicine (*Indigofera* spp.)	Acute to chronic	AUS	B, Cn, E
Iron	Acute	Worldwide	P, E
Karwinskia humboldtiana (coyotillo)	Acute to chronic	MEX, TX	B, Cp, O
Kochia scoparia (fireweed)	Chronic	US	B
Lantana camara	Chronic	AUS, AFR, MEX, US	B, O, E
Mebendazole	Acute	Worldwide	Cn
Moldy hay	Chronic	US	B
Myoporaceae	Acute to chronic	AUS, NZ	B, O
Nolina texana (sacahuiste)	Chronic	US	B, Cp, O
Petroleum	Acute	Worldwide	All
Phalloidin (*Amanita*)	Acute	Worldwide	All
Phomopsin (mycotoxin on lupines)	Acute to chronic	AUS, US	B, O
Phenytoin/primidone	Chronic	Worldwide	Cn
Phosphorus	Acute	Worldwide	All
Pyrrolizidine alkaloids	Chronic	Worldwide	B, Cp, E, O, P
Ricinus communis (Castor bean)	Acute	Worldwide	All
Rubratoxins (*Penicillium rubrum*)	Acute to chronic	Worldwide	B, CN, E, F
Sawfly larvae	Acute	AUS, DENMARK	B, Cp, O
Solanaceae	Acute	AFR, AUS, S AM, US	B, Cp, O
Sporodesmin (*Pithomyces chartarum*)	Chronic	AUS, AFR, NZ	B, O
Tannic acid (*Quercus* spp., oaks)	Acute	Worldwide	B, Cp, E, O
Terminalia oblongata (yellow-wood)	Acute to chronic	AUS	B, O
Tetradymia spp. (rabbit/horsebrush)	Chronic	US	B, Cp, O
Trema aspera (poison peach)	Acute	AUS	B
Trifolium hybridum (Alsike clover)	Chronic	Worldwide	B, E, O
Xanthium strumarium (Cocklebur)	Acute	US	B, P

[a] B, bovine; Cn, canine; Cp, caprine; E, equine; F, feline; O, ovine; P, porcine.

injury in acute disease. Several days following injury, ALT levels may be spuriously low. Therefore, ALT is not helpful in evaluating chronic liver disease. ALT levels also may be elevated by corticosteroid treatment. ALT is not useful in evaluating hepatic disease in the horse, cow, sheep, goat, or pig.

Aspartate aminotransferase (AST) occurs in most cells; however, it is useful in evaluating hepatocellular and muscular injury because of its high activity in these tissues. Elevation of AST is more specific than that of ALT in evaluation of hepatocellular insult in large animals.

Alkaline phosphatase (AP) isoenzymes occur in liver, bone, intestine, kidney, placenta, and leukocytes. Serum AP is of mainly hepatic origin in the dog, cat, and horse. Hepatic isoenzymes have a longer half-life (days) than intestinal, renal, or placental (minutes) isoenzymes. Cholestasis induces hepatic AP. Intrahepatic cholestasis tends to cause progressive increase, whereas extrahepatic obstruction may produce an increase that subsequently plateaus. Increase in serum AP precedes hyperbilirubinemia. Corticosteroids, phenobarbital, chlorinated hydrocarbons, and other compounds may induce hepatic AP. Increased osteoblastic activity in hyperparathyroidism, bone healing, or osteosarcoma may elevate AP. Alkaline phosphatase is insensitive for the detection of cholestasis in ruminants because of the wide variation in the normal activity of this enzyme.

Gamma glutamyltransferase (GGT) is found in many cells; however, the renal convoluted tubular epithelium, canalicular surfaces of hepatocytes, and bile-duct epithelium contain the highest activity. In renal disease GGT is excreted in the urine. Serum GGT is generally of hepatic origin and is elevated during cholestasis. GGT is more sensitive than AP for the detection of cholestatic disease in horses, cattle, sheep, and pigs. Increased serum GGT activity proved to be a sensitive and long-lived indicator of liver insult in cattle exposed to moldy hay (Casteel *et al.*, 1995). Like alkaline phosphatase, GGT appears in serum as a result of increased synthesis, rather than as a result of leakage from cells (Pearson, 1990).

In acute hepatic injury ALT and AST are generally elevated and AP may be normal. In chronic disease with cholestasis AP and GGT often are elevated and ALT and AST may be normal or only slightly elevated (Cornelius, 1989).

High activity of sorbitol dehydrogenase (SDH) is found in hepatocellular cytoplasm. The plasma half-life of this enzyme is very short and serum activities may return to normal within 5 days of hepatocellular insult.

Lactate dehydrogenase (LDH) is a tetrameric enzyme with five isoenzymes that catalyze the reversible conversion of L-lactate to pyruvate in all tissues. All LDH isoenzymes are found in varying concentrations in all tissues. LDH_1 is the principal isoenzyme in cardiac muscle and kidney of most species. It is also found in the liver of cattle and sheep. Unlike the other isoenzymes, it is heat stable at 65°C for 30 minutes. LDH_5 is the principal isoenzyme in skeletal muscle and erythrocytes. Serum LDH activity is tissue nonspecific; however, necrosis of muscle, liver, and hemolysis are the major causes of elevations.

Bilirubin is derived from destruction of damaged or senescent erythrocytes by macrophages of the spleen, liver, and bone marrow. It is noteworthy that bilirubin at physiologic levels is an antioxidant (Stocker *et al.*, 1987). Bilirubin is transported in plasma bound to proteins (albumin, globulin). Hepatic uptake and glucuronide conjugation renders it water soluble. Conjugated bilirubin is secreted into bile canaliculi and transported to the intestine, where the majority is transformed into urobilinogen by intestinal flora and excreted. Direct diazo assay for bilirubin detects conjugated bilirubin. Total bilirubin is measured after addition of alcohol, which allows additional color development. Unconjugated bilirubin is determined by the difference in direct and total bilirubin. Cholestasis results in conjugated hyperbilirubinemia. Bilirubinuria may occur because of "regurgitation" of conjugated bilirubin. Increased AP or GGT precedes hyperbilirubinemia in most species. Hemolysis may result in unconjugated hyperbilirubinemia and elevations of LDH_5.

Sulfobromophthalein (BSP) injected intravenously is removed rapidly from the blood, conjugated by hepatocytes, and excreted in bile. The rate of hepatic blood flow, functional hepatic mass, and patency of the biliary system affect hepatic clearance of this compound. Altered blood flow secondary to cardiotoxicity discussed later may increase BSP retention.

Acute toxic hepatic necrosis increases serum bile acids (cholic and chenodeoxycholic) in the dog, horse, sheep, and cow.

Ammonia is generated by microbial activity within the intestinal tract. It is absorbed from the intestine and transported to the liver by the portal venous system, where it is converted to urea by the healthy liver. Elevations of plasma ammonia during fasting or following ammonia challenge suggests reduction in functional hepatic mass.

Severe hepatic insufficiency may result in hypoproteinemia (Kaneko, 1989a) with reduction of plasma oncotic pressure that promotes tissue edema and effusions which mimic the effects of cardiotoxins.

The clinical signs of acute submassive or massive hepatic necrosis may include anorexia, vomiting, icterus, hepatic encephalopathy, disseminated intravascular coagulopathy, edema, and effusions. Surprisingly, there may be few or no clinical signs in some cases. The activity of ALT and SDH with short half-lives may be elevated, but often fall rapidly. Inducible enzymes such as AP and GGT may increase gradually.

All enzymes may return to normal in the presence of chronic severe liver disease. Hyperbilirubinemia may follow if lesions progress to chronicity and fibrosis.

Chronic hepatotoxicity has sequelae for most organ systems, but especially the nervous (hepatic encephalopathy), integumentary (secondary photosensitization in herbivores), and cardiovascular systems. Cardiotoxins and pneumotoxins may produce enzyme elevations suggestive of hepatic or renal disease due to ischemia/hypoxia.

Ingestion of toxic plants (Table 28.1) tends to be more common in herbivores than carnivores; however, nonherbivorous species are often susceptible if they are willing to ingest them. In addition to hepatotoxins, the blue-green algae (actually classified as cyanobacteria) that contaminate water (Carmichael, 1994) possess neurotoxins that may induce sudden death before any alterations in clinical biochemistry or morphologic changes.

Hepatotoxic plants of the order Compositae include *Asaemia axillaris, Athanasia trifurcata, Helichrysum blandowskianum, Lasiospermum bipinatum,* and *Xanthium* spp. The toxin in these plants has been identified as carboxyatractyloside.

Cycadales contain methylazoxymethanol, which is converted by hepatic microsomal activity to potent alkylating agents.

Hepatotoxic plants of the order Myoporaceae include *Myosporum deserti, M. acuminatum, M. insulare, M. tetramdum,* and *M. laetum.* These plants contain furanosesquiterpenoid oils, the best characterized of which is ngaione.

Chronic intoxication of swine with fumonisins, mycotoxins produced by *Fusarium moniliforme,* is associated with elevations of serum total bilirubin, AP, AST, GGT, and cholesterol (Casteel *et al.*, 1994b). In addition, fumonisins inhibit *N*-acetyl transferase, resulting in loss of complex sphingolipids and accumulation of sphinganine and sphingosine in tissues and serum (Riley *et al.*, 1993).

Iron toxicity from nutritional supplements may produce hepatic necrosis in foals (Acland *et al.*, 1984) and pigs (Kelly, 1993).

Hepatotoxic pyrrolizidine alkaloids are found in the plant species *Amsinckia intermedia, Crotolaria* spp., *Cynoglossum officinale, Echium plantagineum, Heliotropium europeaum, Senecio jacobea, S. vulgaris,* and *S. longilobus.*

Hepatotoxic Solanaceae include *Cestrum parqui, C. laevigatum,* and *C. aurantiacum.* The toxin in these plants has been identified as atractyloside.

Sawfly larvae (*Lophyrotoma interruptus* and *Arge pullata*) infesting eucalyptus and birch trees in Australia and Denmark, respectively, have been reported to induce hepatic necrosis in cattle, sheep, and goats browsing foliage.

Terminalia oblongata, the yellow-wood tree of Australia, contains hepatotoxic tannins and pucicalagin, which induces acute hepatic necrosis (Kelly, 1993).

II. NEPHROTOXICITY

Acute nephrotoxicity (Table 28.2) may initially induce polyuria that is followed by oliguria or anuria. Nephrotoxins affecting approximately 66% of the nephrons will result in inability to concentrate urine to a specific gravity greater than 1.030 in the dog, 1.035 in the cat, and 1.025 in the horse and cow. Chronic toxicity may result in isosthenuria (constant urine osmolality in the range of glomerular filtrate, 1.008–1.012).

The majority of urea is synthesized in the liver from ammonia formed by protein catabolism or intestinal absorption. Urea enters the vascular system and is distributed throughout the total body water compartment by passive diffusion. The urea concentrations of blood and glomerular filtrate are approximately equal. Urea passively diffuses from the tubular lumen back to the blood. Urine flow rate is inversely related to urea reabsorption. Gastrointestinal secretion is inconsequential in monogastrics; however, in ruminants up to 90% of urea in glomerular filtrate may be reabsorbed and enter the rumen via saliva for utilization in amino acid synthesis.

Azotemia, elevation in blood urea nitrogen (BUN), may occur as prerenal, renal, or postrenal. Prerenal azotemia may result from increased protein catabolism (e.g., hepatotoxins, myotoxins, other necrosis) or decreased renal perfusion (e.g., cardiotoxins). Renal azotemia occurs only after approximately 75% of the nephrons have lost function.

Renal disease also results in elevation of serum creatinine. The majority of serum creatinine originates from the endogenous conversion of phosphocreatine in muscle, which occurs at a relatively constant rate. Creatinine is not reutilized. The creatinine pool is modified by muscle mass, conditioning, and muscle disease. Creatinine also is distributed throughout the compartment of total body water. It diffuses more slowly than urea, however, and is not reabsorbed within the tubules after leaving as glomerular filtrate. Creatinine concentration is not affected significantly by diet, protein catabolism, or urinary flow. Reduced renal perfusion affects BUN and creatinine similarly (Finco, 1989). Elevations of BUN and creatinine are not proportional in renal disease of ruminants because of reutilization of urea by the rumen.

TABLE 28.2 Nephrotoxins[a]

Nephrotoxin	Disease onset	Geography	Species affected
Amaranthus retroflexus (pigweed)	Subacute	US	B, P
Antibiotics	Acute to chronic	Worldwide	All
Cantharidin (blister beetle)	Acute	US	B, Cp, E, O
Chlorinated hydrocarbons	Acute	Worldwide	All
Ethylene glycol	Acute	Worldwide	B, Cn, F, P
Fumonisin (*Fusarium moniliforme*)	Acute to chronic	Worldwide	E, O, P
Gossypol (cottonseed)	Chronic	Worldwide	B
Hemoglobin	Acute		All
Hypercalcemia	Chronic	Worldwide	All
Isotropis spp.	Acute	AUS	B, O
Lantana camara	Chronic	US, MEX, AUS, AFR	B, E, O
Menadione	Acute	Worldwide	E
Metals (Cd, Hg, Pb, Tl)	Acute to chronic	Worldwide	All
Myoglobin	Acute		All
Ochratoxin (*Penicillium ochraceus*)	Chronic	Worldwide	P
Oxalates	Acute to chronic	Worldwide	B, Cp, O, P
Paraquat/diquat	Acute	Worldwide	All
Petroleum	Acute	Worldwide	All
Phenothiazine	Acute	Worldwide	E
Phenylbutazone	Acute	Worldwide	E
Phosphorus	Acute	Worldwide	All
Pyrrolizidine alkaloids	Chronic	Worldwide	B, Cp, E, O, P
Tannins (*Quercus* spp., oaks)	Acute	Worldwide	B, Cp, E, O
Terminalia oblongata (yellow-wood)	Acute to chronic	AUS	B, O

[a] B, bovine; Cn, canine; Cp, caprine; E, equine; F, feline; O, ovine; P, porcine.

The heat stable LDH_1 isoenzyme discussed earlier may be elevated in renal disease.

Hyperkalemia may occur in renal failure with oliguria or anuria and acidosis. Hypercalcemia is common in equine renal disease, whereas hypocalcemia and hypochloridemia are common in bovine renal disease.

Proteinuria in the absence of occult blood and cellular sediment is suggestive of renal disease. Glomerular lesions typically result in high protein levels in which albumin is the major constituent. Tubular disease, typical of many nephrotoxins, generally results in lower protein levels containing high levels of globulins. Hypoproteinemia secondary to chronic urinary loss (Kaneko, 1989a) promotes tissue edema and effusions that may mimic cardiotoxicity and hepatotoxicity, as discussed previously.

Bilirubin is considered to be mildly nephrotoxic. Bilirubinuria may occur due to "regurgitation" of conjugated bilirubin resulting from cholestatic hepatotoxins. Myoglobin is also nephrotoxic. Myoglobinuria may occur with toxic necrosis of skeletal and cardiac muscle. Hemoglobin appears to be nephrotoxic in the presence of concurrent dehydration or hypovolemia. Hemoglobinuria may occur with hemolytic toxins.

Dehydration exacerbates the nephrotoxicity of many agents, especially antibiotics in all species and nonsteroidal anti-inflammatory drugs (phenylbutazone) in the horse. Nephrotoxic antibiotics include the aminoglycosides (amikacin, gentamicin, kanamycin, neomycin, streptomycin, and tobramycin), amphotericin B, cephalosporins, polymixins, sulfonamides, and tetracyclines (Maxie, 1993). Elevation of GGT in urine is a sensitive indicator of aminoglycoside toxicity (Gossett *et al.*, 1987).

Hypercalcemia and hyperphosphatemia may result in nephrocalcinosis following iatrogenic hypervitaminosis D or ingestion of cholecalciferol rodenticide (Fooshee and Forrester, 1990) by any species. Ingestion of the toxic plants containing vitamin D-like analogs, including *Cestrum diurnum, Dactylis glomerata,* some *Solanum* species, and *Trisetum flavescens*, by herbivores also may produce hypercalcemia with calcification of soft tissues, including the kidney.

Nephrotoxic metals include arsenic, bismuth, cadmium, lead, mercury, and thallium (Maxie, 1993).

Plants containing toxic concentrations of soluble oxalates include the species *Amaranthus retroflexus* (pigweed), *Halogeton glomeratus, Oxalis* spp., *Rheum rhaponticum* (rhubarb), and *Sarcobatus vermiculatus* (greasewood). Intoxication with ethylene glycol from antifreeze is one of the more common accidental or malicious poisonings encountered in dogs and cats. Birefringent hippurate and oxalate crystals may be observed in urine sediments (Kramer *et al.*, 1984).

Nephrotoxic pyrrolizidine alkaloids include the plant species listed in Section I on hepatotoxicity.

Trees of the genus *Quercus* (oaks) and *Terminalia oblongata* (yellow-wood tree) contain tannins that induce acute tubular necrosis when leaves, buds, or acorns are ingested. *Amaranthus retroflexus* (pigweed), via an unidentified toxic principle, also induces similar renal disease in cattle (Casteel *et al.*, 1994a) and pigs (Osweiler *et al.*, 1969) in the absence of oxalate nephrosis. At postmortem examination, there were consistent elevations of urea and creatinine concentrations in ocular fluid and serum.

III. TOXINS AFFECTING SKELETAL AND CARDIAC MUSCLE

The clinical signs of weakness, dysmetria, and incoordination suggest not only the possibility of neurologic disease, but also skeletal muscular or cardiovascular disease. Acute toxicity of skeletal or cardiac muscle (Table 28.3) can be detected by elevations in serum creatine (phospho)kinase [C(P)K] (Cardinet, 1989). This dimeric enzyme catalyzes the reversible reaction phosphocreatine + ADP $\leftrightarrows$ creatine + ATP and has three isoenzyme types, CK_1, CK_2, and CK_3. CK_1 is found in brain, peripheral nerves, cerebrospinal fluid, and viscera, but is not found in serum during neurologic disease. CK_2 is found in cardiac muscle and in minute amounts in skeletal muscle. CK_3 is found in cardiac and skeletal muscle. CK plasma half-life is short and is considered to be specific for muscle when hemolysis, elevated bilirubin, muscle fluid contamination during venipuncture, and dilution of CK inhibitors during sample processing can be excluded. When injury is not progressive CK elevations maximize within 6–12 hours and return to normal within 24–48 hours. Continuing necrosis results in persistent elevation.

As discussed previously, LDH_5 is the principal isoenzyme in skeletal muscle and erythrocytes. LDH activity is tissue nonspecific, but muscle and liver necrosis and hemolysis are the major sources for elevations of serum activity. In muscle necrosis elevation of serum LDH activity is of lower magnitude, parallels CK elevation, and peaks with 48–72 hours. Elevation of serum AST occurs more slowly than that of CK or LDH and may persist for several days after cessation of muscle injury.

Necrosis of skeletal muscle may result in release of myoglobin and potassium, resulting in myoglobinemia and hyperkalemia. Myoglobinuria, detectable by urinalysis, may induce secondary nephrotoxicity.

Cardiac glycosides disrupt cardiac ion channels, producing sudden dysrhythmias and often death (Cheville, 1988). Plants containing cardiac glycosides include *Bryophyllum tubiflorum, Digitalis* spp. (foxglove), *Homeria* spp. (cape tulip), *Nerium oleander* (oleander), *Thevetia peruviana* (yellow oleander), and *Tylecodon* spp. (Robinson and Maxie, 1993). Most intoxications occur in herbivores. Amphibians producing cardiac glycosides include *Bufo alvarius, B. marinus,* and *Dendrobates* spp. (poison dart frog). Most intoxications occur in cats and dogs that become curious about these animals.

Fluoroacetate containing plants include *Acacia georginae, Dichapetalum cymosum, Gastrolobium* spp., and *Oxylobium* spp. Fluoroacetate (compound 1080) also has been utilized as a rodenticide. It is not directly toxic, but combines with oxaloacetic acid to form fluorocitrate, which inhibits *cis*-aconitase and succinic dehydrogenase of the citric acid cycle, thus reducing ATP generation.

Glycosides and fluoroacetate may produce sudden death that precedes alterations of clinical chemistry or morphologic changes.

Hypercalcemia may induced cardiac calcinosis and nephrotoxicity, as discussed earlier.

Hyperkalemia from myotoxicity (especially gossypol)(Albrecht *et al.*, 1969), nephrotoxicity, or adrenal necrosis (hypoaldosteronism) may exert a dysrhythmogenic (Q–T prolongation and high-amplitude T waves) effect upon the heart. Potassium chloride injection also has been used for lethal poisoning by individuals attempting to circumvent detection by insurance adjusters (Casteel *et al.*, 1989).

Cardiotoxic metals include lithium, cadmium, nickel, barium, lanthanum, manganese, vanadium, lead, and cobalt (Van Vleet and Ferrans, 1986). Iron–dextran toxicity in pigs may produce necrosis of skeletal muscle and hyperkalemia, sparing the myocardium (Kelly, 1993).

Quinolizidine alkaloids in *Lupinus* spp., *Laburnum anagyroides,* and *Thermopsis montana* have been shown to produce skeletal muscle necrosis in cattle with elevations of serum CK and AST in the absence of myoglobi-

TABLE 28.3 Toxins Affecting Cardiac and Skeletal Muscle[a]

Toxins	Disease onset	Geography	Species affected
Cardiac glycosides	Acute	Worldwide	All
Cassia occidentalis (coffee senna)	Acute	Worldwide	B, Cp, O, P
Cantharidin (blister beetle)	Acute	US	E
Catecholamines	Acute		All
Eupatorium rugosum (white snakeroot)	Chronic	US	B, E
Fluoroacetate	Acute	AUS, S AFR	All
Gossypol (cottonseed)	Chronic	Worldwide	P, B, Cn
Hypercalcemia	Chronic	Worldwide	All
Iron	Acute	Worldwide	P
Karwinskia humboldtiana (coyotillo)	Acute to chronic	MEX, TX	Cp, O
Lantana camara	Chronic	AFR, AUS, MEX, US	B, E, O
Metals (As, Bi, Cd, Hg, Pb, Tl)	Acute to chronic	Worldwide	All
Methylxanthines (caffeine, theophylline, theobromine)	Acute	Worldwide	Cn
Monensin	Acute to chronic	Worldwide	B, E, Cp, O, P
Nephrotoxins (uremia; see text)	Chronic		All
Persea americana (avocado)	Acute	Worldwide	B, Cp, E, O
Phalaris	Acute	Worldwide	B, E, O
Phosphorus	Acute	Worldwide	All
Potassium	Acute to chronic	Worldwide	All
Pteridium aquilinum (Bracken fern)	Acute to chronic	Worldwide	B, Cp, E, O, P
Quinolizidine alkaloids (*Lupinus* spp.)	Chronic	Worldwide	B, O
Vicia villosa (hairy vetch)	Chronic	Worldwide	B, E

[a] B, bovine; Cn, canine; Cp, caprine; E, equine; F, feline; O, ovine; P, porcine.

nuria (Keeler and Baker, 1990). Quinolizidine alkaloids are also teratogenic.

The numerous chemotherapeutic agents that have been associated with cardiotoxicity have been reviewed elsewhere (Van Vleet and Ferrans, 1986).

Cardiotoxins may produce serum enzyme elevations suggestive of hepatic or renal disease secondary to ischemia/hypoxia.

IV. TOXINS AFFECTING THE LUNG AND RESPIRATORY TRACT

Disease affecting the respiratory tract is often clinically apparent on the basis of dyspnea. Dyspnea in veterinary medicine is more often the result of pneumonia rather than intoxication. However, when body temperature is normal, the possibility of pulmonary edema induced by toxins affecting the lung and respiratory tract (Table 28.4) or cardiovascular system should be considered. Since, with the exception of the horse, domestic mammals remove excess heat by panting, reduced respiratory capacity secondary to intoxication also may result in elevated body temperature.

Angiotensin-converting enzyme (ACE) is concentrated on the luminal surface of pulmonary endothelial cells. Most circulating ACE originates from the lung; however, many tissues including tubular epithelial and endocrine cells contain this enzyme (Erdos, 1987). Serum ACE activity is altered in chronic and acute pulmonary disease (Hollinger, 1983). Unfortunately, assays for this enzyme are not readily available in most veterinary clinical biochemistry laboratories, and ACE remains primarily a research tool.

Acute pulmonary edema is the typical lesion resulting from toxins affecting the epithelial–endothelial interface of the alveolus. Because CO_2 is approximately 20 times more diffusible than O_2, early pulmonary edema typically results in decreased p_aO_2, while p_aCO_2 remains normal or may decrease with hyperventilation, producing respiratory alkalosis. Severe pulmonary edema may result in elevated p_aCO_2 (hypercapnia and respiratory acidosis) (Carlson, 1989).

Chronic insult to the alveolar epithelial–endothelial interface may progress to pulmonary fibrosis (e.g.,

TABLE 28.4 Toxins Affecting the Lung and Respiratory Tract[a]

Toxins	Disease onset	Geography	Species affected
Eupatorium adenophorum	Chronic	AUS	E
Fumonisin (*Fusarium moniliforme*)	Acute to chronic	Worldwide	P
4-Ipomeanol (*Fusarium solanii*)	Acute	Worldwide	B
Kerosene/petroleum	Acute	Worldwide	All
3-Methylindole	Acute	Worldwide	B
Myoporaceae	Acute	AUS, NZ	B, O
Organobromines/organochlorines	Chronic	Worldwide	All
O_2	Acute to chronic	Worldwide	All
Paraquat	Acute to chronic	Worldwide	All
Perilla frutescens	Acute	US	B, E, O
Pyrrolizidine alkaloids	Chronic	Worldwide	B, Cp, E, O, P

[a] B, bovine; Cn, canine; Cp, caprine; E, equine; F, feline; O, ovine; P, porcine.

paraquat intoxication) and be associated with low p_aO_2 and elevated p_aCO_2. Pulmonary fibrosis impinging upon the pulmonary vasculature may induce pulmonary hypertension and cor pulmonale associated with enzymologic alterations suggestive of hepatoxicity due to passive hepatic congestion.

Cattle are susceptible to several toxins that induce acute pulmonary edema, including ingestion of perilla ketone in *Perilla frutescens,* 4-ipomeanol in sweet potatoes (*Ipomoea batatas*) infected with *Fusarium solani* (Doster *et al.,* 1978), and the generation of 3-methylindole by ruminal *Lactobacillus skatoli* from tryptophan in lush pasture grasses (Breeze and Carlson, 1982).

Some of the pyrrolizidine alkaloids, notably monocrotaline from *Crotolaria spectabilis,* may induce chronic pulmonary arteriopathy resulting in pulmonary hypertension that progresses to right heart failure and elevation of enzymes suggestive of cardiotoxicity or hepatotoxicity. Similar pulmonary arteriopathy occurs in pigs with chronic fumonisin intoxication (Casteel *et al.,* 1994b).

Organochlorines and organobromines include chlorinated naphthalenes, polychlorinated biphenyls (PCBs), polybrominated biphenyls (PBBs), and dibenzofurans. These industrial toxins are cumulative and result in hypovitaminosis A that is associated with squamous metaplasia of columnar epithelium of the respiratory tract and hyperkeratosis, as discussed under integumentary toxins.

The numerous chemotherapeutic agents that have been associated with pulmonary toxicity or hypersensitivity have been reviewed elsewhere (Myers, 1993).

V. TOXINS AFFECTING THE GASTROINTESTINAL TRACT

Disease affecting the gastrointestinal tract is often clinically apparent on the basis of vomiting or diarrhea. In veterinary medicine these are more often the result of infectious disease rather than intoxication. However, the possibility of gastrointestinal toxins (Table 28.5) should be considered, especially in acute outbreaks affecting several animals sharing the same environment.

Vomiting and diarrhea may produce dehydration that results in mild to moderate elevations of BUN, plasma protein, packed cell volume (PCV), and urine specific gravity. BUN elevations are usually of lesser magnitude than those encountered in nephrotoxicity.

Metabolic alkalosis (increased pH, normal or increased pCO_2, increased HCO_3^- and HCO_3^-/H_2CO_3) and hypochloridemia may result from chloride loss associated with vomiting (Tennant and Hornbuckle, 1989).

Metabolic acidosis (decreased pH, normal or decreased pCO_2, decreased HCO_3^- and HCO_3^-/H_2CO_3) may result from secretory loss of bicarbonate in diarrhea.

In the absence of evidence of malnutrition, hepatic, or renal disease, chronic gastrointestinal malabsorption or protein loss should be considered as a potential cause of hypoproteinemia (Kaneko, 1989a).

Peracute intestinal toxicity, such as inorganic arsenic ingestion, may result in transient elevations of serum AP.

Ingestion of strong acids or alkalis may induce immediate and severe damage to the gastrointestinal mucosa.

TABLE 28.5 Toxins Affecting the Gastrointestinal Tract[a]

Toxins	Disease onset	Geography	Species affected
Acids/alkalis	Acute	Worldwide	All
Aleurites fordi (Tung oil tree)	Acute	AUS, US	B
Arsenic	Acute	Worldwide	All
Asclepias spp. (milkweed)	Acute	N AM	B, E, Cp, O
Cantharidin (blister beetle)	Acute	US	All
Copper	Acute	Worldwide	All
Fluoroacetate	Acute	AUS, MEX, S AFR	All
Fluorensia cernua (blackbush/tarbush)	Acute	MEX, US	Cp, O
Hymenoxys odorata (bitterweed)	Acute	US	O
Isotropis spp.	Acute	AUS	B, O
Kerosene/petroleum	Acute	Worldwide	All
Nerium oleander (oleander)	Acute	Worldwide	B, Cp, E, O
Pyrrolizidine alkaloids	Acute	Worldwide	B, Cp, E, O, P
Ricinus communis (Castor bean)	Acute	Worldwide	B, E
Robinia pseudoacacia (black locust)	Acute	N AM	B, E
Solanaceae	Acute	Worldwide	All
Selenium	Acute	Worldwide	B, Cp, E, O, P
Tannins (*Quercus* spp., oaks)	Acute	Worldwide	B, Cp, O
Trichothecenes	Acute	Worldwide	All
Urginea maritima (sea onion)	Acute	Worldwide	Cn, F, P
Zearalenone (*Fusarium roseum*)	Chronic	Worldwide	P
Zinc	Acute	Worldwide	B, Cn, O

[a] B, bovine; Cn, canine; Cp, caprine; E, equine; F, feline; O, ovine; P, porcine.

The seleniferous plants that may be associated with acute gastroenteritis in herbivores are discussed under the integumentary system.

Trichothecenes, especially T-2, produced by *Fusarium* spp., and the macrocyclic trichothecene produced by *Stachybotrys alternans*, are highly irritant and produce acute ulceration and hemorrhage of the gastrointestinal tract. These mycotoxins also produce acute ulcerative and necrotizing lesions of the skin and chronic pancytopenia with hemorrhage.

The estrogenic mycotoxin zearalenone may induce rectal prolapse in pigs in addition to affecting reproductive performance.

VI. TOXINS AFFECTING ERYTHROCYTES AND THE HEMATOPOIETIC SYSTEM

Toxins inducing hemolysis (Table 28.6) may produce elevations of serum LDH, anemia, icterus, hemoglobinemia, hemoglobinuria with secondary nephrotoxicity, and unconjugated bilirubinemia.

Chronic, drug-induced, immune-mediated hemolysis has been associated with *para*-aminosalicylic acid, chlorpromazine, dipyrine, insecticides, penicillin, phenacetin, quinine, quinidine, and sulfonamides (Valli, 1993).

Mellitin is a hemolyzing component of hymenopterous toxins and comprises approximately 50% of bee venom (Cheville, 1988).

Chronic nephrotoxicosis leading to uremia may result in hemolysis due to retention of creatinine and guanidinosuccinic acid. This mechanism is usually insufficient to produce acute hemolytic crises, but may result in anemia due to increased erythrocyte destruction.

The anthelmintic phenothiazine may be acutely hemolytic in sheep and horses. It also may induce primary photosensitization as discussed under integumentary toxins.

The venom of various snakes, including the Crotalidae, Elapidae, Hydrophidae, and Viperidae, contains a mixture of toxins of which phospholipase A_2 (PLA_2) is an important component. PLA_2 is directly lytic for

TABLE 28.6 Toxins Inducing Hemolysis[a]

Toxins	Disease onset	Geography	Species affected
Allium spp. (onion)	Acute	Worldwide	B, Cn, E, O
Acer rubrum (red maple)	Acute	US	E
Brassicae	Acute	Worldwide	B, O
Copper	Acute	Worldwide	Cn, O
Drug-induced immune-mediated anemia	Chronic	Worldwide	All
Mellitin	Acute	Worldwide	All
Methylene blue	Acute	Worldwide	F
Molybdenum	Chronic	Worldwide	B, Cp, O
Naphthalene	Acute	Worldwide	Cn
Nephrotoxins	Chronic		All
Phenothiazine	Acute	Worldwide	E, O
Snake venoms	Acute	Worldwide	All
Zinc	Acute to chronic	Worldwide	B, Cn, O, P

[a] B, bovine; Cn, canine; Cp, caprine; E, equine; F, feline; O, ovine; P, porcine.

erythrocytes and may induce hemolysis, and for platelets and may induce hemorrhage and coagulopathy (Cheville, 1988).

Additional toxins inducing hemorrhage (Table 28.7) include dicumarol derived from moldy sweet clover (*Melilotus* spp.) and the synthetic derivatives used as anticoagulant rodenticides, such as brodifacoum, bromadiolone, diphacinone, fumarin, pindone, and warfarin. These agents are vitamin K antagonists. Prothrombin (factor II), and factors VII, IX, and X require vitamin K for their production. The half-life of factor VII is approximately 4–6 hours, whereas half-lives of factors IX and X are approximately 14–18 hours, and that of prothrombin is 40 hours. Therefore, prolongation of the one-stage prothrombin time (PT) occurs earliest, followed by prolongation of the activated partial thromboplastin test (PTT). Vitamin K antagonists do not affect fibrinogen or platelet numbers initially, but may eventually result in exhaustion of their supplies. Excessive hemorrhage following slight trauma, epistaxis, melena, and hematuria may occur with these intoxications (Dodds, 1989).

Reduced synthesis of antithrombin III and antiplasmin in acute hepatotoxicity may promote thrombosis followed by unchecked fibrinolysis with generation of fibrin split products that interfere with platelet function.

Chronic hepatotoxicity in which hepatic mass is reduced by 70% or more may result in sufficiently inadequate synthesis of both clotting factors and their inhibitors to prolong PT and PTT. Chronic cholestasis with interruption of the enterohepatic circulation of bile salts also may result in malabsorption of fat-soluble vitamin K, producing a syndrome similar to anticoagulant intoxication.

Toxins inducing pancytopenia include estrogen, *Pteridium aquilinum* (bracken fern), and the trichothecene mycotoxins. Insufficient numbers of platelets

TABLE 28.7 Toxins Inducing Hemorrhage/Coagulopathy[a]

Toxins	Disease onset	Geography	Species affected
Anticoagulant rodenticides	Chronic	Worldwide	All
Dicoumarol (moldy *Melilotus alba*)	Chronic	Worldwide	B, E, P
Drugs	Chronic	Worldwide	All
Estrogen	Chronic	Worldwide	All
Hepatotoxins	Chronic	Worldwide	All
Pteridium aquilinum (bracken fern)	Chronic	Worldwide	B, Cp, E, O, P
Trichothecenes	Acute	Worldwide	All
Venoms	Acute	Worldwide	All

[a] B, bovine; Cn, canine; Cp, caprine; E, equine; F, feline; O, ovine; P, porcine.

promote hemorrhage and consumption coagulopathy (Valli, 1993).

VII. TOXINS AFFECTING HEMOGLOBIN AND OXIDATIVE METABOLISM

Lead poisoning interrupts heme synthesis at the level of formation of protoporphyrin and causes accumulation of delta-aminolevulinic acid. Increased urinary excretion of this metabolite is an indication of lead intoxication.

Toxins inducing oxidation of ferrous iron in hemoglobin to ferric iron in methemoglobin (Harvey, 1989) include the herbicide sodium chlorate, the stalk parts of nitrate-accumulating plants such as corn and wheat, hay grown on heavily fertilized soils under drought conditions, fertilizer, or water contaminated by fertilizers or organic material. Ruminants are most susceptible to the nitrate-accumulating plants *Amaranthus* spp. (pigweed), *Avena sativa* (oats), *Chenopodium* spp. (lambsquarter), *Sorghum* spp., and *Triticum aestivum* (wheat) because of the ability of rumen microbes to reduce nitrate to the proximate toxicant, nitrite. Monogastrics and ruminants are equally susceptible to nitrite-based fertilizers (Osweiler *et al.*, 1985). Nitrate/nitrite intoxication produces a brown discoloration of the blood due to methemoglobinemia.

Carbon monoxide (CO) competes with oxygen binding to the heme moiety in hemoglobin and myoglobin. CO affinity for the hemoglobin binding site is approximately 200 times that of O_2, resulting in tightly bound carboxyhemoglobin and decreased blood oxygen transport. Because continuous delivery of O_2 is critical to the heart and brain, carbon monoxide may induce signs of cardiotoxicity or neurotoxicity. Anoxia of the liver, kidney, and muscle may produce elevations of serum enzymes referable to these systems.

CO, cyanide, and H_2S are potent inhibitors of cytochrome oxidase and may produce sudden death due to failure of oxidative metabolism that precedes alterations in clinical biochemistry or morphology.

There are numerous plants containing cyanogenic glycosides that may affect herbivores, especially ruminants. The most common cultivated species include *Cynodon* spp., *Sorghum* spp., and *Prunus* spp. (Jubb and Huxtable, 1993).

VIII. TOXINS AFFECTING THE ENDOCRINE SYSTEM

Carbadox/Mecadox is an antibacterial agent that, with prolonged exposure at levels greater than 25 ppm, induces degeneration of the zona glomerulosa of the adrenal gland associated with reduced plasma aldosterone, hyperkalemia, and hyponatremia (Capen, 1993).

The drug *ortho-,para'*-2,2-bis(2-chlorophenyl-4-chlorophenyl)-1,1-dichloroethane (*o,p'*DDD) is toxic to the zonae fasciculata and reticularis of the adrenal gland and is used as therapy for canine hyperadrenocorticism. This toxin reduces circulating cortisol levels.

Goitrogenic substances induce iodine deficiency or inhibit organification of iodine (Kaneko, 1989b). Thiocyanates, produced by ruminal digestion of cyanogenic glycosides from the toxic plants *Cynodon* spp. and *Trifolium repens,* and goitrin, derived from *Brassica* spp., are goitrogenic. Mimosine (discussed later under integumentary toxins) is metabolized in the rumen to a compound that inhibits organic binding of iodine by the thyroid gland. Thioamides (sulfonamides) inhibit thyroperoxidase. All of these substances may reduce serum T_4 and T_3. Iodine toxicity producing hyperplastic goiter in horses has been associated with feeding kelp.

Hepatic glucuronidation is the rate-limiting step for biliary excretion of T_4. Sulfation by phenol sulfotransferase is the rate-limiting step for excretion of T_3. Induction of hepatic microsomal enzymes may increase T_4/T_3 elimination and disrupt the hypothalamic–pituitary–thyroid axis, resulting in excessive thyroid-stimulating hormone (TSH). Xenobiotics that induce hepatic microsomal enzymes include benzodiazepines, calcium channel blockers, chlorinated hydrocarbons, phenobarbital, PCBs, PBBs, retinoids, and steroids (Capen, 1993).

Toxins inducing hypercalcemia are discussed in the section on nephrotoxicity. Certain species of the Solanaceous produce toxins that may induce chronic atrophy of parathyroid chief cells. Chronic nephrotoxicity, especially in the dog, may result in hypocalcemia and hyperphosphatemia, which stimulates excessive production of parathyroid hormone. Impaired intestinal absorption of calcium and increased mobilization from the skeleton also may occur secondary to insufficient renal production of 1,25-dihydroxycholecalciferol by the kidney (Capen, 1993).

Vicia villosa (hairy vetch) produces angiocentric eosinophilic granulomatous inflammation of the skin, myocardium, kidney, lymph nodes, thyroid, and adrenal glands. The mechanism is unknown. Biochemical alterations suggestive of cardiotoxicity, nephrotoxicity, and depression of serum thyroxine and cortisol levels may occur.

IX. TOXINS AFFECTING THE NERVOUS SYSTEM

Many acute and chronic neurotoxins (Table 28.8) produce illness or death without alterations detectable by routine clinical biochemistry performed on blood

TABLE 28.8 Toxins Affecting the Nervous System[a]

Toxins	Disease onset	Geography	Species affected
Asclepias spp. (milkweed)	Acute	N AM	All
Aspergillus clavatus	Acute	UK, S AFR	B, O
CO	Acute	Worldwide	All
Centaurea spp.	Chronic	US	E
Cyanide	Acute	Worldwide	All
Cycadales	Chronic	AUS, DOM REP, US	B
Eupatorium rugosum (white snakeroot)	Acute to chronic	US	B
Fluoroacetate	Acute	AUS, S AFR	B, Cn, Cp, F, O
Fumonisin (*Fusarium moniliforme*)	Chronic	Worldwide	E
Hepatic encephalopathy	Chronic		All
Hexachlorophene	Chronic	Worldwide	All
Halogenated salicylanilide	Chronic	Worldwide	Cp, O
Helichrysum spp.	Acute to chronic	AUS, S AFR	B, Cp, O
Karwinskia humboldtiana (coyotillo)	Chronic	US	All
Kochia scoparia (fireweed)	Chronic	US	B
Lead	Chronic	Worldwide	All
Lolium perenne (perennial ryegrass)	Acute	Worldwide	B, E, O
Nitrate/nitrite	Acute	Worldwide	B, Cp, O
Organophosphates	Acute to chronic	Worldwide	All
Phalaris	Chronic	AUS, NZ, S AFR, US	B, O
Renal encephalopathy	Chronic		All
Selenium	Acute to chronic	Worldwide	P
Solanum spp.	Chronic	Worldwide	B
Strychnine	Acute	Worldwide	All
Stypandrol	Acute to chronic	ASIA, AUS	B, Cp, O
Swainsonine	Chronic	AUS, N AM	B, Cp, E, O
Thiaminase	Chronic	Worldwide	B, Cn, E, F, O
Trachyandra spp.	Chronic	AUS, S AFR	Cp, E, O, P

[a] B, bovine; Cn, canine; Cp, caprine; E, equine; F, feline; O, ovine; P, porcine.

or serum. Protein, AST, LDH, and CK_1 may be elevated in cerebrospinal fluid samples (Feldman, 1989).

Accumulations of endogenous toxins secondary to hepatotoxicity and nephrotoxicity may produce neurologic dysfunction. Conversely alterations that mimic hepatotoxicity, nephrotoxicity, and muscle, respiratory, and gastrointestinal toxicity may occur secondary to ischemia/anoxia from depression of cardiopulmonary centers, or by affecting sympathetic/parasympathetic balance.

Increased urinary excretion of delta-aminolevulinic acid is a potential indicator of lead intoxication.

Organophosphates induce cholinesterase inhibition, which can be detected as reduction of plasma or whole blood cholinesterase activity.

Subacute selenium intoxication in pigs exposed to complete rations containing 9.7–27 ppm selenium for 45 days is manifested as a central nervous system disorder characterized initially by hindlimb ataxia progressing to posterior paralysis. The clinical syndrome is associated with focal symmetrical poliomyelomalacia of the ventral horns of the cervical and lumbosacral intumescences (Casteel *et al.*, 1985). Hoof separation at the coronary band also occurs. The clinical pathologic alterations are consistent with dehydration from inability to reach water sources. Iatrogenic disease in the dog may result from parenteral administration of selenium preparations (Turk, 1980).

Solanum kwebense, *S. dimidiatum*, and *S. fastigiatum* produce neuronal vacuolation resembling a lysosomal storage disease. The biochemical basis of this lesion is unknown.

Swainsonine is an indolizidine alkaloid produced by certain species of the plants genera *Astragalus*, *Oxy-*

tropis, and *Swainsona*. This toxin inhibits lysosomal alpha-mannosidase, resulting in a lysosomal storage alteration that affects cells in many organs, but is often lethal because of its neurologic effect.

Thiaminases that may induce polioencephalomalacia in herbivores are present in *Equisetum arvense* and *Pteridium aquilinum*. Many uncooked fish species also contain thiaminase, which may produce encephalopathy affecting primarily carnivores. The coccidiostat amprolium is a thiamine antagonist that produces polioencephalomalacia in ruminants. Calves early in the course of polioencephalomalacia may have reduced blood transketolase (which requires thiamine pyrophosphate as a cofactor) and increased pyruvate (Jubb and Huxtable, 1993).

X. TOXINS AFFECTING THE INTEGUMENT

Most toxins affecting the skin (Table 28.9) will induce no or nonspecific alterations in clinical biochemistry. Fortunately, lesions are usually readily visible by physical examination.

Topical exposure to strong acids or alkalis may induce immediate and severe damage of the stratum corneum and epidermis.

Ergot and ergot-like syndromes are produced by the fungi *Claviceps purpurea* in infected rye and other cereal grains, and *Acremonium coenophilaum* in fescue and other pasture grasses. These fungi produce vasoconstrictive alkaloids that are derivatives of lysergic acid, including ergotamine, ergometrine, and ergotoxine (*C. purpurea*) and ergovaline (*A. coenophilaum*). Skin lesions are the result of ischemic necrosis that is usually most impressive in the distal extremities.

Dermatotoxic heavy metals include thallium and arsenic (Yager and Scott, 1993). Thallium is still used as a rodenticide in some countries, but is mainly of historical interest in many developed countries in which it has been banned. Thallium induces parakeratosis and alopecia. The mechanism is unknown, but is speculated to center around alteration of sulfhydryl groups in keratin. Arsenic toxicity exerts similar influences.

Mimosine is a toxic amino acid occurring in *Mimosa pudica* and *Leucaena leucocephala*. This toxin produces alopecia by mechanisms that are incompletely understood, but may involve metal chelation that inhibits metalloenzymes.

Organochlorines and organobromines include chlorinated naphthalenes, PCBs, polybrominated PBBs, and dibenzofurans. These industrial toxins are cumulative and result in alopecia, hyperkeratosis, and squamous metaplasia of columnar epithelium of the respiratory tract.

Molybdenum toxicosis results in a relative copper deficiency in cattle and sheep that produces depigmentation as a consequence of the decreased activity of tyrosinase.

Toxic photosensitization, enhanced susceptibility of the skin to actinic radiation, occurs primarily in lightly pigmented skin of herbivores and may be primary or secondary to chronic hepatotoxicity (Yager and Scott, 1993). Primary disease is due to exogenous photodynamic agents that include treatment with the anthelmintic phenothiazine and grazing of toxic plants such as *Ammi majus* (Bishop's weed, furocoumarin), *Cymopterus watsoni* (spring parsley, furocoumarin) *Fagopyrum* spp. (buckwheat, fagopyrin), *Hypericum perforatum* (St. John's wort, hypericin), and *Thamnosma texana* (Dutchmen's britches, furocoumarin).

TABLE 28.9 Toxins Affecting the Integumentary System[a]

Toxins	Disease onset	Geography	Species affected
Acids/alkalis	Acute	Worldwide	All
Ergotism	Acute to chronic	Worldwide	All
Kerosene	Chronic	Worldwide	B
Metals (As, Tl)	Chronic	Worldwide	All
Mimosine	Chronic	Worldwide	B, E, O, P
Organobromines/organochlorines	Chronic	Worldwide	All
Photosensitization, primary	Acute to chronic	Worldwide	All
Photosensitization, secondary	Acute to chronic	Worldwide	B, O
Selenium	Chronic	Worldwide	All
Trichothecenes	Acute	Worldwide	All
Vicia villosa (hairy vetch)	Chronic	Worldwide	B, E

[a] B, bovine; Cn, canine; Cp, caprine; E, equine; F, feline; O, ovine; P, porcine.

Secondary, or hepatogenous, photosensitization occurs in herbivores with diffuse liver damage that reduces the ability to excrete phylloerythrin. This photodynamic agent is formed from chlorophyll by gastrointestinal flora and is transported by the portal system to the liver where it is normally conjugated and excreted in the bile. When phylloerythrin escapes into the systemic circulation, it is poorly excreted by the kidneys and accumulates in tissues, including the skin.

Plants classified as seleniferous actively concentrate Se. Seleniferous plants can be subclassified as obligate (requiring Se) and facultative (Se not required). *Astragalus* spp. (locoweed), *Machaeranthera* spp., *Oonopsis* spp. (goldenweed), *Stanleya* spp., and *Xylorrhiza* spp. (wood aster) are obligate accumulators of selenium. These species are generally nonpalatable and are consumed only by herbivores with little other available forage. Several plants are facultative accumulators of selenium, including the genera *Aster, Atriplex, Catilleja, Gutierrezia,* and *Sideranthus,* which are more often linked with forage-associated selenium intoxication. In addition, various grasses and crops may accumulate from 1 to 25 ppm selenium when grown on seleniferous soils. The difference in selenium accumulation by these three groups is rather indistinct. Grasses are by far the most important group from the standpoint of sheer numbers of livestock affected.

Grazing of seleniferous plants may result in acute gastroenteritis, but also induces what has historically been called "alkali disease," which manifests as alopecia and dystrophic growth of the hooves, primarily in horses, cattle, and goats. There are conflicting reports regarding the hepatotoxicity, nephrotoxicity, and cardiotoxicity of selenium.

Vicia villosa (hairy vetch) produces angiocentric eosinophilic granulomatous inflammation of the skin, myocardium, kidney, lymph nodes, thyroid, and adrenal glands. The mechanism is unknown. The skin lesions are pruritic.

References

Acland, H. M., Mann, P. C., Robertson, J. L., Divers, T. J., Lichensteiger, C. A., and Whitlock, R. H. (1984). Toxic hepatopathy in foals. *Vet. Pathol.* **21,** 3–9.

Albrecht, J. E., Clawson, A. J., Ulberg, L. C., and Smith, F. H. (1969). Effects of high gossypol cottonseed meal on the electrocardiogram of swine. *J. Anim. Sci.* **27,** 976–980.

Breeze, R. G., and Carlson, J. R. (1982). Chemical-induced lung disease in domestic animals. *Adv. Vet. Sci. Comp. Med.* **26,** 201–232.

Capen, C. C. (1993). The endocrine glands. *In* "Pathology of Domestic Animals" (K. V. F. Jubb, P. C. Kennedy, and N. Palmer, ed.), Vol. 3, pp. 267–347. Academic Press, San Diego.

Cardinet, G. H. III (1989). Skeletal muscle function. *In* "Clinical Biochemistry of Domestic Animals" (J. J. Kaneko, ed.), pp. 462–495. Academic Press, San Diego.

Carlson, G. W. (1989). Fluid, electrolyte, and acid–base balance. *In* "Clinical Biochemistry of Domestic Animals" (J. J. Kaneko, ed.), pp. 544–575. Academic Press, San Diego.

Carmicheal, W. W. (1994). The toxins of cyanobacteria. *Scientific Amer.* **270,** 78–86,1994.

Casteel, S. W., Osweiler, G. D., Cook, W. O., Daniels, G., and Kadlec, R. (1985). Selenium toxicosis in swine. *J. Am. Vet. Med. Assoc.* **186,** 1084–1085.

Casteel, S. W., Thomas, B. R., and South, P. J. (1989). Postmortem diagnosis of potassium poisoning. *J. Eq. Vet. Sci.* **9,** 247–249.

Casteel, S. W., Johnson, G. C., Miller, M. A., Chudomelka, H. J., Cupps, D. E., Haskins, H. E., and Gosser, H. S. (1994). *Amaranthus retroflexus* (redroot pigweed) poisoning in cattle. *J. Am. Vet. Med. Assoc.* **204,** 1068–1070.

Casteel, S. W., Turk, J. R., and Rottinghaus. G. E. (1994b). Chronic effects of dietary fumonisin on the heart and pulmonary vasculature of swine. *Fundam. Appl. Toxicol.* **23,** 518–524.

Casteel, S. W., Rottinghaus, G. E., Johnson, G. C., and Wicklow, D. T. (1995). Liver disease in cattle induced by consumption of moldy hay. *Vet. Human Toxicol.* **37,** 248–251.

Cheville, N. F. (1988). Chemical causes of disease. *In* "Introduction to Veterinary Pathology," pp. 432–454. Iowa State University Press, Ames, Iowa.

Cornelius, C. E. (1989). Liver function. *In* "Clinical Biochemistry of Domestic Animals" (J. J. Kaneko, ed.), pp. 364–397. Academic Press, San Diego.

Dodds, W. J. (1989). Hemostasis. *In* "Clinical Biochemistry of Domestic Animals" (J. J. Kaneko, ed.), pp. 274–315. Academic Press, San Diego.

Doster, A. R., Mitchell, F. E., Farrell, R. L., and Wilson, B. J. (1978). Effects of 4-ipomeanol, a produce from mold-damage sweet potatoes, on the bovine lung. *Vet. Pathol.* **15,** 367–375.

Erdos, E. G. (1987). The angiotensin I-converting enzyme. *Lab. Invest.* **56,** 345–348.

Feldman, B. J. (1989). Cerebrospinal fluid. *In* "Clinical Biochemistry of Domestic Animals" (J. J. Kaneko, ed.), pp. 835–865. Academic Press, San Diego.

Finco, D. R. (1989). Kidney function. *In* "Clinical Biochemistry of Domestic Animals" (J. J. Kaneko, ed.), pp. 496–542. Academic Press, San Diego.

Fooshee, S. K., and Forrester, S. D. (1990). Hypercalcemia secondary to cholecalciferol rodenticide toxicosis in two dogs. *J. Am. Vet. Med. Assoc.* **196,** 1265–1268.

Gossett, K. A., Turnwald, G. H., Kearney, M. T., Greco, D. S., and Cleghorn, G. (1987). Evaluation of gamma-glutamyl transpeptidase-to-creatinine ratio from spot samples of urine supernatant, as an indicator of urinary enzyme excretion. *Am. J. Vet. Res.* **48,** 455–457.

Gram, T. E., Okine, L. K., and Gram, R. A. (1986). The metabolism of xenobiotics by certain extrahepatic organs and its relation to toxicity. *Ann. Rev. Pharmacol. Toxicol.* **26,** 259–291.

Harvey, J. W. (1989). Erythrocyte metabolism. *In* "Clinical Biochemistry of Domestic Animals" (J. J. Kaneko, ed.), pp. 185–234. Academic Press, San Diego.

Hollinger, M. A. (1983). Serum angiotensin-converting enzyme. Status report on its diagnostic significance in pulmonary disease. *Chest* **83,** 589–590.

Jubb, K. V. F., and Huxtable, C. R. (1993). VII. Degeneration in the nervous system. *In* "Pathology of Domestic Animals" (K. V. F. Jubb, P. C. Kennedy, and N. Palmer, eds.), Vol. 1, pp. 334–383. Academic Press, San Diego.

Kaneko, J. J. (1989a). Serum proteins and dysproteinemias. *In* "Clinical Biochemistry of Domestic Animals" (J. J. Kaneko, ed.), pp. 142–165. Academic Press, San Diego.

Kaneko, J. J. (1989b). Thyroid function. *In* "Clinical Biochemistry of Domestic Animals" (J. J. Kaneko, ed.), pp. 630–649. Academic Press, San Diego.

Keeler, R. F., and Baker, D. C. (1990). Myopathy in cattle induced by alkaloid extracts from *Thermopsis montana, Laburnum anagyroides,* and a *Lupinus* sp. *J. Comp. Pathol.* **103,** 169–182.

Kelly, W. R. (1993). XI. Toxic liver disease. *In* "Pathology of Domestic Animals" (K. V. F. Jubb, P. C. Kennedy, and N. Palmer, eds.), Vol. 2, pp. 382–400. Academic Press, San Diego.

Kramer, J. W. (1989). Clinical enzymology. *In* "Clinical Biochemistry of Domestic Animals" (J. J. Kaneko, ed.), pp. 338–363. Academic Press, San Diego.

Kramer, J. W., Bistline, D., Sheridan, P., and Emerson, C. (1984). Identification of hippuric acid crystals in the urine of ethylene glycol-intoxicated dogs and rats. *J. Am. Vet. Med. Assoc.* **184,** 584–585.

Maxie, M. G. (1993). The kidney. *In* "Pathology of Domestic Animals" (K. V. F. Jubb, P. C. Kennedy, and N. Palmer, eds.), Vol. 2, pp. 447–522. Academic Press, San Diego.

Miyai, K. (1991). Structural organization of the liver. *In* "Hepatotoxicology" (R. G. Meeks, S. D. Harrison, and R. J. Bull, eds.), pp. 1–66. CRC Press, Boca Raton, Florida.

Myers, J. L, (1993). Diagnosis of drug reactions in the lung. *In* "The Lung: Current Concepts" (A. Churg and A. A. Katzenstein, eds.), pp. 32–53. Williams and Wilkins, Baltimore.

Monshouwer, M., Witkamp, R. F., Nijmeijer, S. M., van Leengoed, L. A. M. G., Verheijden, J. H. M., and van Miert, A. S. J. P. A. M. (1995). Infection (*Actinobacillus pleuropneumoniae*)–mediated suppression of oxidative hepatic drug metabolism and cytochrome P_{4503a} mRNA levels in pigs. *Drug Metab. Dispos.* **23,** 44–47.

Osweiler, G. D., Buck, W. B., and Bicknell, E. J. (1969). Production of perirenal edema in swine with *Amaranthus retroflexus. Am. J. Vet. Res.* **30,** 557–566.

Osweiler, G. D., Carson, T. L., Buck, W. B., and van Gelder, G. A. (1985). Nitrates, nitrites, and related problems. *In* "Clinical and Diagnostic Veterinary Toxicology," pp. 460–466. Kendall/Hunt Publishing, Dubuque, Iowa.

Pearson, E. G. (1990). Diseases of the hepatobiliary system. *In* "Large Animal Internal Medicine" (B. P. Smith, ed.), pp. 837–843. C. V. Mosby, St. Louis.

Riley, R. T., An, N. H., Showker, J. L., Yoo, H.-S., Norred, W. P., Chamberlain, W. J., Wang, E., Merrill, A. H., Motelin, G., Beasley, V. R., and Haschek, W. M. (1993). Alteration of tissue and serum sphinganine to sphingosine ratio: An early biomarker of exposure to fumonisin-containing feeds. *Toxicol. Appl. Pharmacol.* **118,** 105–112.

Robinson, W. F., and Maxie, M. G. (1993). VII. Myocardium. *In* "Pathology of Domestic Animals" (K. V. F. Jubb, P. C. Kennedy, and N. Palmer, eds.), Vol. 2, pp. 27–36. Academic Press, San Diego.

Snyder, R. (1979). Classes of hepatic microsomal mixed-function oxidase inducers. *Pharmacol. Therap.* **7,** 203–244.

Stocker, R., Yamamoto, Y., McDonagh, A. F., Glazer, A. N., and Ames, B. N. (1987). Bilirubin is an antioxidant of possible physiological importance. *Science* **235,** 1043–1046.

Tennant, B. C., and Hornbuckle, W. E. (1989). Gastrointestinal function. *In* "Clinical Biochemistry of Domestic Animals" (J. J. Kaneko, ed.), pp. 417–461. Academic Press, San Diego.

Turk, J. R. (1980). Chronic parenteral selenium administration in a dog. *Vet. Pathol.* **17,** 493–496.

Valli, V. E. O. (1993). The erythron. *In* "Pathology of Domestic Animals" (K. V. F. Jubb, P. C. Kennedy, and N. Palmer, eds.), Vol. 3, pp. 157–265. Academic Press, San Diego.

Van Vleet, J. F., and Ferrans, V. J. (1986). Myocardial diseases of animals. *Am. J. Pathol.* **124,** 98–174.

Yager, J. A., and Scott, D. W. (1993). VI. Physicochemical diseases of the skin. VII. Actinic diseases of the skin. *In* "Pathology of Domestic Animals" (K. V. F. Jubb, P. C. Kennedy, and N. Palmer, eds.), Vol. 1, pp. 579–597. Academic Press, San Diego.

CHAPTER

29

Clinical Biochemistry of Laboratory Rodents and Rabbits

WALTER F. LOEB

I. INTRODUCTION

Vast numbers of clinical laboratory determinations are performed on sera and other body fluids from rodents and rabbits. The majority of these are performed in fulfillment of objectives of the research in which the animals are utilized, but others are to validate the animals' health status prior to research or occasionally as an aid in the diagnosis of spontaneous laboratory animal diseases.

To enable the results of the determinations to yield optimal and maximal information, several criteria must be met. First, the specimen, as analyzed in the laboratory, should represent the animal's status as validly as possible. Artifacts introduced by specimen collection, handling, and deterioration must be avoided. Second, the analyses performed must be appropriate to the species and the objectives of the determination. The appropriate analyte to determine may vary with the species. For example, in the guinea pig the principal glucocorticoid is cortisol, whereas in the rat and mouse it is corticosterone.

Analytes vary in turnover time with species and disease. The marked elevation of lactate dehydrogenase, isocitrate dehydrogenase, malate dehydrogenase, phosphohexose isomerase, and aspartate aminotransferase in mice infected with Riley's agent is due to impaired reticuloendothelial clearance rather than to increased release (Notkins, 1965). The structure of an analyte, particularly a protein or peptide, may vary from one species to another, which becomes a critical issue in the selection of immunoassay techniques (see Section X, Hormones).

Third, interpretation of the data must be performed with a comprehension of the physiological and pathological factors that influence the results of the analyses under consideration. Appropriate methods of quality control should be utilized to ensure the accuracy and precision of the determined values. Historical reference values appropriate to the species, breed or strain, sex, and age are particularly important when determinations are performed on individual animals for diagnostic purposes. In research studies, the principal decision-making criterion lies in the statistical significance of the difference of group means between treated and appropriate contemporary control animals, but historical reference ranges are useful as an overall quality control measure to evaluate the control animals as well as the analytic procedures (Weingand *et al.*, 1992).

This chapter covers the clinical biochemistry of the rat, mouse, hamster, gerbil, guinea pig, and rabbit and describes some unique features of each of these species. It is not intended to redundantly repeat features common to all or most mammalian species. Among laboratory rodents, rats have been most extensively investigated by the tools of clinical chemistry. Available serum volume may limit the clinical chemical determinations that can be performed on a mouse, particularly on a survival basis. This must be considered in experimental design. Lesser numbers of hamsters and rabbits are utilized in research, generally for selected applications. Guinea pigs represent such a divergence from classic phylogenic pathways that their utilization in research should be carefully evaluated (Wriston, 1984). Despite their convenient body size and pleasant temperament, they are difficult to bleed.

II. SPECIMEN COLLECTION

Blood volume in rodents and rabbits varies with species from 5 to 9% of body weight (Levine, 1995; Morton *et al.*, 1994; McLaughlin and Fish, 1994). Laboratory rats such as CD (cesarian-derived) rats of Sprague–Dawley origin are quite uniform and free from spontaneous disease. In adult rats, 1 ml of blood can be collected every 2 weeks without influencing subsequent results. Ten milliliters can be collected terminally. Mice share uniformity and quality as laboratory animals with rats; however, their small body size constitutes a distinct disadvantage to the investigator who requires repeated or multiple clinical laboratory determinations. Seventy-five microliters of blood may be collected weekly without affecting subsequent results, and approximately 1 ml may be collected terminally from a well-hydrated adult mouse by a skillful animal technician (Loeb, 1987).

Artifacts resulting from techniques of blood sampling must be avoided or minimized. Mice from which approximately 0.4 ml of blood is drawn rapidly develop a marked elevation of alanine aminotransferase, aspartate aminotransferase, and lactate dehydrogenase. This elevation persists for 48 hours. Additional serum samples drawn during this period reflect the effect produced by the prior blood collection, rather than the animal's pathophysiologic status (Goldsmith and Loeb, unpublished data, 1982).

Jugular puncture can be performed easily in rats and is particularly valuable in obtaining blood samples with minimal tissue trauma (Fig. 29.1). The rat is held on its back by an assistant, with the rat's forelegs spread on a bleeding board. The rat's head is covered with a perforated plastic hood, to prevent biting and to calm the animal, and turned to one side. The jugular vein is located in a triangle formed on the opposite side by the rat's neck and shoulder and can be venipunctured with a 23- gauge-$\frac{5}{8}$-inch needle. Three milliliters of blood can readily be collected from an adult rat without evidence of undue stress. Skilled teams of an animal technician and an assistant can bleed rats in this fashion at the rate of one every 30 seconds (Everett, 1982, personal communication).

Other sites from which blood samples may be collected, in some instances only on a terminal basis, include the retroorbital sinus (currently most widely used in the United States), tail vein, heart, aorta, or vena cava. A technique for collection of blood from the tributaries of the dorsal metatarsal vein between the digits of the rat has been reported by Snitily *et al.* (1991). In the United Kingdom, the collection of blood from the retroorbital sinus or plexus is discouraged and is only permitted terminally, under anesthesia (Morton *et al.*, 1994). The site of sampling and anesthetic agent used (or physical restraint without anesthesia) may affect results. Specimens collected from the retroorbital sinus had higher creatine kinase values than those collected from the heart or vena cava, and specimens collected under carbon dioxide anesthesia had higher leukocyte counts than those collected under methoxyflurane, pentobarbital, or ketamine–xylazine (Thompson, 1986; Everett and Harrison, 1983; Friedel *et al.*, 1975; Suber and Kodell, 1985).

In the rabbit, blood samples are most often collected from the ear vein. Though conveniently located, this vein is extremely small and it is difficult to obtain an adequate specimen. Jugular puncture of an anesthetized rabbit with a shaved neck may readily be performed, yielding a more satisfactory sample. In mice, hamsters, and guinea pigs, the retroorbital sinus, heart, aorta, or vena cava are most commonly used for blood collection, the latter two only terminally. In the mouse, small samples may be collected by capillarity by an incision in the tail vein.

When repeated sampling over a short time period is required, appropriate venous or arterial vascular catheters can be implanted and the catheterized animal restrained with an elastic tether. These catheters are kept full of heparinized saline between sampling and must be flushed with the animal's blood immediately prior to sampling (Bodziony and Schwille, 1985). When animals were caged in pairs and removed from the cage for bleeding, stress-affected analyte values such as glucose and corticosteroids were shown to be higher in the second animal bled than in the first (Dunn and Scheuing, 1971).

In collecting serum or plasma, it is essential to remove it from the erythrocytes as promptly as possible. Prolonged exposure of serum to erythrocytes results in a

A

B

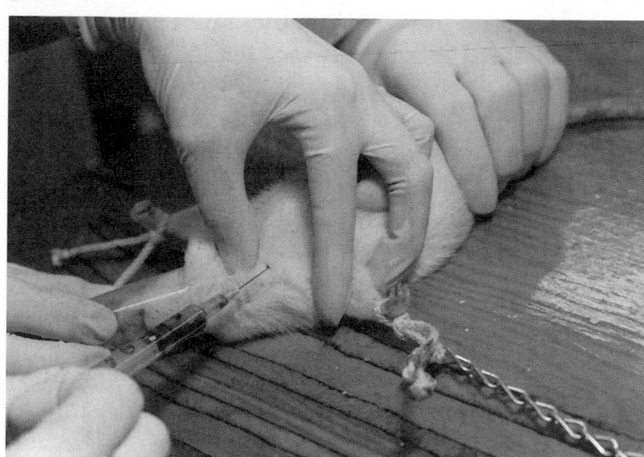

FIGURE 29.1 Technique for jugular venipuncture in the rat. (a) The rat is placed on its back on a rat board by an assistant. The forelegs are spread apart by cords hooked to pins in the sides of the board. The head is covered with a plastic hood made from the bottom of a centrifuge tube and is turned away from the technician. (b) Venipuncture is performed in the triangle formed by the rat's neck and shoulder, using a $\frac{5}{8}$-inch 23-gauge needle.

loss of glucose owing to glycolysis, elevation of lactate dehydrogenase, and, to a lesser degree, elevation of inorganic phosphorus and aspartate aminotransferase. In rodents, potassium-rich erythrocytes leak large quantities of potassium into the serum. Generally, serum should be removed from erythrocytes 20–30 minutes after blood collection. Although serum is generally used in the United States, plasma is more commonly used in Europe. This has advantages: the blood specimen may be centrifuged immediately, precluding leakage of the analytes discussed earlier and consumption of glucose, and the volume yield is better. Concentrations of analytes are generally comparable in serum or plasma, except for globulin and total protein, which are approximately 0.5 g/dl higher in plasma because of the presence of fibrinogen. The anticoagulant used must not interfere with the required analyses. Lithium heparin is generally appropriate. Slaughter and Moen (1991) recommend the use of aqueous solution of lithium heparin in preference to dry lithium heparin, particularly for blood collected from the orbital sinus.

Studies in rodents, particularly rats, have shown that 72 hours of fasting are required to achieve a nonabsorptive state similar to that achieved in dogs, rhesus monkeys, or humans in 12–24 hours (Thompson, 1986). Shorter fasts of 12, 24, or 48 hours do result in a progressive reduction in serum glucose. Overnight fasting in rats decreases the activity of alkaline phosphatase and the concentration of albumin and increases the concentration of free fatty acids, bilirubin, and bile acids. As rats, in particular, are night feeders, depriving rats of food overnight essentially eliminates food consumption for 24 hours. The decision whether or not to deprive rodents of food prior to blood sampling for clinical chemistry should be made on an individual basis depending on the objectives of the study, with the understanding that a nonabsorptive state is not achieved by overnight food deprivation. Valid comparisons can only be made if the protocol with respect to food deprivation is consistent throughout the study (Weingand *et al.*, 1992).

Knudtzon (1984) studied the effects of saline injection, feeding, fasting, transportation, immobilization, and noise on the serum glucose, insulin, glucagon, and free fatty acids of rabbits. Saline injection increased the serum glucose but did not affect the other analytes studied. Serum levels of glucose and fatty acids were elevated for 2 hours after the rabbits were transported. Bleeding the rabbits 44 ml increased all of the analytes. After 16 hours of fasting, plasma levels of glucose and insulin decreased while those of glucagon and fatty acids increased. Immobilization overnight inhibited the hypoglycemia induced by fasting. Impulse noise (96 decibels) did not affect the levels of the analytes.

III. CYCLIC BIORHYTHMS

Cyclic variations occur in the concentrations of many analytes, especially hormones, and must be taken into consideration in experimental design and the interpretation of results. Daily variation, termed circadian, is most often the result of the light–dark cycle. In rats and mice, levels of ACTH and corticosterone are highest early in the dark period, falling rapidly to lowest values in the middle of the dark period (D'Agostino *et al.*, 1982; Ottenweller *et al.*, 1979). These cycles are not present at birth but develop early in life. Takahashi *et al.* (1982) have shown that either natural biological mother or foster mother rats entrain blinded rat pups with their own cycles of corticosterone, which in turn are a function of the light–dark cycle to which the mother rats have been exposed. Cycles less than a day in duration, such as the episodic secretion of growth hormone or luteinizing hormone, are termed ultradian.

In sexually mature female rats, levels of ACTH, corticosterone, and thyroxine vary with the estrous cycle and are highest during proestrus and lowest during diestrus (Buckingham *et al.*, 1978). Similar fluctuations of serum corticosterone were demonstrated in mice (Nichols and Chevins, 1981). In hamsters, both cortisol and corticosterone are present in significant concentrations. Peak levels of both hormones occur during the beginning of the dark period. Corticosterone levels are higher than cortisol in the early part of the light phase, but cortisol levels are higher than corticosterone when both hormones are increasing (Albers *et al.*, 1985; Otterweller *et al.*, 1985).

Other cyclic variations may be seasonal. In both wild and captive environments, rabbits in Tunisia were shown to have seasonal variations in testosterone and thyroxine. Testosterone peaked sharply in October and fell in November and December, with low values observed from January to September. Thyroxine peaked twice yearly, in October and again in spring (Ben Saad and Bayle, 1985).

IV. TRANSGENIC ANIMALS

The development of transgenic technology, incorporating the insertion or deletion of genes, has revolutionized animal research. In effect, each transgenic manipulation creates a new species. Transgenic animals differ from conventional ones with respect to clinical chemistry in that they may express a protein not normally found in that species, have an altered concentration of an analyte, have an aberrant regulatory mechanism, or, in the case of knockout animals, fail to express a protein normally present or lack a receptor. Immunochemical analytic methods utilizing monoclonal antibodies have special applicability in discriminating between the protein present in the conventional animal and that expressed as a result of transgenic manipulation (i.e., human growth hormone vs mouse growth hormone in a transgenic mouse). A comprehensive consideration of the unique aspects of transgenic animals is beyond the scope of this chapter. Rather, several studies in transgenic animals are reviewed to provide examples of differences from conventional animals with respect to clinical biochemistry.

Human renin reacts specifically with human angiotensinogen to release angiotensin I, converted by angiotensin-converting enzyme (ACE) to angiotensin II. Rat or mouse angiotensinogen is very resistant to the action of human renin. Transgenic mice have been produced that express human renin, as well as others that express human angiotensinogen. These animals are normotensive. Cross breeding these mice results in mice that express both human renin and human angiotensinogen. These so-called dual gene strain mice are hypertensive but respond to renin inhibitors, ACE inhibitors, and angiotensin II receptor antagonists. Similar studies have been performed in rats (Wagner *et al.*, 1993).

Transgenic mice that express human growth hormone, ones that express bovine growth hormone, and ones that express rat growth hormone have been developed. Transgenic mice expressing both mouse growth hormone and human growth hormone are of normal size when born. At about 4 weeks of age, growth becomes accelerated because of the stimulation of insulin-like growth factor 1 (somatomedin C) by growth hormone. Adult transgenic mice expressing human growth hormone weigh 50–100% more than their conventional siblings. Males express approximately twice the concentration of human growth hormone as females. The transgenic mice age faster and have shorter lifespans than their conventional siblings. The transgenic females are sterile. Ovarian function and ovulation appear to be normal, but human growth hormone, unlike mouse growth hormone, fails to support the mating-induced prolactin release required for the activation and maintenance of luteal function. If female transgenic mice, expressing human growth hormone, breed and are supported by progesterone administration or prolactin-secreting pituitary implants, they become pregnant and deliver live pups, but generally do not lactate adequately for the survival of the pups. Male transgenic mice expressing human growth hormone are fertile, but have a shortened reproductive lifespan and decreased reproductive aggressiveness with age (Asa *et al.*, 1992; Bartke *et al.*, 1991; Stefaneanu *et al.*, 1993).

Chorionic gonadotropin in the species in which it occurs, FSH, LH, and TSH are glycoprotein hormones having an α- and a β-subunit. The α-subunit is highly species specific and is similar among hormones of this class within a given species of animal, that is, it confers the species specificity on the hormone. The β-subunit conveys the function on the molecule, and β-subunits for a given hormone are similar though not identical among various species. Transgenic mice expressing human FSH-β have been developed. The human FSH-β gene is expressed exclusively in pituitary gonadotrophs, and the mice secrete an interspecies heterodimer consisting of mouse FSH-α and human FSH-β, as well as a much smaller amount of conventional mouse FSH. These mice have elevated pituitary and serum FSH when compared to conventional controls, but their responses to gonadectomy and sex steroid replacement are qualitatively similar to those of conventional animals.

Transgenic mice that express human insulin have been developed. These mice also express mouse insulin I and mouse insulin II in a normal ratio (see Section V, Glucose and Glucoregulatory Hormones). These mice have normal total serum insulin concentrations and normal serum glucose, and they respond normally to challenges to glucose metabolism. These findings document posttranscriptional controls in glucose/insulin regulation (Schnetzler *et al.*, 1993).

The enzyme creatine kinase consists of dimers of two monomers termed m (for muscle) and b (for brain). In the conventional mouse, as in other species, the enzyme in skeletal muscle is mm, that in cardiac muscle is predominantly mm and partially mb, and the enzyme in brain is bb. Because mb is present in significant activity only in cardiac muscle, the finding of this isoenzyme in serum generally characterizes myocardial injury. Transgenic mice microinjected with rat CK b gene overexpress CK b monomere. Their skeletal muscle averages 60% mm, 32% mb, and 8% bb (Brosnan *et al.*, 1993).

These few selected examples have been presented to document some of the unique features that have been reported with respect to clinical chemistry in transgenic animals.

V. GLUCOSE AND GLUCOREGULATORY HORMONES

In both rats and mice, serum glucose decreases with age (Loeb and Caracostas, 1990; Loeb *et al.*, 1996). In mice the rate of decrease is approximately 2 mg/dl/month. This is associated with improved response to glucose challenge and with increased size and insulin content of the islets of Langerhans (Leiter *et al.*, 1988). A similar age-related decrease in serum glucose was observed in spontaneously hypertensive rats (SHRs), though the islet size and insulin contents did not differ between young and aged SHRs (Iwase *et al.*, 1994).

Although serum glucose is affected by numerous factors, it is most extensively studied in the numerous animal models of diabetes mellitus that include both rodent and nonrodent species. In humans, diabetes may be subdivided into Type I, or the insulin-dependent type, also termed the severe, juvenile, or early-onset type, and Type II, or the non-insulin-dependent type, also termed the adult-onset type. The former generally represents an absolute deficiency of insulin, whereas the latter is characterized by hyperglycemia in the presence of hyperinsulinemia. An interesting model of insulin-dependent diabetes mellitus described in the guinea pig has been shown to result from an infectious agent affecting the pancreas. There is profound glycosuria and variable but mild hyperglycemia and ketonuria. There is fatty degeneration of the pancreas with impaired exocrine as well as endocrine function. Diabetic cataracts and diabetic renal glomerular disease in the guinea pig resemble those seen in man (Lang and Munger, 1976).

The C57Bl/6J *ob/ob* obese mouse is a widely studied model of non-insulin-dependent diabetes mellitus. The mouse is obese, hyperglycemic, and hyperinsulinemic, consistent with the corresponding disease in humans (Austin *et al.*, 1984). Using radiolabeled glucose, Wittmers and Haller (1983) have demonstrated that these mice clear glucose from the blood similarly to lean controls, and concluded that the characteristic hyperglycemia is due to accelerated glycogenolysis and gluconeogenesis. Surwitt *et al.* (1985) demonstrated that these obese mice were not consistently hyperglycemic, but were subject to stress hyperglycemia while lean controls were not. Adrenalectomy of these obese mice resulted in normalization of glucose tolerance curves.

The diurnal cycles of serum glucose and insulin as well as tissue enzymes participating in glucoregulation were studied in *db/db* mice, another model of non-insulin-dependent diabetes. Though insulin levels were consistently elevated, glucose levels varied widely, falling to levels approximating those of nondiabetic controls near the end of the light period. Little evidence of circadian rhythm was seen in the glucose or insulin of nondiabetic controls, or the insulin of diabetic mice (Roesler *et al.*, 1985).

The Zucker fatty rat (*fa/fa*) is characterized by obesity and hyperinsulinemia, though it is normoglycemic. Although not truly diabetic, it has some of the characteristics of non-insulin-dependent diabetes. Hyperinsulinemia has been demonstrated at 2 weeks of age, though total pancreatic insulin does not differ from lean (*Fa/Fa*) rats (Blonz *et al.*, 1985).

The LA/N corpulent rat resembles the Zucker *fa/fa* rat in the properties of obesity, hyperinsulinemia, normoglycemia, and hyperlipidemia, though hyperinsulinemia is more marked in the LA/N corpulent rat. Serum corticosterone levels are significantly higher in

LA/N corpulent rats (340 ng/ml) than in lean controls (255 ng/ml) (Elwood *et al.*, 1985).

The BB rat is a widely used model of insulin dependent diabetes mellitus (IDDM, Type I). The incidence of diabetes in this strain may be markedly reduced by treatment with an inhibitor of platelet-activating factor, a potent modulator of inflammation, anaphylaxis, hypersensitivity, and cellular immune responses (Jobe *et al.*, 1993).

Glycosylation of proteins is a nonenzymatic process occurring when specific proteins are exposed to carbohydrates. Glycosylated hemoglobin (in humans, principally hemoglobin A1c) forms continuously during the life of the erythrocyte in relationship to the concentration of serum glucose. Thus, it has proved to be a valuable long-term monitor of glucoregulation, particularly in diabetics. Conventional Sprague–Dawley rats had a glycosylated hemoglobin of 4.22 ± 2.18%, whereas Sprague–Dawley rats made diabetic by treatment with streptozotocin had values of 7.96 ± 3.40 (mean ± 2 SD). The rats made diabetic with streptozotocin also had significantly higher levels of erythrocyte 2,3-diphosphoglycerate (Alder *et al.*, 1992). In the healthy rabbit, Higgins *et al.* (1982) demonstrated a mean hemoglobin A1c of 1.4% and a total glycosylated hemoglobin of 2.3% (of total hemoglobin). In contrast, in humans these values were 4.9 and 6.9%, in the dog 3.3 and 3.3%, and in the pig 0 and 0.3%.

Insulin and glucagon, in common with other peptide hormones, vary in structure among species. Whether an immunoassay having antibody to and competitive hormone of one species will be valid in another species depends on the structural similarity of the molecule, in particular the portion of the molecule to which the antibody is raised. Rats, mice, and hamsters have two distinct nonallelic insulin molecules present in all members of the species. Neither of these is proinsulin or C-peptide. Mouse insulin II is identical to rat insulin II, and mouse insulin I differs from rat insulin I by a single amino acid. Rats have approximately equal quantities of insulin I and insulin II. The assay used to measure total insulin in rats, mice, and hamsters must recognize both molecules equally. The assay for rat insulin may be used to measure mouse insulin; for greatest accuracy, it should then be calibrated with mouse insulin (Markussen 1971; Smith, 1966). The insulin produced by the pancreatic islets of the guinea pig is immunologically quite different from the insulin of either man or rat and cannot be accurately measured by either assay. However, the guinea pig produces another insulin-like molecule from nonpancreatic sources. This molecule, which is not released into the plasma, does not appear to be involved in overall glucoregulation. It reacts with the immunoassay for human insulin. Therefore, using the immunoassay for human insulin on guinea pig serum fails to measure the principal glucoregulatory insulin (Jukes, 1979; Rosenzweig *et al.*, 1980)

Glucagon is a molecule that is generally similar among various animal species. Rat glucagon can be measured with the human immunoassay. The guinea pig, however, has a distinctly different molecule, not measurable by the human assay (Huang *et al.*, 1986).

VI. PROTEINS

Fetal mice have three unique fetal proteins. One of these, α-1-fetoprotein, has also been demonstrated in the sera of mice with transplantable hepatoma. Three specific pregnancy-dependent proteins have been described in the sera of pregnant mice (Quimby, 1988). One of these, pregnancy-associated murine protein-1 (PAMP-1), shows partial immunologic cross reactivity with human pregnancy zone protein. Its concentration varies with the estrus cycle of the mouse; its concentration is highest in diestrus. Its serum level remains low until the 7th day of gestation, when it rises rapidly, peaking at the 11th day and then declining (Hau *et al.*, 1991).

In the Syrian hamster, synthesis of a unique pentraxin termed "female protein" is under the control of sex steroids. Concentrations in females (0.5–3 mg/dl) are 100- to 1000-fold greater than in males. Levels may be experimentally modified by sex steroid treatment. Female protein shares structural similarity with human C-reactive protein and serum amyloid P component and can be demonstrated to be present in hamster tissue amyloid. Although similar to acute-phase reactants of other species, female protein, in acute-phase reactions in the hamster, increases in concentration 5- to 10-fold in males and decreases by 50% in females (Coe and Ross, 1985; Dowton *et al.*, 1985).

In rabbits, experimental infection with the parasite *Setaria cervi* resulted in increases in total proteins and all electrophoretic globulin bands and a decrease in albumin. These changes began within the week after infection, peaked at 3 weeks, and almost returned to control values by 6 weeks. Two precipitin antibodies to the parasite were demonstrated immunoelectrophoretically during the peak hyperglobulinemia (Khatoon *et al.*, 1984).

VII. LIPIDS AND LIPOPROTEINS

Because of the role of the rat in biomedical research, the lipids and lipoproteins of the rat, though differing markedly from those of humans, have been extensively studied. Normal serum cholesterol values in rats fed conventional diets are typically about one-third those of humans. In contrast to humans, 60% of the cholesterol is transported as high-density lipoprotein (HDL) cholesterol. Serum cholesterol levels in the hooded Long–Evans rat are somewhat higher than in Wistar rats. Possibly because of the low level of serum cholesterol and proportionately high level of HDL cholesterol, the rat is quite resistant to the induction of hyperlipidemia and atherosclerosis. Hypothyroidism may result in mild hypercholesterolemia. The Zucker Fatty Rat (*fa/fa*), previously discussed in Section IV, is one of several hypercholesterolemic and hypertriglyceridemic genetic strains, having values approximately three time those of Wistar controls (Carroll and Feldman, 1988). In addition to the lipoproteins recognized in other species, a very high density lipoprotein (VHDL) having very rapid catabolism has been demonstrated in the rat. This lipoprotein has a density of 1.23 g/ml and consists of 85% protein and 15% lipid including cholesterol esters and free cholesterol (Därr *et al.*, 1985).

Sullivan *et al.* (1993) studied the response of gerbils, hamsters, and guinea pigs to the feeding of high-fat/high-cholesterol diets and concluded that among these rodents, the guinea pig is the best rodent model for man. Transgenic rodents expressing human lipoprotein apoproteins appear to be superior to conventional rodents for the study lipid disorders and atherosclerosis (Rubin and Schultz, 1993).

Although reference values for cholesterol in the rabbit resemble those of the rat, and, as in the rat, the principal component is HDL cholesterol, hypercholesterolemia is more readily induced by dietary manipulation. The Watanabe hereditable hyperlipemic rabbit (WHHL) is a model of familial hypercholesterolemia. Homozygotes have mean serum cholesterol levels of 500 mg/dl and may reach 1000 mg/dl. The principal fraction in these animals is low-density lipoprotein cholesterol with only small quantities of HDL. Affected animals develop atherosclerosis by 5 months of age (Tanzawa *et al.*, 1980).

VIII. ENZYMES

The sensitivity and organ specificity of the leakage enzymes in detecting cellular injury depend on several factors, including the comparative activities in various tissues, turnover time, and intracellular distribution. There are two isoenzymes of alanine aminotransferase, one located in the mitochondria, the other in the cytosol. The cytosol isoenzyme is lost when cell membrane permeability increases, whereas loss of the mitochondrial isoenzyme generally requires cell necrosis. Coupling the difference in intracellular distribution with the distribution among various tissues leads to the conclusion that in the rat and mouse, the measurement of alanine aminotransferase is quite sensitive and hepatospecific, whereas in the guinea pig it is insensitive and not hepatospecific. In the rat, the measurement of sorbitol dehydrogenase (L-iditol dehydrogenase) has been shown to be quite hepatospecific and more sensitive in the detection of hepatocellular injury than other enzymes, including alanine aminotransferase (Clampitt and Hart, 1978; Dooley, 1983; Teschke *et al.*, 1983). The article by Boyd (1983) is an excellent resource for the determination of sensitivity and organ specificity of the various enzymes.

The increase in serum lactate dehydrogenase and other enzymes in mice infected with Riley's agent (LDH virus) has been shown to be due to impaired clearance rather than increased release of the enzymes (Notkins, 1965). Serum levels of alkaline phosphatase of hepatobiliary origin are increased in cholestasis and in response to inducer drugs such as barbiturates that result in increased hepatocellular endoplasmic reticulum. Two isoenzymes of alkaline phosphatase are present in the sera of normal adult mice. The inoculation of mice with mouse hepatitis virus results in the appearance of a third isoenzyme of hepatic origin in the serum (Everett and Harrison, 1983). In guinea pigs, dietary deficiencies of zinc or manganese result in an elevation of serum alkaline phosphatase (Alberts *et al.*, 1977; Underwood, 1971).

γ-Glutamyltransferase activity in the sera of healthy rats may be insufficient to permit detection, but it is significantly increased in cholestasis (Ringler and Dabich, 1979).

IX. BILE ACIDS AND BILIRUBIN

Measurement of total serum bile-acid concentration has been shown to be a sensitive and specific indicator of hepatic injury and disease as well as of disorders of enterohepatic circulation in various species (Hardison *et al.*, 1985; Iga and Klaasen, 1982; Gopinath *et al.*, 1980). Thompson *et al.* (1987) have developed and validated a method for the separation and quantification of individual bile acids by high-performance liquid chromatography (HPLC) and demonstrated a greater number of individual bile acids in rat serum than in the sera of several other species studied. Using various chemical and physical methods to induce hepatic disease or enterohepatic circulatory disturbances, they demonstrated that each of these had a unique pattern of changes in the bile-acid profile. This approach shows great potential for the detection and definition of hepatic disease in rodents; it appears to be more sensitive and specific than the measurement of individual analytes utilized for the assessment of hepatic status, and it does not require the extensive handling essential to liver function tests utilizing exogenous dyes.

Several hereditary models of aberrant bilirubin metabolism are available. Gunn rats lack bilirubin UDP-glucuronyltransferase comparable to Crigler–Najjar syndrome of humans. Jansen *et al.* (1985) have described a jaundiced mutant rat of Wistar derivation with conjugated hyperbilirubinemia in which serum bilirubin levels are 100 times those of unaffected controls. Over 90% of the bilirubin is conjugated, predominantly diconjugated. Transport of bromsulfophthalein is impaired. No morphologic lesions are associated with this defect. This may represent an appropriate model for some of the features of the Dubin–Johnson syndrome of humans. These rats have been designated TR− (transport minus) and may be a valuable model for the study of ATP-dependent hepatocanalicular anion transport (Jansen and Elferink, 1993). A hyperbilirubinemic mutant rat of Sprague–Dawley derivation, designated EHBR, or Eisai hyperbilirubinuria rat, has some of the features of human Dubin–Johnson syndrome, and some of Rotor's syndrome. This rat has normal hepatic uptake and storage, and impaired secretion into bile. Affected animals have mesangial granular deposition of IgG, IgM, IgA, and C3, suggestive of IgA nephropathy (Hosokawa *et al.*, 1992).

X. HORMONES

The release of growth hormone and of luteinizing hormone is pulsatile, and valid interpretation requires measurement over time. Single values may be misleading (Fox and Smith, 1984; Watts and Fink, 1984; Bullier-Picard *et al.*, 1986). The growth defect in the Brattleboro rat appears to be the result of decreased hepatic growth hormone receptors rather than a deficiency of growth hormone (Bullier-Picard *et al.*, 1986). The mean growth hormone levels in obese Zucker rats are significantly lower than in lean controls, and pulses tend to be lower and less frequent (Finkelstein *et al.*, 1986). The frequency of pulses of luteinizing hormone decreases with age in mice, though peak values do not change (vom Saal *et al.*, 1994).

Immunoassay procedures depend on the "recognition" of an antigen by an antibody. Species-specific differences in the amino acid sequence of a hormone may result in failure of an antibody raised against the same hormone from another species to recognize, causing a nonsense result, or in "incomplete recognition," underestimation of the concentration of the target hormone. In contrast to TSH, FSH, growth hormone, or prolactin, which are species specific, ACTH can be measured in all mammalian species by a single assay method using an antibody prepared against the first 24 amino acids of the molecule. More than 50 species, including all common laboratory and domestic species, share this same amino acid sequence. Information on cyclic periodicity of hormones of rodents and rabbits is found in Section III.

Thyroxine (T_4) and 3,3′,5-triiodothyronine (T_3), its monodeiodinated, physiologically more active derivative, exist in serum partially free, but predominantly bound to transport proteins. Both the transport proteins and the overall binding affinity vary with species. In the rat, approximately 0.05% of the T4 and 0.25% of the T_3 are free. In the rat and mouse, approximately 80% of thyroid hormones are bound to albumin, and 20% to thyroxine-binding prealbumin. Thyroxine-binding prealbumin is a transport protein that appears to exist in all species, though its electrophoretic migration may vary somewhat with species. Only the free hormone is believed to be physiologically active. The effect of the interaction of obesity and fasting on thyroid hormones was investigated in lean and obese Zucker rats by Young *et al.* (1985). In lean rats 4–7 days of fasting resulted in a marked decrease in total and free thyroxine (T_4) and total and free triiodothyronine (T_3). In obese Zucker rats 4–7 days of fasting produced no change in the serum levels of these hormone, whereas more prolonged fasting resulted in elevation of total hormones from baseline without any change in the free hormones. These alterations were explained by an increase in serum thyroxine-binding globulin, a protein present in low concentration in the rat. In hamsters and rats, serum levels of thyroxine and triiodothyronine decrease significantly as a function of age (Neve *et al.*, 1981; Rao-Rupanagudi *et al.*, 1992). In rats this is associated with a decrease in the height of follicular epithelial cells and a decreased number of follicles.

The chronic administration of ethanol to mice significantly increases hepatic T_3 receptors, which may relate to alcohol-induced hypothyroidism. Experimental autoimmune thyroiditis in mice, a model for Hashimoto's thyroiditis of man, can be induced by the administration of mouse thyroglobulin with Freund's adjuvant. Its experimental production can be blocked by pretreatment with an infusion of thyroid-stimulating hormone.

Calcitonin is produced by the parafollicular cells originating from the neural crest. Calcitonin is a straight-chain polypeptide consisting of 32 amino acids. Its amino acid sequence is highly species specific. Its synthesis is stimulated by calcium ion. The release of calcitonin is stimulated by glucagon, prolactin, TSH, thyroxine, gastrin, secretin, and pancreazymin. In rats serum calcitonin increases with age from undetectable values (less than 90 pg/ml) at 6 weeks of age, to over 14,000 pg/ml at 27 months. It varies with the estrus cycle, being lowest during proestrus. It is higher in feeding than in nonfeeding rats.

The adrenal glucocorticoid axis (consisting of hypothalamus, pituitary, and adrenal cortex, particularly the zona fasciculata), becomes functional in fetal rats between the 18th and 22nd day of gestation. In adult rats, serum corticosterone levels are significantly higher in females than in males. In mice, 18-day female fetuses have significantly higher corticosterone levels than male fetuses, though the corticosterone content of the adrenal is similar in fetuses of both sexes. Stress to or adrenalectomy of the mother eliminates this sex-dependent difference in the fetus (Montano *et al.*, 1993). In the rat and mouse, cleavage of proopiomelanocortin (the precursor of ACTH) yields a large ACTH molecule about twice as long, but containing the same 24-amino-acid sequence, as the 39-amino-acid ACTH molecule of man. This large ACTH molecule is reported not to induce the formation of 17α hydroxylase. This has been proposed to explain the secretion of corticosterone rather than cortisol in rats and mice (Woodman, 1990). The hamster, however, secretes both in ratios that vary with circadian rhythm (see Section III, this chapter). The rabbit also has significant levels of both cortisol and corticosterone. In the serum of the rat, corticosterone occurs 2–19% free, approximately 80% bound to transcortin, and 10% bound to albumin. Serum transcortin levels are reduced by the exogenous administration of ACTH (Amario *et al.*, 1994).

Stress, including very mild stress, elevates the blood level of glucocorticoids. Restraint doubles the corticosterone level of rats in 2 minutes and elevates it 12-fold in 20 minutes (Liu *et al.*, 1996). If rats are housed in pairs and are removed to be bled, the second animal generally has a higher corticosterone value than the first; removal of the cagemate is sufficient stress to elevate the corticosterone. If mice are housed in groups of 8, corticosterone levels are elevated for 2 weeks in comparison with mice housed in pairs or groups of 4. After 2 weeks, corticosterone levels decline to those seen in pairs or groups of 4.

Serum testosterone levels have been shown to be significantly higher in neonatal (1–5 days old) male mice and hamsters than in females of the same age. Pang and Tang (1984) believed this to play an important role in the neural sexual differentiation of rodents.

XI. URINE AND URINARY TRACT EVALUATION

The measurement of serum urea nitrogen (BUN) or serum creatinine constitutes an extremely insensitive indicator of impaired renal function in rodent and rabbits. Serum creatinine is higher in young male rats than in young females, but this difference is abolished by 8 months of age. The use of the endogenous creatinine clearance test is a more sensitive and specific test of impaired glomerular filtration than the measurement of urea nitrogen or creatinine. It requires a timed urine specimen and a serum sample (Ringler and Dabich, 1979).

The validity of studies on the urine of rodents and rabbits may be compromised by artifacts resulting from urine collection. Timed urines must be protected from contamination and evaporation, and kept cold throughout the collection period. A technique of bladder catheterization and urine collection via the catheter passed through a tether has been described by Mandavilli *et al.* (1991).

The urine of the rat (*Rattus norwegicus*) and that of the golden hamster (*Mesocrecetus auratus*) was studied and compared to the urine of the mole rat (*Spalax ehrenbergi*), a subterranean rodent, by Haim *et al.* (1985). These workers showed a higher content of calcium and magnesium in the urine of mole rats and hamsters than in albino laboratory rats. Urine osmolality was highest in rats (388.2 ± 56.7 mOsm/kg), lower in hamsters (330.9 ± 23.6 mOsm/kg), and lowest in mole rats (301.6 ± 27.9 Mosm/kg), and urea and phosphate were significantly lower in mole rats than in hamsters or rats. Haim *et al.* (1985) speculated that these characteristics may adapt the mole rat to subterranian existence and the hamster to burrow dwelling.

Flatt and Carpenter (1971) studied crystals in the normally cloudy urine of the rabbit. They determined that normal rabbit urine contains calcium carbonate monohydrate, anhydrous calcium carbonate, and magnesium ammonium phosphate. A detailed compilation of the urinary analytes of the Sprague–Dawley rat has been reported by Shevock *et al.* (1993).

Proteinuria is commonly present in aged male mice of many strains and less frequently in females. The protein is of lower molecular weight than albumin and is of renal origin (Everett and Harrison, 1983).

A modified water-concentration test may readily be applied to laboratory rodents and appears to be more sensitive than other simple methods of detecting renal disease or impaired renal function. Water is withheld for 12 hours, after which a urine sample is collected for the measurement of specific gravity (Ragan, 1988). The urine concentrating ability of mice in response to water deprivation declines with age (Harrison and Archer, 1983).

Hypertension-prone Sabra rats challenged by acute sodium loading secreted significantly less urine than normotensive controls, including rats of the parent strain. Although this urine contained a higher concentration than that of normal controls, the sodium excretion (per unit time) was significantly lower in the hypertensive rats, demonstrating the role of impaired renal sodium excretion in hypertension (Fregly and Rowland, 1985; Meckler *et al.*, 1985).

The measurement of urinary enzymes constitutes a sensitive assessment of renal injury. Enzymes may be expressed in terms of activity per 24 hours or in a ratio to urine creatinine. Although many enzymes have been studied, the diagnostic utility of *N*-acetyl-β-glucosaminidase (NAG) and of γ-glutamyltransferase (GGT) has been best validated in the rat and mouse. The excretion of NAG, a lysosomal enzyme, is elevated in renal papillary damage. The brush border of the proximal convoluted tubules is very rich in GGT. In the rat, GGT activity in the kidney is 200–300 times as high in the kidney as it is in the liver, though the kidney does not contribute appreciably to the GGT activity of the serum. Factors other than renal injury, such as diuresis, may elevate the urinary excretion of renal enzymes, in particular that of NAG (Stonard *et al.*, 1987; Bomhard *et al.*, 1991, 1994).

References

Albers, H. E., Yogev, L., Todd, R. B., and Goldman, B. D. (1985). *Am. J. Physiol.* **248,** R434.

Alberts, J., Lang, L., Reyes, P., and Briggs, G. (1977). *J. Nutr.* **107,** 1517.

Alder, V. A., Yu, D.-Y., Su, E.-N., and Cringle, S. J. (1992). *Lab. An. Sci.* **42,** 170.

Amario, A., Giralt, M., Marti, O., Gavalda, A., Hidalgo, J., Hsu, B. R.-S., and Kuhn, R. W. (1994). *Endocrine Res.* **20,** 139.

Asa, S. L. *et al.* (1992). *Endocrinol.* **131,** 2083.

Austin, B. P., Garthwaite, T. L., Hagen, T. C., Stevens, J. O., and Menahan, L. A. (1984). *Exp. Gerontol.* **19,** 121.

Bartke, A. *et al.* (1991). *In* "Transgenic animals" (N. L. First and F. P. Huseltine, eds.). Buttersworth-Heinemann, Boston.

Ben Saad, M., and Bayle, J.-D. (1985). *Gen. Comp. Endocrinol.* **57,** 383.

Blonz, E. R., Stern, J. S., and Curry, D. L. (1985) *Am. J. Physiol.* **248,** E188.

Bodziony, J., and Schwille, P. O. (1985). *Z. Versuchstierk.* **27,** 29.

Bomhard, E., Maruhn, D., and Mager, H. (1991). *Eur. J. Clin. Chem. Clin. Biochem.* **29,** 775.

Bomhard, E. M., Falkenberg, F. W., and Loof, I. (1994). *Kidney Int.* **46** (Suppl. 47), S-60.

Boyd, J. W. (1983). *Vet. Clin. Pathol.* **12,** 9.

Brosnan, M. J. *et al.* (1993). *Am. J. Physiol.* **264,** C151.

Buckingham, J. C., Dohler, K. D., and Wilson, C. A. (1978). *J. Endocrinol.* **78,** 359.

Bullier-Picard, F., Wolf, B. A., Hugues, J. N., Durand, D., Voirol, M. J., Charrier, J., Czernichow, P., and Postel-Vinay, M. C. (1986). *Mol. and Cell. Endocrinol.* **45,** 49.

Carroll, R. M., and Feldman, E. B. (1988). *In* "Clinical Chemistry of Laboratory Animals" (W. Loeb and F. Quimby, eds.), p. 95. Pergamon, New York.

Clampitt, R. B., and Hart, R. J. (1978). *J. Comp. Pathol.* **88,** 607.

Coe, J. C., and Ross, M. J. (1985). *J. Clin. Invest.* **76.** 66.

D'Agostino, J. B., Vaeth, G. F., and Henning, S. J. (1982). *Acta Endocrinol.* **100,** 85.

Därr, W. H., Windler, E. E., Stephan, K. U., Walli, A. K., and Greten, H. (1985). *J. Lipid Res.* **26,** 672.

Dooley, J. F. (1983). *Lab. Anim.* **13,** 20.

Dowton, S. B., Woods, D. E., Mantzouranis, E. C., and Colten, H. R. (1985). *Science* **228,** 1206.

Dunn, J., and Scheuing, L. (1971). *J. Endocrinol.* **49,** 347.

Elwood, K. C., Michaelis IV, O. E., Emberland, J. J., and Bhathena, S. J. (1985). *Proc. Soc. Exp. Biol. Med.* **17,** 163.

Everett, R. M., and Harrison, S. D., Jr. (1983). *In* "The Mouse in Biomedical Research" (H. L. Foster, J. D. Small, and J. G. Fox, eds.), Vol. 3, p. 313. Academic Press, New York.

Finkelstein, J. A., Jervois, P., Menadue, M., and Willoughby, J. O. (1986). *Endocrinology* **115,** 1233.

Flatt, R. E. and Carpenter, A. B. (1971). *Am. J. Vet. Res.* **32,** 655.

Fox, S. R., and Smith, M. S. (1984). *Endocrinology* **115,** 2045.

Fregly, M. J., and Rowland, N. E. (1985). *Am. J. Physiol.* **248,** R1.

Friedel, R., Trautschold, I., Gartner, K., Helle-Feldmann, M., and Gaudssuhn, D. (1975). *Z. Klin. Chem. Klin. Biochem.* **13,** 499.

Gopinath, C., Prentice, D. E., Street, A. E., and Crook, D. (1980). *Toxicology* **15,** 113.

Haim, A., Heth, G., and Nevo, E. (1985). *Comp. Biochem. Physiol.* **80A,** 503.

Hardison, W. G. M., Weiner, R. G., Hatoff, D. E., and Miyaki, K. (1985). *Hepatology* **3,** 383.

Harrison, D. E., and Archer, J. R. (1983). *Exp. Aging Res.* **9,** 245.

Hau, J., Anderson, L. L. I., Poulsen, O. M., Krog, N. L., and Worm, K. (1991). *Lab. Animals* **25,** 122.

Liu, J. Y., Diaz, T. G. III, Vadgama, J. V., and Henry, J. P. (1996). *Lab. Anim. Sci.* **46,** 243.

Higgins, P. J., Garlick, R. L., and Bunn, H. F. (1982). *Diabetes* **31,** 743.

Hosokawa, S., Tagaya, O., Mikami, T., Nozaki, Y., Kawaguchi, A., Yamatsu, K., and Shamoto, M. (1992). *Lab. An. Sci.* **42,** 27.

Huang, C. G., Eng, J., Pan, Y. C. E., Hulmer, J. D., and Yalow, R. S. (1986). *Diabetes* **35,** 508.

Iga, T., and Klaasen, C. D. (1982). *Biochem. Pharmacol.* **31,** 205.

Iwase, M., Kodama, T., Himeno, H., Yoshinari, M., Tsutsu, N., Sadoshima, S., and Fujishima, M. (1994). *Clin. and Exp. Hypertension* **16,** 67.

Jansen, P. L. M., and Elferink, R. P. J. O. (1993). *Advances in Vet. Sci. and Comp. Med.* **37,** 175.

Jansen, P. L. M., Peters, W. H., and Lamers, W. H. (1985). *Hepatology* **5,** 573.

Jobe, L. W., Ubungen, R., Goodner, C. J., Baskin, D. G., Braquet, P., and Lernmark, A. (1993). *Autoimmunity* **16,** 259.

Jukes, T. H. (1979). *Can. J. Biochem.* **57,** 455.

Khatoon, H., Wajihullah, A. B., and Ansari, J. A. (1984). *Angew. Parasitol.* **25,** 203.

Knudtzon, J. (1984). *Z. Versuchstierk.* **26,** 123.

Lang, C. M., and Munger, B. L. (1976). *Diabetes* **25,** 434.

Leiter, E. H., Premdas, F., Harrison, D. E., and Lipson, L. G. (1988). *FASEB J.* **2,** 2087.

Levine, B. S. (1995). *In* "CRC Handbook of Toxicology" (M. J. Derelanko and M. A. Hollinger, eds.), p. 516. CRC Press, Boca Raton, Florida.

Loeb, W. F. (1987). *In* "Human Risk Assessment—The Role of Animal Selection and Extrapolation" (M. V. Roloff, ed.), p. 81. Taylor and Francis, New York.

Loeb, W. F., and Caracostas, M. C. (1992). *In* "Pathobiology of the Aging Rat" (U. Mohr, D. L. Dungworth, and C. C. Capen, eds.), p. 7. ILSI Press, Washington, D.C.

Loeb, W. F., Das, S. R., Harbour, L. S., Turturro, A., Bucci, T. J., and Clifford, C. B. (1996). *In* " Pathobiology of the Aging Mouse," (U. Mohr, D. L. Dungworth, C. C. Capen, W. W. Carlton, J. P. Sandberg, and J. M. Ward, eds.). ILSI Press, Washington, D.C.

Mandavilli, U., Schmidt, J., Rattner, D. W., Watson, W. T., and Warshaw, A. L. (1991). *Lab. An. Sci.* **41,** 258.

Markussen, J. (1971). *J. Protein Res.* **3,** 149.

McLaughlin, R. M., and Fish, R. E. (1994). *In* "The Biology of the Laboratory Rabbit" (P. J. Manning, D. H. Ringler, and C. E. Newcomer, eds.). Academic Press, San Diego.

Meckler, J., Yagil, Y., and Ben-Ishay, D. (1985). *Experientia* **41,** 923.

Montano, M. M., Wang, M.-H., and vom Saal, F. S. (1993). *J. Reprod. Fertil.* **99,** 283.

Morton, D. B., Abbot, D., Barclay, R., Close B. S., Ewbank, R., Gask, D., Heath, M., Mattic, S., Poole, T., Seamer, J., Southee, J., Thompson, A., Trussell, B., West, C., and Jennings, M. (1994). *Lab. Animals* **27,** 1.

Neve, P., Authelet, M., and Golstein, J. (1981). *Cell Tissue Res.* **220,** 449.

Nichols, D. J., and Chevins, P. F. D. (1981). *Experientia* **37,** 319.

Notkins, A. L. (1965). *Bacteriol. Rev.* **29,** 143.

Ottenweller, J. E., Meier, A. H., Russo, A. C., and Frenzeke, M. E., (1979). *Acta Endocrinol.* **91,** 150.

Ottenweller, J. E., Tapp, W. N., Burke, J. M., and Natelson, B. H. (1985). *Life Sciences* **37,** 1551.

Pang, S. F., and Tang, F. (1984). *J. Endocrinol.* **100,** 7.

Quimby, F. W. (1988). *In* "Clinical Chemistry of Laboratory Animals" (W. Loeb and F. Quimby, eds.), p. 1. Pergamon, New York.

Ragan, H. A. (1988). *In* "Clinical Chemistry of Laboratory Animals" (W. Loeb and F. Quimby, eds.), p. 321. Pergamon, New York.

Rao-Rupanagudi, S., Heywood, R., and Gopinath, C. (1992). *Vet. Pathol.* **29,** 278.

Ringler, D. H., and Dabich, L. (1979). *In* "The Laboratory Rat" (H. J. Baker, J. R. Lindsey, and S. H. Weisbroth, eds.), Vol. I, p. 105. Academic Press, New York.

Roesler, W. J., Helgason, C., Gulka, M., and Khandel, R. L. (1980). *Horm. Metab. Res.* **17,** 572.

Rosenzweig, J., Haurankoua, J., Lesniak, M., Brownstein, M., and Roth, J. (1980). *Trans. Assoc. Am. Phys.* **40,** 263.

Rubin, E., and Schultz, J. (1993). *In* "Transgenic Animals as Model Systems for Human Disease" (E. F. Wagner and F. Theuring, eds.), p. 25. Springer Verlag, Berlin.

Schnetzler, B. *et al.* (1993). *J. Clin. Invest.* **92,** 272.

Shevock, P. N., Kahn, S. R., and Hackett, R. L. (1993). *Urol. Res.* **21,** 309.

Slaughter, M. R., and Moen, J. S. (1991). *Lab. Animals* **25,** 272.

Smith, L. F. (1966). *Am. J. Med.* **40,** 662.

Snitily, M. U., Gentry, M. J., Mellencamp, M. A., and Preheim, L. C. (1991). *Lab. An. Sci.* **41,** 285.

Stefaneanu, L. *et al.* (1993). *Lab. Invest.* **68,** 584.

Stonard, M. D., Gore, C. W., Oliver, G. J. A., and Smith, I. K. (1987). *Fund. and Appl. Tox.* **9,** 339.

Suber, R. L., and Kodell, R. L. (1985). *Vet. Clin. Pathol.* **14,** 28.

Sullivan, M. P., Cerda, J. J., Robbins, F. L., Burgin, C. W., and Beatty, R. J. (1993). *Lab. An. Sci.* **43,** 575.

Surwitt, R. S., McCubbin, J. A., Livingston, E. G., and Feinglos, M. N. (1985). *Psychosomat. Med.* **47,** 565.

Takahashi, K., Hayafuji, C., and Murakami, N. (1982). *Am. J. Physiol.* **243,** E443.

Tanzawa, K., Shimada, Y., Kuroda, M., Tsujita, Y., and Watanabe, W. (1980). *FEBS Lett.* **118,** 81.

Teschke, R., Vierke, W., and Goldermann, L. (1983). *Toxicol. Lett.* **17,** 175.

Thompson, M. B. (1986). *In* "Managing Conduct and Data Quality of Toxicology Studies" (D. Bristol, ed.), p. 199. Princeton Scientific, Princeton, New Jersey.

Thompson, M. B., Blair, P. C., Morris, R. W., Neptun, D. A., Deyo, D. F., and Popp, J. A. (1987). *Clin. Chem.* **33,** 1856.

Underwood, E. (1971). "Trace Elements in Human and Animal Nutrition," 3rd Ed. Academic Press, New York.

vom Saal, F. S., Finch, C. E., and Nelson, J. F. (1994). *In* "The Physiology of Reproduction" (E. Knobil and J.D. Neill, eds.), p. 1213. Raven Press, New York.

Wagner, J., Paul, M., and Ganter, D. (1993). *In* "Transgenic Animals as Model Systems for Human Disease" (E. Wagner and F. Theuring, eds.), p. 1. Springer-Verlag, Berlin.

Watts, A. G., and Fink, G. (1984). *Endocrinology* **115,** 2251.

Weingand, K., Bloom, J., Carakostas, M., Hall, R., Helfrich, M., Latimer, K., Levine, B., Neptun, D., Rebar, A., Stitzel, K., and Troup, C. (1992). *Tox. Path.* **20,** 539.

Wittmers, L. E., and Haller, E. W. (1983). *Metabolism* **32,** 1093.

Woodman, D. D. (1990). *In* "Proc. of the IVth Congress of the Int. Soc. of An. Clin. Biochem."(J. J. Kaneko, ed.), p. 74.

Wriston, J. C., Jr. (1984). *Comp. Biochem. Physiol.* **B77,** 253.

Young, R. A., Rajatanavin, R., Morin, A. F., and Braverman, L. E. (1985). *Endocrinology* **116,** 1248.

CHAPTER

30

Avian Clinical Biochemistry

J. T. LUMEIJ

I. INTRODUCTION

Avian medicine and surgery has been recognized as a distinct specialty in veterinary medicine on three continents (Europe, Australia, and North America). This has occurred because of the increasing demand for veterinary care for individual birds that have attained a high sentimental and/or economic value. The flock approach that had been practiced for decades in poultry medicine had only limited applicability for individual pet birds, and alternative diagnostic and therapeutic techniques were developed. Clinical signs in birds are often nonspecific and the information gained by physical examination is limited. Earlier demands for large blood sample volumes and limited veterinary involvement in pet bird disease were major obstacles to the diagnostic application of clinical biochemistry in avian medicine. The introduction of micromethods in clinical laboratories and the public demand for veterinary care for individual birds has removed these obstacles. The scientific and clinical work of the past decade in avian clinical biochemistry has led to its development as an indispensable tool in avian medicine.

II. COLLECTION OF BLOOD SAMPLES

A. Size of Blood Samples

An important consideration when taking blood samples from small birds is response to blood loss. Kovách *et al.* (1969) studied the mortality of various avian and mammalian species following blood loss and showed that birds can better tolerate severe blood loss than mammals (Fig. 30.1). This is due to their much greater capacity for extravascular fluid mobilization (Djojosugito *et al.*, 1968; Wyse and Nickerson, 1971). Kovách *et al.* (1969) found that in healthy birds, the amount of blood that can be removed without deleterious effects is 3% of body weight in ducks and pigeons, 2% in chickens, and 1% in crows and pheasants (Fig. 30.1). In patients a maximum of 1% is a safe limit for the amount of blood that can be collected for diagnostic purposes.

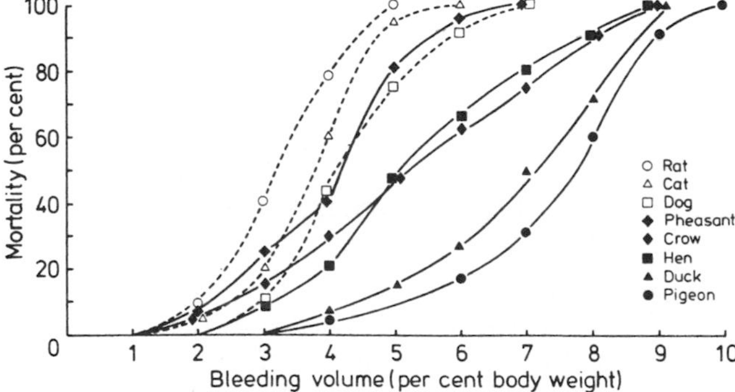

FIGURE 30.1 Mortality after identical blood losses in various avian and mammalian species (abscissa). Every hour 1% of body weight blood was withdrawn from every animal (ordinate). The percentage of animals lost during the hour following bleeding has been recorded and plotted. (After Kovách *et al.*, 1969. Reprinted with permission from Lumeij, 1987a.)

B. Handling of Blood Samples

Nearly all routine hematological and biochemical investigations can be performed when lithium heparin is used. The use of one single sample limits unnecessary blood spillage, which is an important consideration when dealing with small birds. When plasma is used, more plasma can be harvested than serum from the collecting tube.

In mammals, EDTA is regarded as the best anticoagulant for preservation of cellular morphology and good staining characteristics (Schmidt *et al.*, 1963), but this is not necessarily true in avian hematology. There are many avian species where EDTA causes disruption of the red blood cells. Fourie (1977) found heparin to be the most suitable anticoagulant for hematology in pigeons. Hawkey *et al.* (1983) found that EDTA produced progressive hemolysis in blood samples from crowned cranes. Dein (1986) reported a similar reaction in crows, jays, brush turkey, and hornbills. Similar reactions to EDTA are see in blood from crows and magpies (Lumeij, unpublished). Blood smears should be made immediately after collection of the sample to prevent changes in blood cell morphology.

The normal time lag of up to 60 minutes between collection of a blood sample and separation of plasma from cells that is common in human medicine (Laessig *et al.*, 1976) is not acceptable in avian clinical biochemistry. Immediately after collection, plasma and cells should be separated by centrifuging. In pigeon blood at room temperature, a rapid decline (10% in 10 minutes 30% in 30 minutes, up to 65% in 2 hours) of potassium concentrations occurs because of a shift of potassium ions from the plasma into the red blood cells. In chickens, decreases were smaller overall, with a 29% decrease being noted after 2 hours (Lumeij, 1985a). Many reports on blood chemistry in birds are based on serum instead of plasma or on plasma from blood samples that were not centrifuged immediately. Therefore, reported plasma potassium concentrations are often too low.

C. Sampling Procedures

In most species, the right jugular vein is the preferred site for blood sampling. This thick-walled vein is less prone to hematoma formation (Law, 1960; McClure and Cedeno, 1955; Stevens and Ridgeway, 1966). The medial metatarsal vein is especially useful for multiple sampling of small blood volumes in larger birds such as pigeons. Blood can be collected using a needle and syringe or a blood lancet. In the pigeon, the jugular vein is not readily accessible.

The wing vein or ulnar vein is the vein that is traditionally used in poultry. After being swabbed with alcohol, the vein is punctured with a blood lancet (Gratzl and Koehler, 1968). These authors warn against the use of the comb for blood collection in poultry because the risk of exsanguination is high, especially during cold weather. In pet birds, the use of a blood lancet for blood collection from the ulnar vein cannot be recommended because this site is very prone to hematoma formation, often even when a needle is used. The advantage of the cutaneous ulnar vein is that it can be located in all avian species.

In ostriches, it is safer to avoid the clearly visible jugular vein and take blood from the ulnar vein using a sideways approach to the standing animal after it has been hooded and the wing has been lifted upwards by two assistants.

In ducks and geese, the venous occipital sinus is the preferred site for blood sampling (Vuillaume, 1983). It is located at the junction of the base of the skull and the first cervical vertebra.

Cardiac puncture carries the risk of cardiac tamponade, so this technique is not recommended for use in avian clinical practice. Some clip a toenail to obtain a blood sample. Disadvantages of this method are that it is painful to the bird, the sample may become contaminated with tissue fluids, it may cause damage to the nail bed, and the amount of blood that can be obtained is limited. When a toenail is clipped to obtain a blood sample, contamination of the sample with urates from the droppings may give falsely high readings (Rosskopf *et al.*, 1982; Ekstrom and Degernes, 1989). Therefore, the nail should be thoroughly cleaned before clipping. Different bleeding sites (e.g., venous blood vs cardiac blood) may cause variation in hematological or biochemical values (Kern and De Graw, 1978).

A vacuum system greatly facilitates blood sampling from the jugular and cutaneous ulnar veins and from the venous occipital sinus. A 3-ml vacuum tube is sufficient for most cases (Venoject, Omnilabo, Breda, The Netherlands).

III. STARVATION AND POSTPRANDIAL EFFECTS—CIRCADIAN AND CIRCANNUAL RHYTHMS

Some plasma chemical constituents are influenced by starvation or food consumption. Up to 4 days of starvation in pigeons did not result in hypoglycemia, but rather a starvation hyperglycemia occurred after 3 days (Lumeij, 1987b). Constituents that have markedly increased values postprandially are uric acid and total bile acid concentrations (Lumeij, 1991; Lumeij and Remple, 1991, 1992). Furthermore, daily or yearly fluctuations have also been reported for some chemical constituents. In fasted pigeons maintained on a natural daily 17-hour photoperiod, a circadian rhythm was found in plasma glucose concentrations (Lumeij *et al.*, 1987a) with high values early during the photophase (Fig. 30.2). Basal plasma thyroxine concentrations in racing pigeons were significantly higher in July than in September and December (Lumeij and Westerhof, 1988a). Age, sex, altitude, nutritional status and egg laying may also cause variation (Driver, 1981; Kocan and Pits, 1976; McGrath, 1971; Mori and George, 1978; Simkiss, 1967).

IV. PLASMA PROTEINS

A. Introduction

Plasma proteins are important complementary constituents in the diagnosis of gastrointestinal, hepatic, renal, and/or infectious diseases. Determination of plasma proteins seldom leads to a specific diagnosis (e.g., monoclonal gammopathies), but will help the clinician to evaluate the nature, severity, and progress of a disease.

B. Plasma vs Serum

In pigeons, the concentration of total protein (TP) in plasma is about 1.5 g/liter higher than in serum because the former contains fibrinogen: TP_{serum} =

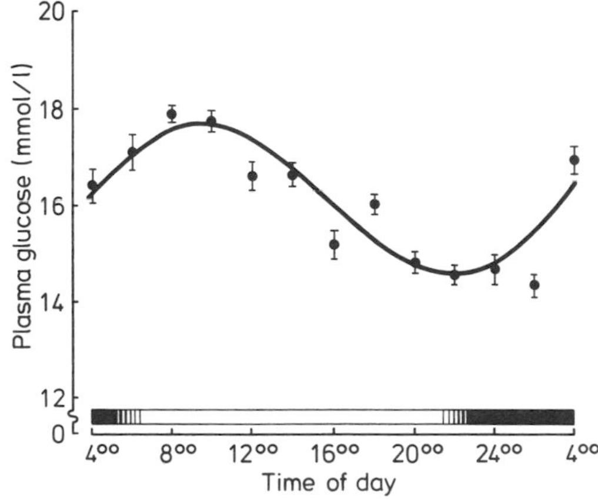

FIGURE 30.2 Mean (± SEM) plasma glucose concentration as a function of time in fasted racing pigeons. A cosines function $y(t)$ is fitted to the data: $y(t) = 16.26 + 1.55 \cdot \cos(0.2618t - 2.4646)$. The relevance of the fit as judged by means of the multiple correlation coefficient was significant ($R = 0.892$, $P < .01$). Reprinted with permission from Lumeij *et al.* (1987a).

$-1.7 + 1.01\ TP_{plasma}$ (Fig. 30.3). The correlation is highly significant ($P < .000001$; $R = 0.99$; $n = 50$) (Lumeij and Maclean, 1996).

C. Physiological Variation in Female Birds

In female birds, a considerable increase in plasma TP concentration occurs just prior to egg laying because of an estrogen-induced increase in the globulin fraction (Griminger, 1976). The proteins are the yolk precursors (e.g., vitellogenin and lipoproteins) that are synthesized in the liver and transported to the ovary, where they are incorporated into the oocyte (Griffin *et al.*, 1984).

D. Refractometry vs the Biuret Method

Lumeij and De Bruijne (1985b) demonstrated that the refractometric method is unreliable for use in avian blood and that this method should not be used in avian practice. The refractometric method consistently yields higher results than TP as determined by the biuret method, and the correlation coefficient between these two methods is low. Even though it has been demonstrated that the refractometric method is unreliable in avian blood, this method is still being used (Clubb *et al.*, 1990). One study suggested that only temperature-compensated refractometers are reliable (Andreasen *et al.*, 1989). In another study in our laboratory (Lumeij and Maclean, 1996) using plasma and serum of 58 pigeons, two types of refractometers were compared with the biuret method. Neither instrument proved to give an accurate measurement of plasma total protein. Both refractometers gave considerably higher values than the biuret method, with the temperature-compensated instrument being consistently higher in readings than the non-temperature-compensated one (Fig. 30.4). It was concluded that a species- and refractometric-specific conversion factor must be applied before refractometric results can be used, and then only as a rough estimate of the TP. Another important consideration is that TP determinations without information on plasma protein electrophoresis have limited value (see later discussion). For avian clinical practice it is advised to establish TP values using the biuret method.

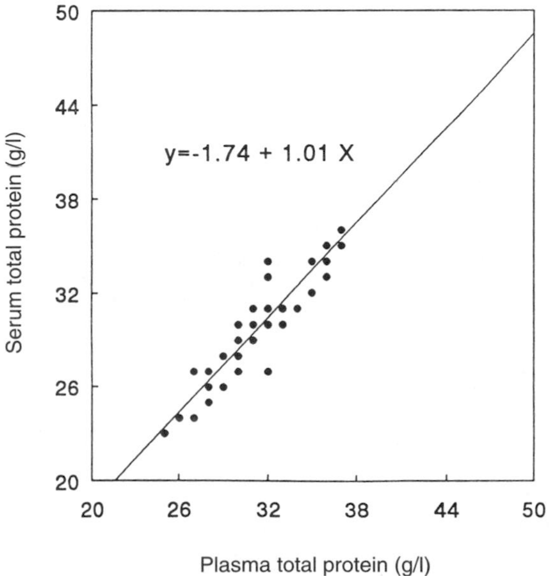

FIGURE 30.3 Relation between plasma and serum total protein by biuret method in racing pigeons (n = 50; g/liter). (From Lumeij and Maclean, 1996).

E. Effect of Protein Standards

Most commercial laboratories use a human standard for TP and albumin (Alb) determinations without validating the method for the species from which the blood sample was obtained. There are significant differences between TP concentrations when different standards are used, such as human, bovine, pigeon, and chicken standards, although there is a high correlation between the results obtained with the various standards (Spano *et al.*, 1988; Lumeij *et al.*, 1990). When

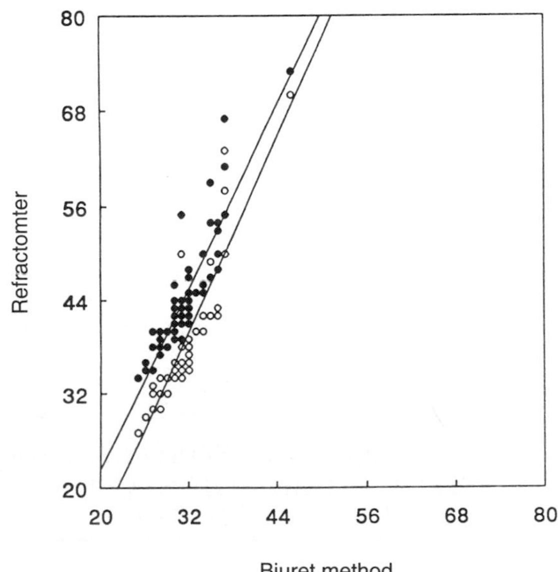

FIGURE 30.4 Relation between plasma total protein concentration by biuret method and two types of refractometers in pigeons (g/liter; $n = 58$). ● = temperature-compensated refractometer ($y = -16.38 + 1.93\,x$; $r = 0.89$); ○ = non-temperature-compensated refractometer ($y = -25.88 + 2.06\,x$; $r = 0.89$). (From Lumeij and Maclean, 1996).

a pigeon standard was used to determine serum TP concentration (TP_p) with the biuret method, values found were significantly higher than values found with the biuret method using the human standard (TP_h), but there was a high correlation (Lumeij *et al.*, 1990):

$$TP_h = 2.94 + 0.83\ TP_p\ (P < .0001;\ r = 0.93).$$

Spano *et al.* (1988) found consistently lower TP values in chicken serum with the use of a chicken standard than with a bovine standard. Because the use of a species-specific standard for all avian species is unrealistic and because a high correlation exists between the results obtained with the various standards, it is recommended that reference values for the various avian species be established using the most commonly used standard, namely the human standard.

F. Plasma Protein Electrophoresis—Albumin/Globulin Ratio

Plasma protein electrophoresis (PPE) on cellulose acetate membranes is used routinely in avian patients in our clinic (Lumeij and De Bruijne, 1985a) and the protein fractions are called albumin and α-, β- and τ-globulin. Often a prealbumin fraction can also be observed (Fig. 30.5). In healthy birds the albumin fraction is the largest protein fraction. In acute or chronic inflammatory conditions a rise in total protein caused by elevated globulin fractions may occur. Often albumin concentrations are decreased in these situations. The combined effect of these changes is a decrease in the albumin/globulin (A/G) ratio. Often the TP concentration is within the reference range, while the A/G ratio is decreased; therefore the A/G ratio is often of greater clinical significance than the TP. Examples of diseases with a decrease in the A/G ratio are egg-related peritonitis, and chronic infectious diseases such as aspergillosis, psittacosis, and tuberculosis. Protein electrophoresis can also be used to monitor response to treatment. In liver failure, extremely low plasma TP can occur in combination with a decreased A/G ratio. Gastrointestinal and renal diseases can also lead to severe hypoproteinemia. In birds, protein malnutrition may lead to hypoproteinemia (Leveille and Sauberlich, 1961). Increased TP with a normal A/G ratio can be expected in dehydrated birds if the primary disease did not cause hypoproteinemia. To calculate the A/G ratio, prealbumin and Alb as determined by plasma protein electrophoresis are combined as A and all globulin fractions as G (Figs. 30.6 through 30.9; Lumeij, 1987e).

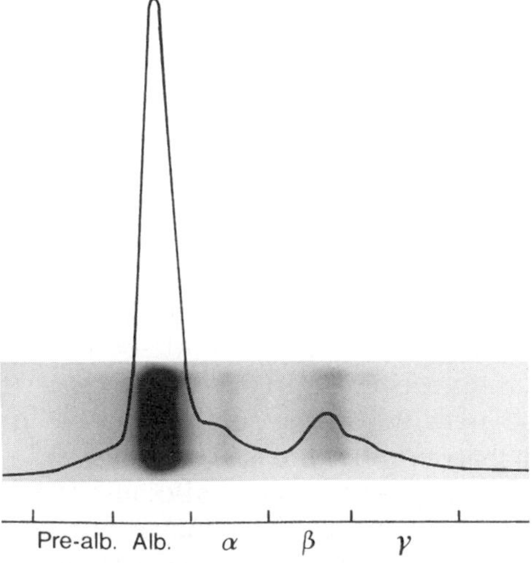

FIGURE 30.5 Densitometer scan and electrophoretic pattern from a representative pigeon serum. Prealbumin, albumin, and α-, β-, and τ-globulin. Reprinted with permission from Lumeij and De Bruijne (1985a).

G. Albumin Methodology

The labor-intensive PPE is not available in every laboratory, and Alb is commonly determined chemically by the bromcresol green (BCG) dye-binding method. Furthermore, most commercial laboratories will use a human standard for TP and Alb determinations, without validation for the species in question, although various standards can be used. The BCG method is unreliable in avian blood. Discrepancies between values obtained by dye-binding techniques and those obtained by electrophoresis have been demonstrated for chicken, duck, turkey, and pigeon (Spano *et al.*, 1988; Lumeij *et al.*, 1990).

Albumin determinations performed with dry methods (e.g., Kodak Ektachem) have not been validated for use in birds.

V. RENAL FUNCTION

A. End Products of Protein Metabolism—Hyperuricemia and Gout

Uric acid (UA) is the major end product of nitrogen (N) metabolism in birds. It constitutes approximately 60 to 80% of the total excreted N in avian urine (Skadhauge, 1983). Uricotelism permits excretion or storage of N waste in a small volume of water. Uric acid is relatively nontoxic when compared to ammonia (NH_3) and is essential for the development of the embryo in the egg of reptiles and birds. Uric acid is synthesized in the liver, and 90% is excreted via tubular secretion, largely independent of the urine flow rate (Skadhauge, 1981). The clearance of UA exceeds the glomerular

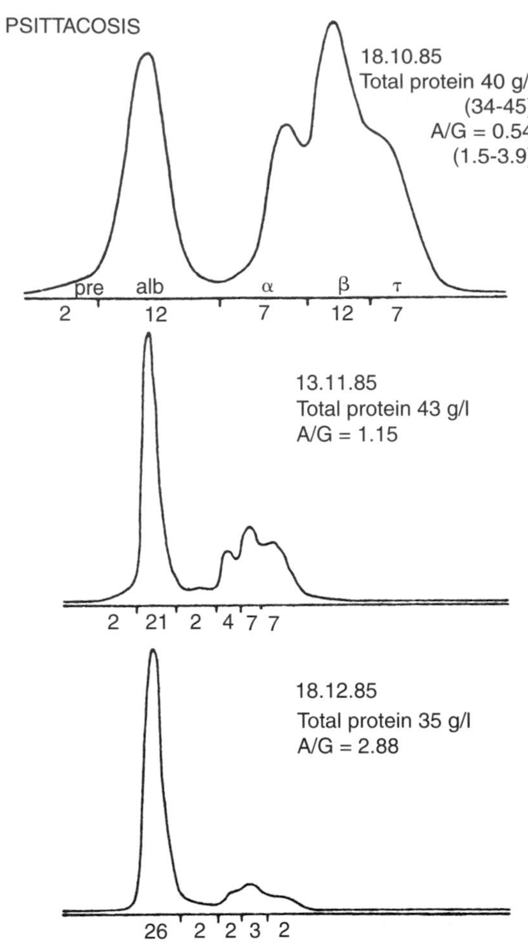

FIGURE 30.6 Plasma protein electrophoresis and the albumin : globulin ratio (A/G) in an African grey parrot (*Psittacus erithacus erithacus*) with psittacosis, before, during, and after treatment with doxycycline. Pre = prealbumin; alb = albumin; α, β, τ = globulin fractions (reference values in parentheses). Reprinted with permission from Lumeij (1987e).

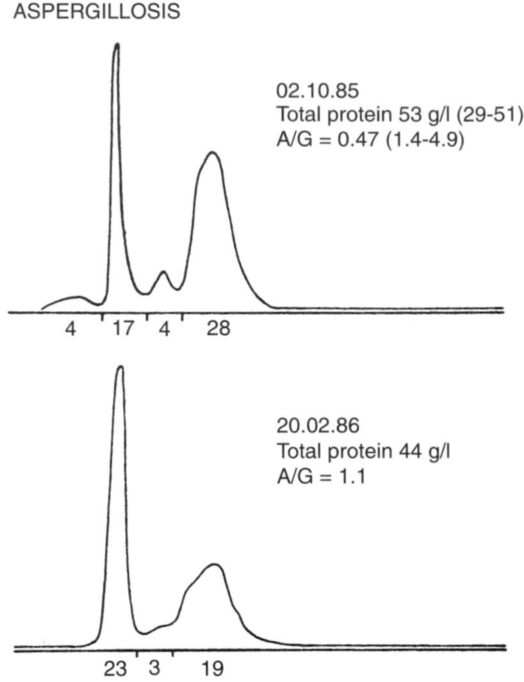

FIGURE 30.7 Plasma protein electrophoresis in an Amazon parrot (*Amazona* sp.) with aspergillosis. At 4 1/2 months after a diagnosis was made and despite treatment with ketoconazole and 5-fluorocytosine, the globulin fraction was still elevated, causing the A/G ration to be decreased (reference values in parentheses). Reprinted with permission from Lumeij (1987e).

filtration rate by a factor of 8–16. The rate of secretion is largely independent of the state of hydration of the bird. Very high concentrations of UA can be found in ureteral urine in dehydrated birds. Renal function disorders can eventually lead to increased plasma UA concentrations. Nonprotein nitrogen (NPN) substances in plasma such as UA, creatinine (Cr), and urea (UR) are increased only when renal function is below 30% of its original capacity.

B. Articular and Visceral Gout

Hyperuricemia can result in the precipitation of monosodium urate monohydrate crystals in joints (articular gout) and on visceral surfaces (visceral gout). The exact mechanism of deposition or the predilection for certain sites is unknown, although lower temperatures at predilection sites have been suggested. Gout should not be regarded as a disease entity, but rather as a sign of a severe renal-function disorder.

When birds are given dietary protein in excess of their requirements, the excess protein is catabolized and the N released is converted to uric acid. The total amount of uric acid formed may surpass the clearing capacity of this substance from the body, and hyperuricemia and articular or visceral gout may result. The use of high-protein poultry pellets as the bulk food in psittacines may result in an increased incidence of gout.

There is no consensus on the different etiologies of articular and visceral gout in birds. The following hypothesis, however, seems to explain all known facts about avian gout. A plasma urate concentration that is slightly above the solubility of sodium urate will lead to urate precipitates in the body. Predilection sites are those areas where the solubility of sodium urate is lower than in other areas. The joints and synovial sheaths may be predilection sites because of a comparatively low temperature, which may reduce urate solubility. Articular gout is a sign of chronic moderate hyperuricemia. Uric acid deposits grow with time and form the typical tophi of articular gout.

If urate crystals precipitate in the tubules or collecting ducts of the kidney or the ureters (e.g., severe

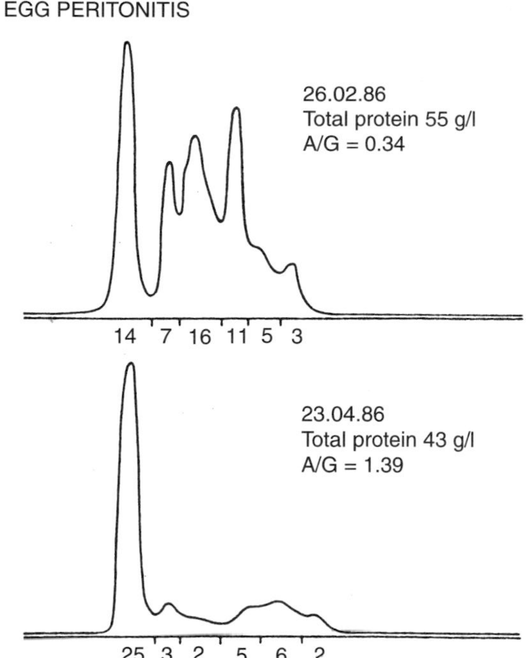

FIGURE 30.8 Serum protein electrophoresis and albumin : globulin (A / G) ratio in an emu (*Dromiceius novaehollandiae*) with an egg-related peritonitis. Two months after surgical treatment and remission of clinical signs a marked increase in the albumin fraction and decrease of the globulin fraction was observed. Reprinted with permission from Lumeij (1987e).

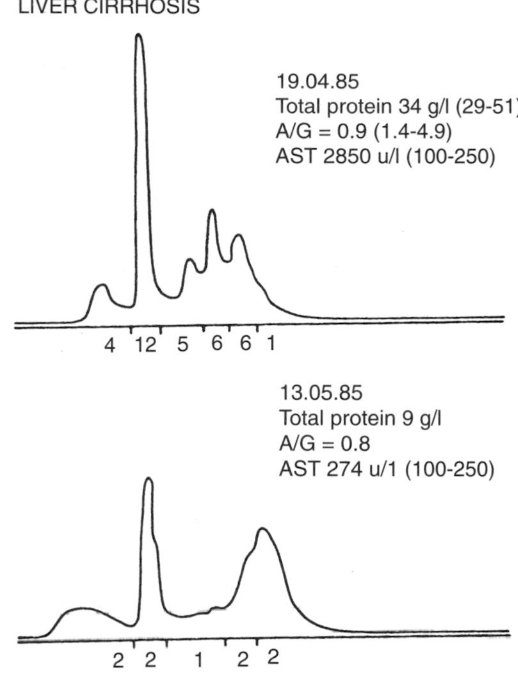

FIGURE 30.9 Total protein concentration, plasma protein electrophoresis, albumin : globulin ratio (A / G), and plasma aspartate aminotransferase (AST) activity in an Amazon parrot (*Amazona* sp.). At the first examination there was a marked elevation of AST and a decreased A / G. Liver cirrhosis was diagnosed by means of histological examination of a liver biopsy. The second protein electrophoresis was made from a plasma sample collected just before the bird was euthanized 1 month later. Reference values in parentheses. Reprinted with permission from Lumeij (1987e).

dehydration of long duration, vitamin A deficiency), this will lead to an acute obstructive uropathy (postrenal obstruction). Anuria or gross oliguria occurs, and tubular secretion of UA is severely compromised. This leads to a rapid and severe increase in plasma UA with precipitation of urates on many visceral surfaces and those predilection sites for articular gout. This condition of visceral gout will rapidly lead to death of the affected animal. This hypothesis is based on the fact that no inflammation or tophi are seen in typical predilection sites for articular gout because the condition has a rapidly fatal course. In this situation, the kidney tubules, collecting ducts, and ureters may contain uric acid deposits. An alternative situation could occur in acute tubular failure. In this condition, visceral gout could develop without urate deposits in the tubules, collecting ducts, and ureters.

C. Acute vs Chronic Renal Failure

Renal function disorders may result either from any progressive destructive condition affecting both kidneys (chronic renal failure) or under conditions where the function of the kidneys is rapidly and severely, but often reversibly compromised (acute renal failure). In acute renal failure, oliguria is common, whereas in chronic renal failure, polyuria is more common. It is important to exclude reversible conditions that can mimic irreversible chronic renal failure, such as prerenal renal failure caused by dehydration, shock, urolithiasis (postrenal renal failure), and renal causes. Appropriate and timely treatment of these conditions can often prevent further damage and in some cases result in improved function. Extrarenal factors such as infection, gastrointestinal hemorrhage, and hypovolemia can disturb an otherwise stable, well-compensated asymptomatic chronic renal patient and precipitate a desperately dangerous condition.

D. Prerenal Azotemia

Prerenal azotemia can be defined as the clinical condition associated with reduced renal arterial tension leading to oliguria and retention of nitrogenous wastes in the blood. It is often seen during shock or severe dehydration. No increased plasma UA concentrations were observed in 4-day dehydrated racing pigeons, whereas plasma urea concentration had a significant 6.5–15.3-fold increase. Plasma urea appeared to be the single most useful variable for early detection of prerenal causes of renal failure (Lumeij, 1987c). This is because urea is excreted by glomerular filtration, whereas tubular reabsorption is dependent on urine flow, which in turn depends on the state of hydration. During hydration almost all of the filtered urea is excreted and during dehydration nearly all of the filtered urea is reabsorbed. The tubular reabsorption of urea in conditions of renal failure accompanied by a low urine flow (e.g., dehydration) in combination with a nearly unchanged excretion of UA causes a disproportionate increase in plasma urea concentration, resulting in an increased urea/uric acid ratio. Urea is normally present in low concentration in avian plasma, and therefore plasma urea concentration has traditionally been considered as an inappropriate variable to evaluate renal function in birds.

When the state of dehydration continues and gets more severe, this may eventually lead to hyperuricemia. This might be caused by reduced tubular blood supply, which leads to reduced urate secretion. Urates may also precipitate in the tubuli when there is active tubular secretion of UA in the absence of urine flow. The latter condition looks very much like the one that is called acute uric acid nephropathy in humans (Watts, 1978).

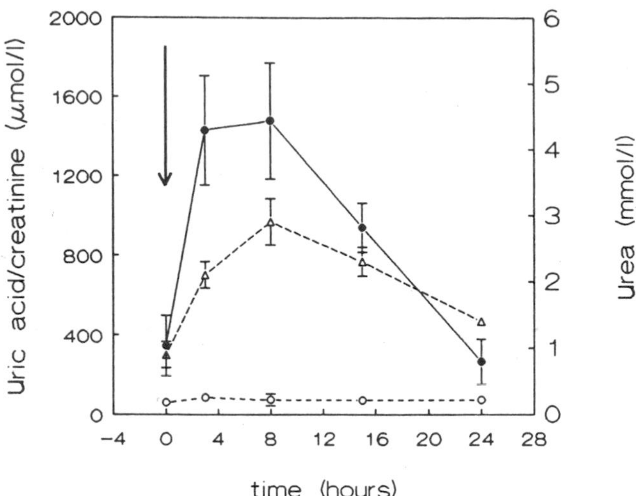

FIGURE 30.10 Fasting and postprandial nonprotein nitrogen substances in plasma of peregrine falcons, *Falco peregrinus* (mean ± SD). ● = [uric acid] (μmol/liter); △ = [urea] (mmol/liter); O = [creatinine] (μmol/liter); ↓ = feeding quail. Reprinted with permission from Lumeij and Remple (1991).

E. Postprandial Effects

In raptors a significant postprandial increase in plasma UA and urea concentration occurs (Lumeij and Remple, 1991; Lumeij and Redig, 1992). Postprandial UA was similar to those in birds with hyperuricemia and gout and was well above the theoretical limit of solubility of urate in plasma. It is not clear why under physiological conditions, no urate deposits occur in raptors, which have hyperuricemia for at least 12 hours after ingesting a natural meal (Fig. 30.10). To avoid physiological food-induced increases in UA and urea that can complicate interpretation of plasma chemistry results in birds, blood samples should be collected after a 24-hour fasting period.

F. Other Changes Associated with Renal Failure

Hyperkalemia is a particular problem in acute renal failure that may lead to severe electrocardiographic changes and eventually to cardiac arrest. Hypocalcemia and hyperphosphatemia are usual in humans with renal failure. The former may lead to hypocalcemic tetany, especially with rapid correction of acidosis. In birds, special attention should be paid to these constituents of renal disease because these changes may have therapeutic implications. Anemia has been documented in birds with chronic renal failure.

G. Murexide Test

Macroscopically, the aspirated urates from articular gout look like toothpaste. The presence of urate can be confirmed by performing the murexide test or by microscopic examination of aspirates of tophi or joint accumulations. The test is performed by mixing a drop of nitric acid with a small amount of the suspected material on a slide. The material is evaporated to dryness over a Bunsen flame and allowed to cool. Then one drop of concentrated ammonia is added. If urates are present, a mauve color will develop. Microscopically sharp needle-shaped crystals can be seen in smears. A polarizing microscope is helpful in identifying the typical crystals.

VI. HEPATOBILIARY DISEASE

A. Clinical Enzymology

Clinical enzymology is described more fully in a separate chapter in this book and only a brief synopsis will be given here. Enzymes occur normally in the cytoplasm (e.g., aspartate aminotransferase [AST], alanine aminotransferase [ALT], lactic dehydrogenase [LDH]), mitochondria (glutamate dehydrogenase [GLDH] and AST), nucleus, or membranes (alkaline phosphatase [AP], gamma glutamyl transferase

[GGT]) of body cells, where they catalyze specific reactions. The distribution of various enzymes is markedly different among various organs and various animal species, which explains the variations in organ and tissue specificities among the various animal species. Generally, increased enzyme concentrations are a measure of recent organ damage rather than decreased organ function. Increased enzyme production has been reported in cholestatic liver disease in mammals (AP and GGT). Sometimes a decreased activity is of diagnostic value (e.g., decreased cholinesterase activity in organophosphate toxicity). Baseline activity of an enzyme in plasma is generally a reflection of the amount and turnover of the tissue that contains this enzyme. For example, the creatine kinase (CK) activity in plasma increases in direct proportion to the increase of skeletal muscle mass as a result of training. Increases in CK may also be seen simply as a result of the excitement of handling. Conversely, in chronic liver disease with fibrosis and a reduction in the number of functional hepatocytes, plasma activities of liver enzymes may be within normal limits despite the presence of severe liver fibrosis. The increase of a particular enzyme also depends on factors such as its rate of release, rate of production, and rate of clearance from plasma. Cytoplasmic enzymes will be released early in cell degeneration, whereas mitochondrial enzymes will be released after advanced cell damage. Enzymes with high tissue concentrations but with short half-lives are of limited value in clinical enzymology because of their rapid disappearance from plasma.

Generally, EDTA samples are not appropriate for enzyme assays because this anticoagulant may chelate metal ions that are required for maximal enzyme activity. Plasma and cells should be separated immediately after sampling to prevent leakage of intracellular enzymes into the plasma. Even if the cellular elements are separated from the plasma, freezing/thawing and refrigeration of plasma samples for several days may severely decrease enzyme activity and therefore should be avoided unless the effects of the storage procedure used are known.

B. Enzyme Activities in Avian Tissues

Enzyme profiles of the various organs have been studied in chickens, mallard ducks, turkeys, racing pigeons, budgerigars, and African grey parrots (Lumeij, 1994f; Figs. 30.11 through 30.13).

C. Clearance of Enzymes from Plasma

The half-life of an enzyme is defined as the time required for its concentration to be reduced by half. When an enzyme is injected into plasma, its clearance from the plasma generally follows a biphasic exponential decline. Initially, there is a very rapid decline, which is the primary mixing or distribution phase, followed by a slower secondary decline phase, which is the actual clearance of the enzyme from the plasma. During this secondary phase, a constant fraction of the enzyme is cleared per unit time; hence, the decline is linear on a semilogarithmic scale. The half-life ($t_{1/2}$) of the enzyme can be calculated from the regression function of the secondary linear phase of the semilogarithmic concentration–time curve and is independent of plasma enzyme activity.

Clearance half-lives of various enzymes considered to be of use for the differential diagnosis of liver and muscle disease in pigeons have been established by studying the disappearance rates of enzymes from plasma after IV injections of supernatants of homologous liver and muscle homogenates (Lumeij *et al.*, 1988a). For AST, ALT, and LDH, the half lives of the respective enzymes from liver and muscle were compared, whereas for GLDH and CK only liver and muscle were used, respectively (Table 30.1).

D. Experimentally Induced Liver and Muscle Damage

Plasma enzyme profiles after experimental or spontaneously occurring liver disease have been studied in a number of avian species. The results of studies in racing pigeons (Lumeij *et al.*, 1988a,b) with two different types of liver disease were compared to plasma chemistry changes after muscle injury. Liver disease was induced by ethylene glycol or D-galactosamine and muscle injury was induced by an intramuscular injection of doxycycline in three groups of six pigeons each. Plasma chemical changes were correlated with histological findings from organ samples taken just after the last blood collection (Fig. 30.14, Table 30.2).

Plasma AST activity and bile acid (BA) concentration were the most sensitive indicators of liver disease in the racing pigeon, followed by ALT, GGT, and LDH. Although all pigeons with histological proven ethylene glycol- or galactosamine-induced liver damage had increased AST activity and BA concentrations in their plasmas, these constituents were not raised at every sampling time. Increased plasma GLDH activities were associated with large necrotic areas in the liver. Moderate necrosis of liver cells resulted in slightly elevated GLDH activities. Degeneration of liver cells and hepatitis with single-cell necrosis did not give rise to elevated plasma GLDH activities. Plasma CK activity was never elevated because of liver damage, whereas GLDH, GGT, and BA were never elevated during muscle damage. Thus, these four constituents are useful for differ-

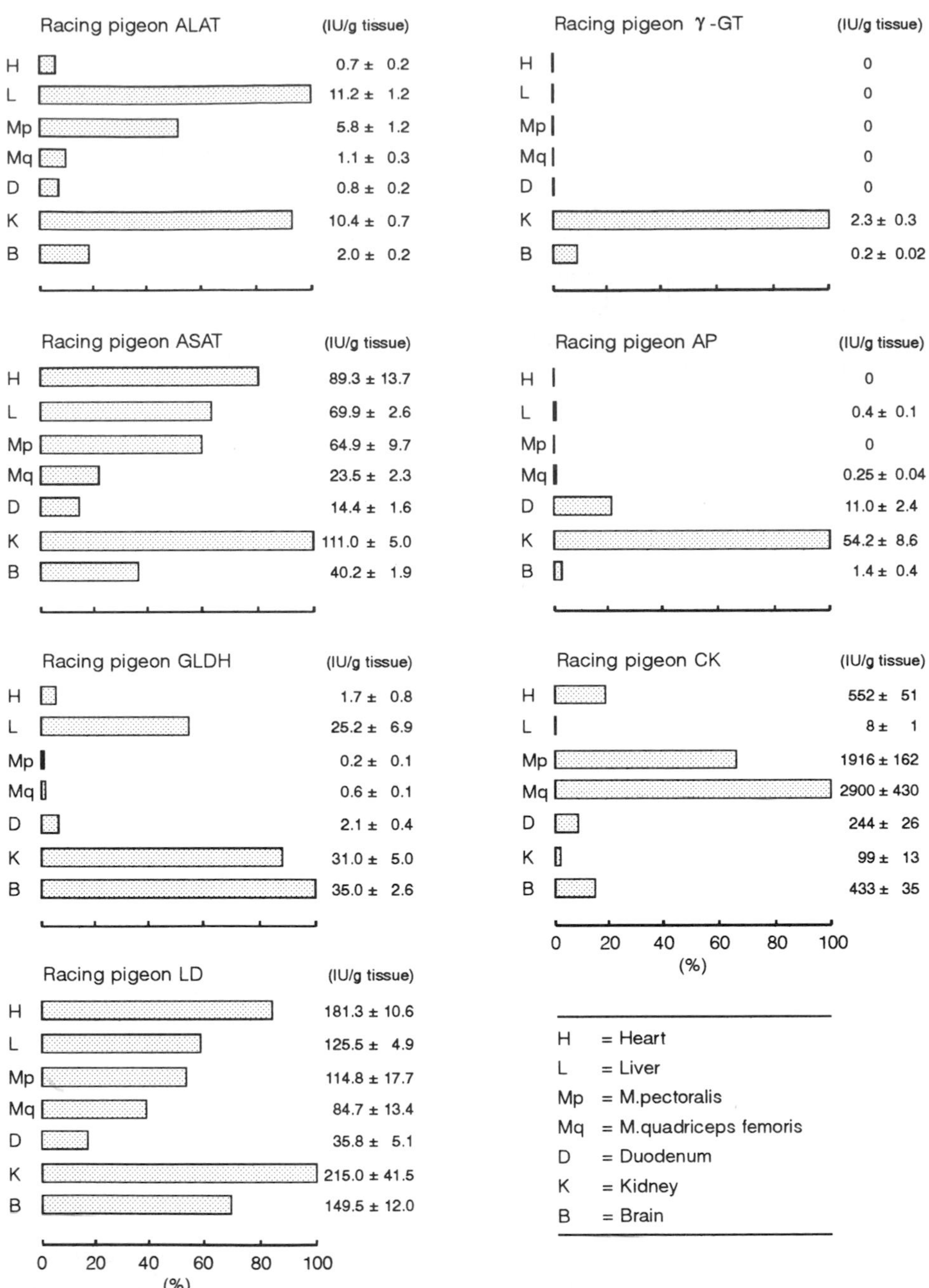

FIGURE 30.11 Relative distribution of various tissue enzymes in supernatants of pigeon tissue homogenates. Mean ± SEM, U/g wet tissue, $n = 6$. Reprinted with permission from Lumeij *et al.* (1988a).

entiating between liver and muscle disease. GLDH is the most liver-specific enzyme in the racing pigeon. Because GLDH is localized within the mitochondria of the liver cells, increased plasma GLDH activities, however, are only observed after liver cell necrosis.

The enzymes that are the most specific indicators of muscle and liver cell damage (CK and GLDH, respectively) have shorter half-lives than AST and ALT, which are not specific indicators of damage to either organ. Thus, after muscle or liver cell damage, AST and ALT activities in plasma can be increased even though CK or GLDH activities have returned to baseline values. The fact that LDH has a shorter half-life than CK makes this enzyme valuable in the differential diagnosis between muscle and liver disease in the pigeon. When plasma LDH activity is increased in the absence of elevated CK activity, muscle damage is unlikely. Enzyme profiles can only serve as rough guides

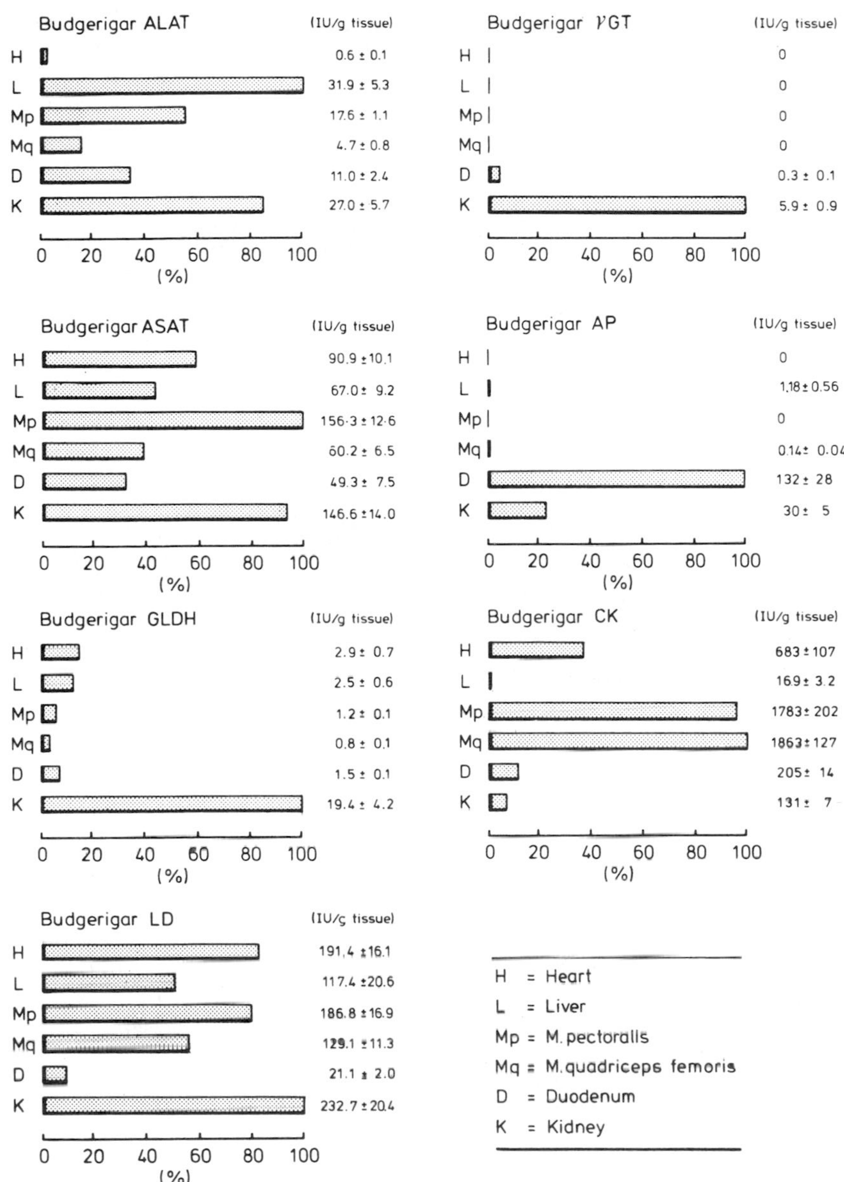

FIGURE 30.12 Relative distribution of various tissue enzymes in supernatants of tissue homogenates from the budgerigar, *Melopsittacus undulatus*. Mean ± SEM, n = 7, U/g wet tissue. Reprinted with permission from Lumeij and Wolfswinkel (1987).

to interpretation of increased plasma enzyme activity and are not characteristic for a particular organ. The most important reason is that the enzyme profile alters after enzymes have entered the circulation because of different removal rates for the various components.

Based on the tissue enzyme studies, it seemed that LDH would be the most sensitive indicator of liver cell damage, whereas ALT, because of its low activity in liver, would be of limited value. It also seemed that LD would be a more sensitive indicator of muscle cell damage than ALT. Experimentally induced liver and muscle damage, however, demonstrated that ALT activity in plasma is increased sooner than LDH activity. The relative increase of ALT was also larger than that of LDH, except in severe liver cell damage when accompanied by large areas of necrosis. Plasma ALT activities were increased for a longer period after organ damage compared to LDH. These findings can be explained by differences between their clearance half-lives. The clearance half-life of LDH is 15 to 30 times less than that of ALT.

Campbell (1986) reported increased AST and ALT activities in 75% of pigeons with aflatoxin B1-induced liver damage and increased LDH activities in 33%. In

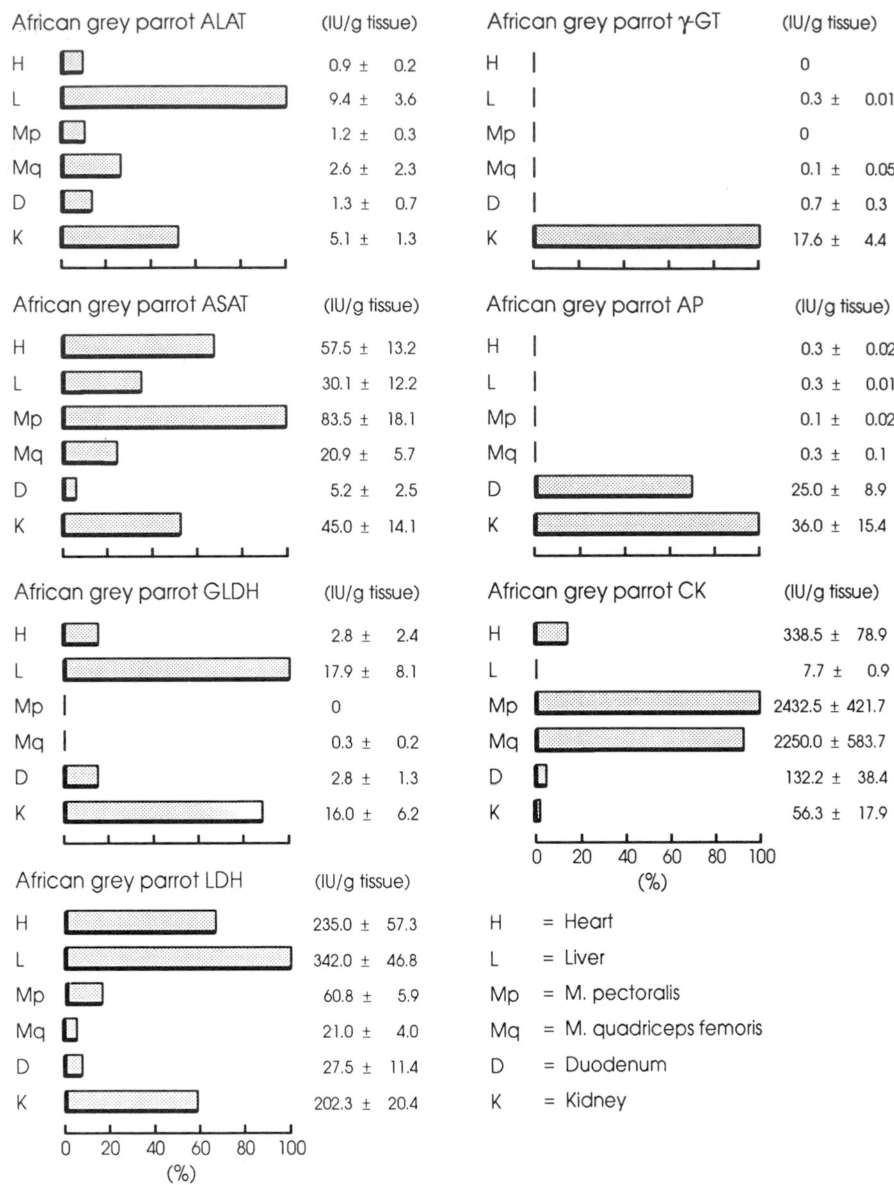

FIGURE 30.13 Relative distribution of various tissue enzymes in supernatants of tissue homogenates from the African grey parrot, *Psittacus erithacus.* Mean ± SEM, U/g wet tissue. Reprinted with permission from Lumeij (1994f).

addition, AP and GGT were not sensitive indicators of liver disease in pigeons, cockatiels, red-tailed hawks, and great horned owls. With the exception of GGT, these findings were confirmed by Lumeij *et al.* (1988a,b). The discrepancy might reflect the difference in the hepatotoxic agents used. In Lumeij *et al.* (1988a,b), GGT activities were increased in the majority of pigeons with experimentally induced liver disease, though no GGT could be detected in liver tissue homogenates. This might be explained by synthesis of GGT during (cholestatic) liver diseases, as has been reported in mammalian species (Kaplan and Righetti, 1969; Kryszewski *et al.*, 1973).

In birds, increased AP activities are predominantly associated with increased osteoblastic activity, such as skeletal growth, nutritional secondary hyperparathyroidism, rickets, and fracture repair osteomyelitis, as well as impending ovulation (Lumeij and Westerhof, 1987). Increased plasma AP associated with liver disease in birds is rare (Ahmed *et al.*, 1974). AP and CK were never elevated after liver cell damage and these enzymes in liver tissue were negligible. Increased ac-

TABLE 30.1 Mean (± SD) $t_{1/2}\beta$ in Plasma of IV-Injected Homologous Enzymes in the Racing Pigeon (n = 6)[a]

Source	Enzyme	$t_{1/2}\beta$ (hours)	Regression function	SD of slope	r
Liver	AST	7.66 ± 1.55	$\log y = 2.6 - 0.04\,x$	0.008	0.9711
	ALT	15.69 ± 1.70	$\log y = 2.1 - 0.02\,x$	0.002	0.9455
	LD	0.71 ± 0.10	$\log y = 3.5 - 0.44\,x$	0.065	0.9901
	GLDH	0.68 ± 0.17	$\log y = 2.3 - 0.46\,x$	0.150	0.9964
Muscle	AST	6.51 ± 0.83	$\log y = 2.7 - 0.05\,x$	0.007	0.9652
	ALT	11.99 ± 1.32	$\log y = 2.7 - 0.03\,x$	0.003	0.9712
	LD	0.48 ± 0.07	$\log y = 4.0 - 0.60\,x$	0.119	0.9882
	CK	3.07 ± 0.59	$\log y = 3.7 - 0.10\,x$	0.019	0.9652

[a] Reprinted with permission from Lumeij *et al.* (1988a).

tivities of liver enzymes in plasma may indicate recent damage to liver cells but do not give information on liver function. In end-stage liver disease (cirrhosis), it is possible to have normal activities of liver enzymes in the plasma, because active damage to liver cells has ceased (e.g., Fig. 30.9).

In psittacine birds, the use of a good anamnesis and the AST is sufficient to make a tentative diagnosis of liver disease. When birds are known to have been recently injected intramuscularly, increased AST and ALT should be interpreted with caution. Including a plasma constituent that specifically gives information on liver function, for example, total bile acids, has proven to be of great value. When liver disease is suspected, a biopsy is essential to establish a definite diagnosis.

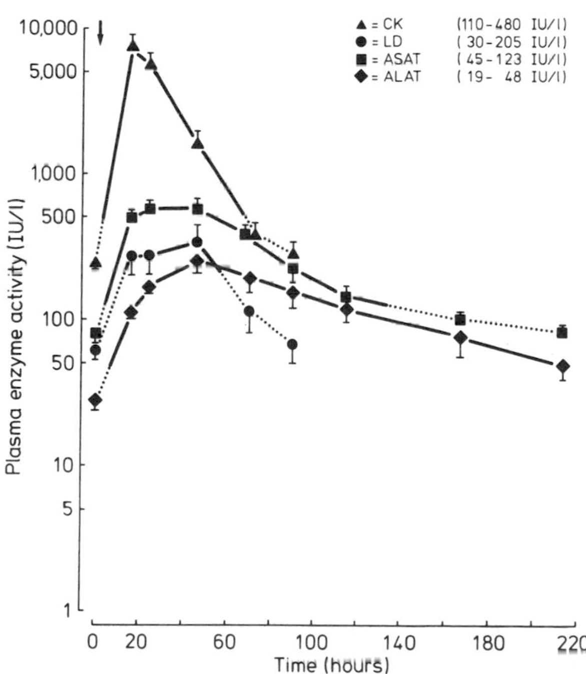

FIGURE 30.14 Mean (± SEM) plasma enzyme activities after intramuscular injection of 75 mg/kg doxycycline (Vibramycin—Pfizer, New York) in 6 racing pigeons. The parts of the curves that are above the reference range are indicated with continuous lines. Variables that did not show elevations above the reference range during the entire experiment (GLDH, AP, and bile acids) are not indicated. On histological examination at t = 215 hr degeneration and necrosis of muscle cells was seen. Reprinted with permission from Lumeij *et al.* (1988b).

E. Bile Pigments

The excretion of green urates is suggestive for liver disease in birds (Galvin, 1980; Lothrop *et al.*, 1986; Steiner and Davis, 1981). This discoloration is caused by biliverdin, which is the most important bile pigment

TABLE 30.2 Summary of Specificity and Sensitivity of Plasma Chemical Variables in Liver and Muscle Disease Based on Experimental Studies in Pigeons by Lumeij *et al.* (1988a,b)

	Liver disease		Muscle disease	
Variable	Specificity	Sensitivity	Specificity	Sensitivity
Bile acids	+++	+++	−	−
γ-GT	+++	+	−	−
AST	−	+++	−	+++
ALT	−	++	−	+++
AP	−	−	−	−
CK	−	−	+++	+++
LD	−	+	−	+
GLDH	+++	(+)[a]	−	−

[a] Elevated GLDH activity is a sign of extensive liver cell necrosis, since GLDH is a mitochondrial and not a cytoplasmatic enzyme. Liver cell degeneration will not cause elevated GLDH activities. In the budgerigar (*Melopsittacus undulatus*) GLDH activity in liver tissue is relatively low when compared to that of man and most of the domestic animals, including cockerel, duck, turkey, and pigeon (Lumeij and Wolfswinkel, 1987). However, in Amazon parrots with extensive liver necrosis due to Pacheco's disease, elevated GLDH activities were observed in plasma, which indicates that this variable is also useful in at least some of the psittacine species (Lumeij, unpublished observations).

in birds (biliverdinuria). Icterus or jaundice, which is caused by a hyperbilirubinemia, is seen infrequently in birds. When in chickens both bile ducts are ligated, the concentration of plasma bile pigments rises immediately but stabilizes after 2 weeks at about 85 μmol/liter, which is a much lower concentration than in mammals with total biliary obstruction. In sera of healthy ducks, low levels of bilirubin may be detected and significantly increased levels have been reported after experimental duck hepatitis virus infection. However, the observed levels of about 17 μmol/liter were well below the serum concentration of 34–51 μmol/liter, which has been mentioned as the level above which icterus becomes apparent in man. The infrequent occurrence of icterus in birds is explained by the absence of biliverdin reductase, which converts biliverdin to bilirubin (Lin *et al.*, 1974; Lind *et al.*, 1967; Tenhunen, 1971). It has been suggested that in birds, biliverdin may be converted to bilirubin by bacteria or nonspecific reducing enzymes (Lewandowski *et al.*, 1986). A yellow discoloration of avian plasma is often caused by the presence of carotenoids, which is often misinterpreted as being icteric.

F. Bile Acids as Indicators of Hepatobiliary Disease

Plasma bile acids and their salts are formed in the liver from cholesterol and are excreted in the intestine, where they assist in digestion of lipids. There is a continuous secretion of bile in birds both with and without a gall bladder. This is slightly increased postprandially because of the intrahepatic effects of intestinal hormones such as secretin, avian vasoactive intestinal peptide (VIP), and/or cholecystokinin (CCK). The sites of increased bile secretion and the regulatory mechanism are unknown (Lumeij, 1991). Via the enterohepatic recirculation, over 90% is reabsorbed in the jejunum and ileum (Hill, 1983). Plasma BA concentrations, including their salts and corresponding glycine and taurine conjugates (BA), are a reflection of the clearing capacity for bile acids of the liver. All liver functions, such as extraction, conjugation, and excretion, are involved in this process, and the BA provides information on the combined effects of these functions. It is now known that BA is a sensitive constituent for evaluating liver function, and it is widely used in man and domestic animals, including birds (De Bruijne and Rothuizen, 1988; Rutgers *et al.*, 1988; Hoffmann *et al.*, 1987; Lumeij, 1988). Reference values for BA have been established for the racing pigeon, the most common psittacine species (Lumeij and Overduin, 1990; Lumeij and Wolfswinkel, 1988), ostriches and peregrine falcons (Table 30.3). Lumeij (1987f) found that BA was the single most useful plasma constituent for detecting liver function disorders in the racing pigeon, where it is both specific and sensitive. The main advantage over AST is that the latter is not liver specific. In experimental liver disease, a 5- to 10-fold increase of BA over the upper limit of the reference range is common.

G. Effect of Feeding on Bile Acids

There is a significant postprandial increase in BA in granivorous birds with and without a gallbladder (Lumeij, 1991) as well as in carnivorous birds (Fig. 30.15; Lumeij and Remple, 1992). Although up to a 4.5-fold postprandial increase of BA was seen in individual birds, the concentrations were never increased more than 1.65-fold over the upper limit of the reference range, while in hepatobiliary disease five to ten-fold increases are common (Lumeij *et al.*, 1988a). Even though postprandial increases might complicate interpretation of BA, differentiation between postprandial increases and increases due to hepatobiliary disease is possible. Experimental findings suggest that values >70 μmol/l in fasted racing pigeons and values >100 μmol/l postprandially should be considered increased and suggestive for hepatobiliary disease.

H. Plasma Ammonia—Hepatoencephalopathy

A tentative diagnosis of hepatoencephalopathy is often made when neurologic signs are seen in birds with liver disease. The syndrome, however, has not been well documented in birds. In mammals, hepatoencephalopathy and hepatic coma are most often seen in portosystemic shunting in which portal blood and its ammonia are shunted away from the liver. Hepatoencephalopathy is not a disease in itself, except for the anatomic anomaly, but a medical condition characterized by neurologic symptoms caused by intoxication of the brain by products of protein digestion, namely NH_3. Most likely, degradation products from protein act as false neurotransmitters. Protein-rich foods often trigger neurologic symptoms in these patients. Fasting plasma NH_3 levels and plasma NH_3 levels 30 minutes after oral loading with NH_4Cl in the form of the NH_3 test (ATT) can be used to test the ability of the liver to convert NH_3 into urea. Fasting plasma NH_3 in healthy psittacines ranged from 36 to 274 μmol/liter, which is well above the fasting concentrations reported in dogs (Lumeij and Peccati, 1993). Furthermore, some avian species will normally have up to an eightfold increase of plasma NH_3 on the ATT using the canine protocol. Therefore, an abnormal ATT

TABLE 30.3 Plasma Chemistry Reference values ($P_{2.5}$–$P_{97.5}$) for Pigeons, Some Psittacine Species, Peregrine Falcon, and Ostrich as Established by the Division of Avian and Exotic Animal Medicine, Utrecht University Department of Clinical Sciences of Companion Animals[a]

Variable	Ostrich (n = 60)	Peregrine falcon (n = 79)	Pigeon (n = 50)	African grey (n = 71)	Amazon (n = 99)	Cockatoo (n = 27)	Macaw[b] (n = 15)
Urea (mmol/liter)	0.5–0.8	0.8–2.9	0.4–0.7	0.7–2.4	0.9–4.6	0.8–2.1	0.3–3.3
Creatinine (μmol/liter)	—	24–64	23–36	23–40	19–33	21–36	20–59
Uric acid (μmol/liter)	357–643	253–996	150–765	93–414	72–312	190–327	109–231
Urea/uric acid ratio	0.9–1.7	1.7–6.4	1.8 ± 1.8[c]	2.4–15.6	4.4–33	2.7–8.9	5–28
Osmolality (mOsmol/kg)	305–328	322–356	297–317	320–347	316–373	317–347	319–378
Sodium (mmol/liter)	147–157	150–170	141–149	154–164	149–164	152–164	150–175
Potassium (mmol/liter)	4.5–5.9	—	3.9–4.7	2.5–3.9	2.3–4.2	3.2–4.9	1.9–4.1
Chloride (mmol/liter)	94–105	114–131	101–113	—	—	—	—
Calcium (mmol/liter)	2.5–4.6	1.9–2.4	1.9–2.6	2.1–2.6	2.0–2.8	2.2–2.7	2.2–2.8
Phosphorus (mmol/liter)	1.3–2.2	0.55–1.53	0.57–1.33	—	—	—	—
Glucose (mmol/liter)	10.4–13.7	16.5–22.0	12.9–20.5	11.4–16.1	12.6–16.9	12.8–17.6	12.0–17.9
LD (IU/liter) EC 1.1.1.27	869–2047	1008–2650	30–205	147–384	46–208	203–442	66–166
GLDH (IU/liter) EC 1.4.1.3	<8	<8	<8	<8	<8	<8	<8
τ-GT (IU/liter) EC 2.3.2.2	<1–1	<1–3	<1–3	1–4	1–10	2–5	<1–5
AST (IU/liter) EC 2.6.1.1	252–401	34–116	45–123	54–155	57–194	52–203	58–206
ALT (IU/liter) EC 2.6.1.2	—	29–90	19–48	12–59	19–98	12–37	22–105
CK (IU/liter) EC 2.7.3.2	1655–4246	120–442	110–480	123–875	45–265	34–204	61–531
Lipase EC 3.1.1.3	—	—	—	—	268–1161	—	—
α-Amylase EC 3.2.1.1	—	—	—	—	571–1987	—	—
Bile acids (μmol/liter)	8–30	5–69	22–60	18–71	19–144	23–70	25–71
Total protein (g/liter)	40–54	24–39	21–35	32–44	33–50	35–44	33–53
Albumin/globulin ratio	0.9–1.4	0.8–2.4	1.5–3.6	1.4–4.7	2.6–7.0	1.5–4.3	1.4–3.9
Thyroxine (nmol/liter) before and 16 hr after stimulation with 2 IU/kg TSH			6–35 100–300	—	—	—	—
Corticosterone (nmol/liter) before and 90 min after stimulation with 250 μg/kg ACTH			6–36[d] 64–324	—	16–39[d] 108–506	—	—
AVT (pg/ml) before and after 24-hr water deprivation			1.7 ± 1.4 6.3 ± 2.2	—	—	—	—

[a] Lumeij, 1987f; Lumeij and Overduin, 1990; Lumeij and Westerhof, 1988a; Lumeij *et al.*, 1987b; and unpublished findings, from Lumeij, Remple, Riddle, 1995; Lumeij, Verstappen, and Bronnenberg, 1995; Van der Horst, 1995; Westerhof, 1995.
[b] Range.
[c] Mean ± SD (n = 6).
[d] Range (n = 6).

is not diagnostic for portosystemic shunting in these species (Lumeij and Peccati, 1993).

VII. MUSCLE DISEASE

Muscle enzyme profiles, half-lives of these enzymes in plasma, and plasma chemistry changes after experimental muscle damage have been reported for racing pigeons (see Section VI.B). Enzyme profiles were studied for pectoral muscle, quadriceps muscle, and heart muscle in pigeons and parrots (Figs. 30.11 through 30.13). Creatine kinase was the most important enzyme in these three muscles, followed by LDH, AST, and ALT. Muscle damage was induced by injection of doxycycline in the pectoral muscle in pigeons (Fig. 30.14). Creatine kinase in plasma was markedly increased (about 20-fold) 16 hours after injection. However, within 66 hours plasma activities of CK had returned to the maximum value of the reference range. LD activities were only slightly elevated (about 2-fold) and only for about 40 hours. AST activities showed a marked increase (about 4-fold) for about 140 hours, whereas ALT showed a marked increase (about 5-fold) for about 214 hours. Despite that ALT activities in individual muscles are low, elevated plasma activities of this enzyme can be seen until 9 days after muscle damage. Plasma CK activities, on the other hand, return to within the reference range within 3 days after muscle damage, despite very high tissue activities. LD ap-

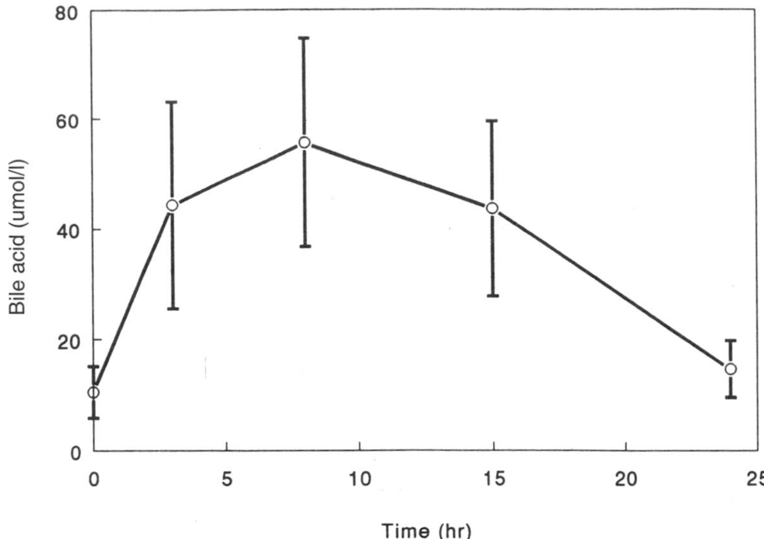

FIGURE 30.15 Plasma bile acid concentrations (mean ± SD) in peregrine falcons, *Falco peregrinus* (n = 6) after a 42-hr fasting period and 3, 8, 15, and 24 hr after birds were fed a complete skinned quail, *Coturnix coturnix* at 0 hr. Values at 3, 8, and 15 hr are significantly different from values at 0 and 24 hr ($P < .05$). Reprinted with permission from Lumeij and Remple (1992).

peared to be a relatively poor indicator of muscle cell damage, despite relatively high activity of this enzyme in muscle (Fig. 30.14). These findings can be explained by the differences in elimination half-lives of the respective enzymes (LD 50 minutes, CK 3 hours, AST 7 hours, ALT 12 hours) (Tables 30.1 and 30.2).

Not all elevated concentrations of muscle enzymes in plasma are an indication for muscle disease. Extreme muscular activity in the period preceding blood sampling is an important cause of elevated enzyme activities in plasma. In dogs plasma CK activities approximately double with exercise (Heffron *et al.*, 1976). Trained persons have plasma CK activities that are twice those of more sedentary people (Okinaka *et al.*, 1964). In humans, elevated activities of plasma CK can persist for about 1 week after exercise (Newham *et al.*, 1983). Chronic elevated plasma CK activities have been reported in certain occupational workers as a result of local muscular strain (Hagberg *et al.*, 1982; Brewster and De Visser, 1988). Plasma CK activity in healthy turkeys is extremely sensitive to physical exercise and stress. With controlled conditions of minimal exercise, stress, and time of handling, however, iatrogenic elevations of plasma CK activities can be prevented (Tripp and Schmitz, 1982). Elevated plasma CK activities can be expected in birds with large muscle mass after capture stress (e.g., ostrich). Intramuscular injections in birds are a well-known cause of elevated activities of plasma enzymes of muscle origin (Fig. 30.14). When physiological or iatrogenic causes of hyperCKemia can be ruled out, primary neuromuscular disease should be considered.

In birds several causes of degenerative myopathy have been reported. In poultry furazolidone and ionophore coccidiostats are well-known causes of myocyte degeneration (Julian, 1991). Ingestion of the beans of coffee senna (*Cassia* spp.) has been suggested as a possible cause of acute myocyte degeneration in birds (Rae, 1992). Two important causes of degenerative myopathy in birds are exertional rhabdomyolysis (capture myopathy) and nutritional myopathy.

One of the signs of a deficiency of selenium and/or vitamin E in birds is muscular degeneration. Some authors believe that exertional rhabdomyolysis is an acute manifestation of nutritional myopathy, although it has been recorded in species with apparently normal vitamin E levels (Spraker, 1980).

Capture myopathy has been reported in flamingos (Young, 1967; Fowler, 1978a,b), cranes (Brannian *et al.*, 1982; Windingstad, 1983; Carpenter *et al.*, 1991), Canada geese (Chalmers and Barrett, 1982), turkeys (Spraker *et al.*, 1987), and ratites (Phalen *et al.*, 1990; Dolensek and Bruning, 1978; Rae, 1992).

Nutrition-related myopathies have been reported in piscivorous birds that have been fed an unsupplemented diet of previously frozen fish, primarily smelt (Campbell and Montali, 1980; Carpenter *et al.*, 1979; Nichols and Montali, 1987; Nichols *et al.*, 1986). Vitamin E deficiency has also been associated with muscle lesions in raptors (Calle *et al.*, 1989; Dierenfeld *et al.*,

1989). Rae (1992) reported that a large percentage of young ratites submitted for necropsy exhibited evidence of degenerative myopathy, and she considered nutritional deficiency of vitamin E and possibly selenium as the most probable cause.

The muscular lesions produced by the various causes cannot be distinguished from each other, and the clinical history is important to establish a diagnosis (Rae, 1992). The use of serum or plasma vitamin E concentrations has been advocated to enable a clinical diagnosis of nutritional myopathy in birds (Rae, 1992). Mean (± SE) plasma concentrations of vitamin E (quantified as α-tocopherol) established in 274 captive cranes were 6.57 ± 0.82 μg/ml. Cranes species that evolved in temperate habitats had higher circulating levels of α-tocopherol than tropical or subtropical species: for example, black crowned crane (*Balearica pavonina*) (n = 10) 2.77 ± 0.23 μg/ml; Siberian crane (*Grus leucogeranus*) (n = 51) 9.41 ± 0.64 μg/ml (Dierenfeld *et al.*, 1993). In peregrine falcons (*Falco peregrinus*) circulating α-tocopherol concentrations <10 μg/ml were considered a reflection of a marginal vitamin E status, whereas plasma concentrations <5 μg/ml were considered deficient (Dierenfeld *et al.*, 1989). Only limited data are available on normal plasma concentrations of α-tocopherol in ratites. In apparently healthy rheas, circulating α-tocopherol concentrations ranged between 9.0 and 14.5 μg/ml, while two rheas, with muscular problems exhibited mean plasma concentrations of 1.34 μg/ml (Dierenfeld, 1989).

For the diagnosis of diseases of the cardiac muscle in birds, plasma chemistry has also been used. Enzymes that have been used include AST, LDH, and CPK. CPK activity in plasma from cardiac muscle origin (CPK-MB isoenzyme) was significantly higher in ducklings with furazolidone-induced cardiotoxicosis than in controls (Webb *et al.*, 1991).

VIII. CALCIUM AND PHOSPHORUS—METABOLIC BONE DISEASE

A. Relation between Total Calcium and Protein in Avian Plasma

About one-third of plasma calcium is protein bound and is biologically inactive. Total calcium concentration is influenced by plasma protein concentrations. The ionized fraction is important with regard to deposition of calcium salts and excitability of nervous tissues. In most laboratories, for technical reasons, only total calcium is measured. Hence, when total plasma calcium is measured it is also important to measure plasma protein concentrations and to make allowances for any deviations from the normal in the latter. A significant linear correlation was found between total calcium and albumin concentration in the plasma of 70 healthy African grey parrots (r = 0.37; P < .05) and a correction formula was derived on the basis of the concentration of albumin: Adjusted calcium (mmol/liter) = calcium (mmol/liter) −0.015 albumin (g/liter) + 0.4. Approximately 14% of the variability of calcium was attributable to the change in the concentration of plasma albumin (R^2 = 0.137) (Lumeij, 1990).

A significant correlation was also found between total calcium and total protein concentration in 124 plasma samples of peregrine falcons (r = 0.65; P < .01). About 42% of the variability in calcium was attributable to the change in the plasma total protein concentration (R^2 = 0.417). The correlation between calcium and albumin was significant (r = 0.33; P < .01), but significantly smaller than the correlation between calcium and total protein (P < .01). Only 11% of the plasma calcium concentration was attributable to difference in concentration of albumin (R^2 = 0.108). An adjustment formula for plasma calcium concentration in the peregrine falcon was derived on the basis of the total protein concentration: Adj. Ca (mmol/liter) = Ca(mmol/liter) −0.02 total protein (g/liter) + 0.67 (Fig. 30.16; Lumeij *et al.*, 1993a). Application of a correction formula in African grey parrots and peregrine falcons is indicated when extremely low or extremely high plasma protein concentrations are found. The correction formulas just mentioned are based on total protein and albumin determinations with the methods outlined before (Section IV).

B. Estrogen-Induced Hypercalcemia

About 4 days before female pigeons are due to ovulate, the blood calcium concentration rises from a normal value of about 2.2 mmol/liter to a value of over 5.0 mmol/liter at the time of ovulation. This rise in calcium is caused by an increase in the protein-bound calcium due to the estrogen-induced transport of yolk proteins to the ovary as calcium complexes, whereby the concentration of ionized calcium remains constant (Simkiss, 1967). Since the total calcium concentration is the sum of biologically active ionized calcium, protein-bound calcium, and calcium chelated to anions, a total calcium concentration should always be interpreted in conjunction with plasma proteins (see Section VIII.A).

C. Physiological Marrow Ossification

Physiological marrow ossification is induced by the *combined* effects of estrogens *and androgens* and can be observed at about the same time as the estrogen-

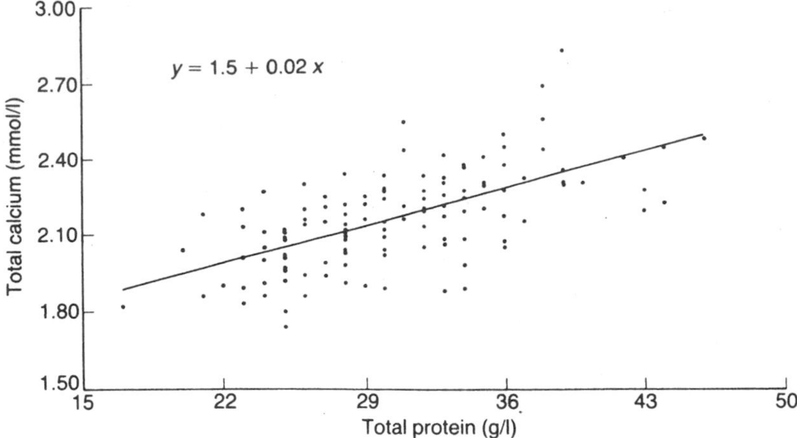

FIGURE 30.16 Significant relationship ($r = 0.65$; $P < 0.01$) between total protein and calcium in plasma of 124 peregrine falcons (*Falco peregrinus*). The least-square regression line is indicated. As the concentration of total protein decreases, there is a concurrent decrease in plasma total calcium. About 42% of the variability in calcium was attributable to the changes in the plasma total protein concentration ($R^2 = 0.417$). Reprinted with permission from Lumeij *et al.* (1993a).

induced hypercalcemia in female birds (Simkiss, 1967). There is a large increase in the quantities of calcium and phosphorus that are retained from the diet and laid down as medullary bone. This medullary bone may completely fill the marrow spaces of the long bones. It is most clearly seen in the limb bones, but occurs in most parts of the skeleton. This period of bone deposition coincides with increased osteoblastic activity. When the bird starts to secrete the eggshell, the medullary bone is resorbed by osteoclastic activity. Calcium is deposited in the eggshell as calcium carbonate and the phosphorus is excreted from the body. Medullary bone might be mistaken for a pathological condition when radiographs are being evaluated.

D. Hypocalcemia Syndrome in African Grey Parrots

A hypocalcemia syndrome known in birds of prey and African grey parrots is characterized by hypocalcemic seizures. A striking feature of this syndrome in African grey parrots which is not known in other birds is that demineralization of the skeleton is not obvious at the moment the seizures occur. The hypocalcemia syndrome is an important differential diagnosis in an African grey parrot that repeatedly falls off its perch. Reference values for plasma calcium concentrations in African grey parrots range from 2.0 to 3.25 mmol/liter (Rosskopf *et al.*, 1982). Lumeij (1990) studying a population of 72 African grey parrots found reference values of 2.1–2.6 mmol (inner limits of the percentiles $P_{2.5}$–$P_{97.5}$, with a probability of 90%) and a range from 2.0 to 3.4 mmol/liter. Hochleithner (1989b) studying 68 African grey parrots and using a dry chemistry system (Kodak Ektachem) reported reference values for Ca of 1.75–2.38 mmol/l (inner limits of the percentiles $P_{2.5}$–$P_{97.5}$). Hochleithner (1989a) reported five cases of hypocalcemia in African grey parrots with plasma calcium concentrations ranging from 0.75 to 1.5 mmol/liter. Rosskopf *et al.* (1985) state that the one consistent finding of the hypocalcemia syndrome is a blood calcium level below 1.5 mmol/liter. Values as low as 0.6 mmol/liter have been reported (Rosskopf *et al.*, 1985). When borderline calcium concentrations are found, the correction formula reported in Section VIII.A should be used. When timely treatment with parenteral calcium and vitamin D_3 preparations is started and sufficient dietary uptake of calcium is taken care of, clinical signs will regress in a short time. It is therefore likely that the disease is caused by calcium and vitamin D_3 deficiency. The higher incidence of the hypocalcemia syndrome in African grey parrots might be associated with lower plasma concentrations of free calcium in this species (Lumeij, 1990).

E. Alkaline Phosphatase in Bone Disease

Alkaline phosphatases (APs) form a group of membrane-bound glycoproteins that hydrolyze monophosphate esters at alkaline pH. Three different isoenzymes have been identified. Although there is a significant activity of AP in various tissues, the physiological role is unclear, except for AP in bone tissue. AP activity in bone reflects the activity of osteoblasts, and this enzyme is involved in the formation and mineralization of the bone matrix. In humans increased AP activity is observed during growth and in osteoproliferative disorders (Savova and Kirev, 1992). Different techniques have been used for identification of fractions responsible for increased plasma activities. The heat inactivation test has been developed to distinguish AP activity of bone origin from that of liver origin (Johnson *et al.*, 1972; Posen *et al.*, 1965). In humans, after heat inactivation at 56°C, residual activities higher than 35% indicate hepatic disease, whereas residual values lower than 25% indicate bone disease with increased osteoblastic activity (Fennely *et al.*, 1969; Fitzgerald *et al.*, 1969; Stolbach, 1969). Using a guinea fowl model with bone tumors induced by osteopetrosis virus, Savova and Kirev (1992) were able to confirm these findings also for an avian species. By comparing the findings with the more sensitive wheat germ lectin method (Rosalki and Foo, 1984; Brixen *et al.*, 1989), they showed that for guinea fowl the AP activity of bone origin can be inactivated at 58°C rather then 56°C. Savova and Kirev (1992) found that AP activity of bone origin in 15-week-old guinea fowl was twice as high as that of 1-year-old birds. They also confirmed the positive correlation between the intensity of virus-induced excessive bone growth and serum AP activity reported previously (Sanger *et al.*,1966; Barnes and Smith, 1977). The presumed high proportion of AP of bone origin was supported by the low values of residual activity after heat inactivation at 58°C (14.7 ± 3.7%) and after precipitation with wheat germ lectin (13 ± 1.2) during the period of active bone tumor formation (Savova and Kirev, 1992).

IX. IRON STORAGE DISEASE

Hemochromatosis is a disease of excessive storage of hemosiderin (hemosiderosis) in various body tissues. Hemosiderin is an iron-containing pigment derived from hemoglobin. The principal site of Fe accumulation is the liver, and many other body tissues are also affected. Hemochromatosis is especially well known in Ramphastidae, Sturnidae, mynahs, quetzals (Worell, 1994), and Psittaciformes (Rosskopf *et al.*, 1992). The etiology is at present unknown, but a high dietary Fe has been suggested. Cardiac complications are often seen in mynahs due to cardiac muscle hemosiderosis. Ramphastidae are generally clinically normal prior to death, but occasionally affected birds are listless 24 hours prior to death. In mynahs, general weakness, dyspnea, and ascites are commonly seen. Radiography may reveal hepatomegaly, cardiomegaly, or ascites. Blood chemistry is indicative of a liver function disorder. A specific diagnosis can be made by histological examination of a liver biopsy after specific staining for Fe (Turnbull blue). Total serum/plasma Fe and total iron binding capacity (TIBC) may be helpful in evaluating the iron status of the animal, although only limited work has been done on the clinical pathology associated with avian hemochromatosis. To measure the Fe status of an individual, three main compartments should be considered: storage Fe, transport Fe, and erythrocyte Fe. A semiquantitative method of establishing the amount of storage Fe is by examining histological sections for stainable Fe in liver biopsies. In humans with hemochromatosis, the urinary Fe excretion at 6 hours following the injection of the iron chelating agent deferoxamine or diethylenetriaminepentaacetic acid (DTPA) is significantly higher than in normal individuals. The concentration of the Fe storage protein ferritin (Ftn) in the serum is directly related to the available storage iron in the liver and is clinically the most useful method for assessing iron stores (Jacobs, 1978).

Transport iron is evaluated by measuring the serum Fe, the total iron binding capacity, and the unbound iron binding capacity (UIBC). The UIBC is calculated as the difference between the serum Fe and the TIBC. The TIBC represents the total amount of the Fe transport protein transferrin (Tf) in the serum, and the UIBC represents the amount of the Tf that is not bound to Fe and is available for Fe binding. Reference values for serum or plasma Fe concentration and TIBC in humans are 10–34 μmol/liter and 45–72 μmol/liter; respectively (Scott, 1978). In pigeons these values were 11–33 μmol/liter and 30–45 μmol/liter, respectively (Lumeij and De Bruijne, 1985a). In normal toucans (Ramphastidae) total serum Fe was below 63 μmol/liter, whereas TIBC was below 100 μmol/liter (Worell, 1991). Clinically normal toucans with bile acids, plasma chemistry, and hematological values within the normal reference limits had histological evidence of hemochromatosis and liver iron levels in excess of the normal 100–300 ppm (Worell, 1993). Currently, antemortem diagnosis of hemochromatosis in clinically normal toucans relies on a hepatic biopsy to determine characteristic histopathologic changes with or without serum Fe (Worell, 1994). Total serum Fe in a mynah bird (*Gracula religiosa*) with hemochromatosis exceeded 160 μmol/liter, whereas two control birds had values < 36 μmol/liter (Morris *et al.*, 1989). In another Balih mynah (*Leucopsar rothschildi*) with hemochromatosis, serum Fe at the time of diagnosis was 88 μmol/liter.

After combined treatment by phlebotomy and deferoxamine mesylate medication, a liver biopsy had minimal stainable Fe and the corresponding serum Fe was 21 μmol/liter (Loomis and Wright, 1993). During treatment of hemochromatosis by repeated phlebotomies, establishment of the various red cell parameters is essential to detect excessive Fe depletion. Erythrocyte Fe is evaluated by red blood cell (RBC) morphology and the various RBC parameters such as the packed cell volume (PCV), hemoglobin (Hb), mean cell volume (MCV), mean cell Hb (MCH), and mean cell Hb concentration (MCHC).

X. DIABETES MELLITUS AND PLASMA GLUCOSE

The fundamental regulation of glucose metabolism in birds is identical to that in mammals, but there is a quantitative difference. Reference ranges for plasma glucose in birds are between 11 and 25 mmol/liter (Rosskopf *et al.*, 1982; Lumeij and Overduin, 1990). Physiological values up to 33 mmol/liter have been observed postprandially in pigeons (Lumeij, 1987b). Under stress, plasma glucose concentrations up to 33 mmol/liter may also be observed (Jenkins, 1994). The insulin content of the pancreas of granivorous birds is about one-sixth that of mammalian pancreata, while the glucagon content is about two to five times greater. Plasma glucagon (1 to 4 ng/ml) is about 10 to 50 times higher in birds than in mammals. Insulin is synthesized in the β cells of the pancreas, while glucagon is synthesized in the A cells.

Spontaneous diabetes mellitus, as characterized by polyuria/polydipsia (PU/PD), glucosuria, persistent hyperglycemia, and loss of weight despite good appetite, has been reported in a number of avian species, including budgerigars, cockatiels, an Amazon parrot, an African grey parrot, toco toucans, a red-tailed hawk, and the pigeon (Altman and Kirmayer, 1976; Candeletta *et al.*, 1993; Lothrop *et al.*, 1986; Lumeij, 1994d; Murphey, 1992; Schlumberger, 1956; Spira, 1981; Wallner-Pendleton *et al.*, 1993; Wiesner, 1971; Woerpel *et al.*, 1987). The species in which diabetes is most commonly encountered are budgerigars, cockatiels, and toco toucans. A tentative diagnosis can be made by finding glucose in urine, and a definitive diagnosis can be made by finding an increased plasma glucose concentration. A PU/PD with glucosuria does not always indicate diabetes mellitus.

In mammals, Fanconi's syndrome is characterized by renal glucosuria, hyperaminoaciduria, and hyperphosphaturia as well as renal loss of potassium, bicarbonate, water, and other substances conserved by the proximal tubule. Fanconi's syndrome should be considered as the final result of any one of many possible primary insults to proximal tubular function. The patient's symptoms reflect the disturbance of tubular function in addition to the primary cause of the syndrome. The syndrome may be inherited or acquired.

There are some striking differences between birds and mammals with regard to pancreatic effects on carbohydrate metabolism. In mammals, pancreatectomy results in diabetes mellitus, whereas effects of pancreatectomy in birds are controversial. In granivorous birds, surgical extirpation or destruction of the pancreas with cytotoxic agents leads to hypoglycemic crisis and death, whereas pancreatectomies in carnivorous birds lead to diabetes mellitus. It is generally accepted that glucagon is more effective in granivorous birds that have a marked insulin insensitivity. The limited data available on spontaneous diabetes mellitus in granivorous birds suggest that in these species, diabetes mellitus is not caused by an insulin deficiency. Birds of prey may be more insulin dependent than granivorous birds (Wallner-Pendleton *et al.*, 1993).

Spontaneous diabetes mellitus in birds has been reported to be successfully treated with daily injections of insulin in dosages comparable to dosages used in dogs. These successful treatments of diabetic birds (disappearance of clinical signs) are surprising considering the relative insulin insensitivity that has been reported to occur in a variety of avian species.

Plasma insulin and glucagon concentrations have been established in three birds with hyperglycemia (Lothrop *et al.*, 1986). In all cases, insulin concentrations were similar to those of controls while glucagon concentrations were extremely high or extremely low. In another case of diabetes in an African grey parrot, Candeletta *et al.* (1993) reported extremely low insulin concentrations. The findings suggest that different etiologies are likely to be involved in the development of avian diabetes. Possible etiologies of diabetes mellitus in birds have been discussed (Lumeij,1994d).

XI. EXOCRINE PANCREATIC DISEASE

Exocrine pancreatic hormones present in the duodenum of birds include amylase, lipase, trypsin, and chymotrypsin. They facilitate degradation of carbohydrates, fats, and proteins, respectively. The inactive precursors of trypsin and chymotrypsin, trypsinogen and chymotrypsinogen, enter the duodenum, where they are activated by intestinal enterokinase. This mechanism prevents autodigestion of pancreatic tissue (Duke, 1986).

There are two basic manifestations of exocrine pancreatic hormone disorders: (1) acute pancreatitis or acute pancreatic necrosis, and (2) chronic pancreatitis resulting in pancreatic fibrosis and pancreatic exocrine insufficiency.

The pathogenesis of acute pancreatitis involves the activation of pancreatic enzymes in and around the pancreas and bloodstream, resulting in coagulation necrosis of the pancreas and necrosis and hemorrhage of peripancreatic and peritoneal adipose tissue. Increased amylase and lipase activities in plasma have been reported in birds with active pancreatitis. Reference values for plasma lipase and amylase have been measured in a population of 87 African grey parrots (Van der Horst and Lumeij, unpublished observations). α-Amylase activity in plasma was determined by a kinetic *p*-nitrophenylmaltoheptaoside method (Sopachem α-Amylase kit # 003-0311-00, Sopar-biochem, 1080 Brussels) at 30°C. Values ranged from 571 to 1987 U/liter (inner limits of $P_{2.5}$–$P_{97.5}$ with a probability of 90%).

Lipase activity was measured at 30°C using a test based on the conversion of triolein by lipase to monoglyceride and oleic acid. The associated decreased turbidity is measured in the UV range (Boehringer Mannheim kit # MPR 3-1442651). Reference values ranged from 268 to 1161 U/liter.

Hochleithner (1989b) reported reference values for plasma amylase in four different psittacine species using a dry chemistry system (Kodak Ektachem, Amylopectin, 25°C; Kodak, 1986). The values were considerably lower than the ones just discussed: budgerigar (n = 50) 187–585 U/liter; African grey parrot (n = 68) 211–519 U/liter; Amazon parrot (n = 30) 106–524 U/liter; macaw 276–594 U/liter.

Chronic pancreatitis results in fibrosis and decreased production of pancreatic hormones. When insufficient pancreatic enzymes are available in the duodenum, maldigestion and a passing of feces with excessive amylum and fat will occur. Affected animals have voluminous, pale or tan greasy feces. Fat can be demonstrated by Sudan staining.

Fecal amylase and proteolytic activity were determined in a population of 87 African grey parrots by Van der Horst and Lumeij (unpublished observations), using radial enzyme diffusion as reported by Westermarck and Sandholm (1980). Reference values (inner limits of $P_{2.5}$–$P_{97.5}$ with a probability of 90%) for fecal amylase were 6 to 18 mm and for fecal trypsin 14–19 mm.

XII. TOXICOLOGY

A. Lead

Lead poisoning is common in birds (Lumeij, 1985b; Dumonceaux and Harrison, 1994). A clinical diagnosis is made by demonstrating increased lead in whole blood or by demonstrating secondary effects of lead on various enzymes involved in heme synthesis. Blood Pb in clinically normal birds and in birds with signs of lead poisoning can be much higher than in mammals (Lumeij, 1985b). Blood Pb in birds without clinical signs and without known exposure to lead ranged between 2.5 and 180 μg/dl. Birds that had been exposed to Pb but had no clinical signs had Pb between 40 and 2000 μg/dl, whereas birds with clinical signs had blood Pb from 52 to 5840 μg/dl. Dieter (1979) proposed that a blood Pb of 20 μg/dl was physiologically detrimental in canvasback ducks, *Aythya valisaneria,* because a significant decrease in δ-aminolevulinic acid activity was observed at these Pb levels. The U.S. Fish and Wildlife Service presumes that blood Pb concentrations >20 μg/dl in 5% of hunter-killed or live-trapped waterfowl is indicative of some type of Pb assimilation in tissues (U.S. Federal Register 29673, 20 August 1986). Draury *et al.* (1993) uses the terms "detrimental" and "deleterious" in association with blood Pb of 20 μg/dl. Although a blood Pb of 20 μg/dl indicates increased exposure to Pb, these levels are not considered harmful to animals. The current clinical view is that blood Pb <50 μg/dl is generally not associated with clinical signs and has a good prognosis even without treatment. Lead between 50 and 100 μg/dl is associated with mild clinical signs and carries a good prognosis for recovery with treatment. Clinical signs and prognosis worsen when blood Pb exceeds 100 μg/dl, and when Pb exceeds 200 μg/dl the prognosis is guarded to poor (Degernes, 1995). When clinical signs are present, blood Pb >35 μg/dl is suggestive of Pb toxicosis (Klein and Galey, 1989). In psittacines blood Pb levels as low as 20 μg/dl are considered suggestive for Pb exposure (Dumonceaux and Harrison, 1994).

Most of the Pb in blood is associated with the red blood cells (Buggiani and Rindi, 1980). The nuclear inclusions that have been observed by electron microscopy in nucleated erythrocytes of pigeons with high blood Pb have led to the assumption that these could serve as storage sites, like the Pb inclusion bodies described in kidneys of Pb-poisoned rats. The capacity of birds to survive high blood Pb without overt toxicosis might be associated with these erythrocytic inclusion bodies (Barthalmus *et al.*, 1977).

Lead interferes with two enzymes in the hemoglobin biosynthetic pathway: δ-aminolevulinic acid dehydratase (ALA-D) and heme synthetase (Fig. 30.17). In humans ALA-D is inhibited even at normal blood Pb levels (McIntire *et al.*, 1973). When there is an increase in Pb, ALA-D is uniformly low. A level >600 IU/dl excludes Pb poisoning (Beeson *et al.*, 1979). A significant negative correlation exists between blood Pb and ALA-D in pigeons, urban-dwelling humans, urban

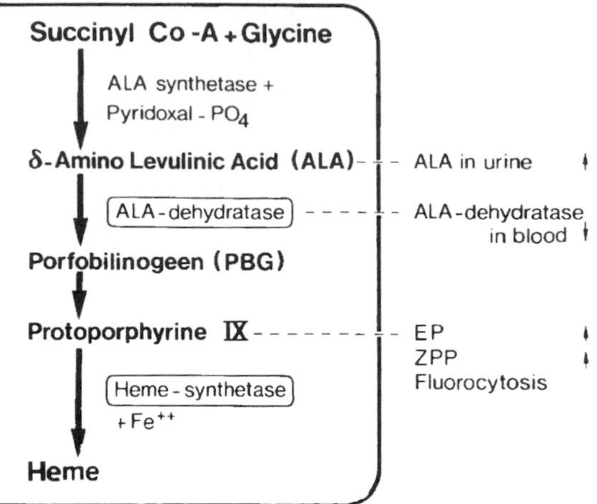

FIGURE 30.17 Schematic representation of heme synthesis. Lead interferes with the enzyme δ-aminolevulinic acid dehydratase (ALA-D), resulting in a lower activity of ALA-D in blood and a higher concentration of δ-aminolevulinic acid in the urine. Interference with the enzyme heme synthetase leads to an accumulation of protoporphyrin IX in the erythrocytes, which can be measured as FEPP or ZPP and leads to fluorocytosis. Modified with permission from Lumeij (1985b).

rats, and Pb-dosed wildfowl as long as the blood Pb is only moderately elevated. If the blood Pb increases above a moderate level, ALA-D does not decrease further. This phenomenon has been observed in pigeons and humans (Hutton, 1980). The inhibition of ALA-D in Pb poisoning leads to accumulation of ALA, and excess ALA is excreted in the urine.

Because Pb inhibits heme synthetase, protoporphyrin IX also accumulates in the erythrocytes. In human Pb poisoning, free erythrocyte protoporphyrin (FEPP) is in the range of 300–3000 μg/dl (reference range 15–60 μg/dl). If a fresh wet film of blood of a patient is examined under UV light (400 nm), 75–100% of the erythrocytes have a red fluorescence, fluorocytes. The accumulated FEPP imparts the red fluorescence to the fluorocytes (Beeson *et al.*, 1979). The fluorocyte test has also been used successfully for the diagnosis of Pb poisoning in rabbits (Roscoe *et al.*, 1975) and waterfowl (Barret and Karstad, 1971).

In humans, FEPP binds to zinc to form the fluorescent compound zinc protoporphyrin (ZPP), which can be measured fluorometrically in a single drop of whole blood (Roscoe *et al.*, 1979). Roscoe *et al.* (1979) found that the blood fluorescence spectra of Pb-poisoned mallard ducks were characteristic of metal-free protoporphyrin IX. They suggested that the reason for chelation of Zn by FEPP in human erythrocytes and not in duck erythrocytes might be due in part to the fact that duck erythrocytes contained only about one-third as much Zn as the human erythrocytes. By changing the factory-installed emission filter in a commercially available fluorometer used to screen humans for Pb intoxication by measurement of ZPP, they made the instrument suitable for measuring FEPP. They found that FEPP levels were at their highest 8 days after Pb-shot ingestion in mallard ducks. The highest value they recorded was 2284 μg/dl. FEPP was rarely elevated (>40 μg/dl) in freshly drawn blood from Pb-poisoned ducks. However, when the same blood was oxygenated and refrigerated prior to testing, FEPP concentrations increased because *in vitro* synthesis that terminated within 2 days. No such increase was manifested by controls. They found that FEPP was directly related to the clinical signs of Pb poisoning in mallard ducks, and those with Pb higher than 500 μg/dl began to show impaired motor function that could seriously affect their survival (Table 30.4).

In raptors, birds showing clinical signs of Pb toxicosis had consistently higher ZPP levels than other Pb-dosed birds with similar blood Pb values (Reiser and Temple, 1981). False-positive FEPP and ZPP elevations occur in humans suffering from iron deficiency anemia or erythrocytic protoporphyria (Wijngaarden and Smith, 1982).

B. Zinc

Zinc poisoning has been reported in birds after they ingest pennies (United States) minted after 1983 (98% Zn) or metal fence clips (96% Zn). Galvanized wire is another well-known source of Zn poisoning in aviary birds (new-wire disease). Clinical signs include

TABLE 30.4 Blood Protoporphyrin IX Concentrations Measured on the Hematofluorometer and Corresponding Clinical Signs of Lead Poisoning in Mallard Ducks[a]

Blood protoporphyrin IX	Clinical signs
> 801 μg/dl > 14.4 μmol/liter	Death Inability to stand, walk, fly Loss of voice Green watery diarrhea
501–800 μg/dl 9.0–14.4 μmol/liter	Muscular weakness Easily fatigued Unsteady gait Slight tail drop Green watery diarrhea
201–500 μg/dl 3.6–9.0 μmol/liter	Hyperexcitability Green watery diarrhea
40–200 μg/dl 0.7–3.6 μmol/liter	Green watery diarrhea
0–39 μg/dl 0–0.7 μmol/liter	No evidence of lead poisoning

[a] From Roscoe *et al.* (1979).

weight loss, depression, anorexia, gastrointestinal signs, and posterior paresis. Pathological lesions are especially seen in the pancreas and include acinar atrophy and proliferation of pancreatic ductules (Howard, 1992; LaBonde, 1995; Lloyd, 1992; Morris, 1985; Reece, 1986; Zdziarski *et al.*, 1994; Wight *et al.*, 1986).

Serum Zn can be used to establish a diagnosis, but extreme care must be taken to exclude contamination by Zn-containing grommets in plastic syringes or rubber stoppers in collecting tubes (Minnick *et al.*, 1982). Serum Zn levels in an affected group of ducks were 1260–1660 μg/dl as compared to a reference group ranging between 184 and 465 μg/dl. In a group of normal cockatiels, the mean serum Zn was 163 μg/dl. In general, blood Zn >1000 μg/dl is considered diagnostic and >200 μg/dl suggestive for Zn poisoning in psittacine birds (LaBonde, 1995).

C. Organophosphate and Carbamate

Organophosphates (OPs) and carbamates (CBMs) are the most common causes of avian insecticide poisoning. Poisoning occurs through inhalation or ingestion. The fundamental mechanism of OP and CBM poisoning is the inhibition of acetylcholinesterase (AChE). The LD-50 of this group of poisons is 10–20 times higher in birds than in mammals. The binding of CBM to AChE is reversible, in contrast to that of OP. Clinical signs include acute anorexia, crop stasis, ptyalism, ataxia, wing twitching, stargazing, weakness, diarrhea, prolapsed nictitans, and muscular tremors or stiffness. Dyspnea and bradycardia may be observed as the toxicity progresses. In severe cases, the birds may be recumbent with varying degrees of paralysis or seizures. An OP ester-induced delayed neuropathy has been reported in mammals and birds. The onset occurs 1 to 3 weeks after exposure and is not associated with AChE inhibition. With aging, metabolites of some OP compounds can affect peripheral axons and myelin sheaths, resulting in sensory and motor neuropathy. Associated clinical signs include weakness, ataxia, and decreased proprioception in the limbs progressing to paralysis (Porter, 1993; LaBonde, 1992, 1995; Lumeij *et al.*, 1993b).

Diagnosis of OP or CBM poisoning can be established by the AChE activity in blood, plasma, or serum. There are a number of different test procedures, but their results are not interchangeable. Besides AChE, another enzyme, pseudocholinesterase (PsChE) or butyrylcholinesterase (BChE, E.C. 3.1.1.8.), is found in sera. Although its physiological role is not defined, it is a useful indicator of exposure to OP and CBM compounds (Sherman *et al.*, 1964; Ludke *et al.*, 1975). Because plasma BChE activity increases with age in nestling passerines, this might partially account for its decreasing sensitivity in older birds to OP and CBM poisoning. Plasma BChE usually is inhibited more rapidly and to a larger degree than brain AChE and may be scavenging the active oxon forms of OP compounds that otherwise might inhibit brain AChE activity. Because of the lack of OP hydrolyzing enzymes in the plasma in many bird species or the low affinity of this class of enzymes for OP compounds in birds, the role of BChE in protecting individuals is important (Gard and Hooper, 1993).

It is important that results of AChE, PsChE, or BChE determinations be compared with those of nonexposed animals of the same species and age. Age-dependent changes in plasma AChE activities have been reported for many avian species. Furthermore, development patterns of plasma AChE differ between altricial and precocial species. In contrast to plasma BChE activities in nestlings of altricial species, plasma AChE and BChE activity decreased significantly with age in precocial species (Bennet and Bennet, 1991; Gard and Hooper, 1993).

Cases of suspected CBM toxicity may have normal AChE activities because of rapid regeneration, and therefore these samples should be run as soon as possible to be accurate. Because CBMs, unlike OP compounds, are reversible AChE inhibitors, AChE inhibition followed by thermal reactivation has been employed to discriminate between these poisonings (Hunt and Hooper 1993; Hunt *et al.*, 1993, 1995; Stansley, 1993).

Acknowledgments

The author is grateful to the publishers of the journals *Avian Diseases, Avian Pathology, Seminars in Avian and Exotic Pet Medicine,* and *The Veterinary Quarterly* for permission to use illustrations that were published previously in these journals, and to Wingers Publishing, Inc. (Lake Worth, Florida) for granting permission to use material previously published in *Avian Medicine: Principles and Application* (B. W. Ritchie, G. J. Harrison, and L. R. Harrison, eds.), 1994. The respective sources have been indicated in the text where appropriate.

References

Ahmed, A. A. S., El-Abdin, Y. Z., Hamza, A., and Saad, F. E. (1974). *Avian Dis.* **19,** 305–309.

Allen, J. L. (1987). *In* "Disorders of Exotic Animals" (E. D. Jacobson and G. Kollias, eds.). Churchill Livingstone, New York.

Altman, R. B., and Kirmayer, A. H. (1976). *J. Am. Anim. Hosp. Assoc.* **12,** 531–537.

Andreasen, C. B., Latimer, K. S., Kircher, I. M., and Brown, J. (1989). *Avian Dis.* **33,** 93–96.

Barnes, A., and Smith R. (1977). *Infect. Immun.* **16,** 876–884.

Baron, H. W. (1980). Veterinary Dissertation. München.

Barret, M. W., and Karstad, L. H. (1971). *J. Wildl. Manage.* 35, 109–118.

Barthalmus, G. T., Leander, J. D., McMillan, D. E., Mushak, P., and Krigman, M. R. (1977). *Toxicol. Appl. Pharmacol.* **42,** 271–284.

Beeson, P. B., McDermott, W., and Wijngaarden, J. B. (1979). "Cecil Textbook of Medicine," 15th ed. Saunders, London.

Bell, D. J., and Sturkie, P. D. (1965). *In* "Avian Physiology" (P. D. Sturkie, ed.), 2nd ed., pp. 32–84. Comstock Publications Associate Division, Cornell University Press, Ithaca, New York.

Bennet, R. S., and Bennet, J. W. (1991). *J. Wildl. Dis.* **27,** 116–118.

Bogin, E., Avidar, Y., and Israeli, B. (1976). *Zentralblatt für Veterinärmedizin A* **23,** 152–157.

Brannian, R. E., Graham, D. L., and Creswell, J. (1982). *Proc. Am. Assoc. Zoo. Vet.,* 21–23.

Brewster, L. M., and De Visser, M. (1988). *Acta Neurol. Scand.* **77,** 60–63.

Brixen, K., Nielsen, H., Eriksen, E., Charles, P., and Mosekilde, L. (1989). *Calcif. Tissue Int.* **44,** 93–98.

Buggiani, S. S., and Rindi, S. (1980). *Bull. Environm. Contam. Toxicol.* **24,** 152–155

Calle, P. P., and Stewart, C. A. (1987). *J. Zoo. Anim. Med.* **18,** 98–99.

Calle, P. P., Dierenfeld, E. S., and Robert, M. E. (1989). *J. Zoo. Wildl. Med.* **20,** 62–67.

Campbell, J. G. (1969). Tumours of the Fowl. J. B. Lippincott, Philadelphia, Pennsylvania.

Campbell, T. W. (1986). *Proc. Ann. Meeting Assoc. Avian Vet., Miami,* pp. 43–51.

Campbell, G., and Montali, R. J. (1980). *J. Zoo Anim. Med.* **11,** 35–40.

Candeletta, S. C., Homer, B. C., Garner, M. M., and Isaza, R. (1993). *J. Assoc. Avian Vet.* **7,** 39–43.

Carpenter, J. W., Spann, J. W., and Norvilla, M. N. (1979). *Proc. Am. Assoc. Zoo Vet.* 51–55.

Carpenter, J. W., Thomas, N. J., and Reeves, S. (1991). *J. Zoo Wildl. Med.* **22,** 488–493.

Chalmers, G. A. and Barret, M. W. (1982). *In* "Noninfectious disease in wildlife" (G. L. Hoff and J. W. Davis, eds.), pp. 84–94. Iowa State University Press, Ames, Iowa.

Chonko, A. M., and Grantham, J. J. (1981). *In* "The Kidney" (B. M. Brenner and F. C. Rector, eds.), pp. 1023–1055. Saunders, Philadelphia, Pennsylvania.

Clarkson, M. J., and Richards, T. G. (1971). *In* "Physiology and Biochemistry of the Domestic Fowl" (D. J. Bell and B. M. Freeman, eds.), pp. 1085–1114. Academic Press, New York.

Clubb, S. L., Schubot, R. M., Joyner, K., Zinkl, J. G., Wolf, S., Escobar, J., Clubb, K. J., and Kabbur, M. B. (1990). *J. Assoc. Avian Vet.* **4,** 218–225.

Clubb, S. L., Schubot, R. M., Joyner, K., Zinkl, J. G., Wolf, S., Escobar, J., and Kabbur, M. B. (1991a). *J. Assoc. Avian Vet.* **5,** 16–26.

Clubb, S. L., Schubot, R. M., Joyner, K., Zinkl, J. G., Wolf, S., Escobar, J., and Kabbur, M. B. (1991b). *J. Assoc. Avian Vet.* **5,** 154–162.

Cooper, J. E., Needham, J. R., Applebee, K., and Jones, C. G. (1988). *Ibis* **130,** 57–64.

De Bruijne, J. J. and Rothuizen, J. (1988). *In* "Animal Clinical Biochemistry, The Future" (D. J. Blackmore, Eckersall, P. D., Evans, G. O., Sommer, H., Stonard, M. D., and Woodman, D. D. eds.), pp. 175–180. Cambridge University Press, Cambridge.

Degernes, L. A. (1995).*Seminars in Avian and Exotic Pet Medicine* **4,** 15–22.

Dein, F. J. (1986a). *In* "Clinical Avian Medicine and Surgery" (G. J. Harrison, and L. R. Harrison, eds.), pp. 174–191. Saunders, Philadelphia.

Dein, F. J. (1986b). *Proc. Annual Meeting Assoc. Avian Vet.,* 41–42.

Dierenfeld, E. S. (1989). *J. Zoo Wildl. Med.* **10,** 3–11.

Dierenfeld, E. S., Sandford, C. E., and Satterfield, W. D. (1989). *J. Wildl. Manage.* **53,** 160–164.

Dierenfeld, E. S., Sheppard, C. D., Langenberg, J., Mirande, C., Spratt, J., and Dein, F. J. (1993). *J. Wildl. Dis.* **29,** 98–102.

Dieter, M. P. (1979). *In* "Animals as Monitors of Environmental Pollutants" (F. W. Nielsen, G. Migaki, and D. G. Scarpelli, eds.), pp. 177–191. National Academy of Sciences, Washington, D.C.

Dieter, M. P., and Wiemeyer, S. N. (1978). *Comp. Biochem. Physiol. C* **61,** 153–155

Djojosugito, A. M., Folkow, B., and Kovach, A. G. B. (1968). *Acta Physiol. Scand.* **74,** 114–122.

Dolensek, E. and Bruning, D. (1978). *In* "Zoo and Wild Animal Medicine" (M. E. Fowler, ed.), pp. 165–180. Saunders, Philadelphia, Pennsylvania.

Draury, R. W., Schwab, F., and Bateman, M. C. (1993). *J. Wildl. Dis.* **29,** 577–581.

Drew, M. L., Joyner, K., and Lobingier, R. (1993). *J. Assoc. Avian. Vet.* 7, 35–38.

Driver, E. A. (1981) *J. Wildl. Dis.* **17,** 413.

Duke, G. E. (1986). *In* "Avian Physiology" (P. D. Sturkie, ed.), 4th ed., pp. 289–302. Springer-Verlag, New York.

Dumonceaux, G., and Harrison, G. J. (1994). *In* "Avian Medicine. Principles and Application" (B. W. Ritchie, G. J. Harrison, and L. R. Harrison, eds.), pp. 1030–1052. Wingers Publishing, Lake Worth, Florida.

Ekstrom, D. D., and Degernes, L. (1989). *Proc. Ann. Meeting Assoc. Avian. Vet, Seattle,* pp. 130–138.

Ettinger, S. J., ed. (1989). "Textbook of Veterinary Internal Medicine," 3rd ed., Vol. 2, pp. 1579–1774. Saunders, Philadelphia, Pennsylvania.

Fairbrother, A. (1990). *J. Wildl. Dis.* **26,** 67–77.

Fennely, J., Dunne, J., and McGeeney, K. (1969). *Ann. NY Acad. Sci.* **166,** 794–807.

Ferrer, M., Garcia-Rodriguez, T., Carrillo, J. C., and Castroviejo, J. (1987). *Comp. Biochem. Physiol. A* **88,** 663–669.

Fitzgerald, M., Fennely, J., and McGeeney, K. (1969). *Am. J. Clin. Path.* **51,** 194–201.

Flammer, K. (1994). *Proc. Annual Meeting Assoc. Avian. Vet.,* 9–12.

Fourie, F. le R. (1977). *Poultry Sci.* **56,** 1842–1846.

Fowler, M. E. (1978a). *In* "Zoo and Wild Animal Medicine" (M. E. Fowler, ed.), p. 215. Saunders, Philadelphia, Pennsylvania.

Fowler, M. E. (1978b). *In* "Zoo and Wild Animal Medicine" (M. E. Fowler, ed.), pp. 33–34. Saunders, Philadelphia, Pennsylvania.

Fowler, M. E. (1978c). "Restraint and Handling of Wild and Domestic Animals," pp. 264–268. Iowa State University Press, Ames, Iowa.

Franson, J. C. (1982). *J. Wildl. Dis.* **18,** 481.

Galvin, C. (1980). *Proc. 47th Ann. Meeting Am. Animal Hosp. Assoc.,* pp. 41–52.

Garcia-Rodriguez, T., Ferrer, M., Recio, F., and Castroviejo, J. (1987). *Comp. Biochem. Physiol. A* **88,** 663–669.

Gard, N. W., and Hooper, M. J. (1993). *J. Wildl. Dis.* **29,** 1-7.

George, J. C. and Berger, A. J. (1966). *In* "Avian Myology," pp. 199–223. Academic Press, New York.

Ghebremeskel, K, Williams, G., Keymer, I. F., Horsley, D., and Gardner, D. A. (1989). *Comp. Biochem. Physiol. A* **92,** 43–47.

Goodman, L. S., and Gilman, A., eds. (1975). "The Pharmacological Basis of Therapeutics," 5th ed. MacMillan Publishing Co., New York.

Gratzl, E., and Koehler, H., eds. (1968). "Spezielle Pathologie und Therapie der Gefluegelkrankheiten," p. 14. Ferdinand Enke Verlag, Stuttgart.

Gremmels, K. (1987). *Zeitschrift für Jagdwissenschaft* **33,** 78–79.

Griffin, H. D., Perry, M. M., and Gilbert, A. B. (1984). *In* "Physiology and Biochemistry of the Domestic Fowl," Vol. 5 (B. M. Freeman, ed.), pp. 345–380. Academic Press, London.

Griminger, P.(1976). *In* "Avian Physiology," 3rd ed. (P. Sturkie, ed.), pp. 233–251. Springer-Verlag, New York.

Hagberg, M., Michaelson, G, and Örtelius, A. (1982). *Int. Arch. Occup. Environ. Health* **50,** 377–386.

Halliwell, W. H. (1981). *In* "Recent Advances in the Study of Raptor Diseases" (J. E. Cooper and A. G. Greenwood, eds.), pp. 111–112. Chiron Publications, Keighly.

Hawkey, C., Samour, J. H., Ashton, D. G., Hart, M. G., Cindery, R. N., Ffinch, J. M., and Jones, D. M. (1983). *Avian Pathol.* **12,** 73–84.

Heffron, J. J. A., Bomzon, L., and Pattinson, R. A. (1976). *Vet. Rec.* **98,** 338–340.

Hendriks, H. J., Haage, A., and De Bruijne, J. J. (1976). *Zentralblatt für Veterinärmedizin A* **23,** 683–687.

Hernandez, M. (1991). *Proc. European Chapter Assoc. Avian. Vet., Vienna,* 411–419.

Hill, K. J. (1983). *In* "Physiology and Biochemistry of the Domestic Fowl," Vol. 4 (B. M. Freeman, ed.), pp. 31–49. Academic Press, London.

Hochleithner, M. (1989a). *Proc. Annual Meeting Assoc. Avian. Vet., Seattle,* 78–81.

Hochleithner, M. (1989b). *J Assoc. Avian Vet.* **3,** 207–209.

Hochleithner, M. (1989c). *Inaugural Dissertation Wien.*

Hoffmann, W. E., Baker, G., Rieser, S., and Dorner, J. L. (1987). *Am. J. Vet. Res.* **48,** 1343–1347.

Howard, B. R. (1992). *J. Am. Vet. Med. Assoc.* **200,** 1667–1674.

Hunt, K. A., and Hooper, M. J. (1993). *Analyt. Biochem.* **212,** 335–343.

Hunt, K. A., Hooper, M. J., and Weisskopf, C. P. (1993). *The Toxicologists* **13,** 371.

Hunt, K. A., Hooper, M. J., and Littrell, E. (1995). *J. Wildl. Dis.* **31,** 186–192.

Hutton, M. (1980). *Environm. Pollut. Ser. A* **22,** 281–293.

Ivins, G. K., Weddle, G. D., and Halliwell, W. H. (1986). *In* "Zoo and Wild Animal Medicine" (M. E. Fowler, ed.), 2nd ed., pp. 434–437. Saunders, Philadelphia, Pennsylvania.

Jacobs, A. (1978). *In* "Price's Textbook of the Practice of Medicine" (R. B. Scott, ed.), pp. 1118–1129. Oxford University Press, Oxford.

Jenkins, J. (1994). *Seminars in Avian and Exotic Pet Medicine* **3,** 25–32.

Johnson, R., Ellingboe, K., and Gibbs, P. (1972). *Clin. Chem.* **19,** 110–115.

Julian, R. J. (1991). *In* "Diseases of Poultry" (B. W. Calnek, Barnes, H. J., Beard, C. W., Reid, W. M., and Yoder, H. W. eds.), 9th ed., pp. 863–884. Iowa State University Press, Ames, Iowa.

Kaplan, M. M., and Righetti, A. (1969). *Biochim. Biophys. Acta* **184,** 667–669.

Kern, M. D., and De Graw, W. A. (1978). *Condor* **80,** 230.

Klein, P. N., and Galey, F. D. (1989). *Proc. Annual Meeting Assoc. Avian. Vet., Seattle,* 139–143.

Kocan, R. M. (1972). *J. Wildl. Dis.* **8,** 115.

Kocan, R. M., and Pits, S. M. (1976). *J. Wildl. Dis.* **12,** 341.

Kodak Company (1986). "Kodak Ektachem DT System, Test Methodologies." Eastman Kodak Company, Rochester, New York.

Kovách, A. G. B., Szasz, E., and E. Pilmayer, N. (1969). *Acta Physiologica Academiae Scientarum Hungaricae* **35,** 109–116.

Kramer, J. W. (1989). *In* "Clinical Biochemistry of Domestic Animals" (J. J. Kaneko, ed.), 4th ed., pp. 338–363. Academic Press, New York.

Kryszewski, A. J., Neale, G., and Whitfield, J. B. (1973). *Clin. Chim. Acta* **47,** 175–182.

LaBonde, J. (1992). *Proc. Annual Meeting Assoc. Avian Vet.,* 113–118.

LaBonde, J. (1995). *Seminars in Avian and Exotic Pet Medicine* **4,** 23–31.

Laessig, R. H., Indriksons, A. A., Hassemer, D. J., Paskey, T. A., and Schwartz, T. H. (1976). *Am. J. Clin. Pathol.* **66,** 598–60.

Lavis, S., Cuenca, R., Marco, I., Velarde, R., and Viñas, L. (1992). *Avian Pathol.* **21,** 711–715.

Law, G. R. J. (1960). *Poultry Sci.* **39,** 1450–1452.

Leveille, G. A., and Sauberlich, H. E. (1961). *J. Nutr.* **74,** 500.

Levy, A., Perelman, B., Waner, T., Van Grevenbroek, M, Van Creveld, C., and Yagil, R. (1989). *Am. J. Vet. Res.* **50,** 1548–1550.

Lewandowski, A. H., Campbell, T. W., and Harrison, G. J. (1986). *In* "Clinical Avian Medicine and Surgery" (G. J. Harrison and L. R. Harrison, eds.), pp. 192–200. Saunders, Philadelphia, Pennsylvania.

Lin, G. L., Himes, J. A., and Cornelius, C. E. (1974). *Am. J. Physiol.* **226,** 881–885.

Lind, G. W., Gronwall, R. R., and Cornelius, C. E. (1967). *Res. Vet. Sci.* **8,** 280–282.

Lloyd, M. (1992). *J. Assoc. Avian Vet.* **6,** 25–29.

Long, S. (1982). *Comp. Biochem. Physiol. A* **71,** 519–526.

Loomis, R, and Wright, J. F. (1993). *Proc. Am. Assoc. Zoo Vet., Saint Louis,* p. 28.

Lothrop, C., Harrison, G., Schultz, D. and Utteridge, T. (1986). *In* "Clinical Avian Medicine and Surgery" (G. J. Harrison and L. R. Harrison, eds.), pp. 525–536. Saunders, Philadelphia, Pennsylvania.

Ludke, J. L., Hill, E. F., and Dieter, M. P. (1975). *Arch. Environm. Contam. Toxicol.* **3,** 1–21.

Lumeij, J. T. (1985a). *Avian Pathol.* **14,** 257–260.

Lumeij, J. T. (1985b). *Vet. Quart.* **7,** 133–136.

Lumeij, J. T. (1987a). *Vet. Quart.* **9,** 249–254.

Lumeij, J. T. (1987b). "A Contribution to Clinical Investigative Methods for Birds, with Special Reference to the Racing Pigeon *Columba livia domestica.*" Ph.D. Thesis, University of Utrecht, Utrecht, pp. 26–30.

Lumeij, J. T. (1987c). *Avian Pathol.* **16,** 377–382.

Lumeij, J. T. (1987d). *Vet. Quart.* **9,** 255–261.

Lumeij, J. T. (1987e). *Vet. Quart.* **9,** 262–268.

Lumeij, J. T. (1987f). "A Contribution to Clinical Investigative Methods for Birds, with Special Reference to the Racing Pigeon, *Columba livia domestica.*" PhD. Thesis, Utrecht University, Utrecht, pp. 35–77.

Lumeij, J. T. (1988). *In* "Animal Clinical Biochemistry, The Future" (D. J. Blackmore, Eckersall, P. D., Evans, G. O., Sommer, H., Stonard, M. D., and Woodman, D. D. eds.), pp. 161–174. Cambridge University Press, Cambridge.

Lumeij, J. T. (1990). *Avian Pathol.* 19, 661-667.

Lumeij, J. T. (1991). *J. Assoc. Avian Vet.* **5,** 197-200.

Lumeij, J. T. (1994b). *In* "Avian Medicine. Principles and Application" (B. W. Ritchie, G. J. Harrison, and L. R. Harrison, eds.), pp. 522–537. Wingers Publishing, Lake Worth, Florida.

Lumeij, J. T. (1994c). *In* "Avian Medicine. Principles and Application" (B. W. Ritchie, G. J. Harrison, and L. R. Harrison, eds.), pp. 538–555. Wingers Publishing, Lake Worth, Florida.

Lumeij, J. T. (1994d). *In* "Avian Medicine. Principles and Application" (B. W. Ritchie, G. J. Harrison, and L. R. Harrison, eds.), pp. 582–606. Wingers Publishing, Lake Worth, Florida.

Lumeij, J. T. (1994f). *Seminars in Avian and Exotic Pet Medicine* **3,** 14–24.

Lumeij, J. T., and De Bruijne, J. J. (1985a). *Avian Pathol.* **14,** 401–408.

Lumeij, J. T., and De Bruijne, J. J. (1985b). *Avian Pathol.* 14, 441–444.

Lumeij, J. T., and Overduin, L. M. (1990). *Avian Pathol.* **19,** 235–244.

Lumeij, J. T. and Maclean, B. (1996). *J. Avian Med. Surg.* **10,** 150–152.

Lumeij, J. T., and Peccati, C. (1993). *Proc. Europ. Chapter Assoc. Avian Vet., Utrecht,* 558–566.

Lumeij, J. T., and Redig, P. T. (1992). "Proceedings VIII. Tagung über Vogelkrankheiten. München 1992," pp. 265–269. Deutsche Veterinärmedizinische Gesellschaft, Giessen.

Lumeij, J. T., and Remple, J. D. (1991). *Avian Pathol.* **20,** 79–83.

Lumeij, J. T., and Remple, J. D. (1992). *Avian Dis.* **36,** 1060–1062.

Lumeij, J. T., and Westerhof, I. (1987). *Vet. Quart.* **9,** 255–261.

Lumeij, J. T., and Westerhof, I. (1988a). *Avian Pathol.* **17,** 63–70.
Lumeij, J. T., and Westerhof, I. (1988b). *Avian Pathol.* **17,** 875–878.
Lumeij, J. T., and Wolfswinkel, J. (1987). *In* "A Contribution to Clinical Investigative Methods for Birds, with Special Reference to the Racing Pigeon *Columba livia domestica.*" Ph.D. Thesis, J. T. Lumeij, University of Utrecht, pp. 71–77.
Lumeij, J. T., and Wolfswinkel, J. (1988). *Avian Pathol.* **17,** 515–517.
Lumeij, J. T., Kantor, A., and Van den Brom, W. E. (1987a). *In* "A Contribution to Clinical Investigative Methods for Birds, with Special Reference to the Racing Pigeon *Columba livia domestica.*" Ph.D. Thesis, J. T. Lumeij), University of Utrecht, pp. 31–34.
Lumeij, J. T., Boschma, Y., Mol, J., De Kloet, E. R., and Van den Brom, W. E. (1987b). *Avian Pathol.* **16,** 199–204.
Lumeij, J. T., De Bruijne, J. J., Slob, A., Wolfswinkel, J., and Rothuizen, J. (1988a). *Avian Pathol.* **17,** 851–864.
Lumeij, J. T., Meidam, M., Wolfswinkel, J., Van der Hage, M. H., and Dorrestein, G. M. (1988b). *Avian Pathol.* **17,** 865–874.
Lumeij, J. T., De Bruijne, J. J., and Kwant, M. M. (1990). *Avian Pathol.* **19,** 255–261.
Lumeij, J. T., Remple, J. D., and Riddle, K. E. (1993a). *Avian Pathol.* **22,** 183–188.
Lumeij, J. T., Westerhof, I, Smit, T., and Spierenburg, T. J. (1993b). *In* "Raptor Biomedicine" (P. T. Redig, J. E. Cooper, D. Remple, and D. B. Hunter, eds.), pp. 233–238. University of Minnesota Press, Minneapolis.
McClure, H. E., and Cedeno, R. (1955). *J. Wildl. Manage.* **19,** 477–478.
McGrath, J. J. (1971). *J. Appl. Physiol.* **31,** 274–276.
McIntire, M. S., Wolf, G. L., and Angle, C. R. (1973). *Clin. Toxicol.* **6,** 183–188.
Minnick, P. D., Braselton, W. E., and Meerdink, G. L. (1982). *Vet. Human. Toxicol.* **24,** 413–414.
Mori, J. G., and George, J. C. (1978). *Comp. Biochem. Physiol. B* **59,** 263.
Morris, P. (1985). *Assoc. Avian Vet. Newsletter* **6,** 75.
Morris, P. J., Avgeris, S. E., and Baumgartner, R. E. (1989). *J. Assoc. Avian Vet.* **3,** 87–92.
Mulley, R. C. (1979). *J. Wildl. Dis.* **15,** 437.
Murphey, J. (1992). *Proc. Annual Meeting Assoc. Avian Vet.,* 165–170.
Newham, D. J., Jones, D. A., and Edwards, R. H. T. (1983). *Muscle Nerve* **6,** 380–385.
Nichols, D. K. and Montali, R. J. (1987). *Proc. Annual Meeting Assoc. Avian Vet., Oahu,* 419–421.
Nichols, D. K., Campbell, V. L. and Montali, R. J. (1986). *J. Am. Vet. Med. Assoc.* **189,** 1110–1112.
Odell, W. D. (1988). *In* "Cecil Textbook of Medicine" (J. B. Wijngaarden and L. H. Smith, eds.), pp. 1100–1104. Saunders, Philadelphia, Pennsylvania.
Okinaka, S., Sugita, H., Momoi, H., Toyokura, Y., Watanabe, T., Ebashi, F., and Ebashi, S. (1964). *J. Lab. Clin. Med.* **64,** 299–305.
Olsen, J. H. (1994). *In* "Avian Medicine. Principles and Application" (B. W. Ritchie, G. J. Harrison, and L. R. Harrison, eds.), pp. 1237-1275. Wingers Publishing, Lake Worth, Florida.
Palomeque, J., and Planas, J. (1978). *Comp. Biochem. Physiol. A* **59,** 413–417.
Palomeque, J., Pintó, D., and Viscor, G. (1991). *J. Wildl. Dis.* **27,** 34–40.
Peinado, V. I., Polo, F. J., and Celdrán, J. F. (1992a). *J. Zoo Wildl. Med.* **23,** 65–71.
Peinado, V. I., Polo, F. J., and Palomeque, J. (1992b). *Avian Pathol.* **21,** 55–64.
Perry, M. C., Obrecht, H. H., Williams, B. K., and Kuenzel, W. J. (1986). *J. Wildl. Manage.* **50,** 435–441.
Phalen, D. N., Ambrus, S., and Graham, D. L. (1990). *Proc. Ann. Meeting Assoc. Avian Vet., Phoenix,* pp. 44–57.
Porter, S. (1993). *In* "Raptor Biomedicine" (P. T. Redig, J. E. Cooper, D. Remple, and D. B. Hunter, eds.), pp. 239–245. University of Minnesota Press, Minneapolis.
Posen, S., Neale, F., and Clubb, J. (1965). *Ann. Int. Med.* **62,** 1234.
Puerta, M. L., Huecas, V., and Garcia del Campo, A. L. (1989). *Comp. Biochem. Physiol. A* **94,** 623–625.
Puerta, M. L., Alonso, J. C., Huecas, V, Alonso, J. A., Abelenda, M., and Muñoz-Pulido, R. (1990). *The Condor* **92,** 210–214.
Puerta, M. L., Garcia del Campo, A. L., Huecas, V., and Abelenda, M. (1991). *Comp. Biochem. Physiol. A* **98,** 393–394.
Quintavalla, F., and Zucca, P. (1995). *Proc. Conf. European Chapter Assoc. Avian Vet., Jerusalem,* 544–551.
Rae, M. (1992). *Proc. Annual Meeting Assoc. Avian. Vet.,* 328–335.
Raphael, B. L. (1981). *Proc. Am. Assoc. Zoo Vet.,* 97–98.
Reece, R. (1986). *Austr. Vet. J.* **63,** 199.
Reiser, M. H., and Temple, S. A. (1981). *In* "Recent Advances in the Study of Raptor Diseases" (J. E. Cooper and A. G. Greenwood, eds.), pp. 21–25. Chiron Publications, Keighly, United Kingdom.
Rivetz, B., Bogin, E., Weisman, Y., and Avidar, J. (1977). *Ref. Vet.* **34,** 57–62.
Rosalki, S., and Foo, A. (1984). *Clin. Chem.* **30,** 1182–1186.
Roscoe, D. E., Nielsen, S. W., Eaton, H. D., and Rousseau, J. E. (1975). *Am. J. Vet. Res.* **26,** 1225–1229.
Roscoe, D. E., Nielsen, S. W., Lamola, A. A., and Zuckerman, D. (1979). *J. Wildl. Dis.* **15,** 127–136.
Rosskopf, W. J., Woerpel, R. W., Rosskopf, G., and Van de Water, D. (1982). *Vet. Med./Small Anim. Clin.* **77,** 1233–1239.
Rosskopf, W. J., Woerpel, R. W., Rosskopf, G., and Van De Water, D. (1984). *Avian/Exotic Practice* **1,** 30–33.
Rosskopf, W. J., Woerpel, R. W., and Lane, R. A. (1985). *Proc. Ann. Meeting Assoc. Avian Vet., Boulder,* 129–131.
Rosskopf, W. J., Woerpel, R. W., and Lane, R. A. (1986). *Proc. Ann. Conf. Assoc. Avian Vet., Miami,* 61–73.
Rosskopf, W. J., Woerpel, R. W., Fudge, A. M., Reavill, D. R., and Rae, M. (1992). *Proc. Ann. Meeting Assoc. Avian Vet.,* 98–107.
Rutgers, H. C., Stradley, R. P., and Johnson, S. E. (1988). *Am. J. Vet. Res.* **49,** 317–320.
Ryan, C. P., Walder, E. J., and Howard, E. B. (1982). *J. Am. Anim. Hosp. Assoc.* **18,** 139–142.
Sanger, V., Burmester, B., and Morril, C. (1986). *Avian Dis.* **10,** 364–371.
Savova, M. N., and Kirev, T. (1992). *Avian Pathol.* **21,** 667–673.
Schlumberger, H. G. (1956). *Cancer Res.* **16,** 149–153.
Schmidt, C. H., Hane, M. E., and Gomez, D. C. (1963). *U.S. Armed Forces Med. J.* **4,** 1556.
Schöpf, A., and Vasicek, L. (1991). *Proc. European Chapter Assoc. Avian Vet., Vienna,* 437–439.
Scott, R. B., ed. (1978). "Price's Textbook of the Practice of Medicine," p. 1458. Oxford University Press, Oxford.
Sherman, M. E., Ross, E., and Chang, M. T. Y. (1964). *Toxicol. Appl. Pharmacol.* **6,** 147–153.
Simkiss, K. (1967). *In* "Calcium in Reproductive Physiology. A Comparative Study of Vertebrate," pp. 155–196. Chapman and Hall, London.
Sitbon, G., and Mialhe, P. (1979). *Horm. Metab. Res.* **11,** 85.
Skadhauge, E. (1981). "Osmoregulation in Birds," pp. 84–91. Springer-Verlag, Berlin.
Spano, J. S., Pedersoli, W. M., Kemppainen, R. J., Krista, L. M., and Young, D. W. (1987). *Avian Dis.* **31,** 800–803.
Spano, J. S., Whiteside, M. S., Pedersoli, W. M., Krista, L. M., and Ravis, W. M. (1988). *Am. J. Vet. Res.* **49,** 325–326.
Sperber, I. (1960). *In* "Biology and Physiology of Birds" (A. J. Marshal, ed.), Vol. I, pp. 469–492. Academic Press, London.
Spira, A. (1981). *Proc. 48th Ann. Meeting Am. Anim. Hosp. Assoc., Atlanta,* 13.
Spraker, T. R. (1980). *In* "The Comparative Pathology of Zoo Animals" (R. J. Montali and G. Migaki eds.), pp. 403–414. Smithsonian Institution Press, Washington, D.C.

Spraker, T. R., Adrian, W. J., and Lange, W. R. (1987). *J. Wildlife Dis.* **23,** 447–453.

Stansley, W. (1993). *Arch. Environm. Contam. Toxicol.* **25,** 315–321.

Steiner, C. V., and Davis, R. B. (1981). "Caged Bird Medicine." Iowa State University Press, Ames, Iowa.

Stevens, R. W. C., and Ridgway, C. J. (1966). *Poultry Sci.* **45,** 204–205.

Stolbach, L. (1969). *Ann. NY Acad. Sci.* **166,** 760–773.

Stoskopf, M. J., Beali, F. B., Ensley, P. K., and Neely, E. (1982). *J. Zoo Wildl. Med.* **13,** 160–168.

Tell, L. A., and Citino, S. B. (1992). *J. Zoo Wildl. Med.* **23,** 62–64.

Tenhunen, R. (1971). *Scand. J. Clin. Lab. Invest.* **116** suppl. 27, p. 9.

Timoshevskaya, G. V., Blinkina, M. S., Korzhuev, P. A., Bevol'skaya, M. V., and Savitskaya, O. N. (1983). *Zhurnal Evolyutsionnoi Biokhimii i Fiziologii* **19,** 226–229.

Trautschold, I., and Werle, E. (1966). "Handbuch der physiologischen-und pathologisch-chemischen Analyse von Hoppe-Seyler/Thierfelder," 10. Auflage, 6. Band/Teil B. Springer-Verlag, Berlin.

Tripp, M. J., and Schmitz, J. A. (1982). *Am. J. Vet. Res.* **43,** 2220–2223.

Van Heerden, J., Daugh, J., Jarvis, M. J. F., Keffen, R. H., Denny, J. E. F. M., Dreyer, M. J., and Kriek, N. P. J. (1985). *J. South Afr. Vet. Assoc.* **56,** 75–79.

Vuillaume, A. (1983). *Avian Pathol.* **12,** 389–391.

Wallner-Pendleton, E. A., Rogers, D., and Epple, A. (1993). *Avian Pathol.* **22,** 631–635.

Watts, R. W. E. (1978). *In* "Price's Textbook of the Practice of Medicine" (R. B. Scott, ed.), 12th ed., pp. 1056–1074. Oxford University Press, Oxford.

Webb, D. M., Denicola, D. B., and Van Vleet, C. R. (1991). *Avian Dis.* **35,** 662–667.

Westermarck, E., and Sandholm, M. (1980). *Res. Vet. Sci.* **28,** 341.

Wiesner, H. (1971). *Der Praktische Tierarzt* **13,** 594–596.

Wight, P. A., Dewar, W. A., and Saunderson, C. L. (1986). *Avian Pathol.* **15,** 23–38.

Wijngaarden, J. B., and Smith, L. H. (1982). "Cecil Textbook of Medicine," 16th ed. Saunders, London.

Windingstad, R. M., Hurley, S. S., and Sileo, L. (1983). *J. Wildl. Dis.* **19,** 289–290.

Woerpel, R. W., Rosskopf, W. J., and Monahan-Brennan, M. (1987). *In* "Companion Bird Medicine" (E. W. Burr, ed.), pp. 180–196. Iowa State University Press, Ames, Iowa.

Worell, A. (1991). *Proc. Ann. Meeting Assoc. Avian Vet., Chicago,* 120–130.

Worell, A. (1993). *Proc. Ann. Meeting Assoc. Avian Vet., Nashville,* 98–107.

Worell, A. (1994). *Seminars in Avian and Exotic Pet Medicine* **3,** 37–39.

Wyse, D. G., and Nickerson, M. (1971). *Canad. J. Physiol. Pharmacol.* **49,** 919–926.

Young, E. (1967). *Int. Zoo Yearb.* **7,** 226–227.

Zdziarski, J. M., Mattix, M., Bush, R. M., and Montalli, R. J. (1994). *J. Zoo Wildl. Med.* **25,** 438–445.

Appendixes

J. JERRY KANEKO, JOHN W. HARVEY, AND MICHAEL L. BRUSS

APPENDIX I. SI UNITS

The Systéme Internationale d'Unites (SI), or the International System of Units, was recommended for use in the health professions by the World Health Assembly (WHA 30.39) in May 1977. The SI is the culmination of more than a century of effort to develop a universally acceptable system of units of measure. In the past decade, the use of SI has been rapidly gaining acceptance, with many nations now mandating its use and many others strongly recommending its use. Furthermore, many scientific journals now require that units be expressed in SI along with the conventional units if used. The following tables in this appendix briefly describe the basis of SI and provide factors for the conversion of conventional units to SI.

TABLE A SI Base Units

Quantity	Name of unit	Symbol
Length	meter (metre)	m
Mass	kilogram	kg
Time	second	s
Electric current	ampere	A
Thermodynamic temperature	kelvin	K
Luminous intensity	candela	cd
Amount of substance	mole	mol

TABLE B SI Derived Units

Quantity	Name of unit	Symbol
Area	square meter	m^2
Volume	cubic meter	m^3
Speed, velocity	meter per second	m/s
Acceleration	meter per second squared	m/s^2
Substance concentration	mole per cubic meter	mol/m^3
Pressure	pascal	Pa
Work, energy	joule	J
Celsius temperature	degree Celsius	°C

TABLE C SI Prefixes

Factor	Prefix	Symbol	Factor	Prefix	Symbol
10^{18}	exa	E	10^{-1}	deci	d
10^{15}	peta	P	10^{-2}	centi	c
10^{12}	tera	T	10^{-3}	milli	m
10^{9}	giga	G	10^{-6}	micro	μ
10^{6}	mega	M	10^{-9}	nano	n
10^{3}	kilo	k	10^{-12}	pico	p
10^{2}	hecto	h	10^{-15}	femto	f
10^{1}	deca	da	10^{-18}	atto	a

TABLE D Non-SI Units Still in General Use

Quantity	Unit	Symbol	Value in SI
Time	minute	m	60 s
	hour	h	3600 s
	day	d	86,400 s
Volume	liter (litre)	l[a]	10^{-3} m^3
Mass	tonne	l	1000 kg
Length	angstrom	Å	10^{-10} m (0.1 nm)
Pressure	bar	bar	10^5 Pa
	atmosphere	atm	101,325 Pa
	mmHg	mm Hg	1.333×10^{-4} Pa
Radioactivity	curie	Ci	3.7×10^{10} Bequerel (Bq)
	roentgen	R	2.58×10^{-4} Ci/kg
	rad	rad	10^{-2} gray (Gy)
	rem	rem	10^{-2} sievert (Sv)

[a] Liter is often spelled out or symbolized by L to avoid confusion.

TABLE E SI Conversion Factors

Chemical constituent	Conventional "old unit"	× Factor	= New SI unit
Acetoacetate	mg/dl	0.098	mmol/l
Acetone	mg/dl	0.172	mmol/l
Albumin	g/dl	10.0	g/l
Ammonia	μg/dl	0.5872	μmol/l
Bicarbonate	mEq/l	1.0	mmol/l
Bilirubin	mg/dl	17.10	μmol/l
Bromosulfonphthalein	mg/dl	11.93	μmol/l
Calcium	mg/dl	0.2495	mmol/l
Carotenes	μg/dl	0.01863	μmol/l
Chloride	mEq/l	1.0	mmol/
Cholesterol	mg/dl	0.02586	mmol/l
Cobalt	μg/dl	0.1697	μmol/l
Coproporphyrin	μg/dl	15.0	nmol/l
Cortisol	μg/dl	27.59	nmol/l
CO_2, total	mEq/l	1.0	mmol/l
CO_2 pressure, pCO_2	mmHg	0.1333	kPa
Copper	μg/dl	0.1574	μmol/l
Creatinine	mg/dl	88.40	μmol/l
Fibrinogen	mg/dl	0.01	g/l
Fructose	mg/dl	55.51	μmol/l
Glucose	mg/dl	0.05551	mmol/l
Haptoglobin	mg/dl	0.01	g/l
Hemoglobin	g/dl	10.0	g/l
β-Hydroxybutyrate	mg/dl	0.096	mmol/l
Iodine	μg/dl	78.8	nmol/l
Insulin	μU/ml	7.175	pmol/l
Iron	μg/dl	0.1791	μmol/l
Lactate	mg/dl	0.1110	mmol/l
Lead	μg/dl	0.04826	μmol/l
Magnesium	mg/dl	0.4114	μmol/l
Manganese	μg/dl	0.1820	μmol/l
Mercury	μg/l	4.985	nmol/l
Methemoglobin	g/dl	10.0	g/l
Molybdenum	μg/dl	0.1042	μmol/l
Myoglobin	mg/dl	0.5848	μmol/l
Nitrogen	mg/dl	0.7138	mmol/l
O_2 pressure, pO_2	mmHg	0.1333	kPa
Phosphorus	g/dl	0.3229	mmol/l
Porphobilinogen	mg/dl	44.20	μmol/l
Potassium	mEq/l	1	mmol/l
Protein	g/dl	10	g/l
Protoporphyrin	μg/dl	0.0178	μmol/l
Pyruvate	mg/dl	113.6	μmol/l
Selenium	μg/dl	0.1266	μmol/l
Sodium	mEq/l	1	mmol/l
Thyroxine	μg/dl	12.87	nmol/l
Transferrin	mg/dl	0.01	g/l
Triglycerides (trioleate)	mg/dl	0.01129	mmol/l
Triiodothyronine	ng/dl	0.01536	nmol/l
Urate	mg/dl	59.48	μmol/l
Urea Nitrogen	mg/dl	0.7140	mmol/l
Urea Nitrogen	mg/dl	0.3570	mmol Urea/l
Urea	mg/dl	0.1665	mmol/l
Urobilinogen	mg/dl	16.90	μmol/l
Uroporphyrin	μg/dl	12.00	nmol/l
Vitamin A	μg/dl	0.03491	μmol/l

(continues)

TABLE E (*Continued*)

Chemical constituent	Conventional "old unit"	× Factor	= New SI unit
Xylose	mg/dl	0.06660	mmol/l
Zinc	μg/dl	0.1530	μmol/l
Enzymes[a]	U/l	16.67	nkat/l

[a] There is yet no general agreement or recommendation for the use of the katal (1 kat = 1 mol/s) in place of the widely used international unit (1 U = 1 μmol/m). The U/l should continue to be used for all enzyme activities.

APPENDIX III. TEMPERATURE CORRECTION FACTORS (Tf) FOR SOME COMMON ENZYMES[a]

Assay temperature (°C)	AlP	CK	LDH	SDH	ALT/AST
20	2.61	2.05	2.10	1.48	2.29
21	2.37	1.82	1.96	1.42	1.85
22	2.15	1.70	1.80	1.37	1.71
23	1.95	1.59	1.67	1.32	1.59
24	1.77	1.49	1.55	1.27	1.45
25	1.61	1.39	1.45	1.22	1.37
26	1.46	1.31	1.33	1.17	1.29
27	1.33	1.23	1.26	1.12	1.21
28	1.21	1.15	1.16	1.08	1.12
29	1.10	1.07	1.07	1.04	1.05
30	1.00	1.00	1.00	1.00	1.00
31	0.90	0.93	0.93	0.96	0.95
32	0.81	0.87	0.86	0.93	0.89
33	0.73	0.81	0.80	0.89	0.85
34	0.66	0.75	0.74	0.85	0.80
35	0.59	0.70	0.68	0.82	0.77
36	0.53	0.65	0.64	0.79	0.73
37	0.48	0.50	0.59	0.76	0.70

[a] Abbreviations: AlP, alkaline phosphatase; CK, creatine kinase; LDH, lactate dehydrogenase; SDH, sorbitol (iditol) dehydrogenase; ALT, alanine aminotransferase; AST, aspartate aminotransferase.

APPENDIX II. CONVERSION FACTORS OF SOME NON-SI SERUM ENZYME UNITS TO SI INTERNATIONAL ENZYME UNITS[a]

Serum enzyme	Non-SI unit	× Factor	= SI
Aldolase (ALD)	Sibley–Lehninger unit (SLU) (Mg DNP/hour/ml)	0.75	U/liter
Amylase (AMYL)	Somogyi unit (SU) (mg G/30 minutes)	1.85	U/liter
Glutamic-oxalacetic transaminase (SGOT, AST)	Sigma–Frankel unit (SFU) Karmen unit (KU) Wroblewski–LaDue unit (WLU) Reitman–Frankel unit (RFU) (0.001 OD/m/ml)	0.48	U/liter
Glutamic pyruvic transaminase (SGPT, ALT)	Sigma–Frankel unit (SFU) Karmen unit (KU) Wroblewski–LaDue unit (WLU) Reitman–Frankel unit (RFU) (0.001 OD/m/ml)	0.48	U/liter
Isocitric dehydrogenase (ICD)	Wolfson–Williams–Ashman unit (WWAU) (nmol/h/ml)	0.0167	U/liter
Lipase	Roe–Byler unit (RBU) (μmol/h/ml)	16.7	U/liter
	Cherry–Crandall unit (CCU) (50 μmol/3h/ml)	2.77	U/liter
Phosphatase, acid	King–Armstrong unit (KAU) (mg phenol P/30 minutes)	1.85	U/liter
Phosphatase, alkaline (A1P)	King–Armstrong unit (KAU) (mg Phenyl P/30 m) Bodansky unit (BU) (mg P/hour)	7.10 5.4	U/liter U/liter
Sorbitol (iditol) dehydrogenase (SDH, IDH)	Sigma–Frankel unit (SFU) (nmol/hour/ml)	0.0167	U/liter

[a] U = 1 international unit = 1 μmol/minute = 16.67 nmol/second = 16.67 nkatal/second = 0.01667 μkatal/second.

Examples of Use

A. To correct an enzyme activity to a standard temperature of 30°C, multiply the assay value at any temperature by the Tf of that temperature: AlP activity at 27°C = 48 U/l; correct it to 30°C

$$48 \times 1.33 = 63.8 \text{ U/l}.$$

B. To correct an enzyme activity at any one temperature to another temperature, multiply the assay value at the first temperature by the ratio of the Tf of the first temperature to the Tf of the second temperature: AlP activity at 22°C = 38 U/l; correct it to 37°C:

$$38 \text{ U/l} \times 2.15/0.48 = 170.2 \text{ U/l at } 37°\text{C}.$$

APPENDIX IV. STABILITY OF SERUM ENZYMES UNDER VARIOUS STORAGE CONDITIONS[a]

Enzyme	25°C	4°C	−25°C
Acid phosphatase	4 hr[b]	3 days[c]	3 days[c]
Alkaline phosphatase	2–3 days[d]	2–3 days	1 month
Aldolase	2 days	2 days	Unstable[e]
Amylase	1 month	7 months	2 months
Cholinesterase	1 week	1 week	1 week
Creatine kinase	2 days	1 week	1 month
Glutamyl transferase	2 days	1 week	1 month
Glutamate dehydrogenase	1 day	2 days	1 day
Aspartate aminotransferase	3 days	1 week	1 month
Alanine aminotransferase	2 days	1 week	Unstable[e]
Hydroxybutyrate dehydrogenase	Unstable	3 days	Unstable[e]
Isocitrate dehydrogenase	5 hr	3 days	3 weeks
Leucine aminopeptidase	1 week	1 week	1 week
Lactate dehydrogenase	1 week	1–3 days[f]	1–3 days[f]
Malate dehydrogenase	Unstable	3 days	3 days
Sorbitol (iditol)dehydrogenase	Unstable	1 day	2 days

[a] No more than 10% of the original activity lost during the specified time.
[b] At pH 5–6.
[c] With added citrate or acetate.
[d] Activity may increase.
[e] Enzyme does not tolerate thawing well.
[f] Depends on the isoenzyme profile.

APPENDIX V. TEMPERATURE CONVERSIONS BETWEEN DEGREES CELSIUS AND DEGREES FAHRENHEIT[a]

°C	°F	°C	°F	°C	°F	°C	°F
−40	−40.0	−4	24.8	32	89.6	68	154.4
−39	−38.2	−3	26.6	33	91.4	69	156.0
−38	−36.4	−2	28.4	34	93.2	70	158.0
−37	−34.6	−1	30.2	35	95.0	71	159.8
−36	−32.8	0	32.0	36	96.8	72	161.6
−35	−31.0	1	33.8	37	98.6	73	163.4
−34	−29.2	2	35.6	38	100.4	74	165.2
−33	−27.4	3	37.4	39	102.2	75	167.0
−32	−25.6	4	39.2	40	104.0	76	168.8
−31	−23.8	5	41.0	41	105.8	77	170.6
−30	−22.0	6	42.8	42	107.6	78	172.4
−29	−20.2	7	44.6	43	109.4	79	174.2
−28	−18.4	8	46.4	44	111.2	80	176.0
−27	−16.6	9	48.2	45	113.0	81	177.8
−26	−14.8	10	50.0	46	114.8	82	179.6
−25	−13.0	11	51.8	47	116.6	83	181.4
−24	−11.2	12	53.6	48	118.4	84	183.2
−23	−9.4	13	55.4	49	120.2	85	185.0
−22	−7.6	14	57.2	50	122.0	86	186.8
−21	−5.8	15	59.0	51	123.8	87	188.6
−20	−4.0	16	60.8	52	125.6	88	190.4
−19	−2.2	17	62.6	53	127.4	89	192.2
−18	−0.4	18	64.4	54	129.2	90	194.0
−17	1.4	19	66.2	55	131.0	91	195.8
−16	3.2	20	68.0	56	132.8	92	197.6
−15	5.0	21	69.8	57	134.6	93	199.4
−14	6.8	22	71.6	58	136.4	94	201.2
−13	8.6	23	73.4	59	138.2	95	203.0
−12	10.4	24	75.2	60	140.0	96	204.8
−11	12.2	25	77.0	61	141.8	97	206.6
−10	14.0	26	78.8	62	143.6	98	208.4
−9	15.8	27	80.6	63	145.4	99	210.2
−8	17.6	28	82.4	64	147.2	100	212.0
−7	19.4	29	84.2	65	149.0	101	213.8
−6	21.2	30	86.0	66	150.8	102	215.6
−5	23.0	31	87.8	67	152.6	103	217.4

[a] $°C = 5/9 \times (°F - 32)$; $°F = (\frac{9}{5} \times °C) + 32$.

APPENDIX VI. NOMOGRAM FOR COMPUTING RELATIVE CENTRIFUGAL FORCES (RCF)

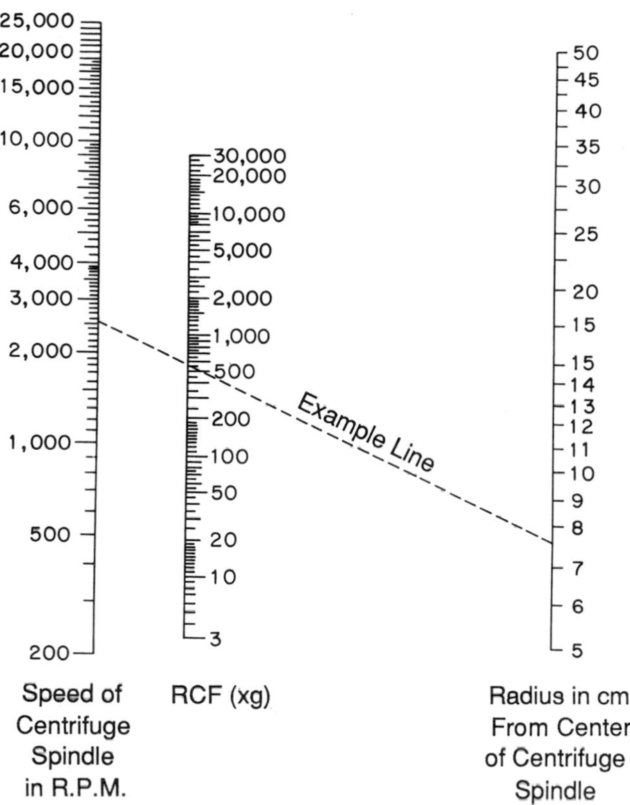

FIGURE VI.1 To calculate the RCF value at any point along the tube, measure the radius in cm from the center of the centrifuge spindle to the point. Draw a line from this radius value on the right hand column to the centrifuge speed on the left hand column. The RCF value is the point where the line intersects the center column.

APPENDIX VII. CONVERSIONS OF BODY WEIGHT TO SQUARE METERS OF BODY SURFACE AREA FOR DOGS AND CATS[a]

kg	m^2	kg	m^2
0.5	0.06		
1.0	0.10	26	0.89
2.0	0.16	27	0.91
3.0	0.21	28	0.93
4.0	0.26	29	0.95
5.0	0.30	30	0.98
6.0	0.33	31	1.00
7.0	0.37	32	1.02
8.0	0.40	33	1.04
9.0	0.44	34	1.06
10.0	0.47	35	1.08
11.0	0.50	36	1.10
12.0	0.53	37	1.12
13.0	0.56	38	1.14
14.0	0.59	39	1.16
15.0	0.62	40	1.18
16.0	0.64	41	1.20
17.0	0.67	42	1.22
18.0	0.69	43	1.24
19.0	0.72	44	1.26
20.0	0.74	45	1.28
21.0	0.77	46	1.30
22.0	0.79	47	1.32
23.0	0.82	48	1.33
24.0	0.84	49	1.35
25.0	0.86	50	1.37

[a] Calculated from

$$A = kW^{0.667} \times 10^{-4},$$

where A = body surface area in square meters (m^2); k = constant 10.1 for dogs; W = body weight in grams. For cats, the constant k is 10.0, so the conversion to body surface area closely approximates that for the dog and may be used interchangeably.

APPENDIX VIII. BLOOD ANALYTE REFERENCE VALUES IN LARGE ANIMALS[a]

Analyte[b]	Unit	Horse	Cow	Sheep	Goat	Llama	Pig
Acetylcholinesterase (AcChE): R	U/liter	450–790	1270–2430	640	270		930
Alanine aminotransferase (ALT, GPT): S, HP	U/liter	3–23	11–40		6–19	6–14	31–58
		(14 ± 11)	(27 ± 14)	(30 ± 4)		(8.8 ± 2.6)	(45 ± 14)
Aldolase (ALD): S, HP	U/liter						
Ammonia (NH_4): S, HP	μmol/liter	7.63–63.4					
		(35.8 ± 17.0)					
	μg/dl	13–108					
		(61 ± 29)					
Amylase (Amyl): S, HP	U/liter	75–150					
Arginase (ARG): S, HP	U/liter	0–14	1–30	0–14			0–14
		(11 ± 18)	(8.3 ± 6)	(5 ± 1)			
Aspartate aminotransferase (AST, GOT): S, HP	U/liter	226–366	78–132	60–280	167–513	216–378	32–84
		(296 ± 70)	(105 ± 27)	(307 ± 43)		(292 ± 50)	(61 ± 26)
Bicarbonate (HCO_3): S, P	mmol/liter	20–28	17–29	20–25			18–27
Bile acids, total (TBA): S	μmol/liter	5–28	20–80				
Bilirubin: S, P, HP							
Conjugated (CB)	μmol/liter	0–6.84	0.68–7.52	0–4.61			0–5.13
		(1.71)	(3.08)	(2.05)			(1.71 ± 1.71)
	mg/dl	0–0.4	0.04–0.44	0–0.27			0–0.3
		(0.1)	(0.18)	(0.12)			(0.1 ± 0.1)
Unconjugated (UCB)	μmol/liter	3.42–34.2	0.51	0–2.05			0–5.13
				(17.1)			
	mg/dl	0.2–2.0	0.03	0–0.12			0–0.3
		(1.0)					
Total (TB)	μmol/liter	7.1–34.2	0.17–8.55	1.71–8.55	0–1.71	0–17.1	0–17.1
		(17.1)	(3.42)	(3.93 ± 1.71)		(3.42 ± 3.42)	(3.42 ± 3.42)
	mg/dl	1–2.0	0.01–0.5	0.1–0.5	0–0.1	0–1.0	0–10
		(1.0)	(0.2)	(0.23 ± 0.1)		(0.2 ± 0.2)	(0.2 ± 0.2)
Butyrylcholinesterase (ButChE): P	U/liter	2000–3100	70	0–70	110		400–430
Calcium (Ca): S, HP	mmol/liter	2.80–3.40	2.43–3.10	2.88–3.20	2.23–2.93	2.20–2.58	1.78–2.90
		(3.10 ± 0.14)	(2.78 ± 0.15)	(3.04 ± 0.07)	(2.58 ± 0.18)	(2.30 ± 0.23)	(2.41 ± 0.25)
	mg/dl	11.2–13.6	9.7–12.4	11.5–12.8	8.9–11.7	8.0–10.3	7.1–11.6
		(12.4 ± 0.58)	(11.08 ± 0.67)	(12.16 ± 0.28)	(10.3 ± 0.7)	(9.2 ± 0.9)	(9.65 ± 0.99)
Carbon dioxide, pressure (pCO_2): S, P	mm Hg	38–46	35–44				
		(42.4 ± 2.0)		(413 ± 4.7)			
Carbon dioxide, total (TCO_2): S, P	mmol/liter	24–32	21.2–32.2	21–28	25.6–29.6		
		(28)	(26.5)	(26.2)	(27.4 ± 1.4)		
Chloride (Cl): S, HP	mmol/liter	99–109	97–111	95–103	99–110.3	102–109	94–106
		(104 ± 2.6)	(104)		(105.1 ± 2.9)	(105 ± 2)	
Cholesterol (Chol): S, P, HP							
Ester	mmol/liter		1.50–2.28				
			(1.89 ± 0.39)				
	mg/dl		58–88				28–48
			(81.1)	(73 ± 15)			

Free	mmol/liter		0.57–1.35				0.72–1.24
		(0.41)	(0.96 ± 0.39)	(1.66 ± 0.31)			
	mg/dl		22–52				5.7–10.9
		(15.7)	(37 ± 15)				
Total	mmol/liter	1.94–3.89	2.07–3.11	1.35–1.97	2.07–3.37	0.91–2.93	0.93–1.40
		(2.88 ± 0.47)		(1.66–0.31)	(1.55 ± 0.67)		
	mg/dl	75–150	80–120	52–76	80–130	35–113	36–54
		(111 ± 18)		(64 ± 12)		(60 ± 26)	
Copper (Cu): S	μmol/liter		5.16–5.54	9.13–25.2			20.9–43.8
							(32.4)
	μg/dl		32.8–35.2	58–160			133–278
Coproporphyrin (COPRO): HB, HP, R	μmol/liter		trace				
	μg/dl		trace				
Cortisol (Cort–RIA): S, HP	nmol/liter	36–81					
			(17 ± 2)	(62 ± 10)	(65 ± 8)		(82 ± 3)
	μg/dl	1.30–2.93					
			(0.61 ± 0.07)	(2.24 ± 0.36)	(2.35 ± 0.29)		(2.97 ± 0.10)
Creatine kinase (CK): S, HP	U/liter	2.4–23.4	4.8–12.1	8.1–12.9	0.8–8.9	17–101	2.4–22.5
		(12.9 ± 5.2)	(7.4 ± 2.4)	(10.3 ± 1.6)	(4.5 ± 2.8)	(40.8 ± 29.9)	(8.9 ± 6.0)
Creatinine (Creat): S, P, HP	μmol/liter	106–168	88.4–177	106–168	88.4–159	97.2–221	141–239
						(150 ± 35.4)	(141 ± 5.3)
Creatinine (Creat): S, P, HP	mg/dl	1.2–1.9	1.0–2.0	1.2–1.9	1.0–1.8	1.1–2.5	1.0–2.7
						(1.7 ± 0.4)	(1.6 ± 0.06)
Fatty acid, free (FFA): HP	mg/liter		30–100				
Fibrinogen (FIbr): P, HP	μmol/liter	2.94–11.8	8.82–20.6	2.94–14.7	2.94–11.8		2.94–14.7
		(7.65 ± 2.35)					
	g/liter	1.0–4.0	3.0–7.0	1.0–5.0	1.0–4.0		1.0–5.0
		(2.6 ± 0.8)					
	mg/dl	100–400	300–700	100–500	100–400		100–500
		(260 ± 80)					
Glucose (GLU): S, P, HP	mmol/liter	4.16–6.39	2.50–4.16	2.78–4.44	2.78–4.16	5.72–8.89	4.72–8.33
		(5.30 ± 0.47)	(3.19 ± 0.38)	(3.80 ± 0.33)	(3.49 ± 0.39)	(7.10 ± 0.89)	(6.61 ± 0.96)
	mg/dl	75–115	45–75	50–80	50–75	103–160	85–150
		(95.6 ± 8.5)	(57.4 ± 6.8)	(68.4 ± 6.0)	(62.8 ± 7.1)	(128 ± 16)	(119 ± 17)
Glutamate dehydrogenase (GD): S, HP	U/liter	0–11.8	31	20			0
		(5.6 ± 4.2)					
Glutamic oxaloacetic transaminase (GOT): see AST							
Glutamic pyruvate transaminase (GPT): see ALT							
γ-Glutamyltransferase (GGT): S, P	U/liter	4.3–13.4	6.1–17.4	20–52	20–56	7–29	10–60
		(7.6 ± 1.5)	(15.7 ± 4.0)	(33.5 ± 4.3)	(38 ± 13)	(15.8 ± 6.4)	(35 ± 21)
Glutathione (GSH): B	mmol/liter		2.47–3.67				
			(2.89 ± 0.46)				
Glutathione (GSH): B	mg/dl		76–113				
			(89 ± 14)				
Glutathione peroxidase (GP_x): H, B	U/100 g Hb	(7931 ± 1620)					
Glutathione reductase (GR): H, B	U/100 g Hb	(33.3 ± 10.5)	(19.5 ± 3.9)	(34.3 ± 7.5)	(98 ± 16)		(68.2 ± 9.2)
Haptoglobin (Hp): S, HP	g/liter						
Hemoglobin (Hb): B	g/liter	110–190	80–150	90–140	80–120	132–205	100–160
		(144 ± 17)	(110)	(115)	(100)	(173)	(130)
Icterus index (II): P, HP	unit	5–20	5–15	2–5	2–5		2–5

(continues)

Analyte[b]	Unit	Horse	Cow	Sheep	Goat	Llama	Pig
Iditol dehydrogenase (ID): see SDH							
Insulin (Ins): S, HP	pmol/liter		0–35.9				
	μU/ml		0–5				
Iodine, total (I): S	nmol/liter	394–946					
	μg/dl	5–12					
Iron (Fe): S	μmol/liter	13.1–25.1	10.2–29.0	29.7–39.7			16.3–35.6
		(19.9 ± 1.97)	(17.4 ± 5.19)	(34.5 ± 1.25)			(21.7 ± 5.91)
	μg/dl	73–140	57–162	166–222			91–199
		(111 ± 11)	(97 ± 29)	(193 ± 7)			(121 ± 33)
Iron binding capacity, total (TIBC): S	μmol/liter	(59.1 ± 5.7)	(41.2 ± 11.6)				(74.6 ± 12.9)
	μg/dl	(330 ± 32)	(230 ± 65)				(417 ± 72)
Iron binding capacity, unbound (UIBC): S	μmol/liter	35.8–46.9	11.3–33.3				
		(39.0 ± 3.8)	(23.5 ± 6.4)				
	μg/dl	200–262	63–186				100–262
		(218 ± 21)	(131 ± 36)				(196 ± 39)
Isocitrate dehydrogenase (ICD): S, HP	U/liter	4.8–18.0	9.4–21.9	0.4–8.0			
		(10.0 ± 3.3)	(16.7 ± 2.8)	(4.7 ± 2.8)			
Ketones (Ket): HP							
Acetoacetate (AcAc):	mmol/liter		0–0.11				
		(0.029 ± 0.003)	(0.043 ± 0)	(0.030 ± 0.002)			
	mg/dl		0–1.1				
		(0.30 ± 0.03)	(0.5)	(0.30 ± 0.02)			
Acetone (Ac):	mmol/liter		0–1.72	0–1.72			
	mg/dl	0–10	0–10				
β-Hydroxybutyrate (β-OHB) or 3-HydroxyButyrate (3-OHB):	mmol/liter	(0.064 ± 0.006)	(0.41 ± 0.03)	(0.55 ± 0.04)			
	mg/dl	(0.67 ± 0.06)	(9.90 ± 1.88)	(5.73 ± 0.42)			
Lactate (Lac): B	mmol/liter	1.11–1.78	0.56–2.22	1.00–1.33			
	mg/dl	10–16	5–20	9–12			
Lactate dehydrogenase (LDH): S, HP	U/liter	162–412	692–1445	238–440	123–392	88–487	380–634
		(252 ± 63)	(1061 ± 222)	(352 ± 59)	(281 ± 71)	(320 ± 116)	(499 ± 75)
LDH isoenzymes: S, P							
LDH-1 (heart, anodal)	%	6.3–18.5	39.8–63.5	45.7–63.6	29.3–51.8		34.1–61.8
		(11.5 ± 4.0)	(49.0 ± 5.4)	(54.3 ± 6.5)	(41.0 ± 8.0)		(50.8 ± 10.1)
LDH-2	%	8.4–20.5	19.7–34.8	0–3.0	0–5.4		5.9–9.2
		(14.8 ± 3.2)	(27.8 ± 3.4)	(0.8 ± 1.2)	(2.4–1.8)		(7.3–1.2)
LDH-3	%	41.0–65.9	11.7–18.1	16.4–29.9	24.4–39.9		5.7–11.7
		(50.2 ± 7.2)	(14.5 ± 1.9)	(23.3 ± 4.0)	(31.2 ± 6.2)		(7.4 ± 1.9)
LDH-4	%	9.5–20.9	0–8.8	4.3–7.3	0–5.5		6.9–15.9
		(16.2 ± 3.8)	(4.4 ± 2.4)	(5.3 ± 1.0)	(2.5 ± 2.5)		(10.9 ± 3.1)
LDH-5 (liver, muscle, cathodal)	%	1.7–16.5	0–12.4	10.5–29.1	14.1–36.8		16.3–35.2
		(7.3 ± 4)	(4.3 ± 3.4)	(16.3 ± 6.2)	(20.9 ± 9.4)		(23.6 ± 6.5)
Lead (Pb): HB	μmol/liter	0.24–1.21	0–1.16	0.24–1.21	0.24–1.21		
			(0.48 ± 0.29)				
	μg/dl	5–25	0–24	5–25	5–25		
			(10 ± 6)				

Lipase (Lip): S	U/liter						
Malate dehydrogenase (MD): S, HP	U/liter						
Magnesium (Mg): S	mmol/liter	0.90–1.15 (1.03 ± 0.13)	0.74–0.95 (0.84 ± 0.10)	0.90–0.31 (1.03 ± 0.12)	0.31–1.48 (1.32 ± 0.14)	0.75–1.55 (0.95 ± 0.10)	1.11–1.52 (1.31 ± 0.20)
	mg/dl	2.2–2.8 (2.5 ± 0.31)	1.8–2.3 (2.05 ± 0.25)	2.2–2.8 (2.5 ± 0.3)	2.8–3.6 (3.2 ± 0.35)	1.82–3.77 (2.31 ± 0.24)	2.7–3.7 (3.2 ± 0.49)
Manganese (Mn): S	μmol/liter						
	μg/dl						
Ornithine carbamyl transferase (OCT): S, HP	U/liter	(3.3 ± 4.2)	(4.7 ± 0.3)				
Oxygen, pressure (pO_2): HB	mm Hg						
pH: HB	unit	7.32 ± 7.44 (7.38 ± 0.03)	7.31–7.53 (7.38)	7.32–7.54 (7.44)			
Phosphatase, acid (AcP): S, HP	U/liter						
Phosphatase, alkaline (AlP): S, HP	U/liter	143–395 (244 ± 101)	0–488 (194 ± 126)	68–387 (178 ± 102)	93–387 (219 ± 76)	41–92 (63 ± 17)	118–395 (194 ± 84)
Phosphate, Inorg (Pi): S, HP	mmol/liter	1.00–1.81	1.81–2.10	1.62–2.36 (2.07 ± 0.06)	(4.62 ± 0.25)	1.00–3.49 (2.06 ± 0.87)	1.71–3.10
	mg/dl	3.1–5.6	5.6–6.5	5.0–7.3 (6.4 ± 0.2)	4.2–9.1 (6.5)	3.1–10.8 (6.4 ± 2.7)	5.3–9.6
Potassium (K): S, HP	mmol/liter	2.4–4.7 (3.51 ± 0.57)	3.9–5.8 (4.8)	3.9–5.4	3.5–6.7 (4.3 ± 0.5)	4.6–7.1 (5.6 ± 0.8)	4.4–6.7
Potassium (K): R	mmol/liter	(88)	10–45 (24 ± 7.0)	(64 or 18)			(100)
Protein (Prot): S							
Total (TP)	g/liter	52.0–79.0 (63.5 ± 5.9)	67.4–74.6 (71.0 ± 1.8)	60.0–79.0 (72.0 ± 5.2)	64.0–70.0 (69.0 ± 4.8)	58.0–75.0 (64.9 ± 4.9)	79.0–89.0 (84.0 ± 5.0)
Protein (Prot): S							
Electrophoresis (SPE), cellulose acetate (CA):							
Albumin	g/liter	26.0–37.0 (30.9 ± 2.8)	30.3–35.5 (32.9 ± 1.3)	24.0–30.0 (27.0 ± 1.9)	27.0–39.0 (33.0 ± 3.3)	36.0–48.0 (42.5 ± 3.9)	19.0–39.0 (25.9 ± 7.1)
Globulin, total	g/liter	26.2–40.4 (33.3 ± 7.1)	30.0–34.8 (32.4 ± 2.4)	35.0–57.0 (44.0 ± 5.3)	27.0–41.0 (36.0 ± 5.0)	16.0–29.0 (22.4 ± 3.9)	52.9–64.3 (58.6 ± 5.7)
α	g/liter		7.5–8.8 (7.9 ± 0.2)	3.0–6.0 (5.0 ± 1.0)	5.0–7.0 (6.0 ± 0.6)	6.0–9.0 (7.7 ± 1.3)	
α_1	g/liter	0.6–7.0 (1.9 ± 2.6)					3.2–4.4 (3.8 ± 0.6)
α_2	g/liter	3.1–13.1 (6.5 ± 1.3)					12.8–15.4 (14.1 ± 1.3)
β	g/liter		8.0–11.2 (9.6 ± 0.8)			10.0–11.0 (10.3 ± 0.5)	
β_1	g/liter	4.0–15.8 (9.2 ± 3.0)		7.0–12.0 10.0 ± 1.4	7.0–12.0 (9.0 ± 1.0)		1.3–3.3 (2.3 ± 1.0)
β_2	g/liter	2.9–8.9 (5.7 ± 1.1)		4.0–14.0 (7.0 ± 2.6)	3.0–6.0 (4.0 ± 0.2)		12.6–16.8 (14.7 ± 2.1)
γ	g/liter	5.5–19.0 (10.0 ± 1.4)	16.9–22.5 (19.7 ± 1.4)		9.0–30.0 (17.0 ± 4.4)	5.0–10.0 (7.0 ± 2.2)	22.4–24.6 (23.5 ± 1.1)
γ_1	g/liter			7.0–22.0 (16.0 ± 4.1)			
γ_2	g/liter			2.0–11.0 (8.0 ± 3.0)			

(continues)

APPENDIX VIII. *(continued)*

Analyte[b]	Unit	Horse	Cow	Sheep	Goat	Llama	Pig
A/G Ratio	—	6.2–14.6	8.4–9.4	4.2–7.6	6.3–12.6	1.31–3.86	3.7–5.1
		(9.6 ± 1.7)	(8.9 ± 0.5)	(6.3 ± 0.9)	(9.5 ± 1.7)	(1.96 ± 0.45)	(4.4 ± 0.7)
Protoporphyrin (PROTO): R	μmol/liter		trace				(2.1)
	μg/dl		trace				118
Pseudocholinesterase (PsChE); see ButChE							
Pyruvate (PYR): R	μmol/liter		(54.0 ± 24.0)				
Sodium (Na): S, HP	mmol/liter	132–146	132–152	139–152	142–155	148–155	135–150
		(139 ± 3.5)	(142)		(150 ± 3.1)	(152 ± 1.9)	
Sorbitol dehydrogenase (SDH): S, HP	U/liter	1.9–5.8	4.3–15.3	5.8–27.9	14.0–23.6	1–17	1.0–5.8
		(3.3 ± 1.3)	(9.2 ± 3.1)	(15.7 ± 7.5)	(19.4 ± 3.6)	(4.9 ± 6.2)	(2.6 ± 1.6)
Thyroxine (T_4-RIA): S	nmol/liter	11.6–36.0	54.0–110.7			131.6–286.4	
		(0.024 ± 0.004)	(82.4)			(185.8 ± 50.3)	
	μg/dl	0.9–2.8	4.2–8.6			10.2–22.2	
		(1.55 ± 0.27)	(6.4)			(14.4 ± 3.9)	
Thyroxine, free (FT_4): S	pmol/liter						
	pg/dl						
Triglyceride, total (TG): S	mmol/liter	0.1–0.5	0–0.2				
	mg/dl	4–44	0–14				
Triiodothyronine (T_3-RIA)	nmol/liter					1.35–4.06	
		(0.85 ± 0.52)				(2.27 ± 0.94)	
	ng/dl					88–264	
		(55.34 ± 33.9)					
Urate (UA): S, P, HP							
	mmol/liter	53.5–65.4	0–119.0	0–113.0	17.8–59.5		
	mg/dl	0.9–1.1	0–2	0–1.9	0.3–1		
Urea (UR): S, P, HP	mmol/liter	3.57–8.57	7.14–10.7	2.86–7.14	3.57–7.14	4.28–12.14	3.57–10.7
					(5.36 ± 0.71)	(9.71 ± 2.61)	
Urea nitrogen (UN): S, P, HP	mg/dl	10–24	20–30	8–20	10–20	12–34	10–30
					(15 ± 2.0)	(27.2 ± 7.3)	
Vitamin A (Vit A):							
Carotene: S	μmol/liter	0.37–3.26	0.47–17.7	0–0.37			
		(1.86)	(0.74)	(18.8)			
	μg/dl	20–175	25–950	0–20			
		(100)	(40)	(10)			
Carotenol: S	μmol/liter	0.17–0.30	0.19–0.56	0.37–0.84			0.19–0.65
		(0.22)	(0.45)				(0.37)
	μg/dl	9–16	10–30	20–45			10–35
		(12)	(24)				(20)

[a] Ranges, with means and standard deviations in parentheses.
[b] B, blood; HB, heparinized blood; HP, heparinized plasma; P, plasma; S, serum; R, erythrocytes.

APPENDIX IX. BLOOD ANALYTE REFERENCE VALUES IN SMALL AND SOME LABORATORY ANIMALS[a]

Analyte[b]	Unit	Dog	Cat	Rat	Mouse	Rabbit	Monkey
Acetylcholinesterase (AcChE): R	U/liter	270	540				
Alanine aminotransferase (ALT, GPT): S, HP	U/liter	21–102	6–83				0–82
		(47 ± 26)	(26 ± 16)	(35.1 ± 13.3)	(19.0)	(79.0)	(27 ± 28)
Aldolase (ALD): S, HP	U/liter						
Ammonia (NH_4): S, HP	μmol/liter	11.2–70.4					
			(31.1 ± 14.7)				
	μg/dl	19–120					
		(53 ± 25)					
Amylase (Amyl): S, HP	U/liter	185–700					
Arginase (ARG): S, HP	U/liter	0–14	0–14				
				(21.3)			
Aspartate aminotransferase (AST, GOT): S, HP	U/liter	23–66	26–43				13–37
		(33 ± 12)	(35 ± 9)	(42.9 ± 10.1)	(37.0)	(47.0)	(22 ± 8)
Bicarbonate (HCO_3): S, P	mmol/liter	18–24	17–21				
Bilirubin: S, P, HP							
Conjugated (CB)	μmol/liter	1.03–2.05					0.68–5.98
							(0.68 ± 0.68)
							0.04–0.35
	mg/dl	0.06–0.12					(0.04 ± 0.04)
Total	μmol/liter	1.71–8.55	2.57–8.55				1.71–8.55
		(3.42 ± 1.71)		(5.13 ± 2.39)	(6.84 ± 8.55)	(6.84 ± 8.55)	(4.28 ± 0.86)
	mg/dl	0.10–0.50	0.15–0.50				0.10–0.50
		(0.20 ± 0.10)		(0.30 ± 0.14)	(0.4 ± 0.5)	(0.40 ± 0.50)	(0.25 ± 0.05)
Unconjugated (UCB)	μmol/liter	0.17–8.38					0–3.76
							(3.42 ± 3.08)
	mg/dl	0.01–0.49					0–0.22
		(0.20 ± 0.18)					
Bile acid, total (TBA): S	μmol/liter	0–5.0	0–5.0				5.0–14.0
		(2.60 ± 0.40)	(1.70 ± 0.30)				(10)
Butyrylcholinesterase (ButChE): P	U/liter	1210–3020	640–1400				523–1711
							(589 ± 260)
Calcium (Ca): S, HP	mmol/liter	2.25–2.83	1.55–2.55	1.50–2.65	1.20–1.86	1.46–3.60	2.28–2.95
		(2.55 ± 0.15)	(2.06–0.24)	(2.00 ± 0.32)	(1.39 ± 0.20)	(2.50 ± 0.56)	(2.55 ± 1.50)
	mg/dl	9.0–11.3	6.2–10.2	6.00–10.6	4.80–7.44	5.84–14.4	9.1–11.8
		(10.2 ± 0.60)	(8.22 ± 0.97)	(8.00 ± 1.28)	(5.56 ± 0.80)	(10.0 ± 2.24)	(10.2 ± 6.0)
Carbon dioxide, pressure (pCO_2): S, P	mm Hg	(38)	(36)				
Carbon dioxide, total (tCO_2): S, P	mmol/liter	17–24	17–24				9.6–25.9
		(21.4)	(20.4)				(18.6 ± 4.0)
Chloride (Cl): S, HP	mmol/liter	105–115	117–123	79.4–111.3	95.6–128.9	85.0–105.3	97.5–113.5
				(96.8 ± 6.4)	(107.6 ± 6.7)	(96.5 ± 6.8)	(105 ± 4.0)
Cholesterol (Chol): S, P, HP							
Ester	mmol/liter	1.04–2.02	1.04–2.23				
		(1.53 ± 0.49)	(1.63 ± 0.60)				
	mg/dl	40–78	40–86				
		(59 ± 19)	(63 ± 23)				

(continues)

Analyte[b]	Unit	Dog	Cat	Rat	Mouse	Rabbit	Monkey
Free	mmol/liter	0.80–1.84	0.52–1.04				0.19–1.08
		(1.32 ± 0.52)	(0.78 ± 0.26)				(0.57 ± 0.26)
	mg/dl	31–71	20–40				7.4–41.7
		(51 ± 20)	(30 ± 10)				(22 ± 10)
Total	mmol/liter	3.50–6.99	2.46–3.37	0.13–1.41	0.74–2.86	0.14–1.86	2.51–4.82
		(4.61 ± 0.98)		(0.73 ± 0.35)	(1.61 ± 0.43)	(0.69 ± 0.41)	(3.81 ± 0.88)
	mg/dl	135–270	95–130	5.1–54.2	28.6–110.4	5.3–71.0	97–186
		(178 ± 38)		(28.3 ± 13.7)	(62.1 ± 16.7)	(26.7 ± 15.9)	(147 ± 34)
Copper (Cu): S	μmol/liter	15.7–31.5					
	μg/dl	100–200					
Cortisol (Cort-RIA): S, HP	nmol/liter	27–188	9–71				(850 ± 224)
	μg/dl	0.96–6.81	0.33–2.57				(30.8 ± 8.1)
Creatine kinase (CK): S, HP	U/liter	1.15–28.40	7.2–28.2				
		(6.25 ± 2.06)	(19.5 ± 6.7)	(183)	(155)	(544)	(125)
Creatinine (Creat): S, P, HP	μmol/liter	44.2–132.6	70.7–159	35.4–331.5	44.2–123.8	70.7–227.2	70.7–205.0
				(140.6 ± 69.8)	(74.2 ± 16.8)	(140.6 ± 30.1)	(124.6 ± 27.4)
	mg/dl	0.5–1.5	0.8–1.8	0.40–3.75	0.5–1.4	0.8–2.57	0.8–2.32
				(1.59 ± 0.79)	(0.84 ± 0.19)	(1.59 ± 0.34)	(1.41 ± 0.31)
Fatty acid, free (FFA): HP	mg/dl						
Fibrinogen (Fibr): P, HP	μmol/liter	5.88–11.8	1.47–8.82				
	g/liter	2.0–4.0	0.5–3.0				
	mg/dl	200–400	50–300				
Fructosamine (FrAm): S, P, HP							
(Sigma method)	mmol/liter	1.70–3.38	21.9–3.47				
		(2.54 ± 0.42)	(2.83 ± 0.32)				
(BMC method)	μmol/liter	170–338	219–347				
		(182.30 ± 30.14)	(203 ± 22.96)				
Glucose (Glu): S, P, HP	mmol/liter	3.61–6.55	3.89–6.11	2.65–5.94	1.74–11.11	2.78–5.18	4.72–7.27
		(5.05 ± 0.67)	(5.05 ± 0.42)	(4.07 ± 1.01)	(5.12 ± 2.49)	(4.08 ± 0.53)	(5.94 ± 0.72)
	mg/dl	65–118	73–134	47.7–107.0	31.4–200	50.0–93.2	85–131
		(91 ± 12)	(91 ± 7.5)	(73.3 ± 18.2)	(92.2 ± 44.9)	(73.4 ± 9.5)	(107 ± 12.9)
Glutamate dehydrogenase (GD): S, HP	U/liter	(3)		(4)	(9)	(16)	(40)
Glutamic oxaloacetic transaminase (GOT): see AST							
Glutamic pyruvate transaminase (GPT): see ALT							
γ-Glutamyltransferase (GGT): S, P	U/liter	1.2–6.4	1.3–5.1				
		(3.5 ± 1.8)				(9)	(62)
Glutathione (GSH): R	mmol/liter	(2.07 ± 0.36)	(1.97 ± 0.19)				
Glutathione peroxidase (GP_x): HB	U/100 gHb	(8921 ± 237)	(12135 ± 616)				
Glutathione reductase (GR): HB	U/100 gHb	(137 ± 7.0)	(405 ± 48)				
Haptoglobin (Hp): S, HP	g/liter						
Hemoglobin (Hb): B	g/liter	12–18	8–14				
Hemoglobin A1c (HbA1c): HB	%	2.3–6.4					
Icterus index (II): P, HP	unit	2–5	2–5				
Iditol dehydrogenase (ID): see SDH							

Insulin (Ins): S, HP	pmol/liter	36–144	0–129				
		(86.1 ± 35.9)					
	μU/ml	5–20	0–18				
		(12 ± 5)					
Iodine, total (I): S	nmol/liter	394–1576					
		(473 ± 276)					
	μg/dl	5–20					
		(6.0 ± 3.5)					
Iron (Fe): S	μmol/liter	5.37–32.2	12.2–38.5				
		(15.5 ± 5.5)	(25.1)	(39.4 ± 22)	(60.1 ± 2.2)	(36.5 ± 33.4)	
	μg/dl	30–180	68–215				
		(86.4 ± 30.8)	(140)	(220 ± 124)	(336 ± 12)	(204 ± 19)	
Iron binding capacity, total (TIBC): S	μmol/liter	29.5–74.9					
		(57.7 ± 7.9)	(51.9)	(65.9)		(48.4)	(79.7)
	μg/dl	165–418					
		(322 ± 44)	(290)	(368)		(270)	(445)
Iron binding capacity, unbound (UIBC): S	μmol/liter	30.4–39.7	18.8–36.7				
		(35.8)	(26.9)				
	μg/dl	170–222	105–205				
		(200)	(150)				
Isocitrate dehydrogenase (ICD): S, HP	U/liter	0.4–7.3	2.0–11.7				
Ketones (Ket): HP		(3.0 ± 1.7)	(5.3 ± 3.2)	(4)	(32)	(137)	(28)
Acetoacetic acid (AcAc)	mmol/liter	(0.018 ± 0.018)					
	mg/dl	(0.18 ± 0.18)					
3-Hydroxybutyric acid (3-OHB)	mmol/liter	(0.030 ± 0.006)					
	mg/dl	(0.30 ± 0.06)					
Lactate (Lac): B	mol/liter	0.22–1.44					
	mg/dl	2–13					
Lactate dehyrogenase (LDH): S, HP	U/liter	45–233	63–273				173–275
		(93 ± 50)	(137 ± 59)	(46.6 ± 22.0)	(366)	(94.3 ± 28.8)	(232 ± 31)
LDH isoenzymes: S, HP							
LDH-1 (heart, anodal)	%	1.7–30.2	0–8.0				2.7–38.2
		(13.9 ± 9.5)	(4.5 ± 2.8)				(17.2 ± 8.4)
LDH-2	%	1.2–11.7	3.3–13.7				4.3–39.7
		(5.5 ± 4.2)	(6.1 ± 3.4)				(19.8 ± 9.4)
LDH-3	%	10.9–25.0	10.2–20.4				12.8–50.4
		(17.1 ± 5.7)	(13.3 ± 3.4)				(24.5 ± 7.2)
LDH-4	%	11.9–15.4	11.6–35.9				0.8–38.0
		(13.0 ± 1.2)	(23.6 ± 8.6)				(17.7 ± 10.6)
LDH-5 (liver, muscle, cathodal)	%	30.0–72.8	40.0–66.3				4.7–36.3
		(50.5 ± 16.9)	(52.5 ± 9.3)				(18.6 ± 8.3)
Lead (Pb): HB	μmol/liter	0–2.42					
	μg/dl	0–50					
Leucine aminopeptidase (LAP): S, HP	U/liter	(13)		(25)	(25)	(46)	(9)
Lipase (Lip): S	U/liter	13–200	0–83				
Magnesium (Mg): S	mmol/liter	0.74–0.99					
		(0.86 ± 0.12)	(0.90)	(1.28 ± 0.17)	(1.28 ± 0.15)	(0.92 ± 0.07)	(0.68 ± 0.13)
	mg/dl	1.8–2.4					
		(2.1 ± 0.3)	(2.2)	(3.12 ± 0.41)	(3.11 ± 0.37)	(2.25 ± 0.16)	(1.65 ± 0.32)

(*continues*)

APPENDIX IX. *(continued)*

Analyte[b]	Unit	Dog	Cat	Rat	Mouse	Rabbit	Monkey
Malate dehydrogenase (MD): S, HP	U/liter	(199)	(132)	(118)	(419)	(1000)	(109)
Ornithine carbamyltransferase (OCT): S, HP	U/liter	(2.7 ± 0.7)	(3.8 ± 1.0)				
Oxygen, pressure (pO_2): HB pH: HB	mm Hg	85–100	78–100				
pH: HB	unit	7.31–7.42	7.24–7.40				
		(7.36)	(7.35)				
Phosphatase, acid (AcP): S, HP	U/liter	5–25	0.5–24				
Phosphatase, alkaline (AlP): S, HP	U/liter	20–156	25–93				100–277
		(66 ± 36)	(50 ± 35)	(133 ± 134)	(66 ± 19)	120 ± 13.8)	(171 ± 55)
Phosphate (Pi): S, HP	mmol/liter	0.84–2.00	1.45–2.62				1.42–1.78
		(1.39 ± 0.29)	(2.00)	(2.29 ± 0.38)	(2.12 ± 0.42)	(1.34 ± 0.15)	(1.62 ± 0.13)
	mg/dl	2.6–6.2	4.5–8.1				4.4–5.5
		(4.3 ± 0.9)	(6.2)	(7.08 ± 1.19)	(6.55 ± 1.30)	(4.16 ± 0.46)	(5.0 ± 0.4)
Potassium (K): S, HP	mmol/liter	4.37–5.35	4.0–4.5				3.5–6.5
		(4.90)	(4.3)	(6.50 ± 1.33)	(5.40 ± 0.15)	(5.3 ± 0.5)	(4.7 ± 0.6)
Protein (Prot): S							
Total (TP)	g/liter	54.0–71.0	54.0–78.0				78.0–96.0
		(61.0 ± 5.2)	(66.0 ± 7.0)	(75.2 ± 2.7)	(62.0 ± 2.0)	(64.5 ± 3.1)	(87.2 ± 7.3)
Electrophoresis (SPE), cellulose acetate (CA):							
Albumin	g/liter	26.0–33.0	21.0–33.0				31.3–53.0
		(29.1 ± 1.9)	(27.0 ± 1.7)	(41.7 ± 2.1)	(34.0 ± 1.0)	(27.3 ± 3.0)	(42.1 ± 2.0)
Globulin, total	g/liter	27.0–44.0	26.0–51.0				30.5–52.2
		(34.0 ± 5.1)	(39.0 ± 6.9)				(41.4–2.0)
α	g/liter						
α_1	g/liter	2.0–5.0	2.0–11.0				1.0–4.9
		(3.0 ± 0.3)	(7.0 ± 0.2)				(2.7 ± 0.3)
α_2	g/liter	3.0–11.0	4.0–9.0				2.5–8.0
		(6.0 ± 2.1)	(7.0 ± 0.2)				(4.7 ± 0.5)
β	g/liter						9.6–27.2
							(18.9 ± 1.7)
β_1	g/liter	7.0–13.0	3.0–9.0				
		(8.2 ± 2.3)	(7.0 ± 0.3)				
β_2	g/liter	6.0–14.0	6.0–10.0				
		(8.9 ± 3.3)	(7.0 ± 0.2)				
γ	g/liter						7.3–28.4
γ_1	g/liter	5.0–13.0	3.0–25.0				
		(8.0 ± 2.5)	(16.0 ± 7.7)				
γ_2	g/liter	4.0–9.0	14.0–19.0				
		(7.0 ± 1.4)	(17.0 ± 3.6)				
A/G ratio	—	0.59–1.11	0.45–1.19				0.72–1.21
		(0.83 ± 0.16)	(0.71 ± 0.20)	(0.59)	(0.62)	(0.58)	(0.94 ± 0.16)
Protoporphyrin (PROTO): R	μmol/liter						
Pseudocholinesterase (PsChE); see ButChE							
Sodium (Na): S, HP	mmol/liter	141–152	147–156				142–160
		(146)	(152)	(146.8 ± 0.93)	(138.0 ± 2.9)	(141.0 ± 4.5)	(149 ± 5)
Sodium (Na): R	mmol/liter	(107)	(104)				

Sorbitol dehydrogenase (SDH): S, HP	U/liter	2.9–8.2	3.9–7.7				
		(4.5 ± 1.9)	(5.4 ± 1.3)	(20.3 ± 4.16)	(29.6 ± 7.4)		
Thyroxine (T_4-RIA): S	nmol/liter	7.7–46.4	1.3–32.3				
		(29.7 ± 10.3)	(12.9 ± 6.5)				
	μg/dl	0.6–3.6	0.1–2.5				
		(2.3 ± 0.8)	(1.0 ± 0.5)				(4.1 ± 0.6)
Thyroxine, free (FT_4): S	pmol/liter	6.35–34.75					
		(45.5 ± 4.4)					
	ng/dl	0.5–2.7					
		(3.5 ± 0.34)					
Triglyceride, total (TG): S	mmol/liter	(0.43)	(0.40)	(1.96 ± 0.29)	(1.53)	(1.38)	(0.75 ± 0.58)
	mg/dl	(38.1)	(35.4)	(173.3 ± 25.9)	(135.4)	(122.0)	(66.6 ± 51.3)
Triiodothyronine (T_3-RIA): S	nmol/liter	1.26–2.13	0.23–1.59				
		(1.65 ± 0.28)					
	ng/dl	82–138	15–104				
		(107 ± 18)					
Triglyceride, total (TG): S	mmol/liter	0.2–1.3	0.1–1.3				
	mg/dl	20–112	10–114				
Urate (UA): S, P, HP	mmol/liter	0–119	0–59.5				
				(90.4 ± 17.8)		(70.2 ± 16.6)	(71.4 ± 16.6)
	mg/dl	0–2	0–1				
					(1.52 ± 0.30)	(1.18 ± 0.28)	(1.20 ± 0.28)
Urea (UR): S, P, HP	mmol/liter	1.67–3.33	3.33–5.00				1.33–3.33
		(2.83 ± 0.67)		(2.82 ± 0.35)	(3.45 ± 0.85)	(2.38 ± 0.50)	(2.50 ± 0.5)
Urea nitrogen (UN): S, P, HP	mg/dl	10–28	20–30				8–20
		(17 ± 4.0)		(16.9 ± 2.1)	(20.7 ± 5.1)	(14.3 ± 3.0)	(15 ± 3.3)
Vitamin A (Vit A):							
Carotenol: S	nmol/liter	0–93	932–3614				
		(56)					
	μg/dl	0–5	50–194				
		(3.0)					
Carotene: S	μmol/liter	652–1677					
			(3502)				
	μg/dl	35–90					
			(188)				
Vitamin B_{12}: S	pmol/liter	125–133					
	pg/ml	170–180					
Zinc (Zn): S	μmol/liter						(12.1 ± 0.6)
	μg/dl						(79 ± 4.0)

[a] Ranges, with means and standard deviations in parentheses.
[b] B, blood; HB, heparinized blood; HP, heparinized plasma; S, serum; R, erythrocytes.

APPENDIX X. BLOOD ANALYTE REFERENCE VALUES IN SELECTED AVIANS—I[a]

Analyte[b]	Unit	Chicken	Budgerigar	Cockatoo	Macaw	Eagle	Hawk
Amylase (Amyl): S, HP	U/liter		185–585				
Aspartate amino transferase (AST, GOT): S, HP	U/liter		150–350	59–1310	40–2408	316–2881	126–500
		(174.8)		(410 ± 452)	(508 ± 950)	(1045 ± 918)	(266 ± 117)
Calcium (Ca): S, HP	mmol/liter			1.30–2.83	1.93–3.73	2.25–3.08	0.90–2.80
		(7.10)		(2.20 ± 0.45)	(2.41 ± 0.66)	(2.58 ± 0.26)	(2.28 ± 0.58)
	mg/dl			5.2–11.3	7.70–14.9	9.0–12.3	3.60–11.2
		(28.4)		(8.81 ± 1.80)	(9.64 ± 2.65)	(10.3 ± 1.05)	(9.13 ± 2.30)
Chloride (Cl): S, HP	mmol/liter					(116)	
Cholesterol (Chol): S, P, HP							
Total	mmol/liter	(4.75)					
	mg/dl	(183.8)					
Creatinine (Creat): S, P, HP	μmol/liter		8.8–35.4	26.5–167.9	35.4–247.5	70.7–132.6	26.5–79.6
				(68.1 ± 52.2)	(64.5 ± 50.4)	(91.9 ± 25.6)	(48.6 ± 15.9)
	mg/dl		0.1–0.4	0.3–1.9	0.4–2.0	0.8–1.5	0.3–0.9
				(0.77 ± 0.59)	(0.73 ± 0.57)	(1.04 ± 0.29)	(0.55 ± 0.18)
Glucose (Glu): S, P, HP	mmol/liter		11.1–22.2	10.2–20.8	11.9–23.2	14.9–32.6	8.5–20.7
		(9.3)		(15.8 ± 5.3)	(16.9 ± 3.2)	(19.9 ± 8.6)	(16.7 ± 4.1)
	mg/dl		200–400	184–375	215–418	268–587	153–373
		(167.8)		(285 ± 95)	(304 ± 57)	(359 ± 154)	(301 ± 74)
Glutamic oxaloacetic transaminase (GOT): see AST							
Lactate dehydrogenase (LDH): S, HP	U/liter		150–450	151–1337	48–831	358–3400	58–708
		(636.0)		(467 ± 435)	(293 ± 269)	(1256 ± 1072)	(301 ± 226)
Osmolality (mOsm): S, P, HP	mOsm/kg			317–347	319–378		
Phosphatase, alkaline (AlP): S, HP	U/liter			36–229	10–239	63–174	6–235
		(482.5)		(109 ± 60)	(88.5 ± 75.0)	(61.8 ± 57.8)	(88.7 ± 84.0)
Phosphate (P_i): S, HP	mmol/liter			0.23–1.91	0.70–3.36	0.68–3.55	1.16–2.16
		(2.52)		(1.0 ± 0.6)	(1.68 ± 0.84)	(1.58 ± 1.00)	(1.58 ± 0.32)
	mg/dl			0.7–5.9	2–10.4	2.1–11.1	3.6–6.7
		(7.81)		(3.1 ± 1.7)	(5.2 ± 2.6)	(4.9 ± 3.1)	(4.9 ± 1.0)

Potassium (K): S, HP	mmol/liter			2.9–11.0	2.2–10.1	2.4–4.4	1.6–4.2
				(6.0 ± 3.2)	(4.7 ± 2.7)	(3.6 ± 0.7)	(3.0 ± 0.9)
Protein (Prot): S							
Total (TP)	g/liter		25.0–45.0	27.0–54.0	22.0–52.0	32.0–49.0	27.0–46.0
		(56.0)		(41.5 ± 7.1)	(35.8 ± 7.3)	(38.7 ± 6.4)	(37.6 ± 6.3)
Electrophoresis (SPE), cellulose acetate (CA):							
Prealbumin	g/liter				5.0–11.1		
				(3.0)	(8.1 ± 5.0)		
Albumin	g/liter		21.0–33.0		11.0–24.0		
		(25.0)	(27.0 ± 1.7)	(23.0)	(17.3 ± 5.3)		
Globulin, total	g/liter		26.0–51.0		8.0–33.0		
		(31.0)	(39.0 ± 6.9)	(16.0)	(19.7 ± 10.0)		
A/G Ratio			0.45–1.19	1.5–4.3	1.40–3.90		
			(0.71 ± 0.20)	(1.74)	(1.96 ± 1.29)		
Sodium (Na): S, HP	mmol/liter			149.0–155.0	138.0–157.0	147–171	154.0–158
				(153.7 ± 2.1)	(148.6 ± 5.3)	(159.0 ± 7.0)	(156.4 ± 1.6)
Urate (UA): S, P, HP	mmol/liter		0.24–0.83	0.10–1.07	0.09–0.88	0.26–2.28	0.37–1.77
				(0.46 ± 0.29)	(0.39 ± 0.27)	(1.07 ± 0.60)	(0.77 ± 0.50)
	mg/dl		4.0–14.0	1.6–18.0	1.5–14.8	4.3–38.4	6.2–29.8
				(7.8 ± 4.8)	(6.6 ± 4.6)	(18.0 ± 10.0)	(13.0 ± 8.4)
Urea (UR): S, P, HP	mmol/liter			0.8–2.1	0.3–3.3		
	mg/dl			4.80–12.6	1.8–19.8		
Urea nitrogen (UN): S, P, HP	mmol/liter			1.60–4.20	0.60–6.60		
	mg/dl			2.24–5.88	0.84–9.24		

[a] Ranges, with means and standard deviations in parentheses.
[b] B, blood; HB, heparinized blood; HP, heparinized plasma; P, plasma; S, serum; R, erythrocytes.

APPENDIX XI. BLOOD ANALYTE REFERENCE VALUES IN SELECTED AVIANS—II[a]

Analyte[b]	Unit	Ostrich	Peregrine falcon	Pigeon	African grey parrot	Amazon parrot
Alanine aminotransferase	U/liter		29–90	19–48	12–59	19–98
Amylase (Amyl): S, HP	U/liter					571–1987
Aspartate aminotransferase (AST, GOT): S, HP	U/liter	252–401	34–116	45–123	54–155	57–194
Bile acids	μmol/liter	8–30	5–69	22–60	18–71	19–144
Calcium (Ca): S, HP	mmol/liter	2.5–4.6	1.9–2.4	1.9–2.6	2.1–2.6	2.0–2.8
	mg/dl	10.0–18.4	7.6–9.6	7.6–10.4	8.4–10.4	8.0–11.2
Chloride (Cl): S, HP	mmol/liter	94–105	114–131	101–113		
Corticosterone: S, P, HP						
Pre—250 μg/kg ACTH	nmol/liter			6–36		16–39
Post—250 μg/kg ACTH	nmol/liter			64–324		108–506
Creatinine (Creat): S, P, HP	μmol/liter		24–64	23–36	23–40	19–33
	mg/dl		0.27–0.72	0.26–0.40	0.26–0.45	0.21–0.37
Creatine kinase (CK): S, P, HP	U/liter	1655–4246	120–442	110–480	123–875	45–265
Glucose (Glu): S, P, HP	mmol/liter	10.4–13.7	16.5–22.0	12.9–20.5	11.4–16.1	12.6–16.9
	mg/dl	187–247	297–396	232–369	205–290	227–304
Glutamate dehydrogenase (GD): S, P, HP	U/liter	<8	<8	<8	<8	<8
Glutamic oxaloacetic transaminase (GOT): see AST						
Glutamic pyruvic transaminase (GPT: see ALT)						
γ-Glutamyltransferase (GGT): S, P, HP	U/liter	0–1	0–3	0–3	0–4	1–10
Lactate dehydrogenase (LD): S, P, HP	U/liter	869–2047	1008–2650	30–205	147–384	46–208
Lipase (Lip): S, P, HP	U/liter					268–1161
Osmolality (mOsm): S, P, HP	mOsm/kg	305–328	322–356	297–317	320–347	316–373
Phosphate (Pi): S, HP	mmol/liter	1.3–2.2	0.55–1.53	0.57–1.33		
	mg/dl	4.0–6.80	1.70–4.74	1.76–4.12		
Potassium (K): S, HP	mmol/liter	4.5–5.9		3.9–4.7	2.5–3.9	2.3–4.2
Protein (Prot): S						
Total (TP)	g/liter	40–54	24–39	21–35	32–44	33–50
Albumin/globulin (A/G) ratio		0.9–1.4	0.8–2.4	1.5–3.6	1.4–4.7	2.6–7.0
Sodium (Na): S, HP	mmol/liter	147–157	150–170	141–149	154–164	149–164
Thyroxine (T_4): S, P, HP						
Pre—2 IU/kg TSH IM	nmol/liter			6–35		
	mg/dl			0.46–2.72		
Post—2 IU/kg TSH IM	nmol/liter			100–300		
	mg/dl			7.77–23.3		
Urate (UA): S, P, HP	μmol/liter	357–643	253–996	150–765	93–414	72–312
	mg/dl	6.00–10.80	4.26–16.75	2.52–12.56	1.56–6.96	1.21–5.25
Urea (UR): S, P, HP	mmol/liter	0.5–0.8	0.8–2.9	0.4–0.7	0.7–2.4	0.9–4.6
	mg/dl	3.0–4.8	4.8–17.4	2.4–4.2	4.2–14.4	5.4–27.6
Urea nitrogen (UN): S, P, HP	mmol/liter	0.99–1.57	1.52–2.07	0.78–1.43	1.43–4.78	1.79–9.20
	mg/dl	1.4–2.2	2.2–8.1	1.1–2.0	2.0–6.7	2.5–27.6

[a] Ranges, with means and standard deviations in parentheses.
[b] B, blood; HB, heparinized blood; HP, heparinized plasma; P, plasma; S, serum; R, erythrocytes.

APPENDIX XII. URINE ANALYTE REFERENCE VALUES IN ANIMALS

Analyte	Units	Horse	Cow	Sheep	Pig	Dog	Cat	Goat
Allantoin	mg/kg/day	5–15	20–60	20–50	20–80	35–45	80	
Arsenic	μg/dl					30–150		
Bicarbonate	mmol/kg/day					0.05–3.2		
Calcium	mg/kg/day		0.10–1.40	2.0		1–3	0.20–0.45	1.0
Chloride	mmol/kg/day		0.10–1.10			0–10.3		
Coproporphyrin	μg/dl		5–14	8.8		16–28		
Creatinine	mg/kg/day		15–20	10	20–90	30–80	12–20	10.0
Cystine	mg/g creatinine					(67 ± 15)		
Hydrogen ion (pH)	unit	7.0–8.0	7.4–8.4	7.4–8.4	5.0–8.0	5.0–7.0	5.0–7.0	7.4–8.4
Lead	μg/dl					20–75		
Lysine	mg/g creatinine					21 ± 6		
Magnesium	mg/kg/day		3.7			1.7–3.0	3–12	
Mercury	μg/dl					1.0–10		
Nitrogen								
Urea N	mg/kg/day		23–28	98	201	140–230	374–1872	107
Total N	mg/kg/day	100–600	40–450	120–350	40–240	250–800	500–1100	120–400
Ammonia N	mg/kg/day		1.0–17.0			30–60	60	3–5
Phosphorus	mg/kg/day			0.2		20–30	108	1.0
Potassium	mmol/kg/day		0.08–0.15			0.1–2.4		
Sodium	mmol/kg/day		0.2–1.1			0.04–13.0		
Specific gravity	units	1.020–1.050	1.025–1.045	1.015–1.045	1.010–1.030	1.015–1.045	1.015–1.065	1.015–1.045
Sulfate	mg/kg/day		3.0–5.0			30–50		
Uric acid	mg/kg/day	1–2	1–4	2–4	1–2			2–5
Urine volume	ml/kg/day	3–18	17–45	10–40	5–30	17–45	10–20	10–40
Uroporphyrin	mg/dl		1.5–7.0	3.8	5.0			

APPENDIX XIII. CEREBROSPINAL FLUID (CSF) ANALYTE REFERENCE VALUES IN LARGE ANIMALS

Analyte	Units	Horse	Cow	Sheep	Goat	Llama	Pig
Rate of formation	μl/m		290	118	164		
Alkaline phosphatase (AIP)	U/liter	0–8 (0.8 ± 0.9)					
Aspartate aminotransferase (AST)	U/liter	15–50 (30.7 ± 6.3)					
Calcium	mg/dl	2.5–6.0 (4.2 ± 0.9)	5.1–6.3	5.1–5.5	4.6		
Chloride	mmol/liter	95–123 (109.2 ± 6.9)	111–123	128–148	116–130	(116–143)	(134 ± 6.5)
Cholesterol (Chol)	mg/dl	020.0 (4.8 ± 5.7)					
Creatine kinase (CK)	U/liter	0–8 (1.1 ± 3.1)					
Glucose	mg/dl	30–70 (48.0 ± 10.0)	37–51 (42.9 ± 1.0)	52–85	70	59–86 (69.3 ± 7.35)	45–87
γ-Glutamyltransferase (GGT)	U/liter	0.8–4.2 (2.6 ± 1.9)					
Hydrogen ion (pH)	units	7.13–7.36	7.22–7.26	7.3–7.4 (7.35)			
Lactate dehydrogenase (LD)	U/liter	12–34 (27.7 ± 8.0)	2–25 (13.94 ± 1.32)			7–24 (13 ± 5.6)	
Lactic acid (LAC)	mg/dl	(2.3 ± 0.2)					
Magnesium	mg/dl	1.1–3.0 (2.0)	1.8–2.1 (1.99 ± 0.03)	2.2–2.8	2.3		
Phosphorus	mg/dl	0.5–1.5 (0.8 ± 0.2)	0.9–2.5	1.2–2.0			
Potassium	mmol/liter	2.5–3.5 (3.0 ± 0.1)	2.7–3.2 (2.96 ± 0.03)	3.0–3.3	3.0	2.9–3.3 (3.19 ± 0.10)	
Pressure	mm H_2O	272–490					
Protein							
Total	mg/dl	40–170 (105 ± 38)	23.4–66.3 (39.1 ± 3.39)	29–42	12	31.2–66.8 (43.1 ± 9.0)	24–29
Albumin	mg/dl	22.6–67.9 (38.6)	8.21–28.71 (15.75 ± 1.53)			11.8–27.1 (17.9 ± 4.45)	17–24
Globulin:	mg/dl	3.4–18.4 (9.3)					5–10
γ-Globulin	mg/dl	3–10 (6.0 ± 2.1)	2.46–8.85			3.4–13.8	
Sodium	mmol/liter	140–150 (144.6 ± 1.9)	132–142 (140 ± 0.78)	145–157	131	134–160 (154 ± 5.8)	134–144
Specific gravity	unit	1.004–1.008	1.005–1.008				
Urea (UR)	mg/dl	0–43.2					
	mmol/liter	0–7.2					
Urea nitrogen (UN)	mg/dl	0–20 (11.8 ± 3.3)	8–11				
	mmol/liter	0–14.3 (0 ± 3.4)					
Viscosity	unit	1.00–1.05	1.019–1.029				
Cells (total WBC)	#/μl	0–6	0.85–3.52	0–5	0–4	0–3	0–7
Cells (total RBC)	#/μl	(195 ± 512)	5–1930			0–1360	

APPENDIX XIV. CEREBROSPINAL FLUID (CSF) ANALYTE REFERENCE VALUES IN SMALL AND SOME LABORATORY ANIMALS

Analyte	Units	Dog	Cat	Rat	Mouse	Rabbit	Monkey
Rate of formation	μl/m	47–66	20–22	2.1–5.4	0.325	10	28.6–41.0
Alaninine aminotransferase (ALT)	U/liter	0.96–15.36 (6.58 ± 0.65)					
Aspartate aminotransferase (AST)	U/liter	4.32–22.08 (9.65 ± 0.65)	0–34 (17 ± 7)				
Calcium	mg/dl	5.13–7.40 (6.56)	5.1–6.3 (6.0 ± 0.24)				
Chloride	mmol/liter	109–126 (130 ± 0.5)	111–123				
Creatine kinase (CK)	U/liter	(23.5 ± 0.19)	2–236 (47 ± 51)				
Glucose	mg/dl	48–57	18.2–130.9 (74.5 ± 23.6)				
Hydrogen ion (pH)	units	7.13–7.36	7.22–7.26				
Lactate dehydrogenase (LD)	U/liter		0–24 (12 ± 5)				
Magnesium	mg/dl	25.8–3.81	(3.24 ± 0.05)				
Phosphorus	mg/dl	2.82–3.47 (3.09)					
Potassium	mmol/liter	2.9–3.2 (3.3 ± 0.04)	(2.69 ± 0.09)				
Pressure	mm H_2O	24–172					
Protein							
Total	mg/dl	18–44 (28.68 ± 5.52)	0–30 (5 ± 9)				
Albumin	mg/dl	7.5–27.6 (17.1 ± 6.7)	19–25 (10.1 ± 12.9)				
Globulin:		14.0–21.1 (17.45 ± 0.83)					
IgA	μg/ml	0–0.2 (0.08)					
IgG	mg/dl	2.5–8.5 (4.68 ± 0.68)	0–5.3 (1.4 ± 0.27)				
IgM	μg/ml	0–5.8 (1.7)					
Sodium	mmol/liter	151.6–155 (153 ± 0.5)	158 (158 ± 4)				
Specific gravity	unit	1.003–1.012 (1.005)					
Urea	mg/dl	10–11	21.5–23.6				
	mmol/liter	1.7–1.8	3.6–3.9				
Urea nitrogen	mg/dl	4.7–5.1	10–11				
	mmol/liter	3.3–3.7	7.1–7.9				
Cells (total WBC)	#/μl	0–25	0–1				
Small mononuclear	%	15–95					
Large mononuclear	%	5–40					
Degenerate	%	0–40					
Cells (total RBC)	#/μl	<1500	<30				

Index

A

B

D

E

H

J

K

L

Q

R

S

U

W

X

Z

ISBN 0-12-396305-2